Be sure to keep up-to-date in 2011!

Register now and we'll notify you by email each quarter when the newest cost index is available free online.

Please provide your name, address, and email below and return this card by mail or Fax: 1-800-632-6732.

To register online:
www.reedconstructiondata.com/rsmeans/chgnotice/842339

Name ______________________________

email ______________________________

Title ______________________________

Company ______________________________

Street ______________________________

City/Town ____________ State/Prov. _____ Zip/Postal Code ________

RS**Means**

AF680840

PLACE
STAMP
HERE

RSMeans
Quarterly Updates
700 Longwater Drive
Suite 100
Norwell, MA 02061
USA

$134.95 per copy (in United States)
Price is subject to change without prior notice.

0151

RSMeans Open Shop Construction Cost Data

27TH ANNUAL EDITION

2011

RSMeans
A division of Reed Construction Data
Construction Publishers & Consultants
700 Longwater Drive
Norwell, MA 02061
USA
1-800-334-3509

Printed in the United States of America
ISSN 0883-8127
ISBN 978-1-936335-15-2

This book is printed using soy-based printing ink.
This book is recyclable.

PRINTED WITH SOY INK™

Related RSMeans Products and Services

This book is aimed primarily at commercial and industrial projects or large multi-family housing projects costing $3,500,000 and up. For civil engineering structures such as bridges, dams, highways, or the like, please refer to *RSMeans Heavy Construction Cost Data.*

The engineers at RSMeans suggest the following products and services as companion information resources to *RSMeans Open Shop Building Construction Cost Data:*

Construction Cost Data Books

Light Commercial Cost Data 2011
Residential Cost Data 2011

Reference Books

Unit Price Estimating Methods
ADA Compliance Pricing Guide
Building Security: Strategies & Costs
Designing & Building with the IBC
Estimating Building Costs
Estimating Handbook
Green Building: Project Planning & Estimating
How to Estimate with Means Data and CostWorks
Plan Reading & Material Takeoff
Project Scheduling and Management for Construction

Seminars and In-House Training

Unit Price Estimating
Means CostWorks® Training
Means Data for Job Order Contracting (JOC)
Plan Reading & Material Takeoff
Scheduling & Project Management
Mechanical & Electrical Estimating

RSMeans on the Internet

Visit RSMeans at **www.reedconstructiondata.com** for a one-stop portal for the most reliable and current resources available on the market. Here you can request or download **FREE** estimating software demos, visit our bookstore for convenient ordering, and learn more about new publications and companion products.

Also visit **www.reedconstructiondata.com** for RSMeans business solutions, seminars, training, and more.

RSMeans Electronic Data

Get the information found in RSMeans cost books electronically on *Means CostWorks®* CD or on the Web at **MeansCostWorks.com**.

RSMeans Business Solutions

Cost engineers and analysts conduct facility life cycle and benchmark research studies and predictive cost modeling, as well as offer consultation services for conceptual estimating, real property management, and job order contracting. Research studies are designed with the application of proprietary cost and project data from Reed Construction Data and RSMeans' extensive North American databases. Analysts offer building product manufacturers qualitative and quantitative market research, as well as time/motion studies for new products, and Web-based dashboards for market opportunity analysis. Clients are from both the public and private sectors, including federal, state, and municipal agencies, corporations, institutions, construction management firms, hospitals, and associations.

RSMeans for Job Order Contracting (JOC)

Best practice job order contracting (JOC) services support the owner in the development of their program. RSMeans develops all required contracting documents, works with stakeholders, and develops qualified and experienced contractor lists. Means JOCWorks™ software is the best of class in the functional tools that meet JOC specific requirements. RSMeans data is organized to meet JOC cost estimating requirements.

Construction Costs for Software Applications

Over 25 unit price and assemblies cost databases are available through a number of leading estimating and facilities management software providers (listed below). For more information see the "Other RSMeans Products" pages at the back of this publication.

MeansData™ is also available to federal, state, and local government agencies as multi-year, multi-seat licenses.

- 4Clicks-Solutions, LLC
- ArenaSoft Estimating
- Beck Technology
- BSD – Building Systems Design, Inc.
- CMS – Construction Management Software
- Corecon Technologies, Inc.
- CorVet Systems
- Hard Dollar Corporation
- MC^2
- Parsons Corporation
- Sage Timberline Office
- UDA Technologies, Inc.
- US Cost, Inc.
- VFA – Vanderweil Facility Advisers
- WinEstimator, Inc.
- R&K Maxwell Solutions
- Maxwell Systems
- Tririga

Table of Contents

Foreword

Our Mission

Since 1942, RSMeans has been actively engaged in construction cost publishing and consulting throughout North America.

Today, more than 65 years after RSMeans began, our primary objective remains the same: to provide you, the construction and facilities professional, with the most current and comprehensive construction cost data possible.

Whether you are a contractor, owner, architect, engineer, facilities manager, or anyone else who needs a reliable construction cost estimate, you'll find this publication to be a highly useful and necessary tool.

With the constant flow of new construction methods and materials today, it's difficult to find the time to look at and evaluate all the different construction cost possibilities. In addition, because labor and material costs keep changing, last year's cost information is not a reliable basis for today's estimate or budget.

That's why so many construction professionals turn to RSMeans. We keep track of the costs for you, along with a wide range of other key information, from city cost indexes . . . to productivity rates . . . to crew composition . . . to contractor's overhead and profit rates.

RSMeans performs these functions by collecting data from all facets of the industry and organizing it in a format that is instantly accessible to you. From the preliminary budget to the detailed unit price estimate, you'll find the data in this book useful for all phases of construction cost determination.

The Staff, the Organization, and Our Services

When you purchase one of RSMeans' publications, you are, in effect, hiring the services of a full-time staff of construction and engineering professionals.

Our thoroughly experienced and highly qualified staff works daily at collecting, analyzing, and disseminating comprehensive cost information for your needs. These staff members have years of practical construction experience and engineering training prior to joining the firm. As a result, you can count on them not only for accurate cost figures, but also for additional background reference information that will help you create a realistic estimate.

The RSMeans organization is always prepared to help you solve construction problems through its variety of data solutions, including online, CD, and print book formats, as well as cost estimating expertise available via our business solutions, training, and seminars.

Besides a full array of construction cost estimating books, RSMeans also publishes a number of other reference works for the construction industry. Subjects include construction estimating and project and business management, special topics such as green building and job order contracting, and a library of facility management references.

In addition, you can access all of our construction cost data electronically using *Means CostWorks®* CD or on the Web using **MeansCostWorks.com.**

What's more, you can increase your knowledge and improve your construction estimating and management performance with an RSMeans construction seminar or in-house training program. These two-day seminar programs offer unparalleled opportunities for everyone in your organization to become updated on a wide variety of construction-related issues.

RSMeans is also a worldwide provider of construction cost management and analysis services for commercial and government owners.

In short, RSMeans can provide you with the tools and expertise for constructing accurate and dependable construction estimates and budgets in a variety of ways.

Robert Snow Means Established a Tradition of Quality That Continues Today

Robert Snow Means spent years building RSMeans, making certain he always delivered a quality product.

Today, at RSMeans, we do more than talk about the quality of our data and the usefulness of our books. We stand behind all of our data, from historical cost indexes to construction materials and techniques to current costs.

If you have any questions about our products or services, please call us toll-free at 1-800-334-3509. Our customer service representatives will be happy to assist you. You can also visit our Web site at **www.rsmeans.com**

How the Book is Built: An Overview

The Construction Specifications Institute (CSI) and Construction Specifications Canada (CSC) have produced the 2010 edition of MasterFormat™, a system of titles and numbers used extensively to organize construction information.

All unit price data in the RSMeans cost data books is now arranged in the 50-division MasterFormat™ 2010 system.

A Powerful Construction Tool

You have in your hands one of the most powerful construction tools available today. A successful project is built on the foundation of an accurate and dependable estimate. This book will enable you to construct just such an estimate.

For the casual user the book is designed to be:

- quickly and easily understood so you can get right to your estimate.
- filled with valuable information so you can understand the necessary factors that go into the cost estimate.

For the regular user, the book is designed to be:

- a handy desk reference that can be quickly referred to for key costs.
- a comprehensive, fully reliable source of current construction costs and productivity rates so you'll be prepared to estimate any project.
- a source book for preliminary project cost, product selections, and alternate materials and methods.

To meet all of these requirements, we have organized the book into the following clearly defined sections.

New: Quick Start

See our new "Quick Start" instructions on the following page to get started right away.

Estimating with RSMeans Unit Price Cost Data

Please refer to these steps for guidance on completing an estimate using RSMeans unit price cost data.

How To Use the Book: The Details

This section contains an in-depth explanation of how the book is arranged . . . and how you can use it to determine a reliable construction cost estimate. It includes information about how we develop our cost figures and how to completely prepare your estimate.

Unit Price Section

All cost data has been divided into the 50 divisions according to the MasterFormat system of classification and numbering.

For a listing of these divisions and an outline of their subdivisions, see the Unit Price Section Table of Contents.

Estimating tips are included at the beginning of each division.

Reference Section

This section includes information on Equipment Rental Costs, Crew Listings, Historical Cost Indexes, City Cost Indexes, Location Factors, Reference Tables, Change Orders, Square Foot Costs, and a listing of Abbreviations.

Equipment Rental Costs: This section contains the average costs to rent and operate hundreds of pieces of construction equipment.

Crew Listings: This section lists all of the crews referenced in the book. For the purposes of this book, a crew is composed of more than one trade classification and/or the addition of power equipment to any trade classification. Power equipment is included in the cost of the crew. Costs are shown both with bare labor rates and with the installing contractor's overhead and profit added. For each, the total crew cost per eight-hour day and the composite cost per labor-hour are listed.

Historical Cost Indexes: These indexes provide you with data to adjust construction costs over time.

City Cost Indexes: All costs in this book are U.S. national averages. Costs vary because of the regional economy. You can adjust costs by CSI Division to over 700 locations throughout the U.S. and Canada by using the data in this section.

Location Factors: You can adjust total project costs to over 900 locations throughout the U.S. and Canada by using the data in this section.

Reference Tables: At the beginning of selected major classifications in the Unit Price section are reference numbers shown in a shaded box. These numbers refer you to related information in the Reference Section. In this section, you'll find reference tables, explanations, estimating information that support how we develop the unit price data, technical data, and estimating procedures.

Change Orders: This section includes information on the factors that influence the pricing of change orders.

Square Foot Costs: This section contains costs for 59 different building types that allow you to make a rough estimate for the overall cost of a project or its major components.

Abbreviations: A listing of abbreviations used throughout this book, along with the terms they represent, is included in this section.

Index

A comprehensive listing of all terms and subjects in this book will help you quickly find what you need when you are not sure where it occurs in MasterFormat.

The Scope of This Book

This book is designed to be as comprehensive and as easy to use as possible. To that end we have made certain assumptions and limited its scope in two key ways:

1. We have established material prices based on a national average.
2. We have computed labor costs based on a 30-city national average of union wage rates.

For a more detailed explanation of how the cost data is developed, see "How To Use the Book: The Details."

Project Size/Type

The material prices in Means data cost books are "contractor's prices." They are the prices that contractors can expect to pay at the lumberyards, suppliers/distributers warehouses, etc. Small orders of speciality items would be higher than the costs shown, while very large orders, such as truckload lots, would be less. The variation would depend on the size, timing, and negotiating power of the contractor. The labor costs are primarily for new construction or major renovation rather than repairs or minor alterations.

With reasonable exercise of judgment, the figures can be used for any building work.

Absolute Essentials for a Quick Start

If you feel you are ready to use this book and don't think you will need the detailed instructions that begin on the following page, this Absolute Essentials for a Quick Start page is for you. These steps will allow you to get started estimating in a matter of minutes.

1 Scope

Think through the project that you will be estimating, and identify the many individual work tasks that will need to be covered in your estimate.

2 Quantify

Determine the number of units that will be required for each work task that you identified.

3 Pricing

Locate individual Unit Price line items that match the work tasks you identified. The Unit Price Section Table of Contents that begins on page 1 and the Index in the back of the book will help you find these line items.

4 Multiply

Multiply the Total Incl O&P cost for a Unit Price line item in the book by your quantity for that item. The price you calculate will be an estimate for a completed item of work performed by a subcontractor. Keep adding line items in this manner to build your estimate.

5 Project Overhead

Include project overhead items in your estimate. These items are needed to make the job run and are typically, but not always, provided by the General Contractor. They can be found in Division 1. An alternate method of estimating project overhead costs is to apply a percentage of the total project cost.

Include rented tools not included in crews, waste, rubbish handling, and cleanup.

6 Estimate Summary

Include General Contractor's markup on subcontractors, General Contractor's office overhead and profit, and sales tax on materials and equipment.

Adjust your estimate to the project's location by using the City Cost Indexes or Location Factors found in the Reference Section.

Editors' Note: We urge you to spend time reading and understanding the supporting material in the front of this book. An accurate estimate requires experience, knowledge, and careful calculation. The more you know about how we at RSMeans developed the data, the more accurate your estimate will be. In addition, it is important to take into consideration the reference material in the back of the book such as Equipment Listings, Crew Listings, City Cost Indexes, Location Factors, and Reference Numbers.

Estimating with RSMeans Unit Price Cost Data

Following these steps will allow you to complete an accurate estimate using RSMeans Unit Price cost data.

1 Scope Out the Project

- Identify the individual work tasks that will need to be covered in your estimate.
- The Unit Price data in this book has been divided into 50 Divisions according to the CSI MasterFormat 2010—their titles are listed on the back cover of your book.
- Think through the project and identify those CSI Divisions needed in your estimate.
- The Unit Price Section Table of Contents on page 1 may also be helpful when scoping out your project.
- Experienced estimators find it helpful to begin with Division 2 and continue through completion. Division 1 can be estimated after the full project scope is known.

2 Quantify

- Determine the number of units required for each work task that you identified.
- Experienced estimators include an allowance for waste in their quantities. (Waste is not included in RSMeans Unit Price line items unless so stated.)

3 Price the Quantities

- Use the Unit Price Table of Contents, and the Index, to locate individual Unit Price line items for your estimate.
- Reference Numbers indicated within a Unit Price section refer to additional information that you may find useful.
- The crew indicates who is performing the work for that task. Crew codes are expanded in the Crew Listings in the Reference Section to include all trades and equipment that comprise the crew.
- The Daily Output is the amount of work the crew is expected to do in one day.
- The Labor-Hours value is the amount of time it will take for the crew to install one unit of work.
- The abbreviated Unit designation indicates the unit of measure upon which the crew, productivity, and prices are based.
- Bare Costs are shown for materials, labor, and equipment needed to complete the Unit Price line item. Bare costs do not include waste, project overhead, payroll insurance, payroll taxes, main office overhead, or profit.
- The Total Incl O&P cost is the billing rate or invoice amount of the installing contractor or subcontractor who performs the work for the Unit Price line item.

4 Multiply

- Multiply the total number of units needed for your project by the Total Incl O&P cost for each Unit Price line item.
- Be careful that your take off unit of measure matches the unit of measure in the Unit column.
- The price you calculate is an estimate for a completed item of work.
- Keep scoping individual tasks, determining the number of units required for those tasks, matching each task with individual Unit Price line items in the book, and multiply quantities by Total Incl O&P costs.
- An estimate completed in this manner is priced as if a subcontractor, or set of subcontractors, is performing the work. The estimate does not yet include Project Overhead or Estimate Summary components such as General Contractor markups on subcontracted work, General Contractor office overhead and profit, contingency, and location factor.

5 Project Overhead

- Include project overhead items from Division 1 – General Requirements.
- These items are needed to make the job run. They are typically, but not always, provided by the General Contractor. Items include, but are not limited to, field personnel, insurance, performance bond, permits, testing, temporary utilities, field office and storage facilities, temporary scaffolding and platforms, equipment mobilization and demobilization, temporary roads and sidewalks, winter protection, temporary barricades and fencing, temporary security, temporary signs, field engineering and layout, final cleaning and commissioning.
- Each item should be quantified, and matched to individual Unit Price line items in Division 1, and then priced and added to your estimate.
- An alternate method of estimating project overhead costs is to apply a percentage of the total project cost, usually 5% to 15% with an average of 10% (see General Conditions, page ix).
- Include other project related expenses in your estimate such as:
  - Rented equipment not itemized in the Crew Listings
  - Rubbish handling throughout the project (see 02 41 19.23)

6 Estimate Summary

- Include sales tax as required by laws of your state or country.
- Include the General Contractor's markup on self-performed work, usually 5% to 15% with an average of 10%.
- Include the General Contractor's markup on subcontracted work, usually 5% to 15% with an average of 10%.
- Include General Contractor's main office overhead and profit.
- RSMeans gives general guidelines on the General Contractor's main office overhead (see section 01 31 13.50 and Reference Number R013113-50).
- RSMeans gives no guidance on the General Contractor's profit.
- Markups will depend on the size of the General Contractor's operations, his projected annual revenue, the level of risk he is taking on, and on the level of competition in the local area and for this project in particular.
- Include a contingency, usually 3% to 5%, if appropriate.
- Adjust your estimate to the project's location by using the City Cost Indexes or the Location Factors in the Reference Section.
- Look at the rules on the pages for How to Use the City Cost Indexes to see how to apply the Indexes for your location.
- When the proper Index or Factor has been identified for the project's location, convert it to a multiplier by dividing it by 100, and then multiply that multiplier by your estimated total cost. The original estimated total cost will now be adjusted up or down from the national average to a total that is appropriate for your location.

Editors' Notes:

1) *We urge you to spend time reading and understanding the supporting material in the front of this book. An accurate estimate requires experience, knowledge, and careful calculation. The more you know about how we at RSMeans developed the data, the more accurate your estimate will be. In addition, it is important to take into consideration the reference material in the back of the book such as Equipment Listings, Crew Listings, City Cost Indexes, Location Factors, and Reference Numbers.*
2) *Contractors who are bidding or are involved in JOC, DOC, SABER, or IDIQ type contracts are cautioned that workers' compensation insurance, federal and state payroll taxes, waste, project supervision, project overhead, main office overhead, and profit are not included in bare costs. The coefficient or multiplier must cover these costs.*

How to Use the Book: The Details

What's Behind the Numbers? The Development of Cost Data

The staff at RSMeans continually monitors developments in the construction industry in order to ensure reliable, thorough, and up-to-date cost information. While overall construction costs may vary relative to general economic conditions, price fluctuations within the industry are dependent upon many factors. Individual price variations may, in fact, be opposite to overall economic trends. Therefore, costs are constantly tracked and complete updates are published yearly. Also, new items are frequently added in response to changes in materials and methods.

Costs—$ (U.S.)

All costs represent U.S. national averages and are given in U.S. dollars. The RSMeans City Cost Indexes can be used to adjust costs to a particular location. The City Cost Indexes for Canada can be used to adjust U.S. national averages to local costs in Canadian dollars. No exchange rate conversion is necessary.

G The processes or products identified by the green symbol in our publications have been determined to be environmentally responsible and/or resource-efficient solely by the RSMeans engineering staff. The inclusion of the green symbol does not represent compliance with any specific industry association or standard.

Material Costs

The RSMeans staff contacts manufacturers, dealers, distributors, and contractors all across the U.S. and Canada to determine national average material costs. If you have access to current material costs for your specific location, you may wish to make adjustments to reflect differences from the national average. Included within material costs are fasteners for a normal installation. RSMeans engineers use manufacturers' recommendations, written specifications, and/or standard construction practice for size and spacing of fasteners. Adjustments to material costs may be required for your specific application or location. Material costs do not include sales tax.

Labor Costs

Labor costs are based on the average of open shop wage rates from 30 major U.S. cities for construction trades for the current year. Rates, along with overhead and profit markups, are listed on the inside back cover of this book.

- If wage rates in your area vary from those used in this book, or if rate increases are expected within a given year, labor costs should be adjusted accordingly.

Labor costs reflect productivity based on actual working conditions. In addition to actual installation, these figures include time spent during a normal weekday on tasks such as, material receiving and handling, mobilization at site, site movement, breaks, and cleanup. Productivity data is developed over an extended period so as not to be influenced by abnormal variations and reflects a typical average.

Equipment Costs

Equipment costs include not only rental, but also operating costs for equipment under normal use. The operating costs include parts and labor for routine servicing such as repair and replacement of pumps, filters, and worn lines. Normal operating expendables, such as fuel, lubricants, tires, and electricity (where applicable), are also included. Extraordinary operating expendables with highly variable wear patterns, such as diamond bits and blades, are excluded. These costs are included under materials. Equipment rental rates are obtained from industry sources throughout North America—contractors, suppliers, dealers, manufacturers, and distributors.

Rental rates can also be treated as reimbursement costs for contractor-owned equipment. Owned equipment costs include depreciation, loan payments, interest, taxes, insurance, storage and major repairs.

Equipment costs do not include operators' wages; nor do they include the cost to move equipment to a job site (mobilization) or from a job site (demobilization).

Equipment Cost/Day——The cost of power equipment required for each crew is included in the Crew Listings in the Reference Section (small tools that are considered as essential everyday tools are not listed out separately). The Crew Listings itemize specialized tools and heavy equipment along with labor trades. The daily cost of itemized equipment included in a crew is based on dividing the weekly bare rental rate by 5 (number of working days per week) and then adding the hourly operating cost times 8 (the number of hours per day). This Equipment Cost/Day is shown in the last column of the Equipment Rental Cost pages in the Reference Section.

Mobilization/Demobilization——The cost to move construction equipment from an equipment yard or rental company to the job site and back again is not included in equipment costs. Mobilization (to the site) and demobilization (from the site) costs can be found in the Unit Price Section. If a piece of equipment is already at the job site, it is not appropriate to utilize mob./demob. costs again in an estimate.

Overhead and Profit

Total Cost including O&P for the *Installing Contractor* is shown in the last column on the Unit Price and/or the Assemblies pages of this book. This figure is the sum of the bare material cost plus 10% for profit, the bare labor cost plus total overhead and profit, and the bare equipment cost plus 10% for profit. Details for the calculation of Overhead and Profit on labor are shown on the inside back cover and in the Reference Section of this book. (See the "How to Use the Unit Price Pages" for an example of this calculation.)

General Conditions

Cost data in this book is presented in two ways: Bare Costs and Total Cost including O&P (Overhead and Profit). General Conditions, or General Requirements, of the contract should also be added to the Total Cost including O&P when applicable. Costs for General Conditions are listed in Division 1 of the Unit Price Section and the Reference

Section of this book. General Conditions for the *Installing Contractor* may range from 0% to 10% of the Total Cost including O&P. For the *General* or *Prime Contractor*, costs for General Conditions may range from 5% to 15% of the Total Cost including O&P, with a figure of 10% as the most typical allowance. If applicable, the Assemblies and Models sections of this book use costs that include the installing contractor's overhead and profit (O&P).

Factors Affecting Costs

Costs can vary depending upon a number of variables. Here's how we have handled the main factors affecting costs.

Quality—The prices for materials and the workmanship upon which productivity is based represent sound construction work. They are also in line with U.S. government specifications.

Overtime—We have made no allowance for overtime. If you anticipate premium time or work beyond normal working hours, be sure to make an appropriate adjustment to your labor costs.

Productivity—The productivity, daily output, and labor-hour figures for each line item are based on working an eight-hour day in daylight hours in moderate temperatures. For work that extends beyond normal work hours or is performed under adverse conditions, productivity may decrease. (See the section in "How To Use the Unit Price Pages" for more on productivity.)

Size of Project—The size, scope of work, and type of construction project will have a significant impact on cost. Economies of scale can reduce costs for large projects. Unit costs can often run higher for small projects.

Location—Material prices in this book are for metropolitan areas. However, in dense urban areas, traffic and site storage limitations may increase costs. Beyond a 20-mile radius of large cities, extra trucking or transportation charges may also increase the material costs slightly. On the other hand, lower wage rates may be in effect. Be sure to consider both of these factors when preparing an estimate, particularly if the job site is located in a central city or remote rural location. In addition, highly specialized subcontract items may require travel and per-diem expenses for mechanics.

Other Factors—

- season of year
- contractor management
- weather conditions
- local union restrictions
- building code requirements
- availability of:
  - adequate energy
  - skilled labor
  - building materials
- owner's special requirements/restrictions
- safety requirements
- environmental considerations

Unpredictable Factors—General business conditions influence "in-place" costs of all items. Substitute materials and construction methods may have to be employed. These may affect the installed cost and/or life cycle costs. Such factors may be difficult to evaluate and cannot necessarily be predicted on the basis of the job's location in a particular section of the country. Thus, where these factors apply, you may find significant but unavoidable cost variations for which you will have to apply a measure of judgment to your estimate.

Rounding of Costs

In general, all unit prices in excess of $5.00 have been rounded to make them easier to use and still maintain adequate precision of the results. The rounding rules we have chosen are in the following table.

Prices from . . .	Rounded to the nearest . . .
$.01 to $5.00	$.01
$5.01 to $20.00	$.05
$20.01 to $100.00	$.50
$100.01 to $300.00	$1.00
$300.01 to $1,000.00	$5.00
$1,000.01 to $10,000.00	$25.00
$10,000.01 to $50,000.00	$100.00
$50,000.01 and above	$500.00

How Subcontracted Items Affect Costs

A considerable portion of all large construction jobs is usually subcontracted. In fact, the percentage done by subcontractors is constantly increasing and may run over 90%. Since the workers employed by these companies do nothing else but install their particular product, they soon become expert in that line. The result is, installation by these firms is accomplished so efficiently that the total in-place cost, even adding the general contractor's overhead and profit, is no more, and often less, than if the principal contractor had handled the installation himself/herself. Companies that deal with construction specialties are anxious to have their product perform well and, consequently, the installation will be the best possible.

Contingencies

The allowance for contingencies generally provides for unforeseen construction difficulties. On alterations or repair jobs, 20% is not too much. If drawings are final and only field contingencies are being considered, 2% or 3% is probably sufficient, and often nothing need be added. Contractually, changes in plans will be covered by extras. The contractor should consider inflationary price trends and possible material shortages during the course of the job. These escalation factors are dependent upon both economic conditions and the anticipated time between the estimate and actual construction. If drawings are not complete or approved, or a budget cost is wanted, it is wise to add 5% to 10%. Contingencies, then, are a matter of judgment. Additional allowances for contingencies are shown in Division 1.

Important Estimating Considerations

The "productivity," or daily output, of each craftsman includes mobilization and cleanup time, break time and plan layout time, as well as an allowance to carry stock from the storage trailer or location on the job site up to 200′ into the building and to the first or second floor. If material has to be transported over greater distances or to higher floors, an additional allowance should be considered by the estimator. An allowance has also been included in the piping and fittings installation time for leak checking and minor tightening. Equipment installation time includes the following applicable items: positioning, leveling and securing the unit in place, connecting all associated piping, ducts, vents, etc., which shall have been estimated separately, connecting to an adjacent power source, filling/bleeding, startup, adjusting the controls up and down to ensure proper response, setting the integral controls/valves/regulators/thermostats for proper operation (does not include external building type control systems, DDC systems, etc.), explaining/training owner's operator, and warranty. A reasonable breakdown of the labor costs is as follows:

1.	Movement into building, installation/setting of equipment	35%
2.	Connection to piping/duct/power, etc.	25%
3.	Filling/flushing/cleaning/touchup, etc.	15%
4.	Startup/running adjustments	5%
5.	Training owner's representative	5%
6.	Warranty/call back/service	15%

Note that cranes or other lifting equipment are not included on any lines in the Mechanical divisions. For example, if a crane is required to lift a heavy piece of pipe into place high above a gym floor, or to put a rooftop unit on the roof of a four-story building, etc., it must be added. Due to the potential for extreme variation—from nothing additional required to a major crane or helicopter—we feel that including a nominal amount for "lifting contingency" would be useless and detract from the accuracy of the estimate. When using equipment rental from RSMeans, remember to include the cost of the operator(s).

Estimating Labor-Hours

The labor-hours expressed in this publication are based on Average Installation time, using an efficiency level of approximately 60%-65% (see item 7), which has been found reasonable and acceptable by many contractors. The book uses this national efficiency average to establish a consistent benchmark. For bid situations, adjustments to this efficiency level should be the responsibility of the contractor bidding the project. The unit labor-hour is divided in the following manner. A typical day for a journeyman might be:

1.	Study Plans	3%	14.4 min.
2.	Material Procurement	3%	14.4 min.
3.	Receiving and Storing	3%	14.4 min
4.	Mobilization	5%	24.0 min.
5.	Site Movement	5%	24.0 min.
6.	Layout and Marking	8%	38.4 min.
7.	Actual Installation	64%	307.2 min.
8.	Cleanup	3%	14.4 min.
9.	Breaks, Nonproductive	6%	28.8 min.
		100%	480.0 min.

If any of the percentages expressed in this breakdown do not apply to the particular work or project situation, then that percentage or a portion of it may be deducted from or added to labor-hours.

Final Checklist

Estimating can be a straightforward process provided you remember the basics. Here's a checklist of some of the steps you should remember to complete before finalizing your estimate. Did you remember to . . .

- factor in the City Cost Index for your locale?
- take into consideration which items have been marked up and by how much?
- mark up the entire estimate sufficiently for your purposes?
- read the background information on techniques and technical matters that could impact your project time span and cost?
- include all components of your project in the final estimate?
- double check your figures for accuracy?
- call RSMeans if you have any questions about your estimate or the data you've found in our publications? Remember, RSMeans stands behind its publications. If you have any questions about your estimate . . . about the costs you've used from our books . . . or even about the technical aspects of the job that may affect your estimate, feel free to call the RSMeans editors at 1-800-334-3509.

Unit Price Section

Table of Contents

Table of Contents (cont.)

Table of Contents (cont.)

How to Use the Unit Price Pages

The following is a detailed explanation of a sample entry in the Unit Price Section. Next to each bold number below is the described item with the appropriate component of the sample entry following in parentheses. Some prices are listed as bare costs; others as costs that include overhead and profit of the installing contractor. In most cases, if the work is to be subcontracted, the general contractor will need to add an additional markup (RSMeans suggests using 10%) to the figures in the column "Total Incl. O&P."

1 Division Number/Title (03 30/Cast-In-Place Concrete)

Use the Unit Price Section Table of Contents to locate specific items. The sections are classified according to the CSI MasterFormat 2004 system.

2 Line Numbers (03 30 53.40 3920)

Each unit price line item has been assigned a unique 12-digit code based on the CSI MasterFormat classification.

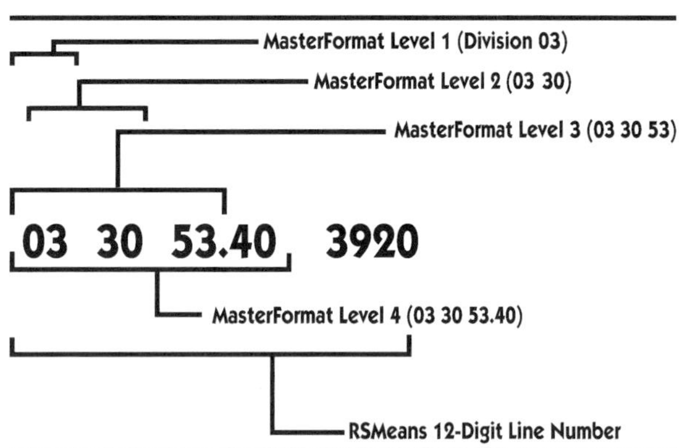

3 Description (Concrete In Place, etc.)

Each line item is described in detail. Sub-items and additional sizes are indented beneath the appropriate line items. The first line or two after the main item (in boldface) may contain descriptive information that pertains to all line items beneath this boldface listing.

4 Reference Number Information

R033053 -50 You'll see reference numbers shown in shaded boxes at the beginning of some sections. These refer to related items in the Reference Section, visually identified by a vertical gray bar on the page edges.

The relation may be an estimating procedure that should be read before estimating, or technical information.

The "R" designates the Reference Section. The numbers refer to the MasterFormat 2010 classification system.

It is strongly recommended that you review all reference numbers that appear within the section in which you are working.

Note: Not all reference numbers appear in all RSMeans publications.

03 30 Cast-In-Place Concrete

03 30 53 – Miscellaneous Cast-In-Place Concrete

03 30 53.40 Concrete In Place			Crew	Daily Output	Labor-Hours	Unit	2011 Bare Costs Material	Labor	Equipment	Total	Total Incl O&P
3825	1 C.Y. to 5 C.Y.			43	2.605		185	83	.55	268.55	345
3850	Over 5 C.Y.	R033105-80		75	1.493		171	47.50	.31	218.81	268
3900	Footings, strip (3000 psi), 18" x 9", unreinforced		C-14L	40	2.400		119	75	.58	194.58	257
3920	18" x 9", reinforced	R033105-85	C-14C	35	3.200		141	102	.67	243.67	330
3925	20" x 10", unreinforced		C-14L	45	2.133		116	66.50	.51	183.01	240
3930	20" x 10", reinforced		C-14C	40	2.800		134	89.50	.59	224.09	298
3935	24" x 12", unreinforced		C-14L	55	1.745		114	54.50	.42	168.92	[illegible]
3940	24" x 12", reinforced		C-14C	48	2.333		132	74.50	.49	206.99	[illegible]
3945	36" x 12", unreinforced		C-14L	[illegible]	1.371		111	43	.33	154.33	194
3950	36" x 12", reinforced		C-14C	[illegible]	1.867		127	59.50	.39	186.89	240
4000	Foundation mat (3000 psi), under 10 C.Y.			38.67	2.896		192	92.50	.61	285.11	365
4050	Over 20 C.Y.			56.40	1.986		169	63.50	.42	232.92	292

5 Crew (C-14C)

The "Crew" column designates the typical trade or crew used to install the item. If an installation can be accomplished by one trade and requires no power equipment, that trade and the number of workers are listed (for example, "2 Carpenters"). If an installation requires a composite crew, a crew code designation is listed (for example, "C-14C"). You'll find full details on all composite crews in the Crew Listings.

- For a complete list of all trades utilized in this book and their abbreviations, see the inside back cover.

Crews

Crew No.	Bare Costs		Incl. Subs O&P		Cost Per Labor-Hour	
Crew C-14C	Hr.	Daily	Hr.	Daily	Bare Costs	Incl. O&P
1 Carpenter Foreman (out)	$35.55	$284.40	$59.50	$476.00	$31.94	$53.62
6 Carpenters	33.55	1610.40	56.15	2695.20		
2 Rodmen (reinf.)	36.30	580.80	63.10	1009.60		
4 Laborers	26.45	846.40	44.30	1417.60		
1 Cement Finisher	31.85	254.80	50.85	406.80		
1 Gas Engine Vibrator		23.20		25.52	0.21	0.23
112 L.H., Daily Totals		$3600.00		$6030.72	$32.14	$53.85

6 Productivity: Daily Output (35.0)/ Labor-Hours (3.20)

The "Daily Output" represents the typical number of units the designated crew will install in a normal 8-hour day. To find out the number of days the given crew would require to complete the installation, divide your quantity by the daily output. For example:

Quantity	÷	Daily Output	=	Duration
100 C.Y.	÷	35.0/ Crew Day	=	2.86 Crew Days

The "Labor-Hours" figure represents the number of labor-hours required to install one unit of work. To find out the number of labor-hours required for your particular task, multiply the quantity of the item times the number of labor-hours shown. For example:

Quantity	x	Productivity Rate	=	Duration
100 C.Y.	x	3.20 Labor-Hours/ C.Y.	=	320 Labor-Hours

7 Unit (C.Y.)

The abbreviated designation indicates the unit of measure upon which the price, production, and crew are based (C.Y. = Cubic Yard). For a complete listing of abbreviations, refer to the Abbreviations Listing in the Reference Section of this book.

8 Bare Costs:

Mat. (Bare Material Cost) (141)

The unit material cost is the "bare" material cost with no overhead or profit included. Costs shown reflect national average material prices for January of the current year and include delivery to the job site. No sales taxes are included.

Labor (102.00)

The unit labor cost is derived by multiplying bare labor-hour costs for Crew C-14C by labor-hour units. The bare labor-hour cost is found in the Crew Section under C-14C. (If a trade is listed, the hourly labor cost—the wage rate—is found on the inside back cover.)

Labor-Hour Cost Crew C-14C	x	Labor-Hour Units	=	Labor
$31.94	x	3.20	=	$102.00

Equip. (Equipment) (.67)

Costs for equipment included in each crew are listed in the Crew Listings in the Reference Section. Costs for small tools that are considered as essential everyday tools are covered in the "Overhead" column on the "Installing Contractor's Overhead and Profit" table that lists labor trades, base rates and markups. The unit equipment cost displayed here is derived by multiplying the bare equipment hourly cost in the Crew Listing by the unit Labor-Hours value.

Equipment Cost Crew C-14C	x	Labor-Hour Units	=	Equip.
$.21	x	3.20	=	$.67

Total (243.67)

The total of the bare costs is the arithmetic total of the three previous columns: mat., labor, and equip.

Material	+	Labor	+	Equip.	=	Total
$141	+	$102.00	+	$.67	=	$243.67

9 Total Costs Including O&P

This figure is the sum of the bare material cost plus 10% for profit; the bare labor cost plus total overhead and profit (per the inside back cover or, if a crew is listed, from the crew listings); and the bare equipment cost plus 10% for profit.

Material is Bare Material Cost + 10% = 141 + 14.10	=	$155.10
Labor for Crew C-14C = Labor-Hour Cost (53.52) x Labor-Hour Units (3.20)	=	$171.26
Equip. is Bare Equip. Cost + 10% = .67 + .07	=	$ .74
Total (Rounded)	=	$ 330

Estimating Tips

01 20 00 Price and Payment Procedures

- Allowances that should be added to estimates to cover contingencies and job conditions that are not included in the national average material and labor costs are shown in section 01 21.
- When estimating historic preservation projects (depending on the condition of the existing structure and the owner's requirements), a 15%–20% contingency or allowance is recommended, regardless of the stage of the drawings.

01 30 00 Administrative Requirements

- Before determining a final cost estimate, it is a good practice to review all the items listed in Subdivisions 01 31 and 01 32 to make final adjustments for items that may need customizing to specific job conditions.
- Requirements for initial and periodic submittals can represent a significant cost to the General Requirements of a job. Thoroughly check the submittal specifications when estimating a project to determine any costs that should be included.

01 40 00 Quality Requirements

- All projects will require some degree of Quality Control. This cost is not included in the unit cost of construction listed in each division. Depending upon the terms of the contract, the various costs of inspection and testing can be the responsibility of either the owner or the contractor. Be sure to include the required costs in your estimate.

01 50 00 Temporary Facilities and Controls

- Barricades, access roads, safety nets, scaffolding, security, and many more requirements for the execution of a safe project are elements of direct cost. These costs can easily be overlooked when preparing an estimate. When looking through the major classifications of this subdivision, determine which items apply to each division in your estimate.
- Construction Equipment Rental Costs can be found in the Reference Section in section 01 54 33. Operators' wages are not included in equipment rental costs.
- Equipment mobilization and demobilization costs are not included in equipment rental costs and must be considered separately in section 01 54 36.50.
- The cost of small tools provided by the installing contractor for his workers is covered in the "Overhead" column on the "Installing Contractor's Overhead and Profit" table that lists labor trades, base rates and markups and, therefore, is included in the "Total Incl. O&P" cost of any Unit Price line item. For those users who are constrained by contract terms to use only bare costs, there are two line items in section 01 54 39.70 for small tools as a percentage of bare labor cost. If some of those users are further constrained by contract terms to refrain from using line items from Division 1, you are advised to cover the cost of small tools within your coefficient or multiplier.

01 70 00 Execution and Closeout Requirements

- When preparing an estimate, thoroughly read the specifications to determine the requirements for Contract Closeout. Final cleaning, record documentation, operation and maintenance data, warranties and bonds, and spare parts and maintenance materials can all be elements of cost for the completion of a contract. Do not overlook these in your estimate.

Reference Numbers

Reference numbers are shown in shaded boxes at the beginning of some major classifications. These numbers refer to related items in the Reference Section. The reference information may be an estimating procedure, an alternate pricing method, or technical information.

Note: Not all subdivisions listed here necessarily appear in this publication.

01 11 Summary of Work

01 11 31 – Professional Consultants

01 11 31.10 Architectural Fees

	01 11 31.10 Architectural Fees		Crew	Daily Output	Labor-Hours	Unit	Material	2011 Bare Costs Labor	Equipment	Total	Total Incl O&P
0010	**ARCHITECTURAL FEES**	R011110-10									
0020	For new construction										
0060	Minimum					Project					4.90%
0090	Maximum										16%
0100	For alteration work, to $500,000, add to new construction fee										50%
0150	Over $500,000, add to new construction fee										25%
2000	For "Greening" of building	G				↓					3%

01 11 31.20 Construction Management Fees

			Crew	Daily Output	Labor-Hours	Unit	Material	Labor	Equipment	Total	Total Incl O&P
0010	**CONSTRUCTION MANAGEMENT FEES**										
0020	$1,000,000 job, minimum					Project					4.50%
0050	Maximum										7.50%
0300	$50,000,000 job, minimum										2.50%
0350	Maximum					↓					4%

01 11 31.30 Engineering Fees

			Crew	Daily Output	Labor-Hours	Unit	Material	Labor	Equipment	Total	Total Incl O&P
0010	**ENGINEERING FEES**	R011110-30									
0020	Educational planning consultant, minimum					Project					.50%
0100	Maximum					"					2.50%
0200	Electrical, minimum					Contrct					4.10%
0300	Maximum										10.10%
0400	Elevator & conveying systems, minimum										2.50%
0500	Maximum										5%
0600	Food service & kitchen equipment, minimum										8%
0700	Maximum										12%
0800	Landscaping & site development, minimum										2.50%
0900	Maximum										6%
1000	Mechanical (plumbing & HVAC), minimum										4.10%
1100	Maximum					↓					10.10%
1200	Structural, minimum					Project					1%
1300	Maximum					"					2.50%

01 11 31.50 Models

			Crew	Daily Output	Labor-Hours	Unit	Material	Labor	Equipment	Total	Total Incl O&P
0010	**MODELS**										
0020	Cardboard & paper, 1 building, minimum					Ea.	710			710	780
0050	Maximum						1,625			1,625	1,775
0100	2 buildings, minimum						945			945	1,050
0150	Maximum					↓	2,150			2,150	2,350
0200	Plexiglass and metal, basic layout					SF Flr.	.06			.06	.07
0210	Including equipment and personnel					"	.31			.31	.34
0300	Site plan layout, minimum					Ea.	1,350			1,350	1,500
0350	Maximum					"	2,275			2,275	2,500

01 11 31.75 Renderings

			Crew	Daily Output	Labor-Hours	Unit	Material	Labor	Equipment	Total	Total Incl O&P
0010	**RENDERINGS** Color, matted, 20" x 30", eye level,										
0020	1 building, minimum					Ea.	1,975			1,975	2,175
0050	Average						2,800			2,800	3,100
0100	Maximum						4,500			4,500	4,950
1000	5 buildings, minimum						3,950			3,950	4,325
1100	Maximum						7,875			7,875	8,675
2000	Aerial perspective, color, 1 building, minimum						2,800			2,800	3,100
2100	Maximum						7,875			7,875	8,675
3000	5 buildings, minimum						5,675			5,675	6,225
3100	Maximum					↓	11,200			11,200	12,400

01 21 Allowances

01 21 16 – Contingency Allowances

01 21 16.50 Contingencies		Crew	Daily Output	Labor-Hours	Unit	2011 Bare Costs Material	Labor	Equipment	Total	Total Incl O&P
0010	**CONTINGENCIES**, Add to estimate									
0020	Conceptual stage				Project					20%
0050	Schematic stage									15%
0100	Preliminary working drawing stage (Design Dev.)									10%
0150	Final working drawing stage				↓					3%

01 21 55 – Job Conditions Allowance

01 21 55.50 Job Conditions		Crew	Daily Output	Labor-Hours	Unit	Material	Labor	Equipment	Total	Total Incl O&P
0010	**JOB CONDITIONS** Modifications to applicable									
0020	cost summaries									
0100	Economic conditions, favorable, deduct				Project				2%	2%
0200	Unfavorable, add								5%	5%
0300	Hoisting conditions, favorable, deduct								2%	2%
0400	Unfavorable, add								5%	5%
0500	General Contractor management, experienced, deduct								2%	2%
0600	Inexperienced, add								10%	10%
0700	Labor availability, surplus, deduct								1%	1%
0800	Shortage, add								10%	10%
0900	Material storage area, available, deduct								1%	1%
1000	Not available, add								2%	2%
1100	Subcontractor availability, surplus, deduct								5%	5%
1200	Shortage, add								12%	12%
1300	Work space, available, deduct								2%	2%
1400	Not available, add				↓				5%	5%

01 21 57 – Overtime Allowance

01 21 57.50 Overtime		Crew	Daily Output	Labor-Hours	Unit	Material	Labor	Equipment	Total	Total Incl O&P
0010	**OVERTIME** for early completion of projects or where R012909-90									
0020	labor shortages exist, add to usual labor, up to				Costs		100%			

01 21 61 – Cost Indexes

01 21 61.10 Construction Cost Index		Crew	Daily Output	Labor-Hours	Unit	Material	Labor	Equipment	Total	Total Incl O&P
0010	**CONSTRUCTION COST INDEX** (Reference) over 930 zip code locations in									
0020	the U.S. and Canada, total bldg. cost, min. (Guymon, OK)				%					67.30%
0050	Average									100%
0100	Maximum (New York, NY)				↓					133%

01 21 61.20 Historical Cost Indexes		Crew	Daily Output	Labor-Hours	Unit	Material	Labor	Equipment	Total	Total Incl O&P
0010	**HISTORICAL COST INDEXES** (See Reference Section)									

01 21 61.30 Labor Index		Crew	Daily Output	Labor-Hours	Unit	Material	Labor	Equipment	Total	Total Incl O&P
0010	**LABOR INDEX** (Reference) For over 930 zip code locations in									
0020	the U.S. and Canada, minimum (Guymon, OK)				%		31.10%			
0050	Average						100%			
0100	Maximum (New York, NY)				↓		166.30%			

01 21 61.50 Material Index		Crew	Daily Output	Labor-Hours	Unit	Material	Labor	Equipment	Total	Total Incl O&P
0010	**MATERIAL INDEX** (Reference) For over 930 zip code locations in									
0020	the U.S. and Canada, minimum (Elizabethtown, KY)				%	91%				
0040	Average					100%				
0060	Maximum (Ketchikan, AK)				↓	136.90%				

01 21 63 – Taxes

01 21 63.10 Taxes		Crew	Daily Output	Labor-Hours	Unit	Material	Labor	Equipment	Total	Total Incl O&P
0010	**TAXES** R012909-80									
0020	Sales tax, State, average				%	5.04%				
0050	Maximum R012909-85				↓	8.25%				

01 21 Allowances

01 21 63 – Taxes

01 21 63.10 Taxes		Crew	Daily Output	Labor-Hours	Unit	Material	2011 Bare Costs Labor	Equipment	Total	Total Incl O&P
0200	Social Security, on first $106,800 of wages				%		7.65%			
0300	Unemployment, combined Federal and State, minimum						.80%			
0350	Average						6.20%			
0400	Maximum						13.16%			

01 31 Project Management and Coordination

01 31 13 – Project Coordination

01 31 13.20 Field Personnel

		Crew	Daily Output	Labor-Hours	Unit	Material	Labor	Equipment	Total	Total Incl O&P
0010	**FIELD PERSONNEL**									
0020	Clerk, average				Week		410		410	630
0100	Field engineer, minimum						975		975	1,500
0120	Average						1,265		1,265	1,950
0140	Maximum						1,450		1,450	2,225
0160	General purpose laborer, average						1,350		1,350	2,075
0180	Project manager, minimum						1,800		1,800	2,750
0200	Average						2,075		2,075	3,175
0220	Maximum						2,375		2,375	3,650
0240	Superintendent, minimum						1,750		1,750	2,675
0260	Average						1,925		1,925	2,950
0280	Maximum						2,200		2,200	3,375
0290	Timekeeper, average						1,130		1,130	1,725

01 31 13.30 Insurance

		Crew	Daily Output	Labor-Hours	Unit	Material	Labor	Equipment	Total	Total Incl O&P
0010	**INSURANCE** R013113-40									
0020	Builders risk, standard, minimum				Job					.24%
0050	Maximum R013113-50									.64%
0200	All-risk type, minimum									.25%
0250	Maximum R013113-60									.62%
0400	Contractor's equipment floater, minimum				Value					.50%
0450	Maximum				"					1.50%
0600	Public liability, average				Job					2.02%
0800	Workers' compensation & employer's liability, average									
0850	by trade, carpentry, general				Payroll		16.07%			
0900	Clerical						.51%			
0950	Concrete						13.33%			
1000	Electrical						5.81%			
1050	Excavation						9.10%			
1100	Glazing						12.89%			
1150	Insulation						12.30%			
1200	Lathing						8.59%			
1250	Masonry						12.66%			
1300	Painting & decorating						11.07%			
1350	Pile driving						18%			
1400	Plastering						11.43%			
1450	Plumbing						7.04%			
1500	Roofing						29.20%			
1550	Sheet metal work (HVAC)						8.94%			
1600	Steel erection, structural						36.30%			
1650	Tile work, interior ceramic						8.45%			
1700	Waterproofing, brush or hand caulking						6.43%			
1800	Wrecking						30.85%			
2000	Range of 35 trades in 50 states, excl. wrecking, min.						2.36%			

01 31 Project Management and Coordination

01 31 13 – Project Coordination

01 31 13.30 Insurance

		Crew	Daily Output	Labor-Hours	Unit	Material	2011 Bare Costs Labor	Equipment	Total	Total Incl O&P
2100	Average				Payroll		14.10%			
2200	Maximum						155.90%			

01 31 13.40 Main Office Expense

		Crew	Daily Output	Labor-Hours	Unit	Material	2011 Bare Costs Labor	Equipment	Total	Total Incl O&P
0010	**MAIN OFFICE EXPENSE** Average for General Contractors R013113-50									
0020	As a percentage of their annual volume									
0125	Annual volume under 1 million dollars				% Vol.				17.50%	
0145	Up to 2.5 million dollars								8%	
0150	Up to 4.0 million dollars								6.80%	
0200	Up to 7.0 million dollars								5.60%	
0250	Up to 10 million dollars								5.10%	
0300	Over 10 million dollars								3.90%	

01 31 13.50 General Contractor's Mark-Up

		Crew	Daily Output	Labor-Hours	Unit	Material	2011 Bare Costs Labor	Equipment	Total	Total Incl O&P
0010	**GENERAL CONTRACTOR'S MARK-UP** on Change Orders									
0200	Extra work, by subcontractors, add				%					10%
0250	By General Contractor, add									15%
0400	Omitted work, by subcontractors, deduct all but									5%
0450	By General Contractor, deduct all but									7.50%
0600	Overtime work, by subcontractors, add									15%
0650	By General Contractor, add									10%

01 31 13.60 Installing Contractor's Main Office Overhead

		Crew	Daily Output	Labor-Hours	Unit	Material	2011 Bare Costs Labor	Equipment	Total	Total Incl O&P
0010	**INSTALLING CONTRACTOR'S MAIN OFFICE OVERHEAD** R013113-50									
0020	As percent of direct costs, minimum				%				5%	
0050	Average								25%	
0100	Maximum								30%	

01 31 13.80 Overhead and Profit

		Crew	Daily Output	Labor-Hours	Unit	Material	2011 Bare Costs Labor	Equipment	Total	Total Incl O&P
0010	**OVERHEAD & PROFIT** Allowance to add to items in this									
0020	book that do not include Subs O&P, average				%				25%	
0100	Allowance to add to items in this book that									
0110	do include Subs O&P, minimum				%					5%
0150	Average									10%
0200	Maximum									15%
0300	Typical, by size of project, under $100,000								30%	
0350	$500,000 project								25%	
0400	$2,000,000 project								20%	
0450	Over $10,000,000 project								15%	

01 31 13.90 Performance Bond

		Crew	Daily Output	Labor-Hours	Unit	Material	2011 Bare Costs Labor	Equipment	Total	Total Incl O&P
0010	**PERFORMANCE BOND** R013113-80									
0020	For buildings, minimum				Job					.60%
0100	Maximum				″					2.50%

01 32 Construction Progress Documentation

01 32 13 – Scheduling of work

01 32 13.50 Scheduling		Crew	Daily Output	Labor-Hours	Unit	Material	2011 Bare Costs Labor	Equipment	Total	Total Incl O&P
0010	**SCHEDULING**									
0020	Critical path, as % of architectural fee, minimum				%					.50%
0100	Maximum				"					1%
0300	Computer-update, micro, no plots, minimum				Ea.				455	500
0400	Including plots, maximum				"				1,450	1,600
0600	Rule of thumb, CPM scheduling, small job ($10 Million)				Job					.05%
0650	Large job ($50 Million +)									.03%
0700	Including cost control, small job									.08%
0750	Large job									.04%

01 32 33 – Photographic Documentation

01 32 33.50 Photographs		Crew	Daily Output	Labor-Hours	Unit	Material	Labor	Equipment	Total	Total Incl O&P
0010	**PHOTOGRAPHS**									
0020	8" x 10", 4 shots, 2 prints ea., std. mounting				Set	475			475	520
0100	Hinged linen mounts					530			530	580
0200	8" x 10", 4 shots, 2 prints each, in color					415			415	460
0300	For I.D. slugs, add to all above					5.30			5.30	5.85
0500	Aerial photos, initial fly-over, 6 shots, 1 print ea., 8" x 10"					845			845	925
0550	11" x 14" prints					1,025			1,025	1,125
0600	16" x 20" prints					1,200			1,200	1,325
0700	For full color prints, add					40%				40%
0750	Add for traffic control area					305			305	335
0900	For over 30 miles from airport, add per				Mile	5.45			5.45	6
1000	Vertical photography, 4 to 6 shots with									
1010	different scales, 1 print each				Set	1,125			1,125	1,225
1500	Time lapse equipment, camera and projector, buy					845			845	930
1550	Rent per month					260			260	286
1700	Cameraman and film, including processing, B.&W.				Day	1,375			1,375	1,525
1720	Color				"	1,375			1,375	1,525

01 41 Regulatory Requirements

01 41 26 – Permit Requirements

01 41 26.50 Permits		Crew	Daily Output	Labor-Hours	Unit	Material	Labor	Equipment	Total	Total Incl O&P
0010	**PERMITS**									
0020	Rule of thumb, most cities, minimum				Job					.50%
0100	Maximum				"					2%

01 45 Quality Control

01 45 23 – Testing and Inspecting Services

01 45 23.50 Testing		Crew	Daily Output	Labor-Hours	Unit	Material	Labor	Equipment	Total	Total Incl O&P
0010	**TESTING** and Inspecting Services									
0015	For concrete building costing $1,000,000, minimum				Project				4,725	5,200
0020	Maximum								38,000	41,800
0050	Steel building, minimum								4,725	5,200
0070	Maximum								14,800	16,300
0100	For building costing, $10,000,000, minimum								30,100	33,100
0150	Maximum								48,200	53,000
0200	Asphalt testing, compressive strength Marshall stability, set of 3				Ea.				145	165
0220	Density, set of 3								86	95
0250	Extraction, individual tests on sample								136	150

01 45 23 – Testing and Inspecting Services

01 45 23.50 Testing		Crew	Daily Output	Labor-Hours	Unit	2011 Bare Costs Material	Labor	Equipment	Total	Total Incl O&P
0300	Penetration				Ea.				41	45
0350	Mix design, 5 specimens								182	200
0360	Additional specimen								36	40
0400	Specific gravity								41	45
0420	Swell test								64	70
0450	Water effect and cohesion, set of 6								182	200
0470	Water effect and plastic flow								64	70
0600	Concrete testing, aggregates, abrasion, ASTM C 131								136	150
0650	Absorption, ASTM C 127								42	46
0800	Petrographic analysis, ASTM C 295								775	850
0900	Specific gravity, ASTM C 127								50	55
1000	Sieve analysis, washed, ASTM C 136								59	65
1050	Unwashed								59	65
1200	Sulfate soundness								114	125
1300	Weight per cubic foot								36	40
1500	Cement, physical tests, ASTM C 150								320	350
1600	Chemical tests, ASTM C 150								245	270
1800	Compressive test, cylinder, delivered to lab, ASTM C 39								12	13
1900	Picked up by lab, minimum								14	15
1950	Average								18	20
2000	Maximum								27	30
2200	Compressive strength, cores (not incl. drilling), ASTM C 42								36	40
2250	Core drilling, 4" diameter (plus technician)				Inch				23	25
2260	Technician for core drilling				Hr.				45	50
2300	Patching core holes				Ea.				22	24
2400	Drying shrinkage at 28 days								236	260
2500	Flexural test beams, ASTM C 78								59	65
2600	Mix design, one batch mix								259	285
2650	Added trial batches								120	132
2800	Modulus of elasticity, ASTM C 469								164	180
2900	Tensile test, cylinders, ASTM C 496								45	50
3000	Water-Cement ratio curve, 3 batches								141	155
3100	4 batches								186	205
3300	Masonry testing, absorption, per 5 brick, ASTM C 67								45	50
3350	Chemical resistance, per 2 brick								50	55
3400	Compressive strength, per 5 brick, ASTM C 67								68	75
3420	Efflorescence, per 5 brick, ASTM C 67								68	75
3440	Imperviousness, per 5 brick								87	96
3470	Modulus of rupture, per 5 brick								86	95
3500	Moisture, block only								32	35
3550	Mortar, compressive strength, set of 3								23	25
4100	Reinforcing steel, bend test								55	61
4200	Tensile test, up to #8 bar								36	40
4220	#9 to #11 bar								41	45
4240	#14 bar and larger								64	70
4400	Soil testing, Atterberg limits, liquid and plastic limits								59	65
4510	Hydrometer analysis								109	120
4530	Specific gravity, ASTM D 354								44	48
4600	Sieve analysis, washed, ASTM D 422								55	60
4700	Unwashed, ASTM D 422								59	65
4710	Consolidation test (ASTM D2435), minimum								250	275
4715	Maximum								430	475
4720	Density and classification of undisturbed sample								73	80

01 45 Quality Control

01 45 23 – Testing and Inspecting Services

01 45 23.50 Testing		Crew	Daily Output	Labor-Hours	Unit	2011 Bare Costs Material	Labor	Equipment	Total	Total Incl O&P
4735	Soil density, nuclear method, ASTM D2922				Ea.				35	38.50
4740	Sand cone method ASTM D1556								27	30
4750	Moisture content, ASTM D 2216								9	10
4780	Permeability test, double ring infiltrometer								500	550
4800	Permeability, var. or constant head, undist., ASTM D 2434								227	250
4850	Recompacted								250	275
4900	Proctor compaction, 4" standard mold, ASTM D 698								123	135
4950	6" modified mold								68	75
5100	Shear tests, triaxial, minimum								410	450
5150	Maximum								545	600
5300	Direct shear, minimum, ASTM D 3080								320	350
5350	Maximum								410	450
5550	Technician for inspection, per day, earthwork								320	350
5650	Bolting								400	440
5750	Roofing								335	370
5790	Welding								370	405
5820	Non-destructive metal testing, dye penetrant				Day				310	340
5840	Magnetic particle								310	340
5860	Radiography								450	495
5880	Ultrasonic								310	340
6000	Welding certification, minimum				Ea.				91	100
6100	Maximum				"				250	275
7000	Underground storage tank									
7500	Volumetric tightness test ,<=12,000 gal.				Ea.				435	480
7510	<=30,000 gal.				"				615	675
7600	Vadose zone (soil gas) sampling, 10-40 samples, min.				Day				1,375	1,500
7610	Maximum				"				2,275	2,500
7700	Ground water monitoring incl. drilling 3 wells, min.				Total				4,550	5,000
7710	Maximum				"				6,375	7,000
8000	X-ray concrete slabs				Ea.				182	200
9000	Thermographic testing, for bldg envelope heat loss, average 2,000 S.F. G				"					500

01 51 Temporary Utilities

01 51 13 – Temporary Electricity

01 51 13.80 Temporary Utilities		Crew	Daily Output	Labor-Hours	Unit	Material	Labor	Equipment	Total	Total Incl O&P
0010	**TEMPORARY UTILITIES**									
0101	Heat, incl. fuel and operation, per week, 12 hrs. per day	1 Clab	100	.080	CSF Flr	27.50	2.12		29.62	34
0201	24 hrs. per day	"	60	.133		52.50	3.53		56.03	64
0350	Lighting, incl. service lamps, wiring & outlets, minimum	1 Elec	34	.235		2.73	9.45		12.18	18.35
0360	Maximum	"	17	.471		5.80	18.95		24.75	37
0400	Power for temp lighting only, per month, min/month 6.6 KWH								.92	1.01
0430	Average/month 11.8 KWH								1.65	1.82
0450	Maximum/month 23.6 KWH								3.30	3.63
0600	Power for job duration incl. elevator, etc., minimum								47	51.50
0650	Maximum								110	121
1000	Toilet, portable, see Equip. Rental 01 54 33 in Reference Section									

01 52 Construction Facilities

01 52 13 – Field Offices and Sheds

01 52 13.20 Office and Storage Space

	01 52 13.20 Office and Storage Space	Crew	Daily Output	Labor-Hours	Unit	2011 Bare Costs Material	Labor	Equipment	Total	Total Incl O&P
0010	**OFFICE AND STORAGE SPACE**									
0020	Trailer, furnished, no hookups, 20′ x 8′, buy	2 Skwk	1	16	Ea.	8,550	545		9,095	10,300
0250	Rent per month					146			146	160
0300	32′ x 8′, buy	2 Skwk	.70	22.857		12,300	780		13,080	14,800
0350	Rent per month					193			193	213
0400	50′ x 10′, buy	2 Skwk	.60	26.667		23,300	910		24,210	27,200
0450	Rent per month					288			288	315
0500	50′ x 12′, buy	2 Skwk	.50	32		26,000	1,100		27,100	30,400
0550	Rent per month					360			360	395
0700	For air conditioning, rent per month, add					41.50			41.50	45.50
0800	For delivery, add per mile				Mile	4.60			4.60	5.05
1000	Portable buildings, prefab, on skids, economy, 8′ x 8′	2 Carp	265	.060	S.F.	85	2.03		87.03	97
1100	Deluxe, 8′ x 12′	″	150	.107	″	95	3.58		98.58	111
1200	Storage boxes, 20′ x 8′, buy	2 Skwk	1.80	8.889	Ea.	2,175	305		2,480	2,875
1250	Rent per month					71.50			71.50	78.50
1300	40′ x 8′, buy	2 Skwk	1.40	11.429		3,025	390		3,415	3,975
1350	Rent per month					94			94	103
5000	Air supported structures, see Section 13 31 13.13									

01 52 13.40 Field Office Expense

	01 52 13.40 Field Office Expense	Crew	Daily Output	Labor-Hours	Unit	Material	Labor	Equipment	Total	Total Incl O&P
0010	**FIELD OFFICE EXPENSE**									
0100	Office equipment rental average				Month	200			200	220
0120	Office supplies, average				″	86			86	94.50
0125	Office trailer rental, see Section 01 52 13.20									
0140	Telephone bill; avg. bill/month incl. long dist.				Month	81			81	89
0160	Lights & HVAC				″	152			152	167

01 54 Construction Aids

01 54 09 – Protection Equipment

01 54 09.50 Personnel Protective Equipment

	01 54 09.50 Personnel Protective Equipment	Crew	Daily Output	Labor-Hours	Unit	Material	Labor	Equipment	Total	Total Incl O&P
0010	**PERSONNEL PROTECTIVE EQUIPMENT**									
0015	Hazardous waste protection									
0020	Respirator mask only, full face, silicone				Ea.	242			242	266
0030	Half face, silicone					36			36	39.50
0040	Respirator cartridges, 2 req'd/mask, dust or asbestos					2.34			2.34	2.57
0050	Chemical vapor					2.66			2.66	2.92
0060	Combination vapor and dust					7			7	7.70
0100	Emergency escape breathing apparatus, 5 min					465			465	510
0110	10 min					500			500	550
0150	Self contained breathing apparatus with full face piece, 30 min					1,750			1,750	1,925
0160	60 min					2,925			2,925	3,225
0200	Encapsulating suits, limited use, level A					950			950	1,050
0210	Level B					244			244	269
0300	Over boots, latex				Pr.	6.35			6.35	7
0310	PVC					21.50			21.50	24
0320	Neoprene					41.50			41.50	46
0400	Gloves, nitrile/PVC					21			21	23.50
0410	Neoprene coated					24			24	26.50

01 54 09.60 Safety Nets

	01 54 09.60 Safety Nets	Crew	Daily Output	Labor-Hours	Unit	Material	Labor	Equipment	Total	Total Incl O&P
0010	**SAFETY NETS**									
0020	No supports, stock sizes, nylon, 4″ mesh				S.F.	1.10			1.10	1.21

01 54 Construction Aids

01 54 09 – Protection Equipment

01 54 09.60 Safety Nets		Crew	Daily Output	Labor-Hours	Unit	2011 Bare Costs Material	Labor	Equipment	Total	Total Incl O&P
0100	Polypropylene, 6″ mesh				S.F.	1.59			1.59	1.75
0200	Small mesh debris nets, 1/4″ & 3/4″ mesh, stock sizes					.74			.74	.81
0220	Combined 4″ mesh and 1/4″ mesh, stock sizes					2.05			2.05	2.26
0300	Monthly rental, 4″ mesh, stock sizes, 1st month					.50			.50	.55
0320	2nd month rental					.25			.25	.28
0340	Maximum rental/year					1.15			1.15	1.27

01 54 16 – Temporary Hoists

01 54 16.50 Weekly Forklift Crew

Line	Description	Crew	Daily Output	Labor-Hours	Unit	Material	Labor	Equipment	Total	Total Incl O&P
0010	**WEEKLY FORKLIFT CREW**									
0100	All-terrain forklift, 45′ lift, 35′ reach, 9000 lb. capacity	A-3P	.20	40	Week		1,350	2,225	3,575	4,650

01 54 19 – Temporary Cranes

01 54 19.50 Daily Crane Crews

Line	Description	Crew	Daily Output	Labor-Hours	Unit	Material	Labor	Equipment	Total	Total Incl O&P
0010	**DAILY CRANE CREWS** for small jobs, portal to portal									
0100	12-ton truck-mounted hydraulic crane	A-3H	1	8	Day		286	870	1,156	1,425
0200	25-ton	A-3I	1	8			286	1,025	1,311	1,600
0300	40-ton	A-3J	1	8			286	1,250	1,536	1,850
0400	55-ton	A-3K	1	16			535	1,750	2,285	2,800
0500	80-ton	A-3L	1	16			535	2,425	2,960	3,550
0600	100-ton	A-3M	1	16			535	2,400	2,935	3,500

01 54 19.60 Monthly Tower Crane Crew

Line	Description	Crew	Daily Output	Labor-Hours	Unit	Material	Labor	Equipment	Total	Total Incl O&P
0010	**MONTHLY TOWER CRANE CREW**, excludes concrete footing									
0100	Static tower crane, 130′ high, 106′ jib, 6200 lb. capacity	A-3N	.05	176	Month		6,300	21,700	28,000	34,200

01 54 23 – Temporary Scaffolding and Platforms

01 54 23.60 Pump Staging

Line	Description	Crew	Daily Output	Labor-Hours	Unit	Material	Labor	Equipment	Total	Total Incl O&P
0010	**PUMP STAGING**, Aluminum R015423-20									
0200	24′ long pole section, buy				Ea.	395			395	435
0300	18′ long pole section, buy					305			305	335
0400	12′ long pole section, buy					206			206	227
0500	6′ long pole section, buy					108			108	119
0600	6′ long splice joint section, buy					80.50			80.50	88.50
0700	Pump jack, buy					139			139	153
0900	Foldable brace, buy					56.50			56.50	62.50
1000	Workbench/back safety rail support, buy					74.50			74.50	82
1100	Scaffolding planks/workbench, 14″ wide x 24′ long, buy					695			695	760
1200	Plank end safety rail, buy					370			370	405
1250	Safety net, 22′ long, buy					335			335	365
1300	System in place, 50′ working height, per use based on 50 uses	2 Carp	84.80	.189	C.S.F.	6.35	6.35		12.70	17.55
1400	100 uses		84.80	.189		3.16	6.35		9.51	14.10
1500	150 uses		84.80	.189		2.12	6.35		8.47	12.95

01 54 23.70 Scaffolding

Line	Description	Crew	Daily Output	Labor-Hours	Unit	Material	Labor	Equipment	Total	Total Incl O&P
0010	**SCAFFOLDING** R015423-10									
0015	Steel tube, regular, no plank, labor only to erect & dismantle									
0091	Building exterior, wall face, 1 to 5 stories, 6′-4″ x 5′ frames	3 Clab	8	3	C.S.F.		79.50		79.50	133
0201	6 to 12 stories	4 Clab	8	4			106		106	177
0310	13 to 20 stories	5 Carp	8	5			168		168	281
0461	Building interior, walls face area, up to 16′ high	3 Clab	12	2			53		53	88.50
0561	16′ to 40′ high		10	2.400			63.50		63.50	106
0801	Building interior floor area, up to 30′ high		150	.160	C.C.F.		4.23		4.23	7.10
0901	Over 30′ high	4 Clab	160	.200	″		5.30		5.30	8.85
0906	Complete system for face of walls, no plank, material only rent/mo				C.S.F.	36			36	39.50

01 54 Construction Aids

01 54 23 – Temporary Scaffolding and Platforms

01 54 23.70 Scaffolding		Crew	Daily Output	Labor-Hours	Unit	2011 Bare Costs Material	Labor	Equipment	Total	Total Incl O&P
0908	Interior spaces, no plank, material only rent/mo				C.C.F.	3.42			3.42	3.76
0910	Steel tubular, heavy duty shoring, buy									
0920	Frames 5′ high 2′ wide				Ea.	93			93	102
0925	5′ high 4′ wide					105			105	116
0930	6′ high 2′ wide					106			106	117
0935	6′ high 4′ wide					124			124	136
0940	Accessories									
0945	Cross braces				Ea.	18			18	19.80
0950	U-head, 8″ x 8″					21.50			21.50	24
0955	J-head, 4″ x 8″					15.80			15.80	17.40
0960	Base plate, 8″ x 8″					17.60			17.60	19.35
0965	Leveling jack					38			38	41.50
1000	Steel tubular, regular, buy									
1100	Frames 3′ high 5′ wide				Ea.	71			71	78
1150	5′ high 5′ wide					83			83	91.50
1200	6′-4″ high 5′ wide					104			104	114
1350	7′-6″ high 6′ wide					179			179	197
1500	Accessories cross braces					20.50			20.50	22.50
1550	Guardrail post					18.60			18.60	20.50
1600	Guardrail 7′ section					9.90			9.90	10.90
1650	Screw jacks & plates					27			27	29.50
1700	Sidearm brackets					43			43	47.50
1750	8″ casters					35.50			35.50	39
1800	Plank 2″ x 10″ x 16′-0″					51			51	56
1900	Stairway section					325			325	360
1910	Stairway starter bar					36.50			36.50	40
1920	Stairway inside handrail					66			66	72.50
1930	Stairway outside handrail					98			98	108
1940	Walk-thru frame guardrail					47			47	51.50
2000	Steel tubular, regular, rent/mo.									
2100	Frames 3′ high 5′ wide				Ea.	5			5	5.50
2150	5′ high 5′ wide					5			5	5.50
2200	6′-4″ high 5′ wide					5.15			5.15	5.65
2250	7′-6″ high 6′ wide					7			7	7.70
2500	Accessories, cross braces					1			1	1.10
2550	Guardrail post					1			1	1.10
2600	Guardrail 7′ section					1			1	1.10
2650	Screw jacks & plates					2			2	2.20
2700	Sidearm brackets					2			2	2.20
2750	8″ casters					8			8	8.80
2800	Outrigger for rolling tower					3			3	3.30
2850	Plank 2″ x 10″ x 16′-0″					6			6	6.60
2900	Stairway section					40			40	44
2940	Walk-thru frame guardrail					2.50			2.50	2.75
3000	Steel tubular, heavy duty shoring, rent/mo.									
3250	5′ high 2′ & 4′ wide				Ea.	5			5	5.50
3300	6′ high 2′ & 4′ wide					5			5	5.50
3500	Accessories, cross braces					1			1	1.10
3600	U - head, 8″ x 8″					1			1	1.10
3650	J - head, 4″ x 8″					1			1	1.10
3700	Base plate, 8″ x 8″					1			1	1.10
3750	Leveling jack					2			2	2.20
5700	Planks, 2″ x 10″ x 16′-0″, labor only to erect & remove to 50′ H	3 Carp	72	.333			11.20		11.20	18.70

01 54 23 – Temporary Scaffolding and Platforms

01 54 23.70 Scaffolding		Crew	Daily Output	Labor-Hours	Unit	2011 Bare Costs Material	Labor	Equipment	Total	Total Incl O&P
5800	Over 50′ high	4 Carp	80	.400	Ea.		13.40		13.40	22.50
6000	Heavy duty shoring for elevated slab forms to 8′-2″ high, floor area									
6101	Labor only to erect & dismantle	4 Clab	16	2	C.S.F.		53		53	88.50
6110	Materials only, rent.mo				″	29.50			29.50	32.50
6500	To 14′-8″ high									
6601	Labor only to erect & dismantle	4 Clab	10	3.200	C.S.F.		84.50		84.50	142
6610	Materials only, rent/mo				″	43			43	47.50

01 54 23.75 Scaffolding Specialties

Line	Description	Crew	Daily Output	Labor-Hours	Unit	Material	Labor	Equipment	Total	Total Incl O&P
0010	**SCAFFOLDING SPECIALTIES**									
1200	Sidewalk bridge, heavy duty steel posts & beams, including									
1210	parapet protection & waterproofing									
1221	8′ to 10′ wide, 2 posts	3 Clab	15	1.600	L.F.	35	42.50		77.50	109
1231	3 posts	″	10	2.400	″	53	63.50		116.50	165
1500	Sidewalk bridge using tubular steel									
1511	scaffold frames, including planking	3 Clab	45	.533	L.F.	5.60	14.10		19.70	29.50
1600	For 2 uses per month, deduct from all above					50%				
1700	For 1 use every 2 months, add to all above					100%				
1900	Catwalks, 20″ wide, no guardrails, 7′ span, buy				Ea.	145			145	160
2000	10′ span, buy					203			203	223
3720	Putlog, standard, 8′ span, with hangers, buy					75			75	82.50
3730	Rent per month					10			10	11
3750	12′ span, buy					113			113	124
3755	Rent per month					15			15	16.50
3760	Trussed type, 16′ span, buy					260			260	286
3770	Rent per month					20			20	22
3790	22′ span, buy					310			310	345
3795	Rent per month					30			30	33
3800	Rolling ladders with handrails, 30″ wide, buy, 2 step					237			237	261
4000	7 step					725			725	800
4050	10 step					1,000			1,000	1,100
4100	Rolling towers, buy, 5′ wide, 7′ long, 10′ high					1,350			1,350	1,500
4200	For 5′ high added sections, to buy, add				↓	207			207	228
4300	Complete incl. wheels, railings, outriggers,									
4350	21′ high, to buy				Ea.	2,250			2,250	2,475
4400	Rent/month = 5% of purchase cost				″	113			113	124

01 54 23.80 Staging Aids

Line	Description	Crew	Daily Output	Labor-Hours	Unit	Material	Labor	Equipment	Total	Total Incl O&P
0010	**STAGING AIDS** and fall protection equipment									
0100	Sidewall staging bracket, tubular, buy				Ea.	47.50			47.50	52
0110	Cost each per day, based on 250 days use				Day	.19			.19	.21
0200	Guard post, buy				Ea.	22.50			22.50	25
0210	Cost each per day, based on 250 days use				Day	.09			.09	.10
0300	End guard chains, buy per pair				Pair	31			31	34
0310	Cost per set per day, based on 250 days use				Day	.19			.19	.21
1000	Roof shingling bracket, steel, buy				Ea.	10.10			10.10	11.10
1010	Cost each per day, based on 250 days use				Day	.04			.04	.04
1100	Wood bracket, buy				Ea.	17.10			17.10	18.80
1110	Cost each per day, based on 250 days use				Day	.07			.07	.08
2000	Ladder jack, aluminum, buy per pair				Pair	108			108	119
2010	Cost per pair per day, based on 250 days use				Day	.43			.43	.48
2100	Steel siderail jack, buy per pair				Pair	122			122	134
2110	Cost per pair per day, based on 250 days use				Day	.49			.49	.54
3000	Laminated wood plank, 2″ x 10″ x 16′, buy				Ea.	51			51	56

01 54 Construction Aids

01 54 23 – Temporary Scaffolding and Platforms

01 54 23.80 Staging Aids		Crew	Daily Output	Labor-Hours	Unit	2011 Bare Costs Material	Labor	Equipment	Total	Total Incl O&P
3010	Cost each per day, based on 250 days use				Day	.20			.20	.22
3100	Aluminum scaffolding plank, 20" wide x 24' long, buy				Ea.	895			895	985
3110	Cost each per day, based on 250 days use				Day	3.58			3.58	3.94
4000	Nylon full body harness, lanyard and rope grab				Ea.	195			195	214
4010	Cost each per day, based on 250 days use				Day	.78			.78	.86
4100	Rope for safety line, 5/8" x 100' nylon, buy				Ea.	70			70	77
4110	Cost each per day, based on 250 days use				Day	.28			.28	.31
4200	Permanent U-Bolt roof anchor, buy				Ea.	35			35	38.50
4300	Temporary (one use) roof ridge anchor, buy				"	28.50			28.50	31
5000	Installation (setup and removal) of staging aids									
5010	Sidewall staging bracket	2 Carp	64	.250	Ea.		8.40		8.40	14.05
5020	Guard post with 2 wood rails	"	64	.250			8.40		8.40	14.05
5030	End guard chains, set	1 Carp	64	.125			4.19		4.19	7
5100	Roof shingling bracket		96	.083			2.80		2.80	4.68
5200	Ladder jack		64	.125			4.19		4.19	7
5300	Wood plank, 2" x 10" x 16'	2 Carp	80	.200			6.70		6.70	11.25
5310	Aluminum scaffold plank, 20" x 24'	"	40	.400			13.40		13.40	22.50
5410	Safety rope	1 Carp	40	.200			6.70		6.70	11.25
5420	Permanent U-Bolt roof anchor (install only)	2 Carp	40	.400			13.40		13.40	22.50
5430	Temporary roof ridge anchor (install only)	1 Carp	64	.125			4.19		4.19	7

01 54 26 – Temporary Swing Staging

01 54 26.50 Swing Staging

Line	Description	Crew	Daily Output	Labor-Hours	Unit	Material	Labor	Equipment	Total	Total Incl O&P
0010	**SWING STAGING**, 500 lb cap., 2' wide to 24' long, hand operated									
0020	steel cable type, with 60' cables, buy				Ea.	4,050			4,050	4,450
0030	Rent per month				"	405			405	445
0600	Lightweight (not for masons) 24' long for 150' height,									
0610	manual type, buy				Ea.	4,150			4,150	4,575
0620	Rent per month					415			415	455
0700	Powered, electric or air, to 150' high, buy					13,300			13,300	14,700
0710	Rent per month					935			935	1,025
0780	To 300' high, buy					13,500			13,500	14,800
0800	Rent per month					945			945	1,025
1000	Bosun's chair or work basket 3' x 3.5', to 300' high, electric, buy					8,625			8,625	9,500
1010	Rent per month					605			605	665
2200	Move swing staging (setup and remove)	E-4	2	16	Move		590	54.50	644.50	1,175

01 54 36 – Equipment Mobilization

01 54 36.50 Mobilization

Line	Description	Crew	Daily Output	Labor-Hours	Unit	Material	Labor	Equipment	Total	Total Incl O&P
0010	**MOBILIZATION** (Use line item again for demobilization) R015433-10									
0015	Up to 25 mi haul dist (50 mi RT for mob/demob crew)									
0020	Dozer, loader, backhoe, excav., grader, paver, roller, 70 to 150 H.P.	B-34N	4	2	Ea.		54.50	116	170.50	218
0100	Above 150 HP	B-34K	3	2.667			72.50	259	331.50	405
0300	Scraper, towed type (incl. tractor), 6 C.Y. capacity		3	2.667			72.50	259	331.50	405
0400	10 C.Y.		2.50	3.200			87	310	397	485
0600	Self-propelled scraper, 15 C.Y.		2.50	3.200			87	310	397	485
0700	24 C.Y.		2	4			109	390	499	610
0900	Shovel or dragline, 3/4 C.Y.		3.60	2.222			60.50	216	276.50	340
1000	1-1/2 C.Y.		3	2.667			72.50	259	331.50	405
1100	Small equipment, placed in rear of, or towed by pickup truck	A-3A	8	1			26.50	14.60	41.10	60
1150	Equip up to 70 HP, on flatbed trailer behind pickup truck	A-3D	4	2			53	57	110	150
2000	Crane, truck-mounted, up to 75 ton, (driver only, one-way)	1 Eqhv	7.20	1.111			40		40	65
2100	Crane, truck-mounted, over 75 ton	A-3E	2.50	6.400			202	46.50	248.50	380
2200	Crawler-mounted, up to 75 ton	A-3F	2	8			252	400	652	850

01 54 Construction Aids

01 54 36 – Equipment Mobilization

01 54 36.50 Mobilization		Crew	Daily Output	Labor-Hours	Unit	Material	2011 Bare Costs Labor	Equipment	Total	Total Incl O&P
2300	Over 75 ton	A-3G	1.50	10.667	Ea.		335	595	930	1,200
2500	For each additional 5 miles haul distance, add						10%	10%		
3000	For large pieces of equipment, allow for assembly/knockdown									
3001	For mob/demob of vibrofloatation equip, see Section 31 45 13.10									
3100	For mob/demob of micro-tunneling equip, see Section 33 05 23.19									
3200	For mob/demob of pile driving equip, see Section 31 62 19.10									
3300	For mob/demob of caisson drilling equip, see Section 31 63 26.13									

01 54 39 – Construction Equipment

01 54 39.70 Small Tools		Crew	Daily Output	Labor-Hours	Unit	Material	Labor	Equipment	Total	Total Incl O&P
0010	**SMALL TOOLS** R013113-50									
0020	As % of contractor's bare labor cost for project, minimum				Total		.50%			
0100	Maximum				"		2%			

01 55 Vehicular Access and Parking

01 55 23 – Temporary Roads

01 55 23.50 Roads and Sidewalks		Crew	Daily Output	Labor-Hours	Unit	Material	Labor	Equipment	Total	Total Incl O&P
0010	**ROADS AND SIDEWALKS** Temporary									
0050	Roads, gravel fill, no surfacing, 4" gravel depth	B-14	715	.067	S.Y.	4	1.88	.45	6.33	8
0100	8" gravel depth	"	615	.078	"	8	2.18	.52	10.70	13
1001	Ramp, 3/4" plywood on 2" x 6" joists, 16" O.C.	2 Clab	300	.053	S.F.	1.34	1.41		2.75	3.84
1101	On 2" x 10" joists, 16" O.C.	"	275	.058	"	1.92	1.54		3.46	4.69

01 56 Temporary Barriers and Enclosures

01 56 13 – Temporary Air Barriers

01 56 13.60 Tarpaulins		Crew	Daily Output	Labor-Hours	Unit	Material	Labor	Equipment	Total	Total Incl O&P
0010	**TARPAULINS**									
0020	Cotton duck, 10 oz. to 13.13 oz. per S.Y., minimum				S.F.	.48			.48	.53
0050	Maximum					.60			.60	.66
0100	Polyvinyl coated nylon, 14 oz. to 18 oz., minimum					.48			.48	.53
0150	Maximum					.68			.68	.75
0200	Reinforced polyethylene 3 mils thick, white					.15			.15	.17
0300	4 mils thick, white, clear or black					.20			.20	.22
0400	5.5 mils thick, clear					.30			.30	.33
0500	White, fire retardant					.35			.35	.39
0600	7.5 mils, oil resistant, fire retardant					.40			.40	.44
0700	8.5 mils, black					.53			.53	.58
0710	Woven polyethylene, 6 mils thick					.35			.35	.39
0730	Polyester reinforced w/integral fastening system 11 mils thick					1.07			1.07	1.18
0740	Mylar polyester, non-reinforced, 7 mils thick				↓	1.17			1.17	1.29

01 56 13.90 Winter Protection		Crew	Daily Output	Labor-Hours	Unit	Material	Labor	Equipment	Total	Total Incl O&P
0010	**WINTER PROTECTION**									
0100	Framing to close openings	2 Clab	750	.021	S.F.	.39	.56		.95	1.37
0200	Tarpaulins hung over scaffolding, 8 uses, not incl. scaffolding		1500	.011		.25	.28		.53	.75
0250	Tarpaulin polyester reinf. w/integral fastening system 11 mils thick		1600	.010		.80	.26		1.06	1.32
0300	Prefab fiberglass panels, steel frame, 8 uses	↓	1200	.013	↓	.85	.35		1.20	1.53

01 56 Temporary Barriers and Enclosures

01 56 16 – Temporary Dust Barriers

01 56 16.10 Dust Barriers, Temporary		Crew	Daily Output	Labor-Hours	Unit	Material	2011 Bare Costs Labor	Equipment	Total	Total Incl O&P
0010	**DUST BARRIERS, TEMPORARY**									
0020	Spring loaded telescoping pole & head, to 12′, erect and dismantle	1 Clab	240	.033	Ea.		.88		.88	1.48
0025	Cost per day (based upon 250 days)				Day	.22			.22	.24
0030	To 21′, erect and dismantle	1 Clab	240	.033	Ea.		.88		.88	1.48
0035	Cost per day (based upon 250 days)				Day	.39			.39	.42
0040	Accessories, caution tape reel, erect and dismantle	1 Clab	480	.017	Ea.		.44		.44	.74
0045	Cost per day (based upon 250 days)				Day	.05			.05	.05
0060	Foam rail and connector, erect and dismantle	1 Clab	240	.033	Ea.		.88		.88	1.48
0065	Cost per day (based upon 250 days)				Day	.10			.10	.11
0070	Caution tape	1 Clab	384	.021	C.L.F.	2.67	.55		3.22	3.85
0080	Zipper, standard duty		60	.133	Ea.	8	3.53		11.53	14.70
0090	Heavy duty		48	.167	″	9.50	4.41		13.91	17.85
0100	Polyethylene sheet, 4 mil		37	.216	Sq.	3.34	5.70		9.04	13.25
0110	6 mil		37	.216	″	5.25	5.70		10.95	15.35
1001	Dust partition, 6 mil polyethylene, 1″ x 3″ frame	2 Clab	2000	.008	S.F.	.45	.21		.66	.84
1081	2″ x 4″ frame	″	2000	.008	″	.30	.21		.51	.68

01 56 23 – Temporary Barricades

01 56 23.10 Barricades		Crew	Daily Output	Labor-Hours	Unit	Material	Labor	Equipment	Total	Total Incl O&P
0011	**BARRICADES**									
0030	5′ high, 3 rail @ 2″ x 8″, fixed	2 Clab	20	.800	L.F.	4.93	21		25.93	41
0200	Precast barrier walls, 10′ sections	B-6	240	.100	″	36	2.88	1.34	40.22	46.50
0300	Stock units, 6′ high, 8′ wide, plain, buy				Ea.	435			435	480
0350	With reflective tape, buy				″	525			525	580
0400	Break-a-way 3″ PVC pipe barricade									
0410	with 3 ea. 1′ x 4′ reflectorized panels, buy				Ea.	305			305	335
0500	Plywood with steel legs, 32″ wide					72			72	79
0600	Telescoping Christmas tree, 9′ high, 5 flags, buy					122			122	134
0800	Traffic cones, PVC, 18″ high					10.55			10.55	11.65
0850	28″ high					17.85			17.85	19.65
1001	Guardrail, wooden, 3′ high, 1″ x 6″ on 2″ x 4″ posts	2 Clab	200	.080	L.F.	1.11	2.12		3.23	4.76
1101	2″ x 6″ on 4″ x 4″ posts	″	165	.097		2.46	2.56		5.02	7
1200	Portable metal with base pads, buy					19.30			19.30	21
1251	Typical installation, assume 10 reuses	2 Clab	600	.027		2.50	.71		3.21	3.93

01 56 26 – Temporary Fencing

01 56 26.50 Temporary Fencing		Crew	Daily Output	Labor-Hours	Unit	Material	Labor	Equipment	Total	Total Incl O&P
0010	**TEMPORARY FENCING**									
0020	Chain link, 11 ga, 5′ high	2 Clab	400	.040	L.F.	4.50	1.06		5.56	6.70
0100	6′ high		300	.053		5.30	1.41		6.71	8.20
0200	Rented chain link, 6′ high, to 1000′ (up to 12 mo.)		400	.040		6.30	1.06		7.36	8.70
0250	Over 1000′ (up to 12 mo.)		300	.053		3.29	1.41		4.70	6
0351	Plywood, painted, 2″ x 4″ frame, 4′ high	3 Clab	135	.178		5.60	4.70		10.30	14.05
0401	4″ x 4″ frame, 8′ high	″	110	.218		10.45	5.75		16.20	21
0501	Wire mesh on 4″ x 4″ posts, 4′ high	2 Clab	100	.160		8.55	4.23		12.78	16.50
0551	8′ high	″	80	.200		13	5.30		18.30	23

01 56 29 – Temporary Protective Walkways

01 56 29.50 Protection		Crew	Daily Output	Labor-Hours	Unit	Material	Labor	Equipment	Total	Total Incl O&P
0010	**PROTECTION**									
0020	Stair tread, 2″ x 12″ planks, 1 use	1 Carp	75	.107	Tread	4.08	3.58		7.66	10.50
0100	Exterior plywood, 1/2″ thick, 1 use		65	.123		1.74	4.13		5.87	8.80
0200	3/4″ thick, 1 use		60	.133		2.45	4.47		6.92	10.20

01 56 Temporary Barriers and Enclosures

01 56 29 – Temporary Protective Walkways

01 56 29.50 Protection		Crew	Daily Output	Labor-Hours	Unit	Material (2011 Bare Costs)	Labor (2011 Bare Costs)	Equipment (2011 Bare Costs)	Total (2011 Bare Costs)	Total Incl O&P
2201	Sidewalks, 2" x 12" planks, 2 uses	1 Clab	350	.023	S.F.	.68	.60		1.28	1.76
2301	Exterior plywood, 2 uses, 1/2" thick		750	.011		.29	.28		.57	.79
2401	5/8" thick		650	.012		.35	.33		.68	.93
2501	3/4" thick		600	.013		.41	.35		.76	1.04

01 56 32 – Temporary Security

01 56 32.50 Watchman

		Crew	Daily Output	Labor-Hours	Unit	Material	Labor	Equipment	Total	Total Incl O&P
0010	**WATCHMAN**									
0020	Service, monthly basis, uniformed person, minimum				Hr.				25	27.50
0100	Maximum								45.50	50
0200	Person and command dog, minimum								31	34
0300	Maximum								54.50	60
0500	Sentry dog, leased, with job patrol (yard dog), 1 dog				Week				290	320
0600	2 dogs				"				390	430
0800	Purchase, trained sentry dog, minimum				Ea.				1,375	1,500
0900	Maximum				"				2,725	3,000

01 58 Project Identification

01 58 13 – Temporary Project Signage

01 58 13.50 Signs

		Crew	Daily Output	Labor-Hours	Unit	Material	Labor	Equipment	Total	Total Incl O&P
0010	**SIGNS**									
0020	High intensity reflectorized, no posts, buy				S.F.	26.50			26.50	29.50

01 71 Examination and Preparation

01 71 23 – Field Engineering

01 71 23.13 Construction Layout

		Crew	Daily Output	Labor-Hours	Unit	Material	Labor	Equipment	Total	Total Incl O&P
0010	**CONSTRUCTION LAYOUT**									
1100	Crew for layout of building, trenching or pipe laying, 2 person crew	A-6	1	16	Day		540	69.50	609.50	965
1200	3 person crew	A-7	1	24			860	69.50	929.50	1,500
1400	Crew for roadway layout, 4 person crew	A-8	1	32			1,125	70	1,195	1,950

01 71 23.19 Surveyor Stakes

		Crew	Daily Output	Labor-Hours	Unit	Material	Labor	Equipment	Total	Total Incl O&P
0010	**SURVEYOR STAKES**									
0020	Hardwood, 1" x 1" x 48" long				C	58			58	64
0100	2" x 2" x 18" long					66			66	72.50
0150	2" x 2" x 24" long					110			110	121
0200	2" x 2" x 30" long					74			74	81.50

01 74 Cleaning and Waste Management

01 74 13 – Progress Cleaning

01 74 13.20 Cleaning Up

		Crew	Daily Output	Labor-Hours	Unit	Material	Labor	Equipment	Total	Total Incl O&P
0010	**CLEANING UP**									
0020	After job completion, allow, minimum				Job					.30%
0040	Maximum				"					1%
0050	Cleanup of floor area, continuous, per day, during const.	A-5	24	.750	M.S.F.	1.70	19.85	1.79	23.34	37
0100	Final by GC at end of job	"	11.50	1.565	"	2.71	41.50	3.73	47.94	76.50
0200	Rubbish removal, see Section 02 41 19.23									

01 76 Protecting Installed Construction

01 76 13 – Temporary Protection of Installed Construction

01 76 13.20 Temporary Protection		Crew	Daily Output	Labor-Hours	Unit	2011 Bare Costs Material	Labor	Equipment	Total	Total Incl O&P
0010	**TEMPORARY PROTECTION**									
0020	Flooring, 1/8" tempered hardboard, taped seams	2 Carp	1500	.011	S.F.	.38	.36		.74	1.02
0030	Peel away carpet protection	1 Clab	3200	.003	"	.11	.07		.18	.23

01 91 Commissioning

01 91 13 – General Commissioning Requirements

01 91 13.50 Building Commissioning		Crew	Daily Output	Labor-Hours	Unit	Material	Labor	Equipment	Total	Total Incl O&P
0010	**BUILDING COMMISSIONING**									
0100	Basic building commissioning, minimum				%					.25%
0150	Maximum									.50%
0200	Enhanced building commissioning, minimum									.50%
0250	Maximum				▼					1%

Division Notes

		CREW	DAILY OUTPUT	LABOR-HOURS	UNIT	2011 BARE COSTS MAT.	LABOR	EQUIP.	TOTAL	TOTAL INCL O&P

Estimating Tips

02 30 00 Subsurface Investigation

In preparing estimates on structures involving earthwork or foundations, all information concerning soil characteristics should be obtained. Look particularly for hazardous waste, evidence of prior dumping of debris, and previous stream beds.

02 41 00 Demolition and Structure Moving

The costs shown for selective demolition do not include rubbish handling or disposal. These items should be estimated separately using RSMeans data or other sources.

- Historic preservation often requires that the contractor remove materials from the existing structure, rehab them, and replace them. The estimator must be aware of any related measures and precautions that must be taken when doing selective demolition and cutting and patching. Requirements may include special handling and storage, as well as security.
- In addition to Subdivision 02 41 00, you can find selective demolition items in each division. Example: Roofing demolition is in Division 7.

02 42 10 Building Deconstruction

This section provides costs for the careful dismantling and recycling of most of low-rise building materials.

02 56 13 Containment of Hazardous Waste

This section addresses on-site hazardous waste disposal costs.

02 80 00 Hazardous Material Disposal/ Remediation

This subdivision includes information on hazardous waste handling, asbestos remediation, lead remediation, and mold remediation. See reference R028213-20 and R028319-60 for further guidance in using these unit price lines.

02 91 10 Monitoring Chemical Sampling, Testing Analysis

This section provides costs for on-site sampling and testing hazardous waste.

Reference Numbers

Reference numbers are shown in shaded boxes at the beginning of some major classifications. These numbers refer to related items in the Reference Section. The reference information may be an estimating procedure, an alternate pricing method, or technical information.

Note: Not all subdivisions listed here necessarily appear in this publication.

02 21 Surveys

02 21 13 – Site Surveys

02 21 13.09 Topographical Surveys		Crew	Daily Output	Labor-Hours	Unit	Material	2011 Bare Costs Labor	Equipment	Total	Total Incl O&P
0010	**TOPOGRAPHICAL SURVEYS**									
0020	Topographical surveying, conventional, minimum	A-7	3.30	7.273	Acre	18.20	261	21	300.20	480
0100	Maximum	A-8	.60	53.333	″	55.50	1,875	116	2,046.50	3,300

02 21 13.13 Boundary and Survey Markers

		Crew	Daily Output	Labor-Hours	Unit	Material	Labor	Equipment	Total	Total Incl O&P
0010	**BOUNDARY AND SURVEY MARKERS**									
0300	Lot location and lines, large quantities, minimum	A-7	2	12	Acre	32.50	430	35	497.50	790
0320	Average	″	1.25	19.200		51.50	690	55.50	797	1,275
0400	Small quantities, maximum	A-8	1	32	↓	68.50	1,125	70	1,263.50	2,025
0600	Monuments, 3′ long	A-7	10	2.400	Ea.	30.50	86	6.95	123.45	184
0800	Property lines, perimeter, cleared land	″	1000	.024	L.F.	.03	.86	.07	.96	1.54
0900	Wooded land	A-8	875	.037	″	.05	1.29	.08	1.42	2.28

02 21 13.16 Aerial Surveys

		Crew	Daily Output	Labor-Hours	Unit	Material	Labor	Equipment	Total	Total Incl O&P
0010	**AERIAL SURVEYS**									
1500	Aerial surveying, including ground control, minimum fee, 10 acres				Total					4,200
1510	100 acres									8,400
1550	From existing photography, deduct				↓					1,400
1600	2′ contours, 10 acres				Acre					155
1650	20 acres									75
1800	50 acres									30
1850	100 acres									25
2000	1000 acres									25
2050	10,000 acres				↓					25
2150	For 1′ contours and									
2160	dense urban areas, add to above				Acre					20%
3000	Inertial guidance system for									
3010	locating coordinates, rent per day				Ea.					4,300

02 32 Geotechnical Investigations

02 32 13 – Subsurface Drilling and Sampling

02 32 13.10 Boring and Exploratory Drilling

		Crew	Daily Output	Labor-Hours	Unit	Material	Labor	Equipment	Total	Total Incl O&P
0010	**BORING AND EXPLORATORY DRILLING**									
0020	Borings, initial field stake out & determination of elevations	A-6	1	16	Day		540	69.50	609.50	965
0100	Drawings showing boring details				Total		300		300	375
0200	Report and recommendations from P.E.						700		700	875
0300	Mobilization and demobilization, minimum	B-55	4	4	↓		106	231	337	430
0350	For over 100 miles, per added mile		450	.036	Mile		.94	2.05	2.99	3.83
0600	Auger holes in earth, no samples, 2-1/2″ diameter		78.60	.204	L.F.		5.40	11.75	17.15	22
0650	4″ diameter		67.50	.237			6.25	13.70	19.95	25.50
0800	Cased borings in earth, with samples, 2-1/2″ diameter		55.50	.288		20.50	7.60	16.65	44.75	53.50
0850	4″ diameter	↓	32.60	.491		32.50	12.95	28.50	73.95	88
1000	Drilling in rock, "BX" core, no sampling	B-56	34.90	.458			12.15	37	49.15	61
1050	With casing & sampling		31.70	.505		20.50	13.35	40.50	74.35	89.50
1200	"NX" core, no sampling		25.92	.617			16.35	49.50	65.85	82
1250	With casing and sampling	↓	25	.640	↓	25	16.95	51.50	93.45	113
1400	Borings, earth, drill rig and crew with truck mounted auger	B-55	1	16	Day		425	925	1,350	1,725
1450	Rock using crawler type drill	B-56	1	16	″		425	1,300	1,725	2,125
1500	For inner city borings add, minimum									10%
1510	Maximum									20%

02 32 Geotechnical Investigations

02 32 19 – Exploratory Excavations

02 32 19.10 Test Pits

	02 32 19.10 Test Pits	Crew	Daily Output	Labor-Hours	Unit	Material	2011 Bare Costs Labor	Equipment	Total	Total Incl O&P
0010	**TEST PITS**									
0020	Hand digging, light soil	1 Clab	4.50	1.778	C.Y.		47		47	79
0100	Heavy soil	"	2.50	3.200			84.50		84.50	142
0120	Loader-backhoe, light soil	B-11M	28	.571			17.55	12.50	30.05	43
0130	Heavy soil	"	20	.800			24.50	17.45	41.95	59.50
1000	Subsurface exploration, mobilization				Mile				6	7.50
1010	Difficult access for rig, add				Hr.				230	290
1020	Auger borings, drill rig, incl. samples				L.F.				20	25
1030	Hand auger								25	30
1050	Drill and sample every 5′, split spoon								25	30
1060	Extra samples				Ea.				30	40

02 41 Demolition

02 41 13 – Selective Site Demolition

02 41 13.15 Hydrodemolition

	02 41 13.15 Hydrodemolition	Crew	Daily Output	Labor-Hours	Unit	Material	2011 Bare Costs Labor	Equipment	Total	Total Incl O&P
0010	**HYDRODEMOLITION** R024119-10									
0015	Hydrodemolition, concrete pavement, 4000 PSI, 2″ depth	B-5	500	.080	S.F.		2.28	2.41	4.69	6.45
0120	4″ depth		450	.089			2.54	2.67	5.21	7.15
0130	6″ depth		400	.100			2.85	3.01	5.86	8.05
0410	6000 PSI, 2″ depth		410	.098			2.78	2.93	5.71	7.85
0420	4″ depth		350	.114			3.26	3.44	6.70	9.25
0430	6″ depth		300	.133			3.81	4.01	7.82	10.75
0510	8000 PSI, 2″ depth		330	.121			3.46	3.65	7.11	9.75
0520	4″ depth		280	.143			4.08	4.30	8.38	11.55
0530	6″ depth		240	.167			4.76	5	9.76	13.40

02 41 13.17 Demolish, Remove Pavement and Curb

	02 41 13.17 Demolish, Remove Pavement and Curb	Crew	Daily Output	Labor-Hours	Unit	Material	2011 Bare Costs Labor	Equipment	Total	Total Incl O&P
0010	**DEMOLISH, REMOVE PAVEMENT AND CURB** R024119-10									
5010	Pavement removal, bituminous roads, 3″ thick	B-38	690	.035	S.Y.		1	.69	1.69	2.42
5050	4″ to 6″ thick		420	.057			1.65	1.14	2.79	3.98
5100	Bituminous driveways		640	.038			1.08	.75	1.83	2.61
5200	Concrete to 6″ thick, hydraulic hammer, mesh reinforced		255	.094			2.71	1.88	4.59	6.55
5300	Rod reinforced		200	.120			3.46	2.39	5.85	8.40
5400	Concrete, 7″ to 24″ thick, plain		33	.727	C.Y.		21	14.50	35.50	51
5500	Reinforced		24	1	"		29	19.95	48.95	70
5600	With hand held air equipment, bituminous, to 6″ thick	B-39	1900	.025	S.F.		.68	.10	.78	1.24
5700	Concrete to 6″ thick, no reinforcing		1600	.030			.80	.12	.92	1.48
5800	Mesh reinforced		1400	.034			.92	.14	1.06	1.69
5900	Rod reinforced		765	.063			1.68	.25	1.93	3.10
6000	Curbs, concrete, plain	B-6	360	.067	L.F.		1.92	.89	2.81	4.17
6100	Reinforced		275	.087			2.52	1.17	3.69	5.45
6200	Granite		360	.067			1.92	.89	2.81	4.17
6300	Bituminous		528	.045			1.31	.61	1.92	2.84

02 41 13.33 Minor Site Demolition

	02 41 13.33 Minor Site Demolition	Crew	Daily Output	Labor-Hours	Unit	Material	2011 Bare Costs Labor	Equipment	Total	Total Incl O&P
0010	**MINOR SITE DEMOLITION** R024119-10									
0015	No hauling, abandon catch basin or manhole	B-6	7	3.429	Ea.		99	46	145	215
0020	Remove existing catch basin or manhole, masonry		4	6			173	80	253	375
0030	Catch basin or manhole frames and covers, stored		13	1.846			53	24.50	77.50	116
0040	Remove and reset		7	3.429			99	46	145	215
0100	Roadside delineators, remove only	B-80	175	.137			3.72	3.54	7.26	10.15
0110	Remove and reset	"	100	.240			6.50	6.20	12.70	17.70

02 41 Demolition

02 41 13 – Selective Site Demolition

02 41 13.33 Minor Site Demolition

	02 41 13.33 Minor Site Demolition	Crew	Daily Output	Labor-Hours	Unit	Material	2011 Bare Costs Labor	Equipment	Total	Total Incl O&P
0800	Guiderail, corrugated steel, remove only	B-80A	100	.240	L.F.		6.35	2.17	8.52	13.05
0850	Remove and reset	″	40	.600	″		15.85	5.40	21.25	32.50
0860	Guide posts, remove only	B-80B	120	.267	Ea.		7.55	1.88	9.43	14.55
0870	Remove and reset	B-55	50	.320			8.45	18.45	26.90	34.50
0901	Hydrants, fire, remove only	2 Skwk	4.70	3.404			116		116	194
0951	Remove and reset	″	1.40	11.429			390		390	650
1000	Masonry walls, block, solid	B-5	1800	.022	C.F.		.63	.67	1.30	1.80
1200	Brick, solid		900	.044			1.27	1.34	2.61	3.58
1400	Stone, with mortar		900	.044			1.27	1.34	2.61	3.58
1500	Dry set		1500	.027			.76	.80	1.56	2.15
1600	Median barrier, precast concrete, remove and store	B-3	430	.112	L.F.		3.17	4.70	7.87	10.40
1610	Remove and reset	″	390	.123			3.50	5.20	8.70	11.50
2900	Pipe removal, sewer/water, no excavation, 12″ diameter	B-6	175	.137			3.95	1.83	5.78	8.55
2930	15″-18″ diameter	B-12Z	150	.160			4.73	10.75	15.48	19.70
2960	21″-24″ diameter		120	.200			5.90	13.45	19.35	24.50
3000	27″-36″ diameter		90	.267			7.90	17.90	25.80	33
3200	Steel, welded connections, 4″ diameter	B-6	160	.150			4.32	2	6.32	9.35
3300	10″ diameter	″	80	.300			8.65	4.01	12.66	18.75
3500	Railroad track removal, ties and track	B-13	330	.145			4.12	2.27	6.39	9.35
3600	Ballast	B-14	500	.096	C.Y.		2.69	.64	3.33	5.15
3700	Remove and re-install, ties & track using new bolts & spikes		50	.960	L.F.		27	6.40	33.40	52
3800	Turnouts using new bolts and spikes		1	48	Ea.		1,350	320	1,670	2,575
4000	Sidewalk removal, bituminous, 2-1/2″ thick	B-6	325	.074	S.Y.		2.13	.99	3.12	4.61
4050	Brick, set in mortar		185	.130			3.74	1.73	5.47	8.10
4100	Concrete, plain, 4″		160	.150			4.32	2	6.32	9.35
4200	Mesh reinforced		150	.160			4.61	2.14	6.75	10
4300	Slab on grade removal, plain	B-5	45	.889	C.Y.		25.50	26.50	52	71.50
4310	Mesh reinforced		33	1.212			34.50	36.50	71	97.50
4320	Rod reinforced		25	1.600			45.50	48	93.50	129
4400	For congested sites or small quantities, add up to								200%	200%
4450	For disposal on site, add	B-11A	232	.069			2.12	4.53	6.65	8.50
4500	To 5 miles, add	B-34D	76	.105			2.86	7.90	10.76	13.45

02 41 13.60 Selective Demolition Fencing

	02 41 13.60 Selective Demolition Fencing	Crew	Daily Output	Labor-Hours	Unit	Material	Labor	Equipment	Total	Total Incl O&P
0010	**SELECTIVE DEMOLITION FENCING** R024119-10									
1600	Fencing, barbed wire, 3 strand	2 Clab	430	.037	L.F.		.98		.98	1.65
1650	5 strand	″	280	.057			1.51		1.51	2.53
1700	Chain link, posts & fabric, 8′ to 10′ high, remove only	B-6	445	.054			1.55	.72	2.27	3.37
1750	Remove and reset	″	70	.343			9.90	4.58	14.48	21.50

02 41 16 – Structure Demolition

02 41 16.13 Building Demolition

	02 41 16.13 Building Demolition	Crew	Daily Output	Labor-Hours	Unit	Material	Labor	Equipment	Total	Total Incl O&P
0010	**BUILDING DEMOLITION** Large urban projects, incl. 20 mi. haul R024119-10									
0011	No foundation or dump fees, C.F. is vol. of building standing									
0020	Steel	B-8	21500	.003	C.F.		.08	.13	.21	.27
0050	Concrete		15300	.004			.11	.18	.29	.38
0080	Masonry		20100	.003			.08	.14	.22	.29
0100	Mixture of types, average		20100	.003			.08	.14	.22	.29
0500	Small bldgs, or single bldgs, no salvage included, steel	B-3	14800	.003			.09	.14	.23	.30
0600	Concrete		11300	.004			.12	.18	.30	.40
0650	Masonry		14800	.003			.09	.14	.23	.30
0700	Wood		14800	.003			.09	.14	.23	.30
0750	For buildings with no interior walls, deduct								50%	
1000	Demoliton single family house, one story, wood 1600 S.F.	B-3	1	48	Ea.		1,375	2,025	3,400	4,500

02 41 Demolition

02 41 16 – Structure Demolition

02 41 16.13 Building Demolition		Crew	Daily Output	Labor-Hours	Unit	Material	2011 Bare Costs Labor	Equipment	Total	Total Incl O&P
1020	3200 S.F.	B-3	.50	96	Ea.		2,725	4,050	6,775	8,975
1200	Demoliton two family house, two story, wood 2400 S.F.		.67	71.964			2,050	3,025	5,075	6,725
1220	4200 S.F.		.38	128			3,650	5,375	9,025	12,000
1300	Demoliton three family house, three story, wood 3200 S.F.		.50	96			2,725	4,050	6,775	8,975
1320	5400 S.F.		.30	160			4,550	6,725	11,275	15,000
5000	For buildings with no interior walls, deduct								50%	

02 41 16.15 Explosive/Implosive Demolition		Crew	Daily Output	Labor-Hours	Unit	Material	Labor	Equipment	Total	Total Incl O&P
0010	**EXPLOSIVE/IMPLOSIVE DEMOLITION** R024119-10									
0011	Large projects,									
0020	No disposal fee based on building volume, steel building	B-5B	16900	.003	C.F.		.09	.13	.22	.29
0100	Concrete building		16900	.003			.09	.13	.22	.29
0200	Masonry building		16900	.003			.09	.13	.22	.29
0400	Disposal of material, minimum	B-3	445	.108	C.Y.		3.07	4.54	7.61	10.10
0500	Maximum	"	365	.132	"		3.74	5.55	9.29	12.30

02 41 16.17 Building Demolition Footings and Foundations		Crew	Daily Output	Labor-Hours	Unit	Material	Labor	Equipment	Total	Total Incl O&P
0010	**BUILDING DEMOLITION FOOTINGS AND FOUNDATIONS** R024119-10									
0200	Floors, concrete slab on grade,									
0240	4" thick, plain concrete	B-9	500	.080	S.F.		2.15	.39	2.54	4.03
0280	Reinforced, wire mesh		470	.085			2.29	.42	2.71	4.29
0300	Rods		400	.100			2.69	.49	3.18	5.05
0400	6" thick, plain concrete		375	.107			2.86	.52	3.38	5.35
0420	Reinforced, wire mesh		340	.118			3.16	.57	3.73	5.95
0440	Rods		300	.133			3.58	.65	4.23	6.70
1000	Footings, concrete, 1' thick, 2' wide	B-5	300	.133	L.F.		3.81	4.01	7.82	10.75
1080	1'-6" thick, 2' wide		250	.160			4.57	4.81	9.38	12.90
1120	3' wide		200	.200			5.70	6	11.70	16.10
1140	2' thick, 3' wide		175	.229			6.50	6.90	13.40	18.40
1200	Average reinforcing, add								10%	10%
1220	Heavy reinforcing, add								20%	20%
2000	Walls, block, 4" thick	1 Clab	180	.044	S.F.		1.18		1.18	1.97
2040	6" thick		170	.047			1.24		1.24	2.08
2080	8" thick		150	.053			1.41		1.41	2.36
2100	12" thick		150	.053			1.41		1.41	2.36
2200	For horizontal reinforcing, add								10%	10%
2220	For vertical reinforcing, add								20%	20%
2400	Concrete, plain concrete, 6" thick	B-9	160	.250			6.70	1.22	7.92	12.60
2420	8" thick		140	.286			7.65	1.39	9.04	14.40
2440	10" thick		120	.333			8.95	1.63	10.58	16.80
2500	12" thick		100	.400			10.75	1.95	12.70	20
2600	For average reinforcing, add								10%	10%
2620	For heavy reinforcing, add								20%	20%
4000	For congested sites or small quantities, add up to								200%	200%
4200	Add for disposal, on site	B-11A	232	.069	C.Y.		2.12	4.53	6.65	8.50
4250	To five miles	B-30	220	.109	"		3.25	8.90	12.15	15.15

02 41 19 – Selective Demolition

02 41 19.13 Selective Building Demolition		Crew	Daily Output	Labor-Hours	Unit	Material	Labor	Equipment	Total	Total Incl O&P
0010	**SELECTIVE BUILDING DEMOLITION**									
0020	Costs related to selective demolition of specific building components									
0025	are included under Common Work Results (XX 05)									
0030	in the component's appropriate division.									

02 41 19.16 Selective Demolition, Cutout

Line	02 41 19.16 Selective Demolition, Cutout	Crew	Daily Output	Labor-Hours	Unit	2011 Bare Costs Material	Labor	Equipment	Total	Total Incl O&P
0010	**SELECTIVE DEMOLITION, CUTOUT** R024119-10									
0020	Concrete, elev. slab, light reinforcement, under 6 C.F.	B-9	65	.615	C.F.		16.50	3	19.50	31
0050	Light reinforcing, over 6 C.F.		75	.533	"		14.30	2.60	16.90	27
0200	Slab on grade to 6" thick, not reinforced, under 8 S.F.		85	.471	S.F.		12.65	2.30	14.95	23.50
0250	8 – 16 S.F.		175	.229	"		6.15	1.12	7.27	11.55
0255	For over 16 S.F. see Line 02 41 16.17 0400									
0600	Walls, not reinforced, under 6 C.F.	B-9	60	.667	C.F.		17.90	3.25	21.15	33.50
0650	6 – 12 C.F.	"	80	.500	"		13.45	2.44	15.89	25
0655	For over 12 C.F. see Line 02 41 16.17 2500									
1000	Concrete, elevated slab, bar reinforced, under 6 C.F.	B-9	45	.889	C.F.		24	4.34	28.34	45
1050	Bar reinforced, over 6 C.F.		50	.800	"		21.50	3.90	25.40	40.50
1200	Slab on grade to 6" thick, bar reinforced, under 8 S.F.		75	.533	S.F.		14.30	2.60	16.90	27
1250	8 – 16 S.F.		150	.267	"		7.15	1.30	8.45	13.45
1255	For over 16 S.F. see Line 02 41 16.17 0440									
1400	Walls, bar reinforced, under 6 C.F.	B-9	50	.800	C.F.		21.50	3.90	25.40	40.50
1450	6 – 12 C.F.	"	70	.571	"		15.35	2.79	18.14	28.50
1455	For over 12 C.F. see Lines 02 41 16.17 2500 and 2600									
2000	Brick, to 4 S.F. opening, not including toothing									
2040	4" thick	B-9	30	1.333	Ea.		36	6.50	42.50	67
2060	8" thick		18	2.222			59.50	10.85	70.35	112
2080	12" thick		10	4			107	19.50	126.50	202
2400	Concrete block, to 4 S.F. opening, 2" thick		35	1.143			30.50	5.60	36.10	57.50
2420	4" thick		30	1.333			36	6.50	42.50	67
2440	8" thick		27	1.481			40	7.25	47.25	74.50
2460	12" thick		24	1.667			45	8.15	53.15	84
2600	Gypsum block, to 4 S.F. opening, 2" thick		80	.500			13.45	2.44	15.89	25
2620	4" thick		70	.571			15.35	2.79	18.14	28.50
2640	8" thick		55	.727			19.55	3.55	23.10	36.50
2800	Terra cotta, to 4 S.F. opening, 4" thick		70	.571			15.35	2.79	18.14	28.50
2840	8" thick		65	.615			16.50	3	19.50	31
2880	12" thick		50	.800			21.50	3.90	25.40	40.50
3000	Toothing masonry cutouts, brick, soft old mortar	1 Brhe	40	.200	V.L.F.		5.45		5.45	8.90
3100	Hard mortar		30	.267			7.25		7.25	11.90
3200	Block, soft old mortar		70	.114			3.10		3.10	5.10
3400	Hard mortar		50	.160			4.34		4.34	7.15
6000	Walls, interior, not including re-framing,									
6010	openings to 5 S.F.									
6100	Drywall to 5/8" thick	1 Clab	24	.333	Ea.		8.80		8.80	14.75
6200	Paneling to 3/4" thick		20	.400			10.60		10.60	17.70
6300	Plaster, on gypsum lath		20	.400			10.60		10.60	17.70
6340	On wire lath		14	.571			15.10		15.10	25.50
7000	Wood frame, not including re-framing, openings to 5 S.F.									
7200	Floors, sheathing and flooring to 2" thick	1 Clab	5	1.600	Ea.		42.50		42.50	71
7310	Roofs, sheathing to 1" thick, not including roofing		6	1.333			35.50		35.50	59
7410	Walls, sheathing to 1" thick, not including siding		7	1.143			30		30	50.50

02 41 19.18 Selective Demolition, Disposal Only

Line	02 41 19.18 Selective Demolition, Disposal Only	Crew	Daily Output	Labor-Hours	Unit	Material	Labor	Equipment	Total	Total Incl O&P
0010	**SELECTIVE DEMOLITION, DISPOSAL ONLY** R024119-10									
0015	Urban bldg w/salvage value allowed									
0020	Including loading and 5 mile haul to dump									
0200	Steel frame	B-3	430	.112	C.Y.		3.17	4.70	7.87	10.40
0300	Concrete frame		365	.132			3.74	5.55	9.29	12.30
0400	Masonry construction		445	.108			3.07	4.54	7.61	10.10

02 41 19.18 Selective Demolition, Disposal Only		Crew	Daily Output	Labor-Hours	Unit	Material	2011 Bare Costs Labor	Equipment	Total	Total Incl O&P
0500	Wood frame	B-3	247	.194	C.Y.		5.55	8.20	13.75	18.20

02 41 19.19 Selective Demolition, Dump Charges

		Crew	Daily Output	Labor-Hours	Unit	Material	Labor	Equipment	Total	Total Incl O&P
0010	**SELECTIVE DEMOLITION, DUMP CHARGES** R024119-10									
0020	Dump charges, typical urban city, tipping fees only									
0100	Building construction materials				Ton	82			82	90
0200	Trees, brush, lumber					70			70	77
0300	Rubbish only					70			70	77
0500	Reclamation station, usual charge					82			82	90

02 41 19.21 Selective Demolition, Gutting

		Crew	Daily Output	Labor-Hours	Unit	Material	Labor	Equipment	Total	Total Incl O&P
0010	**SELECTIVE DEMOLITION, GUTTING** R024119-10									
0020	Building interior, including disposal, dumpster fees not included									
0500	Residential building									
0560	Minimum	B-16	400	.080	SF Flr.		2.17	1.26	3.43	5
0580	Maximum	″	360	.089	″		2.41	1.41	3.82	5.60
0900	Commercial building									
1000	Minimum	B-16	350	.091	SF Flr.		2.48	1.45	3.93	5.75
1020	Maximum	″	250	.128	″		3.47	2.02	5.49	8.05

02 41 19.23 Selective Demolition, Rubbish Handling

		Crew	Daily Output	Labor-Hours	Unit	Material	Labor	Equipment	Total	Total Incl O&P
0010	**SELECTIVE DEMOLITION, RUBBISH HANDLING** R024119-10									
0020	The following are to be added to the demolition prices									
0400	Chute, circular, prefabricated steel, 18″ diameter	B-1	40	.600	L.F.	53	16.25		69.25	86
0440	30″ diameter	″	30	.800	″	47.50	21.50		69	89
0600	Dumpster, weekly rental, 1 dump/week, 6 C.Y. capacity (2 Tons)				Week	300			300	330
0700	10 C.Y. capacity (4 Tons)					475			475	525
0725	20 C.Y. capacity (8 Tons) R024119-20					630			630	695
0800	30 C.Y. capacity (10 Tons)					750			750	825
0840	40 C.Y. capacity (13 Tons)					875			875	965
2000	Load, haul, dump and return, 50′ haul, hand carried	2 Clab	24	.667	C.Y.		17.65		17.65	29.50
2005	Wheeled		37	.432			11.45		11.45	19.15
2040	51′ to 100′ haul, hand carried		16.50	.970			25.50		25.50	43
2045	Wheeled		25	.640			16.95		16.95	28.50
2080	Over 100′ haul, add per 100 L.F., hand carried		35.50	.451			11.90		11.90	19.95
2085	Wheeled		54	.296			7.85		7.85	13.15
2120	In elevators, per 10 floors, add		140	.114			3.02		3.02	5.05
2130	Load, haul, dump and return, up to 50′ haul, incl. up to 5 rsr stair, hand		23	.696			18.40		18.40	31
2135	Wheeled		35	.457			12.10		12.10	20.50
2140	6 – 10 riser stairs, hand carried		22	.727			19.25		19.25	32
2145	Wheeled		34	.471			12.45		12.45	21
2150	11 – 20 riser stairs, hand carried		20	.800			21		21	35.50
2155	Wheeled		31	.516			13.65		13.65	23
2160	21 – 40 riser stairs, hand carried		16	1			26.50		26.50	44.50
2165	Wheeled		24	.667			17.65		17.65	29.50
2170	100′ haul, incl. 5 riser stair, hand carried		15	1.067			28		28	47.50
2175	Wheeled		23	.696			18.40		18.40	31
2180	6 – 10 riser stair, hand carried		14	1.143			30		30	50.50
2185	Wheeled		21	.762			20		20	34
2190	11 – 20 riser stair, hand carried		12	1.333			35.50		35.50	59
2195	Wheeled		18	.889			23.50		23.50	39.50
2200	21 – 40 riser stair, hand carried		8	2			53		53	88.50
2205	Wheeled		12	1.333			35.50		35.50	59
2210	Over 100′ haul, add per 100 L.F., hand carried		35.50	.451			11.90		11.90	19.95
2215	Wheeled		54	.296			7.85		7.85	13.15

02 41 Demolition

02 41 19 – Selective Demolition

02 41 19.23 Selective Demolition, Rubbish Handling		Crew	Daily Output	Labor-Hours	Unit	Material	2011 Bare Costs Labor	Equipment	Total	Total Incl O&P
2220	For each additional flight of stairs, up to 5 risers, add	2 Clab	550	.029	Flight		.77		.77	1.29
2225	6 – 10 risers, add		275	.058			1.54		1.54	2.58
2230	11 – 20 risers, add		138	.116			3.07		3.07	5.15
2235	21 – 40 risers, add		69	.232			6.15		6.15	10.25
3000	Loading & trucking, including 2 mile haul, chute loaded	B-16	45	.711	C.Y.		19.30	11.25	30.55	44.50
3040	Hand loading truck, 50′ haul	"	48	.667			18.10	10.55	28.65	41.50
3080	Machine loading truck	B-17	120	.267			7.60	5.30	12.90	18.40
5000	Haul, per mile, up to 8 C.Y. truck	B-34B	1165	.007			.19	.43	.62	.79
5100	Over 8 C.Y. truck	"	1550	.005			.14	.33	.47	.59

02 41 19.25 Selective Demolition, Saw Cutting

Line	Description	Crew	Daily Output	Labor-Hours	Unit	Material	Labor	Equipment	Total	Total Incl O&P
0010	**SELECTIVE DEMOLITION, SAW CUTTING** R024119-10									
0015	Asphalt, up to 3″ deep	B-89	1050	.015	L.F.	.36	.46	.35	1.17	1.56
0020	Each additional inch of depth	"	1800	.009		.12	.27	.21	.60	.81
1200	Masonry walls, hydraulic saw, brick, per inch of depth	B-89B	300	.053		.44	1.60	2.18	4.22	5.50
1220	Block walls, solid, per inch of depth	"	250	.064		.44	1.92	2.61	4.97	6.50
2000	Brick or masonry w/hand held saw, per inch of depth	A-1	125	.064		.35	1.69	.51	2.55	3.79
5001	Wood sheathing, to 1″ thick on walls	1 Clab	200	.040			1.06		1.06	1.77
5021	On roof	"	250	.032			.85		.85	1.42

02 41 19.27 Selective Demolition, Torch Cutting

Line	Description	Crew	Daily Output	Labor-Hours	Unit	Material	Labor	Equipment	Total	Total Incl O&P
0010	**SELECTIVE DEMOLITION, TORCH CUTTING** R024119-10									
0020	Steel, 1″ thick plate	1 Clab	360	.022	L.F.	.24	.59		.83	1.25
0040	1″ diameter bar	"	210	.038	Ea.		1.01		1.01	1.69
1000	Oxygen lance cutting, reinforced concrete walls									
1040	12″ to 16″ thick walls	1 Clab	10	.800	L.F.		21		21	35.50
1080	24″ thick walls	"	6	1.333	"		35.50		35.50	59

02 42 Removal and Salvage of Construction Materials

02 42 10 – Building Deconstruction

02 42 10.10 Estimated Salvage Value or Savings

Line	Description	Crew	Daily Output	Labor-Hours	Unit	Material	Labor	Equipment	Total	Total Incl O&P
0010	**ESTIMATED SALVAGE VALUE OR SAVINGS**									
0015	Excludes material handling, packaging, container costs and									
0020	transportation for salvage or disposal									
0050	All Items in Section 02 42 10.10 are credit deducts and not costs									
0100	Copper Wire Salvage Value G				Lb.				1.30	1.30
0110	Disposal Savings G								.03	.03
0200	Copper Pipe Salvage Value G								1.34	1.34
0210	Disposal Savings G								.03	.03
0300	Steel Pipe Salvage Value G								.06	.06
0310	Disposal Savings G								.03	.03
0400	Cast Iron Pipe Salvage Value G								.06	.06
0410	Disposal Savings G								.03	.03
0500	Steel Doors or Windows Salvage Value G								.06	.06
0510	Aluminum G								.66	.66
0520	Disposal Savings G								.03	.03
0600	Aluminum Siding Salvage Value G								.49	.49
0630	Disposal Savings G								.03	.03
0640	Wood Siding (no lead or asbestos) G				C.Y.				10.90	10.90
0800	Clean Concrete Disposal Savings G				Ton				60	60
0850	Asphalt Shingles Disposal Savings G				"				60	60
1000	Wood wall framing clean salvage value G				M.B.F.				50	50

02 42 Removal and Salvage of Construction Materials

02 42 10 – Building Deconstruction

02 42 10.10 Estimated Salvage Value or Savings

Line	Description		Crew	Daily Output	Labor-Hours	Unit	Material	Labor (2011 Bare Costs)	Equipment	Total	Total Incl O&P
1010	Painted	G				M.B.F.				40	40
1020	Floor framing	G								50	50
1030	Painted	G								40	40
1050	Roof framing	G								50	50
1060	Painted	G								40	40
1100	Wood beams salvage value	G								50	50
1200	Wood framing and beams disposal savings	G				Ton				60	60
1220	Wood sheating and sub-base flooring	G								60	60
1230	Wood wall paneling (1/4 inch thick)	G								60	60
1300	Wood panel 3/4-1 inch thick low salvage value	G				S.F.				.50	.50
1350	high salvage value	G				"				2	2
1400	Disposal savings	G				Ton				60	60
1500	Flooring tongue and groove 25/32 inch thick low salvage value	G				S.F.				.50	.50
1530	High salvage value	G				"				1	1
1560	Disposal savings	G				Ton				60	60
1600	Drywall or sheet rock salvage value	G								20	20
1650	Disposal savings	G								60	60

02 42 10.20 Deconstruction of Building Components

Line	Description		Crew	Daily Output	Labor-Hours	Unit	Material	Labor (2011 Bare Costs)	Equipment	Total	Total Incl O&P
0010	**DECONSTRUCTION OF BUILDING COMPONENTS**										
0012	Buildings one or two stories only										
0015	Excludes material handling, packaging, container costs and										
0020	transportation for salvage or disposal										
0050	Deconstruction of Plumbing Fixtures										
0100	Wall hung or countertop lavatory	G	2 Clab	16	1	Ea.		26.50		26.50	44.50
0110	Single or double compartment kitchen sink	G		14	1.143			30		30	50.50
0120	Wall hung urinal	G		14	1.143			30		30	50.50
0130	Floor mounted	G		8	2			53		53	88.50
0140	Floor mounted water closet	G		16	1			26.50		26.50	44.50
0150	Wall hung	G		14	1.143			30		30	50.50
0160	Water fountain, free standing	G		16	1			26.50		26.50	44.50
0170	Wall hung or deck mounted	G		12	1.333			35.50		35.50	59
0180	Bathtub, steel or fiberglass	G		10	1.600			42.50		42.50	71
0190	Cast iron	G		8	2			53		53	88.50
0200	Shower, single	G		6	2.667			70.50		70.50	118
0210	Group	G		7	2.286			60.50		60.50	101
0300	Deconstruction of Electrical Fixtures										
0310	Surface mount incandescent fixtures	G	2 Clab	48	.333	Ea.		8.80		8.80	14.75
0320	Fluorescent, 2 lamp	G		32	.500			13.25		13.25	22
0330	4 lamp	G		24	.667			17.65		17.65	29.50
0340	Strip Fluorescent, 1 lamp	G		40	.400			10.60		10.60	17.70
0350	2 lamp	G		32	.500			13.25		13.25	22
0400	Recessed drop-in fluorescent fixture, 2 lamp	G		27	.593			15.65		15.65	26.50
0410	4 lamp	G		18	.889			23.50		23.50	39.50
0500	Deconstruction of appliances										
0510	Cooking stoves	G	2 Clab	26	.615	Ea.		16.30		16.30	27.50
0520	Dishwashers	G	"	26	.615	"		16.30		16.30	27.50
0600	Deconstruction of millwork and trim										
0610	Cabinets, wood	G	2 Carp	40	.400	L.F.		13.40		13.40	22.50
0620	Countertops	G		100	.160	"		5.35		5.35	9
0630	Wall paneling, 1 inch thick	G		500	.032	S.F.		1.07		1.07	1.80
0640	Ceiling trim	G		500	.032	L.F.		1.07		1.07	1.80
0650	Wainscoting	G		500	.032	S.F.		1.07		1.07	1.80

02 42 10 – Building Deconstruction

02 42 10.20 Deconstruction of Building Components			Crew	Daily Output	Labor-Hours	Unit	Material	2011 Bare Costs Labor	Equipment	Total	Total Incl O&P
0660	Base, 3/4" to 1" thick	G	2 Carp	600	.027	L.F.		.89		.89	1.50
0700	Deconstruction of doors and windows										
0710	Doors, wrap, interior, wood, single, no closers	G	2 Carp	21	.762	Ea.	7.50	25.50		33	51.50
0720	Double	G		13	1.231		15	41.50		56.50	85.50
0730	Solid core, single, exterior or interior	G		10	1.600		7.50	53.50		61	98.50
0740	Double	G		8	2		15	67		82	129
0810	Windows, wrap, wood, single	G		21	.762		7.50	25.50		33	51.50
0812	with no casement or cladding										
0820	with casement and/or cladding	G	2 Carp	18	.889	Ea.	7.50	30		37.50	58.50
0900	Deconstruction of interior finishes										
0910	Drywall for recycling	G	2 Clab	1775	.009	S.F.		.24		.24	.40
0920	Plaster wall, first floor	G		1775	.009			.24		.24	.40
0930	Second floor	G		1330	.012			.32		.32	.53
1000	Deconstruction of roofing and accessories										
1010	Built-up roofs	G	2 Clab	570	.028	S.F.		.74		.74	1.24
1020	Gutters, facia and rakes	G	"	1140	.014	L.F.		.37		.37	.62
2000	Deconstruction of wood components										
2010	Roof sheeting	G	2 Clab	570	.028	S.F.		.74		.74	1.24
2020	Main roof framing	G		760	.021	L.F.		.56		.56	.93
2030	Porch roof framing	G		445	.036			.95		.95	1.59
2040	Beams 4" x 8"	G	B-1	375	.064			1.74		1.74	2.91
2050	4" x 10"	G		300	.080			2.17		2.17	3.63
2055	4" x 12"	G		250	.096			2.60		2.60	4.36
2060	6" x 8"	G		250	.096			2.60		2.60	4.36
2065	6" x 10"	G		200	.120			3.25		3.25	5.45
2070	6" x 12"	G		170	.141			3.83		3.83	6.40
2075	8" x 12"	G		126	.190			5.15		5.15	8.65
2080	10" x 12"	G		100	.240			6.50		6.50	10.90
2100	Ceiling joists	G	2 Clab	800	.020			.53		.53	.89
2150	Wall framing, interior	G		1230	.013			.34		.34	.58
2160	Sub-floor	G		2000	.008	S.F.		.21		.21	.35
2170	Floor joists	G		2000	.008	L.F.		.21		.21	.35
2200	Wood siding (no lead or asbestos)	G		1300	.012	S.F.		.33		.33	.55
2300	Wall framing, exterior	G		1600	.010	L.F.		.26		.26	.44
2400	Stair risers	G		53	.302	Ea.		8		8	13.35
2500	Posts	G		800	.020	L.F.		.53		.53	.89
3000	Deconstruction of exterior brick walls										
3010	Exterior brick walls, first floor	G	2 Clab	200	.080	S.F.		2.12		2.12	3.54
3020	Second floor	G		64	.250	"		6.60		6.60	11.10
3030	Brick chiminey	G		100	.160	C.F.		4.23		4.23	7.10
4000	Deconstruction of concrete										
4010	Slab on grade, 4" thick, plain concrete	G	B-9	500	.080	S.F.		2.15	.39	2.54	4.03
4020	Wire mesh reinforced	G		470	.085			2.29	.42	2.71	4.29
4030	Rod reinforced	G		400	.100			2.69	.49	3.18	5.05
4110	Foundation wall, 6" thick, plain concrete	G		160	.250			6.70	1.22	7.92	12.60
4120	8" thick	G		140	.286			7.65	1.39	9.04	14.40
4130	10" thick	G		120	.333			8.95	1.63	10.58	16.80
9000	Deconstruction process, support equipment as needed										
9010	Daily use, portal to portal, 12-ton truck-mounted hydraulic crane crew	G	A-3H	1	8	Day		286	870	1,156	1,425
9020	Daily use, skid steer and operator	G	A-3C	1	8			268	261	529	725
9030	Daily use, backhoe 48 H. P., operator and labor	G	"	1	8			268	261	529	725

02 42 Removal and Salvage of Construction Materials

02 42 10 – Building Deconstruction

02 42 10.30 Deconstruction Material Handling			Crew	Daily Output	Labor-Hours	Unit	2011 Bare Costs Material	Labor	Equipment	Total	Total Incl O&P
0010	**DECONSTRUCTION MATERIAL HANDLING**										
0012	Buildings one or two stories only										
0100	Clean and stack brick on pallet	G	2 Clab	1200	.013	Ea.		.35		.35	.59
0200	Haul 50′ and load rough lumber up to 2″ x 8″ size	G		2000	.008	"		.21		.21	.35
0210	Lumber larger than 2″ x 8″	G		3200	.005	B.F.		.13		.13	.22
0300	Finish wood for recycling stack and wrap per pallet	G		8	2	Ea.	60	53		113	155
0350	Light fixtures			6	2.667		108	70.50		178.50	237
0375	Windows			6	2.667		102	70.50		172.50	230
0400	Miscellaneous materials			8	2		30	53		83	122
1000	See Section 02 41 19.23 for bulk material handling										

02 43 Structure Moving

02 43 13 – Structure Relocation

02 43 13.13 Building Relocation			Crew	Daily Output	Labor-Hours	Unit	Material	Labor	Equipment	Total	Total Incl O&P
0010	**BUILDING RELOCATION**										
0011	One day move, up to 24′ wide										
0020	Reset on new foundation, patch & hook-up, average move					Total					11,200
0040	Wood or steel frame bldg., based on ground floor area	G	B-4	185	.259	S.F.		7	2.32	9.32	14.20
0060	Masonry bldg., based on ground floor area	G	"	137	.350			9.45	3.14	12.59	19.20
0200	For 24′ to 42′ wide, add										15%
0220	For each additional day on road, add	G	B-4	1	48	Day		1,300	430	1,730	2,625
0240	Construct new basement, move building, 1 day										
0300	move, patch & hook-up, based on ground floor area	G	B-3	155	.310	S.F.	9.25	8.80	13.05	31.10	39

02 56 Site containment

02 56 13 – Waste containment

02 56 13.10 Containment of Hazardous Waste		Crew	Daily Output	Labor-Hours	Unit	Material	Labor	Equipment	Total	Total Incl O&P
0010	**CONTAINMENT OF HAZARDOUS WASTE**									
0020	OSHA hazard level C									
0030	OSHA Hazard level D decrease labor and equipment, deduct						-45%	-45%		
0035	OSHA Hazard level B increase labor and equipment, add						22%	22%		
0040	OSHA Hazard level A increase labor and equipment, add						71%	71%		
0100	Excavation of contaminated soil & waste									
0105	Includes one respirator filter and two disposable suits per work day									
0110	3/4 C.Y. excavator to 10 feet deep	B-12F	51	.314	B.C.Y.	1.64	9.75	12.55	23.94	32
0120	Labor crew to 6′ deep	B-2	19	2.105		11	56.50		67.50	107
0130	6′ - 12′ deep	"	12	3.333		17.40	89.50		106.90	169
0200	Move contaminated soil/waste upto 150′ on-site with 2.5 C.Y. loader	B-10T	300	.027	L.C.Y.	.28	.93	1.33	2.54	3.29
0210	300′	"	186	.043	"	.45	1.50	2.14	4.09	5.30
0300	Secure burial cell construction									
0310	Various liner and cover materials									
0400	Very low density polyethelene (VLDPE)									
0410	50 mil top cover	B-47H	4000	.008	S.F.	.37	.28	.05	.70	.93
0420	80 mil liner	"	4000	.008	"	.37	.28	.05	.70	.93
0500	Chlorosulfunated polyethylene									
0510	36 mil hypalon top cover	B-47H	4000	.008	S.F.	1.18	.28	.05	1.51	1.82
0520	45 mil hypalon liner	"	4000	.008	"	1.24	.28	.05	1.57	1.88
0600	Polyvinyl chloride (PVC)									
0610	60 mil top cover	B-47H	4000	.008	S.F.	.65	.28	.05	.98	1.24

02 56 Site containment

02 56 13 – Waste containment

02 56 13.10	Containment of Hazardous Waste	Crew	Daily Output	Labor-Hours	Unit	Material	2011 Bare Costs Labor	Equipment	Total	Total Incl O&P
0620	80 mil liner	B-47H	4000	.008	S.F.	.76	.28	.05	1.09	1.36
0700	Rough textured H.D. polyethylene (HDPE)									
0710	40 mil top cover	B-47H	4000	.008	S.F.	.37	.28	.05	.70	.93
0720	60 mil top cover		4000	.008		.39	.28	.05	.72	.95
0722	60 mil liner		4000	.008		.35	.28	.05	.68	.91
0730	80 mil liner		3800	.008		.42	.29	.06	.77	1.01
1000	3/4" crushed stone, 6" deep ballast around liner	B-6	30	.800	L.C.Y.	38	23	10.70	71.70	92
1100	Hazardous waste, ballast cover with common borrow material	B-63	56	.714		15	18.90	2.71	36.61	51
1110	Mixture of common borrow & topsoil		56	.714		19.50	18.90	2.71	41.11	56
1120	Bank sand		56	.714		15	18.90	2.71	36.61	51
1130	Medium priced clay		44	.909		25	24	3.45	52.45	72
1140	Mixture of common borrow & medium priced clay		56	.714		20	18.90	2.71	41.61	56.50

02 58 Snow Control

02 58 13 – Snow Fencing

02 58 13.10	Snow Fencing System	Crew	Daily Output	Labor-Hours	Unit	Material	Labor	Equipment	Total	Total Incl O&P
0010	**SNOW FENCING SYSTEM**									
7001	Snow fence on steel posts 10′ O.C., 4′ high	B-1	500	.048	L.F.	2.02	1.30		3.32	4.40

02 65 Underground Storage Tank Removal

02 65 10 – Underground Tank and Contaminated Soil Removal

02 65 10.30	Removal of Underground Storage Tanks	Crew	Daily Output	Labor-Hours	Unit	Material	Labor	Equipment	Total	Total Incl O&P
0010	**REMOVAL OF UNDERGROUND STORAGE TANKS** R026510-20									
0011	Petroleum storage tanks, non-leaking									
0100	Excavate & load onto trailer									
0110	3000 gal. to 5000 gal. tank G	B-14	4	12	Ea.		335	80	415	650
0120	6000 gal. to 8000 gal. tank G	B-3A	3	13.333			375	315	690	970
0130	9000 gal. to 12000 gal. tank G	"	2	20			565	475	1,040	1,450
0190	Known leaking tank, add				%				100%	100%
0200	Remove sludge, water and remaining product from tank bottom									
0201	of tank with vacuum truck									
0300	3000 gal. to 5000 gal. tank G	A-13	5	1.600	Ea.		53.50	147	200.50	250
0310	6000 gal. to 8000 gal. tank G		4	2			67	184	251	310
0320	9000 gal. to 12000 gal. tank G		3	2.667			89.50	245	334.50	415
0390	Dispose of sludge off-site, average				Gal.				6	6.60
0400	Insert inert solid CO_2 "dry ice" into tank									
0401	For cleaning/transporting tanks (1.5 lb./100 gal. cap) G	1 Clab	500	.016	Lb.	1.87	.42		2.29	2.77
0403	Insert solid carbon dioxide, 1.5 lb./100 gal. G	"	400	.020	"	1.87	.53		2.40	2.95
0503	Disconnect and remove piping G	1 Plum	160	.050	L.F.		1.98		1.98	3.24
0603	Transfer liquids, 10% of volume G	"	1600	.005	Gal.		.20		.20	.32
0703	Cut accessway into underground storage tank G	1 Clab	5.33	1.501	Ea.		39.50		39.50	66.50
0813	Remove sludge, wash and wipe tank, 500 gal. G	1 Plum	8	1			39.50		39.50	65
0823	3,000 gal. G		6.67	1.199			47.50		47.50	77.50
0833	5,000 gal. G		6.15	1.301			51.50		51.50	84
0843	8,000 gal. G		5.33	1.501			59.50		59.50	97
0853	10,000 gal. G		4.57	1.751			69.50		69.50	113
0863	12,000 gal. G		4.21	1.900			75.50		75.50	123
1020	Haul tank to certified salvage dump, 100 miles round trip									
1023	3000 gal. to 5000 gal. tank				Ea.				710	780

02 65 Underground Storage Tank Removal

02 65 10 – Underground Tank and Contaminated Soil Removal

02 65 10.30	Removal of Underground Storage Tanks		Crew	Daily Output	Labor-Hours	Unit	2011 Bare Costs Material	Labor	Equipment	Total	Total Incl O&P
1026	6000 gal. to 8000 gal. tank					Ea.				810	890
1029	9,000 gal. to 12,000 gal. tank					↓				950	1,050
1100	Disposal of contaminated soil to landfill										
1110	Minimum					C.Y.				125	140
1111	Maximum					"				400	440
1120	Disposal of contaminated soil to										
1121	bituminous concrete batch plant										
1130	Minimum					C.Y.				70	77
1131	Maximum					"				110	120
1203	Excavate, pull, & load tank, backfill hole, 8,000 gal. +	G	B-12C	.50	32	Ea.		995	2,450	3,445	4,350
1213	Haul tank to certified dump, 100 miles rt, 8,000 gal. +	G	B-34K	1	8			218	780	998	1,225
1223	Excavate, pull, & load tank, backfill hole, 500 gal.	G	B-11C	1	16			490	320	810	1,150
1233	Excavate, pull, & load tank, backfill hole, 3,000 – 5,000 gal.	G	B-11M	.50	32			980	700	1,680	2,400
1243	Haul tank to certified dump, 100 miles rt, 500 gal.	G	B-34L	1	8	↓		268	171	439	630
2010	Decontamination of soil on site incl poly tarp on top/bottom										
2011	Soil containment berm, and chemical treatment										
2020	Minimum	G	B-11C	100	.160	C.Y.	7.05	4.91	3.20	15.16	19.40
2021	Maximum	G	"	100	.160		9.15	4.91	3.20	17.26	21.50
2050	Disposal of decontaminated soil, minimum									125	140
2055	Maximum					↓				400	440

02 81 Transportation and Disposal of Hazardous Materials

02 81 20 – Hazardous Waste Handling

02 81 20.10	Hazardous Waste Cleanup/Pickup/Disposal	Crew	Daily Output	Labor-Hours	Unit	Material	Labor	Equipment	Total	Total Incl O&P
0010	**HAZARDOUS WASTE CLEANUP/PICKUP/DISPOSAL**									
0100	For contractor rental equipment, i.e., Dozer,									
0110	Front end loader, Dump truck, etc., see 01 54 33 Reference Section									
1000	Solid pickup									
1100	55 gal. drums				Ea.				225	250
1120	Bulk material, minimum				Ton				170	190
1130	Maximum				"				575	635
1200	Transportation to disposal site									
1220	Truckload = 80 drums or 25 C.Y. or 18 tons									
1260	Minimum				Mile				3.50	3.85
1270	Maximum				"				6.40	7.05
3000	Liquid pickup, vacuum truck, stainless steel tank									
3100	Minimum charge, 4 hours									
3110	1 compartment, 2200 gallon				Hr.				120	135
3120	2 compartment, 5000 gallon				"				170	190
3400	Transportation in 6900 gallon bulk truck				Mile				7	7.70
3410	In teflon lined truck				"				9	9.90
5000	Heavy sludge or dry vacuumable material				Hr.				125	140
6000	Dumpsite disposal charge, minimum				Ton				125	140
6020	Maximum				"				400	440

02 82 Asbestos Remediation

02 82 13 – Asbestos Abatement

02 82 13.39 Asbestos Remediation Plans and Methods		Crew	Daily Output	Labor-Hours	Unit	2011 Bare Costs Material	Labor	Equipment	Total	Total Incl O&P
0010	**ASBESTOS REMEDIATION PLANS AND METHODS**									
0100	Building Survey-Commercial Building				Ea.				2,000	2,200
0200	Asbestos Abatement Remediation Plan				"				1,200	1,325

02 82 13.41 Asbestos Abatement Equip.		Crew	Daily Output	Labor-Hours	Unit	Material	Labor	Equipment	Total	Total Incl O&P
0010	**ASBESTOS ABATEMENT EQUIP.** R028213-20									
0011	Equipment and supplies, buy									
0200	Air filtration device, 2000 C.F.M.				Ea.	870			870	955
0250	Large volume air sampling pump, minimum					355			355	390
0260	Maximum					520			520	570
0300	Airless sprayer unit, 2 gun					4,800			4,800	5,300
0350	Light stand, 500 watt				↓	46			46	50.50
0400	Personal respirators									
0410	Negative pressure, 1/2 face, dual operation, min.				Ea.	25.50			25.50	28
0420	Maximum					31.50			31.50	35
0450	P.A.P.R., full face, minimum					107			107	118
0460	Maximum					195			195	215
0470	Supplied air, full face, incl. air line, minimum					187			187	206
0480	Maximum					385			385	425
0500	Personnel sampling pump, minimum					221			221	244
1500	Power panel, 20 unit, incl. G.F.I.					550			550	610
1600	Shower unit, including pump and filters					1,175			1,175	1,275
1700	Supplied air system (type C)					5,325			5,325	5,875
1750	Vacuum cleaner, HEPA, 16 gal., stainless steel, wet/dry					1,175			1,175	1,300
1760	55 gallon					1,925			1,925	2,100
1800	Vacuum loader, 9 – 18 ton/hr.					90,000			90,000	99,000
1900	Water atomizer unit, including 55 gal. drum					230			230	253
2000	Worker protection, whole body, foot, head cover & gloves, plastic					8.70			8.70	9.60
2500	Respirator, single use					29			29	32
2550	Cartridge for respirator					6.85			6.85	7.55
2570	Glove bag, 7 mil, 50" x 64"					17.25			17.25	19
2580	10 mil, 44" x 60"					5.40			5.40	5.95
3000	HEPA vacuum for work area, minimum					1,375			1,375	1,500
3050	Maximum					1,425			1,425	1,575
6000	Disposable polyethylene bags, 6 mil, 3 C.F.					2.30			2.30	2.53
6300	Disposable fiber drums, 3 C.F.					16.90			16.90	18.60
6400	Pressure sensitive caution labels, 3" x 5"					3.16			3.16	3.48
6450	11" x 17"					7.05			7.05	7.75
6500	Negative air machine, 1800 C.F.M.				↓	820			820	905

02 82 13.42 Preparation of Asbestos Containment Area		Crew	Daily Output	Labor-Hours	Unit	Material	Labor	Equipment	Total	Total Incl O&P
0010	**PREPARATION OF ASBESTOS CONTAINMENT AREA**									
0100	Pre-cleaning, HEPA vacuum and wet wipe, flat surfaces	A-9	12000	.005	S.F.	.02	.19		.21	.33
0200	Protect carpeted area, 2 layers 6 mil poly on 3/4" plywood	"	1000	.064		2	2.22		4.22	5.95
0300	Separation barrier, 2" x 4" @ 16", 1/2" plywood ea. side, 8' high	2 Carp	400	.040		3	1.34		4.34	5.55
0310	12' high		320	.050		3	1.68		4.68	6.10
0320	16' high		200	.080		2	2.68		4.68	6.70
0400	Personnel decontam. chamber, 2" x 4" @ 16", 3/4" ply ea. side		280	.057		4	1.92		5.92	7.60
0450	Waste decontam. chamber, 2" x 4" studs @ 16", 3/4" ply ea. side	↓	360	.044	↓	4	1.49		5.49	6.90
0500	Cover surfaces with polyethylene sheeting									
0501	Including glue and tape									
0550	Floors, each layer, 6 mil	A-9	8000	.008	S.F.	.05	.28		.33	.53
0551	4 mil		9000	.007		.04	.25		.29	.46
0560	Walls, each layer, 6 mil	↓	6000	.011	↓	.05	.37		.42	.69

02 82 13.42 Preparation of Asbestos Containment Area		Crew	Daily Output	Labor-Hours	Unit	2011 Bare Costs Material	Labor	Equipment	Total	Total Incl O&P
0561	4 mil	A-9	7000	.009	S.F.	.04	.32		.36	.58
0570	For heights above 12′, add						20%			
0575	For heights above 20′, add						30%			
0580	For fire retardant poly, add					100%				
0590	For large open areas, deduct					10%	20%			
0600	Seal floor penetrations with foam firestop to 36 sq. in.	2 Carp	200	.080	Ea.	7.60	2.68		10.28	12.90
0610	36 sq. in. to 72 sq. in.		125	.128		15.25	4.29		19.54	24
0615	72 sq. in. to 144 sq. in.		80	.200		30.50	6.70		37.20	45
0620	Wall penetrations, to 36 square inches		180	.089		7.60	2.98		10.58	13.40
0630	36 sq. in. to 72 sq. in.		100	.160		15.25	5.35		20.60	26
0640	72 sq. in. to 144 sq. in.		60	.267		30.50	8.95		39.45	48.50
0800	Caulk seams with latex	1 Carp	230	.035	L.F.	.15	1.17		1.32	2.12

02 82 13.43 Bulk Asbestos Removal

		Crew	Daily Output	Labor-Hours	Unit	Material	Labor	Equipment	Total	Total Incl O&P
0010	**BULK ASBESTOS REMOVAL**									
0020	Includes disposable tools and 2 suits and 1respirator filter/day/worker									
0100	Beams, W 10 x 19	A-9	235	.272	L.F.	.83	9.45		10.28	16.85
0110	W 12 x 22		210	.305		.93	10.60		11.53	18.85
0120	W 14 x 26		180	.356		1.08	12.35		13.43	22
0130	W 16 x 31		160	.400		1.22	13.90		15.12	25
0140	W 18 x 40		140	.457		1.39	15.90		17.29	28.50
0150	W 24 x 55		110	.582		1.77	20		21.77	36
0160	W 30 x 108		85	.753		2.29	26		28.29	46.50
0170	W 36 x 150		72	.889		2.70	31		33.70	55
0200	Boiler insulation		480	.133	S.F.	.48	4.63		5.11	8.30
0210	With metal lath, add				%				50%	
0300	Boiler breeching or flue insulation	A-9	520	.123	S.F.	.37	4.28		4.65	7.60
0310	For active boiler, add				%				100%	
0400	Duct or AHU insulation	A-10B	440	.073	S.F.	.22	2.53		2.75	4.51
0500	Duct vibration isolation joints, up to 24 sq. in. duct	A-9	56	1.143	Ea.	3.47	39.50		42.97	71
0520	25 sq. in. to 48 sq. in. duct		48	1.333		4.05	46.50		50.55	82.50
0530	49 sq. in. to 76 sq. in. duct		40	1.600		4.86	55.50		60.36	99.50
0600	Pipe insulation, air cell type, up to 4″ diameter pipe		900	.071	L.F.	.22	2.47		2.69	4.41
0610	4″ to 8″ diameter pipe		800	.080		.24	2.78		3.02	4.96
0620	10″ to 12″ diameter pipe		700	.091		.28	3.18		3.46	5.65
0630	14″ to 16″ diameter pipe		550	.116		.35	4.04		4.39	7.20
0650	Over 16″ diameter pipe		650	.098	S.F.	.30	3.42		3.72	6.10
0700	With glove bag up to 3″ diameter pipe		200	.320	L.F.	6.05	11.10		17.15	25.50
1000	Pipe fitting insulation up to 4″ diameter pipe		320	.200	Ea.	.61	6.95		7.56	12.35
1100	6″ to 8″ diameter pipe		304	.211		.64	7.30		7.94	13.05
1110	10″ to 12″ diameter pipe		192	.333		1.01	11.60		12.61	20.50
1120	14″ to 16″ diameter pipe		128	.500		1.52	17.40		18.92	31
1130	Over 16″ diameter pipe		176	.364	S.F.	1.11	12.65		13.76	22.50
1200	With glove bag, up to 8″ diameter pipe		75	.853	L.F.	6.55	29.50		36.05	57
2000	Scrape foam fireproofing from flat surface		2400	.027	S.F.	.08	.93		1.01	1.65
2100	Irregular surfaces		1200	.053		.16	1.85		2.01	3.31
3000	Remove cementitious material from flat surface		1800	.036		.11	1.24		1.35	2.20
3100	Irregular surface		1000	.064		.14	2.22		2.36	3.90
4000	Scrape acoustical coating/fireproofing, from ceiling		3200	.020		.06	.70		.76	1.24
5000	Remove VAT and mastic from floor by hand		2400	.027		.08	.93		1.01	1.65
5100	By machine	A-11	4800	.013		.04	.46	.01	.51	.83
5150	For 2 layers, add				%				50%	
6000	Remove contaminated soil from crawl space by hand	A-9	400	.160	C.F.	.49	5.55		6.04	9.95

02 82 Asbestos Remediation

02 82 13 – Asbestos Abatement

02 82 13.43 Bulk Asbestos Removal		Crew	Daily Output	Labor-Hours	Unit	Material	Labor (2011 Bare Costs)	Equipment (2011 Bare Costs)	Total (2011 Bare Costs)	Total Incl O&P
6100	With large production vacuum loader	A-12	700	.091	C.F.	.28	3.18	1.05	4.51	6.80
7000	Radiator backing, not including radiator removal	A-9	1200	.053	S.F.	.16	1.85		2.01	3.31
8000	Cement-asbestos transite board	2 Asbe	1000	.016		.14	.56		.70	1.09
8100	Transite shingle siding	A-10B	750	.043		.21	1.49		1.70	2.74
8200	Shingle roofing	″	2000	.016		.08	.56		.64	1.03
8250	Built-up, no gravel, non-friable	B-2	1400	.029		.08	.77		.85	1.37
8260	Bituminous flashing	1 Rofc	300	.027		.08	.75		.83	1.44
8300	Asbestos millboard	2 Asbe	1000	.016		.08	.56		.64	1.03
9000	For type B (supplied air) respirator equipment, add				%					10%

02 82 13.44 Demolition In Asbestos Contaminated Area

Line	Description	Crew	Daily Output	Labor-Hours	Unit	Material	Labor	Equipment	Total	Total Incl O&P
0010	**DEMOLITION IN ASBESTOS CONTAMINATED AREA**									
0200	Ceiling, including suspension system, plaster and lath	A-9	2100	.030	S.F.	.09	1.06		1.15	1.89
0210	Finished plaster, leaving wire lath		585	.109		.33	3.80		4.13	6.75
0220	Suspended acoustical tile		3500	.018		.06	.64		.70	1.13
0230	Concealed tile grid system		3000	.021		.06	.74		.80	1.32
0240	Metal pan grid system		1500	.043		.13	1.48		1.61	2.64
0250	Gypsum board		2500	.026		.08	.89		.97	1.59
0260	Lighting fixtures up to 2′ x 4′		72	.889	Ea.	2.70	31		33.70	55
0400	Partitions, non load bearing									
0410	Plaster, lath, and studs	A-9	690	.093	S.F.	.85	3.22		4.07	6.40
0450	Gypsum board and studs	″	1390	.046	″	.14	1.60		1.74	2.85
9000	For type B (supplied air) respirator equipment, add				%					10%

02 82 13.45 OSHA Testing

Line	Description	Crew	Daily Output	Labor-Hours	Unit	Material	Labor	Equipment	Total	Total Incl O&P
0010	**OSHA TESTING**									
0100	Certified technician, minimum				Day				200	220
0110	Maximum								300	330
0120	Industrial hygienist, minimum								300	330
0130	Maximum								400	440
0200	Asbestos sampling and PCM analysis, NIOSH 7400, minimum	1 Asbe	8	1	Ea.	2.75	34.50		37.25	61.50
0210	Maximum		4	2		3	69.50		72.50	120
1000	Cleaned area samples		8	1		2.63	34.50		37.13	61.50
1100	PCM air sample analysis, NIOSH 7400, minimum		8	1		30.50	34.50		65	92.50
1110	Maximum		4	2		3.20	69.50		72.70	121
1200	TEM air sample analysis, NIOSH 7402, minimum								90	120
1210	Maximum								360	450

02 82 13.46 Decontamination of Asbestos Containment Area

Line	Description	Crew	Daily Output	Labor-Hours	Unit	Material	Labor	Equipment	Total	Total Incl O&P
0010	**DECONTAMINATION OF ASBESTOS CONTAINMENT AREA**									
0100	Spray exposed substrate with surfactant (bridging)									
0200	Flat surfaces	A-9	6000	.011	S.F.	.36	.37		.73	1.03
0250	Irregular surfaces		4000	.016	″	.31	.56		.87	1.28
0300	Pipes, beams, and columns		2000	.032	L.F.	.56	1.11		1.67	2.50
1000	Spray encapsulate polyethylene sheeting		8000	.008	S.F.	.31	.28		.59	.81
1100	Roll down polyethylene sheeting		8000	.008	″		.28		.28	.47
1500	Bag polyethylene sheeting		400	.160	Ea.	.77	5.55		6.32	10.25
2000	Fine clean exposed substrate, with nylon brush		2400	.027	S.F.		.93		.93	1.56
2500	Wet wipe substrate		4800	.013			.46		.46	.78
2600	Vacuum surfaces, fine brush		6400	.010			.35		.35	.59
3000	Structural demolition									
3100	Wood stud walls	A-9	2800	.023	S.F.		.79		.79	1.34
3500	Window manifolds, not incl. window replacement		4200	.015			.53		.53	.89
3600	Plywood carpet protection		2000	.032			1.11		1.11	1.88
4000	Remove custom decontamination facility	A-10A	8	3	Ea.	15	105		120	193

02 82 Asbestos Remediation

02 82 13 – Asbestos Abatement

02 82 13.46 Decontamination of Asbestos Containment Area		Crew	Daily Output	Labor-Hours	Unit	Material	2011 Bare Costs Labor	Equipment	Total	Total Incl O&P
4100	Remove portable decontamination facility	3 Asbe	12	2	Ea.	12.75	69.50		82.25	131
5000	HEPA vacuum, shampoo carpeting	A-9	4800	.013	S.F.	.05	.46		.51	.84
9000	Final cleaning of protected surfaces	A-10A	8000	.003	″		.10		.10	.18

02 82 13.47 Asbestos Waste Pkg., Handling, and Disp.

Code	Description	Crew	Daily Output	Labor-Hours	Unit	Material	Labor	Equipment	Total	Total Incl O&P
0010	**ASBESTOS WASTE PACKAGING, HANDLING, AND DISPOSAL**									
0100	Collect and bag bulk material, 3 C.F. bags, by hand	A-9	400	.160	Ea.	2.30	5.55		7.85	11.95
0200	Large production vacuum loader	A-12	880	.073		.80	2.53	.83	4.16	6.05
1000	Double bag and decontaminate	A-9	960	.067		2.30	2.32		4.62	6.45
2000	Containerize bagged material in drums, per 3 C.F. drum	″	800	.080		16.90	2.78		19.68	23.50
3000	Cart bags 50′ to dumpster	2 Asbe	400	.040			1.39		1.39	2.34
5000	Disposal charges, not including haul, minimum				C.Y.				60	66
5020	Maximum				″				350	385
5100	Remove refrigerant from system	1 Stpi	40	.200	Lb.		8		8	13.10
9000	For type B (supplied air) respirator equipment, add				%					10%

02 82 13.48 Asbestos Encapsulation With Sealants

Code	Description	Crew	Daily Output	Labor-Hours	Unit	Material	Labor	Equipment	Total	Total Incl O&P
0010	**ASBESTOS ENCAPSULATION WITH SEALANTS**									
0100	Ceilings and walls, minimum	A-9	21000	.003	S.F.	.27	.11		.38	.48
0110	Maximum		10600	.006		.42	.21		.63	.81
0200	Columns and beams, minimum		13300	.005		.27	.17		.44	.58
0210	Maximum		5325	.012		.47	.42		.89	1.22
0300	Pipes to 12″ diameter including minor repairs, minimum		800	.080	L.F.	.37	2.78		3.15	5.10
0310	Maximum		400	.160	″	1.04	5.55		6.59	10.55

02 83 Lead Remediation

02 83 19 – Lead-Based Paint Remediation

02 83 19.21 Lead Paint Remediation Plans and Methods

Code	Description	Crew	Daily Output	Labor-Hours	Unit	Material	Labor	Equipment	Total	Total Incl O&P
0010	**LEAD PAINT REMEDIATION PLANS AND METHODS**									
0100	Building Survey-Commercial Building				Ea.				2,000	2,200
0200	Lead Abatement Remediation Plan								1,200	1,325
0300	Lead Paint Testing, AAS Analysis								50	55
0400	Lead Paint Testing, X-Ray Fluorescence								50	55

02 83 19.23 Encapsulation of Lead-Based Paint

Code	Description	Crew	Daily Output	Labor-Hours	Unit	Material	Labor	Equipment	Total	Total Incl O&P
0010	**ENCAPSULATION OF LEAD-BASED PAINT**									
0020	Interior, brushwork, trim, under 6″	1 Pord	240	.033	L.F.	2.46	.99		3.45	4.31
0030	6″ to 12″ wide		180	.044		3.28	1.32		4.60	5.75
0040	Balustrades		300	.027		1.98	.79		2.77	3.46
0050	Pipe to 4″ diameter		500	.016		1.19	.47		1.66	2.08
0060	To 8″ diameter		375	.021		1.57	.63		2.20	2.76
0070	To 12″ diameter		250	.032		2.36	.95		3.31	4.14
0080	To 16″ diameter		170	.047		3.47	1.40		4.87	6.10
0090	Cabinets, ornate design		200	.040	S.F.	2.97	1.19		4.16	5.20
0100	Simple design		250	.032	″	2.36	.95		3.31	4.14
0110	Doors, 3′ x 7′, both sides, incl. frame & trim									
0120	Flush	1 Pord	6	1.333	Ea.	30.50	39.50		70	97.50
0130	French, 10 – 15 lite		3	2.667		6.05	79		85.05	135
0140	Panel		4	2		36.50	59.50		96	137
0150	Louvered		2.75	2.909		33.50	86.50		120	177
0160	Windows, per interior side, per 15 S.F.									
0170	1 to 6 lite	1 Pord	14	.571	Ea.	21	16.95		37.95	50.50
0180	7 to 10 lite		7.50	1.067		23	31.50		54.50	77

02 83 Lead Remediation

02 83 19 – Lead-Based Paint Remediation

02 83 19.23 Encapsulation of Lead-Based Paint		Crew	Daily Output	Labor-Hours	Unit	Material	2011 Bare Costs Labor	Equipment	Total	Total Incl O&P
0190	12 lite	1 Pord	5.75	1.391	Ea.	31	41.50		72.50	101
0200	Radiators		8	1		73.50	29.50		103	129
0210	Grilles, vents		275	.029	S.F.	2.15	.86		3.01	3.77
0220	Walls, roller, drywall or plaster		1000	.008		.59	.24		.83	1.04
0230	With spunbonded reinforcing fabric		720	.011		.68	.33		1.01	1.28
0240	Wood		800	.010		.74	.30		1.04	1.29
0250	Ceilings, roller, drywall or plaster		900	.009		.68	.26		.94	1.18
0260	Wood		700	.011		.84	.34		1.18	1.47
0270	Exterior, brushwork, gutters and downspouts		300	.027	L.F.	1.98	.79		2.77	3.46
0280	Columns		400	.020	S.F.	1.47	.59		2.06	2.58
0290	Spray, siding		600	.013	"	.99	.40		1.39	1.73
0300	Miscellaneous									
0310	Electrical conduit, brushwork, to 2" diameter	1 Pord	500	.016	L.F.	1.19	.47		1.66	2.08
0320	Brick, block or concrete, spray		500	.016	S.F.	1.19	.47		1.66	2.08
0330	Steel, flat surfaces and tanks to 12"		500	.016		1.19	.47		1.66	2.08
0340	Beams, brushwork		400	.020		1.47	.59		2.06	2.58
0350	Trusses		400	.020		1.47	.59		2.06	2.58

02 83 19.26 Removal of Lead-Based Paint

		Crew	Daily Output	Labor-Hours	Unit	Material	Labor	Equipment	Total	Total Incl O&P
0010	**REMOVAL OF LEAD-BASED PAINT** R028319-60									
0011	By chemicals, per application									
0050	Baseboard, to 6" wide	1 Pord	64	.125	L.F.	.66	3.71		4.37	6.75
0070	To 12" wide		32	.250	"	1.32	7.40		8.72	13.50
0200	Balustrades, one side		28	.286	S.F.	1.32	8.45		9.77	15.20
1400	Cabinets, simple design		32	.250		1.32	7.40		8.72	13.50
1420	Ornate design		25	.320		1.32	9.50		10.82	16.85
1600	Cornice, simple design		60	.133		1.32	3.95		5.27	7.85
1620	Ornate design		20	.400		5.05	11.85		16.90	25
2800	Doors, one side, flush		84	.095		1	2.82		3.82	5.70
2820	Two panel		80	.100		1.32	2.97		4.29	6.25
2840	Four panel		45	.178		1.32	5.25		6.57	10
2880	For trim, one side, add		64	.125	L.F.	.66	3.71		4.37	6.75
3000	Fence, picket, one side		30	.267	S.F.	1.32	7.90		9.22	14.30
3200	Grilles, one side, simple design		30	.267		1.32	7.90		9.22	14.30
3220	Ornate design		25	.320		1.32	9.50		10.82	16.85
4400	Pipes, to 4" diameter		90	.089	L.F.	1	2.64		3.64	5.40
4420	To 8" diameter		50	.160		2	4.74		6.74	9.90
4440	To 12" diameter		36	.222		3	6.60		9.60	14
4460	To 16" diameter		20	.400		4	11.85		15.85	23.50
4500	For hangers, add		40	.200	Ea.	2.54	5.95		8.49	12.45
4800	Siding		90	.089	S.F.	1.17	2.64		3.81	5.55
5000	Trusses, open		55	.145	SF Face	1.85	4.31		6.16	9.05
6200	Windows, one side only, double hung, 1/1 light, 24" x 48" high		4	2	Ea.	25.50	59.50		85	125
6220	30" x 60" high		3	2.667		34	79		113	166
6240	36" x 72" high		2.50	3.200		41	95		136	199
6280	40" x 80" high		2	4		51	119		170	249
6400	Colonial window, 6/6 light, 24" x 48" high		2	4		51	119		170	250
6420	30" x 60" high		1.50	5.333		68	158		226	330
6440	36" x 72" high		1	8		102	237		339	495
6480	40" x 80" high		1	8		102	237		339	495
6600	8/8 light, 24" x 48" high		2	4		51	119		170	250
6620	40" x 80" high		1	8		102	237		339	495
6800	12/12 light, 24" x 48" high		1	8		102	237		339	495

02 83 Lead Remediation

02 83 19 – Lead-Based Paint Remediation

02 83 19.26 Removal of Lead-Based Paint		Crew	Daily Output	Labor-Hours	Unit	Material	2011 Bare Costs Labor	Equipment	Total	Total Incl O&P
6820	40" x 80" high	1 Pord	.75	10.667	Ea.	136	315		451	665
6840	Window frame & trim items, included in pricing above									

02 85 Mold Remediation

02 85 16 – Mold Remediation Preparation and Containment

02 85 16.40 Mold Remediation Plans and Methods

		Crew	Daily Output	Labor-Hours	Unit	Material	Labor	Equipment	Total	Total Incl O&P
0010	**MOLD REMEDIATION PLANS AND METHODS**									
0020	Initial inspection, average 3 bedroom home				Total				250	275
0030	Average 5 bedroom home								400	440
0040	Testing, air sample each								125	138
0050	Swab sample								125	138
0060	Tape sample								125	140
0070	Post remediation air test								125	140
0080	Mold abatement plan, average 3 bedroom home								1,200	1,300
0090	Average 5 bedroom home								1,600	1,750
0100	Packup & removal of contents, average 3 bedroom home, excl storage								8,000	8,800
0110	Average 5 bedroom home, excl storage								15,000	16,500
0610	For personal protection equipment, see Section 02 82 13.41									

02 85 16.50 Preparation of Mold Containment Area

		Crew	Daily Output	Labor-Hours	Unit	Material	Labor	Equipment	Total	Total Incl O&P
0010	**PREPARATION OF MOLD CONTAINMENT AREA**									
0100	Pre-cleaning, HEPA vacuum and wet wipe, flat surfaces	A-9	12000	.005	S.F.	.02	.19		.21	.33
0300	Separation barrier, 2" x 4" @ 16", 1/2" plywood ea. side, 8' high	2 Carp	400	.040		3	1.34		4.34	5.55
0310	12' high		320	.050		3	1.68		4.68	6.10
0320	16' high		200	.080		2	2.68		4.68	6.70
0400	Personnel decontam. chamber, 2" x 4" @ 16", 3/4" ply ea. side		280	.057		4	1.92		5.92	7.60
0450	Waste decontam. chamber, 2" x 4" studs @ 16", 3/4" ply each side		360	.044		4	1.49		5.49	6.90
0500	Cover surfaces with polyethylene sheeting									
0501	Including glue and tape									
0550	Floors, each layer, 6 mil	A-9	8000	.008	S.F.	.06	.28		.34	.54
0551	4 mil		9000	.007		.04	.25		.29	.47
0560	Walls, each layer, 6 mil		6000	.011		.09	.37		.46	.73
0561	4 mil		7000	.009		.07	.32		.39	.62
0570	For heights above 12', add						20%			
0575	For heights above 20', add						30%			
0580	For fire retardant poly, add					100%				
0590	For large open areas, deduct					10%	20%			
0600	Seal floor penetrations with foam firestop to 36 sq. in.	2 Carp	200	.080	Ea.	7.60	2.68		10.28	12.90
0610	36 sq. in. to 72 sq. in.		125	.128		15.25	4.29		19.54	24
0615	72 sq. in. to 144 sq. in.		80	.200		30.50	6.70		37.20	45
0620	Wall penetrations, to 36 square inches		180	.089		7.60	2.98		10.58	13.40
0630	36 sq. in. to 72 sq. in.		100	.160		15.25	5.35		20.60	26
0640	72 sq. in. to 144 sq. in.		60	.267		30.50	8.95		39.45	48.50
0800	Caulk seams with latex caulk	1 Carp	230	.035	L.F.	.15	1.17		1.32	2.12

02 85 33 – Removal and Disposal of Materials with Mold

02 85 33.50 Demolition in Mold Contaminated Area

		Crew	Daily Output	Labor-Hours	Unit	Material	Labor	Equipment	Total	Total Incl O&P
0010	**DEMOLITION IN MOLD CONTAMINATED AREA**									
0200	Ceiling, including suspension system, plaster and lath	A-9	2100	.030	S.F.	.09	1.06		1.15	1.89
0210	Finished plaster, leaving wire lath		585	.109		.33	3.80		4.13	6.75
0220	Suspended acoustical tile		3500	.018		.06	.64		.70	1.13
0230	Concealed tile grid system		3000	.021		.06	.74		.80	1.32

02 85 Mold Remediation

02 85 33 – Removal and Disposal of Materials with Mold

02 85 33.50 Demolition in Mold Contaminated Area		Crew	Daily Output	Labor-Hours	Unit	Material	2011 Bare Costs Labor	Equipment	Total	Total Incl O&P
0240	Metal pan grid system	A-9	1500	.043	S.F.	.13	1.48		1.61	2.64
0250	Gypsum board		2500	.026		.08	.89		.97	1.59
0255	Plywood		2500	.026		.08	.89		.97	1.59
0260	Lighting fixtures up to 2′ x 4′		72	.889	Ea.	2.70	31		33.70	55
0400	Partitions, non load bearing									
0410	Plaster, lath, and studs	A-9	690	.093	S.F.	.85	3.22		4.07	6.40
0450	Gypsum board and studs		1390	.046		.14	1.60		1.74	2.85
0465	Carpet & pad		1390	.046		.14	1.60		1.74	2.85
0600	Pipe insulation, air cell type, up to 4″ diameter pipe		900	.071	L.F.	.22	2.47		2.69	4.41
0610	4″ to 8″ diameter pipe		800	.080		.24	2.78		3.02	4.96
0620	10″ to 12″ diameter pipe		700	.091		.28	3.18		3.46	5.65
0630	14″ to 16″ diameter pipe		550	.116		.35	4.04		4.39	7.20
0650	Over 16″ diameter pipe		650	.098	S.F.	.30	3.42		3.72	6.10
9000	For type B (supplied air) respirator equipment, add				%					10%

02 91 Chemical Sampling, Testing and Analysis

02 91 10 – Monitoring, Sampling, Testing and Analysis

02 91 10.10 Monitoring, Chemical Sampling, Testing and Analysis		Crew	Daily Output	Labor-Hours	Unit	Material	Labor	Equipment	Total	Total Incl O&P
0010	**MONITORING, CHEMICAL SAMPLING, TESTING AND ANALYSIS**									
0015	Field Sampling of waste									
0100	Field samples, sample collection, sludge	1 Skwk	32	.250	Ea.		8.50		8.50	14.25
0110	Contaminated soils	″	32	.250	″		8.50		8.50	14.25
0200	Vials and bottles									
0210	32 oz. clear wide mouth jar (case of 12)				Ea.	41.50			41.50	46
0220	32 oz. Boston round bottle (case of 12)					29			29	32
0230	32 oz. HDPE bottle (case of 12)					17.60			17.60	19.35
0300	Laboratory analytical services									
0310	Laboratory testing 13 metals				Ea.	190			190	209
0312	13 metals + mercury					223			223	245
0314	8 metals					160			160	176
0316	Mercury only					37.50			37.50	41.50
0318	Single metal (only Cs, Li, Sr, Ta)					20			20	22
0320	Single metal (excludes Hg, Cs, Li, Sr, Ta)					17.50			17.50	19.25
0400	Hydrocarbons standard					90			90	99
0410	Hydrocarbons fingerprint					163			163	179
0500	Radioactivity gross alpha					150			150	165
0510	Gross alpha & beta					150			150	165
0520	Radium 226					85			85	93.50
0530	Radium 228					125			125	138
0540	Radon					145			145	160
0550	Uranium					75			75	82.50
0600	Volatile organics without GC/MS					140			140	154
0610	Volatile organics including GC/MS					185			185	204
0630	Synthetic organic compounds					1,000			1,000	1,100
0640	Herbicides					213			213	234
0650	Pesticides					135			135	149
0660	PCB's					135			135	149

Estimating Tips

General

- Carefully check all the plans and specifications. Concrete often appears on drawings other than structural drawings, including mechanical and electrical drawings for equipment pads. The cost of cutting and patching is often difficult to estimate. See Subdivision 03 81 for Concrete Cutting, Subdivision 02 41 19.16 for Cutout Demolition, Subdivision 03 05 05.10 for Concrete Demolition, and Subdivision 02 41 19.23 for Rubbish Handling (handling, loading and hauling of debris).
- Always obtain concrete prices from suppliers near the job site. A volume discount can often be negotiated, depending upon competition in the area. Remember to add for waste, particularly for slabs and footings on grade.

03 10 00 Concrete Forming and Accessories

- A primary cost for concrete construction is forming. Most jobs today are constructed with prefabricated forms. The selection of the forms best suited for the job and the total square feet of forms required for efficient concrete forming and placing are key elements in estimating concrete construction. Enough forms must be available for erection to make efficient use of the concrete placing equipment and crew.
- Concrete accessories for forming and placing depend upon the systems used. Study the plans and specifications to ensure that all special accessory requirements have been included in the cost estimate, such as anchor bolts, inserts, and hangers.
- Included within costs for forms-in-place are all necessary bracing and shoring.

03 20 00 Concrete Reinforcing

- Ascertain that the reinforcing steel supplier has included all accessories, cutting, bending, and an allowance for lapping, splicing, and waste. A good rule of thumb is 10% for lapping, splicing, and waste. Also, 10% waste should be allowed for welded wire fabric.
- The unit price items in the subdivision for Reinforcing In Place include the labor to install accessories such as beam and slab bolsters, high chairs, and bar ties and tie wire. The material cost for these accessories is not included; they may be obtained from the Accessories Division.

03 30 00 Cast-In-Place Concrete

- When estimating structural concrete, pay particular attention to requirements for concrete additives, curing methods, and surface treatments. Special consideration for climate, hot or cold, must be included in your estimate. Be sure to include requirements for concrete placing equipment, and concrete finishing.
- For accurate concrete estimating, the estimator must consider each of the following major components individually: forms, reinforcing steel, ready-mix concrete, placement of the concrete, and finishing of the top surface. For faster estimating, Subdivision 03 30 53.40 for Concrete-In-Place can be used; here, various items of concrete work are presented that include the costs of all five major components (unless specifically stated otherwise).

03 40 00 Precast Concrete
03 50 00 Cast Decks and Underlayment

- The cost of hauling precast concrete structural members is often an important factor. For this reason, it is important to get a quote from the nearest supplier. It may become economically feasible to set up precasting beds on the site if the hauling costs are prohibitive.

Reference Numbers

Reference numbers are shown in shaded boxes at the beginning of some major classifications. These numbers refer to related items in the Reference Section. The reference information may be an estimating procedure, an alternate pricing method, or technical information.

Note: Not all subdivisions listed here necessarily appear in this publication.

03 01 Maintenance of Concrete

03 01 30 – Maintenance of Cast-In-Place Concrete

03 01 30.62 Concrete Patching		Crew	Daily Output	Labor-Hours	Unit	Material	2011 Bare Costs Labor	Equipment	Total	Total Incl O&P
0010	CONCRETE PATCHING									
0100	Floors, 1/4" thick, small areas, regular grout	1 Cefi	170	.047	S.F.	1.15	1.50		2.65	3.65
0150	Epoxy grout	"	100	.080	"	7.65	2.55		10.20	12.45
2000	Walls, including chipping, cleaning and epoxy grout									
2100	1/4" deep	1 Cefi	65	.123	S.F.	7.50	3.92		11.42	14.50
2150	1/2" deep		50	.160		15.05	5.10		20.15	24.50
2200	3/4" deep		40	.200		22.50	6.35		28.85	35

03 05 Common Work Results for Concrete

03 05 05 – Selective Concrete Demolition

03 05 05.10 Selective Demolition, Concrete

Line	Description	Crew	Daily Output	Labor-Hours	Unit	Material	Labor	Equipment	Total	Total Incl O&P
0010	SELECTIVE DEMOLITION, CONCRETE R024119-10									
0012	Excludes saw cutting, torch cutting, loading or hauling									
0050	Break up into small pieces, minimum reinforcing	B-9	24	1.667	C.Y.		45	8.15	53.15	84
0060	Average reinforcing		16	2.500			67	12.20	79.20	125
0070	Maximum reinforcing		8	5			134	24.50	158.50	252
0150	Remove whole pieces, up to 2 tons per piece	E-18	36	1.111	Ea.		40.50	29	69.50	107
0160	2 – 5 tons per piece		30	1.333			48.50	34.50	83	129
0170	5 – 10 tons per piece		24	1.667			61	43	104	161
0180	10 – 15 tons per piece		18	2.222			81	57.50	138.50	213
0250	Precast unit embedded in masonry, up to 1 C.F.	D-1	16	1			30.50		30.50	50
0260	1 – 2 C.F.		12	1.333			40.50		40.50	66.50
0270	2 – 5 C.F.		10	1.600			48.50		48.50	80
0280	5 – 10 C.F.		8	2			61		61	100
0990	For hydrodemolition see Section 02 41 13.15									

03 05 13 – Basic Concrete Materials

03 05 13.20 Concrete Admixtures and Surface Treatments

Line	Description	Crew	Daily Output	Labor-Hours	Unit	Material	Labor	Equipment	Total	Total Incl O&P
0010	CONCRETE ADMIXTURES AND SURFACE TREATMENTS									
0040	Abrasives, aluminum oxide, over 20 tons				Lb.	1.48			1.48	1.63
0050	1 to 20 tons					1.63			1.63	1.80
0070	Under 1 ton					1.70			1.70	1.87
0100	Silicon carbide, black, over 20 tons					2.15			2.15	2.37
0110	1 to 20 tons					2.49			2.49	2.74
0120	Under 1 ton					2.59			2.59	2.85
0200	Air entraining agent, .7 to 1.5 oz. per bag, 55 gallon drum				Gal.	10.55			10.55	11.60
0220	5 gallon pail					14.50			14.50	15.95
0300	Bonding agent, acrylic latex, 250 S.F. per gallon, 5 gallon pail					18.85			18.85	20.50
0320	Epoxy resin, 80 S.F. per gallon, 4 gallon case					55.50			55.50	61.50
0400	Calcium chloride, 50 lb. bags, TL lots				Ton	610			610	670
0420	Less than truckload lots				Bag	19.05			19.05	21
0500	Carbon black, liquid, 2 to 8 lb. per bag of cement				Lb.	6.85			6.85	7.55
0600	Colors, integral, 2 to 10 lb. per bag of cement, minimum					2.52			2.52	2.77
0610	Average					3.46			3.46	3.81
0620	Maximum					5.50			5.50	6.05
0920	Dustproofing compound, 250 S.F./gal., 5 gallon pail				Gal.	7.40			7.40	8.10
1010	Epoxy based, 125 S.F./gal., 5 gallon pail				"	53			53	58.50
1100	Hardeners, metallic, 55 lb. bags, natural (grey)				Lb.	.71			.71	.78
1200	Colors					.87			.87	.96
1300	Non-metallic, 55 lb. bags, natural grey					.45			.45	.50
1320	Colors					.75			.75	.82

03 05 Common Work Results for Concrete

03 05 13 – Basic Concrete Materials

03 05 13.20 Concrete Admixtures and Surface Treatments

	03 05 13.20 Concrete Admixtures and Surface Treatments	Crew	Daily Output	Labor-Hours	Unit	Material	2011 Bare Costs Labor	Equipment	Total	Total Incl O&P
1550	Release agent, for tilt slabs, 5 gallon pail				Gal.	14.90			14.90	16.35
1570	For forms, 5 gallon pail					11.10			11.10	12.20
1590	Concrete release agent for forms, 100% biodegradeable, zero VOC, 5 gal pail G					16.75			16.75	18.45
1595	55 gallon drum G					14			14	15.40
1600	Sealer, hardener and dustproofer, epoxy-based, 125 S.F./gal., 5 gallon, min					53			53	58.50
1620	5 gallon pail, maximum					58.50			58.50	64
1630	Sealer, solvent-based, 250 S.F./gal., 55 gallon drum					16.75			16.75	18.40
1640	5 gallon pail					24			24	26.50
1650	Sealer, water based, 350 S.F., 55 gallon drum					18.75			18.75	20.50
1660	5 gallon pail					20			20	22
1900	Set retarder, 100 S.F./gal., 1 gallon pail					17.80			17.80	19.60
2000	Waterproofing, integral 1 lb. per bag of cement				Lb.	1.60			1.60	1.76
2100	Powdered metallic, 40 lb. per 100 S.F., minimum					1.90			1.90	2.09
2120	Maximum					2.65			2.65	2.92
3000	For integral colors, 2500 psi (5 bag mix)									
3100	Red, yellow or brown, 1.8 lb. per bag, add				C.Y.	22.50			22.50	25
3200	9.4 lb. per bag, add					118			118	130
3400	Black, 1.8 lb. per bag, add					31			31	34.50
3500	7.5 lb. per bag, add					130			130	143
3700	Green, 1.8 lb. per bag, add					49.50			49.50	54.50
3800	7.5 lb. per bag, add					206			206	226
6000	Concrete ready-mix additives, recycled coal fly ash, mixed at plant G				Ton	56			56	61.50
6010	Recycled blast furnace slag, mixed at plant G				″	87			87	95.50

03 05 13.25 Aggregate

	03 05 13.25 Aggregate	Crew	Daily Output	Labor-Hours	Unit	Material	Labor	Equipment	Total	Total Incl O&P
0010	**AGGREGATE** R033105-20									
0100	Lightweight vermiculite or perlite, 4 C.F. bag, C.L. lots G				Bag	19.25			19.25	21
0150	L.C.L. lots R033105-40				″	21.50			21.50	23.50
0250	Sand & stone, loaded at pit, crushed bank gravel				Ton	21.50			21.50	23.50
0350	Sand, washed, for concrete R033105-50					16.60			16.60	18.25
0400	For plaster or brick					16.60			16.60	18.25
0450	Stone, 3/4″ to 1-1/2″					18.60			18.60	20.50
0470	Round, river stone					30			30	33
0500	3/8″ roofing stone & 1/2″ pea stone					24			24	26.50
0550	For trucking 10 miles (5 mile R/T), add to the above	B-34B	117	.068			1.86	4.33	6.19	7.85
0600	30 miles (15 mi R/T), add to the above	″	72	.111			3.02	7.05	10.07	12.75
0850	Sand & stone, loaded at pit, crushed bank gravel				C.Y.	30			30	33
0950	Sand, washed, for concrete					23			23	25.50
1000	For plaster or brick					23			23	25.50
1050	Stone, 3/4″ to 1-1/2″					35.50			35.50	39
1055	Round, river stone					33.50			33.50	37
1100	3/8″ roofing stone & 1/2″ pea stone					39.50			39.50	43.50
1150	For trucking 10 miles (5 mi R/T), add to the above	B-34B	78	.103			2.79	6.50	9.29	11.80
1200	30 miles (15 mi R/T), add to the above	″	48	.167			4.53	10.55	15.08	19.10
1310	Quartz chips, 50 lb. bags				Cwt.	19.20			19.20	21
1330	Silica chips, 50 lb. bags					10.55			10.55	11.60
1410	White marble, 3/8″ to 1/2″, 50 lb. bags					15.60			15.60	17.15
1430	3/4″				Ton	90			90	99

03 05 13.30 Cement

	03 05 13.30 Cement	Crew	Daily Output	Labor-Hours	Unit	Material	Labor	Equipment	Total	Total Incl O&P
0010	**CEMENT** R033105-20									
0240	Portland, Type I/II, TL lots, 94 lb bags				Bag	10.70			10.70	11.80
0250	LTL/LCL lots R033105-30				″	12.45			12.45	13.70
0300	Trucked in bulk, per cwt				Cwt.	7.90			7.90	8.70

03 05 Common Work Results for Concrete

03 05 13 – Basic Concrete Materials

03 05 13.30 Cement			Crew	Daily Output	Labor-Hours	Unit	2011 Bare Costs Material	Labor	Equipment	Total	Total Incl O&P
0400	Type III, high early strength, TL lots, 94 lb bags	R033105-40				Bag	12.50			12.50	13.75
0420	L.T.L. or L.C.L. lots						12.40			12.40	13.65
0500	White, type III, high early strength, T.L. or C.L. lots, bags	R033105-50					25			25	27.50
0520	L.T.L. or L.C.L. lots						25			25	27.50
0600	White, type I, T.L. or C.L. lots, bags						23.50			23.50	25.50
0620	L.T.L. or L.C.L. lots						24.50			24.50	27

03 05 13.80 Waterproofing and Dampproofing

		Crew	Daily Output	Labor-Hours	Unit	Material	Labor	Equipment	Total	Total Incl O&P
0010	**WATERPROOFING AND DAMPPROOFING**									
0050	Integral waterproofing, add to cost of regular concrete				C.Y.	9.60			9.60	10.55

03 05 13.85 Winter Protection

		Crew	Daily Output	Labor-Hours	Unit	Material	Labor	Equipment	Total	Total Incl O&P
0010	**WINTER PROTECTION**									
0012	For heated ready mix, add, minimum				C.Y.	4.25			4.25	4.68
0050	Maximum				"	5.30			5.30	5.85
0100	Temporary heat to protect concrete, 24 hours, minimum	2 Clab	50	.320	M.S.F.	525	8.45		533.45	595
0150	Maximum	"	25	.640	"	685	16.95		701.95	785
0200	Temporary shelter for slab on grade, wood frame/polyethylene sheeting									
0201	Build or remove, minimum	2 Carp	10	1.600	M.S.F.	268	53.50		321.50	385
0210	Maximum	"	3	5.333	"	325	179		504	655
0500	Electrically heated pads, 110 volts, 15 watts per S.F., buy				S.F.	5.90			5.90	6.50
0600	20 watts per S.F., buy					7.85			7.85	8.65
0710	Electrically, heated pads, 15 watts/S.F., 20 uses, minimum					.21			.21	.23
0800	Maximum					.35			.35	.39

03 11 Concrete Forming

03 11 13 – Structural Cast-In-Place Concrete Forming

03 11 13.20 Forms In Place, Beams and Girders

			Crew	Daily Output	Labor-Hours	Unit	Material	Labor	Equipment	Total	Total Incl O&P
0010	**FORMS IN PLACE, BEAMS AND GIRDERS**	R031113-40									
0500	Exterior spandrel, job-built plywood, 12" wide, 1 use	R031113-60	C-2	225	.213	SFCA	2.50	6.40		8.90	13.45
0550	2 use			275	.175		1.31	5.25		6.56	10.20
0600	3 use			295	.163		1	4.88		5.88	9.25
0650	4 use			310	.155		.81	4.64		5.45	8.65
1000	18" wide, 1 use			250	.192		2.19	5.75		7.94	12
1050	2 use			275	.175		1.21	5.25		6.46	10.10
1100	3 use			305	.157		.88	4.72		5.60	8.85
1150	4 use			315	.152		.72	4.57		5.29	8.45
1500	24" wide, 1 use			265	.181		2	5.45		7.45	11.30
1550	2 use			290	.166		1.13	4.96		6.09	9.55
1600	3 use			315	.152		.80	4.57		5.37	8.55
1650	4 use			325	.148		.65	4.43		5.08	8.10
2000	Interior beam, job-built plywood, 12" wide, 1 use			300	.160		3.02	4.80		7.82	11.35
2050	2 use			340	.141		1.46	4.23		5.69	8.70
2100	3 use			364	.132		1.21	3.95		5.16	7.95
2150	4 use			377	.127		.98	3.82		4.80	7.50
2500	24" wide, 1 use			320	.150		2.04	4.50		6.54	9.75
2550	2 use			365	.132		1.15	3.94		5.09	7.85
2600	3 use			385	.125		.81	3.74		4.55	7.15
2650	4 use			395	.122		.66	3.64		4.30	6.80
3000	Encasing steel beam, hung, job-built plywood, 1 use			325	.148		2.43	4.43		6.86	10.05
3050	2 use			390	.123		1.33	3.69		5.02	7.60
3100	3 use			415	.116		.97	3.47		4.44	6.85

03 11 13 – Structural Cast-In-Place Concrete Forming

03 11 13.20 Forms In Place, Beams and Girders

	03 11 13.20 Forms In Place, Beams and Girders	Crew	Daily Output	Labor-Hours	Unit	Material	2011 Bare Costs Labor	Equipment	Total	Total Incl O&P
3150	4 use	C-2	430	.112	SFCA	.79	3.35		4.14	6.45
3500	Bottoms only, to 30" wide, job-built plywood, 1 use		230	.209		3.42	6.25		9.67	14.20
3550	2 use		265	.181		1.91	5.45		7.36	11.20
3600	3 use		280	.171		1.37	5.15		6.52	10.10
3650	4 use		290	.166		1.11	4.96		6.07	9.50
4000	Sides only, vertical, 36" high, job-built plywood, 1 use		335	.143		3.46	4.29		7.75	11
4050	2 use		405	.119		1.91	3.55		5.46	8.05
4100	3 use		430	.112		1.39	3.35		4.74	7.10
4150	4 use		445	.108		1.13	3.23		4.36	6.65
4500	Sloped sides, 36" high, 1 use		305	.157		3.41	4.72		8.13	11.65
4550	2 use		370	.130		1.90	3.89		5.79	8.60
4600	3 use		405	.119		1.36	3.55		4.91	7.45
4650	4 use		425	.113		1.11	3.38		4.49	6.85
5000	Upstanding beams, 36" high, 1 use		225	.213		4.14	6.40		10.54	15.25
5050	2 use		255	.188		2.30	5.65		7.95	12
5100	3 use		275	.175		1.67	5.25		6.92	10.60
5150	4 use	↓	280	.171	↓	1.36	5.15		6.51	10.10

03 11 13.25 Forms In Place, Columns

	03 11 13.25 Forms In Place, Columns	Crew	Daily Output	Labor-Hours	Unit	Material	Labor	Equipment	Total	Total Incl O&P
0010	**FORMS IN PLACE, COLUMNS** R031113-40									
0500	Round fiberglass, 4 use per mo., rent, 12" diameter	C-1	160	.200	L.F.	8.15	5.95		14.10	18.90
0550	16" diameter R031113-60		150	.213		9.70	6.35		16.05	21.50
0600	18" diameter		140	.229		10.85	6.80		17.65	23.50
0650	24" diameter		135	.237		13.50	7.05		20.55	26.50
0700	28" diameter		130	.246		15.10	7.30		22.40	29
0800	30" diameter		125	.256		15.75	7.60		23.35	30
0850	36" diameter		120	.267		21	7.95		28.95	36.50
1500	Round fiber tube, recycled paper, 1 use, 8" diameter G		155	.206		1.42	6.15		7.57	11.80
1550	10" diameter G		155	.206		1.97	6.15		8.12	12.40
1600	12" diameter G		150	.213		2.30	6.35		8.65	13.15
1650	14" diameter G		145	.221		3.33	6.55		9.88	14.60
1700	16" diameter G		140	.229		4.06	6.80		10.86	15.80
1720	18" diameter G		140	.229		4.75	6.80		11.55	16.60
1750	20" diameter G		135	.237		6.55	7.05		13.60	19
1800	24" diameter G		130	.246		8.40	7.30		15.70	21.50
1850	30" diameter G		125	.256		11.40	7.60		19	25.50
1900	36" diameter G		115	.278		15.05	8.25		23.30	30.50
1950	42" diameter G		100	.320		38	9.50		47.50	57.50
2000	48" diameter G	↓	85	.376	↓	44	11.20		55.20	67
2200	For seamless type, add					15%				
3000	Round, steel, 4 use per mo., rent, regular duty, 14" diameter G	C-1	145	.221	L.F.	15.15	6.55		21.70	27.50
3050	16" diameter G		125	.256		15.50	7.60		23.10	30
3100	Heavy duty, 20" diameter G		105	.305		17	9.05		26.05	34
3150	24" diameter G		85	.376		18.65	11.20		29.85	39
3200	30" diameter G		70	.457		21.50	13.60		35.10	46
3250	36" diameter G		60	.533		23	15.85		38.85	52
3300	48" diameter G		50	.640		68	19.05		87.05	107
3350	60" diameter G	↓	45	.711	↓	42	21		63	81.50
4500	For second and succeeding months, deduct					50%				
5000	Job-built plywood, 8" x 8" columns, 1 use	C-1	165	.194	SFCA	2.31	5.75		8.06	12.20
5050	2 use		195	.164		1.32	4.88		6.20	9.60
5100	3 use		210	.152		.92	4.53		5.45	8.60
5150	4 use	↓	215	.149	↓	.76	4.43		5.19	8.25

03 11 13 – Structural Cast-In-Place Concrete Forming

03 11 13.25 Forms In Place, Columns		Crew	Daily Output	Labor-Hours	Unit	Material	Labor	Equipment	Total	Total Incl O&P
						2011 Bare Costs				
5500	12" x 12" columns, 1 use	C-1	180	.178	SFCA	2.19	5.30		7.49	11.25
5550	2 use		210	.152		1.20	4.53		5.73	8.90
5600	3 use		220	.145		.87	4.32		5.19	8.20
5650	4 use		225	.142		.71	4.23		4.94	7.85
6000	16" x 16" columns, 1 use		185	.173		2.17	5.15		7.32	11
6050	2 use		215	.149		1.16	4.43		5.59	8.70
6100	3 use		230	.139		.87	4.14		5.01	7.85
6150	4 use		235	.136		.71	4.05		4.76	7.55
6500	24" x 24" columns, 1 use		190	.168		2.43	5		7.43	11.05
6550	2 use		216	.148		1.33	4.40		5.73	8.80
6600	3 use		230	.139		.97	4.14		5.11	7.95
6650	4 use		238	.134		.79	4		4.79	7.55
7000	36" x 36" columns, 1 use		200	.160		1.77	4.76		6.53	9.90
7050	2 use		230	.139		1.01	4.14		5.15	8
7100	3 use		245	.131		.71	3.88		4.59	7.30
7150	4 use	↓	250	.128	↓	.58	3.81		4.39	7
7400	Steel framed plywood, based on 50 uses of purchased									
7420	forms, and 4 uses of bracing lumber									
7500	8" x 8" column	C-1	340	.094	SFCA	1.81	2.80		4.61	6.65
7550	10" x 10"		350	.091		1.58	2.72		4.30	6.30
7600	12" x 12"		370	.086		1.35	2.57		3.92	5.80
7650	16" x 16"		400	.080		1.05	2.38		3.43	5.15
7700	20" x 20"		420	.076		.93	2.27		3.20	4.81
7750	24" x 24"		440	.073		.67	2.16		2.83	4.35
7755	30" x 30"		440	.073		.86	2.16		3.02	4.56
7760	36" x 36"	↓	460	.070	↓	.75	2.07		2.82	4.29

03 11 13.30 Forms In Place, Culvert

Line	Description	Crew	Daily Output	Labor-Hours	Unit	Material	Labor	Equipment	Total	Total Incl O&P
0010	**FORMS IN PLACE, CULVERT** R031113-40									
0015	5' to 8' square or rectangular, 1 use	C-1	170	.188	SFCA	3.10	5.60		8.70	12.75
0050	2 use R031113-60		180	.178		1.83	5.30		7.13	10.85
0100	3 use		190	.168		1.40	5		6.40	9.95
0150	4 use	↓	200	.160	↓	1.19	4.76		5.95	9.25

03 11 13.35 Forms In Place, Elevated Slabs

Line	Description	Crew	Daily Output	Labor-Hours	Unit	Material	Labor	Equipment	Total	Total Incl O&P
0010	**FORMS IN PLACE, ELEVATED SLABS** R031113-40									
1000	Flat plate, job-built plywood, to 15' high, 1 use R031113-60	C-2	470	.102	S.F.	3.16	3.06		6.22	8.60
1050	2 use		520	.092		1.74	2.77		4.51	6.55
1100	3 use		545	.088		1.26	2.64		3.90	5.80
1150	4 use		560	.086		1.03	2.57		3.60	5.45
1500	15' to 20' high ceilings, 4 use		495	.097		1.03	2.91		3.94	6
1600	21' to 35' high ceilings, 4 use		450	.107		1.29	3.20		4.49	6.75
2000	Flat slab, drop panels, job-built plywood, to 15' high, 1 use		449	.107		3.59	3.20		6.79	9.30
2050	2 use		509	.094		1.97	2.83		4.80	6.90
2100	3 use		532	.090		1.43	2.70		4.13	6.10
2150	4 use		544	.088		1.17	2.64		3.81	5.70
2250	15' to 20' high ceilings, 4 use		480	.100		2.08	3		5.08	7.30
2350	20' to 35' high ceilings, 4 use		435	.110		2.34	3.31		5.65	8.15
3000	Floor slab hung from steel beams, 1 use		485	.099		1.99	2.97		4.96	7.15
3050	2 use		535	.090		1.54	2.69		4.23	6.20
3100	3 use		550	.087		1.40	2.62		4.02	5.90
3150	4 use		565	.085		1.32	2.55		3.87	5.70
3500	Floor slab, with 1-way joist pans, 1 use		415	.116		5.35	3.47		8.82	11.70
3550	2 use	↓	445	.108	↓	3.73	3.23		6.96	9.50

03 11 Concrete Forming

03 11 13 – Structural Cast-In-Place Concrete Forming

03 11 13.35 Forms In Place, Elevated Slabs

	03 11 13.35 Forms In Place, Elevated Slabs	Crew	Daily Output	Labor-Hours	Unit	Material	2011 Bare Costs Labor	Equipment	Total	Total Incl O&P
3600	3 use	C-2	475	.101	S.F.	3.19	3.03		6.22	8.55
3650	4 use		500	.096		2.92	2.88		5.80	8
4500	With 2-way waffle domes, 1 use		405	.119		5.50	3.55		9.05	12
4520	2 use		450	.107		3.86	3.20		7.06	9.60
4530	3 use		460	.104		3.32	3.13		6.45	8.90
4550	4 use		470	.102		3.05	3.06		6.11	8.45
5000	Box out for slab openings, over 16" deep, 1 use		190	.253	SFCA	3.51	7.55		11.06	16.50
5050	2 use		240	.200	"	1.93	6		7.93	12.10
5500	Shallow slab box outs, to 10 S.F.		42	1.143	Ea.	9.65	34.50		44.15	68
5550	Over 10 S.F. (use perimeter)		600	.080	L.F.	1.29	2.40		3.69	5.45
6000	Bulkhead forms for slab, with keyway, 1 use, 2 piece		500	.096		1.67	2.88		4.55	6.65
6100	3 piece (see also edge forms)		460	.104		1.80	3.13		4.93	7.25
6200	Slab bulkhead form, 4-1/2" high, exp metal, w/keyway & stakes G	C-1	1200	.027		1.64	.79		2.43	3.13
6210	5-1/2" high G		1100	.029		1.90	.86		2.76	3.54
6215	7-1/2" high G		960	.033		2.31	.99		3.30	4.20
6220	9-1/2" high G		840	.038		2.40	1.13		3.53	4.53
6500	Curb forms, wood, 6" to 12" high, on elevated slabs, 1 use		180	.178	SFCA	1.34	5.30		6.64	10.35
6550	2 use		205	.156		.74	4.64		5.38	8.55
6600	3 use		220	.145		.54	4.32		4.86	7.85
6650	4 use		225	.142		.44	4.23		4.67	7.55
7000	Edge forms to 6" high, on elevated slab, 4 use		500	.064	L.F.	.17	1.90		2.07	3.37
7500	Depressed area forms to 12" high, 4 use		300	.107		.78	3.17		3.95	6.15
7550	12" to 24" high, 4 use		175	.183		1.06	5.45		6.51	10.25
8000	Perimeter deck and rail for elevated slabs, straight		90	.356		9.70	10.55		20.25	28.50
8050	Curved		65	.492		13.35	14.65		28	39
8500	Void forms, round fiber, 3" diameter G		450	.071		.99	2.11		3.10	4.63
8550	4" diameter G		425	.075		1.46	2.24		3.70	5.35
8600	6" diameter G		400	.080		2.28	2.38		4.66	6.50
8650	8" diameter G		375	.085		3.90	2.54		6.44	8.55
8700	10" diameter G		350	.091		2.52	2.72		5.24	7.30
8750	12" diameter G		300	.107		3.20	3.17		6.37	8.80

03 11 13.40 Forms In Place, Equipment Foundations

	03 11 13.40 Forms In Place, Equipment Foundations	Crew	Daily Output	Labor-Hours	Unit	Material	Labor	Equipment	Total	Total Incl O&P
0010	**FORMS IN PLACE, EQUIPMENT FOUNDATIONS** R031113-40									
0020	1 use	C-2	160	.300	SFCA	2.66	9		11.66	18
0050	2 use R031113-60		190	.253		1.47	7.55		9.02	14.25
0100	3 use		200	.240		1.07	7.20		8.27	13.20
0150	4 use		205	.234		.87	7		7.87	12.70

03 11 13.45 Forms In Place, Footings

	03 11 13.45 Forms In Place, Footings	Crew	Daily Output	Labor-Hours	Unit	Material	Labor	Equipment	Total	Total Incl O&P
0010	**FORMS IN PLACE, FOOTINGS** R031113-40									
0020	Continuous wall, plywood, 1 use	C-1	375	.085	SFCA	5	2.54		7.54	9.75
0050	2 use R031113-60		440	.073		2.76	2.16		4.92	6.65
0100	3 use		470	.068		2	2.02		4.02	5.60
0150	4 use		485	.066		1.63	1.96		3.59	5.10
0500	Dowel supports for footings or beams, 1 use		500	.064	L.F.	.74	1.90		2.64	4
1000	Integral starter wall, to 4" high, 1 use		400	.080		.79	2.38		3.17	4.85
1500	Keyway, 4 use, tapered wood, 2" x 4"	1 Carp	530	.015		.18	.51		.69	1.05
1550	2" x 6"		500	.016		.26	.54		.80	1.19
2000	Tapered plastic		530	.015		.45	.51		.96	1.35
2250	For keyway hung from supports, add		150	.053		.74	1.79		2.53	3.81
3000	Pile cap, square or rectangular, job-built plywood, 1 use	C-1	290	.110	SFCA	2.33	3.28		5.61	8.05
3050	2 use		346	.092		1.28	2.75		4.03	6
3100	3 use		371	.086		.93	2.56		3.49	5.30

03 11 13.45 Forms In Place, Footings

Line	03 11 13.45 Forms In Place, Footings	Crew	Daily Output	Labor-Hours	Unit	Material	Labor	Equipment	Total	Total Incl O&P
3150	4 use	C-1	383	.084	SFCA	.76	2.48		3.24	4.98
4000	Triangular or hexagonal, 1 use		225	.142		2.72	4.23		6.95	10.05
4050	2 use		280	.114		1.50	3.40		4.90	7.35
4100	3 use		305	.105		1.09	3.12		4.21	6.40
4150	4 use		315	.102		.88	3.02		3.90	6
5000	Spread footings, job-built lumber, 1 use		305	.105		1.79	3.12		4.91	7.15
5050	2 use		371	.086		1	2.56		3.56	5.40
5100	3 use		401	.080		.72	2.37		3.09	4.76
5150	4 use		414	.077		.58	2.30		2.88	4.48
6000	Supports for dowels, plinths or templates, 2′ x 2′ footing		25	1.280	Ea.	4.92	38		42.92	69
6050	4′ x 4′ footing		22	1.455		9.85	43		52.85	83.50
6100	8′ x 8′ footing		20	1.600		19.70	47.50		67.20	101
6150	12′ x 12′ footing		17	1.882		24	56		80	120
7000	Plinths, job-built plywood, 1 use		250	.128	SFCA	2.49	3.81		6.30	9.10
7100	4 use		270	.119	"	.82	3.52		4.34	6.80

03 11 13.47 Forms In Place, Gas Station Forms

Line	03 11 13.47 Forms In Place, Gas Station Forms	Crew	Daily Output	Labor-Hours	Unit	Material	Labor	Equipment	Total	Total Incl O&P
0010	**FORMS IN PLACE, GAS STATION FORMS**									
0050	Curb fascia, with template, 12 ga. steel, left in place, 9″ high G	1 Carp	50	.160	L.F.	11.80	5.35		17.15	22
1000	Sign or light bases, 18″ diameter, 9″ high G		9	.889	Ea.	74.50	30		104.50	132
1050	30″ diameter, 13″ high G		8	1		118	33.50		151.50	186
2000	Island forms, 10′ long, 9″ high, 3′- 6″ wide G	C-1	10	3.200		330	95		425	525
2050	4′ wide G		9	3.556		340	106		446	550
2500	20′ long, 9″ high, 4′ wide G		6	5.333		545	159		704	865
2550	5′ wide G		5	6.400		570	190		760	945

03 11 13.50 Forms In Place, Grade Beam

Line	03 11 13.50 Forms In Place, Grade Beam	Crew	Daily Output	Labor-Hours	Unit	Material	Labor	Equipment	Total	Total Incl O&P
0010	**FORMS IN PLACE, GRADE BEAM** R031113-40									
0020	Job-built plywood, 1 use	C-2	530	.091	SFCA	2.37	2.71		5.08	7.15
0050	2 use R031113-60		580	.083		1.30	2.48		3.78	5.60
0100	3 use		600	.080		.95	2.40		3.35	5.05
0150	4 use		605	.079		.77	2.38		3.15	4.83

03 11 13.55 Forms In Place, Mat Foundation

Line	03 11 13.55 Forms In Place, Mat Foundation	Crew	Daily Output	Labor-Hours	Unit	Material	Labor	Equipment	Total	Total Incl O&P
0010	**FORMS IN PLACE, MAT FOUNDATION** R031113-40									
0020	Job-built plywood, 1 use	C-2	290	.166	SFCA	2.44	4.96		7.40	11
0050	2 use R031113-60		310	.155		.99	4.64		5.63	8.85
0100	3 use		330	.145		.65	4.36		5.01	8
0120	4 use		350	.137		.58	4.11		4.69	7.50

03 11 13.65 Forms In Place, Slab On Grade

Line	03 11 13.65 Forms In Place, Slab On Grade	Crew	Daily Output	Labor-Hours	Unit	Material	Labor	Equipment	Total	Total Incl O&P
0010	**FORMS IN PLACE, SLAB ON GRADE** R031113-40									
1000	Bulkhead forms w/keyway, wood, 6″ high, 1 use	C-1	510	.063	L.F.	.86	1.87		2.73	4.06
1050	2 uses R031113-60		400	.080		.47	2.38		2.85	4.50
1100	4 uses		350	.091		.28	2.72		3	4.86
1400	Bulkhead form for slab, 4-1/2″ high, exp metal, incl keyway & stakes G		1200	.027		1.64	.79		2.43	3.13
1410	5-1/2″ high G		1100	.029		1.90	.86		2.76	3.54
1420	7-1/2″ high G		960	.033		2.31	.99		3.30	4.20
1430	9-1/2″ high G		840	.038		2.40	1.13		3.53	4.53
2000	Curb forms, wood, 6″ to 12″ high, on grade, 1 use		215	.149	SFCA	1.80	4.43		6.23	9.40
2050	2 use		250	.128		1	3.81		4.81	7.45
2100	3 use		265	.121		.72	3.59		4.31	6.80
2150	4 use		275	.116		.58	3.46		4.04	6.45
3000	Edge forms, wood, 4 use, on grade, to 6″ high		600	.053	L.F.	.27	1.59		1.86	2.95
3050	7″ to 12″ high		435	.074	SFCA	.59	2.19		2.78	4.31
3500	For depressed slabs, 4 use, to 12″ high		300	.107	L.F.	.60	3.17		3.77	5.95

03 11 13 – Structural Cast-In-Place Concrete Forming

03 11 13.65 Forms In Place, Slab On Grade		Crew	Daily Output	Labor-Hours	Unit	Material	2011 Bare Costs Labor	Equipment	Total	Total Incl O&P
3550	To 24" high	C-1	175	.183	L.F.	.79	5.45		6.24	9.95
4000	For slab blockouts, to 12" high, 1 use		200	.160		.63	4.76		5.39	8.65
4050	To 24" high, 1 use		120	.267		.80	7.95		8.75	14.15
4100	Plastic (extruded), to 6" high, multiple use, on grade		800	.040		7.90	1.19		9.09	10.70
5000	Screed, 24 ga. metal key joint, see Section 03 15 05.25									
5020	Wood, incl. wood stakes, 1" x 3"	C-1	900	.036	L.F.	.68	1.06		1.74	2.52
5050	2" x 4"		900	.036	"	.67	1.06		1.73	2.51
6000	Trench forms in floor, wood, 1 use		160	.200	SFCA	1.55	5.95		7.50	11.65
6050	2 use		175	.183		.85	5.45		6.30	10.05
6100	3 use		180	.178		.62	5.30		5.92	9.55
6150	4 use		185	.173		.50	5.15		5.65	9.15
8760	Void form, corrugated fiberboard, 4" x 12", 4' long G		3000	.011	S.F.	2.39	.32		2.71	3.15
8770	6" x 12", 4' long		3000	.011		2.81	.32		3.13	3.62
8780	1/4" thick hardboard protective cover for void form	2 Carp	1500	.011		.57	.36		.93	1.23

03 11 13.85 Forms In Place, Walls		Crew	Daily Output	Labor-Hours	Unit	Material	Labor	Equipment	Total	Total Incl O&P
0010	**FORMS IN PLACE, WALLS** R031113-10									
0100	Box out for wall openings, to 16" thick, to 10 S.F.	C-2	24	2	Ea.	23	60		83	125
0150	Over 10 S.F. (use perimeter) R031113-40	"	280	.171	L.F.	1.94	5.15		7.09	10.75
0250	Brick shelf, 4" w, add to wall forms, use wall area abv shelf									
0260	1 use R031113-60	C-2	240	.200	SFCA	2.06	6		8.06	12.25
0300	2 use		275	.175		1.13	5.25		6.38	10
0350	4 use		300	.160		.82	4.80		5.62	8.90
0500	Bulkhead, wood with keyway, 1 use, 2 piece		265	.181	L.F.	1.69	5.45		7.14	10.95
0600	Bulkhead forms with keyway, 1 piece expanded metal, 8" wall G	C-1	1000	.032		2.31	.95		3.26	4.13
0610	10" wall G		800	.040		2.40	1.19		3.59	4.63
0620	12" wall G		525	.061		2.88	1.81		4.69	6.20
0700	Buttress, to 8' high, 1 use	C-2	350	.137	SFCA	2.89	4.11		7	10.05
0750	2 use		430	.112		1.59	3.35		4.94	7.35
0800	3 use		460	.104		1.16	3.13		4.29	6.55
0850	4 use		480	.100		.96	3		3.96	6.05
1000	Corbel or haunch, to 12" wide, add to wall forms, 1 use		150	.320	L.F.	1.91	9.60		11.51	18.15
1050	2 use		170	.282		1.05	8.45		9.50	15.30
1100	3 use		175	.274		.76	8.20		8.96	14.60
1150	4 use		180	.267		.62	8		8.62	14.05
2000	Wall, job-built plywood, to 8' high, 1 use		370	.130	SFCA	2.17	3.89		6.06	8.90
2050	2 use		435	.110		1.38	3.31		4.69	7.05
2100	3 use		495	.097		1.01	2.91		3.92	5.95
2150	4 use		505	.095		.82	2.85		3.67	5.65
2400	Over 8' to 16' high, 1 use		280	.171		2.38	5.15		7.53	11.20
2450	2 use		345	.139		1.08	4.17		5.25	8.15
2500	3 use		375	.128		.77	3.84		4.61	7.25
2550	4 use		395	.122		.63	3.64		4.27	6.80
2700	Over 16' high, 1 use		235	.204		2.18	6.10		8.28	12.65
2750	2 use		290	.166		1.20	4.96		6.16	9.60
2800	3 use		315	.152		.87	4.57		5.44	8.60
2850	4 use		330	.145		.71	4.36		5.07	8.10
4000	Radial, smooth curved, job-built plywood, 1 use		245	.196		2.16	5.85		8.01	12.20
4050	2 use		300	.160		1.19	4.80		5.99	9.30
4100	3 use		325	.148		.86	4.43		5.29	8.35
4150	4 use		335	.143		.70	4.29		4.99	7.95
4200	Below grade, job-built plywood, 1 use		225	.213		2.25	6.40		8.65	13.20
4210	2 use		225	.213		1.24	6.40		7.64	12.05

03 11 13 – Structural Cast-In-Place Concrete Forming

03 11 13.85 Forms In Place, Walls		Crew	Daily Output	Labor-Hours	Unit	Material	2011 Bare Costs Labor	Equipment	Total	Total Incl O&P
4220	3 use	C-2	225	.213	SFCA	1.03	6.40		7.43	11.85
4230	4 use		225	.213		.73	6.40		7.13	11.50
4300	Curved, 2′ chords, job-built plywood, 1 use		290	.166		1.76	4.96		6.72	10.25
4350	2 use		355	.135		.97	4.05		5.02	7.85
4400	3 use		385	.125		.70	3.74		4.44	7
4450	4 use		400	.120		.57	3.60		4.17	6.65
4500	Over 8′ high, 1 use		290	.166		.82	4.96		5.78	9.20
4525	2 use		355	.135		.45	4.05		4.50	7.30
4550	3 use		385	.125		.33	3.74		4.07	6.60
4575	4 use		400	.120		.27	3.60		3.87	6.30
4600	Retaining wall, battered, job-built plyw'd, to 8′ high, 1 use		300	.160		1.66	4.80		6.46	9.85
4650	2 use		355	.135		.91	4.05		4.96	7.80
4700	3 use		375	.128		.66	3.84		4.50	7.15
4750	4 use		390	.123		.54	3.69		4.23	6.75
4900	Over 8′ to 16′ high, 1 use		240	.200		1.81	6		7.81	12
4950	2 use		295	.163		.99	4.88		5.87	9.25
5000	3 use		305	.157		.72	4.72		5.44	8.70
5050	4 use	↓	320	.150		.59	4.50		5.09	8.15
5500	For gang wall forming, 192 S.F. sections, deduct					10%	10%			
5550	384 S.F. sections, deduct				↓	20%	20%			
5750	Liners for forms (add to wall forms), A.B.S. plastic									
5800	Aged wood, 4″ wide, 1 use	1 Carp	256	.031	SFCA	2.91	1.05		3.96	4.95
5820	2 use		256	.031		1.60	1.05		2.65	3.51
5830	3 use		256	.031		1.16	1.05		2.21	3.03
5840	4 use		256	.031		.95	1.05		2	2.79
5900	Fractured rope rib, 1 use		192	.042		4.22	1.40		5.62	7
5925	2 use		192	.042		2.32	1.40		3.72	4.89
5950	3 use		192	.042		1.69	1.40		3.09	4.20
6000	4 use		192	.042		1.37	1.40		2.77	3.85
6100	Ribbed, 3/4″ deep x 1-1/2″ O.C., 1 use		224	.036		4.22	1.20		5.42	6.65
6125	2 use		224	.036		2.32	1.20		3.52	4.56
6150	3 use		224	.036		1.69	1.20		2.89	3.87
6200	4 use		224	.036		1.37	1.20		2.57	3.52
6300	Rustic brick pattern, 1 use		224	.036		2.91	1.20		4.11	5.20
6325	2 use		224	.036		1.60	1.20		2.80	3.77
6350	3 use		224	.036		1.16	1.20		2.36	3.29
6400	4 use		224	.036		.95	1.20		2.15	3.05
6500	3/8″ striated, random, 1 use		224	.036		2.91	1.20		4.11	5.20
6525	2 use		224	.036		1.60	1.20		2.80	3.77
6550	3 use		224	.036		1.16	1.20		2.36	3.29
6600	4 use		224	.036		.95	1.20		2.15	3.05
6850	Random vertical rustication, 1 use		384	.021		5.55	.70		6.25	7.25
6900	2 use		384	.021		3.04	.70		3.74	4.52
6925	3 use		384	.021		2.21	.70		2.91	3.60
6950	4 use		384	.021	↓	1.80	.70		2.50	3.15
7050	Wood, beveled edge, 3/4″ deep, 1 use		384	.021	L.F.	.19	.70		.89	1.38
7100	1″ deep, 1 use		384	.021	″	.39	.70		1.09	1.60
7200	4″ wide aged cedar, 1 use		256	.031	SFCA	2.91	1.05		3.96	4.95
7300	4″ variable depth rough cedar		224	.036		4.22	1.20		5.42	6.65
7500	Lintel or sill forms, 1 use		30	.267		2.62	8.95		11.57	17.85
7520	2 use		34	.235		1.44	7.90		9.34	14.80
7540	3 use		36	.222		1.05	7.45		8.50	13.65
7560	4 use	↓	37	.216	↓	.85	7.25		8.10	13.10

03 11 Concrete Forming

03 11 13 – Structural Cast-In-Place Concrete Forming

03 11 13.85 Forms In Place, Walls		Crew	Daily Output	Labor-Hours	Unit	2011 Bare Costs Material	Labor	Equipment	Total	Total Incl O&P
7800	Modular prefabricated plywood, based on 20 uses of purchased									
7820	forms, and 4 uses of bracing lumber									
7860	To 8′ high	C-2	800	.060	SFCA	.89	1.80		2.69	3.99
8060	Over 8′ to 16′ high		600	.080		.93	2.40		3.33	5.05
8600	Pilasters, 1 use		270	.178		2.51	5.35		7.86	11.65
8620	2 use		330	.145		1.38	4.36		5.74	8.80
8640	3 use		370	.130		1	3.89		4.89	7.60
8660	4 use		385	.125		.81	3.74		4.55	7.15
9010	Steel framed plywood, based on 50 uses of purchased									
9020	forms, and 4 uses of bracing lumber									
9060	To 8′ high	C-2	600	.080	SFCA	.55	2.40		2.95	4.62
9260	Over 8′ to 16′ high		450	.107		.55	3.20		3.75	5.95
9460	Over 16′ to 20′ high		400	.120		.55	3.60		4.15	6.60
9475	For elevated walls, add						10%			
9480	For battered walls, 1 side battered, add					10%	10%			
9485	For battered walls, 2 sides battered, add					15%	15%			

03 11 19 – Insulating Concrete Forming

03 11 19.10 Insulating Forms, Left In Place		Crew	Daily Output	Labor-Hours	Unit	Material	Labor	Equipment	Total	Total Incl O&P
0010	**INSULATING FORMS, LEFT IN PLACE**									
0020	S.F. is for exterior face, but includes forms for both faces (total R22)									
2000	4″ wall, straight block, 16″ x 48″ (5.33 S.F.) G	2 Carp	90	.178	Ea.	16.45	5.95		22.40	28
2010	90 corner block, exterior 16″ x 38″ x 22″ (6.67 S.F.) G		75	.213		18.80	7.15		25.95	32.50
2020	45 corner block, exterior 16″ x 34″ x 18″ (5.78 S.F.) G		75	.213		18.95	7.15		26.10	33
2100	6″ wall, straight block, 16″ x 48″ (5.33 S.F.) G		90	.178		16.70	5.95		22.65	28.50
2110	90 corner block, exterior 16″ x 32″ x 24″ (6.22 S.F.) G		75	.213		20.50	7.15		27.65	34.50
2120	45 corner block, exterior 16″ x 26″ x 18″ (4.89 S.F.) G		75	.213		17.80	7.15		24.95	31.50
2130	Brick ledge block, 16″ x 48″ (5.33 S.F.) G		80	.200		21	6.70		27.70	35
2140	Taper top block, 16″ x 48″ (5.33 S.F.) G		80	.200		19.65	6.70		26.35	33
2200	8″ wall, straight block, 16″ x 48″ (5.33 S.F.) G		90	.178		17.25	5.95		23.20	29
2210	90 corner block, exterior 16″ x 34″ x 26″ (6.67 S.F.) G		75	.213		22.50	7.15		29.65	36.50
2220	45 corner block, exterior 16″ x 28″ x 20″ (5.33 S.F.) G		75	.213		19.15	7.15		26.30	33
2230	Brick ledge block, 16″ x 48″ (5.33 S.F.) G		80	.200		22	6.70		28.70	35.50
2240	Taper top block, 16″ x 48″ (5.33 S.F.) G		80	.200		20	6.70		26.70	33.50

03 11 19.60 Roof Deck Form Boards		Crew	Daily Output	Labor-Hours	Unit	Material	Labor	Equipment	Total	Total Incl O&P
0010	**ROOF DECK FORM BOARDS** R051223-50									
0050	Includes bulb tee sub-purlins @ 32-5/8″ O.C.									
0070	Non-asbestos fiber cement, 5/16″ thick	C-13	2950	.008	S.F.	2.49	.29	.04	2.82	3.30
0100	Fiberglass, 1″ thick		2700	.009		3.48	.32	.04	3.84	4.44
0500	Wood fiber, 1″ thick G		2700	.009		2.04	.32	.04	2.40	2.86

03 11 23 – Permanent Stair Forming

03 11 23.75 Forms In Place, Stairs		Crew	Daily Output	Labor-Hours	Unit	Material	Labor	Equipment	Total	Total Incl O&P
0010	**FORMS IN PLACE, STAIRS** R031113-40									
0015	(Slant length x width), 1 use	C-2	165	.291	S.F.	4.65	8.70		13.35	19.70
0050	2 use R031113-60		170	.282		2.65	8.45		11.10	17.05
0100	3 use		180	.267		1.98	8		9.98	15.55
0150	4 use		190	.253		1.65	7.55		9.20	14.45
1000	Alternate pricing method (1.0 L.F./S.F.), 1 use		100	.480	LF Rsr	4.65	14.40		19.05	29
1050	2 use		105	.457		2.65	13.70		16.35	26
1100	3 use		110	.436		1.98	13.10		15.08	24
1150	4 use		115	.417		1.65	12.50		14.15	23
2000	Stairs, cast on sloping ground (length x width), 1 use		220	.218	S.F.	1.88	6.55		8.43	13

03 11 Concrete Forming

03 11 23 – Permanent Stair Forming

03 11 23.75 Forms In Place, Stairs		Crew	Daily Output	Labor-Hours	Unit	Material	Labor	Equipment	Total	Total Incl O&P
						2011 Bare Costs				
2025	2 use	C-2	232	.207	S.F.	1.03	6.20		7.23	11.50
2050	3 use		244	.197		.75	5.90		6.65	10.70
2100	4 use	↓	256	.188	↓	.61	5.60		6.21	10.05

03 15 Concrete Accessories

03 15 05 – Concrete Forming Accessories

03 15 05.02 Anchor Bolt Accessories		Crew	Daily Output	Labor-Hours	Unit	Material	Labor	Equipment	Total	Total Incl O&P
0010	**ANCHOR BOLT ACCESSORIES**									
0015	For anchor bolts set in fresh concrete, see Section 05 05 23.05									
8150	Anchor bolt sleeve, plastic, 1" diam. bolts	1 Carp	60	.133	Ea.	7	4.47		11.47	15.20
8500	1-1/2" diameter		28	.286		12.40	9.60		22	29.50
8600	2" diameter		24	.333		15.45	11.20		26.65	35.50
8650	3" diameter	↓	20	.400	↓	28.50	13.40		41.90	53.50

03 15 05.12 Chamfer Strips		Crew	Daily Output	Labor-Hours	Unit	Material	Labor	Equipment	Total	Total Incl O&P
0010	**CHAMFER STRIPS**									
2000	Polyvinyl chloride, 1/2" wide with leg	1 Carp	535	.015	L.F.	.42	.50		.92	1.30
2200	3/4" wide with leg		525	.015		.69	.51		1.20	1.62
2400	1" radius with leg		515	.016		.30	.52		.82	1.20
2800	2" radius with leg		500	.016		.52	.54		1.06	1.47
5000	Wood, 1/2" wide		535	.015		.13	.50		.63	.98
5200	3/4" wide		525	.015		.19	.51		.70	1.07
5400	1" wide	↓	515	.016	↓	.39	.52		.91	1.30

03 15 05.15 Column Form Accessories		Crew	Daily Output	Labor-Hours	Unit	Material	Labor	Equipment	Total	Total Incl O&P
0010	**COLUMN FORM ACCESSORIES**									
1000	Column clamps, adjustable to 24" x 24", buy G				Set	126			126	139
1100	Rent per month G					11.80			11.80	13
1300	For sizes to 30" x 30", buy G					128			128	141
1400	Rent per month G					12.55			12.55	13.80
1600	For sizes to 36" x 36", buy G					134			134	147
1700	Rent per month G					13.80			13.80	15.15
2000	Bull winch (band iron) 36" x 36", buy G					52			52	57
2100	Rent per month G					5.30			5.30	5.85
2300	48" x 48", buy G					53			53	58.50
2400	Rent per month G					6.35			6.35	7
3000	Chain & wedge type 36" x 36", buy G					72			72	79
3100	Rent per month G					8.10			8.10	8.90
3300	60" x 60", buy G					98			98	108
3400	Rent per month G					11.65			11.65	12.80
4000	Friction collars 2'-6" diam., buy G					735			735	810
4100	Rent per month G					72			72	79
4300	4'-0" diam., buy G					845			845	925
4400	Rent per month G				↓	93			93	102

03 15 05.20 Dovetail Anchor System		Crew	Daily Output	Labor-Hours	Unit	Material	Labor	Equipment	Total	Total Incl O&P
0010	**DOVETAIL ANCHOR SYSTEM**									
0500	Anchor slot, galv., filled, 24 ga. G	1 Carp	425	.019	L.F.	.74	.63		1.37	1.87
0600	20 ga. G		400	.020		1.57	.67		2.24	2.85
0625	12 ga. G		400	.020		1.77	.67		2.44	3.07
0900	26 ga. stainless steel, foam filled G	↓	375	.021	↓	1.18	.72		1.90	2.50
1200	Brick anchor, corr., galv., 3-1/2" long, 16 ga. G	1 Bric	10.50	.762	C	26	25.50		51.50	70.50
1300	12 ga. G	↓	10.50	.762	↓	37	25.50		62.50	82.50

03 15 Concrete Accessories

03 15 05 – Concrete Forming Accessories

03 15 05.20 Dovetail Anchor System			Crew	Daily Output	Labor-Hours	Unit	Material	Labor	Equipment	Total	Total Incl O&P
							2011 Bare Costs				
1500	Flat, galv., 3-1/2" long, 16 ga.	G	1 Bric	10.50	.762	C	38	25.50		63.50	83.50
1600	12 ga.	G		10.50	.762		53	25.50		78.50	101
2000	Cavity wall, corr., galv., 5" long, 16 ga.	G		10.50	.762		33	25.50		58.50	78.50
2100	12 ga.	G		10.50	.762		47	25.50		72.50	94
3000	Furring anchors, corr., galv., 1-1/2" long, 16 ga.	G		10.50	.762		12.80	25.50		38.30	56
3100	12 ga.	G		10.50	.762		22	25.50		47.50	66.50
6000	Stone anchors, 3-1/2" long, galv., 1/8" x 1" wide	G		10.50	.762		82	25.50		107.50	132
6100	1/4" x 1" wide	G		10.50	.762		94.50	25.50		120	146

03 15 05.25 Expansion Joints			Crew	Daily Output	Labor-Hours	Unit	Material	Labor	Equipment	Total	Total Incl O&P
0010	**EXPANSION JOINTS**										
0020	Keyed, cold, 24 ga., incl. stakes, 3-1/2" high	G	1 Carp	200	.040	L.F.	1.41	1.34		2.75	3.80
0050	4-1/2" high	G		200	.040		1.64	1.34		2.98	4.05
0100	5-1/2" high	G		195	.041		1.90	1.38		3.28	4.39
0150	7-1/2" high	G		190	.042		2.31	1.41		3.72	4.90
0160	9-1/2" high	G		185	.043		2.40	1.45		3.85	5.05
0300	Poured asphalt, plain, 1/2" x 1"		1 Clab	450	.018		.65	.47		1.12	1.51
0350	1" x 2"			400	.020		2.60	.53		3.13	3.75
0500	Neoprene, liquid, cold applied, 1/2" x 1"			450	.018		2.22	.47		2.69	3.23
0550	1" x 2"			400	.020		8.85	.53		9.38	10.65
0700	Polyurethane, poured, 2 part, 1/2" x 1"			400	.020		1.30	.53		1.83	2.32
0750	1" x 2"			350	.023		5.20	.60		5.80	6.70
0900	Rubberized asphalt, hot or cold applied, 1/2" x 1"			450	.018		.35	.47		.82	1.18
0950	1" x 2"			400	.020		1.40	.53		1.93	2.43
1100	Hot applied, fuel resistant, 1/2" x 1"			450	.018		.53	.47		1	1.37
1150	1" x 2"			400	.020		2.10	.53		2.63	3.20
2000	Premolded, bituminous fiber, 1/2" x 6"		1 Carp	375	.021		.41	.72		1.13	1.65
2050	1" x 12"			300	.027		1.91	.89		2.80	3.60
2140	Concrete expansion joint, recycled paper and fiber, 1/2" x 6"	G		390	.021		.36	.69		1.05	1.55
2150	1/2" x 12"	G		360	.022		.72	.75		1.47	2.05
2250	Cork with resin binder, 1/2" x 6"			375	.021		1.81	.72		2.53	3.19
2300	1" x 12"			300	.027		7.55	.89		8.44	9.80
2500	Neoprene sponge, closed cell, 1/2" x 6"			375	.021		1.78	.72		2.50	3.16
2550	1" x 12"			300	.027		6.65	.89		7.54	8.80
2750	Polyethylene foam, 1/2" x 6"			375	.021		.62	.72		1.34	1.88
2800	1" x 12"			300	.027		1.99	.89		2.88	3.69
3000	Polyethylene backer rod, 3/8" diameter			460	.017		.05	.58		.63	1.04
3050	3/4" diameter			460	.017		.10	.58		.68	1.09
3100	1" diameter			460	.017		.19	.58		.77	1.19
3500	Polyurethane foam, with polybutylene, 1/2" x 1/2"			475	.017		.95	.57		1.52	2
3550	1" x 1"			450	.018		2.40	.60		3	3.64
3750	Polyurethane foam, regular, closed cell, 1/2" x 6"			375	.021		.70	.72		1.42	1.97
3800	1" x 12"			300	.027		2.50	.89		3.39	4.25
4000	Polyvinyl chloride foam, closed cell, 1/2" x 6"			375	.021		1.90	.72		2.62	3.29
4050	1" x 12"			300	.027		6.55	.89		7.44	8.70
4250	Rubber, gray sponge, 1/2" x 6"			375	.021		1.60	.72		2.32	2.96
4300	1" x 12"			300	.027		5.75	.89		6.64	7.85
4500	Lead wool for joints					Lb.	7.85			7.85	8.60
5000	For installation in walls, add							75%			
5250	For installation in boxouts, add							25%			

03 15 Concrete Accessories

03 15 05 – Concrete Forming Accessories

03 15 05.30 Hangers

			Crew	Daily Output	Labor-Hours	Unit	2011 Bare Costs Material	Labor	Equipment	Total	Total Incl O&P
0010	**HANGERS**										
0020	Slab and beam form										
0500	Banding iron, 3/4" x 22 ga, 14 L.F. per lb or										
0550	1/2" x 14 ga, 7 L.F. per lb.	G				Lb.	1.24			1.24	1.36
1000	Fascia ties, coil type, to 24" long	G				C	320			320	355
1500	Frame ties to 8-1/8"	G					380			380	415
1550	8-1/8" to 10-1/8"	G					400			400	440
5000	Snap tie hanger, to 30" overall length, 3000 #	G					320			320	350
5050	To 36" overall length	G					350			350	385
5100	To 48" overall length	G					410			410	450
5500	Steel beam hanger										
5600	Flange to 8-1/8"	G				C	380			380	415
5650	8-1/8" to 10-1/8"	G					400			400	440
6000	Tie hangers to 30" overall length, 4000 #	G					365			365	400
6100	To 36" overall length	G					395			395	435
6500	Tie back hanger, up to 12-1/8" flange	G					1,025			1,025	1,125
8500	Wire, black annealed, 9 ga	G				Cwt.	200			200	220
8600	16 ga					"	164			164	180

03 15 05.35 Inserts

			Crew	Daily Output	Labor-Hours	Unit	Material	Labor	Equipment	Total	Total Incl O&P
0010	**INSERTS**										
1000	Inserts, slotted nut type for 3/4" bolts, 4" long	G	1 Carp	84	.095	Ea.	16	3.20		19.20	23
2100	6" long	G		84	.095		18	3.20		21.20	25
2150	8" long	G		84	.095		24	3.20		27.20	32
2200	Slotted, strap type, 4" long	G		84	.095		15.70	3.20		18.90	22.50
2250	6" long	G		84	.095		17.70	3.20		20.90	25
2300	8" long	G		84	.095		23.50	3.20		26.70	31.50
2350	Strap for slotted insert, 4" long	G		84	.095		8.60	3.20		11.80	14.80
4100	6" long	G		84	.095		9.30	3.20		12.50	15.60
4150	8" long	G		84	.095		11.60	3.20		14.80	18.10
4200	10" long	G		84	.095		14.10	3.20		17.30	21
7000	Loop ferrule type										
7100	1/4" diameter bolt	G	1 Carp	84	.095	Ea.	1.71	3.20		4.91	7.25
7350	7/8" diameter bolt	G	"	84	.095	"	7.95	3.20		11.15	14.10
9000	Wedge type										
9100	For 3/4" diameter bolt	G	1 Carp	60	.133	Ea.	6.60	4.47		11.07	14.75
9800	Cut washers, black										
9900	3/4" bolt	G				Ea.	.82			.82	.90
9950	For galvanized inserts, add						30%				

03 15 05.70 Shores

			Crew	Daily Output	Labor-Hours	Unit	Material	Labor	Equipment	Total	Total Incl O&P
0010	**SHORES**										
0020	Erect and strip, by hand, horizontal members										
0501	Aluminum joists and stringers	G	L-2	60	.267	Ea.		7.85		7.85	13.15
0602	Steel, adjustable beams	G		45	.356			10.45		10.45	17.50
0701	Wood joists			50	.320			9.40		9.40	15.75
0801	Wood stringers			30	.533			15.70		15.70	26.50
1001	Vertical members to 10' high	G		55	.291			8.55		8.55	14.30
1051	To 13' high	G		50	.320			9.40		9.40	15.75
1101	To 16' high	G		45	.356			10.45		10.45	17.50
1501	Reshoring	G		1400	.011	S.F.	.50	.34		.84	1.11
1600	Flying truss system	G	C-17D	9600	.009	SFCA		.30	.09	.39	.61
1760	Horizontal, aluminum joists, 6-1/4" high x 5' to 21' span, buy	G				L.F.	13.40			13.40	14.70
1770	Beams, 7-1/4" high x 4' to 30' span	G				"	15.65			15.65	17.25

03 15 Concrete Accessories

03 15 05 – Concrete Forming Accessories

03 15 05.70 Shores			Crew	Daily Output	Labor-Hours	Unit	2011 Bare Costs Material	Labor	Equipment	Total	Total Incl O&P
1810	Horizontal, steel beam, W8x10, 7′ span, buy	G				Ea.	47.50			47.50	52
1830	10′ span	G					56			56	61.50
1920	15′ span	G					96			96	106
1940	20′ span	G					135			135	149
1970	Steel stringer, W8x10, 4′ to 16′ span, buy	G				L.F.	5.60			5.60	6.15
3000	Rent for job duration, aluminum joist @ 2′ O.C., per mo	G				SF Flr.	.33			.33	.37
3050	Steel W8x10	G					.14			.14	.15
3060	Steel adjustable	G					.14			.14	.15
3500	#1 post shore, steel, 5′-7″ to 9′-6″ high, 10000# cap., buy	G				Ea.	128			128	141
3550	#2 post shore, 7′-3″ to 12′-10″ high, 7800# capacity	G					148			148	163
3600	#3 post shore, 8′-10″ to 16′-1″ high, 3800# capacity	G					162			162	178
5010	Frame shoring systems, steel, 12000#/leg, buy										
5040	Frame, 2′ wide x 6′ high	G				Ea.	106			106	117
5250	X-brace	G					18			18	19.80
5550	Base plate	G					17.60			17.60	19.35
5600	Screw jack	G					38			38	41.50
5650	U-head, 8″ x 8″	G					21.50			21.50	24

03 15 05.75 Sleeves and Chases			Crew	Daily Output	Labor-Hours	Unit	Material	Labor	Equipment	Total	Total Incl O&P
0010	**SLEEVES AND CHASES**										
0100	Plastic, 1 use, 12″ long, 2″ diameter		1 Carp	100	.080	Ea.	1.74	2.68		4.42	6.40
0150	4″ diameter			90	.089		5.10	2.98		8.08	10.60
0200	6″ diameter			75	.107		9	3.58		12.58	15.90
0250	12″ diameter			60	.133		25.50	4.47		29.97	35.50

03 15 05.80 Snap Ties			Crew	Daily Output	Labor-Hours	Unit	Material	Labor	Equipment	Total	Total Incl O&P
0010	**SNAP TIES**, 8-1/4″ L&W (Lumber and wedge)										
0100	2250 lb., w/flat washer, 8″ wall	G				C	68			68	75
0200	12″ wall	G					79			79	87
0250	16″ wall	G					91			91	100
0300	18″ wall	G					97			97	107
0500	With plastic cone, 8″ wall	G					61			61	67
0600	12″ wall	G					67			67	73.50
0650	16″ wall	G					74			74	81.50
0700	18″ wall	G					77			77	84.50
1000	3350 lb., w/flat washer, 8″ wall	G					96			96	106
1150	12″ wall	G					108			108	119
1200	16″ wall	G					124			124	136
1250	18″ wall	G					129			129	142
1500	With plastic cone, 8″ wall	G					96			96	106
1600	12″ wall	G					108			108	119
1650	16″ wall	G					124			124	136
1700	18″ wall	G					129			129	142

03 15 05.85 Stair Tread Inserts			Crew	Daily Output	Labor-Hours	Unit	Material	Labor	Equipment	Total	Total Incl O&P
0010	**STAIR TREAD INSERTS**										
0015	Cast iron, abrasive, 3″ wide	G	1 Carp	90	.089	L.F.	9.65	2.98		12.63	15.60
0020	4″ wide	G		80	.100		12.80	3.36		16.16	19.70
0040	6″ wide	G		75	.107		19.25	3.58		22.83	27
0050	9″ wide	G		70	.114		29	3.83		32.83	38.50
0100	12″ wide	G		65	.123		38.50	4.13		42.63	49.50
0300	Cast aluminum, compared to cast iron, deduct						10%				
0500	Extruded aluminum safety tread, 3″ wide	G	1 Carp	75	.107		10.95	3.58		14.53	18.05
0550	4″ wide	G		75	.107		14.60	3.58		18.18	22
0600	6″ wide	G		75	.107		22	3.58		25.58	30

03 15 Concrete Accessories

03 15 05 – Concrete Forming Accessories

03 15 05.85 Stair Tread Inserts			Crew	Daily Output	Labor-Hours	Unit	Material	Labor	Equipment	Total	Total Incl O&P
							2011 Bare Costs				
0650	9″ wide to resurface stairs	G	1 Carp	70	.114	L.F.	33	3.83		36.83	42.50
1700	Cement fill for pan-type metal treads, plain		1 Cefi	115	.070	S.F.	1.83	2.22		4.05	5.55
1750	Non-slip		″	100	.080	″	2.02	2.55		4.57	6.30

03 15 05.95 Wall and Foundation Form Accessories

			Crew	Daily Output	Labor-Hours	Unit	Material	Labor	Equipment	Total	Total Incl O&P
0010	**WALL AND FOUNDATION FORM ACCESSORIES**										
2000	Footings, turnbuckle form aligner	G				Ea.	13.50			13.50	14.85
2050	Spreaders for footer, adjustable	G				″	17.90			17.90	19.70
3000	Form oil, coverage varies greatly, minimum					Gal.	6.80			6.80	7.50
3050	Maximum					″	15.90			15.90	17.50
3500	Form patches, 1-3/4″ diameter					C	39			39	43
3550	2-3/4″ diameter					″	68			68	75
4000	Nail stakes, 3/4″ diameter, 18″ long	G				Ea.	2.44			2.44	2.68
4050	24″ long	G					3.16			3.16	3.48
4200	30″ long	G					4.05			4.05	4.46
4250	36″ long	G				↓	4.89			4.89	5.40

03 15 13 – Waterstops

03 15 13.50 Waterstops

		Crew	Daily Output	Labor-Hours	Unit	Material	Labor	Equipment	Total	Total Incl O&P
0010	**WATERSTOPS**, PVC and Rubber									
0020	PVC, ribbed 3/16″ thick, 4″ wide	1 Carp	155	.052	L.F.	.95	1.73		2.68	3.95
0050	6″ wide		145	.055		1.79	1.85		3.64	5.05
0500	With center bulb, 6″ wide, 3/16″ thick		135	.059		1.61	1.99		3.60	5.10
0550	3/8″ thick		130	.062		2.91	2.06		4.97	6.65
0600	9″ wide x 3/8″ thick		125	.064		4.48	2.15		6.63	8.50
0800	Dumbbell type, 6″ wide, 3/16″ thick		150	.053		1.65	1.79		3.44	4.81
0850	3/8″ thick		145	.055		3.27	1.85		5.12	6.70
1000	9″ wide, 3/8″ thick, plain		130	.062		4.86	2.06		6.92	8.80
1050	Center bulb		130	.062		6.95	2.06		9.01	11.05
1250	Ribbed type, split, 3/16″ thick, 6″ wide		145	.055		1.62	1.85		3.47	4.88
1300	3/8″ thick		130	.062		3.43	2.06		5.49	7.25
2000	Rubber, flat dumbbell, 3/8″ thick, 6″ wide		145	.055		8.10	1.85		9.95	12.05
2050	9″ wide		135	.059		12.40	1.99		14.39	17
2500	Flat dumbbell split, 3/8″ thick, 6″ wide		145	.055		1.62	1.85		3.47	4.88
2550	9″ wide		135	.059		3.43	1.99		5.42	7.10
3000	Center bulb, 1/4″ thick, 6″ wide		145	.055		6.95	1.85		8.80	10.75
3050	9″ wide		135	.059		15.55	1.99		17.54	20.50
3500	Center bulb split, 3/8″ thick, 6″ wide		145	.055		10.30	1.85		12.15	14.45
3550	9″ wide	↓	135	.059	↓	17.80	1.99		19.79	23
5000	Waterstop fittings, rubber, flat									
5010	Dumbbell or center bulb, 3/8″ thick,									
5200	Field union, 6″ wide	1 Carp	50	.160	Ea.	21	5.35		26.35	32
5250	9″ wide		50	.160		27	5.35		32.35	39
5500	Flat cross, 6″ wide		30	.267		35	8.95		43.95	53.50
5550	9″ wide		30	.267		51	8.95		59.95	71
6000	Flat tee, 6″ wide		30	.267		36	8.95		44.95	54.50
6050	9″ wide		30	.267		49	8.95		57.95	69
6500	Flat ell, 6″ wide		40	.200		35	6.70		41.70	50
6550	9″ wide		40	.200		45.50	6.70		52.20	61.50
7000	Vertical tee, 6″ wide		25	.320		28	10.75		38.75	48.50
7050	9″ wide		25	.320		39.50	10.75		50.25	61.50
7500	Vertical ell, 6″ wide		35	.229		26	7.65		33.65	41.50
7550	9″ wide	↓	35	.229	↓	35	7.65		42.65	51.50

03 21 05.10 Rebar Accessories			Crew	Daily Output	Labor-Hours	Unit	2011 Bare Costs Material	Labor	Equipment	Total	Total Incl O&P
0010	**REBAR ACCESSORIES**										
0030	Steel & plastic made from recycled materials	G									
0100	Beam bolsters (BB), lower, 1-1/2" high, plain steel	G				C.L.F.	43			43	47.50
0102	Galvanized	G					51.50			51.50	57
0104	Stainless tipped legs	G					345			345	375
0106	Plastic tipped legs	G					47			47	51.50
0108	Epoxy dipped	G					90			90	99
0110	2" high, plain	G					47			47	51.50
0120	Galvanized	G					56			56	61.50
0140	Stainless tipped legs	G					345			345	380
0160	Plastic tipped legs	G					48			48	53
0162	Epoxy dipped	G					100			100	110
0200	Upper (BBU), 1-1/2" high, plain steel	G					84			84	92.50
0210	3" high	G					97			97	107
0300	Beam bolster with plate (BBP), 1-1/2" high, plain steel	G					167			167	184
0310	3" high	G					267			267	294
0500	Slab bolsters, continuous (SB), 1" high, plain steel	G					23			23	25.50
0502	Galvanized	G					40			40	44
0504	Stainless tipped legs	G					325			325	355
0506	Plastic tipped legs	G					30			30	33
0510	2" high, plain steel	G					25			25	27.50
0515	Galvanized	G					51			51	56
0520	Stainless tipped legs	G					325			325	360
0525	Plastic tipped legs	G					35			35	38.50
0530	For bolsters with wire runners (SBR), add	G					32			32	35
0540	For bolsters with plates (SBP), add	G					64			64	70.50
0700	Bag ties, 16 ga., plain, 4" long	G				C	2			2	2.20
0710	5" long	G					3			3	3.30
0720	6" long	G					2			2	2.20
0730	7" long	G					3			3	3.30
1200	High chairs, individual (HC), 3" high, plain steel	G					34			34	37.50
1202	Galvanized	G					41			41	45
1204	Stainless tipped legs	G					335			335	365
1206	Plastic tipped legs	G					46			46	50.50
1210	5" high, plain	G					50			50	55
1212	Galvanized	G					55			55	60.50
1214	Stainless tipped legs	G					350			350	385
1216	Plastic tipped legs	G					63			63	69.50
1220	8" high, plain	G					102			102	112
1222	Galvanized	G					122			122	135
1224	Stainless tipped legs	G					400			400	440
1226	Plastic tipped legs	G					131			131	144
1230	12" high, plain	G					238			238	262
1232	Galvanized	G					285			285	315
1234	Stainless tipped legs	G					540			540	590
1236	Plastic tipped legs	G					238			238	262
1400	Individual high chairs, with plate (HCP), 5" high	G					147			147	162
1410	8" high	G					204			204	224
1500	Bar chair (BC), 1-1/2" high, plain steel	G					22			22	24
1520	Galvanized	G					31			31	34
1530	Stainless tipped legs	G					320			320	355
1540	Plastic tipped legs	G					35			35	38.50
1550	Joist chair (JC), 3/4" x 5", plain steel	G					28			28	31

03 21 05.10 Rebar Accessories			Crew	Daily Output	Labor-Hours	Unit	Material (2011 Bare Costs)	Labor (2011 Bare Costs)	Equipment (2011 Bare Costs)	Total (2011 Bare Costs)	Total Incl O&P
1580	Galvanized	G				C	33.50			33.50	37
1600	Stainless tipped legs	G					330			330	360
1620	Plastic tipped legs	G					47			47	51.50
1630	Epoxy dipped	G					104			104	114
1700	Continuous high chairs (CHC), legs 8" O.C., 4" high, plain steel	G				C.L.F.	41			41	45
1705	Galvanized	G					49			49	54
1710	Stainless tipped legs	G					340			340	375
1715	Plastic tipped legs	G					54			54	59.50
1718	Epoxy dipped	G					75			75	82.50
1720	6" high, plain	G					59			59	65
1725	Galvanized	G					71			71	78
1730	Stainless tipped legs	G					360			360	395
1735	Plastic tipped legs	G					82			82	90
1738	Epoxy dipped	G					99			99	109
1740	8" high, plain	G					87			87	95.50
1745	Galvanized	G					104			104	115
1750	Stainless tipped legs	G					385			385	425
1755	Plastic tipped legs	G					131			131	144
1758	Epoxy dipped	G					150			150	165
1900	For continuous bottom wire runners, add	G					32			32	35
1940	For continuous bottom plate, add	G					151			151	166
2100	Paper tubing, 4' lengths, for #4 bar	G					64			64	70.50
2120	For #6 bar	G					75			75	82.50
2200	Screed chair base, 1/2" coil thread diam., 2-1/2" high, plain steel	G				C	244			244	268
2210	Galvanized	G					257			257	283
2220	5-1/2" high, plain	G					295			295	325
2250	Galvanized	G					310			310	340
2300	3/4" coil thread diam., 2-1/2" high, plain steel	G					310			310	340
2310	Galvanized	G					415			415	455
2320	5-1/2" high, plain steel	G					395			395	435
2350	Galvanized	G					530			530	585
2400	Screed holder, 1/2" coil thread diam. for pipe screed, plain steel, 6" long	G					275			275	305
2420	12" long	G					425			425	465
2500	3/4" coil thread diam. for pipe screed, plain steel, 6" long	G					395			395	435
2520	12" long	G					635			635	700
2700	Screw anchor for bolts, plain steel, 3/4" diameter x 4" long	G					405			405	445
2720	1" diameter x 6" long	G					660			660	730
2740	1-1/2" diameter x 8" long	G					840			840	925
2800	Screw anchor eye bolts, 3/4" x 7" long	G					3,450			3,450	3,800
2820	1" x 9" long	G					6,825			6,825	7,525
2840	1-1/2" x 14" long	G					17,300			17,300	19,000
2900	Screw anchor bolts, 3/4" x 9" long	G					1,200			1,200	1,325
2920	1" x 12" long	G					2,325			2,325	2,575
3001	Slab lifting inserts, single pickup, galv, 3/4" diam., 5" high	G					1,150			1,150	1,250
3010	6" high	G					1,175			1,175	1,275
3030	7" high	G					1,175			1,175	1,300
3100	1" diameter, 5-1/2" high	G					1,200			1,200	1,325
3120	7" high	G					1,250			1,250	1,375
3200	Double pickup lifting inserts, 1" diameter, 5-1/2" high	G					2,325			2,325	2,575
3220	7" high	G					2,575			2,575	2,850
3330	1-1/2" diameter, 5-1/2" high	G					2,825			2,825	3,125
3800	Subgrade chairs, #4 bar head, 3-1/2" high	G					28			28	31
3850	12" high	G					40			40	44

03 21 05 – Reinforcing Steel Accessories

03 21 05.10 Rebar Accessories			Crew	Daily Output	Labor-Hours	Unit	2011 Bare Costs Material	Labor	Equipment	Total	Total Incl O&P
3900	#6 bar head, 3-1/2" high	G				C	28			28	31
3950	12" high	G					40			40	44
4200	Subgrade stakes, no nail holes, 3/4" diameter, 12" long	G					196			196	216
4250	24" long	G					284			284	310
4300	7/8" diameter, 12" long	G					277			277	305
4350	24" long	G					465			465	510
4500	Tie wire, 16 ga. annealed steel	G				Cwt.	164			164	180
4550	Tie wire holder, plastic case	G				Ea.	40			40	44
4600	Aluminum case	G				"	56.50			56.50	62.50

03 21 05.75 Splicing Reinforcing Bars			Crew	Daily Output	Labor-Hours	Unit	Material	Labor	Equipment	Total	Total Incl O&P
0010	**SPLICING REINFORCING BARS** R032110-70										
0020	Including holding bars in place while splicing										
0100	Standard, self-aligning type, taper threaded, #4 bars	G	C-25	190	.168	Ea.	4.55	4.82		9.37	13.50
0105	#5 bars	G		170	.188		5.55	5.40		10.95	15.60
0110	#6 bars	G		150	.213		6.40	6.10		12.50	17.80
0120	#7 bars	G		130	.246		7.45	7.05		14.50	20.50
0300	#8 bars	G		115	.278		12.90	7.95		20.85	28
0305	#9 bars	G	C-5	105	.457		14.10	15.20	7.15	36.45	49.50
0310	#10 bars	G		95	.505		15.70	16.80	7.90	40.40	54.50
0320	#11 bars	G		85	.565		16.70	18.80	8.80	44.30	60
0330	#14 bars	G		65	.738		25	24.50	11.50	61	82
0340	#18 bars	G		45	1.067		38	35.50	16.65	90.15	121
0500	Transition self-aligning, taper threaded, #18-14	G		45	1.067		40.50	35.50	16.65	92.65	123
0510	#18-11	G		45	1.067		41.50	35.50	16.65	93.65	125
0520	#14-11	G		65	.738		29.50	24.50	11.50	65.50	87
0540	#11-10	G		85	.565		23.50	18.80	8.80	51.10	67.50
0550	#10-9	G		95	.505		22	16.80	7.90	46.70	61
0560	#9-8	G	C-25	105	.305		21	8.70		29.70	38.50
0580	#8-7	G		115	.278		19.55	7.95		27.50	35.50
0590	#7-6	G		130	.246		19.05	7.05		26.10	33.50
0600	Position coupler for curved bars, taper threaded, #4 bars	G		160	.200		22	5.70		27.70	34
0610	#5 bars	G		145	.221		23	6.30		29.30	36
0620	#6 bars	G		130	.246		26	7.05		33.05	41.50
0630	#7 bars	G		110	.291		38.50	8.30		46.80	57
0640	#8 bars	G		100	.320		40	9.15		49.15	60
0650	#9 bars	G	C-5	90	.533		43.50	17.75	8.30	69.55	87
0660	#10 bars	G		80	.600		47	19.95	9.35	76.30	96
0670	#11 bars	G		70	.686		49	23	10.70	82.70	105
0680	#14 bars	G		55	.873		61	29	13.60	103.60	132
0690	#18 bars	G		40	1.200		88	40	18.70	146.70	185
0700	Transition position coupler for curved bars, taper threaded, #18-14	G		40	1.200		111	40	18.70	169.70	211
0710	#18-11	G		40	1.200		112	40	18.70	170.70	212
0720	#14-11	G		55	.873		95	29	13.60	137.60	169
0730	#11-10	G		70	.686		54.50	23	10.70	88.20	111
0740	#10-9	G		80	.600		52	19.95	9.35	81.30	101
0750	#9-8	G	C-25	90	.356		48	10.15		58.15	71
0760	#8-7	G		100	.320		44	9.15		53.15	64.50
0770	#7-6	G		110	.291		42.50	8.30		50.80	61.50
0800	Sleeve type w/grout filler, for precast concrete, #6 bars	G		72	.444		23.50	12.70		36.20	48.50
0802	#7 bars	G		64	.500		25.50	14.30		39.80	53
0805	#8 bars	G		56	.571		29	16.35		45.35	61
0807	#9 bars	G		48	.667		31.50	19.05		50.55	68.50

03 21 05 – Reinforcing Steel Accessories

03 21 05.75 Splicing Reinforcing Bars			Crew	Daily Output	Labor-Hours	Unit	Material	2011 Bare Costs Labor	Equipment	Total	Total Incl O&P
0810	#10 bars	G	C-5	40	1.200	Ea.	41	40	18.70	99.70	134
0900	#11 bars	G		32	1.500		49	50	23.50	122.50	165
0920	#14 bars	G		24	2		79.50	66.50	31	177	235
1000	Sleeve type w/ferrous filler, for critical structures, #6 bars	G	C-25	72	.444		41	12.70		53.70	68
1210	#7 bars	G		64	.500		42	14.30		56.30	71
1220	#8 bars	G		56	.571		44	16.35		60.35	77.50
1230	#9 bars	G	C-5	48	1		45	33.50	15.60	94.10	124
1240	#10 bars	G		40	1.200		48	40	18.70	106.70	142
1250	#11 bars	G		32	1.500		58	50	23.50	131.50	175
1260	#14 bars	G		24	2		73	66.50	31	170.50	228
1270	#18 bars	G		16	3		106	100	47	253	340
2000	Weldable half coupler, taper threaded, #4 bars	G	E-16	120	.133		7.10	4.99	.91	13	18.35
2100	#5 bars	G		112	.143		8.40	5.35	.97	14.72	20.50
2200	#6 bars	G		104	.154		13.35	5.75	1.05	20.15	27
2300	#7 bars	G		96	.167		15.50	6.25	1.14	22.89	30
2400	#8 bars	G		88	.182		16.15	6.80	1.24	24.19	32
2500	#9 bars	G		80	.200		17.80	7.50	1.36	26.66	35.50
2600	#10 bars	G		72	.222		18.20	8.30	1.51	28.01	37.50
2700	#11 bars	G		64	.250		19.45	9.35	1.70	30.50	41
2800	#14 bars	G		56	.286		22.50	10.70	1.95	35.15	47
2900	#18 bars	G		48	.333		36.50	12.45	2.27	51.22	66.50

03 21 10 – Uncoated Reinforcing Steel

03 21 10.60 Reinforcing In Place			Crew	Daily Output	Labor-Hours	Unit	Material	Labor	Equipment	Total	Total Incl O&P
0015	**REINFORCING IN PLACE**, 50-60 ton lots, A615 Grade 60 R032110-10										
0020	Includes labor, but not material cost, to install accessories										
0030	Made from recycled materials	G									
0100	Beams & Girders, #3 to #7	G	4 Rodm	1.60	20	Ton	900	725		1,625	2,250
0150	#8 to #18 R032110-20	G		2.70	11.852		900	430		1,330	1,750
0200	Columns, #3 to #7	G		1.50	21.333		900	775		1,675	2,350
0250	#8 to #18	G		2.30	13.913		900	505		1,405	1,875
0300	Spirals, hot rolled, 8″ to 15″ diameter	G		2.20	14.545		1,400	530		1,930	2,475
0320	15″ to 24″ diameter R032110-40	G		2.20	14.545		1,350	530		1,880	2,400
0330	24″ to 36″ diameter	G		2.30	13.913		1,275	505		1,780	2,275
0340	36″ to 48″ diameter R032110-50	G		2.40	13.333		1,225	485		1,710	2,175
0360	48″ to 64″ diameter	G		2.50	12.800		1,350	465		1,815	2,275
0380	64″ to 84″ diameter R032110-70	G		2.60	12.308		1,400	445		1,845	2,325
0390	84″ to 96″ diameter	G		2.70	11.852		1,475	430		1,905	2,375
0400	Elevated slabs, #4 to #7 R032110-80	G		2.90	11.034		955	400		1,355	1,750
0500	Footings, #4 to #7	G		2.10	15.238		855	555		1,410	1,900
0550	#8 to #18	G		3.60	8.889		810	325		1,135	1,450
0600	Slab on grade, #3 to #7	G		2.30	13.913		855	505		1,360	1,825
0700	Walls, #3 to #7	G		3	10.667		855	385		1,240	1,625
0750	#8 to #18	G		4	8		855	290		1,145	1,450
0900	For other than 50 – 60 ton lots										
1000	Under 10 ton job, #3 to #7, add					Ton	25%	10%			
1010	#8 to #18, add						20%	10%			
1050	10 – 50 ton job, #3 to #7, add						10%				
1060	#8 to #18, add						5%				
1100	60 – 100 ton job, #3 to #7, deduct						5%				
1110	#8 to #18, deduct						10%				
1150	Over 100 ton job, #3 to #7, deduct						10%				
1160	#8 to #18, deduct						15%				

03 21 Reinforcing Steel

03 21 10 – Uncoated Reinforcing Steel

03 21 10.60 Reinforcing In Place			Crew	Daily Output	Labor-Hours	Unit	2011 Bare Costs Material	Labor	Equipment	Total	Total Incl O&P
1200	Reinforcing in place, A615 Grade 75, add	G				Ton	82.50			82.50	90.50
1220	Grade 90, add						99			99	109
2000	Unloading & sorting, add to above		C-5	100	.480			15.95	7.50	23.45	35.50
2200	Crane cost for handling, add to above, minimum			135	.356			11.85	5.55	17.40	26
2210	Average			92	.522			17.35	8.15	25.50	38.50
2220	Maximum			35	1.371			45.50	21.50	67	101
2400	Dowels, 2 feet long, deformed, #3	G	2 Rodm	520	.031	Ea.	.37	1.12		1.49	2.35
2410	#4	G		480	.033		.66	1.21		1.87	2.83
2420	#5	G		435	.037		1.03	1.34		2.37	3.46
2430	#6	G		360	.044		1.49	1.61		3.10	4.44
2450	Longer and heavier dowels, add	G		725	.022	Lb.	.50	.80		1.30	1.93
2500	Smooth dowels, 12" long, 1/4" or 3/8" diameter	G		140	.114	Ea.	.72	4.15		4.87	8
2520	5/8" diameter	G		125	.128		1.26	4.65		5.91	9.50
2530	3/4" diameter	G		110	.145		1.57	5.30		6.87	10.90
2600	Dowel sleeves for CIP concrete, 2-part system										
2610	Sleeve base, plastic, for 5/8" smooth dowel sleeve, fasten to edge form		1 Rodm	200	.040	Ea.	.57	1.45		2.02	3.15
2615	Sleeve, plastic, 12" long, for 5/8" smooth dowel, snap onto base			400	.020		1.09	.73		1.82	2.46
2620	Sleeve base, for 3/4" smooth dowel sleeve			175	.046		.54	1.66		2.20	3.47
2625	Sleeve, 12" long, for 3/4" smooth dowel			350	.023		.97	.83		1.80	2.51
2630	Sleeve base, for 1" smooth dowel sleeve			150	.053		.64	1.94		2.58	4.07
2635	Sleeve, 12" long, for 1" smooth dowel			300	.027		1.28	.97		2.25	3.09
2700	Dowel caps, visual warning only, plastic, #3 to #8		2 Rodm	800	.020		.26	.73		.99	1.55
2720	#8 to #18			750	.021		.78	.77		1.55	2.21
2750	Impalement protective, plastic, #4 to #9			800	.020		1.41	.73		2.14	2.81

03 21 10.70 Glass Fiber Reinforced Polymer Bars

		Crew	Daily Output	Labor-Hours	Unit	Material	Labor	Equipment	Total	Total Incl O&P
0010	**GLASS FIBER REINFORCED POLYMER BARS**									
0050	#2 bar, .043 lb./L.F.	4 Rodm	9500	.003	L.F.	.35	.12		.47	.60
0100	#3 bar, .092 lb./L.F.		9300	.003		.48	.12		.60	.75
0150	#4 bar, .160 lb./L.F.		9100	.004		.70	.13		.83	.99
0200	#5 bar, .258 lb./L.F.		8700	.004		1.05	.13		1.18	1.39
0250	#6 bar, .372 lb./L.F.		8300	.004		1.38	.14		1.52	1.76
0300	#7 bar, .497 lb./L.F.		7900	.004		1.73	.15		1.88	2.16
0350	#8 bar, .620 lb./L.F.		7400	.004		2.32	.16		2.48	2.82
0400	#9 bar, .800 lb./L.F.		6800	.005		2.92	.17		3.09	3.51
0450	#10 bar, 1.08 lb./L.F.		5800	.006		3.50	.20		3.70	4.20
0500	For Bends, add per bend				Ea.	1.25			1.25	1.38

03 21 13 – Galvanized Reinforcing Steel

03 21 13.10 Galvanized Reinforcing

		Crew	Daily Output	Labor-Hours	Unit	Material	Labor	Equipment	Total	Total Incl O&P
0010	**GALVANIZED REINFORCING**									
0150	Add to uncoated reinforcing price for galvanizing				Ton	575			575	635

03 21 16 – Epoxy-Coated Reinforcing Steel

03 21 16.10 Epoxy-Coated Reinforcing

		Crew	Daily Output	Labor-Hours	Unit	Material	Labor	Equipment	Total	Total Incl O&P
0010	**EPOXY-COATED REINFORCING**									
0100	Add to uncoated reinforcing price for coating with epoxy				Ton	350			350	385

03 22 Welded Wire Fabric Reinforcing

03 22 05 – Uncoated Welded Wire Fabric

03 22 05.50 Welded Wire Fabric		Crew	Daily Output	Labor-Hours	Unit	Material	2011 Bare Costs Labor	Equipment	Total	Total Incl O&P
0010	**WELDED WIRE FABRIC** ASTM A185 R032205-30									
0030	Made from recycled materials G									
0050	Sheets									
0100	6 x 6 - W1.4 x W1.4 (10 x 10) 21 lb. per C.S.F. G	2 Rodm	35	.457	C.S.F.	12.50	16.60		29.10	43
0200	6 x 6 - W2.1 x W2.1 (8 x 8) 30 lb. per C.S.F. G		31	.516		18.90	18.75		37.65	53.50
0300	6 x 6 - W2.9 x W2.9 (6 x 6) 42 lb. per C.S.F. G		29	.552		21.50	20		41.50	58.50
0400	6 x 6 - W4 x W4 (4 x 4) 58 lb. per C.S.F. G		27	.593		29.50	21.50		51	70
0500	4 x 4 - W1.4 x W1.4 (10 x 10) 31 lb. per C.S.F. G		31	.516		17.90	18.75		36.65	52
0600	4 x 4 - W2.1 x W2.1 (8 x 8) 44 lb. per C.S.F. G		29	.552		22.50	20		42.50	59.50
0650	4 x 4 - W2.9 x W2.9 (6 x 6) 61 lb. per C.S.F. G		27	.593		36	21.50		57.50	77
0700	4 x 4 - W4 x W4 (4 x 4) 85 lb. per C.S.F. G		25	.640		45	23		68	90
0750	Rolls									
0800	2 x 2 - #14 galv., 21 lb./C.S.F., beam & column wrap G	2 Rodm	6.50	2.462	C.S.F.	44	89.50		133.50	204
0900	2 x 2 - #12 galv. for gunite reinforcing G	"	6.50	2.462	"	65	89.50		154.50	227

03 23 Stressing Tendons

03 23 05 – Prestressing Tendons

03 23 05.50 Prestressing Steel		Crew	Daily Output	Labor-Hours	Unit	Material	2011 Bare Costs Labor	Equipment	Total	Total Incl O&P
0010	**PRESTRESSING STEEL** R034136-90									
0100	Grouted strand, post-tensioned in field, 50′ span, 100 kip G	C-3	1200	.053	Lb.	2.28	1.73	.09	4.10	5.55
0150	300 kip G		2700	.024		1.04	.77	.04	1.85	2.50
0300	100′ span, 100 kip G		1700	.038		2.29	1.22	.06	3.57	4.67
0350	300 kip G		3200	.020		1.98	.65	.03	2.66	3.33
0500	200′ span, 100 kip G		2700	.024		2.28	.77	.04	3.09	3.86
0550	300 kip G		3500	.018		1.98	.59	.03	2.60	3.21
0800	Grouted bars, 50′ span, 42 kip G		2600	.025		.91	.80	.04	1.75	2.41
0850	143 kip G		3200	.020		.88	.65	.03	1.56	2.11
1000	75′ span, 42 kip G		3200	.020		.93	.65	.03	1.61	2.17
1050	143 kip G		4200	.015		.78	.50	.03	1.31	1.73
1200	Ungrouted strand, 50′ span, 100 kip G	C-4	1275	.025		.62	.86	.02	1.50	2.20
1250	300 kip G		1475	.022		.62	.74	.02	1.38	1.99
1400	100′ span, 100 kip G		1500	.021		.62	.73	.02	1.37	1.96
1450	300 kip G		1650	.019		.62	.67	.02	1.31	1.85
1600	200′ span, 100 kip G		1500	.021		.62	.73	.02	1.37	1.96
1650	300 kip G		1700	.019		.62	.65	.02	1.29	1.82
1800	Ungrouted bars, 50′ span, 42 kip G		1400	.023		.45	.79	.02	1.26	1.87
1850	143 kip G		1700	.019		.45	.65	.02	1.12	1.64
2000	75′ span, 42 kip G		1800	.018		.45	.61	.02	1.08	1.57
2050	143 kip G		2200	.015		.45	.50	.01	.96	1.37
2220	Ungrouted single strand, 100′ slab, 25 kip G		1200	.027		.62	.92	.02	1.56	2.29
2250	35 kip G		1475	.022		.62	.74	.02	1.38	1.99
3000	Slabs on grade, 0.5-inch diam. non-bonded strands, HDPE sheathed,									
3050	attached dead-end anchors, loose stressing-end anchors									
3100	25′ x 30′ slab, strands @ 36″ O.C., placing	2 Rodm	2940	.005	S.F.	.59	.20		.79	.99
3105	Stressing	C-4A	3750	.004			.16	.01	.17	.28
3110	42″ O.C., placing	2 Rodm	3200	.005		.52	.18		.70	.89
3115	Stressing	C-4A	4040	.004			.14	.01	.15	.26
3120	48″ O.C., placing	2 Rodm	3510	.005		.46	.17		.63	.79
3125	Stressing	C-4A	4390	.004			.13	.01	.14	.24
3150	25′ x 40′ slab, strands @ 36″ O.C., placing	2 Rodm	3370	.005		.57	.17		.74	.93
3155	Stressing	C-4A	4360	.004			.13	.01	.14	.24

03 23 Stressing Tendons

03 23 05 – Prestressing Tendons

03 23 05.50 Prestressing Steel		Crew	Daily Output	Labor-Hours	Unit	2011 Bare Costs Material	Labor	Equipment	Total	Total Incl O&P
3160	42" O.C., placing	2 Rodm	3760	.004	S.F.	.49	.15		.64	.81
3165	Stressing	C-4A	4820	.003			.12	.01	.13	.22
3170	48" O.C., placing	2 Rodm	4090	.004		.44	.14		.58	.73
3175	Stressing	C-4A	5190	.003			.11	.01	.12	.20
3200	30' x 30' slab, strands @ 36" O.C., placing	2 Rodm	3260	.005		.57	.18		.75	.94
3205	Stressing	C-4A	4190	.004			.14	.01	.15	.25
3210	42" O.C., placing	2 Rodm	3530	.005		.51	.16		.67	.86
3215	Stressing	C-4A	4500	.004			.13	.01	.14	.23
3220	48" O.C., placing	2 Rodm	3840	.004		.46	.15		.61	.76
3225	Stressing	C-4A	4850	.003			.12	.01	.13	.22
3230	30' x 40' slab, strands @ 36" O.C., placing	2 Rodm	3780	.004		.55	.15		.70	.88
3235	Stressing	C-4A	4920	.003			.12	.01	.13	.22
3240	42" O.C., placing	2 Rodm	4190	.004		.48	.14		.62	.77
3245	Stressing	C-4A	5410	.003			.11	.01	.12	.20
3250	48" O.C., placing	2 Rodm	4520	.004		.44	.13		.57	.70
3255	Stressing	C-4A	5790	.003			.10	.01	.11	.18
3260	30' x 50' slab, strands @ 36" O.C., placing	2 Rodm	4300	.004		.52	.14		.66	.80
3265	Stressing	C-4A	5650	.003			.10	.01	.11	.19
3270	42" O.C., placing	2 Rodm	4720	.003		.46	.12		.58	.72
3275	Stressing	C-4A	6150	.003			.09	.01	.10	.17
3280	48" O.C., placing	2 Rodm	5240	.003		.41	.11		.52	.64
3285	Stressing	C-4A	6760	.002	↓		.09	.01	.10	.16

03 24 Fibrous Reinforcing

03 24 05 – Reinforcing Fibers

03 24 05.30 Synthetic Fibers

		Crew	Daily Output	Labor-Hours	Unit	Material	Labor	Equipment	Total	Total Incl O&P
0010	**SYNTHETIC FIBERS**									
0100	Synthetic fibers, add to concrete				Lb.	5.05			5.05	5.55
0110	1-1/2 lb. per C.Y.				C.Y.	7.80			7.80	8.60

03 24 05.70 Steel Fibers

		Crew	Daily Output	Labor-Hours	Unit	Material	Labor	Equipment	Total	Total Incl O&P
0010	**STEEL FIBERS**									
0150	Steel fibers, add to concrete G				Lb.	.77			.77	.85
0155	25 lb. per C.Y. G				C.Y.	19.25			19.25	21
0160	50 lb. per C.Y. G					38.50			38.50	42.50
0170	75 lb. per C.Y. G					59.50			59.50	65
0180	100 lb. per C.Y. G				↓	77			77	84.50

03 30 Cast-In-Place Concrete

03 30 53 – Miscellaneous Cast-In-Place Concrete

03 30 53.40 Concrete In Place

			Crew	Daily Output	Labor-Hours	Unit	Material	Labor	Equipment	Total	Total Incl O&P
0010	**CONCRETE IN PLACE**	R033105-10									
0020	Including forms (4 uses), reinforcing steel, concrete, placement,	R033105-20									
0050	and finishing unless otherwise indicated	R033105-50									
0300	Beams (3500 psi), 5 kip per L.F., 10' span	R033105-65	C-14A	15.62	12.804	C.Y.	291	430	49	770	1,100
0350	25' span	R033105-70	"	18.55	10.782		305	360	41	706	985
0500	Chimney foundations (5000 psi), industrial, minimum	R033105-85	C-14C	32.22	3.476		144	111	.73	255.73	345
0510	Maximum		"	23.71	4.724		168	151	.99	319.99	440
0700	Columns, square (4000 psi), 12" x 12", minimum reinforcing		C-14A	11.96	16.722		325	560	64	949	1,375
0720	Average reinforcing		↓	10.13	19.743	↓	510	660	75.50	1,245.50	1,750

	03 30 53.40 Concrete In Place	Crew	Daily Output	Labor-Hours	Unit	2011 Bare Costs Material	Labor	Equipment	Total	Total Incl O&P
0740	Maximum reinforcing	C-14A	9.03	22.148	C.Y.	770	740	84.50	1,594.50	2,200
0800	16" x 16", minimum reinforcing		16.22	12.330		259	415	47	721	1,025
0820	Average reinforcing		12.57	15.911		435	535	61	1,031	1,450
0840	Maximum reinforcing		10.25	19.512		675	655	74.50	1,404.50	1,925
0900	24" x 24", minimum reinforcing		23.66	8.453		221	283	32.50	536.50	755
0920	Average reinforcing		17.71	11.293		390	380	43	813	1,125
0940	Maximum reinforcing		14.15	14.134		620	475	54	1,149	1,550
1000	36" x 36", minimum reinforcing		33.69	5.936		197	199	22.50	418.50	575
1020	Average reinforcing		23.32	8.576		345	287	33	665	900
1040	Maximum reinforcing		17.82	11.223		580	375	43	998	1,325
1100	Columns, round (4000 psi), tied, 12" diameter, minimum reinforcing		20.97	9.537		285	320	36.50	641.50	890
1120	Average reinforcing		15.27	13.098		470	440	50	960	1,300
1140	Maximum reinforcing		12.11	16.515		715	555	63	1,333	1,775
1200	16" diameter, minimum reinforcing		31.49	6.351		262	213	24.50	499.50	670
1220	Average reinforcing		19.12	10.460		450	350	40	840	1,125
1240	Maximum reinforcing		13.77	14.524		675	485	55.50	1,215.50	1,625
1300	20" diameter, minimum reinforcing		41.04	4.873		264	163	18.60	445.60	585
1320	Average reinforcing		24.05	8.316		435	279	32	746	985
1340	Maximum reinforcing		17.01	11.758		675	395	45	1,115	1,450
1400	24" diameter, minimum reinforcing		51.85	3.857		248	129	14.75	391.75	505
1420	Average reinforcing		27.06	7.391		435	248	28	711	925
1440	Maximum reinforcing		18.29	10.935		665	365	42	1,072	1,400
1500	36" diameter, minimum reinforcing		75.04	2.665		247	89.50	10.20	346.70	435
1520	Average reinforcing		37.49	5.335		415	179	20.50	614.50	780
1540	Maximum reinforcing		22.84	8.757		645	293	33.50	971.50	1,250
1900	Elevated slab (4000 psi), flat slab with drops, 125 psf Sup. Load, 20' span	C-14B	38.45	5.410		234	181	19.85	434.85	585
1950	30' span		50.99	4.079		252	136	14.95	402.95	525
2100	Flat plate, 125 psf Sup. Load, 15' span		30.24	6.878		215	230	25	470	650
2150	25' span		49.60	4.194		227	140	15.40	382.40	500
2300	Waffle const., 30" domes, 125 psf Sup. Load, 20' span		37.07	5.611		233	188	20.50	441.50	595
2350	30' span		44.07	4.720		218	158	17.30	393.30	525
2500	One way joists, 30" pans, 125 psf Sup. Load, 15' span		27.38	7.597		278	254	28	560	760
2550	25' span		31.15	6.677		263	223	24.50	510.50	690
2700	One way beam & slab, 125 psf Sup. Load, 15' span		20.59	10.102		231	340	37	608	860
2750	25' span		28.36	7.334		222	245	27	494	685
2900	Two way beam & slab, 125 psf Sup. Load, 15' span		24.04	8.652		224	289	32	545	765
2950	25' span		35.87	5.799		196	194	21.50	411.50	565
3100	Elevated slabs, flat plate, including finish, not									
3110	including forms or reinforcing									
3150	Regular concrete (4000 psi), 4" slab	C-8	2613	.021	S.F.	1.32	.63	.28	2.23	2.80
3200	6" slab		2585	.022		1.96	.64	.29	2.89	3.52
3250	2-1/2" thick floor fill		2685	.021		.85	.62	.28	1.75	2.24
3300	Lightweight, 110# per C.F., 2-1/2" thick floor fill		2585	.022		1.14	.64	.29	2.07	2.63
3400	Cellular concrete, 1-5/8" fill, under 5000 S.F.		2000	.028		.76	.83	.37	1.96	2.61
3450	Over 10,000 S.F.		2200	.025		.73	.75	.34	1.82	2.40
3500	Add per floor for 3 to 6 stories high		31800	.002			.05	.02	.07	.12
3520	For 7 to 20 stories high		21200	.003			.08	.03	.11	.17
3540	Equipment pad (3000 psi), 3' x 3' x 6" thick	C-14H	45	1.067	Ea.	40.50	35	.51	76.01	104
3550	4' x 4' x 6" thick		30	1.600		62	52.50	.77	115.27	157
3560	5' x 5' x 8" thick		18	2.667		111	87.50	1.28	199.78	270
3570	6' x 6' x 8" thick		14	3.429		150	113	1.65	264.65	355
3580	8' x 8' x 10" thick		8	6		320	197	2.88	519.88	690
3590	10' x 10' x 12" thick		5	9.600		550	315	4.61	869.61	1,150

03 30 53.40 Concrete In Place		Crew	Daily Output	Labor-Hours	Unit	2011 Bare Costs Material	Labor	Equipment	Total	Total Incl O&P
3800	Footings (3000 psi), spread under 1 C.Y.	C-14C	28	4	C.Y.	158	128	.84	286.84	390
3825	1 C.Y. to 5 C.Y.		43	2.605		185	83	.55	268.55	345
3850	Over 5 C.Y.		75	1.493		171	47.50	.31	218.81	268
3900	Footings, strip (3000 psi), 18″ x 9″, unreinforced	C-14L	40	2.400		119	75	.58	194.58	257
3920	18″ x 9″, reinforced	C-14C	35	3.200		141	102	.67	243.67	330
3925	20″ x 10″, unreinforced	C-14L	45	2.133		116	66.50	.51	183.01	240
3930	20″ x 10″, reinforced	C-14C	40	2.800		134	89.50	.59	224.09	298
3935	24″ x 12″, unreinforced	C-14L	55	1.745		114	54.50	.42	168.92	217
3940	24″ x 12″, reinforced	C-14C	48	2.333		132	74.50	.49	206.99	271
3945	36″ x 12″, unreinforced	C-14L	70	1.371		111	43	.33	154.33	194
3950	36″ x 12″, reinforced	C-14C	60	1.867		127	59.50	.39	186.89	240
4000	Foundation mat (3000 psi), under 10 C.Y.		38.67	2.896		192	92.50	.61	285.11	365
4050	Over 20 C.Y.		56.40	1.986		169	63.50	.42	232.92	292
4200	Wall, free-standing (3000 psi), 8″ thick, 8′ high	C-14D	45.83	4.364		160	145	16.65	321.65	435
4250	14′ high		27.26	7.337		192	244	28	464	650
4260	12″ thick, 8′ high		64.32	3.109		146	103	11.90	260.90	345
4270	14′ high		40.01	4.999		155	166	19.10	340.10	470
4300	15″ thick, 8′ high		80.02	2.499		140	83	9.55	232.55	305
4350	12′ high		51.26	3.902		140	130	14.90	284.90	385
4500	18′ high		48.85	4.094		156	136	15.65	307.65	415
4520	Handicap access ramp (4000 psi), railing both sides, 3′ wide	C-14H	14.58	3.292	L.F.	278	108	1.58	387.58	490
4525	5′ wide		12.22	3.928		288	129	1.89	418.89	535
4530	With 6″ curb and rails both sides, 3′ wide		8.55	5.614		287	185	2.69	474.69	630
4535	5′ wide		7.31	6.566		292	216	3.15	511.15	685
4650	Slab on grade (3500 psi), not including finish, 4″ thick	C-14E	60.75	1.449	C.Y.	117	47.50	.38	164.88	209
4700	6″ thick	″	92	.957	″	113	31	.25	144.25	177
4701	Thickened slab edge (3500 psi), for slab on grade poured									
4702	monolithically with slab; depth is in addition to slab thickness;									
4703	formed vertical outside edge, earthen bottom and inside slope									
4705	8″ deep x 8″ wide bottom, unreinforced	C-14L	2190	.044	L.F.	3.18	1.37	.01	4.56	5.80
4710	8″ x 8″, reinforced	C-14C	1670	.067		5.30	2.14	.01	7.45	9.45
4715	12″ deep x 12″ wide bottom, unreinforced	C-14L	1800	.053		6.55	1.66	.01	8.22	10
4720	12″ x 12″, reinforced	C-14C	1310	.086		10.40	2.73	.02	13.15	16.05
4725	16″ deep x 16″ wide bottom, unreinforced	C-14L	1440	.067		11.10	2.08	.02	13.20	15.70
4730	16″ x 16″, reinforced	C-14C	1120	.100		15.70	3.19	.02	18.91	22.50
4735	20″ deep x 20″ wide bottom, unreinforced	C-14L	1150	.083		16.85	2.61	.02	19.48	23
4740	20″ x 20″, reinforced	C-14C	920	.122		22.50	3.89	.03	26.42	31.50
4745	24″ deep x 24″ wide bottom, unreinforced	C-14L	930	.103		24	3.22	.02	27.24	31.50
4750	24″ x 24″, reinforced	C-14C	740	.151		31.50	4.83	.03	36.36	42.50
4751	Slab on grade (3500 psi), incl. troweled finish, not incl. forms									
4760	or reinforcing, over 10,000 S.F., 4″ thick	C-14F	3425	.021	S.F.	1.29	.64	.01	1.94	2.46
4820	6″ thick		3350	.021		1.89	.65	.01	2.55	3.14
4840	8″ thick		3184	.023		2.59	.68	.01	3.28	3.97
4900	12″ thick		2734	.026		3.88	.80	.01	4.69	5.55
4950	15″ thick		2505	.029		4.88	.87	.01	5.76	6.75
5000	Slab on grade (3000 psi), incl. textured finish, not incl. forms									
5001	or reinforcing, 4″ thick	C-14G	2873	.019	S.F.	1.29	.58	.01	1.88	2.37
5010	6″ thick		2590	.022		2.01	.64	.01	2.66	3.27
5020	8″ thick		2320	.024		2.62	.72	.01	3.35	4.07
5200	Lift slab in place above the foundation, incl. forms,									
5210	reinforcing, concrete (4000 psi) and columns, minimum	C-14B	2113	.098	S.F.	6.20	3.29	.36	9.85	12.70
5250	Average		1650	.126		6.75	4.21	.46	11.42	14.95
5300	Maximum		1500	.139		7.30	4.64	.51	12.45	16.35

03 30 Cast-In-Place Concrete

03 30 53 – Miscellaneous Cast-In-Place Concrete

03 30 53.40 Concrete In Place		Crew	Daily Output	Labor-Hours	Unit	2011 Bare Costs Material	Labor	Equipment	Total	Total Incl O&P
5500	Lightweight, ready mix, including screed finish only,									
5510	not including forms or reinforcing									
5550	1:4 (2500 psi) for structural roof decks	C-14B	260	.800	C.Y.	136	26.50	2.94	165.44	197
5600	1:6 (3000 psi) for ground slab with radiant heat	C-14F	92	.783		133	23.50	.25	156.75	186
5650	1:3:2 (2000 psi) with sand aggregate, roof deck	C-14B	260	.800		133	26.50	2.94	162.44	194
5700	Ground slab (2000 psi)	C-14F	107	.673		133	20.50	.22	153.72	179
5900	Pile caps (3000 psi), incl. forms and reinf., sq. or rect., under 10 C.Y.	C-14C	54.14	2.069		154	66	.43	220.43	281
5950	Over 10 C.Y.		75	1.493		147	47.50	.31	194.81	241
6000	Triangular or hexagonal, under 10 C.Y.		53	2.113		118	67.50	.44	185.94	243
6050	Over 10 C.Y.		85	1.318		132	42	.28	174.28	216
6200	Retaining walls (3000 psi), gravity, 4′ high see Section 32 32	C-14D	66.20	3.021		136	101	11.55	248.55	330
6250	10′ high		125	1.600		129	53	6.10	188.10	238
6300	Cantilever, level backfill loading, 8′ high		70	2.857		148	95	10.90	253.90	335
6350	16′ high		91	2.198		141	73	8.40	222.40	286
6800	Stairs (3500 psi), not including safety treads, free standing, 3′-6″ wide	C-14H	83	.578	LF Nose	5	19	.28	24.28	38
6850	Cast on ground		125	.384	″	4.28	12.65	.18	17.11	26
7000	Stair landings, free standing		200	.240	S.F.	4.03	7.90	.12	12.05	17.75
7050	Cast on ground		475	.101	″	3.29	3.32	.05	6.66	9.20

03 31 Structural Concrete

03 31 05 – Normal Weight Structural Concrete

03 31 05.30 Concrete, Field Mix

		Crew	Daily Output	Labor-Hours	Unit	Material	Labor	Equipment	Total	Total Incl O&P
0010	**CONCRETE, FIELD MIX** R033105-65									
0015	FOB forms 2250 psi				C.Y.	85			85	93.50
0020	3000 psi				″	92			92	101

03 31 05.35 Normal Weight Concrete, Ready Mix

		Crew	Daily Output	Labor-Hours	Unit	Material	Labor	Equipment	Total	Total Incl O&P
0010	**NORMAL WEIGHT CONCRETE, READY MIX**, delivered R033105-10									
0012	Includes local aggregate, sand, Portland cement, and water									
0015	Excludes all additives and treatments R033105-20									
0020	2000 psi				C.Y.	91.50			91.50	101
0100	2500 psi R033105-30					94			94	103
0150	3000 psi					99			99	109
0200	3500 psi R033105-40					99.50			99.50	110
0300	4000 psi					103			103	113
0350	4500 psi R033105-50					106			106	116
0400	5000 psi					111			111	122
0411	6000 psi					127			127	139
0412	8000 psi					206			206	227
0413	10,000 psi					293			293	320
0414	12,000 psi					355			355	390
1000	For high early strength cement, add					10%				
1010	For structural lightweight with regular sand, add					25%				
1300	For winter concrete (hot water), add					4.25			4.25	4.68
1400	For hot weather concrete (ice), add					9.35			9.35	10.25
1410	For mid-range water reducer, add					4.13			4.13	4.54
1420	For high-range water reducer/superplasticizer, add					6.35			6.35	6.95
1430	For retarder, add					2.71			2.71	2.98
1440	For non-Chloride accelerator, add					4.83			4.83	5.30
1450	For Chloride accelerator, per 1%, add					3.28			3.28	3.61
1460	For fiber reinforcing, synthetic (1 lb./C.Y.), add					6.65			6.65	7.30
1500	For Saturday delivery, add					8.85			8.85	9.70

03 31 05 – Normal Weight Structural Concrete

03 31 05.35 Normal Weight Concrete, Ready Mix		Crew	Daily Output	Labor-Hours	Unit	2011 Bare Costs Material	Labor	Equipment	Total	Total Incl O&P
1510	For truck holding/waiting time past 1st hour per load, add				Hr.	87.50			87.50	96.50
1520	For short load (less than 4 C.Y.), add per load				Ea.	112			112	124
2000	For all lightweight aggregate, add				C.Y.	45%				

03 31 05.70 Placing Concrete		Crew	Daily Output	Labor-Hours	Unit	Material	Labor	Equipment	Total	Total Incl O&P
0010	**PLACING CONCRETE** R033105-70									
0020	Includes labor and equipment to place, strike off and consolidate									
0050	Beams, elevated, small beams, pumped	C-20	60	1.067	C.Y.		30.50	13.10	43.60	65
0100	With crane and bucket	C-7	45	1.600			46	28.50	74.50	107
0200	Large beams, pumped	C-20	90	.711			20	8.75	28.75	43
0250	With crane and bucket	C-7	65	1.108			32	19.65	51.65	74
0400	Columns, square or round, 12" thick, pumped	C-20	60	1.067			30.50	13.10	43.60	65
0450	With crane and bucket	C-7	40	1.800			51.50	32	83.50	121
0600	18" thick, pumped	C-20	90	.711			20	8.75	28.75	43
0650	With crane and bucket	C-7	55	1.309			37.50	23	60.50	87.50
0800	24" thick, pumped	C-20	92	.696			19.80	8.55	28.35	42.50
0850	With crane and bucket	C-7	70	1.029			29.50	18.25	47.75	69
1000	36" thick, pumped	C-20	140	.457			13	5.60	18.60	27.50
1050	With crane and bucket	C-7	100	.720			20.50	12.75	33.25	48
1400	Elevated slabs, less than 6" thick, pumped	C-20	140	.457			13	5.60	18.60	27.50
1450	With crane and bucket	C-7	95	.758			22	13.45	35.45	51
1500	6" to 10" thick, pumped	C-20	160	.400			11.35	4.92	16.27	24.50
1550	With crane and bucket	C-7	110	.655			18.80	11.60	30.40	44
1600	Slabs over 10" thick, pumped	C-20	180	.356			10.10	4.37	14.47	21.50
1650	With crane and bucket	C-7	130	.554			15.90	9.85	25.75	37.50
1900	Footings, continuous, shallow, direct chute	C-6	120	.400			11.05	.39	11.44	18.80
1950	Pumped	C-20	150	.427			12.15	5.25	17.40	26
2000	With crane and bucket	C-7	90	.800			23	14.20	37.20	53.50
2100	Footings, continuous, deep, direct chute	C-6	140	.343			9.50	.33	9.83	16.10
2150	Pumped	C-20	160	.400			11.35	4.92	16.27	24.50
2200	With crane and bucket	C-7	110	.655			18.80	11.60	30.40	44
2400	Footings, spread, under 1 C.Y., direct chute	C-6	55	.873			24	.85	24.85	41
2450	Pumped	C-20	65	.985			28	12.10	40.10	60
2500	With crane and bucket	C-7	45	1.600			46	28.50	74.50	107
2600	Over 5 C.Y., direct chute	C-6	120	.400			11.05	.39	11.44	18.80
2650	Pumped	C-20	150	.427			12.15	5.25	17.40	26
2700	With crane and bucket	C-7	100	.720			20.50	12.75	33.25	48
2900	Foundation mats, over 20 C.Y., direct chute	C-6	350	.137			3.80	.13	3.93	6.45
2950	Pumped	C-20	400	.160			4.55	1.97	6.52	9.70
3000	With crane and bucket	C-7	300	.240			6.90	4.26	11.16	16.10
3200	Grade beams, direct chute	C-6	150	.320			8.85	.31	9.16	15.05
3250	Pumped	C-20	180	.356			10.10	4.37	14.47	21.50
3300	With crane and bucket	C-7	120	.600			17.25	10.65	27.90	40
3500	High rise, for more than 5 stories, pumped, add per story	C-20	2100	.030			.87	.37	1.24	1.85
3510	With crane and bucket, add per story	C-7	2100	.034			.98	.61	1.59	2.30
3700	Pile caps, under 5 C.Y., direct chute	C-6	90	.533			14.75	.52	15.27	25
3750	Pumped	C-20	110	.582			16.55	7.15	23.70	35.50
3800	With crane and bucket	C-7	80	.900			26	15.95	41.95	60.50
3850	Pile cap, 5 C.Y. to 10 C.Y., direct chute	C-6	175	.274			7.60	.27	7.87	12.90
3900	Pumped	C-20	200	.320			9.10	3.94	13.04	19.45
3950	With crane and bucket	C-7	150	.480			13.80	8.50	22.30	32.50
4000	Over 10 C.Y., direct chute	C-6	215	.223			6.20	.22	6.42	10.50
4050	Pumped	C-20	240	.267			7.60	3.28	10.88	16.15

03 31 Structural Concrete

03 31 05 – Normal Weight Structural Concrete

03 31 05.70 Placing Concrete		Crew	Daily Output	Labor-Hours	Unit	Material	2011 Bare Costs Labor	Equipment	Total	Total Incl O&P
4100	With crane and bucket	C-7	185	.389	C.Y.		11.20	6.90	18.10	26
4300	Slab on grade, up to 6" thick, direct chute	C-6	110	.436			12.10	.42	12.52	20.50
4350	Pumped	C-20	130	.492			14	6.05	20.05	29.50
4400	With crane and bucket	C-7	110	.655			18.80	11.60	30.40	44
4600	Over 6" thick, direct chute	C-6	165	.291			8.05	.28	8.33	13.65
4650	Pumped	C-20	185	.346			9.85	4.26	14.11	21
4700	With crane and bucket	C-7	145	.497			14.25	8.80	23.05	33
4900	Walls, 8" thick, direct chute	C-6	90	.533			14.75	.52	15.27	25
4950	Pumped	C-20	100	.640			18.20	7.85	26.05	38.50
5000	With crane and bucket	C-7	80	.900			26	15.95	41.95	60.50
5050	12" thick, direct chute	C-6	100	.480			13.30	.47	13.77	22.50
5100	Pumped	C-20	110	.582			16.55	7.15	23.70	35.50
5200	With crane and bucket	C-7	90	.800			23	14.20	37.20	53.50
5300	15" thick, direct chute	C-6	105	.457			12.65	.44	13.09	21.50
5350	Pumped	C-20	120	.533			15.15	6.55	21.70	32
5400	With crane and bucket	C-7	95	.758			22	13.45	35.45	51
5600	Wheeled concrete dumping, add to placing costs above									
5610	Walking cart, 50' haul, add	C-18	32	.281	C.Y.		7.50	1.72	9.22	14.45
5620	150' haul, add		24	.375			10	2.29	12.29	19.25
5700	250' haul, add		18	.500			13.35	3.06	16.41	26
5800	Riding cart, 50' haul, add	C-19	80	.113			3	1.15	4.15	6.30
5810	150' haul, add		60	.150			4	1.53	5.53	8.40
5900	250' haul, add		45	.200			5.35	2.04	7.39	11.20

03 35 Concrete Finishing

03 35 29 – Tooled Concrete Finishing

03 35 29.30 Finishing Floors		Crew	Daily Output	Labor-Hours	Unit	Material	Labor	Equipment	Total	Total Incl O&P
0010	**FINISHING FLOORS**									
0012	Finishing requires that concrete first be placed, struck off & consolidated									
0015	Basic finishing for various unspecified flatwork									
0100	Bull float only	C-10	4000	.006	S.F.		.18		.18	.29
0125	Bull float & manual float		2000	.012			.36		.36	.58
0150	Bull float, manual float, & broom finish, w/edging & joints		1850	.013			.39		.39	.63
0200	Bull float, manual float & manual steel trowel		1265	.019			.57		.57	.92
0210	For specified Random Access Floors in ACI Classes 1, 2, 3 and 4 to achieve									
0215	Composite Overall Floor Flatness and Levelness values up to F35/F25									
0250	Bull float, machine float & machine trowel (walk-behind)	C-10C	1715	.014	S.F.		.42	.02	.44	.70
0300	Power screed, bull float, machine float & trowel (walk-behind)	C-10D	2400	.010			.30	.04	.34	.54
0350	Power screed, bull float, machine float & trowel (ride-on)	C-10E	4000	.006			.18	.06	.24	.35
0352	For specified Random Access Floors in ACI Classes 5, 6, 7 and 8 to achieve									
0354	Composite Overall Floor Flatness and Levelness values up to F50/F50									
0356	Add for two-dimensional restraightening after power float	C-10	6000	.004	S.F.		.12		.12	.19
0358	For specified Random or Defined Access Floors in ACI Class 9 to achieve									
0360	Composite Overall Floor Flatness and Levelness values up to F100/F100									
0362	Add for two-dimensional restraightening after bull float & power float	C-10	3000	.008	S.F.		.24		.24	.39
0364	For specified Superflat Defined Access Floors in ACI Class 9 to achieve									
0366	Minimum Floor Flatness and Levelness values of F100/F100									
0368	Add for 2-dim'l restraightening after bull float, power float, power trowel	C-10	2000	.012	S.F.		.36		.36	.58
0400	Integral topping and finish, using 1:1:2 mix, 3/16" thick	C-10B	1000	.040		.10	1.14	.23	1.47	2.24
0450	1/2" thick		950	.042		.27	1.20	.24	1.71	2.54
0500	3/4" thick		850	.047		.41	1.35	.27	2.03	2.96

03 35 29 – Tooled Concrete Finishing

03 35 29.30 Finishing Floors		Crew	Daily Output	Labor-Hours	Unit	2011 Bare Costs Material	Labor	Equipment	Total	Total Incl O&P
0600	1″ thick	C-10B	750	.053	S.F.	.55	1.53	.30	2.38	3.43
0800	Granolithic topping, laid after, 1:1:1-1/2 mix, 1/2″ thick		590	.068		.31	1.94	.39	2.64	3.95
0820	3/4″ thick		580	.069		.46	1.97	.39	2.82	4.17
0850	1″ thick		575	.070		.61	1.99	.40	3	4.37
0950	2″ thick		500	.080		1.22	2.29	.46	3.97	5.60
1200	Heavy duty, 1:1:2, 3/4″ thick, preshrunk, gray, 20 M.S.F.		320	.125		.72	3.58	.71	5.01	7.40
1300	100 M.S.F.		380	.105		.41	3.01	.60	4.02	6.05
1600	Exposed local aggregate finish, minimum	1 Cefi	625	.013		.21	.41		.62	.88
1650	Maximum		465	.017		.64	.55		1.19	1.57
1800	Floor abrasives, .25 psf, aluminum oxide		850	.009		.43	.30		.73	.95
1850	Silicon carbide		850	.009		.65	.30		.95	1.19
2000	Floor hardeners, metallic, light service, .50 psf, add		850	.009		.51	.30		.81	1.04
2050	Medium service, .75 psf		750	.011		.76	.34		1.10	1.37
2100	Heavy service, 1.0 psf		650	.012		1.01	.39		1.40	1.74
2150	Extra heavy, 1.5 psf		575	.014		1.52	.44		1.96	2.38
2300	Non-metallic, light service, .50 psf		850	.009		.23	.30		.53	.73
2350	Medium service, .75 psf		750	.011		.34	.34		.68	.92
2400	Heavy service, 1.00 psf		650	.012		.45	.39		.84	1.13
2450	Extra heavy, 1.50 psf		575	.014		.68	.44		1.12	1.46
2800	Trap rock wearing surface for monolithic floors									
2810	2.0 psf	C-10B	1250	.032	S.F.	.02	.92	.18	1.12	1.73
3000	Floor coloring, dusted on, minimum (0.6 psf), add to above	1 Cefi	1300	.006		.35	.20		.55	.69
3050	Maximum (1.0 psf), add to above	″	625	.013		.58	.41		.99	1.28
3100	Colored powder only				Lb.	.58			.58	.63
3600	1/2″ topping using 0.6 psf powdered color	C-10B	590	.068	S.F.	4.41	1.94	.39	6.74	8.45
3650	1/2″ topping using 1.0 psf powdered color	″	590	.068		4.64	1.94	.39	6.97	8.70
3800	Dustproofing, solvent-based, 1 coat	1 Cefi	1900	.004		.14	.13		.27	.36
3850	2 coats		1300	.006		.50	.20		.70	.86
4000	Epoxy-based, 1 coat		1500	.005		.12	.17		.29	.40
4050	2 coats		1500	.005		.23	.17		.40	.52
4400	Stair finish, float		275	.029			.93		.93	1.48
4500	Steel trowel finish		200	.040			1.27		1.27	2.03
4600	Silicon carbide finish, .25 psf		150	.053		.43	1.70		2.13	3.18
03 35 29.35 Control Joints, Saw Cut										
0010	**CONTROL JOINTS, SAW CUT**									
0100	Sawcut control joints in green concrete									
0120	1″ depth	C-27	2000	.008	L.F.	.05	.25	.07	.37	.53
0140	1-1/2″ depth		1800	.009		.07	.28	.07	.42	.61
0160	2″ depth		1600	.010		.10	.32	.08	.50	.71
0180	Sawcut joint reservoir in cured concrete									
0182	3/8″ wide x 3/4″ deep, with single saw blade	C-27	1000	.016	L.F.	.07	.51	.13	.71	1.04
0184	1/2″ wide x 1″ deep, with double saw blades		900	.018		.14	.57	.15	.86	1.22
0186	3/4″ wide x 1-1/2″ deep, with double saw blades		800	.020		.30	.64	.17	1.11	1.53
0190	Water blast joint to wash away laitance, 2 passes	C-29	2500	.003			.08	.02	.10	.17
0200	Air blast joint to blow out debris and air dry, 2 passes	C-28	2000	.004			.13	.01	.14	.21
0300	For backer rod, see Section 07 91 23.10									
0340	For joint sealant, see Sections 03 15 05.25 or 07 92 13.20, if available									
0900	For replacement of joint sealant, see Section 07 01 90.81									
03 35 29.60 Finishing Walls										
0010	**FINISHING WALLS**									
0020	Break ties and patch voids	1 Cefi	540	.015	S.F.	.03	.47		.50	.79
0050	Burlap rub with grout		450	.018		.03	.57		.60	.94

03 35 Concrete Finishing

03 35 29 – Tooled Concrete Finishing

03 35 29.60 Finishing Walls		Crew	Daily Output	Labor-Hours	Unit	Material	2011 Bare Costs Labor	Equipment	Total	Total Incl O&P
0100	Carborundum rub, dry	1 Cefi	270	.030	S.F.		.94		.94	1.51
0150	Wet rub		175	.046			1.46		1.46	2.32
0300	Bush hammer, green concrete	B-39	1000	.048			1.29	.19	1.48	2.36
0350	Cured concrete	"	650	.074			1.98	.30	2.28	3.64
0600	Float finish, 1/16" thick	1 Cefi	300	.027		.29	.85		1.14	1.68
0701	Sandblast, light penetration	E-11	1100	.029		.73	.88	.18	1.79	2.58
0751	Heavy penetration	"	375	.085		1.46	2.58	.52	4.56	6.80
0850	Grind form fins flush	1 Clab	700	.011	L.F.		.30		.30	.51

03 35 33 – Stamped Concrete Finishing

03 35 33.50 Slab Texture Stamping

		Crew	Daily Output	Labor-Hours	Unit	Material	Labor	Equipment	Total	Total Incl O&P
0010	**SLAB TEXTURE STAMPING**									
0050	Stamping requires that concrete first be placed, struck off, consolidated,									
0060	bull floated and free of bleed water. Decorative stamping tasks include:									
0100	Step 1 - first application of dry shake colored hardener	1 Cefi	6400	.001	S.F.	.38	.04		.42	.48
0110	Step 2 - bull float		6400	.001			.04		.04	.06
0130	Step 3 - second application of dry shake colored hardener		6400	.001		.19	.04		.23	.27
0140	Step 4 - bull float, manual float & steel trowel	3 Cefi	1280	.019			.60		.60	.95
0150	Step 5 - application of dry shake colored release agent	1 Cefi	6400	.001		.08	.04		.12	.15
0160	Step 6 - place, tamp & remove mats	3 Cefi	2400	.010		1.41	.32		1.73	2.06
0170	Step 7 - touch up edges, mat joints & simulated grout lines	1 Cefi	1280	.006			.20		.20	.32
0300	Alternate stamping estimating method includes all tasks above	4 Cefi	800	.040		2.07	1.27		3.34	4.31
0400	Step 8 - pressure wash @ 3000 psi after 24 hours	1 Cefi	1600	.005			.16		.16	.25
0500	Step 9 - roll 2 coats cure/seal compound when dry	"	800	.010		.48	.32		.80	1.04

03 35 43 – Polished Concrete Finishing

03 35 43.10 Polished Concrete Floors

		Crew	Daily Output	Labor-Hours	Unit	Material	Labor	Equipment	Total	Total Incl O&P
0010	**POLISHED CONCRETE FLOORS**									
0015	Processing of cured concrete to include grinding, honing,									
0020	and polishing of interior floors with 22" segmented diamond									
0025	planetary floor grinder (2 passes in different directions per grit)									
0100	Removal of pre-existing coatings, dry, with carbide discs using									
0105	dry vacuum pick-up system, final hand sweeping									
0110	Glue, adhesive or tar	J-4	1.60	15	M.S.F.	19.50	450	107	576.50	870
0120	Paint, epoxy, 1 coat		3.60	6.667		19.50	200	47.50	267	400
0130	2 coats		1.80	13.333		19.50	400	95.50	515	775
0200	Grinding and edging, wet, including wet vac pick-up and auto									
0205	scrubbing between grit changes									
0210	40-grit diamond/metal matrix	J-4A	1.60	20	M.S.F.	33	585	204	822	1,200
0220	80-grit diamond/metal matrix		2	16		33	465	163	661	975
0230	120-grit diamond/metal matrix		2.40	13.333		33	390	136	559	820
0240	200-grit diamond/metal matrix		2.80	11.429		33	335	116	484	710
0300	Spray on dye or stain (1 coat)	1 Cefi	16	.500		271	15.95		286.95	325
0400	Spray on densifier/hardener (2 coats)	"	8	1		279	32		311	355
0410	Auto scrubbing after 2nd coat, when dry	J-4B	16	.500			13.25	9.65	22.90	32.50
0500	Honing and edging, wet, including wet vac pick-up and auto									
0505	scrubbing between grit changes									
0510	100-grit diamond/resin matrix	J-4A	2.80	11.429	M.S.F.	33	335	116	484	710
0520	200-grit diamond/resin matrix	"	2.80	11.429	"	33	335	116	484	710
0530	Dry, including dry vacuum pick-up system, final hand sweeping									
0540	400-grit diamond/resin matrix	J-4A	2.80	11.429	M.S.F.	33	335	116	484	710
0600	Polishing and edging, dry, including dry vac pick-up and hand									
0605	sweeping between grit changes									
0610	800-grit diamond/resin matrix	J-4A	2.80	11.429	M.S.F.	33	335	116	484	710

03 35 Concrete Finishing

03 35 43 – Polished Concrete Finishing

03 35 43.10 Polished Concrete Floors		Crew	Daily Output	Labor-Hours	Unit	Material	2011 Bare Costs Labor	Equipment	Total	Total Incl O&P
0620	1500-grit diamond/resin matrix	J-4A	2.80	11.429	M.S.F.	33	335	116	484	710
0630	3000-grit diamond/resin matrix	↓	2.80	11.429		33	335	116	484	710
0700	Auto scrubbing after final polishing step	J-4B	16	.500	↓		13.25	9.65	22.90	32.50

03 37 Specialty Placed Concrete

03 37 13 – Shotcrete

03 37 13.30 Gunite (Dry-Mix)		Crew	Daily Output	Labor-Hours	Unit	Material	Labor	Equipment	Total	Total Incl O&P
0010	**GUNITE (DRY-MIX)**									
0020	Typical in place, 1" layers, no mesh included	C-16	2000	.028	S.F.	.33	.83	.10	1.26	1.83
0100	Mesh for gunite 2 x 2, #12	2 Rodm	800	.020		.65	.73		1.38	1.98
0150	#4 reinforcing bars @ 6" each way	"	500	.032		1.50	1.16		2.66	3.67
0300	Typical in place, including mesh, 2" thick, minimum	C-16	1000	.056		1.31	1.65	.20	3.16	4.37
0350	Maximum		500	.112		1.31	3.30	.40	5.01	7.35
0500	4" thick, minimum		750	.075		1.96	2.20	.27	4.43	6.05
0550	Maximum	↓	350	.160	↓	1.96	4.72	.57	7.25	10.55
0901	Prepare old walls, no scaffolding, minimum	2 Clab	665	.024	S.Y.		.64		.64	1.07
0951	Maximum	"	185	.086	"		2.29		2.29	3.83
1100	For high finish requirement or close tolerance, add, minimum				S.F.		50%			
1150	Maximum				"		110%			

03 37 13.60 Shotcrete (Wet-Mix)		Crew	Daily Output	Labor-Hours	Unit	Material	Labor	Equipment	Total	Total Incl O&P
0010	**SHOTCRETE (WET-MIX)**									
0020	Wet mix, placed @ up to 12 C.Y. per hour, 3000 psi	C-8C	80	.600	C.Y.	106	17.45	5.25	128.70	152
0100	Up to 35 C.Y. per hour	C-8E	240	.200	"	95.50	5.75	2.04	103.29	117
1010	Fiber reinforced, 1" thick	C-8C	1740	.028	S.F.	.79	.80	.24	1.83	2.47
1020	2" thick		900	.053		1.58	1.55	.47	3.60	4.82
1030	3" thick		825	.058		2.37	1.69	.51	4.57	5.95
1040	4" thick	↓	750	.064	↓	3.16	1.86	.56	5.58	7.20

03 39 Concrete Curing

03 39 13 – Water Concrete Curing

03 39 13.50 Water Curing		Crew	Daily Output	Labor-Hours	Unit	Material	Labor	Equipment	Total	Total Incl O&P
0010	**WATER CURING**									
0015	With burlap, 4 uses assumed, 7.5 oz.	2 Clab	55	.291	C.S.F.	8.55	7.70		16.25	22.50
0100	10 oz.	"	55	.291	"	15.40	7.70		23.10	30
0400	Curing blankets, 1" to 2" thick, buy, minimum				S.F.	.32			.32	.35
0450	Maximum				"	.48			.48	.53

03 39 23 – Membrane Concrete Curing

03 39 23.13 Chemical Compound Membrane Concrete Curing		Crew	Daily Output	Labor-Hours	Unit	Material	Labor	Equipment	Total	Total Incl O&P
0010	**CHEMICAL COMPOUND MEMBRANE CONCRETE CURING**									
0300	Sprayed membrane curing compound	2 Clab	95	.168	C.S.F.	5.60	4.45		10.05	13.60
0700	Curing compound, solvent based, 400 S.F./gal., 55 gallon lots				Gal.	16.75			16.75	18.40
0720	5 gallon lots					24			24	26.50
0800	Curing compound, water based, 250 S.F./gal., 55 gallon lots					18.75			18.75	20.50
0820	5 gallon lots				↓	20			20	22

03 39 23.23 Sheet Membrane Concrete Curing		Crew	Daily Output	Labor-Hours	Unit	Material	Labor	Equipment	Total	Total Incl O&P
0010	**SHEET MEMBRANE CONCRETE CURING**									
0200	Curing blanket, burlap/poly, 2-ply	2 Clab	70	.229	C.S.F.	14.85	6.05		20.90	26.50

03 41 Precast Structural Concrete

03 41 05 – Precast Concrete Members

03 41 05.10 Precast Beams

	03 41 05.10 Precast Beams	Crew	Daily Output	Labor-Hours	Unit	Material	2011 Bare Costs Labor	Equipment	Total	Total Incl O&P
0010	**PRECAST BEAMS** R034105-30									
0011	L-shaped, 20′ span, 12″ x 20″	C-11	32	1.750	Ea.	2,000	60.50	59.50	2,120	2,375
1000	Inverted tee beams, add to above, small beams					15%				
1050	Large beams					20%				
1200	Rectangular, 20′ span, 12″ x 20″	C-11	32	1.750		1,500	60.50	59.50	1,620	1,825
1250	18″ x 36″		24	2.333		2,725	80.50	79.50	2,885	3,225
1300	24″ x 44″		22	2.545		3,775	88	86.50	3,949.50	4,425
1400	30′ span, 12″ x 36″		24	2.333		3,425	80.50	79.50	3,585	4,000
1450	18″ x 44″		20	2.800		4,925	97	95.50	5,117.50	5,700
1500	24″ x 52″		16	3.500		6,575	121	119	6,815	7,550
1600	40′ span, 12″ x 52″		20	2.800		6,375	97	95.50	6,567.50	7,275
1650	18″ x 52″		16	3.500		7,575	121	119	7,815	8,650
1700	24″ x 52″		12	4.667		8,775	161	159	9,095	10,100
2000	"T" shaped, 20′ span, 12″ x 20″		32	1.750		2,600	60.50	59.50	2,720	3,025
2050	18″ x 36″		24	2.333		4,675	80.50	79.50	4,835	5,375
2100	24″ x 44″		22	2.545		6,475	88	86.50	6,649.50	7,375
2200	30′ span, 12″ x 36″		24	2.333		5,825	80.50	79.50	5,985	6,625
2250	18″ x 44″		20	2.800		8,375	97	95.50	8,567.50	9,475
2300	24″ x 52″		16	3.500		11,200	121	119	11,440	12,600
2500	40′ span, 12″ x 52″		20	2.800		10,800	97	95.50	10,992.50	12,100
2550	18″ x 52″		16	3.500		12,900	121	119	13,140	14,500
2600	24″ x 52″		12	4.667		14,900	161	159	15,220	16,800

03 41 05.15 Precast Columns

	03 41 05.15 Precast Columns	Crew	Daily Output	Labor-Hours	Unit	Material	Labor	Equipment	Total	Total Incl O&P
0010	**PRECAST COLUMNS** R034105-30									
0020	Rectangular to 12′ high, small columns	C-11	120	.467	L.F.	106	16.15	15.90	138.05	160
0050	Large columns		96	.583		184	20	19.85	223.85	258
0300	24′ high, small columns		192	.292		106	10.10	9.95	126.05	144
0350	Large columns		144	.389		184	13.45	13.25	210.70	239

03 41 05.25 Precast Joists

	03 41 05.25 Precast Joists	Crew	Daily Output	Labor-Hours	Unit	Material	Labor	Equipment	Total	Total Incl O&P
0011	**PRECAST JOISTS**									
0020	40 psf L.L., 6″ deep for 12′ spans	C-11	700	.080	L.F.	17.65	2.77	2.72	23.14	27
0051	8″ deep for 16′ spans		670	.084		29.50	2.89	2.85	35.24	40.50
0101	10″ deep for 20′ spans		640	.088		51.50	3.03	2.98	57.51	65
0151	12″ deep for 24′ spans		610	.092		70.50	3.18	3.12	76.80	86

03 41 13 – Precast Concrete Hollow Core Planks

03 41 13.50 Precast Slab Planks

	03 41 13.50 Precast Slab Planks	Crew	Daily Output	Labor-Hours	Unit	Material	Labor	Equipment	Total	Total Incl O&P
0010	**PRECAST SLAB PLANKS** R034105-30									
0020	Prestressed roof/floor members, grouted, solid, 4″ thick	C-11	2400	.023	S.F.	5.55	.81	.79	7.15	8.30
0050	6″ thick		2800	.020		6.50	.69	.68	7.87	9.05
0100	Hollow, 8″ thick		3200	.018		7.20	.61	.60	8.41	9.55
0150	10″ thick		3600	.016		7.50	.54	.53	8.57	9.75
0200	12″ thick		4000	.014		8	.48	.48	8.96	10.15

03 41 16 – Precast Concrete Slabs

03 41 16.20 Precast Concrete Channel Slabs

	03 41 16.20 Precast Concrete Channel Slabs	Crew	Daily Output	Labor-Hours	Unit	Material	Labor	Equipment	Total	Total Incl O&P
0010	**PRECAST CONCRETE CHANNEL SLABS**									
0335	Lightweight concrete channel slab, long runs, 2-3/4″ thick	C-12	1575	.030	S.F.	7.25	1.01	.42	8.68	10.15
0375	3-3/4″ thick		1550	.031		7.55	1.02	.42	8.99	10.50
0475	4-3/4″ thick		1525	.031		8.25	1.04	.43	9.72	11.30
1275	Short pieces, 2-3/4″ thick		785	.061		10.90	2.02	.84	13.76	16.25
1375	3-3/4″ thick		770	.062		11.35	2.06	.85	14.26	16.85
1475	4-3/4″ thick		762	.063		12.40	2.08	.86	15.34	18

03 41 Precast Structural Concrete

03 41 16 – Precast Concrete Slabs

03 41 16.50 Precast Lightweight Concrete Plank		Crew	Daily Output	Labor-Hours	Unit	Material	2011 Bare Costs Labor	Equipment	Total	Total Incl O&P
0011	**PRECAST LIGHTWEIGHT CONCRETE PLANK**									
0020	Lightweight plank, nailable, T&G, 2" thick	C-11	2100	.027	S.F.	6.65	.92	.91	8.48	9.85
0150	For premium ceiling finish, add				"	10%				
0200	For sloping roofs, slope over 4 in 12, add						25%			
0250	Slope over 6 in 12, add						150%			

03 41 23 – Precast Concrete Stairs

03 41 23.50 Precast Stairs		Crew	Daily Output	Labor-Hours	Unit	Material	Labor	Equipment	Total	Total Incl O&P
0010	**PRECAST STAIRS**									
0020	Precast concrete treads on steel stringers, 3' wide	C-12	75	.640	Riser	129	21	8.75	158.75	187
0300	Front entrance, 5' wide with 48" platform, 2 risers		16	3	Flight	460	99	41	600	715
0350	5 risers		12	4		715	132	54.50	901.50	1,075
0500	6' wide, 2 risers		15	3.200		510	106	43.50	659.50	785
0550	5 risers		11	4.364		790	144	59.50	993.50	1,175
0700	7' wide, 2 risers		14	3.429		660	113	47	820	965
0751	5 risers	B-1	5	4.800		1,075	130		1,205	1,425
1200	Basement entrance stairs, steel bulkhead doors, minimum	B-51	22	2.182		1,350	58.50	7.80	1,416.30	1,600
1250	Maximum	"	11	4.364		2,275	117	15.60	2,407.60	2,700

03 41 33 – Precast Structural Pretensioned Concrete

03 41 33.60 Tees		Crew	Daily Output	Labor-Hours	Unit	Material	Labor	Equipment	Total	Total Incl O&P
0010	**TEES** R034105-30									
0020	Quad tee, short spans, roof	C-11	7200	.008	S.F.	7.90	.27	.26	8.43	9.45
0050	Floor		7200	.008		7.90	.27	.26	8.43	9.45
0200	Double tee, floor members, 60' span		8400	.007		9.40	.23	.23	9.86	11
0250	80' span		8000	.007		12.15	.24	.24	12.63	14
0300	Roof members, 30' span		4800	.012		7.90	.40	.40	8.70	9.80
0350	50' span		6400	.009		8.65	.30	.30	9.25	10.35
0400	Wall members, up to 55' high		3600	.016		11.75	.54	.53	12.82	14.40
0500	Single tee roof members, 40' span		3200	.018		10.70	.61	.60	11.91	13.40
0550	80' span		5120	.011		12.15	.38	.37	12.90	14.40
0600	100' span		6000	.009		17.80	.32	.32	18.44	20.50
0650	120' span		6000	.009		19.50	.32	.32	20.14	22.50
1000	Double tees, floor members									
1100	Lightweight, 20" x 8' wide, 45' span	C-11	20	2.800	Ea.	3,375	97	95.50	3,567.50	3,975
1150	24" x 8' wide, 50' span		18	3.111		3,650	108	106	3,864	4,325
1200	32" x 10' wide, 60' span		16	3.500		6,175	121	119	6,415	7,125
1250	Standard weight, 12" x 8' wide, 20' span		22	2.545		1,250	88	86.50	1,424.50	1,625
1300	16" x 8' wide, 25' span		20	2.800		1,675	97	95.50	1,867.50	2,125
1350	18" x 8' wide, 30' span		20	2.800		2,125	97	95.50	2,317.50	2,625
1400	20" x 8' wide, 45' span		18	3.111		2,400	108	106	2,614	2,925
1450	24" x 8' wide, 50' span		16	3.500		3,100	121	119	3,340	3,725
1500	32" x 10' wide, 60' span		14	4		5,675	138	136	5,949	6,600
2000	Roof members									
2050	Lightweight, 20" x 8' wide, 40' span	C-11	20	2.800	Ea.	2,850	97	95.50	3,042.50	3,400
2100	24" x 8' wide, 50' span		18	3.111		3,800	108	106	4,014	4,475
2150	32" x 10' wide, 60' span		16	3.500		6,275	121	119	6,515	7,225
2200	Standard weight, 12" x 8' wide, 30' span		22	2.545		1,900	88	86.50	2,074.50	2,350
2250	16" x 8' wide, 30' span		20	2.800		2,000	97	95.50	2,192.50	2,475
2300	18" x 8' wide, 30' span		20	2.800		2,225	97	95.50	2,417.50	2,725
2350	20" x 8' wide, 40' span		18	3.111		2,275	108	106	2,489	2,800
2400	24" x 8' wide, 50' span		16	3.500		3,050	121	119	3,290	3,675
2450	32" x 10' wide, 60' span		14	4		5,325	138	136	5,599	6,225

03 41 Precast Structural Concrete

03 41 36 – Precast Structural Post-Tensioned Concrete

03 41 36.50 Post-Tensioned Jobs			Crew	Daily Output	Labor-Hours	Unit	Material	2011 Bare Costs Labor	Equipment	Total	Total Incl O&P
0010	**POST-TENSIONED JOBS**	R034105-30									
0100	Post-tensioned in place, small job	R034136-90	C-17B	8.50	9.647	C.Y.	1,175	335	51	1,561	1,900
0200	Large job		"	10	8.200	"	885	283	43.50	1,211.50	1,500

03 45 Precast Architectural Concrete

03 45 13 – Faced Architectural Precast Concrete

03 45 13.50 Precast Wall Panels

Line	Description	Ref	Crew	Daily Output	Labor-Hours	Unit	Material	Labor	Equipment	Total	Total Incl O&P
0010	**PRECAST WALL PANELS**	R034513-10									
0050	Uninsulated 4" thick, smooth gray										
0150	Low rise, 4' x 8' x 4" thick		C-11	320	.175	S.F.	21.50	6.05	5.95	33.50	40
0210	8' x 8', 4" thick			576	.097		21	3.36	3.31	27.67	32.50
0250	8' x 16' x 4" thick			1024	.055		21	1.89	1.86	24.75	28
0600	High rise, 4' x 8' x 4" thick			288	.194		21.50	6.75	6.60	34.85	42
0650	8' x 8' x 4" thick			512	.109		21	3.78	3.72	28.50	34
0700	8' x 16' x 4" thick			768	.073		21	2.52	2.48	26	30
0750	20' x 10', 6" thick, smooth gray		↓	1400	.040		35.50	1.38	1.36	38.24	43.50
0800	Insulated panel, 2" polystyrene, add						1			1	1.10
0850	2" urethane, add						.79			.79	.87
1200	Finishes, white, add						2.73			2.73	3.01
1250	Exposed aggregate, add						1.77			1.77	1.95
1300	Granite faced, domestic, add						29			29	32
1350	Brick faced, modular, red, add					↓	8.75			8.75	9.65
2200	Fiberglass reinforced cement with urethane core										
2210	R20, 8' x 8', minimum		E-2	750	.064	S.F.	22	2.34	2.16	26.50	31
2220	Maximum		"	600	.080	"	33	2.93	2.70	38.63	44.50

03 47 Site-Cast Concrete

03 47 13 – Tilt-Up Concrete

03 47 13.50 Tilt-Up Wall Panels

Line	Description	Ref	Crew	Daily Output	Labor-Hours	Unit	Material	Labor	Equipment	Total	Total Incl O&P
0010	**TILT-UP WALL PANELS**	R034713-20									
0015	Wall panel construction, walls only, 5-1/2" thick		C-14	1600	.085	S.F.	4.78	2.58	1.06	8.42	10.70
0100	7-1/2" thick			1550	.088		6	2.66	1.09	9.75	12.25
0500	Walls and columns, 5-1/2" thick walls, 12" x 12" columns			1565	.087		7.20	2.63	1.08	10.91	13.50
0550	7-1/2" thick wall, 12" x 12" columns			1370	.099	↓	8.85	3.01	1.24	13.10	16.15
0800	Columns only, site precast, minimum			200	.680	L.F.	11.90	20.50	8.45	40.85	57
0850	Maximum		↓	105	1.295	"	15.05	39.50	16.15	70.70	100

03 48 Precast Concrete Specialties

03 48 43 – Precast Concrete Trim

03 48 43.40 Precast Lintels		Crew	Daily Output	Labor-Hours	Unit	Material	2011 Bare Costs Labor	Equipment	Total	Total Incl O&P
0010	**PRECAST LINTELS**, smooth gray, prestressed, stock units only									
0800	4" wide, 8" high, x 4' long	D-10	28	1.143	Ea.	22.50	38	15.50	76	104
0850	8' long		24	1.333		57.50	44	18.10	119.60	155
1000	6" wide, 8" high, x 4' long		26	1.231		33	40.50	16.70	90.20	121
1050	10' long		22	1.455		88	48	19.75	155.75	198
1200	8" wide, 8" high, x 4' long		24	1.333		41	44	18.10	103.10	137
1250	12' long		20	1.600		134	53	21.50	208.50	259
1275	For custom sizes, types, colors, or finishes of precast lintels, add					150%				

03 48 43.90 Precast Window Sills		Crew	Daily Output	Labor-Hours	Unit	Material	Labor	Equipment	Total	Total Incl O&P
0010	**PRECAST WINDOW SILLS**									
0600	Precast concrete, 4" tapers to 3", 9" wide	D-1	70	.229	L.F.	12.25	6.95		19.20	25
0650	11" wide		60	.267		16	8.10		24.10	31
0700	13" wide, 3 1/2" tapers to 2 1/2", 12" wall		50	.320		16	9.75		25.75	33.50

03 51 Cast Roof Decks

03 51 13 – Cementitious Wood Fiber Decks

03 51 13.50 Cementitious/Wood Fiber Planks		Crew	Daily Output	Labor-Hours	Unit	Material	Labor	Equipment	Total	Total Incl O&P
0010	**CEMENTITIOUS/WOOD FIBER PLANKS** R051223-50									
0050	Plank, beveled edge, 1" thick	2 Carp	1000	.016	S.F.	2.32	.54		2.86	3.45
0100	1-1/2" thick		975	.016		3.02	.55		3.57	4.24
0150	T & G, 2" thick		950	.017		3.54	.57		4.11	4.84
0200	2-1/2" thick		925	.017		4.14	.58		4.72	5.50
0250	3" thick		900	.018		4.75	.60		5.35	6.25
1000	Bulb tee, sub-purlin and grout, 6' span, add	E-1	5000	.003		2.06	.12	.02	2.20	2.50
1100	8' span	"	4200	.004		2.31	.14	.03	2.48	2.83

03 51 16 – Gypsum Concrete Roof Decks

03 51 16.50 Gypsum Roof Deck		Crew	Daily Output	Labor-Hours	Unit	Material	Labor	Equipment	Total	Total Incl O&P
0010	**GYPSUM ROOF DECK**									
1000	Poured gypsum, 2" thick	C-8	6000	.009	S.F.	1.25	.28	.12	1.65	1.96
1100	3" thick	"	4800	.012	"	1.87	.34	.15	2.36	2.80

03 52 Lightweight Concrete Roof Insulation

03 52 16 – Lightweight Insulating Concrete

03 52 16.13 Lightweight Cellular Insulating Concrete		Crew	Daily Output	Labor-Hours	Unit	Material	Labor	Equipment	Total	Total Incl O&P
0010	**LIGHTWEIGHT CELLULAR INSULATING CONCRETE** R035216-10									
0020	Portland cement and foaming agent G	C-8	50	1.120	C.Y.	144	33	14.80	191.80	229

03 52 16.16 Lightweight Aggregate Insulating Concrete		Crew	Daily Output	Labor-Hours	Unit	Material	Labor	Equipment	Total	Total Incl O&P
0010	**LIGHTWEIGHT AGGREGATE INSULATING CONCRETE** R035216-10									
0100	Poured vermiculite or perlite, field mix,									
0110	1:6 field mix G	C-8	50	1.120	C.Y.	222	33	14.80	269.80	315
0200	Ready mix, 1:6 mix, roof fill, 2" thick G		10000	.006	S.F.	1.24	.17	.07	1.48	1.71
0250	3" thick G		7700	.007		1.85	.21	.10	2.16	2.50
0401	Expanded volcanic glass rock, with binder, minimum G	L-2	1500	.011		.40	.31		.71	.97
0451	Maximum G	"	1200	.013		1.19	.39		1.58	1.97

03 54 Cast Underlayment

03 54 13 – Gypsum Cement Underlayment

03 54 13.50 Poured Gypsum Underlayment		Crew	Daily Output	Labor-Hours	Unit	2011 Bare Costs Material	Labor	Equipment	Total	Total Incl O&P
0010	**POURED GYPSUM UNDERLAYMENT**									
0400	Underlayment, gypsum based, self-leveling 2500 psi, pumped, 1/2"	C-8	24000	.002	S.F.	.31	.07	.03	.41	.48
0500	3/4"		20000	.003		.47	.08	.04	.59	.69
0600	1"		16000	.004		.62	.10	.05	.77	.91
1400	Hand placed, 1/2"	C-18	450	.020		.31	.53	.12	.96	1.36
1500	3/4"	"	300	.030		.47	.80	.18	1.45	2.05

03 54 16 – Hydraulic Cement Underlayment

03 54 16.50 Cement Underlayment		Crew	Daily Output	Labor-Hours	Unit	Material	Labor	Equipment	Total	Total Incl O&P
0010	**CEMENT UNDERLAYMENT**									
2510	Underlayment, P.C based self-leveling, 4100 psi, pumped, 1/4"	C-8	20000	.003	S.F.	1.35	.08	.04	1.47	1.67
2520	1/2"		19000	.003		2.71	.09	.04	2.84	3.16
2530	3/4"		18000	.003		4.06	.09	.04	4.19	4.66
2540	1"		17000	.003		5.40	.10	.04	5.54	6.15
2550	1-1/2"		15000	.004		8.10	.11	.05	8.26	9.20
2560	Hand placed, 1/2"	C-18	450	.020		2.71	.53	.12	3.36	4
2610	Topping, P.C. based self-level/dry 6100 psi, pumped, 1/4"	C-8	20000	.003		2.03	.08	.04	2.15	2.42
2620	1/2"		19000	.003		4.06	.09	.04	4.19	4.65
2630	3/4"		18000	.003		6.10	.09	.04	6.23	6.90
2660	1"		17000	.003		8.15	.10	.04	8.29	9.15
2670	1-1/2"		15000	.004		12.20	.11	.05	12.36	13.65
2680	Hand placed, 1/2"	C-18	450	.020		4.06	.53	.12	4.71	5.50

03 62 Non-Shrink Grouting

03 62 13 – Non-Metallic Non-Shrink Grouting

03 62 13.50 Grout, Non-Metallic Non-shrink		Crew	Daily Output	Labor-Hours	Unit	Material	Labor	Equipment	Total	Total Incl O&P
0010	**GROUT, NON-METALLIC NON-SHRINK**									
0300	Non-shrink, non-metallic, 1" deep	1 Cefi	35	.229	S.F.	7.30	7.30		14.60	19.65
0350	2" deep	"	25	.320	"	14.65	10.20		24.85	32.50

03 62 16 – Metallic Non-Shrink Grouting

03 62 16.50 Grout, Metallic Non-Shrink		Crew	Daily Output	Labor-Hours	Unit	Material	Labor	Equipment	Total	Total Incl O&P
0010	**GROUT, METALLIC NON-SHRINK**									
0020	Column & machine bases, non-shrink, metallic, 1" deep	1 Cefi	35	.229	S.F.	8.65	7.30		15.95	21
0050	2" deep	"	25	.320	"	17.35	10.20		27.55	35.50

03 63 Epoxy Grouting

03 63 05 – Grouting of Dowels and Fasteners

03 63 05.10 Epoxy Only		Crew	Daily Output	Labor-Hours	Unit	Material	Labor	Equipment	Total	Total Incl O&P
0010	**EPOXY ONLY**									
1500	Chemical anchoring, epoxy cartridge, excludes layout, drilling, fastener									
1530	For fastener 3/4" diam. x 6" embedment	2 Skwk	72	.222	Ea.	4.95	7.55		12.50	18.10
1535	1" diam. x 8" embedment		66	.242		7.45	8.25		15.70	22
1540	1-1/4" diam. x 10" embedment		60	.267		14.85	9.10		23.95	31.50
1545	1-3/4" diam. x 12" embedment		54	.296		25	10.10		35.10	44
1550	14" embedment		48	.333		29.50	11.35		40.85	51.50
1555	2" diam. x 12" embedment		42	.381		39.50	12.95		52.45	65
1560	18" embedment		32	.500		49.50	17.05		66.55	83

03 81 Concrete Cutting

03 81 13 – Flat Concrete Sawing

03 81 13.50 Concrete Floor/Slab Cutting		Crew	Daily Output	Labor-Hours	Unit	2011 Bare Costs Material	Labor	Equipment	Total	Total Incl O&P
0010	**CONCRETE FLOOR/SLAB CUTTING**									
0300	Saw cut concrete slabs, plain, up to 3" deep	B-89	1060	.015	L.F.	.18	.46	.35	.99	1.34
0320	Each additional inch of depth		3180	.005		.06	.15	.12	.33	.45
0400	Mesh reinforced, up to 3" deep		980	.016		.20	.49	.38	1.07	1.46
0420	Each additional inch of depth		2940	.005		.07	.16	.13	.36	.49
0500	Rod reinforced, up to 3" deep		800	.020		.25	.61	.46	1.32	1.79
0520	Each additional inch of depth		2400	.007		.08	.20	.15	.43	.60

03 81 16 – Track Mounted Concrete Wall Sawing

03 81 16.50 Concrete Wall Cutting		Crew	Daily Output	Labor-Hours	Unit	Material	Labor	Equipment	Total	Total Incl O&P
0010	**CONCRETE WALL CUTTING**									
0800	Concrete walls, hydraulic saw, plain, per inch of depth	B-89B	250	.064	L.F.	.46	1.92	2.61	4.99	6.55
0820	Rod reinforcing, per inch of depth	"	150	.107	"	.65	3.20	4.36	8.21	10.75

03 82 Concrete Boring

03 82 13 – Concrete Core Drilling

03 82 13.10 Core Drilling		Crew	Daily Output	Labor-Hours	Unit	Material	Labor	Equipment	Total	Total Incl O&P
0010	**CORE DRILLING**									
0020	Reinf. conc slab, up to 6" thick, incl. bit, layout & set up									
0100	1" diameter core	B-89A	17	.941	Ea.	3.66	28.50	6.45	38.61	58.50
0150	Each added inch thick in same hole, add		1440	.011		.69	.34	.08	1.11	1.40
0300	3" diameter core		16	1		12.20	30.50	6.85	49.55	71.50
0350	Each added inch thick in same hole, add		720	.022		1.41	.67	.15	2.23	2.85
0500	4" diameter core		15	1.067		15.45	32.50	7.30	55.25	79
0550	Each added inch thick in same hole, add		480	.033		1.73	1.01	.23	2.97	3.84
0700	6" diameter core		14	1.143		20	34.50	7.85	62.35	88.50
0750	Each added inch thick in same hole, add		360	.044		2.74	1.34	.30	4.38	5.60
0900	8" diameter core		13	1.231		33	37	8.45	78.45	108
0950	Each added inch thick in same hole, add		288	.056		3.76	1.68	.38	5.82	7.35
1100	10" diameter core		12	1.333		26.50	40.50	9.15	76.15	107
1150	Each added inch thick in same hole, add		240	.067		3.76	2.02	.46	6.24	8
1300	12" diameter core		11	1.455		31	44	10	85	119
1350	Each added inch thick in same hole, add		206	.078		4.56	2.35	.53	7.44	9.50
1500	14" diameter core		10	1.600		37	48.50	11	96.50	134
1550	Each added inch thick in same hole, add		180	.089		6.20	2.69	.61	9.50	12
1700	18" diameter core		9	1.778		45.50	54	12.20	111.70	153
1750	Each added inch thick in same hole, add		144	.111		8.80	3.36	.76	12.92	16.20
1760	For horizontal holes, add to above						20%	20%		
1770	Prestressed hollow core plank, 8" thick									
1780	1" diameter core	B-89A	17.50	.914	Ea.	2.38	27.50	6.25	36.13	56
1790	Each added inch thick in same hole, add		3840	.004		.48	.13	.03	.64	.77
1800	3" diameter core		17	.941		4.96	28.50	6.45	39.91	60
1810	Each added inch thick in same hole, add		1920	.008		.86	.25	.06	1.17	1.43
1820	4" diameter core		16.50	.970		6.80	29.50	6.65	42.95	64
1830	Each added inch thick in same hole, add		1280	.013		1.26	.38	.09	1.73	2.11
1840	6" diameter core		15.50	1.032		8.35	31	7.10	46.45	69.50
1850	Each added inch thick in same hole, add		960	.017		1.43	.50	.11	2.04	2.54
1860	8" diameter core		15	1.067		11.85	32.50	7.30	51.65	75
1870	Each added inch thick in same hole, add		768	.021		2	.63	.14	2.77	3.42
1880	10" diameter core		14	1.143		15.70	34.50	7.85	58.05	84
1890	Each added inch thick in same hole, add		640	.025		2.08	.76	.17	3.01	3.75

03 82 Concrete Boring

03 82 13 – Concrete Core Drilling

03 82 13.10 Core Drilling		Crew	Daily Output	Labor-Hours	Unit	Material	2011 Bare Costs Labor	Equipment	Total	Total Incl O&P
1900	12" diameter core	B-89A	13.50	1.185	Ea.	18.35	36	8.15	62.50	89
1910	Each added inch thick in same hole, add		548	.029		3.33	.88	.20	4.41	5.35
1999	Drilling, core, minimum labor/equipment charge		4	4	Job		121	27.50	148.50	233
3010	Bits for core drill, diamond, premium, 1" diameter				Ea.	80			80	88
3020	3" diameter					206			206	226
3040	4" diameter					229			229	252
3050	6" diameter					365			365	405
3080	8" diameter					585			585	645
3120	12" diameter					900			900	990
3180	18" diameter					1,400			1,400	1,525
3240	24" diameter					1,700			1,700	1,875

03 82 16 – Concrete Drilling

03 82 16.10 Concrete Impact Drilling		Crew	Daily Output	Labor-Hours	Unit	Material	Labor	Equipment	Total	Total Incl O&P
0010	**CONCRETE IMPACT DRILLING**									
0050	Up to 4" deep in conc/brick floor/wall, incl. bit & layout, no anchor									
0100	Holes, 1/4" diameter	1 Carp	75	.107	Ea.	.05	3.58		3.63	6.05
0150	For each additional inch of depth, add		430	.019		.01	.62		.63	1.05
0200	3/8" diameter		63	.127		.05	4.26		4.31	7.20
0250	For each additional inch of depth, add		340	.024		.01	.79		.80	1.33
0300	1/2" diameter		50	.160		.05	5.35		5.40	9.05
0350	For each additional inch of depth, add		250	.032		.01	1.07		1.08	1.81
0400	5/8" diameter		48	.167		.07	5.60		5.67	9.45
0450	For each additional inch of depth, add		240	.033		.02	1.12		1.14	1.89
0500	3/4" diameter		45	.178		.10	5.95		6.05	10.10
0550	For each additional inch of depth, add		220	.036		.02	1.22		1.24	2.07
0600	7/8" diameter		43	.186		.13	6.25		6.38	10.60
0650	For each additional inch of depth, add		210	.038		.03	1.28		1.31	2.18
0700	1" diameter		40	.200		.14	6.70		6.84	11.40
0750	For each additional inch of depth, add		190	.042		.04	1.41		1.45	2.40
0800	1-1/4" diameter		38	.211		.22	7.05		7.27	12.05
0850	For each additional inch of depth, add		180	.044		.06	1.49		1.55	2.56
0900	1-1/2" diameter		35	.229		.34	7.65		7.99	13.20
0950	For each additional inch of depth, add		165	.048		.08	1.63		1.71	2.81
1000	For ceiling installations, add						40%			

Estimating Tips

04 05 00 Common Work Results for Masonry

- The terms *mortar* and *grout* are often used interchangeably, and incorrectly. Mortar is used to bed masonry units, seal the entry of air and moisture, provide architectural appearance, and allow for size variations in the units. Grout is used primarily in reinforced masonry construction and is used to bond the masonry to the reinforcing steel. Common mortar types are M(2500 psi), S(1800 psi), N(750 psi), and O(350 psi), and conform to ASTM C270. Grout is either fine or coarse and conforms to ASTM C476, and in-place strengths generally exceed 2500 psi. Mortar and grout are different components of masonry construction and are placed by entirely different methods. An estimator should be aware of their unique uses and costs.
- Mortar is included in all assembled masonry line items. The mortar cost, part of the assembled masonry material cost, includes all ingredients, all labor, and all equipment required. Please see reference number R040513-10.
- Waste, specifically the loss/droppings of mortar and the breakage of brick and block, is included in all masonry assemblies in this division. A factor of 25% is added for mortar and 3% for brick and concrete masonry units.
- Scaffolding or staging is not included in any of the Division 4 costs. Refer to Subdivision 01 54 23 for scaffolding and staging costs.

04 20 00 Unit Masonry

- The most common types of unit masonry are brick and concrete masonry. The major classifications of brick are building brick (ASTM C62), facing brick (ASTM C216), glazed brick, fire brick, and pavers. Many varieties of texture and appearance can exist within these classifications, and the estimator would be wise to check local custom and availability within the project area. For repair and remodeling jobs, matching the existing brick may be the most important criteria.
- Brick and concrete block are priced by the piece and then converted into a price per square foot of wall. Openings less than two square feet are generally ignored by the estimator because any savings in units used is offset by the cutting and trimming required.
- It is often difficult and expensive to find and purchase small lots of historic brick. Costs can vary widely. Many design issues affect costs, selection of mortar mix, and repairs or replacement of masonry materials. Cleaning techniques must be reflected in the estimate.
- All masonry walls, whether interior or exterior, require bracing. The cost of bracing walls during construction should be included by the estimator, and this bracing must remain in place until permanent bracing is complete. Permanent bracing of masonry walls is accomplished by masonry itself, in the form of pilasters or abutting wall corners, or by anchoring the walls to the structural frame. Accessories in the form of anchors, anchor slots, and ties are used, but their supply and installation can be by different trades. For instance, anchor slots on spandrel beams and columns are supplied and welded in place by the steel fabricator, but the ties from the slots into the masonry are installed by the bricklayer. Regardless of the installation method, the estimator must be certain that these accessories are accounted for in pricing.

Reference Numbers

Reference numbers are shown in shaded boxes at the beginning of some major classifications. These numbers refer to related items in the Reference Section. The reference information may be an estimating procedure, an alternate pricing method, or technical information.

Note: Not all subdivisions listed here necessarily appear in this publication.

04 01 Maintenance of Masonry

04 01 20 – Maintenance of Unit Masonry

04 01 20.20 Pointing Masonry

		Crew	Daily Output	Labor-Hours	Unit	2011 Bare Costs Material	Labor	Equipment	Total	Total Incl O&P
0010	**POINTING MASONRY**									
0300	Cut and repoint brick, hard mortar, running bond	1 Bric	80	.100	S.F.	.53	3.37		3.90	6.10
0320	Common bond		77	.104		.53	3.50		4.03	6.35
0360	Flemish bond		70	.114		.56	3.85		4.41	6.90
0400	English bond		65	.123		.56	4.14		4.70	7.40
0600	Soft old mortar, running bond		100	.080		.53	2.69		3.22	5
0620	Common bond		96	.083		.53	2.80		3.33	5.20
0640	Flemish bond		90	.089		.56	2.99		3.55	5.50
0680	English bond		82	.098		.56	3.28		3.84	6
0700	Stonework, hard mortar		140	.057	L.F.	.70	1.92		2.62	3.92
0720	Soft old mortar		160	.050	"	.70	1.68		2.38	3.53
1000	Repoint, mask and grout method, running bond		95	.084	S.F.	.70	2.83		3.53	5.40
1020	Common bond		90	.089		.70	2.99		3.69	5.70
1040	Flemish bond		86	.093		.74	3.13		3.87	5.95
1060	English bond		77	.104		.74	3.50		4.24	6.55
2000	Scrub coat, sand grout on walls, minimum		120	.067		2.77	2.24		5.01	6.75
2020	Maximum		98	.082		3.84	2.75		6.59	8.75

04 01 20.30 Pointing CMU

		Crew	Daily Output	Labor-Hours	Unit	Material	Labor	Equipment	Total	Total Incl O&P
0010	**POINTING CMU**									
0300	Cut and repoint block, hard mortar, running bond	1 Bric	190	.042	S.F.	.22	1.42		1.64	2.56
0310	Stacked bond		200	.040		.22	1.35		1.57	2.45
0600	Soft old mortar, running bond		230	.035		.22	1.17		1.39	2.16
0610	Stacked bond		245	.033		.22	1.10		1.32	2.04

04 01 20.40 Sawing Masonry

		Crew	Daily Output	Labor-Hours	Unit	Material	Labor	Equipment	Total	Total Incl O&P
0010	**SAWING MASONRY**									
0050	Brick or block by hand, per inch depth	A-1	125	.064	L.F.		1.69	.51	2.20	3.40

04 01 30 – Unit Masonry Cleaning

04 01 30.40 Masonry Building Cleaning

		Crew	Daily Output	Labor-Hours	Unit	Material	Labor	Equipment	Total	Total Incl O&P
0010	**MASONRY BUILDING CLEANING**									
0020	Steam clean building, not incl scaffolding, minimum	B-9	3000	.013	S.F.		.36	.07	.43	.67
0100	Maximum		1500	.027			.72	.13	.85	1.34
0300	Common face brick		1750	.023			.61	.11	.72	1.15
0400	Wire cut face brick		1250	.032			.86	.16	1.02	1.61

04 01 30.60 Brick Washing

		Crew	Daily Output	Labor-Hours	Unit	Material	Labor	Equipment	Total	Total Incl O&P
0010	**BRICK WASHING** R040130-10									
0012	Acid cleanser, smooth brick surface	1 Bric	560	.014	S.F.	.05	.48		.53	.84
0050	Rough brick		400	.020		.06	.67		.73	1.17
0060	Stone, acid wash		600	.013		.08	.45		.53	.82
1000	Muriatic acid, price per gallon in 5 gallon lots				Gal.	9.55			9.55	10.50

04 05 05 – Selective Masonry Demolition

04 05 05.10 Selective Demolition		Crew	Daily Output	Labor-Hours	Unit	Material	2011 Bare Costs Labor	Equipment	Total	Total Incl O&P
0010	**SELECTIVE DEMOLITION** R024119-10									
0200	Bond beams, 8" block with #4 bar	2 Clab	32	.500	L.F.		13.25		13.25	22
0300	Concrete block walls, unreinforced, 2" thick		1200	.013	S.F.		.35		.35	.59
0310	4" thick		1150	.014			.37		.37	.62
0320	6" thick		1100	.015			.38		.38	.64
0330	8" thick		1050	.015			.40		.40	.68
0340	10" thick		1000	.016			.42		.42	.71
0360	12" thick		950	.017			.45		.45	.75
0380	Reinforced alternate courses, 2" thick		1130	.014			.37		.37	.63
0390	4" thick		1080	.015			.39		.39	.66
0400	6" thick		1035	.015			.41		.41	.68
0410	8" thick		990	.016			.43		.43	.72
0420	10" thick		940	.017			.45		.45	.75
0430	12" thick		890	.018			.48		.48	.80
0440	Reinforced alternate courses & vertically 48" OC, 4" thick		900	.018			.47		.47	.79
0450	6" thick		850	.019			.50		.50	.83
0460	8" thick		800	.020			.53		.53	.89
0480	10" thick		750	.021			.56		.56	.94
0490	12" thick		700	.023			.60		.60	1.01
1000	Chimney, 16" x 16", soft old mortar	1 Clab	55	.145	C.F.		3.85		3.85	6.45
1020	Hard mortar		40	.200			5.30		5.30	8.85
1030	16" x 20", soft old mortar		55	.145			3.85		3.85	6.45
1040	Hard mortar		40	.200			5.30		5.30	8.85
1050	16" x 24", soft old mortar		55	.145			3.85		3.85	6.45
1060	Hard mortar		40	.200			5.30		5.30	8.85
1080	20" x 20", soft old mortar		55	.145			3.85		3.85	6.45
1100	Hard mortar		40	.200			5.30		5.30	8.85
1110	20" x 24", soft old mortar		55	.145			3.85		3.85	6.45
1120	Hard mortar		40	.200			5.30		5.30	8.85
1140	20" x 32", soft old mortar		55	.145			3.85		3.85	6.45
1160	Hard mortar		40	.200			5.30		5.30	8.85
1200	48" x 48", soft old mortar		55	.145			3.85		3.85	6.45
1220	Hard mortar		40	.200			5.30		5.30	8.85
1250	Metal, high temp steel jacket, 24" diameter	E-2	130	.369	V.L.F.		13.50	12.50	26	38.50
1260	60" diameter	"	60	.800			29.50	27	56.50	84
1280	Flue lining, up to 12" x 12"	1 Clab	200	.040			1.06		1.06	1.77
1282	Up to 24" x 24"		150	.053			1.41		1.41	2.36
2000	Columns, 8" x 8", soft old mortar		48	.167			4.41		4.41	7.40
2020	Hard mortar		40	.200			5.30		5.30	8.85
2060	16" x 16", soft old mortar		16	.500			13.25		13.25	22
2100	Hard mortar		14	.571			15.10		15.10	25.50
2140	24" x 24", soft old mortar		8	1			26.50		26.50	44.50
2160	Hard mortar		6	1.333			35.50		35.50	59
2200	36" x 36", soft old mortar		4	2			53		53	88.50
2220	Hard mortar		3	2.667			70.50		70.50	118
2230	Alternate pricing method, soft old mortar		30	.267	C.F.		7.05		7.05	11.80
2240	Hard mortar		23	.348	"		9.20		9.20	15.40
3000	Copings, precast or masonry, to 8" wide									
3020	Soft old mortar	1 Clab	180	.044	L.F.		1.18		1.18	1.97
3040	Hard mortar	"	160	.050	"		1.32		1.32	2.22
3100	To 12" wide									
3120	Soft old mortar	1 Clab	160	.050	L.F.		1.32		1.32	2.22
3140	Hard mortar	"	140	.057	"		1.51		1.51	2.53

04 05 Common Work Results for Masonry

04 05 05 – Selective Masonry Demolition

04 05 05.10 Selective Demolition		Crew	Daily Output	Labor-Hours	Unit	Material	Labor	Equipment	Total	Total Incl O&P
						2011 Bare Costs				
4000	Fireplace, brick, 30" x 24" opening									
4020	Soft old mortar	1 Clab	2	4	Ea.		106		106	177
4040	Hard mortar		1.25	6.400			169		169	284
4100	Stone, soft old mortar		1.50	5.333			141		141	236
4120	Hard mortar		1	8			212		212	355
5000	Veneers, brick, soft old mortar		140	.057	S.F.		1.51		1.51	2.53
5020	Hard mortar		125	.064			1.69		1.69	2.84
5100	Granite and marble, 2" thick		180	.044			1.18		1.18	1.97
5120	4" thick		170	.047			1.24		1.24	2.08
5140	Stone, 4" thick		180	.044			1.18		1.18	1.97
5160	8" thick		175	.046			1.21		1.21	2.03
5400	Alternate pricing method, stone, 4" thick		60	.133	C.F.		3.53		3.53	5.90
5420	8" thick		85	.094	"		2.49		2.49	4.17

04 05 13 – Masonry Mortaring

04 05 13.10 Cement

		Crew	Daily Output	Labor-Hours	Unit	Material	Labor	Equipment	Total	Total Incl O&P
0010	**CEMENT**									
0100	Masonry, 70 lb. bag, T.L. lots				Bag	9.45			9.45	10.35
0150	L.T.L. lots					10			10	11
0200	White, 70 lb. bag, T.L. lots					14.85			14.85	16.35
0250	L.T.L. lots					16.70			16.70	18.35

04 05 13.20 Lime

		Crew	Daily Output	Labor-Hours	Unit	Material	Labor	Equipment	Total	Total Incl O&P
0010	**LIME**									
0020	Masons, hydrated, 50 lb. bag, T.L. lots				Bag	9.10			9.10	10
0050	L.T.L. lots					9.30			9.30	10.25
0200	Finish, double hydrated, 50 lb. bag, T.L. lots					9.90			9.90	10.90
0250	L.T.L. lots					11			11	12.10

04 05 13.23 Surface Bonding Masonry Mortaring

		Crew	Daily Output	Labor-Hours	Unit	Material	Labor	Equipment	Total	Total Incl O&P
0010	**SURFACE BONDING MASONRY MORTARING**									
0020	Gray or white colors, not incl. block work	1 Bric	540	.015	S.F.	.10	.50		.60	.93

04 05 13.30 Mortar

		Crew	Daily Output	Labor-Hours	Unit	Material	Labor	Equipment	Total	Total Incl O&P
0010	**MORTAR** R040513-10									
0020	With masonry cement									
0100	Type M, 1:1:6 mix	1 Brhe	143	.056	C.F.	4.88	1.52		6.40	7.85
0200	Type N, 1:3 mix	"	143	.056	"	4.36	1.52		5.88	7.30
2000	With portland cement and lime									
2100	Type M, 1:1/4:3 mix	1 Brhe	143	.056	C.F.	8.80	1.52		10.32	12.20
2200	Type N, 1:1:6 mix, 750 psi		143	.056		6.80	1.52		8.32	9.95
2300	Type O, 1:2:9 mix (Pointing Mortar)		143	.056		7.50	1.52		9.02	10.75
2650	Pre-mixed, type S or N					4.94			4.94	5.45
2700	Mortar for glass block	1 Brhe	143	.056		10.05	1.52		11.57	13.55
2900	Mortar for Fire Brick, 80 lb. bag, T.L. Lots				Bag	25.50			25.50	28

04 05 13.91 Masonry Restoration Mortaring

		Crew	Daily Output	Labor-Hours	Unit	Material	Labor	Equipment	Total	Total Incl O&P
0010	**MASONRY RESTORATION MORTARING**									
0020	Masonry restoration mix				Lb.	1.97			1.97	2.17
0050	White				"	2.27			2.27	2.50

04 05 13.93 Mortar Pigments

		Crew	Daily Output	Labor-Hours	Unit	Material	Labor	Equipment	Total	Total Incl O&P
0010	**MORTAR PIGMENTS**, 50 lb. bags (2 bags per M bricks) R040513-10									
0020	Color admixture, range 2 to 10 lb. per bag of cement, minimum				Lb.	5.25			5.25	5.80
0050	Average					7.80			7.80	8.60
0100	Maximum					11.10			11.10	12.20

04 05 Common Work Results for Masonry

04 05 13 – Masonry Mortaring

04 05 13.95 Sand

	04 05 13.95 Sand	Crew	Daily Output	Labor-Hours	Unit	Material	Labor (2011 Bare Costs)	Equipment	Total	Total Incl O&P
0010	**SAND**, screened and washed at pit									
0020	For mortar, per ton				Ton	16.60			16.60	18.25
0050	With 10 mile haul					27			27	29.50
0100	With 30 mile haul					36			36	39.50
0200	Screened and washed, at the pit				C.Y.	23			23	25.50
0250	With 10 mile haul					37.50			37.50	41
0300	With 30 mile haul					50			50	55

04 05 13.98 Mortar Admixtures

	04 05 13.98 Mortar Admixtures	Crew	Daily Output	Labor-Hours	Unit	Material	Labor	Equipment	Total	Total Incl O&P
0010	**MORTAR ADMIXTURES**									
0020	Waterproofing admixture, per quart (1 qt. to 2 bags of masonry cement)				Qt.	9.50			9.50	10.45

04 05 16 – Masonry Grouting

04 05 16.30 Grouting

	04 05 16.30 Grouting	Crew	Daily Output	Labor-Hours	Unit	Material	Labor	Equipment	Total	Total Incl O&P
0010	**GROUTING** R040513-10									
0011	Bond beams & lintels, 8″ deep, 6″ thick, 0.15 C.F. per L.F.	D-4	1480	.027	L.F.	.64	.77	.09	1.50	2.05
0020	8″ thick, 0.2 C.F. per L.F.		1400	.029		.99	.81	.09	1.89	2.51
0050	10″ thick, 0.25 C.F. per L.F.		1200	.033		1.06	.94	.11	2.11	2.83
0060	12″ thick, 0.3 C.F. per L.F.		1040	.038		1.27	1.09	.12	2.48	3.32
0200	Concrete block cores, solid, 4″ thk., by hand, 0.067 C.F./S.F. of wall	D-8	1100	.036	S.F.	.28	1.13		1.41	2.16
0210	6″ thick, pumped, 0.175 C.F. per S.F.	D-4	720	.056		.74	1.57	.18	2.49	3.59
0250	8″ thick, pumped, 0.258 C.F. per S.F.		680	.059		1.09	1.67	.19	2.95	4.14
0300	10″ thick, pumped, 0.340 C.F. per S.F.		660	.061		1.44	1.72	.19	3.35	4.62
0350	12″ thick, pumped, 0.422 C.F. per S.F.		640	.063		1.79	1.77	.20	3.76	5.10
0500	Cavity walls, 2″ space, pumped, 0.167 C.F./S.F. of wall		1700	.024		.71	.67	.07	1.45	1.96
0550	3″ space, 0.250 C.F./S.F.		1200	.033		1.06	.94	.11	2.11	2.83
0600	4″ space, 0.333 C.F. per S.F.		1150	.035		1.41	.98	.11	2.50	3.29
0700	6″ space, 0.500 C.F. per S.F.		800	.050		2.12	1.42	.16	3.70	4.83
0800	Door frames, 3′ x 7′ opening, 2.5 C.F. per opening		60	.667	Opng.	10.60	18.85	2.10	31.55	45
0850	6′ x 7′ opening, 3.5 C.F. per opening		45	.889	″	14.80	25	2.80	42.60	61
2000	Grout, C476, for bond beams, lintels and CMU cores		350	.114	C.F.	4.23	3.24	.36	7.83	10.40

04 05 19 – Masonry Anchorage and Reinforcing

04 05 19.05 Anchor Bolts

	04 05 19.05 Anchor Bolts	Crew	Daily Output	Labor-Hours	Unit	Material	Labor	Equipment	Total	Total Incl O&P
0010	**ANCHOR BOLTS**									
0015	Installed in fresh grout in CMU bond beams or filled cores, no templates									
0020	Hooked, with nut and washer, 1/2″ diam., 8″ long	1 Bric	132	.061	Ea.	1.36	2.04		3.40	4.84
0030	12″ long		131	.061		1.51	2.06		3.57	5.05
0040	5/8″ diameter, 8″ long		129	.062		2.96	2.09		5.05	6.70
0050	12″ long		127	.063		3.65	2.12		5.77	7.50
0060	3/4″ diameter, 8″ long		127	.063		3.65	2.12		5.77	7.50
0070	12″ long		125	.064		4.56	2.15		6.71	8.55

04 05 19.16 Masonry Anchors

	04 05 19.16 Masonry Anchors	Crew	Daily Output	Labor-Hours	Unit	Material	Labor	Equipment	Total	Total Incl O&P
0010	**MASONRY ANCHORS**									
0020	For brick veneer, galv., corrugated, 7/8″ x 7″, 22 Ga.	1 Bric	10.50	.762	C	9.45	25.50		34.95	52.50
0100	24 Ga.		10.50	.762		8.90	25.50		34.40	52
0150	16 Ga.		10.50	.762		25	25.50		50.50	69.50
0200	Buck anchors, galv., corrugated, 16 gauge, 2″ bend, 8″ x 2″		10.50	.762		49	25.50		74.50	96
0250	8″ x 3″		10.50	.762		51	25.50		76.50	98
0660	Cavity wall, Z-type, galvanized, 6″ long, 1/8″ diam.		10.50	.762		23	25.50		48.50	67
0670	3/16″ diameter		10.50	.762		21	25.50		46.50	65
0680	1/4″ diameter		10.50	.762		39.50	25.50		65	85.50
0850	8″ long, 3/16″ diameter		10.50	.762		23.50	25.50		49	68

04 05 Common Work Results for Masonry

04 05 19 – Masonry Anchorage and Reinforcing

04 05 19.16 Masonry Anchors

	04 05 19.16 Masonry Anchors	Crew	Daily Output	Labor-Hours	Unit	Material	Labor	Equipment	Total	Total Incl O&P
						2011 Bare Costs				
0855	1/4″ diameter	1 Bric	10.50	.762	C	46.50	25.50		72	93
1000	Rectangular type, galvanized, 1/4″ diameter, 2″ x 6″		10.50	.762		67	25.50		92.50	116
1050	4″ x 6″		10.50	.762		83.50	25.50		109	134
1100	3/16″ diameter, 2″ x 6″		10.50	.762		38	25.50		63.50	83.50
1150	4″ x 6″		10.50	.762		39	25.50		64.50	85
1500	Rigid partition anchors, plain, 8″ long, 1″ x 1/8″		10.50	.762		109	25.50		134.50	162
1550	1″ x 1/4″		10.50	.762		180	25.50		205.50	240
1580	1-1/2″ x 1/8″		10.50	.762		145	25.50		170.50	202
1600	1-1/2″ x 1/4″		10.50	.762		295	25.50		320.50	365
1650	2″ x 1/8″		10.50	.762		188	25.50		213.50	249
1700	2″ x 1/4″		10.50	.762		345	25.50		370.50	420

04 05 19.26 Masonry Reinforcing Bars

	04 05 19.26 Masonry Reinforcing Bars	Crew	Daily Output	Labor-Hours	Unit	Material	Labor	Equipment	Total	Total Incl O&P
0010	**MASONRY REINFORCING BARS** R040519-50									
0015	Steel bars A615, placed horiz., #3 & #4 bars	1 Bric	450	.018	Lb.	.45	.60		1.05	1.48
0020	#5 & #6 bars		800	.010		.45	.34		.79	1.05
0050	Placed vertical, #3 & #4 bars		350	.023		.45	.77		1.22	1.76
0060	#5 & #6 bars		650	.012		.45	.41		.86	1.18
0200	Joint reinforcing, regular truss, to 6″ wide, mill std galvanized		30	.267	C.L.F.	15.45	8.95		24.40	31.50
0250	12″ wide		20	.400		17	13.45		30.45	40.50
0400	Cavity truss with drip section, to 6″ wide		30	.267		14.80	8.95		23.75	31
0450	12″ wide		20	.400		15.20	13.45		28.65	38.50

04 05 23 – Masonry Accessories

04 05 23.13 Masonry Control and Expansion Joints

	04 05 23.13 Masonry Control and Expansion Joints	Crew	Daily Output	Labor-Hours	Unit	Material	Labor	Equipment	Total	Total Incl O&P
0010	**MASONRY CONTROL AND EXPANSION JOINTS**									
0020	Rubber, for double wythe 8″ minimum wall (Brick/CMU)	1 Bric	400	.020	L.F.	1.72	.67		2.39	2.99
0025	"T" shaped		320	.025		1.05	.84		1.89	2.54
0030	Cross-shaped for CMU units		280	.029		1.38	.96		2.34	3.10
0050	PVC, for double wythe 8″ minimum wall (Brick/CMU)		400	.020		1.26	.67		1.93	2.49
0120	"T" shaped		320	.025		.64	.84		1.48	2.08
0160	Cross-shaped for CMU units		280	.029		.88	.96		1.84	2.55

04 05 23.19 Masonry Cavity Drainage, Weepholes, and Vents

	04 05 23.19 Masonry Cavity Drainage, Weepholes, and Vents	Crew	Daily Output	Labor-Hours	Unit	Material	Labor	Equipment	Total	Total Incl O&P
0010	**MASONRY CAVITY DRAINAGE, WEEPHOLES, AND VENTS**									
0020	Extruded aluminum, 4″ deep, 2-3/8″ x 8-1/8″	1 Bric	30	.267	Ea.	30.50	8.95		39.45	48
0050	5″ x 8-1/8″		25	.320		40	10.75		50.75	61.50
0100	2-1/4″ x 25″		25	.320		69.50	10.75		80.25	94
0150	5″ x 16-1/2″		22	.364		58.50	12.25		70.75	84
0200	6″ x 16-1/2″		22	.364		81	12.25		93.25	109
0250	7-3/4″ x 16-1/2″		20	.400		69	13.45		82.45	98
0400	For baked enamel finish, add					35%				
0500	For cast aluminum, painted, add					60%				
1000	Stainless steel ventilators, 6″ x 6″	1 Bric	25	.320		126	10.75		136.75	156
1050	8″ x 8″		24	.333		132	11.20		143.20	164
1100	12″ x 12″		23	.348		153	11.70		164.70	187
1150	12″ x 6″		24	.333		134	11.20		145.20	165
1200	Foundation block vent, galv., 1-1/4″ thk, 8″ high, 16″ long, no damper		30	.267		24.50	8.95		33.45	41.50
1250	For damper, add					8.20			8.20	9

04 05 23.95 Wall Plugs

	04 05 23.95 Wall Plugs	Crew	Daily Output	Labor-Hours	Unit	Material	Labor	Equipment	Total	Total Incl O&P
0010	**WALL PLUGS** (for nailing to brickwork)									
0020	26 ga., galvanized, plain	1 Bric	10.50	.762	C	31	25.50		56.50	76
0050	Wood filled	″	10.50	.762	″	104	25.50		129.50	156

04 21 Clay Unit Masonry

04 21 13 – Brick Masonry

04 21 13.13 Brick Veneer Masonry		Crew	Daily Output	Labor-Hours	Unit	Material	2011 Bare Costs Labor	Equipment	Total	Total Incl O&P
0010	**BRICK VENEER MASONRY**, T.L. lots, excl. scaff., grout & reinforcing R042110-20									
0015	Material costs incl. 3% brick and 25% mortar waste									
0020	Standard, select common, 4" x 2-2/3" x 8" (6.75/S.F.)	D-8	1.50	26.667	M	535	830		1,365	1,950
0050	Red, 4" x 2-2/3" x 8", running bond		1.50	26.667		620	830		1,450	2,025
0100	Full header every 6th course (7.88/S.F.) R042110-50		1.45	27.586		620	855		1,475	2,075
0150	English, full header every 2nd course (10.13/S.F.)		1.40	28.571		615	885		1,500	2,125
0200	Flemish, alternate header every course (9.00/S.F.)		1.40	28.571		620	885		1,505	2,125
0250	Flemish, alt. header every 6th course (7.13/S.F.)		1.45	27.586		620	855		1,475	2,075
0300	Full headers throughout (13.50/S.F.)		1.40	28.571		615	885		1,500	2,125
0350	Rowlock course (13.50/S.F.)		1.35	29.630		615	920		1,535	2,175
0400	Rowlock stretcher (4.50/S.F.)		1.40	28.571		625	885		1,510	2,125
0450	Soldier course (6.75/S.F.)		1.40	28.571		620	885		1,505	2,125
0500	Sailor course (4.50/S.F.)		1.30	30.769		625	955		1,580	2,250
0601	Buff or gray face, running bond, (6.75/S.F.)		1.50	26.667		620	830		1,450	2,025
0700	Glazed face, 4" x 2-2/3" x 8", running bond		1.40	28.571		1,850	885		2,735	3,500
0750	Full header every 6th course (7.88/S.F.)		1.35	29.630		1,775	920		2,695	3,450
1000	Jumbo, 6" x 4" x 12",(3.00/S.F.)		1.30	30.769		1,775	955		2,730	3,525
1051	Norman, 4" x 2-2/3" x 12" (4.50/S.F.)		1.45	27.586		1,100	855		1,955	2,625
1100	Norwegian, 4" x 3-1/5" x 12" (3.75/S.F.)		1.40	28.571		1,150	885		2,035	2,700
1150	Economy, 4" x 4" x 8" (4.50 per S.F.)		1.40	28.571		940	885		1,825	2,475
1201	Engineer, 4" x 3-1/5" x 8", (5.63/S.F.)		1.45	27.586		630	855		1,485	2,100
1251	Roman, 4" x 2" x 12", (6.00/S.F.)		1.50	26.667		985	830		1,815	2,425
1300	S.C.R. 6" x 2-2/3" x 12" (4.50/S.F.)		1.40	28.571		1,175	885		2,060	2,750
1350	Utility, 4" x 4" x 12" (3.00/S.F.)		1.35	29.630		1,550	920		2,470	3,200
1360	For less than truck load lots, add					15			15	16.50
1400	For battered walls, add						30%			
1450	For corbels, add						75%			
1500	For curved walls, add						30%			
1550	For pits and trenches, deduct						20%			
1999	Alternate method of figuring by square foot									
2000	Standard, sel. common, 4" x 2-2/3" x 8", (6.75/S.F.)	D-8	230	.174	S.F.	3.61	5.40		9.01	12.80
2020	Standard, red, 4" x 2-2/3" x 8", running bond (6.75/S.F.)		220	.182		4.19	5.65		9.84	13.85
2050	Full header every 6th course (7.88/S.F.)		185	.216		4.88	6.70		11.58	16.35
2100	English, full header every 2nd course (10.13/S.F.)		140	.286		6.25	8.85		15.10	21.50
2150	Flemish, alternate header every course (9.00/S.F.)		150	.267		5.55	8.30		13.85	19.70
2200	Flemish, alt. header every 6th course (7.13/S.F.)		205	.195		4.42	6.05		10.47	14.80
2250	Full headers throughout (13.50/S.F.)		105	.381		8.30	11.85		20.15	28.50
2300	Rowlock course (13.50/S.F.)		100	.400		8.30	12.40		20.70	29.50
2350	Rowlock stretcher (4.50/S.F.)		310	.129		2.81	4.01		6.82	9.65
2400	Soldier course (6.75/S.F.)		200	.200		4.19	6.20		10.39	14.80
2450	Sailor course (4.50/S.F.)		290	.138		2.81	4.28		7.09	10.15
2600	Buff or gray face, running bond, (6.75/S.F.)		220	.182		4.43	5.65		10.08	14.10
2700	Glazed face brick, running bond		210	.190		11.95	5.90		17.85	23
2750	Full header every 6th course (7.88/S.F.)		170	.235		13.95	7.30		21.25	27.50
3000	Jumbo, 6" x 4" x 12" running bond (3.00/S.F.)		435	.092		4.87	2.86		7.73	10.05
3050	Norman, 4" x 2-2/3" x 12" running bond, (4.5/S.F.)		320	.125		5.70	3.88		9.58	12.60
3100	Norwegian, 4" x 3-1/5" x 12" (3.75/S.F.)		375	.107		4.15	3.31		7.46	10
3150	Economy, 4" x 4" x 8" (4.50/S.F.)		310	.129		4.19	4.01		8.20	11.15
3200	Engineer, 4" x 3-1/5" x 8" (5.63/S.F.)		260	.154		3.54	4.78		8.32	11.75
3250	Roman, 4" x 2" x 12" (6.00/S.F.)		250	.160		5.75	4.97		10.72	14.50
3300	SCR, 6" x 2-2/3" x 12" (4.50/S.F.)		310	.129		5.25	4.01		9.26	12.30
3350	Utility, 4" x 4" x 12" (3.00/S.F.)		450	.089		4.51	2.76		7.27	9.50
3400	For cavity wall construction, add						15%			

04 21 13.13 Brick Veneer Masonry		Crew	Daily Output	Labor-Hours	Unit	2011 Bare Costs Material	Labor	Equipment	Total	Total Incl O&P
3450	For stacked bond, add						10%			
3500	For interior veneer construction, add						15%			
3550	For curved walls, add						30%			

04 21 13.14 Thin Brick Veneer

		Crew	Daily Output	Labor-Hours	Unit	Material	Labor	Equipment	Total	Total Incl O&P
0010	**THIN BRICK VENEER**									
0015	Material costs incl. 3% brick and 25% mortar waste									
0020	On & incl. metal panel support sys, modular, 2-2/3" x 5/8" x 8", red	D-7	92	.174	S.F.	8.75	4.90		13.65	17.50
0100	Closure, 4" x 5/8" x 8"		110	.145		8.75	4.10		12.85	16.20
0110	Norman, 2-2/3" x 5/8" x 12"		110	.145		8.95	4.10		13.05	16.40
0120	Utility, 4" x 5/8" x 12"		125	.128		8.95	3.61		12.56	15.60
0130	Emperor, 4" x 3/4" x 16"		175	.091		10.15	2.58		12.73	15.30
0140	Super emperor, 8" x 3/4" x 16"		195	.082		10.15	2.31		12.46	14.90
0150	For L shaped corners with 4" return, add				L.F.	9.25			9.25	10.20
0200	On masonry/plaster back-up, modular, 2-2/3" x 5/8" x 8", red	D-7	137	.117	S.F.	3.95	3.29		7.24	9.60
0210	Closure, 4" x 5/8" x 8"		165	.097		3.95	2.73		6.68	8.70
0220	Norman, 2-2/3" x 5/8" x 12"		165	.097		4.15	2.73		6.88	8.95
0230	Utility, 4" x 5/8" x 12"		185	.086		4.15	2.44		6.59	8.45
0240	Emperor, 4" x 3/4" x 16"		260	.062		5.35	1.74		7.09	8.65
0250	Super emperor, 8" x 3/4" x 16"		285	.056		5.35	1.58		6.93	8.45
0260	For L shaped corners with 4" return, add				L.F.	9.25			9.25	10.20
0270	For embedment into pre-cast concrete panels, add				S.F.	14.40			14.40	15.85

04 21 13.15 Chimney

		Crew	Daily Output	Labor-Hours	Unit	Material	Labor	Equipment	Total	Total Incl O&P
0010	**CHIMNEY**, excludes foundation, scaffolding, grout and reinforcing									
0100	Brick, 16" x 16", 8" flue	D-1	18.20	.879	V.L.F.	21.50	26.50		48	67.50
0150	16" x 20" with one 8" x 12" flue		16	1		34	30.50		64.50	87.50
0200	16" x 24" with two 8" x 8" flues		14	1.143		50	34.50		84.50	112
0250	20" x 20" with one 12" x 12" flue		13.70	1.168		38.50	35.50		74	101
0300	20" x 24" with two 8" x 12" flues		12	1.333		57	40.50		97.50	129
0350	20" x 32" with two 12" x 12" flues		10	1.600		68	48.50		116.50	155
1800	Metal, high temp. steel jacket, factory lining, 24" diam.	E-2	65	.738		210	27	25	262	310
1900	60" diameter	"	30	1.600		765	58.50	54	877.50	1,000
2100	Poured concrete, brick lining, 200' high x 10' diam.					7,275			7,275	8,000
2800	500' x 20' diameter					12,700			12,700	14,000

04 21 13.18 Columns

		Crew	Daily Output	Labor-Hours	Unit	Material	Labor	Equipment	Total	Total Incl O&P
0010	**COLUMNS**, solid, excludes scaffolding, grout and reinforcing R042110-10									
0050	Brick, 8" x 8", 9 brick per V.L.F.	D-1	56	.286	V.L.F.	5.45	8.70		14.15	20.50
0100	12" x 8", 13.5 brick per V.L.F.		37	.432		8.15	13.15		21.30	30.50
0200	12" x 12", 20 brick per V.L.F.		25	.640		12.10	19.45		31.55	45.50
0300	16" x 12", 27 brick per V.L.F.		19	.842		16.35	25.50		41.85	60
0400	16" x 16", 36 brick per V.L.F.		14	1.143		22	34.50		56.50	81
0500	20" x 16", 45 brick per V.L.F.		11	1.455		27	44		71	103
0600	20" x 20", 56 brick per V.L.F.		9	1.778		34	54		88	126
0700	24" x 20", 68 brick per V.L.F.		7	2.286		41	69.50		110.50	160
0800	24" x 24", 81 brick per V.L.F.		6	2.667		49	81		130	187
1000	36" x 36", 182 brick per V.L.F.		3	5.333		110	162		272	385

04 21 13.30 Oversized Brick

		Crew	Daily Output	Labor-Hours	Unit	Material	Labor	Equipment	Total	Total Incl O&P
0010	**OVERSIZED BRICK**, excludes scaffolding, grout and reinforcing									
0100	Veneer, 4" x 2.25" x 16"	D-8	387	.103	S.F.	6.45	3.21		9.66	12.30
0105	4" x 2.75" x 16"		412	.097		5.80	3.01		8.81	11.30
0110	4" x 4" x 16"		460	.087		5.30	2.70		8	10.30
0120	4" x 8" x 16"		533	.075		6.10	2.33		8.43	10.55
0125	Loadbearing, 6" x 4" x 16", grouted and reinforced		387	.103		9	3.21		12.21	15.15

04 21 13.30 Oversized Brick		Crew	Daily Output	Labor-Hours	Unit	Material	Labor	Equipment	Total	Total Incl O&P
						2011 Bare Costs				
0130	8" x 4" x 16", grouted and reinforced	D-8	327	.122	S.F.	9.40	3.80		13.20	16.60
0135	6" x 8" x 16", grouted and reinforced		440	.091		8.65	2.82		11.47	14.20
0140	8" x 8" x 16", grouted and reinforced		400	.100		9.55	3.11		12.66	15.60
0145	Curtainwall/reinforced veneer, 6" x 4" x 16"		387	.103		13.20	3.21		16.41	19.75
0150	8" x 4" x 16"		327	.122		15.65	3.80		19.45	23.50
0155	6" x 8" x 16"		440	.091		13.30	2.82		16.12	19.30
0160	8" x 8" x 16"		400	.100		15.80	3.11		18.91	22.50
0200	For 1 to 3 slots in face, add					15%				
0210	For 4 to 7 slots in face, add					25%				
0220	For bond beams, add					20%				
0230	For bullnose shapes, add					20%				
0240	For open end knockout, add					10%				
0250	For white or gray color group, add					10%				
0260	For 135 degree corner, add					250%				

04 21 13.35 Common Building Brick

0010	**COMMON BUILDING BRICK**, C62, TL lots, material only R042110-20									
0020	Standard, minimum				M	400			400	440
0050	Average (select)				"	430			430	475

04 21 13.40 Structural Brick

0010	**STRUCTURAL BRICK** C652, Grade SW, incl. mortar, scaffolding not incl.									
0100	Standard unit, 4-5/8" x 2-3/4" x 9-5/8"	D-8	245	.163	S.F.	5.15	5.05		10.20	14
0120	Bond beam		225	.178		9.65	5.50		15.15	19.70
0140	V cut bond beam		225	.178		10.45	5.50		15.95	20.50
0160	Stretcher quoin, 5-5/8" x 2-3/4" x 9-5/8"		245	.163		10.10	5.05		15.15	19.40
0180	Corner quoin		245	.163		11.30	5.05		16.35	21
0200	Corner, 45 deg, 4-5/8" x 2-3/4" x 10-7/16"		235	.170		11.70	5.30		17	21.50

04 21 13.45 Face Brick

0010	**FACE BRICK** Material Only, C216, TL lots R042110-20									
0300	Standard modular, 4" x 2-2/3" x 8", minimum				M	515			515	565
0350	Maximum					685			685	750
0450	Economy, 4" x 4" x 8", minimum					815			815	895
0500	Maximum					1,050			1,050	1,150
0510	Economy, 4" x 4" x 12", minimum					1,200			1,200	1,325
0520	Maximum					1,550			1,550	1,725
0550	Jumbo, 6" x 4" x 12", minimum					1,475			1,475	1,625
0600	Maximum					1,875			1,875	2,075
0610	Jumbo, 8" x 4" x 12", minimum					1,475			1,475	1,625
0620	Maximum					1,875			1,875	2,075
0650	Norwegian, 4" x 3-1/5" x 12", minimum					980			980	1,075
0700	Maximum					1,275			1,275	1,400
0710	Norwegian, 6" x 3-1/5" x 12", minimum					1,425			1,425	1,575
0720	Maximum					1,875			1,875	2,050
0850	Standard glazed, plain colors, 4" x 2-2/3" x 8", minimum					1,625			1,625	1,800
0900	Maximum					2,125			2,125	2,325
1000	Deep trim shades, 4" x 2-2/3" x 8", minimum					1,900			1,900	2,100
1050	Maximum					2,175			2,175	2,400
1080	Jumbo utility, 4" x 4" x 12"					1,350			1,350	1,500
1120	4" x 8" x 8"					1,675			1,675	1,850
1140	4" x 8" x 16"					4,925			4,925	5,400
1260	Engineer, 4" x 3-1/5" x 8", minimum					520			520	570
1270	Maximum					775			775	850
1350	King, 4" x 2-3/4" x 10", minimum					460			460	505

04 21 Clay Unit Masonry

04 21 13 – Brick Masonry

04 21 13.45 Face Brick		Crew	Daily Output	Labor-Hours	Unit	Material	2011 Bare Costs Labor	Equipment	Total	Total Incl O&P
1360	Maximum				M	560			560	620
1400	Norman, 4" x 2-3/4" x 12"					460			460	505
1450	Roman, 4" x 2" x 12"					460			460	505
1500	SCR, 6" x 2-2/3" x 12"					460			460	505
1550	Double, 4" x 5-1/3" x 8"					460			460	505
1600	Triple, 4" x 5-1/3" x 12"					460			460	505
1770	Standard modular, double glazed, 4" x 2-2/3" x 8"					2,275			2,275	2,500
1850	Jumbo, colored glazed ceramic, 6" x 4" x 12"					2,400			2,400	2,650
2050	Jumbo utility, glazed, 4" x 4" x 12"					4,550			4,550	5,000
2100	4" x 8" x 8"					5,325			5,325	5,875
2150	4" x 16" x 8"					6,200			6,200	6,825
2170	For less than truck load lots, add					15			15	16.50
2180	For buff or gray brick, add					16			16	17.60

04 21 26 – Glazed Structural Clay Tile Masonry

04 21 26.10 Structural Facing Tile		Crew	Daily Output	Labor-Hours	Unit	Material	2011 Bare Costs Labor	Equipment	Total	Total Incl O&P
0010	**STRUCTURAL FACING TILE**, std. colors, excl. scaffolding, grout, reinforcing									
0020	6T series, 5-1/3" x 12", 2.3 pieces per S.F., glazed 1 side, 2" thick	D-8	225	.178	S.F.	7.95	5.50		13.45	17.80
0100	4" thick		220	.182		10.95	5.65		16.60	21.50
0150	Glazed 2 sides		195	.205		14.05	6.35		20.40	26
0250	6" thick		210	.190		16.25	5.90		22.15	27.50
0300	Glazed 2 sides		185	.216		19.30	6.70		26	32
0400	8" thick		180	.222		21.50	6.90		28.40	35
0500	Special shapes, group 1		400	.100	Ea.	6.80	3.11		9.91	12.55
0550	Group 2		375	.107		11.05	3.31		14.36	17.65
0600	Group 3		350	.114		14.30	3.55		17.85	21.50
0650	Group 4		325	.123		29.50	3.82		33.32	39
0700	Group 5		300	.133		35.50	4.14		39.64	46
0750	Group 6		275	.145		48.50	4.52		53.02	61
1000	Fire rated, 4" thick, 1 hr. rating		210	.190	S.F.	15.75	5.90		21.65	27
1300	Acoustic, 4" thick		210	.190	"	34	5.90		39.90	47
2000	8W series, 8" x 16", 1.125 pieces per S.F.									
2050	2" thick, glazed 1 side	D-8	360	.111	S.F.	9.30	3.45		12.75	15.85
2100	4" thick, glazed 1 side		345	.116		11.95	3.60		15.55	19.05
2150	Glazed 2 sides		325	.123		13.85	3.82		17.67	21.50
2200	6" thick, glazed 1 side		330	.121		21	3.76		24.76	29
2250	8" thick, glazed 1 side		310	.129		22	4.01		26.01	30.50
2500	Special shapes, group 1		300	.133	Ea.	12.85	4.14		16.99	21
2550	Group 2		280	.143		17.20	4.44		21.64	26
2600	Group 3		260	.154		18.25	4.78		23.03	28
2650	Group 4		250	.160		36.50	4.97		41.47	48
2700	Group 5		240	.167		39	5.20		44.20	51.50
2750	Group 6		230	.174		84.50	5.40		89.90	101
3000	4" thick, glazed 1 side		345	.116	S.F.	11.45	3.60		15.05	18.50
3100	Acoustic, 4" thick		345	.116	"	13.95	3.60		17.55	21.50
3120	4W series, 8" x 8", 2.25 pieces per S.F.									
3125	2" thick, glazed 1 side	D-8	360	.111	S.F.	8	3.45		11.45	14.45
3130	4" thick, glazed 1 side		345	.116		9.65	3.60		13.25	16.50
3135	Glazed 2 sides		325	.123		13.20	3.82		17.02	21
3140	6" thick, glazed 1 side		330	.121		13.80	3.76		17.56	21.50
3150	8" thick, glazed 1 side		310	.129		20	4.01		24.01	28.50
3155	Special shapes, group I		300	.133	Ea.	6.35	4.14		10.49	13.80
3160	Group II		280	.143	"	7.30	4.44		11.74	15.30

04 21 Clay Unit Masonry

04 21 26 – Glazed Structural Clay Tile Masonry

04 21 26.10 Structural Facing Tile		Crew	Daily Output	Labor-Hours	Unit	Material	2011 Bare Costs Labor	Equipment	Total	Total Incl O&P
3200	For designer colors, add					25%				
3300	For epoxy mortar joints, add				S.F.	1.58			1.58	1.74

04 21 29 – Terra Cotta Masonry

04 21 29.10 Terra Cotta Masonry Components		Crew	Daily Output	Labor-Hours	Unit	Material	Labor	Equipment	Total	Total Incl O&P
0010	**TERRA COTTA MASONRY COMPONENTS**									
0020	Coping, split type, not glazed, 9" wide	D-1	90	.178	L.F.	10.40	5.40		15.80	20.50
0100	13" wide		80	.200		15.75	6.10		21.85	27.50
0200	Coping, split type, glazed, 9" wide		90	.178		17.40	5.40		22.80	28
0250	13" wide	↓	80	.200	↓	22.50	6.10		28.60	35
0500	Partition or back-up blocks, scored, in C.L. lots									
0700	Non-load bearing 12" x 12", 3" thick, special order	D-8	550	.073	S.F.	26	2.26		28.26	32
0750	4" thick, standard		500	.080		7.10	2.48		9.58	11.90
0800	6" thick		450	.089		8.70	2.76		11.46	14.10
0850	8" thick		400	.100		11.05	3.11		14.16	17.30
1000	Load bearing, 12" x 12", 4" thick, in walls		500	.080		7.60	2.48		10.08	12.45
1050	In floors		750	.053		7.60	1.66		9.26	11.05
1200	6" thick, in walls		450	.089		8.70	2.76		11.46	14.10
1250	In floors		675	.059		8.70	1.84		10.54	12.55
1400	8" thick, in walls		400	.100		11.05	3.11		14.16	17.30
1450	In floors		575	.070		11.05	2.16		13.21	15.75
1600	10" thick, in walls, special order		350	.114		23.50	3.55		27.05	31.50
1650	In floors, special order		500	.080		23.50	2.48		25.98	29.50
1800	12" thick, in walls, special order		300	.133		28	4.14		32.14	37.50
1850	In floors, special order	↓	450	.089		28	2.76		30.76	35
2000	For reinforcing with steel rods, add to above					15%	5%			
2100	For smooth tile instead of scored, add					3.74			3.74	4.11
2200	For L.C.L. quantities, add				↓	10%	10%			

04 21 29.20 Terra Cotta Tile		Crew	Daily Output	Labor-Hours	Unit	Material	Labor	Equipment	Total	Total Incl O&P
0010	**TERRA COTTA TILE**, on walls, dry set, 1/2" thick									
0100	Square, hexagonal or lattice shapes, unglazed	1 Tilf	135	.059	S.F.	6.35	1.87		8.22	9.95
0300	Glazed, plain colors		130	.062		8.75	1.94		10.69	12.70
0400	Intense colors	↓	125	.064	↓	10.20	2.02		12.22	14.45

04 22 Concrete Unit Masonry

04 22 10 – Concrete Masonry Units

04 22 10.11 Autoclave Aerated Concrete Block		Crew	Daily Output	Labor-Hours	Unit	Material	Labor	Equipment	Total	Total Incl O&P
0010	**AUTOCLAVE AERATED CONCRETE BLOCK**, excl. scaffolding, grout & reinforcing									
0050	Solid, 4" x 12" x 24", incl mortar G	D-8	600	.067	S.F.	1.39	2.07		3.46	4.93
0060	6" x 12" x 24" G		600	.067		2.08	2.07		4.15	5.70
0070	8" x 8" x 24" G		575	.070		2.78	2.16		4.94	6.60
0080	10" x 12" x 24" G		575	.070		3.40	2.16		5.56	7.30
0090	12" x 12" x 24" G	↓	550	.073	↓	4.16	2.26		6.42	8.30

04 22 10.12 Chimney Block		Crew	Daily Output	Labor-Hours	Unit	Material	Labor	Equipment	Total	Total Incl O&P
0010	**CHIMNEY BLOCK**, excludes scaffolding, grout and reinforcing									
0220	1 piece, with 8" x 8" flue, 16" x 16"	D-1	28	.571	V.L.F.	17.50	17.35		34.85	48
0230	2 piece, 16" x 16"		26	.615		15.05	18.70		33.75	47
0240	2 piece, with 8" x 12" flue, 16" x 20"	↓	24	.667	↓	30	20.50		50.50	66.50

04 22 10.14 Concrete Block, Back-Up		Crew	Daily Output	Labor-Hours	Unit	Material	Labor	Equipment	Total	Total Incl O&P
0010	**CONCRETE BLOCK, BACK-UP**, C90, 2000 psi R042210-20									
0020	Normal weight, 8" x 16" units, tooled joint 1 side									

04 22 Concrete Unit Masonry

04 22 10 – Concrete Masonry Units

04 22 10.14 Concrete Block, Back-Up

	04 22 10.14 Concrete Block, Back-Up	Crew	Daily Output	Labor-Hours	Unit	Material	Labor	Equipment	Total	Total Incl O&P
						2011 Bare Costs				
0050	Not-reinforced, 2000 psi, 2" thick	D-8	475	.084	S.F.	1.39	2.61		4	5.80
0200	4" thick		460	.087		1.58	2.70		4.28	6.15
0300	6" thick		440	.091		2.18	2.82		5	7.05
0350	8" thick		400	.100		2.49	3.11		5.60	7.85
0400	10" thick		330	.121		3.12	3.76		6.88	9.60
0450	12" thick	D-9	310	.155		3.52	4.71		8.23	11.60
1000	Reinforced, alternate courses, 4" thick	D-8	450	.089		1.71	2.76		4.47	6.40
1100	6" thick		430	.093		2.31	2.89		5.20	7.30
1150	8" thick		395	.101		2.62	3.14		5.76	8.05
1200	10" thick		320	.125		3.25	3.88		7.13	9.90
1250	12" thick	D-9	300	.160		3.65	4.86		8.51	12

04 22 10.16 Concrete Block, Bond Beam

	04 22 10.16 Concrete Block, Bond Beam	Crew	Daily Output	Labor-Hours	Unit	Material	Labor	Equipment	Total	Total Incl O&P
0010	**CONCRETE BLOCK, BOND BEAM**, C90, 2000 psi									
0020	Not including grout or reinforcing									
0125	Regular block, 6" thick	D-8	584	.068	L.F.	2.32	2.13		4.45	6.05
0130	8" high, 8" thick	"	565	.071		2.50	2.20		4.70	6.35
0150	12" thick	D-9	510	.094		3.50	2.86		6.36	8.55
0525	Lightweight, 6" thick	D-8	592	.068		2.55	2.10		4.65	6.25
0530	8" high, 8" thick	"	575	.070		3.09	2.16		5.25	6.95
0550	12" thick	D-9	520	.092		4.16	2.81		6.97	9.15
2000	Including grout and 2 #5 bars									
2100	Regular block, 8" high, 8" thick	D-8	300	.133	L.F.	4.45	4.14		8.59	11.70
2150	12" thick	D-9	250	.192		6	5.85		11.85	16.20
2500	Lightweight, 8" high, 8" thick	D-8	305	.131		5.05	4.07		9.12	12.25
2550	12" thick	D-9	255	.188		6.65	5.70		12.35	16.70

04 22 10.18 Concrete Block, Column

	04 22 10.18 Concrete Block, Column	Crew	Daily Output	Labor-Hours	Unit	Material	Labor	Equipment	Total	Total Incl O&P
0010	**CONCRETE BLOCK, COLUMN** or pilaster									
0050	Including vertical reinforcing (4-#4 bars) and grout									
0160	1 piece unit, 16" x 16"	D-1	26	.615	V.L.F.	18.05	18.70		36.75	50.50
0170	2 piece units, 16" x 20"		24	.667		23.50	20.50		44	59.50
0180	20" x 20"		22	.727		34.50	22		56.50	74.50
0190	22" x 24"		18	.889		48.50	27		75.50	98
0200	20" x 32"		14	1.143		53	34.50		87.50	116

04 22 10.19 Concrete Block, Insulation Inserts

	04 22 10.19 Concrete Block, Insulation Inserts	Crew	Daily Output	Labor-Hours	Unit	Material	Labor	Equipment	Total	Total Incl O&P
0010	**CONCRETE BLOCK, INSULATION INSERTS**									
0100	Styrofoam, plant installed, add to block prices									
0200	8" x 16" units, 6" thick				S.F.	1.81			1.81	1.99
0250	8" thick					1.81			1.81	1.99
0300	10" thick					2.13			2.13	2.34
0350	12" thick					2.24			2.24	2.46
0500	8" x 8" units, 8" thick					1.49			1.49	1.64
0550	12" thick					1.81			1.81	1.99

04 22 10.23 Concrete Block, Decorative

	04 22 10.23 Concrete Block, Decorative	Crew	Daily Output	Labor-Hours	Unit	Material	Labor	Equipment	Total	Total Incl O&P
0010	**CONCRETE BLOCK, DECORATIVE**, C90, 2000 psi									
0020	Embossed, simulated brick face									
0100	8" x 16" units, 4" thick	D-8	400	.100	S.F.	3.51	3.11		6.62	8.95
0200	8" thick		340	.118		4.83	3.65		8.48	11.30
0250	12" thick		300	.133		6.35	4.14		10.49	13.80
0400	Embossed both sides									
0500	8" thick	D-8	300	.133	S.F.	5.40	4.14		9.54	12.75
0550	12" thick	"	275	.145	"	6.85	4.52		11.37	14.95
1000	Fluted high strength									

04 22 10.23 Concrete Block, Decorative		Crew	Daily Output	Labor-Hours	Unit	2011 Bare Costs Material	Labor	Equipment	Total	Total Incl O&P
1100	8″ x 16″ x 4″ thick, flutes 1 side,	D-8	345	.116	S.F.	4.17	3.60		7.77	10.50
1150	Flutes 2 sides		335	.119		5.05	3.71		8.76	11.65
1200	8″ thick		300	.133		6.60	4.14		10.74	14.05
1250	For special colors, add					.41			.41	.45
1400	Deep grooved, smooth face									
1450	8″ x 16″ x 4″ thick	D-8	345	.116	S.F.	2.72	3.60		6.32	8.90
1500	8″ thick	″	300	.133	″	4.68	4.14		8.82	11.95
2000	Formblock, incl. inserts & reinforcing									
2100	8″ x 16″ x 8″ thick	D-8	345	.116	S.F.	4.76	3.60		8.36	11.15
2150	12″ thick	″	310	.129	″	5.95	4.01		9.96	13.05
2500	Ground face									
2600	8″ x 16″ x 4″ thick	D-8	345	.116	S.F.	3.27	3.60		6.87	9.50
2650	6″ thick		325	.123		4.78	3.82		8.60	11.50
2700	8″ thick		300	.133		6.35	4.14		10.49	13.75
2750	12″ thick	D-9	265	.181		5.70	5.50		11.20	15.30
2900	For special colors, add, minimum					15%				
2950	For special colors, add, maximum					45%				
4000	Slump block									
4100	4″ face height x 16″ x 4″ thick	D-1	165	.097	S.F.	3.94	2.95		6.89	9.20
4150	6″ thick		160	.100		5.45	3.04		8.49	10.95
4200	8″ thick		155	.103		6.05	3.14		9.19	11.80
4250	10″ thick		140	.114		10.90	3.47		14.37	17.70
4300	12″ thick		130	.123		11.40	3.74		15.14	18.70
4400	6″ face height x 16″ x 6″ thick		155	.103		4.75	3.14		7.89	10.35
4450	8″ thick		150	.107		6.50	3.24		9.74	12.45
4500	10″ thick		130	.123		10	3.74		13.74	17.15
4550	12″ thick		120	.133		10.35	4.05		14.40	18.05
5000	Split rib profile units, 1″ deep ribs, 8 ribs									
5100	8″ x 16″ x 4″ thick	D-8	345	.116	S.F.	3.17	3.60		6.77	9.40
5150	6″ thick		325	.123		3.64	3.82		7.46	10.25
5200	8″ thick		300	.133		4.16	4.14		8.30	11.40
5250	12″ thick	D-9	275	.175		4.93	5.30		10.23	14.15
5400	For special deeper colors, 4″ thick, add					1.15			1.15	1.27
5450	12″ thick, add					1.19			1.19	1.31
5600	For white, 4″ thick, add					1.15			1.15	1.27
5650	6″ thick, add					1.14			1.14	1.26
5700	8″ thick, add					1.20			1.20	1.32
5750	12″ thick, add					1.23			1.23	1.35
6000	Split face									
6100	8″ x 16″ x 4″ thick	D-8	350	.114	S.F.	2.82	3.55		6.37	8.90
6150	6″ thick		325	.123		3.28	3.82		7.10	9.85
6200	8″ thick		300	.133		3.73	4.14		7.87	10.90
6250	12″ thick	D-9	270	.178		4.57	5.40		9.97	13.90
6300	For scored, add					.35			.35	.39
6400	For special deeper colors, 4″ thick, add					.67			.67	.73
6450	6″ thick, add					.66			.66	.72
6500	8″ thick, add					.68			.68	.75
6550	12″ thick, add					.70			.70	.77
6650	For white, 4″ thick, add					1.14			1.14	1.26
6700	6″ thick, add					1.13			1.13	1.24
6750	8″ thick, add					1.16			1.16	1.28
6800	12″ thick, add					1.19			1.19	1.31
7000	Scored ground face, 2 to 5 scores									

04 22 Concrete Unit Masonry

04 22 10 – Concrete Masonry Units

04 22 10.23 Concrete Block, Decorative

	04 22 10.23 Concrete Block, Decorative	Crew	Daily Output	Labor-Hours	Unit	2011 Bare Costs Material	Labor	Equipment	Total	Total Incl O&P
7100	8″ x 16″ x 4″ thick	D-8	340	.118	S.F.	4.91	3.65		8.56	11.40
7150	6″ thick		310	.129		5.60	4.01		9.61	12.75
7200	8″ thick		290	.138		6.60	4.28		10.88	14.30
7250	12″ thick	D-9	265	.181		8.20	5.50		13.70	18.10
8000	Hexagonal face profile units, 8″ x 16″ units									
8100	4″ thick, hollow	D-8	340	.118	S.F.	3.12	3.65		6.77	9.45
8200	Solid		340	.118		4.03	3.65		7.68	10.45
8300	6″ thick, hollow		310	.129		3.49	4.01		7.50	10.40
8350	8″ thick, hollow		290	.138		4.73	4.28		9.01	12.25
8500	For stacked bond, add						26%			
8550	For high rise construction, add per story	D-8	67.80	.590	M.S.F.		18.30		18.30	30
8600	For scored block, add					10%				
8650	For honed or ground face, per face, add				Ea.	.38			.38	.42
8700	For honed or ground end, per end, add				"	2.98			2.98	3.28
8750	For bullnose block, add					10%				
8800	For special color, add					13%				

04 22 10.24 Concrete Block, Exterior

	04 22 10.24 Concrete Block, Exterior	Crew	Daily Output	Labor-Hours	Unit	Material	Labor	Equipment	Total	Total Incl O&P
0010	**CONCRETE BLOCK, EXTERIOR**, C90, 2000 psi									
0020	Reinforced alt courses, tooled joints 2 sides									
0100	Normal weight, 8″ x 16″ x 6″ thick	D-8	395	.101	S.F.	2.56	3.14		5.70	7.95
0200	8″ thick		360	.111		3.26	3.45		6.71	9.25
0250	10″ thick		290	.138		4.28	4.28		8.56	11.75
0300	12″ thick	D-9	250	.192		4.29	5.85		10.14	14.30
0500	Lightweight, 8″ x 16″ x 6″ thick	D-8	450	.089		2.73	2.76		5.49	7.55
0600	8″ thick		430	.093		3.47	2.89		6.36	8.55
0650	10″ thick		395	.101		4.32	3.14		7.46	9.90
0700	12″ thick	D-9	350	.137		6.35	4.17		10.52	13.80

04 22 10.26 Concrete Block Foundation Wall

	04 22 10.26 Concrete Block Foundation Wall	Crew	Daily Output	Labor-Hours	Unit	Material	Labor	Equipment	Total	Total Incl O&P
0010	**CONCRETE BLOCK FOUNDATION WALL**, C90/C145									
0050	Normal-weight, cut joints, horiz joint reinf, no vert reinf									
0200	Hollow, 8″ x 16″ x 6″ thick	D-8	455	.088	S.F.	2.62	2.73		5.35	7.35
0250	8″ thick		425	.094		2.94	2.92		5.86	8
0300	10″ thick		350	.114		3.57	3.55		7.12	9.75
0350	12″ thick	D-9	300	.160		3.98	4.86		8.84	12.40
0500	Solid, 8″ x 16″ block, 6″ thick	D-8	440	.091		2.69	2.82		5.51	7.60
0550	8″ thick	"	415	.096		3.80	2.99		6.79	9.10
0600	12″ thick	D-9	350	.137		5.40	4.17		9.57	12.80

04 22 10.28 Concrete Block, High Strength

	04 22 10.28 Concrete Block, High Strength	Crew	Daily Output	Labor-Hours	Unit	Material	Labor	Equipment	Total	Total Incl O&P
0010	**CONCRETE BLOCK, HIGH STRENGTH**									
0050	Hollow, reinforced alternate courses, 8″ x 16″ units									
0200	3500 psi, 4″ thick	D-8	440	.091	S.F.	2.22	2.82		5.04	7.05
0250	6″ thick		395	.101		2.47	3.14		5.61	7.85
0300	8″ thick		360	.111		3.18	3.45		6.63	9.15
0350	12″ thick	D-9	250	.192		4.17	5.85		10.02	14.20
0500	5000 psi, 4″ thick	D-8	440	.091		2.43	2.82		5.25	7.30
0550	6″ thick		395	.101		3.05	3.14		6.19	8.50
0600	8″ thick		360	.111		4.06	3.45		7.51	10.10
0650	12″ thick	D-9	300	.160		6	4.86		10.86	14.60
1000	For 75% solid block, add					30%				
1050	For 100% solid block, add					50%				

04 22 10 – Concrete Masonry Units

04 22 10.30 Concrete Block, Interlocking		Crew	Daily Output	Labor-Hours	Unit	Material	2011 Bare Costs Labor	Equipment	Total	Total Incl O&P
0010	**CONCRETE BLOCK, INTERLOCKING**									
0100	Not including grout or reinforcing									
0200	8" x 16" units, 2,000 psi, 8" thick	D-1	245	.065	S.F.	2.72	1.99		4.71	6.25
0300	12" thick		220	.073		4.03	2.21		6.24	8.05
0350	16" thick		185	.086		6.05	2.63		8.68	10.95
0400	Including grout & reinforcing, 8" thick	D-4	245	.163		7.35	4.62	.51	12.48	16.20
0450	12" thick		220	.182		8.85	5.15	.57	14.57	18.85
0500	16" thick		185	.216		11.05	6.10	.68	17.83	23

04 22 10.32 Concrete Block, Lintels		Crew	Daily Output	Labor-Hours	Unit	Material	Labor	Equipment	Total	Total Incl O&P
0010	**CONCRETE BLOCK, LINTELS**, C90, normal weight									
0100	Including grout and horizontal reinforcing									
0200	8" x 8" x 8", 1 #4 bar	D-4	300	.133	L.F.	4.85	3.77	.42	9.04	12
0250	2 #4 bars		295	.136		5.05	3.84	.43	9.32	12.30
0400	8" x 16" x 8", 1 #4 bar		275	.145		3.33	4.12	.46	7.91	10.95
0450	2 #4 bars		270	.148		3.54	4.19	.47	8.20	11.30
1000	12" x 8" x 8", 1 #4 bar		275	.145		6.80	4.12	.46	11.38	14.80
1100	2 #4 bars		270	.148		7	4.19	.47	11.66	15.10
1150	2 #5 bars		270	.148		7.25	4.19	.47	11.91	15.35
1200	2 #6 bars		265	.151		7.50	4.27	.48	12.25	15.80
1500	12" x 16" x 8", 1 #4 bar		250	.160		10.80	4.53	.50	15.83	19.85
1600	2 #3 bars		245	.163		10.80	4.62	.51	15.93	20
1650	2 #4 bars		245	.163		11	4.62	.51	16.13	20.50
1700	2 #5 bars		240	.167		11.20	4.72	.53	16.45	20.50

04 22 10.34 Concrete Block, Partitions		Crew	Daily Output	Labor-Hours	Unit	Material	Labor	Equipment	Total	Total Incl O&P
0010	**CONCRETE BLOCK, PARTITIONS**, excludes scaffolding R042210-20									
0100	Acoustical slotted block									
0200	4" thick, type A-1	D-8	315	.127	S.F.	3.74	3.94		7.68	10.55
0210	8" thick		275	.145		4.72	4.52		9.24	12.60
0250	8" thick, type Q		275	.145		8.30	4.52		12.82	16.55
0260	4" thick, type RSC		315	.127		5.80	3.94		9.74	12.85
0270	6" thick		295	.136		5.80	4.21		10.01	13.30
0280	8" thick		275	.145		5.80	4.52		10.32	13.80
0290	12" thick		250	.160		5.80	4.97		10.77	14.55
0300	8" thick, type RSR		275	.145		5.80	4.52		10.32	13.80
0400	8" thick, type RSC/RF		275	.145		6.55	4.52		11.07	14.60
0410	10" thick		260	.154		7	4.78		11.78	15.55
0420	12" thick		250	.160		7.55	4.97		12.52	16.45
0430	12" thick, type RSC/RF-4		250	.160		8.95	4.97		13.92	18
0500	NRC .60 type R, 8" thick		265	.151		8	4.69		12.69	16.50
0600	NRC .65 type RR, 8" thick		265	.151		7.35	4.69		12.04	15.75
0700	NRC .65 type 4R-RF, 8" thick		265	.151		8.25	4.69		12.94	16.75
0710	NRC .70 type R, 12" thick		245	.163		8.65	5.05		13.70	17.80
1000	Lightweight block, tooled joints, 2 sides, hollow									
1100	Not reinforced, 8" x 16" x 4" thick	D-8	440	.091	S.F.	1.65	2.82		4.47	6.45
1150	6" thick		410	.098		2.37	3.03		5.40	7.60
1200	8" thick		385	.104		2.88	3.23		6.11	8.45
1250	10" thick		370	.108		3.49	3.36		6.85	9.35
1300	12" thick	D-9	350	.137		4.34	4.17		8.51	11.60
2000	Not reinforced, 8" x 24" x 4" thick, hollow		460	.104		1.21	3.17		4.38	6.55
2100	6" thick		440	.109		1.71	3.32		5.03	7.35
2150	8" thick		415	.116		2.12	3.52		5.64	8.10
2200	10" thick		385	.125		2.56	3.79		6.35	9

04 22 Concrete Unit Masonry

04 22 10 – Concrete Masonry Units

04 22 10.34 Concrete Block, Partitions		Crew	Daily Output	Labor-Hours	Unit	Material	2011 Bare Costs Labor	Equipment	Total	Total Incl O&P
2250	12" thick	D-9	365	.132	S.F.	3.17	4		7.17	10.05
2800	Solid, not reinforced, 8" x 16" x 2" thick	D-8	440	.091		1.46	2.82		4.28	6.25
2900	4" thick		420	.095		2.36	2.96		5.32	7.45
2950	6" thick		390	.103		2.63	3.18		5.81	8.10
3000	8" thick		365	.110		3.38	3.40		6.78	9.30
3050	10" thick		350	.114		4.22	3.55		7.77	10.45
3100	12" thick	D-9	330	.145		6.25	4.42		10.67	14.10
4000	Regular block, tooled joints, 2 sides, hollow									
4100	Not reinforced, 8" x 16" x 4" thick	D-8	430	.093	S.F.	1.49	2.89		4.38	6.40
4150	6" thick		400	.100		2.09	3.11		5.20	7.40
4200	8" thick		375	.107		2.40	3.31		5.71	8.10
4250	10" thick		360	.111		3.03	3.45		6.48	9
4300	12" thick	D-9	340	.141		3.43	4.29		7.72	10.85
4500	Reinforced alternate courses, 8" x 16" x 4" thick	D-8	425	.094		1.62	2.92		4.54	6.55
4550	6" thick		395	.101		2.22	3.14		5.36	7.60
4600	8" thick		370	.108		2.53	3.36		5.89	8.30
4650	10" thick		355	.113		3.43	3.50		6.93	9.50
4700	12" thick	D-9	335	.143		3.56	4.36		7.92	11.05
4900	Solid, not reinforced, 2" thick	D-8	435	.092		1.34	2.86		4.20	6.15
5000	3" thick		430	.093		1.60	2.89		4.49	6.50
5050	4" thick		415	.096		1.83	2.99		4.82	6.90
5100	6" thick		385	.104		2.16	3.23		5.39	7.70
5150	8" thick		360	.111		3.26	3.45		6.71	9.25
5200	12" thick	D-9	325	.148		4.88	4.49		9.37	12.70
5500	Solid, reinforced alternate courses, 4" thick	D-8	420	.095		1.90	2.96		4.86	6.95
5550	6" thick		380	.105		2.25	3.27		5.52	7.85
5600	8" thick		355	.113		3.34	3.50		6.84	9.45
5650	12" thick	D-9	320	.150		4.41	4.56		8.97	12.35

04 22 10.38 Concrete Brick		Crew	Daily Output	Labor-Hours	Unit	Material	Labor	Equipment	Total	Total Incl O&P
0010	**CONCRETE BRICK**, C55, grade N, type 1									
0100	Regular, 4 x 2-1/4 x 8	D-8	660	.061	Ea.	.45	1.88		2.33	3.59
0125	Rusticated, 4 x 2-1/4 x 8		660	.061		.50	1.88		2.38	3.64
0150	Frog, 4 x 2-1/4 x 8		660	.061		.49	1.88		2.37	3.62
0200	Double, 4 x 4-7/8 x 8		535	.075		.79	2.32		3.11	4.68

04 22 10.42 Concrete Block, Screen Block		Crew	Daily Output	Labor-Hours	Unit	Material	Labor	Equipment	Total	Total Incl O&P
0010	**CONCRETE BLOCK, SCREEN BLOCK**									
0200	8" x 16", 4" thick	D-8	330	.121	S.F.	1.97	3.76		5.73	8.30
0300	8" thick		270	.148		2.99	4.60		7.59	10.85
0350	12" x 12", 4" thick		290	.138		2.21	4.28		6.49	9.50
0500	8" thick		250	.160		3.06	4.97		8.03	11.50

04 22 10.44 Glazed Concrete Block		Crew	Daily Output	Labor-Hours	Unit	Material	Labor	Equipment	Total	Total Incl O&P
0010	**GLAZED CONCRETE BLOCK** C744									
0100	Single face, 8" x 16" units, 2" thick	D-8	360	.111	S.F.	8.25	3.45		11.70	14.75
0200	4" thick		345	.116		8.50	3.60		12.10	15.25
0250	6" thick		330	.121		9.15	3.76		12.91	16.20
0300	8" thick		310	.129		9.25	4.01		13.26	16.70
0350	10" thick		295	.136		11.10	4.21		15.31	19.10
0400	12" thick	D-9	280	.171		11.80	5.20		17	21.50
0700	Double face, 8" x 16" units, 4" thick	D-8	340	.118		12.20	3.65		15.85	19.40
0750	6" thick		320	.125		14.50	3.88		18.38	22.50
0800	8" thick		300	.133		15.20	4.14		19.34	23.50
1000	Jambs, bullnose or square, single face, 8" x 16", 2" thick		315	.127	Ea.	17.50	3.94		21.44	25.50

04 22 Concrete Unit Masonry

04 22 10 – Concrete Masonry Units

04 22 10.44 Glazed Concrete Block		Crew	Daily Output	Labor-Hours	Unit	Material	Labor (2011 Bare Costs)	Equipment	Total	Total Incl O&P
1050	4″ thick	D-8	285	.140	Ea.	20.50	4.36		24.86	29.50
1200	Caps, bullnose or square, 8″ x 16″, 2″ thick		420	.095	L.F.	14.15	2.96		17.11	20.50
1250	4″ thick		380	.105		15.95	3.27		19.22	23
1500	Cove base, 8″ x 16″, 2″ thick		315	.127		11	3.94		14.94	18.55
1550	4″ thick		285	.140		7.15	4.36		11.51	15
1600	6″ thick		265	.151		7.70	4.69		12.39	16.15
1650	8″ thick		245	.163		8.10	5.05		13.15	17.25

04 23 Glass Unit Masonry

04 23 13 – Vertical Glass Unit Masonry

04 23 13.10 Glass Block		Crew	Daily Output	Labor-Hours	Unit	Material	Labor	Equipment	Total	Total Incl O&P
0010	**GLASS BLOCK**									
0100	Plain, 4″ thick, under 1,000 S.F., 6″ x 6″	D-8	115	.348	S.F.	25.50	10.80		36.30	45.50
0150	8″ x 8″		160	.250		15	7.75		22.75	29.50
0160	end block		160	.250		50	7.75		57.75	68
0170	90 deg corner		160	.250		49.50	7.75		57.25	67.50
0180	45 deg corner		160	.250		46	7.75		53.75	63.50
0200	12″ x 12″		175	.229		21.50	7.10		28.60	35
0210	4″ x 8″		160	.250		15.30	7.75		23.05	29.50
0220	6″ x 8″		160	.250		14.70	7.75		22.45	29
0300	1,000 to 5,000 S.F., 6″ x 6″		135	.296		25	9.20		34.20	42.50
0350	8″ x 8″		190	.211		14.70	6.55		21.25	27
0400	12″ x 12″		215	.186		21	5.80		26.80	33
0410	4″ x 8″		215	.186		6.05	5.80		11.85	16.15
0420	6″ x 8″		215	.186		5.75	5.80		11.55	15.85
0500	Over 5,000 S.F., 6″ x 6″		145	.276		24.50	8.55		33.05	41
0550	8″ x 8″		215	.186		14.25	5.80		20.05	25
0600	12″ x 12″		240	.167		20.50	5.20		25.70	31
0610	4″ x 8″		240	.167		5.65	5.20		10.85	14.75
0620	6″ x 8″		240	.167		5.75	5.20		10.95	14.85
0700	For solar reflective blocks, add					100%				
1000	Thinline, plain, 3-1/8″ thick, under 1,000 S.F., 6″ x 6″	D-8	115	.348	S.F.	17.50	10.80		28.30	37
1050	8″ x 8″		160	.250		9.80	7.75		17.55	23.50
1200	Over 5,000 S.F., 6″ x 6″		145	.276		16.40	8.55		24.95	32
1250	8″ x 8″		215	.186		9.20	5.80		15	19.60
1400	For cleaning block after installation (both sides), add		1000	.040		.11	1.24		1.35	2.16

04 24 Adobe Unit Masonry

04 24 16 – Manufactured Adobe Unit Masonry

04 24 16.06 Adobe Brick		Crew	Daily Output	Labor-Hours	Unit	Material	Labor	Equipment	Total	Total Incl O&P
0010	**ADOBE BRICK**, Semi-stabilized, with cement mortar									
0060	Brick, 10″ x 4″ x 14″, 2.6/S.F. G	D-8	560	.071	S.F.	3.49	2.22		5.71	7.50
0080	12″ x 4″ x 16″, 2.3/S.F. G		580	.069		4.61	2.14		6.75	8.55
0100	10″ x 4″ x 16″, 2.3/S.F. G		590	.068		4.37	2.11		6.48	8.25
0120	8″ x 4″ x 16″, 2.3/S.F. G		560	.071		3.56	2.22		5.78	7.55
0140	4″ x 4″ x 16″, 2.3/S.F. G		540	.074		2.56	2.30		4.86	6.60
0160	6″ x 4″ x 16″, 2.3/S.F. G		540	.074		2.59	2.30		4.89	6.60
0180	4″ x 4″ x 12″, 3.0/S.F. G		520	.077		2.45	2.39		4.84	6.60
0200	8″ x 4″ x 12″, 3.0/S.F. G		520	.077		3.16	2.39		5.55	7.40

04 25 Unit Masonry Panels

04 25 20 – Pre-Fabricated Masonry Panels

04 25 20.10 Brick and Epoxy Mortar Panels		Crew	Daily Output	Labor-Hours	Unit	Material	2011 Bare Costs Labor	Equipment	Total	Total Incl O&P
0010	**BRICK AND EPOXY MORTAR PANELS**									
0020	Prefabricated brick & epoxy mortar, 4" thick, minimum	C-11	775	.072	S.F.	9.75	2.50	2.46	14.71	17.65
0100	Maximum	"	500	.112		10.75	3.87	3.81	18.43	22.50
0200	For 2" concrete back-up, add					50%				
0300	For 1" urethane & 3" concrete back-up, add				↓	70%				

04 27 Multiple-Wythe Unit Masonry

04 27 10 – Multiple-Wythe Masonry

04 27 10.30 Brick Walls

		Crew	Daily Output	Labor-Hours	Unit	Material	Labor	Equipment	Total	Total Incl O&P
0010	**BRICK WALLS**, including mortar, excludes scaffolding R042110-20									
0020	Estimating by number of brick									
0140	Face brick, 4" thick wall, 6.75 brick/S.F.	D-8	1.45	27.586	M	610	855		1,465	2,075
0150	Common brick, 4" thick wall, 6.75 brick/S.F.		1.60	25		495	775		1,270	1,825
0204	8" thick, 13.50 bricks per S.F.		1.80	22.222		510	690		1,200	1,675
0250	12" thick, 20.25 bricks per S.F.		1.90	21.053		515	655		1,170	1,650
0304	16" thick, 27.00 bricks per S.F.		2	20		520	620		1,140	1,600
0500	Reinforced, face brick, 4" thick wall, 6.75 brick/S.F.		1.40	28.571		635	885		1,520	2,150
0520	Common brick, 4" thick wall, 6.75 brick/S.F.		1.55	25.806		515	800		1,315	1,900
0550	8" thick, 13.50 bricks per S.F.		1.75	22.857		535	710		1,245	1,750
0600	12" thick, 20.25 bricks per S.F.		1.85	21.622		535	670		1,205	1,700
0650	16" thick, 27.00 bricks per S.F.	↓	1.95	20.513	↓	545	635		1,180	1,650
0790	Alternate method of figuring by square foot									
0800	Face brick, 4" thick wall, 6.75 brick/S.F.	D-8	215	.186	S.F.	4.12	5.80		9.92	14.05
0850	Common brick, 4" thick wall, 6.75 brick/S.F.		240	.167		3.33	5.20		8.53	12.15
0900	8" thick, 13.50 bricks per S.F.		135	.296		6.90	9.20		16.10	22.50
1000	12" thick, 20.25 bricks per S.F.		95	.421		10.40	13.05		23.45	33
1050	16" thick, 27.00 bricks per S.F.		75	.533		14.05	16.55		30.60	42.50
1200	Reinforced, face brick, 4" thick wall, 6.75 brick/S.F.		210	.190		4.27	5.90		10.17	14.40
1220	Common brick, 4" thick wall, 6.75 brick/S.F.		235	.170		3.48	5.30		8.78	12.50
1250	8" thick, 13.50 bricks per S.F.		130	.308		7.20	9.55		16.75	23.50
1300	12" thick, 20.25 bricks per S.F.		90	.444		10.85	13.80		24.65	34.50
1350	16" thick, 27.00 bricks per S.F.	↓	70	.571	↓	14.70	17.75		32.45	45

04 27 10.40 Steps

		Crew	Daily Output	Labor-Hours	Unit	Material	Labor	Equipment	Total	Total Incl O&P
0010	**STEPS**									
0012	Entry steps, select common brick	D-1	.30	53.333	M	430	1,625		2,055	3,125

04 41 Dry-Placed Stone

04 41 10 – Dry Placed Stone

04 41 10.10 Rough Stone Wall

		Crew	Daily Output	Labor-Hours	Unit	Material	Labor	Equipment	Total	Total Incl O&P
0011	**ROUGH STONE WALL**, Dry									
0012	Dry laid (no mortar), under 18" thick [G]	D-1	60	.267	C.F.	8.80	8.10		16.90	23
0100	Random fieldstone, under 18" thick [G]	D-12	60	.533		8.80	16.20		25	36
0150	Over 18" thick [G]	"	63	.508	↓	10.55	15.45		26	37
0500	Field stone veneer [G]	D-8	120	.333	S.F.	9.95	10.35		20.30	28
0510	Valley stone veneer [G]		120	.333		9.95	10.35		20.30	28
0520	River stone veneer [G]	↓	120	.333	↓	9.95	10.35		20.30	28
0600	Rubble stone walls, in mortar bed, up to 18" thick [G]	D-11	75	.320	C.F.	10.65	10.05		20.70	28.50

04 43 Stone Masonry

04 43 10 – Masonry with Natural and Processed Stone

04 43 10.05 Ashlar Veneer		Crew	Daily Output	Labor-Hours	Unit	Material	2011 Bare Costs Labor	Equipment	Total	Total Incl O&P
0011	**ASHLAR VENEER** 4" + or - thk, random or random rectangular									
0150	Sawn face, split joints, low priced stone	D-8	140	.286	S.F.	9.50	8.85		18.35	25
0200	Medium priced stone		130	.308		11.30	9.55		20.85	28
0300	High priced stone		120	.333		14.50	10.35		24.85	33
0600	Seam face, split joints, medium price stone		125	.320		7.55	9.95		17.50	24.50
0700	High price stone		120	.333		15	10.35		25.35	33.50
1000	Split or rock face, split joints, medium price stone		125	.320		10.80	9.95		20.75	28
1100	High price stone		120	.333		15	10.35		25.35	33.50
04 43 10.10 Bluestone										
0010	**BLUESTONE**, cut to size									
0500	Sills, natural cleft, 10" wide to 6' long, 1-1/2" thick	D-11	70	.343	L.F.	13.35	10.80		24.15	32.50
0550	2" thick	"	63	.381		14.60	12		26.60	36
1000	Stair treads, natural cleft, 12" wide, 6' long, 1-1/2" thick	D-10	115	.278		30	9.20	3.78	42.98	52.50
1050	2" thick		105	.305		30	10.10	4.14	44.24	54
1100	Smooth finish, 1-1/2" thick		115	.278		15	9.20	3.78	27.98	36
1300	Thermal finish, 1-1/2" thick		115	.278		15	9.20	3.78	27.98	36
1350	2" thick		105	.305		15	10.10	4.14	29.24	37.50
04 43 10.45 Granite										
0010	**GRANITE**, cut to size									
0050	Veneer, polished face, 3/4" to 1-1/2" thick									
0150	Low price, gray, light gray, etc.	D-10	130	.246	S.F.	26.50	8.15	3.34	37.99	46
0220	High price, red, black, etc.	"	130	.246	"	41.50	8.15	3.34	52.99	63
0300	1-1/2" to 2-1/2" thick, veneer									
0350	Low price, gray, light gray, etc.	D-10	130	.246	S.F.	28.50	8.15	3.34	39.99	48
0550	High price, red, black, etc.	"	130	.246	"	51.50	8.15	3.34	62.99	73.50
0700	2-1/2" to 4" thick, veneer									
0750	Low price, gray, light gray, etc.	D-10	110	.291	S.F.	38	9.60	3.95	51.55	62
0950	High price, red, black, etc.	"	110	.291		62.50	9.60	3.95	76.05	88.50
1000	For bush hammered finish, deduct					5%				
1050	Coarse rubbed finish, deduct					10%				
1100	Honed finish, deduct					5%				
1150	Thermal finish, deduct					18%				
2450	For radius under 5', add				L.F.	100%				
2500	Steps, copings, etc., finished on more than one surface									
2550	Minimum	D-10	50	.640	C.F.	91	21	8.70	120.70	144
2600	Maximum	"	50	.640	"	146	21	8.70	175.70	204
2800	Pavers, 4" x 4" x 4" blocks, split face and joints									
2850	Minimum	D-11	80	.300	S.F.	12.95	9.45		22.40	29.50
2900	Maximum	"	80	.300		28.50	9.45		37.95	47
4000	Soffits, 2" thick, minimum	D-13	35	1.371		37.50	44	12.40	93.90	128
4100	Maximum		35	1.371		90	44	12.40	146.40	185
4200	4" thick, minimum		35	1.371		62	44	12.40	118.40	155
4300	Maximum		35	1.371		117	44	12.40	173.40	215
04 43 10.50 Lightweight Natural Stone										
0011	**LIGHTWEIGHT NATURAL STONE** Lava type									
0100	Veneer, rubble face, sawed back, irregular shapes G	D-10	130	.246	S.F.	6.30	8.15	3.34	17.79	24
0200	Sawed face and back, irregular shapes G	"	130	.246	"	6.30	8.15	3.34	17.79	24
04 43 10.55 Limestone										
0010	**LIMESTONE**, cut to size									
0020	Veneer facing panels									
0500	Texture finish, light stick, 4-1/2" thick, 5' x 12'	D-4	300	.133	S.F.	48	3.77	.42	52.19	59.50
0750	5" thick, 5' x 14' panels	D-10	275	.116		52.50	3.85	1.58	57.93	65.50

04 43 10 – Masonry with Natural and Processed Stone

04 43 10.55 Limestone		Crew	Daily Output	Labor-Hours	Unit	Material	2011 Bare Costs Labor	Equipment	Total	Total Incl O&P
1000	Sugarcube finish, 2″ Thick, 3′ x 5′ panels	D-10	275	.116	S.F.	35	3.85	1.58	40.43	46.50
1050	3″ Thick, 4′ x 9′ panels		275	.116		38	3.85	1.58	43.43	50
1200	4″ Thick, 5′ x 11′ panels		275	.116		41	3.85	1.58	46.43	53
1400	Sugarcube, textured finish, 4-1/2″ thick, 5′ x 12′		275	.116		45	3.85	1.58	50.43	57.50
1450	5″ thick, 5′ x 14′ panels		275	.116		53.50	3.85	1.58	58.93	66.50
2000	Coping, sugarcube finish, top & 2 sides		30	1.067	C.F.	68	35.50	14.50	118	149
2100	Sills, lintels, jambs, trim, stops, sugarcube finish, average		20	1.600		68	53	21.50	142.50	186
2150	Detailed		20	1.600		68	53	21.50	142.50	186
2300	Steps, extra hard, 14″ wide, 6″ rise		50	.640	L.F.	41.50	21	8.70	71.20	90
3000	Quoins, plain finish, 6″ x 12″ x 12″	D-12	25	1.280	Ea.	38	39		77	106
3050	6″ x 16″ x 24″	″	25	1.280	″	50.50	39		89.50	120

04 43 10.60 Marble		Crew	Daily Output	Labor-Hours	Unit	Material	Labor	Equipment	Total	Total Incl O&P
0011	**MARBLE**, ashlar, split face, 4″ + or - thick, random									
0040	Lengths 1′ to 4′ & heights 2″ to 7-1/2″, average	D-8	175	.229	S.F.	17.75	7.10		24.85	31
0100	Base, polished, 3/4″ or 7/8″ thick, polished, 6″ high	D-10	65	.492	L.F.	16.15	16.30	6.70	39.15	51.50
0300	Carvings or bas relief, from templates, average		80	.400	S.F.	144	13.20	5.45	162.65	185
0350	Maximum		80	.400	″	335	13.20	5.45	353.65	395
0600	Columns, cornices, mouldings, etc.									
0650	Hand or special machine cut, average	D-10	35	.914	C.F.	144	30	12.40	186.40	221
0700	Maximum	″	35	.914	″	310	30	12.40	352.40	405
1000	Facing, polished finish, cut to size, 3/4″ to 7/8″ thick									
1050	Average	D-10	130	.246	S.F.	23.50	8.15	3.34	34.99	43
1100	Maximum		130	.246		55	8.15	3.34	66.49	77.50
1300	1-1/4″ thick, average		125	.256		34.50	8.45	3.48	46.43	55
1350	Maximum		125	.256		68.50	8.45	3.48	80.43	93
1500	2″ thick, average		120	.267		40	8.80	3.62	52.42	62.50
1550	Maximum		120	.267		69	8.80	3.62	81.42	94.50
2200	Window sills, 6″ x 3/4″ thick	D-1	85	.188	L.F.	8.95	5.70		14.65	19.20
2500	Flooring, polished tiles, 12″ x 12″ x 3/8″ thick									
2510	Thin set, average	D-11	90	.267	S.F.	11.90	8.40		20.30	27
2600	Maximum		90	.267		183	8.40		191.40	215
2700	Mortar bed, average		65	.369		10.65	11.60		22.25	31
2740	Maximum		65	.369		183	11.60		194.60	220
2780	Travertine, 3/8″ thick, average	D-10	130	.246		15.15	8.15	3.34	26.64	33.50
2790	Maximum	″	130	.246		37	8.15	3.34	48.49	58
2800	Patio tile, non-slip, 1/2″ thick, flame finish	D-11	75	.320		15.45	10.05		25.50	33.50
2900	Shower or toilet partitions, 7/8″ thick partitions									
3050	3/4″ or 1-1/4″ thick stiles, polished 2 sides, average	D-11	75	.320	S.F.	46	10.05		56.05	67
3201	Soffits, add to above prices				″	20%	100%			
3210	Stairs, risers, 7/8″ thick x 6″ high	D-10	115	.278	L.F.	15.55	9.20	3.78	28.53	36.50
3360	Treads, 12″ wide x 1-1/4″ thick	″	115	.278	″	22.50	9.20	3.78	35.48	44.50
3500	Thresholds, 3′ long, 7/8″ thick, 4″ to 5″ wide, plain	D-12	24	1.333	Ea.	17.15	40.50		57.65	85.50
3550	Beveled		24	1.333	″	19.90	40.50		60.40	88.50
3700	Window stools, polished, 7/8″ thick, 5″ wide		85	.376	L.F.	15.95	11.45		27.40	36.50

04 43 10.75 Sandstone or Brownstone		Crew	Daily Output	Labor-Hours	Unit	Material	Labor	Equipment	Total	Total Incl O&P
0011	**SANDSTONE OR BROWNSTONE**									
0100	Sawed face veneer, 2-1/2″ thick, to 2′ x 4′ panels	D-10	130	.246	S.F.	21	8.15	3.34	32.49	40
0150	4″ thick, to 3′-6″ x 8′panels		100	.320		21	10.60	4.35	35.95	45
0300	Split face, random sizes		100	.320		14	10.60	4.35	28.95	37.50
0350	Cut stone trim (limestone)									
0360	Ribbon stone, 4″ thick, 5′ pieces	D-8	120	.333	Ea.	172	10.35		182.35	206
0370	Cove stone, 4″ thick, 5′ pieces		105	.381		173	11.85		184.85	209

04 43 10 – Masonry with Natural and Processed Stone

04 43 10.75 Sandstone or Brownstone		Crew	Daily Output	Labor-Hours	Unit	Material	Labor (2011 Bare Costs)	Equipment (2011 Bare Costs)	Total (2011 Bare Costs)	Total Incl O&P
0380	Cornice stone, 10″ to 12″ wide	D-8	90	.444	Ea.	214	13.80		227.80	258
0390	Band stone, 4″ thick, 5′ pieces		145	.276		110	8.55		118.55	135
0410	Window and door trim, 3″ to 4″ wide		160	.250		93.50	7.75		101.25	116
0420	Key stone, 18″ long		60	.667		98.50	20.50		119	142

04 43 10.80 Slate		Crew	Daily Output	Labor-Hours	Unit	Material	Labor	Equipment	Total	Total Incl O&P
0010	**SLATE**									
0040	Pennsylvania - blue gray to black									
0050	Vermont - unfading green, mottled green & purple, gray & purple									
0100	Virginia - blue black									
0200	Exterior paving, natural cleft, 1″ thick									
0500	24″ x 24″, Pennsylvania	D-12	120	.267	S.F.	13	8.10		21.10	27.50
0550	Vermont		120	.267		31	8.10		39.10	47.50
0600	Virginia		120	.267		20.50	8.10		28.60	36
1000	Interior flooring, natural cleft, 1/2″ thick									
1300	24″ x 24″ Pennsylvania	D-12	120	.267	S.F.	7.75	8.10		15.85	22
1350	Vermont		120	.267		25	8.10		33.10	41
1400	Virginia		120	.267		14.75	8.10		22.85	29.50
2000	Facing panels, 1-1/4″ thick, to 4′ x 4′ panels									
2100	Natural cleft finish, Pennsylvania	D-10	180	.178	S.F.	34	5.90	2.41	42.31	49.50
2110	Vermont		180	.178		27	5.90	2.41	35.31	42.50
2120	Virginia		180	.178		33	5.90	2.41	41.31	49
2150	Sand rubbed finish, surface, add					10.15			10.15	11.20
2200	Honed finish, add					7.35			7.35	8.10
2500	Ribbon, natural cleft finish, 1″ thick, to 9 S.F.	D-10	80	.400		13.10	13.20	5.45	31.75	42
2700	1-1/2″ thick		78	.410		17.05	13.55	5.55	36.15	47
2850	2″ thick		76	.421		20.50	13.90	5.70	40.10	52
3500	Stair treads, sand finish, 1″ thick x 12″ wide									
3600	3 L.F. to 6 L.F.	D-10	120	.267	L.F.	24	8.80	3.62	36.42	44.50
3700	Ribbon, sand finish, 1″ thick x 12″ wide									
3750	To 6 L.F.	D-10	120	.267	L.F.	20	8.80	3.62	32.42	40.50
4000	Stools or sills, sand finish, 1″ thick, 6″ wide	D-12	160	.200		11.50	6.10		17.60	22.50
4200	10″ wide		90	.356		17.75	10.80		28.55	37.50
4400	2″ thick, 6″ wide		140	.229		18.50	6.95		25.45	32
4600	10″ wide		90	.356		29	10.80		39.80	50
4800	For lengths over 3′, add					25%				

04 43 10.85 Window Sill		Crew	Daily Output	Labor-Hours	Unit	Material	Labor	Equipment	Total	Total Incl O&P
0010	**WINDOW SILL**									
0020	Bluestone, thermal top, 10″ wide, 1-1/2″ thick	D-1	85	.188	S.F.	16.05	5.70		21.75	27
0050	2″ thick		75	.213	″	18.75	6.50		25.25	31
0100	Cut stone, 5″ x 8″ plain		48	.333	L.F.	11.90	10.15		22.05	29.50
0200	Face brick on edge, brick, 8″ wide		80	.200		2.53	6.10		8.63	12.80
0400	Marble, 9″ wide, 1″ thick		85	.188		8.50	5.70		14.20	18.75
0900	Slate, colored, unfading, honed, 12″ wide, 1″ thick		85	.188		17.10	5.70		22.80	28
0950	2″ thick		70	.229		24	6.95		30.95	37.50

04 51 Flue Liner Masonry

04 51 10 – Clay Flue Lining

04 51 10.10 Flue Lining		Crew	Daily Output	Labor-Hours	Unit	2011 Bare Costs Material	Labor	Equipment	Total	Total Incl O&P
0010	**FLUE LINING**, including mortar									
0020	Clay, 8" x 8"	D-1	125	.128	V.L.F.	4.97	3.89		8.86	11.85
0100	8" x 12"		103	.155		7.55	4.72		12.27	16.05
0200	12" x 12"		93	.172		9.75	5.25		15	19.30
0300	12" x 18"		84	.190		18.90	5.80		24.70	30.50
0400	18" x 18"		75	.213		22.50	6.50		29	35
0500	20" x 20"		66	.242		36	7.35		43.35	51.50
0600	24" x 24"		56	.286		47	8.70		55.70	66
1000	Round, 18" diameter		66	.242		30.50	7.35		37.85	45.50
1100	24" diameter	↓	47	.340	↓	67.50	10.35		77.85	91

04 54 Refractory Brick Masonry

04 54 10 – Refractory Brick Work

04 54 10.10 Fire Brick

		Crew	Daily Output	Labor-Hours	Unit	Material	Labor	Equipment	Total	Total Incl O&P
0010	**FIRE BRICK**									
0012	Low duty, 2000°F, 9" x 2-1/2" x 4-1/2"	D-1	.60	26.667	M	1,375	810		2,185	2,850
0050	High duty, 3000°F	"	.60	26.667	"	2,100	810		2,910	3,650

04 54 10.20 Fire Clay

		Crew	Daily Output	Labor-Hours	Unit	Material	Labor	Equipment	Total	Total Incl O&P
0010	**FIRE CLAY**									
0020	Gray, high duty, 100 lb. bag				Bag	59			59	65
0050	100 lb. drum, premixed (400 brick per drum)				Drum	73			73	80.50

04 57 Masonry Fireplaces

04 57 10 – Brick or Stone Fireplaces

04 57 10.10 Fireplace

		Crew	Daily Output	Labor-Hours	Unit	Material	Labor	Equipment	Total	Total Incl O&P
0010	**FIREPLACE**									
0100	Brick fireplace, not incl. foundations or chimneys									
0110	30" x 29" opening, incl. chamber, plain brickwork	D-1	.40	40	Ea.	535	1,225		1,760	2,600
0200	Fireplace box only (110 brick)	"	2	8	"	176	243		419	595
0300	For elaborate brickwork and details, add					35%	35%			
0400	For hearth, brick & stone, add	D-1	2	8	Ea.	197	243		440	615
0410	For steel angle, damper, cleanouts, add		4	4		138	122		260	350
0600	Plain brickwork, incl. metal circulator		.50	32	↓	1,025	975		2,000	2,725
0800	Face brick only, standard size, 8" x 2-2/3" x 4"	↓	.30	53.333	M	535	1,625		2,160	3,250

04 71 Manufactured Brick Masonry

04 71 10 – Simulated or Manufactured Brick

04 71 10.10 Simulated Brick

		Crew	Daily Output	Labor-Hours	Unit	Material	Labor	Equipment	Total	Total Incl O&P
0010	**SIMULATED BRICK**									
0020	Aluminum, baked on colors	1 Carp	200	.040	S.F.	3.53	1.34		4.87	6.15
0050	Fiberglass panels		200	.040		3.87	1.34		5.21	6.50
0100	Urethane pieces cemented in mastic		150	.053		6.70	1.79		8.49	10.35
0150	Vinyl siding panels	↓	200	.040		2.67	1.34		4.01	5.20
0160	Cement base, brick, incl. mastic	D-1	100	.160	↓	4.22	4.86		9.08	12.65
0170	Corner		50	.320	V.L.F.	10.55	9.75		20.30	27.50
0180	Stone face, incl. mastic		100	.160	S.F.	10.20	4.86		15.06	19.20
0190	Corner	↓	50	.320	V.L.F.	11.25	9.75		21	28.50

04 72 Cast Stone Masonry

04 72 10 – Cast Stone Masonry Features

04 72 10.10 Coping		Crew	Daily Output	Labor-Hours	Unit	Material	2011 Bare Costs Labor	Equipment	Total	Total Incl O&P
0010	**COPING**, stock units									
0050	Precast concrete, 10" wide, 4" tapers to 3-1/2", 8" wall	D-1	75	.213	L.F.	17.55	6.50		24.05	30
0100	12" wide, 3-1/2" tapers to 3", 10" wall		70	.229		17.55	6.95		24.50	30.50
0110	14" wide, 4" tapers to 3-1/2", 12" wall		65	.246		20.50	7.50		28	35
0150	16" wide, 4" tapers to 3-1/2", 14" wall		60	.267		17.10	8.10		25.20	32
0250	Precast concrete corners		40	.400	Ea.	28.50	12.15		40.65	51
0300	Limestone for 12" wall, 4" thick		90	.178	L.F.	15.35	5.40		20.75	26
0350	6" thick		80	.200		23	6.10		29.10	35.50
0500	Marble, to 4" thick, no wash, 9" wide		90	.178		23	5.40		28.40	34.50
0550	12" wide		80	.200		35	6.10		41.10	48.50
0700	Terra cotta, 9" wide		90	.178		5.60	5.40		11	15
0750	12" wide		80	.200		9.20	6.10		15.30	20
0800	Aluminum, for 12" wall		80	.200		13.30	6.10		19.40	24.50

04 72 20 – Cultured Stone Veneer

04 72 20.10 Cultured Stone Veneer Components		Crew	Daily Output	Labor-Hours	Unit	Material	Labor	Equipment	Total	Total Incl O&P
0010	**CULTURED STONE VENEER COMPONENTS**									
0110	On wood frame and sheathing substrate, random sized cobbles, corner stones	D-8	70	.571	V.L.F.	12.10	17.75		29.85	42.50
0120	Field stones		140	.286	S.F.	8.80	8.85		17.65	24
0130	Random sized flats, corner stones		70	.571	V.L.F.	11.90	17.75		29.65	42
0140	Field stones		140	.286	S.F.	10.05	8.85		18.90	25.50
0150	Horizontal lined ledgestones, corner stones		75	.533	V.L.F.	12.10	16.55		28.65	40.50
0160	Field stones		150	.267	S.F.	8.80	8.30		17.10	23.50
0170	Random shaped flats, corner stones		65	.615	V.L.F.	12.10	19.10		31.20	45
0180	Field stones		150	.267	S.F.	8.80	8.30		17.10	23.50
0190	Random shaped/textured face, corner stones		65	.615	V.L.F.	12.10	19.10		31.20	45
0200	Field stones		130	.308	S.F.	8.80	9.55		18.35	25.50
0210	Random shaped river rock, corner stones		65	.615	V.L.F.	12.10	19.10		31.20	45
0220	Field stones		130	.308	S.F.	8.80	9.55		18.35	25.50
0240	On concrete or CMU substrate, random sized cobbles, corner stones		70	.571	V.L.F.	11.50	17.75		29.25	41.50
0250	Field stones		140	.286	S.F.	8.50	8.85		17.35	24
0260	Random sized flats, corner stones		70	.571	V.L.F.	11.30	17.75		29.05	41.50
0270	Field stones		140	.286	S.F.	9.75	8.85		18.60	25.50
0280	Horizontal lined ledgestones, corner stones		75	.533	V.L.F.	11.50	16.55		28.05	39.50
0290	Field stones		150	.267	S.F.	8.50	8.30		16.80	23
0300	Random shaped flats, corner stones		70	.571	V.L.F.	11.50	17.75		29.25	41.50
0310	Field stones		140	.286	S.F.	8.50	8.85		17.35	24
0320	Random shaped/textured face, corner stones		65	.615	V.L.F.	11.50	19.10		30.60	44
0330	Field stones		130	.308	S.F.	8.50	9.55		18.05	25
0340	Random shaped river rock, corner stones		65	.615	V.L.F.	11.50	19.10		30.60	44
0350	Field stones		130	.308	S.F.	8.50	9.55		18.05	25
0360	Cultured stone veneer, #15 felt weather resistant barrier	1 Clab	3700	.002	Sq.	5.65	.06		5.71	6.30
0370	Expanded metal lath, diamond, 2.5 lb./S.Y., galvanized	1 Lath	85	.094	S.Y.	2.63	2.83		5.46	7.40
0390	Water table or window sill, 18" long	1 Bric	80	.100	Ea.	10.60	3.37		13.97	17.15

04 73 Manufactured Stone Masonry

04 73 20 – Simulated or Manufactured Stone

04 73 20.10 Simulated Stone		Crew	Daily Output	Labor-Hours	Unit	2011 Bare Costs Material	Labor	Equipment	Total	Total Incl O&P
0010	**SIMULATED STONE**									
0100	Insulated fiberglass panels, 5/8" ply backer	L-4	200	.080	S.F.	11.40	2.38		13.78	16.45

Estimating Tips

05 05 00 Common Work Results for Metals

- Nuts, bolts, washers, connection angles, and plates can add a significant amount to both the tonnage of a structural steel job and the estimated cost. As a rule of thumb, add 10% to the total weight to account for these accessories.
- Type 2 steel construction, commonly referred to as "simple construction," consists generally of field-bolted connections with lateral bracing supplied by other elements of the building, such as masonry walls or x-bracing. The estimator should be aware, however, that shop connections may be accomplished by welding or bolting. The method may be particular to the fabrication shop and may have an impact on the estimated cost.

05 12 23 Structural Steel

- Steel items can be obtained from two sources: a fabrication shop or a metals service center. Fabrication shops can fabricate items under more controlled conditions than can crews in the field. They are also more efficient and can produce items more economically. Metal service centers serve as a source of long mill shapes to both fabrication shops and contractors.
- Most line items in this structural steel subdivision, and most items in 05 50 00 Metal Fabrications, are indicated as being shop fabricated. The bare material cost for these shop fabricated items is the "Invoice Cost" from the shop and includes the mill base price of steel plus mill extras, transportation to the shop, shop drawings and detailing where warranted, shop fabrication and handling, sandblasting and a shop coat of primer paint, all necessary structural bolts, and delivery to the job site. The bare labor cost and bare equipment cost for these shop fabricated items is for field installation or erection.
- Line items in Subdivision 05 12 23.40 Lightweight Framing, and other items scattered in Division 5, are indicated as being field fabricated. The bare material cost for these field fabricated items is the "Invoice Cost" from the metals service center and includes the mill base price of steel plus mill extras, transportation to the metals service center, material handling, and delivery of long lengths of mill shapes to the job site. Material costs for structural bolts and welding rods should be added to the estimate. The bare labor cost and bare equipment cost for these items is for both field fabrication and field installation or erection, and include time for cutting, welding and drilling in the fabricated metal items. Drilling into concrete and fasteners to fasten field fabricated items to other work are not included and should be added to the estimate.

05 20 00 Steel Joist Framing

- In any given project the total weight of open web steel joists is determined by the loads to be supported and the design. However, economies can be realized in minimizing the amount of labor used to place the joists. This is done by maximizing the joist spacing, and therefore minimizing the number of joists required to be installed on the job. Certain spacings and locations may be required by the design, but in other cases maximizing the spacing and keeping it as uniform as possible will keep the costs down.

05 30 00 Steel Decking

- The takeoff and estimating of metal deck involves more than simply the area of the floor or roof and the type of deck specified or shown on the drawings. Many different sizes and types of openings may exist. Small openings for individual pipes or conduits may be drilled after the floor/roof is installed, but larger openings may require special deck lengths as well as reinforcing or structural support. The estimator should determine who will be supplying this reinforcing. Additionally, some deck terminations are part of the deck package, such as screed angles and pour stops, and others will be part of the steel contract, such as angles attached to structural members and cast-in-place angles and plates. The estimator must ensure that all pieces are accounted for in the complete estimate.

05 50 00 Metal Fabrications

- The most economical steel stairs are those that use common materials, standard details, and most importantly, a uniform and relatively simple method of field assembly. Commonly available A36 channels and plates are very good choices for the main stringers of the stairs, as are angles and tees for the carrier members. Risers and treads are usually made by specialty shops, and it is most economical to use a typical detail in as many places as possible. The stairs should be pre-assembled and shipped directly to the site. The field connections should be simple and straightforward to be accomplished efficiently, and with minimum equipment and labor.

Reference Numbers

Reference numbers are shown in shaded boxes at the beginning of some major classifications. These numbers refer to related items in the Reference Section. The reference information may be an estimating procedure, an alternate pricing method, or technical information. *Note:* Not all subdivisions listed here necessarily appear in this publication.

05 01 Maintenance of Metals

05 01 10 – Maintenance of Structural Metal Framing

05 01 10.51 Cleaning of Structural Metal Framing		Crew	Daily Output	Labor-Hours	Unit	Material	2011 Bare Costs Labor	Equipment	Total	Total Incl O&P
0010	**CLEANING OF STRUCTURAL METAL FRAMING**									
6125	Steel surface treatments, PDCA guidelines									
6171	Wire brush, hand (SSPC-SP2)	1 Pord	400	.020	S.F.	.02	.59		.61	.98
6181	Power tool (SSPC-SP3)		700	.011		.06	.34		.40	.62
6215	Pressure washing, up to 5000 psi, 5000-15,000 S.F./day		10000	.001			.02		.02	.04
6220	Steam cleaning, 600 psi @ 300 F, 1250 – 2500 S.F./day		2000	.004			.12		.12	.19
6225	Water blasting, up to 25,000 psi, 1750 – 3500 S.F./day		2500	.003			.09		.09	.15
6230	Brush-off blast (SSPC-SP7)	E-11	1750	.018		.24	.55	.11	.90	1.38
6235	Com'l blast (SSPC-SP6), loose scale, fine pwder rust, 2.0#/S.F. sand		1200	.027		.49	.81	.16	1.46	2.17
6240	Tight mill scale, little/no rust, 3.0#/S.F. sand		1000	.032		.73	.97	.20	1.90	2.76
6245	Exist coat blistered/pitted, 4.0#/S.F. sand		875	.037		.97	1.11	.22	2.30	3.30
6250	Exist coat badly pitted/nodules, 6.7#/S.F. sand		825	.039		1.63	1.17	.24	3.04	4.15
6255	Near white blast (SSPC-SP10), loose scale, fine rust, 5.6#/S.F. sand		450	.071		1.36	2.15	.44	3.95	5.85
6260	Tight mill scale, little/no rust, 6.9#/S.F. sand		325	.098		1.68	2.98	.60	5.26	7.85
6265	Exist coat blistered/pitted, 9.0#/S.F. sand		225	.142		2.19	4.31	.87	7.37	11.05
6270	Exist coat badly pitted/nodules, 11.3#/S.F. sand		150	.213		2.75	6.45	1.31	10.51	16

05 05 Common Work Results for Metals

05 05 05 – Selective Metals Demolition

05 05 05.10 Selective Demolition, Metals

		Crew	Daily Output	Labor-Hours	Unit	Material	Labor	Equipment	Total	Total Incl O&P
0010	**SELECTIVE DEMOLITION, METALS** R024119-10									
0015	Excludes shores, bracing, cutting, loading, hauling, dumping									
0020	Remove nuts only up to 3/4" diameter	1 Sswk	480	.017	Ea.		.61		.61	1.16
0030	7/8" to 1-1/4" diameter		240	.033			1.21		1.21	2.31
0040	1-3/8" to 2" diameter		160	.050			1.82		1.82	3.47
0060	Unbolt and remove structural bolts up to 3/4" diameter		240	.033			1.21		1.21	2.31
0070	7/8" to 2" diameter		160	.050			1.82		1.82	3.47
0140	Light weight framing members, remove whole or cut up, up to 20 lb		240	.033			1.21		1.21	2.31
0150	21 – 40 lb	2 Sswk	210	.076			2.77		2.77	5.30
0160	41 – 80 lb	3 Sswk	180	.133			4.85		4.85	9.25
0170	81 – 120 lb	4 Sswk	150	.213			7.75		7.75	14.80
0230	Structural members, remove whole or cut up, up to 500 lb	E-19	48	.500			18.05	21.50	39.55	56.50
0240	1/4 – 2 tons	E-18	36	1.111			40.50	29	69.50	107
0250	2 – 5 tons	E-24	30	1.067			38.50	25	63.50	98
0260	5 – 10 tons	E-20	24	2.667			96	52.50	148.50	234
0270	10 – 15 tons	E-2	18	2.667			97.50	90	187.50	281
0340	Fabricated item, remove whole or cut up, up to 20 lb	1 Sswk	96	.083			3.03		3.03	5.80
0350	21 – 40 lb	2 Sswk	84	.190			6.95		6.95	13.20
0360	41 – 80 lb	3 Sswk	72	.333			12.15		12.15	23
0370	81 – 120 lb	4 Sswk	60	.533			19.40		19.40	37
0380	121 – 500 lb	E-19	48	.500			18.05	21.50	39.55	56.50
0390	501 – 1000 lb	"	36	.667			24	29	53	75.50

05 05 13 – Shop-Applied Coatings for Metal

05 05 13.50 Paints and Protective Coatings

		Crew	Daily Output	Labor-Hours	Unit	Material	Labor	Equipment	Total	Total Incl O&P
0010	**PAINTS AND PROTECTIVE COATINGS**									
5900	Galvanizing structural steel in shop, under 1 ton				Ton	460			460	505
5950	1 ton to 20 tons					400			400	440
6000	Over 20 tons					375			375	415

05 05 Common Work Results for Metals

05 05 21 – Fastening Methods for Metal

05 05 21.10 Cutting Steel

	05 05 21.10 Cutting Steel	Crew	Daily Output	Labor-Hours	Unit	2011 Bare Costs Material	Labor	Equipment	Total	Total Incl O&P
0010	**CUTTING STEEL**									
0020	Hand burning, incl. preparation, torch cutting & grinding, no staging									
0100	Steel to 1/2" thick	E-25	320	.025	L.F.		.91	.32	1.23	2.09
0150	3/4" thick		260	.031			1.12	.39	1.51	2.57
0200	1" thick		200	.040			1.46	.50	1.96	3.33

05 05 21.15 Drilling Steel

	05 05 21.15 Drilling Steel	Crew	Daily Output	Labor-Hours	Unit	Material	Labor	Equipment	Total	Total Incl O&P
0010	**DRILLING STEEL**									
1910	Drilling & layout for steel, up to 1/4" deep, no anchor									
1920	Holes, 1/4" diameter	1 Sswk	112	.071	Ea.	.07	2.60		2.67	5.05
1925	For each additional 1/4" depth, add		336	.024		.07	.87		.94	1.72
1930	3/8" diameter		104	.077		.08	2.80		2.88	5.45
1935	For each additional 1/4" depth, add		312	.026		.08	.93		1.01	1.87
1940	1/2" diameter		96	.083		.08	3.03		3.11	5.90
1945	For each additional 1/4" depth, add		288	.028		.08	1.01		1.09	2.02
1950	5/8" diameter		88	.091		.12	3.31		3.43	6.45
1955	For each additional 1/4" depth, add		264	.030		.12	1.10		1.22	2.23
1960	3/4" diameter		80	.100		.15	3.64		3.79	7.10
1965	For each additional 1/4" depth, add		240	.033		.15	1.21		1.36	2.47
1970	7/8" diameter		72	.111		.19	4.04		4.23	7.90
1975	For each additional 1/4" depth, add		216	.037		.19	1.35		1.54	2.78
1980	1" diameter		64	.125		.20	4.55		4.75	8.90
1985	For each additional 1/4" depth, add		192	.042		.20	1.52		1.72	3.11
1990	For drilling up, add						40%			

05 05 21.90 Welding Steel

	05 05 21.90 Welding Steel	Crew	Daily Output	Labor-Hours	Unit	Material	Labor	Equipment	Total	Total Incl O&P
0010	**WELDING STEEL**, Structural R050521-20									
0020	Field welding, 1/8" E6011, cost per welder, no oper. engr	E-14	8	1	Hr.	4.22	36.50	13.65	54.37	89
0200	With 1/2 operating engineer	E-13	8	1.500		4.22	55	13.60	72.82	121
0300	With 1 operating engineer	E-12	8	2		4.22	72	13.60	89.82	148
0500	With no operating engineer, 2# weld rod per ton	E-14	8	1	Ton	4.22	36.50	13.65	54.37	89
0600	8# E6011 per ton	"	2	4		16.85	146	54.50	217.35	355
0800	With one operating engineer per welder, 2# E6011 per ton	E-12	8	2		4.22	72	13.60	89.82	148
0900	8# E6011 per ton	"	2	8		16.85	288	54.50	359.35	590
1200	Continuous fillet, stick welding, incl. equipment									
1300	Single pass, 1/8" thick, 0.1#/L.F.	E-14	150	.053	L.F.	.21	1.94	.73	2.88	4.73
1400	3/16" thick, 0.2#/L.F.		75	.107		.42	3.88	1.45	5.75	9.45
1500	1/4" thick, 0.3#/L.F.		50	.160		.63	5.80	2.18	8.61	14.20
1610	5/16" thick, 0.4#/L.F.		38	.211		.84	7.65	2.87	11.36	18.70
1800	3 passes, 3/8" thick, 0.5#/L.F.		30	.267		1.05	9.70	3.63	14.38	23.50
2010	4 passes, 1/2" thick, 0.7#/L.F.		22	.364		1.48	13.25	4.96	19.69	32
2200	5 to 6 passes, 3/4" thick, 1.3#/L.F.		12	.667		2.74	24.50	9.10	36.34	59.50
2400	8 to 11 passes, 1" thick, 2.4#/L.F.		6	1.333		5.05	48.50	18.15	71.70	118
2600	For all position welding, add, minimum						20%			
2700	Maximum						300%			
2900	For semi-automatic welding, deduct, minimum						5%			
3000	Maximum						15%			
4000	Cleaning and welding plates, bars, or rods									
4010	to existing beams, columns, or trusses	E-14	12	.667	L.F.	1.05	24.50	9.10	34.65	57.50

05 05 23 – Metal Fastenings

05 05 23.05 Anchor Bolts

	05 05 23.05 Anchor Bolts	Crew	Daily Output	Labor-Hours	Unit	Material	Labor	Equipment	Total	Total Incl O&P
0010	**ANCHOR BOLTS**									
0015	Made from recycled materials G									

05 05 23.05 Anchor Bolts			Crew	Daily Output	Labor-Hours	Unit	Material	2011 Bare Costs Labor	Equipment	Total	Total Incl O&P
0025	Single bolts installed in fresh concrete, no templates										
0030	Hooked w/nut and washer, 1/2" diameter, 8" long	G	1 Carp	132	.061	Ea.	1.36	2.03		3.39	4.89
0040	12" long	G		131	.061		1.51	2.05		3.56	5.10
0050	5/8" diameter, 8" long	G		129	.062		2.96	2.08		5.04	6.75
0060	12" long	G		127	.063		3.65	2.11		5.76	7.55
0070	3/4" diameter, 8" long	G		127	.063		3.65	2.11		5.76	7.55
0080	12" long	G		125	.064		4.56	2.15		6.71	8.60
0090	2-bolt pattern, including job-built 2-hole template, per set										
0100	J-type, incl. hex nut & washer, 1/2" diameter x 6" long	G	1 Carp	21	.381	Set	4.50	12.80		17.30	26.50
0110	12" long	G		21	.381		5.10	12.80		17.90	27
0120	18" long	G		21	.381		6	12.80		18.80	28
0130	3/4" diameter x 8" long	G		20	.400		9.40	13.40		22.80	33
0140	12" long	G		20	.400		11.20	13.40		24.60	35
0150	18" long	G		20	.400		13.95	13.40		27.35	38
0160	1" diameter x 12" long	G		19	.421		19.65	14.15		33.80	45
0170	18" long	G		19	.421		23.50	14.15		37.65	49
0180	24" long	G		19	.421		28	14.15		42.15	54
0190	36" long	G		18	.444		37.50	14.90		52.40	66.50
0200	1-1/2" diameter x 18" long	G		17	.471		60	15.80		75.80	92.50
0210	24" long	G		16	.500		71.50	16.80		88.30	107
0300	L-type, incl. hex nut & washer, 3/4" diameter x 12" long	G		20	.400		10.70	13.40		24.10	34.50
0310	18" long	G		20	.400		13.10	13.40		26.50	37
0320	24" long	G		20	.400		15.50	13.40		28.90	39.50
0330	30" long	G		20	.400		19.15	13.40		32.55	43.50
0340	36" long	G		20	.400		21.50	13.40		34.90	46
0350	1" diameter x 12" long	G		19	.421		16.75	14.15		30.90	42
0360	18" long	G		19	.421		20.50	14.15		34.65	46
0370	24" long	G		19	.421		24.50	14.15		38.65	50.50
0380	30" long	G		19	.421		28.50	14.15		42.65	55
0390	36" long	G		18	.444		32.50	14.90		47.40	61
0400	42" long	G		18	.444		39	14.90		53.90	68
0410	48" long	G		18	.444		43.50	14.90		58.40	73
0420	1-1/4" diameter x 18" long	G		18	.444		30	14.90		44.90	58
0430	24" long	G		18	.444		35	14.90		49.90	64
0440	30" long	G		17	.471		40.50	15.80		56.30	71
0450	36" long	G		17	.471		45.50	15.80		61.30	77
0460	42" long	G	2 Carp	32	.500		51.50	16.80		68.30	84.50
0470	48" long	G		32	.500		58.50	16.80		75.30	92
0480	54" long	G		31	.516		68.50	17.30		85.80	104
0490	60" long	G		31	.516		75	17.30		92.30	112
0500	1-1/2" diameter x 18" long	G		33	.485		43.50	16.25		59.75	75
0510	24" long	G		32	.500		50.50	16.80		67.30	83.50
0520	30" long	G		31	.516		57	17.30		74.30	91.50
0530	36" long	G		30	.533		65	17.90		82.90	102
0540	42" long	G		30	.533		74	17.90		91.90	111
0550	48" long	G		29	.552		82.50	18.50		101	122
0560	54" long	G		28	.571		100	19.15		119.15	142
0570	60" long	G		28	.571		110	19.15		129.15	153
0580	1-3/4" diameter x 18" long	G		31	.516		61.50	17.30		78.80	96.50
0590	24" long	G		30	.533		72	17.90		89.90	109
0600	30" long	G		29	.552		83.50	18.50		102	123
0610	36" long	G		28	.571		95	19.15		114.15	136
0620	42" long	G		27	.593		106	19.90		125.90	151

05 05 23.05 Anchor Bolts			Crew	Daily Output	Labor-Hours	Unit	Material	2011 Bare Costs Labor	Equipment	Total	Total Incl O&P
0630	48" long	G	2 Carp	26	.615	Set	116	20.50		136.50	163
0640	54" long	G		26	.615		144	20.50		164.50	193
0650	60" long	G		25	.640		155	21.50		176.50	207
0660	2" diameter x 24" long	G		27	.593		96.50	19.90		116.40	140
0670	30" long	G		27	.593		108	19.90		127.90	153
0680	36" long	G		26	.615		119	20.50		139.50	166
0690	42" long	G		25	.640		132	21.50		153.50	182
0700	48" long	G		24	.667		152	22.50		174.50	205
0710	54" long	G		23	.696		180	23.50		203.50	237
0720	60" long	G		23	.696		194	23.50		217.50	252
0730	66" long	G		22	.727		207	24.50		231.50	269
0740	72" long	G		21	.762		227	25.50		252.50	292
1000	4-bolt pattern, including job-built 4-hole template, per set										
1100	J-type, incl. hex nut & washer, 1/2" diameter x 6" long	G	1 Carp	19	.421	Set	6.90	14.15		21.05	31
1110	12" long	G		19	.421		8.15	14.15		22.30	32.50
1120	18" long	G		18	.444		9.95	14.90		24.85	36
1130	3/4" diameter x 8" long	G		17	.471		16.70	15.80		32.50	45
1140	12" long	G		17	.471		20.50	15.80		36.30	49
1150	18" long	G		17	.471		26	15.80		41.80	55
1160	1" diameter x 12" long	G		16	.500		37.50	16.80		54.30	69
1170	18" long	G		15	.533		44.50	17.90		62.40	79
1180	24" long	G		15	.533		54	17.90		71.90	89
1190	36" long	G		15	.533		73	17.90		90.90	110
1200	1-1/2" diameter x 18" long	G		13	.615		118	20.50		138.50	165
1210	24" long	G		12	.667		140	22.50		162.50	193
1300	L-type, incl. hex nut & washer, 3/4" diameter x 12" long	G		17	.471		19.25	15.80		35.05	47.50
1310	18" long	G		17	.471		24	15.80		39.80	53
1320	24" long	G		17	.471		29	15.80		44.80	58.50
1330	30" long	G		16	.500		36	16.80		52.80	68
1340	36" long	G		16	.500		41	16.80		57.80	73
1350	1" diameter x 12" long	G		16	.500		31.50	16.80		48.30	62.50
1360	18" long	G		15	.533		38.50	17.90		56.40	72.50
1370	24" long	G		15	.533		47	17.90		64.90	82
1380	30" long	G		15	.533		55.50	17.90		73.40	91
1390	36" long	G		15	.533		63	17.90		80.90	99
1400	42" long	G		14	.571		76	19.15		95.15	116
1410	48" long	G		14	.571		85	19.15		104.15	126
1420	1-1/4" diameter x 18" long	G		14	.571		58	19.15		77.15	95.50
1430	24" long	G		14	.571		68.50	19.15		87.65	107
1440	30" long	G		13	.615		79	20.50		99.50	121
1450	36" long	G		13	.615		89.50	20.50		110	133
1460	42" long	G	2 Carp	25	.640		101	21.50		122.50	147
1470	48" long	G		24	.667		115	22.50		137.50	164
1480	54" long	G		23	.696		135	23.50		158.50	187
1490	60" long	G		23	.696		148	23.50		171.50	201
1500	1-1/2" diameter x 18" long	G		25	.640		85	21.50		106.50	130
1510	24" long	G		24	.667		99	22.50		121.50	147
1520	30" long	G		23	.696		112	23.50		135.50	162
1530	36" long	G		22	.727		128	24.50		152.50	182
1540	42" long	G		22	.727		146	24.50		170.50	201
1550	48" long	G		21	.762		163	25.50		188.50	222
1560	54" long	G		20	.800		198	27		225	263
1570	60" long	G		20	.800		217	27		244	284

05 05 Common Work Results for Metals

05 05 23 – Metal Fastenings

05 05 23.05 Anchor Bolts

	05 05 23.05 Anchor Bolts		Crew	Daily Output	Labor-Hours	Unit	Material	2011 Bare Costs Labor	Equipment	Total	Total Incl O&P
1580	1-3/4" diameter x 18" long	G	2 Carp	22	.727	Set	121	24.50		145.50	174
1590	24" long	G		21	.762		142	25.50		167.50	199
1600	30" long	G		21	.762		165	25.50		190.50	224
1610	36" long	G		20	.800		187	27		214	251
1620	42" long	G		19	.842		210	28.50		238.50	279
1630	48" long	G		18	.889		231	30		261	305
1640	54" long	G		18	.889		286	30		316	365
1650	60" long	G		17	.941		310	31.50		341.50	395
1660	2" diameter x 24" long	G		19	.842		191	28.50		219.50	258
1670	30" long	G		18	.889		215	30		245	286
1680	36" long	G		18	.889		236	30		266	310
1690	42" long	G		17	.941		263	31.50		294.50	340
1700	48" long	G		16	1		300	33.50		333.50	385
1710	54" long	G		15	1.067		360	36		396	455
1720	60" long	G		15	1.067		385	36		421	485
1730	66" long	G		14	1.143		410	38.50		448.50	520
1740	72" long	G		14	1.143		450	38.50		488.50	560
1990	For galvanized, add					Ea.	75%				

05 05 23.10 Bolts and Hex Nuts

	05 05 23.10 Bolts and Hex Nuts		Crew	Daily Output	Labor-Hours	Unit	Material	Labor	Equipment	Total	Total Incl O&P
0010	**BOLTS & HEX NUTS**, Steel, A307										
0100	1/4" diameter, 1/2" long	G	1 Sswk	140	.057	Ea.	.07	2.08		2.15	4.05
0200	1" long	G		140	.057		.08	2.08		2.16	4.06
0300	2" long	G		130	.062		.11	2.24		2.35	4.39
0400	3" long	G		130	.062		.16	2.24		2.40	4.45
0500	4" long	G		120	.067		.18	2.43		2.61	4.82
0600	3/8" diameter, 1" long	G		130	.062		.18	2.24		2.42	4.46
0700	2" long	G		130	.062		.22	2.24		2.46	4.51
0800	3" long	G		120	.067		.29	2.43		2.72	4.95
0900	4" long	G		120	.067		.36	2.43		2.79	5.05
1000	5" long	G		115	.070		.45	2.53		2.98	5.30
1100	1/2" diameter, 1-1/2" long	G		120	.067		.43	2.43		2.86	5.10
1200	2" long	G		120	.067		.49	2.43		2.92	5.15
1300	4" long	G		115	.070		.78	2.53		3.31	5.70
1400	6" long	G		110	.073		1.08	2.65		3.73	6.25
1500	8" long	G		105	.076		1.42	2.77		4.19	6.85
1600	5/8" diameter, 1-1/2" long	G		120	.067		1.01	2.43		3.44	5.75
1700	2" long	G		120	.067		1.12	2.43		3.55	5.85
1800	4" long	G		115	.070		1.61	2.53		4.14	6.60
1900	6" long	G		110	.073		2.07	2.65		4.72	7.35
2000	8" long	G		105	.076		3.07	2.77		5.84	8.70
2100	10" long	G		100	.080		3.87	2.91		6.78	9.80
2200	3/4" diameter, 2" long	G		120	.067		1.22	2.43		3.65	5.95
2300	4" long	G		110	.073		1.74	2.65		4.39	6.95
2400	6" long	G		105	.076		2.23	2.77		5	7.75
2500	8" long	G		95	.084		3.35	3.07		6.42	9.55
2600	10" long	G		85	.094		4.39	3.43		7.82	11.40
2700	12" long	G		80	.100		5.15	3.64		8.79	12.60
2800	1" diameter, 3" long	G		105	.076		3.53	2.77		6.30	9.20
2900	6" long	G		90	.089		5.25	3.24		8.49	11.90
3000	12" long	G		75	.107		9.55	3.88		13.43	17.90
3100	For galvanized, add						75%				
3200	For stainless, add						350%				

05 05 23 – Metal Fastenings

05 05 23.15 Chemical Anchors

	05 05 23.15 Chemical Anchors	Crew	Daily Output	Labor-Hours	Unit	2011 Bare Costs Material	Labor	Equipment	Total	Total Incl O&P
0010	**CHEMICAL ANCHORS**									
0020	Includes layout & drilling									
1430	Chemical anchor, w/rod & epoxy cartridge, 3/4" diam. x 9-1/2" long	B-89A	27	.593	Ea.	8.25	17.95	4.07	30.27	43.50
1435	1" diameter x 11-3/4" long		24	.667		16.05	20	4.57	40.62	56.50
1440	1-1/4" diameter x 14" long		21	.762		29.50	23	5.25	57.75	77
1445	1-3/4" diameter x 15" long		20	.800		58	24	5.50	87.50	110
1450	18" long		17	.941		69.50	28.50	6.45	104.45	131
1455	2" diameter x 18" long		16	1		84	30.50	6.85	121.35	151
1460	24" long		15	1.067		109	32.50	7.30	148.80	182

05 05 23.20 Expansion Anchors

	05 05 23.20 Expansion Anchors	Crew	Daily Output	Labor-Hours	Unit	Material	Labor	Equipment	Total	Total Incl O&P
0010	**EXPANSION ANCHORS**									
0100	Anchors for concrete, brick or stone, no layout and drilling									
0200	Expansion shields, zinc, 1/4" diameter, 1-5/16" long, single G	1 Carp	90	.089	Ea.	.42	2.98		3.40	5.45
0300	1-3/8" long, double G		85	.094		.50	3.16		3.66	5.85
0400	3/8" diameter, 1-1/2" long, single G		85	.094		.72	3.16		3.88	6.10
0500	2" long, double G		80	.100		1.08	3.36		4.44	6.80
0600	1/2" diameter, 2-1/16" long, single G		80	.100		1.36	3.36		4.72	7.10
0700	2-1/2" long, double G		75	.107		1.65	3.58		5.23	7.80
0800	5/8" diameter, 2-5/8" long, single G		75	.107		2.46	3.58		6.04	8.70
0900	2-3/4" long, double G		70	.114		2.39	3.83		6.22	9.05
1000	3/4" diameter, 2-3/4" long, single G		70	.114		2.74	3.83		6.57	9.40
1100	3-15/16" long, double G		65	.123		4.65	4.13		8.78	12
2100	Hollow wall anchors for gypsum wall board, plaster or tile									
2300	1/8" diameter, short G	1 Carp	160	.050	Ea.	.19	1.68		1.87	3.02
2400	Long G		150	.053		.22	1.79		2.01	3.23
2500	3/16" diameter, short G		150	.053		.34	1.79		2.13	3.36
2600	Long G		140	.057		.48	1.92		2.40	3.74
2700	1/4" diameter, short G		140	.057		.56	1.92		2.48	3.83
2800	Long G		130	.062		.52	2.06		2.58	4.03
3000	Toggle bolts, bright steel, 1/8" diameter, 2" long G		85	.094		.18	3.16		3.34	5.50
3100	4" long G		80	.100		.23	3.36		3.59	5.85
3200	3/16" diameter, 3" long G		80	.100		.21	3.36		3.57	5.85
3300	6" long G		75	.107		.31	3.58		3.89	6.35
3400	1/4" diameter, 3" long G		75	.107		.27	3.58		3.85	6.30
3500	6" long G		70	.114		.45	3.83		4.28	6.90
3600	3/8" diameter, 3" long G		70	.114		.67	3.83		4.50	7.15
3700	6" long G		60	.133		1.13	4.47		5.60	8.75
3800	1/2" diameter, 4" long G		60	.133		1.67	4.47		6.14	9.35
3900	6" long G		50	.160		2.30	5.35		7.65	11.55
4000	Nailing anchors									
4100	Nylon nailing anchor, 1/4" diameter, 1" long	1 Carp	3.20	2.500	C	12.45	84		96.45	154
4200	1-1/2" long		2.80	2.857		15.85	96		111.85	177
4300	2" long		2.40	3.333		17.95	112		129.95	207
4400	Metal nailing anchor, 1/4" diameter, 1" long G		3.20	2.500		15.20	84		99.20	157
4500	1-1/2" long G		2.80	2.857		19.95	96		115.95	182
4600	2" long G		2.40	3.333		25	112		137	215
5000	Screw anchors for concrete, masonry,									
5100	stone & tile, no layout or drilling included									
5700	Lag screw shields, 1/4" diameter, short G	1 Carp	90	.089	Ea.	.31	2.98		3.29	5.35
5800	Long G		85	.094		.40	3.16		3.56	5.75
5900	3/8" diameter, short G		85	.094		.59	3.16		3.75	5.95
6000	Long G		80	.100		.73	3.36		4.09	6.40

05 05 Common Work Results for Metals

05 05 23 – Metal Fastenings

05 05 23.20 Expansion Anchors

05 05 23.20 Expansion Anchors			Crew	Daily Output	Labor-Hours	Unit	Material	2011 Bare Costs Labor	Equipment	Total	Total Incl O&P
6100	1/2" diameter, short	G	1 Carp	80	.100	Ea.	.99	3.36		4.35	6.70
6200	Long	G		75	.107		1.35	3.58		4.93	7.50
6300	5/8" diameter, short	G		70	.114		1.58	3.83		5.41	8.15
6400	Long	G		65	.123		2.12	4.13		6.25	9.25
6600	Lead, #6 & #8, 3/4" long	G		260	.031		.16	1.03		1.19	1.91
6700	#10 - #14, 1-1/2" long	G		200	.040		.26	1.34		1.60	2.54
6800	#16 & #18, 1-1/2" long	G		160	.050		.32	1.68		2	3.16
6900	Plastic, #6 & #8, 3/4" long			260	.031		.04	1.03		1.07	1.77
7000	#8 & #10, 7/8" long			240	.033		.04	1.12		1.16	1.91
7100	#10 & #12, 1" long			220	.036		.05	1.22		1.27	2.10
7200	#14 & #16, 1-1/2" long			160	.050		.07	1.68		1.75	2.89
8000	Wedge anchors, not including layout or drilling										
8050	Carbon steel, 1/4" diameter, 1-3/4" long	G	1 Carp	150	.053	Ea.	.32	1.79		2.11	3.34
8100	3-1/4" long	G		140	.057		.42	1.92		2.34	3.67
8150	3/8" diameter, 2-1/4" long	G		145	.055		.38	1.85		2.23	3.52
8200	5" long	G		140	.057		.67	1.92		2.59	3.95
8250	1/2" diameter, 2-3/4" long	G		140	.057		.79	1.92		2.71	4.08
8300	7" long	G		125	.064		1.35	2.15		3.50	5.10
8350	5/8" diameter, 3-1/2" long	G		130	.062		1.40	2.06		3.46	5
8400	8-1/2" long	G		115	.070		2.97	2.33		5.30	7.20
8450	3/4" diameter, 4-1/4" long	G		115	.070		2.09	2.33		4.42	6.20
8500	10" long	G		95	.084		4.75	2.83		7.58	10
8550	1" diameter, 6" long	G		100	.080		7.65	2.68		10.33	12.90
8575	9" long	G		85	.094		9.95	3.16		13.11	16.20
8600	12" long	G		75	.107		10.70	3.58		14.28	17.80
8650	1-1/4" diameter, 9" long	G		70	.114		21.50	3.83		25.33	30
8700	12" long	G		60	.133		27.50	4.47		31.97	37.50
8750	For type 303 stainless steel, add						350%				
8800	For type 316 stainless steel, add						450%				
8950	Self-drilling concrete screw, hex washer head, 3/16" diam. x 1-3/4" long	G	1 Carp	300	.027	Ea.	.18	.89		1.07	1.70
8960	2-1/4" long	G		250	.032		.20	1.07		1.27	2.02
8970	Phillips flat head, 3/16" diam. x 1-3/4" long	G		300	.027		.18	.89		1.07	1.70
8980	2-1/4" long	G		250	.032		.19	1.07		1.26	2.01

05 05 23.25 High Strength Bolts

05 05 23.25 High Strength Bolts			Crew	Daily Output	Labor-Hours	Unit	Material	Labor	Equipment	Total	Total Incl O&P
0010	**HIGH STRENGTH BOLTS**	R050523-10									
0020	A325 Type 1, structural steel, bolt-nut-washer set										
0100	1/2" diameter x 1-1/2" long	G	1 Sswk	130	.062	Ea.	.92	2.24		3.16	5.30
0120	2" long	G		125	.064		.99	2.33		3.32	5.55
0150	3" long	G		120	.067		1.38	2.43		3.81	6.15
0170	5/8" diameter x 1-1/2" long	G		125	.064		1.61	2.33		3.94	6.20
0180	2" long	G		120	.067		1.73	2.43		4.16	6.55
0190	3" long	G		115	.070		2.14	2.53		4.67	7.20
0200	3/4" diameter x 2" long	G		120	.067		2.62	2.43		5.05	7.50
0220	3" long	G		115	.070		3.14	2.53		5.67	8.30
0250	4" long	G		110	.073		3.84	2.65		6.49	9.30
0300	6" long	G		105	.076		4.99	2.77		7.76	10.80
0350	8" long	G		95	.084		9.85	3.07		12.92	16.70
0360	7/8" diameter x 2" long	G		115	.070		3.66	2.53		6.19	8.85
0365	3" long	G		110	.073		4.33	2.65		6.98	9.80
0370	4" long	G		105	.076		5.20	2.77		7.97	11.05
0380	6" long	G		100	.080		6.65	2.91		9.56	12.85
0390	8" long	G		90	.089		10.55	3.24		13.79	17.75

05 05 23 – Metal Fastenings

05 05 23.25 High Strength Bolts			Crew	Daily Output	Labor-Hours	Unit	2011 Bare Costs Material	Labor	Equipment	Total	Total Incl O&P
0400	1" diameter x 2" long	G	1 Sswk	105	.076	Ea.	4.67	2.77		7.44	10.45
0420	3" long	G		100	.080		5.25	2.91		8.16	11.35
0450	4" long	G		95	.084		5.95	3.07		9.02	12.40
0500	6" long	G		90	.089		7.90	3.24		11.14	14.85
0550	8" long	G		85	.094		13.70	3.43		17.13	21.50
0600	1-1/4" diameter x 3" long	G		85	.094		10.80	3.43		14.23	18.40
0650	4" long	G		80	.100		2.78	3.64		6.42	10
0700	6" long	G		75	.107		15.25	3.88		19.13	24
0750	8" long	G		70	.114		19.40	4.16		23.56	29.50
1020	A490, bolt-nut-washer set										
1170	5/8" diameter x 1-1/2" long	G	1 Sswk	125	.064	Ea.	2.02	2.33		4.35	6.65
1180	2" long	G		120	.067		2.38	2.43		4.81	7.25
1190	3" long	G		115	.070		2.89	2.53		5.42	8
1200	3/4" diameter x 2" long	G		120	.067		3.47	2.43		5.90	8.45
1220	3" long	G		115	.070		4.11	2.53		6.64	9.35
1250	4" long	G		110	.073		4.79	2.65		7.44	10.30
1300	6" long	G		105	.076		7.05	2.77		9.82	13.05
1350	8" long	G		95	.084		12	3.07		15.07	19.05
1360	7/8" diameter x 2" long	G		115	.070		5.25	2.53		7.78	10.60
1365	3" long	G		110	.073		6.15	2.65		8.80	11.85
1370	4" long	G		105	.076		7.65	2.77		10.42	13.75
1380	6" long	G		100	.080		10.80	2.91		13.71	17.45
1390	8" long	G		90	.089		15.75	3.24		18.99	23.50
1400	1" diameter x 2" long	G		105	.076		6.20	2.77		8.97	12.15
1420	3" long	G		100	.080		7.55	2.91		10.46	13.85
1450	4" long	G		95	.084		8.75	3.07		11.82	15.45
1500	6" long	G		90	.089		11.75	3.24		14.99	19.10
1550	8" long	G		85	.094		18.70	3.43		22.13	27
1600	1-1/4" diameter x 3" long	G		85	.094		13.70	3.43		17.13	21.50
1650	4" long	G		80	.100		15.70	3.64		19.34	24.50
1700	6" long	G		75	.107		21.50	3.88		25.38	31.50
1750	8" long	G		70	.114		28	4.16		32.16	39

05 05 23.30 Lag Screws			Crew	Daily Output	Labor-Hours	Unit	Material	Labor	Equipment	Total	Total Incl O&P
0010	**LAG SCREWS**										
0020	Steel, 1/4" diameter, 2" long	G	1 Carp	200	.040	Ea.	.09	1.34		1.43	2.35
0100	3/8" diameter, 3" long	G		150	.053		.26	1.79		2.05	3.28
0200	1/2" diameter, 3" long	G		130	.062		.48	2.06		2.54	3.99
0300	5/8" diameter, 3" long	G		120	.067		.99	2.24		3.23	4.83

05 05 23.35 Machine Screws			Crew	Daily Output	Labor-Hours	Unit	Material	Labor	Equipment	Total	Total Incl O&P
0010	**MACHINE SCREWS**										
0020	Steel, round head, #8 x 1" long	G	1 Carp	4.80	1.667	C	3.73	56		59.73	97.50
0110	#8 x 2" long	G		2.40	3.333		5.75	112		117.75	193
0200	#10 x 1" long	G		4	2		5.80	67		72.80	118
0300	#10 x 2" long	G		2	4		7.80	134		141.80	234

05 05 23.40 Machinery Anchors			Crew	Daily Output	Labor-Hours	Unit	Material	Labor	Equipment	Total	Total Incl O&P
0010	**MACHINERY ANCHORS**, heavy duty, incl. sleeve, floating base nut,										
0020	lower stud & coupling nut, fiber plug, connecting stud, washer & nut.										
0030	For flush mounted embedment in poured concrete heavy equip. pads.										
0200	Stud & bolt, 1/2" diameter	G	E-16	40	.400	Ea.	68.50	14.95	2.72	86.17	107
0300	5/8" diameter	G		35	.457		76	17.10	3.11	96.21	119
0500	3/4" diameter	G		30	.533		87.50	19.95	3.63	111.08	138
0600	7/8" diameter	G		25	.640		95.50	24	4.36	123.86	155

05 05 Common Work Results for Metals

05 05 23 – Metal Fastenings

05 05 23.40 Machinery Anchors			Crew	Daily Output	Labor-Hours	Unit	Material	2011 Bare Costs Labor	Equipment	Total	Total Incl O&P
0800	1" diameter	G	E-16	20	.800	Ea.	100	30	5.45	135.45	173
0900	1-1/4" diameter	G		15	1.067		133	40	7.25	180.25	231

05 05 23.50 Powder Actuated Tools and Fasteners			Crew	Daily Output	Labor-Hours	Unit	Material	Labor	Equipment	Total	Total Incl O&P
0010	**POWDER ACTUATED TOOLS & FASTENERS**										
0020	Stud driver, .22 caliber, buy, minimum					Ea.	104			104	114
0100	Maximum					"	370			370	405
0300	Powder charges for above, low velocity					C	7			7	7.70
0400	Standard velocity						7.70			7.70	8.50
0600	Drive pins & studs, 1/4" & 3/8" diam., to 3" long, minimum	G	1 Carp	4.80	1.667		4.02	56		60.02	98
0700	Maximum	G	"	4	2		12.85	67		79.85	126

05 05 23.55 Rivets			Crew	Daily Output	Labor-Hours	Unit	Material	Labor	Equipment	Total	Total Incl O&P
0010	**RIVETS**										
0100	Aluminum rivet & mandrel, 1/2" grip length x 1/8" diameter	G	1 Carp	4.80	1.667	C	6.60	56		62.60	101
0200	3/16" diameter	G		4	2		11.20	67		78.20	124
0300	Aluminum rivet, steel mandrel, 1/8" diameter	G		4.80	1.667		8.75	56		64.75	103
0400	3/16" diameter	G		4	2		14	67		81	127
0500	Copper rivet, steel mandrel, 1/8" diameter	G		4.80	1.667		7.70	56		63.70	102
0800	Stainless rivet & mandrel, 1/8" diameter	G		4.80	1.667		24	56		80	120
0900	3/16" diameter	G		4	2		39.50	67		106.50	156
1000	Stainless rivet, steel mandrel, 1/8" diameter	G		4.80	1.667		20.50	56		76.50	116
1100	3/16" diameter	G		4	2		28	67		95	143
1200	Steel rivet and mandrel, 1/8" diameter	G		4.80	1.667		7.50	56		63.50	102
1300	3/16" diameter	G		4	2		11.10	67		78.10	124
1400	Hand riveting tool, minimum					Ea.	55			55	60.50
1500	Maximum						257			257	283
1600	Power riveting tool, minimum						450			450	495
1700	Maximum						2,450			2,450	2,700

05 05 23.70 Structural Blind Bolts			Crew	Daily Output	Labor-Hours	Unit	Material	Labor	Equipment	Total	Total Incl O&P
0010	**STRUCTURAL BLIND BOLTS**										
0100	1/4" diameter x 1/4" grip	G	1 Sswk	240	.033	Ea.	1.24	1.21		2.45	3.67
0150	1/2" grip	G		216	.037		.95	1.35		2.30	3.62
0200	3/8" diameter x 1/2" grip	G		232	.034		1.75	1.26		3.01	4.32
0250	3/4" grip	G		208	.038		1.83	1.40		3.23	4.68
0300	1/2" diameter x 1/2" grip	G		224	.036		5.15	1.30		6.45	8.15
0350	3/4" grip	G		200	.040		5.60	1.46		7.06	8.95
0400	5/8" diameter x 3/4" grip	G		216	.037		8.25	1.35		9.60	11.65
0450	1" grip	G		192	.042		9.05	1.52		10.57	12.85

05 05 23.80 Vibration and Bearing Pads			Crew	Daily Output	Labor-Hours	Unit	Material	Labor	Equipment	Total	Total Incl O&P
0010	**VIBRATION & BEARING PADS**										
0300	Laminated synthetic rubber impregnated cotton duck, 1/2" thick		2 Sswk	24	.667	S.F.	64	24.50		88.50	117
0400	1" thick			20	.800		126	29		155	195
0600	Neoprene bearing pads, 1/2" thick			24	.667		24.50	24.50		49	73.50
0700	1" thick			20	.800		49	29		78	110
0900	Fabric reinforced neoprene, 5000 psi, 1/2" thick			24	.667		10.75	24.50		35.25	58.50
1000	1" thick			20	.800		21.50	29		50.50	79
1200	Felt surfaced vinyl pads, cork and sisal, 5/8" thick			24	.667		27	24.50		51.50	76.50
1300	1" thick			20	.800		49	29		78	110
1500	Teflon bonded to 10 ga. carbon steel, 1/32" layer			24	.667		48	24.50		72.50	99.50
1600	3/32" layer			24	.667		72	24.50		96.50	126
1800	Bonded to 10 ga. stainless steel, 1/32" layer			24	.667		85	24.50		109.50	140
1900	3/32" layer			24	.667		110	24.50		134.50	168
2100	Circular machine leveling pad & stud					Kip	6			6	6.60

05 05 23 – Metal Fastenings

05 05 23.85 Weld Shear Connectors			Crew	Daily Output	Labor-Hours	Unit	Material	2011 Bare Costs Labor	Equipment	Total	Total Incl O&P
0010	**WELD SHEAR CONNECTORS**										
0020	3/4" diameter, 3-3/16" long	G	E-10	960	.025	Ea.	.51	.93	.34	1.78	2.70
0030	3-3/8" long	G		950	.025		.53	.94	.34	1.81	2.76
0200	3-7/8" long	G		945	.025		.57	.94	.34	1.85	2.81
0300	4-3/16" long	G		935	.026		.60	.95	.35	1.90	2.85
0500	4-7/8" long	G		930	.026		.67	.96	.35	1.98	2.95
0600	5-3/16" long	G		920	.026		.70	.97	.35	2.02	3
0800	5-3/8" long	G		910	.026		.70	.98	.36	2.04	3.02
0900	6-3/16" long	G		905	.027		.77	.98	.36	2.11	3.12
1000	7-3/16" long	G		895	.027		.96	.99	.36	2.31	3.36
1100	8-3/16" long	G		890	.027		1.05	1	.37	2.42	3.46
1500	7/8" diameter, 3-11/16" long	G		920	.026		.82	.97	.35	2.14	3.14
1600	4-3/16" long	G		910	.026		.88	.98	.36	2.22	3.22
1700	5-3/16" long	G		905	.027		1	.98	.36	2.34	3.37
1800	6-3/16" long	G		895	.027		1.12	.99	.36	2.47	3.53
1900	7-3/16" long	G		890	.027		1.24	1	.37	2.61	3.68
2000	8-3/16" long	G		880	.027		1.36	1.01	.37	2.74	3.83

05 05 23.87 Weld Studs

			Crew	Daily Output	Labor-Hours	Unit	Material	Labor	Equipment	Total	Total Incl O&P
0010	**WELD STUDS**										
0020	1/4" diameter, 2-11/16" long	G	E-10	1120	.021	Ea.	.33	.79	.29	1.41	2.19
0100	4-1/8" long	G		1080	.022		.31	.82	.30	1.43	2.24
0200	3/8" diameter, 4-1/8" long	G		1080	.022		.36	.82	.30	1.48	2.30
0300	6-1/8" long	G		1040	.023		.47	.86	.31	1.64	2.48
0400	1/2" diameter, 2-1/8" long	G		1040	.023		.34	.86	.31	1.51	2.34
0500	3-1/8" long	G		1025	.023		.41	.87	.32	1.60	2.45
0600	4-1/8" long	G		1010	.024		.48	.88	.32	1.68	2.57
0700	5-5/16" long	G		990	.024		.59	.90	.33	1.82	2.72
0800	6-1/8" long	G		975	.025		.64	.91	.33	1.88	2.81
0900	8-1/8" long	G		960	.025		.90	.93	.34	2.17	3.13
1000	5/8" diameter, 2-11/16" long	G		1000	.024		.58	.89	.33	1.80	2.70
1010	4-3/16" long	G		990	.024		.72	.90	.33	1.95	2.86
1100	6-9/16" long	G		975	.025		.94	.91	.33	2.18	3.14
1200	8-3/16" long	G		960	.025		1.26	.93	.34	2.53	3.53

05 05 23.90 Welding Rod

		Crew	Daily Output	Labor-Hours	Unit	Material	Labor	Equipment	Total	Total Incl O&P
0010	**WELDING ROD**									
0020	Steel, type 6011, 1/8" diam., less than 500#				Lb.	2.11			2.11	2.32
0100	500# to 2,000#					1.90			1.90	2.09
0200	2,000# to 5,000#					1.79			1.79	1.96
0300	5/32" diameter, less than 500#					2.23			2.23	2.45
0310	500# to 2,000#					2.01			2.01	2.21
0320	2,000# to 5,000#					1.89			1.89	2.08
0400	3/16" diam., less than 500#					2.26			2.26	2.49
0500	500# to 2,000#					2.04			2.04	2.24
0600	2,000# to 5,000#					1.92			1.92	2.11
0620	Steel, type 6010, 1/8" diam., less than 500#					2.20			2.20	2.42
0630	500# to 2,000#					1.98			1.98	2.18
0640	2,000# to 5,000#					1.86			1.86	2.05
0650	Steel, type 7018 Low Hydrogen, 1/8" diam., less than 500#					2.12			2.12	2.33
0660	500# to 2,000#					1.91			1.91	2.10
0670	2,000# to 5,000#					1.80			1.80	1.97
0700	Steel, type 7024 Jet Weld, 1/8" diam., less than 500#					2.26			2.26	2.49
0710	500# to 2,000#					2.04			2.04	2.24

05 05 Common Work Results for Metals

05 05 23 – Metal Fastenings

05 05 23.90 Welding Rod			Crew	Daily Output	Labor-Hours	Unit	2011 Bare Costs Material	Labor	Equipment	Total	Total Incl O&P
0720	2,000# to 5,000#					Lb.	1.92			1.92	2.11
1550	Aluminum, type 4043 TIG, 1/8" diam., less than 10#						4.40			4.40	4.84
1560	10# to 60#						3.96			3.96	4.36
1570	Over 60#						3.72			3.72	4.09
1600	Aluminum, type 5356 TIG, 1/8" diam., less than 10#						4.47			4.47	4.92
1610	10# to 60#						4.03			4.03	4.43
1620	Over 60#						3.79			3.79	4.17
1900	Cast iron, type 8 Nickel, 1/8" diam., less than 500#						27.50			27.50	30
1910	500# to 1,000#						24.50			24.50	27
1920	Over 1,000#						23			23	25.50
2000	Stainless steel, type 316/316L, 1/8" diam., less than 500#						7.60			7.60	8.35
2100	500# to 1000#						6.85			6.85	7.55
2220	Over 1000#						6.45			6.45	7.10

05 12 Structural Steel Framing

05 12 23 – Structural Steel for Buildings

05 12 23.05 Canopy Framing			Crew	Daily Output	Labor-Hours	Unit	Material	Labor	Equipment	Total	Total Incl O&P
0010	**CANOPY FRAMING**										
0020	6" and 8" members, shop fabricated	G	E-4	3000	.011	Lb.	1.35	.39	.04	1.78	2.28

05 12 23.10 Ceiling Supports			Crew	Daily Output	Labor-Hours	Unit	Material	Labor	Equipment	Total	Total Incl O&P
0010	**CEILING SUPPORTS**										
1000	Entrance door/folding partition supports, shop fabricated	G	E-4	60	.533	L.F.	22.50	19.70	1.82	44.02	64.50
1100	Linear accelerator door supports	G		14	2.286		102	84.50	7.80	194.30	283
1200	Lintels or shelf angles, hung, exterior hot dipped galv.	G		267	.120		15.35	4.42	.41	20.18	26
1250	Two coats primer paint instead of galv.	G		267	.120		13.30	4.42	.41	18.13	23.50
1400	Monitor support, ceiling hung, expansion bolted	G		4	8	Ea.	355	295	27.50	677.50	985
1450	Hung from pre-set inserts	G		6	5.333		385	197	18.20	600.20	815
1600	Motor supports for overhead doors	G		4	8		181	295	27.50	503.50	795
1700	Partition support for heavy folding partitions, without pocket	G		24	1.333	L.F.	51	49	4.55	104.55	156
1750	Supports at pocket only	G		12	2.667		102	98.50	9.10	209.60	310
2000	Rolling grilles & fire door supports	G		34	.941		44	34.50	3.21	81.71	118
2100	Spider-leg light supports, expansion bolted to ceiling slab	G		8	4	Ea.	146	148	13.65	307.65	455
2150	Hung from pre-set inserts	G		12	2.667	"	158	98.50	9.10	265.60	370
2400	Toilet partition support	G		36	.889	L.F.	51	33	3.03	87.03	122
2500	X-ray travel gantry support	G		12	2.667	"	176	98.50	9.10	283.60	390

05 12 23.15 Columns, Lightweight			Crew	Daily Output	Labor-Hours	Unit	Material	Labor	Equipment	Total	Total Incl O&P
0010	**COLUMNS, LIGHTWEIGHT**										
1000	Lightweight units (lally), 3-1/2" diameter		E-2	780	.062	L.F.	6.65	2.25	2.08	10.98	13.85
1050	4" diameter		"	900	.053	"	8.10	1.95	1.80	11.85	14.50
5800	Adjustable jack post, 8' maximum height, 2-3/4" diameter	G				Ea.	42			42	46.50
5850	4" diameter	G				"	67.50			67.50	74

05 12 23.17 Columns, Structural			Crew	Daily Output	Labor-Hours	Unit	Material	Labor	Equipment	Total	Total Incl O&P
0010	**COLUMNS, STRUCTURAL** R051223-10	G									
0015	Made from recycled materials	G									
0020	Shop fab'd for 100-ton, 1-2 story project, bolted connections										
0800	Steel, concrete filled, extra strong pipe, 3-1/2" diameter		E-2	660	.073	L.F.	37	2.66	2.46	42.12	48.50
0830	4" diameter			780	.062		41.50	2.25	2.08	45.83	52
0890	5" diameter			1020	.047		49.50	1.72	1.59	52.81	59
0930	6" diameter			1200	.040		65.50	1.47	1.35	68.32	76
0940	8" diameter			1100	.044		65.50	1.60	1.47	68.57	76.50

05 12 Structural Steel Framing

05 12 23 – Structural Steel for Buildings

05 12 23.17 Columns, Structural			Crew	Daily Output	Labor-Hours	Unit	2011 Bare Costs Material	Labor	Equipment	Total	Total Incl O&P
1100	For galvanizing, add					Lb.	.20			.20	.22
1300	For web ties, angles, etc., add per added lb.		1 Sswk	945	.008		1.13	.31		1.44	1.83
1500	Steel pipe, extra strong, no concrete, 3" to 5" diameter	G	E-2	16000	.003		1.13	.11	.10	1.34	1.55
1600	6" to 12" diameter	G		14000	.003		1.13	.13	.12	1.38	1.60
1700	Steel pipe, extra strong, no concrete, 3" diameter x 12'-0"	G		60	.800	Ea.	138	29.50	27	194.50	236
1750	4" diameter x 12'-0"	G		58	.828		202	30.50	28	260.50	310
1800	6" diameter x 12'-0"	G		54	.889		385	32.50	30	447.50	520
1850	8" diameter x 14'-0"	G		50	.960		685	35	32.50	752.50	850
1900	10" diameter x 16'-0"	G		48	1		985	36.50	34	1,055.50	1,175
1950	12" diameter x 18'-0"	G		45	1.067		1,325	39	36	1,400	1,575
3300	Structural tubing, square, A500GrB, 4" to 6" square, light section	G		11270	.004	Lb.	1.13	.16	.14	1.43	1.69
3600	Heavy section	G		32000	.002	"	1.13	.05	.05	1.23	1.40
4000	Concrete filled, add					L.F.	4.03			4.03	4.43
4500	Structural tubing, sq, 4" x 4" x 1/4" x 12'-0"	G	E-2	58	.828	Ea.	186	30.50	28	244.50	292
4550	6" x 6" x 1/4" x 12'-0"	G		54	.889		305	32.50	30	367.50	430
4600	8" x 8" x 3/8" x 14'-0"	G		50	.960		660	35	32.50	727.50	825
4650	10" x 10" x 1/2" x 16'-0"	G		48	1		1,225	36.50	34	1,295.50	1,450
5100	Structural tubing, rect, 5" to 6" wide, light section	G		8000	.006	Lb.	1.13	.22	.20	1.55	1.87
5200	Heavy section	G		12000	.004		1.13	.15	.14	1.42	1.66
5300	7" to 10" wide, light section	G		15000	.003		1.13	.12	.11	1.36	1.58
5400	Heavy section	G		18000	.003		1.13	.10	.09	1.32	1.52
5500	Structural tubing, rect, 5" x 3" x 1/4" x 12'-0"	G		58	.828	Ea.	180	30.50	28	238.50	286
5550	6" x 4" x 5/16" x 12'-0"	G		54	.889		281	32.50	30	343.50	405
5600	8" x 4" x 3/8" x 12'-0"	G		54	.889		410	32.50	30	472.50	545
5650	10" x 6" x 3/8" x 14'-0"	G		50	.960		660	35	32.50	727.50	825
5700	12" x 8" x 1/2" x 16'-0"	G		48	1		1,225	36.50	34	1,295.50	1,425
6800	W Shape, A992 steel, 2 tier, W8 x 24	G		1080	.044	L.F.	29.50	1.63	1.50	32.63	37
6850	W8 x 31	G		1080	.044		38.50	1.63	1.50	41.63	46.50
6900	W8 x 48	G		1032	.047		59.50	1.70	1.57	62.77	70.50
6950	W8 x 67	G		984	.049		83	1.79	1.65	86.44	96
7000	W10 x 45	G		1032	.047		55.50	1.70	1.57	58.77	66.50
7050	W10 x 68	G		984	.049		84	1.79	1.65	87.44	97.50
7100	W10 x 112	G		960	.050		139	1.83	1.69	142.52	157
7150	W12 x 50	G		1032	.047		62	1.70	1.57	65.27	73
7200	W12 x 87	G		984	.049		108	1.79	1.65	111.44	123
7250	W12 x 120	G		960	.050		149	1.83	1.69	152.52	168
7300	W12 x 190	G		912	.053		235	1.93	1.78	238.71	265
7350	W14 x 74	G		984	.049		91.50	1.79	1.65	94.94	106
7400	W14 x 120	G		960	.050		149	1.83	1.69	152.52	168
7450	W14 x 176	G		912	.053		218	1.93	1.78	221.71	246
8090	For projects 75 to 99 tons, add					All	10%				
8092	50 to 74 tons, add						20%				
8094	25 to 49 tons, add						30%	10%			
8096	10 to 24 tons, add						50%	25%			
8098	2 to 9 tons, add						75%	50%			
8099	Less than 2 tons, add						100%	100%			

05 12 23.20 Curb Edging

05 12 23.20 Curb Edging			Crew	Daily Output	Labor-Hours	Unit	Material	Labor	Equipment	Total	Total Incl O&P
0010	**CURB EDGING**										
0030	Steel angle w/anchors, shop fabricated, on forms, 1" x 1", 0.8#/L.F.	G	2 Carp	175	.091	L.F.	1.44	3.07		4.51	6.75
0101	2" x 2" angles, 3.92#/L.F.	G		165	.097		5.65	3.25		8.90	11.65
0201	3" x 3" angles, 6.1#/L.F.	G		150	.107		11.05	3.58		14.63	18.15
0301	4" x 4" angles, 8.2#/L.F.	G		140	.114		11.75	3.83		15.58	19.35

05 12 23 – Structural Steel for Buildings

05 12 23.20 Curb Edging			Crew	Daily Output	Labor-Hours	Unit	2011 Bare Costs Material	Labor	Equipment	Total	Total Incl O&P
1002	6" x 4" angles, 12.3#/L.F.	G	2 Carp	125	.128	L.F.	17.30	4.29		21.59	26
1051	Steel channels with anchors, on forms, 3" channel, 5#/L.F.	G		145	.110		7.10	3.70		10.80	14
1101	4" channel, 5.4#/L.F.	G		135	.119		7.65	3.98		11.63	15.05
1201	6" channel, 8.2#/L.F.	G		130	.123		11.75	4.13		15.88	19.85
1301	8" channel, 11.5#/L.F.	G		115	.139		16.20	4.67		20.87	25.50
1401	10" channel, 15.3#/L.F.	G		90	.178		21.50	5.95		27.45	33.50
1501	12" channel, 20.7#/L.F.	G		70	.229		28.50	7.65		36.15	44.50
2000	For curved edging, add						35%	10%			

05 12 23.40 Lightweight Framing			Crew	Daily Output	Labor-Hours	Unit	Material	Labor	Equipment	Total	Total Incl O&P
0010	**LIGHTWEIGHT FRAMING** R051223-35										
0015	Made from recycled materials	G									
0200	For load-bearing steel studs see Section 05 41 13.30										
0400	Angle framing, field fabricated, 4" and larger R051223-45	G	E-3	440	.055	Lb.	.65	2.02	.25	2.92	4.85
0450	Less than 4" angles	G		265	.091	"	.68	3.36	.41	4.45	7.60
0460	1/2" x 1/2" x 1/8"	G		200	.120	L.F.	.14	4.45	.54	5.13	9.25
0462	3/4" x 3/4" x 1/8"	G		160	.150		.38	5.55	.68	6.61	11.75
0464	1" x 1" x 1/8"	G		135	.178		.54	6.60	.81	7.95	14.05
0466	1-1/4" x 1-1/4" x 3/16"	G		115	.209		1	7.75	.95	9.70	16.90
0468	1-1/2" x 1-1/2" x 3/16"	G		100	.240		1.22	8.90	1.09	11.21	19.50
0470	2" x 2" x 1/4"	G		90	.267		2.15	9.90	1.21	13.26	22.50
0472	2-1/2" x 2-1/2" x 1/4"	G		72	.333		2.77	12.35	1.51	16.63	28
0474	3" x 2" x 3/8"	G		65	.369		3.98	13.70	1.68	19.36	32
0476	3" x 3" x 3/8"	G		57	.421		4.86	15.60	1.91	22.37	37.50
0600	Channel framing, field fabricated, 8" and larger	G		500	.048	Lb.	.68	1.78	.22	2.68	4.37
0650	Less than 8" channels	G		335	.072	"	.68	2.66	.33	3.67	6.15
0660	C2 x 1.78	G		115	.209	L.F.	1.20	7.75	.95	9.90	17.10
0662	C3 x 4.1	G		80	.300		2.77	11.10	1.36	15.23	25.50
0664	C4 x 5.4	G		66	.364		3.65	13.50	1.65	18.80	31.50
0666	C5 x 6.7	G		57	.421		4.52	15.60	1.91	22.03	37
0668	C6 x 8.2	G		55	.436		5.35	16.20	1.98	23.53	39
0670	C7 x 9.8	G		40	.600		6.60	22	2.72	31.32	53
0672	C8 x 11.5	G		36	.667		7.75	24.50	3.03	35.28	59
0710	Structural bar tee, field fabricated, 3/4" x 3/4" x 1/8"	G		160	.150		.38	5.55	.68	6.61	11.75
0712	1" x 1" x 1/8"	G		135	.178		.54	6.60	.81	7.95	14.05
0714	1-1/2" x 1-1/2" x 1/4"	G		114	.211		1.58	7.80	.96	10.34	17.70
0716	2" x 2" x 1/4"	G		89	.270		2.15	10	1.22	13.37	23
0718	2-1/2" x 2-1/2" x 3/8"	G		72	.333		3.98	12.35	1.51	17.84	29.50
0720	3" x 3" x 3/8"	G		57	.421		4.86	15.60	1.91	22.37	37.50
0730	Structural zee, field fabricated, 1-1/4" x 1-3/4" x 1-3/4"	G		114	.211		.51	7.80	.96	9.27	16.50
0732	2-11/16" x 3" x 2-11/16"	G		114	.211		1.20	7.80	.96	9.96	17.25
0734	3-1/16" x 4" x 3-1/16"	G		133	.180		1.82	6.70	.82	9.34	15.65
0736	3-1/4" x 5" x 3-1/4"	G		133	.180		2.48	6.70	.82	10	16.35
0738	3-1/2" x 6" x 3-1/2"	G		160	.150		3.73	5.55	.68	9.96	15.45
0740	Junior beam, field fabricated, 3"	G		80	.300		3.85	11.10	1.36	16.31	26.50
0742	4"	G		72	.333		5.20	12.35	1.51	19.06	31
0744	5"	G		67	.358		6.75	13.30	1.63	21.68	34.50
0746	6"	G		62	.387		8.45	14.35	1.76	24.56	38.50
0748	7"	G		57	.421		10.35	15.60	1.91	27.86	43.50
0750	8"	G		53	.453		12.40	16.80	2.06	31.26	48
1000	Continuous slotted channel framing system, shop fab, minimum	G	2 Sswk	2400	.007	Lb.	3.49	.24		3.73	4.30
1200	Maximum	G	"	1600	.010		3.94	.36		4.30	5
1300	Cross bracing, rods, shop fabricated, 3/4" diameter	G	E-3	700	.034		1.35	1.27	.16	2.78	4.08

05 12 23.40 Lightweight Framing

	05 12 23.40 Lightweight Framing		Crew	Daily Output	Labor-Hours	Unit	2011 Bare Costs Material	Labor	Equipment	Total	Total Incl O&P
1310	7/8" diameter	G	E-3	850	.028	Lb.	1.35	1.05	.13	2.53	3.63
1320	1" diameter	G		1000	.024		1.35	.89	.11	2.35	3.31
1330	Angle, 5" x 5" x 3/8"	G		2800	.009		1.35	.32	.04	1.71	2.14
1350	Hanging lintels, shop fabricated, average	G		850	.028		1.35	1.05	.13	2.53	3.63
1380	Roof frames, shop fabricated, 3'-0" square, 5' span	G	E-2	4200	.011		1.35	.42	.39	2.16	2.69
1400	Tie rod, not upset, 1-1/2" to 4" diameter, with turnbuckle	G	2 Sswk	800	.020		1.46	.73		2.19	3
1420	No turnbuckle	G		700	.023		1.41	.83		2.24	3.14
1500	Upset, 1-3/4" to 4" diameter, with turnbuckle	G		800	.020		1.46	.73		2.19	3
1520	No turnbuckle	G		700	.023		1.41	.83		2.24	3.14

05 12 23.45 Lintels

	Description		Crew	Daily Output	Labor-Hours	Unit	Material	Labor	Equipment	Total	Total Incl O&P
0010	**LINTELS**										
0015	Made from recycled materials	G									
0020	Plain steel angles, shop fabricated, under 500 lb.	G	1 Bric	550	.015	Lb.	.87	.49		1.36	1.75
0100	500 to 1000 lb.	G		640	.013		.84	.42		1.26	1.62
0200	1,000 to 2,000 lb.	G		640	.013		.82	.42		1.24	1.59
0300	2,000 to 4,000 lb.	G		640	.013		.80	.42		1.22	1.57
0500	For built-up angles and plates, add to above	G					.28			.28	.31
0700	For engineering, add to above						.11			.11	.12
0900	For galvanizing, add to above, under 500 lb.						.25			.25	.28
0950	500 to 2,000 lb.						.23			.23	.25
1000	Over 2,000 lb.						.20			.20	.22
2000	Steel angles, 3-1/2" x 3", 1/4" thick, 2'-6" long	G	1 Bric	47	.170	Ea.	12.15	5.75		17.90	23
2100	4'-6" long	G		26	.308		22	10.35		32.35	41
2600	4" x 3-1/2", 1/4" thick, 5'-0" long	G		21	.381		28	12.80		40.80	51.50
2700	9'-0" long	G		12	.667		50	22.50		72.50	92

05 12 23.60 Pipe Support Framing

	Description		Crew	Daily Output	Labor-Hours	Unit	Material	Labor	Equipment	Total	Total Incl O&P
0010	**PIPE SUPPORT FRAMING**										
0020	Under 10#/L.F., shop fabricated	G	E-4	3900	.008	Lb.	1.51	.30	.03	1.84	2.27
0200	10.1 to 15#/L.F.	G		4300	.007		1.49	.27	.03	1.79	2.18
0400	15.1 to 20#/L.F.	G		4800	.007		1.46	.25	.02	1.73	2.11
0600	Over 20#/L.F.	G		5400	.006		1.44	.22	.02	1.68	2.02

05 12 23.65 Plates

	Description		Crew	Daily Output	Labor-Hours	Unit	Material	Labor	Equipment	Total	Total Incl O&P
0010	**PLATES** R051223-80										
0015	Made from recycled materials	G									
0020	For connections & stiffener plates, shop fabricated										
0050	1/8" thick (5.1 lb./S.F.)	G				S.F.	5.75			5.75	6.30
0100	1/4" thick (10.2 lb./S.F.)	G					11.50			11.50	12.60
0300	3/8" thick (15.3 lb./S.F.)	G					17.20			17.20	18.95
0400	1/2" thick (20.4 lb./S.F.)	G					23			23	25.50
0450	3/4" thick (30.6 lb./S.F.)	G					34.50			34.50	38
0500	1" thick (40.8 lb./S.F.)	G					46			46	50.50
2000	Steel plate, warehouse prices, no shop fabrication										
2100	1/4" thick (10.2 lb./S.F.)	G				S.F.	7.15			7.15	7.85

05 12 23.70 Stressed Skin Steel Roof and Ceiling System

	Description		Crew	Daily Output	Labor-Hours	Unit	Material	Labor	Equipment	Total	Total Incl O&P
0010	**STRESSED SKIN STEEL ROOF & CEILING SYSTEM**										
0020	Double panel flat roof, spans to 100'	G	E-2	1150	.042	S.F.	9	1.53	1.41	11.94	14.30
0100	Double panel convex roof, spans to 200'	G		960	.050		14.65	1.83	1.69	18.17	21.50
0200	Double panel arched roof, spans to 300'	G		760	.063		22.50	2.31	2.13	26.94	31.50

05 12 23.75 Structural Steel Members

	Description		Crew	Daily Output	Labor-Hours	Unit	Material	Labor	Equipment	Total	Total Incl O&P
0010	**STRUCTURAL STEEL MEMBERS** R051223-10										
0015	Made from recycled materials	G									

05 12 23.75 Structural Steel Members			Crew	Daily Output	Labor-Hours	Unit	2011 Bare Costs Material	Labor	Equipment	Total	Total Incl O&P
0020	Shop fab'd for 100-ton, 1-2 story project, bolted connections										
0100	W 6 x 9	G	E-2	600	.080	L.F.	11.15	2.93	2.70	16.78	20.50
0120	x 15	G		600	.080		18.55	2.93	2.70	24.18	29
0140	x 20	G		600	.080		25	2.93	2.70	30.63	35.50
0300	W 8 x 10	G		600	.080		12.40	2.93	2.70	18.03	22
0320	x 15	G		600	.080		18.55	2.93	2.70	24.18	29
0350	x 21	G		600	.080		26	2.93	2.70	31.63	37
0360	x 24	G		550	.087		29.50	3.20	2.95	35.65	41.50
0370	x 28	G		550	.087		34.50	3.20	2.95	40.65	47
0500	x 31	G		550	.087		38.50	3.20	2.95	44.65	51
0520	x 35	G		550	.087		43.50	3.20	2.95	49.65	56.50
0540	x 48	G		550	.087		59.50	3.20	2.95	65.65	74.50
0600	W 10 x 12	G		600	.080		14.85	2.93	2.70	20.48	25
0620	x 15	G		600	.080		18.55	2.93	2.70	24.18	29
0700	x 22	G		600	.080		27	2.93	2.70	32.63	38.50
0720	x 26	G		600	.080		32	2.93	2.70	37.63	44
0740	x 33	G		550	.087		41	3.20	2.95	47.15	54
0900	x 49	G		550	.087		60.50	3.20	2.95	66.65	75.50
1100	W 12 x 16	G		880	.055		19.80	2	1.84	23.64	28
1300	x 22	G		880	.055		27	2	1.84	30.84	36
1500	x 26	G		880	.055		32	2	1.84	35.84	41.50
1520	x 35	G		810	.059		43.50	2.17	2	47.67	53.50
1560	x 50	G		750	.064		62	2.34	2.16	66.50	75
1580	x 58	G		750	.064		72	2.34	2.16	76.50	86
1700	x 72	G		640	.075		89	2.75	2.53	94.28	106
1740	x 87	G		640	.075		108	2.75	2.53	113.28	126
1900	W 14 x 26	G		990	.048		32	1.78	1.64	35.42	40.50
2100	x 30	G		900	.053		37	1.95	1.80	40.75	46.50
2300	x 34	G		810	.059		42	2.17	2	46.17	52.50
2320	x 43	G		810	.059		53	2.17	2	57.17	64.50
2340	x 53	G		800	.060		65.50	2.20	2.03	69.73	78.50
2360	x 74	G		760	.063		91.50	2.31	2.13	95.94	108
2380	x 90	G		740	.065		111	2.38	2.19	115.57	130
2500	x 120	G		720	.067		149	2.44	2.25	153.69	170
2700	W 16 x 26	G		1000	.048		32	1.76	1.62	35.38	40.50
2900	x 31	G		900	.053		38.50	1.95	1.80	42.25	47.50
3100	x 40	G		800	.060		49.50	2.20	2.03	53.73	61
3120	x 50	G		800	.060		62	2.20	2.03	66.23	74.50
3140	x 67	G		760	.063		83	2.31	2.13	87.44	97.50
3300	W 18 x 35	G	E-5	960	.075		43.50	2.74	1.80	48.04	54.50
3500	x 40	G		960	.075		49.50	2.74	1.80	54.04	61.50
3520	x 46	G		960	.075		57	2.74	1.80	61.54	69.50
3700	x 50	G		912	.079		62	2.89	1.90	66.79	75.50
3900	x 55	G		912	.079		68	2.89	1.90	72.79	82.50
3920	x 65	G		900	.080		80.50	2.92	1.92	85.34	96
3940	x 76	G		900	.080		94	2.92	1.92	98.84	111
3960	x 86	G		900	.080		106	2.92	1.92	110.84	125
3980	x 106	G		900	.080		131	2.92	1.92	135.84	152
4100	W 21 x 44	G		1064	.068		54.50	2.47	1.63	58.60	66.50
4300	x 50	G		1064	.068		62	2.47	1.63	66.10	74.50
4500	x 62	G		1036	.070		76.50	2.54	1.67	80.71	91
4700	x 68	G		1036	.070		84	2.54	1.67	88.21	99
4720	x 83	G		1000	.072		103	2.63	1.73	107.36	120

05 12 23 – Structural Steel for Buildings

05 12 23.75 Structural Steel Members			Crew	Daily Output	Labor-Hours	Unit	Material	2011 Bare Costs Labor	Equipment	Total	Total Incl O&P
4740	x 93	G	E-5	1000	.072	L.F.	115	2.63	1.73	119.36	134
4760	x 101	G		1000	.072		125	2.63	1.73	129.36	144
4780	x 122	G		1000	.072		151	2.63	1.73	155.36	173
4900	W 24 x 55	G		1110	.065		68	2.37	1.56	71.93	81
5100	x 62	G		1110	.065		76.50	2.37	1.56	80.43	90.50
5300	x 68	G		1110	.065		84	2.37	1.56	87.93	98.50
5500	x 76	G		1110	.065		94	2.37	1.56	97.93	109
5700	x 84	G		1080	.067		104	2.44	1.60	108.04	120
5720	x 94	G		1080	.067		116	2.44	1.60	120.04	134
5740	x 104	G		1050	.069		129	2.51	1.65	133.16	149
5760	x 117	G		1050	.069		145	2.51	1.65	149.16	166
5780	x 146	G		1050	.069		181	2.51	1.65	185.16	206
5800	W 27 x 84	G		1190	.061		104	2.21	1.45	107.66	120
5900	x 94	G		1190	.061		116	2.21	1.45	119.66	134
5920	x 114	G		1150	.063		141	2.29	1.51	144.80	161
5940	x 146	G		1150	.063		181	2.29	1.51	184.80	205
5960	x 161	G		1150	.063		199	2.29	1.51	202.80	225
6100	W 30 x 99	G		1200	.060		123	2.19	1.44	126.63	141
6300	x 108	G		1200	.060		134	2.19	1.44	137.63	153
6500	x 116	G		1160	.062		144	2.27	1.49	147.76	164
6520	x 132	G		1160	.062		163	2.27	1.49	166.76	186
6540	x 148	G		1160	.062		183	2.27	1.49	186.76	207
6560	x 173	G		1120	.064		214	2.35	1.55	217.90	242
6580	x 191	G		1120	.064		236	2.35	1.55	239.90	266
6700	W 33 x 118	G		1176	.061		146	2.24	1.47	149.71	167
6900	x 130	G		1134	.063		161	2.32	1.53	164.85	183
7100	x 141	G		1134	.063		174	2.32	1.53	177.85	198
7120	x 169	G		1100	.065		209	2.39	1.57	212.96	236
7140	x 201	G		1100	.065		249	2.39	1.57	252.96	280
7300	W 36 x 135	G		1170	.062		167	2.25	1.48	170.73	190
7500	x 150	G		1170	.062		186	2.25	1.48	189.73	210
7600	x 170	G		1150	.063		210	2.29	1.51	213.80	237
7700	x 194	G		1125	.064		240	2.34	1.54	243.88	270
7900	x 231	G		1125	.064		286	2.34	1.54	289.88	320
7920	x 262	G		1035	.070		325	2.54	1.67	329.21	360
8100	x 302	G		1035	.070		375	2.54	1.67	379.21	415
8490	For projects 75 to 99 tons, add						10%				
8492	50 to 74 tons, add						20%				
8494	25 to 49 tons, add						30%	10%			
8496	10 to 24 tons, add						50%	25%			
8498	2 to 9 tons, add						75%	50%			
8499	Less than 2 tons, add						100%	100%			

05 12 23.77 Structural Steel Projects											
0010	**STRUCTURAL STEEL PROJECTS** R050516-30										
0015	Made from recycled materials	G									
0020	Shop fab'd for 100-ton, 1-2 story project, bolted connections										
0201	Apartments, nursing homes, etc., 1 to 2 stories R050523-10	G	E-2	6.45	7.442	Ton	2,250	273	251	2,774	3,250
0302	3 to 6 stories	G	"	6.30	7.619		2,300	279	257	2,836	3,325
0402	7 to 15 stories R051223-10	G	E-5	8.74	8.238		2,350	300	198	2,848	3,350
0500	Over 15 stories	G	E-6	13.90	8.633		2,425	315	136	2,876	3,400
0701	Offices, hospitals, etc., steel bearing, 1 to 2 stories R051223-20	G	E-2	6.45	7.442		2,250	273	251	2,774	3,250
0801	3 to 6 stories	G	"	6.30	7.619		2,300	279	257	2,836	3,325

05 12 Structural Steel Framing

05 12 23 – Structural Steel for Buildings

05 12 23.77 Structural Steel Projects			Crew	Daily Output	Labor-Hours	Unit	2011 Bare Costs Material	Labor	Equipment	Total	Total Incl O&P
0901	7 to 15 stories R051223-25	G	E-5	8.74	8.238	Ton	2,350	300	198	2,848	3,350
1000	Over 15 stories	G	E-6	13.90	8.633		2,425	315	136	2,876	3,400
1100	For multi-story masonry wall bearing construction, add R051223-30							30%			
1301	Industrial bldgs., 1 story, beams & girders, steel bearing	G	E-2	8.06	5.955		2,250	218	201	2,669	3,100
1401	Masonry bearing	G	″	6.25	7.680		2,250	281	260	2,791	3,275
1500	Industrial bldgs., 1 story, under 10 tons,										
1510	steel from warehouse, trucked	G	E-2	7.50	6.400	Ton	2,700	234	216	3,150	3,650
1601	1 story with roof trusses, steel bearing	G		6.63	7.240		2,650	265	245	3,160	3,700
1701	Masonry bearing	G		6	8		2,650	293	270	3,213	3,775
1901	Monumental structures, banks, stores, etc., minimum	G	E-5	8	9		2,250	330	216	2,796	3,325
2000	Maximum	G	E-6	9	13.333		3,725	485	210	4,420	5,225
2201	Churches, minimum	G	E-2	7.25	6.621		2,100	243	224	2,567	3,000
2300	Maximum	G	E-5	5.20	13.846		2,800	505	335	3,640	4,400
2800	Power stations, fossil fuels, minimum	G	E-6	11	10.909		2,250	395	172	2,817	3,400
2900	Maximum	G		5.70	21.053		3,375	765	330	4,470	5,525
2950	Nuclear fuels, non-safety steel, minimum	G		7	17.143		2,250	620	270	3,140	3,950
3000	Maximum	G		5.50	21.818		3,375	790	345	4,510	5,575
3040	Safety steel, minimum	G		2.50	48		3,275	1,750	755	5,780	7,700
3070	Maximum	G		1.50	80		4,325	2,900	1,250	8,475	11,600
3101	Roof trusses, minimum	G	E-2	8.13	5.904		3,150	216	200	3,566	4,100
3200	Maximum	G	E-5	8.30	8.675		3,825	315	209	4,349	5,025
3211	Schools, minimum	G	E-2	9	5.333		2,250	195	180	2,625	3,050
3220	Maximum	G	E-5	8.30	8.675		3,275	315	209	3,799	4,450
3400	Welded construction, simple commercial bldgs., 1 to 2 stories	G	E-7	7.60	9.474		2,300	345	242	2,887	3,450
3501	7 to 15 stories	G	E-8	6.38	13.793		2,650	505	325	3,480	4,225
3701	Welded rigid frame, 1 story, minimum	G	E-2	9.88	4.858		2,350	178	164	2,692	3,075
3800	Maximum	G	E-7	5.50	13.091		3,050	480	335	3,865	4,625
3810	Fabrication shop costs (included in project material cost, above)										
3820	Mini mill base price, A992	G				Ton	770			770	845
3830	Mill extra for delivery to shop						240			240	264
3840	Shop extra for shop drawings and detailing						270			270	297
3850	Shop fabricating and handling						730			730	805
3860	Shop sandblasting and primer coat of paint						135			135	149
3870	Shop delivery to the job site						105			105	116
3880	Total material cost, shop fabricated, primed, delivered						2,250			2,250	2,475
3900	High strength steel mill spec extras:										
3950	A529, A572 (50 ksi) and A36: same as A992 steel (no extra)										
4000	Add to A992 price for A572 (60, 65 ksi)	G				Ton	100			100	110
4100	A242 and A588 Weathering	G				″	85			85	93.50
4200	Mill size extras for W-Shapes: 0 to 30 plf: no extra charge										
4210	Member sizes 31 to 65 plf, deduct	G				Ton	.01			.01	.01
4220	Member sizes 66 to 100 plf, deduct	G					8.40			8.40	9.25
4230	Member sizes 101 to 387 plf, add	G					58			58	64
4300	Column base plates, light, up to 150 lb	G	2 Sswk	2000	.008	Lb.	1.24	.29		1.53	1.92
4400	Heavy, over 150 lb	G	E-2	7500	.006	″	1.29	.23	.22	1.74	2.10
4600	Castellated beams, light sections, to 50#/L.F., minimum	G		10.70	4.486	Ton	2,375	164	152	2,691	3,075
4700	Maximum	G		7	6.857		2,600	251	232	3,083	3,575
4900	Heavy sections, over 50# per L.F., minimum	G		11.70	4.103		2,475	150	139	2,764	3,150
5000	Maximum	G		7.80	6.154		2,700	225	208	3,133	3,625
5390	For projects 75 to 99 tons, add						10%				
5392	50 to 74 tons, add						20%				
5394	25 to 49 tons, add						30%	10%			
5396	10 to 24 tons, add						50%	25%			

05 12 Structural Steel Framing

05 12 23 – Structural Steel for Buildings

05 12 23.77 Structural Steel Projects			Crew	Daily Output	Labor-Hours	Unit	2011 Bare Costs Material	Labor	Equipment	Total	Total Incl O&P
5398	2 to 9 tons, add					Ton	75%	50%			
5399	Less than 2 tons, add						100%	100%			

05 12 23.80 Subpurlins

			Crew	Daily Output	Labor-Hours	Unit	Material	Labor	Equipment	Total	Total Incl O&P
0010	**SUBPURLINS** R051223-50										
0015	Made from recycled materials	G									
0020	Bulb tees, shop fabricated, painted, 32-5/8" O.C., 40 psf L.L.										
0100	Type 178, max 8'-9" span, 2.15 plf, 2" high x 1-5/8" wide	G	E-1	4200	.004	S.F.	1.56	.14	.03	1.73	2
0200	Type 218, max 10'-2" span, 3.19 plf, 2-1/8" high x 2-1/8" wide	G	"	3100	.005		1.81	.19	.04	2.04	2.39
1420	For 24-5/8" spacing, add						33%	33%			
1430	For 48-5/8" spacing, deduct						50%	50%			

05 14 Structural Aluminum Framing

05 14 23 – Non-Exposed Structural Aluminum Framing

05 14 23.05 Aluminum Shapes

			Crew	Daily Output	Labor-Hours	Unit	Material	Labor	Equipment	Total	Total Incl O&P
0010	**ALUMINUM SHAPES**										
0015	Made from recycled materials	G									
0020	Structural shapes, 1" to 10" members, under 1 ton	G	E-2	1050	.046	Lb.	2.98	1.67	1.54	6.19	8.10
0050	1 to 5 tons	G		1330	.036		2.73	1.32	1.22	5.27	6.80
0100	Over 5 tons	G		1330	.036		2.61	1.32	1.22	5.15	6.65
0300	Extrusions, over 5 tons, stock shapes	G		1330	.036		3.10	1.32	1.22	5.64	7.20
0400	Custom shapes	G		1330	.036		3.10	1.32	1.22	5.64	7.20

05 15 Wire Rope Assemblies

05 15 16 – Steel Wire Rope Assemblies

05 15 16.05 Accessories for Steel Wire Rope

			Crew	Daily Output	Labor-Hours	Unit	Material	Labor	Equipment	Total	Total Incl O&P
0010	**ACCESSORIES FOR STEEL WIRE ROPE**										
0015	Made from recycled materials	G									
1500	Thimbles, heavy duty, 1/4"	G	E-17	160	.100	Ea.	.50	3.74		4.24	7.70
1510	1/2"	G		160	.100		2.21	3.74		5.95	9.60
1520	3/4"	G		105	.152		5	5.70		10.70	16.35
1530	1"	G		52	.308		10.05	11.50		21.55	33
1540	1-1/4"	G		38	.421		15.45	15.75		31.20	47
1550	1-1/2"	G		13	1.231		43.50	46		89.50	136
1560	1-3/4"	G		8	2		89.50	75		164.50	242
1570	2"	G		6	2.667		130	99.50		229.50	335
1580	2-1/4"	G		4	4		177	150		327	480
1600	Clips, 1/4" diameter	G		160	.100		2.54	3.74		6.28	9.95
1610	3/8" diameter	G		160	.100		2.78	3.74		6.52	10.20
1620	1/2" diameter	G		160	.100		4.47	3.74		8.21	12.05
1630	3/4" diameter	G		102	.157		7.25	5.85		13.10	19.15
1640	1" diameter	G		64	.250		12.10	9.35		21.45	31
1650	1-1/4" diameter	G		35	.457		19.80	17.10		36.90	54.50
1670	1-1/2" diameter	G		26	.615		26.50	23		49.50	73.50
1680	1-3/4" diameter	G		16	1		62	37.50		99.50	140
1690	2" diameter	G		12	1.333		69	50		119	171
1700	2-1/4" diameter	G		10	1.600		102	60		162	226
1800	Sockets, open swage, 1/4" diameter	G		160	.100		41	3.74		44.74	52
1810	1/2" diameter	G		77	.208		59	7.75		66.75	80
1820	3/4" diameter	G		19	.842		91.50	31.50		123	161

05 15 16 – Steel Wire Rope Assemblies

05 15 16.05 Accessories for Steel Wire Rope		Crew	Daily Output	Labor-Hours	Unit	Material	Labor	Equipment	Total	Total Incl O&P
						2011 Bare Costs				
1830	1" diameter G	E-17	9	1.778	Ea.	164	66.50		230.50	305
1840	1-1/4" diameter G		5	3.200		227	120		347	480
1850	1-1/2" diameter G		3	5.333		500	199		699	930
1860	1-3/4" diameter G		3	5.333		885	199		1,084	1,350
1870	2" diameter G		1.50	10.667		1,350	400		1,750	2,225
1900	Closed swage, 1/4" diameter G		160	.100		24	3.74		27.74	33.50
1910	1/2" diameter G		104	.154		41.50	5.75		47.25	57
1920	3/4" diameter G		32	.500		62.50	18.70		81.20	104
1930	1" diameter G		15	1.067		109	40		149	196
1940	1-1/4" diameter G		7	2.286		164	85.50		249.50	345
1950	1-1/2" diameter G		4	4		298	150		448	610
1960	1-3/4" diameter G		3	5.333		440	199		639	865
1970	2" diameter G		2	8		855	299		1,154	1,500
2000	Open spelter, galv., 1/4" diameter G		160	.100		52	3.74		55.74	64.50
2010	1/2" diameter G		70	.229		54.50	8.55		63.05	76.50
2020	3/4" diameter G		26	.615		81.50	23		104.50	134
2030	1" diameter G		10	1.600		226	60		286	365
2040	1-1/4" diameter G		5	3.200		325	120		445	585
2050	1-1/2" diameter G		4	4		685	150		835	1,050
2060	1-3/4" diameter G		2	8		1,200	299		1,499	1,900
2070	2" diameter G		1.20	13.333		1,375	500		1,875	2,475
2080	2-1/2" diameter G		1	16		2,525	600		3,125	3,950
2100	Closed spelter, galv., 1/4" diameter G		160	.100		43	3.74		46.74	54.50
2110	1/2" diameter G		88	.182		46	6.80		52.80	63.50
2120	3/4" diameter G		30	.533		70	19.95		89.95	115
2130	1" diameter G		13	1.231		149	46		195	251
2140	1-1/4" diameter G		7	2.286		238	85.50		323.50	425
2150	1-1/2" diameter G		6	2.667		510	99.50		609.50	750
2160	1-3/4" diameter G		2.80	5.714		680	214		894	1,150
2170	2" diameter G		2	8		840	299		1,139	1,500
2200	Jaw & jaw turnbuckles, 1/4" x 4" G		160	.100		16.35	3.74		20.09	25
2250	1/2" x 6" G		96	.167		20.50	6.25		26.75	34.50
2260	1/2" x 9" G		77	.208		27.50	7.75		35.25	45.50
2270	1/2" x 12" G		66	.242		31	9.05		40.05	51.50
2300	3/4" x 6" G		38	.421		40.50	15.75		56.25	74.50
2310	3/4" x 9" G		30	.533		45	19.95		64.95	87.50
2320	3/4" x 12" G		28	.571		57.50	21.50		79	104
2330	3/4" x 18" G		23	.696		69	26		95	126
2350	1" x 6" G		17	.941		78.50	35		113.50	153
2360	1" x 12" G		13	1.231		86	46		132	183
2370	1" x 18" G		10	1.600		129	60		189	256
2380	1" x 24" G		9	1.778		142	66.50		208.50	283
2400	1-1/4" x 12" G		7	2.286		145	85.50		230.50	320
2410	1-1/4" x 18" G		6.50	2.462		179	92		271	375
2420	1-1/4" x 24" G		5.60	2.857		241	107		348	470
2450	1-1/2" x 12" G		5.20	3.077		425	115		540	685
2460	1-1/2" x 18" G		4	4		450	150		600	785
2470	1-1/2" x 24" G		3.20	5		610	187		797	1,025
2500	1-3/4" x 18" G		3.20	5		920	187		1,107	1,350
2510	1-3/4" x 24" G		2.80	5.714		1,050	214		1,264	1,550
2550	2" x 24" G	↓	1.60	10	↓	1,425	375		1,800	2,275

05 15 Wire Rope Assemblies

05 15 16 – Steel Wire Rope Assemblies

05 15 16.50 Steel Wire Rope			Crew	Daily Output	Labor-Hours	Unit	Material	2011 Bare Costs Labor	Equipment	Total	Total Incl O&P
0010	**STEEL WIRE ROPE**										
0015	Made from recycled materials	G									
0020	6 x 19, bright, fiber core, 5000′ rolls, 1/2″ diameter	G				L.F.	.76			.76	.84
0050	Steel core	G					1			1	1.10
0100	Fiber core, 1″ diameter	G					2.56			2.56	2.82
0150	Steel core	G					2.92			2.92	3.21
0300	6 x 19, galvanized, fiber core, 1/2″ diameter	G					1.12			1.12	1.23
0400	Fiber core, 1″ diameter	G					3.28			3.28	3.61
0450	Steel core	G					3.44			3.44	3.78
0500	6 x 7, bright, IPS, fiber core, <500 L.F. w/acc., 1/4″ diameter	G	E-17	6400	.003		1.34	.09		1.43	1.65
0510	1/2″ diameter	G		2100	.008		3.27	.29		3.56	4.14
0520	3/4″ diameter	G		960	.017		5.90	.62		6.52	7.70
0550	6 x 19, bright, IPS, IWRC, <500 L.F. w/acc., 1/4″ diameter	G		5760	.003		.84	.10		.94	1.12
0560	1/2″ diameter	G		1730	.009		1.36	.35		1.71	2.16
0570	3/4″ diameter	G		770	.021		2.36	.78		3.14	4.08
0580	1″ diameter	G		420	.038		4	1.42		5.42	7.10
0590	1-1/4″ diameter	G		290	.055		6.65	2.06		8.71	11.25
0600	1-1/2″ diameter	G		192	.083		8.15	3.12		11.27	14.95
0610	1-3/4″ diameter	G	E-18	240	.167		13	6.10	4.31	23.41	30.50
0620	2″ diameter	G		160	.250		16.70	9.15	6.45	32.30	42.50
0630	2-1/4″ diameter	G		160	.250		22.50	9.15	6.45	38.10	48.50
0650	6 x 37, bright, IPS, IWRC, <500 L.F. w/acc., 1/4″ diameter	G	E-17	6400	.003		1	.09		1.09	1.28
0660	1/2″ diameter	G		1730	.009		1.70	.35		2.05	2.53
0670	3/4″ diameter	G		770	.021		2.75	.78		3.53	4.50
0680	1″ diameter	G		430	.037		4.36	1.39		5.75	7.45
0690	1-1/4″ diameter	G		290	.055		6.60	2.06		8.66	11.20
0700	1-1/2″ diameter	G		190	.084		9.40	3.15		12.55	16.35
0710	1-3/4″ diameter	G	E-18	260	.154		14.95	5.60	3.98	24.53	31
0720	2″ diameter	G		200	.200		19.40	7.30	5.20	31.90	41
0730	2-1/4″ diameter	G		160	.250		25.50	9.15	6.45	41.10	52
0800	6 x 19 & 6 x 37, swaged, 1/2″ diameter	G	E-17	1220	.013		2.20	.49		2.69	3.35
0810	9/16″ diameter	G		1120	.014		2.55	.53		3.08	3.83
0820	5/8″ diameter	G		930	.017		3.03	.64		3.67	4.56
0830	3/4″ diameter	G		640	.025		3.86	.94		4.80	6.05
0840	7/8″ diameter	G		480	.033		4.87	1.25		6.12	7.75
0850	1″ diameter	G		350	.046		5.95	1.71		7.66	9.80
0860	1-1/8″ diameter	G		288	.056		7.30	2.08		9.38	12
0870	1-1/4″ diameter	G		230	.070		8.85	2.60		11.45	14.70
0880	1-3/8″ diameter	G		192	.083		10.20	3.12		13.32	17.20
0890	1-1/2″ diameter	G	E-18	300	.133		12.40	4.87	3.45	20.72	26.50

05 15 16.60 Galvanized Steel Wire Rope and Accessories

			Crew	Daily Output	Labor-Hours	Unit	Material	Labor	Equipment	Total	Total Incl O&P
0010	**GALVANIZED STEEL WIRE ROPE & ACCESSORIES**										
0015	Made from recycled materials	G									
3000	Aircraft cable, galvanized, 7 x 7 x 1/8″	G	E-17	5000	.003	L.F.	.17	.12		.29	.42
3100	Clamps, 1/8″	G	″	125	.128	Ea.	1.81	4.79		6.60	11.15

05 15 16.70 Temporary Cable Safety Railing

			Crew	Daily Output	Labor-Hours	Unit	Material	Labor	Equipment	Total	Total Incl O&P
0010	**TEMPORARY CABLE SAFETY RAILING**, Each 100′ strand incl.										
0020	2 eyebolts, 1 turnbuckle, 100′ cable, 2 thimbles, 6 clips										
0025	Made from recycled materials	G									
0100	One strand using 1/4″ cable & accessories	G	2 Sswk	4	4	C.L.F.	191	146		337	490
0200	1/2″ cable & accessories	G	″	2	8	″	400	291		691	995

05 21 Steel Joist Framing

05 21 13 – Deep Longspan Steel Joist Framing

05 21 13.50 Deep Longspan Joists			Crew	Daily Output	Labor-Hours	Unit	Material	Labor (2011 Bare Costs)	Equipment (2011 Bare Costs)	Total (2011 Bare Costs)	Total Incl O&P
0010	**DEEP LONGSPAN JOISTS**										
3010	DLH series, 40-ton job lots, bolted cross bridging, shop primer										
3015	Made from recycled materials	G									
3020	Spans to 144' (shipped in 2 pieces), minimum	G	E-7	16	4.500	Ton	1,375	165	115	1,655	1,950
3040	Average	G		13	5.538		1,475	202	142	1,819	2,150
3100	Maximum	G		11	6.545		1,800	239	167	2,206	2,600
3200	52DLH11, 26 lb./L.F.	G		2000	.036	L.F.	18.55	1.32	.92	20.79	24
3220	52DLH16, 45 lb./L.F.	G		2000	.036		33.50	1.32	.92	35.74	40
3240	56DLH11, 26 lb./L.F.	G		2000	.036		19.30	1.32	.92	21.54	24.50
3260	56DLH16, 46 lb./L.F.	G		2000	.036		34	1.32	.92	36.24	41
3280	60DLH12, 29 lb./L.F.	G		2000	.036		21.50	1.32	.92	23.74	27
3300	60DLH17, 52 lb./L.F.	G		2000	.036		38.50	1.32	.92	40.74	46
3320	64DLH12, 31 lb./L.F.	G		2200	.033		23	1.20	.84	25.04	28.50
3340	64DLH17, 52 lb./L.F.	G		2200	.033		38.50	1.20	.84	40.54	45.50
3360	68DLH13, 37 lb./L.F.	G		2200	.033		27.50	1.20	.84	29.54	33
3380	68DLH18, 61 lb./L.F.	G		2200	.033		45.50	1.20	.84	47.54	53
3400	72DLH14, 41 lb./L.F.	G		2200	.033		30.50	1.20	.84	32.54	36.50
3420	72DLH19, 70 lb./L.F.	G		2200	.033		52	1.20	.84	54.04	60
3500	For less than 40-ton job lots										
3502	For 30 to 39 tons, add						10%				
3504	20 to 29 tons, add						20%				
3506	10 to 19 tons, add						30%				
3507	5 to 9 tons, add						50%	25%			
3508	1 to 4 tons, add						75%	50%			
3509	Less than 1 ton, add						100%	100%			
4010	SLH series, 40-ton job lots, bolted cross bridging, shop primer										
4020	Spans to 200' (shipped in 3 pieces), minimum	G	E-7	16	4.500	Ton	1,350	165	115	1,630	1,925
4040	Average	G		13	5.538		1,525	202	142	1,869	2,200
4060	Maximum	G		11	6.545		1,825	239	167	2,231	2,625
4200	80SLH15, 40 lb./L.F.	G		1500	.048	L.F.	30.50	1.75	1.23	33.48	38
4220	80SLH20, 75 lb./L.F.	G		1500	.048		57	1.75	1.23	59.98	67.50
4240	88SLH16, 46 lb./L.F.	G		1500	.048		35	1.75	1.23	37.98	43
4260	88SLH21, 89 lb./L.F.	G		1500	.048		68	1.75	1.23	70.98	79
4280	96SLH17, 52 lb./L.F.	G		1500	.048		39.50	1.75	1.23	42.48	48
4300	96SLH22, 102 lb./L.F.	G		1500	.048		78	1.75	1.23	80.98	90
4320	104SLH18, 59 lb./L.F.	G		1800	.040		45	1.46	1.02	47.48	53.50
4340	104SLH23, 109 lb./L.F.	G		1800	.040		83	1.46	1.02	85.48	95.50
4360	112SLH19, 67 lb./L.F.	G		1800	.040		51	1.46	1.02	53.48	60
4380	112SLH24, 131 lb./L.F.	G		1800	.040		100	1.46	1.02	102.48	114
4400	120SLH20, 77 lb./L.F.	G		1800	.040		58.50	1.46	1.02	60.98	68.50
4420	120SLH25, 152 lb./L.F.	G		1800	.040		116	1.46	1.02	118.48	132
6100	For less than 40-ton job lots										
6102	For 30 to 39 tons, add						10%				
6104	20 to 29 tons, add						20%				
6106	10 to 19 tons, add						30%				
6107	5 to 9 tons, add						50%	25%			
6108	1 to 4 tons, add						75%	50%			
6109	Less than 1 ton, add						100%	100%			

05 21 16 – Longspan Steel Joist Framing

05 21 16.50 Longspan Joists											
0010	**LONGSPAN JOISTS**										
2000	LH series, 40-ton job lots, bolted cross bridging, shop primer										

05 21 Steel Joist Framing

05 21 16 – Longspan Steel Joist Framing

05 21 16.50 Longspan Joists			Crew	Daily Output	Labor-Hours	Unit	Material	Labor (2011 Bare Costs)	Equipment	Total	Total Incl O&P
2015	Made from recycled materials	G									
2020	Spans to 96′, minimum	G	E-7	16	4.500	Ton	1,275	165	115	1,555	1,850
2040	Average	G		13	5.538		1,400	202	142	1,744	2,075
2080	Maximum	G		11	6.545		1,675	239	167	2,081	2,450
2200	18LH04, 12 lb./L.F.	G		1400	.051	L.F.	8.45	1.88	1.31	11.64	14.30
2220	18LH08, 19 lb./L.F.	G		1400	.051		13.40	1.88	1.31	16.59	19.75
2240	20LH04, 12 lb./L.F.	G		1400	.051		8.45	1.88	1.31	11.64	14.30
2260	20LH08, 19 lb./L.F.	G		1400	.051		13.40	1.88	1.31	16.59	19.75
2280	24LH05, 13 lb./L.F.	G		1400	.051		9.20	1.88	1.31	12.39	15.10
2300	24LH10, 23 lb./L.F.	G		1400	.051		16.25	1.88	1.31	19.44	23
2320	28LH06, 16 lb./L.F.	G		1800	.040		11.30	1.46	1.02	13.78	16.25
2340	28LH11, 25 lb./L.F.	G		1800	.040		17.65	1.46	1.02	20.13	23.50
2360	32LH08, 17 lb./L.F.	G		1800	.040		12	1.46	1.02	14.48	17.05
2380	32LH13, 30 lb./L.F.	G		1800	.040		21	1.46	1.02	23.48	27.50
2400	36LH09, 21 lb./L.F.	G		1800	.040		14.80	1.46	1.02	17.28	20
2420	36LH14, 36 lb./L.F.	G		1800	.040		25.50	1.46	1.02	27.98	32
2440	40LH10, 21 lb./L.F.	G		2200	.033		14.80	1.20	.84	16.84	19.45
2460	40LH15, 36 lb./L.F.	G		2200	.033		25.50	1.20	.84	27.54	31
2480	44LH11, 22 lb./L.F.	G		2200	.033		15.55	1.20	.84	17.59	20.50
2500	44LH16, 42 lb./L.F.	G		2200	.033		29.50	1.20	.84	31.54	35.50
2520	48LH11, 22 lb./L.F.	G		2200	.033		15.55	1.20	.84	17.59	20.50
2540	48LH16, 42 lb./L.F.	G		2200	.033		29.50	1.20	.84	31.54	35.50
2600	For less than 40-ton job lots										
2602	For 30 to 39 tons, add						10%				
2604	20 to 29 tons, add						20%				
2606	10 to 19 tons, add						30%				
2607	5 to 9 tons, add						50%	25%			
2608	1 to 4 tons, add						75%	50%			
2609	Less than 1 ton, add						100%	100%			
6000	For welded cross bridging, add							30%			

05 21 19 – Open Web Steel Joist Framing

05 21 19.10 Open Web Joists			Crew	Daily Output	Labor-Hours	Unit	Material	Labor (2011 Bare Costs)	Equipment	Total	Total Incl O&P
0010	**OPEN WEB JOISTS**										
0015	Made from recycled materials	G									
0020	K series, 40-ton lots, horiz. bridging, spans to 30′, shop primer, minimum	G	E-7	15	4.800	Ton	1,125	175	123	1,423	1,725
0050	Average	G		12	6		1,275	219	153	1,647	1,975
0080	Maximum	G		9	8		1,525	292	204	2,021	2,450
0130	8K1, 5.1 lb./L.F.	G		1200	.060	L.F.	3.25	2.19	1.53	6.97	9.40
0140	10K1, 5.0 lb./L.F.	G		1200	.060		3.19	2.19	1.53	6.91	9.30
0160	12K3, 5.7 lb./L.F.	G		1500	.048		3.63	1.75	1.23	6.61	8.65
0180	14K3, 6.0 lb./L.F.	G		1500	.048		3.82	1.75	1.23	6.80	8.85
0200	16K3, 6.3 lb./L.F.	G		1800	.040		4.01	1.46	1.02	6.49	8.30
0220	16K6, 8.1 lb./L.F.	G		1800	.040		5.15	1.46	1.02	7.63	9.55
0240	18K5, 7.7 lb./L.F.	G		2000	.036		4.91	1.32	.92	7.15	8.90
0260	18K9, 10.2 lb./L.F.	G		2000	.036		6.50	1.32	.92	8.74	10.65
0410	Span 30′ to 50′, minimum	G		17	4.235	Ton	1,125	155	108	1,388	1,625
0440	Average	G		17	4.235		1,250	155	108	1,513	1,775
0460	Maximum	G		10	7.200		1,325	263	184	1,772	2,150
0500	20K5, 8.2 lb./L.F.	G		2000	.036	L.F.	5.15	1.32	.92	7.39	9.15
0520	20K9, 10.8 lb./L.F.	G		2000	.036		6.75	1.32	.92	8.99	10.95
0540	22K5, 8.8 lb./L.F.	G		2000	.036		5.50	1.32	.92	7.74	9.55
0560	22K9, 11.3 lb./L.F.	G		2000	.036		7.05	1.32	.92	9.29	11.25

05 21 Steel Joist Framing

05 21 19 – Open Web Steel Joist Framing

05 21 19.10 Open Web Joists			Crew	Daily Output	Labor-Hours	Unit	Material	2011 Bare Costs Labor	Equipment	Total	Total Incl O&P
0580	24K6, 9.7 lb./L.F.	G	E-7	2200	.033	L.F.	6.05	1.20	.84	8.09	9.80
0600	24K10, 13.1 lb./L.F.	G		2200	.033		8.20	1.20	.84	10.24	12.15
0620	26K6, 10.6 lb./L.F.	G		2200	.033		6.65	1.20	.84	8.69	10.45
0640	26K10, 13.8 lb./L.F.	G		2200	.033		8.65	1.20	.84	10.69	12.65
0660	28K8, 12.7 lb./L.F.	G		2400	.030		7.95	1.10	.77	9.82	11.65
0680	28K12, 17.1 lb./L.F.	G		2400	.030		10.70	1.10	.77	12.57	14.65
0700	30K8, 13.2 lb./L.F.	G		2400	.030		8.25	1.10	.77	10.12	12
0720	30K12, 17.6 lb./L.F.	G	↓	2400	.030	↓	11	1.10	.77	12.87	15
0800	For less than 40-ton job lots										
0802	For 30 to 39 tons, add						10%				
0804	20 to 29 tons, add						20%				
0806	10 to 19 tons, add						30%				
0807	5 to 9 tons, add						50%	25%			
0808	1 to 4 tons, add						75%	50%			
0809	Less than 1 ton, add						100%	100%			
1010	CS series, 40-ton job lots, horizontal bridging, shop primer										
1020	Spans to 30′, minimum	G	E-7	15	4.800	Ton	1,175	175	123	1,473	1,775
1040	Average	G		12	6		1,300	219	153	1,672	2,025
1060	Maximum	G		9	8	↓	1,550	292	204	2,046	2,475
1100	10CS2, 7.5 lb./L.F.	G		1200	.060	L.F.	4.91	2.19	1.53	8.63	11.20
1120	12CS2, 8.0 lb./L.F.	G		1500	.048		5.25	1.75	1.23	8.23	10.40
1140	14CS2, 8.0 lb./L.F.	G		1500	.048		5.25	1.75	1.23	8.23	10.40
1160	16CS2, 8.5 lb./L.F.	G		1800	.040		5.55	1.46	1.02	8.03	9.95
1180	16CS4, 14.5 lb./L.F.	G		1800	.040		9.50	1.46	1.02	11.98	14.30
1200	18CS2, 9.0 lb./L.F.	G		2000	.036		5.90	1.32	.92	8.14	10
1220	18CS4, 15.0 lb./L.F.	G		2000	.036		9.80	1.32	.92	12.04	14.30
1240	20CS2, 9.5 lb./L.F.	G		2000	.036		6.20	1.32	.92	8.44	10.35
1260	20CS4, 16.5 lb./L.F.	G		2000	.036		10.80	1.32	.92	13.04	15.40
1280	22CS2, 10.0 lb./L.F.	G		2000	.036		6.55	1.32	.92	8.79	10.70
1300	22CS4, 16.5 lb./L.F.	G		2000	.036		10.80	1.32	.92	13.04	15.40
1320	24CS2, 10.0 lb./L.F.	G		2200	.033		6.55	1.20	.84	8.59	10.35
1340	24CS4, 16.5 lb./L.F.	G		2200	.033		10.80	1.20	.84	12.84	15.05
1360	26CS2, 10.0 lb./L.F.	G		2200	.033		6.55	1.20	.84	8.59	10.35
1380	26CS4, 16.5 lb./L.F.	G		2200	.033		10.80	1.20	.84	12.84	15.05
1400	28CS2, 10.5 lb./L.F.	G		2400	.030		6.85	1.10	.77	8.72	10.45
1420	28CS4, 16.5 lb./L.F.	G		2400	.030		10.80	1.10	.77	12.67	14.80
1440	30CS2, 11.0 lb./L.F.	G		2400	.030		7.20	1.10	.77	9.07	10.80
1460	30CS4, 16.5 lb./L.F.	G	↓	2400	.030	↓	10.80	1.10	.77	12.67	14.80
1500	For less than 40-ton job lots										
1502	For 30 to 39 tons, add						10%				
1504	20 to 29 tons, add						20%				
1506	10 to 19 tons, add						30%				
1507	5 to 9 tons, add						50%	25%			
1508	1 to 4 tons, add						75%	50%			
1509	Less than 1 ton, add						100%	100%			
6200	For shop prime paint other than mfrs. standard, add						20%				
6300	For bottom chord extensions, add per chord	G				Ea.	27.50			27.50	30.50
6400	Individual steel bearing plate, 6″ x 6″ x 1/4″ with J-hook	G	1 Bric	160	.050	″	6.75	1.68		8.43	10.20

05 21 23 – Steel Joist Girder Framing

05 21 23.50 Joist Girders			Crew	Daily Output	Labor-Hours	Unit	Material	Labor	Equipment	Total	Total Incl O&P
0010	**JOIST GIRDERS**										
0015	Made from recycled materials	G									

05 21 Steel Joist Framing

05 21 23 – Steel Joist Girder Framing

05 21 23.50 Joist Girders		Crew	Daily Output	Labor-Hours	Unit	Material	2011 Bare Costs Labor	Equipment	Total	Total Incl O&P
7000	Joist girders, 40-ton job lots, shop primer, minimum G	E-5	15	4.800	Ton	1,150	175	115	1,440	1,725
7020	Average G		13	5.538		1,275	202	133	1,610	1,925
7040	Maximum G		11	6.545		1,350	239	157	1,746	2,100
7100	For less than 40-ton job lots									
7102	For 30 to 39 tons, add					10%				
7104	20 to 29 tons, add					20%				
7106	10 to 19 tons, add					30%				
7107	5 to 9 tons, add					50%	25%			
7108	1 to 4 tons, add					75%	50%			
7109	Less than 1 ton, add					100%	100%			
8000	Trusses, 40-ton job lots, shop fabricated WT chords, shop primer, average G	E-5	11	6.545	Ton	4,175	239	157	4,571	5,225
8100	For less than 40-ton job lots									
8102	For 30 to 39 tons, add					10%				
8104	20 to 29 tons, add					20%				
8106	10 to 19 tons, add					30%				
8107	5 to 9 tons, add					50%	25%			
8108	1 to 4 tons, add					75%	50%			
8109	Less than 1 ton, add					100%	100%			

05 31 Steel Decking

05 31 13 – Steel Floor Decking

05 31 13.50 Floor Decking		Crew	Daily Output	Labor-Hours	Unit	Material	Labor	Equipment	Total	Total Incl O&P
0010	**FLOOR DECKING** R053100-10									
0015	Made from recycled materials G									
5100	Non-cellular composite decking, galvanized, 1-1/2" deep, 16 gauge G	E-4	3500	.009	S.F.	2.39	.34	.03	2.76	3.30
5120	18 gauge G		3650	.009		1.92	.32	.03	2.27	2.76
5140	20 gauge G		3800	.008		1.55	.31	.03	1.89	2.32
5200	2" deep, 22 gauge G		3860	.008		1.35	.31	.03	1.69	2.10
5300	20 gauge G		3600	.009		1.49	.33	.03	1.85	2.30
5400	18 gauge G		3380	.009		1.88	.35	.03	2.26	2.78
5500	16 gauge G		3200	.010		2.37	.37	.03	2.77	3.34
5700	3" deep, 22 gauge G		3200	.010		1.47	.37	.03	1.87	2.36
5800	20 gauge G		3000	.011		1.64	.39	.04	2.07	2.59
5900	18 gauge G		2850	.011		2	.41	.04	2.45	3.03
6000	16 gauge G		2700	.012		2.69	.44	.04	3.17	3.83

05 31 23 – Steel Roof Decking

05 31 23.50 Roof Decking		Crew	Daily Output	Labor-Hours	Unit	Material	Labor	Equipment	Total	Total Incl O&P
0010	**ROOF DECKING**									
0015	Made from recycled materials G									
2100	Open type, 1-1/2" deep, Type B, wide rib, galv., 22 gauge, under 50 sq. G	E-4	4500	.007	S.F.	1.42	.26	.02	1.70	2.09
2200	50-500 squares G		4900	.007		1.10	.24	.02	1.36	1.69
2400	Over 500 squares G		5100	.006		1.02	.23	.02	1.27	1.58
2600	20 gauge, under 50 squares G		3865	.008		1.67	.31	.03	2.01	2.45
2650	50-500 squares G		4170	.008		1.34	.28	.03	1.65	2.04
2700	Over 500 squares G		4300	.007		1.21	.27	.03	1.51	1.88
2900	18 gauge, under 50 squares G		3800	.008		2.15	.31	.03	2.49	2.99
2950	50-500 squares G		4100	.008		1.72	.29	.03	2.04	2.47
3000	Over 500 squares G		4300	.007		1.55	.27	.03	1.85	2.25
3050	16 gauge, under 50 squares G		3700	.009		2.90	.32	.03	3.25	3.83
3060	50-500 squares G		4000	.008		2.32	.30	.03	2.65	3.14

05 31 Steel Decking

05 31 23 – Steel Roof Decking

05 31 23.50 Roof Decking			Crew	Daily Output	Labor-Hours	Unit	Material	Labor	Equipment	Total	Total Incl O&P
							2011 Bare Costs				
3100	Over 500 squares	G	E-4	4200	.008	S.F.	2.09	.28	.03	2.40	2.86
3200	3" deep, 22 gauge, under 50 squares	G		3600	.009		2.10	.33	.03	2.46	2.97
3250	50-500 squares	G		3800	.008		1.68	.31	.03	2.02	2.47
3260	over 500 squares	G		4000	.008		1.51	.30	.03	1.84	2.25
3300	20 gauge, under 50 squares	G		3400	.009		2.27	.35	.03	2.65	3.19
3350	50-500 squares	G		3600	.009		1.81	.33	.03	2.17	2.65
3360	over 500 squares	G		3800	.008		1.63	.31	.03	1.97	2.41
3400	18 gauge, under 50 squares	G		3200	.010		2.94	.37	.03	3.34	3.97
3450	50-500 squares	G		3400	.009		2.35	.35	.03	2.73	3.28
3460	over 500 squares	G		3600	.009		2.11	.33	.03	2.47	2.98
3500	16 gauge, under 50 squares	G		3000	.011		3.88	.39	.04	4.31	5.05
3550	50-500 squares	G		3200	.010		3.10	.37	.03	3.50	4.15
3560	over 500 squares	G		3400	.009		2.79	.35	.03	3.17	3.77
3700	4-1/2" deep, 20 gauge, over 50 squares	G		2700	.012		3.64	.44	.04	4.12	4.87
3800	18 gauge	G		2460	.013		4.69	.48	.04	5.21	6.10
3900	16 gauge	G		2350	.014		3.51	.50	.05	4.06	4.87
4100	6" deep, 18 gauge, over 50 squares	G		2000	.016		6.70	.59	.05	7.34	8.60
4200	16 gauge	G		1930	.017		5	.61	.06	5.67	6.75
4300	14 gauge	G		1860	.017		6.45	.63	.06	7.14	8.35
4500	7-1/2" deep, 18 gauge, over 50 squares	G		1690	.019		7.35	.70	.06	8.11	9.50
4600	16 gauge	G		1590	.020		5.50	.74	.07	6.31	7.55
4700	14 gauge	G	↓	1490	.021	↓	7.10	.79	.07	7.96	9.40
4800	For painted instead of galvanized, deduct						5%				
5000	For acoustical perforated with fiberglass insulation, add					S.F.	25%				
5100	For type F intermediate rib instead of type B wide rib, add	G					25%				
5150	For type A narrow rib instead of type B wide rib, add	G				↓	25%				

05 31 33 – Steel Form Decking

05 31 33.50 Form Decking			Crew	Daily Output	Labor-Hours	Unit	Material	Labor	Equipment	Total	Total Incl O&P
0010	**FORM DECKING**										
0015	Made from recycled materials	G									
6100	Slab form, steel, 28 gauge, 9/16" deep, type UFS, uncoated	G	E-4	4000	.008	S.F.	1.07	.30	.03	1.40	1.77
6200	Galvanized	G		4000	.008		.95	.30	.03	1.28	1.64
6220	24 gauge, 1" deep , type UF1X, uncoated	G		3900	.008		1.04	.30	.03	1.37	1.75
6240	Galvanized	G		3900	.008		1.22	.30	.03	1.55	1.95
6300	24 gauge, 1-5/16" deep, type UFX, uncoated	G		3800	.008		1.10	.31	.03	1.44	1.83
6400	Galvanized	G		3800	.008		1.29	.31	.03	1.63	2.04
6500	22 gauge, 1-5/16" deep, uncoated	G		3700	.009		1.38	.32	.03	1.73	2.16
6600	Galvanized	G		3700	.009		1.41	.32	.03	1.76	2.19
6700	22 gauge, 2" deep, uncoated	G		3600	.009		1.82	.33	.03	2.18	2.66
6800	Galvanized	G	↓	3600	.009	↓	1.78	.33	.03	2.14	2.62
7000	Sheet metal edge closure form, 12" wide with 2 bends, galv										
7100	18 gauge	G	E-14	360	.022	L.F.	2.92	.81	.30	4.03	5.10
7200	16 gauge	G	"	360	.022	"	3.97	.81	.30	5.08	6.25

05 35 Raceway Decking Assemblies

05 35 13 – Steel Cellular Decking

05 35 13.50 Cellular Decking			Crew	Daily Output	Labor-Hours	Unit	Material	2011 Bare Costs Labor	Equipment	Total	Total Incl O&P
0010	**CELLULAR DECKING**										
0015	Made from recycled materials	G									
0200	Cellular units, galv, 1-1/2" deep, Type BC, 20-20 gauge, over 15 squares	G	E-4	1460	.022	S.F.	5.65	.81	.07	6.53	7.80
0250	18-20 gauge	G		1420	.023		6.40	.83	.08	7.31	8.70
0300	18-18 gauge	G		1390	.023		6.55	.85	.08	7.48	8.90
0320	16-18 gauge	G		1360	.024		7.80	.87	.08	8.75	10.35
0340	16-16 gauge	G		1330	.024		8.70	.89	.08	9.67	11.40
0400	3" deep, Type NC, galvanized, 20-20 gauge	G		1375	.023		6.20	.86	.08	7.14	8.55
0500	18-20 gauge	G		1350	.024		7.50	.87	.08	8.45	9.95
0600	18-18 gauge	G		1290	.025		7.45	.92	.08	8.45	10.05
0700	16-18 gauge	G		1230	.026		8.40	.96	.09	9.45	11.20
0800	16-16 gauge	G		1150	.028		9.15	1.03	.09	10.27	12.15
1000	4-1/2" deep, Type JC, galvanized, 18-20 gauge	G		1100	.029		8.65	1.07	.10	9.82	11.65
1100	18-18 gauge	G		1040	.031		8.60	1.14	.10	9.84	11.75
1200	16-18 gauge	G		980	.033		9.65	1.20	.11	10.96	13.05
1300	16-16 gauge	G		935	.034		10.55	1.26	.12	11.93	14.15
1500	For acoustical deck, add						15%				
1700	For cells used for ventilation, add						15%				
1900	For multi-story or congested site, add							50%			
8000	Metal deck and trench, 2" thick, 20 gauge, combination										
8010	60% cellular, 40% non-cellular, inserts and trench	G	R-4	1100	.036	S.F.	11.25	1.37	.10	12.72	15.05

05 41 Structural Metal Stud Framing

05 41 13 – Load-Bearing Metal Stud Framing

05 41 13.05 Bracing

			Crew	Daily Output	Labor-Hours	Unit	Material	Labor	Equipment	Total	Total Incl O&P
0010	**BRACING**, shear wall X-bracing, per 10' x 10' bay, one face										
0015	Made of recycled materials	G									
0120	Metal strap, 20 ga x 4" wide	G	2 Carp	18	.889	Ea.	16.65	30		46.65	68.50
0130	6" wide	G		18	.889		25	30		55	77.50
0160	18 ga x 4" wide	G		16	1		26	33.50		59.50	84.50
0170	6" wide	G		16	1		38.50	33.50		72	98.50
0410	Continuous strap bracing, per horizontal row on both faces										
0420	Metal strap, 20 ga x 2" wide, studs 12" O.C.	G	1 Carp	7	1.143	C.L.F.	46	38.50		84.50	115
0430	16" O.C.	G		8	1		46	33.50		79.50	107
0440	24" O.C.	G		10	.800		46	27		73	96
0450	18 ga x 2" wide, studs 12" O.C.	G		6	1.333		64	44.50		108.50	145
0460	16" O.C.	G		7	1.143		64	38.50		102.50	134
0470	24" O.C.	G		8	1		64	33.50		97.50	126

05 41 13.10 Bridging

			Crew	Daily Output	Labor-Hours	Unit	Material	Labor	Equipment	Total	Total Incl O&P
0010	**BRIDGING**, solid between studs w/1-1/4" leg track, per stud bay										
0015	Made from recycled materials	G									
0200	Studs 12" O.C., 18 ga x 2-1/2" wide	G	1 Carp	125	.064	Ea.	.76	2.15		2.91	4.43
0210	3-5/8" wide	G		120	.067		.91	2.24		3.15	4.75
0220	4" wide	G		120	.067		.98	2.24		3.22	4.82
0230	6" wide	G		115	.070		1.26	2.33		3.59	5.30
0240	8" wide	G		110	.073		1.57	2.44		4.01	5.80
0300	16 ga x 2-1/2" wide	G		115	.070		.97	2.33		3.30	4.97
0310	3-5/8" wide	G		110	.073		1.17	2.44		3.61	5.35
0320	4" wide	G		110	.073		1.25	2.44		3.69	5.45
0330	6" wide	G		105	.076		1.60	2.56		4.16	6.05
0340	8" wide	G		100	.080		2.01	2.68		4.69	6.70

05 41 Structural Metal Stud Framing

05 41 13 – Load-Bearing Metal Stud Framing

05 41 13.10 Bridging			Crew	Daily Output	Labor-Hours	Unit	Material	Labor (2011 Bare Costs)	Equipment	Total	Total Incl O&P
1200	Studs 16" O.C., 18 ga x 2-1/2" wide	G	1 Carp	125	.064	Ea.	.97	2.15		3.12	4.66
1210	3-5/8" wide	G		120	.067		1.17	2.24		3.41	5.05
1220	4" wide	G		120	.067		1.25	2.24		3.49	5.10
1230	6" wide	G		115	.070		1.62	2.33		3.95	5.70
1240	8" wide	G		110	.073		2.01	2.44		4.45	6.30
1300	16 ga x 2-1/2" wide	G		115	.070		1.24	2.33		3.57	5.25
1310	3-5/8" wide	G		110	.073		1.50	2.44		3.94	5.75
1320	4" wide	G		110	.073		1.60	2.44		4.04	5.85
1330	6" wide	G		105	.076		2.05	2.56		4.61	6.55
1340	8" wide	G		100	.080		2.57	2.68		5.25	7.30
2200	Studs 24" O.C., 18 ga x 2-1/2" wide	G		125	.064		1.41	2.15		3.56	5.15
2210	3-5/8" wide	G		120	.067		1.69	2.24		3.93	5.60
2220	4" wide	G		120	.067		1.81	2.24		4.05	5.75
2230	6" wide	G		115	.070		2.34	2.33		4.67	6.50
2240	8" wide	G		110	.073		2.91	2.44		5.35	7.30
2300	16 ga x 2-1/2" wide	G		115	.070		1.79	2.33		4.12	5.90
2310	3-5/8" wide	G		110	.073		2.17	2.44		4.61	6.45
2320	4" wide	G		110	.073		2.32	2.44		4.76	6.65
2330	6" wide	G		105	.076		2.96	2.56		5.52	7.55
2340	8" wide	G		100	.080		3.72	2.68		6.40	8.60
3000	Continuous bridging, per row										
3100	16 ga x 1-1/2" channel thru studs 12" O.C.	G	1 Carp	6	1.333	C.L.F.	41	44.50		85.50	120
3110	16" O.C.	G		7	1.143		41	38.50		79.50	109
3120	24" O.C.	G		8.80	.909		41	30.50		71.50	96
4100	2" x 2" angle x 18 ga, studs 12" O.C.	G		7	1.143		64	38.50		102.50	134
4110	16" O.C.	G		9	.889		64	30		94	120
4120	24" O.C.	G		12	.667		64	22.50		86.50	108
4200	16 ga, studs 12" O.C.	G		5	1.600		81.50	53.50		135	180
4210	16" O.C.	G		7	1.143		81.50	38.50		120	154
4220	24" O.C.	G		10	.800		81.50	27		108.50	135

05 41 13.25 Framing, Boxed Headers/Beams			Crew	Daily Output	Labor-Hours	Unit	Material	Labor	Equipment	Total	Total Incl O&P
0010	**FRAMING, BOXED HEADERS/BEAMS**										
0015	Made from recycled materials	G									
0200	Double, 18 ga x 6" deep	G	2 Carp	220	.073	L.F.	4.40	2.44		6.84	8.90
0210	8" deep	G		210	.076		4.84	2.56		7.40	9.65
0220	10" deep	G		200	.080		5.90	2.68		8.58	11
0230	12" deep	G		190	.084		6.45	2.83		9.28	11.80
0300	16 ga x 8" deep	G		180	.089		5.60	2.98		8.58	11.15
0310	10" deep	G		170	.094		6.75	3.16		9.91	12.70
0320	12" deep	G		160	.100		7.30	3.36		10.66	13.65
0400	14 ga x 10" deep	G		140	.114		7.75	3.83		11.58	14.95
0410	12" deep	G		130	.123		8.50	4.13		12.63	16.25
1210	Triple, 18 ga x 8" deep	G		170	.094		7	3.16		10.16	13
1220	10" deep	G		165	.097		8.45	3.25		11.70	14.75
1230	12" deep	G		160	.100		9.25	3.36		12.61	15.80
1300	16 ga x 8" deep	G		145	.110		8.10	3.70		11.80	15.10
1310	10" deep	G		140	.114		9.70	3.83		13.53	17.05
1320	12" deep	G		135	.119		10.55	3.98		14.53	18.30
1400	14 ga x 10" deep	G		115	.139		10.60	4.67		15.27	19.45
1410	12" deep	G		110	.145		11.70	4.88		16.58	21

05 41 Structural Metal Stud Framing

05 41 13 – Load-Bearing Metal Stud Framing

05 41 13.30 Framing, Stud Walls			Crew	Daily Output	Labor-Hours	Unit	2011 Bare Costs Material	Labor	Equipment	Total	Total Incl O&P
0010	**FRAMING, STUD WALLS** w/top & bottom track, no openings,										
0020	Headers, beams, bridging or bracing										
0025	Made from recycled materials	G									
4100	8′ high walls, 18 ga x 2-1/2″ wide, studs 12″ O.C.	G	2 Carp	54	.296	L.F.	7.30	9.95		17.25	24.50
4110	16″ O.C.	G		77	.208		5.85	6.95		12.80	18.05
4120	24″ O.C.	G		107	.150		4.37	5		9.37	13.20
4130	3-5/8″ wide, studs 12″ O.C.	G		53	.302		8.60	10.15		18.75	26.50
4140	16″ O.C.	G		76	.211		6.90	7.05		13.95	19.40
4150	24″ O.C.	G		105	.152		5.20	5.10		10.30	14.25
4160	4″ wide, studs 12″ O.C.	G		52	.308		9.05	10.30		19.35	27.50
4170	16″ O.C.	G		74	.216		7.25	7.25		14.50	20
4180	24″ O.C.	G		103	.155		5.45	5.20		10.65	14.70
4190	6″ wide, studs 12″ O.C.	G		51	.314		11.45	10.55		22	30
4200	16″ O.C.	G		73	.219		9.20	7.35		16.55	22.50
4210	24″ O.C.	G		101	.158		6.90	5.30		12.20	16.50
4220	8″ wide, studs 12″ O.C.	G		50	.320		13.85	10.75		24.60	33
4230	16″ O.C.	G		72	.222		11.15	7.45		18.60	25
4240	24″ O.C.	G		100	.160		8.45	5.35		13.80	18.25
4300	16 ga x 2-1/2″ wide, studs 12″ O.C.	G		47	.340		8.65	11.40		20.05	28.50
4310	16″ O.C.	G		68	.235		6.85	7.90		14.75	21
4320	24″ O.C.	G		94	.170		5.05	5.70		10.75	15.10
4330	3-5/8″ wide, studs 12″ O.C.	G		46	.348		10.20	11.65		21.85	31
4340	16″ O.C.	G		66	.242		8.10	8.15		16.25	22.50
4350	24″ O.C.	G		92	.174		6	5.85		11.85	16.35
4360	4″ wide, studs 12″ O.C.	G		45	.356		10.80	11.95		22.75	32
4370	16″ O.C.	G		65	.246		8.60	8.25		16.85	23.50
4380	24″ O.C.	G		90	.178		6.35	5.95		12.30	17
4390	6″ wide, studs 12″ O.C.	G		44	.364		13.50	12.20		25.70	35.50
4400	16″ O.C.	G		64	.250		10.75	8.40		19.15	26
4410	24″ O.C.	G		88	.182		7.95	6.10		14.05	18.95
4420	8″ wide, studs 12″ O.C.	G		43	.372		16.65	12.50		29.15	39.50
4430	16″ O.C.	G		63	.254		13.25	8.50		21.75	29
4440	24″ O.C.	G		86	.186		9.85	6.25		16.10	21.50
5100	10′ high walls, 18 ga x 2-1/2″ wide, studs 12″ O.C.	G		54	.296		8.75	9.95		18.70	26.50
5110	16″ O.C.	G		77	.208		6.90	6.95		13.85	19.25
5120	24″ O.C.	G		107	.150		5.10	5		10.10	14
5130	3-5/8″ wide, studs 12″ O.C.	G		53	.302		10.35	10.15		20.50	28.50
5140	16″ O.C.	G		76	.211		8.20	7.05		15.25	21
5150	24″ O.C.	G		105	.152		6.05	5.10		11.15	15.20
5160	4″ wide, studs 12″ O.C.	G		52	.308		10.85	10.30		21.15	29.50
5170	16″ O.C.	G		74	.216		8.60	7.25		15.85	21.50
5180	24″ O.C.	G		103	.155		6.35	5.20		11.55	15.70
5190	6″ wide, studs 12″ O.C.	G		51	.314		13.70	10.55		24.25	32.50
5200	16″ O.C.	G		73	.219		10.90	7.35		18.25	24.50
5210	24″ O.C.	G		101	.158		8.05	5.30		13.35	17.75
5220	8″ wide, studs 12″ O.C.	G		50	.320		16.60	10.75		27.35	36
5230	16″ O.C.	G		72	.222		13.20	7.45		20.65	27
5240	24″ O.C.	G		100	.160		9.80	5.35		15.15	19.75
5300	16 ga x 2-1/2″ wide, studs 12″ O.C.	G		47	.340		10.45	11.40		21.85	30.50
5310	16″ O.C.	G		68	.235		8.20	7.90		16.10	22
5320	24″ O.C.	G		94	.170		5.95	5.70		11.65	16.10
5330	3-5/8″ wide, studs 12″ O.C.	G		46	.348		12.35	11.65		24	33

05 41 13.30 Framing, Stud Walls			Crew	Daily Output	Labor-Hours	Unit	Material	2011 Bare Costs Labor	Equipment	Total	Total Incl O&P
5340	16" O.C.	G	2 Carp	66	.242	L.F.	9.70	8.15		17.85	24.50
5350	24" O.C.	G		92	.174		7.05	5.85		12.90	17.50
5360	4" wide, studs 12" O.C.	G		45	.356		13.05	11.95		25	34.50
5370	16" O.C.	G		65	.246		10.25	8.25		18.50	25
5380	24" O.C.	G		90	.178		7.45	5.95		13.40	18.20
5390	6" wide, studs 12" O.C.	G		44	.364		16.30	12.20		28.50	38.50
5400	16" O.C.	G		64	.250		12.85	8.40		21.25	28
5410	24" O.C.	G		88	.182		9.35	6.10		15.45	20.50
5420	8" wide, studs 12" O.C.	G		43	.372		20	12.50		32.50	43
5430	16" O.C.	G		63	.254		15.80	8.50		24.30	31.50
5440	24" O.C.	G		86	.186		11.55	6.25		17.80	23
6190	12' high walls, 18 ga x 6" wide, studs 12" O.C.	G		41	.390		15.95	13.10		29.05	39.50
6200	16" O.C.	G		58	.276		12.55	9.25		21.80	29.50
6210	24" O.C.	G		81	.198		9.20	6.65		15.85	21
6220	8" wide, studs 12" O.C.	G		40	.400		19.30	13.40		32.70	43.50
6230	16" O.C.	G		57	.281		15.25	9.40		24.65	32.50
6240	24" O.C.	G		80	.200		11.15	6.70		17.85	23.50
6390	16 ga x 6" wide, studs 12" O.C.	G		35	.457		19.10	15.35		34.45	46.50
6400	16" O.C.	G		51	.314		14.90	10.55		25.45	34
6410	24" O.C.	G		70	.229		10.75	7.65		18.40	24.50
6420	8" wide, studs 12" O.C.	G		34	.471		23.50	15.80		39.30	52.50
6430	16" O.C.	G		50	.320		18.40	10.75		29.15	38
6440	24" O.C.	G		69	.232		13.25	7.80		21.05	27.50
6530	14 ga x 3-5/8" wide, studs 12" O.C.	G		34	.471		18.20	15.80		34	46.50
6540	16" O.C.	G		48	.333		14.20	11.20		25.40	34.50
6550	24" O.C.	G		65	.246		10.20	8.25		18.45	25
6560	4" wide, studs 12" O.C.	G		33	.485		19.20	16.25		35.45	48
6570	16" O.C.	G		47	.340		15	11.40		26.40	35.50
6580	24" O.C.	G		64	.250		10.80	8.40		19.20	26
6730	12 ga x 3-5/8" wide, studs 12" O.C.	G		31	.516		25	17.30		42.30	56.50
6740	16" O.C.	G		43	.372		19.40	12.50		31.90	42.50
6750	24" O.C.	G		59	.271		13.70	9.10		22.80	30.50
6760	4" wide, studs 12" O.C.	G		30	.533		27	17.90		44.90	59.50
6770	16" O.C.	G		42	.381		20.50	12.80		33.30	44.50
6780	24" O.C.	G		58	.276		14.60	9.25		23.85	31.50
7390	16' high walls, 16 ga x 6" wide, studs 12" O.C.	G		33	.485		24.50	16.25		40.75	54
7400	16" O.C.	G		48	.333		19.10	11.20		30.30	39.50
7410	24" O.C.	G		67	.239		13.50	8		21.50	28.50
7420	8" wide, studs 12" O.C.	G		32	.500		30.50	16.80		47.30	61.50
7430	16" O.C.	G		47	.340		23.50	11.40		34.90	45
7440	24" O.C.	G		66	.242		16.65	8.15		24.80	32
7560	14 ga x 4" wide, studs 12" O.C.	G		31	.516		25	17.30		42.30	56.50
7570	16" O.C.	G		45	.356		19.20	11.95		31.15	41
7580	24" O.C.	G		61	.262		13.60	8.80		22.40	29.50
7590	6" wide, studs 12" O.C.	G		30	.533		31	17.90		48.90	64.50
7600	16" O.C.	G		44	.364		24	12.20		36.20	47
7610	24" O.C.	G		60	.267		17.10	8.95		26.05	34
7760	12 ga x 4" wide, studs 12" O.C.	G		29	.552		35	18.50		53.50	69.50
7770	16" O.C.	G		40	.400		27	13.40		40.40	52
7780	24" O.C.	G		55	.291		18.70	9.75		28.45	37
7790	6" wide, studs 12" O.C.	G		28	.571		44	19.15		63.15	80.50
7800	16" O.C.	G		39	.410		34	13.75		47.75	60.50
7810	24" O.C.	G		54	.296		23.50	9.95		33.45	42.50

05 41 Structural Metal Stud Framing

05 41 13 – Load-Bearing Metal Stud Framing

05 41 13.30 Framing, Stud Walls		Crew	Daily Output	Labor-Hours	Unit	Material	2011 Bare Costs Labor	Equipment	Total	Total Incl O&P
8590	20′ high walls, 14 ga x 6″ wide, studs 12″ O.C. G	2 Carp	29	.552	L.F.	38	18.50		56.50	73
8600	16″ O.C. G		42	.381		29.50	12.80		42.30	54
8610	24″ O.C. G		57	.281		20.50	9.40		29.90	38.50
8620	8″ wide, studs 12″ O.C. G		28	.571		46.50	19.15		65.65	83.50
8630	16″ O.C. G		41	.390		36	13.10		49.10	61.50
8640	24″ O.C. G		56	.286		25	9.60		34.60	43.50
8790	12 ga x 6″ wide, studs 12″ O.C. G		27	.593		54.50	19.90		74.40	93.50
8800	16″ O.C. G		37	.432		41.50	14.50		56	70.50
8810	24″ O.C. G		51	.314		28.50	10.55		39.05	49
8820	8″ wide, studs 12″ O.C. G		26	.615		66	20.50		86.50	107
8830	16″ O.C. G		36	.444		50.50	14.90		65.40	80.50
8840	24″ O.C. G	↓	50	.320	↓	35	10.75		45.75	56.50

05 42 Cold-Formed Metal Joist Framing

05 42 13 – Cold-Formed Metal Floor Joist Framing

05 42 13.05 Bracing		Crew	Daily Output	Labor-Hours	Unit	Material	Labor	Equipment	Total	Total Incl O&P
0010	**BRACING**, continuous, per row, top & bottom									
0015	Made from recycled materials G									
0120	Flat strap, 20 ga x 2″ wide, joists at 12″ O.C. G	1 Carp	4.67	1.713	C.L.F.	48.50	57.50		106	149
0130	16″ O.C. G		5.33	1.501		46.50	50.50		97	136
0140	24″ O.C. G		6.66	1.201		45	40.50		85.50	117
0150	18 ga x 2″ wide, joists at 12″ O.C. G		4	2		63	67		130	182
0160	16″ O.C. G		4.67	1.713		62	57.50		119.50	165
0170	24″ O.C. G	↓	5.33	1.501	↓	61	50.50		111.50	152

05 42 13.10 Bridging		Crew	Daily Output	Labor-Hours	Unit	Material	Labor	Equipment	Total	Total Incl O&P
0010	**BRIDGING**, solid between joists w/1-1/4″ leg track, per joist bay									
0015	Made from recycled materials G									
0230	Joists 12″ O.C., 18 ga track x 6″ wide G	1 Carp	80	.100	Ea.	1.26	3.36		4.62	7
0240	8″ wide G		75	.107		1.57	3.58		5.15	7.75
0250	10″ wide G		70	.114		1.96	3.83		5.79	8.55
0260	12″ wide G		65	.123		2.23	4.13		6.36	9.35
0330	16 ga track x 6″ wide G		70	.114		1.60	3.83		5.43	8.15
0340	8″ wide G		65	.123		2.01	4.13		6.14	9.10
0350	10″ wide G		60	.133		2.48	4.47		6.95	10.25
0360	12″ wide G		55	.145		2.86	4.88		7.74	11.30
0440	14 ga track x 8″ wide G		60	.133		2.51	4.47		6.98	10.25
0450	10″ wide G		55	.145		3.11	4.88		7.99	11.60
0460	12″ wide G		50	.160		3.59	5.35		8.94	12.95
0550	12 ga track x 10″ wide G		45	.178		4.54	5.95		10.49	15
0560	12″ wide G		40	.200		4.66	6.70		11.36	16.35
1230	16″ O.C., 18 ga track x 6″ wide G		80	.100		1.62	3.36		4.98	7.40
1240	8″ wide G		75	.107		2.01	3.58		5.59	8.20
1250	10″ wide G		70	.114		2.51	3.83		6.34	9.15
1260	12″ wide G		65	.123		2.85	4.13		6.98	10.05
1330	16 ga track x 6″ wide G		70	.114		2.05	3.83		5.88	8.65
1340	8″ wide G		65	.123		2.57	4.13		6.70	9.75
1350	10″ wide G		60	.133		3.18	4.47		7.65	11
1360	12″ wide G		55	.145		3.66	4.88		8.54	12.20
1440	14 ga track x 8″ wide G		60	.133		3.22	4.47		7.69	11.05
1450	10″ wide G		55	.145		3.99	4.88		8.87	12.55
1460	12″ wide G	↓	50	.160	↓	4.60	5.35		9.95	14.05

05 42 Cold-Formed Metal Joist Framing

05 42 13 – Cold-Formed Metal Floor Joist Framing

05 42 13.10 Bridging

	05 42 13.10 Bridging		Crew	Daily Output	Labor-Hours	Unit	Material	Labor	Equipment	Total	Total Incl O&P
							2011 Bare Costs				
1550	12 ga track x 10″ wide	G	1 Carp	45	.178	Ea.	5.80	5.95		11.75	16.40
1560	12″ wide	G		40	.200		5.95	6.70		12.65	17.80
2230	24″ O.C., 18 ga track x 6″ wide	G		80	.100		2.34	3.36		5.70	8.15
2240	8″ wide	G		75	.107		2.91	3.58		6.49	9.20
2250	10″ wide	G		70	.114		3.63	3.83		7.46	10.40
2260	12″ wide	G		65	.123		4.13	4.13		8.26	11.45
2330	16 ga track x 6″ wide	G		70	.114		2.96	3.83		6.79	9.65
2340	8″ wide	G		65	.123		3.72	4.13		7.85	11
2350	10″ wide	G		60	.133		4.61	4.47		9.08	12.55
2360	12″ wide	G		55	.145		5.30	4.88		10.18	14
2440	14 ga track x 8″ wide	G		60	.133		4.65	4.47		9.12	12.60
2450	10″ wide	G		55	.145		5.80	4.88		10.68	14.50
2460	12″ wide	G		50	.160		6.65	5.35		12	16.35
2550	12 ga track x 10″ wide	G		45	.178		8.45	5.95		14.40	19.25
2560	12″ wide	G		40	.200		8.65	6.70		15.35	21

05 42 13.25 Framing, Band Joist

	05 42 13.25 Framing, Band Joist		Crew	Daily Output	Labor-Hours	Unit	Material	Labor	Equipment	Total	Total Incl O&P
0010	**FRAMING, BAND JOIST** (track) fastened to bearing wall										
0015	Made from recycled materials	G									
0220	18 ga track x 6″ deep	G	2 Carp	1000	.016	L.F.	1.03	.54		1.57	2.03
0230	8″ deep	G		920	.017		1.28	.58		1.86	2.39
0240	10″ deep	G		860	.019		1.60	.62		2.22	2.80
0320	16 ga track x 6″ deep	G		900	.018		1.30	.60		1.90	2.43
0330	8″ deep	G		840	.019		1.64	.64		2.28	2.87
0340	10″ deep	G		780	.021		2.03	.69		2.72	3.38
0350	12″ deep	G		740	.022		2.33	.73		3.06	3.77
0430	14 ga track x 8″ deep	G		750	.021		2.05	.72		2.77	3.45
0440	10″ deep	G		720	.022		2.54	.75		3.29	4.05
0450	12″ deep	G		700	.023		2.93	.77		3.70	4.50
0540	12 ga track x 10″ deep	G		670	.024		3.71	.80		4.51	5.40
0550	12″ deep	G		650	.025		3.80	.83		4.63	5.55

05 42 13.30 Framing, Boxed Headers/Beams

	05 42 13.30 Framing, Boxed Headers/Beams		Crew	Daily Output	Labor-Hours	Unit	Material	Labor	Equipment	Total	Total Incl O&P
0010	**FRAMING, BOXED HEADERS/BEAMS**										
0015	Made from recycled materials	G									
0200	Double, 18 ga x 6″ deep	G	2 Carp	220	.073	L.F.	4.40	2.44		6.84	8.90
0210	8″ deep	G		210	.076		4.84	2.56		7.40	9.65
0220	10″ deep	G		200	.080		5.90	2.68		8.58	11
0230	12″ deep	G		190	.084		6.45	2.83		9.28	11.80
0300	16 ga x 8″ deep	G		180	.089		5.60	2.98		8.58	11.15
0310	10″ deep	G		170	.094		6.75	3.16		9.91	12.70
0320	12″ deep	G		160	.100		7.30	3.36		10.66	13.65
0400	14 ga x 10″ deep	G		140	.114		7.75	3.83		11.58	14.95
0410	12″ deep	G		130	.123		8.50	4.13		12.63	16.25
0500	12 ga x 10″ deep	G		110	.145		10.25	4.88		15.13	19.40
0510	12″ deep	G		100	.160		11.35	5.35		16.70	21.50
1210	Triple, 18 ga x 8″ deep	G		170	.094		7	3.16		10.16	13
1220	10″ deep	G		165	.097		8.45	3.25		11.70	14.75
1230	12″ deep	G		160	.100		9.25	3.36		12.61	15.80
1300	16 ga x 8″ deep	G		145	.110		8.10	3.70		11.80	15.10
1310	10″ deep	G		140	.114		9.70	3.83		13.53	17.05
1320	12″ deep	G		135	.119		10.55	3.98		14.53	18.30
1400	14 ga x 10″ deep	G		115	.139		11.25	4.67		15.92	20
1410	12″ deep	G		110	.145		12.35	4.88		17.23	22

05 42 13 – Cold-Formed Metal Floor Joist Framing

05 42 13.30 Framing, Boxed Headers/Beams			Crew	Daily Output	Labor-Hours	Unit	Material	Labor	Equipment	Total	Total Incl O&P
1500	12 ga x 10″ deep	G	2 Carp	90	.178	L.F.	15	5.95		20.95	26.50
1510	12″ deep	G		85	.188		16.60	6.30		22.90	29

05 42 13.40 Framing, Joists			Crew	Daily Output	Labor-Hours	Unit	Material	Labor	Equipment	Total	Total Incl O&P
0010	**FRAMING, JOISTS**, no band joists (track), web stiffeners, headers,										
0020	Beams, bridging or bracing										
0025	Made from recycled materials	G									
0030	Joists (2″ flange) and fasteners, materials only										
0220	18 ga x 6″ deep	G				L.F.	1.35			1.35	1.49
0230	8″ deep	G					1.59			1.59	1.74
0240	10″ deep	G					1.87			1.87	2.06
0320	16 ga x 6″ deep	G					1.65			1.65	1.81
0330	8″ deep	G					1.97			1.97	2.17
0340	10″ deep	G					2.31			2.31	2.54
0350	12″ deep	G					2.61			2.61	2.88
0430	14 ga x 8″ deep	G					2.47			2.47	2.71
0440	10″ deep	G					2.86			2.86	3.14
0450	12″ deep	G					3.24			3.24	3.57
0540	12 ga x 10″ deep	G					4.16			4.16	4.57
0550	12″ deep	G					4.73			4.73	5.20
1010	Installation of joists to band joists, beams & headers, labor only										
1220	18 ga x 6″ deep		2 Carp	110	.145	Ea.		4.88		4.88	8.15
1230	8″ deep			90	.178			5.95		5.95	10
1240	10″ deep			80	.200			6.70		6.70	11.25
1320	16 ga x 6″ deep			95	.168			5.65		5.65	9.45
1330	8″ deep			70	.229			7.65		7.65	12.85
1340	10″ deep			60	.267			8.95		8.95	14.95
1350	12″ deep			55	.291			9.75		9.75	16.35
1430	14 ga x 8″ deep			65	.246			8.25		8.25	13.80
1440	10″ deep			45	.356			11.95		11.95	19.95
1450	12″ deep			35	.457			15.35		15.35	25.50
1540	12 ga x 10″ deep			40	.400			13.40		13.40	22.50
1550	12″ deep			30	.533			17.90		17.90	30

05 42 13.45 Framing, Web Stiffeners			Crew	Daily Output	Labor-Hours	Unit	Material	Labor	Equipment	Total	Total Incl O&P
0010	**FRAMING, WEB STIFFENERS** at joist bearing, fabricated from										
0020	Stud piece (1-5/8″ flange) to stiffen joist (2″ flange)										
0025	Made from recycled materials	G									
2120	For 6″ deep joist, with 18 ga x 2-1/2″ stud	G	1 Carp	120	.067	Ea.	1.61	2.24		3.85	5.50
2130	3-5/8″ stud	G		110	.073		1.76	2.44		4.20	6
2140	4″ stud	G		105	.076		1.71	2.56		4.27	6.15
2150	6″ stud	G		100	.080		1.86	2.68		4.54	6.55
2160	8″ stud	G		95	.084		1.90	2.83		4.73	6.80
2220	8″ deep joist, with 2-1/2″ stud	G		120	.067		1.76	2.24		4	5.70
2230	3-5/8″ stud	G		110	.073		1.90	2.44		4.34	6.15
2240	4″ stud	G		105	.076		1.87	2.56		4.43	6.35
2250	6″ stud	G		100	.080		2.04	2.68		4.72	6.75
2260	8″ stud	G		95	.084		2.19	2.83		5.02	7.15
2320	10″ deep joist, with 2-1/2″ stud	G		110	.073		2.48	2.44		4.92	6.80
2330	3-5/8″ stud	G		100	.080		2.71	2.68		5.39	7.45
2340	4″ stud	G		95	.084		2.69	2.83		5.52	7.70
2350	6″ stud	G		90	.089		2.91	2.98		5.89	8.20
2360	8″ stud	G		85	.094		2.93	3.16		6.09	8.55
2420	12″ deep joist, with 2-1/2″ stud	G		110	.073		2.63	2.44		5.07	6.95

05 42 Cold-Formed Metal Joist Framing

05 42 13 – Cold-Formed Metal Floor Joist Framing

05 42 13.45 Framing, Web Stiffeners		Crew	Daily Output	Labor-Hours	Unit	Material	2011 Bare Costs Labor	Equipment	Total	Total Incl O&P
2430	3-5/8" stud G	1 Carp	100	.080	Ea.	2.84	2.68		5.52	7.60
2440	4" stud G		95	.084		2.79	2.83		5.62	7.80
2450	6" stud G		90	.089		3.05	2.98		6.03	8.35
2460	8" stud G		85	.094		3.26	3.16		6.42	8.90
3130	For 6" deep joist, with 16 ga x 3-5/8" stud G		100	.080		1.86	2.68		4.54	6.55
3140	4" stud G		95	.084		1.85	2.83		4.68	6.75
3150	6" stud G		90	.089		2.02	2.98		5	7.20
3160	8" stud G		85	.094		2.14	3.16		5.30	7.65
3230	8" deep joist, with 3-5/8" stud G		100	.080		2.06	2.68		4.74	6.75
3240	4" stud G		95	.084		2.03	2.83		4.86	6.95
3250	6" stud G		90	.089		2.24	2.98		5.22	7.45
3260	8" stud G		85	.094		2.41	3.16		5.57	7.95
3330	10" deep joist, with 3-5/8" stud G		85	.094		2.82	3.16		5.98	8.40
3340	4" stud G		80	.100		2.88	3.36		6.24	8.75
3350	6" stud G		75	.107		3.11	3.58		6.69	9.45
3360	8" stud G		70	.114		3.26	3.83		7.09	10
3430	12" deep joist, with 3-5/8" stud G		85	.094		3.07	3.16		6.23	8.70
3440	4" stud G		80	.100		3.02	3.36		6.38	8.95
3450	6" stud G		75	.107		3.34	3.58		6.92	9.65
3460	8" stud G		70	.114		3.59	3.83		7.42	10.35
4230	For 8" deep joist, with 14 ga x 3-5/8" stud G		90	.089		2.67	2.98		5.65	7.95
4240	4" stud G		85	.094		2.72	3.16		5.88	8.30
4250	6" stud G		80	.100		2.95	3.36		6.31	8.85
4260	8" stud G		75	.107		3.15	3.58		6.73	9.45
4330	10" deep joist, with 3-5/8" stud G		75	.107		3.75	3.58		7.33	10.15
4340	4" stud G		70	.114		3.72	3.83		7.55	10.50
4350	6" stud G		65	.123		4.09	4.13		8.22	11.40
4360	8" stud G		60	.133		4.26	4.47		8.73	12.20
4430	12" deep joist, with 3-5/8" stud G		75	.107		3.99	3.58		7.57	10.40
4440	4" stud G		70	.114		4.06	3.83		7.89	10.85
4450	6" stud G		65	.123		4.40	4.13		8.53	11.75
4460	8" stud G		60	.133		4.71	4.47		9.18	12.70
5330	For 10" deep joist, with 12 ga x 3-5/8" stud G		65	.123		3.96	4.13		8.09	11.25
5340	4" stud G		60	.133		4.06	4.47		8.53	11.95
5350	6" stud G		55	.145		4.48	4.88		9.36	13.10
5360	8" stud G		50	.160		4.90	5.35		10.25	14.40
5430	12" deep joist, with 3-5/8" stud G		65	.123		4.39	4.13		8.52	11.75
5440	4" stud G		60	.133		4.28	4.47		8.75	12.20
5450	6" stud G		55	.145		4.88	4.88		9.76	13.50
5460	8" stud G	↓	50	.160	↓	5.60	5.35		10.95	15.15

05 42 23 – Cold-Formed Metal Roof Joist Framing

05 42 23.05 Framing, Bracing		Crew	Daily Output	Labor-Hours	Unit	Material	Labor	Equipment	Total	Total Incl O&P
0010	**FRAMING, BRACING**									
0015	Made from recycled materials G									
0020	Continuous bracing, per row									
0100	16 ga x 1-1/2" channel thru rafters/trusses @ 16" O.C. G	1 Carp	4.50	1.778	C.L.F.	41	59.50		100.50	145
0120	24" O.C. G		6	1.333		41	44.50		85.50	120
0300	2" x 2" angle x 18 ga, rafters/trusses @ 16" O.C. G		6	1.333		64	44.50		108.50	145
0320	24" O.C. G		8	1		64	33.50		97.50	126
0400	16 ga, rafters/trusses @ 16" O.C. G		4.50	1.778		81.50	59.50		141	190
0420	24" O.C. G	↓	6.50	1.231	↓	81.50	41.50		123	159

05 42 23 – Cold-Formed Metal Roof Joist Framing

05 42 23.10 Framing, Bridging			Crew	Daily Output	Labor-Hours	Unit	2011 Bare Costs Material	Labor	Equipment	Total	Total Incl O&P
0010	**FRAMING, BRIDGING**										
0015	Made from recycled materials	G									
0020	Solid, between rafters w/1-1/4" leg track, per rafter bay										
1200	Rafters 16" O.C., 18 ga x 4" deep	G	1 Carp	60	.133	Ea.	1.25	4.47		5.72	8.90
1210	6" deep	G		57	.140		1.62	4.71		6.33	9.70
1220	8" deep	G		55	.145		2.01	4.88		6.89	10.35
1230	10" deep	G		52	.154		2.51	5.15		7.66	11.40
1240	12" deep	G		50	.160		2.85	5.35		8.20	12.15
2200	24" O.C., 18 ga x 4" deep	G		60	.133		1.81	4.47		6.28	9.50
2210	6" deep	G		57	.140		2.34	4.71		7.05	10.45
2220	8" deep	G		55	.145		2.91	4.88		7.79	11.35
2230	10" deep	G		52	.154		3.63	5.15		8.78	12.65
2240	12" deep	G		50	.160		4.13	5.35		9.48	13.55

05 42 23.50 Framing, Parapets			Crew	Daily Output	Labor-Hours	Unit	Material	Labor	Equipment	Total	Total Incl O&P
0010	**FRAMING, PARAPETS**										
0015	Made from recycled materials	G									
0100	3' high installed on 1st story, 18 ga x 4" wide studs, 12" O.C.	G	2 Carp	100	.160	L.F.	4.56	5.35		9.91	14
0110	16" O.C.	G		150	.107		3.89	3.58		7.47	10.30
0120	24" O.C.	G		200	.080		3.21	2.68		5.89	8
0200	6" wide studs, 12" O.C.	G		100	.160		5.80	5.35		11.15	15.35
0210	16" O.C.	G		150	.107		4.94	3.58		8.52	11.45
0220	24" O.C.	G		200	.080		4.10	2.68		6.78	9
1100	Installed on 2nd story, 18 ga x 4" wide studs, 12" O.C.	G		95	.168		4.56	5.65		10.21	14.45
1110	16" O.C.	G		145	.110		3.89	3.70		7.59	10.50
1120	24" O.C.	G		190	.084		3.21	2.83		6.04	8.25
1200	6" wide studs, 12" O.C.	G		95	.168		5.80	5.65		11.45	15.80
1210	16" O.C.	G		145	.110		4.94	3.70		8.64	11.65
1220	24" O.C.	G		190	.084		4.10	2.83		6.93	9.25
2100	Installed on gable, 18 ga x 4" wide studs, 12" O.C.	G		85	.188		4.56	6.30		10.86	15.55
2110	16" O.C.	G		130	.123		3.89	4.13		8.02	11.20
2120	24" O.C.	G		170	.094		3.21	3.16		6.37	8.85
2200	6" wide studs, 12" O.C.	G		85	.188		5.80	6.30		12.10	16.90
2210	16" O.C.	G		130	.123		4.94	4.13		9.07	12.35
2220	24" O.C.	G		170	.094		4.10	3.16		7.26	9.80

05 42 23.60 Framing, Roof Rafters			Crew	Daily Output	Labor-Hours	Unit	Material	Labor	Equipment	Total	Total Incl O&P
0010	**FRAMING, ROOF RAFTERS**										
0015	Made from recycled materials	G									
0100	Boxed ridge beam, double, 18 ga x 6" deep	G	2 Carp	160	.100	L.F.	4.40	3.36		7.76	10.45
0110	8" deep	G		150	.107		4.84	3.58		8.42	11.35
0120	10" deep	G		140	.114		5.90	3.83		9.73	12.90
0130	12" deep	G		130	.123		6.45	4.13		10.58	13.95
0200	16 ga x 6" deep	G		150	.107		4.96	3.58		8.54	11.45
0210	8" deep	G		140	.114		5.60	3.83		9.43	12.55
0220	10" deep	G		130	.123		6.75	4.13		10.88	14.30
0230	12" deep	G		120	.133		7.30	4.47		11.77	15.55
1100	Rafters, 2" flange, material only, 18 ga x 6" deep	G					1.35			1.35	1.49
1110	8" deep	G					1.59			1.59	1.74
1120	10" deep	G					1.87			1.87	2.06
1130	12" deep	G					2.15			2.15	2.37
1200	16 ga x 6" deep	G					1.65			1.65	1.81
1210	8" deep	G					1.97			1.97	2.17
1220	10" deep	G					2.31			2.31	2.54

05 42 Cold-Formed Metal Joist Framing

05 42 23 – Cold-Formed Metal Roof Joist Framing

05 42 23.60 Framing, Roof Rafters			Crew	Daily Output	Labor-Hours	Unit	2011 Bare Costs Material	Labor	Equipment	Total	Total Incl O&P
1230	12" deep	G				L.F.	2.61			2.61	2.88
2100	Installation only, ordinary rafter to 4:12 pitch, 18 ga x 6" deep		2 Carp	35	.457	Ea.		15.35		15.35	25.50
2110	8" deep			30	.533			17.90		17.90	30
2120	10" deep			25	.640			21.50		21.50	36
2130	12" deep			20	.800			27		27	45
2200	16 ga x 6" deep			30	.533			17.90		17.90	30
2210	8" deep			25	.640			21.50		21.50	36
2220	10" deep			20	.800			27		27	45
2230	12" deep		↓	15	1.067	↓		36		36	60
8100	Add to labor, ordinary rafters on steep roofs							25%			
8110	Dormers & complex roofs							50%			
8200	Hip & valley rafters to 4:12 pitch							25%			
8210	Steep roofs							50%			
8220	Dormers & complex roofs							75%			
8300	Hip & valley jack rafters to 4:12 pitch							50%			
8310	Steep roofs							75%			
8320	Dormers & complex roofs							100%			

05 42 23.70 Framing, Soffits and Canopies			Crew	Daily Output	Labor-Hours	Unit	Material	Labor	Equipment	Total	Total Incl O&P
0010	**FRAMING, SOFFITS & CANOPIES**										
0015	Made from recycled materials	G									
0130	Continuous ledger track @ wall, studs @ 16" O.C., 18 ga x 4" wide	G	2 Carp	535	.030	L.F.	.84	1		1.84	2.60
0140	6" wide	G		500	.032		1.08	1.07		2.15	2.99
0150	8" wide	G		465	.034		1.34	1.15		2.49	3.41
0160	10" wide	G		430	.037		1.67	1.25		2.92	3.93
0230	Studs @ 24" O.C., 18 ga x 4" wide	G		800	.020		.80	.67		1.47	2
0240	6" wide	G		750	.021		1.03	.72		1.75	2.33
0250	8" wide	G		700	.023		1.28	.77		2.05	2.69
0260	10" wide	G	↓	650	.025	↓	1.60	.83		2.43	3.14
1000	Horizontal soffit and canopy members, material only										
1030	1-5/8" flange studs, 18 ga x 4" deep	G				L.F.	1.08			1.08	1.19
1040	6" deep	G					1.36			1.36	1.49
1050	8" deep	G					1.63			1.63	1.80
1140	2" flange joists, 18 ga x 6" deep	G					1.55			1.55	1.70
1150	8" deep	G					1.81			1.81	1.99
1160	10" deep	G				↓	2.14			2.14	2.35
4030	Installation only, 18 ga, 1-5/8" flange x 4" deep		2 Carp	130	.123	Ea.		4.13		4.13	6.90
4040	6" deep			110	.145			4.88		4.88	8.15
4050	8" deep			90	.178			5.95		5.95	10
4140	2" flange, 18 ga x 6" deep			110	.145			4.88		4.88	8.15
4150	8" deep			90	.178			5.95		5.95	10
4160	10" deep		↓	80	.200			6.70		6.70	11.25
6010	Clips to attach facia to rafter tails, 2" x 2" x 18 ga angle	G	1 Carp	120	.067		.75	2.24		2.99	4.57
6020	16 ga angle	G	"	100	.080	↓	.96	2.68		3.64	5.55

05 44 Cold-Formed Metal Trusses

05 44 13 – Cold-Formed Metal Roof Trusses

05 44 13.60 Framing, Roof Trusses		Crew	Daily Output	Labor-Hours	Unit	Material	2011 Bare Costs Labor	Equipment	Total	Total Incl O&P
0010	**FRAMING, ROOF TRUSSES**									
0015	Made from recycled materials G									
0020	Fabrication of trusses on ground, Fink (W) or King Post, to 4:12 pitch									
0120	18 ga x 4" chords, 16' span G	2 Carp	12	1.333	Ea.	50.50	44.50		95	131
0130	20' span G		11	1.455		63	49		112	151
0140	24' span G		11	1.455		75.50	49		124.50	165
0150	28' span G		10	1.600		88	53.50		141.50	187
0160	32' span G		10	1.600		101	53.50		154.50	201
0250	6" chords, 28' span G		9	1.778		111	59.50		170.50	222
0260	32' span G		9	1.778		127	59.50		186.50	239
0270	36' span G		8	2		142	67		209	269
0280	40' span G		8	2		158	67		225	286
1120	5:12 to 8:12 pitch, 18 ga x 4" chords, 16' span G		10	1.600		57.50	53.50		111	154
1130	20' span G		9	1.778		72	59.50		131.50	179
1140	24' span G		9	1.778		86.50	59.50		146	195
1150	28' span G		8	2		101	67		168	223
1160	32' span G		8	2		115	67		182	239
1250	6" chords, 28' span G		7	2.286		127	76.50		203.50	267
1260	32' span G		7	2.286		145	76.50		221.50	287
1270	36' span G		6	2.667		163	89.50		252.50	330
1280	40' span G		6	2.667		181	89.50		270.50	350
2120	9:12 to 12:12 pitch, 18 ga x 4" chords, 16' span G		8	2		72	67		139	191
2130	20' span G		7	2.286		90	76.50		166.50	227
2140	24' span G		7	2.286		108	76.50		184.50	247
2150	28' span G		6	2.667		126	89.50		215.50	289
2160	32' span G		6	2.667		144	89.50		233.50	310
2250	6" chords, 28' span G		5	3.200		158	107		265	355
2260	32' span G		5	3.200		181	107		288	380
2270	36' span G		4	4		203	134		337	450
2280	40' span G		4	4		226	134		360	475
5120	Erection only of roof trusses, to 4:12 pitch, 16' span	F-6	48	.833			26	13.65	39.65	58
5130	20' span		46	.870			27	14.25	41.25	60.50
5140	24' span		44	.909			28.50	14.90	43.40	63.50
5150	28' span		42	.952			29.50	15.60	45.10	66.50
5160	32' span		40	1			31	16.40	47.40	70
5170	36' span		38	1.053			33	17.25	50.25	73.50
5180	40' span		36	1.111			34.50	18.20	52.70	77.50
5220	5:12 to 8:12 pitch, 16' span		42	.952			29.50	15.60	45.10	66.50
5230	20' span		40	1			31	16.40	47.40	70
5240	24' span		38	1.053			33	17.25	50.25	73.50
5250	28' span		36	1.111			34.50	18.20	52.70	77.50
5260	32' span		34	1.176			36.50	19.30	55.80	82
5270	36' span		32	1.250			39	20.50	59.50	87.50
5280	40' span		30	1.333			41.50	22	63.50	93
5320	9:12 to 12:12 pitch, 16' span		36	1.111			34.50	18.20	52.70	77.50
5330	20' span		34	1.176			36.50	19.30	55.80	82
5340	24' span		32	1.250			39	20.50	59.50	87.50
5350	28' span		30	1.333			41.50	22	63.50	93
5360	32' span		28	1.429			44.50	23.50	68	100
5370	36' span		26	1.538			48	25	73	108
5380	40' span		24	1.667			52	27.50	79.50	117

05 51 Metal Stairs

05 51 13 – Metal Pan Stairs

05 51 13.50 Pan Stairs			Crew	Daily Output	Labor-Hours	Unit	Material	2011 Bare Costs Labor	Equipment	Total	Total Incl O&P
0010	**PAN STAIRS**, shop fabricated, steel stringers										
0015	Made from recycled materials	G									
0200	Cement fill metal pan, picket rail, 3′-6″ wide	G	E-4	35	.914	Riser	420	33.50	3.12	456.62	530
0300	4′-0″ wide	G		30	1.067		475	39.50	3.64	518.14	605
0350	Wall rail, both sides, 3′-6″ wide	G		53	.604		320	22.50	2.06	344.56	395
1500	Landing, steel pan, conventional	G		160	.200	S.F.	56	7.40	.68	64.08	76.50
1600	Pre-erected	G		255	.125	″	98.50	4.63	.43	103.56	117
1700	Pre-erected, steel pan tread, 3′-6″ wide, 2 line pipe rail	G	E-2	87	.552	Riser	465	20	18.65	503.65	570

05 51 16 – Metal Floor Plate Stairs

05 51 16.50 Floor Plate Stairs			Crew	Daily Output	Labor-Hours	Unit	Material	Labor	Equipment	Total	Total Incl O&P
0010	**FLOOR PLATE STAIRS**, shop fabricated, steel stringers										
0015	Made from recycled materials										
0400	Cast iron tread and pipe rail, 3′-6″ wide	G	E-4	35	.914	Riser	450	33.50	3.12	486.62	565
0500	Checkered plate tread, industrial, 3′-6″ wide	G		28	1.143		281	42	3.90	326.90	395
0550	Circular, for tanks, 3′-0″ wide	G		33	.970		310	36	3.31	349.31	410
0600	For isolated stairs, add							100%			
0800	Custom steel stairs, 3′-6″ wide, minimum	G	E-4	35	.914		420	33.50	3.12	456.62	530
0810	Average	G		30	1.067		560	39.50	3.64	603.14	695
0900	Maximum	G		20	1.600		700	59	5.45	764.45	890
1100	For 4′ wide stairs, add						5%	5%			
1300	For 5′ wide stairs, add						10%	10%			

05 51 19 – Metal Grating Stairs

05 51 19.50 Grating Stairs			Crew	Daily Output	Labor-Hours	Unit	Material	Labor	Equipment	Total	Total Incl O&P
0010	**GRATING STAIRS**, shop fabricated, steel stringers, safety nosing on treads										
0015	Made from recycled materials	G									
0020	Grating tread and pipe railing, 3′-6″ wide	G	E-4	35	.914	Riser	281	33.50	3.12	317.62	380
0100	4′-0″ wide	G	″	30	1.067	″	365	39.50	3.64	408.14	480

05 51 23 – Metal Fire Escapes

05 51 23.25 Fire Escapes			Crew	Daily Output	Labor-Hours	Unit	Material	Labor	Equipment	Total	Total Incl O&P
0010	**FIRE ESCAPES**, shop fabricated										
0200	2′ wide balcony, 1″ x 1/4″ bars 1-1/2″ O.C.	G	1 Sswk	5	1.600	L.F.	49	58		107	165
0400	1st story cantilevered stair, standard	G		.09	88.889	Ea.	2,050	3,225		5,275	8,425
0700	Platform & fixed stair, 36″ x 40″	G		.17	47.059	Flight	905	1,725		2,630	4,275
0900	For 3′-6″ wide escapes, add to above						100%	150%			

05 51 23.50 Fire Escape Stairs			Crew	Daily Output	Labor-Hours	Unit	Material	Labor	Equipment	Total	Total Incl O&P
0010	**FIRE ESCAPE STAIRS**, shop fabricated										
0020	One story, disappearing, stainless steel	G	2 Sswk	20	.800	V.L.F.	220	29		249	298
0100	Portable ladder					Ea.	120			120	132
1100	Fire escape, galvanized steel, 8′-0″ to 10′-4″ ceiling	G	2 Carp	1	16		1,600	535		2,135	2,650
1110	10′-6″ to 13′-6″ ceiling	G	″	1	16		2,100	535		2,635	3,225

05 51 33 – Metal Ladders

05 51 33.13 Vertical Metal Ladders			Crew	Daily Output	Labor-Hours	Unit	Material	Labor	Equipment	Total	Total Incl O&P
0010	**VERTICAL METAL LADDERS**, shop fabricated										
0015	Made from recycled materials	G									
0020	Steel, 20″ wide, bolted to concrete, with cage	G	E-4	50	.640	V.L.F.	65	23.50	2.18	90.68	119
0100	Without cage	G		85	.376		36	13.90	1.28	51.18	67.50
0300	Aluminum, bolted to concrete, with cage	G		50	.640		111	23.50	2.18	136.68	170
0400	Without cage	G		85	.376		48	13.90	1.28	63.18	81

05 51 Metal Stairs

05 51 33 – Metal Ladders

05 51 33.16 Inclined Metal Ladders			Crew	Daily Output	Labor-Hours	Unit	2011 Bare Costs Material	Labor	Equipment	Total	Total Incl O&P
0010	**INCLINED METAL LADDERS**, shop fabricated										
0015	Made from recycled materials	G									
3900	Industrial ships ladder, steel, 24" W, grating treads, 2 line pipe rail	G	E-4	30	1.067	Riser	187	39.50	3.64	230.14	285
4000	Aluminum	G	"	30	1.067	"	206	39.50	3.64	249.14	305

05 51 33.23 Alternating Tread Ladders

Line	Description		Crew	Daily Output	Labor-Hours	Unit	Material	Labor	Equipment	Total	Total Incl O&P
0010	**ALTERNATING TREAD LADDERS**, shop fabricated										
0015	Made from recycled materials	G									
1350	Alternating tread stair, 56/68°, steel, standard paint color	G	2 Sswk	50	.320	V.L.F.	258	11.65		269.65	305
1360	Non-standard paint color	G		50	.320		292	11.65		303.65	340
1370	Galvanized steel	G		50	.320		300	11.65		311.65	350
1380	Stainless steel	G		50	.320		435	11.65		446.65	495
1390	68°, aluminum	G		50	.320		315	11.65		326.65	370

05 52 Metal Railings

05 52 13 – Pipe and Tube Railings

05 52 13.50 Railings, Pipe

Line	Description		Crew	Daily Output	Labor-Hours	Unit	Material	Labor	Equipment	Total	Total Incl O&P
0010	**RAILINGS, PIPE**, shop fab'd, 3'-6" high, posts @ 5' O.C.										
0015	Made from recycled materials	G									
0020	Aluminum, 2 rail, satin finish, 1-1/4" diameter	G	E-4	160	.200	L.F.	31.50	7.40	.68	39.58	49.50
0030	Clear anodized	G		160	.200		39	7.40	.68	47.08	58
0040	Dark anodized	G		160	.200		44	7.40	.68	52.08	63.50
0080	1-1/2" diameter, satin finish	G		160	.200		37.50	7.40	.68	45.58	56.50
0090	Clear anodized	G		160	.200		42	7.40	.68	50.08	61
0100	Dark anodized	G		160	.200		46.50	7.40	.68	54.58	66
0140	Aluminum, 3 rail, 1-1/4" diam., satin finish	G		137	.234		48	8.60	.80	57.40	70.50
0150	Clear anodized	G		137	.234		60	8.60	.80	69.40	83.50
0160	Dark anodized	G		137	.234		66.50	8.60	.80	75.90	90.50
0200	1-1/2" diameter, satin finish	G		137	.234		57.50	8.60	.80	66.90	81
0210	Clear anodized	G		137	.234		65.50	8.60	.80	74.90	89.50
0220	Dark anodized	G		137	.234		71.50	8.60	.80	80.90	96
0700	Stainless steel, 2 rail, 1-1/4" diam. #4 finish	G		137	.234		94	8.60	.80	103.40	120
0720	High polish	G		137	.234		152	8.60	.80	161.40	184
0740	Mirror polish	G		137	.234		190	8.60	.80	199.40	226
0760	Stainless steel, 3 rail, 1-1/2" diam., #4 finish	G		120	.267		142	9.85	.91	152.76	176
0770	High polish	G		120	.267		235	9.85	.91	245.76	279
0780	Mirror finish	G		120	.267		285	9.85	.91	295.76	335
0900	Wall rail, alum. pipe, 1-1/4" diam., satin finish	G		213	.150		17.95	5.55	.51	24.01	31
0905	Clear anodized	G		213	.150		22	5.55	.51	28.06	35
0910	Dark anodized	G		213	.150		26.50	5.55	.51	32.56	40
0915	1-1/2" diameter, satin finish	G		213	.150		20	5.55	.51	26.06	33
0920	Clear anodized	G		213	.150		25	5.55	.51	31.06	38.50
0925	Dark anodized	G		213	.150		31	5.55	.51	37.06	45
0930	Steel pipe, 1-1/4" diameter, primed	G		213	.150		12.70	5.55	.51	18.76	25
0935	Galvanized	G		213	.150		18.45	5.55	.51	24.51	31.50
0940	1-1/2" diameter	G		176	.182		13.10	6.70	.62	20.42	28
0945	Galvanized	G		213	.150		18.50	5.55	.51	24.56	31.50
0955	Stainless steel pipe, 1-1/2" diam., #4 finish	G		107	.299		75	11.05	1.02	87.07	105
0960	High polish	G		107	.299		153	11.05	1.02	165.07	190
0965	Mirror polish	G		107	.299		181	11.05	1.02	193.07	221
2000	2-line pipe rail (1-1/2" T&B) with 1/2" pickets @ 4-1/2" O.C.,										

05 52 Metal Railings

05 52 13 – Pipe and Tube Railings

05 52 13.50 Railings, Pipe			Crew	Daily Output	Labor-Hours	Unit	Material	2011 Bare Costs Labor	Equipment	Total	Total Incl O&P
2005	attached handrail on brackets										
2010	42" high aluminum, satin finish, straight & level	G	E-4	120	.267	L.F.	175	9.85	.91	185.76	213
2050	42" high steel, primed, straight & level	G	"	120	.267		107	9.85	.91	117.76	138
4000	For curved and level rails, add						10%	10%			
4100	For sloped rails for stairs, add					↓	30%	30%			

05 52 16 – Industrial Railings

05 52 16.50 Railings, Industrial			Crew	Daily Output	Labor-Hours	Unit	Material	Labor	Equipment	Total	Total Incl O&P
0010	**RAILINGS, INDUSTRIAL**, shop fab'd, 3'-6" high, posts @ 5' O.C.										
0020	2 rail, 3'-6" high, 1-1/2" pipe	G	E-4	255	.125	L.F.	23	4.63	.43	28.06	35
0100	2" angle rail	G	"	255	.125		21	4.63	.43	26.06	32.50
0200	For 4" high kick plate, 10 gauge, add	G					4.80			4.80	5.30
0300	1/4" thick, add	G					6.20			6.20	6.80
0500	For curved level rails, add						10%	10%			
0550	For sloped rails for stairs, add					↓	30%	30%			

05 53 Metal Gratings

05 53 13 – Bar Gratings

05 53 13.10 Floor Grating, Aluminum			Crew	Daily Output	Labor-Hours	Unit	Material	Labor	Equipment	Total	Total Incl O&P
0010	**FLOOR GRATING, ALUMINUM**, field fabricated from panels										
0015	Made from recycled materials	G									
0110	Bearing bars @ 1-3/16" O.C., cross bars @ 4" O.C.,										
0111	Up to 300 S.F., 1" x 1/8" bar	G	E-4	900	.036	S.F.	23.50	1.31	.12	24.93	28.50
0112	Over 300 S.F.	G		850	.038		21.50	1.39	.13	23.02	26.50
0113	1-1/4" x 1/8" bar, up to 300 S.F.	G		800	.040		27.50	1.48	.14	29.12	33.50
0114	Over 300 S.F.	G		1000	.032		25	1.18	.11	26.29	30
0122	1-1/4" x 3/16" bar, up to 300 S.F.	G		750	.043		36	1.57	.15	37.72	42.50
0124	Over 300 S.F.	G		1000	.032		32.50	1.18	.11	33.79	38.50
0132	1-1/2" x 3/16" bar, up to 300 S.F.	G		700	.046		41.50	1.69	.16	43.35	49
0134	Over 300 S.F.	G		1000	.032		37.50	1.18	.11	38.79	44
0136	1-3/4" x 3/16" bar, up to 300 S.F.	G		500	.064		50	2.36	.22	52.58	59.50
0138	Over 300 S.F.	G		1000	.032		45.50	1.18	.11	46.79	52.50
0146	2-1/4" x 3/16" bar, up to 300 S.F.	G		600	.053		69	1.97	.18	71.15	79.50
0148	Over 300 S.F.	G		1000	.032		62.50	1.18	.11	63.79	71.50
0162	Cross bars @ 2" O.C., 1" x 1/8", up to 300 S.F.	G		600	.053		30.50	1.97	.18	32.65	37.50
0164	Over 300 S.F.	G		1000	.032		27.50	1.18	.11	28.79	33
0172	1-1/4" x 3/16" bar, up to 300 S.F.	G		600	.053		40.50	1.97	.18	42.65	49
0174	Over 300 S.F.	G		1000	.032		37	1.18	.11	38.29	43
0182	1-1/2" x 3/16" bar, up to 300 S.F.	G		600	.053		45.50	1.97	.18	47.65	54
0184	Over 300 S.F.	G		1000	.032		41.50	1.18	.11	42.79	48
0186	1-3/4" x 3/16" bar, up to 300 S.F.	G		600	.053		55.50	1.97	.18	57.65	65
0188	Over 300 S.F.	G	↓	1000	.032	↓	50.50	1.18	.11	51.79	58
0200	For straight cuts, add					L.F.	3.70			3.70	4.07
0300	For curved cuts, add						4.55			4.55	5
0400	For straight banding, add	G					4.75			4.75	5.25
0500	For curved banding, add	G					5.75			5.75	6.35
0600	For aluminum checkered plate nosings, add	G					6.20			6.20	6.80
0700	For straight toe plate, add	G					9.35			9.35	10.30
0800	For curved toe plate, add	G					10.90			10.90	12
1000	For cast aluminum abrasive nosings, add	G				↓	9.10			9.10	10
1200	Expanded aluminum, .65# per S.F.	G	E-4	1050	.030	S.F.	11.65	1.12	.10	12.87	15.05

05 53 Metal Gratings

05 53 13 – Bar Gratings

05 53 13.10 Floor Grating, Aluminum			Crew	Daily Output	Labor-Hours	Unit	2011 Bare Costs Material	Labor	Equipment	Total	Total Incl O&P
1400	Extruded I bars are 10% less than 3/16″ bars										
1600	Heavy duty, all extruded plank, 3/4″ deep, 1.8 # per S.F.	G	E-4	1100	.029	S.F.	30	1.07	.10	31.17	35
1700	1-1/4″ deep, 2.9# per S.F.	G		1000	.032		35	1.18	.11	36.29	41
1800	1-3/4″ deep, 4.2# per S.F.	G		925	.035		42	1.28	.12	43.40	48.50
1900	2-1/4″ deep, 5.0# per S.F.	G		875	.037		50	1.35	.12	51.47	57.50
2100	For safety serrated surface, add						15%				

05 53 13.70 Floor Grating, Steel			Crew	Daily Output	Labor-Hours	Unit	Material	Labor	Equipment	Total	Total Incl O&P
0010	**FLOOR GRATING, STEEL**, field fabricated from panels										
0015	Made from recycled materials	G									
0300	Platforms, to 12′ high, rectangular	G	E-4	3150	.010	Lb.	3.20	.37	.03	3.60	4.27
0400	Circular	G	″	2300	.014	″	4	.51	.05	4.56	5.45
0410	Painted bearing bars @ 1-3/16″										
0412	Cross bars @ 4″ O.C., 3/4″ x 1/8″ bar, up to 300 S.F.	G	E-2	500	.096	S.F.	7.30	3.52	3.24	14.06	18.15
0414	Over 300 S.F.	G		750	.064		6.65	2.34	2.16	11.15	14.05
0422	1-1/4″ x 3/16″, up to 300 S.F.	G		400	.120		11.20	4.40	4.05	19.65	25
0424	Over 300 S.F.	G		600	.080		10.20	2.93	2.70	15.83	19.60
0432	1-1/2″ x 3/16″, up to 300 S.F.	G		400	.120		12.95	4.40	4.05	21.40	27
0434	Over 300 S.F.	G		600	.080		11.75	2.93	2.70	17.38	21.50
0436	1-3/4″ x 3/16″, up to 300 S.F.	G		400	.120		17.90	4.40	4.05	26.35	32.50
0438	Over 300 S.F.	G		600	.080		16.25	2.93	2.70	21.88	26.50
0452	2-1/4″ x 3/16″, up to 300 S.F.	G		300	.160		21.50	5.85	5.40	32.75	40.50
0454	Over 300 S.F.	G		450	.107		19.55	3.91	3.60	27.06	33
0462	Cross bars @ 2″ O.C., 3/4″ x 1/8″, up to 300 S.F.	G		500	.096		13.65	3.52	3.24	20.41	25
0464	Over 300 S.F.	G		750	.064		11.35	2.34	2.16	15.85	19.25
0472	1-1/4″ x 3/16″, up to 300 S.F.	G		400	.120		17.85	4.40	4.05	26.30	32.50
0474	Over 300 S.F.	G		600	.080		14.85	2.93	2.70	20.48	25
0482	1-1/2″ x 3/16″, up to 300 S.F.	G		400	.120		20	4.40	4.05	28.45	34.50
0484	Over 300 S.F.	G		600	.080		16.75	2.93	2.70	22.38	27
0486	1-3/4″ x 3/16″, up to 300 S.F.	G		400	.120		30.50	4.40	4.05	38.95	46
0488	Over 300 S.F.	G		600	.080		25	2.93	2.70	30.63	36
0502	2-1/4″ x 3/16″, up to 300 S.F.	G		300	.160		29.50	5.85	5.40	40.75	49.50
0504	Over 300 S.F.	G		450	.107		24.50	3.91	3.60	32.01	38.50
0690	For galvanized grating, add						25%				
0800	For straight cuts, add					L.F.	5.35			5.35	5.90
0900	For curved cuts, add						6.75			6.75	7.45
1000	For straight banding, add	G					5.75			5.75	6.35
1100	For curved banding, add	G					7.50			7.50	8.25
1200	For checkered plate nosings, add	G					6.60			6.60	7.25
1300	For straight toe or kick plate, add	G					11.85			11.85	13.05
1400	For curved toe or kick plate, add	G					13.40			13.40	14.75
1500	For abrasive nosings, add	G					8.80			8.80	9.70
1510	For stair treads, see Section 05 55 13.50										
1600	For safety serrated surface, minimum, add						15%				
1700	Maximum, add						25%				
2000	Stainless steel gratings, close spaced, 1″ x 1/8″ bars, up to 300 S.F.	G	E-4	450	.071	S.F.	106	2.62	.24	108.86	122
2100	Standard spacing, 3/4″ x 1/8″ bars	G		500	.064		75.50	2.36	.22	78.08	87.50
2200	1-1/4″ x 3/16″ bars	G		400	.080		116	2.95	.27	119.22	134
2400	Expanded steel grating, at ground, 3.0# per S.F.	G		900	.036		7.30	1.31	.12	8.73	10.70
2500	3.14# per S.F.	G		900	.036		7.85	1.31	.12	9.28	11.30
2600	4.0# per S.F.	G		850	.038		9.70	1.39	.13	11.22	13.45
2650	4.27# per S.F.	G		850	.038		10.40	1.39	.13	11.92	14.25
2700	5.0# per S.F.	G		800	.040		12.35	1.48	.14	13.97	16.55

05 53 Metal Gratings

05 53 13 – Bar Gratings

05 53 13.70 Floor Grating, Steel			Crew	Daily Output	Labor-Hours	Unit	Material	Labor (2011 Bare Costs)	Equipment (2011 Bare Costs)	Total (2011 Bare Costs)	Total Incl O&P
2800	6.25# per S.F.	G	E-4	750	.043	S.F.	16	1.57	.15	17.72	21
2900	7.0# per S.F.	G		700	.046		17.70	1.69	.16	19.55	23
3100	For flattened expanded steel grating, add						8%				
3300	For elevated installation above 15′, add							15%			

05 53 16 – Plank Gratings

05 53 16.50 Grating Planks			Crew	Daily Output	Labor-Hours	Unit	Material	Labor	Equipment	Total	Total Incl O&P
0010	**GRATING PLANKS**, field fabricated from planks										
0020	Aluminum, 9-1/2″ wide, 14 ga., 2″ rib	G	E-4	950	.034	L.F.	28	1.24	.11	29.35	33.50
0200	Galvanized steel, 9-1/2″ wide, 14 ga., 2-1/2″ rib	G		950	.034		15.15	1.24	.11	16.50	19.15
0300	4″ rib	G		950	.034		17.80	1.24	.11	19.15	22
0500	12 gauge, 2-1/2″ rib	G		950	.034		26	1.24	.11	27.35	31
0600	3″ rib	G		950	.034		21	1.24	.11	22.35	25.50
0800	Stainless steel, type 304, 16 ga., 2″ rib	G		950	.034		41	1.24	.11	42.35	47.50
0900	Type 316	G		950	.034		50.50	1.24	.11	51.85	58

05 53 19 – Floor Grating Frame

05 53 19.30 Grating Frame			Crew	Daily Output	Labor-Hours	Unit	Material	Labor	Equipment	Total	Total Incl O&P
0010	**GRATING FRAME**, field fabricated										
0021	Aluminum, for gratings 1″ to 1-1/2″ deep	G	1 Carp	70	.114	L.F.	3.44	3.83		7.27	10.20
0100	For each corner, add	G				Ea.	5.15			5.15	5.70

05 54 Metal Floor Plates

05 54 13 – Floor Plates

05 54 13.20 Checkered Plates			Crew	Daily Output	Labor-Hours	Unit	Material	Labor	Equipment	Total	Total Incl O&P
0010	**CHECKERED PLATES**, steel, field fabricated										
0015	Made from recycled materials	G									
0020	1/4″ & 3/8″, 2000 to 5000 S.F., bolted	G	E-4	2900	.011	Lb.	.66	.41	.04	1.11	1.55
0100	Welded	G		4400	.007	″	.62	.27	.02	.91	1.22
0300	Pit or trench cover and frame, 1/4″ plate, 2′ to 3′ wide	G		100	.320	S.F.	9.10	11.80	1.09	21.99	34
0400	For galvanizing, add	G				Lb.	.29			.29	.32
0500	Platforms, 1/4″ plate, no handrails included, rectangular	G	E-4	4200	.008		3.20	.28	.03	3.51	4.09
0600	Circular	G	″	2500	.013		4	.47	.04	4.51	5.35

05 54 13.70 Trench Covers			Crew	Daily Output	Labor-Hours	Unit	Material	Labor	Equipment	Total	Total Incl O&P
0010	**TRENCH COVERS**, field fabricated										
0021	Cast iron grating with bar stops and angle frame, to 18″ wide	G	1 Carp	20	.400	L.F.	259	13.40		272.40	310
0101	Frame only (both sides of trench), 1″ grating	G		45	.178		1.26	5.95		7.21	11.40
0151	2″ grating	G		35	.229		2.70	7.65		10.35	15.80
0200	Aluminum, stock units, including frames and										
0211	3/8″ plain cover plate, 4″ opening	G	2 Carp	100	.160	L.F.	25.50	5.35		30.85	37
0301	6″ opening	G		90	.178		30	5.95		35.95	43
0401	10″ opening	G		85	.188		37.50	6.30		43.80	52
0501	16″ opening	G		80	.200		50	6.70		56.70	66.50
0700	Add per inch for additional widths to 24″	G					2			2	2.20
0900	For custom fabrication, add						50%				
1100	For 1/4″ plain cover plate, deduct						12%				
1500	For cover recessed for tile, 1/4″ thick, deduct						12%				
1600	3/8″ thick, add						5%				
1800	For checkered plate cover, 1/4″ thick, deduct						12%				
1900	3/8″ thick, add						2%				
2100	For slotted or round holes in cover, 1/4″ thick, add						3%				
2200	3/8″ thick, add						4%				

05 54 Metal Floor Plates

05 54 13 – Floor Plates

05 54 13.70 Trench Covers		Crew	Daily Output	Labor-Hours	Unit	2011 Bare Costs Material	Labor	Equipment	Total	Total Incl O&P
2300	For abrasive cover, add				L.F.	12%				

05 55 Metal Stair Treads and Nosings

05 55 13 – Metal Stair Treads

05 55 13.50 Stair Treads		Crew	Daily Output	Labor-Hours	Unit	Material	Labor	Equipment	Total	Total Incl O&P
0010	**STAIR TREADS**									
0020	Aluminum grating, 3' long, 1-1/2" x 3/16" rect. bars, 6" wide G	1 Sswk	24	.333	Ea.	59	12.15		71.15	88
0100	12" wide G		22	.364		87	13.25		100.25	121
0200	1-1/2" x 3/16" I-bars, 6" wide G		24	.333		49.50	12.15		61.65	77.50
0300	12" wide G		22	.364		80	13.25		93.25	113
0400	For abrasive nosings, add G					15.55			15.55	17.10
0500	For narrow mesh, add					60%				
0600	Stair treads, not incl. stringers, see also Section 03 15 05.85									
0701	Cast aluminum, abrasive, 5/16" thick G	1 Carp	15	.533	Ea.	131	17.90		148.90	174
0801	3/8" thick G		15	.533		140	17.90		157.90	183
0901	1/2" thick G		15	.533		153	17.90		170.90	198
1001	Cast bronze, abrasive, 3/8" thick G		8	1		595	33.50		628.50	710
1101	1/2" thick G		8	1		780	33.50		813.50	915
1201	Cast iron, abrasive, 3/8" thick G		15	.533		116	17.90		133.90	157
1301	1/2" thick G		15	.533		133	17.90		150.90	176
1400	Fiberglass reinforced plastic with safety nosing,									
1501	1-1/2" thick, 12" wide, 24" long	1 Carp	22	.364	Ea.	49.50	12.20		61.70	75
1601	30" long		22	.364		59	12.20		71.20	85.50
1701	36" long		22	.364		69.50	12.20		81.70	97
2000	Steel grating, painted, 3' long, 1-1/4" x 3/16" bars, 6" wide G	1 Sswk	20	.400		48	14.55		62.55	81
2010	12" wide G	"	18	.444		56	16.20		72.20	92.50
2100	Add for abrasive nosing, 3' long, painted G					15.05			15.05	16.55
2200	Galvanized G					20			20	22
2300	Painting 3' treads, in shop, nonstandard paint					2.67			2.67	2.94
2400	Added coats, standard paint					1.14			1.14	1.25
2501	Expanded steel, 2-1/2" deep, 9" x 3' long, 18 gauge G	1 Carp	20	.400		26.50	13.40		39.90	51.50
2601	14 gauge G	"	20	.400		32.50	13.40		45.90	58.50

05 56 Metal Castings

05 56 13 – Metal Construction Castings

05 56 13.50 Construction Castings		Crew	Daily Output	Labor-Hours	Unit	Material	Labor	Equipment	Total	Total Incl O&P
0010	**CONSTRUCTION CASTINGS**									
0020	Manhole covers and frames, see Section 33 44 13.13									
0101	Column bases, cast iron, 16" x 16", approx. 65 lb. G	2 Carp	23	.696	Ea.	138	23.50		161.50	191
0201	32" x 32", approx. 256 lb. G	"	11.50	1.391		515	46.50		561.50	645
0400	Cast aluminum for wood columns, 8" x 8" G	1 Carp	32	.250		41	8.40		49.40	59
0500	12" x 12" G	"	32	.250		89	8.40		97.40	112
0601	Miscellaneous C.I. castings, light sections G	2 Carp	1600	.010	Lb.	2.13	.34		2.47	2.90
1101	Heavy sections G		2100	.008		1.04	.26		1.30	1.57
1301	Special low volume items G		1600	.010		3.70	.34		4.04	4.63
1500	For ductile iron, add					100%				

05 58 Formed Metal Fabrications

05 58 21 – Formed Chain

05 58 21.05 Alloy Steel Chain			Crew	Daily Output	Labor-Hours	Unit	2011 Bare Costs Material	Labor	Equipment	Total	Total Incl O&P
0010	**ALLOY STEEL CHAIN**, Grade 80, for lifting										
0015	Self-colored, cut lengths, 1/4"	G	E-17	4	4	C.L.F.	690	150		840	1,050
0020	3/8"	G		2	8		875	299		1,174	1,525
0030	1/2"	G		1.20	13.333		1,400	500		1,900	2,475
0040	5/8"	G		.72	22.222		2,125	830		2,955	3,925
0050	3/4"	G	E-18	.48	83.333		2,900	3,050	2,150	8,100	11,200
0060	7/8"	G		.40	100		4,850	3,650	2,600	11,100	15,000
0070	1"	G		.35	114		6,550	4,175	2,950	13,675	18,200
0080	1-1/4"	G		.24	166		13,600	6,075	4,325	24,000	31,100
0110	Hook, Grade 80, Clevis slip, 1/4"	G				Ea.	33			33	36.50
0120	3/8"	G					43			43	47.50
0130	1/2"	G					69.50			69.50	76.50
0140	5/8"	G					87.50			87.50	96.50
0150	3/4"	G					143			143	157
0160	Hook, Grade 80, eye/sling w/hammerlock coupling, 15 Ton	G					345			345	380
0170	22 Ton	G					1,375			1,375	1,500
0180	37 Ton	G					2,300			2,300	2,525

05 58 23 – Formed Metal Guards

05 58 23.90 Window Guards			Crew	Daily Output	Labor-Hours	Unit	Material	Labor	Equipment	Total	Total Incl O&P
0010	**WINDOW GUARDS**, shop fabricated										
0015	Expanded metal, steel angle frame, permanent	G	E-4	350	.091	S.F.	19.50	3.37	.31	23.18	28.50
0025	Steel bars, 1/2" x 1/2", spaced 5" O.C.	G	"	290	.110	"	13.50	4.07	.38	17.95	23
0030	Hinge mounted, add	G				Opng.	39			39	43
0040	Removable type, add	G				"	25			25	27.50
0050	For galvanized guards, add					S.F.	35%				
0070	For pivoted or projected type, add						105%	40%			
0100	Mild steel, stock units, economy	G	E-4	405	.079		5.30	2.92	.27	8.49	11.70
0200	Deluxe	G		405	.079		10.85	2.92	.27	14.04	17.80
0400	Woven wire, stock units, 3/8" channel frame, 3' x 5' opening	G		40	.800	Opng.	143	29.50	2.73	175.23	217
0500	4' x 6' opening	G		38	.842		230	31	2.87	263.87	315
0800	Basket guards for above, add	G					196			196	216
1000	Swinging guards for above, add	G					67			67	73.50

05 58 25 – Formed Lamp Posts

05 58 25.40 Lamp Posts			Crew	Daily Output	Labor-Hours	Unit	Material	Labor	Equipment	Total	Total Incl O&P
0010	**LAMP POSTS**										
0020	Aluminum, 7' high, stock units, post only	G	1 Carp	16	.500	Ea.	49	16.80		65.80	82
0100	Mild steel, plain	G	"	16	.500	"	57	16.80		73.80	90.50

05 71 Decorative Metal Stairs

05 71 13 – Fabricated Metal Spiral Stairs

05 71 13.50 Spiral Stairs			Crew	Daily Output	Labor-Hours	Unit	Material	Labor	Equipment	Total	Total Incl O&P
0010	**SPIRAL STAIRS**, shop fabricated										
1810	Spiral aluminum, 5'-0" diameter, stock units	G	E-4	45	.711	Riser	510	26	2.42	538.42	615
1820	Custom units	G		45	.711		965	26	2.42	993.42	1,100
1900	Spiral, cast iron, 4'-0" diameter, ornamental, minimum	G		45	.711		455	26	2.42	483.42	555
1920	Maximum	G		25	1.280		620	47	4.36	671.36	775
2000	Spiral, steel, industrial checkered plate, 4' diameter	G		45	.711		455	26	2.42	483.42	555
2200	Stock units, 6'-0" diameter	G		40	.800		550	29.50	2.73	582.23	665
3110	Spiral steel, stock units, primed, flat metal tread, 3'-6" dia	G	2 Carp	1.60	10	Flight	1,250	335		1,585	1,900
3120	4'-0" dia	G		1.45	11.034		1,425	370		1,795	2,200

05 71 Decorative Metal Stairs

05 71 13 – Fabricated Metal Spiral Stairs

05 71 13.50 Spiral Stairs			Crew	Daily Output	Labor-Hours	Unit	Material	2011 Bare Costs Labor	Equipment	Total	Total Incl O&P
3130	4'-6" dia	G	2 Carp	1.35	11.852	Flight	1,575	400		1,975	2,400
3140	5'-0" dia	G		1.25	12.800		1,700	430		2,130	2,600
3210	Galvanized, 3'-6" dia	G		1.60	10		1,475	335		1,810	2,175
3220	4'-0" dia	G		1.45	11.034		1,700	370		2,070	2,500
3230	4'-6" dia	G		1.35	11.852		1,900	400		2,300	2,750
3240	5'-0" dia	G		1.25	12.800		2,050	430		2,480	2,975
3310	Checkered plate tread, 3'-6" dia	G		1.45	11.034		1,450	370		1,820	2,225
3320	4'-0" dia	G		1.35	11.852		1,650	400		2,050	2,475
3330	4'-6" dia	G		1.25	12.800		1,800	430		2,230	2,700
3340	5'-0" dia	G		1.15	13.913		1,950	465		2,415	2,925
3410	Galvanized, 3'-6" dia	G		1.45	11.034		1,750	370		2,120	2,550
3420	4'-0" dia	G		1.35	11.852		1,975	400		2,375	2,850
3430	4'-6" dia	G		1.25	12.800		2,175	430		2,605	3,100
3440	5'-0" dia	G		1.15	13.913		2,350	465		2,815	3,350
3510	Red oak tread on flat metal, 3'-6" dia			1.35	11.852		2,550	400		2,950	3,500
3520	4'-0" dia			1.25	12.800		2,825	430		3,255	3,825
3530	4'-6" dia			1.15	13.913		3,025	465		3,490	4,125
3540	5'-0" dia		↓	1.05	15.238	↓	3,325	510		3,835	4,500

05 73 Decorative Metal Railings

05 73 16 – Wire Rope Decorative Metal Railings

05 73 16.10 Cable Railings

			Crew	Daily Output	Labor-Hours	Unit	Material	Labor	Equipment	Total	Total Incl O&P
0010	**CABLE RAILINGS**, with 316 stainless steel 1 x 19 cable, 3/16" diameter										
0015	Made from recycled materials	G									
0100	1-3/4" diameter stainless steel posts x 42" high, cables 4" OC	G	2 Sswk	25	.640	L.F.	61.50	23.50		85	113

05 73 23 – Ornamental Railings

05 73 23.50 Railings, Ornamental

			Crew	Daily Output	Labor-Hours	Unit	Material	Labor	Equipment	Total	Total Incl O&P
0010	**RAILINGS, ORNAMENTAL**, 3'-6" high, posts @ 6' O.C.										
0020	Bronze or stainless, hand forged, minimum	G	2 Sswk	24	.667	L.F.	112	24.50		136.50	170
0100	Maximum	G		18	.889		167	32.50		199.50	245
0200	Aluminum, panelized, minimum	G		24	.667		19.85	24.50		44.35	68.50
0300	Maximum	G		18	.889		36	32.50		68.50	101
0400	Wrought iron, hand forged, minimum	G		24	.667		74.50	24.50		99	129
0500	Maximum	G		18	.889		119	32.50		151.50	193
0550	Steel, panelized, minimum	G		24	.667		18.70	24.50		43.20	67
0560	Maximum			18	.889		24	32.50		56.50	87.50
0600	Composite metal/wood/glass, minimum			18	.889		102	32.50		134.50	174
0700	Maximum		↓	12	1.333	↓	204	48.50		252.50	315

05 75 Decorative Formed Metal

05 75 13 – Columns

05 75 13.10 Aluminum Columns			Crew	Daily Output	Labor-Hours	Unit	Material	2011 Bare Costs Labor	Equipment	Total	Total Incl O&P
0010	**ALUMINUM COLUMNS**										
0015	Made from recycled materials	G									
0020	Aluminum, extruded, stock units, no cap or base, 6″ diameter	G	E-4	240	.133	L.F.	9.45	4.92	.45	14.82	20.50
0100	8″ diameter	G		170	.188		12.55	6.95	.64	20.14	28
0200	10″ diameter	G		150	.213		17.15	7.85	.73	25.73	34.50
0300	12″ diameter	G		140	.229		30.50	8.45	.78	39.73	50.50
0400	15″ diameter	G	↓	120	.267	↓	40	9.85	.91	50.76	64
0410	Caps and bases, plain, 6″ diameter	G				Set	22.50			22.50	24.50
0420	8″ diameter	G					26.50			26.50	29.50
0430	10″ diameter	G					37.50			37.50	41
0440	12″ diameter	G					53			53	58.50
0450	15″ diameter	G					102			102	112
0460	Caps, ornamental, minimum	G					278			278	305
0470	Maximum	G				↓	1,350			1,350	1,475
0500	For square columns, add to column prices above					L.F.	50%				
0700	Residential, flat, 8′ high, plain	G	E-4	20	1.600	Ea.	95	59	5.45	159.45	223
0720	Fancy	G		20	1.600		181	59	5.45	245.45	320
0740	Corner type, plain	G		20	1.600		165	59	5.45	229.45	300
0760	Fancy	G	↓	20	1.600	↓	320	59	5.45	384.45	470

05 75 13.20 Columns, Ornamental			Crew	Daily Output	Labor-Hours	Unit	Material	Labor	Equipment	Total	Total Incl O&P
0010	**COLUMNS, ORNAMENTAL**, shop fabricated	R051223-10									
6400	Mild steel, flat, 9″ wide, stock units, painted, plain	G	E-4	160	.200	V.L.F.	8.90	7.40	.68	16.98	24.50
6450	Fancy	G		160	.200		17	7.40	.68	25.08	33.50
6500	Corner columns, painted, plain	G		160	.200		15.50	7.40	.68	23.58	32
6550	Fancy	G	↓	160	.200	↓	30	7.40	.68	38.08	48

Estimating Tips

06 05 00 Common Work Results for Wood, Plastics, and Composites

- Common to any wood-framed structure are the accessory connector items such as screws, nails, adhesives, hangers, connector plates, straps, angles, and hold-downs. For typical wood-framed buildings, such as residential projects, the aggregate total for these items can be significant, especially in areas where seismic loading is a concern. For floor and wall framing, the material cost is based on 10 to 25 lbs. per MBF. Hold-downs, hangers, and other connectors should be taken off by the piece.

  Included with material costs are fasteners for a normal installation. RSMeans engineers use manufacturer's recommendations, written specifications, and/or standard construction practice for size and spacing of fasteners. Prices for various fasteners are shown for informational purposes only. Adjustments should be made if unusual fastening conditions exist.

06 10 00 Carpentry

- Lumber is a traded commodity and therefore sensitive to supply and demand in the marketplace. Even in "budgetary" estimating of wood-framed projects, it is advisable to call local suppliers for the latest market pricing.
- Common quantity units for wood-framed projects are "thousand board feet" (MBF). A board foot is a volume of wood, 1" x 1' x 1', or 144 cubic inches. Board-foot quantities are generally calculated using nominal material dimensions—dressed sizes are ignored. Board foot per lineal foot of any stick of lumber can be calculated by dividing the nominal cross-sectional area by 12. As an example, 2,000 lineal feet of 2 x 12 equates to 4 MBF by dividing the nominal area,2 x 12, by 12, which equals 2, and multiplying by 2,000 to give 4,000 board feet. This simple rule applies to all nominal dimensioned lumber.
- Waste is an issue of concern at the quantity takeoff for any area of construction. Framing lumber is sold in even foot lengths, i.e., 10', 12', 14', 16', and depending on spans, wall heights, and the grade of lumber, waste is inevitable. A rule of thumb for lumber waste is 5%-10% depending on material quality and the complexity of the framing.
- Wood in various forms and shapes is used in many projects, even where the main structural framing is steel, concrete, or masonry. Plywood as a back-up partition material and 2x boards used as blocking and cant strips around roof edges are two common examples. The estimator should ensure that the costs of all wood materials are included in the final estimate.

06 20 00 Finish Carpentry

- It is necessary to consider the grade of workmanship when estimating labor costs for erecting millwork and interior finish. In practice, there are three grades: premium, custom, and economy. The RSMeans daily output for base and case moldings is in the range of 200 to 250 L.F. per carpenter per day. This is appropriate for most average custom-grade projects. For premium projects, an adjustment to productivity of 25%-50% should be made, depending on the complexity of the job.

Reference Numbers

Reference numbers are shown in shaded boxes at the beginning of some major classifications. These numbers refer to related items in the Reference Section. The reference information may be an estimating procedure, an alternate pricing method, or technical information.

Note: Not all subdivisions listed here necessarily appear in this publication.

06 05 05 – Selective Wood and Plastics Demolition

06 05 05.10 Selective Demolition Wood Framing		Crew	Daily Output	Labor-Hours	Unit	Material	2011 Bare Costs Labor	Equipment	Total	Total Incl O&P
0010	SELECTIVE DEMOLITION WOOD FRAMING R024119-10									
0100	Timber connector, nailed, small	1 Clab	96	.083	Ea.		2.20		2.20	3.69
0110	Medium		60	.133			3.53		3.53	5.90
0120	Large		48	.167			4.41		4.41	7.40
0130	Bolted, small		48	.167			4.41		4.41	7.40
0140	Medium		32	.250			6.60		6.60	11.10
0150	Large		24	.333			8.80		8.80	14.75
2958	Beams, 2" x 6"	2 Clab	1100	.015	L.F.		.38		.38	.64
2960	2" x 8"		825	.019			.51		.51	.86
2965	2" x 10"		665	.024			.64		.64	1.07
2970	2" x 12"		550	.029			.77		.77	1.29
2972	2" x 14"		470	.034			.90		.90	1.51
2975	4" x 8"	B-1	413	.058			1.58		1.58	2.64
2980	4" x 10"		330	.073			1.97		1.97	3.30
2985	4" x 12"		275	.087			2.37		2.37	3.96
3000	6" x 8"		275	.087			2.37		2.37	3.96
3040	6" x 10"		220	.109			2.96		2.96	4.95
3080	6" x 12"		185	.130			3.52		3.52	5.90
3120	8" x 12"		140	.171			4.65		4.65	7.80
3160	10" x 12"		110	.218			5.90		5.90	9.90
3162	Alternate pricing method		1.10	21.818	M.B.F.		590		590	990
3170	Blocking, in 16" OC wall framing, 2" x 4"	1 Clab	600	.013	L.F.		.35		.35	.59
3172	2" x 6"		400	.020			.53		.53	.89
3174	In 24" OC wall framing, 2" x 4"		600	.013			.35		.35	.59
3176	2" x 6"		400	.020			.53		.53	.89
3178	Alt method, wood blocking removal from wood framing		.40	20	M.B.F.		530		530	885
3179	Wood blocking removal from steel framing		.36	22.222	"		590		590	985
3180	Bracing, let in, 1" x 3", studs 16" OC		1050	.008	L.F.		.20		.20	.34
3181	Studs 24" OC		1080	.007			.20		.20	.33
3182	1" x 4", studs 16" OC		1050	.008			.20		.20	.34
3183	Studs 24" OC		1080	.007			.20		.20	.33
3184	1" x 6", studs 16" OC		1050	.008			.20		.20	.34
3185	Studs 24" OC		1080	.007			.20		.20	.33
3186	2" x 3", studs 16" OC		800	.010			.26		.26	.44
3187	Studs 24" OC		830	.010			.26		.26	.43
3188	2" x 4", studs 16" OC		800	.010			.26		.26	.44
3189	Studs 24" OC		830	.010			.26		.26	.43
3190	2" x 6", studs 16" OC		800	.010			.26		.26	.44
3191	Studs 24" OC		830	.010			.26		.26	.43
3192	2" x 8", studs 16" OC		800	.010			.26		.26	.44
3193	Studs 24" OC		830	.010			.26		.26	.43
3194	"T" shaped metal bracing, studs at 16" OC		1060	.008			.20		.20	.33
3195	Studs at 24" OC		1200	.007			.18		.18	.30
3196	Metal straps, studs at 16" OC		1200	.007			.18		.18	.30
3197	Studs at 24" OC		1240	.006			.17		.17	.29
3200	Columns, round, 8' to 14' tall		40	.200	Ea.		5.30		5.30	8.85
3202	Dimensional lumber sizes	2 Clab	1.10	14.545	M.B.F.		385		385	645
3250	Blocking, between joists	1 Clab	320	.025	Ea.		.66		.66	1.11
3252	Bridging, metal strap, between joists		320	.025	Pr.		.66		.66	1.11
3254	Wood, between joists		320	.025	"		.66		.66	1.11
3260	Door buck, studs, header & access., 8' high 2" x 4" wall, 3' wide		32	.250	Ea.		6.60		6.60	11.10
3261	4' wide		32	.250			6.60		6.60	11.10
3262	5' wide		32	.250			6.60		6.60	11.10

06 05 05 – Selective Wood and Plastics Demolition

06 05 05.10 Selective Demolition Wood Framing		Crew	Daily Output	Labor-Hours	Unit	Material	2011 Bare Costs Labor	Equipment	Total	Total Incl O&P
3263	6′ wide	1 Clab	32	.250	Ea.		6.60		6.60	11.10
3264	8′ wide		30	.267			7.05		7.05	11.80
3265	10′ wide		30	.267			7.05		7.05	11.80
3266	12′ wide		30	.267			7.05		7.05	11.80
3267	2″ x 6″ wall, 3′ wide		32	.250			6.60		6.60	11.10
3268	4′ wide		32	.250			6.60		6.60	11.10
3269	5′ wide		32	.250			6.60		6.60	11.10
3270	6′ wide		32	.250			6.60		6.60	11.10
3271	8′ wide		30	.267			7.05		7.05	11.80
3272	10′ wide		30	.267			7.05		7.05	11.80
3273	12′ wide		30	.267			7.05		7.05	11.80
3274	Window buck, studs, header & access, 8′ high 2″ x 4″ wall, 2′ wide		24	.333			8.80		8.80	14.75
3275	3′ wide		24	.333			8.80		8.80	14.75
3276	4′ wide		24	.333			8.80		8.80	14.75
3277	5′ wide		24	.333			8.80		8.80	14.75
3278	6′ wide		24	.333			8.80		8.80	14.75
3279	7′ wide		24	.333			8.80		8.80	14.75
3280	8′ wide		22	.364			9.60		9.60	16.10
3281	10′ wide		22	.364			9.60		9.60	16.10
3282	12′ wide		22	.364			9.60		9.60	16.10
3283	2″ x 6″ wall, 2′ wide		24	.333			8.80		8.80	14.75
3284	3′ wide		24	.333			8.80		8.80	14.75
3285	4′ wide		24	.333			8.80		8.80	14.75
3286	5′ wide		24	.333			8.80		8.80	14.75
3287	6′ wide		24	.333			8.80		8.80	14.75
3288	7′ wide		24	.333			8.80		8.80	14.75
3289	8′ wide		22	.364			9.60		9.60	16.10
3290	10′ wide		22	.364			9.60		9.60	16.10
3291	12′ wide		22	.364			9.60		9.60	16.10
3400	Fascia boards, 1″ x 6″		500	.016	L.F.		.42		.42	.71
3440	1″ x 8″		450	.018			.47		.47	.79
3480	1″ x 10″		400	.020			.53		.53	.89
3490	2″ x 6″		450	.018			.47		.47	.79
3500	2″ x 8″		400	.020			.53		.53	.89
3510	2″ x 10″		350	.023			.60		.60	1.01
3610	Furring, on wood walls or ceiling		4000	.002	S.F.		.05		.05	.09
3620	On masonry or concrete walls or ceiling		1200	.007	″		.18		.18	.30
3800	Headers over openings, 2 @ 2″ x 6″		110	.073	L.F.		1.92		1.92	3.22
3840	2 @ 2″ x 8″		100	.080			2.12		2.12	3.54
3880	2 @ 2″ x 10″		90	.089			2.35		2.35	3.94
3885	Alternate pricing method		.26	30.651	M.B.F.		810		810	1,350
3920	Joists, 1″ x 4″		1250	.006	L.F.		.17		.17	.28
3930	1″ x 6″		1135	.007			.19		.19	.31
3940	1″ x 8″		1000	.008			.21		.21	.35
3950	1″ x 10″		895	.009			.24		.24	.40
3960	1″ x 12″		765	.010			.28		.28	.46
4200	2″ x 4″	2 Clab	1000	.016			.42		.42	.71
4230	2″ x 6″		970	.016			.44		.44	.73
4240	2″ x 8″		940	.017			.45		.45	.75
4250	2″ x 10″		910	.018			.47		.47	.78
4280	2″ x 12″		880	.018			.48		.48	.81
4281	2″ x 14″		850	.019			.50		.50	.83
4282	Composite joists, 9-1/2″		960	.017			.44		.44	.74

06 05 05 – Selective Wood and Plastics Demolition

06 05 05.10 Selective Demolition Wood Framing		Crew	Daily Output	Labor-Hours	Unit	2011 Bare Costs Material	Labor	Equipment	Total	Total Incl O&P
4283	11-7/8"	2 Clab	930	.017	L.F.		.45		.45	.76
4284	14"		897	.018			.47		.47	.79
4285	16"		865	.019			.49		.49	.82
4290	Wood joists, alternate pricing method		1.50	10.667	M.B.F.		282		282	475
4500	Open web joist, 12" deep		500	.032	L.F.		.85		.85	1.42
4505	14" deep		475	.034			.89		.89	1.49
4510	16" deep		450	.036			.94		.94	1.58
4520	18" deep		425	.038			1		1	1.67
4530	24" deep		400	.040			1.06		1.06	1.77
4550	Ledger strips, 1" x 2"	1 Clab	1200	.007			.18		.18	.30
4560	1" x 3"		1200	.007			.18		.18	.30
4570	1" x 4"		1200	.007			.18		.18	.30
4580	2" x 2"		1100	.007			.19		.19	.32
4590	2" x 4"		1000	.008			.21		.21	.35
4600	2" x 6"		1000	.008			.21		.21	.35
4601	2" x 8 or 2" x 10"		800	.010			.26		.26	.44
4602	4" x 6"		600	.013			.35		.35	.59
4604	4" x 8"		450	.018			.47		.47	.79
5400	Posts, 4" x 4"	2 Clab	800	.020			.53		.53	.89
5405	4" x 6"		550	.029			.77		.77	1.29
5410	4" x 8"		440	.036			.96		.96	1.61
5425	4" x 10"		390	.041			1.09		1.09	1.82
5430	4" x 12"		350	.046			1.21		1.21	2.03
5440	6" x 6"		400	.040			1.06		1.06	1.77
5445	6" x 8"		350	.046			1.21		1.21	2.03
5450	6" x 10"		320	.050			1.32		1.32	2.22
5455	6" x 12"		290	.055			1.46		1.46	2.44
5480	8" x 8"		300	.053			1.41		1.41	2.36
5500	10" x 10"		240	.067			1.76		1.76	2.95
5660	Tongue and groove floor planks		2	8	M.B.F.		212		212	355
5682	Rafters, ordinary, 16" OC, 2" x 4"		880	.018	S.F.		.48		.48	.81
5683	2" x 6"		840	.019			.50		.50	.84
5684	2" x 8"		820	.020			.52		.52	.86
5685	2" x 10"		820	.020			.52		.52	.86
5686	2" x 12"		810	.020			.52		.52	.87
5687	24" OC, 2" x 4"		1170	.014			.36		.36	.61
5688	2" x 6"		1117	.014			.38		.38	.63
5689	2" x 8"		1091	.015			.39		.39	.65
5690	2" x 10"		1091	.015			.39		.39	.65
5691	2" x 12"		1077	.015			.39		.39	.66
5795	Rafters, ordinary, 2" x 4" (alternate method)		862	.019	L.F.		.49		.49	.82
5800	2" x 6" (alternate method)		850	.019			.50		.50	.83
5840	2" x 8" (alternate method)		837	.019			.51		.51	.85
5855	2" x 10" (alternate method)		825	.019			.51		.51	.86
5865	2" x 12" (alternate method)		812	.020			.52		.52	.87
5870	Sill plate, 2" x 4"	1 Clab	1170	.007			.18		.18	.30
5871	2" x 6"		780	.010			.27		.27	.45
5872	2" x 8"		586	.014			.36		.36	.60
5873	Alternate pricing method		.78	10.256	M.B.F.		271		271	455
5885	Ridge board, 1" x 4"	2 Clab	900	.018	L.F.		.47		.47	.79
5886	1" x 6"		875	.018			.48		.48	.81
5887	1" x 8"		850	.019			.50		.50	.83
5888	1" x 10"		825	.019			.51		.51	.86

06 05 05 – Selective Wood and Plastics Demolition

06 05 05.10 Selective Demolition Wood Framing		Crew	Daily Output	Labor-Hours	Unit	Material	2011 Bare Costs Labor	Equipment	Total	Total Incl O&P
5889	1" x 12"	2 Clab	800	.020	L.F.		.53		.53	.89
5890	2" x 4"		900	.018			.47		.47	.79
5892	2" x 6"		875	.018			.48		.48	.81
5894	2" x 8"		850	.019			.50		.50	.83
5896	2" x 10"		825	.019			.51		.51	.86
5898	2" x 12"		800	.020			.53		.53	.89
6050	Rafter tie, 1" x 4"		1250	.013			.34		.34	.57
6052	1" x 6"		1135	.014			.37		.37	.62
6054	2" x 4"		1000	.016			.42		.42	.71
6056	2" x 6"		970	.016			.44		.44	.73
6070	Sleepers, on concrete, 1" x 2"	1 Clab	4700	.002			.05		.05	.08
6075	1" x 3"		4000	.002			.05		.05	.09
6080	2" x 4"		3000	.003			.07		.07	.12
6085	2" x 6"		2600	.003			.08		.08	.14
6086	Sheathing from roof, 5/16"	2 Clab	1600	.010	S.F.		.26		.26	.44
6088	3/8"		1525	.010			.28		.28	.46
6090	1/2"		1400	.011			.30		.30	.51
6092	5/8"		1300	.012			.33		.33	.55
6094	3/4"		1200	.013			.35		.35	.59
6096	Board sheathing from roof		1400	.011			.30		.30	.51
6100	Sheathing, from walls, 1/4"		1200	.013			.35		.35	.59
6110	5/16"		1175	.014			.36		.36	.60
6120	3/8"		1150	.014			.37		.37	.62
6130	1/2"		1125	.014			.38		.38	.63
6140	5/8"		1100	.015			.38		.38	.64
6150	3/4"		1075	.015			.39		.39	.66
6152	Board sheathing from walls		1500	.011			.28		.28	.47
6158	Subfloor, with boards		1050	.015			.40		.40	.68
6160	Plywood, 1/2" thick		768	.021			.55		.55	.92
6162	5/8" thick		760	.021			.56		.56	.93
6164	3/4" thick		750	.021			.56		.56	.94
6165	1-1/8" thick		720	.022			.59		.59	.98
6166	Underlayment, particle board, 3/8" thick	1 Clab	780	.010			.27		.27	.45
6168	1/2" thick		768	.010			.28		.28	.46
6170	5/8" thick		760	.011			.28		.28	.47
6172	3/4" thick		750	.011			.28		.28	.47
6200	Stairs and stringers, minimum	2 Clab	40	.400	Riser		10.60		10.60	17.70
6240	Maximum	"	26	.615	"		16.30		16.30	27.50
6300	Components, tread	1 Clab	110	.073	Ea.		1.92		1.92	3.22
6320	Riser		80	.100	"		2.65		2.65	4.43
6390	Stringer, 2" x 10"		260	.031	L.F.		.81		.81	1.36
6400	2" x 12"		260	.031			.81		.81	1.36
6410	3" x 10"		250	.032			.85		.85	1.42
6420	3" x 12"		250	.032			.85		.85	1.42
6590	Wood studs, 2" x 3"	2 Clab	3076	.005			.14		.14	.23
6600	2" x 4"		2000	.008			.21		.21	.35
6640	2" x 6"		1600	.010			.26		.26	.44
6720	Wall framing, including studs, plates and blocking, 2" x 4"	1 Clab	600	.013	S.F.		.35		.35	.59
6740	2" x 6"		480	.017	"		.44		.44	.74
6750	Headers, 2" x 4"		1125	.007	L.F.		.19		.19	.32
6755	2" x 6"		1125	.007			.19		.19	.32
6760	2" x 8"		1050	.008			.20		.20	.34
6765	2" x 10"		1050	.008			.20		.20	.34

06 05 05 – Selective Wood and Plastics Demolition

06 05 05.10 Selective Demolition Wood Framing

		Crew	Daily Output	Labor-Hours	Unit	2011 Bare Costs Material	Labor	Equipment	Total	Total Incl O&P
6770	2″ x 12″	1 Clab	1000	.008	L.F.		.21		.21	.35
6780	4″ x 10″		525	.015			.40		.40	.68
6785	4″ x 12″		500	.016			.42		.42	.71
6790	6″ x 8″		560	.014			.38		.38	.63
6795	6″ x 10″		525	.015			.40		.40	.68
6797	6″ x 12″		500	.016			.42		.42	.71
8000	Soffit, T & G wood		520	.015	S.F.		.41		.41	.68
8010	Hardboard, vinyl or aluminum		640	.013			.33		.33	.55
8030	Plywood	2 Carp	315	.051			1.70		1.70	2.85
9500	See Section 02 41 19.23 for rubbish handling									

06 05 05.20 Selective Demolition Millwork and Trim

		Crew	Daily Output	Labor-Hours	Unit	Material	Labor	Equipment	Total	Total Incl O&P
0010	**SELECTIVE DEMOLITION MILLWORK AND TRIM** R024119-10									
1000	Cabinets, wood, base cabinets, per L.F.	2 Clab	80	.200	L.F.		5.30		5.30	8.85
1020	Wall cabinets, per L.F.	″	80	.200	″		5.30		5.30	8.85
1060	Remove and reset, base cabinets	2 Carp	18	.889	Ea.		30		30	50
1070	Wall cabinets	″	20	.800	″		27		27	45
1100	Steel, painted, base cabinets	2 Clab	60	.267	L.F.		7.05		7.05	11.80
1120	Wall cabinets		60	.267	″		7.05		7.05	11.80
1200	Casework, large area		320	.050	S.F.		1.32		1.32	2.22
1220	Selective		200	.080	″		2.12		2.12	3.54
1500	Counter top, minimum		200	.080	L.F.		2.12		2.12	3.54
1510	Maximum		120	.133			3.53		3.53	5.90
1550	Remove and reset, minimum	2 Carp	50	.320			10.75		10.75	17.95
1560	Maximum	″	40	.400			13.40		13.40	22.50
2000	Paneling, 4′ x 8′ sheets	2 Clab	2000	.008	S.F.		.21		.21	.35
2100	Boards, 1″ x 4″		700	.023			.60		.60	1.01
2120	1″ x 6″		750	.021			.56		.56	.94
2140	1″ x 8″		800	.020			.53		.53	.89
3000	Trim, baseboard, to 6″ wide		1200	.013	L.F.		.35		.35	.59
3040	Greater than 6″ and up to 12″ wide		1000	.016			.42		.42	.71
3080	Remove and reset, minimum	2 Carp	400	.040			1.34		1.34	2.25
3090	Maximum	″	300	.053			1.79		1.79	2.99
3100	Ceiling trim	2 Clab	1000	.016			.42		.42	.71
3120	Chair rail		1200	.013			.35		.35	.59
3140	Railings with balusters		240	.067			1.76		1.76	2.95
3160	Wainscoting		700	.023	S.F.		.60		.60	1.01

06 05 23 – Wood, Plastic, and Composite Fastenings

06 05 23.10 Nails

		Crew	Daily Output	Labor-Hours	Unit	Material	Labor	Equipment	Total	Total Incl O&P
0010	**NAILS**, material only, based upon 50# box purchase									
0020	Copper nails, plain				Lb.	10			10	11
0400	Stainless steel, plain					7.80			7.80	8.60
0500	Box, 3d to 20d, bright					1.25			1.25	1.38
0520	Galvanized					1.77			1.77	1.95
0600	Common, 3d to 60d, plain					1.30			1.30	1.43
0700	Galvanized					1.56			1.56	1.72
0800	Aluminum					5.80			5.80	6.40
1000	Annular or spiral thread, 4d to 60d, plain					1.80			1.80	1.98
1200	Galvanized					3			3	3.30
1400	Drywall nails, plain					3.59			3.59	3.95
1600	Galvanized					2.19			2.19	2.41
1800	Finish nails, 4d to 10d, plain					1.36			1.36	1.50
2000	Galvanized					1.79			1.79	1.97

06 05 23 – Wood, Plastic, and Composite Fastenings

06 05 23.10 Nails		Crew	Daily Output	Labor-Hours	Unit	Material (2011 Bare Costs)	Labor (2011 Bare Costs)	Equipment (2011 Bare Costs)	Total (2011 Bare Costs)	Total Incl O&P
2100	Aluminum				Lb.	5.80			5.80	6.40
2300	Flooring nails, hardened steel, 2d to 10d, plain					3.10			3.10	3.41
2400	Galvanized					4			4	4.40
2500	Gypsum lath nails, 1-1/8", 13 ga. flathead, blued					3.59			3.59	3.95
2600	Masonry nails, hardened steel, 3/4" to 3" long, plain					2.20			2.20	2.42
2700	Galvanized					3.75			3.75	4.13
2900	Roofing nails, threaded, galvanized					1.63			1.63	1.79
3100	Aluminum					4.80			4.80	5.30
3300	Compressed lead head, threaded, galvanized					2.47			2.47	2.72
3600	Siding nails, plain shank, galvanized					2			2	2.20
3800	Aluminum					5.80			5.80	6.40
5000	Add to prices above for cement coating					.11			.11	.12
5200	Zinc or tin plating					.14			.14	.15
5500	Vinyl coated sinkers, 8d to 16d				↓	.65			.65	.72

06 05 23.40 Sheet Metal Screws		Crew	Daily Output	Labor-Hours	Unit	Material	Labor	Equipment	Total	Total Incl O&P
0010	**SHEET METAL SCREWS**									
0020	Steel, standard, #8 x 3/4", plain				C	2.92			2.92	3.21
0100	Galvanized					2.92			2.92	3.21
0300	#10 x 1", plain					4.03			4.03	4.43
0400	Galvanized					4.03			4.03	4.43
0600	With washers, #14 x 1", plain					12.35			12.35	13.60
0700	Galvanized					12.35			12.35	13.60
0900	#14 x 2", plain					16.90			16.90	18.60
1000	Galvanized					16.90			16.90	18.60
1500	Self-drilling, with washers, (pinch point) #8 x 3/4", plain					7.80			7.80	8.55
1600	Galvanized					7.80			7.80	8.55
1800	#10 x 3/4", plain					8.90			8.90	9.80
1900	Galvanized					8.90			8.90	9.80
3000	Stainless steel w/aluminum or neoprene washers, #14 x 1", plain					30			30	33
3100	#14 x 2", plain				↓	40			40	44

06 05 23.50 Wood Screws		Crew	Daily Output	Labor-Hours	Unit	Material	Labor	Equipment	Total	Total Incl O&P
0010	**WOOD SCREWS**									
0020	#8 x 1" long, steel				C	3.41			3.41	3.75
0100	Brass					10.50			10.50	11.55
0200	#8, 2" long, steel					5.25			5.25	5.80
0300	Brass					19.95			19.95	22
0400	#10, 1" long, steel					3.65			3.65	4.02
0500	Brass					14.15			14.15	15.60
0600	#10, 2" long, steel					5.85			5.85	6.45
0700	Brass					24.50			24.50	27
0800	#10, 3" long, steel					9.80			9.80	10.75
1000	#12, 2" long, steel					8.05			8.05	8.85
1100	Brass					29.50			29.50	32
1500	#12, 3" long, steel					11.65			11.65	12.80
2000	#12, 4" long, steel				↓	21.50			21.50	23.50

06 05 23.60 Timber Connectors		Crew	Daily Output	Labor-Hours	Unit	Material	Labor	Equipment	Total	Total Incl O&P
0010	**TIMBER CONNECTORS**									
0020	Add up cost of each part for total cost of connection									
0100	Connector plates, steel, with bolts, straight	2 Carp	75	.213	Ea.	28	7.15		35.15	42.50
0110	Tee, 7 gauge		50	.320		32	10.75		42.75	53
0120	T- Strap, 14 gauge, 12" x 8" x 2"		50	.320		32	10.75		42.75	53
0150	Anchor plates, 7 ga, 9" x 7"	↓	75	.213	↓	28	7.15		35.15	42.50

06 05 23 – Wood, Plastic, and Composite Fastenings

06 05 23.60 Timber Connectors		Crew	Daily Output	Labor-Hours	Unit	Material	2011 Bare Costs Labor	Equipment	Total	Total Incl O&P
0200	Bolts, machine, sq. hd. with nut & washer, 1/2" diameter, 4" long	1 Carp	140	.057	Ea.	.78	1.92		2.70	4.07
0300	7-1/2" long		130	.062		1.41	2.06		3.47	5
0500	3/4" diameter, 7-1/2" long		130	.062		3.35	2.06		5.41	7.15
0610	Machine bolts, w/nut, washer, 3/4" diam., 15" L, HD's & beam hangers		95	.084		6.25	2.83		9.08	11.60
0800	Drilling bolt holes in timber, 1/2" diameter		450	.018	Inch		.60		.60	1
0900	1" diameter		350	.023	"		.77		.77	1.28
1100	Framing anchor, angle, 3" x 3" x 1-1/2", 12 ga		175	.046	Ea.	2.40	1.53		3.93	5.20
1150	Framing anchors, 18 gauge, 4-1/2" x 2-3/4"		175	.046		2.40	1.53		3.93	5.20
1160	Framing anchors, 18 gauge, 4-1/2" x 3"		175	.046		2.40	1.53		3.93	5.20
1170	Clip anchors plates, 18 gauge, 12" x 1-1/8"		175	.046		2.40	1.53		3.93	5.20
1250	Holdowns, 3 gauge base, 10 gauge body		8	1		23	33.50		56.50	81
1260	Holdowns, 7 gauge 11-1/16" x 3-1/4"		8	1		23	33.50		56.50	81
1270	Holdowns, 7 gauge 14-3/8" x 3-1/8"		8	1		23	33.50		56.50	81
1275	Holdowns, 12 gauge 8" x 2-1/2"		8	1		23	33.50		56.50	81
1300	Joist and beam hangers, 18 ga. galv., for 2" x 4" joist		175	.046		.67	1.53		2.20	3.31
1400	2" x 6" to 2" x 10" joist		165	.048		1.28	1.63		2.91	4.13
1600	16 ga. galv., 3" x 6" to 3" x 10" joist		160	.050		2.72	1.68		4.40	5.80
1700	3" x 10" to 3" x 14" joist		160	.050		4.52	1.68		6.20	7.80
1800	4" x 6" to 4" x 10" joist		155	.052		2.82	1.73		4.55	6
1900	4" x 10" to 4" x 14" joist		155	.052		4.62	1.73		6.35	8
2000	Two-2" x 6" to two-2" x 10" joists		150	.053		3.85	1.79		5.64	7.25
2100	Two-2" x 10" to two-2" x 14" joists		150	.053		4.30	1.79		6.09	7.70
2300	3/16" thick, 6" x 8" joist		145	.055		59.50	1.85		61.35	68.50
2400	6" x 10" joist		140	.057		62	1.92		63.92	71.50
2500	6" x 12" joist		135	.059		64.50	1.99		66.49	74.50
2700	1/4" thick, 6" x 14" joist		130	.062		67.50	2.06		69.56	77.50
2900	Plywood clips, extruded aluminum H clip, for 3/4" panels					.22			.22	.24
3000	Galvanized 18 ga. back-up clip					.17			.17	.19
3200	Post framing, 16 ga. galv. for 4" x 4" base, 2 piece	1 Carp	130	.062		15.05	2.06		17.11	20
3300	Cap		130	.062		21	2.06		23.06	26.50
3500	Rafter anchors, 18 ga. galv., 1-1/2" wide, 5-1/4" long		145	.055		.45	1.85		2.30	3.60
3600	10-3/4" long		145	.055		1.34	1.85		3.19	4.57
3800	Shear plates, 2-5/8" diameter		120	.067		2.19	2.24		4.43	6.15
3900	4" diameter		115	.070		5.20	2.33		7.53	9.60
4000	Sill anchors, embedded in concrete or block, 25-1/2" long		115	.070		11.65	2.33		13.98	16.75
4100	Spike grids, 3" x 6"		120	.067		.88	2.24		3.12	4.71
4400	Split rings, 2-1/2" diameter		120	.067		1.80	2.24		4.04	5.70
4500	4" diameter		110	.073		2.73	2.44		5.17	7.10
4550	Tie plate, 20 gauge, 7" x 3 1/8"		110	.073		2.73	2.44		5.17	7.10
4560	Tie plate, 20 gauge, 5" x 4 1/8"		110	.073		2.73	2.44		5.17	7.10
4575	Twist straps, 18 gauge, 12" x 1 1/4"		110	.073		2.73	2.44		5.17	7.10
4580	Twist straps, 18 gauge, 16" x 1 1/4"		110	.073		2.73	2.44		5.17	7.10
4600	Strap ties, 20 ga., 2 -1/16" wide, 12 13/16" long		180	.044		.87	1.49		2.36	3.46
4700	Strap ties, 16 ga., 1-3/8" wide, 12" long		180	.044		.87	1.49		2.36	3.46
4800	21-5/8" x 1-1/4"		160	.050		2.72	1.68		4.40	5.80
5000	Toothed rings, 2-5/8" or 4" diameter		90	.089		1.63	2.98		4.61	6.80
5200	Truss plates, nailed, 20 gauge, up to 32' span		17	.471	Truss	11.85	15.80		27.65	39.50
5400	Washers, 2" x 2" x 1/8"				Ea.	.37			.37	.41
5500	3" x 3" x 3/16"				"	.98			.98	1.08
6000	Angles and gussets, painted									
6012	7 ga., 3-1/4" x 3-1/4" x 2-1/2" long	1 Carp	1.90	4.211	C	1,025	141		1,166	1,350
6014	3-1/4" x 3-1/4" x 5" long		1.90	4.211		1,975	141		2,116	2,400
6016	3-1/4" x 3-1/4" x 7-1/2" long		1.85	4.324		3,750	145		3,895	4,375

06 05 23 – Wood, Plastic, and Composite Fastenings

06 05 23.60 Timber Connectors		Crew	Daily Output	Labor-Hours	Unit	Material	2011 Bare Costs Labor	Equipment	Total	Total Incl O&P
6018	5-3/4" x 5-3/4" x 2-1/2" long	1 Carp	1.85	4.324	C	2,425	145		2,570	2,900
6020	5-3/4" x 5-3/4" x 5" long		1.85	4.324		3,850	145		3,995	4,475
6022	5-3/4" x 5-3/4" x 7-1/2" long		1.80	4.444		5,675	149		5,824	6,500
6024	3 ga., 4-1/4" x 4-1/4" x 3" long		1.85	4.324		2,575	145		2,720	3,075
6026	4-1/4" x 4-1/4" x 6" long		1.85	4.324		5,525	145		5,670	6,325
6028	4-1/4" x 4-1/4" x 9" long		1.80	4.444		6,200	149		6,349	7,075
6030	7-1/4" x 7-1/4" x 3" long		1.80	4.444		4,450	149		4,599	5,125
6032	7-1/4" x 7-1/4" x 6" long		1.80	4.444		6,000	149		6,149	6,850
6034	7-1/4" x 7-1/4" x 9" long	↓	1.75	4.571	↓	13,400	153		13,553	15,100
6036	Gussets									
6038	7 ga., 8-1/8" x 8-1/8" x 2-3/4" long	1 Carp	1.80	4.444	C	4,250	149		4,399	4,925
6040	3 ga., 9-3/4" x 9-3/4" x 3-1/4" long	"	1.80	4.444	"	5,875	149		6,024	6,725
6101	Beam hangers, polymer painted									
6102	Bolted, 3 ga., (W x H x L)									
6104	3-1/4" x 9" x 12" top flange	1 Carp	1	8	C	18,000	268		18,268	20,300
6106	5-1/4" x 9" x 12" top flange		1	8		18,800	268		19,068	21,200
6108	5-1/4" x 11" x 11-3/4" top flange		1	8		21,400	268		21,668	24,000
6110	6-7/8" x 9" x 12" top flange		1	8		19,500	268		19,768	21,900
6112	6-7/8" x 11" x 13-1/2" top flange		1	8		22,500	268		22,768	25,200
6114	8-7/8" x 11" x 15-1/2" top flange	↓	1	8	↓	24,000	268		24,268	26,900
6116	Nailed, 3 ga., (W x H x L)									
6118	3-1/4" x 10-1/2" x 10" top flange	1 Carp	1.80	4.444	C	18,800	149		18,949	21,000
6120	3-1/4" x 10-1/2" x 12" top flange		1.80	4.444		18,800	149		18,949	21,000
6122	5-1/4" x 9-1/2" x 10" top flange		1.80	4.444		19,200	149		19,349	21,400
6124	5-1/4" x 9-1/2" x 12" top flange		1.80	4.444		19,200	149		19,349	21,400
6128	6-7/8" x 8-1/2" x 12" top flange	↓	1.80	4.444	↓	19,700	149		19,849	21,900
6134	Saddle hangers, glu-lam (W x H x L)									
6136	3-1/4" x 10-1/2" x 5-1/4" x 6" saddle	1 Carp	.50	16	C	14,100	535		14,635	16,400
6138	3-1/4" x 10-1/2" x 6-7/8" x 6" saddle		.50	16		14,800	535		15,335	17,200
6140	3-1/4" x 10-1/2" x 8-7/8" x 6" saddle		.50	16		15,600	535		16,135	18,000
6142	3-1/4" x 19-1/2" x 5-1/4" x 10-1/8" saddle		.40	20		14,100	670		14,770	16,600
6144	3-1/4" x 19-1/2" x 6-7/8" x 10-1/8" saddle		.40	20		14,800	670		15,470	17,400
6146	3-1/4" x 19-1/2" x 8-7/8" x 10-1/8" saddle		.40	20		15,600	670		16,270	18,200
6148	5-1/4" x 9-1/2" x 5-1/4" x 12" saddle		.50	16		16,800	535		17,335	19,400
6150	5-1/4" x 9-1/2" x 6-7/8" x 9" saddle		.50	16		18,400	535		18,935	21,100
6152	5-1/4" x 10-1/2" x spec x 12" saddle		.50	16		20,000	535		20,535	22,900
6154	5-1/4" x 18" x 5-1/4" x 12-1/8" saddle		.40	20		16,800	670		17,470	19,600
6156	5-1/4" x 18" x 6-7/8" x 12-1/8" saddle		.40	20		18,400	670		19,070	21,300
6158	5-1/4" x 18" x spec x 12-1/8" saddle		.40	20		20,000	670		20,670	23,100
6160	6-7/8" x 8-1/2" x 6-7/8" x 12" saddle		.50	16		20,100	535		20,635	23,000
6162	6-7/8" x 8-1/2" x 8-7/8" x 12" saddle		.50	16		20,800	535		21,335	23,700
6164	6-7/8" x 10-1/2" x spec x 12" saddle		.50	16		20,100	535		20,635	23,000
6166	6-7/8" x 18" x 6-7/8" x 13-3/4" saddle		.40	20		20,100	670		20,770	23,200
6168	6-7/8" x 18" x 8-7/8" x 13-3/4" saddle		.40	20		20,800	670		21,470	23,900
6170	6-7/8" x 18" x spec x 13-3/4" saddle		.40	20		22,400	670		23,070	25,700
6172	8-7/8" x 18" x spec x 15-3/4" saddle	↓	.40	20	↓	35,000	670		35,670	39,600
6201	Beam and purlin hangers, galvanized, 12 ga.									
6202	Purlin or joist size, 3" x 8"	1 Carp	1.70	4.706	C	1,875	158		2,033	2,325
6204	3" x 10"		1.70	4.706		2,000	158		2,158	2,475
6206	3" x 12"		1.65	4.848		2,300	163		2,463	2,800
6208	3" x 14"		1.65	4.848		2,450	163		2,613	2,975
6210	3" x 16"		1.65	4.848		2,600	163		2,763	3,125
6212	4" x 8"	↓	1.65	4.848	↓	1,875	163		2,038	2,325

06 05 Common Work Results for Wood, Plastics, and Composites

06 05 23 – Wood, Plastic, and Composite Fastenings

06 05 23.60 Timber Connectors		Crew	Daily Output	Labor-Hours	Unit	2011 Bare Costs Material	Labor	Equipment	Total	Total Incl O&P
6214	4" x 10"	1 Carp	1.65	4.848	C	2,025	163		2,188	2,500
6216	4" x 12"		1.60	5		2,400	168		2,568	2,900
6218	4" x 14"		1.60	5		2,550	168		2,718	3,075
6220	4" x 16"		1.60	5		2,675	168		2,843	3,225
6222	6" x 8"		1.60	5		2,400	168		2,568	2,925
6224	6" x 10"		1.55	5.161		2,450	173		2,623	3,000
6226	6" x 12"		1.55	5.161		4,200	173		4,373	4,925
6228	6" x 14"		1.50	5.333		4,450	179		4,629	5,200
6230	6" x 16"		1.50	5.333		4,725	179		4,904	5,475
6250	Beam seats									
6252	Beam size, 5-1/4" wide									
6254	5" x 7" x 1/4"	1 Carp	1.80	4.444	C	7,100	149		7,249	8,050
6256	6" x 7" x 3/8"		1.80	4.444		7,950	149		8,099	9,000
6258	7" x 7" x 3/8"		1.80	4.444		8,500	149		8,649	9,600
6260	8" x 7" x 3/8"		1.80	4.444		10,000	149		10,149	11,300
6262	Beam size, 6-7/8" wide									
6264	5" x 9" x 1/4"	1 Carp	1.80	4.444	C	8,475	149		8,624	9,575
6266	6" x 9" x 3/8"		1.80	4.444		11,100	149		11,249	12,500
6268	7" x 9" x 3/8"		1.80	4.444		11,200	149		11,349	12,600
6270	8" x 9" x 3/8"		1.80	4.444		13,200	149		13,349	14,900
6272	Special beams, over 6-7/8" wide									
6274	5" x 10" x 3/8"	1 Carp	1.80	4.444	C	11,500	149		11,649	12,900
6276	6" x 10" x 3/8"		1.80	4.444		13,400	149		13,549	15,000
6278	7" x 10" x 3/8"		1.80	4.444		14,000	149		14,149	15,700
6280	8" x 10" x 3/8"		1.75	4.571		15,000	153		15,153	16,800
6282	5-1/4" x 12" x 5/16"		1.75	4.571		11,600	153		11,753	13,100
6284	6-1/2" x 12" x 3/8"		1.75	4.571		19,200	153		19,353	21,500
6286	5-1/4" x 16" x 5/16"		1.70	4.706		17,100	158		17,258	19,100
6288	6-1/2" x 16" x 3/8"		1.70	4.706		22,300	158		22,458	24,800
6290	5-1/4" x 20" x 5/16"		1.70	4.706		20,100	158		20,258	22,400
6292	6-1/2" x 20" x 3/8"		1.65	4.848		26,200	163		26,363	29,100
6300	Column bases									
6302	4 x 4, 16 ga.	1 Carp	1.80	4.444	C	715	149		864	1,025
6306	7 ga.		1.80	4.444		2,625	149		2,774	3,125
6308	4 x 6, 16 ga.		1.80	4.444		1,675	149		1,824	2,075
6312	7 ga.		1.80	4.444		2,725	149		2,874	3,250
6314	6 x 6, 16 ga.		1.75	4.571		1,900	153		2,053	2,325
6318	7 ga.		1.75	4.571		3,725	153		3,878	4,350
6320	6 x 8, 7 ga.		1.70	4.706		2,900	158		3,058	3,475
6322	6 x 10, 7 ga.		1.70	4.706		3,125	158		3,283	3,700
6324	6 x 12, 7 ga.		1.70	4.706		3,375	158		3,533	4,000
6326	8 x 8, 7 ga.		1.65	4.848		5,700	163		5,863	6,550
6330	8 x 10, 7 ga.		1.65	4.848		6,825	163		6,988	7,800
6332	8 x 12, 7 ga.		1.60	5		7,425	168		7,593	8,425
6334	10 x 10, 3 ga.		1.60	5		7,575	168		7,743	8,600
6336	10 x 12, 3 ga.		1.60	5		8,700	168		8,868	9,850
6338	12 x 12, 3 ga.		1.55	5.161		9,450	173		9,623	10,700
6350	Column caps, painted, 3 ga.									
6352	3-1/4" x 3-5/8"	1 Carp	1.80	4.444	C	9,600	149		9,749	10,900
6354	3-1/4" x 5-1/2"		1.80	4.444		9,600	149		9,749	10,900
6356	3-5/8" x 3-5/8"		1.80	4.444		7,850	149		7,999	8,900
6358	3-5/8" x 5-1/2"		1.80	4.444		7,850	149		7,999	8,900
6360	5-1/4" x 5-1/2"		1.75	4.571		10,300	153		10,453	11,600

06 05 23 – Wood, Plastic, and Composite Fastenings

06 05 23.60 Timber Connectors		Crew	Daily Output	Labor-Hours	Unit	Material	2011 Bare Costs Labor	Equipment	Total	Total Incl O&P
6362	5-1/4" x 7-1/2"	1 Carp	1.75	4.571	C	10,300	153		10,453	11,600
6364	5-1/2" x 3-5/8"		1.75	4.571		11,100	153		11,253	12,500
6366	5-1/2" x 5-1/2"		1.75	4.571		11,100	153		11,253	12,500
6368	5-1/2" x 7-1/2"		1.70	4.706		11,100	158		11,258	12,500
6370	6-7/8" x 5-1/2"		1.70	4.706		11,600	158		11,758	13,000
6372	6-7/8" x 6-7/8"		1.70	4.706		11,600	158		11,758	13,000
6374	6-7/8" x 7-1/2"		1.70	4.706		11,600	158		11,758	13,000
6376	7-1/2" x 5-1/2"		1.65	4.848		12,100	163		12,263	13,600
6378	7-1/2" x 7-1/2"		1.65	4.848		12,100	163		12,263	13,600
6380	8-7/8" x 5-1/2"		1.60	5		12,800	168		12,968	14,300
6382	8-7/8" x 7-1/2"		1.60	5		12,800	168		12,968	14,300
6384	9-1/2" x 5-1/2"	↓	1.60	5	↓	17,300	168		17,468	19,300
6400	Floor tie anchors, polymer paint									
6402	10 ga., 3" x 37-1/2"	1 Carp	1.80	4.444	C	4,600	149		4,749	5,300
6404	3-1/2" x 45-1/2"		1.75	4.571		4,825	153		4,978	5,550
6406	3 ga., 3-1/2" x 56"	↓	1.70	4.706	↓	8,500	158		8,658	9,625
6410	Girder hangers									
6412	6" wall thickness, 4" x 6"	1 Carp	1.80	4.444	C	2,575	149		2,724	3,100
6414	4" x 8"		1.80	4.444		2,875	149		3,024	3,425
6416	8" wall thickness, 4" x 6"		1.80	4.444		3,000	149		3,149	3,525
6418	4" x 8"	↓	1.80	4.444	↓	2,975	149		3,124	3,525
6420	Hinge connections, polymer painted									
6422	3/4" thick top plate									
6424	5-1/4" x 12" w/5" x 5" top	1 Carp	1	8	C	33,000	268		33,268	36,800
6426	5-1/4" x 15" w/6" x 6" top		.80	10		35,100	335		35,435	39,200
6428	5-1/4" x 18" w/7" x 7" top		.70	11.429		36,900	385		37,285	41,100
6430	5-1/4" x 26" w/9" x 9" top	↓	.60	13.333	↓	39,200	445		39,645	44,000
6432	1" thick top plate									
6434	6-7/8" x 14" w/5" x 5" top	1 Carp	.80	10	C	40,200	335		40,535	44,800
6436	6-7/8" x 17" w/6" x 6" top		.80	10		44,600	335		44,935	49,700
6438	6-7/8" x 21" w/7" x 7" top		.70	11.429		48,700	385		49,085	54,000
6440	6-7/8" x 31" w/9" x 9" top	↓	.60	13.333	↓	53,500	445		53,945	59,500
6442	1-1/4" thick top plate									
6444	8-7/8" x 16" w/5" x 5" top	1 Carp	.60	13.333	C	50,000	445		50,445	56,000
6446	8-7/8" x 21" w/6" x 6" top		.50	16		55,000	535		55,535	61,500
6448	8-7/8" x 26" w/7" x 7" top		.40	20		62,500	670		63,170	69,500
6450	8-7/8" x 39" w/9" x 9" top	↓	.30	26.667	↓	78,000	895		78,895	87,000
6460	Holddowns									
6462	Embedded along edge									
6464	26" long, 12 ga.	1 Carp	.90	8.889	C	1,250	298		1,548	1,875
6466	35" long, 12 ga.		.85	9.412		1,675	315		1,990	2,350
6468	35" long, 10 ga.	↓	.85	9.412	↓	1,325	315		1,640	1,975
6470	Embedded away from edge									
6472	Medium duty, 12 ga.									
6474	18-1/2" long	1 Carp	.95	8.421	C	760	283		1,043	1,300
6476	23-3/4" long		.90	8.889		885	298		1,183	1,475
6478	28" long		.85	9.412		905	315		1,220	1,525
6480	35" long	↓	.85	9.412	↓	1,250	315		1,565	1,900
6482	Heavy duty, 10 ga.									
6484	28" long	1 Carp	.85	9.412	C	1,625	315		1,940	2,300
6486	35" long	"	.85	9.412	"	1,775	315		2,090	2,475
6490	Surface mounted (W x H)									
6492	2-1/2" x 5-3/4", 7 ga.	1 Carp	1	8	C	1,850	268		2,118	2,475

06 05 23 – Wood, Plastic, and Composite Fastenings

06 05 23.60 Timber Connectors		Crew	Daily Output	Labor-Hours	Unit	Material	2011 Bare Costs Labor	Equipment	Total	Total Incl O&P
6494	2-1/2" x 8", 12 ga.	1 Carp	1	8	C	1,125	268		1,393	1,700
6496	2-7/8" x 6-3/8", 7 ga.		1	8		4,075	268		4,343	4,925
6498	2-7/8" x 12-1/2", 3 ga.		1	8		4,175	268		4,443	5,050
6500	3-3/16" x 9-3/8", 10 ga.		1	8		2,800	268		3,068	3,525
6502	3-1/2" x 11-5/8", 3 ga.		1	8		5,100	268		5,368	6,050
6504	3-1/2" x 14-3/4", 3 ga.		1	8		3,400	268		3,668	4,175
6506	3-1/2" x 16-1/2", 3 ga.		1	8		7,775	268		8,043	9,000
6508	3-1/2" x 20-1/2", 3 ga.		.90	8.889		7,950	298		8,248	9,250
6510	3-1/2" x 24-1/2", 3 ga.		.90	8.889		10,100	298		10,398	11,600
6512	4-1/4" x 20-3/4", 3 ga.	↓	.90	8.889	↓	6,850	298		7,148	8,025
6520	Joist hangers									
6522	Sloped, field adjustable, 18 ga.									
6524	2" x 6"	1 Carp	1.65	4.848	C	500	163		663	820
6526	2" x 8"		1.65	4.848		900	163		1,063	1,250
6528	2" x 10" and up		1.65	4.848		1,500	163		1,663	1,925
6530	3" x 10" and up		1.60	5		1,125	168		1,293	1,500
6532	4" x 10" and up	↓	1.55	5.161	↓	1,350	173		1,523	1,800
6536	Skewed 45°, 16 ga.									
6538	2" x 4"	1 Carp	1.75	4.571	C	800	153		953	1,125
6540	2" x 6" or 2" x 8"		1.65	4.848		815	163		978	1,175
6542	2" x 10" or 2" x 12"		1.65	4.848		935	163		1,098	1,300
6544	2" x 14" or 2" x 16"		1.60	5		1,675	168		1,843	2,100
6546	(2) 2" x 6" or (2) 2" x 8"		1.60	5		1,500	168		1,668	1,925
6548	(2) 2" x 10" or (2) 2" x 12"		1.55	5.161		1,625	173		1,798	2,075
6550	(2) 2" x 14" or (2) 2" x 16"		1.50	5.333		2,550	179		2,729	3,100
6552	4" x 6" or 4" x 8"		1.60	5		1,250	168		1,418	1,675
6554	4" x 10" or 4" x 12"		1.55	5.161		1,475	173		1,648	1,925
6556	4" x 14" or 4" x 16"	↓	1.55	5.161	↓	2,275	173		2,448	2,800
6560	Skewed 45°, 14 ga.									
6562	(2) 2" x 6" or (2) 2" x 8"	1 Carp	1.60	5	C	1,725	168		1,893	2,175
6564	(2) 2" x 10" or (2) 2" x 12"		1.55	5.161		2,400	173		2,573	2,950
6566	(2) 2" x 14" or (2) 2" x 16"		1.50	5.333		3,525	179		3,704	4,175
6568	4" x 6" or 4" x 8"		1.60	5		2,075	168		2,243	2,550
6570	4" x 10" or 4" x 12"		1.55	5.161		2,175	173		2,348	2,700
6572	4" x 14" or 4" x 16"	↓	1.55	5.161	↓	2,925	173		3,098	3,525
6590	Joist hangers, heavy duty 12 ga., galvanized									
6592	2" x 4"	1 Carp	1.75	4.571	C	1,175	153		1,328	1,525
6594	2" x 6"		1.65	4.848		1,275	163		1,438	1,675
6595	2" x 6", 16 gauge		1.65	4.848		1,275	163		1,438	1,675
6596	2" x 8"		1.65	4.848		1,950	163		2,113	2,400
6597	2" x 8", 16 gauge		1.65	4.848		1,950	163		2,113	2,400
6598	2" x 10"		1.65	4.848		2,000	163		2,163	2,475
6600	2" x 12"		1.65	4.848		2,450	163		2,613	2,975
6602	2" x 14"		1.65	4.848		2,550	163		2,713	3,075
6604	2" x 16"		1.65	4.848		2,700	163		2,863	3,250
6606	3" x 4"		1.65	4.848		1,625	163		1,788	2,050
6608	3" x 6"		1.65	4.848		2,175	163		2,338	2,675
6610	3" x 8"		1.65	4.848		2,200	163		2,363	2,700
6612	3" x 10"		1.60	5		2,550	168		2,718	3,100
6614	3" x 12"		1.60	5		3,075	168		3,243	3,650
6616	3" x 14"		1.60	5		3,600	168		3,768	4,225
6618	3" x 16"		1.60	5		3,950	168		4,118	4,625
6620	(2) 2" x 4"	↓	1.75	4.571	↓	1,900	153		2,053	2,350

06 05 23 – Wood, Plastic, and Composite Fastenings

06 05 23.60 Timber Connectors		Crew	Daily Output	Labor-Hours	Unit	Material	2011 Bare Costs Labor	Equipment	Total	Total Incl O&P
6622	(2) 2" x 6"	1 Carp	1.60	5	C	2,300	168		2,468	2,800
6624	(2) 2" x 8"		1.60	5		2,350	168		2,518	2,850
6626	(2) 2" x 10"		1.55	5.161		2,550	173		2,723	3,100
6628	(2) 2" x 12"		1.55	5.161		3,250	173		3,423	3,875
6630	(2) 2" x 14"		1.50	5.333		3,275	179		3,454	3,900
6632	(2) 2" x 16"		1.50	5.333		3,300	179		3,479	3,950
6634	4" x 4"		1.65	4.848		1,375	163		1,538	1,800
6636	4" x 6"		1.60	5		1,525	168		1,693	1,950
6638	4" x 8"		1.60	5		1,750	168		1,918	2,200
6640	4" x 10"		1.55	5.161		2,150	173		2,323	2,675
6642	4" x 12"		1.55	5.161		2,300	173		2,473	2,825
6644	4" x 14"		1.55	5.161		2,750	173		2,923	3,325
6646	4" x 16"		1.55	5.161		3,000	173		3,173	3,600
6648	(3) 2" x 10"		1.50	5.333		3,325	179		3,504	3,950
6650	(3) 2" x 12"		1.50	5.333		3,700	179		3,879	4,375
6652	(3) 2" x 14"		1.45	5.517		3,975	185		4,160	4,675
6654	(3) 2" x 16"		1.45	5.517		4,050	185		4,235	4,750
6656	6" x 6"		1.60	5		1,775	168		1,943	2,250
6658	6" x 8"		1.60	5		1,850	168		2,018	2,300
6660	6" x 10"		1.55	5.161		2,200	173		2,373	2,725
6662	6" x 12"		1.55	5.161		2,500	173		2,673	3,050
6664	6" x 14"		1.50	5.333		3,150	179		3,329	3,750
6666	6" x 16"	↓	1.50	5.333	↓	3,700	179		3,879	4,375
6690	Knee braces, galvanized, 12 ga.									
6692	Beam depth, 10" x 15" x 5' long	1 Carp	1.80	4.444	C	4,800	149		4,949	5,550
6694	15" x 22-1/2" x 7' long		1.70	4.706		5,525	158		5,683	6,350
6696	22-1/2" x 28-1/2" x 8' long		1.60	5		5,925	168		6,093	6,800
6698	28-1/2" x 36" x 10' long		1.55	5.161		6,175	173		6,348	7,100
6700	36" x 42" x 12' long	↓	1.50	5.333	↓	6,825	179		7,004	7,800
6710	Mudsill anchors									
6714	2" x 4" or 3" x 4"	1 Carp	115	.070	C	1,225	2.33		1,227.33	1,350
6716	2" x 6" or 3" x 6"		115	.070		1,225	2.33		1,227.33	1,350
6718	Block wall, 13-1/4" long		115	.070		850	2.33		852.33	940
6720	21-1/4" long	↓	115	.070	↓	127	2.33		129.33	143
6730	Post bases, 12 ga. galvanized									
6732	Adjustable, 3-9/16" x 3-9/16"	1 Carp	1.30	6.154	C	980	206		1,186	1,425
6734	3-9/16" x 5-1/2"		1.30	6.154		1,275	206		1,481	1,750
6736	4" x 4"		1.30	6.154		715	206		921	1,125
6738	4" x 6"		1.30	6.154		1,675	206		1,881	2,175
6740	5-1/2" x 5-1/2"		1.30	6.154		2,475	206		2,681	3,075
6742	6" x 6"		1.30	6.154		1,900	206		2,106	2,425
6744	Elevated, 3-9/16" x 3-1/4"		1.30	6.154		1,075	206		1,281	1,525
6746	5-1/2" x 3-5/16"		1.30	6.154		1,525	206		1,731	2,025
6748	5-1/2" x 5"		1.30	6.154		2,275	206		2,481	2,850
6750	Regular, 3-9/16" x 3-3/8"		1.30	6.154		815	206		1,021	1,250
6752	4" x 3-3/8"		1.30	6.154		1,150	206		1,356	1,600
6754	18 gauge, 5-1/4" x 3-1/8"		1.30	6.154		1,200	206		1,406	1,675
6755	5-1/2" x 3-3/8"		1.30	6.154		1,200	206		1,406	1,675
6756	5-1/2" x 5-3/8"		1.30	6.154		1,725	206		1,931	2,250
6758	6" x 3-3/8"		1.30	6.154		2,075	206		2,281	2,625
6760	6" x 5-3/8"	↓	1.30	6.154	↓	2,300	206		2,506	2,900
6762	Post combination cap/bases									
6764	3-9/16" x 3-9/16"	1 Carp	1.20	6.667	C	465	224		689	885

06 05 Common Work Results for Wood, Plastics, and Composites

06 05 23 – Wood, Plastic, and Composite Fastenings

06 05 23.60 Timber Connectors		Crew	Daily Output	Labor-Hours	Unit	2011 Bare Costs Material	Labor	Equipment	Total	Total Incl O&P
6766	3-9/16" x 5-1/2"	1 Carp	1.20	6.667	C	1,050	224		1,274	1,525
6768	4" x 4"		1.20	6.667		2,100	224		2,324	2,675
6770	5-1/2" x 5-1/2"		1.20	6.667		1,150	224		1,374	1,650
6772	6" x 6"		1.20	6.667		3,825	224		4,049	4,600
6774	7-1/2" x 7-1/2"		1.20	6.667		4,200	224		4,424	5,000
6776	8" x 8"		1.20	6.667		4,350	224		4,574	5,150
6790	Post-beam connection caps									
6792	Beam size 3-9/16"									
6794	12 ga. post, 4" x 4"	1 Carp	1	8	C	2,650	268		2,918	3,375
6796	4" x 6"		1	8		3,550	268		3,818	4,350
6798	4" x 8"		1	8		5,225	268		5,493	6,200
6800	16 ga. post, 4" x 4"		1	8		1,200	268		1,468	1,775
6802	4" x 6"		1	8		2,000	268		2,268	2,650
6804	4" x 8"		1	8		3,375	268		3,643	4,150
6805	18 ga. post, 2-7/8" x 3"		1	8		3,375	268		3,643	4,150
6806	Beam size 5-1/2"									
6808	12 ga. post, 6" x 4"	1 Carp	1	8	C	3,475	268		3,743	4,275
6810	6" x 6"		1	8		5,475	268		5,743	6,475
6812	6" x 8"		1	8		3,775	268		4,043	4,600
6816	16 ga. post, 6" x 4"		1	8		1,875	268		2,143	2,525
6818	6" x 6"		1	8		2,000	268		2,268	2,650
6820	Beam size 7-1/2"									
6822	12 ga. post, 8" x 4"	1 Carp	1	8	C	4,800	268		5,068	5,750
6824	8" x 6"		1	8		5,050	268		5,318	6,000
6826	8" x 8"		1	8		7,575	268		7,843	8,800
6840	Purlin anchors, embedded									
6842	Heavy duty, 10 ga.									
6844	Straight, 28" long	1 Carp	1.60	5	C	1,325	168		1,493	1,725
6846	35" long		1.50	5.333		1,625	179		1,804	2,100
6848	Twisted, 28" long		1.60	5		1,325	168		1,493	1,725
6850	35" long		1.50	5.333		1,625	179		1,804	2,100
6852	Regular duty, 12 ga.									
6854	Straight, 18-1/2" long	1 Carp	1.80	4.444	C	755	149		904	1,075
6856	23-3/4" long		1.70	4.706		950	158		1,108	1,325
6858	29" long		1.60	5		970	168		1,138	1,350
6860	35" long		1.50	5.333		1,325	179		1,504	1,775
6862	Twisted, 18" long		1.80	4.444		755	149		904	1,075
6866	28" long		1.60	5		900	168		1,068	1,275
6868	35" long		1.50	5.333		1,325	179		1,504	1,775
6870	Straight, plastic coated									
6872	23-1/2" long	1 Carp	1.60	5	C	1,850	168		2,018	2,325
6874	26-7/8" long		1.60	5		2,175	168		2,343	2,675
6876	32-1/2" long		1.50	5.333		2,325	179		2,504	2,850
6878	35-7/8" long		1.50	5.333		2,425	179		2,604	2,950
6890	Purlin hangers, painted									
6892	12 ga., 2" x 6"	1 Carp	1.80	4.444	C	1,700	149		1,849	2,125
6894	2" x 8"		1.80	4.444		1,850	149		1,999	2,300
6896	2" x 10"		1.80	4.444		2,000	149		2,149	2,450
6898	2" x 12"		1.75	4.571		2,150	153		2,303	2,625
6900	2" x 14"		1.75	4.571		2,300	153		2,453	2,775
6902	2" x 16"		1.75	4.571		2,450	153		2,603	2,925
6904	3" x 6"		1.70	4.706		1,725	158		1,883	2,175
6906	3" x 8"		1.70	4.706		1,875	158		2,033	2,325

06 05 23 – Wood, Plastic, and Composite Fastenings

06 05 23.60 Timber Connectors		Crew	Daily Output	Labor-Hours	Unit	Material	Labor	Equipment	Total	Total Incl O&P
						2011 Bare Costs				
6908	3″ x 10″	1 Carp	1.70	4.706	C	2,000	158		2,158	2,475
6910	3″ x 12″		1.65	4.848		2,300	163		2,463	2,800
6912	3″ x 14″		1.65	4.848		2,450	163		2,613	2,975
6914	3″ x 16″		1.65	4.848		2,600	163		2,763	3,125
6916	4″ x 6″		1.65	4.848		1,725	163		1,888	2,175
6918	4″ x 8″		1.65	4.848		1,875	163		2,038	2,325
6920	4″ x 10″		1.65	4.848		2,025	163		2,188	2,500
6922	4″ x 12″		1.60	5		2,400	168		2,568	2,900
6924	4″ x 14″		1.60	5		2,550	168		2,718	3,075
6926	4″ x 16″		1.60	5		2,675	168		2,843	3,225
6928	6″ x 6″		1.60	5		2,275	168		2,443	2,775
6930	6″ x 8″		1.60	5		2,400	168		2,568	2,925
6932	6″ x 10″		1.55	5.161		2,450	173		2,623	3,000
6934	double 2″ x 6″		1.70	4.706		1,875	158		2,033	2,325
6936	double 2″ x 8″		1.70	4.706		2,000	158		2,158	2,475
6938	double 2″ x 10″		1.70	4.706		2,150	158		2,308	2,650
6940	double 2″ x 12″		1.65	4.848		2,300	163		2,463	2,800
6942	double 2″ x 14″		1.65	4.848		2,450	163		2,613	2,975
6944	double 2″ x 16″		1.65	4.848		2,600	163		2,763	3,125
6960	11 ga., 4″ x 6″		1.65	4.848		3,400	163		3,563	4,025
6962	4″ x 8″		1.65	4.848		3,675	163		3,838	4,300
6964	4″ x 10″		1.65	4.848		3,925	163		4,088	4,575
6966	6″ x 6″		1.60	5		3,450	168		3,618	4,075
6968	6″ x 8″		1.60	5		3,700	168		3,868	4,350
6970	6″ x 10″		1.55	5.161		3,950	173		4,123	4,650
6972	6″ x 12″		1.55	5.161		4,200	173		4,373	4,925
6974	6″ x 14″		1.55	5.161		4,450	173		4,623	5,200
6976	6″ x 16″		1.50	5.333		4,725	179		4,904	5,475
6978	7 ga., 8″ x 6″		1.60	5		3,750	168		3,918	4,400
6980	8″ x 8″		1.60	5		4,000	168		4,168	4,675
6982	8″ x 10″		1.55	5.161		4,250	173		4,423	4,975
6984	8″ x 12″		1.55	5.161		4,500	173		4,673	5,250
6986	8″ x 14″		1.50	5.333		4,750	179		4,929	5,525
6988	8″ x 16″	↓	1.50	5.333	↓	5,000	179		5,179	5,800
7000	Strap connectors, galvanized									
7002	12 ga., 2-1/16″ x 36″	1 Carp	1.55	5.161	C	1,075	173		1,248	1,500
7004	2-1/16″ x 47″		1.50	5.333		1,500	179		1,679	1,950
7005	10 ga., 2-1/16″ x 72″		1.50	5.333		1,500	179		1,679	1,950
7006	7 ga., 2-1/16″ x 34″		1.55	5.161		2,700	173		2,873	3,275
7008	2-1/16″ x 45″		1.50	5.333		3,500	179		3,679	4,150
7010	3 ga., 3″ x 32″		1.55	5.161		4,525	173		4,698	5,300
7012	3″ x 41″		1.55	5.161		4,700	173		4,873	5,475
7014	3″ x 50″		1.50	5.333		7,175	179		7,354	8,200
7016	3″ x 59″		1.50	5.333		8,750	179		8,929	9,925
7018	3-1/2″ x 68″	↓	1.45	5.517	↓	8,875	185		9,060	10,100
7030	Tension ties									
7032	19-1/8″ long, 16 ga., 3/4″ anchor bolt	1 Carp	1.80	4.444	C	1,250	149		1,399	1,625
7034	20″ long, 12 ga., 1/2″ anchor bolt		1.80	4.444		1,625	149		1,774	2,025
7036	20″ long, 12 ga., 3/4″ anchor bolt		1.80	4.444		1,625	149		1,774	2,025
7038	27-3/4″ long, 12 ga., 3/4″ anchor bolt	↓	1.75	4.571	↓	2,850	153		3,003	3,400
7050	Truss connectors, galvanized									
7052	Adjustable hanger									
7054	18 ga., 2″ x 6″	1 Carp	1.65	4.848	C	490	163		653	805

06 05 Common Work Results for Wood, Plastics, and Composites

06 05 23 – Wood, Plastic, and Composite Fastenings

06 05 23.60 Timber Connectors		Crew	Daily Output	Labor-Hours	Unit	Material	2011 Bare Costs Labor	Equipment	Total	Total Incl O&P
7056	4″ x 6″	1 Carp	1.65	4.848	C	640	163		803	975
7058	16 ga., 4″ x 10″	↓	1.60	5	↓	940	168		1,108	1,300
7060	(2) 2″ x 10″	↓	1.60	5	↓	940	168		1,108	1,300
7062	Connectors to plate									
7064	16 ga., 2″ x 4″ plate	1 Carp	1.80	4.444	C	480	149		629	780
7066	2″ x 6″ plate	″	1.80	4.444	″	620	149		769	935
7068	Hip jack connector									
7070	14 ga.	1 Carp	1.50	5.333	C	2,525	179		2,704	3,075

06 05 23.70 Rough Hardware

		Crew	Daily Output	Labor-Hours	Unit	Material	Labor	Equipment	Total	Total Incl O&P
0010	**ROUGH HARDWARE**, average percent of carpentry material									
0020	Minimum				Job	.50%				
0200	Maximum				↓	1.50%				
0210	In seismic or hurricane areas, up to				↓	10%				

06 05 23.80 Metal Bracing

		Crew	Daily Output	Labor-Hours	Unit	Material	Labor	Equipment	Total	Total Incl O&P
0010	**METAL BRACING**									
0302	Let-in, "T" shaped, 22 ga. galv. steel, studs at 16″ O.C.	1 Carp	580	.014	L.F.	.75	.46		1.21	1.60
0402	Studs at 24″ O.C.	↓	600	.013	↓	.75	.45		1.20	1.58
0502	16 ga. galv. steel straps, studs at 16″ O.C.	↓	600	.013	↓	.97	.45		1.42	1.82
0602	Studs at 24″ O.C.	↓	620	.013	↓	.97	.43		1.40	1.79

06 11 Wood Framing

06 11 10 – Framing with Dimensional, Engineered or Composite Lumber

06 11 10.01 Forest Stewardship Council Certification

		Crew	Daily Output	Labor-Hours	Unit	Material	Labor	Equipment	Total	Total Incl O&P
0010	**FOREST STEWARDSHIP COUNCIL CERTIFICATION**									
0020	For Forest Stewardship Council (FSC) cert dimension lumber, add [G]					65%				

06 11 10.02 Blocking

		Crew	Daily Output	Labor-Hours	Unit	Material	Labor	Equipment	Total	Total Incl O&P
0010	**BLOCKING**									
2600	Miscellaneous, to wood construction									
2621	2″ x 4″	2 Carp	.34	47.059	M.B.F.	495	1,575		2,070	3,200
2626	Pneumatic nailed	↓	.42	38.095	↓	495	1,275		1,770	2,700
2661	2″ x 8″	↓	.54	29.630	↓	545	995		1,540	2,275
2666	Pneumatic nailed	↓	.66	24.242	↓	545	815		1,360	1,950
2720	To steel construction									
2741	2″ x 4″	2 Carp	.28	57.143	M.B.F.	495	1,925		2,420	3,750
2781	2″ x 8″	″	.42	38.095	″	545	1,275		1,820	2,750

06 11 10.04 Wood Bracing

		Crew	Daily Output	Labor-Hours	Unit	Material	Labor	Equipment	Total	Total Incl O&P
0010	**WOOD BRACING**									
0011	Let-in, with 1″ x 6″ boards, studs 16″ O.C.	1 Carp	1.50	5.333	C.L.F.	67.50	179		246.50	375
0200	Studs @ 24″ O.C.	″	2.30	3.478	″	67.50	117		184.50	269

06 11 10.06 Bridging

		Crew	Daily Output	Labor-Hours	Unit	Material	Labor	Equipment	Total	Total Incl O&P
0010	**BRIDGING**									
0011	Wood, for joists 16″ O.C., 1″ x 3″	1 Carp	1.30	6.154	C.Pr.	51	206		257	400
0015	Pneumatic nailed	↓	1.70	4.706	↓	51	158		209	320
0100	2″ x 3″ bridging	↓	1.30	6.154	↓	49.50	206		255.50	400
0105	Pneumatic nailed	↓	1.70	4.706	↓	49.50	158		207.50	320
0300	Steel, galvanized, 18 ga., for 2″ x 10″ joists at 12″ O.C.	↓	1.30	6.154	↓	172	206		378	535
0400	24″ O.C.	↓	1.40	5.714	↓	185	192		377	525
0600	For 2″ x 14″ joists at 16″ O.C.	↓	1.30	6.154	↓	185	206		391	550
0700	24″ O.C.	↓	1.40	5.714	↓	283	192		475	630
0900	Compression type, 16″ O.C., 2″ x 8″ joists	↓	2	4	↓	178	134		312	420

06 11 10 – Framing with Dimensional, Engineered or Composite Lumber

06 11 10.06 Bridging		Crew	Daily Output	Labor-Hours	Unit	Material	2011 Bare Costs Labor	Equipment	Total	Total Incl O&P
1000	2″ x 12″ joists	1 Carp	2	4	C.Pr.	178	134		312	420

06 11 10.10 Beam and Girder Framing

		Crew	Daily Output	Labor-Hours	Unit	Material	Labor	Equipment	Total	Total Incl O&P
0010	BEAM AND GIRDER FRAMING R061110-30									
3500	Single, 2″ x 6″	2 Carp	.70	22.857	M.B.F.	525	765		1,290	1,850
3505	Pneumatic nailed		.81	19.704		525	660		1,185	1,675
3520	2″ x 8″		.86	18.605		545	625		1,170	1,650
3525	Pneumatic nailed		1	16.048		545	540		1,085	1,500
3540	2″ x 10″		1	16		670	535		1,205	1,625
3545	Pneumatic nailed		1.16	13.793		670	465		1,135	1,500
3560	2″ x 12″		1.10	14.545		680	490		1,170	1,575
3565	Pneumatic nailed		1.28	12.539		680	420		1,100	1,450
3580	2″ x 14″		1.17	13.675		725	460		1,185	1,575
3585	Pneumatic nailed		1.36	11.791		725	395		1,120	1,450
3600	3″ x 8″		1.10	14.545		1,050	490		1,540	2,000
3620	3″ x 10″		1.25	12.800		1,075	430		1,505	1,900
3640	3″ x 12″		1.35	11.852		1,100	400		1,500	1,875
3660	3″ x 14″	↓	1.40	11.429		1,100	385		1,485	1,850
3680	4″ x 8″	F-3	2.66	15.038		1,125	460	246	1,831	2,275
3700	4″ x 10″		3.16	12.658		1,150	390	207	1,747	2,125
3720	4″ x 12″		3.60	11.111		1,375	340	182	1,897	2,300
3740	4″ x 14″	↓	3.96	10.101		1,525	310	166	2,001	2,375
4000	Double, 2″ x 6″	2 Carp	1.25	12.800		525	430		955	1,300
4005	Pneumatic nailed		1.45	11.034		525	370		895	1,200
4020	2″ x 8″		1.60	10		545	335		880	1,150
4025	Pneumatic nailed		1.86	8.621		545	289		834	1,075
4040	2″ x 10″		1.92	8.333		670	280		950	1,200
4045	Pneumatic nailed		2.23	7.185		670	241		911	1,150
4060	2″ x 12″		2.20	7.273		680	244		924	1,150
4065	Pneumatic nailed		2.55	6.275		680	211		891	1,100
4080	2″ x 14″		2.45	6.531		725	219		944	1,175
4085	Pneumatic nailed		2.84	5.634		725	189		914	1,125
5000	Triple, 2″ x 6″		1.65	9.697		525	325		850	1,125
5005	Pneumatic nailed		1.91	8.377		525	281		806	1,050
5020	2″ x 8″		2.10	7.619		545	256		801	1,025
5025	Pneumatic nailed		2.44	6.568		545	220		765	970
5040	2″ x 10″		2.50	6.400		670	215		885	1,100
5045	Pneumatic nailed		2.90	5.517		670	185		855	1,050
5060	2″ x 12″		2.85	5.614		680	188		868	1,075
5065	Pneumatic nailed		3.31	4.840		680	162		842	1,025
5080	2″ x 14″		3.15	5.079		725	170		895	1,075
5085	Pneumatic nailed	↓	3.35	4.770	↓	725	160		885	1,075

06 11 10.12 Ceiling Framing

		Crew	Daily Output	Labor-Hours	Unit	Material	Labor	Equipment	Total	Total Incl O&P
0010	CEILING FRAMING									
6400	Suspended, 2″ x 3″	2 Carp	.50	32	M.B.F.	660	1,075		1,735	2,525
6450	2″ x 4″		.59	27.119		495	910		1,405	2,075
6500	2″ x 6″		.80	20		525	670		1,195	1,700
6550	2″ x 8″	↓	.86	18.605	↓	545	625		1,170	1,650

06 11 10.14 Posts and Columns

		Crew	Daily Output	Labor-Hours	Unit	Material	Labor	Equipment	Total	Total Incl O&P
0010	POSTS AND COLUMNS									
0400	4″ x 4″	2 Carp	.52	30.769	M.B.F.	1,100	1,025		2,125	2,950
0420	4″ x 6″		.55	29.091		1,175	975		2,150	2,925
0440	4″ x 8″	↓	.59	27.119	↓	1,125	910		2,035	2,750

06 11 Wood Framing

06 11 10 – Framing with Dimensional, Engineered or Composite Lumber

06 11 10.14 Posts and Columns

06 11 10.14 Posts and Columns		Crew	Daily Output	Labor-Hours	Unit	Material	Labor	Equipment	Total	Total Incl O&P
						2011 Bare Costs				
0460	6" x 6"	2 Carp	.65	24.615	M.B.F.	1,200	825		2,025	2,700
0480	6" x 8"		.70	22.857		1,200	765		1,965	2,600
0500	6" x 10"		.75	21.333		1,300	715		2,015	2,625

06 11 10.18 Joist Framing

Code	Description	Crew	Daily Output	Labor-Hours	Unit	Material	Labor	Equipment	Total	Total Incl O&P
0010	**JOIST FRAMING** R061110-30									
2650	Joists, 2" x 4"	2 Carp	.83	19.277	M.B.F.	495	645		1,140	1,625
2655	Pneumatic nailed		.96	16.667		495	560		1,055	1,475
2680	2" x 6"		1.25	12.800		525	430		955	1,300
2685	Pneumatic nailed		1.44	11.111		525	375		900	1,200
2700	2" x 8"		1.46	10.959		545	370		915	1,225
2705	Pneumatic nailed		1.68	9.524		545	320		865	1,125
2720	2" x 10"		1.49	10.738		670	360		1,030	1,350
2725	Pneumatic nailed		1.71	9.357		670	315		985	1,250
2740	2" x 12"		1.75	9.143		680	305		985	1,275
2745	Pneumatic nailed		2.01	7.960		680	267		947	1,200
2760	2" x 14"		1.79	8.939		725	300		1,025	1,300
2765	Pneumatic nailed		2.06	7.767		725	261		986	1,225
2780	3" x 6"		1.39	11.511		1,050	385		1,435	1,800
2790	3" x 8"		1.90	8.421		1,050	283		1,333	1,650
2800	3" x 10"		1.95	8.205		1,075	275		1,350	1,625
2820	3" x 12"		1.80	8.889		1,100	298		1,398	1,700
2840	4" x 6"		1.60	10		1,175	335		1,510	1,850
2851	4" x 8"	F-3	4.15	9.639		1,125	296	158	1,579	1,900
2861	4" x 10"		5	8		1,150	246	131	1,527	1,800
2881	4" x 12"		4.50	8.889		1,375	273	146	1,794	2,150
3000	Composite wood joist 9-1/2" deep	2 Carp	.90	17.778	M.L.F.	1,550	595		2,145	2,700
3010	11-1/2" deep		.88	18.182		1,725	610		2,335	2,925
3020	14" deep		.82	19.512		2,400	655		3,055	3,725
3030	16" deep		.78	20.513		2,975	690		3,665	4,400
4000	Open web joist 12" deep		.88	18.182		2,775	610		3,385	4,075
4010	14" deep		.82	19.512		2,975	655		3,630	4,375
4020	16" deep		.78	20.513		2,950	690		3,640	4,400
4030	18" deep		.74	21.622		3,175	725		3,900	4,725

06 11 10.24 Miscellaneous Framing

Code	Description	Crew	Daily Output	Labor-Hours	Unit	Material	Labor	Equipment	Total	Total Incl O&P
0010	**MISCELLANEOUS FRAMING**									
8500	Firestops, 2" x 4"	2 Carp	.51	31.373	M.B.F.	495	1,050		1,545	2,300
8505	Pneumatic nailed		.62	25.806		495	865		1,360	2,000
8520	2" x 6"		.60	26.667		525	895		1,420	2,075
8525	Pneumatic nailed		.73	21.858		525	735		1,260	1,800
8540	2" x 8"		.60	26.667		545	895		1,440	2,100
8560	2" x 12"		.70	22.857		680	765		1,445	2,025
8600	Nailers, treated, wood construction, 2" x 4"		.53	30.189		540	1,025		1,565	2,300
8605	Pneumatic nailed		.64	25.157		540	845		1,385	2,025
8620	2" x 6"		.75	21.333		615	715		1,330	1,875
8625	Pneumatic nailed		.90	17.778		615	595		1,210	1,675
8640	2" x 8"		.93	17.204		715	575		1,290	1,750
8645	Pneumatic nailed		1.12	14.337		715	480		1,195	1,600
8660	Steel construction, 2" x 4"		.50	32		540	1,075		1,615	2,400
8680	2" x 6"		.70	22.857		615	765		1,380	1,950
8700	2" x 8"		.87	18.391		715	615		1,330	1,825
8760	Rough bucks, treated, for doors or windows, 2" x 6"		.40	40		615	1,350		1,965	2,925
8765	Pneumatic nailed		.48	33.333		615	1,125		1,740	2,550

06 11 10 – Framing with Dimensional, Engineered or Composite Lumber

06 11 10.24 Miscellaneous Framing

	06 11 10.24 Miscellaneous Framing	Crew	Daily Output	Labor-Hours	Unit	2011 Bare Costs Material	Labor	Equipment	Total	Total Incl O&P
8780	2" x 8"	2 Carp	.51	31.373	M.B.F.	715	1,050		1,765	2,550
8785	Pneumatic nailed		.61	26.144		715	875		1,590	2,275
8800	Stair stringers, 2" x 10"		.22	72.727		670	2,450		3,120	4,800
8820	2" x 12"		.26	61.538		680	2,075		2,755	4,200
8840	3" x 10"		.31	51.613		1,075	1,725		2,800	4,075
8860	3" x 12"	↓	.38	42.105	↓	1,100	1,425		2,525	3,575

06 11 10.26 Partitions

	06 11 10.26 Partitions	Crew	Daily Output	Labor-Hours	Unit	Material	Labor	Equipment	Total	Total Incl O&P
0010	**PARTITIONS**									
0020	Single bottom and double top plate, no waste, std. & better lumber									
0180	2" x 4" studs, 8' high, studs 12" O.C.	2 Carp	80	.200	L.F.	3.64	6.70		10.34	15.25
0185	12" O.C., pneumatic nailed		96	.167		3.64	5.60		9.24	13.35
0200	16" O.C.		100	.160		2.98	5.35		8.33	12.25
0205	16" O.C., pneumatic nailed		120	.133		2.98	4.47		7.45	10.75
0300	24" O.C.		125	.128		2.31	4.29		6.60	9.75
0305	24" O.C., pneumatic nailed		150	.107		2.31	3.58		5.89	8.55
0380	10' high, studs 12" O.C.		80	.200		4.30	6.70		11	16
0385	12" O.C., pneumatic nailed		96	.167		4.30	5.60		9.90	14.10
0400	16" O.C.		100	.160		3.47	5.35		8.82	12.80
0405	16" O.C., pneumatic nailed		120	.133		3.47	4.47		7.94	11.30
0500	24" O.C.		125	.128		2.65	4.29		6.94	10.10
0505	24" O.C., pneumatic nailed		150	.107		2.65	3.58		6.23	8.90
0580	12' high, studs 12" O.C.		65	.246		4.96	8.25		13.21	19.25
0585	12" O.C., pneumatic nailed		78	.205		4.96	6.90		11.86	16.95
0600	16" O.C.		80	.200		3.97	6.70		10.67	15.60
0605	16" O.C., pneumatic nailed		96	.167		3.97	5.60		9.57	13.70
0700	24" O.C.		100	.160		2.98	5.35		8.33	12.25
0705	24" O.C., pneumatic nailed		120	.133		2.98	4.47		7.45	10.75
0780	2" x 6" studs, 8' high, studs 12" O.C.		70	.229		5.80	7.65		13.45	19.25
0785	12" O.C., pneumatic nailed		84	.190		5.80	6.40		12.20	17.10
0800	16" O.C.		90	.178		4.74	5.95		10.69	15.20
0805	16" O.C., pneumatic nailed		108	.148		4.74	4.97		9.71	13.50
0900	24" O.C.		115	.139		3.69	4.67		8.36	11.85
0905	24" O.C., pneumatic nailed		138	.116		3.69	3.89		7.58	10.55
0980	10' high, studs 12" O.C.		70	.229		6.85	7.65		14.50	20.50
0985	12" O.C., pneumatic nailed		84	.190		6.85	6.40		13.25	18.25
1000	16" O.C.		90	.178		5.55	5.95		11.50	16.10
1005	16" O.C., pneumatic nailed		108	.148		5.55	4.97		10.52	14.40
1100	24" O.C.		115	.139		4.22	4.67		8.89	12.45
1105	24" O.C., pneumatic nailed		138	.116		4.22	3.89		8.11	11.15
1180	12' high, studs 12" O.C.		55	.291		7.90	9.75		17.65	25
1185	12" O.C., pneumatic nailed		66	.242		7.90	8.15		16.05	22.50
1200	16" O.C.		70	.229		6.30	7.65		13.95	19.80
1205	16" O.C., pneumatic nailed		84	.190		6.30	6.40		12.70	17.65
1300	24" O.C.		90	.178		4.74	5.95		10.69	15.20
1305	24" O.C., pneumatic nailed		108	.148		4.74	4.97		9.71	13.50
1400	For horizontal blocking, 2" x 4", add		600	.027		.33	.89		1.22	1.86
1500	2" x 6", add		600	.027		.53	.89		1.42	2.08
1600	For openings, add	↓	250	.064	↓		2.15		2.15	3.59
1700	Headers for above openings, material only, add				M.B.F.	545			545	600

06 11 10.28 Porch or Deck Framing

	06 11 10.28 Porch or Deck Framing	Crew	Daily Output	Labor-Hours	Unit	Material	Labor	Equipment	Total	Total Incl O&P
0010	**PORCH OR DECK FRAMING**									
0100	Treated lumber, posts or columns, 4" x 4"	2 Carp	390	.041	L.F.	1.37	1.38		2.75	3.80

06 11 Wood Framing

06 11 10 – Framing with Dimensional, Engineered or Composite Lumber

06 11 10.28 Porch or Deck Framing		Crew	Daily Output	Labor-Hours	Unit	2011 Bare Costs Material	Labor	Equipment	Total	Total Incl O&P
0110	4" x 6"	2 Carp	275	.058	L.F.	2.46	1.95		4.41	6
0120	4" x 8"		220	.073		4.74	2.44		7.18	9.30
0130	Girder, single, 4" x 4"		675	.024		1.37	.80		2.17	2.83
0140	4" x 6"		600	.027		2.46	.89		3.35	4.21
0150	4" x 8"		525	.030		4.74	1.02		5.76	6.90
0160	Double, 2" x 4"		625	.026		.75	.86		1.61	2.26
0170	2" x 6"		600	.027		1.28	.89		2.17	2.90
0180	2" x 8"		575	.028		1.97	.93		2.90	3.73
0190	2" x 10"		550	.029		2.65	.98		3.63	4.55
0200	2" x 12"		525	.030		3.49	1.02		4.51	5.55
0210	Triple, 2" x 4"		575	.028		1.12	.93		2.05	2.80
0220	2" x 6"		550	.029		1.92	.98		2.90	3.74
0230	2" x 8"		525	.030		2.96	1.02		3.98	4.96
0240	2" x 10"		500	.032		3.98	1.07		5.05	6.20
0250	2" x 12"		475	.034		5.25	1.13		6.38	7.65
0260	Ledger, bolted 4' O.C., 2" x 4"		400	.040		.48	1.34		1.82	2.78
0270	2" x 6"		395	.041		.74	1.36		2.10	3.08
0280	2" x 8"		390	.041		1.07	1.38		2.45	3.48
0300	2" x 12"		380	.042		1.82	1.41		3.23	4.36
0310	Joists, 2" x 4"		1250	.013		.38	.43		.81	1.13
0320	2" x 6"		1250	.013		.64	.43		1.07	1.43
0330	2" x 8"		1100	.015		.99	.49		1.48	1.91
0340	2" x 10"		900	.018		1.33	.60		1.93	2.47
0350	2" x 12"		875	.018		1.60	.61		2.21	2.79
0360	Railings and trim , 1" x 4"	1 Carp	300	.027		.53	.89		1.42	2.08
0370	2" x 2"		300	.027		.35	.89		1.24	1.89
0380	2" x 4"		300	.027		.36	.89		1.25	1.90
0390	2" x 6"		300	.027		.62	.89		1.51	2.18
0400	Decking, 1" x 4"		275	.029	S.F.	1.55	.98		2.53	3.34
0410	2" x 4"		300	.027		1.23	.89		2.12	2.85
0420	2" x 6"		320	.025		1.34	.84		2.18	2.87
0430	5/4" x 6"		320	.025		2.09	.84		2.93	3.70
0440	Balusters, square, 2" x 2"	2 Carp	660	.024	L.F.	.36	.81		1.17	1.75
0450	Turned, 2" x 2"		420	.038		.48	1.28		1.76	2.66
0460	Stair stringer, 2" x 10"		130	.123		1.33	4.13		5.46	8.35
0470	2" x 12"		130	.123		1.60	4.13		5.73	8.65
0480	Stair treads, 1" x 4"		140	.114		1.56	3.83		5.39	8.10
0490	2" x 4"		140	.114		.38	3.83		4.21	6.80
0500	2" x 6"		160	.100		.57	3.36		3.93	6.20
0510	5/4" x 6"		160	.100		.98	3.36		4.34	6.65
0520	Turned handrail post, 4" x 4"		64	.250	Ea.	32	8.40		40.40	49
0530	Lattice panel, 4' x 8', 1/2"		1600	.010	S.F.	.68	.34		1.02	1.31
0535	3/4"		1600	.010	"	1.02	.34		1.36	1.68
0540	Cedar, posts or columns, 4" x 4"		390	.041	L.F.	3.01	1.38		4.39	5.60
0550	4" x 6"		275	.058		5.75	1.95		7.70	9.55
0560	4" x 8"		220	.073		9.70	2.44		12.14	14.80
0800	Decking, 1" x 4"		550	.029		1.49	.98		2.47	3.27
0810	2" x 4"		600	.027		2.98	.89		3.87	4.78
0820	2" x 6"		640	.025		5.25	.84		6.09	7.20
0830	5/4" x 6"		640	.025		3.50	.84		4.34	5.25
0840	Railings and trim, 1" x 4"		600	.027		1.49	.89		2.38	3.14
0860	2" x 4"		600	.027		2.98	.89		3.87	4.78
0870	2" x 6"		600	.027		5.25	.89		6.14	7.30

06 11 10 – Framing with Dimensional, Engineered or Composite Lumber

06 11 10.28 Porch or Deck Framing		Crew	Daily Output	Labor-Hours	Unit	Material	2011 Bare Costs Labor	Equipment	Total	Total Incl O&P
0920	Stair treads, 1" x 4"	2 Carp	140	.114	L.F.	1.49	3.83		5.32	8.05
0930	2" x 4"		140	.114		2.98	3.83		6.81	9.70
0940	2" x 6"		160	.100		5.25	3.36		8.61	11.40
0950	5/4" x 6"		160	.100		3.50	3.36		6.86	9.45
0980	Redwood, posts or columns, 4" x 4"		390	.041		6.40	1.38		7.78	9.35
0990	4" x 6"		275	.058		11.20	1.95		13.15	15.55
1000	4" x 8"		220	.073		20.50	2.44		22.94	26.50
1240	Redwood decking, 1" x 4"	1 Carp	275	.029	S.F.	3.97	.98		4.95	6
1260	2" x 6"		340	.024		7.50	.79		8.29	9.55
1270	5/4" x 6"		320	.025		4.74	.84		5.58	6.60
1280	Railings and trim, 1" x 4"	2 Carp	600	.027	L.F.	1.17	.89		2.06	2.78
1310	2" x 6"		600	.027		7.50	.89		8.39	9.75
1420	Alternative decking, wood/plastic composite, 5/4" x 6" G		640	.025		3	.84		3.84	4.70
1430	Vinyl, 1-1/2" x 5-1/2"		640	.025		4.15	.84		4.99	5.95
1440	1" x 4" square edge fir		550	.029		1.57	.98		2.55	3.35
1450	1" x 4" tongue and groove fir		450	.036		1.41	1.19		2.60	3.55
1460	1" x 4" mahogany		550	.029		2.51	.98		3.49	4.39
1462	5/4" x 6" PVC		550	.029		2.72	.98		3.70	4.62
1465	Framing, porch or deck, alt deck fastening, screws, add	1 Carp	240	.033	S.F.		1.12		1.12	1.87
1470	Accessories, joist hangers, 2" x 4"		160	.050	Ea.	.67	1.68		2.35	3.55
1480	2" x 6" through 2" x 12"		150	.053		1.28	1.79		3.07	4.40
1530	Post footing, incl excav, backfill, tube form & concrete, 4' deep, 8" dia	F-7	12	2.667		10.65	80		90.65	146
1540	10" diameter		11	2.909		15.80	87.50		103.30	163
1550	12" diameter		10	3.200		21	96		117	184

06 11 10.30 Roof Framing		Crew	Daily Output	Labor-Hours	Unit	Material	Labor	Equipment	Total	Total Incl O&P
0010	**ROOF FRAMING**									
6070	Fascia boards, 2" x 8"	2 Carp	.30	53.333	M.B.F.	545	1,800		2,345	3,600
6080	2" x 10"		.30	53.333		670	1,800		2,470	3,725
7000	Rafters, to 4 in 12 pitch, 2" x 6"		1	16		525	535		1,060	1,475
7060	2" x 8"		1.26	12.698		545	425		970	1,325
7300	Hip and valley rafters, 2" x 6"		.76	21.053		525	705		1,230	1,750
7360	2" x 8"		.96	16.667		545	560		1,105	1,525
7540	Hip and valley jacks, 2" x 6"		.60	26.667		525	895		1,420	2,075
7600	2" x 8"		.65	24.615		545	825		1,370	1,975
7780	For slopes steeper than 4 in 12, add						30%			
7790	For dormers or complex roofs, add						50%			
7800	Rafter tie, 1" x 4", #3	2 Carp	.27	59.259	M.B.F.	1,600	2,000		3,600	5,075
7820	Ridge board, #2 or better, 1" x 6"		.30	53.333		1,350	1,800		3,150	4,475
7840	1" x 8"		.37	43.243		1,450	1,450		2,900	4,025
7860	1" x 10"		.42	38.095		1,450	1,275		2,725	3,750
7880	2" x 6"		.50	32		525	1,075		1,600	2,375
7900	2" x 8"		.60	26.667		545	895		1,440	2,100
7920	2" x 10"		.66	24.242		670	815		1,485	2,075
7940	Roof cants, split, 4" x 4"		.86	18.605		1,100	625		1,725	2,275
7960	6" x 6"		1.80	8.889		1,200	298		1,498	1,825
7980	Roof curbs, untreated, 2" x 6"		.52	30.769		525	1,025		1,550	2,300
8000	2" x 12"		.80	20		680	670		1,350	1,875

06 11 10.32 Sill and Ledger Framing		Crew	Daily Output	Labor-Hours	Unit	Material	Labor	Equipment	Total	Total Incl O&P
0010	**SILL AND LEDGER FRAMING**									
4482	Ledgers, nailed, 2" x 4"	2 Carp	.50	32	M.B.F.	495	1,075		1,570	2,350
4484	2" x 6"		.60	26.667		525	895		1,420	2,075
4486	Bolted, not including bolts, 3" x 8"		.65	24.615		1,050	825		1,875	2,550

06 11 Wood Framing

06 11 10 – Framing with Dimensional, Engineered or Composite Lumber

	06 11 10.32 Sill and Ledger Framing	Crew	Daily Output	Labor-Hours	Unit	2011 Bare Costs Material	Labor	Equipment	Total	Total Incl O&P
4488	3″ x 12″	2 Carp	.70	22.857	M.B.F.	1,100	765		1,865	2,475
4490	Mud sills, redwood, construction grade, 2″ x 4″		.59	27.119		3,400	910		4,310	5,250
4492	2″ x 6″		.78	20.513		3,400	690		4,090	4,900
4500	Sills, 2″ x 4″		.40	40		495	1,350		1,845	2,800
4520	2″ x 6″		.55	29.091		525	975		1,500	2,200
4540	2″ x 8″		.67	23.881		545	800		1,345	1,950
4600	Treated, 2″ x 4″		.36	44.444		540	1,500		2,040	3,100
4620	2″ x 6″		.50	32		615	1,075		1,690	2,475
4640	2″ x 8″		.60	26.667		715	895		1,610	2,300
4700	4″ x 4″		.60	26.667		1,000	895		1,895	2,600
4720	4″ x 6″		.70	22.857		1,200	765		1,965	2,600
4740	4″ x 8″		.80	20		1,750	670		2,420	3,050
4760	4″ x 10″		.87	18.391		1,750	615		2,365	2,950

06 11 10.34 Sleepers

		Crew	Daily Output	Labor-Hours	Unit	Material	Labor	Equipment	Total	Total Incl O&P
0010	**SLEEPERS**									
0300	On concrete, treated, 1″ x 2″	2 Carp	.39	41.026	M.B.F.	1,700	1,375		3,075	4,175
0320	1″ x 3″		.50	32		2,100	1,075		3,175	4,100
0340	2″ x 4″		.99	16.162		540	540		1,080	1,500
0360	2″ x 6″		1.30	12.308		615	415		1,030	1,375

06 11 10.36 Soffit and Canopy Framing

		Crew	Daily Output	Labor-Hours	Unit	Material	Labor	Equipment	Total	Total Incl O&P
0010	**SOFFIT AND CANOPY FRAMING**									
1300	Canopy or soffit framing, 1″ x 4″	2 Carp	.30	53.333	M.B.F.	1,600	1,800		3,400	4,750
1340	1″ x 8″		.50	32		1,450	1,075		2,525	3,400
1360	2″ x 4″		.41	39.024		495	1,300		1,795	2,750
1400	2″ x 8″		.67	23.881		545	800		1,345	1,950
1420	3″ x 4″		.50	32		850	1,075		1,925	2,725
1460	3″ x 8″		.60	26.667		1,050	895		1,945	2,675

06 11 10.38 Treated Lumber Framing Material

		Crew	Daily Output	Labor-Hours	Unit	Material	Labor	Equipment	Total	Total Incl O&P
0010	**TREATED LUMBER FRAMING MATERIAL**									
0100	2″ x 4″				M.B.F.	540			540	595
0110	2″ x 6″					615			615	675
0120	2″ x 8″					715			715	790
0130	2″ x 10″					775			775	850
0140	2″ x 12″					850			850	935
0200	4″ x 4″					1,000			1,000	1,100
0210	4″ x 6″					1,200			1,200	1,325
0220	4″ x 8″					1,750			1,750	1,925

06 11 10.40 Wall Framing

		Crew	Daily Output	Labor-Hours	Unit	Material	Labor	Equipment	Total	Total Incl O&P
0010	**WALL FRAMING** R061110-30									
5860	Headers over openings, 2″ x 6″	2 Carp	.36	44.444	M.B.F.	525	1,500		2,025	3,075
5865	2″ x 6″, pneumatic nailed		.43	37.209		525	1,250		1,775	2,675
5880	2″ x 8″		.45	35.556		545	1,200		1,745	2,600
5885	2″ x 8″, pneumatic nailed		.54	29.630		545	995		1,540	2,275
5900	2″ x 10″		.53	30.189		670	1,025		1,695	2,425
5905	2″ x 10″, pneumatic nailed		.67	23.881		670	800		1,470	2,075
5920	2″ x 12″		.60	26.667		680	895		1,575	2,250
5925	2″ x 12″, pneumatic nailed		.72	22.222		680	745		1,425	2,000
5940	4″ x 12″		.76	21.053		1,375	705		2,080	2,700
5945	4″ x 12″, pneumatic nailed		.92	17.391		1,375	585		1,960	2,500
5960	6″ x 12″		.84	19.048		1,700	640		2,340	2,950
5965	6″ x 12″, pneumatic nailed		1.01	15.873		1,700	535		2,235	2,775
6000	Plates, untreated, 2″ x 3″		.43	37.209		660	1,250		1,910	2,825

06 11 10 – Framing with Dimensional, Engineered or Composite Lumber

06 11 10.40 Wall Framing

06 11 10.40 Wall Framing		Crew	Daily Output	Labor-Hours	Unit	Material	Labor	Equipment	Total	Total Incl O&P
						2011 Bare Costs				
6005	2" x 3", pneumatic nailed	2 Carp	.52	30.769	M.B.F.	660	1,025		1,685	2,450
6020	2" x 4"		.53	30.189		495	1,025		1,520	2,250
6025	2" x 4", pneumatic nailed		.67	23.881		495	800		1,295	1,900
6040	2" x 6"		.75	21.333		525	715		1,240	1,775
6045	2" x 6", pneumatic nailed		.90	17.778		525	595		1,120	1,575
6120	Studs, 8' high wall, 2" x 3"		.60	26.667		660	895		1,555	2,225
6125	2" x 3", pneumatic nailed		.72	22.222		660	745		1,405	1,975
6140	2" x 4"		.92	17.391		495	585		1,080	1,525
6145	2" x 4", pneumatic nailed		1.10	14.493		495	485		980	1,350
6160	2" x 6"		1	16		525	535		1,060	1,475
6165	2" x 6", pneumatic nailed		1.20	13.333		525	445		970	1,325
6180	3" x 4"		.80	20		850	670		1,520	2,050
6185	3" x 4", pneumatic nailed	↓	.96	16.667	↓	850	560		1,410	1,875
8200	For 12' high walls, deduct						5%			
8220	For stub wall, 6' high, add						20%			
8240	3' high, add						40%			
8250	For second story & above, add						5%			
8300	For dormer & gable, add						15%			

06 11 10.42 Furring

06 11 10.42 Furring		Crew	Daily Output	Labor-Hours	Unit	Material	Labor	Equipment	Total	Total Incl O&P
0010	**FURRING**									
0012	Wood strips, 1" x 2", on walls, on wood	1 Carp	550	.015	L.F.	.25	.49		.74	1.09
0015	On wood, pneumatic nailed		710	.011		.25	.38		.63	.90
0300	On masonry		495	.016		.25	.54		.79	1.18
0400	On concrete		260	.031		.25	1.03		1.28	2
0600	1" x 3", on walls, on wood		550	.015		.34	.49		.83	1.20
0605	On wood, pneumatic nailed		710	.011		.34	.38		.72	1.01
0700	On masonry		495	.016		.34	.54		.88	1.29
0800	On concrete		260	.031		.34	1.03		1.37	2.11
0850	On ceilings, on wood		350	.023		.34	.77		1.11	1.66
0855	On wood, pneumatic nailed		450	.018		.34	.60		.94	1.38
0900	On masonry		320	.025		.34	.84		1.18	1.78
0950	On concrete	↓	210	.038	↓	.34	1.28		1.62	2.52

06 11 10.44 Grounds

06 11 10.44 Grounds		Crew	Daily Output	Labor-Hours	Unit	Material	Labor	Equipment	Total	Total Incl O&P
0010	**GROUNDS**									
0020	For casework, 1" x 2" wood strips, on wood	1 Carp	330	.024	L.F.	.25	.81		1.06	1.63
0100	On masonry		285	.028		.25	.94		1.19	1.85
0200	On concrete		250	.032		.25	1.07		1.32	2.07
0400	For plaster, 3/4" deep, on wood		450	.018		.25	.60		.85	1.27
0500	On masonry		225	.036		.25	1.19		1.44	2.27
0600	On concrete		175	.046		.25	1.53		1.78	2.84
0700	On metal lath	↓	200	.040	↓	.25	1.34		1.59	2.52

06 12 Structural Panels

06 12 10 – Structural Insulated Panels

06 12 10.10 OSB Faced Panels			Crew	Daily Output	Labor-Hours	Unit	2011 Bare Costs Material	Labor	Equipment	Total	Total Incl O&P
0010	**OSB FACED PANELS**										
0100	Structural insul. panels, 7/16" OSB both faces, EPS insul, 3-5/8" T	G	F-3	2075	.019	S.F.	2.72	.59	.32	3.63	4.32
0110	5-5/8" thick	G		1725	.023		3.05	.71	.38	4.14	4.96
0120	7-3/8" thick	G		1425	.028		3.33	.86	.46	4.65	5.60
0130	9-3/8" thick	G		1125	.036		3.64	1.09	.58	5.31	6.45
0140	7/16" OSB one face, EPS insul, 3-5/8" thick	G		2175	.018		2.90	.56	.30	3.76	4.46
0150	5-5/8" thick	G		1825	.022		3.32	.67	.36	4.35	5.15
0160	7-3/8" thick	G		1525	.026		3.70	.81	.43	4.94	5.90
0170	9-3/8" thick	G		1225	.033		4.12	1	.54	5.66	6.80
0190	7/16" OSB - 1/2" GWB faces , EPS insul, 3-5/8" T	G		2075	.019		2.87	.59	.32	3.78	4.49
0200	5-5/8" thick	G		1725	.023		3.29	.71	.38	4.38	5.20
0210	7-3/8" thick	G		1425	.028		3.66	.86	.46	4.98	5.95
0220	9-3/8" thick	G		1125	.036		4.29	1.09	.58	5.96	7.20
0240	7/16" OSB - 1/2" MRGWB faces , EPS insul, 3-5/8" T	G		2075	.019		2.87	.59	.32	3.78	4.49
0250	5-5/8" thick	G		1725	.023		3.29	.71	.38	4.38	5.20
0260	7-3/8" thick	G		1425	.028		3.66	.86	.46	4.98	5.95
0270	9-3/8" thick	G		1125	.036		4.29	1.09	.58	5.96	7.20
0300	For 1/2" GWB added to OSB skin, add	G					.60			.60	.66
0310	For 1/2" MRGWB added to OSB skin, add	G					.60			.60	.66
0320	For one T1-11 skin, add to OSB-OSB	G					1.39			1.39	1.53
0330	For one 19/32" CDX skin, add to OSB-OSB	G					.70			.70	.77
0500	Structural insulated panel, 7/16" OSB both sides, straw core										
0510	4-3/8" T, walls (w/sill, splines, plates)	G	F-6	2400	.017	S.F.	6.60	.52	.27	7.39	8.40
0520	Floors (w/splines)	G		2400	.017		6.60	.52	.27	7.39	8.40
0530	Roof (w/splines)	G		2400	.017		6.60	.52	.27	7.39	8.40
0550	7-7/8" T, walls (w/sill, splines, plates)	G		2400	.017		9.95	.52	.27	10.74	12.10
0560	Floors (w/splines)	G		2400	.017		9.95	.52	.27	10.74	12.10
0570	Roof (w/splines)	G		2400	.017		9.95	.52	.27	10.74	12.10

06 12 19 – Composite Shearwall Panels

06 12 19.10 Steel and Wood Composite Shearwall Panels		Crew	Daily Output	Labor-Hours	Unit	Material	Labor	Equipment	Total	Total Incl O&P
0010	**STEEL & WOOD COMPOSITE SHEARWALL PANELS**									
0020	Anchor bolts, 36" long (must be placed in wet concrete)	1 Carp	150	.053	Ea.	32.50	1.79		34.29	38.50
0030	On concrete, 2 x 4 & 2 x 6 walls, 7' - 10' high, 360 lb. shear, 12" wide	2 Carp	8	2		278	67		345	415
0040	715 lb. shear, 15" wide		8	2		320	67		387	460
0050	1860 lb. shear, 18" wide		8	2		355	67		422	500
0060	2780 lb. shear, 21" wide		8	2		365	67		432	510
0070	3790 lb. shear, 24" wide		8	2		460	67		527	615
0080	2 x 6 walls, 11' to 13' high, 1180 lb. shear, 18" wide		6	2.667		350	89.50		439.50	535
0090	1555 lb. shear, 21" wide		6	2.667		510	89.50		599.50	710
0100	2280 lb. shear, 24" wide		6	2.667		620	89.50		709.50	830
0110	For installing above on wood floor frame, add									
0120	Coupler nuts, threaded rods, bolts, shear transfer plate kit	1 Carp	16	.500	Ea.	48	16.80		64.80	81
0130	Framing anchors, angle (2 required)	"	96	.083	"	2.40	2.80		5.20	7.30
0140	For blocking see Section 06 11 10.02									
0150	For installing above, first floor to second floor, wood floor frame, add									
0160	Add stack option to first floor wall panel				Ea.	53			53	58.50
0170	Threaded rods, bolts, shear transfer plate kit	1 Carp	16	.500		57.50	16.80		74.30	91.50
0180	Framing anchors, angle (2 required)	"	96	.083		2.40	2.80		5.20	7.30
0200	For installing stacked panels, balloon framing									
0210	Add stack option to first floor wall panel				Ea.	53			53	58.50
0220	Threaded rods, bolts kit	1 Carp	16	.500	"	31	16.80		47.80	62

06 13 Heavy Timber Construction

06 13 23 – Heavy Timber Framing

06 13 23.10 Heavy Framing		Crew	Daily Output	Labor-Hours	Unit	Material	Labor	Equipment	Total	Total Incl O&P
0010	**HEAVY FRAMING**									
0020	Beams, single 6" x 10"	2 Carp	1.10	14.545	M.B.F.	1,400	490		1,890	2,375
0100	Single 8" x 16"		1.20	13.333		1,975	445		2,420	2,925
0200	Built from 2" lumber, multiple 2" x 14"		.90	17.778		725	595		1,320	1,800
0210	Built from 3" lumber, multiple 3" x 6"		.70	22.857		1,050	765		1,815	2,425
0220	Multiple 3" x 8"		.80	20		1,050	670		1,720	2,300
0230	Multiple 3" x 10"		.90	17.778		1,075	595		1,670	2,175
0240	Multiple 3" x 12"		1	16		1,100	535		1,635	2,100
0250	Built from 4" lumber, multiple 4" x 6"		.80	20		1,175	670		1,845	2,425
0260	Multiple 4" x 8"		.90	17.778		1,125	595		1,720	2,225
0270	Multiple 4" x 10"		1	16		1,150	535		1,685	2,150
0280	Multiple 4" x 12"		1.10	14.545		1,375	490		1,865	2,350
0290	Columns, structural grade, 1500f, 4" x 4"		.60	26.667		1,275	895		2,170	2,900
0300	6" x 6"		.65	24.615		1,300	825		2,125	2,800
0400	8" x 8"		.70	22.857		1,350	765		2,115	2,750
0500	10" x 10"		.75	21.333		1,450	715		2,165	2,800
0600	12" x 12"		.80	20		1,525	670		2,195	2,800
0800	Floor planks, 2" thick, T & G, 2" x 6"		1.05	15.238		1,050	510		1,560	2,000
0900	2" x 10"		1.10	14.545		1,050	490		1,540	1,975
1100	3" thick, 3" x 6"		1.05	15.238		1,250	510		1,760	2,225
1200	3" x 10"		1.10	14.545		1,250	490		1,740	2,200
1400	Girders, structural grade, 12" x 12"		.80	20		1,525	670		2,195	2,800
1500	10" x 16"		1	16		2,425	535		2,960	3,575
2300	Roof purlins, 4" thick, structural grade		1.05	15.238		1,125	510		1,635	2,075

06 15 Wood Decking

06 15 16 – Wood Roof Decking

06 15 16.10 Solid Wood Roof Decking

		Crew	Daily Output	Labor-Hours	Unit	Material	Labor	Equipment	Total	Total Incl O&P
0010	**SOLID WOOD ROOF DECKING**									
0400	Cedar planks, 3" thick	2 Carp	320	.050	S.F.	7.70	1.68		9.38	11.30
0500	4" thick		250	.064		10.30	2.15		12.45	14.95
0700	Douglas fir, 3" thick		320	.050		3.60	1.68		5.28	6.75
0800	4" thick		250	.064		4.80	2.15		6.95	8.90
1000	Hemlock, 3" thick		320	.050		3.28	1.68		4.96	6.40
1100	4" thick		250	.064		4.36	2.15		6.51	8.40
1300	Western white spruce, 3" thick		320	.050		3.28	1.68		4.96	6.40
1400	4" thick		250	.064		4.36	2.15		6.51	8.40

06 15 23 – Laminated Wood Decking

06 15 23.10 Laminated Roof Deck

		Crew	Daily Output	Labor-Hours	Unit	Material	Labor	Equipment	Total	Total Incl O&P
0010	**LAMINATED ROOF DECK**									
0020	Pine or hemlock, 3" thick	2 Carp	425	.038	S.F.	3.75	1.26		5.01	6.25
0100	4" thick		325	.049		5	1.65		6.65	8.25
0300	Cedar, 3" thick		425	.038		4.11	1.26		5.37	6.65
0400	4" thick		325	.049		5.25	1.65		6.90	8.50
0600	Fir, 3" thick		425	.038		3.80	1.26		5.06	6.30
0700	4" thick		325	.049		5.05	1.65		6.70	8.35

06 16 Sheathing

06 16 23 – Subflooring

06 16 23.10 Subfloor		Crew	Daily Output	Labor-Hours	Unit	Material	Labor (2011 Bare Costs)	Equipment	Total	Total Incl O&P
0010	**SUBFLOOR** R061636-20									
0011	Plywood, CDX, 1/2" thick	2 Carp	1500	.011	SF Flr.	.58	.36		.94	1.24
0015	Pneumatic nailed		1860	.009		.58	.29		.87	1.12
0100	5/8" thick		1350	.012		.70	.40		1.10	1.44
0105	Pneumatic nailed		1674	.010		.70	.32		1.02	1.31
0200	3/4" thick		1250	.013		.82	.43		1.25	1.62
0205	Pneumatic nailed		1550	.010		.82	.35		1.17	1.48
0300	1-1/8" thick, 2-4-1 including underlayment		1050	.015		1.53	.51		2.04	2.54
0440	With boards, 1" x 6", S4S, laid regular		900	.018		1.47	.60		2.07	2.61
0450	1" x 8", laid regular		1000	.016		1.55	.54		2.09	2.60
0460	Laid diagonal		850	.019		1.55	.63		2.18	2.76
0500	1" x 10", laid regular		1100	.015		1.52	.49		2.01	2.49
0600	Laid diagonal		900	.018		1.52	.60		2.12	2.67
8990	Subfloor adhesive, 3/8" bead	1 Carp	2300	.003	L.F.	.09	.12		.21	.30

06 16 26 – Underlayment

06 16 26.10 Wood Product Underlayment		Crew	Daily Output	Labor-Hours	Unit	Material	Labor (2011 Bare Costs)	Equipment	Total	Total Incl O&P
0010	**WOOD PRODUCT UNDERLAYMENT** R061636-20									
0030	Plywood, underlayment grade, 3/8" thick	2 Carp	1500	.011	SF Flr.	.77	.36		1.13	1.45
0070	Pneumatic nailed		1860	.009		.77	.29		1.06	1.33
0100	1/2" thick		1450	.011		.80	.37		1.17	1.50
0105	Pneumatic nailed		1798	.009		.80	.30		1.10	1.38
0200	5/8" thick		1400	.011		1.01	.38		1.39	1.75
0205	Pneumatic nailed		1736	.009		1.01	.31		1.32	1.63
0300	3/4" thick		1300	.012		1.23	.41		1.64	2.04
0305	Pneumatic nailed		1612	.010		1.23	.33		1.56	1.91
0500	Particle board, 3/8" thick G		1500	.011		.33	.36		.69	.96
0505	Pneumatic nailed G		1860	.009		.33	.29		.62	.84
0600	1/2" thick G		1450	.011		.36	.37		.73	1.02
0605	Pneumatic nailed G		1798	.009		.36	.30		.66	.90
0800	5/8" thick G		1400	.011		.46	.38		.84	1.15
0805	Pneumatic nailed G		1736	.009		.46	.31		.77	1.03
0900	3/4" thick G		1300	.012		.65	.41		1.06	1.41
0905	Pneumatic nailed G		1612	.010		.65	.33		.98	1.28
0955	Particleboard, 100% recycled straw/wheat, 4' x 8' x 1/4" G		1450	.011	S.F.	.26	.37		.63	.91
0960	4' x 8' x 3/8" G		1450	.011		.39	.37		.76	1.05
0965	4' x 8' x 1/2" G		1350	.012		.53	.40		.93	1.25
0970	4' x 8' x 5/8" G		1300	.012		.63	.41		1.04	1.38
0975	4' x 8' x 3/4" G		1250	.013		.73	.43		1.16	1.52
0980	4' x 8' x 1" G		1150	.014		.98	.47		1.45	1.86
0985	4' x 8' x 1-1/4" G		1100	.015		1.15	.49		1.64	2.09
1100	Hardboard, underlayment grade, 4' x 4', .215" thick G		1500	.011	SF Flr.	.57	.36		.93	1.23

06 16 36 – Wood Panel Product Sheathing

06 16 36.10 Sheathing		Crew	Daily Output	Labor-Hours	Unit	Material	Labor (2011 Bare Costs)	Equipment	Total	Total Incl O&P
0010	**SHEATHING** R061110-30									
0012	Plywood on roofs, CDX									
0030	5/16" thick	2 Carp	1600	.010	S.F.	.55	.34		.89	1.16
0035	Pneumatic nailed R061636-20		1952	.008		.55	.28		.83	1.06
0050	3/8" thick		1525	.010		.47	.35		.82	1.11
0055	Pneumatic nailed		1860	.009		.47	.29		.76	1
0100	1/2" thick		1400	.011		.58	.38		.96	1.28
0105	Pneumatic nailed		1708	.009		.58	.31		.89	1.17
0200	5/8" thick		1300	.012		.70	.41		1.11	1.46

06 16 Sheathing

06 16 36 – Wood Panel Product Sheathing

06 16 36.10 Sheathing		Crew	Daily Output	Labor-Hours	Unit	2011 Bare Costs Material	Labor	Equipment	Total	Total Incl O&P
0205	Pneumatic nailed	2 Carp	1586	.010	S.F.	.70	.34		1.04	1.34
0300	3/4" thick		1200	.013		.82	.45		1.27	1.65
0305	Pneumatic nailed		1464	.011		.82	.37		1.19	1.51
0500	Plywood on walls with exterior CDX, 3/8" thick		1200	.013		.47	.45		.92	1.27
0505	Pneumatic nailed		1488	.011		.47	.36		.83	1.12
0600	1/2" thick		1125	.014		.58	.48		1.06	1.44
0605	Pneumatic nailed		1395	.011		.58	.38		.96	1.28
0700	5/8" thick		1050	.015		.70	.51		1.21	1.63
0705	Pneumatic nailed		1302	.012		.70	.41		1.11	1.46
0800	3/4" thick		975	.016		.82	.55		1.37	1.82
0805	Pneumatic nailed		1209	.013		.82	.44		1.26	1.64
0840	Oriented strand board, 7/16" thick G		1400	.011		.33	.38		.71	1
0845	Pneumatic nailed G		1736	.009		.33	.31		.64	.88
0846	1/2" thick G		1325	.012		.33	.41		.74	1.04
0847	Pneumatic nailed G		1643	.010		.33	.33		.66	.91
0852	5/8" thick G		1250	.013		.79	.43		1.22	1.59
0857	Pneumatic nailed G		1550	.010		.79	.35		1.14	1.45
1000	For shear wall construction, add						20%			
1200	For structural 1 exterior plywood, add				S.F.	10%				
1400	With boards, on roof 1" x 6" boards, laid horizontal	2 Carp	725	.022		1.47	.74		2.21	2.85
1500	Laid diagonal		650	.025		1.47	.83		2.30	2.99
1700	1" x 8" boards, laid horizontal		875	.018		1.55	.61		2.16	2.73
1800	Laid diagonal		725	.022		1.55	.74		2.29	2.94
2000	For steep roofs, add						40%			
2200	For dormers, hips and valleys, add					5%	50%			
2400	Boards on walls, 1" x 6" boards, laid regular	2 Carp	650	.025		1.47	.83		2.30	2.99
2500	Laid diagonal		585	.027		1.47	.92		2.39	3.15
2700	1" x 8" boards, laid regular		765	.021		1.55	.70		2.25	2.87
2800	Laid diagonal		650	.025		1.55	.83		2.38	3.08
2850	Gypsum, weatherproof, 1/2" thick		1125	.014		.38	.48		.86	1.22
2900	With embedded glass mats		1100	.015		.75	.49		1.24	1.65
3000	Wood fiber, regular, no vapor barrier, 1/2" thick		1200	.013		.56	.45		1.01	1.37
3100	5/8" thick		1200	.013		.71	.45		1.16	1.53
3300	No vapor barrier, in colors, 1/2" thick		1200	.013		.68	.45		1.13	1.50
3400	5/8" thick		1200	.013		.72	.45		1.17	1.54
3600	With vapor barrier one side, white, 1/2" thick		1200	.013		.55	.45		1	1.36
3700	Vapor barrier 2 sides, 1/2" thick		1200	.013		.77	.45		1.22	1.60
3800	Asphalt impregnated, 25/32" thick		1200	.013		.28	.45		.73	1.06
3850	Intermediate, 1/2" thick		1200	.013		.20	.45		.65	.97

06 17 Shop-Fabricated Structural Wood

06 17 33 – Wood I-Joists

06 17 33.10 Wood and Composite I-Joists		Crew	Daily Output	Labor-Hours	Unit	Material	Labor	Equipment	Total	Total Incl O&P
0010	**WOOD AND COMPOSITE I-JOISTS**									
0100	Plywood webs, incl. bridging & blocking, panels 24" O.C.									
1200	15' to 24' span, 50 psf live load	F-5	2400	.013	SF Flr.	1.76	.39		2.15	2.59
1300	55 psf live load		2250	.014		1.98	.42		2.40	2.88
1400	24' to 30' span, 45 psf live load		2600	.012		2.73	.36		3.09	3.61
1500	55 psf live load		2400	.013		3.38	.39		3.77	4.38

06 17 Shop-Fabricated Structural Wood

06 17 53 – Shop-Fabricated Wood Trusses

06 17 53.10 Roof Trusses		Crew	Daily Output	Labor-Hours	Unit	2011 Bare Costs Material	Labor	Equipment	Total	Total Incl O&P
0010	**ROOF TRUSSES**									
0100	Fink (W) or King post type, 2'-0" O.C.									
0200	Metal plate connected, 4 in 12 slope									
0210	24' to 29' span	F-3	3000	.013	SF Flr.	1.50	.41	.22	2.13	2.57
0300	30' to 43' span		3000	.013		2.18	.41	.22	2.81	3.32
0400	44' to 60' span		3000	.013		2.36	.41	.22	2.99	3.51
0700	Glued and nailed, add					50%				

06 18 Glued-Laminated Construction

06 18 13 – Glued-Laminated Beams

06 18 13.20 Laminated Framing		Crew	Daily Output	Labor-Hours	Unit	2011 Bare Costs Material	Labor	Equipment	Total	Total Incl O&P
0010	**LAMINATED FRAMING**									
0020	30 lb., short term live load, 15 lb. dead load									
0200	Straight roof beams, 20' clear span, beams 8' O.C.	F-3	2560	.016	SF Flr.	1.99	.48	.26	2.73	3.27
0300	Beams 16' O.C.		3200	.013		1.44	.38	.20	2.02	2.45
0500	40' clear span, beams 8' O.C.		3200	.013		3.80	.38	.20	4.38	5.05
0600	Beams 16' O.C.		3840	.010		3.11	.32	.17	3.60	4.14
0800	60' clear span, beams 8' O.C.	F-4	2880	.014		6.55	.43	.42	7.40	8.35
0900	Beams 16' O.C.	"	3840	.010		4.87	.32	.31	5.50	6.25
1100	Tudor arches, 30' to 40' clear span, frames 8' O.C.	F-3	1680	.024		8.50	.73	.39	9.62	11
1200	Frames 16' O.C.	"	2240	.018		6.70	.55	.29	7.54	8.60
1400	50' to 60' clear span, frames 8' O.C.	F-4	2200	.018		9.20	.56	.55	10.31	11.65
1500	Frames 16' O.C.		2640	.015		7.80	.47	.46	8.73	9.85
1700	Radial arches, 60' clear span, frames 8' O.C.		1920	.021		8.60	.64	.63	9.87	11.20
1800	Frames 16' O.C.		2880	.014		6.60	.43	.42	7.45	8.40
2000	100' clear span, frames 8' O.C.		1600	.025		8.90	.77	.76	10.43	11.90
2100	Frames 16' O.C.		2400	.017		7.80	.51	.50	8.81	10
2300	120' clear span, frames 8' O.C.		1440	.028		11.85	.85	.84	13.54	15.35
2400	Frames 16' O.C.		1920	.021		10.80	.64	.63	12.07	13.65
2600	Bowstring trusses, 20' O.C., 40' clear span	F-3	2400	.017		5.35	.51	.27	6.13	7
2700	60' clear span	F-4	3600	.011		4.79	.34	.34	5.47	6.20
2800	100' clear span		4000	.010		6.80	.31	.30	7.41	8.30
2900	120' clear span		3600	.011		7.25	.34	.34	7.93	8.95
3000	For less than 1000 B.F., add					20%				
3050	For over 5000 B.F., deduct					10%				
3100	For premium appearance, add to S.F. prices					5%				
3300	For industrial type, deduct					15%				
3500	For stain and varnish, add					5%				
3900	For 3/4" laminations, add to straight					25%				
4100	Add to curved					15%				
4300	Alternate pricing method: (use nominal footage of									
4310	components). Straight beams, camber less than 6"	F-3	3.50	11.429	M.B.F.	2,950	350	187	3,487	4,050
4400	Columns, including hardware		2	20		3,175	615	330	4,120	4,875
4600	Curved members, radius over 32'		2.50	16		3,250	490	262	4,002	4,675
4700	Radius 10' to 32'		3	13.333		3,225	410	219	3,854	4,450
4900	For complicated shapes, add maximum					100%				
5100	For pressure treating, add to straight					35%				
5200	Add to curved					45%				
6000	Laminated veneer members, southern pine or western species									
6050	1-3/4" wide x 5-1/2" deep	2 Carp	480	.033	L.F.	3.19	1.12		4.31	5.40

06 18 13.20 Laminated Framing		Crew	Daily Output	Labor-Hours	Unit	Material	2011 Bare Costs Labor	Equipment	Total	Total Incl O&P
6100	9-1/2" deep	2 Carp	480	.033	L.F.	4.56	1.12		5.68	6.85
6150	14" deep		450	.036		7.30	1.19		8.49	10.05
6200	18" deep		450	.036		9.90	1.19		11.09	12.85
6300	Parallel strand members, southern pine or western species									
6350	1-3/4" wide x 9-1/4" deep	2 Carp	480	.033	L.F.	4.56	1.12		5.68	6.85
6400	11-1/4" deep		450	.036		5.70	1.19		6.89	8.25
6450	14" deep		400	.040		7.30	1.34		8.64	10.30
6500	3-1/2" wide x 9-1/4" deep		480	.033		15.85	1.12		16.97	19.30
6550	11-1/4" deep		450	.036		19.15	1.19		20.34	23
6600	14" deep		400	.040		23.50	1.34		24.84	28
6650	7" wide x 9-1/4" deep		450	.036		31.50	1.19		32.69	37
6700	11-1/4" deep		420	.038		39.50	1.28		40.78	45.50
6750	14" deep		400	.040		46.50	1.34		47.84	54
8000	Straight beams									
8102	20' span									
8104	3-1/8" x 9"	F-3	30	1.333	Ea.	138	41	22	201	244
8106	X 10-1/2"		30	1.333		161	41	22	224	269
8108	X 12"		30	1.333		184	41	22	247	295
8110	X 13-1/2"		30	1.333		207	41	22	270	320
8112	X 15"		29	1.379		230	42.50	22.50	295	350
8114	5-1/8" x 10-1/2"		30	1.333		265	41	22	328	385
8116	X 12"		30	1.333		300	41	22	363	425
8118	X 13-1/2"		30	1.333		340	41	22	403	465
8120	X 15"		29	1.379		380	42.50	22.50	445	510
8122	X 16-1/2"		29	1.379		415	42.50	22.50	480	550
8124	X 18"		29	1.379		455	42.50	22.50	520	595
8126	X 19-1/2"		29	1.379		490	42.50	22.50	555	635
8128	X 21"		28	1.429		530	44	23.50	597.50	680
8130	X 22-1/2"		28	1.429		565	44	23.50	632.50	725
8132	X 24"		28	1.429		605	44	23.50	672.50	765
8134	6-3/4" x 12"		29	1.379		400	42.50	22.50	465	535
8136	X 13-1/2"		29	1.379		450	42.50	22.50	515	590
8138	X 15"		29	1.379		500	42.50	22.50	565	645
8140	X 16-1/2"		28	1.429		550	44	23.50	617.50	700
8142	X 18"		28	1.429		595	44	23.50	662.50	755
8144	X 19-1/2"		28	1.429		645	44	23.50	712.50	810
8146	X 21"		27	1.481		695	45.50	24.50	765	865
8148	X 22-1/2"		27	1.481		745	45.50	24.50	815	920
8150	X 24"		27	1.481		795	45.50	24.50	865	975
8152	X 25-1/2"		27	1.481		845	45.50	24.50	915	1,025
8154	X 27"		26	1.538		895	47.50	25	967.50	1,100
8156	X 28-1/2"		26	1.538		945	47.50	25	1,017.50	1,150
8158	X 30"		26	1.538		995	47.50	25	1,067.50	1,200
8200	30' span									
8250	3-1/8" x 9"	F-3	30	1.333	Ea.	207	41	22	270	320
8252	X 10-1/2"		30	1.333		242	41	22	305	360
8254	X 12"		30	1.333		277	41	22	340	395
8256	X 13-1/2"		30	1.333		310	41	22	373	430
8258	X 15"		29	1.379		345	42.50	22.50	410	475
8260	5-1/8" x 10-1/2"		30	1.333		395	41	22	458	525
8262	X 12"		30	1.333		455	41	22	518	590
8264	X 13-1/2"		30	1.333		510	41	22	573	650
8266	X 15"		29	1.379		565	42.50	22.50	630	720

06 18 Glued-Laminated Construction

06 18 13 – Glued-Laminated Beams

06 18 13.20 Laminated Framing		Crew	Daily Output	Labor-Hours	Unit	Material	Labor	Equipment	Total	Total Incl O&P
						2011 Bare Costs				
8268	X 16-1/2"	F-3	29	1.379	Ea.	625	42.50	22.50	690	780
8270	X 18"		29	1.379		680	42.50	22.50	745	845
8272	X 19-1/2"		29	1.379		735	42.50	22.50	800	905
8274	X 21"		28	1.429		795	44	23.50	862.50	975
8276	X 22-1/2"		28	1.429		850	44	23.50	917.50	1,025
8278	X 24"		28	1.429		905	44	23.50	972.50	1,100
8280	6-3/4" x 12"		29	1.379		595	42.50	22.50	660	750
8282	X 13-1/2"		29	1.379		670	42.50	22.50	735	835
8284	X 15"		29	1.379		745	42.50	22.50	810	915
8286	X 16-1/2"		28	1.429		820	44	23.50	887.50	1,000
8288	X 18"		28	1.429		895	44	23.50	962.50	1,075
8290	X 19-1/2"		28	1.429		970	44	23.50	1,037.50	1,175
8292	X 21"		27	1.481		1,050	45.50	24.50	1,120	1,250
8294	X 22-1/2"		27	1.481		1,125	45.50	24.50	1,195	1,325
8296	X 24"		27	1.481		1,200	45.50	24.50	1,270	1,425
8298	X 25-1/2"		27	1.481		1,275	45.50	24.50	1,345	1,500
8300	X 27"		26	1.538		1,350	47.50	25	1,422.50	1,575
8302	X 28-1/2"		26	1.538		1,425	47.50	25	1,497.50	1,650
8304	X 30"	↓	26	1.538	↓	1,500	47.50	25	1,572.50	1,750
8400	40' span									
8402	3-1/8" x 9"	F-3	30	1.333	Ea.	277	41	22	340	395
8404	X 10-1/2"		30	1.333		325	41	22	388	445
8406	X 12"		30	1.333		370	41	22	433	495
8408	X 13-1/2"		30	1.333		415	41	22	478	545
8410	X 15"		29	1.379		460	42.50	22.50	525	600
8412	5-1/8" x 10-1/2"		30	1.333		530	41	22	593	670
8414	X 12"		30	1.333		605	41	22	668	755
8416	X 13-1/2"		30	1.333		680	41	22	743	840
8418	X 15"		29	1.379		755	42.50	22.50	820	925
8420	X 16-1/2"		29	1.379		830	42.50	22.50	895	1,000
8422	X 18"		29	1.379		905	42.50	22.50	970	1,100
8424	X 19-1/2"		29	1.379		985	42.50	22.50	1,050	1,175
8426	X 21"		28	1.429		1,050	44	23.50	1,117.50	1,275
8428	X 22-1/2"		28	1.429		1,125	44	23.50	1,192.50	1,350
8430	X 24"		28	1.429		1,200	44	23.50	1,267.50	1,425
8432	6-3/4" x 12"		29	1.379		795	42.50	22.50	860	970
8434	X 13-1/2"		29	1.379		895	42.50	22.50	960	1,075
8436	X 15"		29	1.379		995	42.50	22.50	1,060	1,200
8438	X 16-1/2"		28	1.429		1,100	44	23.50	1,167.50	1,300
8440	X 18"		28	1.429		1,200	44	23.50	1,267.50	1,425
8442	X 19-1/2"		28	1.429		1,300	44	23.50	1,367.50	1,525
8444	X 21"		27	1.481		1,400	45.50	24.50	1,470	1,625
8446	X 22-1/2"		27	1.481		1,500	45.50	24.50	1,570	1,750
8448	X 24"		27	1.481		1,600	45.50	24.50	1,670	1,850
8450	X 25-1/2"		27	1.481		1,700	45.50	24.50	1,770	1,950
8452	X 27"		26	1.538		1,800	47.50	25	1,872.50	2,075
8454	X 28-1/2"		26	1.538		1,900	47.50	25	1,972.50	2,175
8456	X 30"	↓	26	1.538	↓	2,000	47.50	25	2,072.50	2,300

06 22 Millwork

06 22 13 – Standard Pattern Wood Trim

06 22 13.15 Moldings, Base

		Crew	Daily Output	Labor-Hours	Unit	2011 Bare Costs Material	Labor	Equipment	Total	Total Incl O&P
0010	**MOLDINGS, BASE**									
0500	Base, stock pine, 9/16" x 3-1/2"	1 Carp	240	.033	L.F.	2.22	1.12		3.34	4.31
0550	9/16" x 4-1/2"		200	.040		2.46	1.34		3.80	4.96
0561	Base shoe, oak, 3/4" x 1"		240	.033		1.13	1.12		2.25	3.11

06 22 13.30 Moldings, Casings

		Crew	Daily Output	Labor-Hours	Unit	Material	Labor	Equipment	Total	Total Incl O&P
0010	**MOLDINGS, CASINGS**									
0090	Apron, stock pine, 5/8" x 2"	1 Carp	250	.032	L.F.	1.22	1.07		2.29	3.14
0110	5/8" x 3-1/2"		220	.036		1.62	1.22		2.84	3.82
0300	Band, stock pine, 11/16" x 1-1/8"		270	.030		.82	.99		1.81	2.56
0350	11/16" x 1-3/4"		250	.032		1.06	1.07		2.13	2.97
0700	Casing, stock pine, 11/16" x 2-1/2"		240	.033		1.33	1.12		2.45	3.33
0750	11/16" x 3-1/2"		215	.037		1.62	1.25		2.87	3.87

06 22 13.35 Moldings, Ceilings

		Crew	Daily Output	Labor-Hours	Unit	Material	Labor	Equipment	Total	Total Incl O&P
0010	**MOLDINGS, CEILINGS**									
0600	Bed, stock pine, 9/16" x 1-3/4"	1 Carp	270	.030	L.F.	1.02	.99		2.01	2.78
0650	9/16" x 2"		240	.033		1.22	1.12		2.34	3.21
1200	Cornice molding, stock pine, 9/16" x 1-3/4"		330	.024		1.02	.81		1.83	2.48
1300	9/16" x 2-1/4"		300	.027		1.24	.89		2.13	2.86
2400	Cove scotia, stock pine, 9/16" x 1-3/4"		270	.030		.67	.99		1.66	2.40
2500	11/16" x 2-3/4"		255	.031		1.86	1.05		2.91	3.81
2600	Crown, stock pine, 9/16" x 3-5/8"		250	.032		2.30	1.07		3.37	4.33
2700	11/16" x 4-5/8"		220	.036		3.87	1.22		5.09	6.30

06 22 13.40 Moldings, Exterior

		Crew	Daily Output	Labor-Hours	Unit	Material	Labor	Equipment	Total	Total Incl O&P
0010	**MOLDINGS, EXTERIOR**									
1500	Cornice, boards, pine, 1" x 2"	1 Carp	330	.024	L.F.	.25	.81		1.06	1.63
1700	1" x 6"		250	.032		.67	1.07		1.74	2.54
2000	1" x 12"		180	.044		1.46	1.49		2.95	4.11
2200	Three piece, built-up, pine, minimum		80	.100		1.45	3.36		4.81	7.20
2300	Maximum		65	.123		3.64	4.13		7.77	10.90
3000	Corner board, sterling pine, 1" x 4"		200	.040		.79	1.34		2.13	3.12
3100	1" x 6"		200	.040		1.05	1.34		2.39	3.41
3350	Fascia, sterling pine, 1" x 6"		250	.032		1.05	1.07		2.12	2.96
3370	1" x 8"		225	.036		1.43	1.19		2.62	3.57
3400	Trim, back band, 11/16" x 1-1/16"		250	.032		.83	1.07		1.90	2.71
3500	Casing, 11/16" x 4-1/4"		250	.032		2.74	1.07		3.81	4.81
3600	Crown, 11/16" x 4-1/4"		250	.032		3.85	1.07		4.92	6.05
3700	1" x 4" and 1" x 6" railing w/balusters 4" O.C.		22	.364		22.50	12.20		34.70	45
3800	Insect screen framing stock, 1-1/16" x 1-3/4"		395	.020		2.03	.68		2.71	3.37
4100	Verge board, sterling pine, 1" x 4"		200	.040		.79	1.34		2.13	3.12
4200	1" x 6"		200	.040		1.05	1.34		2.39	3.41
4300	2" x 6"		165	.048		2.10	1.63		3.73	5.05
4400	2" x 8"		165	.048		2.86	1.63		4.49	5.85
4700	For redwood trim, add					200%				

06 22 13.45 Moldings, Trim

		Crew	Daily Output	Labor-Hours	Unit	Material	Labor	Equipment	Total	Total Incl O&P
0010	**MOLDINGS, TRIM**									
0200	Astragal, stock pine, 11/16" x 1-3/4"	1 Carp	255	.031	L.F.	1.22	1.05		2.27	3.10
0250	1-5/16" x 2-3/16"		240	.033		2.57	1.12		3.69	4.70
0800	Chair rail, stock pine, 5/8" x 2-1/2"		270	.030		1.43	.99		2.42	3.23
0900	5/8" x 3-1/2"		240	.033		2.26	1.12		3.38	4.36
1000	Closet pole, stock pine, 1-1/8" diameter		200	.040		1.05	1.34		2.39	3.41
1100	Fir, 1-5/8" diameter		200	.040		1.93	1.34		3.27	4.37
3300	Half round, stock pine, 1/4" x 1/2"		270	.030		.24	.99		1.23	1.92

06 22 Millwork

06 22 13 – Standard Pattern Wood Trim

06 22 13.45 Moldings, Trim		Crew	Daily Output	Labor-Hours	Unit	Material	Labor	Equipment	Total	Total Incl O&P
						2011 Bare Costs				
3350	1/2" x 1"	1 Carp	255	.031	L.F.	.62	1.05		1.67	2.44
3400	Handrail, fir, single piece, stock, hardware not included									
3450	1-1/2" x 1-3/4"	1 Carp	80	.100	L.F.	2.03	3.36		5.39	7.85
3470	Pine, 1-1/2" x 1-3/4"		80	.100		2.03	3.36		5.39	7.85
3500	1-1/2" x 2-1/2"		76	.105		3.04	3.53		6.57	9.25
3600	Lattice, stock pine, 1/4" x 1-1/8"		270	.030		.42	.99		1.41	2.12
3700	1/4" x 1-3/4"		250	.032		.65	1.07		1.72	2.52
3800	Miscellaneous, custom, pine, 1" x 1"		270	.030		.35	.99		1.34	2.05
3900	1" x 3"		240	.033		1.06	1.12		2.18	3.04
4100	Birch or oak, nominal 1" x 1"		240	.033		.47	1.12		1.59	2.38
4200	Nominal 1" x 3"		215	.037		1.40	1.25		2.65	3.63
4400	Walnut, nominal 1" x 1"		215	.037		.51	1.25		1.76	2.65
4500	Nominal 1" x 3"		200	.040		1.53	1.34		2.87	3.94
4700	Teak, nominal 1" x 1"		215	.037		2.47	1.25		3.72	4.81
4800	Nominal 1" x 3"		200	.040		7.40	1.34		8.74	10.40
4900	Quarter round, stock pine, 1/4" x 1/4"		275	.029		.22	.98		1.20	1.87
4950	3/4" x 3/4"		255	.031		.60	1.05		1.65	2.42
5600	Wainscot moldings, 1-1/8" x 9/16", 2' high, minimum		76	.105	S.F.	11.15	3.53		14.68	18.20
5700	Maximum		65	.123	"	20	4.13		24.13	29.50
06 22 13.50 Moldings, Window and Door										
0010	**MOLDINGS, WINDOW AND DOOR**									
2800	Door moldings, stock, decorative, 1-1/8" wide, plain	1 Carp	17	.471	Set	45	15.80		60.80	76
2900	Detailed		17	.471	"	87.50	15.80		103.30	123
2960	Clear pine door jamb, no stops, 11/16" x 4-9/16"		240	.033	L.F.	5.15	1.12		6.27	7.50
3150	Door trim set, 1 head and 2 sides, pine, 2-1/2 wide		12	.667	Opng.	22.50	22.50		45	62.50
3170	3-1/2" wide		11	.727	"	27.50	24.50		52	71.50
3250	Glass beads, stock pine, 3/8" x 1/2"		275	.029	L.F.	.31	.98		1.29	1.97
3270	3/8" x 7/8"		270	.030		.41	.99		1.40	2.11
4850	Parting bead, stock pine, 3/8" x 3/4"		275	.029		.36	.98		1.34	2.03
4870	1/2" x 3/4"		255	.031		.45	1.05		1.50	2.26
5000	Stool caps, stock pine, 11/16" x 3-1/2"		200	.040		2.27	1.34		3.61	4.75
5100	1-1/16" x 3-1/4"		150	.053		3.56	1.79		5.35	6.90
5300	Threshold, oak, 3' long, inside, 5/8" x 3-5/8"		32	.250	Ea.	9	8.40		17.40	24
5400	Outside, 1-1/2" x 7-5/8"		16	.500	"	38	16.80		54.80	70
5900	Window trim sets, including casings, header, stops,									
5910	stool and apron, 2-1/2" wide, minimum	1 Carp	13	.615	Opng.	32.50	20.50		53	70
5950	Average		10	.800		38	27		65	87
6000	Maximum		6	1.333		62.50	44.50		107	144
06 22 13.60 Moldings, Soffits										
0010	**MOLDINGS, SOFFITS**									
0200	Soffits, pine, 1" x 4"	2 Carp	420	.038	L.F.	.53	1.28		1.81	2.72
0210	1" x 6"		420	.038		.67	1.28		1.95	2.88
0220	1" x 8"		420	.038		.97	1.28		2.25	3.20
0230	1" x 10"		400	.040		1.21	1.34		2.55	3.58
0240	1" x 12"		400	.040		1.46	1.34		2.80	3.86
0250	STK cedar, 1" x 4"		420	.038		.43	1.28		1.71	2.61
0260	1" x 6"		420	.038		.76	1.28		2.04	2.98
0270	1" x 8"		420	.038		1.18	1.28		2.46	3.44
0280	1" x 10"		400	.040		1.68	1.34		3.02	4.10
0290	1" x 12"		400	.040		2.49	1.34		3.83	4.99
1000	Exterior AC plywood, 1/4" thick		420	.038	S.F.	.62	1.28		1.90	2.82
1050	3/8" thick		420	.038		.77	1.28		2.05	2.99

06 22 Millwork

06 22 13 – Standard Pattern Wood Trim

06 22 13.60 Moldings, Soffits		Crew	Daily Output	Labor-Hours	Unit	Material	Labor	Equipment	Total	Total Incl O&P
						2011 Bare Costs				
1100	1/2" thick	2 Carp	420	.038	S.F.	.80	1.28		2.08	3.02
1150	Polyvinyl chloride, white, solid	1 Carp	230	.035		1.96	1.17		3.13	4.11
1160	Perforated	"	230	.035	↓	1.96	1.17		3.13	4.11

06 25 Prefinished Paneling

06 25 13 – Prefinished Hardboard Paneling

06 25 13.10 Paneling, Hardboard

		Crew	Daily Output	Labor-Hours	Unit	Material	Labor	Equipment	Total	Total Incl O&P
0010	**PANELING, HARDBOARD**									
0050	Not incl. furring or trim, hardboard, tempered, 1/8" thick G	2 Carp	500	.032	S.F.	.38	1.07		1.45	2.22
0100	1/4" thick G		500	.032		.55	1.07		1.62	2.41
0300	Tempered pegboard, 1/8" thick G		500	.032		.45	1.07		1.52	2.30
0400	1/4" thick G		500	.032		.64	1.07		1.71	2.50
0600	Untempered hardboard, natural finish, 1/8" thick G		500	.032		.38	1.07		1.45	2.22
0700	1/4" thick G		500	.032		.46	1.07		1.53	2.31
0900	Untempered pegboard, 1/8" thick G		500	.032		.39	1.07		1.46	2.23
1000	1/4" thick G		500	.032		.42	1.07		1.49	2.26
1200	Plastic faced hardboard, 1/8" thick G		500	.032		.68	1.07		1.75	2.55
1300	1/4" thick G		500	.032		.84	1.07		1.91	2.72
1500	Plastic faced pegboard, 1/8" thick G		500	.032		.62	1.07		1.69	2.48
1600	1/4" thick G		500	.032		.74	1.07		1.81	2.61
1800	Wood grained, plain or grooved, 1/4" thick, minimum G		500	.032		.58	1.07		1.65	2.44
1900	Maximum G		425	.038	↓	1.12	1.26		2.38	3.34
2100	Moldings for hardboard, wood or aluminum, minimum		500	.032	L.F.	.39	1.07		1.46	2.23
2200	Maximum	↓	425	.038	"	1.09	1.26		2.35	3.31

06 25 16 – Prefinished Plywood Paneling

06 25 16.10 Paneling, Plywood

		Crew	Daily Output	Labor-Hours	Unit	Material	Labor	Equipment	Total	Total Incl O&P
0010	**PANELING, PLYWOOD** R061636-20									
2400	Plywood, prefinished, 1/4" thick, 4' x 8' sheets									
2410	with vertical grooves. Birch faced, minimum	2 Carp	500	.032	S.F.	.89	1.07		1.96	2.78
2420	Average		420	.038		1.35	1.28		2.63	3.63
2430	Maximum		350	.046		1.97	1.53		3.50	4.74
2600	Mahogany, African		400	.040		2.51	1.34		3.85	5
2700	Philippine (Lauan)		500	.032		1.08	1.07		2.15	2.99
2900	Oak or Cherry, minimum		500	.032		2.10	1.07		3.17	4.11
3000	Maximum		400	.040		3.22	1.34		4.56	5.80
3200	Rosewood		320	.050		4.58	1.68		6.26	7.85
3400	Teak		400	.040		3.22	1.34		4.56	5.80
3600	Chestnut		375	.043		4.77	1.43		6.20	7.65
3800	Pecan		400	.040		2.06	1.34		3.40	4.52
3900	Walnut, minimum		500	.032		2.75	1.07		3.82	4.83
3950	Maximum		400	.040		5.20	1.34		6.54	8
4000	Plywood, prefinished, 3/4" thick, stock grades, minimum		320	.050		1.24	1.68		2.92	4.17
4100	Maximum		224	.071		5.40	2.40		7.80	9.90
4300	Architectural grade, minimum		224	.071		3.98	2.40		6.38	8.40
4400	Maximum		160	.100		6.10	3.36		9.46	12.30
4600	Plywood, "A" face, birch, V.C., 1/2" thick, natural		450	.036		1.89	1.19		3.08	4.08
4700	Select		450	.036		2.06	1.19		3.25	4.27
4900	Veneer core, 3/4" thick, natural		320	.050		1.99	1.68		3.67	5
5000	Select		320	.050		2.23	1.68		3.91	5.25
5200	Lumber core, 3/4" thick, natural	↓	320	.050	↓	2.99	1.68		4.67	6.10

06 25 Prefinished Paneling

06 25 16 – Prefinished Plywood Paneling

06 25 16.10 Paneling, Plywood		Crew	Daily Output	Labor-Hours	Unit	Material	2011 Bare Costs Labor	Equipment	Total	Total Incl O&P
5500	Plywood, knotty pine, 1/4" thick, A2 grade	2 Carp	450	.036	S.F.	1.63	1.19		2.82	3.79
5600	A3 grade		450	.036		2.06	1.19		3.25	4.27
5800	3/4" thick, veneer core, A2 grade		320	.050		2.11	1.68		3.79	5.15
5900	A3 grade		320	.050		2.38	1.68		4.06	5.45
6100	Aromatic cedar, 1/4" thick, plywood		400	.040		2.08	1.34		3.42	4.54
6200	1/4" thick, particle board	↓	400	.040	↓	1.01	1.34		2.35	3.36

06 26 Board Paneling

06 26 13 – Profile Board Paneling

06 26 13.10 Paneling, Boards

		Crew	Daily Output	Labor-Hours	Unit	Material	Labor	Equipment	Total	Total Incl O&P
0010	**PANELING, BOARDS**									
6400	Wood board paneling, 3/4" thick, knotty pine	2 Carp	300	.053	S.F.	1.89	1.79		3.68	5.05
6500	Rough sawn cedar		300	.053		2.86	1.79		4.65	6.15
6700	Redwood, clear, 1" x 4" boards		300	.053		4.96	1.79		6.75	8.45
6900	Aromatic cedar, closet lining, boards	↓	275	.058	↓	2	1.95		3.95	5.45

06 43 Wood Stairs and Railings

06 43 13 – Wood Stairs

06 43 13.20 Prefabricated Wood Stairs

		Crew	Daily Output	Labor-Hours	Unit	Material	Labor	Equipment	Total	Total Incl O&P
0010	**PREFABRICATED WOOD STAIRS**									
0100	Box stairs, prefabricated, 3'-0" wide									
0110	Oak treads, up to 14 risers	2 Carp	39	.410	Riser	85	13.75		98.75	117
0600	With pine treads for carpet, up to 14 risers	"	39	.410	"	55	13.75		68.75	83.50
1100	For 4' wide stairs, add				Flight	25%				
1550	Stairs, prefabricated stair handrail with balusters	1 Carp	30	.267	L.F.	75	8.95		83.95	97
1700	Basement stairs, prefabricated, pine treads									
1710	Pine risers, 3' wide, up to 14 risers	2 Carp	52	.308	Riser	55	10.30		65.30	78
4000	Residential, wood, oak treads, prefabricated		1.50	10.667	Flight	1,100	360		1,460	1,800
4200	Built in place	↓	.44	36.364	"	1,750	1,225		2,975	3,975
4400	Spiral, oak, 4'-6" diameter, unfinished, prefabricated,									
4500	incl. railing, 9' high	2 Carp	1.50	10.667	Flight	3,625	360		3,985	4,575

06 43 13.40 Wood Stair Parts

		Crew	Daily Output	Labor-Hours	Unit	Material	Labor	Equipment	Total	Total Incl O&P
0010	**WOOD STAIR PARTS**									
0020	Pin top balusters, 1-1/4", oak, 34"	1 Carp	96	.083	Ea.	7.80	2.80		10.60	13.30
0030	38"		96	.083		9.85	2.80		12.65	15.55
0040	42"		96	.083		12.10	2.80		14.90	18
0050	Poplar, 34"		96	.083		3.05	2.80		5.85	8.05
0060	38"		96	.083		3.62	2.80		6.42	8.65
0070	42"		96	.083		8.55	2.80		11.35	14.15
0080	Maple, 34"		96	.083		4.90	2.80		7.70	10.10
0090	38"		96	.083		6.20	2.80		9	11.50
0100	42"		96	.083		7.60	2.80		10.40	13.05
0130	Primed, 34"		96	.083		2.85	2.80		5.65	7.80
0140	38"		96	.083		3.62	2.80		6.42	8.65
0150	42"		96	.083		4.42	2.80		7.22	9.55
0180	Box top balusters, 1-1/4", oak, 34"		60	.133		10.85	4.47		15.32	19.45
0190	38"		60	.133		12.50	4.47		16.97	21.50
0200	42"		60	.133		12.95	4.47		17.42	22
0210	Poplar, 34"	↓	60	.133	↓	8.65	4.47		13.12	17

06 43 13.40 Wood Stair Parts		Crew	Daily Output	Labor-Hours	Unit	2011 Bare Costs Material	Labor	Equipment	Total	Total Incl O&P
0220	38"	1 Carp	60	.133	Ea.	9.50	4.47		13.97	17.95
0230	42"		60	.133		10	4.47		14.47	18.50
0240	Maple, 34"		60	.133		11.90	4.47		16.37	20.50
0250	38"		60	.133		13.80	4.47		18.27	22.50
0260	42"		60	.133		14.20	4.47		18.67	23
0290	Primed, 34"		60	.133		6.95	4.47		11.42	15.15
0300	38"		60	.133		8	4.47		12.47	16.30
0310	42"		60	.133		8.35	4.47		12.82	16.70
0340	Square balusters, cut from lineal stock, pine, 1-1/16" x 1-1/16"		180	.044	L.F.	1.32	1.49		2.81	3.95
0350	1-5/16" x 1-5/16"		180	.044		1.70	1.49		3.19	4.37
0360	1-5/8" x 1-5/8"		180	.044		2.82	1.49		4.31	5.60
0370	Turned newel, oak, 3-1/2" square, 48" high		8	1	Ea.	94.50	33.50		128	160
0380	62" high		8	1		138	33.50		171.50	208
0390	Poplar, 3-1/2" square, 48" high		8	1		52	33.50		85.50	113
0400	62" high		8	1		75	33.50		108.50	139
0410	Maple, 3-1/2" square, 48" high		8	1		66	33.50		99.50	129
0420	62" high		8	1		93	33.50		126.50	158
0430	Square newel, oak, 3-1/2" square, 48" high		8	1		50	33.50		83.50	111
0440	58" high		8	1		70	33.50		103.50	133
0450	Poplar, 3-1/2" square, 48" high		8	1		49	33.50		82.50	110
0460	58" high		8	1		54	33.50		87.50	116
0470	Maple, 3" square, 48" high		8	1		50	33.50		83.50	111
0480	58" high		8	1		62	33.50		95.50	124
0490	Railings, oak, minimum		96	.083	L.F.	10.40	2.80		13.20	16.15
0500	Average		96	.083		12.50	2.80		15.30	18.45
0510	Maximum		96	.083		15.50	2.80		18.30	21.50
0520	Maple, minimum		96	.083		15	2.80		17.80	21
0530	Average		96	.083		15.25	2.80		18.05	21.50
0540	Maximum		96	.083		15.50	2.80		18.30	21.50
0550	Oak, for bending rail, minimum		48	.167		21	5.60		26.60	32.50
0560	Average		48	.167		22	5.60		27.60	33.50
0570	Maximum		48	.167		23	5.60		28.60	35
0580	Maple, for bending rail, minimum		48	.167		23	5.60		28.60	34.50
0590	Average		48	.167		25	5.60		30.60	36.50
0600	Maximum		48	.167		27	5.60		32.60	39
0610	Risers, oak, 3/4" x 8", 36" long		80	.100	Ea.	18	3.36		21.36	25.50
0620	42" long		80	.100		21	3.36		24.36	28.50
0630	48" long		80	.100		24	3.36		27.36	32
0640	54" long		80	.100		27	3.36		30.36	35
0650	60" long		80	.100		30	3.36		33.36	38.50
0660	72" long		80	.100		36	3.36		39.36	45
0670	Poplar, 3/4" x 8", 36" long		80	.100		11.10	3.36		14.46	17.80
0680	42" long		80	.100		12.95	3.36		16.31	19.85
0690	48" long		80	.100		14.80	3.36		18.16	22
0700	54" long		80	.100		16.65	3.36		20.01	24
0710	60" long		80	.100		18.50	3.36		21.86	26
0720	72" long		80	.100		22	3.36		25.36	30
0730	Pine, 1" x 8", 36" long		80	.100		2.90	3.36		6.26	8.80
0740	42" long		80	.100		3.38	3.36		6.74	9.30
0750	48" long		80	.100		3.87	3.36		7.23	9.85
0760	54" long		80	.100		4.35	3.36		7.71	10.40
0770	60" long		80	.100		4.83	3.36		8.19	10.90
0780	72" long		80	.100		5.80	3.36		9.16	12

06 43 Wood Stairs and Railings

06 43 13 – Wood Stairs

06 43 13.40 Wood Stair Parts		Crew	Daily Output	Labor-Hours	Unit	Material	Labor	Equipment	Total	Total Incl O&P
0790	Treads, oak, no returns, 1-1/32" x 11-1/2" x 36" long	1 Carp	32	.250	Ea.	32	8.40		40.40	49
0800	42" long		32	.250		37.50	8.40		45.90	55
0810	48" long		32	.250		42.50	8.40		50.90	61
0820	54" long		32	.250		48	8.40		56.40	67
0830	60" long		32	.250		53.50	8.40		61.90	72.50
0840	72" long		32	.250		64	8.40		72.40	84.50
0850	Mitred return one end, 1-1/32" x 11-1/2" x 36" long		24	.333		46	11.20		57.20	69
0860	42" long		24	.333		53.50	11.20		64.70	77.50
0870	48" long		24	.333		61.50	11.20		72.70	86
0880	54" long		24	.333		69	11.20		80.20	94.50
0890	60" long		24	.333		76.50	11.20		87.70	103
0900	72" long		24	.333		92	11.20		103.20	120
0910	Mitred return two ends, 1-1/32" x 11-1/2" x 36" long		12	.667		49	22.50		71.50	91.50
0920	42" long		12	.667		57	22.50		79.50	101
0930	48" long		12	.667		65.50	22.50		88	110
0940	54" long		12	.667		73.50	22.50		96	119
0950	60" long		12	.667		81.50	22.50		104	128
0960	72" long		12	.667		98	22.50		120.50	146
0970	Starting step, oak, 48", bullnose		8	1		192	33.50		225.50	267
0980	Double end bullnose		8	1	↓	345	33.50		378.50	435
1030	Skirt board, pine, 1" x 10"		55	.145	L.F.	1.21	4.88		6.09	9.50
1040	1" x 12"		52	.154	"	1.46	5.15		6.61	10.25
1050	Oak landing tread, 1-1/16" thick		54	.148	S.F.	8.25	4.97		13.22	17.40
1060	Oak cove molding		96	.083	L.F.	1.45	2.80		4.25	6.30
1070	Oak stringer molding		96	.083	"	2.75	2.80		5.55	7.70
1090	Rail bolt, 5/16" x 3-1/2"		48	.167	Ea.	2.75	5.60		8.35	12.40
1100	5/16" x 4-1/2"		48	.167		2.75	5.60		8.35	12.40
1120	Newel post anchor		16	.500		13.50	16.80		30.30	43
1130	Tapered plug, 1/2"		240	.033		.30	1.12		1.42	2.20
1140	1"	↓	240	.033	↓	.40	1.12		1.52	2.31

06 43 16 – Wood Railings

06 43 16.10 Wood Handrails and Railings		Crew	Daily Output	Labor-Hours	Unit	Material	Labor	Equipment	Total	Total Incl O&P
0010	**WOOD HANDRAILS AND RAILINGS**									
0020	Custom design, architectural grade, hardwood, minimum	1 Carp	38	.211	L.F.	9.50	7.05		16.55	22.50
0100	Maximum		30	.267		50	8.95		58.95	70
0300	Stock interior railing with spindles 4" O.C., 4' long		40	.200		37	6.70		43.70	52
0400	8' long	↓	48	.167	↓	37	5.60		42.60	50

06 44 Ornamental Woodwork

06 44 19 – Wood Grilles

06 44 19.10 Grilles		Crew	Daily Output	Labor-Hours	Unit	Material	Labor	Equipment	Total	Total Incl O&P
0010	**GRILLES** and panels, hardwood, sanded									
0020	2' x 4' to 4' x 8', custom designs, unfinished, minimum	1 Carp	38	.211	S.F.	54	7.05		61.05	71.50
0050	Average		30	.267		62.50	8.95		71.45	84
0100	Maximum	↓	19	.421	↓	71	14.15		85.15	102

06 44 33 – Wood Mantels

06 44 33.10 Fireplace Mantels		Crew	Daily Output	Labor-Hours	Unit	Material	Labor	Equipment	Total	Total Incl O&P
0010	**FIREPLACE MANTELS**									
0015	6" molding, 6' x 3'-6" opening, minimum	1 Carp	5	1.600	Opng.	228	53.50		281.50	340
0100	Maximum	↓	5	1.600	↓	410	53.50		463.50	545

(Material, Labor, Equipment and Total columns: 2011 Bare Costs)

06 44 Ornamental Woodwork

06 44 33 – Wood Mantels

06 44 33.10 Fireplace Mantels

	06 44 33.10 Fireplace Mantels	Crew	Daily Output	Labor-Hours	Unit	Material	Labor (2011 Bare Costs)	Equipment	Total	Total Incl O&P
0300	Prefabricated pine, colonial type, stock, deluxe	1 Carp	2	4	Opng.	1,250	134		1,384	1,625
0400	Economy		3	2.667		530	89.50		619.50	735

06 44 33.20 Fireplace Mantel Beam

	06 44 33.20 Fireplace Mantel Beam	Crew	Daily Output	Labor-Hours	Unit	Material	Labor	Equipment	Total	Total Incl O&P
0010	**FIREPLACE MANTEL BEAM**									
0020	Rough texture wood, 4" x 8"	1 Carp	36	.222	L.F.	8.30	7.45		15.75	21.50
0100	4" x 10"		35	.229	"	10.90	7.65		18.55	25
0300	Laminated hardwood, 2-1/4" x 10-1/2" wide, 6' long		5	1.600	Ea.	110	53.50		163.50	211
0400	8' long		5	1.600	"	150	53.50		203.50	255
0600	Brackets for above, rough sawn		12	.667	Pr.	10	22.50		32.50	48.50
0700	Laminated		12	.667	"	15	22.50		37.50	54

06 44 39 – Wood Posts and Columns

06 44 39.10 Decorative Beams

	06 44 39.10 Decorative Beams	Crew	Daily Output	Labor-Hours	Unit	Material	Labor	Equipment	Total	Total Incl O&P
0010	**DECORATIVE BEAMS**									
0020	Rough sawn cedar, non-load bearing, 4" x 4"	2 Carp	180	.089	L.F.	1.31	2.98		4.29	6.45
0100	4" x 6"		170	.094		1.97	3.16		5.13	7.45
0200	4" x 8"		160	.100		2.63	3.36		5.99	8.50
0300	4" x 10"		150	.107		3.99	3.58		7.57	10.40
0400	4" x 12"		140	.114		5.05	3.83		8.88	12
0500	8" x 8"		130	.123		5.25	4.13		9.38	12.70

06 44 39.20 Columns

	06 44 39.20 Columns	Crew	Daily Output	Labor-Hours	Unit	Material	Labor	Equipment	Total	Total Incl O&P
0010	**COLUMNS**									
0050	Aluminum, round colonial, 6" diameter	2 Carp	80	.200	V.L.F.	16	6.70		22.70	29
0100	8" diameter		62.25	.257		19.30	8.60		27.90	35.50
0200	10" diameter		55	.291		23.50	9.75		33.25	42.50
0250	Fir, stock units, hollow round, 6" diameter		80	.200		26.50	6.70		33.20	40.50
0300	8" diameter		80	.200		31	6.70		37.70	46
0350	10" diameter		70	.229		39.50	7.65		47.15	56.50
0400	Solid turned, to 8' high, 3-1/2" diameter		80	.200		9	6.70		15.70	21
0500	4-1/2" diameter		75	.213		14	7.15		21.15	27.50
0600	5-1/2" diameter		70	.229		18.40	7.65		26.05	33
0800	Square columns, built-up, 5" x 5"		65	.246		15.50	8.25		23.75	31
0900	Solid, 3-1/2" x 3-1/2"		130	.123		9	4.13		13.13	16.80
1600	Hemlock, tapered, T & G, 12" diam., 10' high		100	.160		53	5.35		58.35	67
1700	16' high		65	.246		79.50	8.25		87.75	101
1900	10' high, 14" diameter		100	.160		121	5.35		126.35	142
2000	18' high		65	.246		107	8.25		115.25	132
2200	18" diameter, 12' high		65	.246		153	8.25		161.25	183
2300	20' high		50	.320		148	10.75		158.75	181
2500	20" diameter, 14' high		40	.400		159	13.40		172.40	198
2600	20' high		35	.457		149	15.35		164.35	189
2800	For flat pilasters, deduct					33%				
3000	For splitting into halves, add				Ea.	116			116	128
4000	Rough sawn cedar posts, 4" x 4"	2 Carp	250	.064	V.L.F.	2.98	2.15		5.13	6.85
4100	4" x 6"		235	.068		5.70	2.28		7.98	10.05
4200	6" x 6"		220	.073		10.30	2.44		12.74	15.45
4300	8" x 8"		200	.080		19.30	2.68		21.98	26

06 48 Wood Frames

06 48 13 – Exterior Wood Door Frames

06 48 13.10 Exterior Wood Door Frames and Accessories		Crew	Daily Output	Labor-Hours	Unit	Material	2011 Bare Costs Labor	Equipment	Total	Total Incl O&P
0010	**EXTERIOR WOOD DOOR FRAMES AND ACCESSORIES**									
0400	Exterior frame, incl. ext. trim, pine, 5/4 x 4-9/16" deep	2 Carp	375	.043	L.F.	5.55	1.43		6.98	8.50
0420	5-3/16" deep		375	.043		11.35	1.43		12.78	14.90
0440	6-9/16" deep		375	.043		8.50	1.43		9.93	11.75
0600	Oak, 5/4 x 4-9/16" deep		350	.046		10.70	1.53		12.23	14.35
0620	5-3/16" deep		350	.046		11.80	1.53		13.33	15.50
0640	6-9/16" deep		350	.046		14.50	1.53		16.03	18.50
1000	Sills, 8/4 x 8" deep, oak, no horns		100	.160		17	5.35		22.35	27.50
1020	2" horns		100	.160		17.10	5.35		22.45	28
1040	3" horns		100	.160		17.10	5.35		22.45	28
1100	8/4 x 10" deep, oak, no horns		90	.178		20.50	5.95		26.45	32.50
1120	2" horns		90	.178		21	5.95		26.95	33.50
1140	3" horns		90	.178		21	5.95		26.95	33.50
2000	Exterior, colonial, frame & trim, 3' opng., in-swing, minimum		22	.727	Ea.	315	24.50		339.50	390
2010	Average		21	.762		470	25.50		495.50	560
2020	Maximum		20	.800		1,075	27		1,102	1,225
2100	5'-4" opening, in-swing, minimum		17	.941		360	31.50		391.50	450
2120	Maximum		15	1.067		1,075	36		1,111	1,225
2140	Out-swing, minimum		17	.941		370	31.50		401.50	460
2160	Maximum		15	1.067		1,100	36		1,136	1,275
2400	6'-0" opening, in-swing, minimum		16	1		345	33.50		378.50	435
2420	Maximum		10	1.600		1,100	53.50		1,153.50	1,325
2460	Out-swing, minimum		16	1		375	33.50		408.50	470
2480	Maximum		10	1.600		1,300	53.50		1,353.50	1,525
2600	For two sidelights, add, minimum		30	.533	Opng.	355	17.90		372.90	420
2620	Maximum		20	.800	"	1,125	27		1,152	1,300
2700	Custom birch frame, 3'-0" opening		16	1	Ea.	205	33.50		238.50	282
2750	6'-0" opening		16	1		305	33.50		338.50	390
2900	Exterior, modern, plain trim, 3' opng., in-swing, minimum		26	.615		35.50	20.50		56	73.50
2920	Average		24	.667		42	22.50		64.50	83.50
2940	Maximum		22	.727		50.50	24.50		75	96.50

06 48 16 – Interior Wood Door Frames

06 48 16.10 Interior Wood Door Jamb and Frames		Crew	Daily Output	Labor-Hours	Unit	Material	Labor	Equipment	Total	Total Incl O&P
0010	**INTERIOR WOOD DOOR JAMB AND FRAMES**									
3000	Interior frame, pine, 11/16" x 3-5/8" deep	2 Carp	375	.043	L.F.	5.55	1.43		6.98	8.50
3020	4-9/16" deep		375	.043		6.50	1.43		7.93	9.55
3200	Oak, 11/16" x 3-5/8" deep		350	.046		4.35	1.53		5.88	7.35
3220	4-9/16" deep		350	.046		4.50	1.53		6.03	7.50
3240	5-3/16" deep		350	.046		4.58	1.53		6.11	7.60
3400	Walnut, 11/16" x 3-5/8" deep		350	.046		7.30	1.53		8.83	10.60
3420	4-9/16" deep		350	.046		7.50	1.53		9.03	10.80
3440	5-3/16" deep		350	.046		7.90	1.53		9.43	11.20
3600	Pocket door frame		16	1	Ea.	74.50	33.50		108	138
3800	Threshold, oak, 5/8" x 3-5/8" deep		200	.080	L.F.	3.83	2.68		6.51	8.70
3820	4-5/8" deep		190	.084		4.58	2.83		7.41	9.80
3840	5-5/8" deep		180	.089		7.15	2.98		10.13	12.85

06 49 Wood Screens and Exterior Wood Shutters

06 49 19 – Exterior Wood Shutters

06 49 19.10 Shutters, Exterior		Crew	Daily Output	Labor-Hours	Unit	2011 Bare Costs Material	Labor	Equipment	Total	Total Incl O&P
0010	**SHUTTERS, EXTERIOR**									
0012	Aluminum, louvered, 1′-4″ wide, 3′-0″ long	1 Carp	10	.800	Pr.	61	27		88	113
0400	6′-8″ long		9	.889		123	30		153	185
1000	Pine, louvered, primed, each 1′-2″ wide, 3′-3″ long		10	.800		106	27		133	162
1100	4′-7″ long		10	.800		144	27		171	203
1500	Each 1′-6″ wide, 3′-3″ long		10	.800		113	27		140	169
1620	Hemlock, louvered, 1′-2″ wide, 5′-7″ long		10	.800		174	27		201	236
1630	Each 1′-4″ wide, 2′-2″ long		10	.800		108	27		135	164
1670	4′-3″ long		10	.800		130	27		157	188
1690	5′-11″ long		10	.800		183	27		210	246
1700	Door blinds, 6′-9″ long, each 1′-3″ wide		9	.889		184	30		214	252
1710	1′-6″ wide		9	.889		198	30		228	268
1720	Hemlock, solid raised panel, each 1′-4″ wide, 3′-3″ long		10	.800		175	27		202	238
1740	4′-3″ long		10	.800		222	27		249	289
1770	5′-11″ long		10	.800		297	27		324	370
1800	Door blinds, 6′-9″ long, each 1′-3″ wide		9	.889		335	30		365	415
1900	1′-6″ wide		9	.889		365	30		395	450
2500	Polystyrene, solid raised panel, each 1′-4″ wide, 3′-3″ long		10	.800		41	27		68	90
2700	4′-7″ long		10	.800		51.50	27		78.50	102
4500	Polystyrene, louvered, each 1′-2″ wide, 3′-3″ long		10	.800		28	27		55	75.50
4750	5′-3″ long		10	.800		39.50	27		66.50	88.50
6000	Vinyl, louvered, each 1′-2″ x 4′-7″ long		10	.800		34.50	27		61.50	82.50
6200	Each 1′-4″ x 6′-8″ long		9	.889		48	30		78	103

06 51 Structural Plastic Shapes and Plates

06 51 13 – Plastic Lumber

06 51 13.10 Recycled Plastic Lumber			Crew	Daily Output	Labor-Hours	Unit	Material	Labor	Equipment	Total	Total Incl O&P
0010	**RECYCLED PLASTIC LUMBER**										
4000	Sheeting, recycled plastic, black or white, 4′ x 8′ x 1/8″	G	2 Carp	1100	.015	S.F.	1.56	.49		2.05	2.54
4010	4′ x 8′ x 3/16″	G		1100	.015		2.33	.49		2.82	3.38
4020	4′ x 8′ x 1/4″	G		950	.017		3.11	.57		3.68	4.37
4030	4′ x 8′ x 3/8″	G		950	.017		4.67	.57		5.24	6.10
4040	4′ x 8′ x 1/2″	G		900	.018		6.20	.60		6.80	7.85
4050	4′ x 8′ x 5/8″	G		900	.018		7.70	.60		8.30	9.45
4060	4′ x 8′ x 3/4″	G		850	.019		9.15	.63		9.78	11.15
4070	Add for colors	G				Ea.	5%			5%	6%
8500	100% recycled plastic, var colors, NLB, 2″ x 2″	G				L.F.	1.41			1.41	1.55
8510	2″ x 4″	G					2.95			2.95	3.25
8520	2″ x 6″	G					4.63			4.63	5.10
8530	2″ x 8″	G					6.35			6.35	6.95
8540	2″ x 10″	G					9.25			9.25	10.15
8550	5/4″ x 4″	G					3.02			3.02	3.32
8560	5/4″ x 6″	G					2.99			2.99	3.29
8570	1″ x 6″	G					2.28			2.28	2.51
8580	1/2″ x 8″	G					2.46			2.46	2.71
8590	2″ x 10″ T & G	G					8.60			8.60	9.45
8600	3″ x 10″ T & G	G					14.25			14.25	15.70
8610	Add for premium colors	G					20%			20%	22%

06 52 10 – Fiberglass Structural Assemblies

06 52 10.10 Castings, Fiberglass

	06 52 10.10 Castings, Fiberglass	Crew	Daily Output	Labor-Hours	Unit	2011 Bare Costs Material	Labor	Equipment	Total	Total Incl O&P
0010	**CASTINGS, FIBERGLASS**									
0100	Angle, 1" x 1" x 1/8" thick	2 Sswk	240	.067	L.F.	1.44	2.43		3.87	6.20
0120	3" x 3" x 1/4" thick		200	.080		6	2.91		8.91	12.15
0140	4" x 4" x 1/4" thick		200	.080		7.70	2.91		10.61	14.05
0160	4" x 4" x 3/8" thick		200	.080		11.85	2.91		14.76	18.60
0180	6" x 6" x 1/2" thick		160	.100		23.50	3.64		27.14	33
1000	Flat sheet, 1/8" thick		140	.114	S.F.	5.05	4.16		9.21	13.50
1020	1/4" thick		120	.133		9.90	4.85		14.75	20
1040	3/8" thick		100	.160		16.85	5.80		22.65	29.50
1060	1/2" thick		80	.200		22	7.30		29.30	38.50
2000	Handrail, 42" high, 2" diam. rails pickets 5' O.C.		32	.500	L.F.	51	18.20		69.20	90.50
3000	Round bar, 1/4" diam.		240	.067		.58	2.43		3.01	5.25
3020	1/2" diam.		200	.080		1.01	2.91		3.92	6.65
3040	3/4" diam.		200	.080		2.05	2.91		4.96	7.80
3060	1" diam.		160	.100		3.62	3.64		7.26	10.95
3080	1-1/4" diam.		160	.100		4.13	3.64		7.77	11.50
3100	1-1/2" diam.		140	.114		5.05	4.16		9.21	13.50
3500	Round tube, 1" diam. x 1/8" thick		240	.067		1.97	2.43		4.40	6.80
3520	2" diam. x 1/4" thick		200	.080		6.35	2.91		9.26	12.55
3540	3" diam. x 1/4" thick		160	.100		10.10	3.64		13.74	18.05
4000	Square bar, 1/2" square		240	.067		3.98	2.43		6.41	9
4020	1" square		200	.080		5.10	2.91		8.01	11.15
4040	1-1/2" square		160	.100		10.40	3.64		14.04	18.40
4500	Square tube, 1" x 1" x 1/8" thick		240	.067		2.24	2.43		4.67	7.10
4520	2" x 2" x 1/8" thick		200	.080		4.34	2.91		7.25	10.30
4540	3" x 3" x 1/4" thick		160	.100		11.85	3.64		15.49	20
5000	Threaded rod, 3/8" diam.		320	.050		3.45	1.82		5.27	7.25
5020	1/2" diam.		320	.050		4.06	1.82		5.88	7.95
5040	5/8" diam.		280	.057		4.49	2.08		6.57	8.90
5060	3/4" diam.		280	.057		5.10	2.08		7.18	9.55
6000	Wide flange beam, 4" x 4" x 1/4" thick		120	.133		10.55	4.85		15.40	21
6020	6" x 6" x 1/4" thick		100	.160		19.95	5.80		25.75	33
6040	8" x 8" x 3/8" thick		80	.200		35	7.30		42.30	52.50

06 52 10.20 Fiberglass Stair Treads

	06 52 10.20 Fiberglass Stair Treads	Crew	Daily Output	Labor-Hours	Unit	Material	Labor	Equipment	Total	Total Incl O&P
0010	**FIBERGLASS STAIR TREADS**									
0100	24" wide	2 Sswk	52	.308	Ea.	39.50	11.20		50.70	65
0140	30" wide		52	.308		47	11.20		58.20	73.50
0180	36" wide		52	.308		55.50	11.20		66.70	82.50
0220	42" wide		52	.308		65.50	11.20		76.70	93.50

06 52 10.30 Fiberglass Grating

	06 52 10.30 Fiberglass Grating	Crew	Daily Output	Labor-Hours	Unit	Material	Labor	Equipment	Total	Total Incl O&P
0010	**FIBERGLASS GRATING**									
0100	Molded, green (for mod. corrosive environment)									
0140	1" x 4" mesh, 1" thick	2 Sswk	400	.040	S.F.	11.15	1.46		12.61	15.10
0180	1-1/2" square mesh, 1" thick		400	.040		16.60	1.46		18.06	21
0220	1-1/4" thick		400	.040		13.25	1.46		14.71	17.40
0260	1-1/2" thick		400	.040		24	1.46		25.46	29.50
0300	2" square mesh, 2" thick		320	.050		22.50	1.82		24.32	28.50
1000	Orange (for highly corrosive environment)									
1040	1" x 4" mesh, 1" thick	2 Sswk	400	.040	S.F.	14.05	1.46		15.51	18.25
1080	1-1/2" square mesh, 1" thick		400	.040		18.60	1.46		20.06	23.50
1120	1-1/4" thick		400	.040		19.30	1.46		20.76	24.50
1160	1-1/2" thick		400	.040		21	1.46		22.46	26

06 52 Plastic Structural Assemblies

06 52 10 – Fiberglass Structural Assemblies

06 52 10.30 Fiberglass Grating		Crew	Daily Output	Labor-Hours	Unit	Material	2011 Bare Costs Labor	Equipment	Total	Total Incl O&P
1200	2" square mesh, 2" thick	2 Sswk	320	.050	S.F.	25	1.82		26.82	31
3000	Pultruded, green (for mod. corrosive environment)									
3040	1" O.C. bar spacing, 1" thick	2 Sswk	400	.040	S.F.	16.65	1.46		18.11	21
3080	1-1/2" thick		320	.050		18.50	1.82		20.32	24
3120	1-1/2" O.C. bar spacing, 1" thick		400	.040		13.50	1.46		14.96	17.65
3160	1-1/2" thick		400	.040		15.15	1.46		16.61	19.45
4000	Grating support legs, fixed height, no base				Ea.	43.50			43.50	48
4040	With base					39			39	43
4080	Adjustable to 60"					59			59	65

06 52 10.40 Fiberglass Floor Grating		Crew	Daily Output	Labor-Hours	Unit	Material	Labor	Equipment	Total	Total Incl O&P
0010	**FIBERGLASS FLOOR GRATING**									
0101	Reinforced polyester, fire retardant, 1" x 4" grid, 1" thick	2 Carp	255	.063	S.F.	14.25	2.11		16.36	19.15
0201	1-1/2" x 6" mesh, 1-1/2" thick		250	.064		16.65	2.15		18.80	22
0301	With grit surface, 1-1/2" x 6" grid, 1-1/2" thick		250	.064		17	2.15		19.15	22.50

06 63 Plastic Railings

06 63 10 – Plastic (PVC) Railings

06 63 10.10 Plastic Railings		Crew	Daily Output	Labor-Hours	Unit	Material	Labor	Equipment	Total	Total Incl O&P
0010	**PLASTIC RAILINGS**									
0100	Horizontal PVC handrail with balusters, 3-1/2" wide, 36" high	1 Carp	96	.083	L.F.	24	2.80		26.80	31
0150	42" high		96	.083		27	2.80		29.80	34
0200	Angled PVC handrail with balusters, 3-1/2" wide, 36" high		72	.111		28	3.73		31.73	37
0250	42" high		72	.111		30.50	3.73		34.23	40
0300	Post sleeve for 4 x 4 post		96	.083		12.55	2.80		15.35	18.50
0400	Post cap for 4 x 4 post, flat profile		48	.167	Ea.	11.40	5.60		17	22
0450	Newel post style profile		48	.167		21	5.60		26.60	32.50
0500	Raised corbeled profile		48	.167		23	5.60		28.60	34.50
0550	Post base trim for 4 x 4 post		96	.083		12.80	2.80		15.60	18.80

06 65 Plastic Simulated Wood Trim

06 65 10 – PVC Trim

06 65 10.10 PVC Trim, Exterior		Crew	Daily Output	Labor-Hours	Unit	Material	Labor	Equipment	Total	Total Incl O&P
0010	**PVC TRIM, EXTERIOR**									
0100	Cornerboards, 5/4" x 6" x 6"	1 Carp	240	.033	L.F.	7.30	1.12		8.42	9.90
0110	Door/window casing, 1" x 4"		200	.040		1.42	1.34		2.76	3.81
0120	1" x 6"		200	.040		2.24	1.34		3.58	4.71
0130	1" x 8"		195	.041		2.96	1.38		4.34	5.55
0140	1" x 10"		195	.041		3.74	1.38		5.12	6.40
0150	1" x 12"		190	.042		4.57	1.41		5.98	7.40
0160	5/4" x 4"		195	.041		2.05	1.38		3.43	4.56
0170	5/4" x 6"		195	.041		3.24	1.38		4.62	5.85
0180	5/4" x 8"		190	.042		4.24	1.41		5.65	7
0190	5/4" x 10"		190	.042		5.40	1.41		6.81	8.30
0200	5/4" x 12"		185	.043		6.60	1.45		8.05	9.70
0210	Fascia, 1" x 4"		250	.032		1.42	1.07		2.49	3.36
0220	1" x 6"		250	.032		2.24	1.07		3.31	4.26
0230	1" x 8"		225	.036		2.96	1.19		4.15	5.25
0240	1" x 10"		225	.036		3.74	1.19		4.93	6.10
0250	1" x 12"		200	.040		4.57	1.34		5.91	7.30

06 65 Plastic Simulated Wood Trim

06 65 10 – PVC Trim

06 65 10.10 PVC Trim, Exterior		Crew	Daily Output	Labor-Hours	Unit	Material	2011 Bare Costs Labor	Equipment	Total	Total Incl O&P
0260	5/4" x 4"	1 Carp	240	.033	L.F.	2.05	1.12		3.17	4.13
0270	5/4" x 6"		240	.033		3.24	1.12		4.36	5.45
0280	5/4" x 8"		215	.037		4.24	1.25		5.49	6.75
0290	5/4" x 10"		215	.037		5.40	1.25		6.65	8.05
0300	5/4" x 12"		190	.042		6.60	1.41		8.01	9.60
0310	Frieze, 1" x 4"		250	.032		1.42	1.07		2.49	3.36
0320	1" x 6"		250	.032		2.24	1.07		3.31	4.26
0330	1" x 8"		225	.036		2.96	1.19		4.15	5.25
0340	1" x 10"		225	.036		3.74	1.19		4.93	6.10
0350	1" x 12"		200	.040		4.57	1.34		5.91	7.30
0360	5/4" x 4"		240	.033		2.05	1.12		3.17	4.13
0370	5/4" x 6"		240	.033		3.24	1.12		4.36	5.45
0380	5/4" x 8"		215	.037		4.24	1.25		5.49	6.75
0390	5/4" x 10"		215	.037		5.40	1.25		6.65	8.05
0400	5/4" x 12"		190	.042		6.60	1.41		8.01	9.60
0410	Rake, 1" x 4"		200	.040		1.42	1.34		2.76	3.81
0420	1" x 6"		200	.040		2.24	1.34		3.58	4.71
0430	1" x 8"		190	.042		2.96	1.41		4.37	5.60
0440	1" x 10"		190	.042		3.74	1.41		5.15	6.45
0450	1" x 12"		180	.044		4.57	1.49		6.06	7.55
0460	5/4" x 4"		195	.041		2.05	1.38		3.43	4.56
0470	5/4" x 6"		195	.041		3.24	1.38		4.62	5.85
0480	5/4" x 8"		185	.043		4.24	1.45		5.69	7.10
0490	5/4" x 10"		185	.043		5.40	1.45		6.85	8.40
0500	5/4" x 12"		175	.046		6.60	1.53		8.13	9.80
0510	Rake trim, 1" x 4"		225	.036		1.42	1.19		2.61	3.56
0520	1" x 6"		225	.036		2.24	1.19		3.43	4.46
0560	5/4" x 4"		220	.036		2.05	1.22		3.27	4.30
0570	5/4" x 6"		220	.036		3.24	1.22		4.46	5.60
0610	Soffit, 1" x 4"	2 Carp	420	.038		1.42	1.28		2.70	3.70
0620	1" x 6"		420	.038		2.24	1.28		3.52	4.60
0630	1" x 8"		420	.038		2.96	1.28		4.24	5.40
0640	1" x 10"		400	.040		3.74	1.34		5.08	6.35
0650	1" x 12"		400	.040		4.57	1.34		5.91	7.30
0660	5/4" x 4"		410	.039		2.05	1.31		3.36	4.45
0670	5/4" x 6"		410	.039		3.24	1.31		4.55	5.75
0680	5/4" x 8"		410	.039		4.24	1.31		5.55	6.85
0690	5/4" x 10"		390	.041		5.40	1.38		6.78	8.25
0700	5/4" x 12"		390	.041		6.60	1.38		7.98	9.55

Estimating Tips

07 10 00 Dampproofing and Waterproofing

- Be sure of the job specifications before pricing this subdivision. The difference in cost between waterproofing and dampproofing can be great. Waterproofing will hold back standing water. Dampproofing prevents the transmission of water vapor. Also included in this section are vapor retarding membranes.

07 20 00 Thermal Protection

- Insulation and fireproofing products are measured by area, thickness, volume or R-value. Specifications may give only what the specific R-value should be in a certain situation. The estimator may need to choose the type of insulation to meet that R-value.

07 30 00 Steep Slope Roofing
07 40 00 Roofing and Siding Panels

- Many roofing and siding products are bought and sold by the square. One square is equal to an area that measures 100 square feet.

This simple change in unit of measure could create a large error if the estimator is not observant. Accessories necessary for a complete installation must be figured into any calculations for both material and labor.

07 50 00 Membrane Roofing
07 60 00 Flashing and Sheet Metal
07 70 00 Roofing and Wall Specialties and Accessories

- The items in these subdivisions compose a roofing system. No one component completes the installation, and all must be estimated. Built-up or single-ply membrane roofing systems are made up of many products and installation trades. Wood blocking at roof perimeters or penetrations, parapet coverings, reglets, roof drains, gutters, downspouts, sheet metal flashing, skylights, smoke vents, and roof hatches all need to be considered along with the roofing material. Several different installation trades will need to work together on the roofing system. Inherent difficulties in the scheduling and coordination of various trades must be accounted for when estimating labor costs.

07 90 00 Joint Protection

- To complete the weather-tight shell, the sealants and caulkings must be estimated. Where different materials meet—at expansion joints, at flashing penetrations, and at hundreds of other locations throughout a construction project—they provide another line of defense against water penetration. Often, an entire system is based on the proper location and placement of caulking or sealants. The detailed drawings that are included as part of a set of architectural plans show typical locations for these materials. When caulking or sealants are shown at typical locations, this means the estimator must include them for all the locations where this detail is applicable. Be careful to keep different types of sealants separate, and remember to consider backer rods and primers if necessary.

Reference Numbers

Reference numbers are shown in shaded boxes at the beginning of some major classifications. These numbers refer to related items in the Reference Section. The reference information may be an estimating procedure, an alternate pricing method, or technical information.

Note: Not all subdivisions listed here necessarily appear in this publication.

07 01 Operation and Maint. of Thermal and Moisture Protection

07 01 50 – Maintenance of Membrane Roofing

07 01 50.10 Roof Coatings		Crew	Daily Output	Labor-Hours	Unit	2011 Bare Costs Material	Labor	Equipment	Total	Total Incl O&P
0010	**ROOF COATINGS**									
0012	Asphalt, brush grade, material only				Gal.	7.90			7.90	8.65
0200	Asphalt base, fibered aluminum coating G					10.75			10.75	11.85
0300	Asphalt primer, 5 gallon					6.40			6.40	7.05
0600	Coal tar pitch, 200 lb. barrels				Ton	1,100			1,100	1,200
0700	Tar roof cement, 5 gal. lots				Gal.	14			14	15.40
0800	Glass fibered roof & patching cement, 5 gallon				"	7.40			7.40	8.15
0900	Reinforcing glass membrane, 450 S.F./roll				Ea.	111			111	122
1000	Neoprene roof coating, 5 gal., 2 gal./sq.				Gal.	27.50			27.50	30.50
1100	Roof patch & flashing cement, 5 gallon					8.90			8.90	9.75
1200	Roof resaturant, glass fibered, 3 gal./sq.					5.35			5.35	5.85

07 01 90 – Maintenance of Joint Protection

07 01 90.81 Joint Sealant Replacement		Crew	Daily Output	Labor-Hours	Unit	Material	Labor	Equipment	Total	Total Incl O&P
0010	**JOINT SEALANT REPLACEMENT**									
0050	Control joints in concrete floors/slabs									
0100	Option 1 for joints with hard dry sealant									
0110	Step 1: Sawcut to remove 95% of old sealant									
0112	1/4" wide x 1/2" deep, with single saw blade	C-27	4800	.003	L.F.	.02	.11	.03	.16	.22
0114	3/8" wide x 3/4" deep, with single saw blade		4000	.004		.04	.13	.03	.20	.29
0116	1/2" wide x 1" deep, with double saw blades		3600	.004		.08	.14	.04	.26	.36
0118	3/4" wide x 1-1/2" deep, with double saw blades		3200	.005		.17	.16	.04	.37	.48
0120	Step 2: Water blast joint faces and edges	C-29	2500	.003			.08	.02	.10	.17
0130	Step 3: Air blast joint faces and edges	C-28	2000	.004			.13	.01	.14	.21
0140	Step 4: Sand blast joint faces and edges	E-11	2000	.016			.48	.10	.58	.98
0150	Step 5: Air blast joint faces and edges	C-28	2000	.004			.13	.01	.14	.21
0200	Option 2 for joints with soft pliable sealant									
0210	Step 1: Plow joint with rectangular blade	B-62	2600	.009	L.F.		.27	.06	.33	.50
0220	Step 2: Sawcut to re-face joint faces									
0222	1/4" wide x 1/2" deep, with single saw blade	C-27	2400	.007	L.F.	.03	.21	.06	.30	.43
0224	3/8" wide x 3/4" deep, with single saw blade		2000	.008		.06	.25	.07	.38	.54
0226	1/2" wide x 1" deep, with double saw blades		1800	.009		.11	.28	.07	.46	.65
0228	3/4" wide x 1-1/2" deep, with double saw blades		1600	.010		.22	.32	.08	.62	.84
0230	Step 3: Water blast joint faces and edges	C-29	2500	.003			.08	.02	.10	.17
0240	Step 4: Air blast joint faces and edges	C-28	2000	.004			.13	.01	.14	.21
0250	Step 5: Sand blast joint faces and edges	E-11	2000	.016			.48	.10	.58	.98
0260	Step 6: Air blast joint faces and edges	C-28	2000	.004			.13	.01	.14	.21
0290	For saw cutting new control joints, see Section 03 35 29.35									
8910	For backer rod, see Section 07 91 23.10									
8920	For joint sealant, see Sections 03 15 05.25 or 07 92 13.20, if available									

07 05 Common Work Results for Thermal and Moisture Protection

07 05 05 – Selective Demolition

07 05 05.10 Selective Demo., Thermal and Moist. Protection		Crew	Daily Output	Labor-Hours	Unit	Material	Labor	Equipment	Total	Total Incl O&P
0010	**SELECTIVE DEMO., THERMAL AND MOISTURE PROTECTION**									
0020	Caulking/sealant, to 1" x 1" joint R024119-10	1 Clab	600	.013	L.F.		.35		.35	.59
0120	Downspouts, including hangers		350	.023	"		.60		.60	1.01
0220	Flashing, sheet metal		290	.028	S.F.		.73		.73	1.22
0420	Gutters, aluminum or wood, edge hung		240	.033	L.F.		.88		.88	1.48
0520	Built-in		100	.080	"		2.12		2.12	3.54
0620	Insulation, air/vapor barrier		3500	.002	S.F.		.06		.06	.10

07 05 05 – Selective Demolition

07 05 05.10 Selective Demo., Thermal and Moist. Protection		Crew	Daily Output	Labor-Hours	Unit	Material	2011 Bare Costs Labor	Equipment	Total	Total Incl O&P
0670	Batts or blankets	1 Clab	1400	.006	C.F.		.15		.15	.25
0720	Foamed or sprayed in place	2 Clab	1000	.016	B.F.		.42		.42	.71
0770	Loose fitting	1 Clab	3000	.003	C.F.		.07		.07	.12
0870	Rigid board		3450	.002	B.F.		.06		.06	.10
1120	Roll roofing, cold adhesive		12	.667	Sq.		17.65		17.65	29.50
1170	Roof accessories, adjustable metal chimney flashing		9	.889	Ea.		23.50		23.50	39.50
1325	Plumbing vent flashing		32	.250	"		6.60		6.60	11.10
1375	Ridge vent strip, aluminum		310	.026	L.F.		.68		.68	1.14
1620	Skylight to 10 S.F.		8	1	Ea.		26.50		26.50	44.50
2120	Roof edge, aluminum soffit and fascia		570	.014	L.F.		.37		.37	.62
2170	Concrete coping, up to 12" wide	2 Clab	160	.100			2.65		2.65	4.43
2220	Drip edge	1 Clab	1650	.005			.13		.13	.21
2270	Gravel stop		1650	.005			.13		.13	.21
2370	Sheet metal coping, up to 12" wide		240	.033			.88		.88	1.48
2470	Roof insulation board, over 2" thick	B-2	7800	.005	B.F.		.14		.14	.23
2520	Up to 2" thick	"	3900	.010	S.F.		.28		.28	.46
2620	Roof ventilation, louvered gable vent	1 Clab	16	.500	Ea.		13.25		13.25	22
2670	Remove, roof hatch	G-3	15	2.133			69		69	115
2675	Rafter vents	1 Clab	960	.008			.22		.22	.37
2720	Soffit vent and/or fascia vent		575	.014	L.F.		.37		.37	.62
2775	Soffit vent strip, aluminum, 3" to 4" wide		160	.050			1.32		1.32	2.22
2820	Roofing accessories, shingle moulding, to 1" x 4"		1600	.005			.13		.13	.22
2870	Cant strip	B-2	2000	.020			.54		.54	.90
2920	Concrete block walkway	1 Clab	230	.035			.92		.92	1.54
3070	Roofing, felt paper, 15#		70	.114	Sq.		3.02		3.02	5.05
3125	#30 felt		30	.267	"		7.05		7.05	11.80
3170	Asphalt shingles, 1 layer	B-2	3500	.011	S.F.		.31		.31	.51
3180	2 layers		1750	.023	"		.61		.61	1.03
3370	Modified bitumen		26	1.538	Sq.		41.50		41.50	69
3420	Built-up, no gravel, 3 ply		25	1.600			43		43	72
3470	4 ply		21	1.905			51		51	85.50
3620	5 ply		1600	.025	S.F.		.67		.67	1.12
3720	5 ply, with gravel		890	.045			1.21		1.21	2.02
3725	Gravel removal, minimum		5000	.008			.21		.21	.36
3730	Maximum		2000	.020			.54		.54	.90
3870	Fiberglass sheet		1200	.033			.89		.89	1.50
4120	Slate shingles		1900	.021			.57		.57	.95
4170	Ridge shingles, clay or slate		2000	.020	L.F.		.54		.54	.90
4320	Single ply membrane, attached at seams		52	.769	Sq.		20.50		20.50	34.50
4370	Ballasted		75	.533			14.30		14.30	24
4420	Fully adhered		39	1.026			27.50		27.50	46
4550	Roof hatch, 2'-6" x 3'-0"	1 Clab	10	.800	Ea.		21		21	35.50
4670	Wood shingles	B-2	2200	.018	S.F.		.49		.49	.82
4820	Sheet metal roofing	"	2150	.019			.50		.50	.84
4970	Siding, horizontal wood clapboards	1 Clab	380	.021			.56		.56	.93
5025	Exterior insulation finish system	"	120	.067			1.76		1.76	2.95
5070	Tempered hardboard, remove and reset	1 Carp	380	.021			.71		.71	1.18
5120	Tempered hardboard sheet siding	"	375	.021			.72		.72	1.20
5170	Metal, corner strips	1 Clab	850	.009	L.F.		.25		.25	.42
5225	Horizontal strips		444	.018	S.F.		.48		.48	.80
5320	Vertical strips		400	.020			.53		.53	.89
5520	Wood shingles		350	.023			.60		.60	1.01
5620	Stucco siding		360	.022			.59		.59	.98

07 05 Common Work Results for Thermal and Moisture Protection

07 05 05 – Selective Demolition

07 05 05.10 Selective Demo., Thermal and Moist. Protection		Crew	Daily Output	Labor-Hours	Unit	Material	2011 Bare Costs Labor	Equipment	Total	Total Incl O&P
5670	Textured plywood	1 Clab	725	.011	S.F.		.29		.29	.49
5720	Vinyl siding		510	.016			.42		.42	.70
5770	Corner strips		900	.009	L.F.		.24		.24	.39
5870	Wood, boards, vertical		400	.020	S.F.		.53		.53	.89
5920	Waterproofing, protection/drain board	2 Clab	3900	.004	B.F.		.11		.11	.18
5970	Over 1/2″ thick		1750	.009	S.F.		.24		.24	.40
6020	To 1/2″ thick		2000	.008	″		.21		.21	.35

07 11 Dampproofing

07 11 13 – Bituminous Dampproofing

07 11 13.10 Bituminous Asphalt Coating

Line	Description	Crew	Daily Output	Labor-Hours	Unit	Material	Labor	Equipment	Total	Total Incl O&P
0010	**BITUMINOUS ASPHALT COATING**									
0030	Brushed on, below grade, 1 coat	1 Rofc	665	.012	S.F.	.20	.34		.54	.83
0100	2 coat		500	.016		.39	.45		.84	1.24
0300	Sprayed on, below grade, 1 coat, 25.6 S.F./gal.		830	.010		.20	.27		.47	.71
0400	2 coat, 20.5 S.F./gal.		500	.016		.38	.45		.83	1.23
0500	Asphalt coating, with fibers				Gal.	10.75			10.75	11.85
0600	Troweled on, asphalt with fibers, 1/16″ thick	1 Rofc	500	.016	S.F.	.47	.45		.92	1.32
0700	1/8″ thick		400	.020		.83	.56		1.39	1.93
1000	1/2″ thick		350	.023		2.69	.64		3.33	4.12

07 11 16 – Cementitious Dampproofing

07 11 16.20 Cementitious Parging

Line	Description	Crew	Daily Output	Labor-Hours	Unit	Material	Labor	Equipment	Total	Total Incl O&P
0010	**CEMENTITIOUS PARGING**									
0020	Portland cement, 2 coats, 1/2″ thick	D-1	250	.064	S.F.	.29	1.95		2.24	3.51
0100	Waterproofed Portland cement, 1/2″ thick, 2 coats	″	250	.064	″	3.01	1.95		4.96	6.50

07 12 Built-up Bituminous Waterproofing

07 12 13 – Built-Up Asphalt Waterproofing

07 12 13.20 Membrane Waterproofing

Line	Description	Crew	Daily Output	Labor-Hours	Unit	Material	Labor	Equipment	Total	Total Incl O&P
0010	**MEMBRANE WATERPROOFING**									
0012	On slabs, 1 ply, felt, mopped	G-1	3000	.019	S.F.	.33	.49	.14	.96	1.40
0100	On slabs, 1 ply, glass fiber fabric, mopped		2100	.027		.34	.70	.20	1.24	1.86
0300	On slabs, 2 ply, felt, mopped		2500	.022		.66	.59	.16	1.41	1.98
0400	On slabs, 2 ply, glass fiber fabric, mopped		1650	.034		.77	.89	.25	1.91	2.72
0600	On slabs, 3 ply, felt, mopped		2100	.027		.99	.70	.20	1.89	2.57
0700	On slabs, 3 ply, glass fiber fabric, mopped		1550	.036		1.03	.95	.26	2.24	3.14
0710	Asphaltic hardboard protection board, 1/8″ thick	2 Rofc	500	.032		.40	.90		1.30	2.07
1000	1/4″ EPS membrane protection board		3500	.005		.28	.13		.41	.54
1050	3/8″ thick		3500	.005		.32	.13		.45	.58
1060	1/2″ thick		3500	.005		.36	.13		.49	.62
1070	Fiberglass fabric, black, 20/10 mesh		116	.138	Sq.	18.10	3.88		21.98	27
1080	White, 20/10 mesh		116	.138	″	18.10	3.88		21.98	27

07 13 Sheet Waterproofing

07 13 53 – Elastomeric Sheet Waterproofing

07 13 53.10 Elastomeric Sheet Waterproofing and Access.		Crew	Daily Output	Labor-Hours	Unit	2011 Bare Costs Material	Labor	Equipment	Total	Total Incl O&P
0010	**ELASTOMERIC SHEET WATERPROOFING AND ACCESS.**									
0090	EPDM, plain, 45 mils thick	2 Rofc	580	.028	S.F.	1.30	.78		2.08	2.83
0100	60 mils thick		570	.028		1.37	.79		2.16	2.94
0300	Nylon reinforced sheets, 45 mils thick		580	.028		1.33	.78		2.11	2.86
0400	60 mils thick		570	.028		1.59	.79		2.38	3.18
0600	Vulcanizing splicing tape for above, 2″ wide				C.L.F.	48			48	53
0700	4″ wide				″	110			110	121
0900	Adhesive, bonding, 60 S.F. per gal.				Gal.	21			21	23
1000	Splicing, 75 S.F. per gal.				″	31			31	34
1200	Neoprene sheets, plain, 45 mils thick	2 Rofc	580	.028	S.F.	1.53	.78		2.31	3.08
1300	60 mils thick		570	.028		2.61	.79		3.40	4.30
1500	Nylon reinforced, 45 mils thick		580	.028		1.75	.78		2.53	3.33
1600	60 mils thick		570	.028		2.38	.79		3.17	4.05
1800	120 mils thick		500	.032		4.74	.90		5.64	6.85
1900	Adhesive, splicing, 150 S.F. per gal. per coat				Gal.	31			31	34
2100	Fiberglass reinforced, fluid applied, 1/8″ thick	2 Rofc	500	.032	S.F.	1.90	.90		2.80	3.72
2200	Polyethylene and rubberized asphalt sheets, 1/8″ thick		550	.029		.70	.82		1.52	2.25
2400	Polyvinyl chloride sheets, plain, 10 mils thick		580	.028		.17	.78		.95	1.59
2500	20 mils thick		570	.028		.28	.79		1.07	1.74
2700	30 mils thick		560	.029		.39	.80		1.19	1.88
3000	Adhesives, trowel grade, 40-100 S.F. per gal.				Gal.	21.50			21.50	23.50
3100	Brush grade, 100-250 S.F. per gal.				″	17.50			17.50	19.25
3300	Bitumen modified polyurethane, fluid applied, 55 mils thick	2 Rofc	665	.024	S.F.	.78	.68		1.46	2.08

07 16 Cementitious and Reactive Waterproofing

07 16 16 – Crystalline Waterproofing

07 16 16.20 Cementitious Waterproofing		Crew	Daily Output	Labor-Hours	Unit	Material	Labor	Equipment	Total	Total Incl O&P
0010	**CEMENTITIOUS WATERPROOFING**									
0020	1/8″ application, sprayed on	G-2A	1000	.024	S.F.	2	.60	.59	3.19	3.91
0050	4 coat cementitious metallic slurry	1 Cefi	1.20	6.667	C.S.F.	26	212		238	370

07 17 Bentonite Waterproofing

07 17 13 – Bentonite Panel Waterproofing

07 17 13.10 Bentonite		Crew	Daily Output	Labor-Hours	Unit	Material	Labor	Equipment	Total	Total Incl O&P
0010	**BENTONITE**									
0020	Panels, 4′ x 4′, 3/16″ thick	1 Rofc	625	.013	S.F.	1.10	.36		1.46	1.86
0100	Rolls, 3/8″ thick, with geotextile fabric both sides	″	550	.015	″	1.29	.41		1.70	2.16
0300	Granular bentonite, 50 lb. bags (.625 C.F.)				Bag	9.80			9.80	10.80
0400	3/8″ thick, troweled on	1 Rofc	475	.017	S.F.	.49	.47		.96	1.40
0500	Drain board, expanded polystyrene, 1-1/2″ thick	1 Rohe	1600	.005		.44	.10		.54	.67
0510	2″ thick		1600	.005		.58	.10		.68	.83
0520	3″ thick		1600	.005		.87	.10		.97	1.15
0530	4″ thick		1600	.005		1.16	.10		1.26	1.47
0600	With filter fabric, 1-1/2″ thick		1600	.005		.51	.10		.61	.75
0625	2″ thick		1600	.005		.66	.10		.76	.91
0650	3″ thick		1600	.005		.95	.10		1.05	1.23
0675	4″ thick		1600	.005		1.24	.10		1.34	1.55

07 19 Water Repellents

07 19 19 – Silicone Water Repellents

07 19 19.10 Silicone Based Water Repellents		Crew	Daily Output	Labor-Hours	Unit	2011 Bare Costs Material	Labor	Equipment	Total	Total Incl O&P
0010	**SILICONE BASED WATER REPELLENTS**									
0020	Water base liquid, roller applied	2 Rofc	7000	.002	S.F.	.86	.06		.92	1.07
0200	Silicone or stearate, sprayed on CMU, 1 coat	1 Rofc	4000	.002		.36	.06		.42	.50
0300	2 coats	″	3000	.003		.73	.08		.81	.94

07 21 Thermal Insulation

07 21 13 – Board Insulation

07 21 13.10 Rigid Insulation

Line	Description		Crew	Daily Output	Labor-Hours	Unit	Material	Labor	Equipment	Total	Total Incl O&P
0010	**RIGID INSULATION**, for walls										
0040	Fiberglass, 1.5#/C.F., unfaced, 1″ thick, R4.1	G	1 Carp	1000	.008	S.F.	.40	.27		.67	.89
0060	1-1/2″ thick, R6.2	G		1000	.008		.58	.27		.85	1.09
0080	2″ thick, R8.3	G		1000	.008		.64	.27		.91	1.15
0120	3″ thick, R12.4	G		800	.010		.76	.34		1.10	1.40
0370	3#/C.F., unfaced, 1″ thick, R4.3	G		1000	.008		.46	.27		.73	.96
0390	1-1/2″ thick, R6.5	G		1000	.008		.86	.27		1.13	1.40
0400	2″ thick, R8.7	G		890	.009		.91	.30		1.21	1.50
0420	2-1/2″ thick, R10.9	G		800	.010		.96	.34		1.30	1.62
0440	3″ thick, R13	G		800	.010		1.39	.34		1.73	2.09
0520	Foil faced, 1″ thick, R4.3	G		1000	.008		.83	.27		1.10	1.36
0540	1-1/2″ thick, R6.5	G		1000	.008		1.24	.27		1.51	1.81
0560	2″ thick, R8.7	G		890	.009		1.55	.30		1.85	2.21
0580	2-1/2″ thick, R10.9	G		800	.010		1.82	.34		2.16	2.56
0600	3″ thick, R13	G		800	.010		2	.34		2.34	2.76
1600	Isocyanurate, 4′ x 8′ sheet, foil faced, both sides										
1610	1/2″ thick	G	1 Carp	800	.010	S.F.	.29	.34		.63	.88
1620	5/8″ thick	G		800	.010		.32	.34		.66	.91
1630	3/4″ thick	G		800	.010		.33	.34		.67	.92
1640	1″ thick	G		800	.010		.51	.34		.85	1.12
1650	1-1/2″ thick	G		730	.011		.64	.37		1.01	1.32
1660	2″ thick	G		730	.011		.79	.37		1.16	1.49
1670	3″ thick	G		730	.011		1.78	.37		2.15	2.58
1680	4″ thick	G		730	.011		2	.37		2.37	2.82
1700	Perlite, 1″ thick, R2.77	G		800	.010		.30	.34		.64	.89
1750	2″ thick, R5.55	G		730	.011		.60	.37		.97	1.28
1900	Extruded polystyrene, 25 PSI compressive strength, 1″ thick, R5	G		800	.010		.50	.34		.84	1.11
1940	2″ thick R10	G		730	.011		1	.37		1.37	1.72
1960	3″ thick, R15	G		730	.011		1.44	.37		1.81	2.20
2100	Expanded polystyrene, 1″ thick, R3.85	G		800	.010		.29	.34		.63	.88
2120	2″ thick, R7.69	G		730	.011		.58	.37		.95	1.26
2140	3″ thick, R11.49	G		730	.011		.87	.37		1.24	1.58

07 21 13.13 Foam Board Insulation

Line	Description		Crew	Daily Output	Labor-Hours	Unit	Material	Labor	Equipment	Total	Total Incl O&P
0010	**FOAM BOARD INSULATION**										
0600	Polystyrene, expanded, 1″ thick, R4	G	1 Carp	680	.012	S.F.	.29	.39		.68	.98
0700	2″ thick, R8	G	″	675	.012	″	.58	.40		.98	1.31

07 21 16 – Blanket Insulation

07 21 16.10 Blanket Insulation for Floors/Ceilings

Line	Description		Crew	Daily Output	Labor-Hours	Unit	Material	Labor	Equipment	Total	Total Incl O&P
0010	**BLANKET INSULATION FOR FLOORS/CEILINGS**										
0020	Including spring type wire fasteners										
2000	Fiberglass, blankets or batts, paper or foil backing										
2100	3-1/2″ thick, R13	G	1 Carp	700	.011	S.F.	.32	.38		.70	.99

07 21 16 – Blanket Insulation

07 21 16.10 Blanket Insulation for Floors/Ceilings			Crew	Daily Output	Labor-Hours	Unit	Material	Labor	Equipment	Total	Total Incl O&P
							2011 Bare Costs				
2150	6-1/4" thick, R19	G	1 Carp	600	.013	S.F.	.42	.45		.87	1.21
2210	9-1/2" thick, R30	G		500	.016		.70	.54		1.24	1.67
2220	12" thick, R38	G		475	.017		.85	.57		1.42	1.89
3000	Unfaced, 3-1/2" thick, R13	G		600	.013		.28	.45		.73	1.06
3010	6-1/4" thick, R19	G		500	.016		.44	.54		.98	1.38
3020	9-1/2" thick, R30	G		450	.018		.70	.60		1.30	1.77
3030	12" thick, R38	G	↓	425	.019	↓	.73	.63		1.36	1.86

07 21 16.20 Blanket Insulation for Walls

Line	Description		Crew	Daily Output	Labor-Hours	Unit	Material	Labor	Equipment	Total	Total Incl O&P
0010	**BLANKET INSULATION FOR WALLS**										
0020	Kraft faced fiberglass, 3-1/2" thick, R11, 15" wide	G	1 Carp	1350	.006	S.F.	.28	.20		.48	.64
0030	23" wide	G		1600	.005		.28	.17		.45	.59
0060	R13, 11" wide	G		1150	.007		.27	.23		.50	.69
0080	15" wide	G		1350	.006		.27	.20		.47	.63
0100	23" wide	G		1600	.005		.27	.17		.44	.58
0110	R-15, 11" wide	G		1150	.007		.41	.23		.64	.84
0120	15" wide	G		1350	.006		.41	.20		.61	.78
0130	23" wide	G		1600	.005		.41	.17		.58	.73
0140	6" thick, R19, 11" wide	G		1150	.007		.37	.23		.60	.80
0160	15" wide	G		1350	.006		.37	.20		.57	.74
0180	23" wide	G		1600	.005		.37	.17		.54	.69
0182	R21, 11" wide	G		1150	.007		.51	.23		.74	.95
0184	15" wide	G		1350	.006		.51	.20		.71	.89
0186	23" wide	G		1600	.005		.51	.17		.68	.84
0188	9" thick, R-30, 11" wide	G		985	.008		.71	.27		.98	1.24
0200	15" wide	G		1150	.007		.71	.23		.94	1.17
0220	23" wide	G		1350	.006		.71	.20		.91	1.11
0230	12" thick, R38, 11" wide	G		985	.008		.95	.27		1.22	1.51
0240	15" wide	G		1150	.007		.95	.23		1.18	1.44
0260	23" wide	G		1350	.006		.95	.20		1.15	1.38
0410	Foil faced fiberglass, 3-1/2" thick, R13, 11" wide	G		1150	.007		.54	.23		.77	.98
0420	15" wide	G		1350	.006		.54	.20		.74	.92
0440	23" wide	G		1600	.005		.54	.17		.71	.87
0442	R15, 11" wide	G		1150	.007		.56	.23		.79	1.01
0444	15" wide	G		1350	.006		.56	.20		.76	.95
0446	23" wide	G		1600	.005		.56	.17		.73	.90
0448	6" thick, R19, 11" wide	G		1150	.007		.77	.23		1	1.24
0460	15" wide	G		1350	.006		.77	.20		.97	1.18
0480	23" wide	G		1600	.005		.77	.17		.94	1.13
0482	R-21, 11" wide	G		1150	.007		.82	.23		1.05	1.29
0484	15" wide	G		1350	.006		.82	.20		1.02	1.23
0486	23" wide	G		1600	.005		.82	.17		.99	1.18
0488	9" thick, R-30, 11" wide	G		985	.008		.89	.27		1.16	1.44
0500	9" thick, R30, 15" wide	G		1150	.007		.89	.23		1.12	1.37
0550	23" wide	G		1350	.006		.89	.20		1.09	1.31
0560	12" thick, R-38, 11" wide	G		985	.008		.90	.27		1.17	1.45
0570	15" wide	G		1150	.007		.90	.23		1.13	1.38
0580	23" wide	G		1350	.006		.90	.20		1.10	1.32
0620	Unfaced fiberglass, 3-1/2" thick, R-13, 11" wide	G		1150	.007		.28	.23		.51	.70
0820	15" wide	G		1350	.006		.28	.20		.48	.64
0830	23" wide	G		1600	.005		.28	.17		.45	.59
0832	R15, 11" wide	G		1150	.007		.40	.23		.63	.83
0836	23" wide	G	↓	1600	.005	↓	.40	.17		.57	.72

07 21 Thermal Insulation

07 21 16 – Blanket Insulation

07 21 16.20 Blanket Insulation for Walls			Crew	Daily Output	Labor-Hours	Unit	Material	2011 Bare Costs Labor	Equipment	Total	Total Incl O&P
0838	6″ thick, R19, 11″ wide	G	1 Carp	1150	.007	S.F.	.44	.23		.67	.87
0860	15″ wide	G		1150	.007		.44	.23		.67	.87
0880	23″ wide	G		1350	.006		.44	.20		.64	.81
0882	R-21, 11″ wide	G		1150	.007		.46	.23		.69	.90
0886	15″ wide	G		1350	.006		.46	.20		.66	.84
0888	23″ wide	G		1600	.005		.46	.17		.63	.79
0890	9″ thick, R30, 11″ wide	G		985	.008		.70	.27		.97	1.23
0900	15″ wide	G		1150	.007		.70	.23		.93	1.16
0920	23″ wide	G		1350	.006		.70	.20		.90	1.10
0930	12″ thick, R38, 11″ wide	G		985	.008		.73	.27		1	1.26
0940	15″ wide	G		1000	.008		.73	.27		1	1.25
0960	23″ wide	G		1150	.007		.73	.23		.96	1.19
1300	Mineral fiber batts, kraft faced										
1320	3-1/2″ thick, R12	G	1 Carp	1600	.005	S.F.	.38	.17		.55	.70
1340	6″ thick, R19	G		1600	.005		.42	.17		.59	.74
1380	10″ thick, R30	G		1350	.006		.67	.20		.87	1.07
1850	Friction fit wire insulation supports, 16″ O.C.			960	.008	Ea.	.05	.28		.33	.53

07 21 23 – Loose-Fill Insulation

07 21 23.10 Poured Loose-Fill Insulation			Crew	Daily Output	Labor-Hours	Unit	Material	Labor	Equipment	Total	Total Incl O&P
0010	**POURED LOOSE-FILL INSULATION**										
0020	Cellulose fiber, R3.8 per inch	G	1 Carp	200	.040	C.F.	.56	1.34		1.90	2.87
0040	Ceramic type (perlite), R3.2 per inch	G		200	.040		1.95	1.34		3.29	4.40
0080	Fiberglass wool, R4 per inch	G		200	.040		.42	1.34		1.76	2.71
0100	Mineral wool, R3 per inch	G		200	.040		.39	1.34		1.73	2.68
0300	Polystyrene, R4 per inch	G		200	.040		3.09	1.34		4.43	5.65
0400	Perlite, R2.7 per inch	G		200	.040		1.95	1.34		3.29	4.40
0700	Wood fiber, R3.85 per inch	G		200	.040		.70	1.34		2.04	3.02

07 21 23.20 Masonry Loose-Fill Insulation			Crew	Daily Output	Labor-Hours	Unit	Material	Labor	Equipment	Total	Total Incl O&P
0010	**MASONRY LOOSE-FILL INSULATION**, vermiculite or perlite										
0100	In cores of concrete block, 4″ thick wall, .115 C.F./S.F.	G	D-1	4800	.003	S.F.	.22	.10		.32	.42
0200	6″ thick wall, .175 C.F./S.F.	G		3000	.005		.34	.16		.50	.65
0300	8″ thick wall, .258 C.F./S.F.	G		2400	.007		.50	.20		.70	.88
0400	10″ thick wall, .340 C.F./S.F.	G		1850	.009		.66	.26		.92	1.16
0500	12″ thick wall, .422 C.F./S.F.	G		1200	.013		.82	.41		1.23	1.57
0550	For sand fill, deduct from above	G					70%				
0600	Poured cavity wall, vermiculite or perlite, water repellant	G	D-1	250	.064	C.F.	1.95	1.95		3.90	5.35
0700	Foamed in place, urethane in 2-5/8″ cavity	G	G-2A	1035	.023	S.F.	.49	.58	.57	1.64	2.20
0800	For each 1″ added thickness, add	G	″	2372	.010	″	.12	.25	.25	.62	.86

07 21 26 – Blown Insulation

07 21 26.10 Blown Insulation			Crew	Daily Output	Labor-Hours	Unit	Material	Labor	Equipment	Total	Total Incl O&P
0010	**BLOWN INSULATION** Ceilings, with open access										
0020	Cellulose, 3-1/2″ thick, R13	G	G-4	5000	.005	S.F.	.17	.13	.06	.36	.48
0030	5-3/16″ thick, R19	G		3800	.006		.25	.17	.08	.50	.66
0050	6-1/2″ thick, R22	G		3000	.008		.32	.22	.10	.64	.83
0100	8-11/16″ thick, R30	G		2600	.009		.44	.25	.11	.80	1.03
0120	10-7/8″ thick, R38	G		1800	.013		.56	.36	.16	1.08	1.40
1000	Fiberglass, 5.5″ thick, R11	G		3800	.006		.15	.17	.08	.40	.55
1050	6″ thick, R12	G		3000	.008		.19	.22	.10	.51	.67
1100	8.8″ thick, R19	G		2200	.011		.26	.30	.13	.69	.94
1200	10″ thick, R22	G		1800	.013		.31	.36	.16	.83	1.13
1300	11.5″ thick, R26	G		1500	.016		.37	.43	.20	1	1.36

07 21 Thermal Insulation

07 21 26 – Blown Insulation

07 21 26.10 Blown Insulation			Crew	Daily Output	Labor-Hours	Unit	2011 Bare Costs Material	Labor	Equipment	Total	Total Incl O&P
1350	13" thick, R30	G	G-4	1400	.017	S.F.	.40	.46	.21	1.07	1.45
1450	16" thick, R38	G		1145	.021		.49	.57	.26	1.32	1.77
1500	20" thick, R49	G		920	.026		.62	.71	.32	1.65	2.22

07 21 27 – Reflective Insulation

07 21 27.10 Reflective Insulation Options			Crew	Daily Output	Labor-Hours	Unit	Material	Labor	Equipment	Total	Total Incl O&P
0010	**REFLECTIVE INSULATION OPTIONS**										
0020	Aluminum foil on reinforced scrim	G	1 Carp	19	.421	C.S.F.	14.20	14.15		28.35	39
0100	Reinforced with woven polyolefin	G		19	.421		18	14.15		32.15	43.50
0500	With single bubble air space, R8.8	G		15	.533		28	17.90		45.90	61
0600	With double bubble air space, R9.8	G		15	.533		29	17.90		46.90	61.50

07 21 29 – Sprayed Insulation

07 21 29.10 Sprayed-On Insulation			Crew	Daily Output	Labor-Hours	Unit	Material	Labor	Equipment	Total	Total Incl O&P
0010	**SPRAYED-ON INSULATION**										
0020	Fibrous/cementitious, finished wall, 1" thick, R3.7	G	G-2	2050	.012	S.F.	.24	.33	.06	.63	.87
0100	Attic, 5.2" thick, R19	G		1550	.015	"	.42	.44	.08	.94	1.27
0200	Fiberglass, R4 per inch, vertical	G		2050	.012	B.F.	.15	.33	.06	.54	.78
0210	Horizontal	G		1550	.015	"	.15	.44	.08	.67	.98
0300	Closed cell, spray polyurethane foam, 2 pounds per cubic foot density										
0310	1" thick	G	G-2A	6000	.004	S.F.	.41	.10	.10	.61	.74
0320	2" thick	G		3000	.008		.82	.20	.20	1.22	1.47
0330	3" thick	G		2000	.012		1.23	.30	.30	1.83	2.22
0335	3-1/2" thick	G		1715	.014		1.44	.35	.35	2.14	2.58
0340	4" thick	G		1500	.016		1.64	.40	.40	2.44	2.95
0350	5" thick	G		1200	.020		2.05	.50	.49	3.04	3.69
0355	5-1/2" thick	G		1090	.022		2.26	.55	.54	3.35	4.05
0360	6" thick	G		1000	.024		2.46	.60	.59	3.65	4.42

07 22 Roof and Deck Insulation

07 22 16 – Roof Board Insulation

07 22 16.10 Roof Deck Insulation			Crew	Daily Output	Labor-Hours	Unit	Material	Labor	Equipment	Total	Total Incl O&P
0010	**ROOF DECK INSULATION**										
0020	Fiberboard low density, 1/2" thick R1.39	G	1 Rofc	1000	.008	S.F.	.26	.23		.49	.70
0030	1" thick R2.78	G		800	.010		.46	.28		.74	1.02
0080	1 1/2" thick R4.17	G		800	.010		.70	.28		.98	1.28
0100	2" thick R5.56	G		800	.010		.93	.28		1.21	1.53
0110	Fiberboard high density, 1/2" thick R1.3	G		1000	.008		.24	.23		.47	.67
0120	1" thick R2.5	G		800	.010		.48	.28		.76	1.04
0130	1-1/2" thick R3.8	G		800	.010		.72	.28		1	1.30
0200	Fiberglass, 3/4" thick R2.78	G		1000	.008		.53	.23		.76	.99
0400	15/16" thick R3.70	G		1000	.008		.71	.23		.94	1.19
0460	1-1/16" thick R4.17	G		1000	.008		.89	.23		1.12	1.39
0600	1-5/16" thick R5.26	G		1000	.008		1.20	.23		1.43	1.73
0650	2-1/16" thick R8.33	G		800	.010		1.27	.28		1.55	1.91
0700	2-7/16" thick R10	G		800	.010		1.47	.28		1.75	2.13
1650	Perlite, 1/2" thick R1.32	G		1050	.008		.34	.21		.55	.76
1655	3/4" thick R2.08	G		800	.010		.37	.28		.65	.92
1660	1" thick R2.78	G		800	.010		.46	.28		.74	1.02
1670	1-1/2" thick R4.17	G		800	.010		.48	.28		.76	1.04
1680	2" thick R5.56	G		700	.011		.80	.32		1.12	1.46
1685	2-1/2" thick R6.67	G		700	.011		.95	.32		1.27	1.63

07 22 16 – Roof Board Insulation

07 22 16.10 Roof Deck Insulation			Crew	Daily Output	Labor-Hours	Unit	Material	2011 Bare Costs Labor	Equipment	Total	Total Incl O&P
1690	Tapered for drainage	G	1 Rofc	800	.010	B.F.	.73	.28		1.01	1.31
1700	Polyisocyanurate, 2#/C.F. density, 3/4" thick	G		1500	.005	S.F.	.44	.15		.59	.75
1705	1" thick	G		1400	.006		.51	.16		.67	.85
1715	1-1/2" thick	G		1250	.006		.66	.18		.84	1.06
1725	2" thick	G		1100	.007		.78	.20		.98	1.23
1735	2-1/2" thick	G		1050	.008		1.01	.21		1.22	1.50
1745	3" thick	G		1000	.008		1.22	.23		1.45	1.75
1755	3-1/2" thick	G		1000	.008		1.95	.23		2.18	2.56
1765	Tapered for drainage	G		1400	.006	B.F.	1.95	.16		2.11	2.44
1900	Extruded Polystyrene										
1910	15 PSI compressive strength, 1" thick, R5	G	1 Rofc	1500	.005	S.F.	.44	.15		.59	.75
1920	2" thick, R10	G		1250	.006		.56	.18		.74	.95
1930	3" thick R15	G		1000	.008		1.15	.23		1.38	1.68
1932	4" thick R20	G		1000	.008		1.52	.23		1.75	2.08
1934	Tapered for drainage	G		1500	.005	B.F.	.51	.15		.66	.83
1940	25 PSI compressive strength, 1" thick R5	G		1500	.005	S.F.	.61	.15		.76	.94
1942	2" thick R10	G		1250	.006		1.15	.18		1.33	1.60
1944	3" thick R15	G		1000	.008		1.75	.23		1.98	2.34
1946	4" thick R20	G		1000	.008		2.50	.23		2.73	3.16
1948	Tapered for drainage	G		1500	.005	B.F.	.53	.15		.68	.85
1950	40 psi compressive strength, 1" thick R5	G		1500	.005	S.F.	.47	.15		.62	.79
1952	2" thick R10	G		1250	.006		.88	.18		1.06	1.30
1954	3" thick R15	G		1000	.008		1.28	.23		1.51	1.82
1956	4" thick R20	G		1000	.008		1.70	.23		1.93	2.28
1958	Tapered for drainage	G		1400	.006	B.F.	.67	.16		.83	1.03
1960	60 PSI compressive strength, 1" thick R5	G		1450	.006	S.F.	.65	.16		.81	1
1962	2" thick R10	G		1200	.007		1.25	.19		1.44	1.72
1964	3" thick R15	G		975	.008		2	.23		2.23	2.62
1966	4" thick R20	G		950	.008		2.50	.24		2.74	3.18
1968	Tapered for drainage	G		1400	.006	B.F.	.85	.16		1.01	1.23
2010	Expanded polystyrene, 1#/C.F. density, 3/4" thick R2.89	G		1500	.005	S.F.	.22	.15		.37	.51
2020	1" thick R3.85	G		1500	.005		.29	.15		.44	.59
2100	2" thick R7.69	G		1250	.006		.58	.18		.76	.97
2110	3" thick R11.49	G		1250	.006		.87	.18		1.05	1.29
2120	4" thick R15.38	G		1200	.007		1.16	.19		1.35	1.62
2130	5" thick R19.23	G		1150	.007		1.45	.20		1.65	1.95
2140	6" thick R23.26	G		1150	.007		1.74	.20		1.94	2.26
2150	Tapered for drainage	G		1500	.005	B.F.	.54	.15		.69	.86
2400	Composites with 2" EPS										
2410	1" fiberboard	G	1 Rofc	950	.008	S.F.	1.28	.24		1.52	1.84
2420	7/16" oriented strand board	G		800	.010		1	.28		1.28	1.61
2430	1/2" plywood	G		800	.010		1.20	.28		1.48	1.83
2440	1" perlite	G		800	.010		1.18	.28		1.46	1.81
2450	Composites with 1-1/2" polyisocyanurate										
2460	1" fiberboard	G	1 Rofc	800	.010	S.F.	1.20	.28		1.48	1.83
2470	1" perlite	G		850	.009		1.08	.26		1.34	1.67
2480	7/16" oriented strand board	G		800	.010		.84	.28		1.12	1.43

07 24 Exterior Insulation and Finish Systems

07 24 13 – Polymer-Based Exterior Insulation and Finish System

07 24 13.10 Exterior Insulation and Finish Systems		Crew	Daily Output	Labor-Hours	Unit	Material	2011 Bare Costs Labor	Equipment	Total	Total Incl O&P
0010	**EXTERIOR INSULATION AND FINISH SYSTEMS**									
0095	Field applied, 1" EPS insulation G	J-1	390	.103	S.F.	2.09	3.01	.32	5.42	7.55
0100	With 1/2" cement board sheathing G		268	.149		2.76	4.38	.46	7.60	10.70
0105	2" EPS insulation G		390	.103		2.38	3.01	.32	5.71	7.85
0110	With 1/2" cement board sheathing G		268	.149		3.05	4.38	.46	7.89	11
0115	3" EPS insulation G		390	.103		2.67	3.01	.32	6	8.20
0120	With 1/2" cement board sheathing G		268	.149		3.34	4.38	.46	8.18	11.35
0125	4" EPS insulation G		390	.103		2.96	3.01	.32	6.29	8.50
0130	With 1/2" cement board sheathing G		268	.149		4.31	4.38	.46	9.15	12.40
0140	Premium finish add		1265	.032		.31	.93	.10	1.34	1.96
0150	Heavy duty reinforcement add	↓	914	.044	↓	1.08	1.28	.14	2.50	3.43
0160	2.5#/S.Y. metal lath substrate add	1 Lath	75	.107	S.Y.	2.37	3.21		5.58	7.75
0170	3.4#/S.Y. metal lath substrate add	"	75	.107	"	2.58	3.21		5.79	8
0180	Color or texture change,	J-1	1265	.032	S.F.	.83	.93	.10	1.86	2.53
0190	With substrate leveling base coat	1 Plas	530	.015		.83	.46		1.29	1.66
0210	With substrate sealing base coat	1 Pord	1224	.007	↓	.08	.19		.27	.40
0370	V groove shape in panel face				L.F.	.60			.60	.66
0380	U groove shape in panel face				"	.79			.79	.87
0440	For higher than one story, add						25%			

07 25 Weather Barriers

07 25 10 – Weather Barriers or Wraps

07 25 10.10 Weather Barriers		Crew	Daily Output	Labor-Hours	Unit	Material	Labor	Equipment	Total	Total Incl O&P
0010	**WEATHER BARRIERS**									
0400	Asphalt felt paper, 15#	1 Carp	37	.216	Sq.	5.65	7.25		12.90	18.35
0401	Per square foot	"	3700	.002	S.F.	.06	.07		.13	.18
0450	Housewrap, exterior, spun bonded polypropylene									
0470	Small roll	1 Carp	3800	.002	S.F.	.21	.07		.28	.36
0480	Large roll	"	4000	.002	"	.13	.07		.20	.25
2100	Asphalt felt roof deck vapor barrier, class 1 metal decks	1 Rofc	37	.216	Sq.	19.70	6.10		25.80	32.50
2200	For all other decks	"	37	.216		14.05	6.10		20.15	26.50
2800	Asphalt felt, 50% recycled content, 15 lb, 4 sq per roll	1 Carp	36	.222		5.65	7.45		13.10	18.70
2810	30 lb, 2 sq per roll	"	36	.222	↓	9.45	7.45		16.90	23
3000	Building wrap, spunbonded polyethylene	2 Carp	8000	.002	S.F.	.14	.07		.21	.26

07 26 Vapor Retarders

07 26 10 – Above-Grade Vapor Retarders

07 26 10.10 Vapor Retarders		Crew	Daily Output	Labor-Hours	Unit	Material	Labor	Equipment	Total	Total Incl O&P
0010	**VAPOR RETARDERS**									
0020	Aluminum and kraft laminated, foil 1 side G	1 Carp	37	.216	Sq.	9.35	7.25		16.60	22.50
0100	Foil 2 sides G		37	.216		9.75	7.25		17	23
0600	Polyethylene vapor barrier, standard, .002" thick G		37	.216		1.19	7.25		8.44	13.45
0700	.004" thick G		37	.216		3.34	7.25		10.59	15.80
0900	.006" thick G		37	.216		5.25	7.25		12.50	17.90
1200	.010" thick G		37	.216		6.70	7.25		13.95	19.50
1300	Clear reinforced, fire retardant, .008" thick G		37	.216		10.15	7.25		17.40	23.50
1350	Cross laminated type, .003" thick G		37	.216		7.10	7.25		14.35	19.95
1400	.004" thick G		37	.216		7.80	7.25		15.05	21
1800	Reinf. waterproof, .002" polyethylene backing, 1 side	↓	37	.216	↓	5.65	7.25		12.90	18.35

07 26 Vapor Retarders

07 26 10 – Above-Grade Vapor Retarders

07 26 10.10 Vapor Retarders		Crew	Daily Output	Labor-Hours	Unit	Material	Labor (2011 Bare Costs)	Equipment	Total	Total Incl O&P
1900	2 sides	1 Carp	37	.216	Sq.	7.45	7.25		14.70	20.50
2400	Waterproofed kraft with sisal or fiberglass fibers, minimum		37	.216		6.10	7.25		13.35	18.85
2500	Maximum		37	.216		15.10	7.25		22.35	29

07 31 Shingles and Shakes

07 31 13 – Asphalt Shingles

07 31 13.10 Asphalt Roof Shingles

		Crew	Daily Output	Labor-Hours	Unit	Material	Labor	Equipment	Total	Total Incl O&P
0010	**ASPHALT ROOF SHINGLES**									
0100	Standard strip shingles									
0150	Inorganic, class A, 210-235 lb./sq.	1 Rofc	5.50	1.455	Sq.	74	41		115	156
0155	Pneumatic nailed		7	1.143		74	32		106	140
0200	Organic, class C, 235-240 lb./sq.		5	1.600		72.50	45		117.50	162
0205	Pneumatic nailed		6.25	1.280		72.50	36		108.50	145
0250	Standard, laminated multi-layered shingles									
0300	Class A, 240-260 lb./sq.	1 Rofc	4.50	1.778	Sq.	97.50	50		147.50	198
0305	Pneumatic nailed		5.63	1.422		97.50	40		137.50	180
0350	Class C, 260-300 lb/square, 4 bundles/square		4	2		120	56.50		176.50	234
0355	Pneumatic nailed		5	1.600		120	45		165	214
0400	Premium, laminated multi-layered shingles									
0450	Class A, 260-300 lb, 4 bundles/sq	1 Rofc	3.50	2.286	Sq.	157	64.50		221.50	289
0455	Pneumatic nailed		4.37	1.831		157	51.50		208.50	266
0500	Class C, 300-385 lb/square, 5 bundles/square		3	2.667		150	75		225	300
0505	Pneumatic nailed		3.75	2.133		150	60		210	273
0800	#15 felt underlayment		64	.125		5.65	3.52		9.17	12.55
0825	#30 felt underlayment		58	.138		9.45	3.88		13.33	17.35
0850	Self adhering polyethylene and rubberized asphalt underlayment		22	.364		70	10.25		80.25	95.50
0900	Ridge shingles		330	.024	L.F.	1.90	.68		2.58	3.32
0905	Pneumatic nailed		412.50	.019	″	1.90	.55		2.45	3.08
1000	For steep roofs (7 to 12 pitch or greater), add						50%			

07 31 16 – Metal Shingles

07 31 16.10 Aluminum Shingles

		Crew	Daily Output	Labor-Hours	Unit	Material	Labor	Equipment	Total	Total Incl O&P
0010	**ALUMINUM SHINGLES**									
0020	Mill finish, .019 thick	1 Carp	5	1.600	Sq.	191	53.50		244.50	300
0100	.020″ thick	″	5	1.600		214	53.50		267.50	325
0300	For colors, add					20			20	22
0600	Ridge cap, .024″ thick	1 Carp	170	.047	L.F.	2.18	1.58		3.76	5.05
0700	End wall flashing, .024″ thick		170	.047		1.78	1.58		3.36	4.60
0900	Valley section, .024″ thick		170	.047		2.90	1.58		4.48	5.85
1000	Starter strip, .024″ thick		400	.020		1.41	.67		2.08	2.67
1200	Side wall flashing, .024″ thick		170	.047		1.78	1.58		3.36	4.60
1500	Gable flashing, .024″ thick		400	.020		1.41	.67		2.08	2.67

07 31 16.20 Steel Shingles

		Crew	Daily Output	Labor-Hours	Unit	Material	Labor	Equipment	Total	Total Incl O&P
0010	**STEEL SHINGLES**									
0012	Galvanized, 26 gauge	1 Rots	2.20	3.636	Sq.	250	103		353	460
0201	24 gauge	1 Rofc	2.20	3.636		265	102		367	475
0300	For colored galvanized shingles, add					55			55	60.50
0500	For 1″ factory applied polystyrene insulation, add					40			40	44

07 31 26 – Slate Shingles

07 31 26.10 Slate Roof Shingles		Crew	Daily Output	Labor-Hours	Unit	2011 Bare Costs Material	Labor	Equipment	Total	Total Incl O&P
0010	**SLATE ROOF SHINGLES** R073126-20									
0100	Buckingham Virginia black, 3/16" - 1/4" thick G	1 Rots	1.75	4.571	Sq.	475	129		604	760
0900	Pennsylvania black, Bangor, #1 clear G		1.75	4.571		490	129		619	775
1200	Vermont, unfading, green, mottled green G		1.75	4.571		480	129		609	760
1300	Semi-weathering green & gray G		1.75	4.571		350	129		479	615
1400	Purple G		1.75	4.571		425	129		554	700
1500	Black or gray G		1.75	4.571		460	129		589	740
1600	Red G		1.75	4.571		1,175	129		1,304	1,500
1700	Variegated purple		1.75	4.571		415	129		544	690

07 31 29 – Wood Shingles and Shakes

07 31 29.13 Wood Shingles

Line	Description	Crew	Daily Output	Labor-Hours	Unit	Material	Labor	Equipment	Total	Total Incl O&P
0010	**WOOD SHINGLES**									
0012	16" No. 1 red cedar shingles, 5" exposure, on roof	1 Carp	2.50	3.200	Sq.	215	107		322	415
0015	Pneumatic nailed		3.25	2.462		215	82.50		297.50	375
0200	7-1/2" exposure, on walls		2.05	3.902		143	131		274	375
0205	Pneumatic nailed		2.67	2.996		143	101		244	325
0300	18" No. 1 red cedar perfections, 5-1/2" exposure, on roof		2.75	2.909		220	97.50		317.50	405
0305	Pneumatic nailed		3.57	2.241		220	75		295	370
0500	7-1/2" exposure, on walls		2.25	3.556		162	119		281	380
0505	Pneumatic nailed		2.92	2.740		162	92		254	330
0600	Resquared, and rebutted, 5-1/2" exposure, on roof		3	2.667		238	89.50		327.50	410
0605	Pneumatic nailed		3.90	2.051		238	69		307	375
0900	7-1/2" exposure, on walls		2.45	3.265		175	110		285	375
0905	Pneumatic nailed		3.18	2.516		175	84.50		259.50	335
1000	Add to above for fire retardant shingles					55			55	60.50
1060	Preformed ridge shingles	1 Carp	400	.020	L.F.	3.60	.67		4.27	5.10
2000	White cedar shingles, 16" long, extras, 5" exposure, on roof		2.40	3.333	Sq.	149	112		261	350
2005	Pneumatic nailed		3.12	2.564		149	86		235	310
2050	5" exposure on walls		2	4		149	134		283	390
2055	Pneumatic nailed		2.60	3.077		149	103		252	335
2100	7-1/2" exposure, on walls		2	4		106	134		240	340
2105	Pneumatic nailed		2.60	3.077		106	103		209	290
2150	"B" grade, 5" exposure on walls		2	4		136	134		270	375
2155	Pneumatic nailed		2.60	3.077		136	103		239	320
2300	For 15# organic felt underlayment on roof, 1 layer, add		64	.125		5.65	4.19		9.84	13.20
2400	2 layers, add		32	.250		11.25	8.40		19.65	26.50
2600	For steep roofs (7/12 pitch or greater), add to above						50%			
2700	Panelized systems, No.1 cedar shingles on 5/16" CDX plywood									
2800	On walls, 8' strips, 7" or 14" exposure	2 Carp	700	.023	S.F.	5.55	.77		6.32	7.40
3500	On roofs, 8' strips, 7" or 14" exposure	1 Carp	3	2.667	Sq.	555	89.50		644.50	760
3505	Pneumatic nailed	"	4	2	"	555	67		622	720

07 31 29.16 Wood Shakes

Line	Description	Crew	Daily Output	Labor-Hours	Unit	Material	Labor	Equipment	Total	Total Incl O&P
0010	**WOOD SHAKES**									
1100	Hand-split red cedar shakes, 1/2" thick x 24" long, 10" exp. on roof	1 Carp	2.50	3.200	Sq.	233	107		340	435
1105	Pneumatic nailed		3.25	2.462		233	82.50		315.50	395
1110	3/4" thick x 24" long, 10" exp. on roof		2.25	3.556		233	119		352	455
1115	Pneumatic nailed		2.92	2.740		233	92		325	410
1200	1/2" thick, 18" long, 8-1/2" exp. on roof		2	4		159	134		293	400
1205	Pneumatic nailed		2.60	3.077		159	103		262	350
1210	3/4" thick x 18" long, 8 1/2" exp. on roof		1.80	4.444		159	149		308	425
1215	Pneumatic nailed		2.34	3.419		159	115		274	365

07 31 Shingles and Shakes

07 31 29 – Wood Shingles and Shakes

07 31 29.16 Wood Shakes		Crew	Daily Output	Labor-Hours	Unit	Material	Labor	Equipment	Total	Total Incl O&P
1255	10" exp. on walls	1 Carp	2	4	Sq.	154	134		288	395
1260	10" exposure on walls, pneumatic nailed		2.60	3.077		154	103		257	340
1700	Add to above for fire retardant shakes, 24" long					55			55	60.50
1800	18" long					55			55	60.50
1810	Ridge shakes	1 Carp	350	.023	L.F.	3.60	.77		4.37	5.25

(Material, Labor, Equipment and Total are 2011 Bare Costs.)

07 32 Roof Tiles

07 32 13 – Clay Roof Tiles

07 32 13.10 Clay Tiles

			Crew	Daily Output	Labor-Hours	Unit	Material	Labor	Equipment	Total	Total Incl O&P
0010	**CLAY TILES**										
0200	Lanai tile or Classic tile, 158 pc per sq	G	1 Rots	1.65	4.848	Sq.	200	137		337	470
0300	Americana, 158 pc per sq, most colors	G		1.65	4.848		440	137		577	735
0350	Green, gray or brown	G		1.65	4.848		440	137		577	735
0400	Blue	G		1.65	4.848		440	137		577	735
0600	Spanish tile, 171 pc per sq, red	G		1.80	4.444		325	126		451	585
0800	Buff, green, gray, brown	G		1.80	4.444		550	126		676	830
0900	Glazed white	G		1.80	4.444		630	126		756	920
1100	Mission tile, 192 pc per sq, machine scored finish, red	G		1.15	6.957		740	197		937	1,175
1700	French tile, 133 pc per sq, smooth finish, red	G		1.35	5.926		675	168		843	1,050
1750	Blue or green	G		1.35	5.926		870	168		1,038	1,250
1800	Norman black 317 pc per sq	G		1	8		1,050	226		1,276	1,550
2200	Williamsburg tile, 158 pc per sq, aged cedar	G		1.35	5.926		660	168		828	1,025
2250	Gray or green	G		1.35	5.926		595	168		763	960
2350	Ridge shingles, clay tile	G		200	.040	L.F.	10.70	1.13		11.83	13.85
3010	Clay tile, #15 felt underlayment		1 Rofc	64	.125	Sq.	5.65	3.52		9.17	12.55
3020	Clay tile, #30 felt underlayment			58	.138		9.45	3.88		13.33	17.35
3040	Clay tile, polyethylene and rubberized asph. underlayment			22	.364		70	10.25		80.25	95.50

07 32 16 – Concrete Roof Tiles

07 32 16.10 Concrete Tiles

		Crew	Daily Output	Labor-Hours	Unit	Material	Labor	Equipment	Total	Total Incl O&P
0010	**CONCRETE TILES**									
0020	Corrugated, 13" x 16-1/2", 90 per sq, 950 lb per sq									
0050	Earthtone colors, nailed to wood deck	1 Rots	1.35	5.926	Sq.	97.50	168		265.50	410
0150	Blues		1.35	5.926		101	168		269	415
0200	Greens		1.35	5.926		101	168		269	415
0500	Shakes, 13" x 16-1/2", 90 per sq, 950 lb per sq									
0600	All colors, nailed to wood deck	1 Rots	1.50	5.333	Sq.	260	151		411	560
1500	Accessory pieces, ridge & hip, 10" x 16-1/2", 8 lb. each	"	120	.067	Ea.	3.64	1.89		5.53	7.40
1700	Rake, 6-1/2" x 16-3/4", 9 lb. each					3.64			3.64	4
1800	Mansard hip, 10" x 16-1/2", 9.2 lb. each					3.64			3.64	4
1900	Hip starter, 10" x 16-1/2", 10.5 lb. each					11.45			11.45	12.60
2000	3 or 4 way apex, 10" each side, 11.5 lb. each					13.25			13.25	14.55

07 32 19 – Metal Roof Tiles

07 32 19.10 Metal Roof Tiles

		Crew	Daily Output	Labor-Hours	Unit	Material	Labor	Equipment	Total	Total Incl O&P
0010	**METAL ROOF TILES**									
0020	Accessories included, .032" thick aluminum, mission tile	1 Carp	2.50	3.200	Sq.	815	107		922	1,075
0200	Spanish tiles	"	3	2.667	"	560	89.50		649.50	765

07 33 Natural Roof Coverings

07 33 63 – Vegetated Roofing

07 33 63.10 Green Roof Systems			Crew	Daily Output	Labor-Hours	Unit	2011 Bare Costs Material	Labor	Equipment	Total	Total Incl O&P
0010	**GREEN ROOF SYSTEMS**										
0020	Soil mixture for green roof 30% sand, 55% gravel, 15% soil	G									
0100	Hoist and spread soil mixture 4 inch depth up to five stories tall roof	G	B-13B	4000	.014	S.F.	.24	.40	.30	.94	1.26
0150	6 inch depth	G		2667	.021		.36	.60	.45	1.41	1.89
0200	8 inch depth	G		2000	.028		.48	.80	.60	1.88	2.53
0250	10 inch depth	G		1600	.035		.60	1.01	.76	2.37	3.16
0300	12 inch depth	G		1335	.042		.72	1.21	.90	2.83	3.79
0310	Alt. man-made soil mix, hoist & spread, 4" deep up to 5 stories tall roof	G		4000	.014		1.50	.40	.30	2.20	2.65
0350	Mobilization 55 ton crane to site	G	1 Eqhv	3.60	2.222	Ea.		79.50		79.50	130
0355	Hoisting cost to five stories per day (Avg. 28 picks per day)	G	B-13B	1	56	Day		1,600	1,200	2,800	4,000
0360	Mobilization or demobilization, 100 ton crane to site driver & escort	G	A-3E	2.50	6.400	Ea.		202	46.50	248.50	380
0365	Hoisting cost six to ten stories per day (Avg. 21 picks per day)	G	B-13C	1	56	Day		1,600	1,650	3,250	4,500
0370	Hoist and spread soil mixture 4 inch depth six to ten stories tall roof	G		4000	.014	S.F.	.24	.40	.41	1.05	1.38
0375	6 inch depth	G		2667	.021		.36	.60	.62	1.58	2.07
0380	8 inch depth	G		2000	.028		.48	.80	.83	2.11	2.78
0385	10 inch depth	G		1600	.035		.60	1.01	1.03	2.64	3.47
0390	12 inch depth	G		1335	.042		.72	1.21	1.24	3.17	4.15
0400	Green roof edging treated lumber 4" x 4" no hoisting included	G	2 Carp	400	.040	L.F.	1.37	1.34		2.71	3.75
0410	4" x 6"	G		400	.040		2.46	1.34		3.80	4.96
0420	4" x 8"	G		360	.044		4.74	1.49		6.23	7.70
0430	4" x 6" double stacked	G		300	.053		4.93	1.79		6.72	8.40
0500	Green roof edging redwood lumber 4" x 4" no hoisting included	G		400	.040		6.40	1.34		7.74	9.30
0510	4" x 6"	G		400	.040		11.20	1.34		12.54	14.55
0520	4" x 8"	G		360	.044		20.50	1.49		21.99	25
0530	4" x 6" double stacked	G		300	.053		22.50	1.79		24.29	27.50
0550	Roof membrane and root barrier, not including insulation:	G									
0560	Fluid applied rubber membrane, reinforced, 215 mil thick	G	G-5	350	.114	S.F.	.26	2.93	.48	3.67	6.10
0570	Root barrier	G	2 Rofc	775	.021		.70	.58		1.28	1.82
0580	Moisture retention barrier and reservoir	G	"	900	.018		2.70	.50		3.20	3.87
0600	Planting sedum, light soil, potted, 2-1/4" diameter, two per S.F.	G	1 Clab	420	.019		4.50	.50		5	5.80
0610	one per S.F.	G	"	840	.010		2.25	.25		2.50	2.90
0630	Planting sedum mat per S.F. including shipping (4000 S.F. min)	G	4 Clab	4000	.008		5.40	.21		5.61	6.30
0640	Installation sedum mat system (no soil required) per S.F. (4000 S.F. min)	G	"	4000	.008		7.50	.21		7.71	8.60
0645	Note: pricing of sedum mats shipped in full truck loads (4000-5000 S.F.)	G									

07 41 Roof Panels

07 41 13 – Metal Roof Panels

07 41 13.10 Aluminum Roof Panels		Crew	Daily Output	Labor-Hours	Unit	Material	Labor	Equipment	Total	Total Incl O&P
0010	**ALUMINUM ROOF PANELS**									
0020	Corrugated or ribbed, .0155" thick, natural	G-3	1200	.027	S.F.	.96	.86		1.82	2.49
0300	Painted		1200	.027		1.40	.86		2.26	2.97
0400	Corrugated, .018" thick, on steel frame, natural finish		1200	.027		1.27	.86		2.13	2.83
0600	Painted		1200	.027		1.55	.86		2.41	3.14
0700	Corrugated, on steel frame, natural, .024" thick		1200	.027		1.80	.86		2.66	3.41
0800	Painted		1200	.027		2.19	.86		3.05	3.84
0900	.032" thick, natural		1200	.027		2.32	.86		3.18	3.98
1200	Painted		1200	.027		2.97	.86		3.83	4.70
1300	V-Beam, on steel frame construction, .032" thick, natural		1200	.027		2.45	.86		3.31	4.13
1500	Painted		1200	.027		3.01	.86		3.87	4.74
1600	.040" thick, natural		1200	.027		2.99	.86		3.85	4.72
1800	Painted		1200	.027		3.59	.86		4.45	5.40

07 41 Roof Panels

07 41 13 – Metal Roof Panels

07 41 13.10 Aluminum Roof Panels		Crew	Daily Output	Labor-Hours	Unit	Material	2011 Bare Costs Labor	Equipment	Total	Total Incl O&P
1900	.050″ thick, natural	G-3	1200	.027	S.F.	3.58	.86		4.44	5.35
2100	Painted		1200	.027		4.33	.86		5.19	6.20
2200	For roofing on wood frame, deduct		4600	.007		.08	.23		.31	.46
2400	Ridge cap, .032″ thick, natural		800	.040	L.F.	2.83	1.29		4.12	5.25

07 41 13.20 Steel Roofing Panels

Line	Description	Crew	Daily Output	Labor-Hours	Unit	Material	Labor	Equipment	Total	Total Incl O&P
0010	**STEEL ROOFING PANELS**									
0012	Corrugated or ribbed, on steel framing, 30 ga galv	G-3	1100	.029	S.F.	1.68	.94		2.62	3.41
0100	28 ga		1050	.030		1.79	.99		2.78	3.61
0300	26 ga		1000	.032		1.85	1.03		2.88	3.76
0400	24 ga		950	.034		2.91	1.09		4	5
0600	Colored, 28 ga		1050	.030		1.68	.99		2.67	3.49
0700	26 ga		1000	.032		1.99	1.03		3.02	3.91
0710	Flat profile, 1-3/4″ standing seams, 10″ wide, standard finish, 26 ga		1000	.032		3.89	1.03		4.92	6
0715	24 ga		950	.034		4.51	1.09		5.60	6.75
0720	22 ga		900	.036		5.55	1.15		6.70	8.05
0725	Zinc aluminum alloy finish, 26 ga		1000	.032		3.05	1.03		4.08	5.10
0730	24 ga		950	.034		3.64	1.09		4.73	5.80
0735	22 ga		900	.036		4.17	1.15		5.32	6.50
0740	12″ wide, standard finish, 26 ga		1000	.032		3.88	1.03		4.91	6
0745	24 ga		950	.034		5.10	1.09		6.19	7.40
0750	Zinc aluminum alloy finish, 26 ga		1000	.032		4.41	1.03		5.44	6.55
0755	24 ga		950	.034		3.63	1.09		4.72	5.80
0840	Flat profile, 1″ x 3/8″ batten, 12″ wide, standard finish, 26 ga		1000	.032		3.42	1.03		4.45	5.50
0845	24 ga		950	.034		4.01	1.09		5.10	6.20
0850	22 ga		900	.036		4.80	1.15		5.95	7.20
0855	Zinc aluminum alloy finish, 26 ga		1000	.032		3.28	1.03		4.31	5.35
0860	24 ga		950	.034		3.66	1.09		4.75	5.85
0865	22 ga		900	.036		4.23	1.15		5.38	6.55
0870	16-1/2″ wide, standard finish, 24 ga		950	.034		3.95	1.09		5.04	6.15
0875	22 ga		900	.036		4.42	1.15		5.57	6.75
0880	Zinc aluminum alloy finish, 24 ga		950	.034		3.45	1.09		4.54	5.60
0885	22 ga		900	.036		3.85	1.15		5	6.15
0890	Flat profile, 2″ x 2″ batten, 12″ wide, standard finish, 26 ga		1000	.032		3.93	1.03		4.96	6.05
0895	24 ga		950	.034		4.69	1.09		5.78	6.95
0900	22 ga		900	.036		5.75	1.15		6.90	8.25
0905	Zinc aluminum alloy finish, 26 ga		1000	.032		3.66	1.03		4.69	5.75
0910	24 ga		950	.034		4.18	1.09		5.27	6.40
0915	22 ga		900	.036		4.87	1.15		6.02	7.25
0920	16-1/2″ wide, standard finish, 24 ga		950	.034		4.32	1.09		5.41	6.55
0925	22 ga		900	.036		5.05	1.15		6.20	7.45
0930	Zinc aluminum alloy finish, 24 ga		950	.034		3.91	1.09		5	6.10
0935	22 ga		900	.036		4.45	1.15		5.60	6.80
1200	Ridge, galvanized, 10″ wide		800	.040	L.F.	3.18	1.29		4.47	5.65
1210	20″ wide		750	.043	″	4.93	1.38		6.31	7.70

07 41 33 – Plastic Roof Panels

07 41 33.10 Fiberglass Panels

Line	Description	Crew	Daily Output	Labor-Hours	Unit	Material	Labor	Equipment	Total	Total Incl O&P
0010	**FIBERGLASS PANELS**									
0012	Corrugated panels, roofing, 8 oz per S.F.	G-3	1000	.032	S.F.	1.74	1.03		2.77	3.63
0100	12 oz per S.F.		1000	.032		3.44	1.03		4.47	5.50
0300	Corrugated siding, 6 oz per S.F.		880	.036		1.53	1.18		2.71	3.63
0400	8 oz per S.F.		880	.036		1.74	1.18		2.92	3.86
0500	Fire retardant		880	.036		3.20	1.18		4.38	5.45

07 41 Roof Panels

07 41 33 – Plastic Roof Panels

07 41 33.10 Fiberglass Panels		Crew	Daily Output	Labor-Hours	Unit	2011 Bare Costs Material	Labor	Equipment	Total	Total Incl O&P
0600	12 oz. siding, textured	G-3	880	.036	S.F.	3.31	1.18		4.49	5.60
0700	Fire retardant		880	.036		4.40	1.18		5.58	6.80
0900	Flat panels, 6 oz per S.F., clear or colors		880	.036		1.95	1.18		3.13	4.10
1100	Fire retardant, class A		880	.036		3.18	1.18		4.36	5.45
1300	8 oz per S.F., clear or colors		880	.036		2.15	1.18		3.33	4.32
1700	Sandwich panels, fiberglass, 1-9/16" thick, panels to 20 S.F.		180	.178		23	5.75		28.75	35
1900	As above, but 2-3/4" thick, panels to 100 S.F.	↓	265	.121	↓	17	3.90		20.90	25

07 42 Wall Panels

07 42 13 – Metal Wall Panels

07 42 13.10 Mansard Panels

		Crew	Daily Output	Labor-Hours	Unit	Material	Labor	Equipment	Total	Total Incl O&P
0010	**MANSARD PANELS**									
0600	Aluminum, stock units, straight surfaces	1 Shee	115	.070	S.F.	3.35	2.66		6.01	8.10
0700	Concave or convex surfaces		75	.107	"	3.67	4.07		7.74	10.80
0800	For framing, to 5' high, add		115	.070	L.F.	3.67	2.66		6.33	8.45
0900	Soffits, to 1' wide	↓	125	.064	S.F.	1.82	2.44		4.26	6.05

07 42 13.20 Aluminum Siding Panels

		Crew	Daily Output	Labor-Hours	Unit	Material	Labor	Equipment	Total	Total Incl O&P
0010	**ALUMINUM SIDING PANELS**									
0012	Corrugated, on steel framing, .019 thick, natural finish	G-3	775	.041	S.F.	1.47	1.33		2.80	3.84
0100	Painted		775	.041		1.58	1.33		2.91	3.96
0400	Farm type, .021" thick on steel frame, natural		775	.041		1.48	1.33		2.81	3.85
0600	Painted		775	.041		1.58	1.33		2.91	3.96
0700	Industrial type, corrugated, on steel, .024" thick, mill		775	.041		2.06	1.33		3.39	4.49
0900	Painted		775	.041		2.22	1.33		3.55	4.66
1000	.032" thick, mill		775	.041		2.37	1.33		3.70	4.83
1200	Painted		775	.041		2.89	1.33		4.22	5.40
1300	V-Beam, on steel frame, .032" thick, mill		775	.041		2.67	1.33		4	5.15
1500	Painted		775	.041		3.06	1.33		4.39	5.60
1600	.040" thick, mill		775	.041		3.24	1.33		4.57	5.80
1800	Painted		775	.041		3.80	1.33		5.13	6.40
1900	.050" thick, mill		775	.041		3.83	1.33		5.16	6.45
2100	Painted		775	.041		4.57	1.33		5.90	7.25
2200	Ribbed, 3" profile, on steel frame, .032" thick, natural		775	.041		2.31	1.33		3.64	4.76
2400	Painted		775	.041		2.93	1.33		4.26	5.45
2500	.040" thick, natural		775	.041		2.64	1.33		3.97	5.10
2700	Painted		775	.041		3.08	1.33		4.41	5.60
2750	.050" thick, natural		775	.041		3.04	1.33		4.37	5.55
2760	Painted		775	.041		3.49	1.33		4.82	6.05
3300	For siding on wood frame, deduct from above	↓	2800	.011	↓	.09	.37		.46	.71
3400	Screw fasteners, aluminum, self tapping, neoprene washer, 1"				M	200			200	220
3600	Stitch screws, self tapping, with neoprene washer, 5/8"				"	150			150	165
3630	Flashing, sidewall, .032" thick	G-3	800	.040	L.F.	2.77	1.29		4.06	5.20
3650	End wall, .040" thick		800	.040		3.24	1.29		4.53	5.70
3670	Closure strips, corrugated, .032" thick		800	.040		.82	1.29		2.11	3.05
3680	Ribbed, 4" or 8", .032" thick		800	.040		.82	1.29		2.11	3.05
3690	V-beam, .040" thick	↓	800	.040	↓	1.11	1.29		2.40	3.37
3800	Horizontal, colored clapboard, 8" wide, plain	2 Carp	515	.031	S.F.	2.17	1.04		3.21	4.13
3810	Insulated		515	.031		2.32	1.04		3.36	4.29
4000	Vertical board & batten, colored, non-insulated	↓	515	.031		1.65	1.04		2.69	3.56
4200	For simulated wood design, add				↓	.10			.10	.11
4300	Corners for above, outside	2 Carp	515	.031	V.L.F.	3.19	1.04		4.23	5.25

07 42 Wall Panels

07 42 13 – Metal Wall Panels

07 42 13.20 Aluminum Siding Panels		Crew	Daily Output	Labor-Hours	Unit	2011 Bare Costs Material	Labor	Equipment	Total	Total Incl O&P
4500	Inside corners	2 Carp	515	.031	V.L.F.	1.39	1.04		2.43	3.27
4600	Sandwich panels, 1" insulation, single story	G-3	395	.081	S.F.	7.35	2.62		9.97	12.40
4900	Multi-story	"	345	.093		10.30	3		13.30	16.35
5100	For baked enamel finish 1 side, add				↓	.40			.40	.44

07 42 13.30 Steel Siding

		Crew	Daily Output	Labor-Hours	Unit	Material	Labor	Equipment	Total	Total Incl O&P
0010	**STEEL SIDING**									
0020	Beveled, vinyl coated, 8" wide	1 Carp	265	.030	S.F.	1.97	1.01		2.98	3.87
0050	10" wide	"	275	.029		2.11	.98		3.09	3.95
0080	Galv, corrugated or ribbed, on steel frame, 30 gauge	G-3	800	.040		1.32	1.29		2.61	3.60
0100	28 gauge		795	.040		1.38	1.30		2.68	3.68
0300	26 gauge		790	.041		1.94	1.31		3.25	4.31
0400	24 gauge		785	.041		1.95	1.32		3.27	4.34
0600	22 gauge		770	.042		2.25	1.34		3.59	4.71
0700	Colored, corrugated/ribbed, on steel frame, 10 yr. finish, 28 ga.		800	.040		2.05	1.29		3.34	4.41
0900	26 gauge		795	.040		2.14	1.30		3.44	4.51
1000	24 gauge		790	.041		2.49	1.31		3.80	4.92
1020	20 gauge		785	.041		3.16	1.32		4.48	5.65
1200	Factory sandwich panel, 26 ga., 1" insulation, galvanized		380	.084		5.25	2.72		7.97	10.30
1300	Colored 1 side		380	.084		6.45	2.72		9.17	11.55
1500	Galvanized 2 sides		380	.084		8	2.72		10.72	13.30
1600	Colored 2 sides		380	.084		8.25	2.72		10.97	13.60
1800	Acrylic paint face, regular paint liner	↓	380	.084		6.25	2.72		8.97	11.35
1900	For 2" thick polystyrene, add					.96			.96	1.06
2000	22 ga, galv, 2" insulation, baked enamel exterior	G-3	360	.089		12.20	2.87		15.07	18.20
2100	Polyvinylidene exterior finish	"	360	.089	↓	12.85	2.87		15.72	18.90

07 44 Faced Panels

07 44 73 – Metal Faced Panels

07 44 73.10 Metal Faced Panels and Accessories

		Crew	Daily Output	Labor-Hours	Unit	Material	Labor	Equipment	Total	Total Incl O&P
0010	**METAL FACED PANELS AND ACCESSORIES**									
0400	Textured aluminum, 4' x 8' x 5/16" plywood backing, single face	2 Shee	375	.043	S.F.	3.67	1.63		5.30	6.75
0600	Double face		375	.043		4.99	1.63		6.62	8.20
0700	4' x 10' x 5/16" plywood backing, single face		375	.043		3.93	1.63		5.56	7
0900	Double face		375	.043		5.25	1.63		6.88	8.50
1000	4' x 12' x 5/16" plywood backing, single face		375	.043		3.33	1.63		4.96	6.35
1300	Smooth aluminum, 1/4" plywood panel, fluoropolymer finish, double face		375	.043		5.45	1.63		7.08	8.65
1350	Clear anodized finish, double face		375	.043		9.55	1.63		11.18	13.20
1400	Double face textured aluminum, structural panel, 1" EPS insulation	↓	375	.043	↓	4.92	1.63		6.55	8.10
1500	Accessories, outside corner	1 Shee	175	.046	L.F.	1.90	1.75		3.65	4.97
1600	Inside corner		175	.046		1.34	1.75		3.09	4.35
1800	Batten mounting clip		200	.040		.49	1.53		2.02	3.06
1900	Low profile batten		480	.017		.61	.64		1.25	1.72
2100	High profile batten		480	.017		1.41	.64		2.05	2.60
2200	Water table		200	.040		2.12	1.53		3.65	4.85
2400	Horizontal joint connector		200	.040		1.67	1.53		3.20	4.36
2500	Corner cap		200	.040		1.84	1.53		3.37	4.54
2700	H - moulding	↓	480	.017	↓	1.25	.64		1.89	2.43

07 46 Siding

07 46 23 – Wood Siding

07 46 23.10 Wood Board Siding		Crew	Daily Output	Labor-Hours	Unit	Material	Labor	Equipment	Total	Total Incl O&P
						2011 Bare Costs				
0010	**WOOD BOARD SIDING**									
3200	Wood, cedar bevel, A grade, 1/2" x 6"	1 Carp	295	.027	S.F.	4.39	.91		5.30	6.35
3300	1/2" x 8"		330	.024		5.85	.81		6.66	7.75
3500	3/4" x 10", clear grade		375	.021		6.55	.72		7.27	8.40
3600	"B" grade		375	.021		3.79	.72		4.51	5.35
3800	Cedar, rough sawn, 1" x 4", A grade, natural		220	.036		3.88	1.22		5.10	6.30
3900	Stained		220	.036		4.25	1.22		5.47	6.70
4100	1" x 12", board & batten, #3 & Btr., natural		420	.019		4.02	.64		4.66	5.50
4200	Stained		420	.019		4.25	.64		4.89	5.75
4400	1" x 8" channel siding, #3 & Btr., natural		330	.024		2.92	.81		3.73	4.57
4500	Stained		330	.024		2.67	.81		3.48	4.30
4700	Redwood, clear, beveled, vertical grain, 1/2" x 4"		220	.036		3.85	1.22		5.07	6.30
4800	1/2" x 8"		330	.024		4.10	.81		4.91	5.85
5000	3/4" x 10"		375	.021		4.12	.72		4.84	5.75
5200	Channel siding, 1" x 10", B grade		375	.021		4.12	.72		4.84	5.75
5250	Redwood, T&G boards, B grade, 1" x 4"		220	.036		2.54	1.22		3.76	4.83
5270	1" x 8"		330	.024		4.10	.81		4.91	5.85
5400	White pine, rough sawn, 1" x 8", natural		330	.024		2.20	.81		3.01	3.78
5500	Stained		330	.024		2.60	.81		3.41	4.22

07 46 29 – Plywood Siding

07 46 29.10 Plywood Siding Options		Crew	Daily Output	Labor-Hours	Unit	Material	Labor	Equipment	Total	Total Incl O&P
0010	**PLYWOOD SIDING OPTIONS**									
0900	Plywood, medium density overlaid, 3/8" thick	2 Carp	750	.021	S.F.	1.25	.72		1.97	2.58
1000	1/2" thick		700	.023		1.22	.77		1.99	2.62
1100	3/4" thick		650	.025		1.55	.83		2.38	3.09
1600	Texture 1-11, cedar, 5/8" thick, natural		675	.024		2.43	.80		3.23	4
1700	Factory stained		675	.024		1.95	.80		2.75	3.48
1900	Texture 1-11, fir, 5/8" thick, natural		675	.024		1.37	.80		2.17	2.84
2000	Factory stained		675	.024		1.73	.80		2.53	3.23
2050	Texture 1-11, S.Y.P., 5/8" thick, natural		675	.024		1.19	.80		1.99	2.64
2100	Factory stained		675	.024		1.24	.80		2.04	2.69
2200	Rough sawn cedar, 3/8" thick, natural		675	.024		1.20	.80		2	2.65
2300	Factory stained		675	.024		.96	.80		1.76	2.39
2500	Rough sawn fir, 3/8" thick, natural		675	.024		.76	.80		1.56	2.17
2600	Factory stained		675	.024		.96	.80		1.76	2.39
2800	Redwood, textured siding, 5/8" thick		675	.024		1.92	.80		2.72	3.44
3000	Polyvinyl chloride coated, 3/8" thick		750	.021		1.11	.72		1.83	2.42

07 46 33 – Plastic Siding

07 46 33.10 Vinyl Siding		Crew	Daily Output	Labor-Hours	Unit	Material	Labor	Equipment	Total	Total Incl O&P
0010	**VINYL SIDING**									
3995	Clapboard profile, woodgrain texture, .048 thick, double 4	2 Carp	495	.032	S.F.	.87	1.08		1.95	2.77
4000	Double 5		550	.029		.87	.98		1.85	2.59
4005	Single 8		495	.032		.94	1.08		2.02	2.85
4010	Single 10		550	.029		1.13	.98		2.11	2.87
4015	.044 thick, double 4		495	.032		.82	1.08		1.90	2.72
4020	Double 5		550	.029		.82	.98		1.80	2.54
4025	.042 thick, double 4		495	.032		.82	1.08		1.90	2.72
4030	Double 5		550	.029		.82	.98		1.80	2.54
4035	Cross sawn texture, .040 thick, double 4		495	.032		.64	1.08		1.72	2.52
4040	Double 5		550	.029		.64	.98		1.62	2.34
4045	Smooth texture, .042 thick, double 4		495	.032		.68	1.08		1.76	2.56
4050	Double 5		550	.029		.73	.98		1.71	2.44

07 46 33 – Plastic Siding

07 46 33.10 Vinyl Siding		Crew	Daily Output	Labor-Hours	Unit	Material	2011 Bare Costs Labor	Equipment	Total	Total Incl O&P
4055	Single 8	2 Carp	495	.032	S.F.	.94	1.08		2.02	2.85
4060	Cedar texture, .044 thick, double 4		495	.032		.85	1.08		1.93	2.75
4065	Double 6		600	.027		.94	.89		1.83	2.54
4070	Dutch lap profile, woodgrain texture, .048 thick, double 5		550	.029		.91	.98		1.89	2.64
4075	.044 thick, double 4.5		525	.030		.87	1.02		1.89	2.67
4080	.042 thick, double 4.5		525	.030		.72	1.02		1.74	2.51
4085	.040 thick, double 4.5		525	.030		.65	1.02		1.67	2.43
4090	Shingle profile, random grooves, double 7		400	.040		2.71	1.34		4.05	5.25
4095	Triple 5		400	.040		2.71	1.34		4.05	5.25
4100	Shake profile, 10" wide		400	.040		2.14	1.34		3.48	4.61
4105	Vertical pattern, .046 thick, double 5		550	.029		1.27	.98		2.25	3.03
4110	.044 thick, triple 3		550	.029		1.31	.98		2.29	3.08
4115	.040 thick, triple 4		550	.029		1.25	.98		2.23	3.01
4120	.040 thick, triple 2.66		550	.029		1.41	.98		2.39	3.19
4125	Insulation, fan folded extruded polystyrene, 1/4"		2000	.008		.23	.27		.50	.71
4130	3/8"		2000	.008		.26	.27		.53	.73
4135	Accessories, J channel, 5/8" pocket		700	.023	L.F.	.37	.77		1.14	1.68
4140	3/4" pocket		695	.023		.38	.77		1.15	1.70
4145	1-1/4" pocket		680	.024		.59	.79		1.38	1.97
4150	Flexible, 3/4" pocket		600	.027		1.92	.89		2.81	3.61
4155	Under sill finish trim		500	.032		.39	1.07		1.46	2.23
4160	Vinyl starter strip		700	.023		.39	.77		1.16	1.71
4165	Aluminum starter strip		700	.023		.26	.77		1.03	1.56
4170	Window casing, 2-1/2" wide, 3/4" pocket		510	.031		1.22	1.05		2.27	3.10
4175	Outside corner, woodgrain finish, 4" face, 3/4" pocket		700	.023		1.58	.77		2.35	3.02
4180	5/8" pocket		700	.023		1.66	.77		2.43	3.11
4185	Smooth finish, 4" face, 3/4" pocket		700	.023		1.58	.77		2.35	3.02
4190	7/8" pocket		690	.023		1.46	.78		2.24	2.91
4195	1-1/4" pocket		700	.023		1.05	.77		1.82	2.44
4200	Soffit and fascia, 1' overhang, solid		120	.133		3.38	4.47		7.85	11.20
4205	Vented		120	.133		3.38	4.47		7.85	11.20
4207	18" overhang, solid		110	.145		4.02	4.88		8.90	12.55
4208	Vented		110	.145		4.02	4.88		8.90	12.55
4210	2' overhang, solid		100	.160		4.65	5.35		10	14.10
4215	Vented		100	.160		4.65	5.35		10	14.10
4217	3' overhang, solid		100	.160		5.90	5.35		11.25	15.50
4218	Vented		100	.160		5.90	5.35		11.25	15.50
4220	Colors for siding and soffits, add				S.F.	.16			.16	.18
4225	Colors for accessories and trim, add				L.F.	.32			.32	.35

07 46 46 – Fiber Cement Siding

07 46 46.10 Fiber Cement Siding		Crew	Daily Output	Labor-Hours	Unit	Material	2011 Bare Costs Labor	Equipment	Total	Total Incl O&P
0010	**FIBER CEMENT SIDING**									
0020	Lap siding, 5/16" thick, 6" wide, 4-3/4" exposure, smooth texture	2 Carp	415	.039	S.F.	1.16	1.29		2.45	3.44
0025	Woodgrain texture		415	.039		1.16	1.29		2.45	3.44
0030	7-1/2" wide, 6-1/4" exposure, smooth texture		425	.038		1.04	1.26		2.30	3.25
0035	Woodgrain texture		425	.038		1.04	1.26		2.30	3.25
0040	8" wide, 6-3/4" exposure, smooth texture		425	.038		1.12	1.26		2.38	3.34
0045	Roughsawn texture		425	.038		1.12	1.26		2.38	3.34
0050	9-1/2" wide, 8-1/4" exposure, smooth texture		440	.036		1.16	1.22		2.38	3.32
0055	Woodgrain texture		440	.036		1.16	1.22		2.38	3.32
0060	12" wide, 10-3/8" exposure, smooth texture		455	.035		1.28	1.18		2.46	3.38
0065	Woodgrain texture		455	.035		1.28	1.18		2.46	3.38

07 46 Siding

07 46 46 – Fiber Cement Siding

07 46 46.10 Fiber Cement Siding		Crew	Daily Output	Labor-Hours	Unit	Material	Labor (2011 Bare Costs)	Equipment	Total	Total Incl O&P
0070	Panel siding, 5/16" thick, smooth texture	2 Carp	750	.021	S.F.	1.05	.72		1.77	2.36
0075	Stucco texture		750	.021		1.05	.72		1.77	2.36
0080	Grooved woodgrain texture		750	.021		1.05	.72		1.77	2.36
0085	V - grooved woodgrain texture		750	.021		1.05	.72		1.77	2.36
0090	Wood starter strip		400	.040	L.F.	.53	1.34		1.87	2.83

07 46 73 – Soffit

07 46 73.10 Soffit Options

		Crew	Daily Output	Labor-Hours	Unit	Material	Labor	Equipment	Total	Total Incl O&P
0010	**SOFFIT OPTIONS**									
0012	Aluminum, residential, .020" thick	1 Carp	210	.038	S.F.	1.96	1.28		3.24	4.30
0100	Baked enamel on steel, 16 or 18 gauge		105	.076		5.35	2.56		7.91	10.20
0300	Polyvinyl chloride, white, solid		230	.035		1.96	1.17		3.13	4.11
0400	Perforated		230	.035		1.96	1.17		3.13	4.11
0500	For colors, add					.13			.13	.14

07 51 Built-Up Bituminous Roofing

07 51 13 – Built-Up Asphalt Roofing

07 51 13.10 Built-Up Roofing Components

		Crew	Daily Output	Labor-Hours	Unit	Material	Labor	Equipment	Total	Total Incl O&P
0010	**BUILT-UP ROOFING COMPONENTS**									
0012	Asphalt saturated felt, #30, 2 square per roll	1 Rofc	58	.138	Sq.	9.45	3.88		13.33	17.35
0200	#15, 4 sq per roll, plain or perforated, not mopped		58	.138		5.65	3.88		9.53	13.20
0300	Roll roofing, smooth, #65		15	.533		9.65	15		24.65	37.50
0500	#90		12	.667		34.50	18.75		53.25	72
0520	Mineralized		12	.667		34.50	18.75		53.25	72
0540	D.C. (Double coverage), 19" selvage edge		10	.800		44	22.50		66.50	88.50
0580	Adhesive (lap cement)				Gal.	4.79			4.79	5.25
0600	Steep, flat or dead level asphalt, 10 ton lots, bulk				Ton	415			415	455
0800	Packaged				"	700			700	770

07 51 13.13 Cold-Applied Built-Up Asphalt Roofing

		Crew	Daily Output	Labor-Hours	Unit	Material	Labor	Equipment	Total	Total Incl O&P
0010	**COLD-APPLIED BUILT-UP ASPHALT ROOFING**									
0020	3 ply system, installation only (components listed below)	G-5	50	.800	Sq.		20.50	3.37	23.87	40.50
0100	Spunbond poly. fabric, 1.35 oz./S.Y., 36"W, 10.8 sq./roll				Ea.	175			175	193
0200	49" wide, 14.6 sq./roll					244			244	268
0300	2.10 oz./S.Y., 36" wide, 10.8 sq./roll					266			266	293
0400	49" wide, 14.6 sq./roll					360			360	395
0500	Base & finish coat, 3 gal./sq., 5 gal./can				Gal.	5			5	5.50
0600	Coating, ceramic granules, 1/2 sq./bag				Ea.	16.75			16.75	18.45
0700	Aluminum, 2 gal./sq.				Gal.	13.50			13.50	14.85
0800	Emulsion, fibered or non-fibered, 4 gal./sq.				"	4.70			4.70	5.15

07 51 13.20 Built-Up Roofing Systems

		Crew	Daily Output	Labor-Hours	Unit	Material	Labor	Equipment	Total	Total Incl O&P
0010	**BUILT-UP ROOFING SYSTEMS** R075113-20									
0120	Asphalt flood coat with gravel/slag surfacing, not including									
0140	Insulation, flashing or wood nailers									
0200	Asphalt base sheet, 3 plies #15 asphalt felt, mopped	G-1	22	2.545	Sq.	82.50	67	18.60	168.10	233
0350	On nailable decks		21	2.667		87.50	70.50	19.50	177.50	245
0500	4 plies #15 asphalt felt, mopped		20	2.800		115	74	20.50	209.50	282
0550	On nailable decks		19	2.947		103	77.50	21.50	202	277
0700	Coated glass base sheet, 2 plies glass (type IV), mopped		22	2.545		86	67	18.60	171.60	236
0850	3 plies glass, mopped		20	2.800		103	74	20.50	197.50	269
0950	On nailable decks		19	2.947		97	77.50	21.50	196	270
1100	4 plies glass fiber felt (type IV), mopped		20	2.800		125	74	20.50	219.50	294

07 51 Built-Up Bituminous Roofing

07 51 13 – Built-Up Asphalt Roofing

07 51 13.20 Built-Up Roofing Systems		Crew	Daily Output	Labor-Hours	Unit	Material	2011 Bare Costs Labor	Equipment	Total	Total Incl O&P
1150	On nailable decks	G-1	19	2.947	Sq.	113	77.50	21.50	212	289
1200	Coated & saturated base sheet, 3 plies #15 asph. felt, mopped		20	2.800		92.50	74	20.50	187	258
1250	On nailable decks		19	2.947		86.50	77.50	21.50	185.50	259
1300	4 plies #15 asphalt felt, mopped	↓	22	2.545	↓	108	67	18.60	193.60	261
2000	Asphalt flood coat, smooth surface									
2200	Asphalt base sheet & 3 plies #15 asphalt felt, mopped	G-1	24	2.333	Sq.	87.50	61.50	17.05	166.05	226
2400	On nailable decks		23	2.435		81.50	64	17.80	163.30	226
2600	4 plies #15 asphalt felt, mopped		24	2.333		103	61.50	17.05	181.55	243
2700	On nailable decks	↓	23	2.435	↓	97	64	17.80	178.80	243
2900	Coated glass fiber base sheet, mopped, and 2 plies of									
2910	glass fiber felt (type IV)	G-1	25	2.240	Sq.	80.50	59	16.35	155.85	214
3100	On nailable decks		24	2.333		76	61.50	17.05	154.55	213
3200	3 plies, mopped		23	2.435		97	64	17.80	178.80	243
3300	On nailable decks		22	2.545		91	67	18.60	176.60	242
3800	4 plies glass fiber felt (type IV), mopped		23	2.435		114	64	17.80	195.80	261
3900	On nailable decks		22	2.545		108	67	18.60	193.60	260
4000	Coated & saturated base sheet, 3 plies #15 asph. felt, mopped		24	2.333		86.50	61.50	17.05	165.05	225
4200	On nailable decks		23	2.435		81	64	17.80	162.80	225
4300	4 plies #15 organic felt, mopped	↓	22	2.545	↓	102	67	18.60	187.60	254
4500	Coal tar pitch with gravel/slag surfacing									
4600	4 plies #15 tarred felt, mopped	G-1	21	2.667	Sq.	176	70.50	19.50	266	345
4800	3 plies glass fiber felt (type IV), mopped	"	19	2.947	"	146	77.50	21.50	245	325
5000	Coated glass fiber base sheet, and 2 plies of									
5010	glass fiber felt, (type IV), mopped	G-1	19	2.947	Sq.	150	77.50	21.50	249	330
5300	On nailable decks		18	3.111		131	82	22.50	235.50	320
5600	4 plies glass fiber felt (type IV), mopped		21	2.667		202	70.50	19.50	292	370
5800	On nailable decks	↓	20	2.800	↓	183	74	20.50	277.50	360

07 51 13.30 Cants

		Crew	Daily Output	Labor-Hours	Unit	Material	Labor	Equipment	Total	Total Incl O&P
0010	**CANTS**									
0012	Lumber, treated, 4" x 4" cut diagonally	1 Rofc	325	.025	L.F.	2	.69		2.69	3.45
0100	Foamglass		325	.025		2.23	.69		2.92	3.70
0300	Mineral or fiber, trapezoidal, 1" x 4" x 48"		325	.025		.19	.69		.88	1.46
0400	1-1/2" x 5-5/8" x 48"	↓	325	.025	↓	.31	.69		1	1.59

07 51 13.40 Felts

		Crew	Daily Output	Labor-Hours	Unit	Material	Labor	Equipment	Total	Total Incl O&P
0010	**FELTS**									
0012	Glass fibered roofing felt, #15, not mopped	1 Rofc	58	.138	Sq.	6.90	3.88		10.78	14.55
0300	Base sheet, #45, channel vented		58	.138		32	3.88		35.88	42.50
0400	#50, coated		58	.138		16.45	3.88		20.33	25
0500	Cap, mineral surfaced		58	.138		45	3.88		48.88	56.50
0600	Flashing membrane, #65		16	.500		9.65	14.10		23.75	36
0800	Coal tar fibered, #15, no mopping		58	.138		14	3.88		17.88	22.50
0900	Asphalt felt, #15, 4 sq per roll, no mopping		58	.138		5.65	3.88		9.53	13.20
1100	#30, 2 sq per roll		58	.138		9.45	3.88		13.33	17.35
1200	Double coated, #33		58	.138		9.90	3.88		13.78	17.90
1400	#40, base sheet		58	.138		10.90	3.88		14.78	19
1450	Coated and saturated		58	.138		9.95	3.88		13.83	17.95
1500	Tarred felt, organic, #15, 4 sq rolls		58	.138		11.50	3.88		15.38	19.65
1550	#30, 2 sq roll	↓	58	.138		23	3.88		26.88	32.50
1700	Add for mopping above felts, per ply, asphalt, 24 lb per sq	G-1	192	.292		8.40	7.70	2.13	18.23	25.50
1800	Coal tar mopping, 30 lb per sq		186	.301		16.50	7.95	2.20	26.65	35
1900	Flood coat, with asphalt, 60 lb per sq		60	.933		21	24.50	6.80	52.30	75
2000	With coal tar, 75 lb per sq	↓	56	1	↓	41.50	26.50	7.30	75.30	101

07 51 Built-Up Bituminous Roofing

07 51 13 – Built-Up Asphalt Roofing

	07 51 13.50 Walkways for Built-Up Roofs	Crew	Daily Output	Labor-Hours	Unit	Material	2011 Bare Costs Labor	Equipment	Total	Total Incl O&P
0010	**WALKWAYS FOR BUILT-UP ROOFS**									
0020	Asphalt impregnated, 3′ x 6′ x 1/2″ thick	1 Rofc	400	.020	S.F.	1.72	.56		2.28	2.91
0100	3′ x 3′ x 3/4″ thick	″	400	.020		4.96	.56		5.52	6.45
0300	Concrete patio blocks, 2″ thick, natural	1 Clab	115	.070		3.71	1.84		5.55	7.15
0400	Colors	″	115	.070		3.75	1.84		5.59	7.20

07 52 Modified Bituminous Membrane Roofing

07 52 13 – Atactic-Polypropylene-Modified Bituminous Membrane Roofing

	07 52 13.10 APP Modified Bituminous Membrane	Crew	Daily Output	Labor-Hours	Unit	Material	Labor	Equipment	Total	Total Incl O&P
0010	**APP MODIFIED BITUMINOUS MEMBRANE** R075213-30									
0020	Base sheet, #15 glass fiber felt, nailed to deck	1 Rofc	58	.138	Sq.	7.80	3.88		11.68	15.60
0030	Spot mopped to deck	G-1	295	.190		11.10	5	1.39	17.49	23
0040	Fully mopped to deck	″	192	.292		15.30	7.70	2.13	25.13	33
0050	#15 organic felt, nailed to deck	1 Rofc	58	.138		6.55	3.88		10.43	14.20
0060	Spot mopped to deck	G-1	295	.190		9.85	5	1.39	16.24	21.50
0070	Fully mopped to deck	″	192	.292		14.05	7.70	2.13	23.88	31.50
2100	APP mod., smooth surf. cap sheet, poly. reinf., torched, 160 mils	G-5	2100	.019	S.F.	.48	.49	.08	1.05	1.50
2150	170 mils		2100	.019		.55	.49	.08	1.12	1.58
2200	Granule surface cap sheet, poly. reinf., torched, 180 mils		2000	.020		.59	.51	.08	1.18	1.67
2250	Smooth surface flashing, torched, 160 mils		1260	.032		.48	.81	.13	1.42	2.15
2300	170 mils		1260	.032		.55	.81	.13	1.49	2.23
2350	Granule surface flashing, torched, 180 mils		1260	.032		.59	.81	.13	1.53	2.27
2400	Fibrated aluminum coating	1 Rofc	3800	.002		.11	.06		.17	.23

07 52 16 – Styrene-Butadiene-Styrene Modified Bituminous Membrane Roofing

	07 52 16.10 SBS Modified Bituminous Membrane	Crew	Daily Output	Labor-Hours	Unit	Material	Labor	Equipment	Total	Total Incl O&P
0010	**SBS MODIFIED BITUMINOUS MEMBRANE**									
0080	SBS modified, granule surf cap sheet, polyester rein., mopped									
0600	150 mils	G-1	2000	.028	S.F.	.58	.74	.20	1.52	2.20
1100	160 mils		2000	.028		.77	.74	.20	1.71	2.41
1500	Glass fiber reinforced, mopped, 160 mils		2000	.028		.49	.74	.20	1.43	2.10
1600	Smooth surface cap sheet, mopped, 145 mils		2100	.027		.49	.70	.20	1.39	2.02
1700	Smooth surface flashing, 145 mils		1260	.044		.49	1.17	.32	1.98	3.01
1800	150 mils		1260	.044		.48	1.17	.32	1.97	3
1900	Granular surface flashing, 150 mils		1260	.044		.58	1.17	.32	2.07	3.11
2000	160 mils		1260	.044		.77	1.17	.32	2.26	3.32

07 53 Elastomeric Membrane Roofing

07 53 16 – Chlorosulfonate-Polyethylene Roofing

	07 53 16.10 Chlorosulfonated Polyethylene Roofing	Crew	Daily Output	Labor-Hours	Unit	Material	Labor	Equipment	Total	Total Incl O&P
0010	**CHLOROSULFONATED POLYETHYLENE ROOFING**									
0800	Chlorosulfonated polyethylene-hypalon (CSPE), 45 mils,									
0900	0.29 P.S.F., fully adhered	G-5	26	1.538	Sq.	185	39.50	6.50	231	282
1100	Loose-laid & ballasted with stone (10 P.S.F.)		51	.784		193	20	3.30	216.30	252
1200	Mechanically attached		35	1.143		186	29.50	4.81	220.31	263
1300	Plates with adhesive attachment		35	1.143		183	29.50	4.81	217.31	259

07 53 Elastomeric Membrane Roofing

07 53 23 – Ethylene-Propylene-Diene-Monomer Roofing

07 53 23.20 Ethylene-Propylene-Diene-Monomer Roofing		Crew	Daily Output	Labor-Hours	Unit	2011 Bare Costs Material	Labor	Equipment	Total	Total Incl O&P
0010	**ETHYLENE-PROPYLENE-DIENE-MONOMER ROOFING (E.P.D.M.)**									
3500	Ethylene-propylene-diene-monomer (EPDM), 45 mils, 0.28 P.S.F.									
3600	Loose-laid & ballasted with stone (10 P.S.F.)	G-5	51	.784	Sq.	73.50	20	3.30	96.80	121
3700	Mechanically attached		35	1.143		65.50	29.50	4.81	99.81	130
3800	Fully adhered with adhesive		26	1.538		101	39.50	6.50	147	189
4500	60 mils, 0.40 P.S.F.									
4600	Loose-laid & ballasted with stone (10 P.S.F.)	G-5	51	.784	Sq.	87	20	3.30	110.30	136
4700	Mechanically attached		35	1.143		77.50	29.50	4.81	111.81	143
4800	Fully adhered with adhesive		26	1.538		113	39.50	6.50	159	202
4810	45 mil, .28 PSF, membrane only					39.50			39.50	43.50
4820	60 mil, .40 PSF, membrane only					50.50			50.50	55.50
4850	Seam tape for membrane, 3" x 100' roll				Ea.	48			48	53
4900	Batten strips, 10' sections					3.05			3.05	3.36
4910	Cover tape for batten strips, 6" x 100' roll					110			110	121
4930	Plate anchors				M	105			105	116
4970	Adhesive for fully adhered systems, 60 S.F./gal.				Gal.	21			21	23

07 53 29 – Polyisobutylene Roofing

07 53 29.10 Polyisobutylene Roofing		Crew	Daily Output	Labor-Hours	Unit	Material	Labor	Equipment	Total	Total Incl O&P
0010	**POLYISOBUTYLENE ROOFING**									
7500	Polyisobutylene (PIB), 100 mils, 0.57 P.S.F.									
7600	Loose-laid & ballasted with stone/gravel (10 P.S.F.)	G-5	51	.784	Sq.	181	20	3.30	204.30	239
7700	Partially adhered with adhesive		35	1.143		227	29.50	4.81	261.31	310
7800	Hot asphalt attachment		35	1.143		217	29.50	4.81	251.31	296
7900	Fully adhered with contact cement		26	1.538		234	39.50	6.50	280	335

07 54 Thermoplastic Membrane Roofing

07 54 19 – Polyvinyl-Chloride Roofing

07 54 19.10 Polyvinyl-Chloride Roofing (P.V.C.)		Crew	Daily Output	Labor-Hours	Unit	Material	Labor	Equipment	Total	Total Incl O&P
0010	**POLYVINYL-CHLORIDE ROOFING (P.V.C.)**									
8200	Heat welded seams									
8700	Reinforced, 48 mils, 0.33 P.S.F.									
8750	Loose-laid & ballasted with stone/gravel (12 P.S.F.)	G-5	51	.784	Sq.	102	20	3.30	125.30	152
8800	Mechanically attached		35	1.143		94	29.50	4.81	128.31	161
8850	Fully adhered with adhesive		26	1.538		127	39.50	6.50	173	217
8860	Reinforced, 60 mils, .40 P.S.F.									
8870	Loose-laid & ballasted with stone/gravel (12 P.S.F.)	G-5	51	.784	Sq.	103	20	3.30	126.30	154
8880	Mechanically attached		35	1.143		95.50	29.50	4.81	129.81	163
8890	Fully adhered with adhesive		26	1.538		128	39.50	6.50	174	219

07 54 23 – Thermoplastic-Polyolefin Roofing

07 54 23.10 Thermoplastic Polyolefin Roofing (T.P.O)		Crew	Daily Output	Labor-Hours	Unit	Material	Labor	Equipment	Total	Total Incl O&P
0010	**THERMOPLASTIC POLYOLEFIN ROOFING (T.P.O.)**									
0100	45 mils, loose laid & ballasted with stone(1/2 ton/sq.)	G-5	51	.784	Sq.	83.50	20	3.30	106.80	132
0120	Fully adhered		25	1.600		72.50	41	6.75	120.25	161
0140	Mechanically attached		34	1.176		72	30	4.95	106.95	139
0160	Self adhered		35	1.143		88	29.50	4.81	122.31	155
0180	60 mil membrane, heat welded seams, ballasted		50	.800		89	20.50	3.37	112.87	139
0200	Fully adhered		25	1.600		77	41	6.75	124.75	166
0220	Mechanically attached		34	1.176		80.50	30	4.95	115.45	148
0240	Self adhered		35	1.143		109	29.50	4.81	143.31	178

07 56 Fluid-Applied Roofing

07 56 10 – Fluid-Applied Roofing Elastomers

07 56 10.10 Elastomeric Roofing		Crew	Daily Output	Labor-Hours	Unit	Material	2011 Bare Costs Labor	Equipment	Total	Total Incl O&P
0010	**ELASTOMERIC ROOFING**									
0110	Acrylic rubber, fluid applied, 20 mils thick	G-5	2000	.020	S.F.	2.04	.51	.08	2.63	3.26
0120	50 mils, reinforced		1200	.033		3.16	.85	.14	4.15	5.15
0130	For walking surface, add		900	.044		.96	1.14	.19	2.29	3.33
0300	Hypalon neoprene, fluid applied, 20 mil thick, not-reinforced	G-1	1135	.049		2.57	1.30	.36	4.23	5.60
0600	Non-woven polyester, reinforced		960	.058		2.61	1.54	.43	4.58	6.10
0700	5 coat neoprene deck, 60 mil thick, under 10,000 S.F.		325	.172		5.75	4.54	1.26	11.55	15.90
0900	Over 10,000 S.F.		625	.090		5.35	2.36	.66	8.37	10.90
1300	Vinyl plastic traffic deck, sprayed, 2 to 4 mils thick		625	.090		1.68	2.36	.66	4.70	6.85
1500	Vinyl and neoprene membrane traffic deck		1550	.036		1.77	.95	.26	2.98	3.96

07 57 Coated Foamed Roofing

07 57 13 – Sprayed Polyurethane Foam Roofing

07 57 13.10 Sprayed Polyurethane Foam Roofing (S.P.F.)		Crew	Daily Output	Labor-Hours	Unit	Material	Labor	Equipment	Total	Total Incl O&P
0010	**SPRAYED POLYURETHANE FOAM ROOFING (S.P.F.)**									
0100	Primer for metal substrate (when required)	G-2A	3000	.008	S.F.	.42	.20	.20	.82	1.03
0200	Primer for non-metal substrate (when required)		3000	.008		.16	.20	.20	.56	.75
0300	Closed cell spray, polyurethane foam, 3 lb. per C.F. density, 1", R6.7		15000	.002		.55	.04	.04	.63	.72
0400	2", R13.4		13125	.002		1.10	.05	.05	1.20	1.34
0500	3", R18.6		11485	.002		1.66	.05	.05	1.76	1.97
0550	4", R24.8		10080	.002		2.21	.06	.06	2.33	2.60
0700	Spray-on silicone coating		2500	.010		.92	.24	.24	1.40	1.70
0800	Warranty 5-20 year manufacturer's									.15
0900	Warranty 20 year, no dollar limit									.20

07 58 Roll Roofing

07 58 10 – Asphalt Roll Roofing

07 58 10.10 Roll Roofing		Crew	Daily Output	Labor-Hours	Unit	Material	Labor	Equipment	Total	Total Incl O&P
0010	**ROLL ROOFING**									
0100	Asphalt, mineral surface									
0200	1 ply #15 organic felt, 1 ply mineral surfaced									
0300	Selvage roofing, lap 19", nailed & mopped	G-1	27	2.074	Sq.	60.50	54.50	15.15	130.15	182
0400	3 plies glass fiber felt (type IV), 1 ply mineral surfaced									
0500	Selvage roofing, lapped 19", mopped	G-1	25	2.240	Sq.	98	59	16.35	173.35	233
0600	Coated glass fiber base sheet, 2 plies of glass fiber									
0700	Felt (type IV), 1 ply mineral surfaced selvage									
0800	Roofing, lapped 19", mopped	G-1	25	2.240	Sq.	108	59	16.35	183.35	244
0900	On nailable decks	"	24	2.333	"	99.50	61.50	17.05	178.05	239
1000	3 plies glass fiber felt (type III), 1 ply mineral surfaced									
1100	Selvage roofing, lapped 19", mopped	G-1	25	2.240	Sq.	98	59	16.35	173.35	233

07 61 Sheet Metal Roofing

07 61 13 – Standing Seam Sheet Metal Roofing

07 61 13.10 Standing Seam Sheet Metal Roofing, Field Fab.		Crew	Daily Output	Labor-Hours	Unit	2011 Bare Costs Material	Labor	Equipment	Total	Total Incl O&P
0010	**STANDING SEAM SHEET METAL ROOFING, FIELD FABRICATED**									
0400	Copper standing seam roofing, over 10 squares, 16 oz, 125 lb per sq	1 Shee	1.30	6.154	Sq.	735	235		970	1,200
0600	18 oz, 140 lb per sq		1.20	6.667		825	255		1,080	1,325
0700	20 oz, 150 lb per sq		1.10	7.273		920	278		1,198	1,450
1200	For abnormal conditions or small areas, add					25%	100%			
1300	For lead-coated copper, add					25%				

07 61 16 – Batten Seam Sheet Metal Roofing

07 61 16.10 Batten Seam Sheet Metal Roofing, Field Fabricated

		Crew	Daily Output	Labor-Hours	Unit	Material	Labor	Equipment	Total	Total Incl O&P
0009	**BATTEN SEAM SHEET METAL ROOFING, FIELD FABRICATED**									
0012	Copper batten seam roofing, over 10 sq, 16 oz, 130 lb per sq	1 Shee	1.10	7.273	Sq.	935	278		1,213	1,475
0020	Lead batten seam roofing, 5 lb. per S.F.		1.20	6.667		600	255		855	1,075
0100	Zinc/copper alloy batten seam roofing, .020 thick		1.20	6.667		1,100	255		1,355	1,650
0200	Copper roofing, batten seam, over 10 sq, 18 oz, 145 lb per sq		1	8		1,050	305		1,355	1,650
0300	20 oz, 160 lb per sq		1	8		1,175	305		1,480	1,775
0500	Stainless steel batten seam roofing, type 304, 28 gauge		1.20	6.667		625	255		880	1,100
0600	26 gauge		1.15	6.957		630	266		896	1,125
0800	Zinc, copper alloy roofing, batten seam, .027" thick		1.15	6.957		1,550	266		1,816	2,175
0900	.032" thick		1.10	7.273		1,875	278		2,153	2,525
1000	.040" thick		1.05	7.619		2,100	291		2,391	2,775

07 61 19 – Flat Seam Sheet Metal Roofing

07 61 19.10 Flat Seam Sheet Metal Roofing, Field Fabricated

		Crew	Daily Output	Labor-Hours	Unit	Material	Labor	Equipment	Total	Total Incl O&P
0010	**FLAT SEAM SHEET METAL ROOFING, FIELD FABRICATED**									
0900	Copper flat seam roofing, over 10 squares, 16 oz, 115 lb./sq.	1 Shee	1.20	6.667	Sq.	685	255		940	1,175
1000	20 oz, 145 lb./sq.		1.10	7.273		855	278		1,133	1,400
1100	Lead flat seam roofing, 5 lb. per S.F.		1.30	6.154		600	235		835	1,050

07 65 Flexible Flashing

07 65 10 – Sheet Metal Flashing

07 65 10.10 Sheet Metal Flashing and Counter Flashing

		Crew	Daily Output	Labor-Hours	Unit	Material	Labor	Equipment	Total	Total Incl O&P
0010	**SHEET METAL FLASHING AND COUNTER FLASHING**									
0011	Including up to 4 bends									
0020	Aluminum, mill finish, .013" thick	1 Rofc	145	.055	S.F.	.73	1.55		2.28	3.60
0030	.016" thick		145	.055		.84	1.55		2.39	3.72
0060	.019" thick		145	.055		1.07	1.55		2.62	3.98
0100	.032" thick		145	.055		1.25	1.55		2.80	4.18
0200	.040" thick		145	.055		2.14	1.55		3.69	5.15
0300	.050" thick		145	.055		2.25	1.55		3.80	5.30
0325	Mill finish 5" x 7" step flashing, .016" thick		1920	.004	Ea.	.18	.12		.30	.41
0350	Mill finish 12" x 12" step flashing, .016" thick		1600	.005	"	.62	.14		.76	.93
0400	Painted finish, add				S.F.	.34			.34	.37
1000	Mastic-coated 2 sides, .005" thick	1 Rofc	330	.024		1.67	.68		2.35	3.07
1100	.016" thick		330	.024		1.84	.68		2.52	3.25
1600	Copper, 16 oz, sheets, under 1000 lb.		115	.070		5.90	1.96		7.86	10.05
1700	Over 4000 lb.		155	.052		5.55	1.45		7	8.70
1900	20 oz sheets, under 1000 lb.		110	.073		7.35	2.05		9.40	11.80
2000	Over 4000 lb.		145	.055		7	1.55		8.55	10.50
2200	24 oz sheets, under 1000 lb.		105	.076		9.25	2.14		11.39	14.05
2300	Over 4000 lb.		135	.059		8.80	1.67		10.47	12.65
2500	32 oz sheets, under 1000 lb.		100	.080		11.10	2.25		13.35	16.25
2600	Over 4000 lb.		130	.062		10.55	1.73		12.28	14.75

07 65 Flexible Flashing

07 65 10 – Sheet Metal Flashing

07 65 10.10 Sheet Metal Flashing and Counter Flashing		Crew	Daily Output	Labor-Hours	Unit	2011 Bare Costs Material	Labor	Equipment	Total	Total Incl O&P
2700	W shape for valleys, 16 oz, 24" wide	1 Rofc	100	.080	L.F.	12	2.25		14.25	17.25
5800	Lead, 2.5 lb. per S.F., up to 12" wide		135	.059	S.F.	5.40	1.67		7.07	8.95
5900	Over 12" wide		135	.059		4.43	1.67		6.10	7.90
8900	Stainless steel sheets, 32 ga, .010" thick		155	.052		4	1.45		5.45	7
9000	28 ga, .015" thick		155	.052		5.65	1.45		7.10	8.80
9100	26 ga, .018" thick		155	.052		7.25	1.45		8.70	10.60
9200	24 ga, .025" thick		155	.052		8.90	1.45		10.35	12.40
9290	For mechanically keyed flashing, add					40%				
9320	Steel sheets, galvanized, 20 gauge	1 Rofc	130	.062	S.F.	1.57	1.73		3.30	4.86
9340	30 gauge		160	.050		1.46	1.41		2.87	4.15
9400	Terne coated stainless steel, .015" thick, 28 ga		155	.052		7.35	1.45		8.80	10.70
9500	.018" thick, 26 ga		155	.052		8.30	1.45		9.75	11.70
9600	Zinc and copper alloy (brass), .020" thick		155	.052		11.10	1.45		12.55	14.80
9700	.027" thick		155	.052		15.60	1.45		17.05	19.75
9800	.032" thick		155	.052		18.80	1.45		20.25	23
9900	.040" thick		155	.052		21	1.45		22.45	25.50

07 65 12 – Fabric and Mastic Flashings

07 65 12.10 Fabric and Mastic Flashing and Counter Flashing

		Crew	Daily Output	Labor-Hours	Unit	Material	Labor	Equipment	Total	Total Incl O&P
0010	**FABRIC AND MASTIC FLASHING AND COUNTER FLASHING**									
1300	Asphalt flashing cement, 5 gallon				Gal.	10.80			10.80	11.85
4900	Fabric, asphalt-saturated cotton, specification grade	1 Rofc	35	.229	S.Y.	2.36	6.45		8.81	14.20
5000	Utility grade		35	.229		1.26	6.45		7.71	13
5300	Close-mesh fabric, saturated, 17 oz per S.Y.		35	.229		2.36	6.45		8.81	14.20
5500	Fiberglass, resin-coated		35	.229		1.23	6.45		7.68	12.95
8500	Shower pan, bituminous membrane, 7 oz		155	.052	S.F.	2	1.45		3.45	4.82

07 65 13 – Laminated Sheet Flashing

07 65 13.10 Laminated Sheet Flashing

		Crew	Daily Output	Labor-Hours	Unit	Material	Labor	Equipment	Total	Total Incl O&P
0010	**LAMINATED SHEET FLASHING**, Including up to 4 bends									
0500	Fabric-backed 2 sides, .004" thick	1 Rofc	330	.024	S.F.	1.41	.68		2.09	2.78
0700	.005" thick		330	.024		1.67	.68		2.35	3.07
0750	Mastic-backed, self adhesive		460	.017		3.46	.49		3.95	4.69
0800	Mastic-coated 2 sides, .004" thick		330	.024		1.41	.68		2.09	2.78
2800	Copper, paperbacked 1 side, 2 oz		330	.024		1.25	.68		1.93	2.61
2900	3 oz		330	.024		1.63	.68		2.31	3.02
3100	Paperbacked 2 sides, 2 oz		330	.024		1.26	.68		1.94	2.62
3150	3 oz		330	.024		1.62	.68		2.30	3.01
3200	5 oz		330	.024		2.43	.68		3.11	3.90
3250	7 oz		330	.024		3.96	.68		4.64	5.60
3400	Mastic-backed 2 sides, copper, 2 oz		330	.024		1.49	.68		2.17	2.87
3500	3 oz		330	.024		1.83	.68		2.51	3.24
3700	5 oz		330	.024		2.68	.68		3.36	4.18
3800	Fabric-backed 2 sides, copper, 2 oz		330	.024		1.57	.68		2.25	2.96
4000	3 oz		330	.024		2.04	.68		2.72	3.47
4100	5 oz		330	.024		2.76	.68		3.44	4.27
4300	Copper-clad stainless steel, .015" thick, under 500 lb.		115	.070		5.80	1.96		7.76	9.95
4400	Over 2000 lb.		155	.052		5.55	1.45		7	8.70
4600	.018" thick, under 500 lb.		100	.080		6.55	2.25		8.80	11.25
4700	Over 2000 lb.		145	.055		6.20	1.55		7.75	9.65
6100	Lead-coated copper, fabric-backed, 2 oz		330	.024		2.42	.68		3.10	3.89
6200	5 oz		330	.024		2.78	.68		3.46	4.29
6400	Mastic-backed 2 sides, 2 oz		330	.024		1.89	.68		2.57	3.31
6500	5 oz		330	.024		2.34	.68		3.02	3.80

07 65 Flexible Flashing

07 65 13 – Laminated Sheet Flashing

07 65 13.10 Laminated Sheet Flashing		Crew	Daily Output	Labor-Hours	Unit	Material	2011 Bare Costs Labor	Equipment	Total	Total Incl O&P
6700	Paperbacked 1 side, 2 oz	1 Rofc	330	.024	S.F.	1.63	.68		2.31	3.02
6800	3 oz		330	.024		1.92	.68		2.60	3.34
7000	Paperbacked 2 sides, 2 oz		330	.024		1.69	.68		2.37	3.09
7100	5 oz		330	.024		2.75	.68		3.43	4.26
8550	3 ply copper and fabric, 3 oz		155	.052		2.32	1.45		3.77	5.15
8600	7 oz		155	.052		3.25	1.45		4.70	6.20
8700	Lead on copper and fabric, 5 oz		155	.052		2.78	1.45		4.23	5.70
8800	7 oz		155	.052		5.25	1.45		6.70	8.40
9300	Stainless steel, paperbacked 2 sides, .005" thick	↓	330	.024	↓	3.62	.68		4.30	5.20

07 65 19 – Plastic Sheet Flashing

07 65 19.10 Plastic Sheet Flashing and Counter Flashing		Crew	Daily Output	Labor-Hours	Unit	Material	Labor	Equipment	Total	Total Incl O&P
0010	**PLASTIC SHEET FLASHING AND COUNTER FLASHING**									
7300	Polyvinyl chloride, black, .010" thick	1 Rofc	285	.028	S.F.	.21	.79		1	1.66
7400	.020" thick		285	.028		.30	.79		1.09	1.76
7600	.030" thick		285	.028		.39	.79		1.18	1.86
7700	.056" thick		285	.028		.94	.79		1.73	2.46
7900	Black or white for exposed roofs, .060" thick	↓	285	.028	↓	2	.79		2.79	3.63
8060	PVC tape, 5" x 45 mils, for joint covers, 100 L.F./roll				Ea.	128			128	141
8850	Polyvinyl chloride, .030" thick	1 Rofc	160	.050	S.F.	.46	1.41		1.87	3.05

07 65 23 – Rubber Sheet Flashing

07 65 23.10 Rubber Sheet Flashing and Counterflashing		Crew	Daily Output	Labor-Hours	Unit	Material	Labor	Equipment	Total	Total Incl O&P
0010	**RUBBER SHEET FLASHING AND COUNTERFLASHING**									
8100	Rubber, butyl, 1/32" thick	1 Rofc	285	.028	S.F.	1.75	.79		2.54	3.36
8200	1/16" thick		285	.028		2.38	.79		3.17	4.05
8300	Neoprene, cured, 1/16" thick		285	.028		1.92	.79		2.71	3.54
8400	1/8" thick	↓	285	.028	↓	4	.79		4.79	5.85

07 71 Roof Specialties

07 71 19 – Manufactured Gravel Stops and Fascias

07 71 19.10 Gravel Stop		Crew	Daily Output	Labor-Hours	Unit	Material	Labor	Equipment	Total	Total Incl O&P
0010	**GRAVEL STOP**									
0020	Aluminum, .050" thick, 4" face height, mill finish	1 Shee	145	.055	L.F.	6.10	2.11		8.21	10.25
0080	Duranodic finish		145	.055		5.90	2.11		8.01	10
0100	Painted		145	.055		6.85	2.11		8.96	11
0300	6" face height		135	.059		6.55	2.26		8.81	11
0350	Duranodic finish		135	.059		6.95	2.26		9.21	11.40
0400	Painted		135	.059		8.10	2.26		10.36	12.65
0600	8" face height		125	.064		8.20	2.44		10.64	13.10
0650	Duranodic finish		125	.064		8	2.44		10.44	12.85
0700	Painted		125	.064		8.15	2.44		10.59	13
0900	12" face height, .080 thick, 2 piece		100	.080		10.95	3.06		14.01	17.10
0950	Duranodic finish		100	.080		10.05	3.06		13.11	16.10
1000	Painted		100	.080		11.90	3.06		14.96	18.15
1350	Galv steel, 24 ga., 4" leg, plain, with continuous cleat, 4" face		145	.055		2.65	2.11		4.76	6.40
1360	6" face height		145	.055		4	2.11		6.11	7.90
1500	Polyvinyl chloride, 6" face height		135	.059		4.90	2.26		7.16	9.15
1600	9" face height		125	.064		5.80	2.44		8.24	10.40
1800	Stainless steel, 24 ga., 6" face height		135	.059		12.60	2.26		14.86	17.60
1900	12" face height		100	.080		26.50	3.06		29.56	34
2100	20 ga., 6" face height	↓	135	.059	↓	14.30	2.26		16.56	19.45

07 71 Roof Specialties

07 71 19 – Manufactured Gravel Stops and Fascias

07 71 19.10 Gravel Stop		Crew	Daily Output	Labor-Hours	Unit	2011 Bare Costs Material	Labor	Equipment	Total	Total Incl O&P
2200	12" face height	1 Shee	100	.080	L.F.	30	3.06		33.06	38

07 71 19.30 Fascia

		Crew	Daily Output	Labor-Hours	Unit	Material	Labor	Equipment	Total	Total Incl O&P
0010	**FASCIA**									
0100	Aluminum, reverse board and batten, .032" thick, colored, no furring incl	1 Shee	145	.055	S.F.	5.95	2.11		8.06	10.05
0300	Steel, galv and enameled, stock, no furring, long panels		145	.055		4	2.11		6.11	7.90
0600	Short panels		115	.070		5.45	2.66		8.11	10.40

07 71 23 – Manufactured Gutters and Downspouts

07 71 23.10 Downspouts

		Crew	Daily Output	Labor-Hours	Unit	Material	Labor	Equipment	Total	Total Incl O&P
0010	**DOWNSPOUTS**									
0020	Aluminum 2" x 3", .020" thick, embossed	1 Shee	190	.042	L.F.	1.03	1.61		2.64	3.79
0100	Enameled		190	.042		1.26	1.61		2.87	4.05
0300	Enameled, .024" thick, 2" x 3"		180	.044		1.89	1.70		3.59	4.88
0400	3" x 4"		140	.057		2.75	2.18		4.93	6.65
0600	Round, corrugated aluminum, 3" diameter, .020" thick		190	.042		1.60	1.61		3.21	4.42
0700	4" diameter, .025" thick		140	.057		2.47	2.18		4.65	6.35
0900	Wire strainer, round, 2" diameter		155	.052	Ea.	3.42	1.97		5.39	7
1000	4" diameter		155	.052		6.30	1.97		8.27	10.20
1200	Rectangular, perforated, 2" x 3"		145	.055		2.30	2.11		4.41	6
1300	3" x 4"		145	.055		3.32	2.11		5.43	7.15
1500	Copper, round, 16 oz., stock, 2" diameter		190	.042	L.F.	6.90	1.61		8.51	10.25
1600	3" diameter		190	.042		6.90	1.61		8.51	10.25
1800	4" diameter		145	.055		10.20	2.11		12.31	14.70
1900	5" diameter		130	.062		14.55	2.35		16.90	19.95
2100	Rectangular, corrugated copper, stock, 2" x 3"		190	.042		7.55	1.61		9.16	10.95
2200	3" x 4"		145	.055		9.95	2.11		12.06	14.45
2400	Rectangular, plain copper, stock, 2" x 3"		190	.042		6.70	1.61		8.31	10
2500	3" x 4"		145	.055		9.05	2.11		11.16	13.50
2700	Wire strainers, rectangular, 2" x 3"		145	.055	Ea.	5.50	2.11		7.61	9.55
2800	3" x 4"		145	.055		6.60	2.11		8.71	10.75
3000	Round, 2" diameter		145	.055		6.75	2.11		8.86	10.90
3100	3" diameter		145	.055		6.75	2.11		8.86	10.90
3300	4" diameter		145	.055		12.20	2.11		14.31	16.90
3400	5" diameter		115	.070		16.90	2.66		19.56	23
3600	Lead-coated copper, round, stock, 2" diameter		190	.042	L.F.	34	1.61		35.61	40
3700	3" diameter		190	.042		30	1.61		31.61	35.50
3900	4" diameter		145	.055		31	2.11		33.11	38
4000	5" diameter, corrugated		130	.062		24.50	2.35		26.85	31
4200	6" diameter, corrugated		105	.076		30	2.91		32.91	38
4300	Rectangular, corrugated, stock, 2" x 3"		190	.042		12.70	1.61		14.31	16.60
4500	Plain, stock, 2" x 3"		190	.042		16.35	1.61		17.96	20.50
4600	3" x 4"		145	.055		18.20	2.11		20.31	23.50
4800	Steel, galvanized, round, corrugated, 2" or 3" diameter, 28 gauge		190	.042		1.72	1.61		3.33	4.55
4900	4" diameter, 28 gauge		145	.055		2.35	2.11		4.46	6.05
5100	5" diameter, 28 gauge		130	.062		3.48	2.35		5.83	7.70
5200	26 gauge		130	.062		3.60	2.35		5.95	7.85
5400	6" diameter, 28 gauge		105	.076		4	2.91		6.91	9.20
5500	26 gauge		105	.076		5.20	2.91		8.11	10.55
5700	Rectangular, corrugated, 28 gauge, 2" x 3"		190	.042		1.28	1.61		2.89	4.07
5800	3" x 4"		145	.055		1.64	2.11		3.75	5.30
6000	Rectangular, plain, 28 gauge, galvanized, 2" x 3"		190	.042		1.28	1.61		2.89	4.07
6100	3" x 4"		145	.055		2.33	2.11		4.44	6.05
6300	Epoxy painted, 24 gauge, corrugated, 2" x 3"		190	.042		2	1.61		3.61	4.86

07 71 Roof Specialties

07 71 23 – Manufactured Gutters and Downspouts

07 71 23.10 Downspouts		Crew	Daily Output	Labor-Hours	Unit	Material	Labor	Equipment	Total	Total Incl O&P
						2011 Bare Costs				
6400	3″ x 4″	1 Shee	145	.055	L.F.	2.50	2.11		4.61	6.25
6600	Wire strainers, rectangular, 2″ x 3″		145	.055	Ea.	12	2.11		14.11	16.70
6700	3″ x 4″		145	.055		14	2.11		16.11	18.90
6900	Round strainers, 2″ or 3″ diameter		145	.055		2.80	2.11		4.91	6.55
7000	4″ diameter		145	.055		5.20	2.11		7.31	9.20
7800	Stainless steel tubing, schedule 5, 2″ x 3″ or 3″ diameter		190	.042	L.F.	44.50	1.61		46.11	51.50
7900	3″ x 4″ or 4″ diameter		145	.055		56.50	2.11		58.61	65.50
8100	4″ x 5″ or 5″ diameter		135	.059		117	2.26		119.26	132
8200	Vinyl, rectangular, 2″ x 3″		210	.038		1.94	1.46		3.40	4.53
8300	Round, 2-1/2″		220	.036		1.28	1.39		2.67	3.70

07 71 23.20 Downspout Elbows

Code	Description	Crew	Daily Output	Labor-Hours	Unit	Material	Labor	Equipment	Total	Total Incl O&P
0010	**DOWNSPOUT ELBOWS**									
0020	Aluminum, 2″ x 3″, embossed	1 Shee	100	.080	Ea.	.90	3.06		3.96	6.05
0100	Enameled		100	.080		.96	3.06		4.02	6.10
0200	3″ x 4″, .025″ thick, embossed		100	.080		3.50	3.06		6.56	8.90
0300	Enameled		100	.080		1.75	3.06		4.81	7
0400	Round corrugated, 3″, embossed, .020″ thick		100	.080		2.27	3.06		5.33	7.55
0500	4″, .025″ thick		100	.080		4.74	3.06		7.80	10.25
0600	Copper, 16 oz. round, 2″ diameter		100	.080		8	3.06		11.06	13.85
0700	3″ diameter		100	.080		10.55	3.06		13.61	16.65
0800	4″ diameter		100	.080		12.40	3.06		15.46	18.70
1000	2″ x 3″ corrugated		100	.080		7.25	3.06		10.31	13.05
1100	3″ x 4″ corrugated		100	.080		12.15	3.06		15.21	18.40
1300	Vinyl, 2-1/2″ diameter, 45° or 75°		100	.080		3.98	3.06		7.04	9.45
1400	Tee Y junction		75	.107		12.90	4.07		16.97	21

07 71 23.30 Gutters

Code	Description	Crew	Daily Output	Labor-Hours	Unit	Material	Labor	Equipment	Total	Total Incl O&P
0010	**GUTTERS**									
0012	Aluminum, stock units, 5″ K type, .027″ thick, plain	1 Shee	120	.067	L.F.	2.58	2.55		5.13	7.05
0100	Enameled		120	.067		2.51	2.55		5.06	6.95
0300	5″ K type type, .032″ thick, plain		120	.067		2.74	2.55		5.29	7.20
0400	Enameled		120	.067		3.29	2.55		5.84	7.85
0700	Copper, half round, 16 oz, stock units, 4″ wide		120	.067		8.20	2.55		10.75	13.20
0900	5″ wide		120	.067		8.55	2.55		11.10	13.65
1000	6″ wide		115	.070		9.65	2.66		12.31	15.05
1200	K type, 16 oz, stock, 4″ wide		120	.067		6.85	2.55		9.40	11.75
1300	5″ wide		120	.067		10.65	2.55		13.20	15.95
1500	Lead coated copper, half round, stock, 4″ wide		120	.067		9.45	2.55		12	14.60
1600	6″ wide		115	.070		15.60	2.66		18.26	21.50
1800	K type, stock, 4″ wide		120	.067		11.35	2.55		13.90	16.70
1900	5″ wide		120	.067		12.55	2.55		15.10	18.05
2100	Stainless steel, half round or box, stock, 4″ wide		120	.067		5.90	2.55		8.45	10.65
2200	5″ wide		120	.067		6.35	2.55		8.90	11.15
2400	Steel, galv, half round or box, 28 ga, 5″ wide, plain		120	.067		1.98	2.55		4.53	6.40
2500	Enameled		120	.067		1.51	2.55		4.06	5.85
2700	26 ga, stock, 5″ wide		120	.067		1.59	2.55		4.14	5.95
2800	6″ wide		120	.067		2.79	2.55		5.34	7.30
3000	Vinyl, O.G., 4″ wide	1 Carp	110	.073		1.08	2.44		3.52	5.25
3100	5″ wide		110	.073		1.38	2.44		3.82	5.60
3200	4″ half round, stock units		110	.073		1.08	2.44		3.52	5.25
3250	Joint connectors				Ea.	2.88			2.88	3.17
3300	Wood, clear treated cedar, fir or hemlock, 3″ x 4″	1 Carp	100	.080	L.F.	7.95	2.68		10.63	13.25
3400	4″ x 5″	″	100	.080	″	11.55	2.68		14.23	17.20

07 71 Roof Specialties

07 71 23 – Manufactured Gutters and Downspouts

07 71 23.35 Gutter Guard

	07 71 23.35 Gutter Guard	Crew	Daily Output	Labor-Hours	Unit	2011 Bare Costs Material	Labor	Equipment	Total	Total Incl O&P
0010	**GUTTER GUARD**									
0020	6" wide strip, aluminum mesh	1 Carp	500	.016	L.F.	2.18	.54		2.72	3.30
0100	Vinyl mesh	"	500	.016	"	2.58	.54		3.12	3.74

07 71 26 – Reglets

07 71 26.10 Reglets and Accessories

	07 71 26.10 Reglets and Accessories	Crew	Daily Output	Labor-Hours	Unit	Material	Labor	Equipment	Total	Total Incl O&P
0010	**REGLETS AND ACCESSORIES**									
0020	Aluminum, .025" thick, in concrete parapet	1 Carp	225	.036	L.F.	1.62	1.19		2.81	3.78
0100	Copper, 10 oz.		225	.036		2.62	1.19		3.81	4.88
0300	16 oz.		225	.036		5.10	1.19		6.29	7.65
0400	Galvanized steel, 24 gauge		225	.036		.94	1.19		2.13	3.03
0600	Stainless steel, .020" thick		225	.036		3.15	1.19		4.34	5.45
0700	Zinc and copper alloy, 20 oz.		225	.036		2.62	1.19		3.81	4.88
0900	Counter flashing for above, 12" wide, .032" aluminum	1 Shee	150	.053		1.63	2.04		3.67	5.15
1000	Copper, 10 oz.		150	.053		4.95	2.04		6.99	8.80
1200	16 oz.		150	.053		4.80	2.04		6.84	8.65
1300	Galvanized steel, .020" thick		150	.053		1.17	2.04		3.21	4.66
1500	Stainless steel, .020" thick		150	.053		4.69	2.04		6.73	8.50
1600	Zinc and copper alloy, 20 oz.		150	.053		4.78	2.04		6.82	8.60

07 71 29 – Manufactured Roof Expansion Joints

07 71 29.10 Expansion Joints

	07 71 29.10 Expansion Joints	Crew	Daily Output	Labor-Hours	Unit	Material	Labor	Equipment	Total	Total Incl O&P
0010	**EXPANSION JOINTS**									
0300	Butyl or neoprene center with foam insulation, metal flanges									
0400	Aluminum, .032" thick for openings to 2-1/2"	1 Rofc	165	.048	L.F.	16	1.36		17.36	20
0600	For joint openings to 3-1/2"		165	.048		16	1.36		17.36	20
0610	For joint openings to 5"		165	.048		16.40	1.36		17.76	20.50
0620	For joint openings to 8"		165	.048		17.30	1.36		18.66	21.50
0700	Copper, 16 oz. for openings to 2-1/2"		165	.048		24.50	1.36		25.86	29.50
0900	For joint openings to 3-1/2"		165	.048		24.50	1.36		25.86	29.50
0910	For joint openings to 5"		165	.048		25.50	1.36		26.86	30.50
0920	For joint openings to 8"		165	.048		27	1.36		28.36	32.50
1000	Galvanized steel, 26 ga. for openings to 2-1/2"		165	.048		13	1.36		14.36	16.75
1200	For joint openings to 3-1/2"		165	.048		13	1.36		14.36	16.75
1210	For joint openings to 5"		165	.048		13.90	1.36		15.26	17.75
1220	For joint openings to 8"		165	.048		17.30	1.36		18.66	21.50
1300	Lead-coated copper, 16 oz. for openings to 2-1/2"		165	.048		33.50	1.36		34.86	39.50
1500	For joint openings to 3-1/2"		165	.048		40	1.36		41.36	46.50
1600	Stainless steel, .018", for openings to 2-1/2"		165	.048		17.90	1.36		19.26	22
1800	For joint openings to 3-1/2"		165	.048		17.90	1.36		19.26	22
1810	For joint openings to 5"		165	.048		18.60	1.36		19.96	23
1820	For joint openings to 8"		165	.048		20.50	1.36		21.86	25
1900	Neoprene, double-seal type with thick center, 4-1/2" wide		125	.064		14	1.80		15.80	18.65
1950	Polyethylene bellows, with galv steel flat flanges		100	.080		5.40	2.25		7.65	10
1960	With galvanized angle flanges		100	.080		6	2.25		8.25	10.65
2000	Roof joint with extruded aluminum cover, 2"	1 Shee	115	.070		37.50	2.66		40.16	46
2100	Roof joint, plastic curbs, foam center, standard	1 Rofc	100	.080		13.70	2.25		15.95	19.10
2200	Large		100	.080		18.25	2.25		20.50	24
2300	Transitions, regular, minimum		10	.800	Ea.	120	22.50		142.50	173
2350	Maximum		4	2		150	56.50		206.50	268
2400	Large, minimum		9	.889		173	25		198	236
2450	Maximum		3	2.667		182	75		257	335
2500	Roof to wall joint with extruded aluminum cover	1 Shee	115	.070	L.F.	31.50	2.66		34.16	39.50

07 71 Roof Specialties

07 71 29 – Manufactured Roof Expansion Joints

07 71 29.10 Expansion Joints		Crew	Daily Output	Labor-Hours	Unit	Material	2011 Bare Costs Labor	Equipment	Total	Total Incl O&P
2700	Wall joint, closed cell foam on PVC cover, 9" wide	1 Rofc	125	.064	L.F.	4.90	1.80		6.70	8.65
2800	12" wide	"	115	.070		5.50	1.96		7.46	9.60

07 71 43 – Drip Edge

07 71 43.10 Drip Edge, Rake Edge, Ice Belts		Crew	Daily Output	Labor-Hours	Unit	Material	Labor	Equipment	Total	Total Incl O&P
0010	**DRIP EDGE, RAKE EDGE, ICE BELTS**									
0020	Aluminum, .016" thick, 5" wide, mill finish	1 Carp	400	.020	L.F.	.38	.67		1.05	1.54
0100	White finish		400	.020		.38	.67		1.05	1.54
0200	8" wide, mill finish		400	.020		.46	.67		1.13	1.63
0300	Ice belt, 28" wide, mill finish		100	.080		3.85	2.68		6.53	8.75
0310	Vented, mill finish		400	.020		2.76	.67		3.43	4.16
0320	Painted finish		400	.020		2.86	.67		3.53	4.27
0400	Galvanized, 5" wide		400	.020		.36	.67		1.03	1.52
0500	8" wide, mill finish		400	.020		.69	.67		1.36	1.88
0510	Rake edge, aluminum, 1-1/2" x 1-1/2"		400	.020		.28	.67		.95	1.43
0520	3-1/2" x 1-1/2"		400	.020		.38	.67		1.05	1.54

07 72 Roof Accessories

07 72 23 – Relief Vents

07 72 23.10 Roof Vents		Crew	Daily Output	Labor-Hours	Unit	Material	Labor	Equipment	Total	Total Incl O&P
0010	**ROOF VENTS**									
0020	Mushroom shape, for built-up roofs, aluminum	1 Rofc	30	.267	Ea.	88.50	7.50		96	111
0100	PVC, 6" high	"	30	.267	"	28.50	7.50		36	45

07 72 26 – Ridge Vents

07 72 26.10 Ridge Vents and Accessories		Crew	Daily Output	Labor-Hours	Unit	Material	Labor	Equipment	Total	Total Incl O&P
0010	**RIDGE VENTS AND ACCESSORIES**									
0100	Aluminum strips, mill finish	1 Rofc	160	.050	L.F.	2.62	1.41		4.03	5.40
0150	Painted finish		160	.050	"	3.74	1.41		5.15	6.65
0200	Connectors		48	.167	Ea.	3.68	4.69		8.37	12.50
0300	End caps		48	.167	"	1.37	4.69		6.06	9.95
0400	Galvanized strips		160	.050	L.F.	2	1.41		3.41	4.74
0430	Molded polyethylene, shingles not included		160	.050	"	2.99	1.41		4.40	5.85
0440	End plugs		48	.167	Ea.	1.37	4.69		6.06	9.95
0450	Flexible roll, shingles not included		160	.050	L.F.	2.15	1.41		3.56	4.91
2300	Ridge vent strip, mill finish	1 Shee	155	.052	"	2.73	1.97		4.70	6.25

07 72 33 – Roof Hatches

07 72 33.10 Roof Hatch Options		Crew	Daily Output	Labor-Hours	Unit	Material	Labor	Equipment	Total	Total Incl O&P
0010	**ROOF HATCH OPTIONS**									
0500	2'-6" x 3', aluminum curb and cover	G-3	10	3.200	Ea.	985	103		1,088	1,250
0520	Galvanized steel curb and aluminum cover		10	3.200		595	103		698	825
0540	Galvanized steel curb and cover		10	3.200		550	103		653	775
0600	2'-6" x 4'-6", aluminum curb and cover		9	3.556		850	115		965	1,125
0800	Galvanized steel curb and aluminum cover		9	3.556		600	115		715	850
0900	Galvanized steel curb and cover		9	3.556		825	115		940	1,100
1100	4' x 4' aluminum curb and cover		8	4		1,675	129		1,804	2,050
1120	Galvanized steel curb and aluminum cover		8	4		1,675	129		1,804	2,050
1140	Galvanized steel curb and cover		8	4		1,500	129		1,629	1,875
1200	2'-6" x 8'-0", aluminum curb and cover		6.60	4.848		1,675	157		1,832	2,075
1400	Galvanized steel curb and aluminum cover		6.60	4.848		1,600	157		1,757	2,000
1500	Galvanized steel curb and cover		6.60	4.848		1,200	157		1,357	1,575

07 72 Roof Accessories

07 72 33 – Roof Hatches

07 72 33.10 Roof Hatch Options		Crew	Daily Output	Labor-Hours	Unit	Material	2011 Bare Costs Labor	Equipment	Total	Total Incl O&P
1800	For plexiglass panels, 2'-6" x 3'-0", add to above				Ea.	400			400	440

07 72 36 – Smoke Vents

07 72 36.10 Smoke Hatches

		Crew	Daily Output	Labor-Hours	Unit	Material	Labor	Equipment	Total	Total Incl O&P
0010	**SMOKE HATCHES**									
0200	For 3'-0" long, add to roof hatches from Section 07 72 33.10				Ea.	25%	5%			
0250	For 4'-0" long, add to roof hatches from Section 07 72 33.10					20%	5%			
0300	For 8'-0" long, add to roof hatches from Section 07 72 33.10					10%	5%			

07 72 36.20 Smoke Vent Options

		Crew	Daily Output	Labor-Hours	Unit	Material	Labor	Equipment	Total	Total Incl O&P
0010	**SMOKE VENT OPTIONS**									
0100	4' x 4' aluminum cover and frame	G-3	13	2.462	Ea.	1,900	79.50		1,979.50	2,225
0200	Galvanized steel cover and frame		13	2.462		1,675	79.50		1,754.50	1,950
0300	4' x 8' aluminum cover and frame		8	4		2,600	129		2,729	3,075
0400	Galvanized steel cover and frame		8	4		2,200	129		2,329	2,650

07 72 53 – Snow Guards

07 72 53.10 Snow Guard Options

		Crew	Daily Output	Labor-Hours	Unit	Material	Labor	Equipment	Total	Total Incl O&P
0010	**SNOW GUARD OPTIONS**									
0100	Slate & asphalt shingle roofs, fastened with nails	1 Rofc	160	.050	Ea.	9.40	1.41		10.81	12.90
0200	Standing seam metal roofs, fastened with set screws		48	.167		14.60	4.69		19.29	24.50
0300	Surface mount for metal roofs, fastened with solder		48	.167		6.60	4.69		11.29	15.70
0400	Double rail pipe type, including pipe		130	.062	L.F.	22.50	1.73		24.23	28

07 72 73 – Pitch Pockets

07 72 73.10 Pitch Pockets, Variable Sizes

		Crew	Daily Output	Labor-Hours	Unit	Material	Labor	Equipment	Total	Total Incl O&P
0010	**PITCH POCKETS, VARIABLE SIZES**									
0100	Adjustable, 4" to 7", welded corners, 4" deep	1 Rofc	48	.167	Ea.	12.90	4.69		17.59	22.50
0200	Side extenders, 6"	"	240	.033	"	1.97	.94		2.91	3.86

07 72 80 – Vents

07 72 80.30 Vent Options

		Crew	Daily Output	Labor-Hours	Unit	Material	Labor	Equipment	Total	Total Incl O&P
0010	**VENT OPTIONS**									
0020	Plastic, for insulated decks, 1 per M.S.F., minimum	1 Rofc	40	.200	Ea.	19	5.65		24.65	31
0100	Maximum		20	.400		38	11.25		49.25	62.50
0300	Aluminum		30	.267		24	7.50		31.50	40
0800	Polystyrene baffles, 12" wide for 16" O.C. rafter spacing	1 Carp	90	.089		.40	2.98		3.38	5.45
0900	For 24" O.C. rafter spacing	"	110	.073		.75	2.44		3.19	4.91

07 81 Applied Fireproofing

07 81 16 – Cementitious Fireproofing

07 81 16.10 Sprayed Cementitious Fireproofing

		Crew	Daily Output	Labor-Hours	Unit	Material	Labor	Equipment	Total	Total Incl O&P
0010	**SPRAYED CEMENTITIOUS FIREPROOFING**									
0050	Not incl tamping or canvas protection									
0100	1" thick, on flat plate steel	G-2	3000	.008	S.F.	.53	.23	.04	.80	1
0200	Flat decking		2400	.010		.53	.28	.05	.86	1.10
0400	Beams		1500	.016		.53	.45	.08	1.06	1.41
0500	Corrugated or fluted decks		1250	.019		.79	.54	.10	1.43	1.87
0700	Columns, 1-1/8" thick		1100	.022		.59	.61	.11	1.31	1.79
0800	2-3/16" thick		700	.034		1.13	.97	.18	2.28	3.03
0850	For tamping, add						10%			
0900	For canvas protection, add	G-2	5000	.005	S.F.	.07	.14	.03	.24	.33
1500	Intumescent epoxy fireproofing on wire mesh, 3/16" thick									

07 81 Applied Fireproofing

07 81 16 – Cementitious Fireproofing

07 81 16.10 Sprayed Cementitious Fireproofing		Crew	Daily Output	Labor-Hours	Unit	Material	Labor	Equipment	Total	Total Incl O&P
						2011 Bare Costs				
1550	1 hour rating, exterior use	G-2	136	.176	S.F.	6.90	4.97	.92	12.79	16.70
1600	Magnesium oxychloride, 35# to 40# density, 1/4" thick		3000	.008		1.44	.23	.04	1.71	2
1650	1/2" thick		2000	.012		2.90	.34	.06	3.30	3.82
1700	60# to 70# density, 1/4" thick		3000	.008		1.92	.23	.04	2.19	2.53
1750	1/2" thick		2000	.012		3.85	.34	.06	4.25	4.87
2000	Vermiculite cement, troweled or sprayed, 1/4" thick		3000	.008		1.31	.23	.04	1.58	1.86
2050	1/2" thick	↓	2000	.012	↓	2.60	.34	.06	3	3.49

07 84 Firestopping

07 84 13 – Penetration Firestopping

07 84 13.10 Firestopping

		Crew	Daily Output	Labor-Hours	Unit	Material	Labor	Equipment	Total	Total Incl O&P
0010	**FIRESTOPPING** R078413-30									
0100	Metallic piping, non insulated									
0110	Through walls, 2" diameter	1 Carp	16	.500	Ea.	13.75	16.80		30.55	43
0120	4" diameter		14	.571		21	19.15		40.15	55
0130	6" diameter		12	.667		28.50	22.50		51	68.50
0140	12" diameter		10	.800		50	27		77	100
0150	Through floors, 2" diameter		32	.250		8.35	8.40		16.75	23
0160	4" diameter		28	.286		12	9.60		21.60	29.50
0170	6" diameter		24	.333		15.75	11.20		26.95	36
0180	12" diameter	↓	20	.400	↓	26.50	13.40		39.90	52
0190	Metallic piping, insulated									
0200	Through walls, 2" diameter	1 Carp	16	.500	Ea.	19.50	16.80		36.30	49.50
0210	4" diameter		14	.571		26.50	19.15		45.65	61.50
0220	6" diameter		12	.667		34	22.50		56.50	75
0230	12" diameter		10	.800		55.50	27		82.50	106
0240	Through floors, 2" diameter		32	.250		14.10	8.40		22.50	29.50
0250	4" diameter		28	.286		17.70	9.60		27.30	35.50
0260	6" diameter		24	.333		21.50	11.20		32.70	42
0270	12" diameter	↓	20	.400	↓	26.50	13.40		39.90	52
0280	Non metallic piping, non insulated									
0290	Through walls, 2" diameter	1 Carp	12	.667	Ea.	56.50	22.50		79	100
0300	4" diameter		10	.800		71	27		98	123
0310	6" diameter		8	1		99	33.50		132.50	165
0330	Through floors, 2" diameter		16	.500		44.50	16.80		61.30	77
0340	4" diameter		6	1.333		55	44.50		99.50	136
0350	6" diameter	↓	6	1.333	↓	66	44.50		110.50	148
0370	Ductwork, insulated & non insulated, round									
0380	Through walls, 6" diameter	1 Carp	12	.667	Ea.	29	22.50		51.50	69.50
0390	12" diameter		10	.800		57	27		84	108
0400	18" diameter		8	1		93	33.50		126.50	159
0410	Through floors, 6" diameter		16	.500		15.75	16.80		32.55	45.50
0420	12" diameter		14	.571		29	19.15		48.15	64
0430	18" diameter	↓	12	.667	↓	50	22.50		72.50	92.50
0440	Ductwork, insulated & non insulated, rectangular									
0450	With stiffener/closure angle, through walls, 6" x 12"	1 Carp	8	1	Ea.	24	33.50		57.50	82
0460	12" x 24"		6	1.333		31.50	44.50		76	110
0470	24" x 48"		4	2		90.50	67		157.50	212
0480	With stiffener/closure angle, through floors, 6" x 12"		10	.800		12.95	27		39.95	59
0490	12" x 24"		8	1		23.50	33.50		57	81.50
0500	24" x 48"	↓	6	1.333	↓	45.50	44.50		90	125

07 84 Firestopping

07 84 13 – Penetration Firestopping

07 84 13.10 Firestopping		Crew	Daily Output	Labor-Hours	Unit	Material	2011 Bare Costs Labor	Equipment	Total	Total Incl O&P
0510	Multi trade openings									
0520	Through walls, 6" x 12"	1 Carp	2	4	Ea.	50	134		184	280
0530	12" x 24"	"	1	8		202	268		470	670
0540	24" x 48"	2 Carp	1	16		810	535		1,345	1,800
0550	48" x 96"	"	.75	21.333		3,250	715		3,965	4,775
0560	Through floors, 6" x 12"	1 Carp	2	4		50	134		184	280
0570	12" x 24"	"	1	8		202	268		470	670
0580	24" x 48"	2 Carp	.75	21.333		810	715		1,525	2,100
0590	48" x 96"	"	.50	32	↓	3,250	1,075		4,325	5,375
0600	Structural penetrations, through walls									
0610	Steel beams, W8 x 10	1 Carp	8	1	Ea.	31.50	33.50		65	91
0620	W12 x 14		6	1.333		50	44.50		94.50	130
0630	W21 x 44		5	1.600		100	53.50		153.50	201
0640	W36 x 135		3	2.667		244	89.50		333.50	420
0650	Bar joists, 18" deep		6	1.333		46	44.50		90.50	126
0660	24" deep		6	1.333		57	44.50		101.50	138
0670	36" deep		5	1.600		86	53.50		139.50	185
0680	48" deep	↓	4	2	↓	100	67		167	223
0690	Construction joints, floor slab at exterior wall									
0700	Precast, brick, block or drywall exterior									
0710	2" wide joint	1 Carp	125	.064	L.F.	7.15	2.15		9.30	11.50
0720	4" wide joint	"	75	.107	"	14.30	3.58		17.88	22
0730	Metal panel, glass or curtain wall exterior									
0740	2" wide joint	1 Carp	40	.200	L.F.	17	6.70		23.70	30
0750	4" wide joint	"	25	.320	"	23.50	10.75		34.25	43.50
0760	Floor slab to drywall partition									
0770	Flat joint	1 Carp	100	.080	L.F.	7.05	2.68		9.73	12.25
0780	Fluted joint		50	.160		14.30	5.35		19.65	25
0790	Etched fluted joint	↓	75	.107	↓	9.30	3.58		12.88	16.25
0800	Floor slab to concrete/masonry partition									
0810	Flat joint	1 Carp	75	.107	L.F.	15.75	3.58		19.33	23.50
0820	Fluted joint	"	50	.160	"	18.65	5.35		24	29.50
0830	Concrete/CMU wall joints									
0840	1" wide	1 Carp	100	.080	L.F.	8.60	2.68		11.28	13.95
0850	2" wide		75	.107		15.75	3.58		19.33	23.50
0860	4" wide	↓	50	.160	↓	30	5.35		35.35	42
0870	Concrete/CMU floor joints									
0880	1" wide	1 Carp	200	.040	L.F.	4.31	1.34		5.65	7
0890	2" wide		150	.053		7.90	1.79		9.69	11.65
0900	4" wide	↓	100	.080	↓	15.05	2.68		17.73	21

07 91 Preformed Joint Seals

07 91 13 – Compression Seals

07 91 13.10 Compression Seals		Crew	Daily Output	Labor-Hours	Unit	Material	2011 Bare Costs Labor	Equipment	Total	Total Incl O&P
0010	**COMPRESSION SEALS**									
4900	O-ring type cord, 1/4"	1 Bric	472	.017	L.F.	.47	.57		1.04	1.46
4910	1/2"		440	.018		1.78	.61		2.39	2.96
4920	3/4"		424	.019		3.61	.64		4.25	5
4930	1"		408	.020		6.95	.66		7.61	8.75
4940	1-1/4"		384	.021		13.60	.70		14.30	16.10
4950	1-1/2"		368	.022		17.20	.73		17.93	20
4960	1-3/4"		352	.023		36	.76		36.76	41
4970	2"		344	.023		63.50	.78		64.28	71.50

07 91 16 – Joint Gaskets

07 91 16.10 Joint Gaskets		Crew	Daily Output	Labor-Hours	Unit	Material	Labor	Equipment	Total	Total Incl O&P
0010	**JOINT GASKETS**									
4400	Joint gaskets, neoprene, closed cell w/adh, 1/8" x 3/8"	1 Bric	240	.033	L.F.	.26	1.12		1.38	2.13
4500	1/4" x 3/4"		215	.037		.58	1.25		1.83	2.69
4700	1/2" x 1"		200	.040		1.30	1.35		2.65	3.64
4800	3/4" x 1-1/2"		165	.048		1.44	1.63		3.07	4.26

07 91 23 – Backer Rods

07 91 23.10 Backer Rods		Crew	Daily Output	Labor-Hours	Unit	Material	Labor	Equipment	Total	Total Incl O&P
0010	**BACKER RODS**									
0030	Backer rod, polyethylene, 1/4" diameter	1 Bric	4.60	1.739	C.L.F.	2.83	58.50		61.33	99
0050	1/2" diameter		4.60	1.739		6.60	58.50		65.10	103
0070	3/4" diameter		4.60	1.739		10.10	58.50		68.60	107
0090	1" diameter		4.60	1.739		18.70	58.50		77.20	117

07 91 26 – Joint Fillers

07 91 26.10 Joint Fillers		Crew	Daily Output	Labor-Hours	Unit	Material	Labor	Equipment	Total	Total Incl O&P
0010	**JOINT FILLERS**									
4360	Butyl rubber filler, 1/4" x 1/4"	1 Bric	290	.028	L.F.	.14	.93		1.07	1.68
4365	1/2" x 1/2"		250	.032		.57	1.08		1.65	2.39
4370	1/2" x 3/4"		210	.038		.85	1.28		2.13	3.04
4375	3/4" x 3/4"		230	.035		1.28	1.17		2.45	3.33
4380	1" x 1"		180	.044		1.71	1.50		3.21	4.33
4390	For coloring, add					12%				
4980	Polyethylene joint backing, 1/4" x 2"	1 Bric	2.08	3.846	C.L.F.	12	129		141	225
4990	1/4" x 6"		1.28	6.250	"	36	210		246	385
5600	Silicone, room temp vulcanizing foam seal, 1/4" x 1/2"		1312	.006	L.F.	.29	.21		.50	.66
5610	1/2" x 1/2"		656	.012		.57	.41		.98	1.30
5620	1/2" x 3/4"		442	.018		.86	.61		1.47	1.95
5630	3/4" x 3/4"		328	.024		1.29	.82		2.11	2.77
5640	1/8" x 1"		1312	.006		.29	.21		.50	.66
5650	1/8" x 3"		442	.018		.86	.61		1.47	1.95
5670	1/4" x 3"		295	.027		1.72	.91		2.63	3.39
5680	1/4" x 6"		148	.054		3.44	1.82		5.26	6.75
5690	1/2" x 6"		82	.098		6.90	3.28		10.18	12.95
5700	1/2" x 9"		52.50	.152		10.30	5.15		15.45	19.75
5710	1/2" x 12"		33	.242		13.75	8.15		21.90	28.50

07 92 Joint Sealants

07 92 13 – Elastomeric Joint Sealants

07 92 13.20 Caulking and Sealant Options		Crew	Daily Output	Labor-Hours	Unit	Material	Labor	Equipment	Total	Total Incl O&P
						2011 Bare Costs				
0010	**CAULKING AND SEALANT OPTIONS**									
0050	Latex acrylic based, bulk				Gal.	27			27	29.50
0055	Bulk in place 1/4" x 1/4" bead	1 Bric	300	.027	L.F.	.09	.90		.99	1.56
0060	1/4" x 3/8"		294	.027		.14	.92		1.06	1.66
0065	1/4" x 1/2"		288	.028		.19	.93		1.12	1.74
0075	3/8" x 3/8"		284	.028		.21	.95		1.16	1.79
0080	3/8" x 1/2"		280	.029		.28	.96		1.24	1.89
0085	3/8" x 5/8"		276	.029		.35	.98		1.33	1.99
0095	3/8" x 3/4"		272	.029		.42	.99		1.41	2.09
0100	1/2" x 1/2"		275	.029		.38	.98		1.36	2.03
0105	1/2" x 5/8"		269	.030		.47	1		1.47	2.16
0110	1/2" x 3/4"		263	.030		.57	1.02		1.59	2.31
0115	1/2" x 7/8"		256	.031		.66	1.05		1.71	2.46
0120	1/2" x 1"		250	.032		.76	1.08		1.84	2.60
0125	3/4" x 3/4"		244	.033		.85	1.10		1.95	2.75
0130	3/4" x 1"		225	.036		1.14	1.20		2.34	3.21
0135	1" x 1"	↓	200	.040	↓	1.52	1.35		2.87	3.88
0190	Cartridges				Gal.	22.50			22.50	24.50
0200	11 fl. oz cartridge				Ea.	1.93			1.93	2.12
0500	1/4" x 1/2"	1 Bric	288	.028	L.F.	.16	.93		1.09	1.70
0600	1/2" x 1/2"		275	.029		.32	.98		1.30	1.96
0800	3/4" x 3/4"		244	.033		.71	1.10		1.81	2.59
0900	3/4" x 1"		225	.036		.95	1.20		2.15	3
1000	1" x 1"	↓	200	.040	↓	1.18	1.35		2.53	3.51
1400	Butyl based, bulk				Gal.	26			26	28.50
1500	Cartridges				"	32			32	35
1700	1/4" x 1/2", 154 L.F./gal.	1 Bric	288	.028	L.F.	.17	.93		1.10	1.72
1800	1/2" x 1/2", 77 L.F./gal.	"	275	.029	"	.34	.98		1.32	1.98
2300	Polysulfide compounds, 1 component, bulk				Gal.	56			56	61.50
2600	1 or 2 component, in place, 1/4" x 1/4", 308 L.F./gal.	1 Bric	300	.027	L.F.	.18	.90		1.08	1.67
2700	1/2" x 1/4", 154 L.F./gal.		288	.028		.36	.93		1.29	1.93
2900	3/4" x 3/8", 68 L.F./gal.		272	.029		.82	.99		1.81	2.53
3000	1" x 1/2", 38 L.F./gal.	↓	250	.032	↓	1.47	1.08		2.55	3.39
3200	Polyurethane, 1 or 2 component				Gal.	50			50	55
3500	Bulk, in place, 1/4" x 1/4"	1 Bric	300	.027	L.F.	.16	.90		1.06	1.65
3655	1/2" x 1/4"		288	.028		.32	.93		1.25	1.89
3800	3/4" x 3/8"		272	.029		.74	.99		1.73	2.43
3900	1" x 1/2"	↓	250	.032	↓	1.30	1.08		2.38	3.20
4100	Silicone rubber, bulk				Gal.	41			41	45
4200	Cartridges				"	41			41	45

07 92 16 – Rigid Joint Sealants

07 92 16.10 Rigid Joint Sealants		Crew	Daily Output	Labor-Hours	Unit	Material	Labor	Equipment	Total	Total Incl O&P
0010	**RIGID JOINT SEALANTS**									
5800	Tapes, sealant, P.V.C. foam adhesive, 1/16" x 1/4"				C.L.F.	5.65			5.65	6.20
5900	1/16" x 1/2"					8.35			8.35	9.20
5950	1/16" x 1"					13.85			13.85	15.25
6000	1/8" x 1/2"				↓	9.35			9.35	10.30

07 92 19 – Acoustical Joint Sealants

07 92 19.10 Acoustical Sealant		Crew	Daily Output	Labor-Hours	Unit	Material	Labor	Equipment	Total	Total Incl O&P
0010	**ACOUSTICAL SEALANT**									
0020	Acoustical sealant, elastomeric, cartridges				Ea.	2.30			2.30	2.53
0025	In place, 1/4" x 1/4"	1 Bric	300	.027	L.F.	.09	.90		.99	1.57

07 92 Joint Sealants

07 92 19 – Acoustical Joint Sealants

07 92 19.10 Acoustical Sealant		Crew	Daily Output	Labor-Hours	Unit	Material	2011 Bare Costs Labor	Equipment	Total	Total Incl O&P
0030	1/4" x 1/2"	1 Bric	288	.028	L.F.	.19	.93		1.12	1.74
0035	1/2" x 1/2"		275	.029		.38	.98		1.36	2.02
0040	1/2" x 3/4"		263	.030		.56	1.02		1.58	2.30
0045	3/4" x 3/4"		244	.033		.85	1.10		1.95	2.74
0050	1" x 1"	↓	200	.040	↓	1.50	1.35		2.85	3.86

07 95 Expansion Control

07 95 13 – Expansion Joint Cover Assemblies

07 95 13.50 Expansion Joint Assemblies		Crew	Daily Output	Labor-Hours	Unit	Material	Labor	Equipment	Total	Total Incl O&P
0010	**EXPANSION JOINT ASSEMBLIES**									
0201	Floor cover assemblies, 1" space, aluminum	1 Carp	38	.211	L.F.	18.35	7.05		25.40	32
0301	Bronze or stainless		38	.211		43.50	7.05		50.55	59.50
0501	2" space, aluminum		38	.211		22.50	7.05		29.55	36.50
0601	Bronze or stainless		38	.211		46.50	7.05		53.55	63.50
0801	Wall and ceiling assemblies, 1" space, aluminum		38	.211		10.95	7.05		18	24
0901	Bronze or stainless		38	.211		40.50	7.05		47.55	56.50
1101	2" space, aluminum		38	.211		18.60	7.05		25.65	32.50
1201	Bronze or stainless		38	.211		50	7.05		57.05	67
1401	Floor to wall assemblies, 1" space, aluminum		38	.211		16.70	7.05		23.75	30
1501	Bronze or stainless		38	.211		49	7.05		56.05	66
1701	Gym floor angle covers, aluminum, 3" x 3" angle		46	.174		14.45	5.85		20.30	25.50
1801	3" x 4" angle		46	.174		20	5.85		25.85	32
2001	Roof closures, aluminum, 1" space, flat roof, low profile		57	.140		33.50	4.71		38.21	44.50
2101	High profile		57	.140		48	4.71		52.71	61
2302	Roof to wall, 1" space, low profile		57	.140		18.45	4.71		23.16	28.50
2401	High profile	↓	57	.140	↓	24	4.71		28.71	34.50

Estimating Tips

08 10 00 Doors and Frames

All exterior doors should be addressed for their energy conservation.

- Most metal doors and frames look alike, but there may be significant differences among them. When estimating these items, be sure to choose the line item that most closely compares to the specification or door schedule requirements regarding:
  - — type of metal
  - — metal gauge
  - — door core material
  - — fire rating
  - — finish
- Wood and plastic doors vary considerably in price. The primary determinant is the veneer material. Lauan, birch, and oak are the most common veneers. Other variables include the following:
  - — hollow or solid core
  - — fire rating
  - — flush or raised panel
  - — finish

08 30 00 Specialty Doors and Frames

- There are many varieties of special doors, and they are usually priced per each. Add frames, hardware, or operators required for a complete installation.

08 40 00 Entrances, Storefronts, and Curtain Walls

- Glazed curtain walls consist of the metal tube framing and the glazing material. The cost data in this subdivision is presented for the metal tube framing alone or the composite wall. If your estimate requires a detailed takeoff of the framing, be sure to add the glazing cost.

08 50 00 Windows

- Most metal windows are delivered preglazed. However, some metal windows are priced without glass. Refer to 08 80 00 Glazing for glass pricing. The grade C indicates commercial grade windows, usually ASTM C-35.
- All wood windows and vinyl are priced preglazed. The glazing is insulating glass. Some wood windows may have single pane float glass. Add the cost of screens and grills if required, and not already included.

08 70 00 Hardware

- Hardware costs add considerably to the cost of a door. The most efficient method to determine the hardware requirements for a project is to review the door schedule.
- Door hinges are priced by the pair, with most doors requiring 1-1/2 pairs per door. The hinge prices do not include installation labor because it is included in door installation. Hinges are classified according to the frequency of use.

08 80 00 Glazing

- Different openings require different types of glass. The three most common types are:
  - — float
  - — tempered
  - — insulating
  - — impact-resistant
  - — ballistic-resistant
- Most exterior windows are glazed with insulating glass. Entrance doors and window walls, where the glass is less than 18″ from the floor, are generally glazed with tempered glass. Interior windows and some residential windows are glazed with float glass.
- Coastal communities are requiring the use of impact-resistant glass.
- The insulation or 'u' value is a strong consideration, along with solar heat gain.

Reference Numbers

Reference numbers are shown in shaded boxes at the beginning of some major classifications. These numbers refer to related items in the Reference Section. The reference information may be an estimating procedure, an alternate pricing method, or technical information.

Note: Not all subdivisions listed here necessarily appear in this publication.

08 05 Common Work Results for Openings

08 05 05 – Selective Windows and Doors Demolition

08 05 05.10 Selective Demolition Doors		Crew	Daily Output	Labor-Hours	Unit	Material	2011 Bare Costs Labor	Equipment	Total	Total Incl O&P
0010	**SELECTIVE DEMOLITION DOORS** R024119-10									
0200	Doors, exterior, 1-3/4″ thick, single, 3′ x 7′ high	1 Clab	16	.500	Ea.		13.25		13.25	22
0210	Door demo, doors, exterior, 1-3/4″ thick, single, 3′-0″ x 8′-0″		10	.800			21		21	35.50
0215	Door demo, doors, exterior, double 1-3/4″ thick, 3′-0″ x 8′-0″		6	1.333			35.50		35.50	59
0220	Double, 6′ x 7′ high		12	.667			17.65		17.65	29.50
0500	Interior, 1-3/8″ thick, single, 3′ x 7′ high		20	.400			10.60		10.60	17.70
0520	Double, 6′ x 7′ high		16	.500			13.25		13.25	22
0700	Bi-folding, 3′ x 6′-8″ high		20	.400			10.60		10.60	17.70
0720	6′ x 6′-8″ high		18	.444			11.75		11.75	19.70
0900	Bi-passing, 3′ x 6′-8″ high		16	.500			13.25		13.25	22
0940	6′ x 6′-8″ high		14	.571			15.10		15.10	25.50
1500	Remove and reset, minimum	1 Carp	8	1			33.50		33.50	56
1520	Maximum		6	1.333			44.50		44.50	75
2000	Frames, including trim, metal		8	1			33.50		33.50	56
2201	Alternate pricing method		200	.040	L.F.		1.34		1.34	2.25
3001	Special doors, counter doors	2 Clab	6	2.667	Ea.		70.50		70.50	118
3101	Double acting		10	1.600			42.50		42.50	71
3201	Floor door (trap type)		8	2			53		53	88.50
3301	Glass, sliding, including frames		12	1.333			35.50		35.50	59
3400	Overhead, commercial, 12′ x 12′ high	2 Carp	4	4			134		134	225
3441	20′ x 16′ high	2 Clab	3	5.333			141		141	236
3500	Residential, 9′ x 7′ high	2 Carp	8	2			67		67	112
3540	16′ x 7′ high	″	7	2.286			76.50		76.50	128
3601	Remove and reset, minimum	1 Clab	2	4			106		106	177
3621	Maximum	″	1.25	6.400			169		169	284
3701	Roll-up grille	2 Clab	5	3.200			84.50		84.50	142
3801	Revolving door		2	8			212		212	355
3902	Cafe/bar swing door		8	2			53		53	88.50
6600	Demo flexible transparent strip entrance	3 Shee	115	.209	SF Surf		7.95		7.95	13.15
7100	Remove double swing pneumatic doors, openers and sensors	2 Skwk	.50	32	Opng.		1,100		1,100	1,825
7110	Remove automatic operators, industrial, sliding doors, to 12′ wide	″	.40	40	″		1,350		1,350	2,275

08 05 05.20 Selective Demolition of Windows		Crew	Daily Output	Labor-Hours	Unit	Material	2011 Bare Costs Labor	Equipment	Total	Total Incl O&P
0010	**SELECTIVE DEMOLITION OF WINDOWS** R024119-10									
0200	Aluminum, including trim, to 12 S.F.	1 Clab	16	.500	Ea.		13.25		13.25	22
0240	To 25 S.F.		11	.727			19.25		19.25	32
0280	To 50 S.F.		5	1.600			42.50		42.50	71
0320	Storm windows/screens, to 12 S.F.		27	.296			7.85		7.85	13.15
0360	To 25 S.F.		21	.381			10.10		10.10	16.90
0400	To 50 S.F.		16	.500			13.25		13.25	22
0600	Glass, minimum		200	.040	S.F.		1.06		1.06	1.77
0620	Maximum		150	.053	″		1.41		1.41	2.36
1000	Steel, including trim, to 12 S.F.		13	.615	Ea.		16.30		16.30	27.50
1020	To 25 S.F.		9	.889			23.50		23.50	39.50
1040	To 50 S.F.		4	2			53		53	88.50
2000	Wood, including trim, to 12 S.F.		22	.364			9.60		9.60	16.10
2020	To 25 S.F.		18	.444			11.75		11.75	19.70
2060	To 50 S.F.		13	.615			16.30		16.30	27.50
2065	To 180 S.F.		8	1			26.50		26.50	44.50
4410	Remove skylight, plstc domes, flush/curb mtd	G-3	395	.081	S.F.		2.62		2.62	4.35
5020	Remove and reset window, minimum	1 Carp	6	1.333	Ea.		44.50		44.50	75
5040	Average		4	2			67		67	112
5080	Maximum		2	4			134		134	225

08 11 16 – Aluminum Doors and Frames

08 11 16.10 Entrance Doors		Crew	Daily Output	Labor-Hours	Unit	Material	Labor (2011 Bare Costs)	Equipment	Total	Total Incl O&P
0010	**ENTRANCE DOORS** Aluminum, narrow stile									
0011	Including standard hardware, clear finish, no glass									
0020	3′-0″ x 7′-0″ opening	2 Sswk	2	8	Ea.	760	291		1,051	1,400
0100	3′-0″ x 10′-0″ opening, 3′ high transom		1.80	8.889		1,200	325		1,525	1,925
0200	3′-6″ x 10′-0″ opening, 3′ high transom		1.80	8.889		1,250	325		1,575	2,000
0301	6′-0″ x 7′-0″ opening		1.30	12.308	Pr.	1,150	450		1,600	2,100
0400	6′-0″ x 10′-0″ opening, 3′ high transom		1.10	14.545	″	1,525	530		2,055	2,675
0520	3′-0″ x 7′-0″ opening, wide stile		2	8	Ea.	910	291		1,201	1,550
0540	3′-6″ x 7′-0″ opening		2	8		1,125	291		1,416	1,800
0560	5′-0″ x 7′-0″ opening		2	8		1,550	291		1,841	2,250
0580	6′-0″ x 7′-0″ opening		1.30	12.308	Pr.	1,475	450		1,925	2,475
0600	7′-0″ x 7′-0″ opening		1	16	″	1,625	580		2,205	2,875
1100	For full vision doors, with 1/2″ glass, add				Leaf	55%				
1200	For non-standard size, add					67%				
1300	Light bronze finish, add					36%				
1400	Dark bronze finish, add					18%				
1500	For black finish, add					36%				
1600	Concealed panic device, add					910			910	1,000
1700	Electric striker release, add				Opng.	252			252	277
1800	Floor check, add				Leaf	630			630	695
1900	Concealed closer, add				″	505			505	555

08 11 63 – Metal Screen and Storm Doors and Frames

08 11 63.23 Aluminum Screen and Storm Doors and Frames		Crew	Daily Output	Labor-Hours	Unit	Material	Labor	Equipment	Total	Total Incl O&P
0010	**ALUMINUM SCREEN AND STORM DOORS AND FRAMES**									
0020	Combination storm and screen									
0420	Clear anodic coating, 2′-8″ wide	2 Carp	14	1.143	Ea.	158	38.50		196.50	238
0440	3′-0″ wide	″	14	1.143	″	158	38.50		196.50	238
0500	For 7′ door height, add					5%				
1020	Mill finish, 2′-8″ wide	2 Carp	14	1.143	Ea.	225	38.50		263.50	310
1040	3′-0″ wide	″	14	1.143		246	38.50		284.50	335
1100	For 7′-0″ door, add					5%				
1520	White painted, 2′-8″ wide	2 Carp	14	1.143		281	38.50		319.50	375
1540	3′-0″ wide	″	14	1.143		298	38.50		336.50	390
1600	For 7′-0″ door, add					5%				
2000	Wood door & screen, see Section 08 14 33.20									

08 11 74 – Sliding Metal Grilles

08 11 74.10 Rolling Grille Supports		Crew	Daily Output	Labor-Hours	Unit	Material	Labor	Equipment	Total	Total Incl O&P
0010	**ROLLING GRILLE SUPPORTS**									
0020	Rolling grille supports, overhead framed	E-4	36	.889	L.F.	24.50	33	3.03	60.53	93

08 12 Metal Frames

08 12 13 – Hollow Metal Frames

08 12 13.13 Standard Hollow Metal Frames		Crew	Daily Output	Labor-Hours	Unit	2011 Bare Costs Material	Labor	Equipment	Total	Total Incl O&P
0010	**STANDARD HOLLOW METAL FRAMES**									
0020	16 ga., up to 5-3/4" jamb depth									
0025	6'-8" high, 3'-0" wide, single G	2 Carp	16	1	Ea.	138	33.50		171.50	208
0028	3'-6" wide, single G		16	1		146	33.50		179.50	217
0030	4'-0" wide, single G		16	1		146	33.50		179.50	217
0040	6'-0" wide, double G		14	1.143		195	38.50		233.50	278
0045	8'-0" wide, double G		14	1.143		203	38.50		241.50	287
0100	7'-0" high, 3'-0" wide, single G		16	1		143	33.50		176.50	213
0110	3'-6" wide, single G		16	1		151	33.50		184.50	222
0112	4'-0" wide, single G		16	1		151	33.50		184.50	222
0140	6'-0" wide, double G		14	1.143		185	38.50		223.50	268
0145	8'-0" wide, double G		14	1.143		219	38.50		257.50	305
1000	16 ga., up to 4-7/8" deep, 7'-0" H, 3'-0" W, single G		16	1		157	33.50		190.50	228
1140	6'-0" wide, double G		14	1.143		177	38.50		215.50	259
1200	16 ga., 8-3/4" deep, 7'-0" H, 3'-0" W, single G		16	1		190	33.50		223.50	265
1240	6'-0" wide, double G		14	1.143		223	38.50		261.50	310
2800	14 ga., up to 3-7/8" deep, 7'-0" high, 3'-0" wide, single G		16	1		171	33.50		204.50	244
2840	6'-0" wide, double G		14	1.143		205	38.50		243.50	289
3000	14 ga., up to 5-3/4" deep, 6'-8" high, 3'-0" wide, single G		16	1		145	33.50		178.50	215
3002	3'-6" wide, single G		16	1		188	33.50		221.50	263
3005	4'-0" wide, single G		16	1		188	33.50		221.50	263
3600	up to 5-3/4" jamb depth, 7'-0" high, 4'-0" wide, single G		15	1.067		176	36		212	254
3620	6'-0" wide, double G		12	1.333		224	44.50		268.50	320
3640	8'-0" wide, double G		12	1.333		235	44.50		279.50	335
3700	8'-0" high, 4'-0" wide, single G		15	1.067		224	36		260	305
3740	8'-0" wide, double G		12	1.333		276	44.50		320.50	380
4000	6-3/4" deep, 7'-0" high, 4'-0" wide, single G		15	1.067		206	36		242	287
4020	6'-0" wide, double G		12	1.333		257	44.50		301.50	360
4040	8'-0", wide double G		12	1.333		264	44.50		308.50	365
4100	8'-0" high, 4'-0" wide, single G		15	1.067		263	36		299	350
4140	8'-0" wide, double G		12	1.333		305	44.50		349.50	415
4400	8-3/4" deep, 7'-0" high, 4'-0" wide, single G		15	1.067		241	36		277	325
4440	8'-0" wide, double G		12	1.333		299	44.50		343.50	405
4500	8'-0" high, 4'-0" wide, single G		15	1.067		270	36		306	355
4540	8'-0" wide, double G		12	1.333		335	44.50		379.50	445
4900	For welded frames, add					60			60	66
5400	14 ga., "B" label, up to 5-3/4" deep, 7'-0" high, 4'-0" wide, single G	2 Carp	15	1.067		197	36		233	277
5440	8'-0" wide, double G		12	1.333		258	44.50		302.50	360
5800	6-3/4" deep, 7'-0" high, 4'-0" wide, single G		15	1.067		208	36		244	289
5840	8'-0" wide, double G		12	1.333		365	44.50		409.50	475
6200	8-3/4" deep, 7'-0" high, 4'-0" wide, single G		15	1.067		296	36		332	385
6240	8'-0" wide, double G		12	1.333		340	44.50		384.50	450
6300	For "A" label use same price as "B" label									
6400	For baked enamel finish, add					30%	15%			
6500	For galvanizing, add					15%				
6600	For hospital stop, add				Ea.	282			282	310
6620	For hospital stop, stainless steel add				"	365			365	400
7900	Transom lite frames, fixed, add	2 Carp	155	.103	S.F.	52	3.46		55.46	63
8000	Movable, add	"	130	.123	"	65	4.13		69.13	78.50

08 12 Metal Frames

08 12 13 – Hollow Metal Frames

08 12 13.25 Channel Metal Frames			Crew	Daily Output	Labor-Hours	Unit	2011 Bare Costs Material	Labor	Equipment	Total	Total Incl O&P
0010	**CHANNEL METAL FRAMES**										
0020	Steel channels with anchors and bar stops										
0100	6" channel @ 8.2#/L.F., 3' x 7' door, weighs 150#	G	E-4	13	2.462	Ea.	203	91	8.40	302.40	405
0200	8" channel @ 11.5#/L.F., 6' x 8' door, weighs 275#	G		9	3.556		370	131	12.10	513.10	675
0300	8' x 12' door, weighs 400#	G		6.50	4.923		540	182	16.80	738.80	960
0400	10" channel @ 15.3#/L.F., 10' x 10' door, weighs 500#	G		6	5.333		675	197	18.20	890.20	1,150
0500	12' x 12' door, weighs 600#	G		5.50	5.818		810	215	19.85	1,044.85	1,325
0600	12" channel @ 20.7#/L.F., 12' x 12' door, weighs 825#	G		4.50	7.111		1,125	262	24.50	1,411.50	1,750
0700	12' x 16' door, weighs 1000#	G		4	8		1,350	295	27.50	1,672.50	2,075
0800	For frames without bar stops, light sections, deduct						15%				
0900	Heavy sections, deduct						10%				

08 13 Metal Doors

08 13 13 – Hollow Metal Doors

08 13 13.13 Standard Hollow Metal Doors

			Crew	Daily Output	Labor-Hours	Unit	Material	Labor	Equipment	Total	Total Incl O&P
0010	**STANDARD HOLLOW METAL DOORS** R081313-20										
0015	Flush, full panel, hollow core										
0017	When noted doors are prepared but do not include glass or louvers										
0020	1-3/8" thick, 20 ga., 2'-0" x 6'-8"	G	2 Carp	20	.800	Ea.	320	27		347	395
0040	2'-8" x 6'-8"	G		18	.889		335	30		365	420
0060	3'-0" x 6'-8"	G		17	.941		335	31.50		366.50	425
0100	3'-0" x 7'-0"	G		17	.941		345	31.50		376.50	435
0120	For vision lite, add						90			90	99
0140	For narrow lite, add						97.50			97.50	107
0320	Half glass, 20 ga., 2'-0" x 6'-8"	G	2 Carp	20	.800		470	27		497	560
0340	2'-8" x 6'-8"	G		18	.889		490	30		520	590
0360	3'-0" x 6'-8"	G		17	.941		485	31.50		516.50	590
0400	3'-0" x 7'-0"	G		17	.941		595	31.50		626.50	710
0410	1-3/8" thick, 18 ga., 2'-0" x 6'-8"	G		20	.800		380	27		407	465
0420	3'-0" x 6'-8"	G		17	.941		385	31.50		416.50	480
0425	3'-0" x 7'-0"	G		17	.941		400	31.50		431.50	495
0450	For vision lite, add						90			90	99
0452	For narrow lite, add						97.50			97.50	107
0460	Half glass, 18 ga., 2'-0" x 6'-8"	G	2 Carp	20	.800		530	27		557	630
0465	2'-8" x 6'-8"	G		18	.889		550	30		580	655
0470	3'-0" x 6'-8"	G		17	.941		535	31.50		566.50	645
0475	3'-0" x 7'-0"	G		17	.941		550	31.50		581.50	660
0500	Hollow core, 1-3/4" thick, full panel, 20 ga., 2'-8" x 6'-8"	G		18	.889		400	30		430	490
0520	3'-0" x 6'-8"	G		17	.941		400	31.50		431.50	495
0640	3'-0" x 7'-0"	G		17	.941		420	31.50		451.50	515
0680	4'-0" x 7'-0"	G		15	1.067		605	36		641	725
0700	4'-0" x 8'-0"	G		13	1.231		705	41.50		746.50	845
1000	18 ga., 2'-8" x 6'-8"	G		17	.941		465	31.50		496.50	570
1020	3'-0" x 6'-8"	G		16	1		455	33.50		488.50	555
1120	3'-0" x 7'-0"	G		17	.941		475	31.50		506.50	580
1180	4'-0" x 7'-0"	G		14	1.143		605	38.50		643.50	730
1200	4'-0" x 8'-0"	G		17	.941		705	31.50		736.50	830
1212	For vision lite, add						90			90	99
1214	For narrow lite, add						97.50			97.50	107
1230	Half glass, 20 ga., 2'-8" x 6'-8"	G	2 Carp	20	.800		550	27		577	650

08 13 13 – Hollow Metal Doors

08 13 13.13 Standard Hollow Metal Doors		Crew	Daily Output	Labor-Hours	Unit	Material	2011 Bare Costs Labor	Equipment	Total	Total Incl O&P
1240	3'-0" x 6'-8" G	2 Carp	18	.889	Ea.	550	30		580	655
1260	3'-0" x 7'-0" G		18	.889		565	30		595	670
1320	18 ga., 2'-8" x 6'-8" G		18	.889		610	30		640	720
1340	3'-0" x 6'-8" G		17	.941		605	31.50		636.50	720
1360	3'-0" x 7'-0" G		17	.941		615	31.50		646.50	730
1380	4'-0" x 7'-0" G		15	1.067		750	36		786	885
1400	4'-0" x 8'-0" G		14	1.143		850	38.50		888.50	1,000
1720	Insulated, 1-3/4" thick, full panel, 18 ga., 3'-0" x 6'-8" G		15	1.067		455	36		491	560
1740	2'-8" x 7'-0" G		16	1		480	33.50		513.50	585
1760	3'-0" x 7'-0" G		15	1.067		465	36		501	575
1800	4'-0" x 8'-0" G		13	1.231		705	41.50		746.50	845
1805	For vision lite, add					90			90	99
1810	For narrow lite, add					97.50			97.50	107
1820	Half glass, 18 ga., 3'-0" x 6'-8" G	2 Carp	16	1		605	33.50		638.50	720
1840	2'-8" x 7'-0" G		17	.941		630	31.50		661.50	745
1860	3'-0" x 7'-0" G		16	1		660	33.50		693.50	780
1900	4'-0" x 8'-0" G		14	1.143		705	38.50		743.50	840
2000	For bottom louver, add					264			264	291
2020	For baked enamel finish, add					30%	15%			
2040	For galvanizing, add					15%				

08 13 13.15 Metal Fire Doors		Crew	Daily Output	Labor-Hours	Unit	Material	Labor	Equipment	Total	Total Incl O&P
0010	**METAL FIRE DOORS** R081313-20									
0015	Steel, flush, "B" label, 90 minute									
0020	Full panel, 20 ga., 2'-0" x 6'-8"	2 Carp	20	.800	Ea.	400	27		427	485
0040	2'-8" x 6'-8"		18	.889		415	30		445	510
0060	3'-0" x 6'-8"		17	.941		415	31.50		446.50	515
0080	3'-0" x 7'-0"		17	.941		430	31.50		461.50	530
0140	18 ga., 3'-0" x 6'-8"		16	1		470	33.50		503.50	570
0160	2'-8" x 7'-0"		17	.941		495	31.50		526.50	600
0180	3'-0" x 7'-0"		16	1		480	33.50		513.50	585
0200	4'-0" x 7'-0"		15	1.067		620	36		656	740
0220	For "A" label, 3 hour, 18 ga., use same price as "B" label									
0240	For vision lite, add				Ea.	149			149	164
0520	Flush, "B" label 90 min., egress core, 20 ga., 2'-0" x 6'-8"	2 Carp	18	.889		625	30		655	735
0540	2'-8" x 6'-8"		17	.941		635	31.50		666.50	750
0560	3'-0" x 6'-8"		16	1		635	33.50		668.50	750
0580	3'-0" x 7'-0"		16	1		655	33.50		688.50	775
0640	Flush, "A" label 3 hour, egress core, 18 ga., 3'-0" x 6'-8"		15	1.067		695	36		731	820
0660	2'-8" x 7'-0"		16	1		715	33.50		748.50	840
0680	3'-0" x 7'-0"		15	1.067		705	36		741	835
0700	4'-0" x 7'-0"		14	1.143		840	38.50		878.50	990

08 13 13.20 Residential Steel Doors		Crew	Daily Output	Labor-Hours	Unit	Material	Labor	Equipment	Total	Total Incl O&P
0010	**RESIDENTIAL STEEL DOORS**									
0020	Prehung, insulated, exterior									
0030	Embossed, full panel, 2'-8" x 6'-8" G	2 Carp	17	.941	Ea.	293	31.50		324.50	380
0040	3'-0" x 6'-8" G		15	1.067		242	36		278	325
0060	3'-0" x 7'-0" G		15	1.067		325	36		361	420
0070	5'-4" x 6'-8", double G		8	2		625	67		692	795
0220	Half glass, 2'-8" x 6'-8"		17	.941		300	31.50		331.50	385
0240	3'-0" x 6'-8"		16	1		300	33.50		333.50	385
0260	3'-0" x 7'-0"		16	1		345	33.50		378.50	435
0270	5'-4" x 6'-8", double		8	2		640	67		707	815

08 13 Metal Doors

08 13 13 – Hollow Metal Doors

08 13 13.20 Residential Steel Doors		Crew	Daily Output	Labor-Hours	Unit	Material	2011 Bare Costs Labor	Equipment	Total	Total Incl O&P
1320	Flush face, full panel, 2′-8″ x 6′-8″	2 Carp	16	1	Ea.	230	33.50		263.50	310
1340	3′-0″ x 6′-8″		15	1.067		230	36		266	315
1360	3′-0″ x 7′-0″		15	1.067		290	36		326	380
1380	5′-4″ x 6′-8″, double		8	2		470	67		537	625
1420	Half glass, 2′-8″ x 6′-8″		17	.941		320	31.50		351.50	405
1440	3′-0″ x 6′-8″		16	1		320	33.50		353.50	405
1460	3′-0″ x 7′-0″		16	1		345	33.50		378.50	435
1480	5′-4″ x 6′-8″, double		8	2		635	67		702	805
1500	Sidelight, full lite, 1′-0″ x 6′-8″with grille					236			236	260
1510	1′-0″ x 6′-8″, low e					257			257	282
1520	1′-0″ x 6′-8″, half lite					254			254	280
1530	1′-0″ x 6′-8″, half lite, low e					262			262	289
2300	Interior, residential, closet, bi-fold, 6′-8″ x 2′-0″ wide	2 Carp	16	1		161	33.50		194.50	233
2330	3′-0″ wide		16	1		200	33.50		233.50	276
2360	4′-0″ wide		15	1.067		260	36		296	345
2400	5′-0″ wide		14	1.143		315	38.50		353.50	415
2420	6′-0″ wide		13	1.231		330	41.50		371.50	435

08 13 13.25 Doors Hollow Metal		Crew	Daily Output	Labor-Hours	Unit	Material	Labor	Equipment	Total	Total Incl O&P
0010	**DOORS HOLLOW METAL**									
0500	Exterior, commercial, flush, 20 ga., 1-3/4″ x 7′-0″ x 2′-6″ wide	2 Carp	15	1.067	Ea.	375	36		411	475
0530	2′-8″ wide		15	1.067		415	36		451	520
0560	3′-0″ wide		14	1.143		415	38.50		453.50	525
1000	18 ga., 1-3/4″ x 7′-0″ x 2′-6″ wide		15	1.067		475	36		511	585
1030	2′-8″ wide		15	1.067		480	36		516	590
1060	3′-0″ wide		14	1.143		465	38.50		503.50	580
1500	16 ga., 1-3/4 x 7′-0″ x 2′-6″ wide		15	1.067		540	36		576	655
1530	2′-8″ wide		15	1.067		545	36		581	660
1560	3′-0″ wide		14	1.143		530	38.50		568.50	650
1590	3′-6″ wide		14	1.143		625	38.50		663.50	750
2900	Fire door, "A" label, 1-3/4″ x 7′-0″ x 2′-6″ wide		15	1.067		630	36		666	750
2930	2′-8″ wide		15	1.067		630	36		666	755
2960	3′-0″ wide		14	1.143		620	38.50		658.50	745
2990	3′-6″ wide		14	1.143		705	38.50		743.50	840
3100	"B" label, 2′-6″ wide		15	1.067		565	36		601	680
3130	2′-8″ wide		15	1.067		565	36		601	685
3160	3′-0″ wide		14	1.143		560	38.50		598.50	680

08 13 16 – Aluminum Doors

08 13 16.10 Commercial Aluminum Doors		Crew	Daily Output	Labor-Hours	Unit	Material	Labor	Equipment	Total	Total Incl O&P
0010	**COMMERCIAL ALUMINUM DOORS**, no glazing									
0020	Incl. hinges, push/pull, deadlock, cyl., threshold									
2100	Medium stile, 3′-0″ x 7′-0″, single	2 Carp	3	5.333	Ea.	740	179		919	1,125
2200	Pair of 3′-0″ x 7′-0″		1.70	9.412	Pr.	1,475	315		1,790	2,150
2300	3′-6″ x 7′-0″, single		3	5.333	Ea.	955	179		1,134	1,350
5000	Flush panel doors, pair of 2′-6″ x 7′-0″	2 Sswk	2	8	Pr.	1,375	291		1,666	2,050
5050	3′-0″ x 7′-0″, single		2.50	6.400	Ea.	780	233		1,013	1,300
5100	Pair of 3′-0″ x 7′-0″		2	8	Pr.	1,450	291		1,741	2,150
5150	3′-6″ x 7′-0″, single		2.50	6.400	Ea.	930	233		1,163	1,475

08 13 73 – Sliding Metal Doors

08 13 73.10 Steel Sliding Doors		Crew	Daily Output	Labor-Hours	Unit	Material	Labor	Equipment	Total	Total Incl O&P
0010	**STEEL SLIDING DOORS**									
0021	Up to 50′ x 18′, electric, standard duty, minimum	E-2	360	.133	S.F.	29	4.88	4.51	38.39	45.50

08 13 Metal Doors

08 13 73 – Sliding Metal Doors

08 13 73.10 Steel Sliding Doors		Crew	Daily Output	Labor-Hours	Unit	Material	2011 Bare Costs Labor	Equipment	Total	Total Incl O&P
0101	Maximum	E-2	340	.141	S.F.	45.50	5.15	4.77	55.42	65
0501	Heavy duty, minimum		297	.162		37	5.90	5.45	48.35	57.50
0601	Maximum		277	.173		92.50	6.35	5.85	104.70	120

08 14 Wood Doors

08 14 13 – Carved Wood Doors

08 14 13.10 Types of Wood Doors, Carved

		Crew	Daily Output	Labor-Hours	Unit	Material	Labor	Equipment	Total	Total Incl O&P
0010	**TYPES OF WOOD DOORS, CARVED**									
3000	Solid wood, 1-3/4" thick stile and rail									
3020	Mahogany, 3'-0" x 7'-0", minimum	2 Carp	14	1.143	Ea.	945	38.50		983.50	1,125
3030	Maximum		10	1.600		1,500	53.50		1,553.50	1,750
3040	3'-6" x 8'-0", minimum		10	1.600		1,250	53.50		1,303.50	1,475
3050	Maximum		8	2		1,850	67		1,917	2,150
3100	Pine, 3'-0" x 7'-0", minimum		14	1.143		490	38.50		528.50	600
3110	Maximum		10	1.600		825	53.50		878.50	1,000
3120	3'-6" x 8'-0", minimum		10	1.600		885	53.50		938.50	1,075
3130	Maximum		8	2		1,800	67		1,867	2,075
3200	Red oak, 3'-0" x 7'-0", minimum		14	1.143		1,625	38.50		1,663.50	1,875
3210	Maximum		10	1.600		2,025	53.50		2,078.50	2,325
3220	3'-6" x 8'-0", minimum		10	1.600		2,200	53.50		2,253.50	2,525
3230	Maximum		8	2		3,200	67		3,267	3,625
4000	Hand carved door, mahogany									
4020	3'-0" x 7'-0", minimum	2 Carp	14	1.143	Ea.	1,600	38.50		1,638.50	1,825
4030	Maximum		11	1.455		3,600	49		3,649	4,025
4040	3'-6" x 8'-0", minimum		10	1.600		2,500	53.50		2,553.50	2,850
4050	Maximum		8	2		3,500	67		3,567	3,950
4200	Rose wood, 3'-0" x 7'-0", minimum		14	1.143		5,000	38.50		5,038.50	5,575
4210	Maximum		11	1.455		13,600	49		13,649	15,100
4220	3'-6" x 8'-0", minimum		10	1.600		5,700	53.50		5,753.50	6,375
4225	Maximum		10	1.600		8,700	53.50		8,753.50	9,675
4280	For 6'-8" high door, deduct from 7'-0" door					34			34	37.50
4400	For custom finish, add					400			400	440
4600	Side light, mahogany, 7'-0" x 1'-6" wide, minimum	2 Carp	18	.889		900	30		930	1,050
4610	Maximum		14	1.143		2,625	38.50		2,663.50	2,975
4620	8'-0" x 1'-6" wide, minimum		14	1.143		1,700	38.50		1,738.50	1,950
4630	Maximum		10	1.600		1,950	53.50		2,003.50	2,250
4640	Side light, oak, 7'-0" x 1'-6" wide, minimum		18	.889		1,100	30		1,130	1,250
4650	Maximum		14	1.143		1,900	38.50		1,938.50	2,175
4660	8'-0" x 1'-6" wide, minimum		14	1.143		1,000	38.50		1,038.50	1,175
4670	Maximum		10	1.600		1,900	53.50		1,953.50	2,200

08 14 16 – Flush Wood Doors

08 14 16.09 Smooth Wood Doors

		Crew	Daily Output	Labor-Hours	Unit	Material	Labor	Equipment	Total	Total Incl O&P
0010	**SMOOTH WOOD DOORS**									
0015	Flush, interior, hollow core									
0025	Lauan face, 1-3/8", 3'-0" x 6'-8"	2 Carp	17	.941	Ea.	52.50	31.50		84	111
0030	4'-0" x 6'-8"		16	1		126	33.50		159.50	195
0080	1-3/4", 2'-0" x 6'-8"		17	.941		35	31.50		66.50	91.50
0140	Birch face, 1-3/8", 2'-6" x 6'-8"		17	.941		82.50	31.50		114	144
0180	3'-0" x 6'-8"		17	.941		94	31.50		125.50	156
0200	4'-0" x 6'-8"		16	1		158	33.50		191.50	230

08 14 16.09 Smooth Wood Doors		Crew	Daily Output	Labor-Hours	Unit	2011 Bare Costs Material	Labor	Equipment	Total	Total Incl O&P
0202	1-3/4", 2'-0" x 6'-8"	2 Carp	17	.941	Ea.	43.50	31.50		75	101
0220	Oak face, 1-3/8", 2'-0" x 6'-8"		17	.941		96	31.50		127.50	159
0280	3'-0" x 6'-8"		17	.941		110	31.50		141.50	174
0300	4'-0" x 6'-8"		16	1		134	33.50		167.50	204
0305	1-3/4", 2'-6" x 6'-8"		17	.941		106	31.50		137.50	169
0320	Walnut face, 1-3/8", 2'-0" x 6'-8"		17	.941		178	31.50		209.50	248
0340	2'-6" x 6'-8"		17	.941		184	31.50		215.50	256
0380	3'-0" x 6'-8"		17	.941		192	31.50		223.50	264
0400	4'-0" x 6'-8"		16	1		213	33.50		246.50	290
0430	For 7'-0" high, add					16			16	17.60
0440	For 8'-0" high, add					30			30	33
0480	For prefinishing, clear, add					35			35	38.50
0500	For prefinishing, stain, add					45			45	49.50
1320	M.D. overlay on hardboard, 1-3/8", 2'-0" x 6'-8"	2 Carp	17	.941		110	31.50		141.50	174
1340	2'-6" x 6'-8"		17	.941		110	31.50		141.50	174
1380	3'-0" x 6'-8"		17	.941		120	31.50		151.50	185
1400	4'-0" x 6'-8"		16	1		170	33.50		203.50	243
1420	For 7'-0" high, add					12			12	13.20
1440	For 8'-0" high, add					25			25	27.50
1720	H.P. plastic laminate, 1-3/8", 2'-0" x 6'-8"	2 Carp	16	1		228	33.50		261.50	305
1740	2'-6" x 6'-8"		16	1		228	33.50		261.50	305
1780	3'-0" x 6'-8"		15	1.067		276	36		312	365
1800	4'-0" x 6'-8"		14	1.143		360	38.50		398.50	460
1820	For 7'-0" high, add					11			11	12.10
1840	For 8'-0" high, add					26			26	28.50
2020	Particle core, lauan face, 1-3/8", 2'-6" x 6'-8"	2 Carp	15	1.067		90	36		126	159
2040	3'-0" x 6'-8"		14	1.143		92	38.50		130.50	165
2080	3'-0" x 7'-0"		13	1.231		93.50	41.50		135	172
2085	4'-0" x 7'-0"		12	1.333		120	44.50		164.50	207
2120	Birch face, 1-3/8", 2'-6" x 6'-8"		15	1.067		100	36		136	170
2140	3'-0" x 6'-8"		14	1.143		110	38.50		148.50	185
2180	3'-0" x 7'-0"		13	1.231		120	41.50		161.50	201
2200	4'-0" x 7'-0"		12	1.333		135	44.50		179.50	224
2220	Oak face, 1-3/8", 2'-6" x 6'-8"		15	1.067		113	36		149	184
2240	3'-0" x 6'-8"		14	1.143		125	38.50		163.50	202
2280	3'-0" x 7'-0"		13	1.231		130	41.50		171.50	212
2300	4'-0" x 7'-0"		12	1.333		152	44.50		196.50	242
2320	Walnut face, 1-3/8", 2'-0" x 6'-8"		15	1.067		120	36		156	192
2340	2'-6" x 6'-8"		14	1.143		135	38.50		173.50	213
2380	3'-0" x 6'-8"		13	1.231		152	41.50		193.50	236
2400	4'-0" x 6'-8"		12	1.333		200	44.50		244.50	295
2440	For 8'-0" high, add					35			35	38.50
2460	For 8'-0" high walnut, add					20			20	22
2480	For solid wood core, add					39			39	43
2720	For prefinishing, clear, add					25			25	27.50
2740	For prefinishing, stain, add					52			52	57
3320	M.D. overlay on hardboard, 1-3/8", 2'-6" x 6'-8"	2 Carp	14	1.143		106	38.50		144.50	181
3340	3'-0" x 6'-8"		13	1.231		112	41.50		153.50	192
3380	3'-0" x 7'-0"		12	1.333		114	44.50		158.50	200
3400	4'-0" x 7'-0"		10	1.600		150	53.50		203.50	255
3440	For 8'-0" height, add					35			35	38.50
3460	For solid wood core, add					40			40	44
3720	H.P. plastic laminate, 1-3/8", 2'-6" x 6'-8"	2 Carp	13	1.231		155	41.50		196.50	240

08 14 Wood Doors

08 14 16 – Flush Wood Doors

08 14 16.09 Smooth Wood Doors		Crew	Daily Output	Labor-Hours	Unit	2011 Bare Costs Material	Labor	Equipment	Total	Total Incl O&P
3740	3'-0" x 6'-8"	2 Carp	12	1.333	Ea.	180	44.50		224.50	273
3780	3'-0" x 7'-0"		11	1.455		185	49		234	286
3800	4'-0" x 7'-0"		8	2		222	67		289	355
3840	For 8'-0" height, add					35			35	38.50
3860	For solid wood core, add					40			40	44
4000	Exterior, flush, solid core, birch, 1-3/4" x 2'-6" x 7'-0"	2 Carp	15	1.067		165	36		201	242
4020	2'-8" wide		15	1.067		153	36		189	228
4040	3'-0" wide		14	1.143		195	38.50		233.50	279
4100	Oak faced 1-3/4" x 2'-6" x 7'-0"		15	1.067		185	36		221	264
4120	2'-8" wide		15	1.067		200	36		236	280
4140	3'-0" wide		14	1.143		210	38.50		248.50	295
4180	3'-6" wide	1 Carp	17	.471		375	15.80		390.80	440
4200	Walnut faced, 1-3/4" x 2'-6" x 7'-0"	2 Carp	15	1.067		270	36		306	355
4220	2'-8" wide		15	1.067		285	36		321	375
4240	3'-0" wide		14	1.143		295	38.50		333.50	390
4300	For 6'-8" high door, deduct from 7'-0" door					16			16	17.60

08 14 16.10 Wood Doors Decorator

0010	**WOOD DOORS DECORATOR**									

08 14 16.20 Wood Fire Doors

		Crew	Daily Output	Labor-Hours	Unit	Material	Labor	Equipment	Total	Total Incl O&P
0010	**WOOD FIRE DOORS**									
0020	Particle core, 7 face plys, "B" label,									
0040	1 hour, birch face, 1-3/4" x 2'-6" x 6'-8"	2 Carp	14	1.143	Ea.	385	38.50		423.50	490
0080	3'-0" x 6'-8"		13	1.231		390	41.50		431.50	500
0090	3'-0" x 7'-0"		12	1.333		435	44.50		479.50	550
0100	4'-0" x 7'-0"		12	1.333		510	44.50		554.50	635
0140	Oak face, 2'-6" x 6'-8"		14	1.143		335	38.50		373.50	435
0180	3'-0" x 6'-8"		13	1.231		350	41.50		391.50	455
0190	3'-0" x 7'-0"		12	1.333		365	44.50		409.50	475
0200	4'-0" x 7'-0"		12	1.333		480	44.50		524.50	605
0240	Walnut face, 2'-6" x 6'-8"		14	1.143		450	38.50		488.50	560
0280	3'-0" x 6'-8"		13	1.231		470	41.50		511.50	585
0290	3'-0" x 7'-0"		12	1.333		490	44.50		534.50	615
0300	4'-0" x 7'-0"		12	1.333		610	44.50		654.50	745
0440	M.D. overlay on hardboard, 2'-6" x 6'-8"		15	1.067		300	36		336	390
0480	3'-0" x 6'-8"		14	1.143		315	38.50		353.50	410
0490	3'-0" x 7'-0"		13	1.231		330	41.50		371.50	435
0500	4'-0" x 7'-0"		12	1.333		400	44.50		444.50	515
0740	90 minutes, birch face, 1-3/4" x 2'-6" x 6'-8"		14	1.143		335	38.50		373.50	430
0780	3'-0" x 6'-8"		13	1.231		310	41.50		351.50	410
0790	3'-0" x 7'-0"		12	1.333		375	44.50		419.50	490
0800	4'-0" x 7'-0"		12	1.333		490	44.50		534.50	610
0840	Oak face, 2'-6" x 6'-8"		14	1.143		425	38.50		463.50	530
0880	3'-0" x 6'-8"		13	1.231		435	41.50		476.50	545
0890	3'-0" x 7'-0"		12	1.333		450	44.50		494.50	570
0900	4'-0" x 7'-0"		12	1.333		580	44.50		624.50	715
0940	Walnut face, 2'-6" x 6'-8"		14	1.143		390	38.50		428.50	495
0980	3'-0" x 6'-8"		13	1.231		400	41.50		441.50	510
0990	3'-0" x 7'-0"		12	1.333		460	44.50		504.50	580
1000	4'-0" x 7'-0"		12	1.333		610	44.50		654.50	745
1140	M.D. overlay on hardboard, 2'-6" x 6'-8"		15	1.067		350	36		386	445
1180	3'-0" x 6'-8"		14	1.143		370	38.50		408.50	475
1190	3'-0" x 7'-0"		13	1.231		400	41.50		441.50	510

08 14 Wood Doors

08 14 16 – Flush Wood Doors

08 14 16.20 Wood Fire Doors		Crew	Daily Output	Labor-Hours	Unit	2011 Bare Costs Material	Labor	Equipment	Total	Total Incl O&P
1200	4′-0″ x 7′-0″	2 Carp	12	1.333	Ea.	450	44.50		494.50	570
1240	For 8′-0″ height, add					65			65	71.50
1260	For 8′-0″ height walnut, add					80			80	88
2200	Custom architectural "B" label, flush, 1-3/4″ thick, birch,									
2210	Solid core									
2220	2′-6″ x 7′-0″	2 Carp	15	1.067	Ea.	296	36		332	385
2260	3′-0″ x 7′-0″		14	1.143		305	38.50		343.50	400
2300	4′-0″ x 7′-0″		13	1.231		400	41.50		441.50	505
2420	4′-0″ x 8′-0″		11	1.455		430	49		479	550
2480	For oak veneer, add					50%				
2500	For walnut veneer, add					75%				

08 14 23 – Clad Wood Doors

08 14 23.13 Metal-Faced Wood Doors

Line	Description	Crew	Daily Output	Labor-Hours	Unit	Material	Labor	Equipment	Total	Total Incl O&P
0010	**METAL-FACED WOOD DOORS**									
0020	Interior, flush type, 3′ x 7′	2 Carp	4.30	3.721	Opng.	340	125		465	585

08 14 23.20 Tin Clad Wood Doors

Line	Description	Crew	Daily Output	Labor-Hours	Unit	Material	Labor	Equipment	Total	Total Incl O&P
0010	**TIN CLAD WOOD DOORS**									
0020	3 ply, 6′ x 7′, double sliding, doors only	2 Carp	1	16	Opng.	1,900	535		2,435	3,000
1000	For electric operator, add	1 Elec	2	4	″	3,250	161		3,411	3,825

08 14 33 – Stile and Rail Wood Doors

08 14 33.10 Wood Doors Paneled

Line	Description	Crew	Daily Output	Labor-Hours	Unit	Material	Labor	Equipment	Total	Total Incl O&P
0010	**WOOD DOORS PANELED**									
0020	Interior, six panel, hollow core, 1-3/8″ thick									
0040	Molded hardboard, 2′-0″ x 6′-8″	2 Carp	17	.941	Ea.	60	31.50		91.50	119
0060	2′-6″ x 6′-8″		17	.941		62	31.50		93.50	121
0070	2′-8″ x 6′-8″		17	.941		65	31.50		96.50	125
0080	3′-0″ x 6′-8″		17	.941		68	31.50		99.50	128
0140	Embossed print, molded hardboard, 2′-0″ x 6′-8″		17	.941		62	31.50		93.50	121
0160	2′-6″ x 6′-8″		17	.941		62	31.50		93.50	121
0180	3′-0″ x 6′-8″		17	.941		68	31.50		99.50	128
0540	Six panel, solid, 1-3/8″ thick, pine, 2′-0″ x 6′-8″		15	1.067		150	36		186	225
0560	2′-6″ x 6′-8″		14	1.143		165	38.50		203.50	246
0580	3′-0″ x 6′-8″		13	1.231		140	41.50		181.50	223
1020	Two panel, bored rail, solid, 1-3/8″ thick, pine, 1′-6″ x 6′-8″		16	1		265	33.50		298.50	350
1040	2′-0″ x 6′-8″		15	1.067		350	36		386	445
1060	2′-6″ x 6′-8″		14	1.143		395	38.50		433.50	500
1340	Two panel, solid, 1-3/8″ thick, fir, 2′-0″ x 6′-8″		15	1.067		155	36		191	231
1360	2′-6″ x 6′-8″		14	1.143		205	38.50		243.50	290
1380	3′-0″ x 6′-8″		13	1.231		410	41.50		451.50	520
1740	Five panel, solid, 1-3/8″ thick, fir, 2′-0″ x 6′-8″		15	1.067		275	36		311	365
1760	2′-6″ x 6′-8″		14	1.143		415	38.50		453.50	520
1780	3′-0″ x 6′-8″		13	1.231		415	41.50		456.50	525

08 14 33.20 Wood Doors Residential

Line	Description	Crew	Daily Output	Labor-Hours	Unit	Material	Labor	Equipment	Total	Total Incl O&P
0010	**WOOD DOORS RESIDENTIAL**									
0200	Exterior, combination storm & screen, pine									
0260	2′-8″ wide	2 Carp	10	1.600	Ea.	298	53.50		351.50	420
0280	3′-0″ wide		9	1.778		305	59.50		364.50	435
0300	7′-1″ x 3′-0″ wide		9	1.778		325	59.50		384.50	460
0400	Full lite, 6′-9″ x 2′-6″ wide		11	1.455		310	49		359	420
0420	2′-8″ wide		10	1.600		310	53.50		363.50	430
0440	3′-0″ wide		9	1.778		320	59.50		379.50	450

08 14 33.20 Wood Doors Residential		Crew	Daily Output	Labor-Hours	Unit	2011 Bare Costs Material	Labor	Equipment	Total	Total Incl O&P
0500	7′-1″ x 3′-0″ wide	2 Carp	9	1.778	Ea.	345	59.50		404.50	475
0700	Dutch door, pine, 1-3/4″ x 6′-8″ x 2′-8″ wide, minimum		12	1.333		675	44.50		719.50	820
0720	Maximum		10	1.600		825	53.50		878.50	1,000
0800	3′-0″ wide, minimum		12	1.333		690	44.50		734.50	830
0820	Maximum		10	1.600		900	53.50		953.50	1,075
1000	Entrance door, colonial, 1-3/4″ x 6′-8″ x 2′-8″ wide		16	1		465	33.50		498.50	565
1020	6 panel pine, 3′-0″ wide		15	1.067		435	36		471	540
1100	8 panel pine, 2′-8″ wide		16	1		570	33.50		603.50	685
1120	3′-0″ wide		15	1.067		560	36		596	675
1200	For tempered safety glass lites, (min of 2) add					75.50			75.50	83
1300	Flush, birch, solid core, 1-3/4″ x 6′-8″ x 2′-8″ wide	2 Carp	16	1		105	33.50		138.50	171
1320	3′-0″ wide		15	1.067		115	36		151	186
1350	7′-0″ x 2′-8″ wide		16	1		114	33.50		147.50	182
1360	3′-0″ wide		15	1.067		123	36		159	195
1380	For tempered safety glass lites, add					103			103	113
2700	Interior, closet, bi-fold, w/hardware, no frame or trim incl.									
2720	Flush, birch, 6′-6″ or 6′-8″ x 2′-6″ wide	2 Carp	13	1.231	Ea.	52.50	41.50		94	127
2740	3′-0″ wide		13	1.231		56.50	41.50		98	131
2760	4′-0″ wide		12	1.333		101	44.50		145.50	187
2780	5′-0″ wide		11	1.455		101	49		150	193
2800	6′-0″ wide		10	1.600		127	53.50		180.50	230
3000	Raised panel pine, 6′-6″ or 6′-8″ x 2′-6″ wide		13	1.231		187	41.50		228.50	275
3020	3′-0″ wide		13	1.231		275	41.50		316.50	375
3040	4′-0″ wide		12	1.333		305	44.50		349.50	410
3060	5′-0″ wide		11	1.455		365	49		414	480
3080	6′-0″ wide		10	1.600		400	53.50		453.50	530
3200	Louvered, pine 6′-6″ or 6′-8″ x 2′-6″ wide		13	1.231		137	41.50		178.50	220
3220	3′-0″ wide		13	1.231		197	41.50		238.50	286
3240	4′-0″ wide		12	1.333		222	44.50		266.50	320
3260	5′-0″ wide		11	1.455		256	49		305	365
3280	6′-0″ wide		10	1.600		287	53.50		340.50	405
4400	Bi-passing closet, incl. hardware and frame, no trim incl.									
4420	Flush, lauan, 6′-8″ x 4′-0″ wide	2 Carp	12	1.333	Opng.	169	44.50		213.50	260
4440	5′-0″ wide		11	1.455		186	49		235	286
4460	6′-0″ wide		10	1.600		209	53.50		262.50	320
4600	Flush, birch, 6′-8″ x 4′-0″ wide		12	1.333		220	44.50		264.50	315
4620	5′-0″ wide		11	1.455		212	49		261	315
4640	6′-0″ wide		10	1.600		256	53.50		309.50	370
4800	Louvered, pine, 6′-8″ x 4′-0″ wide		12	1.333		400	44.50		444.50	515
4820	5′-0″ wide		11	1.455		395	49		444	515
4840	6′-0″ wide		10	1.600		515	53.50		568.50	655
5000	Paneled, pine, 6′-8″ x 4′-0″ wide		12	1.333		495	44.50		539.50	620
5020	5′-0″ wide		11	1.455		410	49		459	530
5040	6′-0″ wide		10	1.600		590	53.50		643.50	735
5042	8′-0″ wide		12	1.333		985	44.50		1,029.50	1,150
6100	Folding accordion, closet, including track and frame									
6120	Vinyl, 2 layer, stock	2 Carp	10	1.600	Ea.	62	53.50		115.50	158
6200	Rigid PVC	″	10	1.600	″	34.50	53.50		88	128
6220	For custom partition, add					25%	10%			
7310	Passage doors, flush, no frame included									
7320	Hardboard, hollow core, 1-3/8″ x 6′-8″ x 1′-6″ wide	2 Carp	18	.889	Ea.	42.50	30		72.50	96.50
7330	2′-0″ wide		18	.889		42	30		72	96
7340	2′-6″ wide		18	.889		46.50	30		76.50	101

08 14 Wood Doors

08 14 33 – Stile and Rail Wood Doors

08 14 33.20 Wood Doors Residential		Crew	Daily Output	Labor-Hours	Unit	Material	2011 Bare Costs Labor	Equipment	Total	Total Incl O&P
7350	2′-8″ wide	2 Carp	18	.889	Ea.	48	30		78	103
7360	3′-0″ wide		17	.941		51	31.50		82.50	109
7420	Lauan, hollow core, 1-3/8″ x 6′-8″ x 1′-6″ wide		18	.889		31	30		61	84
7440	2′-0″ wide		18	.889		34	30		64	87
7450	2′-4″ wide		18	.889		37.50	30		67.50	91.50
7460	2′-6″ wide		18	.889		37.50	30		67.50	91.50
7480	2′-8″ wide		18	.889		39	30		69	93
7500	3′-0″ wide		17	.941		41.50	31.50		73	98.50
7700	Birch, hollow core, 1-3/8″ x 6′-8″ x 1′-6″ wide		18	.889		38.50	30		68.50	92.50
7720	2′-0″ wide		18	.889		41	30		71	95
7740	2′-6″ wide		18	.889		49	30		79	104
7760	2′-8″ wide		18	.889		49.50	30		79.50	105
7780	3′-0″ wide		17	.941		51	31.50		82.50	109
8000	Pine louvered, 1-3/8″ x 6′-8″ x 1′-6″ wide		19	.842		101	28.50		129.50	159
8020	2′-0″ wide		18	.889		128	30		158	190
8040	2′-6″ wide		18	.889		139	30		169	203
8060	2′-8″ wide		18	.889		146	30		176	211
8080	3′-0″ wide		17	.941		173	31.50		204.50	243
8300	Pine paneled, 1-3/8″ x 6′-8″ x 1′-6″ wide		19	.842		111	28.50		139.50	170
8320	2′-0″ wide		18	.889		130	30		160	193
8330	2′-4″ wide		18	.889		150	30		180	215
8340	2′-6″ wide		18	.889		158	30		188	223
8360	2′-8″ wide		18	.889		161	30		191	227
8380	3′-0″ wide		17	.941		167	31.50		198.50	236

08 14 40 – Interior Cafe Doors

08 14 40.10 Cafe Style Doors		Crew	Daily Output	Labor-Hours	Unit	Material	Labor	Equipment	Total	Total Incl O&P
0010	**CAFE STYLE DOORS**									
6520	Interior cafe doors, 2′-6″ opening, stock, panel pine	2 Carp	16	1	Ea.	191	33.50		224.50	266
6540	3′-0″ opening	″	16	1	″	199	33.50		232.50	275
6550	Louvered pine									
6560	2′-6″ opening	2 Carp	16	1	Ea.	165	33.50		198.50	238
8000	3′-0″ opening		16	1		176	33.50		209.50	250
8010	2′-6″ opening, hardwood		16	1		242	33.50		275.50	320
8020	3′-0″ opening		16	1		268	33.50		301.50	350

08 16 Composite Doors

08 16 13 – Fiberglass Doors

08 16 13.10 Entrance Doors, Fiberous Glass		Crew	Daily Output	Labor-Hours	Unit	Material	Labor	Equipment	Total	Total Incl O&P
0010	**ENTRANCE DOORS, FIBEROUS GLASS**									
0020	Exterior, fiberglass, door, 2′-8″ wide x 6′-8″ high G	2 Carp	15	1.067	Ea.	252	36		288	335
0040	3′-0″ wide x 6′-8″ high G		15	1.067		252	36		288	335
0060	3′-0″ wide x 7′-0″ high G		15	1.067		455	36		491	560
0080	3′-0″ wide x 6′-8″ high, with two lites G		15	1.067		294	36		330	385
0100	3′-0″ wide x 8′-0″ high, with two lites G		15	1.067		490	36		526	595
0110	Half glass, 3′-0″ wide x 6′-8″ high G		15	1.067		294	36		330	385
0120	3′-0″ wide x 6′-8″ high, low e G		15	1.067		325	36		361	415
0130	3′-0″ wide x 8′-0″ high G		15	1.067		580	36		616	695
0140	3′-0″ wide x 8′-0″ high, low e G		15	1.067		640	36		676	760
0150	Side lights, 1′-0″ wide x 6′-8″ high, G					247			247	271
0160	1′-0″ wide x 6′-8″ high, low e G					256			256	281

08 16 Composite Doors

08 16 13 – Fiberglass Doors

08 16 13.10 Entrance Doors, Fiberous Glass		Crew	Daily Output	Labor-Hours	Unit	2011 Bare Costs Material	Labor	Equipment	Total	Total Incl O&P
0180	1'-0" wide x 6'-8" high, full glass G				Ea.	275			275	305
0190	1'-0" wide x 6'-8" high, low e G					284			284	315

08 16 14 – French Doors

08 16 14.10 Exterior Doors With Glass Lites		Crew	Daily Output	Labor-Hours	Unit	Material	Labor	Equipment	Total	Total Incl O&P
0010	**EXTERIOR DOORS WITH GLASS LITES**									
0020	French, Fir, 1-3/4", 3'-0"wide x 6'-8" high	2 Carp	12	1.333	Ea.	600	44.50		644.50	735
0025	Double		12	1.333		1,200	44.50		1,244.50	1,400
0030	Maple, 1-3/4", 3'-0"wide x 6'-8" high		12	1.333		675	44.50		719.50	820
0035	Double		12	1.333		1,350	44.50		1,394.50	1,550
0040	Cherry, 1-3/4", 3'-0"wide x 6'-8" high		12	1.333		790	44.50		834.50	940
0045	Double		12	1.333		1,575	44.50		1,619.50	1,800
0100	Mahogany, 1-3/4", 3'-0"wide x 8'-0" high		10	1.600		700	53.50		753.50	860
0105	Double		10	1.600		1,400	53.50		1,453.50	1,650
0110	Fir, 1-3/4", 3'-0"wide x 8'-0" high		10	1.600		1,100	53.50		1,153.50	1,300
0115	Double		10	1.600		2,200	53.50		2,253.50	2,525
0120	Oak, 1-3/4", 3'-0"wide x 8'-0" high		10	1.600		1,825	53.50		1,878.50	2,100
0125	Double		10	1.600		3,650	53.50		3,703.50	4,100

08 17 Integrated Door Opening Assemblies

08 17 23 – Integrated Wood Door Opening Assemblies

08 17 23.10 Pre-Hung Doors		Crew	Daily Output	Labor-Hours	Unit	Material	Labor	Equipment	Total	Total Incl O&P
0010	**PRE-HUNG DOORS**									
0300	Exterior, wood, comb. storm & screen, 6'-9" x 2'-6" wide	2 Carp	15	1.067	Ea.	289	36		325	380
0320	2'-8" wide		15	1.067		289	36		325	380
0340	3'-0" wide		15	1.067		296	36		332	385
0360	For 7'-0" high door, add					24			24	26.50
1600	Entrance door, flush, birch, solid core									
1620	4-5/8" solid jamb, 1-3/4" x 6'-8" x 2'-8" wide	2 Carp	16	1	Ea.	285	33.50		318.50	370
1640	3'-0" wide	"	16	1		325	33.50		358.50	415
1680	For 7'-0" high door, add					21			21	23
2000	Entrance door, colonial, 6 panel pine									
2020	4-5/8" solid jamb, 1-3/4" x 6'-8" x 2'-8" wide	2 Carp	16	1	Ea.	635	33.50		668.50	755
2040	3'-0" wide	"	16	1		665	33.50		698.50	790
2060	For 7'-0" high door, add					52.50			52.50	58
2200	For 5-5/8" solid jamb, add					40.50			40.50	44.50
4000	Interior, passage door, 4-5/8" solid jamb									
4400	Lauan, flush, solid core, 1-3/8" x 6'-8" x 2'-6" wide	2 Carp	17	.941	Ea.	184	31.50		215.50	255
4420	2'-8" wide		17	.941		184	31.50		215.50	255
4440	3'-0" wide		16	1		199	33.50		232.50	275
4600	Hollow core, 1-3/8" x 6'-8" x 2'-6" wide		17	.941		124	31.50		155.50	189
4620	2'-8" wide		17	.941		124	31.50		155.50	189
4640	3'-0" wide		16	1		139	33.50		172.50	209
4700	For 7'-0" high door, add					30			30	33
5000	Birch, flush, solid core, 1-3/8" x 6'-8" x 2'-6" wide	2 Carp	17	.941		238	31.50		269.50	315
5020	2'-8" wide		17	.941		199	31.50		230.50	272
5040	3'-0" wide		16	1		270	33.50		303.50	355
5200	Hollow core, 1-3/8" x 6'-8" x 2'-6" wide		17	.941		182	31.50		213.50	253
5220	2'-8" wide		17	.941		158	31.50		189.50	226
5240	3'-0" wide		16	1		195	33.50		228.50	270
5280	For 7'-0" high door, add					25			25	27.50

08 17 Integrated Door Opening Assemblies

08 17 23 – Integrated Wood Door Opening Assemblies

08 17 23.10 Pre-Hung Doors		Crew	Daily Output	Labor-Hours	Unit	Material	2011 Bare Costs Labor	Equipment	Total	Total Incl O&P
5500	Hardboard paneled, 1-3/8" x 6'-8" x 2'-6" wide	2 Carp	17	.941	Ea.	143	31.50		174.50	211
5520	2'-8" wide		17	.941		152	31.50		183.50	220
5540	3'-0" wide		16	1		149	33.50		182.50	220
6000	Pine paneled, 1-3/8" x 6'-8" x 2'-6" wide		17	.941		247	31.50		278.50	325
6020	2'-8" wide		17	.941		267	31.50		298.50	345
6040	3'-0" wide		16	1		272	33.50		305.50	355
6500	For 5-5/8" solid jamb, add					20			20	22
6520	For split jamb, deduct					21.50			21.50	23.50

08 31 Access Doors and Panels

08 31 13 – Access Doors and Frames

08 31 13.10 Types of Framed Access Doors		Crew	Daily Output	Labor-Hours	Unit	Material	Labor	Equipment	Total	Total Incl O&P
0010	**TYPES OF FRAMED ACCESS DOORS**									
1000	Fire rated door with lock									
1100	Metal, 12" x 12"	1 Carp	10	.800	Ea.	158	27		185	219
1150	18" x 18"		9	.889		205	30		235	276
1200	24" x 24"		9	.889		325	30		355	410
1250	24" x 36"		8	1		335	33.50		368.50	420
1300	24" x 48"		8	1		410	33.50		443.50	510
1350	36" x 36"		7.50	1.067		495	36		531	605
1400	48" x 48"		7.50	1.067		635	36		671	755
1600	Stainless steel, 12" x 12"		10	.800		282	27		309	355
1650	18" x 18"		9	.889		410	30		440	500
1700	24" x 24"		9	.889		500	30		530	600
1750	24" x 36"		8	1		640	33.50		673.50	760
2000	Flush door for finishing									
2100	Metal 8" x 8"	1 Carp	10	.800	Ea.	35	27		62	83.50
2150	12" x 12"	"	10	.800	"	40	27		67	89
3000	Recessed door for acoustic tile									
3100	Metal, 12" x 12"	1 Carp	4.50	1.778	Ea.	72	59.50		131.50	179
3150	12" x 24"		4.50	1.778		90	59.50		149.50	199
3200	24" x 24"		4	2		121	67		188	245
3250	24" x 36"		4	2		170	67		237	299
4000	Recessed door for drywall									
4100	Metal 12" x 12"	1 Carp	6	1.333	Ea.	77	44.50		121.50	160
4150	12" x 24"		5.50	1.455		114	49		163	207
4200	24" x 36"		5	1.600		182	53.50		235.50	290
6000	Standard door									
6100	Metal, 8" x 8"	1 Carp	10	.800	Ea.	35	27		62	83.50
6150	12" x 12"		10	.800		40	27		67	89
6200	18" x 18"		9	.889		62	30		92	118
6250	24" x 24"		9	.889		75	30		105	133
6300	24" x 36"		8	1		115	33.50		148.50	183
6350	36" x 36"		8	1		140	33.50		173.50	210
6500	Stainless steel, 8" x 8"		10	.800		80	27		107	133
6550	12" x 12"		10	.800		107	27		134	163
6600	18" x 18"		9	.889		198	30		228	268
6650	24" x 24"		9	.889		260	30		290	335

08 31 13.20 Bulkhead/Cellar Doors		Crew	Daily Output	Labor-Hours	Unit	Material	Labor	Equipment	Total	Total Incl O&P
0010	**BULKHEAD/CELLAR DOORS**									
0020	Steel, not incl. sides, 44" x 62"	1 Carp	5.50	1.455	Ea.	490	49		539	620

08 31 Access Doors and Panels

08 31 13 – Access Doors and Frames

08 31 13.20 Bulkhead/Cellar Doors		Crew	Daily Output	Labor-Hours	Unit	Material	2011 Bare Costs Labor	Equipment	Total	Total Incl O&P
0100	52" x 73"	1 Carp	5.10	1.569	Ea.	670	52.50		722.50	825
0500	With sides and foundation plates, 57" x 45" x 24"		4.70	1.702		700	57		757	865
0600	42" x 49" x 51"	↓	4.30	1.860	↓	790	62.50		852.50	975

08 31 13.30 Commercial Floor Doors

		Crew	Daily Output	Labor-Hours	Unit	Material	Labor	Equipment	Total	Total Incl O&P
0010	**COMMERCIAL FLOOR DOORS**									
0021	Aluminum tile, steel frame, one leaf, 2' x 2' opng.	L-4	3.50	4.571	Opng.	775	136		911	1,075
0051	3'-6" x 3'-6" opening		3.50	4.571		1,475	136		1,611	1,850
0501	Double leaf, 4' x 4' opening		3	5.333		1,575	158		1,733	2,000
0551	5' x 5' opening	↓	3	5.333	↓	2,800	158		2,958	3,350

08 31 13.35 Industrial Floor Doors

		Crew	Daily Output	Labor-Hours	Unit	Material	Labor	Equipment	Total	Total Incl O&P
0010	**INDUSTRIAL FLOOR DOORS**									
0021	Steel 300 psf L.L., single leaf, 2' x 2', 175#	L-4	6	2.667	Opng.	735	79		814	940
0051	3' x 3' opening, 300#		5.50	2.909		1,000	86.50		1,086.50	1,250
0301	Double leaf, 4' x 4' opening, 455#		5	3.200		1,625	95		1,720	1,950
0351	5' x 5' opening, 645#		4.50	3.556		2,450	106		2,556	2,875
1001	Aluminum, 300 psf L.L, single leaf, 2' x 2', 60#		6	2.667		740	79		819	945
1051	3' x 3' opening, 100#		5.50	2.909		1,100	86.50		1,186.50	1,350
1501	Double leaf, 4' x 4' opening, 160#		5	3.200		1,950	95		2,045	2,300
1551	5' x 5' opening, 235#		4.50	3.556		2,675	106		2,781	3,125
2001	Aluminum, 150 psf L.L., single leaf, 2' x 2', 60#		6	2.667		610	79		689	800
2051	3' x 3' opening, 95#		5.50	2.909		920	86.50		1,006.50	1,150
2501	Double leaf, 4' x 4' opening, 150#		5	3.200		1,350	95		1,445	1,625
2551	5' x 5' opening, 230#	↓	4.50	3.556	↓	1,825	106		1,931	2,200

08 31 13.40 Kennel Doors

		Crew	Daily Output	Labor-Hours	Unit	Material	Labor	Equipment	Total	Total Incl O&P
0010	**KENNEL DOORS**									
0020	2 way, swinging type, 13" x 19" opening	2 Carp	11	1.455	Opng.	90	49		139	181
0100	17" x 29" opening		11	1.455		110	49		159	203
0200	9" x 9" opening, electronic with accessories	↓	11	1.455	↓	144	49		193	240

08 32 Sliding Glass Doors

08 32 13 – Sliding Aluminum-Framed Glass Doors

08 32 13.10 Sliding Aluminum Doors		Crew	Daily Output	Labor-Hours	Unit	Material	Labor	Equipment	Total	Total Incl O&P
0010	**SLIDING ALUMINUM DOORS**									
0350	Aluminum, 5/8" tempered insulated glass, 6' wide									
0400	Premium	2 Carp	4	4	Ea.	1,550	134		1,684	1,950
0450	Economy		4	4		885	134		1,019	1,200
0500	8' wide, premium		3	5.333		1,625	179		1,804	2,075
0550	Economy		3	5.333		1,400	179		1,579	1,825
0600	12' wide, premium		2.50	6.400		2,950	215		3,165	3,600
0650	Economy		2.50	6.400		1,375	215		1,590	1,850
0700	Non-insulated, 6' wide		4	4		1,100	134		1,234	1,425
4000	Aluminum, baked on enamel, temp glass, 6'-8" x 10'-0" wide		4	4		1,025	134		1,159	1,350
4020	Insulating glass, 6'-8" x 6'-0" wide		4	4		895	134		1,029	1,200
4040	8'-0" wide		3	5.333		1,050	179		1,229	1,450
4060	10'-0" wide		2	8		1,300	268		1,568	1,875
4080	Anodized, temp glass, 6'-8" x 6'-0" wide		4	4		450	134		584	720
4100	8'-0" wide		3	5.333		570	179		749	930
4120	10'-0" wide		2	8		645	268		913	1,150
4200	Vinyl clad, anodized, temp glass, 6'-8" x 6'-0" wide		4	4		420	134		554	685
4240	8'-0" wide	↓	3	5.333	↓	580	179		759	940

08 32 Sliding Glass Doors

08 32 13 – Sliding Aluminum-Framed Glass Doors

08 32 13.10 Sliding Aluminum Doors		Crew	Daily Output	Labor-Hours	Unit	2011 Bare Costs Material	Labor	Equipment	Total	Total Incl O&P
4260	10'-0" wide	2 Carp	2	8	Ea.	630	268		898	1,150
4280	Insulating glass, 6'-8" x 6'-0' wide		4	4		1,100	134		1,234	1,425
4300	8'-0" wide		3	5.333		920	179		1,099	1,300
4320	10'-0" wide		4	4		1,625	134		1,759	2,000

08 32 19 – Sliding Wood-Framed Glass Doors

08 32 19.15 Sliding Glass Vinyl-Clad Wood Doors		Crew	Daily Output	Labor-Hours	Unit	Material	Labor	Equipment	Total	Total Incl O&P
0010	**SLIDING GLASS VINYL-CLAD WOOD DOORS**									
0025	6'-0" x 6'-10" high G	2 Carp	4	4	Opng.	1,625	134		1,759	2,000
0100	8'-0" x 6'-10" high G		4	4		2,000	134		2,134	2,425
0500	4 leaf, 9'-0" x 6'-10" high G		3	5.333		3,175	179		3,354	3,800
0600	12'-0" x 6'-10" high G		3	5.333		3,800	179		3,979	4,475

08 33 Coiling Doors and Grilles

08 33 13 – Coiling Counter Doors

08 33 13.10 Counter Doors, Coiling Type		Crew	Daily Output	Labor-Hours	Unit	Material	Labor	Equipment	Total	Total Incl O&P
0010	**COUNTER DOORS, COILING TYPE**									
0020	Manual, incl. frm and hdwe, galv. stl., 4' roll-up, 6' long	2 Carp	2	8	Opng.	1,150	268		1,418	1,725
0300	Galvanized steel, UL label		1.80	8.889		1,200	298		1,498	1,825
0600	Stainless steel, 4' high roll-up, 6' long		2	8		2,100	268		2,368	2,750
0700	10' long		1.80	8.889		2,450	298		2,748	3,200
2000	Aluminum, 4' high, 4' long		2.20	7.273		1,325	244		1,569	1,875
2020	6' long		2	8		1,475	268		1,743	2,075
2040	8' long		1.90	8.421		1,675	283		1,958	2,325
2060	10' long		1.80	8.889		1,775	298		2,073	2,450
2080	14' long		1.40	11.429		2,500	385		2,885	3,400
2100	6' high, 4' long		2	8		1,500	268		1,768	2,100
2120	6' long		1.60	10		1,550	335		1,885	2,275
2140	10' long		1.40	11.429		2,050	385		2,435	2,900

08 33 16 – Coiling Counter Grilles

08 33 16.10 Coiling Grilles		Crew	Daily Output	Labor-Hours	Unit	Material	Labor	Equipment	Total	Total Incl O&P
0010	**COILING GRILLES**									
2021	Aluminum, manual operated, mill finish	L-4	82	.195	S.F.	27.50	5.80		33.30	39.50
2041	Bronze anodized		82	.195	"	43.50	5.80		49.30	57.50
2061	Steel, manual operated, 10' x 10' high		1	16	Opng.	2,450	475		2,925	3,500
2081	15' x 8' high		.80	20	"	2,850	595		3,445	4,125
3000	For safety edge bottom bar, electric, add				L.F.	49			49	54
3101	For motor operation, add	L-4	5	3.200	Opng.	1,250	95		1,345	1,525

08 33 23 – Overhead Coiling Doors

08 33 23.10 Rolling Service Doors		Crew	Daily Output	Labor-Hours	Unit	Material	Labor	Equipment	Total	Total Incl O&P
0010	**ROLLING SERVICE DOORS** Steel, manual, 20 ga., incl. hardware									
0051	8' x 8' high	L-4	1.60	10	Ea.	1,075	297		1,372	1,675
0101	10' x 10' high		1.40	11.429		1,775	340		2,115	2,525
0201	20' x 10' high		1	16		3,000	475		3,475	4,100
0301	12' x 12' high		1.20	13.333		1,800	395		2,195	2,650
0401	20' x 12' high		.90	17.778		2,050	530		2,580	3,150
0501	14' x 14' high		.80	20		2,950	595		3,545	4,225
0601	20' x 16' high		.60	26.667		3,325	790		4,115	4,975
0701	10' x 20' high		.50	32		2,350	950		3,300	4,175
1001	12' x 12', crank operated, crank on door side		.80	20		1,600	595		2,195	2,750

08 33 Coiling Doors and Grilles

08 33 23 – Overhead Coiling Doors

08 33 23.10 Rolling Service Doors		Crew	Daily Output	Labor-Hours	Unit	Material	2011 Bare Costs Labor	Equipment	Total	Total Incl O&P
1101	Crank thru wall	L-4	.70	22.857	Ea.	1,850	680		2,530	3,150
1300	For vision panel, add					315			315	350
1400	For 22 ga., deduct				S.F.	1.10			1.10	1.21
1600	3′ x 7′ pass door within rolling steel door, new construction				Ea.	1,725			1,725	1,900
1701	Existing construction	L-4	2	8		1,825	238		2,063	2,400
2001	Class A fire doors, manual, 20 ga., 8′ x 8′ high		1.40	11.429		1,425	340		1,765	2,150
2101	10′ x 10′ high		1.10	14.545		1,950	430		2,380	2,875
2201	20′ x 10′ high		.80	20		3,975	595		4,570	5,375
2301	12′ x 12′ high		1	16		3,050	475		3,525	4,175
2401	20′ x 12′ high		.80	20		4,325	595		4,920	5,750
2501	14′ x 14′ high		.60	26.667		3,350	790		4,140	5,025
2601	20′ x 16′ high		.50	32		5,400	950		6,350	7,550
2701	10′ x 20′ high		.40	40		4,175	1,200		5,375	6,575
3000	For 18 ga. doors, add				S.F.	1.40			1.40	1.54
3300	For enamel finish, add				"	1.65			1.65	1.82
3600	For safety edge bottom bar, pneumatic, add				L.F.	20.50			20.50	22.50
3700	Electric, add					38.50			38.50	42
4000	For weatherstripping, extruded rubber, jambs, add					12.90			12.90	14.20
4100	Hood, add					7.90			7.90	8.70
4200	Sill, add					4.55			4.55	5
4501	Motor operators, to 14′ x 14′ opening	L-4	5	3.200	Ea.	1,025	95		1,120	1,300
4601	Over 14′ x 14′, jack shaft type	"	5	3.200		1,075	95		1,170	1,350
4700	For fire door, additional fusible link, add					25			25	27.50

08 34 Special Function Doors

08 34 13 – Cold Storage Doors

08 34 13.10 Doors for Cold Area Storage

Code	Description	Crew	Daily Output	Labor-Hours	Unit	Material	Labor	Equipment	Total	Total Incl O&P
0010	**DOORS FOR COLD AREA STORAGE**									
0020	Single, 20 ga. galvanized steel									
0300	Horizontal sliding, 5′ x 7′, manual operation, 3.5″ thick	2 Carp	2	8	Ea.	3,175	268		3,443	3,925
0400	4″ thick		2	8		3,150	268		3,418	3,900
0500	6″ thick		2	8		3,300	268		3,568	4,075
0800	5′ x 7′, power operation, 2″ thick		1.90	8.421		5,375	283		5,658	6,375
0900	4″ thick		1.90	8.421		5,500	283		5,783	6,525
1000	6″ thick		1.90	8.421		6,275	283		6,558	7,375
1300	9′ x 10′, manual operation, 2″ insulation		1.70	9.412		4,375	315		4,690	5,325
1400	4″ insulation		1.70	9.412		4,475	315		4,790	5,450
1500	6″ insulation		1.70	9.412		5,375	315		5,690	6,450
1800	Power operation, 2″ insulation		1.60	10		7,400	335		7,735	8,700
1900	4″ insulation		1.60	10		7,650	335		7,985	8,950
2000	6″ insulation		1.70	9.412		8,600	315		8,915	9,975
2300	For stainless steel face, add					22%				
3001	Hinged, lightweight, 3′ x 7′-0″, galvanized, 2″ thick	2 Carp	2	8	Ea.	1,375	268		1,643	1,950
3051	4″ thick		1.90	8.421		1,700	283		1,983	2,350
3301	Aluminum doors, 3′ x 7′-0″, 4″ thick		1.90	8.421		1,300	283		1,583	1,900
3351	6″ thick		1.40	11.429		2,275	385		2,660	3,150
3601	Stainless steel, 3′ x 7′-0″, 4″ thick		1.90	8.421		1,675	283		1,958	2,300
3651	6″ thick		1.40	11.429		2,850	385		3,235	3,775
3901	Painted, 3′ x 7′-0″, 4″ thick		1.90	8.421		1,225	283		1,508	1,825
3951	6″ thick		1.40	11.429		2,275	385		2,660	3,150
5000	Bi-parting, electric operated									

08 34 Special Function Doors

08 34 13 – Cold Storage Doors

08 34 13.10 Doors for Cold Area Storage		Crew	Daily Output	Labor-Hours	Unit	Material	2011 Bare Costs Labor	Equipment	Total	Total Incl O&P
5011	6′ x 8′ opening, galv. faces, 4″ thick for cooler	2 Carp	.80	20	Opng.	7,000	670		7,670	8,825
5051	For freezer		.80	20		7,725	670		8,395	9,625
5301	For door buck framing and door protection, add		2.50	6.400		620	215		835	1,050
6001	Galvanized batten door, galvanized hinges, 4′ x 7′		2	8		1,750	268		2,018	2,375
6051	6′ x 8′		1.80	8.889		2,400	298		2,698	3,125
6501	Fire door, 3 hr., 6′ x 8′, single slide		.80	20		8,025	670		8,695	9,950
6551	Double, bi-parting		.70	22.857		12,200	765		12,965	14,700

08 34 16 – Hangar Doors

08 34 16.10 Aircraft Hangar Doors		Crew	Daily Output	Labor-Hours	Unit	Material	Labor	Equipment	Total	Total Incl O&P
0010	**AIRCRAFT HANGAR DOORS**									
0020	Bi-fold, ovhd., 20 PSF wind load, incl. elec. oper.									
0101	12′ high x 40′	L-4	240	.067	S.F.	15.45	1.98		17.43	20.50
0201	16′ high x 60′		230	.070		18.50	2.07		20.57	24
0301	20′ high x 80′		220	.073		20	2.16		22.16	25.50

08 34 36 – Darkroom Doors

08 34 36.10 Various Types of Darkroom Doors		Crew	Daily Output	Labor-Hours	Unit	Material	Labor	Equipment	Total	Total Incl O&P
0010	**VARIOUS TYPES OF DARKROOM DOORS**									
0015	Revolving, standard, 2 way, 36″ diameter	2 Carp	3.10	5.161	Opng.	2,250	173		2,423	2,775
0021	41″ diameter		3.10	5.161		2,700	173		2,873	3,275
0051	3 way, 51″ diameter		1.40	11.429		3,275	385		3,660	4,250
1001	4 way, 49″ diameter		1.40	11.429		3,900	385		4,285	4,925
2001	Hinged safety, 2 way, 41″ diameter		2.30	6.957		3,050	233		3,283	3,750
2501	3 way, 51″ diameter		1.40	11.429		3,900	385		4,285	4,950
3001	Pop out safety, 2 way, 41″ diameter		3.10	5.161		3,600	173		3,773	4,250
4001	3 way, 51″ diameter		1.40	11.429		3,550	385		3,935	4,550
5001	Wheelchair-type, pop out, 51″ diameter		1.40	11.429		4,125	385		4,510	5,200
9300	For complete darkrooms, see Section 13 21 53.50									

08 34 53 – Security Doors and Frames

08 34 53.20 Steel Door		Crew	Daily Output	Labor-Hours	Unit	Material	Labor	Equipment	Total	Total Incl O&P
0010	**STEEL DOOR** with ballistic core and welded frame both 14 gauge									
0050	Flush, UL 752 Level 3, 1-3/4″, 3′-0″ x 6′-8″	2 Carp	1.50	10.667	Opng.	2,150	360		2,510	2,950
0055	1-3/4″, 3′-6″ x 6′-8″		1.50	10.667		2,250	360		2,610	3,075
0060	1-3/4″, 4′-0″ x 6′-8″		1.20	13.333		2,425	445		2,870	3,425
0100	UL 752 Level 8, 1-3/4″, 3′-0″ x 6′-8″		1.50	10.667		9,475	360		9,835	11,000
0105	1-3/4″, 3′-6″ x 6′-8″		1.50	10.667		9,575	360		9,935	11,100
0110	1-3/4″, 4′-0″ x 6′-8″		1.20	13.333		10,900	445		11,345	12,800
0120	UL 752 Level 3, 1-3/4″, 3′-0″ x 7′-0″		1.50	10.667		2,175	360		2,535	2,975
0125	1-3/4″, 3′-6″ x 7′-0″		1.50	10.667		2,275	360		2,635	3,100
0130	1-3/4″, 4′-0″ x 7′-0″		1.20	13.333		2,475	445		2,920	3,450
0150	UL 752 Level 8, 1-3/4″, 3′-0″ x 7′-0″		1.50	10.667		9,500	360		9,860	11,100
0155	1-3/4″, 3′-6″ x 7′-0″		1.50	10.667		9,600	360		9,960	11,200
0160	1-3/4″, 4′-0″ x 7′-0″		1.20	13.333		10,900	445		11,345	12,800
1000	Safe Room sliding door and hardware, 1-3/4″, 3′-0″ x 7′-0″ UL 752 Level 3		.50	32		22,700	1,075		23,775	26,700
1050	Safe Room swinging door and hardware, 1-3/4″, 3′-0″ x 7′-0″ UL 752 Level 3		.50	32		26,800	1,075		27,875	31,300

08 34 53.30 Wood Ballistic Doors		Crew	Daily Output	Labor-Hours	Unit	Material	Labor	Equipment	Total	Total Incl O&P
0010	**WOOD BALLISTIC DOORS** with frames and hardware									
0050	Wood, 1-3/4″, 3′-0″ x 7′-0″ UL 752 Level 3	2 Carp	1.50	10.667	Opng.	2,050	360		2,410	2,875

08 34 Special Function Doors

08 34 59 – Vault Doors and Day Gates

08 34 59.10 Secure Storage Doors

	08 34 59.10 Secure Storage Doors	Crew	Daily Output	Labor-Hours	Unit	2011 Bare Costs Material	Labor	Equipment	Total	Total Incl O&P
0010	**SECURE STORAGE DOORS**									
0020	Door and frame, 32" x 78", clear opening									
0101	1 hour test, weighs 750 lb.	L-4	1.50	10.667	Opng.	2,750	315		3,065	3,550
0201	2 hour test, 32" door, weighs 950 lb.		1.30	12.308		2,950	365		3,315	3,850
0251	40" door, weighs 1130 lb.		1	16		3,350	475		3,825	4,475
0301	4 hour test, 32" door, weighs 1025 lb.		1.20	13.333		3,250	395		3,645	4,225
0351	40" door, weighs 1140 lb.		.90	17.778		3,775	530		4,305	5,025
0500	For stainless steel front, including frame, add to above					1,900			1,900	2,100
0550	Back, add					1,900			1,900	2,100
0600	For time lock, two movement, add	1 Elec	2	4	Ea.	1,525	161		1,686	1,925
0650	Three movement, add	"	2	4		2,200	161		2,361	2,675
0801	Day gate, painted, wire mesh, 32" wide	L-4	1.50	10.667		1,850	315		2,165	2,575
0851	40" wide		1.40	11.429		2,050	340		2,390	2,825
0901	Aluminum, 32" wide		1.50	10.667		2,950	315		3,265	3,775
0951	40" wide		1.40	11.429		3,275	340		3,615	4,175
2051	Security vault door, class I	E-24	.19	166	Opng.	15,200	6,000	3,900	25,100	32,100
2101	Class II		.19	166		17,700	6,000	3,900	27,600	34,800
2151	Class III		.13	250		22,200	9,000	5,850	37,050	47,400

08 34 63 – Detention Doors and Frames

08 34 63.13 Steel Detention Doors and Frames

	08 34 63.13 Steel Detention Doors and Frames	Crew	Daily Output	Labor-Hours	Unit	Material	Labor	Equipment	Total	Total Incl O&P
0010	**STEEL DETENTION DOORS AND FRAMES**									
0500	Rolling cell door, bar front, 7/8" bars, 4" O.C., 7' H, 5' W, with hardware G	E-4	2	16	Ea.	5,025	590	54.50	5,669.50	6,700
0550	Actuator for rolling cell door, bar front	2 Skwk	2	8		4,200	272		4,472	5,075
1000	Doors & frames, 3' x 7', complete, with hardware, single plate	E-4	4	8		4,200	295	27.50	4,522.50	5,225
1650	Double plate	"	4	8		5,175	295	27.50	5,497.50	6,300

08 34 73 – Sound Control Door Assemblies

08 34 73.10 Acoustical Doors

	08 34 73.10 Acoustical Doors	Crew	Daily Output	Labor-Hours	Unit	Material	Labor	Equipment	Total	Total Incl O&P
0010	**ACOUSTICAL DOORS**									
0020	Including framed seals, 3' x 7', wood, 27 STC rating	2 Carp	1.50	10.667	Ea.	900	360		1,260	1,600
0100	Steel, 40 STC rating		1.50	10.667		2,900	360		3,260	3,775
0200	45 STC rating		1.50	10.667		3,400	360		3,760	4,350
0300	48 STC rating		1.50	10.667		3,925	360		4,285	4,900
0400	52 STC rating		1.50	10.667		4,425	360		4,785	5,475

08 36 Panel Doors

08 36 13 – Sectional Doors

08 36 13.10 Overhead Commercial Doors

	08 36 13.10 Overhead Commercial Doors	Crew	Daily Output	Labor-Hours	Unit	Material	Labor	Equipment	Total	Total Incl O&P
0010	**OVERHEAD COMMERCIAL DOORS**									
1000	Stock, sectional, heavy duty, wood, 1-3/4" thick, 8' x 8' high	2 Carp	2	8	Ea.	810	268		1,078	1,350
1100	10' x 10' high		1.80	8.889		1,175	298		1,473	1,800
1200	12' x 12' high		1.50	10.667		1,700	360		2,060	2,450
1300	Chain hoist, 14' x 14' high		1.30	12.308		2,600	415		3,015	3,575
1400	12' x 16' high		1	16		2,575	535		3,110	3,725
1500	20' x 8' high		1.30	12.270		2,325	410		2,735	3,275
1600	20' x 16' high		.65	24.615		4,700	825		5,525	6,550
1800	Center mullion openings, 8' high		4	4		1,200	134		1,334	1,550
1900	20' high		2	8		2,000	268		2,268	2,650
2100	For medium duty custom door, deduct					5%	5%			
2150	For medium duty stock doors, deduct					10%	5%			

08 36 Panel Doors

08 36 13 – Sectional Doors

08 36 13.10 Overhead Commercial Doors		Crew	Daily Output	Labor-Hours	Unit	2011 Bare Costs Material	Labor	Equipment	Total	Total Incl O&P
2300	Fiberglass and aluminum, heavy duty, sectional, 12′ x 12′ high	2 Carp	1.50	10.667	Ea.	2,625	360		2,985	3,475
2450	Chain hoist, 20′ x 20′ high		.50	32		6,550	1,075		7,625	9,000
2600	Steel, 24 ga. sectional, manual, 8′ x 8′ high		2	8		805	268		1,073	1,325
2650	10′ x 10′ high		1.80	8.889		1,025	298		1,323	1,625
2700	12′ x 12′ high		1.50	10.667		1,250	360		1,610	1,975
2800	Chain hoist, 20′ x 14′ high		.70	22.857		3,450	765		4,215	5,075
2850	For 1-1/4″ rigid insulation and 26 ga. galv.									
2860	back panel, add				S.F.	4.50			4.50	4.95
2900	For electric trolley operator, 1/3 H.P., to 12′ x 12′, add	1 Carp	2	4	Ea.	905	134		1,039	1,225
2950	Over 12′ x 12′, 1/2 H.P., add	″	1	8	″	1,050	268		1,318	1,600

08 36 13.20 Residential Garage Doors		Crew	Daily Output	Labor-Hours	Unit	Material	Labor	Equipment	Total	Total Incl O&P
0010	**RESIDENTIAL GARAGE DOORS**									
0051	Hinged, wood, custom, double door, 9′ x 7′	2 Carp	3	5.333	Ea.	640	179		819	1,000
0071	16′ x 7′		2	8		1,000	268		1,268	1,550
0201	Overhead, sectional, incl. hardware, fiberglass, 9′ x 7′, standard		8	2		855	67		922	1,050
0221	Deluxe		8	2		1,075	67		1,142	1,300
0301	16′ x 7′, standard		6	2.667		1,550	89.50		1,639.50	1,850
0321	Deluxe		6	2.667		2,100	89.50		2,189.50	2,450
0501	Hardboard, 9′ x 7′, standard		8	2		575	67		642	745
0521	Deluxe		8	2		800	67		867	990
0601	16′ x 7′, standard		6	2.667		1,200	89.50		1,289.50	1,475
0621	Deluxe		6	2.667		1,400	89.50		1,489.50	1,700
0701	Metal, 9′ x 7′, standard		8	2		625	67		692	800
0721	Deluxe		8	2		860	67		927	1,050
0801	16′ x 7′, standard		6	2.667		760	89.50		849.50	985
0821	Deluxe		6	2.667		1,375	89.50		1,464.50	1,650
0901	Wood, 9′ x 7′, standard		8	2		900	67		967	1,100
0921	Deluxe		8	2		2,100	67		2,167	2,400
1001	16′ x 7′, standard		6	2.667		1,600	89.50		1,689.50	1,900
1021	Deluxe		6	2.667		2,975	89.50		3,064.50	3,425
1800	Door hardware, sectional	1 Carp	4	2		320	67		387	460
1810	Door tracks only		4	2		139	67		206	265
1820	One side only		7	1.143		110	38.50		148.50	185
3001	Swing-up, including hardware, fiberglass, 9′ x 7′, standard	2 Carp	8	2		980	67		1,047	1,175
3021	Deluxe		8	2		1,000	67		1,067	1,200
3101	16′ x 7′ standard		6	2.667		1,225	89.50		1,314.50	1,500
3121	Deluxe		6	2.667		1,500	89.50		1,589.50	1,800
3201	Hardboard, 9′ x 7′, standard		8	2		500	67		567	660
3221	Deluxe		8	2		600	67		667	770
3301	16′ x 7′, standard		6	2.667		620	89.50		709.50	830
3321	Deluxe		6	2.667		800	89.50		889.50	1,025
3401	Metal, 9′ x 7′, standard		8	2		420	67		487	575
3421	Deluxe		8	2		900	67		967	1,100
3501	16′ x 7′, standard		6	2.667		685	89.50		774.50	905
3522	Deluxe		6	2.667		1,000	89.50		1,089.50	1,250
3601	Wood, 9′ x 7′, standard		8	2		600	67		667	770
3621	Deluxe		8	2		1,025	67		1,092	1,225
3701	16′ x 7′, standard		6	2.667		800	89.50		889.50	1,025
3721	Deluxe		6	2.667		2,000	89.50		2,089.50	2,350
3900	Door hardware only, swing up	1 Carp	4	2		150	67		217	277
3920	One side only		7	1.143		80	38.50		118.50	152
4000	For electric operator, economy, add		8	1		410	33.50		443.50	505

08 36 Panel Doors

08 36 13 – Sectional Doors

08 36 13.20 Residential Garage Doors		Crew	Daily Output	Labor-Hours	Unit	2011 Bare Costs Material	Labor	Equipment	Total	Total Incl O&P
4100	Deluxe, including remote control	1 Carp	8	1	Ea.	600	33.50		633.50	715
4500	For transmitter/receiver control , add to operator				Total	110			110	121
4600	Transmitters, additional				"	55			55	60.50

08 36 19 – Multi-Leaf Vertical Lift Doors

08 36 19.10 Sectional Vertical Lift Doors

		Crew	Daily Output	Labor-Hours	Unit	Material	Labor	Equipment	Total	Total Incl O&P
0010	SECTIONAL VERTICAL LIFT DOORS									
0020	Motorized, 14 ga. stl, incl., frm and ctrl pnl									
0050	16′ x 16′ high	L-10	.50	48	Ea.	21,500	1,775	1,300	24,575	28,400
0100	10′ x 20′ high		1.30	18.462		31,500	680	505	32,685	36,500
0120	15′ x 20′ high		1.30	18.462		39,000	680	505	40,185	44,700
0140	20′ x 20′ high		1	24		45,400	885	655	46,940	52,500
0160	25′ x 20′ high		1	24		52,000	885	655	53,540	59,500
0170	32′ x 24′ high		.75	32		46,400	1,175	875	48,450	54,000
0180	20′ x 25′ high		1	24		53,000	885	655	54,540	61,000
0200	25′ x 25′ high		.70	34.286		62,000	1,275	935	64,210	72,000
0220	25′ x 30′ high		.70	34.286		67,000	1,275	935	69,210	77,500
0240	30′ x 30′ high		.70	34.286		78,000	1,275	935	80,210	89,500
0260	35′ x 30′ high		.70	34.286		88,000	1,275	935	90,210	100,500

08 36 23 – Telescoping Vertical Lift Doors

08 36 23.10 Telescoping Steel Doors

		Crew	Daily Output	Labor-Hours	Unit	Material	Labor	Equipment	Total	Total Incl O&P
0010	TELESCOPING STEEL DOORS									
1000	Overhead, .03″ thick, electric operated, 10′ x 10′	E-3	.80	30	Ea.	14,400	1,100	136	15,636	18,100
2000	20′ x 10′		.60	40		18,500	1,475	182	20,157	23,400
3000	20′ x 16′		.40	60		19,000	2,225	272	21,497	25,500

08 38 Traffic Doors

08 38 13 – Flexible Strip Doors

08 38 13.10 Flexible Transparent Strip Doors

		Crew	Daily Output	Labor-Hours	Unit	Material	Labor	Equipment	Total	Total Incl O&P
0010	FLEXIBLE TRANSPARENT STRIP DOORS									
0100	12″ strip width, 2/3 overlap	3 Shee	135	.178	SF Surf	7.50	6.80		14.30	19.45
0200	Full overlap		115	.209		9.40	7.95		17.35	23.50
0220	8″ strip width, 1/2 overlap		140	.171		6	6.55		12.55	17.40
0240	Full overlap		120	.200		7.60	7.65		15.25	21
0300	Add for suspension system, header mount				L.F.	8.90			8.90	9.80
0400	Wall mount				"	9			9	9.90

08 38 19 – Rigid Traffic Doors

08 38 19.20 Double Acting Swing Doors

		Crew	Daily Output	Labor-Hours	Unit	Material	Labor	Equipment	Total	Total Incl O&P
0010	DOUBLE ACTING SWING DOORS									
1000	.063″ aluminum, 7′-0″ high, 4′-0″ wide	2 Carp	4.20	3.810	Pr.	1,900	128		2,028	2,325
1050	6′-8″ wide	"	4	4	"	2,300	134		2,434	2,750
2000	Solid core wood, 3/4″ thick, metal frame, stainless steel									
2010	base plate, 7′ high opening, 4′ wide	2 Carp	4	4	Pr.	2,200	134		2,334	2,650
2050	7′ wide	"	3.80	4.211	"	2,500	141		2,641	2,975

08 38 19.30 Shock Absorbing Doors

		Crew	Daily Output	Labor-Hours	Unit	Material	Labor	Equipment	Total	Total Incl O&P
0010	SHOCK ABSORBING DOORS									
0021	Rigid, no frame, insulated, 1-13/16″ thick, 5′ x 7′	L-4	1.90	8.421	Opng.	1,400	250		1,650	1,975
0101	8′ x 8′		1.80	8.889		2,000	264		2,264	2,650
0501	Flexible, frame not incl., 5′ x 7′ opening, economy		2	8		1,750	238		1,988	2,325
0601	Deluxe		1.90	8.421		2,625	250		2,875	3,300

08 38 Traffic Doors

08 38 19 – Rigid Traffic Doors

08 38 19.30 Shock Absorbing Doors		Crew	Daily Output	Labor-Hours	Unit	Material	2011 Bare Costs Labor	Equipment	Total	Total Incl O&P
1001	8′ x 8′ opening, economy	L-4	2	8	Opng.	2,750	238		2,988	3,425
1101	Deluxe		1.90	8.421		3,500	250		3,750	4,275

08 41 Entrances and Storefronts

08 41 13 – Aluminum-Framed Entrances and Storefronts

08 41 13.20 Tube Framing

		Crew	Daily Output	Labor-Hours	Unit	Material	Labor	Equipment	Total	Total Incl O&P
0010	**TUBE FRAMING**, For window walls and store fronts, aluminum stock									
0050	Plain tube frame, mill finish, 1-3/4″ x 1-3/4″	2 Glaz	103	.155	L.F.	9.05	5.15		14.20	18.40
0150	1-3/4″ x 4″		98	.163		12.50	5.40		17.90	22.50
0200	1-3/4″ x 4-1/2″		95	.168		14.35	5.60		19.95	25
0250	2″ x 6″		89	.180		21.50	5.95		27.45	33.50
0350	4″ x 4″		87	.184		24.50	6.10		30.60	36.50
0400	4-1/2″ x 4-1/2″		85	.188		25.50	6.25		31.75	38.50
0450	Glass bead		240	.067		2.73	2.21		4.94	6.65
1000	Flush tube frame, mill finish, 1/4″ glass, 1-3/4″ x 4″, open header		80	.200		12.05	6.65		18.70	24
1050	Open sill		82	.195		9.75	6.50		16.25	21.50
1100	Closed back header		83	.193		17.20	6.40		23.60	29.50
1150	Closed back sill		85	.188		16.40	6.25		22.65	28.50
1160	Tube fmg, spandrel cover both sides, alum 1″ wide	1 Sswk	85	.094	S.F.	91.50	3.43		94.93	107
1170	Tube fmg, spandrel cover both sides, alum 2″ wide	″	85	.094	″	35	3.43		38.43	45
1200	Vertical mullion, one piece	2 Glaz	75	.213	L.F.	18	7.10		25.10	31.50
1250	Two piece		73	.219		19.15	7.30		26.45	33
1300	90° or 180° vertical corner post		75	.213		29	7.10		36.10	43.50
1400	1-3/4″ x 4-1/2″, open header		80	.200		14.60	6.65		21.25	27
1450	Open sill		82	.195		12.20	6.50		18.70	24
1500	Closed back header		83	.193		17.45	6.40		23.85	29.50
1550	Closed back sill		85	.188		17.15	6.25		23.40	29
1600	Vertical mullion, one piece		75	.213		19.20	7.10		26.30	32.50
1650	Two piece		73	.219		20	7.30		27.30	34.50
1700	90° or 180° vertical corner post		75	.213		20.50	7.10		27.60	34.50
2000	Flush tube frame, mil fin.,ins. glass w/thml brk, 2″ x 4-1/2″, open header		75	.213		15	7.10		22.10	28
2050	Open sill		77	.208		12.60	6.90		19.50	25
2100	Closed back header		78	.205		14.15	6.80		20.95	27
2150	Closed back sill		80	.200		15.15	6.65		21.80	27.50
2200	Vertical mullion, one piece		70	.229		15.70	7.60		23.30	30
2250	Two piece		68	.235		16.95	7.80		24.75	31.50
2300	90° or 180° vertical corner post		70	.229		16.20	7.60		23.80	30.50
5000	Flush tube frame, mill fin., thermal brk., 2-1/4″ x 4-1/2″, open header		74	.216		15.60	7.20		22.80	29
5050	Open sill		75	.213		13.90	7.10		21	27
5100	Vertical mullion, one piece		69	.232		17.05	7.70		24.75	31.50
5150	Two piece		67	.239		19.75	7.95		27.70	34.50
5200	90° or 180° vertical corner post		69	.232		17.40	7.70		25.10	32
6980	Door stop (snap in)		380	.042		3	1.40		4.40	5.60
7000	For joints, 90°, clip type, add				Ea.	23.50			23.50	26
7050	Screw spline joint, add					22			22	24
7100	For joint other than 90°, add					45.50			45.50	50
8000	For bronze anodized aluminum, add					15%				
8020	For black finish, add					27%				
8050	For stainless steel materials, add					350%				
8100	For monumental grade, add					52%				
8150	For steel stiffener, add	2 Glaz	200	.080	L.F.	10.50	2.66		13.16	15.90

08 41 Entrances and Storefronts

08 41 13 – Aluminum-Framed Entrances and Storefronts

08 41 13.20 Tube Framing		Crew	Daily Output	Labor-Hours	Unit	2011 Bare Costs Material	Labor	Equipment	Total	Total Incl O&P
8200	For 2 to 5 stories, add per story				Story		6%			

08 41 19 – Stainless-Steel-Framed Entrances and Storefronts

08 41 19.10 Stainless-Steel and Glass Entrance Unit

		Crew	Daily Output	Labor-Hours	Unit	Material	Labor	Equipment	Total	Total Incl O&P
0010	**STAINLESS-STEEL AND GLASS ENTRANCE UNIT**, narrow stiles									
0021	3′ x 7′ opening, including hardware, minimum	L-4	1.60	10	Opng.	6,400	297		6,697	7,550
0051	Average		1.40	11.429		6,800	340		7,140	8,050
0101	Maximum		1.20	13.333		7,350	395		7,745	8,725
1000	For solid bronze entrance units, statuary finish, add					62%				
1100	Without statuary finish, add					45%				
2001	Balanced doors, 3′ x 7′, economy	L-4	.90	17.778	Ea.	8,600	530		9,130	10,300
2101	Premium	"	.70	22.857	"	14,800	680		15,480	17,400

08 41 26 – All-Glass Entrances and Storefronts

08 41 26.10 Window Walls Aluminum, Stock

		Crew	Daily Output	Labor-Hours	Unit	Material	Labor	Equipment	Total	Total Incl O&P
0010	**WINDOW WALLS ALUMINUM, STOCK**, including glazing									
0020	Minimum	H-2	160	.150	S.F.	42	4.64		46.64	54
0050	Average		140	.171		58	5.30		63.30	72.50
0100	Maximum		110	.218		162	6.75		168.75	189
0500	For translucent sandwich wall systems, see Section 07 41 33.10									
0850	Cost of the above walls depends on material,									
0860	finish, repetition, and size of units.									
0870	The larger the opening, the lower the S.F. cost									
1200	Double glazed acoustical window wall for airports,									
1220	including 1″ thick glass with 2″ x 4-1/2″ tube frame	H-2	40	.600	S.F.	112	18.55		130.55	154

08 42 Entrances

08 42 26 – All-Glass Entrances

08 42 26.10 Swinging Glass Doors

		Crew	Daily Output	Labor-Hours	Unit	Material	Labor	Equipment	Total	Total Incl O&P
0010	**SWINGING GLASS DOORS**									
0020	Including hardware, 1/2″ thick, tempered, 3′ x 7′ opening	2 Glaz	2	8	Opng.	2,100	266		2,366	2,750
0100	6′ x 7′ opening	"	1.40	11.429	"	4,025	380		4,405	5,050

08 42 29 – Automatic Entrances

08 42 29.23 Sliding Automatic Entrances

		Crew	Daily Output	Labor-Hours	Unit	Material	Labor	Equipment	Total	Total Incl O&P
0010	**SLIDING AUTOMATIC ENTRANCES** 12′ x 7′-6″ opng., 5′ x 7′ door, 2 way traffic									
0020	Mat or electronic activated, panic pushout, incl. operator & hardware,									
0030	not including glass or glazing	2 Glaz	.70	22.857	Opng.	8,150	760		8,910	10,200

08 42 33 – Revolving Door Entrances

08 42 33.10 Circular Rotating Entrance Doors

		Crew	Daily Output	Labor-Hours	Unit	Material	Labor	Equipment	Total	Total Incl O&P
0010	**CIRCULAR ROTATING ENTRANCE DOORS**, Aluminum									
0021	6′-10″ to 7′ high, stock units, minimum	L-4	.38	42.105	Opng.	19,100	1,250		20,350	23,100
0051	Average		.30	53.333		23,100	1,575		24,675	28,100
0101	Maximum		.23	69.565		39,700	2,075		41,775	47,100
1001	Stainless steel		.15	106		39,000	3,175		42,175	48,200
1101	Solid bronze		.07	228		46,000	6,800		52,800	62,000
1500	For automatic controls, add	2 Elec	2	8		14,500	320		14,820	16,400

08 42 36 – Balanced Door Entrances

08 42 36.10 Balanced Entrance Doors

		Crew	Daily Output	Labor-Hours	Unit	Material	Labor	Equipment	Total	Total Incl O&P
0010	**BALANCED ENTRANCE DOORS**									
0021	Hardware & frame, alum. & glass, 3′ x 7′, econ.	L-4	.90	17.778	Ea.	6,050	530		6,580	7,525

08 42 Entrances

08 42 36 – Balanced Door Entrances

08 42 36.10 Balanced Entrance Doors		Crew	Daily Output	Labor-Hours	Unit	Material	2011 Bare Costs Labor	Equipment	Total	Total Incl O&P
0151	Premium	L-4	.70	22.857	Ea.	7,400	680		8,080	9,275

08 43 Storefronts

08 43 13 – Aluminum-Framed Storefronts

08 43 13.10 Aluminum-Framed Entrance Doors		Crew	Daily Output	Labor-Hours	Unit	Material	Labor	Equipment	Total	Total Incl O&P
0010	**ALUMINUM-FRAMED ENTRANCE DOORS**									
0015	Standard hardware and glass stops but no glazing									
0021	Entrance, 3′ x 7′ opening, clear finish	2 Sswk	7	2.286	Opng.	355	83		438	555
0040	Bronze finish		7	2.286		375	83		458	575
0060	Black finish		7	2.286		420	83		503	620
0500	6′ x 7′ opening, clear finish		6	2.667		720	97		817	975
0520	Bronze finish		6	2.667		755	97		852	1,025
0540	Black finish		6	2.667		845	97		942	1,100
1001	With 3′ high transoms, 3′ x 10′ opening, clear finish		5.50	2.909		415	106		521	660
1051	Bronze finish		5.50	2.909		440	106		546	685
1101	Black finish		5.50	2.909		475	106		581	720
1501	With 3′ high transoms, 6′ x 10′ opening, clear finish		5.50	2.909		505	106		611	755
1551	Bronze finish		5.50	2.909		530	106		636	785
1601	Black finish	↓	5.50	2.909	↓	590	106		696	845

08 43 13.20 Storefront Systems		Crew	Daily Output	Labor-Hours	Unit	Material	Labor	Equipment	Total	Total Incl O&P
0010	**STOREFRONT SYSTEMS**, aluminum frame clear 3/8″ plate glass									
0020	incl. 3′ x 7′ door with hardware (400 sq. ft. max. wall)									
0500	Wall height to 12′ high, commercial grade	2 Glaz	150	.107	S.F.	20	3.54		23.54	28
0600	Institutional grade		130	.123		25	4.09		29.09	34
0700	Monumental grade		115	.139		38	4.62		42.62	49
1000	6′ x 7′ door with hardware, commercial grade		135	.119		20	3.93		23.93	28.50
1100	Institutional grade		115	.139		28	4.62		32.62	38
1200	Monumental grade	↓	100	.160		51	5.30		56.30	65
1500	For bronze anodized finish, add					15%				
1600	For black anodized finish, add					35%				
1700	For stainless steel framing, add to monumental				↓	76%				

08 43 29 – Sliding Storefronts

08 43 29.10 Sliding Panels		Crew	Daily Output	Labor-Hours	Unit	Material	Labor	Equipment	Total	Total Incl O&P
0010	**SLIDING PANELS**									
0020	Mall fronts, aluminum & glass, 15′ x 9′ high	2 Glaz	1.30	12.308	Opng.	3,250	410		3,660	4,250
0100	24′ x 9′ high		.70	22.857		4,625	760		5,385	6,325
0200	48′ x 9′ high, with fixed panels	↓	.90	17.778		8,600	590		9,190	10,400
0500	For bronze finish, add				↓	17%				

08 44 Curtain Wall and Glazed Assemblies

08 44 13 – Glazed Aluminum Curtain Walls

08 44 13.10 Glazed Curtain Walls		Crew	Daily Output	Labor-Hours	Unit	Material	Labor	Equipment	Total	Total Incl O&P
0010	**GLAZED CURTAIN WALLS**, aluminum, stock, including glazing									
0020	Minimum	H-1	205	.156	S.F.	33.50	5.45		38.95	46
0050	Average, single glazed		195	.164		48	5.70		53.70	63
0150	Average, double glazed		180	.178		62	6.20		68.20	79.50
0200	Maximum	↓	160	.200	↓	169	6.95		175.95	198

08 45 Translucent Wall and Roof Assemblies

08 45 10 – Translucent Roof Assemblies

08 45 10.10 Skyroofs		Crew	Daily Output	Labor-Hours	Unit	Material	2011 Bare Costs Labor	Equipment	Total	Total Incl O&P
0010	**SKYROOFS**, Translucent panels, 2-3/4" thick									
0020	Under 5000 S.F.	G-3	395	.081	SF Hor.	33	2.62		35.62	41
0100	Over 5000 S.F.		465	.069		31	2.23		33.23	37.50
0300	Continuous vaulted, semi-circular, to 8' wide, double glazed G		145	.221		51.50	7.15		58.65	68.50
0400	Single glazed		160	.200		34.50	6.45		40.95	49
0600	To 20' wide, single glazed		175	.183		45	5.90		50.90	59.50
0700	Over 20' wide, single glazed		200	.160		44	5.15		49.15	57
0900	Motorized opening type, single glazed, 1/3 opening		145	.221		45.50	7.15		52.65	62
1000	Full opening		130	.246		51.50	7.95		59.45	70
1200	Pyramid type units, self-supporting, to 30' clear opening,									
1300	square or circular, single glazed, minimum	G-3	200	.160	SF Hor.	26.50	5.15		31.65	37.50
1310	Average		165	.194		36	6.25		42.25	50.50
1400	Maximum		130	.246		52	7.95		59.95	70
1500	Grid type, 4' to 10' modules, single glass glazed, minimum		200	.160		35.50	5.15		40.65	47.50
1550	Maximum		128	.250		55	8.10		63.10	74
1600	Preformed acrylic, minimum		300	.107		42.50	3.45		45.95	52.50
1650	Maximum		175	.183		58	5.90		63.90	74
1800	Skyroofs, dome type, self-supporting, to 100' clear opening, circular									
1900	Rise to span ratio = 0.20									
1920	Minimum	G-3	197	.162	SF Hor.	15.50	5.25		20.75	26
1950	Maximum		113	.283		52	9.15		61.15	72
2100	Rise to span ratio = 0.33, minimum		169	.189		31	6.10		37.10	44
2150	Maximum		101	.317		61.50	10.25		71.75	84.50
2200	Rise to span ratio = 0.50, minimum		148	.216		47	7		54	63
2250	Maximum		87	.368		68	11.90		79.90	95
2400	Ridge units, continuous, to 8' wide, double G		130	.246		119	7.95		126.95	144
2500	Single G		200	.160		82	5.15		87.15	98.50
2700	Ridge and furrow units, over 4' O.C., double, minimum G		200	.160		25.50	5.15		30.65	36.50
2750	Maximum G		120	.267		50.50	8.60		59.10	70
2800	Single, minimum		214	.150		22	4.83		26.83	32
2850	Maximum		153	.209		46	6.75		52.75	62.50
3000	Rolling roof, translucent panels, flat roof, minimum		253	.126	S.F.	19.80	4.09		23.89	29
3030	Maximum G		160	.200		39.50	6.45		45.95	54.50
3100	Lean-to skyroof, long span, double, minimum G		197	.162		26.50	5.25		31.75	37.50
3150	Maximum G		101	.317		53	10.25		63.25	75
3300	Single, minimum		321	.100		20	3.22		23.22	27.50
3350	Maximum		160	.200		33	6.45		39.45	47.50

08 51 Metal Windows

08 51 13 – Aluminum Windows

08 51 13.10 Aluminum Sash

		Crew	Daily Output	Labor-Hours	Unit	Material	Labor	Equipment	Total	Total Incl O&P
0010	**ALUMINUM SASH**									
0021	Stock, grade C, glaze & trim not incl., csmt.	L-4	200	.080	S.F.	37.50	2.38		39.88	45
0051	Double hung		200	.080		38	2.38		40.38	45.50
0101	Fixed casement		200	.080		16.55	2.38		18.93	22
0151	Picture window		200	.080		17.55	2.38		19.93	23.50
0201	Projected window		200	.080		34	2.38		36.38	41
0251	Single hung		200	.080		15.80	2.38		18.18	21.50
0301	Sliding		200	.080		20.50	2.38		22.88	26.50
1001	Mullions for above, tubular		240	.067	L.F.	5.90	1.98		7.88	9.75
2001	Custom aluminum sash, grade 3, glazing not included, minimum		200	.080	S.F.	38	2.38		40.38	46

08 51 Metal Windows

08 51 13 – Aluminum Windows

08 51 13.10 Aluminum Sash		Crew	Daily Output	Labor-Hours	Unit	Material	2011 Bare Costs Labor	Equipment	Total	Total Incl O&P
2101	Maximum	L-4	85	.188	S.F.	49.50	5.60		55.10	64

08 51 13.20 Aluminum Windows		Crew	Daily Output	Labor-Hours	Unit	Material	Labor	Equipment	Total	Total Incl O&P
0010	**ALUMINUM WINDOWS**, incl. frame and glazing, commercial grade									
1001	Stock units, casement, 3′-1″ x 3′-2″ opening	L-4	10	1.600	Ea.	355	47.50		402.50	470
1050	Add for storms					115			115	126
1601	Projected, with screen, 3′-1″ x 3′-2″ opening	L-4	10	1.600		335	47.50		382.50	450
1700	Add for storms					112			112	123
2001	4′-5″ x 5′-3″ opening	L-4	8	2		380	59.50		439.50	520
2100	Add for storms					120			120	132
2501	Enamel finish windows, 3′-1″ x 3′-2″	L-4	10	1.600		340	47.50		387.50	455
2601	4′-5″ x 5′-3″		8	2		385	59.50		444.50	525
3001	Single hung, 2′ x 3′ opening, enameled, standard glazed		10	1.600		198	47.50		245.50	298
3101	Insulating glass		10	1.600		240	47.50		287.50	345
3301	2′-8″ x 6′-8″ opening, standard glazed		8	2		350	59.50		409.50	485
3401	Insulating glass		8	2		450	59.50		509.50	595
3701	3′-4″ x 5′-0″ opening, standard glazed		9	1.778		286	53		339	405
3801	Insulating glass		9	1.778		315	53		368	440
3891	Awning type, 3′ x 3′ opening standard glass		14	1.143		405	34		439	500
3901	Insulating glass		14	1.143		430	34		464	530
3911	3′ x 4′ opening, standard glass		10	1.600		470	47.50		517.50	600
3921	Insulating glass		10	1.600		545	47.50		592.50	675
3931	3′ x 5′-4″ opening, standard glass		10	1.600		570	47.50		617.50	705
3941	Insulating glass		10	1.600		670	47.50		717.50	815
3951	4′ x 5′-4″ opening, standard glass		9	1.778		625	53		678	775
3961	Insulating glass		9	1.778		750	53		803	915
4001	Sliding aluminum, 3′ x 2′ opening, standard glazed		10	1.600		209	47.50		256.50	310
4101	Insulating glass		10	1.600		224	47.50		271.50	325
4301	5′ x 3′ opening, standard glazed		9	1.778		320	53		373	440
4401	Insulating glass		9	1.778		370	53		423	500
4601	8′ x 4′ opening, standard glazed		6	2.667		335	79		414	500
4701	Insulating glass		6	2.667		545	79		624	725
5001	9′ x 5′ opening, standard glazed		4	4		510	119		629	760
5101	Insulating glass		4	4		820	119		939	1,100
5501	Sliding, with thermal barrier and screen, 6′ x 4′, 2 track		8	2		695	59.50		754.50	865
5701	4 track	↓	8	2		880	59.50		939.50	1,075
6000	For above units with bronze finish, add					12%				
6200	For installation in concrete openings, add				↓	6%				

08 51 23 – Steel Windows

08 51 23.10 Steel Sash		Crew	Daily Output	Labor-Hours	Unit	Material	Labor	Equipment	Total	Total Incl O&P
0010	**STEEL SASH** Custom units, glazing and trim not included R085123-10									
0103	Casement, 100% vented	L-4	200	.080	S.F.	62	2.38		64.38	72
0201	50% vented		200	.080		51	2.38		53.38	60
0301	Fixed		200	.080		27.50	2.38		29.88	34
1001	Projected, commercial, 40% vented		200	.080		48.50	2.38		50.88	57.50
1101	Intermediate, 50% vented		200	.080		54.50	2.38		56.88	64
1501	Industrial, horizontally pivoted		200	.080		50.50	2.38		52.88	59.50
1601	Fixed		200	.080		29.50	2.38		31.88	36
2001	Industrial security sash, 50% vented		200	.080		54.50	2.38		56.88	64
2101	Fixed		200	.080		44.50	2.38		46.88	53
2501	Picture window		200	.080		28.50	2.38		30.88	35.50
3001	Double hung		200	.080	↓	56	2.38		58.38	65.50
5001	Mullions for above, open interior face	↓	240	.067	L.F.	9.80	1.98		11.78	14.05

08 51 Metal Windows

08 51 23 – Steel Windows

08 51 23.10 Steel Sash		Crew	Daily Output	Labor-Hours	Unit	Material	2011 Bare Costs Labor	Equipment	Total	Total Incl O&P
5101	With interior cover	L-4	240	.067	L.F.	16.35	1.98		18.33	21.50

08 51 23.20 Steel Windows

		Crew	Daily Output	Labor-Hours	Unit	Material	Labor	Equipment	Total	Total Incl O&P
0010	**STEEL WINDOWS** Stock, including frame, trim and insul. glass R085123-10									
0020	See Section 13 34 19.50									
1001	Custom units, double hung, 2′-8″ x 4′-6″ opening	L-4	12	1.333	Ea.	680	39.50		719.50	815
1101	2′-4″ x 3′-9″ opening		12	1.333		565	39.50		604.50	685
1501	Commercial projected, 3′-9″ x 5′-5″ opening		10	1.600		1,200	47.50		1,247.50	1,375
1601	6′-9″ x 4′-1″ opening		7	2.286		1,575	68		1,643	1,850
2001	Intermediate projected, 2′-9″ 4′-1″ opening		12	1.333		670	39.50		709.50	800
2101	4′-1″ x 5′-5″ opening		10	1.600		1,350	47.50		1,397.50	1,550

08 51 66 – Metal Window Screens

08 51 66.10 Screens

		Crew	Daily Output	Labor-Hours	Unit	Material	Labor	Equipment	Total	Total Incl O&P
0010	**SCREENS**									
0021	For metal sash, aluminum or bronze mesh, flat screen	L-4	1200	.013	S.F.	4.10	.40		4.50	5.15
0500	Wicket screen, inside window	2 Sswk	1000	.016		6.30	.58		6.88	8.05
0800	Security screen, aluminum frame with stainless steel cloth	"	1200	.013		22.50	.49		22.99	26
0901	Steel grate, painted, on steel frame	L-4	1600	.010		11	.30		11.30	12.60
1001	For solar louvers, add	"	160	.100		26	2.97		28.97	33.50
4000	See Section 05 58 27.90									

08 52 Wood Windows

08 52 10 – Plain Wood Windows

08 52 10.20 Awning Window

		Crew	Daily Output	Labor-Hours	Unit	Material	Labor	Equipment	Total	Total Incl O&P
0010	**AWNING WINDOW**, Including frame, screens and grilles									
0100	Average quality, 34″ x 22″, double insulated glass	1 Carp	10	.800	Ea.	256	27		283	325
0200	Low E glass		10	.800		265	27		292	335
0300	40″ x 28″, double insulated glass		9	.889		310	30		340	390
0400	Low E Glass		9	.889		335	30		365	420
0500	48″ x 36″, double insulated glass		8	1		460	33.50		493.50	560
0600	Low E glass		8	1		485	33.50		518.50	585
1000	Vinyl clad, 34″ x 22″		10	.800		260	27		287	330
1100	40″ x 22″		10	.800		284	27		311	355
1200	36″ x 28″		9	.889		300	30		330	380
1300	36″ x 36″		9	.889		335	30		365	420
1400	48″ x 28″		8	1		360	33.50		393.50	455
1500	60″ x 36″		8	1		510	33.50		543.50	620
2200	Metal clad, 36″ x 25″		9	.889		269	30		299	345
2300	40″ x 30″		9	.889		335	30		365	420
2400	48″ x 28″		8	1		345	33.50		378.50	435
2500	60″ x 36″		8	1		365	33.50		398.50	460
4000	Impact windows, minimum, add					60%				
4010	Impact windows, maximum, add					160%				

08 52 10.40 Casement Window

		Crew	Daily Output	Labor-Hours	Unit	Material	Labor	Equipment	Total	Total Incl O&P
0010	**CASEMENT WINDOW**, including frame, screen and grilles									
0100	Avg. quality, bldrs. model, 2′-0″ x 3′-0″ H, dbl. insulated glass G	1 Carp	10	.800	Ea.	276	27		303	350
0150	Low E glass G		10	.800		225	27		252	293
0200	2′-0″ x 4′-6″ high, double insulated glass G		9	.889		345	30		375	430
0250	Low E glass G		9	.889		263	30		293	340
0260	Casement 4′-2″ x 4′-2″ double insulated glass G		11	.727		855	24.50		879.50	980
0270	4′-0″ x 4′-0″ Low E glass G		11	.727		465	24.50		489.50	555

08 52 Wood Windows

08 52 10 – Plain Wood Windows

08 52 10.40 Casement Window

	08 52 10.40 Casement Window		Crew	Daily Output	Labor-Hours	Unit	2011 Bare Costs Material	Labor	Equipment	Total	Total Incl O&P
0290	6′-4″ x 5′-7″ Low E glass	G	1 Carp	9	.889	Ea.	1,100	30		1,130	1,250
0300	2′-4″ x 6′-0″ high, double insulated glass	G		8	1		400	33.50		433.50	495
0350	Low E glass	G		8	1		465	33.50		498.50	570
0522	Vinyl clad, premium, double insulated glass, 2′-0″ x 3′-0″	G		10	.800		266	27		293	335
0524	2′-0″ x 4′-0″	G		9	.889		310	30		340	390
0525	2′-0″ x 5′-0″	G		8	1		355	33.50		388.50	445
0528	2′-0″ x 6′-0″	G		8	1		355	33.50		388.50	445
0600	3′-0″ x 5′-0″			8	1		650	33.50		683.50	770
0700	4′-0″ x 3′-0″			8	1		715	33.50		748.50	845
0710	4′-0″ x 4′-0″	G		8	1		610	33.50		643.50	730
0720	4′-8″ x 4′-0″			8	1		675	33.50		708.50	800
0730	4′-8″ x 5′-0″			6	1.333		770	44.50		814.50	925
0740	4′-8″ x 6′-0″			6	1.333		870	44.50		914.50	1,025
0750	6′-0″ x 4′-0″			6	1.333		790	44.50		834.50	940
0800	6′-0″ x 5′-0″			6	1.333		885	44.50		929.50	1,050
0900	5′-6″ x 5′-6″	G	2 Carp	15	1.067		1,400	36		1,436	1,575
2000	Bay, casement units, 8′ x 5′, w/screens, dbl. insul. glass			2.50	6.400	Opng.	1,575	215		1,790	2,075
2100	Low E glass			2.50	6.400	"	1,650	215		1,865	2,150
8100	Metal clad, deluxe, dbl. insul. glass, 2′-0″ x 3′-0″ high	G	1 Carp	10	.800	Ea.	272	27		299	345
8120	2′-0″ x 4′-0″ high	G		9	.889		305	30		335	385
8140	2′-0″ x 5′-0″ high	G		8	1		320	33.50		353.50	405
8160	2′-0″ x 6′-0″ high	G		8	1		355	33.50		388.50	445
8190	For installation, add per leaf							15%			
8200	For multiple leaf units, deduct for stationary sash										
8220	2′ high					Ea.	22.50			22.50	24.50
8240	4′-6″ high						25.50			25.50	28
8260	6′ high						33.50			33.50	37
8300	Impact windows, minimum, add						60%				
8310	Impact windows, maximum, add						160%				

08 52 10.50 Double Hung

	08 52 10.50 Double Hung		Crew	Daily Output	Labor-Hours	Unit	Material	Labor	Equipment	Total	Total Incl O&P
0010	**DOUBLE HUNG**, Including frame, screens and grilles	R085216-10									
0100	Avg. quality, bldrs. model, 2′-0″ x 3′-0″ high, low E insul. glass	G	1 Carp	10	.800	Ea.	208	27		235	274
0200	3′-0″ x 4′-0″ high, double insulated glass	G		9	.889		264	30		294	340
0300	4′-0″ x 4′-6″ high, low E insulated glass	G		8	1		315	33.50		348.50	400
1000	Vinyl clad, premium, double insulated glass, 2′-6″ x 3′-0″	G		10	.800		300	27		327	375
1005	2′-6″ x 4′-0″	G		10	.800		355	27		382	435
1100	3′-0″ x 3′-6″	G		10	.800		330	27		357	405
1200	3′-0″ x 4′-0″	G		9	.889		375	30		405	465
1300	3′-0″ x 4′-6″	G		9	.889		395	30		425	485
1400	3′-0″ x 5′-0″	G		8	1		435	33.50		468.50	535
1490	3′-4″ x 5′-0″	G		8	1		445	33.50		478.50	545
1500	3′-6″ x 6′-0″	G		8	1		485	33.50		518.50	585
1520	4′-0″ x 5′-0″	G		7	1.143		550	38.50		588.50	670
1530	4′-0″ x 6′-0″	G		7	1.143		685	38.50		723.50	820
2000	Metal clad, deluxe, dbl. insul. glass, 2′-6″ x 3′-0″ high	G		10	.800		265	27		292	335
2100	3′-0″ x 3′-6″ high	G		10	.800		305	27		332	380
2200	3′-0″ x 4′-0″ high	G		9	.889		320	30		350	400
2300	3′-0″ x 4′-6″ high	G		9	.889		335	30		365	420
2400	3′-0″ x 5′-0″ high	G		8	1		365	33.50		398.50	455
2500	3′-6″ x 6′-0″ high	G		8	1		440	33.50		473.50	535
8000	Impact windows, minimum, add						60%				
8010	Impact windows, maximum, add						160%				

08 52 Wood Windows

08 52 10 – Plain Wood Windows

08 52 10.55 Picture Window		Crew	Daily Output	Labor-Hours	Unit	Material	Labor	Equipment	Total	Total Incl O&P
						2011 Bare Costs				
0010	**PICTURE WINDOW**, Including frame and grilles									
0100	Average quality, bldrs. model, 3′-6″ x 4′-0″ high, dbl. insulated glass	2 Carp	12	1.333	Ea.	395	44.50		439.50	510
0150	Low E glass		12	1.333		410	44.50		454.50	525
0200	4′-0″ x 4′-6″ high, double insulated glass		11	1.455		490	49		539	620
0250	Low E glass		11	1.455		520	49		569	650
0300	5′-0″ x 4′-0″ high, double insulated glass		11	1.455		570	49		619	705
0350	Low E glass		11	1.455		590	49		639	730
0400	6′-0″ x 4′-6″ high, double insulated glass		10	1.600		610	53.50		663.50	765
0450	Low E glass		10	1.600		620	53.50		673.50	775

08 52 10.65 Wood Sash

Line	Description	Crew	Daily Output	Labor-Hours	Unit	Material	Labor	Equipment	Total	Total Incl O&P
0010	**WOOD SASH**, Including glazing but not trim									
0050	Custom, 5′-0″ x 4′-0″, 1″ dbl. glazed, 3/16″ thick lites	2 Carp	3.20	5	Ea.	225	168		393	530
0100	1/4″ thick lites		5	3.200		240	107		347	445
0200	1″ thick, triple glazed		5	3.200		405	107		512	625
0300	7′-0″ x 4′-6″ high, 1″ double glazed, 3/16″ thick lites		4.30	3.721		415	125		540	665
0400	1/4″ thick lites		4.30	3.721		465	125		590	720
0500	1″ thick, triple glazed		4.30	3.721		530	125		655	795
0600	8′-6″ x 5′-0″ high, 1″ double glazed, 3/16″ thick lites		3.50	4.571		555	153		708	865
0700	1/4″ thick lites		3.50	4.571		605	153		758	920
0800	1″ thick, triple glazed		3.50	4.571		610	153		763	930
0900	Window frames only, based on perimeter length				L.F.	3.94			3.94	4.33
1200	Window sill, stock, per lineal foot					8.45			8.45	9.30
1250	Casing, stock					3.23			3.23	3.55

08 52 10.70 Sliding Windows

Line	Description		Crew	Daily Output	Labor-Hours	Unit	Material	Labor	Equipment	Total	Total Incl O&P
0010	**SLIDING WINDOWS**										
0100	Average quality, bldrs. model, 3′-0″ x 3′-0″ high, double insulated	G	1 Carp	10	.800	Ea.	272	27		299	345
0120	Low E glass	G		10	.800		293	27		320	365
0200	4′-0″ x 3′-6″ high, double insulated	G		9	.889		325	30		355	410
0220	Low E glass	G		9	.889		350	30		380	435
0300	6′-0″ x 5′-0″ high, double insulated	G		8	1		460	33.50		493.50	565
0320	Low E glass	G		8	1		500	33.50		533.50	605

08 52 13 – Metal-Clad Wood Windows

08 52 13.10 Awning Windows, Metal-Clad

Line	Description	Crew	Daily Output	Labor-Hours	Unit	Material	Labor	Equipment	Total	Total Incl O&P
0010	**AWNING WINDOWS, METAL-CLAD**									
2000	Metal clad, awning deluxe, double insulated glass, 34″ x 22″	1 Carp	10	.800	Ea.	243	27		270	310
2100	40″ x 22″	″	10	.800	″	286	27		313	360

08 52 13.35 Picture and Sliding Windows Metal-Clad

Line	Description		Crew	Daily Output	Labor-Hours	Unit	Material	Labor	Equipment	Total	Total Incl O&P
0010	**PICTURE AND SLIDING WINDOWS METAL-CLAD**										
2000	Metal clad, dlx picture, dbl. insul. glass, 4′-0″ x 4′-0″ high		2 Carp	12	1.333	Ea.	365	44.50		409.50	480
2100	4′-0″ x 6′-0″ high			11	1.455		535	49		584	670
2200	5′-0″ x 6′-0″ high			10	1.600		595	53.50		648.50	745
2300	6′-0″ x 6′-0″ high			10	1.600		680	53.50		733.50	840
2400	Metal clad, dlx sliding, double insulated glass, 3′-0″ x 3′-0″ high	G	1 Carp	10	.800		330	27		357	405
2420	4′-0″ x 3′-6″ high	G		9	.889		400	30		430	490
2440	5′-0″ x 4′-0″ high	G		9	.889		480	30		510	575
2460	6′-0″ x 5′-0″ high	G		8	1		705	33.50		738.50	830

08 52 16 – Plastic-Clad Wood Windows

08 52 16.10 Bow Window

Line	Description	Crew	Daily Output	Labor-Hours	Unit	Material	Labor	Equipment	Total	Total Incl O&P
0010	**BOW WINDOW** Including frames, screens, and grilles									
0020	End panels operable									

08 52 Wood Windows

08 52 16 – Plastic-Clad Wood Windows

08 52 16.10 Bow Window		Crew	Daily Output	Labor-Hours	Unit	2011 Bare Costs Material	Labor	Equipment	Total	Total Incl O&P
1001	Bow type, casement, wood, bldrs mdl, 8′ x 5′ dbl. insltd glass, 4 panel	2 Carp	10	1.600	Ea.	1,500	53.50		1,553.50	1,750
1050	Low E glass		10	1.600		1,325	53.50		1,378.50	1,550
1100	10′-0″ x 5′-0″, double insulated glass, 6 panels		6	2.667		1,350	89.50		1,439.50	1,650
1200	Low E glass, 6 panels		6	2.667		1,450	89.50		1,539.50	1,750
1300	Vinyl clad, bldrs model, double insulated glass, 6′-0″ x 4′-0″, 3 panel		10	1.600		1,025	53.50		1,078.50	1,225
1340	9′-0″ x 4′-0″, 4 panel		8	2		1,350	67		1,417	1,600
1380	10′-0″ x 6′-0″, 5 panels		7	2.286		2,250	76.50		2,326.50	2,600
1420	12′-0″ x 6′-0″, 6 panels		6	2.667		2,925	89.50		3,014.50	3,375
1601	Metal clad, bldrs model, 6′-0″ x 4′-0″ dbl. insltd glass, 3 panels		10	1.600		1,200	53.50		1,253.50	1,425
1640	9′-0″ x 4′-0″, 4 panels		8	2		1,550	67		1,617	1,800
1680	10′-0″ x 5′-0″, 5 panels		7	2.286		2,100	76.50		2,176.50	2,450
1720	12′-0″ x 6′-0″, 6 panels		6	2.667		2,700	89.50		2,789.50	3,100
2000	Bay window, 8′ x 5′, dbl. insul glass		10	1.600		1,825	53.50		1,878.50	2,125
2050	Low E glass		10	1.600		2,225	53.50		2,278.50	2,550
2100	12′-0″ x 6′-0″, double insulated glass, 6 panels		6	2.667		2,300	89.50		2,389.50	2,675
2200	Low E glass		6	2.667		2,350	89.50		2,439.50	2,750
2280	6′-0″ x 4′-0″		11	1.455		1,225	49		1,274	1,425
2300	Vinyl clad, premium, double insulated glass, 8′-0″ x 5′-0″		10	1.600		1,750	53.50		1,803.50	2,025
2340	10′-0″ x 5′-0″		8	2		2,275	67		2,342	2,600
2380	10′-0″ x 6′-0″		7	2.286		2,600	76.50		2,676.50	2,975
2420	12′-0″ x 6′-0″		6	2.667		3,200	89.50		3,289.50	3,675
2600	Metal clad, deluxe, dbl. insul. glass, 8′-0″ x 5′-0″ high, 4 panels		10	1.600		1,625	53.50		1,678.50	1,875
2640	10′-0″ x 5′-0″ high, 5 panels		8	2		1,750	67		1,817	2,025
2680	10′-0″ x 6′-0″ high, 5 panels		7	2.286		2,050	76.50		2,126.50	2,400
2720	12′-0″ x 6′-0″ high, 6 panels		6	2.667		2,850	89.50		2,939.50	3,300
3000	Double hung, bldrs. model, bay, 8′ x 4′ high, dbl. insulated glass		10	1.600		1,300	53.50		1,353.50	1,525
3050	Low E glass		10	1.600		1,375	53.50		1,428.50	1,625
3101	9′-0″ x 5′-0″ high, 4 panels, double insulated glass		6	2.667		1,400	89.50		1,489.50	1,675
3200	Low E glass		6	2.667		1,475	89.50		1,564.50	1,775
3300	Vinyl clad, premium, double insulated glass, 7′-0″ x 4′-6″		10	1.600		1,325	53.50		1,378.50	1,575
3340	8′-0″ x 4′-6″		8	2		1,375	67		1,442	1,600
3380	8′-0″ x 5′-0″		7	2.286		1,425	76.50		1,501.50	1,700
3420	9′-0″ x 5′-0″		6	2.667		1,475	89.50		1,564.50	1,775
3601	Metal clad, deluxe, dbl. insul. glass, 7′-0″ x 4′-0″ high, 3 panels		10	1.600		1,225	53.50		1,278.50	1,450
3641	8′-0″ x 4′-0″ high, 4 panels		8	2		1,275	67		1,342	1,500
3681	8′-0″ x 5′-0″ high, 4 panels		7	2.286		1,325	76.50		1,401.50	1,575
3721	9′-0″ x 5′-0″ high, 4 panels		6	2.667		1,400	89.50		1,489.50	1,700

08 52 16.30 Palladian Windows		Crew	Daily Output	Labor-Hours	Unit	Material	Labor	Equipment	Total	Total Incl O&P
0010	**PALLADIAN WINDOWS**									
0020	Vinyl clad, double insulated glass, including frame and grilles									
0040	3′-2″ x 2′-6″ high	2 Carp	11	1.455	Ea.	1,250	49		1,299	1,450
0060	3′-2″ x 4′-10″		11	1.455		1,750	49		1,799	1,975
0080	3′-2″ x 6′-4″		10	1.600		1,700	53.50		1,753.50	1,975
0100	4′-0″ x 4′-0″		10	1.600		1,525	53.50		1,578.50	1,775
0120	4′-0″ x 5′-4″	3 Carp	10	2.400		1,875	80.50		1,955.50	2,175
0140	4′-0″ x 6′-0″		9	2.667		1,950	89.50		2,039.50	2,300
0160	4′-0″ x 7′-4″		9	2.667		2,125	89.50		2,214.50	2,475
0180	5′-5″ x 4′-10″		9	2.667		2,275	89.50		2,364.50	2,650
0200	5′-5″ x 6′-10″		9	2.667		2,575	89.50		2,664.50	3,000
0220	5′-5″ x 7′-9″		9	2.667		2,800	89.50		2,889.50	3,225
0240	6′-0″ x 7′-11″		8	3		3,475	101		3,576	4,000
0260	8′-0″ x 6′-0″		8	3		3,075	101		3,176	3,550

08 52 Wood Windows

08 52 16 – Plastic-Clad Wood Windows

08 52 16.40 Transom Windows		Crew	Daily Output	Labor-Hours	Unit	Material	2011 Bare Costs Labor	Equipment	Total	Total Incl O&P
0010	**TRANSOM WINDOWS**									
1000	Vinyl clad, premium, dbl. insul. glass, 4′-0″ x 4′-0″	2 Carp	12	1.333	Ea.	480	44.50		524.50	605
1100	4′-0″ x 6′-0″		11	1.455		910	49		959	1,075
1200	5′-0″ x 6′-0″		10	1.600		1,000	53.50		1,053.50	1,225
1300	6′-0″ x 6′-0″		10	1.600		1,025	53.50		1,078.50	1,225

08 52 16.70 Vinyl Clad, Premium, Dbl. Insulated Glass		Crew	Daily Output	Labor-Hours	Unit	Material	Labor	Equipment	Total	Total Incl O&P
0010	**VINYL CLAD, PREMIUM, DBL. INSULATED GLASS**									
1000	Sliding, 3′-0″ x 3′-0″ G	1 Carp	10	.800	Ea.	575	27		602	675
1050	4′-0″ x 3′-6″ G		9	.889		670	30		700	785
1100	5′-0″ x 4′-0″ G		9	.889		860	30		890	995
1150	6′-0″ x 5′-0″ G		8	1		1,075	33.50		1,108.50	1,250

08 52 50 – Window Accessories

08 52 50.10 Window Grille or Muntin		Crew	Daily Output	Labor-Hours	Unit	Material	Labor	Equipment	Total	Total Incl O&P
0010	**WINDOW GRILLE OR MUNTIN**, snap in type									
0020	Standard pattern interior grilles									
2000	Wood, awning window, glass size 28″ x 16″ high	1 Carp	30	.267	Ea.	25.50	8.95		34.45	43
2060	44″ x 24″ high		32	.250		37	8.40		45.40	54.50
2100	Casement, glass size, 20″ x 36″ high		30	.267		29.50	8.95		38.45	47.50
2180	20″ x 56″ high		32	.250		41	8.40		49.40	59
2200	Double hung, glass size, 16″ x 24″ high		24	.333	Set	49	11.20		60.20	72
2280	32″ x 32″ high		34	.235	″	128	7.90		135.90	154
2500	Picture, glass size, 48″ x 48″ high		30	.267	Ea.	114	8.95		122.95	140
2580	60″ x 68″ high		28	.286	″	178	9.60		187.60	211
2600	Sliding, glass size, 14″ x 36″ high		24	.333	Set	34	11.20		45.20	55.50
2680	36″ x 36″ high		22	.364	″	41.50	12.20		53.70	66

08 52 66 – Wood Window Screens

08 52 66.10 Wood Screens		Crew	Daily Output	Labor-Hours	Unit	Material	Labor	Equipment	Total	Total Incl O&P
0010	**WOOD SCREENS**									
0021	Over 3 S.F., 3/4″ frames	2 Carp	375	.043	S.F.	4.66	1.43		6.09	7.55
0101	1-1/8″ frames	″	375	.043	″	7.85	1.43		9.28	11

08 52 69 – Wood Storm Windows

08 52 69.10 Storm Windows		Crew	Daily Output	Labor-Hours	Unit	Material	Labor	Equipment	Total	Total Incl O&P
0010	**STORM WINDOWS**, aluminum residential									
0300	Basement, mill finish, incl. fiberglass screen									
0320	1′-10″ x 1′-0″ high G	2 Carp	30	.533	Ea.	35	17.90		52.90	68.50
0340	2′-9″ x 1′-6″ high G	″	30	.533	″	38	17.90		55.90	72
1600	Double-hung, combination, storm & screen									
2000	Average quality, clear anodic coating, 2′-0″ x 3′-5″ high G	2 Carp	30	.533	Ea.	85	17.90		102.90	124
2020	2′-6″ x 5′-0″ high G		28	.571		107	19.15		126.15	149
2040	4′-0″ x 6′-0″ high G		25	.640		126	21.50		147.50	175
2400	White painted, 2′-0″ x 3′-5″ high G		30	.533		85	17.90		102.90	124
2420	2′-6″ x 5′-0″ high G		28	.571		90	19.15		109.15	131
2440	4′-0″ x 6′-0″ high G		25	.640		98	21.50		119.50	144
2600	Mill finish, 2′-0″ x 3′-5″ high G		30	.533		75	17.90		92.90	113
2620	2′-6″ x 5′-0″ high G		28	.571		85	19.15		104.15	126
2640	4′-0″ x 6′-8″ high G		25	.640		99	21.50		120.50	145

08 53 Plastic Windows

08 53 13 – Vinyl Windows

08 53 13.20 Vinyl Single Hung Windows

	08 53 13.20 Vinyl Single Hung Windows		Crew	Daily Output	Labor-Hours	Unit	2011 Bare Costs Material	Labor	Equipment	Total	Total Incl O&P
0010	**VINYL SINGLE HUNG WINDOWS**, insulated glass										
0100	Grids, low E, J fin, ext. jambs, 21" x 53"	G	2 Carp	18	.889	Ea.	198	30		228	268
0110	21" x 57"	G		17	.941		195	31.50		226.50	268
0120	21" x 65"	G		16	1		200	33.50		233.50	276
0130	25" x 41"	G		20	.800		188	27		215	252
0140	25" x 49"	G		18	.889		198	30		228	268
0150	25" x 57"	G		17	.941		201	31.50		232.50	274
0160	25" x 65"	G		16	1		207	33.50		240.50	284
0170	29" x 41"	G		18	.889		192	30		222	261
0180	29" x 53"	G		18	.889		202	30		232	272
0190	29" x 57"	G		17	.941		206	31.50		237.50	280
0200	29" x 65"	G		16	1		212	33.50		245.50	289
0210	33" x 41"	G		20	.800		198	27		225	263
0220	33" x 53"	G		18	.889		209	30		239	280
0230	33" x 57"	G		17	.941		212	31.50		243.50	286
0240	33" x 65"	G		16	1		218	33.50		251.50	296
0250	37" x 41"	G		20	.800		206	27		233	272
0260	37" x 53"	G		18	.889		218	30		248	290
0270	37" x 57"	G		17	.941		221	31.50		252.50	296
0280	37" x 65"	G		16	1		232	33.50		265.50	310

08 53 13.30 Vinyl Double Hung Windows

	08 53 13.30 Vinyl Double Hung Windows		Crew	Daily Output	Labor-Hours	Unit	Material	Labor	Equipment	Total	Total Incl O&P
0010	**VINYL DOUBLE HUNG WINDOWS**, insulated glass										
0100	Grids, low E, J fin, ext. jambs, 21" x 53"	G	2 Carp	18	.889	Ea.	215	30		245	287
0102	21" x 37"	G		18	.889		197	30		227	267
0104	21" x 41"	G		18	.889		201	30		231	271
0106	21" x 49"	G		18	.889		221	30		251	293
0110	21" x 57"	G		17	.941		218	31.50		249.50	293
0120	21" x 65"	G		16	1		228	33.50		261.50	305
0128	25" x 37"	G		20	.800		206	27		233	272
0130	25" x 41"	G		20	.800		209	27		236	275
0140	25" x 49"	G		18	.889		213	30		243	284
0145	25" x 53"	G		18	.889		221	30		251	293
0150	25" x 57"	G		17	.941		221	31.50		252.50	296
0160	25" x 65"	G		16	1		231	33.50		264.50	310
0162	25" x 69"	G		16	1		238	33.50		271.50	320
0164	25" x 77"	G		16	1		248	33.50		281.50	330
0168	29" x 37"	G		18	.889		211	30		241	282
0170	29" x 41"	G		18	.889		214	30		244	285
0172	29" x 49"	G		18	.889		221	30		251	293
0180	29" x 53"	G		18	.889		225	30		255	298
0190	29" x 57"	G		17	.941		229	31.50		260.50	305
0200	29" x 65"	G		16	1		236	33.50		269.50	315
0202	29" x 69"	G		16	1		243	33.50		276.50	325
0205	29" x 77"	G		16	1		254	33.50		287.50	335
0208	33" x 37"	G		20	.800		215	27		242	282
0210	33" x 41"	G		20	.800		219	27		246	286
0215	33" x 49"	G		20	.800		227	27		254	295
0220	33" x 53"	G		18	.889		231	30		261	305
0230	33" x 57"	G		17	.941		235	31.50		266.50	310
0240	33" x 65"	G		16	1		239	33.50		272.50	320
0242	33" x 69"	G		16	1		245	33.50		278.50	325
0246	33" x 77"	G		16	1		251	33.50		284.50	330

08 53 Plastic Windows

08 53 13 – Vinyl Windows

08 53 13.30 Vinyl Double Hung Windows		Crew	Daily Output	Labor-Hours	Unit	Material	2011 Bare Costs Labor	Equipment	Total	Total Incl O&P
0250	37″ x 41″ G	2 Carp	20	.800	Ea.	223	27		250	290
0255	37″ x 49″ G		20	.800		231	27		258	299
0260	37″ x 53″ G		18	.889		239	30		269	315
0270	37″ x 57″ G		17	.941		243	31.50		274.50	320
0280	37″ x 65″ G		16	1		255	33.50		288.50	335
0282	37″ x 69″ G		16	1		262	33.50		295.50	345
0286	37″ x 77″ G		16	1		270	33.50		303.50	355
0300	Solid vinyl, average quality, double insulated glass, 2′-0″ x 3′-0″ G	1 Carp	10	.800		287	27		314	360
0310	3′-0″ x 4′-0″ G		9	.889		198	30		228	267
0320	4′-0″ x 4′-6″ G		8	1		298	33.50		331.50	385
0330	Premium, double insulated glass, 2′-6″ x 3′-0″ G		10	.800		230	27		257	298
0340	3′-0″ x 3′-6″ G		9	.889		254	30		284	330
0350	3′-0″ x 4″-0″ G		9	.889		274	30		304	350
0360	3′-0″ x 4′-6″ G		9	.889		266	30		296	340
0370	3′-0″ x 5′-0″ G		8	1		286	33.50		319.50	370
0380	3′-6″ x 6″-0″ G		8	1		320	33.50		353.50	405

08 53 13.40 Vinyl Casement Windows		Crew	Daily Output	Labor-Hours	Unit	Material	Labor	Equipment	Total	Total Incl O&P
0010	**VINYL CASEMENT WINDOWS**, insulated glass									
0015	Grids, low E, J fin, extension jambs, screens									
0100	One lite, 21″ x 41″ G	2 Carp	20	.800	Ea.	271	27		298	345
0110	21″ x 47″ G		20	.800		294	27		321	370
0120	21″ x 53″ G		20	.800		315	27		342	395
0128	24″ x 35″ G		19	.842		263	28.50		291.50	335
0130	24″ x 41″ G		19	.842		283	28.50		311.50	360
0140	24″ x 47″ G		19	.842		305	28.50		333.50	385
0150	24″ x 53″ G		19	.842		325	28.50		353.50	410
0158	28″ x 35″ G		19	.842		278	28.50		306.50	355
0160	28″ x 41″ G		19	.842		298	28.50		326.50	380
0170	28″ x 47″ G		19	.842		320	28.50		348.50	400
0180	28″ x 53″ G		19	.842		350	28.50		378.50	435
0184	28″ x 59″ G		19	.842		350	28.50		378.50	435
0188	Two lites, 33″ x 35″ G		18	.889		430	30		460	520
0190	33″ x 41″ G		18	.889		455	30		485	555
0200	33″ x 47″ G		18	.889		490	30		520	585
0210	33″ x 53″ G		18	.889		520	30		550	620
0212	33″ x 59″ G		18	.889		550	30		580	655
0215	33″ x 72″ G		18	.889		570	30		600	675
0220	41″ x 41″ G		18	.889		495	30		525	595
0230	41″ x 47″ G		18	.889		530	30		560	630
0240	41″ x 53″ G		17	.941		560	31.50		591.50	670
0242	41″ x 59″ G		17	.941		590	31.50		621.50	700
0246	41″ x 72″ G		17	.941		615	31.50		646.50	730
0250	47″ x 41″ G		17	.941		500	31.50		531.50	605
0260	47″ x 47″ G		17	.941		530	31.50		561.50	640
0270	47″ x 53″ G		17	.941		560	31.50		591.50	675
0272	47″ x 59″ G		17	.941		615	31.50		646.50	730
0280	56″ x 41″ G		15	1.067		535	36		571	650
0290	56″ x 47″ G		15	1.067		560	36		596	680
0300	56″ x 53″ G		15	1.067		610	36		646	730
0302	56″ x 59″ G		15	1.067		635	36		671	760
0310	56″ x 72″ G		15	1.067		690	36		726	820
0340	Solid vinyl, premium, double insulated glass, 2′-0″ x 3′-0″ high G	1 Carp	10	.800		259	27		286	330

08 53 Plastic Windows

08 53 13 – Vinyl Windows

08 53 13.40 Vinyl Casement Windows			Crew	Daily Output	Labor-Hours	Unit	2011 Bare Costs Material	Labor	Equipment	Total	Total Incl O&P
0360	2'-0" x 4'-0" high	G	1 Carp	9	.889	Ea.	289	30		319	370
0380	2'-0" x 5'-0" high	G		8	1		295	33.50		328.50	380

08 53 13.50 Vinyl Picture Windows

Line	Description	Crew	Daily Output	Labor-Hours	Unit	Material	Labor	Equipment	Total	Total Incl O&P
0010	**VINYL PICTURE WINDOWS**, insulated glass									
0100	Grids, low E, J fin, ext. jambs, 33" x 47"	2 Carp	12	1.333	Ea.	238	44.50		282.50	335
0110	35" x 71"		12	1.333		310	44.50		354.50	415
0120	41" x 47"		12	1.333		252	44.50		296.50	350
0130	41" x 71"		12	1.333		320	44.50		364.50	430
0140	47" x 47"		12	1.333		277	44.50		321.50	380
0150	47" x 71"		11	1.455		350	49		399	465
0160	53" x 47"		11	1.455		300	49		349	410
0170	53" x 71"		11	1.455		315	49		364	425
0180	59" x 47"		11	1.455		340	49		389	455
0190	59" x 71"		11	1.455		365	49		414	480
0200	71" x 47"		10	1.600		375	53.50		428.50	505
0210	71" x 71"		10	1.600		395	53.50		448.50	525

08 54 Composite Windows

08 54 13 – Fiberglass Windows

08 54 13.10 Fiberglass Single Hung Windows

Line	Description		Crew	Daily Output	Labor-Hours	Unit	Material	Labor	Equipment	Total	Total Incl O&P
0010	**FIBERGLASS SINGLE HUNG WINDOWS**										
0100	Grids, low E, 18" x 24"	G	2 Carp	18	.889	Ea.	335	30		365	420
0110	18" x 40"	G		17	.941		300	31.50		331.50	385
0130	24" x 40"	G		20	.800		330	27		357	410
0230	36" x 36"	G		17	.941		370	31.50		401.50	460
0250	36" x 48"	G		20	.800		405	27		432	490
0260	36" x 60"	G		18	.889		445	30		475	540
0280	36" x 72"	G		16	1		470	33.50		503.50	570
0290	48" x 40"	G		16	1		470	33.50		503.50	570

08 56 Special Function Windows

08 56 63 – Detention Windows

08 56 63.13 Visitor Cubicle Windows

Line	Description	Crew	Daily Output	Labor-Hours	Unit	Material	Labor	Equipment	Total	Total Incl O&P
0010	**VISITOR CUBICLE WINDOWS**									
4000	Visitor cubicle, vision panel, no intercom	E-4	2	16	Ea.	2,925	590	54.50	3,569.50	4,400

08 62 Unit Skylights

08 62 13 – Domed Unit Skylights

08 62 13.20 Skylights

Line	Description		Crew	Daily Output	Labor-Hours	Unit	Material	Labor	Equipment	Total	Total Incl O&P
0010	**SKYLIGHTS**, Plastic domes, flush or curb mounted										
2120	Ventilating insulated plexiglass dome with										
2130	curb mounting, 36" x 36"	G	G-3	12	2.667	Ea.	460	86		546	650
2150	52" x 52"	G		12	2.667		640	86		726	850
2160	28" x 52"	G		10	3.200		470	103		573	685
2170	36" x 52"	G		10	3.200		520	103		623	740
2180	For electric opening system, add	G					300			300	330
2300	Insulated safety glass with aluminum frame	G	G-3	160	.200	S.F.	90	6.45		96.45	110

08 62 Unit Skylights

08 62 13 – Domed Unit Skylights

08 62 13.20 Skylights		Crew	Daily Output	Labor-Hours	Unit	Material (2011 Bare Costs)	Labor (2011 Bare Costs)	Equipment (2011 Bare Costs)	Total (2011 Bare Costs)	Total Incl O&P
2400	Sandwich panels, fiberglass, for walls, 1-9/16" thick, to 250 S.F. G	G-3	200	.160	S.F.	16.20	5.15		21.35	26.50
2700	As above, but for roofs, 2-3/4" thick, to 250 S.F. G	↓	295	.108	↓	23.50	3.51		27.01	31.50
2800	250 S.F. and up G	↓	330	.097	↓	19	3.14		22.14	26

08 63 Metal-Framed Skylights

08 63 23 – Ridge Metal-Framed Skylights

08 63 23.10 Prefabricated		Crew	Daily Output	Labor-Hours	Unit	Material	Labor	Equipment	Total	Total Incl O&P
0010	**PREFABRICATED** glass block with metal frame									
0020	Minimum G	G-3	265	.121	S.F.	56	3.90		59.90	68
0100	Maximum G	"	160	.200	"	108	6.45		114.45	130

08 71 Door Hardware

08 71 13 – Automatic Door Operators

08 71 13.10 Automatic Openers Commercial		Crew	Daily Output	Labor-Hours	Unit	Material	Labor	Equipment	Total	Total Incl O&P
0010	**AUTOMATIC OPENERS COMMERCIAL**									
0020	Pneumatic, incl opener, motion sens, control box, tubing, compressor									
0050	For single swing door, per opening	2 Skwk	.80	20	Ea.	4,300	680		4,980	5,875
0100	Pair, per opening	↓	.50	32	Opng.	7,000	1,100		8,100	9,525
1000	For single sliding door, per opening	↓	.60	26.667	↓	4,700	910		5,610	6,700
1300	Bi-parting pair	↓	.50	32	↓	7,100	1,100		8,200	9,625
1420	Electronic door opener incl motion sens, 12 V control box, motor									
1450	For single swing door, per opening	2 Skwk	.80	20	Opng.	3,500	680		4,180	5,000
1500	Pair, per opening	↓	.50	32	↓	5,700	1,100		6,800	8,100
1600	For single sliding door, per opening	↓	.60	26.667	↓	3,800	910		4,710	5,700
1700	Bi-parting pair	↓	.50	32	↓	5,800	1,100		6,900	8,200
1750	Handicap actuator buttons, 2, including 12 V DC wiring, add	1 Carp	1.50	5.333	Pr.	430	179		609	775

08 71 13.20 Automatic Openers Industrial		Crew	Daily Output	Labor-Hours	Unit	Material	Labor	Equipment	Total	Total Incl O&P
0010	**AUTOMATIC OPENERS INDUSTRIAL**									
0015	Sliding doors up to 6' wide	2 Skwk	.60	26.667	Opng.	5,500	910		6,410	7,575
0200	To 12' wide	"	.40	40	"	6,600	1,350		7,950	9,525
0400	Over 12' wide, add per L.F. of excess				L.F.	750			750	825
1000	Swing doors, to 5' wide	2 Skwk	.80	20	Ea.	3,200	680		3,880	4,675
1860	Add for controls, wall pushbutton, 3 button	↓	4	4	↓	215	136		351	465
1870	Ceiling pull cord	↓	4.30	3.721	↓	175	127		302	405

08 71 20 – Hardware

08 71 20.10 Bolts, Flush		Crew	Daily Output	Labor-Hours	Unit	Material	Labor	Equipment	Total	Total Incl O&P
0010	**BOLTS, FLUSH**									
0020	Standard, concealed	1 Carp	7	1.143	Ea.	23	38.50		61.50	89
0800	Automatic fire exit	"	5	1.600	↓	299	53.50		352.50	420
1600	For electric release, add	1 Elec	3	2.667	↓	125	107		232	310
3000	Barrel, brass, 2" long	1 Carp	40	.200	↓	5.20	6.70		11.90	16.95
3020	4" long	↓	40	.200	↓	5.30	6.70		12	17.10
3060	6" long	↓	40	.200	↓	10.40	6.70		17.10	22.50

08 71 20.15 Hardware		Crew	Daily Output	Labor-Hours	Unit	Material	Labor	Equipment	Total	Total Incl O&P
0009	**HARDWARE**									
0010	Average percentage for hardware, total job cost									
0025	Minimum				Job					1%
0050	Maximum				↓					4%

08 71 Door Hardware

08 71 20 – Hardware

08 71 20.15 Hardware		Crew	Daily Output	Labor-Hours	Unit	Material	Labor	Equipment	Total	Total Incl O&P
						2011 Bare Costs				
0500	Total hardware for building, average distribution				Job	85%	15%			
1000	Door hardware, apartment, interior	1 Carp	4	2	Door	510	67		577	670
1300	Average, door hardware, motel/hotel interior, with access card		4	2		625	67		692	800
1500	Hospital bedroom, minimum		4	2		595	67		662	765
2000	Maximum		3	2.667		705	89.50		794.50	925
2100	Pocket door		6	1.333	Ea.	134	44.50		178.50	222
2250	School, single exterior, incl. lever, incl. panic device		3	2.667	Door	1,100	89.50		1,189.50	1,375
2500	Single interior, regular use, lever included		3	2.667		595	89.50		684.50	800
2550	Average, door hdwe, school, classroom, ANSI F84, lever handle		3	2.667		750	89.50		839.50	975
2600	Average, door hdwe, school, classroom, ANSI F88, incl. lever		3	2.667		790	89.50		879.50	1,025
2850	Stairway, single interior		3	2.667		525	89.50		614.50	725
3100	Double exterior, with panic device		2	4	Pr.	2,075	134		2,209	2,500
3600	Toilet, public, single interior				Door	210			210	231
6020	Add for fire alarm door holder, electro-magnetic	1 Elec	4	2	Ea.	99	80.50		179.50	240
08 71 20.20 Door Protectors										
0010	**DOOR PROTECTORS**									
0020	1-3/4" x 3/4" U channel	2 Carp	80	.200	L.F.	25	6.70		31.70	39
0021	1-3/4" x 1-1/4" U channel		80	.200	"	15	6.70		21.70	28
1000	Tear drop, spring-stl, 8" high x 19" long		15	1.067	Ea.	92	36		128	161
1010	8" high x 32" long		15	1.067		114	36		150	185
1100	Tear drop, stainless stl, 8" high x 19" long		15	1.067		240	36		276	325
1200	8" high x 32" long		15	1.067		300	36		336	390
08 71 20.30 Door Closers										
0010	**DOOR CLOSERS**									
0020	Adjustable backcheck, 3 way mount, all sizes, regular arm	1 Carp	6	1.333	Ea.	182	44.50		226.50	275
0040	Hold open arm		6	1.333		230	44.50		274.50	330
0100	Fusible link		6.50	1.231		155	41.50		196.50	240
0200	Non sized, regular arm		6	1.333		165	44.50		209.50	257
0240	Hold open arm		6	1.333		200	44.50		244.50	295
0400	4 way mount, non sized, regular arm		6	1.333		220	44.50		264.50	315
0440	Hold open arm		6	1.333		230	44.50		274.50	330
2000	Backcheck and adjustable power, hinge face mount									
2010	All sizes, regular arm	1 Carp	6.50	1.231	Ea.	210	41.50		251.50	300
2040	Hold open arm		6.50	1.231		245	41.50		286.50	340
2400	Top jamb mount, all sizes, regular arm		6	1.333		200	44.50		244.50	295
2440	Hold open arm		6	1.333		210	44.50		254.50	305
2800	Top face mount, all sizes, regular arm		6.50	1.231		195	41.50		236.50	284
2840	Hold open arm		6.50	1.231		215	41.50		256.50	305
4000	Backcheck, overhead concealed, all sizes, regular arm		5.50	1.455		216	49		265	320
4040	Concealed arm		5	1.600		227	53.50		280.50	340
4400	Compact overhead, concealed, all sizes, regular arm		5.50	1.455		345	49		394	460
4440	Concealed arm		5	1.600		380	53.50		433.50	510
4800	Concealed in door, all sizes, regular arm		5.50	1.455		355	49		404	470
4840	Concealed arm		5	1.600		375	53.50		428.50	505
4900	Floor concealed, all sizes, single acting		2.20	3.636		380	122		502	625
4940	Double acting		2.20	3.636		320	122		442	555
5000	For cast aluminum cylinder, deduct					35			35	38.50
5040	For delayed action, add					45			45	49.50
5080	For fusible link arm, add					30			30	33
5120	For shock absorbing arm, add					50			50	55
5160	For spring power adjustment, add					40			40	44
6000	Closer-holder, hinge face mount, all sizes, exposed arm	1 Carp	6.50	1.231		165	41.50		206.50	251

08 71 Door Hardware

08 71 20 – Hardware

08 71 20.30 Door Closers		Crew	Daily Output	Labor-Hours	Unit	2011 Bare Costs Material	Labor	Equipment	Total	Total Incl O&P
7000	Electronic closer-holder, hinge facemount, concealed arm	1 Carp	5	1.600	Ea.	270	53.50		323.50	385
7400	With built-in detector		5	1.600		665	53.50		718.50	820
8000	Surface mounted, stand. duty, parallel arm, primed, traditional		6	1.333		187	44.50		231.50	281
8030	Light duty		6	1.333		123	44.50		167.50	210
8050	Heavy duty		6	1.333		213	44.50		257.50	310
8100	Standard duty, parallel arm, modern		6	1.333		213	44.50		257.50	310
8150	Heavy duty	↓	6	1.333	↓	228	44.50		272.50	325

08 71 20.31 Door Closers		Crew	Daily Output	Labor-Hours	Unit	Material	Labor	Equipment	Total	Total Incl O&P
0010	**DOOR CLOSERS**									
0015	Door closer, rack and pinion	1 Carp	6.50	1.231	Ea.	168	41.50		209.50	253
1520	Door, single acting, standard arm		1	8		199	268		467	670
1522	Hold open arm		1	8		216	268		484	690
1526	Frame, single acting, standard arm		1	8		335	268		603	820
1530	Hold open arm		1	8		355	268		623	840
1534	Double acting, standard arm		1	8		450	268		718	945
1536	Hold open arm		1	8		460	268		728	955
1554	Floor, center hung, single acting, bottom arm		1	8		320	268		588	800
1558	Double acting		1	8		330	268		598	815
1560	Offset hung, single acting, bottom arm	↓	1	8	↓	370	268		638	860
6500	Electro magnetic closer/holder									
6510	Single point, no detector	1 Carp	1	8	Ea.	350	268		618	835
6515	Including detector		1	8		670	268		938	1,200
6520	Multi-point, no detector		1	8		770	268		1,038	1,300
6524	Including detector	↓	1	8	↓	780	268		1,048	1,300
6550	Electric automatic operators									
6555	Operator	1 Carp	1	8	Ea.	1,525	268		1,793	2,125
6570	Wall plate actuator		1	8		195	268		463	665
8010	Light duty, regular arm		1	8		111	268		379	570
8032	Parallel arm		1	8		115	268		383	575
8034	Hold open arm		1	8		121	268		389	585
8036	Fusible link arm		1	8		176	268		444	645
8040	Medium duty, regular arm		1	8		140	268		408	605
8042	Extra duty parallel arm		1	8		150	268		418	615
8044	Hold open arm		1	8		157	268		425	620
8046	Positive stop arm		1	8		170	268		438	635
8052	Heavy duty, regular arm		1	8		172	268		440	640
8054	Top jamb mount		1	8		172	268		440	640
8056	Extra duty parallel arm		1	8		172	268		440	640
8058	Hold open arm		1	8		187	268		455	655
8060	Positive stop arm		1	8		189	268		457	660
8062	Fusible link arm		1	8		217	268		485	690
8080	Universal heavy duty, regular arm		1	8		186	268		454	655
8084	Parallel arm		1	8		186	268		454	655
8088	Extra duty, parallel arm		1	8		191	268		459	660
8090	Hold open arm		1	8		272	268		540	750
8094	Positive stop arm		1	8		277	268		545	755
8098	For delayed action add	↓	1	8	↓	30	268		298	485

08 71 20.35 Panic Devices		Crew	Daily Output	Labor-Hours	Unit	Material	Labor	Equipment	Total	Total Incl O&P
0010	**PANIC DEVICES**									
0015	For rim locks, single door exit only	1 Carp	6	1.333	Ea.	530	44.50		574.50	660
0020	Outside key and pull		5	1.600		575	53.50		628.50	720
0200	Bar and vertical rod, exit only	↓	5	1.600	↓	735	53.50		788.50	900

08 71 20.35 Panic Devices		Crew	Daily Output	Labor-Hours	Unit	2011 Bare Costs Material	Labor	Equipment	Total	Total Incl O&P
0210	Outside key and pull	1 Carp	4	2	Ea.	810	67		877	1,000
0400	Bar and concealed rod		4	2		595	67		662	765
0600	Touch bar, exit only		6	1.333		505	44.50		549.50	630
0610	Outside key and pull		5	1.600		600	53.50		653.50	755
0700	Touch bar and vertical rod, exit only		5	1.600		670	53.50		723.50	825
0710	Outside key and pull		4	2		780	67		847	965
1000	Mortise, bar, exit only		4	2		550	67		617	715
1600	Touch bar, exit only		4	2		620	67		687	795
2000	Narrow stile, rim mounted, bar, exit only		6	1.333		650	44.50		694.50	790
2010	Outside key and pull		5	1.600		700	53.50		753.50	860
2200	Bar and vertical rod, exit only		5	1.600		680	53.50		733.50	835
2210	Outside key and pull		4	2		680	67		747	855
2400	Bar and concealed rod, exit only		3	2.667		765	89.50		854.50	990
3000	Mortise, bar, exit only		4	2		610	67		677	780
3600	Touch bar, exit only		4	2		880	67		947	1,075

08 71 20.40 Lockset

		Crew	Daily Output	Labor-Hours	Unit	Material	Labor	Equipment	Total	Total Incl O&P
0010	**LOCKSET**, Standard duty									
0020	Non-keyed, passage, w/sect.trim	1 Carp	12	.667	Ea.	54.50	22.50		77	97
0100	Privacy		12	.667		62	22.50		84.50	106
0400	Keyed, single cylinder function		10	.800		92.50	27		119.50	147
0420	Hotel (see also Section 08 71 20.15)		8	1		135	33.50		168.50	205
0500	Lever handled, keyed, single cylinder function		10	.800		159	27		186	220
1000	Heavy duty with sectional trim, non-keyed, passages		12	.667		145	22.50		167.50	198
1100	Privacy		12	.667		185	22.50		207.50	242
1400	Keyed, single cylinder function		10	.800		220	27		247	287
1420	Hotel		8	1		325	33.50		358.50	415
1600	Communicating		10	.800		250	27		277	320
1690	For re-core cylinder, add					40			40	44
1700	Residential, interior door, minimum	1 Carp	16	.500		25	16.80		41.80	55.50
1720	Maximum		8	1		70	33.50		103.50	133
1800	Exterior, minimum		14	.571		45	19.15		64.15	81.50
1810	Average		8	1		72	33.50		105.50	136
1820	Maximum		8	1		170	33.50		203.50	243

08 71 20.41 Dead Locks

		Crew	Daily Output	Labor-Hours	Unit	Material	Labor	Equipment	Total	Total Incl O&P
0010	**DEAD LOCKS**									
0011	Mortise heavy duty outside key (security item)	1 Carp	9	.889	Ea.	269	30		299	345
0020	Double cylinder		9	.889		265	30		295	340
0100	Medium duty, outside key		10	.800		140	27		167	199
0110	Double cylinder		10	.800		155	27		182	216
1000	Tubular, standard duty, outside key		10	.800		65	27		92	117
1010	Double cylinder		10	.800		82	27		109	135
1200	Night latch, outside key		10	.800		81	27		108	134

08 71 20.42 Mortise Locksets

		Crew	Daily Output	Labor-Hours	Unit	Material	Labor	Equipment	Total	Total Incl O&P
0010	**MORTISE LOCKSETS**, Comm., wrought knobs & full escutcheon trim									
0020	Non-keyed, passage, minimum	1 Carp	9	.889	Ea.	190	30		220	259
0030	Maximum		8	1		300	33.50		333.50	385
0040	Privacy, minimum		9	.889		200	30		230	270
0050	Maximum		8	1		325	33.50		358.50	415
0100	Keyed, office/entrance/apartment, minimum		8	1		230	33.50		263.50	310
0110	Maximum		7	1.143		385	38.50		423.50	490
0120	Single cylinder, typical, minimum		8	1		198	33.50		231.50	274
0130	Maximum		7	1.143		360	38.50		398.50	460

08 71 Door Hardware

08 71 20 – Hardware

08 71 20.42 Mortise Locksets		Crew	Daily Output	Labor-Hours	Unit	Material	2011 Bare Costs Labor	Equipment	Total	Total Incl O&P
0200	Hotel, minimum (see also Section 08 71 20.15)	1 Carp	7	1.143	Ea.	230	38.50		268.50	315
0210	Maximum (see also Section 08 71 20.15)		6	1.333		375	44.50		419.50	490
0300	Communication, double cylinder, minimum		8	1		225	33.50		258.50	305
0310	Maximum		7	1.143		300	38.50		338.50	395
1000	Wrought knobs and sectional trim, non-keyed, passage, minimum		10	.800		130	27		157	188
1010	Maximum		9	.889		240	30		270	315
1040	Privacy, minimum		10	.800		145	27		172	205
1050	Maximum		9	.889		260	30		290	335
1100	Keyed, entrance, office/apartment, minimum		9	.889		220	30		250	292
1110	Maximum		8	1		310	33.50		343.50	395
1120	Single cylinder, typical, minimum		9	.889		215	30		245	287
1130	Maximum		8	1		295	33.50		328.50	380
2000	Cast knobs and full escutcheon trim									
2010	Non-keyed, passage, minimum	1 Carp	9	.889	Ea.	260	30		290	335
2020	Maximum		8	1		410	33.50		443.50	505
2040	Privacy, minimum		9	.889		310	30		340	390
2050	Maximum		8	1		445	33.50		478.50	545
2120	Keyed, single cylinder, typical, minimum		8	1		320	33.50		353.50	405
2130	Maximum		7	1.143		490	38.50		528.50	605
3000	Cast knob and sectional trim, non-keyed, passage, minimum		10	.800		210	27		237	276
3010	Maximum		10	.800		400	27		427	485
3040	Privacy, minimum		10	.800		225	27		252	293
3050	Maximum		10	.800		400	27		427	485
3100	Keyed, office/entrance/apartment, minimum		9	.889		255	30		285	330
3110	Maximum		9	.889		400	30		430	490
3120	Single cylinder, typical, minimum		9	.889		255	30		285	330
3130	Maximum		9	.889		490	30		520	590
3190	For re-core cylinder, add					35			35	38.50
3980	Keyless, pushbutton type									
4000	Residential/light commercial, deadbolt, standard	1 Carp	9	.889	Ea.	130	30		160	193
4010	Heavy duty		9	.889		150	30		180	215
4020	Industrial, heavy duty, with deadbolt		9	.889		284	30		314	365
4030	Key override		9	.889		320	30		350	400
4040	Lever activated handle		9	.889		335	30		365	420
4050	Key override		9	.889		345	30		375	430
4060	Double sided pushbutton type		8	1		695	33.50		728.50	815
4070	Key override		8	1		690	33.50		723.50	810

08 71 20.45 Peepholes

		Crew	Daily Output	Labor-Hours	Unit	Material	Labor	Equipment	Total	Total Incl O&P
0010	**PEEPHOLES**									
2010	Peephole	1 Carp	32	.250	Ea.	15.80	8.40		24.20	31.50

08 71 20.50 Door Stops

		Crew	Daily Output	Labor-Hours	Unit	Material	Labor	Equipment	Total	Total Incl O&P
0010	**DOOR STOPS**									
0020	Holder & bumper, floor or wall	1 Carp	32	.250	Ea.	35.50	8.40		43.90	53
1300	Wall bumper, 4" diameter, with rubber pad, aluminum		32	.250		10.95	8.40		19.35	26
1600	Door bumper, floor type, aluminum		32	.250		6.70	8.40		15.10	21.50
1900	Plunger type, door mounted		32	.250		25.50	8.40		33.90	42

08 71 20.55 Push-Pull Plates

		Crew	Daily Output	Labor-Hours	Unit	Material	Labor	Equipment	Total	Total Incl O&P
0010	**PUSH-PULL PLATES**									
0100	Push plate, .050 thick, 4" x 16", aluminum	1 Carp	12	.667	Ea.	10.20	22.50		32.70	49
0500	Bronze		12	.667		21.50	22.50		44	61
1500	Pull handle and push bar, aluminum		11	.727		131	24.50		155.50	185
2000	Bronze		10	.800		169	27		196	230

08 71 20.55 Push-Pull Plates		Crew	Daily Output	Labor-Hours	Unit	2011 Bare Costs Material	Labor	Equipment	Total	Total Incl O&P
3000	Push plate both sides, aluminum	1 Carp	14	.571	Ea.	19.95	19.15		39.10	54
3500	Bronze		13	.615		44	20.50		64.50	83
4000	Door pull, designer style, cast aluminum, minimum		12	.667		71.50	22.50		94	116
5000	Maximum		8	1		370	33.50		403.50	460
6000	Cast bronze, minimum		12	.667		87	22.50		109.50	133
7000	Maximum		8	1		405	33.50		438.50	500
8000	Walnut, minimum		12	.667		67	22.50		89.50	111
9000	Maximum		8	1		370	33.50		403.50	460

08 71 20.60 Entrance Locks

		Crew	Daily Output	Labor-Hours	Unit	Material	Labor	Equipment	Total	Total Incl O&P
0010	**ENTRANCE LOCKS**									
0015	Cylinder, grip handle deadlocking latch	1 Carp	9	.889	Ea.	132	30		162	195
0020	Deadbolt		8	1		160	33.50		193.50	232
0100	Push and pull plate, dead bolt		8	1		150	33.50		183.50	221
0900	For handicapped lever, add					170			170	187

08 71 20.65 Thresholds

		Crew	Daily Output	Labor-Hours	Unit	Material	Labor	Equipment	Total	Total Incl O&P
0010	**THRESHOLDS**									
0011	Threshold 3′ long saddles aluminum	1 Carp	48	.167	L.F.	4.93	5.60		10.53	14.75
0100	Aluminum, 8″ wide, 1/2″ thick		12	.667	Ea.	39	22.50		61.50	80.50
0500	Bronze		60	.133	L.F.	41.50	4.47		45.97	53
0600	Bronze, panic threshold, 5″ wide, 1/2″ thick		12	.667	Ea.	67.50	22.50		90	112
0700	Rubber, 1/2″ thick, 5-1/2″ wide		20	.400		40	13.40		53.40	66.50
0800	2-3/4″ wide		20	.400		29	13.40		42.40	54.50

08 71 20.70 Floor Checks

		Crew	Daily Output	Labor-Hours	Unit	Material	Labor	Equipment	Total	Total Incl O&P
0010	**FLOOR CHECKS**									
0020	For over 3′ wide doors single acting	1 Carp	2.50	3.200	Ea.	520	107		627	750
0500	Double acting	″	2.50	3.200	″	670	107		777	915

08 71 20.75 Door Hardware Accessories

		Crew	Daily Output	Labor-Hours	Unit	Material	Labor	Equipment	Total	Total Incl O&P
0010	**DOOR HARDWARE ACCESSORIES**									
0050	Door closing coordinator, 36″ (for paired openings up to 56″)	1 Carp	8	1	Ea.	94	33.50		127.50	159
0060	48″ (for paired openings up to 84″)		8	1		101	33.50		134.50	167
0070	56″ (for paired openings up to 96″)		8	1		111	33.50		144.50	178

08 71 20.80 Hasps

		Crew	Daily Output	Labor-Hours	Unit	Material	Labor	Equipment	Total	Total Incl O&P
0010	**HASPS**, steel assembly									
0015	3″	1 Carp	26	.308	Ea.	4.71	10.30		15.01	22.50
0020	4-1/2″		13	.615		6.50	20.50		27	41.50
0040	6″		12.50	.640		9.25	21.50		30.75	46

08 71 20.90 Hinges

		Crew	Daily Output	Labor-Hours	Unit	Material	Labor	Equipment	Total	Total Incl O&P
0010	**HINGES** R087120-10									
0012	Full mortise, avg. freq., steel base, USP, 4-1/2″ x 4-1/2″				Pr.	27			27	29.50
0100	5″ x 5″, USP					43			43	47.50
0200	6″ x 6″, USP					91			91	100
0400	Brass base, 4-1/2″ x 4-1/2″, US10					53.50			53.50	59
0500	5″ x 5″, US10					77			77	85
0600	6″ x 6″, US10					130			130	143
0800	Stainless steel base, 4-1/2″ x 4-1/2″, US32					79.50			79.50	87.50
0900	For non removable pin, add (security item)				Ea.	4.72			4.72	5.20
0910	For floating pin, driven tips, add					3.25			3.25	3.58
0930	For hospital type tip on pin, add					14.10			14.10	15.50
0940	For steeple type tip on pin, add					13			13	14.30
0950	Full mortise, high frequency, steel base, 3-1/2″ x 3-1/2″, US26D				Pr.	27			27	29.50
1000	4-1/2″ x 4-1/2″, USP					62			62	68.50

08 71 Door Hardware

08 71 20 – Hardware

08 71 20.90 Hinges		Crew	Daily Output	Labor-Hours	Unit	Material	2011 Bare Costs Labor	Equipment	Total	Total Incl O&P
1100	5" x 5", USP				Pr.	58.50			58.50	64
1200	6" x 6", USP					142			142	156
1400	Brass base, 3-1/2" x 3-1/2", US4					48			48	52.50
1430	4-1/2" x 4-1/2", US10					84			84	92
1500	5" x 5", US10					123			123	135
1600	6" x 6", US10					178			178	196
1800	Stainless steel base, 4-1/2" x 4-1/2", US32					127			127	140
1810	5" x 4-1/2", US32					182			182	200
1930	For hospital type tip on pin, add				Ea.	12.60			12.60	13.85
1950	Full mortise, low frequency, steel base, 3-1/2" x 3-1/2", US26D				Pr.	11.55			11.55	12.75
2000	4-1/2" x 4-1/2", USP					12.80			12.80	14.05
2100	5" x 5", USP					31.50			31.50	34.50
2200	6" x 6", USP					63.50			63.50	69.50
2300	4-1/2" x 4-1/2", US3					18.40			18.40	20
2310	5" x 5", US3					44			44	48.50
2400	Brass bass, 4-1/2" x 4-1/2", US10					44.50			44.50	49
2500	5" x 5", US10					67			67	74
2800	Stainless steel base, 4-1/2" x 4-1/2", US32					75.50			75.50	83

08 71 20.91 Special Hinges		Crew	Daily Output	Labor-Hours	Unit	Material	Labor	Equipment	Total	Total Incl O&P
0010	**SPECIAL HINGES**									
0015	Paumelle, high frequency									
0020	Steel base, 6" x 4-1/2", US10				Pr.	145			145	160
0100	Brass base, 5" x 4-1/2", US10				Ea.	205			205	226
0200	Paumelle, average frequency, steel base, 4-1/2" x 3-1/2", US10				Pr.	98.50			98.50	108
0400	Olive knuckle, low frequency, brass base, 6" x 4-1/2", US10				Ea.	152			152	167
1000	Electric hinge with concealed conductor, average frequency									
1010	Steel base, 4-1/2" x 4-1/2", US26D				Pr.	315			315	345
1100	Bronze base, 4-1/2" x 4-1/2", US26D				"	320			320	350
1200	Electric hinge with concealed conductor, high frequency									
1210	Steel base, 4-1/2" x 4-1/2", US26D				Pr.	241			241	265
1600	Double weight, 800 lb., steel base, removable pin, 5" x 6", USP					132			132	145
1700	Steel base-welded pin, 5" x 6", USP					163			163	179
1800	Triple weight, 2000 lb., steel base, welded pin, 5" x 6", USP					151			151	166
2000	Pivot reinf., high frequency, steel base, 7-3/4" door plate, USP					187			187	206
2200	Bronze base, 7-3/4" door plate, US10					232			232	255
3000	Swing clear, full mortise, full or half surface, high frequency,									
3010	Steel base, 5" high, USP				Pr.	159			159	175
3200	Swing clear, full mortise, average frequency									
3210	Steel base, 4-1/2" high, USP				Pr.	127			127	140
4000	Wide throw, average frequency, steel base, 4-1/2" x 6", USP					97			97	107
4200	High frequency, steel base, 4-1/2" x 6", USP					148			148	163
4600	Spring hinge, single acting, 6" flange, steel				Ea.	54.50			54.50	60
4700	Brass					95.50			95.50	105
4900	Double acting, 6" flange, steel					96.50			96.50	106
4950	Brass					158			158	173
9000	Continuous hinge, steel, full mortise, heavy duty 501-85	2 Carp	64	.250	L.F.	71	8.40		79.40	92

08 71 20.95 Kick Plates		Crew	Daily Output	Labor-Hours	Unit	Material	Labor	Equipment	Total	Total Incl O&P
0010	**KICK PLATES**									
0020	Stainless steel, .050, 16 ga, 8" x 28", US32	1 Carp	15	.533	Ea.	35	17.90		52.90	68.50
0030	8" x 30"		15	.533		36	17.90		53.90	69.50
0040	8" x 34"		15	.533		40	17.90		57.90	74
0050	10" x 28"		15	.533		67	17.90		84.90	104

08 71 20.95 Kick Plates		Crew	Daily Output	Labor-Hours	Unit	Material	2011 Bare Costs Labor	Equipment	Total	Total Incl O&P
0060	10" x 30"	1 Carp	15	.533	Ea.	72	17.90		89.90	109
0070	10" x 34"		15	.533		81	17.90		98.90	119
0080	Mop/Kick, 4" x 28"		15	.533		30	17.90		47.90	63
0090	4" x 30"		15	.533		32	17.90		49.90	65
0100	4" x 34"		15	.533		36	17.90		53.90	69.50
0110	6" x 28"		15	.533		38	17.90		55.90	72
0120	6" x 30"		15	.533		41	17.90		58.90	75
0130	6" x 34"		15	.533		47	17.90		64.90	81.50
0500	Bronze, .050", 8" x 28"		15	.533		47	17.90		64.90	81.50
0510	8" x 30"		15	.533		63	17.90		80.90	99.50
0520	8" x 34"		15	.533		71	17.90		88.90	108
0530	10" x 28"		15	.533		73	17.90		90.90	111
0540	10" x 30"		15	.533		78	17.90		95.90	116
0550	10" x 34"		15	.533		89	17.90		106.90	128
0560	Mop/Kick, 4" x 28"		15	.533		31	17.90		48.90	64
0570	4" x 30"		15	.533		34	17.90		51.90	67.50
0580	4" x 34"		15	.533		35	17.90		52.90	68.50
0590	6" x 28"		15	.533		44	17.90		61.90	78.50
0600	6" x 30"		15	.533		50	17.90		67.90	85
0610	6" x 34"		15	.533		54	17.90		71.90	89.50
1000	Acrylic, .125", 8" x 26"		15	.533		28	17.90		45.90	61
1010	8" x 36"		15	.533		38	17.90		55.90	72
1020	8" x 42"		15	.533		45	17.90		62.90	79.50
1030	10" x 26"		15	.533		35	17.90		52.90	68.50
1040	10" x 36"		15	.533		48	17.90		65.90	83
1050	10" x 42"		15	.533		68.50	17.90		86.40	106
1060	Mop/Kick, 4" x 26"		15	.533		17	17.90		34.90	48.50
1070	4" x 36"		15	.533		24	17.90		41.90	56.50
1080	4" x 42"		15	.533		27	17.90		44.90	59.50
1090	6" x 26"		15	.533		23	17.90		40.90	55.50
1100	6" x 36"		15	.533		34	17.90		51.90	67.50
1110	6" x 42"		15	.533		39	17.90		56.90	73
1220	Brass, .050", 8" x 26"		15	.533		56	17.90		73.90	91.50
1230	8" x 36"		15	.533		75	17.90		92.90	113
1240	8" x 42"		15	.533		86	17.90		103.90	125
1250	10" x 26"		15	.533		71	17.90		88.90	108
1260	10" x 36"		15	.533		91	17.90		108.90	130
1270	10" x 42"		15	.533		105	17.90		122.90	146
1320	Mop/Kick, 4" x 26"		15	.533		28	17.90		45.90	61
1330	4" x 36"		15	.533		39	17.90		56.90	73
1340	4" x 42"		15	.533		44	17.90		61.90	78.50
1350	6" x 26"		15	.533		38	17.90		55.90	72
1360	6" x 36"		15	.533		48	17.90		65.90	83
1370	6" x 42"		15	.533		55	17.90		72.90	90.50
1800	Aluminum, .050", 8" x 26"		15	.533		22	17.90		39.90	54
1810	8" x 36"		15	.533		30.50	17.90		48.40	63.50
1820	8" x 42"		15	.533		35.50	17.90		53.40	69
1830	10" x 26"		15	.533		27.50	17.90		45.40	60.50
1840	10" x 36"		15	.533		38	17.90		55.90	72
1850	10" x 42"		15	.533		44.50	17.90		62.40	79
1860	Mop/Kick, 4" x 26"		15	.533		11	17.90		28.90	42
1870	4" x 36"		15	.533		15.50	17.90		33.40	47
1880	4" x 42"		15	.533		17	17.90		34.90	48.50

08 71 Door Hardware

08 71 20 – Hardware

08 71 20.95 Kick Plates		Crew	Daily Output	Labor-Hours	Unit	Material	2011 Bare Costs Labor	Equipment	Total	Total Incl O&P
1890	6″ x 26″	1 Carp	15	.533	Ea.	16.50	17.90		34.40	48
1900	6″ x 36″		15	.533		23	17.90		40.90	55.50
1910	6″ x 42″		15	.533		27	17.90		44.90	59.50

08 71 21 – Astragals

08 71 21.10 Exterior Mouldings, Astragals

		Crew	Daily Output	Labor-Hours	Unit	Material	Labor	Equipment	Total	Total Incl O&P
0010	**EXTERIOR MOULDINGS, ASTRAGALS**									
0400	One piece, overlapping cadmium plated steel, flat, 3/16″ x 2″	1 Carp	90	.089	L.F.	3.60	2.98		6.58	8.95
0600	Prime coated steel, flat, 1/8″ x 3″		90	.089		4.75	2.98		7.73	10.25
0800	Stainless steel, flat, 3/32″ x 1-5/8″		90	.089		17.95	2.98		20.93	24.50
1000	Aluminum, flat, 1/8″ x 2″		90	.089		3.60	2.98		6.58	8.95
1200	Nail on, "T" extrusion		120	.067		1	2.24		3.24	4.84
1300	Vinyl bulb insert		105	.076		1.40	2.56		3.96	5.80
1600	Screw on, "T" extrusion		90	.089		5.20	2.98		8.18	10.70
1700	Vinyl insert		75	.107		3.50	3.58		7.08	9.85
2000	"L" extrusion, neoprene bulbs		75	.107		2.10	3.58		5.68	8.30
2100	Neoprene sponge insert		75	.107		6.05	3.58		9.63	12.65
2200	Magnetic		75	.107		10	3.58		13.58	17
2400	Spring hinged security seal, with cam		75	.107		6.45	3.58		10.03	13.10
2600	Spring loaded locking bolt, vinyl insert		45	.178		8.70	5.95		14.65	19.55
2800	Neoprene sponge strip, "Z" shaped, aluminum		60	.133		4.25	4.47		8.72	12.20
2900	Solid neoprene strip, nail on aluminum strip		90	.089		3.50	2.98		6.48	8.85
3000	One piece stile protection									
3020	Neoprene fabric loop, nail on aluminum strips	1 Carp	60	.133	L.F.	.90	4.47		5.37	8.50
3110	Flush mounted aluminum extrusion, 1/2″ x 1-1/4″		60	.133		3.35	4.47		7.82	11.20
3140	3/4″ x 1-3/8″		60	.133		4.25	4.47		8.72	12.20
3160	1-1/8″ x 1-3/4″		60	.133		6.70	4.47		11.17	14.85
3300	Mortise, 9/16″ x 3/4″		60	.133		3.65	4.47		8.12	11.50
3320	13/16″ x 1-3/8″		60	.133		3.90	4.47		8.37	11.80
3600	Spring bronze strip, nail on type		105	.076		3.15	2.56		5.71	7.75
3620	Screw on, with retainer		75	.107		2.55	3.58		6.13	8.80
3800	Flexible stainless steel housing, pile insert, 1/2″ door		105	.076		6.85	2.56		9.41	11.85
3820	3/4″ door		105	.076		7.70	2.56		10.26	12.75
4000	Extruded aluminum retainer, flush mount, pile insert		105	.076		2.45	2.56		5.01	7
4080	Mortise, felt insert		90	.089		4.30	2.98		7.28	9.70
4160	Mortise with spring, pile insert		90	.089		3.40	2.98		6.38	8.75
4400	Rigid vinyl retainer, mortise, pile insert		105	.076		2.45	2.56		5.01	7
4600	Wool pile filler strip, aluminum backing		105	.076		2.50	2.56		5.06	7.05
5000	Two piece overlapping astragal, extruded aluminum retainer									
5010	Pile insert	1 Carp	60	.133	L.F.	3.15	4.47		7.62	10.95
5020	Vinyl bulb insert		60	.133		2.10	4.47		6.57	9.80
5040	Vinyl flap insert		60	.133		6.25	4.47		10.72	14.40
5060	Solid neoprene flap insert		60	.133		6.20	4.47		10.67	14.30
5080	Hypalon rubber flap insert		60	.133		6.30	4.47		10.77	14.45
5090	Snap on cover, pile insert		60	.133		7.20	4.47		11.67	15.40
5400	Magnetic aluminum, surface mounted		60	.133		24.50	4.47		28.97	34.50
5500	Interlocking aluminum, 5/8″ x 1″ neoprene bulb insert		45	.178		3.95	5.95		9.90	14.35
5600	Adjustable aluminum, 9/16″ x 21/32″, pile insert		45	.178		18.25	5.95		24.20	30
5790	For vinyl bulb, deduct					.65			.65	.72
5800	Magnetic, adjustable, 9/16″ x 21/32″	1 Carp	45	.178		23.50	5.95		29.45	36
6000	Two piece stile protection									
6010	Cloth backed rubber loop, 1″ gap, nail on aluminum strips	1 Carp	45	.178	L.F.	4.10	5.95		10.05	14.50
6040	Screw on aluminum strips		45	.178		6.20	5.95		12.15	16.80

08 71 Door Hardware

08 71 21 – Astragals

08 71 21.10 Exterior Mouldings, Astragals		Crew	Daily Output	Labor-Hours	Unit	2011 Bare Costs Material	Labor	Equipment	Total	Total Incl O&P
6100	1-1/2″ gap, screw on aluminum extrusion	1 Carp	45	.178	L.F.	5.55	5.95		11.50	16.10
6240	Vinyl fabric loop, slotted aluminum extrusion, 1″ gap		45	.178		2.05	5.95		8	12.25
6300	1-1/4″ gap		45	.178		5.85	5.95		11.80	16.45

08 71 25 – Weatherstripping

08 71 25.10 Mechanical Seals, Weatherstripping		Crew	Daily Output	Labor-Hours	Unit	Material	Labor	Equipment	Total	Total Incl O&P
0010	**MECHANICAL SEALS, WEATHERSTRIPPING**									
1000	Doors, wood frame, interlocking, for 3′ x 7′ door, zinc	1 Carp	3	2.667	Opng.	16	89.50		105.50	168
1100	Bronze		3	2.667		25	89.50		114.50	178
1300	6′ x 7′ opening, zinc		2	4		18	134		152	245
1400	Bronze		2	4		33	134		167	262
1700	Wood frame, spring type, bronze									
1800	3′ x 7′ door	1 Carp	7.60	1.053	Opng.	19.35	35.50		54.85	80.50
1900	6′ x 7′ door	″	7	1.143	″	28	38.50		66.50	94.50
2200	Metal frame, spring type, bronze									
2300	3′ x 7′ door	1 Carp	3	2.667	Opng.	37	89.50		126.50	191
2400	6′ x 7′ door	″	2.50	3.200	″	47	107		154	232
2500	For stainless steel, spring type, add					133%				
2700	Metal frame, extruded sections, 3′ x 7′ door, aluminum	1 Carp	3	2.667	Opng.	46	89.50		135.50	201
2800	Bronze		3	2.667		120	89.50		209.50	282
3100	6′ x 7′ door, aluminum		1.50	5.333		58	179		237	365
3200	Bronze		1.50	5.333		137	179		316	450
3500	Threshold weatherstripping									
3650	Door sweep, flush mounted, aluminum	1 Carp	25	.320	Ea.	13.55	10.75		24.30	33
3700	Vinyl		25	.320		16	10.75		26.75	35.50
5000	Garage door bottom weatherstrip, 12′ aluminum, clear		14	.571		21.50	19.15		40.65	56
5010	Bronze		14	.571		84	19.15		103.15	125
5050	Bottom protection, Rubber		14	.571		24.50	19.15		43.65	58.50
5100	Threshold		14	.571		91.50	19.15		110.65	133

08 74 Access Control Hardware

08 74 13 – Card Key Access Control Hardware

08 74 13.50 Card Key Access		Crew	Daily Output	Labor-Hours	Unit	Material	Labor	Equipment	Total	Total Incl O&P
0010	**CARD KEY ACCESS**, does not include door hardware									
0020	Card type, 1 time zone, minimum				Ea.	665			665	730
0040	Maximum					1,775			1,775	1,950
0060	3 time zones, minimum					2,200			2,200	2,425
0080	Maximum					2,450			2,450	2,700
0100	System with printer, and control console, 3 zones				Total	9,150			9,150	10,100
0120	6 zones				″	12,000			12,000	13,200
0140	For each door, minimum, add				Ea.	1,400			1,400	1,550
0160	Maximum, add				″	2,000			2,000	2,200

08 74 19 – Biometric Identity Access Control Hardware

08 74 19.50 Biometric Identity Access		Crew	Daily Output	Labor-Hours	Unit	Material	Labor	Equipment	Total	Total Incl O&P
0010	**BIOMETRIC IDENTITY ACCESS**									
0220	Hand geometry scanner, mem of 512 users, excl striker/powr	1 Elec	3	2.667	Ea.	2,025	107		2,132	2,400
0230	Memory upgrade for, adds 9,700 user profiles		8	1		264	40.50		304.50	355
0240	Adds 32,500 user profiles		8	1		580	40.50		620.50	705
0250	Prison type, memory of 256 users, excl striker, power		3	2.667		2,600	107		2,707	3,025
0260	Memory upgrade for, adds 3,300 user profiles		8	1		211	40.50		251.50	298
0270	Adds 9,700 user profiles		8	1		425	40.50		465.50	530

08 74 Access Control Hardware

08 74 19 – Biometric Identity Access Control Hardware

08 74 19.50 Biometric Identity Access		Crew	Daily Output	Labor-Hours	Unit	2011 Bare Costs Material	Labor	Equipment	Total	Total Incl O&P
0280	Adds 27,900 user profiles	1 Elec	8	1	Ea.	580	40.50		620.50	705
0290	All weather, mem of 512 users, excl striker/pwr		3	2.667		3,900	107		4,007	4,450
0300	Facial & fingerprint scanner, combination unit, excl striker/power		3	2.667		4,200	107		4,307	4,800
0310	Access for, for initial setup, excl striker/power	↓	3	2.667	↓	1,000	107		1,107	1,275

08 75 Window Hardware

08 75 30 – Weatherstripping

08 75 30.10 Mechanical Weather Seals		Crew	Daily Output	Labor-Hours	Unit	Material	Labor	Equipment	Total	Total Incl O&P
0010	**MECHANICAL WEATHER SEALS**, Window, double hung, 3' X 5'									
0020	Zinc	1 Carp	7.20	1.111	Opng.	12.80	37.50		50.30	76.50
0100	Bronze		7.20	1.111		28	37.50		65.50	93.50
0500	As above but heavy duty, zinc		4.60	1.739		18	58.50		76.50	117
0600	Bronze	↓	4.60	1.739	↓	32	58.50		90.50	133

08 79 Hardware Accessories

08 79 13 – Key Storage Equipment

08 79 13.10 Key Cabinets		Crew	Daily Output	Labor-Hours	Unit	Material	Labor	Equipment	Total	Total Incl O&P
0010	**KEY CABINETS**									
0020	Wall mounted, 60 key capacity	1 Carp	20	.400	Ea.	95.50	13.40		108.90	128
0200	Drawer type, 600 key capacity	1 Clab	15	.533		725	14.10		739.10	825
0300	2,400 key capacity		20	.400		4,200	10.60		4,210.60	4,650
0400	Tray type, 20 key capacity		50	.160		41.50	4.23		45.73	52.50
0500	50 key capacity	↓	40	.200	↓	92.50	5.30		97.80	111

08 79 20 – Door Accessories

08 79 20.10 Door Hardware Accessories		Crew	Daily Output	Labor-Hours	Unit	Material	Labor	Equipment	Total	Total Incl O&P
0010	**DOOR HARDWARE ACCESSORIES**									
0140	Door bolt, surface, 4"	1 Carp	32	.250	Ea.	9	8.40		17.40	24
0160	Door latch	"	12	.667	"	8.45	22.50		30.95	47
0200	Sliding closet door									
0220	Track and hanger, single	1 Carp	10	.800	Ea.	54	27		81	105
0240	Double		8	1		75	33.50		108.50	139
0260	Door guide, single		48	.167		25	5.60		30.60	37
0280	Double		48	.167		34	5.60		39.60	47
0600	Deadbolt and lock cover plate, brass or stainless steel		30	.267		28	8.95		36.95	46
0620	Hole cover plate, brass or chrome		35	.229		7.20	7.65		14.85	21
2240	Mortise lockset, passage, lever handle		9	.889		198	30		228	268
4000	Security chain, standard	↓	18	.444	↓	7.90	14.90		22.80	33.50

08 81 Glass Glazing

08 81 10 – Float Glass

08 81 10.10 Various Types and Thickness of Float Glass		Crew	Daily Output	Labor-Hours	Unit	2011 Bare Costs Material	Labor	Equipment	Total	Total Incl O&P
0010	**VARIOUS TYPES AND THICKNESS OF FLOAT GLASS** R088110-10									
0020	3/16″ Plain	2 Glaz	130	.123	S.F.	4.63	4.09		8.72	11.80
0200	Tempered, clear		130	.123		6.45	4.09		10.54	13.80
0300	Tinted		130	.123		7.35	4.09		11.44	14.75
0600	1/4″ thick, clear, plain		120	.133		5.40	4.43		9.83	13.20
0700	Tinted		120	.133		8	4.43		12.43	16.05
0800	Tempered, clear		120	.133		7.85	4.43		12.28	15.90
0900	Tinted		120	.133		10.05	4.43		14.48	18.30
1600	3/8″ thick, clear, plain		75	.213		9.45	7.10		16.55	22
1700	Tinted		75	.213		14.80	7.10		21.90	28
1800	Tempered, clear		75	.213		15.60	7.10		22.70	29
1900	Tinted		75	.213		17.40	7.10		24.50	31
2200	1/2″ thick, clear, plain		55	.291		16	9.65		25.65	33.50
2300	Tinted		55	.291		26	9.65		35.65	44.50
2400	Tempered, clear		55	.291		23	9.65		32.65	41.50
2500	Tinted		55	.291		24	9.65		33.65	42.50
2800	5/8″ thick, clear, plain		45	.356		26	11.80		37.80	48
2900	Tempered, clear		45	.356		29.50	11.80		41.30	52
3200	3/4″ thick, clear, plain		35	.457		33.50	15.20		48.70	61.50
3300	Tempered, clear		35	.457		39	15.20		54.20	68
3600	1″ thick, clear, plain		30	.533		55.50	17.70		73.20	90
8900	For low emissivity coating for 3/16″ & 1/4″ only, add to above					16%				

08 81 13 – Decorative Glass Glazing

08 81 13.10 Beveled Glass		Crew	Daily Output	Labor-Hours	Unit	Material	Labor	Equipment	Total	Total Incl O&P
0010	**BEVELED GLASS**, with design patterns									
0020	Minimum	2 Glaz	150	.107	S.F.	58	3.54		61.54	70
0050	Average		125	.128		129	4.25		133.25	149
0100	Maximum		100	.160		229	5.30		234.30	260
0102	Maximum, per L.F. for 30″ cabinet	1 Glaz	40	.200	L.F.	570	6.65		576.65	640

08 81 13.20 Faceted Glass		Crew	Daily Output	Labor-Hours	Unit	Material	Labor	Equipment	Total	Total Incl O&P
0010	**FACETED GLASS**, Color tinted 3/4″ thick									
0020	Minimum	2 Glaz	95	.168	S.F.	46.50	5.60		52.10	60
0100	Maximum	″	75	.213	″	81.50	7.10		88.60	101

08 81 13.30 Sandblasted Glass		Crew	Daily Output	Labor-Hours	Unit	Material	Labor	Equipment	Total	Total Incl O&P
0010	**SANDBLASTED GLASS**, float glass									
0020	1/8″ thick	2 Glaz	160	.100	S.F.	10.40	3.32		13.72	16.90
0100	3/16″ thick		130	.123		11.50	4.09		15.59	19.35
0500	1/4″ thick		120	.133		12	4.43		16.43	20.50
0600	3/8″ thick		75	.213		12.90	7.10		20	26

08 81 20 – Vision Panels

08 81 20.10 Full Vision		Crew	Daily Output	Labor-Hours	Unit	Material	Labor	Equipment	Total	Total Incl O&P
0010	**FULL VISION**, window system with 3/4″ glass mullions									
0020	Up to 10′ high	H-2	130	.185	S.F.	59.50	5.70		65.20	75
0100	10′ to 20′ high, minimum		110	.218		63	6.75		69.75	80
0150	Average		100	.240		68	7.45		75.45	87.50
0200	Maximum		80	.300		76.50	9.30		85.80	99.50

08 81 25 – Glazing Variables

08 81 25.10 Applications of Glazing		Crew	Daily Output	Labor-Hours	Unit	Material	Labor	Equipment	Total	Total Incl O&P
0010	**APPLICATIONS OF GLAZING** R088110-10									
0500	For high rise glazing, exterior, add per S.F. per story				S.F.					.40

08 81 Glass Glazing

08 81 25 – Glazing Variables

08 81 25.10 Applications of Glazing		Crew	Daily Output	Labor-Hours	Unit	Material	2011 Bare Costs Labor	Equipment	Total	Total Incl O&P
0600	For glass replacement, add				S.F.		100%			
0700	For gasket settings, add				L.F.	5.55			5.55	6.10
0900	For sloped glazing, add				S.F.		25%			
2000	Fabrication, polished edges, 1/4″ thick				Inch	.50			.50	.55
2100	1/2″ thick					1.23			1.23	1.35
2500	Mitered edges, 1/4″ thick					1.23			1.23	1.35
2600	1/2″ thick					2			2	2.20

08 81 30 – Insulating Glass

08 81 30.10 Reduce Heat Transfer Glass

Line	Description	Crew	Daily Output	Labor-Hours	Unit	Material	Labor	Equipment	Total	Total Incl O&P
0010	**REDUCE HEAT TRANSFER GLASS** R088110-10									
0015	2 lites 1/8″ float, 1/2″ thk under 15 S.F.									
0020	Clear G	2 Glaz	95	.168	S.F.	9.40	5.60		15	19.55
0100	Tinted G		95	.168		13.10	5.60		18.70	23.50
0200	2 lites 3/16″ float, for 5/8″ thk unit, 15 to 30 S.F., clear G		90	.178		12.95	5.90		18.85	24
0300	Tinted G		90	.178		13.30	5.90		19.20	24.50
0400	1″ thk, dbl. glazed, 1/4″ float, 30-70 S.F., clear G		75	.213		15.65	7.10		22.75	29
0500	Tinted G		75	.213		22	7.10		29.10	36
0600	1″ thick double glazed, 1/4″ float, 1/4″ wire		75	.213		22	7.10		29.10	35.50
0700	1/4″ float, 1/4″ tempered		75	.213		18.55	7.10		25.65	32
0800	1/4″ wire, 1/4″ tempered		75	.213		27.50	7.10		34.60	42
0900	Both lites, 1/4″ wire		75	.213		35	7.10		42.10	50
2000	Both lites, light & heat reflective G		85	.188		29.50	6.25		35.75	43
2500	Heat reflective, film inside, 1″ thick unit, clear G		85	.188		26	6.25		32.25	39
2600	Tinted G		85	.188		27	6.25		33.25	40
3000	Film on weatherside, clear, 1/2″ thick unit G		95	.168		18.60	5.60		24.20	29.50
3100	5/8″ thick unit G		90	.178		18.60	5.90		24.50	30
3200	1″ thick unit G		85	.188		25.50	6.25		31.75	38.50

08 81 35 – Translucent Glass

08 81 35.10 Obscure Glass

Line	Description	Crew	Daily Output	Labor-Hours	Unit	Material	Labor	Equipment	Total	Total Incl O&P
0010	**OBSCURE GLASS**									
0020	1/8″ thick, Minimum	2 Glaz	140	.114	S.F.	11	3.79		14.79	18.35
0100	Maximum		125	.128		13	4.25		17.25	21.50
0300	7/32″ thick, minimum		120	.133		12	4.43		16.43	20.50
0400	Maximum		105	.152		15.10	5.05		20.15	25

08 81 35.20 Patterned Glass

Line	Description	Crew	Daily Output	Labor-Hours	Unit	Material	Labor	Equipment	Total	Total Incl O&P
0010	**PATTERNED GLASS**, colored									
0020	1/8″ thick, Minimum	2 Glaz	140	.114	S.F.	9	3.79		12.79	16.15
0100	Maximum		125	.128		15	4.25		19.25	23.50
0300	7/32″ thick, minimum		120	.133		11.05	4.43		15.48	19.45
0400	Maximum		105	.152		16	5.05		21.05	26

08 81 45 – Sheet Glass

08 81 45.10 Window Glass, Sheet

Line	Description	Crew	Daily Output	Labor-Hours	Unit	Material	Labor	Equipment	Total	Total Incl O&P
0010	**WINDOW GLASS, SHEET** gray									
0020	1/8″ thick	2 Glaz	160	.100	S.F.	5.40	3.32		8.72	11.40
0200	1/4″ thick	″	130	.123	″	6.60	4.09		10.69	13.95

08 81 50 – Spandrel Glass

08 81 50.10 Glass for Non Vision Areas

Line	Description	Crew	Daily Output	Labor-Hours	Unit	Material	Labor	Equipment	Total	Total Incl O&P
0010	**GLASS FOR NON VISION AREAS**, 1/4″ thick standard colors									
0020	Up to 1000 S.F.	2 Glaz	110	.145	S.F.	15.55	4.83		20.38	25
0200	1,000 to 2,000 S.F.	″	120	.133	″	14.45	4.43		18.88	23

08 81 Glass Glazing

08 81 50 – Spandrel Glass

08 81 50.10 Glass for Non Vision Areas		Crew	Daily Output	Labor-Hours	Unit	Material	Labor	Equipment	Total	Total Incl O&P
						2011 Bare Costs				
0300	For custom colors, add				Total	10%				
0500	For 3/8" thick, add				S.F.	10.70			10.70	11.75
1000	For double coated, 1/4" thick, add					4			4	4.40
1200	For insulation on panels, add					6.60			6.60	7.25
2000	Panels, insulated, with aluminum backed fiberglass, 1" thick	2 Glaz	120	.133		15.90	4.43		20.33	25
2100	2" thick	"	120	.133		18.90	4.43		23.33	28.50
2500	With galvanized steel backing, add					5.50			5.50	6.05

08 81 55 – Window Glass

08 81 55.10 Sheet Glass		Crew	Daily Output	Labor-Hours	Unit	Material	Labor	Equipment	Total	Total Incl O&P
0010	**SHEET GLASS** (window), clear float, stops, putty bed									
0015	1/8" thick, clear float	2 Glaz	480	.033	S.F.	3.40	1.11		4.51	5.55
0500	3/16" thick, clear		480	.033		5.40	1.11		6.51	7.75
0600	Tinted		480	.033		7.15	1.11		8.26	9.65
0700	Tempered		480	.033		8.65	1.11		9.76	11.30

08 81 65 – Wire Glass

08 81 65.10 Glass Reinforced With Wire		Crew	Daily Output	Labor-Hours	Unit	Material	Labor	Equipment	Total	Total Incl O&P
0010	**GLASS REINFORCED WITH WIRE**									
0012	1/4" thick rough obscure	2 Glaz	135	.119	S.F.	17.75	3.93		21.68	26
1000	Polished wire, 1/4" thick, diamond, clear		135	.119		20.50	3.93		24.43	29
1500	Pinstripe, obscure		135	.119		22	3.93		25.93	30.50

08 83 Mirrors

08 83 13 – Mirrored Glass Glazing

08 83 13.10 Mirrors		Crew	Daily Output	Labor-Hours	Unit	Material	Labor	Equipment	Total	Total Incl O&P
0010	**MIRRORS**, No frames, wall type, 1/4" plate glass, polished edge									
0100	Up to 5 S.F.	2 Glaz	125	.128	S.F.	9.35	4.25		13.60	17.30
0200	Over 5 S.F.		160	.100		9.05	3.32		12.37	15.40
0500	Door type, 1/4" plate glass, up to 12 S.F.		160	.100		8.20	3.32		11.52	14.45
1000	Float glass, up to 10 S.F., 1/8" thick		160	.100		5.45	3.32		8.77	11.40
1100	3/16" thick		150	.107		6.65	3.54		10.19	13.10
1500	12" x 12" wall tiles, square edge, clear		195	.082		1.90	2.72		4.62	6.55
1600	Veined		195	.082		4.86	2.72		7.58	9.80
2000	1/4" thick, stock sizes, one way transparent		125	.128		18.10	4.25		22.35	27
2010	Bathroom, unframed, laminated		160	.100		12.60	3.32		15.92	19.30

08 83 13.15 Reflective Glass		Crew	Daily Output	Labor-Hours	Unit	Material	Labor	Equipment	Total	Total Incl O&P
0010	**REFLECTIVE GLASS**									
0100	1/4" float with fused metallic oxide fixed G	2 Glaz	115	.139	S.F.	16	4.62		20.62	25
0500	1/4" float glass with reflective applied coating G	"	115	.139	"	13	4.62		17.62	22

08 84 Plastic Glazing

08 84 10 – Plexiglass Glazing

08 84 10.10 Plexiglass Acrylic		Crew	Daily Output	Labor-Hours	Unit	2011 Bare Costs Material	Labor	Equipment	Total	Total Incl O&P
0010	**PLEXIGLASS ACRYLIC**, clear, masked,									
0020	1/8" thick, cut sheets	2 Glaz	170	.094	S.F.	5.30	3.12		8.42	11
0200	Full sheets		195	.082		2.75	2.72		5.47	7.50
0500	1/4" thick, cut sheets		165	.097		9.20	3.22		12.42	15.40
0600	Full sheets		185	.086		5	2.87		7.87	10.20
0900	3/8" thick, cut sheets		155	.103		16.85	3.43		20.28	24
1000	Full sheets		180	.089		9.10	2.95		12.05	14.85
1300	1/2" thick, cut sheets		135	.119		19.40	3.93		23.33	28
1400	Full sheets		150	.107		18.80	3.54		22.34	26.50
1700	3/4" thick, cut sheets		115	.139		69	4.62		73.62	83.50
1800	Full sheets		130	.123		40	4.09		44.09	50.50
2100	1" thick, cut sheets		105	.152		77.50	5.05		82.55	94
2200	Full sheets		125	.128		48	4.25		52.25	60
3000	Colored, 1/8" thick, cut sheets		170	.094		16	3.12		19.12	23
3200	Full sheets		195	.082		10.30	2.72		13.02	15.80
3500	1/4" thick, cut sheets		165	.097		18	3.22		21.22	25
3600	Full sheets		185	.086		12.20	2.87		15.07	18.10
4000	Mirrors, untinted, cut sheets, 1/8" thick		185	.086		7.50	2.87		10.37	12.95
4200	1/4" thick		180	.089		12.10	2.95		15.05	18.15

08 84 20 – Polycarbonate

08 84 20.10 Thermoplastic		Crew	Daily Output	Labor-Hours	Unit	Material	Labor	Equipment	Total	Total Incl O&P
0010	**THERMOPLASTIC**, clear, masked, cut sheets									
0020	1/8" thick	2 Glaz	170	.094	S.F.	9.70	3.12		12.82	15.80
0500	3/16" thick		165	.097		11.10	3.22		14.32	17.50
1000	1/4" thick		155	.103		12.50	3.43		15.93	19.40
1500	3/8" thick		150	.107		22	3.54		25.54	30.50

08 84 30 – Vinyl Glass

08 84 30.10 Vinyl Sheets		Crew	Daily Output	Labor-Hours	Unit	Material	Labor	Equipment	Total	Total Incl O&P
0010	**VINYL SHEETS**, Steel mesh reinforced, stock sizes									
0020	.090" thick	2 Glaz	170	.094	S.F.	13.10	3.12		16.22	19.55
0500	.120" thick		170	.094		16	3.12		19.12	23
1000	.250" thick		155	.103		21.50	3.43		24.93	29
1500	For non-standard sizes, add					18%				

08 87 Glazing Surface Films

08 87 13 – Solar Control Films

08 87 13.10 Solar Films On Glass		Crew	Daily Output	Labor-Hours	Unit	Material	Labor	Equipment	Total	Total Incl O&P
0010	**SOLAR FILMS ON GLASS** (glass not included)									
2000	Minimum G	2 Glaz	180	.089	S.F.	6.50	2.95		9.45	12
2050	Maximum G	"	225	.071	"	14.50	2.36		16.86	19.85

08 87 23 – Safety and Security Films

08 87 23.13 Safety Films		Crew	Daily Output	Labor-Hours	Unit	Material	Labor	Equipment	Total	Total Incl O&P
0010	**SAFETY FILMS**									
2010	Window protection film, controls blast damage G	2 Glaz	80	.200	S.F.	6.50	6.65		13.15	18.05

08 87 23.16 Security Films		Crew	Daily Output	Labor-Hours	Unit	Material	Labor	Equipment	Total	Total Incl O&P
0010	**SECURITY FILMS**, clear, 32000 psi tensile strength, adhered to glass R088110-10									
0100	.002" thick, daylight installation	H-2	950	.025	S.F.	2.20	.78		2.98	3.71
0150	.004" thick, daylight installation		800	.030		2.30	.93		3.23	4.06
0200	.006" thick, daylight installation		700	.034		2.45	1.06		3.51	4.45

08 87 Glazing Surface Films

08 87 23 – Safety and Security Films

08 87 23.16 Security Films		Crew	Daily Output	Labor-Hours	Unit	Material	Labor	Equipment	Total	Total Incl O&P
						2011 Bare Costs				
0210	Install for anchorage	H-2	600	.040	S.F.	2.72	1.24		3.96	5.05
0400	.007" thick, daylight installation		600	.040		2.65	1.24		3.89	4.96
0410	Install for anchorage		500	.048		2.94	1.49		4.43	5.70
0500	.008" thick, daylight installation		500	.048		2.85	1.49		4.34	5.60
0510	Install for anchorage		500	.048		3.17	1.49		4.66	5.95
0600	.015" thick, daylight installation		400	.060		4.75	1.86		6.61	8.30
0610	Install for anchorage		400	.060		3.17	1.86		5.03	6.55
0900	Security film anchorage, mechanical attachment and cover plate	H-3	370	.043	L.F.	9.05	1.27		10.32	12.05
0950	Security film anchorage, wet glaze structural caulking	1 Glaz	225	.036	"	.83	1.18		2.01	2.85
1000	Adhered security film removal	1 Clab	275	.029	S.F.		.77		.77	1.29

08 88 Special Function Glazing

08 88 40 – Acoustical Glass Units

08 88 40.10 Sound Reduction Units

		Crew	Daily Output	Labor-Hours	Unit	Material	Labor	Equipment	Total	Total Incl O&P
0010	**SOUND REDUCTION UNITS**, 1lite at 3/8", 1 lite at 3/16"									
0020	For 1" thick	2 Glaz	100	.160	S.F.	32	5.30		37.30	44
0100	For 4" thick	"	80	.200	"	55	6.65		61.65	71.50

08 88 56 – Ballistics-Resistant Glazing

08 88 56.10 Laminated Glass

		Crew	Daily Output	Labor-Hours	Unit	Material	Labor	Equipment	Total	Total Incl O&P
0010	**LAMINATED GLASS**									
0020	Clear float .03" vinyl 1/4"	2 Glaz	90	.178	S.F.	11.55	5.90		17.45	22.50
0100	3/8" thick		78	.205		21	6.80		27.80	34
0200	.06" vinyl, 1/2" thick		65	.246		24	8.15		32.15	40
1000	5/8" thick		90	.178		28	5.90		33.90	40.50
2000	Bullet-resisting, 1-3/16" thick, to 15 S.F.		16	1		79.50	33		112.50	142
2100	Over 15 S.F.		16	1		96	33		129	160
2500	2-1/4" thick, to 15 S.F.		12	1.333		93	44.50		137.50	175
2600	Over 15 S.F.		12	1.333		83	44.50		127.50	164
2700	Level 2 (.357 magnum)		12	1.333		65	44.50		109.50	144
2750	Level 3A (.44 magnum)		12	1.333		70	44.50		114.50	150
2800	Level 4 (AK-47)		12	1.333		91.50	44.50		136	174
2850	Level 5 (M-16)		12	1.333		102	44.50		146.50	185
2900	Level 3 (7.62 Armor Piercing)		12	1.333		122	44.50		166.50	207

08 91 Louvers

08 91 19 – Fixed Louvers

08 91 19.10 Aluminum Louvers

		Crew	Daily Output	Labor-Hours	Unit	Material	Labor	Equipment	Total	Total Incl O&P
0010	**ALUMINUM LOUVERS**									
0020	Aluminum with screen, residential, 8" x 8"	1 Carp	38	.211	Ea.	10.55	7.05		17.60	23.50
0100	12" x 12"		38	.211		15	7.05		22.05	28.50
0200	12" x 18"		35	.229		18.85	7.65		26.50	33.50
0250	14" x 24"		30	.267		26	8.95		34.95	43.50
0300	18" x 24"		27	.296		29	9.95		38.95	48.50
0500	24" x 30"		24	.333		32	11.20		43.20	53.50
0700	Triangle, adjustable, small		20	.400		32	13.40		45.40	57.50
0800	Large		15	.533		41	17.90		58.90	75
1200	Extruded aluminum, see Section 23 37 15.40									
2100	Midget, aluminum, 3/4" deep, 1" diameter	1 Carp	85	.094	Ea.	.69	3.16		3.85	6.05
2150	3" diameter		60	.133		1.87	4.47		6.34	9.55

08 91 Louvers

08 91 19 – Fixed Louvers

08 91 19.10 Aluminum Louvers		Crew	Daily Output	Labor-Hours	Unit	Material	Labor	Equipment	Total	Total Incl O&P
						2011 Bare Costs				
2200	4" diameter	1 Carp	50	.160	Ea.	3.64	5.35		8.99	13
2250	6" diameter	↓	30	.267	↓	3.74	8.95		12.69	19.05

08 91 26 – Door Louvers

08 91 26.10 Steel Louvers, 18 Gauge, Fixed Blade		Crew	Daily Output	Labor-Hours	Unit	Material	Labor	Equipment	Total	Total Incl O&P
0010	**STEEL LOUVERS, 18 GAUGE, FIXED BLADE**									
0050	12" x 12", with enamel or powder coat	1 Carp	20	.400	Ea.	67.50	13.40		80.90	97
0055	18" x 12"		20	.400		76	13.40		89.40	106
0060	18" x 18"		20	.400		83	13.40		96.40	114
0065	24" x 12"		20	.400		91.50	13.40		104.90	124
0070	24" x 18"		20	.400		126	13.40		139.40	162
0075	24" x 24"		20	.400		130	13.40		143.40	166
0100	12" x 12", galvanized		20	.400		58	13.40		71.40	86.50
0105	18" x 12"		20	.400		65.50	13.40		78.90	94.50
0115	24" x 12"		20	.400		82	13.40		95.40	113
0125	24" x 24"	↓	20	.400	↓	119	13.40		132.40	153

08 95 Vents

08 95 13 – Soffit Vents

08 95 13.10 Wall Louvers		Crew	Daily Output	Labor-Hours	Unit	Material	Labor	Equipment	Total	Total Incl O&P
0010	**WALL LOUVERS**									
2330	Soffit vent, continuous, 3" wide, aluminum, mill finish	1 Carp	200	.040	L.F.	.48	1.34		1.82	2.78
2340	Baked enamel finish		200	.040	"	.52	1.34		1.86	2.82
2400	Under eaves vent, aluminum, mill finish, 16" x 4"		48	.167	Ea.	1.37	5.60		6.97	10.85
2500	16" x 8"	↓	48	.167	"	1.93	5.60		7.53	11.45

08 95 16 – Wall Vents

08 95 16.10 Louvers		Crew	Daily Output	Labor-Hours	Unit	Material	Labor	Equipment	Total	Total Incl O&P
0010	**LOUVERS**									
0020	Redwood, 2'-0" diameter, full circle	1 Carp	16	.500	Ea.	169	16.80		185.80	214
0100	Half circle		16	.500		189	16.80		205.80	236
0200	Octagonal		16	.500		129	16.80		145.80	170
0300	Triangular, 5/12 pitch, 5'-0" at base		16	.500		189	16.80		205.80	236
7000	Vinyl gable vent, 8" x 8"		38	.211		11.55	7.05		18.60	24.50
7020	12" x 12"		38	.211		22.50	7.05		29.55	37
7080	12" x 18"		35	.229		29.50	7.65		37.15	45.50
7200	18" x 24"	↓	30	.267	↓	36.50	8.95		45.45	55

Estimating Tips

General

- Room Finish Schedule: A complete set of plans should contain a room finish schedule. If one is not available, it would be well worth the time and effort to obtain one.

09 20 00 Plaster and Gypsum Board

- Lath is estimated by the square yard plus a 5% allowance for waste. Furring, channels, and accessories are measured by the linear foot. An extra foot should be allowed for each accessory miter or stop.
- Plaster is also estimated by the square yard. Deductions for openings vary by preference, from zero deduction to 50% of all openings over 2 feet in width. The estimator should allow one extra square foot for each linear foot of horizontal interior or exterior angle located below the ceiling level. Also, double the areas of small radius work.
- Drywall accessories, studs, track, and acoustical caulking are all measured by the linear foot. Drywall taping is figured by the square foot. Gypsum wallboard is estimated by the square foot. No material deductions should be made for door or window openings under 32 S.F.

09 60 00 Flooring

- Tile and terrazzo areas are taken off on a square foot basis. Trim and base materials are measured by the linear foot. Accent tiles are listed per each. Two basic methods of installation are used. Mud set is approximately 30% more expensive than thin set. In terrazzo work, be sure to include the linear footage of embedded decorative strips, grounds, machine rubbing, and power cleanup.
- Wood flooring is available in strip, parquet, or block configuration. The latter two types are set in adhesives with quantities estimated by the square foot. The laying pattern will influence labor costs and material waste. In addition to the material and labor for laying wood floors, the estimator must make allowances for sanding and finishing these areas unless the flooring is prefinished.
- Sheet flooring is measured by the square yard. Roll widths vary, so consideration should be given to use the most economical width, as waste must be figured into the total quantity. Consider also the installation methods available, direct glue down or stretched.

09 70 00 Wall Finishes

- Wall coverings are estimated by the square foot. The area to be covered is measured, length by height of wall above baseboards, to calculate the square footage of each wall. This figure is divided by the number of square feet in the single roll which is being used. Deduct, in full, the areas of openings such as doors and windows. Where a pattern match is required allow 25%–30% waste.

09 80 00 Acoustic Treatment

- Acoustical systems fall into several categories. The takeoff of these materials should be by the square foot of area with a 5% allowance for waste. Do not forget about scaffolding, if applicable, when estimating these systems.

09 90 00 Painting and Coating

- A major portion of the work in painting involves surface preparation. Be sure to include cleaning, sanding, filling, and masking costs in the estimate.
- Protection of adjacent surfaces is not included in painting costs. When considering the method of paint application, an important factor is the amount of protection and masking required. These must be estimated separately and may be the determining factor in choosing the method of application.

Reference Numbers

Reference numbers are shown in shaded boxes at the beginning of some major classifications. These numbers refer to related items in the Reference Section. The reference information may be an estimating procedure, an alternate pricing method, or technical information.

Note: Not all subdivisions listed here necessarily appear in this publication.

09 01 Maintenance of Finishes

09 01 60 – Maintenance of Flooring

09 01 60.10 Carpet Maintenance		Crew	Daily Output	Labor-Hours	Unit	Material	2011 Bare Costs Labor	Equipment	Total	Total Incl O&P
0010	**CARPET MAINTENANCE**									
0020	Steam clean, per cleaning, minimum	1 Clab	3000	.003	S.F.	.07	.07		.14	.20
0500	Maximum	"	2000	.004	"	.12	.11		.23	.31

09 01 70 – Maintenance of Wall Finishes

09 01 70.10 Gypsum Wallboard Repairs

		Crew	Daily Output	Labor-Hours	Unit	Material	Labor	Equipment	Total	Total Incl O&P
0010	**GYPSUM WALLBOARD REPAIRS**									
0100	Fill and sand, pin/nail holes	1 Carp	960	.008	Ea.		.28		.28	.47
0110	Screw head pops		480	.017			.56		.56	.94
0120	Dents, up to 2" square		48	.167		.01	5.60		5.61	9.35
0130	2" to 4" square		24	.333		.03	11.20		11.23	18.75
0140	Cut square, patch, sand and finish, holes, up to 2" square		12	.667		.04	22.50		22.54	37.50
0150	2" to 4" square		11	.727		.09	24.50		24.59	41
0160	4" to 8" square		10	.800		.22	27		27.22	45
0170	8" to 12" square		8	1		.43	33.50		33.93	56.50
0180	12" to 32" square		6	1.333		1.39	44.50		45.89	76.50
0210	16" by 48"		5	1.600		2.33	53.50		55.83	92.50
0220	32" by 48"		4	2		3.30	67		70.30	116
0230	48" square		3.50	2.286		4.62	76.50		81.12	133
0240	60" square		3.20	2.500		8.30	84		92.30	149
0500	Skim coat surface with joint compound		1600	.005	S.F.	.03	.17		.20	.31
0510	Prepare, retape and refinish joints		60	.133	L.F.	.58	4.47		5.05	8.15

09 05 Common Work Results for Finishes

09 05 05 – Selective Finishes Demolition

09 05 05.10 Selective Demolition, Ceilings

		Crew	Daily Output	Labor-Hours	Unit	Material	Labor	Equipment	Total	Total Incl O&P
0010	**SELECTIVE DEMOLITION, CEILINGS** R024119-10									
0200	Ceiling, drywall, furred and nailed or screwed	2 Clab	800	.020	S.F.		.53		.53	.89
0220	On metal frame		760	.021			.56		.56	.93
0240	On suspension system, including system		720	.022			.59		.59	.98
1000	Plaster, lime and horse hair, on wood lath, incl. lath		700	.023			.60		.60	1.01
1020	On metal lath		570	.028			.74		.74	1.24
1100	Gypsum, on gypsum lath		720	.022			.59		.59	.98
1120	On metal lath		500	.032			.85		.85	1.42
1200	Suspended ceiling, mineral fiber, 2' x 2' or 2' x 4'		1500	.011			.28		.28	.47
1250	On suspension system, incl. system		1200	.013			.35		.35	.59
1500	Tile, wood fiber, 12" x 12", glued		900	.018			.47		.47	.79
1540	Stapled		1500	.011			.28		.28	.47
1580	On suspension system, incl. system		760	.021			.56		.56	.93
2000	Wood, tongue and groove, 1" x 4"		1000	.016			.42		.42	.71
2040	1" x 8"		1100	.015			.38		.38	.64
2400	Plywood or wood fiberboard, 4' x 8' sheets		1200	.013			.35		.35	.59

09 05 05.20 Selective Demolition, Flooring

		Crew	Daily Output	Labor-Hours	Unit	Material	Labor	Equipment	Total	Total Incl O&P
0010	**SELECTIVE DEMOLITION, FLOORING** R024119-10									
0200	Brick with mortar	2 Clab	475	.034	S.F.		.89		.89	1.49
0400	Carpet, bonded, including surface scraping		2000	.008			.21		.21	.35
0440	Scrim applied		8000	.002			.05		.05	.09
0480	Tackless		9000	.002			.05		.05	.08
0550	Carpet tile, releasable adhesive		5000	.003			.08		.08	.14
0560	Permanent adhesive		1850	.009			.23		.23	.38
0601	Composition	1 Clab	200	.040			1.06		1.06	1.77

09 05 Common Work Results for Finishes

09 05 05 – Selective Finishes Demolition

09 05 05.20 Selective Demolition, Flooring		Crew	Daily Output	Labor-Hours	Unit	Material	2011 Bare Costs Labor	Equipment	Total	Total Incl O&P
0800	Resilient, sheet goods	2 Clab	1400	.011	S.F.		.30		.30	.51
0820	For gym floors	″	900	.018			.47		.47	.79
0850	Vinyl or rubber cove base	1 Clab	1000	.008	L.F.		.21		.21	.35
0860	Vinyl or rubber cove base, molded corner	″	1000	.008	Ea.		.21		.21	.35
0870	For glued and caulked installation, add to labor						50%			
0900	Vinyl composition tile, 12″ x 12″	2 Clab	1000	.016	S.F.		.42		.42	.71
2000	Tile, ceramic, thin set		675	.024			.63		.63	1.05
2020	Mud set		625	.026			.68		.68	1.13
2200	Marble, slate, thin set		675	.024			.63		.63	1.05
2220	Mud set		625	.026			.68		.68	1.13
2600	Terrazzo, thin set		450	.036			.94		.94	1.58
2620	Mud set		425	.038			1		1	1.67
2630	Grout removal, carbide tipped, rotary grinder	1 Clab	240	.033	L.F.		.88		.88	1.48
2640	Terrazzo, cast in place	2 Clab	300	.053	S.F.		1.41		1.41	2.36
3000	Wood, block, on end	1 Carp	400	.020			.67		.67	1.12
3200	Parquet		450	.018			.60		.60	1
3400	Strip flooring, interior, 2-1/4″ x 25/32″ thick		325	.025			.83		.83	1.38
3500	Exterior, porch flooring, 1″ x 4″		220	.036			1.22		1.22	2.04
3800	Subfloor, tongue and groove, 1″ x 6″		325	.025			.83		.83	1.38
3820	1″ x 8″		430	.019			.62		.62	1.04
3840	1″ x 10″		520	.015			.52		.52	.86
4000	Plywood, nailed		600	.013			.45		.45	.75
4100	Glued and nailed		400	.020			.67		.67	1.12
4200	Hardboard, 1/4″ thick		760	.011			.35		.35	.59
8000	Remove flooring, bead blast, minimum	A-1A	1000	.008			.27	.22	.49	.70
8100	Maximum		400	.020			.68	.54	1.22	1.73
8150	Mastic only		1500	.005			.18	.14	.32	.46

09 05 05.30 Selective Demolition, Walls and Partitions

		Crew	Daily Output	Labor-Hours	Unit	Material	Labor	Equipment	Total	Total Incl O&P
0010	**SELECTIVE DEMOLITION, WALLS AND PARTITIONS** R024119-10									
0020	Walls, concrete, reinforced	B-39	120	.400	C.F.		10.70	1.62	12.32	19.75
0025	Plain	″	160	.300			8.05	1.22	9.27	14.80
0100	Brick, 4″ to 12″ thick	B-9	220	.182			4.88	.89	5.77	9.15
0200	Concrete block, 4″ thick		1000	.040	S.F.		1.07	.20	1.27	2.01
0280	8″ thick		810	.049			1.33	.24	1.57	2.48
0300	Exterior stucco 1″ thick over mesh		3200	.013			.34	.06	.40	.63
1000	Drywall, nailed or screwed	1 Clab	1000	.008			.21		.21	.35
1010	2 layers		400	.020			.53		.53	.89
1020	Glued and nailed		900	.009			.24		.24	.39
1500	Fiberboard, nailed		900	.009			.24		.24	.39
1520	Glued and nailed		800	.010			.26		.26	.44
1568	Plenum barrier, sheet lead		300	.027			.71		.71	1.18
2000	Movable walls, metal, 5′ high		300	.027			.71		.71	1.18
2020	8′ high		400	.020			.53		.53	.89
2200	Metal or wood studs, finish 2 sides, fiberboard	B-1	520	.046			1.25		1.25	2.10
2250	Lath and plaster		260	.092			2.50		2.50	4.19
2300	Plasterboard (drywall)		520	.046			1.25		1.25	2.10
2350	Plywood		450	.053			1.45		1.45	2.42
2800	Paneling, 4′ x 8′ sheets	1 Clab	475	.017			.45		.45	.75
3000	Plaster, lime and horsehair, on wood lath		400	.020			.53		.53	.89
3020	On metal lath		335	.024			.63		.63	1.06
3400	Gypsum or perlite, on gypsum lath		410	.020			.52		.52	.86
3420	On metal lath		300	.027			.71		.71	1.18

09 05 Common Work Results for Finishes

09 05 05 – Selective Finishes Demolition

09 05 05.30 Selective Demolition, Walls and Partitions		Crew	Daily Output	Labor-Hours	Unit	2011 Bare Costs Material	Labor	Equipment	Total	Total Incl O&P
3450	Plaster, interior gypsum, acoustic, or cement	1 Clab	60	.133	S.Y.		3.53		3.53	5.90
3500	Stucco, on masonry		145	.055			1.46		1.46	2.44
3510	Commercial 3-coat		80	.100			2.65		2.65	4.43
3520	Interior stucco		25	.320			8.45		8.45	14.20
3600	Plywood, one side	B-1	1500	.016	S.F.		.43		.43	.73
3750	Terra cotta block and plaster, to 6″ thick	″	175	.137			3.72		3.72	6.25
3760	Tile, ceramic, on walls, thin set	1 Clab	300	.027			.71		.71	1.18
3765	Mud set		250	.032			.85		.85	1.42
3800	Toilet partitions, slate or marble		5	1.600	Ea.		42.50		42.50	71
3820	Metal or plastic		8	1	″		26.50		26.50	44.50

09 21 Plaster and Gypsum Board Assemblies

09 21 13 – Plaster Assemblies

09 21 13.10 Plaster Partition Wall

Line	Description	Crew	Daily Output	Labor-Hours	Unit	Material	Labor	Equipment	Total	Total Incl O&P
0010	**PLASTER PARTITION WALL**									
0400	Stud walls, 3.4 lb. metal lath, 3 coat gypsum plaster, 2 sides									
0600	2″ x 4″ wood studs, 16″ O.C.	J-2	315	.152	S.F.	3.31	4.49	.39	8.19	11.35
0700	2-1/2″ metal studs, 25 ga., 12″ O.C.		325	.148		3.06	4.35	.38	7.79	10.85
0800	3-5/8″ metal studs, 25 ga., 16″ O.C.		320	.150		3.08	4.42	.39	7.89	10.95
0900	Gypsum lath, 2 coat vermiculite plaster, 2 sides									
1000	2″ x 4″ wood studs, 16″ O.C.	J-2	355	.135	S.F.	3.69	3.99	.35	8.03	10.90
1200	2-1/2″ metal studs, 25 ga., 12″ O.C.		365	.132		3.27	3.88	.34	7.49	10.25
1300	3-5/8″ metal studs, 25 ga., 16″ O.C.		360	.133		3.36	3.93	.34	7.63	10.50

09 21 16 – Gypsum Board Assemblies

09 21 16.23 Gypsum Board Shaft Wall Assemblies

Line	Description	Crew	Daily Output	Labor-Hours	Unit	Material	Labor	Equipment	Total	Total Incl O&P
0010	**GYPSUM BOARD SHAFT WALL ASSEMBLIES**									
0030	1″ thick coreboard wall liner on shaft side									
0040	2-hour assembly with double layer									
0060	5/8″ fire rated gypsum board on room side	2 Carp	220	.073	S.F.	1.68	2.44		4.12	5.95
0100	3-hour assembly with triple layer									
0300	5/8″ fire rated gypsum board on room side	2 Carp	180	.089	S.F.	1.40	2.98		4.38	6.55
0400	4-hour assembly, 1″ coreboard, 5/8″ fire rated gypsum board									
0600	and 3/4″ galv. metal furring channels, 24″ O.C., with									
0700	Double layer 5/8″ fire rated gypsum board on room side	2 Carp	110	.145	S.F.	1.30	4.88		6.18	9.60
0900	For taping & finishing, add per side	1 Carp	1050	.008	″	.04	.26		.30	.48
1000	For insulation, see Section 07 21									
5200	For work over 8′ high, add	2 Carp	3060	.005	S.F.		.18		.18	.29
5300	For distribution cost over 3 stories high, add per story	″	6100	.003	″		.09		.09	.15

09 21 16.33 Partition Wall

Line	Description	Crew	Daily Output	Labor-Hours	Unit	Material	Labor	Equipment	Total	Total Incl O&P
0010	**PARTITION WALL** Stud wall, 8′ to 12′ high									
0050	1/2″, interior, gypsum board, std, tape & finish 2 sides									
0500	Installed on and incl., 2″ x 4″ wood studs, 16″ O.C.	2 Carp	310	.052	S.F.	.92	1.73		2.65	3.91
1000	Metal studs, NLB, 25 ga., 16″ O.C., 3-5/8″ wide		350	.046		.86	1.53		2.39	3.51
1200	6″ wide		330	.048		.99	1.63		2.62	3.81
1400	Water resistant, on 2″ x 4″ wood studs, 16″ O.C.		310	.052		1.18	1.73		2.91	4.20
1600	Metal studs, NLB, 25 ga., 16″ O.C., 3-5/8″ wide		350	.046		1.12	1.53		2.65	3.80
1800	6″ wide		330	.048		1.25	1.63		2.88	4.09
2000	Fire res., 2 layers, 1-1/2 hr., on 2″ x 4″ wood studs, 16″ O.C.		210	.076		1.58	2.56		4.14	6
2200	Metal studs, NLB, 25 ga., 16″ O.C., 3-5/8″ wide		250	.064		1.52	2.15		3.67	5.25
2400	6″ wide		230	.070		1.65	2.33		3.98	5.70

09 21 Plaster and Gypsum Board Assemblies

09 21 16 – Gypsum Board Assemblies

09 21 16.33 Partition Wall		Crew	Daily Output	Labor-Hours	Unit	2011 Bare Costs Material	Labor	Equipment	Total	Total Incl O&P
2600	Fire & water res., 2 layers, 1-1/2 hr., 2" x 4" studs, 16" O.C.	2 Carp	210	.076	S.F.	1.58	2.56		4.14	6
2800	Metal studs, NLB, 25 ga., 16" O.C., 3-5/8" wide		250	.064		1.52	2.15		3.67	5.25
3000	6" wide		230	.070		1.65	2.33		3.98	5.70
3200	5/8", interior, gypsum board, std, tape & finish 2 sides									
3400	Installed on and including 2" x 4" wood studs, 16" O.C.	2 Carp	300	.053	S.F.	1.02	1.79		2.81	4.11
3600	24" O.C.		330	.048		.94	1.63		2.57	3.75
3800	Metal studs, NLB, 25 ga., 16" O.C., 3-5/8" wide		340	.047		.96	1.58		2.54	3.69
4000	6" wide		320	.050		1.09	1.68		2.77	4.01
4200	24" O.C., 3-5/8" wide		360	.044		.88	1.49		2.37	3.46
4400	6" wide		340	.047		.97	1.58		2.55	3.71
4800	Water resistant, on 2" x 4" wood studs, 16" O.C.		300	.053		1.26	1.79		3.05	4.38
5000	24" O.C.		330	.048		1.18	1.63		2.81	4.02
5200	Metal studs, NLB, 25 ga. 16" O.C., 3-5/8" wide		340	.047		1.20	1.58		2.78	3.96
5400	6" wide		320	.050		1.33	1.68		3.01	4.27
5600	24" O.C., 3-5/8" wide		360	.044		1.12	1.49		2.61	3.73
5800	6" wide		340	.047		1.21	1.58		2.79	3.97
6000	Fire resistant, 2 layers, 2 hr., on 2" x 4" wood studs, 16" O.C.		205	.078		1.50	2.62		4.12	6.05
6200	24" O.C.		235	.068		1.50	2.28		3.78	5.45
6400	Metal studs, NLB, 25 ga., 16" O.C., 3-5/8" wide		245	.065		1.55	2.19		3.74	5.35
6600	6" wide		225	.071		1.65	2.39		4.04	5.80
6800	24" O.C., 3-5/8" wide		265	.060		1.44	2.03		3.47	4.97
7000	6" wide		245	.065		1.53	2.19		3.72	5.35
7200	Fire & water resistant, 2 layers, 2 hr., 2" x 4" studs, 16" O.C.		205	.078		1.58	2.62		4.20	6.10
7400	24" O.C.		235	.068		1.50	2.28		3.78	5.45
7600	Metal studs, NLB, 25 ga., 16" O.C., 3-5/8" wide		245	.065		1.52	2.19		3.71	5.35
7800	6" wide		225	.071		1.65	2.39		4.04	5.80
8000	24" O.C., 3-5/8" wide		265	.060		1.44	2.03		3.47	4.97
8200	6" wide		245	.065		1.53	2.19		3.72	5.35
8600	1/2" blueboard, mesh tape both sides									
8620	Installed on and including 2" x 4" wood studs, 16" O.C.	2 Carp	300	.053	S.F.	1.02	1.79		2.81	4.11
8640	Metal studs, NLB, 25 ga., 16" O.C., 3-5/8" wide		340	.047		.96	1.58		2.54	3.69
8660	6" wide		320	.050		1.09	1.68		2.77	4.01
9000	Exterior, 1/2" gypsum sheathing, 1/2" gypsum finished, interior,									
9100	including foil faced insulation, metal studs, 20 ga.									
9200	16" O.C., 3-5/8" wide	2 Carp	290	.055	S.F.	1.62	1.85		3.47	4.89
9400	6" wide		270	.059		1.79	1.99		3.78	5.30
9600	Partitions, for work over 8' high, add		1530	.010			.35		.35	.59

09 22 Supports for Plaster and Gypsum Board

09 22 03 – Fastening Methods for Finishes

09 22 03.20 Drilling Plaster/Drywall		Crew	Daily Output	Labor-Hours	Unit	Material	Labor	Equipment	Total	Total Incl O&P
0010	**DRILLING PLASTER/DRYWALL**									
1100	Drilling & layout for drywall/plaster walls, up to 1" deep, no anchor									
1200	Holes, 1/4" diameter	1 Carp	150	.053	Ea.	.01	1.79		1.80	3
1300	3/8" diameter		140	.057		.01	1.92		1.93	3.22
1400	1/2" diameter		130	.062		.01	2.06		2.07	3.47
1500	3/4" diameter		120	.067		.01	2.24		2.25	3.75
1600	1" diameter		110	.073		.02	2.44		2.46	4.10
1700	1-1/4" diameter		100	.080		.03	2.68		2.71	4.52
1800	1-1/2" diameter		90	.089		.04	2.98		3.02	5.05
1900	For ceiling installations, add						40%			

09 22 Supports for Plaster and Gypsum Board

09 22 13 – Metal Furring

09 22 13.13 Metal Channel Furring		Crew	Daily Output	Labor-Hours	Unit	Material	2011 Bare Costs Labor	Equipment	Total	Total Incl O&P
0010	**METAL CHANNEL FURRING**									
0030	Beams and columns, 7/8" channels, galvanized, 12" O.C.	1 Lath	155	.052	S.F.	.31	1.55		1.86	2.82
0050	16" O.C.		170	.047		.25	1.41		1.66	2.54
0070	24" O.C.		185	.043		.17	1.30		1.47	2.27
0100	Ceilings, on steel, 7/8" channels, galvanized, 12" O.C.		210	.038		.28	1.14		1.42	2.14
0300	16" O.C.		290	.028		.25	.83		1.08	1.61
0400	24" O.C.		420	.019		.17	.57		.74	1.11
0600	1-5/8" channels, galvanized, 12" O.C.		190	.042		.38	1.27		1.65	2.44
0700	16" O.C.		260	.031		.34	.92		1.26	1.85
0900	24" O.C.		390	.021		.23	.62		.85	1.24
0930	7/8" channels with sound isolation clips, 12" O.C.		120	.067		1.35	2		3.35	4.68
0940	16" O.C.		100	.080		1.88	2.40		4.28	5.90
0950	24" O.C.		165	.048		1.23	1.46		2.69	3.69
0960	1-5/8" channels, galvanized, 12" O.C.		110	.073		1.44	2.19		3.63	5.10
0970	16" O.C.		100	.080		1.97	2.40		4.37	6
0980	24" O.C.		155	.052		1.29	1.55		2.84	3.90
1000	Walls, 7/8" channels, galvanized, 12" O.C.		235	.034		.28	1.02		1.30	1.95
1200	16" O.C.		265	.030		.25	.91		1.16	1.73
1300	24" O.C.		350	.023		.17	.69		.86	1.29
1500	1-5/8" channels, galvanized, 12" O.C.		210	.038		.38	1.14		1.52	2.25
1600	16" O.C.		240	.033		.34	1		1.34	1.97
1800	24" O.C.		305	.026		.23	.79		1.02	1.51
1920	7/8" channels with sound isolation clips, 12" O.C.		125	.064		1.35	1.92		3.27	4.56
1940	16" O.C.		100	.080		1.88	2.40		4.28	5.90
1950	24" O.C.		150	.053		1.23	1.60		2.83	3.92
1960	1-5/8" channels, galvanized, 12" O.C.		115	.070		1.44	2.09		3.53	4.93
1970	16" O.C.		95	.084		1.97	2.53		4.50	6.20
1980	24" O.C.		140	.057		1.29	1.72		3.01	4.17

09 22 16 – Non-Structural Metal Framing

09 22 16.13 Non-Structural Metal Stud Framing		Crew	Daily Output	Labor-Hours	Unit	Material	Labor	Equipment	Total	Total Incl O&P
0010	**NON-STRUCTURAL METAL STUD FRAMING**									
1600	Non-load bearing, galv, 8' high, 25 ga. 1-5/8" wide, 16" O.C.	1 Carp	619	.013	S.F.	.24	.43		.67	1
1610	24" O.C.		950	.008		.18	.28		.46	.67
1620	2-1/2" wide, 16" O.C.		613	.013		.28	.44		.72	1.04
1630	24" O.C.		938	.009		.21	.29		.50	.71
1640	3-5/8" wide, 16" O.C.		600	.013		.32	.45		.77	1.11
1650	24" O.C.		925	.009		.24	.29		.53	.76
1660	4" wide, 16" O.C.		594	.013		.37	.45		.82	1.16
1670	24" O.C.		925	.009		.27	.29		.56	.79
1680	6" wide, 16" O.C.		588	.014		.46	.46		.92	1.27
1690	24" O.C.		906	.009		.35	.30		.65	.88
1700	20 ga. studs, 1-5/8" wide, 16" O.C.		494	.016		.37	.54		.91	1.31
1710	24" O.C.		763	.010		.27	.35		.62	.89
1720	2-1/2" wide, 16" O.C.		488	.016		.42	.55		.97	1.38
1730	24" O.C.		750	.011		.32	.36		.68	.95
1740	3-5/8" wide, 16" O.C.		481	.017		.45	.56		1.01	1.43
1750	24" O.C.		738	.011		.34	.36		.70	.98
1760	4" wide, 16" O.C.		475	.017		.54	.57		1.11	1.54
1770	24" O.C.		738	.011		.40	.36		.76	1.05
1780	6" wide, 16" O.C.		469	.017		.63	.57		1.20	1.65
1790	24" O.C.		725	.011		.47	.37		.84	1.14
2000	Non-load bearing, galv, 10' high, 25 ga. 1-5/8" wide, 16" O.C.		495	.016		.23	.54		.77	1.16

09 22 16 – Non-Structural Metal Framing

	09 22 16.13 Non-Structural Metal Stud Framing	Crew	Daily Output	Labor-Hours	Unit	2011 Bare Costs Material	Labor	Equipment	Total	Total Incl O&P
2100	24" O.C.	1 Carp	760	.011	S.F.	.17	.35		.52	.78
2200	2-1/2" wide, 16" O.C.		490	.016		.27	.55		.82	1.21
2250	24" O.C.		750	.011		.20	.36		.56	.81
2300	3-5/8" wide, 16" O.C.		480	.017		.31	.56		.87	1.28
2350	24" O.C.		740	.011		.23	.36		.59	.86
2400	4" wide, 16" O.C.		475	.017		.35	.57		.92	1.33
2450	24" O.C.		740	.011		.25	.36		.61	.89
2500	6" wide, 16" O.C.		470	.017		.44	.57		1.01	1.44
2550	24" O.C.		725	.011		.32	.37		.69	.97
2600	20 ga. studs, 1-5/8" wide, 16" O.C.		395	.020		.35	.68		1.03	1.52
2650	24" O.C.		610	.013		.25	.44		.69	1.02
2700	2-1/2" wide, 16" O.C.		390	.021		.40	.69		1.09	1.59
2750	24" O.C.		600	.013		.29	.45		.74	1.07
2800	3-5/8" wide, 16" OC		385	.021		.43	.70		1.13	1.64
2850	24" O.C.		590	.014		.32	.45		.77	1.11
2900	4" wide, 16" O.C.		380	.021		.51	.71		1.22	1.74
2950	24" O.C.		590	.014		.37	.45		.82	1.17
3000	6" wide, 16" O.C.		375	.021		.60	.72		1.32	1.86
3050	24" O.C.		580	.014		.44	.46		.90	1.25
3060	Non-load bearing, galv, 12' high, 25 ga. 1-5/8" wide, 16" O.C.		413	.019		.22	.65		.87	1.33
3070	24" O.C.		633	.013		.16	.42		.58	.89
3080	2-1/2" wide, 16" O.C.		408	.020		.25	.66		.91	1.38
3090	24" O.C.		625	.013		.18	.43		.61	.92
3100	3-5/8" wide, 16" O.C.		400	.020		.29	.67		.96	1.44
3110	24" O.C.		617	.013		.21	.44		.65	.97
3120	4" wide, 16" O.C.		396	.020		.33	.68		1.01	1.50
3130	24" O.C.		617	.013		.24	.44		.68	1
3140	6" wide, 16" O.C.		392	.020		.42	.68		1.10	1.61
3150	24" O.C.		604	.013		.30	.44		.74	1.07
3160	20 ga. studs, 1-5/8" wide, 16" O.C.		329	.024		.33	.82		1.15	1.73
3170	24" O.C.		508	.016		.24	.53		.77	1.15
3180	2-1/2" wide, 16" O.C.		325	.025		.38	.83		1.21	1.80
3190	24" O.C.		500	.016		.28	.54		.82	1.20
3200	3-5/8" wide, 16" O.C.		321	.025		.41	.84		1.25	1.85
3210	24" O.C.		492	.016		.30	.55		.85	1.24
3220	4" wide, 16" O.C.		317	.025		.49	.85		1.34	1.96
3230	24" O.C.		492	.016		.35	.55		.90	1.30
3240	6" wide, 16" O.C.		313	.026		.57	.86		1.43	2.07
3250	24" O.C.	↓	483	.017	↓	.42	.56		.98	1.39
5000	Load bearing studs, see Section 05 41 13.30									

09 22 26 – Suspension Systems

	09 22 26.13 Ceiling Suspension Systems	Crew	Daily Output	Labor-Hours	Unit	Material	Labor	Equipment	Total	Total Incl O&P
0010	**CEILING SUSPENSION SYSTEMS** For gypsum board or plaster									
8000	Suspended ceilings, including carriers									
8200	1-1/2" carriers, 24" O.C. with:									
8300	7/8" channels, 16" O.C.	1 Lath	275	.029	S.F.	.43	.87		1.30	1.87
8320	24" O.C.		310	.026		.34	.78		1.12	1.62
8400	1-5/8" channels, 16" O.C.		205	.039		.51	1.17		1.68	2.44
8420	24" O.C.	↓	250	.032	↓	.40	.96		1.36	1.98
8600	2" carriers, 24" O.C. with:									
8700	7/8" channels, 16" O.C.	1 Lath	250	.032	S.F.	.50	.96		1.46	2.10
8720	24" O.C.	↓	285	.028	↓	.42	.84		1.26	1.81

09 22 Supports for Plaster and Gypsum Board

09 22 26 – Suspension Systems

09 22 26.13 Ceiling Suspension Systems		Crew	Daily Output	Labor-Hours	Unit	Material	Labor (2011 Bare Costs)	Equipment (2011 Bare Costs)	Total (2011 Bare Costs)	Total Incl O&P
8800	1-5/8" channels, 16" O.C.	1 Lath	190	.042	S.F.	.59	1.27		1.86	2.67
8820	24" O.C.	↓	225	.036	↓	.48	1.07		1.55	2.23

09 22 36 – Lath

09 22 36.13 Gypsum Lath

		Crew	Daily Output	Labor-Hours	Unit	Material	Labor	Equipment	Total	Total Incl O&P
0010	**GYPSUM LATH** R092000-50									
0020	Plain or perforated, nailed, 3/8" thick	1 Lath	85	.094	S.Y.	4.95	2.83		7.78	9.95
0100	1/2" thick		80	.100		4.86	3.01		7.87	10.15
0300	Clipped to steel studs, 3/8" thick		75	.107		4.95	3.21		8.16	10.60
0400	1/2" thick		70	.114		4.86	3.43		8.29	10.85
1500	For ceiling installations, add		216	.037			1.11		1.11	1.78
1600	For columns and beams, add	↓	170	.047	↓		1.41		1.41	2.26

09 22 36.23 Metal Lath

		Crew	Daily Output	Labor-Hours	Unit	Material	Labor	Equipment	Total	Total Incl O&P
0010	**METAL LATH** R092000-50									
0020	Diamond, expanded, 2.5 lb. per S.Y., painted				S.Y.	3.62			3.62	3.98
0100	Galvanized					2.63			2.63	2.89
0300	3.4 lb. per S.Y., painted					4.38			4.38	4.82
0400	Galvanized					4			4	4.40
0600	For 15# asphalt sheathing paper, add					.51			.51	.56
0900	Flat rib, 1/8" high, 2.75 lb., painted					3.14			3.14	3.45
1000	Foil backed					2.80			2.80	3.08
1200	3.4 lb. per S.Y., painted					4			4	4.40
1300	Galvanized					3.48			3.48	3.83
1500	For 15# asphalt sheathing paper, add					.51			.51	.56
1800	High rib, 3/8" high, 3.4 lb. per S.Y., painted					4.84			4.84	5.30
1900	Galvanized				↓	4.84			4.84	5.30
2400	3/4" high, painted, .60 lb. per S.F.				S.F.	.50			.50	.55
2500	.75 lb. per S.F.				"	1.08			1.08	1.19
2800	Stucco mesh, painted, 3.6 lb.				S.Y.	3.03			3.03	3.33
3000	K-lath, perforated, absorbent paper, regular					4.21			4.21	4.63
3100	Heavy duty					4.97			4.97	5.45
3300	Waterproof, heavy duty, grade B backing					4.86			4.86	5.35
3400	Fire resistant backing					5.40			5.40	5.90
3600	2.5 lb. diamond painted, on wood framing, on walls	1 Lath	85	.094		3.62	2.83		6.45	8.50
3700	On ceilings		75	.107		3.62	3.21		6.83	9.15
3900	3.4 lb. diamond painted, on wood framing, on walls		80	.100		4	3.01		7.01	9.20
4000	On ceilings		70	.114		4	3.43		7.43	9.90
4200	3.4 lb. diamond painted, wired to steel framing		75	.107		4	3.21		7.21	9.55
4300	On ceilings		60	.133		4	4.01		8.01	10.80
4500	Columns and beams, wired to steel		40	.200		4	6		10	14
4600	Cornices, wired to steel		35	.229		4	6.85		10.85	15.40
4800	Screwed to steel studs, 2.5 lb.		80	.100		3.62	3.01		6.63	8.80
4900	3.4 lb.		75	.107		4.38	3.21		7.59	9.95
5100	Rib lath, painted, wired to steel, on walls, 2.5 lb.		75	.107		3.14	3.21		6.35	8.60
5200	3.4 lb.		70	.114		4.84	3.43		8.27	10.80
5400	4.0 lb.	↓	65	.123		5.45	3.70		9.15	11.85
5500	For self-furring lath, add					.10			.10	.11
5700	Suspended ceiling system, incl. 3.4 lb. diamond lath, painted	1 Lath	15	.533		3.35	16.05		19.40	29
5800	Galvanized	"	15	.533	↓	6.50	16.05		22.55	32.50
6000	Hollow metal stud partitions, 3.4 lb. painted lath both sides									
6010	Non-load bearing, 25 ga., w/rib lath 2-1/2" studs, 12" O.C.	1 Lath	20.30	.394	S.Y.	12.70	11.85		24.55	33
6300	16" O.C.		21.10	.379		12.05	11.40		23.45	31.50
6350	24" O.C.	↓	22.70	.352	↓	11.45	10.60		22.05	29.50

09 22 Supports for Plaster and Gypsum Board

09 22 36 – Lath

09 22 36.23 Metal Lath		Crew	Daily Output	Labor-Hours	Unit	Material	2011 Bare Costs Labor	Equipment	Total	Total Incl O&P
6400	3-5/8" studs, 16" O.C.	1 Lath	19.50	.410	S.Y.	12.45	12.35		24.80	33.50
6600	24" O.C.		20.40	.392		11.70	11.80		23.50	32
6700	4" studs, 16" O.C.		20.40	.392		12.80	11.80		24.60	33
6900	24" O.C.		21.60	.370		11.95	11.15		23.10	31
7000	6" studs, 16" O.C.		19.50	.410		13.60	12.35		25.95	34.50
7100	24" O.C.		21.10	.379		12.60	11.40		24	32
7200	L.B. partitions, 16 ga., w/rib lath, 2-1/2" studs, 16" O.C.		20	.400		12.30	12		24.30	33
7300	3-5/8" studs, 16 ga.		19.70	.406		13.70	12.20		25.90	34.50
7500	4" studs, 16 ga.		19.50	.410		14.20	12.35		26.55	35.50
7600	6" studs, 16 ga.		18.70	.428		16.50	12.85		29.35	38.50

09 22 36.43 Security Mesh		Crew	Daily Output	Labor-Hours	Unit	Material	Labor	Equipment	Total	Total Incl O&P
0010	**SECURITY MESH**, expanded metal, flat, screwed to framing									
0100	On walls, 3/4", 1.76 lb./S.F.	2 Carp	1500	.011	S.F.	1.49	.36		1.85	2.24
0110	1-1/2", 1.14 lb./S.F.		1600	.010		1.21	.34		1.55	1.89
0200	On ceilings, 3/4", 1.76 lb./S.F.		1350	.012		1.49	.40		1.89	2.31
0210	1-1/2", 1.14 lb./S.F.		1450	.011		1.21	.37		1.58	1.95

09 22 36.83 Accessories, Plaster		Crew	Daily Output	Labor-Hours	Unit	Material	Labor	Equipment	Total	Total Incl O&P
0010	**ACCESSORIES, PLASTER**									
0020	Casing bead, expanded flange, galvanized	1 Lath	2.70	2.963	C.L.F.	43.50	89		132.50	190
0200	Foundation weep screed, galvanized	"	2.70	2.963		58	89		147	206
0900	Channels, cold rolled, 16 ga., 3/4" deep, galvanized					28.50			28.50	31
1200	1-1/2" deep, 16 ga., galvanized					38			38	41.50
1620	Corner bead, expanded bullnose, 3/4" radius, #10, galvanized	1 Lath	2.60	3.077		30.50	92.50		123	182
1650	#1, galvanized		2.55	3.137		48	94.50		142.50	204
1670	Expanded wing, 2-3/4" wide, #1, galvanized		2.65	3.019		33.50	90.50		124	182
1700	Inside corner (corner rite), 3" x 3", painted		2.60	3.077		28.50	92.50		121	180
1750	Strip-ex, 4" wide, painted		2.55	3.137		29	94.50		123.50	183
1800	Expansion joint, 3/4" grounds, limited expansion, galv., 1 piece		2.70	2.963		98	89		187	250
2100	Extreme expansion, galvanized, 2 piece		2.60	3.077		157	92.50		249.50	320

09 23 Gypsum Plastering

09 23 13 – Acoustical Gypsum Plastering

09 23 13.10 Perlite or Vermiculite Plaster		Crew	Daily Output	Labor-Hours	Unit	Material	Labor	Equipment	Total	Total Incl O&P
0010	**PERLITE OR VERMICULITE PLASTER** R092000-50									
0020	In 100 lb. bags, under 200 bags				Bag	14.40			14.40	15.80
0100	Over 200 bags				"	13.05			13.05	14.35
0300	2 coats, no lath included, on walls	J-1	92	.435	S.Y.	3.78	12.75	1.34	17.87	26.50
0400	On ceilings	"	79	.506		3.78	14.85	1.56	20.19	30
0600	On and incl. 3/8" gypsum lath, on metal studs	J-2	84	.571		8.60	16.85	1.47	26.92	38.50
0700	On ceilings	"	70	.686		8.60	20	1.77	30.37	44.50
0900	3 coats, no lath included, on walls	J-1	74	.541		6.30	15.85	1.67	23.82	35
1000	On ceilings	"	63	.635		6.30	18.65	1.96	26.91	39.50
1200	On and incl. painted metal lath, on metal studs	J-2	72	.667		11.15	19.65	1.72	32.52	46
1300	On ceilings		61	.787		11.15	23	2.03	36.18	52
1500	On and incl. suspended metal lath ceiling		37	1.297		9.65	38	3.35	51	76.50
1700	For irregular or curved surfaces, add to above						30%			
1800	For columns and beams, add to above						50%			
1900	For soffits, add to ceiling prices						40%			

09 23 Gypsum Plastering

09 23 20 – Gypsum Plaster

09 23 20.10 Gypsum Plaster On Walls and Ceilings

	09 23 20.10 Gypsum Plaster On Walls and Ceilings	Crew	Daily Output	Labor-Hours	Unit	Material	2011 Bare Costs Labor	Equipment	Total	Total Incl O&P
0010	**GYPSUM PLASTER ON WALLS AND CEILINGS** R092000-50									
0020	80# bag, less than 1 ton				Bag	17.30			17.30	19
0100	Over 1 ton				″	15.40			15.40	16.95
0300	2 coats, no lath included, on walls	J-1	105	.381	S.Y.	3.87	11.20	1.18	16.25	24
0400	On ceilings	″	92	.435		3.87	12.75	1.34	17.96	26.50
0600	On and incl. 3/8″ gypsum lath on steel, on walls	J-2	97	.495		8.80	14.60	1.28	24.68	34.50
0700	On ceilings	″	83	.578		8.80	17.05	1.49	27.34	39
0900	3 coats, no lath included, on walls	J-1	87	.460		5.55	13.50	1.42	20.47	29.50
1000	On ceilings	″	78	.513		5.55	15.05	1.58	22.18	32.50
1200	On and including painted metal lath, on wood studs	J-2	86	.558		10.50	16.45	1.44	28.39	39.50
1300	On ceilings	″	76.50	.627		10.50	18.50	1.62	30.62	43.50
1600	For irregular or curved surfaces, add						30%			
1800	For columns & beams, add						50%			

09 23 20.20 Gauging Plaster

	09 23 20.20 Gauging Plaster	Crew	Daily Output	Labor-Hours	Unit	Material	Labor	Equipment	Total	Total Incl O&P
0010	**GAUGING PLASTER** R092000-50									
0020	100 lb. bags, less than 1 ton				Bag	18.10			18.10	19.90
0100	Over 1 ton				″	15.85			15.85	17.45

09 23 20.30 Keenes Cement

	09 23 20.30 Keenes Cement	Crew	Daily Output	Labor-Hours	Unit	Material	Labor	Equipment	Total	Total Incl O&P
0010	**KEENES CEMENT** R092000-50									
0020	In 100 lb. bags, less than 1 ton				Bag	26			26	28.50
0100	Over 1 ton				″	24.50			24.50	27
0300	Finish only, add to plaster prices, standard	J-1	215	.186	S.Y.	2.36	5.45	.57	8.38	12.15
0400	High quality	″	144	.278	″	2.39	8.15	.86	11.40	16.80

09 24 Cement Plastering

09 24 23 – Cement Stucco

09 24 23.40 Stucco

	09 24 23.40 Stucco	Crew	Daily Output	Labor-Hours	Unit	Material	Labor	Equipment	Total	Total Incl O&P
0010	**STUCCO** R092000-50									
0015	3 coats 1″ thick, float finish, with mesh, on wood frame	J-2	63	.762	S.Y.	5.30	22.50	1.97	29.77	44.50
0100	On masonry construction, no mesh incl.	J-1	67	.597		2.28	17.55	1.84	21.67	33
0300	For trowel finish, add	1 Plas	170	.047			1.44		1.44	2.35
0400	For 3/4″ thick, on masonry, deduct	J-1	880	.045		.61	1.33	.14	2.08	2.99
0600	For coloring and special finish, add, minimum		685	.058		.39	1.71	.18	2.28	3.42
0700	Maximum		200	.200		1.36	5.85	.62	7.83	11.75
0900	For soffits, add	J-2	155	.310		2.11	9.15	.80	12.06	18
1000	Exterior stucco, with bonding agent, 3 coats, on walls, no mesh incl.	J-1	200	.200		3.54	5.85	.62	10.01	14.10
1200	Ceilings		180	.222		3.54	6.50	.69	10.73	15.25
1300	Beams		80	.500		3.54	14.70	1.55	19.79	29.50
1500	Columns		100	.400		3.54	11.75	1.24	16.53	24.50
1600	Mesh, painted, nailed to wood, 1.8 lb.	1 Lath	60	.133		5	4.01		9.01	11.90
1800	3.6 lb.		55	.145		3.03	4.37		7.40	10.35
1900	Wired to steel, painted, 1.8 lb.		53	.151		5	4.54		9.54	12.75
2100	3.6 lb.		50	.160		3.03	4.81		7.84	11.05

09 26 Veneer Plastering

09 26 13 – Gypsum Veneer Plastering

09 26 13.20 Blueboard

	09 26 13.20 Blueboard	Crew	Daily Output	Labor-Hours	Unit	Material	Labor	Equipment	Total	Total Incl O&P
						2011 Bare Costs				
0010	**BLUEBOARD** For use with thin coat									
0100	plaster application (see Section 09 26 13.80)									
1000	3/8" thick, on walls or ceilings, standard, no finish included	2 Carp	1900	.008	S.F.	.26	.28		.54	.76
1100	With thin coat plaster finish		875	.018		.36	.61		.97	1.42
1400	On beams, columns, or soffits, standard, no finish included		675	.024		.30	.80		1.10	1.66
1450	With thin coat plaster finish		475	.034		.40	1.13		1.53	2.32
3000	1/2" thick, on walls or ceilings, standard, no finish included		1900	.008		.29	.28		.57	.79
3100	With thin coat plaster finish		875	.018		.39	.61		1	1.45
3300	Fire resistant, no finish included		1900	.008		.29	.28		.57	.79
3400	With thin coat plaster finish		875	.018		.39	.61		1	1.45
3450	On beams, columns, or soffits, standard, no finish included		675	.024		.33	.80		1.13	1.70
3500	With thin coat plaster finish		475	.034		.43	1.13		1.56	2.36
3700	Fire resistant, no finish included		675	.024		.33	.80		1.13	1.70
3800	With thin coat plaster finish		475	.034		.43	1.13		1.56	2.36
5000	5/8" thick, on walls or ceilings, fire resistant, no finish included		1900	.008		.30	.28		.58	.80
5100	With thin coat plaster finish		875	.018		.40	.61		1.01	1.47
5500	On beams, columns, or soffits, no finish included		675	.024		.35	.80		1.15	1.71
5600	With thin coat plaster finish		475	.034		.44	1.13		1.57	2.38
6000	For high ceilings, over 8' high, add		3060	.005			.18		.18	.29
6500	For over 3 stories high, add per story		6100	.003			.09		.09	.15

09 26 13.80 Thin Coat Plaster

	09 26 13.80 Thin Coat Plaster	Crew	Daily Output	Labor-Hours	Unit	Material	Labor	Equipment	Total	Total Incl O&P
0010	**THIN COAT PLASTER** R092000-50									
0012	1 coat veneer, not incl. lath	J-1	3600	.011	S.F.	.10	.33	.03	.46	.68
1000	In 50 lb. bags				Bag	12.95			12.95	14.25

09 28 Backing Boards and Underlayments

09 28 13 – Cementitious Backing Boards

09 28 13.10 Cementitious Backerboard

	09 28 13.10 Cementitious Backerboard	Crew	Daily Output	Labor-Hours	Unit	Material	Labor	Equipment	Total	Total Incl O&P
0010	**CEMENTITIOUS BACKERBOARD**									
0070	Cementitious backerboard, on floor, 3' x 4' x 1/2" sheets	2 Carp	525	.030	S.F.	.73	1.02		1.75	2.51
0080	3' x 5' x 1/2" sheets		525	.030		.71	1.02		1.73	2.49
0090	3' x 6' x 1/2" sheets		525	.030		.68	1.02		1.70	2.46
0100	3' x 4' x 5/8" sheets		525	.030		.96	1.02		1.98	2.76
0110	3' x 5' x 5/8" sheets		525	.030		.97	1.02		1.99	2.78
0120	3' x 6' x 5/8" sheets		525	.030		.82	1.02		1.84	2.62
0150	On wall, 3' x 4' x 1/2" sheets		350	.046		.73	1.53		2.26	3.37
0160	3' x 5' x 1/2" sheets		350	.046		.71	1.53		2.24	3.35
0170	3' x 6' x 1/2" sheets		350	.046		.68	1.53		2.21	3.32
0180	3' x 4' x 5/8" sheets		350	.046		.96	1.53		2.49	3.62
0190	3' x 5' x 5/8" sheets		350	.046		.97	1.53		2.50	3.64
0200	3' x 6' x 5/8" sheets		350	.046		.82	1.53		2.35	3.48
0250	On counter, 3' x 4' x 1/2" sheets		180	.089		.73	2.98		3.71	5.80
0260	3' x 5' x 1/2" sheets		180	.089		.71	2.98		3.69	5.75
0270	3' x 6' x 1/2" sheets		180	.089		.68	2.98		3.66	5.75
0300	3' x 4' x 5/8" sheets		180	.089		.96	2.98		3.94	6.05
0310	3' x 5' x 5/8" sheets		180	.089		.97	2.98		3.95	6.05
0320	3' x 6' x 5/8" sheets		180	.089		.82	2.98		3.80	5.90

09 29 10 – Gypsum Board Panels

09 29 10.30 Gypsum Board		Crew	Daily Output	Labor-Hours	Unit	Material	2011 Bare Costs Labor	Equipment	Total	Total Incl O&P
0010	**GYPSUM BOARD** on walls & ceilings R092910-10									
0100	Nailed or screwed to studs unless otherwise noted									
0150	3/8" thick, on walls, standard, no finish included	2 Carp	2000	.008	S.F.	.24	.27		.51	.71
0200	On ceilings, standard, no finish included		1800	.009		.24	.30		.54	.76
0250	On beams, columns, or soffits, no finish included		675	.024		.24	.80		1.04	1.59
0300	1/2" thick, on walls, standard, no finish included		2000	.008		.23	.27		.50	.70
0350	Taped and finished (level 4 finish)		965	.017		.27	.56		.83	1.23
0390	With compound skim coat (level 5 finish)		775	.021		.32	.69		1.01	1.52
0400	Fire resistant, no finish included		2000	.008		.28	.27		.55	.76
0450	Taped and finished (level 4 finish)		965	.017		.32	.56		.88	1.29
0490	With compound skim coat (level 5 finish)		775	.021		.37	.69		1.06	1.57
0500	Water resistant, no finish included		2000	.008		.36	.27		.63	.85
0550	Taped and finished (level 4 finish)		965	.017		.40	.56		.96	1.38
0590	With compound skim coat (level 5 finish)		775	.021		.45	.69		1.14	1.66
0600	Prefinished, vinyl, clipped to studs		900	.018		.46	.60		1.06	1.51
0700	Mold resistant, no finish included		2000	.008		.37	.27		.64	.86
0710	Taped and finished (level 4 finish)		965	.017		.41	.56		.97	1.39
0720	With compound skim coat (level 5 finish)		775	.021		.46	.69		1.15	1.67
1000	On ceilings, standard, no finish included		1800	.009		.23	.30		.53	.75
1050	Taped and finished (level 4 finish)		765	.021		.27	.70		.97	1.47
1090	With compound skim coat (level 5 finish)		610	.026		.32	.88		1.20	1.83
1100	Fire resistant, no finish included		1800	.009		.28	.30		.58	.81
1150	Taped and finished (level 4 finish)		765	.021		.32	.70		1.02	1.53
1195	With compound skim coat (level 5 finish)		610	.026		.37	.88		1.25	1.88
1200	Water resistant, no finish included		1800	.009		.36	.30		.66	.90
1250	Taped and finished (level 4 finish)		765	.021		.40	.70		1.10	1.62
1290	With compound skim coat (level 5 finish)		610	.026		.45	.88		1.33	1.97
1310	Mold resistant, no finish included		1800	.009		.37	.30		.67	.91
1320	Taped and finished (level 4 finish)		765	.021		.41	.70		1.11	1.63
1330	With compound skim coat (level 5 finish)		610	.026		.46	.88		1.34	1.98
1500	On beams, columns, or soffits, standard, no finish included		675	.024		.26	.80		1.06	1.62
1550	Taped and finished (level 4 finish)		475	.034		.27	1.13		1.40	2.19
1590	With compound skim coat (level 5 finish)		540	.030		.32	.99		1.31	2.02
1600	Fire resistant, no finish included		675	.024		.28	.80		1.08	1.64
1650	Taped and finished (level 4 finish)		475	.034		.32	1.13		1.45	2.25
1690	With compound skim coat (level 5 finish)		540	.030		.37	.99		1.36	2.07
1700	Water resistant, no finish included		675	.024		.41	.80		1.21	1.79
1750	Taped and finished (level 4 finish)		475	.034		.40	1.13		1.53	2.34
1790	With compound skim coat (level 5 finish)		540	.030		.45	.99		1.44	2.16
1800	Mold resistant, no finish included		675	.024		.43	.80		1.23	1.80
1810	Taped and finished (level 4 finish)		475	.034		.41	1.13		1.54	2.35
1820	With compound skim coat (level 5 finish)		540	.030		.46	.99		1.45	2.17
2000	5/8" thick, on walls, standard, no finish included		2000	.008		.28	.27		.55	.76
2050	Taped and finished (level 4 finish)		965	.017		.32	.56		.88	1.29
2090	With compound skim coat (level 5 finish)		775	.021		.37	.69		1.06	1.57
2100	Fire resistant, no finish included		2000	.008		.28	.27		.55	.76
2150	Taped and finished (level 4 finish)		965	.017		.32	.56		.88	1.29
2195	With compound skim coat (level 5 finish)		775	.021		.37	.69		1.06	1.57
2200	Water resistant, no finish included		2000	.008		.40	.27		.67	.89
2250	Taped and finished (level 4 finish)		965	.017		.44	.56		1	1.42
2290	With compound skim coat (level 5 finish)		775	.021		.49	.69		1.18	1.70
2300	Prefinished, vinyl, clipped to studs		900	.018		.72	.60		1.32	1.79
2510	Mold resistant, no finish included		2000	.008		.42	.27		.69	.91

09 29 Gypsum Board

09 29 10 – Gypsum Board Panels

09 29 10.30 Gypsum Board		Crew	Daily Output	Labor-Hours	Unit	2011 Bare Costs Material	Labor	Equipment	Total	Total Incl O&P
2520	Taped and finished (level 4 finish)	2 Carp	965	.017	S.F.	.46	.56		1.02	1.44
2530	With compound skim coat (level 5 finish)		775	.021		.51	.69		1.20	1.72
3000	On ceilings, standard, no finish included		1800	.009		.28	.30		.58	.81
3050	Taped and finished (level 4 finish)		765	.021		.32	.70		1.02	1.53
3090	With compound skim coat (level 5 finish)		615	.026		.37	.87		1.24	1.87
3100	Fire resistant, no finish included		1800	.009		.28	.30		.58	.81
3150	Taped and finished (level 4 finish)		765	.021		.32	.70		1.02	1.53
3190	With compound skim coat (level 5 finish)		615	.026		.37	.87		1.24	1.87
3200	Water resistant, no finish included		1800	.009		.40	.30		.70	.94
3250	Taped and finished (level 4 finish)		765	.021		.44	.70		1.14	1.66
3290	With compound skim coat (level 5 finish)		615	.026		.49	.87		1.36	2
3300	Mold resistant, no finish included		1800	.009		.42	.30		.72	.96
3310	Taped and finished (level 4 finish)		765	.021		.46	.70		1.16	1.68
3320	With compound skim coat (level 5 finish)		615	.026		.51	.87		1.38	2.02
3500	On beams, columns, or soffits, no finish included		675	.024		.32	.80		1.12	1.68
3550	Taped and finished (level 4 finish)		475	.034		.37	1.13		1.50	2.30
3590	With compound skim coat (level 5 finish)		380	.042		.43	1.41		1.84	2.83
3600	Fire resistant, no finish included		675	.024		.32	.80		1.12	1.68
3650	Taped and finished (level 4 finish)		475	.034		.37	1.13		1.50	2.30
3690	With compound skim coat (level 5 finish)		380	.042		.37	1.41		1.78	2.77
3700	Water resistant, no finish included		675	.024		.46	.80		1.26	1.84
3750	Taped and finished (level 4 finish)		475	.034		.51	1.13		1.64	2.45
3790	With compound skim coat (level 5 finish)		380	.042		.49	1.41		1.90	2.90
3800	Mold resistant, no finish included		675	.024		.48	.80		1.28	1.86
3810	Taped and finished (level 4 finish)		475	.034		.53	1.13		1.66	2.48
3820	With compound skim coat (level 5 finish)		380	.042		.51	1.41		1.92	2.92
4000	Fireproofing, beams or columns, 2 layers, 1/2″ thick, incl finish		330	.048		.60	1.63		2.23	3.39
4010	Mold resistant		330	.048		.78	1.63		2.41	3.58
4050	5/8″ thick		300	.053		.65	1.79		2.44	3.70
4060	Mold resistant		300	.053		.93	1.79		2.72	4.01
4100	3 layers, 1/2″ thick		225	.071		.88	2.39		3.27	4.96
4110	Mold resistant		225	.071		1.15	2.39		3.54	5.25
4150	5/8″ thick		210	.076		.97	2.56		3.53	5.35
4160	Mold resistant		210	.076		1.39	2.56		3.95	5.80
5050	For 1″ thick coreboard on columns		480	.033		.56	1.12		1.68	2.49
5100	For foil-backed board, add					.14			.14	.15
5200	For work over 8′ high, add	2 Carp	3060	.005			.18		.18	.29
5270	For textured spray, add	2 Lath	1600	.010		.05	.30		.35	.54
5300	For distribution cost over 3 stories high, add per story	2 Carp	6100	.003			.09		.09	.15
5350	For finishing inner corners, add		950	.017	L.F.	.10	.57		.67	1.06
5355	For finishing outer corners, add		1250	.013		.21	.43		.64	.96
5500	For acoustical sealant, add per bead	1 Carp	500	.016		.04	.54		.58	.95
5550	Sealant, 1 quart tube				Ea.	7.05			7.05	7.80
5600	Sound deadening board, 1/4″ gypsum	2 Carp	1800	.009	S.F.	.34	.30		.64	.87
5650	1/2″ wood fiber	″	1800	.009	″	.34	.30		.64	.87

09 29 10.50 High Abuse Gypsum Board		Crew	Daily Output	Labor-Hours	Unit	Material	Labor	Equipment	Total	Total Incl O&P
0010	**HIGH ABUSE GYPSUM BOARD**, fiber reinforced, nailed or									
0100	screwed to studs unless otherwise noted									
0110	1/2″ thick, on walls, no finish included	2 Carp	1800	.009	S.F.	.86	.30		1.16	1.45
0120	Taped and finished (level 4 finish)		870	.018		.90	.62		1.52	2.03
0130	With compound skim coat (level 5 finish)		700	.023		.95	.77		1.72	2.33
0150	On ceilings, no finish included		1620	.010		.86	.33		1.19	1.50

09 29 10 – Gypsum Board Panels

09 29 10.50 High Abuse Gypsum Board		Crew	Daily Output	Labor-Hours	Unit	Material	Labor	Equipment	Total	Total Incl O&P
						2011 Bare Costs				
0160	Taped and finished (level 4 finish)	2 Carp	690	.023	S.F.	.90	.78		1.68	2.30
0170	With compound skim coat (level 5 finish)		550	.029		.95	.98		1.93	2.68
0210	5/8" thick, on walls, no finish included		1800	.009		.90	.30		1.20	1.49
0220	Taped and finished (level 4 finish)		870	.018		.94	.62		1.56	2.07
0230	With compound skim coat (level 5 finish)		700	.023		.99	.77		1.76	2.37
0250	On ceilings, no finish included		1620	.010		.90	.33		1.23	1.54
0260	Taped and finished (level 4 finish)		690	.023		.94	.78		1.72	2.34
0270	With compound skim coat (level 5 finish)		550	.029		.99	.98		1.97	2.72
0310	5/8" thick, on walls, very high impact, no finish included		1800	.009		.98	.30		1.28	1.58
0320	Taped and finished (level 4 finish)		870	.018		1.02	.62		1.64	2.16
0330	With compound skim coat (level 5 finish)		700	.023		1.07	.77		1.84	2.46
0350	On ceilings, no finish included		1620	.010		.98	.33		1.31	1.63
0360	Taped and finished (level 4 finish)		690	.023		1.02	.78		1.80	2.43
0370	With compound skim coat (level 5 finish)	↓	550	.029	↓	1.07	.98		2.05	2.81
0400	High abuse, gypsum core, paper face									
0410	1/2" thick, on walls, no finish included	2 Carp	1800	.009	S.F.	.79	.30		1.09	1.37
0420	Taped and finished (level 4 finish)		870	.018		.83	.62		1.45	1.95
0430	With compound skim coat (level 5 finish)		700	.023		.88	.77		1.65	2.25
0450	On ceilings, no finish included		1620	.010		.79	.33		1.12	1.42
0460	Taped and finished (level 4 finish)		690	.023		.83	.78		1.61	2.22
0470	With compound skim coat (level 5 finish)		550	.029		.88	.98		1.86	2.60
0510	5/8" thick, on walls, no finish included		1800	.009		.79	.30		1.09	1.37
0520	Taped and finished (level 4 finish)		870	.018		.83	.62		1.45	1.95
0530	With compound skim coat (level 5 finish)		700	.023		.88	.77		1.65	2.25
0550	On ceilings, no finish included		1620	.010		.79	.33		1.12	1.42
0560	Taped and finished (level 4 finish)		690	.023		.83	.78		1.61	2.22
0570	With compound skim coat (level 5 finish)		550	.029		.88	.98		1.86	2.60
1000	For high ceilings, over 8' high, add		2750	.006			.20		.20	.33
1010	For distribution cost over 3 stories high, add per story	↓	5500	.003	↓		.10		.10	.16

09 29 15 – Gypsum Board Accessories

09 29 15.10 Accessories, Gypsum Board		Crew	Daily Output	Labor-Hours	Unit	Material	Labor	Equipment	Total	Total Incl O&P
0010	**ACCESSORIES, GYPSUM BOARD**									
0020	Casing bead, galvanized steel	1 Carp	2.90	2.759	C.L.F.	27	92.50		119.50	185
0100	Vinyl		3	2.667		23	89.50		112.50	176
0300	Corner bead, galvanized steel, 1" x 1"		4	2		14.35	67		81.35	128
0400	1-1/4" x 1-1/4"		3.50	2.286		15.30	76.50		91.80	145
0600	Vinyl		4	2		16.75	67		83.75	130
0900	Furring channel, galv. steel, 7/8" deep, standard		2.60	3.077		24.50	103		127.50	200
1000	Resilient		2.55	3.137		27	105		132	206
1100	J trim, galvanized steel, 1/2" wide		3	2.667		29	89.50		118.50	182
1120	5/8" wide		2.95	2.712		28	91		119	183
1140	L trim, galvanized		3	2.667		23	89.50		112.50	175
1150	U trim, galvanized	↓	2.95	2.712	↓	22	91		113	176
1160	Screws #6 x 1" A				M	8.45			8.45	9.30
1170	#6 x 1-5/8" A				"	12.40			12.40	13.65
1200	For stud partitions, see Section 05 41 13.30 and 09 22 16.13									
1500	Z stud, galvanized steel, 1-1/2" wide	1 Carp	2.60	3.077	C.L.F.	38.50	103		141.50	216
1600	2" wide	"	2.55	3.137	"	64.50	105		169.50	247

09 30 13 – Ceramic Tiling

09 30 13.10 Ceramic Tile		Crew	Daily Output	Labor-Hours	Unit	Material	Labor	Equipment	Total	Total Incl O&P
						2011 Bare Costs				
0010	**CERAMIC TILE**									
0020	Backsplash, thinset, average grade tiles	1 Tilf	50	.160	S.F.	2.32	5.05		7.37	10.60
0022	Custom grade tiles		50	.160		4.65	5.05		9.70	13.15
0024	Luxury grade tiles		50	.160		9.30	5.05		14.35	18.30
0026	Economy grade tiles		50	.160		2.12	5.05		7.17	10.40
0050	Base, using 1′ x 4″ high pc. with 1″ x 1″ tiles, mud set	D-7	82	.195	L.F.	4.81	5.50		10.31	14.10
0100	Thin set	"	128	.125		4.46	3.53		7.99	10.55
0300	For 6″ high base, 1″ x 1″ tile face, add					.72			.72	.79
0400	For 2″ x 2″ tile face, add to above					.38			.38	.42
0600	Cove base, 4-1/4″ x 4-1/4″ high, mud set	D-7	91	.176		3.75	4.96		8.71	12.05
0700	Thin set		128	.125		3.81	3.53		7.34	9.85
0900	6″ x 4-1/4″ high, mud set		100	.160		3.20	4.51		7.71	10.70
1000	Thin set		137	.117		3.07	3.29		6.36	8.65
1200	Sanitary cove base, 6″ x 4-1/4″ high, mud set		93	.172		3.85	4.85		8.70	12
1300	Thin set		124	.129		3.75	3.64		7.39	9.95
1500	6″ x 6″ high, mud set		84	.190		4.34	5.35		9.69	13.40
1600	Thin set		117	.137		4.21	3.86		8.07	10.80
1800	Bathroom accessories, average (soap dish, tooth brush holder)		82	.195	Ea.	11	5.50		16.50	21
1900	Bathtub, 5′, rec. 4-1/4″ x 4-1/4″ tile wainscot, adhesive set 6′ high		2.90	5.517		156	156		312	420
2100	7′ high wainscot		2.50	6.400		179	180		359	485
2200	8′ high wainscot		2.20	7.273		190	205		395	540
2400	Bullnose trim, 4-1/4″ x 4-1/4″, mud set		82	.195	L.F.	3.52	5.50		9.02	12.65
2500	Thin set		128	.125		3.37	3.53		6.90	9.35
2700	6″ x 4-1/4″ bullnose trim, mud set		84	.190		2.70	5.35		8.05	11.55
2800	Thin set		124	.129		2.61	3.64		6.25	8.65
3000	Floors, natural clay, random or uniform, thin set, color group 1		183	.087	S.F.	3.83	2.47		6.30	8.15
3100	Color group 2		183	.087		5.15	2.47		7.62	9.60
3255	Floors, glazed, thin set, 6″ x 6″, color group 1		200	.080		4.43	2.26		6.69	8.50
3260	8″ x 8″ tile		250	.064		4.43	1.80		6.23	7.75
3270	12″ x 12″ tile		325	.049		5.25	1.39		6.64	7.95
3280	16″ x 16″ tile		550	.029		6.05	.82		6.87	7.95
3281	18″ x 18″ tile		600	.027		6.05	.75		6.80	7.85
3283	24″ x 24″ tile		650	.025		6.05	.69		6.74	7.75
3285	Border, 6″ x 12″ tile		275	.058		11.55	1.64		13.19	15.35
3290	3″ x 12″ tile		200	.080		34	2.26		36.26	40.50
3300	Porcelain type, 1 color, color group 2, 1″ x 1″		183	.087		4.39	2.47		6.86	8.75
3310	2″ x 2″ or 2″ x 1″, thin set		190	.084		4.26	2.37		6.63	8.50
3350	For random blend, 2 colors, add					.88			.88	.97
3360	4 colors, add					1.24			1.24	1.36
3370	For color group 3, add					.50			.50	.55
3380	For abrasive non-slip tile, add					.49			.49	.54
4300	Specialty tile, 4-1/4″ x 4-1/4″ x 1/2″, decorator finish	D-7	183	.087		9.90	2.47		12.37	14.85
4500	Add for epoxy grout, 1/16″ joint, 1″ x 1″ tile		800	.020		.62	.56		1.18	1.58
4600	2″ x 2″ tile		820	.020		.59	.55		1.14	1.53
4800	Pregrouted sheets, walls, 4-1/4″ x 4-1/4″, 6″ x 4-1/4″									
4810	and 8-1/2″ x 4-1/4″, 4 S.F. sheets, silicone grout	D-7	240	.067	S.F.	4.73	1.88		6.61	8.20
5100	Floors, unglazed, 2 S.F. sheets,									
5110	Urethane adhesive	D-7	180	.089	S.F.	4.71	2.51		7.22	9.20
5400	Walls, interior, thin set, 4-1/4″ x 4-1/4″ tile		190	.084		2.14	2.37		4.51	6.15
5500	6″ x 4-1/4″ tile		190	.084		2.73	2.37		5.10	6.80
5700	8-1/2″ x 4-1/4″ tile		190	.084		3.80	2.37		6.17	7.95
5800	6″ x 6″ tile		200	.080		2.38	2.26		4.64	6.20
5810	8″ x 8″ tile		225	.071		4.08	2.01		6.09	7.70

09 30 Tiling

09 30 13 – Ceramic Tiling

09 30 13.10 Ceramic Tile

Line	09 30 13.10 Ceramic Tile	Crew	Daily Output	Labor-Hours	Unit	Material	2011 Bare Costs Labor	Equipment	Total	Total Incl O&P
5820	12" x 12" tile	D-7	300	.053	S.F.	3.53	1.50		5.03	6.30
5830	16" x 16" tile		500	.032		4.13	.90		5.03	6
6000	Decorated wall tile, 4-1/4" x 4-1/4", minimum		270	.059		2.64	1.67		4.31	5.60
6100	Maximum		180	.089		44	2.51		46.51	52
6300	Exterior walls, frostproof, mud set, 4-1/4" x 4-1/4"		102	.157		7	4.42		11.42	14.75
6400	1-3/8" x 1-3/8"		93	.172		4.65	4.85		9.50	12.85
6600	Crystalline glazed, 4-1/4" x 4-1/4", mud set, plain		100	.160		3.78	4.51		8.29	11.35
6700	4-1/4" x 4-1/4", scored tile		100	.160		4.95	4.51		9.46	12.65
6900	6" x 6" plain		93	.172		5.15	4.85		10	13.40
7000	For epoxy grout, 1/16" joints, 4-1/4" tile, add		800	.020		.39	.56		.95	1.33
7200	For tile set in dry mortar, add		1735	.009			.26		.26	.42
7300	For tile set in Portland cement mortar, add		290	.055		.16	1.56		1.72	2.67
9000	Regrout tile 4-1/2 x 4-1/2, or larger, wall	1 Tilf	100	.080		.14	2.52		2.66	4.18
9220	Floor	"	125	.064		.15	2.02		2.17	3.40
9300	Ceramic tiles, recycled glass, standard colors, 2" x 2" thru 6" x 6" G	D-7	190	.084		17	2.37		19.37	22.50
9310	6" x 6" G		200	.080		17	2.26		19.26	22.50
9320	8" x 8" G		225	.071		18.45	2.01		20.46	23.50
9330	12" x 12" G		300	.053		18.45	1.50		19.95	23
9340	Earthtones, 2" x 2" to 4" x 8" G		190	.084		21.50	2.37		23.87	27.50
9350	6" x 6" G		200	.080		21.50	2.26		23.76	27
9360	8" x 8" G		225	.071		23	2.01		25.01	28
9370	12" x 12" G		300	.053		23	1.50		24.50	27.50
9380	Deep colors, 2" x 2" to 4" x 8" G		190	.084		31	2.37		33.37	38
9390	6" x 6" G		200	.080		31	2.26		33.26	37.50
9400	8" x 8" G		225	.071		32	2.01		34.01	38.50
9410	12" x 12" G		300	.053		32	1.50		33.50	38

09 30 13.45 Ceramic Tile Accessories

Line	09 30 13.45 Ceramic Tile Accessories	Crew	Daily Output	Labor-Hours	Unit	Material	Labor	Equipment	Total	Total Incl O&P
0010	**CERAMIC TILE ACCESSORIES**									
0100	Spacers, 1/8"				Ea.	1.44			1.44	1.58
1310	Sealer, installed	1 Tilf	650	.012	S.F.	.38	.39		.77	1.04

09 30 16 – Quarry Tiling

09 30 16.10 Quarry Tile

Line	09 30 16.10 Quarry Tile	Crew	Daily Output	Labor-Hours	Unit	Material	Labor	Equipment	Total	Total Incl O&P
0010	**QUARRY TILE**									
0100	Base, cove or sanitary, mud set, to 5" high, 1/2" thick	D-7	110	.145	L.F.	5.30	4.10		9.40	12.35
0300	Bullnose trim, red, mud set, 6" x 6" x 1/2" thick		120	.133		4.34	3.76		8.10	10.80
0400	4" x 4" x 1/2" thick		110	.145		4.82	4.10		8.92	11.85
0600	4" x 8" x 1/2" thick, using 8" as edge		130	.123		4.26	3.47		7.73	10.25
0700	Floors, mud set, 1,000 S.F. lots, red, 4" x 4" x 1/2" thick		120	.133	S.F.	6.10	3.76		9.86	12.70
0900	6" x 6" x 1/2" thick		140	.114		5.50	3.22		8.72	11.20
1000	4" x 8" x 1/2" thick		130	.123		6.10	3.47		9.57	12.25
1300	For waxed coating, add					.71			.71	.78
1500	For non-standard colors, add					.42			.42	.46
1600	For abrasive surface, add					.49			.49	.54
1800	Brown tile, imported, 6" x 6" x 3/4"	D-7	120	.133		7.30	3.76		11.06	14
1900	8" x 8" x 1"		110	.145		7.95	4.10		12.05	15.30
2100	For thin set mortar application, deduct		700	.023			.64		.64	1.03
2200	For epoxy grout & mortar, 6" x 6" x 1/2", add		350	.046		1.94	1.29		3.23	4.19
2700	Stair tread, 6" x 6" x 3/4", plain		50	.320		5.15	9		14.15	20
2800	Abrasive		47	.340		5.60	9.60		15.20	21.50
3000	Wainscot, 6" x 6" x 1/2", thin set, red		105	.152		4.31	4.30		8.61	11.60
3100	Non-standard colors		105	.152		4.78	4.30		9.08	12.10
3300	Window sill, 6" wide, 3/4" thick		90	.178	L.F.	5.20	5		10.20	13.75

09 30 Tiling

09 30 16 – Quarry Tiling

09 30 16.10 Quarry Tile		Crew	Daily Output	Labor-Hours	Unit	2011 Bare Costs Material	Labor	Equipment	Total	Total Incl O&P
3400	Corners	D-7	80	.200	Ea.	6	5.65		11.65	15.60

09 30 23 – Glass Mosaic Tiling

09 30 23.10 Glass Mosaics

		Crew	Daily Output	Labor-Hours	Unit	Material	Labor	Equipment	Total	Total Incl O&P
0010	**GLASS MOSAICS** 3/4" tile on 12" sheets, standard grout									
0300	Color group 1 & 2	D-7	73	.219	S.F.	19	6.20		25.20	31
0350	Color group 3		73	.219		19.30	6.20		25.50	31.50
0400	Color group 4		73	.219		25	6.20		31.20	37.50
0450	Color group 5		73	.219		28.50	6.20		34.70	41.50
0500	Color group 6		73	.219		40.50	6.20		46.70	54.50
0600	Color group 7		73	.219		43.50	6.20		49.70	58
0700	Color group 8, golds, silvers & specialties		64	.250		62	7.05		69.05	80

09 30 29 – Metal Tiling

09 30 29.10 Metal Tile

		Crew	Daily Output	Labor-Hours	Unit	Material	Labor	Equipment	Total	Total Incl O&P
0010	**METAL TILE** 4' x 4' sheet, 24 ga., tile pattern, nailed									
0200	Stainless steel	2 Carp	512	.031	S.F.	25.50	1.05		26.55	30
0400	Aluminized steel	"	512	.031	"	13.80	1.05		14.85	16.95

09 34 Waterproofing-Membrane Tiling

09 34 13 – Waterproofing-Membrane Ceramic Tiling

09 34 13.10 Ceramic Tile Waterproofing Membrane

		Crew	Daily Output	Labor-Hours	Unit	Material	Labor	Equipment	Total	Total Incl O&P
0010	**CERAMIC TILE WATERPROOFING MEMBRANE**									
0020	On floors, including thinset									
0030	Fleece laminated polyethylene grid, 1/8" thick	D-7	250	.064	S.F.	2.24	1.80		4.04	5.35
0040	5/16" thick	"	250	.064	"	2.60	1.80		4.40	5.75
0050	On walls, including thinset									
0060	Fleece laminated polyethylene sheet, 8 mil thick	D-7	480	.033	S.F.	2.24	.94		3.18	3.96
0070	Accessories, including thinset									
0080	Joint and corner sheet, 4 mils thick, 5" wide	1 Tilf	240	.033	L.F.	1.45	1.05		2.50	3.27
0090	7-1/4" wide		180	.044		1.84	1.40		3.24	4.26
0100	10" wide		120	.067		2.24	2.10		4.34	5.80
0110	Pre-formed corners, inside		32	.250	Ea.	3.93	7.90		11.83	16.90
0120	Outside		32	.250		7.40	7.90		15.30	21
0130	2" flanged floor drain with 6" stainless steel grate		16	.500		355	15.80		370.80	415
0140	EPS, sloped shower floor		480	.017	S.F.	4.82	.53		5.35	6.15
0150	Curb		32	.250	L.F.	13.55	7.90		21.45	27.50

09 51 Acoustical Ceilings

09 51 23 – Acoustical Tile Ceilings

09 51 23.10 Suspended Acoustic Ceiling Tiles

		Crew	Daily Output	Labor-Hours	Unit	Material	Labor	Equipment	Total	Total Incl O&P
0010	**SUSPENDED ACOUSTIC CEILING TILES**, not including									
0100	suspension system									
0300	Fiberglass boards, film faced, 2' x 2' or 2' x 4', 5/8" thick	1 Carp	625	.013	S.F.	.72	.43		1.15	1.51
0400	3/4" thick		600	.013		1.66	.45		2.11	2.58
0500	3" thick, thermal, R11		450	.018		1.53	.60		2.13	2.68
0600	Glass cloth faced fiberglass, 3/4" thick		500	.016		1.49	.54		2.03	2.54
0700	1" thick		485	.016		2.18	.55		2.73	3.33
0820	1-1/2" thick, nubby face		475	.017		2.66	.57		3.23	3.88
1110	Mineral fiber tile, lay-in, 2' x 2' or 2' x 4', 5/8" thick, fine texture		625	.013		.59	.43		1.02	1.37

09 51 Acoustical Ceilings

09 51 23 – Acoustical Tile Ceilings

09 51 23.10 Suspended Acoustic Ceiling Tiles		Crew	Daily Output	Labor-Hours	Unit	2011 Bare Costs Material	Labor	Equipment	Total	Total Incl O&P
1115	Rough textured	1 Carp	625	.013	S.F.	1.44	.43		1.87	2.30
1125	3/4" thick, fine textured		600	.013		1.50	.45		1.95	2.40
1130	Rough textured		600	.013		1.79	.45		2.24	2.72
1135	Fissured		600	.013		2.45	.45		2.90	3.45
1150	Tegular, 5/8" thick, fine textured		470	.017		1.45	.57		2.02	2.56
1155	Rough textured		470	.017		1.90	.57		2.47	3.05
1165	3/4" thick, fine textured		450	.018		2.08	.60		2.68	3.29
1170	Rough textured		450	.018		2.35	.60		2.95	3.59
1175	Fissured		450	.018		3.63	.60		4.23	4.99
1180	For aluminum face, add					6.05			6.05	6.65
1185	For plastic film face, add					.94			.94	1.03
1190	For fire rating, add					.45			.45	.50
1300	Mirror tile, acrylic, 2' x 2'	1 Carp	500	.016		8.70	.54		9.24	10.45
1900	Eggcrate, acrylic, 1/2" x 1/2" x 1/2" cubes		500	.016		1.82	.54		2.36	2.90
2100	Polystyrene eggcrate, 3/8" x 3/8" x 1/2" cubes		510	.016		1.53	.53		2.06	2.56
2200	1/2" x 1/2" x 1/2" cubes		500	.016		1.75	.54		2.29	2.83
2400	Luminous panels, prismatic, acrylic		400	.020		2.22	.67		2.89	3.56
2500	Polystyrene		400	.020		1.14	.67		1.81	2.37
2700	Flat white acrylic		400	.020		3.86	.67		4.53	5.35
2800	Polystyrene		400	.020		2.65	.67		3.32	4.04
3000	Drop pan, white, acrylic		400	.020		5.65	.67		6.32	7.35
3100	Polystyrene		400	.020		4.73	.67		5.40	6.30
3600	Perforated aluminum sheets, .024" thick, corrugated, painted		490	.016		2.25	.55		2.80	3.40
3700	Plain		500	.016		3.76	.54		4.30	5.05
3720	Mineral fiber, 24" x 24" or 48", reveal edge, painted, 5/8" thick		600	.013		1.38	.45		1.83	2.27
3740	3/4" thick		575	.014		2.24	.47		2.71	3.24
5020	66 – 78% recycled content, 3/4" thick G		600	.013		1.82	.45		2.27	2.75
5040	Mylar, 42% recycled content, 3/4" thick G		600	.013		4.28	.45		4.73	5.45

09 51 23.30 Suspended Ceilings, Complete		Crew	Daily Output	Labor-Hours	Unit	Material	Labor	Equipment	Total	Total Incl O&P
0010	**SUSPENDED CEILINGS, COMPLETE** Including standard									
0100	suspension system but not incl. 1-1/2" carrier channels									
0600	Fiberglass ceiling board, 2' x 4' x 5/8", plain faced	1 Carp	500	.016	S.F.	1.37	.54		1.91	2.41
0700	Offices, 2' x 4' x 3/4"		380	.021		2.31	.71		3.02	3.72
0800	Mineral fiber, on 15/16" T bar susp. 2' x 2' x 3/4" lay-in board		345	.023		2.31	.78		3.09	3.84
0810	2' x 4' x 5/8" tile		380	.021		1.24	.71		1.95	2.54
0820	Tegular, 2' x 2' x 5/8" tile on 9/16" grid		250	.032		2.42	1.07		3.49	4.46
0830	2' x 4' x 3/4" tile		275	.029		2.89	.98		3.87	4.81
0900	Luminous panels, prismatic, acrylic		255	.031		2.87	1.05		3.92	4.92
1200	Metal pan with acoustic pad, steel		75	.107		4.18	3.58		7.76	10.60
1300	Painted aluminum		75	.107		2.90	3.58		6.48	9.20
1500	Aluminum, degreased finish		75	.107		4.82	3.58		8.40	11.30
1600	Stainless steel		75	.107		9.05	3.58		12.63	15.95
1800	Tile, Z bar suspension, 5/8" mineral fiber tile		150	.053		2.16	1.79		3.95	5.35
1900	3/4" mineral fiber tile		150	.053		2.31	1.79		4.10	5.55
2400	For strip lighting, see Section 26 51 13.50									
2500	For rooms under 500 S.F., add				S.F.		25%			

09 51 53 – Direct-Applied Acoustical Ceilings

09 51 53.10 Ceiling Tile		Crew	Daily Output	Labor-Hours	Unit	Material	Labor	Equipment	Total	Total Incl O&P
0010	**CEILING TILE**, Stapled or cemented									
0100	12" x 12" or 12" x 24", not including furring									
0600	Mineral fiber, vinyl coated, 5/8" thick	1 Carp	300	.027	S.F.	1.79	.89		2.68	3.47
0700	3/4" thick		300	.027		1.91	.89		2.80	3.60

09 51 Acoustical Ceilings

09 51 53 – Direct-Applied Acoustical Ceilings

09 51 53.10 Ceiling Tile		Crew	Daily Output	Labor-Hours	Unit	2011 Bare Costs Material	Labor	Equipment	Total	Total Incl O&P
0900	Fire rated, 3/4" thick, plain faced	1 Carp	300	.027	S.F.	1.40	.89		2.29	3.04
1000	Plastic coated face		300	.027		1.18	.89		2.07	2.80
1200	Aluminum faced, 5/8" thick, plain		300	.027		1.51	.89		2.40	3.16
3700	Wall application of above, add		1000	.008			.27		.27	.45
3900	For ceiling primer, add					.13			.13	.14
4000	For ceiling cement, add					.37			.37	.41

09 53 Acoustical Ceiling Suspension Assemblies

09 53 23 – Metal Acoustical Ceiling Suspension Assemblies

09 53 23.30 Ceiling Suspension Systems

Code	Description	Crew	Daily Output	Labor-Hours	Unit	Material	Labor	Equipment	Total	Total Incl O&P
0010	**CEILING SUSPENSION SYSTEMS** for boards and tile									
0050	Class A suspension system, 15/16" T bar, 2' x 4' grid	1 Carp	800	.010	S.F.	.65	.34		.99	1.28
0300	2' x 2' grid		650	.012		.81	.41		1.22	1.58
0310	25% recycled steel, 2' x 4' grid G		800	.010		.74	.34		1.08	1.37
0320	2' x 2' grid G		650	.012		.93	.41		1.34	1.71
0350	For 9/16" grid, add					.16			.16	.18
0360	For fire rated grid, add					.09			.09	.10
0370	For colored grid, add					.21			.21	.23
0400	Concealed Z bar suspension system, 12" module	1 Carp	520	.015		.66	.52		1.18	1.59
0600	1-1/2" carrier channels, 4' O.C., add	"	470	.017		.10	.57		.67	1.06
0700	Carrier channels for ceilings with									
0900	recessed lighting fixtures, add	1 Carp	460	.017	S.F.	.17	.58		.75	1.17
1040	Hanging wire, 12 ga., 4' long		65	.123	C.S.F.	7.55	4.13		11.68	15.20
1080	8' long		65	.123	"	15.15	4.13		19.28	23.50
1200	Seismic bracing, 2-1/2" compression post with four bracing wires		50	.160	Ea.	2.74	5.35		8.09	12

09 54 Specialty Ceilings

09 54 33 – Decorative Panel Ceilings

09 54 33.20 Metal Panel Ceilings

Code	Description	Crew	Daily Output	Labor-Hours	Unit	Material	Labor	Equipment	Total	Total Incl O&P
0010	**METAL PANEL CEILINGS**									
0020	Lay-in or screwed to furring, not including grid									
0100	Tin ceilings, 2' x 2' or 2' x 4', bare steel finish	2 Carp	300	.053	S.F.	2.37	1.79		4.16	5.60
0120	Painted white finish		300	.053	"	3.66	1.79		5.45	7
0140	Copper, chrome or brass finish		300	.053	L.F.	6.40	1.79		8.19	10.05
0200	Cornice molding, 2-1/2" to 3-1/2" wide, 4' long, bare steel finish		200	.080	S.F.	2.25	2.68		4.93	6.95
0220	Painted white finish		200	.080		2.94	2.68		5.62	7.70
0240	Copper, chrome or brass finish		200	.080		3.78	2.68		6.46	8.65
0320	5" to 6-1/2" wide, 4' long, bare steel finish		150	.107		3.19	3.58		6.77	9.50
0340	Painted white finish		150	.107		4.20	3.58		7.78	10.60
0360	Copper, chrome or brass finish		150	.107		6.40	3.58		9.98	13.05
0420	Flat molding, 3-1/2" to 5" wide, 4' long, bare steel finish		250	.064		3.43	2.15		5.58	7.35
0440	Painted white finish		250	.064		4.09	2.15		6.24	8.10
0460	Copper, chrome or brass finish		250	.064		7.35	2.15		9.50	11.70

09 62 Specialty Flooring

09 62 19 – Laminate Flooring

09 62 19.10 Floating Floor			Crew	Daily Output	Labor-Hours	Unit	2011 Bare Costs Material	Labor	Equipment	Total	Total Incl O&P
0010	**FLOATING FLOOR**										
8300	Floating floor, laminate, wood pattern strip, complete		1 Clab	133	.060	S.F.	4.57	1.59		6.16	7.70
8310	Components, T & G wood composite strips						3.66			3.66	4.03
8320	Film						.15			.15	.16
8330	Foam						.24			.24	.27
8340	Adhesive						.50			.50	.55
8350	Installation kit						.17			.17	.19
8360	Trim, 2" wide x 3' long					L.F.	3.85			3.85	4.24
8370	Reducer moulding					"	5.20			5.20	5.70

09 62 23 – Bamboo Flooring

09 62 23.10 Flooring, Bamboo			Crew	Daily Output	Labor-Hours	Unit	2011 Bare Costs Material	Labor	Equipment	Total	Total Incl O&P
0010	**FLOORING, BAMBOO**										
8600	Flooring, wood, bamboo strips, unfinished, 5/8" x 4" x 3'	G	1 Carp	255	.031	S.F.	4.44	1.05		5.49	6.65
8610	5/8" x 4" x 4'	G		275	.029		4.61	.98		5.59	6.70
8620	5/8" x 4" x 6'	G		295	.027		5.05	.91		5.96	7.05
8630	Finished, 5/8" x 4" x 3'	G		255	.031		4.89	1.05		5.94	7.15
8640	5/8" x 4" x 4'	G		275	.029		5.10	.98		6.08	7.30
8650	5/8" x 4" x 6'	G		295	.027		4.74	.91		5.65	6.70
8660	Stair treads, unfinished, 1-1/16" x 11-1/2" x 4'	G		18	.444	Ea.	42.50	14.90		57.40	72
8670	Finished, 1-1/16" x 11-1/2" x 4'	G		18	.444		71.50	14.90		86.40	104
8680	Stair risers, unfinished, 5/8" x 7-1/2" x 4'	G		18	.444		15.75	14.90		30.65	42.50
8690	Finished, 5/8" x 7-1/2" x 4'	G		18	.444		30	14.90		44.90	58
8700	Stair nosing, unfinished, 6' long	G		16	.500		35	16.80		51.80	66.50
8710	Finished, 6' long	G		16	.500		39.50	16.80		56.30	71.50

09 63 Masonry Flooring

09 63 13 – Brick Flooring

09 63 13.10 Miscellaneous Brick Flooring			Crew	Daily Output	Labor-Hours	Unit	Material	Labor	Equipment	Total	Total Incl O&P
0010	**MISCELLANEOUS BRICK FLOORING**										
0020	Acid-proof shales, red, 8" x 3-3/4" x 1-1/4" thick		D-7	.43	37.209	M	860	1,050		1,910	2,625
0050	2-1/4" thick		D-1	.40	40		1,000	1,225		2,225	3,100
0200	Acid-proof clay brick, 8" x 3-3/4" x 2-1/4" thick	G		.40	40		895	1,225		2,120	2,975
0250	9" x 4-1/2" x 3"	G		95	.168	S.F.	3.91	5.10		9.01	12.70
0260	Cast ceramic, pressed, 4" x 8" x 1/2", unglazed		D-7	100	.160		6.25	4.51		10.76	14.10
0270	Glazed			100	.160		8.35	4.51		12.86	16.35
0280	Hand molded flooring, 4" x 8" x 3/4", unglazed			95	.168		8.25	4.75		13	16.70
0290	Glazed			95	.168		10.35	4.75		15.10	19
0300	8" hexagonal, 3/4" thick, unglazed			85	.188		9.05	5.30		14.35	18.45
0310	Glazed			85	.188		16.35	5.30		21.65	26.50
0400	Heavy duty industrial, cement mortar bed, 2" thick, not incl. brick		D-1	80	.200		.83	6.10		6.93	10.90
0450	Acid-proof joints, 1/4" wide		"	65	.246		1.44	7.50		8.94	13.90
0500	Pavers, 8" x 4", 1" to 1-1/4" thick, red		D-7	95	.168		3.64	4.75		8.39	11.60
0510	Ironspot		"	95	.168		5.15	4.75		9.90	13.25
0540	1-3/8" to 1-3/4" thick, red		D-1	95	.168		3.51	5.10		8.61	12.25
0560	Ironspot			95	.168		5.10	5.10		10.20	14
0580	2-1/4" thick, red			90	.178		3.57	5.40		8.97	12.80
0590	Ironspot			90	.178		5.55	5.40		10.95	14.95
0700	Paver, adobe brick, 6" x 12", 1/2" joint	G		42	.381		.84	11.60		12.44	19.90
0710	Mexican red, 12" x 12"	G	1 Tilf	48	.167		1.38	5.25		6.63	9.90
0720	Saltillo, 12" x 12"	G	"	48	.167		1.38	5.25		6.63	9.90

09 63 Masonry Flooring

09 63 13 – Brick Flooring

09 63 13.10 Miscellaneous Brick Flooring		Crew	Daily Output	Labor-Hours	Unit	2011 Bare Costs Material	Labor	Equipment	Total	Total Incl O&P
0800	For sidewalks and patios with pavers, see Section 32 14 16.10									
0870	For epoxy joints, add	D-1	600	.027	S.F.	2.72	.81		3.53	4.32
0880	For Furan underlayment, add	"	600	.027		2.25	.81		3.06	3.81
0890	For waxed surface, steam cleaned, add	A-1H	1000	.008		.20	.21	.06	.47	.64

09 63 40 – Stone Flooring

09 63 40.10 Marble		Crew	Daily Output	Labor-Hours	Unit	Material	Labor	Equipment	Total	Total Incl O&P
0010	**MARBLE**									
0020	Thin gauge tile, 12" x 6", 3/8", white Carara	D-7	60	.267	S.F.	11.25	7.50		18.75	24.50
0100	Travertine		60	.267		12.30	7.50		19.80	25.50
0200	12" x 12" x 3/8", thin set, floors		60	.267		6.05	7.50		13.55	18.65
0300	On walls		52	.308		9.50	8.70		18.20	24.50
1000	Marble threshold, 4" wide x 36" long x 5/8" thick, white		60	.267	Ea.	6.05	7.50		13.55	18.65

09 63 40.20 Slate Tile		Crew	Daily Output	Labor-Hours	Unit	Material	Labor	Equipment	Total	Total Incl O&P
0010	**SLATE TILE**									
0020	Vermont, 6" x 6" x 1/4" thick, thin set	D-7	180	.089	S.F.	6.80	2.51		9.31	11.45
0200	See also Section 32 14 40.10									

09 64 Wood Flooring

09 64 16 – Wood Block Flooring

09 64 16.10 End Grain Block Flooring		Crew	Daily Output	Labor-Hours	Unit	Material	Labor	Equipment	Total	Total Incl O&P
0010	**END GRAIN BLOCK FLOORING**									
0020	End grain flooring, coated, 2" thick	1 Carp	295	.027	S.F.	3.29	.91		4.20	5.15
0400	Natural finish, 1" thick, fir		125	.064		3.40	2.15		5.55	7.35
0600	1-1/2" thick, pine		125	.064		3.34	2.15		5.49	7.25
0700	2" thick, pine		125	.064		4.64	2.15		6.79	8.70

09 64 19 – Wood Composition Flooring

09 64 19.10 Wood Composition		Crew	Daily Output	Labor-Hours	Unit	Material	Labor	Equipment	Total	Total Incl O&P
0010	**WOOD COMPOSITION** Gym floors									
0100	2-1/4" x 6-7/8" x 3/8", on 2" grout setting bed	D-7	150	.107	S.F.	5.85	3.01		8.86	11.25
0200	Thin set, on concrete	"	250	.064		5.35	1.80		7.15	8.80
0300	Sanding and finishing, add	1 Carp	200	.040		.79	1.34		2.13	3.12

09 64 23 – Wood Parquet Flooring

09 64 23.10 Wood Parquet		Crew	Daily Output	Labor-Hours	Unit	Material	Labor	Equipment	Total	Total Incl O&P
0010	**WOOD PARQUET** flooring									
5200	Parquetry, standard, 5/16" thick, not incl. finish, oak, minimum	1 Carp	160	.050	S.F.	4.55	1.68		6.23	7.80
5300	Maximum		100	.080		5.55	2.68		8.23	10.60
5500	Teak, minimum		160	.050		4.86	1.68		6.54	8.15
5600	Maximum		100	.080		8.50	2.68		11.18	13.85
5650	13/16" thick, select grade oak, minimum		160	.050		9.35	1.68		11.03	13.10
5700	Maximum		100	.080		14.20	2.68		16.88	20
5800	Custom parquetry, including finish, minimum		100	.080		15.65	2.68		18.33	21.50
5900	Maximum		50	.160		21	5.35		26.35	32
6700	Parquetry, prefinished white oak, 5/16" thick, minimum		160	.050		3.71	1.68		5.39	6.90
6800	Maximum		100	.080		5.50	2.68		8.18	10.55
7000	Walnut or teak, parquetry, minimum		160	.050		5.35	1.68		7.03	8.70
7100	Maximum		100	.080		9.35	2.68		12.03	14.80
7200	Acrylic wood parquet blocks, 12" x 12" x 5/16",									
7210	Irradiated, set in epoxy	1 Carp	160	.050	S.F.	7.80	1.68		9.48	11.40

09 64 Wood Flooring

09 64 29 – Wood Strip and Plank Flooring

09 64 29.10 Wood		Crew	Daily Output	Labor-Hours	Unit	2011 Bare Costs Material	Labor	Equipment	Total	Total Incl O&P
0010	**WOOD**									
0020	Fir, vertical grain, 1" x 4", not incl. finish, grade B & better	1 Carp	255	.031	S.F.	2.74	1.05		3.79	4.77
0100	C grade & better		255	.031		2.58	1.05		3.63	4.60
4000	Maple, strip, 25/32" x 2-1/4", not incl. finish, select		170	.047		4.37	1.58		5.95	7.45
4100	#2 & better		170	.047		3.49	1.58		5.07	6.50
4300	33/32" x 3-1/4", not incl. finish, #1 grade		170	.047		4.19	1.58		5.77	7.25
4400	#2 & better		170	.047		3.73	1.58		5.31	6.75
4600	Oak, white or red, 25/32" x 2-1/4", not incl. finish									
4700	#1 common	1 Carp	170	.047	S.F.	2.49	1.58		4.07	5.40
4900	Select quartered, 2-1/4" wide		170	.047		3.09	1.58		4.67	6.05
5000	Clear		170	.047		3.95	1.58		5.53	7
6100	Prefinished, white oak, prime grade, 2-1/4" wide		170	.047		4.49	1.58		6.07	7.60
6200	3-1/4" wide		185	.043		4.99	1.45		6.44	7.95
6400	Ranch plank		145	.055		6.75	1.85		8.60	10.50
6500	Hardwood blocks, 9" x 9", 25/32" thick		160	.050		5.75	1.68		7.43	9.15
7400	Yellow pine, 3/4" x 3-1/8", T & G, C & better, not incl. finish		200	.040		1.49	1.34		2.83	3.89
7500	Refinish wood floor, sand, 2 cts poly, wax, soft wood, min.	1 Clab	400	.020		.78	.53		1.31	1.75
7600	Hard wood, max		130	.062		1.16	1.63		2.79	4.01
7800	Sanding and finishing, 2 coats polyurethane		295	.027		.78	.72		1.50	2.06
7900	Subfloor and underlayment, see Section 06 16									
8015	Transition molding, 2 1/4" wide, 5' long	1 Carp	19.20	.417	Ea.	15	14		29	40

09 64 66 – Wood Athletic Flooring

09 64 66.10 Gymnasium Flooring		Crew	Daily Output	Labor-Hours	Unit	Material	Labor	Equipment	Total	Total Incl O&P
0010	**GYMNASIUM FLOORING**									
0600	Gym floor, in mastic, over 2 ply felt, #2 & better									
0700	25/32" thick maple	1 Carp	100	.080	S.F.	4.14	2.68		6.82	9.05
0900	33/32" thick maple		98	.082		4.99	2.74		7.73	10.10
1000	For 1/2" corkboard underlayment, add		750	.011		.90	.36		1.26	1.59
1300	For #1 grade maple, add					.47			.47	.52
1600	Maple flooring, over sleepers, #2 & better									
1700	25/32" thick	1 Carp	85	.094	S.F.	4.32	3.16		7.48	10.05
1900	33/32" thick	"	83	.096		5.05	3.23		8.28	10.95
2000	For #1 grade, add					.50			.50	.55
2200	For 3/4" subfloor, add	1 Carp	350	.023		1.12	.77		1.89	2.51
2300	With two 1/2" subfloors, 25/32" thick	"	69	.116		5.40	3.89		9.29	12.45
2500	Maple, incl. finish, #2 & btr., 25/32" thick, on rubber									
2600	Sleepers, with two 1/2" subfloors	1 Carp	76	.105	S.F.	5.80	3.53		9.33	12.30
2800	With steel spline, double connection to channels	"	73	.110		6.20	3.68		9.88	12.95
2900	For 33/32" maple, add					.67			.67	.74
3100	For #1 grade maple, add					.50			.50	.55
3500	For termite proofing all of the above, add					.27			.27	.30
3700	Portable hardwood, prefinished panels	1 Carp	83	.096		7.75	3.23		10.98	13.90
3720	Insulated with polystyrene, 1" thick, add		165	.048		.67	1.63		2.30	3.46
3750	Running tracks, Sitka spruce surface, 25/32" x 2-1/4"		62	.129		14.15	4.33		18.48	23
3770	3/4" plywood surface, finished		100	.080		3.36	2.68		6.04	8.20

09 65 Resilient Flooring

09 65 10 – Resilient Tile Underlayment

09 65 10.10 Latex Underlayment		Crew	Daily Output	Labor-Hours	Unit	Material	Labor (2011 Bare Costs)	Equipment (2011 Bare Costs)	Total (2011 Bare Costs)	Total Incl O&P
0010	**LATEX UNDERLAYMENT**									
3600	Latex underlayment, 1/8" thk., cementitious for resilient flooring	1 Tilf	160	.050	S.F.	.82	1.58		2.40	3.42
4000	Liquid, fortified				Gal.	40.50			40.50	44.50

09 65 13 – Resilient Base and Accessories

09 65 13.13 Resilient Base

Line	Description	Crew	Daily Output	Labor-Hours	Unit	Material	Labor	Equipment	Total	Total Incl O&P
0010	**RESILIENT BASE**									
0800	Base, cove, rubber or vinyl									
1100	Standard colors, 0.080" thick, 2-1/2" high	1 Tilf	315	.025	L.F.	1	.80		1.80	2.38
1150	4" high		315	.025		1.08	.80		1.88	2.47
1200	6" high		315	.025		1.50	.80		2.30	2.93
1450	1/8" thick, 2-1/2" high		315	.025		1	.80		1.80	2.38
1500	4" high		315	.025		.90	.80		1.70	2.27
1550	6" high		315	.025		1.45	.80		2.25	2.88
1600	Corners, 2-1/2" high		315	.025	Ea.	1.60	.80		2.40	3.04
1630	4" high		315	.025		3	.80		3.80	4.58
1660	6" high		315	.025		3	.80		3.80	4.58

09 65 13.23 Resilient Stair Treads and Risers

Line	Description	Crew	Daily Output	Labor-Hours	Unit	Material	Labor	Equipment	Total	Total Incl O&P
0010	**RESILIENT STAIR TREADS AND RISERS**									
0300	Rubber, molded tread, 12" wide, 5/16" thick, black	1 Tilf	115	.070	L.F.	12.75	2.19		14.94	17.55
0400	Colors		115	.070		13.35	2.19		15.54	18.20
0600	1/4" thick, black		115	.070		12.50	2.19		14.69	17.25
0700	Colors		115	.070		12.75	2.19		14.94	17.55
0900	Grip strip safety tread, colors, 5/16" thick		115	.070		16.95	2.19		19.14	22
1000	3/16" thick		120	.067		11.75	2.10		13.85	16.25
1200	Landings, smooth sheet rubber, 1/8" thick		120	.067	S.F.	6.30	2.10		8.40	10.30
1300	3/16" thick		120	.067	"	7.95	2.10		10.05	12.10
1500	Nosings, 3" wide, 3/16" thick, black		140	.057	L.F.	3.67	1.80		5.47	6.90
1600	Colors		140	.057		3.67	1.80		5.47	6.90
1800	Risers, 7" high, 1/8" thick, flat		250	.032		4.55	1.01		5.56	6.60
1900	Coved		250	.032		5.95	1.01		6.96	8.15
2100	Vinyl, molded tread, 12" wide, colors, 1/8" thick		115	.070		6.25	2.19		8.44	10.40
2200	1/4" thick		115	.070		8	2.19		10.19	12.30
2300	Landing material, 1/8" thick		200	.040	S.F.	6.25	1.26		7.51	8.90
2400	Riser, 7" high, 1/8" thick, coved		175	.046	L.F.	3.50	1.44		4.94	6.15
2500	Tread and riser combined, 1/8" thick		80	.100	"	9.50	3.16		12.66	15.50

09 65 16 – Resilient Sheet Flooring

09 65 16.10 Rubber and Vinyl Sheet Flooring

Line	Description	Crew	Daily Output	Labor-Hours	Unit	Material	Labor	Equipment	Total	Total Incl O&P
0010	**RUBBER AND VINYL SHEET FLOORING**									
5500	Linoleum, sheet goods G	1 Tilf	360	.022	S.F.	4.28	.70		4.98	5.85
5900	Rubber, sheet goods, 36" wide, 1/8" thick		120	.067		6.85	2.10		8.95	10.90
5950	3/16" thick		100	.080		9.25	2.52		11.77	14.25
6000	1/4" thick		90	.089		10.95	2.80		13.75	16.55
8000	Vinyl sheet goods, backed, .065" thick, minimum		250	.032		3.75	1.01		4.76	5.75
8050	Maximum		200	.040		3.69	1.26		4.95	6.10
8100	.080" thick, minimum		230	.035		3.80	1.10		4.90	5.95
8150	Maximum		200	.040		5.40	1.26		6.66	7.95
8200	.125" thick, minimum		230	.035		4.35	1.10		5.45	6.55
8250	Maximum		200	.040		6.50	1.26		7.76	9.15
8700	Adhesive cement, 1 gallon per 200 to 300 S.F.				Gal.	25			25	27.50
8800	Asphalt primer, 1 gallon per 300 S.F.					12.75			12.75	14.05
8900	Emulsion, 1 gallon per 140 S.F.					16.25			16.25	17.90

09 65 Resilient Flooring

09 65 19 – Resilient Tile Flooring

09 65 19.10 Miscellaneous Resilient Tile Flooring			Crew	Daily Output	Labor-Hours	Unit	2011 Bare Costs Material	Labor	Equipment	Total	Total Incl O&P
0010	**MISCELLANEOUS RESILIENT TILE FLOORING**										
2200	Cork tile, standard finish, 1/8" thick	G	1 Tilf	315	.025	S.F.	6.25	.80		7.05	8.20
2250	3/16" thick	G		315	.025		6.30	.80		7.10	8.20
2300	5/16" thick	G		315	.025		7.75	.80		8.55	9.85
2350	1/2" thick	G		315	.025		9.20	.80		10	11.40
2500	Urethane finish, 1/8" thick	G		315	.025		7.45	.80		8.25	9.45
2550	3/16" thick	G		315	.025		7.80	.80		8.60	9.90
2600	5/16" thick	G		315	.025		10	.80		10.80	12.30
2650	1/2" thick	G		315	.025		13.50	.80		14.30	16.15
6050	Rubber tile, marbleized colors, 12" x 12", 1/8" thick			400	.020		6.65	.63		7.28	8.35
6100	3/16" thick			400	.020		10	.63		10.63	12
6300	Special tile, plain colors, 1/8" thick			400	.020		10	.63		10.63	12
6350	3/16" thick			400	.020		10	.63		10.63	12
6410	Raised, radial or square, minimum			400	.020		9.25	.63		9.88	11.20
6430	Maximum			400	.020		12.50	.63		13.13	14.75
6450	For golf course, skating rink, etc., 1/4" thick			275	.029		12.50	.92		13.42	15.20
6700	Synthetic turf, 3/8" thick			90	.089		6.20	2.80		9	11.30
6750	Interlocking 2' x 2' squares, 1/2" thick, not										
6810	cemented, for playgrounds, minimum		1 Tilf	210	.038	S.F.	4.75	1.20		5.95	7.15
6850	Maximum			190	.042		12	1.33		13.33	15.30
7000	Vinyl composition tile, 12" x 12", 1/16" thick			500	.016		.93	.50		1.43	1.83
7050	Embossed			500	.016		2.17	.50		2.67	3.20
7100	Marbleized			500	.016		2.17	.50		2.67	3.20
7150	Solid			500	.016		2.79	.50		3.29	3.88
7200	3/32" thick, embossed			500	.016		1.35	.50		1.85	2.30
7250	Marbleized			500	.016		2.49	.50		2.99	3.55
7300	Solid			500	.016		2.31	.50		2.81	3.35
7350	1/8" thick, marbleized			500	.016		1.64	.50		2.14	2.61
7400	Solid			500	.016		1.73	.50		2.23	2.71
7450	Conductive			500	.016		5.95	.50		6.45	7.35
7500	Vinyl tile, 12" x 12", .050" thick, minimum			500	.016		3	.50		3.50	4.11
7550	Maximum			500	.016		6	.50		6.50	7.40
7600	1/8" thick, minimum			500	.016		4.50	.50		5	5.75
7650	Solid colors			500	.016		6.20	.50		6.70	7.60
7700	Marbleized or Travertine pattern			500	.016		5.65	.50		6.15	7
7750	Florentine pattern			500	.016		6	.50		6.50	7.40
7800	Maximum			500	.016		13.50	.50		14	15.65

09 65 33 – Conductive Resilient Flooring

09 65 33.10 Conductive Rubber and Vinyl Flooring		Crew	Daily Output	Labor-Hours	Unit	Material	Labor	Equipment	Total	Total Incl O&P
0010	**CONDUCTIVE RUBBER AND VINYL FLOORING**									
1700	Conductive flooring, rubber tile, 1/8" thick	1 Tilf	315	.025	S.F.	6	.80		6.80	7.90
1800	Homogeneous vinyl tile, 1/8" thick	"	315	.025	"	8	.80		8.80	10.10

09 66 Terrazzo Flooring

09 66 13 – Portland Cement Terrazzo Flooring

09 66 13.10 Portland Cement Terrazzo		Crew	Daily Output	Labor-Hours	Unit	Material	Labor	Equipment	Total	Total Incl O&P
						2011 Bare Costs				
0010	**PORTLAND CEMENT TERRAZZO**, cast-in-place R096613-10									
0101	Curb, 6" high and 6" wide	J-3	12	1.333	L.F.	5.55	38.50	20	64.05	89
0300	Divider strip for floors, 14 ga., 1-1/4" deep, zinc	1 Mstz	375	.021		1.39	.67		2.06	2.60
0400	Brass		375	.021		2.43	.67		3.10	3.74
0600	Heavy top strip 1/4" thick, 1-1/4" deep, zinc		300	.027		1.90	.83		2.73	3.42
0900	Galv. bottoms, brass		300	.027		3.13	.83		3.96	4.77
1200	For thin set floors, 16 ga., 1/2" x 1/2", zinc		350	.023		.82	.72		1.54	2.04
1300	Brass		350	.023		2.27	.72		2.99	3.64
1500	Floor, bonded to concrete, 1-3/4" thick, gray cement	J-3	75	.213	S.F.	2.92	6.10	3.22	12.24	16.55
1600	White cement, mud set		75	.213		3.32	6.10	3.22	12.64	17
1800	Not bonded, 3" total thickness, gray cement		70	.229		3.65	6.55	3.45	13.65	18.30
1900	White cement, mud set		70	.229		3.98	6.55	3.45	13.98	18.70
2100	For Venetian terrazzo, 1" topping, add					50%	50%			
2200	For heavy duty abrasive terrazzo, add					50%	50%			
2700	Monolithic terrazzo, 1/2" thick									
2710	10' panels	J-3	125	.128	S.F.	2.65	3.67	1.93	8.25	10.90
3000	Stairs, cast in place, pan filled treads		30	.533	L.F.	2.71	15.30	8.05	26.06	36.50
3100	Treads and risers		14	1.143	"	5.55	33	17.25	55.80	77.50
3300	For stair landings, add to floor prices						50%			
3400	Stair stringers and fascia	J-3	30	.533	S.F.	4.44	15.30	8.05	27.79	38
3600	For abrasive metal nosings on stairs, add		150	.107	L.F.	8.60	3.06	1.61	13.27	16.10
3700	For abrasive surface finish, add		600	.027	S.F.	1.39	.77	.40	2.56	3.19
3900	For raised abrasive strips, add		150	.107	L.F.	1.13	3.06	1.61	5.80	7.90
4000	Wainscot, bonded, 1-1/2" thick		30	.533	S.F.	3.50	15.30	8.05	26.85	37
4200	Epoxy terrazzo, 1/4" thick		40	.400	"	4.96	11.50	6.05	22.51	30.50
4300	Stone chips, onyx gemstone, per 50 lb. bag				Bag	16.30			16.30	17.95

09 66 13.30 Terrazzo, Precast		Crew	Daily Output	Labor-Hours	Unit	Material	Labor	Equipment	Total	Total Incl O&P
0010	**TERRAZZO, PRECAST**									
0020	Base, 6" high, straight	1 Mstz	70	.114	L.F.	10.55	3.58		14.13	17.35
0100	Cove		60	.133		12.15	4.17		16.32	20
0300	8" high, straight		60	.133		10.85	4.17		15.02	18.55
0400	Cove		50	.160		15.95	5		20.95	25.50
0600	For white cement, add					.43			.43	.47
0700	For 16 ga. zinc toe strip, add					1.63			1.63	1.79
0900	Curbs, 4" x 4" high	1 Mstz	40	.200		30.50	6.25		36.75	43.50
1000	8" x 8" high	"	30	.267		35.50	8.35		43.85	52.50
1200	Floor tiles, non-slip, 1" thick, 12" x 12"	D-1	60	.267	S.F.	18.60	8.10		26.70	34
1300	1-1/4" thick, 12" x 12"		60	.267		19.60	8.10		27.70	35
1500	16" x 16"		50	.320		21.50	9.75		31.25	39.50
1600	1-1/2" thick, 16" x 16"		45	.356		19.45	10.80		30.25	39.50
1800	For Venetian terrazzo, add					5.85			5.85	6.40
1900	For white cement, add					.55			.55	.61
2401	Stair treads, 1-1/2" thick, non-slip, three line pattern	2 Mstz	70	.229	L.F.	39.50	7.15		46.65	55
2501	Nosing and two lines	J-3	70	.229		39.50	6.55	3.45	49.50	58
2701	2" thick treads, straight	2 Mstz	60	.267		46	8.35		54.35	64
2801	Curved		50	.320		61.50	10		71.50	84
3001	Stair risers, straight sections, 1" thick, to 6" high		60	.267		9.80	8.35		18.15	24
3101	Cove		50	.320		15.20	10		25.20	33
3301	Curved, 1" thick, to 6" high, vertical		48	.333		25.50	10.45		35.95	44.50
3402	Cove		38	.421		49.50	13.20		62.70	75.50
3601	Stair tread and riser, single piece, straight, minimum		60	.267		50	8.35		58.35	68.50
3701	Maximum		40	.400		65	12.50		77.50	91

09 66 Terrazzo Flooring

09 66 13 – Portland Cement Terrazzo Flooring

09 66 13.30 Terrazzo, Precast		Crew	Daily Output	Labor-Hours	Unit	Material	2011 Bare Costs Labor	Equipment	Total	Total Incl O&P
3901	Curved tread and riser, minimum	J-3	40	.400	L.F.	70	11.50	6.05	87.55	102
4001	Maximum	"	32	.500		88.50	14.35	7.55	110.40	128
4201	Stair stringers, notched, 1" thick	2 Mstz	25	.640		29.50	20		49.50	64.50
4301	2" thick		22	.727		34.50	23		57.50	74
4501	Stair landings, structural, non-slip, 1-1/2" thick		85	.188	S.F.	27.50	5.90		33.40	39.50
4601	3" thick		14	1.143		34	36		70	94
4800	Wainscot, 12" x 12" x 1" tiles	1 Mstz	12	.667		6.50	21		27.50	40.50
4900	16" x 16" x 1-1/2" tiles	"	8	1		13.70	31.50		45.20	65

09 66 16 – Terrazzo Floor Tile

09 66 16.10 Tile or Terrazzo Base

		Crew	Daily Output	Labor-Hours	Unit	Material	Labor	Equipment	Total	Total Incl O&P
0010	**TILE OR TERRAZZO BASE**									
0020	Scratch coat only	1 Mstz	150	.053	S.F.	.41	1.67		2.08	3.12
0500	Scratch and brown coat only	"	75	.107	"	.78	3.34		4.12	6.20

09 66 33 – Conductive Terrazzo Flooring

09 66 33.10 Conductive Terrazzo

		Crew	Daily Output	Labor-Hours	Unit	Material	Labor	Equipment	Total	Total Incl O&P
0010	**CONDUCTIVE TERRAZZO**									
2400	Bonded conductive floor for hospitals	J-3	90	.178	S.F.	3.98	5.10	2.68	11.76	15.50
2500	Epoxy terrazzo, 1/4" thick, minimum		100	.160		4.50	4.59	2.42	11.51	14.95
2550	Average		75	.213		4.96	6.10	3.22	14.28	18.80
2600	Maximum		60	.267		5.45	7.65	4.03	17.13	22.50

09 67 Fluid-Applied Flooring

09 67 20 – Epoxy-Marble Chip Flooring

09 67 20.13 Elastomeric Liquid Flooring

		Crew	Daily Output	Labor-Hours	Unit	Material	Labor	Equipment	Total	Total Incl O&P
0010	**ELASTOMERIC LIQUID FLOORING**									
0020	Cementitious acrylic, 1/4" thick	C-6	520	.092	S.F.	1.47	2.56	.09	4.12	5.95
0100	3/8" thick	"	450	.107		1.91	2.95	.10	4.96	7.10
0200	Methyl methachrylate, 1/4" thick	C-8A	3000	.016		5.40	.46		5.86	6.70
0210	1/8" thick	"	3000	.016		2.97	.46		3.43	4.02
0300	Cupric oxychloride, on bond coat, minimum	C-6	480	.100		3.25	2.77	.10	6.12	8.30
0400	Maximum		420	.114		5.40	3.16	.11	8.67	11.30
2400	Mastic, hot laid, 2 coat, 1-1/2" thick, standard, minimum		690	.070		3.77	1.93	.07	5.77	7.40
2500	Maximum		520	.092		4.84	2.56	.09	7.49	9.65
2700	Acid-proof, minimum		605	.079		4.84	2.20	.08	7.12	9.05
2800	Maximum		350	.137		6.70	3.80	.13	10.63	13.80
3000	Neoprene, trowelled on, 1/4" thick, minimum		545	.088		3.71	2.44	.09	6.24	8.20
3100	Maximum		430	.112		5.10	3.09	.11	8.30	10.85
4300	Polyurethane, with suspended vinyl chips, minimum		1065	.045		7.20	1.25	.04	8.49	10.05
4500	Maximum		860	.056		10.40	1.54	.05	11.99	14.05

09 67 20.16 Epoxy Terrazzo

		Crew	Daily Output	Labor-Hours	Unit	Material	Labor	Equipment	Total	Total Incl O&P
0010	**EPOXY TERRAZZO**									
1800	Epoxy terrazzo, 1/4" thick, chemical resistant, minimum	J-3	200	.080	S.F.	5.15	2.30	1.21	8.66	10.70
1900	Maximum		150	.107		7.85	3.06	1.61	12.52	15.30
2100	Conductive, minimum		210	.076		7.05	2.19	1.15	10.39	12.55
2200	Maximum		160	.100		9.10	2.87	1.51	13.48	16.25

09 67 20.19 Polyacrylate Terrazzo

		Crew	Daily Output	Labor-Hours	Unit	Material	Labor	Equipment	Total	Total Incl O&P
0010	**POLYACRYLATE TERRAZZO**									
3150	Polyacrylate, 1/4" thick, minimum	C-6	735	.065	S.F.	3.07	1.81	.06	4.94	6.45
3170	Maximum		480	.100		3.91	2.77	.10	6.78	9

09 67 Fluid-Applied Flooring

09 67 20 – Epoxy-Marble Chip Flooring

09 67 20.19 Polyacrylate Terrazzo		Crew	Daily Output	Labor-Hours	Unit	2011 Bare Costs Material	Labor	Equipment	Total	Total Incl O&P
3200	3/8" thick, minimum	C-6	620	.077	S.F.	4	2.14	.08	6.22	8.05
3220	Maximum		480	.100		5.65	2.77	.10	8.52	10.95
3300	Conductive, 1/4" thick, minimum		450	.107		6.40	2.95	.10	9.45	12.05
3330	Maximum		305	.157		8.80	4.36	.15	13.31	17.10
3350	3/8" thick, minimum		365	.132		8.40	3.64	.13	12.17	15.45
3370	Maximum		255	.188		11.25	5.20	.18	16.63	21.50
3450	Granite, conductive, 1/4" thick, minimum		695	.069		8.15	1.91	.07	10.13	12.20
3470	Maximum		420	.114		10.50	3.16	.11	13.77	16.90
3500	3/8" thick, minimum		695	.069		11.90	1.91	.07	13.88	16.35
3520	Maximum		380	.126		14.40	3.50	.12	18.02	22

09 67 20.26 Quartz Flooring

Line	Description	Crew	Daily Output	Labor-Hours	Unit	Material	Labor	Equipment	Total	Total Incl O&P
0010	**QUARTZ FLOORING**									
0600	Epoxy, with colored quartz chips, broadcast, minimum	C-6	675	.071	S.F.	2.57	1.97	.07	4.61	6.20
0700	Maximum		490	.098		3.12	2.71	.10	5.93	8.05
0900	Trowelled, minimum		560	.086		3.31	2.37	.08	5.76	7.65
1000	Maximum		480	.100		4.84	2.77	.10	7.71	10
1200	Heavy duty epoxy topping, 1/4" thick,									
1300	500 to 1,000 S.F.	C-6	420	.114	S.F.	4.89	3.16	.11	8.16	10.75
1500	1,000 to 2,000 S.F.		450	.107		4.06	2.95	.10	7.11	9.50
1600	Over 10,000 S.F.		480	.100		3.83	2.77	.10	6.70	8.90
3600	Polyester, with colored quartz chips, 1/16" thick, minimum		1065	.045		2.74	1.25	.04	4.03	5.15
3700	Maximum		560	.086		3.73	2.37	.08	6.18	8.15
3900	1/8" thick, minimum		810	.059		3.19	1.64	.06	4.89	6.30
4000	Maximum		675	.071		4.23	1.97	.07	6.27	8
4200	Polyester, heavy duty, compared to epoxy, add		2590	.019		1.31	.51	.02	1.84	2.31

09 67 66 – Fluid-Applied Athletic Flooring

09 67 66.10 Polyurethane

Line	Description	Crew	Daily Output	Labor-Hours	Unit	Material	Labor	Equipment	Total	Total Incl O&P
0010	**POLYURETHANE**									
4400	Thermoset, prefabricated in place, indoor									
4500	3/8" thick for basketball, gyms, etc.	1 Tilf	100	.080	S.F.	4.95	2.52		7.47	9.50
4600	1/2" thick for professional sports		95	.084		6.65	2.66		9.31	11.55
4700	Outdoor, 1/4" thick, smooth, for tennis		100	.080		5.15	2.52		7.67	9.70
4800	Rough, for track, 3/8" thick		95	.084		5.95	2.66		8.61	10.80
5000	Poured in place, indoor, with finish, 1/4" thick		80	.100		3.66	3.16		6.82	9.10
5050	3/8" thick		65	.123		4.45	3.88		8.33	11.10
5100	1/2" thick		50	.160		5.85	5.05		10.90	14.50

09 68 Carpeting

09 68 05 – Carpet Accessories

09 68 05.11 Flooring Transition Strip

Line	Description	Crew	Daily Output	Labor-Hours	Unit	Material	Labor	Equipment	Total	Total Incl O&P
0010	**FLOORING TRANSITION STRIP**									
0107	Clamp down brass divider, 12' strip, vinyl to carpet	1 Tilf	31.25	.256	Ea.	10.45	8.10		18.55	24.50
0117	Vinyl to hard surface	"	31.25	.256	"	10.45	8.10		18.55	24.50

09 68 10 – Carpet Pad

09 68 10.10 Commercial Grade Carpet Pad

Line	Description	Crew	Daily Output	Labor-Hours	Unit	Material	Labor	Equipment	Total	Total Incl O&P
0010	**COMMERCIAL GRADE CARPET PAD**									
9000	Sponge rubber pad, minimum	1 Tilf	150	.053	S.Y.	4.06	1.68		5.74	7.15
9100	Maximum		150	.053		10	1.68		11.68	13.70
9200	Felt pad, minimum		150	.053		4.14	1.68		5.82	7.25
9300	Maximum		150	.053		7.10	1.68		8.78	10.50

09 68 Carpeting

09 68 10 – Carpet Pad

09 68 10.10 Commercial Grade Carpet Pad		Crew	Daily Output	Labor-Hours	Unit	Material	2011 Bare Costs Labor	Equipment	Total	Total Incl O&P
9400	Bonded urethane pad, minimum	1 Tilf	150	.053	S.Y.	4.57	1.68		6.25	7.75
9500	Maximum		150	.053		7.80	1.68		9.48	11.30
9600	Prime urethane pad, minimum		150	.053		3	1.68		4.68	6
9700	Maximum		150	.053		5	1.68		6.68	8.20

09 68 13 – Tile Carpeting

09 68 13.10 Carpet Tile

Code	Description	Crew	Daily Output	Labor-Hours	Unit	Material	Labor	Equipment	Total	Total Incl O&P
0010	**CARPET TILE**									
0100	Tufted nylon, 18" x 18", hard back, 20 oz.	1 Tilf	80	.100	S.Y.	22	3.16		25.16	29.50
0110	26 oz.		80	.100		38	3.16		41.16	46.50
0200	Cushion back, 20 oz.		80	.100		28	3.16		31.16	35.50
0210	26 oz.		80	.100		43.50	3.16		46.66	53
1100	Tufted, 24" x 24", hard back, 24 oz. nylon		80	.100		29	3.16		32.16	36.50
1180	35 oz.		80	.100		34	3.16		37.16	42.50
5060	42 oz.		80	.100		45.50	3.16		48.66	55

09 68 16 – Sheet Carpeting

09 68 16.10 Sheet Carpet

Code	Description	Crew	Daily Output	Labor-Hours	Unit	Material	Labor	Equipment	Total	Total Incl O&P
0010	**SHEET CARPET**									
0700	Nylon, level loop, 26 oz., light to medium traffic	1 Tilf	75	.107	S.Y.	28	3.37		31.37	36.50
0720	28 oz., light to medium traffic		75	.107		29	3.37		32.37	37.50
0900	32 oz., medium traffic		75	.107		35	3.37		38.37	43.50
1100	40 oz., medium to heavy traffic		75	.107		51	3.37		54.37	61.50
2920	Nylon plush, 30 oz., medium traffic		57	.140		27	4.43		31.43	36.50
3000	36 oz., medium traffic		75	.107		35.50	3.37		38.87	44.50
3100	42 oz., medium to heavy traffic		70	.114		40	3.61		43.61	50
3200	46 oz., medium to heavy traffic		70	.114		46.50	3.61		50.11	57.50
3300	54 oz., heavy traffic		70	.114		51	3.61		54.61	62
3340	60 oz., heavy traffic		70	.114		70	3.61		73.61	83
3665	Olefin, 24 oz., light to medium traffic		75	.107		16	3.37		19.37	23
3670	26 oz., medium traffic		75	.107		17.35	3.37		20.72	24.50
3680	28 oz., medium to heavy traffic		75	.107		18.95	3.37		22.32	26.50
3700	32 oz., medium to heavy traffic		75	.107		24	3.37		27.37	32
3730	42 oz., heavy traffic		70	.114		25.50	3.61		29.11	34
4110	Wool, level loop, 40 oz., medium traffic		75	.107		119	3.37		122.37	136
4500	50 oz., medium to heavy traffic		75	.107		120	3.37		123.37	137
4700	Patterned, 32 oz., medium to heavy traffic		70	.114		119	3.61		122.61	137
4900	48 oz., heavy traffic		70	.114		122	3.61		125.61	140
5000	For less than full roll (approx. 1500 S.F.), add					25%				
5100	For small rooms, less than 12' wide, add						25%			
5200	For large open areas (no cuts), deduct						25%			
5600	For bound carpet baseboard, add	1 Tilf	300	.027	L.F.	2.75	.84		3.59	4.37
5610	For stairs, not incl. price of carpet, add	"	30	.267	Riser		8.40		8.40	13.45
5620	For borders and patterns, add to labor						18%			
8950	For tackless, stretched installation, add padding from 09 68 10.10 to above									
9850	For brand-named specific fiber, add				S.Y.	25%				

09 68 20 – Athletic Carpet

09 68 20.10 Indoor Athletic Carpet

Code	Description	Crew	Daily Output	Labor-Hours	Unit	Material	Labor	Equipment	Total	Total Incl O&P
0010	**INDOOR ATHLETIC CARPET**									
3700	Polyethylene, in rolls, no base incl., landscape surfaces	1 Tilf	275	.029	S.F.	3.30	.92		4.22	5.10
3800	Nylon action surface, 1/8" thick		275	.029		3.46	.92		4.38	5.30
3900	1/4" thick		275	.029		4.99	.92		5.91	6.95
4000	3/8" thick		275	.029		6.25	.92		7.17	8.35

09 68 Carpeting

09 68 20 – Athletic Carpet

09 68 20.10 Indoor Athletic Carpet		Crew	Daily Output	Labor-Hours	Unit	2011 Bare Costs Material	Labor	Equipment	Total	Total Incl O&P
4100	Golf tee surface with foam back	1 Tilf	235	.034	S.F.	6.20	1.07		7.27	8.50
4200	Practice putting, knitted nylon surface		235	.034		5.25	1.07		6.32	7.50
5500	Polyvinyl chloride, sheet goods for gyms, 1/4″ thick		80	.100		8.25	3.16		11.41	14.15
5600	3/8″ thick	↓	60	.133	↓	9.25	4.21		13.46	16.90

09 69 Access Flooring

09 69 13 – Rigid-Grid Access Flooring

09 69 13.10 Access Floors		Crew	Daily Output	Labor-Hours	Unit	Material	Labor	Equipment	Total	Total Incl O&P
0010	**ACCESS FLOORS**									
0015	Access floor package including panel, pedestal, stringers & laminate cover									
0100	Computer room, greater than 6,000 S.F.	4 Carp	750	.043	S.F.	7	1.43		8.43	10.10
0110	Less than 6,000 S.F.	2 Carp	375	.043		7.65	1.43		9.08	10.80
0120	Office, greater than 6,000 S.F.	4 Carp	1050	.030		5	1.02		6.02	7.20
0250	Panels, particle board or steel, 1250# load, no covering, under 6,000 S.F.	2 Carp	600	.027		3.50	.89		4.39	5.35
0300	Over 6,000 S.F.		640	.025		3	.84		3.84	4.70
0400	Aluminum, 24″ panels	↓	500	.032		31.50	1.07		32.57	37
0600	For carpet covering, add					8.55			8.55	9.40
0700	For vinyl floor covering, add					6.70			6.70	7.35
0900	For high pressure laminate covering, add					5.45			5.45	6
0910	For snap on stringer system, add	2 Carp	1000	.016	↓	1.44	.54		1.98	2.48
0950	Office applications, steel or concrete panels,									
0960	no covering, over 6,000 S.F.	2 Carp	960	.017	S.F.	10.10	.56		10.66	12.05
1000	Machine cutouts after initial installation	1 Carp	50	.160	Ea.	5.05	5.35		10.40	14.55
1050	Pedestals, 6″ to 12″	2 Carp	85	.188		7.95	6.30		14.25	19.30
1100	Air conditioning grilles, 4″ x 12″	1 Carp	17	.471		66	15.80		81.80	99
1150	4″ x 18″	″	14	.571	↓	90.50	19.15		109.65	132
1200	Approach ramps, minimum	2 Carp	60	.267	S.F.	24.50	8.95		33.45	42
1300	Maximum	″	40	.400	″	33.50	13.40		46.90	59
1500	Handrail, 2 rail, aluminum	1 Carp	15	.533	L.F.	94.50	17.90		112.40	134

09 72 Wall Coverings

09 72 23 – Wallpapering

09 72 23.10 Wallpaper		Crew	Daily Output	Labor-Hours	Unit	Material	Labor	Equipment	Total	Total Incl O&P
0010	**WALLPAPER** including sizing; add 10-30 percent waste @ takeoff R097223-10									
0050	Aluminum foil	1 Pape	275	.029	S.F.	1.02	.87		1.89	2.53
0100	Copper sheets, .025″ thick, vinyl backing		240	.033		5.45	1		6.45	7.60
0300	Phenolic backing		240	.033		7.05	1		8.05	9.40
0600	Cork tiles, light or dark, 12″ x 12″ x 3/16″		240	.033		4.47	1		5.47	6.55
0700	5/16″ thick		235	.034		3.80	1.02		4.82	5.85
0900	1/4″ basketweave		240	.033		5.85	1		6.85	8.05
1000	1/2″ natural, non-directional pattern		240	.033		7.85	1		8.85	10.25
1200	Granular surface, 12″ x 36″, 1/2″ thick		385	.021		1.26	.62		1.88	2.40
1300	1″ thick		370	.022		1.63	.65		2.28	2.84
1500	Polyurethane coated, 12″ x 12″ x 3/16″ thick		240	.033		3.95	1		4.95	5.95
1600	5/16″ thick		235	.034		5.65	1.02		6.67	7.85
1800	Cork wallpaper, paperbacked, natural		480	.017		1.99	.50		2.49	3
1900	Colors		480	.017		2.78	.50		3.28	3.87
2100	Flexible wood veneer, 1/32″ thick, plain woods		100	.080		2.36	2.39		4.75	6.50
2200	Exotic woods	↓	95	.084	↓	3.58	2.52		6.10	8.05

09 72 Wall Coverings

09 72 23 – Wallpapering

09 72 23.10 Wallpaper		Crew	Daily Output	Labor-Hours	Unit	2011 Bare Costs Material	Labor	Equipment	Total	Total Incl O&P
2400	Gypsum-based, fabric-backed, fire resistant									
2501	resistant for masonry walls, minimum	1 Pord	800	.010	S.F.	.82	.30		1.12	1.38
2700	Maximum (small quantities)	1 Pape	640	.013		1.37	.37		1.74	2.12
2750	Acrylic, modified, semi-rigid PVC, .028" thick	2 Carp	330	.048		1.13	1.63		2.76	3.96
2800	.040" thick	"	320	.050		1.49	1.68		3.17	4.45
3001	Vinyl wall covering, fabric-backed, lightweight	1 Pord	640	.013		.69	.37		1.06	1.36
3301	Medium weight		480	.017		1.13	.49		1.62	2.04
3401	Heavy weight		435	.018		1.53	.55		2.08	2.57
3600	Adhesive, 5 gal. lots (18 S.Y./gal.)				Gal.	10.40			10.40	11.45
3700	Wallpaper, average workmanship, solid pattern, low cost paper	1 Pape	640	.013	S.F.	.37	.37		.74	1.02
3900	basic patterns (matching required), avg. cost paper		535	.015		.78	.45		1.23	1.59
4000	Paper at $85 per double roll, quality workmanship		435	.018		2.47	.55		3.02	3.61
4100	Linen wall covering, paper backed									
4150	Flame treatment, minimum				S.F.	1.04			1.04	1.14
4180	Maximum					1.55			1.55	1.71
4190	Grass cloth, natural fabric G	1 Pape	400	.020		2.99	.60		3.59	4.26
4200	Grass cloths with lining paper, minimum G		400	.020		.78	.60		1.38	1.83
4300	Maximum G		350	.023		2.49	.68		3.17	3.85

09 77 Special Wall Surfacing

09 77 33 – Fiberglass Reinforced Panels

09 77 33.10 Fiberglass Reinforced Plastic Panels		Crew	Daily Output	Labor-Hours	Unit	Material	Labor	Equipment	Total	Total Incl O&P
0010	**FIBERGLASS REINFORCED PLASTIC PANELS**, .090" thick									
0020	On walls, adhesive mounted, embossed surface	2 Carp	640	.025	S.F.	1.23	.84		2.07	2.75
0030	Smooth surface		640	.025		1.51	.84		2.35	3.06
0040	Fire rated, embossed surface		640	.025		2.19	.84		3.03	3.81
0050	Nylon rivet mounted, on drywall, embossed surface		480	.033		1.15	1.12		2.27	3.14
0060	Smooth surface		480	.033		1.43	1.12		2.55	3.44
0070	Fire rated, embossed surface		480	.033		2.11	1.12		3.23	4.19
0080	On masonry, embossed surface		320	.050		1.15	1.68		2.83	4.08
0090	Smooth surface		320	.050		1.43	1.68		3.11	4.38
0100	Fire rated, embossed surface		320	.050		2.11	1.68		3.79	5.15
0110	Nylon rivet and adhesive mounted, on drywall, embossed surface		240	.067		1.30	2.24		3.54	5.15
0120	Smooth surface		240	.067		1.58	2.24		3.82	5.50
0130	Fire rated, embossed surface		240	.067		2.26	2.24		4.50	6.25
0140	On masonry, embossed surface		190	.084		1.30	2.83		4.13	6.15
0150	Smooth surface		190	.084		1.58	2.83		4.41	6.45
0160	Fire rated, embossed surface		190	.084		2.26	2.83		5.09	7.20
0170	For moldings add	1 Carp	250	.032	L.F.	.28	1.07		1.35	2.11
0180	On ceilings, for lay in grid system, embossed surface		400	.020	S.F.	1.23	.67		1.90	2.47
0190	Smooth surface		400	.020		1.51	.67		2.18	2.78
0200	Fire rated, embossed surface		400	.020		2.19	.67		2.86	3.53

09 81 Acoustic Insulation

09 81 16 – Acoustic Blanket Insulation

09 81 16.10 Sound Attenuation Blanket		Crew	Daily Output	Labor-Hours	Unit	Material	Labor (2011 Bare Costs)	Equipment	Total	Total Incl O&P
0010	**SOUND ATTENUATION BLANKET**									
0020	Blanket, 1" thick	1 Carp	925	.009	S.F.	.25	.29		.54	.77
0500	1-1/2" thick		920	.009		.25	.29		.54	.77
1000	2" thick		915	.009		.34	.29		.63	.86
1500	3" thick		910	.009		.50	.29		.79	1.04
3000	Thermal or acoustical batt above ceiling, 2" thick		900	.009		.49	.30		.79	1.04
3100	3" thick		900	.009		.74	.30		1.04	1.31
3200	4" thick		900	.009		.92	.30		1.22	1.51
3400	Urethane plastic foam, open cell, on wall, 2" thick	2 Carp	2050	.008		3.05	.26		3.31	3.80
3500	3" thick		1550	.010		4.05	.35		4.40	5.05
3600	4" thick		1050	.015		5.70	.51		6.21	7.10
3700	On ceiling, 2" thick		1700	.009		3.04	.32		3.36	3.87
3800	3" thick		1300	.012		4.05	.41		4.46	5.15
3900	4" thick		900	.018		5.70	.60		6.30	7.25
4000	Nylon matting 0.4" thick, with carbon black spinerette									
4010	plus polyester fabric, on floor	D-7	1600	.010	S.F.	1.10	.28		1.38	1.66
4200	Fiberglass reinf. backer board underlayment, 7/16" thick, on floor	"	1500	.011	"	1.90	.30		2.20	2.57

09 84 Acoustic Room Components

09 84 13 – Fixed Sound-Absorptive Panels

09 84 13.10 Fixed Panels		Crew	Daily Output	Labor-Hours	Unit	Material	Labor	Equipment	Total	Total Incl O&P
0010	**FIXED PANELS** Perforated steel facing, painted with									
0100	Fiberglass or mineral filler, no backs, 2-1/4" thick, modular									
0200	space units, ceiling or wall hung, white or colored	1 Carp	100	.080	S.F.	7.35	2.68		10.03	12.60
0300	Fiberboard sound deadening panels, 1/2" thick	"	600	.013	"	.32	.45		.77	1.10
0500	Fiberglass panels, 4' x 8' x 1" thick, with									
0600	glass cloth face for walls, cemented	1 Carp	155	.052	S.F.	6.50	1.73		8.23	10.05
0700	1-1/2" thick, dacron covered, inner aluminum frame,									
0710	wall mounted	1 Carp	300	.027	S.F.	8.75	.89		9.64	11.10
0900	Mineral fiberboard panels, fabric covered, 30" x 108",									
1000	3/4" thick, concealed spline, wall mounted	1 Carp	150	.053	S.F.	6.25	1.79		8.04	9.90

09 84 36 – Sound-Absorbing Ceiling Units

09 84 36.10 Barriers		Crew	Daily Output	Labor-Hours	Unit	Material	Labor	Equipment	Total	Total Incl O&P
0010	**BARRIERS** Plenum									
0600	Aluminum foil, fiberglass reinf., parallel with joists	1 Carp	275	.029	S.F.	.92	.98		1.90	2.64
0700	Perpendicular to joists		180	.044		.92	1.49		2.41	3.51
0900	Aluminum mesh, kraft paperbacked		275	.029		.77	.98		1.75	2.48
0970	Fiberglass batts, kraft faced, 3-1/2" thick		1400	.006		.36	.19		.55	.72
0980	6" thick		1300	.006		.65	.21		.86	1.07
1000	Sheet lead, 1 lb., 1/64" thick, perpendicular to joists		150	.053		3.47	1.79		5.26	6.80
1100	Vinyl foam reinforced, 1/8" thick, 1.0 lb. per S.F.		150	.053		4	1.79		5.79	7.40

09 91 Painting

09 91 03 – Paint Restoration

09 91 03.20 Sanding		Crew	Daily Output	Labor-Hours	Unit	Material	2011 Bare Costs Labor	Equipment	Total	Total Incl O&P
0010	**SANDING** and puttying interior trim, compared to									
0100	Painting 1 coat, on quality work				L.F.		100%			
0300	Medium work						50%			
0400	Industrial grade						25%			
0500	Surface protection, placement and removal									
0510	Basic drop cloths	1 Pord	6400	.001	S.F.		.04		.04	.06
0520	Masking with paper		800	.010		.05	.30		.35	.54
0530	Volume cover up (using plastic sheathing, or building paper)		16000	.001			.01		.01	.02

09 91 03.30 Exterior Surface Preparation		Crew	Daily Output	Labor-Hours	Unit	Material	Labor	Equipment	Total	Total Incl O&P
0010	**EXTERIOR SURFACE PREPARATION**									
0015	Doors, per side, not incl. frames or trim									
0020	Scrape & sand									
0030	Wood, flush	1 Pord	616	.013	S.F.		.39		.39	.63
0040	Wood, detail		496	.016			.48		.48	.78
0050	Wood, louvered		280	.029			.85		.85	1.38
0060	Wood, overhead		616	.013			.39		.39	.63
0070	Wire brush									
0080	Metal, flush	1 Pord	640	.013	S.F.		.37		.37	.60
0090	Metal, detail		520	.015			.46		.46	.74
0100	Metal, louvered		360	.022			.66		.66	1.07
0110	Metal or fibr., overhead		640	.013			.37		.37	.60
0120	Metal, roll up		560	.014			.42		.42	.69
0130	Metal, bulkhead		640	.013			.37		.37	.60
0140	Power wash, based on 2500 lb. operating pressure									
0150	Metal, flush	A-1H	2240	.004	S.F.		.09	.03	.12	.19
0160	Metal, detail		2120	.004			.10	.03	.13	.20
0170	Metal, louvered		2000	.004			.11	.03	.14	.21
0180	Metal or fibr., overhead		2400	.003			.09	.03	.12	.18
0190	Metal, roll up		2400	.003			.09	.03	.12	.18
0200	Metal, bulkhead		2200	.004			.10	.03	.13	.19
0400	Windows, per side, not incl. trim									
0410	Scrape & sand									
0420	Wood, 1-2 lite	1 Pord	320	.025	S.F.		.74		.74	1.20
0430	Wood, 3-6 lite		280	.029			.85		.85	1.38
0440	Wood, 7-10 lite		240	.033			.99		.99	1.60
0450	Wood, 12 lite		200	.040			1.19		1.19	1.93
0460	Wood, Bay/Bow		320	.025			.74		.74	1.20
0470	Wire brush									
0480	Metal, 1-2 lite	1 Pord	480	.017	S.F.		.49		.49	.80
0490	Metal, 3-6 lite		400	.020			.59		.59	.96
0500	Metal, Bay/Bow		480	.017			.49		.49	.80
0510	Power wash, based on 2500 lb. operating pressure									
0520	1-2 lite	A-1H	4400	.002	S.F.		.05	.01	.06	.10
0530	3-6 lite		4320	.002			.05	.01	.06	.10
0540	7-10 lite		4240	.002			.05	.01	.06	.10
0550	12 lite		4160	.002			.05	.01	.06	.11
0560	Bay/Bow		4400	.002			.05	.01	.06	.10
0600	Siding, scrape and sand, light=10-30%, med.=30-70%									
0610	Heavy=70-100% of surface to sand									
0650	Texture 1-11, light	1 Pord	480	.017	S.F.		.49		.49	.80
0660	Med.		440	.018			.54		.54	.88
0670	Heavy		360	.022			.66		.66	1.07

09 91 03 – Paint Restoration

09 91 03.30 Exterior Surface Preparation		Crew	Daily Output	Labor-Hours	Unit	Material	2011 Bare Costs Labor	Equipment	Total	Total Incl O&P
0680	Wood shingles, shakes, light	1 Pord	440	.018	S.F.		.54		.54	.88
0690	Med.		360	.022			.66		.66	1.07
0700	Heavy		280	.029			.85		.85	1.38
0710	Clapboard, light		520	.015			.46		.46	.74
0720	Med.		480	.017			.49		.49	.80
0730	Heavy	↓	400	.020	↓		.59		.59	.96
0740	Wire brush									
0750	Aluminum, light	1 Pord	600	.013	S.F.		.40		.40	.64
0760	Med.		520	.015			.46		.46	.74
0770	Heavy	↓	440	.018	↓		.54		.54	.88
0780	Pressure wash, based on 2500 lb. operating pressure									
0790	Stucco	A-1H	3080	.003	S.F.		.07	.02	.09	.14
0800	Aluminum or vinyl		3200	.003			.07	.02	.09	.13
0810	Siding, masonry, brick & block	↓	2400	.003	↓		.09	.03	.12	.18
1300	Miscellaneous, wire brush									
1310	Metal, pedestrian gate	1 Pord	100	.080	S.F.		2.37		2.37	3.85
8020	For sand blasting, see Section 03 35 29.60 and 05 01 10.51									

09 91 03.40 Interior Surface Preparation		Crew	Daily Output	Labor-Hours	Unit	Material	Labor	Equipment	Total	Total Incl O&P
0010	**INTERIOR SURFACE PREPARATION**									
0020	Doors, per side, not incl. frames or trim									
0030	Scrape & sand									
0040	Wood, flush	1 Pord	616	.013	S.F.		.39		.39	.63
0050	Wood, detail		496	.016			.48		.48	.78
0060	Wood, louvered	↓	280	.029	↓		.85		.85	1.38
0070	Wire brush									
0080	Metal, flush	1 Pord	640	.013	S.F.		.37		.37	.60
0090	Metal, detail		520	.015			.46		.46	.74
0100	Metal, louvered	↓	360	.022	↓		.66		.66	1.07
0110	Hand wash									
0120	Wood, flush	1 Pord	2160	.004	S.F.		.11		.11	.18
0130	Wood, detailed		2000	.004			.12		.12	.19
0140	Wood, louvered		1360	.006			.17		.17	.28
0150	Metal, flush		2160	.004			.11		.11	.18
0160	Metal, detail		2000	.004			.12		.12	.19
0170	Metal, louvered	↓	1360	.006	↓		.17		.17	.28
0400	Windows, per side, not incl. trim									
0410	Scrape & sand									
0420	Wood, 1-2 lite	1 Pord	360	.022	S.F.		.66		.66	1.07
0430	Wood, 3-6 lite		320	.025			.74		.74	1.20
0440	Wood, 7-10 lite		280	.029			.85		.85	1.38
0450	Wood, 12 lite		240	.033			.99		.99	1.60
0460	Wood, Bay/Bow	↓	360	.022	↓		.66		.66	1.07
0470	Wire brush									
0480	Metal, 1-2 lite	1 Pord	520	.015	S.F.		.46		.46	.74
0490	Metal, 3-6 lite		440	.018			.54		.54	.88
0500	Metal, Bay/Bow	↓	520	.015	↓		.46		.46	.74
0600	Walls, sanding, light=10-30%, medium - 30-70%,									
0610	heavy=70-100% of surface to sand									
0650	Walls, sand									
0660	Gypsum board or plaster, light	1 Pord	3077	.003	S.F.		.08		.08	.13
0670	Gypsum board or plaster, medium		2160	.004			.11		.11	.18
0680	Gypsum board or plaster, heavy	↓	923	.009	↓		.26		.26	.42

09 91 Painting

09 91 03 – Paint Restoration

09 91 03.40 Interior Surface Preparation		Crew	Daily Output	Labor-Hours	Unit	Material	2011 Bare Costs Labor	Equipment	Total	Total Incl O&P
0690	Wood, T&G, light	1 Pord	2400	.003	S.F.		.10		.10	.16
0700	Wood, T&G, med.		1600	.005			.15		.15	.24
0710	Wood, T&G, heavy		800	.010			.30		.30	.48
0720	Walls, wash									
0730	Gypsum board or plaster	1 Pord	3200	.003	S.F.		.07		.07	.12
0740	Wood, T&G		3200	.003			.07		.07	.12
0750	Masonry, brick & block, smooth		2800	.003			.08		.08	.14
0760	Masonry, brick & block, coarse		2000	.004			.12		.12	.19
8020	For sand blasting, see Section 03 35 29.60 and 05 01 10.51									

09 91 13 – Exterior Painting

09 91 13.30 Fences

		Crew	Daily Output	Labor-Hours	Unit	Material	Labor	Equipment	Total	Total Incl O&P
0010	**FENCES** R099100-20									
0100	Chain link or wire metal, one side, water base									
0110	Roll & brush, first coat	1 Pord	960	.008	S.F.	.06	.25		.31	.47
0120	Second coat		1280	.006		.06	.19		.25	.36
0130	Spray, first coat		2275	.004		.06	.10		.16	.24
0140	Second coat		2600	.003		.06	.09		.15	.22
0150	Picket, water base									
0160	Roll & brush, first coat	1 Pord	865	.009	S.F.	.06	.27		.33	.52
0170	Second coat		1050	.008		.06	.23		.29	.44
0180	Spray, first coat		2275	.004		.06	.10		.16	.24
0190	Second coat		2600	.003		.06	.09		.15	.22
0200	Stockade, water base									
0210	Roll & brush, first coat	1 Pord	1040	.008	S.F.	.06	.23		.29	.44
0220	Second coat		1200	.007		.06	.20		.26	.39
0230	Spray, first coat		2275	.004		.06	.10		.16	.24
0240	Second coat		2600	.003		.06	.09		.15	.22

09 91 13.42 Miscellaneous, Exterior

		Crew	Daily Output	Labor-Hours	Unit	Material	Labor	Equipment	Total	Total Incl O&P
0010	**MISCELLANEOUS, EXTERIOR** R099100-20									
0015	For painting metals, see Section 09 97 13.23									
0100	Railing, ext., decorative wood, incl. cap & baluster									
0110	Newels & spindles @ 12" O.C.									
0120	Brushwork, stain, sand, seal & varnish									
0130	First coat	1 Pord	90	.089	L.F.	.69	2.64		3.33	5.05
0140	Second coat	"	120	.067	"	.69	1.98		2.67	3.97
0150	Rough sawn wood, 42" high, 2" x 2" verticals, 6" O.C.									
0160	Brushwork, stain, each coat	1 Pord	90	.089	L.F.	.22	2.64		2.86	4.53
0170	Wrought iron, 1" rail, 1/2" sq. verticals									
0180	Brushwork, zinc chromate, 60" high, bars 6" O.C.									
0190	Primer	1 Pord	130	.062	L.F.	.59	1.82		2.41	3.61
0200	Finish coat		130	.062		.18	1.82		2	3.16
0210	Additional coat		190	.042		.21	1.25		1.46	2.26
0220	Shutters or blinds, single panel, 2' x 4', paint all sides									
0230	Brushwork, primer	1 Pord	20	.400	Ea.	.58	11.85		12.43	19.90
0240	Finish coat, exterior latex		20	.400		.49	11.85		12.34	19.80
0250	Primer & 1 coat, exterior latex		13	.615		.94	18.25		19.19	30.50
0260	Spray, primer		35	.229		.85	6.80		7.65	11.95
0270	Finish coat, exterior latex		35	.229		1.04	6.80		7.84	12.15
0280	Primer & 1 coat, exterior latex		20	.400		.91	11.85		12.76	20.50
0290	For louvered shutters, add				S.F.	10%				
0300	Stair stringers, exterior, metal									
0310	Roll & brush, zinc chromate, to 14", each coat	1 Pord	320	.025	L.F.	.06	.74		.80	1.27

09 91 13.42 Miscellaneous, Exterior		Crew	Daily Output	Labor-Hours	Unit	Material	2011 Bare Costs Labor	Equipment	Total	Total Incl O&P
0320	Rough sawn wood, 4" x 12"									
0330	Roll & brush, exterior latex, each coat	1 Pord	215	.037	L.F.	.07	1.10		1.17	1.87
0340	Trellis/lattice, 2" x 2" @ 3" O.C. with 2" x 8" supports									
0350	Spray, latex, per side, each coat	1 Pord	475	.017	S.F.	.07	.50		.57	.89
0450	Decking, ext., sealer, alkyd, brushwork, sealer coat		1140	.007		.09	.21		.30	.44
0460	1st coat		1140	.007		.09	.21		.30	.43
0470	2nd coat		1300	.006		.06	.18		.24	.37
0500	Paint, alkyd, brushwork, primer coat		1140	.007		.08	.21		.29	.42
0510	1st coat		1140	.007		.16	.21		.37	.51
0520	2nd coat		1300	.006		.12	.18		.30	.43
0600	Sand paint, alkyd, brushwork, 1 coat		150	.053		.12	1.58		1.70	2.70

09 91 13.60 Siding Exterior		Crew	Daily Output	Labor-Hours	Unit	Material	Labor	Equipment	Total	Total Incl O&P
0010	**SIDING EXTERIOR**, Alkyd (oil base)									
0450	Steel siding, oil base, paint 1 coat, brushwork	2 Pord	2015	.008	S.F.	.13	.24		.37	.52
0500	Spray		4550	.004		.19	.10		.29	.38
0800	Paint 2 coats, brushwork		1300	.012		.25	.37		.62	.87
1000	Spray		2750	.006		.17	.17		.34	.47
1200	Stucco, rough, oil base, paint 2 coats, brushwork		1300	.012		.25	.37		.62	.87
1400	Roller		1625	.010		.27	.29		.56	.76
1600	Spray		2925	.005		.28	.16		.44	.57
1800	Texture 1-11 or clapboard, oil base, primer coat, brushwork		1300	.012		.10	.37		.47	.70
2000	Spray		4550	.004		.10	.10		.20	.28
2400	Paint 2 coats, brushwork		810	.020		.37	.59		.96	1.35
2600	Spray		2600	.006		.41	.18		.59	.75
3400	Stain 2 coats, brushwork		950	.017		.15	.50		.65	.97
4000	Spray		3050	.005		.17	.16		.33	.43
4200	Wood shingles, oil base primer coat, brushwork		1300	.012		.09	.37		.46	.69
4400	Spray		3900	.004		.09	.12		.21	.29
5000	Paint 2 coats, brushwork		810	.020		.31	.59		.90	1.29
5200	Spray		2275	.007		.29	.21		.50	.66
6500	Stain 2 coats, brushwork		950	.017		.15	.50		.65	.97
7000	Spray		2660	.006		.21	.18		.39	.52
8000	For latex paint, deduct					10%				
8100	For work over 12' H, from pipe scaffolding, add						15%			
8200	For work over 12' H, from extension ladder, add						25%			
8300	For work over 12' H, from swing staging, add						35%			

09 91 13.62 Siding, Misc.		Crew	Daily Output	Labor-Hours	Unit	Material	Labor	Equipment	Total	Total Incl O&P
0010	**SIDING, MISC.**, latex paint R099100-10									
0100	Aluminum siding									
0110	Brushwork, primer	2 Pord	2275	.007	S.F.	.05	.21		.26	.40
0120	Finish coat, exterior latex		2275	.007		.04	.21		.25	.39
0130	Primer & 1 coat exterior latex		1300	.012		.10	.37		.47	.70
0140	Primer & 2 coats exterior latex		975	.016		.14	.49		.63	.95
0150	Mineral fiber shingles									
0160	Brushwork, primer	2 Pord	1495	.011	S.F.	.10	.32		.42	.63
0170	Finish coat, industrial enamel		1495	.011		.15	.32		.47	.68
0180	Primer & 1 coat enamel		810	.020		.25	.59		.84	1.22
0190	Primer & 2 coats enamel		540	.030		.40	.88		1.28	1.86
0200	Roll, primer		1625	.010		.11	.29		.40	.59
0210	Finish coat, industrial enamel		1625	.010		.16	.29		.45	.65
0220	Primer & 1 coat enamel		975	.016		.27	.49		.76	1.09
0230	Primer & 2 coats enamel		650	.025		.43	.73		1.16	1.66

09 91 13.62 Siding, Misc.		Crew	Daily Output	Labor-Hours	Unit	2011 Bare Costs Material	Labor	Equipment	Total	Total Incl O&P
0240	Spray, primer	2 Pord	3900	.004	S.F.	.09	.12		.21	.29
0250	Finish coat, industrial enamel		3900	.004		.13	.12		.25	.35
0260	Primer & 1 coat enamel		2275	.007		.22	.21		.43	.58
0270	Primer & 2 coats enamel		1625	.010		.35	.29		.64	.86
0280	Waterproof sealer, first coat		4485	.004		.07	.11		.18	.24
0290	Second coat		5235	.003		.06	.09		.15	.22
0300	Rough wood incl. shingles, shakes or rough sawn siding									
0310	Brushwork, primer	2 Pord	1280	.013	S.F.	.12	.37		.49	.73
0320	Finish coat, exterior latex		1280	.013		.07	.37		.44	.68
0330	Primer & 1 coat exterior latex		960	.017		.19	.49		.68	1.01
0340	Primer & 2 coats exterior latex		700	.023		.26	.68		.94	1.39
0350	Roll, primer		2925	.005		.16	.16		.32	.43
0360	Finish coat, exterior latex		2925	.005		.09	.16		.25	.36
0370	Primer & 1 coat exterior latex		1790	.009		.24	.27		.51	.70
0380	Primer & 2 coats exterior latex		1300	.012		.33	.37		.70	.95
0390	Spray, primer		3900	.004		.13	.12		.25	.35
0400	Finish coat, exterior latex		3900	.004		.07	.12		.19	.27
0410	Primer & 1 coat exterior latex		2600	.006		.20	.18		.38	.52
0420	Primer & 2 coats exterior latex		2080	.008		.27	.23		.50	.66
0430	Waterproof sealer, first coat		4485	.004		.12	.11		.23	.30
0440	Second coat		4485	.004		.07	.11		.18	.24
0450	Smooth wood incl. butt, T&G, beveled, drop or B&B siding									
0460	Brushwork, primer	2 Pord	2325	.007	S.F.	.08	.20		.28	.42
0470	Finish coat, exterior latex		1280	.013		.07	.37		.44	.68
0480	Primer & 1 coat exterior latex		800	.020		.16	.59		.75	1.13
0490	Primer & 2 coats exterior latex		630	.025		.23	.75		.98	1.47
0500	Roll, primer		2275	.007		.09	.21		.30	.44
0510	Finish coat, exterior latex		2275	.007		.08	.21		.29	.43
0520	Primer & 1 coat exterior latex		1300	.012		.17	.37		.54	.78
0530	Primer & 2 coats exterior latex		975	.016		.25	.49		.74	1.07
0540	Spray, primer		4550	.004		.07	.10		.17	.25
0550	Finish coat, exterior latex		4550	.004		.07	.10		17	.24
0560	Primer & 1 coat exterior latex		2600	.006		.14	.18		.32	.45
0570	Primer & 2 coats exterior latex		1950	.008		.21	.24		.45	.63
0580	Waterproof sealer, first coat		5230	.003		.07	.09		.16	.22
0590	Second coat		5980	.003		.07	.08		.15	.20
0600	For oil base paint, add					10%				

09 91 13.70 Doors and Windows, Exterior		Crew	Daily Output	Labor-Hours	Unit	Material	Labor	Equipment	Total	Total Incl O&P
0010	**DOORS AND WINDOWS, EXTERIOR**									
0100	Door frames & trim, only									
0110	Brushwork, primer R099100-20	1 Pord	512	.016	L.F.	.05	.46		.51	.81
0120	Finish coat, exterior latex		512	.016		.06	.46		.52	.82
0130	Primer & 1 coat, exterior latex		300	.027		.11	.79		.90	1.41
0140	Primer & 2 coats, exterior latex		265	.030		.18	.90		1.08	1.64
0150	Doors, flush, both sides, incl. frame & trim									
0160	Roll & brush, primer	1 Pord	10	.800	Ea.	4.03	23.50		27.53	43
0170	Finish coat, exterior latex		10	.800		4.65	23.50		28.15	43.50
0180	Primer & 1 coat, exterior latex		7	1.143		8.70	34		42.70	64.50
0190	Primer & 2 coats, exterior latex		5	1.600		13.30	47.50		60.80	91.50
0200	Brushwork, stain, sealer & 2 coats polyurethane		4	2		24	59.50		83.50	123
0210	Doors, French, both sides, 10-15 lite, incl. frame & trim									
0220	Brushwork, primer	1 Pord	6	1.333	Ea.	2.02	39.50		41.52	66

09 91 13.70 Doors and Windows, Exterior

	09 91 13.70 Doors and Windows, Exterior	Crew	Daily Output	Labor-Hours	Unit	Material	Labor (2011 Bare Costs)	Equipment (2011 Bare Costs)	Total (2011 Bare Costs)	Total Incl O&P
0230	Finish coat, exterior latex	1 Pord	6	1.333	Ea.	2.32	39.50		41.82	66.50
0240	Primer & 1 coat, exterior latex		3	2.667		4.34	79		83.34	133
0250	Primer & 2 coats, exterior latex		2	4		6.55	119		125.55	200
0260	Brushwork, stain, sealer & 2 coats polyurethane		2.50	3.200		8.70	95		103.70	164
0270	Doors, louvered, both sides, incl. frame & trim									
0280	Brushwork, primer	1 Pord	7	1.143	Ea.	4.03	34		38.03	59.50
0290	Finish coat, exterior latex		7	1.143		4.65	34		38.65	60
0300	Primer & 1 coat, exterior latex		4	2		8.70	59.50		68.20	106
0310	Primer & 2 coats, exterior latex		3	2.667		13.05	79		92.05	142
0320	Brushwork, stain, sealer & 2 coats polyurethane		4.50	1.778		24	52.50		76.50	112
0330	Doors, panel, both sides, incl. frame & trim									
0340	Roll & brush, primer	1 Pord	6	1.333	Ea.	4.03	39.50		43.53	68.50
0350	Finish coat, exterior latex		6	1.333		4.65	39.50		44.15	69
0360	Primer & 1 coat, exterior latex		3	2.667		8.70	79		87.70	138
0370	Primer & 2 coats, exterior latex		2.50	3.200		13.05	95		108.05	168
0380	Brushwork, stain, sealer & 2 coats polyurethane		3	2.667		24	79		103	154
0400	Windows, per ext. side, based on 15 S.F.									
0410	1 to 6 lite									
0420	Brushwork, primer	1 Pord	13	.615	Ea.	.80	18.25		19.05	30.50
0430	Finish coat, exterior latex		13	.615		.92	18.25		19.17	30.50
0440	Primer & 1 coat, exterior latex		8	1		1.71	29.50		31.21	50
0450	Primer & 2 coats, exterior latex		6	1.333		2.58	39.50		42.08	67
0460	Stain, sealer & 1 coat varnish		7	1.143		3.44	34		37.44	59
0470	7 to 10 lite									
0480	Brushwork, primer	1 Pord	11	.727	Ea.	.80	21.50		22.30	36
0490	Finish coat, exterior latex		11	.727		.92	21.50		22.42	36
0500	Primer & 1 coat, exterior latex		7	1.143		1.71	34		35.71	57
0510	Primer & 2 coats, exterior latex		5	1.600		2.58	47.50		50.08	80
0520	Stain, sealer & 1 coat varnish		6	1.333		3.44	39.50		42.94	68
0530	12 lite									
0540	Brushwork, primer	1 Pord	10	.800	Ea.	.80	23.50		24.30	39.50
0550	Finish coat, exterior latex		10	.800		.92	23.50		24.42	39.50
0560	Primer & 1 coat, exterior latex		6	1.333		1.71	39.50		41.21	66
0570	Primer & 2 coats, exterior latex		5	1.600		2.58	47.50		50.08	80
0580	Stain, sealer & 1 coat varnish		6	1.333		3.37	39.50		42.87	67.50
0590	For oil base paint, add					10%				

09 91 13.80 Trim, Exterior

	09 91 13.80 Trim, Exterior	Crew	Daily Output	Labor-Hours	Unit	Material	Labor (2011 Bare Costs)	Equipment (2011 Bare Costs)	Total (2011 Bare Costs)	Total Incl O&P
0010	**TRIM, EXTERIOR** R099100-10									
0100	Door frames & trim (see Doors, interior or exterior)									
0110	Fascia, latex paint, one coat coverage									
0120	1" x 4", brushwork	1 Pord	640	.013	L.F.	.02	.37		.39	.62
0130	Roll		1280	.006		.02	.19		.21	.32
0140	Spray		2080	.004		.02	.11		.13	.21
0150	1" x 6" to 1" x 10", brushwork		640	.013		.06	.37		.43	.67
0160	Roll		1230	.007		.07	.19		.26	.39
0170	Spray		2100	.004		.05	.11		.16	.24
0180	1" x 12", brushwork		640	.013		.06	.37		.43	.67
0190	Roll		1050	.008		.07	.23		.30	.45
0200	Spray		2200	.004		.05	.11		.16	.24
0210	Gutters & downspouts, metal, zinc chromate paint									
0220	Brushwork, gutters, 5", first coat	1 Pord	640	.013	L.F.	.06	.37		.43	.67
0230	Second coat		960	.008		.06	.25		.31	.47

09 91 13 – Exterior Painting

09 91 13.80 Trim, Exterior		Crew	Daily Output	Labor-Hours	Unit	Material	2011 Bare Costs Labor	Equipment	Total	Total Incl O&P
0240	Third coat	1 Pord	1280	.006	L.F.	.05	.19		.24	.35
0250	Downspouts, 4", first coat		640	.013		.06	.37		.43	.67
0260	Second coat		960	.008		.06	.25		.31	.47
0270	Third coat		1280	.006		.05	.19		.24	.35
0280	Gutters & downspouts, wood									
0290	Brushwork, gutters, 5", primer	1 Pord	640	.013	L.F.	.05	.37		.42	.66
0300	Finish coat, exterior latex		640	.013		.05	.37		.42	.66
0310	Primer & 1 coat exterior latex		400	.020		.11	.59		.70	1.09
0320	Primer & 2 coats exterior latex		325	.025		.18	.73		.91	1.38
0330	Downspouts, 4", primer		640	.013		.05	.37		.42	.66
0340	Finish coat, exterior latex		640	.013		.05	.37		.42	.66
0350	Primer & 1 coat exterior latex		400	.020		.11	.59		.70	1.09
0360	Primer & 2 coats exterior latex		325	.025		.09	.73		.82	1.29
0370	Molding, exterior, up to 14" wide									
0380	Brushwork, primer	1 Pord	640	.013	L.F.	.06	.37		.43	.67
0390	Finish coat, exterior latex		640	.013		.07	.37		.44	.67
0400	Primer & 1 coat exterior latex		400	.020		.14	.59		.73	1.11
0410	Primer & 2 coats exterior latex		315	.025		.14	.75		.89	1.37
0420	Stain & fill		1050	.008		.09	.23		.32	.47
0430	Shellac		1850	.004		.09	.13		.22	.31
0440	Varnish		1275	.006		.10	.19		.29	.41

09 91 13.90 Walls, Masonry (CMU), Exterior		Crew	Daily Output	Labor-Hours	Unit	Material	Labor	Equipment	Total	Total Incl O&P
0350	**WALLS, MASONRY (CMU), EXTERIOR**									
0360	Concrete masonry units (CMU), smooth surface									
0370	Brushwork, latex, first coat	1 Pord	640	.013	S.F.	.06	.37		.43	.66
0380	Second coat		960	.008		.05	.25		.30	.45
0390	Waterproof sealer, first coat		736	.011		.25	.32		.57	.79
0400	Second coat		1104	.007		.25	.22		.47	.62
0410	Roll, latex, paint, first coat		1465	.005		.07	.16		.23	.34
0420	Second coat		1790	.004		.05	.13		.18	.28
0430	Waterproof sealer, first coat		1680	.005		.25	.14		.39	.50
0440	Second coat		2060	.004		.25	.12		.37	.46
0450	Spray, latex, paint, first coat		1950	.004		.05	.12		.17	.26
0460	Second coat		2600	.003		.04	.09		.13	.20
0470	Waterproof sealer, first coat		2245	.004		.25	.11		.36	.44
0480	Second coat		2990	.003		.25	.08		.33	.40
0490	Concrete masonry unit (CMU), porous									
0500	Brushwork, latex, first coat	1 Pord	640	.013	S.F.	.12	.37		.49	.73
0510	Second coat		960	.008		.06	.25		.31	.46
0520	Waterproof sealer, first coat		736	.011		.25	.32		.57	.79
0530	Second coat		1104	.007		.25	.22		.47	.62
0540	Roll latex, first coat		1465	.005		.09	.16		.25	.36
0550	Second coat		1790	.004		.06	.13		.19	.28
0560	Waterproof sealer, first coat		1680	.005		.25	.14		.39	.50
0570	Second coat		2060	.004		.25	.12		.37	.46
0580	Spray latex, first coat		1950	.004		.06	.12		.18	.27
0590	Second coat		2600	.003		.04	.09		.13	.20
0600	Waterproof sealer, first coat		2245	.004		.25	.11		.36	.44
0610	Second coat		2990	.003		.25	.08		.33	.40

09 91 Painting

09 91 23 – Interior Painting

09 91 23.20 Cabinets and Casework		Crew	Daily Output	Labor-Hours	Unit	Material	2011 Bare Costs Labor	Equipment	Total	Total Incl O&P
0010	CABINETS AND CASEWORK R099100-10									
1000	Primer coat, oil base, brushwork	1 Pord	650	.012	S.F.	.06	.37		.43	.66
2000	Paint, oil base, brushwork, 1 coat		650	.012		.08	.37		.45	.68
3000	Stain, brushwork, wipe off		650	.012		.07	.37		.44	.67
4000	Shellac, 1 coat, brushwork		650	.012		.07	.37		.44	.67
4500	Varnish, 3 coats, brushwork, sand after 1st coat		325	.025		.24	.73		.97	1.45
5000	For latex paint, deduct					10%				

09 91 23.33 Doors and Windows, Interior Alkyd (Oil Base)		Crew	Daily Output	Labor-Hours	Unit	Material	Labor	Equipment	Total	Total Incl O&P
0010	DOORS AND WINDOWS, INTERIOR ALKYD (OIL BASE)									
0500	Flush door & frame, 3' x 7', oil, primer, brushwork	1 Pord	10	.800	Ea.	3.04	23.50		26.54	42
1000	Paint, 1 coat		10	.800		3.62	23.50		27.12	42.50
1400	Stain, brushwork, wipe off		18	.444		1.57	13.20		14.77	23
1600	Shellac, 1 coat, brushwork		25	.320		1.54	9.50		11.04	17.10
1800	Varnish, 3 coats, brushwork, sand after 1st coat		9	.889		5	26.50		31.50	48.50
2000	Panel door & frame, 3' x 7', oil, primer, brushwork		6	1.333		2.31	39.50		41.81	66.50
2200	Paint, 1 coat		6	1.333		3.62	39.50		43.12	68
2600	Stain, brushwork, panel door, 3' x 7', not incl. frame		16	.500		1.57	14.85		16.42	25.50
2800	Shellac, 1 coat, brushwork		22	.364		1.54	10.80		12.34	19.20
3000	Varnish, 3 coats, brushwork, sand after 1st coat		7.50	1.067		5	31.50		36.50	57
4400	Windows, including frame and trim, per side									
4600	Colonial type, 6/6 lites, 2' x 3', oil, primer, brushwork	1 Pord	14	.571	Ea.	.36	16.95		17.31	28
5800	Paint, 1 coat		14	.571		.57	16.95		17.52	28
6200	3' x 5' opening, 6/6 lites, primer coat, brushwork		12	.667		.91	19.75		20.66	33
6400	Paint, 1 coat		12	.667		1.43	19.75		21.18	33.50
6800	4' x 8' opening, 6/6 lites, primer coat, brushwork		8	1		1.95	29.50		31.45	50
7000	Paint, 1 coat		8	1		3.05	29.50		32.55	51.50
8000	Single lite type, 2' x 3', oil base, primer coat, brushwork		33	.242		.36	7.20		7.56	12.05
8200	Paint, 1 coat		33	.242		.57	7.20		7.77	12.30
8600	3' x 5' opening, primer coat, brushwork		20	.400		.91	11.85		12.76	20.50
8800	Paint, 1 coat		20	.400		1.43	11.85		13.28	21
9200	4' x 8' opening, primer coat, brushwork		14	.571		1.95	16.95		18.90	29.50
9400	Paint, 1 coat		14	.571		3.05	16.95		20	31

09 91 23.35 Doors and Windows, Interior Latex		Crew	Daily Output	Labor-Hours	Unit	Material	Labor	Equipment	Total	Total Incl O&P
0010	DOORS & WINDOWS, INTERIOR LATEX R099100-10									
0100	Doors, flush, both sides, incl. frame & trim									
0110	Roll & brush, primer	1 Pord	10	.800	Ea.	3.73	23.50		27.23	42.50
0120	Finish coat, latex		10	.800		4.13	23.50		27.63	43
0130	Primer & 1 coat latex		7	1.143		7.85	34		41.85	63.50
0140	Primer & 2 coats latex		5	1.600		11.75	47.50		59.25	90
0160	Spray, both sides, primer		20	.400		3.93	11.85		15.78	23.50
0170	Finish coat, latex		20	.400		4.33	11.85		16.18	24
0180	Primer & 1 coat latex		11	.727		8.30	21.50		29.80	44
0190	Primer & 2 coats latex		8	1		12.45	29.50		41.95	61.50
0200	Doors, French, both sides, 10-15 lite, incl. frame & trim									
0210	Roll & brush, primer	1 Pord	6	1.333	Ea.	1.87	39.50		41.37	66
0220	Finish coat, latex		6	1.333		2.06	39.50		41.56	66.50
0230	Primer & 1 coat latex		3	2.667		3.93	79		82.93	132
0240	Primer & 2 coats latex		2	4		5.90	119		124.90	199
0260	Doors, louvered, both sides, incl. frame & trim									
0270	Roll & brush, primer	1 Pord	7	1.143	Ea.	3.73	34		37.73	59
0280	Finish coat, latex		7	1.143		4.13	34		38.13	59.50
0290	Primer & 1 coat, latex		4	2		7.65	59.50		67.15	105

09 91 23.35 Doors and Windows, Interior Latex

	09 91 23.35 Doors and Windows, Interior Latex	Crew	Daily Output	Labor-Hours	Unit	2011 Bare Costs Material	Labor	Equipment	Total	Total Incl O&P
0300	Primer & 2 coats, latex	1 Pord	3	2.667	Ea.	12	79		91	141
0320	Spray, both sides, primer		20	.400		3.93	11.85		15.78	23.50
0330	Finish coat, latex		20	.400		4.33	11.85		16.18	24
0340	Primer & 1 coat, latex		11	.727		8.30	21.50		29.80	44
0350	Primer & 2 coats, latex		8	1		12.70	29.50		42.20	62
0360	Doors, panel, both sides, incl. frame & trim									
0370	Roll & brush, primer	1 Pord	6	1.333	Ea.	3.93	39.50		43.43	68.50
0380	Finish coat, latex		6	1.333		4.13	39.50		43.63	68.50
0390	Primer & 1 coat, latex		3	2.667		7.85	79		86.85	137
0400	Primer & 2 coats, latex		2.50	3.200		12	95		107	167
0420	Spray, both sides, primer		10	.800		3.93	23.50		27.43	43
0430	Finish coat, latex		10	.800		4.33	23.50		27.83	43.50
0440	Primer & 1 coat, latex		5	1.600		8.30	47.50		55.80	86
0450	Primer & 2 coats, latex		4	2		12.70	59.50		72.20	110
0460	Windows, per interior side, based on 15 S.F.									
0470	1 to 6 lite									
0480	Brushwork, primer	1 Pord	13	.615	Ea.	.74	18.25		18.99	30.50
0490	Finish coat, enamel		13	.615		.81	18.25		19.06	30.50
0500	Primer & 1 coat enamel		8	1		1.55	29.50		31.05	49.50
0510	Primer & 2 coats enamel		6	1.333		2.37	39.50		41.87	66.50
0530	7 to 10 lite									
0540	Brushwork, primer	1 Pord	11	.727	Ea.	.74	21.50		22.24	36
0550	Finish coat, enamel		11	.727		.81	21.50		22.31	36
0560	Primer & 1 coat enamel		7	1.143		1.55	34		35.55	56.50
0570	Primer & 2 coats enamel		5	1.600		2.37	47.50		49.87	79.50
0590	12 lite									
0600	Brushwork, primer	1 Pord	10	.800	Ea.	.74	23.50		24.24	39.50
0610	Finish coat, enamel		10	.800		.81	23.50		24.31	39.50
0620	Primer & 1 coat enamel		6	1.333		1.55	39.50		41.05	65.50
0630	Primer & 2 coats enamel		5	1.600		2.37	47.50		49.87	79.50
0650	For oil base paint, add					10%				

09 91 23.39 Doors and Windows, Interior Latex, Zero Voc

	09 91 23.39 Doors and Windows, Interior Latex, Zero Voc	Crew	Daily Output	Labor-Hours	Unit	2011 Bare Costs Material	Labor	Equipment	Total	Total Incl O&P
0010	**DOORS & WINDOWS, INTERIOR LATEX, ZERO VOC**									
0100	Doors flush, both sides, incl. frame & trim									
0110	Roll & brush, primer G	1 Pord	10	.800	Ea.	5	23.50		28.50	44
0120	Finish coat, latex G		10	.800		5.25	23.50		28.75	44.50
0130	Primer & 1 coat latex G		7	1.143		10.25	34		44.25	66.50
0140	Primer & 2 coats latex G		5	1.600		15.20	47.50		62.70	94
0160	Spray, both sides, primer G		20	.400		5.25	11.85		17.10	25
0170	Finish coat, latex G		20	.400		5.50	11.85		17.35	25.50
0180	Primer & 1 coat latex G		11	.727		10.85	21.50		32.35	47
0190	Primer & 2 coats latex G		8	1		16.10	29.50		45.60	65.50
0200	Doors, French, both sides, 10-15 lite, incl. frame & trim									
0210	Roll & brush, primer G	1 Pord	6	1.333	Ea.	2.50	39.50		42	67
0220	Finish coat, latex G		6	1.333		2.63	39.50		42.13	67
0230	Primer & 1 coat latex G		3	2.667		5.15	79		84.15	134
0240	Primer & 2 coats latex G		2	4		7.60	119		126.60	201
0360	Doors, panel, both sides, incl. frame & trim									
0370	Roll & brush, primer G	1 Pord	6	1.333	Ea.	5.25	39.50		44.75	70
0380	Finish coat, latex G		6	1.333		5.25	39.50		44.75	70
0390	Primer & 1 coat, latex G		3	2.667		10.25	79		89.25	139
0400	Primer & 2 coats, latex G		2.50	3.200		15.50	95		110.50	171

09 91 23 – Interior Painting

09 91 23.39 Doors and Windows, Interior Latex, Zero Voc		Crew	Daily Output	Labor-Hours	Unit	2011 Bare Costs Material	Labor	Equipment	Total	Total Incl O&P
0420	Spray, both sides, primer G	1 Pord	10	.800	Ea.	5.25	23.50		28.75	44.50
0430	Finish coat, latex G		10	.800		5.50	23.50		29	44.50
0440	Primer & 1 coat, latex G		5	1.600		10.85	47.50		58.35	89
0450	Primer & 2 coats, latex G		4	2		16.45	59.50		75.95	115
0460	Windows, per interior side, based on 15 S.F.									
0470	1 to 6 lite									
0480	Brushwork, primer G	1 Pord	13	.615	Ea.	.99	18.25		19.24	30.50
0490	Finish coat, enamel G		13	.615		1.04	18.25		19.29	30.50
0500	Primer & 1 coat enamel G		8	1		2.02	29.50		31.52	50
0510	Primer & 2 coats enamel G		6	1.333		3.06	39.50		42.56	67.50

09 91 23.40 Floors, Interior

Line	Description	Crew	Daily Output	Labor-Hours	Unit	Material	Labor	Equipment	Total	Total Incl O&P
0010	**FLOORS, INTERIOR**									
0100	Concrete paint, latex									
0110	Brushwork									
0120	1st coat	1 Pord	975	.008	S.F.	.15	.24		.39	.57
0130	2nd coat		1150	.007		.10	.21		.31	.45
0140	3rd coat		1300	.006		.08	.18		.26	.39
0150	Roll									
0160	1st coat	1 Pord	2600	.003	S.F.	.20	.09		.29	.37
0170	2nd coat		3250	.002		.12	.07		.19	.25
0180	3rd coat		3900	.002		.09	.06		.15	.20
0190	Spray									
0200	1st coat	1 Pord	2600	.003	S.F.	.17	.09		.26	.34
0210	2nd coat		3250	.002		.09	.07		.16	.22
0220	3rd coat		3900	.002		.08	.06		.14	.18
0300	Acid stain and sealer									
0310	Stain, one coat	1 Pord	650	.012	S.F.	.11	.37		.48	.71
0320	Two coats		570	.014		.22	.42		.64	.92
0330	Acrylic sealer, one coat		2600	.003		.18	.09		.27	.35
0340	Two coats		1400	.006		.36	.17		.53	.66

09 91 23.52 Miscellaneous, Interior

Line	Description	Crew	Daily Output	Labor-Hours	Unit	Material	Labor	Equipment	Total	Total Incl O&P
0010	**MISCELLANEOUS, INTERIOR**									
2400	Floors, conc./wood, oil base, primer/sealer coat, brushwork	2 Pord	1950	.008	S.F.	.06	.24		.30	.47
2450	Roller		5200	.003		.06	.09		.15	.22
2600	Spray		6000	.003		.06	.08		.14	.20
2650	Paint 1 coat, brushwork		1950	.008		.10	.24		.34	.51
2800	Roller		5200	.003		.10	.09		.19	.26
2850	Spray		6000	.003		.11	.08		.19	.25
3000	Stain, wood floor, brushwork, 1 coat		4550	.004		.07	.10		.17	.25
3200	Roller		5200	.003		.08	.09		.17	.24
3250	Spray		6000	.003		.08	.08		.16	.22
3400	Varnish, wood floor, brushwork		4550	.004		.08	.10		.18	.26
3450	Roller		5200	.003		.08	.09		.17	.24
3600	Spray		6000	.003		.09	.08		.17	.23
3650	For dust proofing or anti skid, see Section 03 35 29.30									
3800	Grilles, per side, oil base, primer coat, brushwork	1 Pord	520	.015	S.F.	.12	.46		.58	.87
3850	Spray		1140	.007		.13	.21		.34	.48
3920	Paint 2 coats, brushwork		325	.025		.37	.73		1.10	1.60
3940	Spray		650	.012		.42	.37		.79	1.06
5000	Pipe, 1" - 4" diameter, primer or sealer coat, oil base, brushwork	2 Pord	1250	.013	L.F.	.06	.38		.44	.69
5100	Spray		2165	.007		.06	.22		.28	.43
5350	Paint 2 coats, brushwork		775	.021		.18	.61		.79	1.19

09 91 Painting

09 91 23 – Interior Painting

09 91 23.52 Miscellaneous, Interior

	09 91 23.52 Miscellaneous, Interior	Crew	Daily Output	Labor-Hours	Unit	2011 Bare Costs Material	Labor	Equipment	Total	Total Incl O&P
5400	Spray	2 Pord	1240	.013	L.F.	.20	.38		.58	.84
6300	13″ - 16″ diameter, primer or sealer coat, brushwork		310	.052		.26	1.53		1.79	2.78
6350	Spray		540	.030		.29	.88		1.17	1.75
6500	Paint 2 coats, brushwork		195	.082		.72	2.43		3.15	4.74
6550	Spray		310	.052		.80	1.53		2.33	3.37
7000	Trim, wood, incl. puttying, under 6″ wide									
7200	Primer coat, oil base, brushwork	1 Pord	650	.012	L.F.	.03	.37		.40	.62
7250	Paint, 1 coat, brushwork		650	.012		.05	.37		.42	.64
7450	3 coats		325	.025		.14	.73		.87	1.34
7500	Over 6″ wide, primer coat, brushwork		650	.012		.06	.37		.43	.66
7550	Paint, 1 coat, brushwork		650	.012		.10	.37		.47	.69
7650	3 coats		325	.025		.27	.73		1	1.49
8000	Cornice, simple design, primer coat, oil base, brushwork		650	.012	S.F.	.06	.37		.43	.66
8250	Paint, 1 coat		650	.012		.10	.37		.47	.69
8350	Ornate design, primer coat		350	.023		.06	.68		.74	1.17
8400	Paint, 1 coat		350	.023		.10	.68		.78	1.20
8600	Balustrades, primer coat, oil base, brushwork		520	.015		.06	.46		.52	.81
8650	Paint, 1 coat		520	.015		.10	.46		.56	.84
8900	Trusses and wood frames, primer coat, oil base, brushwork		800	.010		.06	.30		.36	.55
8950	Spray		1200	.007		.06	.20		.26	.39
9220	Paint 2 coats, brushwork		500	.016		.19	.47		.66	.97
9240	Spray		600	.013		.21	.40		.61	.87
9260	Stain, brushwork, wipe off		600	.013		.07	.40		.47	.72
9280	Varnish, 3 coats, brushwork		275	.029		.24	.86		1.10	1.66
9350	For latex paint, deduct					10%				

09 91 23.62 Electrostatic Painting

	09 91 23.62 Electrostatic Painting	Crew	Daily Output	Labor-Hours	Unit	Material	Labor	Equipment	Total	Total Incl O&P
0010	**ELECTROSTATIC PAINTING**									
0100	In Shop									
0200	Flat Surfaces (lockers, casework, elev doors. etc)									
0300	One coat	1 Pord	200	.040	S.F.	.46	1.19		1.65	2.44
0400	Two coats	″	120	.067	″	.67	1.98		2.65	3.95
0500	Irregular Surfaces (furniture, door frames, etc)									
0600	One coat	1 Pord	150	.053	S.F.	.46	1.58		2.04	3.08
0700	Two coats	″	100	.080	″	.67	2.37		3.04	4.59
0800	On Site									
0900	Flat Surfaces (lockers, casework, elev doors. etc)									
1000	One coat	1 Pord	150	.053	S.F.	.46	1.58		2.04	3.08
1100	Two coats	″	100	.080	″	.67	2.37		3.04	4.59
1200	Irregular Surfaces (furniture, door frames, etc)									
1300	One coat	1 Pord	115	.070	S.F.	.46	2.06		2.52	3.86
1400	Two coats	″	70	.114	″	.67	3.39		4.06	6.25

09 91 23.72 Walls and Ceilings, Interior

	09 91 23.72 Walls and Ceilings, Interior	Crew	Daily Output	Labor-Hours	Unit	Material	Labor	Equipment	Total	Total Incl O&P
0010	**WALLS AND CEILINGS, INTERIOR** R099100-10									
0100	Concrete, drywall or plaster, latex, primer or sealer coat R099100-20									
0200	Smooth finish, brushwork	1 Pord	1150	.007	S.F.	.06	.21		.27	.40
0240	Roller		1350	.006		.06	.18		.24	.35
0280	Spray		2750	.003		.05	.09		.14	.19
0300	Sand finish, brushwork		975	.008		.06	.24		.30	.46
0340	Roller		1150	.007		.06	.21		.27	.40
0380	Spray		2275	.004		.05	.10		.15	.22
0800	Paint 2 coats, smooth finish, brushwork		680	.012		.11	.35		.46	.69
0840	Roller		800	.010		.11	.30		.41	.60

09 91 23 – Interior Painting

09 91 23.72 Walls and Ceilings, Interior		Crew	Daily Output	Labor-Hours	Unit	Material	Labor	Equipment	Total	Total Incl O&P
						2011 Bare Costs				
0880	Spray	1 Pord	1625	.005	S.F.	.10	.15		.25	.35
0900	Sand finish, brushwork		605	.013		.11	.39		.50	.76
0940	Roller		1020	.008		.11	.23		.34	.50
0980	Spray		1700	.005		.10	.14		.24	.34
1200	Paint 3 coats, smooth finish, brushwork		510	.016		.16	.47		.63	.94
1240	Roller		650	.012		.16	.37		.53	.77
1280	Spray		1625	.005		.15	.15		.30	.41
1600	Glaze coating, 2 coats, spray, clear		1200	.007		.49	.20		.69	.86
1640	Multicolor		1200	.007		.95	.20		1.15	1.37
1660	Painting walls, complete, including surface prep, primer &									
1670	2 coats finish, on drywall or plaster, with roller	1 Pord	325	.025	S.F.	.17	.73		.90	1.38
1700	For oil base paint, add					10%				
1800	For ceiling installations, add						25%			
2000	Masonry or concrete block, primer/sealer, latex paint									
2100	Primer, smooth finish, brushwork	1 Pord	1000	.008	S.F.	.10	.24		.34	.50
2110	Roller		1150	.007		.10	.21		.31	.45
2180	Spray		2400	.003		.09	.10		.19	.26
2200	Sand finish, brushwork		850	.009		.10	.28		.38	.56
2210	Roller		975	.008		.10	.24		.34	.51
2280	Spray		2050	.004		.09	.12		.21	.29
2800	Primer plus one finish coat, smooth brush		525	.015		.27	.45		.72	1.03
2810	Roller		615	.013		.17	.39		.56	.82
2880	Spray		1200	.007		.15	.20		.35	.49
2900	Sand finish, brushwork		450	.018		.17	.53		.70	1.05
2910	Roller		515	.016		.17	.46		.63	.94
2980	Spray		1025	.008		.15	.23		.38	.55
3600	Glaze coating, 3 coats, spray, clear		900	.009		.70	.26		.96	1.20
3620	Multicolor		900	.009		1.10	.26		1.36	1.64
4000	Block filler, 1 coat, brushwork		425	.019		.13	.56		.69	1.06
4100	Silicone, water repellent, 2 coats, spray		2000	.004		.26	.12		.38	.47
4120	For oil base paint, add					10%				
8200	For work 8' - 15' H, add						10%			
8300	For work over 15' H, add						20%			
8400	For light textured surfaces, add						10%			
8410	Heavy textured, add						25%			

09 91 23.74 Walls and Ceilings, Interior, Zero VOC Latex		Crew	Daily Output	Labor-Hours	Unit	Material	Labor	Equipment	Total	Total Incl O&P
0010	**WALLS AND CEILINGS, INTERIOR, ZERO VOC LATEX**									
0100	Concrete, dry wall or plaster, latex, primer or sealer coat									
0200	Smooth finish, brushwork G	1 Pord	1150	.007	S.F.	.07	.21		.28	.41
0240	Roller G		1350	.006		.06	.18		.24	.36
0280	Spray G		2750	.003		.05	.09		.14	.19
0300	Sand finish, brushwork G		975	.008		.06	.24		.30	.47
0340	Roller G		1150	.007		.07	.21		.28	.42
0380	Spray G		2275	.004		.05	.10		.15	.23
0800	Paint 2 coats, smooth finish, brushwork G		680	.012		.13	.35		.48	.71
0840	Roller G		800	.010		.13	.30		.43	.62
0880	Spray G		1625	.005		.11	.15		.26	.36
0900	Sand finish, brushwork G		605	.013		.12	.39		.51	.78
0940	Roller G		1020	.008		.13	.23		.36	.52
0980	Spray G		1700	.005		.11	.14		.25	.35
1200	Paint 3 coats, smooth finish, brushwork G		510	.016		.18	.47		.65	.96
1240	Roller G		650	.012		.19	.37		.56	.80

09 91 Painting

09 91 23 – Interior Painting

09 91 23.74 Walls and Ceilings, Interior, Zero VOC Latex		Crew	Daily Output	Labor-Hours	Unit	Material	Labor	Equipment	Total	Total Incl O&P
						2011 Bare Costs				
1280	Spray G	1 Pord	1625	.005	S.F.	.17	.15		.32	.42
1800	For ceiling installations, add G						25%			
8200	For work 8' - 15' H, add						10%			
8300	For work over 15' H, add						20%			

09 91 23.75 Dry Fall Painting		Crew	Daily Output	Labor-Hours	Unit	Material	Labor	Equipment	Total	Total Incl O&P
0010	**DRY FALL PAINTING**									
0100	Sprayed on walls, gypsum board or plaster									
0220	One coat	1 Pord	2600	.003	S.F.	.05	.09		.14	.20
0250	Two coats		1560	.005		.10	.15		.25	.36
0280	Concrete or textured plaster, one coat		1560	.005		.05	.15		.20	.30
0310	Two coats		1300	.006		.10	.18		.28	.41
0340	Concrete block, one coat		1560	.005		.05	.15		.20	.30
0370	Two coats		1300	.006		.10	.18		.28	.41
0400	Wood, one coat		877	.009		.05	.27		.32	.49
0430	Two coats		650	.012		.10	.37		.47	.70
0440	On ceilings, gypsum board or plaster									
0470	One coat	1 Pord	1560	.005	S.F.	.05	.15		.20	.30
0500	Two coats		1300	.006		.10	.18		.28	.41
0530	Concrete or textured plaster, one coat		1560	.005		.05	.15		.20	.30
0560	Two coats		1300	.006		.10	.18		.28	.41
0570	Structural steel, bar joists or metal deck, one coat		1560	.005		.05	.15		.20	.30
0580	Two coats		1040	.008		.10	.23		.33	.48

09 93 Staining and Transparent Finishing

09 93 23 – Interior Staining and Finishing

09 93 23.10 Varnish		Crew	Daily Output	Labor-Hours	Unit	Material	Labor	Equipment	Total	Total Incl O&P
0010	**VARNISH**									
0012	1 coat + sealer, on wood trim, brush, no sanding included	1 Pord	400	.020	S.F.	.07	.59		.66	1.04
0100	Hardwood floors, 2 coats, no sanding included, roller	"	1890	.004	"	.15	.13		.28	.37

09 96 High-Performance Coatings

09 96 23 – Graffiti-Resistant Coatings

09 96 23.10 Graffiti Resistant Treatments		Crew	Daily Output	Labor-Hours	Unit	Material	Labor	Equipment	Total	Total Incl O&P
0010	**GRAFFITI RESISTANT TREATMENTS**, sprayed on walls									
0100	Non-sacrificial, permanent non-stick coating, clear, on metals	1 Pord	2000	.004	S.F.	2.06	.12		2.18	2.46
0200	Concrete		2000	.004		2.35	.12		2.47	2.77
0300	Concrete block		2000	.004		3.03	.12		3.15	3.53
0400	Brick		2000	.004		3.44	.12		3.56	3.97
0500	Stone		2000	.004		3.44	.12		3.56	3.97
0600	Unpainted wood		2000	.004		3.97	.12		4.09	4.56
2000	Semi-permanent cross linking polymer primer, on metals		2000	.004		.40	.12		.52	.63
2100	Concrete		2000	.004		.48	.12		.60	.72
2200	Concrete block		2000	.004		.60	.12		.72	.85
2300	Brick		2000	.004		.48	.12		.60	.72
2400	Stone		2000	.004		.48	.12		.60	.72
2500	Unpainted wood		2000	.004		.67	.12		.79	.92
3000	Top coat, on metals		2000	.004		.43	.12		.55	.66
3100	Concrete		2000	.004		.49	.12		.61	.72
3200	Concrete block		2000	.004		.68	.12		.80	.94

09 96 High-Performance Coatings

09 96 23 – Graffiti-Resistant Coatings

09 96 23.10 Graffiti Resistant Treatments		Crew	Daily Output	Labor-Hours	Unit	Material	Labor	Equipment	Total	Total Incl O&P
						2011 Bare Costs				
3300	Brick	1 Pord	2000	.004	S.F.	.57	.12		.69	.81
3400	Stone		2000	.004		.57	.12		.69	.81
3500	Unpainted wood		2000	.004		.68	.12		.80	.94
5000	Sacrificial, water based, on metal		2000	.004		.24	.12		.36	.46
5100	Concrete		2000	.004		.24	.12		.36	.46
5200	Concrete block		2000	.004		.24	.12		.36	.46
5300	Brick		2000	.004		.24	.12		.36	.46
5400	Stone		2000	.004		.24	.12		.36	.46
5500	Unpainted wood		2000	.004		.24	.12		.36	.46
8000	Cleaner for use after treatment									
8100	Towels or wipes, per package of 30				Ea.	.63			.63	.70
8200	Aerosol spray, 24 oz. can				"	18			18	19.80

09 96 56 – Epoxy Coatings

09 96 56.20 Wall Coatings		Crew	Daily Output	Labor-Hours	Unit	Material	Labor	Equipment	Total	Total Incl O&P
0010	**WALL COATINGS**									
0100	Acrylic glazed coatings, minimum	1 Pord	525	.015	S.F.	.29	.45		.74	1.05
0200	Maximum		305	.026		.62	.78		1.40	1.94
0300	Epoxy coatings, minimum		525	.015		.38	.45		.83	1.15
0400	Maximum		170	.047		1.15	1.40		2.55	3.54
0600	Exposed aggregate, troweled on, 1/16" to 1/4", minimum		235	.034		.58	1.01		1.59	2.28
0700	Maximum (epoxy or polyacrylate)		130	.062		1.24	1.82		3.06	4.32
0900	1/2" to 5/8" aggregate, minimum		130	.062		1.14	1.82		2.96	4.21
1000	Maximum		80	.100		1.96	2.97		4.93	7
1500	Exposed aggregate, sprayed on, 1/8" aggregate, minimum		295	.027		.53	.80		1.33	1.89
1600	Maximum		145	.055		.99	1.64		2.63	3.75
1800	High build epoxy, 50 mil, minimum		390	.021		.64	.61		1.25	1.69
1900	Maximum		95	.084		1.09	2.50		3.59	5.25
2100	Laminated epoxy with fiberglass, minimum		295	.027		.69	.80		1.49	2.07
2200	Maximum		145	.055		1.23	1.64		2.87	4.01
2400	Sprayed perlite or vermiculite, 1/16" thick, minimum		2935	.003		.25	.08		.33	.41
2500	Maximum		640	.013		.70	.37		1.07	1.37
2700	Vinyl plastic wall coating, minimum		735	.011		.31	.32		.63	.86
2800	Maximum		240	.033		.77	.99		1.76	2.45
3000	Urethane on smooth surface, 2 coats, minimum		1135	.007		.25	.21		.46	.62
3100	Maximum		665	.012		.55	.36		.91	1.19
3600	Ceramic-like glazed coating, cementitious, minimum		440	.018		.45	.54		.99	1.38
3700	Maximum		345	.023		.76	.69		1.45	1.96
3900	Resin base, minimum		640	.013		.31	.37		.68	.94
4000	Maximum		330	.024		.50	.72		1.22	1.72

09 97 Special Coatings

09 97 13 – Steel Coatings

09 97 13.23 Exterior Steel Coatings		Crew	Daily Output	Labor-Hours	Unit	Material	Labor	Equipment	Total	Total Incl O&P
0010	**EXTERIOR STEEL COATINGS** R050516-30									
6101	Cold galvanizing, brush in field	1 Pord	1100	.007	S.F.	.07	.22		.29	.42
6510	Paints & protective coatings, sprayed in field									
6520	Alkyds, primer	2 Psst	3600	.004	S.F.	.06	.14		.20	.33
6540	Gloss topcoats		3200	.005		.07	.15		.22	.37
6560	Silicone alkyd		3200	.005		.12	.15		.27	.42
6610	Epoxy, primer		3000	.005		.24	.16		.40	.57

09 97 Special Coatings

09 97 13 – Steel Coatings

09 97 13.23 Exterior Steel Coatings		Crew	Daily Output	Labor-Hours	Unit	Material	Labor	Equipment	Total	Total Incl O&P
						2011 Bare Costs				
6630	Intermediate or topcoat	2 Psst	2800	.006	S.F.	.21	.17		.38	.58
6650	Enamel coat		2800	.006		.27	.17		.44	.63
6700	Epoxy ester, primer		2800	.006		.52	.17		.69	.91
6720	Topcoats		2800	.006		.18	.17		.35	.54
6810	Latex primer		3600	.004		.05	.14		.19	.32
6830	Topcoats		3200	.005		.06	.15		.21	.35
6910	Universal primers, one part, phenolic, modified alkyd		2000	.008		.11	.24		.35	.59
6940	Two part, epoxy spray		2000	.008		.28	.24		.52	.78
7000	Zinc rich primers, self cure, spray, inorganic		1800	.009		.59	.27		.86	1.17
7010	Epoxy, spray, organic	▼	1800	.009	▼	.21	.27		.48	.76
7020	Above one story, spray painting simple structures, add						25%			
7030	Intricate structures, add						50%			

Estimating Tips

General

- The items in this division are usually priced per square foot or each.
- Many items in Division 10 require some type of support system or special anchors that are not usually furnished with the item. The required anchors must be added to the estimate in the appropriate division.
- Some items in Division 10, such as lockers, may require assembly before installation. Verify the amount of assembly required. Assembly can often exceed installation time.

10 20 00 Interior Specialties

- Support angles and blocking are not included in the installation of toilet compartments, shower/dressing compartments, or cubicles. Appropriate line items from Divisions 5 or 6 may need to be added to support the installations.
- Toilet partitions are priced by the stall. A stall consists of a side wall, pilaster, and door with hardware. Toilet tissue holders and grab bars are extra.
- The required acoustical rating of a folding partition can have a significant impact on costs. Verify the sound transmission coefficient rating of the panel priced to the specification requirements.
- Grab bar installation does not include supplemental blocking or backing to support the required load. When grab bars are installed at an existing facility, provisions must be made to attach the grab bars to solid structure.

Reference Numbers

Reference numbers are shown in shaded boxes at the beginning of some major classifications. These numbers refer to related items in the Reference Section. The reference information may be an estimating procedure, an alternate pricing method, or technical information.

Note: Not all subdivisions listed here necessarily appear in this publication.

10 05 Common Work Results for Specialties

10 05 05 – Selective Specialties Demolition

10 05 05.10 Selective Demolition, Specialties		Crew	Daily Output	Labor-Hours	Unit	Material	2011 Bare Costs Labor	Equipment	Total	Total Incl O&P
0010	**SELECTIVE DEMOLITION, SPECIALTIES**									
1100	Boards and panels, wall mounted	2 Clab	15	1.067	Ea.		28		28	47.50
1200	Cases, for directory and/or bulletin boards, incl doors		24	.667			17.65		17.65	29.50
1850	Shower partitions, cabinet or stall, incl base and door		8	2			53		53	88.50
1855	Shower receptor, terrazzo or concrete	1 Clab	14	.571			15.10		15.10	25.50
1900	Curtain track or rod, hospital type, ceiling mounted or suspended	"	220	.036	L.F.		.96		.96	1.61
1910	Toilet Cubicles, remove	2 Clab	8	2	Ea.		53		53	88.50
1930	Urinal Screen, remove	1 Clab	12	.667	"		17.65		17.65	29.50
2650	Wall guard, misc. wall or corner protection	"	320	.025	L.F.		.66		.66	1.11
2750	Access floor, metal panel system, incl pedestals, covering	2 Clab	850	.019	S.F.		.50		.50	.83
3050	Fireplace, prefab, freestanding or wall hung, incl hood and screen	1 Clab	2	4	Ea.		106		106	177
3054	Chimney top, simulated brick, 4′ high	"	15	.533			14.10		14.10	23.50
3200	Stove, woodburning, cast iron	2 Clab	2	8			212		212	355
3440	Weathervane, residential	1 Clab	12	.667			17.65		17.65	29.50
3500	Flagpole, groundset, to 70′ high, excl base/fndn	K-1	1	16			480	217	697	1,050
3555	To 30′ high	"	2.50	6.400			192	87	279	415
4300	Letter, signs or plaques, exterior on wall	1 Clab	20	.400			10.60		10.60	17.70
4310	Signs, street, reflective aluminum, incl post and bracket		60	.133			3.53		3.53	5.90
4320	Door signs interior on door 6″ x 6″, selective demolition		20	.400			10.60		10.60	17.70
4550	Turnstiles, manual or electric	2 Clab	2	8			212		212	355
5050	Lockers	1 Clab	15	.533	Opng.		14.10		14.10	23.50
5250	Cabinets, recessed	Q-12	12	1.333	Ea.		47		47	77
5260	Mail boxes, Horiz., Key Lock, front loading, Remove	1 Carp	34	.235	"		7.90		7.90	13.20
5350	Awning, fabric, including frame	2 Clab	100	.160	S.F.		4.23		4.23	7.10
6050	Partition, woven wire		1400	.011			.30		.30	.51
6100	Folding gate, security, door or window		500	.032			.85		.85	1.42
6580	Acoustic air wall		650	.025			.65		.65	1.09
7550	Telephone enclosure, exterior, post mounted		3	5.333	Ea.		141		141	236
8850	Scale, platform, excludes foundation or pit		.25	64	"		1,700		1,700	2,825

10 11 Visual Display Units

10 11 13 – Chalkboards

10 11 13.13 Fixed Chalkboards

10 11 13.13 Fixed Chalkboards		Crew	Daily Output	Labor-Hours	Unit	Material	Labor	Equipment	Total	Total Incl O&P
0010	**FIXED CHALKBOARDS** Porcelain enamel steel									
3900	Wall hung									
4000	Aluminum frame and chalktrough									
4200	3′ x 4′	2 Carp	16	1	Ea.	185	33.50		218.50	260
4300	3′ x 5′		15	1.067		248	36		284	335
4500	4′ x 8′		14	1.143		355	38.50		393.50	455
4600	4′ x 12′		13	1.231		515	41.50		556.50	635
4700	Wood frame and chalktrough									
4800	3′ x 4′	2 Carp	16	1	Ea.	188	33.50		221.50	263
5000	3′ x 5′		15	1.067		233	36		269	315
5100	4′ x 5′		14	1.143		248	38.50		286.50	335
5300	4′ x 8′		13	1.231		350	41.50		391.50	455
5400	Liquid chalk, white porcelain enamel, wall hung									
5420	Deluxe units, aluminum trim and chalktrough									
5450	4′ x 4′	2 Carp	16	1	Ea.	219	33.50		252.50	297
5500	4′ x 8′		14	1.143		335	38.50		373.50	435
5550	4′ x 12′		12	1.333		495	44.50		539.50	620
5700	Wood trim and chalktrough									

10 11 Visual Display Units

10 11 13 – Chalkboards

10 11 13.13 Fixed Chalkboards		Crew	Daily Output	Labor-Hours	Unit	Material	2011 Bare Costs Labor	Equipment	Total	Total Incl O&P
5900	4′ x 4′	2 Carp	16	1	Ea.	550	33.50		583.50	660
6000	4′ x 6′		15	1.067		655	36		691	780
6200	4′ x 8′		14	1.143		760	38.50		798.50	900
6300	Liquid chalk, felt tip markers					1.52			1.52	1.67
6500	Erasers					1.54			1.54	1.69
6600	Board cleaner, 8 oz. bottle					4.49			4.49	4.94

10 11 13.23 Modular-Support-Mounted Chalkboards		Crew	Daily Output	Labor-Hours	Unit	Material	Labor	Equipment	Total	Total Incl O&P
0010	**MODULAR-SUPPORT-MOUNTED CHALKBOARDS**									
0400	Sliding chalkboards									
0450	Vertical, one sliding board with back panel, wall mounted									
0500	8′ x 4′	2 Carp	8	2	Ea.	2,050	67		2,117	2,350
0520	8′ x 8′		7.50	2.133		2,875	71.50		2,946.50	3,300
0540	8′ x 12′		7	2.286		3,500	76.50		3,576.50	3,975
0600	Two sliding boards, with back panel									
0620	8′ x 4′	2 Carp	8	2	Ea.	3,300	67		3,367	3,725
0640	8′ x 8′		7.50	2.133		4,900	71.50		4,971.50	5,500
0660	8′ x 12′		7	2.286		7,925	76.50		8,001.50	8,850
0700	Horizontal, two track									
0800	4′ x 8′, 2 sliding panels	2 Carp	8	2	Ea.	1,825	67		1,892	2,125
0820	4′ x 12′, 2 sliding panels		7.50	2.133		2,400	71.50		2,471.50	2,750
0840	4′ x 16′, 4 sliding panels		7	2.286		3,225	76.50		3,301.50	3,675
0900	Four track, four sliding panels									
0920	4′ x 8′	2 Carp	8	2	Ea.	2,975	67		3,042	3,350
0940	4′ x 12′		7.50	2.133		3,875	71.50		3,946.50	4,375
0960	4′ x 16′		7	2.286		5,050	76.50		5,126.50	5,675
1200	Vertical, motor operated									
1400	One sliding panel with back panel									
1450	10′ x 4′	2 Carp	4	4	Ea.	5,325	134		5,459	6,100
1500	10′ x 10′		3.75	4.267		6,425	143		6,568	7,325
1550	10′ x 16′		3.50	4.571		7,575	153		7,728	8,600
1700	Two sliding panels with back panel									
1750	10′ x 4′	2 Carp	4	4	Ea.	9,475	134		9,609	10,600
1800	10′ x 10′		3.75	4.267		11,000	143		11,143	12,300
1850	10′ x 16′		3.50	4.571		12,600	153		12,753	14,200
2000	Three sliding panels with back panel									
2100	10′ x 4′	2 Carp	4	4	Ea.	13,200	134		13,334	14,700
2150	10′ x 10′		3.75	4.267		14,500	143		14,643	16,100
2200	10′ x 16′		3.50	4.571		17,500	153		17,653	19,500
2400	For projection screen, glass beaded, add				S.F.	7.65			7.65	8.40
2500	For remote control, 1 panel control, add				Ea.	360			360	395
2600	2 panel control, add				″	620			620	680
2800	For units without back panels, deduct				S.F.	4.84			4.84	5.30
2850	For liquid chalk porcelain panels, add				″	5.15			5.15	5.70
3000	Swing leaf, any comb. of chalkboard & cork, aluminum frame									
3100	Floor style, 6 panels									
3150	30″ x 40″ panels				Ea.	1,575			1,575	1,750
3200	48″ x 40″ panels				″	2,650			2,650	2,925
3300	Wall mounted, 6 panels									
3400	30″ x 40″ panels	2 Carp	16	1	Ea.	1,500	33.50		1,533.50	1,700
3450	48″ x 40″ panels	″	16	1	″	1,850	33.50		1,883.50	2,075
3600	Extra panels for swing leaf units									
3700	30″ x 40″ panels				Ea.	299			299	330

10 11 Visual Display Units

10 11 13 – Chalkboards

10 11 13.23 Modular-Support-Mounted Chalkboards

	10 11 13.23 Modular-Support-Mounted Chalkboards	Crew	Daily Output	Labor-Hours	Unit	Material	2011 Bare Costs Labor	Equipment	Total	Total Incl O&P
3750	48" x 40" panels				Ea.	370			370	405

10 11 13.43 Portable Chalkboards

	10 11 13.43 Portable Chalkboards	Crew	Daily Output	Labor-Hours	Unit	Material	Labor	Equipment	Total	Total Incl O&P
0010	**PORTABLE CHALKBOARDS**									
0100	Freestanding, reversible									
0120	Economy, wood frame, 4' x 6'									
0140	Chalkboard both sides				Ea.	525			525	575
0160	Chalkboard one side, cork other side				"	500			500	555
0200	Standard, lightweight satin finished aluminum, 4' x 6'									
0220	Chalkboard both sides				Ea.	565			565	620
0240	Chalkboard one side, cork other side				"	595			595	655
0300	Deluxe, heavy duty extruded aluminum, 4' x 6'									
0320	Chalkboard both sides				Ea.	990			990	1,100
0340	Chalkboard one side, cork other side				"	890			890	980

10 11 23 – Tackboards

10 11 23.10 Fixed Tackboards

	10 11 23.10 Fixed Tackboards	Crew	Daily Output	Labor-Hours	Unit	Material	Labor	Equipment	Total	Total Incl O&P
0010	**FIXED TACKBOARDS**									
0020	Cork sheets, unbacked, no frame, 1/4" thick	2 Carp	290	.055	S.F.	1.94	1.85		3.79	5.25
0100	1/2" thick		290	.055		3.85	1.85		5.70	7.35
0300	Fabric-face, no frame, on 7/32" cork underlay		290	.055		6	1.85		7.85	9.70
0400	On 1/4" cork on 1/4" hardboard		290	.055		7.85	1.85		9.70	11.75
0600	With edges wrapped		290	.055		8.45	1.85		10.30	12.35
0700	On 7/16" fire retardant core		290	.055		7.60	1.85		9.45	11.45
0900	With edges wrapped		290	.055		12.30	1.85		14.15	16.60
1000	Designer fabric only, cut to size					2.07			2.07	2.28
1200	1/4" vinyl cork, on 1/4" hardboard, no frame	2 Carp	290	.055		8.40	1.85		10.25	12.30
1300	On 1/4" coreboard		290	.055		7.95	1.85		9.80	11.85
2000	For map and display rail, economy, add		385	.042	L.F.	2.52	1.39		3.91	5.10
2100	Deluxe, add		350	.046	"	4.44	1.53		5.97	7.45
2120	Prefabricated, 1/4" cork, 3' x 5' with aluminum frame		16	1	Ea.	113	33.50		146.50	180
2140	Wood frame		16	1		143	33.50		176.50	213
2160	4' x 4' with aluminum frame		16	1		98.50	33.50		132	164
2180	Wood frame		16	1		128	33.50		161.50	197
2200	4' x 8' with aluminum frame		14	1.143		177	38.50		215.50	258
2210	With wood frame		14	1.143		177	38.50		215.50	258
2220	4' x 12' with aluminum frame		12	1.333		263	44.50		307.50	365
2230	Bulletin board case, single glass door, with lock									
2240	36" x 24", economy	2 Carp	12	1.333	Ea.	297	44.50		341.50	400
2250	Deluxe		12	1.333		520	44.50		564.50	645
2260	42" x 30", economy		12	1.333		340	44.50		384.50	450
2270	Deluxe		12	1.333		300	44.50		344.50	410
2300	Glass enclosed cabinets, alum., cork panel, hinged doors									
2400	3' x 3', 1 door	2 Carp	12	1.333	Ea.	570	44.50		614.50	705
2500	4' x 4', 2 door		11	1.455		955	49		1,004	1,125
2600	4' x 7', 3 door		10	1.600		1,650	53.50		1,703.50	1,925
2800	4' x 10', 4 door		8	2		2,200	67		2,267	2,500
2900	For lights, add per door opening	1 Elec	13	.615		131	25		156	185
3100	Horizontal sliding units, 4 doors, 4' x 8', 8' x 4'	2 Carp	9	1.778		1,825	59.50		1,884.50	2,125
3200	4' x 12'		7	2.286		2,400	76.50		2,476.50	2,750
3400	8 doors, 4' x 16'		5	3.200		3,250	107		3,357	3,750
3500	4' x 24'		4	4		4,375	134		4,509	5,050

10 11 Visual Display Units

10 11 23 – Tackboards

10 11 23.20 Control Boards		Crew	Daily Output	Labor-Hours	Unit	Material	2011 Bare Costs Labor	Equipment	Total	Total Incl O&P
0010	**CONTROL BOARDS**									
0020	Magnetic, porcelain finish, 18″ x 24″, framed	2 Carp	8	2	Ea.	199	67		266	330
0100	24″ x 36″		7.50	2.133		295	71.50		366.50	445
0200	36″ x 48″		7	2.286		390	76.50		466.50	560
0300	48″ x 72″		6	2.667		860	89.50		949.50	1,100
0400	48″ x 96″		5	3.200		1,075	107		1,182	1,350

10 13 Directories

10 13 10 – Building Directories

10 13 10.10 Directory Boards		Crew	Daily Output	Labor-Hours	Unit	Material	Labor	Equipment	Total	Total Incl O&P
0010	**DIRECTORY BOARDS**									
0050	Plastic, glass covered, 30″ x 20″	2 Carp	3	5.333	Ea.	260	179		439	585
0100	36″ x 48″		2	8		745	268		1,013	1,275
0300	Grooved cork, 30″ x 20″		3	5.333		375	179		554	710
0400	36″ x 48″		2	8		505	268		773	1,000
0600	Black felt, 30″ x 20″		3	5.333		227	179		406	550
0700	36″ x 48″		2	8		440	268		708	935
0900	Outdoor, weatherproof, black plastic, 36″ x 24″		2	8		720	268		988	1,250
1000	36″ x 36″		1.50	10.667		830	360		1,190	1,525
1800	Indoor, economy, open face, 18″ x 24″		7	2.286		134	76.50		210.50	276
1900	24″ x 36″		7	2.286		124	76.50		200.50	264
2000	36″ x 24″		6	2.667		141	89.50		230.50	305
2100	36″ x 48″		6	2.667		218	89.50		307.50	390
2400	Building directory, alum., black felt panels, 1 door, 24″ x 18″		4	4		300	134		434	555
2500	36″ x 24″		3.50	4.571		355	153		508	645
2600	48″ x 32″		3	5.333		575	179		754	935
2700	2 door, 36″ x 48″		2.50	6.400		630	215		845	1,050
2800	36″ x 60″		2	8		825	268		1,093	1,350
2900	48″ x 60″		1	16		915	535		1,450	1,900
3100	For bronze enamel finish, add					15%				
3200	For bronze anodized finish, add					25%				
3400	For illuminated directory, single door unit, add					110			110	120
3500	For 6″ header panel, 6 letters per foot, add				L.F.	20.50			20.50	22.50
6050	Building directory, electronic display, alum. frame, wall mounted	2 Carp	32	.500	S.F.	550	16.80		566.80	635
6100	Free standing	″	60	.267	″	655	8.95		663.95	735

10 14 Signage

10 14 19 – Dimensional Letter Signage

10 14 19.10 Exterior Signs		Crew	Daily Output	Labor-Hours	Unit	Material	Labor	Equipment	Total	Total Incl O&P
0010	**EXTERIOR SIGNS**									
0020	Letters, 2″ high, 3/8″ deep, cast bronze	1 Carp	24	.333	Ea.	22.50	11.20		33.70	43.50
0140	1/2″ deep, cast aluminum		18	.444		22.50	14.90		37.40	50
0160	Cast bronze		32	.250		36	8.40		44.40	53.50
0300	6″ high, 5/8″ deep, cast aluminum		24	.333		28.50	11.20		39.70	49.50
0400	Cast bronze		24	.333		59	11.20		70.20	83.50
0600	8″ high, 3/4″ deep, cast aluminum		14	.571		34.50	19.15		53.65	70
0700	Cast bronze		20	.400		84.50	13.40		97.90	116
0900	10″ high, 1″ deep, cast aluminum		18	.444		52	14.90		66.90	82
1000	Bronze		18	.444		114	14.90		128.90	150

10 14 19 – Dimensional Letter Signage

10 14 19.10 Exterior Signs		Crew	Daily Output	Labor-Hours	Unit	2011 Bare Costs Material	Labor	Equipment	Total	Total Incl O&P
1200	12" high, 1-1/4" deep, cast aluminum	1 Carp	12	.667	Ea.	55	22.50		77.50	98
1500	Cast bronze		18	.444		129	14.90		143.90	167
1600	14" high, 2-5/16" deep, cast aluminum		12	.667		79	22.50		101.50	125
1800	Fabricated stainless steel, 6" high, 2" deep		20	.400		53	13.40		66.40	80.50
1900	12" high, 3" deep		18	.444		118	14.90		132.90	154
2100	18" high, 3" deep		12	.667		228	22.50		250.50	288
2200	24" high, 4" deep		10	.800		385	27		412	465
2700	Acrylic, on high density foam, 12" high, 2" deep		20	.400		22.50	13.40		35.90	47
2800	18" high, 2" deep		18	.444		61	14.90		75.90	92
3900	Plaques, custom, 20" x 30", for up to 450 letters, cast aluminum	2 Carp	4	4		1,150	134		1,284	1,475
4000	Cast bronze		4	4		1,675	134		1,809	2,075
4200	30" x 36", up to 900 letters cast aluminum		3	5.333		1,925	179		2,104	2,425
4300	Cast bronze		3	5.333		3,475	179		3,654	4,125
4500	36" x 48", for up to 1300 letters, cast bronze		2	8		4,250	268		4,518	5,150
4800	Signs, reflective alum. directional signs, dbl. face, 2-way, w/bracket		30	.533		92.50	17.90		110.40	132
4900	4-way		30	.533		148	17.90		165.90	193
5100	Exit signs, 24 ga. alum., 14" x 12" surface mounted	1 Carp	30	.267		41.50	8.95		50.45	60.50
5200	10" x 7"		20	.400		26.50	13.40		39.90	52
5400	Bracket mounted, double face, 12" x 10"		30	.267		42	8.95		50.95	61.50
5500	Sticky back, stock decals, 14" x 10"	1 Clab	50	.160		19.60	4.23		23.83	28.50
6000	Interior elec., wall mount, fiberglass panels, 2 lamps, 6"	1 Elec	8	1		121	40.50		161.50	199
6100	8"	"	8	1		95.50	40.50		136	171
6400	Replacement sign faces, 6" or 8"	1 Clab	50	.160		54.50	4.23		58.73	66.50

10 14 23 – Panel Signage

10 14 23.13 Engraved Panel Signage		Crew	Daily Output	Labor-Hours	Unit	Material	Labor	Equipment	Total	Total Incl O&P
0010	**ENGRAVED PANEL SIGNAGE**, Interior									
1010	Flexible door sign, adhesive back, w/Braille, 5/8" letters, 4" x 4"	1 Clab	32	.250	Ea.	29	6.60		35.60	43
1050	6" x 6"		32	.250		42.50	6.60		49.10	58
1100	8" x 2"		32	.250		29.50	6.60		36.10	43.50
1150	8" x 4"		32	.250		37.50	6.60		44.10	52.50
1200	8" x 8"		32	.250		59.50	6.60		66.10	76.50
1250	12" x 2"		32	.250		35	6.60		41.60	49.50
1300	12" x 6"		32	.250		50	6.60		56.60	66
1350	12" x 12"		32	.250		119	6.60		125.60	142
1500	Graphic symbols, 2" x 2"		32	.250		12	6.60		18.60	24.50
1550	6" x 6"		32	.250		31	6.60		37.60	45
1600	8" x 8"		32	.250		31	6.60		37.60	45

10 14 53 – Traffic Signage

10 14 53.20 Traffic Signs		Crew	Daily Output	Labor-Hours	Unit	Material	Labor	Equipment	Total	Total Incl O&P
0010	**TRAFFIC SIGNS**									
0012	Stock, 24" x 24", no posts, .080" alum. reflectorized	B-80	70	.343	Ea.	78.50	9.30	8.85	96.65	112
0100	High intensity		70	.343		78.50	9.30	8.85	96.65	112
0300	30" x 30", reflectorized		70	.343		161	9.30	8.85	179.15	202
0400	High intensity		70	.343		161	9.30	8.85	179.15	202
0600	Guide and directional signs, 12" x 18", reflectorized		70	.343		34.50	9.30	8.85	52.65	63.50
0700	High intensity		70	.343		49.50	9.30	8.85	67.65	80
0900	18" x 24", stock signs, reflectorized		70	.343		52.50	9.30	8.85	70.65	83
1000	High intensity		70	.343		52.50	9.30	8.85	70.65	83
1200	24" x 24", stock signs, reflectorized		70	.343		64.50	9.30	8.85	82.65	96.50
1300	High intensity		70	.343		64.50	9.30	8.85	82.65	96.50
1500	Add to above for steel posts, galvanized, 10'-0" upright, bolted		200	.120		35	3.25	3.09	41.34	47
1600	12'-0" upright, bolted		140	.171		33	4.65	4.42	42.07	49

10 14 Signage

10 14 53 – Traffic Signage

10 14 53.20 Traffic Signs		Crew	Daily Output	Labor-Hours	Unit	2011 Bare Costs Material	Labor	Equipment	Total	Total Incl O&P
1800	Highway road signs, aluminum, over 20 S. F., reflectorized	B-80	350	.069	S.F.	28	1.86	1.77	31.63	36
2000	High intensity		350	.069		28	1.86	1.77	31.63	36
2200	Highway, suspended over road, 80 S.F. min., reflectorized		165	.145		28	3.94	3.75	35.69	41.50
2300	High intensity		165	.145		28	3.94	3.75	35.69	41.50

10 17 Telephone Specialties

10 17 16 – Telephone Enclosures

10 17 16.10 Commercial Telephone Enclosures		Crew	Daily Output	Labor-Hours	Unit	Material	Labor	Equipment	Total	Total Incl O&P
0010	**COMMERCIAL TELEPHONE ENCLOSURES**									
0300	Shelf type, wall hung, minimum	2 Carp	5	3.200	Ea.	1,050	107		1,157	1,325
0400	Maximum		5	3.200		2,350	107		2,457	2,750
0600	Booth type, painted steel, indoor or outdoor, minimum		1.50	10.667		3,475	360		3,835	4,425
0700	Maximum (stainless steel)		1.50	10.667		11,600	360		11,960	13,300
1300	Outdoor, acoustical, on post		3	5.333		1,525	179		1,704	1,975
1400	Phone carousel, pedestal mounted with dividers		.60	26.667		5,800	895		6,695	7,875
1900	Outdoor, drive-up type, wall mounted		4	4		930	134		1,064	1,250
2000	Post mounted, stainless steel posts		3	5.333		1,425	179		1,604	1,875
2200	Directory shelf, wall mounted, stainless steel									
2300	3 binders	2 Carp	8	2	Ea.	1,100	67		1,167	1,325
2500	4 binders		7	2.286		1,275	76.50		1,351.50	1,525
2600	5 binders		6	2.667		1,625	89.50		1,714.50	1,950
2800	Table type, stainless steel, 4 binders		8	2		1,300	67		1,367	1,550
2900	7 binders		7	2.286		1,625	76.50		1,701.50	1,925

10 21 Compartments and Cubicles

10 21 13 – Toilet Compartments

10 21 13.13 Metal Toilet Compartments		Crew	Daily Output	Labor-Hours	Unit	Material	Labor	Equipment	Total	Total Incl O&P
0010	**METAL TOILET COMPARTMENTS**									
0110	Cubicles, ceiling hung									
0200	Powder coated steel	2 Carp	4	4	Ea.	480	134		614	755
0500	Stainless steel	"	4	4		755	134		889	1,050
0600	For handicap units, incl. 52" grab bars, add					410			410	450
0900	Floor and ceiling anchored									
1000	Powder coated steel	2 Carp	5	3.200	Ea.	590	107		697	825
1300	Stainless steel	"	5	3.200		1,050	107		1,157	1,325
1400	For handicap units, incl. 52" grab bars, add					287			287	315
1610	Floor anchored									
1700	Powder coated steel	2 Carp	7	2.286	Ea.	555	76.50		631.50	740
2000	Stainless steel	"	7	2.286		1,250	76.50		1,326.50	1,500
2100	For handicap units, incl. 52" grab bars, add					285			285	315
2200	For juvenile units, deduct					41.50			41.50	45.50
2450	Floor anchored, headrail braced									
2500	Powder coated steel	2 Carp	6	2.667	Ea.	485	89.50		574.50	685
2804	Stainless steel	"	4.60	3.478		1,175	117		1,292	1,500
2900	For handicap units, incl. 52" grab bars, add					330			330	365
3000	Wall hung partitions, powder coated steel	2 Carp	7	2.286		625	76.50		701.50	815
3300	Stainless steel	"	7	2.286		1,650	76.50		1,726.50	1,950
3400	For handicap units, incl. 52" grab bars, add					330			330	365
4000	Screens, entrance, floor mounted, 58" high, 48" wide									

10 21 13 – Toilet Compartments

10 21 13.13 Metal Toilet Compartments		Crew	Daily Output	Labor-Hours	Unit	Material	2011 Bare Costs Labor	Equipment	Total	Total Incl O&P
4200	Powder coated steel	2 Carp	15	1.067	Ea.	238	36		274	320
4500	Stainless steel	"	15	1.067	"	895	36		931	1,050
4650	Urinal screen, 18" wide									
4704	Powder coated steel	2 Carp	6.15	2.602	Ea.	220	87.50		307.50	390
5004	Stainless steel	"	6.15	2.602	"	580	87.50		667.50	785
5100	Floor mounted, head rail braced									
5300	Powder coated steel	2 Carp	8	2	Ea.	244	67		311	380
5600	Stainless steel	"	8	2	"	575	67		642	740
5750	Pilaster, flush									
5800	Powder coated steel	2 Carp	10	1.600	Ea.	276	53.50		329.50	395
6100	Stainless steel		10	1.600		670	53.50		723.50	825
6300	Post braced, powder coated steel		10	1.600		175	53.50		228.50	283
6600	Stainless steel		10	1.600		450	53.50		503.50	585
6700	Wall hung, bracket supported									
6800	Powder coated steel	2 Carp	10	1.600	Ea.	305	53.50		358.50	425
7100	Stainless steel		10	1.600		214	53.50		267.50	325
7400	Flange supported, powder coated steel		10	1.600		106	53.50		159.50	206
7700	Stainless steel		10	1.600		310	53.50		363.50	430
7800	Wedge type, powder coated steel		10	1.600		134	53.50		187.50	238
8100	Stainless steel		10	1.600		565	53.50		618.50	710

10 21 13.14 Metal Toilet Compartment Components		Crew	Daily Output	Labor-Hours	Unit	Material	Labor	Equipment	Total	Total Incl O&P
0010	**METAL TOILET COMPARTMENT COMPONENTS**									
0100	Pilasters									
0110	Overhead braced, powder coated steel, 7" wide x 82" high	2 Carp	22.20	.721	Ea.	82.50	24		106.50	131
0120	Stainless steel		22.20	.721		168	24		192	226
0130	Floor braced, powder coated steel, 7" wide x 70" high		23.30	.687		122	23		145	174
0140	Stainless steel		23.30	.687		232	23		255	294
0150	Ceiling hung, powder coated steel, 7" wide x 83" high		13.30	1.203		129	40.50		169.50	210
0160	Stainless steel		13.30	1.203		258	40.50		298.50	350
0170	Wall hung, powder coated steel, 3" wide x 58" high		18.90	.847		123	28.50		151.50	183
0180	Stainless steel		18.90	.847		171	28.50		199.50	236
0200	Panels									
0210	Powder coated steel, 31" wide x 58" high	2 Carp	18.90	.847	Ea.	136	28.50		164.50	197
0220	Stainless steel		18.90	.847		345	28.50		373.50	425
0230	Powder coated steel, 53" wide x 58" high		18.90	.847		244	28.50		272.50	315
0240	Stainless steel		18.90	.847		455	28.50		483.50	550
0250	Powder coated steel, 63" wide x 58" high		18.90	.847		211	28.50		239.50	280
0260	Stainless steel		18.90	.847		515	28.50		543.50	615
0300	Doors									
0310	Powder coated steel, 24" wide x 58" high	2 Carp	14.10	1.135	Ea.	149	38		187	228
0320	Stainless steel		14.10	1.135		335	38		373	435
0330	Powder coated steel, 26" wide x 58" high		14.10	1.135		148	38		186	227
0340	Stainless steel		14.10	1.135		340	38		378	440
0350	Powder coated steel, 28" wide x 58" high		14.10	1.135		157	38		195	237
0360	Stainless steel		14.10	1.135		360	38		398	460
0370	Powder coated steel, 36" wide x 58" high		14.10	1.135		182	38		220	264
0380	Stainless steel		14.10	1.135		405	38		443	510
0400	Headrails									
0410	For powder coated steel, 62" long	2 Carp	65	.246	Ea.	24	8.25		32.25	40.50
0420	Stainless steel		65	.246		24	8.25		32.25	40.50
0430	For powder coated steel, 84" long		50	.320		32.50	10.75		43.25	54
0440	Stainless steel		50	.320		31	10.75		41.75	52

10 21 Compartments and Cubicles

10 21 13 – Toilet Compartments

10 21 13.14 Metal Toilet Compartment Components		Crew	Daily Output	Labor-Hours	Unit	2011 Bare Costs Material	Labor	Equipment	Total	Total Incl O&P
0450	For powder coated steel, 120" long	2 Carp	30	.533	Ea.	44.50	17.90		62.40	79
0460	Stainless steel		30	.533		44.50	17.90		62.40	79

10 21 13.16 Plastic-Laminate-Clad Toilet Compartments

		Crew	Daily Output	Labor-Hours	Unit	Material	Labor	Equipment	Total	Total Incl O&P
0010	**PLASTIC-LAMINATE-CLAD TOILET COMPARTMENTS**									
0110	Cubicles, ceiling hung									
0300	Plastic laminate on particle board	2 Carp	4	4	Ea.	575	134		709	855
0600	For handicap units, incl. 52" grab bars, add				"	410			410	450
0900	Floor and ceiling anchored									
1100	Plastic laminate on particle board	2 Carp	5	3.200	Ea.	750	107		857	1,000
1400	For handicap units, incl. 52" grab bars, add				"	287			287	315
1610	Floor mounted									
1800	Plastic laminate on particle board	2 Carp	7	2.286	Ea.	535	76.50		611.50	715
2450	Floor mounted, headrail braced									
2600	Plastic laminate on particle board	2 Carp	6	2.667	Ea.	800	89.50		889.50	1,025
3400	For handicap units, incl. 52" grab bars, add					330			330	365
4300	Entrance screen, floor mtd., plas. lam., 58" high, 48" wide	2 Carp	15	1.067		580	36		616	700
4800	Urinal screen, 18" wide, ceiling braced, plastic laminate		8	2		174	67		241	305
5400	Floor mounted, headrail braced		8	2		180	67		247	310
5900	Pilaster, flush, plastic laminate		10	1.600		485	53.50		538.50	620
6400	Post braced, plastic laminate		10	1.600		285	53.50		338.50	405
6700	Wall hung, bracket supported									
6900	Plastic laminate on particle board	2 Carp	10	1.600	Ea.	89.50	53.50		143	189
7450	Flange supported									
7500	Plastic laminate on particle board	2 Carp	10	1.600	Ea.	221	53.50		274.50	335

10 21 13.17 Plastic-Laminate Clad Toilet Compartment Components

		Crew	Daily Output	Labor-Hours	Unit	Material	Labor	Equipment	Total	Total Incl O&P
0010	**PLASTIC-LAMINATE CLAD TOILET COMPARTMENT COMPONENTS**									
0100	Pilasters									
0110	Overhead braced, 7" wide x 82" high	2 Carp	22.20	.721	Ea.	98	24		122	149
0130	Floor anchored, 7" wide x 70" high		23.30	.687		93	23		116	141
0150	Ceiling hung, 7" wide x 83" high		13.30	1.203		97.50	40.50		138	175
0180	Wall hung, 3" wide x 58" high		18.90	.847		87.50	28.50		116	144
0200	Panels									
0210	31" wide x 58" high	2 Carp	18.90	.847	Ea.	143	28.50		171.50	206
0230	51" wide x 58" high		18.90	.847		193	28.50		221.50	260
0250	63" wide x 58" high		18.90	.847		223	28.50		251.50	293
0300	Doors									
0310	24" wide x 58" high	2 Carp	14.10	1.135	Ea.	133	38		171	211
0330	26" wide x 58" high		14.10	1.135		138	38		176	215
0350	28" wide x 58" high		14.10	1.135		143	38		181	221
0370	36" wide x 58" high		14.10	1.135		170	38		208	251
0400	Headrails									
0410	62" long	2 Carp	65	.246	Ea.	25.50	8.25		33.75	42
0430	84" long		60	.267		31	8.95		39.95	49
0450	120" long		30	.533		44.50	17.90		62.40	79

10 21 13.19 Plastic Toilet Compartments

		Crew	Daily Output	Labor-Hours	Unit	Material	Labor	Equipment	Total	Total Incl O&P
0010	**PLASTIC TOILET COMPARTMENTS**									
0110	Cubicles, ceiling hung									
0250	Phenolic	2 Carp	4	4	Ea.	805	134		939	1,100
0600	For handicap units, incl. 52" grab bars, add				"	410			410	450
0900	Floor and ceiling anchored									
1050	Phenolic	2 Carp	5	3.200	Ea.	805	107		912	1,075
1400	For handicap units, incl. 52" grab bars, add				"	287			287	315

10 21 Compartments and Cubicles

10 21 13 – Toilet Compartments

10 21 13.19 Plastic Toilet Compartments		Crew	Daily Output	Labor-Hours	Unit	2011 Bare Costs Material	Labor	Equipment	Total	Total Incl O&P
1610	Floor mounted									
1750	Phenolic	2 Carp	7	2.286	Ea.	1,350	76.50		1,426.50	1,600
2100	For handicap units, incl. 52″ grab bars, add					285			285	315
2200	For juvenile units, deduct					41.50			41.50	45.50
2450	Floor mounted, headrail braced									
2550	Phenolic	2 Carp	6	2.667	Ea.	895	89.50		984.50	1,125

10 21 13.20 Plastic Toilet Compartment Components

Line	Description	Crew	Daily Output	Labor-Hours	Unit	Material	Labor	Equipment	Total	Total Incl O&P
0010	**PLASTIC TOILET COMPARTMENT COMPONENTS**									
0100	Pilasters									
0110	Overhead braced, polymer plastic, 7″ wide x 82″ high	2 Carp	22.20	.721	Ea.	128	24		152	182
0120	Phenolic		22.20	.721		136	24		160	191
0130	Floor braced, polymer plastic, 7″ wide x 70″ high		23.30	.687		170	23		193	226
0140	Phenolic		23.30	.687		128	23		151	180
0150	Ceiling hung, polymer plastic, 7″ wide x 83″ high		13.30	1.203		151	40.50		191.50	234
0160	Phenolic		13.30	1.203		145	40.50		185.50	228
0180	Wall hung, phenolic, 3″ wide x 58″ high		18.90	.847		90.50	28.50		119	147
0200	Panels									
0210	Polymer plastic, 31″ high x 55″ high	2 Carp	18.90	.847	Ea.	289	28.50		317.50	370
0220	Phenolic, 31″ wide x 58″ high		18.90	.847		251	28.50		279.50	325
0230	Polymer plastic, 51″ wide x 55″ high		18.90	.847		435	28.50		463.50	530
0240	Phenolic, 51″ wide x 58″ high		18.90	.847		375	28.50		403.50	460
0250	Polymer plastic, 63″ wide x 55″ high		18.90	.847		520	28.50		548.50	620
0260	Phenolic, 63″ wide x 58″ high		18.90	.847		445	28.50		473.50	540
0300	Doors									
0310	Polymer plastic, 24″ wide x 55″ high	2 Carp	14.10	1.135	Ea.	200	38		238	284
0320	Phenolic, 24″ wide x 58″ high		14.10	1.135		274	38		312	365
0330	Polymer plastic, 26″ high x 55″ high		14.10	1.135		213	38		251	298
0340	Phenolic, 26″ wide x 58″ high		14.10	1.135		291	38		329	385
0350	Polymer plastic, 28″ wide x 55″ high		14.10	1.135		225	38		263	310
0360	Phenolic, 28″ wide x 58″ high		14.10	1.135		310	38		348	405
0370	Polymer plastic, 36″ wide x 55″ high		14.10	1.135		285	38		323	380
0380	Phenolic, 36″ wide x 58″ high		14.10	1.135		380	38		418	485
0400	Headrails									
0410	For polymer plastic, 62″ long	2 Carp	65	.246	Ea.	32	8.25		40.25	49.50
0420	Phenolic		65	.246		23	8.25		31.25	39.50
0430	For polymer plastic, 84″ long		50	.320		36	10.75		46.75	58
0440	Phenolic		50	.320		29	10.75		39.75	50
0450	For polymer plastic, 120″ long		30	.533		51	17.90		68.90	86
0460	Phenolic		30	.533		39	17.90		56.90	73

10 21 13.40 Stone Toilet Compartments

Line	Description	Crew	Daily Output	Labor-Hours	Unit	Material	Labor	Equipment	Total	Total Incl O&P
0010	**STONE TOILET COMPARTMENTS**									
0100	Cubicles, ceiling hung, marble	2 Marb	2	8	Ea.	1,450	254		1,704	2,025
0600	For handicap units, incl. 52″ grab bars, add					410			410	450
0800	Floor & ceiling anchored, marble	2 Marb	2.50	6.400		1,825	204		2,029	2,325
1400	For handicap units, incl. 52″ grab bars, add					287			287	315
1600	Floor mounted, marble	2 Marb	3	5.333		1,075	170		1,245	1,450
2401	Floor mounted, headrail braced, marble	2 Ston	3	5.333		1,025	175		1,200	1,400
2900	For handicap units, incl. 52″ grab bars, add					330			330	365
4100	Entrance screen, floor mounted marble, 58″ high, 48″ wide	2 Marb	9	1.778		675	56.50		731.50	840
4600	Urinal screen, 18″ wide, ceiling braced, marble	D-1	6	2.667		675	81		756	880
5100	Floor mounted, head rail braced									
5200	Marble	D-1	6	2.667	Ea.	575	81		656	765

10 21 Compartments and Cubicles

10 21 13 – Toilet Compartments

10 21 13.40 Stone Toilet Compartments		Crew	Daily Output	Labor-Hours	Unit	2011 Bare Costs Material	Labor	Equipment	Total	Total Incl O&P
5700	Pilaster, flush, marble	D-1	9	1.778	Ea.	750	54		804	915
6200	Post braced, marble		9	1.778		745	54		799	910

10 21 16 – Shower and Dressing Compartments

10 21 16.10 Partitions, Shower		Crew	Daily Output	Labor-Hours	Unit	2011 Bare Costs Material	Labor	Equipment	Total	Total Incl O&P
0010	**PARTITIONS, SHOWER** Floor mounted, no plumbing									
0400	Cabinet, one piece, fiberglass, 32″ x 32″	2 Carp	5	3.200	Ea.	475	107		582	700
0420	36″ x 36″		5	3.200		530	107		637	760
0440	36″ x 48″		5	3.200		1,375	107		1,482	1,700
0460	Acrylic, 32″ x 32″		5	3.200		325	107		432	535
0480	36″ x 36″		5	3.200		980	107		1,087	1,250
0500	36″ x 48″		5	3.200		2,100	107		2,207	2,500
0520	Shower door for above, clear plastic, 24″ wide	1 Carp	8	1		162	33.50		195.50	234
0540	28″ wide		8	1		184	33.50		217.50	258
0560	Tempered glass, 24″ wide		8	1		175	33.50		208.50	248
0580	28″ wide		8	1		197	33.50		230.50	273
2400	Glass stalls, with doors, no receptors, chrome on brass	2 Shee	3	5.333		1,525	204		1,729	2,000
2700	Anodized aluminum	"	4	4		1,050	153		1,203	1,400
2901	Marble shower stall, stock design, with shower door	2 Ston	1.20	13.333		2,275	440		2,715	3,225
3001	With curtain	"	1.30	12.308		2,000	405		2,405	2,875
3200	Receptors, precast terrazzo, 32″ x 32″	2 Marb	14	1.143		147	36.50		183.50	222
3300	48″ x 34″	"	9.50	1.684		203	53.50		256.50	310
3501	Plastic, simulated terrazzo receptor, 32″ x 32″	2 Ston	14	1.143		120	37.50		157.50	194
3600	32″ x 48″	2 Marb	12	1.333		206	42.50		248.50	296
3801	Precast concrete, colors, 32″ x 32″	2 Ston	14	1.143		225	37.50		262.50	310
3900	48″ x 48″	2 Marb	8	2		400	63.50		463.50	545
4100	Shower doors, economy plastic, 24″ wide	1 Shee	9	.889		132	34		166	201
4200	Tempered glass door, economy		8	1		218	38		256	305
4400	Folding, tempered glass, aluminum frame		6	1.333		395	51		446	520
4500	Sliding, tempered glass, 48″ opening		6	1.333		263	51		314	375
4700	Deluxe, tempered glass, chrome on brass frame, minimum		8	1		235	38		273	320
4800	Maximum		1	8		910	305		1,215	1,500
4850	On anodized aluminum frame, minimum		2	4		148	153		301	415
4900	Maximum		1	8		535	305		840	1,100
5100	Shower enclosure, tempered glass, anodized alum. frame									
5120	2 panel & door, corner unit, 32″ x 32″	1 Shee	2	4	Ea.	520	153		673	825
5140	Neo-angle corner unit, 16″ x 24″ x 16″	"	2	4		960	153		1,113	1,300
5200	Shower surround, 3 wall, polypropylene, 32″ x 32″	1 Carp	4	2		385	67		452	535
5220	PVC, 32″ x 32″		4	2		350	67		417	495
5240	Fiberglass		4	2		385	67		452	530
5250	2 wall, polypropylene, 32″ x 32″		4	2		290	67		357	430
5270	PVC		4	2		360	67		427	505
5290	Fiberglass		4	2		365	67		432	515
5300	Tub doors, tempered glass & frame, minimum	1 Shee	8	1		210	38		248	294
5400	Maximum		6	1.333		490	51		541	625
5600	Chrome plated, brass frame, minimum		8	1		278	38		316	370
5700	Maximum		6	1.333		680	51		731	835
5900	Tub/shower enclosure, temp. glass, alum. frame, minimum		2	4		375	153		528	665
6200	Maximum		1.50	5.333		785	204		989	1,200
6500	On chrome-plated brass frame, minimum		2	4		520	153		673	820
6600	Maximum		1.50	5.333		1,100	204		1,304	1,550
6800	Tub surround, 3 wall, polypropylene	1 Carp	4	2		235	67		302	370
6900	PVC		4	2		355	67		422	505

10 21 Compartments and Cubicles

10 21 16 – Shower and Dressing Compartments

10 21 16.10 Partitions, Shower		Crew	Daily Output	Labor-Hours	Unit	Material	Labor	Equipment	Total	Total Incl O&P
						2011 Bare Costs				
7000	Fiberglass, minimum	1 Carp	4	2	Ea.	370	67		437	515
7100	Maximum	↓	3	2.667	↓	625	89.50		714.50	835

10 21 23 – Cubicle Curtains and Track

10 21 23.16 Cubicle Track and Hardware

		Crew	Daily Output	Labor-Hours	Unit	Material	Labor	Equipment	Total	Total Incl O&P
0010	**CUBICLE TRACK AND HARDWARE**									
0020	Curtain track, box channel, ceiling mounted	1 Carp	135	.059	L.F.	6	1.99		7.99	9.95
0100	Suspended	"	100	.080	"	7.60	2.68		10.28	12.85
0300	Curtains, nylon mesh tops, fire resistant, 11 oz. per lineal yard									
0310	Polyester oxford cloth, 9′ ceiling height	1 Carp	425	.019	L.F.	12	.63		12.63	14.25
0500	8′ ceiling height		425	.019		10.15	.63		10.78	12.20
0700	Designer oxford cloth	↓	425	.019	↓	25.50	.63		26.13	29.50
0800	I.V. track systems									
0820	I.V. track, oval	1 Carp	135	.059	L.F.	5.25	1.99		7.24	9.15
0830	I.V. trolley		32	.250	Ea.	36	8.40		44.40	54
0840	I.V. pendent, (tree, 5 hook)	↓	32	.250	"	129	8.40		137.40	156

10 22 Partitions

10 22 13 – Wire Mesh Partitions

10 22 13.10 Partitions, Woven Wire

		Crew	Daily Output	Labor-Hours	Unit	Material	Labor	Equipment	Total	Total Incl O&P
0010	**PARTITIONS, WOVEN WIRE** For tool or stockroom enclosures									
0100	Channel frame, 1-1/2″ diamond mesh, 10 ga. wire, painted									
0300	Wall panels, 4′-0″ wide, 7′ high	2 Carp	25	.640	Ea.	126	21.50		147.50	175
0400	8′ high		23	.696		147	23.50		170.50	201
0600	10′ high	↓	18	.889		173	30		203	240
0700	For 5′ wide panels, add					5%				
0900	Ceiling panels, 10′ long, 2′ wide	2 Carp	25	.640		130	21.50		151.50	179
1000	4′ wide		15	1.067		164	36		200	240
1200	Panel with service window & shelf, 5′ wide, 7′ high		20	.800		405	27		432	490
1300	8′ high		15	1.067		400	36		436	500
1500	Sliding doors, full height, 3′ wide, 7′ high		6	2.667		420	89.50		509.50	610
1600	10′ high		5	3.200		375	107		482	590
1800	6′ wide sliding door, 7′ full height		5	3.200		575	107		682	810
1900	10′ high		4	4		710	134		844	1,000
2100	Swinging doors, 3′ wide, 7′ high, no transom		6	2.667		265	89.50		354.50	440
2200	7′ high, 3′ transom	↓	5	3.200	↓	320	107		427	530

10 22 16 – Folding Gates

10 22 16.10 Security Gates

		Crew	Daily Output	Labor-Hours	Unit	Material	Labor	Equipment	Total	Total Incl O&P
0010	**SECURITY GATES** For roll up type, see Section 08 33 13.10									
0300	Scissors type folding gate, ptd. steel, single, 6-1/2′ high, 5-1/2′wide	2 Sswk	4	4	Opng.	196	146		342	495
0350	6-1/2′ wide		4	4		218	146		364	515
0400	7-1/2′ wide		4	4		200	146		346	500
0600	Double gate, 8′ high, 8′ wide		2.50	6.400		345	233		578	825
0650	10′ wide		2.50	6.400		375	233		608	860
0700	12′ wide		2	8		550	291		841	1,150
0750	14′ wide		2	8		630	291		921	1,250
0900	Door gate, folding steel, 4′ wide, 61″ high		4	4		121	146		267	410
1000	71″ high		4	4		109	146		255	400
1200	81″ high		4	4		118	146		264	410
1300	Window gates, 2′ to 4′ wide, 31″ high		4	4		59.50	146		205.50	345
1500	55″ high	↓	3.75	4.267	↓	93	155		248	400

10 22 Partitions

10 22 16 – Folding Gates

10 22 16.10 Security Gates		Crew	Daily Output	Labor-Hours	Unit	Material	Labor (2011 Bare Costs)	Equipment	Total	Total Incl O&P
1600	79″ high	2 Sswk	3.50	4.571	Opng.	108	166		274	435

10 22 19 – Demountable Partitions

10 22 19.43 Demountable Composite Partitions

		Crew	Daily Output	Labor-Hours	Unit	Material	Labor	Equipment	Total	Total Incl O&P
0010	**DEMOUNTABLE COMPOSITE PARTITIONS**, add for doors									
0100	Do not deduct door openings from total L.F.									
0900	Demountable gypsum system on 2″ to 2-1/2″									
1000	steel studs, 9′ high, 3″ to 3-3/4″ thick									
1200	Vinyl clad gypsum	2 Carp	48	.333	L.F.	58	11.20		69.20	82
1300	Fabric clad gypsum		44	.364		144	12.20		156.20	179
1500	Steel clad gypsum		40	.400		156	13.40		169.40	195
1600	1.75 system, aluminum framing, vinyl clad hardboard,									
1800	paper honeycomb core panel, 1-3/4″ to 2-1/2″ thick									
1900	9′ high	2 Carp	48	.333	L.F.	97	11.20		108.20	126
2100	7′ high		60	.267		87	8.95		95.95	110
2200	5′ high		80	.200		73.50	6.70		80.20	92.50
2250	Unitized gypsum system									
2300	Unitized panel, 9′ high, 2″ to 2-1/2″ thick									
2350	Vinyl clad gypsum	2 Carp	48	.333	L.F.	125	11.20		136.20	156
2400	Fabric clad gypsum	″	44	.364	″	206	12.20		218.20	247
2500	Unitized mineral fiber system									
2510	Unitized panel, 9′ high, 2-1/4″ thick, aluminum frame									
2550	Vinyl clad mineral fiber	2 Carp	48	.333	L.F.	124	11.20		135.20	156
2600	Fabric clad mineral fiber	″	44	.364	″	185	12.20		197.20	225
2800	Movable steel walls, modular system									
2900	Unitized panels, 9′ high, 48″ wide									
3100	Baked enamel, pre-finished	2 Carp	60	.267	L.F.	141	8.95		149.95	170
3200	Fabric clad steel		56	.286	″	203	9.60		212.60	240
5310	Trackless wall, cork finish, semi-acoustic, 1-5/8″ thick, minimum		325	.049	S.F.	38.50	1.65		40.15	45.50
5320	Maximum		190	.084		36.50	2.83		39.33	44.50
5330	Acoustic, 2″ thick, minimum		305	.052		31	1.76		32.76	37
5340	Maximum		225	.071		53.50	2.39		55.89	63
5500	For acoustical partitions, add, minimum					2.22			2.22	2.44
5550	Maximum					10.35			10.35	11.40
5700	For doors, see Sections 08 11 & 08 16									
5800	For door hardware, see Section 08 71									
6100	In-plant modular office system, w/prehung hollow core door									
6200	3″ thick polystyrene core panels									
6250	12′ x 12′, 2 wall	2 Clab	3.80	4.211	Ea.	3,700	111		3,811	4,250
6300	4 wall		1.90	8.421		4,700	223		4,923	5,550
6350	16′ x 16′, 2 wall		3.60	4.444		4,925	118		5,043	5,600
6400	4 wall		1.80	8.889		6,625	235		6,860	7,675

10 22 23 – Portable Partitions, Screens, and Panels

10 22 23.13 Wall Screens

		Crew	Daily Output	Labor-Hours	Unit	Material	Labor	Equipment	Total	Total Incl O&P
0010	**WALL SCREENS**, divider panels, free standing, fiber core									
0020	Fabric face straight									
0100	3′-0″ long, 4′-0″ high	2 Carp	100	.160	L.F.	146	5.35		151.35	170
0200	5′-0″ high		90	.178		103	5.95		108.95	123
0500	6′-0″ high		75	.213		120	7.15		127.15	144
0900	5′-0″ long, 4′-0″ high		175	.091		90.50	3.07		93.57	105
1000	5′-0″ high		150	.107		98.50	3.58		102.08	114
1500	6″-0″ high		125	.128		114	4.29		118.29	133
1600	6′-0″ long, 5′-0″ high		162	.099		95.50	3.31		98.81	111

10 22 Partitions

10 22 23 – Portable Partitions, Screens, and Panels

10 22 23.13 Wall Screens		Crew	Daily Output	Labor-Hours	Unit	Material	2011 Bare Costs Labor	Equipment	Total	Total Incl O&P
3200	Economical panels, fabric face, 4'-0" long, 5'-0" high	2 Carp	132	.121	L.F.	51.50	4.07		55.57	63.50
3250	6'-0" high		112	.143		54	4.79		58.79	67
3300	5'-0" long, 5'-0" high		150	.107		44	3.58		47.58	54.50
3350	6'-0" high		125	.128		46	4.29		50.29	57.50
3450	Acoustical panels, 60 to 90 NRC, 3'-0" long, 5'-0" high		90	.178		89.50	5.95		95.45	109
3550	6'-0" high		75	.213		99.50	7.15		106.65	121
3600	5'-0" long, 5'-0" high		150	.107		55.50	3.58		59.08	67.50
3650	6'-0" high		125	.128		68	4.29		72.29	81.50
3700	6'-0" long, 5'-0" high		162	.099		51.50	3.31		54.81	62
3750	6'-0" high		138	.116		56	3.89		59.89	68
3800	Economy acoustical panels, 40 N.R.C., 4'-0" long, 5'-0" high		132	.121		51.50	4.07		55.57	63.50
3850	6'-0" high		112	.143		54	4.79		58.79	67
3900	5'-0" long, 6'-0" high		125	.128		46	4.29		50.29	57.50
3950	6'-0" long, 5'-0" high		162	.099		40.50	3.31		43.81	50
4000	Metal chalkboard, 6'-6" high, chalkboard, 1 side		125	.128		106	4.29		110.29	123
4100	Metal chalkboard, 2 sides		120	.133		121	4.47		125.47	141
4300	Tackboard, both sides		123	.130		95.50	4.36		99.86	112

10 22 33 – Accordion Folding Partitions

10 22 33.10 Partitions, Accordion Folding		Crew	Daily Output	Labor-Hours	Unit	Material	Labor	Equipment	Total	Total Incl O&P
0010	**PARTITIONS, ACCORDION FOLDING**									
0100	Vinyl covered, over 150 S.F., frame not included									
0300	Residential, 1.25 lb. per S.F., 8' maximum height	2 Carp	300	.053	S.F.	19.75	1.79		21.54	24.50
0400	Commercial, 1.75 lb. per S.F., 8' maximum height		225	.071		22.50	2.39		24.89	29
0600	2 lb. per S.F., 17' maximum height		150	.107		23.50	3.58		27.08	31.50
0700	Industrial, 4 lb. per S.F., 20' maximum height		75	.213		33.50	7.15		40.65	48.50
0900	Acoustical, 3 lb. per S.F., 17' maximum height		100	.160		26	5.35		31.35	38
1200	5 lb. per S.F., 20' maximum height		95	.168		36	5.65		41.65	49.50
1300	5.5 lb. per S.F., 17' maximum height		90	.178		42.50	5.95		48.45	57
1400	Fire rated, 4.5 psf, 20' maximum height		160	.100		42.50	3.36		45.86	52.50
1500	Vinyl clad wood or steel, electric operation, 5.0 psf		160	.100		49.50	3.36		52.86	59.50
1900	Wood, non-acoustic, birch or mahogany, to 10' high		300	.053		26.50	1.79		28.29	32

10 22 39 – Folding Panel Partitions

10 22 39.10 Partitions, Folding Panel		Crew	Daily Output	Labor-Hours	Unit	Material	Labor	Equipment	Total	Total Incl O&P
0010	**PARTITIONS, FOLDING PANEL**, acoustic, wood									
0100	Vinyl faced, to 18' high, 6 psf, minimum	2 Carp	60	.267	S.F.	51.50	8.95		60.45	71.50
0150	Average		45	.356		61.50	11.95		73.45	87.50
0200	Maximum		30	.533		79.50	17.90		97.40	118
0400	Formica or hardwood finish, minimum		60	.267		53	8.95		61.95	73.50
0500	Maximum		30	.533		56.50	17.90		74.40	92.50
0600	Wood, low acoustical type, 4.5 psf, to 14' high		50	.320		38.50	10.75		49.25	60.50
1100	Steel, acoustical, 9 to 12 lb. per S.F., vinyl faced, minimum		60	.267		55	8.95		63.95	75.50
1200	Maximum		30	.533		67	17.90		84.90	104
1700	Aluminum framed, acoustical, to 12' high, 5.5 psf, minimum		60	.267		37	8.95		45.95	55.50
1800	Maximum		30	.533		44.50	17.90		62.40	79
2000	6.5 lb. per S.F., minimum		60	.267		39	8.95		47.95	58
2100	Maximum		30	.533		48	17.90		65.90	83

10 22 43 – Sliding Partitions

10 22 43.10 Partitions, Sliding		Crew	Daily Output	Labor-Hours	Unit	Material	Labor	Equipment	Total	Total Incl O&P
0010	**PARTITIONS, SLIDING**									
0020	Acoustic air wall, 1-5/8" thick, minimum	2 Carp	375	.043	S.F.	31.50	1.43		32.93	37
0100	Maximum		365	.044		54	1.47		55.47	62

10 22 Partitions

10 22 43 – Sliding Partitions

10 22 43.10 Partitions, Sliding		Crew	Daily Output	Labor-Hours	Unit	2011 Bare Costs Material	Labor	Equipment	Total	Total Incl O&P
0300	2-1/4" thick, minimum	2 Carp	360	.044	S.F.	35.50	1.49		36.99	41.50
0400	Maximum	↓	330	.048	↓	62	1.63		63.63	70.50
0600	For track type, add to above				L.F.	115			115	127
0700	Overhead track type, acoustical, 3" thick, 11 psf, minimum	2 Carp	350	.046	S.F.	79.50	1.53		81.03	90
0800	Maximum	"	300	.053	"	95.50	1.79		97.29	108

10 26 Wall and Door Protection

10 26 13 – Corner Guards

10 26 13.10 Metal Corner Guards

Code	Description	Crew	Daily Output	Labor-Hours	Unit	Material	Labor	Equipment	Total	Total Incl O&P
0010	**METAL CORNER GUARDS**									
0020	Steel angle w/anchors, 1" x 1" x 1/4", 1.5#/L.F.	2 Carp	160	.100	L.F.	6.85	3.36		10.21	13.15
0101	2" x 2" x 1/4" angles, 3.2#/L.F.		150	.107		9.60	3.58		13.18	16.55
0201	3" x 3" x 5/16" angles, 6.1#/L.F.		140	.114		16.30	3.83		20.13	24.50
0301	4" x 4" x 5/16" angles, 8.2#/L.F.	↓	120	.133		14.85	4.47		19.32	24
0350	For angles drilled and anchored to masonry, add					15%	120%			
0370	Drilled and anchored to concrete, add					20%	170%			
0400	For galvanized angles, add					35%				
0450	For stainless steel angles, add				↓	100%				
0501	Steel door track/wheel guard, 4' - 0" high	2 Carp	12	1.333	Ea.	153	44.50		197.50	243
0801	Pipe bumper for truck doors, conc. filled, 8' long, 6" diam.		10	1.600		665	53.50		718.50	820
0901	8" diameter	↓	10	1.600	↓	980	53.50		1,033.50	1,175

10 26 13.20 Corner Protection

Code	Description	Crew	Daily Output	Labor-Hours	Unit	Material	Labor	Equipment	Total	Total Incl O&P
0010	**CORNER PROTECTION**									
0101	Stainless steel, 16 ga., adhesive mount, 3-1/2" leg	1 Carp	80	.100	L.F.	18.60	3.36		21.96	26
0201	12 ga. stainless, adhesive mount	"	80	.100		21	3.36		24.36	28.50
0300	For screw mount, add						10%			
0500	Vinyl acrylic, adhesive mount, 3" leg	1 Carp	128	.063		7.75	2.10		9.85	12
0550	1-1/2" leg		160	.050		4.38	1.68		6.06	7.65
0600	Screw mounted, 3" leg		80	.100		8.30	3.36		11.66	14.75
0650	1-1/2" leg		100	.080		4.12	2.68		6.80	9
0700	Clear plastic, screw mounted, 2-1/2"		60	.133		4.55	4.47		9.02	12.50
1000	Vinyl cover, alum. retainer, surface mount, 3" x 3"		48	.167		9.75	5.60		15.35	20
1050	2" x 2"		48	.167		9.30	5.60		14.90	19.60
1100	Flush mounted, 3" x 3"		32	.250		19.25	8.40		27.65	35
1150	2" x 2"	↓	32	.250	↓	15.65	8.40		24.05	31.50

10 26 16 – Bumper Guards

10 26 16.10 Wallguard

Code	Description	Crew	Daily Output	Labor-Hours	Unit	Material	Labor	Equipment	Total	Total Incl O&P
0010	**WALLGUARD**									
0400	Rub rail, vinyl, adhesive mounted	1 Carp	185	.043	L.F.	7.70	1.45		9.15	10.90
0500	Neoprene, aluminum backing, 1-1/2" x 2"		110	.073		8.65	2.44		11.09	13.60
1000	Trolley rail, PVC, clipped to wall, 5" high		185	.043		8.65	1.45		10.10	11.95
1050	8" high		180	.044	↓	10.30	1.49		11.79	13.80
1200	Bed bumper, vinyl acrylic, alum. retainer, 21" long		10	.800	Ea.	39.50	27		66.50	88.50
1300	53" long with aligner		9	.889	"	100	30		130	160
1400	Bumper, vinyl cover, alum. retain., cush. mnt., 1-1/2" x 2-3/4"		80	.100	L.F.	13.95	3.36		17.31	21
1500	2" x 4-1/4"		80	.100		20.50	3.36		23.86	28
1600	Surface mounted, 1-3/4" x 3-5/8"		80	.100		11.70	3.36		15.06	18.50
2000	Crash rail, vinyl cover, alum. retainer, 1" x 4"		110	.073		10.80	2.44		13.24	16
2100	1" x 8"		90	.089		16.90	2.98		19.88	23.50
2150	Vinyl inserts, aluminum plate, 1" x 2-1/2"	↓	110	.073	↓	14.60	2.44		17.04	20

10 26 Wall and Door Protection

10 26 16 – Bumper Guards

10 26 16.10 Wallguard		Crew	Daily Output	Labor-Hours	Unit	Material	Labor	Equipment	Total	Total Incl O&P
						2011 Bare Costs				
2200	1" x 5"	1 Carp	90	.089	L.F.	22.50	2.98		25.48	30
3000	Handrail/bumper, vinyl cover, alum. retainer									
3010	Bracket mounted, flat rail, 5-1/2"	1 Carp	80	.100	L.F.	17.55	3.36		20.91	25
3100	6-1/2"		80	.100		22	3.36		25.36	29.50
3200	Bronze bracket, 1-3/4" diam. rail		80	.100		15.85	3.36		19.21	23

10 28 Toilet, Bath, and Laundry Accessories

10 28 13 – Toilet Accessories

10 28 13.13 Commercial Toilet Accessories

Line	Description	Crew	Daily Output	Labor-Hours	Unit	Material	Labor	Equipment	Total	Total Incl O&P
0010	**COMMERCIAL TOILET ACCESSORIES**									
0200	Curtain rod, stainless steel, 5' long, 1" diameter	1 Carp	13	.615	Ea.	34.50	20.50		55	72.50
0300	1-1/4" diameter		13	.615		23.50	20.50		44	60.50
0350	Chrome, 1" diameter		13	.615		25	20.50		45.50	62
0360	For vinyl curtain, add		13	.615	S.F.	.76	20.50		21.26	35.50
0400	Diaper changing station, horizontal, wall mounted, plastic		10	.800	Ea.	224	27		251	291
0500	Dispenser units, combined soap & towel dispensers,									
0510	mirror and shelf, flush mounted	1 Carp	10	.800	Ea.	310	27		337	385
0600	Towel dispenser and waste receptacle,									
0610	18 gallon capacity	1 Carp	10	.800	Ea.	263	27		290	335
0800	Grab bar, straight, 1-1/4" diameter, stainless steel, 18" long		24	.333		29	11.20		40.20	50.50
0900	24" long		23	.348		29.50	11.65		41.15	52
1000	30" long		22	.364		31.50	12.20		43.70	55
1100	36" long		20	.400		39	13.40		52.40	65.50
1105	42" long		20	.400		42	13.40		55.40	69
1200	1-1/2" diameter, 24" long		23	.348		29.50	11.65		41.15	52
1300	36" long		20	.400		34	13.40		47.40	60
1310	42" long		18	.444		43	14.90		57.90	72.50
1500	Tub bar, 1-1/4" diameter, 24" x 36"		14	.571		86	19.15		105.15	127
1600	Plus vertical arm		12	.667		85.50	22.50		108	132
1900	End tub bar, 1" diameter, 90° angle, 16" x 32"		12	.667		96	22.50		118.50	143
2010	Tub/shower/toilet, 2-wall, 36" x 24"		12	.667		91	22.50		113.50	138
2300	Hand dryer, surface mounted, electric, 115 volt, 20 amp		4	2		490	67		557	650
2400	230 volt, 10 amp		4	2		655	67		722	830
2600	Hat and coat strip, stainless steel, 4 hook, 36" long		24	.333		71	11.20		82.20	97
2700	6 hook, 60" long		20	.400		112	13.40		125.40	146
3000	Mirror, with stainless steel 3/4" square frame, 18" x 24"		20	.400		47.50	13.40		60.90	74.50
3100	36" x 24"		15	.533		101	17.90		118.90	141
3200	48" x 24"		10	.800		151	27		178	211
3300	72" x 24"		6	1.333		196	44.50		240.50	290
3500	With 5" stainless steel shelf, 18" x 24"		20	.400		188	13.40		201.40	230
3600	36" x 24"		15	.533		195	17.90		212.90	245
3700	48" x 24"		10	.800		214	27		241	281
3800	72" x 24"		6	1.333		264	44.50		308.50	365
4100	Mop holder strip, stainless steel, 5 holders, 48" long		20	.400		78	13.40		91.40	109
4200	Napkin/tampon dispenser, recessed		15	.533		525	17.90		542.90	605
4250	Napkin receptacle, recessed		6.50	1.231		143	41.50		184.50	226
4300	Robe hook, single, regular		36	.222		11.30	7.45		18.75	25
4400	Heavy duty, concealed mounting		36	.222		14.75	7.45		22.20	28.50
4600	Soap dispenser, chrome, surface mounted, liquid		20	.400		42	13.40		55.40	69
4700	Powder		20	.400		50.50	13.40		63.90	78
5000	Recessed stainless steel, liquid		10	.800		123	27		150	180

10 28 Toilet, Bath, and Laundry Accessories

10 28 13 – Toilet Accessories

10 28 13.13 Commercial Toilet Accessories		Crew	Daily Output	Labor-Hours	Unit	Material	Labor	Equipment	Total	Total Incl O&P
						2011 Bare Costs				
5100	Powder	1 Carp	10	.800	Ea.	200	27		227	265
5300	Soap tank, stainless steel, 1 gallon		10	.800		200	27		227	265
5400	5 gallon		5	1.600		275	53.50		328.50	390
5600	Shelf, stainless steel, 5" wide, 18 ga., 24" long		24	.333		53.50	11.20		64.70	77.50
5700	48" long		16	.500		108	16.80		124.80	147
5800	8" wide shelf, 18 ga., 24" long		22	.364		61.50	12.20		73.70	88
5900	48" long		14	.571		119	19.15		138.15	162
6000	Toilet seat cover dispenser, stainless steel, recessed		20	.400		160	13.40		173.40	199
6050	Surface mounted		15	.533		30	17.90		47.90	63
6100	Toilet tissue dispenser, surface mounted, SS, single roll		30	.267		16.95	8.95		25.90	33.50
6200	Double roll		24	.333		22.50	11.20		33.70	43.50
6240	Plastic, twin/jumbo dbl. roll		24	.333		27.50	11.20		38.70	49
6400	Towel bar, stainless steel, 18" long		23	.348		39.50	11.65		51.15	63
6500	30" long		21	.381		111	12.80		123.80	145
6700	Towel dispenser, stainless steel, surface mounted		16	.500		33.50	16.80		50.30	64.50
6800	Flush mounted, recessed		10	.800		233	27		260	300
6900	Plastic, touchless, battery operated		16	.500		76	16.80		92.80	112
7000	Towel holder, hotel type, 2 guest size		20	.400		43.50	13.40		56.90	70.50
7200	Towel shelf, stainless steel, 24" long, 8" wide		20	.400		56	13.40		69.40	84
7400	Tumbler holder, tumbler only		30	.267		41.50	8.95		50.45	60.50
7500	Soap, tumbler & toothbrush		30	.267		18.30	8.95		27.25	35
7700	Wall urn ash receiver, surface mount, 11" long		12	.667		72	22.50		94.50	117
7800	7-1/2", long		18	.444		104	14.90		118.90	139
8000	Waste receptacles, stainless steel, with top, 13 gallon		10	.800		270	27		297	340
8100	36 gallon	↓	8	1	↓	325	33.50		358.50	415

10 28 16 – Bath Accessories

10 28 16.20 Medicine Cabinets		Crew	Daily Output	Labor-Hours	Unit	Material	Labor	Equipment	Total	Total Incl O&P
0010	**MEDICINE CABINETS**									
0020	With mirror, st. st. frame, 16" x 22", unlighted	1 Carp	14	.571	Ea.	64.50	19.15		83.65	103
0100	Wood frame		14	.571		126	19.15		145.15	171
0300	Sliding mirror doors, 20" x 16" x 4-3/4", unlighted		7	1.143		110	38.50		148.50	185
0400	24" x 19" x 8-1/2", lighted		5	1.600		159	53.50		212.50	265
0600	Triple door, 30" x 32", unlighted, plywood body		7	1.143		280	38.50		318.50	375
0700	Steel body		7	1.143		310	38.50		348.50	405
0900	Oak door, wood body, beveled mirror, single door		7	1.143		180	38.50		218.50	262
1000	Double door		6	1.333		375	44.50		419.50	485
1200	Hotel cabinets, stainless, with lower shelf, unlighted		10	.800		202	27		229	267
1300	Lighted	↓	5	1.600	↓	300	53.50		353.50	420

10 28 23 – Laundry Accessories

10 28 23.13 Built-In Ironing Boards		Crew	Daily Output	Labor-Hours	Unit	Material	Labor	Equipment	Total	Total Incl O&P
0010	**BUILT-IN IRONING BOARDS**									
0020	Including cabinet, board & light, minimum	1 Carp	2	4	Ea.	315	134		449	570
0100	Maximum, see also Section 11 23 13.13	"	1.50	5.333	"	525	179		704	880

10 31 Manufactured Fireplaces

10 31 13 – Manufactured Fireplace Chimneys

10 31 13.10 Fireplace Chimneys		Crew	Daily Output	Labor-Hours	Unit	Material	2011 Bare Costs Labor	Equipment	Total	Total Incl O&P
0010	**FIREPLACE CHIMNEYS**									
0500	Chimney dbl. wall, all stainless, over 8′-6″, 7″ diam., add to fireplace	1 Carp	33	.242	V.L.F.	72	8.15		80.15	92.50
0600	10″ diameter, add to fireplace		32	.250		106	8.40		114.40	131
0700	12″ diameter, add to fireplace		31	.258		139	8.65		147.65	167
0800	14″ diameter, add to fireplace		30	.267		183	8.95		191.95	216
1000	Simulated brick chimney top, 4′ high, 16″ x 16″		10	.800	Ea.	258	27		285	330
1100	24″ x 24″		7	1.143	″	480	38.50		518.50	595

10 31 13.20 Chimney Accessories

		Crew	Daily Output	Labor-Hours	Unit	Material	Labor	Equipment	Total	Total Incl O&P
0010	**CHIMNEY ACCESSORIES**									
0020	Chimney screens, galv., 13″ x 13″ flue	1 Bric	8	1	Ea.	51	33.50		84.50	111
0050	24″ x 24″ flue		5	1.600		118	54		172	219
0200	Stainless steel, 13″ x 13″ flue		8	1		315	33.50		348.50	400
0250	20″ x 20″ flue		5	1.600		425	54		479	560
2400	Squirrel and bird screens, galvanized, 8″ x 8″ flue		16	.500		44.50	16.85		61.35	76.50
2450	13″ x 13″ flue		12	.667		48	22.50		70.50	90

10 31 16 – Manufactured Fireplace Forms

10 31 16.10 Fireplace Forms

		Crew	Daily Output	Labor-Hours	Unit	Material	Labor	Equipment	Total	Total Incl O&P
0010	**FIREPLACE FORMS**									
1800	Fireplace forms, no accessories, 32″ opening	1 Bric	3	2.667	Ea.	650	89.50		739.50	860
1900	36″ opening		2.50	3.200		830	108		938	1,075
2000	40″ opening		2	4		1,100	135		1,235	1,425
2100	78″ opening		1.50	5.333		1,600	179		1,779	2,050

10 31 23 – Prefabricated Fireplaces

10 31 23.10 Fireplace, Prefabricated

		Crew	Daily Output	Labor-Hours	Unit	Material	Labor	Equipment	Total	Total Incl O&P
0010	**FIREPLACE, PREFABRICATED**, free standing or wall hung									
0100	With hood & screen, minimum	1 Carp	1.30	6.154	Ea.	1,200	206		1,406	1,650
0150	Average		1	8		1,675	268		1,943	2,300
0200	Maximum		.90	8.889		3,475	298		3,773	4,325
1500	Simulated logs, gas fired, 40,000 BTU, 2′ long, minimum		7	1.143	Set	445	38.50		483.50	555
1600	Maximum		6	1.333		1,025	44.50		1,069.50	1,200
1700	Electric, 1,500 BTU, 1′-6″ long, minimum		7	1.143		175	38.50		213.50	257
1800	11,500 BTU, maximum		6	1.333		395	44.50		439.50	505
2000	Fireplace, built-in, 36″ hearth, radiant		1.30	6.154	Ea.	565	206		771	965
2100	Recirculating, small fan		1	8		900	268		1,168	1,450
2150	Large fan		.90	8.889		1,775	298		2,073	2,475
2200	42″ hearth, radiant		1.20	6.667		800	224		1,024	1,250
2300	Recirculating, small fan		.90	8.889		1,075	298		1,373	1,700
2350	Large fan		.80	10		1,775	335		2,110	2,525
2400	48″ hearth, radiant		1.10	7.273		1,675	244		1,919	2,225
2500	Recirculating, small fan		.80	10		2,075	335		2,410	2,825
2550	Large fan		.70	11.429		3,200	385		3,585	4,175
3000	See through, including doors		.80	10		2,725	335		3,060	3,550
3200	Corner (2 wall)		1	8		2,725	268		2,993	3,450

10 32 Fireplace Specialties

10 32 13 – Fireplace Dampers

10 32 13.10 Dampers		Crew	Daily Output	Labor-Hours	Unit	Material	2011 Bare Costs Labor	Equipment	Total	Total Incl O&P
0010	**DAMPERS**									
0800	Damper, rotary control, steel, 30" opening	1 Bric	6	1.333	Ea.	75.50	45		120.50	157
0850	Cast iron, 30" opening		6	1.333		84	45		129	166
0880	36" opening		6	1.333		91.50	45		136.50	175
0900	48" opening		6	1.333		131	45		176	218
0920	60" opening		6	1.333		292	45		337	395
0950	72" opening		5	1.600		350	54		404	475
1000	84" opening, special order		5	1.600		755	54		809	920
1050	96" opening, special order		4	2		765	67.50		832.50	950
1200	Steel plate, poker control, 60" opening		8	1		267	33.50		300.50	350
1250	84" opening, special order		5	1.600		485	54		539	625
1400	"Universal" type, chain operated, 32" x 20" opening		8	1		207	33.50		240.50	283
1450	48" x 24" opening	↓	5	1.600	↓	310	54		364	430

10 32 23 – Fireplace Doors

10 32 23.10 Doors		Crew	Daily Output	Labor-Hours	Unit	Material	Labor	Equipment	Total	Total Incl O&P
0010	**DOORS**									
0400	Cleanout doors and frames, cast iron, 8" x 8"	1 Bric	12	.667	Ea.	35.50	22.50		58	76
0450	12" x 12"		10	.800		43.50	27		70.50	91.50
0500	18" x 24"		8	1		124	33.50		157.50	192
0550	Cast iron frame, steel door, 24" x 30"		5	1.600		269	54		323	385
1600	Dutch Oven door and frame, cast iron, 12" x 15" opening		13	.615		109	20.50		129.50	154
1650	Copper plated, 12" x 15" opening	↓	13	.615	↓	213	20.50		233.50	268

10 35 Stoves

10 35 13 – Heating Stoves

10 35 13.10 Woodburning Stoves		Crew	Daily Output	Labor-Hours	Unit	Material	Labor	Equipment	Total	Total Incl O&P
0010	**WOODBURNING STOVES**									
0015	Cast iron, minimum	2 Carp	1.30	12.308	Ea.	1,200	415		1,615	2,025
0020	Average		1	16		1,600	535		2,135	2,650
0030	Maximum	↓	.80	20		2,875	670		3,545	4,275
0050	For gas log lighter, add				↓	42			42	46.50

10 44 Fire Protection Specialties

10 44 13 – Fire Protection Cabinets

10 44 13.53 Fire Equipment Cabinets		Crew	Daily Output	Labor-Hours	Unit	Material	Labor	Equipment	Total	Total Incl O&P
0010	**FIRE EQUIPMENT CABINETS**, not equipped, 20 ga. steel box									
0040	recessed, D.S. glass in door, box size given									
1000	Portable extinguisher, single, 8" x 12" x 27", alum. door & frame	Q-12	8	2	Ea.	124	71		195	253
1100	Steel door and frame	"	8	2	"	93	71		164	218
3000	Hose rack assy., 1-1/2" valve & 100' hose, 24" x 40" x 5-1/2"									
3100	Aluminum door and frame	Q-12	6	2.667	Ea.	305	94.50		399.50	490
3200	Steel door and frame		6	2.667		370	94.50		464.50	565
3300	Stainless steel door and frame	↓	6	2.667	↓	400	94.50		494.50	595
4000	Hose rack assy., 2-1/2" x 1-1/2" valve, 100' hose, 24" x 40" x 8"									
4100	Aluminum door and frame	Q-12	6	2.667	Ea.	305	94.50		399.50	490
4200	Steel door and frame		6	2.667		208	94.50		302.50	385
4300	Stainless steel door and frame	↓	6	2.667	↓	410	94.50		504.50	605
5000	Hose rack assy., 2-1/2" x 1-1/2" valve, 100' hose									
5010	and extinguisher, 30" x 40" x 8"									

10 44 Fire Protection Specialties

10 44 13 – Fire Protection Cabinets

10 44 13.53 Fire Equipment Cabinets		Crew	Daily Output	Labor-Hours	Unit	Material	2011 Bare Costs Labor	Equipment	Total	Total Incl O&P
5100	Aluminum door and frame	Q-12	5	3.200	Ea.	390	113		503	615
5200	Steel door and frame		5	3.200		229	113		342	435
5300	Stainless steel door and frame		5	3.200		465	113		578	700
8000	Valve cabinet for 2-1/2" FD angle valve, 18" x 18" x 8"									
8100	Aluminum door and frame	Q-12	12	1.333	Ea.	135	47		182	226
8200	Steel door and frame		12	1.333		111	47		158	200
8300	Stainless steel door and frame		12	1.333		182	47		229	277

10 44 16 – Fire Extinguishers

10 44 16.13 Portable Fire Extinguishers

		Crew	Daily Output	Labor-Hours	Unit	Material	Labor	Equipment	Total	Total Incl O&P
0010	**PORTABLE FIRE EXTINGUISHERS**									
0140	CO_2, with hose and "H" horn, 10 lb.				Ea.	191			191	210
0160	15 lb.					250			250	276
0180	20 lb.					277			277	305
1000	Dry chemical, pressurized									
1040	Standard type, portable, painted, 2-1/2 lb.				Ea.	31.50			31.50	34.50
1060	5 lb.					44			44	48.50
1080	10 lb.					69.50			69.50	76.50
1100	20 lb.					113			113	124
1120	30 lb.					375			375	415
1300	Standard type, wheeled, 150 lb.					1,875			1,875	2,075
2000	ABC all purpose type, portable, 2-1/2 lb.					21			21	23.50
2060	5 lb.					26.50			26.50	29
2080	9-1/2 lb.					44			44	48
2100	20 lb.					79.50			79.50	87.50
5000	Pressurized water, 2-1/2 gallon, stainless steel					97.50			97.50	107
5060	With anti-freeze					106			106	116
9400	Installation of extinguishers, 12 or more, on nailable surface	1 Carp	30	.267			8.95		8.95	14.95
9420	On masonry or concrete	"	15	.533			17.90		17.90	30

10 44 16.16 Wheeled Fire Extinguisher Units

		Crew	Daily Output	Labor-Hours	Unit	Material	Labor	Equipment	Total	Total Incl O&P
0010	**WHEELED FIRE EXTINGUISHER UNITS**									
0350	CO_2, portable, with swivel horn									
0360	Wheeled type, cart mounted, 50 lb.				Ea.	940			940	1,025
0400	100 lb.				"	2,750			2,750	3,025
2200	ABC all purpose type									
2300	Wheeled, 45 lb.				Ea.	735			735	810
2360	150 lb.				"	1,850			1,850	2,025

10 51 Lockers

10 51 13 – Metal Lockers

10 51 13.10 Lockers

		Crew	Daily Output	Labor-Hours	Unit	Material	Labor	Equipment	Total	Total Incl O&P
0011	**LOCKERS** Steel, baked enamel, pre-assembled									
0110	Single tier box locker, 12" x 15" x 72"	1 Shee	20	.400	Ea.	213	15.30		228.30	260
0120	18" x 15" x 72"		20	.400		236	15.30		251.30	285
0130	12" x 18" x 72"		20	.400		243	15.30		258.30	292
0140	18" x 18" x 72"		20	.400		247	15.30		262.30	296
0410	Double tier, 12" x 15" x 36"		30	.267		227	10.20		237.20	267
0420	18" x 15" x 36"		30	.267		263	10.20		273.20	305
0430	12" x 18" x 36"		30	.267		259	10.20		269.20	300
0440	18" x 18" x 36"		30	.267		241	10.20		251.20	282
0500	Two person, 18" x 15" x 72"		20	.400		264	15.30		279.30	315

10 51 Lockers

10 51 13 – Metal Lockers

10 51 13.10 Lockers		Crew	Daily Output	Labor-Hours	Unit	Material	Labor	Equipment	Total	Total Incl O&P
						2011 Bare Costs				
0510	18" x 18" x 72"	1 Shee	20	.400	Ea.	266	15.30		281.30	320
0520	Duplex, 15" x 15" x 72"		20	.400		291	15.30		306.30	345
0530	15" x 21" x 72"		20	.400		315	15.30		330.30	375
0600	5 tier box lockers, minimum		30	.267	Opng.	43.50	10.20		53.70	65
0700	Maximum		24	.333		51.50	12.75		64.25	77.50
0900	6 tier box lockers, minimum		36	.222		36	8.50		44.50	54
1000	Maximum		30	.267		47	10.20		57.20	69
1100	Wire meshed wardrobe, floor. mtd., open front varsity type		7.50	1.067	Ea.	236	41		277	330
2400	16-person locker unit with clothing rack									
2500	72 wide x 15" deep x 72" high	1 Shee	15	.533	Ea.	425	20.50		445.50	500
2550	18" deep	"	15	.533	"	595	20.50		615.50	690
3000	Wall mounted lockers, 4 person, with coat bar									
3100	48" wide x 18" deep x 12" high	1 Shee	20	.400	Ea.	300	15.30		315.30	355
3250	Rack w/24 wire mesh baskets		1.50	5.333	Set	365	204		569	735
3260	30 baskets		1.25	6.400		330	244		574	770
3270	36 baskets		.95	8.421		360	320		680	925
3280	42 baskets		.80	10		415	380		795	1,075
3300	For built-in lock with 2 keys, add				Ea.	12.90			12.90	14.20
3600	For hanger rods, add					1.86			1.86	2.05
3650	For number plate kit, 100 plates #1 - #100, add	1 Shee	4	2		61.50	76.50		138	194
3700	For locker base, closed front panel		90	.089		6.45	3.40		9.85	12.70
3710	End panel, bolted		36	.222		7.85	8.50		16.35	22.50
3800	For sloping top, 12" wide		24	.333		24.50	12.75		37.25	48
3810	15" wide		24	.333		31.50	12.75		44.25	56
3820	18" wide		24	.333		34	12.75		46.75	58.50
3850	Sloping top end panel, 12" deep		72	.111		10.35	4.24		14.59	18.40
3860	15" deep		72	.111		10.50	4.24		14.74	18.55
3870	18" deep		72	.111		10.95	4.24		15.19	19.05
3900	For finish end panels, steel, 60" high, 15" deep		12	.667		50	25.50		75.50	97
3910	72" high, 12" deep		12	.667		50	25.50		75.50	97
3920	18" deep		12	.667		58.50	25.50		84	107
5000	For 'ready to assemble' lockers,									
5010	Add to labor						75%			
5020	Deduct from material					20%				

10 51 26 – Plastic Lockers

10 51 26.13 Recycled Plastic Lockers		Crew	Daily Output	Labor-Hours	Unit	Material	Labor	Equipment	Total	Total Incl O&P
0011	**RECYCLED PLASTIC LOCKERS**, 30% recycled									
0110	Single tier box locker, 12" x 12" x 72" G	1 Shee	8	1	Ea.	435	38		473	545
0120	12" x 15" x 72" G		8	1		420	38		458	525
0130	12" x 18" x 72" G		8	1		350	38		388	450
0410	Double tier, 12" x 12" x 72" G		21	.381		445	14.55		459.55	515
0420	12" x 15" x 72" G		21	.381		400	14.55		414.55	465
0430	12" x 18" x 72" G		21	.381		400	14.55		414.55	465

10 51 53 – Locker Room Benches

10 51 53.10 Benches		Crew	Daily Output	Labor-Hours	Unit	Material	Labor	Equipment	Total	Total Incl O&P
0010	**BENCHES**									
2100	Locker bench, laminated maple, top only	1 Shee	100	.080	L.F.	14.75	3.06		17.81	21.50
2200	Pedestals, steel pipe		25	.320	Ea.	40	12.20		52.20	64
2250	Plastic, 9.5" top with PVC pedestals		80	.100	L.F.	40.50	3.82		44.32	51

10 55 Postal Specialties

10 55 23 – Mail Boxes

10 55 23.10 Commercial Mail Boxes		Crew	Daily Output	Labor-Hours	Unit	2011 Bare Costs Material	Labor	Equipment	Total	Total Incl O&P
0010	**COMMERCIAL MAIL BOXES**									
0020	Horiz., key lock, 5"H x 6"W x 15"D, alum., rear load	1 Carp	34	.235	Ea.	35	7.90		42.90	51.50
0100	Front loading		34	.235		35	7.90		42.90	51.50
0200	Double, 5"H x 12"W x 15"D, rear loading		26	.308		71.50	10.30		81.80	96.50
0300	Front loading		26	.308		66	10.30		76.30	90
0500	Quadruple, 10"H x 12"W x 15"D, rear loading		20	.400		91	13.40		104.40	123
0600	Front loading		20	.400		95.50	13.40		108.90	128
0800	Vertical, front load, 15"H x 5"W x 6"D, alum., per compartment		34	.235		30	7.90		37.90	46
0900	Bronze, duranodic finish		34	.235		35	7.90		42.90	51.50
1000	Steel, enameled		34	.235		30	7.90		37.90	46
1700	Alphabetical directories, 120 names		10	.800		127	27		154	185
1800	Letter collection box		6	1.333		660	44.50		704.50	805
1830	Lobby collection boxes, aluminum	2 Shee	5	3.200		2,025	122		2,147	2,425
1840	Bronze or stainless	"	4.50	3.556		2,500	136		2,636	2,975
1900	Letter slot, residential	1 Carp	20	.400		89.50	13.40		102.90	121
2000	Post office type		8	1		105	33.50		138.50	172
2200	Post office counter window, with grille		2	4		550	134		684	830
2250	Key keeper, single key, aluminum		26	.308		37	10.30		47.30	58.50
2300	Steel, enameled		26	.308		107	10.30		117.30	135

10 56 Storage Assemblies

10 56 13 – Metal Storage Shelving

10 56 13.10 Shelving		Crew	Daily Output	Labor-Hours	Unit	Material	Labor	Equipment	Total	Total Incl O&P
0010	**SHELVING**									
0020	Metal, industrial, cross-braced, 3' wide, 12" deep	1 Sswk	175	.046	SF Shlf	6.65	1.66		8.31	10.50
0100	24" deep		330	.024		4.56	.88		5.44	6.70
0300	4' wide, 12" deep		185	.043		6	1.57		7.57	9.60
0400	24" deep		380	.021		4.44	.77		5.21	6.35
1200	Enclosed sides, cross-braced back, 3' wide, 12" deep		175	.046		11.85	1.66		13.51	16.15
1300	24" deep		290	.028		9.60	1		10.60	12.45
1500	Fully enclosed, sides and back, 3' wide, 12" deep		150	.053		12.95	1.94		14.89	17.95
1600	24" deep		255	.031		8.60	1.14		9.74	11.65
1800	4' wide, 12" deep		150	.053		9.95	1.94		11.89	14.65
1900	24" deep		290	.028		7.20	1		8.20	9.80
2200	Wide span, 1600 lb. capacity per shelf, 6' wide, 24" deep		380	.021		6.70	.77		7.47	8.85
2400	36" deep		440	.018		6.40	.66		7.06	8.30
2600	8' wide, 24" deep		440	.018		6.70	.66		7.36	8.60
2800	36" deep		520	.015		5.20	.56		5.76	6.75
4000	Pallet racks, steel frame 5,000 lb. capacity, 8' long, 36" deep	2 Sswk	450	.036		8.80	1.29		10.09	12.15
4200	42" deep		500	.032		8	1.16		9.16	11
4400	48" deep		520	.031		7.15	1.12		8.27	10

10 56 13.20 Parts Bins		Crew	Daily Output	Labor-Hours	Unit	Material	Labor	Equipment	Total	Total Incl O&P
0010	**PARTS BINS** Metal, gray baked enamel finish									
0100	6'-3" high, 3' wide									
0300	12 bins, 18" wide x 12" high, 12" deep	2 Clab	10	1.600	Ea.	300	42.50		342.50	400
0400	24" deep		10	1.600		415	42.50		457.50	525
0600	72 bins, 6" wide x 6" high, 12" deep		8	2		500	53		553	640
0700	18" deep		8	2		750	53		803	915
1000	7'-3" high, 3' wide									
1200	14 bins, 18" wide x 12" high, 12" deep	2 Clab	10	1.600	Ea.	310	42.50		352.50	410
1300	24" deep		10	1.600		405	42.50		447.50	515

10 56 Storage Assemblies

10 56 13 – Metal Storage Shelving

10 56 13.20 Parts Bins		Crew	Daily Output	Labor-Hours	Unit	Material	2011 Bare Costs Labor	Equipment	Total	Total Incl O&P
1500	84 bins, 6" wide x 6" high, 12" deep	2 Clab	8	2	Ea.	730	53		783	895
1600	24" deep	↓	8	2	↓	905	53		958	1,100

10 57 Wardrobe and Closet Specialties

10 57 13 – Hat and Coat Racks

10 57 13.10 Coat Racks and Wardrobes

		Crew	Daily Output	Labor-Hours	Unit	Material	Labor	Equipment	Total	Total Incl O&P
0010	**COAT RACKS AND WARDROBES**									
0020	Hat & coat rack, floor model, 6 hangers									
0050	Standing, beech wood, 21" x 21" x 72", chrome				Ea.	237			237	260
0100	18 gauge tubular steel, 21" x 21" x 69", wood walnut				"	281			281	310
0500	16 gauge steel frame, 22 gauge steel shelves									
0650	Single pedestal, 30" x 18" x 63"				Ea.	253			253	278
0800	Single face rack, 29" x 18-1/2" x 62"					254			254	279
0900	51" x 18-1/2" x 70"					385			385	420
0910	Double face rack, 39" x 26" x 70"					340			340	375
0920	63" x 26" x 70"				↓	410			410	455
0940	For 2" ball casters, add				Set	88			88	97
1400	Utility hook strips, 3/8" x 2-1/2" x 18", 6 hooks	1 Carp	48	.167	Ea.	57	5.60		62.60	72.50
1500	34" long, 12 hooks	"	48	.167	"	51	5.60		56.60	65.50
1650	Wall mounted racks, 16 gauge steel frame, 22 gauge steel shelves									
1850	12" x 15" x 26", 6 hangers	1 Carp	32	.250	Ea.	155	8.40		163.40	184
2000	12" x 15" x 50", 12 hangers	"	32	.250	"	181	8.40		189.40	213
2150	Wardrobe cabinet, steel, baked enamel finish									
2300	36" x 21" x 78", incl. top shelf & hanger rod				Ea.	282			282	310

10 57 23 – Closet and Utility Shelving

10 57 23.19 Wood Closet and Utility Shelving

		Crew	Daily Output	Labor-Hours	Unit	Material	Labor	Equipment	Total	Total Incl O&P
0010	**WOOD CLOSET AND UTILITY SHELVING**									
0020	Pine, clear grade, no edge band, 1" x 8"	1 Carp	115	.070	L.F.	2.83	2.33		5.16	7
0100	1" x 10"		110	.073		3.52	2.44		5.96	7.95
0200	1" x 12"		105	.076		4.24	2.56		6.80	8.95
0600	Plywood, 3/4" thick with lumber edge, 12" wide		75	.107		1.58	3.58		5.16	7.75
0700	24" wide		70	.114	↓	2.81	3.83		6.64	9.50
0900	Bookcase, clear grade pine, shelves 12" O.C., 8" deep, per S.F. shelf		70	.114	S.F.	9.20	3.83		13.03	16.50
1000	12" deep shelves		65	.123	"	13.80	4.13		17.93	22
1200	Adjustable closet rod and shelf, 12" wide, 3' long		20	.400	Ea.	8.25	13.40		21.65	31.50
1300	8' long		15	.533	"	17.15	17.90		35.05	49
1500	Prefinished shelves with supports, stock, 8" wide		75	.107	L.F.	3.14	3.58		6.72	9.45
1600	10" wide	↓	70	.114	"	4.27	3.83		8.10	11.10

10 73 Protective Covers

10 73 13 – Awnings

10 73 13.10 Awnings, Fabric		Crew	Daily Output	Labor-Hours	Unit	2011 Bare Costs Material	Labor	Equipment	Total	Total Incl O&P
0010	**AWNINGS, FABRIC**									
0020	Including acrylic canvas and frame, standard design									
0100	Door and window, slope, 3′ high, 4′ wide	1 Carp	4.50	1.778	Ea.	690	59.50		749.50	860
0110	6′ wide		3.50	2.286		890	76.50		966.50	1,100
0120	8′ wide		3	2.667		1,100	89.50		1,189.50	1,350
0200	Quarter round convex, 4′ wide		3	2.667		1,075	89.50		1,164.50	1,325
0210	6′ wide		2.25	3.556		1,400	119		1,519	1,725
0220	8′ wide		1.80	4.444		1,700	149		1,849	2,125
0300	Dome, 4′ wide		7.50	1.067		415	36		451	515
0310	6′ wide		3.50	2.286		935	76.50		1,011.50	1,150
0320	8′ wide		2	4		1,650	134		1,784	2,050
0350	Elongated dome, 4′ wide		1.33	6.015		1,550	202		1,752	2,050
0360	6′ wide		1.11	7.207		1,850	242		2,092	2,450
0370	8′ wide	↓	1	8		2,175	268		2,443	2,850
1000	Entry or walkway, peak, 12′ long, 4′ wide	2 Carp	.90	17.778		4,925	595		5,520	6,400
1010	6′ wide		.60	26.667		7,575	895		8,470	9,825
1020	8′ wide		.40	40		10,500	1,350		11,850	13,800
1100	Radius with dome end, 4′ wide		1.10	14.545		3,725	490		4,215	4,925
1110	6′ wide		.70	22.857		6,000	765		6,765	7,875
1120	8′ wide	↓	.50	32	↓	8,525	1,075		9,600	11,200
2000	Retractable lateral arm awning, manual									
2010	To 12′ wide, 8′ - 6″ projection	2 Carp	1.70	9.412	Ea.	1,100	315		1,415	1,750
2020	To 14′ wide, 8′ - 6″ projection		1.10	14.545		1,300	490		1,790	2,250
2030	To 19′ wide, 8′ - 6″ projection		.85	18.824		1,750	630		2,380	2,975
2040	To 24′ wide, 8′ - 6″ projection	↓	.67	23.881		2,225	800		3,025	3,800
2050	Motor for above, add	1 Carp	2.67	3	↓	955	101		1,056	1,225
3000	Patio/deck canopy with frame									
3010	12′ wide, 12′ projection	2 Carp	2	8	Ea.	1,575	268		1,843	2,175
3020	16′ wide, 14′ projection	″	1.20	13.333		2,450	445		2,895	3,425
9000	For fire retardant canvas, add					7%				
9010	For lettering or graphics, add					35%				
9020	For painted or coated acrylic canvas, deduct					8%				
9030	For translucent or opaque vinyl canvas, add					10%				
9040	For 6 or more units, deduct				↓	20%	15%			

10 73 16 – Canopies

10 73 16.20 Metal Canopies		Crew	Daily Output	Labor-Hours	Unit	Material	Labor	Equipment	Total	Total Incl O&P
0010	**METAL CANOPIES**									
0020	Wall hung, .032″, aluminum, prefinished, 8′ x 10′	K-2	1.30	18.462	Ea.	2,150	625	167	2,942	3,675
0300	8′ x 20′		1.10	21.818		4,275	735	197	5,207	6,275
0500	10′ x 10′		1.30	18.462		2,400	625	167	3,192	3,950
0700	10′ x 20′		1.10	21.818		4,450	735	197	5,382	6,475
1000	12′ x 20′		1	24		5,475	810	217	6,502	7,775
1360	12′ x 30′		.80	30		8,050	1,000	271	9,321	11,000
1700	12′ x 40′	↓	.60	40		9,800	1,350	360	11,510	13,700
1900	For free standing units, add				↓	20%	10%			
2300	Aluminum entrance canopies, flat soffit, .032″									
2500	3′-6″ x 4′-0″, clear anodized	2 Carp	4	4	Ea.	900	134		1,034	1,225
2700	Bronze anodized		4	4		1,575	134		1,709	1,975
3000	Polyurethane painted		4	4		1,275	134		1,409	1,625
3300	4′-6″ x 10′-0″, clear anodized		2	8		2,475	268		2,743	3,175
3500	Bronze anodized		2	8		3,175	268		3,443	3,925
3700	Polyurethane painted	↓	2	8	↓	2,650	268		2,918	3,375

10 73 Protective Covers

10 73 16 – Canopies

10 73 16.20 Metal Canopies		Crew	Daily Output	Labor-Hours	Unit	2011 Bare Costs Material	Labor	Equipment	Total	Total Incl O&P
4000	Wall downspout, 10 L.F., clear anodized	1 Carp	7	1.143	Ea.	154	38.50		192.50	234
4300	Bronze anodized		7	1.143		269	38.50		307.50	360
4500	Polyurethane painted		7	1.143		231	38.50		269.50	320
7000	Carport, baked vinyl finish, .032", 20' x 10', no foundations, min.	K-2	4	6	Car	3,850	202	54	4,106	4,650
7250	Maximum		2	12	"	7,000	405	108	7,513	8,575
7500	Walkway cover, to 12' wide, stl., vinyl finish, .032",no fndtns., min.		250	.096	S.F.	22.50	3.24	.87	26.61	32
7750	Maximum		200	.120	"	24.50	4.05	1.08	29.63	35.50

10 74 Manufactured Exterior Specialties

10 74 23 – Cupolas

10 74 23.10 Wood Cupolas

		Crew	Daily Output	Labor-Hours	Unit	Material	Labor	Equipment	Total	Total Incl O&P
0010	**WOOD CUPOLAS**									
0020	Stock units, pine, painted, 18" sq., 28" high, alum. roof	1 Carp	4.10	1.951	Ea.	162	65.50		227.50	288
0100	Copper roof		3.80	2.105		208	70.50		278.50	345
0300	23" square, 33" high, aluminum roof		3.70	2.162		385	72.50		457.50	545
0400	Copper roof		3.30	2.424		465	81.50		546.50	645
0600	30" square, 37" high, aluminum roof		3.70	2.162		495	72.50		567.50	665
0700	Copper roof		3.30	2.424		590	81.50		671.50	785
0900	Hexagonal, 31" wide, 46" high, copper roof		4	2		730	67		797	915
1000	36" wide, 50" high, copper roof		3.50	2.286		1,475	76.50		1,551.50	1,750
1200	For deluxe stock units, add to above					25%				
1400	For custom built units, add to above					50%	50%			

10 74 29 – Steeples

10 74 29.10 Prefabricated Steeples

		Crew	Daily Output	Labor-Hours	Unit	Material	Labor	Equipment	Total	Total Incl O&P
0010	**PREFABRICATED STEEPLES**									
4000	Steeples, translucent fiberglass, 30" square, 15' high	F-3	2	20	Ea.	5,825	615	330	6,770	7,800
4150	25' high		1.80	22.222		6,800	685	365	7,850	9,000
4350	Opaque fiberglass, 24" square, 14' high		2	20		4,850	615	330	5,795	6,725
4500	28' high		1.80	22.222		5,000	685	365	6,050	7,025
4600	Aluminum, baked finish, 16" square, 14' high					4,400			4,400	4,850
4620	20' high, 3'-6" base					7,400			7,400	8,150
4640	35' high, 8' base					28,600			28,600	31,500
4660	60' high, 14' base					61,500			61,500	67,500
4680	152' high, custom					512,500			512,500	564,000
4700	Porcelain enamel steeples, custom, 40' high	F-3	.50	80		14,100	2,450	1,300	17,850	21,100
4800	60' high	"	.30	133		24,500	4,100	2,175	30,775	36,200

10 74 33 – Weathervanes

10 74 33.10 Residential Weathervanes

		Crew	Daily Output	Labor-Hours	Unit	Material	Labor	Equipment	Total	Total Incl O&P
0010	**RESIDENTIAL WEATHERVANES**									
0020	Residential types, minimum	1 Carp	8	1	Ea.	151	33.50		184.50	222
0100	Maximum	"	2	4	"	1,700	134		1,834	2,100

10 74 46 – Window Wells

10 74 46.10 Area Window Wells

		Crew	Daily Output	Labor-Hours	Unit	Material	Labor	Equipment	Total	Total Incl O&P
0010	**AREA WINDOW WELLS**, Galvanized steel									
0020	20 ga., 3'-2" wide, 1' deep	1 Sswk	29	.276	Ea.	14	10.05		24.05	34.50
0100	2' deep		23	.348		25	12.65		37.65	51.50
0300	16 ga., 3'-2" wide, 1' deep		29	.276		20	10.05		30.05	41
0400	3' deep		23	.348		39	12.65		51.65	67
0600	Welded grating for above, 15 lb., painted		45	.178		85	6.45		91.45	106

10 74 Manufactured Exterior Specialties

10 74 46 – Window Wells

10 74 46.10 Area Window Wells		Crew	Daily Output	Labor-Hours	Unit	Material	2011 Bare Costs Labor	Equipment	Total	Total Incl O&P
0700	Galvanized	1 Sswk	45	.178	Ea.	114	6.45		120.45	137
0900	Translucent plastic cap for above	↓	60	.133	↓	16	4.85		20.85	27

10 75 Flagpoles

10 75 16 – Ground-Set Flagpoles

10 75 16.10 Flagpoles		Crew	Daily Output	Labor-Hours	Unit	Material	2011 Bare Costs Labor	Equipment	Total	Total Incl O&P
0010	**FLAGPOLES**, Ground set									
0050	Not including base or foundation									
0100	Aluminum, tapered, ground set 20′ high	K-1	2	8	Ea.	880	240	108	1,228	1,500
0200	25′ high		1.70	9.412		1,275	282	128	1,685	2,000
0300	30′ high		1.50	10.667		1,225	320	145	1,690	2,050
0400	35′ high		1.40	11.429		1,650	345	155	2,150	2,575
0500	40′ high		1.20	13.333		2,350	400	181	2,931	3,450
0600	50′ high		1	16		2,850	480	217	3,547	4,200
0700	60′ high		.90	17.778		5,225	535	241	6,001	6,875
0800	70′ high		.80	20		6,600	600	271	7,471	8,550
1100	Counterbalanced, internal halyard, 20′ high		1.80	8.889		2,350	266	121	2,737	3,150
1200	30′ high		1.50	10.667		2,600	320	145	3,065	3,550
1300	40′ high		1.30	12.308		5,675	370	167	6,212	7,050
1400	50′ high		1	16		7,900	480	217	8,597	9,725
2820	Aluminum, electronically operated, 30′ high		1.40	11.429		4,300	345	155	4,800	5,475
2840	35′ high		1.30	12.308		4,625	370	167	5,162	5,900
2860	39′ high		1.10	14.545		5,050	435	197	5,682	6,500
2880	45′ high		1	16		6,450	480	217	7,147	8,150
2900	50′ high		.90	17.778		6,875	535	241	7,651	8,700
3000	Fiberglass, tapered, ground set, 23′ high		2	8		1,100	240	108	1,448	1,725
3100	29′-7″ high		1.50	10.667		1,450	320	145	1,915	2,275
3200	36′-1″ high		1.40	11.429		1,900	345	155	2,400	2,850
3300	39′-5″ high		1.20	13.333		2,225	400	181	2,806	3,325
3400	49′-2″ high		1	16		3,975	480	217	4,672	5,425
3500	59′ high	↓	.90	17.778	↓	4,800	535	241	5,576	6,425
4300	Steel, direct imbedded installation									
4400	Internal halyard, 20′ high	K-1	2.50	6.400	Ea.	1,550	192	87	1,829	2,150
4500	25′ high		2.50	6.400		1,750	192	87	2,029	2,325
4600	30′ high		2.30	6.957		2,050	209	94.50	2,353.50	2,700
4700	40′ high		2.10	7.619		3,050	228	103	3,381	3,850
4800	50′ high		1.90	8.421		4,500	252	114	4,866	5,500
5000	60′ high		1.80	8.889		6,800	266	121	7,187	8,050
5100	70′ high		1.60	10		8,300	300	136	8,736	9,775
5200	80′ high		1.40	11.429		10,900	345	155	11,400	12,700
5300	90′ high		1.20	13.333		15,000	400	181	15,581	17,400
5500	100′ high	↓	1	16	↓	15,500	480	217	16,197	18,100
6400	Wood poles, tapered, clear vertical grain fir with tilting									
6410	base, not incl. foundation, 4″ butt, 25′ high	K-1	1.90	8.421	Ea.	1,375	252	114	1,741	2,075
6800	6″ butt, 30′ high	″	1.30	12.308	″	1,925	370	167	2,462	2,925
7300	Foundations for flagpoles, including									
7400	excavation and concrete, to 35′ high poles	C-1	10	3.200	Ea.	540	95		635	755
7600	40′ to 50′ high		3.50	9.143		985	272		1,257	1,525
7700	Over 60′ high	↓	2	16	↓	1,100	475		1,575	2,000

10 75 Flagpoles

10 75 23 – Wall-Mounted Flagpoles

10 75 23.10 Flagpoles		Crew	Daily Output	Labor-Hours	Unit	Material	2011 Bare Costs Labor	Equipment	Total	Total Incl O&P
0010	**FLAGPOLES**, Structure mounted									
0100	Fiberglass, vertical wall set, 19′-8″ long	K-1	1.50	10.667	Ea.	1,300	320	145	1,765	2,125
0200	23′ long		1.40	11.429		1,375	345	155	1,875	2,250
0300	26′-3″ long		1.30	12.308		1,775	370	167	2,312	2,750
0800	19′-8″ long outrigger		1.30	12.308		1,325	370	167	1,862	2,250
1300	Aluminum, vertical wall set, tapered, with base, 20′ high		1.20	13.333		1,050	400	181	1,631	2,025
1400	29′-6″ high		1	16		2,250	480	217	2,947	3,525
2400	Outrigger poles with base, 12′ long		1.30	12.308		930	370	167	1,467	1,825
2500	14′ long		1	16		1,425	480	217	2,122	2,625

10 81 Pest Control Devices

10 81 13 – Bird Control Devices

10 81 13.10 Bird Control Netting		Crew	Daily Output	Labor-Hours	Unit	Material	Labor	Equipment	Total	Total Incl O&P
0010	**BIRD CONTROL NETTING**									
0020	1/8″ square mesh	4 Clab	4000	.008	S.F.	.06	.21		.27	.42
0100	1/4″ square mesh		4000	.008		.07	.21		.28	.43
0120	1/2″ square mesh		4000	.008		.08	.21		.29	.44
0140	5/8″ x 3/4″ mesh		4000	.008		.08	.21		.29	.44
0160	1-1/4″ x 1-1/2″ mesh		4000	.008		.10	.21		.31	.46
0200	4″ square mesh		4000	.008		.11	.21		.32	.47
1000	Poly clips				Ea.	.02			.02	.02

10 88 Scales

10 88 05 – Commercial Scales

10 88 05.10 Scales		Crew	Daily Output	Labor-Hours	Unit	Material	Labor	Equipment	Total	Total Incl O&P
0010	**SCALES**									
0700	Truck scales, incl. steel weigh bridge,									
0800	not including foundation, pits									
1550	Digital, electronic, 100 ton capacity, steel deck 12′ x 10′ platform	3 Carp	.20	120	Ea.	12,500	4,025		16,525	20,500
1600	40′ x 10′ platform		.14	171		25,300	5,750		31,050	37,500
1640	60′ x 10′ platform		.13	184		31,400	6,200		37,600	44,900
1680	70′ x 10′ platform		.12	200		35,000	6,700		41,700	49,700
2000	For standard automatic printing device, add					2,525			2,525	2,775
2100	For remote reading electronic system, add					2,275			2,275	2,500
2300	Concrete foundation pits for above, 8′ x 6′, 5 C.Y. required	C-1	.50	64		955	1,900		2,855	4,225
2400	14′ x 6′ platform, 10 C.Y. required		.35	91.429		1,400	2,725		4,125	6,075
2600	50′ x 10′ platform, 30 C.Y. required		.25	128		1,875	3,800		5,675	8,450
2700	70′ x 10′ platform, 40 C.Y. required		.15	213		4,125	6,350		10,475	15,100
2750	Crane scales, dial, 1 ton capacity					1,025			1,025	1,125
2780	5 ton capacity					1,400			1,400	1,525
2800	Digital, 1 ton capacity					1,775			1,775	1,950
2850	10 ton capacity					4,475			4,475	4,900
2900	Low profile electronic warehouse scale,									
3000	not incl. printer, 4′ x 4′ platform, 10,000 lb. capacity	2 Carp	.30	53.333	Ea.	1,425	1,800		3,225	4,550
3300	5′ x 7′ platform, 10,000 lb. capacity		.25	64		4,125	2,150		6,275	8,125
3400	20,000 lb. capacity		.20	80		5,300	2,675		7,975	10,400
3500	For printers, incl. time, date & numbering, add					1,325			1,325	1,450
3800	Portable, beam type, capacity 1000#, platform 18″ x 24″					745			745	820
3900	Dial type, capacity 2000#, platform 24″ x 24″					1,325			1,325	1,450

10 88 Scales

10 88 05 – Commercial Scales

10 88 05.10 Scales		Crew	Daily Output	Labor-Hours	Unit	Material	Labor	Equipment	Total	Total Incl O&P
						2011 Bare Costs				
4000	Digital type, capacity 1000#, platform 24" x 30"				Ea.	2,125			2,125	2,325
4100	Portable contractor truck scales, 50 ton cap., 40' x 10' platform					32,000			32,000	35,200
4200	60' x 10' platform				↓	29,500			29,500	32,500

Estimating Tips

General

- The items in this division are usually priced per square foot or each. Many of these items are purchased by the owner for installation by the contractor. Check the specifications for responsibilities and include time for receiving, storage, installation, and mechanical and electrical hookups in the appropriate divisions.
- Many items in Division 11 require some type of support system that is not usually furnished with the item. Examples of these systems include blocking for the attachment of casework and support angles for ceiling-hung projection screens. The required blocking or supports must be added to the estimate in the appropriate division.
- Some items in Division 11 may require assembly or electrical hookups. Verify the amount of assembly required or the need for a hard electrical connection and add the appropriate costs.

Reference Numbers

Reference numbers are shown in shaded boxes at the beginning of some major classifications. These numbers refer to related items in the Reference Section. The reference information may be an estimating procedure, an alternate pricing method, or technical information.

Note: Not all subdivisions listed here necessarily appear in this publication.

11 05 05 – Selective Equipment Demolition

11 05 05.10 Selective Demolition		Crew	Daily Output	Labor-Hours	Unit	Material	2011 Bare Costs Labor	Equipment	Total	Total Incl O&P
0010	**SELECTIVE DEMOLITION**									
0130	Central vacuum, motor unit, residential or commercial	1 Clab	2	4	Ea.		106		106	177
0210	Vault door and frame	2 Skwk	2	8			272		272	455
0215	Day gate, for vault	″	3	5.333			182		182	305
0380	Bank equipment, teller window, bullet resistant	1 Clab	1.20	6.667			176		176	295
0381	Counter	2 Clab	1.50	10.667	Station		282		282	475
0382	Drive-up window, including drawer and glass		1.50	10.667	″		282		282	475
0383	Thru-wall boxes and chests, selective demolition		2.50	6.400	Ea.		169		169	284
0384	Bullet resistant partitions		20	.800	L.F.		21		21	35.50
0385	Pneumatic tube system, 2 lane drive-up	L-3	.45	22.222	Ea.		775		775	1,300
0386	Safety deposit box	1 Clab	50	.160	Opng.		4.23		4.23	7.10
0387	Surveillance system, video, complete	2 Elec	2	8	Ea.		320		320	520
0410	Church equipment, misc moveable fixtures	2 Clab	1	16			425		425	710
0412	Steeple, to 28′ high	F-3	3	13.333			410	219	629	920
0414	40′ to 60′ high	″	.80	50			1,525	820	2,345	3,450
0510	Library equipment, bookshelves, wood, to 90″ high	1 Clab	20	.400	L.F.		10.60		10.60	17.70
0515	Carrels, hardwood, 36″ x 24″	″	9	.889	Ea.		23.50		23.50	39.50
0630	Stage equipment, light control panel	1 Elec	1	8	″		320		320	520
0632	Border lights		40	.200	L.F.		8.05		8.05	13.05
0634	Spotlights		8	1	Ea.		40.50		40.50	65.50
0636	Telescoping platforms and risers	2 Clab	175	.091	SF Stg.		2.42		2.42	4.05
1020	Barber equipment, hydraulic chair	1 Clab	40	.200	Ea.		5.30		5.30	8.85
1030	Checkout counter, supermarket or warehouse conveyor	2 Clab	18	.889			23.50		23.50	39.50
1040	Food cases, refrigerated or frozen	Q-5	6	2.667			96		96	157
1190	Laundry equipment, commercial	L-6	3	4			159		159	260
1360	Movie equipment, lamphouse, to 4000 watt, incl rectifier	1 Elec	4	2			80.50		80.50	131
1365	Sound system, incl amplifier	″	1.25	6.400			258		258	420
1410	Air compressor, to 5 H.P.	2 Clab	2.50	6.400			169		169	284
1412	Lubrication equipment, automotive, 3 reel type, incl pump, excl piping	L-4	1	16	Set		475		475	795
1414	Booth, spray paint, complete, to 26′ long	″	.80	20	″		595		595	995
1560	Parking equipment, cashier booth	B-22	2	15	Ea.		435	97	532	825
1600	Loading dock equipment, dock bumpers, rubber	1 Clab	50	.160	″		4.23		4.23	7.10
1610	Door seal for door perimeter	″	50	.160	L.F.		4.23		4.23	7.10
1620	Platform lifter, fixed, 6′ x 8′, 5000 lb capacity	E-16	1.50	10.667	Ea.		400	72.50	472.50	840
1630	Dock leveller	″	2	8			299	54.50	353.50	630
1640	Lights, single or double arm	1 Elec	8	1			40.50		40.50	65.50
1650	Shelter, fabric, truck or train	1 Clab	1.50	5.333			141		141	236
1790	Waste handling equipment, commercial compactor	L-4	2	8			238		238	395
1792	Commercial or municipal incinerator, gas	″	2	8			238		238	395
1795	Crematory, excluding building	Q-3	.25	128			4,575		4,575	7,450
1910	Detection equipment, cell bar front	E-4	4	8			295	27.50	322.50	595
1912	Cell door and frame		8	4			148	13.65	161.65	296
1914	Prefab cell, 4′ to 5′ wide, 7′ to 8′ high, 7′ deep		8	4			148	13.65	161.65	296
1916	Cot, bolted, single		40	.800			29.50	2.73	32.23	59.50
1918	Visitor cubicle		4	8			295	27.50	322.50	595
2850	Hydraulic gates, canal, flap, knife, slide or sluice, to 18″ diameter	L-5A	8	4			147	70	217	350
2852	19″ to 36″ diameter		6	5.333			196	93.50	289.50	465
2854	37″ to 48″ diameter		2	16			590	280	870	1,375
2856	49″ to 60″ diameter		1	32			1,175	560	1,735	2,800
2858	Over 60″ diameter		.30	106			3,925	1,875	5,800	9,275
3100	Sewage pumping system, prefabricated, to 1000 GPM	C-17D	.20	420			14,500	4,225	18,725	28,900
3110	Sewage treatment, holding tank for recirc chemical water closet	1 Plum	8	1			39.50		39.50	65
3900	Wastewater treatment system, to 1500 gallons	B-21	2	14			395	64.50	459.50	730

11 05 Common Work Results for Equipment

11 05 05 – Selective Equipment Demolition

11 05 05.10 Selective Demolition

11 05 05.10 Selective Demolition		Crew	Daily Output	Labor-Hours	Unit	Material	2011 Bare Costs Labor	Equipment	Total	Total Incl O&P
4050	Food storage equipment, walk-in refrigerator/freezer	2 Clab	64	.250	S.F.		6.60		6.60	11.10
4052	Shelving, stainless steel, 4 tier or dunnage rack	1 Clab	12	.667	Ea.		17.65		17.65	29.50
4100	Food preparation equipment, small countertop		18	.444			11.75		11.75	19.70
4150	Food delivery carts, heated cabinets		18	.444			11.75		11.75	19.70
4200	Cooking equipment, commercial range	Q-1	12	1.333			47.50		47.50	77.50
4250	Hood and ventilation equipment, kitchen exhaust hood, excl fire prot	1 Clab	3	2.667			70.50		70.50	118
4255	Fire protection system	Q-1	3	5.333			190		190	310
4300	Food dispensing equipment, countertop items	1 Clab	15	.533			14.10		14.10	23.50
4310	Serving counter	"	65	.123	L.F.		3.26		3.26	5.45
4350	Ice machine, ice cube maker, flakers and storage bins, to 2000 lb/day	Q-1	1.60	10	Ea.		355		355	585
4400	Cleaning and disposal, commercial dishwasher, to 50 racks per hour	L-6	1	12			480		480	780
4405	To 275 racks per hour	L-4	1	16			475		475	795
4410	Dishwasher hood	2 Clab	5	3.200			84.50		84.50	142
4420	Garbage disposal, commercial, to 5 H.P.	L-1	8	1.250			49.50		49.50	81
4540	Water heater, residential, to 80 gal/day	"	5	2			79.50		79.50	130
4542	Water softener, automatic	2 Plum	10	1.600			63.50		63.50	104
4544	Disappearing stairway, to 15′ floor height	2 Clab	6	2.667			70.50		70.50	118
4710	Darkroom equipment, light	L-7	10	2.600			75.50		75.50	126
4712	Heavy	"	1.50	17.333			505		505	840
4720	Doors	2 Clab	3.50	4.571	Opng.		121		121	203
4830	Bowling alley, complete, incl pinsetter, scorer, counters, misc supplies	4 Clab	.40	80	Lane		2,125		2,125	3,550
4840	Health club equipment, circuit training apparatus	2 Clab	2	8	Set		212		212	355
4842	Squat racks	"	10	1.600	Ea.		42.50		42.50	71
4860	School equipment, basketball backstop	L-2	2	8			236		236	395
4862	Table and benches, folding, in wall, 14′ long	L-4	4	4			119		119	199
4864	Bleachers, telescoping, to 30 tier	F-5	120	.267	Seat		7.85		7.85	13.15
4866	Boxing ring, elevated	L-4	.20	80	Ea.		2,375		2,375	3,975
4867	Boxing ring, floor level	"	2	8			238		238	395
4868	Exercise equipment	1 Clab	6	1.333			35.50		35.50	59
4870	Gym divider	L-4	1000	.016	S.F.		.48		.48	.79
4875	Scoreboard	R-3	2	10	Ea.		395	64.50	459.50	710
4880	Shooting range, incl bullet traps, targets, excl structure	L-9	1	36	Point		1,125		1,125	1,875
5200	Vocational shop equipment	2 Clab	8	2	Ea.		53		53	88.50
6200	Fume hood, incl countertop, excl HVAC	"	6	2.667	L.F.		70.50		70.50	118
7100	Medical sterilizing, distiller, water, steam heated, 50 gal. capacity	1 Plum	2.80	2.857	Ea.		113		113	185
7200	Medical equipment, surgery table, minor	1 Clab	1	8			212		212	355
7210	Surgical lights, doctors office, single or double arm	2 Elec	3	5.333			215		215	350
7300	Physical therapy, table	2 Clab	4	4			106		106	177
7310	Whirlpool bath, fixed, incl mixing valves	1 Plum	4	2			79.50		79.50	130
7400	Dental equipment, chair, electric or hydraulic	1 Clab	.75	10.667			282		282	475
7410	Central suction system	1 Plum	2	4			159		159	259
7420	Drill console with accessories	1 Clab	3.20	2.500			66		66	111
7430	X-ray unit	"	4	2			53		53	88.50
7440	X-ray developer	1 Plum	10	.800			31.50		31.50	52

11 05 10 – Equipment Installation

11 05 10.10 Industrial Equipment Installation

11 05 10.10 Industrial Equipment Installation		Crew	Daily Output	Labor-Hours	Unit	Material	Labor	Equipment	Total	Total Incl O&P
0010	**INDUSTRIAL EQUIPMENT INSTALLATION**									
0020	Industrial equipment, minimum	E-2	12	4	Ton		147	135	282	420
0200	Maximum	"	2	24	"		880	810	1,690	2,525

11 11 Vehicle Service Equipment

11 11 13 – Compressed-Air Vehicle Service Equipment

11 11 13.10 Compressed Air Equipment		Crew	Daily Output	Labor-Hours	Unit	2011 Bare Costs Material	Labor	Equipment	Total	Total Incl O&P
0010	COMPRESSED AIR EQUIPMENT									
0030	Compressors, electric, 1-1/2 H.P., standard controls	L-4	1.50	10.667	Ea.	410	315		725	985
0550	Dual controls		1.50	10.667		740	315		1,055	1,350
0600	5 H.P., 115/230 volt, standard controls		1	16		2,175	475		2,650	3,200
0650	Dual controls	↓	1	16	↓	2,350	475		2,825	3,375

11 11 19 – Vehicle Lubrication Equipment

11 11 19.10 Lubrication Equipment		Crew	Daily Output	Labor-Hours	Unit	Material	Labor	Equipment	Total	Total Incl O&P
0010	LUBRICATION EQUIPMENT									
3000	Lube equipment, 3 reel type, with pumps, not including piping	L-4	.50	32	Set	7,900	950		8,850	10,300

11 11 33 – Vehicle Spray Painting Equipment

11 11 33.10 Spray Painting Equipment		Crew	Daily Output	Labor-Hours	Unit	Material	Labor	Equipment	Total	Total Incl O&P
0010	SPRAY PAINTING EQUIPMENT									
4000	Spray painting booth, 26′ long, complete	L-4	.40	40	Ea.	15,900	1,200		17,100	19,400

11 12 Parking Control Equipment

11 12 13 – Parking Key and Card Control Units

11 12 13.10 Parking Control Units		Crew	Daily Output	Labor-Hours	Unit	Material	Labor	Equipment	Total	Total Incl O&P
0010	PARKING CONTROL UNITS									
5100	Card reader	1 Elec	2	4	Ea.	1,875	161		2,036	2,300
5120	Proximity with customer display	2 Elec	1	16		5,500	645		6,145	7,100
6000	Parking control software, minimum	1 Elec	.50	16		22,600	645		23,245	25,900
6020	Maximum	"	.20	40	↓	98,500	1,600		100,100	111,000

11 12 16 – Parking Ticket Dispensers

11 12 16.10 Ticket Dispensers		Crew	Daily Output	Labor-Hours	Unit	Material	Labor	Equipment	Total	Total Incl O&P
0010	TICKET DISPENSERS									
5900	Ticket spitter with time/date stamp, standard	2 Elec	2	8	Ea.	6,150	320		6,470	7,275
5920	Mag stripe encoding	"	2	8	"	18,600	320		18,920	20,900

11 12 26 – Parking Fee Collection Equipment

11 12 26.13 Parking Fee Coin Collection Equipment		Crew	Daily Output	Labor-Hours	Unit	Material	Labor	Equipment	Total	Total Incl O&P
0010	PARKING FEE COIN COLLECTION EQUIPMENT									
5200	Cashier booth, average	B-22	1	30	Ea.	9,875	865	194	10,934	12,600
5300	Collector station, pay on foot	2 Elec	.20	80		109,500	3,225		112,725	125,500
5320	Credit card only	"	.50	32	↓	20,200	1,300		21,500	24,400

11 12 26.23 Fee Equipment		Crew	Daily Output	Labor-Hours	Unit	Material	Labor	Equipment	Total	Total Incl O&P
0010	FEE EQUIPMENT									
5600	Fee computer	1 Elec	1.50	5.333	Ea.	13,600	215		13,815	15,400

11 12 33 – Parking Gates

11 12 33.13 Parking Gate Equipment		Crew	Daily Output	Labor-Hours	Unit	Material	Labor	Equipment	Total	Total Incl O&P
0010	PARKING GATE EQUIPMENT									
5000	Barrier gate with programmable controller	2 Elec	3	5.333	Ea.	3,425	215		3,640	4,125
5020	Industrial		3	5.333		4,725	215		4,940	5,550
5500	Exit verifier	↓	1	16		17,400	645		18,045	20,300
5700	Full sign, 4″ letters	1 Elec	2	4		1,225	161		1,386	1,575
5800	Inductive loop	2 Elec	4	4		169	161		330	445
5950	Vehicle detector, microprocessor based	1 Elec	3	2.667	↓	395	107		502	610

11 13 Loading Dock Equipment

11 13 13 – Loading Dock Bumpers

11 13 13.10 Dock Bumpers		Crew	Daily Output	Labor-Hours	Unit	Material	Labor	Equipment	Total	Total Incl O&P
						2011 Bare Costs				
0010	**DOCK BUMPERS** Bolts not included									
0020	2″ x 6″ to 4″ x 8″, average	1 Carp	.30	26.667	M.B.F.	1,325	895		2,220	2,975
0050	Bumpers, rubber blocks 4-1/2″ thick, 10″ high, 14″ long		26	.308	Ea.	50.50	10.30		60.80	73
0200	24″ long		22	.364		95	12.20		107.20	125
0300	36″ long		17	.471		96	15.80		111.80	133
0500	12″ high, 14″ long		25	.320		100	10.75		110.75	128
0550	24″ long		20	.400		111	13.40		124.40	145
0600	36″ long		15	.533		123	17.90		140.90	166
0800	Rubber blocks 6″ thick, 10″ high, 14″ long		22	.364		93	12.20		105.20	123
0850	24″ long		18	.444		121	14.90		135.90	158
0900	36″ long		13	.615		147	20.50		167.50	197
0910	20″ high, 11″ long		13	.615		147	20.50		167.50	196
0920	Extruded rubber bumpers, T section, 22″ x 22″ x 3″ thick		41	.195		55	6.55		61.55	71.50
0940	Molded rubber bumpers, 24″ x 12″ x 3″ thick		20	.400		54	13.40		67.40	81.50
1000	Welded installation of above bumpers	E-14	8	1		3.74	36.50	13.65	53.89	88.50
1100	For drilled anchors, add per anchor	1 Carp	36	.222		6.70	7.45		14.15	19.85
1300	Steel bumpers, see Section 10 26 13.10									

11 13 16 – Loading Dock Seals and Shelters

11 13 16.10 Dock Seals and Shelters		Crew	Daily Output	Labor-Hours	Unit	Material	Labor	Equipment	Total	Total Incl O&P
0010	**DOCK SEALS AND SHELTERS**									
3600	Door seal for door perimeter, 12″ x 12″, vinyl covered	1 Carp	26	.308	L.F.	30.50	10.30		40.80	51
3900	Folding gates, see Section 10 22 16.10									
6200	Shelters, fabric, for truck or train, scissor arms, minimum	1 Carp	1	8	Ea.	1,575	268		1,843	2,200
6300	Maximum	″	.50	16	″	2,050	535		2,585	3,150

11 13 19 – Stationary Loading Dock Equipment

11 13 19.10 Dock Equipment		Crew	Daily Output	Labor-Hours	Unit	Material	Labor	Equipment	Total	Total Incl O&P
0010	**DOCK EQUIPMENT**									
2200	Dock boards, heavy duty, 60″ x 60″, aluminum, 5,000 lb. capacity				Ea.	1,200			1,200	1,325
2700	9,000 lb. capacity					1,375			1,375	1,525
3200	15,000 lb. capacity					1,625			1,625	1,775
4200	Platform lifter, 6′ x 6′, portable, 3,000 lb. capacity					9,150			9,150	10,100
4250	4,000 lb. capacity					11,300			11,300	12,400
4400	Fixed, 6′ x 8′, 5,000 lb. capacity	E-16	.70	22.857		9,850	855	156	10,861	12,600
4500	Levelers, hinged for trucks, 10 ton capacity, 6′ x 8′		1.08	14.815		4,975	555	101	5,631	6,625
4650	7′ x 8′		1.08	14.815		5,350	555	101	6,006	7,025
4670	Air bag power operated, 10 ton cap., 6′ x 8′		1.08	14.815		5,775	555	101	6,431	7,500
4680	7′ x 8′		1.08	14.815		5,800	555	101	6,456	7,525
4700	Hydraulic, 10 ton capacity, 6′ x 8′		1.08	14.815		8,650	555	101	9,306	10,700
4800	7′ x 8′		1.08	14.815		9,325	555	101	9,981	11,400
5800	Loading dock safety restraints, manual style		1.08	14.815		3,175	555	101	3,831	4,625
5900	Automatic style		1.08	14.815		5,325	555	101	5,981	7,000

11 13 26 – Loading Dock Lights

11 13 26.10 Dock Lights		Crew	Daily Output	Labor-Hours	Unit	Material	Labor	Equipment	Total	Total Incl O&P
0010	**DOCK LIGHTS**									
5000	Lights for loading docks, single arm, 24″ long	1 Elec	3.80	2.105	Ea.	139	84.50		223.50	290
5700	Double arm, 60″ long	″	3.80	2.105	″	184	84.50		268.50	340

11 14 Pedestrian Control Equipment

11 14 13 – Pedestrian Gates

11 14 13.13 Portable Posts and Railings		Crew	Daily Output	Labor-Hours	Unit	Material	2011 Bare Costs Labor	Equipment	Total	Total Incl O&P
0010	**PORTABLE POSTS AND RAILINGS**									
0020	Portable for pedestrian traffic control, standard, minimum				Ea.	122			122	135
0100	Maximum					181			181	199
0300	Deluxe posts, minimum					172			172	189
0400	Maximum					325			325	355
0600	Ropes for above posts, plastic covered, 1-1/2" diameter				L.F.	15.55			15.55	17.10
0700	Chain core				"	10.90			10.90	12
1500	Portable security or safety barrier, black with 7' yellow strap				Ea.	250			250	275
1510	12' yellow strap					290			290	320
1550	Sign holder, standard design					135			135	148

11 14 13.19 Turnstiles		Crew	Daily Output	Labor-Hours	Unit	Material	Labor	Equipment	Total	Total Incl O&P
0010	**TURNSTILES**									
0020	One way, 4 arm, 46" diameter, economy, manual	2 Carp	5	3.200	Ea.	805	107		912	1,075
0100	Electric		1.20	13.333		1,300	445		1,745	2,200
0300	High security, galv., 5'-5" diameter, 7' high, manual		1	16		3,975	535		4,510	5,275
0350	Electric		.60	26.667		4,500	895		5,395	6,450
0420	Three arm, 24" opening, light duty, manual		2	8		1,250	268		1,518	1,850
0450	Heavy duty		1.50	10.667		3,125	360		3,485	4,050
0460	Manual, with registering & controls, light duty		2	8		2,725	268		2,993	3,450
0470	Heavy duty		1.50	10.667		3,300	360		3,660	4,225
0480	Electric, heavy duty		1.10	14.545		3,825	490		4,315	5,025
0500	For coin or token operating, add					610			610	670
1200	One way gate with horizontal bars, 5'-5" diameter									
1300	7' high, recreation or transit type	2 Carp	.80	20	Ea.	4,425	670		5,095	5,975
1500	For electronic counter, add				"	210			210	231

11 16 Vault Equipment

11 16 16 – Safes

11 16 16.10 Commercial Safes		Crew	Daily Output	Labor-Hours	Unit	Material	Labor	Equipment	Total	Total Incl O&P
0010	**COMMERCIAL SAFES**									
0200	Office, 1 hr. rating, 30" x 18" x 18"				Ea.	2,175			2,175	2,400
0250	40" x 18" x 18"					4,100			4,100	4,525
0300	60" x 36" x 18", double door					8,150			8,150	8,975
0600	Data, 1 hr. rating, 27" x 19" x 16"					7,600			7,600	8,350
0700	63" x 34" x 16"					14,500			14,500	15,900
0750	Diskette, 1 hr., 14" x 12" x 11", inside					4,200			4,200	4,625
0800	Money, "B" label, 9" x 14" x 14"					525			525	580
0900	Tool resistive, 24" x 24" x 20"					3,575			3,575	3,925
1050	Tool and torch resistive, 24" x 24" x 20"					9,150			9,150	10,100
1150	Jewelers, 23" x 20" x 18"					11,700			11,700	12,800
1200	63" x 25" x 18"					22,400			22,400	24,700
1300	For handling into building, add, minimum	A-2	8.50	2.824			74.50	20	94.50	147
1400	Maximum	"	.78	30.769			815	220	1,035	1,600

11 17 Teller and Service Equipment

11 17 13 – Teller Equipment Systems

11 17 13.10 Bank Equipment		Crew	Daily Output	Labor-Hours	Unit	2011 Bare Costs Material	Labor	Equipment	Total	Total Incl O&P
0010	**BANK EQUIPMENT**									
0020	Alarm system, police	2 Elec	1.60	10	Ea.	4,675	405		5,080	5,800
0100	With vault alarm	"	.40	40		18,600	1,600		20,200	23,000
0400	Bullet resistant teller window, 44" x 60"	1 Glaz	.60	13.333		3,525	445		3,970	4,600
0500	48" x 60"	"	.60	13.333		4,700	445		5,145	5,900
3000	Counters for banks, frontal only	2 Carp	1	16	Station	1,750	535		2,285	2,825
3100	Complete with steel undercounter	"	.50	32	"	3,400	1,075		4,475	5,550
4600	Door and frame, bullet-resistant, with vision panel, minimum	2 Sswk	1.10	14.545	Ea.	5,000	530		5,530	6,500
4700	Maximum		1.10	14.545		6,775	530		7,305	8,475
4800	Drive-up window, drawer & mike, not incl. glass, minimum		1	16		5,800	580		6,380	7,475
4900	Maximum		.50	32		8,875	1,175		10,050	12,000
5000	Night depository, with chest, minimum		1	16		7,250	580		7,830	9,075
5100	Maximum		.50	32		10,300	1,175		11,475	13,500
5200	Package receiver, painted		3.20	5		1,300	182		1,482	1,775
5300	Stainless steel		3.20	5		2,225	182		2,407	2,775
5400	Partitions, bullet-resistant, 1-3/16" glass, 8' high	2 Carp	10	1.600	L.F.	189	53.50		242.50	298
5450	Acrylic	"	10	1.600	"	360	53.50		413.50	485
5500	Pneumatic tube systems, 2 lane drive-up, complete	L-3	.25	40	Total	24,600	1,400		26,000	29,400
5550	With T.V. viewer	"	.20	50	"	47,600	1,750		49,350	55,500
5570	Safety deposit boxes, minimum	1 Sswk	44	.182	Opng.	54.50	6.60		61.10	72.50
5580	Maximum, 10" x 15" opening		19	.421		116	15.35		131.35	156
5590	Teller locker, average		15	.533		1,500	19.40		1,519.40	1,675
5600	Pass thru, bullet-res. window, painted steel, 24" x 36"	2 Sswk	1.60	10	Ea.	2,375	365		2,740	3,325
5700	48" x 48"		1.20	13.333		2,550	485		3,035	3,725
5800	72" x 40"		.80	20		3,725	730		4,455	5,500
5900	For stainless steel frames, add					20%				
6100	Surveillance system, video camera, complete	2 Elec	1	16	Ea.	9,700	645		10,345	11,800
6110	For each additional camera, add				"	945			945	1,050
6120	CCTV system, see Section 27 41 19.10									
6200	Twenty-four hour teller, single unit,									
6300	automated deposit, cash and memo	L-3	.25	40	Ea.	42,800	1,400		44,200	49,300
7000	Vault front, see Section 08 34 59.10									

11 19 Detention Equipment

11 19 30 – Detention Cell Equipment

11 19 30.10 Cell Equipment		Crew	Daily Output	Labor-Hours	Unit	Material	Labor	Equipment	Total	Total Incl O&P
0010	**CELL EQUIPMENT**									
3000	Toilet apparatus including wash basin, average	L-8	1.50	13.333	Ea.	3,350	420		3,770	4,400

11 21 Mercantile and Service Equipment

11 21 13 – Cash Registers and Checking Equipment

11 21 13.10 Checkout Counter		Crew	Daily Output	Labor-Hours	Unit	Material	Labor	Equipment	Total	Total Incl O&P
0010	**CHECKOUT COUNTER**									
0020	Supermarket conveyor, single belt	2 Clab	10	1.600	Ea.	2,900	42.50		2,942.50	3,275
0100	Double belt, power take-away		9	1.778		4,175	47		4,222	4,675
0400	Double belt, power take-away, incl. side scanning		7	2.286		4,900	60.50		4,960.50	5,500
0800	Warehouse or bulk type		6	2.667		5,800	70.50		5,870.50	6,500
1000	Scanning system, 2 lanes, w/registers, scan gun & memory				System	15,200			15,200	16,700
1100	10 lanes, single processor, full scan, with scales				"	144,500			144,500	158,500

11 21 Mercantile and Service Equipment

11 21 13 – Cash Registers and Checking Equipment

11 21 13.10 Checkout Counter		Crew	Daily Output	Labor-Hours	Unit	2011 Bare Costs Material	Labor	Equipment	Total	Total Incl O&P
2000	Register, restaurant, minimum				Ea.	675			675	745
2100	Maximum					2,850			2,850	3,125
2150	Store, minimum					675			675	745
2200	Maximum					2,850			2,850	3,125

11 21 53 – Barber and Beauty Shop Equipment

11 21 53.10 Barber Equipment		Crew	Daily Output	Labor-Hours	Unit	Material	Labor	Equipment	Total	Total Incl O&P
0010	**BARBER EQUIPMENT**									
0020	Chair, hydraulic, movable, minimum	1 Carp	24	.333	Ea.	460	11.20		471.20	525
0050	Maximum	″	16	.500		2,975	16.80		2,991.80	3,300
0200	Wall hung styling station with mirrors, minimum	L-2	8	2		395	59		454	535
0300	Maximum	″	4	4		1,975	118		2,093	2,375
0500	Sink, hair washing basin, rough plumbing not incl.	1 Plum	8	1		425	39.50		464.50	530
1000	Sterilizer, liquid solution for tools					138			138	151
1100	Total equipment, rule of thumb, per chair, minimum	L-8	1	20		1,725	630		2,355	2,950
1150	Maximum	″	1	20		4,400	630		5,030	5,900

11 23 Commercial Laundry and Dry Cleaning Equipment

11 23 13 – Dry Cleaning Equipment

11 23 13.13 Dry Cleaners		Crew	Daily Output	Labor-Hours	Unit	Material	Labor	Equipment	Total	Total Incl O&P
0010	**DRY CLEANERS** Not incl. rough-in									
2000	Dry cleaners, electric, 20 lb. capacity	L-1	.20	50	Ea.	33,300	2,000		35,300	39,900
2050	25 lb. capacity		.17	58.824		45,800	2,350		48,150	54,500
2100	30 lb. capacity		.15	66.667		48,300	2,650		50,950	57,500
2150	60 lb. capacity		.09	111		73,000	4,425		77,425	87,000

11 23 16 – Drying and Conditioning Equipment

11 23 16.13 Dryers		Crew	Daily Output	Labor-Hours	Unit	Material	Labor	Equipment	Total	Total Incl O&P
0010	**DRYERS**, Not including rough-in									
1500	Industrial, 30 lb. capacity	1 Plum	2	4	Ea.	2,975	159		3,134	3,525
1600	50 lb. capacity	″	1.70	4.706		3,200	187		3,387	3,825
4700	Lint collector, ductwork not included, 8,000 to 10,000 C.F.M.	Q-10	.30	80		8,650	2,850		11,500	14,200

11 23 19 – Finishing Equipment

11 23 19.13 Folders and Spreaders		Crew	Daily Output	Labor-Hours	Unit	Material	Labor	Equipment	Total	Total Incl O&P
0010	**FOLDERS AND SPREADERS**									
3500	Folders, blankets & sheets, minimum	1 Elec	.17	47.059	Ea.	31,100	1,900		33,000	37,300
3700	King size with automatic stacker		.10	80		56,500	3,225		59,725	67,000
3800	For conveyor delivery, add		.45	17.778		6,000	715		6,715	7,750

11 23 23 – Commercial Ironing Equipment

11 23 23.13 Irons and Pressers		Crew	Daily Output	Labor-Hours	Unit	Material	Labor	Equipment	Total	Total Incl O&P
0010	**IRONS AND PRESSERS**									
4500	Ironers, institutional, 110″, single roll	1 Elec	.20	40	Ea.	30,200	1,600		31,800	35,800

11 23 26 – Commercial Washers and Extractors

11 23 26.13 Washers and Extractors		Crew	Daily Output	Labor-Hours	Unit	Material	Labor	Equipment	Total	Total Incl O&P
0010	**WASHERS AND EXTRACTORS**, not including rough-in									
6000	Combination washer/extractor, 20 lb. capacity	L-6	1.50	8	Ea.	5,400	320		5,720	6,475
6100	30 lb. capacity		.80	15		8,550	600		9,150	10,400
6200	50 lb. capacity		.68	17.647		10,000	705		10,705	12,200
6300	75 lb. capacity		.30	40		18,700	1,600		20,300	23,200
6350	125 lb. capacity		.16	75		24,900	3,000		27,900	32,300

11 23 Commercial Laundry and Dry Cleaning Equipment

11 23 33 – Coin-Operated Laundry Equipment

11 23 33.13 Coin Operated Washers and Dryers		Crew	Daily Output	Labor-Hours	Unit	Material	2011 Bare Costs Labor	Equipment	Total	Total Incl O&P
0010	**COIN OPERATED WASHERS AND DRYERS**									
0990	Dryer, gas fired									
1000	Commercial, 30 lb. capacity, coin operated, single	1 Plum	3	2.667	Ea.	3,100	106		3,206	3,575
1100	Double stacked	"	2	4	"	6,275	159		6,434	7,150
5290	Clothes washer									
5300	Commercial, coin operated, average	1 Plum	3	2.667	Ea.	1,125	106		1,231	1,425

11 24 Maintenance Equipment

11 24 19 – Vacuum Cleaning Systems

11 24 19.10 Vacuum Cleaning

		Crew	Daily Output	Labor-Hours	Unit	Material	Labor	Equipment	Total	Total Incl O&P
0010	**VACUUM CLEANING**									
0020	Central, 3 inlet, residential	1 Skwk	.90	8.889	Total	1,025	305		1,330	1,650
0200	Commercial		.70	11.429		1,225	390		1,615	2,000
0400	5 inlet system, residential		.50	16		1,450	545		1,995	2,500
0600	7 inlet system, commercial		.40	20		1,675	680		2,355	3,000
0800	9 inlet system, residential	↓	.30	26.667		3,600	910		4,510	5,475
4010	Rule of thumb: First 1200 S.F., installed				↓				1,325	1,450
4020	For each additional S.F., add				S.F.					.24

11 26 Unit Kitchens

11 26 13 – Metal Unit Kitchens

11 26 13.10 Commercial Unit Kitchens

		Crew	Daily Output	Labor-Hours	Unit	Material	Labor	Equipment	Total	Total Incl O&P
0010	**COMMERCIAL UNIT KITCHENS**									
1500	Combination range, refrigerator and sink, 30" wide, minimum	L-1	2	5	Ea.	925	199		1,124	1,350
1550	Maximum		1	10		3,675	400		4,075	4,700
1570	60" wide, average		1.40	7.143		3,225	284		3,509	4,000
1590	72" wide, average		1.20	8.333		4,200	330		4,530	5,175
1600	Office model, 48" wide		2	5		2,650	199		2,849	3,225
1620	Refrigerator and sink only	↓	2.40	4.167	↓	2,725	166		2,891	3,250
1640	Combination range, refrigerator, sink, microwave									
1660	Oven and ice maker	L-1	.80	12.500	Ea.	4,450	495		4,945	5,700

11 27 Photographic Processing Equipment

11 27 13 – Darkroom Processing Equipment

11 27 13.10 Darkroom Equipment

		Crew	Daily Output	Labor-Hours	Unit	Material	Labor	Equipment	Total	Total Incl O&P
0010	**DARKROOM EQUIPMENT**									
0020	Developing sink, 5" deep, 24" x 48"	Q-1	2	8	Ea.	5,625	285		5,910	6,650
0050	48" x 52"		1.70	9.412		5,675	335		6,010	6,775
0200	10" deep, 24" x 48"		1.70	9.412		7,775	335		8,110	9,100
0250	24" x 108"	↓	1.50	10.667		10,200	380		10,580	11,900
0500	Dryers, dehumidified filtered air, 36" x 25" x 68" high	L-7	6	4.333		5,075	126		5,201	5,775
0550	48" x 25" x 68" high		5	5.200		10,000	151		10,151	11,300
2000	Processors, automatic, color print, minimum		4	6.500		15,700	189		15,889	17,600
2050	Maximum		.60	43.333		24,700	1,250		25,950	29,300
2300	Black and white print, minimum		2	13		11,900	375		12,275	13,700
2350	Maximum		.80	32.500		61,500	945		62,445	69,000
2600	Manual processor, 16" x 20" maximum print size	↓	2	13	↓	9,450	375		9,825	11,000

11 27 Photographic Processing Equipment

11 27 13 – Darkroom Processing Equipment

11 27 13.10 Darkroom Equipment		Crew	Daily Output	Labor-Hours	Unit	Material	Labor	Equipment	Total	Total Incl O&P
						2011 Bare Costs				
2650	20″ x 24″ maximum print size	L-7	1	26	Ea.	8,975	755		9,730	11,100
3000	Viewing lights, 20″ x 24″		6	4.333		380	126		506	625
3100	20″ x 24″ with color correction		6	4.333		535	126		661	800
3500	Washers, round, minimum sheet 11″ x 14″	Q-1	2	8		3,400	285		3,685	4,225
3550	Maximum sheet 20″ x 24″		1	16		3,475	570		4,045	4,750
3800	Square, minimum sheet 20″ x 24″		1	16		3,125	570		3,695	4,350
3900	Maximum sheet 50″ x 56″		.80	20		4,850	715		5,565	6,500
4500	Combination tank sink, tray sink, washers, with									
4510	Dry side tables, average	Q-1	.45	35.556	Ea.	10,600	1,275		11,875	13,800

11 31 Residential Appliances

11 31 13 – Residential Kitchen Appliances

11 31 13.13 Cooking Equipment

		Crew	Daily Output	Labor-Hours	Unit	Material	Labor	Equipment	Total	Total Incl O&P
0010	**COOKING EQUIPMENT**									
0020	Cooking range, 30″ free standing, 1 oven, minimum	2 Clab	10	1.600	Ea.	505	42.50		547.50	625
0050	Maximum		4	4		1,700	106		1,806	2,050
0150	2 oven, minimum		10	1.600		1,050	42.50		1,092.50	1,225
0200	Maximum		10	1.600		1,100	42.50		1,142.50	1,275
0350	Built-in, 30″ wide, 1 oven, minimum	1 Elec	6	1.333		605	53.50		658.50	750
0400	Maximum	2 Carp	2	8		1,150	268		1,418	1,700
0500	2 oven, conventional, minimum		4	4		1,350	134		1,484	1,700
0550	1 conventional, 1 microwave, maximum		2	8		1,600	268		1,868	2,225
0700	Free-standing, 1 oven, 21″ wide range, minimum	2 Clab	10	1.600		330	42.50		372.50	430
0750	21″ wide, maximum	″	4	4		335	106		441	545
0900	Countertop cooktops, 4 burner, standard, minimum	1 Elec	6	1.333		284	53.50		337.50	395
0950	Maximum		3	2.667		1,100	107		1,207	1,375
1050	As above, but with grill and griddle attachment, minimum		6	1.333		1,225	53.50		1,278.50	1,425
1100	Maximum		3	2.667		3,125	107		3,232	3,625
1200	Induction cooktop, 30″ wide		3	2.667		1,125	107		1,232	1,425
1250	Microwave oven, minimum		4	2		110	80.50		190.50	252
1300	Maximum		2	4		495	161		656	805

11 31 13.23 Refrigeration Equipment

		Crew	Daily Output	Labor-Hours	Unit	Material	Labor	Equipment	Total	Total Incl O&P
0010	**REFRIGERATION EQUIPMENT**									
2000	Deep freeze, 15 to 23 C.F., minimum	2 Clab	10	1.600	Ea.	550	42.50		592.50	675
2050	Maximum		5	3.200		730	84.50		814.50	945
2200	30 C.F., minimum		8	2		800	53		853	970
2250	Maximum		3	5.333		850	141		991	1,175
5200	Icemaker, automatic, 20 lb. per day	1 Plum	7	1.143		880	45.50		925.50	1,050
5350	51 lb. per day	″	2	4		1,250	159		1,409	1,625
5500	Refrigerator, no frost, 10 C.F. to 12 C.F., minimum	2 Clab	10	1.600		415	42.50		457.50	525
5600	Maximum		6	2.667		505	70.50		575.50	675
5750	14 C.F. to 16 C.F., minimum		9	1.778		480	47		527	610
5800	Maximum		5	3.200		635	84.50		719.50	840
5950	18 C.F. to 20 C.F., minimum		8	2		610	53		663	760
6000	Maximum		4	4		1,175	106		1,281	1,475
6150	21 C.F. to 29 C.F., minimum		7	2.286		950	60.50		1,010.50	1,150
6200	Maximum		3	5.333		3,175	141		3,316	3,700
6790	Energy-star qualified, 18 C.F., minimum G	2 Carp	4	4		600	134		734	885
6795	Maximum G		2	8		1,300	268		1,568	1,875
6797	21.7 C.F., minimum G		4	4		920	134		1,054	1,225
6799	Maximum G		4	4		1,500	134		1,634	1,875

11 31 Residential Appliances

11 31 13 – Residential Kitchen Appliances

11 31 13.33 Kitchen Cleaning Equipment

	11 31 13.33 Kitchen Cleaning Equipment	Crew	Daily Output	Labor-Hours	Unit	Material	2011 Bare Costs Labor	Equipment	Total	Total Incl O&P
0010	**KITCHEN CLEANING EQUIPMENT**									
2750	Dishwasher, built-in, 2 cycles, minimum	L-1	4	2.500	Ea.	221	99.50		320.50	405
2800	Maximum		2	5		400	199		599	765
2950	4 or more cycles, minimum		4	2.500		325	99.50		424.50	515
2960	Average		4	2.500		430	99.50		529.50	635
3000	Maximum		2	5		1,075	199		1,274	1,525
3100	Energy-star qualified, minimum G		4	2.500		284	99.50		383.50	470
3110	Maximum G		2	5		1,300	199		1,499	1,750

11 31 13.43 Waste Disposal Equipment

	11 31 13.43 Waste Disposal Equipment	Crew	Daily Output	Labor-Hours	Unit	Material	Labor	Equipment	Total	Total Incl O&P
0010	**WASTE DISPOSAL EQUIPMENT**									
1750	Compactor, residential size, 4 to 1 compaction, minimum	1 Carp	5	1.600	Ea.	535	53.50		588.50	675
1800	Maximum	"	3	2.667		850	89.50		939.50	1,075
3300	Garbage disposal, sink type, minimum	L-1	10	1		72.50	40		112.50	145
3350	Maximum	"	10	1		191	40		231	275

11 31 13.53 Kitchen Ventilation Equipment

	11 31 13.53 Kitchen Ventilation Equipment	Crew	Daily Output	Labor-Hours	Unit	Material	Labor	Equipment	Total	Total Incl O&P
0010	**KITCHEN VENTILATION EQUIPMENT**									
4150	Hood for range, 2 speed, vented, 30" wide, minimum	L-3	5	2	Ea.	62.50	70		132.50	185
4200	Maximum		3	3.333		1,000	116		1,116	1,300
4300	42" wide, minimum		5	2		188	70		258	320
4330	Custom		5	2		1,650	70		1,720	1,950
4350	Maximum		3	3.333		2,025	116		2,141	2,425
4500	For ventless hood, 2 speed, add					17.05			17.05	18.75
4650	For vented 1 speed, deduct from maximum					45.50			45.50	50

11 31 23 – Residential Laundry Appliances

11 31 23.13 Washers

	11 31 23.13 Washers	Crew	Daily Output	Labor-Hours	Unit	Material	Labor	Equipment	Total	Total Incl O&P
0010	**WASHERS**									
5000	Residential, 4 cycle, average	1 Plum	3	2.667	Ea.	800	106		906	1,050
6650	Washing machine, automatic, minimum		3	2.667		400	106		506	615
6700	Maximum		1	8		1,300	315		1,615	1,950
6750	Energy star qualified, front loading, minimum G		3	2.667		620	106		726	860
6760	Maximum G		1	8		1,275	315		1,590	1,925
6764	Top loading, minimum G		3	2.667		370	106		476	585
6766	Maximum G		3	2.667		1,075	106		1,181	1,350

11 31 23.23 Dryers

	11 31 23.23 Dryers	Crew	Daily Output	Labor-Hours	Unit	Material	Labor	Equipment	Total	Total Incl O&P
0010	**DRYERS**									
0500	Gas fired residential, 16 lb. capacity, average	1 Plum	3	2.667	Ea.	635	106		741	875
6770	Electric, front loading, energy-star qualified, minimum G	L-2	3	5.333		500	157		657	815
6780	Maximum G	"	2	8		1,600	236		1,836	2,150
7450	Vent kits for dryers	1 Carp	10	.800		37.50	27		64.50	86

11 31 33 – Miscellaneous Residential Appliances

11 31 33.13 Sump Pumps

	11 31 33.13 Sump Pumps	Crew	Daily Output	Labor-Hours	Unit	Material	Labor	Equipment	Total	Total Incl O&P
0010	**SUMP PUMPS**									
6400	Cellar drainer, pedestal, 1/3 H.P., molded PVC base	1 Plum	3	2.667	Ea.	119	106		225	305
6450	Solid brass	"	2	4	"	263	159		422	550
6460	Sump pump, see also Section 22 14 29.16									

11 31 33.23 Water Heaters

	11 31 33.23 Water Heaters	Crew	Daily Output	Labor-Hours	Unit	Material	Labor	Equipment	Total	Total Incl O&P
0010	**WATER HEATERS**									
6900	Electric, glass lined, 30 gallon, minimum	L-1	5	2	Ea.	520	79.50		599.50	700
6950	Maximum		3	3.333		720	133		853	1,000
7100	80 gallon, minimum		2	5		935	199		1,134	1,350

11 31 Residential Appliances

11 31 33 – Miscellaneous Residential Appliances

11 31 33.23 Water Heaters		Crew	Daily Output	Labor-Hours	Unit	Material	2011 Bare Costs Labor	Equipment	Total	Total Incl O&P
7150	Maximum	L-1	1	10	Ea.	1,300	400		1,700	2,075
7180	Gas, glass lined, 30 gallon, minimum	2 Plum	5	3.200		805	127		932	1,100
7220	Maximum		3	5.333		1,125	211		1,336	1,575
7260	50 gallon, minimum		2.50	6.400		845	254		1,099	1,350
7300	Maximum		1.50	10.667		1,175	425		1,600	1,975
7310	Water heater, see also Section 22 33 30.13									

11 31 33.43 Air Quality		Crew	Daily Output	Labor-Hours	Unit	Material	Labor	Equipment	Total	Total Incl O&P
0010	**AIR QUALITY**									
2450	Dehumidifier, portable, automatic, 15 pint				Ea.	154			154	170
2550	40 pint					171			171	188
3550	Heater, electric, built-in, 1250 watt, ceiling type, minimum	1 Elec	4	2		98.50	80.50		179	240
3600	Maximum		3	2.667		161	107		268	350
3700	Wall type, minimum		4	2		159	80.50		239.50	305
3750	Maximum		3	2.667		171	107		278	360
3900	1500 watt wall type, with blower		4	2		159	80.50		239.50	305
3950	3000 watt		3	2.667		325	107		432	530
4850	Humidifier, portable, 8 gallons per day					165			165	182
5000	15 gallons per day					198			198	218

11 33 Retractable Stairs

11 33 10 – Disappearing Stairs

11 33 10.10 Disappearing Stairway		Crew	Daily Output	Labor-Hours	Unit	Material	Labor	Equipment	Total	Total Incl O&P
0010	**DISAPPEARING STAIRWAY** No trim included									
0100	Custom grade, pine, 8′-6″ ceiling, minimum	1 Carp	4	2	Ea.	142	67		209	268
0150	Average		3.50	2.286		217	76.50		293.50	365
0200	Maximum		3	2.667		289	89.50		378.50	470
0500	Heavy duty, pivoted, from 7′-7″ to 12′-10″ floor to floor		3	2.667		685	89.50		774.50	900
0600	16′-0″ ceiling		2	4		1,400	134		1,534	1,750
0800	Economy folding, pine, 8′-6″ ceiling		4	2		160	67		227	288
0900	9′-6″ ceiling		4	2		160	67		227	288
1100	Automatic electric, aluminum, floor to floor height, 8′ to 9′	2 Carp	1	16		8,675	535		9,210	10,500
1400	11′ to 12′		.90	17.778		9,325	595		9,920	11,300
1700	14′ to 15′		.70	22.857		9,950	765		10,715	12,200

11 41 Foodservice Storage Equipment

11 41 13 – Refrigerated Food Storage Cases

11 41 13.10 Refrigerated Food Cases		Crew	Daily Output	Labor-Hours	Unit	Material	Labor	Equipment	Total	Total Incl O&P
0010	**REFRIGERATED FOOD CASES**									
0030	Dairy, multi-deck, 12′ long	Q-5	3	5.333	Ea.	10,700	192		10,892	12,100
0100	For rear sliding doors, add					1,525			1,525	1,700
0200	Delicatessen case, service deli, 12′ long, single deck	Q-5	3.90	4.103		7,225	148		7,373	8,200
0300	Multi-deck, 18 S.F. shelf display		3	5.333		6,450	192		6,642	7,400
0400	Freezer, self-contained, chest-type, 30 C.F.		3.90	4.103		7,400	148		7,548	8,400
0500	Glass door, upright, 78 C.F.		3.30	4.848		10,000	175		10,175	11,300
0600	Frozen food, chest type, 12′ long		3.30	4.848		7,250	175		7,425	8,250
0700	Glass door, reach-in, 5 door		3	5.333		13,900	192		14,092	15,600
0800	Island case, 12′ long, single deck		3.30	4.848		8,225	175		8,400	9,300
0900	Multi-deck		3	5.333		17,400	192		17,592	19,400
1000	Meat case, 12′ long, single deck		3.30	4.848		5,975	175		6,150	6,825

11 41 Foodservice Storage Equipment

11 41 13 – Refrigerated Food Storage Cases

11 41 13.10 Refrigerated Food Cases		Crew	Daily Output	Labor-Hours	Unit	2011 Bare Costs Material	Labor	Equipment	Total	Total Incl O&P
1050	Multi-deck	Q-5	3.10	5.161	Ea.	10,300	186		10,486	11,600
1100	Produce, 12′ long, single deck		3.30	4.848		7,900	175		8,075	8,975
1200	Multi-deck		3.10	5.161		8,625	186		8,811	9,775
11 41 13.20 Refrigerated Food Storage Equipment										
0010	**REFRIGERATED FOOD STORAGE EQUIPMENT**									
2350	Cooler, reach-in, beverage, 6′ long	Q-1	6	2.667	Ea.	4,625	95		4,720	5,225
4300	Freezers, reach-in, 44 C.F.		4	4		2,700	143		2,843	3,200
4500	68 C.F.		3	5.333		3,650	190		3,840	4,300
4600	Freezer, pre-fab, 8′ x 8′ w/refrigeration	2 Carp	.45	35.556		10,800	1,200		12,000	13,900
4620	8′ x 12′		.35	45.714		12,800	1,525		14,325	16,700
4640	8′ x 16′		.25	64		15,800	2,150		17,950	21,000
4660	8′ x 20′		.17	94.118		19,100	3,150		22,250	26,300
4680	Reach-in, 1 compartment	Q-1	4	4		2,275	143		2,418	2,725
4685	Energy star rated G	R-18	7.80	3.333		2,225	104		2,329	2,625
4700	2 compartment	Q-1	3	5.333		3,600	190		3,790	4,250
4705	Energy star rated G	R-18	6.20	4.194		3,200	131		3,331	3,750
4710	3 compartment	Q-1	3	5.333		4,875	190		5,065	5,675
4715	Energy star rated G	R-18	5.60	4.643		4,025	144		4,169	4,675
8320	Refrigerator, reach-in, 1 compartment		7.80	3.333		2,175	104		2,279	2,575
8325	Energy star rated G		7.80	3.333		2,225	104		2,329	2,625
8330	2 compartment		6.20	4.194		3,275	131		3,406	3,825
8335	Energy star rated G		6.20	4.194		3,200	131		3,331	3,750
8340	3 compartment		5.60	4.643		4,000	144		4,144	4,650
8345	Energy star rated G		5.60	4.643		4,025	144		4,169	4,675
8350	Pre-fab, with refrigeration, 8′ x 8′	2 Carp	.45	35.556		6,350	1,200		7,550	8,975
8360	8′ x 12′		.35	45.714		7,525	1,525		9,050	10,900
8370	8′ x 16′		.25	64		11,200	2,150		13,350	15,900
8380	8′ x 20′		.17	94.118		14,200	3,150		17,350	20,900
8390	Pass-thru/roll-in, 1 compartment	R-18	7.80	3.333		4,075	104		4,179	4,650
8400	2 compartment		6.24	4.167		5,800	130		5,930	6,600
8410	3 compartment		5.60	4.643		7,925	144		8,069	8,950
8420	Walk-in, alum, door & floor only, no refrig, 6′ x 6′ x 7′-6″	2 Carp	1.40	11.429		8,725	385		9,110	10,200
8430	10′ x 6′ x 7′-6″		.55	29.091		12,500	975		13,475	15,300
8440	12′ x 14′ x 7′-6″		.25	64		17,500	2,150		19,650	22,800
8450	12′ x 20′ x 7′-6″		.17	94.118		21,300	3,150		24,450	28,700
8460	Refrigerated cabinets, mobile					3,650			3,650	4,025
8470	Refrigerator/freezer, reach-in, 1 compartment	R-18	5.60	4.643		4,825	144		4,969	5,550
8480	2 compartment	″	4.80	5.417		6,775	169		6,944	7,750
11 41 13.30 Wine Cellar										
0010	**WINE CELLAR**, refrigerated, Redwood interior, carpeted, walk-in type									
0020	6′-8″ high, including racks									
0200	80″ W x 48″ D for 900 bottles	2 Carp	1.50	10.667	Ea.	3,275	360		3,635	4,200
0250	80″ W x 72″ D for 1300 bottles		1.33	12.030		4,125	405		4,530	5,225
0300	80″ W x 94″ D for 1900 bottles		1.17	13.675		4,900	460		5,360	6,150
0400	80″ W x 124″ D for 2500 bottles		1	16		6,000	535		6,535	7,500
0600	Portable cabinets, red oak, reach-in temp.& humidity controlled									
0650	26-5/8″W x 26-1/2″D x 68″H for 235 bottles				Ea.	2,675			2,675	2,950
0660	32″W x 21-1/2″D x 73-1/2″H for 144 bottles					2,625			2,625	2,900
0670	32″W x 29-1/2″D x 73-1/2″H for 288 bottles					2,700			2,700	2,975
0680	39-1/2″W x 29-1/2″D x 86-1/2″H for 440 bottles					4,025			4,025	4,425
0690	52-1/2″W x 29-1/2″D x 73-1/2″H for 468 bottles					3,750			3,750	4,100
0700	52-1/2″W x 29-1/2″D x 86-1/2″H for 572 bottles					3,850			3,850	4,250

11 41 Foodservice Storage Equipment

11 41 13 – Refrigerated Food Storage Cases

11 41 13.30 Wine Cellar		Crew	Daily Output	Labor-Hours	Unit	Material	Labor	Equipment	Total	Total Incl O&P
						2011 Bare Costs				
0730	Portable, red oak, can be built-in with glass door									
0750	23-7/8"W x 24"D x 34-1/2"H for 50 bottles				Ea.	965			965	1,050

11 41 33 – Foodservice Shelving

11 41 33.20 Metal Food Storage Shelving

No.	Description	Crew	Daily Output	Labor-Hours	Unit	Material	Labor	Equipment	Total	Total Incl O&P
0010	**METAL FOOD STORAGE SHELVING**									
8600	Stainless steel shelving, louvered 4-tier, 20" x 3'	1 Clab	6	1.333	Ea.	1,250	35.50		1,285.50	1,425
8605	20" x 4'		6	1.333		1,800	35.50		1,835.50	2,025
8610	20" x 6'		6	1.333		2,600	35.50		2,635.50	2,900
8615	24" x 3'		6	1.333		1,650	35.50		1,685.50	1,850
8620	24" x 4'		6	1.333		1,925	35.50		1,960.50	2,150
8625	24" x 6'		6	1.333		2,775	35.50		2,810.50	3,100
8630	Flat 4-tier, 20" x 3'		6	1.333		1,125	35.50		1,160.50	1,275
8635	20" x 4'		6	1.333		1,350	35.50		1,385.50	1,550
8640	20" x 5'		6	1.333		1,575	35.50		1,610.50	1,775
8645	24" x 3'		6	1.333		1,275	35.50		1,310.50	1,450
8650	24" x 4'		6	1.333		2,100	35.50		2,135.50	2,350
8655	24" x 6'		6	1.333		2,500	35.50		2,535.50	2,800
8700	Galvanized shelving, louvered 4-tier, 20" x 3'		6	1.333		640	35.50		675.50	765
8705	20" x 4'		6	1.333		715	35.50		750.50	845
8710	20" x 6'		6	1.333		870	35.50		905.50	1,025
8715	24" x 3'		6	1.333		565	35.50		600.50	685
8720	24" x 4'		6	1.333		765	35.50		800.50	900
8725	24" x 6'		6	1.333		1,100	35.50		1,135.50	1,250
8730	Flat 4-tier, 20" x 3'		6	1.333		565	35.50		600.50	680
8735	20" x 4'		6	1.333		620	35.50		655.50	745
8740	20" x 6'		6	1.333		895	35.50		930.50	1,050
8745	24" x 3'		6	1.333		625	35.50		660.50	750
8750	24" x 4'		6	1.333		635	35.50		670.50	760
8755	24" x 6'		6	1.333		930	35.50		965.50	1,075
8760	Stainless steel dunnage rack, 24" x 3'		8	1		320	26.50		346.50	395
8765	24" x 4'		8	1		405	26.50		431.50	490
8770	Galvanized dunnage rack, 24" x 3'		8	1		162	26.50		188.50	223
8775	24" x 4'		8	1		182	26.50		208.50	246

11 42 Food Preparation Equipment

11 42 10 – Commercial Food Preparation Equipment

11 42 10.10 Choppers, Mixers and Misc. Equipment

No.	Description	Crew	Daily Output	Labor-Hours	Unit	Material	Labor	Equipment	Total	Total Incl O&P
0010	**CHOPPERS, MIXERS AND MISC. EQUIPMENT**									
1700	Choppers, 5 pounds	R-18	7	3.714	Ea.	1,825	116		1,941	2,200
1720	16 pounds		5	5.200		2,200	162		2,362	2,700
1740	35 to 40 pounds		4	6.500		3,300	202		3,502	3,950
1840	Coffee brewer, 5 burners	1 Plum	3	2.667		1,250	106		1,356	1,550
1850	Coffee urn, twin 6 gallon urns		2	4		2,725	159		2,884	3,250
1860	Single, 3 gallon		3	2.667		1,975	106		2,081	2,350
3000	Fast food equipment, total package, minimum	6 Skwk	.08	600		192,500	20,400		212,900	246,000
3100	Maximum	"	.07	685		262,500	23,300		285,800	327,500
3800	Food mixers, 20 quarts	L-7	7	3.714		2,525	108		2,633	2,950
3850	40 quarts		5.40	4.815		6,525	140		6,665	7,400
3900	60 quarts		5	5.200		9,925	151		10,076	11,200
4040	80 quarts		3.90	6.667		11,700	193		11,893	13,200

11 42 Food Preparation Equipment

11 42 10 – Commercial Food Preparation Equipment

11 42 10.10 Choppers, Mixers and Misc. Equipment		Crew	Daily Output	Labor-Hours	Unit	Material	2011 Bare Costs Labor	Equipment	Total	Total Incl O&P
4080	130 quarts	L-7	2.20	11.818	Ea.	17,100	345		17,445	19,400
4100	Floor type, 20 quarts		15	1.733		3,625	50.50		3,675.50	4,075
4120	60 quarts		14	1.857		9,050	54		9,104	10,000
4140	80 quarts		12	2.167		12,000	63		12,063	13,200
4160	140 quarts		8.60	3.023		18,700	87.50		18,787.50	20,700
6700	Peelers, small	R-18	8	3.250		1,600	101		1,701	1,925
6720	Large	"	6	4.333		4,750	135		4,885	5,450
6800	Pulper/extractor, close coupled, 5 HP	1 Plum	1.90	4.211		3,600	167		3,767	4,225
8580	Slicer with table	R-18	9	2.889		4,925	90		5,015	5,575

11 43 Food Delivery Carts and Conveyors

11 43 13 – Food Delivery Carts

11 43 13.10 Mobile Carts, Racks and Trays		Crew	Daily Output	Labor-Hours	Unit	Material	Labor	Equipment	Total	Total Incl O&P
0010	**MOBILE CARTS, RACKS AND TRAYS**									
1650	Cabinet, heated, 1 compartment, reach-in	R-18	5.60	4.643	Ea.	3,625	144		3,769	4,250
1655	Pass-thru roll-in		5.60	4.643		4,950	144		5,094	5,700
1660	2 compartment, reach-in		4.80	5.417		7,625	169		7,794	8,675
1670	Mobile					2,925			2,925	3,225
6850	Mobile rack w/pan slide					1,300			1,300	1,425
9180	Tray and silver dispenser, mobile	1 Clab	16	.500		855	13.25		868.25	960

11 44 Food Cooking Equipment

11 44 13 – Commercial Ranges

11 44 13.10 Cooking Equipment		Crew	Daily Output	Labor-Hours	Unit	Material	Labor	Equipment	Total	Total Incl O&P
0010	**COOKING EQUIPMENT**									
0020	Bake oven, gas, one section	Q-1	8	2	Ea.	4,650	71.50		4,721.50	5,250
0300	Two sections		7	2.286		9,300	81.50		9,381.50	10,300
0600	Three sections		6	2.667		13,500	95		13,595	15,000
0900	Electric convection, single deck	L-7	4	6.500		6,475	189		6,664	7,450
1300	Broiler, without oven, standard	Q-1	8	2		3,550	71.50		3,621.50	4,025
1550	Infrared	L-7	4	6.500		9,325	189		9,514	10,600
4750	Fryer, with twin baskets, modular model	Q-1	7	2.286		1,575	81.50		1,656.50	1,850
5000	Floor model, on 6" legs	"	5	3.200		1,975	114		2,089	2,350
5100	Extra single basket, large					102			102	112
5170	Energy star rated, 50 lb. capacity [G]	R-18	4	6.500		3,950	202		4,152	4,650
5175	85 lb. capacity [G]	"	4	6.500		7,450	202		7,652	8,500
5300	Griddle, SS, 24" plate, w/4" legs, elec, 208 V, 3 phase, 3' long	Q-1	7	2.286		1,300	81.50		1,381.50	1,550
5550	4' long	"	6	2.667		2,225	95		2,320	2,600
6200	Iced tea brewer	1 Plum	3.44	2.326		700	92		792	925
6350	Kettle, w/steam jacket, tilting, w/positive lock, SS, 20 gallons	L-7	7	3.714		7,400	108		7,508	8,300
6600	60 gallons	"	6	4.333		9,800	126		9,926	11,000
6900	Range, restaurant type, 6 burners and 1 standard oven, 36" wide	Q-1	7	2.286		2,825	81.50		2,906.50	3,225
6950	Convection		7	2.286		5,325	81.50		5,406.50	5,975
7150	2 standard ovens, 24" griddle, 60" wide		6	2.667		4,000	95		4,095	4,550
7200	1 standard, 1 convection oven		6	2.667		8,150	95		8,245	9,100
7450	Heavy duty, single 34" standard oven, open top		5	3.200		4,975	114		5,089	5,650
7500	Convection oven		5	3.200		4,650	114		4,764	5,275
7700	Griddle top		6	2.667		2,525	95		2,620	2,950
7750	Convection oven		6	2.667		7,000	95		7,095	7,850

11 44 Food Cooking Equipment

11 44 13 – Commercial Ranges

11 44 13.10 Cooking Equipment		Crew	Daily Output	Labor-Hours	Unit	Material	2011 Bare Costs Labor	Equipment	Total	Total Incl O&P
8850	Steamer, electric 27 KW	L-7	7	3.714	Ea.	8,125	108		8,233	9,125
9100	Electric, 10 KW or gas 100,000 BTU	″	5	5.200		5,975	151		6,126	6,825
9150	Toaster, conveyor type, 16-22 slices per minute					975			975	1,075
9160	Pop-up, 2 slot					600			600	660
9200	For deluxe models of above equipment, add					75%				
9400	Rule of thumb: Equipment cost based									
9410	on kitchen work area									
9420	Office buildings, minimum	L-7	77	.338	S.F.	85.50	9.80		95.30	111
9450	Maximum		58	.448		145	13		158	181
9550	Public eating facilities, minimum		77	.338		112	9.80		121.80	140
9600	Maximum		46	.565		183	16.40		199.40	229
9750	Hospitals, minimum		58	.448		115	13		128	149
9800	Maximum		39	.667		213	19.35		232.35	266

11 46 Food Dispensing Equipment

11 46 16 – Service Line Equipment

11 46 16.10 Commercial Food Dispensing Equipment		Crew	Daily Output	Labor-Hours	Unit	Material	2011 Bare Costs Labor	Equipment	Total	Total Incl O&P
0010	**COMMERCIAL FOOD DISPENSING EQUIPMENT**									
1050	Butter pat dispenser	1 Clab	13	.615	Ea.	925	16.30		941.30	1,050
1100	Bread dispenser, counter top		13	.615		885	16.30		901.30	1,000
1900	Cup and glass dispenser, drop in		4	2		960	53		1,013	1,150
1920	Disposable cup, drop in		16	.500		435	13.25		448.25	500
2650	Dish dispenser, drop in, 12″		11	.727		2,025	19.25		2,044.25	2,250
2660	Mobile		10	.800		2,450	21		2,471	2,725
3300	Food warmer, counter, 1.2 KW					480			480	530
3550	1.6 KW					1,975			1,975	2,175
3600	Well, hot food, built-in, rectangular, 12″ x 20″	R-30	10	2.600		640	83		723	835
3610	Circular, 7 qt.		10	2.600		360	83		443	535
3620	Refrigerated, 2 compartments		10	2.600		2,725	83		2,808	3,125
3630	3 compartments		9	2.889		3,325	92		3,417	3,825
3640	4 compartments		8	3.250		3,900	104		4,004	4,475
4720	Frost cold plate		9	2.889		17,400	92		17,492	19,400
5700	Hot chocolate dispenser	1 Plum	4	2		1,075	79.50		1,154.50	1,300
6250	Jet spray dispenser	R-18	4.50	5.778		3,100	180		3,280	3,700
6300	Juice dispenser, concentrate	″	4.50	5.778		2,150	180		2,330	2,650
6690	Milk dispenser, bulk, 2 flavor	R-30	8	3.250		1,525	104		1,629	1,850
6695	3 flavor	″	8	3.250		2,100	104		2,204	2,475
8800	Serving counter, straight	1 Carp	40	.200	L.F.	805	6.70		811.70	895
8820	Curved section	″	30	.267	″	1,000	8.95		1,008.95	1,125
8825	Solid surface, see Section 12 36 61.16									
8830	Soft serve ice cream machine, medium	R-18	11	2.364	Ea.	11,700	73.50		11,773.50	12,900
8840	Large	″	9	2.889	″	19,500	90		19,590	21,600

11 47 Ice Machines

11 47 10 – Commercial Ice Machines

11 47 10.10 Commercial Ice Equipment		Crew	Daily Output	Labor-Hours	Unit	2011 Bare Costs Material	Labor	Equipment	Total	Total Incl O&P
0010	**COMMERCIAL ICE EQUIPMENT**									
5800	Ice cube maker, 50 pounds per day	Q-1	6	2.667	Ea.	1,475	95		1,570	1,775
5900	250 pounds per day		1.20	13.333		2,600	475		3,075	3,625
6050	500 pounds per day		4	4		2,825	143		2,968	3,325
6060	With bin		1.20	13.333		2,900	475		3,375	3,975
6090	1000 pounds per day, with bin		1	16		4,725	570		5,295	6,100
6100	Ice flakers, 300 pounds per day		1.60	10		2,700	355		3,055	3,550
6120	600 pounds per day		.95	16.842		3,850	600		4,450	5,200
6130	1000 pounds per day		.75	21.333		4,325	760		5,085	6,000
6140	2000 pounds per day		.65	24.615		20,400	880		21,280	23,800
6160	Ice storage bin, 500 pound capacity	Q-5	1	16		1,000	575		1,575	2,050
6180	1000 pound	"	.56	28.571		2,425	1,025		3,450	4,350

11 48 Cleaning and Disposal Equipment

11 48 13 – Commercial Dishwashers

11 48 13.10 Dishwashers		Crew	Daily Output	Labor-Hours	Unit	Material	Labor	Equipment	Total	Total Incl O&P
0010	**DISHWASHERS**									
2700	Dishwasher, commercial, rack type									
2720	10 to 12 racks per hour	Q-1	3.20	5	Ea.	4,300	178		4,478	5,025
2730	Energy star rated, 35 to 40 racks/hour G		1.30	12.308		4,650	440		5,090	5,825
2740	50 to 60 racks/hour G		1.30	12.308		10,100	440		10,540	11,800
2800	Automatic, 190 to 230 racks per hour	L-6	.35	34.286		14,300	1,375		15,675	17,900
2820	235 to 275 racks per hour		.25	48		30,300	1,925		32,225	36,400
2840	8,750 to 12,500 dishes per hour		.10	120		49,500	4,775		54,275	62,500
2950	Dishwasher hood, canopy type	L-3A	10	1.200	L.F.	800	43.50		843.50	955
2960	Pant leg type	"	2.50	4.800	Ea.	8,225	175		8,400	9,350
5200	Garbage disposal 1.5 HP, 100 GPH	L-1	4.80	2.083		1,950	83		2,033	2,250
5210	3 HP, 120 GPH		4.60	2.174		2,075	86.50		2,161.50	2,425
5220	5 HP, 250 GPH		4.50	2.222		3,275	88.50		3,363.50	3,750
6750	Pot sink, 3 compartment	1 Plum	7.25	1.103	L.F.	1,000	44		1,044	1,175
6760	Pot washer, low temp wash/rinse		1.60	5	Ea.	3,825	198		4,023	4,525
6770	High pressure wash, high temperature rinse		1.20	6.667		36,900	264		37,164	41,000
9170	Trash compactor, small, up to 125 lb. compacted weight	L-4	4	4		22,800	119		22,919	25,300
9175	Large, up to 175 lb. compacted weight	"	3	5.333		27,900	158		28,058	31,000

11 52 Audio-Visual Equipment

11 52 13 – Projection Screens

11 52 13.10 Projection Screens, Wall or Ceiling Hung		Crew	Daily Output	Labor-Hours	Unit	Material	Labor	Equipment	Total	Total Incl O&P
0010	**PROJECTION SCREENS, WALL OR CEILING HUNG**, matte white									
0100	Manually operated, economy	2 Carp	500	.032	S.F.	5.85	1.07		6.92	8.25
0300	Intermediate		450	.036		6.80	1.19		7.99	9.50
0400	Deluxe		400	.040		9.45	1.34		10.79	12.60
0600	Electric operated, matte white, 25 S.F., economy		5	3.200	Ea.	855	107		962	1,125
0700	Deluxe		4	4		1,775	134		1,909	2,175
0900	50 S.F., economy		3	5.333		750	179		929	1,125
1000	Deluxe		2	8		1,975	268		2,243	2,625
1200	Heavy duty, electric operated, 200 S.F.		1.50	10.667		3,900	360		4,260	4,900
1300	400 S.F.		1	16		4,800	535		5,335	6,200
1500	Rigid acrylic in wall, for rear projection, 1/4" thick	2 Glaz	30	.533	S.F.	47	17.70		64.70	81

11 52 Audio-Visual Equipment

11 52 13 – Projection Screens

11 52 13.10 Projection Screens, Wall or Ceiling Hung		Crew	Daily Output	Labor-Hours	Unit	2011 Bare Costs Material	Labor	Equipment	Total	Total Incl O&P
1600	1/2″ thick (maximum size 10′ x 20′)	2 Glaz	25	.640	S.F.	83.50	21.50		105	127

11 52 16 – Projectors

11 52 16.10 Movie Equipment

		Crew	Daily Output	Labor-Hours	Unit	Material	Labor	Equipment	Total	Total Incl O&P
0010	**MOVIE EQUIPMENT**									
0020	Changeover, minimum				Ea.	465			465	510
0100	Maximum					905			905	995
0400	Film transport, incl. platters and autowind, minimum					5,025			5,025	5,525
0500	Maximum					14,300			14,300	15,700
0800	Lamphouses, incl. rectifiers, xenon, 1,000 watt	1 Elec	2	4		6,650	161		6,811	7,550
0900	1,600 watt		2	4		7,100	161		7,261	8,050
1000	2,000 watt		1.50	5.333		7,625	215		7,840	8,725
1100	4,000 watt		1.50	5.333		9,425	215		9,640	10,800
1400	Lenses, anamorphic, minimum					1,275			1,275	1,400
1500	Maximum					2,875			2,875	3,150
1800	Flat 35 mm, minimum					1,100			1,100	1,225
1900	Maximum					1,725			1,725	1,900
2200	Pedestals, for projectors					1,475			1,475	1,625
2300	Console type					10,700			10,700	11,800
2600	Projector mechanisms, incl. soundhead, 35 mm, minimum					11,100			11,100	12,200
2700	Maximum					15,200			15,200	16,700
3000	Projection screens, rigid, in wall, acrylic, 1/4″ thick	2 Glaz	195	.082	S.F.	42	2.72		44.72	50.50
3100	1/2″ thick	″	130	.123	″	48.50	4.09		52.59	60
3300	Electric operated, heavy duty, 400 S.F.	2 Carp	1	16	Ea.	2,950	535		3,485	4,150
3320	Theater projection screens, matte white, including frames	″	200	.080	S.F.	6.65	2.68		9.33	11.80
3400	Also see Section 11 52 13.10									
3700	Sound systems, incl. amplifier, mono, minimum	1 Elec	.90	8.889	Ea.	3,300	360		3,660	4,200
3800	Dolby/Super Sound, maximum		.40	20		18,100	805		18,905	21,200
4100	Dual system, 2 channel, front surround, minimum		.70	11.429		4,625	460		5,085	5,850
4200	Dolby/Super Sound, 4 channel, maximum		.40	20		16,500	805		17,305	19,500
4500	Sound heads, 35 mm					5,300			5,300	5,825
4900	Splicer, tape system, minimum					740			740	815
5000	Tape type, maximum					1,325			1,325	1,450
5300	Speakers, recessed behind screen, minimum	1 Elec	2	4		1,050	161		1,211	1,425
5400	Maximum	″	1	8		3,100	320		3,420	3,925
5700	Seating, painted steel, upholstered, minimum	2 Carp	35	.457		132	15.35		147.35	171
5800	Maximum	″	28	.571		420	19.15		439.15	495
6100	Rewind tables, minimum					2,650			2,650	2,900
6200	Maximum					4,725			4,725	5,175
7000	For automation, varying sophistication, minimum	1 Elec	1	8	System	2,375	320		2,695	3,150
7100	Maximum	2 Elec	.30	53.333	″	5,575	2,150		7,725	9,600

11 52 16.20 Movie Equipment- Digital

		Crew	Daily Output	Labor-Hours	Unit	Material	Labor	Equipment	Total	Total Incl O&P
0010	**MOVIE EQUIPMENT- DIGITAL**									
1000	Digital 2K projection system, 98″ DMD	1 Elec	2	4	Ea.	35,800	161		35,961	39,600
1100	OEM lens		2	4		5,125	161		5,286	5,875
2000	Pedestal with power distribution		2	4		1,950	161		2,111	2,400
3000	Software		2	4		1,600	161		1,761	2,000

11 53 Laboratory Equipment

11 53 03 – Laboratory Test Equipment

11 53 03.13 Test Equipment		Crew	Daily Output	Labor-Hours	Unit	Material	Labor	Equipment	Total	Total Incl O&P
						2011 Bare Costs				
0010	**TEST EQUIPMENT**									
1700	Thermometer, electric, portable				Ea.	365			365	400
1800	Titration unit, four 2000 ml reservoirs				"	5,500			5,500	6,050

11 53 13 – Laboratory Fume Hoods

11 53 13.13 Recirculating Laboratory Fume Hoods

Code	Description	Crew	Daily Output	Labor-Hours	Unit	Material	Labor	Equipment	Total	Total Incl O&P
0010	**RECIRCULATING LABORATORY FUME HOODS**									
0600	Fume hood, with countertop & base, not including HVAC									
0610	Simple, minimum	2 Carp	5.40	2.963	L.F.	635	99.50		734.50	865
0620	Complex, including fixtures		2.40	6.667		1,800	224		2,024	2,350
0630	Special, maximum		1.70	9.412		1,825	315		2,140	2,525
0670	Service fixtures, average				Ea.	287			287	315
0680	For sink assembly with hot and cold water, add	1 Plum	1.40	5.714		705	227		932	1,150
0750	Glove box, fiberglass, bacteriological					16,900			16,900	18,600
0760	Controlled atmosphere					18,900			18,900	20,700
0770	Radioisotope					16,900			16,900	18,600
0780	Carcinogenic					16,900			16,900	18,600

11 53 13.23 Exhaust Hoods

Code	Description	Crew	Daily Output	Labor-Hours	Unit	Material	Labor	Equipment	Total	Total Incl O&P
0010	**EXHAUST HOODS**									
0650	Ductwork, minimum	2 Shee	1	16	Hood	3,625	610		4,235	5,000
0660	Maximum	"	.50	32	"	5,575	1,225		6,800	8,175

11 53 16 – Laboratory Incubators

11 53 16.13 Incubators

Code	Description	Crew	Daily Output	Labor-Hours	Unit	Material	Labor	Equipment	Total	Total Incl O&P
0010	**INCUBATORS**									
1000	Incubators, minimum				Ea.	3,400			3,400	3,750
1010	Maximum				"	20,100			20,100	22,100

11 53 19 – Laboratory Sterilizers

11 53 19.13 Sterilizers

Code	Description	Crew	Daily Output	Labor-Hours	Unit	Material	Labor	Equipment	Total	Total Incl O&P
0010	**STERILIZERS**									
0700	Glassware washer, undercounter, minimum	L-1	1.80	5.556	Ea.	5,850	221		6,071	6,800
0710	Maximum	"	1	10		11,700	400		12,100	13,500
1850	Utensil washer-sanitizer	1 Plum	2	4		10,900	159		11,059	12,300

11 53 23 – Laboratory Refrigerators

11 53 23.13 Refrigerators

Code	Description	Crew	Daily Output	Labor-Hours	Unit	Material	Labor	Equipment	Total	Total Incl O&P
0010	**REFRIGERATORS**									
1200	Blood bank, 28.6 C.F. emergency signal				Ea.	8,675			8,675	9,550
1210	Reach-in, 16.9 C.F.				"	8,175			8,175	9,000

11 53 33 – Emergency Safety Appliances

11 53 33.13 Emergency Equipment

Code	Description	Crew	Daily Output	Labor-Hours	Unit	Material	Labor	Equipment	Total	Total Incl O&P
0010	**EMERGENCY EQUIPMENT**									
1400	Safety equipment, eye wash, hand held				Ea.	405			405	445
1450	Deluge shower				"	740			740	815

11 53 43 – Service Fittings and Accessories

11 53 43.13 Fittings

Code	Description	Crew	Daily Output	Labor-Hours	Unit	Material	Labor	Equipment	Total	Total Incl O&P
0010	**FITTINGS**									
1600	Sink, one piece plastic, flask wash, hose, free standing	1 Plum	1.60	5	Ea.	1,800	198		1,998	2,300
1610	Epoxy resin sink, 25" x 16" x 10"	"	2	4	"	198	159		357	475
1950	Utility table, acid resistant top with drawers	2 Carp	30	.533	L.F.	147	17.90		164.90	191
8000	Alternate pricing method: as percent of lab furniture									
8050	Installation, not incl. plumbing & duct work				% Furn.				20%	22%

11 53 Laboratory Equipment

11 53 43 – Service Fittings and Accessories

11 53 43.13 Fittings		Crew	Daily Output	Labor-Hours	Unit	2011 Bare Costs Material	Labor	Equipment	Total	Total Incl O&P
8100	Plumbing, final connections, simple system				% Furn.				9.09%	10%
8110	Moderately complex system								13.64%	15%
8120	Complex system								18.18%	20%
8150	Electrical, simple system								9.09%	10%
8160	Moderately complex system								18.18%	20%
8170	Complex system								31.80%	35%

11 57 Vocational Shop Equipment

11 57 10 – Shop Equipment

11 57 10.10 Vocational School Shop Equipment

		Crew	Daily Output	Labor-Hours	Unit	Material	Labor	Equipment	Total	Total Incl O&P
0010	**VOCATIONAL SCHOOL SHOP EQUIPMENT**									
0020	Benches, work, wood, average	2 Carp	5	3.200	Ea.	525	107		632	755
0100	Metal, average		5	3.200		475	107		582	700
0400	Combination belt & disc sander, 6"		4	4		1,400	134		1,534	1,750
0700	Drill press, floor mounted, 12", 1/2 H.P.		4	4		425	134		559	690
0800	Dust collector, not incl. ductwork, 6" diameter	1 Shee	1.10	7.273		4,500	278		4,778	5,400
0810	Dust collector bag, 20" diameter	"	5	1.600		310	61		371	440
1000	Grinders, double wheel, 1/2 H.P.	2 Carp	5	3.200		254	107		361	460
1300	Jointer, 4", 3/4 H.P.		4	4		1,325	134		1,459	1,675
1600	Kilns, 16 C.F., to 2000°		4	4		2,275	134		2,409	2,725
1900	Lathe, woodworking, 10", 1/2 H.P.		4	4		735	134		869	1,025
2200	Planer, 13" x 6"		4	4		1,000	134		1,134	1,325
2500	Potter's wheel, motorized		4	4		1,175	134		1,309	1,525
2800	Saws, band, 14", 3/4 H.P.		4	4		865	134		999	1,175
3100	Metal cutting band saw, 14"		4	4		2,200	134		2,334	2,625
3400	Radial arm saw, 10", 2 H.P.		4	4		1,325	134		1,459	1,675
3700	Scroll saw, 24"		4	4		590	134		724	875
4000	Table saw, 10", 3 H.P.		4	4		2,150	134		2,284	2,575
4300	Welder AC arc, 30 amp capacity		4	4		2,100	134		2,234	2,550

11 61 Theater and Stage Equipment

11 61 23 – Folding and Portable Stages

11 61 23.10 Portable Stages

		Crew	Daily Output	Labor-Hours	Unit	Material	Labor	Equipment	Total	Total Incl O&P
0010	**PORTABLE STAGES**									
1500	Flooring, portable oak parquet, 3' x 3' sections				S.F.	13.40			13.40	14.70
1600	Cart to carry 225 S.F. of flooring				Ea.	400			400	440
5000	Stages, portable with steps, folding legs, stock, 8" high				SF Stg.	30.50			30.50	33.50
5100	16" high					23.50			23.50	26
5200	32" high					41.50			41.50	46
5300	40" high					59.50			59.50	65
6000	Telescoping platforms, extruded alum., straight, minimum	4 Carp	157	.204		33	6.85		39.85	47.50
6100	Maximum		77	.416		44.50	13.95		58.45	72.50
6500	Pie-shaped, minimum		150	.213		71	7.15		78.15	90.50
6600	Maximum		70	.457		79.50	15.35		94.85	113
6800	For 3/4" plywood covered deck, deduct					4.05			4.05	4.46
7000	Band risers, steel frame, plywood deck, minimum	4 Carp	275	.116		30.50	3.90		34.40	40
7100	Maximum	"	138	.232		67	7.80		74.80	86.50
7500	Chairs for above, self-storing, minimum	2 Carp	43	.372	Ea.	107	12.50		119.50	139
7600	Maximum	"	40	.400	"	189	13.40		202.40	231

11 61 Theater and Stage Equipment

11 61 33 – Rigging Systems and Controls

11 61 33.10 Controls		Crew	Daily Output	Labor-Hours	Unit	2011 Bare Costs Material	Labor	Equipment	Total	Total Incl O&P
0010	**CONTROLS**									
0050	Control boards with dimmers and breakers, minimum	1 Elec	1	8	Ea.	12,300	320		12,620	14,000
0100	Average		.50	16		38,600	645		39,245	43,500
0150	Maximum		.20	40		126,000	1,600		127,600	141,000
8000	Rule of thumb: total stage equipment, minimum	4 Carp	100	.320	SF Stg.	97	10.75		107.75	125
8100	Maximum	"	25	1.280	"	545	43		588	670

11 61 43 – Stage Curtains

11 61 43.10 Curtains		Crew	Daily Output	Labor-Hours	Unit	Material	Labor	Equipment	Total	Total Incl O&P
0010	**CURTAINS**									
0500	Curtain track, straight, light duty	2 Carp	20	.800	L.F.	27	27		54	74.50
0600	Heavy duty		18	.889		61	30		91	117
0700	Curved sections		12	1.333		173	44.50		217.50	265
1000	Curtains, velour, medium weight		600	.027	S.F.	7.90	.89		8.79	10.20
1150	Silica based yarn, inherently fire retardant		50	.320	"	15.20	10.75		25.95	34.50

11 62 Musical Equipment

11 62 16 – Carillons

11 62 16.10 Bell Tower Equipment		Crew	Daily Output	Labor-Hours	Unit	Material	Labor	Equipment	Total	Total Incl O&P
0010	**BELL TOWER EQUIPMENT**									
0300	Carillon, 4 octave (48 bells), with keyboard				System	937,000			937,000	1,031,000
0320	2 octave (24 bells)					441,000			441,000	485,000
0340	3 to 4 bell peal, minimum					110,500			110,500	121,500
0360	Maximum					661,500			661,500	727,500
0380	Cast bronze bell, average				Ea.	99,000			99,000	109,000
0400	Electronic, digital, minimum					16,500			16,500	18,200
0410	With keyboard, maximum					82,500			82,500	91,000

11 66 Athletic Equipment

11 66 13 – Exercise Equipment

11 66 13.10 Physical Training Equipment		Crew	Daily Output	Labor-Hours	Unit	Material	Labor	Equipment	Total	Total Incl O&P
0010	**PHYSICAL TRAINING EQUIPMENT**									
0020	Abdominal rack, 2 board capacity				Ea.	485			485	535
0050	Abdominal board, upholstered					605			605	665
0200	Bicycle trainer, minimum					400			400	440
0300	Deluxe, electric					3,800			3,800	4,175
0400	Barbell set, chrome plated steel, 25 lb.					281			281	310
0420	100 lb.					420			420	460
0450	200 lb.					685			685	755
0500	Weight plates, cast iron, per lb.				Lb.	5.20			5.20	5.70
0520	Storage rack, 10 station				Ea.	900			900	990
0600	Circuit training apparatus, 12 machines minimum	2 Clab	1.25	12.800	Set	28,700	340		29,040	32,100
0700	Average		1	16		35,200	425		35,625	39,400
0800	Maximum		.75	21.333		41,700	565		42,265	46,800
0820	Dumbbell set, cast iron, with rack and 5 pair					605			605	670
0900	Squat racks	2 Clab	5	3.200	Ea.	890	84.50		974.50	1,125
1200	Multi-station gym machine, 5 station					5,000			5,000	5,500
1250	9 station					10,000			10,000	11,000
1280	Rowing machine, hydraulic					1,700			1,700	1,875

11 66 Athletic Equipment

11 66 13 – Exercise Equipment

11 66 13.10 Physical Training Equipment		Crew	Daily Output	Labor-Hours	Unit	2011 Bare Costs Material	Labor	Equipment	Total	Total Incl O&P
1300	Treadmill, manual				Ea.	1,000			1,000	1,100
1320	Motorized					3,625			3,625	4,000
1340	Electronic					3,700			3,700	4,075
1360	Cardio-testing					4,000			4,000	4,400
1400	Treatment/massage tables, minimum					590			590	650
1420	Deluxe, with accessories					715			715	785
4150	Exercise equipment, bicycle trainer					595			595	655
4180	Chinning bar, adjustable, wall mounted	1 Carp	5	1.600		300	53.50		353.50	420
4200	Exercise ladder, 16′ x 1′-7″, suspended	L-2	3	5.333		1,525	157		1,682	1,975
4210	High bar, floor plate attached	1 Carp	4	2		1,150	67		1,217	1,375
4240	Parallel bars, adjustable		4	2		2,125	67		2,192	2,450
4270	Uneven parallel bars, adjustable		4	2		2,775	67		2,842	3,150
4280	Wall mounted, adjustable	L-2	1.50	10.667	Set	820	315		1,135	1,425
4300	Rope, ceiling mounted, 18′ long	1 Carp	3.66	2.186	Ea.	268	73.50		341.50	420
4330	Side horse, vaulting		5	1.600		1,775	53.50		1,828.50	2,050
4360	Treadmill, motorized, deluxe, training type		5	1.600		3,000	53.50		3,053.50	3,400
4390	Weight lifting multi-station, minimum	2 Clab	1	16		450	425		875	1,200
4450	Maximum	″	.50	32		14,700	845		15,545	17,600

11 66 23 – Gymnasium Equipment

11 66 23.13 Basketball Equipment

		Crew	Daily Output	Labor-Hours	Unit	Material	Labor	Equipment	Total	Total Incl O&P
0010	**BASKETBALL EQUIPMENT**									
1000	Backstops, wall mtd., 6′ extended, fixed, minimum	L-2	1	16	Ea.	1,175	470		1,645	2,100
1100	Maximum		1	16		1,750	470		2,220	2,725
1200	Swing up, minimum		1	16		1,425	470		1,895	2,375
1250	Maximum		1	16		5,575	470		6,045	6,925
1300	Portable, manual, heavy duty, spring operated		1.90	8.421		13,800	248		14,048	15,600
1400	Ceiling suspended, stationary, minimum		.78	20.513		3,550	605		4,155	4,900
1450	Fold up, with accessories, maximum		.40	40		6,350	1,175		7,525	8,950
1600	For electrically operated, add	1 Elec	1	8		2,075	320		2,395	2,825
5800	Wall pads, 1-1/2″ thick, standard (not fire rated)	2 Carp	640	.025	S.F.	6.95	.84		7.79	9.05

11 66 23.19 Boxing Ring

		Crew	Daily Output	Labor-Hours	Unit	Material	Labor	Equipment	Total	Total Incl O&P
0010	**BOXING RING**									
4100	Elevated, 22′ x 22′	L-4	.10	160	Ea.	7,725	4,750		12,475	16,500
4110	For cellular plastic foam padding, add		.10	160		900	4,750		5,650	8,950
4120	Floor level, including posts and ropes only, 20′ x 20′		.80	20		3,750	595		4,345	5,125
4130	Canvas, 30′ x 30′		5	3.200		1,200	95		1,295	1,475

11 66 23.47 Gym Mats

		Crew	Daily Output	Labor-Hours	Unit	Material	Labor	Equipment	Total	Total Incl O&P
0010	**GYM MATS**									
5500	2″ thick, naugahyde covered				S.F.	3.77			3.77	4.15
5600	Vinyl/nylon covered					6.35			6.35	7
6000	Wrestling mats, 1″ thick, heavy duty					5.55			5.55	6.10

11 66 43 – Interior Scoreboards

11 66 43.10 Scoreboards

		Crew	Daily Output	Labor-Hours	Unit	Material	Labor	Equipment	Total	Total Incl O&P
0010	**SCOREBOARDS**									
7000	Baseball, minimum	R-3	1.30	15.385	Ea.	4,000	610	99	4,709	5,500
7200	Maximum		.05	400		17,500	15,800	2,575	35,875	47,800
7300	Football, minimum		.86	23.256		5,175	920	150	6,245	7,350
7400	Maximum		.20	100		15,400	3,950	645	19,995	24,100
7500	Basketball (one side), minimum		2.07	9.662		2,325	380	62.50	2,767.50	3,250
7600	Maximum		.30	66.667		3,400	2,625	430	6,455	8,500
7700	Hockey-basketball (four sides), minimum		.25	80		5,500	3,175	515	9,190	11,800

11 66 Athletic Equipment

11 66 43 – Interior Scoreboards

11 66 43.10 Scoreboards		Crew	Daily Output	Labor-Hours	Unit	2011 Bare Costs Material	Labor	Equipment	Total	Total Incl O&P
7800	Maximum	R-3	.15	133	Ea.	5,575	5,275	860	11,710	15,600

11 66 53 – Gymnasium Dividers

11 66 53.10 Divider Curtains

		Crew	Daily Output	Labor-Hours	Unit	Material	Labor	Equipment	Total	Total Incl O&P
0010	**DIVIDER CURTAINS**									
4500	Gym divider curtain, mesh top, vinyl bottom, manual	L-4	500	.032	S.F.	8.70	.95		9.65	11.20
4700	Electric roll up	L-7	400	.065	"	10.65	1.89		12.54	14.85

11 67 Recreational Equipment

11 67 13 – Bowling Alley Equipment

11 67 13.10 Bowling Alleys

		Crew	Daily Output	Labor-Hours	Unit	Material	Labor	Equipment	Total	Total Incl O&P
0010	**BOWLING ALLEYS** Including alley, pinsetter, scorer,									
0020	Counters and misc. supplies, minimum	4 Carp	.20	160	Lane	41,300	5,375		46,675	54,500
0150	Average		.19	168		45,400	5,650		51,050	59,500
0300	Maximum		.18	177		52,000	5,975		57,975	67,500
0400	Combo table ball rack, add					1,150			1,150	1,250
0600	For automatic scorer, add, minimum					8,225			8,225	9,050
0700	Maximum					9,875			9,875	10,900

11 67 23 – Shooting Range Equipment

11 67 23.10 Shooting Range

		Crew	Daily Output	Labor-Hours	Unit	Material	Labor	Equipment	Total	Total Incl O&P
0010	**SHOOTING RANGE** Incl. bullet traps, target provisions, controls,									
0100	Separators, ceiling system, etc. Not incl. structural shell									
0200	Commercial	L-9	.64	56.250	Point	17,500	1,750		19,250	22,200
0300	Law enforcement		.28	128		33,000	4,025		37,025	43,000
0400	National Guard armories		.71	50.704		21,000	1,600		22,600	25,800
0500	Reserve training centers		.71	50.704		15,000	1,600		16,600	19,200
0600	Schools and colleges		.32	112		31,000	3,525		34,525	40,000
0700	Major academies		.19	189		46,000	5,925		51,925	60,500
0800	For acoustical treatment, add					10%	10%			
0900	For lighting, add					28%	25%			
1000	For plumbing, add					5%	5%			
1100	For ventilating system, add, minimum					40%	40%			
1200	Add, average					25%	25%			
1300	Add, maximum					35%	35%			

11 68 Play Field Equipment and Structures

11 68 13 – Playground Equipment

11 68 13.10 Free-Standing Playground Equipment

		Crew	Daily Output	Labor-Hours	Unit	Material	Labor	Equipment	Total	Total Incl O&P
0010	**FREE-STANDING PLAYGROUND EQUIPMENT** See also individual items									
0200	Bike rack, 10′ long, permanent G	B-1	12	2	Ea.	555	54		609	700
0400	Horizontal monkey ladder, 14′ long, 6′ high		4	6		940	163		1,103	1,300
0590	Parallel bars, 10′ long		4	6		273	163		436	575
0600	Posts, tether ball set, 2-3/8″ O.D.		12	2		300	54		354	420
0800	Poles, multiple purpose, 10′-6″ long		12	2	Pr.	160	54		214	267
1000	Ground socket for movable posts, 2-3/8″ post		10	2.400		102	65		167	221
1100	3-1/2″ post		10	2.400		139	65		204	262
1300	See-saw, spring, steel, 2 units		6	4	Ea.	760	108		868	1,025
1400	4 units		4	6		1,375	163		1,538	1,775
1500	6 units		3	8		1,675	217		1,892	2,200

11 68 Play Field Equipment and Structures

11 68 13 – Playground Equipment

11 68 13.10 Free-Standing Playground Equipment		Crew	Daily Output	Labor-Hours	Unit	2011 Bare Costs Material	Labor	Equipment	Total	Total Incl O&P
1700	Shelter, fiberglass golf tee, 3 person	B-1	4.60	5.217	Ea.	2,675	142		2,817	3,175
1900	Slides, stainless steel bed, 12′ long, 6′ high		3	8		3,525	217		3,742	4,250
2000	20′ long, 10′ high		2	12		4,250	325		4,575	5,225
2200	Swings, plain seats, 8′ high, 4 seats		2	12		1,025	325		1,350	1,675
2300	8 seats		1.30	18.462		1,900	500		2,400	2,950
2500	12′ high, 4 seats		2	12		1,975	325		2,300	2,725
2600	8 seats		1.30	18.462		3,450	500		3,950	4,625
2800	Whirlers, 8′ diameter		3	8		2,900	217		3,117	3,575
2900	10′ diameter		3	8		6,050	217		6,267	7,025

11 68 13.20 Modular Playground		Crew	Daily Output	Labor-Hours	Unit	Material	Labor	Equipment	Total	Total Incl O&P
0010	**MODULAR PLAYGROUND** Basic components									
0100	Deck, square, steel, 48″ x 48″	B-1	1	24	Ea.	525	650		1,175	1,675
0110	Recycled polyurethane		1	24		525	650		1,175	1,675
0120	Triangular, steel, 48″ side		1	24		630	650		1,280	1,800
0130	Post, steel, 5″ square		18	1.333	L.F.	35	36		71	99
0140	Aluminum, 2-3/8″ square		20	1.200	″	30	32.50		62.50	87.50
0160	Roof, square poly, 54″ side		18	1.333	Ea.	1,150	36		1,186	1,325
0170	Wheelchair transfer module, for 3′ high deck		3	8	″	2,725	217		2,942	3,375
0180	Guardrail, pipe, 36″ high		60	.400	L.F.	202	10.85		212.85	240
0190	Steps, deck-to-deck, 3 – 8″ steps		8	3	Ea.	990	81.50		1,071.50	1,200
0200	Activity panel, crawl through panel		2	12		365	325		690	950
0210	Alphabet/spelling panel		2	12		495	325		820	1,100
0360	With guardrails		3	8		1,625	217		1,842	2,150
0370	Crawl tunnel, straight, 56″ long		4	6		925	163		1,088	1,300
0380	90°, 4′ long		4	6		965	163		1,128	1,350
1200	Slide, tunnel, for 56″ high deck		8	3		1,275	81.50		1,356.50	1,550
1210	Straight, poly		8	3		2,325	81.50		2,406.50	2,675
1220	Stainless steel, 54″ high deck		6	4		520	108		628	755
1230	Curved, poly, 40″ high deck		6	4		610	108		718	855
1240	Spiral slide, 56″ - 72″ high		5	4.800		3,225	130		3,355	3,775
1300	Ladder, vertical, for 24″ - 72″ high deck		5	4.800		600	130		730	880
1310	Horizontal, 8′ long		5	4.800		580	130		710	855
1320	Corkscrew climber, 6′ high		3	8		595	217		812	1,025
1330	Fire pole for 72″ high deck		6	4		550	108		658	785
1340	Bridge, ring climber, 8′ long		4	6		1,475	163		1,638	1,900
1350	Suspension		4	6	L.F.	1,025	163		1,188	1,400

11 68 16 – Play Structures

11 68 16.10 Handball/Squash Court		Crew	Daily Output	Labor-Hours	Unit	Material	Labor	Equipment	Total	Total Incl O&P
0010	**HANDBALL/SQUASH COURT**, outdoor									
0900	Handball or squash court, outdoor, wood	2 Carp	.50	32	Ea.	5,250	1,075		6,325	7,600
1000	Masonry handball/squash court	D-1	.30	53.333	″	24,000	1,625		25,625	29,000

11 68 16.30 Platform/Paddle Tennis Court		Crew	Daily Output	Labor-Hours	Unit	Material	Labor	Equipment	Total	Total Incl O&P
0010	**PLATFORM/PADDLE TENNIS COURT** Complete with lighting, etc.									
0100	Aluminum slat deck with aluminum frame	B-1	.08	300	Court	49,500	8,125		57,625	68,000
0500	Aluminum slat deck with wood frame	C-1	.12	266		49,500	7,925		57,425	68,000
0800	Aluminum deck heater, add	B-1	1.18	20.339		4,900	550		5,450	6,325
0900	Douglas fir planking with wood frame 2″ x 6″ x 30′	C-1	.12	266		46,500	7,925		54,425	64,500
1000	Plywood deck with steel frame		.12	266		46,500	7,925		54,425	64,500
1100	Steel slat deck with wood frame		.12	266		33,000	7,925		40,925	49,600

11 68 Play Field Equipment and Structures

11 68 33 – Athletic Field Equipment

11 68 33.13 Football Field Equipment		Crew	Daily Output	Labor-Hours	Unit	2011 Bare Costs Material	Labor	Equipment	Total	Total Incl O&P
0010	**FOOTBALL FIELD EQUIPMENT**									
0020	Goal posts, steel, football, double post	B-1	1.50	16	Pr.	3,375	435		3,810	4,450
0100	Deluxe, single post		1.50	16		2,425	435		2,860	3,375
0300	Football, convertible to soccer		1.50	16		2,025	435		2,460	2,950
0500	Soccer, regulation	↓	2	12	↓	1,075	325		1,400	1,725

11 71 Medical Sterilizing Equipment

11 71 10 – Medical Sterilizers & Distillers

11 71 10.10 Sterilizers and Distillers		Crew	Daily Output	Labor-Hours	Unit	Material	Labor	Equipment	Total	Total Incl O&P
0010	**STERILIZERS AND DISTILLERS**									
0700	Distiller, water, steam heated, 50 gal. capacity	1 Plum	1.40	5.714	Ea.	18,500	227		18,727	20,700
5600	Sterilizers, floor loading, 26" x 62" x 42", single door, steam					158,500			158,500	174,500
5650	Double door, steam					204,000			204,000	224,000
5800	General purpose, 20" x 20" x 38", single door					15,300			15,300	16,800
6000	Portable, counter top, steam, minimum					3,800			3,800	4,175
6020	Maximum					5,950			5,950	6,550
6050	Portable, counter top, gas, 17" x 15" x 32-1/2"					39,300			39,300	43,200
6150	Manual washer/sterilizer, 16" x 16" x 26"	1 Plum	2	4	↓	54,000	159		54,159	59,500
6200	Steam generators, electric 10 kW to 180 kW, freestanding									
6250	Minimum	1 Elec	3	2.667	Ea.	8,000	107		8,107	8,975
6300	Maximum	"	.70	11.429		28,800	460		29,260	32,400
8200	Bed pan washer-sanitizer	1 Plum	2	4	↓	7,125	159		7,284	8,100

11 72 Examination and Treatment Equipment

11 72 13 – Examination Equipment

11 72 13.13 Examination Equipment		Crew	Daily Output	Labor-Hours	Unit	Material	Labor	Equipment	Total	Total Incl O&P
0010	**EXAMINATION EQUIPMENT**									
0300	Blood pressure unit, mercurial, wall				Ea.	198			198	218
0400	Diagnostic set, wall					650			650	715
4400	Scale, physician's, with height rod				↓	286			286	315

11 72 53 – Treatment Equipment

11 72 53.13 Medical Treatment Equipment		Crew	Daily Output	Labor-Hours	Unit	Material	Labor	Equipment	Total	Total Incl O&P
0010	**MEDICAL TREATMENT EQUIPMENT**									
6500	Surgery table, minor minimum	1 Sswk	.70	11.429	Ea.	13,900	415		14,315	16,000
6520	Maximum	"	.50	16		26,100	580		26,680	29,800
6700	Surgical lights, doctor's office, single arm	2 Elec	2	8		1,225	320		1,545	1,875
6750	Dual arm	"	1	16	↓	5,800	645		6,445	7,450

11 73 Patient Care Equipment

11 73 10 – Patient Treatment Equipment

11 73 10.10 Treatment Equipment		Crew	Daily Output	Labor-Hours	Unit	2011 Bare Costs Material	Labor	Equipment	Total	Total Incl O&P
0010	**TREATMENT EQUIPMENT**									
0750	Exam room furnishings, average per room				Ea.	6,450			6,450	7,100
1800	Heat therapy unit, humidified, 26″ x 78″ x 28″				″	3,400			3,400	3,725
2100	Hubbard tank with accessories, stainless steel,									
2110	125 GPM at 45 psi water pressure				Ea.	26,300			26,300	29,000
2150	For electric overhead hoist, add					2,875			2,875	3,150
2900	K-Module for heat therapy, 20 oz. capacity, 75°F to 110°F					405			405	445
3600	Paraffin bath, 126°F, auto controlled					1,475			1,475	1,600
3900	Parallel bars for walking training, 12′-0″					1,200			1,200	1,325
4600	Station, dietary, medium, with ice					16,000			16,000	17,600
4700	Medicine					7,250			7,250	7,975
7000	Tables, physical therapy, walk off, electric	2 Carp	3	5.333		3,200	179		3,379	3,825
7150	Standard, vinyl top with base cabinets, minimum		3	5.333		1,300	179		1,479	1,725
7200	Maximum		2	8		3,850	268		4,118	4,700
8400	Whirlpool bath, mobile, sst, 18″ x 24″ x 60″					4,500			4,500	4,950
8450	Fixed, incl. mixing valves	1 Plum	2	4		9,200	159		9,359	10,400

11 74 Dental Equipment

11 74 10 – Dental Office Equipment

11 74 10.10 Diagnostic and Treatment Equipment		Crew	Daily Output	Labor-Hours	Unit	Material	Labor	Equipment	Total	Total Incl O&P
0010	**DIAGNOSTIC AND TREATMENT EQUIPMENT**									
0020	Central suction system, minimum	1 Plum	1.20	6.667	Ea.	1,550	264		1,814	2,125
0100	Maximum	″	.90	8.889		4,250	350		4,600	5,250
0300	Air compressor, minimum	1 Skwk	.80	10		3,075	340		3,415	3,950
0400	Maximum		.50	16		7,200	545		7,745	8,825
0600	Chair, electric or hydraulic, minimum		.50	16		2,625	545		3,170	3,775
0700	Maximum		.25	32		14,400	1,100		15,500	17,700
0800	Doctor's/assistant's stool, minimum					119			119	131
0850	Maximum					810			810	890
1000	Drill console with accessories, minimum	1 Skwk	1.60	5		1,575	170		1,745	2,000
1100	Maximum		1.60	5		5,375	170		5,545	6,200
2000	Light, ceiling mounted, minimum		8	1		1,225	34		1,259	1,400
2100	Maximum		8	1		1,625	34		1,659	1,850
2200	Unit light, minimum	2 Skwk	5.33	3.002		600	102		702	830
2210	Maximum		5.33	3.002		1,625	102		1,727	1,975
2220	Track light, minimum		3.20	5		1,525	170		1,695	1,950
2230	Maximum		3.20	5		3,725	170		3,895	4,375
2300	Sterilizers, steam portable, minimum					1,850			1,850	2,050
2350	Maximum					10,400			10,400	11,400
2600	Steam, institutional					4,250			4,250	4,675
2650	Dry heat, electric, portable, 3 trays					1,225			1,225	1,325
2700	Ultra-sonic cleaner, portable, minimum					239			239	263
2750	Maximum (institutional)					1,525			1,525	1,675
3000	X-ray unit, wall, minimum	1 Skwk	4	2		2,300	68		2,368	2,650
3010	Maximum		4	2		5,125	68		5,193	5,775
3100	Panoramic unit		.60	13.333		14,600	455		15,055	16,800
3105	Deluxe, minimum	2 Skwk	1.60	10		16,800	340		17,140	19,100
3110	Maximum	″	1.60	10		54,000	340		54,340	60,000
3500	Developers, X-ray, average	1 Plum	5.33	1.501		4,950	59.50		5,009.50	5,550
3600	Maximum	″	5.33	1.501		8,100	59.50		8,159.50	9,000

11 76 Operating Room Equipment

11 76 10 – Operating Room Equipment

11 76 10.10 Surgical Equipment		Crew	Daily Output	Labor-Hours	Unit	2011 Bare Costs Material	Labor	Equipment	Total	Total Incl O&P
0010	**SURGICAL EQUIPMENT**									
5000	Scrub, surgical, stainless steel, single station, minimum	1 Plum	3	2.667	Ea.	4,200	106		4,306	4,800
5100	Maximum					7,875			7,875	8,650
6550	Major surgery table, minimum	1 Sswk	.50	16		31,400	580		31,980	35,700
6570	Maximum	"	.50	16		103,500	580		104,080	115,000
6800	Surgical lights, major operating room, dual head, minimum	2 Elec	1	16		17,500	645		18,145	20,300
6850	Maximum	"	1	16	↓	29,100	645		29,745	33,100

11 77 Radiology Equipment

11 77 10 – Radiology Equipment

11 77 10.10 X-Ray Equipment		Crew	Daily Output	Labor-Hours	Unit	Material	Labor	Equipment	Total	Total Incl O&P
0010	**X-RAY EQUIPMENT**									
8700	X-ray, mobile, minimum				Ea.	13,400			13,400	14,700
8750	Maximum					75,500			75,500	83,000
8900	Stationary, minimum					41,800			41,800	46,000
8950	Maximum					217,500			217,500	239,500
9150	Developing processors, minimum					11,600			11,600	12,800
9200	Maximum				↓	30,600			30,600	33,600

11 78 Mortuary Equipment

11 78 13 – Mortuary Refrigerators

11 78 13.10 Mortuary and Autopsy Equipment		Crew	Daily Output	Labor-Hours	Unit	Material	Labor	Equipment	Total	Total Incl O&P
0010	**MORTUARY AND AUTOPSY EQUIPMENT**									
0015	Autopsy table, standard	1 Plum	1	8	Ea.	8,850	315		9,165	10,300
0020	Deluxe	"	.60	13.333		15,600	530		16,130	18,100
3200	Mortuary refrigerator, end operated, 2 capacity					12,800			12,800	14,100
3300	6 capacity				↓	23,900			23,900	26,300

11 78 16 – Crematorium Equipment

11 78 16.10 Crematory		Crew	Daily Output	Labor-Hours	Unit	Material	Labor	Equipment	Total	Total Incl O&P
0010	**CREMATORY**									
1500	Crematory, not including building, 1 place	Q-3	.20	160	Ea.	63,000	5,700		68,700	79,000
1750	2 place	"	.10	320	"	90,000	11,400		101,400	117,500

11 82 Solid Waste Handling Equipment

11 82 19 – Packaged Incinerators

11 82 19.10 Packaged Gas Fired Incinerators		Crew	Daily Output	Labor-Hours	Unit	Material	Labor	Equipment	Total	Total Incl O&P
0010	**PACKAGED GAS FIRED INCINERATORS**									
4400	Incinerator, gas, not incl. chimney, elec. or pipe, 50#/hr., minimum	Q-3	.80	40	Ea.	11,500	1,425		12,925	15,000
4420	Maximum		.70	45.714		37,100	1,625		38,725	43,500
4440	200 lb. per hr., minimum (batch type)		.60	53.333		63,500	1,900		65,400	73,000
4460	Maximum (with feeder)		.50	64		90,000	2,275		92,275	102,500
4480	400 lb. per hr., minimum (batch type)		.30	106		70,500	3,800		74,300	83,500
4500	Maximum (with feeder)		.25	128		95,500	4,575		100,075	112,500
4520	800 lb. per hr., with feeder, minimum		.20	160		115,500	5,700		121,200	136,500
4540	Maximum		.17	188		159,000	6,725		165,725	186,000
4560	1,200 lb. per hr., with feeder, minimum		.15	213		155,000	7,600		162,600	183,000
4580	Maximum	↓	.11	290	↓	185,000	10,400		195,400	220,500

11 82 Solid Waste Handling Equipment

11 82 19 – Packaged Incinerators

11 82 19.10 Packaged Gas Fired Incinerators		Crew	Daily Output	Labor-Hours	Unit	Material	2011 Bare Costs Labor	Equipment	Total	Total Incl O&P
4600	2,000 lb. per hr., with feeder, minimum	Q-3	.10	320	Ea.	379,000	11,400		390,400	435,500
4620	Maximum		.05	640		519,000	22,800		541,800	608,500
4700	For heat recovery system, add, minimum		.25	128		73,000	4,575		77,575	88,000
4710	Add, maximum		.11	290		234,000	10,400		244,400	274,500
4720	For automatic ash conveyer, add		.50	64		30,700	2,275		32,975	37,500
4750	Large municipal incinerators, incl. stack, minimum		.25	128	Ton/day	18,700	4,575		23,275	28,100
4850	Maximum		.10	320	"	49,700	11,400		61,100	73,000

11 82 26 – Facility Waste Compactors

11 82 26.10 Compactors		Crew	Daily Output	Labor-Hours	Unit	Material	Labor	Equipment	Total	Total Incl O&P
0010	**COMPACTORS**									
0020	Compactors, 115 volt, 250#/hr., chute fed	L-4	1	16	Ea.	11,600	475		12,075	13,500
0100	Hand fed		2.40	6.667		8,450	198		8,648	9,625
0300	Multi-bag, 230 volt, 600#/hr, chute fed		1	16		10,500	475		10,975	12,400
0400	Hand fed		1	16		9,025	475		9,500	10,700
0500	Containerized, hand fed, 2 to 6 C.Y. containers, 250#/hr.		1	16		11,100	475		11,575	13,100
0550	For chute fed, add per floor		1	16		1,275	475		1,750	2,200
1000	Heavy duty industrial compactor, 0.5 C.Y. capacity		1	16		7,525	475		8,000	9,075
1050	1.0 C.Y. capacity		1	16		11,500	475		11,975	13,400
1100	3.0 C.Y. capacity		.50	32		15,700	950		16,650	18,800
1150	5.0 C.Y. capacity		.50	32		19,400	950		20,350	22,900
1200	Combination shredder/compactor (5,000 lb./hr.)		.50	32		38,300	950		39,250	43,800
1400	For handling hazardous waste materials, 55 gallon drum packer, std.					17,800			17,800	19,600
1410	55 gallon drum packer w/HEPA filter					22,200			22,200	24,400
1420	55 gallon drum packer w/charcoal & HEPA filter					29,700			29,700	32,600
1430	All of the above made explosion proof, add					13,500			13,500	14,900
5500	Shredder, municipal use, 35 ton per hour					278,000			278,000	306,000
5600	60 ton per hour					592,500			592,500	651,500
5750	Shredder & baler, 50 ton per day					555,500			555,500	611,000
5800	Shredder, industrial, minimum					21,900			21,900	24,100
5850	Maximum					117,500			117,500	129,000
5900	Baler, industrial, minimum					8,775			8,775	9,650
5950	Maximum					512,000			512,000	563,500
6000	Transfer station compactor, with power unit									
6050	and pedestal, not including pit, 50 ton per hour				Ea.	175,500			175,500	193,000

11 91 Religious Equipment

11 91 13 – Baptisteries

11 91 13.10 Baptistry		Crew	Daily Output	Labor-Hours	Unit	Material	Labor	Equipment	Total	Total Incl O&P
0010	**BAPTISTRY**									
0150	Fiberglass, 3′-6″ deep, x 13′-7″ long,									
0160	steps at both ends, incl. plumbing, minimum	L-8	1	20	Ea.	4,100	630		4,730	5,550
0200	Maximum	"	.70	28.571		6,675	900		7,575	8,850
0250	Add for filter, heater and lights					1,475			1,475	1,600

11 91 23 – Sanctuary Equipment

11 91 23.10 Sanctuary Furnishings		Crew	Daily Output	Labor-Hours	Unit	Material	Labor	Equipment	Total	Total Incl O&P
0010	**SANCTUARY FURNISHINGS**									
0020	Altar, wood, custom design, plain	1 Carp	1.40	5.714	Ea.	2,000	192		2,192	2,525
0050	Deluxe	"	.20	40		9,700	1,350		11,050	13,000
0071	Granite or marble, average	2 Ston	.50	32		11,200	1,050		12,250	14,000
0091	Deluxe	"	.20	80		31,900	2,625		34,525	39,400

11 91 Religious Equipment

11 91 23 – Sanctuary Equipment

11 91 23.10 Sanctuary Furnishings		Crew	Daily Output	Labor-Hours	Unit	2011 Bare Costs Material	Labor	Equipment	Total	Total Incl O&P
0100	Arks, prefabricated, plain	2 Carp	.80	20	Ea.	8,425	670		9,095	10,400
0130	Deluxe, maximum	"	.20	80		119,500	2,675		122,175	136,000
0500	Reconciliation room, wood, prefabricated, single, plain	1 Carp	.60	13.333		2,550	445		2,995	3,550
0550	Deluxe		.40	20		7,025	670		7,695	8,850
0650	Double, plain		.40	20		5,100	670		5,770	6,750
0700	Deluxe		.20	40		15,900	1,350		17,250	19,800
1000	Lecterns, wood, plain		5	1.600		770	53.50		823.50	940
1100	Deluxe		2	4		5,375	134		5,509	6,150
2000	Pulpits, hardwood, prefabricated, plain		2	4		1,275	134		1,409	1,625
2100	Deluxe		1.60	5		8,725	168		8,893	9,875
2500	Railing, hardwood, average		25	.320	L.F.	175	10.75		185.75	210
3000	Seating, individual, oak, contour, laminated		21	.381	Person	154	12.80		166.80	191
3100	Cushion seat		21	.381		139	12.80		151.80	175
3200	Fully upholstered		21	.381		136	12.80		148.80	172
3300	Combination, self-rising		21	.381		350	12.80		362.80	405
3500	For cherry, add					30%				
5000	Wall cross, aluminum, extruded, 2" x 2" section	1 Carp	34	.235	L.F.	146	7.90		153.90	173
5150	4" x 4" section		29	.276		209	9.25		218.25	246
5300	Bronze, extruded, 1" x 2" section		31	.258		287	8.65		295.65	330
5350	2-1/2" x 2-1/2" section		34	.235		435	7.90		442.90	495
5450	Solid bar stock, 1/2" x 3" section		29	.276		570	9.25		579.25	640
5600	Fiberglass, stock		34	.235		118	7.90		125.90	143
5700	Stainless steel, 4" deep, channel section		29	.276		460	9.25		469.25	520
5800	4" deep box section		29	.276		625	9.25		634.25	705

Division Notes

	CREW	DAILY OUTPUT	LABOR-HOURS	UNIT	2011 BARE COSTS MAT.	LABOR	EQUIP.	TOTAL	TOTAL INCL O&P

Estimating Tips

General

- The items in this division are usually priced per square foot or each. Most of these items are purchased by the owner and placed by the contractor.
  Do not assume the items in Division 12 will be purchased and installed by the contractor. Check the specifications for responsibilities and include receiving, storage, installation, and mechanical and electrical hookups in the appropriate divisions.
- Some items in this division require some type of support system that is not usually furnished with the item. Examples of these systems include blocking for the attachment of casework and heavy drapery rods. The required blocking must be added to the estimate in the appropriate division.

Reference Numbers

Reference numbers are shown in shaded boxes at the beginning of some major classifications. These numbers refer to related items in the Reference Section. The reference information may be an estimating procedure, an alternate pricing method, or technical information.

Note: Not all subdivisions listed here necessarily appear in this publication.

12 21 Window Blinds

12 21 13 – Horizontal Louver Blinds

12 21 13.13 Metal Horizontal Louver Blinds		Crew	Daily Output	Labor-Hours	Unit	Material	2011 Bare Costs Labor	Equipment	Total	Total Incl O&P
0010	**METAL HORIZONTAL LOUVER BLINDS**									
0020	Horizontal, 1" aluminum slats, solid color, stock	1 Carp	590	.014	S.F.	4.75	.45		5.20	6
0090	Custom, minimum		590	.014		5.25	.45		5.70	6.55
0100	Maximum		440	.018		7.15	.61		7.76	8.85
0250	2" aluminum slats, custom, minimum		590	.014		8.10	.45		8.55	9.70
0350	Maximum		440	.018		13.15	.61		13.76	15.50
0450	Stock, minimum		590	.014		4.86	.45		5.31	6.10
0500	Maximum		440	.018		7.90	.61		8.51	9.70
0600	2" steel slats, stock, minimum		590	.014		1.84	.45		2.29	2.78
0630	Maximum		440	.018		5.15	.61		5.76	6.70
0750	Custom, minimum		590	.014		1.67	.45		2.12	2.60
0850	Maximum	↓	400	.020	↓	8.25	.67		8.92	10.20

12 21 13.33 Vinyl Horizontal Louver Blinds		Crew	Daily Output	Labor-Hours	Unit	Material	Labor	Equipment	Total	Total Incl O&P
0010	**VINYL HORIZONTAL LOUVER BLINDS**									
0100	2" custom composite, 48" wide, 48" high	1 Carp	30	.267	Ea.	89	8.95		97.95	113
0120	72" high		29	.276		126	9.25		135.25	155
0140	96" high		28	.286		161	9.60		170.60	193
0200	60" wide, 60" high		27	.296		93.50	9.95		103.45	120
0220	72" high		25	.320		108	10.75		118.75	136
0240	96" high		24	.333		151	11.20		162.20	185
0300	72" wide, 72" high		25	.320		169	10.75		179.75	203
0320	96" high		23	.348		249	11.65		260.65	294
0400	96" wide, 96" high		20	.400		295	13.40		308.40	350
1000	2" custom faux wood, 48" wide, 48" high		30	.267		55.50	8.95		64.45	76
1020	72" high		29	.276		74	9.25		83.25	97
1040	96" high		28	.286		92.50	9.60		102.10	118
1300	72" wide, 72" high		25	.320		115	10.75		125.75	144
1320	96" high		23	.348		164	11.65		175.65	201
1400	96" wide, 96" high	↓	20	.400	↓	180	13.40		193.40	221

12 21 16 – Vertical Louver Blinds

12 21 16.13 Metal Vertical Louver Blinds		Crew	Daily Output	Labor-Hours	Unit	Material	Labor	Equipment	Total	Total Incl O&P
0010	**METAL VERTICAL LOUVER BLINDS**									
1500	Vertical, 3" to 5" PVC or cloth strips, minimum	1 Carp	460	.017	S.F.	5.95	.58		6.53	7.55
1600	Maximum		400	.020		24	.67		24.67	27.50
1800	4" aluminum slats, minimum		460	.017		7.65	.58		8.23	9.40
1900	Maximum		400	.020		9.10	.67		9.77	11.10
1950	Mylar mirror-finish strips, to 8" wide, minimum		460	.017		4.62	.58		5.20	6.10
1970	Maximum	↓	400	.020	↓	10.50	.67		11.17	12.65

12 22 Curtains and Drapes

12 22 16 – Drapery Track and Accessories

12 22 16.10 Drapery Hardware		Crew	Daily Output	Labor-Hours	Unit	Material	Labor	Equipment	Total	Total Incl O&P
0010	**DRAPERY HARDWARE**									
0030	Standard traverse, per foot, minimum	1 Carp	59	.136	L.F.	3.42	4.55		7.97	11.35
0100	Maximum		51	.157	"	8.05	5.25		13.30	17.65
4000	Traverse rods, adjustable, 28" to 48"		22	.364	Ea.	21	12.20		33.20	43.50
4020	48" to 84"		20	.400		29.50	13.40		42.90	55
4040	66" to 120"		18	.444		31.50	14.90		46.40	59.50
4060	84" to 156"		16	.500		33	16.80		49.80	64
4080	100" to 180"	↓	14	.571	↓	42.50	19.15		61.65	78.50

12 22 Curtains and Drapes

12 22 16 – Drapery Track and Accessories

12 22 16.10 Drapery Hardware		Crew	Daily Output	Labor-Hours	Unit	Material	2011 Bare Costs Labor	Equipment	Total	Total Incl O&P
4100	228″ to 312″	1 Carp	13	.615	Ea.	59.50	20.50		80	100
4600	Valance, pinch pleated fabric, 12″ deep, up to 54″ long, minimum					36.50			36.50	40.50
4610	Maximum					91.50			91.50	101
4620	Up to 77″ long, minimum					56.50			56.50	62
4630	Maximum					148			148	163
5000	Stationary rods, first 2′					9.35			9.35	10.25
5020	Each additional foot, add				L.F.	3.64			3.64	4

12 22 16.20 Blast Curtains		Crew	Daily Output	Labor-Hours	Unit	Material	Labor	Equipment	Total	Total Incl O&P
0010	**BLAST CURTAINS** per L.F. horizontal opening width, off-white or gray fabric									
0100	Blast curtains, drapery system, complete, including hardware, minimum	1 Carp	10.25	.780	L.F.	170	26		196	231
0120	Average		10.25	.780		200	26		226	264
0140	Maximum		10.25	.780		230	26		256	297

12 23 Interior Shutters

12 23 10 – Wood Interior Shutters

12 23 10.10 Wood Interior Shutters		Crew	Daily Output	Labor-Hours	Unit	Material	Labor	Equipment	Total	Total Incl O&P
0010	**WOOD INTERIOR SHUTTERS**, louvered									
0200	Two panel, 27″ wide, 36″ high	1 Carp	5	1.600	Set	125	53.50		178.50	228
0300	33″ wide, 36″ high		5	1.600		164	53.50		217.50	270
0500	47″ wide, 36″ high		5	1.600		219	53.50		272.50	330
1000	Four panel, 27″ wide, 36″ high		5	1.600		148	53.50		201.50	253
1100	33″ wide, 36″ high		5	1.600		181	53.50		234.50	290
1300	47″ wide, 36″ high		5	1.600		258	53.50		311.50	375

12 23 10.13 Wood Panels		Crew	Daily Output	Labor-Hours	Unit	Material	Labor	Equipment	Total	Total Incl O&P
0010	**WOOD PANELS**									
3000	Wood folding panels with movable louvers, 7″ x 20″ each	1 Carp	17	.471	Pr.	50	15.80		65.80	81.50
3300	8″ x 28″ each		17	.471		64	15.80		79.80	97
3450	9″ x 36″ each		17	.471		84	15.80		99.80	119
3600	10″ x 40″ each		17	.471		90	15.80		105.80	126
4000	Fixed louver type, stock units, 8″ x 20″ each		17	.471		58	15.80		73.80	90.50
4150	10″ x 28″ each		17	.471		68	15.80		83.80	102
4300	12″ x 36″ each		17	.471		86	15.80		101.80	121
4450	18″ x 40″ each		17	.471		122	15.80		137.80	161
5000	Insert panel type, stock, 7″ x 20″ each		17	.471		19.65	15.80		35.45	48
5150	8″ x 28″ each		17	.471		36	15.80		51.80	66
5300	9″ x 36″ each		17	.471		46	15.80		61.80	77
5450	10″ x 40″ each		17	.471		49	15.80		64.80	80.50
5600	Raised panel type, stock, 10″ x 24″ each		17	.471		80	15.80		95.80	115
5650	12″ x 26″ each		17	.471		86	15.80		101.80	121
5700	14″ x 30″ each		17	.471		258	15.80		273.80	310
5750	16″ x 36″ each		17	.471		286	15.80		301.80	340
6000	For custom built pine, add					22%				
6500	For custom built hardwood blinds, add					42%				

12 24 Window Shades

12 24 13 – Roller Window Shades

12 24 13.10 Shades		Crew	Daily Output	Labor-Hours	Unit	2011 Bare Costs Material	Labor	Equipment	Total	Total Incl O&P
0010	**SHADES**									
0020	Basswood, roll-up, stain finish, 3/8" slats	1 Carp	300	.027	S.F.	13.95	.89		14.84	16.85
0200	7/8" slats		300	.027		13.15	.89		14.04	16
0300	Vertical side slide, stain finish, 3/8" slats		300	.027		18.05	.89		18.94	21.50
0400	7/8" slats		300	.027		18.05	.89		18.94	21.50
0500	For fire retardant finishes, add					16%				
0600	For "B" rated finishes, add					20%				
0900	Mylar, single layer, non-heat reflective	1 Carp	685	.012		6.70	.39		7.09	8
1000	Double layered, heat reflective		685	.012		9.25	.39		9.64	10.85
1100	Triple layered, heat reflective		685	.012		10.80	.39		11.19	12.50
1200	For metal roller instead of wood, add per				Shade	4.20			4.20	4.62
1300	Vinyl coated cotton, standard	1 Carp	685	.012	S.F.	3.43	.39		3.82	4.43
1400	Lightproof decorator shades		685	.012		2.15	.39		2.54	3.03
1500	Vinyl, lightweight, 4 gauge		685	.012		.56	.39		.95	1.28
1600	Heavyweight, 6 gauge		685	.012		1.70	.39		2.09	2.53
1700	Vinyl laminated fiberglass, 6 ga., translucent		685	.012		2.41	.39		2.80	3.31
1800	Lightproof		685	.012		3.91	.39		4.30	4.96
3000	Woven aluminum, 3/8" thick, lightproof and fireproof		350	.023		5.50	.77		6.27	7.35

12 32 Manufactured Wood Casework

12 32 16 – Manufactured Plastic-Laminate-Clad Casework

12 32 16.20 Plastic Laminate Casework Doors

		Crew	Daily Output	Labor-Hours	Unit	Material	Labor	Equipment	Total	Total Incl O&P
0010	**PLASTIC LAMINATE CASEWORK DOORS**									
1000	For casework frames, see Section 12 32 23.15									
1100	For casework hardware, see Section 12 32 23.35									
6000	Plastic laminate on particle board									
6100	12" wide, 18" high	1 Carp	25	.320	Ea.	10.65	10.75		21.40	29.50
6140	30" high		23	.348		17.70	11.65		29.35	39
6500	18" wide, 18" high		24	.333		15.95	11.20		27.15	36.50
6600	30" high		22	.364		26.50	12.20		38.70	49.50

12 32 16.25 Plastic Laminate Drawer Fronts

		Crew	Daily Output	Labor-Hours	Unit	Material	Labor	Equipment	Total	Total Incl O&P
0010	**PLASTIC LAMINATE DRAWER FRONTS**									
2800	Plastic laminate on particle board front									
3000	4" high, 12" wide	1 Carp	17	.471	Ea.	5.05	15.80		20.85	32
3200	18" wide	"	16	.500	"	7.60	16.80		24.40	36.50

12 32 23 – Hardwood Casework

12 32 23.10 Manufactured Wood Casework, Stock Units

		Crew	Daily Output	Labor-Hours	Unit	Material	Labor	Equipment	Total	Total Incl O&P
0010	**MANUFACTURED WOOD CASEWORK, STOCK UNITS**									
0300	Built-in drawer units, pine, 18" deep, 32" high, unfinished									
0400	Minimum	2 Carp	53	.302	L.F.	145	10.15		155.15	177
0500	Maximum	"	40	.400	"	180	13.40		193.40	221
0700	Kitchen base cabinets, hardwood, not incl. counter tops,									
0710	24" deep, 35" high, prefinished									
0800	One top drawer, one door below, 12" wide	2 Carp	24.80	.645	Ea.	229	21.50		250.50	288
0840	18" wide		23.30	.687		259	23		282	325
0880	24" wide		22.30	.717		310	24		334	385
1000	Four drawers, 12" wide		24.80	.645		240	21.50		261.50	300
1040	18" wide		23.30	.687		268	23		291	335
1060	24" wide		22.30	.717		305	24		329	375
1200	Two top drawers, two doors below, 27" wide		22	.727		330	24.50		354.50	405

12 32 23.10 Manufactured Wood Casework, Stock Units		Crew	Daily Output	Labor-Hours	Unit	Material	2011 Bare Costs Labor	Equipment	Total	Total Incl O&P
1260	36″ wide	2 Carp	20.30	.788	Ea.	400	26.50		426.50	485
1300	48″ wide		18.90	.847		460	28.50		488.50	555
1500	Range or sink base, two doors below, 30″ wide		21.40	.748		305	25		330	375
1540	36″ wide		20.30	.788		345	26.50		371.50	425
1580	48″ wide		18.90	.847		385	28.50		413.50	475
1800	For sink front units, deduct					145			145	160
2000	Corner base cabinets, 36″ wide, standard	2 Carp	18	.889		540	30		570	645
2100	Lazy Susan with revolving door	″	16.50	.970		725	32.50		757.50	850
4000	Kitchen wall cabinets, hardwood, 12″ deep with two doors									
4050	12″ high, 30″ wide	2 Carp	24.80	.645	Ea.	205	21.50		226.50	261
4100	36″ wide		24	.667		240	22.50		262.50	300
4400	15″ high, 30″ wide		24	.667		165	22.50		187.50	219
4440	36″ wide		22.70	.705		240	23.50		263.50	305
4700	24″ high, 30″ wide		23.30	.687		275	23		298	345
4720	36″ wide		22.70	.705		305	23.50		328.50	375
5000	30″ high, one door, 12″ wide		22	.727		185	24.50		209.50	245
5040	18″ wide		20.90	.766		229	25.50		254.50	295
5060	24″ wide		20.30	.788		267	26.50		293.50	340
5300	Two doors, 27″ wide		19.80	.808		294	27		321	370
5340	36″ wide		18.80	.851		350	28.50		378.50	435
5380	48″ wide		18.40	.870		430	29		459	525
6000	Corner wall, 30″ high, 24″ wide		18	.889		233	30		263	305
6050	30″ wide		17.20	.930		264	31		295	340
6100	36″ wide		16.50	.970		295	32.50		327.50	380
6500	Revolving Lazy Susan		15.20	1.053		415	35.50		450.50	515
7000	Broom cabinet, 84″ high, 24″ deep, 18″ wide		10	1.600		575	53.50		628.50	725
7500	Oven cabinets, 84″ high, 24″ deep, 27″ wide		8	2		860	67		927	1,050
7750	Valance board trim		396	.040	L.F.	12.40	1.36		13.76	15.90
9000	For deluxe models of all cabinets, add					40%				
9500	For custom built in place, add					25%	10%			
9558	Rule of thumb, kitchen cabinets not including									
9560	appliances & counter top, minimum	2 Carp	30	.533	L.F.	146	17.90		163.90	191
9600	Maximum	″	25	.640	″	310	21.50		331.50	380
9610	For metal cabinets, see Section 12 35 70.13									

12 32 23.15 Manufactured Wood Casework Frames		Crew	Daily Output	Labor-Hours	Unit	Material	Labor	Equipment	Total	Total Incl O&P
0010	**MANUFACTURED WOOD CASEWORK FRAMES**									
0050	Base cabinets, counter storage, 36″ high									
0100	One bay, 18″ wide	1 Carp	2.70	2.963	Ea.	134	99.50		233.50	315
0400	Two bay, 36″ wide		2.20	3.636		204	122		326	430
1100	Three bay, 54″ wide		1.50	5.333		243	179		422	565
2800	Bookcases, one bay, 7′ high, 18″ wide		2.40	3.333		157	112		269	360
3500	Two bay, 36″ wide		1.60	5		228	168		396	530
4100	Three bay, 54″ wide		1.20	6.667		375	224		599	790
5100	Coat racks, one bay, 7′ high, 24″ wide		4.50	1.778		157	59.50		216.50	273
5300	Two bay, 48″ wide		2.75	2.909		219	97.50		316.50	405
5800	Three bay, 72″ wide		2.10	3.810		320	128		448	570
6100	Wall mounted cabinet, one bay, 24″ high, 18″ wide		3.60	2.222		86.50	74.50		161	220
6800	Two bay, 36″ wide		2.20	3.636		126	122		248	345
7400	Three bay, 54″ wide		1.70	4.706		157	158		315	435
8400	30″ high, one bay, 18″ wide		3.60	2.222		94	74.50		168.50	228
9000	Two bay, 36″ wide		2.15	3.721		125	125		250	345
9400	Three bay, 54″ wide		1.60	5		156	168		324	450

12 32 Manufactured Wood Casework

12 32 23 – Hardwood Casework

12 32 23.15 Manufactured Wood Casework Frames		Crew	Daily Output	Labor-Hours	Unit	Material	2011 Bare Costs Labor	Equipment	Total	Total Incl O&P
9800	Wardrobe, 7′ high, single, 24″ wide	1 Carp	2.70	2.963	Ea.	173	99.50		272.50	355
9880	Partition & adjustable shelves, 48″ wide		1.70	4.706		220	158		378	505
9950	Partition, adjustable shelves & drawers, 48″ wide		1.40	5.714		330	192		522	685

12 32 23.20 Manufactured Hardwood Casework Doors

Code	Description	Crew	Daily Output	Labor-Hours	Unit	Material	Labor	Equipment	Total	Total Incl O&P
0010	**MANUFACTURED HARDWOOD CASEWORK DOORS**									
2000	Glass panel, hardwood frame									
2200	12″ wide, 18″ high	1 Carp	34	.235	Ea.	22	7.90		29.90	37
2600	30″ high		32	.250		36.50	8.40		44.90	54
4450	18″ wide, 18″ high		32	.250		32.50	8.40		40.90	50
4550	30″ high		29	.276		54.50	9.25		63.75	75.50
5000	Hardwood, raised panel									
5100	12″ wide, 18″ high	1 Carp	16	.500	Ea.	26	16.80		42.80	56.50
5200	30″ high		15	.533		43.50	17.90		61.40	77.50
5500	18″ wide, 18″ high		15	.533		39	17.90		56.90	73
5600	30″ high		14	.571		65	19.15		84.15	104

12 32 23.25 Manufactured Wood Casework Drawer Fronts

Code	Description	Crew	Daily Output	Labor-Hours	Unit	Material	Labor	Equipment	Total	Total Incl O&P
0010	**MANUFACTURED WOOD CASEWORK DRAWER FRONTS**									
0100	Solid hardwood front									
1000	4″ high, 12″ wide	1 Carp	17	.471	Ea.	3.33	15.80		19.13	30
1200	18″ wide	"	16	.500	"	4.95	16.80		21.75	33.50

12 32 23.30 Manufactured Wood Casework Vanities

Code	Description	Crew	Daily Output	Labor-Hours	Unit	Material	Labor	Equipment	Total	Total Incl O&P
0010	**MANUFACTURED WOOD CASEWORK VANITIES**									
8000	Vanity bases, 2 doors, 30″ high, 21″ deep, 24″ wide	2 Carp	20	.800	Ea.	227	27		254	295
8050	30″ wide		16	1		262	33.50		295.50	345
8100	36″ wide		13.33	1.200		350	40.50		390.50	455
8150	48″ wide		11.43	1.400		415	47		462	540
9000	For deluxe models of all vanities, add to above					40%				
9500	For custom built in place, add to above					25%	10%			

12 32 23.35 Manufactured Wood Casework Hardware

Code	Description	Crew	Daily Output	Labor-Hours	Unit	Material	Labor	Equipment	Total	Total Incl O&P
0010	**MANUFACTURED WOOD CASEWORK HARDWARE**									
1000	Catches, minimum	1 Carp	235	.034	Ea.	1	1.14		2.14	3.01
1040	Maximum	"	80	.100	"	6.25	3.36		9.61	12.45
2000	Door/drawer pulls, handles									
2200	Handles and pulls, projecting, metal, minimum	1 Carp	48	.167	Ea.	4.52	5.60		10.12	14.30
2240	Maximum		24	.333		9.55	11.20		20.75	29.50
2300	Wood, minimum		48	.167		4.76	5.60		10.36	14.60
2340	Maximum		24	.333		8.75	11.20		19.95	28.50
2600	Flush, metal, minimum		48	.167		4.76	5.60		10.36	14.60
2640	Maximum		24	.333		8.75	11.20		19.95	28.50
3000	Drawer tracks/glides, minimum		48	.167	Pr.	8.05	5.60		13.65	18.25
3040	Maximum		24	.333		23.50	11.20		34.70	44.50
4000	Cabinet hinges, minimum		160	.050		2.74	1.68		4.42	5.80
4040	Maximum		68	.118		10.35	3.95		14.30	18

12 35 Specialty Casework

12 35 50 – Educational/Library Casework

12 35 50.13 Educational Casework		Crew	Daily Output	Labor-Hours	Unit	2011 Bare Costs Material	Labor	Equipment	Total	Total Incl O&P
0010	**EDUCATIONAL CASEWORK**									
5000	School, 24″ deep, metal, 84″ high units	2 Carp	15	1.067	L.F.	385	36		421	485
5150	Counter height units		20	.800		258	27		285	330
5450	Wood, custom fabricated, 32″ high counter		20	.800		215	27		242	281
5600	Add for counter top		56	.286		23	9.60		32.60	41
5800	84″ high wall units		15	1.067		420	36		456	520
6000	Laminated plastic finish is same price as wood									

12 35 53 – Laboratory Casework

12 35 53.13 Metal Laboratory Casework		Crew	Daily Output	Labor-Hours	Unit	Material	Labor	Equipment	Total	Total Incl O&P
0010	**METAL LABORATORY CASEWORK**									
0020	Cabinets, base, door units, metal	2 Carp	18	.889	L.F.	202	30		232	272
0300	Drawer units		18	.889		450	30		480	545
0700	Tall storage cabinets, open, 7′ high		20	.800		430	27		457	520
0900	With glazed doors		20	.800		545	27		572	645
1300	Wall cabinets, metal, 12-1/2″ deep, open		20	.800		142	27		169	201
1500	With doors		20	.800		295	27		322	370
6300	Rule of thumb: lab furniture including installation & connection									
6320	High school				S.F.				33.50	37
6340	College								49.50	54.50
6360	Clinical, health care								43	47
6380	Industrial								69	75.50

12 35 59 – Display Casework

12 35 59.10 Display Cases		Crew	Daily Output	Labor-Hours	Unit	Material	Labor	Equipment	Total	Total Incl O&P
0010	**DISPLAY CASES** Free standing, all glass									
0020	Aluminum frame, 42″ high x 36″ x 12″ deep	2 Carp	8	2	Ea.	1,250	67		1,317	1,450
0100	70″ high x 48″ x 18″ deep	″	6	2.667		3,075	89.50		3,164.50	3,525
0500	For wood bases, add					9%				
0600	For hardwood frames, deduct					8%				
0700	For bronze, baked enamel finish, add					10%				
2000	Wall mounted, glass front, aluminum frame									
2010	Non-illuminated, one section 3′ x 4′ x 1′-4″	2 Carp	5	3.200	Ea.	2,150	107		2,257	2,525
2100	5′ x 4′ x 1′-4″		5	3.200		2,475	107		2,582	2,900
2200	6′ x 4′ x 1′-4″		4	4		2,975	134		3,109	3,500
2500	Two sections, 8′ x 4′ x 1′-4″		2	8		1,900	268		2,168	2,550
2600	10′ x 4′ x 1′-4″		2	8		2,250	268		2,518	2,925
3000	Three sections, 16′ x 4′ x 1′-4″		1.50	10.667		3,900	360		4,260	4,900
3500	For fluorescent lights, add				Section	330			330	365
4000	Table exhibit cases, 2′ wide, 3′ high, 4′ long, flat top	2 Carp	5	3.200	Ea.	1,150	107		1,257	1,450
4100	3′ wide, 3′ high, 4′ long, sloping top	″	3	5.333	″	1,850	179		2,029	2,325

12 35 70 – Healthcare Casework

12 35 70.13 Hospital Casework		Crew	Daily Output	Labor-Hours	Unit	Material	Labor	Equipment	Total	Total Incl O&P
0010	**HOSPITAL CASEWORK**									
0500	Base cabinets, laminated plastic	2 Carp	10	1.600	L.F.	246	53.50		299.50	360
1000	Stainless steel	″	10	1.600		455	53.50		508.50	590
1200	For all drawers, add					26			26	28.50
1300	Cabinet base trim, 4″ high, enameled steel	2 Carp	200	.080		41.50	2.68		44.18	50
1400	Stainless steel		200	.080		82.50	2.68		85.18	95.50
1450	Countertop, laminated plastic, no backsplash		40	.400		42.50	13.40		55.90	69.50
1650	With backsplash		40	.400		53	13.40		66.40	80.50
1800	For sink cutout, add		12.20	1.311	Ea.		44		44	73.50
1900	Stainless steel counter top		40	.400	L.F.	138	13.40		151.40	174

12 35 Specialty Casework

12 35 70 – Healthcare Casework

12 35 70.13 Hospital Casework

	12 35 70.13 Hospital Casework	Crew	Daily Output	Labor-Hours	Unit	Material	Labor	Equipment	Total	Total Incl O&P
						2011 Bare Costs				
2000	For drop-in stainless 43" x 21" sink, add				Ea.	905			905	995
2500	Wall cabinets, laminated plastic	2 Carp	15	1.067	L.F.	184	36		220	263
2600	Enameled steel		15	1.067		227	36		263	310
2700	Stainless steel		15	1.067		450	36		486	555
2800	For glass doors, add					31			31	34

12 35 70.16 Nurse Station Casework

	12 35 70.16 Nurse Station Casework	Crew	Daily Output	Labor-Hours	Unit	Material	Labor	Equipment	Total	Total Incl O&P
0010	**NURSE STATION CASEWORK**									
2100	Door type, laminated plastic	2 Carp	10	1.600	L.F.	286	53.50		339.50	405
2200	Enameled steel		10	1.600		273	53.50		326.50	390
2300	Stainless steel		10	1.600		545	53.50		598.50	690
2400	For drawer type, add					232			232	255

12 35 80 – Commercial Kitchen Casework

12 35 80.13 Metal Kitchen Casework

	12 35 80.13 Metal Kitchen Casework	Crew	Daily Output	Labor-Hours	Unit	Material	Labor	Equipment	Total	Total Incl O&P
0010	**METAL KITCHEN CASEWORK**									
3500	Base cabinets, metal, minimum	2 Carp	30	.533	L.F.	66.50	17.90		84.40	103
3600	Maximum		25	.640		169	21.50		190.50	222
3700	Wall cabinets, metal, minimum		30	.533		66.50	17.90		84.40	103
3800	Maximum		25	.640		153	21.50		174.50	204

12 36 Countertops

12 36 16 – Metal Countertops

12 36 16.10 Stainless Steel Countertops

	12 36 16.10 Stainless Steel Countertops	Crew	Daily Output	Labor-Hours	Unit	Material	Labor	Equipment	Total	Total Incl O&P
0010	**STAINLESS STEEL COUNTERTOPS**									
3200	Stainless steel, custom	1 Carp	24	.333	S.F.	145	11.20		156.20	179

12 36 19 – Wood Countertops

12 36 19.10 Maple Countertops

	12 36 19.10 Maple Countertops	Crew	Daily Output	Labor-Hours	Unit	Material	Labor	Equipment	Total	Total Incl O&P
0010	**MAPLE COUNTERTOPS**									
2900	Solid, laminated, 1-1/2" thick, no splash	1 Carp	28	.286	L.F.	74	9.60		83.60	97.50
3000	With square splash		28	.286	"	88	9.60		97.60	113
3400	Recessed cutting block with trim, 16" x 20" x 1"		8	1	Ea.	74	33.50		107.50	138

12 36 23 – Plastic Countertops

12 36 23.13 Plastic-Laminate-Clad Countertops

	12 36 23.13 Plastic-Laminate-Clad Countertops	Crew	Daily Output	Labor-Hours	Unit	Material	Labor	Equipment	Total	Total Incl O&P
0010	**PLASTIC-LAMINATE-CLAD COUNTERTOPS**									
0020	Stock, 24" wide w/backsplash, minimum	1 Carp	30	.267	L.F.	9.85	8.95		18.80	26
0100	Maximum		25	.320		21	10.75		31.75	41
0300	Custom plastic, 7/8" thick, aluminum molding, no splash		30	.267		22.50	8.95		31.45	40
0400	Cove splash		30	.267		29	8.95		37.95	47
0600	1-1/4" thick, no splash		28	.286		31.50	9.60		41.10	50.50
0700	Square splash		28	.286		32	9.60		41.60	51
0900	Square edge, plastic face, 7/8" thick, no splash		30	.267		28	8.95		36.95	46
1000	With splash		30	.267		36	8.95		44.95	54.50
1200	For stainless channel edge, 7/8" thick, add					3			3	3.30
1300	1-1/4" thick, add					3.50			3.50	3.85
1500	For solid color suede finish, add					3			3	3.30
1700	For end splash, add				Ea.	18			18	19.80
1900	For cut outs, standard, add, minimum	1 Carp	32	.250		4	8.40		12.40	18.45
2000	Maximum		8	1		6	33.50		39.50	62.50
2100	Postformed, including backsplash and front edge		30	.267	L.F.	11	8.95		19.95	27
2110	Mitred, add		12	.667	Ea.		22.50		22.50	37.50

12 36 Countertops

12 36 23 – Plastic Countertops

12 36 23.13 Plastic-Laminate-Clad Countertops		Crew	Daily Output	Labor-Hours	Unit	Material	2011 Bare Costs Labor	Equipment	Total	Total Incl O&P
2200	Built-in place, 25″ wide, plastic laminate	1 Carp	25	.320	L.F.	30	10.75		40.75	51

12 36 33 – Tile Countertops

12 36 33.10 Ceramic Tile Countertops		Crew	Daily Output	Labor-Hours	Unit	Material	Labor	Equipment	Total	Total Incl O&P
0010	**CERAMIC TILE COUNTERTOPS**									
2300	Ceramic tile mosaic	1 Carp	25	.320	L.F.	33	10.75		43.75	54.50

12 36 40 – Stone Countertops

12 36 40.10 Natural Stone Countertops		Crew	Daily Output	Labor-Hours	Unit	Material	Labor	Equipment	Total	Total Incl O&P
0010	**NATURAL STONE COUNTERTOPS**									
2500	Marble, stock, with splash, 1/2″ thick, minimum	1 Bric	17	.471	L.F.	41	15.85		56.85	71
2700	3/4″ thick, maximum		13	.615		103	20.50		123.50	147
2800	Granite, average, 1-1/4″ thick, 24″ wide, no splash	↓	13.01	.615	↓	135	20.50		155.50	183

12 36 53 – Laboratory Countertops

12 36 53.10 Laboratory Countertops and Sinks		Crew	Daily Output	Labor-Hours	Unit	Material	Labor	Equipment	Total	Total Incl O&P
0010	**LABORATORY COUNTERTOPS AND SINKS**									
0020	Countertops, not incl. base cabinets, acid-proof, minimum	2 Carp	82	.195	S.F.	34.50	6.55		41.05	49
0030	Maximum		70	.229		44.50	7.65		52.15	62
0040	Stainless steel	↓	82	.195	↓	110	6.55		116.55	131

12 36 61 – Simulated Stone Countertops

12 36 61.16 Solid Surface Countertops		Crew	Daily Output	Labor-Hours	Unit	Material	Labor	Equipment	Total	Total Incl O&P
0010	**SOLID SURFACE COUNTERTOPS**, Acrylic polymer									
0020	Pricing for orders of 100 L.F. or greater									
0100	25″ wide, solid colors	2 Carp	28	.571	L.F.	44.50	19.15		63.65	81
0200	Patterned colors		28	.571		56.50	19.15		75.65	94
0300	Premium patterned colors		28	.571		70.50	19.15		89.65	110
0400	With silicone attached 4″ backsplash, solid colors		27	.593		49	19.90		68.90	87.50
0500	Patterned colors		27	.593		62	19.90		81.90	102
0600	Premium patterned colors		27	.593		77	19.90		96.90	119
0700	With hard seam attached 4″ backsplash, solid colors		23	.696		49	23.50		72.50	93
0800	Patterned colors		23	.696		62	23.50		85.50	107
0900	Premium patterned colors	↓	23	.696	↓	77	23.50		100.50	124
1000	Pricing for order of 51 – 99 L.F.									
1100	25″ wide, solid colors	2 Carp	24	.667	L.F.	51.50	22.50		74	94
1200	Patterned colors		24	.667		65	22.50		87.50	109
1300	Premium patterned colors		24	.667		81.50	22.50		104	127
1400	With silicone attached 4″ backsplash, solid colors		23	.696		56.50	23.50		80	101
1500	Patterned colors		23	.696		71.50	23.50		95	118
1600	Premium patterned colors		23	.696		89	23.50		112.50	137
1700	With hard seam attached 4″ backsplash, solid colors		20	.800		56.50	27		83.50	107
1800	Patterned colors		20	.800		71.50	27		98.50	124
1900	Premium patterned colors	↓	20	.800	↓	89	27		116	143
2000	Pricing for order of 1 – 50 L.F.									
2100	25″ wide, solid colors	2 Carp	20	.800	L.F.	60	27		87	111
2200	Patterned colors		20	.800		76.50	27		103.50	129
2300	Premium patterned colors		20	.800		95.50	27		122.50	150
2400	With silicone attached 4″ backsplash, solid colors		19	.842		66	28.50		94.50	120
2500	Patterned colors		19	.842		83.50	28.50		112	140
2600	Premium patterned colors		19	.842		104	28.50		132.50	163
2700	With hard seam attached 4″ backsplash, solid colors		15	1.067		66	36		102	133
2800	Patterned colors		15	1.067		83.50	36		119.50	152
2900	Premium patterned colors	↓	15	1.067	↓	104	36		140	175

12 36 Countertops

12 36 61 – Simulated Stone Countertops

12 36 61.16 Solid Surface Countertops		Crew	Daily Output	Labor-Hours	Unit	Material	2011 Bare Costs Labor	Equipment	Total	Total Incl O&P
3000	Sinks, pricing for order of 100 or greater units									
3100	Single bowl, hard seamed, solid colors, 13" x 17"	1 Carp	3	2.667	Ea.	300	89.50		389.50	480
3200	10" x 15"		7	1.143		139	38.50		177.50	217
3300	Cutouts for sinks		8	1			33.50		33.50	56
3400	Sinks, pricing for order of 51 – 99 units									
3500	Single bowl, hard seamed, solid colors, 13" x 17"	1 Carp	2.55	3.137	Ea.	345	105		450	555
3600	10" x 15"		6	1.333		160	44.50		204.50	251
3700	Cutouts for sinks		7	1.143			38.50		38.50	64
3800	Sinks, pricing for order of 1 – 50 units									
3900	Single bowl, hard seamed, solid colors, 13" x 17"	1 Carp	2	4	Ea.	405	134		539	670
4000	10" x 15"		4.55	1.758		188	59		247	305
4100	Cutouts for sinks		5.25	1.524			51		51	85.50
4200	Cooktop cutouts, pricing for 100 or greater units		4	2		22	67		89	137
4300	51 – 99 units		3.40	2.353		25.50	79		104.50	160
4400	1 – 50 units		3	2.667		30	89.50		119.50	183
12 36 61.17 Solid Surface Vanity Tops										
0010	**SOLID SURFACE VANITY TOPS**									
0015	Solid surface, center bowl, 17" x 19"	1 Carp	12	.667	Ea.	189	22.50		211.50	246
0020	19" x 25"		12	.667		229	22.50		251.50	290
0030	19" x 31"		12	.667		278	22.50		300.50	345
0040	19" x 37"		12	.667		325	22.50		347.50	395
0050	22" x 25"		10	.800		202	27		229	267
0060	22" x 31"		10	.800		236	27		263	305
0070	22" x 37"		10	.800		274	27		301	345
0080	22" x 43"		10	.800		315	27		342	390
0090	22" x 49"		10	.800		345	27		372	425
0110	22" x 55"		8	1		395	33.50		428.50	490
0120	22" x 61"		8	1		450	33.50		483.50	550
0220	Double bowl, 22" x 61"		8	1		505	33.50		538.50	615
0230	Double bowl, 22" x 73"		8	1		705	33.50		738.50	830
0240	For aggregate colors, add					35%				
0250	For faucets and fittings, see Section 22 41 39.10									
12 36 61.19 Engineered Stone Countertops										
0010	**ENGINEERED STONE COUNTERTOPS**									
0100	25" wide, 4" backsplash, color group A, minimum	2 Carp	15	1.067	L.F.	54	36		90	120
0110	Maximum		15	1.067		75	36		111	143
0120	Color group B, minimum		15	1.067		58	36		94	124
0130	Maximum		15	1.067		80	36		116	148
0140	Color group C, minimum		15	1.067		65	36		101	132
0150	Maximum		15	1.067		91	36		127	160
0160	Color group D, minimum		15	1.067		71	36		107	138
0170	Maximum		15	1.067		98	36		134	168

12 46 Furnishing Accessories

12 46 13 – Ash Receptacles

12 46 13.10 Ash/Trash Receivers		Crew	Daily Output	Labor-Hours	Unit	2011 Bare Costs Material	Labor	Equipment	Total	Total Incl O&P
0010	**ASH/TRASH RECEIVERS**									
1000	Ash urn, cylindrical metal									
1020	8″ diameter, 20″ high	1 Clab	60	.133	Ea.	85	3.53		88.53	99.50
1060	10″ diameter, 26″ high	″	60	.133	″	100	3.53		103.53	116
2000	Combination ash/trash urn, metal									
2020	8″ diameter, 20″ high	1 Clab	60	.133	Ea.	85	3.53		88.53	99.50
2050	10″ diameter, 26″ high	″	60	.133	″	105	3.53		108.53	121

12 46 19 – Clocks

12 46 19.50 Wall Clocks		Crew	Daily Output	Labor-Hours	Unit	Material	Labor	Equipment	Total	Total Incl O&P
0010	**WALL CLOCKS**									
0080	12″ diameter, single face	1 Elec	8	1	Ea.	139	40.50		179.50	218
0100	Double face	″	6.20	1.290	″	350	52		402	470

12 46 33 – Waste Receptacles

12 46 33.13 Trash Receptacles		Crew	Daily Output	Labor-Hours	Unit	Material	Labor	Equipment	Total	Total Incl O&P
0010	**TRASH RECEPTACLES**									
4000	Trash receptacle, metal									
4020	8″ diameter, 15″ high	1 Clab	60	.133	Ea.	67	3.53		70.53	79.50
4040	10″ diameter, 18″ high		60	.133		135	3.53		138.53	154
5040	16″ x 8″ x 14″ high	↓	60	.133	↓	21	3.53		24.53	29
5500	Plastic, with lid									
5520	35 gallon	1 Clab	60	.133	Ea.	135	3.53		138.53	154
5540	45 gallon		60	.133		220	3.53		223.53	248
5550	Plastic recycling barrel, w/lid & wheels, 32 gal. G		60	.133		75	3.53		78.53	88.50
5560	65 gal. G		60	.133		135	3.53		138.53	154
5570	95 gal. G	↓	60	.133	↓	220	3.53		223.53	248

12 48 Rugs and Mats

12 48 13 – Entrance Floor Mats and Frames

12 48 13.13 Entrance Floor Mats		Crew	Daily Output	Labor-Hours	Unit	Material	Labor	Equipment	Total	Total Incl O&P
0010	**ENTRANCE FLOOR MATS**									
0020	Recessed, in-laid black rubber, 3/8″ thick, solid	1 Clab	155	.052	S.F.	19.85	1.37		21.22	24.50
0050	Perforated		155	.052		13.80	1.37		15.17	17.50
0100	1/2″ thick, solid		155	.052		16.60	1.37		17.97	20.50
0150	Perforated		155	.052		22	1.37		23.37	26.50
0200	In colors, 3/8″ thick, solid		155	.052		17.85	1.37		19.22	22
0250	Perforated		155	.052		18.45	1.37		19.82	23
0300	1/2″ thick, solid		155	.052		23	1.37		24.37	28
0350	Perforated		155	.052		23.50	1.37		24.87	28.50
2000	Recycled rubber tire tile, 12″ x 12″ x 3/8″ thick G		125	.064		8.50	1.69		10.19	12.20
2510	Natural cocoa fiber, 1/2″ thick G		125	.064		8.10	1.69		9.79	11.80
2520	3/4″ thick G		125	.064		6.25	1.69		7.94	9.70
2530	1″ thick G	↓	125	.064	↓	6.90	1.69		8.59	10.45

12 51 Office Furniture

12 51 16 – Case Goods

12 51 16.13 Metal Case Goods		Crew	Daily Output	Labor-Hours	Unit	2011 Bare Costs Material	Labor	Equipment	Total	Total Incl O&P
0010	**METAL CASE GOODS**									
0020	Desks, 29" high, double pedestal, 30" x 60", metal, minimum				Ea.	500			500	550
0030	Maximum					1,275			1,275	1,400
0600	Desks, single pedestal, 30" x 60", metal, minimum					475			475	520
0620	Maximum					1,050			1,050	1,150
0720	Desks, secretarial, 30" x 60", metal, minimum					435			435	480
0730	Maximum					750			750	825
0740	Return, 20" x 42", minimum					325			325	355
0750	Maximum					490			490	540
0940	59" x 12" x 23" high, steel, minimum					286			286	315
0960	Maximum					350			350	385

12 51 16.16 Wood Case Goods

Line	Description	Crew	Daily Output	Labor-Hours	Unit	Material	Labor	Equipment	Total	Total Incl O&P
0010	**WOOD CASE GOODS**									
0150	Desk, 29" high, double pedestal, 30" x 60"									
0160	Wood, minimum				Ea.	330			330	365
0180	Maximum				"	2,475			2,475	2,725
0630	Single pedestal, 30" x 60"									
0640	Wood, minimum				Ea.	560			560	615
0650	Maximum					780			780	855
0670	Executive return, 24" x 42", with box, file, wood, minimum					395			395	430
0680	Maximum					780			780	855
0790	Desk, 29" high, secretarial, 30" x 60"									
0800	Wood, minimum				Ea.	495			495	545
0810	Maximum					2,525			2,525	2,775
0820	Return, 20" x 42", minimum					290			290	320
0830	Maximum					995			995	1,100
0900	Desktop organizer, 72" x 14" x 36" high, wood, minimum					169			169	186
0920	Maximum					400			400	440
1110	Furniture, credenza, 29" high, 18" to 22" x 60" to 72"									

12 51 23 – Office Tables

12 51 23.33 Conference Tables

Line	Description	Crew	Daily Output	Labor-Hours	Unit	Material	Labor	Equipment	Total	Total Incl O&P
0010	**CONFERENCE TABLES**									
6050	Boat, 96" x 42", minimum				Ea.	735			735	810
6150	Maximum					3,075			3,075	3,375
6720	Rectangle, 96" x 42", minimum					1,075			1,075	1,175
6740	Maximum					3,075			3,075	3,375

12 52 Seating

12 52 23 – Office Seating

12 52 23.13 Office Chairs

Line	Description	Crew	Daily Output	Labor-Hours	Unit	Material	Labor	Equipment	Total	Total Incl O&P
0010	**OFFICE CHAIRS**									
2000	Standard office chair, executive, minimum				Ea.	330			330	360
2150	Maximum					1,800			1,800	2,000
2200	Management, minimum					206			206	226
2250	Maximum					2,300			2,300	2,525
2280	Task, minimum					136			136	150
2290	Maximum					575			575	630
2300	Arm kit, minimum					61			61	67
2320	Maximum					91			91	100

12 54 Hospitality Furniture

12 54 13 – Hotel and Motel Furniture

12 54 13.10 Hotel Furniture		Crew	Daily Output	Labor-Hours	Unit	2011 Bare Costs Material	Labor	Equipment	Total	Total Incl O&P
0010	**HOTEL FURNITURE**									
0020	Standard quality set, minimum				Room	2,325			2,325	2,550
0200	Maximum				"	8,300			8,300	9,125

12 54 16 – Restaurant Furniture

12 54 16.10 Tables, Folding		Crew	Daily Output	Labor-Hours	Unit	Material	Labor	Equipment	Total	Total Incl O&P
0010	**TABLES, FOLDING** Laminated plastic tops									
1000	Tubular steel legs with glides									
1020	18" x 60", minimum				Ea.	236			236	260
1040	Maximum					1,300			1,300	1,425
1840	36" x 96", minimum					272			272	299
1860	Maximum					2,600			2,600	2,875
2000	Round, wood stained, plywood top, 60" diameter, minimum					385			385	420
2020	Maximum					1,400			1,400	1,525

12 54 16.20 Furniture, Restaurant		Crew	Daily Output	Labor-Hours	Unit	Material	Labor	Equipment	Total	Total Incl O&P
0010	**FURNITURE, RESTAURANT**									
0020	Bars, built-in, front bar	1 Carp	5	1.600	L.F.	250	53.50		303.50	365
0200	Back bar	"	5	1.600	"	182	53.50		235.50	290
0300	Booth seating, see Section 12 54 16.70									
2000	Chair, bentwood side chair, metal, minimum				Ea.	83.50			83.50	92
2020	Maximum					98			98	108
2600	Upholstered seat & back, arms, minimum					136			136	150
2620	Maximum					385			385	425

12 54 16.70 Booths		Crew	Daily Output	Labor-Hours	Unit	Material	Labor	Equipment	Total	Total Incl O&P
0010	**BOOTHS**									
1000	Banquet, upholstered seat and back, custom									
1500	Straight, minimum	2 Carp	40	.400	L.F.	169	13.40		182.40	209
1520	Maximum		36	.444		330	14.90		344.90	385
1600	"L" or "U" shape, minimum		35	.457		172	15.35		187.35	216
1620	Maximum		30	.533		305	17.90		322.90	365
1800	Upholstered outside finished backs for									
1810	single booths and custom banquets									
1820	Minimum	2 Carp	44	.364	L.F.	19.50	12.20		31.70	42
1840	Maximum	"	40	.400	"	58.50	13.40		71.90	87
3000	Fixed seating, one piece plastic chair and									
3010	plastic laminate table top									
3100	Two seat, 24" x 24" table, minimum	F-7	30	1.067	Ea.	680	32		712	805
3120	Maximum		26	1.231		970	37		1,007	1,125
3200	Four seat, 24" x 48" table, minimum		28	1.143		675	34.50		709.50	805
3220	Maximum		24	1.333		1,150	40		1,190	1,350
5000	Mount in floor, wood fiber core with									
5010	plastic laminate face, single booth									
5050	24" wide	F-7	30	1.067	Ea.	261	32		293	340
5100	48" wide	"	28	1.143	"	330	34.50		364.50	425

12 55 Detention Furniture

12 55 13 – Detention Bunks

12 55 13.13 Cots		Crew	Daily Output	Labor-Hours	Unit	Material	2011 Bare Costs Labor	Equipment	Total	Total Incl O&P
0010	**COTS**									
2500	Bolted, single, painted steel	E-4	20	1.600	Ea.	330	59	5.45	394.45	485
2700	Stainless steel	"	20	1.600	"	960	59	5.45	1,024.45	1,175

12 56 Institutional Furniture

12 56 33 – Classroom Furniture

12 56 33.10 Furniture, School

		Crew	Daily Output	Labor-Hours	Unit	Material	Labor	Equipment	Total	Total Incl O&P
0010	**FURNITURE, SCHOOL**									
0500	Classroom, movable chair & desk type, minimum				Set				67	74
0600	Maximum				"				141	155
1000	Chair, molded plastic									
1100	Integral tablet arm, minimum				Ea.	77.50			77.50	85
1150	Maximum					151			151	166
2000	Desk, single pedestal, top book compartment, minimum					55			55	60.50
2020	Maximum					67.50			67.50	74.50
2200	Flip top, minimum					95.50			95.50	105
2220	Maximum					122			122	134

12 56 43 – Dormitory Furniture

12 56 43.10 Dormitory Furnishings

		Crew	Daily Output	Labor-Hours	Unit	Material	Labor	Equipment	Total	Total Incl O&P
0010	**DORMITORY FURNISHINGS**									
0300	Bunkable bed, twin, minimum				Ea.	300			300	330
0320	Maximum					685			685	750
1000	Chest, four drawer, minimum					440			440	485
1020	Maximum					580			580	635
1050	Built-in, minimum	2 Carp	13	1.231	L.F.	145	41.50		186.50	229
1150	Maximum		10	1.600		268	53.50		321.50	385
1200	Desk top, built-in, laminated plastic, 24" deep, minimum		50	.320		39.50	10.75		50.25	61.50
1300	Maximum		40	.400		118	13.40		131.40	153
1450	30" deep, minimum		50	.320		50.50	10.75		61.25	73.50
1550	Maximum		40	.400		221	13.40		234.40	266
1750	Dressing unit, built-in, minimum		12	1.333		218	44.50		262.50	315
1850	Maximum		8	2		655	67		722	830
8000	Rule of thumb: total cost for furniture, minimum				Student				2,300	2,525
8050	Maximum				"				4,425	4,875

12 56 51 – Library Furniture

12 56 51.10 Library Furnishings

		Crew	Daily Output	Labor-Hours	Unit	Material	Labor	Equipment	Total	Total Incl O&P
0010	**LIBRARY FURNISHINGS**									
0100	Attendant desk, 36" x 62" x 29" high	1 Carp	16	.500	Ea.	2,700	16.80		2,716.80	3,000
0200	Book display, "A" frame display, both sides, 42" x 42" x 60" high		16	.500		1,500	16.80		1,516.80	1,675
0220	Table with bulletin board, 42" x 24" x 49" high		16	.500		1,325	16.80		1,341.80	1,475
0800	Card catalogue, 30 tray unit		16	.500		2,850	16.80		2,866.80	3,150
0840	60 tray unit		16	.500		5,225	16.80		5,241.80	5,775
0880	72 tray unit	2 Carp	16	1		6,400	33.50		6,433.50	7,100
1000	Carrels, single face, initial unit	1 Carp	16	.500		720	16.80		736.80	825
1500	Double face, initial unit	2 Carp	16	1		1,150	33.50		1,183.50	1,300
1710	Carrels, hardwood, 36" x 24", minimum	1 Carp	5	1.600		520	53.50		573.50	665
1720	Maximum	"	4	2		1,275	67		1,342	1,500
2700	Card catalog file, 60 trays, complete					5,600			5,600	6,150
2720	Alternate method: each tray					93.50			93.50	103
3800	Charging desk, built-in, with counter, plastic laminated top	1 Carp	7	1.143	L.F.	410	38.50		448.50	520

12 56 Institutional Furniture

12 56 51 – Library Furniture

12 56 51.10 Library Furnishings		Crew	Daily Output	Labor-Hours	Unit	Material	2011 Bare Costs Labor	Equipment	Total	Total Incl O&P
4000	Dictionary stand, stationary	1 Carp	16	.500	Ea.	880	16.80		896.80	1,000
4020	Revolving		16	.500		345	16.80		361.80	410
4200	Exhibit case, table style, 60" x 28" x 36"		11	.727		2,800	24.50		2,824.50	3,150
6010	Bookshelf, metal, 90" high, 10" shelf, double face		11.50	.696	L.F.	118	23.50		141.50	169
6020	Single face		12	.667	"	113	22.50		135.50	162
6050	For 8" shelving, subtract from above					10%				
6060	For 12" shelving, add to above					10%				
6070	For 42" high with countertop, subtract from above					20%				
6100	Mobile compacted shelving, hand crank, 9'-0" high									
6110	Double face, including track, 3' section				Ea.	1,125			1,125	1,250
6150	For electrical operation, add					25%				
6200	Magazine shelving, 82" high, 12" deep, single face	1 Carp	11.50	.696	L.F.	92.50	23.50		116	141
6210	Double face		11.50	.696	"	119	23.50		142.50	169
7000	Tables, card catalog reference, 24" x 60" x 42"		16	.500	Ea.	975	16.80		991.80	1,100
7200	Reading table, laminated top, 60" x 36"				"	720			720	790

12 56 70 – Healthcare Furniture

12 56 70.10 Furniture, Hospital		Crew	Daily Output	Labor-Hours	Unit	Material	Labor	Equipment	Total	Total Incl O&P
0010	**FURNITURE, HOSPITAL**									
0020	Beds, manual, minimum				Ea.	755			755	830
0100	Maximum					2,575			2,575	2,825
0300	Manual and electric beds, minimum					1,050			1,050	1,150
0400	Maximum					2,975			2,975	3,275
0600	All electric hospital beds, minimum					1,325			1,325	1,450
0700	Maximum					4,300			4,300	4,750
0900	Manual, nursing home beds, minimum					850			850	935
1000	Maximum					1,775			1,775	1,950
1020	Overbed table, laminated top, minimum					360			360	400
1040	Maximum					900			900	990
1100	Patient wall systems, not incl. plumbing, minimum				Room	1,075			1,075	1,200
1200	Maximum				"	2,000			2,000	2,200
2000	Geriatric chairs, minimum				Ea.	420			420	465
2020	Maximum				"	720			720	790

12 61 Fixed Audience Seating

12 61 13 – Upholstered Audience Seating

12 61 13.13 Auditorium Chairs		Crew	Daily Output	Labor-Hours	Unit	Material	Labor	Equipment	Total	Total Incl O&P
0010	**AUDITORIUM CHAIRS**									
2000	All veneer construction	2 Carp	22	.727	Ea.	175	24.50		199.50	234
2200	Veneer back, padded seat		22	.727		221	24.50		245.50	284
2350	Fully upholstered, spring seat		22	.727		211	24.50		235.50	273
2450	For tablet arms, add					58			58	63.50
2500	For fire retardancy, CATB-133, add					25			25	27.50

12 61 13.23 Lecture Hall Seating		Crew	Daily Output	Labor-Hours	Unit	Material	Labor	Equipment	Total	Total Incl O&P
0010	**LECTURE HALL SEATING**									
1000	Pedestal type, minimum	2 Carp	22	.727	Ea.	165	24.50		189.50	223
1200	Maximum	"	14.50	1.103	"	555	37		592	670

12 63 Stadium and Arena Seating

12 63 13 – Stadium and Arena Bench Seating

12 63 13.13 Bleachers		Crew	Daily Output	Labor-Hours	Unit	Material	Labor	Equipment	Total	Total Incl O&P
						2011 Bare Costs				
0010	**BLEACHERS**									
3000	Telescoping, manual to 15 tier, minimum	F-5	65	.492	Seat	86	14.50		100.50	119
3100	Maximum		60	.533		129	15.70		144.70	169
3300	16 to 20 tier, minimum		60	.533		207	15.70		222.70	255
3400	Maximum		55	.582		259	17.15		276.15	315
3600	21 to 30 tier, minimum		50	.640		215	18.85		233.85	269
3700	Maximum		40	.800		259	23.50		282.50	325
3900	For integral power operation, add, minimum	2 Elec	300	.053		43	2.15		45.15	51
4000	Maximum	"	250	.064		69	2.58		71.58	80
5000	Benches, folding, in wall, 14' table, 2 benches	L-4	2	8	Set	730	238		968	1,200

12 67 Pews and Benches

12 67 13 – Pews

12 67 13.13 Sanctuary Pews

		Crew	Daily Output	Labor-Hours	Unit	Material	Labor	Equipment	Total	Total Incl O&P
0010	**SANCTUARY PEWS**									
1500	Bench type, hardwood, minimum	1 Carp	20	.400	L.F.	77.50	13.40		90.90	108
1550	Maximum	"	15	.533		154	17.90		171.90	199
1570	For kneeler, add					19.05			19.05	21

12 92 Interior Planters and Artificial Plants

12 92 33 – Interior Planters

12 92 33.10 Planters

		Crew	Daily Output	Labor-Hours	Unit	Material	Labor	Equipment	Total	Total Incl O&P
0010	**PLANTERS**									
1000	Fiberglass, hanging, 12" diameter, 7" high				Ea.	105			105	115
1500	Rectangular, 48" long, 16" high x 15" wide					690			690	760
1650	60" long, 30" high, 28" wide					1,050			1,050	1,150
2000	Round, 12" diameter, 13" high					119			119	131
2050	25" high					135			135	148
5000	Square, 10" side, 20" high					158			158	173
5100	14" side, 15" high					190			190	209
6000	Metal bowl, 32" diameter, 8" high, minimum					480			480	525
6050	Maximum					620			620	680
8750	Wood, fiberglass liner, square									
8780	14" square, 15" high, minimum				Ea.	300			300	330
8800	Maximum					370			370	410
9400	Plastic cylinder, molded, 10" diameter, 10" high					12.25			12.25	13.45
9500	11" diameter, 11" high					39			39	42.50

12 93 Site Furnishings

12 93 23 – Trash and Litter Receptors

12 93 23.10 Trash Receptacles		Crew	Daily Output	Labor-Hours	Unit	Material	2011 Bare Costs Labor	Equipment	Total	Total Incl O&P
0010	**TRASH RECEPTACLES**									
0020	Fiberglass, 2′ square, 18″ high	2 Clab	30	.533	Ea.	525	14.10		539.10	600
0100	2′ square, 2′-6″ high		30	.533		525	14.10		539.10	600
0300	Circular , 2′ diameter, 18″ high		30	.533		515	14.10		529.10	590
0400	2′ diameter, 2′-6″ high		30	.533		595	14.10		609.10	680
0500	Recycled plastic, var colors, round, 32 gal., 28″ x 38″ H G		5	3.200		390	84.50		474.50	565
0510	32 gal., 31″ x 32″ H G		5	3.200		430	84.50		514.50	610

12 93 23.20 Trash Closure		Crew	Daily Output	Labor-Hours	Unit	Material	Labor	Equipment	Total	Total Incl O&P
0010	**TRASH CLOSURE**									
0020	Steel with pullover cover, 2′-3″ wide, 4′-7″ high, 6′-2″ long	2 Clab	5	3.200	Ea.	1,375	84.50		1,459.50	1,675
0100	10′-1″ long		4	4		1,950	106		2,056	2,325
0300	Wood, 10′ wide, 6′ high, 10′ long		1.20	13.333		1,450	355		1,805	2,200

12 93 33 – Manufactured Planters

12 93 33.10 Planters		Crew	Daily Output	Labor-Hours	Unit	Material	Labor	Equipment	Total	Total Incl O&P
0010	**PLANTERS**									
0012	Concrete, sandblasted, precast, 48″ diameter, 24″ high	2 Clab	15	1.067	Ea.	630	28		658	745
0100	Fluted, precast, 7′ diameter, 36″ high		10	1.600		1,300	42.50		1,342.50	1,500
0300	Fiberglass, circular, 36″ diameter, 24″ high		15	1.067		600	28		628	710
0320	36″ diameter, 27″ high		12	1.333		710	35.50		745.50	840
0330	33″ high		15	1.067		760	28		788	885
0335	24″ diameter, 36″ high		15	1.067		485	28		513	585
0340	60″ diameter, 39″ high		8	2		1,225	53		1,278	1,450
0400	60″ diameter, 24″ high		10	1.600		705	42.50		747.50	845
0600	Square, 24″ side, 36″ high		15	1.067		615	28		643	730
0610	24″ side, 27″ high		12	1.333		615	35.50		650.50	740
0620	24″ side, 16″ high		20	.800		330	21		351	395
0700	48″ side, 36″ high		15	1.067		900	28		928	1,050
0900	Planter/bench, 72″ square, 36″ high		5	3.200		1,300	84.50		1,384.50	1,575
1000	96″ square, 27″ high		5	3.200		1,975	84.50		2,059.50	2,325
1200	Wood, square, 48″ side, 24″ high		15	1.067		1,225	28		1,253	1,375
1300	Circular, 48″ diameter, 30″ high		10	1.600		910	42.50		952.50	1,075
1500	72″ diameter, 30″ high		10	1.600		1,425	42.50		1,467.50	1,625
1600	Planter/bench, 72″		5	3.200		3,150	84.50		3,234.50	3,625

12 93 43 – Site Seating and Tables

12 93 43.13 Site Seating		Crew	Daily Output	Labor-Hours	Unit	Material	Labor	Equipment	Total	Total Incl O&P
0010	**SITE SEATING**									
0012	Seating, benches, park, precast conc, w/backs, wood rails, 4′ long	2 Clab	5	3.200	Ea.	470	84.50		554.50	655
0100	8′ long		4	4		930	106		1,036	1,200
0300	Fiberglass, without back, one piece, 4′ long		10	1.600		605	42.50		647.50	735
0400	8′ long		7	2.286		805	60.50		865.50	985
0500	Steel barstock pedestals w/backs, 2″ x 3″ wood rails, 4′ long		10	1.600		1,075	42.50		1,117.50	1,250
0510	8′ long		7	2.286		1,300	60.50		1,360.50	1,525
0520	3″ x 8″ wood plank, 4′ long		10	1.600		1,075	42.50		1,117.50	1,250
0530	8′ long		7	2.286		1,300	60.50		1,360.50	1,525
0540	Backless, 4″ x 4″ wood plank, 4′ square		10	1.600		1,075	42.50		1,117.50	1,275
0550	8′ long		7	2.286		835	60.50		895.50	1,025
0600	Aluminum pedestals, with backs, aluminum slats, 8′ long		8	2		335	53		388	460
0610	15′ long		5	3.200		400	84.50		484.50	580
0620	Portable, aluminum slats, 8′ long		8	2		340	53		393	460
0630	15′ long		5	3.200		435	84.50		519.50	620
0800	Cast iron pedestals, back & arms, wood slats, 4′ long		8	2		320	53		373	445

12 93 Site Furnishings

12 93 43 – Site Seating and Tables

12 93 43.13 Site Seating		Crew	Daily Output	Labor-Hours	Unit	Material	Labor	Equipment	Total	Total Incl O&P
						2011 Bare Costs				
0820	8′ long	2 Clab	5	3.200	Ea.	890	84.50		974.50	1,125
0840	Backless, wood slats, 4′ long		8	2		395	53		448	525
0860	8′ long		5	3.200		925	84.50		1,009.50	1,175
1700	Steel frame, fir seat, 10′ long	↓	10	1.600	↓	330	42.50		372.50	430

Estimating Tips

General

- The items and systems in this division are usually estimated, purchased, supplied, and installed as a unit by one or more subcontractors. The estimator must ensure that all parties are operating from the same set of specifications and assumptions, and that all necessary items are estimated and will be provided. Many times the complex items and systems are covered, but the more common ones, such as excavation or a crane, are overlooked for the very reason that everyone assumes nobody could miss them. The estimator should be the central focus and be able to ensure that all systems are complete.
- Another area where problems can develop in this division is at the interface between systems. The estimator must ensure, for instance, that anchor bolts, nuts, and washers are estimated and included for the air-supported structures and pre-engineered buildings to be bolted to their foundations. Utility supply is a common area where essential items or pieces of equipment can be missed or overlooked due to the fact that each subcontractor may feel it is another's responsibility. The estimator should also be aware of certain items which may be supplied as part of a package but installed by others, and ensure that the installing contractor's estimate includes the cost of installation. Conversely, the estimator must also ensure that items are not costed by two different subcontractors, resulting in an inflated overall estimate.

13 30 00 Special Structures

- The foundations and floor slab, as well as rough mechanical and electrical, should be estimated, as this work is required for the assembly and erection of the structure. Generally, as noted in the book, the pre-engineered building comes as a shell. Pricing is based on the size and structural design parameters stated in the reference section. Additional features, such as windows and doors with their related structural framing, must also be included by the estimator. Here again, the estimator must have a clear understanding of the scope of each portion of the work and all the necessary interfaces.

Reference Numbers

Reference numbers are shown in shaded boxes at the beginning of some major classifications. These numbers refer to related items in the Reference Section. The reference information may be an estimating procedure, an alternate pricing method, or technical information.

Note: Not all subdivisions listed here necessarily appear in this publication.

13 05 Common Work Results for Special Construction

13 05 05 – Selective Special Construction Demolition

13 05 05.10 Selective Demolition, Air Supported Structures

		Crew	Daily Output	Labor-Hours	Unit	Material	2011 Bare Costs Labor	Equipment	Total	Total Incl O&P
0010	**SELECTIVE DEMOLITION, AIR SUPPORTED STRUCTURES**									
0020	Tank covers, scrim, dbl. layer, vinyl poly w/hdw, blower & controls									
0050	Round and rectangular R024119-10	B-2	9000	.004	S.F.		.12		.12	.20
0100	Warehouse structures									
0120	Poly/vinyl fabric, 28 oz, incl tension cables & inflation system	4 Clab	9000	.004	SF Flr.		.09		.09	.16
0150	Reinforced vinyl, 12 oz., 3000 S.F.	"	5000	.006			.17		.17	.28
0200	12,000 to 24,000 S.F.	8 Clab	20000	.003			.08		.08	.14
0250	Tedlar vinyl fabric, 28 oz. w/liner, to 3000 S.F.	4 Clab	5000	.006			.17		.17	.28
0300	12,000 to 24,000 S.F.	8 Clab	20000	.003			.08		.08	.14
0350	Greenhouse/shelter, woven polyethylene with liner									
0400	3000 S.F.	4 Clab	5000	.006	SF Flr.		.17		.17	.28
0450	12,000 to 24,000 S.F.	8 Clab	20000	.003			.08		.08	.14
0500	Tennis/gymnasium, poly/vinyl fabric, 28 oz., incl thermal liner	4 Clab	9000	.004			.09		.09	.16
0600	Stadium/convention center, teflon coated fiberglass, incl thermal liner	9 Clab	40000	.002			.05		.05	.08
0700	Doors, air lock, 15′ long, 10′ x 10′	2 Carp	1.50	10.667	Ea.		360		360	600
0720	15′ x 15′		.80	20			670		670	1,125
0750	Revolving personnel door, 6′ dia. x 6′-6″ high		1.50	10.667			360		360	600

13 05 05.20 Selective Demolition, Garden Houses

		Crew	Daily Output	Labor-Hours	Unit	Material	Labor	Equipment	Total	Total Incl O&P
0010	**SELECTIVE DEMOLITION, GARDEN HOUSES** R024119-10									
0020	Prefab, wood, excl foundation, average	2 Clab	400	.040	SF Flr.		1.06		1.06	1.77

13 05 05.25 Selective Demolition, Geodesic Domes

		Crew	Daily Output	Labor-Hours	Unit	Material	Labor	Equipment	Total	Total Incl O&P
0010	**SELECTIVE DEMOLITION, GEODESIC DOMES**									
0050	Shell only, interlocking plywood panels, 30′ diameter	F-5	3.20	10	Ea.		295		295	490
0060	34′ diameter		2.30	13.913			410		410	685
0070	39′ diameter		2	16			470		470	790
0080	45′ diameter	F-3	2.20	18.182			560	298	858	1,250
0090	55′ diameter		2	20			615	330	945	1,375
0100	60′ diameter		2	20			615	330	945	1,375
0110	65′ diameter		1.60	25			770	410	1,180	1,725

13 05 05.30 Selective Demolition, Greenhouses

		Crew	Daily Output	Labor-Hours	Unit	Material	Labor	Equipment	Total	Total Incl O&P
0010	**SELECTIVE DEMOLITION, GREENHOUSES** R024119-10									
0020	Resi-type, free standing, excl foundations, 9′ long x 8′ wide	2 Clab	160	.100	SF Flr.		2.65		2.65	4.43
0030	9′ long x 11′ wide		170	.094			2.49		2.49	4.17
0040	9′ long x 14′ wide		220	.073			1.92		1.92	3.22
0050	9′ long x 17′ wide		320	.050			1.32		1.32	2.22
0060	Lean-to type, 4′ wide		64	.250			6.60		6.60	11.10
0070	7′ wide		120	.133			3.53		3.53	5.90
0080	Geodesic hemisphere, 1/8″ plexiglass glazing, 8′ diam.		4	4	Ea.		106		106	177
0090	24′ dia		.80	20			530		530	885
0100	48′ dia		.40	40			1,050		1,050	1,775

13 05 05.35 Selective Demolition, Hangars

		Crew	Daily Output	Labor-Hours	Unit	Material	Labor	Equipment	Total	Total Incl O&P
0010	**SELECTIVE DEMOLITION, HANGARS**									
0020	T type hangars, prefab, steel , galv roof & walls, incl doors, excl fndtn	E-2	2550	.019	SF Flr.		.69	.64	1.33	1.98
0030	Circular type, prefab, steel frame, plastic skin, incl foundation, 80′ diam	"	.50	96	Total		3,525	3,250	6,775	10,100

13 05 05.45 Selective Demolition, Lightning Protection

		Crew	Daily Output	Labor-Hours	Unit	Material	Labor	Equipment	Total	Total Incl O&P
0010	**SELECTIVE DEMOLITION, LIGHTNING PROTECTION**									
0020	Air terminal & base, copper, 3/8″ diam. x 10″, to 75′ h	1 Clab	16	.500	Ea.		13.25		13.25	22
0030	1/2″ diam. x 12″, over 75′ h		16	.500			13.25		13.25	22
0050	Aluminum, 1/2″ diam. x 12″, to 75′ h		16	.500			13.25		13.25	22
0060	5/8″ diam. x 12″, over 75′ h		16	.500			13.25		13.25	22
0070	Cable, copper, 220 lb per thousand feet, to 75′ high		640	.013	L.F.		.33		.33	.55

13 05 Common Work Results for Special Construction

13 05 05 – Selective Special Construction Demolition

13 05 05.45 Selective Demolition, Lightning Protection

		Crew	Daily Output	Labor-Hours	Unit	2011 Bare Costs Material	Labor	Equipment	Total	Total Incl O&P
0080	375 lb per thousand feet, over 75′ high	1 Clab	460	.017	L.F.		.46		.46	.77
0090	Aluminum, 101 lb per thousand feet, to 75′ high		560	.014			.38		.38	.63
0100	199 lb per thousand feet, over 75′ high		480	.017			.44		.44	.74
0110	Arrester, 175 V AC, to ground		16	.500	Ea.		13.25		13.25	22
0120	650 V AC, to ground		13	.615	"		16.30		16.30	27.50

13 05 05.50 Selective Demolition, Pre-Engineered Steel Buildings

		Crew	Daily Output	Labor-Hours	Unit	Material	Labor	Equipment	Total	Total Incl O&P
0010	**SELECTIVE DEMOLITION, PRE-ENGINEERED STEEL BUILDINGS**									
0500	Pre-engd steel bldgs, rigid frame, clear span & multi post, excl salvage									
0550	3,500 to 7,500 S.F.	L-10	1000	.024	SF Flr.		.88	.66	1.54	2.33
0600	7,501 to 12,500 S.F.		1500	.016			.59	.44	1.03	1.55
0650	12,500 S.F. or greater		1650	.015			.54	.40	.94	1.42
0700	Pre-engd steel building components									
0710	Entrance canopy, including frame 4′ x 4′	E-24	8	4	Ea.		144	93.50	237.50	370
0720	4′ x 8′	"	7	4.571			165	107	272	425
0730	H.M. doors, self framing, single leaf	2 Skwk	8	2			68		68	114
0740	Double leaf		5	3.200			109		109	182
0760	Gutter, eave type		600	.027	L.F.		.91		.91	1.52
0770	Sash, single slide, double slide or fixed		24	.667	Ea.		22.50		22.50	38
0780	Skylight, fiberglass, to 30 S.F.		16	1			34		34	57
0785	Roof vents, circular, 12″ to 24″ diameter		12	1.333			45.50		45.50	76
0790	Continuous, 10′ long		8	2			68		68	114
0900	Shelters, aluminum frame									
0910	Acrylic glazing, 3′ x 9′ x 8′ high	2 Skwk	2	8	Ea.		272		272	455
0920	9′ x 12′ x 8′ high	"	1.50	10.667	"		365		365	610

13 05 05.60 Selective Demolition, Silos

		Crew	Daily Output	Labor-Hours	Unit	Material	Labor	Equipment	Total	Total Incl O&P
0010	**SELECTIVE DEMOLITION, SILOS**									
0020	Conc stave, indstrl, conical/sloping bott, excl fndtn, 12′ diam., 35′ h	E-24	.18	177	Ea.		6,400	4,150	10,550	16,400
0030	16′ diam., 45′ h		.12	266			9,600	6,225	15,825	24,600
0040	25′ dia, 75′ h		.08	400			14,400	9,350	23,750	36,800
0050	Steel, factory fabricated, 30,000 gal. cap, painted or epoxy lined	L-5	2	28			1,025	375	1,400	2,325

13 05 05.65 Selective Demolition, Sound Control

		Crew	Daily Output	Labor-Hours	Unit	Material	Labor	Equipment	Total	Total Incl O&P
0010	**SELECTIVE DEMOLITION, SOUND CONTROL** R024119-10									
0120	Acoustical enclosure, 4″ thick walls & ceiling panels, 8 lb./S.F.	3 Carp	144	.167	SF Surf		5.60		5.60	9.35
0130	10.5 lb./S.F.		128	.188			6.30		6.30	10.55
0140	Reverb chamber, parallel walls, 4″ thick		120	.200			6.70		6.70	11.25
0150	Skewed walls, parallel roof, 4″ thick		110	.218			7.30		7.30	12.25
0160	Skewed walls/roof, 4″ layer/air space		96	.250			8.40		8.40	14.05
0170	Sound-absorbing panels, painted metal, 2′-6″ x 8′, under 1,000 S.F.		430	.056			1.87		1.87	3.13
0180	Over 1,000 S.F.		480	.050			1.68		1.68	2.81
0190	Flexible transparent curtain, clear	3 Shee	430	.056			2.13		2.13	3.52
0192	50% clear, 50% foam		430	.056			2.13		2.13	3.52
0194	25% clear, 75% foam		430	.056			2.13		2.13	3.52
0196	100% foam		430	.056			2.13		2.13	3.52
0200	Audio-masking sys, incl speakers, amplfr, signal gnrtr, clng mntd									
0205	Ceiling mounted, 5,000 S.F.	2 Elec	4800	.003	S.F.		.13		.13	.22
0210	10,000 S.F.		5600	.003			.12		.12	.19
0220	Plenum mounted, 5,000 S.F.		7600	.002			.08		.08	.14
0230	10,000 S.F.		8800	.002			.07		.07	.12

13 05 05.70 Selective Demolition, Special Purpose Rooms

		Crew	Daily Output	Labor-Hours	Unit	Material	Labor	Equipment	Total	Total Incl O&P
0010	**SELECTIVE DEMOLITION, SPECIAL PURPOSE ROOMS** R024119-10									
0100	Audiometric rooms, under 500 S.F. surface	4 Carp	200	.160	SF Surf		5.35		5.35	9
0110	Over 500 S.F. surface	"	240	.133	"		4.47		4.47	7.50

13 05 05 – Selective Special Construction Demolition

13 05 05.70 Selective Demolition, Special Purpose Rooms		Crew	Daily Output	Labor-Hours	Unit	Material	2011 Bare Costs Labor	Equipment	Total	Total Incl O&P
0200	Clean rooms, 12′ x 12′ soft wall, class 100	1 Carp	.30	26.667	Ea.		895		895	1,500
0210	Class 1000		.30	26.667			895		895	1,500
0220	Class 10,000		.35	22.857			765		765	1,275
0230	Class 100,000		.35	22.857			765		765	1,275
0300	Darkrooms, shell complete, 8′ high	2 Carp	220	.073	SF Flr.		2.44		2.44	4.08
0310	12′ high		110	.145	″		4.88		4.88	8.15
0350	Darkrooms doors, mini-cylindrical, revolving		4	4	Ea.		134		134	225
0400	Music room, practice modular		140	.114	SF Surf		3.83		3.83	6.40
0500	Refrigeration structures and finishes									
0510	Wall finish, 2 coat portland cement plaster, 1/2″ thick	1 Clab	200	.040	S.F.		1.06		1.06	1.77
0520	Fiberglass panels, 1/8″ thick		400	.020			.53		.53	.89
0530	Ceiling finish, polystyrene plastic, 1″ to 2″ thick		500	.016			.42		.42	.71
0540	4″ thick		450	.018			.47		.47	.79
0550	Refrigerator, prefab aluminum walk-in, 7′-6″ high, 6′ x 6′ OD	2 Carp	100	.160	SF Flr.		5.35		5.35	9
0560	10′ x 10′ OD		160	.100			3.36		3.36	5.60
0570	Over 150 S.F.		200	.080			2.68		2.68	4.49
0600	Sauna, prefabricated, including heater & controls, 7′ high, to 30 S.F.		120	.133			4.47		4.47	7.50
0610	To 40 S.F.		140	.114			3.83		3.83	6.40
0620	To 60 S.F.		175	.091			3.07		3.07	5.15
0630	To 100 S.F.		220	.073			2.44		2.44	4.08
0640	To 130 S.F.		250	.064			2.15		2.15	3.59
0650	Steam bath, heater, timer, head, single, to 140 C.F.	1 Plum	2.20	3.636	Ea.		144		144	235
0660	To 300 C.F.		2.20	3.636			144		144	235
0670	Steam bath, comm. size, w/blow-down assembly, to 800 C.F.		1.80	4.444			176		176	288
0680	To 2500 C.F.		1.60	5			198		198	325
0690	Steam bath, comm. size, multiple, for motels, apts, 500 C.F., 2 baths		2	4			159		159	259
0700	1,000 C.F., 4 baths		1.40	5.714			227		227	370

13 05 05.75 Selective Demolition, Storage Tanks		Crew	Daily Output	Labor-Hours	Unit	Material	Labor	Equipment	Total	Total Incl O&P
0010	**SELECTIVE DEMOLITION, STORAGE TANKS**									
0500	Steel tank, single wall, above ground, not incl fdn, pumps or piping									
0510	Single wall, 275 gallon R024119-10	Q-1	3	5.333	Ea.		190		190	310
0520	550 thru 2,000 gallon	B-34P	2	12			405	269	674	960
0530	5,000 thru 10,000 gallon	B-34Q	2	12			410	590	1,000	1,325
0540	15,000 thru 30,000 gallon	B-34S	2	16			570	1,775	2,345	2,900
0600	Steel tank, double wall, above ground not incl fdn, pumps & piping									
0620	500 thru 2,000 gallon	B-34P	2	12	Ea.		405	269	674	960

13 05 05.85 Selective Demolition, Swimming Pool Equip		Crew	Daily Output	Labor-Hours	Unit	Material	Labor	Equipment	Total	Total Incl O&P
0010	**SELECTIVE DEMOLITION, SWIMMING POOL EQUIP**									
0020	Diving stand, stainless steel, 3 meter	2 Clab	3	5.333	Ea.		141		141	236
0030	1 meter		5	3.200			84.50		84.50	142
0040	Diving board, 16′ long, aluminum		5.40	2.963			78.50		78.50	131
0050	Fiberglass		5.40	2.963			78.50		78.50	131
0070	Ladders, heavy duty, stainless steel, 2 tread		14	1.143			30		30	50.50
0080	4 tread		12	1.333			35.50		35.50	59
0090	Lifeguard chair, stainless steel, fixed		5	3.200			84.50		84.50	142
0100	Slide, tubular, fiberglass, aluminum handrails & ladder, 5′, straight		4	4			106		106	177
0110	8′, curved		6	2.667			70.50		70.50	118
0120	10′, curved		3	5.333			141		141	236
0130	12′ straight, with platform		2.50	6.400			169		169	284
0140	Removable access ramp, stainless steel		4	4			106		106	177
0150	Removable stairs, stainless steel, collapsible		4	4			106		106	177

13 05 Common Work Results for Special Construction

13 05 05 – Selective Special Construction Demolition

13 05 05.90 Selective Demolition, Tension Structures		Crew	Daily Output	Labor-Hours	Unit	Material	2011 Bare Costs Labor	Equipment	Total	Total Incl O&P
0010	**SELECTIVE DEMOLITION, TENSION STRUCTURES**									
0020	Steel/alum frame, fabric shell, 60′ clear span, 6,000 S.F.	B-41	2000	.022	SF Flr.		.60	.14	.74	1.17
0030	12,000 S.F.		2200	.020			.55	.13	.68	1.06
0040	80′ clear span, 20,800 S.F.		2440	.018			.49	.12	.61	.96
0050	100′ clear span, 10,000 S.F.	L-5	4350	.013			.47	.17	.64	1.07
0060	26,000 S.F.	″	4600	.012			.45	.16	.61	1.01

13 05 05.95 Selective Demo, X-Ray/Radio Freq Protection

13 05 05.95 Selective Demo, X-Ray/Radio Freq Protection		Crew	Daily Output	Labor-Hours	Unit	Material	Labor	Equipment	Total	Total Incl O&P
0010	**SELECTIVE DEMO, X-RAY/RADIO FREQ PROTECTION**									
0020	Shielding lead, lined door frame, excl hdwe, 1/16″ thick	1 Clab	4.80	1.667	Ea.		44		44	74
0030	Lead sheets, 1/16″ thick	2 Clab	270	.059	S.F.		1.57		1.57	2.63
0040	1/8″ thick		240	.067			1.76		1.76	2.95
0050	Lead shielding, 1/4″ thick		270	.059			1.57		1.57	2.63
0060	1/2″ thick		240	.067			1.76		1.76	2.95
0070	Lead glass, 1/4″ thick, 2.0 mm LE, 12″ x 16″	2 Glaz	16	1	Ea.		33		33	54.50
0080	24″ x 36″		8	2			66.50		66.50	109
0090	36″ x 60″		4	4			133		133	218
0100	Lead glass window frame, with 1/16″ lead & voice passage, 36″ x 60″		4	4			133		133	218
0110	Lead glass window frame, 24″ x 36″		8	2			66.50		66.50	109
0120	Lead gypsum board, 5/8″ thick with 1/16″ lead	2 Clab	320	.050	S.F.		1.32		1.32	2.22
0130	1/8″ lead		280	.057			1.51		1.51	2.53
0140	1/32″ lead		400	.040			1.06		1.06	1.77
0150	Butt joints, 1/8″ lead or thicker, 2″ x 7′ long batten strip		480	.033	Ea.		.88		.88	1.48
0160	X-ray protection, average radiography room, up to 300 S.F., 1/16″ lead, min		.50	32	Total		845		845	1,425
0170	Maximum		.30	53.333			1,400		1,400	2,375
0180	Deep therapy X-ray room, 250 kV cap, up to 300 S.F., 1/4″ lead, min		.20	80			2,125		2,125	3,550
0190	Maximum		.12	133			3,525		3,525	5,900
0880	Radio frequency shielding, prefab or screen-type copper or steel, minimum		360	.044	SF Surf		1.18		1.18	1.97
0890	Average		310	.052			1.37		1.37	2.29
0895	Maximum		290	.055			1.46		1.46	2.44

13 11 Swimming Pools

13 11 13 – Below-Grade Swimming Pools

13 11 13.50 Swimming Pools

13 11 13.50 Swimming Pools		Crew	Daily Output	Labor-Hours	Unit	Material	Labor	Equipment	Total	Total Incl O&P
0010	**SWIMMING POOLS** Residential in-ground, vinyl lined, concrete									
0020	Sides including equipment, sand bottom	B-52	300	.187	SF Surf	15.95	5.40	1.68	23.03	28.50
0100	Metal or polystyrene sides R131113-20	B-14	410	.117		13.35	3.27	.78	17.40	21
0200	Add for vermiculite bottom					1.02			1.02	1.12
0500	Gunite bottom and sides, white plaster finish									
0600	12′ x 30′ pool	B-52	145	.386	SF Surf	29.50	11.20	3.48	44.18	55
0720	16′ x 32′ pool		155	.361		27	10.50	3.25	40.75	50.50
0750	20′ x 40′ pool		250	.224		24	6.50	2.02	32.52	39.50
0810	Concrete bottom and sides, tile finish									
0820	12′ x 30′ pool	B-52	80	.700	SF Surf	30	20.50	6.30	56.80	74
0830	16′ x 32′ pool		95	.589		25	17.15	5.30	47.45	62
0840	20′ x 40′ pool		130	.431		19.70	12.50	3.88	36.08	47
1100	Motel, gunite with plaster finish, incl. medium									
1150	capacity filtration & chlorination	B-52	115	.487	SF Surf	36.50	14.15	4.38	55.03	68.50
1200	Municipal, gunite with plaster finish, incl. high									
1250	capacity filtration & chlorination	B-52	100	.560	SF Surf	47.50	16.25	5.05	68.80	85
1350	Add for formed gutters				L.F.	69.50			69.50	76.50

13 11 Swimming Pools

13 11 13 – Below-Grade Swimming Pools

13 11 13.50 Swimming Pools		Crew	Daily Output	Labor-Hours	Unit	2011 Bare Costs Material	Labor	Equipment	Total	Total Incl O&P
1360	Add for stainless steel gutters				L.F.	206			206	226
1600	For water heating system, see Section 23 52 28.10									
1700	Filtration and deck equipment only, as % of total				Total				20%	20%
1800	Deck equipment, rule of thumb, 20′ x 40′ pool				SF Pool				1.18	1.30
1900	5000 S.F. pool				"				1.73	1.90
3000	Painting pools, preparation + 3 coats, 20′ x 40′ pool, epoxy	2 Pord	.33	48.485	Total	1,700	1,450		3,150	4,175
3100	Rubber base paint, 18 gallons	"	.33	48.485		1,100	1,450		2,550	3,525
3500	42′ x 82′ pool, 75 gallons, epoxy paint	3 Pord	.14	171		7,150	5,075		12,225	16,100
3600	Rubber base paint	"	.14	171		4,500	5,075		9,575	13,200

13 11 46 – Swimming Pool Accessories

13 11 46.50 Swimming Pool Equipment		Crew	Daily Output	Labor-Hours	Unit	Material	Labor	Equipment	Total	Total Incl O&P
0010	**SWIMMING POOL EQUIPMENT**									
0020	Diving stand, stainless steel, 3 meter	2 Carp	.40	40	Ea.	11,000	1,350		12,350	14,400
0300	1 meter		2.70	5.926		6,375	199		6,574	7,325
0600	Diving boards, 16′ long, aluminum		2.70	5.926		3,550	199		3,749	4,225
0700	Fiberglass		2.70	5.926		2,625	199		2,824	3,225
1100	Gutter system, stainless steel, with grating, stock,									
1110	contains supply and drainage system	E-1	20	.800	L.F.	268	29	5.45	302.45	355
1120	Integral gutter and 5′ high wall system, stainless steel	"	10	1.600	"	360	58	10.90	428.90	520
1200	Ladders, heavy duty, stainless steel, 2 tread	2 Carp	7	2.286	Ea.	635	76.50		711.50	825
1500	4 tread		6	2.667		835	89.50		924.50	1,075
1800	Lifeguard chair, stainless steel, fixed		2.70	5.926		3,050	199		3,249	3,675
1900	Portable					2,750			2,750	3,025
2100	Lights, underwater, 12 volt, with transformer, 300 watt	1 Elec	1	8		263	320		583	810
2200	110 volt, 500 watt, standard		1	8		209	320		529	750
2400	Low water cutoff type		1	8		254	320		574	800
2800	Heaters, see Section 23 52 28.10									
3000	Pool covers, reinforced vinyl	3 Clab	1800	.013	S.F.	.35	.35		.70	.98
3050	Automatic, electric								8.75	9.65
3100	Vinyl water tube	3 Clab	3200	.008		.50	.20		.70	.88
3200	Maximum	"	3000	.008		.63	.21		.84	1.04
3250	Sealed air bubble polyethylene solar blanket, 16 mils					.36			.36	.40
3300	Slides, tubular, fiberglass, aluminum handrails & ladder, 5′-0″, straight	2 Carp	1.60	10	Ea.	3,075	335		3,410	3,925
3320	8′-0″, curved		3	5.333		6,575	179		6,754	7,525
3400	10′-0″, curved		1	16		28,700	535		29,235	32,400
3420	12′-0″, straight with platform		1.20	13.333		13,000	445		13,445	15,100
4500	Hydraulic lift, movable pool bottom, single ram									
4520	Under 1,000 S.F. area	L-9	72	.500	S.F.	130	15.65		145.65	169
4600	Four ram lift, over 1,000 S.F.	"	109	.330	"	105	10.35		115.35	133
5000	Removable access ramp, stainless steel	2 Clab	2	8	Ea.	10,500	212		10,712	11,900
5500	Removable stairs, stainless steel, collapsible	"	2	8	"	4,450	212		4,662	5,225

13 17 Tubs and Pools

13 17 13 – Hot Tubs

	13 17 13.10 Redwood Hot Tub System	Crew	Daily Output	Labor-Hours	Unit	Material	2011 Bare Costs Labor	Equipment	Total	Total Incl O&P
0010	**REDWOOD HOT TUB SYSTEM**									
7050	4′ diameter x 4′ deep	Q-1	1	16	Ea.	1,900	570		2,470	3,025
7100	5′ diameter x 4′ deep		1	16		2,475	570		3,045	3,650
7150	6′ diameter x 4′ deep		.80	20		2,950	715		3,665	4,425
7200	8′ diameter x 4′ deep		.80	20		4,100	715		4,815	5,675

13 17 33 – Whirlpool Tubs

	13 17 33.10 Whirlpool Bath	Crew	Daily Output	Labor-Hours	Unit	Material	Labor	Equipment	Total	Total Incl O&P
0010	**WHIRLPOOL BATH**									
6000	Whirlpool, bath with vented overflow, molded fiberglass									
6100	66″ x 48″ x 24″	Q-1	1	16	Ea.	3,525	570		4,095	4,800

13 18 Ice Rinks

13 18 13 – Ice Rink Floor Systems

	13 18 13.50 Ice Skating	Crew	Daily Output	Labor-Hours	Unit	Material	Labor	Equipment	Total	Total Incl O&P
0010	**ICE SKATING** Equipment incl. refrigeration, plumbing & cooling									
0020	coils & concrete slab, 85′ x 200′ rink									
0300	55° system, 5 mos., 100 ton				Total	575,000			575,000	632,500
0700	90° system, 12 mos., 135 ton				″	600,000			600,000	660,000
1200	Subsoil heating system (recycled from compressor), 85′ x 200′	Q-7	.27	118	Ea.	25,000	4,275		29,275	34,500
1300	Subsoil insulation, 2 lb. polystyrene with vapor barrier, 85′ x 200′	2 Carp	.14	114	″	30,000	3,825		33,825	39,400

13 18 16 – Ice Rink Dasher Boards

	13 18 16.50 Ice Rink Dasher Boards	Crew	Daily Output	Labor-Hours	Unit	Material	Labor	Equipment	Total	Total Incl O&P
0010	**ICE RINK DASHER BOARDS**									
1000	Dasher boards, 1/2″ H.D. polyethylene faced steel frame, 3′ acrylic									
1020	screen at sides, 5′ acrylic ends, 85′ x 200′	F-5	.06	533	Ea.	125,000	15,700		140,700	164,000
1100	Fiberglass & aluminum construction, same sides and ends	″	.06	533	″	125,000	15,700		140,700	164,000

13 21 Controlled Environment Rooms

13 21 13 – Clean Rooms

	13 21 13.50 Clean Room Components	Crew	Daily Output	Labor-Hours	Unit	Material	Labor	Equipment	Total	Total Incl O&P
0010	**CLEAN ROOM COMPONENTS**									
1100	Clean room, soft wall, 12′ x 12′, Class 100	1 Carp	.18	44.444	Ea.	15,500	1,500		17,000	19,600
1110	Class 1,000		.18	44.444		12,600	1,500		14,100	16,400
1120	Class 10,000		.21	38.095		11,100	1,275		12,375	14,400
1130	Class 100,000		.21	38.095		10,400	1,275		11,675	13,700
2800	Ceiling grid support, slotted channel struts 4′-0″ O.C., ea. way				S.F.				5.90	6.50
3000	Ceiling panel, vinyl coated foil on mineral substrate									
3020	Sealed, non-perforated				S.F.				1.27	1.40
4000	Ceiling panel seal, silicone sealant, 150 L.F./gal.	1 Carp	150	.053	L.F.	.30	1.79		2.09	3.32
4100	Two sided adhesive tape	″	240	.033	″	.13	1.12		1.25	2.01
4200	Clips, one per panel				Ea.	1.04			1.04	1.14
6000	HEPA filter, 2′ x 4′, 99.97% eff., 3″ dp beveled frame (silicone seal)					365			365	400
6040	6″ deep skirted frame (channel seal)					390			390	430
6100	99.99% efficient, 3″ deep beveled frame (silicone seal)					380			380	420
6140	6″ deep skirted frame (channel seal)					405			405	445
6200	99.999% efficient, 3″ deep beveled frame (silicone seal)					395			395	435
6240	6″ deep skirted frame (channel seal)					420			420	465
7000	Wall panel systems, including channel strut framing									

13 21 Controlled Environment Rooms

13 21 13 – Clean Rooms

13 21 13.50 Clean Room Components		Crew	Daily Output	Labor-Hours	Unit	2011 Bare Costs Material	Labor	Equipment	Total	Total Incl O&P
7020	Polyester coated aluminum, particle board				S.F.				18.20	20
7100	Porcelain coated aluminum, particle board								32	35
7400	Wall panel support, slotted channel struts, to 12′ high								16.35	18

13 21 26 – Cold Storage Rooms

13 21 26.50 Refrigeration

13 21 26.50 Refrigeration		Crew	Daily Output	Labor-Hours	Unit	Material	Labor	Equipment	Total	Total Incl O&P
0010	**REFRIGERATION**									
0020	Curbs, 12″ high, 4″ thick, concrete	2 Carp	58	.276	L.F.	4.37	9.25		13.62	20.50
1000	Doors, see Section 08 34 13.10									
2400	Finishes, 2 coat portland cement plaster, 1/2″ thick	1 Plas	48	.167	S.F.	1.20	5.10		6.30	9.65
2500	For galvanized reinforcing mesh, add	1 Lath	335	.024		.80	.72		1.52	2.03
2700	3/16″ thick latex cement	1 Plas	88	.091		2.05	2.79		4.84	6.80
2900	For glass cloth reinforced ceilings, add	″	450	.018		.49	.55		1.04	1.43
3100	Fiberglass panels, 1/8″ thick	1 Carp	149.45	.054		2.71	1.80		4.51	6
3200	Polystyrene, plastic finish ceiling, 1″ thick		274	.029		2.50	.98		3.48	4.39
3400	2″ thick		274	.029		2.85	.98		3.83	4.78
3500	4″ thick		219	.037		3.17	1.23		4.40	5.55
3800	Floors, concrete, 4″ thick	1 Cefi	93	.086		1.29	2.74		4.03	5.80
3900	6″ thick	″	85	.094		2.01	3		5.01	7
4000	Insulation, 1″ to 6″ thick, cork				B.F.	1.16			1.16	1.28
4100	Urethane					.51			.51	.56
4300	Polystyrene, regular					.50			.50	.55
4400	Bead board					.29			.29	.32
4600	Installation of above, add per layer	2 Carp	657.60	.024	S.F.	.38	.82		1.20	1.79
4700	Wall and ceiling juncture		298.90	.054	L.F.	1.87	1.80		3.67	5.05
4900	Partitions, galvanized sandwich panels, 4″ thick, stock		219.20	.073	S.F.	7.85	2.45		10.30	12.70
5000	Aluminum or fiberglass		219.20	.073	″	8.55	2.45		11	13.55
5200	Prefab walk-in, 7′-6″ high, aluminum, incl. door & floors,									
5210	not incl. partitions or refrigeration, 6′ x 6′ O.D. nominal	2 Carp	54.80	.292	SF Flr.	140	9.80		149.80	170
5500	10′ x 10′ O.D. nominal		82.20	.195		113	6.55		119.55	135
5700	12′ x 14′ O.D. nominal		109.60	.146		101	4.90		105.90	119
5800	12′ x 20′ O.D. nominal		109.60	.146		88	4.90		92.90	105
6100	For 8′-6″ high, add					5%				
6300	Rule of thumb for complete units, w/o doors & refrigeration, cooler	2 Carp	146	.110		127	3.68		130.68	146
6400	Freezer		109.60	.146		150	4.90		154.90	173
6600	Shelving, plated or galvanized, steel wire type		360	.044	SF Hor.	11.10	1.49		12.59	14.75
6700	Slat shelf type		375	.043		13.70	1.43		15.13	17.50
6900	For stainless steel shelving, add					300%				
7000	Vapor barrier, on wood walls	2 Carp	1644	.010	S.F.	.16	.33		.49	.73
7200	On masonry walls	″	1315	.012	″	.42	.41		.83	1.14
7500	For air curtain doors, see Section 23 34 33.10									

13 21 48 – Sound-Conditioned Rooms

13 21 48.10 Anechoic Chambers

13 21 48.10 Anechoic Chambers		Crew	Daily Output	Labor-Hours	Unit	Material	Labor	Equipment	Total	Total Incl O&P
0010	**ANECHOIC CHAMBERS** Standard units, 7′ ceiling heights									
0100	Area for pricing is net inside dimensions									
0300	200 cycles per second cutoff, 25 S.F. floor area				SF Flr.	1,600			1,600	1,750
0400	50 S.F.								1,050	1,150
0600	75 S.F.								1,000	1,100
0700	100 S.F.					1,200			1,200	1,325
0900	For 150 cycles per second cutoff, add to 100 S.F. room									30%
1000	For 100 cycles per second cutoff, add to 100 S.F. room									45%

13 21 Controlled Environment Rooms

13 21 48 – Sound-Conditioned Rooms

13 21 48.15 Audiometric Rooms		Crew	Daily Output	Labor-Hours	Unit	Material	Labor	Equipment	Total	Total Incl O&P
						2011 Bare Costs				
0010	**AUDIOMETRIC ROOMS**									
0020	Under 500 S.F. surface	4 Carp	98	.327	SF Surf	51.50	10.95		62.45	75
0100	Over 500 S.F. surface	"	120	.267	"	49	8.95		57.95	68.50

13 21 53 – Darkrooms

13 21 53.50 Darkrooms		Crew	Daily Output	Labor-Hours	Unit	Material	Labor	Equipment	Total	Total Incl O&P
0010	**DARKROOMS**									
0020	Shell, complete except for door, 64 S.F., 8′ high	2 Carp	128	.125	SF Flr.	53.50	4.19		57.69	65.50
0100	12′ high		64	.250		68.50	8.40		76.90	89
0500	120 S.F. floor, 8′ high		120	.133		37.50	4.47		41.97	48.50
0600	12′ high		60	.267		51.50	8.95		60.45	71.50
0800	240 S.F. floor, 8′ high		120	.133		27	4.47		31.47	37
0900	12′ high		60	.267		38	8.95		46.95	56.50
1200	Mini-cylindrical, revolving, unlined, 4′ diameter		3.50	4.571	Ea.	2,300	153		2,453	2,775
1400	5′-6″ diameter		2.50	6.400		3,950	215		4,165	4,700
1600	Add for lead lining, inner cylinder, 1/32″ thick					1,475			1,475	1,625
1700	1/16″ thick					3,975			3,975	4,375
1800	Add for lead lining, inner and outer cylinder, 1/32″ thick					2,725			2,725	3,000
1900	1/16″ thick					6,100			6,100	6,700
2000	For darkroom door, see Section 08 34 36.10									

13 21 56 – Music Rooms

13 21 56.50 Music Rooms		Crew	Daily Output	Labor-Hours	Unit	Material	Labor	Equipment	Total	Total Incl O&P
0010	**MUSIC ROOMS**									
0020	Practice room, modular, perforated steel, under 500 S.F.	2 Carp	70	.229	SF Surf	31.50	7.65		39.15	47.50
0100	Over 500 S.F.	"	80	.200	"	26.50	6.70		33.20	41

13 24 Special Activity Rooms

13 24 16 – Saunas

13 24 16.50 Saunas and Heaters		Crew	Daily Output	Labor-Hours	Unit	Material	Labor	Equipment	Total	Total Incl O&P
0010	**SAUNAS AND HEATERS**									
0020	Prefabricated, incl. heater & controls, 7′ high, 6′ x 4′, C/C	L-7	2.20	11.818	Ea.	5,125	345		5,470	6,225
0050	6′ x 4′, C/P		2	13		4,300	375		4,675	5,350
0400	6′ x 5′, C/C		2	13		5,225	375		5,600	6,375
0450	6′ x 5′, C/P		2	13		4,450	375		4,825	5,525
0600	6′ x 6′, C/C		1.80	14.444		5,550	420		5,970	6,800
0650	6′ x 6′, C/P		1.80	14.444		4,750	420		5,170	5,925
0800	6′ x 9′, C/C		1.60	16.250		6,875	470		7,345	8,325
0850	6′ x 9′, C/P		1.60	16.250		6,425	470		6,895	7,850
1000	8′ x 12′, C/C		1.10	23.636		11,200	685		11,885	13,500
1050	8′ x 12′, C/P		1.10	23.636		9,900	685		10,585	12,100
1200	8′ x 8′, C/C		1.40	18.571		8,525	540		9,065	10,300
1250	8′ x 8′, C/P		1.40	18.571		7,750	540		8,290	9,425
1400	8′ x 10′, C/C		1.20	21.667		9,450	630		10,080	11,500
1450	8′ x 10′, C/P		1.20	21.667		8,500	630		9,130	10,400
1600	10′ x 12′, C/C		1	26		11,800	755		12,555	14,300
1650	10′ x 12′, C/P		1	26		10,300	755		11,055	12,700
1700	Door only, cedar, 2′ x 6′, with tempered insulated glass window	2 Carp	3.40	4.706		620	158		778	945
1800	Prehung, incl. jambs, pulls & hardware	"	12	1.333		620	44.50		664.50	760
2500	Heaters only (incl. above), wall mounted, to 200 C.F.					615			615	675
2750	To 300 C.F.					795			795	875

13 24 Special Activity Rooms

13 24 16 – Saunas

13 24 16.50 Saunas and Heaters		Crew	Daily Output	Labor-Hours	Unit	Material	2011 Bare Costs Labor	Equipment	Total	Total Incl O&P
3000	Floor standing, to 720 C.F., 10,000 watts, w/controls	1 Elec	3	2.667	Ea.	2,100	107		2,207	2,475
3250	To 1,000 C.F., 16,000 watts	"	3	2.667		2,700	107		2,807	3,125

13 24 26 – Steam Baths

13 24 26.50 Steam Baths and Components

		Crew	Daily Output	Labor-Hours	Unit	Material	Labor	Equipment	Total	Total Incl O&P
0010	**STEAM BATHS AND COMPONENTS**									
0020	Heater, timer & head, single, to 140 C.F.	1 Plum	1.20	6.667	Ea.	2,100	264		2,364	2,725
0500	To 300 C.F.		1.10	7.273		2,425	288		2,713	3,150
1000	Commercial size, with blow-down assembly, to 800 C.F.		.90	8.889		7,800	350		8,150	9,150
1500	To 2500 C.F.		.80	10		8,775	395		9,170	10,300
2000	Multiple, motels, apts., 2 baths, w/blow-down assm., 500 C.F.	Q-1	1.30	12.308		5,900	440		6,340	7,225
2500	4 baths	"	.70	22.857		7,450	815		8,265	9,525
2700	Conversion unit for residential tub, including door					4,100			4,100	4,500

13 28 Athletic and Recreational Special Construction

13 28 33 – Athletic and Recreational Court Walls

13 28 33.50 Sport Court

		Crew	Daily Output	Labor-Hours	Unit	Material	Labor	Equipment	Total	Total Incl O&P
0010	**SPORT COURT**									
0020	Floors, No. 2 & better maple, 25/32" thick				SF Flr.				6.05	6.65
0100	Walls, laminated plastic bonded to galv. steel studs				SF Wall				7.70	8.45
0150	Laminated fiberglass surfacing, minimum								1.91	2.10
0180	Maximum								2.15	2.37
0300	Squash, regulation court in existing building, minimum				Court	36,800			36,800	40,400
0400	Maximum				"	41,000			41,000	45,000
0450	Rule of thumb for components:									
0470	Walls	3 Carp	.15	160	Court	11,000	5,375		16,375	21,100
0500	Floor	"	.25	96		8,725	3,225		11,950	15,000
0550	Lighting	2 Elec	.60	26.667		2,100	1,075		3,175	4,050
0600	Handball, racquetball court in existing building, minimum	C-1	.20	160		39,800	4,750		44,550	52,000
0800	Maximum	"	.10	320		43,100	9,525		52,625	63,500
0900	Rule of thumb for components: walls	3 Carp	.12	200		12,600	6,700		19,300	25,100
1000	Floor		.25	96		8,725	3,225		11,950	15,000
1100	Ceiling		.33	72.727		4,200	2,450		6,650	8,700
1200	Lighting	2 Elec	.60	26.667		2,200	1,075		3,275	4,175

13 31 Fabric Structures

13 31 13 – Air-Supported Fabric Structures

13 31 13.09 Air Supported Tank Covers

		Crew	Daily Output	Labor-Hours	Unit	Material	Labor	Equipment	Total	Total Incl O&P
0010	**AIR SUPPORTED TANK COVERS**, vinyl polyester									
0100	Scrim, double layer, with hardware, blower, standby & controls									
0200	Round, 75' diameter	B-2	4500	.009	S.F.	10	.24		10.24	11.40
0300	100' diameter		5000	.008		9.10	.21		9.31	10.35
0400	150' diameter		5000	.008		7.20	.21		7.41	8.30
0500	Rectangular, 20' x 20'		4500	.009		23	.24		23.24	25.50
0600	30' x 40'		4500	.009		23	.24		23.24	25.50
0700	50' x 60'		4500	.009		23	.24		23.24	25.50
0800	For single wall construction, deduct, minimum					.77			.77	.85
0900	Maximum					2.29			2.29	2.52
1000	For maximum resistance to atmosphere or cold, add					1.12			1.12	1.23
1100	For average shipping charges, add				Total	1,925			1,925	2,125

13 31 Fabric Structures

13 31 13 – Air-Supported Fabric Structures

13 31 13.13 Single-Walled Air-Supported Structures		Crew	Daily Output	Labor-Hours	Unit	2011 Bare Costs Material	Labor	Equipment	Total	Total Incl O&P
0010	**SINGLE-WALLED AIR-SUPPORTED STRUCTURES** R133113-10									
0020	Site preparation, incl. anchor placement and utilities	B-11B	1000	.016	SF Flr.	1.14	.48	.25	1.87	2.31
0030	For concrete, see Section 03 30 53.40									
0050	Warehouse, polyester/vinyl fabric, 28 oz., over 10 yr. life, welded									
0060	Seams, tension cables, primary & auxiliary inflation system,									
0070	airlock, personnel doors and liner									
0100	5,000 S.F.	4 Clab	5000	.006	SF Flr.	24.50	.17		24.67	27.50
0250	12,000 S.F.	″	6000	.005		17.65	.14		17.79	19.65
0400	24,000 S.F.	8 Clab	12000	.005		12.40	.14		12.54	13.90
0500	50,000 S.F.	″	12500	.005	↓	11.50	.14		11.64	12.90
0700	12 oz. reinforced vinyl fabric, 5 yr. life, sewn seams,									
0710	accordian door, including liner									
0750	3000 S.F.	4 Clab	3000	.011	SF Flr.	11.75	.28		12.03	13.35
0800	12,000 S.F.	″	6000	.005		10	.14		10.14	11.25
0850	24,000 S.F.	8 Clab	12000	.005		8.45	.14		8.59	9.55
0950	Deduct for single layer					1.01			1.01	1.11
1000	Add for welded seams					1.01			1.01	1.11
1050	Add for double layer, welded seams included				↓	2.08			2.08	2.29
1250	Tedlar/vinyl fabric, 28 oz., with liner, over 10 yr. life,									
1260	incl. overhead and personnel doors									
1300	3000 S.F.	4 Clab	3000	.011	SF Flr.	24	.28		24.28	27
1450	12,000 S.F.	″	6000	.005		16.85	.14		16.99	18.80
1550	24,000 S.F.	8 Clab	12000	.005		13.05	.14		13.19	14.60
1700	Deduct for single layer				↓	1.44			1.44	1.58
2250	Greenhouse/shelter, woven polyethylene with liner, 2 yr. life,									
2260	sewn seams, including doors									
2300	3000 S.F.	4 Clab	3000	.011	SF Flr.	6.60	.28		6.88	7.70
2350	12,000 S.F.	″	6000	.005		6.70	.14		6.84	7.65
2450	24,000 S.F.	8 Clab	12000	.005		5.60	.14		5.74	6.40
2550	Deduct for single layer				↓	.63			.63	.69
2600	Tennis/gymnasium, polyester/vinyl fabric, 28 oz., over 10 yr. life,									
2610	including thermal liner, heat and lights									
2650	7,200 S.F.	4 Clab	6000	.005	SF Flr.	23	.14		23.14	25.50
2750	13,000 S.F.	″	6500	.005		17.65	.13		17.78	19.60
2850	Over 24,000 S.F.	8 Clab	12000	.005		16.10	.14		16.24	18
2860	For low temperature conditions, add				↓	1.12			1.12	1.23
2870	For average shipping charges, add				Total	5,475			5,475	6,000
2900	Thermal liner, translucent reinforced vinyl				SF Flr.	1.12			1.12	1.23
2950	Metalized mylar fabric and mesh, double liner				″	2.29			2.29	2.52
3050	Stadium/convention center, teflon coated fiberglass, heavy weight,									
3060	over 20 yr. life, incl. thermal liner and heating system									
3100	Minimum	9 Clab	26000	.003	SF Flr.	56	.07		56.07	62
3110	Maximum	″	19000	.004	″	67	.10		67.10	73.50
3400	Doors, air lock, 15′ long, 10′ x 10′	2 Carp	.80	20	Ea.	20,100	670		20,770	23,200
3600	15′ x 15′	″	.50	32		29,700	1,075		30,775	34,500
3700	For each added 5′ length, add					5,300			5,300	5,850
3900	Revolving personnel door, 6′ diameter, 6′-6″ high	2 Carp	.80	20	↓	14,900	670		15,570	17,500
4200	Double wall, self supporting, shell only, minimum				SF Flr.				19.10	21
4300	Maximum				″				35.50	39

13 31 Fabric Structures

13 31 23 – Tensioned Fabric Structures

13 31 23.50 Tension Structures		Crew	Daily Output	Labor-Hours	Unit	Material	2011 Bare Costs Labor	Equipment	Total	Total Incl O&P
0010	**TENSION STRUCTURES** Rigid steel/alum. frame, vinyl coated poly									
0100	Fabric shell, 60′ clear span, not incl. foundations or floors									
0200	6,000 S.F.	B-41	1000	.044	SF Flr.	14.65	1.21	.28	16.14	18.45
0300	12,000 S.F.		1100	.040		13.55	1.10	.26	14.91	17
0400	80′ clear span, 20,800 S.F.		1220	.036		13.45	.99	.23	14.67	16.65
0410	100′ clear span, 10,000 S.F.	L-5	2175	.026		14.80	.94	.34	16.08	18.45
0430	26,000 S.F.		2300	.024		13.95	.89	.33	15.17	17.40
0450	36,000 S.F.		2500	.022		13.60	.82	.30	14.72	16.80
0460	120′ clear span, 24,000 S.F.		3000	.019		15.55	.68	.25	16.48	18.65
0470	150′ clear span, 30,000 S.F.		6000	.009		16.70	.34	.12	17.16	19.15
0480	200′ clear span, 40,000 S.F.	E-6	8000	.015		19.05	.54	.24	19.83	22.50
0500	For roll-up door, 12′ x 14′, add	L-2	1	16	Ea.	6,075	470		6,545	7,475
0600	For personnel doors, add, minimum				SF Flr.	5%				
0700	Add, maximum					15%				
0800	For site work, simple foundation, etc., add, minimum								1.25	1.95
0900	Add, maximum								2.75	3.05

13 34 Fabricated Engineered Structures

13 34 13 – Glazed Structures

13 34 13.13 Greenhouses		Crew	Daily Output	Labor-Hours	Unit	Material	Labor	Equipment	Total	Total Incl O&P
0010	**GREENHOUSES**, Shell only, stock units, not incl. 2′ stub walls,									
0020	foundation, floors, heat or compartments									
0300	Residential type, free standing, 8′-6″ long x 7′-6″ wide	2 Carp	59	.271	SF Flr.	30.50	9.10		39.60	49
0400	10′-6″ wide		85	.188		26	6.30		32.30	39
0600	13′-6″ wide		108	.148		20.50	4.97		25.47	31.50
0700	17′-0″ wide		160	.100		25	3.36		28.36	33
0900	Lean-to type, 3′-10″ wide		34	.471		33.50	15.80		49.30	63.50
1000	6′-10″ wide		58	.276		41	9.25		50.25	60.50
1500	Commercial, custom, truss frame, incl. equip., plumbing, elec.,									
1510	benches and controls, under 2,000 S.F., minimum				SF Flr.				27.50	30
1550	Maximum					14.45			14.45	15.90
1700	Over 5,000 S.F., minimum					14.45			14.45	15.90
1750	Maximum								27.50	30
2000	Institutional, custom, rigid frame, including compartments and									
2010	multi-controls, under 500 S.F., minimum				SF Flr.	45			45	49.50
2050	Maximum					32			32	35.50
2150	Over 2,000 S.F., minimum					11.70			11.70	12.90
2200	Maximum					59			59	65
2400	Concealed rigid frame, under 500 S.F., minimum								82	90
2450	Maximum								100	110
2550	Over 2,000 S.F., minimum								62	68
2600	Maximum								72	79
2800	Lean-to type, under 500 S.F., minimum								72.50	80
2850	Maximum								109	120
3000	Over 2,000 S.F., minimum								40	44
3050	Maximum								66.50	73
3600	For 1/4″ clear plate glass, add				SF Surf	1.86			1.86	2.05
3700	For 1/4″ tempered glass, add				″	4.21			4.21	4.63
3900	For cooling, add, minimum				SF Flr.	2.81			2.81	3.09
4000	Maximum					6.95			6.95	7.65

13 34 Fabricated Engineered Structures

13 34 13 – Glazed Structures

13 34 13.13 Greenhouses		Crew	Daily Output	Labor-Hours	Unit	2011 Bare Costs Material	Labor	Equipment	Total	Total Incl O&P
4200	For heaters, 13.6 MBH, add				SF Flr.	5.35			5.35	5.85
4300	60 MBH, add					2			2	2.20
4500	For benches, 2′ x 3′-6″, add				SF Hor.	24			24	26.50
4600	3′ x 10′, add				S.F.	12.95			12.95	14.25
4800	For controls, add, minimum				Total	2,400			2,400	2,650
4900	Maximum				"	14,300			14,300	15,700
5100	For humidification equipment, add				M.C.F.	6.15			6.15	6.75
5200	For vinyl shading, add				S.F.	1.29			1.29	1.42
6000	Geodesic hemisphere, 1/8″ plexiglass glazing									
6050	8′ diameter	2 Carp	2	8	Ea.	2,700	268		2,968	3,425
6150	24′ diameter		.35	45.714		13,900	1,525		15,425	17,900
6250	48′ diameter		.20	80		37,600	2,675		40,275	45,900

13 34 13.19 Swimming Pool Enclosures		Crew	Daily Output	Labor-Hours	Unit	Material	Labor	Equipment	Total	Total Incl O&P
0010	**SWIMMING POOL ENCLOSURES** Translucent, free standing									
0020	not including foundations, heat or light									
0200	Economy, minimum	2 Carp	200	.080	SF Hor.	14.80	2.68		17.48	20.50
0300	Maximum		100	.160		45.50	5.35		50.85	59
0400	Deluxe, minimum		100	.160		50.50	5.35		55.85	64.50
0600	Maximum		70	.229		227	7.65		234.65	263
0700	For motorized roof, 40% opening, solid roof, add					11.70			11.70	12.85
0800	Skylight type roof, add					9.35			9.35	10.30
0900	Air-inflated, including blowers and heaters, minimum								3.18	3.50
1000	Maximum								6.10	6.70

13 34 16 – Grandstands and Bleachers

13 34 16.13 Grandstands		Crew	Daily Output	Labor-Hours	Unit	Material	Labor	Equipment	Total	Total Incl O&P
0010	**GRANDSTANDS** Permanent, municipal, including foundation									
0050	Steel understructure w/aluminum closed deck, minimum				Seat				118	130
0100	Maximum								205	225
0300	Steel, minimum					55			55	60.50
0400	Maximum					110			110	121
0600	Aluminum, extruded, stock design, minimum								63.50	70
0700	Maximum								105	115
0900	Composite, steel, wood and plastic, stock design, minimum					130			130	143
1000	Maximum					260			260	286

13 34 16.53 Bleachers		Crew	Daily Output	Labor-Hours	Unit	Material	Labor	Equipment	Total	Total Incl O&P
0010	**BLEACHERS**									
0020	Bleachers, outdoor, portable, 3 to 5 tiers, to 300′ long, min	2 Sswk	120	.133	Seat	48	4.85		52.85	62
0100	Maximum, less than 15′ long, prefabricated		80	.200		78	7.30		85.30	100
0200	6 to 20 tiers, minimum, up to 300′ long		120	.133		66.50	4.85		71.35	83
0300	Max., under 15′, (highly prefabricated, on wheels)		80	.200		85.50	7.30		92.80	108
0500	Permanent grandstands, wood seat, steel frame, 24″ row									
0600	3 to 15 tiers, minimum	2 Sswk	60	.267	Seat	103	9.70		112.70	132
0700	Maximum		48	.333		123	12.15		135.15	158
0900	16 to 30 tiers, minimum		60	.267		189	9.70		198.70	227
0950	Average		55	.291		211	10.60		221.60	252
1000	Maximum		48	.333		240	12.15		252.15	287
1200	Seat backs only, 30″ row, fiberglass		160	.100		40	3.64		43.64	51
1300	Steel and wood		160	.100		62	3.64		65.64	75
1400	NOTE: average seating is 1.5′ in width									

	13 34 19.50 Pre-Engineered Steel Buildings	Crew	Daily Output	Labor-Hours	Unit	Material	2011 Bare Costs Labor	Equipment	Total	Total Incl O&P
0010	**PRE-ENGINEERED STEEL BUILDINGS** R133419-10									
0100	Clear span rigid frame, 26 ga. colored roofing and siding									
0150	20′ to 29′ wide, 10′ eave height	E-2	425	.113	SF Flr.	8.85	4.14	3.82	16.81	21.50
0160	14′ eave height		350	.137		9.35	5	4.63	18.98	25
0170	16′ eave height		320	.150		9.80	5.50	5.05	20.35	26.50
0180	20′ eave height		275	.175		10.85	6.40	5.90	23.15	30.50
0190	24′ eave height		240	.200		12.40	7.35	6.75	26.50	35
0200	30′ to 40′ wide, 10′ eave height		535	.090		7.25	3.29	3.03	13.57	17.40
0300	14′ eave height		450	.107		7.60	3.91	3.60	15.11	19.60
0400	16′ eave height		415	.116		8	4.24	3.91	16.15	21
0500	20′ eave height		360	.133		8.75	4.88	4.51	18.14	23.50
0600	24′ eave height		320	.150		9.90	5.50	5.05	20.45	26.50
0700	50′ to 100′ wide, 10′ eave height		770	.062		6.20	2.28	2.11	10.59	13.40
0800	14′ eave height		650	.074		6.50	2.71	2.50	11.71	14.95
0900	16′ eave height		600	.080		7	2.93	2.70	12.63	16.10
1000	20′ eave height		490	.098		7.50	3.59	3.31	14.40	18.60
1100	24′ eave height		435	.110		8.25	4.04	3.73	16.02	20.50
1200	Clear span tapered beam frame, 26 ga. colored roofing/siding									
1300	30′ to 39′ wide, 10′ eave height	E-2	535	.090	SF Flr.	7.95	3.29	3.03	14.27	18.20
1400	14′ eave height		450	.107		8.75	3.91	3.60	16.26	21
1500	16′ eave height		415	.116		9.40	4.24	3.91	17.55	22.50
1600	20′ eave height		360	.133		10.35	4.88	4.51	19.74	25.50
1700	40′ wide, 10′ eave height		600	.080		7.30	2.93	2.70	12.93	16.45
1800	14′ eave height		510	.094		7.95	3.45	3.18	14.58	18.65
1900	16′ eave height		475	.101		8.30	3.70	3.41	15.41	19.75
2000	20′ eave height		415	.116		9.10	4.24	3.91	17.25	22
2100	50′ to 79′ wide, 10′ eave height		770	.062		6.75	2.28	2.11	11.14	14
2200	14′ eave height		675	.071		7.35	2.60	2.40	12.35	15.60
2300	16′ eave height		635	.076		7.65	2.77	2.55	12.97	16.40
2400	20′ eave height		490	.098		8.35	3.59	3.31	15.25	19.55
2410	80′ to 100′ wide, 10′ eave height		935	.051		6	1.88	1.73	9.61	12
2420	14′ eave height		750	.064		6.60	2.34	2.16	11.10	14
2430	16′ eave height		685	.070		6.90	2.57	2.37	11.84	15
2440	20′ eave height		560	.086		7.40	3.14	2.90	13.44	17.15
2460	101′ to 120′ wide, 10′ eave height		950	.051		5.50	1.85	1.71	9.06	11.40
2470	14′ eave height		770	.062		6.10	2.28	2.11	10.49	13.25
2480	16′ eave height		675	.071		6.50	2.60	2.40	11.50	14.65
2490	20′ eave height		560	.086		6.95	3.14	2.90	12.99	16.70
2500	Single post 2-span frame, 26 ga. colored roofing and siding									
2600	80′ wide, 14′ eave height	E-2	740	.065	SF Flr.	5.95	2.38	2.19	10.52	13.40
2700	16′ eave height		695	.069		6.30	2.53	2.33	11.16	14.25
2800	20′ eave height		625	.077		6.90	2.81	2.60	12.31	15.70
2900	24′ eave height		570	.084		7.45	3.08	2.85	13.38	17.10
3000	100′ wide, 14′ eave height		835	.057		5.75	2.11	1.94	9.80	12.40
3100	16′ eave height		795	.060		5.70	2.21	2.04	9.95	12.60
3200	20′ eave height		730	.066		6.65	2.41	2.22	11.28	14.25
3300	24′ eave height		670	.072		7.15	2.62	2.42	12.19	15.45
3400	120′ wide, 14′ eave height		870	.055		5.70	2.02	1.86	9.58	12.10
3500	16′ eave height		830	.058		5.95	2.12	1.95	10.02	12.65
3600	20′ eave height		765	.063		6.45	2.30	2.12	10.87	13.65
3700	24′ eave height		705	.068		7	2.49	2.30	11.79	14.90
3800	Double post 3-span frame, 26 ga. colored roofing and siding									

13 34 19.50 Pre-Engineered Steel Buildings		Crew	Daily Output	Labor-Hours	Unit	2011 Bare Costs Material	Labor	Equipment	Total	Total Incl O&P
3900	150′ wide, 14′ eave height	E-2	925	.052	SF Flr.	4.92	1.90	1.75	8.57	10.85
4000	16′ eave height		890	.054		5.05	1.98	1.82	8.85	11.30
4100	20′ eave height		820	.059		5.35	2.14	1.98	9.47	12.05
4200	24′ eave height		765	.063		6.40	2.30	2.12	10.82	13.60
4300	Triple post 4-span frame, 26 ga. colored roofing and siding									
4400	160′ wide, 14′ eave height	E-2	970	.049	SF Flr.	4.77	1.81	1.67	8.25	10.45
4500	16′ eave height		930	.052		4.99	1.89	1.74	8.62	10.95
4600	20′ eave height		870	.055		4.99	2.02	1.86	8.87	11.30
4700	24′ eave height		815	.059		5.95	2.16	1.99	10.10	12.75
4800	200′ wide, 14′ eave height		1030	.047		4.58	1.71	1.57	7.86	9.95
4900	16′ eave height		995	.048		4.70	1.77	1.63	8.10	10.25
5000	20′ eave height		935	.051		5.10	1.88	1.73	8.71	11
5100	24′ eave height		885	.054		6	1.99	1.83	9.82	12.30
5200	Accessory items: add to the basic building cost above									
5250	Eave overhang, 2′ wide, 26 ga., with soffit	E-2	360	.133	L.F.	25	4.88	4.51	34.39	41.50
5300	4′ wide, without soffit		300	.160		24.50	5.85	5.40	35.75	44
5350	With soffit		250	.192		34.50	7.05	6.50	48.05	58.50
5400	6′ wide, without soffit		250	.192		33.50	7.05	6.50	47.05	57.50
5450	With soffit		200	.240		43.50	8.80	8.10	60.40	73.50
5500	Entrance canopy, incl. frame, 4′ x 4′		25	1.920	Ea.	350	70.50	65	485.50	590
5550	4′ x 8′		19	2.526	"	480	92.50	85.50	658	795
5600	End wall roof overhang, 4′ wide, without soffit		850	.056	L.F.	16.40	2.07	1.91	20.38	24
5650	With soffit		500	.096	"	24	3.52	3.24	30.76	36.50
5700	Doors, H.M. self-framing, incl. butts, lockset and trim									
5750	Single leaf, 3070 (3′ x 7′), economy	2 Sswk	5	3.200	Opng.	560	116		676	835
5800	Deluxe		4	4		610	146		756	950
5825	Glazed		4	4		710	146		856	1,050
5850	3670 (3′-6″ x 7′)		4	4		780	146		926	1,125
5900	4070 (4′ x 7′)		3	5.333		845	194		1,039	1,300
5950	Double leaf, 6070 (6′ x 7′)		2	8		1,050	291		1,341	1,700
6000	Glazed		2	8		1,325	291		1,616	2,025
6050	Framing only, for openings, 3′ x 7′		4	4		157	146		303	450
6100	10′ x 10′		3	5.333		515	194		709	940
6150	For windows below, 2020 (2′ x 2′)		6	2.667		166	97		263	370
6200	4030 (4′ x 3′)		5	3.200		203	116		319	445
6250	Flashings, 26 ga., corner or eave, painted		240	.067	L.F.	4.08	2.43		6.51	9.10
6300	Galvanized		240	.067		3.50	2.43		5.93	8.50
6350	Rake flashing, painted		240	.067		4.37	2.43		6.80	9.45
6400	Galvanized		240	.067		3.84	2.43		6.27	8.85
6450	Ridge flashing, 18″ wide, painted		240	.067		5.35	2.43		7.78	10.55
6500	Galvanized		240	.067		4.95	2.43		7.38	10.10
6550	Gutter, eave type, 26 ga., painted		320	.050		5.60	1.82		7.42	9.60
6600	Galvanized		320	.050		3.54	1.82		5.36	7.35
6650	Valley type, between buildings, painted		120	.133		9.35	4.85		14.20	19.55
6710	Insulation, rated .6 lb density, unfaced 4″ thick, R13	2 Carp	2300	.007	S.F.	.30	.23		.53	.72
6720	6″ thick, R-19		2300	.007		.42	.23		.65	.85
6730	10″ thick, R-30		2300	.007		.72	.23		.95	1.18
6750	Insulation, rated .6 lb density, poly/scrim/foil (PSF) faced									
6801	1-1/2″ thick, R5	2 Clab	2300	.007	S.F.	.27	.18		.45	.61
6851	3″ thick, R10		2300	.007		.29	.18		.47	.63
6901	4″ thick, R13		2300	.007		.42	.18		.60	.77
6930	10″ thick, R30	2 Carp	2300	.007		.92	.23		1.15	1.40
6951	Foil faced, 1-1/2″ thick, R5	2 Clab	2300	.007		.29	.18		.47	.63

13 34 19 – Metal Building Systems

13 34 19.50 Pre-Engineered Steel Buildings		Crew	Daily Output	Labor-Hours	Unit	Material	2011 Bare Costs Labor	Equipment	Total	Total Incl O&P
7001	2" thick, R6	2 Clab	2300	.007	S.F.	.37	.18		.55	.72
7051	3" thick, R10		2300	.007		.38	.18		.56	.73
7101	4" thick, R13		2300	.007		.47	.18		.65	.83
7110	6" thick, R19	2 Carp	2300	.007		.53	.23		.76	.97
7120	10" thick, R30	"	2300	.007		.78	.23		1.01	1.25
7151	Metalized polyester facing, 1-1/2" thick, R5	2 Clab	2300	.007		.47	.18		.65	.83
7201	2" thick, R6		2300	.007		.55	.18		.73	.92
7251	3" thick, R11		2300	.007		.52	.18		.70	.88
7301	4" thick, R13		2300	.007		.63	.18		.81	1
7310	6" thick, R19	2 Carp	2300	.007		.77	.23		1	1.24
7351	Vinyl, scrim foil (SD), 1-1/2" thick, R5	2 Clab	2300	.007		.41	.18		.59	.76
7401	2" thick, R6		2300	.007		.52	.18		.70	.88
7451	3" thick, R10		2300	.007		.58	.18		.76	.95
7501	4" thick, R13		2300	.007		.67	.18		.85	1.05
7510	Vinyl/scrim/vinyl (VSV) 4" thick, R13	2 Carp	2300	.007		.47	.23		.70	.91
7520	6" thick, R19		2300	.007		.62	.23		.85	1.07
7530	10" thick, R30		2300	.007		.83	.23		1.06	1.30
7540	Polyprop/scrim/polyester (PSP), 4", R13		2300	.007		.40	.23		.63	.83
7550	6" thick, R19		2300	.007		.54	.23		.77	.98
7555	10" thick, R30		2300	.007		.99	.23		1.22	1.48
7560	Polyprop/scrim/kraft/polyester (PSKP), 4" thick, R13		2300	.007		.42	.23		.65	.85
7570	6" thick, R19		2300	.007		.55	.23		.78	1
7580	10" thick, R19		2300	.007		1.05	.23		1.28	1.55
7585	Vinyl/scrim/polyester (VSP), 4" thick, R13		2300	.007		.43	.23		.66	.86
7590	6" thick, R19		2300	.007		.57	.23		.80	1.02
7600	10" thick, R30		2300	.007		1.03	.23		1.26	1.52
7635	Insulation installation, over the purlin, second layer, up to 4" thick, add						90%			
7640	Insulation installation, between the purlins, up to 4" thick, add						100%			
7650	Sash, single slide, glazed, with screens, 2020 (2' x 2')	E-1	22	.727	Opng.	118	26.50	4.95	149.45	186
7700	3030 (3' x 3')		14	1.143		266	41.50	7.80	315.30	380
7750	4030 (4' x 3')		13	1.231		355	45	8.40	408.40	485
7800	6040 (6' x 4')		12	1.333		710	48.50	9.10	767.60	885
7850	Double slide sash, 3030 (3' x 3')		14	1.143		207	41.50	7.80	256.30	315
7900	6040 (6' x 4')		12	1.333		550	48.50	9.10	607.60	710
7950	Fixed glass, no screens, 3030 (3' x 3')		14	1.143		220	41.50	7.80	269.30	330
8000	6040 (6' x 4')		12	1.333		585	48.50	9.10	642.60	750
8050	Prefinished storm sash, 3030 (3' x 3')		70	.229		72	8.30	1.56	81.86	96.50
8100	Siding and roofing, see Sections 07 41 13.00 & 07 42 13.00									
8200	Skylight, fiberglass panels, to 30 S.F.	E-1	10	1.600	Ea.	103	58	10.90	171.90	237
8250	Larger sizes, add for excess over 30 S.F.	"	300	.053	S.F.	3.44	1.94	.36	5.74	7.90
8300	Roof vents, circular with damper, birdscreen									
8350	and operator hardware, painted									
8400	26 ga., 12" diameter	1 Sswk	4	2	Ea.	69	73		142	215
8450	20" diameter		3	2.667		142	97		239	340
8500	24 ga., 24" diameter		2	4		282	146		428	590
8550	Galvanized		2	4		236	146		382	540
8600	Continuous, 26 ga., 10' long, 9" wide	2 Sswk	4	4		20	146		166	300
8650	12" wide	"	4	4		24	146		170	305

13 34 23 – Fabricated Structures

13 34 23.10 Comfort Stations

	13 34 23.10 Comfort Stations	Crew	Daily Output	Labor-Hours	Unit	2011 Bare Costs Material	Labor	Equipment	Total	Total Incl O&P
0010	**COMFORT STATIONS** Prefab., stock, w/doors, windows & fixt.									
0100	Not incl. interior finish or electrical									
0300	Mobile, on steel frame, minimum				S.F.	46.50			46.50	51
0350	Maximum					75			75	82.50
0400	Permanent, including concrete slab, minimum	B-12J	50	.320		199	9.95	17.15	226.10	254
0500	Maximum	″	43	.372		288	11.60	19.95	319.55	355
0600	Alternate pricing method, mobile, minimum				Fixture	2,150			2,150	2,375
0650	Maximum					3,175			3,175	3,500
0700	Permanent, minimum	B-12J	.70	22.857		12,200	710	1,225	14,135	16,000
0750	Maximum	″	.50	32		20,400	995	1,725	23,120	25,900

13 34 23.15 Domes

	13 34 23.15 Domes	Crew	Daily Output	Labor-Hours	Unit	Material	Labor	Equipment	Total	Total Incl O&P
0010	**DOMES**									
0020	Domes, rev alum, elec drive, for astronomy obsv shell only, stock units									
0600	10′-0″ diameter, 800#, dome	2 Carp	.25	64	Ea.	9,875	2,150		12,025	14,500
0700	Base		.67	23.881		3,425	800		4,225	5,100
0900	18′-0″ diameter, 2,500#, dome		.17	94.118		28,900	3,150		32,050	37,100
1000	Base		.33	48.485		9,575	1,625		11,200	13,200
1200	24′-0″ diameter, 4,500#, dome		.08	200		53,500	6,700		60,200	70,000
1300	Base		.25	64		17,200	2,150		19,350	22,600
1500	Domes, bulk storage, shell only, dual radius hemisphere, arch, steel									
1600	framing, corrugated steel covering, 150′ diameter	E-2	550	.087	SF Flr.	28	3.20	2.95	34.15	39.50
1700	400′ diameter	″	720	.067		22.50	2.44	2.25	27.19	32
1800	Wood framing, wood decking, to 400′ diameter	F-4	400	.100		20.50	3.07	3.02	26.59	31
1900	Radial framed wood (2″ x 6″), 1/2″ thick									
2000	plywood, asphalt shingles, 50′ diameter	F-3	2000	.020	SF Flr.	40.50	.61	.33	41.44	46
2100	60′ diameter		1900	.021		31	.65	.35	32	35.50
2200	72′ diameter		1800	.022		27.50	.68	.36	28.54	32
2300	116′ diameter		1730	.023		24.50	.71	.38	25.59	28.50
2400	150′ diameter		1500	.027		21	.82	.44	22.26	25

13 34 23.16 Fabricated Control Booths

	13 34 23.16 Fabricated Control Booths	Crew	Daily Output	Labor-Hours	Unit	Material	Labor	Equipment	Total	Total Incl O&P
0010	**FABRICATED CONTROL BOOTHS**									
0100	Guard House, prefab conc w/bullet resistant doors & windows, roof & wiring									
0110	8′ x 8′, Level III	L-10	1	24	Ea.	24,200	885	655	25,740	28,900
0120	8′ x 8′, Level IV	″	1	24	″	30,300	885	655	31,840	35,600

13 34 23.25 Garage Costs

	13 34 23.25 Garage Costs	Crew	Daily Output	Labor-Hours	Unit	Material	Labor	Equipment	Total	Total Incl O&P
0010	**GARAGE COSTS**									
0020	Public parking, average				Car				16,600	18,300
0100	See also Square Foot Costs in Reference Section									
0300	Residential, prefab shell, stock, wood, single car, minimum	2 Carp	1	16	Total	5,100	535		5,635	6,500
0350	Maximum		.67	23.881		5,875	800		6,675	7,825
0400	Two car, minimum		.67	23.881		10,200	800		11,000	12,600
0450	Maximum		.50	32		11,800	1,075		12,875	14,700

13 34 23.30 Garden House

	13 34 23.30 Garden House	Crew	Daily Output	Labor-Hours	Unit	Material	Labor	Equipment	Total	Total Incl O&P
0010	**GARDEN HOUSE** Prefab wood, no floors or foundations									
0100	32 to 200 S.F., minimum	2 Carp	200	.080	SF Flr.	27.50	2.68		30.18	34.50
0300	Maximum	″	48	.333	″	47	11.20		58.20	70

13 34 23.35 Geodesic Domes

	13 34 23.35 Geodesic Domes	Crew	Daily Output	Labor-Hours	Unit	Material	Labor	Equipment	Total	Total Incl O&P
0010	**GEODESIC DOMES** Shell only, interlocking plywood panels R133423-30									
0400	30′ diameter	F-5	1.60	20	Ea.	17,700	590		18,290	20,500
0500	34′ diameter		1.14	28.070		21,800	825		22,625	25,300
0600	39′ diameter		1	32		25,300	940		26,240	29,500

13 34 Fabricated Engineered Structures

13 34 23 – Fabricated Structures

13 34 23.35 Geodesic Domes		Crew	Daily Output	Labor-Hours	Unit	Material	2011 Bare Costs Labor	Equipment	Total	Total Incl O&P
0700	45′ diameter	F-3	1.13	35.556	Ea.	30,500	1,100	585	32,185	36,100
0750	55′ diameter		1	40		45,400	1,225	655	47,280	52,500
0800	60′ diameter		1	40		59,000	1,225	655	60,880	68,000
0850	65′ diameter		.80	50		70,000	1,525	820	72,345	80,500
1100	Aluminum panel, with 6″ insulation									
1200	100′ diameter				SF Flr.	37.50			37.50	41
1300	500′ diameter				″	32			32	35.50
1600	Aluminum framed, plexiglass closure panels									
1700	40′ diameter				SF Flr.	95			95	104
1800	200′ diameter				″	75.50			75.50	83.50
2100	Aluminum framed, aluminum closure panels									
2200	40′ diameter				SF Flr.	30			30	33
2300	100′ diameter					22.50			22.50	24.50
2400	200′ diameter					21.50			21.50	23.50
2500	For VRP faced bonded fiberglass insulation, add									10
2700	Aluminum framed, fiberglass sandwich panel closure									
2800	6′ diameter	2 Carp	150	.107	SF Flr.	33.50	3.58		37.08	42.50
2900	28′ diameter	″	350	.046	″	30.50	1.53		32.03	36
13 34 23.45 Kiosks										
0010	**KIOSKS**									
0020	Round, 5′ diameter, 8′ high, 1/4″ fiberglass wall				Ea.	5,800			5,800	6,375
0100	1″ insulated double wall, fiberglass					6,600			6,600	7,250
0500	Rectangular, 5′ x 9′, 7′-6″ high, 1/4″ fiberglass wall					8,450			8,450	9,300
0600	1″ insulated double wall, fiberglass					10,000			10,000	11,000
13 34 23.60 Portable Booths										
0010	**PORTABLE BOOTHS** Prefab aluminum with doors, windows, ext. roof									
0100	lights wiring & insulation, 15 S.F. building, O.D., painted, minimum				S.F.	300			300	330
0300	30 S.F. building, minimum					254			254	279
0400	50 S.F. building, minimum					189			189	208
0600	80 S.F. building, minimum					154			154	169
0700	100 S.F. building, minimum					144			144	158
0900	Acoustical booth, 27 Db @ 1,000 Hz, 15 S.F. floor				Ea.	4,350			4,350	4,775
1000	7′ x 7′-6″, including light & ventilation					8,900			8,900	9,800
1200	Ticket booth, galv. steel, not incl. foundations., 4′ x 4′					5,625			5,625	6,200
1300	4′ x 6′					7,225			7,225	7,950
13 34 23.70 Shelters										
0010	**SHELTERS**									
0020	Aluminum frame, acrylic glazing, 3′ x 9′ x 8′ high	2 Sswk	1.14	14.035	Ea.	2,250	510		2,760	3,450
0100	9′ x 12′ x 8′ high	″	.73	21.918	″	4,775	800		5,575	6,800

13 34 43 – Aircraft Hangars

13 34 43.50 Hangars		Crew	Daily Output	Labor-Hours	Unit	Material	Labor	Equipment	Total	Total Incl O&P
0010	**HANGARS** Prefabricated steel T hangars, Galv. steel roof &									
0100	walls, incl. electric bi-folding doors, 4 or more units,									
0110	not including floors or foundations, minimum	E-2	1275	.038	SF Flr.	13.20	1.38	1.27	15.85	18.50
0130	Maximum		1063	.045		14.65	1.65	1.53	17.83	21
0900	With bottom rolling doors, minimum		1386	.035		12.85	1.27	1.17	15.29	17.80
1000	Maximum		966	.050		14.25	1.82	1.68	17.75	21
1200	Alternate pricing method:									
1300	Galv. roof and walls, electric bi-folding doors, minimum	E-2	1.06	45.283	Plane	17,100	1,650	1,525	20,275	23,600
1500	Maximum		.91	52.747		20,400	1,925	1,775	24,100	28,100
1600	With bottom rolling doors, minimum		1.25	38.400		12,900	1,400	1,300	15,600	18,300

13 34 Fabricated Engineered Structures

13 34 43 – Aircraft Hangars

13 34 43.50 Hangars		Crew	Daily Output	Labor-Hours	Unit	Material	2011 Bare Costs Labor	Equipment	Total	Total Incl O&P
1800	Maximum	E-2	.97	49.485	Plane	15,000	1,825	1,675	18,500	21,700
2000	Circular type, prefab., steel frame, plastic skin, electric									
2010	door, including foundations, 80′ diameter,									
2020	for up to 5 light planes, minimum	E-2	.50	96	Total	72,500	3,525	3,250	79,275	90,000
2200	Maximum	"	.25	192	"	82,000	7,025	6,500	95,525	110,000

13 34 53 – Agricultural Structures

13 34 53.50 Silos		Crew	Daily Output	Labor-Hours	Unit	Material	Labor	Equipment	Total	Total Incl O&P
0010	**SILOS** Concrete stave industrial, conical or sloping bottom									
0100	excluding foundation, 12′ diameter, 35′ high	E-24	.09	355	Ea.	42,400	12,800	8,325	63,525	79,500
0200	16′ diameter, 45′ high		.06	533		69,000	19,200	12,500	100,700	125,000
0400	25′ diameter, 75′ high		.04	800		159,000	28,800	18,700	206,500	248,500
0500	Steel, factory fab., 30,000 gallon cap., painted, minimum	L-5	1	56		21,000	2,050	750	23,800	27,800
0700	Maximum		.50	112		33,400	4,100	1,500	39,000	46,100
0800	Epoxy lined, minimum		1	56		34,300	2,050	750	37,100	42,500
1000	Maximum		.50	112		43,600	4,100	1,500	49,200	57,000

13 34 63 – Natural Fiber Construction

13 34 63.50 Straw Bale Construction		Crew	Daily Output	Labor-Hours	Unit	Material	Labor	Equipment	Total	Total Incl O&P
0010	**STRAW BALE CONSTRUCTION**									
2000	Straw bales wall, incl labor & material complete, minimum G				SF Flr.					130
2010	Maximum G				"					160
2020	Straw bales in walls w/modified post and beam frame G	2 Carp	320	.050	S.F.	4.10	1.68		5.78	7.30

13 36 Towers

13 36 13 – Metal Towers

13 36 13.50 Control Towers		Crew	Daily Output	Labor-Hours	Unit	Material	Labor	Equipment	Total	Total Incl O&P
0010	**CONTROL TOWERS**									
0020	Modular 12′ x 10′, incl. instruments, min.				Ea.	670,000			670,000	737,000
0100	Maximum								927,500	1,020,000
0500	With standard 40′ tower, average					935,000			935,000	1,028,500
1000	Temporary portable control towers, 8′ x 12′,									
1010	complete with one position communications, minimum				Ea.				266,000	293,000
2000	For fixed facilities, depending on height, minimum								57,000	63,000
2010	Maximum								115,000	127,000

13 42 Building Modules

13 42 63 – Detention Cell Modules

13 42 63.16 Steel Detention Cell Modules		Crew	Daily Output	Labor-Hours	Unit	Material	Labor	Equipment	Total	Total Incl O&P
0010	**STEEL DETENTION CELL MODULES**									
2000	Cells, prefab., 5′ to 6′ wide, 7′ to 8′ high, 7′ to 8′ deep,									
2010	bar front, cot, not incl. plumbing	E-4	1.50	21.333	Ea.	9,650	785	73	10,508	12,200

13 48 Sound, Vibration, and Seismic Control

13 48 13 – Manufactured Sound and Vibration Control Components

13 48 13.50 Audio Masking		Crew	Daily Output	Labor-Hours	Unit	2011 Bare Costs Material	Labor	Equipment	Total	Total Incl O&P
0010	**AUDIO MASKING**, acoustical enclosure, 4" thick wall and ceiling									
0020	8# per S.F., up to 12' span	3 Carp	72	.333	SF Surf	32	11.20		43.20	54
0300	Better quality panels, 10.5# per S.F.		64	.375		36.50	12.60		49.10	61
0400	Reverb-chamber, 4" thick, parallel walls		60	.400		45.50	13.40		58.90	72.50
0600	Skewed wall, parallel roof, 4" thick panels		55	.436		52	14.65		66.65	81.50
0700	Skewed walls, skewed roof, 4" layers, 4" air space		48	.500		58.50	16.80		75.30	92.50
0900	Sound-absorbing panels, pntd mtl, 2'-6" x 8', under 1,000 S.F.		215	.112		12.10	3.75		15.85	19.55
1100	Over 1000 S.F.		240	.100		11.65	3.36		15.01	18.40
1200	Fabric faced		240	.100		9.45	3.36		12.81	15.95
1500	Flexible transparent curtain, clear	3 Shee	215	.112		7.30	4.26		11.56	15.05
1600	50% foam		215	.112		10.20	4.26		14.46	18.25
1700	75% foam		215	.112		10.20	4.26		14.46	18.25
1800	100% foam		215	.112		10.20	4.26		14.46	18.25
3100	Audio masking system, including speakers, amplification									
3110	and signal generator									
3200	Ceiling mounted, 5,000 S.F.	2 Elec	2400	.007	S.F.	1.25	.27		1.52	1.82
3300	10,000 S.F.		2800	.006		1.01	.23		1.24	1.48
3400	Plenum mounted, 5,000 S.F.		3800	.004		1.06	.17		1.23	1.44
3500	10,000 S.F.		4400	.004		.72	.15		.87	1.03

13 49 Radiation Protection

13 49 13 – Integrated X-Ray Shielding Assemblies

13 49 13.50 Lead Sheets		Crew	Daily Output	Labor-Hours	Unit	Material	Labor	Equipment	Total	Total Incl O&P
0010	**LEAD SHEETS**									
0300	Lead sheets, 1/16" thick	2 Lath	135	.119	S.F.	7.15	3.56		10.71	13.60
0400	1/8" thick		120	.133		14.45	4.01		18.46	22.50
0500	Lead shielding, 1/4" thick		135	.119		34	3.56		37.56	43
0550	1/2" thick		120	.133		59.50	4.01		63.51	72
0950	Lead headed nails (average 1 lb. per sheet)				Lb.	7.15			7.15	7.90
1000	Butt joints in 1/8" lead or thicker, 2" batten strip x 7' long	2 Lath	240	.067	Ea.	16.95	2		18.95	22
1200	X-ray protection, average radiography or fluoroscopy									
1210	room, up to 300 S.F. floor, 1/16" lead, minimum	2 Lath	.25	64	Total	8,075	1,925		10,000	12,000
1500	Maximum, 7'-0" walls	"	.15	106	"	9,700	3,200		12,900	15,800
1600	Deep therapy X-ray room, 250 kV capacity,									
1800	up to 300 S.F. floor, 1/4" lead, minimum	2 Lath	.08	200	Total	22,600	6,000		28,600	34,400
1900	Maximum, 7'-0" walls	"	.06	266	"	27,800	8,025		35,825	43,400

13 49 19 – Lead-Lined Materials

13 49 19.50 Shielding Lead		Crew	Daily Output	Labor-Hours	Unit	Material	Labor	Equipment	Total	Total Incl O&P
0010	**SHIELDING LEAD**									
0100	Laminated lead in wood doors, 1/16" thick, no hardware				S.F.	50			50	55
0200	Lead lined door frame, not incl. hardware,									
0210	1/16" thick lead, butt prepared for hardware	1 Lath	2.40	3.333	Ea.	575	100		675	790
0850	Window frame with 1/16" lead and voice passage, 36" x 60"	2 Glaz	2	8		1,275	266		1,541	1,825
0870	24" x 36" frame		4	4		1,025	133		1,158	1,350
0900	Lead gypsum board, 5/8" thick with 1/16" lead		160	.100	S.F.	7.10	3.32		10.42	13.30
0910	1/8" lead		140	.114		14.45	3.79		18.24	22
0930	1/32" lead	2 Lath	200	.080		4.43	2.40		6.83	8.70

13 49 Radiation Protection

13 49 21 – Lead Glazing

13 49 21.50 Lead Glazing		Crew	Daily Output	Labor-Hours	Unit	2011 Bare Costs Material	Labor	Equipment	Total	Total Incl O&P
0010	**LEAD GLAZING**									
0600	Lead glass, 1/4″ thick, 2.0 mm LE, 12″ x 16″	2 Glaz	13	1.231	Ea.	265	41		306	360
0700	24″ x 36″		8	2		980	66.50		1,046.50	1,175
0800	36″ x 60″		2	8		4,125	266		4,391	4,950
2000	X-ray viewing panels, clear lead plastic									
2010	7 mm thick, 0.3 mm LE, 2.3 lb./S.F.	H-3	139	.115	S.F.	179	3.37		182.37	203
2020	12 mm thick, 0.5 mm LE, 3.9 lb./S.F.		82	.195		242	5.70		247.70	275
2030	18 mm thick, 0.8 mm LE, 5.9 lb./S.F.		54	.296		263	8.70		271.70	305
2040	22 mm thick, 1.0 mm LE, 7.2 lb./S.F.		44	.364		269	10.65		279.65	315
2050	35 mm thick, 1.5 mm LE, 11.5 lb./S.F.		28	.571		300	16.75		316.75	360
2060	46 mm thick, 2.0 mm LE, 15.0 lb./S.F.		21	.762		395	22.50		417.50	470
2090	For panels 12 S.F. to 48 S.F., add crating charge				Ea.					50

13 49 23 – Integrated RFI/EMI Shielding Assemblies

13 49 23.50 Modular Shielding Partitions		Crew	Daily Output	Labor-Hours	Unit	Material	Labor	Equipment	Total	Total Incl O&P
0010	**MODULAR SHIELDING PARTITIONS**									
4000	X-ray barriers, modular, panels mounted within framework for									
4002	attaching to floor, wall or ceiling, upper portion is clear lead									
4005	plastic window panels 48″H, lower portion is opaque leaded									
4008	steel panels 36″H, structural supports not incl.									
4010	1-section barrier, 36″W x 84″H overall									
4020	0.5 mm LE panels	H-3	6.40	2.500	Ea.	5,650	73		5,723	6,350
4030	0.8 mm LE panels		6.40	2.500		6,025	73		6,098	6,750
4040	1.0 mm LE panels		5.33	3.002		6,150	88		6,238	6,900
4050	1.5 mm LE panels		5.33	3.002		6,550	88		6,638	7,350
4060	2-section barrier, 72″W x 84″H overall									
4070	0.5 mm LE panels	H-3	4	4	Ea.	11,600	117		11,717	12,900
4080	0.8 mm LE panels		4	4		12,200	117		12,317	13,600
4090	1.0 mm LE panels		3.56	4.494		12,500	132		12,632	13,900
5000	1.5 mm LE panels		3.20	5		14,200	146		14,346	15,800
5010	3-section barrier, 108″W x 84″H overall									
5020	0.5 mm LE panels	H-3	3.20	5	Ea.	17,400	146		17,546	19,300
5030	0.8 mm LE panels		3.20	5		18,300	146		18,446	20,300
5040	1.0 mm LE panels		2.67	5.993		18,600	175		18,775	20,800
5050	1.5 mm LE panels		2.46	6.504		20,200	190		20,390	22,600
7000	X-ray barriers, mobile, mounted within framework w/casters on									
7005	bottom, clear lead plastic window panels on upper portion,									
7010	opaque on lower, 30″W x 75″H overall, incl. framework									
7020	24″H upper w/0.5 mm LE, 48″H lower w/0.8 mm LE	1 Carp	16	.500	Ea.	3,025	16.80		3,041.80	3,350
7030	48″W x 75″H overall, incl. framework									
7040	36″H upper w/0.5 mm LE, 36″H lower w/0.8 mm LE	1 Carp	16	.500	Ea.	5,525	16.80		5,541.80	6,100
7050	36″H upper w/1.0 mm LE, 36″H lower w/1.5 mm LE	″	16	.500	″	6,550	16.80		6,566.80	7,250
7060	72″W x 75″H overall, incl. framework									
7070	36″H upper w/0.5 mm LE, 36″H lower w/0.8 mm LE	1 Carp	16	.500	Ea.	6,550	16.80		6,566.80	7,225
7080	36″H upper w/1.0 mm LE, 36″H lower w/1.5 mm LE	″	16	.500	″	8,175	16.80		8,191.80	9,025

13 49 33 – Radio Frequency Shielding

13 49 33.50 Shielding, Radio Frequency		Crew	Daily Output	Labor-Hours	Unit	Material	Labor	Equipment	Total	Total Incl O&P
0010	**SHIELDING, RADIO FREQUENCY**									
0020	Prefabricated or screen-type copper or steel, minimum	2 Carp	180	.089	SF Surf	30	2.98		32.98	38.50
0100	Average		155	.103		32	3.46		35.46	41.50
0150	Maximum		145	.110		34	3.70		37.70	43.50

13 53 Meteorological Instrumentation

13 53 09 – Weather Instrumentation

13 53 09.50 Weather Station		Crew	Daily Output	Labor-Hours	Unit	2011 Bare Costs Material	Labor	Equipment	Total	Total Incl O&P
0010	**WEATHER STATION**									
0020	Remote recording, solar powered, with rain gauge & display, 400 ft range				Ea.	1,375			1,375	1,500
0100	1 mile range				"	1,650			1,650	1,800

Estimating Tips

General

- Many products in Division 14 will require some type of support or blocking for installation not included with the item itself. Examples are supports for conveyors or tube systems, attachment points for lifts, and footings for hoists or cranes. Add these supports in the appropriate division.

14 10 00 Dumbwaiters
14 20 00 Elevators

- Dumbwaiters and elevators are estimated and purchased in a method similar to buying a car. The manufacturer has a base unit with standard features. Added to this base unit price will be whatever options the owner or specifications require. Increased load capacity, additional vertical travel, additional stops, higher speed, and cab finish options are items to be considered. When developing an estimate for dumbwaiters and elevators, remember that some items needed by the installers may have to be included as part of the general contract.

  Examples are:
  - — shaftway
  - — rail support brackets
  - — machine room
  - — electrical supply
  - — sill angles
  - — electrical connections
  - — pits
  - — roof penthouses
  - — pit ladders

Check the job specifications and drawings before pricing.

- Installation of elevators and handicapped lifts in historic structures can require significant additional costs. The associated structural requirements may involve cutting into and repairing finishes, mouldings, flooring, etc. The estimator must account for these special conditions.

14 30 00 Escalators and Moving Walks

- Escalators and moving walks are specialty items installed by specialty contractors. There are numerous options associated with these items. For specific options, contact a manufacturer or contractor. In a method similar to estimating dumbwaiters and elevators, you should verify the extent of general contract work and add items as necessary.

14 40 00 Lifts
14 90 00 Other Conveying Equipment

- Products such as correspondence lifts, chutes, and pneumatic tube systems, as well as other items specified in this subdivision, may require trained installers. The general contractor might not have any choice as to who will perform the installation or when it will be performed. Long lead times are often required for these products, making early decisions in scheduling necessary.

Reference Numbers

Reference numbers are shown in shaded boxes at the beginning of some major classifications. These numbers refer to related items in the Reference Section. The reference information may be an estimating procedure, an alternate pricing method, or technical information.

Note: Not all subdivisions listed here necessarily appear in this publication.

14 11 Manual Dumbwaiters

14 11 10 – Hand Operated Dumbwaiters

14 11 10.20 Manual Dumbwaiters		Crew	Daily Output	Labor-Hours	Unit	2011 Bare Costs Material	Labor	Equipment	Total	Total Incl O&P
0010	**MANUAL DUMBWAITERS**									
0020	2 stop, minimum	2 Elev	.75	21.333	Ea.	2,925	1,075		4,000	4,975
0100	Maximum		.50	32	"	6,650	1,625		8,275	9,950
0300	For each additional stop, add		.75	21.333	Stop	1,050	1,075		2,125	2,925

14 12 Electric Dumbwaiters

14 12 10 – Dumbwaiters

14 12 10.10 Electric Dumbwaiters		Crew	Daily Output	Labor-Hours	Unit	Material	Labor	Equipment	Total	Total Incl O&P
0010	**ELECTRIC DUMBWAITERS**									
0020	2 stop, minimum	2 Elev	.13	123	Ea.	7,375	6,225		13,600	18,200
0100	Maximum		.11	145	"	22,200	7,350		29,550	36,300
0600	For each additional stop, add		.54	29.630	Stop	3,250	1,500		4,750	6,000
0750	Correspondence lift, 1 floor, 2 stop, 45 lb capacity	2 Elec	.20	80	Ea.	8,350	3,225		11,575	14,400

14 21 Electric Traction Elevators

14 21 13 – Electric Traction Freight Elevators

14 21 13.10 Electric Traction Freight Elevators and Options		Crew	Daily Output	Labor-Hours	Unit	Material	Labor	Equipment	Total	Total Incl O&P
0010	**ELECTRIC TRACTION FREIGHT ELEVATORS AND OPTIONS** R142000-10									
0425	Electric freight, base unit, 4000 lb, 200 fpm, 4 stop, std. fin.	2 Elev	.05	320	Ea.	96,500	16,200		112,700	132,500
0450	For 5000 lb capacity, add					5,450			5,450	6,000
0500	For 6000 lb capacity, add					16,900			16,900	18,600
0525	For 7000 lb capacity, add					20,700			20,700	22,800
0550	For 8000 lb capacity, add					28,000			28,000	30,800
0575	For 10000 lb capacity, add					31,500			31,500	34,700
0600	For 12000 lb capacity, add					38,200			38,200	42,000
0625	For 16000 lb capacity, add					48,700			48,700	53,500
0650	For 20000 lb capacity, add					58,000			58,000	63,500
0675	For increased speed, 250 fpm, add					21,800			21,800	23,900
0700	300 fpm, geared electric, add					27,600			27,600	30,300
0725	350 fpm, geared electric, add					33,300			33,300	36,700
0750	400 fpm, geared electric, add					38,800			38,800	42,600
0775	500 fpm, gearless electric, add					46,400			46,400	51,000
0800	600 fpm, gearless electric, add					56,000			56,000	62,000
0825	700 fpm, gearless electric, add					67,000			67,000	74,000
0850	800 fpm, gearless electric, add					77,500			77,500	85,500
0875	For class "B" loading, add					6,375			6,375	7,000
0900	For class "C-1" loading, add					8,800			8,800	9,700
0925	For class "C-2" loading, add					10,000			10,000	11,000
0950	For class "C-3" loading, add					12,700			12,700	13,900
0975	For travel over 40 V.L.F., add	2 Elev	7.25	2.207	V.L.F.	540	112		652	775
1000	For number of stops over 4, add	"	.27	59.259	Stop	3,850	3,000		6,850	9,125

14 21 23 – Electric Traction Passenger Elevators

14 21 23.10 Electric Traction Passenger Elevators and Options		Crew	Daily Output	Labor-Hours	Unit	Material	Labor	Equipment	Total	Total Incl O&P
0010	**ELECTRIC TRACTION PASSENGER ELEVATORS AND OPTIONS**									
1625	Electric pass., base unit, 2000 lb, 200 fpm, 4 stop, std. fin.	2 Elev	.05	320	Ea.	83,500	16,200		99,700	118,500
1650	For 2500 lb capacity, add					3,475			3,475	3,825
1675	For 3000 lb capacity, add					4,150			4,150	4,575
1700	For 3500 lb capacity, add					6,950			6,950	7,625
1725	For 4000 lb capacity, add					6,525			6,525	7,175

14 21 Electric Traction Elevators

14 21 23 – Electric Traction Passenger Elevators

14 21 23.10 Electric Traction Passenger Elevators and Options		Crew	Daily Output	Labor-Hours	Unit	2011 Bare Costs Material	Labor	Equipment	Total	Total Incl O&P
1750	For 4500 lb capacity, add				Ea.	8,250			8,250	9,075
1775	For 5000 lb capacity, add					9,925			9,925	10,900
1800	For increased speed, 250 fpm, geared electric, add					6,600			6,600	7,250
1825	300 fpm, geared electric, add					9,975			9,975	11,000
1850	350 fpm, geared electric, add					11,300			11,300	12,400
1875	400 fpm, geared electric, add					14,400			14,400	15,800
1900	500 fpm, gearless electric, add					29,500			29,500	32,500
1925	600 fpm, gearless electric, add					32,700			32,700	36,000
1950	700 fpm, gearless electric, add					37,500			37,500	41,300
1975	800 fpm, gearless electric, add					42,800			42,800	47,100
2000	For travel over 40 V.L.F., add	2 Elev	7.25	2.207	V.L.F.	650	112		762	895
2025	For number of stops over 4, add		.27	59.259	Stop	7,400	3,000		10,400	13,000
2400	Electric hospital, base unit, 4000 lb, 200 fpm, 4 stop, std fin.		.05	320	Ea.	72,000	16,200		88,200	105,500
2425	For 4500 lb capacity, add					5,125			5,125	5,650
2450	For 5000 lb capacity, add					6,175			6,175	6,775
2475	For increased speed, 250 fpm, geared electric, add					6,625			6,625	7,300
2500	300 fpm, geared electric, add					9,825			9,825	10,800
2525	350 fpm, geared electric, add					11,200			11,200	12,300
2550	400 fpm, geared electric, add					14,200			14,200	15,600
2575	500 fpm, gearless electric, add					29,200			29,200	32,100
2600	600 fpm, gearless electric, add					33,400			33,400	36,800
2625	700 fpm, gearless electric, add					37,900			37,900	41,700
2650	800 fpm, gearless electric, add					43,700			43,700	48,000
2675	For travel over 40 V.L.F., add	2 Elev	7.25	2.207	V.L.F.	635	112		747	875
2700	For number of stops over 4, add	"	.27	59.259	Stop	3,825	3,000		6,825	9,100

14 21 33 – Electric Traction Residential Elevators

14 21 33.20 Residential Elevators		Crew	Daily Output	Labor-Hours	Unit	Material	Labor	Equipment	Total	Total Incl O&P
0010	**RESIDENTIAL ELEVATORS**									
7000	Residential, cab type, 1 floor, 2 stop, minimum	2 Elev	.20	80	Ea.	11,000	4,050		15,050	18,700
7100	Maximum		.10	160		18,600	8,100		26,700	33,600
7200	2 floor, 3 stop, minimum		.12	133		16,400	6,750		23,150	28,900
7300	Maximum		.06	266		26,700	13,500		40,200	51,500

14 24 Hydraulic Elevators

14 24 13 – Hydraulic Freight Elevators

14 24 13.10 Hydraulic Freight Elevators and Options		Crew	Daily Output	Labor-Hours	Unit	Material	Labor	Equipment	Total	Total Incl O&P
0010	**HYDRAULIC FREIGHT ELEVATORS AND OPTIONS**									
1025	Hydraulic freight, base unit, 2000 lb, 50 fpm, 2 stop, std. fin.	2 Elev	.10	160	Ea.	75,000	8,100		83,100	95,500
1050	For 2500 lb capacity, add					4,725			4,725	5,200
1075	For 3000 lb capacity, add					8,375			8,375	9,225
1100	For 3500 lb capacity, add					15,200			15,200	16,700
1125	For 4000 lb capacity, add					16,200			16,200	17,800
1150	For 4500 lb capacity, add					19,000			19,000	20,900
1175	For 5000 lb capacity, add					23,900			23,900	26,300
1200	For 6000 lb capacity, add					29,900			29,900	32,900
1225	For 7000 lb capacity, add					35,900			35,900	39,500
1250	For 8000 lb capacity, add					39,500			39,500	43,500
1275	For 10000 lb capacity, add					50,000			50,000	55,000
1300	For 12000 lb capacity, add					65,500			65,500	72,000
1325	For 16000 lb capacity, add					90,000			90,000	99,000

14 24 Hydraulic Elevators

14 24 13 – Hydraulic Freight Elevators

14 24 13.10 Hydraulic Freight Elevators and Options		Crew	Daily Output	Labor-Hours	Unit	2011 Bare Costs Material	Labor	Equipment	Total	Total Incl O&P
1350	For 20000 lb capacity, add				Ea.	109,500			109,500	120,500
1375	For increased speed, 100 fpm, add					2,100			2,100	2,325
1400	125 fpm, add					4,525			4,525	4,975
1425	150 fpm, add					5,400			5,400	5,925
1450	175 fpm, add					9,200			9,200	10,100
1475	For class "B" loading, add					6,350			6,350	6,975
1500	For class "C-1" loading, add					8,775			8,775	9,650
1525	For class "C-2" loading, add					10,000			10,000	11,000
1550	For class "C-3" loading, add					12,600			12,600	13,900
1575	For travel over 20 V.L.F., add	2 Elev	7.25	2.207	V.L.F.	1,650	112		1,762	2,000
1600	For number of stops over 2, add	"	.27	59.259	Stop	4,450	3,000		7,450	9,775

14 24 23 – Hydraulic Passenger Elevators

14 24 23.10 Hydraulic Passenger Elevators and Options		Crew	Daily Output	Labor-Hours	Unit	Material	Labor	Equipment	Total	Total Incl O&P
0010	**HYDRAULIC PASSENGER ELEVATORS AND OPTIONS**									
2050	Hyd. pass., base unit, 1500 lb, 100 fpm, 2 stop, std. fin.	2 Elev	.10	160	Ea.	34,700	8,100		42,800	51,500
2075	For 2000 lb capacity, add					1,900			1,900	2,100
2100	For 2500 lb. capacity, add					2,975			2,975	3,250
2125	For 3000 lb capacity, add					3,900			3,900	4,300
2150	For 3500 lb capacity, add					5,650			5,650	6,225
2175	For 4000 lb capacity, add					6,950			6,950	7,650
2200	For 4500 lb capacity, add					8,175			8,175	9,000
2225	For 5000 lb capacity, add					10,800			10,800	11,900
2250	For increased speed, 125 fpm, add					2,800			2,800	3,075
2275	150 fpm, add					3,525			3,525	3,875
2300	175 fpm, add					5,275			5,275	5,800
2325	200 fpm, add					7,300			7,300	8,025
2350	For travel over 12 V.L.F., add	2 Elev	7.25	2.207	V.L.F.	865	112		977	1,125
2375	For number of stops over 2, add		.27	59.259	Stop	2,450	3,000		5,450	7,550
2725	Hydraulic hospital, base unit, 4000 lb, 100 fpm, 2 stop, std. fin.		.10	160	Ea.	56,500	8,100		64,600	75,000
2775	For 4500 lb capacity, add					6,400			6,400	7,050
2800	For 5000 lb capacity, add					8,475			8,475	9,325
2825	For increased speed, 125 fpm, add					3,225			3,225	3,550
2850	150 fpm, add					3,875			3,875	4,275
2875	175 fpm, add					5,625			5,625	6,200
2900	200 fpm, add					6,800			6,800	7,475
2925	For travel over 12 V.L.F., add	2 Elev	7.25	2.207	V.L.F.	715	112		827	965
2950	For number of stops over 2, add	"	.27	59.259	Stop	3,850	3,000		6,850	9,100

14 27 Custom Elevator Cabs and Doors

14 27 13 – Custom Elevator Cab Finishes

14 27 13.10 Cab Finishes		Crew	Daily Output	Labor-Hours	Unit	Material	Labor	Equipment	Total	Total Incl O&P
0010	**CAB FINISHES**									
3325	Passenger elevator cab finishes (based on 3500 lb cab size)									
3350	Acrylic panel ceiling				Ea.	705			705	775
3375	Aluminum eggcrate ceiling					1,150			1,150	1,275
3400	Stainless steel doors					2,850			2,850	3,125
3425	Carpet flooring					570			570	630
3450	Epoxy flooring					435			435	480
3475	Quarry tile flooring					810			810	890
3500	Slate flooring					1,425			1,425	1,550

14 27 Custom Elevator Cabs and Doors

14 27 13 – Custom Elevator Cab Finishes

14 27 13.10 Cab Finishes		Crew	Daily Output	Labor-Hours	Unit	Material	2011 Bare Costs Labor	Equipment	Total	Total Incl O&P
3525	Textured rubber flooring				Ea.	188			188	207
3550	Stainless steel walls					4,975			4,975	5,475
3575	Stainless steel returns at door					1,150			1,150	1,275
4450	Hospital elevator cab finishes (based on 3500 lb cab size)									
4475	Aluminum eggcrate ceiling				Ea.	1,150			1,150	1,250
4500	Stainless steel doors					2,575			2,575	2,825
4525	Epoxy flooring					350			350	385
4550	Quarry tile flooring					820			820	900
4575	Textured rubber flooring					625			625	685
4600	Stainless steel walls					4,975			4,975	5,475
4625	Stainless steel returns at door					1,150			1,150	1,275

14 28 Elevator Equipment and Controls

14 28 10 – Elevator Equipment and Control Options

14 28 10.10 Elevator Controls and Doors

		Crew	Daily Output	Labor-Hours	Unit	Material	Labor	Equipment	Total	Total Incl O&P
0010	**ELEVATOR CONTROLS AND DOORS**									
2975	Passenger elevator options									
3000	2 car group automatic controls	2 Elev	.66	24.242	Ea.	6,275	1,225		7,500	8,900
3025	3 car group automatic controls		.44	36.364		12,400	1,850		14,250	16,600
3050	4 car group automatic controls		.33	48.485		25,200	2,450		27,650	31,700
3075	5 car group automatic controls		.26	61.538		46,200	3,100		49,300	56,000
3100	6 car group automatic controls		.22	72.727		95,000	3,675		98,675	110,500
3125	Intercom service		3	5.333		1,325	270		1,595	1,925
3150	Duplex car selective collective		.66	24.242		11,700	1,225		12,925	14,800
3175	Center opening 1 speed doors		2	8		1,725	405		2,130	2,550
3200	Center opening 2 speed doors		2	8		2,425	405		2,830	3,325
3225	Rear opening doors (opposite front)		2	8		2,875	405		3,280	3,825
3250	Side opening 2 speed doors		2	8		4,525	405		4,930	5,625
3275	Automatic emergency power switching		.66	24.242		1,100	1,225		2,325	3,200
3300	Manual emergency power switching		8	2		480	101		581	695
3625	Hall finishes, stainless steel doors					1,925			1,925	2,125
3650	Stainless steel frames					1,400			1,400	1,550
3675	12 month maintenance contract								3,600	3,950
3700	Signal devices, hall lanterns	2 Elev	8	2		525	101		626	745
3725	Position indicators, up to 3		9.40	1.702		615	86		701	815
3750	Position indicators, per each over 3		32	.500		595	25.50		620.50	695
3775	High speed heavy duty door opener					4,050			4,050	4,450
3800	Variable voltage, O.H. gearless machine, min.	2 Elev	.16	100		66,500	5,050		71,550	81,000
3815	Maximum		.07	228		98,000	11,600		109,600	127,000
3825	Basement installed geared machine		.33	48.485		69,000	2,450		71,450	80,000
3850	Freight elevator options									
3875	Doors, bi-parting	2 Elev	.66	24.242	Ea.	9,525	1,225		10,750	12,500
3900	Power operated door and gate	"	.66	24.242		22,100	1,225		23,325	26,300
3925	Finishes, steel plate floor					2,375			2,375	2,625
3950	14 ga. 1/4" x 4' steel plate walls					1,800			1,800	1,975
3975	12 month maintenance contract								3,600	3,950
4000	Signal devices, hall lanterns	2 Elev	8	2		520	101		621	735
4025	Position indicators, up to 3		9.40	1.702		625	86		711	825
4050	Position indicators, per each over 3		32	.500		600	25.50		625.50	700
4075	Variable voltage basement installed geared machine		.66	24.242		18,200	1,225		19,425	22,100
4100	Hospital elevator options									

14 28 Elevator Equipment and Controls

14 28 10 – Elevator Equipment and Control Options

14 28 10.10 Elevator Controls and Doors		Crew	Daily Output	Labor-Hours	Unit	Material	Labor	Equipment	Total	Total Incl O&P
						2011 Bare Costs				
4125	2 car group automatic controls	2 Elev	.66	24.242	Ea.	6,275	1,225		7,500	8,900
4150	3 car group automatic controls		.44	36.364		12,400	1,850		14,250	16,600
4175	4 car group automatic controls		.33	48.485		25,200	2,450		27,650	31,700
4200	5 car group automatic controls		.26	61.538		46,000	3,100		49,100	55,500
4225	6 car group automatic controls		.22	72.727		94,500	3,675		98,175	110,000
4250	Intercom service		3	5.333		1,325	270		1,595	1,900
4275	Duplex car selective collective		.66	24.242		11,600	1,225		12,825	14,800
4300	Center opening 1 speed doors		2	8		1,725	405		2,130	2,550
4325	Center opening 2 speed doors		2	8		2,350	405		2,755	3,250
4350	Rear opening doors (opposite front)		2	8		2,875	405		3,280	3,825
4375	Side opening 2 speed doors		2	8		3,600	405		4,005	4,600
4400	Automatic emergency power switching		.66	24.242		1,075	1,225		2,300	3,175
4425	Manual emergency power switching		8	2		470	101		571	680
4675	Hall finishes, stainless steel doors					1,925			1,925	2,125
4700	Stainless steel frames					1,400			1,400	1,550
4725	12 month maintenance contract								3,600	3,950
4750	Signal devices, hall lanterns	2 Elev	8	2		515	101		616	730
4775	Position indicators, up to 3		9.40	1.702		615	86		701	815
4800	Position indicators, per each over 3		32	.500		595	25.50		620.50	695
4825	High speed heavy duty door opener					4,050			4,050	4,450
4850	Variable voltage, O.H. gearless machine, min.	2 Elev	.16	100		66,000	5,050		71,050	80,500
4865	Maximum		.07	228		97,500	11,600		109,100	126,500
4875	Basement installed geared machine		.33	48.485		25,000	2,450		27,450	31,500
5000	Drilling for piston, casing included, 18" diameter	B-48	80	.600	V.L.F.	47	17	35	99	119

14 31 Escalators

14 31 10 – Glass and Steel Escalators

14 31 10.10 Escalators		Crew	Daily Output	Labor-Hours	Unit	Material	Labor	Equipment	Total	Total Incl O&P
0010	**ESCALATORS** R143110-10									
1000	Glass, 32" wide x 10' floor to floor height	M-1	.07	457	Ea.	80,000	22,000	700	102,700	124,500
1010	48" wide x 10' floor to floor height		.07	457		86,000	22,000	700	108,700	131,500
1020	32" wide x 15' floor to floor height		.06	533		84,000	25,600	815	110,415	135,000
1030	48" wide x 15' floor to floor height		.06	533		89,000	25,600	815	115,415	140,500
1040	32" wide x 20' floor to floor height		.05	653		88,500	31,400	1,000	120,900	149,000
1050	48" wide x 20' floor to floor height		.05	653		96,000	31,400	1,000	128,400	157,500
1060	32" wide x 25' floor to floor height		.04	800		96,000	38,400	1,225	135,625	169,500
1070	48" wide x 25' floor to floor height		.04	800		111,500	38,400	1,225	151,125	186,500
1080	Enameled steel, 32" wide x 10' floor to floor height		.07	457		85,500	22,000	700	108,200	130,500
1090	48" wide x 10' floor to floor height		.07	457		92,500	22,000	700	115,200	138,500
1110	32" wide x 15' floor to floor height		.06	533		92,500	25,600	815	118,915	144,500
1120	48" wide x 15' floor to floor height		.06	533		92,500	25,600	815	118,915	144,500
1130	32" wide x 20' floor to floor height		.05	653		94,500	31,400	1,000	126,900	156,000
1140	48" wide x 20' floor to floor height		.05	653		101,500	31,400	1,000	133,900	163,500
1150	32" wide x 25' floor to floor height		.04	800		101,500	38,400	1,225	141,125	175,500
1160	48" wide x 25' floor to floor height		.04	800		101,500	38,400	1,225	141,125	175,500
1170	Stainless steel, 32" wide x 10' floor to floor height		.07	457		101,500	22,000	700	124,200	148,000
1180	48" wide x 10' floor to floor height		.07	457		101,500	22,000	700	124,200	148,000
1500	32" wide x 15' floor to floor height		.06	533		93,000	25,600	815	119,415	144,500
1700	48" wide x 15' floor to floor height		.06	533		94,000	25,600	815	120,415	146,000
2300	32" wide x 25' floor to floor height		.04	800		101,500	38,400	1,225	141,125	176,000
2500	48" wide x 25' floor to floor height		.04	800		123,000	38,400	1,225	162,625	199,000

14 32 Moving Walks

14 32 10 – Moving Walkways

14 32 10.10 Moving Walks		Crew	Daily Output	Labor-Hours	Unit	Material	2011 Bare Costs Labor	Equipment	Total	Total Incl O&P
0010	MOVING WALKS R143210-20									
0020	Walk, 27" tread width, minimum	M-1	6.50	4.923	L.F.	850	236	7.55	1,093.55	1,325
0100	300' to 500', maximum		4.43	7.223		1,175	345	11.05	1,531.05	1,875
0300	48" tread width walk, minimum		4.43	7.223		1,925	345	11.05	2,281.05	2,700
0400	100' to 350', maximum		3.82	8.377		2,250	400	12.80	2,662.80	3,150
0600	Ramp, 12° incline, 36" tread width, minimum		5.27	6.072		1,575	292	9.30	1,876.30	2,200
0700	70' to 90' maximum		3.82	8.377		2,250	400	12.80	2,662.80	3,150
0900	48" tread width, minimum		3.57	8.964		2,300	430	13.70	2,743.70	3,250
1000	40' to 70', maximum		2.91	10.997		2,900	530	16.80	3,446.80	4,075

14 42 Wheelchair Lifts

14 42 13 – Inclined Wheelchair Lifts

14 42 13.10 Inclined Wheelchair Lifts and Stairclimbers		Crew	Daily Output	Labor-Hours	Unit	Material	Labor	Equipment	Total	Total Incl O&P
0010	INCLINED WHEELCHAIR LIFTS AND STAIRCLIMBERS									
7700	Stair climber (chair lift), single seat, minimum	2 Elev	1	16	Ea.	5,325	810		6,135	7,200
7800	Maximum		.20	80		7,350	4,050		11,400	14,700
8700	Stair lift, minimum		1	16		14,500	810		15,310	17,200
8900	Maximum		.20	80		22,900	4,050		26,950	31,800

14 42 16 – Vertical Wheelchair Lifts

14 42 16.10 Wheelchair Lifts		Crew	Daily Output	Labor-Hours	Unit	Material	Labor	Equipment	Total	Total Incl O&P
0010	WHEELCHAIR LIFTS									
8000	Wheelchair lift, minimum	2 Elev	1	16	Ea.	7,325	810		8,135	9,375
8500	Maximum	"	.50	32	"	17,300	1,625		18,925	21,600

14 45 Vehicle Lifts

14 45 10 – Hydraulic Vehicle Lifts

14 45 10.10 Hydraulic Lifts		Crew	Daily Output	Labor-Hours	Unit	Material	Labor	Equipment	Total	Total Incl O&P
0010	HYDRAULIC LIFTS									
0200	Double post, 9000 lb frame	L-4	1.15	13.913	Ea.	5,525	415		5,940	6,775
2200	Hoists, single post, 8000 lb capacity, swivel arms		.40	40		4,975	1,200		6,175	7,450
2400	Two posts, adjustable frames, 11,000 lb capacity		.25	64		6,400	1,900		8,300	10,200
2500	24,000 lb capacity		.15	106		8,550	3,175		11,725	14,700
2700	7500 lb capacity, frame supports		.50	32		7,125	950		8,075	9,425
2800	Four post, roll on ramp		.50	32		6,400	950		7,350	8,650
2810	Hydraulic lifts, above ground, 2 post, clear floor, 6000 lb cap		2.67	5.993		7,025	178		7,203	8,025
2815	9000 lb capacity		2.29	6.987		16,700	208		16,908	18,600
2820	15,000 lb capacity		2	8		18,800	238		19,038	21,100
2825	30,000 lb capacity		1.60	10		41,600	297		41,897	46,300
2830	4 post, ramp style, 25,000 lb capacity		2	8		16,300	238		16,538	18,300
2835	35,000 lb capacity		1	16		76,000	475		76,475	84,500
2840	50,000 lb capacity		1	16		85,000	475		85,475	94,500
2845	75,000 lb capacity		1	16		98,500	475		98,975	109,500
2850	For drive thru tracks, add, minimum					1,050			1,050	1,150
2855	Maximum					1,800			1,800	1,975
2860	Ramp extensions, 3' (set of 2)					860			860	945
2865	Rolling jack platform					3,000			3,000	3,300
2870	Elec/hyd jacking beam					8,000			8,000	8,800
2880	Scissor lift, portable, 6000 lb capacity					7,850			7,850	8,625

14 51 Correspondence and Parcel Lifts

14 51 10 – Electric Correspondence and Parcel Lifts

14 51 10.10 Correspondence Lifts		Crew	Daily Output	Labor-Hours	Unit	Material	Labor	Equipment	Total	Total Incl O&P
						2011 Bare Costs				
0010	**CORRESPONDENCE LIFTS**									
0020	1 floor, 2 stop, 25 lb capacity, electric	2 Elev	.20	80	Ea.	5,975	4,050		10,025	13,200
0100	Hand, 5 lb capacity	"	.20	80	"	2,250	4,050		6,300	9,075
14 51 10.20 Parcel Lifts										
0010	**PARCEL LIFTS**									
0020	20" x 20", 100 lb capacity, electric, per floor	2 Mill	.25	64	Ea.	9,350	2,250		11,600	13,900

14 91 Facility Chutes

14 91 33 – Laundry and Linen Chutes

14 91 33.10 Chutes		Crew	Daily Output	Labor-Hours	Unit	Material	Labor	Equipment	Total	Total Incl O&P
0011	**CHUTES**, linen, trash or refuse									
0050	Aluminized steel, 16 ga., 18" diameter	2 Shee	3.50	4.571	Floor	1,625	175		1,800	2,100
0100	24" diameter		3.20	5		1,725	191		1,916	2,200
0200	30" diameter		3	5.333		2,050	204		2,254	2,575
0300	36" diameter		2.80	5.714		2,475	218		2,693	3,075
0400	Galvanized steel, 16 ga., 18" diameter		3.50	4.571		1,000	175		1,175	1,400
0500	24" diameter		3.20	5		1,125	191		1,316	1,575
0600	30" diameter		3	5.333		1,275	204		1,479	1,725
0700	36" diameter		2.80	5.714		1,500	218		1,718	2,000
0800	Stainless steel, 18" diameter		3.50	4.571		2,725	175		2,900	3,300
0900	24" diameter		3.20	5		2,950	191		3,141	3,575
1000	30" diameter		3	5.333		3,525	204		3,729	4,200
1005	36" diameter		2.80	5.714		3,825	218		4,043	4,550
1200	Linen chute bottom collector, aluminized steel		4	4	Ea.	1,400	153		1,553	1,800
1300	Stainless steel		4	4		1,800	153		1,953	2,225
1500	Refuse, bottom hopper, aluminized steel, 18" diameter		3	5.333		1,025	204		1,229	1,450
1600	24" diameter		3	5.333		1,275	204		1,479	1,725
1800	36" diameter		3	5.333		2,375	204		2,579	2,925

14 91 82 – Trash Chutes

14 91 82.10 Trash Chutes and Accessories		Crew	Daily Output	Labor-Hours	Unit	Material	Labor	Equipment	Total	Total Incl O&P
0010	**TRASH CHUTES AND ACCESSORIES**									
2900	Package chutes, spiral type, minimum	2 Shee	4.50	3.556	Floor	2,425	136		2,561	2,900
3000	Maximum	"	1.50	10.667	"	6,325	405		6,730	7,650

14 92 Pneumatic Tube Systems

14 92 10 – Conventional, Automatic and Computer Controlled Pneumatic Tube Systems

14 92 10.10 Pneumatic Tube Systems		Crew	Daily Output	Labor-Hours	Unit	Material	Labor	Equipment	Total	Total Incl O&P
0010	**PNEUMATIC TUBE SYSTEMS**									
0020	100' long, single tube, 2 stations, stock									
0100	3" diameter	2 Stpi	.12	133	Total	6,600	5,350		11,950	16,000
0300	4" diameter	"	.09	177	"	7,450	7,125		14,575	19,800
0400	Twin tube, two stations or more, conventional system									
0600	2-1/2" round	2 Stpi	62.50	.256	L.F.	13.30	10.25		23.55	31.50
0700	3" round		46	.348		15.25	13.95		29.20	40
0900	4" round		49.60	.323		16.90	12.90		29.80	39.50
1000	4" x 7" oval		37.60	.426		24.50	17.05		41.55	55
1050	Add for blower		2	8	System	5,200	320		5,520	6,250
1110	Plus for each round station, add		7.50	2.133	Ea.	585	85.50		670.50	785

14 92 Pneumatic Tube Systems

14 92 10 – Conventional, Automatic and Computer Controlled Pneumatic Tube Systems

14 92 10.10 Pneumatic Tube Systems		Crew	Daily Output	Labor-Hours	Unit	2011 Bare Costs Material	Labor	Equipment	Total	Total Incl O&P
1150	Plus for each oval station, add	2 Stpi	7.50	2.133	Ea.	585	85.50		670.50	785
1200	Alternate pricing method: base cost, minimum		.75	21.333	Total	6,375	855		7,230	8,425
1300	Maximum		.25	64	"	12,700	2,575		15,275	18,100
1500	Plus total system length, add, minimum		93.40	.171	L.F.	8.20	6.85		15.05	20.50
1600	Maximum		37.60	.426	"	24.50	17.05		41.55	54.50
1800	Completely automatic system, 4" round, 15 to 50 stations		.29	55.172	Station	20,100	2,200		22,300	25,700
2200	51 to 144 stations		.32	50		15,600	2,000		17,600	20,500
2400	6" round or 4" x 7" oval, 15 to 50 stations		.24	66.667		25,200	2,675		27,875	32,100
2800	51 to 144 stations	▼	.23	69.565	▼	21,100	2,775		23,875	27,800

Division Notes

		CREW	DAILY OUTPUT	LABOR-HOURS	UNIT	2011 BARE COSTS MAT.	LABOR	EQUIP.	TOTAL	TOTAL INCL O&P

Estimating Tips

Pipe for fire protection and all uses is located in Subdivision 22 11 13.

The labor adjustment factors listed in Subdivision 22 01 02.20 also apply to Division 21.

Many, but not all, areas in the U.S. require backflow protection in the fire system. It is advisable to check local building codes for specific requirements.

For your reference, the following is a list of the most applicable Fire Codes and Standards which may be purchased from the NFPA, 1 Batterymarch Park, Quincy, MA 02169-7471.

NFPA 1: Uniform Fire Code
NFPA 10: Portable Fire Extinguishers
NFPA 11: Low-, Medium-, and High-Expansion Foam
NFPA 12: Carbon Dioxide Extinguishing Systems (Also companion 12A)
NFPA 13: Installation of Sprinkler Systems (Also companion 13D, 13E, and 13R)
NFPA 14: Installation of Standpipe and Hose Systems
NFPA 15: Water Spray Fixed Systems for Fire Protection
NFPA 16: Installation of Foam-Water Sprinkler and Foam-Water Spray Systems
NFPA 17: Dry Chemical Extinguishing Systems (Also companion 17A)
NFPA 18: Wetting Agents
NFPA 20: Installation of Stationary Pumps for Fire Protection
NFPA 22: Water Tanks for Private Fire Protection
NFPA 24: Installation of Private Fire Service Mains and their Appurtenances
NFPA 25: Inspection, Testing and Maintenance of Water-Based Fire Protection

Reference Numbers

Reference numbers are shown in shaded boxes at the beginning of some major classifications. These numbers refer to related items in the Reference Section. The reference information may be an estimating procedure, an alternate pricing method, or technical information.

Note: Not all subdivisions listed here necessarily appear in this publication.

Note: **Trade Service,** *in part, has been used as a reference source for some of the material prices used in Division 21.*

21 05 Common Work Results for Fire Suppression

21 05 23 – General-Duty Valves for Water-Based Fire-Suppression Piping

21 05 23.50 General-Duty Valves for Water-Based Fire Supp.		Crew	Daily Output	Labor-Hours	Unit	2011 Bare Costs Material	Labor	Equipment	Total	Total Incl O&P
0010	**GENERAL-DUTY VALVES FOR WATER-BASED FIRE SUPPRESSION PIPING**									
6200	Valves and components									
6500	Check, swing, C.I. body, brass fittings, auto. ball drip									
6520	4" size	Q-12	3	5.333	Ea.	270	189		459	605
6800	Check, wafer, butterfly type, C.I. body, bronze fittings									
6820	4" size	Q-12	4	4	Ea.	298	142		440	560

21 11 Facility Fire-Suppression Water-Service Piping

21 11 16 – Facility Fire Hydrants

21 11 16.50 Fire Hydrants for Buildings

		Crew	Daily Output	Labor-Hours	Unit	Material	Labor	Equipment	Total	Total Incl O&P
0010	**FIRE HYDRANTS FOR BUILDINGS**									
3750	Hydrants, wall, w/caps, single, flush, polished brass									
3800	2-1/2" x 2-1/2"	Q-12	5	3.200	Ea.	175	113		288	380
3840	2-1/2" x 3"	"	5	3.200	"	355	113		468	575
3950	Double, flush, polished brass									
4000	2-1/2" x 2-1/2" x 4"	Q-12	5	3.200	Ea.	470	113		583	705
4040	2-1/2" x 2-1/2" x 6"	"	4.60	3.478	↓	675	123		798	945
4200	For polished chrome, add				↓	10%				
4350	Double, projecting, polished brass									
4400	2-1/2" x 2-1/2" x 4"	Q-12	5	3.200	Ea.	209	113		322	415
4450	2-1/2" x 2-1/2" x 6"	"	4.60	3.478	"	425	123		548	670
4460	Valve control, dbl. flush/projecting hydrant, cap &									
4470	chain, extension rod & cplg., escutcheon, polished brass	Q-12	8	2	Ea.	249	71		320	390

21 11 19 – Fire-Department Connections

21 11 19.50 Connections for the Fire-Department

		Crew	Daily Output	Labor-Hours	Unit	Material	Labor	Equipment	Total	Total Incl O&P
0010	**CONNECTIONS FOR THE FIRE-DEPARTMENT**									
7140	Standpipe connections, wall, w/plugs & chains									
7160	Single, flush, brass, 2-1/2" x 2-1/2"	Q-12	5	3.200	Ea.	123	113		236	320
7180	2-1/2" x 3"	"	5	3.200	"	135	113		248	335
7240	For polished chrome, add					15%				
7280	Double, flush, polished brass									
7300	2-1/2" x 2-1/2" x 4"	Q-12	5	3.200	Ea.	425	113		538	655
7330	2-1/2" x 2-1/2" x 6"	"	4.60	3.478	"	595	123		718	855
7400	For polished chrome, add					15%				
7440	For sill cock combination, add				Ea.	50.50			50.50	56
7900	Three way, flush, polished brass									
7920	2-1/2" (3) x 4"	Q-12	4.80	3.333	Ea.	960	118		1,078	1,250
7930	2-1/2" (3) x 6"	"	4.80	3.333	↓	1,150	118		1,268	1,475
8000	For polished chrome, add				↓	9%				
8020	Three way, projecting, polished brass									
8040	2-1/2"(3) x 4"	Q-12	4.80	3.333	Ea.	985	118		1,103	1,275

21 12 Fire-Suppression Standpipes

21 12 13 – Fire-Suppression Hoses and Nozzles

21 12 13.50 Fire Hoses and Nozzles		Crew	Daily Output	Labor-Hours	Unit	2011 Bare Costs Material	Labor	Equipment	Total	Total Incl O&P
0010	**FIRE HOSES AND NOZZLES**									
0200	Adapters, rough brass, straight hose threads									
0220	One piece, female to male, rocker lugs									
0240	1" x 1"				Ea.	39.50			39.50	43.50
2200	Hose, less couplings									
2260	Synthetic jacket, lined, 300 lb. test, 1-1/2" diameter	Q-12	2600	.006	L.F.	2.39	.22		2.61	2.99
2280	2-1/2" diameter		2200	.007		3.98	.26		4.24	4.80
2360	High strength, 500 lb. test, 1-1/2" diameter		2600	.006		2.47	.22		2.69	3.08
2380	2-1/2" diameter	↓	2200	.007	↓	4.29	.26		4.55	5.15
5600	Nozzles, brass									
5620	Adjustable fog, 3/4" booster line				Ea.	91			91	100
5630	1" booster line					109			109	119
5640	1-1/2" leader line					66			66	72.50
5660	2-1/2" direct connection				↓	172			172	189
5780	For chrome plated, add					8%				
5850	Electrical fire, adjustable fog, no shock									
5900	1-1/2"				Ea.	340			340	375
5920	2-1/2"					455			455	505
5980	For polished chrome, add				↓	6%				
6200	Heavy duty, comb. adj. fog and str. stream, with handle									
6210	1" booster line				Ea.	325			325	360

21 12 19 – Fire-Suppression Hose Racks

21 12 19.50 Fire Hose Racks		Crew	Daily Output	Labor-Hours	Unit	Material	Labor	Equipment	Total	Total Incl O&P
0010	**FIRE HOSE RACKS**									
2600	Hose rack, swinging, for 1-1/2" diameter hose,									
2620	Enameled steel, 50' & 75' lengths of hose	Q-12	20	.800	Ea.	48.50	28.50		77	99
2640	100' and 125' lengths of hose	"	20	.800	"	49	28.50		77.50	100

21 12 23 – Fire-Suppression Hose Valves

21 12 23.70 Fire Hose Valves		Crew	Daily Output	Labor-Hours	Unit	Material	Labor	Equipment	Total	Total Incl O&P
0010	**FIRE HOSE VALVES**									
0080	Wheel handle, 300 lb., 1-1/2"	1 Spri	12	.667	Ea.	74.50	26		100.50	125
0090	2-1/2"	"	7	1.143	"	60	45		105	140
0100	For polished brass, add					35%				
0110	For polished chrome, add					50%				

21 13 Fire-Suppression Sprinkler Systems

21 13 13 – Wet-Pipe Sprinkler Systems

21 13 13.50 Wet-Pipe Sprinkler System Components		Crew	Daily Output	Labor-Hours	Unit	Material	Labor	Equipment	Total	Total Incl O&P
0010	**WET-PIPE SPRINKLER SYSTEM COMPONENTS**									
2600	Sprinkler heads, not including supply piping									
3700	Standard spray, pendent or upright, brass, 135°F to 286°F									
3730	1/2" NPT, 7/16" orifice	1 Spri	16	.500	Ea.	12	19.65		31.65	45
3740	1/2" NPT, 1/2" orifice	"	16	.500		7.80	19.65		27.45	40.50
3860	For wax and lead coating, add					25.50			25.50	28
3880	For wax coating, add					15.25			15.25	16.80
3900	For lead coating, add				↓	16.50			16.50	18.15
3920	For 360°F, same cost									
3930	For 400°F, add				Ea.	58.50			58.50	64
3940	For 500°F, add				"	58.50			58.50	64
4500	Sidewall, horizontal, brass, 135°F to 286°F									

21 13 Fire-Suppression Sprinkler Systems

21 13 13 – Wet-Pipe Sprinkler Systems

21 13 13.50 Wet-Pipe Sprinkler System Components		Crew	Daily Output	Labor-Hours	Unit	2011 Bare Costs Material	Labor	Equipment	Total	Total Incl O&P
4520	1/2" NPT, 1/2" orifice	1 Spri	16	.500	Ea.	15	19.65		34.65	48.50
4540	For 360°F, same cost									
4800	Recessed pendent, brass, 135°F to 286°F									
4820	1/2" NPT, 3/8" orifice	1 Spri	10	.800	Ea.	35	31.50		66.50	90
4830	1/2" NPT, 7/16" orifice	↓	10	.800	↓	19.70	31.50		51.20	73
4840	1/2" NPT, 1/2" orifice	↓	10	.800	↓	14.30	31.50		45.80	67.50

21 13 16 – Dry-Pipe Sprinkler Systems

21 13 16.50 Dry-Pipe Sprinkler System Components		Crew	Daily Output	Labor-Hours	Unit	Material	Labor	Equipment	Total	Total Incl O&P
0010	**DRY-PIPE SPRINKLER SYSTEM COMPONENTS**									
0600	Accelerator	1 Spri	8	1	Ea.	595	39.50		634.50	720
2600	Sprinkler heads, not including supply piping									
2640	Dry, pendent, 1/2" orifice, 3/4" or 1" NPT									
2700	15-1/4" to 18" length	1 Spri	14	.571	Ea.	112	22.50		134.50	160
2710	18-1/4" to 21" length		13	.615		116	24		140	167
2720	21-1/4" to 24" length		13	.615		120	24		144	172
2730	24-1/4" to 27" length	↓	13	.615	↓	125	24		149	177
6330	Valves and components									

21 13 26 – Deluge Fire-Suppression Sprinkler Systems

21 13 26.50 Deluge Fire-Suppression Sprinkler Sys. Comp.		Crew	Daily Output	Labor-Hours	Unit	Material	Labor	Equipment	Total	Total Incl O&P
0010	**DELUGE FIRE-SUPPRESSION SPRINKLER SYSTEM COMPONENTS**									
6200	Valves and components									
7000	Deluge, assembly, incl. trim, pressure									
7020	operated relief, emergency release, gauges									
7040	2" size	Q-12	2	8	Ea.	2,900	283		3,183	3,650
7060	3" size	"	1.50	10.667	"	3,250	375		3,625	4,200

21 13 39 – Foam-Water Systems

21 13 39.50 Foam-Water System Components		Crew	Daily Output	Labor-Hours	Unit	Material	Labor	Equipment	Total	Total Incl O&P
0010	**FOAM-WATER SYSTEM COMPONENTS**									
2600	Sprinkler heads, not including supply piping									
3600	Foam-water, pendent or upright, 1/2" NPT	1 Spri	12	.667	Ea.	136	26		162	193

21 21 Carbon-Dioxide Fire-Extinguishing Systems

21 21 16 – Carbon-Dioxide Fire-Extinguishing Equipment

21 21 16.50 CO2 Fire Extinguishing System		Crew	Daily Output	Labor-Hours	Unit	Material	Labor	Equipment	Total	Total Incl O&P
0010	**CO_2 FIRE EXTINGUISHING SYSTEM**									
0042	For detectors and control stations, see Section 28 31 23.50									
1000	Dispersion nozzle, CO_2, 3" x 5"	1 Plum	18	.444	Ea.	56	17.60		73.60	91
2000	Extinguisher, CO_2 system, high pressure, 75 lb. cylinder	Q-1	6	2.667		1,075	95		1,170	1,325
3000	Electro/mechanical release	L-1	4	2.500		140	99.50		239.50	315
3400	Manual pull station	1 Plum	6	1.333	↓	50.50	53		103.50	142

21 22 Clean-Agent Fire-Extinguishing Systems

21 22 16 – Clean-Agent Fire-Extinguishing Equipment

21 22 16.50 FM200 Fire Extinguishing System		Crew	Daily Output	Labor-Hours	Unit	Material	2011 Bare Costs Labor	Equipment	Total	Total Incl O&P
0010	**FM200 FIRE EXTINGUISHING SYSTEM**									
1100	Dispersion nozzle FM200, 1-1/2"	1 Plum	14	.571	Ea.	56	22.50		78.50	99
2400	Extinguisher, FM200 system, filled, with mounting bracket									
2540	196 lb. container	Q-1	4	4	Ea.	6,525	143		6,668	7,400
6000	Average FM200 system, minimum				C.F.	1.48			1.48	1.63
6020	Maximum				"	2.94			2.94	3.23

21 31 Centrifugal Fire Pumps

21 31 13 – Electric-Drive, Centrifugal Fire Pumps

21 31 13.50 Electric-Drive Fire Pumps

		Crew	Daily Output	Labor-Hours	Unit	Material	Labor	Equipment	Total	Total Incl O&P
0010	**ELECTRIC-DRIVE FIRE PUMPS** Including controller, fittings and relief valve									
3100	250 GPM, 55 psi, 15 HP, 3,550 RPM, 2" pump	Q-13	.70	45.714	Ea.	14,000	1,625		15,625	18,100
3200	500 GPM, 50 psi, 27 HP, 1,770 RPM, 4" pump		.68	47.059		14,700	1,675		16,375	18,800
3350	750 GPM, 50 psi, 44 HP, 1,770 RPM, 5" pump		.64	50		15,200	1,775		16,975	19,600
3400	750 GPM, 100 psi, 66 HP, 3,550 RPM, 4" pump	↓	.58	55.172		17,300	1,950		19,250	22,300
5000	For jockey pump 1", 3 HP, with control, add	Q-12	2	8	↓	2,125	283		2,408	2,800

21 31 16 – Diesel-Drive, Centrifugal Fire Pumps

21 31 16.50 Diesel-Drive Fire Pumps

		Crew	Daily Output	Labor-Hours	Unit	Material	Labor	Equipment	Total	Total Incl O&P
0010	**DIESEL-DRIVE FIRE PUMPS** Including controller, fittings and relief valve									
0050	500 GPM, 50 psi, 27 HP, 4" pump	Q-13	.64	50	Ea.	26,400	1,775		28,175	31,900
0200	750 GPM, 50 psi, 44 HP, 5" pump		.60	53.333		27,300	1,875		29,175	33,100
0400	1,000 GPM, 100 psi, 89 HP, 4" pump		.56	57.143		31,100	2,025		33,125	37,500
0700	2,000 GPM, 100 psi, 167 HP, 6" pump		.34	94.118		34,200	3,325		37,525	43,100
0950	3,500 GPM, 100 psi, 300 HP, 10" pump	↓	.24	133	↓	63,500	4,725		68,225	77,500

Division Notes

	CREW	DAILY OUTPUT	LABOR-HOURS	UNIT	2011 BARE COSTS MAT.	LABOR	EQUIP.	TOTAL	TOTAL INCL O&P

Estimating Tips

22 10 00 Plumbing Piping and Pumps

This subdivision is primarily basic pipe and related materials. The pipe may be used by any of the mechanical disciplines, i.e., plumbing, fire protection, heating, and air conditioning.

- The labor adjustment factors listed in Subdivision 22 01 02.20 apply throughout Divisions 21, 22, and 23. CAUTION: the correct percentage may vary for the same items. For example, the percentage add for the basic pipe installation should be based on the maximum height that the craftsman must install for that particular section. If the pipe is to be located 14′ above the floor but it is suspended on threaded rod from beams, the bottom flange of which is 18′ high (4′ rods), then the height is actually 18′ and the add is 20%. The pipe coverer, however, does not have to go above the 14′, and so his or her add should be 10%.
- Most pipe is priced first as straight pipe with a joint (coupling, weld, etc.) every 10′ and a hanger usually every 10′. There are exceptions with hanger spacing such as for cast iron pipe (5′) and plastic pipe (3 per 10′). Following each type of pipe there are several lines listing sizes and the amount to be subtracted to delete couplings and hangers. This is for pipe that is to be buried or supported together on trapeze hangers. The reason that the couplings are deleted is that these runs are usually long, and frequently longer lengths of pipe are used. By deleting the couplings, the estimator is expected to look up and add back the correct reduced number of couplings.
- When preparing an estimate, it may be necessary to approximate the fittings. Fittings usually run between 25% and 50% of the cost of the pipe. The lower percentage is for simpler runs, and the higher number is for complex areas, such as mechanical rooms.
- For historic restoration projects, the systems must be as invisible as possible, and pathways must be sought for pipes, conduit, and ductwork. While installations in accessible spaces (such as basements and attics) are relatively straightforward to estimate, labor costs may be more difficult to determine when delivery systems must be concealed.

22 40 00 Plumbing Fixtures

- Plumbing fixture costs usually require two lines: the fixture itself and its "rough-in, supply, and waste."
- In the Assemblies Section (Plumbing D2010) for the desired fixture, the System Components Group at the center of the page shows the fixture on the first line. The rest of the list (fittings, pipe, tubing, etc.) will total up to what we refer to in the Unit Price section as "Rough-in, supply, waste, and vent." Note that for most fixtures we allow a nominal 5′ of tubing to reach from the fixture to a main or riser.
- Remember that gas- and oil-fired units need venting.

Reference Numbers

Reference numbers are shown in shaded boxes at the beginning of some major classifications. These numbers refer to related items in the Reference Section. The reference information may be an estimating procedure, an alternate pricing method, or technical information.

Note: Not all subdivisions listed here necessarily appear in this publication.

Note: **Trade Service,** *in part, has been used as a reference source for some of the material prices used in Division 22.*

22 01 02 – Labor Adjustments

22 01 02.10 Boilers, General		Crew	Daily Output	Labor-Hours	Unit	Material	2011 Bare Costs Labor	Equipment	Total	Total Incl O&P
0010	**BOILERS, GENERAL**, Prices do not include flue piping, elec. wiring,									
0020	gas or oil piping, boiler base, pad, or tankless unless noted									
0100	Boiler H.P.: 10 KW = 34 lb./steam/hr. = 33,475 BTU/hr.									
0150	To convert SFR to BTU rating: Hot water, 150 x SFR; R235000-50									
0160	Forced hot water, 180 x SFR; steam, 240 x SFR									

22 01 02.20 Labor Adjustment Factors		Crew	Daily Output	Labor-Hours	Unit	Material	Labor	Equipment	Total	Total Incl O&P
0010	**LABOR ADJUSTMENT FACTORS**, (For Div. 21, 22 and 23)									
1000	Add to labor for elevated installation (Above floor level)									
1080	10′ to 14.5′ high						10%			
1100	15′ to 19.5′ high						20%			
1120	20′ to 24.5′ high						25%			
1140	25′ to 29.5′ high						35%			
1160	30′ to 34.5′ high						40%			
1180	35′ to 39.5′ high						50%			
1200	40′ and higher						55%			
2000	Add to labor for crawl space									
2100	3′ high						40%			
2140	4′ high						30%			
3000	Add to labor for multi-story building									
3100	Add per floor for floors 3 thru 19						2%			
3140	Add per floor for floors 20 and up						4%			
4000	Add to labor for working in existing occupied buildings									
4100	Hospital						35%			
4140	Office building						25%			
4180	School						20%			
4220	Factory or warehouse						15%			
4260	Multi dwelling						15%			
5000	Add to labor, miscellaneous									
5100	Cramped shaft						35%			
5140	Congested area						15%			
5180	Excessive heat or cold						30%			
9000	Labor factors, The above are reasonable suggestions, however									
9010	each project should be evaluated for its own peculiarities.									
9100	Other factors to be considered are:									
9140	Movement of material and equipment through finished areas									
9180	Equipment room									
9220	Attic space									
9260	No service road									
9300	Poor unloading/storage area									
9340	Congested site area/heavy traffic									

22 05 Common Work Results for Plumbing

22 05 05 – Selective Plumbing Demolition

22 05 05.10 Plumbing Demolition		Crew	Daily Output	Labor-Hours	Unit	Material	2011 Bare Costs Labor	Equipment	Total	Total Incl O&P
0010	**PLUMBING DEMOLITION**									
1020	Fixtures, including 10′ piping									
1101	Bathtubs, cast iron	1 Clab	4	2	Ea.		53		53	88.50
1121	Fiberglass		6	1.333			35.50		35.50	59
1141	Steel		5	1.600			42.50		42.50	71
1201	Lavatory, wall hung		10	.800			21		21	35.50
1221	Counter top		16	.500			13.25		13.25	22
1301	Sink, single compartment		16	.500			13.25		13.25	22
1321	Double		10	.800			21		21	35.50
1401	Water closet, floor mounted		16	.500			13.25		13.25	22
1421	Wall mounted		7	1.143			30		30	50.50
1501	Urinal, floor mounted		4	2			53		53	88.50
1521	Wall mounted		7	1.143			30		30	50.50
1601	Water fountains, free standing		8	1			26.50		26.50	44.50
1621	Wall or deck mounted		6	1.333			35.50		35.50	59
2001	Piping, metal, to 1-1/2″ diameter		200	.040	L.F.		1.06		1.06	1.77
2051	2″ thru 3-1/2″ diameter		150	.053			1.41		1.41	2.36
2101	4″ thru 6″ diameter		100	.080			2.12		2.12	3.54
2151	16″ thru 20″ diameter		50	.160			4.23		4.23	7.10
2212	Deduct for salvage, aluminum scrap				Ton				700	770
2214	Brass scrap								2,450	2,675
2216	Copper scrap								3,200	3,525
2218	Lead scrap								520	570
2220	Steel scrap								180	200

22 05 23 – General-Duty Valves for Plumbing Piping

22 05 23.10 Valves, Brass		Crew	Daily Output	Labor-Hours	Unit	Material	2011 Bare Costs Labor	Equipment	Total	Total Incl O&P
0010	**VALVES, BRASS**									
0500	Gas cocks, threaded									
0530	1/2″	1 Plum	24	.333	Ea.	8.90	13.20		22.10	31.50
0540	3/4″		22	.364		11.20	14.40		25.60	36
0550	1″		19	.421		17.65	16.70		34.35	47
0560	1-1/4″		15	.533		38	21		59	76.50

22 05 23.20 Valves, Bronze		Crew	Daily Output	Labor-Hours	Unit	Material	2011 Bare Costs Labor	Equipment	Total	Total Incl O&P
0010	**VALVES, BRONZE**									
1020	Angle, 150 lb., rising stem, threaded									
1030	1/8″	1 Plum	24	.333	Ea.	102	13.20		115.20	134
1040	1/4″		24	.333		102	13.20		115.20	134
1050	3/8″		24	.333		102	13.20		115.20	134
1060	1/2″		22	.364		102	14.40		116.40	136
1070	3/4″		20	.400		139	15.85		154.85	179
1080	1″		19	.421		200	16.70		216.70	248
1100	1-1/2″		13	.615		335	24.50		359.50	410
1110	2″		11	.727		540	29		569	640
1380	Ball, 150 psi, threaded									
1400	1/4″	1 Plum	24	.333	Ea.	11.15	13.20		24.35	34
1430	3/8″		24	.333		11.15	13.20		24.35	34
1450	1/2″		22	.364		11.30	14.40		25.70	36
1460	3/4″		20	.400		18.60	15.85		34.45	46.50
1470	1″		19	.421		23.50	16.70		40.20	53.50
1480	1-1/4″		15	.533		39.50	21		60.50	78
1490	1-1/2″		13	.615		55	24.50		79.50	101
1500	2″		11	.727		66.50	29		95.50	120

22 05 23 – General-Duty Valves for Plumbing Piping

22 05 23.20 Valves, Bronze		Crew	Daily Output	Labor-Hours	Unit	Material	2011 Bare Costs Labor	Equipment	Total	Total Incl O&P
1750	Check, swing, class 150, regrinding disc, threaded									
1800	1/8"	1 Plum	24	.333	Ea.	47.50	13.20		60.70	73.50
1830	1/4"		24	.333		47.50	13.20		60.70	73.50
1840	3/8"		24	.333		47.50	13.20		60.70	73.50
1850	1/2"		24	.333		49.50	13.20		62.70	76
1860	3/4"		20	.400		72	15.85		87.85	105
1870	1"		19	.421		104	16.70		120.70	142
1880	1-1/4"		15	.533		149	21		170	199
1890	1-1/2"		13	.615		174	24.50		198.50	231
1900	2"		11	.727		256	29		285	330
1910	2-1/2"	Q-1	15	1.067		575	38		613	690
2000	For 200 lb, add					5%	10%			
2040	For 300 lb, add					15%	15%			
2850	Gate, N.R.S., soldered, 125 psi									
2900	3/8"	1 Plum	24	.333	Ea.	42	13.20		55.20	67.50
2920	1/2"		24	.333		35	13.20		48.20	60
2940	3/4"		20	.400		40.50	15.85		56.35	70.50
2950	1"		19	.421		57	16.70		73.70	90
2960	1-1/4"		15	.533		91	21		112	135
2970	1-1/2"		13	.615		97	24.50		121.50	147
2980	2"		11	.727		138	29		167	199
2990	2-1/2"	Q-1	15	1.067		345	38		383	440
3000	3"	"	13	1.231		450	44		494	565
3850	Rising stem, soldered, 300 psi									
3950	1"	1 Plum	19	.421	Ea.	136	16.70		152.70	178
3980	2"	"	11	.727		360	29		389	440
4000	3"	Q-1	13	1.231		1,150	44		1,194	1,325
4250	Threaded, class 150									
4310	1/4"	1 Plum	24	.333	Ea.	53.50	13.20		66.70	80
4320	3/8"		24	.333		53.50	13.20		66.70	80
4330	1/2"		24	.333		49.50	13.20		62.70	76
4340	3/4"		20	.400		57.50	15.85		73.35	89.50
4350	1"		19	.421		77	16.70		93.70	112
4360	1-1/4"		15	.533		104	21		125	150
4370	1-1/2"		13	.615		132	24.50		156.50	185
4380	2"		11	.727		174	29		203	239
4390	2-1/2"	Q-1	15	1.067		415	38		453	515
4400	3"	"	13	1.231		580	44		624	705
4500	For 300 psi, threaded, add					100%	15%			
4540	For chain operated type, add					15%				
4850	Globe, class 150, rising stem, threaded									
4920	1/4"	1 Plum	24	.333	Ea.	76	13.20		89.20	105
4940	3/8" size		24	.333		76	13.20		89.20	105
4950	1/2"		24	.333		76	13.20		89.20	105
4960	3/4"		20	.400		102	15.85		117.85	138
4970	1"		19	.421		160	16.70		176.70	204
4980	1-1/4"		15	.533		254	21		275	315
4990	1-1/2"		13	.615		310	24.50		334.50	380
5000	2"		11	.727		465	29		494	555
5010	2-1/2"	Q-1	15	1.067		935	38		973	1,075
5020	3"	"	13	1.231		1,325	44		1,369	1,525
5120	For 300 lb threaded, add					50%	15%			
5600	Relief, pressure & temperature, self-closing, ASME, threaded									

22 05 23 – General-Duty Valves for Plumbing Piping

22 05 23.20 Valves, Bronze		Crew	Daily Output	Labor-Hours	Unit	Material	2011 Bare Costs Labor	Equipment	Total	Total Incl O&P
5640	3/4"	1 Plum	28	.286	Ea.	132	11.35		143.35	165
5650	1"		24	.333		202	13.20		215.20	244
5660	1-1/4"		20	.400		390	15.85		405.85	450
5670	1-1/2"		18	.444		745	17.60		762.60	845
5680	2"		16	.500		790	19.85		809.85	905
5950	Pressure, poppet type, threaded									
6000	1/2"	1 Plum	30	.267	Ea.	35	10.55		45.55	56
6040	3/4"	"	28	.286	"	37.50	11.35		48.85	59.50
6400	Pressure, water, ASME, threaded									
6440	3/4"	1 Plum	28	.286	Ea.	74.50	11.35		85.85	100
6450	1"		24	.333		158	13.20		171.20	196
6460	1-1/4"		20	.400		251	15.85		266.85	300
6470	1-1/2"		18	.444		345	17.60		362.60	410
6480	2"		16	.500		500	19.85		519.85	585
6490	2-1/2"		15	.533		1,900	21		1,921	2,125
6900	Reducing, water pressure									
6920	300 psi to 25-75 psi, threaded or sweat									
6940	1/2"	1 Plum	24	.333	Ea.	239	13.20		252.20	285
6950	3/4"		20	.400		239	15.85		254.85	289
6960	1"		19	.421		370	16.70		386.70	440
6970	1-1/4"		15	.533		665	21		686	765
6980	1-1/2"		13	.615		1,000	24.50		1,024.50	1,150
8350	Tempering, water, sweat connections									
8400	1/2"	1 Plum	24	.333	Ea.	98	13.20		111.20	130
8440	3/4"	"	20	.400	"	114	15.85		129.85	151
8650	Threaded connections									
8700	1/2"	1 Plum	24	.333	Ea.	141	13.20		154.20	177
8740	3/4"		20	.400		445	15.85		460.85	510
8750	1"		19	.421		500	16.70		516.70	580
8760	1-1/4"		15	.533		770	21		791	880
8770	1-1/2"		13	.615		845	24.50		869.50	965
8780	2"		11	.727		1,275	29		1,304	1,450

22 05 23.60 Valves, Plastic		Crew	Daily Output	Labor-Hours	Unit	Material	Labor	Equipment	Total	Total Incl O&P
0010	**VALVES, PLASTIC**									
1100	Angle, PVC, threaded									
1110	1/4"	1 Plum	26	.308	Ea.	69.50	12.20		81.70	96.50
1120	1/2"		26	.308		69.50	12.20		81.70	96.50
1130	3/4"		25	.320		82.50	12.70		95.20	112
1140	1"		23	.348		100	13.80		113.80	133
1150	Ball, PVC, socket or threaded, single union									
1230	1/2"	1 Plum	26	.308	Ea.	26.50	12.20		38.70	49
1240	3/4"		25	.320		30	12.70		42.70	53.50
1250	1"		23	.348		38	13.80		51.80	64
1260	1-1/4"		21	.381		51	15.10		66.10	80.50
1270	1-1/2"		20	.400		60.50	15.85		76.35	92.50
1280	2"		17	.471		87.50	18.65		106.15	127
1360	For PVC, flanged, add					100%	15%			
1650	CPVC, socket or threaded, single union									
1700	1/2"	1 Plum	26	.308	Ea.	50	12.20		62.20	75
1720	3/4"		25	.320		62.50	12.70		75.20	89.50
1730	1"		23	.348		75	13.80		88.80	105
1750	1-1/4"		21	.381		120	15.10		135.10	156

22 05 23 – General-Duty Valves for Plumbing Piping

22 05 23.60 Valves, Plastic		Crew	Daily Output	Labor-Hours	Unit	Material	2011 Bare Costs Labor	Equipment	Total	Total Incl O&P
1760	1-1/2"	1 Plum	20	.400	Ea.	120	15.85		135.85	157
1840	For CPVC, flanged, add					65%	15%			
1880	For true union, socket or threaded, add				↓	50%	5%			
2050	Polypropylene, threaded									
2100	1/4"	1 Plum	26	.308	Ea.	38.50	12.20		50.70	62
2120	3/8"		26	.308		38.50	12.20		50.70	62
2130	1/2"		26	.308		38.50	12.20		50.70	62
2140	3/4"		25	.320		46.50	12.70		59.20	71.50
2150	1"		23	.348		53.50	13.80		67.30	81.50
2160	1-1/4"		21	.381		72	15.10		87.10	104
2170	1-1/2"		20	.400		88	15.85		103.85	123
2180	2"	↓	17	.471	↓	119	18.65		137.65	162
4850	Foot valve, PVC, socket or threaded									
4900	1/2"	1 Plum	34	.235	Ea.	66.50	9.35		75.85	88.50
4930	3/4"		32	.250		75	9.90		84.90	98.50
4940	1"		28	.286		97.50	11.35		108.85	126
4950	1-1/4"		27	.296		187	11.75		198.75	225
4960	1-1/2"	↓	26	.308	↓	187	12.20		199.20	226
6350	Y sediment strainer, PVC, socket or threaded									
6400	1/2"	1 Plum	26	.308	Ea.	44	12.20		56.20	68.50
6440	3/4"		24	.333		47	13.20		60.20	73.50
6450	1"		23	.348		57	13.80		70.80	85
6460	1-1/4"		21	.381		94.50	15.10		109.60	129
6470	1-1/2"	↓	20	.400	↓	94.50	15.85		110.35	130

22 05 48 – Vibration and Seismic Controls for Plumbing Piping and Equipment

22 05 48.10 Seismic Bracing Supports		Crew	Daily Output	Labor-Hours	Unit	Material	Labor	Equipment	Total	Total Incl O&P
0010	**SEISMIC BRACING SUPPORTS**									
0020	Clamps									
0030	C-clamp, for mounting on steel beam									
0040	3/8" threaded rod	1 Skwk	160	.050	Ea.	3.15	1.70		4.85	6.30
0050	1/2" threaded rod		160	.050		3.91	1.70		5.61	7.15
0060	5/8" threaded rod		160	.050		6	1.70		7.70	9.45
0070	3/4" threaded rod	↓	160	.050	↓	8.05	1.70		9.75	11.75
0100	Brackets									
0110	Beam side or wall mallable iron									
0120	3/8" threaded rod	1 Skwk	48	.167	Ea.	3.34	5.70		9.04	13.15
0130	1/2" threaded rod		48	.167		5.80	5.70		11.50	15.90
0140	5/8" threaded rod		48	.167		9.10	5.70		14.80	19.50
0150	3/4" threaded rod		48	.167		10.95	5.70		16.65	21.50
0160	7/8" threaded rod	↓	48	.167	↓	12.85	5.70		18.55	23.50
0170	For concrete installation, add						30%			
0180	Wall, welded steel									
0190	0 size 12" wide 18" deep	1 Skwk	34	.235	Ea.	234	8		242	270
0200	1 size 18" wide 24" deep		34	.235		278	8		286	320
0210	2 size 24" wide 30" deep	↓	34	.235	↓	365	8		373	420
0300	Rod, carbon steel									
0310	Continuous thread									
0320	1/4" thread	1 Skwk	144	.056	L.F.	1.35	1.89		3.24	4.66
0330	3/8" thread		144	.056		1.44	1.89		3.33	4.75
0340	1/2" thread		144	.056		2.26	1.89		4.15	5.65
0350	5/8" thread		144	.056		3.21	1.89		5.10	6.70
0360	3/4" thread	↓	144	.056	↓	5.65	1.89		7.54	9.40

22 05 48 – Vibration and Seismic Controls for Plumbing Piping and Equipment

22 05 48.10 Seismic Bracing Supports		Crew	Daily Output	Labor-Hours	Unit	Material	Labor	Equipment	Total	Total Incl O&P
						2011 Bare Costs				
0370	7/8" thread	1 Skwk	144	.056	L.F.	7.10	1.89		8.99	10.95
0380	For galvanized, add					30%				
0400	Channel, steel									
0410	3/4" x 1-1/2"	1 Skwk	80	.100	L.F.	6.10	3.41		9.51	12.40
0420	1-1/2" x 1-1/2"		70	.114		8.05	3.89		11.94	15.40
0430	1-7/8" x 1-1/2"		60	.133		26	4.54		30.54	36
0440	3" x 1-1/2"	↓	50	.160	↓	30	5.45		35.45	42
0450	Spring nuts									
0460	3/8"	1 Skwk	100	.080	Ea.	1.85	2.72		4.57	6.60
0470	1/2"	"	80	.100	"	2	3.41		5.41	7.90
0500	Welding, field									
0510	Cleaning and welding plates, bars, or rods									
0520	To existing beams, columns, or trusses									
0530	1" weld	1 Skwk	144	.056	Ea.	.20	1.89		2.09	3.39
0540	2" weld		72	.111		.35	3.78		4.13	6.75
0550	3" weld		54	.148		.54	5.05		5.59	9.05
0560	4" weld		36	.222		.75	7.55		8.30	13.50
0570	5" weld		30	.267		.94	9.10		10.04	16.25
0580	6" weld	↓	24	.333	↓	1.07	11.35		12.42	20
0600	Vibration absorbers									
0610	Hangers, neoprene flex									
0620	10-120 lb. capacity	1 Skwk	8	1	Ea.	22.50	34		56.50	82
0630	75-550 lb. capacity		8	1		30.50	34		64.50	90.50
0640	250-1100 lb. capacity		6	1.333		63	45.50		108.50	146
0650	1000-4000 lb. capacity	↓	6	1.333	↓	100	45.50		145.50	186

22 05 76 – Facility Drainage Piping Cleanouts

22 05 76.10 Cleanouts		Crew	Daily Output	Labor-Hours	Unit	Material	Labor	Equipment	Total	Total Incl O&P
0010	**CLEANOUTS**									
0060	Floor type									
0080	Round or square, scoriated nickel bronze top									
0100	2" pipe size	1 Plum	10	.800	Ea.	155	31.50		186.50	222
0120	3" pipe size		8	1		232	39.50		271.50	320
0140	4" pipe size	↓	6	1.333	↓	232	53		285	345
0980	Round top, recessed for terrazzo									
1000	2" pipe size	1 Plum	9	.889	Ea.	155	35		190	228
1080	3" pipe size		6	1.333		232	53		285	345
1100	4" pipe size	↓	4	2		232	79.50		311.50	385
1120	5" pipe size	Q-1	6	2.667	↓	295	95		390	480

22 05 76.20 Cleanout Tees		Crew	Daily Output	Labor-Hours	Unit	Material	Labor	Equipment	Total	Total Incl O&P
0010	**CLEANOUT TEES**									
0100	Cast iron, B&S, with countersunk plug									
0200	2" pipe size	1 Plum	4	2	Ea.	211	79.50		290.50	360
0220	3" pipe size		3.60	2.222		230	88		318	395
0240	4" pipe size	↓	3.30	2.424		286	96		382	470
0280	6" pipe size	Q-1	5	3.200	↓	770	114		884	1,025
0500	For round smooth access cover, same price									
4000	Plastic, tees and adapters. Add plugs									
4010	ABS, DWV									
4020	Cleanout tee, 1-1/2" pipe size	1 Plum	15	.533	Ea.	7.25	21		28.25	42.50
4030	2" pipe size	Q-1	27	.593		7.95	21		28.95	43.50
4040	3" pipe size		21	.762		14.90	27		41.90	61
4050	4" pipe size	↓	16	1	↓	32	35.50		67.50	93.50

22 05 Common Work Results for Plumbing

22 05 76 – Facility Drainage Piping Cleanouts

22 05 76.20 Cleanout Tees		Crew	Daily Output	Labor-Hours	Unit	Material	2011 Bare Costs Labor	Equipment	Total	Total Incl O&P
4100	Cleanout plug, 1-1/2" pipe size	1 Plum	32	.250	Ea.	1.26	9.90		11.16	17.60
4110	2" pipe size	Q-1	56	.286		1.68	10.20		11.88	18.50
4120	3" pipe size		36	.444		2.66	15.85		18.51	29
4130	4" pipe size		30	.533		4.65	19.05		23.70	36
4180	Cleanout adapter fitting, 1-1/2" pipe size	1 Plum	32	.250		1.98	9.90		11.88	18.40
4190	2" pipe size	Q-1	56	.286		3.04	10.20		13.24	20
4200	3" pipe size		36	.444		7.80	15.85		23.65	34.50
4210	4" pipe size		30	.533		14.35	19.05		33.40	47
5000	PVC, DWV									
5010	Cleanout tee, 1-1/2" pipe size	1 Plum	15	.533	Ea.	5.35	21		26.35	40.50
5020	2" pipe size	Q-1	27	.593		6.20	21		27.20	41.50
5030	3" pipe size		21	.762		12.05	27		39.05	58
5040	4" pipe size		16	1		21.50	35.50		57	82.50
5090	Cleanout plug, 1-1/2" pipe size	1 Plum	32	.250		1.27	9.90		11.17	17.60
5100	2" pipe size	Q-1	56	.286		1.41	10.20		11.61	18.20
5110	3" pipe size		36	.444		2.52	15.85		18.37	29
5120	4" pipe size		30	.533		3.72	19.05		22.77	35
5130	6" pipe size		24	.667		12.35	24		36.35	52.50
5170	Cleanout adapter fitting, 1-1/2" pipe size	1 Plum	32	.250		1.69	9.90		11.59	18.05
5180	2" pipe size	Q-1	56	.286		2.18	10.20		12.38	19.05
5190	3" pipe size		36	.444		6.15	15.85		22	33
5200	4" pipe size		30	.533		10.10	19.05		29.15	42
5210	6" pipe size		24	.667		29.50	24		53.50	71.50

22 07 Plumbing Insulation

22 07 19 – Plumbing Piping Insulation

22 07 19.10 Piping Insulation			Crew	Daily Output	Labor-Hours	Unit	Material	Labor	Equipment	Total	Total Incl O&P
0010	**PIPING INSULATION**										
0100	Rule of thumb, as a percentage of total mechanical costs					Job				10%	
0110	Insulation req'd is based on the surface size/area to be covered										
4000	Pipe covering (price copper tube one size less than IPS)										
6600	Fiberglass, with all service jacket										
6840	1" wall, 1/2" iron pipe size	G	Q-14	240	.067	L.F.	.95	2.08		3.03	4.56
6870	1" iron pipe size	G		220	.073		1.12	2.27		3.39	5.05
6900	2" iron pipe size	G		200	.080		1.41	2.50		3.91	5.75
6920	3" iron pipe size	G		180	.089		1.72	2.78		4.50	6.55
6940	4" iron pipe size	G		150	.107		2.27	3.33		5.60	8.10
7320	2" wall, 1/2" iron pipe size	G		220	.073		2.81	2.27		5.08	6.90
7440	6" iron pipe size	G		100	.160		5.65	5		10.65	14.60
7460	8" iron pipe size	G		80	.200		6.85	6.25		13.10	18.05
7480	10" iron pipe size	G		70	.229		8.20	7.15		15.35	21
7490	12" iron pipe size	G		65	.246		9.20	7.70		16.90	23
7800	For fiberglass with standard canvas jacket, deduct						5%				
7802	For fittings, add 3 L.F. for each fitting										
7804	plus 4 L.F. for each flange of the fitting										
7810	Finishes										
7811	Finishes, for .010" aluminum jacket, add	G	Q-14	200	.080	S.F.	.77	2.50		3.27	5.05
7812	For .016" aluminum jacket, add	G		200	.080		1.12	2.50		3.62	5.45
7813	For .010" stainless steel, add	G		160	.100		3.11	3.12		6.23	8.65
7814	For single layer of felt, add						10%	10%			
7816	For roofing paper, 45 lb to 55 lb, add						25%	10%			

22 07 Plumbing Insulation

22 07 19 – Plumbing Piping Insulation

22 07 19.10 Piping Insulation			Crew	Daily Output	Labor-Hours	Unit	Material	2011 Bare Costs Labor	Equipment	Total	Total Incl O&P
7879	Rubber tubing, flexible closed cell foam										
7880	3/8" wall, 1/4" iron pipe size	G	1 Asbe	120	.067	L.F.	.55	2.31		2.86	4.51
7910	1/2" iron pipe size	G		115	.070		.70	2.41		3.11	4.84
7920	3/4" iron pipe size	G		115	.070		.85	2.41		3.26	5
7930	1" iron pipe size	G		110	.073		.98	2.52		3.50	5.35
7950	1-1/2" iron pipe size	G		110	.073		1.30	2.52		3.82	5.70
8100	1/2" wall, 1/4" iron pipe size	G		90	.089		.50	3.08		3.58	5.75
8130	1/2" iron pipe size	G		89	.090		.61	3.12		3.73	5.90
8140	3/4" iron pipe size	G		89	.090		.68	3.12		3.80	6
8150	1" iron pipe size	G		88	.091		.75	3.15		3.90	6.15
8170	1-1/2" iron pipe size	G		87	.092		1.06	3.19		4.25	6.55
8180	2" iron pipe size	G		86	.093		1.38	3.23		4.61	6.95
8200	3" iron pipe size	G		85	.094		1.95	3.27		5.22	7.65
8300	3/4" wall, 1/4" iron pipe size	G		90	.089		.77	3.08		3.85	6.05
8330	1/2" iron pipe size	G		89	.090		1.01	3.12		4.13	6.35
8340	3/4" iron pipe size	G		89	.090		1.23	3.12		4.35	6.60
8350	1" iron pipe size	G		88	.091		1.41	3.15		4.56	6.85
8370	1-1/2" iron pipe size	G		87	.092		2.15	3.19		5.34	7.75
8380	2" iron pipe size	G		86	.093		2.52	3.23		5.75	8.20
8400	3" iron pipe size	G		85	.094		3.85	3.27		7.12	9.75
8444	1" wall, 1/2" iron pipe size	G		86	.093		1.93	3.23		5.16	7.55
8445	3/4" iron pipe size	G		84	.095		2.34	3.30		5.64	8.10
8446	1" iron pipe size	G		84	.095		2.72	3.30		6.02	8.55
8447	1-1/4" iron pipe size	G		82	.098		3.08	3.39		6.47	9.10
8448	1-1/2" iron pipe size	G		82	.098		3.58	3.39		6.97	9.65
8449	2" iron pipe size	G		80	.100		4.78	3.47		8.25	11.10
8450	2-1/2" iron pipe size	G	↓	80	.100	↓	6.25	3.47		9.72	12.70
8456	Rubber insulation tape, 1/8" x 2" x 30'	G				Ea.	11.40			11.40	12.50

22 11 Facility Water Distribution

22 11 13 – Facility Water Distribution Piping

22 11 13.14 Pipe, Brass		Crew	Daily Output	Labor-Hours	Unit	Material	Labor	Equipment	Total	Total Incl O&P
0010	**PIPE, BRASS**, Plain end									
0900	Field threaded, coupling & clevis hanger assembly 10' O.C.									
0920	Regular weight									
1120	1/2" diameter	1 Plum	48	.167	L.F.	6.50	6.60		13.10	17.95
1140	3/4" diameter		46	.174		8.60	6.90		15.50	20.50
1160	1" diameter	↓	43	.186		12.50	7.40		19.90	26
1180	1-1/4" diameter	Q-1	72	.222		18.90	7.95		26.85	34
1200	1-1/2" diameter		65	.246		22.50	8.80		31.30	39
1220	2" diameter	↓	53	.302	↓	31	10.75		41.75	51.50

22 11 13.23 Pipe/Tube, Copper		Crew	Daily Output	Labor-Hours	Unit	Material	Labor	Equipment	Total	Total Incl O&P
0010	**PIPE/TUBE, COPPER**, Solder joints									
1000	Type K tubing, couplings & clevis hanger assemblies 10' O.C.									
1100	1/4" diameter	1 Plum	84	.095	L.F.	4.41	3.78		8.19	11
1200	1" diameter		66	.121		11.80	4.81		16.61	21
1260	2" diameter	↓	40	.200	↓	28.50	7.95		36.45	44.50
2000	Type L tubing, couplings & clevis hanger assemblies 10' O.C.									
2100	1/4" diameter	1 Plum	88	.091	L.F.	2.71	3.60		6.31	8.90
2120	3/8" diameter		84	.095		3.65	3.78		7.43	10.15
2140	1/2" diameter	↓	81	.099	↓	3.97	3.92		7.89	10.75

22 11 13.23 Pipe/Tube, Copper

	22 11 13.23 Pipe/Tube, Copper	Crew	Daily Output	Labor-Hours	Unit	2011 Bare Costs Material	Labor	Equipment	Total	Total Incl O&P
2160	5/8" diameter	1 Plum	79	.101	L.F.	5.85	4.02		9.87	13
2180	3/4" diameter		76	.105		5.95	4.17		10.12	13.35
2200	1" diameter		68	.118		9	4.66		13.66	17.55
2220	1-1/4" diameter		58	.138		12.45	5.45		17.90	22.50
2240	1-1/2" diameter		52	.154		15.70	6.10		21.80	27.50
2260	2" diameter		42	.190		24	7.55		31.55	39
2280	2-1/2" diameter	Q-1	62	.258		37	9.20		46.20	56
2300	3" diameter		56	.286		50	10.20		60.20	71.50
2320	3-1/2" diameter		43	.372		69.50	13.30		82.80	98
2340	4" diameter		39	.410		85.50	14.65		100.15	118
2360	5" diameter		34	.471		183	16.80		199.80	229
2380	6" diameter	Q-2	40	.600		258	20.50		278.50	315
2400	8" diameter	"	36	.667		430	23		453	510
2410	For other than full hard temper, add					21%				
2590	For silver solder, add						15%			
4000	Type DWV tubing, couplings & clevis hanger assemblies 10' O.C.									
4100	1-1/4" diameter	1 Plum	60	.133	L.F.	11.05	5.30		16.35	21
4120	1-1/2" diameter		54	.148		13.65	5.85		19.50	24.50
4140	2" diameter		44	.182		18.30	7.20		25.50	32
4160	3" diameter	Q-1	58	.276		33.50	9.85		43.35	53
4180	4" diameter		40	.400		59.50	14.25		73.75	89
4200	5" diameter		36	.444		167	15.85		182.85	209
4220	6" diameter	Q-2	42	.571		239	19.65		258.65	295

22 11 13.44 Pipe, Steel

	22 11 13.44 Pipe, Steel	Crew	Daily Output	Labor-Hours	Unit	Material	Labor	Equipment	Total	Total Incl O&P
0010	**PIPE, STEEL** R221113-50									
0020	All pipe sizes are to Spec. A-53 unless noted otherwise									
0050	Schedule 40, threaded, with couplings, and clevis hanger									
0060	assemblies sized for covering, 10' O.C.									
0540	Black, 1/4" diameter	1 Plum	66	.121	L.F.	6.50	4.81		11.31	15
0550	3/8" diameter		65	.123		7.25	4.88		12.13	15.95
0560	1/2" diameter		63	.127		3.80	5.05		8.85	12.40
0570	3/4" diameter		61	.131		4.37	5.20		9.57	13.30
0580	1" diameter		53	.151		6.15	6		12.15	16.50
0590	1-1/4" diameter	Q-1	89	.180		7.55	6.40		13.95	18.75
0600	1-1/2" diameter		80	.200		8.70	7.15		15.85	21.50
0610	2" diameter		64	.250		11.45	8.90		20.35	27
0620	2-1/2" diameter		50	.320		17.30	11.40		28.70	37.50
0630	3" diameter		43	.372		22	13.30		35.30	45.50
0640	3-1/2" diameter		40	.400		30	14.25		44.25	56
0650	4" diameter		36	.444		32.50	15.85		48.35	61.50
1280	All pipe sizes are to Spec. A-53 unless noted otherwise									
1281	Schedule 40, threaded, with couplings and clevis hanger									
1282	assemblies sized for covering, 10' O. C.									
1290	Galvanized, 1/4" diameter	1 Plum	66	.121	L.F.	8.60	4.81		13.41	17.35
1300	3/8" diameter		65	.123		9.45	4.88		14.33	18.35
1310	1/2" diameter		63	.127		5.15	5.05		10.20	13.85
1320	3/4" diameter		61	.131		10.25	5.20		15.45	19.80
1330	1" diameter		53	.151		8.10	6		14.10	18.65
1340	1-1/4" diameter	Q-1	89	.180		10.15	6.40		16.55	21.50
1350	1-1/2" diameter		80	.200		11.80	7.15		18.95	24.50
1360	2" diameter		64	.250		15.60	8.90		24.50	31.50
1370	2-1/2" diameter		50	.320		25	11.40		36.40	46.50

22 11 13.44 Pipe, Steel		Crew	Daily Output	Labor-Hours	Unit	Material	2011 Bare Costs Labor	Equipment	Total	Total Incl O&P
1380	3″ diameter	Q-1	43	.372	L.F.	32	13.30		45.30	56.50
1390	3-1/2″ diameter		40	.400		40.50	14.25		54.75	68
1400	4″ diameter		36	.444		46	15.85		61.85	76.50
2000	Welded, sch. 40, on yoke & roll hanger assy's, sized for covering, 10′ O.C.									
2040	Black, 1″ diameter	Q-15	93	.172	L.F.	4.64	6.15	.35	11.14	15.50
2070	2″ diameter		61	.262		7.20	9.35	.53	17.08	24
2090	3″ diameter		43	.372		11.40	13.30	.76	25.46	35
2110	4″ diameter		37	.432		15.80	15.45	.88	32.13	43.50
2120	5″ diameter		32	.500		21	17.85	1.02	39.87	53
2130	6″ diameter	Q-16	36	.667		25	24.50	.90	50.40	69
2140	8″ diameter		29	.828		36.50	30.50	1.12	68.12	91
2150	10″ diameter		24	1		51	37	1.35	89.35	118
2160	12″ diameter		19	1.263		87.50	46.50	1.71	135.71	174

22 11 13.48 Pipe, Fittings and Valves, Steel, Grooved-Joint

		Crew	Daily Output	Labor-Hours	Unit	Material	Labor	Equipment	Total	Total Incl O&P
0010	**PIPE, FITTINGS AND VALVES, STEEL, GROOVED-JOINT**									
0012	Fittings are ductile iron. Steel fittings noted.									
0020	Pipe includes coupling & clevis type hanger assemblies, 10′ O.C.									
1000	Schedule 40, black									
1040	3/4″ diameter	1 Plum	71	.113	L.F.	5.40	4.47		9.87	13.25
1050	1″ diameter		63	.127		4.09	5.05		9.14	12.70
1060	1-1/4″ diameter		58	.138		5.10	5.45		10.55	14.60
1070	1-1/2″ diameter		51	.157		5.75	6.20		11.95	16.45
1080	2″ diameter		40	.200		6.60	7.95		14.55	20
1090	2-1/2″ diameter	Q-1	57	.281		9.25	10		19.25	26.50
1100	3″ diameter		50	.320		11.30	11.40		22.70	31
1110	4″ diameter		45	.356		15.75	12.70		28.45	38
1120	5″ diameter		37	.432		21.50	15.45		36.95	48.50
1130	6″ diameter	Q-2	42	.571		27.50	19.65		47.15	62
1800	Galvanized									
1840	3/4″ diameter	1 Plum	71	.113	L.F.	11.25	4.47		15.72	19.70
1850	1″ diameter		63	.127		8.05	5.05		13.10	17.05
1860	1-1/4″ diameter		58	.138		10.15	5.45		15.60	20
1870	1-1/2″ diameter		51	.157		11.70	6.20		17.90	23
1880	2″ diameter		40	.200		14.85	7.95		22.80	29.50
1890	2-1/2″ diameter	Q-1	57	.281		22	10		32	41
1900	3″ diameter		50	.320		28	11.40		39.40	49.50
1910	4″ diameter		45	.356		39	12.70		51.70	63.50
1920	5″ diameter		37	.432		46	15.45		61.45	75.50
1930	6″ diameter	Q-2	42	.571		51.50	19.65		71.15	88.50
3990	Fittings: coupling material required at joints not incl. in fitting price.									
3994	Add 1 selected coupling, material only, per joint for installed price.									
4000	Elbow, 90° or 45°, painted									
4030	3/4″ diameter	1 Plum	50	.160	Ea.	43	6.35		49.35	58
4040	1″ diameter		50	.160		23	6.35		29.35	35.50
4050	1-1/4″ diameter		40	.200		23	7.95		30.95	38
4060	1-1/2″ diameter		33	.242		23	9.60		32.60	40.50
4070	2″ diameter		25	.320		23	12.70		35.70	45.50
4080	2-1/2″ diameter	Q-1	40	.400		23	14.25		37.25	48.50
4090	3″ diameter		33	.485		40.50	17.30		57.80	72.50
4100	4″ diameter		25	.640		44	23		67	86
4110	5″ diameter		20	.800		105	28.50		133.50	163
4120	6″ diameter	Q-2	25	.960		124	33		157	190

22 11 13.48 Pipe, Fittings and Valves, Steel, Grooved-Joint

	22 11 13.48 Pipe, Fittings and Valves, Steel, Grooved-Joint	Crew	Daily Output	Labor-Hours	Unit	2011 Bare Costs Material	Labor	Equipment	Total	Total Incl O&P
4250	For galvanized elbows, add				Ea.	26%				
4690	Tee, painted									
4700	3/4" diameter	1 Plum	38	.211	Ea.	46	8.35		54.35	64.50
4740	1" diameter		33	.242		35.50	9.60		45.10	54.50
4750	1-1/4" diameter		27	.296		35.50	11.75		47.25	58
4760	1-1/2" diameter		22	.364		35.50	14.40		49.90	62.50
4770	2" diameter		17	.471		35.50	18.65		54.15	69.50
4780	2-1/2" diameter	Q-1	27	.593		35.50	21		56.50	73.50
4790	3" diameter		22	.727		49	26		75	96
4800	4" diameter		17	.941		74.50	33.50		108	137
4810	5" diameter		13	1.231		174	44		218	263
4820	6" diameter	Q-2	17	1.412		201	48.50		249.50	300
4900	For galvanized tees, add					24%				
4906	Couplings, rigid style, painted									
4908	1" diameter	1 Plum	100	.080	Ea.	17.20	3.17		20.37	24
4909	1-1/4" diameter		100	.080		17.20	3.17		20.37	24
4910	1-1/2" diameter		67	.119		17.20	4.73		21.93	26.50
4912	2" diameter		50	.160		18.80	6.35		25.15	31
4914	2-1/2" diameter	Q-1	80	.200		22	7.15		29.15	35.50
4916	3" diameter		67	.239		25.50	8.50		34	42
4918	4" diameter		50	.320		35.50	11.40		46.90	57.50
4920	5" diameter		40	.400		46	14.25		60.25	74.50
4922	6" diameter	Q-2	50	.480		61	16.50		77.50	94
4924	8" diameter		42	.571		96	19.65		115.65	138
4926	10" diameter		35	.686		159	23.50		182.50	213
4928	12" diameter		32	.750		177	26		203	237
4930	14" diameter		24	1		204	34.50		238.50	281
4931	16" diameter		20	1.200		266	41		307	360
4932	18" diameter		18	1.333		310	46		356	415
4933	20" diameter		16	1.500		420	51.50		471.50	545
4934	24" diameter		13	1.846		540	63.50		603.50	700
4940	Flexible, standard, painted									
4950	3/4" diameter	1 Plum	100	.080	Ea.	12.40	3.17		15.57	18.85
4960	1" diameter		100	.080		12.40	3.17		15.57	18.85
4970	1-1/4" diameter		80	.100		16.40	3.97		20.37	24.50
4980	1-1/2" diameter		67	.119		18	4.73		22.73	27.50
4990	2" diameter		50	.160		19	6.35		25.35	31.50
5000	2-1/2" diameter	Q-1	80	.200		22	7.15		29.15	36
5010	3" diameter		67	.239		25	8.50		33.50	41.50
5020	3-1/2" diameter		57	.281		36	10		46	56
5030	4" diameter		50	.320		36	11.40		47.40	58
5040	5" diameter		40	.400		54.50	14.25		68.75	83.50
5050	6" diameter	Q-2	50	.480		64	16.50		80.50	97.50

22 11 13.64 Pipe, Stainless Steel

	22 11 13.64 Pipe, Stainless Steel	Crew	Daily Output	Labor-Hours	Unit	Material	Labor	Equipment	Total	Total Incl O&P
0010	**PIPE, STAINLESS STEEL**									
3500	Threaded, couplings and clevis hanger assemblies, 10' O.C.									
3520	Schedule 40, type 304									
3540	1/4" diameter	1 Plum	54	.148	L.F.	6.45	5.85		12.30	16.70
3550	3/8" diameter		53	.151		8.15	6		14.15	18.70
3560	1/2" diameter		52	.154		9.60	6.10		15.70	20.50
3580	1" diameter		45	.178		15.65	7.05		22.70	28.50
3610	2" diameter	Q-1	57	.281		34	10		44	54

22 11 13 – Facility Water Distribution Piping

22 11 13.64 Pipe, Stainless Steel		Crew	Daily Output	Labor-Hours	Unit	Material	2011 Bare Costs Labor	Equipment	Total	Total Incl O&P
3640	4" diameter	Q-2	51	.471	L.F.	97.50	16.15		113.65	134
3740	For small quantities, add				↓	10%				
4250	Schedule 40, type 316									
4290	1/4" diameter	1 Plum	54	.148	L.F.	7.75	5.85		13.60	18.10
4300	3/8" diameter		53	.151		8.75	6		14.75	19.35
4310	1/2" diameter		52	.154		11.65	6.10		17.75	23
4320	3/4" diameter		51	.157		14.20	6.20		20.40	26
4330	1" diameter	↓	45	.178		20	7.05		27.05	33.50
4360	2" diameter	Q-1	57	.281		44.50	10		54.50	65
4390	4" diameter	Q-2	51	.471		145	16.15		161.15	186
4490	For small quantities, add				↓	10%				

22 11 13.74 Pipe, Plastic		Crew	Daily Output	Labor-Hours	Unit	Material	Labor	Equipment	Total	Total Incl O&P
0010	**PIPE, PLASTIC**									
0020	Fiberglass reinforced, couplings 10' O.C., clevis hanger assy's, 3 per 10'									
1800	PVC, couplings 10' O.C., clevis hanger assemblies, 3 per 10'									
1820	Schedule 40									
1860	1/2" diameter	1 Plum	54	.148	L.F.	3.53	5.85		9.38	13.50
1870	3/4" diameter		51	.157		3.81	6.20		10.01	14.35
1880	1" diameter		46	.174		4.89	6.90		11.79	16.65
1890	1-1/4" diameter		42	.190		5.70	7.55		13.25	18.60
1900	1-1/2" diameter	↓	36	.222		5.75	8.80		14.55	20.50
1910	2" diameter	Q-1	59	.271		6.75	9.70		16.45	23.50
1920	2-1/2" diameter		56	.286		9.05	10.20		19.25	26.50
1930	3" diameter		53	.302		11.10	10.75		21.85	30
1940	4" diameter		48	.333		14.20	11.90		26.10	35
1950	5" diameter		43	.372		19.20	13.30		32.50	42.50
1960	6" diameter	↓	39	.410	↓	24.50	14.65		39.15	51
4100	DWV type, schedule 40, couplings 10' O.C., clevis hanger assy's, 3 per 10'									
4210	ABS, schedule 40, foam core type									
4212	Plain end black									
4214	1-1/2" diameter	1 Plum	39	.205	L.F.	4.48	8.15		12.63	18.20
4216	2" diameter	Q-1	62	.258		4.93	9.20		14.13	20.50
4218	3" diameter		56	.286		7.45	10.20		17.65	25
4220	4" diameter		51	.314		9.70	11.20		20.90	29
4222	6" diameter	↓	42	.381	↓	17.85	13.60		31.45	41.50
4240	To delete coupling & hangers, subtract									
4244	1-1/2" diam. to 6" diam.					43%	48%			
4400	PVC									
4410	1-1/4" diameter	1 Plum	42	.190	L.F.	4.65	7.55		12.20	17.45
4420	1-1/2" diameter	"	36	.222		4.53	8.80		13.33	19.40
4460	2" diameter	Q-1	59	.271		4.97	9.70		14.67	21.50
4470	3" diameter		53	.302		7.50	10.75		18.25	26
4480	4" diameter		48	.333		9.50	11.90		21.40	30
4490	6" diameter	↓	39	.410	↓	16.10	14.65		30.75	41.50
5300	CPVC, socket joint, couplings 10' O.C., clevis hanger assemblies, 3 per 10'									
5302	Schedule 40									
5304	1/2" diameter	1 Plum	54	.148	L.F.	4.22	5.85		10.07	14.25
5305	3/4" diameter		51	.157		4.75	6.20		10.95	15.40
5306	1" diameter		46	.174		6.45	6.90		13.35	18.30
5307	1-1/4" diameter		42	.190		7.65	7.55		15.20	21
5308	1-1/2" diameter	↓	36	.222		8.55	8.80		17.35	24
5309	2" diameter	Q-1	59	.271	↓	10.05	9.70		19.75	27

22 11 Facility Water Distribution

22 11 13 – Facility Water Distribution Piping

22 11 13.74 Pipe, Plastic		Crew	Daily Output	Labor-Hours	Unit	2011 Bare Costs Material	Labor	Equipment	Total	Total Incl O&P
5310	2-1/2" diameter	Q-1	56	.286	L.F.	15.85	10.20		26.05	34
5311	3" diameter		53	.302		18.65	10.75		29.40	38
5360	CPVC, threaded, couplings 10' O.C., clevis hanger assemblies, 3 per 10'									
5380	Schedule 40									
5460	1/2" diameter	1 Plum	54	.148	L.F.	4.93	5.85		10.78	15.05
5470	3/4" diameter		51	.157		6	6.20		12.20	16.75
5480	1" diameter		46	.174		7.70	6.90		14.60	19.75
5490	1-1/4" diameter		42	.190		8.65	7.55		16.20	22
5500	1-1/2" diameter		36	.222		9.40	8.80		18.20	25
5510	2" diameter	Q-1	59	.271		11.15	9.70		20.85	28
5520	2-1/2" diameter		56	.286		16.95	10.20		27.15	35.50
5530	3" diameter		53	.302		20.50	10.75		31.25	40

22 11 19 – Domestic Water Piping Specialties

22 11 19.10 Flexible Connectors

		Crew	Daily Output	Labor-Hours	Unit	Material	Labor	Equipment	Total	Total Incl O&P
0010	**FLEXIBLE CONNECTORS**, Corrugated, 7/8" O.D., 1/2" I.D.									
0050	Gas, seamless brass, steel fittings									
0200	12" long	1 Plum	36	.222	Ea.	15.10	8.80		23.90	31
0220	18" long		36	.222		18.70	8.80		27.50	35
0240	24" long		34	.235		22	9.35		31.35	40
0280	36" long		32	.250		26.50	9.90		36.40	45
0340	60" long		30	.267		40	10.55		50.55	61.50
2000	Water, copper tubing, dielectric separators									
2100	12" long	1 Plum	36	.222	Ea.	14	8.80		22.80	30
2260	24" long	"	34	.235	"	20.50	9.35		29.85	38

22 11 19.14 Flexible Metal Hose

		Crew	Daily Output	Labor-Hours	Unit	Material	Labor	Equipment	Total	Total Incl O&P
0010	**FLEXIBLE METAL HOSE**, Connectors, standard lengths									
0100	Bronze braided, bronze ends									
0120	3/8" diameter x 12"	1 Stpi	26	.308	Ea.	18.85	12.30		31.15	41
0160	3/4" diameter x 12"		20	.400		26.50	16		42.50	55
0180	1" diameter x 18"		19	.421		34	16.85		50.85	65
0200	1-1/2" diameter x 18"		13	.615		53.50	24.50		78	99.50
0220	2" diameter x 18"		11	.727		64.50	29		93.50	119

22 11 19.26 Pressure Regulators

		Crew	Daily Output	Labor-Hours	Unit	Material	Labor	Equipment	Total	Total Incl O&P
0010	**PRESSURE REGULATORS**									
3000	Steam, high capacity, bronze body, stainless steel trim									
3020	Threaded, 1/2" diameter	1 Stpi	24	.333	Ea.	1,450	13.35		1,463.35	1,625
3030	3/4" diameter		24	.333		1,450	13.35		1,463.35	1,625
3040	1" diameter		19	.421		1,625	16.85		1,641.85	1,800
3060	1-1/4" diameter		15	.533		1,800	21.50		1,821.50	2,000
3080	1-1/2" diameter		13	.615		2,050	24.50		2,074.50	2,300
3100	2" diameter		11	.727		2,500	29		2,529	2,800
3120	2-1/2" diameter	Q-5	12	1.333		3,125	48		3,173	3,525
3140	3" diameter	"	11	1.455		3,575	52.50		3,627.50	4,025
3500	Flanged connection, iron body, 125 lb. W.S.P.									
3520	3" diameter	Q-5	11	1.455	Ea.	3,925	52.50		3,977.50	4,400
3540	4" diameter	"	5	3.200	"	4,950	115		5,065	5,650

22 11 19.38 Water Supply Meters

		Crew	Daily Output	Labor-Hours	Unit	Material	Labor	Equipment	Total	Total Incl O&P
0010	**WATER SUPPLY METERS**									
2000	Domestic/commercial, bronze									
2020	Threaded									
2060	5/8" diameter, to 20 GPM	1 Plum	16	.500	Ea.	43.50	19.85		63.35	80.50

22 11 19 – Domestic Water Piping Specialties

22 11 19.38 Water Supply Meters		Crew	Daily Output	Labor-Hours	Unit	2011 Bare Costs Material	Labor	Equipment	Total	Total Incl O&P
2080	3/4" diameter, to 30 GPM	1 Plum	14	.571	Ea.	79.50	22.50		102	125
2100	1" diameter, to 50 GPM		12	.667		121	26.50		147.50	176
2300	Threaded/flanged									
2340	1-1/2" diameter, to 100 GPM	1 Plum	8	1	Ea.	295	39.50		334.50	390
2360	2" diameter, to 160 GPM	"	6	1.333	"	400	53		453	525
2600	Flanged, compound									
2640	3" diameter, 320 GPM	Q-1	3	5.333	Ea.	2,725	190		2,915	3,300
2660	4" diameter, to 500 GPM		1.50	10.667		4,350	380		4,730	5,425
2680	6" diameter, to 1,000 GPM		1	16		6,950	570		7,520	8,575
2700	8" diameter, to 1,800 GPM		.80	20		10,900	715		11,615	13,200

22 11 19.42 Backflow Preventers		Crew	Daily Output	Labor-Hours	Unit	Material	Labor	Equipment	Total	Total Incl O&P
0010	**BACKFLOW PREVENTERS**, Includes valves									
0020	and four test cocks, corrosion resistant, automatic operation									
4000	Reduced pressure principle									
4100	Threaded, bronze, valves are ball									
4120	3/4" pipe size	1 Plum	16	.500	Ea.	345	19.85		364.85	415
4140	1" pipe size		14	.571		375	22.50		397.50	445
4150	1-1/4" pipe size		12	.667		640	26.50		666.50	750
4160	1-1/2" pipe size		10	.800		700	31.50		731.50	825
4180	2" pipe size		7	1.143		615	45.50		660.50	750
5000	Flanged, bronze, valves are OS&Y									
5060	2-1/2" pipe size	Q-1	5	3.200	Ea.	5,600	114		5,714	6,325
5080	3" pipe size		4.50	3.556		6,525	127		6,652	7,375
5100	4" pipe size		3	5.333		7,650	190		7,840	8,725
5120	6" pipe size	Q-2	3	8		12,200	275		12,475	13,900
5600	Flanged, iron, valves are OS&Y									
5660	2-1/2" pipe size	Q-1	5	3.200	Ea.	2,625	114		2,739	3,075
5680	3" pipe size		4.50	3.556		2,775	127		2,902	3,250
5700	4" pipe size		3	5.333		3,475	190		3,665	4,125
5720	6" pipe size	Q-2	3	8		5,025	275		5,300	5,975
5740	8" pipe size		2	12		10,300	410		10,710	12,000
5760	10" pipe size		1	24		11,800	825		12,625	14,400

22 11 19.50 Vacuum Breakers		Crew	Daily Output	Labor-Hours	Unit	Material	Labor	Equipment	Total	Total Incl O&P
0010	**VACUUM BREAKERS**									
0013	See also backflow preventers Section 22 11 19.42									
1000	Anti-siphon continuous pressure type									
1010	Max. 150 PSI - 210°F									
1020	Bronze body									
1030	1/2" size	1 Stpi	24	.333	Ea.	160	13.35		173.35	198
1040	3/4" size		20	.400		160	16		176	202
1050	1" size		19	.421		165	16.85		181.85	210
1060	1-1/4" size		15	.533		325	21.50		346.50	395
1070	1-1/2" size		13	.615		400	24.50		424.50	480
1080	2" size		11	.727		415	29		444	505
1200	Max. 125 PSI with atmospheric vent									
1210	Brass, in-line construction									
1220	1/4" size	1 Stpi	24	.333	Ea.	63	13.35		76.35	91.50
1230	3/8" size	"	24	.333		63	13.35		76.35	91.50
1260	For polished chrome finish, add					13%				
2000	Anti-siphon, non-continuous pressure type									
2010	Hot or cold water 125 PSI - 210°F									
2020	Bronze body									

22 11 19 – Domestic Water Piping Specialties

22 11 19.50 Vacuum Breakers

Line	22 11 19.50 Vacuum Breakers	Crew	Daily Output	Labor-Hours	Unit	2011 Bare Costs Material	Labor	Equipment	Total	Total Incl O&P
2030	1/4" size	1 Stpi	24	.333	Ea.	40.50	13.35		53.85	66.50
2040	3/8" size		24	.333		40.50	13.35		53.85	66.50
2050	1/2" size		24	.333		46	13.35		59.35	72.50
2060	3/4" size		20	.400		54	16		70	85.50
2070	1" size		19	.421		84.50	16.85		101.35	121
2080	1-1/4" size		15	.533		148	21.50		169.50	198
2090	1-1/2" size		13	.615		174	24.50		198.50	232
2100	2" size		11	.727		271	29		300	345
2110	2-1/2" size		8	1		780	40		820	920
2120	3" size		6	1.333		1,025	53.50		1,078.50	1,225
2150	For polished chrome finish, add					50%				

22 11 19.54 Water Hammer Arresters/Shock Absorbers

Line	22 11 19.54 Water Hammer Arresters/Shock Absorbers	Crew	Daily Output	Labor-Hours	Unit	Material	Labor	Equipment	Total	Total Incl O&P
0010	**WATER HAMMER ARRESTERS/SHOCK ABSORBERS**									
0490	Copper									
0500	3/4" male I.P.S. For 1 to 11 fixtures	1 Plum	12	.667	Ea.	20.50	26.50		47	65.50
0600	1" male I.P.S. For 12 to 32 fixtures		8	1		41.50	39.50		81	111
0700	1-1/4" male I.P.S. For 33 to 60 fixtures		8	1		44	39.50		83.50	114
0800	1-1/2" male I.P.S. For 61 to 113 fixtures		8	1		63	39.50		102.50	135
0900	2" male I.P.S. For 114 to 154 fixtures		8	1		92	39.50		131.50	166
1000	2-1/2" male I.P.S. For 155 to 330 fixtures		4	2		285	79.50		364.50	445

22 11 19.64 Hydrants

Line	22 11 19.64 Hydrants	Crew	Daily Output	Labor-Hours	Unit	Material	Labor	Equipment	Total	Total Incl O&P
0010	**HYDRANTS**									
0050	Wall type, moderate climate, bronze, encased									
0200	3/4" IPS connection	1 Plum	16	.500	Ea.	600	19.85		619.85	695
0300	1" IPS connection	"	14	.571		690	22.50		712.50	795
0500	Anti-siphon type, 3/4" connection					520			520	570
1000	Non-freeze, bronze, exposed									
1100	3/4" IPS connection, 4" to 9" thick wall	1 Plum	14	.571	Ea.	405	22.50		427.50	480
1120	10" to 14" thick wall		12	.667		520	26.50		546.50	615
1140	15" to 19" thick wall		12	.667		520	26.50		546.50	615
1160	20" to 24" thick wall		10	.800		530	31.50		561.50	635
1200	For 1" IPS connection, add					15%	10%			
1240	For 3/4" adapter type vacuum breaker, add				Ea.	51			51	56
1280	For anti-siphon type, add				"	106			106	117
2000	Non-freeze bronze, encased, anti-siphon type									
2100	3/4" IPS connection, 5" to 9" thick wall	1 Plum	14	.571	Ea.	1,025	22.50		1,047.50	1,150
2140	15" to 19" thick wall	"	12	.667	"	1,100	26.50		1,126.50	1,275
3000	Ground box type, bronze frame, 3/4" IPS connection									
3080	Non-freeze, all bronze, polished face, set flush									
3100	2 feet depth of bury	1 Plum	8	1	Ea.	765	39.50		804.50	910
3180	6 feet depth of bury		7	1.143		1,000	45.50		1,045.50	1,175
3220	8 feet depth of bury		5	1.600		1,125	63.50		1,188.50	1,325
3400	For 1" IPS connection, add					15%	10%			
3550	For 2" connection, add					445%	24%			
3600	For tapped drain port in box, add					69.50			69.50	76
5000	Moderate climate, all bronze, polished face									
5020	and scoriated cover, set flush									
5100	3/4" IPS connection	1 Plum	16	.500	Ea.	530	19.85		549.85	620
5120	1" IPS connection	"	14	.571		655	22.50		677.50	755
5200	For tapped drain port in box, add					69.50			69.50	76

22 11 Facility Water Distribution

22 11 23 – Domestic Water Pumps

22 11 23.10 General Utility Pumps		Crew	Daily Output	Labor-Hours	Unit	Material	Labor	Equipment	Total	Total Incl O&P
						2011 Bare Costs				
0010	**GENERAL UTILITY PUMPS**									
2000	Single stage									
3000	Double suction,									
3190	75 HP, to 2500 GPM	Q-3	.28	114	Ea.	16,800	4,075		20,875	25,200
3220	100 HP, to 3000 GPM		.26	123		21,200	4,400		25,600	30,600
3240	150 HP, to 4000 GPM		.24	133		32,800	4,750		37,550	43,900

22 13 Facility Sanitary Sewerage

22 13 16 – Sanitary Waste and Vent Piping

22 13 16.20 Pipe, Cast Iron

No.	Description	Crew	Daily Output	Labor-Hours	Unit	Material	Labor	Equipment	Total	Total Incl O&P
0010	**PIPE, CAST IRON**, Soil, on clevis hanger assemblies, 5′ O.C.									
0020	Single hub, service wt., lead & oakum joints 10′ O.C.									
2120	2″ diameter	Q-1	63	.254	L.F.	9	9.05		18.05	24.50
2140	3″ diameter		60	.267		12.20	9.50		21.70	29
2160	4″ diameter		55	.291		15.50	10.40		25.90	34
2180	5″ diameter	Q-2	76	.316		20.50	10.85		31.35	40.50
2200	6″ diameter	″	73	.329		25.50	11.30		36.80	46.50
2220	8″ diameter	Q-3	59	.542		39	19.35		58.35	74.50
2240	10″ diameter		54	.593		62.50	21		83.50	104
2260	12″ diameter		48	.667		88	24		112	136
2320	For service weight, double hub, add					10%				
2340	For extra heavy, single hub, add					48%	4%			
2360	For extra heavy, double hub, add					71%	4%			
2400	Lead for caulking, (1#/dia in)	Q-1	160	.100	Lb.	1.04	3.57		4.61	7
2420	Oakum for caulking, (1/8#/dia in)	″	40	.400	″	4.88	14.25		19.13	29
4000	No hub, couplings 10′ O.C.									
4100	1-1/2″ diameter	Q-1	71	.225	L.F.	9	8.05		17.05	23
4120	2″ diameter		67	.239		9.25	8.50		17.75	24
4160	4″ diameter		58	.276		15.80	9.85		25.65	33.50

22 13 16.50 Shower Drains

No.	Description	Crew	Daily Output	Labor-Hours	Unit	Material	Labor	Equipment	Total	Total Incl O&P
0010	**SHOWER DRAINS**									
2780	Shower, with strainer, uniform diam. trap, bronze top									
2800	2″ and 3″ pipe size	Q-1	8	2	Ea.	335	71.50		406.50	480
2820	4″ pipe size	″	7	2.286		370	81.50		451.50	540
2840	For galvanized body, add					148			148	163

22 13 16.60 Traps

No.	Description	Crew	Daily Output	Labor-Hours	Unit	Material	Labor	Equipment	Total	Total Incl O&P
0010	**TRAPS**									
0030	Cast iron, service weight									
0050	Running P trap, without vent									
1100	2″	Q-1	16	1	Ea.	113	35.50		148.50	183
1140	3″		14	1.143		113	41		154	191
1150	4″		13	1.231		113	44		157	196
1160	6″	Q-2	17	1.412		490	48.50		538.50	620
1180	Running trap, single hub, with vent									
2080	3″ pipe size, 3″ vent	Q-1	14	1.143	Ea.	89.50	41		130.50	165
2120	4″ pipe size, 4″ vent	″	13	1.231		121	44		165	206
2300	For double hub, vent, add					10%	20%			
3000	P trap, B&S, 2″ pipe size	Q-1	16	1		26.50	35.50		62	88
3040	3″ pipe size	″	14	1.143		40	41		81	111
3350	Deep seal trap, B&S									

22 13 Facility Sanitary Sewerage

22 13 16 – Sanitary Waste and Vent Piping

22 13 16.60 Traps

	22 13 16.60 Traps	Crew	Daily Output	Labor-Hours	Unit	Material	Labor (2011 Bare Costs)	Equipment (2011 Bare Costs)	Total (2011 Bare Costs)	Total Incl O&P
3400	1-1/4" pipe size	Q-1	14	1.143	Ea.	42.50	41		83.50	114
3410	1-1/2" pipe size		14	1.143		42.50	41		83.50	114
3420	2" pipe size		14	1.143		36	41		77	106
3440	3" pipe size		12	1.333		49	47.50		96.50	132
4700	Copper, drainage, drum trap									
4800	3" x 5" solid, 1-1/2" pipe size	1 Plum	16	.500	Ea.	177	19.85		196.85	227
4840	3" x 6" swivel, 1-1/2" pipe size	"	16	.500	"	280	19.85		299.85	345
5100	P trap, standard pattern									
5200	1-1/4" pipe size	1 Plum	18	.444	Ea.	130	17.60		147.60	172
5240	1-1/2" pipe size		17	.471		126	18.65		144.65	170
5260	2" pipe size		15	.533		195	21		216	249
5280	3" pipe size		11	.727		465	29		494	560
5340	With cleanout, swivel joint and slip joint									
5360	1-1/4" pipe size	1 Plum	18	.444	Ea.	92	17.60		109.60	130
5400	1-1/2" pipe size	"	17	.471	"	192	18.65		210.65	242

22 13 16.80 Vent Flashing and Caps

	22 13 16.80 Vent Flashing and Caps	Crew	Daily Output	Labor-Hours	Unit	Material	Labor	Equipment	Total	Total Incl O&P
0010	**VENT FLASHING AND CAPS**									
0120	Vent caps									
0140	Cast iron									
0180	2-1/2" - 3-5/8" pipe	1 Plum	21	.381	Ea.	40	15.10		55.10	68.50
0190	4" - 4-1/8" pipe	"	19	.421	"	48	16.70		64.70	80.50
0900	Vent flashing									
1000	Aluminum with lead ring									
1020	1-1/4" pipe	1 Plum	20	.400	Ea.	10.70	15.85		26.55	38
1030	1-1/2" pipe		20	.400		10.95	15.85		26.80	38
1040	2" pipe		18	.444		11.45	17.60		29.05	41.50
1050	3" pipe		17	.471		12.70	18.65		31.35	44.50
1060	4" pipe		16	.500		15.30	19.85		35.15	49.50
1350	Copper with neoprene ring									
1400	1-1/4" pipe	1 Plum	20	.400	Ea.	19.95	15.85		35.80	48
1430	1-1/2" pipe		20	.400		19.95	15.85		35.80	48
1440	2" pipe		18	.444		21	17.60		38.60	52
1450	3" pipe		17	.471		24.50	18.65		43.15	57.50
1460	4" pipe		16	.500		27	19.85		46.85	62.50

22 13 19 – Sanitary Waste Piping Specialties

22 13 19.13 Sanitary Drains

	22 13 19.13 Sanitary Drains	Crew	Daily Output	Labor-Hours	Unit	Material	Labor	Equipment	Total	Total Incl O&P
0010	**SANITARY DRAINS**									
0400	Deck, auto park, C.I., 13" top									
0440	3", 4", 5", and 6" pipe size	Q-1	8	2	Ea.	1,125	71.50		1,196.50	1,375
0480	For galvanized body, add				"	610			610	675
2000	Floor, medium duty, C.I., deep flange, 7" dia top									
2040	2" and 3" pipe size	Q-1	12	1.333	Ea.	158	47.50		205.50	252
2080	For galvanized body, add					75.50			75.50	83
2120	For polished bronze top, add					89			89	98
2400	Heavy duty, with sediment bucket, C.I., 12" dia. loose grate									
2420	2", 3", 4", 5", and 6" pipe size	Q-1	9	1.778	Ea.	545	63.50		608.50	705
2460	For polished bronze top, add				"	225			225	248
2500	Heavy duty, cleanout & trap w/bucket, C.I., 15" top									
2540	2", 3", and 4" pipe size	Q-1	6	2.667	Ea.	5,175	95		5,270	5,825
2560	For galvanized body, add					1,325			1,325	1,450
2580	For polished bronze top, add					570			570	630

22 13 Facility Sanitary Sewerage

22 13 23 – Sanitary Waste Interceptors

22 13 23.10 Interceptors		Crew	Daily Output	Labor-Hours	Unit	Material	2011 Bare Costs Labor	Equipment	Total	Total Incl O&P
0010	**INTERCEPTORS**									
0150	Grease, fabricated steel, 4 GPM, 8 lb. fat capacity	1 Plum	4	2	Ea.	935	79.50		1,014.50	1,150
0200	7 GPM, 14 lb. fat capacity		4	2		1,300	79.50		1,379.50	1,550
1000	10 GPM, 20 lb. fat capacity		4	2		1,525	79.50		1,604.50	1,800
1040	15 GPM, 30 lb. fat capacity		4	2		2,250	79.50		2,329.50	2,600
1060	20 GPM, 40 lb. fat capacity		3	2.667		2,775	106		2,881	3,225
1160	100 GPM, 200 lb. fat capacity	Q-1	2	8		11,500	285		11,785	13,200
1580	For seepage pan, add					7%				
3000	Hair, cast iron, 1-1/4" and 1-1/2" pipe connection	1 Plum	8	1	Ea.	350	39.50		389.50	450
3100	For chrome-plated cast iron, add					215			215	236
4000	Oil, fabricated steel, 10 GPM, 2" pipe size	1 Plum	4	2		2,075	79.50		2,154.50	2,425
4100	15 GPM, 2" or 3" pipe size		4	2		2,850	79.50		2,929.50	3,275
4120	20 GPM, 2" or 3" pipe size		3	2.667		3,450	106		3,556	3,975
4220	100 GPM, 3" pipe size	Q-1	2	8		11,500	285		11,785	13,200
6000	Solids, precious metals recovery, C.I., 1-1/4" to 2" pipe	1 Plum	4	2		530	79.50		609.50	710
6100	Dental Lab., large, C.I., 1-1/2" to 2" pipe	"	3	2.667		1,850	106		1,956	2,200

22 13 29 – Sanitary Sewerage Pumps

22 13 29.13 Wet-Pit-Mounted, Vertical Sewerage Pumps		Crew	Daily Output	Labor-Hours	Unit	Material	Labor	Equipment	Total	Total Incl O&P
0010	**WET-PIT-MOUNTED, VERTICAL SEWERAGE PUMPS**									
0020	Controls incl. alarm/disconnect panel w/wire. Excavation not included									
0260	Simplex, 9 GPM at 60 PSIG, 91 gal. tank				Ea.	2,975			2,975	3,275
0300	Unit with manway, 26" I.D., 18" high					3,375			3,375	3,700
0340	26" I.D., 36" high					3,500			3,500	3,850
0380	43" I.D., 4' high					3,575			3,575	3,925
3000	Indoor residential type installation									
3020	Simplex, 9 GPM at 60 PSIG, 91 gal. HDPE tank				Ea.	2,950			2,950	3,225

22 13 29.14 Sewage Ejector Pumps		Crew	Daily Output	Labor-Hours	Unit	Material	Labor	Equipment	Total	Total Incl O&P
0010	**SEWAGE EJECTOR PUMPS**, With operating and level controls									
0100	Simplex system incl. tank, cover, pump 15' head									
0500	37 gal. PE tank, 12 GPM, 1/2 HP, 2" discharge	Q-1	3.20	5	Ea.	440	178		618	775
0510	3" discharge		3.10	5.161		480	184		664	825
0530	87 GPM, .7 HP, 2" discharge		3.20	5		670	178		848	1,025
0540	3" discharge		3.10	5.161		725	184		909	1,100
0600	45 gal. coated stl tank, 12 GPM, 1/2 HP, 2" discharge		3	5.333		780	190		970	1,175
0610	3" discharge		2.90	5.517		810	197		1,007	1,200
0630	87 GPM, .7 HP, 2" discharge		3	5.333		1,000	190		1,190	1,400
0640	3" discharge		2.90	5.517		1,050	197		1,247	1,475
0660	134 GPM, 1 HP, 2" discharge		2.80	5.714		1,075	204		1,279	1,525
0680	3" discharge		2.70	5.926		1,150	211		1,361	1,600
0700	70 gal. PE tank, 12 GPM, 1/2 HP, 2" discharge		2.60	6.154		855	220		1,075	1,300
0710	3" discharge		2.40	6.667		910	238		1,148	1,400
0730	87 GPM, 0.7 HP, 2" discharge		2.50	6.400		1,100	228		1,328	1,575
0740	3" discharge		2.30	6.957		1,150	248		1,398	1,675
0760	134 GPM, 1 HP, 2" discharge		2.20	7.273		1,200	259		1,459	1,725
0770	3" discharge		2	8		1,275	285		1,560	1,875

22 14 Facility Storm Drainage

22 14 23 – Storm Drainage Piping Specialties

22 14 23.33 Backwater Valves		Crew	Daily Output	Labor-Hours	Unit	Material	2011 Bare Costs Labor	Equipment	Total	Total Incl O&P
0010	**BACKWATER VALVES**, C.I. Body									
6980	Bronze gate and automatic flapper valves									
7000	3" and 4" pipe size	Q-1	13	1.231	Ea.	1,600	44		1,644	1,850
7100	5" and 6" pipe size	"	13	1.231	"	2,475	44		2,519	2,775
7240	Bronze flapper valve, bolted cover									
7260	2" pipe size	Q-1	16	1	Ea.	470	35.50		505.50	575
7300	4" pipe size	"	13	1.231		905	44		949	1,075
7340	6" pipe size	Q-2	17	1.412	↓	1,300	48.50		1,348.50	1,500

22 14 26 – Facility Storm Drains

22 14 26.13 Roof Drains

		Crew	Daily Output	Labor-Hours	Unit	Material	Labor	Equipment	Total	Total Incl O&P
0010	**ROOF DRAINS**									
0140	Cornice, C.I., 45° or 90° outlet									
0200	3" and 4" pipe size	Q-1	12	1.333	Ea.	259	47.50		306.50	365
0260	For galvanized body, add					59.50			59.50	65.50
0280	For polished bronze dome, add				↓	46			46	51
3860	Roof, flat metal deck, C.I. body, 12" C.I. dome									
3890	3" pipe size	Q-1	14	1.143	Ea.	310	41		351	405
3920	6" pipe size	"	10	1.600	"	540	57		597	685
4620	Main, all aluminum, 12" low profile dome									
4640	2", 3" and 4" pipe size	Q-1	14	1.143	Ea.	345	41		386	445

22 14 26.16 Facility Area Drains

		Crew	Daily Output	Labor-Hours	Unit	Material	Labor	Equipment	Total	Total Incl O&P
0010	**FACILITY AREA DRAINS**									
4980	Scupper floor, oblique strainer, C.I.									
5000	6" x 7" top, 2", 3" and 4" pipe size	Q-1	16	1	Ea.	223	35.50		258.50	305
5100	8" x 12" top, 5" and 6" pipe size	"	14	1.143		435	41		476	540
5160	For galvanized body, add					40%				
5200	For polished bronze strainer, add				↓	85%				

22 14 26.19 Facility Trench Drains

		Crew	Daily Output	Labor-Hours	Unit	Material	Labor	Equipment	Total	Total Incl O&P
0010	**FACILITY TRENCH DRAINS**									
5980	Trench, floor, hvy duty, modular, C.I., 12" x 12" top									
6000	2", 3", 4", 5", & 6" pipe size	Q-1	8	2	Ea.	705	71.50		776.50	890
6100	For unit with polished bronze top	"	8	2	"	1,050	71.50		1,121.50	1,275
6600	Trench, floor, for cement concrete encasement									
6610	Not including trenching or concrete									
6640	Polyester polymer concrete									
6650	4" internal width, with grate									
6660	Light duty steel grate	Q-1	120	.133	L.F.	26.50	4.76		31.26	37
6670	Medium duty steel grate		115	.139		33	4.96		37.96	44.50
6680	Heavy duty iron grate	↓	110	.145	↓	51	5.20		56.20	64.50
6700	12" internal width, with grate									
6770	Heavy duty galvanized grate	Q-1	80	.200	L.F.	128	7.15		135.15	153
6800	Fiberglass									
6810	8" internal width, with grate									
6820	Medium duty galvanized grate	Q-1	115	.139	L.F.	89	4.96		93.96	106
6830	Heavy duty iron grate	"	110	.145	"	86	5.20		91.20	103

22 14 29 – Sump Pumps

22 14 29.13 Wet-Pit-Mounted, Vertical Sump Pumps

		Crew	Daily Output	Labor-Hours	Unit	Material	Labor	Equipment	Total	Total Incl O&P
0010	**WET-PIT-MOUNTED, VERTICAL SUMP PUMPS**									
0400	Molded PVC base, 21 GPM at 15' head, 1/3 HP	1 Plum	5	1.600	Ea.	119	63.50		182.50	235
0800	Iron base, 21 GPM at 15' head, 1/3 HP		5	1.600		150	63.50		213.50	269
1200	Solid brass, 21 GPM at 15' head, 1/3 HP	↓	5	1.600	↓	263	63.50		326.50	395

22 14 Facility Storm Drainage

22 14 29 – Sump Pumps

22 14 29.16 Submersible Sump Pumps		Crew	Daily Output	Labor-Hours	Unit	Material	Labor	Equipment	Total	Total Incl O&P
						2011 Bare Costs				
0010	**SUBMERSIBLE SUMP PUMPS**									
7000	Sump pump, automatic									
7100	Plastic, 1-1/4" discharge, 1/4 HP	1 Plum	6	1.333	Ea.	148	53		201	249
7140	1/3 HP		5	1.600		181	63.50		244.50	305
7160	1/2 HP		5	1.600		221	63.50		284.50	345
7180	1-1/2" discharge, 1/2 HP		4	2		246	79.50		325.50	400
7500	Cast iron, 1-1/4" discharge, 1/4 HP		6	1.333		170	53		223	274
7540	1/3 HP		6	1.333		200	53		253	305
7560	1/2 HP	↓	5	1.600	↓	242	63.50		305.50	370

22 31 Domestic Water Softeners

22 31 13 – Residential Domestic Water Softeners

22 31 13.10 Residential Water Softeners

Line	Description	Crew	Daily Output	Labor-Hours	Unit	Material	Labor	Equipment	Total	Total Incl O&P
0010	**RESIDENTIAL WATER SOFTENERS**									
7350	Water softener, automatic, to 30 grains per gallon	2 Plum	5	3.200	Ea.	380	127		507	625
7400	To 100 grains per gallon	"	4	4	"	750	159		909	1,075

22 31 16 – Commercial Domestic Water Softeners

22 31 16.10 Water Softeners

Line	Description	Crew	Daily Output	Labor-Hours	Unit	Material	Labor	Equipment	Total	Total Incl O&P
0010	**WATER SOFTENERS**									
5800	Softener systems, automatic, intermediate sizes									
5820	available, may be used in multiples.									
6000	Hardness capacity between regenerations and flow									
6100	150,000 grains, 37 GPM cont., 51 GPM peak	Q-1	1.20	13.333	Ea.	3,900	475		4,375	5,050
6200	300,000 grains, 81 GPM cont., 113 GPM peak		1	16		8,675	570		9,245	10,500
6300	750,000 grains, 160 GPM cont., 230 GPM peak		.80	20		11,300	715		12,015	13,700
6400	900,000 grains, 185 GPM cont., 270 GPM peak	↓	.70	22.857	↓	18,200	815		19,015	21,300

22 33 Electric Domestic Water Heaters

22 33 13 – Instantaneous Electric Domestic Water Heaters

22 33 13.10 Hot Water Dispensers

Line	Description	Crew	Daily Output	Labor-Hours	Unit	Material	Labor	Equipment	Total	Total Incl O&P
0010	**HOT WATER DISPENSERS**									
0160	Commercial, 100 cup, 11.3 amp	1 Plum	14	.571	Ea.	450	22.50		472.50	530
3180	Household, 60 cup	"	14	.571	"	238	22.50		260.50	299

22 33 30 – Residential, Electric Domestic Water Heaters

22 33 30.13 Residential, Small-Capacity Elec. Water Heaters

Line	Description	Crew	Daily Output	Labor-Hours	Unit	Material	Labor	Equipment	Total	Total Incl O&P
0010	**RESIDENTIAL, SMALL-CAPACITY ELECTRIC DOMESTIC WATER HEATERS**									
1000	Residential, electric, glass lined tank, 5 yr., 10 gal., single element	1 Plum	2.30	3.478	Ea.	385	138		523	650
1040	20 gallon, single element		2.20	3.636		440	144		584	720
1060	30 gallon, double element		2.20	3.636		575	144		719	870
1080	40 gallon, double element		2	4		615	159		774	935
1100	52 gallon, double element		2	4		690	159		849	1,025
1180	120 gallon, double element	↓	1.40	5.714	↓	1,450	227		1,677	1,975

22 33 33 – Light-Commercial Electric Domestic Water Heaters

22 33 33.10 Commercial Electric Water Heaters

Line	Description	Crew	Daily Output	Labor-Hours	Unit	Material	Labor	Equipment	Total	Total Incl O&P
0010	**COMMERCIAL ELECTRIC WATER HEATERS**									
4000	Commercial, 100° rise. NOTE: for each size tank, a range of									
4010	heaters between the ones shown are available									

22 33 Electric Domestic Water Heaters

22 33 33 – Light-Commercial Electric Domestic Water Heaters

22 33 33.10 Commercial Electric Water Heaters		Crew	Daily Output	Labor-Hours	Unit	Material	Labor	Equipment	Total	Total Incl O&P
						2011 Bare Costs				
4020	Electric									
4100	5 gal., 3 kW, 12 GPH, 208 volt	1 Plum	2	4	Ea.	2,350	159		2,509	2,825
4120	10 gal., 6 kW, 25 GPH, 208 volt		2	4		2,600	159		2,759	3,125
4140	50 gal., 9 kW, 37 GPH, 208 volt		1.80	4.444		3,275	176		3,451	3,900
4160	50 gal., 36 kW, 148 GPH, 208 volt		1.80	4.444		5,475	176		5,651	6,325
4300	200 gal., 15 kW, 61 GPH, 480 volt	Q-1	1.70	9.412		16,900	335		17,235	19,100
4320	200 gal., 120 kW , 490 GPH, 480 volt		1.70	9.412		23,000	335		23,335	25,900
4460	400 gal., 30 kW, 123 GPH, 480 volt		1	16		22,900	570		23,470	26,100
5400	Modulating step control, 2-5 steps	1 Elec	5.30	1.509		810	61		871	990
5440	6-10 steps		3.20	2.500		1,025	101		1,126	1,300
5460	11-15 steps		2.70	2.963		1,475	119		1,594	1,825
5480	16-20 steps		1.60	5		2,175	201		2,376	2,700

22 34 Fuel-Fired Domestic Water Heaters

22 34 13 – Instantaneous, Tankless, Gas Domestic Water Heaters

22 34 13.10 Instantaneous, Tankless, Gas Water Heaters

		Crew	Daily Output	Labor-Hours	Unit	Material	Labor	Equipment	Total	Total Incl O&P
0010	**INSTANTANEOUS, TANKLESS, GAS WATER HEATERS**									
9410	Natural gas/propane, incl vent, 3.2 GPM G	1 Plum	2	4	Ea.	535	159		694	850

22 34 30 – Residential Gas Domestic Water Heaters

22 34 30.13 Residential, Atmos, Gas Domestic Wtr Heaters

		Crew	Daily Output	Labor-Hours	Unit	Material	Labor	Equipment	Total	Total Incl O&P
0010	**RESIDENTIAL, ATMOSPHERIC, GAS DOMESTIC WATER HEATERS**									
2000	Gas fired, foam lined tank, 10 yr., vent not incl.									
2040	30 gallon	1 Plum	2	4	Ea.	895	159		1,054	1,250
2100	75 gallon		1.50	5.333		1,225	211		1,436	1,700
2120	100 gallon		1.30	6.154		1,525	244		1,769	2,075

22 34 36 – Commercial Gas Domestic Water Heaters

22 34 36.13 Commercial, Atmos., Gas Domestic Water Htrs.

		Crew	Daily Output	Labor-Hours	Unit	Material	Labor	Equipment	Total	Total Incl O&P
0010	**COMMERCIAL, ATMOSPHERIC, GAS DOMESTIC WATER HEATERS**									
6000	Gas fired, flush jacket, std. controls, vent not incl.									
6040	75 MBH input, 73 GPH	1 Plum	1.40	5.714	Ea.	1,575	227		1,802	2,100
6060	98 MBH input, 95 GPH		1.40	5.714		4,575	227		4,802	5,425
6080	120 MBH input, 110 GPH		1.20	6.667		4,975	264		5,239	5,875
6180	200 MBH input, 192 GPH		.60	13.333		7,650	530		8,180	9,275
6200	250 MBH input, 245 GPH		.50	16		7,975	635		8,610	9,800
6900	For low water cutoff, add		8	1		275	39.50		314.50	365
6960	For bronze body hot water circulator, add		4	2		1,750	79.50		1,829.50	2,050

22 34 46 – Oil-Fired Domestic Water Heaters

22 34 46.10 Residential Oil-Fired Water Heaters

		Crew	Daily Output	Labor-Hours	Unit	Material	Labor	Equipment	Total	Total Incl O&P
0010	**RESIDENTIAL OIL-FIRED WATER HEATERS**									
3000	Oil fired, glass lined tank, 5 yr., vent not included, 30 gallon	1 Plum	2	4	Ea.	665	159		824	990
3040	50 gallon		1.80	4.444		1,050	176		1,226	1,450
3060	70 gallon		1.50	5.333		1,275	211		1,486	1,750
3080	85 gallon		1.40	5.714		7,125	227		7,352	8,200

22 34 46.20 Commercial Oil-Fired Water Heaters

		Crew	Daily Output	Labor-Hours	Unit	Material	Labor	Equipment	Total	Total Incl O&P
0010	**COMMERCIAL OIL-FIRED WATER HEATERS**									
8000	Oil fired, glass lined, UL listed, std. controls, vent not incl.									
8060	140 gal., 140 MBH input, 134 GPH	Q-1	2.13	7.512	Ea.	14,900	268		15,168	16,800
8080	140 gal., 199 MBH input, 191 GPH		2	8		15,400	285		15,685	17,400
8100	140 gal., 255 MBH input, 247 GPH		1.60	10		15,900	355		16,255	18,000

22 34 Fuel-Fired Domestic Water Heaters

22 34 46 – Oil-Fired Domestic Water Heaters

22 34 46.20 Commercial Oil-Fired Water Heaters		Crew	Daily Output	Labor-Hours	Unit	2011 Bare Costs Material	Labor	Equipment	Total	Total Incl O&P
8160	140 gal., 540 MBH input, 519 GPH	Q-1	.96	16.667	Ea.	21,000	595		21,595	24,100
8180	140 gal., 720 MBH input, 691 GPH		.92	17.391		21,400	620		22,020	24,500
8280	201 gal., 1250 MBH input, 1200 GPH	Q-2	1.22	19.672		32,900	675		33,575	37,200
8300	201 gal., 1500 MBH input, 1441 GPH	"	1.16	20.690		35,700	710		36,410	40,500
8900	For low water cutoff, add	1 Plum	8	1		300	39.50		339.50	395
8960	For bronze body hot water circulator, add	"	4	2		710	79.50		789.50	910

22 35 Domestic Water Heat Exchangers

22 35 30 – Water Heating by Steam

22 35 30.10 Water Heating Transfer Package

Line	Description	Crew	Daily Output	Labor-Hours	Unit	Material	Labor	Equipment	Total	Total Incl O&P
0010	**WATER HEATING TRANSFER PACKAGE**, Complete controls,									
0020	expansion tank, converter, air separator									
1000	Hot water, 180°F enter, 200°F leaving, 15# steam									
1010	One pump system, 28 GPM	Q-6	.75	32	Ea.	16,700	1,100		17,800	20,100
1020	35 GPM		.70	34.286		18,100	1,200		19,300	21,900
1040	55 GPM		.65	36.923		21,300	1,275		22,575	25,600
1060	130 GPM		.55	43.636		27,000	1,525		28,525	32,200
1080	255 GPM		.40	60		35,300	2,075		37,375	42,300
1100	550 GPM		.30	80		44,800	2,775		47,575	53,500

22 41 Residential Plumbing Fixtures

22 41 06 – Plumbing Fixtures General

22 41 06.10 Plumbing Fixture Notes

Line	Description	Crew	Daily Output	Labor-Hours	Unit	Material	Labor	Equipment	Total	Total Incl O&P
0010	**PLUMBING FIXTURE NOTES**, Incl. trim fittings unless otherwise noted R224000-40									
0080	For rough-in, supply, waste, and vent, see add for each type									
0122	For electric water coolers, see Section 22 47 16.10									
0160	For color, unless otherwise noted, add				Ea.	20%				

22 41 13 – Residential Water Closets, Urinals, and Bidets

22 41 13.40 Water Closets

Line	Description	Crew	Daily Output	Labor-Hours	Unit	Material	Labor	Equipment	Total	Total Incl O&P
0010	**WATER CLOSETS** R224000-40									
0032	For automatic flush, see Line 22 42 39.10 0972									
0150	Tank type, vitreous china, incl. seat, supply pipe w/stop, 1.6 gpf or noted									
0200	Wall hung									
0400	Two piece, close coupled	Q-1	5.30	3.019	Ea.	590	108		698	825
0960	For rough-in, supply, waste, vent and carrier	"	2.73	5.861	"	910	209		1,119	1,350
0999	Floor mounted									
1100	Two piece, close coupled	Q-1	5.30	3.019	Ea.	240	108		348	440
1102	Economy		5.30	3.019		134	108		242	325
1110	Two piece, close coupled, dual flush		5.30	3.019		256	108		364	460
1140	Two piece, close coupled, 1.28 gpf, ADA G		5.30	3.019		295	108		403	500
1960	For color, add					30%				
1980	For rough-in, supply, waste and vent	Q-1	3.05	5.246	Ea.	305	187		492	640

22 41 16 – Residential Lavatories and Sinks

22 41 16.10 Lavatories

Line	Description	Crew	Daily Output	Labor-Hours	Unit	Material	Labor	Equipment	Total	Total Incl O&P
0010	**LAVATORIES**, With trim, white unless noted otherwise R224000-40									
0500	Vanity top, porcelain enamel on cast iron									
0600	20" x 18"	Q-1	6.40	2.500	Ea.	296	89		385	470
0640	33" x 19" oval		6.40	2.500		605	89		694	810

22 41 Residential Plumbing Fixtures

22 41 16 – Residential Lavatories and Sinks

22 41 16.10 Lavatories		Crew	Daily Output	Labor-Hours	Unit	Material	Labor	Equipment	Total	Total Incl O&P
						2011 Bare Costs				
0720	19" round	Q-1	6.40	2.500	Ea.	228	89		317	395
0860	For color, add					25%				
1000	Cultured marble, 19" x 17", single bowl	Q-1	6.40	2.500	Ea.	211	89		300	380
1040	25" x 19", single bowl	"	6.40	2.500	"	242	89		331	410
1580	For color, same price									
1900	Stainless steel, self-rimming, 25" x 22", single bowl, ledge	Q-1	6.40	2.500	Ea.	360	89		449	540
1960	17" x 22", single bowl		6.40	2.500		350	89		439	530
2600	Steel, enameled, 20" x 17", single bowl		5.80	2.759		186	98.50		284.50	365
2660	19" round		5.80	2.759		161	98.50		259.50	340
2900	Vitreous china, 20" x 16", single bowl		5.40	2.963		260	106		366	460
2960	20" x 17", single bowl		5.40	2.963		176	106		282	365
3020	19" round, single bowl		5.40	2.963		174	106		280	365
3200	22" x 13", single bowl		5.40	2.963		267	106		373	465
3560	For color, add					50%				
3580	Rough-in, supply, waste and vent for all above lavatories	Q-1	2.30	6.957	Ea.	320	248		568	755
4000	Wall hung									
4040	Porcelain enamel on cast iron, 16" x 14", single bowl	Q-1	8	2	Ea.	445	71.50		516.50	605
4180	20" x 18", single bowl		8	2		345	71.50		416.50	495
4240	22" x 19", single bowl		8	2		595	71.50		666.50	770
4580	For color, add					30%				
6000	Vitreous china, 18" x 15", single bowl with backsplash	Q-1	7	2.286	Ea.	231	81.50		312.50	385
6500	For color, add					30%				
6960	Rough-in, supply, waste and vent for above lavatories	Q-1	1.66	9.639	Ea.	420	345		765	1,025
7000	Pedestal type									
7600	Vitreous china, 27" x 21", white	Q-1	6.60	2.424	Ea.	545	86.50		631.50	740
7610	27" x 21", colored		6.60	2.424		655	86.50		741.50	860
7620	27" x 21", premium color		6.60	2.424		720	86.50		806.50	930
7660	26" x 20", white		6.60	2.424		485	86.50		571.50	675
7670	26" x 20", colored		6.60	2.424		580	86.50		666.50	775
7680	26" x 20", premium color		6.60	2.424		635	86.50		721.50	840
7700	24" x 18", white		6.60	2.424		470	86.50		556.50	660
7710	24" x 18", colored		6.60	2.424		560	86.50		646.50	760
7720	24" x 18", premium color		6.60	2.424		655	86.50		741.50	860
7760	21" x 18", white		6.60	2.424		215	86.50		301.50	375
7770	21" x 18", colored		6.60	2.424		232	86.50		318.50	395
7990	Rough-in, supply, waste and vent for pedestal lavatories		1.66	9.639		420	345		765	1,025

22 41 16.30 Sinks

Line	Description	Crew	Daily Output	Labor-Hours	Unit	Material	Labor	Equipment	Total	Total Incl O&P
0010	**SINKS**, With faucets and drain R224000-40									
2000	Kitchen, counter top style, P.E. on C.I., 24" x 21" single bowl	Q-1	5.60	2.857	Ea.	257	102		359	450
2100	31" x 22" single bowl		5.60	2.857		500	102		602	715
2200	32" x 21" double bowl		4.80	3.333		277	119		396	500
2300	42" x 21" double bowl		4.80	3.333		1,025	119		1,144	1,325
3000	Stainless steel, self rimming, 19" x 18" single bowl		5.60	2.857		580	102		682	800
3100	25" x 22" single bowl		5.60	2.857		645	102		747	875
4000	Steel, enameled, with ledge, 24" x 21" single bowl		5.60	2.857		420	102		522	625
4100	32" x 21" double bowl		4.80	3.333		400	119		519	635
4960	For color sinks except stainless steel, add					10%				
4980	For rough-in, supply, waste and vent, counter top sinks	Q-1	2.14	7.477		365	267		632	835
5000	Kitchen, raised deck, P.E. on C.I.									
5100	32" x 21", dual level, double bowl	Q-1	2.60	6.154	Ea.	330	220		550	725
5700	For color, add					20%				
5790	For rough-in, supply, waste & vent, sinks	Q-1	1.85	8.649		365	310		675	905

22 41 19 – Residential Bathtubs

22 41 19.10 Baths

	22 41 19.10 Baths	Crew	Daily Output	Labor-Hours	Unit	Material	2011 Bare Costs Labor	Equipment	Total	Total Incl O&P
0010	**BATHS**									
0100	Tubs, recessed porcelain enamel on cast iron, with trim									
0180	48" x 42"	Q-1	4	4	Ea.	2,225	143		2,368	2,650
0220	72" x 36"		3	5.333		2,275	190		2,465	2,825
2000	Enameled formed steel, 4'-6" long		5.80	2.759		415	98.50		513.50	615
4000	Soaking, acrylic with pop-up drain, 60" x 32" x 21" deep		5.50	2.909		1,025	104		1,129	1,300
4100	60" x 48" x 18-1/2" deep		5	3.200		960	114		1,074	1,225
9600	Rough-in, supply, waste and vent, for all above tubs, add		2.07	7.729		395	276		671	880

22 41 23 – Residential Showers

22 41 23.20 Showers

	22 41 23.20 Showers	Crew	Daily Output	Labor-Hours	Unit	Material	Labor	Equipment	Total	Total Incl O&P
0010	**SHOWERS**									
1500	Stall, with drain only. Add for valve and door/curtain									
3000	Fiberglass, one piece, with 3 walls, 32" x 32" square	Q-1	5.50	2.909	Ea.	545	104		649	765
3100	36" x 36" square		5.50	2.909		555	104		659	785
3250	64" x 65-3/4" x 81-1/2" fold. seat, whlchr.		3.80	4.211		2,050	150		2,200	2,525
4000	Polypropylene, stall only, w/molded-stone floor, 30" x 30"		2	8		500	285		785	1,025
4200	Rough-in, supply, waste and vent for above showers		2.05	7.805		485	278		763	990

22 41 23.40 Shower System Components

	22 41 23.40 Shower System Components	Crew	Daily Output	Labor-Hours	Unit	Material	Labor	Equipment	Total	Total Incl O&P
0010	**SHOWER SYSTEM COMPONENTS**									
4500	Receptor only									
4510	For tile, 36" x 36"	1 Plum	4	2	Ea.	325	79.50		404.50	490
4520	Fiberglass receptor only, 32" x 32"		8	1		107	39.50		146.50	183
4530	34" x 34"		7.80	1.026		126	40.50		166.50	206
4540	36" x 36"		7.60	1.053		125	41.50		166.50	206
4600	Rectangular									
4620	32" x 48"	1 Plum	7.40	1.081	Ea.	149	43		192	233
4630	32" x 54"		7.20	1.111		196	44		240	287
4640	32" x 60"		7	1.143		208	45.50		253.50	300
5000	Built-in, head, arm, 2.5 GPM valve		4	2		80	79.50		159.50	218
5200	Head, arm, by-pass, integral stops, handles		3.60	2.222		216	88		304	380

22 41 36 – Residential Laundry Trays

22 41 36.10 Laundry Sinks

	22 41 36.10 Laundry Sinks	Crew	Daily Output	Labor-Hours	Unit	Material	Labor	Equipment	Total	Total Incl O&P
0010	**LAUNDRY SINKS**, With trim									
0020	Porcelain enamel on cast iron, black iron frame									
0050	24" x 21", single compartment	Q-1	6	2.667	Ea.	465	95		560	665
0100	26" x 21", single compartment	"	6	2.667	"	450	95		545	650
2000	Molded stone, on wall hanger or legs									
2020	22" x 23", single compartment	Q-1	6	2.667	Ea.	138	95		233	305
2100	45" x 21", double compartment	"	5	3.200	"	259	114		373	470
3000	Plastic, on wall hanger or legs									
3020	18" x 23", single compartment	Q-1	6.50	2.462	Ea.	145	88		233	300
3300	40" x 24", double compartment		5.50	2.909		257	104		361	450
5000	Stainless steel, counter top, 22" x 17" single compartment		6	2.667		61.50	95		156.50	223
5200	33" x 22", double compartment		5	3.200		73.50	114		187.50	267
9600	Rough-in, supply, waste and vent, for all laundry sinks		2.14	7.477		365	267		632	835

22 41 39 – Residential Faucets, Supplies and Trim

22 41 39.10 Faucets and Fittings

	22 41 39.10 Faucets and Fittings	Crew	Daily Output	Labor-Hours	Unit	Material	Labor	Equipment	Total	Total Incl O&P
0010	**FAUCETS AND FITTINGS**									
0150	Bath, faucets, diverter spout combination, sweat	1 Plum	8	1	Ea.	86.50	39.50		126	160
0200	For integral stops, IPS unions, add					109			109	120

22 41 Residential Plumbing Fixtures

22 41 39 – Residential Faucets, Supplies and Trim

22 41 39.10 Faucets and Fittings		Crew	Daily Output	Labor-Hours	Unit	2011 Bare Costs Material	Labor	Equipment	Total	Total Incl O&P
0420	Bath, press-bal mix valve w/diverter, spout, shower hd, arm/flange	1 Plum	8	1	Ea.	155	39.50		194.50	236
0810	Bidet									
0812	Fitting, over the rim, swivel spray/pop-up drain	1 Plum	8	1	Ea.	172	39.50		211.50	254
1000	Kitchen sink faucets, top mount, cast spout		10	.800		61.50	31.50		93	120
1100	For spray, add		24	.333		16.15	13.20		29.35	39.50
1300	Single control lever handle									
1310	With pull out spray									
1320	Polished chrome	1 Plum	10	.800	Ea.	171	31.50		202.50	241
2000	Laundry faucets, shelf type, IPS or copper unions		12	.667		49.50	26.50		76	97.50
2100	Lavatory faucet, centerset, without drain		10	.800		44.50	31.50		76	101
2210	Porcelain cross handles and pop-up drain									
2220	Polished chrome	1 Plum	6.66	1.201	Ea.	128	47.50		175.50	219
2230	Polished brass	″	6.66	1.201	″	193	47.50		240.50	290
2260	Single lever handle and pop-up drain									
2270	Black nickel	1 Plum	6.66	1.201	Ea.	220	47.50		267.50	320
2280	Polished brass		6.66	1.201		220	47.50		267.50	320
2290	Polished chrome		6.66	1.201		172	47.50		219.50	267
2810	Automatic sensor and operator, with faucet head G		6.15	1.301		370	51.50		421.50	495
4000	Shower by-pass valve with union		18	.444		68.50	17.60		86.10	104
4200	Shower thermostatic mixing valve, concealed, with shower head trim kit		8	1		305	39.50		344.50	400
4220	Shower pressure balancing mixing valve,									
4230	With shower head, arm, flange and diverter tub spout									
4240	Chrome	1 Plum	6.14	1.303	Ea.	222	51.50		273.50	330
4250	Polished brass		6.14	1.303		279	51.50		330.50	390
4260	Satin		6.14	1.303		279	51.50		330.50	390
4270	Polished chrome/brass		6.14	1.303		229	51.50		280.50	335
5000	Sillcock, compact, brass, IPS or copper to hose		24	.333		10.60	13.20		23.80	33

22 42 Commercial Plumbing Fixtures

22 42 13 – Commercial Water Closets, Urinals, and Bidets

22 42 13.30 Urinals

		Crew	Daily Output	Labor-Hours	Unit	Material	Labor	Equipment	Total	Total Incl O&P
0010	**URINALS**									
3000	Wall hung, vitreous china, with hanger & self-closing valve									
3100	Siphon jet type	Q-1	3	5.333	Ea.	289	190		479	625
3120	Blowout type		3	5.333		405	190		595	755
3140	Water saving .5 gpf G		3	5.333		645	190		835	1,025
3300	Rough-in, supply, waste & vent		2.83	5.654		525	202		727	910
5000	Stall type, vitreous china, includes valve		2.50	6.400		640	228		868	1,075
6980	Rough-in, supply, waste and vent		1.99	8.040		465	287		752	980

22 42 13.40 Water Closets

		Crew	Daily Output	Labor-Hours	Unit	Material	Labor	Equipment	Total	Total Incl O&P
0010	**WATER CLOSETS**									
3000	Bowl only, with flush valve, seat, 1.6 gpf unless noted									
3100	Wall hung	Q-1	5.80	2.759	Ea.	445	98.50		543.50	650
3200	For rough-in, supply, waste and vent, single WC		2.56	6.250		975	223		1,198	1,450
3300	Floor mounted		5.80	2.759		315	98.50		413.50	505
3350	With wall outlet		5.80	2.759		445	98.50		543.50	650
3360	With floor outlet, 1.28 gpf G		5.80	2.759		555	98.50		653.50	770
3362	With floor outlet, 1.28 gpf, ADA G		5.80	2.759		595	98.50		693.50	815
3400	For rough-in, supply, waste and vent, single WC		2.84	5.634		365	201		566	735

22 42 Commercial Plumbing Fixtures

22 42 16 – Commercial Lavatories and Sinks

22 42 16.10 Handwasher-Dryer Module		Crew	Daily Output	Labor-Hours	Unit	2011 Bare Costs Material	Labor	Equipment	Total	Total Incl O&P
0010	**HANDWASHER-DRYER MODULE**									
0030	Wall mounted									
0040	With electric dryer									
0050	Sensor operated	Q-1	8	2	Ea.	3,650	71.50		3,721.50	4,125
0110	Sensor operated (ADA)		8	2		3,650	71.50		3,721.50	4,150
0140	Sensor operated (ADA), surface mounted		8	2		4,400	71.50		4,471.50	4,975
0150	With paper towels									
0180	Sensor operated	Q-1	8	2	Ea.	3,550	71.50		3,621.50	4,025

22 42 16.14 Lavatories

0010	**LAVATORIES**, With trim, white unless noted otherwise									
0020	Commercial lavatories same as residential. See Section 22 41 16.10									

22 42 16.40 Service Sinks

0010	**SERVICE SINKS**									
6650	Service, floor, corner, P.E. on C.I., 28" x 28"	Q-1	4.40	3.636	Ea.	845	130		975	1,150
6790	For rough-in, supply, waste & vent, floor service sinks		1.64	9.756		1,025	350		1,375	1,725
7000	Service, wall, P.E. on C.I., roll rim, 22" x 18"		4	4		645	143		788	945
7100	24" x 20"		4	4		715	143		858	1,025
8600	Vitreous china, 22" x 20"		4	4		530	143		673	820
8960	For stainless steel rim guard, front or side, add					38.50			38.50	42
8980	For rough-in, supply, waste & vent, wall service sinks	Q-1	1.30	12.308		1,300	440		1,740	2,150

22 42 23 – Commercial Showers

22 42 23.30 Group Showers

0010	**GROUP SHOWERS**									
6000	Group, w/pressure balancing valve, rough-in and rigging not included									
6800	Column, 6 heads, no receptors, less partitions	Q-1	3	5.333	Ea.	3,000	190		3,190	3,600
6900	With stainless steel partitions		1	16		9,250	570		9,820	11,100
7600	5 heads, no receptors, less partitions		3	5.333		2,475	190		2,665	3,025
7620	4 heads (1 handicap) no receptors, less partitions		3	5.333		3,975	190		4,165	4,675
7700	With stainless steel partitions		1	16		7,600	570		8,170	9,275
8000	Wall, 2 heads, no receptors, less partitions		4	4		1,375	143		1,518	1,750
8100	With stainless steel partitions		2	8		3,225	285		3,510	4,025

22 42 33 – Wash Fountains

22 42 33.20 Commercial Wash Fountains

0010	**COMMERCIAL WASH FOUNTAINS**									
1900	Group, foot control									
2000	Precast terrazzo, circular, 36" diam., 5 or 6 persons	Q-2	3	8	Ea.	4,200	275		4,475	5,050
2100	54" diameter for 8 or 10 persons		2.50	9.600		5,225	330		5,555	6,275
2400	Semi-circular, 36" diam. for 3 persons		3	8		3,850	275		4,125	4,700
2500	54" diam. for 4 or 5 persons		2.50	9.600		4,650	330		4,980	5,675
2700	Quarter circle (corner), 54" for 3 persons		3.50	6.857		4,525	236		4,761	5,350
3000	Stainless steel, circular, 36" diameter		3.50	6.857		4,100	236		4,336	4,900
3100	54" diameter		2.80	8.571		5,300	294		5,594	6,300
3400	Semi-circular, 36" diameter		3.50	6.857		3,575	236		3,811	4,300
3500	54" diameter		2.80	8.571		4,575	294		4,869	5,500
5610	Group, infrared control, barrier free									
5614	Precast terrazzo									
5620	Semi-circular 36" diam. for 3 persons	Q-2	3	8	Ea.	5,875	275		6,150	6,925
5630	46" diam. for 4 persons		2.80	8.571		6,425	294		6,719	7,525
5640	Circular, 54" diam. for 8 persons, button control		2.50	9.600		7,700	330		8,030	9,025
5700	Rough-in, supply, waste and vent for above wash fountains	Q-1	1.82	8.791		525	315		840	1,100

22 42 Commercial Plumbing Fixtures

22 42 33 – Wash Fountains

22 42 33.20 Commercial Wash Fountains		Crew	Daily Output	Labor-Hours	Unit	2011 Bare Costs Material	Labor	Equipment	Total	Total Incl O&P
6200	Duo for small washrooms, stainless steel	Q-1	2	8	Ea.	2,725	285		3,010	3,450
6500	Rough-in, supply, waste & vent for duo fountains		2.02	7.921		310	283		593	800

22 42 39 – Commercial Faucets, Supplies, and Trim

22 42 39.10 Faucets and Fittings		Crew	Daily Output	Labor-Hours	Unit	Material	Labor	Equipment	Total	Total Incl O&P
0010	**FAUCETS AND FITTINGS**									
0840	Flush valves, with vacuum breaker									
0850	Water closet									
0860	Exposed, rear spud	1 Plum	8	1	Ea.	137	39.50		176.50	216
0870	Top spud		8	1		126	39.50		165.50	204
0880	Concealed, rear spud		8	1		180	39.50		219.50	263
0890	Top spud		8	1		149	39.50		188.50	229
0900	Wall hung		8	1		159	39.50		198.50	240
0920	Urinal									
0930	Exposed, stall	1 Plum	8	1	Ea.	126	39.50		165.50	204
0940	Wall, (washout)		8	1		126	39.50		165.50	204
0950	Pedestal, top spud		8	1		129	39.50		168.50	207
0960	Concealed, stall		8	1		145	39.50		184.50	224
0970	Wall (washout)		8	1		156	39.50		195.50	237
0971	Automatic flush sensor and operator for									
0972	urinals or water closets, standard	1 Plum	8	1	Ea.	415	39.50		454.50	525
0980	High efficiency water saving									
0984	Water closets, 1.28 gpf	1 Plum	8	1	Ea.	395	39.50		434.50	500
0988	Urinals, .5 gpf	"	8	1	"	395	39.50		434.50	500
2790	Faucets for lavatories									
2800	Self-closing, center set	1 Plum	10	.800	Ea.	131	31.50		162.50	196
2810	Automatic sensor and operator, with faucet head		6.15	1.301		370	51.50		421.50	495
3000	Service sink faucet, cast spout, pail hook, hose end		14	.571		80	22.50		102.50	125

22 42 39.30 Carriers and Supports		Crew	Daily Output	Labor-Hours	Unit	Material	Labor	Equipment	Total	Total Incl O&P
0010	**CARRIERS AND SUPPORTS**, For plumbing fixtures									
0500	Drinking fountain, wall mounted									
0600	Plate type with studs, top back plate	1 Plum	7	1.143	Ea.	70.50	45.50		116	152
0700	Top front and back plate		7	1.143		87	45.50		132.50	170
0800	Top & bottom, front & back plates, w/bearing jacks		7	1.143		126	45.50		171.50	213
3000	Lavatory, concealed arm									
3050	Floor mounted, single									
3100	High back fixture	1 Plum	6	1.333	Ea.	395	53		448	515
3200	Flat slab fixture		6	1.333		340	53		393	460
3220	Paraplegic		6	1.333		440	53		493	570
3250	Floor mounted, back to back									
3300	High back fixtures	1 Plum	5	1.600	Ea.	560	63.50		623.50	720
3400	Flat slab fixtures		5	1.600		690	63.50		753.50	865
3430	Paraplegic		5	1.600		640	63.50		703.50	810
3500	Wall mounted, in stud or masonry									
3600	High back fixture	1 Plum	6	1.333	Ea.	232	53		285	345
3700	Flat slab fixture	"	6	1.333	"	201	53		254	310
4600	Sink, floor mounted									
4650	Exposed arm system									
4700	Single heavy fixture	1 Plum	5	1.600	Ea.	655	63.50		718.50	830
4750	Single heavy sink with slab		5	1.600		810	63.50		873.50	995
4800	Back to back, standard fixtures		5	1.600		480	63.50		543.50	635
4850	Back to back, heavy fixtures		5	1.600		690	63.50		753.50	865
4900	Back to back, heavy sink with slab		5	1.600		690	63.50		753.50	865

22 42 39.30 Carriers and Supports		Crew	Daily Output	Labor-Hours	Unit	2011 Bare Costs Material	Labor	Equipment	Total	Total Incl O&P
4950	Exposed offset arm system									
5000	Single heavy deep fixture	1 Plum	5	1.600	Ea.	625	63.50		688.50	795
5100	Plate type system									
5200	With bearing jacks, single fixture	1 Plum	5	1.600	Ea.	720	63.50		783.50	895
5300	With exposed arms, single heavy fixture		5	1.600		920	63.50		983.50	1,125
5400	Wall mounted, exposed arms, single heavy fixture		5	1.600		335	63.50		398.50	475
6000	Urinal, floor mounted, 2" or 3" coupling, blowout type		6	1.333		415	53		468	545
6100	With fixture or hanger bolts, blowout or washout		6	1.333		298	53		351	415
6200	With bearing plate		6	1.333		330	53		383	450
6300	Wall mounted, plate type system		6	1.333		257	53		310	370
6980	Water closet, siphon jet									
7000	Horizontal, adjustable, caulk									
7040	Single, 4" pipe size	1 Plum	5.33	1.501	Ea.	755	59.50		814.50	925
7050	4" pipe size, paraplegic		5.33	1.501		755	59.50		814.50	925
7060	5" pipe size		5.33	1.501		955	59.50		1,014.50	1,150
7100	Double, 4" pipe size		5	1.600		1,100	63.50		1,163.50	1,325
7110	4" pipe size, paraplegic		5	1.600		1,100	63.50		1,163.50	1,325
7120	5" pipe size		5	1.600		1,600	63.50		1,663.50	1,850
7160	Horizontal, adjustable, extended, caulk									
7180	Single, 4" pipe size	1 Plum	5.33	1.501	Ea.	755	59.50		814.50	925
7200	5" pipe size		5.33	1.501		955	59.50		1,014.50	1,150
7240	Double, 4" pipe size		5	1.600		1,375	63.50		1,438.50	1,600
7260	5" pipe size		5	1.600		1,600	63.50		1,663.50	1,850
7400	Vertical, adjustable, caulk or thread									
7440	Single, 4" pipe size	1 Plum	5.33	1.501	Ea.	765	59.50		824.50	940
7460	5" pipe size		5.33	1.501		945	59.50		1,004.50	1,150
7480	6" pipe size		5	1.600		1,075	63.50		1,138.50	1,300
7520	Double, 4" pipe size		5	1.600		1,150	63.50		1,213.50	1,350
7540	5" pipe size		5	1.600		1,325	63.50		1,388.50	1,550
7560	6" pipe size		4	2		1,450	79.50		1,529.50	1,725
7600	Vertical, adjustable, extended, caulk									
7620	Single, 4" pipe size	1 Plum	5.33	1.501	Ea.	765	59.50		824.50	940
7640	5" pipe size		5.33	1.501		945	59.50		1,004.50	1,150
7680	6" pipe size		5	1.600		1,075	63.50		1,138.50	1,300
7720	Double, 4" pipe size		5	1.600		1,150	63.50		1,213.50	1,350
7740	5" pipe size		5	1.600		1,325	63.50		1,388.50	1,550
7760	6" pipe size		4	2		1,450	79.50		1,529.50	1,725
7780	Water closet, blow out									
7800	Vertical offset, caulk or thread									
7820	Single, 4" pipe size	1 Plum	5.33	1.501	Ea.	565	59.50		624.50	720
7840	Double, 4" pipe size	"	5	1.600	"	970	63.50		1,033.50	1,175
7880	Vertical offset, extended, caulk									
7900	Single, 4" pipe size	1 Plum	5.33	1.501	Ea.	710	59.50		769.50	875
7920	Double, 4" pipe size	"	5	1.600	"	1,125	63.50		1,188.50	1,325
7960	Vertical, for floor mounted back-outlet									
7980	Single, 4" thread, 2" vent	1 Plum	5.33	1.501	Ea.	500	59.50		559.50	645
8000	Double, 4" thread, 2" vent	"	6	1.333	"	1,475	53		1,528	1,675
8040	Vertical, for floor mounted back-outlet, extended									
8060	Single, 4" caulk, 2" vent	1 Plum	6	1.333	Ea.	500	53		553	635
8080	Double, 4" caulk, 2" vent	"	6	1.333	"	1,475	53		1,528	1,675
8200	Water closet, residential									
8220	Vertical centerline, floor mount									
8240	Single, 3" caulk, 2" or 3" vent	1 Plum	6	1.333	Ea.	455	53		508	585

22 42 Commercial Plumbing Fixtures

22 42 39 – Commercial Faucets, Supplies, and Trim

22 42 39.30 Carriers and Supports		Crew	Daily Output	Labor-Hours	Unit	Material	Labor (2011 Bare Costs)	Equipment	Total	Total Incl O&P
8260	4″ caulk, 2″ or 4″ vent	1 Plum	6	1.333	Ea.	575	53		628	720
8280	3″ copper sweat, 3″ vent		6	1.333		405	53		458	530
8300	4″ copper sweat, 4″ vent		6	1.333		490	53		543	625
8400	Vertical offset, floor mount									
8420	Single, 3″ or 4″ caulk, vent	1 Plum	4	2	Ea.	565	79.50		644.50	755
8440	3″ or 4″ copper sweat, vent		5	1.600		565	63.50		628.50	730
8460	Double, 3″ or 4″ caulk, vent		4	2		970	79.50		1,049.50	1,200
8480	3″ or 4″ copper sweat, vent		5	1.600		970	63.50		1,033.50	1,175
9000	Water cooler (electric), floor mounted									
9100	Plate type with bearing plate, single	1 Plum	6	1.333	Ea.	286	53		339	400

22 45 Emergency Plumbing Fixtures

22 45 13 – Emergency Showers

22 45 13.10 Emergency Showers		Crew	Daily Output	Labor-Hours	Unit	Material	Labor	Equipment	Total	Total Incl O&P
0010	**EMERGENCY SHOWERS**, Rough-in not included									
5000	Shower, single head, drench, ball valve, pull, freestanding	Q-1	4	4	Ea.	340	143		483	610
5200	Horizontal or vertical supply		4	4		490	143		633	775
6000	Multi-nozzle, eye/face wash combination		4	4		595	143		738	890
6400	Multi-nozzle, 12 spray, shower only		4	4		1,600	143		1,743	1,975
6600	For freeze-proof, add		6	2.667		365	95		460	560

22 45 16 – Eyewash Equipment

22 45 16.10 Eyewash Safety Equipment		Crew	Daily Output	Labor-Hours	Unit	Material	Labor	Equipment	Total	Total Incl O&P
0010	**EYEWASH SAFETY EQUIPMENT**, Rough-in not included									
1000	Eye wash fountain									
1400	Plastic bowl, pedestal mounted	Q-1	4	4	Ea.	250	143		393	510
1600	Unmounted		4	4		186	143		329	435
1800	Wall mounted		4	4		380	143		523	650
2000	Stainless steel, pedestal mounted		4	4		330	143		473	600
2200	Unmounted		4	4		248	143		391	505
2400	Wall mounted		4	4		244	143		387	500

22 45 19 – Self-Contained Eyewash Equipment

22 45 19.10 Self-Contained Eyewash Safety Equipment		Crew	Daily Output	Labor-Hours	Unit	Material	Labor	Equipment	Total	Total Incl O&P
0010	**SELF-CONTAINED EYEWASH SAFETY EQUIPMENT**									
3000	Eye wash, portable, self-contained				Ea.	865			865	950

22 45 26 – Eye/Face Wash Equipment

22 45 26.10 Eye/Face Wash Safety Equipment		Crew	Daily Output	Labor-Hours	Unit	Material	Labor	Equipment	Total	Total Incl O&P
0010	**EYE/FACE WASH SAFETY EQUIPMENT**, Rough-in not included									
4000	Eye and face wash, combination fountain									
4200	Stainless steel, pedestal mounted	Q-1	4	4	Ea.	490	143		633	775
4400	Unmounted		4	4		345	143		488	615
4600	Wall mounted		4	4		224	143		367	480

22 47 Drinking Fountains and Water Coolers

22 47 13 – Drinking Fountains

22 47 13.10 Drinking Water Fountains		Crew	Daily Output	Labor-Hours	Unit	Material	2011 Bare Costs Labor	Equipment	Total	Total Incl O&P
0010	**DRINKING WATER FOUNTAINS,** For connection to cold water supply									
1000	Wall mounted, non-recessed									
1400	Bronze, with no back	1 Plum	4	2	Ea.	930	79.50		1,009.50	1,150
1800	Cast aluminum, enameled, for correctional institutions		4	2		1,375	79.50		1,454.50	1,625
2000	Fiberglass, 12" back, single bubbler unit		4	2		1,750	79.50		1,829.50	2,050
2040	Dual bubbler		3.20	2.500		1,950	99		2,049	2,300
2400	Precast stone, no back		4	2		760	79.50		839.50	965
2700	Stainless steel, single bubbler, no back		4	2		920	79.50		999.50	1,125
2740	With back		4	2		635	79.50		714.50	830
2780	Dual handle & wheelchair projection type		4	2		655	79.50		734.50	850
2820	Dual level for handicapped type		3.20	2.500		1,325	99		1,424	1,625
3300	Vitreous china									
3340	7" back	1 Plum	4	2	Ea.	575	79.50		654.50	765
3940	For vandal-resistant bottom plate, add					59.50			59.50	65.50
3960	For freeze-proof valve system, add	1 Plum	2	4		550	159		709	865
3980	For rough-in, supply and waste, add	"	2.21	3.620		254	144		398	515
4000	Wall mounted, semi-recessed									
4200	Poly-marble, single bubbler	1 Plum	4	2	Ea.	925	79.50		1,004.50	1,150
4600	Stainless steel, satin finish, single bubbler		4	2		995	79.50		1,074.50	1,225
4900	Vitreous china, single bubbler		4	2		830	79.50		909.50	1,050
5980	For rough-in, supply and waste, add		1.83	4.372		254	173		427	565
6000	Wall mounted, fully recessed									
6400	Poly-marble, single bubbler	1 Plum	4	2	Ea.	1,525	79.50		1,604.50	1,800
6800	Stainless steel, single bubbler		4	2		1,350	79.50		1,429.50	1,600
7560	For freeze-proof valve system, add		2	4		700	159		859	1,025
7580	For rough-in, supply and waste, add		1.83	4.372		254	173		427	565
7600	Floor mounted, pedestal type									
7700	Aluminum, architectural style, C.I. base	1 Plum	2	4	Ea.	2,200	159		2,359	2,650
7780	Wheelchair handicap unit		2	4		1,500	159		1,659	1,900
8400	Stainless steel, architectural style		2	4		2,650	159		2,809	3,175
8600	Enameled iron, heavy duty service, 2 bubblers		2	4		2,500	159		2,659	3,025
8660	4 bubblers		2	4		3,825	159		3,984	4,450
8880	For freeze-proof valve system, add		2	4		415	159		574	715
8900	For rough-in, supply and waste, add		1.83	4.372		254	173		427	565
9100	Deck mounted									
9500	Stainless steel, circular receptor	1 Plum	4	2	Ea.	370	79.50		449.50	540
9760	White enameled steel, 14" x 9" receptor		4	2		320	79.50		399.50	485
9860	White enameled cast iron, 24" x 16" receptor		3	2.667		315	106		421	525
9980	For rough-in, supply and waste, add		1.83	4.372		254	173		427	565

22 47 16 – Pressure Water Coolers

22 47 16.10 Electric Water Coolers		Crew	Daily Output	Labor-Hours	Unit	Material	2011 Bare Costs Labor	Equipment	Total	Total Incl O&P
0010	**ELECTRIC WATER COOLERS** R224000-50									
0100	Wall mounted, non-recessed									
0140	4 GPH	Q-1	4	4	Ea.	615	143		758	910
0160	8 GPH, barrier free, sensor operated		4	4		890	143		1,033	1,200
0180	8.2 G.P.H.		4	4		680	143		823	985
0600	8 GPH hot and cold water		4	4		1,175	143		1,318	1,525
0640	For stainless steel cabinet, add					84			84	92
1000	Dual height, 8.2 GPH	Q-1	3.80	4.211		1,975	150		2,125	2,425
1040	14.3 GPH	"	3.80	4.211		1,125	150		1,275	1,500
1240	For stainless steel cabinet, add					150			150	165
2600	Wheelchair type, 8 GPH	Q-1	4	4		975	143		1,118	1,300

22 47 Drinking Fountains and Water Coolers

22 47 16 – Pressure Water Coolers

22 47 16.10 Electric Water Coolers		Crew	Daily Output	Labor-Hours	Unit	2011 Bare Costs Material	Labor	Equipment	Total	Total Incl O&P
3300	Semi-recessed, 8.1 GPH	Q-1	4	4	Ea.	630	143		773	925
3320	12 GPH	↓	4	4	↓	725	143		868	1,025
4600	Floor mounted, flush-to-wall									
4640	4 GPH	1 Plum	3	2.667	Ea.	620	106		726	860
4680	8.2 GPH		3	2.667		655	106		761	895
4720	14.3 GPH		3	2.667		750	106		856	1,000
4960	14 GPH hot and cold water	↓	3	2.667		1,050	106		1,156	1,325
4980	For stainless steel cabinet, add					128			128	141
5000	Dual height, 8.2 GPH	1 Plum	2	4		1,000	159		1,159	1,350
5040	14.3 GPH	"	2	4		1,050	159		1,209	1,400
5120	For stainless steel cabinet, add					168			168	185
9800	For supply, waste & vent, all coolers	1 Plum	2.21	3.620	↓	254	144		398	515

22 51 Swimming Pool Plumbing Systems

22 51 19 – Swimming Pool Water Treatment Equipment

22 51 19.50 Swimming Pool Filtration Equipment		Crew	Daily Output	Labor-Hours	Unit	Material	Labor	Equipment	Total	Total Incl O&P
0010	**SWIMMING POOL FILTRATION EQUIPMENT**									
0900	Filter system, sand or diatomite type, incl. pump, 6,000 gal./hr.	2 Plum	1.80	8.889	Total	1,750	350		2,100	2,500
1020	Add for chlorination system, 800 S.F. pool		3	5.333	Ea.	103	211		314	460
1040	5,000 S.F. pool	↓	3	5.333	"	1,875	211		2,086	2,425

22 52 Fountain Plumbing Systems

22 52 16 – Fountain Pumps

22 52 16.10 Fountain Water Pumps		Crew	Daily Output	Labor-Hours	Unit	Material	Labor	Equipment	Total	Total Incl O&P
0010	**FOUNTAIN WATER PUMPS**									
0100	Pump w/controls									
0200	Single phase, 100′ cord, 1/2 H.P. pump	2 Skwk	4.40	3.636	Ea.	1,200	124		1,324	1,525
0300	3/4 H.P. pump		4.30	3.721		2,050	127		2,177	2,450
0400	1 H.P. pump		4.20	3.810		2,225	130		2,355	2,675
0500	1-1/2 H.P. pump		4.10	3.902		2,650	133		2,783	3,125
0600	2 H.P. pump		4	4		3,575	136		3,711	4,150
0700	Three phase, 200′ cord, 5 H.P. pump		3.90	4.103		4,900	140		5,040	5,600
0800	7-1/2 H.P. pump		3.80	4.211		8,500	143		8,643	9,600
0900	10 H.P. pump		3.70	4.324		9,275	147		9,422	10,400
1000	15 H.P. pump	↓	3.60	4.444	↓	11,200	151		11,351	12,600
2000	DESIGN NOTE: Use two horsepower per surface acre.									

22 52 33 – Fountain Ancillary

22 52 33.10 Fountain Miscellaneous		Crew	Daily Output	Labor-Hours	Unit	Material	Labor	Equipment	Total	Total Incl O&P
0010	**FOUNTAIN MISCELLANEOUS**									
1100	Nozzles, minimum	2 Skwk	8	2	Ea.	145	68		213	274
1200	Maximum		8	2		310	68		378	455
1300	Lights w/mounting kits, 200 watt		18	.889		1,025	30.50		1,055.50	1,175
1400	300 watt		18	.889		1,225	30.50		1,255.50	1,400
1500	500 watt		18	.889		1,400	30.50		1,430.50	1,575
1600	Color blender	↓	12	1.333	↓	540	45.50		585.50	670

22 66 53 – Laboratory Chemical-Waste and Vent Piping

22 66 53.30 Glass Pipe		Crew	Daily Output	Labor-Hours	Unit	Material	2011 Bare Costs Labor	Equipment	Total	Total Incl O&P
0010	**GLASS PIPE**, Borosilicate, couplings & clevis hanger assemblies, 10' O.C.									
0020	Drainage									
1100	1-1/2" diameter	Q-1	52	.308	L.F.	10.20	11		21.20	29
1120	2" diameter		44	.364		13.20	12.95		26.15	35.50
1140	3" diameter		39	.410		17.65	14.65		32.30	43.50
1160	4" diameter		30	.533		31	19.05		50.05	65
1180	6" diameter	↓	26	.615	↓	57	22		79	98.50

22 66 53.60 Corrosion Resistant Pipe		Crew	Daily Output	Labor-Hours	Unit	Material	Labor	Equipment	Total	Total Incl O&P
0010	**CORROSION RESISTANT PIPE**, No couplings or hangers									
0020	Iron alloy, drain, mechanical joint									
1000	1-1/2" diameter	Q-1	70	.229	L.F.	46	8.15		54.15	64
1100	2" diameter		66	.242		47	8.65		55.65	65.50
1120	3" diameter		60	.267		60.50	9.50		70	82
1140	4" diameter	↓	52	.308	↓	77.50	11		88.50	103
2980	Plastic, epoxy, fiberglass filament wound, B&S joint									
3000	2" diameter	Q-1	62	.258	L.F.	11.40	9.20		20.60	27.50
3100	3" diameter		51	.314		13.30	11.20		24.50	33
3120	4" diameter		45	.356		19	12.70		31.70	41.50
3140	6" diameter	↓	32	.500	↓	26.50	17.85		44.35	58.50
3980	Polyester, fiberglass filament wound, B&S joint									
4000	2" diameter	Q-1	62	.258	L.F.	12.35	9.20		21.55	28.50
4100	3" diameter		51	.314		16.15	11.20		27.35	36
4120	4" diameter		45	.356		24	12.70		36.70	46.50
4140	6" diameter	↓	32	.500	↓	34	17.85		51.85	66.50
4980	Polypropylene, acid resistant, fire retardant, schedule 40									
5000	1-1/2" diameter	Q-1	68	.235	L.F.	6.05	8.40		14.45	20.50
5100	2" diameter		62	.258		8.20	9.20		17.40	24
5120	3" diameter		51	.314		16.80	11.20		28	37
5140	4" diameter	↓	45	.356	↓	21.50	12.70		34.20	44
5980	Proxylene, fire retardant, Schedule 40									
6000	1-1/2" diameter	Q-1	68	.235	L.F.	9.40	8.40		17.80	24
6100	2" diameter		62	.258		12.90	9.20		22.10	29.50
6120	3" diameter		51	.314		23.50	11.20		34.70	44.50
6140	4" diameter	↓	45	.356	↓	33	12.70		45.70	57

Division Notes

		CREW	DAILY OUTPUT	LABOR-HOURS	UNIT	2011 BARE COSTS MAT.	LABOR	EQUIP.	TOTAL	TOTAL INCL O&P

Estimating Tips

The labor adjustment factors listed in Subdivision 22 01 02.20 also apply to Division 23.

23 10 00 Facility Fuel Systems

- The prices in this subdivision for above- and below-ground storage tanks do not include foundations or hold-down slabs, unless noted. The estimator should refer to Divisions 3 and 31 for foundation system pricing. In addition to the foundations, required tank accessories, such as tank gauges, leak detection devices, and additional manholes and piping, must be added to the tank prices.

23 50 00 Central Heating Equipment

- When estimating the cost of an HVAC system, check to see who is responsible for providing and installing the temperature control system. It is possible to overlook controls, assuming that they would be included in the electrical estimate.
- When looking up a boiler, be careful on specified capacity. Some manufacturers rate their products on output while others use input.
- Include HVAC insulation for pipe, boiler, and duct (wrap and liner).
- Be careful when looking up mechanical items to get the correct pressure rating and connection type (thread, weld, flange).

23 70 00 Central HVAC Equipment

- Combination heating and cooling units are sized by the air conditioning requirements. (See Reference No. R236000-20 for preliminary sizing guide.)
- A ton of air conditioning is nominally 400 CFM.
- Rectangular duct is taken off by the linear foot for each size, but its cost is usually estimated by the pound. Remember that SMACNA standards now base duct on internal pressure.
- Prefabricated duct is estimated and purchased like pipe: straight sections and fittings.
- Note that cranes or other lifting equipment are not included on any lines in Division 23. For example, if a crane is required to lift a heavy piece of pipe into place high above a gym floor, or to put a rooftop unit on the roof of a four-story building, etc., it must be added. Due to the potential for extreme variation—from nothing additional required to a major crane or helicopter—we feel that including a nominal amount for "lifting contingency" would be useless and detract from the accuracy of the estimate. When using equipment rental cost data from RSMeans, do not forget to include the cost of the operator(s).

Reference Numbers

Reference numbers are shown in shaded boxes at the beginning of some major classifications. These numbers refer to related items in the Reference Section. The reference information may be an estimating procedure, an alternate pricing method, or technical information.

Note: Not all subdivisions listed here necessarily appear in this publication.

Note: **Trade Service,** *in part, has been used as a reference source for some of the material prices used in Division 23.*

23 05 Common Work Results for HVAC

23 05 02 – HVAC General

23 05 02.10 Air Conditioning, General		Crew	Daily Output	Labor-Hours	Unit	Material	2011 Bare Costs Labor	Equipment	Total	Total Incl O&P
0010	**AIR CONDITIONING, GENERAL** Prices are for standard efficiencies (SEER 13)									
0020	for upgrade to SEER 14 add					20%				

23 05 05 – Selective HVAC Demolition

23 05 05.10 HVAC Demolition

23 05 05.10 HVAC Demolition		Crew	Daily Output	Labor-Hours	Unit	Material	2011 Bare Costs Labor	Equipment	Total	Total Incl O&P
0010	**HVAC DEMOLITION**									
0100	Air conditioner, split unit, 3 ton	Q-5	2	8	Ea.		288		288	470
0150	Package unit, 3 ton	Q-6	3	8	″		278		278	455
0298	Boilers									
0300	Electric, up thru 148 kW	Q-19	2	12	Ea.		450		450	730
0310	150 thru 518 kW	″	1	24			900		900	1,475
0320	550 thru 2000 kW	Q-21	.40	80			3,050		3,050	4,975
0330	2070 kW and up	″	.30	106			4,075		4,075	6,625
0340	Gas and/or oil, up thru 150 MBH	Q-7	2.20	14.545			525		525	855
0350	160 thru 2000 MBH		.80	40			1,450		1,450	2,350
0360	2100 thru 4500 MBH		.50	64			2,300		2,300	3,775
0370	4600 thru 7000 MBH		.30	106			3,850		3,850	6,275
0380	7100 thru 12,000 MBH		.16	200			7,200		7,200	11,800
0390	12,200 thru 25,000 MBH		.12	266			9,625		9,625	15,700
1000	Ductwork, 4″ high, 8″ wide	1 Clab	200	.040	L.F.		1.06		1.06	1.77
1100	6″ high, 8″ wide		165	.048			1.28		1.28	2.15
1200	10″ high, 12″ wide		125	.064			1.69		1.69	2.84
1300	12″-14″ high, 16″-18″ wide		85	.094			2.49		2.49	4.17
1400	18″ high, 24″ wide		67	.119			3.16		3.16	5.30
1500	30″ high, 36″ wide		56	.143			3.78		3.78	6.35
1540	72″ wide		50	.160			4.23		4.23	7.10
3000	Mechanical equipment, light items. Unit is weight, not cooling.	Q-5	.90	17.778	Ton		640		640	1,050
3600	Heavy items	″	1.10	14.545	″		525		525	855
3700	Deduct for salvage (when applicable), minimum				Job				73	80
3710	Maximum				″				455	500

23 05 23 – General-Duty Valves for HVAC Piping

23 05 23.30 Valves, Iron Body

23 05 23.30 Valves, Iron Body		Crew	Daily Output	Labor-Hours	Unit	Material	2011 Bare Costs Labor	Equipment	Total	Total Incl O&P
0010	**VALVES, IRON BODY**									
1020	Butterfly, wafer type, gear actuator, 200 lb.									
1030	2″	1 Plum	14	.571	Ea.	167	22.50		189.50	220
1040	2-1/2″	Q-1	9	1.778		172	63.50		235.50	293
1050	3″		8	2		178	71.50		249.50	315
1060	4″		5	3.200		202	114		316	410
1070	5″	Q-2	5	4.800		269	165		434	565
1080	6″	″	5	4.800		305	165		470	605
1650	Gate, 125 lb., N.R.S.									
2150	Flanged									
2200	2″	1 Plum	5	1.600	Ea.	550	63.50		613.50	710
2240	2-1/2″	Q-1	5	3.200		565	114		679	805
2260	3″		4.50	3.556		630	127		757	900
2280	4″		3	5.333		905	190		1,095	1,300
2300	6″	Q-2	3	8		1,550	275		1,825	2,150
3550	OS&Y, 125 lb., flanged									
3600	2″	1 Plum	5	1.600	Ea.	350	63.50		413.50	490
3660	3″	Q-1	4.50	3.556		395	127		522	640
3680	4″	″	3	5.333		565	190		755	930
3700	6″	Q-2	3	8		930	275		1,205	1,475
3900	For 175 lb., flanged, add					200%	10%			

23 05 Common Work Results for HVAC

23 05 23 – General-Duty Valves for HVAC Piping

23 05 23.30 Valves, Iron Body		Crew	Daily Output	Labor-Hours	Unit	2011 Bare Costs Material	Labor	Equipment	Total	Total Incl O&P
5450	Swing check, 125 lb., threaded									
5470	1"	1 Plum	13	.615	Ea.	575	24.50		599.50	675
5500	2"	"	11	.727		705	29		734	820
5540	2-1/2"	Q-1	15	1.067		675	38		713	805
5550	3"		13	1.231		750	44		794	895
5560	4"		10	1.600		1,300	57		1,357	1,525
5950	Flanged									
6000	2"	1 Plum	5	1.600	Ea.	284	63.50		347.50	415
6040	2-1/2"	Q-1	5	3.200		286	114		400	500
6050	3"		4.50	3.556		305	127		432	540
6060	4"		3	5.333		480	190		670	840
6070	6"	Q-2	3	8		820	275		1,095	1,350

23 05 23.80 Valves, Steel		Crew	Daily Output	Labor-Hours	Unit	Material	Labor	Equipment	Total	Total Incl O&P
0010	**VALVES, STEEL**									
0800	Cast									
1350	Check valve, swing type, 150 lb., flanged									
1370	1"	1 Plum	10	.800	Ea.	545	31.50		576.50	650
1400	2"	"	8	1		660	39.50		699.50	790
1440	2-1/2"	Q-1	5	3.200		765	114		879	1,025
1450	3"		4.50	3.556		780	127		907	1,050
1460	4"		3	5.333		1,125	190		1,315	1,525
1540	For 300 lb., flanged, add					50%	15%			
1548	For 600 lb., flanged, add					110%	20%			
1950	Gate valve, 150 lb., flanged									
2000	2"	1 Plum	8	1	Ea.	710	39.50		749.50	845
2040	2-1/2"	Q-1	5	3.200		1,000	114		1,114	1,275
2050	3"		4.50	3.556		1,000	127		1,127	1,300
2060	4"		3	5.333		1,250	190		1,440	1,675
2070	6"	Q-2	3	8		1,850	275		2,125	2,475
3650	Globe valve, 150 lb., flanged									
3700	2"	1 Plum	8	1	Ea.	890	39.50		929.50	1,050
3740	2-1/2"	Q-1	5	3.200		1,125	114		1,239	1,425
3750	3"		4.50	3.556		1,125	127		1,252	1,450
3760	4"		3	5.333		1,650	190		1,840	2,125
3770	6"	Q-2	3	8		2,600	275		2,875	3,300
5150	Forged									
5650	Check valve, class 800, horizontal, socket									
5698	Threaded									
5700	1/4"	1 Plum	24	.333	Ea.	88	13.20		101.20	118
5720	3/8"		24	.333		88	13.20		101.20	118
5730	1/2"		24	.333		88	13.20		101.20	118
5740	3/4"		20	.400		94.50	15.85		110.35	130
5750	1"		19	.421		111	16.70		127.70	150
5760	1-1/4"		15	.533		217	21		238	274

23 05 93 – Testing, Adjusting, and Balancing for HVAC

23 05 93.10 Balancing, Air		Crew	Daily Output	Labor-Hours	Unit	Material	Labor	Equipment	Total	Total Incl O&P
0010	**BALANCING, AIR** (Subcontractor's quote incl. material and labor)									
0900	Heating and ventilating equipment									
1000	Centrifugal fans, utility sets				Ea.				340	375
1100	Heating and ventilating unit								510	560
1200	In-line fan								510	560
1300	Propeller and wall fan								96.50	106

23 05 Common Work Results for HVAC

23 05 93 – Testing, Adjusting, and Balancing for HVAC

23 05 93.10 Balancing, Air		Crew	Daily Output	Labor-Hours	Unit	Material	Labor	Equipment	Total	Total Incl O&P
						2011 Bare Costs				
1400	Roof exhaust fan				Ea.				227	250
2000	Air conditioning equipment, central station								740	810
2100	Built-up low pressure unit								680	750
2200	Built-up high pressure unit								795	875
2500	Multi-zone A.C. and heating unit								510	560
2600	For each zone over one, add								114	125
2700	Package A.C. unit								285	310
2800	Rooftop heating and cooling unit								400	440
3000	Supply, return, exhaust, registers & diffusers, avg. height ceiling								68	75
3100	High ceiling								102	112
3200	Floor height								57	62

23 05 93.20 Balancing, Water

Line	Description	Crew	Daily Output	Labor-Hours	Unit	Material	Labor	Equipment	Total	Total Incl O&P
0010	**BALANCING, WATER** (Subcontractor's quote incl. material and labor)									
0050	Air cooled condenser				Ea.				206	226
0080	Boiler								415	455
0100	Cabinet unit heater								70.50	77.50
0200	Chiller								500	550
0300	Convector								59	64.50
0500	Cooling tower								380	420
0600	Fan coil unit, unit ventilator								106	116
0700	Fin tube and radiant panels								118	129
0800	Main and duct re-heat coils								109	120
0810	Heat exchanger								109	120
1000	Pumps								259	284
1100	Unit heater								82.50	90.50

23 07 HVAC Insulation

23 07 13 – Duct Insulation

23 07 13.10 Duct Thermal Insulation

Line	Description	Crew	Daily Output	Labor-Hours	Unit	Material	Labor	Equipment	Total	Total Incl O&P
0010	**DUCT THERMAL INSULATION**									
0100	Rule of thumb, as a percentage of total mechanical costs				Job				10%	
0110	Insulation req'd is based on the surface size/area to be covered									
3000	Ductwork									
3020	Blanket type, fiberglass, flexible									
3140	FSK vapor barrier wrap, .75 lb. density									
3160	1" thick G	Q-14	350	.046	S.F.	.20	1.43		1.63	2.63
3170	1-1/2" thick G		320	.050		.23	1.56		1.79	2.88
3180	2" thick G		300	.053		.28	1.67		1.95	3.12
3190	3" thick G		260	.062		.40	1.92		2.32	3.68
3200	4" thick G		242	.066		.60	2.06		2.66	4.14
3212	Vinyl Jacket, .75 lb. density, 1-1/2" thick G		320	.050		.23	1.56		1.79	2.88
3280	Unfaced, 1 lb. density									
3310	1" thick G	Q-14	360	.044	S.F.	.19	1.39		1.58	2.55
3320	1-1/2" thick G		330	.048		.23	1.51		1.74	2.80
3330	2" thick G		310	.052		.31	1.61		1.92	3.06
3400	FSK facing, 1 lb. density									
3420	1-1/2" thick G	Q-14	310	.052	S.F.	.18	1.61		1.79	2.92
3430	2" thick G	"	300	.053	"	.20	1.67		1.87	3.03
3450	FSK facing, 1.5 lb. density									
3470	1-1/2" thick G	Q-14	300	.053	S.F.	.27	1.67		1.94	3.11
3480	2" thick G	"	290	.055	"	.33	1.72		2.05	3.26

23 07 HVAC Insulation

23 07 13 – Duct Insulation

23 07 13.10 Duct Thermal Insulation			Crew	Daily Output	Labor-Hours	Unit	2011 Bare Costs Material	Labor	Equipment	Total	Total Incl O&P
3795	Finishes										
3800	Stainless steel woven mesh		Q-14	100	.160	S.F.	.84	5		5.84	9.30
3820	18 oz. fiberglass cloth, pasted on			170	.094		.77	2.94		3.71	5.80
3900	8 oz. canvas, pasted on			180	.089		.49	2.78		3.27	5.20

23 07 16 – HVAC Equipment Insulation

23 07 16.10 HVAC Equipment Thermal Insulation

Line	Description		Crew	Daily Output	Labor-Hours	Unit	Material	Labor	Equipment	Total	Total Incl O&P
0010	**HVAC EQUIPMENT THERMAL INSULATION**										
0100	Rule of thumb, as a percentage of total mechanical costs					Job				10%	
0110	Insulation req'd is based on the surface size/area to be covered										
1000	Boiler, 1-1/2" calcium silicate only	G	Q-14	110	.145	S.F.	2.92	4.54		7.46	10.85
1020	Plus 2" fiberglass	G	"	80	.200	"	3.82	6.25		10.07	14.75
2000	Breeching, 2" calcium silicate										
2020	Rectangular	G	Q-14	42	.381	S.F.	5.70	11.90		17.60	26.50
2040	Round	G	"	38.70	.413	"	5.95	12.90		18.85	28.50

23 09 Instrumentation and Control for HVAC

23 09 33 – Electric and Electronic Control System for HVAC

23 09 33.10 Electronic Control Systems

Line	Description		Crew	Daily Output	Labor-Hours	Unit	Material	Labor	Equipment	Total	Total Incl O&P
0010	**ELECTRONIC CONTROL SYSTEMS**										
0020	For electronic costs, add to Section 23 09 43.10					Ea.					15%

23 09 43 – Pneumatic Control System for HVAC

23 09 43.10 Pneumatic Control Systems

Line	Description		Crew	Daily Output	Labor-Hours	Unit	Material	Labor	Equipment	Total	Total Incl O&P
0010	**PNEUMATIC CONTROL SYSTEMS**										
0011	Including a nominal 50 ft. of tubing. Add control panelboard if req'd.										
0100	Heating and ventilating, split system										
0200	Mixed air control, economizer cycle, panel readout, tubing										
0220	Up to 10 tons	G	Q-19	.68	35.294	Ea.	3,675	1,325		5,000	6,200
0240	For 10 to 20 tons	G		.63	37.915		3,925	1,425		5,350	6,650
0260	For over 20 tons	G		.58	41.096		4,275	1,550		5,825	7,200
0300	Heating coil, hot water, 3 way valve,										
0320	Freezestat, limit control on discharge, readout		Q-5	.69	23.088	Ea.	2,725	830		3,555	4,350
0500	Cooling coil, chilled water, room										
0520	Thermostat, 3 way valve		Q-5	2	8	Ea.	1,225	288		1,513	1,825
0600	Cooling tower, fan cycle, damper control,										
0620	Control system including water readout in/out at panel		Q-19	.67	35.821	Ea.	4,850	1,350		6,200	7,500
1000	Unit ventilator, day/night operation,										
1100	freezestat, ASHRAE, cycle 2		Q-19	.91	26.374	Ea.	2,675	990		3,665	4,550
2000	Compensated hot water from boiler, valve control,										
2100	readout and reset at panel, up to 60 GPM		Q-19	.55	43.956	Ea.	5,000	1,650		6,650	8,175
2120	For 120 GPM			.51	47.059		5,350	1,750		7,100	8,775
2140	For 240 GPM			.49	49.180		5,600	1,850		7,450	9,175
3000	Boiler room combustion air, damper to 5 S.F., controls			1.37	17.582		2,425	660		3,085	3,725
3500	Fan coil, heating and cooling valves, 4 pipe control system			3	8		1,100	300		1,400	1,700
3600	Heat exchanger system controls			.86	27.907		2,350	1,050		3,400	4,275
4000	Pneumatic thermostat, including controlling room radiator valve		Q-5	2.43	6.593		730	238		968	1,200
4060	Pump control system		Q-19	3	8		1,125	300		1,425	1,725
4500	Air supply for pneumatic control system										
4600	Tank mounted duplex compressor, starter, alternator,										
4620	piping, dryer, PRV station and filter										
4630	1/2 HP		Q-19	.68	35.139	Ea.	8,975	1,325		10,300	12,000

23 09 Instrumentation and Control for HVAC

23 09 43 – Pneumatic Control System for HVAC

23 09 43.10 Pneumatic Control Systems		Crew	Daily Output	Labor-Hours	Unit	Material	2011 Bare Costs Labor	Equipment	Total	Total Incl O&P
4660	1-1/2 HP	Q-19	.58	41.739	Ea.	11,000	1,575		12,575	14,700
4690	5 HP	↓	.42	57.143	↓	26,100	2,150		28,250	32,200

23 13 Facility Fuel-Storage Tanks

23 13 13 – Facility Underground Fuel-Oil, Storage Tanks

23 13 13.09 Single-Wall Steel Fuel-Oil Tanks

Line	Description	Crew	Daily Output	Labor-Hours	Unit	Material	Labor	Equipment	Total	Total Incl O&P
0010	**SINGLE-WALL STEEL FUEL-OIL TANKS**									
5000	Steel underground, sti-P3, set in place, not incl. hold-down bars.									
5500	Excavation, pad, pumps and piping not included									
5510	Single wall, 500 gallon capacity, 7 gauge shell	Q-5	2.70	5.926	Ea.	2,050	214		2,264	2,600
5520	1,000 gallon capacity, 7 gauge shell	"	2.50	6.400		3,150	231		3,381	3,850
5530	2,000 gallon capacity, 1/4" thick shell	Q-7	4.60	6.957		5,125	251		5,376	6,025
5535	2,500 gallon capacity, 7 gauge shell	Q-5	3	5.333		5,650	192		5,842	6,550
5540	5,000 gallon capacity, 1/4" thick shell	Q-7	3.20	10		7,300	360		7,660	8,650
5580	15,000 gallon capacity, 5/16" thick shell		1.70	18.824		17,400	680		18,080	20,200
5600	20,000 gallon capacity, 5/16" thick shell		1.50	21.333		23,000	770		23,770	26,600
5610	25,000 gallon capacity, 3/8" thick shell		1.30	24.615		31,100	885		31,985	35,700
5620	30,000 gallon capacity, 3/8" thick shell		1.10	29.091		31,700	1,050		32,750	36,600
5630	40,000 gallon capacity, 3/8" thick shell		.90	35.556		35,600	1,275		36,875	41,300
5640	50,000 gallon capacity, 3/8" thick shell	↓	.80	40	↓	39,500	1,450		40,950	45,900

23 13 13.23 Glass-Fiber-Reinfcd-Plastic, Fuel-Oil, Storage

Line	Description	Crew	Daily Output	Labor-Hours	Unit	Material	Labor	Equipment	Total	Total Incl O&P
0010	**GLASS-FIBER-REINFCD-PLASTIC, UNDERGRND FUEL-OIL, STORAGE**									
0210	Fiberglass, underground, single wall, U.L. listed, not including									
0220	manway or hold-down strap									
0240	2,000 gallon capacity	Q-7	4.57	7.002	Ea.	6,075	252		6,327	7,100
0250	4,000 gallon capacity		3.55	9.014		8,650	325		8,975	10,100
0260	6,000 gallon capacity		2.67	11.985		9,600	430		10,030	11,300
0280	10,000 gallon capacity		2	16		12,900	575		13,475	15,100
0284	15,000 gallon capacity		1.68	19.048		23,900	685		24,585	27,400
0290	20,000 gallon capacity	↓	1.45	22.069		25,800	795		26,595	29,700
0500	For manway, fittings and hold-downs, add				↓	20%	15%			
1020	Fiberglass, underground, double wall, U.L. listed									
1030	includes manways, not incl. hold-down straps									
1040	600 gallon capacity	Q-5	2.42	6.612	Ea.	6,925	238		7,163	8,025
1050	1,000 gallon capacity	"	2.25	7.111		8,650	256		8,906	9,950
1060	2,500 gallon capacity	Q-7	4.16	7.692		13,400	277		13,677	15,300
1070	3,000 gallon capacity		3.90	8.205		14,500	296		14,796	16,500
1080	4,000 gallon capacity		3.64	8.791		16,000	315		16,315	18,200
1090	6,000 gallon capacity		2.42	13.223		18,000	475		18,475	20,600
1100	8,000 gallon capacity		2.08	15.385		21,300	555		21,855	24,400
1110	10,000 gallon capacity		1.82	17.582		24,000	635		24,635	27,400
1120	12,000 gallon capacity	↓	1.70	18.824	↓	29,600	680		30,280	33,600
2210	Fiberglass, underground, single wall, U.L. listed, including									
2220	hold-down straps, no manways									
2240	2,000 gallon capacity	Q-7	3.55	9.014	Ea.	6,525	325		6,850	7,700
2250	4,000 gallon capacity		2.90	11.034		9,100	400		9,500	10,700
2260	6,000 gallon capacity		2	16		10,500	575		11,075	12,400
2280	10,000 gallon capacity		1.60	20		13,800	720		14,520	16,400
2284	15,000 gallon capacity		1.39	23.022		24,800	830		25,630	28,700
2290	20,000 gallon capacity	↓	1.14	28.070	↓	27,100	1,000		28,100	31,600
3020	Fiberglass, underground, double wall, U.L. listed									

23 13 Facility Fuel-Storage Tanks

23 13 13 – Facility Underground Fuel-Oil, Storage Tanks

23 13 13.23	Glass-Fiber-Reinfcd-Plastic, Fuel-Oil, Storage	Crew	Daily Output	Labor-Hours	Unit	Material	Labor	Equipment	Total	Total Incl O&P
						2011 Bare Costs				
3030	includes manways and hold-down straps									
3040	600 gallon capacity	Q-5	1.86	8.602	Ea.	7,375	310		7,685	8,600
3050	1,000 gallon capacity	"	1.70	9.412		9,100	340		9,440	10,600
3060	2,500 gallon capacity	Q-7	3.29	9.726		13,900	350		14,250	15,900
3070	3,000 gallon capacity		3.13	10.224		15,000	370		15,370	17,100
3080	4,000 gallon capacity		2.93	10.922		16,500	395		16,895	18,700
3090	6,000 gallon capacity		1.86	17.204		18,900	620		19,520	21,800
3100	8,000 gallon capacity		1.65	19.394		22,200	700		22,900	25,600
3110	10,000 gallon capacity		1.48	21.622		24,900	780		25,680	28,700
3120	12,000 gallon capacity		1.40	22.857		30,500	825		31,325	34,900

23 13 23 – Facility Aboveground Fuel-Oil, Storage Tanks

23 13 23.13 Vertical, Steel, Abvground Fuel-Oil, Stor. Tanks

Line	Description	Crew	Daily Output	Labor-Hours	Unit	Material	Labor	Equipment	Total	Total Incl O&P
0010	**VERTICAL, STEEL, ABOVEGROUND FUEL-OIL, STORAGE TANKS**									
4000	Fixed roof oil storage tanks, steel, (1 BBL=42 gal. w/foundation 3'D x 1'W)									
4200	5,000 barrels				Ea.				194,000	213,500
4300	24,000 barrels								333,500	367,000
4500	56,000 barrels								729,000	802,000
4600	110,000 barrels								1,060,000	1,166,000
4800	143,000 barrels								1,250,000	1,375,000
4900	224,000 barrels								1,360,000	1,496,000
5100	Floating roof gasoline tanks, steel, 5,000 barrels (w/foundation 3'D x 1'W)								204,000	225,000
5200	25,000 barrels								381,000	419,000
5400	55,000 barrels								839,000	923,000
5500	100,000 barrels								1,253,000	1,379,000
5700	150,000 barrels								1,532,000	1,685,000
5800	225,000 barrels								2,300,000	2,783,000

23 13 23.16 Horizontal, Stl, Abvgrd Fuel-Oil, Storage Tanks

Line	Description	Crew	Daily Output	Labor-Hours	Unit	Material	Labor	Equipment	Total	Total Incl O&P
0010	**HORIZONTAL, STEEL, ABOVEGROUND FUEL-OIL, STORAGE TANKS**									
3000	Steel, storage, above ground, including cradles, coating,									
3020	fittings, not including foundation, pumps or piping									
3040	Single wall, 275 gallon	Q-5	5	3.200	Ea.	405	115		520	635
3060	550 gallon	"	2.70	5.926		2,800	214		3,014	3,450
3080	1,000 gallon	Q-7	5	6.400		3,175	231		3,406	3,875
3100	1,500 gallon		4.75	6.737		5,025	243		5,268	5,950
3120	2,000 gallon		4.60	6.957		5,775	251		6,026	6,750
3140	5,000 gallon		3.20	10		6,075	360		6,435	7,300
3150	10,000 gallon		2	16		26,400	575		26,975	29,900
3160	15,000 gallon		1.70	18.824		24,800	680		25,480	28,400
3170	20,000 gallon		1.45	22.069		31,700	795		32,495	36,100
3180	25,000 gallon		1.30	24.615		32,500	885		33,385	37,200
3190	30,000 gallon		1.10	29.091		42,300	1,050		43,350	48,200
3320	Double wall, 500 gallon capacity	Q-5	2.40	6.667		2,200	240		2,440	2,825
3330	2000 gallon capacity	Q-7	4.15	7.711		8,250	278		8,528	9,525
3340	4000 gallon capacity		3.60	8.889		14,700	320		15,020	16,700
3350	6000 gallon capacity		2.40	13.333		17,400	480		17,880	19,900
3360	8000 gallon capacity		2	16		22,300	575		22,875	25,400
3370	10000 gallon capacity		1.80	17.778		25,000	640		25,640	28,600
3380	15000 gallon capacity		1.50	21.333		37,900	770		38,670	43,000
3390	20000 gallon capacity		1.30	24.615		43,200	885		44,085	49,000
3400	25000 gallon capacity		1.15	27.826		52,500	1,000		53,500	59,000
3410	30000 gallon capacity		1	32		57,500	1,150		58,650	65,500

23 21 Hydronic Piping and Pumps

23 21 20 – Hydronic HVAC Piping Specialties

23 21 20.10 Air Control

	23 21 20.10 Air Control	Crew	Daily Output	Labor-Hours	Unit	2011 Bare Costs Material	Labor	Equipment	Total	Total Incl O&P
0010	**AIR CONTROL**									
0030	Air separator, with strainer									
0040	2″ diameter	Q-5	6	2.667	Ea.	865	96		961	1,100
0080	2-1/2″ diameter		5	3.200		980	115		1,095	1,275
0100	3″ diameter		4	4		1,500	144		1,644	1,850
0120	4″ diameter		3	5.333		2,150	192		2,342	2,700
0130	5″ diameter	Q-6	3.60	6.667		2,750	231		2,981	3,400
0140	6″ diameter	″	3.40	7.059		3,275	245		3,520	4,000

23 21 20.18 Automatic Air Vent

	23 21 20.18 Automatic Air Vent	Crew	Daily Output	Labor-Hours	Unit	Material	Labor	Equipment	Total	Total Incl O&P
0010	**AUTOMATIC AIR VENT**									
0020	Cast iron body, stainless steel internals, float type									
0060	1/2″ NPT inlet, 300 psi	1 Stpi	12	.667	Ea.	90.50	26.50		117	143
0220	3/4″ NPT inlet, 250 psi	″	10	.800		289	32		321	370
0340	1-1/2″ NPT inlet, 250 psi	Q-5	12	1.333		895	48		943	1,075

23 21 20.42 Expansion Joints

	23 21 20.42 Expansion Joints	Crew	Daily Output	Labor-Hours	Unit	Material	Labor	Equipment	Total	Total Incl O&P
0010	**EXPANSION JOINTS**									
0100	Bellows type, neoprene cover, flanged spool									
0140	6″ face to face, 1-1/4″ diameter	1 Stpi	11	.727	Ea.	248	29		277	320
0160	1-1/2″ diameter	″	10.60	.755		248	30		278	325
0180	2″ diameter	Q-5	13.30	1.203		251	43.50		294.50	345
0190	2-1/2″ diameter		12.40	1.290		260	46.50		306.50	360
0200	3″ diameter		11.40	1.404		291	50.50		341.50	405
0480	10″ face to face, 2″ diameter		13	1.231		360	44.50		404.50	470
0500	2-1/2″ diameter		12	1.333		380	48		428	500
0520	3″ diameter		11	1.455		390	52.50		442.50	510
0540	4″ diameter		8	2		440	72		512	605
0560	5″ diameter		7	2.286		525	82.50		607.50	715
0580	6″ diameter		6	2.667		545	96		641	755

23 21 20.46 Expansion Tanks

	23 21 20.46 Expansion Tanks	Crew	Daily Output	Labor-Hours	Unit	Material	Labor	Equipment	Total	Total Incl O&P
0010	**EXPANSION TANKS**									
1507	Underground fuel-oil storage tanks, see Section 23 13 13									
1512	Tank leak detection systems, see Section 28 33 33.50									
2000	Steel, liquid expansion, ASME, painted, 15 gallon capacity	Q-5	17	.941	Ea.	555	34		589	670
2020	24 gallon capacity		14	1.143		595	41		636	720
2040	30 gallon capacity		12	1.333		625	48		673	765
2060	40 gallon capacity		10	1.600		730	57.50		787.50	895
2080	60 gallon capacity		8	2		875	72		947	1,075
2100	80 gallon capacity		7	2.286		935	82.50		1,017.50	1,150
2120	100 gallon capacity		6	2.667		1,275	96		1,371	1,550
3000	Steel ASME expansion, rubber diaphragm, 19 gal. cap. accept.		12	1.333		2,225	48		2,273	2,525
3020	31 gallon capacity		8	2		2,475	72		2,547	2,850
3040	61 gallon capacity		6	2.667		3,475	96		3,571	3,975
3080	119 gallon capacity		4	4		3,750	144		3,894	4,350
3100	158 gallon capacity		3.80	4.211		5,200	152		5,352	5,975
3140	317 gallon capacity		2.80	5.714		7,850	206		8,056	8,975
3180	528 gallon capacity		2.40	6.667		12,800	240		13,040	14,400

23 21 20.58 Hydronic Heating Control Valves

	23 21 20.58 Hydronic Heating Control Valves	Crew	Daily Output	Labor-Hours	Unit	Material	Labor	Equipment	Total	Total Incl O&P
0010	**HYDRONIC HEATING CONTROL VALVES**									
0100	Radiator supply, 1/2″ diameter	1 Stpi	24	.333	Ea.	58	13.35		71.35	85.50
0120	3/4″ diameter		20	.400		59.50	16		75.50	91
0140	1″ diameter		19	.421		74.50	16.85		91.35	109
0160	1-1/4″ diameter		15	.533		107	21.50		128.50	152

23 21 20 – Hydronic HVAC Piping Specialties

23 21 20.58 Hydronic Heating Control Valves		Crew	Daily Output	Labor-Hours	Unit	Material	2011 Bare Costs Labor	Equipment	Total	Total Incl O&P
0500	For low pressure steam, add					25%				
23 21 20.70 Steam Traps										
0010	**STEAM TRAPS**									
0030	Cast iron body, threaded									
0040	Inverted bucket									
0050	1/2" pipe size	1 Stpi	12	.667	Ea.	131	26.50		157.50	188
0070	3/4" pipe size		10	.800		228	32		260	305
0100	1" pipe size		9	.889		350	35.50		385.50	445
0120	1-1/4" pipe size	↓	8	1	↓	530	40		570	650
1000	Float & thermostatic, 15 psi									
1010	3/4" pipe size	1 Stpi	16	.500	Ea.	105	20		125	149
1020	1" pipe size		15	.533		127	21.50		148.50	174
1040	1-1/2" pipe size		9	.889		224	35.50		259.50	305
1060	2" pipe size	↓	6	1.333	↓	410	53.50		463.50	540
23 21 20.76 Strainers, Y Type, Bronze Body										
0010	**STRAINERS, Y TYPE, BRONZE BODY**									
0050	Screwed, 150 lb., 1/4" pipe size	1 Stpi	24	.333	Ea.	17.30	13.35		30.65	41
0070	3/8" pipe size		24	.333		23	13.35		36.35	47
0100	1/2" pipe size		20	.400		23	16		39	51
0140	1" pipe size		17	.471		30.50	18.85		49.35	65
0160	1-1/2" pipe size		14	.571		66	23		89	111
0180	2" pipe size		13	.615		88.50	24.50		113	138
0182	3" pipe size	↓	12	.667		670	26.50		696.50	780
0220	3" pipe size	Q-5	16	1		835	36		871	980
0240	4" pipe size	"	15	1.067	↓	1,900	38.50		1,938.50	2,175
0500	For 300 lb. rating 1/4" thru 2", add					15%				
1000	Flanged, 150 lb., 1-1/2" pipe size	1 Stpi	11	.727	Ea.	435	29		464	530
1020	2" pipe size	"	8	1		590	40		630	715
1030	2-1/2" pipe size	Q-5	5	3.200		875	115		990	1,150
1040	3" pipe size		4.50	3.556		1,075	128		1,203	1,375
1060	4" pipe size	↓	3	5.333		1,625	192		1,817	2,125
1100	6" pipe size	Q-6	3	8		3,125	278		3,403	3,875
1106	8" pipe size	"	2.60	9.231	↓	3,425	320		3,745	4,300
1500	For 300 lb. rating, add					40%				
23 21 20.78 Strainers, Y Type, Iron Body										
0010	**STRAINERS, Y TYPE, IRON BODY**									
0050	Screwed, 250 lb., 1/4" pipe size	1 Stpi	20	.400	Ea.	9.60	16		25.60	36.50
0070	3/8" pipe size		20	.400		9.60	16		25.60	36.50
0100	1/2" pipe size		20	.400		9.60	16		25.60	36.50
0140	1" pipe size		16	.500		15.85	20		35.85	50
0160	1-1/2" pipe size		12	.667		26	26.50		52.50	72
0180	2" pipe size	↓	8	1		39	40		79	109
0220	3" pipe size	Q-5	11	1.455		251	52.50		303.50	360
0240	4" pipe size	"	5	3.200	↓	425	115		540	655
0500	For galvanized body, add					50%				
1000	Flanged, 125 lb., 1-1/2" pipe size	1 Stpi	11	.727	Ea.	135	29		164	196
1020	2" pipe size	"	8	1		142	40		182	222
1040	3" pipe size	Q-5	4.50	3.556		133	128		261	355
1060	4" pipe size	"	3	5.333		242	192		434	580
1080	5" pipe size	Q-6	3.40	7.059		380	245		625	815
1100	6" pipe size	"	3	8	↓	460	278		738	965
1500	For 250 lb. rating, add					20%				

23 21 Hydronic Piping and Pumps

23 21 20 – Hydronic HVAC Piping Specialties

23 21 20.78 Strainers, Y Type, Iron Body		Crew	Daily Output	Labor-Hours	Unit	Material	2011 Bare Costs Labor	Equipment	Total	Total Incl O&P
2000	For galvanized body, add					50%				
2500	For steel body, add					40%				

23 21 20.88 Venturi Flow

		Crew	Daily Output	Labor-Hours	Unit	Material	Labor	Equipment	Total	Total Incl O&P
0010	**VENTURI FLOW**, Measuring device									
0050	1/2" diameter	1 Stpi	24	.333	Ea.	281	13.35		294.35	330
0120	1" diameter		19	.421		276	16.85		292.85	335
0140	1-1/4" diameter		15	.533		340	21.50		361.50	410
0160	1-1/2" diameter		13	.615		355	24.50		379.50	430
0180	2" diameter		11	.727		365	29		394	455
0220	3" diameter	Q-5	14	1.143		515	41		556	640
0240	4" diameter	"	11	1.455		775	52.50		827.50	935
0280	6" diameter	Q-6	3.50	6.857		1,125	238		1,363	1,650
0500	For meter, add					2,125			2,125	2,350

23 21 23 – Hydronic Pumps

23 21 23.13 In-Line Centrifugal Hydronic Pumps

		Crew	Daily Output	Labor-Hours	Unit	Material	Labor	Equipment	Total	Total Incl O&P
0010	**IN-LINE CENTRIFUGAL HYDRONIC PUMPS**									
0600	Bronze, sweat connections, 1/40 HP, in line									
0640	3/4" size	Q-1	16	1	Ea.	192	35.50		227.50	270
1000	Flange connection, 3/4" to 1-1/2" size									
1040	1/12 HP	Q-1	6	2.667	Ea.	495	95		590	700
1060	1/8 HP		6	2.667		855	95		950	1,100
1100	1/3 HP		6	2.667		955	95		1,050	1,200
1140	2" size, 1/6 HP		5	3.200		1,025	114		1,139	1,300
1180	2-1/2" size, 1/4 HP		5	3.200		1,525	114		1,639	1,875
2000	Cast iron, flange connection									
2040	3/4" to 1-1/2" size, in line, 1/12 HP	Q-1	6	2.667	Ea.	320	95		415	505
2101	Pumps, circulating, 3/4" to 1-1/2" size, 1/3 HP		6	2.667		695	95		790	920
2140	2" size, 1/6 HP		5	3.200		655	114		769	905
2180	2-1/2" size, 1/4 HP		5	3.200		830	114		944	1,100
2220	3" size, 1/4 HP		4	4		840	143		983	1,150
2600	For non-ferrous impeller, add					3%				

23 21 29 – Automatic Condensate Pump Units

23 21 29.10 Condensate Removal Pump System

		Crew	Daily Output	Labor-Hours	Unit	Material	Labor	Equipment	Total	Total Incl O&P
0010	**CONDENSATE REMOVAL PUMP SYSTEM**									
0020	Pump with 1 gal. ABS tank G									
0100	115 V G									
0120	1/50 HP, 200 GPH G	1 Stpi	12	.667	Ea.	176	26.50		202.50	238
0140	1/18 HP, 270 GPH G		10	.800		183	32		215	254
0160	1/5 HP, 450 GPH G		8	1		420	40		460	530
0200	230 V G									
0240	1/18 HP, 270 GPH G	1 Stpi	10	.800	Ea.	195	32		227	268
0260	1/5 HP, 450 GPH G	"	8	1	"	465	40		505	580

23 22 Steam and Condensate Piping and Pumps

23 22 13 – Steam and Condensate Heating Piping

23 22 13.23 Aboveground Steam and Condensate Piping		Crew	Daily Output	Labor-Hours	Unit	2011 Bare Costs Material	Labor	Equipment	Total	Total Incl O&P
0010	**ABOVEGROUND STEAM AND CONDENSATE HEATING PIPING**									
0020	Condensate meter									
0100	500 lb. per hour	1 Stpi	14	.571	Ea.	2,350	23		2,373	2,650
0140	1500 lb. per hour	"	7	1.143	"	2,550	46		2,596	2,875

23 22 23 – Steam Condensate Pumps

23 22 23.10 Condensate Return System

		Crew	Daily Output	Labor-Hours	Unit	Material	Labor	Equipment	Total	Total Incl O&P
0010	**CONDENSATE RETURN SYSTEM**									
2000	Simplex									
2010	With pump, motor, CI receiver, float switch									
2020	3/4 HP, 15 GPM	Q-1	1.80	8.889	Ea.	5,550	315		5,865	6,625
2100	Duplex									
2110	With 2 pumps and motors, CI receiver, float switch, alternator									
2120	3/4 HP, 15 GPM, 15 gal. CI rcvr	Q-1	1.40	11.429	Ea.	6,050	410		6,460	7,325
2130	1 HP, 25 GPM		1.20	13.333		7,250	475		7,725	8,750
2140	1-1/2 HP, 45 GPM		1	16		8,425	570		8,995	10,200
2150	1-1/2 HP, 60 GPM	↓	1	16	↓	9,450	570		10,020	11,300

23 31 HVAC Ducts and Casings

23 31 13 – Metal Ducts

23 31 13.13 Rectangular Metal Ducts

		Crew	Daily Output	Labor-Hours	Unit	Material	Labor	Equipment	Total	Total Incl O&P
0010	**RECTANGULAR METAL DUCTS** R233100-20									
0020	Fabricated rectangular, includes fittings, joints, supports,									
0030	allowance for flexible connections, no insulation									
0031	NOTE: Fabrication and installation are combined									
0040	as LABOR cost. Approx. 25% fittings assumed.									
0100	Aluminum, alloy 3003-H14, under 100 lb.	Q-10	75	.320	Lb.	3.51	11.40		14.91	22.50
0110	100 to 500 lb.		80	.300		2.31	10.70		13.01	20
0120	500 to 1,000 lb.		95	.253		2.19	9		11.19	17.25
0140	1,000 to 2,000 lb.		120	.200		2.11	7.15		9.26	14.10
0150	2,000 to 5,000 lb.		130	.185		2.11	6.60		8.71	13.15
0160	Over 5,000 lb.		145	.166		2.11	5.90		8.01	12.05
0500	Galvanized steel, under 200 lb.		235	.102		.66	3.64		4.30	6.75
0520	200 to 500 lb.		245	.098		.65	3.49		4.14	6.45
0540	500 to 1,000 lb.		255	.094		.64	3.36		4	6.25
0560	1,000 to 2,000 lb.		265	.091		.63	3.23		3.86	6.05
0580	Over 5,000 lb.		285	.084		.62	3		3.62	5.65
1000	Stainless steel, type 304, under 100 lb.		165	.145		2.40	5.20		7.60	11.20
1020	100 to 500 lb.		175	.137		4.02	4.89		8.91	12.45
1030	500 to 1,000 lb.		190	.126		3.68	4.50		8.18	11.50
1040	1,000 to 2,000 lb.		200	.120		3.65	4.28		7.93	11.05
1050	2,000 to 5,000 lb.		225	.107		3.31	3.80		7.11	9.95
1060	Over 5,000 lb.	↓	235	.102		3.26	3.64		6.90	9.60
1100	For medium pressure ductwork, add						15%			
1200	For high pressure ductwork, add						40%			
1210	For welded ductwork, add						85%			
1220	For 30% fittings, add						11%			
1224	For 40% fittings, add						34%			
1228	For 50% fittings, add						56%			
1232	For 60% fittings, add						79%			
1236	For 70% fittings, add				↓		101%			

23 31 HVAC Ducts and Casings

23 31 13 – Metal Ducts

23 31 13.13 Rectangular Metal Ducts		Crew	Daily Output	Labor-Hours	Unit	Material	2011 Bare Costs Labor	Equipment	Total	Total Incl O&P
1240	For 80% fittings, add				Lb.		124%			
1244	For 90% fittings, add						147%			
1248	For 100% fittings, add						169%			
1252	Note: Fittings add includes time for detailing and installation.									

23 31 13.19 Metal Duct Fittings		Crew	Daily Output	Labor-Hours	Unit	Material	Labor	Equipment	Total	Total Incl O&P
0010	**METAL DUCT FITTINGS**									
0050	Air extractors, 12" x 4"	1 Shee	24	.333	Ea.	16.10	12.75		28.85	39
0100	8" x 6"		22	.364		16.10	13.90		30	41
0200	20" x 8"		16	.500		36.50	19.10		55.60	71.50
0280	24" x 12"		10	.800		49.50	30.50		80	105
2000	Fabrics for flexible connections, with metal edge		100	.080	L.F.	2.13	3.06		5.19	7.40
2100	Without metal edge		160	.050	"	1.53	1.91		3.44	4.84

23 31 16 – Nonmetal Ducts

23 31 16.13 Fibrous-Glass Ducts		Crew	Daily Output	Labor-Hours	Unit	Material	Labor	Equipment	Total	Total Incl O&P
0010	**FIBROUS-GLASS DUCTS** R233100-20									
3490	Rigid fiberglass duct board, foil reinf. kraft facing									
3500	Rectangular, 1" thick, alum. faced, (FRK), std. weight	Q-10	350	.069	SF Surf	.77	2.44		3.21	4.89

23 33 Air Duct Accessories

23 33 13 – Dampers

23 33 13.13 Volume-Control Dampers		Crew	Daily Output	Labor-Hours	Unit	Material	Labor	Equipment	Total	Total Incl O&P
0010	**VOLUME-CONTROL DAMPERS**									
5990	Multi-blade dampers, opposed blade, 8" x 6"	1 Shee	24	.333	Ea.	20	12.75		32.75	43
5994	8" x 8"		22	.364		21	13.90		34.90	46
5996	10" x 10"		21	.381		24.50	14.55		39.05	51
6000	12" x 12"		21	.381		27.50	14.55		42.05	54
6020	12" x 18"		18	.444		36.50	17		53.50	68.50
6030	14" x 10"		20	.400		26.50	15.30		41.80	54.50
6031	14" x 14"		17	.471		32.50	18		50.50	65
6033	16" x 12"		17	.471		32.50	18		50.50	65
6035	16" x 16"		16	.500		40.50	19.10		59.60	76
6037	18" x 16"		15	.533		44.50	20.50		65	82.50
6038	18" x 18"		15	.533		48	20.50		68.50	86.50
6070	20" x 16"		14	.571		48	22		70	89
6072	20" x 20"		13	.615		57.50	23.50		81	103
6074	22" x 18"		14	.571		57.50	22		79.50	99.50
6076	24" x 16"		11	.727		57	28		85	109
6078	24" x 20"		8	1		67	38		105	137
6080	24" x 24"		8	1		78.50	38		116.50	150
6110	26" x 26"		6	1.333		86.50	51		137.50	179
6133	30" x 30"	Q-9	6.60	2.424		124	83.50		207.50	274
6135	32" x 32"		6.40	2.500		144	86		230	300
6180	48" x 36"		5.60	2.857		237	98		335	425
8000	Multi-blade dampers, parallel blade									
8100	8" x 8"	1 Shee	24	.333	Ea.	73	12.75		85.75	102
8140	16" x 10"		20	.400		94	15.30		109.30	128
8200	24" x 16"		11	.727		120	28		148	177
8260	30" x 18"		7	1.143		164	43.50		207.50	252

23 33 Air Duct Accessories

23 33 13 – Dampers

23 33 13.16 Fire Dampers		Crew	Daily Output	Labor-Hours	Unit	Material	Labor (2011 Bare Costs)	Equipment	Total	Total Incl O&P
0010	**FIRE DAMPERS**									
3000	Fire damper, curtain type, 1-1/2 hr. rated, vertical, 6" x 6"	1 Shee	24	.333	Ea.	21.50	12.75		34.25	45
3020	8" x 6"		22	.364		21.50	13.90		35.40	47
3240	16" x 14"		18	.444		29.50	17		46.50	60.50
3400	24" x 20"		8	1		42	38		80	109

23 33 13.28 Splitter Damper Assembly		Crew	Daily Output	Labor-Hours	Unit	Material	Labor	Equipment	Total	Total Incl O&P
0009	**SPLITTER DAMPER ASSEMBLY**									
0010	Self locking, 1' rod	1 Shee	24	.333	Ea.	21.50	12.75		34.25	45
7020	3' rod		22	.364		28	13.90		41.90	54
7040	4' rod		20	.400		31.50	15.30		46.80	59.50
7060	6' rod		18	.444		38	17		55	70

23 33 19 – Duct Silencers

23 33 19.10 Duct Silencers		Crew	Daily Output	Labor-Hours	Unit	Material	Labor	Equipment	Total	Total Incl O&P
0009	**DUCT SILENCERS**									
0010	Silencers, noise control for air flow, duct				MCFM	55.50			55.50	61

23 33 33 – Duct-Mounting Access Doors

23 33 33.13 Duct Access Doors		Crew	Daily Output	Labor-Hours	Unit	Material	Labor	Equipment	Total	Total Incl O&P
0010	**DUCT ACCESS DOORS**									
1000	Duct access door, insulated, 6" x 6"	1 Shee	14	.571	Ea.	15	22		37	52.50
1020	10" x 10"		11	.727		18.15	28		46.15	66
1040	12" x 12"		10	.800		19.40	30.50		49.90	72
1050	12" x 18"		9	.889		34.50	34		68.50	94
1070	18" x 18"		8	1		34.50	38		72.50	101
1074	24" x 18"		8	1		41	38		79	109

23 33 46 – Flexible Ducts

23 33 46.10 Flexible Air Ducts		Crew	Daily Output	Labor-Hours	Unit	Material	Labor	Equipment	Total	Total Incl O&P
0010	**FLEXIBLE AIR DUCTS** R233100-20									
1300	Flexible, coated fiberglass fabric on corr. resist. metal helix									
1400	pressure to 12" (WG) UL-181									
1500	Non-insulated, 3" diameter	Q-9	400	.040	L.F.	1.02	1.38		2.40	3.39
1540	5" diameter		320	.050		1.25	1.72		2.97	4.22
1561	Ductwork, flexible, non-insulated, 6" diameter		280	.057		1.53	1.96		3.49	4.92
1580	7" diameter		240	.067		1.82	2.29		4.11	5.80
1600	8" diameter		200	.080		2.10	2.75		4.85	6.85
1640	10" diameter		160	.100		2.61	3.44		6.05	8.55
1660	12" diameter		120	.133		3.17	4.58		7.75	11.05
1900	Insulated, 1" thick, PE jacket, 3" diameter G		380	.042		2.15	1.45		3.60	4.76
1910	4" diameter G		340	.047		2.15	1.62		3.77	5.05
1920	5" diameter G		300	.053		2.15	1.83		3.98	5.40
1940	6" diameter G		260	.062		2.38	2.12		4.50	6.10
1960	7" diameter G		220	.073		2.78	2.50		5.28	7.20
1980	8" diameter G		180	.089		3	3.06		6.06	8.35
2020	10" diameter G		140	.114		3.51	3.93		7.44	10.35
2040	12" diameter G		100	.160		4.36	5.50		9.86	13.90

23 33 53 – Duct Liners

23 33 53.10 Duct Liner Board		Crew	Daily Output	Labor-Hours	Unit	Material	Labor	Equipment	Total	Total Incl O&P
0010	**DUCT LINER BOARD**									
3490	Board type, fiberglass liner, 3 lb. density									
3500	Fire resistant, black pigmented, 1 side									
3520	1" thick G	Q-14	150	.107	S.F.	.77	3.33		4.10	6.45

23 33 Air Duct Accessories

23 33 53 – Duct Liners

23 33 53.10 Duct Liner Board			Crew	Daily Output	Labor-Hours	Unit	Material	2011 Bare Costs Labor	Equipment	Total	Total Incl O&P
3540	1-1/2" thick	G	Q-14	130	.123	S.F.	.97	3.84		4.81	7.55
3560	2" thick	G	↓	120	.133	↓	1.18	4.16		5.34	8.30
3600	FSK vapor barrier										
3620	1" thick	G	Q-14	150	.107	S.F.	.77	3.33		4.10	6.45
3630	1-1/2" thick	G		130	.123		.97	3.84		4.81	7.55
3640	2" thick	G	↓	120	.133	↓	1.18	4.16		5.34	8.30
3680	No finish										
3700	1" thick	G	Q-14	170	.094	S.F.	.40	2.94		3.34	5.40
3710	1-1/2" thick	G		140	.114		.60	3.57		4.17	6.65
3720	2" thick	G	↓	130	.123	↓	.81	3.84		4.65	7.40
3940	Board type, non-fibrous foam										
3950	Temperature, bacteria and fungi resistant										
3960	1" thick	G	Q-14	150	.107	S.F.	2.25	3.33		5.58	8.10
3970	1-1/2" thick	G		130	.123		3	3.84		6.84	9.80
3980	2" thick	G	↓	120	.133	↓	3.60	4.16		7.76	10.95

23 34 HVAC Fans

23 34 13 – Axial HVAC Fans

23 34 13.10 Axial Flow HVAC Fans			Crew	Daily Output	Labor-Hours	Unit	Material	Labor	Equipment	Total	Total Incl O&P
0010	**AXIAL FLOW HVAC FANS**	R233400-10									
0020	Air conditioning and process air handling										
0030	Axial flow, compact, low sound, 2.5" S.P.										
0050	3,800 CFM, 5 HP		Q-20	3.40	5.882	Ea.	4,625	209		4,834	5,450
0080	6,400 CFM, 5 HP			2.80	7.143		5,175	254		5,429	6,125
0100	10,500 CFM, 7-1/2 HP			2.40	8.333		6,450	296		6,746	7,575
0120	15,600 CFM, 10 HP			1.60	12.500		8,125	445		8,570	9,650
1500	Vaneaxial, low pressure, 2000 CFM, 1/2 HP			3.60	5.556		1,975	198		2,173	2,500
1520	4,000 CFM, 1 HP			3.20	6.250		2,275	222		2,497	2,875
1540	8,000 CFM, 2 HP		↓	2.80	7.143	↓	2,900	254		3,154	3,625

23 34 14 – Blower HVAC Fans

23 34 14.10 Blower Type HVAC Fans			Crew	Daily Output	Labor-Hours	Unit	Material	Labor	Equipment	Total	Total Incl O&P
0010	**BLOWER TYPE HVAC FANS**										
2500	Ceiling fan, right angle, extra quiet, 0.10" S.P.										
2520	95 CFM		Q-20	20	1	Ea.	252	35.50		287.50	335
2540	210 CFM			19	1.053		298	37.50		335.50	385
2560	385 CFM			18	1.111		380	39.50		419.50	480
2580	885 CFM			16	1.250		745	44.50		789.50	895
2600	1,650 CFM			13	1.538		1,025	54.50		1,079.50	1,225
2620	2,960 CFM		↓	11	1.818		1,375	64.50		1,439.50	1,600
2640	For wall or roof cap, add		1 Shee	16	.500		252	19.10		271.10	310
2660	For straight thru fan, add						10%				
2680	For speed control switch, add		1 Elec	16	.500	↓	137	20		157	184
7500	Utility set, steel construction, pedestal, 1/4" S.P.										
7520	Direct drive, 150 CFM, 1/8 HP		Q-20	6.40	3.125	Ea.	775	111		886	1,050
7540	485 CFM, 1/6 HP			5.80	3.448		980	123		1,103	1,275
7560	1950 CFM, 1/2 HP			4.80	4.167		1,150	148		1,298	1,500
7580	2410 CFM, 3/4 HP			4.40	4.545		2,125	162		2,287	2,600
7600	3328 CFM, 1-1/2 HP		↓	3	6.667	↓	2,350	237		2,587	3,000
7680	V-belt drive, drive cover, 3 phase										
7700	800 CFM, 1/4 HP		Q-20	6	3.333	Ea.	855	119		974	1,125

23 34 14 – Blower HVAC Fans

23 34 14.10 Blower Type HVAC Fans		Crew	Daily Output	Labor-Hours	Unit	2011 Bare Costs Material	Labor	Equipment	Total	Total Incl O&P
7720	1,300 CFM, 1/3 HP	Q-20	5	4	Ea.	895	142		1,037	1,225
7740	2,000 CFM, 1 HP		4.60	4.348		1,050	155		1,205	1,425
7760	2,900 CFM, 3/4 HP	↓	4.20	4.762	↓	1,425	169		1,594	1,850

23 34 16 – Centrifugal HVAC Fans

23 34 16.10 Centrifugal Type HVAC Fans		Crew	Daily Output	Labor-Hours	Unit	Material	Labor	Equipment	Total	Total Incl O&P
0010	**CENTRIFUGAL TYPE HVAC FANS**									
0200	In-line centrifugal, supply/exhaust booster									
0220	aluminum wheel/hub, disconnect switch, 1/4" S.P.									
0240	500 CFM, 10" diameter connection	Q-20	3	6.667	Ea.	1,150	237		1,387	1,675
0260	1,380 CFM, 12" diameter connection		2	10		1,225	355		1,580	1,925
0280	1,520 CFM, 16" diameter connection		2	10		1,325	355		1,680	2,025
0300	2,560 CFM, 18" diameter connection		1	20		1,450	710		2,160	2,775
0320	3,480 CFM, 20" diameter connection		.80	25		1,725	890		2,615	3,325
0326	5,080 CFM, 20" diameter connection	↓	.75	26.667	↓	1,875	950		2,825	3,600
3500	Centrifugal, airfoil, motor and drive, complete									
3520	1000 CFM, 1/2 HP	Q-20	2.50	8	Ea.	1,675	284		1,959	2,325
3540	2,000 CFM, 1 HP		2	10		1,875	355		2,230	2,625
3560	4,000 CFM, 3 HP		1.80	11.111		2,375	395		2,770	3,250
3580	8,000 CFM, 7-1/2 HP		1.40	14.286		3,675	510		4,185	4,875
3600	12,000 CFM, 10 HP	↓	1	20	↓	4,425	710		5,135	6,050
4500	Corrosive fume resistant, plastic									
4600	roof ventilators, centrifugal, V belt drive, motor									
4620	1/4" S.P., 250 CFM, 1/4 HP	Q-20	6	3.333	Ea.	3,975	119		4,094	4,575
4640	895 CFM, 1/3 HP		5	4		4,325	142		4,467	4,975
4660	1630 CFM, 1/2 HP		4	5		5,125	178		5,303	5,925
4680	2240 CFM, 1 HP	↓	3	6.667	↓	5,325	237		5,562	6,275
5000	Utility set, centrifugal, V belt drive, motor									
5020	1/4" S.P., 1200 CFM, 1/4 HP	Q-20	6	3.333	Ea.	1,550	119		1,669	1,925
5040	1520 CFM, 1/3 HP		5	4		1,725	142		1,867	2,125
5060	1850 CFM, 1/2 HP		4	5		1,725	178		1,903	2,200
5080	2180 CFM, 3/4 HP		3	6.667		2,050	237		2,287	2,650
5100	1/2" S.P., 3600 CFM, 1 HP		2	10		2,350	355		2,705	3,175
5120	4250 CFM, 1-1/2 HP		1.60	12.500		2,575	445		3,020	3,575
5140	4800 CFM, 2 HP	↓	1.40	14.286	↓	2,900	510		3,410	4,025
7000	Roof exhauster, centrifugal, aluminum housing, 12" galvanized									
7020	curb, bird screen, back draft damper, 1/4" S.P.									
7100	Direct drive, 320 CFM, 11" sq. damper	Q-20	7	2.857	Ea.	610	102		712	835
7120	600 CFM, 11" sq. damper		6	3.333		760	119		879	1,025
7140	815 CFM, 13" sq. damper		5	4		760	142		902	1,075
7160	1450 CFM, 13" sq. damper		4.20	4.762		1,100	169		1,269	1,475
7180	2050 CFM, 16" sq. damper		4	5		1,450	178		1,628	1,900
7200	V-belt drive, 1650 CFM, 12" sq. damper		6	3.333		1,100	119		1,219	1,400
7220	2750 CFM, 21" sq. damper		5	4		1,300	142		1,442	1,650
7230	3500 CFM, 21" sq. damper		4.50	4.444		1,475	158		1,633	1,875
7240	4910 CFM, 23" sq. damper		4	5		1,775	178		1,953	2,275
7260	8525 CFM, 28" sq. damper		3	6.667		2,400	237		2,637	3,050
7280	13,760 CFM, 35" sq. damper		2	10		3,225	355		3,580	4,125
7300	20,558 CFM, 43" sq. damper	↓	1	20		6,550	710		7,260	8,400
7320	For 2 speed winding, add					15%				
7340	For explosionproof motor, add					495			495	545
7360	For belt driven, top discharge, add				↓	15%				
8500	Wall exhausters, centrifugal, auto damper, 1/8" S.P.									

23 34 HVAC Fans

23 34 16 – Centrifugal HVAC Fans

23 34 16.10 Centrifugal Type HVAC Fans		Crew	Daily Output	Labor-Hours	Unit	Material	2011 Bare Costs Labor	Equipment	Total	Total Incl O&P
8520	Direct drive, 610 CFM, 1/20 HP	Q-20	14	1.429	Ea.	365	51		416	485
8540	796 CFM, 1/12 HP		13	1.538		735	54.50		789.50	895
8560	822 CFM, 1/6 HP		12	1.667		925	59.50		984.50	1,125
8580	1,320 CFM, 1/4 HP	↓	12	1.667	↓	1,100	59.50		1,159.50	1,300
9500	V-belt drive, 3 phase									
9520	2,800 CFM, 1/4 HP	Q-20	9	2.222	Ea.	1,650	79		1,729	1,950
9540	3,740 CFM, 1/2 HP	"	8	2.500	"	1,725	89		1,814	2,050

23 34 23 – HVAC Power Ventilators

23 34 23.10 HVAC Power Circulators and Ventilators

		Crew	Daily Output	Labor-Hours	Unit	Material	Labor	Equipment	Total	Total Incl O&P
0010	**HVAC POWER CIRCULATORS AND VENTILATORS**									
3000	Paddle blade air circulator, 3 speed switch									
3020	42", 5,000 CFM high, 3000 CFM low G	1 Elec	2.40	3.333	Ea.	154	134		288	385
3040	52", 6,500 CFM high, 4000 CFM low G	"	2.20	3.636	"	154	146		300	405
3100	For antique white motor, same cost									
3200	For brass plated motor, same cost									
3300	For light adaptor kit, add G				Ea.	38.50			38.50	42.50
6000	Propeller exhaust, wall shutter, 1/4" S.P.									
6020	Direct drive, two speed									
6100	375 CFM, 1/10 HP	Q-20	10	2	Ea.	430	71		501	590
6120	730 CFM, 1/7 HP		9	2.222		480	79		559	655
6140	1000 CFM, 1/8 HP		8	2.500		665	89		754	880
6200	4720 CFM, 1 HP	↓	5	4	↓	1,075	142		1,217	1,400
6300	V-belt drive, 3 phase									
6320	6175 CFM, 3/4 HP	Q-20	5	4	Ea.	870	142		1,012	1,200
6340	7500 CFM, 3/4 HP		5	4		960	142		1,102	1,275
6360	10,100 CFM, 1 HP		4.50	4.444		1,050	158		1,208	1,400
6380	14,300 CFM, 1-1/2 HP	↓	4	5	↓	1,200	178		1,378	1,625
6650	Residential, bath exhaust, grille, back draft damper									
6660	50 CFM	Q-20	24	.833	Ea.	47.50	29.50		77	101
6670	110 CFM		22	.909		79	32.50		111.50	140
6680	Light combination, squirrel cage, 100 watt, 70 CFM	↓	24	.833	↓	85.50	29.50		115	143
6700	Light/heater combination, ceiling mounted									
6710	70 CFM, 1450 watt	Q-20	24	.833	Ea.	127	29.50		156.50	189
6800	Heater combination, recessed, 70 CFM		24	.833		51	29.50		80.50	105
6820	With 2 infrared bulbs		23	.870		79.50	31		110.50	139
6900	Kitchen exhaust, grille, complete, 160 CFM		22	.909		87.50	32.50		120	150
6910	180 CFM		20	1		74.50	35.50		110	140
6920	270 CFM		18	1.111		134	39.50		173.50	213
6930	350 CFM	↓	16	1.250	↓	105	44.50		149.50	188
6940	Residential roof jacks and wall caps									
6944	Wall cap with back draft damper									
6946	3" & 4" dia. round duct	1 Shee	11	.727	Ea.	20.50	28		48.50	69
6948	6" dia. round duct	"	11	.727	"	52.50	28		80.50	104
6958	Roof jack with bird screen and back draft damper									
6960	3" & 4" dia. round duct	1 Shee	11	.727	Ea.	21.50	28		49.50	69.50
6962	3-1/4" x 10" rectangular duct	"	10	.800	"	39	30.50		69.50	93
6980	Transition									
6982	3-1/4" x 10" to 6" dia. round	1 Shee	20	.400	Ea.	23	15.30		38.30	50.50

23 34 HVAC Fans

23 34 33 – Air Curtains

	23 34 33.10 Air Barrier Curtains	Crew	Daily Output	Labor-Hours	Unit	Material	2011 Bare Costs Labor	Equipment	Total	Total Incl O&P
0010	**AIR BARRIER CURTAINS**, Incl. motor starters, transformers,									
0050	door switches & temperature controls									
0100	Shipping and receiving doors, unheated, minimal wind stoppage									
0150	8′ high, multiples of 3′ wide	2 Shee	6	2.667	L.F.	244	102		346	435
0160	5′ wide		10	1.600		235	61		296	360
0210	10′ high, multiples of 4′ wide		8	2		194	76.50		270.50	340
0250	12′ high, 3′-6″ wide		7	2.286		210	87.50		297.50	375
0260	12′ wide		6	2.667		225	102		327	415
0350	16′ high, 3′-6″ wide		7	2.286		210	87.50		297.50	375
0360	12′ wide		6	2.667		225	102		327	415
0500	Maximum wind stoppage									
0550	10′ high, multiples of 4′ wide	2 Shee	8	2	L.F.	194	76.50		270.50	340
0650	14′ high, multiples of 4′ wide		8	2		585	76.50		661.50	770
0750	20′ high, multiples of 8′ wide		6	2.667		600	102		702	825
1100	Heated, maximum wind stoppage, steam heat									
1150	10′ high, multiples of 4′ wide	2 Shee	6	2.667	L.F.	930	102		1,032	1,200
1250	14′ high, multiples of 4′ wide		6	2.667		930	102		1,032	1,200
1350	20′ high, multiples of 8′ wide		4	4		790	153		943	1,125
1500	Customer entrance doors, unheated, minimal wind stoppage									
1550	10′ high, multiples of 3′ wide	2 Shee	6	2.667	L.F.	220	102		322	410
1560	5′ wide		10	1.600		235	61		296	360
1650	Maximum wind stoppage, 12′ high, multiples of 4′ wide		8	2		350	76.50		426.50	510
1700	Heated, minimal wind stoppage, electric heat									
1750	8′ high, multiples of 3′ wide	2 Shee	6	2.667	L.F.	244	102		346	435
1760	Multiples of 5′ wide		10	1.600		235	61		296	360
1850	10′ high, multiples of 3′ wide		6	2.667		220	102		322	410
1860	Multiples of 5′ wide		10	1.600		510	61		571	660
1950	Maximum wind stoppage, steam heat									
1960	12′ high, multiples of 4′ wide	2 Shee	8	2	L.F.	930	76.50		1,006.50	1,150
2000	Walk-in coolers and freezers, ambient air, minimal wind stoppage									
2050	8′ high, multiples of 3′ wide	2 Shee	6	2.667	L.F.	220	102		322	410
2060	Multiples of 5′ wide		10	1.600		510	61		571	660
2250	Maximum wind stoppage, 12′ high, multiples of 3′ wide		6	2.667		220	102		322	410
2450	Conveyor openings or service windows, unheated, 5′ high		5	3.200		172	122		294	390
2460	Heated, electric, 5′ high, 2′-6″ wide		5	3.200		460	122		582	705

23 37 Air Outlets and Inlets

23 37 13 – Diffusers, Registers, and Grilles

	23 37 13.10 Diffusers	Crew	Daily Output	Labor-Hours	Unit	Material	Labor	Equipment	Total	Total Incl O&P
0010	**DIFFUSERS**, Aluminum, opposed blade damper unless noted									
0100	Ceiling, linear, also for sidewall									
0500	Perforated, 24″ x 24″ lay-in panel size, 6″ x 6″	1 Shee	16	.500	Ea.	57.50	19.10		76.60	95
0520	8″ x 8″		15	.533		59	20.50		79.50	98.50
0530	9″ x 9″		14	.571		63	22		85	106
0540	10″ x 10″		14	.571		65	22		87	108
0560	12″ x 12″		12	.667		66	25.50		91.50	115
0590	16″ x 16″		11	.727		109	28		137	166
0600	18″ x 18″		10	.800		115	30.50		145.50	178
0610	20″ x 20″		10	.800		133	30.50		163.50	197
0620	24″ x 24″		9	.889		165	34		199	238

23 37 Air Outlets and Inlets

23 37 13 – Diffusers, Registers, and Grilles

23 37 13.10 Diffusers		Crew	Daily Output	Labor-Hours	Unit	Material	Labor	Equipment	Total	Total Incl O&P
						2011 Bare Costs				
1000	Rectangular, 1 to 4 way blow, 6″ x 6″	1 Shee	16	.500	Ea.	46.50	19.10		65.60	82.50
1010	8″ x 8″		15	.533		55.50	20.50		76	94.50
1014	9″ x 9″		15	.533		61	20.50		81.50	101
1016	10″ x 10″		15	.533		75.50	20.50		96	117
1020	12″ x 6″		15	.533		67.50	20.50		88	108
1040	12″ x 9″		14	.571		73.50	22		95.50	117
1060	12″ x 12″		12	.667		78.50	25.50		104	129
1070	14″ x 6″		13	.615		73	23.50		96.50	120
1074	14″ x 14″		12	.667		121	25.50		146.50	175
1150	18″ x 18″		9	.889		132	34		166	201
1160	21″ x 21″		8	1		210	38		248	294
1170	24″ x 12″		10	.800		160	30.50		190.50	227
1500	Round, butterfly damper, steel, diffuser size, 6″ diameter		18	.444		24	17		41	54.50
1520	8″ diameter		16	.500		25.50	19.10		44.60	59.50
1540	10″ diameter		14	.571		31.50	22		53.50	71
1560	12″ diameter		12	.667		41.50	25.50		67	87.50
1580	14″ diameter		10	.800		51.50	30.50		82	107
2000	T bar mounting, 24″ x 24″ lay-in frame, 6″ x 6″		16	.500		44	19.10		63.10	80
2020	8″ x 8″		14	.571		46	22		68	86.50
2040	12″ x 12″		12	.667		54.50	25.50		80	102
2060	16″ x 16″		11	.727		73	28		101	127
2080	18″ x 18″		10	.800		81	30.50		111.50	140
6000	For steel diffusers instead of aluminum, deduct					10%				

23 37 13.30 Grilles		Crew	Daily Output	Labor-Hours	Unit	Material	Labor	Equipment	Total	Total Incl O&P
0010	**GRILLES**									
0020	Aluminum									
1000	Air return, 6″ x 6″	1 Shee	26	.308	Ea.	17.55	11.75		29.30	38.50
1020	10″ x 6″		24	.333		17.55	12.75		30.30	40.50
1080	16″ x 8″		22	.364		22	13.90		35.90	47.50
1100	12″ x 12″		22	.364		24	13.90		37.90	49
1120	24″ x 12″		18	.444		33	17		50	64
1220	24″ x 18″		16	.500		41.50	19.10		60.60	77
1280	36″ x 24″		14	.571		71	22		93	114
3000	Filter grille with filter, 12″ x 12″		24	.333		47	12.75		59.75	72.50
3020	18″ x 12″		20	.400		65.50	15.30		80.80	97.50
3040	24″ x 18″		18	.444		77	17		94	113
3060	24″ x 24″		16	.500		89	19.10		108.10	130
6000	For steel grilles instead of aluminum in above, deduct					10%				

23 37 13.60 Registers		Crew	Daily Output	Labor-Hours	Unit	Material	Labor	Equipment	Total	Total Incl O&P
0010	**REGISTERS**									
0980	Air supply									
1000	Ceiling/wall, O.B. damper, anodized aluminum									
1010	One or two way deflection, adj. curved face bars									
1020	8″ x 4″	1 Shee	26	.308	Ea.	34	11.75		45.75	56.50
1120	12″ x 12″		18	.444		62	17		79	96
1240	20″ x 6″		18	.444		55	17		72	88.50
1340	24″ x 8″		13	.615		74.50	23.50		98	121
1350	24″ x 18″		12	.667		139	25.50		164.50	194
2700	Above registers in steel instead of aluminum, deduct					10%				
4000	Floor, toe operated damper, enameled steel									
4020	4″ x 8″	1 Shee	32	.250	Ea.	26.50	9.55		36.05	45.50
4100	8″ x 10″		22	.364		32.50	13.90		46.40	59

23 37 13 – Diffusers, Registers, and Grilles

23 37 13.60 Registers		Crew	Daily Output	Labor-Hours	Unit	2011 Bare Costs Material	Labor	Equipment	Total	Total Incl O&P
4140	10″ x 10″	1 Shee	20	.400	Ea.	39	15.30		54.30	68
4220	14″ x 14″		16	.500		137	19.10		156.10	183
4240	14″ x 20″		15	.533		183	20.50		203.50	235
4980	Air return									
5000	Ceiling or wall, fixed 45° face blades									
5010	Adjustable O.B. damper, anodized aluminum									
5020	4″ x 8″	1 Shee	26	.308	Ea.	24	11.75		35.75	45.50
5060	6″ x 10″		19	.421		27	16.10		43.10	56.50
5280	24″ x 24″		11	.727		92	28		120	147
5300	24″ x 36″		8	1		144	38		182	222
6000	For steel construction instead of aluminum, deduct					10%				

23 37 15 – Louvers

23 37 15.40 HVAC Louvers		Crew	Daily Output	Labor-Hours	Unit	Material	Labor	Equipment	Total	Total Incl O&P
0010	**HVAC LOUVERS**									
0100	Aluminum, extruded, with screen, mill finish									
1002	Brick vent, see also Section 04 05 23.19									
1100	Standard, 4″ deep, 8″ wide, 5″ high	1 Shee	24	.333	Ea.	34.50	12.75		47.25	58.50
1200	Modular, 4″ deep, 7-3/4″ wide, 5″ high		24	.333		36	12.75		48.75	60.50
1300	Speed brick, 4″ deep, 11-5/8″ wide, 3-7/8″ high		24	.333		36	12.75		48.75	60.50
1400	Fuel oil brick, 4″ deep, 8″ wide, 5″ high		24	.333		61.50	12.75		74.25	89
2000	Cooling tower and mechanical equip., screens, light weight		40	.200	S.F.	15.40	7.65		23.05	29.50
2020	Standard weight		35	.229		41	8.75		49.75	59.50
2500	Dual combination, automatic, intake or exhaust		20	.400		56	15.30		71.30	86.50
2520	Manual operation		20	.400		41.50	15.30		56.80	71
2540	Electric or pneumatic operation		20	.400		41.50	15.30		56.80	71
2560	Motor, for electric or pneumatic		14	.571	Ea.	480	22		502	565
3000	Fixed blade, continuous line									
3100	Mullion type, stormproof	1 Shee	28	.286	S.F.	41.50	10.90		52.40	64
3200	Stormproof		28	.286		41.50	10.90		52.40	64
3300	Vertical line		28	.286		49.50	10.90		60.40	72
3500	For damper to use with above, add					50%	30%			
3520	Motor, for damper, electric or pneumatic	1 Shee	14	.571	Ea.	480	22		502	565
4000	Operating, 45°, manual, electric or pneumatic		24	.333	S.F.	50	12.75		62.75	76
4100	Motor, for electric or pneumatic		14	.571	Ea.	480	22		502	565
4200	Penthouse, roof		56	.143	S.F.	24.50	5.45		29.95	36
4300	Walls		40	.200		58	7.65		65.65	76.50
5000	Thinline, under 4″ thick, fixed blade		40	.200		24	7.65		31.65	39
5010	Finishes, applied by mfr. at additional cost, available in colors									
5020	Prime coat only, add				S.F.	3.30			3.30	3.63
5040	Baked enamel finish coating, add					6.10			6.10	6.70
5060	Anodized finish, add					6.60			6.60	7.25
5080	Duranodic finish, add					12			12	13.20
5100	Fluoropolymer finish coating, add					18.90			18.90	21
9980	For small orders (under 10 pieces), add					25%				

23 37 23 – HVAC Gravity Ventilators

23 37 23.10 HVAC Gravity Air Ventilators		Crew	Daily Output	Labor-Hours	Unit	Material	Labor	Equipment	Total	Total Incl O&P
0010	**HVAC GRAVITY AIR VENTILATORS**, Incl. base and damper R233400-10									
1280	Rotary ventilators, wind driven, galvanized									
1300	4″ neck diameter	Q-9	20	.800	Ea.	50	27.50		77.50	101
1340	6″ neck diameter		16	1		56	34.50		90.50	119
1400	12″ neck diameter		10	1.600		81	55		136	181
1500	24″ neck diameter, 3,100 CFM		8	2		236	69		305	375

23 37 Air Outlets and Inlets

23 37 23 – HVAC Gravity Ventilators

23 37 23.10 HVAC Gravity Air Ventilators		Crew	Daily Output	Labor-Hours	Unit	Material	2011 Bare Costs Labor	Equipment	Total	Total Incl O&P
1540	36" neck diameter, 5,500 CFM	Q-9	6	2.667	Ea.	695	91.50		786.50	915
2000	Stationary, gravity, syphon, galvanized									
2160	6" neck diameter, 66 CFM	Q-9	16	1	Ea.	53	34.50		87.50	116
2240	12" neck diameter, 160 CFM		10	1.600		91	55		146	191
2340	24" neck diameter, 900 CFM		8	2		248	69		317	385
2380	36" neck diameter, 2,000 CFM		6	2.667		775	91.50		866.50	1,000
4200	Stationary mushroom, aluminum, 16" orifice diameter		10	1.600		535	55		590	680
4220	26" orifice diameter		6.15	2.602		790	89.50		879.50	1,025
4230	30" orifice diameter		5.71	2.802		1,150	96.50		1,246.50	1,425
4240	38" orifice diameter		5	3.200		1,650	110		1,760	2,000
4250	42" orifice diameter		4.70	3.404		2,200	117		2,317	2,625
4260	50" orifice diameter		4.44	3.604		2,625	124		2,749	3,075
5000	Relief vent									
5500	Rectangular, aluminum, galvanized curb									
5510	intake/exhaust, 0.033" SP									
5580	500 CFM, 12" x 12"	Q-9	8.60	1.860	Ea.	625	64		689	795
5600	600 CFM, 12" x 16"		8	2		705	69		774	890
5640	1000 CFM, 12" x 24"		6.60	2.424		795	83.50		878.50	1,025
5680	3000 CFM, 20" x 42"		4	4		1,400	138		1,538	1,750
5880	Size is throat area, volume is at 500 fpm									

23 38 Ventilation Hoods

23 38 13 – Commercial-Kitchen Hoods

23 38 13.10 Hood and Ventilation Equipment

		Crew	Daily Output	Labor-Hours	Unit	Material	Labor	Equipment	Total	Total Incl O&P
0010	**HOOD AND VENTILATION EQUIPMENT**									
2970	Exhaust hood, sst, gutter on all sides, 4' x 4' x 2'	1 Carp	1.80	4.444	Ea.	4,525	149		4,674	5,225
2980	4' x 4' x 7'	"	1.60	5		7,100	168		7,268	8,100
7950	Hood fire protection system, minimum	Q-1	3	5.333		4,400	190		4,590	5,150
8050	Maximum	"	1	16		35,300	570		35,870	39,800

23 41 Particulate Air Filtration

23 41 13 – Panel Air Filters

23 41 13.10 Panel Type Air Filters

		Crew	Daily Output	Labor-Hours	Unit	Material	Labor	Equipment	Total	Total Incl O&P
0010	**PANEL TYPE AIR FILTERS**									
2950	Mechanical media filtration units									
3000	High efficiency type, with frame, non-supported G				MCFM	45			45	49.50
3100	Supported type G				"	60			60	66
5500	Throwaway glass or paper media type				Ea.	3.35			3.35	3.69

23 41 16 – Renewable-Media Air Filters

23 41 16.10 Disposable Media Air Filters

		Crew	Daily Output	Labor-Hours	Unit	Material	Labor	Equipment	Total	Total Incl O&P
0010	**DISPOSABLE MEDIA AIR FILTERS**									
5000	Renewable disposable roll				MCFM	19.35			19.35	21.50

23 41 19 – Washable Air Filters

23 41 19.10 Permanent Air Filters

		Crew	Daily Output	Labor-Hours	Unit	Material	Labor	Equipment	Total	Total Incl O&P
0010	**PERMANENT AIR FILTERS**									
4500	Permanent washable G				MCFM	20			20	22

23 41 Particulate Air Filtration

23 41 23 – Extended Surface Filters

23 41 23.10 Expanded Surface Filters		Crew	Daily Output	Labor-Hours	Unit	Material	Labor	Equipment	Total	Total Incl O&P
0010	**EXPANDED SURFACE FILTERS**									
4000	Medium efficiency, extended surface G				MCFM	5.50			5.50	6.05

23 42 Gas-Phase Air Filtration

23 42 13 – Activated-Carbon Air Filtration

23 42 13.10 Charcoal Type Air Filtration		Crew	Daily Output	Labor-Hours	Unit	Material	Labor	Equipment	Total	Total Incl O&P
0010	**CHARCOAL TYPE AIR FILTRATION**									
0050	Activated charcoal type, full flow				MCFM	600			600	660
0060	Full flow, impregnated media 12" deep					225			225	248
0070	HEPA filter & frame for field erection					300			300	330
0080	HEPA filter-diffuser, ceiling install.				↓	300			300	330

23 51 Breechings, Chimneys, and Stacks

23 51 13 – Draft Control Devices

23 51 13.13 Draft-Induction Fans		Crew	Daily Output	Labor-Hours	Unit	Material	Labor	Equipment	Total	Total Incl O&P
0010	**DRAFT-INDUCTION FANS**									
1000	Breeching installation									
1800	Hot gas, 600°F, variable pitch pulley and motor									
1860	8" diam. inlet, 1/4 H.P., 1phase, 1120 CFM	Q-9	4	4	Ea.	2,675	138		2,813	3,175
1900	12" diam. inlet, 3/4 H.P., 3 phase, 2960 CFM		3	5.333		3,650	183		3,833	4,325
1980	24" diam. inlet, 7-1/2 H.P., 3 phase, 17,760 CFM	↓	.80	20		10,500	690		11,190	12,600
2300	For multi-blade damper at fan inlet, add				↓	20%				

23 51 23 – Gas Vents

23 51 23.10 Gas Chimney Vents		Crew	Daily Output	Labor-Hours	Unit	Material	Labor	Equipment	Total	Total Incl O&P
0010	**GAS CHIMNEY VENTS,** Prefab metal, U.L. listed									
0020	Gas, double wall, galvanized steel									
0080	3" diameter	Q-9	72	.222	V.L.F.	5.55	7.65		13.20	18.75
0100	4" diameter		68	.235		6.95	8.10		15.05	21
0120	5" diameter		64	.250		8.10	8.60		16.70	23
0140	6" diameter		60	.267		9.45	9.15		18.60	25.50
0160	7" diameter		56	.286		13.85	9.80		23.65	31.50
0180	8" diameter		52	.308		15.45	10.60		26.05	34.50
0200	10" diameter		48	.333		35	11.45		46.45	57.50
0220	12" diameter		44	.364		47	12.50		59.50	72
0260	16" diameter	↓	40	.400		107	13.75		120.75	141
0300	20" diameter	Q-10	36	.667		145	24		169	199
0480	38" diameter	"	24	1	↓	520	35.50		555.50	630

23 51 26 – All-Fuel Vent Chimneys

23 51 26.30 All-Fuel Vent Chimneys, Double Wall, St. Stl.		Crew	Daily Output	Labor-Hours	Unit	Material	Labor	Equipment	Total	Total Incl O&P
0010	**ALL-FUEL VENT CHIMNEYS, DOUBLE WALL, STAINLESS STEEL**									
7800	All fuel, double wall, stainless steel, 6" diameter	Q-9	60	.267	V.L.F.	58.50	9.15		67.65	79.50
7802	7" diameter		56	.286		76	9.80		85.80	99.50
7804	8" diameter		52	.308		90	10.60		100.60	116
7806	10" diameter		48	.333		130	11.45		141.45	162
7808	12" diameter		44	.364		174	12.50		186.50	212
7810	14" diameter	↓	42	.381	↓	228	13.10		241.10	273

Costs are 2011 Bare Costs (Material, Labor, Equipment, Total).

23 51 33 – Insulated Sectional Chimneys

23 51 33.10 Prefabricated Insulated Sectional Chimneys		Crew	Daily Output	Labor-Hours	Unit	2011 Bare Costs Material	Labor	Equipment	Total	Total Incl O&P
0010	**PREFABRICATED INSULATED SECTIONAL CHIMNEYS**									
9000	High temp. (2000°F), steel jacket, acid resistant refractory lining									
9010	11 ga. galvanized jacket, U.L. listed									
9020	Straight section, 48" long, 10" diameter	Q-10	13.30	1.805	Ea.	400	64.50		464.50	545
9040	18" diameter		7.40	3.243		615	116		731	865
9070	36" diameter		2.70	8.889		1,325	315		1,640	1,975
9090	48" diameter	Q-11	2.70	11.852		1,925	405		2,330	2,775
9120	Tee section, 10" diameter	Q-10	4.40	5.455		855	194		1,049	1,250
9140	18" diameter		2.40	10		1,225	355		1,580	1,950
9170	36" diameter		.80	30		3,975	1,075		5,050	6,150
9190	48" diameter	Q-11	.90	35.556		6,775	1,225		8,000	9,475
9220	Cleanout pier section, 10" diameter	Q-10	3.50	6.857		655	244		899	1,125
9240	18" diameter		1.90	12.632		885	450		1,335	1,725
9270	36" diameter		.75	32		1,625	1,150		2,775	3,650
9290	48" diameter	Q-11	.75	42.667		2,475	1,475		3,950	5,150
9320	For drain, add					53%				
9330	Elbow, 30° and 45°, 10" diameter	Q-10	6.60	3.636		555	130		685	825
9350	18" diameter		3.70	6.486		900	231		1,131	1,375
9380	36" diameter		1.30	18.462		2,225	660		2,885	3,525
9400	48" diameter	Q-11	1.35	23.704		3,900	815		4,715	5,625
9430	For 60° and 90° elbow, add					112%				
9440	End cap, 10" diameter	Q-10	26	.923		955	33		988	1,100
9460	18" diameter		15	1.600		1,350	57		1,407	1,575
9490	36" diameter		5	4.800		2,350	171		2,521	2,875
9510	48" diameter	Q-11	5.30	6.038		3,475	208		3,683	4,150
9540	Increaser (1 diameter), 10" diameter	Q-10	6.60	3.636		475	130		605	735
9560	18" diameter		3.70	6.486		740	231		971	1,200
9590	36" diameter		1.30	18.462		1,450	660		2,110	2,675
9592	42" diameter	Q-11	1.60	20		1,600	690		2,290	2,875
9594	48" diameter		1.35	23.704		1,925	815		2,740	3,450
9596	54" diameter		1.10	29.091		2,325	1,000		3,325	4,200
9600	For expansion joints, add to straight section					7%				
9610	For 1/4" hot rolled steel jacket, add					157%				
9620	For 2950°F very high temperature, add					89%				
9630	26 ga. aluminized jacket, straight section, 48" long									
9640	10" diameter	Q-10	15.30	1.569	V.L.F.	237	56		293	355
9660	18" diameter		8.50	2.824		365	101		466	570
9690	36" diameter		3.10	7.742		845	276		1,121	1,375
9810	Draw band, galv. stl., 11 gauge, 10" diameter		32	.750	Ea.	65	26.50		91.50	116
9820	12" diameter		30	.800		69	28.50		97.50	123
9830	18" diameter		26	.923		86	33		119	149
9860	36" diameter		18	1.333		146	47.50		193.50	239
9880	48" diameter	Q-11	20	1.600		251	55		306	370

23 52 Heating Boilers

23 52 13 – Electric Boilers

23 52 13.10 Electric Boilers, ASME		Crew	Daily Output	Labor-Hours	Unit	Material	2011 Bare Costs Labor	Equipment	Total	Total Incl O&P
0010	**ELECTRIC BOILERS, ASME**, Standard controls and trim									
1000	Steam, 6 KW, 20.5 MBH	Q-19	1.20	20	Ea.	3,625	750		4,375	5,225
1160	60 KW, 205 MBH		1	24		6,250	900		7,150	8,350
1220	112 KW, 382 MBH		.75	32		8,550	1,200		9,750	11,400
1280	222 KW, 758 MBH		.55	43.636		13,500	1,625		15,125	17,600
1380	518 KW, 1768 MBH	Q-21	.36	88.889		22,800	3,375		26,175	30,600
1480	814 KW, 2778 MBH		.25	128		31,900	4,875		36,775	43,100
1600	2,340 KW, 7984 MBH		.16	200		65,500	7,625		73,125	84,500
2000	Hot water, 7.5 KW, 25.6 MBH	Q-19	1.30	18.462		3,825	690		4,515	5,325
2100	90 KW, 307 MBH		1.10	21.818		5,575	815		6,390	7,450
2220	296 KW, 1010 MBH		.55	43.636		11,900	1,625		13,525	15,800
2500	1036 KW, 3536 MBH	Q-21	.34	94.118		27,700	3,575		31,275	36,400
2680	2400 KW, 8191 MBH		.25	128		58,000	4,875		62,875	71,500
2820	3600 KW, 12,283 MBH		.16	200		81,500	7,625		89,125	102,000

23 52 23 – Cast-Iron Boilers

23 52 23.20 Gas-Fired Boilers

Line	Description	Crew	Daily Output	Labor-Hours	Unit	Material	Labor	Equipment	Total	Total Incl O&P
0010	**GAS-FIRED BOILERS,** Natural or propane, standard controls, packaged									
1000	Cast iron, with insulated jacket									
2000	Steam, gross output, 81 MBH	Q-7	1.40	22.857	Ea.	2,025	825		2,850	3,575
2080	203 MBH		.90	35.556		3,225	1,275		4,500	5,650
2180	400 MBH		.56	56.838		5,400	2,050		7,450	9,300
2240	765 MBH		.43	74.419		10,600	2,675		13,275	16,100
2320	1,875 MBH		.30	106		19,000	3,850		22,850	27,200
2440	4,720 MBH		.15	207		70,000	7,500		77,500	89,000
2480	6,100 MBH		.13	246		77,500	8,875		86,375	99,500
2540	6,970 MBH		.10	320		84,500	11,500		96,000	112,000
3000	Hot water, gross output, 80 MBH		1.46	21.918		1,850	790		2,640	3,350
3140	320 MBH		.80	40		4,375	1,450		5,825	7,175
3260	1,088 MBH		.40	80		12,900	2,875		15,775	18,900
3360	2,856 MBH		.20	160		22,400	5,775		28,175	34,000
3380	3,264 MBH		.18	179		23,900	6,475		30,375	36,900
3480	6,100 MBH		.13	250		77,500	9,025		86,525	99,500
3540	6,970 MBH		.09	359		84,500	13,000		97,500	114,000
7000	For tankless water heater, add					10%				

23 52 23.30 Gas/Oil Fired Boilers

Line	Description	Crew	Daily Output	Labor-Hours	Unit	Material	Labor	Equipment	Total	Total Incl O&P
0010	**GAS/OIL FIRED BOILERS,** Combination with burners and controls, packaged									
1000	Cast iron with insulated jacket									
2000	Steam, gross output, 720 MBH	Q-7	.43	74.074	Ea.	14,400	2,675		17,075	20,300
2080	1,600 MBH		.30	107		17,700	3,850		21,550	25,700
2140	2,700 MBH		.19	165		28,600	5,975		34,575	41,300
2280	5,520 MBH		.14	235		87,000	8,475		95,475	110,000
2340	6,390 MBH		.11	296		93,500	10,700		104,200	120,000
2380	6,970 MBH		.09	372		99,000	13,400		112,400	131,000
2900	Hot water, gross output									
2910	200 MBH	Q-6	.62	39.024	Ea.	8,275	1,350		9,625	11,300
2920	300 MBH		.49	49.080		8,275	1,700		9,975	11,900
2930	400 MBH		.41	57.971		9,650	2,025		11,675	13,900
2940	500 MBH		.36	67.039		10,400	2,325		12,725	15,300
3000	584 MBH	Q-7	.44	72.072		11,500	2,600		14,100	17,000
3060	1,460 MBH		.28	113		23,900	4,075		27,975	33,000
3160	4,088 MBH		.16	195		48,300	7,025		55,325	64,500
3300	13,500 MBH, 403.3 BHP		.04	727		149,500	26,200		175,700	207,500

23 52 Heating Boilers

23 52 23 – Cast-Iron Boilers

23 52 23.40 Oil-Fired Boilers		Crew	Daily Output	Labor-Hours	Unit	Material	2011 Bare Costs Labor	Equipment	Total	Total Incl O&P
0010	**OIL-FIRED BOILERS,** Standard controls, flame retention burner, packaged									
1000	Cast iron, with insulated flush jacket									
2000	Steam, gross output, 109 MBH	Q-7	1.20	26.667	Ea.	2,050	960		3,010	3,825
2060	207 MBH		.90	35.556		2,800	1,275		4,075	5,200
2180	1,084 MBH		.38	85.106		9,400	3,075		12,475	15,300
2280	3,000 MBH		.19	170		21,600	6,125		27,725	33,700
2380	5,520 MBH		.14	235		68,500	8,475		76,975	89,500
2460	6,970 MBH		.09	363		88,000	13,100		101,100	118,500
3000	Hot water, same price as steam									
4000	For tankless coil in smaller sizes, add				Ea.	15%				

23 52 26 – Steel Boilers

23 52 26.40 Oil-Fired Boilers		Crew	Daily Output	Labor-Hours	Unit	Material	Labor	Equipment	Total	Total Incl O&P
0010	**OIL-FIRED BOILERS**, Standard controls, flame retention burner									
5000	Steel, with insulated flush jacket									
7000	Hot water, gross output, 103 MBH	Q-6	1.60	15	Ea.	1,725	520		2,245	2,750
7120	420 MBH		.70	34.483		5,625	1,200		6,825	8,125
7320	3,150 MBH		.13	184		39,300	6,400		45,700	53,500
7340	For tankless coil in steam or hot water, add					7%				

23 52 28 – Swimming Pool Boilers

23 52 28.10 Swimming Pool Heaters		Crew	Daily Output	Labor-Hours	Unit	Material	Labor	Equipment	Total	Total Incl O&P
0010	**SWIMMING POOL HEATERS**, Not including wiring, external									
0020	piping, base or pad,									
0160	Gas fired, input, 155 MBH	Q-6	1.50	16	Ea.	1,625	555		2,180	2,675
0200	199 MBH		1	24		1,725	835		2,560	3,250
0280	500 MBH		.40	60		7,325	2,075		9,400	11,500
0400	1,800 MBH		.14	171		15,900	5,950		21,850	27,200
2000	Electric, 12 KW, 4,800 gallon pool	Q-19	3	8		2,025	300		2,325	2,725
2020	15 KW, 7,200 gallon pool		2.80	8.571		2,175	320		2,495	2,900
2040	24 KW, 9,600 gallon pool		2.40	10		2,375	375		2,750	3,200
2100	57 KW, 24,000 gallon pool		1.20	20		3,475	750		4,225	5,050

23 54 Furnaces

23 54 13 – Electric-Resistance Furnaces

23 54 13.10 Electric Furnaces		Crew	Daily Output	Labor-Hours	Unit	Material	Labor	Equipment	Total	Total Incl O&P
0010	**ELECTRIC FURNACES**, Hot air, blowers, std. controls									
0011	not including gas, oil or flue piping									
1000	Electric, UL listed									
1020	10.2 MBH	Q-20	5	4	Ea.	475	142		617	755
1040	17.1 MBH		4.80	4.167		495	148		643	790
1060	27.3 MBH		4.60	4.348		600	155		755	915
1100	34.1 MBH		4.40	4.545		635	162		797	965

23 54 16 – Fuel-Fired Furnaces

23 54 16.13 Gas-Fired Furnaces		Crew	Daily Output	Labor-Hours	Unit	Material	Labor	Equipment	Total	Total Incl O&P
0010	**GAS-FIRED FURNACES**									
3000	Gas, AGA certified, upflow, direct drive models									
3020	45 MBH input	Q-9	4	4	Ea.	575	138		713	860
3040	60 MBH input		3.80	4.211		615	145		760	915
3060	75 MBH input		3.60	4.444		685	153		838	1,000
3100	100 MBH input		3.20	5		690	172		862	1,050

23 54 Furnaces

23 54 16 – Fuel-Fired Furnaces

23 54 16.16 Oil-Fired Furnaces		Crew	Daily Output	Labor-Hours	Unit	Material	2011 Bare Costs Labor	Equipment	Total	Total Incl O&P
0010	**OIL-FIRED FURNACES**									
6000	Oil, UL listed, atomizing gun type burner									
6020	56 MBH output	Q-9	3.60	4.444	Ea.	1,475	153		1,628	1,875
6030	84 MBH output		3.50	4.571		1,850	157		2,007	2,300
6040	95 MBH output		3.40	4.706		1,875	162		2,037	2,350
6060	134 MBH output		3.20	5		2,175	172		2,347	2,675
6080	151 MBH output		3	5.333		2,250	183		2,433	2,775

23 55 Fuel-Fired Heaters

23 55 13 – Fuel-Fired Duct Heaters

23 55 13.16 Gas-Fired Duct Heaters		Crew	Daily Output	Labor-Hours	Unit	Material	Labor	Equipment	Total	Total Incl O&P
0010	**GAS-FIRED DUCT HEATERS**, Includes burner, controls, stainless steel									
0020	heat exchanger. Gas fired, electric ignition									
0030	Indoor installation									
0100	120 MBH output	Q-5	4	4	Ea.	2,125	144		2,269	2,575
0130	200 MBH output		2.70	5.926		2,850	214		3,064	3,475
0140	240 MBH output		2.30	6.957		2,975	251		3,226	3,650
0180	320 MBH output		1.60	10		3,550	360		3,910	4,525
0300	For powered venter and adapter, add					460			460	510
0502	For required flue pipe, see Section 23 51 23.10									
1000	Outdoor installation, with power venter									
1020	75 MBH output	Q-5	4	4	Ea.	2,775	144		2,919	3,275
1060	120 MBH output		4	4		3,525	144		3,669	4,125
1100	187 MBH output		3	5.333		5,050	192		5,242	5,875
1140	300 MBH output		1.80	8.889		6,100	320		6,420	7,225
1180	450 MBH output		1.40	11.429		7,450	410		7,860	8,875

23 55 33 – Fuel-Fired Unit Heaters

23 55 33.13 Oil-Fired Unit Heaters		Crew	Daily Output	Labor-Hours	Unit	Material	Labor	Equipment	Total	Total Incl O&P
0010	**OIL-FIRED UNIT HEATERS**, Cabinet, grilles, fan, ctrl, burner, no piping									
6000	Oil fired, suspension mounted, 94 MBH output	Q-5	4	4	Ea.	3,400	144		3,544	3,950
6040	140 MBH output		3	5.333		3,625	192		3,817	4,300
6060	184 MBH output		3	5.333		3,900	192		4,092	4,625

23 55 33.16 Gas-Fired Unit Heaters		Crew	Daily Output	Labor-Hours	Unit	Material	Labor	Equipment	Total	Total Incl O&P
0010	**GAS-FIRED UNIT HEATERS**, Cabinet, grilles, fan, ctrls., burner, no piping									
0022	thermostat, no piping. For flue see Section 23 51 23.10									
1000	Gas fired, floor mounted									
1100	60 MBH output	Q-5	10	1.600	Ea.	805	57.50		862.50	980
1140	100 MBH output		8	2		890	72		962	1,100
1180	180 MBH output		6	2.667		1,275	96		1,371	1,550
2000	Suspension mounted, propeller fan, 20 MBH output		8.50	1.882		985	68		1,053	1,175
2040	60 MBH output		7	2.286		1,200	82.50		1,282.50	1,450
2060	80 MBH output		6	2.667		1,225	96		1,321	1,500
2100	130 MBH output		5	3.200		1,500	115		1,615	1,850
2240	320 MBH output		2	8		2,650	288		2,938	3,375
2500	For powered venter and adapter, add					355			355	390
5000	Wall furnace, 17.5 MBH output	Q-5	6	2.667		715	96		811	940
5020	24 MBH output		5	3.200		740	115		855	1,000
5040	35 MBH output		4	4		895	144		1,039	1,225

23 56 Solar Energy Heating Equipment

23 56 16 – Packaged Solar Heating Equipment

23 56 16.40 Solar Heating Systems		Crew	Daily Output	Labor-Hours	Unit	Material	2011 Bare Costs Labor	Equipment	Total	Total Incl O&P
0010	**SOLAR HEATING SYSTEMS** R235616-60									
0020	System/Package prices, not including connecting									
0030	pipe, insulation, or special heating/plumbing fixtures									
0500	Hot water, standard package, low temperature									
0540	1 collector, circulator, fittings, 65 gal. tank G	Q-1	.50	32	Ea.	1,550	1,150		2,700	3,575
0580	2 collectors, circulator, fittings, 120 gal. tank G		.40	40		2,250	1,425		3,675	4,800
0620	3 collectors, circulator, fittings, 120 gal. tank G		.34	47.059		3,075	1,675		4,750	6,125
0700	Medium temperature package									
0720	1 collector, circulator, fittings, 80 gal. tank G	Q-1	.50	32	Ea.	2,250	1,150		3,400	4,350
0740	2 collectors, circulator, fittings, 120 gal. tank G		.40	40		3,275	1,425		4,700	5,925
0780	3 collectors, circulator, fittings, 120 gal. tank G		.30	53.333		4,250	1,900		6,150	7,775
0980	For each additional 120 gal. tank, add G					1,475			1,475	1,625

23 56 19 – Solar Heating Components

23 56 19.50 Solar Heating Ancillary

23 56 19.50 Solar Heating Ancillary		Crew	Daily Output	Labor-Hours	Unit	Material	Labor	Equipment	Total	Total Incl O&P
0010	**SOLAR HEATING ANCILLARY**									
2300	Circulators, air									
2310	Blowers									
2400	Reversible fan, 20" diameter, 2 speed G	Q-9	18	.889	Ea.	108	30.50		138.50	170
2870	1/12 HP, 30 GPM G	Q-1	10	1.600	"	330	57		387	460
3000	Collector panels, air with aluminum absorber plate									
3010	Wall or roof mount									
3040	Flat black, plastic glazing									
3080	4' x 8' G	Q-9	6	2.667	Ea.	635	91.50		726.50	845
3200	Flush roof mount, 10' to 16' x 22" wide G	"	96	.167	L.F.	450	5.75		455.75	505
3300	Collector panels, liquid with copper absorber plate									
3330	Alum. frame, 4' x 8', 5/32" single glazing G	Q-1	9.50	1.684	Ea.	945	60		1,005	1,150
3390	Alum. frame, 4' x 10', 5/32" single glazing G		6	2.667		1,075	95		1,170	1,350
3450	Flat black, alum. frame, 3.5' x 7.5' G		9	1.778		690	63.50		753.50	865
3500	4' x 8' G		5.50	2.909		865	104		969	1,125
3520	4' x 10' G		10	1.600		970	57		1,027	1,175
3540	4' x 12.5' G		5	3.200		1,175	114		1,289	1,450
3600	Liquid, full wetted, plastic, alum. frame, 3' x 10' G		5	3.200		234	114		348	445
3650	Collector panel mounting, flat roof or ground rack G		7	2.286		217	81.50		298.50	370
3670	Roof clamps G		70	.229	Set	2.35	8.15		10.50	15.90
3700	Roof strap, teflon G	1 Plum	205	.039	L.F.	19.95	1.55		21.50	24.50
3900	Differential controller with two sensors									
3930	Thermostat, hard wired G	1 Plum	8	1	Ea.	79.50	39.50		119	152
4100	Five station with digital read-out G	"	3	2.667	"	218	106		324	410
4300	Heat exchanger									
4580	Fluid to fluid package includes two circulating pumps									
4590	expansion tank, check valve, relief valve									
4600	controller, high temperature cutoff and sensors G	Q-1	2.50	6.400	Ea.	695	228		923	1,150
4650	Heat transfer fluid									
4700	Propylene glycol, inhibited anti-freeze G	1 Plum	28	.286	Gal.	12.05	11.35		23.40	32
8250	Water storage tank with heat exchanger and electric element									
8300	80 gal. with 2" x 2 lb. density insulation G	1 Plum	1.60	5	Ea.	1,100	198		1,298	1,525
8380	120 gal. with 2" x 2 lb. density insulation G		1.40	5.714		1,225	227		1,452	1,725
8400	120 gal. with 2" x 2 lb. density insul., 40 S.F. heat coil G		1.40	5.714		1,575	227		1,802	2,100

23 57 Heat Exchangers for HVAC

23 57 16 – Steam-to-Water Heat Exchangers

23 57 16.10 Shell/Tube Type Steam-to-Water Heat Exch.		Crew	Daily Output	Labor-Hours	Unit	Material	Labor (2011 Bare Costs)	Equipment (2011 Bare Costs)	Total (2011 Bare Costs)	Total Incl O&P
0010	**SHELL AND TUBE TYPE STEAM-TO-WATER HEAT EXCHANGERS**									
0016	Shell & tube type, 2 or 4 pass, 3/4" O.D. copper tubes,									
0020	C.I. heads, C.I. tube sheet, steel shell									
0100	Hot water 40°F to 180°F, by steam at 10 PSI									
0120	8 GPM	Q-5	6	2.667	Ea.	1,825	96		1,921	2,150
0140	10 GPM		5	3.200		2,750	115		2,865	3,225
0160	40 GPM		4	4		4,250	144		4,394	4,900
0180	64 GPM		2	8		6,525	288		6,813	7,650
0200	96 GPM		1	16		8,725	575		9,300	10,500
0220	120 GPM	Q-6	1.50	16		11,400	555		11,955	13,500

23 57 19 – Liquid-to-Liquid Heat Exchangers

23 57 19.13 Plate-Type, Liquid-to-Liquid Heat Exchangers

		Crew	Daily Output	Labor-Hours	Unit	Material	Labor	Equipment	Total	Total Incl O&P
0010	**PLATE-TYPE, LIQUID-TO-LIQUID HEAT EXCHANGERS**									
3000	Plate type,									
3100	400 GPM	Q-6	.80	30	Ea.	32,300	1,050		33,350	37,200
3120	800 GPM	"	.50	48		55,500	1,675		57,175	63,500
3140	1200 GPM	Q-7	.34	94.118		82,500	3,400		85,900	96,500
3160	1800 GPM	"	.24	133		110,000	4,800		114,800	129,000

23 57 19.16 Shell-Type, Liquid-to-Liquid Heat Exchangers

		Crew	Daily Output	Labor-Hours	Unit	Material	Labor	Equipment	Total	Total Incl O&P
0010	**SHELL-TYPE, LIQUID-TO-LIQUID HEAT EXCHANGERS**									
1000	Hot water 40°F to 140°F, by water at 200°F									
1020	7 GPM	Q-5	6	2.667	Ea.	2,225	96		2,321	2,600
1040	16 GPM		5	3.200		3,175	115		3,290	3,700
1060	34 GPM		4	4		4,800	144		4,944	5,500
1100	74 GPM		1.50	10.667		8,700	385		9,085	10,200

23 63 Refrigerant Condensers

23 63 13 – Air-Cooled Refrigerant Condensers

23 63 13.10 Air-Cooled Refrig. Condensers

		Crew	Daily Output	Labor-Hours	Unit	Material	Labor	Equipment	Total	Total Incl O&P
0010	**AIR-COOLED REFRIG. CONDENSERS**, Ratings for 30°F TD, R-22.									
0080	Air cooled, belt drive, propeller fan									
0240	50 ton	Q-6	.69	34.985	Ea.	9,975	1,225		11,200	13,000
0280	59 ton		.58	41.308		12,000	1,425		13,425	15,600
0320	73 ton		.47	51.173		15,600	1,775		17,375	20,100
0360	86 ton		.40	60.302		18,100	2,100		20,200	23,300
0380	88 ton		.39	61.697		19,400	2,150		21,550	24,900
1550	Air cooled, direct drive, propeller fan									
1590	1 ton	Q-5	3.80	4.211	Ea.	685	152		837	1,000
1600	1-1/2 ton		3.60	4.444		840	160		1,000	1,175
1620	2 ton		3.20	5		900	180		1,080	1,275
1640	5 ton		2	8		1,825	288		2,113	2,500
1660	10 ton		1.40	11.429		3,025	410		3,435	4,000
1690	16 ton		1.10	14.545		4,150	525		4,675	5,425
1720	26 ton		.84	19.002		5,400	685		6,085	7,075
1760	41 ton	Q-6	.77	31.008		9,700	1,075		10,775	12,500
1800	63 ton	"	.55	44.037		15,300	1,525		16,825	19,300

23 64 Packaged Water Chillers

23 64 13 – Absorption Water Chillers

23 64 13.16 Indirect-Fired Absorption Water Chillers		Crew	Daily Output	Labor-Hours	Unit	2011 Bare Costs Material	Labor	Equipment	Total	Total Incl O&P
0010	**INDIRECT-FIRED ABSORPTION WATER CHILLERS**									
0020	Steam or hot water, water cooled									
0400	420 ton	Q-7	.10	323	Ea.	335,000	11,700		346,700	388,000

23 64 16 – Centrifugal Water Chillers

23 64 16.10 Centrifugal Type Water Chillers

		Crew	Daily Output	Labor-Hours	Unit	Material	Labor	Equipment	Total	Total Incl O&P
0010	**CENTRIFUGAL TYPE WATER CHILLERS**, With standard controls									
0020	Centrifugal liquid chiller, water cooled									
0030	not including water tower									
0100	2000 ton (twin 1000 ton units)	Q-7	.07	477	Ea.	847,500	17,200		864,700	960,500

23 64 19 – Reciprocating Water Chillers

23 64 19.10 Reciprocating Type Water Chillers

		Crew	Daily Output	Labor-Hours	Unit	Material	Labor	Equipment	Total	Total Incl O&P
0010	**RECIPROCATING TYPE WATER CHILLERS**, With standard controls									
0494	Water chillers, integral air cooled condenser									
0600	100 ton cooling	Q-7	.25	129	Ea.	71,500	4,650		76,150	86,000
0980	Water cooled, multiple compressor, semi-hermetic, tower not incl.									
1000	15 ton cooling	Q-6	.36	65.934	Ea.	17,500	2,300		19,800	23,100
1020	25 ton cooling	Q-7	.41	78.049		18,900	2,825		21,725	25,400
1060	35 ton cooling		.31	101		22,600	3,675		26,275	30,800
1100	50 ton cooling		.28	113		32,300	4,100		36,400	42,200
1160	100 ton cooling		.18	179		58,500	6,475		64,975	75,000
1180	125 ton cooling		.16	196		66,500	7,075		73,575	84,500
1200	145 ton cooling		.16	202		77,500	7,300		84,800	97,500
1451	Water cooled, dual compressors, semi-hermetic, tower not incl.									
1500	80 ton cooling	Q-7	.14	222	Ea.	27,400	8,000		35,400	43,200
1520	100 ton cooling		.14	228		37,000	8,250		45,250	54,000
1540	120 ton cooling		.14	231		44,800	8,350		53,150	63,000

23 64 23 – Scroll Water Chillers

23 64 23.10 Scroll Water Chillers

		Crew	Daily Output	Labor-Hours	Unit	Material	Labor	Equipment	Total	Total Incl O&P
0010	**SCROLL WATER CHILLERS,** With standard controls									
0490	Packaged w/integral air cooled condenser, 15 ton cool	Q-7	.37	86.486	Ea.	20,400	3,125		23,525	27,600
0500	20 ton cooling		.34	94.118		26,400	3,400		29,800	34,600
0520	40 ton cooling		.30	108		34,100	3,900		38,000	43,900
0680	Scroll water cooled, single compressor, hermetic, tower not incl.									
0700	2 ton cooling	Q-5	.57	28.070	Ea.	3,225	1,000		4,225	5,200
0710	5 ton cooling		.57	28.070		3,850	1,000		4,850	5,900
0740	8 ton cooling		.31	52.117		5,325	1,875		7,200	8,925
0760	10 ton cooling	Q-6	.36	67.039		6,125	2,325		8,450	10,500
0800	20 ton cooling	Q-7	.38	83.990		10,900	3,025		13,925	17,000
0820	30 ton cooling	"	.33	96.096		12,200	3,475		15,675	19,200

23 64 26 – Rotary-Screw Water Chillers

23 64 26.10 Rotary-Screw Type Water Chillers

		Crew	Daily Output	Labor-Hours	Unit	Material	Labor	Equipment	Total	Total Incl O&P
0010	**ROTARY-SCREW TYPE WATER CHILLERS**, With standard controls									
0110	Screw, liquid chiller, air cooled, insulated evaporator									
0120	130 ton	Q-7	.14	228	Ea.	83,500	8,250		91,750	105,500
0124	160 ton		.13	246		103,000	8,875		111,875	127,500
0128	180 ton		.13	250		115,500	9,025		124,525	141,500
0132	210 ton		.12	258		127,500	9,300		136,800	155,000
0136	270 ton		.12	266		145,500	9,625		155,125	175,500
0140	320 ton		.12	275		183,000	9,950		192,950	217,000
1450	Water cooled, tower not included									

23 64 Packaged Water Chillers

23 64 26 – Rotary-Screw Water Chillers

23 64 26.10 Rotary-Screw Type Water Chillers		Crew	Daily Output	Labor-Hours	Unit	Material	Labor	Equipment	Total	Total Incl O&P
						2011 Bare Costs				
1580	150 ton cooling, screw compressors	Q-7	.13	240	Ea.	63,500	8,675		72,175	84,000
1620	200 ton cooling, screw compressors		.13	250		87,500	9,025		96,525	111,000
1660	291 ton cooling, screw compressors	↓	.12	260	↓	91,500	9,375		100,875	116,500

23 65 Cooling Towers

23 65 13 – Forced-Draft Cooling Towers

23 65 13.10 Forced-Draft Type Cooling Towers		Crew	Daily Output	Labor-Hours	Unit	Material	Labor	Equipment	Total	Total Incl O&P
0010	**FORCED-DRAFT TYPE COOLING TOWERS**, Packaged units									
0070	Galvanized steel									
0080	Induced draft, crossflow									
0100	Vertical, belt drive, 61 tons	Q-6	90	.267	TonAC	171	9.25		180.25	203
0150	100 ton		100	.240		163	8.35		171.35	194
0200	115 ton		109	.220		153	7.65		160.65	181
0250	131 ton		120	.200		151	6.95		157.95	177
0260	162 ton	↓	132	.182	↓	149	6.30		155.30	174
1500	Induced air, double flow									
1900	Vertical, gear drive, 167 ton	Q-6	126	.190	TonAC	145	6.60		151.60	170
2000	297 ton		129	.186		90.50	6.45		96.95	110
2100	582 ton		132	.182		73	6.30		79.30	91
2150	849 ton		142	.169		70.50	5.85		76.35	87.50
2200	1016 ton	↓	150	.160	↓	65.50	5.55		71.05	81
3000	For higher capacities, use multiples									
3500	For pumps and piping, add	Q-6	38	.632	TonAC	85.50	22		107.50	130
4000	For absorption systems, add				"	75%	75%			
5000	Fiberglass tower on galvanized steel support structure									
5010	Draw thru									
5100	100 ton	Q-6	1.40	17.143	Ea.	10,100	595		10,695	12,100
5120	120 ton		1.20	20		11,700	695		12,395	14,000
5140	140 ton		1	24		12,700	835		13,535	15,400
5160	160 ton		.80	30		14,100	1,050		15,150	17,200
5180	180 ton		.65	36.923		16,000	1,275		17,275	19,700
5200	200 ton	↓	.48	50	↓	18,100	1,725		19,825	22,700
5300	For stainless steel support structure, add					30%				
5360	For higher capacities, use multiples of each size									
6000	Stainless steel									
6010	Induced draft, crossflow, horizontal, belt drive									
6100	57 ton	Q-6	1.50	16	Ea.	24,100	555		24,655	27,400
6120	91 ton		.99	24.242		29,000	840		29,840	33,400
6140	111 ton		.43	55.814		37,000	1,950		38,950	43,900
6160	126 ton	↓	.22	109	↓	37,000	3,800		40,800	46,900
6170	Induced draft, crossflow, vertical, gear drive									
6172	167 ton	Q-6	.75	32	Ea.	47,300	1,100		48,400	54,000
6174	297 ton		.43	55.814		52,500	1,950		54,450	61,000
6176	582 ton		.23	104		83,000	3,625		86,625	97,500
6178	849 ton		.17	141		117,500	4,900		122,400	137,000
6180	1016 ton	↓	.15	160	↓	141,000	5,550		146,550	164,500

23 73 Indoor Central-Station Air-Handling Units

23 73 39 – Indoor, Direct Gas-Fired Heating and Ventilating Units

23 73 39.10 Make-Up Air Unit		Crew	Daily Output	Labor-Hours	Unit	Material	Labor (2011 Bare Costs)	Equipment (2011 Bare Costs)	Total (2011 Bare Costs)	Total Incl O&P
0010	**MAKE-UP AIR UNIT**									
0020	Indoor suspension, natural/LP gas, direct fired,									
0032	standard control. For flue see Section 23 51 23.10									
0040	70°F temperature rise, MBH is input									
0100	75 MBH input	Q-6	3.60	6.667	Ea.	2,400	231		2,631	3,025
0160	150 MBH input		3	8		2,850	278		3,128	3,575
0220	225 MBH input		2.40	10		3,300	345		3,645	4,200
0300	400 MBH input		1.60	15		7,450	520		7,970	9,025
0600	For discharge louver assembly, add					5%				
0700	For filters, add					10%				
0800	For air shut-off damper section, add					30%				

23 74 Packaged Outdoor HVAC Equipment

23 74 33 – Dedicated Outdoor-Air Units

23 74 33.10 Roof Top Air Conditioners		Crew	Daily Output	Labor-Hours	Unit	Material	Labor (2011 Bare Costs)	Equipment (2011 Bare Costs)	Total (2011 Bare Costs)	Total Incl O&P
0010	**ROOF TOP AIR CONDITIONERS**, Standard controls, curb, economizer									
1000	Single zone, electric cool, gas heat									
1100	3 ton cooling, 60 MBH heating R236000-20	Q-5	.70	22.857	Ea.	3,600	825		4,425	5,300
1120	4 ton cooling, 95 MBH heating		.61	26.403		4,250	950		5,200	6,225
1140	5 ton cooling, 112 MBH heating		.56	28.521		4,975	1,025		6,000	7,150
1145	6 ton cooling, 140 MBH heating		.52	30.769		5,825	1,100		6,925	8,225
1150	7.5 ton cooling, 170 MBH heating		.50	32.258		7,650	1,175		8,825	10,300
1160	10 ton cooling, 200 MBH heating	Q-6	.67	35.982		10,600	1,250		11,850	13,700
1170	12.5 ton cooling, 230 MBH heating		.63	37.975		13,400	1,325		14,725	16,900
1190	17.5 ton cooling, 330 MBH heating		.52	45.889		18,800	1,600		20,400	23,300
1200	20 ton cooling, 360 MBH heating	Q-7	.67	47.976		21,900	1,725		23,625	26,900
1210	25 ton cooling, 450 MBH heating		.56	57.554		25,000	2,075		27,075	30,900
1220	30 ton cooling, 540 MBH heating		.47	68.376		25,500	2,475		27,975	32,100
1240	40 ton cooling, 675 MBH heating		.35	91.168		33,400	3,275		36,675	42,100
2000	Multizone, electric cool, gas heat, economizer									
2100	15 ton cooling, 360 MBH heating	Q-7	.61	52.545	Ea.	57,000	1,900		58,900	65,500
2120	20 ton cooling, 360 MBH heating		.53	60.038		61,000	2,175		63,175	71,000
2200	40 ton cooling, 540 MBH heating		.28	113		107,000	4,100		111,100	124,500
2210	50 ton cooling, 540 MBH heating		.23	142		133,500	5,125		138,625	155,000
2220	70 ton cooling, 1500 MBH heating		.16	198		144,500	7,175		151,675	170,500
2240	80 ton cooling, 1500 MBH heating		.14	228		165,000	8,250		173,250	195,000
2260	90 ton cooling, 1500 MBH heating		.13	256		173,000	9,225		182,225	205,000
2280	105 ton cooling, 1500 MBH heating		.11	290		190,500	10,500		201,000	226,500
2400	For hot water heat coil, deduct					5%				
2500	For steam heat coil, deduct					2%				
2600	For electric heat, deduct					3%	5%			

23 81 Decentralized Unitary HVAC Equipment

23 81 13 – Packaged Terminal Air-Conditioners

23 81 13.10 Packaged Cabinet Type Air-Conditioners		Crew	Daily Output	Labor-Hours	Unit	2011 Bare Costs Material	Labor	Equipment	Total	Total Incl O&P
0010	**PACKAGED CABINET TYPE AIR-CONDITIONERS**, Cabinet, wall sleeve,									
0100	louver, electric heat, thermostat, manual changeover, 208 V									
0200	6,000 BTUH cooling, 8800 BTU heat	Q-5	6	2.667	Ea.	715	96		811	940
0220	9,000 BTUH cooling, 13,900 BTU heat		5	3.200		810	115		925	1,075
0240	12,000 BTUH cooling, 13,900 BTU heat		4	4		895	144		1,039	1,225
0260	15,000 BTUH cooling, 13,900 BTU heat		3	5.333		985	192		1,177	1,400
0500	For hot water coil, increase heat by 10%, add					5%	10%			
1000	For steam, increase heat output by 30%, add					8%	10%			

23 81 19 – Self-Contained Air-Conditioners

23 81 19.20 Self-Contained Single Package		Crew	Daily Output	Labor-Hours	Unit	Material	Labor	Equipment	Total	Total Incl O&P
0010	**SELF-CONTAINED SINGLE PACKAGE**									
0100	Air cooled, for free blow or duct, not incl. remote condenser									
0200	3 ton cooling	Q-5	1	16	Ea.	3,475	575		4,050	4,775
0220	5 ton cooling	Q-6	1.20	20		4,150	695		4,845	5,700
0240	10 ton cooling	Q-7	1	32		7,000	1,150		8,150	9,575
0260	20 ton cooling		.90	35.556		15,600	1,275		16,875	19,300
0280	30 ton cooling		.80	40		23,600	1,450		25,050	28,400
0340	60 ton cooling	Q-8	.40	80		53,500	3,050	81	56,631	64,000
0490	For duct mounting no price change									
0500	For steam heating coils, add				Ea.	10%	10%			
1000	Water cooled for free blow or duct, not including tower									
1010	Constant volume									
1100	3 ton cooling	Q-6	1	24	Ea.	3,475	835		4,310	5,175
1120	5 ton cooling	″	1	24		4,550	835		5,385	6,375
1140	10 ton cooling	Q-7	.90	35.556		8,550	1,275		9,825	11,500
1160	20 ton cooling		.80	40		26,800	1,450		28,250	31,900
1180	30 ton cooling		.70	45.714		35,500	1,650		37,150	41,800

23 81 23 – Computer-Room Air-Conditioners

23 81 23.10 Computer Room Units		Crew	Daily Output	Labor-Hours	Unit	Material	Labor	Equipment	Total	Total Incl O&P
0010	**COMPUTER ROOM UNITS**									
1000	Air cooled, includes remote condenser but not									
1020	interconnecting tubing or refrigerant									
1080	3 ton	Q-5	.50	32	Ea.	17,000	1,150		18,150	20,600
1120	5 ton		.45	35.556		18,100	1,275		19,375	22,100
1160	6 ton		.30	53.333		32,000	1,925		33,925	38,400
1200	8 ton		.27	59.259		33,900	2,125		36,025	40,800
1240	10 ton		.25	64		35,400	2,300		37,700	42,800
1260	12 ton		.24	66.667		36,700	2,400		39,100	44,300
1280	15 ton		.22	72.727		39,000	2,625		41,625	47,200
1290	18 ton		.20	80		44,400	2,875		47,275	53,500
1300	20 ton	Q-6	.26	92.308		46,700	3,200		49,900	56,500
1320	22 ton		.24	100		47,300	3,475		50,775	57,500
1360	30 ton		.21	114		58,000	3,975		61,975	70,500
2200	Chilled water, for connection to									
2220	existing chiller system of adequate capacity									
2260	5 ton	Q-5	.74	21.622	Ea.	13,000	780		13,780	15,600

23 81 43 – Air-Source Unitary Heat Pumps

23 81 43.10 Air-Source Heat Pumps		Crew	Daily Output	Labor-Hours	Unit	Material	Labor	Equipment	Total	Total Incl O&P
0010	**AIR-SOURCE HEAT PUMPS**, Not including interconnecting tubing									
1000	Air to air, split system, not including curbs, pads, fan coil and ductwork									
1012	Outside condensing unit only, for fan coil see Section 23 82 19.10									

23 81 Decentralized Unitary HVAC Equipment

23 81 43 – Air-Source Unitary Heat Pumps

23 81 43.10 Air-Source Heat Pumps		Crew	Daily Output	Labor-Hours	Unit	2011 Bare Costs Material	Labor	Equipment	Total	Total Incl O&P
1020	2 ton cooling, 8.5 MBH heat @ 0°F	Q-5	2	8	Ea.	2,125	288		2,413	2,800
1060	5 ton cooling, 27 MBH heat @ 0°F		.50	32		3,425	1,150		4,575	5,650
1080	7.5 ton cooling, 33 MBH heat @ 0°F		.45	35.556		5,925	1,275		7,200	8,600
1100	10 ton cooling, 50 MBH heat @ 0°F	Q-6	.64	37.500		8,250	1,300		9,550	11,200
1120	15 ton cooling, 64 MBH heat @ 0°F		.50	48		12,200	1,675		13,875	16,100
1130	20 ton cooling, 85 MBH heat @ 0°F		.35	68.571		16,000	2,375		18,375	21,500
1140	25 ton cooling, 119 MBH heat @ 0°F		.25	96		19,300	3,325		22,625	26,700
1500	Single package, not including curbs, pads, or plenums									
1520	2 ton cooling, 6.5 MBH heat @ 0°F	Q-5	1.50	10.667	Ea.	2,750	385		3,135	3,650
1580	4 ton cooling, 13 MBH heat @ 0°F		.96	16.667		3,850	600		4,450	5,225
1640	7.5 ton cooling, 35 MBH heat @ 0°F		.40	40		6,850	1,450		8,300	9,875

23 81 46 – Water-Source Unitary Heat Pumps

23 81 46.10 Water Source Heat Pumps

		Crew	Daily Output	Labor-Hours	Unit	Material	Labor	Equipment	Total	Total Incl O&P
0010	**WATER SOURCE HEAT PUMPS**, Not incl. connecting tubing or water source									
2000	Water source to air, single package									
2100	1 ton cooling, 13 MBH heat @ 75°F	Q-5	2	8	Ea.	1,700	288		1,988	2,325
2140	2 ton cooling, 19 MBH heat @ 75°F		1.70	9.412		1,875	340		2,215	2,625
2220	5 ton cooling, 29 MBH heat @ 75°F		.90	17.778		2,825	640		3,465	4,150
3960	For supplementary heat coil, add					10%				
4000	For increase in capacity thru use									
4020	of solar collector, size boiler at 60%									

23 82 Convection Heating and Cooling Units

23 82 16 – Air Coils

23 82 16.10 Flanged Coils

		Crew	Daily Output	Labor-Hours	Unit	Material	Labor	Equipment	Total	Total Incl O&P
0010	**FLANGED COILS**									
0500	Chilled water cooling, 6 rows, 24" x 48"	Q-5	3.20	5	Ea.	3,725	180		3,905	4,400
1000	Direct expansion cooling, 6 rows, 24" x 48"		2.80	5.714		4,025	206		4,231	4,750
1500	Hot water heating, 1 row, 24" x 48"		4	4		1,475	144		1,619	1,850
2000	Steam heating, 1 row, 24" x 48"		3.06	5.229		2,100	189		2,289	2,600

23 82 16.20 Duct Heaters

		Crew	Daily Output	Labor-Hours	Unit	Material	Labor	Equipment	Total	Total Incl O&P
0010	**DUCT HEATERS**, Electric, 480 V, 3 Ph.									
0020	Finned tubular insert, 500°F									
0100	8" wide x 6" high, 4.0 kW	Q-20	16	1.250	Ea.	705	44.50		749.50	850
0120	12" high, 8.0 kW		15	1.333		1,175	47.50		1,222.50	1,350
0140	18" high, 12.0 kW		14	1.429		1,650	51		1,701	1,875
0160	24" high, 16.0 kW		13	1.538		2,100	54.50		2,154.50	2,425
0180	30" high, 20.0 kW		12	1.667		2,575	59.50		2,634.50	2,925
0300	12" wide x 6" high, 6.7 kW		15	1.333		750	47.50		797.50	905
0320	12" high, 13.3 kW		14	1.429		1,200	51		1,251	1,400
0340	18" high, 20.0 kW		13	1.538		1,700	54.50		1,754.50	1,975
0360	24" high, 26.7 kW		12	1.667		2,175	59.50		2,234.50	2,500
0700	24" wide x 6" high, 17.8 kW		13	1.538		890	54.50		944.50	1,075
0760	24" high, 71.1 kW		10	2		2,725	71		2,796	3,125
8000	To obtain BTU multiply kW by 3413									

23 82 19 – Fan Coil Units

23 82 19.10 Fan Coil Air Conditioning

		Crew	Daily Output	Labor-Hours	Unit	Material	Labor	Equipment	Total	Total Incl O&P
0010	**FAN COIL AIR CONDITIONING** Cabinet mounted, filters, controls									
0100	Chilled water, 1/2 ton cooling	Q-5	8	2	Ea.	710	72		782	900
0120	1 ton cooling		6	2.667		815	96		911	1,050

23 82 Convection Heating and Cooling Units

23 82 19 – Fan Coil Units

23 82 19.10 Fan Coil Air Conditioning		Crew	Daily Output	Labor-Hours	Unit	Material	Labor	Equipment	Total	Total Incl O&P
						2011 Bare Costs				
0180	3 ton cooling	Q-5	4	4	Ea.	1,725	144		1,869	2,125
0262	For hot water coil, add					40%	10%			
0940	Direct expansion, for use w/air cooled condensing unit, 1.5 ton cooling	Q-5	5	3.200	Ea.	690	115		805	950
1000	5 ton cooling	"	3	5.333		1,325	192		1,517	1,775
1040	10 ton cooling	Q-6	2.60	9.231		2,975	320		3,295	3,800
1060	20 ton cooling	"	.70	34.286		5,575	1,200		6,775	8,100

23 82 19.20 Heating and Ventilating Units

		Crew	Daily Output	Labor-Hours	Unit	Material	Labor	Equipment	Total	Total Incl O&P
0010	**HEATING AND VENTILATING UNITS**, Classroom units									
0020	Includes filter, heating/cooling coils, standard controls									
0080	750 CFM, 2 tons cooling	Q-6	2	12	Ea.	3,775	415		4,190	4,825
0120	1250 CFM, 3 tons cooling		1.40	17.143		4,600	595		5,195	6,025
0140	1500 CFM, 4 tons cooling		.80	30		4,925	1,050		5,975	7,100
0500	For electric heat, add					35%				
1000	For no cooling, deduct					25%	10%			

23 82 27 – Infrared Units

23 82 27.10 Infrared Type Heating Units

		Crew	Daily Output	Labor-Hours	Unit	Material	Labor	Equipment	Total	Total Incl O&P
0010	**INFRARED TYPE HEATING UNITS**									
0020	Gas fired, unvented, electric ignition, 100% shutoff.									
0030	Piping and wiring not included									
0060	Input, 15 MBH	Q-5	7	2.286	Ea.	505	82.50		587.50	690
0120	45 MBH		5	3.200		1,500	115		1,615	1,850
0160	60 MBH		4	4		1,600	144		1,744	1,975
0240	120 MBH		2	8		1,975	288		2,263	2,625

23 82 29 – Radiators

23 82 29.10 Hydronic Heating

		Crew	Daily Output	Labor-Hours	Unit	Material	Labor	Equipment	Total	Total Incl O&P
0010	**HYDRONIC HEATING**, Terminal units, not incl. main supply pipe									
1000	Radiation									
1100	Panel, baseboard, C.I., including supports, no covers	Q-5	46	.348	L.F.	32.50	12.55		45.05	56
3000	Radiators, cast iron									
3100	Free standing or wall hung, 6 tube, 25" high	Q-5	96	.167	Section	36	6		42	49.50
3200	4 tube, 19" high	"	96	.167	"	25	6		31	37.50
3250	Adj. brackets, 2 per wall radiator up to 30 sections	1 Stpi	32	.250	Ea.	28	10		38	47
9500	To convert SFR to BTU rating: Hot water, 150 x SFR									
9510	Forced hot water, 180 x SFR; steam, 240 x SFR									

23 82 33 – Convectors

23 82 33.10 Convector Units

		Crew	Daily Output	Labor-Hours	Unit	Material	Labor	Equipment	Total	Total Incl O&P
0010	**CONVECTOR UNITS,** Terminal units, not incl. main supply pipe									
2204	Convector, multifin, 2 pipe w/cabinet									
2210	17" H x 24" L	Q-5	10	1.600	Ea.	98.50	57.50		156	202
2214	17" H x 36" L		8.60	1.860		148	67		215	272
2218	17" H x 48" L		7.40	2.162		197	78		275	345
2222	21" H x 24" L		9	1.778		96.50	64		160.50	211
2226	21" H x 36" L		8.20	1.951		145	70.50		215.50	274
2228	21" H x 48" L		6.80	2.353		193	85		278	350
2240	For knob operated damper, add					140%				
2241	For metal trim strips, add	Q-5	64	.250	Ea.	19.25	9		28.25	35.50
2243	For snap-on inlet grille, add					10%	10%			
2245	For hinged access door, add	Q-5	64	.250	Ea.	35.50	9		44.50	53.50
2246	For air chamber, auto-venting, add	"	58	.276	"	6.35	9.95		16.30	23.50

23 82 Convection Heating and Cooling Units

23 82 36 – Finned-Tube Radiation Heaters

23 82 36.10 Finned Tube Radiation		Crew	Daily Output	Labor-Hours	Unit	2011 Bare Costs Material	Labor	Equipment	Total	Total Incl O&P
0010	**FINNED TUBE RADIATION**, Terminal units, not incl. main supply pipe									
1150	Fin tube, wall hung, 14" slope top cover, with damper									
1200	1-1/4" copper tube, 4-1/4" alum. fin	Q-5	38	.421	L.F.	40.50	15.20		55.70	70
1250	1-1/4" steel tube, 4-1/4" steel fin	"	36	.444	"	38.50	16		54.50	68.50
1500	Note: fin tube may also require corners, caps, etc.									

23 82 39 – Unit Heaters

23 82 39.16 Propeller Unit Heaters		Crew	Daily Output	Labor-Hours	Unit	Material	Labor	Equipment	Total	Total Incl O&P
0010	**PROPELLER UNIT HEATERS**									
3950	Unit heaters, propeller, 115 V 2 psi steam, 60°F entering air									
4000	Horizontal, 12 MBH	Q-5	12	1.333	Ea.	335	48		383	450
4060	43.9 MBH		8	2		505	72		577	675
4140	96.8 MBH		6	2.667		735	96		831	965
4180	157.6 MBH		4	4		980	144		1,124	1,300
4240	286.9 MBH		2	8		1,600	288		1,888	2,225
4250	326.0 MBH		1.90	8.421		1,825	305		2,130	2,500
4260	364 MBH		1.80	8.889		1,900	320		2,220	2,625
4270	404 MBH		1.60	10		2,500	360		2,860	3,350
4300	For vertical diffuser, add					174			174	191
4310	Vertical flow, 40 MBH	Q-5	11	1.455		500	52.50		552.50	635
4314	58.5 MBH		8	2		520	72		592	690
4326	131.0 MBH		4	4		780	144		924	1,100
4346	297.0 MBH		1.80	8.889		1,500	320		1,820	2,175
4354	420 MBH,(460 V)	Q-6	1.80	13.333		2,000	465		2,465	2,950
4358	500 MBH, (460 V)		1.71	14.035		2,650	485		3,135	3,700
4362	570 MBH, (460 V)		1.40	17.143		3,725	595		4,320	5,075
4366	620 MBH, (460 V)		1.30	18.462		4,275	640		4,915	5,750
4370	960 MBH, (460 V)		1.10	21.818		7,200	760		7,960	9,125

23 83 Radiant Heating Units

23 83 33 – Electric Radiant Heaters

23 83 33.10 Electric Heating		Crew	Daily Output	Labor-Hours	Unit	Material	Labor	Equipment	Total	Total Incl O&P
0010	**ELECTRIC HEATING**, not incl. conduit or feed wiring									
1100	Rule of thumb: Baseboard units, including control	1 Elec	4.40	1.818	kW	95.50	73		168.50	224
1300	Baseboard heaters, 2' long, 375 watt		8	1	Ea.	40	40.50		80.50	110
1400	3' long, 500 watt		8	1		46	40.50		86.50	116
1600	4' long, 750 watt		6.70	1.194		54	48		102	138
1800	5' long, 935 watt		5.70	1.404		68	56.50		124.50	167
2000	6' long, 1125 watt		5	1.600		74	64.50		138.50	186
2400	8' long, 1500 watt		4	2		92	80.50		172.50	232
2950	Wall heaters with fan, 120 to 277 volt									
3600	Thermostats, integral	1 Elec	16	.500	Ea.	30	20		50	65.50
3800	Line voltage, 1 pole	"	8	1	"	31	40.50		71.50	99.50

23 84 13.10 Humidifier Units		Crew	Daily Output	Labor-Hours	Unit	2011 Bare Costs Material	Labor	Equipment	Total	Total Incl O&P
0010	**HUMIDIFIER UNITS**									
0520	Steam, room or duct, filter, regulators, auto. controls, 220 V									
0540	11 lb. per hour	Q-5	6	2.667	Ea.	2,450	96		2,546	2,850
0560	22 lb. per hour		5	3.200		2,700	115		2,815	3,175
0580	33 lb. per hour		4	4		2,775	144		2,919	3,275
0600	50 lb. per hour		4	4		3,400	144		3,544	3,975
0620	100 lb. per hour	↓	3	5.333	↓	4,075	192		4,267	4,800

Division Notes

	CREW	DAILY OUTPUT	LABOR-HOURS	UNIT	2011 BARE COSTS MAT.	LABOR	EQUIP.	TOTAL	TOTAL INCL O&P

Estimating Tips

26 05 00 Common Work Results for Electrical

- Conduit should be taken off in three main categories—power distribution, branch power, and branch lighting—so the estimator can concentrate on systems and components, therefore making it easier to ensure all items have been accounted for.
- For cost modifications for elevated conduit installation, add the percentages to labor according to the height of installation, and only to the quantities exceeding the different height levels, not to the total conduit quantities.
- Remember that aluminum wiring of equal ampacity is larger in diameter than copper and may require larger conduit.
- If more than three wires at a time are being pulled, deduct percentages from the labor hours of that grouping of wires.
- When taking off grounding systems, identify separately the type and size of wire, and list each unique type of ground connection.
- The estimator should take the weights of materials into consideration when completing a takeoff. Topics to consider include: How will the materials be supported? What methods of support are available? How high will the support structure have to reach? Will the final support structure be able to withstand the total burden? Is the support material included or separate from the fixture, equipment, and material specified?
- Do not overlook the costs for equipment used in the installation. If scaffolding or highlifts are available in the field, contractors may use them in lieu of the proposed ladders and rolling staging.

26 20 00 Low-Voltage Electrical Transmission

- Supports and concrete pads may be shown on drawings for the larger equipment, or the support system may be only a piece of plywood for the back of a panelboard. In either case, it must be included in the costs.

26 40 00 Electrical and Cathodic Protection

- When taking off cathodic protections systems, identify the type and size of cable, and list each unique type of anode connection.

26 50 00 Lighting

- Fixtures should be taken off room by room, using the fixture schedule, specifications, and the ceiling plan. For large concentrations of lighting fixtures in the same area, deduct the percentages from labor hours.

Reference Numbers

Reference numbers are shown in shaded boxes at the beginning of some major classifications. These numbers refer to related items in the Reference Section. The reference information may be an estimating procedure, an alternate pricing method, or technical information.

Note: Not all subdivisions listed here necessarily appear in this publication.

Note: **Trade Service,** *in part, has been used as a reference source for some of the material prices used in Division 26.*

26 05 05 – Selective Electrical Demolition

26 05 05.10 Electrical Demolition		Crew	Daily Output	Labor-Hours	Unit	Material	2011 Bare Costs Labor	Equipment	Total	Total Incl O&P
0010	**ELECTRICAL DEMOLITION**									
0020	Conduit to 15′ high, including fittings & hangers									
0100	Rigid galvanized steel, 1/2″ to 1″ diameter	1 Elec	242	.033	L.F.		1.33		1.33	2.16
0120	1-1/4″ to 2″	"	200	.040			1.61		1.61	2.61
0140	2-1/2″ to 3-1/2″	2 Elec	302	.053			2.13		2.13	3.46
0160	4″ to 6″	"	160	.100			4.03		4.03	6.55
0200	Electric metallic tubing (EMT), 1/2″ to 1″	1 Elec	394	.020			.82		.82	1.32
0220	1-1/4″ to 1-1/2″	"	326	.025			.99		.99	1.60
0260	3-1/2″ to 4″	2 Elec	310	.052			2.08		2.08	3.37
0270	Armored cable, (BX) avg. 50′ runs									
0280	#14, 2 wire	1 Elec	690	.012	L.F.		.47		.47	.76
0290	#14, 3 wire		571	.014			.56		.56	.91
0300	#12, 2 wire		605	.013			.53		.53	.86
0310	#12, 3 wire		514	.016			.63		.63	1.02
0320	#10, 2 wire		514	.016			.63		.63	1.02
0330	#10, 3 wire		425	.019			.76		.76	1.23
0340	#8, 3 wire		342	.023			.94		.94	1.53
0350	Non metallic sheathed cable (Romex)									
0360	#14, 2 wire	1 Elec	720	.011	L.F.		.45		.45	.72
0370	#14, 3 wire		657	.012			.49		.49	.79
0380	#12, 2 wire		629	.013			.51		.51	.83
0390	#10, 3 wire		450	.018			.72		.72	1.16
0400	Wiremold raceway, including fittings & hangers									
0420	No. 3000	1 Elec	250	.032	L.F.		1.29		1.29	2.09
0440	No. 4000		217	.037			1.48		1.48	2.41
0460	No. 6000		166	.048			1.94		1.94	3.14
0465	Telephone/power pole		12	.667	Ea.		27		27	43.50
0470	Non-metallic, straight section		480	.017	L.F.		.67		.67	1.09
0500	Channels, steel, including fittings & hangers									
0520	3/4″ x 1-1/2″	1 Elec	308	.026	L.F.		1.05		1.05	1.69
0540	1-1/2″ x 1-1/2″		269	.030			1.20		1.20	1.94
0560	1-1/2″ x 1-7/8″		229	.035			1.41		1.41	2.28
0600	Copper bus duct, indoor, 3 phase									
0610	Including hangers & supports									
0620	225 amp	2 Elec	135	.119	L.F.		4.77		4.77	7.75
0640	400 amp		106	.151			6.10		6.10	9.85
0660	600 amp		86	.186			7.50		7.50	12.15
0680	1000 amp		60	.267			10.75		10.75	17.40
0700	1600 amp		40	.400			16.10		16.10	26
0720	3000 amp		10	1.600			64.50		64.50	104
1300	Transformer, dry type, 1 phase, incl. removal of									
1320	supports, wire & conduit terminations									
1340	1 kVA	1 Elec	7.70	1.039	Ea.		42		42	68
1420	75 kVA	2 Elec	2.50	6.400	"		258		258	420
1440	3 phase to 600V, primary									
1460	3 kVA	1 Elec	3.85	2.078	Ea.		83.50		83.50	136
1520	75 kVA	2 Elec	2.70	5.926			239		239	385
1550	300 kVA	R-3	1.80	11.111			440	71.50	511.50	795
1570	750 kVA	"	1.10	18.182			720	117	837	1,300
1800	Wire, THW-THWN-THHN, removed from									
1810	in place conduit, to 15′ high									
1830	#14	1 Elec	65	.123	C.L.F.		4.95		4.95	8.05
1840	#12		55	.145			5.85		5.85	9.50

26 05 05 – Selective Electrical Demolition

26 05 05.10 Electrical Demolition		Crew	Daily Output	Labor-Hours	Unit	Material	2011 Bare Costs Labor	Equipment	Total	Total Incl O&P
1850	#10	1 Elec	45.50	.176	C.L.F.		7.10		7.10	11.45
1860	#8		40.40	.198			7.95		7.95	12.90
1870	#6		32.60	.245			9.90		9.90	16
1880	#4	2 Elec	53	.302			12.15		12.15	19.70
1890	#3		50	.320			12.90		12.90	21
1900	#2		44.60	.359			14.45		14.45	23.50
1910	1/0		33.20	.482			19.40		19.40	31.50
1920	2/0		29.20	.548			22		22	36
1930	3/0		25	.640			26		26	42
1940	4/0		22	.727			29.50		29.50	47.50
1950	250 kcmil		20	.800			32		32	52
1960	300 kcmil		19	.842			34		34	55
1970	350 kcmil		18	.889			36		36	58
1980	400 kcmil		17	.941			38		38	61.50
1990	500 kcmil		16.20	.988			40		40	64.50
2000	Interior fluorescent fixtures, incl. supports									
2010	& whips, to 15' high									
2100	Recessed drop-in 2' x 2', 2 lamp	2 Elec	35	.457	Ea.		18.40		18.40	30
2120	2' x 4', 2 lamp		33	.485			19.50		19.50	31.50
2140	2' x 4', 4 lamp		30	.533			21.50		21.50	35
2160	4' x 4', 4 lamp		20	.800			32		32	52
2180	Surface mount, acrylic lens & hinged frame									
2200	1' x 4', 2 lamp	2 Elec	44	.364	Ea.		14.65		14.65	23.50
2220	2' x 2', 2 lamp		44	.364			14.65		14.65	23.50
2260	2' x 4', 4 lamp		33	.485			19.50		19.50	31.50
2280	4' x 4', 4 lamp		23	.696			28		28	45.50
2300	Strip fixtures, surface mount									
2320	4' long, 1 lamp	2 Elec	53	.302	Ea.		12.15		12.15	19.70
2340	4' long, 2 lamp		50	.320			12.90		12.90	21
2360	8' long, 1 lamp		42	.381			15.35		15.35	25
2380	8' long, 2 lamp		40	.400			16.10		16.10	26
2400	Pendant mount, industrial, incl. removal									
2410	of chain or rod hangers, to 15' high									
2420	4' long, 2 lamp	2 Elec	35	.457	Ea.		18.40		18.40	30
2440	8' long, 2 lamp	"	27	.593	"		24		24	38.50

26 05 13 – Medium-Voltage Cables

26 05 13.16 Medium-Voltage, Single Cable		Crew	Daily Output	Labor-Hours	Unit	Material	Labor	Equipment	Total	Total Incl O&P
0010	**MEDIUM-VOLTAGE, SINGLE CABLE** Splicing & terminations not included									
0040	Copper, XLP shielding, 5 kV, #6	2 Elec	4.40	3.636	C.L.F.	139	146		285	390
0050	#4		4.40	3.636		181	146		327	435
0100	#2		4	4		214	161		375	495
0200	#1		4	4		246	161		407	530
0400	1/0		3.80	4.211		271	169		440	575
0600	2/0		3.60	4.444		340	179		519	665
0800	4/0		3.20	5		445	201		646	815
1000	250 kcmil	3 Elec	4.50	5.333		550	215		765	955
1200	350 kcmil		3.90	6.154		705	248		953	1,175
1400	500 kcmil		3.60	6.667		865	268		1,133	1,375
1600	15 kV, ungrounded neutral, #1	2 Elec	4	4		299	161		460	590
1800	1/0		3.80	4.211		355	169		524	670
2000	2/0		3.60	4.444		405	179		584	735
2200	4/0		3.20	5		540	201		741	920

26 05 13 – Medium-Voltage Cables

26 05 13.16 Medium-Voltage, Single Cable		Crew	Daily Output	Labor-Hours	Unit	Material	2011 Bare Costs Labor	Equipment	Total	Total Incl O&P
2400	250 kcmil	3 Elec	4.50	5.333	C.L.F.	600	215		815	1,000
2600	350 kcmil		3.90	6.154		755	248		1,003	1,225
2800	500 kcmil		3.60	6.667		930	268		1,198	1,450

26 05 19 – Low-Voltage Electrical Power Conductors and Cables

26 05 19.20 Armored Cable

Line	Description	Crew	Daily Output	Labor-Hours	Unit	Material	Labor	Equipment	Total	Total Incl O&P
0010	**ARMORED CABLE**									
0050	600 volt, copper (BX), #14, 2 conductor, solid	1 Elec	2.40	3.333	C.L.F.	71.50	134		205.50	297
0100	3 conductor, solid		2.20	3.636		111	146		257	360
0150	#12, 2 conductor, solid		2.30	3.478		72.50	140		212.50	305
0200	3 conductor, solid		2	4		117	161		278	390
0250	#10, 2 conductor, solid		2	4		133	161		294	405
0300	3 conductor, solid		1.60	5		184	201		385	525
0350	#8, 3 conductor, solid		1.30	6.154		345	248		593	780
0400	3 conductor with PVC jacket, in cable tray, #6		3.10	2.581		385	104		489	595
0450	#4	2 Elec	5.40	2.963		495	119		614	740
0500	#2		4.60	3.478		660	140		800	950
0550	#1		4	4		895	161		1,056	1,250
0600	1/0		3.60	4.444		1,050	179		1,229	1,450
0650	2/0		3.40	4.706		1,275	189		1,464	1,700
0700	3/0		3.20	5		1,500	201		1,701	1,975
0750	4/0		3	5.333		1,700	215		1,915	2,225
0800	250 kcmil	3 Elec	3.60	6.667		1,925	268		2,193	2,550
0850	350 kcmil		3.30	7.273		2,575	293		2,868	3,300
0900	500 kcmil		3	8		3,350	320		3,670	4,200
1050	5 kV, copper, 3 conductor with PVC jacket,									
1060	non-shielded, in cable tray, #4	2 Elec	380	.042	L.F.	7.60	1.69		9.29	11.10
1100	#2		360	.044		9.85	1.79		11.64	13.75
1200	#1		300	.053		12.55	2.15		14.70	17.30
1400	1/0		290	.055		14.50	2.22		16.72	19.55
1600	2/0		260	.062		16.75	2.48		19.23	22.50
2000	4/0		240	.067		22.50	2.68		25.18	29
2100	250 kcmil	3 Elec	330	.073		30.50	2.93		33.43	38.50
2150	350 kcmil		315	.076		37.50	3.07		40.57	46.50
2200	500 kcmil		270	.089		50	3.58		53.58	61
2400	15 kV, copper, 3 conductor with PVC jacket galv steel armored									
2500	grounded neutral, in cable tray, #2	2 Elec	300	.053	L.F.	15.95	2.15		18.10	21
2600	#1		280	.057		16.95	2.30		19.25	22.50
2800	1/0		260	.062		19.50	2.48		21.98	25.50
2900	2/0		220	.073		26	2.93		28.93	33.50
3000	4/0		190	.084		29.50	3.39		32.89	37.50
3100	250 kcmil	3 Elec	270	.089		32.50	3.58		36.08	42
3150	350 kcmil		240	.100		38.50	4.03		42.53	48.50
3200	500 kcmil		210	.114		51.50	4.60		56.10	64.50
3400	15 kV, copper, 3 conductor with PVC jacket,									
3450	ungrounded neutral, in cable tray, #2	2 Elec	260	.062	L.F.	17.20	2.48		19.68	23
3500	#1		230	.070		19.05	2.80		21.85	25.50
3600	1/0		200	.080		22	3.22		25.22	29
3700	2/0		190	.084		27	3.39		30.39	35
3800	4/0		160	.100		32.50	4.03		36.53	42
4000	250 kcmil	3 Elec	210	.114		37.50	4.60		42.10	49
4050	350 kcmil		195	.123		49.50	4.95		54.45	62.50
4100	500 kcmil		180	.133		60.50	5.35		65.85	75

26 05 19.20 Armored Cable

26 05 19.20 Armored Cable		Crew	Daily Output	Labor-Hours	Unit	Material	2011 Bare Costs Labor	Equipment	Total	Total Incl O&P
9010	600 volt, copper (MC) steel clad, #14, 2 wire	1 Elec	2.40	3.333	C.L.F.	72.50	134		206.50	298
9020	3 wire		2.20	3.636		111	146		257	360
9040	#12, 2 wire		2.30	3.478		73	140		213	310
9050	3 wire		2	4		122	161		283	395
9070	#10, 2 wire		2	4		152	161		313	430
9080	3 wire		1.60	5		211	201		412	555
9100	#8, 2 wire, stranded		1.80	4.444		278	179		457	595
9110	3 wire, stranded		1.30	6.154		430	248		678	875
9200	600 volt, copper (MC) aluminum clad, #14, 2 wire		2.65	3.019		72.50	122		194.50	277
9210	3 wire		2.45	3.265		111	131		242	335
9220	4 wire		2.20	3.636		151	146		297	405
9230	#12, 2 wire		2.55	3.137		73	126		199	286
9240	3 wire		2.20	3.636		122	146		268	370
9250	4 wire		2	4		162	161		323	440
9260	#10, 2 wire		2.20	3.636		152	146		298	405
9270	3 wire		1.80	4.444		211	179		390	520
9280	4 wire	↓	1.55	5.161	↓	325	208		533	695

26 05 19.35 Cable Terminations

26 05 19.35 Cable Terminations		Crew	Daily Output	Labor-Hours	Unit	Material	Labor	Equipment	Total	Total Incl O&P
0010	**CABLE TERMINATIONS**									
0015	Wire connectors, screw type, #22 to #14	1 Elec	260	.031	Ea.	.08	1.24		1.32	2.10
0020	#18 to #12		240	.033		.10	1.34		1.44	2.28
0025	#18 to #10		240	.033		.16	1.34		1.50	2.35
0030	Screw-on connectors, insulated, #18 to #12		240	.033		.12	1.34		1.46	2.30
0035	#16 to #10		230	.035		.17	1.40		1.57	2.46
0040	#14 to #8		210	.038		.38	1.53		1.91	2.91
0045	#12 to #6		180	.044		.63	1.79		2.42	3.59
0050	Terminal lugs, solderless, #16 to #10		50	.160		.60	6.45		7.05	11.10
0100	#8 to #4		30	.267		.93	10.75		11.68	18.40
0150	#2 to #1		22	.364		1.25	14.65		15.90	25
0200	1/0 to 2/0		16	.500		2.22	20		22.22	35
0250	3/0		12	.667		4.30	27		31.30	48
0300	4/0		11	.727		4.30	29.50		33.80	52
0350	250 kcmil		9	.889		4.30	36		40.30	62.50
0400	350 kcmil		7	1.143		5.50	46		51.50	80.50
0450	500 kcmil	↓	6	1.333	↓	10.70	53.50		64.20	99
1600	Crimp 1 hole lugs, copper or aluminum, 600 volt									
1620	#14	1 Elec	60	.133	Ea.	.43	5.35		5.78	9.15
1630	#12		50	.160		.58	6.45		7.03	11.10
1640	#10		45	.178		.58	7.15		7.73	12.25
1780	#8		36	.222		2.44	8.95		11.39	17.20
1800	#6		30	.267		3.20	10.75		13.95	21
2000	#4		27	.296		4.36	11.95		16.31	24
2200	#2		24	.333		6.80	13.40		20.20	29.50
2400	#1		20	.400		7.05	16.10		23.15	34
2600	2/0		15	.533		8.70	21.50		30.20	44.50
2800	3/0		12	.667		10.25	27		37.25	55
3000	4/0		11	.727		11.45	29.50		40.95	60
3200	250 kcmil		9	.889		13.45	36		49.45	73
3400	300 kcmil		8	1		15.50	40.50		56	82.50
3500	350 kcmil		7	1.143		15.95	46		61.95	92
3600	400 kcmil		6.50	1.231		19.45	49.50		68.95	102
3800	500 kcmil	↓	6	1.333	↓	22.50	53.50		76	112

26 05 19 – Low-Voltage Electrical Power Conductors and Cables

26 05 19.50 Mineral Insulated Cable

	26 05 19.50 Mineral Insulated Cable	Crew	Daily Output	Labor-Hours	Unit	Material	Labor (2011 Bare Costs)	Equipment	Total	Total Incl O&P
0010	**MINERAL INSULATED CABLE** 600 volt									
0100	1 conductor, #12	1 Elec	1.60	5	C.L.F.	370	201		571	735
0200	#10		1.60	5		480	201		681	850
0400	#8		1.50	5.333		570	215		785	980
0500	#6		1.40	5.714		645	230		875	1,075
0600	#4	2 Elec	2.40	6.667		890	268		1,158	1,400
0800	#2		2.20	7.273		1,200	293		1,493	1,775
0900	#1		2.10	7.619		1,350	305		1,655	2,000
1000	1/0		2	8		1,550	320		1,870	2,250
1100	2/0		1.90	8.421		1,850	340		2,190	2,600
1200	3/0		1.80	8.889		2,200	360		2,560	3,000
1400	4/0		1.60	10		2,525	405		2,930	3,425
1410	250 kcmil	3 Elec	2.40	10		2,875	405		3,280	3,800
1420	350 kcmil		1.95	12.308		3,225	495		3,720	4,325
1430	500 kcmil		1.95	12.308		4,025	495		4,520	5,225

26 05 19.55 Non-Metallic Sheathed Cable

	26 05 19.55 Non-Metallic Sheathed Cable	Crew	Daily Output	Labor-Hours	Unit	Material	Labor (2011 Bare Costs)	Equipment	Total	Total Incl O&P
0010	**NON-METALLIC SHEATHED CABLE** 600 volt									
0100	Copper with ground wire, (Romex)									
0150	#14, 2 conductor	1 Elec	2.70	2.963	C.L.F.	25.50	119		144.50	221
0200	3 conductor		2.40	3.333		35.50	134		169.50	257
0250	#12, 2 conductor		2.50	3.200		39	129		168	252
0300	3 conductor		2.20	3.636		55	146		201	298
0350	#10, 2 conductor		2.20	3.636		61	146		207	305
0400	3 conductor		1.80	4.444		86.50	179		265.50	385
0450	#8, 3 conductor		1.50	5.333		138	215		353	500
0500	#6, 3 conductor		1.40	5.714		223	230		453	620
0550	SE type SER aluminum cable, 3 RHW and									
0600	1 bare neutral, 3 #8 & 1 #8	1 Elec	1.60	5	C.L.F.	119	201		320	455
0650	3 #6 & 1 #6	"	1.40	5.714		134	230		364	525
0700	3 #4 & 1 #6	2 Elec	2.40	6.667		151	268		419	600
0750	3 #2 & 1 #4		2.20	7.273		222	293		515	720
0800	3 #1/0 & 1 #2		2	8		350	320		670	905
0850	3 #2/0 & 1 #1		1.80	8.889		395	360		755	1,025
0900	3 #4/0 & 1 #2/0		1.60	10		565	405		970	1,275

26 05 19.90 Wire

	26 05 19.90 Wire	Crew	Daily Output	Labor-Hours	Unit	Material	Labor (2011 Bare Costs)	Equipment	Total	Total Incl O&P
0010	**WIRE** R260519-92									
0020	600 volt, copper type THW, solid, #14	1 Elec	13	.615	C.L.F.	7.85	25		32.85	48.50
0030	#12 R260533-22		11	.727		11.95	29.50		41.45	60.50
0040	#10		10	.800		18.70	32		50.70	72.50
0050	Stranded, #14		13	.615		8.95	25		33.95	50
0100	#12		11	.727		12.90	29.50		42.40	61.50
0120	#10		10	.800		19.90	32		51.90	74
0140	#8		8	1		32.50	40.50		73	101
0160	#6		6.50	1.231		55	49.50		104.50	141
0180	#4	2 Elec	10.60	1.509		86.50	61		147.50	194
0200	#3		10	1.600		109	64.50		173.50	224
0220	#2		9	1.778		137	71.50		208.50	266
0240	#1		8	2		173	80.50		253.50	320
0260	1/0		6.60	2.424		216	97.50		313.50	395
0280	2/0		5.80	2.759		271	111		382	480
0300	3/0		5	3.200		340	129		469	585
0350	4/0		4.40	3.636		425	146		571	705

26 05 19 – Low-Voltage Electrical Power Conductors and Cables

26 05 19.90 Wire		Crew	Daily Output	Labor-Hours	Unit	Material	2011 Bare Costs Labor	Equipment	Total	Total Incl O&P
0400	250 kcmil	3 Elec	6	4	C.L.F.	505	161		666	815
0420	300 kcmil		5.70	4.211		605	169		774	940
0450	350 kcmil		5.40	4.444		705	179		884	1,075
0480	400 kcmil		5.10	4.706		805	189		994	1,200
0490	500 kcmil		4.80	5		1,000	201		1,201	1,425
0540	600 volt, aluminum type THHN, stranded, #6	1 Elec	8	1		27	40.50		67.50	95.50
0560	#4	2 Elec	13	1.231		33.50	49.50		83	118
0580	#2		10.60	1.509		46	61		107	149
0600	#1		9	1.778		67	71.50		138.50	190
0620	1/0		8	2		80.50	80.50		161	220
0640	2/0		7.20	2.222		95	89.50		184.50	249
0680	3/0		6.60	2.424		118	97.50		215.50	288
0700	4/0		6.20	2.581		131	104		235	310
0720	250 kcmil	3 Elec	8.70	2.759		160	111		271	355
0740	300 kcmil		8.10	2.963		221	119		340	435
0760	350 kcmil		7.50	3.200		225	129		354	455
0780	400 kcmil		6.90	3.478		263	140		403	515
0800	500 kcmil		6	4		290	161		451	580
0850	600 kcmil		5.70	4.211		365	169		534	680
0880	700 kcmil		5.10	4.706		425	189		614	770
0900	750 kcmil		4.80	5		430	201		631	795
0920	600 volt, copper type THWN-THHN, solid, #14	1 Elec	13	.615		7.85	25		32.85	48.50
0940	#12		11	.727		11.95	29.50		41.45	60.50
0960	#10		10	.800		18.70	32		50.70	72.50
1000	Stranded, #14		13	.615		8.95	25		33.95	50
1200	#12		11	.727		12.90	29.50		42.40	61.50
1250	#10		10	.800		19.90	32		51.90	74
1300	#8		8	1		32.50	40.50		73	101
1350	#6		6.50	1.231		55	49.50		104.50	141
1400	#4	2 Elec	10.60	1.509		86.50	61		147.50	194

26 05 23 – Control-Voltage Electrical Power Cables

26 05 23.10 Control Cable

		Crew	Daily Output	Labor-Hours	Unit	Material	Labor	Equipment	Total	Total Incl O&P
0010	**CONTROL CABLE**									
0020	600 volt, copper, #14 THWN wire with PVC jacket, 2 wires	1 Elec	9	.889	C.L.F.	19.60	36		55.60	79.50
0100	4 wires		7	1.143		33.50	46		79.50	112
0200	6 wires		6	1.333		63	53.50		116.50	157
0300	8 wires		5.30	1.509		77.50	61		138.50	184
0400	10 wires		4.80	1.667		91	67		158	209
0500	12 wires		4.30	1.860		106	75		181	238
0600	14 wires		3.80	2.105		127	84.50		211.50	277
0700	16 wires		3.50	2.286		130	92		222	292
0800	18 wires		3.30	2.424		144	97.50		241.50	315
0900	20 wires		3	2.667		175	107		282	365
1000	22 wires		2.80	2.857		180	115		295	385

26 05 26 – Grounding and Bonding for Electrical Systems

26 05 26.80 Grounding

		Crew	Daily Output	Labor-Hours	Unit	Material	Labor	Equipment	Total	Total Incl O&P
0010	**GROUNDING**									
0030	Rod, copper clad, 8′ long, 1/2″ diameter	1 Elec	5.50	1.455	Ea.	18.65	58.50		77.15	116
0050	3/4″ diameter		5.30	1.509		32.50	61		93.50	134
0080	10′ long, 1/2″ diameter		4.80	1.667		21.50	67		88.50	133
0100	3/4″ diameter		4.40	1.818		38.50	73		111.50	162
0130	15′ long, 3/4″ diameter		4	2		83.50	80.50		164	223

26 05 26 – Grounding and Bonding for Electrical Systems

26 05 26.80 Grounding		Crew	Daily Output	Labor-Hours	Unit	2011 Bare Costs Material	Labor	Equipment	Total	Total Incl O&P
0390	Bare copper wire, stranded, #8	1 Elec	11	.727	C.L.F.	30.50	29.50		60	81
0400	#6		10	.800		53.50	32		85.50	111
0600	#2	2 Elec	10	1.600		135	64.50		199.50	253
0800	3/0		6.60	2.424		320	97.50		417.50	510
1000	4/0		5.70	2.807		400	113		513	625
1200	250 kcmil	3 Elec	7.20	3.333		475	134		609	745
1800	Water pipe ground clamps, heavy duty									
2000	Bronze, 1/2" to 1" diameter	1 Elec	8	1	Ea.	26	40.50		66.50	94
2100	1-1/4" to 2" diameter		8	1		34	40.50		74.50	103
2200	2-1/2" to 3" diameter		6	1.333		59.50	53.50		113	152
2800	Brazed connections, #6 wire		12	.667		13.65	27		40.65	58.50
3000	#2 wire		10	.800		18.30	32		50.30	72
3100	3/0 wire		8	1		27.50	40.50		68	95.50
3200	4/0 wire		7	1.143		31.50	46		77.50	109
3400	250 kcmil wire		5	1.600		36.50	64.50		101	144
3600	500 kcmil wire		4	2		45	80.50		125.50	181

26 05 33 – Raceway and Boxes for Electrical Systems

26 05 33.13 Conduit		Crew	Daily Output	Labor-Hours	Unit	Material	Labor	Equipment	Total	Total Incl O&P
0010	**CONDUIT** To 15' high, includes 2 terminations, 2 elbows, R260533-22									
0020	11 beam clamps, and 11 couplings per 100 L.F.									
0300	Aluminum, 1/2" diameter	1 Elec	100	.080	L.F.	2.26	3.22		5.48	7.70
0500	3/4" diameter		90	.089		3.06	3.58		6.64	9.15
0700	1" diameter		80	.100		4.03	4.03		8.06	11
1000	1-1/4" diameter		70	.114		5.40	4.60		10	13.40
1030	1-1/2" diameter		65	.123		6.50	4.95		11.45	15.20
1050	2" diameter		60	.133		8.85	5.35		14.20	18.45
1070	2-1/2" diameter		50	.160		14.45	6.45		20.90	26.50
1100	3" diameter	2 Elec	90	.178		20	7.15		27.15	33.50
1130	3-1/2" diameter		80	.200		26.50	8.05		34.55	42
1140	4" diameter		70	.229		32	9.20		41.20	50
1750	Rigid galvanized steel, 1/2" diameter	1 Elec	90	.089		2.55	3.58		6.13	8.60
1770	3/4" diameter		80	.100		2.70	4.03		6.73	9.50
1800	1" diameter		65	.123		3.62	4.95		8.57	12.05
1830	1-1/4" diameter		60	.133		5.05	5.35		10.40	14.25
1850	1-1/2" diameter		55	.145		5.80	5.85		11.65	15.90
1870	2" diameter		45	.178		7.55	7.15		14.70	19.90
1900	2-1/2" diameter		35	.229		14.40	9.20		23.60	31
1930	3" diameter	2 Elec	50	.320		17.65	12.90		30.55	40.50
1950	3-1/2" diameter		44	.364		21.50	14.65		36.15	47
1970	4" diameter		40	.400		25.50	16.10		41.60	54.50
2500	Steel, intermediate conduit (IMC), 1/2" diameter	1 Elec	100	.080		1.73	3.22		4.95	7.10
2530	3/4" diameter		90	.089		2.07	3.58		5.65	8.10
2550	1" diameter		70	.114		2.84	4.60		7.44	10.55
2570	1-1/4" diameter		65	.123		3.72	4.95		8.67	12.15
2600	1-1/2" diameter		60	.133		4.47	5.35		9.82	13.60
2630	2" diameter		50	.160		5.65	6.45		12.10	16.70
2650	2-1/2" diameter		40	.200		10.85	8.05		18.90	25
2670	3" diameter	2 Elec	60	.267		14.75	10.75		25.50	33.50
2700	3-1/2" diameter		54	.296		18.80	11.95		30.75	40
2730	4" diameter		50	.320		21	12.90		33.90	44
5000	Electric metallic tubing (EMT), 1/2" diameter	1 Elec	170	.047		.64	1.89		2.53	3.77
5020	3/4" diameter		130	.062		1	2.48		3.48	5.10

26 05 Common Work Results for Electrical

26 05 33 – Raceway and Boxes for Electrical Systems

26 05 33.13 Conduit		Crew	Daily Output	Labor-Hours	Unit	2011 Bare Costs Material	Labor	Equipment	Total	Total Incl O&P
5040	1" diameter	1 Elec	115	.070	L.F.	1.70	2.80		4.50	6.40
5060	1-1/4" diameter		100	.080		2.81	3.22		6.03	8.30
5080	1-1/2" diameter		90	.089		3.66	3.58		7.24	9.80
5100	2" diameter		80	.100		4.74	4.03		8.77	11.75
5120	2-1/2" diameter		60	.133		11.35	5.35		16.70	21
5140	3" diameter	2 Elec	100	.160		13.40	6.45		19.85	25
5160	3-1/2" diameter		90	.178		16.90	7.15		24.05	30
5180	4" diameter		80	.200		18.75	8.05		26.80	33.50
9900	Add to labor for higher elevated installation									
9910	15' to 20' high, add						10%			
9920	20' to 25' high, add						20%			
9930	25' to 30' high, add						25%			
9940	30' to 35' high, add						30%			
9950	35' to 40' high, add						35%			
9960	Over 40' high, add						40%			
9980	Allow. for cond. ftngs., 5% min.-20% max.									

26 05 33.16 Outlet Boxes		Crew	Daily Output	Labor-Hours	Unit	Material	Labor	Equipment	Total	Total Incl O&P
0010	**OUTLET BOXES**									
0020	Pressed steel, octagon, 4"	1 Elec	20	.400	Ea.	3.63	16.10		19.73	30
0060	Covers, blank		64	.125		1.50	5.05		6.55	9.80
0100	Extension rings		40	.200		6	8.05		14.05	19.65
0150	Square, 4"		20	.400		2.48	16.10		18.58	28.50
0200	Extension rings		40	.200		6.10	8.05		14.15	19.75
0250	Covers, blank		64	.125		1.65	5.05		6.70	9.95
0300	Plaster rings		64	.125		3.36	5.05		8.41	11.85
0650	Switchbox		27	.296		5.80	11.95		17.75	26
1100	Concrete, floor, 1 gang		5.30	1.509		89.50	61		150.50	197
2000	Poke-thru fitting, fire rated, for 3-3/4" floor		6.80	1.176		119	47.50		166.50	208
2040	For 7" floor		6.80	1.176		119	47.50		166.50	208
2100	Pedestal, 15 amp, duplex receptacle & blank plate		5.25	1.524		131	61.50		192.50	245
2120	Duplex receptacle and telephone plate		5.25	1.524		131	61.50		192.50	245
2140	Pedestal, 20 amp, duplex recept. & phone plate		5	1.600		132	64.50		196.50	250
2200	Abandonment plate		32	.250		37.50	10.05		47.55	57.50

26 05 33.18 Pull Boxes		Crew	Daily Output	Labor-Hours	Unit	Material	Labor	Equipment	Total	Total Incl O&P
0010	**PULL BOXES**									
0100	Sheet metal, pull box, NEMA 1, type SC, 6" W x 6" H x 4" D	1 Elec	8	1	Ea.	14.90	40.50		55.40	82
0200	8" W x 8" H x 4" D		8	1		18.70	40.50		59.20	86
0300	10" W x 12" H x 6" D		5.30	1.509		38	61		99	140
0400	16" W x 20" H x 8" D		4	2		130	80.50		210.50	274
0500	20" W x 24" H x 8" D		3.20	2.500		141	101		242	320
0600	24" W x 36" H x 8" D		2.70	2.963		217	119		336	430
0650	Pull box, hinged , NEMA 1, 6" W x 6" H x 4" D		8	1		14.90	40.50		55.40	82
0800	12" W x 16" H x 6" D		4.70	1.702		48.50	68.50		117	165
1000	20" W x 20" H x 6" D		3.60	2.222		97	89.50		186.50	252
1200	20" W x 20" H x 8" D		3.20	2.500		194	101		295	375
1400	24" W x 36" H x 8" D		2.70	2.963		310	119		429	535
1600	24" W x 42" H x 8" D		2	4		470	161		631	780
2100	Pull box, NEMA 3R, type SC, raintight & weatherproof									
2150	6" L x 6" W x 6" D	1 Elec	10	.800	Ea.	25	32		57	79.50
2200	8" L x 6" W x 6" D		8	1		31.50	40.50		72	100
2250	10" L x 6" W x 6" D		7	1.143		41.50	46		87.50	120
2300	12" L x 12" W x 6" D		5	1.600		59	64.50		123.50	169

26 05 33 – Raceway and Boxes for Electrical Systems

26 05 33.18 Pull Boxes		Crew	Daily Output	Labor-Hours	Unit	Material	Labor	Equipment	Total	Total Incl O&P
						2011 Bare Costs				
2350	16" L x 16" W x 6" D	1 Elec	4.50	1.778	Ea.	117	71.50		188.50	245
2400	20" L x 20" W x 6" D		4	2		160	80.50		240.50	305
2450	24" L x 18" W x 8" D		3	2.667		166	107		273	355
2500	24" L x 24" W x 10" D		2.50	3.200		213	129		342	445
2550	30" L x 24" W x 12" D		2	4		430	161		591	730
2600	36" L x 36" W x 12" D	↓	1.50	5.333	↓	565	215		780	970
2800	Cast iron, pull boxes for surface mounting									
3000	NEMA 4, watertight & dust tight									
3050	6" L x 6" W x 6" D	1 Elec	4	2	Ea.	233	80.50		313.50	385
3100	8" L x 6" W x 6" D		3.20	2.500		253	101		354	440
3150	10" L x 6" W x 6" D		2.50	3.200		315	129		444	555
3200	12" L x 12" W x 6" D		2.30	3.478		555	140		695	840
3250	16" L x 16" W x 6" D		1.30	6.154		1,450	248		1,698	2,000
3300	20" L x 20" W x 6" D		.80	10		2,700	405		3,105	3,625
3350	24" L x 18" W x 8" D		.70	11.429		2,825	460		3,285	3,875
3400	24" L x 24" W x 10" D		.50	16		4,350	645		4,995	5,825
3450	30" L x 24" W x 12" D		.40	20		7,275	805		8,080	9,300
3500	36" L x 36" W x 12" D	↓	.20	40	↓	9,000	1,600		10,600	12,500
4000	NEMA 7, explosionproof									
4050	6" L x 6" W x 6" D	1 Elec	2	4	Ea.	800	161		961	1,150
4100	8" L x 6" W x 6" D		1.80	4.444		955	179		1,134	1,350
4150	10" L x 6" W x 6" D		1.60	5		1,300	201		1,501	1,750
4200	12" L x 12" W x 6" D		1	8		2,150	320		2,470	2,900
4250	16" L x 14" W x 6" D		.60	13.333		2,650	535		3,185	3,800
4300	18" L x 18" W x 8" D		.50	16		4,700	645		5,345	6,225
4350	24" L x 18" W x 8" D		.40	20		6,275	805		7,080	8,200
4400	24" L x 24" W x 10" D		.30	26.667		8,150	1,075		9,225	10,700
4450	30" L x 24" W x 12" D	↓	.20	40	↓	12,400	1,600		14,000	16,300
6000	J.I.C. wiring boxes, NEMA 12, dust tight & drip tight									
6050	6" L x 8" W x 4" D	1 Elec	10	.800	Ea.	59	32		91	117
6100	8" L x 10" W x 4" D		8	1		74	40.50		114.50	147
6150	12" L x 14" W x 6" D		5.30	1.509		124	61		185	235
6200	14" L x 16" W x 6" D		4.70	1.702		155	68.50		223.50	282
6250	16" L x 20" W x 6" D		4.40	1.818		305	73		378	455
6300	24" L x 30" W x 6" D		3.20	2.500		450	101		551	660
6350	24" L x 30" W x 8" D		2.90	2.759		480	111		591	710
6400	24" L x 36" W x 8" D		2.70	2.963		535	119		654	780
6450	24" L x 42" W x 8" D		2.30	3.478		585	140		725	865
6500	24" L x 48" W x 8" D	↓	2	4	↓	635	161		796	960

26 05 33.23 Wireway		Crew	Daily Output	Labor-Hours	Unit	Material	Labor	Equipment	Total	Total Incl O&P
0010	**WIREWAY** to 15' high									
0100	Screw cover, NEMA 1 w/fittings and supports, 2-1/2" x 2-1/2"	1 Elec	45	.178	L.F.	11.45	7.15		18.60	24
0200	4" x 4"	"	40	.200		12.65	8.05		20.70	27
0400	6" x 6"	2 Elec	60	.267		19.70	10.75		30.45	39
0600	8" x 8"	"	40	.400		33	16.10		49.10	62.50
4475	Screw cover, NEMA 3R w/fittings and supports, 4" x 4"	1 Elec	36	.222		27	8.95		35.95	44
4480	6" x 6"	2 Elec	55	.291		34	11.70		45.70	56.50
4485	8" x 8"		36	.444		55	17.90		72.90	89.50
4490	12" x 12"	↓	18	.889	↓	77	36		113	143
9980	Allow. for wireway ftngs., 5% min.-20% max.									

26 05 33 – Raceway and Boxes for Electrical Systems

26 05 33.35 Flexible Metallic Conduit		Crew	Daily Output	Labor-Hours	Unit	2011 Bare Costs Material	Labor	Equipment	Total	Total Incl O&P
0010	**FLEXIBLE METALLIC CONDUIT**									
0050	Steel, 3/8" diameter	1 Elec	200	.040	L.F.	.60	1.61		2.21	3.27
0100	1/2" diameter		200	.040		.68	1.61		2.29	3.36
0200	3/4" diameter		160	.050		.93	2.01		2.94	4.28
0250	1" diameter		100	.080		1.68	3.22		4.90	7.05
0300	1-1/4" diameter		70	.114		2.17	4.60		6.77	9.85
0350	1-1/2" diameter		50	.160		3.54	6.45		9.99	14.35
0370	2" diameter		40	.200		4.34	8.05		12.39	17.80
0420	Connectors, plain, 3/8" diameter		100	.080	Ea.	3.06	3.22		6.28	8.55
0430	1/2" diameter		80	.100		3.57	4.03		7.60	10.50
0440	3/4" diameter		70	.114		4.08	4.60		8.68	11.95
0450	1" diameter		50	.160		8.65	6.45		15.10	19.95
0452	1-1/4" diameter		45	.178		10.65	7.15		17.80	23.50
0454	1-1/2" diameter		40	.200		15.65	8.05		23.70	30.50
0456	2" diameter		28	.286		23.50	11.50		35	44.50
0490	Insulated, 1" diameter		40	.200		9.40	8.05		17.45	23.50
0500	1-1/4" diameter		40	.200		16.85	8.05		24.90	31.50
0550	1-1/2" diameter		32	.250		24	10.05		34.05	43
0600	2" diameter		23	.348		43.50	14		57.50	70.50

26 05 36 – Cable Trays for Electrical Systems

26 05 36.10 Cable Tray Ladder Type

Line	Description	Crew	Daily Output	Labor-Hours	Unit	Material	Labor	Equipment	Total	Total Incl O&P
0010	**CABLE TRAY LADDER TYPE** w/ftngs. & supports, 4" dp., to 15' elev.									
0160	Galvanized steel tray									
0170	4" rung spacing, 6" wide	2 Elec	98	.163	L.F.	14.10	6.55		20.65	26
0200	12" wide		86	.186		16.90	7.50		24.40	31
0400	18" wide		82	.195		19.65	7.85		27.50	34.50
0600	24" wide		78	.205		22.50	8.25		30.75	38.50
3200	Aluminum tray, 4" deep, 6" rung spacing, 6" wide		134	.119		17.70	4.81		22.51	27.50
3220	12" wide		124	.129		19.80	5.20		25	30.50
3230	18" wide		114	.140		22	5.65		27.65	33.50
3240	24" wide		106	.151		25.50	6.10		31.60	38
9980	Allow. for tray ftngs., 5% min.-20% max.									

26 05 39 – Underfloor Raceways for Electrical Systems

26 05 39.30 Conduit In Concrete Slab

Line	Description	Crew	Daily Output	Labor-Hours	Unit	Material	Labor	Equipment	Total	Total Incl O&P
0010	**CONDUIT IN CONCRETE SLAB** Including terminations,									
0020	fittings and supports									
3230	PVC, schedule 40, 1/2" diameter	1 Elec	270	.030	L.F.	.72	1.19		1.91	2.72
3250	3/4" diameter		230	.035		.86	1.40		2.26	3.22
3270	1" diameter		200	.040		1.14	1.61		2.75	3.87
3300	1-1/4" diameter		170	.047		1.61	1.89		3.50	4.84
3330	1-1/2" diameter		140	.057		1.91	2.30		4.21	5.85
3350	2" diameter		120	.067		2.45	2.68		5.13	7.05
4350	Rigid galvanized steel, 1/2" diameter		200	.040		2.27	1.61		3.88	5.10
4400	3/4" diameter		170	.047		2.42	1.89		4.31	5.75
4450	1" diameter		130	.062		3.34	2.48		5.82	7.70
4500	1-1/4" diameter		110	.073		4.48	2.93		7.41	9.65
4600	1-1/2" diameter		100	.080		5.25	3.22		8.47	11
4800	2" diameter		90	.089		6.75	3.58		10.33	13.25

26 05 39 – Underfloor Raceways for Electrical Systems

26 05 39.40 Conduit In Trench

	26 05 39.40 Conduit In Trench	Crew	Daily Output	Labor-Hours	Unit	2011 Bare Costs Material	Labor	Equipment	Total	Total Incl O&P
0010	**CONDUIT IN TRENCH** Includes terminations and fittings									
0020	Does not include excavation or backfill, see Section 31 23 16.00									
0200	Rigid galvanized steel, 2″ diameter	1 Elec	150	.053	L.F.	6.45	2.15		8.60	10.55
0400	2-1/2″ diameter	″	100	.080		12.30	3.22		15.52	18.70
0600	3″ diameter	2 Elec	160	.100		15.45	4.03		19.48	23.50
0800	3-1/2″ diameter		140	.114		19.95	4.60		24.55	29.50
1000	4″ diameter		100	.160		22.50	6.45		28.95	35
1200	5″ diameter		80	.200		46	8.05		54.05	64
1400	6″ diameter		60	.267		67.50	10.75		78.25	92

26 05 43 – Underground Ducts and Raceways for Electrical Systems

26 05 43.10 Trench Duct

	26 05 43.10 Trench Duct	Crew	Daily Output	Labor-Hours	Unit	Material	Labor	Equipment	Total	Total Incl O&P
0010	**TRENCH DUCT** Steel with cover									
0020	Standard adjustable, depths to 4″									
0100	Straight, single compartment, 9″ wide	2 Elec	40	.400	L.F.	111	16.10		127.10	148
0200	12″ wide		32	.500		126	20		146	171
0400	18″ wide		26	.615		167	25		192	224
0600	24″ wide		22	.727		215	29.50		244.50	284
0800	30″ wide		20	.800		259	32		291	335
1000	36″ wide		16	1		310	40.50		350.50	410
1200	Horizontal elbow, 9″ wide		5.40	2.963	Ea.	410	119		529	645
1400	12″ wide		4.60	3.478		475	140		615	745
1600	18″ wide		4	4		605	161		766	925
1800	24″ wide		3.20	5		850	201		1,051	1,250
2000	30″ wide		2.60	6.154		1,125	248		1,373	1,650
2200	36″ wide		2.40	6.667		1,500	268		1,768	2,050
2400	Vertical elbow, 9″ wide		5.40	2.963		143	119		262	350
2600	12″ wide		4.60	3.478		155	140		295	400
2800	18″ wide		4	4		178	161		339	455
3000	24″ wide		3.20	5		221	201		422	570
3200	30″ wide		2.60	6.154		244	248		492	670
3400	36″ wide		2.40	6.667		269	268		537	730
3600	Cross, 9″ wide		4	4		670	161		831	1,000
3800	12″ wide		3.20	5		710	201		911	1,100
4000	18″ wide		2.60	6.154		850	248		1,098	1,325
4200	24″ wide		2.20	7.273		1,075	293		1,368	1,650
4400	30″ wide		2	8		1,400	320		1,720	2,075
4600	36″ wide		1.80	8.889		1,750	360		2,110	2,500
4800	End closure, 9″ wide		14.40	1.111		42	44.50		86.50	119
5000	12″ wide		12	1.333		48	53.50		101.50	140
5200	18″ wide		10	1.600		73.50	64.50		138	185
5400	24″ wide		8	2		97	80.50		177.50	238
5600	30″ wide		6.60	2.424		122	97.50		219.50	292
5800	36″ wide		5.80	2.759		144	111		255	340
6000	Tees, 9″ wide		4	4		410	161		571	710
6200	12″ wide		3.60	4.444		475	179		654	810
6400	18″ wide		3.20	5		605	201		806	990
6600	24″ wide		3	5.333		870	215		1,085	1,300
6800	30″ wide		2.60	6.154		1,125	248		1,373	1,650
7000	36″ wide		2	8		1,500	320		1,820	2,150
7200	Riser, and cabinet connector, 9″ wide		5.40	2.963		178	119		297	390
7400	12″ wide		4.60	3.478		208	140		348	455
7600	18″ wide		4	4		255	161		416	540

26 05 43 – Underground Ducts and Raceways for Electrical Systems

26 05 43.10 Trench Duct		Crew	Daily Output	Labor-Hours	Unit	Material	Labor	Equipment	Total	Total Incl O&P
						2011 Bare Costs				
7800	24" wide	2 Elec	3.20	5	Ea.	310	201		511	665
8000	30" wide		2.60	6.154		355	248		603	790
8200	36" wide		2	8		415	320		735	975
8400	Insert assembly, cell to conduit adapter, 1-1/4"	1 Elec	16	.500		70.50	20		90.50	110

26 05 43.20 Underfloor Duct		Crew	Daily Output	Labor-Hours	Unit	Material	Labor	Equipment	Total	Total Incl O&P
0010	**UNDERFLOOR DUCT**									
0100	Duct, 1-3/8" x 3-1/8" blank, standard	2 Elec	160	.100	L.F.	13.50	4.03		17.53	21.50
0200	1-3/8" x 7-1/4" blank, super duct		120	.133		27	5.35		32.35	38
0400	7/8" or 1-3/8" insert type, 24" O.C., 1-3/8" x 3-1/8", std.		140	.114		18.10	4.60		22.70	27.50
0600	1-3/8" x 7-1/4", super duct		100	.160		31.50	6.45		37.95	45.50
0800	Junction box, single duct, 1 level, 3-1/8"	1 Elec	4	2	Ea.	405	80.50		485.50	575
1000	Junction box, single duct, 1 level, 7-1/4"		2.70	2.963		475	119		594	715
1200	1 level, 2 duct, 3-1/8"		3.20	2.500		540	101		641	760
1400	Junction box, 1 level, 2 duct, 7-1/4"		2.30	3.478		1,400	140		1,540	1,750
1600	Triple duct, 3-1/8"		2.30	3.478		920	140		1,060	1,250
1800	Insert to conduit adapter, 3/4" & 1"		32	.250		33	10.05		43.05	53
2000	Support, single cell		27	.296		50	11.95		61.95	74.50
2200	Super duct		16	.500		50	20		70	87.50
2400	Double cell		16	.500		50	20		70	87.50
2600	Triple cell		11	.727		50	29.50		79.50	103
2800	Vertical elbow, standard duct		10	.800		89.50	32		121.50	151
3000	Super duct		8	1		89.50	40.50		130	164
3200	Cabinet connector, standard duct		32	.250		67.50	10.05		77.55	91
3400	Super duct		27	.296		67.50	11.95		79.45	94
3600	Conduit adapter, 1" to 1-1/4"		32	.250		67.50	10.05		77.55	91
3800	2" to 1-1/4"		27	.296		81	11.95		92.95	108
4000	Outlet, low tension (tele, computer, etc.)		8	1		94.50	40.50		135	170
4200	High tension, receptacle (120 volt)		8	1		94.50	40.50		135	170

26 05 80 – Wiring Connections

26 05 80.10 Motor Connections		Crew	Daily Output	Labor-Hours	Unit	Material	Labor	Equipment	Total	Total Incl O&P
0010	**MOTOR CONNECTIONS**									
0020	Flexible conduit and fittings, 115 volt, 1 phase, up to 1 HP motor	1 Elec	8	1	Ea.	9.70	40.50		50.20	76
0200	25 HP motor		2.70	2.963		34	119		153	230
0400	50 HP motor		2.20	3.636		83	146		229	330
0600	100 HP motor		1.50	5.333		201	215		416	570

26 05 90 – Residential Applications

26 05 90.10 Residential Wiring		Crew	Daily Output	Labor-Hours	Unit	Material	Labor	Equipment	Total	Total Incl O&P
0010	**RESIDENTIAL WIRING**									
0020	20' avg. runs and #14/2 wiring incl. unless otherwise noted									
1000	Service & panel, includes 24' SE-AL cable, service eye, meter,									
1010	Socket, panel board, main bkr., ground rod, 15 or 20 amp									
1020	1-pole circuit breakers, and misc. hardware									
1100	100 amp, with 10 branch breakers	1 Elec	1.19	6.723	Ea.	560	271		831	1,050
1110	With PVC conduit and wire		.92	8.696		630	350		980	1,250
1120	With RGS conduit and wire		.73	10.959		790	440		1,230	1,575
1150	150 amp, with 14 branch breakers		1.03	7.767		865	315		1,180	1,450
1170	With PVC conduit and wire		.82	9.756		1,000	395		1,395	1,725
1180	With RGS conduit and wire		.67	11.940		1,300	480		1,780	2,225
1200	200 amp, with 18 branch breakers	2 Elec	1.80	8.889		1,150	360		1,510	1,825
1220	With PVC conduit and wire		1.46	10.959		1,275	440		1,715	2,125
1230	With RGS conduit and wire		1.24	12.903		1,700	520		2,220	2,725

26 05 90 – Residential Applications

26 05 90.10 Residential Wiring		Crew	Daily Output	Labor-Hours	Unit	Material	2011 Bare Costs Labor	Equipment	Total	Total Incl O&P
1800	Lightning surge suppressor for above services, add	1 Elec	32	.250	Ea.	55	10.05		65.05	77
2000	Switch devices									
2100	Single pole, 15 amp, Ivory, with a 1-gang box, cover plate,									
2110	Type NM (Romex) cable	1 Elec	17.10	.468	Ea.	10.70	18.85		29.55	42.50
2120	Type MC (BX) cable		14.30	.559		28.50	22.50		51	68
2130	EMT & wire		5.71	1.401		34	56.50		90.50	129
2150	3-way, #14/3, type NM cable		14.55	.550		14.85	22		36.85	52.50
2170	Type MC cable		12.31	.650		38.50	26		64.50	85
2180	EMT & wire		5	1.600		37.50	64.50		102	145
2200	4-way, #14/3, type NM cable		14.55	.550		31.50	22		53.50	71
2220	Type MC cable		12.31	.650		55.50	26		81.50	104
2230	EMT & wire		5	1.600		54.50	64.50		119	164
2250	S.P., 20 amp, #12/2, type NM cable		13.33	.600		18.05	24		42.05	59
2270	Type MC cable		11.43	.700		33.50	28		61.50	82
2280	EMT & wire		4.85	1.649		42.50	66.50		109	155
2290	S.P. rotary dimmer, 600W, no wiring		17	.471		21	18.95		39.95	53.50
2300	S.P. rotary dimmer, 600W, type NM cable		14.55	.550		26	22		48	65
2320	Type MC cable		12.31	.650		44	26		70	91
2330	EMT & wire		5	1.600		51	64.50		115.50	160
2350	3-way rotary dimmer, type NM cable		13.33	.600		33.50	24		57.50	75.50
2370	Type MC cable		11.43	.700		51	28		79	102
2380	EMT & wire	↓	4.85	1.649	↓	58	66.50		124.50	172
2400	Interval timer wall switch, 20 amp, 1-30 min., #12/2									
2410	Type NM cable	1 Elec	14.55	.550	Ea.	53	22		75	94.50
2420	Type MC cable		12.31	.650		61	26		87	110
2430	EMT & wire	↓	5	1.600	↓	77.50	64.50		142	189
2500	Decorator style									
2510	S.P., 15 amp, type NM cable	1 Elec	17.10	.468	Ea.	15.55	18.85		34.40	47.50
2520	Type MC cable		14.30	.559		33.50	22.50		56	73
2530	EMT & wire		5.71	1.401		38.50	56.50		95	134
2550	3-way, #14/3, type NM cable		14.55	.550		19.70	22		41.70	57.50
2570	Type MC cable		12.31	.650		43.50	26		69.50	90
2580	EMT & wire		5	1.600		42.50	64.50		107	151
2600	4-way, #14/3, type NM cable		14.55	.550		36.50	22		58.50	76.50
2620	Type MC cable		12.31	.650		60.50	26		86.50	109
2630	EMT & wire		5	1.600		59.50	64.50		124	169
2650	S.P., 20 amp, #12/2, type NM cable		13.33	.600		23	24		47	64
2670	Type MC cable		11.43	.700		38	28		66	87.50
2680	EMT & wire		4.85	1.649		47.50	66.50		114	160
2700	S.P., slide dimmer, type NM cable		17.10	.468		33.50	18.85		52.35	67.50
2720	Type MC cable		14.30	.559		51.50	22.50		74	93
2730	EMT & wire		5.71	1.401		58.50	56.50		115	156
2750	S.P., touch dimmer, type NM cable		17.10	.468		28.50	18.85		47.35	62
2770	Type MC cable		14.30	.559		46.50	22.50		69	87.50
2780	EMT & wire		5.71	1.401		53	56.50		109.50	150
2800	3-way touch dimmer, type NM cable		13.33	.600		48	24		72	91.50
2820	Type MC cable		11.43	.700		65.50	28		93.50	118
2830	EMT & wire	↓	4.85	1.649	↓	72.50	66.50		139	188
3000	Combination devices									
3100	S.P. switch/15 amp recpt., Ivory, 1-gang box, plate									
3110	Type NM cable	1 Elec	11.43	.700	Ea.	22.50	28		50.50	70.50
3120	Type MC cable		10	.800		40.50	32		72.50	96.50
3130	EMT & wire	↓	4.40	1.818	↓	47	73		120	171

26 05 90.10 Residential Wiring		Crew	Daily Output	Labor-Hours	Unit	Material	2011 Bare Costs Labor	Equipment	Total	Total Incl O&P
3150	S.P. switch/pilot light, type NM cable	1 Elec	11.43	.700	Ea.	23.50	28		51.50	71.50
3170	Type MC cable		10	.800		41.50	32		73.50	97.50
3180	EMT & wire		4.43	1.806		48	72.50		120.50	171
3190	2-S.P. switches, 2-#14/2, no wiring		14	.571		8.40	23		31.40	46.50
3200	2-S.P. switches, 2-#14/2, type NM cables		10	.800		26.50	32		58.50	81.50
3220	Type MC cable		8.89	.900		53.50	36		89.50	118
3230	EMT & wire		4.10	1.951		51	78.50		129.50	183
3250	3-way switch/15 amp recpt., #14/3, type NM cable		10	.800		30.50	32		62.50	85.50
3270	Type MC cable		8.89	.900		54.50	36		90.50	119
3280	EMT & wire		4.10	1.951		53.50	78.50		132	186
3300	2-3 way switches, 2-#14/3, type NM cables		8.89	.900		40.50	36		76.50	103
3320	Type MC cable		8	1		79.50	40.50		120	153
3330	EMT & wire		4	2		61	80.50		141.50	198
3350	S.P. switch/20 amp recpt., #12/2, type NM cable		10	.800		34.50	32		66.50	90
3370	Type MC cable		8.89	.900		42.50	36		78.50	106
3380	EMT & wire		4.10	1.951		59	78.50		137.50	192
3400	Decorator style									
3410	S.P. switch/15 amp recpt., type NM cable	1 Elec	11.43	.700	Ea.	27.50	28		55.50	75.50
3420	Type MC cable		10	.800		45.50	32		77.50	102
3430	EMT & wire		4.40	1.818		52	73		125	177
3450	S.P. switch/pilot light, type NM cable		11.43	.700		28.50	28		56.50	76.50
3470	Type MC cable		10	.800		46	32		78	103
3480	EMT & wire		4.40	1.818		53	73		126	177
3500	2-S.P. switches, 2-#14/2, type NM cables		10	.800		31.50	32		63.50	86.50
3520	Type MC cable		8.89	.900		58.50	36		94.50	123
3530	EMT & wire		4.10	1.951		56	78.50		134.50	189
3550	3-way/15 amp recpt., #14/3, type NM cable		10	.800		35.50	32		67.50	91
3570	Type MC cable		8.89	.900		59	36		95	124
3580	EMT & wire		4.10	1.951		58	78.50		136.50	191
3650	2-3 way switches, 2-#14/3, type NM cables		8.89	.900		45.50	36		81.50	109
3670	Type MC cable		8	1		84	40.50		124.50	158
3680	EMT & wire		4	2		65.50	80.50		146	204
3700	S.P. switch/20 amp recpt., #12/2, type NM cable		10	.800		39.50	32		71.50	95.50
3720	Type MC cable		8.89	.900		47.50	36		83.50	111
3730	EMT & wire		4.10	1.951		64	78.50		142.50	198
4000	Receptacle devices									
4010	Duplex outlet, 15 amp recpt., Ivory, 1-gang box, plate									
4015	Type NM cable	1 Elec	14.55	.550	Ea.	9.10	22		31.10	46
4020	Type MC cable		12.31	.650		27	26		53	72
4030	EMT & wire		5.33	1.501		32	60.50		92.50	134
4050	With #12/2, type NM cable		12.31	.650		11.80	26		37.80	55.50
4070	Type MC cable		10.67	.750		27	30		57	79
4080	EMT & wire		4.71	1.699		36	68.50		104.50	151
4100	20 amp recpt., #12/2, type NM cable		12.31	.650		21	26		47	65.50
4120	Type MC cable		10.67	.750		36.50	30		66.50	89
4130	EMT & wire		4.71	1.699		45.50	68.50		114	161
4140	For GFI see Section 26 05 90.10 line 4300 below									
4150	Decorator style, 15 amp recpt., type NM cable	1 Elec	14.55	.550	Ea.	13.95	22		35.95	51.50
4170	Type MC cable		12.31	.650		32	26		58	77.50
4180	EMT & wire		5.33	1.501		37	60.50		97.50	139
4200	With #12/2, type NM cable		12.31	.650		16.65	26		42.65	61
4220	Type MC cable		10.67	.750		32	30		62	84
4230	EMT & wire		4.71	1.699		41	68.50		109.50	156

26 05 Common Work Results for Electrical

26 05 90 – Residential Applications

26 05 90.10 Residential Wiring		Crew	Daily Output	Labor-Hours	Unit	Material	Labor (2011 Bare Costs)	Equipment	Total	Total Incl O&P
4250	20 amp recpt. #12/2, type NM cable	1 Elec	12.31	.650	Ea.	26	26		52	71
4270	Type MC cable		10.67	.750		41	30		71	94.50
4280	EMT & wire		4.71	1.699		50.50	68.50		119	167
4300	GFI, 15 amp recpt., type NM cable		12.31	.650		43.50	26		69.50	90.50
4320	Type MC cable		10.67	.750		61.50	30		91.50	117
4330	EMT & wire		4.71	1.699		66.50	68.50		135	185
4350	GFI with #12/2, type NM cable		10.67	.750		46.50	30		76.50	100
4370	Type MC cable		9.20	.870		61.50	35		96.50	125
4380	EMT & wire		4.21	1.900		71	76.50		147.50	202
4400	20 amp recpt., #12/2 type NM cable		10.67	.750		48.50	30		78.50	102
4420	Type MC cable		9.20	.870		63.50	35		98.50	127
4430	EMT & wire		4.21	1.900		73	76.50		149.50	204
4500	Weather-proof cover for above receptacles, add		32	.250		5.25	10.05		15.30	22
4550	Air conditioner outlet, 20 amp-240 volt recpt.									
4560	30′ of #12/2, 2 pole circuit breaker									
4570	Type NM cable	1 Elec	10	.800	Ea.	61	32		93	119
4580	Type MC cable		9	.889		79.50	36		115.50	146
4590	EMT & wire		4	2		85	80.50		165.50	225
4600	Decorator style, type NM cable		10	.800		65.50	32		97.50	124
4620	Type MC cable		9	.889		84	36		120	151
4630	EMT & wire		4	2		89.50	80.50		170	230
4650	Dryer outlet, 30 amp-240 volt recpt., 20′ of #10/3									
4660	2 pole circuit breaker									
4670	Type NM cable	1 Elec	6.41	1.248	Ea.	67.50	50		117.50	156
4680	Type MC cable		5.71	1.401		80	56.50		136.50	180
4690	EMT & wire		3.48	2.299		80.50	92.50		173	239
4700	Range outlet, 50 amp-240 volt recpt., 30′ of #8/3									
4710	Type NM cable	1 Elec	4.21	1.900	Ea.	96.50	76.50		173	230
4720	Type MC cable		4	2		165	80.50		245.50	315
4730	EMT & wire		2.96	2.703		110	109		219	297
4750	Central vacuum outlet, Type NM cable		6.40	1.250		65	50.50		115.50	153
4770	Type MC cable		5.71	1.401		91	56.50		147.50	192
4780	EMT & wire		3.48	2.299		90	92.50		182.50	249
4800	30 amp-110 volt locking recpt., #10/2 circ. bkr.									
4810	Type NM cable	1 Elec	6.20	1.290	Ea.	75.50	52		127.50	167
4820	Type MC cable		5.40	1.481		105	59.50		164.50	213
4830	EMT & wire		3.20	2.500		100	101		201	273
4900	Low voltage outlets									
4910	Telephone recpt., 20′ of 4/C phone wire	1 Elec	26	.308	Ea.	11	12.40		23.40	32
4920	TV recpt., 20′ of RG59U coax wire, F type connector	″	16	.500	″	19.25	20		39.25	53.50
4950	Door bell chime, transformer, 2 buttons, 60′ of bellwire									
4970	Economy model	1 Elec	11.50	.696	Ea.	77.50	28		105.50	131
4980	Custom model		11.50	.696		121	28		149	179
4990	Luxury model, 3 buttons		9.50	.842		325	34		359	410
6000	Lighting outlets									
6050	Wire only (for fixture), type NM cable	1 Elec	32	.250	Ea.	7.15	10.05		17.20	24
6070	Type MC cable		24	.333		17.25	13.40		30.65	41
6080	EMT & wire		10	.800		21	32		53	75
6100	Box (4″), and wire (for fixture), type NM cable		25	.320		17.70	12.90		30.60	40.50
6120	Type MC cable		20	.400		28	16.10		44.10	56.50
6130	EMT & wire		11	.727		31.50	29.50		61	82.50
6200	Fixtures (use with lines 6050 or 6100 above)									
6210	Canopy style, economy grade	1 Elec	40	.200	Ea.	34.50	8.05		42.55	51

26 05 90 – Residential Applications

26 05 90.10 Residential Wiring		Crew	Daily Output	Labor-Hours	Unit	2011 Bare Costs Material	Labor	Equipment	Total	Total Incl O&P
6220	Custom grade	1 Elec	40	.200	Ea.	53	8.05		61.05	71.50
6250	Dining room chandelier, economy grade		19	.421		86	16.95		102.95	122
6260	Custom grade		19	.421		253	16.95		269.95	305
6270	Luxury grade		15	.533		570	21.50		591.50	660
6310	Kitchen fixture (fluorescent), economy grade		30	.267		64	10.75		74.75	88
6320	Custom grade		25	.320		175	12.90		187.90	214
6350	Outdoor, wall mounted, economy grade		30	.267		32.50	10.75		43.25	53.50
6360	Custom grade		30	.267		112	10.75		122.75	140
6370	Luxury grade		25	.320		253	12.90		265.90	299
6410	Outdoor PAR floodlights, 1 lamp, 150 watt		20	.400		35	16.10		51.10	64.50
6420	2 lamp, 150 watt each		20	.400		58	16.10		74.10	90
6425	Motion sensing, 2 lamp, 150 watt each		20	.400		70	16.10		86.10	103
6430	For infrared security sensor, add		32	.250		117	10.05		127.05	145
6450	Outdoor, quartz-halogen, 300 watt flood		20	.400		42	16.10		58.10	72.50
6600	Recessed downlight, round, pre-wired, 50 or 75 watt trim		30	.267		43.50	10.75		54.25	65
6610	With shower light trim		30	.267		53.50	10.75		64.25	76.50
6620	With wall washer trim		28	.286		64.50	11.50		76	89.50
6630	With eye-ball trim		28	.286		64.50	11.50		76	89.50
6640	For direct contact with insulation, add					2.30			2.30	2.53
6700	Porcelain lamp holder	1 Elec	40	.200		4.50	8.05		12.55	18
6710	With pull switch		40	.200		5.25	8.05		13.30	18.85
6750	Fluorescent strip, 1-20 watt tube, wrap around diffuser, 24"		24	.333		54	13.40		67.40	81.50
6760	1-34 watt tube, 48"		24	.333		69.50	13.40		82.90	98.50
6770	2-34 watt tubes, 48"		20	.400		82.50	16.10		98.60	117
6780	With residential ballast		20	.400		92.50	16.10		108.60	128
6800	Bathroom heat lamp, 1-250 watt		28	.286		47.50	11.50		59	71
6810	2-250 watt lamps		28	.286		73.50	11.50		85	99.50
6820	For timer switch, see Section 26 05 90.10 line 2400									
6900	Outdoor post lamp, incl. post, fixture, 35' of #14/2									
6910	Type NMC cable	1 Elec	3.50	2.286	Ea.	203	92		295	370
6920	Photo-eye, add		27	.296		35	11.95		46.95	58
6950	Clock dial time switch, 24 hr., w/enclosure, type NM cable		11.43	.700		67	28		95	119
6970	Type MC cable		11	.727		84.50	29.50		114	141
6980	EMT & wire		4.85	1.649		90	66.50		156.50	207
7000	Alarm systems									
7050	Smoke detectors, box, #14/3, type NM cable	1 Elec	14.55	.550	Ea.	35	22		57	74.50
7070	Type MC cable		12.31	.650		51.50	26		77.50	99
7080	EMT & wire		5	1.600		50.50	64.50		115	160
7090	For relay output to security system, add					13			13	14.30
8000	Residential equipment									
8050	Disposal hook-up, incl. switch, outlet box, 3' of flex									
8060	20 amp-1 pole circ. bkr., and 25' of #12/2									
8070	Type NM cable	1 Elec	10	.800	Ea.	31.50	32		63.50	86.50
8080	Type MC cable		8	1		48.50	40.50		89	119
8090	EMT & wire		5	1.600		58	64.50		122.50	168
8100	Trash compactor or dishwasher hook-up, incl. outlet box,									
8110	3' of flex, 15 amp-1 pole circ. bkr., and 25' of #14/2									
8120	Type NM cable	1 Elec	10	.800	Ea.	23	32		55	77
8130	Type MC cable		8	1		43	40.50		83.50	113
8140	EMT & wire		5	1.600		50	64.50		114.50	159
8150	Hot water sink dispensor hook-up, use line 8100									
8200	Vent/exhaust fan hook-up, type NM cable	1 Elec	32	.250	Ea.	7.15	10.05		17.20	24
8220	Type MC cable		24	.333		17.25	13.40		30.65	41

26 05 90 – Residential Applications

26 05 90.10 Residential Wiring		Crew	Daily Output	Labor-Hours	Unit	Material	2011 Bare Costs Labor	Equipment	Total	Total Incl O&P
8230	EMT & wire	1 Elec	10	.800	Ea.	21	32		53	75
8250	Bathroom vent fan, 50 CFM (use with above hook-up)									
8260	Economy model	1 Elec	15	.533	Ea.	25	21.50		46.50	62.50
8270	Low noise model		15	.533		34	21.50		55.50	72.50
8280	Custom model	↓	12	.667	↓	117	27		144	173
8300	Bathroom or kitchen vent fan, 110 CFM									
8310	Economy model	1 Elec	15	.533	Ea.	69.50	21.50		91	112
8320	Low noise model	"	15	.533	"	86	21.50		107.50	130
8350	Paddle fan, variable speed (w/o lights)									
8360	Economy model (AC motor)	1 Elec	10	.800	Ea.	115	32		147	179
8362	With light kit		10	.800		153	32		185	221
8370	Custom model (AC motor)		10	.800		182	32		214	252
8372	With light kit		10	.800		220	32		252	295
8380	Luxury model (DC motor)		8	1		360	40.50		400.50	460
8382	With light kit		8	1		400	40.50		440.50	505
8390	Remote speed switch for above, add	↓	12	.667	↓	30	27		57	76.50
8500	Whole house exhaust fan, ceiling mount, 36", variable speed									
8510	Remote switch, incl. shutters, 20 amp-1 pole circ. bkr.									
8520	30' of #12/2, type NM cable	1 Elec	4	2	Ea.	1,125	80.50		1,205.50	1,375
8530	Type MC cable		3.50	2.286		1,150	92		1,242	1,425
8540	EMT & wire	↓	3	2.667	↓	1,150	107		1,257	1,450
8600	Whirlpool tub hook-up, incl. timer switch, outlet box									
8610	3' of flex, 20 amp-1 pole GFI circ. bkr.									
8620	30' of #12/2, type NM cable	1 Elec	5	1.600	Ea.	128	64.50		192.50	244
8630	Type MC cable		4.20	1.905		139	76.50		215.50	277
8640	EMT & wire	↓	3.40	2.353	↓	148	94.50		242.50	315
8650	Hot water heater hook-up, incl. 1-2 pole circ. bkr., box;									
8660	3' of flex, 20' of #10/2, type NM cable	1 Elec	5	1.600	Ea.	33	64.50		97.50	140
8670	Type MC cable		4.20	1.905		56	76.50		132.50	186
8680	EMT & wire	↓	3.40	2.353	↓	53	94.50		147.50	212
9000	Heating/air conditioning									
9050	Furnace/boiler hook-up, incl. firestat, local on-off switch									
9060	Emergency switch, and 40' of type NM cable	1 Elec	4	2	Ea.	56.50	80.50		137	193
9070	Type MC cable		3.50	2.286		83.50	92		175.50	241
9080	EMT & wire	↓	1.50	5.333	↓	92.50	215		307.50	450
9100	Air conditioner hook-up, incl. local 60 amp disc. switch									
9110	3' sealtite, 40 amp, 2 pole circuit breaker									
9130	40' of #8/2, type NM cable	1 Elec	3.50	2.286	Ea.	225	92		317	395
9140	Type MC cable		3	2.667		325	107		432	535
9150	EMT & wire	↓	1.30	6.154	↓	259	248		507	685
9200	Heat pump hook-up, 1-40 & 1-100 amp 2 pole circ. bkr.									
9210	Local disconnect switch, 3' sealtite									
9220	40' of #8/2 & 30' of #3/2									
9230	Type NM cable	1 Elec	1.30	6.154	Ea.	570	248		818	1,025
9240	Type MC cable		1.08	7.407		755	298		1,053	1,325
9250	EMT & wire	↓	.94	8.511	↓	620	345		965	1,225
9500	Thermostat hook-up, using low voltage wire									
9520	Heating only, 25' of #18-3	1 Elec	24	.333	Ea.	9.70	13.40		23.10	32.50
9530	Heating/cooling, 25' of #18-4	"	20	.400	"	12.25	16.10		28.35	39.50

26 09 Instrumentation and Control for Electrical Systems

26 09 13 – Electrical Power Monitoring

26 09 13.10 Switchboard Instruments		Crew	Daily Output	Labor-Hours	Unit	2011 Bare Costs Material	Labor	Equipment	Total	Total Incl O&P
0010	**SWITCHBOARD INSTRUMENTS** 3 phase, 4 wire									
0100	AC indicating, ammeter & switch	1 Elec	8	1	Ea.	2,025	40.50		2,065.50	2,300
0200	Voltmeter & switch		8	1		2,025	40.50		2,065.50	2,300
0300	Wattmeter		8	1		4,000	40.50		4,040.50	4,475
0400	AC recording, ammeter		4	2		7,150	80.50		7,230.50	7,975
0500	Voltmeter		4	2		7,150	80.50		7,230.50	7,975
0600	Ground fault protection, zero sequence		2.70	2.963		6,300	119		6,419	7,125
0700	Ground return path		2.70	2.963		6,300	119		6,419	7,125
0800	3 current transformers, 5 to 800 amp		2	4		2,925	161		3,086	3,475
0900	1000 to 1500 amp		1.30	6.154		4,225	248		4,473	5,050
1200	2000 to 4000 amp		1	8		4,975	320		5,295	6,000
1300	Fused potential transformer, maximum 600 volt		8	1		1,100	40.50		1,140.50	1,300

26 12 Medium-Voltage Transformers

26 12 19 – Pad-Mounted, Liquid-Filled, Medium-Voltage Transformers

26 12 19.10 Transformer, Oil-Filled

		Crew	Daily Output	Labor-Hours	Unit	Material	Labor	Equipment	Total	Total Incl O&P
0010	**TRANSFORMER, OIL-FILLED** primary delta or Y,									
0050	Pad mounted 5 kV or 15 kV, with taps, 277/480 V secondary, 3 phase									
0100	150 kVA	R-3	.65	30.769	Ea.	9,850	1,225	198	11,273	13,000
0200	300 kVA		.45	44.444		12,700	1,750	287	14,737	17,200
0300	500 kVA		.40	50		19,700	1,975	325	22,000	25,300
0400	750 kVA		.38	52.632		22,900	2,075	340	25,315	29,000
0500	1000 kVA		.26	76.923		27,200	3,050	495	30,745	35,400
0600	1500 kVA		.23	86.957		32,200	3,450	560	36,210	41,700
0700	2000 kVA		.20	100		40,700	3,950	645	45,295	52,000
0800	3750 kVA		.16	125		76,500	4,950	805	82,255	93,000

26 22 Low-Voltage Transformers

26 22 13 – Low-Voltage Distribution Transformers

26 22 13.10 Transformer, Dry-Type

		Crew	Daily Output	Labor-Hours	Unit	Material	Labor	Equipment	Total	Total Incl O&P
0010	**TRANSFORMER, DRY-TYPE**									
0050	Single phase, 240/480 volt primary, 120/240 volt secondary									
0100	1 kVA	1 Elec	2	4	Ea.	282	161		443	570
0300	2 kVA		1.60	5		425	201		626	790
0500	3 kVA		1.40	5.714		525	230		755	950
0700	5 kVA		1.20	6.667		720	268		988	1,225
0900	7.5 kVA	2 Elec	2.20	7.273		1,000	293		1,293	1,575
1100	10 kVA		1.60	10		1,275	405		1,680	2,050
1300	15 kVA		1.20	13.333		1,750	535		2,285	2,800
1500	25 kVA		1	16		2,400	645		3,045	3,675
1700	37.5 kVA		.80	20		3,100	805		3,905	4,700
1900	50 kVA		.70	22.857		3,675	920		4,595	5,550
2100	75 kVA		.65	24.615		4,875	990		5,865	6,950
2190	480 V primary 120/240 V secondary, nonvent., 15 kVA		1.20	13.333		1,775	535		2,310	2,825
2200	25 kVA		.90	17.778		2,600	715		3,315	4,025
2210	37 kVA		.75	21.333		3,100	860		3,960	4,800
2220	50 kVA		.65	24.615		3,675	990		4,665	5,650
2300	3 phase, 480 volt primary 120/208 volt secondary									
2310	Ventilated, 3 kVA	1 Elec	1	8	Ea.	915	320		1,235	1,525

26 22 Low-Voltage Transformers

26 22 13 – Low-Voltage Distribution Transformers

26 22 13.10 Transformer, Dry-Type		Crew	Daily Output	Labor-Hours	Unit	2011 Bare Costs Material	Labor	Equipment	Total	Total Incl O&P
2700	6 kVA	1 Elec	.80	10	Ea.	1,250	405		1,655	2,025
2900	9 kVA		.70	11.429		1,425	460		1,885	2,325
3100	15 kVA	2 Elec	1.10	14.545		1,900	585		2,485	3,050
3300	30 kVA		.90	17.778		2,225	715		2,940	3,600
3500	45 kVA		.80	20		2,675	805		3,480	4,250
3700	75 kVA		.70	22.857		4,025	920		4,945	5,925
3900	112.5 kVA	R-3	.90	22.222		5,350	880	143	6,373	7,450
4100	150 kVA		.85	23.529		6,975	930	152	8,057	9,350
4300	225 kVA		.65	30.769		9,475	1,225	198	10,898	12,600
4500	300 kVA		.55	36.364		12,000	1,450	235	13,685	15,800
4700	500 kVA		.45	44.444		19,800	1,750	287	21,837	25,000
4800	750 kVA		.35	57.143		34,700	2,250	370	37,320	42,300

26 24 Switchboards and Panelboards

26 24 13 – Switchboards

26 24 13.10 Incoming Switchboards

Line	Description	Crew	Daily Output	Labor-Hours	Unit	Material	Labor	Equipment	Total	Total Incl O&P
0010	**INCOMING SWITCHBOARDS** main service section									
0100	Aluminum bus bars, not including CT's or PT's									
0200	No main disconnect, includes CT compartment									
0300	120/208 volt, 4 wire, 600 amp	2 Elec	1	16	Ea.	4,025	645		4,670	5,475
0400	800 amp		.88	18.182		4,025	730		4,755	5,600
0500	1000 amp		.80	20		4,850	805		5,655	6,625
0600	1200 amp		.72	22.222		4,850	895		5,745	6,775
0700	1600 amp		.66	24.242		4,850	975		5,825	6,900
0800	2000 amp		.62	25.806		5,225	1,050		6,275	7,400
1000	3000 amp		.56	28.571		6,875	1,150		8,025	9,450
2000	Fused switch & CT compartment									
2100	120/208 volt, 4 wire, 400 amp	2 Elec	1.12	14.286	Ea.	3,875	575		4,450	5,175
2200	600 amp		.94	17.021		4,575	685		5,260	6,150
2300	800 amp		.84	19.048		10,900	765		11,665	13,300
2400	1200 amp		.68	23.529		14,200	945		15,145	17,100
2900	Pressure switch & CT compartment									
3000	120/208 volt, 4 wire, 800 amp	2 Elec	.80	20	Ea.	9,800	805		10,605	12,100
3100	1200 amp		.66	24.242		19,000	975		19,975	22,400
3200	1600 amp		.62	25.806		20,100	1,050		21,150	23,900
3300	2000 amp		.56	28.571		21,400	1,150		22,550	25,500
4400	Circuit breaker, molded case & CT compartment									
4600	3 pole, 4 wire, 600 amp	2 Elec	.94	17.021	Ea.	8,425	685		9,110	10,400
4800	800 amp		.84	19.048		10,100	765		10,865	12,400
5000	1200 amp		.68	23.529		13,800	945		14,745	16,600
5100	Copper bus bars, not incl. CT's or PT's, add, minimum					15%				

26 24 13.30 Distribution Switchboards Section

Line	Description	Crew	Daily Output	Labor-Hours	Unit	Material	Labor	Equipment	Total	Total Incl O&P
0010	**DISTRIBUTION SWITCHBOARDS SECTION**									
0100	Aluminum bus bars, not including breakers									
0200	120/208 or 277/480 volt, 4 wire, 600 amp	2 Elec	1	16	Ea.	2,300	645		2,945	3,575
0300	800 amp		.88	18.182		3,050	730		3,780	4,525
0400	1000 amp		.80	20		3,050	805		3,855	4,650
0500	1200 amp		.72	22.222		3,775	895		4,670	5,600
0600	1600 amp		.66	24.242		4,325	975		5,300	6,325
0700	2000 amp		.62	25.806		5,075	1,050		6,125	7,250
0800	2500 amp		.60	26.667		5,725	1,075		6,800	8,025

26 24 13 – Switchboards

26 24 13.30 Distribution Switchboards Section		Crew	Daily Output	Labor-Hours	Unit	Material	2011 Bare Costs Labor	Equipment	Total	Total Incl O&P
0900	3000 amp	2 Elec	.56	28.571	Ea.	6,925	1,150		8,075	9,500
0950	4000 amp		.52	30.769		10,100	1,250		11,350	13,100

26 24 13.40 Switchboards Feeder Section

Code	Description	Crew	Daily Output	Labor-Hours	Unit	Material	Labor	Equipment	Total	Total Incl O&P
0010	**SWITCHBOARDS FEEDER SECTION** group mounted devices									
0030	Circuit breakers									
0160	FA frame, 15 to 60 amp, 240 volt, 1 pole	1 Elec	8	1	Ea.	115	40.50		155.50	192
0280	FA frame, 70 to 100 amp, 240 volt, 1 pole		7	1.143		151	46		197	242
0420	KA frame, 70 to 225 amp		3.20	2.500		1,275	101		1,376	1,575
0430	LA frame, 125 to 400 amp		2.30	3.478		2,875	140		3,015	3,375
0460	MA frame, 450 to 600 amp		1.60	5		4,725	201		4,926	5,525
0470	700 to 800 amp		1.30	6.154		6,150	248		6,398	7,175
0480	MAL frame, 1000 amp		1	8		6,375	320		6,695	7,550
0490	PA frame, 1200 amp		.80	10		13,000	405		13,405	15,000
0500	Branch circuit, fusible switch, 600 volt, double 30/30 amp		4	2		1,200	80.50		1,280.50	1,450
0550	60/60 amp		3.20	2.500		1,225	101		1,326	1,525
0600	100/100 amp		2.70	2.963		1,725	119		1,844	2,100
0650	Single, 30 amp		5.30	1.509		740	61		801	915
0700	60 amp		4.70	1.702		740	68.50		808.50	925
0750	100 amp		4	2		1,175	80.50		1,255.50	1,400
0800	200 amp		2.70	2.963		1,700	119		1,819	2,075
0850	400 amp		2.30	3.478		3,200	140		3,340	3,750
0900	600 amp		1.80	4.444		3,800	179		3,979	4,500
0950	800 amp		1.30	6.154		6,225	248		6,473	7,250
1000	1200 amp		.80	10		7,425	405		7,830	8,825

26 24 16 – Panelboards

26 24 16.20 Panelboard and Load Center Circuit Breakers

Code	Description	Crew	Daily Output	Labor-Hours	Unit	Material	Labor	Equipment	Total	Total Incl O&P
0010	**PANELBOARD AND LOAD CENTER CIRCUIT BREAKERS**									
0050	Bolt-on, 10,000 amp IC, 120 volt, 1 pole									
0100	15 to 50 amp	1 Elec	10	.800	Ea.	14.80	32		46.80	68.50
0200	60 amp		8	1		14.80	40.50		55.30	82
0300	70 amp		8	1		28	40.50		68.50	96.50
0350	240 volt, 2 pole									
0400	15 to 50 amp	1 Elec	8	1	Ea.	32.50	40.50		73	102
0500	60 amp		7.50	1.067		32.50	43		75.50	106
0600	80 to 100 amp		5	1.600		83.50	64.50		148	196
0700	3 pole, 15 to 60 amp		6.20	1.290		103	52		155	197
0800	70 amp		5	1.600		130	64.50		194.50	246
0900	80 to 100 amp		3.60	2.222		147	89.50		236.50	305
1000	22,000 amp I.C., 240 volt, 2 pole, 70 – 225 amp		2.70	2.963		635	119		754	890
1100	3 pole, 70 – 225 amp		2.30	3.478		705	140		845	1,000
1200	14,000 amp I.C., 277 volts, 1 pole, 15 – 30 amp		8	1		39	40.50		79.50	109
1300	22,000 amp I.C., 480 volts, 2 pole, 70 – 225 amp		2.70	2.963		635	119		754	890
1400	3 pole, 70 – 225 amp		2.30	3.478		780	140		920	1,075

26 24 16.30 Panelboards Commercial Applications

Code	Description	Crew	Daily Output	Labor-Hours	Unit	Material	Labor	Equipment	Total	Total Incl O&P
0010	**PANELBOARDS COMMERCIAL APPLICATIONS**									
0050	NQOD, w/20 amp 1 pole bolt-on circuit breakers									
0100	3 wire, 120/240 volts, 100 amp main lugs									
0150	10 circuits	1 Elec	1	8	Ea.	545	320		865	1,125
0200	14 circuits		.88	9.091		630	365		995	1,300
0250	18 circuits		.75	10.667		685	430		1,115	1,450
0300	20 circuits		.65	12.308		770	495		1,265	1,650
0350	225 amp main lugs, 24 circuits	2 Elec	1.20	13.333		870	535		1,405	1,825

26 24 Switchboards and Panelboards

26 24 16 – Panelboards

26 24 16.30 Panelboards Commercial Applications		Crew	Daily Output	Labor-Hours	Unit	Material	2011 Bare Costs Labor	Equipment	Total	Total Incl O&P
0400	30 circuits	2 Elec	.90	17.778	Ea.	1,000	715		1,715	2,250
0450	36 circuits		.80	20		1,150	805		1,955	2,550
0500	38 circuits		.72	22.222		1,225	895		2,120	2,800
0550	42 circuits		.66	24.242		1,275	975		2,250	2,975
0600	4 wire, 120/208 volts, 100 amp main lugs, 12 circuits	1 Elec	1	8		610	320		930	1,200
0650	16 circuits		.75	10.667		700	430		1,130	1,475
0700	20 circuits		.65	12.308		810	495		1,305	1,700
0750	24 circuits		.60	13.333		875	535		1,410	1,825
0800	30 circuits		.53	15.094		1,000	610		1,610	2,075
0850	225 amp main lugs, 32 circuits	2 Elec	.90	17.778		1,125	715		1,840	2,375
0900	34 circuits		.84	19.048		1,150	765		1,915	2,525
0950	36 circuits		.80	20		1,175	805		1,980	2,600
1000	42 circuits		.68	23.529		1,325	945		2,270	2,975
1200	NEHB,w/20 amp, 1 pole bolt-on circuit breakers									
1250	4 wire, 277/480 volts, 100 amp main lugs, 12 circuits	1 Elec	.88	9.091	Ea.	1,125	365		1,490	1,850
1300	20 circuits	"	.60	13.333		1,675	535		2,210	2,725
1350	225 amp main lugs, 24 circuits	2 Elec	.90	17.778		1,900	715		2,615	3,250
1400	30 circuits		.80	20		2,275	805		3,080	3,825
1450	36 circuits		.72	22.222		2,650	895		3,545	4,375
1600	NQOD panel, w/20 amp, 1 pole, circuit breakers									
1650	3 wire, 120/240 volt with main circuit breaker									
1700	100 amp main, 12 circuits	1 Elec	.80	10	Ea.	740	405		1,145	1,475
1750	20 circuits	"	.60	13.333		940	535		1,475	1,900
1800	225 amp main, 30 circuits	2 Elec	.68	23.529		1,775	945		2,720	3,475
1850	42 circuits		.52	30.769		2,050	1,250		3,300	4,250
1900	400 amp main, 30 circuits		.54	29.630		2,425	1,200		3,625	4,600
1950	42 circuits		.50	32		2,700	1,300		4,000	5,075
2000	4 wire, 120/208 volts with main circuit breaker									
2050	100 amp main, 24 circuits	1 Elec	.47	17.021	Ea.	1,100	685		1,785	2,300
2100	30 circuits	"	.40	20		1,225	805		2,030	2,650
2200	225 amp main, 32 circuits	2 Elec	.72	22.222		2,050	895		2,945	3,700
2250	42 circuits		.56	28.571		2,250	1,150		3,400	4,350
2300	400 amp main, 42 circuits		.48	33.333		3,000	1,350		4,350	5,475
2350	600 amp main, 42 circuits		.40	40		4,450	1,600		6,050	7,500
2400	NEHB, with 20 amp, 1 pole circuit breaker									
2450	4 wire, 277/480 volts with main circuit breaker									
2500	100 amp main, 24 circuits	1 Elec	.42	19.048	Ea.	2,200	765		2,965	3,675
2550	30 circuits	"	.38	21.053		2,575	845		3,420	4,200
2600	225 amp main, 30 circuits	2 Elec	.72	22.222		3,225	895		4,120	5,000
2650	42 circuits	"	.56	28.571		3,975	1,150		5,125	6,250

26 24 19 – Motor-Control Centers

26 24 19.40 Motor Starters and Controls		Crew	Daily Output	Labor-Hours	Unit	Material	Labor	Equipment	Total	Total Incl O&P
0010	**MOTOR STARTERS AND CONTROLS**									
0050	Magnetic, FVNR, with enclosure and heaters, 480 volt									
0100	5 HP, size 0	1 Elec	2.30	3.478	Ea.	255	140		395	510
0200	10 HP, size 1	"	1.60	5		287	201		488	640
0300	25 HP, size 2	2 Elec	2.20	7.273		540	293		833	1,075
0400	50 HP, size 3		1.80	8.889		880	360		1,240	1,550
0500	100 HP, size 4		1.20	13.333		1,950	535		2,485	3,025
0600	200 HP, size 5		.90	17.778		4,550	715		5,265	6,175
0700	Combination, with motor circuit protectors, 5 HP, size 0	1 Elec	1.80	4.444		830	179		1,009	1,200
0800	10 HP, size 1	"	1.30	6.154		860	248		1,108	1,350

26 24 Switchboards and Panelboards

26 24 19 – Motor-Control Centers

26 24 19.40 Motor Starters and Controls		Crew	Daily Output	Labor-Hours	Unit	Material	2011 Bare Costs Labor	Equipment	Total	Total Incl O&P
0900	25 HP, size 2	2 Elec	2	8	Ea.	1,200	320		1,520	1,850
1000	50 HP, size 3		1.32	12.121		1,750	490		2,240	2,725
1200	100 HP, size 4		.80	20		3,800	805		4,605	5,475
1400	Combination, with fused switch, 5 HP, size 0	1 Elec	1.80	4.444		635	179		814	985
1600	10 HP, size 1	"	1.30	6.154		675	248		923	1,150
1800	25 HP, size 2	2 Elec	2	8		1,100	320		1,420	1,725
2000	50 HP, size 3		1.32	12.121		1,850	490		2,340	2,850
2200	100 HP, size 4		.80	20		3,250	805		4,055	4,875

26 25 Enclosed Bus Assemblies

26 25 13 – Bus Duct/Busway and Fittings

26 25 13.40 Copper Bus Duct		Crew	Daily Output	Labor-Hours	Unit	Material	Labor	Equipment	Total	Total Incl O&P
0010	**COPPER BUS DUCT** 10 ft. long									
0050	Indoor 3 pole 4 wire, plug-in, straight section, 225 amp	2 Elec	40	.400	L.F.	190	16.10		206.10	235
1000	400 amp		32	.500		190	20		210	242
1500	600 amp		26	.615		190	25		215	249
2400	800 amp		20	.800		226	32		258	300
2450	1000 amp		18	.889		250	36		286	335
2500	1350 amp		16	1		355	40.50		395.50	460
2510	1600 amp		12	1.333		405	53.50		458.50	530
2520	2000 amp		10	1.600		510	64.50		574.50	670
2550	Feeder, 600 amp		28	.571		167	23		190	221
2600	800 amp		22	.727		202	29.50		231.50	271
2700	1000 amp		20	.800		226	32		258	300
2800	1350 amp		18	.889		335	36		371	425
2900	1600 amp		14	1.143		380	46		426	495
3000	2000 amp		12	1.333		490	53.50		543.50	620
3100	Elbows, 225 amp		4	4	Ea.	1,125	161		1,286	1,500
3200	400 amp		3.60	4.444		1,125	179		1,304	1,550
3300	600 amp		3.20	5		1,125	201		1,326	1,575
3400	800 amp		2.80	5.714		1,225	230		1,455	1,725
3500	1000 amp		2.60	6.154		1,375	248		1,623	1,900
3600	1350 amp		2.40	6.667		1,600	268		1,868	2,200
3700	1600 amp		2.20	7.273		1,750	293		2,043	2,400
3800	2000 amp		1.80	8.889		2,150	360		2,510	2,950
4000	End box, 225 amp		34	.471		152	18.95		170.95	199
4100	400 amp		32	.500		152	20		172	201
4200	600 amp		28	.571		152	23		175	206
4300	800 amp		26	.615		152	25		177	208
4400	1000 amp		24	.667		152	27		179	212
4500	1350 amp		22	.727		152	29.50		181.50	216
4600	1600 amp		20	.800		152	32		184	220
4700	2000 amp		18	.889		187	36		223	264
4800	Cable tap box end, 225 amp		3.20	5		1,250	201		1,451	1,675
5000	400 amp		2.60	6.154		1,250	248		1,498	1,750
5100	600 amp		2.20	7.273		1,250	293		1,543	1,825
5200	800 amp		2	8		1,425	320		1,745	2,100
5300	1000 amp		1.60	10		1,525	405		1,930	2,325
5400	1350 amp		1.40	11.429		1,900	460		2,360	2,850
5500	1600 amp		1.20	13.333		2,125	535		2,660	3,225
5600	2000 amp		1	16		2,375	645		3,020	3,675

26 25 Enclosed Bus Assemblies

26 25 13 – Bus Duct/Busway and Fittings

26 25 13.40 Copper Bus Duct		Crew	Daily Output	Labor-Hours	Unit	2011 Bare Costs Material	Labor	Equipment	Total	Total Incl O&P
5700	Switchboard stub, 225 amp	2 Elec	5.40	2.963	Ea.	1,150	119		1,269	1,450
5800	400 amp		4.60	3.478		1,150	140		1,290	1,475
5900	600 amp		4	4		1,150	161		1,311	1,500
6000	800 amp		3.20	5		1,375	201		1,576	1,850
6100	1000 amp		3	5.333		1,600	215		1,815	2,100
6200	1350 amp		2.60	6.154		2,050	248		2,298	2,675
6300	1600 amp		2.40	6.667		2,325	268		2,593	2,975
6400	2000 amp		2	8		2,825	320		3,145	3,625
6490	Tee fittings, 225 amp		2.40	6.667		1,550	268		1,818	2,125
6500	400 amp		2	8		1,550	320		1,870	2,225
6600	600 amp		1.80	8.889		1,550	360		1,910	2,275
6700	800 amp		1.60	10		1,800	405		2,205	2,625
6800	1350 amp		1.20	13.333		2,575	535		3,110	3,700
7000	1600 amp		1	16		2,925	645		3,570	4,275
7100	2000 amp		.80	20		3,475	805		4,280	5,125
7200	Plug-in fusible switches w/3 fuses, 600 volt, 3 pole, 30 amp	1 Elec	4	2		695	80.50		775.50	895
7300	60 amp		3.60	2.222		780	89.50		869.50	1,000
7400	100 amp		2.70	2.963		1,200	119		1,319	1,500
7500	200 amp	2 Elec	3.20	5		2,125	201		2,326	2,675
7600	400 amp		1.40	11.429		6,250	460		6,710	7,600
7700	600 amp		.90	17.778		7,075	715		7,790	8,925
7800	800 amp		.66	24.242		9,925	975		10,900	12,500
7900	1200 amp		.50	32		18,600	1,300		19,900	22,600
8000	Plug-in circuit breakers, molded case, 15 to 50 amp	1 Elec	4.40	1.818		660	73		733	845
8100	70 to 100 amp	″	3.10	2.581		730	104		834	975
8200	150 to 225 amp	2 Elec	3.40	4.706		1,975	189		2,164	2,475
8300	250 to 400 amp		1.40	11.429		3,475	460		3,935	4,575
8400	500 to 600 amp		1	16		4,700	645		5,345	6,225
8500	700 to 800 amp		.64	25		5,800	1,000		6,800	8,000
8600	900 to 1000 amp		.56	28.571		8,275	1,150		9,425	11,000
8700	1200 amp		.44	36.364		9,950	1,475		11,425	13,400

26 27 Low-Voltage Distribution Equipment

26 27 16 – Electrical Cabinets and Enclosures

26 27 16.10 Cabinets

		Crew	Daily Output	Labor-Hours	Unit	Material	Labor	Equipment	Total	Total Incl O&P
0010	**CABINETS**									
7000	Cabinets, current transformer									
7050	Single door, 24″ H x 24″ W x 10″ D	1 Elec	1.60	5	Ea.	206	201		407	550
7100	30″ H x 24″ W x 10″ D		1.30	6.154		210	248		458	630
7150	36″ H x 24″ W x 10″ D		1.10	7.273		215	293		508	710
7200	30″ H x 30″ W x 10″ D		1	8		225	320		545	770
7250	36″ H x 30″ W x 10″ D		.90	8.889		305	360		665	915
7300	36″ H x 36″ W x 10″ D		.80	10		325	405		730	1,025
7500	Double door, 48″ H x 36″ W x 10″ D		.60	13.333		670	535		1,205	1,600
7550	24″ H x 24″ W x 12″ D		1	8		274	320		594	820

26 27 23 – Indoor Service Poles

26 27 23.40 Surface Raceway

		Crew	Daily Output	Labor-Hours	Unit	Material	Labor	Equipment	Total	Total Incl O&P
0010	**SURFACE RACEWAY**									
0090	Metal, straight section									
0100	No. 500	1 Elec	100	.080	L.F.	1.03	3.22		4.25	6.35

26 27 Low-Voltage Distribution Equipment

26 27 23 – Indoor Service Poles

26 27 23.40 Surface Raceway		Crew	Daily Output	Labor-Hours	Unit	Material	Labor	Equipment	Total	Total Incl O&P
						2011 Bare Costs				
0110	No. 700	1 Elec	100	.080	L.F.	1.16	3.22		4.38	6.50
0400	No. 1500, small pancake		90	.089		2.10	3.58		5.68	8.10
0600	No. 2000, base & cover, blank		90	.089		2.15	3.58		5.73	8.15
0800	No. 3000, base & cover, blank		75	.107		4.10	4.29		8.39	11.45
1000	No. 4000, base & cover, blank		65	.123		6.70	4.95		11.65	15.40
1200	No. 6000, base & cover, blank		50	.160		11.10	6.45		17.55	22.50
2400	Fittings, elbows, No. 500		40	.200	Ea.	1.85	8.05		9.90	15.10
2800	Elbow cover, No. 2000		40	.200		3.50	8.05		11.55	16.90
2880	Tee, No. 500		42	.190		3.60	7.65		11.25	16.40
2900	No. 2000		27	.296		11.70	11.95		23.65	32
3000	Switch box, No. 500		16	.500		12.40	20		32.40	46
3400	Telephone outlet, No. 1500		16	.500		14.50	20		34.50	48.50
3600	Junction box, No. 1500		16	.500		9.90	20		29.90	43.50
3800	Plugmold wired sections, No. 2000									
4000	1 circuit, 6 outlets, 3 ft. long	1 Elec	8	1	Ea.	36	40.50		76.50	105
4100	2 circuits, 8 outlets, 6 ft. long	″	5.30	1.509	″	59.50	61		120.50	164

26 27 26 – Wiring Devices

26 27 26.10 Low Voltage Switching

Code	Description	Crew	Daily Output	Labor-Hours	Unit	Material	Labor	Equipment	Total	Total Incl O&P
0010	**LOW VOLTAGE SWITCHING**									
3600	Relays, 120 V or 277 V standard	1 Elec	12	.667	Ea.	38.50	27		65.50	86
3800	Flush switch, standard		40	.200		11.30	8.05		19.35	25.50
4000	Interchangeable		40	.200		14.75	8.05		22.80	29.50
4100	Surface switch, standard		40	.200		8	8.05		16.05	22
4200	Transformer 115 V to 25 V		12	.667		128	27		155	185
4400	Master control, 12 circuit, manual		4	2		118	80.50		198.50	261
4500	25 circuit, motorized		4	2		131	80.50		211.50	275
4600	Rectifier, silicon		12	.667		42.50	27		69.50	90.50
4800	Switchplates, 1 gang, 1, 2 or 3 switch, plastic		80	.100		4.67	4.03		8.70	11.70
5000	Stainless steel		80	.100		10.60	4.03		14.63	18.20
5400	2 gang, 3 switch, stainless steel		53	.151		20	6.10		26.10	32
5500	4 switch, plastic		53	.151		9.65	6.10		15.75	20.50
5800	3 gang, 9 switch, stainless steel		32	.250		64	10.05		74.05	87

26 27 26.20 Wiring Devices Elements

Code	Description	Crew	Daily Output	Labor-Hours	Unit	Material	Labor	Equipment	Total	Total Incl O&P
0010	**WIRING DEVICES ELEMENTS**									
0200	Toggle switch, quiet type, single pole, 15 amp	1 Elec	40	.200	Ea.	5.40	8.05		13.45	18.95
0600	3 way, 15 amp		23	.348		7.90	14		21.90	31
0900	4 way, 15 amp		15	.533		26	21.50		47.50	63.50
1650	Dimmer switch, 120 volt, incandescent, 600 watt, 1 pole G		16	.500		13.60	20		33.60	47.50
2460	Receptacle, duplex, 120 volt, grounded, 15 amp		40	.200		1.37	8.05		9.42	14.55
2470	20 amp		27	.296		10.55	11.95		22.50	31
2490	Dryer, 30 amp		15	.533		6.55	21.50		28.05	42
2500	Range, 50 amp		11	.727		14.60	29.50		44.10	63.50
2600	Wall plates, stainless steel, 1 gang		80	.100		2.50	4.03		6.53	9.30
2800	2 gang		53	.151		4.25	6.10		10.35	14.55
3200	Lampholder, keyless		26	.308		12.80	12.40		25.20	34
3400	Pullchain with receptacle		22	.364		13.60	14.65		28.25	38.50

26 27 73 – Door Chimes

26 27 73.10 Doorbell System

Code	Description	Crew	Daily Output	Labor-Hours	Unit	Material	Labor	Equipment	Total	Total Incl O&P
0010	**DOORBELL SYSTEM**, incl. transformer, button & signal									
0100	6″ bell	1 Elec	4	2	Ea.	114	80.50		194.50	257
0200	Buzzer	″	4	2	″	90.50	80.50		171	231

26 28 Low-Voltage Circuit Protective Devices

26 28 16 – Enclosed Switches and Circuit Breakers

26 28 16.10 Circuit Breakers		Crew	Daily Output	Labor-Hours	Unit	Material	2011 Bare Costs Labor	Equipment	Total	Total Incl O&P
0010	**CIRCUIT BREAKERS** (in enclosure)									
0100	Enclosed (NEMA 1), 600 volt, 3 pole, 30 amp	1 Elec	3.20	2.500	Ea.	600	101		701	820
0200	60 amp		2.80	2.857		600	115		715	840
0400	100 amp		2.30	3.478		685	140		825	975
0600	225 amp		1.50	5.333		1,575	215		1,790	2,100
0700	400 amp	2 Elec	1.60	10		2,700	405		3,105	3,625
0800	600 amp		1.20	13.333		3,900	535		4,435	5,175
1000	800 amp		.94	17.021		5,100	685		5,785	6,700

26 28 16.20 Safety Switches		Crew	Daily Output	Labor-Hours	Unit	Material	Labor	Equipment	Total	Total Incl O&P
0010	**SAFETY SWITCHES**									
0100	General duty 240 volt, 3 pole NEMA 1, fusible, 30 amp	1 Elec	3.20	2.500	Ea.	119	101		220	294
0200	60 amp		2.30	3.478		202	140		342	450
0300	100 amp		1.90	4.211		345	169		514	655
0400	200 amp		1.30	6.154		745	248		993	1,225
0500	400 amp	2 Elec	1.80	8.889		1,875	360		2,235	2,650
0600	600 amp	"	1.20	13.333		3,500	535		4,035	4,725
2900	Heavy duty, 240 volt, 3 pole NEMA 1 fusible									
2910	30 amp	1 Elec	3.20	2.500	Ea.	192	101		293	375
3000	60 amp		2.30	3.478		325	140		465	585
3300	100 amp		1.90	4.211		515	169		684	840
3500	200 amp		1.30	6.154		880	248		1,128	1,375
3700	400 amp	2 Elec	1.80	8.889		2,250	360		2,610	3,075
3900	600 amp	"	1.20	13.333		3,900	535		4,435	5,150

26 29 Low-Voltage Controllers

26 29 13 – Enclosed Controllers

26 29 13.20 Control Stations		Crew	Daily Output	Labor-Hours	Unit	Material	Labor	Equipment	Total	Total Incl O&P
0010	**CONTROL STATIONS**									
0050	NEMA 1, heavy duty, stop/start	1 Elec	8	1	Ea.	137	40.50		177.50	216
0100	Stop/start, pilot light		6.20	1.290		186	52		238	289
0200	Hand/off/automatic		6.20	1.290		101	52		153	195
0400	Stop/start/reverse		5.30	1.509		184	61		245	300

26 32 Packaged Generator Assemblies

26 32 13 – Engine Generators

26 32 13.13 Diesel-Engine-Driven Generator Sets		Crew	Daily Output	Labor-Hours	Unit	Material	Labor	Equipment	Total	Total Incl O&P
0010	**DIESEL-ENGINE-DRIVEN GENERATOR SETS**									
2000	Diesel engine, including battery, charger,									
2010	muffler, automatic transfer switch & day tank, 30 kW	R-3	.55	36.364	Ea.	21,000	1,450	235	22,685	25,700
2100	50 kW		.42	47.619		24,900	1,875	305	27,080	30,800
2200	75 kW		.35	57.143		32,300	2,250	370	34,920	39,700
2300	100 kW		.31	64.516		36,000	2,550	415	38,965	44,100
2400	125 kW		.29	68.966		37,900	2,725	445	41,070	46,600
2500	150 kW		.26	76.923		43,500	3,050	495	47,045	53,500
2600	175 kW		.25	80		47,300	3,175	515	50,990	57,500
2700	200 kW		.24	83.333		48,800	3,300	540	52,640	59,500
2800	250 kW		.23	86.957		57,500	3,450	560	61,510	69,000
2900	300 kW		.22	90.909		62,000	3,600	585	66,185	75,000
3000	350 kW		.20	100		70,000	3,950	645	74,595	84,000

26 32 Packaged Generator Assemblies

26 32 13 – Engine Generators

26 32 13.13 Diesel-Engine-Driven Generator Sets		Crew	Daily Output	Labor-Hours	Unit	Material	2011 Bare Costs Labor	Equipment	Total	Total Incl O&P
3100	400 kW	R-3	.19	105	Ea.	86,500	4,175	680	91,355	103,000
3200	500 kW	↓	.18	111	↓	109,000	4,400	715	114,115	128,000

26 32 13.16 Gas-Engine-Driven Generator Sets

		Crew	Daily Output	Labor-Hours	Unit	Material	Labor	Equipment	Total	Total Incl O&P
0010	**GAS-ENGINE-DRIVEN GENERATOR SETS**									
0020	Gas or gasoline operated, includes battery,									
0050	charger, muffler & transfer switch									
0200	3 phase 4 wire, 277/480 volt, 7.5 kW	R-3	.83	24.096	Ea.	7,925	955	155	9,035	10,400
0300	11.5 kW		.71	28.169		11,200	1,125	182	12,507	14,400
0400	20 kW		.63	31.746		13,200	1,250	205	14,655	16,900
0500	35 kW		.55	36.364		15,800	1,450	235	17,485	19,900
0600	80 kW		.40	50		25,900	1,975	325	28,200	32,000
0700	100 kW		.33	60.606		28,300	2,400	390	31,090	35,500
0800	125 kW		.28	71.429		58,000	2,825	460	61,285	69,000
0900	185 kW	↓	.25	80	↓	76,500	3,175	515	80,190	90,000

26 35 Power Filters and Conditioners

26 35 13 – Capacitors

26 35 13.10 Capacitors Indoor

		Crew	Daily Output	Labor-Hours	Unit	Material	Labor	Equipment	Total	Total Incl O&P
0010	**CAPACITORS INDOOR**									
0020	240 volts, single & 3 phase, 0.5 kVAR	1 Elec	2.70	2.963	Ea.	395	119		514	630
0100	1.0 kVAR		2.70	2.963		475	119		594	720
0150	2.5 kVAR		2	4		535	161		696	850
0200	5.0 kVAR		1.80	4.444		640	179		819	995
0250	7.5 kVAR		1.60	5		745	201		946	1,150
0300	10 kVAR		1.50	5.333		900	215		1,115	1,325
0350	15 kVAR		1.30	6.154		1,225	248		1,473	1,750
0400	20 kVAR		1.10	7.273		1,550	293		1,843	2,175
0450	25 kVAR		1	8		1,800	320		2,120	2,525
1000	480 volts, single & 3 phase, 1 kVAR		2.70	2.963		360	119		479	590
1050	2 kVAR		2.70	2.963		415	119		534	650
1100	5 kVAR		2	4		520	161		681	835
1150	7.5 kVAR		2	4		560	161		721	880
1200	10 kVAR		2	4		630	161		791	950
1250	15 kVAR		2	4		775	161		936	1,125
1300	20 kVAR		1.60	5		865	201		1,066	1,275
1350	30 kVAR		1.50	5.333		1,075	215		1,290	1,550
1400	40 kVAR		1.20	6.667		1,375	268		1,643	1,950
1450	50 kVAR	↓	1.10	7.273	↓	1,575	293		1,868	2,200

26 51 13.50 Interior Lighting Fixtures		Crew	Daily Output	Labor-Hours	Unit	2011 Bare Costs Material	Labor	Equipment	Total	Total Incl O&P
0010	**INTERIOR LIGHTING FIXTURES** Including lamps, mounting									
0030	hardware and connections									
0100	Fluorescent, C.W. lamps, troffer, recess mounted in grid, RS									
0130	Grid ceiling mount									
0200	Acrylic lens, 1'W x 4'L, two 40 watt	1 Elec	5.70	1.404	Ea.	54	56.50		110.50	151
0210	1'W x 4'L, three 40 watt		5.40	1.481		62.50	59.50		122	166
0300	2'W x 2'L, two U40 watt		5.70	1.404		59.50	56.50		116	157
0400	2'W x 4'L, two 40 watt		5.30	1.509		58	61		119	163
0500	2'W x 4'L, three 40 watt		5	1.600		64	64.50		128.50	175
0600	2'W x 4'L, four 40 watt		4.70	1.702		67	68.50		135.50	185
0700	4'W x 4'L, four 40 watt	2 Elec	6.40	2.500		340	101		441	540
0800	4'W x 4'L, six 40 watt		6.20	2.581		350	104		454	555
0900	4'W x 4'L, eight 40 watt		5.80	2.759		365	111		476	580
0910	Acrylic lens, 1'W x 4'L, two 32 watt T8 G	1 Elec	5.70	1.404		67	56.50		123.50	165
0930	2'W x 2'L, two U32 watt T8 G		5.70	1.404		80	56.50		136.50	180
0940	2'W x 4'L, two 32 watt T8 G		5.30	1.509		72.50	61		133.50	179
0950	2'W x 4'L, three 32 watt T8 G		5	1.600		76.50	64.50		141	188
0960	2'W x 4'L, four 32 watt T8 G		4.70	1.702		78.50	68.50		147	197
1000	Surface mounted, RS									
1030	Acrylic lens with hinged & latched door frame									
1100	1'W x 4'L, two 40 watt	1 Elec	7	1.143	Ea.	74	46		120	156
1110	1'W x 4'L, three 40 watt		6.70	1.194		76.50	48		124.50	162
1200	2'W x 2'L, two U40 watt		7	1.143		79.50	46		125.50	162
1300	2'W x 4'L, two 40 watt		6.20	1.290		91	52		143	184
1400	2'W x 4'L, three 40 watt		5.70	1.404		92	56.50		148.50	193
1500	2'W x 4'L, four 40 watt		5.30	1.509		94.50	61		155.50	203
1600	4'W x 4'L, four 40 watt	2 Elec	7.20	2.222		465	89.50		554.50	655
1700	4'W x 4'L, six 40 watt		6.60	2.424		505	97.50		602.50	715
1800	4'W x 4'L, eight 40 watt		6.20	2.581		525	104		629	750
1900	2'W x 8'L, four 40 watt		6.40	2.500		185	101		286	365
2000	2'W x 8'L, eight 40 watt		6.20	2.581		199	104		303	385
2100	Strip fixture									
2130	Surface mounted									
2200	4' long, one 40 watt, RS	1 Elec	8.50	.941	Ea.	31.50	38		69.50	96
2300	4' long, two 40 watt, RS		8	1		38	40.50		78.50	107
2400	4' long, one 40 watt, SL		8	1		46	40.50		86.50	116
2500	4' long, two 40 watt, SL		7	1.143		62.50	46		108.50	144
2600	8' long, one 75 watt, SL	2 Elec	13.40	1.194		47	48		95	130
2700	8' long, two 75 watt, SL	"	12.40	1.290		57	52		109	147
2800	4' long, two 60 watt, HO	1 Elec	6.70	1.194		91.50	48		139.50	179
2900	8' long, two 110 watt, HO	2 Elec	10.60	1.509		96.50	61		157.50	205
3000	Strip, pendent mounted, industrial, white porcelain enamel									
3100	4' long, two 40 watt, RS	1 Elec	5.70	1.404	Ea.	52	56.50		108.50	149
3200	4' long, two 60 watt, HO	"	5	1.600		81	64.50		145.50	193
3300	8' long, two 75 watt, SL	2 Elec	8.80	1.818		96	73		169	225
3400	8' long, two 110 watt, HO	"	8	2		123	80.50		203.50	266
3470	Troffer, air handling, 2'W x 4'L with four 32 watt T8 G	1 Elec	4	2		110	80.50		190.50	252
3480	2'W x 2'L with two U32 watt T8 G		5.50	1.455		106	58.50		164.50	212
3490	Air connector insulated, 5" diameter		20	.400		66	16.10		82.10	99
3500	6" diameter		20	.400		67.50	16.10		83.60	100
4450	Incandescent, high hat can, round alzak reflector, prewired									
4470	100 watt	1 Elec	8	1	Ea.	70	40.50		110.50	143
4480	150 watt		8	1		102	40.50		142.50	178

26 51 Interior Lighting

26 51 13 – Interior Lighting Fixtures, Lamps, and Ballasts

26 51 13.50 Interior Lighting Fixtures			Crew	Daily Output	Labor-Hours	Unit	Material	2011 Bare Costs Labor	Equipment	Total	Total Incl O&P
4500	300 watt		1 Elec	6.70	1.194	Ea.	236	48		284	340
4600	Square glass lens with metal trim, prewired										
4630	100 watt		1 Elec	6.70	1.194	Ea.	54	48		102	138
4700	200 watt			6.70	1.194		95	48		143	183
4800	300 watt			5.70	1.404		142	56.50		198.50	248
4900	Ceiling/wall, surface mounted, metal cylinder, 75 watt			10	.800		55	32		87	113
4920	150 watt			10	.800		79	32		111	139
5200	Ceiling, surface mounted, opal glass drum										
5300	8", one 60 watt lamp		1 Elec	10	.800	Ea.	43.50	32		75.50	100
5400	10", two 60 watt lamps			8	1		49	40.50		89.50	120
5500	12", four 60 watt lamps			6.70	1.194		69	48		117	154
6010	Vapor tight, incandescent, ceiling mounted, 200 watt			6.20	1.290		81	52		133	173
6100	Fluorescent, surface mounted, 2 lamps, 4'L, RS, 40 watt			3.20	2.500		113	101		214	287
6850	Vandalproof, surface mounted, fluorescent, two 32 watt T8	G		3.20	2.500		248	101		349	435
6860	Incandescent, one 150 watt			8	1		84	40.50		124.50	158
7500	Ballast replacement, by weight of ballast, to 15' high										
7520	Indoor fluorescent, less than 2 lb.		1 Elec	10	.800	Ea.	25.50	32		57.50	80
7540	Two 40W, watt reducer, 2 to 5 lb.			9.40	.851		40.50	34.50		75	100
7560	Two F96 slimline, over 5 lb.			8	1		75.50	40.50		116	149
7580	Vaportite ballast, less than 2 lb.			9.40	.851		25.50	34.50		60	83.50
7600	2 lb. to 5 lb.			8.90	.899		40.50	36		76.50	103
7620	Over 5 lb.			7.60	1.053		75.50	42.50		118	152
7630	Electronic ballast for two tubes			8	1		36.50	40.50		77	106
7640	Dimmable ballast one lamp	G		8	1		104	40.50		144.50	180
7650	Dimmable ballast two-lamp	G		7.60	1.053		102	42.50		144.50	181

26 52 Emergency Lighting

26 52 13 – Emergency Lighting Equipments

26 52 13.10 Emergency Lighting and Battery Units

			Crew	Daily Output	Labor-Hours	Unit	Material	Labor	Equipment	Total	Total Incl O&P
0010	**EMERGENCY LIGHTING AND BATTERY UNITS**										
0300	Emergency light units, battery operated										
0350	Twin sealed beam light, 25 watt, 6 volt each										
0500	Lead battery operated		1 Elec	4	2	Ea.	137	80.50		217.50	281
0700	Nickel cadmium battery operated			4	2		575	80.50		655.50	765
0900	Self-contained fluorescent lamp pack			10	.800		157	32		189	225

26 53 Exit Signs

26 53 13 – Exit Lighting

26 53 13.10 Exit Lighting Fixtures

			Crew	Daily Output	Labor-Hours	Unit	Material	Labor	Equipment	Total	Total Incl O&P
0010	**EXIT LIGHTING FIXTURES**										
0080	Exit light ceiling or wall mount, incandescent, single face		1 Elec	8	1	Ea.	42.50	40.50		83	113
0100	Double face			6.70	1.194		49	48		97	132
0200	L.E.D. standard, single face	G		8	1		74.50	40.50		115	148
0220	Double face	G		6.70	1.194		74.50	48		122.50	160
0230	L.E.D. vandal-resistant, single face	G		7.27	1.100		223	44.50		267.50	315
0240	L.E.D. w/battery unit, single face	G		4.40	1.818		120	73		193	251
0260	Double face	G		4	2		137	80.50		217.50	282
0262	L.E.D. w/battery unit, vandal-resistant, single face	G		4.40	1.818		370	73		443	530
0270	Combination emergency light units and exit sign			4	2		184	80.50		264.50	335

26 54 Classified Location Lighting

26 54 13 – Classified Lighting

26 54 13.20 Explosionproof		Crew	Daily Output	Labor-Hours	Unit	Material	2011 Bare Costs Labor	Equipment	Total	Total Incl O&P
0010	**EXPLOSIONPROOF**, incl lamps, mounting hardware and connections									
6510	Incandescent, ceiling mounted, 200 watt	1 Elec	4	2	Ea.	960	80.50		1,040.50	1,175
6600	Fluorescent, RS, 4′ long, ceiling mounted, two 40 watt	″	2.70	2.963	″	2,550	119		2,669	3,000

26 55 Special Purpose Lighting

26 55 61 – Theatrical Lighting

26 55 61.10 Lights		Crew	Daily Output	Labor-Hours	Unit	Material	Labor	Equipment	Total	Total Incl O&P
0010	**LIGHTS**									
2000	Lights, border, quartz, reflector, vented,									
2100	colored or white	1 Elec	20	.400	L.F.	173	16.10		189.10	216
2500	Spotlight, follow spot, with transformer, 2,100 watt	″	4	2	Ea.	3,075	80.50		3,155.50	3,500
2600	For no transformer, deduct					910			910	1,000
3000	Stationary spot, fresnel quartz, 6″ lens	1 Elec	4	2		131	80.50		211.50	275
3100	8″ lens		4	2		231	80.50		311.50	385
3500	Ellipsoidal quartz, 1,000W, 6″ lens		4	2		335	80.50		415.50	500
3600	12″ lens		4	2		600	80.50		680.50	790
4000	Strobe light, 1 to 15 flashes per second, quartz		3	2.667		750	107		857	1,000
4500	Color wheel, portable, five hole, motorized		4	2		194	80.50		274.50	345

26 56 Exterior Lighting

26 56 13 – Lighting Poles and Standards

26 56 13.10 Lighting Poles		Crew	Daily Output	Labor-Hours	Unit	Material	Labor	Equipment	Total	Total Incl O&P
0010	**LIGHTING POLES**									
2800	Light poles, anchor base									
2820	not including concrete bases									
2840	Aluminum pole, 8′ high	1 Elec	4	2	Ea.	700	80.50		780.50	900
3000	20′ high	R-3	2.90	6.897		930	273	44.50	1,247.50	1,525
3200	30′ high		2.60	7.692		1,675	305	49.50	2,029.50	2,400
3400	35′ high		2.30	8.696		1,900	345	56	2,301	2,725
3600	40′ high		2	10		2,175	395	64.50	2,634.50	3,100
3800	Bracket arms, 1 arm	1 Elec	8	1		125	40.50		165.50	203
4000	2 arms		8	1		247	40.50		287.50	335
4200	3 arms		5.30	1.509		380	61		441	520
4400	4 arms		5.30	1.509		510	61		571	660
4500	Steel pole, galvanized, 8′ high		3.80	2.105		625	84.50		709.50	820
4600	20′ high	R-3	2.60	7.692		1,100	305	49.50	1,454.50	1,775
4800	30′ high		2.30	8.696		1,200	345	56	1,601	1,925
5000	35′ high		2.20	9.091		1,400	360	58.50	1,818.50	2,200
5200	40′ high		1.70	11.765		1,750	465	76	2,291	2,775
5400	Bracket arms, 1 arm	1 Elec	8	1		176	40.50		216.50	260
5600	2 arms		8	1		273	40.50		313.50	365
5800	3 arms		5.30	1.509		295	61		356	425
6000	4 arms		5.30	1.509		410	61		471	555

26 56 19 – Roadway Lighting

26 56 19.20 Roadway Area Luminaire		Crew	Daily Output	Labor-Hours	Unit	Material	Labor	Equipment	Total	Total Incl O&P
0010	**ROADWAY AREA LUMINAIRE**									
2650	Roadway area luminaire, low pressure sodium, 135 watt	1 Elec	2	4	Ea.	645	161		806	970
2700	180 watt	″	2	4		675	161		836	1,000
2750	Metal halide, 400 watt	2 Elec	4.40	3.636		530	146		676	820

26 56 Exterior Lighting

26 56 19 – Roadway Lighting

26 56 19.20 Roadway Area Luminaire		Crew	Daily Output	Labor-Hours	Unit	2011 Bare Costs Material	Labor	Equipment	Total	Total Incl O&P
2760	1000 watt	2 Elec	4	4	Ea.	600	161		761	920
2780	High pressure sodium, 400 watt		4.40	3.636		550	146		696	840
2790	1000 watt		4	4		630	161		791	950

26 56 23 – Area Lighting

26 56 23.10 Exterior Fixtures

Line	Description	Crew	Daily Output	Labor-Hours	Unit	Material	Labor	Equipment	Total	Total Incl O&P
0010	**EXTERIOR FIXTURES** With lamps									
0200	Wall mounted, incandescent, 100 watt	1 Elec	8	1	Ea.	33.50	40.50		74	103
0400	Quartz, 500 watt		5.30	1.509		56	61		117	161
1100	Wall pack, low pressure sodium, 35 watt		4	2		234	80.50		314.50	390
1150	55 watt		4	2		278	80.50		358.50	435

26 56 36 – Flood Lighting

26 56 36.20 Floodlights

Line	Description	Crew	Daily Output	Labor-Hours	Unit	Material	Labor	Equipment	Total	Total Incl O&P
0010	**FLOODLIGHTS** with ballast and lamp,									
1400	pole mounted, pole not included									
1950	Metal halide, 175 watt	1 Elec	2.70	2.963	Ea.	325	119		444	555
2000	400 watt	2 Elec	4.40	3.636		405	146		551	680
2200	1000 watt	"	4	4		555	161		716	870
2340	High pressure sodium, 70 watt	1 Elec	2.70	2.963		245	119		364	465
2400	400 watt	2 Elec	4.40	3.636		365	146		511	635
2600	1000 watt	"	4	4		625	161		786	950

26 61 Lighting Systems and Accessories

26 61 23 – Lamps Applications

26 61 23.10 Lamps

Line	Description	Crew	Daily Output	Labor-Hours	Unit	Material	Labor	Equipment	Total	Total Incl O&P
0010	**LAMPS**									
0080	Fluorescent, rapid start, cool white, 2' long, 20 watt	1 Elec	1	8	C	355	320		675	910
0100	4' long, 40 watt		.90	8.889		300	360		660	910
0200	Slimline, 4' long, 40 watt		.90	8.889		895	360		1,255	1,550
0210	4' long, 30 watt energy saver G		.90	8.889		895	360		1,255	1,550
0400	High output, 4' long, 60 watt		.90	8.889		1,025	360		1,385	1,700
0410	8' long, 95 watt energy saver G		.80	10		1,025	405		1,430	1,800
0500	8' long, 110 watt		.80	10		1,025	405		1,430	1,800
0512	2' long, T5, 14 watt energy saver G		1	8		1,400	320		1,720	2,075
0514	3' long, T5, 21 watt energy saver G		.90	8.889		1,400	360		1,760	2,125
0516	4' long, T5, 28 watt energy saver G		.90	8.889		1,225	360		1,585	1,900
0517	4' long, T5, 54 watt energy saver G		.90	8.889		1,775	360		2,135	2,525
0560	Twin tube compact lamp G		.90	8.889		585	360		945	1,225
0570	Double twin tube compact lamp G		.80	10		1,300	405		1,705	2,100
0600	Mercury vapor, mogul base, deluxe white, 100 watt		.30	26.667		3,100	1,075		4,175	5,175
0700	250 watt		.30	26.667		4,800	1,075		5,875	7,050
0800	400 watt		.30	26.667		3,875	1,075		4,950	6,000
0900	1000 watt		.20	40		9,025	1,600		10,625	12,500
1000	Metal halide, mogul base, 175 watt		.30	26.667		4,125	1,075		5,200	6,300
1200	400 watt		.30	26.667		4,425	1,075		5,500	6,625
1300	1000 watt		.20	40		10,700	1,600		12,300	14,400
1350	High pressure sodium, 70 watt		.30	26.667		4,800	1,075		5,875	7,025
1380	250 watt		.30	26.667		5,475	1,075		6,550	7,775
1400	400 watt		.30	26.667		5,625	1,075		6,700	7,925

26 61 Lighting Systems and Accessories

26 61 23 – Lamps Applications

26 61 23.10 Lamps		Crew	Daily Output	Labor-Hours	Unit	2011 Bare Costs Material	Labor	Equipment	Total	Total Incl O&P
1450	1000 watt	1 Elec	.20	40	C	15,500	1,600		17,100	19,700
3000	Guards, fluorescent lamp, 4′ long		1	8		1,175	320		1,495	1,800
3200	8′ long	↓	.90	8.889	↓	2,325	360		2,685	3,150

Estimating Tips

27 20 00 Data Communications
27 30 00 Voice Communications
27 40 00 Audio-Video Communications

When estimating material costs for special systems, it is always prudent to obtain manufacturers' quotations for equipment prices and special installation requirements which will affect the total costs.

Reference Numbers

Reference numbers are shown in shaded boxes at the beginning of some major classifications. These numbers refer to related items in the Reference Section. The reference information may be an estimating procedure, an alternate pricing method, or technical information.

Note: Not all subdivisions listed here necessarily appear in this publication.

Note: **Trade Service,** *in part, has been used as a reference source for some of the material prices used in Division 27.*

27 13 Communications Backbone Cabling

27 13 23 – Communications Optical Fiber Backbone Cabling

	27 13 23.13 Communications Optical Fiber	Crew	Daily Output	Labor-Hours	Unit	Material	2011 Bare Costs Labor	Equipment	Total	Total Incl O&P
0010	**COMMUNICATIONS OPTICAL FIBER**									
0040	Specialized tools & techniques cause installation costs to vary.									
0070	Fiber optic, cable, bulk simplex, minimum	1 Elec	8	1	C.L.F.	23	40.50		63.50	91
0080	Bulk plenum quad, maximum	″	2.29	3.493	″	68	141		209	305
0150	Jumper				Ea.	34.50			34.50	38
0200	Pigtail					32			32	35
0300	Connector	1 Elec	24	.333		20	13.40		33.40	44
0350	Finger splice		32	.250		32	10.05		42.05	51.50
0400	Transceiver (low cost bi-directional)		8	1		410	40.50		450.50	520
0450	Rack housing, 4 rack spaces, 12 panels (144 fibers)		2	4		375	161		536	675
0500	Patch panel, 12 ports		6	1.333		242	53.50		295.50	355

27 41 Audio-Video Systems

27 41 19 – Portable Audio-Video Equipment

	27 41 19.10 T.V. Systems	Crew	Daily Output	Labor-Hours	Unit	Material	Labor	Equipment	Total	Total Incl O&P
0010	**T.V. SYSTEMS**, not including rough-in wires, cables & conduits									
0100	Master TV antenna system									
0200	VHF reception & distribution, 12 outlets	1 Elec	6	1.333	Outlet	205	53.50		258.50	310
0400	30 outlets		10	.800		135	32		167	200
0600	100 outlets		13	.615		140	25		165	194
0800	VHF & UHF reception & distribution, 12 outlets		6	1.333		203	53.50		256.50	310
1000	30 outlets		10	.800		135	32		167	200
1200	100 outlets		13	.615		137	25		162	191
1400	School and deluxe systems, 12 outlets		2.40	3.333		268	134		402	515
1600	30 outlets		4	2		235	80.50		315.50	390
1800	80 outlets		5.30	1.509		226	61		287	345

27 51 Distributed Audio-Video Communications Systems

27 51 16 – Public Address and Mass Notification Systems

	27 51 16.10 Public Address System	Crew	Daily Output	Labor-Hours	Unit	Material	Labor	Equipment	Total	Total Incl O&P
0010	**PUBLIC ADDRESS SYSTEM**									
0100	Conventional, office	1 Elec	5.33	1.501	Speaker	129	60.50		189.50	239
0200	Industrial	″	2.70	2.963	″	248	119		367	465

27 51 19 – Sound Masking Systems

	27 51 19.10 Sound System	Crew	Daily Output	Labor-Hours	Unit	Material	Labor	Equipment	Total	Total Incl O&P
0010	**SOUND SYSTEM**, not including rough-in wires, cables & conduits									
0100	Components, projector outlet	1 Elec	8	1	Ea.	55	40.50		95.50	126
0200	Microphone		4	2		73	80.50		153.50	212
0400	Speakers, ceiling or wall		8	1		112	40.50		152.50	189
0600	Trumpets		4	2		209	80.50		289.50	360
0800	Privacy switch		8	1		83	40.50		123.50	157
1000	Monitor panel		4	2		370	80.50		450.50	540
1200	Antenna, AM/FM		4	2		132	80.50		212.50	276
1400	Volume control		8	1		80	40.50		120.50	154
1600	Amplifier, 250 watts		1	8		1,175	320		1,495	1,800
1800	Cabinets		1	8		810	320		1,130	1,400
2000	Intercom, 30 station capacity, master station	2 Elec	2	8		2,300	320		2,620	3,050
2200	Remote station	1 Elec	8	1		156	40.50		196.50	237
2400	Intercom outlets		8	1		91.50	40.50		132	167

27 51 Distributed Audio-Video Communications Systems

27 51 19 – Sound Masking Systems

27 51 19.10 Sound System		Crew	Daily Output	Labor-Hours	Unit	2011 Bare Costs Material	Labor	Equipment	Total	Total Incl O&P
2600	Handset	1 Elec	4	2	Ea.	305	80.50		385.50	465
2800	Emergency call system, 12 zones, annunciator		1.30	6.154		915	248		1,163	1,400
3000	Bell		5.30	1.509		94.50	61		155.50	203
3200	Light or relay		8	1		47	40.50		87.50	118
3400	Transformer		4	2		207	80.50		287.50	360
3600	House telephone, talking station		1.60	5		445	201		646	815
3800	Press to talk, release to listen		5.30	1.509		104	61		165	213
4000	System-on button					62			62	68
4200	Door release	1 Elec	4	2		111	80.50		191.50	253
4400	Combination speaker and microphone		8	1		189	40.50		229.50	274
4600	Termination box		3.20	2.500		59.50	101		160.50	229
4800	Amplifier or power supply		5.30	1.509		680	61		741	850
5000	Vestibule door unit		16	.500	Name	125	20		145	171
5200	Strip cabinet		27	.296	Ea.	237	11.95		248.95	280
5400	Directory		16	.500	"	112	20		132	156

27 52 Healthcare Communications and Monitoring Systems

27 52 23 – Nurse Call/Code Blue Systems

27 52 23.10 Nurse Call Systems		Crew	Daily Output	Labor-Hours	Unit	Material	Labor	Equipment	Total	Total Incl O&P
0010	**NURSE CALL SYSTEMS**									
0100	Single bedside call station	1 Elec	8	1	Ea.	227	40.50		267.50	315
0200	Ceiling speaker station		8	1		65.50	40.50		106	138
0400	Emergency call station		8	1		114	40.50		154.50	191
0600	Pillow speaker		8	1		172	40.50		212.50	256
0800	Double bedside call station		4	2		204	80.50		284.50	355
1000	Duty station		4	2		138	80.50		218.50	283
1200	Standard call button		8	1		106	40.50		146.50	183
1400	Lights, corridor, dome or zone indicator		8	1		61	40.50		101.50	133
1600	Master control station for 20 stations	2 Elec	.65	24.615	Total	3,675	990		4,665	5,650

27 53 Distributed Systems

27 53 13 – Clock Systems

27 53 13.50 Clock Equipments		Crew	Daily Output	Labor-Hours	Unit	Material	Labor	Equipment	Total	Total Incl O&P
0010	**CLOCK EQUIPMENTS**, not including wires & conduits									
0100	Time system components, master controller	1 Elec	.33	24.242	Ea.	1,825	975		2,800	3,575
0200	Program bell		8	1		85	40.50		125.50	159
0400	Combination clock & speaker		3.20	2.500		210	101		311	395
0600	Frequency generator		2	4		2,350	161		2,511	2,850
0800	Job time automatic stamp recorder		4	2		460	80.50		540.50	635
1600	Master time clock system, clocks & bells, 20 room	4 Elec	.20	160		6,300	6,450		12,750	17,300
1800	50 room	"	.08	400		13,000	16,100		29,100	40,400
2000	Time clock, 100 cards in & out, 1 color	1 Elec	3.20	2.500		1,325	101		1,426	1,625
2200	2 colors		3.20	2.500		1,250	101		1,351	1,575
2800	Metal rack for 25 cards		7	1.143		38.50	46		84.50	117

Division Notes

		CREW	DAILY OUTPUT	LABOR-HOURS	UNIT	2011 BARE COSTS MAT.	LABOR	EQUIP.	TOTAL	TOTAL INCL O&P

Estimating Tips

- When estimating material costs for electronic safety and security systems, it is always prudent to obtain manufacturers' quotations for equipment prices and special installation requirements that affect the total cost.
- Fire alarm systems consist of control panels, annunciator panels, battery with rack, charger, and fire alarm actuating and indicating devices. Some fire alarm systems include speakers, telephone lines, door closer controls, and other components. Be careful not to overlook the costs related to installation for these items. Also be aware of costs for integrated automation instrumentation and terminal devices, control equipment, control wiring, and programming.
- Security equipment includes items such as CCTV, access control, and other detection and identification systems to perform alert and alarm functions. Be sure to consider the costs related to installation for this security equipment, such as for integrated automation instrumentation and terminal devices, control equipment, control wiring, and programming.

Reference Numbers

Reference numbers are shown in shaded boxes at the beginning of some major classifications. These numbers refer to related items in the Reference Section. The reference information may be an estimating procedure, an alternate pricing method, or technical information.

Note: Not all subdivisions listed here necessarily appear in this publication.

Division 28 - Electronic Safety and Security

28 13 Access Control

28 13 53 – Security Access Detection

28 13 53.13 Security Access Metal Detectors		Crew	Daily Output	Labor-Hours	Unit	Material	2011 Bare Costs Labor	Equipment	Total	Total Incl O&P
0010	**SECURITY ACCESS METAL DETECTORS**									
0240	Metal detector, hand-held, wand type, unit only				Ea.	89			89	98
0250	Metal detector, walk through portal type, single zone	1 Elec	2	4		2,750	161		2,911	3,275
0260	Multi-zone	″	2	4		3,500	161		3,661	4,100

28 13 53.16 Security Access X-Ray Equipment		Crew	Daily Output	Labor-Hours	Unit	Material	Labor	Equipment	Total	Total Incl O&P
0010	**SECURITY ACCESS X-RAY EQUIPMENT**									
0290	X-ray machine, desk top, for mail/small packages/letters	1 Elec	4	2	Ea.	3,000	80.50		3,080.50	3,425
0300	Conveyor type, incl monitor, minimum		2	4		14,000	161		14,161	15,700
0310	Maximum		2	4		25,000	161		25,161	27,800
0320	X-ray machine, large unit, for airports, incl monitor, min	2 Elec	1	16		35,000	645		35,645	39,600
0330	Maximum	″	.50	32		60,000	1,300		61,300	68,000

28 13 53.23 Security Access Explosive Detection Equipment		Crew	Daily Output	Labor-Hours	Unit	Material	Labor	Equipment	Total	Total Incl O&P
0010	**SECURITY ACCESS EXPLOSIVE DETECTION EQUIPMENT**									
0270	Explosives detector, walk through portal type	1 Elec	2	4	Ea.	3,500	161		3,661	4,100
0280	Hand-held, battery operated				″				25,500	28,100

28 16 Intrusion Detection

28 16 16 – Intrusion Detection Systems Infrastructure

28 16 16.50 Intrusion Detection		Crew	Daily Output	Labor-Hours	Unit	Material	Labor	Equipment	Total	Total Incl O&P
0010	**INTRUSION DETECTION**, not including wires & conduits									
0100	Burglar alarm, battery operated, mechanical trigger	1 Elec	4	2	Ea.	272	80.50		352.50	430
0200	Electrical trigger		4	2		325	80.50		405.50	485
0400	For outside key control, add		8	1		82	40.50		122.50	156
0600	For remote signaling circuitry, add		8	1		122	40.50		162.50	201
0800	Card reader, flush type, standard		2.70	2.963		910	119		1,029	1,200
1000	Multi-code		2.70	2.963		1,175	119		1,294	1,500
1200	Door switches, hinge switch		5.30	1.509		57.50	61		118.50	162
1400	Magnetic switch		5.30	1.509		67.50	61		128.50	173
1600	Exit control locks, horn alarm		4	2		266	80.50		346.50	425
1800	Flashing light alarm		4	2		297	80.50		377.50	455
2000	Indicating panels, 1 channel		2.70	2.963		360	119		479	590
2200	10 channel	2 Elec	3.20	5		1,225	201		1,426	1,675
2400	20 channel		2	8		2,400	320		2,720	3,175
2600	40 channel		1.14	14.035		4,375	565		4,940	5,725
2800	Ultrasonic motion detector, 12 volt	1 Elec	2.30	3.478		225	140		365	475
3000	Infrared photoelectric detector	″	4	2		186	80.50		266.50	335

28 23 Video Surveillance

28 23 13 – Video Surveillance Control and Management Systems

28 23 13.10 Closed Circuit Television System		Crew	Daily Output	Labor-Hours	Unit	Material	Labor	Equipment	Total	Total Incl O&P
0010	**CLOSED CIRCUIT TELEVISION SYSTEM**									
2000	Surveillance, one station (camera & monitor)	2 Elec	2.60	6.154	Total	1,300	248		1,548	1,825
2200	For additional camera stations, add	1 Elec	2.70	2.963	Ea.	725	119		844	995
2400	Industrial quality, one station (camera & monitor)	2 Elec	2.60	6.154	Total	2,700	248		2,948	3,350
2600	For additional camera stations, add	1 Elec	2.70	2.963	Ea.	1,650	119		1,769	2,025
2610	For low light, add		2.70	2.963		1,325	119		1,444	1,650
2620	For very low light, add		2.70	2.963		9,750	119		9,869	10,900
2800	For weatherproof camera station, add		1.30	6.154		1,025	248		1,273	1,525
3000	For pan and tilt, add		1.30	6.154		2,625	248		2,873	3,300

28 23 Video Surveillance

28 23 13 – Video Surveillance Control and Management Systems

28 23 13.10 Closed Circuit Television System		Crew	Daily Output	Labor-Hours	Unit	2011 Bare Costs Material	Labor	Equipment	Total	Total Incl O&P
3200	For zoom lens - remote control, add, minimum	1 Elec	2	4	Ea.	2,425	161		2,586	2,925
3400	Maximum		2	4		8,850	161		9,011	9,975
3410	For automatic iris for low light, add	↓	2	4	↓	2,125	161		2,286	2,575
3600	Educational T.V. studio, basic 3 camera system, black & white,									
3800	electrical & electronic equip. only, minimum	4 Elec	.80	40	Total	12,600	1,600		14,200	16,400
4000	Maximum (full console)		.28	114		53,500	4,600		58,100	66,500
4100	As above, but color system, minimum		.28	114		71,000	4,600		75,600	85,500
4120	Maximum	↓	.12	266	↓	308,500	10,700		319,200	356,500
4200	For film chain, black & white, add	1 Elec	1	8	Ea.	14,400	320		14,720	16,300
4250	Color, add		.25	32		17,500	1,300		18,800	21,400
4400	For video recorders, add, minimum	↓	1	8		3,025	320		3,345	3,850
4600	Maximum	4 Elec	.40	80	↓	25,200	3,225		28,425	32,900

28 23 23 – Video Surveillance Systems Infrastructure

28 23 23.50 Video Surveillance Equipments		Crew	Daily Output	Labor-Hours	Unit	Material	Labor	Equipment	Total	Total Incl O&P
0010	**VIDEO SURVEILLANCE EQUIPMENTS**									
0200	Video cameras, wireless, hidden in exit signs, clocks, etc, incl receiver	1 Elec	3	2.667	Ea.	143	107		250	330
0210	Accessories for video recorder, single camera		3	2.667		260	107		367	460
0220	For multiple cameras		3	2.667		1,750	107		1,857	2,100
0230	Video cameras, wireless, for under vehicle searching, complete	↓	2	4	↓	9,975	161		10,136	11,300

28 31 Fire Detection and Alarm

28 31 23 – Fire Detection and Alarm Annunciation Panels and Fire Stations

28 31 23.50 Alarm Panels and Devices		Crew	Daily Output	Labor-Hours	Unit	Material	Labor	Equipment	Total	Total Incl O&P
0010	**ALARM PANELS AND DEVICES**, not including wires & conduits									
3594	Fire, alarm control panel									
3600	4 zone	2 Elec	2	8	Ea.	390	320		710	945
3800	8 zone		1	16		760	645		1,405	1,875
4000	12 zone	↓	.67	23.988		2,300	965		3,265	4,100
4020	Alarm device	1 Elec	8	1		229	40.50		269.50	320
4050	Actuating device		8	1		325	40.50		365.50	420
4200	Battery and rack		4	2		395	80.50		475.50	565
4400	Automatic charger		8	1		560	40.50		600.50	685
4600	Signal bell		8	1		70.50	40.50		111	143
4800	Trouble buzzer or manual station		8	1		80.50	40.50		121	154
5600	Strobe and horn		5.30	1.509		147	61		208	260
5800	Fire alarm horn		6.70	1.194		58.50	48		106.50	143
6000	Door holder, electro-magnetic		4	2		99	80.50		179.50	240
6200	Combination holder and closer		3.20	2.500		119	101		220	294
6600	Drill switch		8	1		360	40.50		400.50	460
6800	Master box		2.70	2.963		6,175	119		6,294	7,000
7000	Break glass station		8	1		65	40.50		105.50	137
7800	Remote annunciator, 8 zone lamp	↓	1.80	4.444		201	179		380	510
8000	12 zone lamp	2 Elec	2.60	6.154		325	248		573	755
8200	16 zone lamp	"	2.20	7.273	↓	405	293		698	920

28 31 43 – Fire Detection Sensors

28 31 43.50 Fire and Heat Detectors		Crew	Daily Output	Labor-Hours	Unit	Material	Labor	Equipment	Total	Total Incl O&P
0010	**FIRE & HEAT DETECTORS**									
5000	Detector, rate of rise	1 Elec	8	1	Ea.	49	40.50		89.50	120

28 31 Fire Detection and Alarm

28 31 46 – Smoke Detection Sensors

28 31 46.50 Smoke Detectors		Crew	Daily Output	Labor-Hours	Unit	2011 Bare Costs Material	Labor	Equipment	Total	Total Incl O&P
0010	**SMOKE DETECTORS**									
5200	Smoke detector, ceiling type	1 Elec	6.20	1.290	Ea.	126	52		178	223
5400	Duct type	"	3.20	2.500	"	315	101		416	510

28 33 Fuel-Gas Detection and Alarm

28 33 33 – Fuel-Gas Detection Sensors

28 33 33.50 Tank Leak Detection Systems		Crew	Daily Output	Labor-Hours	Unit	Material	Labor	Equipment	Total	Total Incl O&P
0010	**TANK LEAK DETECTION SYSTEMS** Liquid and vapor									
0100	For hydrocarbons and hazardous liquids/vapors									
0120	Controller, data acquisition, incl. printer, modem, RS232 port									
0140	24 channel, for use with all probes				Ea.	3,500			3,500	3,850
0160	9 channel, for external monitoring				"	770			770	845
0170	Integrated control panel with monitoring system and printer									
0180	1 to 2 tanks, double wall	2 Elec	4	4	Ea.	1,750	161		1,911	2,175
0181	1 to 8 tanks, single wall		4	4		1,750	161		1,911	2,175
0190	2 to 8 tanks, double wall		2.50	6.400		4,925	258		5,183	5,850
0200	Probes									
0210	Well monitoring									
0220	Liquid phase detection				Ea.	400			400	440
0230	Hydrocarbon vapor, fixed position					400			400	440
0240	Hydrocarbon vapor, float mounted					400			400	440
0250	Both liquid and vapor hydrocarbon					400			400	440
0300	Secondary containment, liquid phase									
0310	Pipe trench/manway sump				Ea.	690			690	760
0320	Double wall pipe and manual sump					690			690	760
0330	Double wall fiberglass annular space					272			272	299
0340	Double wall steel tank annular space					282			282	310
0500	Accessories									
0510	Modem, non-dedicated phone line				Ea.	255			255	280
0600	Monitoring, internal									
0610	Automatic tank gauge, incl. overfill				Ea.	965			965	1,075
0620	Product line				"	965			965	1,075
0700	Monitoring, special									
0710	Cathodic protection				Ea.	605			605	670
0720	Annular space chemical monitor				"	825			825	910

28 39 Mass Notification Systems

28 39 10 – Notification Systems

28 39 10.10 Mass Notification System		Crew	Daily Output	Labor-Hours	Unit	Material	Labor	Equipment	Total	Total Incl O&P
0010	**MASS NOTIFICATION SYSTEM**									
0100	Wireless command center, 10,000 devices	2 Elec	1.33	12.030	Ea.	3,000	485		3,485	4,075
0200	Option, email notification					1,300			1,300	1,425
0210	Remote device supervision & monitor					1,800			1,800	1,975
0300	Antenna VHF or UHF, for medium range	1 Elec	4	2		95	80.50		175.50	236
0310	For high-power transmitter		2	4		495	161		656	805
0400	Transmitter, 25 watt		4	2		1,500	80.50		1,580.50	1,775
0410	40 watt		2.66	3.008		2,000	121		2,121	2,400

28 39 Mass Notification Systems

28 39 10 – Notification Systems

28 39 10.10 Mass Notification System		Crew	Daily Output	Labor-Hours	Unit	Material	Labor	Equipment	Total	Total Incl O&P
						2011 Bare Costs				
0420	100 watt	1 Elec	1.33	6.015	Ea.	5,000	242		5,242	5,900
0500	Wireless receiver/control module for speaker		8	1		195	40.50		235.50	281
0600	Desktop paging controller, stand alone	↓	4	2	↓	245	80.50		325.50	400

Division Notes

		CREW	DAILY OUTPUT	LABOR-HOURS	UNIT	2011 BARE COSTS MAT.	LABOR	EQUIP.	TOTAL	TOTAL INCL O&P

Estimating Tips

31 05 00 Common Work Results for Earthwork

- Estimating the actual cost of performing earthwork requires careful consideration of the variables involved. This includes items such as type of soil, whether water will be encountered, dewatering, whether banks need bracing, disposal of excavated earth, and length of haul to fill or spoil sites, etc. If the project has large quantities of cut or fill, consider raising or lowering the site to reduce costs, while paying close attention to the effect on site drainage and utilities.
- If the project has large quantities of fill, creating a borrow pit on the site can significantly lower the costs.
- It is very important to consider what time of year the project is scheduled for completion. Bad weather can create large cost overruns from dewatering, site repair, and lost productivity from cold weather.

Reference Numbers

Reference numbers are shown in shaded boxes at the beginning of some major classifications. These numbers refer to related items in the Reference Section. The reference information may be an estimating procedure, an alternate pricing method, or technical information.

Note: Not all subdivisions listed here necessarily appear in this publication.

31 05 Common Work Results for Earthwork

31 05 13 – Soils for Earthwork

31 05 13.10 Borrow		Crew	Daily Output	Labor-Hours	Unit	Material	2011 Bare Costs Labor	Equipment	Total	Total Incl O&P
0010	**BORROW** R312316-40									
0020	Spread, 200 H.P. dozer, no compaction, 2 mi. RT haul									
0200	Common borrow	B-15	600	.047	C.Y.	15	1.37	3.44	19.81	22.50
0700	Screened loam		600	.047		33	1.37	3.44	37.81	42.50
0800	Topsoil, weed free		600	.047		24	1.37	3.44	28.81	32.50
0900	For 5 mile haul, add	B-34B	200	.040			1.09	2.53	3.62	4.58

31 05 16 – Aggregates for Earthwork

31 05 16.10 Borrow		Crew	Daily Output	Labor-Hours	Unit	Material	Labor	Equipment	Total	Total Incl O&P
0010	**BORROW** R312316-40									
0020	Spread, with 200 H.P. dozer, no compaction, 2 mi. RT haul									
0100	Bank run gravel	B-15	600	.047	C.Y.	22	1.37	3.44	26.81	30
0300	Crushed stone (1.40 tons per CY), 1-1/2"		600	.047		38	1.37	3.44	42.81	48
0320	3/4"		600	.047		38	1.37	3.44	42.81	48
0340	1/2"		600	.047		47	1.37	3.44	51.81	58
0360	3/8"		600	.047		32	1.37	3.44	36.81	41.50
0400	Sand, washed, concrete		600	.047		41	1.37	3.44	45.81	51
0500	Dead or bank sand		600	.047		15	1.37	3.44	19.81	22.50
0600	Select structural fill		600	.047		20	1.37	3.44	24.81	28
0900	For 5 mile haul, add	B-34B	200	.040			1.09	2.53	3.62	4.58

31 05 23 – Cement and Concrete for Earthwork

31 05 23.30 Plant Mixed Bituminous Concrete		Crew	Daily Output	Labor-Hours	Unit	Material	Labor	Equipment	Total	Total Incl O&P
0010	**PLANT MIXED BITUMINOUS CONCRETE**									
0020	Asphaltic concrete plant mix (145 lb. per C.F.)				Ton	64			64	70.50
0040	Asphaltic concrete less than 300 tons add trucking costs									
0050	See Section 31 23 23.20 for hauling costs									
0200	All weather patching mix, hot				Ton	65			65	71.50
0250	Cold patch					65			65	71.50
0300	Berm mix					65			65	71.50

31 06 Schedules for Earthwork

31 06 60 – Schedules for Special Foundations and Load Bearing Elements

31 06 60.14 Piling Special Costs		Crew	Daily Output	Labor-Hours	Unit	Material	Labor	Equipment	Total	Total Incl O&P
0010	**PILING SPECIAL COSTS**									
0011	Piling special costs, pile caps, see Section 03 30 53.40									
0500	Cutoffs, concrete piles, plain	1 Pile	5.50	1.455	Ea.		47.50		47.50	83
0600	With steel thin shell, add		38	.211			6.90		6.90	12
0700	Steel pile or "H" piles		19	.421			13.80		13.80	24
0800	Wood piles		38	.211			6.90		6.90	12
0900	Pre-augering up to 30' deep, average soil, 24" diameter	B-43	180	.222	L.F.		5.95	13.80	19.75	25
0920	36" diameter		115	.348			9.35	21.50	30.85	39.50
0960	48" diameter		70	.571			15.35	35.50	50.85	64.50
0980	60" diameter		50	.800			21.50	49.50	71	90.50
1000	Testing, any type piles, test load is twice the design load									
1050	50 ton design load, 100 ton test				Ea.				14,000	15,500
1100	100 ton design load, 200 ton test								20,000	22,000
1150	150 ton design load, 300 ton test								26,000	28,500
1200	200 ton design load, 400 ton test								28,000	31,000
1250	400 ton design load, 800 ton test								32,000	35,000
1500	Wet conditions, soft damp ground									
1600	Requiring mats for crane, add								40%	40%

31 06 Schedules for Earthwork

31 06 60 – Schedules for Special Foundations and Load Bearing Elements

31 06 60.14 Piling Special Costs		Crew	Daily Output	Labor-Hours	Unit	Material	2011 Bare Costs Labor	Equipment	Total	Total Incl O&P
1700	Barge mounted driving rig, add								30%	30%

31 06 60.15 Mobilization

		Crew	Daily Output	Labor-Hours	Unit	Material	Labor	Equipment	Total	Total Incl O&P
0010	**MOBILIZATION**									
0020	Set up & remove, air compressor, 600 C.F.M.	A-5	3.30	5.455	Ea.		144	13	157	255
0100	1200 C.F.M.	″	2.20	8.182			216	19.45	235.45	380
0200	Crane, with pile leads and pile hammer, 75 ton	B-19	.60	93.333			3,050	3,075	6,125	8,600
0300	150 ton	″	.36	155			5,075	5,125	10,200	14,400
0500	Drill rig, for caissons, to 36″, minimum	B-43	2	20			535	1,250	1,785	2,275
0600	Up to 84″	″	1	40			1,075	2,475	3,550	4,525
0800	Auxiliary boiler, for steam small	A-5	1.66	10.843			287	26	313	510
0900	Large	″	.83	21.687			575	51.50	626.50	1,025
1100	Rule of thumb: complete pile driving set up, small	B-19	.45	124			4,050	4,100	8,150	11,500
1200	Large	″	.27	207			6,750	6,850	13,600	19,100
1500	Mobilization, barge, by tug boat	B-83	25	.640	Mile		19.65	25.50	45.15	60.50

31 11 Clearing and Grubbing

31 11 10 – Clearing and Grubbing Land

31 11 10.10 Clear and Grub Site

		Crew	Daily Output	Labor-Hours	Unit	Material	Labor	Equipment	Total	Total Incl O&P
0010	**CLEAR AND GRUB SITE**									
0020	Cut & chip light trees to 6″ diam.	B-7	1	48	Acre		1,350	1,300	2,650	3,700
0150	Grub stumps and remove	B-30	2	12			355	980	1,335	1,675
0200	Cut & chip medium, trees to 12″ diam.	B-7	.70	68.571			1,925	1,875	3,800	5,275
0250	Grub stumps and remove	B-30	1	24			715	1,950	2,665	3,325
0300	Cut & chip heavy, trees to 24″ diam.	B-7	.30	160			4,500	4,350	8,850	12,300
0350	Grub stumps and remove	B-30	.50	48			1,425	3,925	5,350	6,650
0400	If burning is allowed, reduce cut & chip									40%
3000	Chipping stumps, to 18″ deep, 12″ diam.	B-86	20	.400	Ea.		13.95	8.45	22.40	32.50
3040	18″ diameter		16	.500			17.45	10.60	28.05	40
3080	24″ diameter		14	.571			19.95	12.10	32.05	46
3100	30″ diameter		12	.667			23.50	14.10	37.60	53.50
3120	36″ diameter		10	.800			28	16.95	44.95	64
3160	48″ diameter		8	1			35	21	56	80.50
5000	Tree thinning, feller buncher, conifer									
5080	Up to 8″ diameter	B-93	240	.033	Ea.		1.16	2.35	3.51	4.48
5120	12″ diameter		160	.050			1.75	3.53	5.28	6.75
5240	Hardwood, up to 4″ diameter		240	.033			1.16	2.35	3.51	4.48
5280	8″ diameter		180	.044			1.55	3.13	4.68	6
5320	12″ diameter		120	.067			2.33	4.70	7.03	8.95
7000	Tree removal, congested area, aerial lift truck									
7040	8″ diameter	B-85	7	5.714	Ea.		162	123	285	405
7080	12″ diameter		6	6.667			189	143	332	475
7120	18″ diameter		5	8			226	172	398	565
7160	24″ diameter		4	10			283	215	498	705
7240	36″ diameter		3	13.333			375	287	662	940
7280	48″ diameter		2	20			565	430	995	1,425

31 13 Selective Tree and Shrub Removal and Trimming

31 13 13 – Selective Tree and Shrub Removal

31 13 13.10 Selective Clearing		Crew	Daily Output	Labor-Hours	Unit	Material	2011 Bare Costs Labor	Equipment	Total	Total Incl O&P
0010	**SELECTIVE CLEARING**									
0020	Clearing brush with brush saw	A-1C	.25	32	Acre		845	103	948	1,550
0100	By hand	1 Clab	.12	66.667			1,775		1,775	2,950
0300	With dozer, ball and chain, light clearing	B-11A	2	8			245	525	770	985
0400	Medium clearing		1.50	10.667			325	700	1,025	1,300
0500	With dozer and brush rake, light		10	1.600			49	105	154	197
0550	Medium brush to 4" diameter		8	2			61.50	131	192.50	246
0600	Heavy brush to 4" diameter		6.40	2.500			76.50	164	240.50	310
1000	Brush mowing, tractor w/rotary mower, no removal									
1020	Light density	B-84	2	4	Acre		140	147	287	390
1040	Medium density		1.50	5.333			186	196	382	520
1080	Heavy density		1	8			279	294	573	780

31 13 13.20 Selective Tree Removal		Crew	Daily Output	Labor-Hours	Unit	Material	Labor	Equipment	Total	Total Incl O&P
0010	**SELECTIVE TREE REMOVAL**									
0011	With tractor, large tract, firm									
0020	level terrain, no boulders, less than 12" diam. trees									
0300	300 HP dozer, up to 400 trees/acre, 0 to 25% hardwoods	B-10M	.75	10.667	Acre		370	2,150	2,520	2,950
0340	25% to 50% hardwoods		.60	13.333			465	2,675	3,140	3,700
0370	75% to 100% hardwoods		.45	17.778			620	3,575	4,195	4,950
0400	500 trees/acre, 0% to 25% hardwoods		.60	13.333			465	2,675	3,140	3,700
0440	25% to 50% hardwoods		.48	16.667			580	3,350	3,930	4,625
0470	75% to 100% hardwoods		.36	22.222			775	4,475	5,250	6,175
0500	More than 600 trees/acre, 0 to 25% hardwoods		.52	15.385			535	3,100	3,635	4,275
0540	25% to 50% hardwoods		.42	19.048			665	3,825	4,490	5,275
0570	75% to 100% hardwoods		.31	25.806			900	5,175	6,075	7,175
0900	Large tract clearing per tree									
1500	300 HP dozer, to 12" diameter, softwood	B-10M	320	.025	Ea.		.87	5	5.87	6.95
1550	Hardwood		100	.080			2.79	16.05	18.84	22.50
1600	12" to 24" diameter, softwood		200	.040			1.40	8.05	9.45	11.15
1650	Hardwood		80	.100			3.49	20	23.49	27.50
1700	24" to 36" diameter, softwood		100	.080			2.79	16.05	18.84	22.50
1750	Hardwood		50	.160			5.60	32	37.60	44.50
1800	36" to 48" diameter, softwood		70	.114			3.99	23	26.99	32
1850	Hardwood		35	.229			8	46	54	63.50
2000	Stump removal on site by hydraulic backhoe, 1-1/2 C.Y.									
2040	4" to 6" diameter	B-17	60	.533	Ea.		15.15	10.60	25.75	36.50
2050	8" to 12" diameter	B-30	33	.727			21.50	59.50	81	101
2100	14" to 24" diameter		25	.960			28.50	78.50	107	133
2150	26" to 36" diameter		16	1.500			44.50	122	166.50	209
3000	Remove selective trees, on site using chain saws and chipper,									
3050	not incl. stumps, up to 6" diameter	B-7	18	2.667	Ea.		75	72.50	147.50	205
3100	8" to 12" diameter		12	4			113	109	222	310
3150	14" to 24" diameter		10	4.800			135	131	266	370
3200	26" to 36" diameter		8	6			169	164	333	460
3300	Machine load, 2 mile haul to dump, 12" diam. tree	A-3B	8	2			62	113	175	226

31 14 Earth Stripping and Stockpiling

31 14 13 – Soil Stripping and Stockpiling

31 14 13.23 Topsoil Stripping and Stockpiling		Crew	Daily Output	Labor-Hours	Unit	Material	2011 Bare Costs Labor	Equipment	Total	Total Incl O&P
0010	**TOPSOIL STRIPPING AND STOCKPILING**									
0020	200 H.P. dozer, ideal conditions	B-10B	2300	.003	C.Y.		.12	.46	.58	.70
0100	Adverse conditions	"	1150	.007			.24	.91	1.15	1.41
0200	300 H.P. dozer, ideal conditions	B-10M	3000	.003			.09	.54	.63	.74
0300	Adverse conditions	"	1650	.005			.17	.97	1.14	1.35
0400	400 H.P. dozer, ideal conditions	B-10X	3900	.002			.07	.54	.61	.71
0500	Adverse conditions	"	2000	.004			.14	1.05	1.19	1.39
0600	Clay, dry and soft, 200 H.P. dozer, ideal conditions	B-10B	1600	.005			.17	.66	.83	1.01
0700	Adverse conditions	"	800	.010			.35	1.31	1.66	2.02
1000	Medium hard, 300 H.P. dozer, ideal conditions	B-10M	2000	.004			.14	.80	.94	1.11
1100	Adverse conditions	"	1100	.007			.25	1.46	1.71	2.02
1200	Very hard, 400 H.P. dozer, ideal conditions	B-10X	2600	.003			.11	.81	.92	1.07
1300	Adverse conditions	"	1340	.006			.21	1.57	1.78	2.07
1400	Loam or topsoil, remove and stockpile on site									
1420	6″ deep, 200′ haul	B-10B	865	.009	C.Y.		.32	1.22	1.54	1.87
1430	300′ haul		520	.015			.54	2.02	2.56	3.10
1440	500′ haul		225	.036			1.24	4.67	5.91	7.20
1450	Alternate method: 6″ deep, 200′ haul		5090	.002	S.Y.		.05	.21	.26	.32
1460	500′ haul		1325	.006	"		.21	.79	1	1.21

31 22 Grading

31 22 16 – Fine Grading

31 22 16.10 Finish Grading		Crew	Daily Output	Labor-Hours	Unit	Material	Labor	Equipment	Total	Total Incl O&P
0010	**FINISH GRADING**									
0012	Finish grading area to be paved with grader, small area	B-11L	400	.040	S.Y.		1.23	1.36	2.59	3.52
0100	Large area		2000	.008			.25	.27	.52	.71
1100	Fine grade for slab on grade, machine		1040	.015			.47	.52	.99	1.35
1150	Hand grading	B-18	700	.034			.93	.06	.99	1.62
3500	Finish grading lagoon bottoms	B-11L	4	4	M.S.F.		123	136	259	350

31 23 Excavation and Fill

31 23 16 – Excavation

31 23 16.13 Excavating, Trench		Crew	Daily Output	Labor-Hours	Unit	Material	Labor	Equipment	Total	Total Incl O&P
0010	**EXCAVATING, TRENCH**									
0011	Or continuous footing									
0020	Common earth with no sheeting or dewatering included									
0050	1′ to 4′ deep, 3/8 C.Y. excavator	B-11C	150	.107	B.C.Y.		3.27	2.14	5.41	7.75
0060	1/2 C.Y. excavator	B-11M	200	.080			2.45	1.75	4.20	5.95
0090	4′ to 6′ deep, 1/2 C.Y. excavator	"	200	.080			2.45	1.75	4.20	5.95
0101	5/8 C.Y. hydraulic excavator	B-12Q	250	.064			1.99	2.12	4.11	5.60
0300	1/2 C.Y. excavator, truck mounted	B-12J	200	.080			2.49	4.29	6.78	8.85
0500	6′ to 10′ deep, 3/4 C.Y. excavator	B-12F	225	.071			2.21	2.85	5.06	6.80
0510	1 C.Y. excavator	B-12A	400	.040			1.25	1.82	3.07	4.07
0600	1 C.Y. excavator, truck mounted	B-12K	400	.040			1.25	2.50	3.75	4.81
0610	1-1/2 C.Y. excavator	B-12B	600	.027			.83	1.58	2.41	3.11
0900	10′ to 14′ deep, 3/4 C.Y. excavator	B-12F	200	.080			2.49	3.20	5.69	7.65
0910	1 C.Y. excavator	B-12A	360	.044			1.38	2.03	3.41	4.51
1000	1-1/2 C.Y. excavator	B-12B	540	.030			.92	1.75	2.67	3.45
1300	14′ to 20′ deep, 1 C.Y. excavator	B-12A	320	.050			1.56	2.28	3.84	5.10

31 23 16.13 Excavating, Trench		Crew	Daily Output	Labor-Hours	Unit	Material	2011 Bare Costs Labor	Equipment	Total	Total Incl O&P
1310	1-1/2 C.Y. excavator	B-12B	480	.033	B.C.Y.		1.04	1.97	3.01	3.88
1320	2-1/2 C.Y. excavator	B-12S	765	.021			.65	2.11	2.76	3.40
1340	20' to 24' deep, 1 C.Y. excavator	B-12A	288	.056			1.73	2.53	4.26	5.65
1342	1-1/2 C.Y. excavator	B-12B	432	.037			1.15	2.19	3.34	4.31
1344	2-1/2 C.Y. excavator	B-12S	685	.023			.73	2.35	3.08	3.79
1352	4' to 6' deep, 1/2 C.Y. excavator w/trench box	B-13H	188	.085			2.65	5.15	7.80	10.05
1362	6' to 10' deep, 3/4 C.Y. excavator w/trench box	B-13G	212	.075			2.35	3.56	5.91	7.80
1370	1 C.Y. excavator	B-13D	376	.043			1.32	2.24	3.56	4.66
1371	1-1/2 C.Y. excavator	B-13E	564	.028			.88	1.88	2.76	3.53
1374	10' to 14' deep, 3/4 C.Y. excavator w/trench box	B-13G	188	.085			2.65	4.01	6.66	8.80
1375	1 C.Y. excavator	B-13D	338	.047			1.47	2.49	3.96	5.15
1376	1-1/2 C.Y. excavator	B-13E	508	.032			.98	2.09	3.07	3.92
1381	14' to 20' deep, 1 C.Y. excavator w/trench box	B-13D	301	.053			1.65	2.80	4.45	5.80
1382	1-1/2 C.Y. excavator	B-13E	451	.035			1.10	2.35	3.45	4.41
1383	2-1/2 C.Y. excavator	B-13J	720	.022			.69	2.40	3.09	3.78
1386	20' to 24' deep, 1 C.Y. excavator w/trench box	B-13D	271	.059			1.84	3.11	4.95	6.45
1387	1-1/2 C.Y. excavator	B-13E	406	.039			1.23	2.61	3.84	4.90
1388	2-1/2 C.Y. excavator	B-13J	645	.025			.77	2.68	3.45	4.22
1400	By hand with pick and shovel 2' to 6' deep, light soil	1 Clab	8	1			26.50		26.50	44.50
1500	Heavy soil	″	4	2			53		53	88.50
1700	For tamping backfilled trenches, air tamp, add	A-1G	100	.080	E.C.Y.		2.12	.50	2.62	4.09
1900	Vibrating plate, add	B-18	180	.133	″		3.62	.22	3.84	6.30
2100	Trim sides and bottom for concrete pours, common earth		1500	.016	S.F.		.43	.03	.46	.76
2300	Hardpan		600	.040	″		1.08	.07	1.15	1.89
2400	Pier and spread footing excavation, add to above				B.C.Y.				30%	30%
3000	Backfill trench, F.E. loader, wheel mtd., 1 C.Y. bucket									
3020	Minimal haul	B-10R	400	.020	L.C.Y.		.70	.59	1.29	1.79
3040	100' haul	″	200	.040			1.40	1.18	2.58	3.58
3080	2-1/4 C.Y. bucket, minimum haul	B-10T	600	.013			.47	.66	1.13	1.49
3090	100' haul	″	300	.027			.93	1.33	2.26	2.98
5020	Loam & Sandy clay with no sheeting or dewatering included									
5050	1' to 4' deep, 3/8 C.Y. tractor loader/backhoe	B-11C	162	.099	B.C.Y.		3.03	1.98	5.01	7.20
5060	1/2 C.Y. excavator	B-11M	216	.074			2.27	1.62	3.89	5.55
5080	4' to 6' deep, 1/2 C.Y. excavator	″	216	.074			2.27	1.62	3.89	5.55
5130	1/2 C.Y. excavator, truck mounted	B-12J	216	.074			2.31	3.97	6.28	8.20
5140	6' to 10' deep, 3/4 C.Y. excavator	B-12F	243	.066			2.05	2.64	4.69	6.30
5150	1 C.Y. excavator	B-12A	432	.037			1.15	1.69	2.84	3.76
5160	1 C.Y. excavator, truck mounted	B-12K	432	.037			1.15	2.32	3.47	4.45
5170	1-1/2 C.Y. excavator	B-12B	648	.025			.77	1.46	2.23	2.88
5190	10' to 14' deep, 3/4 C.Y. excavator	B-12F	216	.074			2.31	2.97	5.28	7.05
5200	1 C.Y. excavator	B-12A	389	.041			1.28	1.88	3.16	4.17
5210	1-1/2 C.Y. excavator	B-12B	583	.027			.85	1.62	2.47	3.20
5250	14' to 20' deep, 1 C.Y. excavator	B-12A	346	.046			1.44	2.11	3.55	4.70
5260	1-1/2 C.Y. excavator	B-12B	518	.031			.96	1.83	2.79	3.60
5270	2-1/2 C.Y. excavator	B-12S	826	.019			.60	1.95	2.55	3.15
5300	20' to 24' deep, 1 C.Y. excavator	B-12A	311	.051			1.60	2.35	3.95	5.20
5310	1-1/2 C.Y. excavator	B-12B	467	.034			1.07	2.03	3.10	3.99
5320	2-1/2 C.Y. excavator	B-12S	740	.022			.67	2.18	2.85	3.51
5352	4' to 6' deep, 1/2 C.Y. excavator w/trench box	B-13H	205	.078			2.43	4.74	7.17	9.20
5362	6' to 10' deep, 3/4 C.Y. excavator w/trench box	B-13G	231	.069			2.16	3.26	5.42	7.15
5364	1 C.Y. excavator	B-13D	410	.039			1.21	2.06	3.27	4.27
5366	1-1/2 C.Y. excavator	B-13E	616	.026			.81	1.72	2.53	3.22
5370	10' to 14' deep, 3/4 C.Y. excavator w/trench box	B-13G	205	.078			2.43	3.68	6.11	8.05

31 23 16.13 Excavating, Trench		Crew	Daily Output	Labor-Hours	Unit	Material	2011 Bare Costs Labor	Equipment	Total	Total Incl O&P
5372	1 C.Y. excavator	B-13D	370	.043	B.C.Y.		1.35	2.28	3.63	4.73
5374	1-1/2 C.Y. excavator	B-13E	554	.029			.90	1.91	2.81	3.59
5382	14' to 20' deep, 1 C.Y. excavator w/trench box	B-13D	329	.049			1.51	2.56	4.07	5.30
5384	1-1/2 C.Y. excavator	B-13E	492	.033			1.01	2.16	3.17	4.04
5386	2-1/2 C.Y. excavator	B-13J	780	.021			.64	2.21	2.85	3.48
5392	20' to 24' deep, 1 C.Y. excavator w/trench box	B-13D	295	.054			1.69	2.86	4.55	5.95
5394	1-1/2 C.Y. excavator	B-13E	444	.036			1.12	2.39	3.51	4.48
5396	2-1/2 C.Y. excavator	B-13J	695	.023			.72	2.48	3.20	3.91
6020	Sand & gravel with no sheeting or dewatering included									
6050	1' to 4' deep, 3/8 C.Y. excavator	B-11C	165	.097	B.C.Y.		2.98	1.94	4.92	7.05
6060	1/2 C.Y. excavator	B-11M	220	.073			2.23	1.59	3.82	5.45
6080	4' to 6' deep, 1/2 C.Y. excavator	"	220	.073			2.23	1.59	3.82	5.45
6130	1/2 C.Y. excavator, truck mounted	B-12J	220	.073			2.26	3.90	6.16	8.05
6140	6' to 10' deep, 3/4 C.Y. excavator	B-12F	248	.065			2.01	2.58	4.59	6.15
6150	1 C.Y. excavator	B-12A	440	.036			1.13	1.66	2.79	3.69
6160	1 C.Y. excavator, truck mounted	B-12K	440	.036			1.13	2.27	3.40	4.37
6170	1-1/2 C.Y. excavator	B-12B	660	.024			.75	1.44	2.19	2.83
6190	10' to 14' deep, 3/4 C.Y. excavator	B-12F	220	.073			2.26	2.91	5.17	6.95
6200	1 C.Y. excavator	B-12A	396	.040			1.26	1.84	3.10	4.11
6210	1-1/2 C.Y. excavator	B-12B	594	.027			.84	1.60	2.44	3.13
6250	14' to 20' deep, 1 C.Y. excavator	B-12A	352	.045			1.41	2.07	3.48	4.62
6260	1-1/2 C.Y. excavator	B-12B	528	.030			.94	1.79	2.73	3.53
6270	2-1/2 C.Y. excavator	B-12S	840	.019			.59	1.92	2.51	3.09
6300	20' to 24' deep, 1 C.Y. excavator	B-12A	317	.050			1.57	2.30	3.87	5.10
6310	1-1/2 C.Y. excavator	B-12B	475	.034			1.05	1.99	3.04	3.92
6320	2-1/2 C.Y. excavator	B-12S	755	.021			.66	2.14	2.80	3.44
6352	4' to 6' deep, 1/2 C.Y. excavator w/trench box	B-13H	209	.077			2.38	4.65	7.03	9.05
6362	6' to 10' deep, 3/4 C.Y. excavator w/trench box	B-13G	236	.068			2.11	3.19	5.30	7
6364	1 C.Y. excavator	B-13D	418	.038			1.19	2.02	3.21	4.19
6366	1-1/2 C.Y. excavator	B-13E	627	.026			.79	1.69	2.48	3.17
6370	10' to 14' deep, 3/4 C.Y. excavator w/trench box	B-13G	209	.077			2.38	3.61	5.99	7.90
6372	1 C.Y. excavator	B-13D	376	.043			1.32	2.24	3.56	4.66
6374	1-1/2 C.Y. excavator	B-13E	564	.028			.88	1.88	2.76	3.53
6382	14' to 20' deep, 1 C.Y. excavator w/trench box	B-13D	334	.048			1.49	2.52	4.01	5.25
6384	1-1/2 C.Y. excavator	B-13E	502	.032			.99	2.11	3.10	3.96
6386	2-1/2 C.Y. excavator	B-13J	790	.020			.63	2.19	2.82	3.44
6392	20' to 24' deep, 1 C.Y. excavator w/trench box	B-13D	301	.053			1.65	2.80	4.45	5.80
6394	1-1/2 C.Y. excavator	B-13E	452	.035			1.10	2.35	3.45	4.40
6396	2-1/2 C.Y. excavator	B-13J	710	.023			.70	2.43	3.13	3.84
7020	Dense hard clay with no sheeting or dewatering included									
7050	1' to 4' deep, 3/8 C.Y. excavator	B-11C	132	.121	B.C.Y.		3.72	2.43	6.15	8.80
7060	1/2 C.Y. excavator	B-11M	176	.091			2.79	1.99	4.78	6.80
7080	4' to 6' deep, 1/2 C.Y. excavator	"	176	.091			2.79	1.99	4.78	6.80
7130	1/2 C.Y. excavator, truck mounted	B-12J	176	.091			2.83	4.87	7.70	10
7140	6' to 10' deep, 3/4 C.Y. excavator	B-12F	198	.081			2.52	3.23	5.75	7.70
7150	1 C.Y. excavator	B-12A	352	.045			1.41	2.07	3.48	4.62
7160	1 C.Y. excavator, truck mounted	B-12K	352	.045			1.41	2.84	4.25	5.45
7170	1-1/2 C.Y. excavator	B-12B	528	.030			.94	1.79	2.73	3.53
7190	10' to 14' deep, 3/4 C.Y. excavator	B-12F	176	.091			2.83	3.64	6.47	8.65
7200	1 C.Y. excavator	B-12A	317	.050			1.57	2.30	3.87	5.10
7210	1-1/2 C.Y. excavator	B-12B	475	.034			1.05	1.99	3.04	3.92
7250	14' to 20' deep, 1 C.Y. excavator	B-12A	282	.057			1.77	2.59	4.36	5.75
7260	1-1/2 C.Y. excavator	B-12B	422	.038			1.18	2.24	3.42	4.42

31 23 Excavation and Fill

31 23 16 – Excavation

31 23 16.13 Excavating, Trench

	31 23 16.13 Excavating, Trench	Crew	Daily Output	Labor-Hours	Unit	Material	2011 Bare Costs Labor	Equipment	Total	Total Incl O&P
7270	2-1/2 C.Y. excavator	B-12S	675	.024	B.C.Y.		.74	2.39	3.13	3.85
7300	20′ to 24′ deep, 1 C.Y. excavator	B-12A	254	.063			1.96	2.87	4.83	6.40
7310	1-1/2 C.Y. excavator	B-12B	380	.042			1.31	2.49	3.80	4.90
7320	2-1/2 C.Y. excavator	B-12S	605	.026			.82	2.67	3.49	4.29

31 23 16.14 Excavating, Utility Trench

	31 23 16.14 Excavating, Utility Trench	Crew	Daily Output	Labor-Hours	Unit	Material	Labor	Equipment	Total	Total Incl O&P
0010	**EXCAVATING, UTILITY TRENCH**									
0011	Common earth									
0050	Trenching with chain trencher, 12 H.P., operator walking									
0100	4″ wide trench, 12″ deep	B-53	800	.010	L.F.		.26	.07	.33	.52
0150	18″ deep		750	.011			.28	.08	.36	.55
0200	24″ deep		700	.011			.30	.08	.38	.60
0300	6″ wide trench, 12″ deep		650	.012			.33	.09	.42	.65
0350	18″ deep		600	.013			.35	.10	.45	.70
0400	24″ deep		550	.015			.38	.10	.48	.75
0450	36″ deep		450	.018			.47	.13	.60	.93
0600	8″ wide trench, 12″ deep		475	.017			.45	.12	.57	.88
0650	18″ deep		400	.020			.53	.14	.67	1.05
0700	24″ deep		350	.023			.60	.16	.76	1.19
0750	36″ deep		300	.027			.71	.19	.90	1.39
1000	Backfill by hand including compaction, add									
1050	4″ wide trench, 12″ deep	A-1G	800	.010	L.F.		.26	.06	.32	.51
1100	18″ deep		530	.015			.40	.09	.49	.77
1150	24″ deep		400	.020			.53	.12	.65	1.03
1300	6″ wide trench, 12″ deep		540	.015			.39	.09	.48	.76
1350	18″ deep		405	.020			.52	.12	.64	1
1400	24″ deep		270	.030			.78	.18	.96	1.51
1450	36″ deep		180	.044			1.18	.28	1.46	2.27
1600	8″ wide trench, 12″ deep		400	.020			.53	.12	.65	1.03
1650	18″ deep		265	.030			.80	.19	.99	1.55
1700	24″ deep		200	.040			1.06	.25	1.31	2.04
1750	36″ deep		135	.059			1.57	.37	1.94	3.03
2000	Chain trencher, 40 H.P. operator riding									
2050	6″ wide trench and backfill, 12″ deep	B-54	1200	.007	L.F.		.22	.25	.47	.64
2100	18″ deep		1000	.008			.27	.30	.57	.77
2150	24″ deep		975	.008			.28	.30	.58	.79
2200	36″ deep		900	.009			.30	.33	.63	.85
2250	48″ deep		750	.011			.36	.40	.76	1.02
2300	60″ deep		650	.012			.41	.46	.87	1.17
2400	8″ wide trench and backfill, 12″ deep		1000	.008			.27	.30	.57	.77
2450	18″ deep		950	.008			.28	.31	.59	.80
2500	24″ deep		900	.009			.30	.33	.63	.85
2550	36″ deep		800	.010			.34	.37	.71	.96
2600	48″ deep		650	.012			.41	.46	.87	1.17
2700	12″ wide trench and backfill, 12″ deep		975	.008			.28	.30	.58	.79
2750	18″ deep		860	.009			.31	.35	.66	.89
2800	24″ deep		800	.010			.34	.37	.71	.96
2850	36″ deep		725	.011			.37	.41	.78	1.05
3000	16″ wide trench and backfill, 12″ deep		835	.010			.32	.36	.68	.92
3050	18″ deep		750	.011			.36	.40	.76	1.02
3100	24″ deep		700	.011			.38	.42	.80	1.10
3200	Compaction with vibratory plate, add								35%	35%
5100	Hand excavate and trim for pipe bells after trench excavation									

31 23 16.14 Excavating, Utility Trench		Crew	Daily Output	Labor-Hours	Unit	Material	2011 Bare Costs Labor	Equipment	Total	Total Incl O&P
5200	8″ pipe	1 Clab	155	.052	L.F.		1.37		1.37	2.29
5300	18″ pipe	″	130	.062	″		1.63		1.63	2.73

31 23 16.16 Structural Excavation for Minor Structures

		Crew	Daily Output	Labor-Hours	Unit	Material	Labor	Equipment	Total	Total Incl O&P
0010	**STRUCTURAL EXCAVATION FOR MINOR STRUCTURES** R312316-40									
0015	Hand, pits to 6′ deep, sandy soil	1 Clab	8	1	B.C.Y.		26.50		26.50	44.50
0100	Heavy soil or clay		4	2			53		53	88.50
0300	Pits 6′ to 12′ deep, sandy soil		5	1.600			42.50		42.50	71
0500	Heavy soil or clay		3	2.667			70.50		70.50	118
0700	Pits 12′ to 18′ deep, sandy soil		4	2			53		53	88.50
0900	Heavy soil or clay		2	4			106		106	177
1100	Hand loading trucks from stock pile, sandy soil		12	.667			17.65		17.65	29.50
1300	Heavy soil or clay	↓	8	1	↓		26.50		26.50	44.50
1500	For wet or muck hand excavation, add to above				%				50%	50%
6000	Machine excavation, for spread and mat footings, elevator pits,									
6001	and small building foundations									
6030	Common earth, hydraulic backhoe, 1/2 C.Y. bucket	B-12E	55	.291	B.C.Y.		9.05	6.65	15.70	22.50
6035	3/4 C.Y. bucket	B-12F	90	.178			5.55	7.10	12.65	17
6040	1 C.Y. bucket	B-12A	108	.148			4.61	6.75	11.36	15.05
6050	1-1/2 C.Y. bucket	B-12B	144	.111			3.46	6.60	10.06	12.95
6060	2 C.Y. bucket	B-12C	200	.080			2.49	6.10	8.59	10.85
6070	Sand and gravel, 3/4 C.Y. bucket	B-12F	100	.160			4.98	6.40	11.38	15.25
6080	1 C.Y. bucket	B-12A	120	.133			4.15	6.10	10.25	13.55
6090	1-1/2 C.Y. bucket	B-12B	160	.100			3.11	5.90	9.01	11.65
6100	2 C.Y. bucket	B-12C	220	.073			2.26	5.55	7.81	9.85
6110	Clay, till, or blasted rock, 3/4 C.Y. bucket	B-12F	80	.200			6.25	8	14.25	19.10
6120	1 C.Y. bucket	B-12A	95	.168			5.25	7.70	12.95	17.10
6130	1-1/2 C.Y. bucket	B-12B	130	.123			3.83	7.30	11.13	14.35
6140	2 C.Y. bucket	B-12C	175	.091			2.85	7	9.85	12.40
6230	Sandy clay & loam, hydraulic backhoe, 1/2 C.Y. bucket	B-12E	60	.267			8.30	6.10	14.40	20.50
6235	3/4 C.Y. bucket	B-12F	98	.163			5.10	6.55	11.65	15.60
6240	1 C.Y. bucket	B-12A	116	.138			4.29	6.30	10.59	14
6250	1-1/2 C.Y. bucket	B-12B	156	.103	↓		3.19	6.05	9.24	11.95
9010	For mobilization or demobilization, see Section 01 54 36.50									
9020	For dewatering, see Section 31 23 19.20									
9022	For larger structures, see Bulk Excavation, Section 31 23 16.42									
9024	For loading onto trucks, add								15%	15%
9026	For hauling, see Section 31 23 23.20									
9030	For sheeting or soldier bms/lagging, see Section 31 52 16.10									
9040	For trench excavation of strip ftgs, see Section 31 23 16.13									

31 23 16.26 Rock Removal

		Crew	Daily Output	Labor-Hours	Unit	Material	Labor	Equipment	Total	Total Incl O&P
0010	**ROCK REMOVAL** R312316-40									
0015	Drilling only rock, 2″ hole for rock bolts	B-47	316	.051	L.F.		1.39	4.13	5.52	6.85
0800	2-1/2″ hole for pre-splitting		600	.027			.73	2.17	2.90	3.62
4600	Quarry operations, 2-1/2″ to 3-1/2″ diameter	↓	715	.022	↓		.61	1.82	2.43	3.04

31 23 16.30 Drilling and Blasting Rock

		Crew	Daily Output	Labor-Hours	Unit	Material	Labor	Equipment	Total	Total Incl O&P
0010	**DRILLING AND BLASTING ROCK**									
0020	Rock, open face, under 1500 C.Y.	B-47	225	.071	B.C.Y.	2.82	1.95	5.80	10.57	12.75
0100	Over 1500 C.Y.		300	.053		2.82	1.46	4.35	8.63	10.35
0200	Areas where blasting mats are required, under 1500 C.Y.		175	.091		2.82	2.51	7.45	12.78	15.50
0250	Over 1500 C.Y.	↓	250	.064		2.82	1.76	5.20	9.78	11.80
0300	Bulk drilling and blasting, can vary greatly, average								9.50	12
0500	Pits, average				↓				25	31

31 23 Excavation and Fill

31 23 16 – Excavation

31 23 16.30 Drilling and Blasting Rock		Crew	Daily Output	Labor-Hours	Unit	2011 Bare Costs Material	Labor	Equipment	Total	Total Incl O&P
1300	Deep hole method, up to 1500 C.Y.	B-47	50	.320	B.C.Y.	2.82	8.80	26	37.62	46.50
1400	Over 1500 C.Y.		66	.242		2.82	6.65	19.75	29.22	36
1900	Restricted areas, up to 1500 C.Y.		13	1.231		2.82	34	100	136.82	170
2000	Over 1500 C.Y.		20	.800		2.82	22	65	89.82	112
2200	Trenches, up to 1500 C.Y.		22	.727		8.20	19.95	59.50	87.65	108
2300	Over 1500 C.Y.		26	.615		8.20	16.90	50	75.10	92.50
2500	Pier holes, up to 1500 C.Y.		22	.727		2.82	19.95	59.50	82.27	102
2600	Over 1500 C.Y.		31	.516		2.82	14.15	42	58.97	73
2800	Boulders under 1/2 C.Y., loaded on truck, no hauling	B-100	80	.100			3.49	9.65	13.14	16.35
2900	Boulders, drilled, blasted	B-47	100	.160		2.82	4.39	13.05	20.26	25
3100	Jackhammer operators with foreman compressor, air tools	B-9	1	40	Day		1,075	195	1,270	2,025
3300	Track drill, compressor, operator and foreman	B-47	1	16	"		440	1,300	1,740	2,150
3500	Blasting caps				Ea.	5.25			5.25	5.75
3700	Explosives					.38			.38	.41
3900	Blasting mats, rent, for first day					123			123	135
4000	Per added day					41			41	45
4200	Preblast survey for 6 room house, individual lot, minimum	A-6	2.40	6.667			224	29	253	400
4300	Maximum	"	1.35	11.852			400	51.50	451.50	715
4500	City block within zone of influence, minimum	A-8	25200	.001	S.F.		.04		.04	.07
4600	Maximum	"	15100	.002	"		.07		.07	.13

31 23 16.42 Excavating, Bulk Bank Measure		Crew	Daily Output	Labor-Hours	Unit	Material	Labor	Equipment	Total	Total Incl O&P
0010	**EXCAVATING, BULK BANK MEASURE** R312316-40									
0011	Common earth piled									
0020	For loading onto trucks, add								15%	15%
0050	For mobilization and demobilization, see Section 01 54 36.50 R312316-45									
0100	For hauling, see Section 31 23 23.20									
0200	Excavator, hydraulic, crawler mtd., 1 C.Y. cap. = 100 C.Y./hr.	B-12A	800	.020	B.C.Y.		.62	.91	1.53	2.03
0250	1-1/2 C.Y. cap. = 125 C.Y./hr.	B-12B	1000	.016			.50	.95	1.45	1.86
0260	2 C.Y. cap. = 165 C.Y./hr.	B-12C	1320	.012			.38	.93	1.31	1.64
0300	3 C.Y. cap. = 260 C.Y./hr.	B-12D	2080	.008			.24	1.04	1.28	1.55
0305	3-1/2 C.Y. cap. = 300 C.Y./hr.	"	2400	.007			.21	.91	1.12	1.34
0310	Wheel mounted, 1/2 C.Y. cap. = 40 C.Y./hr.	B-12E	320	.050			1.56	1.14	2.70	3.82
0360	3/4 C.Y. cap. = 60 C.Y./hr.	B-12F	480	.033			1.04	1.33	2.37	3.18
0500	Clamshell, 1/2 C.Y. cap. = 20 C.Y./hr.	B-12G	160	.100			3.11	4.35	7.46	9.95
0550	1 C.Y. cap. = 35 C.Y./hr.	B-12H	280	.057			1.78	4.18	5.96	7.55
0950	Dragline, 1/2 C.Y. cap. = 30 C.Y./hr.	B-12I	240	.067			2.08	3.60	5.68	7.40
1000	3/4 C.Y. cap. = 35 C.Y./hr.	"	280	.057			1.78	3.08	4.86	6.35
1050	1-1/2 C.Y. cap. = 65 C.Y./hr.	B-12P	520	.031			.96	2.24	3.20	4.04
1200	Front end loader, track mtd., 1-1/2 C.Y. cap. = 70 C.Y./hr.	B-10N	560	.014			.50	.83	1.33	1.74
1250	2-1/2 C.Y. cap. = 95 C.Y./hr.	B-10O	760	.011			.37	1.02	1.39	1.72
1300	3 C.Y. cap. = 130 C.Y./hr.	B-10P	1040	.008			.27	.97	1.24	1.51
1350	5 C.Y. cap. = 160 C.Y./hr.	B-10Q	1280	.006			.22	1.12	1.34	1.59
1500	Wheel mounted, 3/4 C.Y. cap. = 45 C.Y./hr.	B-10R	360	.022			.78	.66	1.44	1.99
1550	1-1/2 C.Y. cap. = 80 C.Y./hr.	B-10S	640	.013			.44	.50	.94	1.26
1600	2-1/4 C.Y. cap. = 100 C.Y./hr.	B-10T	800	.010			.35	.50	.85	1.12
1650	5 C.Y. cap. = 185 C.Y./hr.	B-10U	1480	.005			.19	.63	.82	1
1800	Hydraulic excavator, truck mtd. 1/2 C.Y. = 30 C.Y./hr.	B-12J	240	.067			2.08	3.57	5.65	7.35
1850	48 inch bucket, 1 C.Y. = 45 C.Y./hr.	B-12K	360	.044			1.38	2.78	4.16	5.35
3700	Shovel, 1/2 C.Y. capacity = 55 C.Y./hr.	B-12L	440	.036			1.13	1.62	2.75	3.65
3750	3/4 C.Y. capacity = 85 C.Y./hr.	B-12M	680	.024			.73	1.33	2.06	2.67
3800	1 C.Y. capacity = 120 C.Y./hr.	B-12N	960	.017			.52	1.24	1.76	2.22
3850	1-1/2 C.Y. capacity = 160 C.Y./hr.	B-12O	1280	.013			.39	.94	1.33	1.68

31 23 16 – Excavation

31 23 16.42 Excavating, Bulk Bank Measure		Crew	Daily Output	Labor-Hours	Unit	Material	2011 Bare Costs Labor	Equipment	Total	Total Incl O&P
3900	3 C.Y. cap. = 250 C.Y./hr.	B-12T	2000	.008	B.C.Y.		.25	.77	1.02	1.26
4000	For soft soil or sand, deduct								15%	15%
4100	For heavy soil or stiff clay, add								60%	60%
4200	For wet excavation with clamshell or dragline, add								100%	100%
4250	All other equipment, add								50%	50%
4400	Clamshell in sheeting or cofferdam, minimum	B-12H	160	.100			3.11	7.30	10.41	13.20
4450	Maximum	"	60	.267			8.30	19.50	27.80	35
5000	Excavating, bulk bank measure, sandy clay & loam piled									
5020	For loading onto trucks, add								15%	15%
5100	Excavator, hydraulic, crawler mtd., 1 C.Y. cap. = 120 C.Y./hr.	B-12A	960	.017	B.C.Y.		.52	.76	1.28	1.70
5150	1-1/2 C.Y. cap. = 150 C.Y./hr.	B-12B	1200	.013			.42	.79	1.21	1.56
5300	2 C.Y. cap. = 195 C.Y./hr.	B-12C	1560	.010			.32	.78	1.10	1.39
5400	3 C.Y. cap. = 300 C.Y./hr.	B-12D	2400	.007			.21	.91	1.12	1.34
5500	3.5 C.Y. cap. = 350 C.Y./hr.	"	2800	.006			.18	.77	.95	1.14
5610	Wheel mounted, 1/2 C.Y. cap. = 44 C.Y./hr.	B-12E	352	.045			1.41	1.04	2.45	3.48
5660	3/4 C.Y. cap. = 66 C.Y./hr.	B-12F	528	.030			.94	1.21	2.15	2.89
8000	For hauling excavated material, see Section 31 23 23.20									

31 23 16.46 Excavating, Bulk, Dozer		Crew	Daily Output	Labor-Hours	Unit	Material	Labor	Equipment	Total	Total Incl O&P
0010	**EXCAVATING, BULK, DOZER**									
0011	Open site									
2000	80 H.P., 50' haul, sand & gravel	B-10L	460	.017	B.C.Y.		.61	.90	1.51	1.98
2010	Sandy clay & loam		440	.018			.63	.94	1.57	2.07
2020	Common earth		400	.020			.70	1.03	1.73	2.27
2040	Clay		250	.032			1.12	1.65	2.77	3.65
2200	150' haul, sand & gravel		230	.035			1.21	1.79	3	3.95
2210	Sandy clay & loam		220	.036			1.27	1.88	3.15	4.13
2220	Common earth		200	.040			1.40	2.06	3.46	4.55
2240	Clay		125	.064			2.23	3.30	5.53	7.30
2400	300' haul, sand & gravel		120	.067			2.33	3.44	5.77	7.60
2440	Clay		65	.123			4.30	6.35	10.65	14
3000	105 H.P., 50' haul, sand & gravel	B-10W	700	.011			.40	.72	1.12	1.44
3010	Sandy clay & loam		680	.012			.41	.74	1.15	1.49
3020	Common earth		610	.013			.46	.83	1.29	1.66
3040	Clay		385	.021			.73	1.31	2.04	2.63
3200	150' haul, sand & gravel		310	.026			.90	1.63	2.53	3.26
3210	Sandy clay & loam		300	.027			.93	1.69	2.62	3.37
3220	Common earth		270	.030			1.03	1.87	2.90	3.75
3240	Clay		170	.047			1.64	2.97	4.61	5.95
3300	300' haul, sand & gravel		140	.057			1.99	3.61	5.60	7.25
3310	Sandy clay & loam		135	.059			2.07	3.74	5.81	7.50
3320	Common earth		120	.067			2.33	4.21	6.54	8.45
3340	Clay		100	.080			2.79	5.05	7.84	10.10
4000	200 H.P., 50' haul, sand & gravel	B-10B	1400	.006			.20	.75	.95	1.16
4010	Sandy clay & loam		1360	.006			.21	.77	.98	1.19
4020	Common earth		1230	.007			.23	.85	1.08	1.31
4040	Clay		770	.010			.36	1.37	1.73	2.09
4200	150' haul, sand & gravel		595	.013			.47	1.77	2.24	2.71
4210	Sandy clay & loam		580	.014			.48	1.81	2.29	2.78
4220	Common earth		516	.016			.54	2.04	2.58	3.12
4240	Clay		325	.025			.86	3.23	4.09	4.96
4400	300' haul, sand & gravel		310	.026			.90	3.39	4.29	5.20
4410	Sandy clay & loam		300	.027			.93	3.50	4.43	5.35

31 23 16.46 Excavating, Bulk, Dozer

	31 23 16.46 Excavating, Bulk, Dozer	Crew	Daily Output	Labor-Hours	Unit	Material	2011 Bare Costs Labor	Equipment	Total	Total Incl O&P
4420	Common earth	B-10B	270	.030	B.C.Y.		1.03	3.89	4.92	5.95
4440	Clay		170	.047			1.64	6.20	7.84	9.50
5000	300 H.P., 50′ haul, sand & gravel	B-10M	1900	.004			.15	.85	1	1.17
5010	Sandy clay & loam		1850	.004			.15	.87	1.02	1.20
5020	Common earth		1650	.005			.17	.97	1.14	1.35
5040	Clay		1025	.008			.27	1.57	1.84	2.17
5200	150′ haul, sand & gravel		920	.009			.30	1.75	2.05	2.42
5210	Sandy clay & loam		895	.009			.31	1.80	2.11	2.49
5220	Common earth		800	.010			.35	2.01	2.36	2.78
5240	Clay		500	.016			.56	3.21	3.77	4.45
5400	300′ haul, sand & gravel		470	.017			.59	3.42	4.01	4.73
5410	Sandy clay & loam		455	.018			.61	3.53	4.14	4.88
5420	Common earth		410	.020			.68	3.92	4.60	5.40
5440	Clay		250	.032			1.12	6.45	7.57	8.90
5500	460 H.P., 50′ haul, sand & gravel	B-10X	1930	.004			.14	1.09	1.23	1.44
5506	Sandy clay & loam		1880	.004			.15	1.12	1.27	1.47
5510	Common earth		1680	.005			.17	1.25	1.42	1.65
5520	Clay		1050	.008			.27	2.01	2.28	2.64
5530	150′ haul, sand & gravel		1290	.006			.22	1.63	1.85	2.14
5535	Sandy clay & loam		1250	.006			.22	1.68	1.90	2.22
5540	Common earth		1120	.007			.25	1.88	2.13	2.48
5550	Clay		700	.011			.40	3.01	3.41	3.96
5560	300′ haul, sand & gravel		660	.012			.42	3.19	3.61	4.20
5565	Sandy clay & loam		640	.013			.44	3.29	3.73	4.33
5570	Common earth		575	.014			.49	3.66	4.15	4.82
5580	Clay		350	.023			.80	6	6.80	7.90
6000	700 H.P., 50′ haul, sand & gravel	B-10V	3500	.002			.08	1.19	1.27	1.44
6006	Sandy clay & loam		3400	.002			.08	1.22	1.30	1.47
6010	Common earth		3035	.003			.09	1.37	1.46	1.66
6020	Clay		1925	.004			.15	2.16	2.31	2.61
6030	150′ haul, sand & gravel		2025	.004			.14	2.05	2.19	2.48
6035	Sandy clay & loam		1960	.004			.14	2.12	2.26	2.56
6040	Common earth		1750	.005			.16	2.37	2.53	2.87
6050	Clay		1100	.007			.25	3.77	4.02	4.56
6060	300′ haul, sand & gravel		1030	.008			.27	4.03	4.30	4.87
6065	Sandy clay & loam		1005	.008			.28	4.13	4.41	4.99
6070	Common earth		900	.009			.31	4.61	4.92	5.55
6080	Clay		550	.015			.51	7.55	8.06	9.15

31 23 16.50 Excavation, Bulk, Scrapers

	31 23 16.50 Excavation, Bulk, Scrapers	Crew	Daily Output	Labor-Hours	Unit	Material	Labor	Equipment	Total	Total Incl O&P
0010	**EXCAVATION, BULK, SCRAPERS** R312316-40									
0100	Elev. scraper 11 C.Y., sand & gravel 1500′ haul, 1/4 dozer	B-33F	690	.014	B.C.Y.		.51	2.16	2.67	3.21
0150	3000′ haul		610	.016			.57	2.44	3.01	3.63
0200	5000′ haul		505	.020			.69	2.95	3.64	4.38
0300	Common earth, 1500′ haul		600	.017			.58	2.49	3.07	3.68
0350	3000′ haul		530	.019			.66	2.81	3.47	4.17
0400	5000′ haul		440	.023			.79	3.39	4.18	5.05
0410	Sandy clay & loam, 1500′ haul		648	.015			.54	2.30	2.84	3.41
0420	3000′ haul		572	.017			.61	2.61	3.22	3.87
0430	5000′ haul		475	.021			.73	3.14	3.87	4.65
0500	Clay, 1500′ haul		375	.027			.93	3.98	4.91	5.90
0550	3000′ haul		330	.030			1.06	4.52	5.58	6.70
0600	5000′ haul		275	.036			1.27	5.40	6.67	8

31 23 16.50 Excavation, Bulk, Scrapers		Crew	Daily Output	Labor-Hours	Unit	2011 Bare Costs Material	Labor	Equipment	Total	Total Incl O&P
1000	Self propelled scraper, 14 C.Y. 1/4 push dozer, sand									
1050	Sand and gravel, 1500' haul	B-33D	920	.011	B.C.Y.		.38	2.08	2.46	2.91
1100	3000' haul		805	.012			.43	2.38	2.81	3.32
1200	5000' haul		645	.016			.54	2.97	3.51	4.14
1300	Common earth, 1500' haul		800	.013			.44	2.39	2.83	3.34
1350	3000' haul		700	.014			.50	2.73	3.23	3.83
1400	5000' haul		560	.018			.62	3.42	4.04	4.78
1420	Sandy clay & loam, 1500' haul		864	.012			.40	2.21	2.61	3.10
1430	3000' haul		786	.013			.44	2.43	2.87	3.41
1440	5000' haul		605	.017			.58	3.16	3.74	4.42
1500	Clay, 1500' haul		500	.020			.70	3.83	4.53	5.35
1550	3000' haul		440	.023			.79	4.35	5.14	6.10
1600	5000' haul		350	.029			1	5.45	6.45	7.65
2000	21 C.Y., 1/4 push dozer, sand & gravel, 1500' haul	B-33E	1180	.008			.30	1.93	2.23	2.60
2100	3000' haul		910	.011			.38	2.50	2.88	3.38
2200	5000' haul		750	.013			.47	3.03	3.50	4.09
2300	Common earth, 1500' haul		1030	.010			.34	2.21	2.55	2.98
2350	3000' haul		790	.013			.44	2.88	3.32	3.89
2400	5000' haul		650	.015			.54	3.50	4.04	4.73
2420	Sandy clay & loam, 1500' haul		1112	.009			.31	2.04	2.35	2.76
2430	3000' haul		854	.012			.41	2.66	3.07	3.60
2440	5000' haul		702	.014			.50	3.24	3.74	4.37
2500	Clay, 1500' haul		645	.016			.54	3.52	4.06	4.76
2550	3000' haul		495	.020			.71	4.59	5.30	6.20
2600	5000' haul		405	.025			.86	5.60	6.46	7.60
2700	Towed, 10 C.Y., 1/4 push dozer, sand & gravel, 1500' haul	B-33B	560	.018			.62	3.84	4.46	5.25
2720	3000' haul		450	.022			.78	4.77	5.55	6.50
2730	5000' haul		365	.027			.96	5.90	6.86	8.05
2750	Common earth, 1500' haul		420	.024			.83	5.10	5.93	7
2770	3000' haul		400	.025			.87	5.35	6.22	7.35
2780	5000' haul		310	.032			1.13	6.95	8.08	9.50
2785	Sandy clay & Loam, 1500' haul		454	.022			.77	4.73	5.50	6.45
2790	3000' haul		432	.023			.81	4.97	5.78	6.75
2795	5000' haul		340	.029			1.03	6.30	7.33	8.65
2800	Clay, 1500' haul		315	.032			1.11	6.80	7.91	9.30
2820	3000' haul		300	.033			1.16	7.15	8.31	9.80
2840	5000' haul		225	.044			1.55	9.55	11.10	13.05
2900	15 C.Y., 1/4 push dozer, sand & gravel, 1500' haul	B-33C	800	.013			.44	2.71	3.15	3.69
2920	3000' haul		640	.016			.55	3.39	3.94	4.62
2940	5000' haul		520	.019			.67	4.17	4.84	5.70
2960	Common earth, 1500' haul		600	.017			.58	3.61	4.19	4.92
2980	3000' haul		560	.018			.62	3.87	4.49	5.30
3000	5000' haul		440	.023			.79	4.93	5.72	6.70
3005	Sandy clay & Loam, 1500' haul		648	.015			.54	3.34	3.88	4.56
3010	3000' haul		605	.017			.58	3.58	4.16	4.88
3015	5000' haul		475	.021			.73	4.56	5.29	6.20
3020	Clay, 1500' haul		450	.022			.78	4.82	5.60	6.55
3040	3000' haul		420	.024			.83	5.15	5.98	7.05
3060	5000' haul		320	.031			1.09	6.75	7.84	9.25

31 23 19.20 Dewatering Systems		Crew	Daily Output	Labor-Hours	Unit	2011 Bare Costs Material	Labor	Equipment	Total	Total Incl O&P
0010	**DEWATERING SYSTEMS**									
0020	Excavate drainage trench, 2′ wide, 2′ deep	B-11C	90	.178	C.Y.		5.45	3.56	9.01	12.90
0100	2′ wide, 3′ deep, with backhoe loader	″	135	.119			3.64	2.37	6.01	8.60
0200	Excavate sump pits by hand, light soil	1 Clab	7.10	1.127			30		30	50
0300	Heavy soil	″	3.50	2.286			60.50		60.50	101
0500	Pumping 8 hr., attended 2 hrs. per day, including 20 L.F.									
0550	of suction hose & 100 L.F. discharge hose									
0601	2″ diaphragm pump used for 8 hours	B-94A	4	2	Day		53	17.60	70.60	108
0652	4″ diaphragm pump used for 8 hours	B-94B	4	2			53	29.50	82.50	121
0801	8 hrs. attended, 2″ diaphragm pump	B-94A	1	8			212	70.50	282.50	435
0901	3″ centrifugal pump	B-10J	1	8			279	77.50	356.50	540
1001	4″ diaphragm pump	B-10I	1	8			279	117	396	585
1101	6″ centrifugal pump	B-10K	1	8			279	315	594	805
1300	CMP, incl. excavation 3′ deep, 12″ diameter	B-6	115	.209	L.F.	8.65	6	2.79	17.44	22.50
1400	18″ diameter		100	.240	″	14.35	6.90	3.20	24.45	30.50
1600	Sump hole construction, incl. excavation and gravel, pit		1250	.019	C.F.	.99	.55	.26	1.80	2.29
1700	With 12″ gravel collar, 12″ pipe, corrugated, 16 ga.		70	.343	L.F.	23	9.90	4.58	37.48	47
1800	15″ pipe, corrugated, 16 ga.		55	.436		29.50	12.60	5.85	47.95	60
1900	18″ pipe, corrugated, 16 ga.		50	.480		34.50	13.85	6.40	54.75	68
2000	24″ pipe, corrugated, 14 ga.		40	.600		41.50	17.30	8	66.80	83
2200	Wood lining, up to 4′ x 4′, add		300	.080	SFCA	17	2.31	1.07	20.38	23.50
9950	See Section 31 23 19.40 for wellpoints									
9960	See Section 31 23 19.30 for deep well systems									

31 23 19.30 Wells		Crew	Daily Output	Labor-Hours	Unit	Material	Labor	Equipment	Total	Total Incl O&P
0010	**WELLS**									
0011	For dewatering 10′ to 20′ deep, 2′ diameter									
0020	with steel casing, minimum	B-6	165	.145	V.L.F.	35	4.19	1.94	41.13	47.50
0050	Average		98	.245		40	7.05	3.27	50.32	59.50
0100	Maximum		49	.490		45	14.10	6.55	65.65	80
0300	For dewatering pumps see 01 54 33 in Reference Section									
0500	For domestic water wells, see Section 33 21 13.10									

31 23 19.40 Wellpoints		Crew	Daily Output	Labor-Hours	Unit	Material	Labor	Equipment	Total	Total Incl O&P
0010	**WELLPOINTS** R312319-90									
0011	For equipment rental, see 01 54 33 in Reference Section									
0100	Installation and removal of single stage system									
0110	Labor only, .75 labor-hours per L.F., minimum	1 Clab	10.70	.748	LF Hdr		19.80		19.80	33
0200	2.0 labor-hours per L.F., maximum	″	4	2	″		53		53	88.50
0400	Pump operation, 4 @ 6 hr. shifts									
0411	Per 24 hour day	4 Clab	1.27	25.197	Day		665		665	1,125
0501	Per 168 hour week, 160 hr. straight, 8 hr. double time		.18	177	Week		4,700		4,700	7,875
0551	Per 4.3 week month		.04	800	Month		21,200		21,200	35,400
0600	Complete installation, operation, equipment rental, fuel &									
0610	removal of system with 2″ wellpoints 5′ O.C.									
0700	100′ long header, 6″ diameter, first month	4 Eqlt	3.23	9.907	LF Hdr	157	330		487	715
0800	Thereafter, per month		4.13	7.748		125	260		385	565
1000	200′ long header, 8″ diameter, first month		6	5.333		142	179		321	450
1100	Thereafter, per month		8.39	3.814		70.50	128		198.50	287
1300	500′ long header, 8″ diameter, first month		10.63	3.010		55	101		156	226
1400	Thereafter, per month		20.91	1.530		39	51.50		90.50	127
1600	1,000′ long header, 10″ diameter, first month		11.62	2.754		47	92.50		139.50	203
1700	Thereafter, per month		41.81	.765		23.50	25.50		49	68
1900	Note: above figures include pumping 168 hrs. per week									

31 23 19 – Dewatering

31 23 19.40 Wellpoints		Crew	Daily Output	Labor-Hours	Unit	Material	Labor	Equipment	Total	Total Incl O&P
						2011 Bare Costs				
1910	and include the pump operator and one stand-by pump.									

31 23 23 – Fill

31 23 23.13 Backfill

		Crew	Daily Output	Labor-Hours	Unit	Material	Labor	Equipment	Total	Total Incl O&P
0010	**BACKFILL** R312323-30									
0015	By hand, no compaction, light soil	1 Clab	14	.571	L.C.Y.		15.10		15.10	25.50
0100	Heavy soil		11	.727	"		19.25		19.25	32
0300	Compaction in 6" layers, hand tamp, add to above		20.60	.388	E.C.Y.		10.25		10.25	17.20
0400	Roller compaction operator walking, add	B-10A	100	.080			2.79	1.47	4.26	6.20
0500	Air tamp, add	B-9D	190	.211			5.65	1.20	6.85	10.75
0600	Vibrating plate, add	A-1D	60	.133			3.53	.51	4.04	6.45
0800	Compaction in 12" layers, hand tamp, add to above	1 Clab	34	.235			6.20		6.20	10.40
0900	Roller compaction operator walking, add	B-10A	150	.053			1.86	.98	2.84	4.12
1000	Air tamp, add	B-9	285	.140			3.77	.68	4.45	7.05
1100	Vibrating plate, add	A-1E	90	.089			2.35	.43	2.78	4.42
1300	Dozer backfilling, bulk, up to 300' haul, no compaction	B-10B	1200	.007	L.C.Y.		.23	.88	1.11	1.34
1400	Air tamped, add	B-11B	80	.200	E.C.Y.		6	3.11	9.11	13.30
1600	Compacting backfill, 6" to 12" lifts, vibrating roller	B-10C	800	.010			.35	1.81	2.16	2.56
1700	Sheepsfoot roller	B-10D	750	.011			.37	2	2.37	2.81
1900	Dozer backfilling, trench, up to 300' haul, no compaction	B-10B	900	.009	L.C.Y.		.31	1.17	1.48	1.79
2000	Air tamped, add	B-11B	80	.200	E.C.Y.		6	3.11	9.11	13.30
2200	Compacting backfill, 6" to 12" lifts, vibrating roller	B-10C	700	.011			.40	2.07	2.47	2.92
2300	Sheepsfoot roller	B-10D	650	.012			.43	2.31	2.74	3.24

31 23 23.16 Fill By Borrow and Utility Bedding

		Crew	Daily Output	Labor-Hours	Unit	Material	Labor	Equipment	Total	Total Incl O&P
0010	**FILL BY BORROW AND UTILITY BEDDING**									
0015	Fill by borrow, load, 1 mile haul, spread with dozer									
0020	for embankments	B-15	1200	.023	L.C.Y.	15	.68	1.72	17.40	19.50
0035	Select fill for shoulders & embankments	"	1200	.023	"	20	.68	1.72	22.40	25
0040	Fill,for hauling over 1 mile,add to above per C.Y., see Section 31 23 23.20				Mile				1.29	1.55
0049	Utility bedding, for pipe & conduit, not incl. compaction									
0050	Crushed or screened bank run gravel	B-6	150	.160	L.C.Y.	25	4.61	2.14	31.75	37.50
0100	Crushed stone 3/4" to 1/2"		150	.160		38	4.61	2.14	44.75	52
0200	Sand, dead or bank		150	.160		15	4.61	2.14	21.75	26.50
0500	Compacting bedding in trench	A-1D	90	.089	E.C.Y.		2.35	.34	2.69	4.32
0600	If material source exceeds 2 miles, add for extra mileage.									
0610	See Section 31 23 23.20 for hauling mileage add.									

31 23 23.17 General Fill

		Crew	Daily Output	Labor-Hours	Unit	Material	Labor	Equipment	Total	Total Incl O&P
0010	**GENERAL FILL**									
0011	Spread dumped material, no compaction									
0020	By dozer, no compaction	B-10B	1000	.008	L.C.Y.		.28	1.05	1.33	1.62
0100	By hand	1 Clab	12	.667	"		17.65		17.65	29.50
0500	Gravel fill, compacted, under floor slabs, 4" deep	B-37	10000	.005	S.F.	.40	.13	.01	.54	.68
0600	6" deep		8600	.006		.60	.16	.02	.78	.94
0700	9" deep		7200	.007		1	.19	.02	1.21	1.43
0800	12" deep		6000	.008		1.40	.22	.02	1.64	1.94
1000	Alternate pricing method, 4" deep		120	.400	E.C.Y.	30	11.20	1.21	42.41	53
1100	6" deep		160	.300		30	8.40	.91	39.31	48
1200	9" deep		200	.240		30	6.70	.73	37.43	45
1300	12" deep		220	.218		30	6.10	.66	36.76	44
1500	For fill under exterior paving, see Section 32 11 23.23									

31 23 23 – Fill

31 23 23.20 Hauling		Crew	Daily Output	Labor-Hours	Unit	Material	2011 Bare Costs Labor	Equipment	Total	Total Incl O&P
0010	**HAULING**									
0011	Excavated or borrow, loose cubic yards									
0012	no loading equipment, including hauling,waiting, loading/dumping									
0013	time per cycle (wait, load, travel, unload or dump & return)									
0014	8 C.Y. truck, 15 MPH ave, cycle 0.5 miles, 10 min. wait/Ld./Uld.	B-34A	320	.025	L.C.Y.		.68	.99	1.67	2.21
0016	cycle 1 mile		272	.029			.80	1.16	1.96	2.61
0018	cycle 2 miles		208	.038			1.05	1.52	2.57	3.40
0020	cycle 4 miles		144	.056			1.51	2.19	3.70	4.92
0022	cycle 6 miles		112	.071			1.94	2.82	4.76	6.30
0024	cycle 8 miles		88	.091			2.47	3.58	6.05	8.05
0026	20 MPH ave,cycle 0.5 mile		336	.024			.65	.94	1.59	2.10
0028	cycle 1 mile		296	.027			.74	1.07	1.81	2.39
0030	cycle 2 miles		240	.033			.91	1.31	2.22	2.95
0032	cycle 4 miles		176	.045			1.24	1.79	3.03	4.02
0034	cycle 6 miles		136	.059			1.60	2.32	3.92	5.20
0036	cycle 8 miles		112	.071			1.94	2.82	4.76	6.30
0044	25 MPH ave, cycle 4 miles		192	.042			1.13	1.64	2.77	3.69
0046	cycle 6 miles		160	.050			1.36	1.97	3.33	4.43
0048	cycle 8 miles		128	.063			1.70	2.46	4.16	5.55
0050	30 MPH ave, cycle 4 miles		216	.037			1.01	1.46	2.47	3.28
0052	cycle 6 miles		176	.045			1.24	1.79	3.03	4.02
0054	cycle 8 miles		144	.056			1.51	2.19	3.70	4.92
0114	15 MPH ave, cycle 0.5 mile, 15 min. wait/Ld./Uld.		224	.036			.97	1.41	2.38	3.16
0116	cycle 1 mile		200	.040			1.09	1.58	2.67	3.53
0118	cycle 2 miles		168	.048			1.30	1.88	3.18	4.22
0120	cycle 4 miles		120	.067			1.81	2.63	4.44	5.90
0122	cycle 6 miles		96	.083			2.27	3.29	5.56	7.35
0124	cycle 8 miles		80	.100			2.72	3.94	6.66	8.85
0126	20 MPH ave, cycle 0.5 mile		232	.034			.94	1.36	2.30	3.06
0128	cycle 1 mile		208	.038			1.05	1.52	2.57	3.40
0130	cycle 2 miles		184	.043			1.18	1.71	2.89	3.85
0132	cycle 4 miles		144	.056			1.51	2.19	3.70	4.92
0134	cycle 6 miles		112	.071			1.94	2.82	4.76	6.30
0136	cycle 8 miles		96	.083			2.27	3.29	5.56	7.35
0144	25 MPH ave, cycle 4 miles		152	.053			1.43	2.08	3.51	4.65
0146	cycle 6 miles		128	.063			1.70	2.46	4.16	5.55
0148	cycle 8 miles		112	.071			1.94	2.82	4.76	6.30
0150	30 MPH ave, cycle 4 miles		168	.048			1.30	1.88	3.18	4.22
0152	cycle 6 miles		144	.056			1.51	2.19	3.70	4.92
0154	cycle 8 miles		120	.067			1.81	2.63	4.44	5.90
0214	15 MPH ave, cycle 0.5 mile, 20 min wait/Ld./Uld.		176	.045			1.24	1.79	3.03	4.02
0216	cycle 1 mile		160	.050			1.36	1.97	3.33	4.43
0218	cycle 2 miles		136	.059			1.60	2.32	3.92	5.20
0220	cycle 4 miles		104	.077			2.09	3.03	5.12	6.80
0222	cycle 6 miles		88	.091			2.47	3.58	6.05	8.05
0224	cycle 8 miles		72	.111			3.02	4.38	7.40	9.80
0226	20 MPH ave, cycle 0.5 mile		176	.045			1.24	1.79	3.03	4.02
0228	cycle 1 mile		168	.048			1.30	1.88	3.18	4.22
0230	cycle 2 miles		144	.056			1.51	2.19	3.70	4.92
0232	cycle 4 miles		120	.067			1.81	2.63	4.44	5.90
0234	cycle 6 miles		96	.083			2.27	3.29	5.56	7.35
0236	cycle 8 miles		88	.091			2.47	3.58	6.05	8.05

31 23 23.20 Hauling		Crew	Daily Output	Labor-Hours	Unit	Material	2011 Bare Costs Labor	Equipment	Total	Total Incl O&P
0244	25 MPH ave, cycle 4 miles	B-34A	128	.063	L.C.Y.		1.70	2.46	4.16	5.55
0246	cycle 6 miles		112	.071			1.94	2.82	4.76	6.30
0248	cycle 8 miles		96	.083			2.27	3.29	5.56	7.35
0250	30 MPH ave, cycle 4 miles		136	.059			1.60	2.32	3.92	5.20
0252	cycle 6 miles		120	.067			1.81	2.63	4.44	5.90
0254	cycle 8 miles		104	.077			2.09	3.03	5.12	6.80
0314	15 MPH ave, cycle 0.5 mile, 25 min wait/Ld./Uld.		144	.056			1.51	2.19	3.70	4.92
0316	cycle 1 mile		128	.063			1.70	2.46	4.16	5.55
0318	cycle 2 miles		112	.071			1.94	2.82	4.76	6.30
0320	cycle 4 miles		96	.083			2.27	3.29	5.56	7.35
0322	cycle 6 miles		80	.100			2.72	3.94	6.66	8.85
0324	cycle 8 miles		64	.125			3.40	4.93	8.33	11.05
0326	20 MPH ave, cycle 0.5 mile		144	.056			1.51	2.19	3.70	4.92
0328	cycle 1 mile		136	.059			1.60	2.32	3.92	5.20
0330	cycle 2 miles		120	.067			1.81	2.63	4.44	5.90
0332	cycle 4 miles		104	.077			2.09	3.03	5.12	6.80
0334	cycle 6 miles		88	.091			2.47	3.58	6.05	8.05
0336	cycle 8 miles		80	.100			2.72	3.94	6.66	8.85
0344	25 MPH ave, cycle 4 miles		112	.071			1.94	2.82	4.76	6.30
0346	cycle 6 miles		96	.083			2.27	3.29	5.56	7.35
0348	cycle 8 miles		88	.091			2.47	3.58	6.05	8.05
0350	30 MPH ave, cycle 4 miles		112	.071			1.94	2.82	4.76	6.30
0352	cycle 6 miles		104	.077			2.09	3.03	5.12	6.80
0354	cycle 8 miles		96	.083			2.27	3.29	5.56	7.35
0414	15 MPH ave, cycle 0.5 mile, 30 min wait/Ld./Uld.		120	.067			1.81	2.63	4.44	5.90
0416	cycle 1 mile		112	.071			1.94	2.82	4.76	6.30
0418	cycle 2 miles		96	.083			2.27	3.29	5.56	7.35
0420	cycle 4 miles		80	.100			2.72	3.94	6.66	8.85
0422	cycle 6 miles		72	.111			3.02	4.38	7.40	9.80
0424	cycle 8 miles		64	.125			3.40	4.93	8.33	11.05
0426	20 MPH ave, cycle 0.5 mile		120	.067			1.81	2.63	4.44	5.90
0428	cycle 1 mile		112	.071			1.94	2.82	4.76	6.30
0430	cycle 2 miles		104	.077			2.09	3.03	5.12	6.80
0432	cycle 4 miles		88	.091			2.47	3.58	6.05	8.05
0434	cycle 6 miles		80	.100			2.72	3.94	6.66	8.85
0436	cycle 8 miles		72	.111			3.02	4.38	7.40	9.80
0444	25 MPH ave, cycle 4 miles		96	.083			2.27	3.29	5.56	7.35
0446	cycle 6 miles		88	.091			2.47	3.58	6.05	8.05
0448	cycle 8 miles		80	.100			2.72	3.94	6.66	8.85
0450	30 MPH ave, cycle 4 miles		96	.083			2.27	3.29	5.56	7.35
0452	cycle 6 miles		88	.091			2.47	3.58	6.05	8.05
0454	cycle 8 miles		80	.100			2.72	3.94	6.66	8.85
0514	15 MPH ave, cycle 0.5 mile, 35 min wait/Ld./Uld.		104	.077			2.09	3.03	5.12	6.80
0516	cycle 1 mile		96	.083			2.27	3.29	5.56	7.35
0518	cycle 2 miles		88	.091			2.47	3.58	6.05	8.05
0520	cycle 4 miles		72	.111			3.02	4.38	7.40	9.80
0522	cycle 6 miles		64	.125			3.40	4.93	8.33	11.05
0524	cycle 8 miles		56	.143			3.89	5.65	9.54	12.65
0526	20 MPH ave, cycle 0.5 mile		104	.077			2.09	3.03	5.12	6.80
0528	cycle 1 mile		96	.083			2.27	3.29	5.56	7.35
0530	cycle 2 miles		96	.083			2.27	3.29	5.56	7.35
0532	cycle 4 miles		80	.100			2.72	3.94	6.66	8.85
0534	cycle 6 miles	↓	72	.111	↓		3.02	4.38	7.40	9.80

31 23 23.20 Hauling		Crew	Daily Output	Labor-Hours	Unit	Material	2011 Bare Costs Labor	Equipment	Total	Total Incl O&P
0536	cycle 8 miles	B-34A	64	.125	L.C.Y.		3.40	4.93	8.33	11.05
0544	25 MPH ave, cycle 4 miles		88	.091			2.47	3.58	6.05	8.05
0546	cycle 6 miles		80	.100			2.72	3.94	6.66	8.85
0548	cycle 8 miles		72	.111			3.02	4.38	7.40	9.80
0550	30 MPH ave, cycle 4 miles		88	.091			2.47	3.58	6.05	8.05
0552	cycle 6 miles		80	.100			2.72	3.94	6.66	8.85
0554	cycle 8 miles		72	.111			3.02	4.38	7.40	9.80
1014	12 C.Y. truck, cycle 0.5 mile, 15 MPH ave, 15 min. wait/Ld./Uld.	B-34B	336	.024			.65	1.51	2.16	2.73
1016	cycle 1 mile		300	.027			.73	1.69	2.42	3.06
1018	cycle 2 miles		252	.032			.86	2.01	2.87	3.64
1020	cycle 4 miles		180	.044			1.21	2.81	4.02	5.10
1022	cycle 6 miles		144	.056			1.51	3.51	5.02	6.40
1024	cycle 8 miles		120	.067			1.81	4.22	6.03	7.65
1025	cycle 10 miles		96	.083			2.27	5.25	7.52	9.55
1026	20 MPH ave, cycle 0.5 mile		348	.023			.63	1.45	2.08	2.64
1028	cycle 1 mile		312	.026			.70	1.62	2.32	2.94
1030	cycle 2 miles		276	.029			.79	1.83	2.62	3.33
1032	cycle 4 miles		216	.037			1.01	2.34	3.35	4.25
1034	cycle 6 miles		168	.048			1.30	3.01	4.31	5.45
1036	cycle 8 miles		144	.056			1.51	3.51	5.02	6.40
1038	cycle 10 miles		120	.067			1.81	4.22	6.03	7.65
1040	25 MPH ave, cycle 4 miles		228	.035			.95	2.22	3.17	4.02
1042	cycle 6 miles		192	.042			1.13	2.64	3.77	4.78
1044	cycle 8 miles		168	.048			1.30	3.01	4.31	5.45
1046	cycle 10 miles		144	.056			1.51	3.51	5.02	6.40
1050	30 MPH ave, cycle 4 miles		252	.032			.86	2.01	2.87	3.64
1052	cycle 6 miles		216	.037			1.01	2.34	3.35	4.25
1054	cycle 8 miles		180	.044			1.21	2.81	4.02	5.10
1056	cycle 10 miles		156	.051			1.39	3.24	4.63	5.90
1060	35 MPH ave, cycle 4 miles		264	.030			.82	1.92	2.74	3.48
1062	cycle 6 miles		228	.035			.95	2.22	3.17	4.02
1064	cycle 8 miles		204	.039			1.07	2.48	3.55	4.50
1066	cycle 10 miles		180	.044			1.21	2.81	4.02	5.10
1068	cycle 20 miles		120	.067			1.81	4.22	6.03	7.65
1069	cycle 30 miles		84	.095			2.59	6	8.59	10.95
1070	cycle 40 miles		72	.111			3.02	7.05	10.07	12.75
1072	40 MPH ave, cycle 6 miles		240	.033			.91	2.11	3.02	3.82
1074	cycle 8 miles		216	.037			1.01	2.34	3.35	4.25
1076	cycle 10 miles		192	.042			1.13	2.64	3.77	4.78
1078	cycle 20 miles		120	.067			1.81	4.22	6.03	7.65
1080	cycle 30 miles		96	.083			2.27	5.25	7.52	9.55
1082	cycle 40 miles		72	.111			3.02	7.05	10.07	12.75
1084	cycle 50 miles		60	.133			3.63	8.45	12.08	15.30
1094	45 MPH ave, cycle 8 miles		216	.037			1.01	2.34	3.35	4.25
1096	cycle 10 miles		204	.039			1.07	2.48	3.55	4.50
1098	cycle 20 miles		132	.061			1.65	3.83	5.48	6.95
1100	cycle 30 miles		108	.074			2.01	4.68	6.69	8.50
1102	cycle 40 miles		84	.095			2.59	6	8.59	10.95
1104	cycle 50 miles		72	.111			3.02	7.05	10.07	12.75
1106	50 MPH ave, cycle 10 miles		216	.037			1.01	2.34	3.35	4.25
1108	cycle 20 miles		144	.056			1.51	3.51	5.02	6.40
1110	cycle 30 miles		108	.074			2.01	4.68	6.69	8.50
1112	cycle 40 miles		84	.095			2.59	6	8.59	10.95

31 23 23.20 Hauling		Crew	Daily Output	Labor-Hours	Unit	2011 Bare Costs Material	Labor	Equipment	Total	Total Incl O&P
1114	cycle 50 miles	B-34B	72	.111	L.C.Y.		3.02	7.05	10.07	12.75
1214	15 MPH ave, cycle 0.5 mile, 20 min. wait/Ld./Uld.		264	.030			.82	1.92	2.74	3.48
1216	cycle 1 mile		240	.033			.91	2.11	3.02	3.82
1218	cycle 2 miles		204	.039			1.07	2.48	3.55	4.50
1220	cycle 4 miles		156	.051			1.39	3.24	4.63	5.90
1222	cycle 6 miles		132	.061			1.65	3.83	5.48	6.95
1224	cycle 8 miles		108	.074			2.01	4.68	6.69	8.50
1225	cycle 10 miles		96	.083			2.27	5.25	7.52	9.55
1226	20 MPH ave, cycle 0.5 mile		264	.030			.82	1.92	2.74	3.48
1228	cycle 1 mile		252	.032			.86	2.01	2.87	3.64
1230	cycle 2 miles		216	.037			1.01	2.34	3.35	4.25
1232	cycle 4 miles		180	.044			1.21	2.81	4.02	5.10
1234	cycle 6 miles		144	.056			1.51	3.51	5.02	6.40
1236	cycle 8 miles		132	.061			1.65	3.83	5.48	6.95
1238	cycle 10 miles		108	.074			2.01	4.68	6.69	8.50
1240	25 MPH ave, cycle 4 miles		192	.042			1.13	2.64	3.77	4.78
1242	cycle 6 miles		168	.048			1.30	3.01	4.31	5.45
1244	cycle 8 miles		144	.056			1.51	3.51	5.02	6.40
1246	cycle 10 miles		132	.061			1.65	3.83	5.48	6.95
1250	30 MPH ave, cycle 4 miles		204	.039			1.07	2.48	3.55	4.50
1252	cycle 6 miles		180	.044			1.21	2.81	4.02	5.10
1254	cycle 8 miles		156	.051			1.39	3.24	4.63	5.90
1256	cycle 10 miles		144	.056			1.51	3.51	5.02	6.40
1260	35 MPH ave, cycle 4 miles		216	.037			1.01	2.34	3.35	4.25
1262	cycle 6 miles		192	.042			1.13	2.64	3.77	4.78
1264	cycle 8 miles		168	.048			1.30	3.01	4.31	5.45
1266	cycle 10 miles		156	.051			1.39	3.24	4.63	5.90
1268	cycle 20 miles		108	.074			2.01	4.68	6.69	8.50
1269	cycle 30 miles		72	.111			3.02	7.05	10.07	12.75
1270	cycle 40 miles		60	.133			3.63	8.45	12.08	15.30
1272	40 MPH ave, cycle 6 miles		192	.042			1.13	2.64	3.77	4.78
1274	cycle 8 miles		180	.044			1.21	2.81	4.02	5.10
1276	cycle 10 miles		156	.051			1.39	3.24	4.63	5.90
1278	cycle 20 miles		108	.074			2.01	4.68	6.69	8.50
1280	cycle 30 miles		84	.095			2.59	6	8.59	10.95
1282	cycle 40 miles		72	.111			3.02	7.05	10.07	12.75
1284	cycle 50 miles		60	.133			3.63	8.45	12.08	15.30
1294	45 MPH ave, cycle 8 miles		180	.044			1.21	2.81	4.02	5.10
1296	cycle 10 miles		168	.048			1.30	3.01	4.31	5.45
1298	cycle 20 miles		120	.067			1.81	4.22	6.03	7.65
1300	cycle 30 miles		96	.083			2.27	5.25	7.52	9.55
1302	cycle 40 miles		72	.111			3.02	7.05	10.07	12.75
1304	cycle 50 miles		60	.133			3.63	8.45	12.08	15.30
1306	50 MPH ave, cycle 10 miles		180	.044			1.21	2.81	4.02	5.10
1308	cycle 20 miles		132	.061			1.65	3.83	5.48	6.95
1310	cycle 30 miles		96	.083			2.27	5.25	7.52	9.55
1312	cycle 40 miles		84	.095			2.59	6	8.59	10.95
1314	cycle 50 miles		72	.111			3.02	7.05	10.07	12.75
1414	15 MPH ave, cycle 0.5 mile, 25 min. wait/Ld./Uld.		204	.039			1.07	2.48	3.55	4.50
1416	cycle 1 mile		192	.042			1.13	2.64	3.77	4.78
1418	cycle 2 miles		168	.048			1.30	3.01	4.31	5.45
1420	cycle 4 miles		132	.061			1.65	3.83	5.48	6.95
1422	cycle 6 miles	↓	120	.067	↓		1.81	4.22	6.03	7.65

	31 23 23.20 Hauling	Crew	Daily Output	Labor-Hours	Unit	Material	2011 Bare Costs Labor	Equipment	Total	Total Incl O&P
1424	cycle 8 miles	B-34B	96	.083	L.C.Y.		2.27	5.25	7.52	9.55
1425	cycle 10 miles		84	.095			2.59	6	8.59	10.95
1426	20 MPH ave, cycle 0.5 mile		216	.037			1.01	2.34	3.35	4.25
1428	cycle 1 mile		204	.039			1.07	2.48	3.55	4.50
1430	cycle 2 miles		180	.044			1.21	2.81	4.02	5.10
1432	cycle 4 miles		156	.051			1.39	3.24	4.63	5.90
1434	cycle 6 miles		132	.061			1.65	3.83	5.48	6.95
1436	cycle 8 miles		120	.067			1.81	4.22	6.03	7.65
1438	cycle 10 miles		96	.083			2.27	5.25	7.52	9.55
1440	25 MPH ave, cycle 4 miles		168	.048			1.30	3.01	4.31	5.45
1442	cycle 6 miles		144	.056			1.51	3.51	5.02	6.40
1444	cycle 8 miles		132	.061			1.65	3.83	5.48	6.95
1446	cycle 10 miles		108	.074			2.01	4.68	6.69	8.50
1450	30 MPH ave, cycle 4 miles		168	.048			1.30	3.01	4.31	5.45
1452	cycle 6 miles		156	.051			1.39	3.24	4.63	5.90
1454	cycle 8 miles		132	.061			1.65	3.83	5.48	6.95
1456	cycle 10 miles		120	.067			1.81	4.22	6.03	7.65
1460	35 MPH ave, cycle 4 miles		180	.044			1.21	2.81	4.02	5.10
1462	cycle 6 miles		156	.051			1.39	3.24	4.63	5.90
1464	cycle 8 miles		144	.056			1.51	3.51	5.02	6.40
1466	cycle 10 miles		132	.061			1.65	3.83	5.48	6.95
1468	cycle 20 miles		96	.083			2.27	5.25	7.52	9.55
1469	cycle 30 miles		72	.111			3.02	7.05	10.07	12.75
1470	cycle 40 miles		60	.133			3.63	8.45	12.08	15.30
1472	40 MPH ave, cycle 6 miles		168	.048			1.30	3.01	4.31	5.45
1474	cycle 8 miles		156	.051			1.39	3.24	4.63	5.90
1476	cycle 10 miles		144	.056			1.51	3.51	5.02	6.40
1478	cycle 20 miles		96	.083			2.27	5.25	7.52	9.55
1480	cycle 30 miles		84	.095			2.59	6	8.59	10.95
1482	cycle 40 miles		60	.133			3.63	8.45	12.08	15.30
1484	cycle 50 miles		60	.133			3.63	8.45	12.08	15.30
1494	45 MPH ave, cycle 8 miles		156	.051			1.39	3.24	4.63	5.90
1496	cycle 10 miles		144	.056			1.51	3.51	5.02	6.40
1498	cycle 20 miles		108	.074			2.01	4.68	6.69	8.50
1500	cycle 30 miles		84	.095			2.59	6	8.59	10.95
1502	cycle 40 miles		72	.111			3.02	7.05	10.07	12.75
1504	cycle 50 miles		60	.133			3.63	8.45	12.08	15.30
1506	50 MPH ave, cycle 10 miles		156	.051			1.39	3.24	4.63	5.90
1508	cycle 20 miles		120	.067			1.81	4.22	6.03	7.65
1510	cycle 30 miles		96	.083			2.27	5.25	7.52	9.55
1512	cycle 40 miles		72	.111			3.02	7.05	10.07	12.75
1514	cycle 50 miles		60	.133			3.63	8.45	12.08	15.30
1614	15 MPH, cycle 0.5 mile, 30 min. wait/Ld./Uld.		180	.044			1.21	2.81	4.02	5.10
1616	cycle 1 mile		168	.048			1.30	3.01	4.31	5.45
1618	cycle 2 miles		144	.056			1.51	3.51	5.02	6.40
1620	cycle 4 miles		120	.067			1.81	4.22	6.03	7.65
1622	cycle 6 miles		108	.074			2.01	4.68	6.69	8.50
1624	cycle 8 miles		84	.095			2.59	6	8.59	10.95
1625	cycle 10 miles		84	.095			2.59	6	8.59	10.95
1626	20 MPH ave, cycle 0.5 mile		180	.044			1.21	2.81	4.02	5.10
1628	cycle 1 mile		168	.048			1.30	3.01	4.31	5.45
1630	cycle 2 miles		156	.051			1.39	3.24	4.63	5.90
1632	cycle 4 miles		132	.061			1.65	3.83	5.48	6.95

31 23 23.20 Hauling		Crew	Daily Output	Labor-Hours	Unit	Material	2011 Bare Costs Labor	Equipment	Total	Total Incl O&P
1634	cycle 6 miles	B-34B	120	.067	L.C.Y.		1.81	4.22	6.03	7.65
1636	cycle 8 miles		108	.074			2.01	4.68	6.69	8.50
1638	cycle 10 miles		96	.083			2.27	5.25	7.52	9.55
1640	25 MPH ave, cycle 4 miles		144	.056			1.51	3.51	5.02	6.40
1642	cycle 6 miles		132	.061			1.65	3.83	5.48	6.95
1644	cycle 8 miles		108	.074			2.01	4.68	6.69	8.50
1646	cycle 10 miles		108	.074			2.01	4.68	6.69	8.50
1650	30 MPH ave, cycle 4 miles		144	.056			1.51	3.51	5.02	6.40
1652	cycle 6 miles		132	.061			1.65	3.83	5.48	6.95
1654	cycle 8 miles		120	.067			1.81	4.22	6.03	7.65
1656	cycle 10 miles		108	.074			2.01	4.68	6.69	8.50
1660	35 MPH ave, cycle 4 miles		156	.051			1.39	3.24	4.63	5.90
1662	cycle 6 miles		144	.056			1.51	3.51	5.02	6.40
1664	cycle 8 miles		132	.061			1.65	3.83	5.48	6.95
1666	cycle 10 miles		120	.067			1.81	4.22	6.03	7.65
1668	cycle 20 miles		84	.095			2.59	6	8.59	10.95
1669	cycle 30 miles		72	.111			3.02	7.05	10.07	12.75
1670	cycle 40 miles		60	.133			3.63	8.45	12.08	15.30
1672	40 MPH, cycle 6 miles		144	.056			1.51	3.51	5.02	6.40
1674	cycle 8 miles		132	.061			1.65	3.83	5.48	6.95
1676	cycle 10 miles		120	.067			1.81	4.22	6.03	7.65
1678	cycle 20 miles		96	.083			2.27	5.25	7.52	9.55
1680	cycle 30 miles		72	.111			3.02	7.05	10.07	12.75
1682	cycle 40 miles		60	.133			3.63	8.45	12.08	15.30
1684	cycle 50 miles		48	.167			4.53	10.55	15.08	19.10
1694	45 MPH ave, cycle 8 miles		144	.056			1.51	3.51	5.02	6.40
1696	cycle 10 miles		132	.061			1.65	3.83	5.48	6.95
1698	cycle 20 miles		96	.083			2.27	5.25	7.52	9.55
1700	cycle 30 miles		84	.095			2.59	6	8.59	10.95
1702	cycle 40 miles		60	.133			3.63	8.45	12.08	15.30
1704	cycle 50 miles		60	.133			3.63	8.45	12.08	15.30
1706	50 MPH ave, cycle 10 miles		132	.061			1.65	3.83	5.48	6.95
1708	cycle 20 miles		108	.074			2.01	4.68	6.69	8.50
1710	cycle 30 miles		84	.095			2.59	6	8.59	10.95
1712	cycle 40 miles		72	.111			3.02	7.05	10.07	12.75
1714	cycle 50 miles		60	.133			3.63	8.45	12.08	15.30
2000	Hauling, 8 C.Y. truck, small project cost per hour	B-34A	8	1	Hr.		27	39.50	66.50	88.50
2100	12 C.Y. Truck	B-34B	8	1			27	63.50	90.50	115
2150	16.5 C.Y. Truck	B-34C	8	1			27	73	100	126
2175	18 C.Y. 8 wheel Truck	B-34I	8	1			27	82.50	109.50	136
2200	20 C.Y. Truck	B-34D	8	1			27	75	102	128
2300	Grading at dump, or embankment if required, by dozer	B-10B	1000	.008	L.C.Y.		.28	1.05	1.33	1.62
2310	Spotter at fill or cut, if required	1 Clab	8	1	Hr.		26.50		26.50	44.50
9014	18 C.Y. truck, 8 wheels,15 min. wait/Ld./Uld.,15 MPH, cycle 0.5 mi.	B-34I	504	.016	L.C.Y.		.43	1.31	1.74	2.16
9016	cycle 1 mile		450	.018			.48	1.47	1.95	2.42
9018	cycle 2 miles		378	.021			.58	1.75	2.33	2.87
9020	cycle 4 miles		270	.030			.81	2.45	3.26	4.03
9022	cycle 6 miles		216	.037			1.01	3.06	4.07	5.05
9024	cycle 8 miles		180	.044			1.21	3.67	4.88	6.05
9025	cycle 10 miles		144	.056			1.51	4.59	6.10	7.55
9026	20 MPH ave, cycle 0.5 mile		522	.015			.42	1.27	1.69	2.08
9028	cycle 1 mile		468	.017			.46	1.41	1.87	2.32
9030	cycle 2 miles		414	.019			.53	1.60	2.13	2.63

31 23 23.20 Hauling		Crew	Daily Output	Labor-Hours	Unit	Material	2011 Bare Costs Labor	Equipment	Total	Total Incl O&P
9032	cycle 4 miles	B-34I	324	.025	L.C.Y.		.67	2.04	2.71	3.35
9034	cycle 6 miles		252	.032			.86	2.62	3.48	4.31
9036	cycle 8 miles		216	.037			1.01	3.06	4.07	5.05
9038	cycle 10 miles		180	.044			1.21	3.67	4.88	6.05
9040	25 MPH ave, cycle 4 miles		342	.023			.64	1.93	2.57	3.18
9042	cycle 6 miles		288	.028			.76	2.29	3.05	3.77
9044	cycle 8 miles		252	.032			.86	2.62	3.48	4.31
9046	cycle 10 miles		216	.037			1.01	3.06	4.07	5.05
9050	30 MPH ave, cycle 4 miles		378	.021			.58	1.75	2.33	2.87
9052	cycle 6 miles		324	.025			.67	2.04	2.71	3.35
9054	cycle 8 miles		270	.030			.81	2.45	3.26	4.03
9056	cycle 10 miles		234	.034			.93	2.82	3.75	4.65
9060	35 MPH ave, cycle 4 miles		396	.020			.55	1.67	2.22	2.75
9062	cycle 6 miles		342	.023			.64	1.93	2.57	3.18
9064	cycle 8 miles		288	.028			.76	2.29	3.05	3.77
9066	cycle 10 miles		270	.030			.81	2.45	3.26	4.03
9068	cycle 20 miles		162	.049			1.34	4.08	5.42	6.70
9070	cycle 30 miles		126	.063			1.73	5.25	6.98	8.60
9072	cycle 40 miles		90	.089			2.42	7.35	9.77	12.10
9074	40 MPH ave, cycle 6 miles		360	.022			.60	1.84	2.44	3.02
9076	cycle 8 miles		324	.025			.67	2.04	2.71	3.35
9078	cycle 10 miles		288	.028			.76	2.29	3.05	3.77
9080	cycle 20 miles		180	.044			1.21	3.67	4.88	6.05
9082	cycle 30 miles		144	.056			1.51	4.59	6.10	7.55
9084	cycle 40 miles		108	.074			2.01	6.10	8.11	10.10
9086	cycle 50 miles		90	.089			2.42	7.35	9.77	12.10
9094	45 MPH ave, cycle 8 miles		324	.025			.67	2.04	2.71	3.35
9096	cycle 10 miles		306	.026			.71	2.16	2.87	3.56
9098	cycle 20 miles		198	.040			1.10	3.34	4.44	5.50
9100	cycle 30 miles		144	.056			1.51	4.59	6.10	7.55
9102	cycle 40 miles		126	.063			1.73	5.25	6.98	8.60
9104	cycle 50 miles		108	.074			2.01	6.10	8.11	10.10
9106	50 MPH ave, cycle 10 miles		324	.025			.67	2.04	2.71	3.35
9108	cycle 20 miles		216	.037			1.01	3.06	4.07	5.05
9110	cycle 30 miles		162	.049			1.34	4.08	5.42	6.70
9112	cycle 40 miles		126	.063			1.73	5.25	6.98	8.60
9114	cycle 50 miles		108	.074			2.01	6.10	8.11	10.10
9214	20 min. wait/Ld./Uld.,15 MPH, cycle 0.5 mi.		396	.020			.55	1.67	2.22	2.75
9216	cycle 1 mile		360	.022			.60	1.84	2.44	3.02
9218	cycle 2 miles		306	.026			.71	2.16	2.87	3.56
9220	cycle 4 miles		234	.034			.93	2.82	3.75	4.65
9222	cycle 6 miles		198	.040			1.10	3.34	4.44	5.50
9224	cycle 8 miles		162	.049			1.34	4.08	5.42	6.70
9225	cycle 10 miles		144	.056			1.51	4.59	6.10	7.55
9226	20 MPH ave, cycle 0.5 mile		396	.020			.55	1.67	2.22	2.75
9228	cycle 1 mile		378	.021			.58	1.75	2.33	2.87
9230	cycle 2 miles		324	.025			.67	2.04	2.71	3.35
9232	cycle 4 miles		270	.030			.81	2.45	3.26	4.03
9234	cycle 6 miles		216	.037			1.01	3.06	4.07	5.05
9236	cycle 8 miles		198	.040			1.10	3.34	4.44	5.50
9238	cycle 10 miles		162	.049			1.34	4.08	5.42	6.70
9240	25 MPH ave, cycle 4 miles		288	.028			.76	2.29	3.05	3.77
9242	cycle 6 miles		252	.032			.86	2.62	3.48	4.31

31 23 23.20 Hauling		Crew	Daily Output	Labor-Hours	Unit	Material	2011 Bare Costs Labor	Equipment	Total	Total Incl O&P
9244	cycle 8 miles	B-34I	216	.037	L.C.Y.		1.01	3.06	4.07	5.05
9246	cycle 10 miles		198	.040			1.10	3.34	4.44	5.50
9250	30 MPH ave, cycle 4 miles		306	.026			.71	2.16	2.87	3.56
9252	cycle 6 miles		270	.030			.81	2.45	3.26	4.03
9254	cycle 8 miles		234	.034			.93	2.82	3.75	4.65
9256	cycle 10 miles		216	.037			1.01	3.06	4.07	5.05
9260	35 MPH ave, cycle 4 miles		324	.025			.67	2.04	2.71	3.35
9262	cycle 6 miles		288	.028			.76	2.29	3.05	3.77
9264	cycle 8 miles		252	.032			.86	2.62	3.48	4.31
9266	cycle 10 miles		234	.034			.93	2.82	3.75	4.65
9268	cycle 20 miles		162	.049			1.34	4.08	5.42	6.70
9270	cycle 30 miles		108	.074			2.01	6.10	8.11	10.10
9272	cycle 40 miles		90	.089			2.42	7.35	9.77	12.10
9274	40 MPH ave, cycle 6 miles		288	.028			.76	2.29	3.05	3.77
9276	cycle 8 miles		270	.030			.81	2.45	3.26	4.03
9278	cycle 10 miles		234	.034			.93	2.82	3.75	4.65
9280	cycle 20 miles		162	.049			1.34	4.08	5.42	6.70
9282	cycle 30 miles		126	.063			1.73	5.25	6.98	8.60
9284	cycle 40 miles		108	.074			2.01	6.10	8.11	10.10
9286	cycle 50 miles		90	.089			2.42	7.35	9.77	12.10
9294	45 MPH ave, cycle 8 miles		270	.030			.81	2.45	3.26	4.03
9296	cycle 10 miles		252	.032			.86	2.62	3.48	4.31
9298	cycle 20 miles		180	.044			1.21	3.67	4.88	6.05
9300	cycle 30 miles		144	.056			1.51	4.59	6.10	7.55
9302	cycle 40 miles		108	.074			2.01	6.10	8.11	10.10
9304	cycle 50 miles		90	.089			2.42	7.35	9.77	12.10
9306	50 MPH ave, cycle 10 miles		270	.030			.81	2.45	3.26	4.03
9308	cycle 20 miles		198	.040			1.10	3.34	4.44	5.50
9310	cycle 30 miles		144	.056			1.51	4.59	6.10	7.55
9312	cycle 40 miles		126	.063			1.73	5.25	6.98	8.60
9314	cycle 50 miles		108	.074			2.01	6.10	8.11	10.10
9414	25 min. wait/Ld./Uld.,15 MPH, cycle 0.5 mi.		306	.026			.71	2.16	2.87	3.56
9416	cycle 1 mile		288	.028			.76	2.29	3.05	3.77
9418	cycle 2 miles		252	.032			.86	2.62	3.48	4.31
9420	cycle 4 miles		198	.040			1.10	3.34	4.44	5.50
9422	cycle 6 miles		180	.044			1.21	3.67	4.88	6.05
9424	cycle 8 miles		144	.056			1.51	4.59	6.10	7.55
9425	cycle 10 miles		126	.063			1.73	5.25	6.98	8.60
9426	20 MPH ave, cycle 0.5 mile		324	.025			.67	2.04	2.71	3.35
9428	cycle 1 mile		306	.026			.71	2.16	2.87	3.56
9430	cycle 2 miles		270	.030			.81	2.45	3.26	4.03
9432	cycle 4 miles		234	.034			.93	2.82	3.75	4.65
9434	cycle 6 miles		198	.040			1.10	3.34	4.44	5.50
9436	cycle 8 miles		180	.044			1.21	3.67	4.88	6.05
9438	cycle 10 miles		144	.056			1.51	4.59	6.10	7.55
9440	25 MPH ave, cycle 4 miles		252	.032			.86	2.62	3.48	4.31
9442	cycle 6 miles		216	.037			1.01	3.06	4.07	5.05
9444	cycle 8 miles		198	.040			1.10	3.34	4.44	5.50
9446	cycle 10 miles		180	.044			1.21	3.67	4.88	6.05
9450	30 MPH ave, cycle 4 miles		252	.032			.86	2.62	3.48	4.31
9452	cycle 6 miles		234	.034			.93	2.82	3.75	4.65
9454	cycle 8 miles		198	.040			1.10	3.34	4.44	5.50
9456	cycle 10 miles	↓	180	.044	↓		1.21	3.67	4.88	6.05

31 23 23.20 Hauling

	31 23 23.20 Hauling	Crew	Daily Output	Labor-Hours	Unit	Material	2011 Bare Costs Labor	Equipment	Total	Total Incl O&P
9460	35 MPH ave, cycle 4 miles	B-34I	270	.030	L.C.Y.		.81	2.45	3.26	4.03
9462	cycle 6 miles		234	.034			.93	2.82	3.75	4.65
9464	cycle 8 miles		216	.037			1.01	3.06	4.07	5.05
9466	cycle 10 miles		198	.040			1.10	3.34	4.44	5.50
9468	cycle 20 miles		144	.056			1.51	4.59	6.10	7.55
9470	cycle 30 miles		108	.074			2.01	6.10	8.11	10.10
9472	cycle 40 miles		90	.089			2.42	7.35	9.77	12.10
9474	40 MPH ave, cycle 6 miles		252	.032			.86	2.62	3.48	4.31
9476	cycle 8 miles		234	.034			.93	2.82	3.75	4.65
9478	cycle 10 miles		216	.037			1.01	3.06	4.07	5.05
9480	cycle 20 miles		144	.056			1.51	4.59	6.10	7.55
9482	cycle 30 miles		126	.063			1.73	5.25	6.98	8.60
9484	cycle 40 miles		90	.089			2.42	7.35	9.77	12.10
9486	cycle 50 miles		90	.089			2.42	7.35	9.77	12.10
9494	45 MPH ave, cycle 8 miles		234	.034			.93	2.82	3.75	4.65
9496	cycle 10 miles		216	.037			1.01	3.06	4.07	5.05
9498	cycle 20 miles		162	.049			1.34	4.08	5.42	6.70
9500	cycle 30 miles		126	.063			1.73	5.25	6.98	8.60
9502	cycle 40 miles		108	.074			2.01	6.10	8.11	10.10
9504	cycle 50 miles		90	.089			2.42	7.35	9.77	12.10
9506	50 MPH ave, cycle 10 miles		234	.034			.93	2.82	3.75	4.65
9508	cycle 20 miles		180	.044			1.21	3.67	4.88	6.05
9510	cycle 30 miles		144	.056			1.51	4.59	6.10	7.55
9512	cycle 40 miles		108	.074			2.01	6.10	8.11	10.10
9514	cycle 50 miles		90	.089			2.42	7.35	9.77	12.10
9614	30 min. wait/Ld./Uld.,15 MPH, cycle 0.5 mi.		270	.030			.81	2.45	3.26	4.03
9616	cycle 1 mile		252	.032			.86	2.62	3.48	4.31
9618	cycle 2 miles		216	.037			1.01	3.06	4.07	5.05
9620	cycle 4 miles		180	.044			1.21	3.67	4.88	6.05
9622	cycle 6 miles		162	.049			1.34	4.08	5.42	6.70
9624	cycle 8 miles		126	.063			1.73	5.25	6.98	8.60
9625	cycle 10 miles		126	.063			1.73	5.25	6.98	8.60
9626	20 MPH ave, cycle 0.5 mile		270	.030			.81	2.45	3.26	4.03
9628	cycle 1 mile		252	.032			.86	2.62	3.48	4.31
9630	cycle 2 miles		234	.034			.93	2.82	3.75	4.65
9632	cycle 4 miles		198	.040			1.10	3.34	4.44	5.50
9634	cycle 6 miles		180	.044			1.21	3.67	4.88	6.05
9636	cycle 8 miles		162	.049			1.34	4.08	5.42	6.70
9638	cycle 10 miles		144	.056			1.51	4.59	6.10	7.55
9640	25 MPH ave, cycle 4 miles		216	.037			1.01	3.06	4.07	5.05
9642	cycle 6 miles		198	.040			1.10	3.34	4.44	5.50
9644	cycle 8 miles		180	.044			1.21	3.67	4.88	6.05
9646	cycle 10 miles		162	.049			1.34	4.08	5.42	6.70
9650	30 MPH ave, cycle 4 miles		216	.037			1.01	3.06	4.07	5.05
9652	cycle 6 miles		198	.040			1.10	3.34	4.44	5.50
9654	cycle 8 miles		180	.044			1.21	3.67	4.88	6.05
9656	cycle 10 miles		162	.049			1.34	4.08	5.42	6.70
9660	35 MPH ave, cycle 4 miles		234	.034			.93	2.82	3.75	4.65
9662	cycle 6 miles		216	.037			1.01	3.06	4.07	5.05
9664	cycle 8 miles		198	.040			1.10	3.34	4.44	5.50
9666	cycle 10 miles		180	.044			1.21	3.67	4.88	6.05
9668	cycle 20 miles		126	.063			1.73	5.25	6.98	8.60
9670	cycle 30 miles		108	.074			2.01	6.10	8.11	10.10

31 23 Excavation and Fill

31 23 23 – Fill

31 23 23.20 Hauling		Crew	Daily Output	Labor-Hours	Unit	Material	2011 Bare Costs Labor	Equipment	Total	Total Incl O&P
9672	cycle 40 miles	B-34I	90	.089	L.C.Y.		2.42	7.35	9.77	12.10
9674	40 MPH ave, cycle 6 miles		216	.037			1.01	3.06	4.07	5.05
9676	cycle 8 miles		198	.040			1.10	3.34	4.44	5.50
9678	cycle 10 miles		180	.044			1.21	3.67	4.88	6.05
9680	cycle 20 miles		144	.056			1.51	4.59	6.10	7.55
9682	cycle 30 miles		108	.074			2.01	6.10	8.11	10.10
9684	cycle 40 miles		90	.089			2.42	7.35	9.77	12.10
9686	cycle 50 miles		72	.111			3.02	9.20	12.22	15.10
9694	45 MPH ave, cycle 8 miles		216	.037			1.01	3.06	4.07	5.05
9696	cycle 10 miles		198	.040			1.10	3.34	4.44	5.50
9698	cycle 20 miles		144	.056			1.51	4.59	6.10	7.55
9700	cycle 30 miles		126	.063			1.73	5.25	6.98	8.60
9702	cycle 40 miles		108	.074			2.01	6.10	8.11	10.10
9704	cycle 50 miles		90	.089			2.42	7.35	9.77	12.10
9706	50 MPH ave, cycle 10 miles		198	.040			1.10	3.34	4.44	5.50
9708	cycle 20 miles		162	.049			1.34	4.08	5.42	6.70
9710	cycle 30 miles		126	.063			1.73	5.25	6.98	8.60
9712	cycle 40 miles		108	.074			2.01	6.10	8.11	10.10
9714	cycle 50 miles		90	.089			2.42	7.35	9.77	12.10

31 23 23.24 Compaction, Structural

		Crew	Daily Output	Labor-Hours	Unit	Material	Labor	Equipment	Total	Total Incl O&P
0010	**COMPACTION, STRUCTURAL** R312323-30									
0020	Steel wheel tandem roller, 5 tons	B-10E	8	1	Hr.		35	18.15	53.15	77
0100	10 tons	B-10F	8	1	"		35	28	63	88
0300	Sheepsfoot or wobbly wheel roller, 8" lifts, common fill	B-10G	1300	.006	E.C.Y.		.21	.84	1.05	1.28
0400	Select fill	"	1500	.005			.19	.73	.92	1.10
0600	Vibratory plate, 8" lifts, common fill	A-1D	200	.040			1.06	.15	1.21	1.94
0700	Select fill	"	216	.037			.98	.14	1.12	1.80

31 25 Erosion and Sedimentation Controls

31 25 14 – Stabilization Measures for Erosion and Sedimentation Control

31 25 14.16 Rolled Erosion Control Mats and Blankets

		Crew	Daily Output	Labor-Hours	Unit	Material	Labor	Equipment	Total	Total Incl O&P
0010	**ROLLED EROSION CONTROL MATS AND BLANKETS**									
0020	Jute mesh, 100 S.Y. per roll, 4' wide, stapled G	B-80A	2400	.010	S.Y.	.76	.26	.09	1.11	1.38
0100	Plastic netting, stapled, 2" x 1" mesh, 20 mil G	B-1	2500	.010		.36	.26		.62	.84
0200	Polypropylene mesh, stapled, 6.5 oz./S.Y. G		2500	.010		1.34	.26		1.60	1.91
0300	Tobacco netting, or jute mesh #2, stapled G		2500	.010		.11	.26		.37	.56
1000	Silt fence, polypropylene, 3' high, ideal conditions G	2 Clab	1600	.010	L.F.	.41	.26		.67	.89
1100	Adverse conditions G	"	950	.017	"	.41	.45		.86	1.20
1200	Place and remove hay bales G	A-2	3	8	Ton	164	211	57	432	600
1250	Hay bales, staked G	"	2500	.010	L.F.	6.55	.25	.07	6.87	7.70

31 31 Soil Treatment

31 31 16 – Termite Control

31 31 16.13 Chemical Termite Control		Crew	Daily Output	Labor-Hours	Unit	2011 Bare Costs Material	Labor	Equipment	Total	Total Incl O&P
0010	**CHEMICAL TERMITE CONTROL**									
0020	Slab and walls, residential	1 Skwk	1200	.007	SF Flr.	.32	.23		.55	.73
0100	Commercial, minimum		2496	.003		.33	.11		.44	.54
0200	Maximum		1645	.005		.50	.17		.67	.82
0400	Insecticides for termite control, minimum		14.20	.563	Gal.	12.75	19.20		31.95	46
0500	Maximum		11	.727	″	22	25		47	65.50

31 32 Soil Stabilization

31 32 13 – Soil Mixing Stabilization

31 32 13.30 Calcium Chloride		Crew	Daily Output	Labor-Hours	Unit	Material	Labor	Equipment	Total	Total Incl O&P
0010	**CALCIUM CHLORIDE**									
0020	Calcium chloride delivered, 100 lb. bags, truckload lots				Ton	650			650	715
0030	Solution, 4 lb. flake per gallon, tank truck delivery				Gal.	1.22			1.22	1.34

31 33 Rock Stabilization

31 33 13 – Rock Bolting and Grouting

31 33 13.10 Rock Bolting		Crew	Daily Output	Labor-Hours	Unit	Material	Labor	Equipment	Total	Total Incl O&P
0010	**ROCK BOLTING**									
2020	Hollow core, prestressable anchor, 1″ diameter, 5′ long	2 Skwk	32	.500	Ea.	89.50	17.05		106.55	127
2025	10′ long		24	.667		179	22.50		201.50	235
2060	2″ diameter, 5′ long		32	.500		340	17.05		357.05	405
2065	10′ long		24	.667		670	22.50		692.50	775
2100	Super high-tensile, 3/4″ diameter, 5′ long		32	.500		20.50	17.05		37.55	51
2105	10′ long		24	.667		40.50	22.50		63	82.50
2160	2″ diameter, 5′ long		32	.500		181	17.05		198.05	228
2165	10′ long		24	.667		360	22.50		382.50	440
4400	Drill hole for rock bolt, 1-3/4″ diam., 5′ long (for 3/4″ bolt)	B-56	17	.941			25	76	101	125
4405	10′ long		9	1.778			47	143	190	237
4420	2″ diameter, 5′ long (for 1″ bolt)		13	1.231			32.50	99	131.50	164
4425	10′ long		7	2.286			60.50	184	244.50	305
4460	3-1/2″ diameter, 5′ long (for 2″ bolt)		10	1.600			42.50	129	171.50	213
4465	10′ long		5	3.200			84.50	258	342.50	425

31 36 Gabions

31 36 13 – Gabion Boxes

31 36 13.10 Gabion Box Systems		Crew	Daily Output	Labor-Hours	Unit	Material	Labor	Equipment	Total	Total Incl O&P
0010	**GABION BOX SYSTEMS**									
0400	Gabions, galvanized steel mesh mats or boxes, stone filled, 6″ deep	B-13	200	.240	S.Y.	26.50	6.80	3.74	37.04	44.50
0500	9″ deep		163	.294		33	8.35	4.59	45.94	55.50
0600	12″ deep		153	.314		48	8.90	4.89	61.79	73
0700	18″ deep		102	.471		63.50	13.35	7.35	84.20	100
0800	36″ deep		60	.800		102	22.50	12.50	137	164

31 37 Riprap

31 37 13 – Machined Riprap

31 37 13.10 Riprap and Rock Lining		Crew	Daily Output	Labor-Hours	Unit	2011 Bare Costs Material	Labor	Equipment	Total	Total Incl O&P
0010	**RIPRAP AND ROCK LINING**									
0011	Random, broken stone									
0100	Machine placed for slope protection	B-12G	62	.258	L.C.Y.	31.50	8.05	11.20	50.75	60
0110	3/8 to 1/4 C.Y. pieces, grouted	B-13	80	.600	S.Y.	65.50	17	9.35	91.85	111
0200	18" minimum thickness, not grouted	"	53	.906	"	19.70	25.50	14.15	59.35	80
0300	Dumped, 50 lb. average	B-11A	800	.020	Ton	30.50	.61	1.31	32.42	36
0350	100 lb. average		700	.023		30.50	.70	1.50	32.70	36.50
0370	300 lb. average		600	.027		30.50	.82	1.75	33.07	37

31 41 Shoring

31 41 13 – Timber Shoring

31 41 13.10 Building Shoring		Crew	Daily Output	Labor-Hours	Unit	Material	Labor	Equipment	Total	Total Incl O&P
0010	**BUILDING SHORING**									
0020	Shoring, existing building, with timber, no salvage allowance	B-51	2.20	21.818	M.B.F.	680	585	78	1,343	1,800
1000	On cribbing with 35 ton screw jacks, per box and jack	"	3.60	13.333	Jack	65	355	47.50	467.50	720

31 41 16 – Sheet Piling

31 41 16.10 Sheet Piling Systems		Crew	Daily Output	Labor-Hours	Unit	Material	Labor	Equipment	Total	Total Incl O&P
0010	**SHEET PILING SYSTEMS**									
0020	Sheet piling steel, not incl. wales, 22 psf, 15' excav., left in place	B-40	10.81	5.180	Ton	1,550	169	300	2,019	2,325
0100	Drive, extract & salvage R314116-40		6	9.333		545	305	545	1,395	1,725
0300	20' deep excavation, 27 psf, left in place		12.95	4.324		1,550	141	252	1,943	2,225
0400	Drive, extract & salvage R314116-45		6.55	8.550		545	278	500	1,323	1,625
0600	25' deep excavation, 38 psf, left in place		19	2.947		1,550	96	172	1,818	2,050
0700	Drive, extract & salvage		10.50	5.333		545	174	310	1,029	1,225
0900	40' deep excavation, 38 psf, left in place		21.20	2.642		1,550	86	154	1,790	2,025
1000	Drive, extract & salvage		12.25	4.571		545	149	267	961	1,150
1200	15' deep excavation, 22 psf, left in place		983	.057	S.F.	18	1.86	3.32	23.18	26.50
1300	Drive, extract & salvage		545	.103		6.10	3.35	6	15.45	19.05
1500	20' deep excavation, 27 psf, left in place		960	.058		22.50	1.90	3.40	27.80	32
1600	Drive, extract & salvage		485	.115		7.90	3.76	6.75	18.41	22.50
1800	25' deep excavation, 38 psf, left in place		1000	.056		33.50	1.82	3.27	38.59	43
1900	Drive, extract & salvage		553	.101		10.85	3.30	5.90	20.05	24
2100	Rent steel sheet piling and wales, first month				Ton	310			310	340
2200	Per added month					31			31	34
2300	Rental piling left in place, add to rental					1,250			1,250	1,375
2500	Wales, connections & struts, 2/3 salvage					495			495	545
2700	High strength piling, 50,000 psi, add					71.50			71.50	78.50
2800	55,000 psi, add					81			81	89
3000	Tie rod, not upset, 1-1/2" to 4" diameter with turnbuckle					2,250			2,250	2,475
3100	No turnbuckle					1,750			1,750	1,925
3300	Upset, 1-3/4" to 4" diameter with turnbuckle					2,550			2,550	2,825
3400	No turnbuckle					2,200			2,200	2,425
3600	Lightweight, 18" to 28" wide, 7 ga., 9.22 psf, and									
3610	9 ga., 8.6 psf, minimum				Lb.	.93			.93	1.02
3700	Average					1.01			1.01	1.11
3750	Maximum					1.15			1.15	1.27
3900	Wood, solid sheeting, incl. wales, braces and spacers,									
3910	drive, extract & salvage, 8' deep excavation	B-31	330	.121	S.F.	1.51	3.25	.55	5.31	7.70
4000	10' deep, 50 S.F./hr. in & 150 S.F./hr. out		300	.133		1.55	3.58	.60	5.73	8.35
4100	12' deep, 45 S.F./hr. in & 135 S.F./hr. out		270	.148		1.60	3.98	.67	6.25	9.15

31 41 Shoring

31 41 16 – Sheet Piling

31 41 16.10 Sheet Piling Systems		Crew	Daily Output	Labor-Hours	Unit	Material	2011 Bare Costs Labor	Equipment	Total	Total Incl O&P
4200	14′ deep, 42 S.F./hr. in & 126 S.F./hr. out	B-31	250	.160	S.F.	1.65	4.30	.72	6.67	9.80
4300	16′ deep, 40 S.F./hr. in & 120 S.F./hr. out		240	.167		1.70	4.48	.76	6.94	10.20
4400	18′ deep, 38 S.F./hr. in & 114 S.F./hr. out		230	.174		1.75	4.67	.79	7.21	10.60
4500	20′ deep, 35 S.F./hr. in & 105 S.F./hr. out		210	.190		1.81	5.10	.86	7.77	11.50
4520	Left in place, 8′ deep, 55 S.F./hr.		440	.091		2.72	2.44	.41	5.57	7.55
4540	10′ deep, 50 S.F./hr.		400	.100		2.86	2.69	.45	6	8.15
4560	12′ deep, 45 S.F./hr.		360	.111		3.02	2.98	.50	6.50	8.85
4565	14′ deep, 42 S.F./hr.		335	.119		3.20	3.21	.54	6.95	9.45
4570	16′ deep, 40 S.F./hr.		320	.125		3.40	3.36	.57	7.33	9.95
4580	18′ deep, 38 S.F./hr.		305	.131		3.63	3.52	.59	7.74	10.55
4590	20′ deep, 35 S.F./hr.		280	.143		3.89	3.84	.65	8.38	11.40
4700	Alternate pricing, left in place, 8′ deep		1.76	22.727	M.B.F.	610	610	103	1,323	1,800
4800	Drive, extract and salvage, 8′ deep		1.32	30.303	"	545	815	137	1,497	2,125
5000	For treated lumber add cost of treatment to lumber									

31 43 Concrete Raising

31 43 13 – Pressure Grouting

31 43 13.13 Concrete Pressure Grouting		Crew	Daily Output	Labor-Hours	Unit	Material	Labor	Equipment	Total	Total Incl O&P
0010	**CONCRETE PRESSURE GROUTING**									
0020	Grouting, pressure, cement & sand, 1:1 mix, minimum	B-61	124	.323	Bag	12.40	8.65	2.58	23.63	31
0100	Maximum		51	.784	"	12.40	21	6.25	39.65	56
0200	Cement and sand, 1:1 mix, minimum		250	.160	C.F.	25	4.30	1.28	30.58	36
0300	Maximum		100	.400		37	10.75	3.20	50.95	62.50
0400	Epoxy cement grout, minimum		137	.292		525	7.85	2.33	535.18	595
0500	Maximum		57	.702		525	18.85	5.60	549.45	620
0700	Alternate pricing method: (Add for materials)									
0710	5 person crew and equipment	B-61	1	40	Day		1,075	320	1,395	2,150

31 45 Vibroflotation and Densification

31 45 13 – Vibroflotation

31 45 13.10 Vibroflotation Densification		Crew	Daily Output	Labor-Hours	Unit	Material	Labor	Equipment	Total	Total Incl O&P
0010	**VIBROFLOTATION DENSIFICATION** R314513-90									
0900	Vibroflotation compacted sand cylinder, minimum	B-60	750	.064	V.L.F.		1.89	2.85	4.74	6.25
0950	Maximum		325	.148			4.36	6.55	10.91	14.50
1100	Vibro replacement compacted stone cylinder, minimum		500	.096			2.83	4.27	7.10	9.40
1150	Maximum		250	.192			5.65	8.55	14.20	18.80
1300	Mobilization and demobilization, minimum		.47	102	Total		3,025	4,550	7,575	10,000
1400	Maximum		.14	342	"		10,100	15,300	25,400	33,600

31 46 Needle Beams

31 46 13 – Cantilever Needle Beams

31 46 13.10 Needle Beams		Crew	Daily Output	Labor-Hours	Unit	Material	2011 Bare Costs Labor	Equipment	Total	Total Incl O&P
0010	**NEEDLE BEAMS**									
0011	Incl. wood shoring 10′ x 10′ opening									
0400	Block, concrete, 8″ thick	B-9	7.10	5.634	Ea.	43	151	27.50	221.50	330
0420	12″ thick		6.70	5.970		51.50	160	29	240.50	355
0800	Brick, 4″ thick with 8″ backup block		5.70	7.018		51.50	188	34.50	274	410
1000	Brick, solid, 8″ thick		6.20	6.452		43	173	31.50	247.50	370
1040	12″ thick		4.90	8.163		51.50	219	40	310.50	465
1080	16″ thick		4.50	8.889		67.50	239	43.50	350	520
2000	Add for additional floors of shoring	B-1	6	4		43	108		151	230

31 48 Underpinning

31 48 13 – Underpinning Piers

31 48 13.10 Underpinning Foundations

		Crew	Daily Output	Labor-Hours	Unit	Material	Labor	Equipment	Total	Total Incl O&P
0010	**UNDERPINNING FOUNDATIONS**									
0011	Including excavation,									
0020	forming, reinforcing, concrete and equipment									
0100	5′ to 16′ below grade, 100 to 500 C.Y.	B-52	2.30	24.348	C.Y.	278	710	219	1,207	1,725
0200	Over 500 C.Y.		2.50	22.400		250	650	202	1,102	1,600
0400	16′ to 25′ below grade, 100 to 500 C.Y.		2	28		305	815	252	1,372	1,975
0500	Over 500 C.Y.		2.10	26.667		289	775	240	1,304	1,875
0700	26′ to 40′ below grade, 100 to 500 C.Y.		1.60	35		335	1,025	315	1,675	2,400
0800	Over 500 C.Y.		1.80	31.111		305	905	280	1,490	2,175
0900	For under 50 C.Y., add					10%	40%			
1000	For 50 C.Y. to 100 C.Y., add					5%	20%			

31 52 Cofferdams

31 52 16 – Timber Cofferdams

31 52 16.10 Cofferdams

		Crew	Daily Output	Labor-Hours	Unit	Material	Labor	Equipment	Total	Total Incl O&P
0010	**COFFERDAMS**									
0011	Incl. mobilization and temporary sheeting									
0080	Soldier beams & lagging H piles with 3″ wood sheeting									
0090	horizontal between piles, including removal of wales & braces									
0100	No hydrostatic head, 15′ deep, 1 line of braces, minimum	B-50	545	.191	S.F.	6.80	5.85	4.24	16.89	22
0200	Maximum		495	.210		7.55	6.45	4.67	18.67	24.50
0400	15′ to 22′ deep with 2 lines of braces, 10″ H, minimum		360	.289		8	8.85	6.40	23.25	31
0500	Maximum		330	.315		9.05	9.70	7	25.75	34
0700	23′ to 35′ deep with 3 lines of braces, 12″ H, minimum		325	.320		10.45	9.85	7.10	27.40	36
0800	Maximum		295	.353		11.35	10.85	7.85	30.05	39.50
1000	36′ to 45′ deep with 4 lines of braces, 14″ H, minimum		290	.359		11.70	11	7.95	30.65	40.50
1100	Maximum		265	.392		12.35	12.05	8.70	33.10	43.50
1300	No hydrostatic head, left in place, 15′ dp., 1 line of braces, min.		635	.164		9.05	5.05	3.64	17.74	22.50
1400	Maximum		575	.181		9.70	5.55	4.02	19.27	24.50
1600	15′ to 22′ deep with 2 lines of braces, minimum		455	.229		13.60	7	5.10	25.70	32.50
1700	Maximum		415	.251		15.10	7.70	5.55	28.35	36
1900	23′ to 35′ deep with 3 lines of braces, minimum		420	.248		16.20	7.60	5.50	29.30	37
2000	Maximum		380	.274		17.90	8.40	6.10	32.40	41
2200	36′ to 45′ deep with 4 lines of braces, minimum		385	.270		19.45	8.30	6	33.75	42.50
2300	Maximum		350	.297		22.50	9.15	6.60	38.25	48
2350	Lagging only, 3″ thick wood between piles 8′ O.C., minimum	B-46	400	.120		1.51	3.59	.10	5.20	7.90

31 52 Cofferdams

31 52 16 – Timber Cofferdams

31 52 16.10 Cofferdams		Crew	Daily Output	Labor-Hours	Unit	2011 Bare Costs Material	Labor	Equipment	Total	Total Incl O&P
2370	Maximum	B-46	250	.192	S.F.	2.27	5.75	.17	8.19	12.50
2400	Open sheeting no bracing, for trenches to 10′ deep, min.		1736	.028		.68	.83	.02	1.53	2.20
2450	Maximum		1510	.032		.76	.95	.03	1.74	2.49
2500	Tie-back method, add to open sheeting, add, minimum								20%	20%
2550	Maximum								60%	60%
2700	Tie-backs only, based on tie-backs total length, minimum	B-46	86.80	.553	L.F.	12.55	16.55	.48	29.58	43
2750	Maximum		38.50	1.247	"	22	37.50	1.08	60.58	89.50
3500	Tie-backs only, typical average, 25′ long		2	24	Ea.	555	720	21	1,296	1,850
3600	35′ long		1.58	30.380	"	735	910	26.50	1,671.50	2,400

31 56 Slurry Walls

31 56 23 – Lean Concrete Slurry Walls

31 56 23.20 Slurry Trench		Crew	Daily Output	Labor-Hours	Unit	Material	Labor	Equipment	Total	Total Incl O&P
0010	**SLURRY TRENCH**									
0011	Excavated slurry trench in wet soils									
0020	backfilled with 3000 PSI concrete, no reinforcing steel									
0050	Minimum	C-7	333	.216	C.F.	7.35	6.20	3.84	17.39	22.50
0100	Maximum		200	.360	"	12.30	10.35	6.40	29.05	37.50
0200	Alternate pricing method, minimum		150	.480	S.F.	14.65	13.80	8.50	36.95	48.50
0300	Maximum		120	.600		22	17.25	10.65	49.90	64
0500	Reinforced slurry trench, minimum	B-48	177	.271		11	7.70	15.80	34.50	42.50
0600	Maximum	"	69	.696		36.50	19.70	40.50	96.70	118
0800	Haul for disposal, 2 mile haul, excavated material, add	B-34B	99	.081	C.Y.		2.20	5.10	7.30	9.25
0900	Haul bentonite castings for disposal, add	"	40	.200	"		5.45	12.65	18.10	23

31 62 Driven Piles

31 62 13 – Concrete Piles

31 62 13.23 Prestressed Concrete Piles		Crew	Daily Output	Labor-Hours	Unit	Material	Labor	Equipment	Total	Total Incl O&P
0010	**PRESTRESSED CONCRETE PILES**, 200 piles									
0020	Unless specified otherwise, not incl. pile caps or mobilization									
2200	Precast, prestressed, 50′ long, 12″ diam., 2-3/8″ wall	B-19	720	.078	V.L.F.	22.50	2.53	2.57	27.60	31.50
2300	14″ diameter, 2-1/2″ wall		680	.082		30	2.68	2.72	35.40	40.50
2500	16″ diameter, 3″ wall		640	.088		42.50	2.85	2.89	48.24	55
2600	18″ diameter, 3″ wall	B-19A	600	.107		53.50	3.58	3.75	60.83	69
2800	20″ diameter, 3-1/2″ wall		560	.114		64.50	3.83	4.02	72.35	82
2900	24″ diameter, 3-1/2″ wall		520	.123		73	4.13	4.33	81.46	92.50
3100	Precast, prestressed, 40′ long, 10″ thick, square	B-19	700	.080		16.35	2.61	2.64	21.60	25.50
3200	12″ thick, square		680	.082		20.50	2.68	2.72	25.90	30
3400	14″ thick, square		600	.093		24	3.04	3.08	30.12	35
3500	Octagonal		640	.088		33	2.85	2.89	38.74	44.50
3700	16″ thick, square		560	.100		38	3.26	3.30	44.56	51
3800	Octagonal		600	.093		39.50	3.04	3.08	45.62	52
4000	18″ thick, square	B-19A	520	.123		46	4.13	4.33	54.46	62.50
4100	Octagonal	B-19	560	.100		47	3.26	3.30	53.56	60.50
4300	20″ thick, square	B-19A	480	.133		57	4.47	4.69	66.16	75.50
4400	Octagonal	B-19	520	.108		52	3.51	3.55	59.06	67
4600	24″ thick, square	B-19A	440	.145		81	4.88	5.10	90.98	103
4700	Octagonal	B-19	480	.117		74.50	3.80	3.85	82.15	93
4730	Precast, prestressed, 60′ long, 10″ thick, square		700	.080		17.15	2.61	2.64	22.40	26

31 62 Driven Piles

31 62 13 – Concrete Piles

31 62 13.23 Prestressed Concrete Piles		Crew	Daily Output	Labor-Hours	Unit	Material	Labor (2011 Bare Costs)	Equipment (2011 Bare Costs)	Total (2011 Bare Costs)	Total Incl O&P
4740	12" thick, square (60' long)	B-19	680	.082	V.L.F.	21.50	2.68	2.72	26.90	31
4750	Mobilization for 10,000 L.F. pile job, add		3300	.017			.55	.56	1.11	1.57
4800	25,000 L.F. pile job, add		8500	.007			.21	.22	.43	.61

31 62 16 – Steel Piles

31 62 16.13 Steel Piles

		Crew	Daily Output	Labor-Hours	Unit	Material	Labor	Equipment	Total	Total Incl O&P
0010	**STEEL PILES**									
0100	Step tapered, round, concrete filled									
0110	8" tip, 60 ton capacity, 30' depth	B-19	760	.074	V.L.F.	10	2.40	2.43	14.83	17.80
0120	60' depth		740	.076		11.30	2.46	2.50	16.26	19.45
0130	80' depth		700	.080		11.70	2.61	2.64	16.95	20
0150	10" tip, 90 ton capacity, 30' depth		700	.080		12	2.61	2.64	17.25	20.50
0160	60' depth		690	.081		12.35	2.64	2.68	17.67	21
0170	80' depth		670	.084		13.30	2.72	2.76	18.78	22.50
0190	12" tip, 120 ton capacity, 30' depth		660	.085		16.30	2.76	2.80	21.86	26
0200	60' depth, 12" diameter		630	.089		16.20	2.90	2.93	22.03	26
0210	80' depth		590	.095		14.95	3.09	3.13	21.17	25
0250	"H" Sections, 50' long, HP8 x 36		640	.088		17.85	2.85	2.89	23.59	27.50
0400	HP10 X 42		610	.092		21	2.99	3.03	27.02	31.50
0500	HP10 X 57		610	.092		28	2.99	3.03	34.02	39.50
0700	HP12 X 53		590	.095		26.50	3.09	3.13	32.72	38
0800	HP12 X 74	B-19A	590	.108		37.50	3.64	3.81	44.95	51.50
1000	HP14 X 73		540	.119		36.50	3.98	4.17	44.65	52
1100	HP14 X 89		540	.119		45.50	3.98	4.17	53.65	61.50
1300	HP14 X 102		510	.125		51.50	4.21	4.41	60.12	69
1400	HP14 X 117		510	.125		59.50	4.21	4.41	68.12	77.50
1601	Splice on standard points, not in leads, 8" or 10"	1 Pile	5	1.600	Ea.	117	52.50		169.50	221
1701	12" or 14"		4	2		171	65.50		236.50	300
1901	Heavy duty points, not in leads, 10" wide		4	2		181	65.50		246.50	315
2101	14" wide		3.50	2.286		235	75		310	390

31 62 19 – Timber Piles

31 62 19.10 Wood Piles

		Crew	Daily Output	Labor-Hours	Unit	Material	Labor	Equipment	Total	Total Incl O&P
0010	**WOOD PILES**									
0011	Friction or end bearing, not including									
0050	mobilization or demobilization									
0100	Untreated piles, up to 30' long, 12" butts, 8" points	B-19	625	.090	V.L.F.	8	2.92	2.96	13.88	17.05
0200	30' to 39' long, 12" butts, 8" points		700	.080		8	2.61	2.64	13.25	16.20
0300	40' to 49' long, 12" butts, 7" points		720	.078		8	2.53	2.57	13.10	15.95
0400	50' to 59' long, 13"butts, 7" points		800	.070		8	2.28	2.31	12.59	15.25
0500	60' to 69' long, 13" butts, 7" points		840	.067		9	2.17	2.20	13.37	16.05
0600	70' to 80' long, 13" butts, 6" points		840	.067		10	2.17	2.20	14.37	17.15
0800	Treated piles, 12 lb. per C.F.,									
0810	friction or end bearing, ASTM class B									
1000	Up to 30' long, 12" butts, 8" points	B-19	625	.090	V.L.F.	10.90	2.92	2.96	16.78	20.50
1100	30' to 39' long, 12" butts, 8" points		700	.080		10.65	2.61	2.64	15.90	19.10
1200	40' to 49' long, 12" butts, 7" points		720	.078		11.40	2.53	2.57	16.50	19.70
1300	50' to 59' long, 13" butts, 7" points		800	.070		12.60	2.28	2.31	17.19	20.50
1400	60' to 69' long, 13" butts, 6" points	B-19A	840	.076		17.20	2.56	2.68	22.44	26
1500	70' to 80' long, 13" butts, 6" points	"	840	.076		21.50	2.56	2.68	26.74	31.50
1600	Treated piles, C.C.A., 2.5# per C.F.									
1610	8" butts, 10' long	B-19	400	.140	V.L.F.	7.95	4.56	4.62	17.13	21.50
1620	11' to 16' long		500	.112		7.95	3.65	3.69	15.29	19.05
1630	17' to 20' long		575	.097		7.95	3.17	3.21	14.33	17.75

31 62 Driven Piles

31 62 19 – Timber Piles

31 62 19.10 Wood Piles		Crew	Daily Output	Labor-Hours	Unit	2011 Bare Costs Material	Labor	Equipment	Total	Total Incl O&P
1640	10" butts, 10' to 16' long	B-19	500	.112	V.L.F.	9.05	3.65	3.69	16.39	20.50
1650	17' to 20' long		575	.097		9.05	3.17	3.21	15.43	18.95
1660	21' to 40' long		700	.080		9.05	2.61	2.64	14.30	17.35
1670	12" butts, 10' to 20' long		575	.097		9.80	3.17	3.21	16.18	19.75
1680	21' to 35' long		650	.086		9.80	2.81	2.84	15.45	18.70
1690	36' to 40' long		700	.080		9.80	2.61	2.64	15.05	18.15
1695	14" butts. to 40' long		700	.080		14	2.61	2.64	19.25	23
1700	Boot for pile tip, minimum	1 Pile	27	.296	Ea.	22.50	9.70		32.20	41.50
1800	Maximum		21	.381		67	12.50		79.50	95.50
2000	Point for pile tip, minimum		20	.400		22.50	13.10		35.60	47.50
2100	Maximum		15	.533		80.50	17.45		97.95	119
2300	Splice for piles over 50' long, minimum	B-46	35	1.371		56	41	1.19	98.19	133
2400	Maximum		20	2.400		67	72	2.09	141.09	199
2600	Concrete encasement with wire mesh and tube		331	.145	V.L.F.	10.50	4.34	.13	14.97	19.15
2700	Mobilization for 10,000 L.F. pile job, add	B-19	3300	.017			.55	.56	1.11	1.57
2800	25,000 L.F. pile job, add	"	8500	.007			.21	.22	.43	.61

31 62 23 – Composite Piles

31 62 23.13 Concrete-Filled Steel Piles		Crew	Daily Output	Labor-Hours	Unit	Material	Labor	Equipment	Total	Total Incl O&P
0010	**CONCRETE-FILLED STEEL PILES** no mobilization or demobilization									
2600	Pipe piles, 50' lg. 8" diam., 29 lb. per L.F., no concrete	B-19	500	.112	V.L.F.	19.50	3.65	3.69	26.84	32
2700	Concrete filled		460	.122		20.50	3.97	4.02	28.49	33.50
2900	10" diameter, 34 lb. per L.F., no concrete		500	.112		24.50	3.65	3.69	31.84	37.50
3000	Concrete filled		450	.124		26.50	4.05	4.11	34.66	40.50
3200	12" diameter, 44 lb. per L.F., no concrete		475	.118		29.50	3.84	3.89	37.23	43.50
3300	Concrete filled		415	.135		30.50	4.40	4.45	39.35	46
3500	14" diameter, 46 lb. per L.F., no concrete		430	.130		32	4.24	4.30	40.54	47
3600	Concrete filled		355	.158		33.50	5.15	5.20	43.85	51.50
3800	16" diameter, 52 lb. per L.F., no concrete		385	.145		35.50	4.74	4.80	45.04	52.50
3900	Concrete filled		335	.167		38	5.45	5.50	48.95	57.50
4100	18" diameter, 59 lb. per L.F., no concrete		355	.158		46.50	5.15	5.20	56.85	65.50
4200	Concrete filled		310	.181		52.50	5.90	5.95	64.35	74
4400	Splices for pipe piles, not in leads, 8" diameter	1 Sswl	4.67	1.713	Ea.	86.50	62.50		149	215
4500	14" diameter		3.79	2.111		114	77		191	271
4600	16" diameter		3.03	2.640		140	96		236	335
4800	Points, standard, 8" diameter		4.61	1.735		142	63		205	276
4900	14" diameter		4.05	1.975		197	72		269	355
5000	16" diameter		3.37	2.374		240	86.50		326.50	430
5200	Points, heavy duty, 10" diameter		2.89	2.768		281	101		382	500
5300	14" or 16" diameter		2.02	3.960		445	144		589	765
5500	For reinforcing steel, add		1150	.007	Lb.	.78	.25		1.03	1.34
5700	For thick wall sections, add				"	.89			.89	.98

	31 63 26.13 Fixed End Caisson Piles	Crew	Daily Output	Labor-Hours	Unit	2011 Bare Costs Material	Labor	Equipment	Total	Total Incl O&P
0010	**FIXED END CAISSON PILES**									
0015	Including excavation, concrete, 50 lb. reinforcing									
0020	per C.Y., not incl. mobilization, boulder removal, disposal									
0100	Open style, machine drilled, to 50′ deep, in stable ground, no R316326-60									
0110	casings or ground water, 18″ diam., 0.065 C.Y./L.F.	B-43	200	.200	V.L.F.	7.90	5.35	12.45	25.70	31.50
0200	24″ diameter, 0.116 C.Y./L.F.		190	.211		14.10	5.65	13.10	32.85	39.50
0300	30″ diameter, 0.182 C.Y./L.F.		150	.267		22	7.15	16.55	45.70	55
0400	36″ diameter, 0.262 C.Y./L.F.		125	.320		32	8.60	19.90	60.50	71.50
0500	48″ diameter, 0.465 C.Y./L.F.		100	.400		56.50	10.75	25	92.25	108
0600	60″ diameter, 0.727 C.Y./L.F.		90	.444		88.50	11.95	27.50	127.95	148
0700	72″ diameter, 1.05 C.Y./L.F.		80	.500		128	13.45	31	172.45	197
0800	84″ diameter, 1.43 C.Y./L.F.		75	.533		174	14.30	33	221.30	252
1000	For bell excavation and concrete, add									
1020	4′ bell diameter, 24″ shaft, 0.444 C.Y.	B-43	20	2	Ea.	44	53.50	124	221.50	276
1040	6′ bell diameter, 30″ shaft, 1.57 C.Y.		5.70	7.018		155	188	435	778	965
1060	8′ bell diameter, 36″ shaft, 3.72 C.Y.		2.40	16.667		370	450	1,025	1,845	2,300
1080	9′ bell diameter, 48″ shaft, 4.48 C.Y.		2	20		445	535	1,250	2,230	2,775
1100	10′ bell diameter, 60″ shaft, 5.24 C.Y.		1.70	23.529		520	630	1,450	2,600	3,225
1120	12′ bell diameter, 72″ shaft, 8.74 C.Y.		1	40		865	1,075	2,475	4,415	5,475
1140	14′ bell diameter, 84″ shaft, 13.6 C.Y.		.70	57.143		1,350	1,525	3,550	6,425	7,950
1200	Open style, machine drilled, to 50′ deep, in wet ground, pulled									
1300	casing and pumping, 18″ diameter, 0.065 C.Y./L.F.	B-48	160	.300	V.L.F.	7.90	8.50	17.50	33.90	42
1400	24″ diameter, 0.116 C.Y./L.F.		125	.384		14.10	10.90	22.50	47.50	58
1500	30″ diameter, 0.182 C.Y./L.F.		85	.565		22	16	33	71	87
1600	36″ diameter, 0.262 C.Y./L.F.		60	.800		32	22.50	46.50	101	125
1700	48″ diameter, 0.465 C.Y./L.F.	B-49	55	1.309		56.50	38	64.50	159	198
1800	60″ diameter, 0.727 C.Y./L.F.		35	2.057		88.50	60	101	249.50	310
1900	72″ diameter, 1.05 C.Y./L.F.		30	2.400		128	70	118	316	390
2000	84″ diameter, 1.43 C.Y./L.F.		25	2.880		174	84	142	400	490
2100	For bell excavation and concrete, add									
2120	4′ bell diameter, 24″ shaft, 0.444 C.Y.	B-48	19.80	2.424	Ea.	44	68.50	141	253.50	320
2140	6′ bell diameter, 30″ shaft, 1.57 C.Y.		5.70	8.421		155	239	490	884	1,100
2160	8′ bell diameter, 36″ shaft, 3.72 C.Y.		2.40	20		370	565	1,175	2,110	2,625
2180	9′ bell diameter, 48″ shaft, 4.48 C.Y.	B-49	3.30	21.818		445	635	1,075	2,155	2,750
2200	10′ bell diameter, 60″ shaft, 5.24 C.Y.		2.80	25.714		520	750	1,275	2,545	3,225
2220	12′ bell diameter, 72″ shaft, 8.74 C.Y.		1.60	45		865	1,300	2,225	4,390	5,575
2240	14′ bell diameter, 84″ shaft, 13.6 C.Y.		1	72		1,350	2,100	3,550	7,000	8,900
2300	Open style, machine drilled, to 50′ deep, in soft rocks and									
2400	medium hard shales, 18″ diameter, 0.065 C.Y./L.F.	B-49	50	1.440	V.L.F.	7.90	42	71	120.90	157
2500	24″ diameter, 0.116 C.Y./L.F.		30	2.400		14.10	70	118	202.10	264
2600	30″ diameter, 0.182 C.Y./L.F.		20	3.600		22	105	177	304	395
2700	36″ diameter, 0.262 C.Y./L.F.		15	4.800		32	140	236	408	530
2800	48″ diameter, 0.465 C.Y./L.F.		10	7.200		56.50	210	355	621.50	805
2900	60″ diameter, 0.727 C.Y./L.F.		7	10.286		88.50	299	505	892.50	1,150
3000	72″ diameter, 1.05 C.Y./L.F.		6	12		128	350	590	1,068	1,375
3100	84″ diameter, 1.43 C.Y./L.F.		5	14.400		174	420	710	1,304	1,675
3200	For bell excavation and concrete, add									
3220	4′ bell diameter, 24″ shaft, 0.444 C.Y.	B-49	10.90	6.606	Ea.	44	192	325	561	735
3240	6′ bell diameter, 30″ shaft, 1.57 C.Y.		3.10	23.226		155	675	1,150	1,980	2,575
3260	8′ bell diameter, 36″ shaft, 3.72 C.Y.		1.30	55.385		370	1,600	2,725	4,695	6,125
3280	9′ bell diameter, 48″ shaft, 4.48 C.Y.		1.10	65.455		445	1,900	3,225	5,570	7,275
3300	10′ bell diameter, 60″ shaft, 5.24 C.Y.		.90	80		520	2,325	3,950	6,795	8,825
3320	12′ bell diameter, 72″ shaft, 8.74 C.Y.		.60	120		865	3,500	5,900	10,265	13,400

31 63 Bored Piles

31 63 26 – Drilled Caissons

	31 63 26.13 Fixed End Caisson Piles	Crew	Daily Output	Labor-Hours	Unit	Material	Labor	Equipment	Total	Total Incl O&P
						2011 Bare Costs				
3340	14' bell diameter, 84" shaft, 13.6 C.Y.	B-49	.40	180	Ea.	1,350	5,250	8,850	15,450	20,100
3600	For rock excavation, sockets, add, minimum		120	.600	C.F.		17.45	29.50	46.95	62
3650	Average		95	.758			22	37.50	59.50	78
3700	Maximum		48	1.500			43.50	74	117.50	155
3900	For 50' to 100' deep, add				V.L.F.				7%	7%
4000	For 100' to 150' deep, add								25%	25%
4100	For 150' to 200' deep, add								30%	30%
4200	For casings left in place, add				Lb.	.99			.99	1.09
4300	For other than 50 lb. reinf. per C.Y., add or deduct				"	1.05			1.05	1.16
4400	For steel "I" beam cores, add	B-49	8.30	8.675	Ton	1,800	253	425	2,478	2,875
4500	Load and haul excess excavation, 2 miles	B-34B	178	.045	L.C.Y.		1.22	2.84	4.06	5.15
4600	For mobilization, 50 mile radius, rig to 36"	B-43	2	20	Ea.		535	1,250	1,785	2,275
4650	Rig to 84"	B-48	1.75	27.429			775	1,600	2,375	3,050
4700	For low headroom, add								50%	
5000	Bottom inspection	1 Skwk	1.20	6.667			227		227	380

31 63 26.16 Concrete Caissons for Marine Construction

	Description	Crew	Daily Output	Labor-Hours	Unit	Material	Labor	Equipment	Total	Total Incl O&P
0010	**CONCRETE CAISSONS FOR MARINE CONSTRUCTION**									
0100	Caissons, incl. mobilization and demobilization, up to 50 miles									
0200	Uncased shafts, 30 to 80 tons cap., 17" diam., 10' depth	B-44	88	.727	V.L.F.	19.80	23	20.50	63.30	84
0300	25' depth		165	.388		14.15	12.35	11	37.50	48.50
0400	80-150 ton capacity, 22" diameter, 10' depth		80	.800		25	25.50	22.50	73	95.50
0500	20' depth		130	.492		19.80	15.65	13.95	49.40	64.50
0700	Cased shafts, 10 to 30 ton capacity, 10-5/8" diam., 20' depth		175	.366		14.15	11.65	10.40	36.20	47
0800	30' depth		240	.267		13.20	8.50	7.55	29.25	37.50
0850	30 to 60 ton capacity, 12" diameter, 20' depth		160	.400		19.80	12.70	11.35	43.85	56.50
0900	40' depth		230	.278		15.25	8.85	7.90	32	40.50
1000	80 to 100 ton capacity, 16" diameter, 20' depth		160	.400		28.50	12.70	11.35	52.55	65.50
1100	40' depth		230	.278		26.50	8.85	7.90	43.25	53
1200	110 to 140 ton capacity, 17-5/8" diameter, 20' depth		160	.400		30.50	12.70	11.35	54.55	68
1300	40' depth		230	.278		28.50	8.85	7.90	45.25	55
1400	140 to 175 ton capacity, 19" diameter, 20' depth		130	.492		33	15.65	13.95	62.60	79
1500	40' depth		210	.305		30.50	9.70	8.65	48.85	59.50
1700	Over 30' long, L.F. cost tends to be lower									
1900	Maximum depth is about 90'									

31 63 29 – Drilled Concrete Piers and Shafts

31 63 29.13 Uncased Drilled Concrete Piers

	Description	Crew	Daily Output	Labor-Hours	Unit	Material	Labor	Equipment	Total	Total Incl O&P
0010	**UNCASED DRILLED CONCRETE PIERS**									
0020	Unless specified otherwise, not incl. pile caps or mobilization									
0050	Cast in place augered piles, no casing or reinforcing									
0060	8" diameter	B-43	540	.074	V.L.F.	3.95	1.99	4.60	10.54	12.75
0065	10" diameter		480	.083		6.30	2.24	5.20	13.74	16.35
0070	12" diameter		420	.095		8.85	2.56	5.90	17.31	20.50
0075	14" diameter		360	.111		11.95	2.98	6.90	21.83	25.50
0080	16" diameter		300	.133		16.10	3.58	8.30	27.98	33
0085	18" diameter		240	.167		19.90	4.48	10.35	34.73	41
0100	Cast in place, thin wall shell pile, straight sided,									
0110	not incl. reinforcing, 8" diam., 16 ga., 5.8 lb./L.F.	B-19	700	.080	V.L.F.	9.15	2.61	2.64	14.40	17.45
0200	10" diameter, 16 ga. corrugated, 7.3 lb./L.F.		650	.086		12	2.81	2.84	17.65	21
0300	12" diameter, 16 ga. corrugated, 8.7 lb./L.F.		600	.093		15.55	3.04	3.08	21.67	25.50
0400	14" diameter, 16 ga. corrugated, 10.0 lb./L.F.		550	.102		18.30	3.32	3.36	24.98	29.50
0500	16" diameter, 16 ga. corrugated, 11.6 lb./L.F.		500	.112		22.50	3.65	3.69	29.84	35
0800	Cast in place friction pile, 50' long, fluted,									

31 63 Bored Piles

31 63 29 – Drilled Concrete Piers and Shafts

31 63 29.13 Uncased Drilled Concrete Piers		Crew	Daily Output	Labor-Hours	Unit	Material	2011 Bare Costs Labor	Equipment	Total	Total Incl O&P
0810	tapered steel, 4000 psi concrete, no reinforcing									
0900	12" diameter, 7 ga.	B-19	600	.093	V.L.F.	27.50	3.04	3.08	33.62	38.50
1000	14" diameter, 7 ga.		560	.100		30	3.26	3.30	36.56	41.50
1100	16" diameter, 7 ga.		520	.108		35	3.51	3.55	42.06	48.50
1200	18" diameter, 7 ga.	↓	480	.117	↓	41	3.80	3.85	48.65	56
1300	End bearing, fluted, constant diameter,									
1320	4000 psi concrete, no reinforcing									
1340	12" diameter, 7 ga.	B-19	600	.093	V.L.F.	28.50	3.04	3.08	34.62	40
1360	14" diameter, 7 ga.		560	.100		35.50	3.26	3.30	42.06	48
1380	16" diameter, 7 ga.		520	.108		41.50	3.51	3.55	48.56	55.50
1400	18" diameter, 7 ga.	↓	480	.117	↓	45.50	3.80	3.85	53.15	61

31 63 29.20 Cast In Place Piles, Adds		Crew	Daily Output	Labor-Hours	Unit	Material	Labor	Equipment	Total	Total Incl O&P
0010	**CAST IN PLACE PILES, ADDS**									
1500	For reinforcing steel, add				Lb.	.90			.90	.99
1700	For ball or pedestal end, add	B-19	11	5.091	C.Y.	147	166	168	481	630
1900	For lengths above 60', concrete, add	"	11	5.091	"	152	166	168	486	640
2000	For steel thin shell, pipe only				Lb.	1.11			1.11	1.22

Division Notes

	CREW	DAILY OUTPUT	LABOR-HOURS	UNIT	2011 BARE COSTS				TOTAL INCL O&P
					MAT.	LABOR	EQUIP.	TOTAL	

Estimating Tips

32 01 00 Operations and Maintenance of Exterior Improvements

- Recycling of asphalt pavement is becoming very popular and is an alternative to removal and replacement. It can be a good value engineering proposal if removed pavement can be recycled, either at the project site or at another site that is reasonably close to the project site. Two new sections on repair of flexible and rigid pavement have been added.

32 10 00 Bases, Ballasts, and Paving

- When estimating paving, keep in mind the project schedule. If an asphaltic paving project is in a colder climate and runs through to the spring, consider placing the base course in the autumn and then topping it in the spring, just prior to completion. This could save considerable costs in spring repair. Keep in mind that prices for asphalt and concrete are generally higher in the cold seasons. New lines have been added for paving intersections and bridge approaches.

32 90 00 Planting

- The timing of planting and guarantee specifications often dictate the costs for establishing tree and shrub growth and a stand of grass or ground cover. Establish the work performance schedule to coincide with the local planting season. Maintenance and growth guarantees can add from 20%-100% to the total landscaping cost. The cost to replace trees and shrubs can be as high as 5% of the total cost, depending on the planting zone, soil conditions, and time of year.

Reference Numbers

Reference numbers are shown in shaded boxes at the beginning of some major classifications. These numbers refer to related items in the Reference Section. The reference information may be an estimating procedure, an alternate pricing method, or technical information.

Note: Not all subdivisions listed here necessarily appear in this publication.

32 01 Operation and Maintenance of Exterior Improvements

32 01 13 – Flexible Paving Surface Treatment

32 01 13.61 Slurry Seal (Latex Modified)		Crew	Daily Output	Labor-Hours	Unit	Material	2011 Bare Costs Labor	Equipment	Total	Total Incl O&P
0010	SLURRY SEAL (LATEX MODIFIED)									
3780	Rubberized asphalt (latex) seal	B-45	5000	.003	S.Y.	2.35	.09	.06	2.50	2.80

32 01 13.64 Sand Seal		Crew	Daily Output	Labor-Hours	Unit	Material	Labor	Equipment	Total	Total Incl O&P
0010	SAND SEAL									
2080	Sand sealing, sharp sand, asphalt emulsion, small area	B-91	10000	.006	S.Y.	1.35	.20	.21	1.76	2.05
2120	Roadway or large area	"	18000	.004	"	1.16	.11	.11	1.38	1.59

32 01 13.66 Fog Seal		Crew	Daily Output	Labor-Hours	Unit	Material	Labor	Equipment	Total	Total Incl O&P
0010	FOG SEAL									
0012	Sealcoating, 2 coat coal tar pitch emulsion over 10,000 S.Y.	B-45	5000	.003	S.Y.	1.20	.09	.06	1.35	1.53
0030	1000 to 10,000 S.Y.	"	3000	.005		1.20	.14	.10	1.44	1.67
0100	Under 1000 S.Y.	B-1	1050	.023		1.20	.62		1.82	2.36
0300	Petroleum resistant, over 10,000 S.Y.	B-45	5000	.003		1.40	.09	.06	1.55	1.75
0320	1000 to 10,000 S.Y.	"	3000	.005		1.40	.14	.10	1.64	1.89
0400	Under 1000 S.Y.	B-1	1050	.023		1.40	.62		2.02	2.58
0600	Non-skid pavement renewal, over 10,000 S.Y.	B-45	5000	.003		1.55	.09	.06	1.70	1.92
0620	1000 to 10,000 S.Y.	"	3000	.005		1.55	.14	.10	1.79	2.06
0700	Under 1000 S.Y.	B-1	1050	.023		1.55	.62		2.17	2.75
0800	Prepare and clean surface for above	A-2	8545	.003	↓		.07	.02	.09	.14
1000	Hand seal asphalt curbing	B-1	4420	.005	L.F.	.85	.15		1	1.19
1900	Asphalt surface treatment, single course, small area									
1901	0.30 gal/S.Y. asphalt material, 20#/S.Y. aggregate	B-91	5000	.013	S.Y.	1.78	.40	.41	2.59	3.07
1910	Roadway or large area		10000	.006		1.64	.20	.21	2.05	2.36
1950	Asphalt surface treatment, dbl. course for small area		3000	.021		3.56	.66	.68	4.90	5.75
1960	Roadway or large area		6000	.011		3.20	.33	.34	3.87	4.45
1980	Asphalt surface treatment, single course, for shoulders	↓	7500	.009	↓	1.88	.26	.27	2.41	2.81

32 06 Schedules for Exterior Improvements

32 06 10 – Schedules for Bases, Ballasts, and Paving

32 06 10.10 Sidewalks, Driveways and Patios		Crew	Daily Output	Labor-Hours	Unit	Material	Labor	Equipment	Total	Total Incl O&P
0010	SIDEWALKS, DRIVEWAYS AND PATIOS No base									
0020	Asphaltic concrete, 2" thick	B-37	720	.067	S.Y.	6.80	1.86	.20	8.86	10.80
0100	2-1/2" thick	"	660	.073	"	8.60	2.03	.22	10.85	13.10
0300	Concrete, 3000 psi, CIP, 6 x 6 - W1.4 x W1.4 mesh,									
0311	broomed finish, no base, 4" thick	3 Clab	600	.040	S.F.	1.65	1.06		2.71	3.59
0351	5" thick		545	.044		2.20	1.16		3.36	4.37
0401	6" thick	↓	510	.047		2.57	1.24		3.81	4.90
0450	For bank run gravel base, 4" thick, add	B-18	2500	.010		.51	.26	.02	.79	1.02
0520	8" thick, add	"	1600	.015		1.03	.41	.02	1.46	1.85
0550	Exposed aggregate finish, add to above, minimum	B-24	1875	.013		.13	.39		.52	.79
0600	Maximum	"	455	.053		.43	1.62		2.05	3.13
1000	Crushed stone, 1" thick, white marble	2 Clab	1700	.009		.25	.25		.50	.70
1050	Bluestone	"	1700	.009		.18	.25		.43	.62
1700	Redwood, prefabricated, 4' x 4' sections	2 Carp	316	.051		4.74	1.70		6.44	8.05
1750	Redwood planks, 1" thick, on sleepers	"	240	.067	↓	4.74	2.24		6.98	8.95
2250	Stone dust, 4" thick	B-62	900	.027	S.Y.	3.33	.77	.17	4.27	5.15

32 06 10.20 Steps		Crew	Daily Output	Labor-Hours	Unit	Material	Labor	Equipment	Total	Total Incl O&P
0010	STEPS									
0011	Incl. excav., borrow & concrete base as required									
0100	Brick steps	B-24	35	.686	LF Riser	12.85	21		33.85	48.50
0300	Bluestone treads, 12" x 2" or 12" x 1-1/2"	"	30	.800	"	28	24.50		52.50	71.50

32 06 Schedules for Exterior Improvements

32 06 10 – Schedules for Bases, Ballasts, and Paving

32 06 10.20 Steps		Crew	Daily Output	Labor-Hours	Unit	Material	2011 Bare Costs Labor	Equipment	Total	Total Incl O&P
0500	Concrete, cast in place, see Section 03 30 53.40									
0600	Precast concrete, see Section 03 41 23.50									

32 11 Base Courses

32 11 23 – Aggregate Base Courses

32 11 23.23 Base Course Drainage Layers

		Crew	Daily Output	Labor-Hours	Unit	Material	Labor	Equipment	Total	Total Incl O&P
0010	**BASE COURSE DRAINAGE LAYERS**									
0011	For roadways and large areas									
0050	Crushed 3/4" stone base, compacted, 3" deep	B-36C	5200	.008	S.Y.	3.71	.25	.66	4.62	5.20
0100	6" deep		5000	.008		7.40	.26	.69	8.35	9.35
0200	9" deep		4600	.009		11.15	.28	.75	12.18	13.55
0300	12" deep		4200	.010		14.85	.31	.82	15.98	17.75
0301	Crushed 1-1/2" stone base, compacted to 4" deep	B-36B	6000	.011		5.10	.33	.67	6.10	6.90
0302	6" deep		5400	.012		7.65	.37	.74	8.76	9.85
0303	8" deep		4500	.014		10.20	.44	.89	11.53	12.95
0304	12" deep		3800	.017		15.30	.52	1.06	16.88	18.85
0350	Bank run gravel, spread and compacted									
0370	6" deep	B-32	6000	.005	S.Y.	4.29	.17	.30	4.76	5.35
0390	9" deep		4900	.007		6.45	.21	.37	7.03	7.80
0400	12" deep		4200	.008		8.55	.25	.43	9.23	10.35
6000	Stabilization fabric, polypropylene, 6 oz./S.Y.	B-6	10000	.002		1.48	.07	.03	1.58	1.78
6900	For small and irregular areas, add						50%	50%		
7000	Prepare and roll sub-base, small areas to 2500 S.Y.	B-32A	1500	.016	S.Y.		.51	.76	1.27	1.68
8000	Large areas over 2500 S.Y.	"	3500	.007			.22	.33	.55	.72
8050	For roadways	B-32	4000	.008			.26	.45	.71	.93

32 11 26 – Asphaltic Base Courses

32 11 26.19 Bituminous-Stabilized Base Courses

		Crew	Daily Output	Labor-Hours	Unit	Material	Labor	Equipment	Total	Total Incl O&P
0010	**BITUMINOUS-STABILIZED BASE COURSES**									
0020	And large paved areas									
0700	Liquid application to gravel base, asphalt emulsion	B-45	6000	.003	Gal.	5.90	.07	.05	6.02	6.70
0800	Prime and seal, cut back asphalt		6000	.003	"	6.95	.07	.05	7.07	7.85
1000	Macadam penetration crushed stone, 2 gal. per S.Y., 4" thick		6000	.003	S.Y.	11.80	.07	.05	11.92	13.20
1100	6" thick, 3 gal. per S.Y.		4000	.004		17.70	.11	.08	17.89	19.70
1200	8" thick, 4 gal. per S.Y.		3000	.005		23.50	.14	.10	23.74	26.50
8900	For small and irregular areas, add						50%	50%		

32 12 Flexible Paving

32 12 16 – Asphalt Paving

32 12 16.13 Plant-Mix Asphalt Paving

		Crew	Daily Output	Labor-Hours	Unit	Material	Labor	Equipment	Total	Total Incl O&P
0010	**PLANT-MIX ASPHALT PAVING**									
0020	And large paved areas with no hauling included									
0025	See Section 31 23 23.20 for hauling costs									
0080	Binder course, 1-1/2" thick	B-25	7725	.011	S.Y.	5.25	.33	.32	5.90	6.65
0120	2" thick		6345	.014		6.95	.40	.39	7.74	8.75
0130	2-1/2" thick		5620	.016		8.70	.45	.44	9.59	10.80
0160	3" thick		4905	.018		10.45	.52	.51	11.48	12.90
0170	3-1/2" thick		4520	.019		12.20	.56	.55	13.31	14.95
0200	4" thick		4140	.021		13.90	.62	.60	15.12	17
0300	Wearing course, 1" thick	B-25B	10575	.009		3.45	.27	.26	3.98	4.52

32 12 16 – Asphalt Paving

32 12 16.13 Plant-Mix Asphalt Paving		Crew	Daily Output	Labor-Hours	Unit	2011 Bare Costs Material	Labor	Equipment	Total	Total Incl O&P
0340	1-1/2" thick	B-25B	7725	.012	S.Y.	5.80	.37	.35	6.52	7.40
0380	2" thick		6345	.015		7.80	.45	.43	8.68	9.75
0420	2-1/2" thick		5480	.018		9.60	.52	.49	10.61	12
0460	3" thick		4900	.020		11.45	.58	.55	12.58	14.15
0470	3-1/2" thick		4520	.021		13.45	.63	.60	14.68	16.50
0480	4" thick		4140	.023		15.35	.68	.65	16.68	18.75
0500	Open graded friction course	B-25C	5000	.010		3.76	.28	.43	4.47	5.10
0800	Alternate method of figuring paving costs									
0810	Binder course, 1-1/2" thick	B-25	630	.140	Ton	64	4.04	3.94	71.98	81.50
0811	2" thick		690	.128		64	3.69	3.60	71.29	80.50
0812	3" thick		800	.110		64	3.18	3.10	70.28	79
0813	4" thick		900	.098		64	2.83	2.76	69.59	78
0850	Wearing course, 1" thick	B-25B	575	.167		65	4.91	4.71	74.62	85
0851	1-1/2" thick		630	.152		65	4.48	4.30	73.78	83.50
0852	2" thick		690	.139		65	4.09	3.92	73.01	82.50
0853	2-1/2" thick		765	.125		65	3.69	3.54	72.23	81.50
0854	3" thick		800	.120		65	3.53	3.38	71.91	81
1000	Pavement replacement over trench, 2" thick	B-37	90	.533	S.Y.	7.20	14.90	1.62	23.72	34.50
1050	4" thick		70	.686		14.20	19.20	2.08	35.48	50
1080	6" thick		55	.873		22.50	24.50	2.64	49.64	68.50

32 12 16.14 Asphaltic Concrete Paving		Crew	Daily Output	Labor-Hours	Unit	Material	Labor	Equipment	Total	Total Incl O&P
0011	**ASPHALTIC CONCRETE PAVING**, parking lots & driveways									
0015	No asphalt hauling included									
0018	Use 6.05 C. Y. per inch per M.S.F. for hauling									
0020	6" stone base, 2" binder course, 1" topping	B-25C	9000	.005	S.F.	1.85	.16	.24	2.25	2.57
0025	2" binder course, 2" topping		9000	.005		2.25	.16	.24	2.65	3
0030	3" binder course, 2" topping		9000	.005		2.64	.16	.24	3.04	3.43
0035	4" binder course, 2" topping		9000	.005		3.02	.16	.24	3.42	3.85
0040	1.5" binder course, 1" topping		9000	.005		1.66	.16	.24	2.06	2.36
0042	3" binder course, 1" topping		9000	.005		2.24	.16	.24	2.64	3
0045	3" binder course, 3" topping		9000	.005		3.03	.16	.24	3.43	3.87
0050	4" binder course, 3" topping		9000	.005		3.42	.16	.24	3.82	4.29
0055	4" binder course, 4" topping		9000	.005		3.81	.16	.24	4.21	4.72
0300	Binder course, 1-1/2" thick		35000	.001		.58	.04	.06	.68	.78
0400	2" thick		25000	.002		.75	.06	.09	.90	1.02
0500	3" thick		15000	.003		1.16	.09	.14	1.39	1.60
0600	4" thick		10800	.004		1.52	.13	.20	1.85	2.12
0800	Sand finish course, 3/4" thick		41000	.001		.31	.03	.05	.39	.46
0900	1" thick		34000	.001		.38	.04	.06	.48	.56
1000	Fill pot holes, hot mix, 2" thick	B-16	4200	.008		.80	.21	.12	1.13	1.36
1100	4" thick		3500	.009		1.17	.25	.14	1.56	1.86
1120	6" thick		3100	.010		1.58	.28	.16	2.02	2.38
1140	Cold patch, 2" thick	B-51	3000	.016		.81	.43	.06	1.30	1.67
1160	4" thick		2700	.018		1.55	.48	.06	2.09	2.57
1180	6" thick		1900	.025		2.41	.68	.09	3.18	3.88

32 13 Rigid Paving

32 13 13 – Concrete Paving

	32 13 13.23 Concrete Paving Surface Treatment	Crew	Daily Output	Labor-Hours	Unit	2011 Bare Costs Material	Labor	Equipment	Total	Total Incl O&P
0010	**CONCRETE PAVING SURFACE TREATMENT**									
0015	Including joints, finishing and curing									
0020	Fixed form, 12′ pass, unreinforced, 6″ thick	B-26	3000	.029	S.Y.	21	.87	1.01	22.88	25.50
0100	8″ thick		2750	.032		31.50	.95	1.10	33.55	38
0110	8″ thick, small area		1375	.064		31.50	1.89	2.20	35.59	40.50
0200	9″ thick		2500	.035		35.50	1.04	1.21	37.75	42
0300	10″ thick		2100	.042		39	1.24	1.44	41.68	46
0310	10″ thick, small area		1050	.084		39	2.48	2.88	44.36	50
0400	12″ thick		1800	.049		44	1.44	1.68	47.12	53
0410	Conc pavement, w/jt,fnsh&curing,fix form,24′ pass,unreinforced,6″T		6000	.015		19.35	.43	.50	20.28	23
0430	8″ thick		5500	.016		28.50	.47	.55	29.52	32.50
0440	9″ thick		5000	.018		32.50	.52	.61	33.63	37
0450	10″ thick		4200	.021		35.50	.62	.72	36.84	41
0460	12″ thick		3600	.024		41	.72	.84	42.56	47
0470	15″ thick		3000	.029		53.50	.87	1.01	55.38	61.50
0500	Fixed form 12′ pass 15″ thick		1500	.059		57	1.73	2.02	60.75	67.50
0510	For small irregular areas, add				%	10%	100%	100%		
0520	Welded wire fabric, sheets for rigid paving 2.33 lb./S.Y.	2 Rodm	389	.041	S.Y.	1.13	1.49		2.62	3.84
0530	Reinforcing steel for rigid paving 12 lb./S.Y.		666	.024		5.70	.87		6.57	7.75
0540	Reinforcing steel for rigid paving 18 lb./S.Y.		444	.036		8.50	1.31		9.81	11.60
0620	Slip form, 12′ pass, unreinforced, 6″ thick	B-26A	5600	.016		20.50	.46	.56	21.52	24
0624	8″ thick		5300	.017		31	.49	.60	32.09	35.50
0626	9″ thick		4820	.018		34.50	.54	.66	35.70	39.50
0628	10″ thick		4050	.022		38	.64	.78	39.42	43.50
0630	12″ thick		3470	.025		43	.75	.91	44.66	50
0632	15″ thick		2890	.030		55.50	.90	1.09	57.49	63.50
0640	Slip form, 24′ pass, unreinforced, 6″ thick		11200	.008		19.65	.23	.28	20.16	22
0644	8″ thick		10600	.008		28	.25	.30	28.55	31
0646	9″ thick		9640	.009		32	.27	.33	32.60	36
0648	10″ thick		8100	.011		35	.32	.39	35.71	39.50
0650	12″ thick		6940	.013		40.50	.37	.46	41.33	45.50
0652	15″ thick		5780	.015		51	.45	.55	52	58
0700	Finishing, broom finish small areas	2 Cefi	120	.133			4.25		4.25	6.80
1000	Curing, with sprayed membrane by hand	2 Clab	1500	.011		.50	.28		.78	1.02
1650	For integral coloring, see Section 03 05 13.20									

32 14 Unit Paving

32 14 13 – Precast Concrete Unit Paving

	32 14 13.13 Interlocking Precast Concrete Unit Paving	Crew	Daily Output	Labor-Hours	Unit	Material	Labor	Equipment	Total	Total Incl O&P
0010	**INTERLOCKING PRECAST CONCRETE UNIT PAVING**									
0020	"V" blocks for retaining soil	D-1	205	.078	S.F.	9.90	2.37		12.27	14.80
	32 14 13.16 Precast Concrete Unit Paving Slabs									
0010	**PRECAST CONCRETE UNIT PAVING SLABS**									
0750	Exposed local aggregate, natural	2 Bric	250	.064	S.F.	7.25	2.15		9.40	11.55
0800	Colors		250	.064		7.25	2.15		9.40	11.55
0850	Exposed granite or limestone aggregate		250	.064		7.50	2.15		9.65	11.80
0900	Exposed white tumblestone aggregate		250	.064		5.55	2.15		7.70	9.65
	32 14 13.18 Precast Concrete Plantable Pavers									
0010	**PRECAST CONCRETE PLANTABLE PAVERS** (50% grass)									
0300	3/4″ crushed stone base for plantable pavers, 6 inch depth	B-62	1000	.024	S.Y.	7.40	.69	.15	8.24	9.45

32 14 Unit Paving

32 14 13 – Precast Concrete Unit Paving

32 14 13.18 Precast Concrete Plantable Pavers		Crew	Daily Output	Labor-Hours	Unit	2011 Bare Costs Material	Labor	Equipment	Total	Total Incl O&P
0400	8 inch depth	B-62	900	.027	S.Y.	9.95	.77	.17	10.89	12.40
0500	10 inch depth		800	.030		12.35	.86	.19	13.40	15.25
0600	12 inch depth		700	.034		14.85	.99	.22	16.06	18.25
0700	Hydro seeding plantable pavers	B-81A	20	.800	M.S.F.	14.85	21	16.50	52.35	69.50
0800	Apply fertilizer and seed to plantable pavers	1 Clab	8	1	"	13.40	26.50		39.90	59.50

32 14 16 – Brick Unit Paving

32 14 16.10 Brick Paving

		Crew	Daily Output	Labor-Hours	Unit	Material	Labor	Equipment	Total	Total Incl O&P
0010	**BRICK PAVING**									
0012	4" x 8" x 1-1/2", without joints (4.5 brick/S.F.)	D-1	110	.145	S.F.	3.27	4.42		7.69	10.85
0100	Grouted, 3/8" joint (3.9 brick/S.F.)		90	.178		3.38	5.40		8.78	12.55
0200	4" x 8" x 2-1/4", without joints (4.5 bricks/S.F.)		110	.145		3.74	4.42		8.16	11.35
0300	Grouted, 3/8" joint (3.9 brick/S.F.)		90	.178		3.45	5.40		8.85	12.65
0500	Bedding, asphalt, 3/4" thick	B-25	5130	.017		.64	.50	.48	1.62	2.05
0540	Course washed sand bed, 1" thick	B-18	5000	.005		.30	.13	.01	.44	.56
0580	Mortar, 1" thick	D-1	300	.053		.66	1.62		2.28	3.38
0620	2" thick		200	.080		1.31	2.43		3.74	5.45
1500	Brick on 1" thick sand bed laid flat, 4.5 per S.F.		100	.160		3.38	4.86		8.24	11.70
2000	Brick pavers, laid on edge, 7.2 per S.F.		70	.229		3.16	6.95		10.11	14.85
2500	For 4" thick concrete bed and joints, add		595	.027		1.12	.82		1.94	2.58
2800	For steam cleaning, add	A-1H	950	.008		.07	.22	.06	.35	.52

32 14 23 – Asphalt Unit Paving

32 14 23.10 Asphalt Blocks

		Crew	Daily Output	Labor-Hours	Unit	Material	Labor	Equipment	Total	Total Incl O&P
0010	**ASPHALT BLOCKS**									
0020	Rectangular, 6" x 12" x 1-1/4", w/bed & neopr. adhesive	D-1	135	.119	S.F.	7.85	3.60		11.45	14.55
0100	3" thick		130	.123		11	3.74		14.74	18.25
0300	Hexagonal tile, 8" wide, 1-1/4" thick		135	.119		7.85	3.60		11.45	14.55
0400	2" thick		130	.123		11	3.74		14.74	18.25
0500	Square, 8" x 8", 1-1/4" thick		135	.119		7.85	3.60		11.45	14.55
0600	2" thick		130	.123		11	3.74		14.74	18.25
0900	For exposed aggregate (ground finish) add					.55			.55	.61
0910	For colors, add					.42			.42	.46

32 14 40 – Stone Paving

32 14 40.10 Stone Pavers

		Crew	Daily Output	Labor-Hours	Unit	Material	Labor	Equipment	Total	Total Incl O&P
0010	**STONE PAVERS**									
1100	Flagging, bluestone, irregular, 1" thick,	D-1	81	.198	S.F.	5.50	6		11.50	15.90
1150	Snapped random rectangular, 1" thick		92	.174		8.35	5.30		13.65	17.80
1200	1-1/2" thick		85	.188		10	5.70		15.70	20.50
1250	2" thick		83	.193		11.65	5.85		17.50	22.50
1300	Slate, natural cleft, irregular, 3/4" thick		92	.174		7	5.30		12.30	16.35
1350	Random rectangular, gauged, 1/2" thick		105	.152		15.15	4.63		19.78	24.50
1400	Random rectangular, butt joint, gauged, 1/4" thick		150	.107		16.30	3.24		19.54	23.50
1500	For interior setting, add								25%	25%
1550	Granite blocks, 3-1/2" x 3-1/2" x 3-1/2"	D-1	92	.174	S.F.	10.10	5.30		15.40	19.75
1600	4" to 12" long, 3" to 5" wide, 3" to 5" thick		98	.163		8.40	4.96		13.36	17.40
1650	6" to 15" long, 3" to 6" wide, 3" to 5" thick		105	.152		4.49	4.63		9.12	12.55

32 16 13 – Curbs and Gutters

32 16 13.13 Cast-in-Place Concrete Curbs and Gutters

	32 16 13.13 Cast-in-Place Concrete Curbs and Gutters	Crew	Daily Output	Labor-Hours	Unit	2011 Bare Costs Material	Labor	Equipment	Total	Total Incl O&P
0010	**CAST-IN-PLACE CONCRETE CURBS AND GUTTERS**									
0290	Forms only, no concrete									
0300	Concrete, wood forms, 6″ x 18″, straight	C-2	500	.096	L.F.	2.03	2.88		4.91	7.05
0400	6″ x 18″, radius	″	200	.240	″	2.13	7.20		9.33	14.40
0402	Forms and concrete complete									
0404	Concrete, wood forms, 6″ x 18″, straight & concrete	C-2A	500	.096	L.F.	4.80	3.11		7.91	10.45
0406	6″ x 18″, radius	″	200	.240		4.90	7.80		12.70	18.30
0415	Machine formed, 6″ x 18″, straight	B-69A	2000	.024		6.10	.70	.37	7.17	8.25
0416	6″ x 18″, radius	″	900	.053		6.25	1.55	.83	8.63	10.30
0421	Curb and gutter, straight									
0422	with 6″ high curb and 6″ thick gutter, wood forms									
0430	24″ wide, .055 C.Y. per L.F.	C-2A	375	.128	L.F.	12.90	4.15		17.05	21
0435	30″ wide, .066 C.Y. per L.F.		340	.141		14.35	4.58		18.93	23.50
0440	Steel forms, 24″ wide, straight		700	.069		10.95	2.22		13.17	15.75
0441	Radius		300	.160		10.95	5.20		16.15	20.50
0442	30″ wide, straight		700	.069		12.90	2.22		15.12	17.90
0443	Radius		300	.160		12.90	5.20		18.10	23
0445	Machine formed, 24″ wide, straight	B-69A	2000	.024		5.55	.70	.37	6.62	7.65
0446	Radius		900	.053		5.55	1.55	.83	7.93	9.55
0447	30″ wide, straight		2000	.024		6.45	.70	.37	7.52	8.65
0448	Radius		900	.053		6.45	1.55	.83	8.83	10.55

32 16 13.23 Precast Concrete Curbs and Gutters

		Crew	Daily Output	Labor-Hours	Unit	Material	Labor	Equipment	Total	Total Incl O&P
0010	**PRECAST CONCRETE CURBS AND GUTTERS**									
0550	Precast, 6″ x 18″, straight	B-29	700	.069	L.F.	8	1.94	1.22	11.16	13.40
0600	6″ x 18″, radius	″	325	.148	″	20	4.19	2.64	26.83	32

32 16 13.33 Asphalt Curbs

		Crew	Daily Output	Labor-Hours	Unit	Material	Labor	Equipment	Total	Total Incl O&P
0010	**ASPHALT CURBS**									
0012	Curbs, asphaltic, machine formed, 8″ wide, 6″ high, 40 L.F./ton	B-27	1000	.032	L.F.	1.63	.86	.28	2.77	3.53
0100	8″ wide, 8″ high, 30 L.F. per ton		900	.036		2.17	.96	.31	3.44	4.34
0150	Asphaltic berm, 12″ W, 3″-6″ H, 35 L.F./ton, before pavement		700	.046		.04	1.23	.40	1.67	2.54
0200	12″ W, 1-1/2″ to 4″ H, 60 L.F. per ton, laid with pavement	B-2	1050	.038		.02	1.02		1.04	1.73

32 16 13.43 Stone Curbs

		Crew	Daily Output	Labor-Hours	Unit	Material	Labor	Equipment	Total	Total Incl O&P
0010	**STONE CURBS**									
1000	Granite, split face, straight, 5″ x 16″	D-13	500	.096	L.F.	13.50	3.09	.87	17.46	21
1100	6″ x 18″	″	450	.107		17.75	3.43	.97	22.15	26
1300	Radius curbing, 6″ x 18″, over 10′ radius	B-29	260	.185		21.50	5.25	3.30	30.05	36.50
1400	Corners, 2′ radius		80	.600	Ea.	73	17	10.70	100.70	120
1600	Edging, 4-1/2″ x 12″, straight		300	.160	L.F.	6.75	4.53	2.86	14.14	18.15
1800	Curb inlets, (guttermouth) straight		41	1.171	Ea.	162	33	21	216	257
2000	Indian granite (belgian block)									
2100	Jumbo, 10-1/2″ x 7-1/2″ x 4″, grey	D-1	150	.107	L.F.	5.70	3.24		8.94	11.55
2150	Pink		150	.107		5.80	3.24		9.04	11.70
2200	Regular, 9″ x 4-1/2″ x 4-1/2″, grey		160	.100		5.45	3.04		8.49	11
2250	Pink		160	.100		5.80	3.04		8.84	11.35
2300	Cubes, 4″ x 4″ x 4″, grey		175	.091		3.18	2.78		5.96	8.05
2350	Pink		175	.091		3.15	2.78		5.93	8.05
2400	6″ x 6″ x 6″, pink		155	.103		10.40	3.14		13.54	16.60
2500	Alternate pricing method for indian granite									
2550	Jumbo, 10-1/2″ x 7-1/2″ x 4″ (30 lb), grey				Ton	325			325	360
2600	Pink					340			340	370
2650	Regular, 9″ x 4-1/2″ x 4-1/2″ (20 lb), grey					385			385	425
2700	Pink					405			405	445

32 16 Curbs, Gutters, Sidewalks, and Driveways

32 16 13 – Curbs and Gutters

32 16 13.43 Stone Curbs		Crew	Daily Output	Labor-Hours	Unit	2011 Bare Costs Material	Labor	Equipment	Total	Total Incl O&P
2750	Cubes, 4" x 4" x 4" (5 lb), grey				Ton	385			385	425
2800	Pink					405			405	445
2850	6" x 6" x 6" (25 lb), pink					405			405	445
2900	For pallets, add					20			20	22

32 17 Paving Specialties

32 17 13 – Parking Bumpers

32 17 13.13 Metal Parking Bumpers

		Crew	Daily Output	Labor-Hours	Unit	Material	Labor	Equipment	Total	Total Incl O&P
0010	**METAL PARKING BUMPERS**									
0015	Bumper rails for garages, 12 Ga. rail, 6" wide, with steel									
0020	posts 12'-6" O.C., minimum	E-4	190	.168	L.F.	17.75	6.20	.57	24.52	32
0030	Average		165	.194		22	7.15	.66	29.81	39
0100	Maximum		140	.229		26.50	8.45	.78	35.73	46.50
0300	12" channel rail, minimum		160	.200		22	7.40	.68	30.08	39.50
0400	Maximum		120	.267		33.50	9.85	.91	44.26	56.50
1300	Pipe bollards, conc filled/paint, 8' L x 4' D hole, 6" diam.	B-6	20	1.200	Ea.	640	34.50	16	690.50	780
1400	8" diam.		15	1.600		940	46	21.50	1,007.50	1,125
1500	12" diam.		12	2		1,050	57.50	26.50	1,134	1,275
1592	Bollards, steel, 3' H, retractable, incl hydraulic controls, min		4	6		40,000	173	80	40,253	44,400
1594	Max		2	12		45,000	345	160	45,505	50,500
2030	Folding with individual padlocks	B-2	50	.800		630	21.50		651.50	725
8000	Parking lot control, see Section 11 12 13.10									

32 17 13.16 Plastic Parking Bumpers

		Crew	Daily Output	Labor-Hours	Unit	Material	Labor	Equipment	Total	Total Incl O&P
0010	**PLASTIC PARKING BUMPERS**									
1200	Thermoplastic, 6" x 10" x 6'-0"	B-2	120	.333	Ea.	159	8.95		167.95	190

32 17 13.19 Precast Concrete Parking Bumpers

		Crew	Daily Output	Labor-Hours	Unit	Material	Labor	Equipment	Total	Total Incl O&P
0010	**PRECAST CONCRETE PARKING BUMPERS**									
1000	Wheel stops, precast concrete incl. dowels, 6" x 10" x 6'-0"	B-2	120	.333	Ea.	60	8.95		68.95	81
1100	8" x 13" x 6'-0"	"	120	.333	"	86	8.95		94.95	110

32 17 13.26 Wood Parking Bumpers

		Crew	Daily Output	Labor-Hours	Unit	Material	Labor	Equipment	Total	Total Incl O&P
0010	**WOOD PARKING BUMPERS**									
0020	Parking barriers, timber w/saddles, treated type									
0100	4" x 4" for cars	B-2	520	.077	L.F.	2.21	2.07		4.28	5.90
0200	6" x 6" for trucks		520	.077	"	4.53	2.07		6.60	8.45
0600	Flexible fixed stanchion, 2' high, 3" diameter		100	.400	Ea.	40	10.75		50.75	62

32 17 23 – Pavement Markings

32 17 23.13 Painted Pavement Markings

		Crew	Daily Output	Labor-Hours	Unit	Material	Labor	Equipment	Total	Total Incl O&P
0010	**PAINTED PAVEMENT MARKINGS**									
0020	Acrylic waterborne, white or yellow, 4" wide	B-78	20000	.002	L.F.	.19	.05	.02	.26	.33
0200	6" wide		11000	.004		.25	.10	.04	.39	.49
0500	8" wide		10000	.004		.30	.11	.05	.46	.56
0600	12" wide		4000	.010		.46	.27	.12	.85	1.08
0620	Arrows or gore lines		2300	.017	S.F.	.15	.47	.20	.82	1.17
0640	Temporary paint, white or yellow		15000	.003	L.F.	.05	.07	.03	.15	.21
0660	Removal	1 Clab	300	.027			.71		.71	1.18
0680	Temporary tape	2 Clab	1500	.011		.40	.28		.68	.91
0710	Thermoplastic, white or yellow, 4" wide	B-79	15000	.002		.20	.06	.07	.33	.40
0730	6" wide		14000	.002		.29	.06	.08	.43	.51
0740	8" wide		12000	.003		.39	.07	.09	.55	.65
0750	12" wide		6000	.005		.58	.14	.18	.90	1.08

32 17 Paving Specialties

32 17 23 – Pavement Markings

32 17 23.13 Painted Pavement Markings		Crew	Daily Output	Labor-Hours	Unit	2011 Bare Costs Material	Labor	Equipment	Total	Total Incl O&P
0760	Arrows	B-79	660	.048	S.F.	.40	1.31	1.64	3.35	4.43
0770	Gore lines		2500	.013		.40	.35	.43	1.18	1.50
0780	Letters		660	.048		.40	1.31	1.64	3.35	4.43
1000	Airport painted markings									
1100	Painting, white or yellow, taxiway markings	B-79	4000	.008	S.F.	.19	.22	.27	.68	.87
1200	Runway markings		3500	.009		.19	.25	.31	.75	.96
1300	Pavement location or direction signs		2500	.013		.19	.35	.43	.97	1.27
1350	Mobilization airport pavement painting		4	8	Ea.		216	270	486	655
1400	Paint markings or pavement signs removal daytime	B-78B	400	.045	S.F.		1.23	.78	2.01	2.91
1500	Removal nighttime		335	.054	"		1.46	.93	2.39	3.47
1600	Mobilization pavement paint removal		4	4.500	Ea.		123	78.50	201.50	291

32 17 23.14 Pavement Parking Markings		Crew	Daily Output	Labor-Hours	Unit	Material	Labor	Equipment	Total	Total Incl O&P
0010	**PAVEMENT PARKING MARKINGS**									
0790	Layout of pavement marking	A-2	25000	.001	L.F.		.03	.01	.04	.05
0800	Lines on pvmt, parking stall, paint, white, 4" wide	B-78	440	.091	Stall	4.50	2.44	1.07	8.01	10.20
1000	Street letters and numbers	"	1600	.025	S.F.	.65	.67	.29	1.61	2.16

32 18 Athletic and Recreational Surfacing

32 18 13 – Synthetic Grass Surfacing

32 18 13.10 Artificial Grass Surfacing		Crew	Daily Output	Labor-Hours	Unit	Material	Labor	Equipment	Total	Total Incl O&P
0010	**ARTIFICIAL GRASS SURFACING**									
0015	Not including asphalt base or drainage,									
0020	but including cushion pad, over 50,000 S.F.									
0200	1/2" pile and 5/16" cushion pad, standard	C-17	3200	.025	S.F.	9.50	.86		10.36	11.90
0300	Deluxe		2560	.031		11.25	1.08		12.33	14.20
0500	1/2" pile and 5/8" cushion pad, standard		2844	.028		13.75	.97		14.72	16.75
0600	Deluxe		2327	.034		15.15	1.18		16.33	18.65
0800	For asphaltic concrete base, 2-1/2" thick,									
0900	with 6" crushed stone sub-base, add	B-25	12000	.007	S.F.	1.77	.21	.21	2.19	2.53

32 18 23 – Athletic Surfacing

32 18 23.33 Running Track Surfacing		Crew	Daily Output	Labor-Hours	Unit	Material	Labor	Equipment	Total	Total Incl O&P
0010	**RUNNING TRACK SURFACING**									
0020	Running track, asphalt, incl base, 3" thick	B-37	300	.160	S.Y.	21	4.48	.48	25.96	31
0100	Surface, latex rubber system, 3/8" thick, black	B-20	125	.192		9.35	5.20		14.55	19
0150	Colors		125	.192		16.65	5.20		21.85	27
0300	Urethane rubber system, 3/8" thick, black		120	.200		25	5.40		30.40	36.50
0400	Color coating		115	.209		30.50	5.65		36.15	43

32 18 23.53 Tennis Court Surfacing		Crew	Daily Output	Labor-Hours	Unit	Material	Labor	Equipment	Total	Total Incl O&P
0010	**TENNIS COURT SURFACING**									
0020	Tennis court, asphalt, incl. base, 2-1/2" thick, one court	B-37	450	.107	S.Y.	33.50	2.98	.32	36.80	42.50
0200	Two courts		675	.071		15.10	1.99	.22	17.31	20
0300	Clay courts		360	.133		37.50	3.73	.40	41.63	48
0400	Pulverized natural greenstone with 4" base, fast dry		250	.192		36	5.35	.58	41.93	49
0800	Rubber-acrylic base resilient pavement		600	.080		51	2.24	.24	53.48	60
1000	Colored sealer, acrylic emulsion, 3 coats	2 Clab	800	.020		5.25	.53		5.78	6.70
1100	3 coat, 2 colors	"	900	.018		7.40	.47		7.87	8.90
1200	For preparing old courts, add	1 Clab	825	.010			.26		.26	.43
1400	Posts for nets, 3-1/2" diameter with eye bolts	B-1	3.40	7.059	Pr.	212	191		403	555
1500	With pulley & reel		3.40	7.059	"	300	191		491	650
1700	Net, 42' long, nylon thread with binder		50	.480	Ea.	279	13		292	325

32 18 Athletic and Recreational Surfacing

32 18 23 – Athletic Surfacing

32 18 23.53 Tennis Court Surfacing		Crew	Daily Output	Labor-Hours	Unit	Material	2011 Bare Costs Labor	Equipment	Total	Total Incl O&P
1800	All metal	B-1	6.50	3.692	Ea.	490	100		590	710
2001	Paint markings on asphalt, 2 coat	1 Clab	2.50	3.200	Court	131	84.50		215.50	286
2200	Complete court with fence, etc., asphaltic conc., minimum	B-37	.20	240		25,000	6,725	725	32,450	39,500
2300	Maximum		.16	300		50,000	8,400	910	59,310	70,000
2800	Clay courts, minimum		.20	240		25,000	6,725	725	32,450	39,500
2900	Maximum		.16	300		40,000	8,400	910	49,310	59,000

32 31 Fences and Gates

32 31 13 – Chain Link Fences and Gates

32 31 13.20 Fence, Chain Link Industrial

		Crew	Daily Output	Labor-Hours	Unit	Material	Labor	Equipment	Total	Total Incl O&P
0010	**FENCE, CHAIN LINK INDUSTRIAL**									
0011	Schedule 40, including concrete									
0020	3 strands barb wire, 2" post @ 10' O.C., set in concrete, 6' H									
0200	9 ga. wire, galv. steel, in concrete	B-80C	240	.100	L.F.	16	2.64	.74	19.38	23
0248	Fence, add for vinyl coated fabric				S.F.	.65			.65	.72
0300	Aluminized steel	B-80C	240	.100	L.F.	17	2.64	.74	20.38	24
0500	6 ga. wire, galv. steel		240	.100		18	2.64	.74	21.38	25
0600	Aluminized steel		240	.100		21.50	2.64	.74	24.88	28.50
0800	6 ga. wire, 6' high but omit barbed wire, galv. steel		250	.096		17.20	2.54	.71	20.45	24
0900	Aluminized steel, in concrete		250	.096		20.50	2.54	.71	23.75	27.50
0920	8' H, 6 ga. wire, 2-1/2" line post, galv. steel, in concrete		180	.133		27	3.52	.99	31.51	37
0940	Aluminized steel, in concrete		180	.133		33	3.52	.99	37.51	43.50
1100	Add for corner posts, 3" diam., galv. steel, in concrete		40	.600	Ea.	80	15.85	4.46	100.31	119
1200	Aluminized steel, in concrete		40	.600		92.50	15.85	4.46	112.81	133
1300	Add for braces, galv. steel		80	.300		27.50	7.95	2.23	37.68	45.50
1350	Aluminized steel		80	.300		36.50	7.95	2.23	46.68	55.50
1400	Gate for 6' high fence, 1-5/8" frame, 3' wide, galv. steel		10	2.400		147	63.50	17.85	228.35	288
1500	Aluminized steel, in concrete		10	2.400		248	63.50	17.85	329.35	400
2000	5'-0" high fence, 9 ga., no barbed wire, 2" line post, in concrete									
2010	10' O.C., 1-5/8" top rail, in concrete									
2100	Galvanized steel, in concrete	B-80C	300	.080	L.F.	16.30	2.11	.60	19.01	22
2200	Aluminized steel, in concrete		300	.080	"	17	2.11	.60	19.71	23
2400	Gate, 4' wide, 5' high, 2" frame, galv. steel, in concrete		10	2.400	Ea.	143	63.50	17.85	224.35	283
2500	Aluminized steel, in concrete		10	2.400	"	153	63.50	17.85	234.35	294
3100	Overhead slide gate, chain link, 6' high, to 18' wide, in concrete		38	.632	L.F.	159	16.70	4.70	180.40	208
3110	Cantilever type, in concrete	B-80	48	.500		137	13.55	12.90	163.45	187
3120	8' high, in concrete		24	1		141	27	26	194	230
3130	10' high, in concrete		18	1.333		156	36	34.50	226.50	270
5000	Double swing gates, incl. posts & hardware, in concrete									
5010	5' high, 12' opening, in concrete	B-80C	3.40	7.059	Opng.	550	187	52.50	789.50	975
5020	20' opening, in concrete		2.80	8.571		865	227	64	1,156	1,400
5060	6' high, 12' opening, in concrete		3.20	7.500		485	198	56	739	925
5070	20' opening, in concrete		2.60	9.231		725	244	68.50	1,037.50	1,275
5080	8' high, 12' opening, in concrete	B-80	2.13	11.252		515	305	290	1,110	1,400
5090	20' opening, in concrete		1.45	16.552		790	450	425	1,665	2,075
5100	10' high, 12' opening, in concrete		1.31	18.321		995	495	470	1,960	2,450
5110	20' opening, in concrete		1.03	23.301		1,375	630	600	2,605	3,200
5120	12' high, 12' opening, in concrete		1.05	22.857		1,350	620	590	2,560	3,175
5130	20' opening, in concrete		.85	28.235		1,900	765	730	3,395	4,175
5190	For aluminized steel add					20%				
7075	Fence, for small jobs 100 L.F. or less fence w/or wo gate, add				S.F.	20%			20%	

	32 31 13.25 Fence, Chain Link Residential	Crew	Daily Output	Labor-Hours	Unit	2011 Bare Costs Material	Labor	Equipment	Total	Total Incl O&P
0010	**FENCE, CHAIN LINK RESIDENTIAL**									
0011	Schedule 20, 11 gauge wire, 1-5/8" post									
0020	10' O.C., 1-3/8" top rail, 2" corner post, galv. stl. 3' high	B-80C	500	.048	L.F.	2.01	1.27	.36	3.64	4.72
0050	4' high		400	.060		4.72	1.59	.45	6.76	8.35
0100	6' high		200	.120		5.95	3.17	.89	10.01	12.85
0150	Add for gate 3' wide, 1-3/8" frame, 3' high		12	2	Ea.	82	53	14.90	149.90	195
0170	4' high		10	2.400		88	63.50	17.85	169.35	223
0190	6' high		10	2.400		94	63.50	17.85	175.35	230
0200	Add for gate 4' wide, 1-3/8" frame, 3' high		9	2.667		88	70.50	19.85	178.35	237
0220	4' high		9	2.667		92.50	70.50	19.85	182.85	242
0240	6' high		8	3		101	79.50	22.50	203	269
0350	Aluminized steel, 11 ga. wire, 3' high		500	.048	L.F.	5.50	1.27	.36	7.13	8.55
0380	4' high		400	.060		7.20	1.59	.45	9.24	11.05
0400	6' high		200	.120		9.85	3.17	.89	13.91	17.15
0450	Add for gate 3' wide, 1-3/8" frame, 3' high		12	2	Ea.	120	53	14.90	187.90	237
0470	4' high		10	2.400		126	63.50	17.85	207.35	264
0490	6' high		10	2.400		137	63.50	17.85	218.35	277
0500	Add for gate 4' wide, 1-3/8" frame, 3' high		10	2.400		129	63.50	17.85	210.35	268
0520	4' high		9	2.667		135	70.50	19.85	225.35	289
0540	6' high		8	3		144	79.50	22.50	246	315
0620	Vinyl covered, 9 ga. wire, 3' high		500	.048	L.F.	4.58	1.27	.36	6.21	7.55
0640	4' high		400	.060		5.10	1.59	.45	7.14	8.75
0660	6' high		200	.120		6.65	3.17	.89	10.71	13.60
0720	Add for gate 3' wide, 1-3/8" frame, 3' high		12	2	Ea.	120	53	14.90	187.90	237
0740	4' high		10	2.400		126	63.50	17.85	207.35	265
0760	6' high		10	2.400		127	63.50	17.85	208.35	266
0780	Add for gate 4' wide, 1-3/8" frame, 3' high		10	2.400		129	63.50	17.85	210.35	268
0800	4' high		9	2.667		135	70.50	19.85	225.35	289
0820	6' high		8	3		144	79.50	22.50	246	315
7076	Fence, for small jobs 100 L.F. fence or less w/or wo gate, add				S.F.	20%			20%	

	32 31 13.26 Tennis Court Fences and Gates	Crew	Daily Output	Labor-Hours	Unit	Material	Labor	Equipment	Total	Total Incl O&P
0010	**TENNIS COURT FENCES AND GATES**									
0860	Tennis courts, 11 ga. wire, 2-1/2" post set									
0870	in concrete, 10' O.C., 1-5/8" top rail									
0900	10' high	B-80	190	.126	L.F.	9.35	3.43	3.26	16.04	19.60
0920	12' high		170	.141	"	11.70	3.83	3.64	19.17	23.50
1000	Add for gate 4' wide, 1-5/8" frame 7' high		10	2.400	Ea.	140	65	62	267	330
1040	Aluminized steel, 11 ga. wire 10' high		190	.126	L.F.	17.30	3.43	3.26	23.99	28.50
1100	12' high		170	.141	"	21.50	3.83	3.64	28.97	34
1140	Add for gate 4' wide, 1-5/8" frame, 7' high		10	2.400	Ea.	161	65	62	288	355
1250	Vinyl covered, 9 ga. wire, 10' high		190	.126	L.F.	10.25	3.43	3.26	16.94	20.50
1300	12' high		170	.141	"	12.40	3.83	3.64	19.87	24
1310	Fence, CL, tennis ct, transom gate, single, galv, 4' x 7' x 3'	B-80A	8.72	2.752	Ea.	435	73	25	533	630
1400	Add for gate 4' wide, 1-5/8" frame, 7' high	B-80	10	2.400	"	262	65	62	389	465

	32 31 13.33 Chain Link Backstops	Crew	Daily Output	Labor-Hours	Unit	Material	Labor	Equipment	Total	Total Incl O&P
0010	**CHAIN LINK BACKSTOPS**									
0015	Backstops, baseball, prefabricated, 30' wide, 12' high & 1 overhang	B-1	1	24	Ea.	2,400	650		3,050	3,750
0100	40' wide, 12' high & 2 overhangs	"	.75	32		5,225	870		6,095	7,200
0300	Basketball, steel, single goal	B-13	3.04	15.789		1,375	445	246	2,066	2,550
0400	Double goal	"	1.92	25		990	710	390	2,090	2,700
0600	Tennis, wire mesh with pair of ends	B-1	2.48	9.677	Set	2,250	262		2,512	2,925
0700	Enclosed court	"	1.30	18.462	Ea.	9,700	500		10,200	11,500

32 31 Fences and Gates

32 31 19 – Decorative Metal Fences and Gates

32 31 19.10 Decorative Fence		Crew	Daily Output	Labor-Hours	Unit	Material	2011 Bare Costs Labor	Equipment	Total	Total Incl O&P
0010	**DECORATIVE FENCE**									
5300	Tubular picket, steel, 6′ sections, 1-9/16″ posts, 4′ high	B-80C	300	.080	L.F.	27.50	2.11	.60	30.21	34
5400	2″ posts, 5′ high		240	.100		41	2.64	.74	44.38	50.50
5600	2″ posts, 6′ high		200	.120		47.50	3.17	.89	51.56	59
5700	Staggered picket 1-9/16″ posts, 4′ high		300	.080		27.50	2.11	.60	30.21	34
5800	2″ posts, 5′ high		240	.100		41	2.64	.74	44.38	50.50
5900	2″ posts, 6′ high		200	.120		47.50	3.17	.89	51.56	59
6200	Gates, 4′ high, 3′ wide	B-1	10	2.400	Ea.	235	65		300	370
6300	5′ high, 3′ wide		10	2.400		248	65		313	380
6400	6′ high, 3′ wide		10	2.400		298	65		363	440
6500	4′ wide		10	2.400		325	65		390	465

32 31 26 – Wire Fences and Gates

32 31 26.10 Fences, Misc. Metal		Crew	Daily Output	Labor-Hours	Unit	Material	Labor	Equipment	Total	Total Incl O&P
0010	**FENCES, MISC. METAL**									
0012	Chicken wire, posts @ 4′, 1″ mesh, 4′ high	B-80C	410	.059	L.F.	2.46	1.55	.44	4.45	5.75
0100	2″ mesh, 6′ high		350	.069		2.38	1.81	.51	4.70	6.20
0200	Galv. steel, 12 ga., 2″ x 4″ mesh, posts 5′ O.C., 3′ high		300	.080		2.44	2.11	.60	5.15	6.85
0300	5′ high		300	.080		3.74	2.11	.60	6.45	8.30
0400	14 ga., 1″ x 2″ mesh, 3′ high		300	.080		2.76	2.11	.60	5.47	7.25
0500	5′ high		300	.080		4.34	2.11	.60	7.05	8.95
1000	Kennel fencing, 1-1/2″ mesh, 6′ long, 3′-6″ wide, 6′-2″ high	2 Clab	4	4	Ea.	460	106		566	680
1050	12′ long		4	4		600	106		706	835
1200	Top covers, 1-1/2″ mesh, 6′ long		15	1.067		153	28		181	217
1250	12′ long		12	1.333		200	35.50		235.50	279
1300	For kennel doors, see Section 08 31 13.40									
4500	Security fence, prison grade, set in concrete, 12′ high	B-80	25	.960	L.F.	50	26	25	101	126
4600	16′ high	″	20	1.200	″	68	32.50	31	131.50	164

32 31 26.20 Wire Fencing, General		Crew	Daily Output	Labor-Hours	Unit	Material	Labor	Equipment	Total	Total Incl O&P
0010	**WIRE FENCING, GENERAL**									
0015	Barbed wire, galvanized, domestic steel, hi-tensile 15-1/2 ga.				M.L.F.	117			117	128
0020	Standard, 12-3/4 ga.					133			133	147
0210	Barbless wire, 2-strand galvanized, 12-1/2 ga.					133			133	147
0500	Helical razor ribbon, stainless steel, 18″ dia x 18″ spacing				C.L.F.	370			370	405
0600	Hardware cloth galv., 1/4″ mesh, 23 ga., 2′ wide				C.S.F.	74.50			74.50	82
0700	3′ wide					74.50			74.50	82
0900	1/2″ mesh, 19 ga., 2′ wide					74.50			74.50	82
1000	4′ wide					56			56	61.50
1200	Chain link fabric, steel, 2″ mesh, 6 ga, galvanized					200			200	220
1300	9 ga, galvanized					76.50			76.50	84
1350	Vinyl coated					85			85	93.50
1360	Aluminized					88			88	96.50
1400	2-1/4″ mesh, 11.5 ga, galvanized					48			48	52.50
1600	1-3/4″ mesh (tennis courts), 11.5 ga (core), vinyl coated					48			48	52.50
1700	9 ga, galvanized					73			73	80.50
2100	Welded wire fabric, galvanized, 1″ x 2″, 14 ga.					54.50			54.50	60
2200	2″ x 4″, 12-1/2 ga.					56			56	61.50

32 31 29 – Wood Fences and Gates

32 31 29.20 Fence, Wood Rail		Crew	Daily Output	Labor-Hours	Unit	Material	Labor	Equipment	Total	Total Incl O&P
0010	**FENCE, WOOD RAIL**									
0012	Picket, No. 2 cedar, Gothic, 2 rail, 3′ high	B-1	160	.150	L.F.	6.10	4.07		10.17	13.50
0050	Gate, 3′-6″ wide	B-80C	9	2.667	Ea.	52	70.50	19.85	142.35	197

32 31 Fences and Gates

32 31 29 – Wood Fences and Gates

32 31 29.20 Fence, Wood Rail		Crew	Daily Output	Labor-Hours	Unit	2011 Bare Costs Material	Labor	Equipment	Total	Total Incl O&P
0400	3 rail, 4′ high	B-80C	150	.160	L.F.	7	4.23	1.19	12.42	16.05
0500	Gate, 3′-6″ wide		9	2.667	Ea.	63	70.50	19.85	153.35	210
0600	Open rail, rustic, No. 1 cedar, 2 rail, 3′ high		160	.150	L.F.	5.40	3.96	1.12	10.48	13.80
0650	Gate, 3′ wide		9	2.667	Ea.	60.50	70.50	19.85	150.85	207
0700	3 rail, 4′ high		150	.160	L.F.	6.20	4.23	1.19	11.62	15.15
0900	Gate, 3′ wide		9	2.667	Ea.	78	70.50	19.85	168.35	226
1200	Stockade, No. 2 cedar, treated wood rails, 6′ high		160	.150	L.F.	7.55	3.96	1.12	12.63	16.15
1250	Gate, 3′ wide		9	2.667	Ea.	62	70.50	19.85	152.35	208
1300	No. 1 cedar, 3-1/4″ cedar rails, 6′ high		160	.150	L.F.	18.55	3.96	1.12	23.63	28.50
1500	Gate, 3′ wide		9	2.667	Ea.	151	70.50	19.85	241.35	305
1520	Open rail, split, No. 1 cedar, 2 rail, 3′ high		160	.150	L.F.	5.45	3.96	1.12	10.53	13.85
1540	3 rail, 4′-0″ high		150	.160		6.60	4.23	1.19	12.02	15.60
3300	Board, shadow box, 1″ x 6″, treated pine, 6′ high		160	.150		10.25	3.96	1.12	15.33	19.15
3400	No. 1 cedar, 6′ high		150	.160		20	4.23	1.19	25.42	30.50
3900	Basket weave, No. 1 cedar, 6′ high		160	.150		28	3.96	1.12	33.08	38.50
3950	Gate, 3′-6″ wide	B-1	8	3	Ea.	159	81.50		240.50	310
4000	Treated pine, 6′ high		150	.160	L.F.	14.90	4.34		19.24	23.50
4200	Gate, 3′-6″ wide		9	2.667	Ea.	82	72.50		154.50	211
5000	Fence rail, redwood, 2″ x 4″, merch grade 8′		2400	.010	L.F.	1.80	.27		2.07	2.43
5050	Select grade, 8′		2400	.010	″	4	.27		4.27	4.85
6000	Fence post, select redwood, earthpacked & treated, 4″ x 4″ x 6′		96	.250	Ea.	11.30	6.80		18.10	24
6010	4″ x 4″ x 8′		96	.250		13.40	6.80		20.20	26
6020	Set in concrete, 4″ x 4″ x 6′		50	.480		14.50	13		27.50	38
6030	4″ x 4″ x 8′		50	.480		17.30	13		30.30	41
6040	Wood post, 4′ high, set in concrete, incl. concrete		50	.480		8.10	13		21.10	31
6050	Earth packed		96	.250		5.55	6.80		12.35	17.45
6060	6′ high, set in concrete, incl. concrete		50	.480		10.35	13		23.35	33.50
6070	Earth packed		96	.250		7.55	6.80		14.35	19.65

32 32 Retaining Walls

32 32 13 – Cast-in-Place Concrete Retaining Walls

32 32 13.10 Retaining Walls, Cast Concrete		Crew	Daily Output	Labor-Hours	Unit	Material	Labor	Equipment	Total	Total Incl O&P
0010	**RETAINING WALLS, CAST CONCRETE**									
1800	Concrete gravity wall with vertical face including excavation & backfill									
1850	No reinforcing									
1900	6′ high, level embankment	C-17C	36	2.306	L.F.	70.50	79.50	17.65	167.65	230
2000	33° slope embankment		32	2.594		66	89.50	19.85	175.35	245
2200	8′ high, no surcharge		27	3.074		90	106	23.50	219.50	300
2300	33° slope embankment		24	3.458		109	119	26.50	254.50	350
2500	10′ high, level embankment		19	4.368		129	151	33.50	313.50	430
2600	33° slope embankment		18	4.611		178	159	35.50	372.50	500
2800	Reinforced concrete cantilever, incl. excavation, backfill & reinf.									
2900	6′ high, 33° slope embankment	C-17C	35	2.371	L.F.	66	82	18.15	166.15	229
3000	8′ high, 33° slope embankment		29	2.862		76	98.50	22	196.50	273
3100	10′ high, 33° slope embankment		20	4.150		99	143	32	274	385
3200	20′ high, 500 lb. per L.F. surcharge		7.50	11.067		297	380	84.50	761.50	1,050
3500	Concrete cribbing, incl. excavation and backfill									
3700	12′ high, open face	B-13	210	.229	S.F.	33	6.50	3.57	43.07	51
3900	Closed face	″	210	.229	″	31	6.50	3.57	41.07	48.50
4100	Concrete filled slurry trench, see Section 31 56 23.20									

32 32 Retaining Walls

32 32 23 – Segmental Retaining Walls

32 32 23.13 Segmental Conc. Unit Masonry Retaining Walls		Crew	Daily Output	Labor-Hours	Unit	Material	2011 Bare Costs Labor	Equipment	Total	Total Incl O&P
0010	**SEGMENTAL CONC. UNIT MASONRY RETAINING WALLS**									
7100	Segmental Retaining Wall system, incl pins, and void fill									
7120	base not included									
7140	Large unit, 8″ high x 18″ wide x 20″ deep, 3 plane split	B-62	300	.080	S.F.	10.20	2.31	.51	13.02	15.60
7150	Straight split		300	.080		10.10	2.31	.51	12.92	15.50
7160	Medium, ltwt, 8″ high x 18″ wide x 12″ deep, 3 plane split		400	.060		8	1.73	.38	10.11	12.10
7170	Straight split		400	.060		7.90	1.73	.38	10.01	12
7180	Small unit, 4″ x 18″ x 10″ deep, 3 plane split		400	.060		12	1.73	.38	14.11	16.50
7190	Straight split		400	.060		12.20	1.73	.38	14.31	16.70
7200	Cap unit, 3 plane split		300	.080		11.80	2.31	.51	14.62	17.40
7210	Cap unit, straight split	↓	300	.080		11.80	2.31	.51	14.62	17.40
7260	For reinforcing, add				↓				4.28	4.71
8000	For higher walls, add components as necessary									

32 32 26 – Metal Crib Retaining Walls

32 32 26.10 Metal Bin Retaining Walls		Crew	Daily Output	Labor-Hours	Unit	Material	Labor	Equipment	Total	Total Incl O&P
0010	**METAL BIN RETAINING WALLS**									
0011	Aluminized steel bin, excavation									
0020	and backfill not included, 10′ wide									
0100	4′ high, 5.5′ deep	B-13	650	.074	S.F.	23	2.09	1.15	26.24	30.50
0200	8′ high, 5.5′ deep		615	.078		26.50	2.21	1.22	29.93	34
0300	10′ high, 7.7′ deep		580	.083		28	2.35	1.29	31.64	36.50
0400	12′ high, 7.7′ deep		530	.091		30	2.57	1.41	33.98	39.50
0500	16′ high, 7.7′ deep		515	.093		32	2.64	1.45	36.09	41
0600	16′ high, 9.9′ deep		500	.096		33.50	2.72	1.50	37.72	43
0700	20′ high, 9.9′ deep		470	.102		38	2.89	1.59	42.48	48
0800	20′ high, 12.1′ deep		460	.104		40.50	2.96	1.63	45.09	51
0900	24′ high, 12.1′ deep		455	.105		43	2.99	1.65	47.64	54.50
1000	24′ high, 14.3′ deep		450	.107		44.50	3.02	1.66	49.18	56
1100	28′ high, 14.3′ deep	↓	440	.109	↓	46.50	3.09	1.70	51.29	58
1300	For plain galvanized bin type walls, deduct					10%				

32 32 36 – Gabion Retaining Walls

32 32 36.10 Stone Gabion Retaining Walls		Crew	Daily Output	Labor-Hours	Unit	Material	Labor	Equipment	Total	Total Incl O&P
0010	**STONE GABION RETAINING WALLS**									
4300	Stone filled gabions, not incl. excavation,									
4310	Stone, delivered, 3′ wide									
4350	Galvanized, 6′ high, 33° slope embankment	B-13	49	.980	L.F.	22	28	15.30	65.30	88
4500	Highway surcharge		27	1.778		80	50.50	27.50	158	203
4600	9′ high, up to 33° slope embankment		24	2		50	56.50	31	137.50	184
4700	Highway surcharge		16	3		140	85	47	272	350
4900	12′ high, up to 33° slope embankment		14	3.429		77.50	97	53.50	228	305
5000	Highway surcharge	↓	11	4.364		200	124	68	392	500
5950	For PVC coating, add				↓	20%				

32 32 60 – Stone Retaining Walls

32 32 60.10 Retaining Walls, Stone		Crew	Daily Output	Labor-Hours	Unit	Material	Labor	Equipment	Total	Total Incl O&P
0010	**RETAINING WALLS, STONE**									
0015	Including excavation, concrete footing and									
0020	stone 3′ below grade. Price is exposed face area.									
0200	Decorative random stone, to 6′ high, 1′-6″ thick, dry set	D-1	35	.457	S.F.	50	13.90		63.90	78
0300	Mortar set		40	.400		55	12.15		67.15	80.50
0500	Cut stone, to 6′ high, 1′-6″ thick, dry set	↓	35	.457	↓	50	13.90		63.90	78

32 32 Retaining Walls

32 32 60 – Stone Retaining Walls

32 32 60.10 Retaining Walls, Stone		Crew	Daily Output	Labor-Hours	Unit	2011 Bare Costs Material	Labor	Equipment	Total	Total Incl O&P
0600	Mortar set	D-1	40	.400	S.F.	55	12.15		67.15	80.50
0800	Random stone, 6′ to 10′ high, 2′ thick, dry set		45	.356		55	10.80		65.80	78.50
0900	Mortar set		50	.320		60	9.75		69.75	82
1100	Cut stone, 6′ to 10′ high, 2′ thick, dry set		45	.356		55	10.80		65.80	78.50
1200	Mortar set		50	.320		60	9.75		69.75	82

32 34 Fabricated Bridges

32 34 20 – Fabricated Pedestrian Bridges

32 34 20.10 Bridges, Pedestrian		Crew	Daily Output	Labor-Hours	Unit	Material	Labor	Equipment	Total	Total Incl O&P
0010	**BRIDGES, PEDESTRIAN**									
0011	Spans over streams, roadways, etc.									
0020	including erection, not including foundations									
0050	Precast concrete, complete in place, 8′ wide, 60′ span	E-2	215	.223	S.F.	105	8.20	7.55	120.75	139
0100	100′ span		185	.259		114	9.50	8.75	132.25	153
0150	120′ span		160	.300		124	11	10.15	145.15	169
0200	150′ span		145	.331		129	12.15	11.20	152.35	177
0300	Steel, trussed or arch spans, compl. in place, 8′ wide, 40′ span		320	.150		99	5.50	5.05	109.55	125
0400	50′ span		395	.122		89	4.45	4.11	97.56	111
0500	60′ span		465	.103		89	3.78	3.49	96.27	109
0600	80′ span		570	.084		106	3.08	2.85	111.93	126
0700	100′ span		465	.103		149	3.78	3.49	156.27	175
0800	120′ span		365	.132		188	4.82	4.44	197.26	221
0900	150′ span		310	.155		200	5.65	5.25	210.90	236
1000	160′ span		255	.188		200	6.90	6.35	213.25	240
1100	10′ wide, 80′ span		640	.075		110	2.75	2.53	115.28	129
1200	120′ span		415	.116		143	4.24	3.91	151.15	169
1300	150′ span		445	.108		160	3.95	3.64	167.59	187
1400	200′ span		205	.234		171	8.60	7.90	187.50	213
1600	Wood, laminated type, complete in place, 80′ span	C-12	203	.236		65.50	7.80	3.23	76.53	88.50
1700	130′ span	"	153	.314		68	10.40	4.29	82.69	97

32 35 Screening Devices

32 35 16 – Sound Barriers

32 35 16.10 Traffic Barriers, Highway Sound Barriers		Crew	Daily Output	Labor-Hours	Unit	Material	Labor	Equipment	Total	Total Incl O&P
0010	**TRAFFIC BARRIERS, HIGHWAY SOUND BARRIERS**									
0020	Highway sound barriers, not including footing									
0100	Precast concrete, concrete columns @ 30′ OC, 8″ T, 8′ H	C-12	400	.120	L.F.	128	3.97	1.64	133.61	149
0110	12′ H		265	.181		192	6	2.47	200.47	224
0120	16′ H		200	.240		256	7.95	3.28	267.23	299
0130	20′ H		160	.300		320	9.90	4.10	334	370
0400	Lt. Wt. composite panel, cementitious face, St. posts @ 12′ OC, 8′ H	B-80B	190	.168		143	4.75	1.19	148.94	167
0410	12′ H		125	.256		215	7.25	1.80	224.05	251
0420	16′ H		95	.337		287	9.50	2.37	298.87	335
0430	20′ H		75	.427		360	12.05	3.01	375.06	420

32 84 Planting Irrigation

32 84 23 – Underground Sprinklers

32 84 23.10 Sprinkler Irrigation System		Crew	Daily Output	Labor-Hours	Unit	2011 Bare Costs Material	Labor	Equipment	Total	Total Incl O&P
0010	**SPRINKLER IRRIGATION SYSTEM**									
0011	For lawns									
0100	Golf course with fully automatic system	C-17	.05	1600	9 holes	100,000	55,000		155,000	202,500
0200	24′ diam. head at 15′ O.C incl. piping, auto oper., minimum	B-20	70	.343	Head	21	9.30		30.30	38.50
0300	Maximum		40	.600		52	16.25		68.25	84.50
0500	60′ diam. head at 40′ O.C. incl. piping, auto oper., minimum		28	.857		66	23.50		89.50	112
0600	Maximum		23	1.043		183	28.50		211.50	249
0800	Residential system, custom, 1″ supply		2000	.012	S.F.	.32	.33		.65	.90
0900	1-1/2″ supply		1800	.013	″	.37	.36		.73	1.02
1020	Pop up spray head w/risers, hi-pop, full circle pattern, 4″	2 Skwk	76	.211	Ea.	4.24	7.15		11.39	16.65
1030	1/2 circle pattern, 4″		76	.211		5.15	7.15		12.30	17.65
1040	6″, full circle pattern		76	.211		10.20	7.15		17.35	23
1050	1/2 circle pattern, 6″		76	.211		11.05	7.15		18.20	24
1060	12″, full circle pattern		76	.211		15.70	7.15		22.85	29.50
1070	1/2 circle pattern, 12″		76	.211		17.35	7.15		24.50	31
1080	Pop up bubbler head w/risers, hi-pop bubbler head, 4″		76	.211		5.15	7.15		12.30	17.70
1090	6″		76	.211		11.45	7.15		18.60	24.50
1100	12″		76	.211		15.35	7.15		22.50	29
1110	Impact full/part circle sprinklers, 28′-54′ 25-60 PSI		37	.432		15.15	14.70		29.85	41
1120	Spaced 37′-49′ @ 25-50 PSI		37	.432		29	14.70		43.70	56
1130	Spaced 43′-61′ @ 30-60 PSI		37	.432		56.50	14.70		71.20	86.50
1140	Spaced 54′-78′ @ 40-80 PSI		37	.432		105	14.70		119.70	140
1145	Impact rotor pop-up full/part commercial circle sprinklers									
1150	Spaced 42′-65′ 35-80 PSI	2 Skwk	25	.640	Ea.	13	22		35	51
1160	Spaced 48′-76′ 45-85 PSI	″	25	.640	″	14.05	22		36.05	52
1165	Impact rotor pop-up part. circle comm., 53′-75′, 55-100 PSI, w/accessories									
1170	Plastic case, metal cover	2 Skwk	25	.640	Ea.	120	22		142	168
1180	Rubber cover		25	.640		132	22		154	182
1190	Iron case, metal cover		22	.727		144	25		169	200
1200	Rubber cover		22	.727		166	25		191	224
1250	Plastic case, 2 nozzle, metal cover		25	.640		143	22		165	194
1260	Rubber cover		25	.640		150	22		172	201
1270	Iron case, 2 nozzle, metal cover		22	.727		213	25		238	276
1280	Rubber cover		22	.727		210	25		235	273
1282	Impact rotor pop-up full circle commercial, 39′-99′, 30-100 PSI									
1284	Plastic case, metal cover	2 Skwk	25	.640	Ea.	150	22		172	202
1286	Rubber cover		25	.640		169	22		191	223
1288	Iron case, metal cover		22	.727		218	25		243	281
1290	Rubber cover		22	.727		226	25		251	290
1292	Plastic case, 2 nozzle, metal cover		22	.727		164	25		189	222
1294	Rubber cover		22	.727		166	25		191	225
1296	Iron case, 2 nozzle, metal cover		20	.800		208	27		235	274
1298	Rubber cover		20	.800		221	27		248	289
1305	Electric remote control valve, plastic, 3/4″		18	.889		19	30.50		49.50	71.50
1310	1″		18	.889		22.50	30.50		53	75.50
1320	1-1/2″		18	.889		117	30.50		147.50	180
1330	2″		18	.889		128	30.50		158.50	191
1335	Quick coupling valves, brass, locking cover									
1340	Inlet coupling valve, 3/4″	2 Skwk	18.75	.853	Ea.	44	29		73	97
1350	1″		18.75	.853		43	29		72	96
1360	Controller valve boxes, 6″ round boxes		18.75	.853		4.98	29		33.98	54
1370	10″ round boxes		14.25	1.123		12.90	38		50.90	78
1380	12″ square box		9.75	1.641		26	56		82	122

32 84 Planting Irrigation

32 84 23 – Underground Sprinklers

32 84 23.10 Sprinkler Irrigation System		Crew	Daily Output	Labor-Hours	Unit	2011 Bare Costs Material	Labor	Equipment	Total	Total Incl O&P
1388	Electromech. control, 14 day 3-60 min, auto start to 23/day									
1390	4 station	2 Skwk	1.04	15.385	Ea.	159	525		684	1,050
1400	7 station		.64	25		315	850		1,165	1,775
1410	12 station		.40	40		330	1,350		1,680	2,650
1420	Dual programs, 18 station		.24	66.667		1,100	2,275		3,375	5,000
1430	23 station		.16	100		1,850	3,400		5,250	7,725
1435	Backflow preventer, bronze, 0-175 PSI, w/valves, test cocks									
1440	3/4"	2 Skwk	2	8	Ea.	136	272		408	605
1450	1"		2	8		159	272		431	630
1460	1-1/2"		2	8		242	272		514	720
1470	2"		2	8		266	272		538	750
1475	Pressure vacuum breaker, brass, 15-150 PSI									
1480	3/4"	2 Skwk	2	8	Ea.	81	272		353	545
1490	1"		2	8		104	272		376	570
1500	1-1/2"		2	8		360	272		632	850
1510	2"		2	8		410	272		682	905

32 91 Planting Preparation

32 91 13 – Soil Preparation

32 91 13.16 Mulching

Line	Description	Crew	Daily Output	Labor-Hours	Unit	Material	Labor	Equipment	Total	Total Incl O&P
0010	**MULCHING**									
0100	Aged barks, 3" deep, hand spread	1 Clab	100	.080	S.Y.	3.02	2.12		5.14	6.85
0150	Skid steer loader	B-63	13.50	2.963	M.S.F.	335	78.50	11.25	424.75	515
0200	Hay, 1" deep, hand spread	1 Clab	475	.017	S.Y.	.50	.45		.95	1.30
0250	Power mulcher, small	B-64	180	.089	M.S.F.	55.50	2.35	1.68	59.53	67
0350	Large	B-65	530	.030	"	55.50	.80	.81	57.11	63
0400	Humus peat, 1" deep, hand spread	1 Clab	700	.011	S.Y.	3.99	.30		4.29	4.90
0450	Push spreader	"	2500	.003	"	3.99	.08		4.07	4.53
0550	Tractor spreader	B-66	700	.011	M.S.F.	445	.38	.32	445.70	490
0600	Oat straw, 1" deep, hand spread	1 Clab	475	.017	S.Y.	.50	.45		.95	1.30
0650	Power mulcher, small	B-64	180	.089	M.S.F.	55.50	2.35	1.68	59.53	67
0700	Large	B-65	530	.030	"	55.50	.80	.81	57.11	63
0750	Add for asphaltic emulsion	B-45	1770	.009	Gal.	2.14	.24	.17	2.55	2.94
0800	Peat moss, 1" deep, hand spread	1 Clab	900	.009	S.Y.	2.98	.24		3.22	3.67
0850	Push spreader	"	2500	.003	"	2.98	.08		3.06	3.42
0950	Tractor spreader	B-66	700	.011	M.S.F.	330	.38	.32	330.70	365
1000	Polyethylene film, 6 mil.	2 Clab	2000	.008	S.Y.	.23	.21		.44	.60
1100	Redwood nuggets, 3" deep, hand spread	1 Clab	150	.053	"	3	1.41		4.41	5.65
1150	Skid steer loader	B-63	13.50	2.963	M.S.F.	335	78.50	11.25	424.75	510
1200	Stone mulch, hand spread, ceramic chips, economy	1 Clab	125	.064	S.Y.	6.95	1.69		8.64	10.50
1250	Deluxe	"	95	.084	"	10.75	2.23		12.98	15.60
1300	Granite chips	B-1	10	2.400	C.Y.	35	65		100	148
1400	Marble chips		10	2.400		130	65		195	252
1600	Pea gravel		28	.857		69	23.50		92.50	115
1700	Quartz		10	2.400		170	65		235	296
1800	Tar paper, 15 lb. felt	1 Clab	800	.010	S.Y.	.51	.26		.77	1
1900	Wood chips, 2" deep, hand spread	"	220	.036	"	1.95	.96		2.91	3.76
1950	Skid steer loader	B-63	20.30	1.970	M.S.F.	217	52	7.50	276.50	335

32 91 13.26 Planting Beds

Line	Description	Crew	Daily Output	Labor-Hours	Unit	Material	Labor	Equipment	Total	Total Incl O&P
0010	**PLANTING BEDS**									
0100	Backfill planting pit, by hand, on site topsoil	2 Clab	18	.889	C.Y.		23.50		23.50	39.50

32 91 Planting Preparation

32 91 13 – Soil Preparation

32 91 13.26 Planting Beds		Crew	Daily Output	Labor-Hours	Unit	Material	2011 Bare Costs Labor	Equipment	Total	Total Incl O&P
0200	Prepared planting mix, by hand	2 Clab	24	.667	C.Y.		17.65		17.65	29.50
0300	Skid steer loader, on site topsoil	B-62	340	.071			2.03	.45	2.48	3.86
0400	Prepared planting mix	"	410	.059			1.69	.37	2.06	3.21
1000	Excavate planting pit, by hand, sandy soil	2 Clab	16	1			26.50		26.50	44.50
1100	Heavy soil or clay	"	8	2			53		53	88.50
1200	1/2 C.Y. backhoe, sandy soil	B-11C	150	.107			3.27	2.14	5.41	7.75
1300	Heavy soil or clay	"	115	.139			4.27	2.79	7.06	10.10
2000	Mix planting soil, incl. loam, manure, peat, by hand	2 Clab	60	.267		42	7.05		49.05	58
2100	Skid steer loader	B-62	150	.160		42	4.61	1.01	47.62	55
3000	Pile sod, skid steer loader	"	2800	.009	S.Y.		.25	.05	.30	.47
3100	By hand	2 Clab	400	.040			1.06		1.06	1.77
4000	Remove sod, F.E. loader	B-10S	2000	.004			.14	.16	.30	.41
4100	Sod cutter	B-12K	3200	.005			.16	.31	.47	.60
4200	By hand	2 Clab	240	.067			1.76		1.76	2.95

32 91 19 – Landscape Grading

32 91 19.13 Topsoil Placement and Grading

Line	Description	Crew	Daily Output	Labor-Hours	Unit	Material	Labor	Equipment	Total	Total Incl O&P
0010	**TOPSOIL PLACEMENT AND GRADING**									
0400	Spread from pile to rough finish grade, F.E. loader, 1.5 C.Y.	B-10S	200	.040	C.Y.		1.40	1.60	3	4.04
0500	Up to 200' radius, by hand	1 Clab	14	.571			15.10		15.10	25.50
0600	Top dress by hand, 1 C.Y. for 600 S.F.	"	11.50	.696		33	18.40		51.40	67.50
0700	Furnish and place, truck dumped, screened, 4" deep	B-10S	1300	.006	S.Y.	4.13	.21	.25	4.59	5.15
0800	6" deep	"	820	.010	"	5.30	.34	.39	6.03	6.80

32 92 Turf and Grasses

32 92 19 – Seeding

32 92 19.13 Mechanical Seeding

Line	Description	Crew	Daily Output	Labor-Hours	Unit	Material	Labor	Equipment	Total	Total Incl O&P
0010	**MECHANICAL SEEDING** R329219-50									
0020	Mechanical seeding, 215 lb./acre	B-66	1.50	5.333	Acre	570	179	149	898	1,075
0100	44 lb./M.S.Y.	"	2500	.003	S.Y.	.20	.11	.09	.40	.50
0300	Fine grading and seeding incl. lime, fertilizer & seed,									
0310	with equipment	B-14	1000	.048	S.Y.	.48	1.34	.32	2.14	3.12
0600	Limestone hand push spreader, 50 lb. per M.S.F.	1 Clab	180	.044	M.S.F.	3.95	1.18		5.13	6.30
0800	Grass seed hand push spreader, 4.5 lb. per M.S.F.	"	180	.044	"	22	1.18		23.18	26.50
1000	Hydro or air seeding for large areas, incl. seed and fertilizer	B-81	8900	.002	S.Y.	.20	.05	.06	.31	.37
1100	With wood fiber mulch added	"	8900	.002	"	.34	.05	.06	.45	.52
1300	Seed only, over 100 lb., field seed, minimum				Lb.	1.15			1.15	1.27
1400	Maximum					2.75			2.75	3.03
1500	Lawn seed, minimum					.80			.80	.88
1600	Maximum					1.40			1.40	1.54
1800	Aerial operations, seeding only, field seed	B-58	50	.480	Acre	375	13.85	58	446.85	495
1900	Lawn seed		50	.480		260	13.85	58	331.85	375
2100	Seed and liquid fertilizer, field seed		50	.480		465	13.85	58	536.85	595
2200	Lawn seed		50	.480		350	13.85	58	421.85	470

32 92 23 – Sodding

32 92 23.10 Sodding Systems

Line	Description	Crew	Daily Output	Labor-Hours	Unit	Material	Labor	Equipment	Total	Total Incl O&P
0010	**SODDING SYSTEMS**									
0020	Sodding, 1" deep, bluegrass sod, on level ground, over 8 M.S.F.	B-63	22	1.818	M.S.F.	215	48	6.90	269.90	325
0200	4 M.S.F.		17	2.353		275	62	8.95	345.95	420
0300	1000 S.F.		13.50	2.963		300	78.50	11.25	389.75	475
0500	Sloped ground, over 8 M.S.F.		6	6.667		215	176	25.50	416.50	560

32 92 Turf and Grasses

32 92 23 – Sodding

32 92 23.10 Sodding Systems		Crew	Daily Output	Labor-Hours	Unit	Material	2011 Bare Costs Labor	Equipment	Total	Total Incl O&P
0600	4 M.S.F.	B-63	5	8	M.S.F.	275	212	30.50	517.50	695
0700	1000 S.F.		4	10		300	265	38	603	815
1000	Bent grass sod, on level ground, over 6 M.S.F.		20	2		530	53	7.60	590.60	675
1100	3 M.S.F.		18	2.222		610	59	8.45	677.45	780
1200	Sodding 1000 S.F. or less		14	2.857		695	75.50	10.85	781.35	905
1500	Sloped ground, over 6 M.S.F.		15	2.667		530	70.50	10.15	610.65	710
1600	3 M.S.F.		13.50	2.963		610	78.50	11.25	699.75	815
1700	1000 S.F.		12	3.333		695	88	12.65	795.65	925

32 93 Plants

32 93 10 – General Planting Costs

32 93 10.12 Travel		Crew	Daily Output	Labor-Hours	Unit	Material	Labor	Equipment	Total	Total Incl O&P
0010	**TRAVEL** to all nursery items, for 10 to 20 miles, add									
0015	10 to 20 miles one way, add				All					5%
0100	30 to 50 miles one way, add				"					10%

32 93 13 – Ground Covers

32 93 13.10 Ground Cover Plants		Crew	Daily Output	Labor-Hours	Unit	Material	Labor	Equipment	Total	Total Incl O&P
0010	**GROUND COVER PLANTS**									
0012	Plants, pachysandra, in prepared beds	B-1	15	1.600	C	30	43.50		73.50	106
0200	Vinca minor, 1 yr., bare root, in prepared beds		12	2	"	43	54		97	139
0600	Stone chips, in 50 lb. bags, Georgia marble		520	.046	Bag	6	1.25		7.25	8.70
0700	Onyx gemstone		260	.092		23	2.50		25.50	29
0800	Quartz		260	.092		8.60	2.50		11.10	13.65
0900	Pea gravel, truckload lots		28	.857	Ton	30	23.50		53.50	72

32 93 33 – Shrubs

32 93 33.10 Shrubs and Trees		Crew	Daily Output	Labor-Hours	Unit	Material	Labor	Equipment	Total	Total Incl O&P
0010	**SHRUBS AND TREES**									
0011	Evergreen, in prepared beds, B & B									
0100	Arborvitae pyramidal, 4′-5′	B-17	30	1.067	Ea.	49	30.50	21	100.50	128
0150	Globe, 12″-15″	B-1	96	.250		20.50	6.80		27.30	34
0300	Cedar, blue, 8′-10′	B-17	18	1.778		210	50.50	35.50	296	355
0500	Hemlock, canadian, 2-1/2′-3′	B-1	36	.667		32	18.10		50.10	65.50
0550	Holly, Savannah, 8′ - 10′ H		9.68	2.479		375	67		442	525
0600	Juniper, andorra, 18″-24″		80	.300		16.80	8.15		24.95	32
0620	Wiltoni, 15″-18″		80	.300		17.25	8.15		25.40	32.50
0640	Skyrocket, 4-1/2′-5′	B-17	55	.582		68.50	16.55	11.55	96.60	116
0660	Blue pfitzer, 2′-2-1/2′	B-1	44	.545		32.50	14.80		47.30	60.50
0680	Ketleerie, 2-1/2′-3′		50	.480		24	13		37	48
0700	Pine, black, 2-1/2′-3′		50	.480		50.50	13		63.50	77.50
0720	Mugo, 18″-24″		60	.400		40	10.85		50.85	62
0740	White, 4′-5′	B-17	75	.427		57	12.10	8.50	77.60	92
0800	Spruce, blue, 18″-24″	B-1	60	.400		41	10.85		51.85	63
0840	Norway, 4′-5′	B-17	75	.427		65	12.10	8.50	85.60	101
0900	Yew, denisforma, 12″-15″	B-1	60	.400		30.50	10.85		41.35	51.50
1000	Capitata, 18″-24″		30	.800		25	21.50		46.50	63.50
1100	Hicksi, 2′-2-1/2′		30	.800		31.50	21.50		53	71.50

32 93 33.20 Shrubs		Crew	Daily Output	Labor-Hours	Unit	Material	Labor	Equipment	Total	Total Incl O&P
0010	**SHRUBS**									
0011	Broadleaf Evergreen, planted in prepared beds									
0100	Andromeda, 15″-18″, container	B-1	96	.250	Ea.	28.50	6.80		35.30	42.50

32 93 Plants

32 93 33 – Shrubs

32 93 33.20 Shrubs		Crew	Daily Output	Labor-Hours	Unit	2011 Bare Costs Material	Labor	Equipment	Total	Total Incl O&P
0200	Azalea, 15" - 18", container	B-1	96	.250	Ea.	21	6.80		27.80	34.50
0300	Barberry, 9"-12", container		130	.185		12.30	5		17.30	22
0400	Boxwood, 15"-18", B&B		96	.250		32	6.80		38.80	46.50
0500	Euonymus, emerald gaiety, 12" to 15", container		115	.209		22	5.65		27.65	33.50
0600	Holly, 15"-18", B & B		96	.250		17	6.80		23.80	30
0900	Mount laurel, 18" - 24", B & B		80	.300		60	8.15		68.15	79.50
1000	Paxistema, 9 – 12" high		130	.185		19	5		24	29.50
1100	Rhododendron, 18"-24", container		48	.500		28	13.55		41.55	53.50
1200	Rosemary, 1 gal. container		600	.040		63	1.08		64.08	71.50
2000	Deciduous, planted in prepared beds, amelanchier, 2'-3', B & B		57	.421		89	11.40		100.40	117
2100	Azalea, 15"-18", B & B		96	.250		26.50	6.80		33.30	40.50
2300	Bayberry, 2'-3', B & B		57	.421		24.50	11.40		35.90	46
2600	Cotoneaster, 15"-18", B & B		80	.300		22.50	8.15		30.65	38.50
2800	Dogwood, 3'-4', B & B	B-17	40	.800		27.50	22.50	15.90	65.90	85.50
2900	Euonymus, alatus compacta, 15" to 18", container	B-1	80	.300		22.50	8.15		30.65	38.50
3200	Forsythia, 2'-3', container	"	60	.400		19.25	10.85		30.10	39
3300	Hibiscus, 3'-4', B & B	B-17	75	.427		39.50	12.10	8.50	60.10	73
3400	Honeysuckle, 3'-4', B & B	B-1	60	.400		21.50	10.85		32.35	41.50
3500	Hydrangea, 2'-3', B & B	"	57	.421		29	11.40		40.40	51
3600	Lilac, 3'-4', B & B	B-17	40	.800		23.50	22.50	15.90	61.90	81
3900	Privet, bare root, 18"-24"	B-1	80	.300		12.55	8.15		20.70	27.50
4100	Quince, 2'-3', B & B	"	57	.421		25	11.40		36.40	46.50
4200	Russian olive, 3'-4', B & B	B-17	75	.427		25.50	12.10	8.50	46.10	57.50
4400	Spirea, 3'-4', B & B	B-1	70	.343		17.25	9.30		26.55	34.50
4500	Viburnum, 3'-4', B & B	B-17	40	.800		28	22.50	15.90	66.40	86

32 93 43 – Trees

32 93 43.20 Trees		Crew	Daily Output	Labor-Hours	Unit	Material	Labor	Equipment	Total	Total Incl O&P
0010	**TREES**									
0011	Deciduous, in prep. beds, balled & burlapped (B&B)									
0100	Ash, 2" caliper G	B-17	8	4	Ea.	79.50	114	79.50	273	365
0200	Beech, 5'-6' G		50	.640		95	18.20	12.70	125.90	149
0300	Birch, 6'-8', 3 stems G		20	1.600		112	45.50	32	189.50	234
0500	Crabapple, 6'-8' G		20	1.600		125	45.50	32	202.50	249
0600	Dogwood, 4'-5' G		40	.800		122	22.50	15.90	160.40	189
0700	Eastern redbud 4'-5' G		40	.800		146	22.50	15.90	184.40	215
0800	Elm, 8'-10' G		20	1.600		119	45.50	32	196.50	242
0900	Ginkgo, 6'-7' G		24	1.333		180	38	26.50	244.50	290
1000	Hawthorn, 8'-10', 1" caliper G		20	1.600		141	45.50	32	218.50	266
1100	Honeylocust, 10'-12', 1-1/2" caliper G		10	3.200		153	91	63.50	307.50	390
1300	Larch, 8' G		32	1		108	28.50	19.85	156.35	187
1400	Linden, 8'-10', 1" caliper G		20	1.600		113	45.50	32	190.50	235
1500	Magnolia, 4'-5' G		20	1.600		81.50	45.50	32	159	201
1600	Maple, red, 8'-10', 1-1/2" caliper G		10	3.200		255	91	63.50	409.50	500
1700	Mountain ash, 8'-10', 1" caliper G		16	2		228	57	39.50	324.50	390
1800	Oak, 2-1/2"-3" caliper G		6	5.333		315	152	106	573	715
2100	Planetree, 9'-11', 1-1/4" caliper G		10	3.200		125	91	63.50	279.50	360
2200	Plum, 6'-8', 1" caliper G		20	1.600		95.50	45.50	32	173	216
2300	Poplar, 9'-11', 1-1/4" caliper G		10	3.200		95	91	63.50	249.50	325
2500	Sumac, 2'-3' G		75	.427		21	12.10	8.50	41.60	52.50
2700	Tulip, 5'-6' G		40	.800		53.50	22.50	15.90	91.90	114
2800	Willow, 6'-8', 1" caliper G		20	1.600		57.50	45.50	32	135	174

Stop guessing...
Estimate with the Experts
Stay up-to-date on localized construction materials, labor, and equipment costs with RSMeans cost data.
We've Got You Covered...
▪ commercial
▪ residential
▪ new construction
▪ renovation and remodeling
▪ green construction
▪ facility maintenance
With Comprehensive Data...
▪ detailed costs
▪ assemblies costs
▪ square foot costs
▪ materials and equipment
▪ labor and productivity rates
▪ overhead and markups
And Powerful Estimating Tools
▪ search and find costs with ease
▪ create, customize and save estimates
▪ share estimates with team members
▪ export estimates to Excel
▪ access your estimates from anywhere on the web
Shop for more RSMeans cost data in online, book, or CD format at www.rsmeans.com
RSMeans

32 94 Planting Accessories

32 94 50 – Tree Guying

32 94 50.10 Tree Guying Systems		Crew	Daily Output	Labor-Hours	Unit	Material	2011 Bare Costs Labor	Equipment	Total	Total Incl O&P
0010	**TREE GUYING SYSTEMS**									
0015	Tree guying Including stakes, guy wire and wrap									
0100	Less than 3" caliper, 2 stakes	2 Clab	35	.457	Ea.	18	12.10		30.10	40.50
0200	3" to 4" caliper, 3 stakes	"	21	.762	"	20.50	20		40.50	56.50
1000	Including arrowhead anchor, cable, turnbuckles and wrap									
1100	Less than 3" caliper, 3" anchors	2 Clab	20	.800	Ea.	51	21		72	91.50
1200	3" to 6" caliper, 4" anchors		15	1.067		51.50	28		79.50	104
1300	6" caliper, 6" anchors		12	1.333		96	35.50		131.50	165
1400	8" caliper, 8" anchors	↓	9	1.778	↓	99	47		146	188

32 96 Transplanting

32 96 23 – Plant and Bulb Transplanting

32 96 23.23 Planting

		Crew	Daily Output	Labor-Hours	Unit	Material	Labor	Equipment	Total	Total Incl O&P
0010	**PLANTING**									
0011	Moving shrubs on site, 12" ball	3 Clab	28	.857	Ea.		22.50		22.50	38
0100	24" ball	B-62	22	1.091	"		31.50	6.90	38.40	59.50

32 96 23.43 Moving Trees

		Crew	Daily Output	Labor-Hours	Unit	Material	Labor	Equipment	Total	Total Incl O&P
0290	**MOVING TREES**, On site									
0300	Moving trees on site, 36" ball	B-6	3.75	6.400	Ea.		184	85.50	269.50	400
0400	60" ball	"	1	24	"		690	320	1,010	1,500

Division Notes

		CREW	DAILY OUTPUT	LABOR-HOURS	UNIT	2011 BARE COSTS MAT.	LABOR	EQUIP.	TOTAL	TOTAL INCL O&P

Estimating Tips

33 10 00 Water Utilities
33 30 00 Sanitary Sewerage Utilities
33 40 00 Storm Drainage Utilities

- Never assume that the water, sewer, and drainage lines will go in at the early stages of the project. Consider the site access needs before dividing the site in half with open trenches, loose pipe, and machinery obstructions. Always inspect the site to establish that the site drawings are complete. Check off all existing utilities on your drawings as you locate them. Be especially careful with underground utilities because appurtenances are sometimes buried during regrading or repaving operations. If you find any discrepancies, mark up the site plan for further research. Differing site conditions can be very costly if discovered later in the project.
- See also Section 33 01 00 for restoration of pipe where removal/replacement may be undesirable. Use of new types of piping materials can reduce the overall project cost. Owners/design engineers should consider the installing contractor as a valuable source of current information on utility products and local conditions that could lead to significant cost savings.

Reference Numbers

Reference numbers are shown in shaded boxes at the beginning of some major classifications. These numbers refer to related items in the Reference Section. The reference information may be an estimating procedure, an alternate pricing method, or technical information.

Note: Not all subdivisions listed here necessarily appear in this publication.

Note: **Trade Service,** *in part, has been used as a reference source for some of the material prices used in Division 33.*

33 01 Operation and Maintenance of Utilities

33 01 10 – Operation and Maintenance of Water Utilities

33 01 10.10 Corrosion Resistance		Crew	Daily Output	Labor-Hours	Unit	Material	2011 Bare Costs Labor	Equipment	Total	Total Incl O&P
0010	**CORROSION RESISTANCE**									
0012	Wrap & coat, add to pipe, 4″ diameter				L.F.	2.15			2.15	2.37
0040	6″ diameter					3.23			3.23	3.55
0060	8″ diameter					4.19			4.19	4.61
0100	12″ diameter					6.35			6.35	6.95
0200	24″ diameter					12.20			12.20	13.45
0500	Coating, bituminous, per diameter inch, 1 coat, add					.46			.46	.51
0540	3 coat					1.37			1.37	1.51
0560	Coal tar epoxy, per diameter inch, 1 coat, add					.27			.27	.30
0600	3 coat					.55			.55	.61

33 01 30 – Operation and Maintenance of Sewer Utilities

33 01 30.72 Relining Sewers		Crew	Daily Output	Labor-Hours	Unit	Material	Labor	Equipment	Total	Total Incl O&P
0010	**RELINING SEWERS**									
0011	With cement incl. bypass & cleaning									
0020	Less than 10,000 L.F., urban, 6″ to 10″	C-17E	130	.615	L.F.	9	21	.68	30.68	46
0200	24″ to 36″		90	.889		14.35	30.50	.99	45.84	68.50
0300	48″ to 72″		80	1		23	34.50	1.11	58.61	83.50

33 05 Common Work Results for Utilities

33 05 16 – Utility Structures

33 05 16.13 Precast Concrete Utility Boxes		Crew	Daily Output	Labor-Hours	Unit	Material	Labor	Equipment	Total	Total Incl O&P
0010	**PRECAST CONCRETE UTILITY BOXES**, 6″ thick									
0050	5′ x 10′ x 6′ high, I.D.	B-13	2	24	Ea.	2,425	680	375	3,480	4,200
0100	6′ x 10′ x 6′ high, I.D.		2	24		2,500	680	375	3,555	4,275
0150	5′ x 12′ x 6′ high, I.D.		2	24		2,675	680	375	3,730	4,450
0200	6′ x 12′ x 6′ high, I.D.		1.80	26.667		2,975	755	415	4,145	4,975
0250	6′ x 13′ x 6′ high, I.D.		1.50	32		3,925	905	500	5,330	6,350
0300	8′ x 14′ x 7′ high, I.D.		1	48		4,225	1,350	750	6,325	7,750
0350	Hand hole, precast concrete, 1-1/2″ thick									
0400	1′-0″ x 2′-0″ x 1′-9″, I.D., light duty	B-1	4	6	Ea.	375	163		538	690
0450	4′-6″ x 3′-2″ x 2′-0″, O.D., heavy duty	B-6	3	8	″	1,100	231	107	1,438	1,700

33 05 23 – Trenchless Utility Installation

33 05 23.19 Microtunneling		Crew	Daily Output	Labor-Hours	Unit	Material	Labor	Equipment	Total	Total Incl O&P
0010	**MICROTUNNELING**									
0011	Not including excavation, backfill, shoring,									
0020	or dewatering, average 50′/day, slurry method									
0100	24″ to 48″ outside diameter, minimum				L.F.				790	870
0110	Adverse conditions, add				%				50%	50%
1000	Rent microtunneling machine, average monthly lease				Month				88,000	97,000
1010	Operating technician				Day				585	650
1100	Mobilization and demobilization, minimum				Job				38,000	42,000
1110	Maximum				″				400,000	440,000

33 05 23.20 Horizontal Boring		Crew	Daily Output	Labor-Hours	Unit	Material	Labor	Equipment	Total	Total Incl O&P
0010	**HORIZONTAL BORING**									
0011	Casing only, 100′ minimum,									
0020	not incl. jacking pits or dewatering									
0100	Roadwork, 1/2″ thick wall, 24″ diameter casing	B-42	20	2.800	L.F.	142	84	65	291	365
0200	36″ diameter		16	3.500		214	105	81	400	500
0300	48″ diameter		15	3.733		285	112	86.50	483.50	595
0500	Railroad work, 24″ diameter		15	3.733		142	112	86.50	340.50	440

33 05 Common Work Results for Utilities

33 05 23 – Trenchless Utility Installation

33 05 23.20 Horizontal Boring		Crew	Daily Output	Labor-Hours	Unit	Material	2011 Bare Costs Labor	Equipment	Total	Total Incl O&P
0600	36″ diameter	B-42	14	4	L.F.	214	120	93	427	535
0700	48″ diameter		12	4.667		285	140	108	533	665
0900	For ledge, add									20%

33 05 26 – Utility Identification

33 05 26.10 Utility Accessories		Crew	Daily Output	Labor-Hours	Unit	Material	Labor	Equipment	Total	Total Incl O&P
0010	**UTILITY ACCESSORIES**									
0400	Underground tape, detectable, reinforced, alum. foil core, 2″	1 Clab	150	.053	C.L.F.	2.25	1.41		3.66	4.84
0500	6″	″	140	.057	″	6.55	1.51		8.06	9.75

33 11 Water Utility Distribution Piping

33 11 13 – Public Water Utility Distribution Piping

33 11 13.15 Water Supply, Ductile Iron Pipe		Crew	Daily Output	Labor-Hours	Unit	Material	Labor	Equipment	Total	Total Incl O&P
0010	**WATER SUPPLY, DUCTILE IRON PIPE** R331113-80									
0020	Not including excavation or backfill									
2000	Pipe, class 50 water piping, 18′ lengths									
2020	Mechanical joint, 4″ diameter	B-21A	200	.200	L.F.	12.50	6.50	2.17	21.17	27
2040	6″ diameter		160	.250		14.60	8.10	2.72	25.42	32.50
2060	8″ diameter		133.33	.300		21	9.70	3.26	33.96	42.50
2080	10″ diameter		114.29	.350		26.50	11.35	3.80	41.65	52
2100	12″ diameter		105.26	.380		32	12.30	4.13	48.43	60
2120	14″ diameter		100	.400		48.50	12.95	4.34	65.79	80
2140	16″ diameter		72.73	.550		50	17.80	5.95	73.75	91
2160	18″ diameter		68.97	.580		53	18.80	6.30	78.10	96.50
2170	20″ diameter		57.14	.700		56	22.50	7.60	86.10	107
2180	24″ diameter		47.06	.850		61	27.50	9.25	97.75	123
3000	Tyton, push-on joint, 4″ diameter		400	.100		10.80	3.24	1.09	15.13	18.45
3020	6″ diameter		333.33	.120		15.20	3.89	1.30	20.39	24.50
3040	8″ diameter		200	.200		21.50	6.50	2.17	30.17	37
3060	10″ diameter		181.82	.220		29.50	7.15	2.39	39.04	47
3080	12″ diameter		160	.250		31	8.10	2.72	41.82	50.50
3100	14″ diameter		133.33	.300		34	9.70	3.26	46.96	57
3120	16″ diameter		114.29	.350		37	11.35	3.80	52.15	63.50
3140	18″ diameter		100	.400		41	12.95	4.34	58.29	71.50
3160	20″ diameter		88.89	.450		42.50	14.60	4.89	61.99	76
3180	24″ diameter		76.92	.520		44.50	16.85	5.65	67	83
8000	Fittings, mechanical joint									
8006	90° bend, 4″ diameter	B-20A	16	2	Ea.	248	63		311	375
8020	6″ diameter		12.80	2.500		370	79		449	535
8040	8″ diameter		10.67	2.999		530	94.50		624.50	740
8060	10″ diameter	B-21A	11.43	3.500		810	113	38	961	1,125
8080	12″ diameter		10.53	3.799		1,075	123	41.50	1,239.50	1,450
8100	14″ diameter		10	4		1,525	130	43.50	1,698.50	1,925
8120	16″ diameter		7.27	5.502		1,925	178	60	2,163	2,475
8140	18″ diameter		6.90	5.797		2,725	188	63	2,976	3,375
8160	20″ diameter		5.71	7.005		3,400	227	76	3,703	4,200
8180	24″ diameter		4.70	8.511		5,975	276	92.50	6,343.50	7,100
8200	Wye or tee, 4″ diameter	B-20A	10.67	2.999		360	94.50		454.50	550
8220	6″ diameter		8.53	3.751		530	118		648	780
8240	8″ diameter		7.11	4.501		785	142		927	1,100
8260	10″ diameter	B-21A	7.62	5.249		1,325	170	57	1,552	1,800

33 11 Water Utility Distribution Piping

33 11 13 – Public Water Utility Distribution Piping

33 11 13.15 Water Supply, Ductile Iron Pipe		Crew	Daily Output	Labor-Hours	Unit	Material	Labor (2011 Bare Costs)	Equipment	Total	Total Incl O&P
8280	12″ diameter	B-21A	7.02	5.698	Ea.	1,750	185	62	1,997	2,300
8300	14″ diameter		6.67	5.997		1,575	194	65	1,834	2,150
8320	16″ diameter		4.85	8.247		2,925	267	89.50	3,281.50	3,775
8340	18″ diameter		4.60	8.696		4,100	282	94.50	4,476.50	5,075
8360	20″ diameter		3.81	10.499		5,325	340	114	5,779	6,525
8380	24″ diameter		3.14	12.739		9,525	415	138	10,078	11,300
8450	Decreaser, 6″ x 4″ diameter	B-20A	14.22	2.250		260	71		331	405
8460	8″ x 6″ diameter	″	11.64	2.749		380	87		467	565
8470	10″ x 6″ diameter	B-21A	13.33	3.001		455	97.50	32.50	585	695
8480	12″ x 6″ diameter		12.70	3.150		595	102	34	731	860
8490	16″ x 6″ diameter		10	4		1,175	130	43.50	1,348.50	1,550
8500	20″ x 6″ diameter		8.42	4.751		2,600	154	51.50	2,805.50	3,150
8550	Piping, butterfly valves, cast iron									
8560	4″ diameter	B-20	6	4	Ea.	795	108		903	1,050
8570	6″ diameter	″	5	4.800		845	130		975	1,150
8580	8″ diameter	B-21	4	7		1,125	199	32	1,356	1,600
8590	10″ diameter		3.50	8		1,675	227	37	1,939	2,275
8600	12″ diameter		3	9.333		2,425	265	43	2,733	3,125
8610	14″ diameter		2	14		4,075	395	64.50	4,534.50	5,200
8620	16″ diameter		2	14		4,775	395	64.50	5,234.50	5,975

33 11 13.25 Water Supply, Polyvinyl Chloride Pipe		Crew	Daily Output	Labor-Hours	Unit	Material	Labor	Equipment	Total	Total Incl O&P
0010	**WATER SUPPLY, POLYVINYL CHLORIDE PIPE**									
0020	Not including excavation or backfill, unless specified									
2100	PVC pipe, Class 150, 1-1/2″ diameter	Q-1A	750	.013	L.F.	.32	.53		.85	1.22
2120	2″ diameter		686	.015		.49	.58		1.07	1.49
2140	2-1/2″ diameter		500	.020		.73	.80		1.53	2.11
2160	3″ diameter	B-20	430	.056		1.11	1.51		2.62	3.75
3010	AWWA C905, PR 100, DR 25									
3030	14″ diameter	B-20A	213	.150	L.F.	10.70	4.74		15.44	19.60
3040	16″ diameter		200	.160		13.90	5.05		18.95	23.50
3050	18″ diameter		160	.200		21	6.30		27.30	33.50
3060	20″ diameter		133	.241		21	7.60		28.60	35.50
3070	24″ diameter		107	.299		29.50	9.45		38.95	48
3080	30″ diameter		80	.400		59	12.60		71.60	86
3090	36″ diameter		80	.400		91	12.60		103.60	121
3100	42″ diameter		60	.533		123	16.85		139.85	163
3200	48″ diameter		60	.533		161	16.85		177.85	205
4520	Pressure pipe Class 150, SDR 18, AWWA C900, 4″ diameter		380	.084		1.76	2.66		4.42	6.35
4530	6″ diameter		316	.101		5.15	3.20		8.35	10.95
4540	8″ diameter		264	.121		8	3.83		11.83	15.10
4550	10″ diameter		220	.145		12.60	4.59		17.19	21.50
4560	12″ diameter		186	.172		17.20	5.45		22.65	28
8000	Fittings with rubber gasket									
8003	Class 150, D.R. 18									
8006	90° Bend , 4″ diameter	B-20	100	.240	Ea.	54	6.50		60.50	70.50
8020	6″ diameter		90	.267		97	7.25		104.25	119
8040	8″ diameter		80	.300		185	8.15		193.15	217
8060	10″ diameter		50	.480		282	13		295	330
8080	12″ diameter		30	.800		430	21.50		451.50	510
8100	Tee, 4″ diameter		90	.267		57	7.25		64.25	75
8120	6″ diameter		80	.300		154	8.15		162.15	183
8140	8″ diameter		70	.343		215	9.30		224.30	252

33 11 Water Utility Distribution Piping

33 11 13 – Public Water Utility Distribution Piping

33 11 13.25 Water Supply, Polyvinyl Chloride Pipe		Crew	Daily Output	Labor-Hours	Unit	2011 Bare Costs Material	Labor	Equipment	Total	Total Incl O&P
8160	10″ diameter	B-20	40	.600	Ea.	455	16.25		471.25	530
8180	12″ diameter		20	1.200		635	32.50		667.50	755
8200	45° Bend, 4″ diameter		100	.240		53.50	6.50		60	70
8220	6″ diameter		90	.267		96.50	7.25		103.75	118
8240	8″ diameter		50	.480		182	13		195	222
8260	10″ diameter		50	.480		286	13		299	335
8280	12″ diameter		30	.800		380	21.50		401.50	455
8300	Reducing tee 6″ x 4″		100	.240		131	6.50		137.50	155
8320	8″ x 6″		90	.267		266	7.25		273.25	305
8330	10″ x 6″		90	.267		310	7.25		317.25	350
8340	10″ x 8″		90	.267		330	7.25		337.25	370
8350	12″ x 6″		90	.267		415	7.25		422.25	465
8360	12″ x 8″		90	.267		460	7.25		467.25	515
8400	Tapped service tee (threaded type) 6″ x 6″ x 3/4″		100	.240		86.50	6.50		93	106
8430	6″ x 6″ x 1″		90	.267		86.50	7.25		93.75	107
8440	6″ x 6″ x 1-1/2″		90	.267		86.50	7.25		93.75	107
8450	6″ x 6″ x 2″		90	.267		86.50	7.25		93.75	107
8460	8″ x 8″ x 3/4″		90	.267		128	7.25		135.25	152
8470	8″ x 8″ x 1″		90	.267		128	7.25		135.25	152
8480	8″ x 8″ x 1-1/2″		90	.267		128	7.25		135.25	152
8490	8″ x 8″ x 2″		90	.267		128	7.25		135.25	152
8500	Repair coupling 4″		100	.240		91	6.50		97.50	111
8520	6″ diameter		90	.267		114	7.25		121.25	138
8540	8″ diameter		50	.480		133	13		146	169
8560	10″ diameter		50	.480		185	13		198	226
8580	12″ diameter		50	.480		223	13		236	268
8600	Plug end 4″		100	.240		34	6.50		40.50	48
8620	6″ diameter		90	.267		50	7.25		57.25	67
8640	8″ diameter		50	.480		76	13		89	106
8660	10″ diameter		50	.480		105	13		118	138
8680	12″ diameter		50	.480		129	13		142	164

33 12 Water Utility Distribution Equipment

33 12 19 – Water Utility Distribution Fire Hydrants

33 12 19.10 Fire Hydrants		Crew	Daily Output	Labor-Hours	Unit	Material	Labor	Equipment	Total	Total Incl O&P
0010	**FIRE HYDRANTS**									
0020	Mechanical joints unless otherwise noted									
1000	Fire hydrants, two way; excavation and backfill not incl.									
1100	4-1/2″ valve size, depth 2′-0″	B-21	10	2.800	Ea.	1,350	79.50	12.90	1,442.40	1,650
1120	2′-6″		10	2.800		1,450	79.50	12.90	1,542.40	1,750
1140	3′-0″		10	2.800		1,550	79.50	12.90	1,642.40	1,850
1300	7′-0″		6	4.667		1,775	132	21.50	1,928.50	2,200
2400	Lower barrel extensions with stems, 1′-0″	B-20	14	1.714		375	46.50		421.50	495
2480	3′-0″	″	12	2		730	54		784	895

33 16 Water Utility Storage Tanks

33 16 13 – Aboveground Water Utility Storage Tanks

33 16 13.13 Steel Water Storage Tanks

	33 16 13.13 Steel Water Storage Tanks	Crew	Daily Output	Labor-Hours	Unit	Material	Labor	Equipment	Total	Total Incl O&P
						2011 Bare Costs				
0010	**STEEL WATER STORAGE TANKS**									
0910	Steel, ground level, ht/diam. less than 1, not incl. fdn., 100,000 gallons				Ea.				178,000	196,000
1000	250,000 gallons								237,000	260,000
1200	500,000 gallons								335,000	368,000
1250	750,000 gallons								432,000	475,000
1300	1,000,000 gallons								530,000	583,000
1500	2,000,000 gallons								921,000	1,013,000
1600	4,000,000 gallons								1,703,000	1,873,000
1800	6,000,000 gallons								2,485,000	2,733,000
1850	8,000,000 gallons								3,266,000	3,593,000
1910	10,000,000 gallons								4,055,000	4,460,000
2100	Steel standpipes, hgt/diam. more than 1, 100′ to overflow, no fdn									
2200	500,000 gallons				Ea.				439,000	482,000
2400	750,000 gallons								580,000	638,000
2500	1,000,000 gallons								568,000	625,000
2700	1,500,000 gallons								1,404,000	1,544,000
2800	2,000,000 gallons								1,049,000	1,154,000

33 16 13.16 Prestressed Conc. Aboveground Water Utility Storage Tanks

	33 16 13.16 Prestressed Conc. Aboveground Water Utility Storage Tanks	Crew	Daily Output	Labor-Hours	Unit	Material	Labor	Equipment	Total	Total Incl O&P
0010	**PRESTRESSED CONC. ABOVEGROUND WATER UTILITY STORAGE TANKS**									
0020	Not including fdn., pipe or pumps, 250,000 gallons				Ea.				270,000	297,000
0100	500,000 gallons								441,000	485,000
0300	1,000,000 gallons								639,000	703,000
0400	2,000,000 gallons								970,000	1,067,000
0600	4,000,000 gallons								1,543,000	1,697,000
0700	6,000,000 gallons								2,050,000	2,255,000
0750	8,000,000 gallons								2,645,000	2,909,000
0800	10,000,000 gallons								3,196,000	3,519,000

33 16 13.23 Plastic-Coated Fabric Pillow Water Tanks

	33 16 13.23 Plastic-Coated Fabric Pillow Water Tanks	Crew	Daily Output	Labor-Hours	Unit	Material	Labor	Equipment	Total	Total Incl O&P
0010	**PLASTIC-COATED FABRIC PILLOW WATER TANKS**									
7000	Water tanks, vinyl coated fabric pillow tanks, freestanding, 5,000 gallons	4 Clab	4	8	Ea.	9,500	212		9,712	10,900
7100	Supporting embankment not included, 25,000 gallons	6 Clab	2	24		14,900	635		15,535	17,500
7200	50,000 gallons	8 Clab	1.50	42.667		27,600	1,125		28,725	32,300
7300	100,000 gallons	9 Clab	.90	80		60,000	2,125		62,125	69,500
7400	150,000 gallons		.50	144		82,000	3,800		85,800	96,500
7500	200,000 gallons		.40	180		100,500	4,750		105,250	118,500
7600	250,000 gallons		.30	240		143,000	6,350		149,350	168,000

33 16 19 – Elevated Water Utility Storage Tanks

33 16 19.50 Elevated Water Storage Tanks

	33 16 19.50 Elevated Water Storage Tanks	Crew	Daily Output	Labor-Hours	Unit	Material	Labor	Equipment	Total	Total Incl O&P
0010	**ELEVATED WATER STORAGE TANKS**									
0011	Not incl. pipe, pumps or foundation									
3000	Elevated water tanks, 100′ to bottom capacity line, incl. painting									
3010	50,000 gallons				Ea.				182,000	200,000
3300	100,000 gallons								273,000	300,000
3400	250,000 gallons								767,000	844,000
3600	500,000 gallons								1,364,000	1,500,000
3700	750,000 gallons								1,655,000	1,820,000
3900	1,000,000 gallons								2,364,000	2,600,000

33 21 Water Supply Wells

33 21 13 – Public Water Supply Wells

33 21 13.10 Wells and Accessories		Crew	Daily Output	Labor-Hours	Unit	2011 Bare Costs Material	Labor	Equipment	Total	Total Incl O&P
0010	**WELLS & ACCESSORIES**									
0011	Domestic									
0100	Drilled, 4" to 6" diameter	B-23	120	.333	L.F.		8.95	22.50	31.45	40
0200	8" diameter	"	95.20	.420	"		11.30	28.50	39.80	50
0400	Gravel pack well, 40' deep, incl. gravel & casing, complete									
0500	24" diameter casing x 18" diameter screen	B-23	.13	307	Total	33,800	8,250	20,800	62,850	74,000
0600	36" diameter casing x 18" diameter screen		.12	333	"	35,900	8,950	22,500	67,350	79,500
0800	Observation wells, 1-1/4" riser pipe		163	.245	V.L.F.	17.70	6.60	16.60	40.90	49
0900	For flush Buffalo roadway box, add	1 Skwk	16.60	.482	Ea.	48.50	16.40		64.90	80.50
1200	Test well, 2-1/2" diameter, up to 50' deep (15 to 50 GPM)	B-23	1.51	26.490	"	725	710	1,800	3,235	3,975
1300	Over 50' deep, add	"	121.80	.328	L.F.	19.25	8.80	22	50.05	60.50
1500	Pumps, installed in wells to 100' deep, 4" submersible									
1510	1/2 H.P.	Q-1	3.22	4.969	Ea.	395	177		572	725
1520	3/4 H.P.		2.66	6.015		485	215		700	885
1600	1 H.P.		2.29	6.987		570	249		819	1,025
1700	1-1/2 H.P.	Q-22	1.60	10		810	355	410	1,575	1,925
1800	2 H.P.		1.33	12.030		1,125	430	495	2,050	2,500
1900	3 H.P.		1.14	14.035		1,325	500	575	2,400	2,900
2000	5 H.P.		1.14	14.035		2,275	500	575	3,350	3,950
2050	Remove and install motor only, 4 H.P.		1.14	14.035		2,350	500	575	3,425	4,050
3000	Pump, 6" submersible, 25' to 150' deep, 25 H.P., 249 to 297 GPM		.89	17.978		6,850	640	735	8,225	9,400
3100	25' to 500' deep, 30 H.P., 100 to 300 GPM		.73	21.918		8,425	780	900	10,105	11,500
9950	See Section 31 23 19.40 for wellpoints									
9960	See Section 31 23 19.30 for drainage wells									

33 21 13.20 Water Supply Wells, Pumps		Crew	Daily Output	Labor-Hours	Unit	Material	Labor	Equipment	Total	Total Incl O&P
0010	**WATER SUPPLY WELLS, PUMPS**									
0011	With pressure control									
1000	Deep well, jet, 42 gal. galvanized tank									
1040	3/4 HP	1 Plum	.80	10	Ea.	1,450	395		1,845	2,250
3000	Shallow well, jet, 30 gal. galvanized tank									
3040	1/2 HP	1 Plum	2	4	Ea.	1,200	159		1,359	1,575

33 31 Sanitary Utility Sewerage Piping

33 31 13 – Public Sanitary Utility Sewerage Piping

33 31 13.15 Sewage Collection, Concrete Pipe		Crew	Daily Output	Labor-Hours	Unit	Material	Labor	Equipment	Total	Total Incl O&P
0010	**SEWAGE COLLECTION, CONCRETE PIPE**									
0020	See Section 33 41 13.60 for sewage/drainage collection, concrete pipe									

33 31 13.25 Sewage Collection, Polyvinyl Chloride Pipe		Crew	Daily Output	Labor-Hours	Unit	Material	Labor	Equipment	Total	Total Incl O&P
0010	**SEWAGE COLLECTION, POLYVINYL CHLORIDE PIPE**									
0020	Not including excavation or backfill									
2000	20' lengths, S.D.R. 35, B&S, 4" diameter	B-20	375	.064	L.F.	1.85	1.74		3.59	4.95
2040	6" diameter		350	.069		3.68	1.86		5.54	7.15
2080	13' lengths , S.D.R. 35, B&S, 8" diameter		335	.072		7.60	1.94		9.54	11.60
2120	10" diameter	B-21	330	.085		11.30	2.41	.39	14.10	16.90
2160	12" diameter		320	.088		13.05	2.48	.40	15.93	18.95
2200	15" diameter		240	.117		14.35	3.31	.54	18.20	22
4000	Piping, DWV PVC, no exc/bkfill, 10' L, Sch 40, 4" diameter	B-20	375	.064		3.85	1.74		5.59	7.15
4010	6" diameter		350	.069		7.65	1.86		9.51	11.55
4020	8" diameter		335	.072		13.40	1.94		15.34	17.95

33 36 Utility Septic Tanks

33 36 13 – Utility Septic Tank and Effluent Wet Wells

33 36 13.10 Septic Tanks		Crew	Daily Output	Labor-Hours	Unit	Material	2011 Bare Costs Labor	Equipment	Total	Total Incl O&P
0010	**SEPTIC TANKS**									
0011	Not including excavation or piping									
0015	Septic tanks, precast, 1,000 gallon	B-21	8	3.500	Ea.	870	99.50	16.10	985.60	1,150
0060	1,500 gallon		7	4		1,475	113	18.40	1,606.40	1,825
0100	2,000 gallon		5	5.600		1,825	159	26	2,010	2,300
0200	5,000 gallon	B-13	3.50	13.714		9,675	390	214	10,279	11,500
0300	15,000 gallon, 4 piece	B-13B	1.70	32.941		21,500	945	710	23,155	26,000
0400	25,000 gallon, 4 piece		1.10	50.909		45,300	1,475	1,100	47,875	53,500
0500	40,000 gallon, 4 piece		.80	70		53,500	2,000	1,500	57,000	64,000
0520	50,000 gallon, 5 piece	B-13C	.60	93.333		61,500	2,675	2,750	66,925	75,000
0540	75,000 gallon, cast in place	C-14C	.25	448		75,000	14,300	94	89,394	106,500
0560	100,000 gallon	"	.15	746		92,500	23,800	157	116,457	142,000
0600	High density polyethylene, 1,000 gallon	B-21	6	4.667		1,050	132	21.50	1,203.50	1,400
0700	1,500 gallon	"	4	7		1,375	199	32	1,606	1,900
1000	Distribution boxes, concrete, 7 outlets	2 Clab	16	1		81	26.50		107.50	134
1100	9 outlets	"	8	2		455	53		508	590
1150	Leaching field chambers, 13′ x 3′-7″ x 1′-4″, standard	B-13	16	3		505	85	47	637	750
1200	Heavy duty, 8′ x 4′ x 1′-6″		14	3.429		280	97	53.50	430.50	530
1300	13′ x 3′-9″ x 1′-6″		12	4		1,025	113	62.50	1,200.50	1,375
1350	20′ x 4′ x 1′-6″		5	9.600		1,075	272	150	1,497	1,825
1400	Leaching pit, precast concrete, 3′ diameter, 3′ deep	B-21	8	3.500		640	99.50	16.10	755.60	890
1500	6′ diameter, 3′ section		4.70	5.957		870	169	27.50	1,066.50	1,275
2000	Velocity reducing pit, precast conc., 6′ diameter, 3′ deep		4.70	5.957		1,525	169	27.50	1,721.50	1,975
2200	Excavation for septic tank, 3/4 C.Y. backhoe	B-12F	145	.110	C.Y.		3.43	4.42	7.85	10.50
2400	4′ trench for disposal field, 3/4 C.Y. backhoe	"	335	.048	L.F.		1.49	1.91	3.40	4.55
2600	Gravel fill, run of bank	B-6	150	.160	C.Y.	20	4.61	2.14	26.75	32
2800	Crushed stone, 3/4″	"	150	.160	"	38	4.61	2.14	44.75	51.50
3000	Effluent filter, 4″ dia.	1 Skwk	8	1	Ea.	47.50	34		81.50	110

33 41 Storm Utility Drainage Piping

33 41 13 – Public Storm Utility Drainage Piping

33 41 13.40 Piping, Storm Drainage, Corrugated Metal		Crew	Daily Output	Labor-Hours	Unit	Material	Labor	Equipment	Total	Total Incl O&P
0010	**PIPING, STORM DRAINAGE, CORRUGATED METAL**									
0020	Not including excavation or backfill									
2000	Corrugated metal pipe, galvanized									
2020	Bituminous coated with paved invert, 20′ lengths									
2040	8″ diameter, 16 ga.	B-14	330	.145	L.F.	10.70	4.07	.97	15.74	19.60
2060	10″ diameter, 16 ga.		260	.185		11.15	5.15	1.23	17.53	22
2080	12″ diameter, 16 ga.		210	.229		11.50	6.40	1.53	19.43	25
2100	15″ diameter, 16 ga.		200	.240		18.75	6.70	1.60	27.05	33.50
2120	18″ diameter, 16 ga.		190	.253		19.10	7.05	1.69	27.84	34.50
2140	24″ diameter, 14 ga.		160	.300		22	8.40	2	32.40	40
2160	30″ diameter, 14 ga.	B-13	120	.400		26	11.35	6.25	43.60	54.50
2180	36″ diameter, 12 ga.		120	.400		31.50	11.35	6.25	49.10	60.50
2200	48″ diameter, 12 ga.		100	.480		52.50	13.60	7.50	73.60	88.50
2220	60″ diameter, 10 ga.	B-13B	75	.747		79	21.50	16.10	116.60	140
2240	72″ diameter, 8 ga.	"	45	1.244		90.50	36	27	153.50	189
2500	Galvanized, uncoated, 20′ lengths									
2520	8″ diameter, 16 ga.	B-14	355	.135	L.F.	8.25	3.78	.90	12.93	16.40
2540	10″ diameter, 16 ga.		280	.171		8.75	4.79	1.15	14.69	18.90
2560	12″ diameter, 16 ga.		220	.218		9	6.10	1.46	16.56	21.50

33 41 Storm Utility Drainage Piping

33 41 13 – Public Storm Utility Drainage Piping

33 41 13.40 Piping, Storm Drainage, Corrugated Metal		Crew	Daily Output	Labor-Hours	Unit	2011 Bare Costs Material	Labor	Equipment	Total	Total Incl O&P
2580	15″ diameter, 16 ga.	B-14	220	.218	L.F.	13	6.10	1.46	20.56	26
2600	18″ diameter, 16 ga.		205	.234		14.75	6.55	1.56	22.86	29
2620	24″ diameter, 14 ga.		175	.274		18.75	7.65	1.83	28.23	35.50
2640	30″ diameter, 14 ga.	B-13	130	.369		23	10.45	5.75	39.20	49
2660	36″ diameter, 12 ga.		130	.369		27.50	10.45	5.75	43.70	54.50
2680	48″ diameter, 12 ga.		110	.436		45.50	12.35	6.80	64.65	78
2690	60″ diameter, 10 ga.	B-13B	78	.718		69	20.50	15.50	105	128
2780	End sections, 8″ diameter	B-14	24	2	Ea.	34	56	13.35	103.35	145
2785	10″ diameter		22	2.182		44.50	61	14.55	120.05	167
2790	12″ diameter		35	1.371		78	38.50	9.15	125.65	160
2800	18″ diameter		30	1.600		108	45	10.70	163.70	204
2810	24″ diameter	B-13	25	1.920		143	54.50	30	227.50	281
2820	30″ diameter		25	1.920		246	54.50	30	330.50	395
2825	36″ diameter		20	2.400		365	68	37.50	470.50	560
2830	48″ diameter		10	4.800		930	136	75	1,141	1,325
2835	60″ diameter	B-13B	5	11.200		985	320	242	1,547	1,875
2840	72″ diameter	″	4	14		1,575	400	300	2,275	2,725

33 41 13.60 Sewage/Drainage Collection, Concrete Pipe		Crew	Daily Output	Labor-Hours	Unit	Material	Labor	Equipment	Total	Total Incl O&P
0010	**SEWAGE/DRAINAGE COLLECTION, CONCRETE PIPE**									
0020	Not including excavation or backfill									
1000	Non-reinforced pipe, extra strength, B&S or T&G joints									
1010	6″ diameter	B-14	265.04	.181	L.F.	6.25	5.05	1.21	12.51	16.70
1020	8″ diameter		224	.214		6.90	6	1.43	14.33	19.15
1030	10″ diameter		216	.222		7.65	6.20	1.48	15.33	20.50
1040	12″ diameter		200	.240		9.35	6.70	1.60	17.65	23.50
1050	15″ diameter		180	.267		12.80	7.45	1.78	22.03	28.50
1060	18″ diameter		144	.333		15.10	9.30	2.23	26.63	34.50
1070	21″ diameter		112	.429		18.60	12	2.86	33.46	43.50
1080	24″ diameter		100	.480		25	13.45	3.21	41.66	53.50
2000	Reinforced culvert, class 3, no gaskets									
2010	12″ diameter	B-14	150	.320	L.F.	11.50	8.95	2.14	22.59	30
2020	15″ diameter		150	.320		15.70	8.95	2.14	26.79	34.50
2030	18″ diameter		132	.364		19.70	10.15	2.43	32.28	41
2035	21″ diameter		120	.400		23.50	11.20	2.67	37.37	47.50
2040	24″ diameter		100	.480		27.50	13.45	3.21	44.16	56.50
2045	27″ diameter	B-13	92	.522		40	14.80	8.15	62.95	77.50
2050	30″ diameter		88	.545		48	15.45	8.50	71.95	88.50
2060	36″ diameter		72	.667		58	18.90	10.40	87.30	106
2070	42″ diameter	B-13B	72	.778		76.50	22.50	16.80	115.80	140
2080	48″ diameter		64	.875		95.50	25	18.85	139.35	168
2090	60″ diameter		48	1.167		123	33.50	25	181.50	218
2100	72″ diameter		40	1.400		200	40	30	270	320
2120	84″ diameter		32	1.750		288	50.50	38	376.50	440
2140	96″ diameter		24	2.333		345	67	50.50	462.50	545
2200	With gaskets, class 3, 12″ diameter	B-21	168	.167		12.65	4.73	.77	18.15	22.50
2220	15″ diameter		160	.175		17.25	4.96	.81	23.02	28
2230	18″ diameter		152	.184		21.50	5.20	.85	27.55	33.50
2240	24″ diameter		136	.206		33.50	5.85	.95	40.30	48
2260	30″ diameter	B-13	88	.545		55	15.45	8.50	78.95	96
2270	36″ diameter	″	72	.667		66.50	18.90	10.40	95.80	116
2290	48″ diameter	B-13B	64	.875		107	25	18.85	150.85	181
2310	72″ diameter	″	40	1.400		217	40	30	287	340

33 41 Storm Utility Drainage Piping

33 41 13 – Public Storm Utility Drainage Piping

33 41 13.60 Sewage/Drainage Collection, Concrete Pipe		Crew	Daily Output	Labor-Hours	Unit	Material	2011 Bare Costs Labor	Equipment	Total	Total Incl O&P
2330	Flared ends, 6′-1″ long, 12″ diameter	B-21	190	.147	L.F.	36.50	4.18	.68	41.36	48
2340	15″ diameter		155	.181		40	5.10	.83	45.93	53
2400	6′-2″ long, 18″ diameter		122	.230		51	6.50	1.06	58.56	68.50
2420	24″ diameter		88	.318		55	9	1.46	65.46	77
2440	36″ diameter	B-13	60	.800		127	22.50	12.50	162	192
3080	Radius pipe, add to pipe prices, 12″ to 60″ diameter					50%				
3090	Over 60″ diameter, add					20%				
3500	Reinforced elliptical, 8′ lengths, C507 class 3									
3520	14″ x 23″ inside, round equivalent 18″ diameter	B-21	82	.341	L.F.	54.50	9.70	1.57	65.77	78
3530	24″ x 38″ inside, round equivalent 30″ diameter	B-13	58	.828		81	23.50	12.90	117.40	142
3540	29″ x 45″ inside, round equivalent 36″ diameter		52	.923		97	26	14.40	137.40	166
3550	38″ x 60″ inside, round equivalent 48″ diameter		38	1.263		161	36	19.70	216.70	258
3560	48″ x 76″ inside, round equivalent 60″ diameter		26	1.846		204	52.50	29	285.50	345
3570	58″ x 91″ inside, round equivalent 72″ diameter		22	2.182		305	62	34	401	480
3780	Concrete slotted pipe, class 4 mortar joint									
3800	12″ diameter	B-21	168	.167	L.F.	35	4.73	.77	40.50	47
3840	18″ diameter	″	152	.184	″	40	5.20	.85	46.05	53.50
3900	Concrete slotted pipe, Class 4 O-ring joint									
3940	12″ diameter	B-21	168	.167	L.F.	35	4.73	.77	40.50	47
3960	18″ diameter	″	152	.184	″	40	5.20	.85	46.05	53.50

33 42 Culverts

33 42 16 – Concrete Culverts

33 42 16.15 Oval Arch Culverts

		Crew	Daily Output	Labor-Hours	Unit	Material	Labor	Equipment	Total	Total Incl O&P
0010	**OVAL ARCH CULVERTS**									
3000	Corrugated galvanized or aluminum, coated & paved									
3020	17″ x 13″, 16 ga., 15″ equivalent	B-14	200	.240	L.F.	17.45	6.70	1.60	25.75	32
3040	21″ x 15″, 16 ga., 18″ equivalent		150	.320		24.50	8.95	2.14	35.59	44.50
3060	28″ x 20″, 14 ga., 24″ equivalent		125	.384		28	10.75	2.57	41.32	51.50
3080	35″ x 24″, 14 ga., 30″ equivalent		100	.480		37	13.45	3.21	53.66	66.50
3100	42″ x 29″, 12 ga., 36″ equivalent	B-13	100	.480		46	13.60	7.50	67.10	82
3120	49″ x 33″, 12 ga., 42″ equivalent		90	.533		62.50	15.10	8.30	85.90	103
3140	57″ x 38″, 12 ga., 48″ equivalent		75	.640		75.50	18.15	10	103.65	125
3160	Steel, plain oval arch culverts, plain									
3180	17″ x 13″, 16 ga., 15″ equivalent	B-14	225	.213	L.F.	10.25	5.95	1.43	17.63	23
3200	21″ x 15″, 16 ga., 18″ equivalent		175	.274		12.40	7.65	1.83	21.88	28.50
3220	28″ x 20″, 14 ga., 24″ equivalent		150	.320		16.60	8.95	2.14	27.69	35.50
3240	35″ x 24″, 14 ga., 30″ equivalent	B-13	108	.444		21.50	12.60	6.95	41.05	52.50
3260	42″ x 29″, 12 ga., 36″ equivalent		108	.444		27	12.60	6.95	46.55	58.50
3280	49″ x 33″, 12 ga., 42″ equivalent		92	.522		31	14.80	8.15	53.95	67.50
3300	57″ x 38″, 12 ga., 48″ equivalent		75	.640		47.50	18.15	10	75.65	93
3320	End sections, 17″ x 13″		22	2.182	Ea.	105	62	34	201	257
3340	42″ x 29″		17	2.824	″	292	80	44	416	500
3360	Multi-plate arch, steel	B-20	1690	.014	Lb.	1.20	.39		1.59	1.97

33 44 Storm Utility Water Drains

33 44 13 – Utility Area Drains

33 44 13.13 Catchbasins		Crew	Daily Output	Labor-Hours	Unit	2011 Bare Costs Material	Labor	Equipment	Total	Total Incl O&P
0010	**CATCHBASINS**									
0011	Not including footing & excavation									
1600	Frames & covers, C.I., 24" square, 500 lb.	B-6	7.80	3.077	Ea.	345	88.50	41	474.50	570
1700	26" D shape, 600 lb.		7	3.429		510	99	46	655	775
1800	Light traffic, 18" diameter, 100 lb.		10	2.400		154	69	32	255	320
1900	24" diameter, 300 lb.		8.70	2.759		236	79.50	37	352.50	435
2000	36" diameter, 900 lb.		5.80	4.138		510	119	55	684	820
2100	Heavy traffic, 24" diameter, 400 lb.		7.80	3.077		220	88.50	41	349.50	435
2200	36" diameter, 1150 lb.		3	8		830	231	107	1,168	1,425
2300	Mass. State standard, 26" diameter, 475 lb.		7	3.429		250	99	46	395	490
2400	30" diameter, 620 lb.		7	3.429		340	99	46	485	590
2500	Watertight, 24" diameter, 350 lb.		7.80	3.077		235	88.50	41	364.50	450
2600	26" diameter, 500 lb.		7	3.429		395	99	46	540	650
2700	32" diameter, 575 lb.		6	4		800	115	53.50	968.50	1,125
2800	3 piece cover & frame, 10" deep,									
2900	1200 lb., for heavy equipment	B-6	3	8	Ea.	1,100	231	107	1,438	1,725
3000	Raised for paving 1-1/4" to 2" high									
3100	4 piece expansion ring									
3200	20" to 26" diameter	1 Clab	3	2.667	Ea.	143	70.50		213.50	275
3300	30" to 36" diameter	"	3	2.667	"	199	70.50		269.50	335
3320	Frames and covers, existing, raised for paving, 2", including									
3340	row of brick, concrete collar, up to 12" wide frame	B-6	18	1.333	Ea.	45	38.50	17.80	101.30	133
3360	20" to 26" wide frame		11	2.182		67.50	63	29	159.50	210
3380	30" to 36" wide frame		9	2.667		83.50	77	35.50	196	258
3400	Inverts, single channel brick	D-1	3	5.333		97	162		259	375
3500	Concrete		5	3.200		75.50	97.50		173	243
3600	Triple channel, brick		2	8		148	243		391	565
3700	Concrete		3	5.333		128	162		290	405

33 46 Subdrainage

33 46 16 – Subdrainage Piping

33 46 16.25 Piping, Subdrainage, Corrugated Metal		Crew	Daily Output	Labor-Hours	Unit	Material	Labor	Equipment	Total	Total Incl O&P
0010	**PIPING, SUBDRAINAGE, CORRUGATED METAL**									
0021	Not including excavation and backfill									
2010	Aluminum, perforated									
2020	6" diameter, 18 ga.	B-14	380	.126	L.F.	7.55	3.53	.84	11.92	15.15
2200	8" diameter, 16 ga.		370	.130		10.90	3.63	.87	15.40	19
2220	10" diameter, 16 ga.		360	.133		13.65	3.73	.89	18.27	22
2240	12" diameter, 16 ga.		285	.168		15.30	4.71	1.13	21.14	26
2260	18" diameter, 16 ga.		205	.234		23	6.55	1.56	31.11	37.50
3000	Uncoated galvanized, perforated									
3020	6" diameter, 18 ga.	B-20	380	.063	L.F.	7.55	1.71		9.26	11.15
3200	8" diameter, 16 ga.	"	370	.065		10.90	1.76		12.66	14.95
3220	10" diameter, 16 ga.	B-21	360	.078		16.40	2.21	.36	18.97	22
3240	12" diameter, 16 ga.		285	.098		17.15	2.79	.45	20.39	24
3260	18" diameter, 16 ga.		205	.137		26	3.87	.63	30.50	36
4000	Steel, perforated, asphalt coated									
4020	6" diameter 18 ga.	B-20	380	.063	L.F.	7.55	1.71		9.26	11.15
4030	8" diameter 18 ga.	"	370	.065		10.90	1.76		12.66	14.95
4040	10" diameter 16 ga.	B-21	360	.078		13.25	2.21	.36	15.82	18.65
4050	12" diameter 16 ga.		285	.098		15.10	2.79	.45	18.34	22

33 46 Subdrainage

33 46 16 – Subdrainage Piping

33 46 16.25 Piping, Subdrainage, Corrugated Metal		Crew	Daily Output	Labor-Hours	Unit	Material	2011 Bare Costs Labor	Equipment	Total	Total Incl O&P
4060	18" diameter 16 ga.	B-21	205	.137	L.F.	21	3.87	.63	25.50	30

33 46 16.40 Piping, Subdrainage, Polyvinyl Chloride

		Crew	Daily Output	Labor-Hours	Unit	Material	Labor	Equipment	Total	Total Incl O&P
0010	**PIPING, SUBDRAINAGE, POLYVINYL CHLORIDE**									
0020	Perforated, price as solid pipe, Section 33 31 13.25									

33 49 Storm Drainage Structures

33 49 13 – Storm Drainage Manholes, Frames, and Covers

33 49 13.10 Storm Drainage Manholes, Frames and Covers

		Crew	Daily Output	Labor-Hours	Unit	Material	Labor	Equipment	Total	Total Incl O&P
0010	**STORM DRAINAGE MANHOLES, FRAMES & COVERS**									
0020	Excludes footing, excavation, backfill (See line items for frame & cover)									
0050	Brick, 4' inside diameter, 4' deep	D-1	1	16	Ea.	445	485		930	1,300
0100	6' deep		.70	22.857		625	695		1,320	1,825
0150	8' deep		.50	32		805	975		1,780	2,475
0200	For depths over 8', add		4	4	V.L.F.	155	122		277	370
0400	Concrete blocks (radial), 4' I.D., 4' deep		1.50	10.667	Ea.	370	325		695	940
0500	6' deep		1	16		495	485		980	1,350
0600	8' deep		.70	22.857		625	695		1,320	1,825
0700	For depths over 8', add		5.50	2.909	V.L.F.	65.50	88.50		154	217
0800	Concrete, cast in place, 4' x 4', 8" thick, 4' deep	C-14H	2	24	Ea.	430	790	11.50	1,231.50	1,825
0900	6' deep		1.50	32		615	1,050	15.35	1,680.35	2,450
1000	8' deep		1	48		895	1,575	23	2,493	3,650
1100	For depths over 8', add		8	6	V.L.F.	98	197	2.88	297.88	440
1110	Precast, 4' I.D., 4' deep	B-22	4.10	7.317	Ea.	690	211	47	948	1,150
1120	6' deep		3	10		795	289	64.50	1,148.50	1,425
1130	8' deep		2	15		990	435	97	1,522	1,925
1140	For depths over 8', add		16	1.875	V.L.F.	101	54	12.10	167.10	214
1150	5' I.D., 4' deep	B-6	3	8	Ea.	1,400	231	107	1,738	2,050
1160	6' deep		2	12		1,700	345	160	2,205	2,625
1170	8' deep		1.50	16		1,950	460	214	2,624	3,150
1180	For depths over 8', add		12	2	V.L.F.	180	57.50	26.50	264	320
1190	6' I.D., 4' deep		2	12	Ea.	2,000	345	160	2,505	2,950
1200	6' deep		1.50	16		2,700	460	214	3,374	3,975
1210	8' deep		1	24		3,250	690	320	4,260	5,100
1220	For depths over 8', add		8	3	V.L.F.	252	86.50	40	378.50	465
1250	Slab tops, precast, 8" thick									
1300	4' diameter manhole	B-6	8	3	Ea.	234	86.50	40	360.50	445
1400	5' diameter manhole		7.50	3.200		400	92	42.50	534.50	640
1500	6' diameter manhole		7	3.429		615	99	46	760	890
3800	Steps, heavyweight cast iron, 7" x 9"	1 Bric	40	.200		16.70	6.75		23.45	29.50
3900	8" x 9"		40	.200		25	6.75		31.75	38.50
3928	12" x 10-1/2"		40	.200		23.50	6.75		30.25	36.50
4000	Standard sizes, galvanized steel		40	.200		22	6.75		28.75	35.50
4100	Aluminum		40	.200		22	6.75		28.75	35.50

33 51 Natural-Gas Distribution

33 51 13 – Natural-Gas Piping

33 51 13.10 Piping, Gas Service and Distribution, P.E.		Crew	Daily Output	Labor-Hours	Unit	Material	Labor	Equipment	Total	Total Incl O&P
						2011 Bare Costs				
0010	**PIPING, GAS SERVICE AND DISTRIBUTION, POLYETHYLENE**									
0020	Not including excavation or backfill									
1000	60 psi coils, compression coupling @ 100′, 1/2″ diameter, SDR 11	B-20A	608	.053	L.F.	.75	1.66		2.41	3.57
1010	1″ diameter, SDR 11		544	.059		1.95	1.86		3.81	5.20
1040	1-1/4″ diameter, SDR 11		544	.059		2.39	1.86		4.25	5.70
1100	2″ diameter, SDR 11		488	.066		3.26	2.07		5.33	7
1160	3″ diameter, SDR 11		408	.078		4.31	2.48		6.79	8.85
1500	60 PSI 40′ joints with coupling, 3″ diameter, SDR 11	B-21A	408	.098		5.35	3.18	1.06	9.59	12.30
1540	4″ diameter, SDR 11		352	.114		7.75	3.68	1.23	12.66	15.90
1600	6″ diameter, SDR 11		328	.122		27	3.95	1.32	32.27	37.50
1640	8″ diameter, SDR 11		272	.147		32.50	4.77	1.60	38.87	45

33 52 Liquid Fuel Distribution

33 52 16 – Gasoline Distribution

33 52 16.13 Gasoline Piping

Line	Description	Crew	Daily Output	Labor-Hours	Unit	Material	Labor	Equipment	Total	Total Incl O&P
0010	**GASOLINE PIPING**									
0020	Primary containment pipe, fiberglass-reinforced									
0030	Plastic pipe 15′ & 30′ lengths									
0040	2″ diameter	Q-6	425	.056	L.F.	6.25	1.96		8.21	10.05
0050	3″ diameter		400	.060		6.65	2.08		8.73	10.75
0060	4″ diameter		375	.064		8.60	2.22		10.82	13.10
0100	Fittings									
0110	Elbows, 90° & 45°, bell-ends, 2″	Q-6	24	1	Ea.	51.50	34.50		86	113
0120	3″ diameter		22	1.091		54	38		92	122
0130	4″ diameter		20	1.200		69	41.50		110.50	144
0200	Tees, bell ends, 2″		21	1.143		51	39.50		90.50	122
0210	3″ diameter		18	1.333		61	46.50		107.50	143
0220	4″ diameter		15	1.600		153	55.50		208.50	259
0230	Flanges bell ends, 2″		24	1		36	34.50		70.50	96
0240	3″ diameter		22	1.091		47	38		85	114
0250	4″ diameter		20	1.200		57.50	41.50		99	132
0260	Sleeve couplings, 2″		21	1.143		21.50	39.50		61	89
0270	3″ diameter		18	1.333		25	46.50		71.50	103
0280	4″ diameter		15	1.600		29.50	55.50		85	123
0290	Threaded adapters 2″		21	1.143		18	39.50		57.50	85
0300	3″ diameter		18	1.333		31.50	46.50		78	110
0310	4″ diameter		15	1.600		42.50	55.50		98	138
0320	Reducers, 2″		27	.889		38.50	31		69.50	93
0330	3″ diameter		22	1.091		50.50	38		88.50	118
0340	4″ diameter		20	1.200		68.50	41.50		110	143
1010	Gas station product line for secondary containment (double wall)									
1100	Fiberglass reinforced plastic pipe 25′ lengths									
1120	Pipe, plain end, 3″ diameter	Q-6	375	.064	L.F.	13.40	2.22		15.62	18.35
1130	4″ diameter		350	.069		15.30	2.38		17.68	20.50
1140	5″ diameter		325	.074		20	2.56		22.56	26
1150	6″ diameter		300	.080		20.50	2.78		23.28	27.50
1200	Fittings									
1230	Elbows, 90° & 45°, 3″ diameter	Q-6	18	1.333	Ea.	117	46.50		163.50	204
1240	4″ diameter		16	1.500		123	52		175	220
1250	5″ diameter		14	1.714		282	59.50		341.50	405
1260	6″ diameter		12	2		285	69.50		354.50	430

33 52 Liquid Fuel Distribution

33 52 16 – Gasoline Distribution

33 52 16.13 Gasoline Piping		Crew	Daily Output	Labor-Hours	Unit	Material	2011 Bare Costs Labor	Equipment	Total	Total Incl O&P
1270	Tees, 3″ diameter	Q-6	15	1.600	Ea.	121	55.50		176.50	224
1280	4″ diameter		12	2		142	69.50		211.50	269
1290	5″ diameter		9	2.667		290	92.50		382.50	470
1300	6″ diameter		6	4		305	139		444	560
1310	Couplings, 3″ diameter		18	1.333		95	46.50		141.50	181
1320	4″ diameter		16	1.500		125	52		177	223
1330	5″ diameter		14	1.714		260	59.50		319.50	385
1340	6″ diameter		12	2		270	69.50		339.50	410
1350	Cross-over nipples, 3″ diameter		18	1.333		16.50	46.50		63	93.50
1360	4″ diameter		16	1.500		19.50	52		71.50	107
1370	5″ diameter		14	1.714		25	59.50		84.50	125
1380	6″ diameter		12	2		30.50	69.50		100	147
1400	Telescoping, reducers, concentric 4″ x 3″		18	1.333		168	46.50		214.50	261
1410	5″ x 4″		17	1.412		168	49		217	265
1420	6″ x 5″		16	1.500		168	52		220	270

33 71 Electrical Utility Transmission and Distribution

33 71 16 – Electrical Utility Poles

33 71 16.33 Wood Electrical Utility Poles

		Crew	Daily Output	Labor-Hours	Unit	Material	Labor	Equipment	Total	Total Incl O&P
0010	**WOOD ELECTRICAL UTILITY POLES**									
0011	Excludes excavation, backfill and cast-in-place concrete									
6200	Electric & tel sitework, 20′ high, treated wd, see Section 26 56 13.10	R-3	3.10	6.452	Ea.	243	255	41.50	539.50	730
6400	25′ high		2.90	6.897		297	273	44.50	614.50	820
6600	30′ high		2.60	7.692		385	305	49.50	739.50	970
6800	35′ high		2.40	8.333		475	330	54	859	1,125
7000	40′ high		2.30	8.696		685	345	56	1,086	1,375
7200	45′ high		1.70	11.765		825	465	76	1,366	1,750
7400	Cross arms with hardware & insulators									
7600	4′ long	1 Elec	2.50	3.200	Ea.	130	129		259	350
7800	5′ long		2.40	3.333		151	134		285	385
8000	6′ long		2.20	3.636		174	146		320	430

33 71 19 – Electrical Underground Ducts and Manholes

33 71 19.17 Electric and Telephone Underground

		Crew	Daily Output	Labor-Hours	Unit	Material	Labor	Equipment	Total	Total Incl O&P
0010	**ELECTRIC AND TELEPHONE UNDERGROUND**									
0011	Not including excavation									
0200	backfill and cast in place concrete									
0400	Hand holes, precast concrete, with concrete cover									
0600	2′ x 2′ x 3′ deep	R-3	2.40	8.333	Ea.	380	330	54	764	1,000
0800	3′ x 3′ x 3′ deep		1.90	10.526		490	415	68	973	1,300
1000	4′ x 4′ x 4′ deep		1.40	14.286		1,200	565	92	1,857	2,350
1200	Manholes, precast with iron racks & pulling irons, C.I. frame									
1400	and cover, 4′ x 6′ x 7′ deep	B-13	2	24	Ea.	2,075	680	375	3,130	3,800
1600	6′ x 8′ x 7′ deep		1.90	25.263		2,550	715	395	3,660	4,425
1800	6′ x 10′ x 7′ deep		1.80	26.667		2,850	755	415	4,020	4,850
4200	Underground duct, banks ready for concrete fill, min. of 7.5″									
4400	between conduits, center to center									
4580	PVC, type EB, 1 @ 2″ diameter	2 Elec	480	.033	L.F.	.36	1.34		1.70	2.56
4600	2 @ 2″ diameter		240	.067		.72	2.68		3.40	5.15
4800	4 @ 2″ diameter		120	.133		1.43	5.35		6.78	10.30
5000	2 @ 3″ diameter		200	.080		1	3.22		4.22	6.30

33 71 Electrical Utility Transmission and Distribution

33 71 19 – Electrical Underground Ducts and Manholes

33 71 19.17 Electric and Telephone Underground		Crew	Daily Output	Labor-Hours	Unit	Material	2011 Bare Costs Labor	Equipment	Total	Total Incl O&P
5200	4 @ 3″ diameter	2 Elec	100	.160	L.F.	2	6.45		8.45	12.65
5400	2 @ 4″ diameter		160	.100		1.47	4.03		5.50	8.15
5600	4 @ 4″ diameter		80	.200		2.93	8.05		10.98	16.30
5800	6 @ 4″ diameter		54	.296		4.40	11.95		16.35	24
6200	Rigid galvanized steel, 2 @ 2″ diameter		180	.089		12.75	3.58		16.33	19.85
6400	4 @ 2″ diameter		90	.178		25.50	7.15		32.65	39.50
6800	2 @ 3″ diameter		100	.160		30	6.45		36.45	43.50
7000	4 @ 3″ diameter		50	.320		60.50	12.90		73.40	87.50
7200	2 @ 4″ diameter		70	.229		43.50	9.20		52.70	62.50
7400	4 @ 4″ diameter		34	.471		87	18.95		105.95	126
7600	6 @ 4″ diameter		22	.727		130	29.50		159.50	191

33 81 Communications Structures

33 81 13 – Communications Transmission Towers

33 81 13.10 Radio Towers		Crew	Daily Output	Labor-Hours	Unit	Material	Labor	Equipment	Total	Total Incl O&P
0010	**RADIO TOWERS**									
0020	Guyed, 50′ H, 40 lb. sec., 70 MPH basic wind spd.	2 Sswk	1	16	Ea.	2,600	580		3,180	3,950
0100	Wind load 90 MPH basic wind speed	″	1	16		3,225	580		3,805	4,650
0300	190′ high, 40 lb. section, wind load 70 MPH basic wind speed	K-2	.33	72.727		10,000	2,450	655	13,105	16,300
0400	200′ high, 70 lb. section, wind load 90 MPH basic wind speed		.33	72.727		17,800	2,450	655	20,905	24,900
0600	300′ high, 70 lb. section, wind load 70 MPH basic wind speed		.20	120		25,000	4,050	1,075	30,125	36,200
0700	270′ high, 90 lb. section, wind load 90 MPH basic wind speed		.20	120		27,000	4,050	1,075	32,125	38,400
0800	400′ high, 100 lb. section, wind load 70 MPH basic wind speed		.14	171		37,000	5,775	1,550	44,325	53,000
0900	Self-supporting, 60′ high, wind load 70 MPH basic wind speed		.80	30		5,500	1,000	271	6,771	8,225
0910	60′ high, wind load 90 MPH basic wind speed		.45	53.333		6,600	1,800	480	8,880	11,100
1000	120′ high, wind load 70 MPH basic wind speed		.40	60		12,000	2,025	540	14,565	17,500
1200	190′ high, wind load 90 MPH basic wind speed		.20	120		31,000	4,050	1,075	36,125	42,800
2000	For states west of Rocky Mountains, add for shipping					10%				

Division Notes

	CREW	DAILY OUTPUT	LABOR-HOURS	UNIT	2011 BARE COSTS MAT.	LABOR	EQUIP.	TOTAL	TOTAL INCL O&P

Estimating Tips

34 11 00 Rail Tracks

This subdivision includes items that may involve either repair of existing, or construction of new, railroad tracks. Additional preparation work, such as the roadbed earthwork, would be found in Division 31. Additional new construction siding and turnouts are found in Subdivision 34 72. Maintenance of railroads is found under 34 01 23 Operation and Maintenance of Railways.

34 41 13 Traffic Signals

This subdivision includes traffic signal systems. Other traffic control devices such as traffic signs are found in Subdivision 10 14 53 Traffic Signage.

34 71 13 Vehicle Barriers

This subdivision includes security vehicle barriers, guide and guard rails, crash barriers, and delineators. The actual maintenance and construction of concrete and asphalt pavement is found in Division 32.

Reference Numbers

Reference numbers are shown in shaded boxes at the beginning of some major classifications. These numbers refer to related items in the Reference Section. The reference information may be an estimating procedure, an alternate pricing method, or technical information.

Note: Not all subdivisions listed here necessarily appear in this publication.

34 01 Operation and Maintenance of Transportation

34 01 23 – Operation and Maintenance of Railways

34 01 23.51 Maintenance of Railroads		Crew	Daily Output	Labor-Hours	Unit	Material	2011 Bare Costs Labor	Equipment	Total	Total Incl O&P
0010	**MAINTENANCE OF RAILROADS**									
0400	Resurface and realign existing track	B-14	200	.240	L.F.		6.70	1.60	8.30	12.95
0600	For crushed stone ballast, add	"	500	.096	"	15.35	2.69	.64	18.68	22

34 11 Rail Tracks

34 11 13 – Track Rails

34 11 13.23 Heavy Rail Track

		Crew	Daily Output	Labor-Hours	Unit	Material	Labor	Equipment	Total	Total Incl O&P
0010	**HEAVY RAIL TRACK** R347216-10									
1000	Rail, 100 lb. prime grade				L.F.	29.50			29.50	32.50
1500	Relay rail				"	14.70			14.70	16.15

34 11 33 – Track Cross Ties

34 11 33.13 Concrete Track Cross Ties

		Crew	Daily Output	Labor-Hours	Unit	Material	Labor	Equipment	Total	Total Incl O&P
0010	**CONCRETE TRACK CROSS TIES**									
1400	Ties, concrete, 8'-6" long, 30" O.C.	B-14	80	.600	Ea.	145	16.80	4.01	165.81	192

34 11 33.16 Timber Track Cross Ties

		Crew	Daily Output	Labor-Hours	Unit	Material	Labor	Equipment	Total	Total Incl O&P
0010	**TIMBER TRACK CROSS TIES**									
1600	Wood, pressure treated, 6" x 8" x 8'-6", C.L. lots	B-14	90	.533	Ea.	43.50	14.90	3.56	61.96	77
1700	L.C.L. lots		90	.533		46	14.90	3.56	64.46	79.50
1900	Heavy duty, 7" x 9" x 8'-6", C.L. lots		70	.686		48	19.20	4.58	71.78	89.50
2000	L.C.L. lots		70	.686		48	19.20	4.58	71.78	89.50

34 11 33.17 Timber Switch Ties

		Crew	Daily Output	Labor-Hours	Unit	Material	Labor	Equipment	Total	Total Incl O&P
0010	**TIMBER SWITCH TIES**									
1200	Switch timber, for a #8 switch, pressure treated	B-14	3.70	12.973	M.B.F.	2,625	365	86.50	3,076.50	3,575
1300	Complete set of timbers, 3.7 M.B.F. for #8 switch	"	1	48	Total	10,100	1,350	320	11,770	13,700

34 11 93 – Track Appurtenances and Accessories

34 11 93.50 Track Accessories

		Crew	Daily Output	Labor-Hours	Unit	Material	Labor	Equipment	Total	Total Incl O&P
0010	**TRACK ACCESSORIES**									
0020	Car bumpers, test	B-14	2	24	Ea.	3,275	670	160	4,105	4,900
0100	Heavy duty R347216-20		2	24		6,200	670	160	7,030	8,125
0200	Derails hand throw (sliding)		10	4.800		1,025	134	32	1,191	1,375
0300	Hand throw with standard timbers, open stand & target		8	6		1,125	168	40	1,333	1,550
2400	Wheel stops, fixed		18	2.667	Pr.	810	74.50	17.80	902.30	1,025
2450	Hinged		14	3.429	"	1,100	96	23	1,219	1,375

34 11 93.60 Track Material

		Crew	Daily Output	Labor-Hours	Unit	Material	Labor	Equipment	Total	Total Incl O&P
0010	**TRACK MATERIAL**									
0020	Track bolts				Ea.	3.36			3.36	3.70
0100	Joint bars				Pr.	78			78	85.50
0200	Spikes				Ea.	1.11			1.11	1.22
0300	Tie plates				"	13.05			13.05	14.35

34 41 Roadway Signaling and Control Equipment

34 41 13 – Traffic Signals

34 41 13.10 Traffic Signals Systems		Crew	Daily Output	Labor-Hours	Unit	Material	2011 Bare Costs Labor	Equipment	Total	Total Incl O&P
0010	**TRAFFIC SIGNALS SYSTEMS**									
0020	Mid block pedestrian crosswalk with pushbutton and mast arms	R-11	.30	186	Total	86,000	7,050	2,700	95,750	109,000
0600	Traffic signals, school flashing system, solar powered, remote controlled	″	1	56	Signal	19,800	2,100	810	22,710	26,100
1000	Intersection traffic signals, LED, mast, programmable, no lane control									
1010	Includes all labor, material and equip. for complete installation	R-11	1	56	Signal	200,000	2,100	810	202,910	224,500
1100	Intersection traffic signals, LED, mast, programmable, R/L lane control									
1110	Includes all labor, material and equip. for complete installation	R-11	1	56	Signal	262,000	2,100	810	264,910	293,000
1200	Add protective/permissive left turns to existing traffic light									
1210	Includes all labor, material and equip. for complete installation	R-11	1	56	Signal	80,000	2,100	810	82,910	92,500
1300	Replace existing light heads with LED heads wire hung									
1310	Includes all labor, material and equip. for complete installation	R-11	1	56	Signal	54,000	2,100	810	56,910	64,000
1400	Replace existing light heads with LED heads mast arm hung									
1410	Includes all labor, material and equip. for complete installation	R-11	1	56	Signal	100,000	2,100	810	102,910	114,500

34 71 Roadway Construction

34 71 13 – Vehicle Barriers

34 71 13.26 Vehicle Guide Rails		Crew	Daily Output	Labor-Hours	Unit	Material	Labor	Equipment	Total	Total Incl O&P
0010	**VEHICLE GUIDE RAILS**									
0012	Corrugated stl, galv. stl posts, 6′-3″ O.C.	B-80	850	.028	L.F.	22	.77	.73	23.50	26.50
0200	End sections, galvanized, flared		50	.480	Ea.	90	13	12.35	115.35	135
0300	Wrap around end		50	.480	″	136	13	12.35	161.35	186
0400	Timber guide rail, 4″ x 8″ with 6″ x 8″ wood posts, treated		960	.025	L.F.	25.50	.68	.64	26.82	30
0600	Cable guide rail, 3 at 3/4″ cables, steel posts, single face		900	.027		10.35	.72	.69	11.76	13.30
0700	Wood posts		950	.025		12.35	.69	.65	13.69	15.40
0900	Guide rail, steel box beam, 6″ x 6″		120	.200		32	5.40	5.15	42.55	50
1100	Median barrier, steel box beam, 6″ x 8″		215	.112		42	3.03	2.88	47.91	54
1400	Resilient guide fence and light shield, 6′ high	B-2	130	.308		35	8.25		43.25	52.50
1500	Concrete posts, individual, 6′-5″, triangular	B-80	110	.218	Ea.	65	5.90	5.60	76.50	87.50
1550	Square	″	110	.218	″	69.50	5.90	5.60	81	92.50
2000	Median, precast concrete, 3′-6″ high, 2′ wide, single face	B-29	380	.126	L.F.	50.50	3.58	2.26	56.34	64
2200	Double face	″	340	.141	″	58	4	2.52	64.52	73.50
2400	Speed bumps, thermoplastic, 10-1/2″ x 2-1/4″ x 48″ long	B-2	120	.333	Ea.	140	8.95		148.95	169
3030	Impact barrier, UTMCD, barrel type	B-16	30	1.067	″	470	29	16.85	515.85	580

34 71 19 – Vehicle Delineators

34 71 19.13 Fixed Vehicle Delineators		Crew	Daily Output	Labor-Hours	Unit	Material	Labor	Equipment	Total	Total Incl O&P
0010	**FIXED VEHICLE DELINEATORS**									
0020	Crash barriers									
0100	Traffic channelizing pavement markers, layout only	A-7	2000	.012	Ea.		.43	.03	.46	.75
0110	13″ x 7-1/2″ x 2-1/2″ high, non-plowable install	2 Clab	96	.167		26.50	4.41		30.91	37
0200	8″ x 8″ x 3-1/4″ high, non-plowable, install		96	.167		23.50	4.41		27.91	33.50
0230	4″ x 4″ x 3/4″ high, non-plowable, install		120	.133		1.41	3.53		4.94	7.45
0240	9-1/4″ x 5-7/8″ x 1/4″ high, plowable, concrete pav't	A-2A	70	.343		20	9.05	4.36	33.41	42
0250	9-1/4″ x 5-7/8″ x 1/4″ high, plowable, asphalt pav't	″	120	.200		4.25	5.30	2.55	12.10	16.35
0300	Barrier and curb delineators, reflectorized, 2″ x 4″	2 Clab	150	.107		2.30	2.82		5.12	7.25
0310	3″ x 5″	″	150	.107		4.50	2.82		7.32	9.70
0500	Rumble strip, polycarbonate									
0510	24″ x 3-1/2″ x 1/2″ high	2 Clab	50	.320	Ea.	8.10	8.45		16.55	23

34 72 Railway Construction

34 72 16 – Railway Siding

34 72 16.50 Railroad Sidings		Crew	Daily Output	Labor-Hours	Unit	2011 Bare Costs Material	Labor	Equipment	Total	Total Incl O&P
0010	**RAILROAD SIDINGS** R347216-10									
0800	Siding, yard spur, level grade									
0820	100 lb. new rail	B-14	57	.842	L.F.	114	23.50	5.65	143.15	172
1020	100 lb. new rail	"	22	2.182	"	156	61	14.55	231.55	290

34 72 16.60 Railroad Turnouts		Crew	Daily Output	Labor-Hours	Unit	Material	Labor	Equipment	Total	Total Incl O&P
0010	**RAILROAD TURNOUTS**									
2200	Turnout, #8 complete, w/rails, plates, bars, frog, switch point,									
2250	timbers, and ballast to 6" below bottom of ties									
2280	90 lb. rails	B-13	.25	192	Ea.	33,700	5,450	3,000	42,150	49,400
2290	90 lb. relay rails		.25	192		22,600	5,450	3,000	31,050	37,300
2300	100 lb. rails		.25	192		37,500	5,450	3,000	45,950	53,500
2310	100 lb. relay rails		.25	192		25,200	5,450	3,000	33,650	40,100
2320	110 lb. rails		.25	192		41,300	5,450	3,000	49,750	58,000
2330	110 lb. relay rails		.25	192		27,800	5,450	3,000	36,250	43,000
2340	115 lb. rails		.25	192		45,300	5,450	3,000	53,750	62,000
2350	115 lb. relay rails		.25	192		30,400	5,450	3,000	38,850	45,900
2360	132 lb. rails		.25	192		52,000	5,450	3,000	60,450	69,500
2370	132 lb. relay rails	↓	.25	192	↓	34,300	5,450	3,000	42,750	50,000

Estimating Tips

35 20 16 Hydraulic Gates

This subdivision includes various types of gates that are commonly used in waterway and canal construction. Various earthwork items and structural support is found in Division 31, and concrete work in Division 3.

35 20 23 Dredging

This subdivision includes barge and shore dredging systems for rivers, canals, and channels.

35 31 00 Shoreline Protection

This subdivision includes breakwaters, bulkheads, and revetments for ocean and river inlets. Additional earthwork may be required from Division 31, and concrete work from Division 3.

35 41 00 Levees

Information on levee construction, including estimated cost of clay cone material.

35 49 00 Waterway Structures

This subdivision includes breakwaters and bulkheads for canals.

35 51 00 Floating Construction

This section includes floating piers, docks, and dock accessories. Fixed Pier Timber Construction is found in 06 13 33. Driven piles are found in Division 31, as well as sheet piling, cofferdams, and riprap.

Reference Numbers

Reference numbers are shown in shaded boxes at the beginning of some major classifications. These numbers refer to related items in the Reference Section. The reference information may be an estimating procedure, an alternate pricing method, or technical information.

Note: Not all subdivisions listed here necessarily appear in this publication.

35 20 Waterway and Marine Construction and Equipment

35 20 23 – Dredging

35 20 23.13 Mechanical Dredging		Crew	Daily Output	Labor-Hours	Unit	Material	2011 Bare Costs Labor	Equipment	Total	Total Incl O&P
0010	**MECHANICAL DREDGING**									
0020	Dredging mobilization and demobilization, add to below, minimum	B-8	.53	105	Total		3,100	5,225	8,325	10,900
0100	Maximum	"	.10	560	"		16,400	27,700	44,100	57,500
0300	Barge mounted clamshell excavation into scows									
0310	Dumped 20 miles at sea, minimum	B-57	310	.129	B.C.Y.		3.71	7.80	11.51	14.75
0400	Maximum	"	213	.188	"		5.40	11.35	16.75	21.50
0500	Barge mounted dragline or clamshell, hopper dumped,									
0510	pumped 1000' to shore dump, minimum	B-57	340	.118	B.C.Y.		3.38	7.10	10.48	13.45
0525	All pumping uses 2000 gallons of water per cubic yard									
0600	Maximum	B-57	243	.165	B.C.Y.		4.73	9.95	14.68	18.80

35 20 23.23 Hydraulic Dredging		Crew	Daily Output	Labor-Hours	Unit	Material	Labor	Equipment	Total	Total Incl O&P
0010	**HYDRAULIC DREDGING**									
1000	Hydraulic method, pumped 1000' to shore dump, minimum	B-57	460	.087	B.C.Y.		2.50	5.25	7.75	9.95
1100	Maximum		310	.129			3.71	7.80	11.51	14.75
1400	Into scows dumped 20 miles, minimum		425	.094			2.70	5.70	8.40	10.75
1500	Maximum	↓	243	.165			4.73	9.95	14.68	18.80
1600	For inland rivers and canals in South, deduct				↓				30%	30%

35 51 Floating Construction

35 51 13 – Floating Piers

35 51 13.23 Floating Wood Piers		Crew	Daily Output	Labor-Hours	Unit	Material	Labor	Equipment	Total	Total Incl O&P
0010	**FLOATING WOOD PIERS**									
0020	Polyethylene encased polystyrene, no pilings included	F-3	330	.121	S.F.	40	3.72	1.99	45.71	52.50
0200	Pile supported, shore constructed, bare, 3" decking		130	.308		26.50	9.45	5.05	41	50.50
0250	4" decking		120	.333		25.50	10.25	5.45	41.20	51
0400	Floating, small boat, prefab, no shore facilities, minimum		250	.160		13.35	4.92	2.62	20.89	25.50
0500	Maximum		150	.267	↓	45	8.20	4.37	57.57	68
0700	Per slip, minimum (180 S.F. each)		1.59	25.157	Ea.	2,900	775	410	4,085	4,925
0800	Maximum	↓	1.40	28.571	"	8,550	880	470	9,900	11,400

Estimating Tips

Products such as conveyors, material handling cranes and hoists, as well as other items specified in this division, require trained installers. The general contractor may not have any choice as to who will perform the installation or when it will be performed. Long lead times are often required for these products, making early decisions in purchasing and scheduling necessary.

The installation of this type of equipment may require the embedment of mounting hardware during construction of floors, structural walls, or interior walls/partitions. Electrical connections will require coordination with the electrical contractor.

Reference Numbers

Reference numbers are shown in shaded boxes at the beginning of some major classifications. These numbers refer to related items in the Reference Section. The reference information may be an estimating procedure, an alternate pricing method, or technical information.

Note: Not all subdivisions listed here necessarily appear in this publication.

41 21 Conveyors

41 21 23 – Piece Material Conveyors

41 21 23.16 Container Piece Material Conveyors		Crew	Daily Output	Labor-Hours	Unit	Material	2011 Bare Costs Labor	Equipment	Total	Total Incl O&P
0010	**CONTAINER PIECE MATERIAL CONVEYORS**									
0020	Gravity fed, 2" rollers, 3" O.C.									
0050	10' sections with 2 supports, 600 lb. capacity, 18" wide				Ea.	450			450	495
0100	24" wide					515			515	570
0150	1400 lb. capacity, 18" wide					415			415	455
0200	24" wide					460			460	505
0350	Horizontal belt, center drive and takeup, 60 fpm									
0400	16" belt, 26.5' length	2 Mill	.50	32	Ea.	3,150	1,125		4,275	5,275
0450	24" belt, 41.5' length		.40	40		4,900	1,425		6,325	7,625
0500	61.5' length		.30	53.333		6,525	1,875		8,400	10,200
0600	Inclined belt, 10' rise with horizontal loader and									
0620	End idler assembly, 27.5' length, 18" belt	2 Mill	.30	53.333	Ea.	9,200	1,875		11,075	13,100
0700	24" belt	"	.15	106	"	10,300	3,775		14,075	17,400
3600	Monorail, overhead, manual, channel type									
3700	125 lb. per L.F.	1 Mill	26	.308	L.F.	10.25	10.90		21.15	28.50
3900	500 lb. per L.F.	"	21	.381	"	9.25	13.45		22.70	31.50
4000	Trolleys for above, 2 wheel, 125 lb. capacity				Ea.	77			77	84.50
4200	4 wheel, 250 lb. capacity					207			207	227
4300	8 wheel, 1,000 lb. capacity					380			380	420

41 22 Cranes and Hoists

41 22 13 – Cranes

41 22 13.10 Crane Rail

		Crew	Daily Output	Labor-Hours	Unit	Material	Labor	Equipment	Total	Total Incl O&P
0010	**CRANE RAIL**									
0020	Box beam bridge, no equipment included	E-4	3400	.009	Lb.	1.19	.35	.03	1.57	2.01
0200	Running track only, 104 lb per yard	"	5600	.006	"	.62	.21	.02	.85	1.10

41 22 13.13 Bridge Cranes

		Crew	Daily Output	Labor-Hours	Unit	Material	Labor	Equipment	Total	Total Incl O&P
0010	**BRIDGE CRANES**									
0100	1 girder, 20' span, 3 ton	M-3	1	34	Ea.	23,800	1,350	128	25,278	28,500
0125	5 ton		1	34		26,000	1,350	128	27,478	30,900
0150	7.5 ton		1	34		31,000	1,350	128	32,478	36,400
0175	10 ton		.80	42.500		41,000	1,675	160	42,835	47,900
0200	15 ton		.80	42.500		52,500	1,675	160	54,335	61,000
0225	30' span, 3 ton		1	34		24,700	1,350	128	26,178	29,500
0250	5 ton		1	34		27,200	1,350	128	28,678	32,200
0275	7.5 ton		1	34		32,600	1,350	128	34,078	38,100
0300	10 ton		.80	42.500		42,400	1,675	160	44,235	49,500
0325	15 ton		.80	42.500		55,000	1,675	160	56,835	63,500
0350	2 girder, 40' span, 3 ton	M-4	.50	72		40,900	2,825	330	44,055	49,900
0375	5 ton		.50	72		42,800	2,825	330	45,955	52,000
0400	7.5 ton		.50	72		46,900	2,825	330	50,055	56,500
0425	10 ton		.40	90		55,000	3,525	410	58,935	66,500
0450	15 ton		.40	90		74,500	3,525	410	78,435	88,000
0475	25 ton		.30	120		88,000	4,700	550	93,250	105,500
0500	50' span, 3 ton		.50	72		46,500	2,825	330	49,655	56,000
0525	5 ton		.50	72		48,400	2,825	330	51,555	58,000
0550	7.5 ton		.50	72		52,000	2,825	330	55,155	62,000
0575	10 ton		.40	90		59,500	3,525	410	63,435	71,500
0600	15 ton		.40	90		78,000	3,525	410	81,935	92,000
0625	25 ton		.30	120		92,500	4,700	550	97,750	110,000

41 22 Cranes and Hoists

41 22 23 – Hoists

41 22 23.10 Material Handling		Crew	Daily Output	Labor-Hours	Unit	Material	2011 Bare Costs Labor	Equipment	Total	Total Incl O&P
0010	**MATERIAL HANDLING**, cranes, hoists and lifts									
1500	Cranes, portable hydraulic, floor type, 2,000 lb. capacity				Ea.	2,475			2,475	2,700
1600	4,000 lb. capacity					4,150			4,150	4,575
1800	Movable gantry type, 12′ to 15′ range, 2,000 lb. capacity					3,600			3,600	3,975
1900	6,000 lb. capacity					5,225			5,225	5,750
2100	Hoists, electric overhead, chain, hook hung, 15′ lift, 1 ton cap.					2,250			2,250	2,475
2200	3 ton capacity					2,625			2,625	2,900
2500	5 ton capacity				↓	6,625			6,625	7,275
2600	For hand-pushed trolley, add					15%				
2700	For geared trolley, add					30%				
2800	For motor trolley, add					75%				
3000	For lifts over 15′, 1 ton, add				L.F.	23.50			23.50	25.50
3100	5 ton, add				"	52			52	57
3300	Lifts, scissor type, portable, electric, 36″ high, 2,000 lb.				Ea.	4,275			4,275	4,700
3400	48″ high, 4,000 lb.				"	4,675			4,675	5,150

Division Notes

		CREW	DAILY OUTPUT	LABOR-HOURS	UNIT	2011 BARE COSTS MAT.	LABOR	EQUIP.	TOTAL	TOTAL INCL O&P

Estimating Tips

This new division contains information about water and wastewater equipment and systems, which was formerly located in Division 44.

The main areas of focus are total wastewater treatment plants and components of wastewater treatment plants. In addition, there are assemblies such as sewage treatment lagoons that can be found in some publications under G30 Site Mechnical Utilities.

Also included in this section are oil/water separators for wastewater treatment.

Reference Numbers

Reference numbers are shown in shaded boxes at the beginning of some major classifications. These numbers refer to related items in the Reference Section. The reference information may be an estimating procedure, an alternate pricing method, or technical information.

Note: Not all subdivisions listed here necessarily appear in this publication.

46 07 Packaged Water and Wastewater Treatment Equipment

46 07 53 – Packaged Wastewater Treatment Equipment

46 07 53.10 Biological Packaged Water Treatment Plants		Crew	Daily Output	Labor-Hours	Unit	2011 Bare Costs Material	Labor	Equipment	Total	Total Incl O&P
0010	BIOLOGICAL PACKAGED WATER TREATMENT PLANTS									
0011	Not including fencing or external piping									
0020	Steel packaged, blown air aeration plants									
0100	1,000 GPD				Gal.				95	109
0200	5,000 GPD								57	66
0300	15,000 GPD								48	55
0400	30,000 GPD								43	50
0500	50,000 GPD								38	44
0600	100,000 GPD								33	38
0700	200,000 GPD								28	32
0800	500,000 GPD				↓				24	28
1000	Concrete, extended aeration, primary and secondary treatment									
1010	10,000 GPD				Gal.				48	55
1100	30,000 GPD								43	50
1200	50,000 GPD								38	44
1400	100,000 GPD								33	38
1500	500,000 GPD				↓				24	28
1700	Municipal wastewater treatment facility									
1720	1.0 MGD				Gal.				9.70	11
1740	1.5 MGD								9.20	10.60
1760	2.0 MGD								8.75	10
1780	3.0 MGD								6.80	7.80
1800	5.0 MGD				↓				5.80	6.70
2000	Holding tank system, not incl. excavation or backfill									
2010	Recirculating chemical water closet	2 Plum	4	4	Ea.	460	159		619	765
2100	For voltage converter, add	"	16	1		250	39.50		289.50	340
2200	For high level alarm, add	1 Plum	7.80	1.026	↓	143	40.50		183.50	224

46 07 53.20 Wastewater Treatment System		Crew	Daily Output	Labor-Hours	Unit	Material	Labor	Equipment	Total	Total Incl O&P
0010	WASTEWATER TREATMENT SYSTEM									
0020	Fiberglass, 1,000 gallon	B-21	1.29	21.705	Ea.	3,450	615	100	4,165	4,925
0100	1,500 gallon	B-21	1.03	27.184	Ea.	7,550	770	125	8,445	9,750

Reference Section

All the reference information is in one section, making it easy to find what you need to know . . . and easy to use the book on a daily basis. This section is visually identified by a vertical gray bar on the page edges.

In this Reference Section, we've included Equipment Rental Costs, a listing of rental and operating costs; Crew Listings, a full listing of all crews and equipment, and their costs; Historical Cost Indexes for cost comparisons over time; City Cost Indexes and Location Factors for adjusting costs to the region you are in; Reference Tables, where you will find explanations, estimating information and procedures, or technical data; Change Orders, information on pricing changes to contract documents; Square Foot Costs that allow you to make a rough estimate for the overall cost of a project; and an explanation of all the Abbreviations in the book.

Table of Contents

Estimating Tips

- This section contains the average costs to rent and operate hundreds of pieces of construction equipment. This is useful information when estimating the time and material requirements of any particular operation in order to establish a unit or total cost. Equipment costs include not only rental, but also operating costs for equipment under normal use.

Rental Costs

- Equipment rental rates are obtained from industry sources throughout North America–contractors, suppliers, dealers, manufacturers, and distributors.
- Rental rates vary throughout the country, with larger cities generally having lower rates. Lease plans for new equipment are available for periods in excess of six months, with a percentage of payments applying toward purchase.
- Monthly rental rates vary from 2% to 5% of the purchase price of the equipment depending on the anticipated life of the equipment and its wearing parts.
- Weekly rental rates are about 1/3 the monthly rates, and daily rental rates are about 1/3 the weekly rate.
- Rental rates can also be treated as reimbursement costs for contractor-owned equipment. Owned equipment costs include depreciation, loan payments, interest, taxes, insurance, storage, and major repairs.

Operating Costs

- The operating costs include parts and labor for routine servicing, such as repair and replacement of pumps, filters and worn lines. Normal operating expendables, such as fuel, lubricants, tires and electricity (where applicable), are also included.
- Extraordinary operating expendables with highly variable wear patterns, such as diamond bits and blades, are excluded. These costs can be found as material costs in the Unit Price section.
- The hourly operating costs listed do not include the operator's wages.

Equipment Cost/Day

- Any power equipment required by a crew is shown in the Crew Listings with a daily cost.
- The daily cost of equipment needed by a crew is based on dividing the weekly rental rate by 5 (number of working days in the week), and then adding the hourly operating cost times 8 (the number of hours in a day). This "Equipment Cost/Day" is shown in the far right column of the Equipment Rental pages.
- If equipment is needed for only one or two days, it is best to develop your own cost by including components for daily rent and hourly operating cost. This is important when the listed Crew for a task does not contain the equipment needed, such as a crane for lifting mechanical heating/cooling equipment up onto a roof.
- If the quantity of work is less than the crew's Daily Output shown for a Unit Price line item that includes a bare unit equipment cost, it is recommended to estimate one day's rental cost and operating cost for equipment shown in the Crew Listing for that line item.

Mobilization/Demobilization

- The cost to move construction equipment from an equipment yard or rental company to the jobsite and back again is not included in equipment rental costs listed in the Reference section, nor in the bare equipment cost of any Unit Price line item, nor in any equipment costs shown in the Crew listings.
- Mobilization (to the site) and demobilization (from the site) costs can be found in the Unit Price section.
- If a piece of equipment is already at the jobsite, it is not appropriate to utilize mobil./demob. costs again in an estimate.

01 54 33 | Equipment Rental

		01 54 33 | Equipment Rental	Unit	Hourly Oper. Cost	Rent per Day	Rent per Week	Rent per Month	Equipment Cost/Day	
10	0010	**CONCRETE EQUIPMENT RENTAL** without operators (R015433-10)							10
	0200	Bucket, concrete lightweight, 1/2 C.Y.	Ea.	.75	23	69	207	19.80	
	0300	1 C.Y.		.80	28	84	252	23.20	
	0400	1-1/2 C.Y.		1.00	38.50	115	345	31	
	0500	2 C.Y.		1.10	45	135	405	35.80	
	0580	8 C.Y.		5.70	265	795	2,375	204.60	
	0600	Cart, concrete, self-propelled, operator walking, 10 C.F.		2.50	58.50	175	525	55	
	0700	Operator riding, 18 C.F.		4.20	96.50	290	870	91.60	
	0800	Conveyer for concrete, portable, gas, 16" wide, 26' long		9.20	127	380	1,150	149.60	
	0900	46' long		9.55	153	460	1,375	168.40	
	1000	56' long		9.70	162	485	1,450	174.60	
	1100	Core drill, electric, 2-1/2 H.P., 1" to 8" bit diameter		1.40	59.50	179	535	47	
	1150	11 H.P., 8" to 18" cores		5.10	115	345	1,025	109.80	
	1200	Finisher, concrete floor, gas, riding trowel, 96" wide		8.55	148	445	1,325	157.40	
	1300	Gas, walk-behind, 3 blade, 36" trowel		1.40	20.50	62	186	23.60	
	1400	4 blade, 48" trowel		2.70	28	84	252	38.40	
	1500	Float, hand-operated (Bull float) 48" wide		.08	13.65	41	123	8.85	
	1570	Curb builder, 14 H.P., gas, single screw		11.95	253	760	2,275	247.60	
	1590	Double screw		12.60	293	880	2,650	276.80	
	1600	Floor grinder, concrete and terrazzo, electric, 22" path		1.62	102	305	915	73.95	
	1700	Edger, concrete, electric, 7" path		.90	51.50	155	465	38.20	
	1750	Vacuum pick-up system for floor grinders, wet/dry		1.35	81.50	245	735	59.80	
	1800	Mixer, powered, mortar and concrete, gas, 6 C.F., 18 H.P.		6.45	120	360	1,075	123.60	
	1900	10 C.F., 25 H.P.		7.90	147	440	1,325	151.20	
	2000	16 C.F.		8.20	170	510	1,525	167.60	
	2100	Concrete, stationary, tilt drum, 2 C.Y.		6.45	238	715	2,150	194.60	
	2120	Pump, concrete, truck mounted 4" line 80' boom		23.25	925	2,775	8,325	741	
	2140	5" line, 110' boom		30.70	1,250	3,770	11,300	999.60	
	2160	Mud jack, 50 C.F. per hr.		6.10	128	385	1,150	125.80	
	2180	225 C.F. per hr		8.15	147	440	1,325	153.20	
	2190	Shotcrete pump rig, 12 C.Y./hr.		14.15	227	680	2,050	249.20	
	2200	35 C.Y./hr.		16.75	243	730	2,200	280	
	2600	Saw, concrete, manual, gas, 18 H.P.		4.65	43.50	130	390	63.20	
	2650	Self-propelled, gas, 30 H.P.		8.90	105	315	945	134.20	
	2700	Vibrators, concrete, electric, 60 cycle, 2 H.P.		.31	9	27	81	7.90	
	2800	3 H.P.		.43	12	36	108	10.65	
	2900	Gas engine, 5 H.P.		1.30	16.35	49	147	20.20	
	3000	8 H.P.		1.75	15.35	46	138	23.20	
	3050	Vibrating screed, gas engine, 8 H.P.		2.81	71.50	215	645	65.50	
	3120	Concrete transit mixer, 6 x 4, 250 H.P., 8 C.Y., rear discharge		43.95	590	1,765	5,300	704.60	
	3200	Front discharge		51.35	725	2,175	6,525	845.80	
	3300	6 x 6, 285 H.P., 12 C.Y., rear discharge		50.45	685	2,050	6,150	813.60	
	3400	Front discharge	↓	52.65	735	2,200	6,600	861.20	
20	0010	**EARTHWORK EQUIPMENT RENTAL** without operators (R015433-10)							20
	0040	Aggregate spreader, push type 8' to 12' wide	Ea.	2.25	26	78	234	33.60	
	0045	Tailgate type, 8' wide		2.20	33.50	100	300	37.60	
	0055	Earth auger, truck-mounted, for fence & sign posts, utility poles		11.60	515	1,545	4,625	401.80	
	0060	For borings and monitoring wells		38.45	665	1,995	5,975	706.60	
	0070	Portable, trailer mounted		2.45	32.50	98	294	39.20	
	0075	Truck-mounted, for caissons, water wells		83.90	3,025	9,075	27,200	2,486	
	0080	Horizontal boring machine, 12" to 36" diameter, 45 H.P.		20.85	200	600	1,800	286.80	
	0090	12" to 48" diameter, 65 H.P.		28.90	350	1,055	3,175	442.20	
	0095	Auger, for fence posts, gas engine, hand held		.45	6	18	54	7.20	
	0100	Excavator, diesel hydraulic, crawler mounted, 1/2 C.Y. cap.		16.70	385	1,155	3,475	364.60	
	0120	5/8 C.Y. capacity		24.50	555	1,670	5,000	530	
	0140	3/4 C.Y. capacity		30.55	660	1,980	5,950	640.40	
	0150	1 C.Y. capacity	↓	36.35	730	2,195	6,575	729.80	

	01 54 33 | Equipment Rental		UNIT	HOURLY OPER. COST	RENT PER DAY	RENT PER WEEK	RENT PER MONTH	EQUIPMENT COST/DAY	
20	0200	1-1/2 C.Y. capacity	Ea.	44.55	985	2,955	8,875	947.40	20
	0300	2 C.Y. capacity		58.25	1,275	3,790	11,400	1,224	
	0320	2-1/2 C.Y. capacity		77.80	1,650	4,955	14,900	1,613	
	0325	3-1/2 C.Y. capacity		104.95	2,225	6,655	20,000	2,171	
	0330	4-1/2 C.Y. capacity		128.05	2,750	8,225	24,700	2,669	
	0335	6 C.Y. capacity		158.65	3,050	9,160	27,500	3,101	
	0340	7 C.Y. capacity		160.90	3,150	9,460	28,400	3,179	
	0342	Excavator attachments, bucket thumbs		3.00	245	735	2,200	171	
	0345	Grapples		2.70	213	640	1,925	149.60	
	0347	Hydraulic hammer for boom mounting, 5000 ft lbs		12.65	420	1,255	3,775	352.20	
	0349	11,000 ft lbs		21.50	750	2,250	6,750	622	
	0350	Gradall type, truck mounted, 3 ton @ 15' radius, 5/8 C.Y.		36.80	940	2,815	8,450	857.40	
	0370	1 C.Y. capacity		41.00	1,125	3,365	10,100	1,001	
	0400	Backhoe-loader, 40 to 45 H.P., 5/8 C.Y. capacity		11.20	223	670	2,000	223.60	
	0450	45 H.P. to 60 H.P., 3/4 C.Y. capacity		16.30	315	950	2,850	320.40	
	0460	80 H.P., 1-1/4 C.Y. capacity		17.80	345	1,035	3,100	349.40	
	0470	112 H.P., 1-1/2 C.Y. capacity		30.25	610	1,830	5,500	608	
	0482	Backhoe-loader attachment, compactor, 20,000 lb.		5.40	145	435	1,300	130.20	
	0485	Hydraulic hammer, 750 ft lbs		3.10	103	310	930	86.80	
	0486	Hydraulic hammer, 1200 ft lbs		5.50	190	570	1,700	158	
	0500	Brush chipper, gas engine, 6" cutter head, 35 H.P.		8.10	110	330	990	130.80	
	0550	12" cutter head, 130 H.P.		13.25	185	555	1,675	217	
	0600	15" cutter head, 165 H.P.		19.45	222	665	2,000	288.60	
	0750	Bucket, clamshell, general purpose, 3/8 C.Y.		1.20	38.50	115	345	32.60	
	0800	1/2 C.Y.		1.30	45	135	405	37.40	
	0850	3/4 C.Y.		1.45	56.50	170	510	45.60	
	0900	1 C.Y.		1.50	60	180	540	48	
	0950	1-1/2 C.Y.		2.40	81.50	245	735	68.20	
	1000	2 C.Y.		2.50	91.50	275	825	75	
	1010	Bucket, dragline, medium duty, 1/2 C.Y.		.70	24.50	73	219	20.20	
	1020	3/4 C.Y.		.70	25.50	77	231	21	
	1030	1 C.Y.		.75	27	81	243	22.20	
	1040	1-1/2 C.Y.		1.15	41.50	125	375	34.20	
	1050	2 C.Y.		1.20	46.50	140	420	37.60	
	1070	3 C.Y.		1.85	63.50	190	570	52.80	
	1200	Compactor, manually guided 2-drum vibratory smooth roller, 7.5 H.P.		5.85	167	500	1,500	146.80	
	1250	Rammer/tamper, gas, 8"		2.20	43.50	130	390	43.60	
	1260	15"		2.45	50	150	450	49.60	
	1300	Vibratory plate, gas, 18" plate, 3000 lb. blow		2.10	23.50	70	210	30.80	
	1350	21" plate, 5000 lb. blow		2.60	30.50	91	273	39	
	1370	Curb builder/extruder, 14 H.P., gas, single screw		11.95	253	760	2,275	247.60	
	1390	Double screw		12.60	293	880	2,650	276.80	
	1500	Disc harrow attachment, for tractor		.41	68.50	205	615	44.30	
	1810	Feller buncher, shearing & accumulating trees, 100 H.P.		27.75	570	1,710	5,125	564	
	1860	Grader, self-propelled, 25,000 lb.		22.90	470	1,405	4,225	464.20	
	1910	30,000 lb.		26.35	555	1,660	4,975	542.80	
	1920	40,000 lb.		43.05	1,050	3,155	9,475	975.40	
	1930	55,000 lb.		55.95	1,425	4,275	12,800	1,303	
	1950	Hammer, pavement breaker, self-propelled, diesel, 1000 to 1250 lb.		22.90	360	1,080	3,250	399.20	
	2000	1300 to 1500 lb.		34.35	720	2,160	6,475	706.80	
	2050	Pile driving hammer, steam or air, 4150 ft lbs @ 225 bpm		9.80	500	1,495	4,475	377.40	
	2100	8750 ft lbs @ 145 bpm		11.80	700	2,095	6,275	513.40	
	2150	15,000 ft lbs @ 60 bpm		13.40	845	2,535	7,600	614.20	
	2200	24,450 ft lbs @ 111 bpm		14.40	940	2,820	8,450	679.20	
	2250	Leads, 60' high for pile driving hammers up to 20,000 ft lbs		3.10	82	246	740	74	
	2300	90' high for hammers over 20,000 ft lbs		4.70	149	448	1,350	127.20	
	2350	Diesel type hammer, 22,400 ft lbs		24.80	655	1,970	5,900	592.40	
	2400	41,300 ft lbs	↓	31.25	710	2,135	6,400	677	

01 54 33 | Equipment Rental

			UNIT	HOURLY OPER. COST	RENT PER DAY	RENT PER WEEK	RENT PER MONTH	EQUIPMENT COST/DAY	
20	2450	141,000 ft lbs	Ea.	47.65	1,225	3,660	11,000	1,113	20
	2500	Vib. elec. hammer/extractor, 200 kW diesel generator, 34 H.P.		36.95	680	2,045	6,125	704.60	
	2550	80 H.P.		64.95	1,000	3,025	9,075	1,125	
	2600	150 H.P.		120.90	1,950	5,865	17,600	2,140	
	2700	Extractor, steam or air, 700 ft lbs		17.75	520	1,560	4,675	454	
	2750	1000 ft lbs		20.20	635	1,910	5,725	543.60	
	2800	Log chipper, up to 22" diameter, 600 H.P.		35.35	455	1,360	4,075	554.80	
	2850	Logger, for skidding & stacking logs, 150 H.P.		41.75	850	2,550	7,650	844	
	2860	Mulcher, diesel powered, trailer mounted		15.60	223	670	2,000	258.80	
	2900	Rake, spring tooth, with tractor		12.66	294	882	2,650	277.70	
	3000	Roller, vibratory, tandem, smooth drum, 20 H.P.		7.40	143	430	1,300	145.20	
	3050	35 H.P.		9.40	248	745	2,225	224.20	
	3100	Towed type vibratory compactor, smooth drum, 50 H.P.		22.45	360	1,075	3,225	394.60	
	3150	Sheepsfoot, 50 H.P.		24.75	420	1,255	3,775	449	
	3170	Landfill compactor, 220 H.P.		68.75	1,425	4,290	12,900	1,408	
	3200	Pneumatic tire roller, 80 H.P.		12.65	345	1,040	3,125	309.20	
	3250	120 H.P.		19.15	600	1,800	5,400	513.20	
	3300	Sheepsfoot vibratory roller, 240 H.P.		55.10	1,100	3,265	9,800	1,094	
	3320	340 H.P.		73.05	1,675	5,025	15,100	1,589	
	3350	Smooth drum vibratory roller, 75 H.P.		20.60	555	1,670	5,000	498.80	
	3400	125 H.P.		25.60	655	1,960	5,875	596.80	
	3410	Rotary mower, brush, 60", with tractor		16.60	268	805	2,425	293.80	
	3420	Rototiller, walk-behind, gas, 5 H.P.		2.23	67.50	203	610	58.45	
	3422	8 H.P.		3.47	103	310	930	89.75	
	3440	Scrapers, towed type, 7 C.Y. capacity		5.25	110	330	990	108	
	3450	10 C.Y. capacity		6.00	153	460	1,375	140	
	3500	15 C.Y. capacity		6.45	178	535	1,600	158.60	
	3525	Self-propelled, single engine, 14 C.Y. capacity		77.50	1,475	4,460	13,400	1,512	
	3550	Dual engine, 21 C.Y. capacity		107.80	1,675	5,050	15,200	1,872	
	3600	31 C.Y. capacity		152.05	2,600	7,805	23,400	2,777	
	3640	44 C.Y. capacity		193.60	3,600	10,780	32,300	3,705	
	3650	Elevating type, single engine, 11 C.Y. capacity		54.35	1,100	3,270	9,800	1,089	
	3700	22 C.Y. capacity		99.85	2,150	6,440	19,300	2,087	
	3710	Screening plant 110 H.P. w/5' x 10' screen		30.50	420	1,260	3,775	496	
	3720	5' x 16' screen		32.65	520	1,560	4,675	573.20	
	3850	Shovel, crawler-mounted, front-loading, 7 C.Y. capacity		168.30	2,950	8,815	26,400	3,109	
	3855	12 C.Y. capacity		225.05	3,750	11,230	33,700	4,046	
	3860	Shovel/backhoe bucket, 1/2 C.Y.		2.25	63.50	190	570	56	
	3870	3/4 C.Y.		2.30	71.50	215	645	61.40	
	3880	1 C.Y.		2.40	80	240	720	67.20	
	3890	1-1/2 C.Y.		2.55	95	285	855	77.40	
	3910	3 C.Y.		2.85	130	390	1,175	100.80	
	3950	Stump chipper, 18" deep, 30 H.P.		7.17	187	560	1,675	169.35	
	4110	Dozer, crawler, torque converter, diesel 80 H.P.		21.70	400	1,195	3,575	412.60	
	4150	105 H.P.		25.80	500	1,495	4,475	505.40	
	4200	140 H.P.		37.35	800	2,400	7,200	778.80	
	4260	200 H.P.		50.75	1,075	3,225	9,675	1,051	
	4310	300 H.P.		71.80	1,725	5,165	15,500	1,607	
	4360	410 H.P.		97.20	2,200	6,635	19,900	2,105	
	4370	500 H.P.		130.15	3,000	9,020	27,100	2,845	
	4380	700 H.P.		191.35	4,375	13,095	39,300	4,150	
	4400	Loader, crawler, torque conv., diesel, 1-1/2 C.Y., 80 H.P.		22.70	475	1,425	4,275	466.60	
	4450	1-1/2 to 1-3/4 C.Y., 95 H.P.		25.75	600	1,795	5,375	565	
	4510	1-3/4 to 2-1/4 C.Y., 130 H.P.		33.70	840	2,520	7,550	773.60	
	4530	2-1/2 to 3-1/4 C.Y., 190 H.P.		46.40	1,050	3,185	9,550	1,008	
	4560	3-1/2 to 5 C.Y., 275 H.P.		64.35	1,525	4,585	13,800	1,432	
	4610	Front end loader, 4WD, articulated frame, 1 to 1-1/4 C.Y., 70 H.P.		13.35	215	645	1,925	235.80	
	4620	1-1/2 to 1-3/4 C.Y., 95 H.P.	↓	17.30	300	905	2,725	319.40	

01 54 33 | Equipment Rental

		01 54 33 Equipment Rental	UNIT	HOURLY OPER. COST	RENT PER DAY	RENT PER WEEK	RENT PER MONTH	EQUIPMENT COST/DAY
20	4650	1-3/4 to 2 C.Y., 130 H.P.	Ea.	19.90	370	1,105	3,325	380.20
	4710	2-1/2 to 3-1/2 C.Y., 145 H.P.		21.20	380	1,140	3,425	397.60
	4730	3 to 4-1/2 C.Y., 185 H.P.		28.95	610	1,835	5,500	598.60
	4760	5-1/4 to 5-3/4 C.Y., 270 H.P.		47.05	930	2,785	8,350	933.40
	4810	7 to 9 C.Y., 475 H.P.		83.30	1,900	5,725	17,200	1,811
	4870	9 - 11 C.Y., 620 H.P.		108.70	2,675	8,030	24,100	2,476
	4880	Skid steer loader, wheeled, 10 C.F., 30 H.P. gas		7.85	148	445	1,325	151.80
	4890	1 C.Y., 78 H.P., diesel		13.90	250	750	2,250	261.20
	4892	Skid-steer attachment, auger		.53	88	264	790	57.05
	4893	Backhoe		.71	119	356	1,075	76.90
	4894	Broom		.74	123	370	1,100	79.90
	4895	Forks		.24	40.50	122	365	26.30
	4896	Grapple		.55	91.50	274	820	59.20
	4897	Concrete hammer		1.07	178	534	1,600	115.35
	4898	Tree spade		.78	129	388	1,175	83.85
	4899	Trencher		.76	127	380	1,150	82.10
	4900	Trencher, chain, boom type, gas, operator walking, 12 H.P.		3.55	48.50	145	435	57.40
	4910	Operator riding, 40 H.P.		14.25	305	915	2,750	297
	5000	Wheel type, diesel, 4' deep, 12" wide		61.50	850	2,545	7,625	1,001
	5100	6' deep, 20" wide		77.45	1,975	5,910	17,700	1,802
	5150	Chain type, diesel, 5' deep, 8" wide		27.10	600	1,805	5,425	577.80
	5200	Diesel, 8' deep, 16" wide		114.45	2,650	7,950	23,900	2,506
	5202	Rock trencher, wheel type, 6" wide x 18" deep		17.35	325	970	2,900	332.80
	5206	Chain type, 18" wide x 7' deep		95.95	2,950	8,850	26,600	2,538
	5210	Tree spade, self-propelled		13.73	267	800	2,400	269.85
	5250	Truck, dump, 2-axle, 12 ton, 8 C.Y. payload, 220 H.P.		21.80	235	705	2,125	315.40
	5300	Three axle dump, 16 ton, 12 CY payload, 400 H.P.		38.00	335	1,010	3,025	506
	5310	Four axle dump, 25 ton, 18 C.Y. payload, 450 H.P.		45.60	495	1,480	4,450	660.80
	5350	Dump trailer only, rear dump, 16-1/2 C.Y.		5.10	142	425	1,275	125.80
	5400	20 C.Y.		5.50	160	480	1,450	140
	5450	Flatbed, single axle, 1-1/2 ton rating		16.05	71.50	215	645	171.40
	5500	3 ton rating		19.50	102	305	915	217
	5550	Off highway rear dump, 25 ton capacity		57.25	1,350	4,030	12,100	1,264
	5600	35 ton capacity		52.80	1,200	3,565	10,700	1,135
	5610	50 ton capacity		71.20	1,625	4,895	14,700	1,549
	5620	65 ton capacity		72.90	1,625	4,840	14,500	1,551
	5630	100 ton capacity		93.85	2,100	6,290	18,900	2,009
	6000	Vibratory plow, 25 H.P., walking		5.95	63.50	190	570	85.60
40	0010	**GENERAL EQUIPMENT RENTAL** without operators R015433 -10						
	0150	Aerial lift, scissor type, to 15' high, 1000 lb. cap., electric	Ea.	2.90	56.50	170	510	57.20
	0160	To 25' high, 2000 lb. capacity		3.30	71.50	215	645	69.40
	0170	Telescoping boom to 40' high, 500 lb. capacity, gas		15.85	320	955	2,875	317.80
	0180	To 45' high, 500 lb. capacity		16.95	370	1,105	3,325	356.60
	0190	To 60' high, 600 lb. capacity		19.20	480	1,445	4,325	442.60
	0195	Air compressor, portable, 6.5 CFM, electric		.33	14	42	126	11.05
	0196	Gasoline		.80	21	63	189	19
	0200	Towed type, gas engine, 60 CFM		10.15	48.50	145	435	110.20
	0300	160 CFM		11.75	51.50	155	465	125
	0400	Diesel engine, rotary screw, 250 CFM		12.35	107	320	960	162.80
	0500	365 CFM		16.40	132	395	1,175	210.20
	0550	450 CFM		20.70	165	495	1,475	264.60
	0600	600 CFM		35.95	228	685	2,050	424.60
	0700	750 CFM		36.10	238	715	2,150	431.80
	0800	For silenced models, small sizes, add to rent		3%	5%	5%	5%	
	0900	Large sizes, add to rent		5%	7%	7%	7%	
	0930	Air tools, breaker, pavement, 60 lb.	Ea.	.50	9.65	29	87	9.80
	0940	80 lb.		.50	10.35	31	93	10.20
	0950	Drills, hand (jackhammer) 65 lb.		.55	17.35	52	156	14.80

01 54 33 | Equipment Rental

			UNIT	HOURLY OPER. COST	RENT PER DAY	RENT PER WEEK	RENT PER MONTH	EQUIPMENT COST/DAY
40	0960	Track or wagon, swing boom, 4" drifter	Ea.	49.45	755	2,270	6,800	849.60
	0970	5" drifter		59.40	815	2,450	7,350	965.20
	0975	Track mounted quarry drill, 6" diameter drill		82.30	1,250	3,725	11,200	1,403
	0980	Dust control per drill		.58	18.65	56	168	15.85
	0990	Hammer, chipping, 12 lb.		.50	24.50	73	219	18.60
	1000	Hose, air with couplings, 50' long, 3/4" diameter		.03	4.67	14	42	3.05
	1100	1" diameter		.04	6.35	19	57	4.10
	1200	1-1/2" diameter		.06	9.65	29	87	6.30
	1300	2" diameter		.07	12	36	108	7.75
	1400	2-1/2" diameter		.11	19	57	171	12.30
	1410	3" diameter		.14	23	69	207	14.90
	1450	Drill, steel, 7/8" x 2'		.05	8.65	26	78	5.60
	1460	7/8" x 6'		.05	9	27	81	5.80
	1520	Moil points		.02	4	12	36	2.55
	1525	Pneumatic nailer w/accessories		.49	32.50	98	294	23.50
	1530	Sheeting driver for 60 lb. breaker		.04	6	18	54	3.90
	1540	For 90 lb. breaker		.12	8	24	72	5.75
	1550	Spade, 25 lb.		.45	7.35	22	66	8
	1560	Tamper, single, 35 lb.		.55	36.50	109	325	26.20
	1570	Triple, 140 lb.		.82	54.50	164	490	39.35
	1580	Wrenches, impact, air powered, up to 3/4" bolt		.40	10	30	90	9.20
	1590	Up to 1-1/4" bolt		.50	23.50	71	213	18.20
	1600	Barricades, barrels, reflectorized, 1 to 50 barrels		.03	4.60	13.80	41.50	3
	1610	100 to 200 barrels		.02	3.53	10.60	32	2.30
	1620	Barrels with flashers, 1 to 50 barrels		.03	5.25	15.80	47.50	3.40
	1630	100 to 200 barrels		.03	4.20	12.60	38	2.75
	1640	Barrels with steady burn type C lights		.04	7	21	63	4.50
	1650	Illuminated board, trailer mounted, with generator		.75	122	365	1,100	79
	1670	Portable barricade, stock, with flashers, 1 to 6 units		.03	5.25	15.80	47.50	3.40
	1680	25 to 50 units		.03	4.90	14.70	44	3.20
	1685	Butt fusion machine, wheeled, 1.5 HP electric, 2" - 8" diam. pipe		2.56	167	500	1,500	120.50
	1690	Tracked, 20 HP diesel, 4"-12" diam. pipe		10.43	490	1,465	4,400	376.45
	1695	83 HP diesel, 8" - 24" diam. pipe		27.52	975	2,930	8,800	806.15
	1700	Carts, brick, hand powered, 1000 lb. capacity		.44	72.50	218	655	47.10
	1800	Gas engine, 1500 lb., 7-1/2' lift		4.04	115	345	1,025	101.30
	1822	Dehumidifier, medium, 6 lb/hr, 150 CFM		.93	60	180	540	43.45
	1824	Large, 18 lb/hr, 600 CFM		1.89	122	366	1,100	88.30
	1830	Distributor, asphalt, trailer mounted, 2000 gal., 38 H.P. diesel		9.60	340	1,020	3,050	280.80
	1840	3000 gal., 38 H.P. diesel		11.00	370	1,105	3,325	309
	1850	Drill, rotary hammer, electric		.82	26.50	80	240	22.55
	1860	Carbide bit, 1-1/2" diameter, add to electric rotary hammer		.02	6.25	18.80	56.50	3.90
	1865	Rotary, crawler, 250 H.P.		122.20	2,050	6,125	18,400	2,203
	1870	Emulsion sprayer, 65 gal., 5 H.P. gas engine		2.89	99	297	890	82.50
	1880	200 gal., 5 H.P. engine		6.85	165	495	1,475	153.80
	1900	Floor auto-scrubbing machine, walk-behind, 28" path		3.26	213	640	1,925	154.10
	1930	Floodlight, mercury vapor, or quartz, on tripod, 1000 watt		.27	14	42	126	10.55
	1940	2000 watt		.50	27	81	243	20.20
	1950	Floodlights, trailer mounted with generator, 1 - 300 watt light		3.10	73.50	220	660	68.80
	1960	2 - 1000 watt lights		3.95	98.50	295	885	90.60
	2000	4 - 300 watt lights		3.75	93.50	280	840	86
	2005	Foam spray rig, incl. box trailer, compressor, generator, proportioner		23.78	495	1,490	4,475	488.25
	2020	Forklift, straight mast, 12' lift, 5000 lb., 2 wheel drive, gas		19.80	202	605	1,825	279.40
	2040	21' lift, 5000 lb., 4 wheel drive, diesel		16.20	248	745	2,225	278.60
	2050	For rough terrain, 42' lift, 35' reach, 9000 lb., 110 H.P.		22.30	450	1,345	4,025	447.40
	2060	For plant, 4 ton capacity, 80 H.P., 2 wheel drive, gas		12.05	102	305	915	157.40
	2080	10 ton capacity, 120 H.P., 2 wheel drive, diesel		17.70	177	530	1,600	247.60
	2100	Generator, electric, gas engine, 1.5 kW to 3 kW		2.80	11.35	34	102	29.20
	2200	5 kW	↓	3.55	15.35	46	138	37.60

01 54 33 | Equipment Rental

			UNIT	HOURLY OPER. COST	RENT PER DAY	RENT PER WEEK	RENT PER MONTH	EQUIPMENT COST/DAY	
40	2300	10 kW	Ea.	6.65	38.50	115	345	76.20	40
	2400	25 kW		7.80	88.50	265	795	115.40	
	2500	Diesel engine, 20 kW		9.00	75	225	675	117	
	2600	50 kW		15.85	105	315	945	189.80	
	2700	100 kW		30.25	145	435	1,300	329	
	2800	250 kW		59.65	253	760	2,275	629.20	
	2850	Hammer, hydraulic, for mounting on boom, to 500 ft lbs		2.40	76.50	230	690	65.20	
	2860	1000 ft lbs		4.05	127	380	1,150	108.40	
	2900	Heaters, space, oil or electric, 50 MBH		1.84	8	24	72	19.50	
	3000	100 MBH		3.31	10.65	32	96	32.90	
	3100	300 MBH		10.62	36.50	110	330	106.95	
	3150	500 MBH		21.42	48.50	145	435	200.35	
	3200	Hose, water, suction with coupling, 20' long, 2" diameter		.02	3	9	27	1.95	
	3210	3" diameter		.03	4.67	14	42	3.05	
	3220	4" diameter		.03	5	15	45	3.25	
	3230	6" diameter		.11	18.35	55	165	11.90	
	3240	8" diameter		.20	33.50	100	300	21.60	
	3250	Discharge hose with coupling, 50' long, 2" diameter		.01	1.33	4	12	.90	
	3260	3" diameter		.02	2.67	8	24	1.75	
	3270	4" diameter		.02	3.67	11	33	2.35	
	3280	6" diameter		.06	9.65	29	87	6.30	
	3290	8" diameter		.20	33.50	100	300	21.60	
	3295	Insulation blower		.20	6	18	54	5.20	
	3300	Ladders, extension type, 16' to 36' long		.15	25.50	76	228	16.40	
	3400	40' to 60' long		.22	36	108	325	23.35	
	3405	Lance for cutting concrete		2.17	104	313	940	79.95	
	3407	Lawn mower, rotary, 22", 5 H.P.		2.14	63.50	190	570	55.10	
	3408	48" self propelled		2.88	94.50	283	850	79.65	
	3410	Level, electronic, automatic, with tripod and leveling rod		1.45	96.50	290	870	69.60	
	3430	Laser type, for pipe and sewer line and grade		.76	50.50	152	455	36.50	
	3440	Rotating beam for interior control		1.16	77.50	232	695	55.70	
	3460	Builder's optical transit, with tripod and rod		.09	14.65	44	132	9.50	
	3500	Light towers, towable, with diesel generator, 2000 watt		3.75	93.50	280	840	86	
	3600	4000 watt		3.95	98.50	295	885	90.60	
	3700	Mixer, powered, plaster and mortar, 6 C.F., 7 H.P.		1.85	30	90	270	32.80	
	3800	10 C.F., 9 H.P.		2.00	31.50	95	285	35	
	3850	Nailer, pneumatic		.49	32.50	98	294	23.50	
	3900	Paint sprayers complete, 8 CFM		.73	48.50	145	435	34.85	
	4000	17 CFM		1.35	89.50	269	805	64.60	
	4020	Pavers, bituminous, rubber tires, 8' wide, 50 H.P., diesel		38.00	1,050	3,165	9,500	937	
	4030	10' wide, 150 H.P.		80.45	1,700	5,120	15,400	1,668	
	4050	Crawler, 8' wide, 100 H.P., diesel		78.80	1,800	5,415	16,200	1,713	
	4060	10' wide, 150 H.P.		90.60	2,125	6,405	19,200	2,006	
	4070	Concrete paver, 12' to 24' wide, 250 H.P.		78.20	1,625	4,875	14,600	1,601	
	4080	Placer-spreader-trimmer, 24' wide, 300 H.P.		111.00	2,650	7,975	23,900	2,483	
	4100	Pump, centrifugal gas pump, 1-1/2" diam., 65 GPM		3.50	48.50	145	435	57	
	4200	2" diameter, 130 GPM		4.50	53.50	160	480	68	
	4300	3" diameter, 250 GPM		4.75	55	165	495	71	
	4400	6" diameter, 1500 GPM		22.95	180	540	1,625	291.60	
	4500	Submersible electric pump, 1-1/4" diameter, 55 GPM		.31	16.35	49	147	12.30	
	4600	1-1/2" diameter, 83 GPM		.35	18.65	56	168	14	
	4700	2" diameter, 120 GPM		1.30	23.50	70	210	24.40	
	4800	3" diameter, 300 GPM		2.20	41.50	125	375	42.60	
	4900	4" diameter, 560 GPM		9.50	163	490	1,475	174	
	5000	6" diameter, 1590 GPM		13.95	220	660	1,975	243.60	
	5100	Diaphragm pump, gas, single, 1-1/2" diameter		1.15	49.50	148	445	38.80	
	5200	2" diameter		3.70	61.50	185	555	66.60	
	5300	3" diameter		3.70	61.50	185	555	66.60	

01 54 33 | Equipment Rental

			UNIT	HOURLY OPER. COST	RENT PER DAY	RENT PER WEEK	RENT PER MONTH	EQUIPMENT COST/DAY	
40	5400	Double, 4″ diameter	Ea.	5.55	108	325	975	109.40	40
	5450	Pressure washer 5 GPM, 3000 psi		3.85	46.50	140	420	58.80	
	5500	Trash pump, self-priming, gas, 2″ diameter		3.70	21	63	189	42.20	
	5600	Diesel, 4″ diameter		9.10	58.50	175	525	107.80	
	5650	Diesel, 6″ diameter		29.10	130	390	1,175	310.80	
	5655	Grout Pump		14.75	100	300	900	178	
	5700	Salamanders, L.P. gas fired, 100,000 Btu		3.35	16.35	49	147	36.60	
	5705	50,000 Btu		2.50	10.35	31	93	26.20	
	5720	Sandblaster, portable, open top, 3 C.F. capacity		.55	27	81	243	20.60	
	5730	6 C.F. capacity		.90	40	120	360	31.20	
	5740	Accessories for above		.12	20.50	61	183	13.15	
	5750	Sander, floor		.60	20	60	180	16.80	
	5760	Edger		.61	24.50	74	222	19.70	
	5800	Saw, chain, gas engine, 18″ long		1.75	19.65	59	177	25.80	
	5900	Hydraulic powered, 36″ long		.70	60	180	540	41.60	
	5950	60″ long		.70	61.50	185	555	42.60	
	6000	Masonry, table mounted, 14″ diameter, 5 H.P.		1.25	56.50	170	510	44	
	6050	Portable cut-off, 8 H.P.		1.90	33	99	297	35	
	6100	Circular, hand held, electric, 7-1/4″ diameter		.14	4.67	14	42	3.90	
	6200	12″ diameter		.21	7.65	23	69	6.30	
	6250	Wall saw, w/hydraulic power, 10 H.P.		6.50	61.50	185	555	89	
	6275	Shot blaster, walk-behind, 20″ wide		4.65	298	895	2,675	216.20	
	6280	Sidewalk broom, walk-behind		2.09	60.50	182	545	53.10	
	6300	Steam cleaner, 100 gallons per hour		3.15	60.50	182	545	61.60	
	6310	200 gallons per hour		4.25	73.50	220	660	78	
	6340	Tar Kettle/Pot, 400 gallons		4.97	81.50	245	735	88.75	
	6350	Torch, cutting, acetylene-oxygen, 150′ hose		.50	23.50	70	210	18	
	6360	Hourly operating cost includes tips and gas		10.35				82.80	
	6410	Toilet, portable chemical		.12	19.65	59	177	12.75	
	6420	Recycle flush type		.14	23.50	71	213	15.30	
	6430	Toilet, fresh water flush, garden hose,		.17	27.50	83	249	17.95	
	6440	Hoisted, non-flush, for high rise		.14	23.50	70	210	15.10	
	6450	Toilet, trailers, minimum		.24	40.50	122	365	26.30	
	6460	Maximum		.74	124	372	1,125	80.30	
	6465	Tractor, farm with attachment		15.50	305	910	2,725	306	
	6500	Trailers, platform, flush deck, 2 axle, 25 ton capacity		5.15	117	350	1,050	111.20	
	6600	40 ton capacity		6.65	162	485	1,450	150.20	
	6700	3 axle, 50 ton capacity		7.20	180	540	1,625	165.60	
	6800	75 ton capacity		8.95	237	710	2,125	213.60	
	6810	Trailer mounted cable reel for high voltage line work		5.04	240	720	2,150	184.30	
	6820	Trailer mounted cable tensioning rig		10.01	475	1,430	4,300	366.10	
	6830	Cable pulling rig		67.52	2,675	8,020	24,100	2,144	
	6900	Water tank trailer, engine driven discharge, 5000 gallons		6.65	155	465	1,400	146.20	
	6925	10,000 gallons		9.15	218	655	1,975	204.20	
	6950	Water truck, off highway, 6000 gallons		64.15	860	2,585	7,750	1,030	
	7010	Tram car for high voltage line work, powered, 2 conductor		4.39	130	391	1,175	113.30	
	7020	Transit (builder's level) with tripod		.09	14.65	44	132	9.50	
	7030	Trench box, 3000 lbs. 6′ x 8′		.56	93	279	835	60.30	
	7040	7200 lbs. 6′ x 20′		1.05	175	525	1,575	113.40	
	7050	8000 lbs., 8′ x 16′		1.05	175	525	1,575	113.40	
	7060	9500 lbs., 8′ x 20′		1.19	199	596	1,800	128.70	
	7065	11,000 lbs., 8′ x 24′		1.25	209	627	1,875	135.40	
	7070	12,000 lbs., 10′ x 20′		1.36	227	680	2,050	146.90	
	7100	Truck, pickup, 3/4 ton, 2 wheel drive		8.70	58.50	175	525	104.60	
	7200	4 wheel drive		8.95	75	225	675	116.60	
	7250	Crew carrier, 9 passenger		12.30	90	270	810	152.40	
	7290	Flat bed truck, 20,000 lbs. GVW		13.75	133	400	1,200	190	
	7300	Tractor, 4 x 2, 220 H.P.	↓	19.40	207	620	1,850	279.20	

01 54 33 | Equipment Rental

		01 54 33 Equipment Rental	UNIT	HOURLY OPER. COST	RENT PER DAY	RENT PER WEEK	RENT PER MONTH	EQUIPMENT COST/DAY	
40	7410	330 H.P.	Ea.	28.45	283	850	2,550	397.60	40
	7500	6 x 4, 380 H.P.		32.55	330	995	2,975	459.40	
	7600	450 H.P.		39.85	400	1,200	3,600	558.80	
	7610	Tractor, with A frame, boom and winch, 225 H.P.		22.10	283	850	2,550	346.80	
	7620	Vacuum truck, hazardous material, 2500 gallons		11.35	315	945	2,825	279.80	
	7625	5,000 gallons		13.89	440	1,320	3,950	375.10	
	7650	Vacuum, HEPA, 16 gallon, wet/dry		.36	18	54	162	13.70	
	7655	55 gallon, wet/dry		.59	27	81	243	20.90	
	7660	Water tank, portable		.17	28.50	85.50	257	18.45	
	7690	Sewer/catch basin vacuum, 14 C.Y., 1500 gallons		18.51	660	1,980	5,950	544.10	
	7700	Welder, electric, 200 amp		1.70	17.35	52	156	24	
	7800	300 amp		2.48	21	63	189	32.45	
	7900	Gas engine, 200 amp		10.15	25.50	76	228	96.40	
	8000	300 amp		11.60	27	81	243	109	
	8100	Wheelbarrow, any size		.08	12.65	38	114	8.25	
	8200	Wrecking ball, 4000 lbs.	↓	2.20	73.50	220	660	61.60	
50	0010	**HIGHWAY EQUIPMENT RENTAL** without operators R015433-10							50
	0050	Asphalt batch plant, portable drum mixer, 100 ton/hr	Ea.	68.35	1,475	4,400	13,200	1,427	
	0060	200 ton/hr		76.45	1,550	4,655	14,000	1,543	
	0070	300 ton/hr		88.65	1,825	5,485	16,500	1,806	
	0100	Backhoe attachment, long stick, up to 185 H.P., 10.5' long		.33	22	66	198	15.85	
	0140	Up to 250 H.P., 12' long		.36	23.50	71	213	17.10	
	0180	Over 250 H.P., 15' long		.49	32.50	97	291	23.30	
	0200	Special dipper arm, up to 100 H.P., 32' long		1.00	66.50	199	595	47.80	
	0240	Over 100 H.P., 33' long		1.24	82.50	248	745	59.50	
	0280	Catch basin/sewer cleaning truck, 3 ton, 9 C.Y., 1000 gal		30.85	415	1,250	3,750	496.80	
	0300	Concrete batch plant, portable, electric, 200 C.Y./hr.		18.20	530	1,585	4,750	462.60	
	0520	Grader/dozer attachment, ripper/scarifier, rear mounted, up to 135 H.P.		3.30	68.50	205	615	67.40	
	0540	Up to 180 H.P.		3.90	86.50	260	780	83.20	
	0580	Up to 250 H.P.		4.30	100	300	900	94.40	
	0700	Pvmt. removal bucket, for hyd. excavator, up to 90 H.P.		1.75	51.50	155	465	45	
	0740	Up to 200 H.P.		1.95	73.50	220	660	59.60	
	0780	Over 200 H.P.		2.10	85	255	765	67.80	
	0900	Aggregate spreader, self-propelled, 187 H.P.		40.05	715	2,140	6,425	748.40	
	1000	Chemical spreader, 3 C.Y.		2.75	45	135	405	49	
	1900	Hammermill, traveling, 250 H.P.		67.55	1,925	5,760	17,300	1,692	
	2000	Horizontal borer, 3" diameter, 13 H.P. gas driven		5.25	56.50	170	510	76	
	2150	Horizontal directional drill, 20,000 lb. thrust, 78 H.P. diesel		26.20	690	2,065	6,200	622.60	
	2160	30,000 lb. thrust, 115 H.P.		32.30	1,050	3,150	9,450	888.40	
	2170	50,000 lb. thrust, 170 H.P.		46.00	1,350	4,030	12,100	1,174	
	2190	Mud trailer for HDD, 1500 gallons, 175 H.P., gas		21.10	162	485	1,450	265.80	
	2200	Hydromulcher, diesel, 3000 gallon, for truck mounting		15.00	263	790	2,375	278	
	2300	Gas, 600 gallon		6.40	103	310	930	113.20	
	2400	Joint & crack cleaner, walk behind, 25 H.P.		2.75	51.50	155	465	53	
	2500	Filler, trailer mounted, 400 gallons, 20 H.P.		7.70	222	665	2,000	194.60	
	3000	Paint striper, self-propelled, double line, 30 H.P.		6.10	165	495	1,475	147.80	
	3200	Post drivers, 6" I-Beam frame, for truck mounting		11.45	440	1,325	3,975	356.60	
	3400	Road sweeper, self-propelled, 8' wide, 90 H.P.		31.10	595	1,780	5,350	604.80	
	3450	Road sweeper, vacuum assisted, 4 C.Y., 220 gallons		51.45	650	1,945	5,825	800.60	
	4000	Road mixer, self-propelled, 130 H.P.		41.45	785	2,350	7,050	801.60	
	4100	310 H.P.		71.70	2,200	6,580	19,700	1,890	
	4220	Cold mix paver, incl. pug mill and bitumen tank, 165 H.P.		90.60	2,150	6,455	19,400	2,016	
	4250	Paver, asphalt, wheel or crawler, 130 H.P., diesel		89.10	2,050	6,180	18,500	1,949	
	4300	Paver, road widener, gas 1' to 6', 67 H.P.		42.15	885	2,650	7,950	867.20	
	4400	Diesel, 2' to 14', 88 H.P.		53.00	1,075	3,215	9,650	1,067	
	4600	Slipform pavers, curb and gutter, 2 track, 75 H.P.		35.60	765	2,300	6,900	744.80	
	4700	4 track, 165 H.P.		44.65	820	2,455	7,375	848.20	
	4800	Median barrier, 215 H.P.	↓	45.40	850	2,550	7,650	873.20	

01 54 33 | Equipment Rental

				UNIT	HOURLY OPER. COST	RENT PER DAY	RENT PER WEEK	RENT PER MONTH	EQUIPMENT COST/DAY	
50	4901	Trailer, low bed, 75 ton capacity		Ea.	9.65	237	710	2,125	219.20	50
	5000	Road planer, walk behind, 10" cutting width, 10 H.P.			2.70	33.50	100	300	41.60	
	5100	Self-propelled, 12" cutting width, 64 H.P.			7.00	115	345	1,025	125	
	5120	Traffic line remover, metal ball blaster, truck mounted, 115 H.P.			47.15	765	2,300	6,900	837.20	
	5140	Grinder, truck mounted, 115 H.P.			50.10	835	2,505	7,525	901.80	
	5160	Walk-behind, 11 H.P.			3.05	55	165	495	57.40	
	5200	Pavement profiler, 4' to 6' wide, 450 H.P.			210.55	3,425	10,285	30,900	3,741	
	5300	8' to 10' wide, 750 H.P.			332.90	4,950	14,815	44,400	5,626	
	5400	Roadway plate, steel, 1" x 8' x 20'			.07	12	36	108	7.75	
	5600	Stabilizer, self-propelled, 150 H.P.			39.35	615	1,850	5,550	684.80	
	5700	310 H.P.			66.85	1,325	3,955	11,900	1,326	
	5800	Striper, thermal, truck mounted 120 gallon paint, 150 H.P.			42.85	520	1,555	4,675	653.80	
	6000	Tar kettle, 330 gallon, trailer mounted			4.58	63.50	190	570	74.65	
	7000	Tunnel locomotive, diesel, 8 to 12 ton			27.45	585	1,755	5,275	570.60	
	7005	Electric, 10 ton			24.75	670	2,005	6,025	599	
	7010	Muck cars, 1/2 C.Y. capacity			1.90	23.50	71	213	29.40	
	7020	1 C.Y. capacity			2.15	32.50	98	294	36.80	
	7030	2 C.Y. capacity			2.25	36.50	110	330	40	
	7040	Side dump, 2 C.Y. capacity			2.45	45	135	405	46.60	
	7050	3 C.Y. capacity			3.30	51.50	155	465	57.40	
	7060	5 C.Y. capacity			4.70	65	195	585	76.60	
	7100	Ventilating blower for tunnel, 7-1/2 H.P.			1.02	51.50	155	465	39.15	
	7110	10 H.P.			1.11	51.50	155	465	39.90	
	7120	20 H.P.			1.67	67.50	202	605	53.75	
	7140	40 H.P.			2.77	96.50	290	870	80.15	
	7160	60 H.P.			4.26	152	455	1,375	125.10	
	7175	75 H.P.			5.51	202	607	1,825	165.50	
	7180	200 H.P.			11.15	305	910	2,725	271.20	
	7800	Windrow loader, elevating			49.40	1,350	4,075	12,200	1,210	
60	0010	**LIFTING AND HOISTING EQUIPMENT RENTAL** without operators	R015433-10							60
	0120	Aerial lift truck, 2 person, to 80'		Ea.	24.90	730	2,185	6,550	636.20	
	0140	Boom work platform, 40' snorkel	R015433-15		14.25	278	835	2,500	281	
	0150	Crane, flatbed mounted, 3 ton capacity			13.45	197	590	1,775	225.60	
	0200	Crane, climbing, 106' jib, 6000 lb. capacity, 410 fpm	R312316-45		29.50	1,525	4,580	13,700	1,152	
	0300	101' jib, 10,250 lb. capacity, 270 fpm			35.60	1,925	5,800	17,400	1,445	
	0500	Tower, static, 130' high, 106' jib, 6200 lb. capacity at 400 fpm			33.05	1,775	5,290	15,900	1,322	
	0600	Crawler mounted, lattice boom, 1/2 C.Y., 15 tons at 12' radius			30.74	685	2,060	6,175	657.90	
	0700	3/4 C.Y., 20 tons at 12' radius			40.99	855	2,570	7,700	841.90	
	0800	1 C.Y., 25 tons at 12' radius			54.65	1,150	3,430	10,300	1,123	
	0900	1-1/2 C.Y., 40 tons at 12' radius			54.65	1,150	3,455	10,400	1,128	
	1000	2 C.Y., 50 tons at 12' radius			58.55	1,350	4,055	12,200	1,279	
	1100	3 C.Y., 75 tons at 12' radius			63.05	1,575	4,710	14,100	1,446	
	1200	100 ton capacity, 60' boom			71.50	1,800	5,405	16,200	1,653	
	1300	165 ton capacity, 60' boom			89.20	2,100	6,325	19,000	1,979	
	1400	200 ton capacity, 70' boom			111.15	2,675	7,995	24,000	2,488	
	1500	350 ton capacity, 80' boom			157.20	4,050	12,160	36,500	3,690	
	1600	Truck mounted, lattice boom, 6 x 4, 20 tons at 10' radius			37.39	1,225	3,680	11,000	1,035	
	1700	25 tons at 10' radius			40.40	1,325	4,000	12,000	1,123	
	1800	8 x 4, 30 tons at 10' radius			43.77	1,425	4,260	12,800	1,202	
	1900	40 tons at 12' radius			46.65	1,475	4,450	13,400	1,263	
	2000	60 tons at 15' radius			52.35	1,575	4,710	14,100	1,361	
	2050	82 tons at 15' radius			58.46	1,675	5,030	15,100	1,474	
	2100	90 tons at 15' radius			65.56	1,825	5,490	16,500	1,622	
	2200	115 tons at 15' radius			73.91	2,050	6,130	18,400	1,817	
	2300	150 tons at 18' radius			76.90	2,150	6,455	19,400	1,906	
	2350	165 tons at 18' radius			86.63	2,275	6,840	20,500	2,061	
	2400	Truck mounted, hydraulic, 12 ton capacity			41.45	540	1,620	4,850	655.60	

	01 54 33 | Equipment Rental		UNIT	HOURLY OPER. COST	RENT PER DAY	RENT PER WEEK	RENT PER MONTH	EQUIPMENT COST/DAY	
60	2500	25 ton capacity	Ea.	43.70	665	1,995	5,975	748.60	60
	2550	33 ton capacity		44.25	680	2,045	6,125	763	
	2560	40 ton capacity		57.20	800	2,405	7,225	938.60	
	2600	55 ton capacity		75.80	1,000	3,010	9,025	1,208	
	2700	80 ton capacity		97.75	1,525	4,555	13,700	1,693	
	2720	100 ton capacity		91.65	1,525	4,585	13,800	1,650	
	2740	120 ton capacity		106.15	1,625	4,910	14,700	1,831	
	2760	150 ton capacity		124.30	2,225	6,710	20,100	2,336	
	2800	Self-propelled, 4 x 4, with telescoping boom, 5 ton		14.60	235	705	2,125	257.80	
	2900	12-1/2 ton capacity		25.80	380	1,140	3,425	434.40	
	3000	15 ton capacity		26.45	405	1,210	3,625	453.60	
	3050	20 ton capacity		28.25	475	1,425	4,275	511	
	3100	25 ton capacity		29.80	535	1,610	4,825	560.40	
	3150	40 ton capacity		36.95	605	1,815	5,450	658.60	
	3200	Derricks, guy, 20 ton capacity, 60' boom, 75' mast		23.35	375	1,122	3,375	411.20	
	3300	100' boom, 115' mast		36.76	645	1,930	5,800	680.10	
	3400	Stiffleg, 20 ton capacity, 70' boom, 37' mast		25.72	485	1,460	4,375	497.75	
	3500	100' boom, 47' mast		39.56	775	2,330	7,000	782.50	
	3550	Helicopter, small, lift to 1250 lbs. maximum, w/pilot		95.92	3,025	9,060	27,200	2,579	
	3600	Hoists, chain type, overhead, manual, 3/4 ton		.15	1	3	9	1.80	
	3900	10 ton		.70	7	21	63	9.80	
	4000	Hoist and tower, 5000 lb. cap., portable electric, 40' high		3.89	215	646	1,950	160.30	
	4100	For each added 10' section, add		.10	17	51	153	11	
	4200	Hoist and single tubular tower, 5000 lb. electric, 100' high		5.34	300	902	2,700	223.10	
	4300	For each added 6'-6" section, add		.17	29	87	261	18.75	
	4400	Hoist and double tubular tower, 5000 lb., 100' high		5.79	330	993	2,975	244.90	
	4500	For each added 6'-6" section, add		.19	32.50	97	291	20.90	
	4550	Hoist and tower, mast type, 6000 lb., 100' high		6.14	345	1,030	3,100	255.10	
	4570	For each added 10' section, add		.12	20.50	61	183	13.15	
	4600	Hoist and tower, personnel, electric, 2000 lb., 100' @ 125 fpm		14.36	915	2,740	8,225	662.90	
	4700	3000 lb., 100' @ 200 fpm		16.38	1,025	3,110	9,325	753.05	
	4800	3000 lb., 150' @ 300 fpm		18.23	1,150	3,480	10,400	841.85	
	4900	4000 lb., 100' @ 300 fpm		18.74	1,175	3,550	10,700	859.90	
	5000	6000 lb., 100' @ 275 fpm	↓	19.92	1,250	3,720	11,200	903.35	
	5100	For added heights up to 500', add	L.F.	.01	1.67	5	15	1.10	
	5200	Jacks, hydraulic, 20 ton	Ea.	.05	2	6	18	1.60	
	5500	100 ton		.35	11.65	35	105	9.80	
	6100	Jacks, hydraulic, climbing w/50' jackrods, control console, 30 ton cap.		1.86	124	371	1,125	89.10	
	6150	For each added 10' jackrod section, add		.05	3.33	10	30	2.40	
	6300	50 ton capacity		2.99	199	597	1,800	143.30	
	6350	For each added 10' jackrod section, add		.06	4	12	36	2.90	
	6500	125 ton capacity		7.85	525	1,570	4,700	376.80	
	6550	For each added 10' jackrod section, add		.54	35.50	107	320	25.70	
	6600	Cable jack, 10 ton capacity with 200' cable		1.56	104	311	935	74.70	
	6650	For each added 50' of cable, add	↓	.18	12	36	108	8.65	
70	0010	**WELLPOINT EQUIPMENT RENTAL** without operators	R015433 -10						70
	0020	Based on 2 months rental							
	0100	Combination jetting & wellpoint pump, 60 H.P. diesel	Ea.	15.74	305	920	2,750	309.90	
	0200	High pressure gas jet pump, 200 H.P., 300 psi	"	36.50	262	786	2,350	449.20	
	0300	Discharge pipe, 8" diameter	L.F.	.01	.50	1.50	4.50	.40	
	0350	12" diameter		.01	.73	2.20	6.60	.50	
	0400	Header pipe, flows up to 150 GPM, 4" diameter		.01	.45	1.35	4.05	.35	
	0500	400 GPM, 6" diameter		.01	.53	1.60	4.80	.40	
	0600	800 GPM, 8" diameter		.01	.73	2.20	6.60	.50	
	0700	1500 GPM, 10" diameter		.01	.77	2.31	6.95	.55	
	0800	2500 GPM, 12" diameter		.02	1.46	4.38	13.15	1.05	
	0900	4500 GPM, 16" diameter	↓	.03	1.87	5.61	16.85	1.35	

01 54 33 | Equipment Rental

			UNIT	HOURLY OPER. COST	RENT PER DAY	RENT PER WEEK	RENT PER MONTH	EQUIPMENT COST/DAY	
70	0950	For quick coupling aluminum and plastic pipe, add	L.F.	.03	1.93	5.80	17.40	1.40	70
	1100	Wellpoint, 25' long, with fittings & riser pipe, 1-1/2" or 2" diameter	Ea.	.06	3.86	11.58	34.50	2.80	
	1200	Wellpoint pump, diesel powered, 4" suction, 20 H.P.		6.81	177	530	1,600	160.50	
	1300	6" suction, 30 H.P.		9.26	219	658	1,975	205.70	
	1400	8" suction, 40 H.P.		12.51	300	902	2,700	280.50	
	1500	10" suction, 75 H.P.		19.00	350	1,054	3,150	362.80	
	1600	12" suction, 100 H.P.		27.26	560	1,680	5,050	554.10	
	1700	12" suction, 175 H.P.	↓	40.08	615	1,850	5,550	690.65	
80	0010	**MARINE EQUIPMENT RENTAL** without operators R015433-10							80
	0200	Barge, 400 Ton, 30' wide x 90' long	Ea.	16.05	1,150	3,460	10,400	820.40	
	0240	800 Ton, 45' wide x 90' long		19.50	1,400	4,190	12,600	994	
	2000	Tugboat, diesel, 100 H.P.		25.55	215	645	1,925	333.40	
	2040	250 H.P.		49.40	395	1,185	3,550	632.20	
	2080	380 H.P.	↓	108.20	1,175	3,505	10,500	1,567	

Crew No.	Bare Costs		Incl. Subs O&P		Cost Per Labor-Hour	

Crew A-1	Hr.	Daily	Hr.	Daily	Bare Costs	Incl. O&P
1 Building Laborer	$26.45	$211.60	$44.30	$354.40	$26.45	$44.30
1 Concrete saw, gas manual		63.20		69.52	7.90	8.69
8 L.H., Daily Totals		$274.80		$423.92	$34.35	$52.99

Crew A-1A	Hr.	Daily	Hr.	Daily	Bare Costs	Incl. O&P
1 Skilled Worker	$34.05	$272.40	$57.00	$456.00	$34.05	$57.00
1 Shot Blaster, 20"		216.20		237.82	27.02	29.73
8 L.H., Daily Totals		$488.60		$693.82	$61.08	$86.73

Crew A-1B	Hr.	Daily	Hr.	Daily	Bare Costs	Incl. O&P
1 Building Laborer	$26.45	$211.60	$44.30	$354.40	$26.45	$44.30
1 Concrete Saw		134.20		147.62	16.77	18.45
8 L.H., Daily Totals		$345.80		$502.02	$43.23	$62.75

Crew A-1C	Hr.	Daily	Hr.	Daily	Bare Costs	Incl. O&P
1 Building Laborer	$26.45	$211.60	$44.30	$354.40	$26.45	$44.30
1 Chain Saw, gas, 18"		25.80		28.38	3.23	3.55
8 L.H., Daily Totals		$237.40		$382.78	$29.68	$47.85

Crew A-1D	Hr.	Daily	Hr.	Daily	Bare Costs	Incl. O&P
1 Building Laborer	$26.45	$211.60	$44.30	$354.40	$26.45	$44.30
1 Vibrating plate, gas, 18"		30.80		33.88	3.85	4.24
8 L.H., Daily Totals		$242.40		$388.28	$30.30	$48.53

Crew A-1E	Hr.	Daily	Hr.	Daily	Bare Costs	Incl. O&P
1 Building Laborer	$26.45	$211.60	$44.30	$354.40	$26.45	$44.30
1 Vibratory Plate, gas, 21"		39.00		42.90	4.88	5.36
8 L.H., Daily Totals		$250.60		$397.30	$31.32	$49.66

Crew A-1F	Hr.	Daily	Hr.	Daily	Bare Costs	Incl. O&P
1 Building Laborer	$26.45	$211.60	$44.30	$354.40	$26.45	$44.30
1 Rammer/tamper, gas, 8"		43.60		47.96	5.45	6.00
8 L.H., Daily Totals		$255.20		$402.36	$31.90	$50.30

Crew A-1G	Hr.	Daily	Hr.	Daily	Bare Costs	Incl. O&P
1 Building Laborer	$26.45	$211.60	$44.30	$354.40	$26.45	$44.30
1 Rammer/tamper, gas, 15"		49.60		54.56	6.20	6.82
8 L.H., Daily Totals		$261.20		$408.96	$32.65	$51.12

Crew A-1H	Hr.	Daily	Hr.	Daily	Bare Costs	Incl. O&P
1 Building Laborer	$26.45	$211.60	$44.30	$354.40	$26.45	$44.30
1 Exterior Steam Cleaner		61.60		67.76	7.70	8.47
8 L.H., Daily Totals		$273.20		$422.16	$34.15	$52.77

Crew A-1J	Hr.	Daily	Hr.	Daily	Bare Costs	Incl. O&P
1 Building Laborer	$26.45	$211.60	$44.30	$354.40	$26.45	$44.30
1 Cultivator, Walk-Behind, 5 H.P.		58.45		64.30	7.31	8.04
8 L.H., Daily Totals		$270.05		$418.69	$33.76	$52.34

Crew A-1K	Hr.	Daily	Hr.	Daily	Bare Costs	Incl. O&P
1 Building Laborer	$26.45	$211.60	$44.30	$354.40	$26.45	$44.30
1 Cultivator, Walk-Behind, 8 H.P.		89.75		98.72	11.22	12.34
8 L.H., Daily Totals		$301.35		$453.13	$37.67	$56.64

Crew A-1M	Hr.	Daily	Hr.	Daily	Bare Costs	Incl. O&P
1 Building Laborer	$26.45	$211.60	$44.30	$354.40	$26.45	$44.30
1 Snow Blower, Walk-Behind		53.10		58.41	6.64	7.30
8 L.H., Daily Totals		$264.70		$412.81	$33.09	$51.60

Crew No.	Bare Costs		Incl. Subs O&P		Cost Per Labor-Hour	

Crew A-2	Hr.	Daily	Hr.	Daily	Bare Costs	Incl. O&P
2 Laborers	$26.45	$423.20	$44.30	$708.80	$26.43	$44.13
1 Truck Driver (light)	26.40	211.20	43.80	350.40		
1 Flatbed Truck, Gas, 1.5 Ton		171.40		188.54	7.14	7.86
24 L.H., Daily Totals		$805.80		$1247.74	$33.58	$51.99

Crew A-2A	Hr.	Daily	Hr.	Daily	Bare Costs	Incl. O&P
2 Laborers	$26.45	$423.20	$44.30	$708.80	$26.43	$44.13
1 Truck Driver (light)	26.40	211.20	43.80	350.40		
1 Flatbed Truck, Gas, 1.5 Ton		171.40		188.54		
1 Concrete Saw		134.20		147.62	12.73	14.01
24 L.H., Daily Totals		$940.00		$1395.36	$39.17	$58.14

Crew A-2B	Hr.	Daily	Hr.	Daily	Bare Costs	Incl. O&P
1 Truck Driver (light)	$26.40	$211.20	$43.80	$350.40	$26.40	$43.80
1 Flatbed Truck, Gas, 1.5 Ton		171.40		188.54	21.43	23.57
8 L.H., Daily Totals		$382.60		$538.94	$47.83	$67.37

Crew A-3A	Hr.	Daily	Hr.	Daily	Bare Costs	Incl. O&P
1 Truck Driver (light)	$26.40	$211.20	$43.80	$350.40	$26.40	$43.80
1 Pickup truck, 4 x 4, 3/4 ton		116.60		128.26	14.57	16.03
8 L.H., Daily Totals		$327.80		$478.66	$40.98	$59.83

Crew A-3B	Hr.	Daily	Hr.	Daily	Bare Costs	Incl. O&P
1 Equip. Oper. (med.)	$34.90	$279.20	$57.05	$456.40	$31.05	$51.08
1 Truck Driver (heav.)	27.20	217.60	45.10	360.80		
1 Dump Truck, 12 C.Y., 400 H.P.		506.00		556.60		
1 F.E. Loader, W.M.,2.5 C.Y.		397.60		437.36	56.48	62.12
16 L.H., Daily Totals		$1400.40		$1811.16	$87.53	$113.20

Crew A-3C	Hr.	Daily	Hr.	Daily	Bare Costs	Incl. O&P
1 Equip. Oper. (light)	$33.55	$268.40	$54.80	$438.40	$33.55	$54.80
1 Loader, Skid Steer, 78 H.P.		261.20		287.32	32.65	35.91
8 L.H., Daily Totals		$529.60		$725.72	$66.20	$90.72

Crew A-3D	Hr.	Daily	Hr.	Daily	Bare Costs	Incl. O&P
1 Truck Driver, Light	$26.40	$211.20	$43.80	$350.40	$26.40	$43.80
1 Pickup truck, 4 x 4, 3/4 ton		116.60		128.26		
1 Flatbed Trailer, 25 Ton		111.20		122.32	28.48	31.32
8 L.H., Daily Totals		$439.00		$600.98	$54.88	$75.12

Crew A-3E	Hr.	Daily	Hr.	Daily	Bare Costs	Incl. O&P
1 Equip. Oper. (crane)	$35.80	$286.40	$58.50	$468.00	$31.50	$51.80
1 Truck Driver (heavy)	27.20	217.60	45.10	360.80		
1 Pickup truck, 4 x 4, 3/4 ton		116.60		128.26	7.29	8.02
16 L.H., Daily Totals		$620.60		$957.06	$38.79	$59.82

Crew A-3F	Hr.	Daily	Hr.	Daily	Bare Costs	Incl. O&P
1 Equip. Oper. (crane)	$35.80	$286.40	$58.50	$468.00	$31.50	$51.80
1 Truck Driver (heavy)	27.20	217.60	45.10	360.80		
1 Pickup truck, 4 x 4, 3/4 ton		116.60		128.26		
1 Truck Tractor, 6x4, 380 H.P.		459.40		505.34		
1 Lowbed Trailer, 75 Ton		219.20		241.12	49.70	54.67
16 L.H., Daily Totals		$1299.20		$1703.52	$81.20	$106.47

Crews

Crew No.	Bare Costs		Incl. Subs O&P		Cost Per Labor-Hour	

Crew A-3G	Hr.	Daily	Hr.	Daily	Bare Costs	Incl. O&P
1 Equip. Oper. (crane)	$35.80	$286.40	$58.50	$468.00	$31.50	$51.80
1 Truck Driver (heavy)	27.20	217.60	45.10	360.80		
1 Pickup truck, 4 x 4, 3/4 ton		116.60		128.26		
1 Truck Tractor, 6x4, 450 H.P.		558.80		614.68		
1 Lowbed Trailer, 75 Ton		219.20		241.12	55.91	61.50
16 L.H., Daily Totals		$1398.60		$1812.86	$87.41	$113.30

Crew A-3H	Hr.	Daily	Hr.	Daily	Bare Costs	Incl. O&P
1 Equip. Oper. (crane)	$35.80	$286.40	$58.50	$468.00	$35.80	$58.50
1 Hyd. crane, 12 Ton (daily)		871.60		958.76	108.95	119.85
8 L.H., Daily Totals		$1158.00		$1426.76	$144.75	$178.35

Crew A-3I	Hr.	Daily	Hr.	Daily	Bare Costs	Incl. O&P
1 Equip. Oper. (crane)	$35.80	$286.40	$58.50	$468.00	$35.80	$58.50
1 Hyd. crane, 25 Ton (daily)		1015.00		1116.50	126.88	139.56
8 L.H., Daily Totals		$1301.40		$1584.50	$162.68	$198.06

Crew A-3J	Hr.	Daily	Hr.	Daily	Bare Costs	Incl. O&P
1 Equip. Oper. (crane)	$35.80	$286.40	$58.50	$468.00	$35.80	$58.50
1 Hyd. crane, 40 Ton (daily)		1258.00		1383.80	157.25	172.97
8 L.H., Daily Totals		$1544.40		$1851.80	$193.05	$231.47

Crew A-3K	Hr.	Daily	Hr.	Daily	Bare Costs	Incl. O&P
1 Equip. Oper. (crane)	$35.80	$286.40	$58.50	$468.00	$33.42	$54.63
1 Equip. Oper. Oiler	31.05	248.40	50.75	406.00		
1 Hyd. crane, 55 Ton (daily)		1611.00		1772.10		
1 P/U Truck, 3/4 Ton (daily)		129.60		142.56	108.79	119.67
16 L.H., Daily Totals		$2275.40		$2788.66	$142.21	$174.29

Crew A-3L	Hr.	Daily	Hr.	Daily	Bare Costs	Incl. O&P
1 Equip. Oper. (crane)	$35.80	$286.40	$58.50	$468.00	$33.42	$54.63
1 Equip. Oper. Oiler	31.05	248.40	50.75	406.00		
1 Hyd. crane, 80 Ton (daily)		2302.00		2532.20		
1 P/U Truck, 3/4 Ton (daily)		129.60		142.56	151.97	167.17
16 L.H., Daily Totals		$2966.40		$3548.76	$185.40	$221.80

Crew A-3M	Hr.	Daily	Hr.	Daily	Bare Costs	Incl. O&P
1 Equip. Oper. (crane)	$35.80	$286.40	$58.50	$468.00	$33.42	$54.63
1 Equip. Oper. Oiler	31.05	248.40	50.75	406.00		
1 Hyd. crane, 100 Ton (daily)		2263.00		2489.30		
1 P/U Truck, 3/4 Ton (daily)		129.60		142.56	149.54	164.49
16 L.H., Daily Totals		$2927.40		$3505.86	$182.96	$219.12

Crew A-3N	Hr.	Daily	Hr.	Daily	Bare Costs	Incl. O&P
1 Equip. Oper. (crane)	$35.80	$286.40	$58.50	$468.00	$35.80	$58.50
1 Tower crane (monthly)		987.20		1085.92	123.40	135.74
8 L.H., Daily Totals		$1273.60		$1553.92	$159.20	$194.24

Crew A-3P	Hr.	Daily	Hr.	Daily	Bare Costs	Incl. O&P
1 Equip. Oper., Light	$33.55	$268.40	$54.80	$438.40	$33.55	$54.80
1 A.T. Forklift, 42' lift		447.40		492.14	55.92	61.52
8 L.H., Daily Totals		$715.80		$930.54	$89.47	$116.32

Crew A-4	Hr.	Daily	Hr.	Daily	Bare Costs	Incl. O&P
2 Carpenters	$33.55	$536.80	$56.15	$898.40	$32.25	$53.48
1 Painter, Ordinary	29.65	237.20	48.15	385.20		
24 L.H., Daily Totals		$774.00		$1283.60	$32.25	$53.48

Crew A-5	Hr.	Daily	Hr.	Daily	Bare Costs	Incl. O&P
2 Laborers	$26.45	$423.20	$44.30	$708.80	$26.44	$44.24
.25 Truck Driver (light)	26.40	52.80	43.80	87.60		
.25 Flatbed Truck, Gas, 1.5 Ton		42.85		47.13	2.38	2.62
18 L.H., Daily Totals		$518.85		$843.53	$28.82	$46.86

Crew A-6	Hr.	Daily	Hr.	Daily	Bare Costs	Incl. O&P
1 Instrument Man	$34.05	$272.40	$57.00	$456.00	$33.63	$55.75
1 Rodman/Chainman	33.20	265.60	54.50	436.00		
1 Level, electronic		69.60		76.56	4.35	4.79
16 L.H., Daily Totals		$607.60		$968.56	$37.98	$60.53

Crew A-7	Hr.	Daily	Hr.	Daily	Bare Costs	Incl. O&P
1 Chief Of Party	$40.30	$322.40	$67.15	$537.20	$35.85	$59.55
1 Instrument Man	34.05	272.40	57.00	456.00		
1 Rodman/Chainman	33.20	265.60	54.50	436.00		
1 Level, electronic		69.60		76.56	2.90	3.19
24 L.H., Daily Totals		$930.00		$1505.76	$38.75	$62.74

Crew A-8	Hr.	Daily	Hr.	Daily	Bare Costs	Incl. O&P
1 Chief of Party	$40.30	$322.40	$67.15	$537.20	$35.19	$58.29
1 Instrument Man	34.05	272.40	57.00	456.00		
2 Rodmen/Chainmen	33.20	531.20	54.50	872.00		
1 Level, electronic		69.60		76.56	2.17	2.39
32 L.H., Daily Totals		$1195.60		$1941.76	$37.36	$60.68

Crew A-9	Hr.	Daily	Hr.	Daily	Bare Costs	Incl. O&P
1 Asbestos Foreman	$35.20	$281.60	$59.35	$474.80	$34.76	$58.61
7 Asbestos Workers	34.70	1943.20	58.50	3276.00		
64 L.H., Daily Totals		$2224.80		$3750.80	$34.76	$58.61

Crew A-10A	Hr.	Daily	Hr.	Daily	Bare Costs	Incl. O&P
1 Asbestos Foreman	$35.20	$281.60	$59.35	$474.80	$34.87	$58.78
2 Asbestos Workers	34.70	555.20	58.50	936.00		
24 L.H., Daily Totals		$836.80		$1410.80	$34.87	$58.78

Crew A-10B	Hr.	Daily	Hr.	Daily	Bare Costs	Incl. O&P
1 Asbestos Foreman	$35.20	$281.60	$59.35	$474.80	$34.83	$58.71
3 Asbestos Workers	34.70	832.80	58.50	1404.00		
32 L.H., Daily Totals		$1114.40		$1878.80	$34.83	$58.71

Crew A-10C	Hr.	Daily	Hr.	Daily	Bare Costs	Incl. O&P
3 Asbestos Workers	$34.70	$832.80	$58.50	$1404.00	$34.70	$58.50
1 Flatbed Truck, Gas, 1.5 Ton		171.40		188.54	7.14	7.86
24 L.H., Daily Totals		$1004.20		$1592.54	$41.84	$66.36

Crew A-10D	Hr.	Daily	Hr.	Daily	Bare Costs	Incl. O&P
2 Asbestos Workers	$34.70	$555.20	$58.50	$936.00	$34.06	$56.56
1 Equip. Oper. (crane)	35.80	286.40	58.50	468.00		
1 Equip. Oper. Oiler	31.05	248.40	50.75	406.00		
1 Hydraulic Crane, 33 Ton		763.00		839.30	23.84	26.23
32 L.H., Daily Totals		$1853.00		$2649.30	$57.91	$82.79

Crew A-11	Hr.	Daily	Hr.	Daily	Bare Costs	Incl. O&P
1 Asbestos Foreman	$35.20	$281.60	$59.35	$474.80	$34.76	$58.61
7 Asbestos Workers	34.70	1943.20	58.50	3276.00		
2 Chip. Hammers, 12 Lb., Elec.		37.20		40.92	0.58	0.64
64 L.H., Daily Totals		$2262.00		$3791.72	$35.34	$59.25

Crew No.	Bare Costs		Incl. Subs O&P		Cost Per Labor-Hour	

Crew A-12	Hr.	Daily	Hr.	Daily	Bare Costs	Incl. O&P
1 Asbestos Foreman	$35.20	$281.60	$59.35	$474.80	$34.76	$58.61
7 Asbestos Workers	34.70	1943.20	58.50	3276.00		
1 Trk-mtd vac, 14 CY, 1500 Gal.		544.10		598.51		
1 Flatbed Truck, 20,000 GVW		190.00		209.00	11.47	12.62
64 L.H., Daily Totals		$2958.90		$4558.31	$46.23	$71.22

Crew A-13	Hr.	Daily	Hr.	Daily	Bare Costs	Incl. O&P
1 Equip. Oper. (light)	$33.55	$268.40	$54.80	$438.40	$33.55	$54.80
1 Trk-mtd vac, 14 CY, 1500 Gal.		544.10		598.51		
1 Flatbed Truck, 20,000 GVW		190.00		209.00	91.76	100.94
8 L.H., Daily Totals		$1002.50		$1245.91	$125.31	$155.74

Crew B-1	Hr.	Daily	Hr.	Daily	Bare Costs	Incl. O&P
1 Labor Foreman (outside)	$28.45	$227.60	$47.65	$381.20	$27.12	$45.42
2 Laborers	26.45	423.20	44.30	708.80		
24 L.H., Daily Totals		$650.80		$1090.00	$27.12	$45.42

Crew B-1A	Hr.	Daily	Hr.	Daily	Bare Costs	Incl. O&P
1 Laborer Foreman	$28.45	$227.60	$47.65	$381.20	$27.12	$45.42
2 Laborers	26.45	423.20	44.30	708.80		
2 Cutting Torches		36.00		39.60		
2 Set of Gases		165.60		182.16	8.40	9.24
24 L.H., Daily Totals		$852.40		$1311.76	$35.52	$54.66

Crew B-1B	Hr.	Daily	Hr.	Daily	Bare Costs	Incl. O&P
1 Laborer Foreman	$28.45	$227.60	$47.65	$381.20	$29.29	$48.69
2 Laborers	26.45	423.20	44.30	708.80		
1 Equip. Oper. (crane)	35.80	286.40	58.50	468.00		
2 Cutting Torches		36.00		39.60		
2 Set of Gases		165.60		182.16		
1 Hyd. Crane, 12 Ton		655.60		721.16	26.79	29.47
32 L.H., Daily Totals		$1794.40		$2500.92	$56.08	$78.15

Crew B-2	Hr.	Daily	Hr.	Daily	Bare Costs	Incl. O&P
1 Labor Foreman (outside)	$28.45	$227.60	$47.65	$381.20	$26.85	$44.97
4 Laborers	26.45	846.40	44.30	1417.60		
40 L.H., Daily Totals		$1074.00		$1798.80	$26.85	$44.97

Crew B-3	Hr.	Daily	Hr.	Daily	Bare Costs	Incl. O&P
1 Labor Foreman (outside)	$28.45	$227.60	$47.65	$381.20	$28.44	$47.25
2 Laborers	26.45	423.20	44.30	708.80		
1 Equip. Oper. (med.)	34.90	279.20	57.05	456.40		
2 Truck Drivers (heavy)	27.20	435.20	45.10	721.60		
1 Crawler Loader, 3 C.Y.		1008.00		1108.80		
2 Dump Trucks 12 C.Y., 400 H.P.		1012.00		1113.20	42.08	46.29
48 L.H., Daily Totals		$3385.20		$4490.00	$70.53	$93.54

Crew B-3A	Hr.	Daily	Hr.	Daily	Bare Costs	Incl. O&P
4 Laborers	$26.45	$846.40	$44.30	$1417.60	$28.14	$46.85
1 Equip. Oper. (med.)	34.90	279.20	57.05	456.40		
1 Hyd. Excavator, 1.5 C.Y.		947.40		1042.14	23.68	26.05
40 L.H., Daily Totals		$2073.00		$2916.14	$51.83	$72.90

Crew B-3B	Hr.	Daily	Hr.	Daily	Bare Costs	Incl. O&P
2 Laborers	$26.45	$423.20	$44.30	$708.80	$28.75	$47.69
1 Equip. Oper. (med.)	34.90	279.20	57.05	456.40		
1 Truck Driver (heavy)	27.20	217.60	45.10	360.80		
1 Backhoe Loader, 80 H.P.		349.40		384.34		
1 Dump Truck, 12 C.Y., 400 H.P.		506.00		556.60	26.73	29.40
32 L.H., Daily Totals		$1775.40		$2466.94	$55.48	$77.09

Crew B-3C	Hr.	Daily	Hr.	Daily	Bare Costs	Incl. O&P
3 Laborers	$26.45	$634.80	$44.30	$1063.20	$28.56	$47.49
1 Equip. Oper. (med.)	34.90	279.20	57.05	456.40		
1 Crawler Loader, 4 C.Y.		1432.00		1575.20	44.75	49.23
32 L.H., Daily Totals		$2346.00		$3094.80	$73.31	$96.71

Crew B-4	Hr.	Daily	Hr.	Daily	Bare Costs	Incl. O&P
1 Labor Foreman (outside)	$28.45	$227.60	$47.65	$381.20	$26.91	$44.99
4 Laborers	26.45	846.40	44.30	1417.60		
1 Truck Driver (heavy)	27.20	217.60	45.10	360.80		
1 Truck Tractor, 220 H.P.		279.20		307.12		
1 Flatbed Trailer, 40 Ton		150.20		165.22	8.95	9.84
48 L.H., Daily Totals		$1721.00		$2631.94	$35.85	$54.83

Crew B-5	Hr.	Daily	Hr.	Daily	Bare Costs	Incl. O&P
1 Labor Foreman (outside)	$28.45	$227.60	$47.65	$381.20	$28.54	$47.52
3 Laborers	26.45	634.80	44.30	1063.20		
1 Equip. Oper. (med.)	34.90	279.20	57.05	456.40		
1 Air Compressor, 250 cfm		162.80		179.08		
2 Breakers, Pavement, 60 lb.		19.60		21.56		
2 -50' Air Hoses, 1.5"		12.60		13.86		
1 Crawler Loader, 3 C.Y.		1008.00		1108.80	30.07	33.08
40 L.H., Daily Totals		$2344.60		$3224.10	$58.62	$80.60

Crew B-5A	Hr.	Daily	Hr.	Daily	Bare Costs	Incl. O&P
1 Foreman	$28.45	$227.60	$47.65	$381.20	$28.74	$47.71
6 Laborers	26.45	1269.60	44.30	2126.40		
2 Equip. Oper. (med.)	34.90	558.40	57.05	912.80		
1 Equip. Oper. (light)	33.55	268.40	54.80	438.40		
2 Truck Drivers (heavy)	27.20	435.20	45.10	721.60		
1 Air Compressor, 365 cfm		210.20		231.22		
2 Breakers, Pavement, 60 lb.		19.60		21.56		
8 -50' Air Hoses, 1"		32.80		36.08		
2 Dump Trucks, 8 C.Y., 220 H.P.		630.80		693.88	9.31	10.24
96 L.H., Daily Totals		$3652.60		$5563.14	$38.05	$57.95

Crew B-5B	Hr.	Daily	Hr.	Daily	Bare Costs	Incl. O&P
1 Powderman	$34.05	$272.40	$57.00	$456.00	$30.91	$51.07
2 Equip. Oper. (med.)	34.90	558.40	57.05	912.80		
3 Truck Drivers (heavy)	27.20	652.80	45.10	1082.40		
1 F.E. Loader, W.M.,2.5 C.Y.		397.60		437.36		
3 Dump Trucks, 12 C.Y., 400 H.P.		1518.00		1669.80		
1 Air Compressor, 365 cfm		210.20		231.22	44.29	48.72
48 L.H., Daily Totals		$3609.40		$4789.58	$75.20	$99.78

Crew B-5C	Hr.	Daily	Hr.	Daily	Bare Costs	Incl. O&P
3 Laborers	$26.45	$634.80	$44.30	$1063.20	$29.44	$48.67
1 Equip. Oper. (med.)	34.90	279.20	57.05	456.40		
2 Truck Drivers (heav.)	27.20	435.20	45.10	721.60		
1 Equip. Oper. (crane)	35.80	286.40	58.50	468.00		
1 Equip. Oper. Oiler	31.05	248.40	50.75	406.00		
2 Dump Trucks, 12 C.Y., 400 H.P.		1012.00		1113.20		
1 Crawler Loader, 4 C.Y.		1432.00		1575.20		
1 S.P. Crane, 4x4, 25 Ton		560.40		616.44	46.94	51.64
64 L.H., Daily Totals		$4888.40		$6420.04	$76.38	$100.31

Crew B-6	Hr.	Daily	Hr.	Daily	Bare Costs	Incl. O&P
2 Laborers	$26.45	$423.20	$44.30	$708.80	$28.82	$47.80
1 Equip. Oper. (light)	33.55	268.40	54.80	438.40		
1 Backhoe Loader, 48 H.P.		320.40		352.44	13.35	14.69
24 L.H., Daily Totals		$1012.00		$1499.64	$42.17	$62.48

Crew No.	Bare Costs		Incl. Subs O&P		Cost Per Labor-Hour	
Crew B-6B	**Hr.**	**Daily**	**Hr.**	**Daily**	**Bare Costs**	**Incl. O&P**
2 Labor Foremen (out)	$28.45	$455.20	$47.65	$762.40	$27.12	$45.42
4 Laborers	26.45	846.40	44.30	1417.60		
1 S.P. Crane, 4x4, 5 Ton		257.80		283.58		
1 Flatbed Truck, Gas, 1.5 Ton		171.40		188.54		
1 Butt Fusion Mach., 4"-12" diam.		376.45		414.10	16.78	18.46
48 L.H., Daily Totals		$2107.25		$3066.22	$43.90	$63.88
Crew B-6C	**Hr.**	**Daily**	**Hr.**	**Daily**	**Bare Costs**	**Incl. O&P**
2 Labor Foremen (out)	$28.45	$455.20	$47.65	$762.40	$27.12	$45.42
4 Laborers	26.45	846.40	44.30	1417.60		
1 S.P. Crane, 4x4, 12 Ton		434.40		477.84		
1 Flatbed Truck, Gas, 3 Ton		217.00		238.70		
1 Butt Fusion Mach., 8"-24" diam.		806.15		886.76	30.37	33.40
48 L.H., Daily Totals		$2759.15		$3783.30	$57.48	$78.82
Crew B-7	**Hr.**	**Daily**	**Hr.**	**Daily**	**Bare Costs**	**Incl. O&P**
1 Labor Foreman (outside)	$28.45	$227.60	$47.65	$381.20	$28.19	$46.98
4 Laborers	26.45	846.40	44.30	1417.60		
1 Equip. Oper. (med.)	34.90	279.20	57.05	456.40		
1 Brush Chipper, 12", 130 H.P.		217.00		238.70		
1 Crawler Loader, 3 C.Y.		1008.00		1108.80		
2 Chain Saws, gas, 36" Long		83.20		91.52	27.25	29.98
48 L.H., Daily Totals		$2661.40		$3694.22	$55.45	$76.96
Crew B-7A	**Hr.**	**Daily**	**Hr.**	**Daily**	**Bare Costs**	**Incl. O&P**
2 Laborers	$26.45	$423.20	$44.30	$708.80	$28.82	$47.80
1 Equip. Oper. (light)	33.55	268.40	54.80	438.40		
1 Rake w/Tractor		277.70		305.47		
2 Chain Saws, gas, 18"		51.60		56.76	13.72	15.09
24 L.H., Daily Totals		$1020.90		$1509.43	$42.54	$62.89
Crew B-8	**Hr.**	**Daily**	**Hr.**	**Daily**	**Bare Costs**	**Incl. O&P**
1 Labor Foreman (outside)	$28.45	$227.60	$47.65	$381.20	$29.36	$48.65
2 Laborers	26.45	423.20	44.30	708.80		
2 Equip. Oper. (med.)	34.90	558.40	57.05	912.80		
2 Truck Drivers (heavy)	27.20	435.20	45.10	721.60		
1 Hyd. Crane, 25 Ton		748.60		823.46		
1 Crawler Loader, 3 C.Y.		1008.00		1108.80		
2 Dump Trucks, 12 C.Y., 400 H.P.		1012.00		1113.20	49.44	54.38
56 L.H., Daily Totals		$4413.00		$5769.86	$78.80	$103.03
Crew B-9	**Hr.**	**Daily**	**Hr.**	**Daily**	**Bare Costs**	**Incl. O&P**
1 Labor Foreman (outside)	$28.45	$227.60	$47.65	$381.20	$26.85	$44.97
4 Laborers	26.45	846.40	44.30	1417.60		
1 Air Compressor, 250 cfm		162.80		179.08		
2 Breakers, Pavement, 60 lb.		19.60		21.56		
2 -50' Air Hoses, 1.5"		12.60		13.86	4.88	5.36
40 L.H., Daily Totals		$1269.00		$2013.30	$31.73	$50.33
Crew B-9A	**Hr.**	**Daily**	**Hr.**	**Daily**	**Bare Costs**	**Incl. O&P**
2 Laborers	$26.45	$423.20	$44.30	$708.80	$26.70	$44.57
1 Truck Driver (heavy)	27.20	217.60	45.10	360.80		
1 Water Tank Trailer, 5000 Gal.		146.20		160.82		
1 Truck Tractor, 220 H.P.		279.20		307.12		
2 -50' Discharge Hoses, 3"		3.50		3.85	17.87	19.66
24 L.H., Daily Totals		$1069.70		$1541.39	$44.57	$64.22

Crew No.	Bare Costs		Incl. Subs O&P		Cost Per Labor-Hour	
Crew B-9B	**Hr.**	**Daily**	**Hr.**	**Daily**	**Bare Costs**	**Incl. O&P**
2 Laborers	$26.45	$423.20	$44.30	$708.80	$26.70	$44.57
1 Truck Driver (heavy)	27.20	217.60	45.10	360.80		
2 -50' Discharge Hoses, 3"		3.50		3.85		
1 Water Tank Trailer, 5000 Gal.		146.20		160.82		
1 Truck Tractor, 220 H.P.		279.20		307.12		
1 Pressure Washer		58.80		64.68	20.32	22.35
24 L.H., Daily Totals		$1128.50		$1606.07	$47.02	$66.92
Crew B-9D	**Hr.**	**Daily**	**Hr.**	**Daily**	**Bare Costs**	**Incl. O&P**
1 Labor Foreman (Outside)	$28.45	$227.60	$47.65	$381.20	$26.85	$44.97
4 Common Laborers	26.45	846.40	44.30	1417.60		
1 Air Compressor, 250 cfm		162.80		179.08		
2 -50' Air Hoses, 1.5"		12.60		13.86		
2 Air Powered Tampers		52.40		57.64	5.70	6.26
40 L.H., Daily Totals		$1301.80		$2049.38	$32.55	$51.23
Crew B-10	**Hr.**	**Daily**	**Hr.**	**Daily**	**Bare Costs**	**Incl. O&P**
1 Equip. Oper. (med.)	$34.90	$279.20	$57.05	$456.40	$34.90	$57.05
8 L.H., Daily Totals		$279.20		$456.40	$34.90	$57.05
Crew B-10A	**Hr.**	**Daily**	**Hr.**	**Daily**	**Bare Costs**	**Incl. O&P**
1 Equip. Oper. (med.)	$34.90	$279.20	$57.05	$456.40	$34.90	$57.05
1 Roller, 2-Drum, W.B., 7.5 H.P.		146.80		161.48	18.35	20.18
8 L.H., Daily Totals		$426.00		$617.88	$53.25	$77.23
Crew B-10B	**Hr.**	**Daily**	**Hr.**	**Daily**	**Bare Costs**	**Incl. O&P**
1 Equip. Oper. (med.)	$34.90	$279.20	$57.05	$456.40	$34.90	$57.05
1 Dozer, 200 H.P.		1051.00		1156.10	131.38	144.51
8 L.H., Daily Totals		$1330.20		$1612.50	$166.28	$201.56
Crew B-10C	**Hr.**	**Daily**	**Hr.**	**Daily**	**Bare Costs**	**Incl. O&P**
1 Equip. Oper. (med.)	$34.90	$279.20	$57.05	$456.40	$34.90	$57.05
1 Dozer, 200 H.P.		1051.00		1156.10		
1 Vibratory Roller, Towed, 23 Ton		394.60		434.06	180.70	198.77
8 L.H., Daily Totals		$1724.80		$2046.56	$215.60	$255.82
Crew B-10D	**Hr.**	**Daily**	**Hr.**	**Daily**	**Bare Costs**	**Incl. O&P**
1 Equip. Oper. (med.)	$34.90	$279.20	$57.05	$456.40	$34.90	$57.05
1 Dozer, 200 H.P.		1051.00		1156.10		
1 Sheepsft. Roller, Towed		449.00		493.90	187.50	206.25
8 L.H., Daily Totals		$1779.20		$2106.40	$222.40	$263.30
Crew B-10E	**Hr.**	**Daily**	**Hr.**	**Daily**	**Bare Costs**	**Incl. O&P**
1 Equip. Oper. (med.)	$34.90	$279.20	$57.05	$456.40	$34.90	$57.05
1 Tandem Roller, 5 Ton		145.20		159.72	18.15	19.97
8 L.H., Daily Totals		$424.40		$616.12	$53.05	$77.02
Crew B-10F	**Hr.**	**Daily**	**Hr.**	**Daily**	**Bare Costs**	**Incl. O&P**
1 Equip. Oper. (med.)	$34.90	$279.20	$57.05	$456.40	$34.90	$57.05
1 Tandem Roller, 10 Ton		224.20		246.62	28.02	30.83
8 L.H., Daily Totals		$503.40		$703.02	$62.92	$87.88
Crew B-10G	**Hr.**	**Daily**	**Hr.**	**Daily**	**Bare Costs**	**Incl. O&P**
1 Equip. Oper. (med.)	$34.90	$279.20	$57.05	$456.40	$34.90	$57.05
1 Sheepsft. Roll., 240 H.P.		1094.00		1203.40	136.75	150.43
8 L.H., Daily Totals		$1373.20		$1659.80	$171.65	$207.47

Crew No.	Bare Costs		Incl. Subs O&P		Cost Per Labor-Hour	
Crew B-10H	Hr.	Daily	Hr.	Daily	Bare Costs	Incl. O&P
1 Equip. Oper. (med.)	$34.90	$279.20	$57.05	$456.40	$34.90	$57.05
1 Diaphragm Water Pump, 2"		66.60		73.26		
1 -20' Suction Hose, 2"		1.95		2.15		
2 -50' Discharge Hoses, 2"		1.80		1.98	8.79	9.67
8 L.H., Daily Totals		$349.55		$533.78	$43.69	$66.72
Crew B-10I	Hr.	Daily	Hr.	Daily	Bare Costs	Incl. O&P
1 Equip. Oper. (med.)	$34.90	$279.20	$57.05	$456.40	$34.90	$57.05
1 Diaphragm Water Pump, 4"		109.40		120.34		
1 -20' Suction Hose, 4"		3.25		3.58		
2 -50' Discharge Hoses, 4"		4.70		5.17	14.67	16.14
8 L.H., Daily Totals		$396.55		$585.49	$49.57	$73.19
Crew B-10J	Hr.	Daily	Hr.	Daily	Bare Costs	Incl. O&P
1 Equip. Oper. (med.)	$34.90	$279.20	$57.05	$456.40	$34.90	$57.05
1 Centrifugal Water Pump, 3"		71.00		78.10		
1 -20' Suction Hose, 3"		3.05		3.36		
2 -50' Discharge Hoses, 3"		3.50		3.85	9.69	10.66
8 L.H., Daily Totals		$356.75		$541.71	$44.59	$67.71
Crew B-10K	Hr.	Daily	Hr.	Daily	Bare Costs	Incl. O&P
1 Equip. Oper. (med.)	$34.90	$279.20	$57.05	$456.40	$34.90	$57.05
1 Centr. Water Pump, 6"		291.60		320.76		
1 -20' Suction Hose, 6"		11.90		13.09		
2 -50' Discharge Hoses, 6"		12.60		13.86	39.51	43.46
8 L.H., Daily Totals		$595.30		$804.11	$74.41	$100.51
Crew B-10L	Hr.	Daily	Hr.	Daily	Bare Costs	Incl. O&P
1 Equip. Oper. (med.)	$34.90	$279.20	$57.05	$456.40	$34.90	$57.05
1 Dozer, 80 H.P.		412.60		453.86	51.58	56.73
8 L.H., Daily Totals		$691.80		$910.26	$86.47	$113.78
Crew B-10M	Hr.	Daily	Hr.	Daily	Bare Costs	Incl. O&P
1 Equip. Oper. (med.)	$34.90	$279.20	$57.05	$456.40	$34.90	$57.05
1 Dozer, 300 H.P.		1607.00		1767.70	200.88	220.96
8 L.H., Daily Totals		$1886.20		$2224.10	$235.78	$278.01
Crew B-10N	Hr.	Daily	Hr.	Daily	Bare Costs	Incl. O&P
1 Equip. Oper. (med.)	$34.90	$279.20	$57.05	$456.40	$34.90	$57.05
1 F.E. Loader, T.M., 1.5 C.Y		466.60		513.26	58.33	64.16
8 L.H., Daily Totals		$745.80		$969.66	$93.22	$121.21
Crew B-10O	Hr.	Daily	Hr.	Daily	Bare Costs	Incl. O&P
1 Equip. Oper. (med.)	$34.90	$279.20	$57.05	$456.40	$34.90	$57.05
1 F.E. Loader, T.M., 2.25 C.Y.		773.60		850.96	96.70	106.37
8 L.H., Daily Totals		$1052.80		$1307.36	$131.60	$163.42
Crew B-10P	Hr.	Daily	Hr.	Daily	Bare Costs	Incl. O&P
1 Equip. Oper. (med.)	$34.90	$279.20	$57.05	$456.40	$34.90	$57.05
1 Crawler Loader, 3 C.Y.		1008.00		1108.80	126.00	138.60
8 L.H., Daily Totals		$1287.20		$1565.20	$160.90	$195.65
Crew B-10Q	Hr.	Daily	Hr.	Daily	Bare Costs	Incl. O&P
1 Equip. Oper. (med.)	$34.90	$279.20	$57.05	$456.40	$34.90	$57.05
1 Crawler Loader, 4 C.Y.		1432.00		1575.20	179.00	196.90
8 L.H., Daily Totals		$1711.20		$2031.60	$213.90	$253.95

Crew No.	Bare Costs		Incl. Subs O&P		Cost Per Labor-Hour	
Crew B-10R	Hr.	Daily	Hr.	Daily	Bare Costs	Incl. O&P
1 Equip. Oper. (med.)	$34.90	$279.20	$57.05	$456.40	$34.90	$57.05
1 F.E. Loader, W.M., 1 C.Y.		235.80		259.38	29.48	32.42
8 L.H., Daily Totals		$515.00		$715.78	$64.38	$89.47
Crew B-10S	Hr.	Daily	Hr.	Daily	Bare Costs	Incl. O&P
1 Equip. Oper. (med.)	$34.90	$279.20	$57.05	$456.40	$34.90	$57.05
1 F.E. Loader, W.M., 1.5 C.Y.		319.40		351.34	39.92	43.92
8 L.H., Daily Totals		$598.60		$807.74	$74.83	$100.97
Crew B-10T	Hr.	Daily	Hr.	Daily	Bare Costs	Incl. O&P
1 Equip. Oper. (med.)	$34.90	$279.20	$57.05	$456.40	$34.90	$57.05
1 F.E. Loader, W.M.,2.5 C.Y.		397.60		437.36	49.70	54.67
8 L.H., Daily Totals		$676.80		$893.76	$84.60	$111.72
Crew B-10U	Hr.	Daily	Hr.	Daily	Bare Costs	Incl. O&P
1 Equip. Oper. (med.)	$34.90	$279.20	$57.05	$456.40	$34.90	$57.05
1 F.E. Loader, W.M., 5.5 C.Y.		933.40		1026.74	116.68	128.34
8 L.H., Daily Totals		$1212.60		$1483.14	$151.57	$185.39
Crew B-10V	Hr.	Daily	Hr.	Daily	Bare Costs	Incl. O&P
1 Equip. Oper. (med.)	$34.90	$279.20	$57.05	$456.40	$34.90	$57.05
1 Dozer, 700 H.P.		4150.00		4565.00	518.75	570.63
8 L.H., Daily Totals		$4429.20		$5021.40	$553.65	$627.67
Crew B-10W	Hr.	Daily	Hr.	Daily	Bare Costs	Incl. O&P
1 Equip. Oper. (med.)	$34.90	$279.20	$57.05	$456.40	$34.90	$57.05
1 Dozer, 105 H.P.		505.40		555.94	63.17	69.49
8 L.H., Daily Totals		$784.60		$1012.34	$98.08	$126.54
Crew B-10X	Hr.	Daily	Hr.	Daily	Bare Costs	Incl. O&P
1 Equip. Oper. (med.)	$34.90	$279.20	$57.05	$456.40	$34.90	$57.05
1 Dozer, 410 H.P.		2105.00		2315.50	263.13	289.44
8 L.H., Daily Totals		$2384.20		$2771.90	$298.02	$346.49
Crew B-10Y	Hr.	Daily	Hr.	Daily	Bare Costs	Incl. O&P
1 Equip. Oper. (med.)	$34.90	$279.20	$57.05	$456.40	$34.90	$57.05
1 Vibr. Roller, Towed, 12 Ton		498.80		548.68	62.35	68.58
8 L.H., Daily Totals		$778.00		$1005.08	$97.25	$125.64
Crew B-11A	Hr.	Daily	Hr.	Daily	Bare Costs	Incl. O&P
1 Equipment Oper. (med.)	$34.90	$279.20	$57.05	$456.40	$30.68	$50.67
1 Laborer	26.45	211.60	44.30	354.40		
1 Dozer, 200 H.P.		1051.00		1156.10	65.69	72.26
16 L.H., Daily Totals		$1541.80		$1966.90	$96.36	$122.93
Crew B-11B	Hr.	Daily	Hr.	Daily	Bare Costs	Incl. O&P
1 Equipment Oper. (light)	$33.55	$268.40	$54.80	$438.40	$30.00	$49.55
1 Laborer	26.45	211.60	44.30	354.40		
1 Air Powered Tamper		26.20		28.82		
1 Air Compressor, 365 cfm		210.20		231.22		
2 -50' Air Hoses, 1.5"		12.60		13.86	15.56	17.12
16 L.H., Daily Totals		$729.00		$1066.70	$45.56	$66.67
Crew B-11C	Hr.	Daily	Hr.	Daily	Bare Costs	Incl. O&P
1 Equipment Oper. (med.)	$34.90	$279.20	$57.05	$456.40	$30.68	$50.67
1 Laborer	26.45	211.60	44.30	354.40		
1 Backhoe Loader, 48 H.P.		320.40		352.44	20.02	22.03
16 L.H., Daily Totals		$811.20		$1163.24	$50.70	$72.70

Crew B-11K	Bare Costs Hr.	Bare Costs Daily	Incl. Subs O&P Hr.	Incl. Subs O&P Daily	Cost Per Labor-Hour Bare Costs	Cost Per Labor-Hour Incl. O&P
1 Equipment Oper. (med.)	$34.90	$279.20	$57.05	$456.40	$30.68	$50.67
1 Laborer	26.45	211.60	44.30	354.40		
1 Trencher, Chain Type, 8' D		2506.00		2756.60	156.63	172.29
16 L.H., Daily Totals		$2996.80		$3567.40	$187.30	$222.96

Crew B-11L	Hr.	Daily	Hr.	Daily	Bare Costs	Incl. O&P
1 Equipment Oper. (med.)	$34.90	$279.20	$57.05	$456.40	$30.68	$50.67
1 Laborer	26.45	211.60	44.30	354.40		
1 Grader, 30,000 Lbs.		542.80		597.08	33.92	37.32
16 L.H., Daily Totals		$1033.60		$1407.88	$64.60	$87.99

Crew B-11M	Hr.	Daily	Hr.	Daily	Bare Costs	Incl. O&P
1 Equipment Oper. (med.)	$34.90	$279.20	$57.05	$456.40	$30.68	$50.67
1 Laborer	26.45	211.60	44.30	354.40		
1 Backhoe Loader, 80 H.P.		349.40		384.34	21.84	24.02
16 L.H., Daily Totals		$840.20		$1195.14	$52.51	$74.70

Crew B-11W	Hr.	Daily	Hr.	Daily	Bare Costs	Incl. O&P
1 Equipment Operator (med.)	$34.90	$279.20	$57.05	$456.40	$27.78	$46.03
1 Common Laborer	26.45	211.60	44.30	354.40		
10 Truck Drivers (hvy.)	27.20	2176.00	45.10	3608.00		
1 Dozer, 200 H.P.		1051.00		1156.10		
1 Vibratory Roller, Towed, 23 Ton		394.60		434.06		
10 Dump Trucks, 8 C.Y., 220 H.P.		3154.00		3469.40	47.91	52.70
96 L.H., Daily Totals		$7266.40		$9478.36	$75.69	$98.73

Crew B-11Y	Hr.	Daily	Hr.	Daily	Bare Costs	Incl. O&P
1 Labor Foreman (Outside)	$28.45	$227.60	$47.65	$381.20	$29.49	$48.92
5 Common Laborers	26.45	1058.00	44.30	1772.00		
3 Equipment Operators (med.)	34.90	837.60	57.05	1369.20		
1 Dozer, 80 H.P.		412.60		453.86		
2 Roller, 2-Drum, W.B., 7.5 H.P.		293.60		322.96		
4 Vibratory Plates, gas, 21"		156.00		171.60	11.98	13.17
72 L.H., Daily Totals		$2985.40		$4470.82	$41.46	$62.09

Crew B-12A	Hr.	Daily	Hr.	Daily	Bare Costs	Incl. O&P
1 Equip. Oper. (crane)	$35.80	$286.40	$58.50	$468.00	$31.13	$51.40
1 Laborer	26.45	211.60	44.30	354.40		
1 Hyd. Excavator, 1 C.Y.		729.80		802.78	45.61	50.17
16 L.H., Daily Totals		$1227.80		$1625.18	$76.74	$101.57

Crew B-12B	Hr.	Daily	Hr.	Daily	Bare Costs	Incl. O&P
1 Equip. Oper. (crane)	$35.80	$286.40	$58.50	$468.00	$31.13	$51.40
1 Laborer	26.45	211.60	44.30	354.40		
1 Hyd. Excavator, 1.5 C.Y.		947.40		1042.14	59.21	65.13
16 L.H., Daily Totals		$1445.40		$1864.54	$90.34	$116.53

Crew B-12C	Hr.	Daily	Hr.	Daily	Bare Costs	Incl. O&P
1 Equip. Oper. (crane)	$35.80	$286.40	$58.50	$468.00	$31.13	$51.40
1 Laborer	26.45	211.60	44.30	354.40		
1 Hyd. Excavator, 2 C.Y.		1224.00		1346.40	76.50	84.15
16 L.H., Daily Totals		$1722.00		$2168.80	$107.63	$135.55

Crew B-12D	Hr.	Daily	Hr.	Daily	Bare Costs	Incl. O&P
1 Equip. Oper. (crane)	$35.80	$286.40	$58.50	$468.00	$31.13	$51.40
1 Laborer	26.45	211.60	44.30	354.40		
1 Hyd. Excavator, 3.5 C.Y.		2171.00		2388.10	135.69	149.26
16 L.H., Daily Totals		$2669.00		$3210.50	$166.81	$200.66

Crew B-12E	Hr.	Daily	Hr.	Daily	Bare Costs	Incl. O&P
1 Equip. Oper. (crane)	$35.80	$286.40	$58.50	$468.00	$31.13	$51.40
1 Laborer	26.45	211.60	44.30	354.40		
1 Hyd. Excavator, .5 C.Y.		364.60		401.06	22.79	25.07
16 L.H., Daily Totals		$862.60		$1223.46	$53.91	$76.47

Crew B-12F	Hr.	Daily	Hr.	Daily	Bare Costs	Incl. O&P
1 Equip. Oper. (crane)	$35.80	$286.40	$58.50	$468.00	$31.13	$51.40
1 Laborer	26.45	211.60	44.30	354.40		
1 Hyd. Excavator, .75 C.Y.		640.40		704.44	40.02	44.03
16 L.H., Daily Totals		$1138.40		$1526.84	$71.15	$95.43

Crew B-12G	Hr.	Daily	Hr.	Daily	Bare Costs	Incl. O&P
1 Equip. Oper. (crane)	$35.80	$286.40	$58.50	$468.00	$31.13	$51.40
1 Laborer	26.45	211.60	44.30	354.40		
1 Crawler Crane, 15 Ton		657.90		723.69		
1 Clamshell Bucket, .5 C.Y.		37.40		41.14	43.46	47.80
16 L.H., Daily Totals		$1193.30		$1587.23	$74.58	$99.20

Crew B-12H	Hr.	Daily	Hr.	Daily	Bare Costs	Incl. O&P
1 Equip. Oper. (crane)	$35.80	$286.40	$58.50	$468.00	$31.13	$51.40
1 Laborer	26.45	211.60	44.30	354.40		
1 Crawler Crane, 25 Ton		1123.00		1235.30		
1 Clamshell Bucket, 1 C.Y.		48.00		52.80	73.19	80.51
16 L.H., Daily Totals		$1669.00		$2110.50	$104.31	$131.91

Crew B-12I	Hr.	Daily	Hr.	Daily	Bare Costs	Incl. O&P
1 Equip. Oper. (crane)	$35.80	$286.40	$58.50	$468.00	$31.13	$51.40
1 Laborer	26.45	211.60	44.30	354.40		
1 Crawler Crane, 20 Ton		841.90		926.09		
1 Dragline Bucket, .75 C.Y.		21.00		23.10	53.93	59.32
16 L.H., Daily Totals		$1360.90		$1771.59	$85.06	$110.72

Crew B-12J	Hr.	Daily	Hr.	Daily	Bare Costs	Incl. O&P
1 Equip. Oper. (crane)	$35.80	$286.40	$58.50	$468.00	$31.13	$51.40
1 Laborer	26.45	211.60	44.30	354.40		
1 Gradall, 5/8 C.Y.		857.40		943.14	53.59	58.95
16 L.H., Daily Totals		$1355.40		$1765.54	$84.71	$110.35

Crew B-12K	Hr.	Daily	Hr.	Daily	Bare Costs	Incl. O&P
1 Equip. Oper. (crane)	$35.80	$286.40	$58.50	$468.00	$31.13	$51.40
1 Laborer	26.45	211.60	44.30	354.40		
1 Gradall, 3 Ton, 1 C.Y.		1001.00		1101.10	62.56	68.82
16 L.H., Daily Totals		$1499.00		$1923.50	$93.69	$120.22

Crew B-12L	Hr.	Daily	Hr.	Daily	Bare Costs	Incl. O&P
1 Equip. Oper. (crane)	$35.80	$286.40	$58.50	$468.00	$31.13	$51.40
1 Laborer	26.45	211.60	44.30	354.40		
1 Crawler Crane, 15 Ton		657.90		723.69		
1 F.E. Attachment, .5 C.Y.		56.00		61.60	44.62	49.08
16 L.H., Daily Totals		$1211.90		$1607.69	$75.74	$100.48

Crew B-12M	Hr.	Daily	Hr.	Daily	Bare Costs	Incl. O&P
1 Equip. Oper. (crane)	$35.80	$286.40	$58.50	$468.00	$31.13	$51.40
1 Laborer	26.45	211.60	44.30	354.40		
1 Crawler Crane, 20 Ton		841.90		926.09		
1 F.E. Attachment, .75 C.Y.		61.40		67.54	56.46	62.10
16 L.H., Daily Totals		$1401.30		$1816.03	$87.58	$113.50

Crew No.	Bare Costs		Incl. Subs O&P		Cost Per Labor-Hour	
Crew B-12N	**Hr.**	**Daily**	**Hr.**	**Daily**	**Bare Costs**	**Incl. O&P**
1 Equip. Oper. (crane)	$35.80	$286.40	$58.50	$468.00	$31.13	$51.40
1 Laborer	26.45	211.60	44.30	354.40		
1 Crawler Crane, 25 Ton		1123.00		1235.30		
1 F.E. Attachment, 1 C.Y.		67.20		73.92	74.39	81.83
16 L.H., Daily Totals		$1688.20		$2131.62	$105.51	$133.23
Crew B-12O	**Hr.**	**Daily**	**Hr.**	**Daily**	**Bare Costs**	**Incl. O&P**
1 Equip. Oper. (crane)	$35.80	$286.40	$58.50	$468.00	$31.13	$51.40
1 Laborer	26.45	211.60	44.30	354.40		
1 Crawler Crane, 40 Ton		1128.00		1240.80		
1 F.E. Attachment, 1.5 C.Y.		77.40		85.14	75.34	82.87
16 L.H., Daily Totals		$1703.40		$2148.34	$106.46	$134.27
Crew B-12P	**Hr.**	**Daily**	**Hr.**	**Daily**	**Bare Costs**	**Incl. O&P**
1 Equip. Oper. (crane)	$35.80	$286.40	$58.50	$468.00	$31.13	$51.40
1 Laborer	26.45	211.60	44.30	354.40		
1 Crawler Crane, 40 Ton		1128.00		1240.80		
1 Dragline Bucket, 1.5 C.Y.		34.20		37.62	72.64	79.90
16 L.H., Daily Totals		$1660.20		$2100.82	$103.76	$131.30
Crew B-12Q	**Hr.**	**Daily**	**Hr.**	**Daily**	**Bare Costs**	**Incl. O&P**
1 Equip. Oper. (crane)	$35.80	$286.40	$58.50	$468.00	$31.13	$51.40
1 Laborer	26.45	211.60	44.30	354.40		
1 Hyd. Excavator, 5/8 C.Y.		530.00		583.00	33.13	36.44
16 L.H., Daily Totals		$1028.00		$1405.40	$64.25	$87.84
Crew B-12S	**Hr.**	**Daily**	**Hr.**	**Daily**	**Bare Costs**	**Incl. O&P**
1 Equip. Oper. (crane)	$35.80	$286.40	$58.50	$468.00	$31.13	$51.40
1 Laborer	26.45	211.60	44.30	354.40		
1 Hyd. Excavator, 2.5 C.Y.		1613.00		1774.30	100.81	110.89
16 L.H., Daily Totals		$2111.00		$2596.70	$131.94	$162.29
Crew B-12T	**Hr.**	**Daily**	**Hr.**	**Daily**	**Bare Costs**	**Incl. O&P**
1 Equip. Oper. (crane)	$35.80	$286.40	$58.50	$468.00	$31.13	$51.40
1 Laborer	26.45	211.60	44.30	354.40		
1 Crawler Crane, 75 Ton		1446.00		1590.60		
1 F.E. Attachment, 3 C.Y.		100.80		110.88	96.67	106.34
16 L.H., Daily Totals		$2044.80		$2523.88	$127.80	$157.74
Crew B-12V	**Hr.**	**Daily**	**Hr.**	**Daily**	**Bare Costs**	**Incl. O&P**
1 Equip. Oper. (crane)	$35.80	$286.40	$58.50	$468.00	$31.13	$51.40
1 Laborer	26.45	211.60	44.30	354.40		
1 Crawler Crane, 75 Ton		1446.00		1590.60		
1 Dragline Bucket, 3 C.Y.		52.80		58.08	93.67	103.04
16 L.H., Daily Totals		$1996.80		$2471.08	$124.80	$154.44
Crew B-12Y	**Hr.**	**Daily**	**Hr.**	**Daily**	**Bare Costs**	**Incl. O&P**
1 Equip. Oper. (crane)	$35.80	$286.40	$58.50	$468.00	$29.57	$49.03
2 Laborers	26.45	423.20	44.30	708.80		
1 Hyd. Excavator, 3.5 C.Y.		2171.00		2388.10	90.46	99.50
24 L.H., Daily Totals		$2880.60		$3564.90	$120.03	$148.54
Crew B-12Z	**Hr.**	**Daily**	**Hr.**	**Daily**	**Bare Costs**	**Incl. O&P**
1 Equip. Oper. (crane)	$35.80	$286.40	$58.50	$468.00	$29.57	$49.03
2 Laborers	26.45	423.20	44.30	708.80		
1 Hyd. Excavator, 2.5 C.Y.		1613.00		1774.30	67.21	73.93
24 L.H., Daily Totals		$2322.60		$2951.10	$96.78	$122.96

Crew No.	Bare Costs		Incl. Subs O&P		Cost Per Labor-Hour	
Crew B-13	**Hr.**	**Daily**	**Hr.**	**Daily**	**Bare Costs**	**Incl. O&P**
1 Labor Foreman (outside)	$28.45	$227.60	$47.65	$381.20	$28.34	$47.23
4 Laborers	26.45	846.40	44.30	1417.60		
1 Equip. Oper. (crane)	35.80	286.40	58.50	468.00		
1 Hyd. Crane, 25 Ton		748.60		823.46	15.60	17.16
48 L.H., Daily Totals		$2109.00		$3090.26	$43.94	$64.38
Crew B-13A	**Hr.**	**Daily**	**Hr.**	**Daily**	**Bare Costs**	**Incl. O&P**
1 Foreman	$28.45	$227.60	$47.65	$381.20	$29.36	$48.65
2 Laborers	26.45	423.20	44.30	708.80		
2 Equipment Operators (med.)	34.90	558.40	57.05	912.80		
2 Truck Drivers (heavy)	27.20	435.20	45.10	721.60		
1 Crawler Crane, 75 Ton		1446.00		1590.60		
1 Crawler Loader, 4 C.Y.		1432.00		1575.20		
2 Dump Trucks, 8 C.Y., 220 H.P.		630.80		693.88	62.66	68.92
56 L.H., Daily Totals		$5153.20		$6584.08	$92.02	$117.57
Crew B-13B	**Hr.**	**Daily**	**Hr.**	**Daily**	**Bare Costs**	**Incl. O&P**
1 Labor Foreman (outside)	$28.45	$227.60	$47.65	$381.20	$28.73	$47.73
4 Laborers	26.45	846.40	44.30	1417.60		
1 Equip. Oper. (crane)	35.80	286.40	58.50	468.00		
1 Equip. Oper. Oiler	31.05	248.40	50.75	406.00		
1 Hyd. Crane, 55 Ton		1208.00		1328.80	21.57	23.73
56 L.H., Daily Totals		$2816.80		$4001.60	$50.30	$71.46
Crew B-13C	**Hr.**	**Daily**	**Hr.**	**Daily**	**Bare Costs**	**Incl. O&P**
1 Labor Foreman (outside)	$28.45	$227.60	$47.65	$381.20	$28.73	$47.73
4 Laborers	26.45	846.40	44.30	1417.60		
1 Equip. Oper. (crane)	35.80	286.40	58.50	468.00		
1 Equip. Oper. Oiler	31.05	248.40	50.75	406.00		
1 Crawler Crane, 100 Ton		1653.00		1818.30	29.52	32.47
56 L.H., Daily Totals		$3261.80		$4491.10	$58.25	$80.20
Crew B-13D	**Hr.**	**Daily**	**Hr.**	**Daily**	**Bare Costs**	**Incl. O&P**
1 Laborer	$26.45	$211.60	$44.30	$354.40	$31.13	$51.40
1 Equip. Oper. (crane)	35.80	286.40	58.50	468.00		
1 Hyd. Excavator, 1 C.Y.		729.80		802.78		
1 Trench Box		113.40		124.74	52.70	57.97
16 L.H., Daily Totals		$1341.20		$1749.92	$83.83	$109.37
Crew B-13E	**Hr.**	**Daily**	**Hr.**	**Daily**	**Bare Costs**	**Incl. O&P**
1 Laborer	$26.45	$211.60	$44.30	$354.40	$31.13	$51.40
1 Equip. Oper. (crane)	35.80	286.40	58.50	468.00		
1 Hyd. Excavator, 1.5 C.Y.		947.40		1042.14		
1 Trench Box		113.40		124.74	66.30	72.93
16 L.H., Daily Totals		$1558.80		$1989.28	$97.42	$124.33
Crew B-13F	**Hr.**	**Daily**	**Hr.**	**Daily**	**Bare Costs**	**Incl. O&P**
1 Laborer	$26.45	$211.60	$44.30	$354.40	$31.13	$51.40
1 Equip. Oper. (crane)	35.80	286.40	58.50	468.00		
1 Hyd. Excavator, 3.5 C.Y.		2171.00		2388.10		
1 Trench Box		113.40		124.74	142.78	157.05
16 L.H., Daily Totals		$2782.40		$3335.24	$173.90	$208.45
Crew B-13G	**Hr.**	**Daily**	**Hr.**	**Daily**	**Bare Costs**	**Incl. O&P**
1 Laborer	$26.45	$211.60	$44.30	$354.40	$31.13	$51.40
1 Equip. Oper. (crane)	35.80	286.40	58.50	468.00		
1 Hyd. Excavator, .75 C.Y.		640.40		704.44		
1 Trench Box		113.40		124.74	47.11	51.82
16 L.H., Daily Totals		$1251.80		$1651.58	$78.24	$103.22

Crew No.	Bare Costs		Incl. Subs O&P		Cost Per Labor-Hour	
Crew B-13H	**Hr.**	**Daily**	**Hr.**	**Daily**	**Bare Costs**	**Incl. O&P**
1 Laborer	$26.45	$211.60	$44.30	$354.40	$31.13	$51.40
1 Equip. Oper. (crane)	35.80	286.40	58.50	468.00		
1 Gradall, 5/8 C.Y.		857.40		943.14		
1 Trench Box		113.40		124.74	60.67	66.74
16 L.H., Daily Totals		$1468.80		$1890.28	$91.80	$118.14
Crew B-13I	**Hr.**	**Daily**	**Hr.**	**Daily**	**Bare Costs**	**Incl. O&P**
1 Laborer	$26.45	$211.60	$44.30	$354.40	$31.13	$51.40
1 Equip. Oper. (crane)	35.80	286.40	58.50	468.00		
1 Gradall, 3 Ton, 1 C.Y.		1001.00		1101.10		
1 Trench Box		113.40		124.74	69.65	76.61
16 L.H., Daily Totals		$1612.40		$2048.24	$100.78	$128.01
Crew B-13J	**Hr.**	**Daily**	**Hr.**	**Daily**	**Bare Costs**	**Incl. O&P**
1 Laborer	$26.45	$211.60	$44.30	$354.40	$31.13	$51.40
1 Equip. Oper. (crane)	35.80	286.40	58.50	468.00		
1 Hyd. Excavator, 2.5 C.Y.		1613.00		1774.30		
1 Trench Box		113.40		124.74	107.90	118.69
16 L.H., Daily Totals		$2224.40		$2721.44	$139.03	$170.09
Crew B-14	**Hr.**	**Daily**	**Hr.**	**Daily**	**Bare Costs**	**Incl. O&P**
1 Labor Foreman (outside)	$28.45	$227.60	$47.65	$381.20	$27.97	$46.61
4 Laborers	26.45	846.40	44.30	1417.60		
1 Equip. Oper. (light)	33.55	268.40	54.80	438.40		
1 Backhoe Loader, 48 H.P.		320.40		352.44	6.67	7.34
48 L.H., Daily Totals		$1662.80		$2589.64	$34.64	$53.95
Crew B-14A	**Hr.**	**Daily**	**Hr.**	**Daily**	**Bare Costs**	**Incl. O&P**
1 Equip. Oper. (crane)	$35.80	$286.40	$58.50	$468.00	$32.68	$53.77
.5 Laborer	26.45	105.80	44.30	177.20		
1 Hyd. Excavator, 4.5 C.Y.		2669.00		2935.90	222.42	244.66
12 L.H., Daily Totals		$3061.20		$3581.10	$255.10	$298.43
Crew B-14B	**Hr.**	**Daily**	**Hr.**	**Daily**	**Bare Costs**	**Incl. O&P**
1 Equip. Oper. (crane)	$35.80	$286.40	$58.50	$468.00	$32.68	$53.77
.5 Laborer	26.45	105.80	44.30	177.20		
1 Hyd. Excavator, 6 C.Y.		3101.00		3411.10	258.42	284.26
12 L.H., Daily Totals		$3493.20		$4056.30	$291.10	$338.02
Crew B-14C	**Hr.**	**Daily**	**Hr.**	**Daily**	**Bare Costs**	**Incl. O&P**
1 Equip. Oper. (crane)	$35.80	$286.40	$58.50	$468.00	$32.68	$53.77
.5 Laborer	26.45	105.80	44.30	177.20		
1 Hyd. Excavator, 7 C.Y.		3179.00		3496.90	264.92	291.41
12 L.H., Daily Totals		$3571.20		$4142.10	$297.60	$345.18
Crew B-14F	**Hr.**	**Daily**	**Hr.**	**Daily**	**Bare Costs**	**Incl. O&P**
1 Equip. Oper. (crane)	$35.80	$286.40	$58.50	$468.00	$32.68	$53.77
.5 Laborer	26.45	105.80	44.30	177.20		
1 Hyd. Shovel, 7 C.Y.		3109.00		3419.90	259.08	284.99
12 L.H., Daily Totals		$3501.20		$4065.10	$291.77	$338.76
Crew B-14G	**Hr.**	**Daily**	**Hr.**	**Daily**	**Bare Costs**	**Incl. O&P**
1 Equip. Oper. (crane)	$35.80	$286.40	$58.50	$468.00	$32.68	$53.77
.5 Laborer	26.45	105.80	44.30	177.20		
1 Hyd. Shovel, 12 C.Y.		4046.00		4450.60	337.17	370.88
12 L.H., Daily Totals		$4438.20		$5095.80	$369.85	$424.65

Crew No.	Bare Costs		Incl. Subs O&P		Cost Per Labor-Hour	
Crew B-14J	**Hr.**	**Daily**	**Hr.**	**Daily**	**Bare Costs**	**Incl. O&P**
1 Equip. Oper. (med.)	$34.90	$279.20	$57.05	$456.40	$32.08	$52.80
.5 Laborer	26.45	105.80	44.30	177.20		
1 F.E. Loader, 8 C.Y.		1811.00		1992.10	150.92	166.01
12 L.H., Daily Totals		$2196.00		$2625.70	$183.00	$218.81
Crew B-14K	**Hr.**	**Daily**	**Hr.**	**Daily**	**Bare Costs**	**Incl. O&P**
1 Equip. Oper. (med.)	$34.90	$279.20	$57.05	$456.40	$32.08	$52.80
.5 Laborer	26.45	105.80	44.30	177.20		
1 F.E. Loader, 10 C.Y.		2476.00		2723.60	206.33	226.97
12 L.H., Daily Totals		$2861.00		$3357.20	$238.42	$279.77
Crew B-15	**Hr.**	**Daily**	**Hr.**	**Daily**	**Bare Costs**	**Incl. O&P**
1 Equipment Oper. (med.)	$34.90	$279.20	$57.05	$456.40	$29.29	$48.40
.5 Laborer	26.45	105.80	44.30	177.20		
2 Truck Drivers (heavy)	27.20	435.20	45.10	721.60		
2 Dump Trucks, 12 C.Y., 400 H.P.		1012.00		1113.20		
1 Dozer, 200 H.P.		1051.00		1156.10	73.68	81.05
28 L.H., Daily Totals		$2883.20		$3624.50	$102.97	$129.45
Crew B-16	**Hr.**	**Daily**	**Hr.**	**Daily**	**Bare Costs**	**Incl. O&P**
1 Labor Foreman (outside)	$28.45	$227.60	$47.65	$381.20	$27.14	$45.34
2 Laborers	26.45	423.20	44.30	708.80		
1 Truck Driver (heavy)	27.20	217.60	45.10	360.80		
1 Dump Truck, 12 C.Y., 400 H.P.		506.00		556.60	15.81	17.39
32 L.H., Daily Totals		$1374.40		$2007.40	$42.95	$62.73
Crew B-17	**Hr.**	**Daily**	**Hr.**	**Daily**	**Bare Costs**	**Incl. O&P**
2 Laborers	$26.45	$423.20	$44.30	$708.80	$28.41	$47.13
1 Equip. Oper. (light)	33.55	268.40	54.80	438.40		
1 Truck Driver (heavy)	27.20	217.60	45.10	360.80		
1 Backhoe Loader, 48 H.P.		320.40		352.44		
1 Dump Truck, 8 C.Y., 220 H.P.		315.40		346.94	19.87	21.86
32 L.H., Daily Totals		$1545.00		$2207.38	$48.28	$68.98
Crew B-17A	**Hr.**	**Daily**	**Hr.**	**Daily**	**Bare Costs**	**Incl. O&P**
2 Laborer Foremen	$28.45	$455.20	$47.65	$762.40	$28.57	$47.84
6 Laborers	26.45	1269.60	44.30	2126.40		
1 Skilled Worker Foreman	36.05	288.40	60.35	482.80		
1 Skilled Worker	34.05	272.40	57.00	456.00		
80 L.H., Daily Totals		$2285.60		$3827.60	$28.57	$47.84
Crew B-18	**Hr.**	**Daily**	**Hr.**	**Daily**	**Bare Costs**	**Incl. O&P**
1 Labor Foreman (outside)	$28.45	$227.60	$47.65	$381.20	$27.12	$45.42
2 Laborers	26.45	423.20	44.30	708.80		
1 Vibratory Plate, gas, 21"		39.00		42.90	1.63	1.79
24 L.H., Daily Totals		$689.80		$1132.90	$28.74	$47.20
Crew B-19	**Hr.**	**Daily**	**Hr.**	**Daily**	**Bare Costs**	**Incl. O&P**
1 Pile Driver Foreman	$34.75	$278.00	$60.55	$484.40	$32.57	$55.96
4 Pile Drivers	32.75	1048.00	57.10	1827.20		
1 Equip. Oper. (crane)	35.80	286.40	58.50	468.00		
1 Building Laborer	26.45	211.60	44.30	354.40		
1 Crawler Crane, 40 Ton		1128.00		1240.80		
1 Lead, 90' high		127.20		139.92		
1 Hammer, Diesel, 22k ft-lb		592.40		651.64	32.99	36.29
56 L.H., Daily Totals		$3671.60		$5166.36	$65.56	$92.26

Crew B-19A	Bare Costs Hr.	Bare Costs Daily	Incl. Subs O&P Hr.	Incl. Subs O&P Daily	Cost Per Labor-Hour Bare Costs	Cost Per Labor-Hour Incl. O&P
1 Pile Driver Foreman	$34.75	$278.00	$60.55	$484.40	$33.55	$57.09
4 Pile Drivers	32.75	1048.00	57.10	1827.20		
2 Equip. Oper. (crane)	35.80	572.80	58.50	936.00		
1 Equip. Oper. Oiler	31.05	248.40	50.75	406.00		
1 Crawler Crane, 75 Ton		1446.00		1590.60		
1 Lead, 90' high		127.20		139.92		
1 Hammer, Diesel, 41k ft-lb		677.00		744.70	35.16	38.68
64 L.H., Daily Totals		$4397.40		$6128.82	$68.71	$95.76

Crew B-20	Hr.	Daily	Hr.	Daily	Bare Costs	Incl. O&P
1 Labor Foreman (out)	$28.45	$227.60	$47.65	$381.20	$27.12	$45.42
2 Laborers	26.45	423.20	44.30	708.80		
24 L.H., Daily Totals		$650.80		$1090.00	$27.12	$45.42

Crew B-20A	Hr.	Daily	Hr.	Daily	Bare Costs	Incl. O&P
1 Labor Foreman	$28.45	$227.60	$47.65	$381.20	$31.56	$52.11
1 Laborer	26.45	211.60	44.30	354.40		
1 Plumber	39.65	317.20	64.75	518.00		
1 Plumber Apprentice	31.70	253.60	51.75	414.00		
32 L.H., Daily Totals		$1010.00		$1667.60	$31.56	$52.11

Crew B-21	Hr.	Daily	Hr.	Daily	Bare Costs	Incl. O&P
1 Labor Foreman (out)	$28.45	$227.60	$47.65	$381.20	$28.36	$47.29
2 Laborers	26.45	423.20	44.30	708.80		
.5 Equip. Oper. (crane)	35.80	143.20	58.50	234.00		
.5 S.P. Crane, 4x4, 5 Ton		128.90		141.79	4.60	5.06
28 L.H., Daily Totals		$922.90		$1465.79	$32.96	$52.35

Crew B-21A	Hr.	Daily	Hr.	Daily	Bare Costs	Incl. O&P
1 Labor Foreman	$28.45	$227.60	$47.65	$381.20	$32.41	$53.39
1 Laborer	26.45	211.60	44.30	354.40		
1 Plumber	39.65	317.20	64.75	518.00		
1 Plumber Apprentice	31.70	253.60	51.75	414.00		
1 Equip. Oper. (crane)	35.80	286.40	58.50	468.00		
1 S.P. Crane, 4x4, 12 Ton		434.40		477.84	10.86	11.95
40 L.H., Daily Totals		$1730.80		$2613.44	$43.27	$65.34

Crew B-21B	Hr.	Daily	Hr.	Daily	Bare Costs	Incl. O&P
1 Laborer Foreman	$28.45	$227.60	$47.65	$381.20	$28.72	$47.81
3 Laborers	26.45	634.80	44.30	1063.20		
1 Equip. Oper. (crane)	35.80	286.40	58.50	468.00		
1 Hyd. Crane, 12 Ton		655.60		721.16	16.39	18.03
40 L.H., Daily Totals		$1804.40		$2633.56	$45.11	$65.84

Crew B-21C	Hr.	Daily	Hr.	Daily	Bare Costs	Incl. O&P
1 Laborer Foreman	$28.45	$227.60	$47.65	$381.20	$28.73	$47.73
4 Laborers	26.45	846.40	44.30	1417.60		
1 Equip. Oper. (crane)	35.80	286.40	58.50	468.00		
1 Equip. Oper. Oiler	31.05	248.40	50.75	406.00		
2 Cutting Torches		36.00		39.60		
2 Set of Gases		165.60		182.16		
1 Lattice Boom Crane, 90 Ton		1622.00		1784.20	32.56	35.82
56 L.H., Daily Totals		$3432.40		$4678.76	$61.29	$83.55

Crew B-22	Hr.	Daily	Hr.	Daily	Bare Costs	Incl. O&P
1 Labor Foreman (out)	$28.45	$227.60	$47.65	$381.20	$28.85	$48.03
2 Laborers	26.45	423.20	44.30	708.80		
.75 Equip. Oper. (crane)	35.80	214.80	58.50	351.00		
.75 S.P. Crane, 4x4, 5 Ton		193.35		212.69	6.45	7.09
30 L.H., Daily Totals		$1058.95		$1653.68	$35.30	$55.12

Crew B-22A	Hr.	Daily	Hr.	Daily	Bare Costs	Incl. O&P
1 Labor Foreman (out)	$28.45	$227.60	$47.65	$381.20	$30.24	$50.35
1 Skilled Worker	34.05	272.40	57.00	456.00		
2 Laborers	26.45	423.20	44.30	708.80		
1 Equipment Oper. (crane)	35.80	286.40	58.50	468.00		
1 S.P. Crane, 4x4, 5 Ton		257.80		283.58		
1 Butt Fusion Mach., 4"-12" diam.		376.45		414.10	15.86	17.44
40 L.H., Daily Totals		$1843.85		$2711.68	$46.10	$67.79

Crew B-22B	Hr.	Daily	Hr.	Daily	Bare Costs	Incl. O&P
1 Labor Foreman (out)	$28.45	$227.60	$47.65	$381.20	$30.24	$50.35
1 Skilled Worker	34.05	272.40	57.00	456.00		
2 Laborers	26.45	423.20	44.30	708.80		
1 Equip. Oper. (crane)	35.80	286.40	58.50	468.00		
1 S.P. Crane, 4x4, 5 Ton		257.80		283.58		
1 Butt Fusion Mach., 8"-24" diam.		806.15		886.76	26.60	29.26
40 L.H., Daily Totals		$2273.55		$3184.34	$56.84	$79.61

Crew B-22C	Hr.	Daily	Hr.	Daily	Bare Costs	Incl. O&P
1 Skilled Worker	$34.05	$272.40	$57.00	$456.00	$30.25	$50.65
1 Laborer	26.45	211.60	44.30	354.40		
1 Butt Fusion Mach., 2"-8" diam.		120.50		132.55	7.53	8.28
16 L.H., Daily Totals		$604.50		$942.95	$37.78	$58.93

Crew B-23	Hr.	Daily	Hr.	Daily	Bare Costs	Incl. O&P
1 Labor Foreman (outside)	$28.45	$227.60	$47.65	$381.20	$26.85	$44.97
4 Laborers	26.45	846.40	44.30	1417.60		
1 Drill Rig, Truck-Mounted		2486.00		2734.60		
1 Flatbed Truck, Gas, 3 Ton		217.00		238.70	67.58	74.33
40 L.H., Daily Totals		$3777.00		$4772.10	$94.42	$119.30

Crew B-23A	Hr.	Daily	Hr.	Daily	Bare Costs	Incl. O&P
1 Labor Foreman (outside)	$28.45	$227.60	$47.65	$381.20	$29.93	$49.67
1 Laborer	26.45	211.60	44.30	354.40		
1 Equip. Operator (med.)	34.90	279.20	57.05	456.40		
1 Drill Rig, Truck-Mounted		2486.00		2734.60		
1 Pickup Truck, 3/4 Ton		104.60		115.06	107.94	118.74
24 L.H., Daily Totals		$3309.00		$4041.66	$137.88	$168.40

Crew B-23B	Hr.	Daily	Hr.	Daily	Bare Costs	Incl. O&P
1 Labor Foreman (outside)	$28.45	$227.60	$47.65	$381.20	$29.93	$49.67
1 Laborer	26.45	211.60	44.30	354.40		
1 Equip. Operator (med.)	34.90	279.20	57.05	456.40		
1 Drill Rig, Truck-Mounted		2486.00		2734.60		
1 Pickup Truck, 3/4 Ton		104.60		115.06		
1 Centr. Water Pump, 6"		291.60		320.76	120.09	132.10
24 L.H., Daily Totals		$3600.60		$4362.42	$150.03	$181.77

Crew B-24	Hr.	Daily	Hr.	Daily	Bare Costs	Incl. O&P
1 Cement Finisher	$31.85	$254.80	$50.85	$406.80	$30.62	$50.43
1 Laborer	26.45	211.60	44.30	354.40		
1 Carpenter	33.55	268.40	56.15	449.20		
24 L.H., Daily Totals		$734.80		$1210.40	$30.62	$50.43

Crew B-25	Hr.	Daily	Hr.	Daily	Bare Costs	Incl. O&P
1 Labor Foreman	$28.45	$227.60	$47.65	$381.20	$28.94	$48.08
7 Laborers	26.45	1481.20	44.30	2480.80		
3 Equip. Oper. (med.)	34.90	837.60	57.05	1369.20		
1 Asphalt Paver, 130 H.P.		1949.00		2143.90		
1 Tandem Roller, 10 Ton		224.20		246.62		
1 Roller, Pneum. Whl, 12 Ton		309.20		340.12	28.21	31.03
88 L.H., Daily Totals		$5028.80		$6961.84	$57.15	$79.11

Crew No.	Bare Costs		Incl. Subs O&P		Cost Per Labor-Hour	
Crew B-25B	**Hr.**	**Daily**	**Hr.**	**Daily**	**Bare Costs**	**Incl. O&P**
1 Labor Foreman	$28.45	$227.60	$47.65	$381.20	$29.43	$48.83
7 Laborers	26.45	1481.20	44.30	2480.80		
4 Equip. Oper. (med.)	34.90	1116.80	57.05	1825.60		
1 Asphalt Paver, 130 H.P.		1949.00		2143.90		
2 Tandem Rollers, 10 Ton		448.40		493.24		
1 Roller, Pneum. Whl, 12 Ton		309.20		340.12	28.19	31.01
96 L.H., Daily Totals		$5532.20		$7664.86	$57.63	$79.84

Crew B-25C	Hr.	Daily	Hr.	Daily	Bare Costs	Incl. O&P
1 Labor Foreman	$28.45	$227.60	$47.65	$381.20	$29.60	$49.11
3 Laborers	26.45	634.80	44.30	1063.20		
2 Equip. Oper. (med.)	34.90	558.40	57.05	912.80		
1 Asphalt Paver, 130 H.P.		1949.00		2143.90		
1 Tandem Roller, 10 Ton		224.20		246.62	45.27	49.80
48 L.H., Daily Totals		$3594.00		$4747.72	$74.88	$98.91

Crew B-26	Hr.	Daily	Hr.	Daily	Bare Costs	Incl. O&P
1 Labor Foreman (outside)	$28.45	$227.60	$47.65	$381.20	$29.55	$49.23
6 Laborers	26.45	1269.60	44.30	2126.40		
2 Equip. Oper. (med.)	34.90	558.40	57.05	912.80		
1 Rodman (reinf.)	36.30	290.40	63.10	504.80		
1 Cement Finisher	31.85	254.80	50.85	406.80		
1 Grader, 30,000 Lbs.		542.80		597.08		
1 Paving Mach. & Equip.		2483.00		2731.30	34.38	37.82
88 L.H., Daily Totals		$5626.60		$7660.38	$63.94	$87.05

Crew B-26A	Hr.	Daily	Hr.	Daily	Bare Costs	Incl. O&P
1 Labor Foreman (outside)	$28.45	$227.60	$47.65	$381.20	$29.55	$49.23
6 Laborers	26.45	1269.60	44.30	2126.40		
2 Equip. Oper. (med.)	34.90	558.40	57.05	912.80		
1 Rodman (reinf.)	36.30	290.40	63.10	504.80		
1 Cement Finisher	31.85	254.80	50.85	406.80		
1 Grader, 30,000 Lbs.		542.80		597.08		
1 Paving Mach. & Equip.		2483.00		2731.30		
1 Concrete Saw		134.20		147.62	35.91	39.50
88 L.H., Daily Totals		$5760.80		$7808.00	$65.46	$88.73

Crew B-26B	Hr.	Daily	Hr.	Daily	Bare Costs	Incl. O&P
1 Labor Foreman (outside)	$28.45	$227.60	$47.65	$381.20	$30.00	$49.88
6 Laborers	26.45	1269.60	44.30	2126.40		
3 Equip. Oper. (med.)	34.90	837.60	57.05	1369.20		
1 Rodman (reinf.)	36.30	290.40	63.10	504.80		
1 Cement Finisher	31.85	254.80	50.85	406.80		
1 Grader, 30,000 Lbs.		542.80		597.08		
1 Paving Mach. & Equip.		2483.00		2731.30		
1 Concrete Pump, 110' Boom		999.60		1099.56	41.93	46.12
96 L.H., Daily Totals		$6905.40		$9216.34	$71.93	$96.00

Crew B-27	Hr.	Daily	Hr.	Daily	Bare Costs	Incl. O&P
1 Labor Foreman (outside)	$28.45	$227.60	$47.65	$381.20	$26.95	$45.14
3 Laborers	26.45	634.80	44.30	1063.20		
1 Berm Machine		276.80		304.48	8.65	9.52
32 L.H., Daily Totals		$1139.20		$1748.88	$35.60	$54.65

Crew B-28	Hr.	Daily	Hr.	Daily	Bare Costs	Incl. O&P
2 Carpenters	$33.55	$536.80	$56.15	$898.40	$31.18	$52.20
1 Laborer	26.45	211.60	44.30	354.40		
24 L.H., Daily Totals		$748.40		$1252.80	$31.18	$52.20

Crew B-29	Hr.	Daily	Hr.	Daily	Bare Costs	Incl. O&P
1 Labor Foreman (outside)	$28.45	$227.60	$47.65	$381.20	$28.34	$47.23
4 Laborers	26.45	846.40	44.30	1417.60		
1 Equip. Oper. (crane)	35.80	286.40	58.50	468.00		
1 Gradall, 5/8 C.Y.		857.40		943.14	17.86	19.65
48 L.H., Daily Totals		$2217.80		$3209.94	$46.20	$66.87

Crew B-30	Hr.	Daily	Hr.	Daily	Bare Costs	Incl. O&P
1 Equip. Oper. (med.)	$34.90	$279.20	$57.05	$456.40	$29.77	$49.08
2 Truck Drivers (heavy)	27.20	435.20	45.10	721.60		
1 Hyd. Excavator, 1.5 C.Y.		947.40		1042.14		
2 Dump Trucks, 12 C.Y., 400 H.P.		1012.00		1113.20	81.64	89.81
24 L.H., Daily Totals		$2673.80		$3333.34	$111.41	$138.89

Crew B-31	Hr.	Daily	Hr.	Daily	Bare Costs	Incl. O&P
1 Labor Foreman (outside)	$28.45	$227.60	$47.65	$381.20	$26.85	$44.97
4 Laborers	26.45	846.40	44.30	1417.60		
1 Air Compressor, 250 cfm		162.80		179.08		
1 Sheeting Driver		5.75		6.33		
2 -50' Air Hoses, 1.5"		12.60		13.86	4.53	4.98
40 L.H., Daily Totals		$1255.15		$1998.07	$31.38	$49.95

Crew B-32	Hr.	Daily	Hr.	Daily	Bare Costs	Incl. O&P
1 Laborer	$26.45	$211.60	$44.30	$354.40	$32.79	$53.86
3 Equip. Oper. (med.)	34.90	837.60	57.05	1369.20		
1 Grader, 30,000 Lbs.		542.80		597.08		
1 Tandem Roller, 10 Ton		224.20		246.62		
1 Dozer, 200 H.P.		1051.00		1156.10	56.81	62.49
32 L.H., Daily Totals		$2867.20		$3723.40	$89.60	$116.36

Crew B-32A	Hr.	Daily	Hr.	Daily	Bare Costs	Incl. O&P
1 Laborer	$26.45	$211.60	$44.30	$354.40	$32.08	$52.80
2 Equip. Oper. (med.)	34.90	558.40	57.05	912.80		
1 Grader, 30,000 Lbs.		542.80		597.08		
1 Roller, Vibratory, 25 Ton		596.80		656.48	47.48	52.23
24 L.H., Daily Totals		$1909.60		$2520.76	$79.57	$105.03

Crew B-32B	Hr.	Daily	Hr.	Daily	Bare Costs	Incl. O&P
1 Laborer	$26.45	$211.60	$44.30	$354.40	$32.08	$52.80
2 Equip. Oper. (med.)	34.90	558.40	57.05	912.80		
1 Dozer, 200 H.P.		1051.00		1156.10		
1 Roller, Vibratory, 25 Ton		596.80		656.48	68.66	75.52
24 L.H., Daily Totals		$2417.80		$3079.78	$100.74	$128.32

Crew B-32C	Hr.	Daily	Hr.	Daily	Bare Costs	Incl. O&P
1 Labor Foreman	$28.45	$227.60	$47.65	$381.20	$31.01	$51.23
2 Laborers	26.45	423.20	44.30	708.80		
3 Equip. Oper. (med.)	34.90	837.60	57.05	1369.20		
1 Grader, 30,000 Lbs.		542.80		597.08		
1 Tandem Roller, 10 Ton		224.20		246.62		
1 Dozer, 200 H.P.		1051.00		1156.10	37.88	41.66
48 L.H., Daily Totals		$3306.40		$4459.00	$68.88	$92.90

Crew B-33A	Hr.	Daily	Hr.	Daily	Bare Costs	Incl. O&P
1 Equip. Oper. (med.)	$34.90	$279.20	$57.05	$456.40	$34.90	$57.05
.25 Equip. Oper. (med.)	34.90	69.80	57.05	114.10		
1 Scraper, Towed, 7 C.Y.		108.00		118.80		
1.250 Dozers, 300 H.P.		2008.75		2209.63	211.68	232.84
10 L.H., Daily Totals		$2465.75		$2898.93	$246.57	$289.89

Crew No.	Bare Costs		Incl. Subs O&P		Cost Per Labor-Hour	
Crew B-33B	**Hr.**	**Daily**	**Hr.**	**Daily**	**Bare Costs**	**Incl. O&P**
1 Equip. Oper. (med.)	$34.90	$279.20	$57.05	$456.40	$34.90	$57.05
.25 Equip. Oper. (med.)	34.90	69.80	57.05	114.10		
1 Scraper, Towed, 10 C.Y.		140.00		154.00		
1.250 Dozers, 300 H.P.		2008.75		2209.63	214.88	236.36
10 L.H., Daily Totals		$2497.75		$2934.13	$249.78	$293.41

Crew B-33C	Hr.	Daily	Hr.	Daily	Bare Costs	Incl. O&P
1 Equip. Oper. (med.)	$34.90	$279.20	$57.05	$456.40	$34.90	$57.05
.25 Equip. Oper. (med.)	34.90	69.80	57.05	114.10		
1 Scraper, Towed, 15 C.Y.		158.60		174.46		
1.250 Dozers, 300 H.P.		2008.75		2209.63	216.74	238.41
10 L.H., Daily Totals		$2516.35		$2954.59	$251.63	$295.46

Crew B-33D	Hr.	Daily	Hr.	Daily	Bare Costs	Incl. O&P
1 Equip. Oper. (med.)	$34.90	$279.20	$57.05	$456.40	$34.90	$57.05
.25 Equip. Oper. (med.)	34.90	69.80	57.05	114.10		
1 S.P. Scraper, 14 C.Y.		1512.00		1663.20		
.25 Dozer, 300 H.P.		401.75		441.93	191.38	210.51
10 L.H., Daily Totals		$2262.75		$2675.63	$226.28	$267.56

Crew B-33E	Hr.	Daily	Hr.	Daily	Bare Costs	Incl. O&P
1 Equip. Oper. (med.)	$34.90	$279.20	$57.05	$456.40	$34.90	$57.05
.25 Equip. Oper. (med.)	34.90	69.80	57.05	114.10		
1 S.P. Scraper, 21 C.Y.		1872.00		2059.20		
.25 Dozer, 300 H.P.		401.75		441.93	227.38	250.11
10 L.H., Daily Totals		$2622.75		$3071.63	$262.27	$307.16

Crew B-33F	Hr.	Daily	Hr.	Daily	Bare Costs	Incl. O&P
1 Equip. Oper. (med.)	$34.90	$279.20	$57.05	$456.40	$34.90	$57.05
.25 Equip. Oper. (med.)	34.90	69.80	57.05	114.10		
1 Elev. Scraper, 11 C.Y.		1089.00		1197.90		
.25 Dozer, 300 H.P.		401.75		441.93	149.07	163.98
10 L.H., Daily Totals		$1839.75		$2210.32	$183.97	$221.03

Crew B-33G	Hr.	Daily	Hr.	Daily	Bare Costs	Incl. O&P
1 Equip. Oper. (med.)	$34.90	$279.20	$57.05	$456.40	$34.90	$57.05
.25 Equip. Oper. (med.)	34.90	69.80	57.05	114.10		
1 Elev. Scraper, 22 C.Y.		2087.00		2295.70		
.25 Dozer, 300 H.P.		401.75		441.93	248.88	273.76
10 L.H., Daily Totals		$2837.75		$3308.13	$283.77	$330.81

Crew B-33K	Hr.	Daily	Hr.	Daily	Bare Costs	Incl. O&P
1 Equipment Operator (med.)	$34.90	$279.20	$57.05	$456.40	$32.49	$53.41
.25 Equipment Operator (med.)	34.90	69.80	57.05	114.10		
.5 Laborer	26.45	105.80	44.30	177.20		
1 S.P. Scraper, 31 C.Y.		2777.00		3054.70		
.25 Dozer, 410 H.P.		526.25		578.88	235.95	259.54
14 L.H., Daily Totals		$3758.05		$4381.27	$268.43	$312.95

Crew B-34A	Hr.	Daily	Hr.	Daily	Bare Costs	Incl. O&P
1 Truck Driver (heavy)	$27.20	$217.60	$45.10	$360.80	$27.20	$45.10
1 Dump Truck, 8 C.Y., 220 H.P.		315.40		346.94	39.42	43.37
8 L.H., Daily Totals		$533.00		$707.74	$66.63	$88.47

Crew B-34B	Hr.	Daily	Hr.	Daily	Bare Costs	Incl. O&P
1 Truck Driver (heavy)	$27.20	$217.60	$45.10	$360.80	$27.20	$45.10
1 Dump Truck, 12 C.Y., 400 H.P.		506.00		556.60	63.25	69.58
8 L.H., Daily Totals		$723.60		$917.40	$90.45	$114.68

Crew B-34C	Hr.	Daily	Hr.	Daily	Bare Costs	Incl. O&P
1 Truck Driver (heavy)	$27.20	$217.60	$45.10	$360.80	$27.20	$45.10
1 Truck Tractor, 6x4, 380 H.P.		459.40		505.34		
1 Dump Trailer, 16.5 C.Y.		125.80		138.38	73.15	80.47
8 L.H., Daily Totals		$802.80		$1004.52	$100.35	$125.57

Crew B-34D	Hr.	Daily	Hr.	Daily	Bare Costs	Incl. O&P
1 Truck Driver (heavy)	$27.20	$217.60	$45.10	$360.80	$27.20	$45.10
1 Truck Tractor, 6x4, 380 H.P.		459.40		505.34		
1 Dump Trailer, 20 C.Y.		140.00		154.00	74.92	82.42
8 L.H., Daily Totals		$817.00		$1020.14	$102.13	$127.52

Crew B-34E	Hr.	Daily	Hr.	Daily	Bare Costs	Incl. O&P
1 Truck Driver (heavy)	$27.20	$217.60	$45.10	$360.80	$27.20	$45.10
1 Dump Truck, Off Hwy., 25 Ton		1264.00		1390.40	158.00	173.80
8 L.H., Daily Totals		$1481.60		$1751.20	$185.20	$218.90

Crew B-34F	Hr.	Daily	Hr.	Daily	Bare Costs	Incl. O&P
1 Truck Driver (heavy)	$27.20	$217.60	$45.10	$360.80	$27.20	$45.10
1 Dump Truck, Off Hwy., 35 Ton		1135.00		1248.50	141.88	156.06
8 L.H., Daily Totals		$1352.60		$1609.30	$169.07	$201.16

Crew B-34G	Hr.	Daily	Hr.	Daily	Bare Costs	Incl. O&P
1 Truck Driver (heavy)	$27.20	$217.60	$45.10	$360.80	$27.20	$45.10
1 Dump Truck, Off Hwy., 50 Ton		1549.00		1703.90	193.63	212.99
8 L.H., Daily Totals		$1766.60		$2064.70	$220.82	$258.09

Crew B-34H	Hr.	Daily	Hr.	Daily	Bare Costs	Incl. O&P
1 Truck Driver (heavy)	$27.20	$217.60	$45.10	$360.80	$27.20	$45.10
1 Dump Truck, Off Hwy., 65 Ton		1551.00		1706.10	193.88	213.26
8 L.H., Daily Totals		$1768.60		$2066.90	$221.07	$258.36

Crew B-34I	Hr.	Daily	Hr.	Daily	Bare Costs	Incl. O&P
1 Truck Driver (heavy)	$27.20	$217.60	$45.10	$360.80	$27.20	$45.10
1 Dump Truck, 18 C.Y., 450 H.P.		660.80		726.88	82.60	90.86
8 L.H., Daily Totals		$878.40		$1087.68	$109.80	$135.96

Crew B-34J	Hr.	Daily	Hr.	Daily	Bare Costs	Incl. O&P
1 Truck Driver (heavy)	$27.20	$217.60	$45.10	$360.80	$27.20	$45.10
1 Dump Truck, Off Hwy., 100 Ton		2009.00		2209.90	251.13	276.24
8 L.H., Daily Totals		$2226.60		$2570.70	$278.32	$321.34

Crew B-34K	Hr.	Daily	Hr.	Daily	Bare Costs	Incl. O&P
1 Truck Driver (heavy)	$27.20	$217.60	$45.10	$360.80	$27.20	$45.10
1 Truck Tractor, 6x4, 450 H.P.		558.80		614.68		
1 Lowbed Trailer, 75 Ton		219.20		241.12	97.25	106.97
8 L.H., Daily Totals		$995.60		$1216.60	$124.45	$152.07

Crew B-34L	Hr.	Daily	Hr.	Daily	Bare Costs	Incl. O&P
1 Equip. Oper. (light)	$33.55	$268.40	$54.80	$438.40	$33.55	$54.80
1 Flatbed Truck, Gas, 1.5 Ton		171.40		188.54	21.43	23.57
8 L.H., Daily Totals		$439.80		$626.94	$54.98	$78.37

Crew B-34N	Hr.	Daily	Hr.	Daily	Bare Costs	Incl. O&P
1 Truck Driver (heavy)	$27.20	$217.60	$45.10	$360.80	$27.20	$45.10
1 Dump Truck, 8 C.Y., 220 H.P.		315.40		346.94		
1 Flatbed Trailer, 40 Ton		150.20		165.22	58.20	64.02
8 L.H., Daily Totals		$683.20		$872.96	$85.40	$109.12

Crew No.	Bare Costs		Incl. Subs O&P		Cost Per Labor-Hour	

Crew B-34P	Hr.	Daily	Hr.	Daily	Bare Costs	Incl. O&P
1 Pipe Fitter	$40.05	$320.40	$65.40	$523.20	$33.78	$55.42
1 Truck Driver (light)	26.40	211.20	43.80	350.40		
1 Equip. Oper. (med.)	34.90	279.20	57.05	456.40		
1 Flatbed Truck, Gas, 3 Ton		217.00		238.70		
1 Backhoe Loader, 48 H.P.		320.40		352.44	22.39	24.63
24 L.H., Daily Totals		$1348.20		$1921.14	$56.17	$80.05

Crew B-34Q	Hr.	Daily	Hr.	Daily	Bare Costs	Incl. O&P
1 Pipe Fitter	$40.05	$320.40	$65.40	$523.20	$34.08	$55.90
1 Truck Driver (light)	26.40	211.20	43.80	350.40		
1 Eqip. Oper. (crane)	35.80	286.40	58.50	468.00		
1 Flatbed Trailer, 25 Ton		111.20		122.32		
1 Dump Truck, 8 C.Y., 220 H.P.		315.40		346.94		
1 Hyd. Crane, 25 Ton		748.60		823.46	48.97	53.86
24 L.H., Daily Totals		$1993.20		$2634.32	$83.05	$109.76

Crew B-34R	Hr.	Daily	Hr.	Daily	Bare Costs	Incl. O&P
1 Pipe Fitter	$40.05	$320.40	$65.40	$523.20	$34.08	$55.90
1 Truck Driver (light)	26.40	211.20	43.80	350.40		
1 Eqip. Oper. (crane)	35.80	286.40	58.50	468.00		
1 Flatbed Trailer, 25 Ton		111.20		122.32		
1 Dump Truck, 8 C.Y., 220 H.P.		315.40		346.94		
1 Hyd. Crane, 25 Ton		748.60		823.46		
1 Hyd. Excavator, 1 C.Y.		729.80		802.78	79.38	87.31
24 L.H., Daily Totals		$2723.00		$3437.10	$113.46	$143.21

Crew B-34S	Hr.	Daily	Hr.	Daily	Bare Costs	Incl. O&P
2 Pipe Fitters	$40.05	$640.80	$65.40	$1046.40	$35.77	$58.60
1 Truck Driver (heavy)	27.20	217.60	45.10	360.80		
1 Eqip. Oper. (crane)	35.80	286.40	58.50	468.00		
1 Flatbed Trailer, 40 Ton		150.20		165.22		
1 Truck Tractor, 6x4, 380 H.P.		459.40		505.34		
1 Hyd. Crane, 80 Ton		1693.00		1862.30		
1 Hyd. Excavator, 2 C.Y.		1224.00		1346.40	110.21	121.23
32 L.H., Daily Totals		$4671.40		$5754.46	$145.98	$179.83

Crew B-34T	Hr.	Daily	Hr.	Daily	Bare Costs	Incl. O&P
2 Pipe Fitters	$40.05	$640.80	$65.40	$1046.40	$35.77	$58.60
1 Truck Driver (heavy)	27.20	217.60	45.10	360.80		
1 Eqip. Oper. (crane)	35.80	286.40	58.50	468.00		
1 Flatbed Trailer, 40 Ton		150.20		165.22		
1 Truck Tractor, 6x4, 380 H.P.		459.40		505.34		
1 Hyd. Crane, 80 Ton		1693.00		1862.30	71.96	79.15
32 L.H., Daily Totals		$3447.40		$4408.06	$107.73	$137.75

Crew B-35	Hr.	Daily	Hr.	Daily	Bare Costs	Incl. O&P
1 Laborer Foreman (out)	$28.45	$227.60	$47.65	$381.20	$32.88	$54.44
1 Skilled Worker	34.05	272.40	57.00	456.00		
1 Welder (plumber)	39.65	317.20	64.75	518.00		
1 Laborer	26.45	211.60	44.30	354.40		
1 Equip. Oper. (crane)	35.80	286.40	58.50	468.00		
1 Welder, electric, 300 amp		32.45		35.70		
1 Hyd. Excavator, .75 C.Y.		640.40		704.44	16.82	18.50
40 L.H., Daily Totals		$1988.05		$2917.74	$49.70	$72.94

Crew B-35A	Hr.	Daily	Hr.	Daily	Bare Costs	Incl. O&P
1 Laborer Foreman (out)	$28.45	$227.60	$47.65	$381.20	$31.70	$52.46
2 Laborers	26.45	423.20	44.30	708.80		
1 Skilled Worker	34.05	272.40	57.00	456.00		
1 Welder (plumber)	39.65	317.20	64.75	518.00		
1 Equip. Oper. (crane)	35.80	286.40	58.50	468.00		
1 Equip. Oper. Oiler	31.05	248.40	50.75	406.00		
1 Welder, gas engine, 300 amp		109.00		119.90		
1 Crawler Crane, 75 Ton		1446.00		1590.60	27.77	30.54
56 L.H., Daily Totals		$3330.20		$4648.50	$59.47	$83.01

Crew B-36	Hr.	Daily	Hr.	Daily	Bare Costs	Incl. O&P
1 Labor Foreman (outside)	$28.45	$227.60	$47.65	$381.20	$30.23	$50.07
2 Laborers	26.45	423.20	44.30	708.80		
2 Equip. Oper. (med.)	34.90	558.40	57.05	912.80		
1 Dozer, 200 H.P.		1051.00		1156.10		
1 Aggregate Spreader		33.60		36.96		
1 Tandem Roller, 10 Ton		224.20		246.62	32.72	35.99
40 L.H., Daily Totals		$2518.00		$3442.48	$62.95	$86.06

Crew B-36A	Hr.	Daily	Hr.	Daily	Bare Costs	Incl. O&P
1 Labor Foreman (outside)	$28.45	$227.60	$47.65	$381.20	$31.56	$52.06
2 Laborers	26.45	423.20	44.30	708.80		
4 Equip. Oper. (med.)	34.90	1116.80	57.05	1825.60		
1 Dozer, 200 H.P.		1051.00		1156.10		
1 Aggregate Spreader		33.60		36.96		
1 Tandem Roller, 10 Ton		224.20		246.62		
1 Roller, Pneum. Whl, 12 Ton		309.20		340.12	28.89	31.78
56 L.H., Daily Totals		$3385.60		$4695.40	$60.46	$83.85

Crew B-36B	Hr.	Daily	Hr.	Daily	Bare Costs	Incl. O&P
1 Labor Foreman (outside)	$28.45	$227.60	$47.65	$381.20	$31.02	$51.19
2 Laborers	26.45	423.20	44.30	708.80		
4 Equip. Oper. (med.)	34.90	1116.80	57.05	1825.60		
1 Truck Driver, Heavy	27.20	217.60	45.10	360.80		
1 Grader, 30,000 Lbs.		542.80		597.08		
1 F.E. Loader, crl, 1.5 C.Y.		565.00		621.50		
1 Dozer, 300 H.P.		1607.00		1767.70		
1 Roller, Vibratory, 25 Ton		596.80		656.48		
1 Truck Tractor, 6x4, 450 H.P.		558.80		614.68		
1 Water Tank Trailer, 5000 Gal.		146.20		160.82	62.76	69.04
64 L.H., Daily Totals		$6001.80		$7694.66	$93.78	$120.23

Crew B-36C	Hr.	Daily	Hr.	Daily	Bare Costs	Incl. O&P
1 Labor Foreman (outside)	$28.45	$227.60	$47.65	$381.20	$32.07	$52.78
3 Equip. Oper. (med.)	34.90	837.60	57.05	1369.20		
1 Truck Driver, Heavy	27.20	217.60	45.10	360.80		
1 Grader, 30,000 Lbs.		542.80		597.08		
1 Dozer, 300 H.P.		1607.00		1767.70		
1 Roller, Vibratory, 25 Ton		596.80		656.48		
1 Truck Tractor, 6x4, 450 H.P.		558.80		614.68		
1 Water Tank Trailer, 5000 Gal.		146.20		160.82	86.29	94.92
40 L.H., Daily Totals		$4734.40		$5907.96	$118.36	$147.70

Crew No.	Bare Costs		Incl. Subs O&P		Cost Per Labor-Hour	
Crew B-36E	**Hr.**	**Daily**	**Hr.**	**Daily**	**Bare Costs**	**Incl. O&P**
1 Labor Foreman (outside)	$28.45	$227.60	$47.65	$381.20	$32.54	$53.49
4 Equip. Oper. (medium)	34.90	1116.80	57.05	1825.60		
1 Truck Driver, Heavy	27.20	217.60	45.10	360.80		
1 Grader, 30,000 Lbs.		542.80		597.08		
1 Dozer, 300 H.P.		1607.00		1767.70		
1 Roller, Vibratory, 25 Ton		596.80		656.48		
1 Truck Tractor, 6x4, 380 H.P.		459.40		505.34		
1 Dist. Tanker, 3000 Gallon		309.00		339.90	73.23	80.55
48 L.H., Daily Totals		$5077.00		$6434.10	$105.77	$134.04

Crew B-37	Hr.	Daily	Hr.	Daily	Bare Costs	Incl. O&P
1 Labor Foreman (outside)	$28.45	$227.60	$47.65	$381.20	$27.97	$46.61
4 Laborers	26.45	846.40	44.30	1417.60		
1 Equip. Oper. (light)	33.55	268.40	54.80	438.40		
1 Tandem Roller, 5 Ton		145.20		159.72	3.02	3.33
48 L.H., Daily Totals		$1487.60		$2396.92	$30.99	$49.94

Crew B-37A	Hr.	Daily	Hr.	Daily	Bare Costs	Incl. O&P
2 Laborers	$26.45	$423.20	$44.30	$708.80	$26.43	$44.13
1 Truck Driver (light)	26.40	211.20	43.80	350.40		
1 Flatbed Truck, Gas, 1.5 Ton		171.40		188.54		
1 Tar Kettle, T.M.		74.65		82.11	10.25	11.28
24 L.H., Daily Totals		$880.45		$1329.86	$36.69	$55.41

Crew B-37B	Hr.	Daily	Hr.	Daily	Bare Costs	Incl. O&P
3 Laborers	$26.45	$634.80	$44.30	$1063.20	$26.44	$44.17
1 Truck Driver (light)	26.40	211.20	43.80	350.40		
1 Flatbed Truck, Gas, 1.5 Ton		171.40		188.54		
1 Tar Kettle, T.M.		74.65		82.11	7.69	8.46
32 L.H., Daily Totals		$1092.05		$1684.26	$34.13	$52.63

Crew B-37C	Hr.	Daily	Hr.	Daily	Bare Costs	Incl. O&P
2 Laborers	$26.45	$423.20	$44.30	$708.80	$26.43	$44.05
2 Truck Drivers (light)	26.40	422.40	43.80	700.80		
2 Flatbed Truck, Gas, 1.5 Ton		342.80		377.08		
1 Tar Kettle, T.M.		74.65		82.11	13.05	14.35
32 L.H., Daily Totals		$1263.05		$1868.80	$39.47	$58.40

Crew B-37D	Hr.	Daily	Hr.	Daily	Bare Costs	Incl. O&P
1 Laborer	$26.45	$211.60	$44.30	$354.40	$26.43	$44.05
1 Truck Driver (light)	26.40	211.20	43.80	350.40		
1 Pickup Truck, 3/4 Ton		104.60		115.06	6.54	7.19
16 L.H., Daily Totals		$527.40		$819.86	$32.96	$51.24

Crew B-37E	Hr.	Daily	Hr.	Daily	Bare Costs	Incl. O&P
3 Laborers	$26.45	$634.80	$44.30	$1063.20	$28.66	$47.48
1 Equip. Oper. (light)	33.55	268.40	54.80	438.40		
1 Equip. Oper. (medium)	34.90	279.20	57.05	456.40		
2 Truck Drivers (light)	26.40	422.40	43.80	700.80		
4 Barrel w/ Flasher		13.60		14.96		
1 Concrete Saw		134.20		147.62		
1 Rotary Hammer Drill		22.55		24.81		
1 Hammer Drill Bit		3.90		4.29		
1 Loader, Skid Steer, 30 H.P., gas		151.80		166.98		
1 Conc. Hammer Attach.		115.35		126.89		
1 Vibrating plate, gas, 18"		30.80		33.88		
2 Flatbed Truck, Gas, 1.5 Ton		342.80		377.08	14.55	16.01
56 L.H., Daily Totals		$2419.80		$3555.30	$43.21	$63.49

Crew B-37F	Hr.	Daily	Hr.	Daily	Bare Costs	Incl. O&P
3 Laborers	$26.45	$634.80	$44.30	$1063.20	$26.44	$44.17
1 Truck Driver (light)	26.40	211.20	43.80	350.40		
4 Barrel w/ Flasher		13.60		14.96		
1 Concrete Mixer, 10 C.F.		151.20		166.32		
1 Air Compressor, 60 cfm		110.20		121.22		
1 -50' Air Hose, 3/4"		3.05		3.36		
1 Spade (chipper)		8.00		8.80		
1 Flatbed Truck, Gas, 1.5 Ton		171.40		188.54	14.30	15.72
32 L.H., Daily Totals		$1303.45		$1916.80	$40.73	$59.90

Crew B-38	Hr.	Daily	Hr.	Daily	Bare Costs	Incl. O&P
2 Laborers	$26.45	$423.20	$44.30	$708.80	$28.82	$47.80
1 Equip. Oper. (light)	33.55	268.40	54.80	438.40		
1 Backhoe Loader, 48 H.P.		320.40		352.44		
1 Hyd.Hammer, (1200 lb.)		158.00		173.80	19.93	21.93
24 L.H., Daily Totals		$1170.00		$1673.44	$48.75	$69.73

Crew B-39	Hr.	Daily	Hr.	Daily	Bare Costs	Incl. O&P
1 Labor Foreman (outside)	$28.45	$227.60	$47.65	$381.20	$26.78	$44.86
5 Laborers	26.45	1058.00	44.30	1772.00		
1 Air Compressor, 250 cfm		162.80		179.08		
2 Breakers, Pavement, 60 lb.		19.60		21.56		
2 -50' Air Hoses, 1.5"		12.60		13.86	4.06	4.47
48 L.H., Daily Totals		$1480.60		$2367.70	$30.85	$49.33

Crew B-40	Hr.	Daily	Hr.	Daily	Bare Costs	Incl. O&P
1 Pile Driver Foreman (out)	$34.75	$278.00	$60.55	$484.40	$32.57	$55.96
4 Pile Drivers	32.75	1048.00	57.10	1827.20		
1 Building Laborer	26.45	211.60	44.30	354.40		
1 Equip. Oper. (crane)	35.80	286.40	58.50	468.00		
1 Crawler Crane, 40 Ton		1128.00		1240.80		
1 Vibratory Hammer & Gen.		2140.00		2354.00	58.36	64.19
56 L.H., Daily Totals		$5092.00		$6728.80	$90.93	$120.16

Crew B-40B	Hr.	Daily	Hr.	Daily	Bare Costs	Incl. O&P
1 Laborer Foreman	$28.45	$227.60	$47.65	$381.20	$29.11	$48.30
3 Laborers	26.45	634.80	44.30	1063.20		
1 Equip. Oper. (crane)	35.80	286.40	58.50	468.00		
1 Equip. Oper. Oiler	31.05	248.40	50.75	406.00		
1 Lattice Boom Crane, 40 Ton		1263.00		1389.30	26.31	28.94
48 L.H., Daily Totals		$2660.20		$3707.70	$55.42	$77.24

Crew B-41	Hr.	Daily	Hr.	Daily	Bare Costs	Incl. O&P
1 Labor Foreman (outside)	$28.45	$227.60	$47.65	$381.20	$27.45	$45.85
4 Laborers	26.45	846.40	44.30	1417.60		
.25 Equip. Oper. (crane)	35.80	71.60	58.50	117.00		
.25 Equip. Oper. Oiler	31.05	62.10	50.75	101.50		
.25 Crawler Crane, 40 Ton		282.00		310.20	6.41	7.05
44 L.H., Daily Totals		$1489.70		$2327.50	$33.86	$52.90

Crew B-42	Hr.	Daily	Hr.	Daily	Bare Costs	Incl. O&P
1 Labor Foreman (outside)	$28.45	$227.60	$47.65	$381.20	$29.96	$49.73
4 Laborers	26.45	846.40	44.30	1417.60		
1 Equip. Oper. (crane)	35.80	286.40	58.50	468.00		
1 Welder	39.65	317.20	64.75	518.00		
1 Hyd. Crane, 25 Ton		748.60		823.46		
1 Welder, gas engine, 300 amp		109.00		119.90		
1 Horz. Boring Csg. Mch.		442.20		486.42	23.21	25.53
56 L.H., Daily Totals		$2977.40		$4214.58	$53.17	$75.26

Crew B-43	Bare Costs Hr.	Bare Costs Daily	Incl. Subs O&P Hr.	Incl. Subs O&P Daily	Cost Per Labor-Hour Bare Costs	Cost Per Labor-Hour Incl. O&P
1 Labor Foreman (outside)	$28.45	$227.60	$47.65	$381.20	$26.85	$44.97
4 Laborers	26.45	846.40	44.30	1417.60		
1 Drill Rig, Truck-Mounted		2486.00		2734.60	62.15	68.36
40 L.H., Daily Totals		$3560.00		$4533.40	$89.00	$113.33

Crew B-44	Hr.	Daily	Hr.	Daily	Bare Costs	Incl. O&P
1 Pile Driver Foreman	$34.75	$278.00	$60.55	$484.40	$31.81	$54.51
4 Pile Drivers	32.75	1048.00	57.10	1827.20		
1 Equip. Oper. (crane)	35.80	286.40	58.50	468.00		
2 Laborers	26.45	423.20	44.30	708.80		
1 Crawler Crane, 40 Ton		1128.00		1240.80		
1 Lead, 60' high		74.00		81.40		
1 Hammer, diesel, 15K ft.-lbs.		614.20		675.62	28.38	31.22
64 L.H., Daily Totals		$3851.80		$5486.22	$60.18	$85.72

Crew B-45	Hr.	Daily	Hr.	Daily	Bare Costs	Incl. O&P
1 Building Laborer	$26.45	$211.60	$44.30	$354.40	$26.82	$44.70
1 Truck Driver (heavy)	27.20	217.60	45.10	360.80		
1 Dist. Tanker, 3000 Gallon		309.00		339.90	19.31	21.24
16 L.H., Daily Totals		$738.20		$1055.10	$46.14	$65.94

Crew B-46	Hr.	Daily	Hr.	Daily	Bare Costs	Incl. O&P
1 Pile Driver Foreman	$34.75	$278.00	$60.55	$484.40	$29.93	$51.27
2 Pile Drivers	32.75	524.00	57.10	913.60		
3 Laborers	26.45	634.80	44.30	1063.20		
1 Chain Saw, gas, 36" Long		41.60		45.76	0.87	0.95
48 L.H., Daily Totals		$1478.40		$2506.96	$30.80	$52.23

Crew B-47	Hr.	Daily	Hr.	Daily	Bare Costs	Incl. O&P
1 Blast Foreman	$28.45	$227.60	$47.65	$381.20	$27.45	$45.98
1 Driller	26.45	211.60	44.30	354.40		
1 Air Track Drill, 4"		849.60		934.56		
1 Air Compressor, 600 cfm		424.60		467.06		
2 -50' Air Hoses, 3"		29.80		32.78	81.50	89.65
16 L.H., Daily Totals		$1743.20		$2170.00	$108.95	$135.63

Crew B-47A	Hr.	Daily	Hr.	Daily	Bare Costs	Incl. O&P
1 Drilling Foreman	$28.45	$227.60	$47.65	$381.20	$31.77	$52.30
1 Equip. Oper. (heavy)	35.80	286.40	58.50	468.00		
1 Oiler	31.05	248.40	50.75	406.00		
1 Air Track Drill, 5"		965.20		1061.72	40.22	44.24
24 L.H., Daily Totals		$1727.60		$2316.92	$71.98	$96.54

Crew B-47C	Hr.	Daily	Hr.	Daily	Bare Costs	Incl. O&P
1 Laborer	$26.45	$211.60	$44.30	$354.40	$30.00	$49.55
1 Equip. Oper. (light)	33.55	268.40	54.80	438.40		
1 Air Compressor, 750 cfm		431.80		474.98		
2 -50' Air Hoses, 3"		29.80		32.78		
1 Air Track Drill, 4"		849.60		934.56	81.95	90.14
16 L.H., Daily Totals		$1791.20		$2235.12	$111.95	$139.69

Crew B-47E	Hr.	Daily	Hr.	Daily	Bare Costs	Incl. O&P
1 Laborer Foreman	$28.45	$227.60	$47.65	$381.20	$26.95	$45.14
3 Laborers	26.45	634.80	44.30	1063.20		
1 Flatbed Truck, Gas, 3 Ton		217.00		238.70	6.78	7.46
32 L.H., Daily Totals		$1079.40		$1683.10	$33.73	$52.60

Crew B-47G	Hr.	Daily	Hr.	Daily	Bare Costs	Incl. O&P
1 Laborer Foreman	$28.45	$227.60	$47.65	$381.20	$27.12	$45.42
2 Laborers	26.45	423.20	44.30	708.80		
1 Air Track Drill, 4"		849.60		934.56		
1 Air Compressor, 600 cfm		424.60		467.06		
2 -50' Air Hoses, 3"		29.80		32.78		
1 Gunnite Pump rig		178.00		195.80	61.75	67.92
24 L.H., Daily Totals		$2132.80		$2720.20	$88.87	$113.34

Crew B-47H	Hr.	Daily	Hr.	Daily	Bare Costs	Incl. O&P
1 Skilled Worker Foreman	$36.05	$288.40	$60.35	$482.80	$34.55	$57.84
3 Skilled Workers	34.05	817.20	57.00	1368.00		
1 Flatbed Truck, Gas, 3 Ton		217.00		238.70	6.78	7.46
32 L.H., Daily Totals		$1322.60		$2089.50	$41.33	$65.30

Crew B-48	Hr.	Daily	Hr.	Daily	Bare Costs	Incl. O&P
1 Labor Foreman (outside)	$28.45	$227.60	$47.65	$381.20	$28.34	$47.23
4 Laborers	26.45	846.40	44.30	1417.60		
1 Equip. Oper. (crane)	35.80	286.40	58.50	468.00		
1 Centr. Water Pump, 6"		291.60		320.76		
1 -20' Suction Hose, 6"		11.90		13.09		
1 -50' Discharge Hose, 6"		6.30		6.93		
1 Drill Rig, Truck-Mounted		2486.00		2734.60	58.25	64.07
48 L.H., Daily Totals		$4156.20		$5342.18	$86.59	$111.30

Crew B-49	Hr.	Daily	Hr.	Daily	Bare Costs	Incl. O&P
1 Labor Foreman (outside)	$28.45	$227.60	$47.65	$381.20	$29.11	$49.09
5 Laborers	26.45	1058.00	44.30	1772.00		
1 Equip. Oper. (crane)	35.80	286.40	58.50	468.00		
2 Pile Drivers	32.75	524.00	57.10	913.60		
1 Hyd. Crane, 25 Ton		748.60		823.46		
1 Centr. Water Pump, 6"		291.60		320.76		
1 -20' Suction Hose, 6"		11.90		13.09		
1 -50' Discharge Hose, 6"		6.30		6.93		
1 Drill Rig, Truck-Mounted		2486.00		2734.60	49.23	54.15
72 L.H., Daily Totals		$5640.40		$7433.64	$78.34	$103.25

Crew B-50	Hr.	Daily	Hr.	Daily	Bare Costs	Incl. O&P
1 Pile Driver Foreman	$34.75	$278.00	$60.55	$484.40	$30.72	$52.55
6 Pile Drivers	32.75	1572.00	57.10	2740.80		
1 Equip. Oper. (crane)	35.80	286.40	58.50	468.00		
5 Laborers	26.45	1058.00	44.30	1772.00		
1 Crawler Crane, 40 Ton		1128.00		1240.80		
1 Lead, 60' high		74.00		81.40		
1 Hammer, diesel, 15K ft.-lbs.		614.20		675.62		
1 Air Compressor, 600 cfm		424.60		467.06		
2 -50' Air Hoses, 3"		29.80		32.78		
1 Chain Saw, gas, 36" Long		41.60		45.76	22.23	24.46
104 L.H., Daily Totals		$5506.60		$8008.62	$52.95	$77.01

Crew B-51	Hr.	Daily	Hr.	Daily	Bare Costs	Incl. O&P
1 Labor Foreman (outside)	$28.45	$227.60	$47.65	$381.20	$26.77	$44.77
4 Laborers	26.45	846.40	44.30	1417.60		
1 Truck Driver (light)	26.40	211.20	43.80	350.40		
1 Flatbed Truck, Gas, 1.5 Ton		171.40		188.54	3.57	3.93
48 L.H., Daily Totals		$1456.60		$2337.74	$30.35	$48.70

Crew No.	Bare Costs		Incl. Subs O&P		Cost Per Labor-Hour	
Crew B-52	**Hr.**	**Daily**	**Hr.**	**Daily**	**Bare Costs**	**Incl. O&P**
1 Labor Foreman	$28.45	$227.60	$47.65	$381.20	$29.06	$48.73
1 Carpenter	33.55	268.40	56.15	449.20		
4 Laborers	26.45	846.40	44.30	1417.60		
.5 Rodman (reinf.)	36.30	145.20	63.10	252.40		
.5 Equip. Oper. (med.)	34.90	139.60	57.05	228.20		
.5 Crawler Loader, 3 C.Y.		504.00		554.40	9.00	9.90
56 L.H., Daily Totals		$2131.20		$3283.00	$38.06	$58.63
Crew B-53	**Hr.**	**Daily**	**Hr.**	**Daily**	**Bare Costs**	**Incl. O&P**
1 Building Laborer	$26.45	$211.60	$44.30	$354.40	$26.45	$44.30
1 Trencher, Chain, 12 H.P.		57.40		63.14	7.17	7.89
8 L.H., Daily Totals		$269.00		$417.54	$33.63	$52.19
Crew B-54	**Hr.**	**Daily**	**Hr.**	**Daily**	**Bare Costs**	**Incl. O&P**
1 Equip. Oper. (light)	$33.55	$268.40	$54.80	$438.40	$33.55	$54.80
1 Trencher, Chain, 40 H.P.		297.00		326.70	37.13	40.84
8 L.H., Daily Totals		$565.40		$765.10	$70.67	$95.64
Crew B-54A	**Hr.**	**Daily**	**Hr.**	**Daily**	**Bare Costs**	**Incl. O&P**
.17 Labor Foreman (outside)	$28.45	$38.69	$47.65	$64.80	$33.96	$55.68
1 Equipment Operator (med.)	34.90	279.20	57.05	456.40		
1 Wheel Trencher, 67 H.P.		1001.00		1101.10	106.94	117.64
9.36 L.H., Daily Totals		$1318.89		$1622.30	$140.91	$173.32
Crew B-54B	**Hr.**	**Daily**	**Hr.**	**Daily**	**Bare Costs**	**Incl. O&P**
.25 Labor Foreman (outside)	$28.45	$56.90	$47.65	$95.30	$33.61	$55.17
1 Equipment Operator (med.)	34.90	279.20	57.05	456.40		
1 Wheel Trencher, 150 H.P.		1802.00		1982.20	180.20	198.22
10 L.H., Daily Totals		$2138.10		$2533.90	$213.81	$253.39
Crew B-54D	**Hr.**	**Daily**	**Hr.**	**Daily**	**Bare Costs**	**Incl. O&P**
1 Laborer	$26.45	$211.60	$44.30	$354.40	$30.68	$50.67
1 Equipment Operator (med.)	34.90	279.20	57.05	456.40		
1 Rock trencher, 6" width		332.80		366.08	20.80	22.88
16 L.H., Daily Totals		$823.60		$1176.88	$51.48	$73.56
Crew B-54E	**Hr.**	**Daily**	**Hr.**	**Daily**	**Bare Costs**	**Incl. O&P**
1 Laborer	$26.45	$211.60	$44.30	$354.40	$30.68	$50.67
1 Equipment Operator (med.)	34.90	279.20	57.05	456.40		
1 Rock trencher, 18" width		2538.00		2791.80	158.63	174.49
16 L.H., Daily Totals		$3028.80		$3602.60	$189.30	$225.16
Crew B-55	**Hr.**	**Daily**	**Hr.**	**Daily**	**Bare Costs**	**Incl. O&P**
1 Laborer	$26.45	$211.60	$44.30	$354.40	$26.43	$44.05
1 Truck Driver (light)	26.40	211.20	43.80	350.40		
1 Truck-mounted earth auger		706.60		777.26		
1 Flatbed Truck, Gas, 3 Ton		217.00		238.70	57.73	63.50
16 L.H., Daily Totals		$1346.40		$1720.76	$84.15	$107.55
Crew B-56	**Hr.**	**Daily**	**Hr.**	**Daily**	**Bare Costs**	**Incl. O&P**
2 Laborers	$26.45	$423.20	$44.30	$708.80	$26.45	$44.30
1 Air Track Drill, 4"		849.60		934.56		
1 Air Compressor, 600 cfm		424.60		467.06		
1 -50' Air Hose, 3"		14.90		16.39	80.57	88.63
16 L.H., Daily Totals		$1712.30		$2126.81	$107.02	$132.93

Crew No.	Bare Costs		Incl. Subs O&P		Cost Per Labor-Hour	
Crew B-57	**Hr.**	**Daily**	**Hr.**	**Daily**	**Bare Costs**	**Incl. O&P**
1 Labor Foreman (outside)	$28.45	$227.60	$47.65	$381.20	$28.72	$47.81
3 Laborers	26.45	634.80	44.30	1063.20		
1 Equip. Oper. (crane)	35.80	286.40	58.50	468.00		
1 Barge, 400 Ton		820.40		902.44		
1 Crawler Crane, 25 Ton		1123.00		1235.30		
1 Clamshell Bucket, 1 C.Y.		48.00		52.80		
1 Centr. Water Pump, 6"		291.60		320.76		
1 -20' Suction Hose, 6"		11.90		13.09		
20 -50' Discharge Hoses, 6"		126.00		138.60	60.52	66.57
40 L.H., Daily Totals		$3569.70		$4575.39	$89.24	$114.38
Crew B-58	**Hr.**	**Daily**	**Hr.**	**Daily**	**Bare Costs**	**Incl. O&P**
2 Laborers	$26.45	$423.20	$44.30	$708.80	$28.82	$47.80
1 Equip. Oper. (light)	33.55	268.40	54.80	438.40		
1 Backhoe Loader, 48 H.P.		320.40		352.44		
1 Small Helicopter, w/pilot		2579.00		2836.90	120.81	132.89
24 L.H., Daily Totals		$3591.00		$4336.54	$149.63	$180.69
Crew B-59	**Hr.**	**Daily**	**Hr.**	**Daily**	**Bare Costs**	**Incl. O&P**
1 Truck Driver (heavy)	$27.20	$217.60	$45.10	$360.80	$27.20	$45.10
1 Truck Tractor, 220 H.P.		279.20		307.12		
1 Water Tank Trailer, 5000 Gal.		146.20		160.82	53.17	58.49
8 L.H., Daily Totals		$643.00		$828.74	$80.38	$103.59
Crew B-60	**Hr.**	**Daily**	**Hr.**	**Daily**	**Bare Costs**	**Incl. O&P**
1 Labor Foreman (outside)	$28.45	$227.60	$47.65	$381.20	$29.52	$48.98
3 Laborers	26.45	634.80	44.30	1063.20		
1 Equip. Oper. (crane)	35.80	286.40	58.50	468.00		
1 Equip. Oper. (light)	33.55	268.40	54.80	438.40		
1 Crawler Crane, 40 Ton		1128.00		1240.80		
1 Lead, 60' high		74.00		81.40		
1 Hammer, diesel, 15K ft.-lbs.		614.20		675.62		
1 Backhoe Loader, 48 H.P.		320.40		352.44	44.51	48.96
48 L.H., Daily Totals		$3553.80		$4701.06	$74.04	$97.94
Crew B-61	**Hr.**	**Daily**	**Hr.**	**Daily**	**Bare Costs**	**Incl. O&P**
1 Labor Foreman (outside)	$28.45	$227.60	$47.65	$381.20	$26.85	$44.97
4 Laborers	26.45	846.40	44.30	1417.60		
1 Cement Mixer, 2 C.Y.		194.60		214.06		
1 Air Compressor, 160 cfm		125.00		137.50	7.99	8.79
40 L.H., Daily Totals		$1393.60		$2150.36	$34.84	$53.76
Crew B-62	**Hr.**	**Daily**	**Hr.**	**Daily**	**Bare Costs**	**Incl. O&P**
2 Laborers	$26.45	$423.20	$44.30	$708.80	$28.82	$47.80
1 Equip. Oper. (light)	33.55	268.40	54.80	438.40		
1 Loader, Skid Steer, 30 H.P., gas		151.80		166.98	6.33	6.96
24 L.H., Daily Totals		$843.40		$1314.18	$35.14	$54.76
Crew B-63	**Hr.**	**Daily**	**Hr.**	**Daily**	**Bare Costs**	**Incl. O&P**
5 Laborers	$26.45	$1058.00	$44.30	$1772.00	$26.45	$44.30
1 Loader, Skid Steer, 30 H.P., gas		151.80		166.98	3.79	4.17
40 L.H., Daily Totals		$1209.80		$1938.98	$30.25	$48.47
Crew B-64	**Hr.**	**Daily**	**Hr.**	**Daily**	**Bare Costs**	**Incl. O&P**
1 Laborer	$26.45	$211.60	$44.30	$354.40	$26.43	$44.05
1 Truck Driver (light)	26.40	211.20	43.80	350.40		
1 Power Mulcher (small)		130.80		143.88		
1 Flatbed Truck, Gas, 1.5 Ton		171.40		188.54	18.89	20.78
16 L.H., Daily Totals		$725.00		$1037.22	$45.31	$64.83

Crew No.	Bare Costs		Incl. Subs O&P		Cost Per Labor-Hour	
Crew B-65	**Hr.**	**Daily**	**Hr.**	**Daily**	**Bare Costs**	**Incl. O&P**
1 Laborer	$26.45	$211.60	$44.30	$354.40	$26.43	$44.05
1 Truck Driver (light)	26.40	211.20	43.80	350.40		
1 Power Mulcher (large)		258.80		284.68		
1 Flatbed Truck, Gas, 1.5 Ton		171.40		188.54	26.89	29.58
16 L.H., Daily Totals		$853.00		$1178.02	$53.31	$73.63
Crew B-66	**Hr.**	**Daily**	**Hr.**	**Daily**	**Bare Costs**	**Incl. O&P**
1 Equip. Oper. (light)	$33.55	$268.40	$54.80	$438.40	$33.55	$54.80
1 Loader-Backhoe		223.60		245.96	27.95	30.75
8 L.H., Daily Totals		$492.00		$684.36	$61.50	$85.55
Crew B-67	**Hr.**	**Daily**	**Hr.**	**Daily**	**Bare Costs**	**Incl. O&P**
1 Millwright	$35.35	$282.80	$56.45	$451.60	$34.45	$55.63
1 Equip. Oper. (light)	33.55	268.40	54.80	438.40		
1 Forklift, R/T, 4,000 Lb.		278.60		306.46	17.41	19.15
16 L.H., Daily Totals		$829.80		$1196.46	$51.86	$74.78
Crew B-68	**Hr.**	**Daily**	**Hr.**	**Daily**	**Bare Costs**	**Incl. O&P**
2 Millwrights	$35.35	$565.60	$56.45	$903.20	$34.75	$55.90
1 Equip. Oper. (light)	33.55	268.40	54.80	438.40		
1 Forklift, R/T, 4,000 Lb.		278.60		306.46	11.61	12.77
24 L.H., Daily Totals		$1112.60		$1648.06	$46.36	$68.67
Crew B-69	**Hr.**	**Daily**	**Hr.**	**Daily**	**Bare Costs**	**Incl. O&P**
1 Labor Foreman (outside)	$28.45	$227.60	$47.65	$381.20	$29.11	$48.30
3 Laborers	26.45	634.80	44.30	1063.20		
1 Equip Oper. (crane)	35.80	286.40	58.50	468.00		
1 Equip Oper. Oiler	31.05	248.40	50.75	406.00		
1 Hyd. Crane, 80 Ton		1693.00		1862.30	35.27	38.80
48 L.H., Daily Totals		$3090.20		$4180.70	$64.38	$87.10
Crew B-69A	**Hr.**	**Daily**	**Hr.**	**Daily**	**Bare Costs**	**Incl. O&P**
1 Labor Foreman	$28.45	$227.60	$47.65	$381.20	$29.09	$48.08
3 Laborers	26.45	634.80	44.30	1063.20		
1 Equip. Oper. (med.)	34.90	279.20	57.05	456.40		
1 Concrete Finisher	31.85	254.80	50.85	406.80		
1 Curb/Gutter Paver, 2-Track		744.80		819.28	15.52	17.07
48 L.H., Daily Totals		$2141.20		$3126.88	$44.61	$65.14
Crew B-69B	**Hr.**	**Daily**	**Hr.**	**Daily**	**Bare Costs**	**Incl. O&P**
1 Labor Foreman	$28.45	$227.60	$47.65	$381.20	$29.09	$48.08
3 Laborers	26.45	634.80	44.30	1063.20		
1 Equip. Oper. (med.)	34.90	279.20	57.05	456.40		
1 Cement Finisher	31.85	254.80	50.85	406.80		
1 Curb/Gutter Paver, 4-Track		848.20		933.02	17.67	19.44
48 L.H., Daily Totals		$2244.60		$3240.62	$46.76	$67.51
Crew B-70	**Hr.**	**Daily**	**Hr.**	**Daily**	**Bare Costs**	**Incl. O&P**
1 Labor Foreman (outside)	$28.45	$227.60	$47.65	$381.20	$30.36	$50.24
3 Laborers	26.45	634.80	44.30	1063.20		
3 Equip. Oper. (med.)	34.90	837.60	57.05	1369.20		
1 Grader, 30,000 Lbs.		542.80		597.08		
1 Ripper, beam & 1 shank		83.20		91.52		
1 Road Sweeper, S.P., 8' wide		604.80		665.28		
1 F.E. Loader, W.M., 1.5 C.Y.		319.40		351.34	27.68	30.45
56 L.H., Daily Totals		$3250.20		$4518.82	$58.04	$80.69

Crew No.	Bare Costs		Incl. Subs O&P		Cost Per Labor-Hour	
Crew B-71	**Hr.**	**Daily**	**Hr.**	**Daily**	**Bare Costs**	**Incl. O&P**
1 Labor Foreman (outside)	$28.45	$227.60	$47.65	$381.20	$30.36	$50.24
3 Laborers	26.45	634.80	44.30	1063.20		
3 Equip. Oper. (med.)	34.90	837.60	57.05	1369.20		
1 Pvmt. Profiler, 750 H.P.		5626.00		6188.60		
1 Road Sweeper, S.P., 8' wide		604.80		665.28		
1 F.E. Loader, W.M., 1.5 C.Y.		319.40		351.34	116.97	128.66
56 L.H., Daily Totals		$8250.20		$10018.82	$147.32	$178.91
Crew B-72	**Hr.**	**Daily**	**Hr.**	**Daily**	**Bare Costs**	**Incl. O&P**
1 Labor Foreman (outside)	$28.45	$227.60	$47.65	$381.20	$30.93	$51.09
3 Laborers	26.45	634.80	44.30	1063.20		
4 Equip. Oper. (med.)	34.90	1116.80	57.05	1825.60		
1 Pvmt. Profiler, 750 H.P.		5626.00		6188.60		
1 Hammermill, 250 H.P.		1692.00		1861.20		
1 Windrow Loader		1210.00		1331.00		
1 Mix Paver 165 H.P.		2016.00		2217.60		
1 Roller, Pneum. Whl, 12 Ton		309.20		340.12	169.58	186.54
64 L.H., Daily Totals		$12832.40		$15208.52	$200.51	$237.63
Crew B-73	**Hr.**	**Daily**	**Hr.**	**Daily**	**Bare Costs**	**Incl. O&P**
1 Labor Foreman (outside)	$28.45	$227.60	$47.65	$381.20	$31.98	$52.69
2 Laborers	26.45	423.20	44.30	708.80		
5 Equip. Oper. (med.)	34.90	1396.00	57.05	2282.00		
1 Road Mixer, 310 H.P.		1890.00		2079.00		
1 Tandem Roller, 10 Ton		224.20		246.62		
1 Hammermill, 250 H.P.		1692.00		1861.20		
1 Grader, 30,000 Lbs.		542.80		597.08		
.5 F.E. Loader, W.M., 1.5 C.Y.		159.70		175.67		
.5 Truck Tractor, 220 H.P.		139.60		153.56		
.5 Water Tank Trailer, 5000 Gal.		73.10		80.41	73.77	81.15
64 L.H., Daily Totals		$6768.20		$8565.54	$105.75	$133.84
Crew B-74	**Hr.**	**Daily**	**Hr.**	**Daily**	**Bare Costs**	**Incl. O&P**
1 Labor Foreman (outside)	$28.45	$227.60	$47.65	$381.20	$31.11	$51.29
1 Laborer	26.45	211.60	44.30	354.40		
4 Equip. Oper. (med.)	34.90	1116.80	57.05	1825.60		
2 Truck Drivers (heavy)	27.20	435.20	45.10	721.60		
1 Grader, 30,000 Lbs.		542.80		597.08		
1 Ripper, beam & 1 shank		83.20		91.52		
2 Stabilizers, 310 H.P.		2652.00		2917.20		
1 Flatbed Truck, Gas, 3 Ton		217.00		238.70		
1 Chem. Spreader, Towed		49.00		53.90		
1 Roller, Vibratory, 25 Ton		596.80		656.48		
1 Water Tank Trailer, 5000 Gal.		146.20		160.82		
1 Truck Tractor, 220 H.P.		279.20		307.12	71.35	78.48
64 L.H., Daily Totals		$6557.40		$8305.62	$102.46	$129.78
Crew B-75	**Hr.**	**Daily**	**Hr.**	**Daily**	**Bare Costs**	**Incl. O&P**
1 Labor Foreman (outside)	$28.45	$227.60	$47.65	$381.20	$31.67	$52.18
1 Laborer	26.45	211.60	44.30	354.40		
4 Equip. Oper. (med.)	34.90	1116.80	57.05	1825.60		
1 Truck Driver (heavy)	27.20	217.60	45.10	360.80		
1 Grader, 30,000 Lbs.		542.80		597.08		
1 Ripper, beam & 1 shank		83.20		91.52		
2 Stabilizers, 310 H.P.		2652.00		2917.20		
1 Dist. Tanker, 3000 Gallon		309.00		339.90		
1 Truck Tractor, 6x4, 380 H.P.		459.40		505.34		
1 Roller, Vibratory, 25 Ton		596.80		656.48	82.91	91.21
56 L.H., Daily Totals		$6416.80		$8029.52	$114.59	$143.38

Crews

Crew No.	Bare Costs		Incl. Subs O&P		Cost Per Labor-Hour	
Crew B-76	**Hr.**	**Daily**	**Hr.**	**Daily**	**Bare Costs**	**Incl. O&P**
1 Dock Builder Foreman	$34.75	$278.00	$60.55	$484.40	$33.46	$57.09
5 Dock Builders	32.75	1310.00	57.10	2284.00		
2 Equip. Oper. (crane)	35.80	572.80	58.50	936.00		
1 Equip. Oper. Oiler	31.05	248.40	50.75	406.00		
1 Crawler Crane, 50 Ton		1279.00		1406.90		
1 Barge, 400 Ton		820.40		902.44		
1 Hammer, diesel, 15K ft.-lbs.		614.20		675.62		
1 Lead, 60' high		74.00		81.40		
1 Air Compressor, 600 cfm		424.60		467.06		
2 -50' Air Hoses, 3"		29.80		32.78	45.03	49.53
72 L.H., Daily Totals		$5651.20		$7676.60	$78.49	$106.62

Crew B-76A	Hr.	Daily	Hr.	Daily	Bare Costs	Incl. O&P
1 Laborer Foreman	$28.45	$227.60	$47.65	$381.20	$28.44	$47.30
5 Laborers	26.45	1058.00	44.30	1772.00		
1 Equip. Oper. (crane)	35.80	286.40	58.50	468.00		
1 Equip. Oper. Oiler	31.05	248.40	50.75	406.00		
1 Crawler Crane, 50 Ton		1279.00		1406.90		
1 Barge, 400 Ton		820.40		902.44	32.80	36.08
64 L.H., Daily Totals		$3919.80		$5336.54	$61.25	$83.38

Crew B-77	Hr.	Daily	Hr.	Daily	Bare Costs	Incl. O&P
1 Labor Foreman	$28.45	$227.60	$47.65	$381.20	$26.84	$44.87
3 Laborers	26.45	634.80	44.30	1063.20		
1 Truck Driver (light)	26.40	211.20	43.80	350.40		
1 Crack Cleaner, 25 H.P.		53.00		58.30		
1 Crack Filler, Trailer Mtd.		194.60		214.06		
1 Flatbed Truck, Gas, 3 Ton		217.00		238.70	11.62	12.78
40 L.H., Daily Totals		$1538.20		$2305.86	$38.45	$57.65

Crew B-78	Hr.	Daily	Hr.	Daily	Bare Costs	Incl. O&P
1 Labor Foreman	$28.45	$227.60	$47.65	$381.20	$26.85	$44.97
4 Laborers	26.45	846.40	44.30	1417.60		
1 Paint Striper, S.P.		147.80		162.58		
1 Flatbed Truck, Gas, 3 Ton		217.00		238.70		
1 Pickup Truck, 3/4 Ton		104.60		115.06	11.73	12.91
40 L.H., Daily Totals		$1543.40		$2315.14	$38.59	$57.88

Crew B-78B	Hr.	Daily	Hr.	Daily	Bare Costs	Incl. O&P
2 Laborers	$26.45	$423.20	$44.30	$708.80	$27.24	$45.47
.25 Equip. Oper. (light)	33.55	67.10	54.80	109.60		
1 Pickup Truck, 3/4 Ton		104.60		115.06		
1 Line Rem., 11 H.P., walk behind		57.40		63.14		
.25 Road Sweeper, S.P., 8' wide		151.20		166.32	17.40	19.14
18 L.H., Daily Totals		$803.50		$1162.92	$44.64	$64.61

Crew B-79	Hr.	Daily	Hr.	Daily	Bare Costs	Incl. O&P
1 Labor Foreman	$28.45	$227.60	$47.65	$381.20	$26.95	$45.14
3 Laborers	26.45	634.80	44.30	1063.20		
1 Thermo. Striper, T.M.		653.80		719.18		
1 Flatbed Truck, Gas, 3 Ton		217.00		238.70		
2 Pickup Trucks, 3/4 Ton		209.20		230.12	33.75	37.13
32 L.H., Daily Totals		$1942.40		$2632.40	$60.70	$82.26

Crew B-80	Hr.	Daily	Hr.	Daily	Bare Costs	Incl. O&P
1 Labor Foreman	$28.45	$227.60	$47.65	$381.20	$27.12	$45.42
2 Laborers	26.45	423.20	44.30	708.80		
1 Flatbed Truck, Gas, 3 Ton		217.00		238.70		
1 Earth Auger, Truck-Mtd.		401.80		441.98	25.78	28.36
24 L.H., Daily Totals		$1269.60		$1770.68	$52.90	$73.78

Crew No.	Bare Costs		Incl. Subs O&P		Cost Per Labor-Hour	
Crew B-80A	**Hr.**	**Daily**	**Hr.**	**Daily**	**Bare Costs**	**Incl. O&P**
3 Laborers	$26.45	$634.80	$44.30	$1063.20	$26.45	$44.30
1 Flatbed Truck, Gas, 3 Ton		217.00		238.70	9.04	9.95
24 L.H., Daily Totals		$851.80		$1301.90	$35.49	$54.25

Crew B-80B	Hr.	Daily	Hr.	Daily	Bare Costs	Incl. O&P
3 Laborers	$26.45	$634.80	$44.30	$1063.20	$28.23	$46.92
1 Equip. Oper. (light)	33.55	268.40	54.80	438.40		
1 Crane, Flatbed Mounted, 3 Ton		225.60		248.16	7.05	7.75
32 L.H., Daily Totals		$1128.80		$1749.76	$35.27	$54.68

Crew B-80C	Hr.	Daily	Hr.	Daily	Bare Costs	Incl. O&P
2 Laborers	$26.45	$423.20	$44.30	$708.80	$26.43	$44.13
1 Truck Driver (light)	26.40	211.20	43.80	350.40		
1 Flatbed Truck, Gas, 1.5 Ton		171.40		188.54		
1 Manual fence post auger, gas		7.20		7.92	7.44	8.19
24 L.H., Daily Totals		$813.00		$1255.66	$33.88	$52.32

Crew B-81	Hr.	Daily	Hr.	Daily	Bare Costs	Incl. O&P
1 Laborer	$26.45	$211.60	$44.30	$354.40	$26.82	$44.70
1 Truck Driver (heavy)	27.20	217.60	45.10	360.80		
1 Hydromulcher, T.M., 3000 Gal.		278.00		305.80		
1 Truck Tractor, 220 H.P.		279.20		307.12	34.83	38.31
16 L.H., Daily Totals		$986.40		$1328.12	$61.65	$83.01

Crew B-81A	Hr.	Daily	Hr.	Daily	Bare Costs	Incl. O&P
1 Laborer	$26.45	$211.60	$44.30	$354.40	$26.43	$44.05
1 Truck Driver (light)	26.40	211.20	43.80	350.40		
1 Hydromulcher, T.M., 600 Gal.		113.20		124.52		
1 Flatbed Truck, Gas, 3 Ton		217.00		238.70	20.64	22.70
16 L.H., Daily Totals		$753.00		$1068.02	$47.06	$66.75

Crew B-82	Hr.	Daily	Hr.	Daily	Bare Costs	Incl. O&P
1 Laborer	$26.45	$211.60	$44.30	$354.40	$30.00	$49.55
1 Equip. Oper. (light)	33.55	268.40	54.80	438.40		
1 Horiz. Borer, 6 H.P.		76.00		83.60	4.75	5.22
16 L.H., Daily Totals		$556.00		$876.40	$34.75	$54.77

Crew B-82A	Hr.	Daily	Hr.	Daily	Bare Costs	Incl. O&P
1 Laborer	$26.45	$211.60	$44.30	$354.40	$30.00	$49.55
1 Equip. Oper. (light)	33.55	268.40	54.80	438.40		
1 Flatbed Truck, Gas, 3 Ton		217.00		238.70		
1 Flatbed Trailer, 25 Ton		111.20		122.32		
1 Horiz. Dir. Drill, 20k lb. thrust		622.60		684.86	59.42	65.37
16 L.H., Daily Totals		$1430.80		$1838.68	$89.42	$114.92

Crew B-82B	Hr.	Daily	Hr.	Daily	Bare Costs	Incl. O&P
2 Laborers	$26.45	$423.20	$44.30	$708.80	$28.82	$47.80
1 Equip. Oper. (light)	33.55	268.40	54.80	438.40		
1 Flatbed Truck, Gas, 3 Ton		217.00		238.70		
1 Flatbed Trailer, 25 Ton		111.20		122.32		
1 Horiz. Dir. Drill, 30k lb. thrust		888.40		977.24	50.69	55.76
24 L.H., Daily Totals		$1908.20		$2485.46	$79.51	$103.56

Crew B-82C	Hr.	Daily	Hr.	Daily	Bare Costs	Incl. O&P
2 Laborers	$26.45	$423.20	$44.30	$708.80	$28.82	$47.80
1 Equip. Oper. (light)	33.55	268.40	54.80	438.40		
1 Flatbed Truck, Gas, 3 Ton		217.00		238.70		
1 Flatbed Trailer, 25 Ton		111.20		122.32		
1 Horiz. Dir. Drill, 50k lb. thrust		1174.00		1291.40	62.59	68.85
24 L.H., Daily Totals		$2193.80		$2799.62	$91.41	$116.65

Crew No.	Bare Costs		Incl. Subs O&P		Cost Per Labor-Hour	
Crew B-82D	**Hr.**	**Daily**	**Hr.**	**Daily**	**Bare Costs**	**Incl. O&P**
1 Equip. Oper. (light)	$33.55	$268.40	$54.80	$438.40	$33.55	$54.80
1 Mud Trailer for HDD, 1500 gallon		265.80		292.38	33.23	36.55
8 L.H., Daily Totals		$534.20		$730.78	$66.78	$91.35

Crew B-83	Hr.	Daily	Hr.	Daily	Bare Costs	Incl. O&P
1 Tugboat Captain	$34.90	$279.20	$57.05	$456.40	$30.68	$50.67
1 Tugboat Hand	26.45	211.60	44.30	354.40		
1 Tugboat, 250 H.P.		632.20		695.42	39.51	43.46
16 L.H., Daily Totals		$1123.00		$1506.22	$70.19	$94.14

Crew B-84	Hr.	Daily	Hr.	Daily	Bare Costs	Incl. O&P
1 Equip. Oper. (med.)	$34.90	$279.20	$57.05	$456.40	$34.90	$57.05
1 Rotary Mower/Tractor		293.80		323.18	36.73	40.40
8 L.H., Daily Totals		$573.00		$779.58	$71.63	$97.45

Crew B-85	Hr.	Daily	Hr.	Daily	Bare Costs	Incl. O&P
3 Laborers	$26.45	$634.80	$44.30	$1063.20	$28.29	$47.01
1 Equip. Oper. (med.)	34.90	279.20	57.05	456.40		
1 Truck Driver (heavy)	27.20	217.60	45.10	360.80		
1 Aerial Lift Truck, 80'		636.20		699.82		
1 Brush Chipper, 12", 130 H.P.		217.00		238.70		
1 Pruning Saw, Rotary		6.30		6.93	21.49	23.64
40 L.H., Daily Totals		$1991.10		$2825.85	$49.78	$70.65

Crew B-86	Hr.	Daily	Hr.	Daily	Bare Costs	Incl. O&P
1 Equip. Oper. (med.)	$34.90	$279.20	$57.05	$456.40	$34.90	$57.05
1 Stump Chipper, S.P.		169.35		186.29	21.17	23.29
8 L.H., Daily Totals		$448.55		$642.68	$56.07	$80.34

Crew B-86A	Hr.	Daily	Hr.	Daily	Bare Costs	Incl. O&P
1 Equip. Oper. (med.)	$34.90	$279.20	$57.05	$456.40	$34.90	$57.05
1 Grader, 30,000 Lbs.		542.80		597.08	67.85	74.64
8 L.H., Daily Totals		$822.00		$1053.48	$102.75	$131.69

Crew B-86B	Hr.	Daily	Hr.	Daily	Bare Costs	Incl. O&P
1 Equip. Oper. (med.)	$34.90	$279.20	$57.05	$456.40	$34.90	$57.05
1 Dozer, 200 H.P.		1051.00		1156.10	131.38	144.51
8 L.H., Daily Totals		$1330.20		$1612.50	$166.28	$201.56

Crew B-87	Hr.	Daily	Hr.	Daily	Bare Costs	Incl. O&P
1 Laborer	$26.45	$211.60	$44.30	$354.40	$33.21	$54.50
4 Equip. Oper. (med.)	34.90	1116.80	57.05	1825.60		
2 Feller Bunchers, 100 H.P.		1128.00		1240.80		
1 Log Chipper, 22" Tree		554.80		610.28		
1 Dozer, 105 H.P.		505.40		555.94		
1 Chain Saw, gas, 36" Long		41.60		45.76	55.74	61.32
40 L.H., Daily Totals		$3558.20		$4632.78	$88.95	$115.82

Crew B-88	Hr.	Daily	Hr.	Daily	Bare Costs	Incl. O&P
1 Laborer	$26.45	$211.60	$44.30	$354.40	$33.69	$55.23
6 Equip. Oper. (med.)	34.90	1675.20	57.05	2738.40		
2 Feller Bunchers, 100 H.P.		1128.00		1240.80		
1 Log Chipper, 22" Tree		554.80		610.28		
2 Log Skidders, 50 H.P.		1688.00		1856.80		
1 Dozer, 105 H.P.		505.40		555.94		
1 Chain Saw, gas, 36" Long		41.60		45.76	69.96	76.96
56 L.H., Daily Totals		$5804.60		$7402.38	$103.65	$132.19

Crew B-89	Hr.	Daily	Hr.	Daily	Bare Costs	Incl. O&P
1 Skilled Worker	$34.05	$272.40	$57.00	$456.00	$30.25	$50.65
1 Building Laborer	26.45	211.60	44.30	354.40		
1 Flatbed Truck, Gas, 3 Ton		217.00		238.70		
1 Concrete Saw		134.20		147.62		
1 Water Tank, 65 Gal.		18.45		20.30	23.10	25.41
16 L.H., Daily Totals		$853.65		$1217.02	$53.35	$76.06

Crew B-89A	Hr.	Daily	Hr.	Daily	Bare Costs	Incl. O&P
1 Skilled Worker	$34.05	$272.40	$57.00	$456.00	$30.25	$50.65
1 Laborer	26.45	211.60	44.30	354.40		
1 Core Drill (large)		109.80		120.78	6.86	7.55
16 L.H., Daily Totals		$593.80		$931.18	$37.11	$58.20

Crew B-89B	Hr.	Daily	Hr.	Daily	Bare Costs	Incl. O&P
1 Equip. Oper. (light)	$33.55	$268.40	$54.80	$438.40	$29.98	$49.30
1 Truck Driver, Light	26.40	211.20	43.80	350.40		
1 Wall Saw, Hydraulic, 10 H.P.		89.00		97.90		
1 Generator, Diesel, 100 kW		329.00		361.90		
1 Water Tank, 65 Gal.		18.45		20.30		
1 Flatbed Truck, Gas, 3 Ton		217.00		238.70	40.84	44.92
16 L.H., Daily Totals		$1133.05		$1507.60	$70.82	$94.22

Crew B-90	Hr.	Daily	Hr.	Daily	Bare Costs	Incl. O&P
1 Labor Foreman (outside)	$28.45	$227.60	$47.65	$381.20	$28.66	$47.54
3 Laborers	26.45	634.80	44.30	1063.20		
2 Equip. Oper. (light)	33.55	536.80	54.80	876.80		
2 Truck Drivers (heavy)	27.20	435.20	45.10	721.60		
1 Road Mixer, 310 H.P.		1890.00		2079.00		
1 Dist. Truck, 2000 Gal.		280.80		308.88	33.92	37.31
64 L.H., Daily Totals		$4005.20		$5430.68	$62.58	$84.85

Crew B-90A	Hr.	Daily	Hr.	Daily	Bare Costs	Incl. O&P
1 Labor Foreman	$28.45	$227.60	$47.65	$381.20	$31.56	$52.06
2 Laborers	26.45	423.20	44.30	708.80		
4 Equip. Oper. (med.)	34.90	1116.80	57.05	1825.60		
2 Graders, 30,000 Lbs.		1085.60		1194.16		
1 Tandem Roller, 10 Ton		224.20		246.62		
1 Roller, Pneum. Whl, 12 Ton		309.20		340.12	28.91	31.80
56 L.H., Daily Totals		$3386.60		$4696.50	$60.48	$83.87

Crew B-90B	Hr.	Daily	Hr.	Daily	Bare Costs	Incl. O&P
1 Labor Foreman	$28.45	$227.60	$47.65	$381.20	$31.01	$51.23
2 Laborers	26.45	423.20	44.30	708.80		
3 Equip. Oper. (med.)	34.90	837.60	57.05	1369.20		
1 Tandem Roller, 10 Ton		224.20		246.62		
1 Roller, Pneum. Whl, 12 Ton		309.20		340.12		
1 Road Mixer, 310 H.P.		1890.00		2079.00	50.49	55.54
48 L.H., Daily Totals		$3911.80		$5124.94	$81.50	$106.77

Crew B-91	Hr.	Daily	Hr.	Daily	Bare Costs	Incl. O&P
1 Labor Foreman (outside)	$28.45	$227.60	$47.65	$381.20	$31.02	$51.19
2 Laborers	26.45	423.20	44.30	708.80		
4 Equip. Oper. (med.)	34.90	1116.80	57.05	1825.60		
1 Truck Driver (heavy)	27.20	217.60	45.10	360.80		
1 Dist. Tanker, 3000 Gallon		309.00		339.90		
1 Truck Tractor, 6x4, 380 H.P.		459.40		505.34		
1 Aggreg. Spreader, S.P.		748.40		823.24		
1 Roller, Pneum. Whl, 12 Ton		309.20		340.12		
1 Tandem Roller, 10 Ton		224.20		246.62	32.03	35.24
64 L.H., Daily Totals		$4035.40		$5531.62	$63.05	$86.43

Crew No.	Bare Costs		Incl. Subs O&P		Cost Per Labor-Hour	
Crew B-92	**Hr.**	**Daily**	**Hr.**	**Daily**	**Bare Costs**	**Incl. O&P**
1 Labor Foreman (outside)	$28.45	$227.60	$47.65	$381.20	$26.95	$45.14
3 Laborers	26.45	634.80	44.30	1063.20		
1 Crack Cleaner, 25 H.P.		53.00		58.30		
1 Air Compressor, 60 cfm		110.20		121.22		
1 Tar Kettle, T.M.		74.65		82.11		
1 Flatbed Truck, Gas, 3 Ton		217.00		238.70	14.21	15.64
32 L.H., Daily Totals		$1317.25		$1944.73	$41.16	$60.77

Crew B-93	Hr.	Daily	Hr.	Daily	Bare Costs	Incl. O&P
1 Equip. Oper. (med.)	$34.90	$279.20	$57.05	$456.40	$34.90	$57.05
1 Feller Buncher, 100 H.P.		564.00		620.40	70.50	77.55
8 L.H., Daily Totals		$843.20		$1076.80	$105.40	$134.60

Crew B-94A	Hr.	Daily	Hr.	Daily	Bare Costs	Incl. O&P
1 Laborer	$26.45	$211.60	$44.30	$354.40	$26.45	$44.30
1 Diaphragm Water Pump, 2"		66.60		73.26		
1 -20' Suction Hose, 2"		1.95		2.15		
2 -50' Discharge Hoses, 2"		1.80		1.98	8.79	9.67
8 L.H., Daily Totals		$281.95		$431.79	$35.24	$53.97

Crew B-94B	Hr.	Daily	Hr.	Daily	Bare Costs	Incl. O&P
1 Laborer	$26.45	$211.60	$44.30	$354.40	$26.45	$44.30
1 Diaphragm Water Pump, 4"		109.40		120.34		
1 -20' Suction Hose, 4"		3.25		3.58		
2 -50' Discharge Hoses, 4"		4.70		5.17	14.67	16.14
8 L.H., Daily Totals		$328.95		$483.49	$41.12	$60.44

Crew B-94C	Hr.	Daily	Hr.	Daily	Bare Costs	Incl. O&P
1 Laborer	$26.45	$211.60	$44.30	$354.40	$26.45	$44.30
1 Centrifugal Water Pump, 3"		71.00		78.10		
1 -20' Suction Hose, 3"		3.05		3.36		
2 -50' Discharge Hoses, 3"		3.50		3.85	9.69	10.66
8 L.H., Daily Totals		$289.15		$439.70	$36.14	$54.96

Crew B-94D	Hr.	Daily	Hr.	Daily	Bare Costs	Incl. O&P
1 Laborer	$26.45	$211.60	$44.30	$354.40	$26.45	$44.30
1 Centr. Water Pump, 6"		291.60		320.76		
1 -20' Suction Hose, 6"		11.90		13.09		
2 -50' Discharge Hoses, 6"		12.60		13.86	39.51	43.46
8 L.H., Daily Totals		$527.70		$702.11	$65.96	$87.76

Crew C-1	Hr.	Daily	Hr.	Daily	Bare Costs	Incl. O&P
2 Carpenters	$33.55	$536.80	$56.15	$898.40	$29.73	$49.73
1 Carpenter Helper	25.35	202.80	42.30	338.40		
1 Laborer	26.45	211.60	44.30	354.40		
32 L.H., Daily Totals		$951.20		$1591.20	$29.73	$49.73

Crew C-2	Hr.	Daily	Hr.	Daily	Bare Costs	Incl. O&P
1 Carpenter Foreman (out)	$35.55	$284.40	$59.50	$476.00	$29.97	$50.12
2 Carpenters	33.55	536.80	56.15	898.40		
2 Carpenter Helpers	25.35	405.60	42.30	676.80		
1 Laborer	26.45	211.60	44.30	354.40		
48 L.H., Daily Totals		$1438.40		$2405.60	$29.97	$50.12

Crew C-2A	Hr.	Daily	Hr.	Daily	Bare Costs	Incl. O&P
1 Carpenter Foreman (out)	$35.55	$284.40	$59.50	$476.00	$32.42	$53.85
3 Carpenters	33.55	805.20	56.15	1347.60		
1 Cement Finisher	31.85	254.80	50.85	406.80		
1 Laborer	26.45	211.60	44.30	354.40		
48 L.H., Daily Totals		$1556.00		$2584.80	$32.42	$53.85

Crew C-3	Hr.	Daily	Hr.	Daily	Bare Costs	Incl. O&P
1 Rodman Foreman	$38.30	$306.40	$66.55	$532.40	$32.51	$55.44
3 Rodmen (reinf.)	36.30	871.20	63.10	1514.40		
1 Equip. Oper. (light)	33.55	268.40	54.80	438.40		
3 Laborers	26.45	634.80	44.30	1063.20		
3 Stressing Equipment		29.40		32.34		
.5 Grouting Equipment		76.60		84.26	1.66	1.82
64 L.H., Daily Totals		$2186.80		$3665.00	$34.17	$57.27

Crew C-4	Hr.	Daily	Hr.	Daily	Bare Costs	Incl. O&P
1 Rodman Foreman	$38.30	$306.40	$66.55	$532.40	$34.34	$59.26
2 Rodmen (reinf.)	36.30	580.80	63.10	1009.60		
1 Building Laborer	26.45	211.60	44.30	354.40		
3 Stressing Equipment		29.40		32.34	0.92	1.01
32 L.H., Daily Totals		$1128.20		$1928.74	$35.26	$60.27

Crew C-4A	Hr.	Daily	Hr.	Daily	Bare Costs	Incl. O&P
2 Rodmen (reinf.)	$36.30	$580.80	$63.10	$1009.60	$36.30	$63.10
4 Stressing Equipment		39.20		43.12	2.45	2.69
16 L.H., Daily Totals		$620.00		$1052.72	$38.75	$65.80

Crew C-5	Hr.	Daily	Hr.	Daily	Bare Costs	Incl. O&P
1 Rodman Foreman	$38.30	$306.40	$66.55	$532.40	$33.27	$56.64
2 Rodmen (reinf.)	36.30	580.80	63.10	1009.60		
1 Equip. Oper. (crane)	35.80	286.40	58.50	468.00		
2 Building Laborers	26.45	423.20	44.30	708.80		
1 Hyd. Crane, 25 Ton		748.60		823.46	15.60	17.16
48 L.H., Daily Totals		$2345.40		$3542.26	$48.86	$73.80

Crew C-6	Hr.	Daily	Hr.	Daily	Bare Costs	Incl. O&P
1 Labor Foreman (outside)	$28.45	$227.60	$47.65	$381.20	$27.68	$45.95
4 Laborers	26.45	846.40	44.30	1417.60		
1 Cement Finisher	31.85	254.80	50.85	406.80		
2 Gas Engine Vibrators		46.40		51.04	0.97	1.06
48 L.H., Daily Totals		$1375.20		$2256.64	$28.65	$47.01

Crew C-7	Hr.	Daily	Hr.	Daily	Bare Costs	Incl. O&P
1 Labor Foreman (outside)	$28.45	$227.60	$47.65	$381.20	$28.72	$47.53
5 Laborers	26.45	1058.00	44.30	1772.00		
1 Cement Finisher	31.85	254.80	50.85	406.80		
1 Equip. Oper. (med.)	34.90	279.20	57.05	456.40		
1 Equip. Oper. (oiler)	31.05	248.40	50.75	406.00		
2 Gas Engine Vibrators		46.40		51.04		
1 Concrete Bucket, 1 C.Y.		23.20		25.52		
1 Hyd. Crane, 55 Ton		1208.00		1328.80	17.74	19.52
72 L.H., Daily Totals		$3345.60		$4827.76	$46.47	$67.05

Crew C-8	Hr.	Daily	Hr.	Daily	Bare Costs	Incl. O&P
1 Labor Foreman (outside)	$28.45	$227.60	$47.65	$381.20	$29.49	$48.47
3 Laborers	26.45	634.80	44.30	1063.20		
2 Cement Finishers	31.85	509.60	50.85	813.60		
1 Equip. Oper. (med.)	34.90	279.20	57.05	456.40		
1 Concrete Pump (small)		741.00		815.10	13.23	14.56
56 L.H., Daily Totals		$2392.20		$3529.50	$42.72	$63.03

Crew C-8A	Hr.	Daily	Hr.	Daily	Bare Costs	Incl. O&P
1 Labor Foreman (outside)	$28.45	$227.60	$47.65	$381.20	$28.58	$47.04
3 Laborers	26.45	634.80	44.30	1063.20		
2 Cement Finishers	31.85	509.60	50.85	813.60		
48 L.H., Daily Totals		$1372.00		$2258.00	$28.58	$47.04

Crew No.	Bare Costs		Incl. Subs O&P		Cost Per Labor-Hour	
Crew C-8B	**Hr.**	**Daily**	**Hr.**	**Daily**	**Bare Costs**	**Incl. O&P**
1 Labor Foreman (outside)	$28.45	$227.60	$47.65	$381.20	$28.54	$47.52
3 Laborers	26.45	634.80	44.30	1063.20		
1 Equip. Oper. (med.)	34.90	279.20	57.05	456.40		
1 Vibrating Power Screed		65.50		72.05		
1 Roller, Vibratory, 25 Ton		596.80		656.48		
1 Dozer, 200 H.P.		1051.00		1156.10	42.83	47.12
40 L.H., Daily Totals		$2854.90		$3785.43	$71.37	$94.64

Crew C-8C	Hr.	Daily	Hr.	Daily	Bare Costs	Incl. O&P
1 Labor Foreman (outside)	$28.45	$227.60	$47.65	$381.20	$29.09	$48.08
3 Laborers	26.45	634.80	44.30	1063.20		
1 Cement Finisher	31.85	254.80	50.85	406.80		
1 Equip. Oper. (med.)	34.90	279.20	57.05	456.40		
1 Shotcrete Rig, 12 C.Y./hr		249.20		274.12		
1 Air Compressor, 160 cfm		125.00		137.50		
4 -50' Air Hoses, 1"		16.40		18.04		
4 -50' Air Hoses, 2"		31.00		34.10	8.78	9.66
48 L.H., Daily Totals		$1818.00		$2771.36	$37.88	$57.74

Crew C-8D	Hr.	Daily	Hr.	Daily	Bare Costs	Incl. O&P
1 Labor Foreman (outside)	$28.45	$227.60	$47.65	$381.20	$30.07	$49.40
1 Laborer	26.45	211.60	44.30	354.40		
1 Cement Finisher	31.85	254.80	50.85	406.80		
1 Equipment Oper. (light)	33.55	268.40	54.80	438.40		
1 Air Compressor, 250 cfm		162.80		179.08		
2 -50' Air Hoses, 1"		8.20		9.02	5.34	5.88
32 L.H., Daily Totals		$1133.40		$1768.90	$35.42	$55.28

Crew C-8E	Hr.	Daily	Hr.	Daily	Bare Costs	Incl. O&P
1 Labor Foreman (outside)	$28.45	$227.60	$47.65	$381.20	$28.87	$47.70
3 Laborers	26.45	634.80	44.30	1063.20		
1 Cement Finisher	31.85	254.80	50.85	406.80		
1 Equipment Oper. (light)	33.55	268.40	54.80	438.40		
1 Shotcrete Rig, 35 C.Y./hr		280.00		308.00		
1 Air Compressor, 250 cfm		162.80		179.08		
4 -50' Air Hoses, 1"		16.40		18.04		
4 -50' Air Hoses, 2"		31.00		34.10	10.21	11.23
48 L.H., Daily Totals		$1875.80		$2828.82	$39.08	$58.93

Crew C-10	Hr.	Daily	Hr.	Daily	Bare Costs	Incl. O&P
1 Laborer	$26.45	$211.60	$44.30	$354.40	$30.05	$48.67
2 Cement Finishers	31.85	509.60	50.85	813.60		
24 L.H., Daily Totals		$721.20		$1168.00	$30.05	$48.67

Crew C-10B	Hr.	Daily	Hr.	Daily	Bare Costs	Incl. O&P
3 Laborers	$26.45	$634.80	$44.30	$1063.20	$28.61	$46.92
2 Cement Finishers	31.85	509.60	50.85	813.60		
1 Concrete Mixer, 10 C.F.		151.20		166.32		
2 Trowels, 48" Walk-Behind		76.80		84.48	5.70	6.27
40 L.H., Daily Totals		$1372.40		$2127.60	$34.31	$53.19

Crew C-10C	Hr.	Daily	Hr.	Daily	Bare Costs	Incl. O&P
1 Laborer	$26.45	$211.60	$44.30	$354.40	$30.05	$48.67
2 Cement Finishers	31.85	509.60	50.85	813.60		
1 Trowel, 48" Walk-Behind		38.40		42.24	1.60	1.76
24 L.H., Daily Totals		$759.60		$1210.24	$31.65	$50.43

Crew C-10D	Hr.	Daily	Hr.	Daily	Bare Costs	Incl. O&P
1 Laborer	$26.45	$211.60	$44.30	$354.40	$30.05	$48.67
2 Cement Finishers	31.85	509.60	50.85	813.60		
1 Vibrating Power Screed		65.50		72.05		
1 Trowel, 48" Walk-Behind		38.40		42.24	4.33	4.76
24 L.H., Daily Totals		$825.10		$1282.29	$34.38	$53.43

Crew C-10E	Hr.	Daily	Hr.	Daily	Bare Costs	Incl. O&P
1 Laborer	$26.45	$211.60	$44.30	$354.40	$30.05	$48.67
2 Cement Finishers	31.85	509.60	50.85	813.60		
1 Vibrating Power Screed		65.50		72.05		
1 Cement Trowel, 96" Ride-On		157.40		173.14	9.29	10.22
24 L.H., Daily Totals		$944.10		$1413.19	$39.34	$58.88

Crew C-11	Hr.	Daily	Hr.	Daily	Bare Costs	Incl. O&P
1 Skilled Worker Foreman	$36.05	$288.40	$60.35	$482.80	$34.59	$57.69
5 Skilled Workers	34.05	1362.00	57.00	2280.00		
1 Equip. Oper. (crane)	35.80	286.40	58.50	468.00		
1 Lattice Boom Crane, 150 Ton		1906.00		2096.60	34.04	37.44
56 L.H., Daily Totals		$3842.80		$5327.40	$68.62	$95.13

Crew C-12	Hr.	Daily	Hr.	Daily	Bare Costs	Incl. O&P
1 Carpenter Foreman (out)	$35.55	$284.40	$59.50	$476.00	$33.08	$55.13
3 Carpenters	33.55	805.20	56.15	1347.60		
1 Laborer	26.45	211.60	44.30	354.40		
1 Equip. Oper. (crane)	35.80	286.40	58.50	468.00		
1 Hyd. Crane, 12 Ton		655.60		721.16	13.66	15.02
48 L.H., Daily Totals		$2243.20		$3367.16	$46.73	$70.15

Crew C-13	Hr.	Daily	Hr.	Daily	Bare Costs	Incl. O&P
2 Struc. Steel Workers	$36.40	$582.40	$69.40	$1110.40	$35.45	$64.98
1 Carpenter	33.55	268.40	56.15	449.20		
1 Welder, gas engine, 300 amp		109.00		119.90	4.54	5.00
24 L.H., Daily Totals		$959.80		$1679.50	$39.99	$69.98

Crew C-14	Hr.	Daily	Hr.	Daily	Bare Costs	Incl. O&P
1 Carpenter Foreman (out)	$35.55	$284.40	$59.50	$476.00	$30.32	$50.63
3 Carpenters	33.55	805.20	56.15	1347.60		
2 Carpenter Helpers	25.35	405.60	42.30	676.80		
4 Laborers	26.45	846.40	44.30	1417.60		
2 Rodmen (reinf.)	36.30	580.80	63.10	1009.60		
2 Rodman Helpers	25.35	405.60	42.30	676.80		
2 Cement Finishers	31.85	509.60	50.85	813.60		
1 Equip. Oper. (crane)	35.80	286.40	58.50	468.00		
1 Hyd. Crane, 80 Ton		1693.00		1862.30	12.45	13.69
136 L.H., Daily Totals		$5817.00		$8748.30	$42.77	$64.33

Crew C-14A	Hr.	Daily	Hr.	Daily	Bare Costs	Incl. O&P
1 Carpenter Foreman (out)	$35.55	$284.40	$59.50	$476.00	$33.49	$56.27
16 Carpenters	33.55	4294.40	56.15	7187.20		
4 Rodmen (reinf.)	36.30	1161.60	63.10	2019.20		
2 Laborers	26.45	423.20	44.30	708.80		
1 Cement Finisher	31.85	254.80	50.85	406.80		
1 Equip. Oper. (med.)	34.90	279.20	57.05	456.40		
1 Gas Engine Vibrator		23.20		25.52		
1 Concrete Pump (small)		741.00		815.10	3.82	4.20
200 L.H., Daily Totals		$7461.80		$12095.02	$37.31	$60.48

Crew No.	Bare Costs		Incl. Subs O&P		Cost Per Labor-Hour	
Crew C-14B	**Hr.**	**Daily**	**Hr.**	**Daily**	**Bare Costs**	**Incl. O&P**
1 Carpenter Foreman (out)	$35.55	$284.40	$59.50	$476.00	$33.42	$56.06
16 Carpenters	33.55	4294.40	56.15	7187.20		
4 Rodmen (reinf.)	36.30	1161.60	63.10	2019.20		
2 Laborers	26.45	423.20	44.30	708.80		
2 Cement Finishers	31.85	509.60	50.85	813.60		
1 Equip. Oper. (med.)	34.90	279.20	57.05	456.40		
1 Gas Engine Vibrator		23.20		25.52		
1 Concrete Pump (small)		741.00		815.10	3.67	4.04
208 L.H., Daily Totals		$7716.60		$12501.82	$37.10	$60.10
Crew C-14C	**Hr.**	**Daily**	**Hr.**	**Daily**	**Bare Costs**	**Incl. O&P**
1 Carpenter Foreman (out)	$35.55	$284.40	$59.50	$476.00	$31.94	$53.62
6 Carpenters	33.55	1610.40	56.15	2695.20		
2 Rodmen (reinf.)	36.30	580.80	63.10	1009.60		
4 Laborers	26.45	846.40	44.30	1417.60		
1 Cement Finisher	31.85	254.80	50.85	406.80		
1 Gas Engine Vibrator		23.20		25.52	0.21	0.23
112 L.H., Daily Totals		$3600.00		$6030.72	$32.14	$53.85
Crew C-14D	**Hr.**	**Daily**	**Hr.**	**Daily**	**Bare Costs**	**Incl. O&P**
1 Carpenter Foreman (out)	$35.55	$284.40	$59.50	$476.00	$33.27	$55.72
18 Carpenters	33.55	4831.20	56.15	8085.60		
2 Rodmen (reinf.)	36.30	580.80	63.10	1009.60		
2 Laborers	26.45	423.20	44.30	708.80		
1 Cement Finisher	31.85	254.80	50.85	406.80		
1 Equip. Oper. (med.)	34.90	279.20	57.05	456.40		
1 Gas Engine Vibrator		23.20		25.52		
1 Concrete Pump (small)		741.00		815.10	3.82	4.20
200 L.H., Daily Totals		$7417.80		$11983.82	$37.09	$59.92
Crew C-14E	**Hr.**	**Daily**	**Hr.**	**Daily**	**Bare Costs**	**Incl. O&P**
1 Carpenter Foreman (out)	$35.55	$284.40	$59.50	$476.00	$32.64	$55.27
2 Carpenters	33.55	536.80	56.15	898.40		
4 Rodmen (reinf.)	36.30	1161.60	63.10	2019.20		
3 Laborers	26.45	634.80	44.30	1063.20		
1 Cement Finisher	31.85	254.80	50.85	406.80		
1 Gas Engine Vibrator		23.20		25.52	0.26	0.29
88 L.H., Daily Totals		$2895.60		$4889.12	$32.90	$55.56
Crew C-14F	**Hr.**	**Daily**	**Hr.**	**Daily**	**Bare Costs**	**Incl. O&P**
1 Laborer Foreman (out)	$28.45	$227.60	$47.65	$381.20	$30.27	$49.04
2 Laborers	26.45	423.20	44.30	708.80		
6 Cement Finishers	31.85	1528.80	50.85	2440.80		
1 Gas Engine Vibrator		23.20		25.52	0.32	0.35
72 L.H., Daily Totals		$2202.80		$3556.32	$30.59	$49.39
Crew C-14G	**Hr.**	**Daily**	**Hr.**	**Daily**	**Bare Costs**	**Incl. O&P**
1 Laborer Foreman (out)	$28.45	$227.60	$47.65	$381.20	$29.82	$48.52
2 Laborers	26.45	423.20	44.30	708.80		
4 Cement Finishers	31.85	1019.20	50.85	1627.20		
1 Gas Engine Vibrator		23.20		25.52	0.41	0.46
56 L.H., Daily Totals		$1693.20		$2742.72	$30.24	$48.98
Crew C-14H	**Hr.**	**Daily**	**Hr.**	**Daily**	**Bare Costs**	**Incl. O&P**
1 Carpenter Foreman (out)	$35.55	$284.40	$59.50	$476.00	$32.88	$55.01
2 Carpenters	33.55	536.80	56.15	898.40		
1 Rodman (reinf.)	36.30	290.40	63.10	504.80		
1 Laborer	26.45	211.60	44.30	354.40		
1 Cement Finisher	31.85	254.80	50.85	406.80		
1 Gas Engine Vibrator		23.20		25.52	0.48	0.53
48 L.H., Daily Totals		$1601.20		$2665.92	$33.36	$55.54

Crew No.	Bare Costs		Incl. Subs O&P		Cost Per Labor-Hour	
Crew C-14L	**Hr.**	**Daily**	**Hr.**	**Daily**	**Bare Costs**	**Incl. O&P**
1 Carpenter Foreman (out)	$35.55	$284.40	$59.50	$476.00	$31.21	$52.04
6 Carpenters	33.55	1610.40	56.15	2695.20		
4 Laborers	26.45	846.40	44.30	1417.60		
1 Cement Finisher	31.85	254.80	50.85	406.80		
1 Gas Engine Vibrator		23.20		25.52	0.24	0.27
96 L.H., Daily Totals		$3019.20		$5021.12	$31.45	$52.30
Crew C-15	**Hr.**	**Daily**	**Hr.**	**Daily**	**Bare Costs**	**Incl. O&P**
1 Carpenter Foreman (out)	$35.55	$284.40	$59.50	$476.00	$31.33	$52.17
2 Carpenters	33.55	536.80	56.15	898.40		
3 Laborers	26.45	634.80	44.30	1063.20		
2 Cement Finishers	31.85	509.60	50.85	813.60		
1 Rodman (reinf.)	36.30	290.40	63.10	504.80		
72 L.H., Daily Totals		$2256.00		$3756.00	$31.33	$52.17
Crew C-16	**Hr.**	**Daily**	**Hr.**	**Daily**	**Bare Costs**	**Incl. O&P**
1 Labor Foreman (outside)	$28.45	$227.60	$47.65	$381.20	$29.49	$48.47
3 Laborers	26.45	634.80	44.30	1063.20		
2 Cement Finishers	31.85	509.60	50.85	813.60		
1 Equip. Oper. (med.)	34.90	279.20	57.05	456.40		
1 Gunite Pump Rig		178.00		195.80		
2 -50' Air Hoses, 3/4"		6.10		6.71		
2 -50' Air Hoses, 2"		15.50		17.05	3.56	3.92
56 L.H., Daily Totals		$1850.80		$2933.96	$33.05	$52.39
Crew C-17	**Hr.**	**Daily**	**Hr.**	**Daily**	**Bare Costs**	**Incl. O&P**
2 Skilled Worker Foremen	$36.05	$576.80	$60.35	$965.60	$34.45	$57.67
8 Skilled Workers	34.05	2179.20	57.00	3648.00		
80 L.H., Daily Totals		$2756.00		$4613.60	$34.45	$57.67
Crew C-17A	**Hr.**	**Daily**	**Hr.**	**Daily**	**Bare Costs**	**Incl. O&P**
2 Skilled Worker Foremen	$36.05	$576.80	$60.35	$965.60	$34.47	$57.68
8 Skilled Workers	34.05	2179.20	57.00	3648.00		
.125 Equip. Oper. (crane)	35.80	35.80	58.50	58.50		
.125 Hyd. Crane, 80 Ton		211.63		232.79	2.61	2.87
81 L.H., Daily Totals		$3003.43		$4904.89	$37.08	$60.55
Crew C-17B	**Hr.**	**Daily**	**Hr.**	**Daily**	**Bare Costs**	**Incl. O&P**
2 Skilled Worker Foremen	$36.05	$576.80	$60.35	$965.60	$34.48	$57.69
8 Skilled Workers	34.05	2179.20	57.00	3648.00		
.25 Equip. Oper. (crane)	35.80	71.60	58.50	117.00		
.25 Hyd. Crane, 80 Ton		423.25		465.57		
.25 Trowel, 48" Walk-Behind		9.60		10.56	5.28	5.81
82 L.H., Daily Totals		$3260.45		$5206.73	$39.76	$63.50
Crew C-17C	**Hr.**	**Daily**	**Hr.**	**Daily**	**Bare Costs**	**Incl. O&P**
2 Skilled Worker Foremen	$36.05	$576.80	$60.35	$965.60	$34.50	$57.70
8 Skilled Workers	34.05	2179.20	57.00	3648.00		
.375 Equip. Oper. (crane)	35.80	107.40	58.50	175.50		
.375 Hyd. Crane, 80 Ton		634.88		698.36	7.65	8.41
83 L.H., Daily Totals		$3498.28		$5487.46	$42.15	$66.11
Crew C-17D	**Hr.**	**Daily**	**Hr.**	**Daily**	**Bare Costs**	**Incl. O&P**
2 Skilled Worker Foremen	$36.05	$576.80	$60.35	$965.60	$34.51	$57.71
8 Skilled Workers	34.05	2179.20	57.00	3648.00		
.5 Equip. Oper. (crane)	35.80	143.20	58.50	234.00		
.5 Hyd. Crane, 80 Ton		846.50		931.15	10.08	11.09
84 L.H., Daily Totals		$3745.70		$5778.75	$44.59	$68.79

Crew C-17E	Hr.	Daily	Hr.	Daily	Bare Costs	Incl. O&P
2 Skilled Worker Foremen	$36.05	$576.80	$60.35	$965.60	$34.45	$57.67
8 Skilled Workers	34.05	2179.20	57.00	3648.00		
1 Hyd. Jack with Rods		89.10		98.01	1.11	1.23
80 L.H., Daily Totals		$2845.10		$4711.61	$35.56	$58.90

Crew C-18	Hr.	Daily	Hr.	Daily	Bare Costs	Incl. O&P
.125 Labor Foreman (out)	$28.45	$28.45	$47.65	$47.65	$26.67	$44.67
1 Laborer	26.45	211.60	44.30	354.40		
1 Concrete Cart, 10 C.F.		55.00		60.50	6.11	6.72
9 L.H., Daily Totals		$295.05		$462.55	$32.78	$51.39

Crew C-19	Hr.	Daily	Hr.	Daily	Bare Costs	Incl. O&P
.125 Labor Foreman (out)	$28.45	$28.45	$47.65	$47.65	$26.67	$44.67
1 Laborer	26.45	211.60	44.30	354.40		
1 Concrete Cart, 18 C.F.		91.60		100.76	10.18	11.20
9 L.H., Daily Totals		$331.65		$502.81	$36.85	$55.87

Crew C-20	Hr.	Daily	Hr.	Daily	Bare Costs	Incl. O&P
1 Labor Foreman (outside)	$28.45	$227.60	$47.65	$381.20	$28.43	$47.13
5 Laborers	26.45	1058.00	44.30	1772.00		
1 Cement Finisher	31.85	254.80	50.85	406.80		
1 Equip. Oper. (med.)	34.90	279.20	57.05	456.40		
2 Gas Engine Vibrators		46.40		51.04		
1 Concrete Pump (small)		741.00		815.10	12.30	13.53
64 L.H., Daily Totals		$2607.00		$3882.54	$40.73	$60.66

Crew C-21	Hr.	Daily	Hr.	Daily	Bare Costs	Incl. O&P
1 Labor Foreman (outside)	$28.45	$227.60	$47.65	$381.20	$28.43	$47.13
5 Laborers	26.45	1058.00	44.30	1772.00		
1 Cement Finisher	31.85	254.80	50.85	406.80		
1 Equip. Oper. (med.)	34.90	279.20	57.05	456.40		
2 Gas Engine Vibrators		46.40		51.04		
1 Concrete Conveyer		174.60		192.06	3.45	3.80
64 L.H., Daily Totals		$2040.60		$3259.50	$31.88	$50.93

Crew C-22	Hr.	Daily	Hr.	Daily	Bare Costs	Incl. O&P
1 Rodman Foreman	$38.30	$306.40	$66.55	$532.40	$36.54	$63.35
4 Rodmen (reinf.)	36.30	1161.60	63.10	2019.20		
.125 Equip. Oper. (crane)	35.80	35.80	58.50	58.50		
.125 Equip. Oper. Oiler	31.05	31.05	50.75	50.75		
.125 Hyd. Crane, 25 Ton		93.58		102.93	2.23	2.45
42 L.H., Daily Totals		$1628.43		$2763.78	$38.77	$65.80

Crew C-23	Hr.	Daily	Hr.	Daily	Bare Costs	Incl. O&P
2 Skilled Worker Foremen	$36.05	$576.80	$60.35	$965.60	$34.33	$57.20
6 Skilled Workers	34.05	1634.40	57.00	2736.00		
1 Equip. Oper. (crane)	35.80	286.40	58.50	468.00		
1 Equip. Oper. Oiler	31.05	248.40	50.75	406.00		
1 Lattice Boom Crane, 90 Ton		1622.00		1784.20	20.27	22.30
80 L.H., Daily Totals		$4368.00		$6359.80	$54.60	$79.50

Crew C-24	Hr.	Daily	Hr.	Daily	Bare Costs	Incl. O&P
2 Skilled Worker Foremen	$36.05	$576.80	$60.35	$965.60	$34.33	$57.20
6 Skilled Workers	34.05	1634.40	57.00	2736.00		
1 Equip. Oper. (crane)	35.80	286.40	58.50	468.00		
1 Equip. Oper. Oiler	31.05	248.40	50.75	406.00		
1 Lattice Boom Crane, 150 Ton		1906.00		2096.60	23.82	26.21
80 L.H., Daily Totals		$4652.00		$6672.20	$58.15	$83.40

Crew C-25	Hr.	Daily	Hr.	Daily	Bare Costs	Incl. O&P
2 Rodmen (reinf.)	$36.30	$580.80	$63.10	$1009.60	$28.60	$50.40
2 Rodmen Helpers	20.90	334.40	37.70	603.20		
32 L.H., Daily Totals		$915.20		$1612.80	$28.60	$50.40

Crew C-27	Hr.	Daily	Hr.	Daily	Bare Costs	Incl. O&P
2 Cement Finishers	$31.85	$509.60	$50.85	$813.60	$31.85	$50.85
1 Concrete Saw		134.20		147.62	8.39	9.23
16 L.H., Daily Totals		$643.80		$961.22	$40.24	$60.08

Crew C-28	Hr.	Daily	Hr.	Daily	Bare Costs	Incl. O&P
1 Cement Finisher	$31.85	$254.80	$50.85	$406.80	$31.85	$50.85
1 Portable Air Compressor, Gas		19.00		20.90	2.38	2.61
8 L.H., Daily Totals		$273.80		$427.70	$34.23	$53.46

Crew C-29	Hr.	Daily	Hr.	Daily	Bare Costs	Incl. O&P
1 Laborer	$26.45	$211.60	$44.30	$354.40	$26.45	$44.30
1 Pressure Washer		58.80		64.68	7.35	8.09
8 L.H., Daily Totals		$270.40		$419.08	$33.80	$52.38

Crew D-1	Hr.	Daily	Hr.	Daily	Bare Costs	Incl. O&P
1 Bricklayer	$33.65	$269.20	$55.20	$441.60	$30.40	$49.88
1 Bricklayer Helper	27.15	217.20	44.55	356.40		
16 L.H., Daily Totals		$486.40		$798.00	$30.40	$49.88

Crew D-2	Hr.	Daily	Hr.	Daily	Bare Costs	Incl. O&P
3 Bricklayers	$33.65	$807.60	$55.20	$1324.80	$31.05	$50.94
2 Bricklayer Helpers	27.15	434.40	44.55	712.80		
40 L.H., Daily Totals		$1242.00		$2037.60	$31.05	$50.94

Crew D-3	Hr.	Daily	Hr.	Daily	Bare Costs	Incl. O&P
3 Bricklayers	$33.65	$807.60	$55.20	$1324.80	$31.17	$51.19
2 Bricklayer Helpers	27.15	434.40	44.55	712.80		
.25 Carpenter	33.55	67.10	56.15	112.30		
42 L.H., Daily Totals		$1309.10		$2149.90	$31.17	$51.19

Crew D-4	Hr.	Daily	Hr.	Daily	Bare Costs	Incl. O&P
1 Bricklayer	$33.65	$269.20	$55.20	$441.60	$28.31	$46.63
3 Bricklayer Helpers	27.15	651.60	44.55	1069.20		
1 Building Laborer	26.45	211.60	44.30	354.40		
1 Grout Pump, 50 C.F./hr.		125.80		138.38	3.15	3.46
40 L.H., Daily Totals		$1258.20		$2003.58	$31.45	$50.09

Crew D-5	Hr.	Daily	Hr.	Daily	Bare Costs	Incl. O&P
1 Block Mason Helper	$27.15	$217.20	$44.55	$356.40	$27.15	$44.55
8 L.H., Daily Totals		$217.20		$356.40	$27.15	$44.55

Crew D-6	Hr.	Daily	Hr.	Daily	Bare Costs	Incl. O&P
3 Bricklayers	$33.65	$807.60	$55.20	$1324.80	$30.40	$49.88
3 Bricklayer Helpers	27.15	651.60	44.55	1069.20		
48 L.H., Daily Totals		$1459.20		$2394.00	$30.40	$49.88

Crew D-7	Hr.	Daily	Hr.	Daily	Bare Costs	Incl. O&P
1 Tile Layer	$31.55	$252.40	$50.40	$403.20	$28.20	$45.05
1 Tile Layer Helper	24.85	198.80	39.70	317.60		
16 L.H., Daily Totals		$451.20		$720.80	$28.20	$45.05

Crew No.	Bare Costs		Incl. Subs O&P		Cost Per Labor-Hour	
Crew D-8	**Hr.**	**Daily**	**Hr.**	**Daily**	**Bare Costs**	**Incl. O&P**
3 Bricklayers	$33.65	$807.60	$55.20	$1324.80	$31.05	$50.94
2 Bricklayer Helpers	27.15	434.40	44.55	712.80		
40 L.H., Daily Totals		$1242.00		$2037.60	$31.05	$50.94

Crew D-9	Hr.	Daily	Hr.	Daily	Bare Costs	Incl. O&P
3 Bricklayers	$33.65	$807.60	$55.20	$1324.80	$30.40	$49.88
3 Bricklayer Helpers	27.15	651.60	44.55	1069.20		
48 L.H., Daily Totals		$1459.20		$2394.00	$30.40	$49.88

Crew D-10	Hr.	Daily	Hr.	Daily	Bare Costs	Incl. O&P
1 Bricklayer Foreman	$35.65	$285.20	$58.45	$467.60	$33.06	$54.17
1 Bricklayer	33.65	269.20	55.20	441.60		
1 Bricklayer Helper	27.15	217.20	44.55	356.40		
1 Equip. Oper. (crane)	35.80	286.40	58.50	468.00		
1 S.P. Crane, 4x4, 12 Ton		434.40		477.84	13.57	14.93
32 L.H., Daily Totals		$1492.40		$2211.44	$46.64	$69.11

Crew D-11	Hr.	Daily	Hr.	Daily	Bare Costs	Incl. O&P
2 Bricklayers	$33.65	$538.40	$55.20	$883.20	$31.48	$51.65
1 Bricklayer Helper	27.15	217.20	44.55	356.40		
24 L.H., Daily Totals		$755.60		$1239.60	$31.48	$51.65

Crew D-12	Hr.	Daily	Hr.	Daily	Bare Costs	Incl. O&P
2 Bricklayers	$33.65	$538.40	$55.20	$883.20	$30.40	$49.88
2 Bricklayer Helpers	27.15	434.40	44.55	712.80		
32 L.H., Daily Totals		$972.80		$1596.00	$30.40	$49.88

Crew D-13	Hr.	Daily	Hr.	Daily	Bare Costs	Incl. O&P
1 Bricklayer Foreman	$35.65	$285.20	$58.45	$467.60	$32.17	$52.74
2 Bricklayers	33.65	538.40	55.20	883.20		
2 Bricklayer Helpers	27.15	434.40	44.55	712.80		
1 Equip. Oper. (crane)	35.80	286.40	58.50	468.00		
1 S.P. Crane, 4x4, 12 Ton		434.40		477.84	9.05	9.96
48 L.H., Daily Totals		$1978.80		$3009.44	$41.23	$62.70

Crew E-1	Hr.	Daily	Hr.	Daily	Bare Costs	Incl. O&P
2 Struc. Steel Workers	$36.40	$582.40	$69.40	$1110.40	$36.40	$69.40
1 Welder, gas engine, 300 amp		109.00		119.90	6.81	7.49
16 L.H., Daily Totals		$691.40		$1230.30	$43.21	$76.89

Crew E-2	Hr.	Daily	Hr.	Daily	Bare Costs	Incl. O&P
1 Struc. Steel Foreman	$38.40	$307.20	$73.20	$585.60	$36.63	$68.22
4 Struc. Steel Workers	36.40	1164.80	69.40	2220.80		
1 Equip. Oper. (crane)	35.80	286.40	58.50	468.00		
1 Lattice Boom Crane, 90 Ton		1622.00		1784.20	33.79	37.17
48 L.H., Daily Totals		$3380.40		$5058.60	$70.42	$105.39

Crew E-3	Hr.	Daily	Hr.	Daily	Bare Costs	Incl. O&P
1 Struc. Steel Foreman	$38.40	$307.20	$73.20	$585.60	$37.07	$70.67
2 Struc. Steel Workers	36.40	582.40	69.40	1110.40		
1 Welder, gas engine, 300 amp		109.00		119.90	4.54	5.00
24 L.H., Daily Totals		$998.60		$1815.90	$41.61	$75.66

Crew E-4	Hr.	Daily	Hr.	Daily	Bare Costs	Incl. O&P
1 Struc. Steel Foreman	$38.40	$307.20	$73.20	$585.60	$36.90	$70.35
3 Struc. Steel Workers	36.40	873.60	69.40	1665.60		
1 Welder, gas engine, 300 amp		109.00		119.90	3.41	3.75
32 L.H., Daily Totals		$1289.80		$2371.10	$40.31	$74.10

Crew E-5	Hr.	Daily	Hr.	Daily	Bare Costs	Incl. O&P
1 Struc. Steel Foreman	$38.40	$307.20	$73.20	$585.60	$36.56	$68.61
7 Struc. Steel Workers	36.40	2038.40	69.40	3886.40		
1 Equip. Oper. (crane)	35.80	286.40	58.50	468.00		
1 Lattice Boom Crane, 90 Ton		1622.00		1784.20		
1 Welder, gas engine, 300 amp		109.00		119.90	24.04	26.45
72 L.H., Daily Totals		$4363.00		$6844.10	$60.60	$95.06

Crew E-6	Hr.	Daily	Hr.	Daily	Bare Costs	Incl. O&P
1 Struc. Steel Foreman	$38.40	$307.20	$73.20	$585.60	$36.30	$67.95
12 Struc. Steel Workers	36.40	3494.40	69.40	6662.40		
1 Equip. Oper. (crane)	35.80	286.40	58.50	468.00		
1 Equip. Oper. (light)	33.55	268.40	54.80	438.40		
1 Lattice Boom Crane, 90 Ton		1622.00		1784.20		
1 Welder, gas engine, 300 amp		109.00		119.90		
1 Air Compressor, 160 cfm		125.00		137.50		
2 Impact Wrenches		36.40		40.04	15.77	17.35
120 L.H., Daily Totals		$6248.80		$10236.04	$52.07	$85.30

Crew E-7	Hr.	Daily	Hr.	Daily	Bare Costs	Incl. O&P
1 Struc. Steel Foreman	$38.40	$307.20	$73.20	$585.60	$36.56	$68.61
7 Struc. Steel Workers	36.40	2038.40	69.40	3886.40		
1 Equip. Oper. (crane)	35.80	286.40	58.50	468.00		
1 Lattice Boom Crane, 90 Ton		1622.00		1784.20		
2 Welders, gas engine, 300 amp		218.00		239.80	25.56	28.11
72 L.H., Daily Totals		$4472.00		$6964.00	$62.11	$96.72

Crew E-8	Hr.	Daily	Hr.	Daily	Bare Costs	Incl. O&P
1 Struc. Steel Foreman	$38.40	$307.20	$73.20	$585.60	$36.53	$68.75
9 Struc. Steel Workers	36.40	2620.80	69.40	4996.80		
1 Equip. Oper. (crane)	35.80	286.40	58.50	468.00		
1 Lattice Boom Crane, 90 Ton		1622.00		1784.20		
4 Welders, gas engine, 300 amp		436.00		479.60	23.39	25.73
88 L.H., Daily Totals		$5272.40		$8314.20	$59.91	$94.48

Crew E-9	Hr.	Daily	Hr.	Daily	Bare Costs	Incl. O&P
2 Struc. Steel Foremen	$38.40	$614.40	$73.20	$1171.20	$36.23	$67.35
5 Struc. Steel Workers	36.40	1456.00	69.40	2776.00		
1 Welder Foreman	38.40	307.20	73.20	585.60		
5 Welders	36.40	1456.00	69.40	2776.00		
1 Equip. Oper. (crane)	35.80	286.40	58.50	468.00		
1 Equip. Oper. Oiler	31.05	248.40	50.75	406.00		
1 Equip. Oper. (light)	33.55	268.40	54.80	438.40		
1 Lattice Boom Crane, 90 Ton		1622.00		1784.20		
5 Welders, gas engine, 300 amp		545.00		599.50	16.93	18.62
128 L.H., Daily Totals		$6803.80		$11004.90	$53.15	$85.98

Crew E-10	Hr.	Daily	Hr.	Daily	Bare Costs	Incl. O&P
1 Struc. Steel Foreman	$38.40	$307.20	$73.20	$585.60	$37.07	$70.67
2 Struc. Steel Workers	36.40	582.40	69.40	1110.40		
1 Welder, gas engine, 300 amp		109.00		119.90		
1 Flatbed Truck, Gas, 3 Ton		217.00		238.70	13.58	14.94
24 L.H., Daily Totals		$1215.60		$2054.60	$50.65	$85.61

Crew E-11	Hr.	Daily	Hr.	Daily	Bare Costs	Incl. O&P
2 Painters, Struc. Steel	$30.55	$488.80	$58.95	$943.20	$30.27	$54.25
1 Building Laborer	26.45	211.60	44.30	354.40		
1 Equip. Oper. (light)	33.55	268.40	54.80	438.40		
1 Air Compressor, 250 cfm		162.80		179.08		
1 Sandblaster, portable, 3 C.F.		20.60		22.66		
1 Set Sand Blasting Accessories		13.15		14.47	6.14	6.76
32 L.H., Daily Totals		$1165.35		$1952.20	$36.42	$61.01

Crew No.	Bare Costs		Incl. Subs O&P		Cost Per Labor-Hour	
Crew E-12	**Hr.**	**Daily**	**Hr.**	**Daily**	**Bare Costs**	**Incl. O&P**
1 Welder Foreman	$38.40	$307.20	$73.20	$585.60	$35.98	$64.00
1 Equip. Oper. (light)	33.55	268.40	54.80	438.40		
1 Welder, gas engine, 300 amp		109.00		119.90	6.81	7.49
16 L.H., Daily Totals		$684.60		$1143.90	$42.79	$71.49
Crew E-13	**Hr.**	**Daily**	**Hr.**	**Daily**	**Bare Costs**	**Incl. O&P**
1 Welder Foreman	$38.40	$307.20	$73.20	$585.60	$36.78	$67.07
.5 Equip. Oper. (light)	33.55	134.20	54.80	219.20		
1 Welder, gas engine, 300 amp		109.00		119.90	9.08	9.99
12 L.H., Daily Totals		$550.40		$924.70	$45.87	$77.06
Crew E-14	**Hr.**	**Daily**	**Hr.**	**Daily**	**Bare Costs**	**Incl. O&P**
1 Struc. Steel Worker	$36.40	$291.20	$69.40	$555.20	$36.40	$69.40
1 Welder, gas engine, 300 amp		109.00		119.90	13.63	14.99
8 L.H., Daily Totals		$400.20		$675.10	$50.02	$84.39
Crew E-16	**Hr.**	**Daily**	**Hr.**	**Daily**	**Bare Costs**	**Incl. O&P**
1 Welder Foreman	$38.40	$307.20	$73.20	$585.60	$37.40	$71.30
1 Welder	36.40	291.20	69.40	555.20		
1 Welder, gas engine, 300 amp		109.00		119.90	6.81	7.49
16 L.H., Daily Totals		$707.40		$1260.70	$44.21	$78.79
Crew E-17	**Hr.**	**Daily**	**Hr.**	**Daily**	**Bare Costs**	**Incl. O&P**
1 Structural Steel Foreman	$38.40	$307.20	$73.20	$585.60	$37.40	$71.30
1 Structural Steel Worker	36.40	291.20	69.40	555.20		
16 L.H., Daily Totals		$598.40		$1140.80	$37.40	$71.30
Crew E-18	**Hr.**	**Daily**	**Hr.**	**Daily**	**Bare Costs**	**Incl. O&P**
1 Structural Steel Foreman	$38.40	$307.20	$73.20	$585.60	$36.50	$67.69
3 Structural Steel Workers	36.40	873.60	69.40	1665.60		
1 Equipment Operator (med.)	34.90	279.20	57.05	456.40		
1 Lattice Boom Crane, 20 Ton		1035.00		1138.50	25.88	28.46
40 L.H., Daily Totals		$2495.00		$3846.10	$62.38	$96.15
Crew E-19	**Hr.**	**Daily**	**Hr.**	**Daily**	**Bare Costs**	**Incl. O&P**
1 Structural Steel Worker	$36.40	$291.20	$69.40	$555.20	$36.12	$65.80
1 Structural Steel Foreman	38.40	307.20	73.20	585.60		
1 Equip. Oper. (light)	33.55	268.40	54.80	438.40		
1 Lattice Boom Crane, 20 Ton		1035.00		1138.50	43.13	47.44
24 L.H., Daily Totals		$1901.80		$2717.70	$79.24	$113.24
Crew E-20	**Hr.**	**Daily**	**Hr.**	**Daily**	**Bare Costs**	**Incl. O&P**
1 Structural Steel Foreman	$38.40	$307.20	$73.20	$585.60	$35.91	$66.18
5 Structural Steel Workers	36.40	1456.00	69.40	2776.00		
1 Equip. Oper. (crane)	35.80	286.40	58.50	468.00		
1 Oiler	31.05	248.40	50.75	406.00		
1 Lattice Boom Crane, 40 Ton		1263.00		1389.30	19.73	21.71
64 L.H., Daily Totals		$3561.00		$5624.90	$55.64	$87.89
Crew E-22	**Hr.**	**Daily**	**Hr.**	**Daily**	**Bare Costs**	**Incl. O&P**
1 Skilled Worker Foreman	$36.05	$288.40	$60.35	$482.80	$34.72	$58.12
2 Skilled Workers	34.05	544.80	57.00	912.00		
24 L.H., Daily Totals		$833.20		$1394.80	$34.72	$58.12
Crew E-24	**Hr.**	**Daily**	**Hr.**	**Daily**	**Bare Costs**	**Incl. O&P**
3 Structural Steel Workers	$36.40	$873.60	$69.40	$1665.60	$36.02	$66.31
1 Equipment Operator (med.)	34.90	279.20	57.05	456.40		
1 Hyd. Crane, 25 Ton		748.60		823.46	23.39	25.73
32 L.H., Daily Totals		$1901.40		$2945.46	$59.42	$92.05

Crew No.	Bare Costs		Incl. Subs O&P		Cost Per Labor-Hour	
Crew E-25	**Hr.**	**Daily**	**Hr.**	**Daily**	**Bare Costs**	**Incl. O&P**
1 Welder	$36.40	$291.20	$69.40	$555.20	$36.40	$69.40
1 Cutting Torch		18.00		19.80		
1 Set of Gases		82.80		91.08	12.60	13.86
8 L.H., Daily Totals		$392.00		$666.08	$49.00	$83.26
Crew F-3	**Hr.**	**Daily**	**Hr.**	**Daily**	**Bare Costs**	**Incl. O&P**
2 Carpenters	$33.55	$536.80	$56.15	$898.40	$30.72	$51.08
2 Carpenter Helpers	25.35	405.60	42.30	676.80		
1 Equip. Oper. (crane)	35.80	286.40	58.50	468.00		
1 Hyd. Crane, 12 Ton		655.60		721.16	16.39	18.03
40 L.H., Daily Totals		$1884.40		$2764.36	$47.11	$69.11
Crew F-4	**Hr.**	**Daily**	**Hr.**	**Daily**	**Bare Costs**	**Incl. O&P**
2 Carpenters	$33.55	$536.80	$56.15	$898.40	$30.72	$51.08
2 Carpenter Helpers	25.35	405.60	42.30	676.80		
1 Equip. Oper. (crane)	35.80	286.40	58.50	468.00		
1 Hyd. Crane, 55 Ton		1208.00		1328.80	30.20	33.22
40 L.H., Daily Totals		$2436.80		$3372.00	$60.92	$84.30
Crew F-5	**Hr.**	**Daily**	**Hr.**	**Daily**	**Bare Costs**	**Incl. O&P**
2 Carpenters	$33.55	$536.80	$56.15	$898.40	$29.45	$49.23
2 Carpenter Helpers	25.35	405.60	42.30	676.80		
32 L.H., Daily Totals		$942.40		$1575.20	$29.45	$49.23
Crew F-6	**Hr.**	**Daily**	**Hr.**	**Daily**	**Bare Costs**	**Incl. O&P**
2 Carpenters	$33.55	$536.80	$56.15	$898.40	$31.16	$51.88
2 Building Laborers	26.45	423.20	44.30	708.80		
1 Equip. Oper. (crane)	35.80	286.40	58.50	468.00		
1 Hyd. Crane, 12 Ton		655.60		721.16	16.39	18.03
40 L.H., Daily Totals		$1902.00		$2796.36	$47.55	$69.91
Crew F-7	**Hr.**	**Daily**	**Hr.**	**Daily**	**Bare Costs**	**Incl. O&P**
2 Carpenters	$33.55	$536.80	$56.15	$898.40	$30.00	$50.23
2 Building Laborers	26.45	423.20	44.30	708.80		
32 L.H., Daily Totals		$960.00		$1607.20	$30.00	$50.23
Crew G-1	**Hr.**	**Daily**	**Hr.**	**Daily**	**Bare Costs**	**Incl. O&P**
1 Roofer Foreman	$30.15	$241.20	$54.40	$435.20	$26.36	$47.57
4 Roofers, Composition	28.15	900.80	50.80	1625.60		
2 Roofer Helpers	20.90	334.40	37.70	603.20		
1 Application Equipment		168.40		185.24		
1 Tar Kettle/Pot		88.75		97.63		
1 Crew Truck		152.40		167.64	7.31	8.04
56 L.H., Daily Totals		$1885.95		$3114.51	$33.68	$55.62
Crew G-2	**Hr.**	**Daily**	**Hr.**	**Daily**	**Bare Costs**	**Incl. O&P**
1 Plasterer	$30.70	$245.60	$49.95	$399.60	$28.17	$46.25
1 Plasterer Helper	27.35	218.80	44.50	356.00		
1 Building Laborer	26.45	211.60	44.30	354.40		
1 Grout Pump, 50 C.F./hr		125.80		138.38	5.24	5.77
24 L.H., Daily Totals		$801.80		$1248.38	$33.41	$52.02
Crew G-2A	**Hr.**	**Daily**	**Hr.**	**Daily**	**Bare Costs**	**Incl. O&P**
1 Roofer, composition	$28.15	$225.20	$50.80	$406.40	$25.17	$44.27
1 Roofer Helper	20.90	167.20	37.70	301.60		
1 Building Laborer	26.45	211.60	44.30	354.40		
1 Foam spray rig, trailer-mtd.		488.25		537.08		
1 Pickup Truck, 3/4 Ton		104.60		115.06	24.70	27.17
24 L.H., Daily Totals		$1196.85		$1714.54	$49.87	$71.44

Crew No.	Bare Costs		Incl. Subs O&P		Cost Per Labor-Hour	
Crew G-3	**Hr.**	**Daily**	**Hr.**	**Daily**	**Bare Costs**	**Incl. O&P**
2 Sheet Metal Workers	$38.20	$611.20	$63.10	$1009.60	$32.33	$53.70
2 Building Laborers	26.45	423.20	44.30	708.80		
32 L.H., Daily Totals		$1034.40		$1718.40	$32.33	$53.70
Crew G-4	**Hr.**	**Daily**	**Hr.**	**Daily**	**Bare Costs**	**Incl. O&P**
1 Labor Foreman (outside)	$28.45	$227.60	$47.65	$381.20	$27.12	$45.42
2 Building Laborers	26.45	423.20	44.30	708.80		
1 Flatbed Truck, Gas, 1.5 Ton		171.40		188.54		
1 Air Compressor, 160 cfm		125.00		137.50	12.35	13.59
24 L.H., Daily Totals		$947.20		$1416.04	$39.47	$59.00
Crew G-5	**Hr.**	**Daily**	**Hr.**	**Daily**	**Bare Costs**	**Incl. O&P**
1 Roofer Foreman	$30.15	$241.20	$54.40	$435.20	$25.65	$46.28
2 Roofers, Composition	28.15	450.40	50.80	812.80		
2 Roofer Helpers	20.90	334.40	37.70	603.20		
1 Application Equipment		168.40		185.24	4.21	4.63
40 L.H., Daily Totals		$1194.40		$2036.44	$29.86	$50.91
Crew G-6A	**Hr.**	**Daily**	**Hr.**	**Daily**	**Bare Costs**	**Incl. O&P**
2 Roofers Composition	$28.15	$450.40	$50.80	$812.80	$28.15	$50.80
1 Small Compressor, Electric		11.05		12.15		
2 Pneumatic Nailers		47.00		51.70	3.63	3.99
16 L.H., Daily Totals		$508.45		$876.65	$31.78	$54.79
Crew G-7	**Hr.**	**Daily**	**Hr.**	**Daily**	**Bare Costs**	**Incl. O&P**
1 Carpenter	$33.55	$268.40	$56.15	$449.20	$33.55	$56.15
1 Small Compressor, Electric		11.05		12.15		
1 Pneumatic Nailer		23.50		25.85	4.32	4.75
8 L.H., Daily Totals		$302.95		$487.20	$37.87	$60.90
Crew H-1	**Hr.**	**Daily**	**Hr.**	**Daily**	**Bare Costs**	**Incl. O&P**
2 Glaziers	$33.20	$531.20	$54.50	$872.00	$34.80	$61.95
2 Struc. Steel Workers	36.40	582.40	69.40	1110.40		
32 L.H., Daily Totals		$1113.60		$1982.40	$34.80	$61.95
Crew H-2	**Hr.**	**Daily**	**Hr.**	**Daily**	**Bare Costs**	**Incl. O&P**
2 Glaziers	$33.20	$531.20	$54.50	$872.00	$30.95	$51.10
1 Building Laborer	26.45	211.60	44.30	354.40		
24 L.H., Daily Totals		$742.80		$1226.40	$30.95	$51.10
Crew H-3	**Hr.**	**Daily**	**Hr.**	**Daily**	**Bare Costs**	**Incl. O&P**
1 Glazier	$33.20	$265.60	$54.50	$436.00	$29.27	$48.40
1 Helper	25.35	202.80	42.30	338.40		
16 L.H., Daily Totals		$468.40		$774.40	$29.27	$48.40
Crew H-4	**Hr.**	**Daily**	**Hr.**	**Daily**	**Bare Costs**	**Incl. O&P**
1 Carpenter	$33.55	$268.40	$56.15	$449.20	$31.61	$52.43
1 Carpenter Helper	25.35	202.80	42.30	338.40		
.5 Electrician	40.25	161.00	65.25	261.00		
20 L.H., Daily Totals		$632.20		$1048.60	$31.61	$52.43
Crew J-1	**Hr.**	**Daily**	**Hr.**	**Daily**	**Bare Costs**	**Incl. O&P**
3 Plasterers	$30.70	$736.80	$49.95	$1198.80	$29.36	$47.77
2 Plasterer Helpers	27.35	437.60	44.50	712.00		
1 Mixing Machine, 6 C.F.		123.60		135.96	3.09	3.40
40 L.H., Daily Totals		$1298.00		$2046.76	$32.45	$51.17

Crew No.	Bare Costs		Incl. Subs O&P		Cost Per Labor-Hour	
Crew J-2	**Hr.**	**Daily**	**Hr.**	**Daily**	**Bare Costs**	**Incl. O&P**
3 Plasterers	$30.70	$736.80	$49.95	$1198.80	$29.48	$47.82
2 Plasterer Helpers	27.35	437.60	44.50	712.00		
1 Lather	30.05	240.40	48.05	384.40		
1 Mixing Machine, 6 C.F.		123.60		135.96	2.58	2.83
48 L.H., Daily Totals		$1538.40		$2431.16	$32.05	$50.65
Crew J-3	**Hr.**	**Daily**	**Hr.**	**Daily**	**Bare Costs**	**Incl. O&P**
1 Terrazzo Worker	$31.30	$250.40	$50.00	$400.00	$28.70	$45.85
1 Terrazzo Helper	26.10	208.80	41.70	333.60		
1 Floor Grinder, 22" path		73.95		81.34		
1 Terrazzo Mixer		167.60		184.36	15.10	16.61
16 L.H., Daily Totals		$700.75		$999.30	$43.80	$62.46
Crew J-4	**Hr.**	**Daily**	**Hr.**	**Daily**	**Bare Costs**	**Incl. O&P**
2 Cement Finishers	$31.85	$509.60	$50.85	$813.60	$30.05	$48.67
1 Laborer	26.45	211.60	44.30	354.40		
1 Floor Grinder, 22" path		73.95		81.34		
1 Floor Edger, 7" path		38.20		42.02		
1 Vacuum Pick-Up System		59.80		65.78	7.16	7.88
24 L.H., Daily Totals		$893.15		$1357.15	$37.21	$56.55
Crew J-4A	**Hr.**	**Daily**	**Hr.**	**Daily**	**Bare Costs**	**Incl. O&P**
2 Cement Finishers	$31.85	$509.60	$50.85	$813.60	$29.15	$47.58
2 Laborers	26.45	423.20	44.30	708.80		
1 Floor Grinder, 22" path		73.95		81.34		
1 Floor Edger, 7" path		38.20		42.02		
1 Vacuum Pick-Up System		59.80		65.78		
1 Floor Auto Scrubber		154.10		169.51	10.19	11.21
32 L.H., Daily Totals		$1258.85		$1881.06	$39.34	$58.78
Crew J-4B	**Hr.**	**Daily**	**Hr.**	**Daily**	**Bare Costs**	**Incl. O&P**
1 Laborer	$26.45	$211.60	$44.30	$354.40	$26.45	$44.30
1 Floor Auto Scrubber		154.10		169.51	19.26	21.19
8 L.H., Daily Totals		$365.70		$523.91	$45.71	$65.49
Crew K-1	**Hr.**	**Daily**	**Hr.**	**Daily**	**Bare Costs**	**Incl. O&P**
1 Carpenter	$33.55	$268.40	$56.15	$449.20	$29.98	$49.98
1 Truck Driver (light)	26.40	211.20	43.80	350.40		
1 Flatbed Truck, Gas, 3 Ton		217.00		238.70	13.56	14.92
16 L.H., Daily Totals		$696.60		$1038.30	$43.54	$64.89
Crew K-2	**Hr.**	**Daily**	**Hr.**	**Daily**	**Bare Costs**	**Incl. O&P**
1 Struc. Steel Foreman	$38.40	$307.20	$73.20	$585.60	$33.73	$62.13
1 Struc. Steel Worker	36.40	291.20	69.40	555.20		
1 Truck Driver (light)	26.40	211.20	43.80	350.40		
1 Flatbed Truck, Gas, 3 Ton		217.00		238.70	9.04	9.95
24 L.H., Daily Totals		$1026.60		$1729.90	$42.77	$72.08
Crew L-1	**Hr.**	**Daily**	**Hr.**	**Daily**	**Bare Costs**	**Incl. O&P**
.25 Electrician	$40.25	$80.50	$65.25	$130.50	$39.77	$64.85
1 Plumber	39.65	317.20	64.75	518.00		
10 L.H., Daily Totals		$397.70		$648.50	$39.77	$64.85
Crew L-2	**Hr.**	**Daily**	**Hr.**	**Daily**	**Bare Costs**	**Incl. O&P**
1 Carpenter	$33.55	$268.40	$56.15	$449.20	$29.45	$49.23
1 Carpenter Helper	25.35	202.80	42.30	338.40		
16 L.H., Daily Totals		$471.20		$787.60	$29.45	$49.23

Crew No.	Bare Costs		Incl. Subs O&P		Cost Per Labor-Hour	

Crew L-3	Hr.	Daily	Hr.	Daily	Bare Costs	Incl. O&P
1 Carpenter	$33.55	$268.40	$56.15	$449.20	$34.89	$57.97
.25 Electrician	40.25	80.50	65.25	130.50		
10 L.H., Daily Totals		$348.90		$579.70	$34.89	$57.97

Crew L-3A	Hr.	Daily	Hr.	Daily	Bare Costs	Incl. O&P
1 Carpenter Foreman (outside)	$35.55	$284.40	$59.50	$476.00	$36.43	$60.70
.5 Sheet Metal Worker	38.20	152.80	63.10	252.40		
12 L.H., Daily Totals		$437.20		$728.40	$36.43	$60.70

Crew L-4	Hr.	Daily	Hr.	Daily	Bare Costs	Incl. O&P
1 Skilled Worker	$34.05	$272.40	$57.00	$456.00	$29.70	$49.65
1 Helper	25.35	202.80	42.30	338.40		
16 L.H., Daily Totals		$475.20		$794.40	$29.70	$49.65

Crew L-5	Hr.	Daily	Hr.	Daily	Bare Costs	Incl. O&P
1 Struc. Steel Foreman	$38.40	$307.20	$73.20	$585.60	$36.60	$68.39
5 Struc. Steel Workers	36.40	1456.00	69.40	2776.00		
1 Equip. Oper. (crane)	35.80	286.40	58.50	468.00		
1 Hyd. Crane, 25 Ton		748.60		823.46	13.37	14.70
56 L.H., Daily Totals		$2798.20		$4653.06	$49.97	$83.09

Crew L-5A	Hr.	Daily	Hr.	Daily	Bare Costs	Incl. O&P
1 Structural Steel Foreman	$38.40	$307.20	$73.20	$585.60	$36.75	$67.63
2 Structural Steel Workers	36.40	582.40	69.40	1110.40		
1 Equip. Oper. (crane)	35.80	286.40	58.50	468.00		
1 S.P. Crane, 4x4, 25 Ton		560.40		616.44	17.51	19.26
32 L.H., Daily Totals		$1736.40		$2780.44	$54.26	$86.89

Crew L-5B	Hr.	Daily	Hr.	Daily	Bare Costs	Incl. O&P
1 Structural Steel Foreman	$38.40	$307.20	$73.20	$585.60	$37.12	$64.01
2 Structural Steel Workers	36.40	582.40	69.40	1110.40		
2 Electricians	40.25	644.00	65.25	1044.00		
2 Steamfitters/Pipefitters	40.05	640.80	65.40	1046.40		
1 Equip. Oper. (crane)	35.80	286.40	58.50	468.00		
1 Common Building Laborer	26.45	211.60	44.30	354.40		
1 Hyd. Crane, 80 Ton		1693.00		1862.30	23.51	25.87
72 L.H., Daily Totals		$4365.40		$6471.10	$60.63	$89.88

Crew L-6	Hr.	Daily	Hr.	Daily	Bare Costs	Incl. O&P
1 Plumber	$39.65	$317.20	$64.75	$518.00	$39.85	$64.92
.5 Electrician	40.25	161.00	65.25	261.00		
12 L.H., Daily Totals		$478.20		$779.00	$39.85	$64.92

Crew L-7	Hr.	Daily	Hr.	Daily	Bare Costs	Incl. O&P
1 Carpenter	$33.55	$268.40	$56.15	$449.20	$29.02	$48.33
2 Carpenter Helpers	25.35	405.60	42.30	676.80		
.25 Electrician	40.25	80.50	65.25	130.50		
26 L.H., Daily Totals		$754.50		$1256.50	$29.02	$48.33

Crew L-8	Hr.	Daily	Hr.	Daily	Bare Costs	Incl. O&P
1 Carpenter	$33.55	$268.40	$56.15	$449.20	$31.49	$52.33
1 Carpenter Helper	25.35	202.80	42.30	338.40		
.5 Plumber	39.65	158.60	64.75	259.00		
20 L.H., Daily Totals		$629.80		$1046.60	$31.49	$52.33

Crew L-9	Hr.	Daily	Hr.	Daily	Bare Costs	Incl. O&P
1 Skilled Worker Foreman	$36.05	$288.40	$60.35	$482.80	$31.32	$52.13
1 Skilled Worker	34.05	272.40	57.00	456.00		
2 Helpers	25.35	405.60	42.30	676.80		
.5 Electrician	40.25	161.00	65.25	261.00		
36 L.H., Daily Totals		$1127.40		$1876.60	$31.32	$52.13

Crew L-10	Hr.	Daily	Hr.	Daily	Bare Costs	Incl. O&P
1 Structural Steel Foreman	$38.40	$307.20	$73.20	$585.60	$36.87	$67.03
1 Structural Steel Worker	36.40	291.20	69.40	555.20		
1 Equip. Oper. (crane)	35.80	286.40	58.50	468.00		
1 Hyd. Crane, 12 Ton		655.60		721.16	27.32	30.05
24 L.H., Daily Totals		$1540.40		$2329.96	$64.18	$97.08

Crew L-11	Hr.	Daily	Hr.	Daily	Bare Costs	Incl. O&P
2 Wreckers	$26.45	$423.20	$48.20	$771.20	$30.56	$52.42
1 Equip. Oper. (crane)	35.80	286.40	58.50	468.00		
1 Equip. Oper. (light)	33.55	268.40	54.80	438.40		
1 Hyd. Excavator, 2.5 C.Y.		1613.00		1774.30		
1 Loader, Skid Steer, 78 H.P.		261.20		287.32	58.57	64.43
32 L.H., Daily Totals		$2852.20		$3739.22	$89.13	$116.85

Crew M-1	Hr.	Daily	Hr.	Daily	Bare Costs	Incl. O&P
3 Elevator Constructors	$50.55	$1213.20	$82.10	$1970.40	$48.02	$78.00
1 Elevator Apprentice	40.45	323.60	65.70	525.60		
5 Hand Tools		49.00		53.90	1.53	1.68
32 L.H., Daily Totals		$1585.80		$2549.90	$49.56	$79.68

Crew M-3	Hr.	Daily	Hr.	Daily	Bare Costs	Incl. O&P
1 Electrician Foreman (out)	$42.25	$338.00	$68.50	$548.00	$39.63	$64.67
1 Common Laborer	26.45	211.60	44.30	354.40		
.25 Equipment Operator, Med.	34.90	69.80	57.05	114.10		
1 Elevator Constructor	50.55	404.40	82.10	656.80		
1 Elevator Apprentice	40.45	323.60	65.70	525.60		
.25 S.P. Crane, 4x4, 20 Ton		127.75		140.53	3.76	4.13
34 L.H., Daily Totals		$1475.15		$2339.43	$43.39	$68.81

Crew M-4	Hr.	Daily	Hr.	Daily	Bare Costs	Incl. O&P
1 Electrician Foreman (out)	$42.25	$338.00	$68.50	$548.00	$39.20	$63.98
1 Common Laborer	26.45	211.60	44.30	354.40		
.25 Equipment Operator, Crane	35.80	71.60	58.50	117.00		
.25 Equipment Operator, Oiler	31.05	62.10	50.75	101.50		
1 Elevator Constructor	50.55	404.40	82.10	656.80		
1 Elevator Apprentice	40.45	323.60	65.70	525.60		
.25 S.P. Crane, 4x4, 40 Ton		164.65		181.12	4.57	5.03
36 L.H., Daily Totals		$1575.95		$2484.42	$43.78	$69.01

Crew Q-1	Hr.	Daily	Hr.	Daily	Bare Costs	Incl. O&P
1 Plumber	$39.65	$317.20	$64.75	$518.00	$35.67	$58.25
1 Plumber Apprentice	31.70	253.60	51.75	414.00		
16 L.H., Daily Totals		$570.80		$932.00	$35.67	$58.25

Crew Q-1A	Hr.	Daily	Hr.	Daily	Bare Costs	Incl. O&P
.25 Plumber Foreman (out)	$41.65	$83.30	$68.00	$136.00	$40.05	$65.40
1 Plumber	39.65	317.20	64.75	518.00		
10 L.H., Daily Totals		$400.50		$654.00	$40.05	$65.40

Crew No.	Bare Costs		Incl. Subs O&P		Cost Per Labor-Hour	
Crew Q-1C	**Hr.**	**Daily**	**Hr.**	**Daily**	**Bare Costs**	**Incl. O&P**
1 Plumber	$39.65	$317.20	$64.75	$518.00	$35.42	$57.85
1 Plumber Apprentice	31.70	253.60	51.75	414.00		
1 Equip. Oper. (medium)	34.90	279.20	57.05	456.40		
1 Trencher, Chain Type, 8' D		2506.00		2756.60	104.42	114.86
24 L.H., Daily Totals		$3356.00		$4145.00	$139.83	$172.71
Crew Q-2	**Hr.**	**Daily**	**Hr.**	**Daily**	**Bare Costs**	**Incl. O&P**
1 Plumber	$39.65	$317.20	$64.75	$518.00	$34.35	$56.08
2 Plumber Apprentices	31.70	507.20	51.75	828.00		
24 L.H., Daily Totals		$824.40		$1346.00	$34.35	$56.08
Crew Q-3	**Hr.**	**Daily**	**Hr.**	**Daily**	**Bare Costs**	**Incl. O&P**
2 Plumbers	$39.65	$634.40	$64.75	$1036.00	$35.67	$58.25
2 Plumber Apprentices	31.70	507.20	51.75	828.00		
32 L.H., Daily Totals		$1141.60		$1864.00	$35.67	$58.25
Crew Q-4	**Hr.**	**Daily**	**Hr.**	**Daily**	**Bare Costs**	**Incl. O&P**
2 Plumbers	$39.65	$634.40	$64.75	$1036.00	$37.66	$61.50
1 Welder (plumber)	39.65	317.20	64.75	518.00		
1 Plumber Apprentice	31.70	253.60	51.75	414.00		
1 Welder, electric, 300 amp		32.45		35.70	1.01	1.12
32 L.H., Daily Totals		$1237.65		$2003.69	$38.68	$62.62
Crew Q-5	**Hr.**	**Daily**	**Hr.**	**Daily**	**Bare Costs**	**Incl. O&P**
1 Steamfitter	$40.05	$320.40	$65.40	$523.20	$36.05	$58.88
1 Steamfitter Apprentice	32.05	256.40	52.35	418.80		
16 L.H., Daily Totals		$576.80		$942.00	$36.05	$58.88
Crew Q-6	**Hr.**	**Daily**	**Hr.**	**Daily**	**Bare Costs**	**Incl. O&P**
1 Steamfitter	$40.05	$320.40	$65.40	$523.20	$34.72	$56.70
2 Steamfitter Apprentices	32.05	512.80	52.35	837.60		
24 L.H., Daily Totals		$833.20		$1360.80	$34.72	$56.70
Crew Q-7	**Hr.**	**Daily**	**Hr.**	**Daily**	**Bare Costs**	**Incl. O&P**
2 Steamfitters	$40.05	$640.80	$65.40	$1046.40	$36.05	$58.88
2 Steamfitter Apprentices	32.05	512.80	52.35	837.60		
32 L.H., Daily Totals		$1153.60		$1884.00	$36.05	$58.88
Crew Q-8	**Hr.**	**Daily**	**Hr.**	**Daily**	**Bare Costs**	**Incl. O&P**
2 Steamfitters	$40.05	$640.80	$65.40	$1046.40	$38.05	$62.14
1 Welder (steamfitter)	40.05	320.40	65.40	523.20		
1 Steamfitter Apprentice	32.05	256.40	52.35	418.80		
1 Welder, electric, 300 amp		32.45		35.70	1.01	1.12
32 L.H., Daily Totals		$1250.05		$2024.10	$39.06	$63.25
Crew Q-9	**Hr.**	**Daily**	**Hr.**	**Daily**	**Bare Costs**	**Incl. O&P**
1 Sheet Metal Worker	$38.20	$305.60	$63.10	$504.80	$34.38	$56.77
1 Sheet Metal Apprentice	30.55	244.40	50.45	403.60		
16 L.H., Daily Totals		$550.00		$908.40	$34.38	$56.77
Crew Q-10	**Hr.**	**Daily**	**Hr.**	**Daily**	**Bare Costs**	**Incl. O&P**
2 Sheet Metal Workers	$38.20	$611.20	$63.10	$1009.60	$35.65	$58.88
1 Sheet Metal Apprentice	30.55	244.40	50.45	403.60		
24 L.H., Daily Totals		$855.60		$1413.20	$35.65	$58.88
Crew Q-11	**Hr.**	**Daily**	**Hr.**	**Daily**	**Bare Costs**	**Incl. O&P**
2 Sheet Metal Workers	$38.20	$611.20	$63.10	$1009.60	$34.38	$56.77
2 Sheet Metal Apprentices	30.55	488.80	50.45	807.20		
32 L.H., Daily Totals		$1100.00		$1816.80	$34.38	$56.77

Crew No.	Bare Costs		Incl. Subs O&P		Cost Per Labor-Hour	
Crew Q-12	**Hr.**	**Daily**	**Hr.**	**Daily**	**Bare Costs**	**Incl. O&P**
1 Sprinkler Installer	$39.30	$314.40	$64.20	$513.60	$35.38	$57.80
1 Sprinkler Apprentice	31.45	251.60	51.40	411.20		
16 L.H., Daily Totals		$566.00		$924.80	$35.38	$57.80
Crew Q-13	**Hr.**	**Daily**	**Hr.**	**Daily**	**Bare Costs**	**Incl. O&P**
2 Sprinkler Installers	$39.30	$628.80	$64.20	$1027.20	$35.38	$57.80
2 Sprinkler Apprentices	31.45	503.20	51.40	822.40		
32 L.H., Daily Totals		$1132.00		$1849.60	$35.38	$57.80
Crew Q-14	**Hr.**	**Daily**	**Hr.**	**Daily**	**Bare Costs**	**Incl. O&P**
1 Asbestos Worker	$34.70	$277.60	$58.50	$468.00	$31.23	$52.65
1 Asbestos Apprentice	27.75	222.00	46.80	374.40		
16 L.H., Daily Totals		$499.60		$842.40	$31.23	$52.65
Crew Q-15	**Hr.**	**Daily**	**Hr.**	**Daily**	**Bare Costs**	**Incl. O&P**
1 Plumber	$39.65	$317.20	$64.75	$518.00	$35.67	$58.25
1 Plumber Apprentice	31.70	253.60	51.75	414.00		
1 Welder, electric, 300 amp		32.45		35.70	2.03	2.23
16 L.H., Daily Totals		$603.25		$967.70	$37.70	$60.48
Crew Q-16	**Hr.**	**Daily**	**Hr.**	**Daily**	**Bare Costs**	**Incl. O&P**
2 Plumbers	$39.65	$634.40	$64.75	$1036.00	$37.00	$60.42
1 Plumber Apprentice	31.70	253.60	51.75	414.00		
1 Welder, electric, 300 amp		32.45		35.70	1.35	1.49
24 L.H., Daily Totals		$920.45		$1485.69	$38.35	$61.90
Crew Q-17	**Hr.**	**Daily**	**Hr.**	**Daily**	**Bare Costs**	**Incl. O&P**
1 Steamfitter	$40.05	$320.40	$65.40	$523.20	$36.05	$58.88
1 Steamfitter Apprentice	32.05	256.40	52.35	418.80		
1 Welder, electric, 300 amp		32.45		35.70	2.03	2.23
16 L.H., Daily Totals		$609.25		$977.70	$38.08	$61.11
Crew Q-17A	**Hr.**	**Daily**	**Hr.**	**Daily**	**Bare Costs**	**Incl. O&P**
1 Steamfitter	$40.05	$320.40	$65.40	$523.20	$35.97	$58.75
1 Steamfitter Apprentice	32.05	256.40	52.35	418.80		
1 Equip. Oper. (crane)	35.80	286.40	58.50	468.00		
1 Hyd. Crane, 12 Ton		655.60		721.16		
1 Welder, electric, 300 amp		32.45		35.70	28.67	31.54
24 L.H., Daily Totals		$1551.25		$2166.86	$64.64	$90.29
Crew Q-18	**Hr.**	**Daily**	**Hr.**	**Daily**	**Bare Costs**	**Incl. O&P**
2 Steamfitters	$40.05	$640.80	$65.40	$1046.40	$37.38	$61.05
1 Steamfitter Apprentice	32.05	256.40	52.35	418.80		
1 Welder, electric, 300 amp		32.45		35.70	1.35	1.49
24 L.H., Daily Totals		$929.65		$1500.90	$38.74	$62.54
Crew Q-19	**Hr.**	**Daily**	**Hr.**	**Daily**	**Bare Costs**	**Incl. O&P**
1 Steamfitter	$40.05	$320.40	$65.40	$523.20	$37.45	$61.00
1 Steamfitter Apprentice	32.05	256.40	52.35	418.80		
1 Electrician	40.25	322.00	65.25	522.00		
24 L.H., Daily Totals		$898.80		$1464.00	$37.45	$61.00
Crew Q-20	**Hr.**	**Daily**	**Hr.**	**Daily**	**Bare Costs**	**Incl. O&P**
1 Sheet Metal Worker	$38.20	$305.60	$63.10	$504.80	$35.55	$58.47
1 Sheet Metal Apprentice	30.55	244.40	50.45	403.60		
.5 Electrician	40.25	161.00	65.25	261.00		
20 L.H., Daily Totals		$711.00		$1169.40	$35.55	$58.47

Crew No.	Bare Costs		Incl. Subs O&P		Cost Per Labor-Hour	
Crew Q-21	**Hr.**	**Daily**	**Hr.**	**Daily**	**Bare Costs**	**Incl. O&P**
2 Steamfitters	$40.05	$640.80	$65.40	$1046.40	$38.10	$62.10
1 Steamfitter Apprentice	32.05	256.40	52.35	418.80		
1 Electrician	40.25	322.00	65.25	522.00		
32 L.H., Daily Totals		$1219.20		$1987.20	$38.10	$62.10
Crew Q-22	**Hr.**	**Daily**	**Hr.**	**Daily**	**Bare Costs**	**Incl. O&P**
1 Plumber	$39.65	$317.20	$64.75	$518.00	$35.67	$58.25
1 Plumber Apprentice	31.70	253.60	51.75	414.00		
1 Hyd. Crane, 12 Ton		655.60		721.16	40.98	45.07
16 L.H., Daily Totals		$1226.40		$1653.16	$76.65	$103.32
Crew Q-22A	**Hr.**	**Daily**	**Hr.**	**Daily**	**Bare Costs**	**Incl. O&P**
1 Plumber	$39.65	$317.20	$64.75	$518.00	$33.40	$54.83
1 Plumber Apprentice	31.70	253.60	51.75	414.00		
1 Laborer	26.45	211.60	44.30	354.40		
1 Equip. Oper. (crane)	35.80	286.40	58.50	468.00		
1 Hyd. Crane, 12 Ton		655.60		721.16	20.49	22.54
32 L.H., Daily Totals		$1724.40		$2475.56	$53.89	$77.36
Crew Q-23	**Hr.**	**Daily**	**Hr.**	**Daily**	**Bare Costs**	**Incl. O&P**
1 Plumber Foreman	$41.65	$333.20	$68.00	$544.00	$38.73	$63.27
1 Plumber	39.65	317.20	64.75	518.00		
1 Equip. Oper. (med.)	34.90	279.20	57.05	456.40		
1 Lattice Boom Crane, 20 Ton		1035.00		1138.50	43.13	47.44
24 L.H., Daily Totals		$1964.60		$2656.90	$81.86	$110.70
Crew R-1	**Hr.**	**Daily**	**Hr.**	**Daily**	**Bare Costs**	**Incl. O&P**
1 Electrician Foreman	$40.75	$326.00	$66.05	$528.40	$35.37	$57.73
3 Electricians	40.25	966.00	65.25	1566.00		
2 Helpers	25.35	405.60	42.30	676.80		
48 L.H., Daily Totals		$1697.60		$2771.20	$35.37	$57.73
Crew R-1A	**Hr.**	**Daily**	**Hr.**	**Daily**	**Bare Costs**	**Incl. O&P**
1 Electrician	$40.25	$322.00	$65.25	$522.00	$32.80	$53.77
1 Helper	25.35	202.80	42.30	338.40		
16 L.H., Daily Totals		$524.80		$860.40	$32.80	$53.77
Crew R-2	**Hr.**	**Daily**	**Hr.**	**Daily**	**Bare Costs**	**Incl. O&P**
1 Electrician Foreman	$40.75	$326.00	$66.05	$528.40	$35.43	$57.84
3 Electricians	40.25	966.00	65.25	1566.00		
2 Helpers	25.35	405.60	42.30	676.80		
1 Equip. Oper. (crane)	35.80	286.40	58.50	468.00		
1 S.P. Crane, 4x4, 5 Ton		257.80		283.58	4.60	5.06
56 L.H., Daily Totals		$2241.80		$3522.78	$40.03	$62.91
Crew R-3	**Hr.**	**Daily**	**Hr.**	**Daily**	**Bare Costs**	**Incl. O&P**
1 Electrician Foreman	$40.75	$326.00	$66.05	$528.40	$39.56	$64.22
1 Electrician	40.25	322.00	65.25	522.00		
.5 Equip. Oper. (crane)	35.80	143.20	58.50	234.00		
.5 S.P. Crane, 4x4, 5 Ton		128.90		141.79	6.45	7.09
20 L.H., Daily Totals		$920.10		$1426.19	$46.01	$71.31
Crew R-4	**Hr.**	**Daily**	**Hr.**	**Daily**	**Bare Costs**	**Incl. O&P**
1 Struc. Steel Foreman	$38.40	$307.20	$73.20	$585.60	$37.57	$69.33
3 Struc. Steel Workers	36.40	873.60	69.40	1665.60		
1 Electrician	40.25	322.00	65.25	522.00		
1 Welder, gas engine, 300 amp		109.00		119.90	2.73	3.00
40 L.H., Daily Totals		$1611.80		$2893.10	$40.30	$72.33

Crew No.	Bare Costs		Incl. Subs O&P		Cost Per Labor-Hour	
Crew R-5	**Hr.**	**Daily**	**Hr.**	**Daily**	**Bare Costs**	**Incl. O&P**
1 Electrician Foreman	$40.75	$326.00	$66.05	$528.40	$34.88	$56.98
4 Electrician Linemen	40.25	1288.00	65.25	2088.00		
2 Electrician Operators	40.25	644.00	65.25	1044.00		
4 Electrician Groundmen	25.35	811.20	42.30	1353.60		
1 Crew Truck		152.40		167.64		
1 Flatbed Truck, 20,000 GVW		190.00		209.00		
1 Pickup Truck, 3/4 Ton		104.60		115.06		
.2 Hyd. Crane, 55 Ton		241.60		265.76		
.2 Hyd. Crane, 12 Ton		131.12		144.23		
.2 Earth Auger, Truck-Mtd.		80.36		88.40		
1 Tractor w/Winch		346.80		381.48	14.17	15.59
88 L.H., Daily Totals		$4316.08		$6385.57	$49.05	$72.56
Crew R-6	**Hr.**	**Daily**	**Hr.**	**Daily**	**Bare Costs**	**Incl. O&P**
1 Electrician Foreman	$40.75	$326.00	$66.05	$528.40	$34.88	$56.98
4 Electrician Linemen	40.25	1288.00	65.25	2088.00		
2 Electrician Operators	40.25	644.00	65.25	1044.00		
4 Electrician Groundmen	25.35	811.20	42.30	1353.60		
1 Crew Truck		152.40		167.64		
1 Flatbed Truck, 20,000 GVW		190.00		209.00		
1 Pickup Truck, 3/4 Ton		104.60		115.06		
.2 Hyd. Crane, 55 Ton		241.60		265.76		
.2 Hyd. Crane, 12 Ton		131.12		144.23		
.2 Earth Auger, Truck-Mtd.		80.36		88.40		
1 Tractor w/Winch		346.80		381.48		
3 Cable Trailers		552.90		608.19		
.5 Tensioning Rig		183.05		201.35		
.5 Cable Pulling Rig		1072.00		1179.20	34.71	38.19
88 L.H., Daily Totals		$6124.03		$8374.31	$69.59	$95.16
Crew R-7	**Hr.**	**Daily**	**Hr.**	**Daily**	**Bare Costs**	**Incl. O&P**
1 Electrician Foreman	$40.75	$326.00	$66.05	$528.40	$27.92	$46.26
5 Electrician Groundmen	25.35	1014.00	42.30	1692.00		
1 Crew Truck		152.40		167.64	3.17	3.49
48 L.H., Daily Totals		$1492.40		$2388.04	$31.09	$49.75
Crew R-8	**Hr.**	**Daily**	**Hr.**	**Daily**	**Bare Costs**	**Incl. O&P**
1 Electrician Foreman	$40.75	$326.00	$66.05	$528.40	$35.37	$57.73
3 Electrician Linemen	40.25	966.00	65.25	1566.00		
2 Electrician Groundmen	25.35	405.60	42.30	676.80		
1 Pickup Truck, 3/4 Ton		104.60		115.06		
1 Crew Truck		152.40		167.64	5.35	5.89
48 L.H., Daily Totals		$1954.60		$3053.90	$40.72	$63.62
Crew R-9	**Hr.**	**Daily**	**Hr.**	**Daily**	**Bare Costs**	**Incl. O&P**
1 Electrician Foreman	$40.75	$326.00	$66.05	$528.40	$32.86	$53.88
1 Electrician Lineman	40.25	322.00	65.25	522.00		
2 Electrician Operators	40.25	644.00	65.25	1044.00		
4 Electrician Groundmen	25.35	811.20	42.30	1353.60		
1 Pickup Truck, 3/4 Ton		104.60		115.06		
1 Crew Truck		152.40		167.64	4.02	4.42
64 L.H., Daily Totals		$2360.20		$3730.70	$36.88	$58.29
Crew R-10	**Hr.**	**Daily**	**Hr.**	**Daily**	**Bare Costs**	**Incl. O&P**
1 Electrician Foreman	$40.75	$326.00	$66.05	$528.40	$37.85	$61.56
4 Electrician Linemen	40.25	1288.00	65.25	2088.00		
1 Electrician Groundman	25.35	202.80	42.30	338.40		
1 Crew Truck		152.40		167.64		
3 Tram Cars		339.90		373.89	10.26	11.28
48 L.H., Daily Totals		$2309.10		$3496.33	$48.11	$72.84

Crew R-11	Bare Costs Hr.	Bare Costs Daily	Incl. Subs O&P Hr.	Incl. Subs O&P Daily	Cost Per Labor-Hour Bare Costs	Cost Per Labor-Hour Incl. O&P
1 Electrician Foreman	$40.75	$326.00	$66.05	$528.40	$37.71	$61.41
4 Electricians	40.25	1288.00	65.25	2088.00		
1 Equip. Oper. (crane)	35.80	286.40	58.50	468.00		
1 Common Laborer	26.45	211.60	44.30	354.40		
1 Crew Truck		152.40		167.64		
1 Hyd. Crane, 12 Ton		655.60		721.16	14.43	15.87
56 L.H., Daily Totals		$2920.00		$4327.60	$52.14	$77.28

Crew R-12	Hr.	Daily	Hr.	Daily	Bare Costs	Incl. O&P
1 Carpenter Foreman	$34.05	$272.40	$57.00	$456.00	$31.40	$53.20
4 Carpenters	33.55	1073.60	56.15	1796.80		
4 Common Laborers	26.45	846.40	44.30	1417.60		
1 Equip. Oper. (med.)	34.90	279.20	57.05	456.40		
1 Steel Worker	36.40	291.20	69.40	555.20		
1 Dozer, 200 H.P.		1051.00		1156.10		
1 Pickup Truck, 3/4 Ton		104.60		115.06	13.13	14.45
88 L.H., Daily Totals		$3918.40		$5953.16	$44.53	$67.65

Crew R-15	Hr.	Daily	Hr.	Daily	Bare Costs	Incl. O&P
1 Electrician Foreman	$40.75	$326.00	$66.05	$528.40	$39.22	$63.64
4 Electricians	40.25	1288.00	65.25	2088.00		
1 Equipment Oper. (light)	33.55	268.40	54.80	438.40		
1 Aerial Lift Truck		317.80		349.58	6.62	7.28
48 L.H., Daily Totals		$2200.20		$3404.38	$45.84	$70.92

Crew R-15A	Hr.	Daily	Hr.	Daily	Bare Costs	Incl. O&P
1 Electrician Foreman	$40.75	$326.00	$66.05	$528.40	$34.62	$56.66
2 Electricians	40.25	644.00	65.25	1044.00		
2 Common Laborers	26.45	423.20	44.30	708.80		
1 Equipment Operator	33.55	268.40	54.80	438.40		
1 Aerial Lift Truck		317.80		349.58	6.62	7.28
48 L.H., Daily Totals		$1979.40		$3069.18	$41.24	$63.94

Crew R-18	Hr.	Daily	Hr.	Daily	Bare Costs	Incl. O&P
.25 Electrician Foreman	$40.75	$81.50	$66.05	$132.10	$31.12	$51.19
1 Electrician	40.25	322.00	65.25	522.00		
2 Helpers	25.35	405.60	42.30	676.80		
26 L.H., Daily Totals		$809.10		$1330.90	$31.12	$51.19

Crew R-19	Hr.	Daily	Hr.	Daily	Bare Costs	Incl. O&P
.5 Electrician Foreman	$40.75	$163.00	$66.05	$264.20	$40.35	$65.41
2 Electricians	40.25	644.00	65.25	1044.00		
20 L.H., Daily Totals		$807.00		$1308.20	$40.35	$65.41

Crew R-21	Hr.	Daily	Hr.	Daily	Bare Costs	Incl. O&P
1 Electrician Foreman	$40.75	$326.00	$66.05	$528.40	$40.24	$65.25
3 Electricians	40.25	966.00	65.25	1566.00		
.1 Equip. Oper. (med.)	34.90	27.92	57.05	45.64		
.1 S.P. Crane, 4x4, 25 Ton		56.04		61.64	1.71	1.88
32.8 L.H., Daily Totals		$1375.96		$2201.68	$41.95	$67.12

Crew R-22	Hr.	Daily	Hr.	Daily	Bare Costs	Incl. O&P
.66 Electrician Foreman	$40.75	$215.16	$66.05	$348.74	$33.93	$55.51
2 Helpers	25.35	405.60	42.30	676.80		
2 Electricians	40.25	644.00	65.25	1044.00		
37.28 L.H., Daily Totals		$1264.76		$2069.54	$33.93	$55.51

Crew R-30	Bare Costs Hr.	Bare Costs Daily	Incl. Subs O&P Hr.	Incl. Subs O&P Daily	Cost Per Labor-Hour Bare Costs	Cost Per Labor-Hour Incl. O&P
.25 Electrician Foreman (out)	$42.25	$84.50	$68.50	$137.00	$31.91	$52.61
1 Electrician	40.25	322.00	65.25	522.00		
2 Laborers, (Semi-Skilled)	26.45	423.20	44.30	708.80		
26 L.H., Daily Totals		$829.70		$1367.80	$31.91	$52.61

Historical Cost Indexes

The table below lists both the RSMeans Historical Cost Index based on Jan. 1, 1993 = 100 as well as the computed value of an index based on Jan. 1, 2011 costs. Since the Jan. 1, 2011 figure is estimated, space is left to write in the actual index figures as they become available through either the quarterly *Means Construction Cost Indexes* or as printed in the *Engineering News-Record.* To compute the actual index based on Jan. 1, 2011 = 100, divide the Historical Cost Index for a particular year by the actual Jan. 1, 2011 Construction Cost Index. Space has been left to advance the index figures as the year progresses.

Year	Historical Cost Index Jan. 1, 1993 = 100		Current Index Based on Jan. 1, 2011 = 100		Year	Historical Cost Index Jan. 1, 1993 = 100	Current Index Based on Jan. 1, 2011 = 100		Year	Historical Cost Index Jan. 1, 1993 = 100	Current Index Based on Jan. 1, 2011 = 100	
	Est.	Actual	Est.	Actual		Actual	Est.	Actual		Actual	Est.	Actual
Oct 2011*					July 1996	110.2	59.6		July 1978	53.5	28.9	
July 2011*					1995	107.6	58.2		1977	49.5	26.8	
April 2011*					1994	104.4	56.4		1976	46.9	25.4	
Jan 2011*	185.0		100.0	100.0	1993	101.7	55.0		1975	44.8	24.2	
July 2010		183.5	99.2		1992	99.4	53.7		1974	41.4	22.4	
2009		180.1	97.4		1991	96.8	52.3		1973	37.7	20.4	
2008		180.4	97.5		1990	94.3	51.0		1972	34.8	18.8	
2007		169.4	91.6		1989	92.1	49.8		1971	32.1	17.4	
2006		162.0	87.6		1988	89.9	48.6		1970	28.7	15.5	
2005		151.6	81.9		1987	87.7	47.4		1969	26.9	14.5	
2004		143.7	77.7		1986	84.2	45.5		1968	24.9	13.5	
2003		132.0	71.4		1985	82.6	44.7		1967	23.5	12.7	
2002		128.7	69.6		1984	82.0	44.3		1966	22.7	12.3	
2001		125.1	67.6		1983	80.2	43.3		1965	21.7	11.7	
2000		120.9	65.4		1982	76.1	41.2		1964	21.2	11.5	
1999		117.6	63.6		1981	70.0	37.8		1963	20.7	11.2	
1998		115.1	62.2		1980	62.9	34.0		1962	20.2	10.9	
1997		112.8	61.0		1979	57.8	31.2		1961	19.8	10.7	

Adjustments to Costs

The Historical Cost Index can be used to convert National Average building costs at a particular time to the approximate building costs for some other time.

Time Adjustment using the Historical Cost Indexes:

$$\frac{\text{Index for Year A}}{\text{Index for Year B}} \text{ x Cost in Year B = Cost in Year A}$$

Example:

Estimate and compare construction costs for different years in the same city.

To estimate the National Average construction cost of a building in 1970, knowing that it cost $900,000 in 2011:

INDEX in 1970 = 28.7

INDEX in 2011 = 185.0

$$\frac{\text{INDEX 1970}}{\text{INDEX 2011}} \text{ x Cost 2011 = Cost 1970}$$

$$\frac{28.7}{185.0} \text{ x \$900,000 = .155 x \$900,000 = \$139,500}$$

The construction cost of the building in 1970 is $139,500.

Note: The City Cost Indexes for Canada can be used to convert U.S. National averages to local costs in Canadian dollars.

*Historical Cost Index updates and other resources are provided on the following website.
http://www.reedconstructiondata.com/rsmeans/chgnotice/842339

How to Use the City Cost Indexes

What you should know before you begin

RSMeans City Cost Indexes (CCI) are an extremely useful tool to use when you want to compare costs from city to city and region to region.

This publication contains average construction cost indexes for 731 U.S. and Canadian cities covering over 930 three-digit zip code locations, as listed directly under each city.

Keep in mind that a City Cost Index number is a percentage ratio of a specific city's cost to the national average cost of the same item at a stated time period.

In other words, these index figures represent relative construction factors (or, if you prefer, multipliers) for Material and Installation costs, as well as the weighted average for Total In Place costs for each CSI MasterFormat division. Installation costs include both labor and equipment rental costs. When estimating equipment rental rates only, for a specific location, use 015433 CONTRACTOR EQUIPMENT index.

The 30 City Average Index is the average of 30 major U.S. cities and serves as a National Average.

Index figures for both material and installation are based on the 30 major city average of 100 and represent the cost relationship as of July 1, 2010. The index for each division is computed from representative material and labor quantities for that division. The weighted average for each city is a weighted total of the components listed above it, but does not include relative productivity between trades or cities.

As changes occur in local material prices, labor rates, and equipment rental rates, (including fuel costs) the impact of these changes should be accurately measured by the change in the City Cost Index for each particular city (as compared to the 30 City Average).

Therefore, if you know (or have estimated) building costs in one city today, you can easily convert those costs to expected building costs in another city.

In addition, by using the Historical Cost Index, you can easily convert National Average building costs at a particular time to the approximate building costs for some other time. The City Cost Indexes can then be applied to calculate the costs for a particular city.

Quick Calculations

Location Adjustment Using the City Cost Indexes:

$$\frac{\text{Index for City A}}{\text{Index for City B}} \times \text{Cost in City B} = \text{Cost in City A}$$

Time Adjustment for the National Average Using the Historical Cost Index:

$$\frac{\text{Index for Year A}}{\text{Index for Year B}} \times \text{Cost in Year B} = \text{Cost in Year A}$$

Adjustment from the National Average:

$$\frac{\text{Index for City A}}{100} \times \text{National Average Cost} = \text{Cost in City A}$$

Since each of the other RSMeans publications contains many different items, any *one* item multiplied by the particular city index may give incorrect results. However, the larger the number of items compiled, the closer the results should be to actual costs for that particular city.

The City Cost Indexes for Canadian cities are calculated using Canadian material and equipment prices and labor rates, in Canadian dollars. Therefore, indexes for Canadian cities can be used to convert U.S. National Average prices to local costs in Canadian dollars.

How to use this section

1. Compare costs from city to city.

In using the RSMeans Indexes, remember that an index number is not a fixed number but a ratio: It's a percentage ratio of a building component's cost at any stated time to the National Average cost of that same component at the same time period. Put in the form of an equation:

$$\frac{\text{Specific City Cost}}{\text{National Average Cost}} \times 100 = \text{City Index Number}$$

Therefore, when making cost comparisons between cities, do not subtract one city's index number from the index number of another city and read the result as a percentage difference. Instead, divide one city's index number by that of the other city. The resulting number may then be used as a multiplier to calculate cost differences from city to city.

The formula used to find cost differences between cities for the purpose of comparison is as follows:

$$\frac{\text{City A Index}}{\text{City B Index}} \times \text{City B Cost (Known)} = \text{City A Cost (Unknown)}$$

In addition, you can use RSMeans CCI to calculate and compare costs division by division between cities using the same basic formula. (Just be sure that you're comparing similar divisions.)

2. Compare a specific city's construction costs with the National Average.

When you're studying construction location feasibility, it's advisable to compare a prospective project's cost index with an index of the National Average cost.

For example, divide the weighted average index of construction costs of a specific city by that of the 30 City Average, which = 100.

$$\frac{\text{City Index}}{100} = \text{\% of National Average}$$

As a result, you get a ratio that indicates the relative cost of construction in that city in comparison with the National Average.

3. Convert U.S. National Average to actual costs in Canadian City.

$$\frac{\text{Index for Canadian City}}{100} \times \text{National Average Cost} = \text{Cost in Canadian City in \$ CAN}$$

4. **Adjust construction cost data based on a National Average.**

When you use a source of construction cost data which is based on a National Average (such as RSMeans cost data publications), it is necessary to adjust those costs to a specific location.

$$\frac{\text{City Index}}{100} \times \begin{array}{c}\text{"Book" Cost Based on}\\ \text{National Average Costs}\end{array} = \begin{array}{c}\text{City Cost}\\ \text{(Unknown)}\end{array}$$

5. **When applying the City Cost Indexes to demolition projects, use the appropriate division installation index.** For example, for removal of existing doors and windows, use Division 8 (Openings) index.

What you might like to know about how we developed the Indexes

The information presented in the CCI is organized according to the Construction Specifications Institute (CSI) MasterFormat 2010.

To create a reliable index, RSMeans researched the building type most often constructed in the United States and Canada. Because it was concluded that no one type of building completely represented the building construction industry, nine different types of buildings were combined to create a composite model.

The exact material, labor, and equipment quantities are based on detailed analysis of these nine building types, then each quantity is weighted in proportion to expected usage. These various material items, labor hours, and equipment rental rates are thus combined to form a composite building representing as closely as possible the actual usage of materials, labor and equipment used in the North American Building Construction Industry.

The following structures were chosen to make up that composite model:

1. Factory, 1 story
2. Office, 2–4 story
3. Store, Retail
4. Town Hall, 2–3 story
5. High School, 2–3 story
6. Hospital, 4–8 story
7. Garage, Parking
8. Apartment, 1–3 story
9. Hotel/Motel, 2–3 story

For the purposes of ensuring the timeliness of the data, the components of the index for the composite model have been streamlined. They currently consist of:

- specific quantities of 66 commonly used construction materials;
- specific labor-hours for 21 building construction trades; and
- specific days of equipment rental for 6 types of construction equipment (normally used to install the 66 material items by the 21 trades.) Fuel costs and routine maintenance costs are included in the equipment cost.

A sophisticated computer program handles the updating of all costs for each city on a quarterly basis. Material and equipment price quotations are gathered quarterly from 731 cities in the United States and Canada. These prices and the latest negotiated labor wage rates for 21 different building trades are used to compile the quarterly update of the City Cost Index.

The 30 major U.S. cities used to calculate the National Average are:

Atlanta, GA	Memphis, TN
Baltimore, MD	Milwaukee, WI
Boston, MA	Minneapolis, MN
Buffalo, NY	Nashville, TN
Chicago, IL	New Orleans, LA
Cincinnati, OH	New York, NY
Cleveland, OH	Philadelphia, PA
Columbus, OH	Phoenix, AZ
Dallas, TX	Pittsburgh, PA
Denver, CO	St. Louis, MO
Detroit, MI	San Antonio, TX
Houston, TX	San Diego, CA
Indianapolis, IN	San Francisco, CA
Kansas City, MO	Seattle, WA
Los Angeles, CA	Washington, DC

What the CCI does not indicate

The weighted average for each city is a total of the divisional components weighted to reflect typical usage, but it does not include the productivity variations between trades or cities.

In addition, the CCI does not take into consideration factors such as the following:

- managerial efficiency
- competitive conditions
- automation
- restrictive union practices
- unique local requirements
- regional variations due to specific building codes

DIVISION		UNITED STATES			ALABAMA															
		30 CITY			ANNISTON			BIRMINGHAM			BUTLER			DECATUR			DOTHAN		
		AVERAGE			362			350 - 352			369			356			363		
		MAT.	INST.	TOTAL	MAT.	INST.	TOTAL	MAT.	INST.	TOTAL	MAT.	INST.	TOTAL	MAT.	INST.	TOTAL	MAT.	INST.	TOTAL
015433	CONTRACTOR EQUIPMENT		100.0	100.0		100.8	100.8		100.9	100.9		97.8	97.8		100.8	100.8		97.8	97.8
0241, 31 - 34	SITE & INFRASTRUCTURE, DEMOLITION	100.0	100.0	100.0	87.3	93.5	91.6	85.2	94.1	91.4	101.5	87.2	91.6	80.7	92.1	88.5	99.0	87.1	90.8
0310	Concrete Forming & Accessories	100.0	100.0	100.0	92.6	50.9	56.4	92.8	73.9	76.4	89.1	46.1	51.7	96.8	49.0	55.3	97.9	45.5	52.4
0320	Concrete Reinforcing	100.0	100.0	100.0	88.9	84.6	86.7	88.9	84.9	86.9	93.9	49.5	71.1	88.9	79.3	84.0	93.9	48.9	70.8
0330	Cast-in-Place Concrete	100.0	100.0	100.0	92.1	52.5	77.0	98.3	67.9	86.7	89.9	58.2	77.7	91.6	66.1	81.9	89.9	48.8	74.1
03	CONCRETE	100.0	100.0	100.0	91.5	59.4	76.2	90.8	74.9	83.2	92.1	52.7	73.3	87.6	62.2	75.5	91.6	49.1	71.3
04	MASONRY	100.0	100.0	100.0	85.7	72.1	77.5	88.1	80.9	83.7	89.5	52.4	67.0	86.6	55.4	67.7	90.7	42.9	61.7
05	METALS	100.0	100.0	100.0	93.6	92.1	93.1	94.5	93.9	94.3	92.6	79.2	88.0	95.8	88.5	93.3	92.7	77.9	87.6
06	WOOD, PLASTICS & COMPOSITES	100.0	100.0	100.0	94.5	46.2	66.8	96.4	72.5	82.7	89.6	45.9	64.5	99.5	46.5	69.1	101.1	45.9	69.4
07	THERMAL & MOISTURE PROTECTION	100.0	100.0	100.0	91.9	56.8	77.9	93.9	83.4	89.8	91.9	60.3	79.4	91.7	62.1	79.9	91.9	52.6	76.3
08	OPENINGS	100.0	100.0	100.0	93.9	52.1	83.3	97.6	74.8	91.8	94.0	47.7	82.3	97.9	53.6	86.7	94.0	47.7	82.3
0920	Plaster & Gypsum Board	100.0	100.0	100.0	89.7	45.1	58.3	98.4	72.2	79.9	87.5	44.8	57.4	94.6	45.4	60.0	97.7	44.8	60.4
0950, 0980	Ceilings & Acoustic Treatment	100.0	100.0	100.0	89.7	45.1	60.2	95.6	72.2	80.0	89.7	44.8	59.9	94.4	45.4	61.9	89.7	44.8	59.9
0960	Flooring	100.0	100.0	100.0	103.2	42.5	85.8	105.9	63.7	93.8	109.1	59.2	94.8	105.9	53.5	90.8	116.1	29.1	91.1
0970, 0990	Wall Finishes & Painting/Coating	100.0	100.0	100.0	96.5	33.9	58.5	96.5	59.8	74.3	96.5	57.4	72.8	96.5	59.9	74.3	96.5	57.4	72.8
09	FINISHES	100.0	100.0	100.0	96.4	46.3	68.8	99.5	70.0	83.2	100.0	49.2	72.0	98.2	50.1	71.7	103.1	43.2	70.0
COVERS	DIVS. 10 - 14, 25, 28, 41, 43, 44, 46	100.0	100.0	100.0	100.0	76.6	95.1	100.0	88.5	97.6	100.0	46.1	88.8	100.0	48.2	89.2	100.0	46.2	88.8
21, 22, 23	FIRE SUPPRESSION, PLUMBING & HVAC	100.0	100.0	100.0	99.0	60.8	83.1	99.8	67.4	86.3	96.9	36.5	71.7	99.8	44.7	76.9	96.9	36.4	71.7
26, 27, 3370	ELECTRICAL, COMMUNICATIONS & UTIL.	100.0	100.0	100.0	96.4	62.2	78.2	105.6	62.7	82.8	98.6	43.6	69.4	98.2	66.2	81.2	97.3	62.3	78.7
MF2010	WEIGHTED AVERAGE	100.0	100.0	100.0	95.1	65.5	81.9	97.0	75.6	87.4	95.7	52.6	76.4	95.8	60.8	80.1	95.8	52.6	76.5

DIVISION		ALABAMA																	
		EVERGREEN			GADSDEN			HUNTSVILLE			JASPER			MOBILE			MONTGOMERY		
		364			359			357 - 358			355			365 - 366			360 - 361		
		MAT.	INST.	TOTAL	MAT.	INST.	TOTAL	MAT.	INST.	TOTAL	MAT.	INST.	TOTAL	MAT.	INST.	TOTAL	MAT.	INST.	TOTAL
015433	CONTRACTOR EQUIPMENT		97.8	97.8		100.8	100.8		100.8	100.8		100.8	100.8		97.8	97.8		97.8	97.8
0241, 31 - 34	SITE & INFRASTRUCTURE, DEMOLITION	101.9	87.4	91.9	86.0	93.0	90.8	80.5	93.1	89.2	85.5	92.8	90.5	93.6	88.2	89.9	92.3	88.1	89.4
0310	Concrete Forming & Accessories	85.6	47.9	52.8	89.1	44.6	50.5	96.8	65.6	69.7	94.3	37.1	44.6	96.9	58.0	63.1	94.7	46.7	53.1
0320	Concrete Reinforcing	94.0	49.5	71.1	94.1	83.7	88.8	88.9	79.8	84.2	88.9	83.7	86.2	91.8	83.0	87.2	91.8	83.6	87.5
0330	Cast-in-Place Concrete	89.9	61.0	78.8	91.6	51.6	76.3	89.2	64.4	79.7	101.5	46.3	80.4	94.2	70.7	85.2	96.2	49.2	78.2
03	CONCRETE	92.3	54.4	74.2	91.3	56.2	74.5	86.5	69.1	78.1	94.9	51.0	73.9	89.4	68.5	79.4	90.2	56.3	74.0
04	MASONRY	89.5	57.4	70.0	84.8	46.4	61.5	87.7	61.6	71.9	81.2	44.5	59.0	88.4	56.9	69.3	86.8	42.5	59.9
05	METALS	92.6	79.2	88.0	93.7	91.2	92.8	95.8	90.1	93.8	93.6	90.5	92.5	94.7	92.2	93.8	94.5	90.8	93.2
06	WOOD, PLASTICS & COMPOSITES	85.7	45.9	62.9	90.2	41.9	62.5	99.5	66.5	80.6	96.4	34.1	60.7	99.6	57.3	75.4	96.0	45.9	67.3
07	THERMAL & MOISTURE PROTECTION	91.9	61.2	79.7	91.8	67.0	82.0	91.6	75.9	85.3	91.8	48.6	74.6	91.6	74.5	84.8	91.1	65.8	81.1
08	OPENINGS	94.0	47.7	82.3	93.9	50.9	83.0	97.9	64.4	89.4	93.9	49.8	82.7	97.9	61.9	88.8	97.9	55.4	87.1
0920	Plaster & Gypsum Board	85.8	44.8	56.9	87.1	40.7	54.4	94.6	66.0	74.5	90.2	32.7	49.7	94.6	56.5	67.8	98.4	44.8	60.6
0950, 0980	Ceilings & Acoustic Treatment	89.7	44.8	59.9	89.7	40.7	57.2	96.5	66.0	76.3	89.7	32.7	51.9	96.5	56.5	70.0	95.6	44.8	61.9
0960	Flooring	106.5	59.2	92.9	101.4	47.5	85.9	105.9	57.1	91.9	104.1	42.5	86.4	115.1	59.2	99.0	113.5	31.3	89.9
0970, 0990	Wall Finishes & Painting/Coating	96.5	57.4	72.8	96.5	65.4	77.6	96.5	62.9	76.1	96.5	41.5	63.2	100.0	58.4	74.8	96.5	57.4	72.8
09	FINISHES	99.1	50.4	72.3	95.4	45.7	68.0	98.7	63.4	79.2	96.5	37.1	63.7	103.4	57.2	77.9	102.7	43.5	70.0
COVERS	DIVS. 10 - 14, 25, 28, 41, 43, 44, 46	100.0	47.8	89.2	100.0	79.9	95.8	100.0	84.8	96.8	100.0	44.2	88.4	100.0	81.1	96.1	100.0	79.8	95.8
21, 22, 23	FIRE SUPPRESSION, PLUMBING & HVAC	96.9	39.0	72.8	101.4	39.4	75.6	99.8	56.8	81.9	101.4	57.7	83.2	99.8	63.6	84.7	99.9	37.2	73.8
26, 27, 3370	ELECTRICAL, COMMUNICATIONS & UTIL.	95.7	43.6	68.0	98.2	62.7	79.3	99.4	66.2	81.8	97.7	62.7	79.1	99.4	62.0	79.5	100.5	66.1	82.2
MF2010	WEIGHTED AVERAGE	95.3	54.1	76.9	95.6	58.2	78.9	95.8	69.0	83.8	96.0	58.2	79.1	96.8	67.9	83.9	96.8	57.2	79.1

DIVISION		ALABAMA									ALASKA								
		PHENIX CITY			SELMA			TUSCALOOSA			ANCHORAGE			FAIRBANKS			JUNEAU		
		368			367			354			995 - 996			997			998		
		MAT.	INST.	TOTAL	MAT.	INST.	TOTAL	MAT.	INST.	TOTAL	MAT.	INST.	TOTAL	MAT.	INST.	TOTAL	MAT.	INST.	TOTAL
015433	CONTRACTOR EQUIPMENT		97.8	97.8		97.8	97.8		100.8	100.8		115.9	115.9		115.9	115.9		115.9	115.9
0241, 31 - 34	SITE & INFRASTRUCTURE, DEMOLITION	105.9	88.1	93.6	98.8	88.0	91.4	80.9	93.2	89.4	129.3	131.2	130.6	121.2	131.2	128.1	129.0	131.2	130.5
0310	Concrete Forming & Accessories	92.5	41.4	48.2	90.3	45.9	51.8	96.7	50.5	56.6	137.3	118.9	121.3	140.5	118.9	121.8	136.5	118.7	121.1
0320	Concrete Reinforcing	93.8	66.1	79.6	93.9	83.4	88.5	88.9	83.7	86.2	153.8	107.1	129.8	148.1	107.1	127.0	115.7	107.1	111.3
0330	Cast-in-Place Concrete	89.9	49.8	74.5	89.9	49.0	74.2	92.9	51.6	77.1	144.8	117.2	134.2	136.9	117.5	129.5	144.7	117.1	134.1
03	CONCRETE	94.6	50.8	73.7	91.1	55.8	74.2	88.3	58.8	74.2	132.9	115.4	124.5	120.4	115.5	118.1	123.5	115.3	119.6
04	MASONRY	89.4	44.3	62.1	92.8	41.7	61.8	86.7	49.6	64.2	200.7	120.0	151.8	203.5	120.0	152.9	185.9	120.0	146.0
05	METALS	92.6	84.2	89.7	92.6	89.7	91.6	94.9	91.1	93.6	122.1	101.5	115.0	119.1	101.6	113.1	119.1	101.4	113.0
06	WOOD, PLASTICS & COMPOSITES	94.2	38.3	62.1	91.2	45.9	65.2	99.5	48.1	70.0	112.6	118.7	116.1	117.6	118.7	118.2	112.6	118.7	116.1
07	THERMAL & MOISTURE PROTECTION	92.2	65.7	81.7	91.8	55.3	77.3	91.7	69.3	82.8	192.8	116.0	162.3	181.6	114.8	155.0	183.0	116.0	156.4
08	OPENINGS	93.9	47.0	82.1	93.9	55.4	84.2	97.9	61.5	88.7	120.1	114.0	118.5	116.6	114.0	116.0	115.9	114.0	115.4
0920	Plaster & Gypsum Board	91.1	37.0	53.0	88.9	44.8	57.8	94.6	47.0	61.1	146.7	119.0	127.2	153.1	119.0	129.1	134.6	119.0	123.6
0950, 0980	Ceilings & Acoustic Treatment	89.7	37.0	54.7	89.7	44.8	59.9	96.5	47.0	63.7	140.9	119.0	126.4	126.0	119.0	121.4	120.3	119.0	119.5
0960	Flooring	111.7	31.3	88.6	109.8	30.1	86.9	105.9	39.6	86.9	162.0	125.2	151.4	152.3	125.2	144.5	152.7	125.2	144.8
0970, 0990	Wall Finishes & Painting/Coating	96.5	57.4	72.8	96.5	57.4	72.8	96.5	49.4	68.0	150.8	117.3	130.5	149.8	122.3	133.1	150.8	117.3	130.5
09	FINISHES	101.7	39.3	67.3	100.1	43.3	68.8	98.7	47.1	70.2	152.6	120.4	134.8	145.2	120.9	131.8	142.4	120.4	130.2
COVERS	DIVS. 10 - 14, 25, 28, 41, 43, 44, 46	100.0	79.3	95.7	100.0	46.1	88.8	100.0	81.8	96.2	100.0	112.3	102.6	100.0	112.3	102.6	100.0	112.3	102.6
21, 22, 23	FIRE SUPPRESSION, PLUMBING & HVAC	96.9	37.6	72.2	96.9	36.7	71.8	99.8	37.0	73.7	100.4	105.9	102.7	100.3	108.0	103.5	100.4	96.9	98.9
26, 27, 3370	ELECTRICAL, COMMUNICATIONS & UTIL.	98.0	74.2	85.4	97.0	43.6	68.6	98.8	62.7	79.6	141.8	116.3	128.2	147.2	116.3	130.8	137.5	116.3	126.3
MF2010	WEIGHTED AVERAGE	96.2	56.3	78.4	95.5	52.5	76.3	95.8	59.2	79.4	127.2	114.5	121.5	124.4	115.0	120.2	122.8	112.6	118.3

DIVISION		ALASKA			ARIZONA															
		KETCHIKAN			CHAMBERS			FLAGSTAFF			GLOBE			KINGMAN			MESA/TEMPE		
		999			865			860			855			864			852		
		MAT.	INST.	TOTAL	MAT.	INST.	TOTAL	MAT.	INST.	TOTAL	MAT.	INST.	TOTAL	MAT.	INST.	TOTAL	MAT.	INST.	TOTAL
015433	CONTRACTOR EQUIPMENT		115.9	115.9		94.8	94.8		94.8	94.8		97.8	97.8		94.8	94.8		97.8	97.8
0241, 31 - 34	SITE & INFRASTRUCTURE, DEMOLITION	180.9	131.2	146.6	61.0	97.5	86.2	77.3	99.3	92.5	88.8	100.9	97.2	60.9	99.5	87.5	80.3	102.5	95.6
0310	Concrete Forming & Accessories	132.2	118.7	120.5	96.6	46.4	53.0	101.7	69.1	73.4	98.1	46.2	53.1	94.9	68.6	72.1	100.8	68.5	72.8
0320	Concrete Reinforcing	116.3	107.1	111.5	96.5	79.2	87.6	96.4	81.3	88.6	102.4	79.1	90.4	96.7	81.3	88.8	103.1	81.3	91.8
0330	Cast-in-Place Concrete	277.0	117.1	215.8	92.7	55.2	78.3	92.8	75.7	86.3	91.2	55.5	77.5	92.4	73.6	85.2	91.9	73.9	85.0
03	CONCRETE	205.7	115.3	162.4	96.9	56.0	77.3	116.0	73.6	95.7	109.6	56.0	84.0	96.5	72.7	85.1	101.0	72.8	87.5
04	MASONRY	212.0	120.0	156.3	93.1	42.1	62.2	93.1	63.7	75.3	107.8	42.1	67.9	93.1	62.6	74.6	107.9	62.6	80.5
05	METALS	119.3	101.4	113.2	94.6	69.0	85.8	95.1	75.0	88.2	92.4	69.3	84.5	95.2	74.9	88.2	92.7	75.3	86.7
06	WOOD, PLASTICS & COMPOSITES	108.8	118.7	114.5	97.1	44.3	66.8	102.9	69.8	83.9	91.7	44.4	64.5	93.1	69.8	79.7	94.9	69.9	80.6
07	THERMAL & MOISTURE PROTECTION	184.5	116.0	157.3	96.6	51.9	78.8	98.5	66.7	85.8	104.5	49.7	82.7	96.6	62.8	83.2	103.6	63.8	87.8
08	OPENINGS	118.3	114.0	117.2	102.7	51.8	89.8	102.8	72.1	95.1	97.4	51.9	85.9	102.9	72.1	95.1	97.4	72.2	91.0
0920	Plaster & Gypsum Board	141.1	119.0	125.6	94.9	42.6	58.1	99.2	68.9	77.8	101.8	42.6	60.1	89.0	68.9	74.8	104.4	68.9	79.4
0950, 0980	Ceilings & Acoustic Treatment	120.7	119.0	119.6	122.0	42.6	69.4	122.9	68.9	87.1	113.0	42.6	66.3	122.9	68.9	87.1	113.0	68.9	83.7
0960	Flooring	152.3	125.2	144.5	92.8	41.8	78.1	95.1	42.0	79.8	98.8	41.8	82.4	91.9	56.7	81.8	100.2	52.3	86.4
0970, 0990	Wall Finishes & Painting/Coating	149.8	117.3	130.1	91.5	39.7	60.1	91.5	58.9	71.8	96.5	39.7	62.1	91.5	58.9	71.8	96.5	58.9	73.7
09	FINISHES	147.6	120.4	132.6	99.5	43.6	68.7	102.7	62.4	80.5	100.2	43.7	69.1	98.7	64.7	80.0	100.0	63.9	80.1
COVERS	DIVS. 10 - 14, 25, 28, 41, 43, 44, 46	100.0	112.3	102.6	100.0	77.5	95.3	100.0	82.7	96.4	100.0	77.8	95.4	100.0	82.4	96.3	100.0	82.7	96.4
21, 22, 23	FIRE SUPPRESSION, PLUMBING & HVAC	98.9	96.9	98.1	96.8	73.1	86.9	100.2	84.5	93.7	94.9	68.0	83.7	96.8	83.9	91.4	100.3	76.0	90.2
26, 27, 3370	ELECTRICAL, COMMUNICATIONS & UTIL.	147.1	116.3	130.8	101.4	70.0	84.7	100.2	63.6	80.7	97.0	65.1	80.1	101.4	63.6	81.3	93.5	63.6	77.6
MF2010	WEIGHTED AVERAGE	136.9	112.6	126.0	96.8	63.5	81.9	100.7	74.4	88.9	98.5	62.0	82.2	96.7	74.2	86.7	98.2	72.8	86.8

DIVISION		ARIZONA												ARKANSAS					
		PHOENIX			PRESCOTT			SHOW LOW			TUCSON			BATESVILLE			CAMDEN		
		850,853			863			859			856 - 857			725			717		
		MAT.	INST.	TOTAL	MAT.	INST.	TOTAL	MAT.	INST.	TOTAL	MAT.	INST.	TOTAL	MAT.	INST.	TOTAL	MAT.	INST.	TOTAL
015433	CONTRACTOR EQUIPMENT		98.4	98.4		94.8	94.8		97.8	97.8		97.8	97.8		87.8	87.8		87.8	87.8
0241, 31 - 34	SITE & INFRASTRUCTURE, DEMOLITION	80.7	103.0	96.1	66.6	97.5	87.9	90.8	100.9	97.8	76.3	102.7	94.5	73.8	84.1	80.9	74.6	83.7	80.9
0310	Concrete Forming & Accessories	101.7	71.5	75.5	98.1	46.2	53.0	104.5	46.5	54.2	101.1	71.1	75.0	83.6	48.6	53.2	83.4	32.9	39.6
0320	Concrete Reinforcing	101.5	81.4	91.1	96.4	79.4	87.6	103.1	79.3	90.8	85.1	81.3	83.1	99.3	65.5	81.9	100.6	65.3	82.4
0330	Cast-in-Place Concrete	92.0	76.1	85.9	92.7	55.1	78.3	91.2	55.6	77.6	94.6	76.0	87.5	81.3	47.8	68.5	83.4	40.1	66.8
03	CONCRETE	100.7	74.9	88.3	102.1	55.9	80.0	111.8	56.3	85.2	99.6	74.6	87.6	84.0	52.2	68.8	85.7	42.6	65.1
04	MASONRY	95.0	64.3	76.4	93.1	47.4	65.4	107.8	43.6	68.9	93.0	63.6	75.2	105.1	44.0	68.1	113.4	33.9	65.2
05	METALS	94.3	76.3	88.1	95.1	69.2	86.2	92.2	69.8	84.5	93.5	75.3	87.3	97.5	67.0	87.0	97.6	66.4	86.9
06	WOOD, PLASTICS & COMPOSITES	95.8	72.8	82.6	98.4	44.3	67.3	99.2	44.4	67.8	95.2	72.8	82.3	86.7	49.3	65.2	86.4	32.1	55.3
07	THERMAL & MOISTURE PROTECTION	103.3	67.0	88.8	97.1	51.2	78.9	104.7	52.1	83.8	104.8	64.1	88.6	95.8	46.8	76.3	95.6	37.9	72.7
08	OPENINGS	98.6	73.7	92.3	102.8	51.8	89.9	96.4	51.9	85.1	93.7	73.8	88.7	95.9	47.8	83.7	92.0	42.1	79.3
0920	Plaster & Gypsum Board	107.0	71.8	82.2	95.1	42.6	58.2	107.1	42.6	61.7	111.0	71.8	83.4	75.4	48.3	56.3	75.4	30.6	43.8
0950, 0980	Ceilings & Acoustic Treatment	120.5	71.8	88.2	121.0	42.6	69.1	113.0	42.6	66.3	113.9	71.8	86.0	86.6	48.3	61.2	86.6	30.6	49.5
0960	Flooring	100.4	54.6	87.3	93.7	41.8	78.8	101.9	47.6	86.3	91.0	42.0	76.9	101.4	62.4	90.2	101.3	41.8	84.2
0970, 0990	Wall Finishes & Painting/Coating	96.5	59.1	73.8	91.5	39.7	60.1	96.5	39.7	62.1	97.0	58.9	73.9	97.1	44.3	65.1	97.1	50.9	69.1
09	FINISHES	102.0	66.6	82.5	100.1	43.6	68.9	102.0	44.7	70.4	98.1	64.2	79.4	88.1	50.4	67.3	88.1	35.7	59.2
COVERS	DIVS. 10 - 14, 25, 28, 41, 43, 44, 46	100.0	83.4	96.5	100.0	77.5	95.3	100.0	77.8	95.4	100.0	83.3	96.5	100.0	44.0	88.4	100.0	39.5	87.4
21, 22, 23	FIRE SUPPRESSION, PLUMBING & HVAC	100.2	84.6	93.7	100.2	72.8	88.8	94.9	73.2	85.8	100.2	72.0	88.5	94.7	54.9	78.1	94.7	52.1	77.0
26, 27, 3370	ELECTRICAL, COMMUNICATIONS & UTIL.	100.4	64.9	81.5	99.9	63.5	80.6	93.9	65.0	78.5	95.9	60.2	76.9	100.2	64.7	81.3	96.1	61.7	77.8
MF2010	WEIGHTED AVERAGE	98.6	75.9	88.5	98.4	63.0	82.6	98.6	63.5	82.9	96.9	72.0	85.8	94.3	56.9	77.6	94.2	50.7	74.8

DIVISION		ARKANSAS																	
		FAYETTEVILLE			FORT SMITH			HARRISON			HOT SPRINGS			JONESBORO			LITTLE ROCK		
		727			729			726			719			724			720 - 722		
		MAT.	INST.	TOTAL	MAT.	INST.	TOTAL	MAT.	INST.	TOTAL	MAT.	INST.	TOTAL	MAT.	INST.	TOTAL	MAT.	INST.	TOTAL
015433	CONTRACTOR EQUIPMENT		87.8	87.8		87.8	87.8		87.8	87.8		87.8	87.8		106.7	106.7		87.8	87.8
0241, 31 - 34	SITE & INFRASTRUCTURE, DEMOLITION	73.2	85.7	81.8	77.9	85.6	83.2	78.5	84.1	82.4	78.0	85.1	82.9	98.8	99.4	99.2	85.3	85.6	85.5
0310	Concrete Forming & Accessories	79.5	41.6	46.6	97.2	59.4	64.4	87.7	48.5	53.7	81.0	38.5	44.1	87.3	52.1	56.8	89.6	63.9	67.3
0320	Concrete Reinforcing	99.3	70.4	84.4	100.4	71.2	85.4	98.8	65.4	81.6	98.7	65.5	81.6	96.1	67.8	81.5	100.6	66.2	82.9
0330	Cast-in-Place Concrete	81.3	52.3	70.2	93.0	67.7	83.3	90.2	45.7	73.2	85.3	40.6	68.2	88.6	58.3	77.0	93.4	67.8	83.6
03	CONCRETE	83.8	51.6	68.4	92.0	64.9	79.0	91.2	51.5	72.2	89.0	45.3	68.1	88.7	58.6	74.3	91.7	66.0	79.4
04	MASONRY	94.5	44.7	64.3	99.8	53.1	71.5	105.5	42.1	67.1	84.5	36.2	55.2	95.9	45.3	65.2	96.6	53.1	70.2
05	METALS	97.5	69.1	87.8	99.9	70.9	89.9	98.7	66.7	87.6	97.5	67.0	87.0	94.4	80.5	89.6	96.1	69.3	86.8
06	WOOD, PLASTICS & COMPOSITES	82.3	38.4	57.1	102.6	61.9	79.3	91.9	49.3	67.5	83.6	38.6	57.8	90.1	52.3	68.5	97.2	67.8	80.4
07	THERMAL & MOISTURE PROTECTION	96.6	48.4	77.4	96.9	56.7	81.0	96.1	45.4	76.0	95.8	40.2	73.7	100.4	52.5	81.4	96.0	57.4	80.7
08	OPENINGS	95.9	47.3	83.5	96.7	58.1	87.0	96.7	48.5	84.5	91.9	44.0	79.8	98.8	56.5	88.1	96.7	61.4	87.8
0920	Plaster & Gypsum Board	72.3	37.1	47.5	81.4	61.3	67.2	80.0	48.3	57.7	73.7	37.3	48.1	91.9	51.1	63.2	86.8	67.4	73.2
0950, 0980	Ceilings & Acoustic Treatment	86.6	37.1	53.8	90.3	61.3	71.1	90.3	48.3	62.5	86.6	37.3	53.9	90.9	51.1	64.5	92.2	67.4	75.8
0960	Flooring	98.4	62.4	88.0	109.0	64.2	96.1	104.1	62.4	92.1	100.0	62.4	89.2	72.6	56.4	68.0	110.3	64.2	97.1
0970, 0990	Wall Finishes & Painting/Coating	97.1	31.6	57.4	97.1	57.9	73.4	97.1	44.3	65.1	97.1	50.9	69.1	85.9	52.1	65.4	97.1	59.5	74.3
09	FINISHES	86.9	43.1	62.7	92.1	60.4	74.6	90.6	50.4	68.4	87.9	43.2	63.3	85.2	52.3	67.1	93.8	64.1	77.4
COVERS	DIVS. 10 - 14, 25, 28, 41, 43, 44, 46	100.0	54.4	90.5	100.0	77.3	95.3	100.0	54.9	90.6	100.0	40.6	87.7	100.0	50.3	89.7	100.0	78.0	95.4
21, 22, 23	FIRE SUPPRESSION, PLUMBING & HVAC	94.8	53.4	77.5	100.2	52.7	80.4	94.7	52.5	77.2	94.7	51.6	76.8	100.4	55.8	81.8	100.1	60.6	83.7
26, 27, 3370	ELECTRICAL, COMMUNICATIONS & UTIL.	93.1	55.6	73.2	97.1	70.1	82.7	98.5	40.3	67.5	98.6	68.6	82.7	103.9	64.8	83.1	101.6	71.2	85.5
MF2010	WEIGHTED AVERAGE	92.9	54.9	75.9	97.0	63.4	82.0	95.7	53.0	76.6	93.4	53.6	75.6	96.7	61.6	81.0	97.0	65.9	83.1

DIVISION		ARKANSAS												CALIFORNIA					
		PINE BLUFF			RUSSELLVILLE			TEXARKANA			WEST MEMPHIS			ALHAMBRA			ANAHEIM		
		716			728			718			723			917 - 918			928		
		MAT.	INST.	TOTAL	MAT.	INST.	TOTAL	MAT.	INST.	TOTAL	MAT.	INST.	TOTAL	MAT.	INST.	TOTAL	MAT.	INST.	TOTAL
015433	CONTRACTOR EQUIPMENT		87.8	87.8		87.8	87.8		88.5	88.5		106.7	106.7		99.1	99.1		101.6	101.6
0241, 31 - 34	SITE & INFRASTRUCTURE, DEMOLITION	79.7	85.6	83.8	75.1	84.0	81.2	93.3	86.1	88.3	106.1	99.4	101.5	99.6	109.4	106.4	98.7	109.2	105.9
0310	Concrete Forming & Accessories	80.7	63.8	66.0	84.2	57.7	61.2	86.7	43.4	49.1	92.2	52.3	57.6	121.7	119.9	120.2	105.6	123.1	120.7
0320	Concrete Reinforcing	100.5	66.2	82.9	100.0	65.3	82.1	100.1	65.5	82.3	96.1	67.8	81.5	120.7	119.9	120.3	99.1	119.9	109.8
0330	Cast-in-Place Concrete	85.3	67.8	78.6	85.2	47.8	70.9	93.0	45.6	74.9	92.9	58.4	79.7	90.3	119.0	101.3	93.7	121.6	104.3
03	CONCRETE	89.7	65.9	78.4	87.2	56.2	72.4	88.6	49.2	69.8	95.5	58.7	77.9	103.7	118.6	110.8	101.9	120.9	111.0
04	MASONRY	122.1	53.1	80.3	99.1	40.3	63.5	99.8	35.8	61.0	81.4	45.3	59.5	113.5	116.9	115.6	82.6	115.1	102.3
05	METALS	98.4	69.1	88.3	97.5	66.5	86.9	91.7	67.0	83.2	93.5	80.7	89.1	86.6	102.1	91.9	102.7	103.1	102.8
06	WOOD, PLASTICS & COMPOSITES	83.1	67.8	74.4	88.0	61.9	73.0	91.4	45.5	65.1	95.9	52.3	70.9	93.3	119.2	108.1	93.6	123.1	110.5
07	THERMAL & MOISTURE PROTECTION	95.9	57.4	80.6	96.8	46.6	76.9	96.6	47.1	77.0	100.8	52.5	81.6	102.5	115.3	107.6	102.6	119.4	109.3
08	OPENINGS	91.9	61.4	84.2	95.9	57.4	86.1	97.2	48.5	84.9	98.8	56.5	88.1	91.0	119.0	98.1	103.2	123.3	108.3
0920	Plaster & Gypsum Board	73.2	67.4	69.1	75.4	61.3	65.5	78.3	44.4	54.4	94.6	51.1	64.0	108.3	119.8	116.4	107.5	123.7	118.9
0950, 0980	Ceilings & Acoustic Treatment	86.6	67.4	73.9	86.6	61.3	69.8	92.2	44.4	60.5	88.8	51.1	63.8	98.6	119.8	112.7	111.4	123.7	119.5
0960	Flooring	99.8	64.2	89.6	101.1	62.4	90.0	102.1	53.9	88.3	75.1	56.4	69.7	98.7	105.9	100.7	110.3	112.4	110.9
0970, 0990	Wall Finishes & Painting/Coating	97.1	59.5	74.3	97.1	37.7	61.1	97.1	32.6	58.0	85.9	52.1	65.4	105.9	109.0	107.8	102.5	109.0	106.4
09	FINISHES	87.9	64.1	74.7	88.2	57.1	71.0	91.0	43.5	64.8	86.5	52.3	67.7	99.4	116.3	108.7	105.9	119.7	113.5
COVERS	DIVS. 10 - 14, 25, 28, 41, 43, 44, 46	100.0	78.0	95.4	100.0	45.4	88.7	100.0	36.2	86.7	100.0	50.3	89.7	100.0	109.5	102.0	100.0	110.3	102.1
21, 22, 23	FIRE SUPPRESSION, PLUMBING & HVAC	100.2	55.2	81.4	94.8	54.7	78.1	100.2	58.6	82.9	95.0	67.8	83.7	94.8	107.8	100.2	100.1	112.6	105.3
26, 27, 3370	ELECTRICAL, COMMUNICATIONS & UTIL.	96.3	71.2	83.0	97.0	48.5	71.2	98.5	39.1	66.9	105.8	61.6	82.3	116.1	114.5	115.3	93.8	107.7	101.2
MF2010	WEIGHTED AVERAGE	96.7	64.8	82.4	94.2	56.1	77.2	95.9	51.9	76.3	95.9	63.7	81.5	98.6	112.7	104.9	99.9	113.7	106.1

DIVISION		CALIFORNIA																	
		BAKERSFIELD			BERKELEY			EUREKA			FRESNO			INGLEWOOD			LONG BEACH		
		932 - 933			947			955			936 - 938			903 - 905			906 - 908		
		MAT.	INST.	TOTAL	MAT.	INST.	TOTAL	MAT.	INST.	TOTAL	MAT.	INST.	TOTAL	MAT.	INST.	TOTAL	MAT.	INST.	TOTAL
015433	CONTRACTOR EQUIPMENT		99.4	99.4		98.4	98.4		99.1	99.1		99.4	99.4		98.8	98.8		98.8	98.8
0241, 31 - 34	SITE & INFRASTRUCTURE, DEMOLITION	101.1	106.6	104.9	127.7	104.4	111.6	112.6	103.7	106.5	104.1	106.5	105.7	91.0	106.2	101.5	98.4	106.2	103.8
0310	Concrete Forming & Accessories	103.7	122.6	120.1	119.8	146.4	142.9	115.6	129.3	127.5	101.6	130.2	126.4	109.9	119.9	118.6	105.1	119.9	118.0
0320	Concrete Reinforcing	111.6	119.8	115.8	93.0	121.1	107.5	108.3	120.6	114.7	94.5	120.3	107.8	114.3	119.9	117.2	113.3	119.9	116.7
0330	Cast-in-Place Concrete	95.4	120.6	105.1	126.7	143.6	133.2	101.6	116.6	107.3	98.7	117.3	105.8	89.7	119.8	101.2	102.1	119.8	108.9
03	CONCRETE	102.2	120.4	110.9	115.4	138.8	126.6	112.8	122.0	117.2	103.5	122.7	112.7	99.3	119.0	108.8	109.4	119.0	114.0
04	MASONRY	102.5	114.5	109.8	139.6	131.5	134.7	108.3	130.1	121.5	105.8	121.8	115.5	75.0	117.0	100.5	83.6	117.0	103.9
05	METALS	98.0	102.3	99.5	105.9	103.7	105.1	102.6	101.7	102.3	102.6	103.0	102.7	93.4	103.5	96.9	93.3	103.6	96.8
06	WOOD, PLASTICS & COMPOSITES	87.8	123.2	108.1	115.3	151.1	135.8	106.6	132.6	121.5	99.0	132.6	118.3	92.1	119.1	107.6	85.8	119.1	104.9
07	THERMAL & MOISTURE PROTECTION	101.3	112.0	105.5	115.6	136.1	123.7	106.7	117.0	110.8	96.1	115.6	103.8	101.4	116.7	107.5	101.7	117.0	107.7
08	OPENINGS	100.7	118.3	105.2	101.2	140.0	111.0	102.0	112.8	104.7	102.5	123.4	107.8	92.8	119.0	99.4	92.8	119.0	99.4
0920	Plaster & Gypsum Board	99.3	123.7	116.5	110.5	152.1	139.8	112.6	133.4	127.3	101.9	133.4	124.1	104.9	119.8	115.4	100.5	119.8	114.1
0950, 0980	Ceilings & Acoustic Treatment	112.9	123.7	120.0	116.4	152.1	140.0	117.0	133.4	127.9	114.5	133.4	127.0	98.9	119.8	112.7	98.9	119.8	112.7
0960	Flooring	110.6	98.3	107.0	119.8	128.0	122.2	115.0	123.1	117.3	117.1	141.0	124.0	100.6	105.9	102.1	97.7	105.9	100.1
0970, 0990	Wall Finishes & Painting/Coating	104.4	95.2	98.8	106.2	143.2	128.7	103.7	49.8	71.0	120.9	107.8	113.0	99.3	109.0	105.2	99.3	109.0	105.2
09	FINISHES	107.2	116.0	112.0	115.3	144.1	131.2	112.3	121.7	117.5	111.5	131.2	122.4	101.1	116.3	109.5	100.2	116.3	109.1
COVERS	DIVS. 10 - 14, 25, 28, 41, 43, 44, 46	100.0	110.5	102.2	100.0	125.5	105.3	100.0	122.1	104.6	100.0	122.0	104.6	100.0	109.4	102.0	100.0	109.4	102.0
21, 22, 23	FIRE SUPPRESSION, PLUMBING & HVAC	100.1	106.0	102.6	94.9	145.0	115.7	94.7	105.8	99.4	100.2	113.7	105.8	94.8	107.8	100.2	94.8	107.8	100.2
26, 27, 3370	ELECTRICAL, COMMUNICATIONS & UTIL.	105.8	103.4	104.5	104.0	138.6	122.4	101.7	119.4	111.1	93.2	99.4	96.5	106.2	114.5	110.6	105.9	114.5	110.5
MF2010	WEIGHTED AVERAGE	101.4	110.4	105.4	106.8	133.6	118.8	103.1	115.2	108.5	101.7	115.3	107.8	96.1	112.6	103.5	97.8	112.6	104.5

DIVISION		CALIFORNIA																	
		LOS ANGELES			MARYSVILLE			MODESTO			MOJAVE			OAKLAND			OXNARD		
		900 - 902			959			953			935			946			930		
		MAT.	INST.	TOTAL	MAT.	INST.	TOTAL	MAT.	INST.	TOTAL	MAT.	INST.	TOTAL	MAT.	INST.	TOTAL	MAT.	INST.	TOTAL
015433	CONTRACTOR EQUIPMENT		100.5	100.5		99.1	99.1		99.1	99.1		99.4	99.4		98.4	98.4		98.2	98.2
0241, 31 - 34	SITE & INFRASTRUCTURE, DEMOLITION	97.5	108.6	105.2	108.5	105.9	106.7	102.5	106.0	104.9	97.5	106.6	103.8	134.5	104.3	113.7	104.3	104.6	104.5
0310	Concrete Forming & Accessories	107.2	123.0	120.9	105.9	130.1	126.9	102.5	130.3	126.6	114.8	117.3	116.9	109.0	146.2	141.3	106.4	123.0	120.8
0320	Concrete Reinforcing	114.3	120.0	117.2	108.3	120.1	114.4	112.3	120.4	116.4	113.7	119.7	116.8	95.1	121.1	108.5	111.6	119.8	115.8
0330	Cast-in-Place Concrete	97.7	120.2	106.3	113.5	117.4	115.0	101.6	117.5	107.7	85.4	120.5	98.9	120.1	121.3	120.5	99.4	120.9	107.6
03	CONCRETE	105.9	120.5	112.9	114.6	122.6	118.5	105.3	122.8	113.7	95.6	117.9	106.3	115.2	131.1	122.8	104.4	120.6	112.2
04	MASONRY	88.9	117.8	106.4	109.3	118.1	114.6	107.1	122.6	116.5	103.9	113.8	109.9	147.9	131.5	138.0	107.6	113.2	111.0
05	METALS	100.7	103.9	101.8	102.0	103.0	102.4	99.2	103.6	100.7	99.6	102.0	100.5	100.1	103.6	101.3	97.8	102.6	99.5
06	WOOD, PLASTICS & COMPOSITES	88.6	122.9	108.3	94.1	132.6	116.2	90.2	132.6	114.5	99.2	116.2	109.0	102.6	151.1	130.4	94.0	123.2	110.7
07	THERMAL & MOISTURE PROTECTION	101.4	119.0	108.4	106.3	118.1	111.0	105.9	117.1	110.4	102.1	110.9	105.6	113.1	136.1	122.2	105.8	114.9	109.4
08	OPENINGS	99.1	123.2	105.2	101.2	124.3	107.0	100.0	124.5	106.2	96.6	113.2	100.8	101.3	140.0	111.1	99.6	123.3	105.6
0920	Plaster & Gypsum Board	104.9	123.7	118.1	104.8	133.4	125.0	108.0	133.4	125.9	115.5	116.5	116.2	103.2	152.1	137.6	107.5	123.7	118.9
0950, 0980	Ceilings & Acoustic Treatment	108.2	123.7	118.5	116.0	133.4	127.6	111.4	133.4	126.0	116.0	116.5	116.3	119.5	152.1	141.1	116.6	123.7	121.3
0960	Flooring	99.1	112.4	102.9	110.0	111.5	110.4	110.6	117.7	112.6	117.3	98.3	111.8	114.3	128.0	118.2	109.0	112.4	109.9
0970, 0990	Wall Finishes & Painting/Coating	99.3	109.0	105.2	103.7	117.7	112.2	103.7	118.5	112.7	103.7	94.6	98.2	106.2	143.2	128.7	103.7	100.7	101.9
09	FINISHES	102.9	119.6	112.1	109.2	127.2	119.1	108.1	128.4	119.3	111.8	111.7	111.7	114.0	144.1	130.6	108.7	118.9	114.3
COVERS	DIVS. 10 - 14, 25, 28, 41, 43, 44, 46	100.0	109.9	102.1	100.0	122.0	104.6	100.0	122.1	104.6	100.0	109.7	102.0	100.0	125.5	105.3	100.0	110.5	102.2
21, 22, 23	FIRE SUPPRESSION, PLUMBING & HVAC	100.1	112.6	105.3	94.7	107.5	100.0	100.1	113.7	105.8	94.7	101.3	97.5	100.3	144.9	118.9	100.1	112.6	105.3
26, 27, 3370	ELECTRICAL, COMMUNICATIONS & UTIL.	102.8	118.2	111.0	97.8	103.9	101.0	100.6	109.9	105.6	92.0	100.4	96.5	103.1	146.7	126.3	97.7	109.8	104.1
MF2010	WEIGHTED AVERAGE	100.6	115.4	107.2	102.4	113.8	107.5	101.9	116.6	108.5	98.2	107.7	102.4	107.5	133.7	119.2	101.4	113.1	106.6

DIVISION		CALIFORNIA																	
		PALM SPRINGS			PALO ALTO			PASADENA			REDDING			RICHMOND			RIVERSIDE		
		922			943			910 - 912			960			948			925		
		MAT.	INST.	TOTAL	MAT.	INST.	TOTAL	MAT.	INST.	TOTAL	MAT.	INST.	TOTAL	MAT.	INST.	TOTAL	MAT.	INST.	TOTAL
015433	CONTRACTOR EQUIPMENT		100.3	100.3		98.4	98.4		99.1	99.1		99.1	99.1		98.4	98.4		100.3	100.3
0241, 31 - 34	SITE & INFRASTRUCTURE, DEMOLITION	90.2	107.2	101.9	123.1	104.3	110.1	96.4	109.4	105.4	108.3	105.9	106.7	134.2	104.3	113.6	96.9	107.2	104.0
0310	Concrete Forming & Accessories	102.2	120.1	117.7	107.1	139.7	135.4	111.1	119.9	118.8	105.7	130.1	126.9	122.2	145.8	142.6	105.9	123.0	120.7
0320	Concrete Reinforcing	113.9	119.8	116.9	93.0	120.8	107.3	121.7	119.9	120.8	108.3	120.1	114.4	93.0	120.8	107.3	110.6	119.8	115.4
0330	Cast-in-Place Concrete	89.3	121.5	101.6	107.1	121.0	112.4	85.7	119.0	98.5	112.6	117.4	114.4	123.2	121.1	122.4	97.1	121.6	106.5
03	CONCRETE	96.8	119.5	107.7	103.7	128.1	115.4	99.2	118.6	108.5	113.6	122.6	117.9	117.2	130.8	123.7	103.1	120.9	111.6
04	MASONRY	80.4	114.0	100.7	116.6	127.6	123.3	97.8	116.9	109.4	109.3	118.1	114.6	139.3	127.6	132.2	81.3	114.0	101.1
05	METALS	103.3	102.7	103.1	97.4	102.8	99.3	86.6	102.1	91.9	102.4	103.0	102.6	97.5	102.8	99.3	102.9	102.9	102.9
06	WOOD, PLASTICS & COMPOSITES	88.3	119.3	106.1	100.0	142.9	124.6	80.5	119.2	102.7	93.8	132.6	116.1	119.0	151.1	137.4	93.6	123.1	110.5
07	THERMAL & MOISTURE PROTECTION	102.4	116.5	108.0	112.7	129.5	119.4	102.2	115.3	107.4	106.3	118.1	111.0	113.2	132.5	120.9	102.9	118.3	109.0
08	OPENINGS	98.7	119.1	103.8	101.3	134.6	109.7	91.0	119.0	98.1	102.2	124.3	107.8	101.3	139.0	110.8	101.7	123.3	107.2
0920	Plaster & Gypsum Board	102.1	119.8	114.6	101.0	143.6	131.0	101.3	119.8	114.3	105.1	133.4	125.0	112.5	152.1	140.4	107.0	123.7	118.7
0950, 0980	Ceilings & Acoustic Treatment	109.5	119.8	116.3	117.4	143.6	134.8	98.6	119.8	112.7	122.2	133.4	129.6	117.4	152.1	140.4	114.7	123.7	120.7
0960	Flooring	112.5	105.9	110.6	113.3	121.9	115.8	93.2	105.9	96.8	110.0	111.5	110.4	121.8	121.9	121.8	114.4	112.4	113.8
0970, 0990	Wall Finishes & Painting/Coating	100.6	119.8	112.2	106.2	138.0	125.5	105.9	109.0	107.8	103.7	117.7	112.2	106.2	138.0	125.5	100.6	109.0	105.7
09	FINISHES	104.7	117.6	111.8	112.1	137.7	126.2	96.7	116.3	107.5	110.4	127.2	119.7	117.0	142.5	131.0	107.4	119.7	114.2
COVERS	DIVS. 10 - 14, 25, 28, 41, 43, 44, 46	100.0	109.8	102.0	100.0	124.6	105.1	100.0	109.5	102.0	100.0	122.0	104.6	100.0	125.5	105.3	100.0	110.3	102.1
21, 22, 23	FIRE SUPPRESSION, PLUMBING & HVAC	94.7	107.8	100.2	94.9	138.9	113.2	94.8	107.8	100.2	100.1	107.5	103.2	94.9	135.4	111.8	100.1	112.6	105.3
26, 27, 3370	ELECTRICAL, COMMUNICATIONS & UTIL.	97.8	106.7	102.5	103.0	140.8	123.1	112.6	114.5	113.6	101.2	103.9	102.6	103.6	133.0	119.2	93.9	106.7	100.7
MF2010	WEIGHTED AVERAGE	97.5	111.5	103.8	102.3	129.4	114.4	96.5	112.7	103.7	104.1	113.8	108.4	106.1	128.9	116.3	100.0	113.3	105.9

DIVISION		CALIFORNIA																	
		SACRAMENTO			SALINAS			SAN BERNARDINO			SAN DIEGO			SAN FRANCISCO			SAN JOSE		
		942,956 - 958			939			923 - 924			919 - 921			940 - 941			951		
		MAT.	INST.	TOTAL	MAT.	INST.	TOTAL	MAT.	INST.	TOTAL	MAT.	INST.	TOTAL	MAT.	INST.	TOTAL	MAT.	INST.	TOTAL
015433	CONTRACTOR EQUIPMENT		98.0	98.0		99.4	99.4		100.3	100.3		99.1	99.1		107.7	107.7		99.8	99.8
0241, 31 - 34	SITE & INFRASTRUCTURE, DEMOLITION	105.8	109.9	108.6	119.0	106.6	110.5	75.9	107.1	97.5	102.2	103.1	102.8	136.9	110.1	118.4	138.8	100.8	112.6
0310	Concrete Forming & Accessories	107.1	132.2	128.9	109.1	133.6	130.3	109.3	120.0	118.6	108.5	111.9	111.5	108.5	146.9	141.8	108.4	146.0	141.0
0320	Concrete Reinforcing	88.0	120.4	104.7	112.3	120.8	116.7	110.6	119.8	115.3	117.9	119.7	118.8	108.3	121.4	115.0	98.7	121.1	110.2
0330	Cast-in-Place Concrete	101.6	118.1	107.9	98.2	117.8	105.7	67.1	121.4	87.9	93.9	107.6	99.2	123.2	122.6	122.9	118.1	120.8	119.1
03	CONCRETE	104.7	123.5	113.7	114.2	124.5	119.1	76.8	119.5	97.2	104.3	111.3	107.6	118.6	132.4	125.2	112.0	131.3	121.2
04	MASONRY	118.6	122.6	121.0	103.5	128.0	118.4	88.2	114.0	103.8	89.7	111.1	102.7	148.4	134.7	140.1	137.9	131.5	134.1
05	METALS	95.7	98.3	96.6	102.2	104.7	103.1	102.8	102.6	102.8	102.2	103.5	102.6	106.5	111.1	108.1	97.7	109.3	101.7
06	WOOD, PLASTICS & COMPOSITES	96.2	135.0	118.4	98.3	135.9	119.9	97.2	119.3	109.9	93.4	109.6	102.7	102.6	151.2	130.5	101.6	150.9	129.9
07	THERMAL & MOISTURE PROTECTION	119.3	120.7	119.9	102.7	125.2	111.7	101.6	116.5	107.5	108.2	104.6	106.8	115.3	137.5	124.1	102.2	137.3	116.2
08	OPENINGS	113.7	125.8	116.8	101.0	131.8	108.8	98.7	119.1	103.9	98.8	112.8	102.3	105.7	140.1	114.4	92.9	139.9	104.8
0920	Plaster & Gypsum Board	97.4	135.7	124.3	108.8	136.8	128.5	108.4	119.8	116.4	108.7	109.7	109.4	105.6	152.1	138.3	102.5	152.1	137.4
0950, 0980	Ceilings & Acoustic Treatment	117.4	135.7	129.5	116.0	136.8	129.8	111.4	119.8	117.0	106.1	109.7	108.5	125.8	152.1	143.2	108.2	152.1	137.3
0960	Flooring	113.5	111.7	113.0	111.3	128.0	116.1	116.6	105.9	113.5	101.5	112.4	104.6	114.3	128.0	118.2	102.8	128.0	110.0
0970, 0990	Wall Finishes & Painting/Coating	103.6	118.5	112.7	104.8	143.2	128.1	100.6	106.1	103.9	102.6	109.0	106.5	106.2	153.2	134.7	104.2	143.2	127.9
09	FINISHES	110.4	128.8	120.5	111.3	134.8	124.3	105.8	116.1	111.4	102.4	111.4	107.4	115.8	145.4	132.1	106.6	144.0	127.2
COVERS	DIVS. 10 - 14, 25, 28, 41, 43, 44, 46	100.0	122.8	104.7	100.0	122.5	104.7	100.0	109.8	102.0	100.0	108.4	101.8	100.0	126.0	105.4	100.0	125.1	105.2
21, 22, 23	FIRE SUPPRESSION, PLUMBING & HVAC	100.1	114.9	106.3	94.7	120.6	105.5	94.7	107.8	100.2	100.1	111.7	105.0	100.2	165.7	127.5	100.1	144.4	118.6
26, 27, 3370	ELECTRICAL, COMMUNICATIONS & UTIL.	98.5	111.4	105.3	93.2	128.6	112.0	97.9	105.8	102.1	99.1	100.3	99.8	103.2	152.8	129.6	104.0	149.7	128.3
MF2010	WEIGHTED AVERAGE	103.9	117.2	109.8	102.0	123.0	111.4	95.2	111.2	102.3	100.6	108.2	104.0	109.6	140.8	123.6	104.6	134.3	117.9

DIVISION		CALIFORNIA																	
		SAN LUIS OBISPO			SAN MATEO			SAN RAFAEL			SANTA ANA			SANTA BARBARA			SANTA CRUZ		
		934			944			949			926 - 927			931			950		
		MAT.	INST.	TOTAL	MAT.	INST.	TOTAL	MAT.	INST.	TOTAL	MAT.	INST.	TOTAL	MAT.	INST.	TOTAL	MAT.	INST.	TOTAL
015433	CONTRACTOR EQUIPMENT		99.4	99.4		98.4	98.4		98.5	98.5		100.3	100.3		99.4	99.4		99.8	99.8
0241, 31 - 34	SITE & INFRASTRUCTURE, DEMOLITION	110.4	106.6	107.8	131.3	104.3	112.7	120.4	110.0	113.2	88.5	107.1	101.4	104.3	106.6	105.9	137.9	100.7	112.2
0310	Concrete Forming & Accessories	116.4	119.8	119.3	112.8	145.8	141.5	115.5	145.6	141.7	109.6	120.0	118.6	107.4	122.9	120.9	108.4	133.7	130.4
0320	Concrete Reinforcing	113.7	119.7	116.8	93.0	121.0	107.4	93.6	121.0	107.7	114.4	119.8	117.2	111.6	119.8	115.8	122.3	120.8	121.5
0330	Cast-in-Place Concrete	105.5	120.6	111.3	119.3	121.1	120.0	138.9	120.1	131.7	85.7	121.5	99.4	99.1	120.7	107.3	117.3	119.7	118.2
03	CONCRETE	112.4	119.1	115.6	113.4	130.9	121.8	136.7	130.3	133.6	94.4	119.5	106.4	104.3	120.5	112.1	115.1	125.4	120.0
04	MASONRY	105.5	113.8	110.5	139.0	130.8	134.0	113.9	130.8	124.1	77.4	114.3	99.8	104.1	113.8	110.0	142.1	128.1	133.7
05	METALS	100.3	102.2	100.9	97.3	103.1	99.3	98.9	100.2	99.3	102.9	102.7	102.8	98.2	102.5	99.7	104.6	108.2	105.8
06	WOOD, PLASTICS & COMPOSITES	101.5	119.4	111.8	107.7	151.1	132.6	104.5	150.8	131.1	98.8	119.3	110.6	94.0	123.2	110.7	101.6	136.0	121.3
07	THERMAL & MOISTURE PROTECTION	102.9	113.6	107.1	113.0	132.5	120.8	117.6	132.2	123.4	102.7	116.6	108.2	102.3	114.0	106.9	102.1	127.9	112.4
08	OPENINGS	98.8	115.0	102.9	101.2	139.0	110.8	111.0	138.9	118.0	97.9	119.1	103.3	100.7	123.3	106.4	94.1	131.8	103.7
0920	Plaster & Gypsum Board	117.2	119.8	119.0	106.8	152.1	138.6	109.2	152.1	139.4	109.3	119.8	116.7	107.5	123.7	118.9	108.3	136.8	128.3
0950, 0980	Ceilings & Acoustic Treatment	116.0	119.8	118.5	117.4	152.1	140.4	123.9	152.1	142.6	111.4	119.8	117.0	116.6	123.7	121.3	113.2	136.8	128.8
0960	Flooring	118.3	98.3	112.6	116.3	121.9	117.9	127.4	121.9	125.8	117.2	105.9	113.9	110.6	98.3	107.0	107.3	128.0	113.2
0970, 0990	Wall Finishes & Painting/Coating	103.7	100.7	101.9	106.2	138.0	125.5	102.3	138.0	124.0	100.6	106.1	103.9	103.7	100.7	101.9	104.5	143.2	128.0
09	FINISHES	113.4	113.4	113.4	114.4	142.1	129.7	116.4	142.3	130.7	107.3	116.1	112.2	109.4	116.5	113.3	109.7	134.9	123.6
COVERS	DIVS. 10 - 14, 25, 28, 41, 43, 44, 46	100.0	120.7	104.3	100.0	125.5	105.3	100.0	124.8	105.2	100.0	109.8	102.0	100.0	110.5	102.2	100.0	122.9	104.7
21, 22, 23	FIRE SUPPRESSION, PLUMBING & HVAC	94.7	107.8	100.2	94.9	138.5	113.0	94.8	149.0	117.4	94.7	107.8	100.2	100.1	112.6	105.3	100.1	120.7	108.7
26, 27, 3370	ELECTRICAL, COMMUNICATIONS & UTIL.	92.0	98.7	95.6	103.0	144.2	124.9	99.9	118.8	109.9	97.9	107.2	102.8	90.9	110.5	101.3	102.8	128.6	116.5
MF2010	WEIGHTED AVERAGE	101.2	109.8	105.0	105.1	131.4	116.9	107.4	130.3	117.6	97.2	111.4	103.6	100.6	113.1	106.2	106.4	123.1	113.9

	DIVISION	CALIFORNIA															COLORADO		
		SANTA ROSA			STOCKTON			SUSANVILLE			VALLEJO			VAN NUYS			ALAMOSA		
		954			952			961			945			913 - 916			811		
		MAT.	INST.	TOTAL	MAT.	INST.	TOTAL	MAT.	INST.	TOTAL	MAT.	INST.	TOTAL	MAT.	INST.	TOTAL	MAT.	INST.	TOTAL
015433	CONTRACTOR EQUIPMENT		99.6	99.6		99.1	99.1		99.1	99.1		98.5	98.5		99.1	99.1		96.4	96.4
0241, 31 - 34	SITE & INFRASTRUCTURE, DEMOLITION	103.1	106.1	105.1	102.2	106.0	104.8	115.8	105.9	109.0	105.0	110.0	108.4	114.3	109.4	111.0	122.3	92.7	101.9
0310	Concrete Forming & Accessories	103.1	145.5	139.9	106.3	130.3	127.1	106.9	130.1	127.0	106.2	145.1	139.9	117.2	119.9	119.6	106.0	72.8	77.2
0320	Concrete Reinforcing	109.3	121.2	115.4	112.3	120.4	116.4	108.3	120.1	114.4	94.7	121.2	108.3	121.7	119.9	120.8	104.4	79.6	91.6
0330	Cast-in-Place Concrete	111.5	119.1	114.4	99.0	117.5	106.1	102.4	117.4	108.2	110.6	119.3	113.9	90.4	119.0	101.3	103.4	80.6	94.7
03	CONCRETE	114.2	130.3	121.9	104.3	122.8	113.2	114.9	122.6	118.6	110.1	129.8	119.5	113.2	118.6	115.8	116.0	77.0	97.4
04	MASONRY	106.8	133.4	122.9	107.2	122.6	116.6	108.2	118.1	114.2	84.5	133.3	114.1	113.5	116.9	115.6	126.1	77.2	96.5
05	METALS	103.3	106.4	104.4	99.4	103.6	100.8	101.5	103.0	102.0	98.8	100.4	99.4	85.6	102.1	91.3	96.5	82.3	91.6
06	WOOD, PLASTICS & COMPOSITES	89.7	150.6	124.6	95.6	132.6	116.8	95.7	132.6	116.9	93.4	150.8	126.3	88.4	119.2	106.0	95.7	73.4	82.9
07	THERMAL & MOISTURE PROTECTION	103.1	134.9	115.8	106.0	118.5	110.9	106.8	118.1	111.3	115.2	134.6	122.9	103.1	115.3	108.0	106.3	81.2	96.3
08	OPENINGS	99.6	139.7	109.8	100.0	124.5	106.2	102.1	124.3	107.7	113.3	139.8	120.1	90.8	119.0	98.0	93.9	78.9	90.1
0920	Plaster & Gypsum Board	103.5	152.1	137.7	108.0	133.4	125.9	106.2	133.4	125.3	102.6	152.1	137.4	106.0	119.8	115.7	85.5	72.3	76.2
0950, 0980	Ceilings & Acoustic Treatment	111.4	152.1	138.3	116.6	133.4	127.7	116.0	133.4	127.6	126.0	152.1	143.3	96.8	119.8	112.0	118.2	72.3	87.8
0960	Flooring	113.1	121.4	115.5	110.6	117.7	112.6	110.6	111.5	110.8	122.3	128.0	123.9	96.4	105.9	99.1	112.3	57.9	96.6
0970, 0990	Wall Finishes & Painting/Coating	100.6	143.2	126.5	103.7	118.5	112.7	103.7	117.7	112.2	103.3	143.2	127.5	105.9	109.0	107.8	107.9	25.4	57.8
09	FINISHES	106.5	142.4	126.3	109.2	128.4	119.8	110.4	127.2	119.7	112.8	143.6	129.8	99.2	116.3	108.6	108.6	64.7	84.4
COVERS	DIVS. 10 - 14, 25, 28, 41, 43, 44, 46	100.0	124.0	105.0	100.0	122.1	104.6	100.0	122.0	104.6	100.0	124.4	105.1	100.0	109.5	102.0	100.0	89.7	97.9
21, 22, 23	FIRE SUPPRESSION, PLUMBING & HVAC	94.7	164.9	123.9	100.1	113.7	105.8	94.7	107.5	100.0	100.2	129.0	112.2	94.8	107.8	100.2	94.6	69.2	84.1
26, 27, 3370	ELECTRICAL, COMMUNICATIONS & UTIL.	98.2	123.5	111.7	100.6	114.7	108.1	101.6	103.9	102.8	95.6	128.3	113.0	112.6	114.5	113.6	96.0	76.7	85.7
MF2010	WEIGHTED AVERAGE	101.7	134.9	116.5	101.9	117.3	108.8	103.1	113.8	107.9	102.9	127.8	114.0	99.7	112.7	105.5	102.1	76.1	90.5

	DIVISION	COLORADO																	
		BOULDER			COLORADO SPRINGS			DENVER			DURANGO			FORT COLLINS			FORT MORGAN		
		803			808 - 809			800 - 802			813			805			807		
		MAT.	INST.	TOTAL	MAT.	INST.	TOTAL	MAT.	INST.	TOTAL	MAT.	INST.	TOTAL	MAT.	INST.	TOTAL	MAT.	INST.	TOTAL
015433	CONTRACTOR EQUIPMENT		96.2	96.2		94.8	94.8		99.2	99.2		96.4	96.4		96.2	96.2		96.2	96.2
0241, 31 - 34	SITE & INFRASTRUCTURE, DEMOLITION	88.7	97.2	94.6	90.0	94.8	93.3	89.1	102.9	98.6	116.5	92.7	100.1	99.6	96.4	97.4	90.9	96.8	94.9
0310	Concrete Forming & Accessories	105.5	81.6	84.7	96.1	81.8	83.7	101.6	82.0	84.6	111.6	72.8	77.9	102.7	78.0	81.2	105.8	81.0	84.3
0320	Concrete Reinforcing	96.3	79.7	87.7	95.5	84.8	90.0	95.5	84.8	90.0	104.4	79.5	91.6	96.3	78.4	87.1	96.5	79.6	87.8
0330	Cast-in-Place Concrete	105.4	80.8	96.0	108.3	81.4	98.0	102.5	81.1	94.3	118.9	80.6	104.2	119.0	75.5	102.3	103.4	79.6	94.3
03	CONCRETE	103.3	81.2	92.7	106.6	82.5	95.0	100.9	82.4	92.1	119.1	77.0	98.9	114.1	77.4	96.5	101.8	80.4	91.5
04	MASONRY	91.6	77.8	83.2	93.4	77.8	83.9	95.3	77.8	84.7	113.6	77.2	91.6	108.0	49.1	72.3	106.3	77.4	88.8
05	METALS	98.9	85.7	94.3	102.0	88.3	97.3	104.5	88.3	98.9	96.5	82.3	91.6	100.1	80.6	93.4	98.6	82.4	93.0
06	WOOD, PLASTICS & COMPOSITES	99.8	83.9	90.7	90.3	84.1	86.7	97.7	84.0	89.8	103.9	73.4	86.4	97.5	83.9	89.7	100.1	83.9	90.8
07	THERMAL & MOISTURE PROTECTION	103.5	79.4	94.0	104.3	83.2	95.9	102.7	77.5	92.7	106.2	81.2	96.2	104.0	66.0	88.9	103.5	82.5	95.1
08	OPENINGS	98.0	84.6	94.6	102.5	86.2	98.4	102.6	86.1	98.4	101.7	78.9	95.9	98.0	84.6	94.6	98.0	84.6	94.6
0920	Plaster & Gypsum Board	108.6	83.7	91.0	88.7	83.7	85.1	104.7	83.7	89.9	97.8	72.3	79.9	102.9	83.7	89.3	109.5	83.7	91.3
0950, 0980	Ceilings & Acoustic Treatment	106.7	83.7	91.4	114.2	83.7	93.9	113.2	83.7	93.7	118.2	72.3	87.8	106.7	83.7	91.4	106.7	83.7	91.4
0960	Flooring	109.9	86.6	103.2	101.3	72.1	92.9	106.4	89.6	101.6	117.0	57.9	100.0	106.3	57.9	92.4	110.4	57.9	95.3
0970, 0990	Wall Finishes & Painting/Coating	103.6	71.7	84.2	103.3	43.6	67.1	103.6	71.7	84.2	107.9	25.4	57.8	103.6	43.3	67.0	103.6	57.7	75.7
09	FINISHES	106.6	81.8	92.9	103.3	76.1	88.3	106.4	82.4	93.2	110.7	64.7	85.3	105.4	71.0	86.4	106.8	74.4	88.9
COVERS	DIVS. 10 - 14, 25, 28, 41, 43, 44, 46	100.0	89.9	97.9	100.0	90.3	98.0	100.0	90.0	97.9	100.0	89.7	97.9	100.0	87.4	97.4	100.0	90.0	97.9
21, 22, 23	FIRE SUPPRESSION, PLUMBING & HVAC	94.6	81.1	89.0	100.1	75.9	90.0	99.9	82.1	92.5	94.6	66.9	83.1	100.0	77.3	90.5	94.6	81.0	89.0
26, 27, 3370	ELECTRICAL, COMMUNICATIONS & UTIL.	98.3	85.8	91.7	102.2	86.5	93.9	103.9	85.8	94.3	95.3	74.5	84.3	98.3	85.8	91.6	98.7	85.8	91.8
MF2010	WEIGHTED AVERAGE	98.5	83.7	91.8	101.3	82.4	92.8	101.4	84.9	94.0	102.6	75.3	90.4	102.3	77.1	91.0	99.1	82.3	91.6

	DIVISION	COLORADO																	
		GLENWOOD SPRINGS			GOLDEN			GRAND JUNCTION			GREELEY			MONTROSE			PUEBLO		
		816			804			815			806			814			810		
		MAT.	INST.	TOTAL	MAT.	INST.	TOTAL	MAT.	INST.	TOTAL	MAT.	INST.	TOTAL	MAT.	INST.	TOTAL	MAT.	INST.	TOTAL
015433	CONTRACTOR EQUIPMENT		99.1	99.1		96.2	96.2		99.1	99.1		96.2	96.2		97.7	97.7		96.4	96.4
0241, 31 - 34	SITE & INFRASTRUCTURE, DEMOLITION	130.5	99.6	109.2	100.7	97.1	98.2	115.8	99.3	104.4	87.3	95.2	92.8	124.6	95.8	104.7	108.1	94.5	98.7
0310	Concrete Forming & Accessories	103.2	80.8	83.7	98.6	81.1	83.4	110.3	80.3	84.3	100.9	78.8	81.7	102.8	80.5	83.4	107.9	82.0	85.5
0320	Concrete Reinforcing	103.3	79.5	91.1	96.5	79.6	87.8	103.6	79.5	91.2	96.3	78.2	86.9	103.1	79.5	91.0	99.9	84.8	92.1
0330	Cast-in-Place Concrete	103.4	79.9	94.4	103.5	79.7	94.4	114.5	79.2	101.0	99.4	52.4	81.4	103.4	79.9	94.4	102.6	82.5	94.9
03	CONCRETE	121.1	80.4	101.6	111.3	80.4	96.5	115.3	79.9	98.3	98.4	69.9	84.7	112.3	80.2	96.9	105.1	83.0	94.5
04	MASONRY	100.9	77.4	86.7	109.0	77.5	89.9	133.6	77.2	99.4	102.1	36.8	62.5	106.7	77.2	88.9	97.3	77.8	85.5
05	METALS	96.2	82.8	91.6	98.8	82.6	93.2	97.9	81.5	92.2	100.1	79.5	93.0	95.5	81.8	90.8	99.4	89.1	95.9
06	WOOD, PLASTICS & COMPOSITES	90.8	84.1	86.9	92.7	83.9	87.7	101.4	84.1	91.4	94.9	90.2	92.2	92.1	84.2	87.6	98.1	84.4	90.2
07	THERMAL & MOISTURE PROTECTION	106.1	82.3	96.7	104.4	77.3	93.7	105.0	71.9	91.8	103.4	57.5	85.2	106.3	82.3	96.8	104.6	83.1	96.0
08	OPENINGS	100.9	84.7	96.8	98.0	84.6	94.6	101.7	84.7	97.4	98.0	88.0	95.5	101.8	84.8	97.5	95.7	86.3	93.3
0920	Plaster & Gypsum Board	112.5	83.7	92.2	100.2	83.7	88.5	124.9	83.7	95.9	101.1	90.1	93.4	83.7	83.7	83.7	90.5	83.7	85.7
0950, 0980	Ceilings & Acoustic Treatment	117.3	83.7	95.0	106.7	83.7	91.4	117.3	83.7	95.0	106.7	90.1	95.7	118.2	83.7	95.3	128.5	83.7	98.8
0960	Flooring	111.4	53.3	94.7	104.1	57.9	90.8	116.4	57.9	99.6	105.2	57.9	91.6	114.5	47.9	95.4	113.2	89.6	106.4
0970, 0990	Wall Finishes & Painting/Coating	107.8	57.7	77.4	103.6	71.7	84.2	107.8	71.7	85.9	103.6	27.0	57.1	107.9	25.4	57.8	107.9	40.4	66.9
09	FINISHES	111.7	73.7	90.8	105.2	76.7	89.5	112.9	76.1	92.6	104.1	70.7	85.7	109.0	69.0	87.0	109.9	79.2	93.0
COVERS	DIVS. 10 - 14, 25, 28, 41, 43, 44, 46	100.0	90.2	98.0	100.0	90.0	97.9	100.0	90.1	98.0	100.0	85.1	96.9	100.0	90.5	98.0	100.0	91.0	98.1
21, 22, 23	FIRE SUPPRESSION, PLUMBING & HVAC	94.6	66.7	83.0	94.6	80.7	88.8	99.9	65.5	85.6	100.0	72.8	88.7	94.6	66.7	83.0	99.9	69.4	87.2
26, 27, 3370	ELECTRICAL, COMMUNICATIONS & UTIL.	92.5	74.5	83.0	98.7	85.8	91.8	95.0	58.5	75.6	98.3	85.8	91.6	95.0	58.5	75.6	96.0	76.8	85.8
MF2010	WEIGHTED AVERAGE	102.1	78.0	91.3	100.6	82.5	92.5	104.7	75.2	91.5	99.6	73.6	88.0	101.2	74.7	89.4	100.8	80.2	91.6

DIVISION		COLORADO			CONNECTICUT														
		SALIDA			BRIDGEPORT			BRISTOL			HARTFORD			MERIDEN			NEW BRITAIN		
		812			066			060			061			064			060		
		MAT.	INST.	TOTAL	MAT.	INST.	TOTAL	MAT.	INST.	TOTAL	MAT.	INST.	TOTAL	MAT.	INST.	TOTAL	MAT.	INST.	TOTAL
015433	CONTRACTOR EQUIPMENT		97.7	97.7		100.3	100.3		100.3	100.3		100.3	100.3		100.8	100.8		100.3	100.3
0241, 31 - 34	SITE & INFRASTRUCTURE, DEMOLITION	116.2	96.1	102.3	103.5	103.8	103.7	102.7	103.8	103.5	96.7	103.8	101.6	101.0	104.6	103.5	102.9	103.8	103.5
0310	Concrete Forming & Accessories	110.2	80.7	84.6	96.5	118.3	115.4	96.5	121.8	118.5	94.5	121.8	118.2	96.2	121.8	118.4	96.8	121.8	118.5
0320	Concrete Reinforcing	102.9	79.5	90.9	108.1	127.0	117.8	108.1	127.0	117.8	108.1	127.0	117.8	108.1	127.0	117.8	108.1	127.0	117.8
0330	Cast-in-Place Concrete	118.4	79.9	103.7	101.4	124.5	110.3	95.0	124.5	106.3	100.7	124.5	109.8	91.4	124.5	104.1	96.6	124.5	107.3
03	CONCRETE	114.2	80.3	98.0	109.2	121.9	115.3	106.0	123.5	114.4	108.7	123.5	115.8	104.2	123.5	113.4	106.8	123.5	114.8
04	MASONRY	135.2	77.2	100.1	108.3	127.9	120.2	100.2	127.9	117.0	100.7	127.9	117.2	99.9	127.9	116.9	102.2	127.9	117.8
05	METALS	95.1	82.0	90.6	99.8	124.3	108.3	99.8	124.3	108.2	104.5	124.3	111.3	97.1	124.3	106.4	96.2	124.3	105.9
06	WOOD, PLASTICS & COMPOSITES	99.1	84.2	90.6	98.1	116.2	108.5	98.1	121.0	111.3	96.6	121.0	110.6	98.1	121.0	111.3	98.1	121.0	111.3
07	THERMAL & MOISTURE PROTECTION	105.1	82.3	96.1	96.4	125.4	107.9	96.5	123.6	107.3	97.7	123.6	108.0	96.5	123.6	107.2	96.5	123.6	107.3
08	OPENINGS	94.0	84.8	91.6	105.6	127.2	111.1	105.6	129.9	111.8	106.3	129.9	112.3	108.5	129.9	113.9	105.6	129.9	111.8
0920	Plaster & Gypsum Board	85.0	83.7	84.1	99.4	116.0	111.1	99.4	121.0	114.6	97.5	121.0	114.0	100.9	121.0	115.1	99.4	121.0	114.6
0950, 0980	Ceilings & Acoustic Treatment	118.2	83.7	95.3	99.4	116.0	110.4	99.4	121.0	113.7	97.5	121.0	113.1	102.9	121.0	114.9	99.4	121.0	113.7
0960	Flooring	119.5	47.9	99.0	95.8	120.8	103.0	95.8	120.8	103.0	95.8	120.8	103.0	95.8	120.8	103.0	95.8	120.8	103.0
0970, 0990	Wall Finishes & Painting/Coating	107.9	25.4	57.8	92.3	114.5	105.8	92.3	116.9	107.2	92.3	116.9	107.2	92.3	116.9	107.2	92.3	116.9	107.2
09	FINISHES	109.7	69.0	87.3	100.1	118.0	110.0	100.1	121.1	111.7	98.6	121.1	111.0	101.1	121.1	112.1	100.1	121.1	111.7
COVERS	DIVS. 10 - 14, 25, 28, 41, 43, 44, 46	100.0	90.6	98.1	100.0	108.0	101.7	100.0	108.6	101.8	100.0	108.6	101.8	100.0	108.6	101.8	100.0	108.6	101.8
21, 22, 23	FIRE SUPPRESSION, PLUMBING & HVAC	94.6	69.2	84.0	100.2	112.8	105.4	100.2	112.8	105.4	100.2	112.8	105.5	94.8	112.8	102.3	100.2	112.8	105.4
26, 27, 3370	ELECTRICAL, COMMUNICATIONS & UTIL.	95.3	76.7	85.4	100.8	105.9	103.5	100.8	110.4	105.9	99.7	110.4	105.4	100.8	109.5	105.4	100.9	110.4	105.9
MF2010	WEIGHTED AVERAGE	102.0	77.8	91.2	102.2	116.4	108.6	101.4	117.8	108.7	102.1	117.8	109.1	99.7	117.7	107.8	101.0	117.8	108.5

DIVISION		CONNECTICUT																	
		NEW HAVEN			NEW LONDON			NORWALK			STAMFORD			WATERBURY			WILLIMANTIC		
		065			063			068			069			067			062		
		MAT.	INST.	TOTAL	MAT.	INST.	TOTAL	MAT.	INST.	TOTAL	MAT.	INST.	TOTAL	MAT.	INST.	TOTAL	MAT.	INST.	TOTAL
015433	CONTRACTOR EQUIPMENT		100.8	100.8		100.8	100.8		100.3	100.3		100.3	100.3		100.3	100.3		100.3	100.3
0241, 31 - 34	SITE & INFRASTRUCTURE, DEMOLITION	102.7	104.6	104.0	95.4	104.6	101.7	103.3	103.8	103.7	103.9	103.8	103.9	103.2	103.8	103.6	103.4	103.8	103.7
0310	Concrete Forming & Accessories	96.2	121.8	118.4	96.2	121.8	118.4	96.5	118.3	115.4	96.4	118.6	115.7	96.5	121.8	118.5	96.5	121.8	118.5
0320	Concrete Reinforcing	108.1	127.0	117.8	84.8	127.0	106.5	108.1	127.1	117.9	108.1	127.1	117.9	108.1	127.0	117.8	108.1	127.0	117.8
0330	Cast-in-Place Concrete	98.3	124.5	108.3	83.7	124.5	99.3	99.8	125.9	109.8	101.4	126.0	110.9	101.4	124.5	110.3	94.7	124.5	106.1
03	CONCRETE	123.0	123.5	123.2	93.7	123.5	108.0	108.4	122.4	115.1	109.2	122.6	115.6	109.2	123.5	116.0	105.9	123.5	114.3
04	MASONRY	100.5	127.9	117.1	98.9	127.9	116.5	99.9	129.5	117.8	100.7	129.5	118.2	100.7	127.9	117.2	100.2	127.9	117.0
05	METALS	96.4	124.3	106.0	96.1	124.3	105.8	99.8	124.4	108.3	99.8	124.8	108.4	99.8	124.3	108.2	99.6	124.3	108.1
06	WOOD, PLASTICS & COMPOSITES	98.1	121.0	111.3	98.1	121.0	111.3	98.1	116.2	108.5	98.1	116.2	108.5	98.1	121.0	111.3	98.1	121.0	111.3
07	THERMAL & MOISTURE PROTECTION	96.6	123.6	107.3	96.4	123.6	107.2	96.5	126.1	108.3	96.5	126.1	108.2	96.5	123.6	107.3	96.7	123.6	107.4
08	OPENINGS	105.6	129.9	111.8	108.9	129.9	114.2	105.6	127.2	111.1	105.6	127.2	111.1	105.6	129.9	111.8	108.9	129.9	114.2
0920	Plaster & Gypsum Board	99.4	121.0	114.6	99.4	121.0	114.6	99.4	116.0	111.1	99.4	116.0	111.1	99.4	121.0	114.6	99.4	121.0	114.6
0950, 0980	Ceilings & Acoustic Treatment	99.4	121.0	113.7	97.3	121.0	113.0	99.4	116.0	110.4	99.4	116.0	110.4	99.4	121.0	113.7	97.3	121.0	113.0
0960	Flooring	95.8	120.8	103.0	95.8	120.8	103.0	95.8	120.8	103.0	95.8	120.8	103.0	95.8	120.8	103.0	95.8	120.8	103.0
0970, 0990	Wall Finishes & Painting/Coating	92.3	114.5	105.8	92.3	116.9	107.2	92.3	114.5	105.8	92.3	114.5	105.8	92.3	116.9	107.2	92.3	116.9	107.2
09	FINISHES	100.2	120.9	111.6	99.1	121.1	111.3	100.1	118.0	110.0	100.2	118.0	110.0	100.0	121.1	111.6	99.8	121.1	111.6
COVERS	DIVS. 10 - 14, 25, 28, 41, 43, 44, 46	100.0	108.6	101.8	100.0	108.6	101.8	100.0	108.0	101.7	100.0	108.2	101.7	100.0	108.6	101.8	100.0	108.6	101.8
21, 22, 23	FIRE SUPPRESSION, PLUMBING & HVAC	100.2	112.8	105.4	94.8	112.8	102.3	100.2	112.8	105.4	100.2	112.9	105.5	100.2	112.8	105.4	100.2	112.8	105.4
26, 27, 3370	ELECTRICAL, COMMUNICATIONS & UTIL.	100.8	109.5	105.4	97.2	109.5	103.7	100.8	104.4	102.7	100.8	151.0	127.5	100.2	109.8	105.3	100.8	110.4	105.9
MF2010	WEIGHTED AVERAGE	102.9	117.7	109.5	97.7	117.7	106.6	101.6	116.5	108.3	101.8	123.0	111.3	101.7	117.7	108.9	101.6	117.8	108.9

DIVISION		D.C.			DELAWARE									FLORIDA					
		WASHINGTON			DOVER			NEWARK			WILMINGTON			DAYTONA BEACH			FORT LAUDERDALE		
		200 - 205			199			197			198			321			333		
		MAT.	INST.	TOTAL	MAT.	INST.	TOTAL	MAT.	INST.	TOTAL	MAT.	INST.	TOTAL	MAT.	INST.	TOTAL	MAT.	INST.	TOTAL
015433	CONTRACTOR EQUIPMENT		103.6	103.6		117.7	117.7		117.7	117.7		117.8	117.8		97.8	97.8		90.5	90.5
0241, 31 - 34	SITE & INFRASTRUCTURE, DEMOLITION	113.8	92.3	99.0	94.0	112.2	106.6	94.2	112.2	106.6	85.3	112.5	104.0	113.2	89.3	96.7	98.7	77.3	83.9
0310	Concrete Forming & Accessories	98.7	80.3	82.7	100.2	99.8	99.9	102.4	99.8	100.1	103.3	99.8	100.3	97.5	73.6	76.8	96.2	74.3	77.2
0320	Concrete Reinforcing	94.6	89.6	92.0	92.7	103.6	98.3	93.5	103.6	98.7	93.5	103.6	98.7	91.8	78.2	84.8	91.8	77.8	84.6
0330	Cast-in-Place Concrete	129.7	87.2	113.5	92.1	102.1	95.9	82.7	102.1	90.1	89.2	102.1	94.2	91.4	75.5	85.3	95.9	82.9	90.9
03	CONCRETE	116.2	85.7	101.6	97.1	102.3	99.6	92.7	102.3	97.3	96.0	102.3	99.0	88.1	76.2	82.4	90.3	79.0	84.9
04	MASONRY	102.2	81.9	89.9	104.4	97.6	100.3	102.7	97.6	99.6	108.7	97.6	102.0	87.1	70.9	77.3	87.4	73.8	79.1
05	METALS	93.5	107.2	98.2	104.6	118.0	109.2	104.6	118.0	109.2	101.6	118.0	107.2	96.1	93.0	95.1	96.0	93.9	95.3
06	WOOD, PLASTICS & COMPOSITES	96.8	79.2	86.7	100.4	98.1	99.1	103.4	98.1	100.3	105.8	98.1	101.4	100.5	75.1	86.0	96.3	72.9	82.9
07	THERMAL & MOISTURE PROTECTION	99.1	84.9	93.4	97.5	113.9	104.0	97.4	113.9	104.0	97.9	113.9	104.3	93.8	79.4	88.1	93.8	87.3	91.2
08	OPENINGS	104.2	88.8	100.3	95.8	109.0	99.2	95.8	109.0	99.2	95.6	109.0	99.0	100.5	71.9	93.2	97.9	70.9	91.1
0920	Plaster & Gypsum Board	104.3	78.6	86.2	105.2	97.9	100.1	106.6	97.9	100.5	107.1	97.9	100.6	94.6	74.9	80.7	93.7	72.6	78.8
0950, 0980	Ceilings & Acoustic Treatment	107.6	78.6	88.4	109.0	97.9	101.6	109.0	97.9	101.6	104.2	97.9	100.0	96.5	74.9	82.2	96.5	72.6	80.7
0960	Flooring	120.1	96.6	113.4	90.3	106.2	94.8	90.5	106.2	95.0	90.1	106.2	94.7	119.4	77.8	107.4	119.4	73.2	106.1
0970, 0990	Wall Finishes & Painting/Coating	121.0	88.5	101.3	94.7	101.6	98.9	94.7	101.6	98.9	94.7	101.6	98.9	111.3	79.8	92.2	105.9	75.6	87.5
09	FINISHES	108.4	83.5	94.7	99.8	100.3	100.1	100.3	100.3	100.3	99.1	100.3	99.8	107.6	74.8	89.5	104.9	73.5	87.6
COVERS	DIVS. 10 - 14, 25, 28, 41, 43, 44, 46	100.0	98.1	99.6	100.0	96.5	99.3	100.0	96.5	99.3	100.0	96.5	99.3	100.0	81.2	96.1	100.0	88.0	97.5
21, 22, 23	FIRE SUPPRESSION, PLUMBING & HVAC	100.2	94.9	98.0	99.9	116.1	106.6	100.1	116.1	106.8	100.1	116.1	106.8	99.8	78.8	91.1	99.8	69.4	87.1
26, 27, 3370	ELECTRICAL, COMMUNICATIONS & UTIL.	101.4	100.2	100.7	101.2	112.5	107.2	99.6	112.5	106.4	103.5	112.5	108.3	98.1	59.6	77.6	98.1	75.9	86.3
MF2010	WEIGHTED AVERAGE	102.8	91.9	97.9	99.9	108.6	103.8	99.3	108.6	103.4	99.6	108.6	103.6	98.0	76.4	88.3	97.3	76.7	88.1

DIVISION		FLORIDA																	
		FORT MYERS			GAINESVILLE			JACKSONVILLE			LAKELAND			MELBOURNE			MIAMI		
		339,341			326,344			320,322			338			329			330 - 332,340		
		MAT.	INST.	TOTAL	MAT.	INST.	TOTAL	MAT.	INST.	TOTAL	MAT.	INST.	TOTAL	MAT.	INST.	TOTAL	MAT.	INST.	TOTAL
015433	CONTRACTOR EQUIPMENT		97.8	97.8		97.8	97.8		97.8	97.8		97.8	97.8		97.8	97.8		90.5	90.5
0241, 31 - 34	SITE & INFRASTRUCTURE, DEMOLITION	110.5	88.6	95.4	123.5	88.3	99.2	113.2	88.6	96.3	112.6	88.8	96.2	121.3	88.6	98.7	100.5	77.1	84.3
0310	Concrete Forming & Accessories	91.5	81.1	82.5	92.4	58.6	63.1	97.3	59.0	64.0	87.9	81.8	82.6	93.7	75.4	77.8	100.7	74.8	78.2
0320	Concrete Reinforcing	92.8	100.1	96.6	97.4	68.6	82.6	91.8	68.7	79.9	95.1	101.0	98.1	92.8	78.2	85.3	91.8	77.9	84.6
0330	Cast-in-Place Concrete	100.0	73.1	89.7	105.0	66.5	90.2	92.3	72.8	84.9	102.3	74.5	91.6	110.2	78.1	97.9	97.8	84.0	92.5
03	CONCRETE	91.2	82.9	87.2	98.8	64.7	82.5	88.5	67.0	78.2	93.1	83.8	88.6	99.6	77.9	89.2	91.5	79.5	85.8
04	MASONRY	80.7	66.9	72.3	100.2	66.0	79.5	87.0	66.1	74.3	95.7	81.9	87.3	84.0	75.3	78.8	85.9	77.7	80.9
05	METALS	98.1	101.5	99.3	95.1	87.7	92.5	94.7	88.0	92.4	98.0	102.3	99.5	104.0	93.1	100.3	100.7	92.7	98.0
06	WOOD, PLASTICS & COMPOSITES	93.2	84.1	88.0	94.4	56.1	72.5	100.5	56.1	75.1	88.9	84.1	86.1	96.0	75.1	84.0	102.3	72.9	85.4
07	THERMAL & MOISTURE PROTECTION	93.7	86.3	90.7	94.1	65.4	82.7	94.0	66.1	82.9	93.6	91.3	92.7	94.2	82.0	89.4	102.4	81.1	93.9
08	OPENINGS	99.0	80.6	94.4	98.4	56.7	87.9	100.5	60.1	90.2	99.0	81.3	94.5	99.6	75.4	93.4	97.9	70.9	91.1
0920	Plaster & Gypsum Board	88.9	84.1	85.5	89.7	55.3	65.5	94.6	55.3	66.9	86.6	84.1	84.9	89.7	74.9	79.3	92.0	72.6	78.4
0950, 0980	Ceilings & Acoustic Treatment	89.7	84.1	86.0	89.7	55.3	66.9	96.5	55.3	69.2	89.7	84.1	86.0	91.8	74.9	80.6	93.7	72.6	79.7
0960	Flooring	115.7	58.2	99.2	116.2	45.6	95.9	119.4	68.2	104.7	112.9	59.7	97.6	116.4	77.8	105.3	127.0	77.1	112.7
0970, 0990	Wall Finishes & Painting/Coating	111.3	65.4	83.5	111.3	65.4	83.5	111.3	66.7	84.3	111.3	65.4	83.5	111.3	99.0	103.8	105.9	75.6	87.5
09	FINISHES	104.2	75.5	88.3	105.7	55.8	78.2	107.7	60.5	81.6	103.2	75.8	88.1	105.9	78.1	90.5	106.5	74.7	89.0
COVERS	DIVS. 10 - 14, 25, 28, 41, 43, 44, 46	100.0	81.5	96.2	100.0	79.4	95.7	100.0	77.4	95.3	100.0	81.5	96.2	100.0	82.8	96.4	100.0	88.6	97.6
21, 22, 23	FIRE SUPPRESSION, PLUMBING & HVAC	96.9	59.7	81.4	98.9	69.1	86.5	99.8	66.7	86.0	96.9	88.7	93.5	99.8	81.0	92.0	99.8	73.0	88.6
26, 27, 3370	ELECTRICAL, COMMUNICATIONS & UTIL.	100.6	66.9	82.7	98.4	77.9	87.5	97.8	67.6	81.7	98.4	66.8	81.6	99.4	69.5	83.5	103.9	78.5	90.4
MF2010	WEIGHTED AVERAGE	97.3	75.4	87.5	99.5	70.6	86.6	97.8	69.7	85.2	98.0	83.5	91.5	100.4	79.6	91.1	99.1	78.1	89.7

DIVISION		FLORIDA																	
		ORLANDO			PANAMA CITY			PENSACOLA			SARASOTA			ST. PETERSBURG			TALLAHASSEE		
		327 - 328,347			324			325			342			337			323		
		MAT.	INST.	TOTAL	MAT.	INST.	TOTAL	MAT.	INST.	TOTAL	MAT.	INST.	TOTAL	MAT.	INST.	TOTAL	MAT.	INST.	TOTAL
015433	CONTRACTOR EQUIPMENT		97.8	97.8		97.8	97.8		97.8	97.8		97.8	97.8		97.8	97.8		97.8	97.8
0241, 31 - 34	SITE & INFRASTRUCTURE, DEMOLITION	113.8	88.5	96.3	127.5	87.5	99.9	127.0	88.0	100.1	114.7	88.6	96.6	114.0	88.1	96.1	110.0	87.7	94.6
0310	Concrete Forming & Accessories	97.1	78.0	80.6	96.6	47.1	53.7	94.2	56.1	61.1	97.5	81.3	83.4	95.5	55.1	60.4	99.4	47.7	54.5
0320	Concrete Reinforcing	91.8	77.5	84.4	95.8	78.2	86.8	98.3	78.8	88.3	91.8	100.9	96.5	95.1	86.4	90.6	91.8	68.5	79.8
0330	Cast-in-Place Concrete	114.6	75.4	99.6	97.0	60.2	82.9	119.8	70.3	100.9	104.6	74.2	93.0	103.4	68.5	90.1	99.8	59.8	84.5
03	CONCRETE	100.9	78.0	89.9	96.2	59.2	78.5	107.1	66.8	87.8	94.7	83.4	89.3	94.5	67.1	81.4	92.4	57.5	75.7
04	MASONRY	90.4	70.9	78.6	91.4	51.0	66.9	109.7	58.7	78.8	88.1	81.9	84.3	133.0	52.8	84.4	89.8	58.1	70.6
05	METALS	101.6	92.3	98.4	95.9	89.7	93.8	97.1	91.6	95.2	98.7	101.7	99.8	99.0	94.1	97.3	89.9	87.4	89.1
06	WOOD, PLASTICS & COMPOSITES	100.1	81.7	89.5	99.3	45.2	68.3	97.4	55.9	73.6	100.5	84.1	91.1	97.8	54.3	72.9	97.9	44.7	67.4
07	THERMAL & MOISTURE PROTECTION	95.0	80.0	89.0	94.3	60.8	81.0	94.2	67.0	83.4	93.8	91.3	92.8	93.8	62.1	81.2	99.7	77.0	90.7
08	OPENINGS	100.5	74.4	93.8	98.0	50.3	85.9	97.9	61.6	88.7	100.4	80.3	95.3	99.0	64.2	90.2	100.9	50.9	88.3
0920	Plaster & Gypsum Board	105.4	81.6	88.7	92.4	44.1	58.4	92.9	55.0	66.3	94.6	84.1	87.2	91.5	53.4	64.7	110.8	43.6	63.5
0950, 0980	Ceilings & Acoustic Treatment	98.4	81.6	87.3	91.8	44.1	60.2	91.8	55.0	67.4	94.4	84.1	87.6	91.8	53.4	66.4	98.4	43.6	62.0
0960	Flooring	119.6	77.8	107.6	118.9	45.4	97.8	113.6	66.5	100.1	119.4	61.1	102.6	118.0	58.3	100.8	119.4	65.5	103.9
0970, 0990	Wall Finishes & Painting/Coating	111.3	75.6	89.6	111.3	67.7	84.9	111.3	67.7	84.9	111.3	65.4	83.5	111.3	65.4	83.5	111.3	67.7	84.9
09	FINISHES	109.1	78.2	92.1	107.7	48.1	74.8	105.8	58.5	79.7	107.2	76.0	90.0	105.9	55.8	78.3	109.4	51.8	77.6
COVERS	DIVS. 10 - 14, 25, 28, 41, 43, 44, 46	100.0	81.9	96.2	100.0	51.5	89.9	100.0	51.6	89.9	100.0	81.5	96.2	100.0	62.4	92.2	100.0	73.8	94.6
21, 22, 23	FIRE SUPPRESSION, PLUMBING & HVAC	99.7	62.0	84.0	99.8	41.3	75.4	99.8	57.5	82.2	99.7	60.2	83.3	99.8	58.6	82.6	99.8	42.7	76.0
26, 27, 3370	ELECTRICAL, COMMUNICATIONS & UTIL.	104.1	63.6	82.6	96.7	58.5	76.4	100.9	58.4	78.3	98.1	66.8	81.5	98.4	66.8	81.6	104.1	58.6	79.9
MF2010	WEIGHTED AVERAGE	101.2	74.2	89.1	99.2	57.5	80.6	101.9	65.1	85.4	99.2	77.3	89.4	101.3	66.1	85.6	98.6	59.7	81.2

DIVISION		FLORIDA						GEORGIA											
		TAMPA			WEST PALM BEACH			ALBANY			ATHENS			ATLANTA			AUGUSTA		
		335 - 336,346			334,349			317,398			306			300 - 303,399			308 - 309		
		MAT.	INST.	TOTAL	MAT.	INST.	TOTAL	MAT.	INST.	TOTAL	MAT.	INST.	TOTAL	MAT.	INST.	TOTAL	MAT.	INST.	TOTAL
015433	CONTRACTOR EQUIPMENT		97.8	97.8		90.5	90.5		91.3	91.3		92.7	92.7		93.3	93.3		92.7	92.7
0241, 31 - 34	SITE & INFRASTRUCTURE, DEMOLITION	114.4	88.7	96.7	95.6	77.4	83.0	98.9	78.5	84.8	101.2	92.3	95.0	97.2	94.1	95.1	93.8	92.0	92.6
0310	Concrete Forming & Accessories	98.3	81.9	84.1	99.5	73.9	77.3	96.8	46.6	53.2	94.8	44.3	51.0	98.2	74.1	77.3	95.7	68.4	72.0
0320	Concrete Reinforcing	91.8	101.0	96.5	94.4	77.6	85.7	91.4	84.5	87.8	99.4	84.0	91.4	98.8	85.9	92.2	99.7	76.6	87.8
0330	Cast-in-Place Concrete	101.1	74.5	90.9	91.2	79.7	86.8	97.6	48.4	78.8	110.0	53.6	88.4	110.0	72.3	95.5	103.9	50.4	83.4
03	CONCRETE	93.0	83.8	88.6	87.3	77.7	82.7	91.0	55.9	74.2	101.0	55.8	79.3	98.9	75.8	87.9	94.5	64.1	79.9
04	MASONRY	88.3	81.9	84.4	86.9	71.9	77.9	89.7	45.6	63.0	74.6	58.1	64.6	88.0	69.2	76.6	88.2	42.2	60.3
05	METALS	98.0	102.4	99.5	95.2	93.4	94.6	95.4	87.3	92.6	91.1	75.5	85.8	92.0	79.6	87.8	90.8	73.0	84.7
06	WOOD, PLASTICS & COMPOSITES	101.8	84.1	91.7	101.1	72.9	84.9	99.5	41.1	66.0	89.3	38.0	59.9	93.3	76.1	83.4	90.7	74.3	81.3
07	THERMAL & MOISTURE PROTECTION	94.0	91.3	92.9	93.6	75.6	86.4	91.5	61.7	79.7	94.5	54.3	78.6	94.4	74.7	86.6	94.1	58.3	79.9
08	OPENINGS	100.4	83.8	96.2	97.0	70.9	90.4	97.9	47.4	85.1	91.9	46.1	80.3	97.8	72.4	91.3	91.9	65.7	85.3
0920	Plaster & Gypsum Board	94.6	84.1	87.2	97.3	72.6	79.9	94.3	39.8	56.0	100.5	36.5	55.5	102.7	75.8	83.7	101.4	73.9	82.0
0950, 0980	Ceilings & Acoustic Treatment	96.5	84.1	88.3	89.7	72.6	78.4	95.6	39.8	58.6	100.6	36.5	58.1	100.6	75.8	84.1	101.6	73.9	83.2
0960	Flooring	119.4	59.7	102.2	121.8	70.2	107.0	119.4	37.9	96.0	89.5	56.9	80.1	90.9	68.9	84.6	89.9	41.5	76.0
0970, 0990	Wall Finishes & Painting/Coating	111.3	65.4	83.5	105.9	75.6	87.5	105.9	55.5	75.3	99.0	37.7	61.8	99.0	86.4	91.3	99.0	39.4	62.8
09	FINISHES	107.7	75.8	90.1	104.3	72.9	87.0	104.7	43.9	71.2	96.1	44.3	67.5	96.3	74.4	84.2	95.8	61.0	76.6
COVERS	DIVS. 10 - 14, 25, 28, 41, 43, 44, 46	100.0	82.1	96.3	100.0	88.0	97.5	100.0	80.2	95.9	100.0	43.5	88.3	100.0	84.7	96.8	100.0	55.4	90.7
21, 22, 23	FIRE SUPPRESSION, PLUMBING & HVAC	99.8	88.7	95.2	96.9	68.8	85.2	99.8	69.8	87.3	94.6	74.5	86.2	99.9	74.4	89.3	99.9	67.2	86.3
26, 27, 3370	ELECTRICAL, COMMUNICATIONS & UTIL.	98.2	66.8	81.5	99.4	75.9	86.9	93.5	59.8	75.6	100.2	74.2	86.4	99.5	77.1	87.6	100.8	67.1	82.9
MF2010	WEIGHTED AVERAGE	99.0	83.6	92.1	96.0	75.7	86.9	96.9	62.0	81.3	94.9	64.9	81.5	97.2	76.8	88.1	95.9	65.5	82.3

	DIVISION	GEORGIA																	
		COLUMBUS			DALTON			GAINESVILLE			MACON			SAVANNAH			STATESBORO		
		318 - 319			307			305			310 - 312			313 - 314			304		
		MAT.	INST.	TOTAL	MAT.	INST.	TOTAL	MAT.	INST.	TOTAL	MAT.	INST.	TOTAL	MAT.	INST.	TOTAL	MAT.	INST.	TOTAL
015433	**CONTRACTOR EQUIPMENT**		91.3	91.3		104.7	104.7		92.7	92.7		102.0	102.0		92.3	92.3		92.2	92.2
0241, 31 - 34	**SITE & INFRASTRUCTURE, DEMOLITION**	98.8	78.7	84.9	102.0	96.2	98.0	100.8	92.3	94.9	99.4	93.9	95.6	99.8	79.4	85.7	102.5	76.4	84.5
0310	Concrete Forming & Accessories	96.7	58.2	63.3	88.2	49.9	54.9	97.9	46.1	53.0	96.0	55.3	60.7	96.8	51.6	57.6	82.8	43.2	48.4
0320	Concrete Reinforcing	91.8	85.3	88.4	98.9	78.9	88.6	99.2	84.0	91.3	92.9	84.6	88.6	92.7	76.8	84.5	98.6	44.6	70.8
0330	Cast-in-Place Concrete	97.2	57.1	81.9	106.8	47.3	84.0	115.6	53.4	91.8	96.0	50.7	78.6	107.0	49.8	85.1	109.8	62.7	91.7
03	**CONCRETE**	90.9	64.3	78.1	99.8	56.1	78.9	103.1	56.6	80.8	90.4	60.7	76.1	95.9	57.2	77.4	100.4	51.9	77.2
04	**MASONRY**	89.2	59.7	71.3	75.5	34.6	50.7	83.6	58.5	68.4	101.6	40.3	64.4	88.6	52.7	66.8	77.4	42.5	56.2
05	**METALS**	94.9	89.1	92.9	91.9	85.3	89.6	90.4	75.2	85.1	91.2	88.1	90.2	91.6	84.6	89.2	95.4	72.4	87.5
06	**WOOD, PLASTICS & COMPOSITES**	99.5	56.5	74.8	73.3	51.4	60.7	93.1	41.0	63.2	109.0	55.5	78.3	115.1	48.5	76.9	66.9	42.6	53.0
07	**THERMAL & MOISTURE PROTECTION**	91.5	67.0	81.8	96.4	52.5	78.9	94.5	57.4	79.7	90.1	63.4	79.5	93.7	58.4	79.7	95.1	50.8	77.5
08	**OPENINGS**	97.9	57.7	87.7	93.6	47.9	82.0	91.9	43.2	79.5	96.1	56.2	86.0	99.1	49.4	86.5	94.3	34.1	79.1
0920	Plaster & Gypsum Board	94.3	55.7	67.1	85.0	50.4	60.6	102.7	39.6	58.3	102.9	54.6	68.9	93.3	47.4	61.0	85.7	41.3	54.5
0950, 0980	Ceilings & Acoustic Treatment	95.6	55.7	69.1	115.0	50.4	72.2	100.6	39.6	60.2	90.2	54.6	66.6	98.4	47.4	64.6	110.1	41.3	64.5
0960	Flooring	119.4	53.3	100.4	90.3	37.9	75.2	90.9	37.9	75.6	92.9	38.3	77.2	119.9	49.0	99.5	106.9	47.4	89.8
0970, 0990	Wall Finishes & Painting/Coating	105.9	68.9	83.4	89.7	57.5	70.2	99.0	37.7	61.8	108.2	55.5	76.2	107.3	58.4	77.6	97.0	35.3	59.6
09	**FINISHES**	104.6	57.4	78.6	105.9	47.8	73.9	96.5	41.6	66.2	91.8	51.0	69.3	105.5	50.9	75.4	108.4	42.4	72.0
COVERS	**DIVS. 10 - 14, 25, 28, 41, 43, 44, 46**	100.0	82.0	96.3	100.0	24.7	84.4	100.0	43.8	88.3	100.0	79.9	95.8	100.0	54.3	90.5	100.0	40.4	87.6
21, 22, 23	**FIRE SUPPRESSION, PLUMBING & HVAC**	99.8	51.0	79.5	94.6	59.0	79.8	94.6	74.5	86.2	99.8	67.8	86.5	99.8	60.5	83.4	95.0	55.5	78.5
26, 27, 3370	**ELECTRICAL, COMMUNICATIONS & UTIL.**	93.7	74.3	83.4	109.9	73.5	90.6	100.2	74.2	86.4	92.7	63.4	77.1	97.7	58.7	77.0	100.3	58.7	78.2
MF2010	**WEIGHTED AVERAGE**	96.8	65.3	82.7	96.8	60.5	80.6	95.5	64.7	81.7	95.6	65.0	81.9	97.7	60.8	81.2	96.8	54.1	77.7

	DIVISION	GEORGIA						HAWAII									IDAHO		
		VALDOSTA			WAYCROSS			HILO			HONOLULU			STATES & POSS., GUAM			BOISE		
		316			315			967			968			969			836 - 837		
		MAT.	INST.	TOTAL	MAT.	INST.	TOTAL	MAT.	INST.	TOTAL	MAT.	INST.	TOTAL	MAT.	INST.	TOTAL	MAT.	INST.	TOTAL
015433	**CONTRACTOR EQUIPMENT**		91.3	91.3		91.3	91.3		99.9	99.9		99.9	99.9		165.5	165.5		100.6	100.6
0241, 31 - 34	**SITE & INFRASTRUCTURE, DEMOLITION**	108.8	78.5	87.9	105.7	77.2	86.1	127.4	108.0	114.0	129.9	108.0	114.8	169.4	105.6	125.4	74.1	99.4	91.6
0310	Concrete Forming & Accessories	86.3	45.7	51.1	88.2	67.8	70.5	108.6	134.5	131.1	115.7	134.5	132.0	110.8	68.4	74.0	100.7	75.0	78.4
0320	Concrete Reinforcing	93.6	72.5	82.8	93.6	78.4	85.8	112.7	120.1	116.5	111.6	120.1	116.0	196.9	32.5	112.2	98.5	76.0	86.9
0330	Cast-in-Place Concrete	95.6	52.4	79.0	108.0	51.0	86.2	189.4	125.4	164.9	159.8	125.4	146.6	164.2	110.7	143.7	93.6	76.4	87.0
03	**CONCRETE**	94.4	54.7	75.4	98.8	64.9	82.6	149.9	127.5	139.2	135.3	127.5	131.6	149.7	76.5	114.7	101.1	75.8	89.0
04	**MASONRY**	94.8	52.2	69.0	95.7	42.4	63.4	126.2	125.4	125.7	118.0	125.4	122.5	179.4	46.9	99.1	122.6	63.9	87.0
05	**METALS**	94.6	83.5	90.7	93.6	80.2	89.0	109.4	107.8	108.8	130.1	107.8	122.4	140.9	78.5	119.4	100.4	75.5	91.8
06	**WOOD, PLASTICS & COMPOSITES**	86.8	40.1	60.0	89.1	75.7	81.4	97.4	137.6	120.4	108.2	137.6	125.0	109.3	72.5	88.2	90.4	77.3	82.9
07	**THERMAL & MOISTURE PROTECTION**	91.7	64.8	81.0	91.5	52.9	76.1	105.8	123.9	113.0	115.7	123.9	118.9	121.8	69.3	100.9	96.2	68.0	85.0
08	**OPENINGS**	93.3	43.5	80.7	93.3	60.4	84.9	99.5	131.3	107.6	104.7	131.3	111.5	104.3	58.1	92.6	94.4	71.9	88.7
0920	Plaster & Gypsum Board	85.8	38.8	52.7	87.1	75.5	78.9	108.0	138.4	129.5	139.1	138.4	138.6	207.0	60.5	103.9	90.8	76.5	80.7
0950, 0980	Ceilings & Acoustic Treatment	91.8	38.8	56.7	89.7	75.5	80.3	113.2	138.4	130.0	118.5	138.4	131.7	224.9	60.5	115.9	122.2	76.5	91.9
0960	Flooring	112.0	49.1	94.0	113.4	31.6	89.9	134.9	131.0	133.8	158.6	131.0	150.7	158.3	49.0	126.9	97.3	67.1	88.6
0970, 0990	Wall Finishes & Painting/Coating	105.9	55.5	75.3	105.9	42.9	67.7	98.4	144.0	126.1	102.7	144.0	127.8	103.7	38.7	64.2	98.5	42.5	64.5
09	**FINISHES**	101.6	45.7	70.8	101.3	59.5	78.2	117.4	135.9	127.6	130.2	135.9	133.4	191.5	63.0	120.7	101.9	71.1	84.9
COVERS	**DIVS. 10 - 14, 25, 28, 41, 43, 44, 46**	100.0	53.4	90.3	100.0	44.0	88.4	100.0	116.5	103.4	100.0	116.5	103.4	100.0	82.7	96.4	100.0	64.5	92.6
21, 22, 23	**FIRE SUPPRESSION, PLUMBING & HVAC**	99.8	69.7	87.3	96.3	59.4	80.9	100.1	108.6	103.7	100.2	108.6	103.7	102.3	40.9	76.7	100.0	67.2	86.3
26, 27, 3370	**ELECTRICAL, COMMUNICATIONS & UTIL.**	91.6	58.2	73.8	96.2	61.8	77.9	108.4	126.1	117.9	111.1	126.1	119.1	160.4	44.7	98.9	95.5	74.6	84.4
MF2010	**WEIGHTED AVERAGE**	96.7	61.2	80.9	96.7	61.8	81.1	111.9	120.6	115.8	115.1	120.6	117.5	133.3	61.7	101.3	99.5	73.3	87.8

	DIVISION	IDAHO															ILLINOIS		
		COEUR D'ALENE			IDAHO FALLS			LEWISTON			POCATELLO			TWIN FALLS			BLOOMINGTON		
		838			834			835			832			833			617		
		MAT.	INST.	TOTAL	MAT.	INST.	TOTAL	MAT.	INST.	TOTAL	MAT.	INST.	TOTAL	MAT.	INST.	TOTAL	MAT.	INST.	TOTAL
015433	**CONTRACTOR EQUIPMENT**		94.6	94.6		100.6	100.6		94.6	94.6		100.6	100.6		100.6	100.6		101.0	101.0
0241, 31 - 34	**SITE & INFRASTRUCTURE, DEMOLITION**	74.0	94.7	88.3	72.6	98.5	90.5	79.9	95.2	90.5	75.6	99.4	92.0	81.6	97.6	92.7	98.8	96.8	97.4
0310	Concrete Forming & Accessories	117.8	58.5	66.3	95.2	42.1	49.1	122.5	60.0	68.3	100.9	74.7	78.2	101.7	30.1	39.6	88.1	118.8	114.8
0320	Concrete Reinforcing	105.8	94.9	100.1	100.4	74.9	87.3	105.8	95.1	100.3	98.9	76.1	87.1	100.7	75.2	87.5	99.6	104.8	102.2
0330	Cast-in-Place Concrete	101.2	88.7	96.4	89.1	53.0	75.3	105.2	90.0	99.3	96.2	76.3	88.6	98.7	40.3	76.4	99.7	114.5	105.4
03	**CONCRETE**	108.2	76.2	92.9	93.3	53.0	74.0	112.0	77.3	95.4	100.5	75.6	88.6	108.1	43.3	77.1	96.7	114.6	105.3
04	**MASONRY**	124.1	81.7	98.4	117.8	35.3	67.8	124.5	84.6	100.3	119.9	65.8	87.1	122.7	33.3	68.5	113.8	119.1	117.0
05	**METALS**	94.8	84.7	91.3	108.1	72.8	95.9	94.2	85.8	91.3	108.1	75.3	96.8	108.2	72.2	95.8	95.0	110.1	100.2
06	**WOOD, PLASTICS & COMPOSITES**	95.6	51.6	70.3	84.5	40.6	59.3	100.3	51.6	72.4	90.4	77.3	82.9	91.2	28.8	55.4	84.1	117.8	103.4
07	**THERMAL & MOISTURE PROTECTION**	148.4	74.6	119.1	95.8	45.4	75.8	148.6	75.3	119.5	96.2	62.5	82.8	97.0	39.4	74.1	98.4	113.3	104.3
08	**OPENINGS**	114.3	57.1	99.8	98.7	46.8	85.6	114.3	59.8	100.5	95.2	66.5	88.0	98.8	38.2	83.4	90.3	105.5	94.1
0920	Plaster & Gypsum Board	155.8	50.1	81.4	83.9	38.7	52.1	156.7	50.1	81.6	87.9	76.5	79.9	88.3	26.5	44.8	91.2	118.2	110.2
0950, 0980	Ceilings & Acoustic Treatment	160.9	50.1	87.5	119.1	38.7	65.8	160.9	50.1	87.5	128.5	76.5	94.0	122.0	26.5	58.7	89.5	118.2	108.5
0960	Flooring	136.5	48.5	111.2	97.1	45.5	82.3	139.5	90.3	125.3	100.4	67.1	90.8	101.6	45.5	85.5	95.1	113.2	100.3
0970, 0990	Wall Finishes & Painting/Coating	117.4	63.2	84.5	98.5	34.9	59.9	117.4	63.2	84.5	98.4	44.2	65.5	98.5	26.1	54.5	96.2	122.3	112.0
09	**FINISHES**	172.0	55.6	107.8	99.9	41.3	67.6	173.1	64.4	113.2	103.8	71.3	85.9	103.5	31.4	63.8	94.9	118.8	108.1
COVERS	**DIVS. 10 - 14, 25, 28, 41, 43, 44, 46**	100.0	53.9	90.4	100.0	40.7	87.7	100.0	67.3	93.2	100.0	64.5	92.6	100.0	36.6	86.8	100.0	105.2	101.1
21, 22, 23	**FIRE SUPPRESSION, PLUMBING & HVAC**	99.8	77.4	90.5	100.7	60.8	84.1	100.8	78.8	91.6	99.9	67.2	86.3	99.9	56.9	82.0	94.7	106.5	99.6
26, 27, 3370	**ELECTRICAL, COMMUNICATIONS & UTIL.**	88.0	79.7	83.6	87.7	74.6	80.7	85.9	82.5	84.1	92.9	74.6	83.2	89.2	64.3	76.0	93.2	95.5	94.5
MF2010	**WEIGHTED AVERAGE**	108.5	75.4	93.7	99.0	59.0	81.1	109.3	78.3	95.4	100.4	73.0	88.2	101.7	53.1	80.0	96.1	108.7	101.7

DIVISION		ILLINOIS																	
		CARBONDALE			CENTRALIA			CHAMPAIGN			CHICAGO			DECATUR			EAST ST. LOUIS		
		629			628			618 - 619			606 - 608			625			620 - 622		
		MAT.	INST.	TOTAL	MAT.	INST.	TOTAL	MAT.	INST.	TOTAL	MAT.	INST.	TOTAL	MAT.	INST.	TOTAL	MAT.	INST.	TOTAL
015433	CONTRACTOR EQUIPMENT		107.4	107.4		107.4	107.4		101.7	101.7		93.4	93.4		101.7	101.7		107.4	107.4
0241, 31 - 34	SITE & INFRASTRUCTURE, DEMOLITION	107.1	97.0	100.1	107.4	98.3	101.1	108.5	97.5	100.9	115.1	94.0	100.6	92.2	97.5	95.9	109.2	98.4	101.7
0310	Concrete Forming & Accessories	96.5	103.7	102.7	98.3	111.4	109.7	94.0	116.6	113.6	100.3	152.2	145.3	98.2	116.0	113.7	94.1	111.7	109.3
0320	Concrete Reinforcing	98.5	91.8	95.0	98.5	106.1	102.4	99.6	101.9	100.8	98.8	157.4	129.0	96.5	101.7	99.2	98.4	106.1	102.4
0330	Cast-in-Place Concrete	93.3	101.4	96.4	93.8	115.1	102.0	115.7	110.7	113.8	112.8	147.0	125.9	101.9	112.1	105.8	95.4	115.3	103.0
03	CONCRETE	87.8	101.5	94.3	88.2	112.3	99.8	109.3	111.7	110.5	104.4	150.3	126.3	98.0	111.9	104.7	89.2	112.5	100.4
04	MASONRY	76.0	103.0	92.4	76.1	117.2	101.0	136.9	114.1	123.1	96.0	154.3	131.3	72.8	113.2	97.3	76.2	117.3	101.1
05	METALS	94.3	110.4	99.8	94.3	119.0	102.8	95.0	107.1	99.2	95.6	133.4	108.6	97.9	107.4	101.1	95.5	119.4	103.7
06	WOOD, PLASTICS & COMPOSITES	98.0	102.2	100.4	100.5	108.8	105.3	90.7	116.0	105.2	100.2	149.9	128.7	97.9	116.0	108.3	95.3	108.8	103.1
07	THERMAL & MOISTURE PROTECTION	93.7	100.1	96.2	93.7	113.0	101.4	99.0	112.5	104.4	102.4	144.6	119.2	99.1	108.1	102.7	93.7	112.5	101.2
08	OPENINGS	86.4	107.1	91.7	86.4	114.4	93.5	91.0	111.2	96.1	100.9	152.9	114.1	96.8	111.1	100.4	86.5	114.4	93.6
0920	Plaster & Gypsum Board	95.6	102.2	100.3	97.0	109.0	105.4	94.3	116.4	109.9	96.3	151.3	135.0	98.5	116.4	111.1	94.3	109.0	104.7
0950, 0980	Ceilings & Acoustic Treatment	89.5	102.2	97.9	89.5	109.0	102.4	89.5	116.4	107.3	104.8	151.3	135.6	95.1	116.4	109.2	89.5	109.0	102.4
0960	Flooring	113.4	126.9	117.3	114.5	119.6	116.0	98.2	126.9	106.5	93.1	149.9	109.4	101.8	114.0	105.3	112.2	119.6	114.4
0970, 0990	Wall Finishes & Painting/Coating	104.0	98.4	100.6	104.0	108.0	106.4	96.2	114.7	107.4	94.9	156.2	132.1	96.2	110.7	105.0	104.0	105.8	105.1
09	FINISHES	98.4	106.5	102.8	98.9	112.3	106.3	97.0	118.8	109.0	98.0	152.8	128.2	99.0	116.0	108.4	97.9	112.3	105.9
COVERS	DIVS. 10 - 14, 25, 28, 41, 43, 44, 46	100.0	104.0	100.8	100.0	106.9	101.4	100.0	104.1	100.9	100.0	125.5	105.3	100.0	103.8	100.8	100.0	106.9	101.4
21, 22, 23	FIRE SUPPRESSION, PLUMBING & HVAC	94.6	98.9	96.4	94.6	93.3	94.1	94.7	104.4	98.7	100.1	136.5	115.3	100.1	100.3	100.2	100.0	95.4	98.1
26, 27, 3370	ELECTRICAL, COMMUNICATIONS & UTIL.	92.0	100.6	96.6	93.5	103.9	99.0	96.8	95.8	96.3	95.3	134.8	116.3	96.3	91.3	93.6	93.0	103.9	98.8
MF2010	WEIGHTED AVERAGE	92.9	102.3	97.1	93.2	106.9	99.3	99.8	107.3	103.1	99.7	138.9	117.2	97.0	105.3	100.7	94.6	107.5	100.4

DIVISION		ILLINOIS																	
		EFFINGHAM			GALESBURG			JOLIET			KANKAKEE			LA SALLE			NORTH SUBURBAN		
		624			614			604			609			613			600 - 603		
		MAT.	INST.	TOTAL	MAT.	INST.	TOTAL	MAT.	INST.	TOTAL	MAT.	INST.	TOTAL	MAT.	INST.	TOTAL	MAT.	INST.	TOTAL
015433	CONTRACTOR EQUIPMENT		101.7	101.7		101.0	101.0		91.5	91.5		91.5	91.5		101.0	101.0		91.5	91.5
0241, 31 - 34	SITE & INFRASTRUCTURE, DEMOLITION	100.1	97.1	98.0	101.4	96.7	98.1	114.8	93.2	99.9	107.9	92.3	97.2	100.8	97.8	98.7	113.8	92.5	99.1
0310	Concrete Forming & Accessories	103.1	110.2	109.2	93.9	114.5	111.8	102.4	154.7	147.8	96.0	130.8	126.2	107.2	119.9	118.2	101.8	148.3	142.1
0320	Concrete Reinforcing	99.6	97.0	98.3	99.1	104.5	101.9	98.8	145.2	122.7	99.5	138.3	119.5	99.2	135.4	117.9	98.8	152.0	126.2
0330	Cast-in-Place Concrete	101.6	107.0	103.6	102.7	105.5	103.8	112.7	144.7	125.0	105.1	127.3	113.6	102.6	117.5	108.3	112.8	142.1	124.0
03	CONCRETE	98.8	106.7	102.6	99.5	109.6	104.3	104.5	148.2	125.4	98.4	130.3	113.6	100.2	122.1	110.7	104.5	145.7	124.2
04	MASONRY	80.2	106.7	96.2	114.0	116.2	115.3	98.7	149.5	129.5	95.5	129.0	115.8	114.0	119.6	117.4	96.0	145.6	126.0
05	METALS	94.9	104.5	98.3	95.0	109.1	99.9	93.3	126.3	104.7	93.3	121.8	103.1	95.1	127.6	106.3	94.6	128.6	106.3
06	WOOD, PLASTICS & COMPOSITES	100.3	111.1	106.5	90.6	112.8	103.4	101.8	155.4	132.6	94.8	130.5	115.3	104.7	118.1	112.4	100.2	149.4	128.4
07	THERMAL & MOISTURE PROTECTION	98.7	106.7	101.8	98.5	107.4	102.1	102.2	141.5	117.8	101.4	130.1	112.8	98.7	114.4	104.9	102.7	135.8	115.8
08	OPENINGS	90.8	107.2	94.9	90.3	106.5	94.4	98.9	152.6	112.5	90.4	137.4	102.3	90.3	121.6	98.2	98.9	151.2	112.2
0920	Plaster & Gypsum Board	98.7	111.3	107.6	94.3	113.1	107.6	93.2	157.0	138.1	90.6	131.4	119.3	101.0	118.6	113.4	96.3	150.8	134.7
0950, 0980	Ceilings & Acoustic Treatment	89.5	111.3	103.9	89.5	113.1	105.2	104.8	157.0	139.4	104.8	131.4	122.4	89.5	118.6	108.8	104.8	150.8	135.3
0960	Flooring	103.0	126.9	109.9	98.1	113.2	102.5	92.7	139.3	106.1	89.8	139.3	104.1	105.2	117.4	108.7	93.1	139.3	106.4
0970, 0990	Wall Finishes & Painting/Coating	96.2	103.5	100.6	96.2	95.9	96.0	93.0	146.1	125.3	93.0	122.3	110.8	96.2	122.3	112.0	94.9	146.1	126.0
09	FINISHES	98.3	113.6	106.8	96.3	113.0	105.5	97.4	152.3	127.7	95.7	129.4	114.3	99.2	118.6	109.9	97.9	147.7	125.4
COVERS	DIVS. 10 - 14, 25, 28, 41, 43, 44, 46	100.0	74.6	94.7	100.0	104.3	100.9	100.0	124.8	105.2	100.0	116.3	103.4	100.0	104.4	100.9	100.0	121.2	104.4
21, 22, 23	FIRE SUPPRESSION, PLUMBING & HVAC	94.7	100.9	97.3	94.7	101.9	97.7	100.2	136.2	115.2	94.8	122.1	106.2	94.7	122.9	106.4	100.1	123.0	109.6
26, 27, 3370	ELECTRICAL, COMMUNICATIONS & UTIL.	94.0	100.6	97.5	94.2	89.6	91.8	94.6	138.5	117.9	89.2	137.6	114.9	91.1	137.6	115.8	94.4	121.2	108.6
MF2010	WEIGHTED AVERAGE	95.1	103.6	98.9	96.8	104.8	100.4	99.2	137.7	116.4	95.2	125.3	108.7	96.9	121.4	107.8	99.3	131.0	113.4

DIVISION		ILLINOIS																	
		PEORIA			QUINCY			ROCK ISLAND			ROCKFORD			SOUTH SUBURBAN			SPRINGFIELD		
		615 - 616			623			612			610 - 611			605			626 - 627		
		MAT.	INST.	TOTAL	MAT.	INST.	TOTAL	MAT.	INST.	TOTAL	MAT.	INST.	TOTAL	MAT.	INST.	TOTAL	MAT.	INST.	TOTAL
015433	CONTRACTOR EQUIPMENT		101.0	101.0		101.7	101.7		101.0	101.0		101.0	101.0		91.5	91.5		101.7	101.7
0241, 31 - 34	SITE & INFRASTRUCTURE, DEMOLITION	101.5	96.6	98.1	98.8	97.0	97.6	99.4	96.1	97.1	100.5	97.9	98.7	113.8	92.5	99.1	100.2	97.6	98.4
0310	Concrete Forming & Accessories	96.4	115.0	112.5	100.9	112.5	110.9	95.4	100.3	99.6	101.1	129.5	125.8	101.8	148.3	142.1	99.2	116.5	114.3
0320	Concrete Reinforcing	96.5	104.6	100.7	99.2	101.3	100.3	99.1	101.3	100.2	91.2	140.2	116.4	98.8	152.0	126.2	96.5	102.4	99.5
0330	Cast-in-Place Concrete	99.6	114.8	105.4	101.7	98.4	100.5	100.5	98.1	99.6	102.0	116.4	107.5	112.8	142.1	124.0	91.6	108.4	98.1
03	CONCRETE	96.7	113.0	104.5	98.5	105.6	101.9	97.6	100.1	98.8	97.4	126.9	111.5	104.5	145.7	124.2	92.9	111.0	101.6
04	MASONRY	114.1	115.4	114.9	99.9	100.7	100.4	113.8	97.6	104.0	88.9	139.1	119.3	96.0	145.6	126.0	77.6	114.3	99.9
05	METALS	97.9	110.0	102.0	95.0	106.2	98.8	95.1	106.9	99.2	97.8	129.2	108.6	94.6	128.6	106.3	100.2	107.7	102.8
06	WOOD, PLASTICS & COMPOSITES	97.9	112.8	106.5	97.7	116.0	108.2	92.1	99.1	96.1	97.9	127.5	114.8	100.2	149.4	128.4	98.9	116.0	108.7
07	THERMAL & MOISTURE PROTECTION	99.2	111.5	104.1	98.7	103.2	100.5	98.5	98.1	98.4	101.7	129.6	112.8	102.7	135.8	115.8	98.9	111.6	103.9
08	OPENINGS	96.9	110.5	100.4	91.6	111.1	96.6	90.3	98.4	92.3	96.9	130.6	105.5	98.9	151.2	112.2	97.9	111.3	101.3
0920	Plaster & Gypsum Board	98.5	113.1	108.8	97.0	116.4	110.7	94.3	99.0	97.6	98.5	128.2	119.4	96.3	150.8	134.7	100.3	116.4	111.6
0950, 0980	Ceilings & Acoustic Treatment	95.1	113.1	107.0	89.5	116.4	107.3	89.5	99.0	95.8	95.1	128.2	117.0	104.8	150.8	135.3	94.2	116.4	108.9
0960	Flooring	101.8	113.2	105.1	101.8	104.9	102.7	99.3	95.5	98.2	101.8	117.4	106.3	93.1	139.3	106.4	101.8	108.4	103.7
0970, 0990	Wall Finishes & Painting/Coating	96.2	122.3	112.0	96.2	105.6	101.9	96.2	99.0	97.9	96.2	129.6	116.4	94.9	146.1	126.0	96.2	110.2	104.7
09	FINISHES	99.1	115.7	108.3	97.7	111.8	105.4	96.6	99.2	98.0	99.2	127.6	114.8	97.9	147.7	125.4	98.9	115.1	107.8
COVERS	DIVS. 10 - 14, 25, 28, 41, 43, 44, 46	100.0	104.3	100.9	100.0	76.8	95.2	100.0	98.2	99.6	100.0	96.8	99.3	100.0	121.2	104.4	100.0	104.2	100.9
21, 22, 23	FIRE SUPPRESSION, PLUMBING & HVAC	100.1	104.2	101.8	94.7	96.3	95.3	94.7	98.4	96.3	100.1	114.3	106.0	100.1	123.0	109.6	99.9	102.5	101.0
26, 27, 3370	ELECTRICAL, COMMUNICATIONS & UTIL.	95.4	95.7	95.5	91.2	83.4	87.0	85.7	96.5	91.5	95.7	127.1	112.4	94.4	121.2	108.6	98.9	94.4	96.5
MF2010	WEIGHTED AVERAGE	99.2	107.3	102.8	95.8	99.6	97.5	95.7	99.0	97.2	98.1	122.7	109.1	99.3	131.0	113.4	97.5	106.2	101.4

	DIVISION	INDIANA																	
		ANDERSON			BLOOMINGTON			COLUMBUS			EVANSVILLE			FORT WAYNE			GARY		
		460			474			472			476 - 477			467 - 468			463 - 464		
		MAT.	INST.	TOTAL	MAT.	INST.	TOTAL	MAT.	INST.	TOTAL	MAT.	INST.	TOTAL	MAT.	INST.	TOTAL	MAT.	INST.	TOTAL
015433	CONTRACTOR EQUIPMENT		96.3	96.3		87.3	87.3		87.3	87.3		117.4	117.4		96.3	96.3		96.3	96.3
0241, 31 - 34	SITE & INFRASTRUCTURE, DEMOLITION	89.3	94.5	92.9	77.4	94.7	89.4	74.4	94.5	88.3	82.2	125.1	111.8	90.6	94.4	93.2	89.9	98.1	95.5
0310	Concrete Forming & Accessories	93.4	79.7	81.5	100.1	80.2	82.8	94.5	77.9	80.1	94.0	79.8	81.7	92.0	73.8	76.2	93.5	111.5	109.1
0320	Concrete Reinforcing	93.9	80.4	86.9	90.0	80.1	84.9	90.4	80.1	85.1	98.6	77.7	87.8	93.9	79.0	86.2	93.9	104.8	99.5
0330	Cast-in-Place Concrete	101.6	80.4	93.5	98.2	76.8	90.0	97.7	75.3	89.1	93.8	86.7	91.1	108.0	76.4	95.9	106.1	108.8	107.2
03	CONCRETE	94.1	80.6	87.6	100.8	78.8	90.3	100.2	77.3	89.2	101.0	82.0	91.9	97.2	76.3	87.2	96.4	109.2	102.5
04	MASONRY	84.2	82.2	83.0	88.4	78.5	82.4	88.3	78.4	82.3	84.2	83.2	83.6	87.7	78.9	82.4	85.5	112.7	102.0
05	METALS	93.4	88.3	91.7	94.5	77.4	88.6	94.6	76.6	88.4	88.5	84.6	87.2	93.4	87.4	91.4	93.4	105.6	97.6
06	WOOD, PLASTICS & COMPOSITES	109.6	79.7	92.4	120.4	80.1	97.2	115.1	77.1	93.3	98.9	78.0	86.9	109.3	72.6	88.3	107.0	110.7	109.1
07	THERMAL & MOISTURE PROTECTION	99.4	79.5	91.5	94.2	81.4	89.1	93.9	81.6	89.0	98.1	84.0	92.5	99.2	81.6	92.2	98.2	109.2	102.6
08	OPENINGS	97.3	80.6	93.1	105.3	80.9	99.1	100.5	79.2	95.1	97.3	79.0	92.6	97.3	73.8	91.4	97.3	114.0	101.5
0920	Plaster & Gypsum Board	99.5	79.5	85.4	94.0	80.1	84.2	90.0	77.0	80.9	88.5	76.8	80.2	98.6	72.2	80.1	92.0	111.5	105.7
0950, 0980	Ceilings & Acoustic Treatment	94.3	79.5	84.5	80.5	80.1	80.2	80.5	77.0	78.2	85.6	76.8	79.8	94.3	72.2	79.7	94.3	111.5	105.7
0960	Flooring	97.0	82.0	92.7	104.8	79.2	97.5	99.8	79.2	93.9	98.6	85.2	94.7	97.0	79.2	91.9	97.0	122.5	104.4
0970, 0990	Wall Finishes & Painting/Coating	102.9	71.5	83.9	92.2	85.1	87.9	92.2	85.1	87.9	97.6	84.5	89.7	102.9	76.0	86.6	102.9	118.6	112.4
09	FINISHES	97.6	79.3	87.5	96.3	80.5	87.6	94.3	78.7	85.7	94.8	81.3	87.3	97.3	75.0	85.0	96.6	114.4	106.4
COVERS	DIVS. 10 - 14, 25, 28, 41, 43, 44, 46	100.0	91.9	98.3	100.0	89.8	97.9	100.0	89.5	97.8	100.0	77.8	95.4	100.0	90.3	98.0	100.0	106.4	101.3
21, 22, 23	FIRE SUPPRESSION, PLUMBING & HVAC	99.9	75.7	89.8	99.7	79.4	91.2	94.3	79.9	88.3	99.9	80.9	92.0	99.9	76.0	90.0	99.9	105.9	102.4
26, 27, 3370	ELECTRICAL, COMMUNICATIONS & UTIL.	85.2	88.2	86.8	100.0	83.8	91.4	99.1	88.1	93.3	96.1	86.3	90.9	85.9	75.7	80.5	97.5	107.4	102.8
MF2010	WEIGHTED AVERAGE	95.3	82.8	89.7	98.1	81.5	90.7	95.9	81.6	89.5	95.9	86.0	91.5	95.9	79.2	88.5	96.7	108.1	101.8

	DIVISION	INDIANA																	
		INDIANAPOLIS			KOKOMO			LAFAYETTE			LAWRENCEBURG			MUNCIE			NEW ALBANY		
		461 - 462			469			479			470			473			471		
		MAT.	INST.	TOTAL	MAT.	INST.	TOTAL	MAT.	INST.	TOTAL	MAT.	INST.	TOTAL	MAT.	INST.	TOTAL	MAT.	INST.	TOTAL
015433	CONTRACTOR EQUIPMENT		91.8	91.8		96.3	96.3		87.3	87.3		106.1	106.1		96.1	96.1		95.8	95.8
0241, 31 - 34	SITE & INFRASTRUCTURE, DEMOLITION	88.8	97.8	95.0	86.6	94.5	92.0	75.4	94.5	88.6	73.4	111.9	100.0	77.5	94.9	89.5	69.6	98.0	89.2
0310	Concrete Forming & Accessories	94.1	86.5	87.5	96.5	76.2	78.9	92.4	81.2	82.6	91.0	74.9	77.0	92.6	79.5	81.2	88.2	72.7	74.8
0320	Concrete Reinforcing	93.3	83.6	88.3	85.0	80.2	82.5	90.0	80.3	85.0	89.3	79.5	84.2	99.6	80.3	89.7	90.7	77.7	84.0
0330	Cast-in-Place Concrete	97.7	82.9	92.1	100.6	83.3	94.0	98.3	82.7	92.3	91.9	74.3	85.2	103.2	79.6	94.1	94.9	73.4	86.7
03	CONCRETE	95.7	84.4	90.3	91.1	80.0	85.8	100.4	81.3	91.2	93.6	76.2	85.2	99.9	80.2	90.5	98.7	74.3	87.0
04	MASONRY	91.7	84.6	87.4	83.9	80.0	81.6	93.5	82.0	86.6	73.7	72.0	72.7	90.2	82.2	85.3	80.6	71.5	75.1
05	METALS	95.6	80.6	90.4	89.9	88.0	89.3	93.0	77.2	87.6	89.9	86.2	88.6	96.4	88.5	93.7	91.7	81.9	88.3
06	WOOD, PLASTICS & COMPOSITES	106.3	86.9	95.2	112.7	74.7	90.9	112.4	81.1	94.4	97.7	74.0	84.1	114.7	79.5	94.5	99.4	73.1	84.3
07	THERMAL & MOISTURE PROTECTION	97.4	84.7	92.4	99.1	78.2	90.7	93.8	83.1	89.6	97.4	78.2	89.8	96.8	79.5	89.9	85.9	71.1	80.0
08	OPENINGS	105.3	85.4	100.3	91.6	77.9	88.2	98.8	81.4	94.4	98.8	76.6	93.1	100.1	80.5	95.1	96.8	75.8	91.5
0920	Plaster & Gypsum Board	97.2	86.8	89.9	104.8	74.4	83.4	87.2	81.2	83.0	74.0	73.8	73.8	88.5	79.5	82.2	86.8	72.6	76.8
0950, 0980	Ceilings & Acoustic Treatment	98.0	86.8	90.6	94.3	74.4	81.1	76.8	81.2	79.7	91.3	73.8	79.7	81.5	79.5	80.2	85.6	72.6	77.0
0960	Flooring	96.8	88.2	94.4	101.4	89.6	98.0	98.7	89.9	96.2	73.4	88.2	77.7	98.3	82.0	93.6	96.7	66.4	88.0
0970, 0990	Wall Finishes & Painting/Coating	102.9	85.1	92.1	102.9	77.2	87.3	92.2	82.3	86.2	93.2	68.6	78.3	92.2	71.5	79.6	97.6	69.5	80.6
09	FINISHES	98.4	87.0	92.1	99.3	78.5	87.9	92.9	82.9	87.4	86.6	77.0	81.3	93.6	79.1	85.6	94.3	71.3	81.6
COVERS	DIVS. 10 - 14, 25, 28, 41, 43, 44, 46	100.0	94.0	98.8	100.0	89.7	97.9	100.0	91.7	98.3	100.0	47.1	89.0	100.0	91.4	98.2	100.0	46.6	88.9
21, 22, 23	FIRE SUPPRESSION, PLUMBING & HVAC	99.9	81.3	92.1	94.5	80.0	88.5	94.3	79.7	88.2	95.0	75.8	87.0	99.7	75.7	89.7	94.5	76.1	86.8
26, 27, 3370	ELECTRICAL, COMMUNICATIONS & UTIL.	99.1	88.2	93.3	90.1	79.0	84.2	98.5	84.5	91.0	93.8	76.1	84.4	91.5	80.8	85.8	94.3	78.0	85.6
MF2010	WEIGHTED AVERAGE	98.3	85.7	92.7	93.1	81.7	88.0	95.6	82.8	89.9	92.5	78.9	86.4	96.8	81.7	90.1	93.6	76.4	85.9

	DIVISION	INDIANA									IOWA								
		SOUTH BEND			TERRE HAUTE			WASHINGTON			BURLINGTON			CARROLL			CEDAR RAPIDS		
		465 - 466			478			475			526			514			522 - 524		
		MAT.	INST.	TOTAL	MAT.	INST.	TOTAL	MAT.	INST.	TOTAL	MAT.	INST.	TOTAL	MAT.	INST.	TOTAL	MAT.	INST.	TOTAL
015433	CONTRACTOR EQUIPMENT		105.0	105.0		117.4	117.4		117.4	117.4		99.8	99.8		99.8	99.8		96.4	96.4
0241, 31 - 34	SITE & INFRASTRUCTURE, DEMOLITION	91.5	94.9	93.9	83.9	125.3	112.5	84.1	123.2	111.1	98.3	96.7	97.2	86.2	96.8	93.5	100.3	96.1	97.4
0310	Concrete Forming & Accessories	95.9	78.0	80.4	94.9	81.6	83.3	95.8	78.7	80.9	95.4	78.8	81.0	85.0	53.8	57.9	100.2	83.4	85.7
0320	Concrete Reinforcing	93.9	77.0	85.2	98.6	80.5	89.3	91.2	52.5	71.2	95.9	83.6	89.6	96.7	83.0	89.7	96.7	86.1	91.3
0330	Cast-in-Place Concrete	98.8	80.5	91.8	90.8	88.6	89.9	98.8	88.4	94.8	109.1	57.9	89.5	109.1	63.0	91.5	109.4	82.4	99.1
03	CONCRETE	89.0	80.0	84.7	103.5	84.0	94.1	109.2	77.1	93.8	104.0	73.3	89.3	103.0	63.8	84.3	104.1	84.2	94.6
04	MASONRY	84.0	81.1	82.2	91.4	82.2	85.8	84.4	84.5	84.4	103.2	69.8	83.0	105.6	76.6	88.0	109.6	82.1	92.9
05	METALS	93.4	99.2	95.4	89.1	87.1	88.4	83.8	70.5	79.2	89.2	92.5	90.3	89.2	92.1	90.2	91.6	94.6	92.6
06	WOOD, PLASTICS & COMPOSITES	106.8	77.4	89.9	100.9	81.1	89.5	101.3	77.9	87.9	97.4	79.6	87.2	85.5	48.6	64.4	103.0	82.4	91.2
07	THERMAL & MOISTURE PROTECTION	93.8	83.6	89.8	98.2	83.1	92.2	98.3	84.2	92.7	105.7	78.4	94.9	106.1	72.9	92.9	106.6	82.1	96.9
08	OPENINGS	90.6	77.6	87.3	97.8	81.4	93.7	94.4	69.0	87.9	90.7	72.6	86.1	95.9	57.3	86.1	96.8	83.4	93.4
0920	Plaster & Gypsum Board	90.8	77.1	81.2	88.5	79.9	82.5	88.3	76.7	80.1	99.2	79.0	85.0	94.3	47.2	61.1	103.3	82.2	88.4
0950, 0980	Ceilings & Acoustic Treatment	88.7	77.1	81.0	85.6	79.9	81.8	80.0	76.7	77.8	104.7	79.0	87.7	104.7	47.2	66.5	108.4	82.2	91.0
0960	Flooring	95.4	84.1	92.1	98.6	88.2	95.6	99.6	80.7	94.2	116.1	39.9	94.2	109.0	35.1	87.7	134.7	82.7	119.8
0970, 0990	Wall Finishes & Painting/Coating	102.4	85.7	92.3	97.6	87.0	91.2	97.6	85.1	90.0	101.4	84.5	91.1	101.4	80.0	88.4	104.3	73.9	85.9
09	FINISHES	94.8	79.9	86.6	94.8	83.3	88.5	94.2	80.4	86.6	107.0	72.2	87.8	102.4	51.2	74.2	113.9	82.2	96.4
COVERS	DIVS. 10 - 14, 25, 28, 41, 43, 44, 46	100.0	73.1	94.4	100.0	94.5	98.9	100.0	77.7	95.4	100.0	90.6	98.1	100.0	72.4	94.3	100.0	91.7	98.3
21, 22, 23	FIRE SUPPRESSION, PLUMBING & HVAC	99.8	78.0	90.7	99.9	80.0	91.6	94.5	80.1	88.5	94.8	79.7	88.5	94.8	72.7	85.6	100.2	83.8	93.4
26, 27, 3370	ELECTRICAL, COMMUNICATIONS & UTIL.	102.7	86.0	93.8	94.0	83.8	88.6	94.6	86.3	90.2	98.6	73.4	85.2	99.4	84.2	91.3	96.0	82.7	89.0
MF2010	WEIGHTED AVERAGE	95.3	83.3	89.9	96.6	86.7	92.2	94.6	83.3	89.5	97.5	78.6	89.1	97.3	73.6	86.7	100.5	85.5	93.8

DIVISION		IOWA																	
		COUNCIL BLUFFS			CRESTON			DAVENPORT			DECORAH			DES MOINES			DUBUQUE		
		515			508			527 - 528			521			500 - 503,509			520		
		MAT.	INST.	TOTAL	MAT.	INST.	TOTAL	MAT.	INST.	TOTAL	MAT.	INST.	TOTAL	MAT.	INST.	TOTAL	MAT.	INST.	TOTAL
015433	CONTRACTOR EQUIPMENT		96.1	96.1		99.8	99.8		99.8	99.8		99.8	99.8		101.5	101.5		95.1	95.1
0241, 31 - 34	SITE & INFRASTRUCTURE, DEMOLITION	108.5	92.7	97.6	86.2	96.4	93.2	97.9	99.1	98.7	96.8	96.2	96.4	87.1	100.1	96.1	98.1	93.2	94.7
0310	Concrete Forming & Accessories	84.5	73.5	75.0	85.5	70.2	72.2	99.8	89.9	91.2	93.3	47.4	53.5	101.1	80.1	82.8	85.6	72.4	74.2
0320	Concrete Reinforcing	98.6	79.1	88.5	96.1	83.1	89.4	96.7	97.2	97.0	95.9	82.6	89.1	96.7	83.3	89.8	95.3	86.0	90.5
0330	Cast-in-Place Concrete	113.8	78.4	100.2	108.8	67.2	92.9	105.4	88.9	99.1	106.2	59.5	88.3	91.7	92.9	92.2	107.1	90.0	100.6
03	CONCRETE	106.8	77.1	92.5	102.8	72.5	88.3	102.0	91.5	97.0	101.9	59.7	81.7	94.1	85.8	90.1	100.9	81.8	91.8
04	MASONRY	110.9	77.5	90.7	105.6	83.8	92.4	106.0	87.7	94.9	125.3	72.7	93.4	99.8	84.9	90.8	110.6	78.3	91.0
05	METALS	96.7	90.9	94.7	89.1	92.4	90.2	91.6	102.4	95.3	89.3	90.2	89.6	91.7	94.7	92.8	90.2	94.0	91.5
06	WOOD, PLASTICS & COMPOSITES	84.5	73.0	77.9	86.0	66.9	75.0	103.0	89.0	94.9	94.7	40.1	63.4	102.8	77.8	88.4	86.1	68.6	76.1
07	THERMAL & MOISTURE PROTECTION	106.1	71.1	92.2	106.8	80.6	96.4	106.1	87.7	98.8	105.9	55.8	86.0	107.5	81.7	97.2	106.4	73.6	93.4
08	OPENINGS	95.9	78.2	91.4	108.5	67.5	98.1	96.8	90.8	95.3	94.2	52.6	83.7	96.8	83.8	93.5	95.9	77.7	91.3
0920	Plaster & Gypsum Board	94.3	72.4	78.9	94.3	66.0	74.4	103.3	88.7	93.0	97.8	38.4	56.0	95.4	77.2	82.6	94.7	68.0	75.9
0950, 0980	Ceilings & Acoustic Treatment	104.7	72.4	83.3	104.7	66.0	79.0	108.4	88.7	95.3	104.7	38.4	60.7	105.6	77.2	86.7	104.7	68.0	80.3
0960	Flooring	108.0	85.2	101.5	109.2	35.1	87.9	118.9	87.7	110.0	115.6	49.8	96.7	118.6	85.3	109.0	123.6	80.5	111.2
0970, 0990	Wall Finishes & Painting/Coating	98.4	69.0	80.5	101.4	80.0	88.4	101.4	90.5	94.7	101.4	35.4	61.4	101.4	84.2	91.0	103.3	68.6	82.3
09	FINISHES	103.6	74.9	87.8	102.5	63.8	81.1	109.1	89.4	98.2	106.6	44.8	72.5	106.0	81.0	92.2	108.5	72.6	88.7
COVERS	DIVS. 10 - 14, 25, 28, 41, 43, 44, 46	100.0	84.8	96.8	100.0	77.0	95.2	100.0	93.7	98.7	100.0	85.0	96.9	100.0	91.6	98.3	100.0	89.6	97.8
21, 22, 23	FIRE SUPPRESSION, PLUMBING & HVAC	100.2	76.3	90.2	94.8	82.6	89.7	100.2	91.7	96.7	94.8	75.0	86.5	100.0	83.2	93.0	100.2	78.2	91.0
26, 27, 3370	ELECTRICAL, COMMUNICATIONS & UTIL.	101.7	83.8	92.2	93.8	84.2	88.7	93.8	88.5	91.0	96.0	48.1	70.6	107.0	84.2	94.9	100.2	80.8	89.9
MF2010	WEIGHTED AVERAGE	101.4	80.3	91.9	98.0	80.2	90.0	99.4	92.0	96.1	98.4	66.7	84.3	98.9	86.2	93.2	99.7	81.2	91.4

DIVISION		IOWA																	
		FORT DODGE			MASON CITY			OTTUMWA			SHENANDOAH			SIBLEY			SIOUX CITY		
		505			504			525			516			512			510 - 511		
		MAT.	INST.	TOTAL	MAT.	INST.	TOTAL	MAT.	INST.	TOTAL	MAT.	INST.	TOTAL	MAT.	INST.	TOTAL	MAT.	INST.	TOTAL
015433	CONTRACTOR EQUIPMENT		99.8	99.8		99.8	99.8		95.1	95.1		96.1	96.1		99.8	99.8		99.8	99.8
0241, 31 - 34	SITE & INFRASTRUCTURE, DEMOLITION	93.7	95.1	94.7	93.7	96.3	95.5	99.0	91.3	93.7	106.7	91.7	96.3	108.9	95.0	99.3	110.0	95.9	100.2
0310	Concrete Forming & Accessories	86.0	48.5	53.4	90.0	48.2	53.7	91.6	60.6	64.7	85.9	60.6	64.0	86.3	40.4	46.5	100.2	67.2	71.6
0320	Concrete Reinforcing	96.1	65.4	80.3	95.9	82.5	89.0	95.9	83.9	89.7	98.6	69.7	83.7	98.6	65.3	81.5	96.7	65.7	80.7
0330	Cast-in-Place Concrete	102.1	46.6	80.8	102.1	60.2	86.1	109.8	57.3	89.7	110.0	63.3	92.1	107.8	47.1	84.5	108.5	58.3	89.3
03	CONCRETE	98.0	52.5	76.2	98.3	60.2	80.1	103.7	65.1	85.2	104.1	64.4	85.1	103.0	49.0	77.2	103.4	64.9	84.9
04	MASONRY	103.8	40.9	65.7	117.9	71.5	89.7	106.9	60.7	78.9	110.5	77.5	90.5	131.1	41.2	76.6	103.0	57.6	75.5
05	METALS	89.2	81.7	86.6	89.2	90.6	89.7	89.1	92.1	90.1	95.7	85.5	92.2	89.4	80.4	86.3	91.6	83.2	88.7
06	WOOD, PLASTICS & COMPOSITES	86.5	48.9	65.0	90.6	40.1	61.6	92.5	62.1	75.0	86.1	57.2	69.6	86.8	38.8	59.3	103.0	67.3	82.5
07	THERMAL & MOISTURE PROTECTION	106.2	60.7	88.1	105.7	66.3	90.0	106.5	65.8	90.4	105.4	69.3	91.1	105.7	51.4	84.2	106.1	66.4	90.4
08	OPENINGS	101.1	51.4	88.5	91.7	52.5	81.8	95.8	67.6	88.7	85.6	58.1	78.6	92.0	45.9	80.3	96.8	63.2	88.3
0920	Plaster & Gypsum Board	94.7	47.5	61.5	95.2	38.4	55.2	96.1	61.2	71.5	94.7	56.3	67.6	94.7	37.0	54.1	103.3	66.4	77.3
0950, 0980	Ceilings & Acoustic Treatment	104.7	47.5	66.7	104.7	38.4	60.7	104.7	61.2	75.9	104.7	56.3	72.6	104.7	37.0	59.8	108.4	66.4	80.5
0960	Flooring	111.0	49.8	93.4	113.8	49.8	95.4	127.4	53.0	106.0	108.9	36.2	88.0	110.0	35.9	88.7	119.3	56.7	101.3
0970, 0990	Wall Finishes & Painting/Coating	101.4	63.3	78.3	101.4	32.4	59.5	103.3	82.9	90.9	98.4	69.0	80.5	101.4	63.3	78.3	101.8	63.3	78.5
09	FINISHES	104.6	49.3	74.1	105.5	44.6	71.9	109.9	60.6	82.7	103.8	55.8	77.3	106.4	40.6	70.1	111.0	64.7	85.5
COVERS	DIVS. 10 - 14, 25, 28, 41, 43, 44, 46	100.0	80.7	96.0	100.0	85.4	97.0	100.0	83.1	96.5	100.0	73.7	94.5	100.0	79.7	95.8	100.0	87.4	97.4
21, 22, 23	FIRE SUPPRESSION, PLUMBING & HVAC	94.8	68.4	83.8	94.8	76.1	87.0	94.8	74.2	86.2	94.8	73.8	86.0	94.8	64.2	82.0	100.2	70.9	88.0
26, 27, 3370	ELECTRICAL, COMMUNICATIONS & UTIL.	101.2	45.5	71.6	100.2	60.4	79.0	98.4	69.4	83.0	96.0	83.8	89.5	96.0	45.0	68.9	96.0	76.0	85.4
MF2010	WEIGHTED AVERAGE	97.7	60.6	81.1	97.5	68.9	84.7	98.4	71.9	86.6	97.9	73.6	87.1	98.9	57.3	80.3	100.1	71.9	87.5

DIVISION		IOWA						KANSAS											
		SPENCER			WATERLOO			BELLEVILLE			COLBY			DODGE CITY			EMPORIA		
		513			506 - 507			669			677			678			668		
		MAT.	INST.	TOTAL	MAT.	INST.	TOTAL	MAT.	INST.	TOTAL	MAT.	INST.	TOTAL	MAT.	INST.	TOTAL	MAT.	INST.	TOTAL
015433	CONTRACTOR EQUIPMENT		99.8	99.8		99.8	99.8		102.7	102.7		102.7	102.7		102.7	102.7		101.0	101.0
0241, 31 - 34	SITE & INFRASTRUCTURE, DEMOLITION	109.0	95.0	99.3	98.4	96.3	96.9	116.8	93.7	100.9	114.1	93.7	100.0	116.3	93.3	100.5	107.6	91.5	96.5
0310	Concrete Forming & Accessories	91.5	40.4	47.2	100.6	50.7	57.3	96.9	57.8	63.0	103.1	62.4	67.8	96.4	62.3	66.8	88.7	70.0	72.4
0320	Concrete Reinforcing	98.6	65.3	81.5	96.7	85.8	91.1	104.5	56.4	79.7	107.1	56.4	81.0	104.5	56.5	79.8	103.0	57.0	79.3
0330	Cast-in-Place Concrete	107.8	47.1	84.5	109.4	57.4	89.5	120.8	58.8	97.1	116.2	58.8	94.2	118.2	58.4	95.3	116.8	49.0	90.8
03	CONCRETE	103.3	49.0	77.3	104.1	61.1	83.5	116.9	59.2	89.3	114.1	61.2	88.8	115.3	61.0	89.3	109.6	61.4	86.5
04	MASONRY	131.1	41.2	76.6	104.5	77.5	88.2	97.9	59.7	74.8	99.3	59.7	75.3	106.3	62.3	79.6	103.6	70.2	83.4
05	METALS	89.3	80.4	86.3	91.6	93.5	92.2	93.2	79.0	88.3	93.5	79.0	88.5	95.1	77.9	89.2	92.9	80.3	88.5
06	WOOD, PLASTICS & COMPOSITES	92.3	38.8	61.6	103.5	40.1	67.1	92.3	56.6	71.8	99.2	62.8	78.3	91.7	62.8	75.1	84.3	72.4	77.5
07	THERMAL & MOISTURE PROTECTION	106.6	51.4	84.7	105.9	73.5	93.0	99.9	60.2	84.1	99.9	60.8	84.4	99.9	61.8	84.7	98.5	75.6	89.4
08	OPENINGS	105.2	45.9	90.2	92.6	55.2	83.1	96.6	51.6	85.2	96.7	55.0	86.2	96.7	55.0	86.1	94.3	60.2	85.6
0920	Plaster & Gypsum Board	96.1	37.0	54.5	103.3	38.4	57.6	96.5	55.2	67.4	101.0	61.6	73.3	95.6	61.6	71.7	92.5	71.5	77.7
0950, 0980	Ceilings & Acoustic Treatment	104.7	37.0	59.8	108.4	38.4	62.0	89.5	55.2	66.7	89.5	61.6	71.0	89.5	61.6	71.0	89.5	71.5	77.5
0960	Flooring	113.3	35.9	91.1	120.7	79.7	108.9	101.4	41.4	84.2	105.1	41.4	86.8	101.2	41.4	84.0	96.3	38.5	79.7
0970, 0990	Wall Finishes & Painting/Coating	101.4	63.3	78.3	101.8	83.7	90.8	96.2	41.9	63.2	96.2	41.9	63.2	96.2	41.9	63.2	96.2	41.9	63.2
09	FINISHES	107.5	40.6	70.6	109.6	57.4	80.8	98.9	52.8	73.5	100.4	56.5	76.2	98.7	56.5	75.4	95.9	61.5	76.9
COVERS	DIVS. 10 - 14, 25, 28, 41, 43, 44, 46	100.0	79.7	95.8	100.0	86.8	97.3	100.0	44.9	88.6	100.0	45.7	88.7	100.0	45.7	88.7	100.0	44.6	88.5
21, 22, 23	FIRE SUPPRESSION, PLUMBING & HVAC	94.8	64.2	82.0	100.2	83.1	93.1	94.7	72.4	85.4	94.7	69.0	84.0	100.1	69.0	87.1	94.7	72.7	85.5
26, 27, 3370	ELECTRICAL, COMMUNICATIONS & UTIL.	97.9	45.0	69.7	96.0	60.4	77.1	98.8	68.7	82.8	98.8	68.7	82.8	95.3	75.3	84.7	95.7	75.2	84.8
MF2010	WEIGHTED AVERAGE	100.6	57.3	81.3	99.4	73.4	87.8	99.6	66.5	84.8	99.5	66.8	84.9	101.0	67.8	86.2	97.8	70.8	85.7

DIVISION		KANSAS																	
		FORT SCOTT			HAYS			HUTCHINSON			INDEPENDENCE			KANSAS CITY			LIBERAL		
		667			676			675			673			660 - 662			679		
		MAT.	INST.	TOTAL	MAT.	INST.	TOTAL	MAT.	INST.	TOTAL	MAT.	INST.	TOTAL	MAT.	INST.	TOTAL	MAT.	INST.	TOTAL
015433	CONTRACTOR EQUIPMENT		101.9	101.9		102.7	102.7		102.7	102.7		102.7	102.7		99.5	99.5		102.7	102.7
0241, 31 - 34	SITE & INFRASTRUCTURE, DEMOLITION	103.9	92.0	95.7	120.5	93.7	102.0	96.5	93.8	94.7	117.9	93.9	101.3	96.3	92.2	93.4	120.4	93.8	102.1
0310	Concrete Forming & Accessories	103.7	82.5	85.3	100.6	62.4	67.5	91.1	62.2	66.1	111.5	71.0	76.3	100.1	99.3	99.4	96.9	62.3	66.8
0320	Concrete Reinforcing	102.3	90.4	96.2	104.5	56.4	79.7	104.5	56.4	79.7	103.9	61.4	82.0	99.3	95.8	97.5	106.0	56.4	80.5
0330	Cast-in-Place Concrete	108.3	57.5	88.8	91.6	58.8	79.0	84.8	48.4	70.9	118.7	48.6	91.9	92.8	98.5	95.0	91.6	48.4	75.1
03	CONCRETE	104.5	76.1	90.9	104.5	61.2	83.8	88.0	57.6	73.5	116.5	62.5	90.7	96.0	98.6	97.2	106.5	57.6	83.1
04	MASONRY	104.3	60.7	77.9	107.0	59.7	78.3	98.6	61.9	76.4	96.4	64.6	77.1	106.8	99.6	102.5	104.1	61.9	78.5
05	METALS	92.8	92.0	92.5	93.1	79.0	88.2	92.9	77.8	87.7	92.8	80.4	88.6	100.6	99.9	100.4	93.4	77.8	88.0
06	WOOD, PLASTICS & COMPOSITES	101.2	90.8	95.2	96.5	62.8	77.2	86.8	62.8	73.0	108.8	73.9	88.8	97.4	99.8	98.8	92.3	62.8	75.4
07	THERMAL & MOISTURE PROTECTION	99.2	72.1	88.5	100.2	60.8	84.5	98.9	61.7	84.1	99.9	74.2	89.7	98.6	99.3	98.9	100.3	61.7	85.0
08	OPENINGS	94.3	82.2	91.2	96.6	55.0	86.1	96.6	55.0	86.1	94.3	62.2	86.1	95.7	92.6	94.9	96.7	55.0	86.1
0920	Plaster & Gypsum Board	99.2	90.4	93.0	99.2	61.6	72.7	94.3	61.6	71.3	107.6	73.0	83.2	91.2	99.7	97.2	96.5	61.6	71.9
0950, 0980	Ceilings & Acoustic Treatment	89.5	90.4	90.1	89.5	61.6	71.0	89.5	61.6	71.0	89.5	73.0	78.6	89.5	99.7	96.3	89.5	61.6	71.0
0960	Flooring	112.4	41.6	92.1	103.9	41.4	85.9	98.1	41.4	81.8	109.8	41.4	90.1	90.0	99.2	92.7	101.4	41.4	84.2
0970, 0990	Wall Finishes & Painting/Coating	98.1	41.9	64.0	96.2	41.9	63.2	96.2	41.9	63.2	96.2	41.9	63.2	104.0	72.9	85.1	96.2	41.9	63.2
09	FINISHES	101.7	72.1	85.3	100.4	56.5	76.2	95.9	56.5	74.2	102.7	63.0	80.8	96.2	95.6	95.9	99.6	56.5	75.8
COVERS	DIVS. 10 - 14, 25, 28, 41, 43, 44, 46	100.0	51.4	89.9	100.0	45.7	88.7	100.0	45.7	88.7	100.0	47.0	89.0	100.0	68.8	93.5	100.0	45.7	88.7
21, 22, 23	FIRE SUPPRESSION, PLUMBING & HVAC	94.7	70.7	84.7	94.7	69.0	84.0	94.7	69.0	84.0	94.7	71.0	84.8	100.0	101.3	100.6	94.7	69.0	84.0
26, 27, 3370	ELECTRICAL, COMMUNICATIONS & UTIL.	94.9	75.2	84.4	97.6	68.7	82.3	92.0	68.7	79.6	94.5	75.1	84.2	101.4	98.0	99.6	95.3	71.1	82.5
MF2010	WEIGHTED AVERAGE	97.6	75.2	87.6	98.7	66.8	84.4	94.5	66.4	82.0	99.2	70.6	86.4	99.2	97.3	98.3	98.5	66.8	84.3

DIVISION		KANSAS									KENTUCKY								
		SALINA			TOPEKA			WICHITA			ASHLAND			BOWLING GREEN			CAMPTON		
		674			664 - 666			670 - 672			411 - 412			421 - 422			413 - 414		
		MAT.	INST.	TOTAL	MAT.	INST.	TOTAL	MAT.	INST.	TOTAL	MAT.	INST.	TOTAL	MAT.	INST.	TOTAL	MAT.	INST.	TOTAL
015433	CONTRACTOR EQUIPMENT		102.7	102.7		101.0	101.0		102.7	102.7		98.4	98.4		95.8	95.8		102.3	102.3
0241, 31 - 34	SITE & INFRASTRUCTURE, DEMOLITION	105.8	93.8	97.5	97.8	90.5	92.8	97.6	92.4	94.0	106.2	87.0	93.0	69.6	98.8	89.8	78.9	99.0	92.8
0310	Concrete Forming & Accessories	93.1	58.2	62.8	100.3	43.5	51.1	99.4	47.4	54.3	88.7	102.1	100.4	84.6	81.1	81.5	89.3	48.6	54.0
0320	Concrete Reinforcing	103.9	65.3	84.0	96.5	94.7	95.6	96.5	77.0	86.4	92.1	100.7	96.6	89.4	81.4	85.2	90.2	65.1	77.3
0330	Cast-in-Place Concrete	102.6	52.1	83.2	92.4	48.1	75.4	90.2	51.2	75.3	86.3	102.2	92.4	85.6	101.7	91.8	95.5	60.0	81.9
03	CONCRETE	102.2	58.7	81.4	93.3	56.4	75.7	92.2	55.7	74.7	94.5	102.4	98.3	92.7	88.3	90.6	96.8	56.6	77.5
04	MASONRY	121.3	59.7	84.0	100.6	58.0	74.8	94.4	54.3	70.1	89.9	105.3	99.3	91.3	74.1	80.9	89.8	40.7	60.0
05	METALS	94.9	82.8	90.7	100.2	95.8	98.7	100.2	84.7	94.9	90.9	108.2	96.8	92.5	83.2	89.3	91.7	74.1	85.6
06	WOOD, PLASTICS & COMPOSITES	88.2	56.6	70.1	93.5	40.3	63.0	95.0	46.1	66.9	80.5	101.5	92.5	94.6	80.5	86.5	93.2	47.1	66.8
07	THERMAL & MOISTURE PROTECTION	99.4	60.2	83.8	100.4	64.2	86.1	98.8	55.4	81.5	90.4	98.8	93.7	85.8	82.5	84.5	98.5	51.5	79.8
08	OPENINGS	96.6	52.5	85.4	96.9	56.8	86.8	96.9	54.1	86.1	95.7	96.7	95.9	96.8	76.2	91.6	98.5	56.9	87.9
0920	Plaster & Gypsum Board	94.3	55.2	66.7	93.4	38.5	54.7	93.2	44.4	58.8	60.6	101.6	89.5	82.8	80.2	80.9	82.8	44.9	56.1
0950, 0980	Ceilings & Acoustic Treatment	89.5	55.2	66.7	90.4	38.5	56.0	89.5	44.4	59.6	80.4	101.6	94.5	85.6	80.2	82.0	85.6	44.9	58.6
0960	Flooring	99.4	65.4	89.6	102.8	77.8	95.6	101.6	66.6	91.5	78.5	105.9	86.4	94.6	80.1	90.4	96.1	43.6	81.0
0970, 0990	Wall Finishes & Painting/Coating	96.2	41.9	63.2	96.2	56.6	72.1	96.2	51.6	69.1	99.8	90.7	94.3	97.6	71.3	81.7	97.6	37.2	61.0
09	FINISHES	97.2	57.5	75.3	97.6	49.5	71.1	97.2	50.8	71.6	82.7	101.9	93.3	93.1	79.9	85.8	93.7	45.3	67.0
COVERS	DIVS. 10 - 14, 25, 28, 41, 43, 44, 46	100.0	56.0	90.9	100.0	52.7	90.2	100.0	52.0	90.0	100.0	102.6	100.6	100.0	63.2	92.3	100.0	54.5	90.5
21, 22, 23	FIRE SUPPRESSION, PLUMBING & HVAC	100.1	69.1	87.2	100.1	58.5	82.8	99.9	65.1	85.4	94.4	90.0	92.6	99.9	83.0	92.9	94.6	66.3	82.8
26, 27, 3370	ELECTRICAL, COMMUNICATIONS & UTIL.	95.0	75.3	84.5	105.4	75.2	89.3	102.1	75.3	87.8	91.8	99.4	95.8	94.7	82.5	88.2	91.7	71.8	81.1
MF2010	WEIGHTED AVERAGE	99.7	68.0	85.5	99.2	65.2	84.0	98.3	65.1	83.4	93.2	98.5	95.6	94.7	82.8	89.4	94.4	62.6	80.2

DIVISION		KENTUCKY																	
		CORBIN			COVINGTON			ELIZABETHTOWN			FRANKFORT			HAZARD			HENDERSON		
		407 - 409			410			427			406			417 - 418			424		
		MAT.	INST.	TOTAL	MAT.	INST.	TOTAL	MAT.	INST.	TOTAL	MAT.	INST.	TOTAL	MAT.	INST.	TOTAL	MAT.	INST.	TOTAL
015433	CONTRACTOR EQUIPMENT		102.3	102.3		106.1	106.1		95.8	95.8		102.3	102.3		102.3	102.3		117.4	117.4
0241, 31 - 34	SITE & INFRASTRUCTURE, DEMOLITION	76.4	99.0	92.0	74.7	114.2	102.0	65.0	98.7	88.2	80.9	99.9	94.0	76.9	100.6	93.3	73.6	125.7	109.5
0310	Concrete Forming & Accessories	86.5	48.2	53.3	85.3	98.4	96.7	80.2	80.1	80.1	96.6	59.0	64.0	86.5	42.0	47.9	92.3	77.9	79.8
0320	Concrete Reinforcing	89.8	65.0	77.1	88.8	97.3	93.2	89.8	91.5	90.7	91.1	79.7	85.2	90.7	65.8	77.8	89.5	88.7	89.1
0330	Cast-in-Place Concrete	90.4	59.8	78.7	91.4	96.3	93.3	77.4	75.3	76.6	87.4	66.9	79.6	91.9	87.6	90.2	75.7	90.6	81.4
03	CONCRETE	92.0	56.3	74.9	95.1	97.4	96.2	85.1	80.8	83.0	91.4	66.3	79.4	93.6	63.1	79.0	89.8	84.5	87.2
04	MASONRY	89.5	41.4	60.4	103.2	108.8	106.6	76.0	80.1	78.5	87.1	73.2	78.6	86.7	50.1	64.6	95.3	93.1	94.0
05	METALS	91.6	73.7	85.4	89.9	94.1	91.3	91.6	87.9	90.3	98.4	83.3	93.2	91.7	73.9	85.6	83.6	89.0	85.4
06	WOOD, PLASTICS & COMPOSITES	90.2	47.1	65.5	91.1	91.4	91.3	89.7	80.5	84.4	102.0	56.9	76.2	90.2	34.6	58.3	96.9	73.7	83.6
07	THERMAL & MOISTURE PROTECTION	98.3	52.6	80.1	97.6	99.6	98.4	85.5	79.3	83.0	98.1	70.0	87.0	98.4	51.0	79.5	97.8	92.4	95.7
08	OPENINGS	97.7	49.6	85.5	99.9	90.2	97.4	96.9	83.8	93.6	97.0	67.8	89.6	98.9	40.6	84.1	94.9	79.3	90.9
0920	Plaster & Gypsum Board	81.4	44.9	55.7	70.1	91.7	85.3	81.0	80.2	80.4	93.1	55.0	66.3	81.4	32.0	46.6	85.2	72.3	76.1
0950, 0980	Ceilings & Acoustic Treatment	85.6	44.9	58.6	88.5	91.7	90.6	85.6	80.2	82.0	87.5	55.0	66.0	85.6	32.0	50.1	80.0	72.3	74.9
0960	Flooring	94.4	43.6	79.8	71.0	88.9	76.1	92.0	70.8	85.9	98.9	57.5	87.0	94.4	44.3	80.0	97.8	90.9	95.8
0970, 0990	Wall Finishes & Painting/Coating	97.6	37.2	61.0	93.2	83.1	87.1	97.6	78.7	86.1	97.6	50.6	69.1	97.6	30.5	56.9	97.6	95.8	96.5
09	FINISHES	92.8	45.3	66.6	85.0	94.9	90.5	91.7	78.0	84.1	96.2	56.7	74.5	92.9	39.7	63.6	92.5	81.6	86.5
COVERS	DIVS. 10 - 14, 25, 28, 41, 43, 44, 46	100.0	49.1	89.4	100.0	105.0	101.0	100.0	93.4	98.6	100.0	67.0	93.1	100.0	56.2	90.9	100.0	70.6	93.9
21, 22, 23	FIRE SUPPRESSION, PLUMBING & HVAC	94.5	65.8	82.5	95.0	93.0	94.2	94.9	81.5	89.3	100.0	67.3	86.4	94.6	43.1	73.2	94.9	76.3	87.1
26, 27, 3370	ELECTRICAL, COMMUNICATIONS & UTIL.	92.0	71.8	81.2	96.1	80.5	87.8	91.5	82.5	86.7	103.1	94.6	98.6	91.7	54.5	71.9	94.0	82.4	87.8
MF2010	WEIGHTED AVERAGE	93.5	62.0	79.4	94.4	96.0	95.1	91.0	83.4	87.6	97.1	74.3	86.9	93.7	55.7	76.7	92.3	86.3	89.6

DIVISION		KENTUCKY																	
		LEXINGTON			LOUISVILLE			OWENSBORO			PADUCAH			PIKEVILLE			SOMERSET		
		403 - 405			400 - 402			423			420			415 - 416			425 - 426		
		MAT.	INST.	TOTAL	MAT.	INST.	TOTAL	MAT.	INST.	TOTAL	MAT.	INST.	TOTAL	MAT.	INST.	TOTAL	MAT.	INST.	TOTAL
015433	**CONTRACTOR EQUIPMENT**		102.3	102.3		95.8	95.8		117.4	117.4		117.4	117.4		98.4	98.4		102.3	102.3
0241, 31 - 34	**SITE & INFRASTRUCTURE, DEMOLITION**	78.0	101.1	93.9	67.8	98.7	89.1	82.2	125.3	112.0	76.2	125.1	109.9	116.2	85.6	95.1	70.7	99.5	90.6
0310	Concrete Forming & Accessories	97.6	74.2	77.3	93.0	80.1	81.8	90.9	69.8	72.6	88.8	80.3	81.4	96.8	67.4	71.3	87.2	46.7	52.1
0320	Concrete Reinforcing	98.6	91.0	94.7	98.6	91.5	95.0	89.5	90.3	89.9	90.0	85.5	87.7	92.6	89.6	91.1	89.8	78.7	84.1
0330	Cast-in-Place Concrete	92.5	84.1	89.3	94.4	75.3	87.1	88.3	97.4	91.8	80.6	88.7	83.7	94.9	92.4	94.0	75.7	101.9	85.7
03	**CONCRETE**	95.1	81.1	88.4	95.8	80.8	88.6	101.5	83.5	92.9	94.1	84.3	89.4	107.6	81.4	95.1	80.4	72.5	76.6
04	**MASONRY**	88.3	63.7	73.4	89.4	80.1	83.8	88.2	69.3	76.7	91.2	90.5	90.8	88.0	103.0	97.1	82.2	60.8	69.2
05	**METALS**	94.0	88.3	92.0	100.7	87.9	96.3	85.1	88.3	86.2	82.0	86.9	83.7	90.8	99.1	93.6	91.6	80.3	87.7
06	**WOOD, PLASTICS & COMPOSITES**	104.6	74.6	87.4	102.0	80.5	89.7	95.0	66.2	78.5	92.7	79.4	85.1	89.4	59.7	72.3	90.8	39.9	61.6
07	**THERMAL & MOISTURE PROTECTION**	98.5	84.6	93.0	92.5	79.3	87.3	98.2	74.7	88.8	97.9	82.4	91.7	91.0	69.5	82.5	97.8	57.0	81.6
08	**OPENINGS**	97.8	77.9	92.8	97.8	83.8	94.3	94.9	76.7	90.3	94.0	78.5	90.1	96.5	68.4	89.4	97.7	56.3	87.3
0920	Plaster & Gypsum Board	91.4	73.2	78.6	84.4	80.2	81.4	83.9	64.5	70.2	82.5	78.2	79.5	64.2	58.5	60.2	81.9	37.5	50.6
0950, 0980	Ceilings & Acoustic Treatment	91.3	73.2	79.3	87.5	80.2	82.6	80.0	64.5	69.7	80.0	78.2	78.8	80.4	58.5	65.9	85.6	37.5	53.7
0960	Flooring	99.3	58.6	87.6	97.8	70.8	90.0	97.0	80.1	92.2	95.8	62.7	86.3	82.9	51.8	74.0	94.6	43.6	79.9
0970, 0990	Wall Finishes & Painting/Coating	97.6	67.6	79.4	97.6	78.7	86.1	97.6	95.6	96.4	97.6	80.6	87.3	99.8	87.4	92.3	97.6	51.3	69.5
09	**FINISHES**	96.7	70.0	82.0	94.1	78.0	85.2	92.7	73.6	82.2	91.8	77.0	83.6	85.3	65.4	74.3	92.3	44.8	66.1
COVERS	**DIVS. 10 - 14, 25, 28, 41, 43, 44, 46**	100.0	93.4	98.6	100.0	93.4	98.6	100.0	79.8	95.8	100.0	67.4	93.2	100.0	59.0	91.5	100.0	58.3	91.3
21, 22, 23	**FIRE SUPPRESSION, PLUMBING & HVAC**	99.9	62.9	84.5	100.0	81.5	92.3	99.9	66.6	86.0	94.9	82.0	89.5	94.4	76.1	86.8	94.9	62.4	81.4
26, 27, 3370	**ELECTRICAL, COMMUNICATIONS & UTIL.**	95.2	71.3	82.5	106.2	82.5	93.6	94.1	82.4	87.9	96.8	82.3	89.0	95.2	79.6	86.9	92.2	71.2	81.0
MF2010	**WEIGHTED AVERAGE**	96.3	75.3	86.9	97.7	83.4	91.3	95.1	80.2	88.4	92.6	86.0	89.7	95.7	80.3	88.8	91.6	66.6	80.4

DIVISION		LOUISIANA																	
		ALEXANDRIA			BATON ROUGE			HAMMOND			LAFAYETTE			LAKE CHARLES			MONROE		
		713 - 714			707 - 708			704			705			706			712		
		MAT.	INST.	TOTAL	MAT.	INST.	TOTAL	MAT.	INST.	TOTAL	MAT.	INST.	TOTAL	MAT.	INST.	TOTAL	MAT.	INST.	TOTAL
015433	**CONTRACTOR EQUIPMENT**		88.5	88.5		88.3	88.3		89.0	89.0		89.0	89.0		88.3	88.3		88.5	88.5
0241, 31 - 34	**SITE & INFRASTRUCTURE, DEMOLITION**	100.0	85.4	90.0	108.0	85.1	92.2	109.7	86.4	93.6	110.3	86.3	93.7	110.9	84.9	93.0	100.0	84.8	89.5
0310	Concrete Forming & Accessories	82.6	36.8	42.9	94.2	63.1	67.2	79.1	49.3	53.2	95.6	54.1	59.6	97.1	58.4	63.5	82.1	37.4	43.3
0320	Concrete Reinforcing	102.0	58.3	79.5	100.6	60.1	79.7	99.2	61.9	80.0	100.6	59.7	79.5	100.6	59.1	79.2	100.8	58.1	78.8
0330	Cast-in-Place Concrete	97.2	39.4	75.1	95.1	58.4	81.0	97.9	45.2	77.7	97.4	45.7	77.6	102.5	67.6	89.2	97.2	56.6	81.6
03	**CONCRETE**	94.0	43.1	69.6	92.4	61.4	77.6	92.7	51.2	72.8	93.7	53.0	74.2	96.3	62.3	80.0	93.8	49.1	72.4
04	**MASONRY**	117.5	48.8	75.9	105.9	50.6	72.4	105.5	52.3	73.3	105.5	43.9	68.2	104.4	57.9	76.2	113.0	46.8	72.9
05	**METALS**	90.2	71.7	83.9	102.7	70.0	91.5	98.5	70.5	88.8	97.5	69.3	87.8	97.5	70.2	88.1	90.2	69.7	83.1
06	**WOOD, PLASTICS & COMPOSITES**	86.3	35.8	57.3	100.7	69.3	82.7	85.8	50.9	65.8	107.1	58.2	79.0	105.2	62.4	80.6	85.7	36.1	57.2
07	**THERMAL & MOISTURE PROTECTION**	97.0	47.9	77.5	95.7	57.2	80.4	97.5	56.8	81.3	98.0	52.0	79.7	97.3	59.2	82.1	97.0	50.7	78.6
08	**OPENINGS**	99.0	42.3	84.6	102.2	60.8	91.7	97.7	55.4	87.0	102.2	54.8	90.2	102.2	57.9	91.0	98.9	46.3	85.6
0920	Plaster & Gypsum Board	74.5	34.4	46.2	94.6	68.7	76.4	88.1	49.8	61.2	98.2	57.3	69.4	98.2	61.6	72.4	74.0	34.7	46.4
0950, 0980	Ceilings & Acoustic Treatment	89.4	34.4	52.9	103.7	68.7	80.5	100.7	49.8	67.0	98.9	57.3	71.3	99.9	61.6	74.5	89.4	34.7	53.2
0960	Flooring	100.0	64.7	89.9	103.9	64.7	92.6	94.8	48.9	81.6	104.1	36.6	84.7	104.1	77.0	96.3	99.6	60.3	88.3
0970, 0990	Wall Finishes & Painting/Coating	97.1	34.3	59.0	100.7	39.3	63.4	100.7	52.0	71.1	100.7	39.3	63.4	100.7	36.2	61.6	97.1	41.0	63.1
09	**FINISHES**	89.8	40.8	62.8	98.9	61.7	78.4	94.8	49.4	69.8	98.4	49.2	71.2	98.6	59.8	77.2	89.6	41.0	62.8
COVERS	**DIVS. 10 - 14, 25, 28, 41, 43, 44, 46**	100.0	52.7	90.2	100.0	79.2	95.7	100.0	50.6	89.7	100.0	77.6	95.4	100.0	78.5	95.5	100.0	48.3	89.3
21, 22, 23	**FIRE SUPPRESSION, PLUMBING & HVAC**	100.2	52.3	80.3	100.0	60.4	83.5	94.7	46.9	74.8	100.1	60.6	83.7	100.1	61.1	83.9	100.2	51.3	79.8
26, 27, 3370	**ELECTRICAL, COMMUNICATIONS & UTIL.**	95.2	57.7	75.3	114.6	62.7	87.0	100.3	57.8	77.7	101.7	68.4	84.0	101.2	68.4	83.8	97.3	58.9	76.9
MF2010	**WEIGHTED AVERAGE**	97.2	53.8	77.8	101.5	63.6	84.5	97.4	56.1	78.9	99.7	60.5	82.2	100.0	65.0	84.3	97.2	54.3	78.0

DIVISION		LOUISIANA									MAINE								
		NEW ORLEANS			SHREVEPORT			THIBODAUX			AUGUSTA			BANGOR			BATH		
		700 - 701			710 - 711			703			043			044			045		
		MAT.	INST.	TOTAL	MAT.	INST.	TOTAL	MAT.	INST.	TOTAL	MAT.	INST.	TOTAL	MAT.	INST.	TOTAL	MAT.	INST.	TOTAL
015433	**CONTRACTOR EQUIPMENT**		89.2	89.2		88.5	88.5		89.0	89.0		100.3	100.3		100.3	100.3		100.3	100.3
0241, 31 - 34	**SITE & INFRASTRUCTURE, DEMOLITION**	113.7	88.0	95.9	102.6	84.7	90.2	112.4	86.3	94.4	82.5	100.6	95.0	84.7	98.4	94.1	83.1	100.6	95.2
0310	Concrete Forming & Accessories	98.9	64.1	68.7	96.7	39.5	47.1	91.7	62.9	66.7	95.2	99.3	98.7	90.8	89.7	89.8	87.0	99.2	97.6
0320	Concrete Reinforcing	100.6	62.2	80.8	100.4	58.6	78.9	99.2	61.9	80.0	88.3	99.8	94.3	88.0	99.9	94.1	87.0	99.7	93.5
0330	Cast-in-Place Concrete	101.5	71.0	89.8	96.7	45.3	77.0	105.1	48.4	83.4	89.7	62.6	79.3	78.6	107.0	89.5	78.6	62.5	72.5
03	**CONCRETE**	95.9	66.7	81.9	93.8	46.2	71.0	98.0	58.3	79.0	105.1	86.2	96.1	98.8	96.8	97.9	98.9	86.1	92.8
04	**MASONRY**	105.7	61.8	79.1	102.7	48.4	69.8	133.7	49.5	82.7	99.2	61.2	76.2	113.7	61.0	81.7	120.3	58.1	82.6
05	**METALS**	110.6	73.9	98.0	87.0	69.7	81.1	98.5	70.7	88.9	87.4	83.9	86.2	87.0	80.0	84.6	85.4	83.6	84.8
06	**WOOD, PLASTICS & COMPOSITES**	101.6	64.1	80.1	101.9	39.2	65.9	94.2	68.5	79.5	96.7	110.4	104.6	91.6	96.8	94.6	87.1	110.4	100.5
07	**THERMAL & MOISTURE PROTECTION**	97.7	62.2	83.6	96.2	49.6	77.7	97.7	55.2	80.8	96.1	64.8	83.7	96.6	66.2	84.6	96.6	64.0	83.7
08	**OPENINGS**	103.0	64.9	93.3	96.7	44.4	83.5	103.2	64.9	93.5	102.3	90.0	99.2	102.3	82.4	97.2	102.2	89.7	99.1
0920	Plaster & Gypsum Board	96.4	63.4	73.2	81.4	37.9	50.8	92.1	67.9	75.1	96.8	110.0	106.1	94.2	96.0	95.5	91.5	110.0	104.6
0950, 0980	Ceilings & Acoustic Treatment	98.9	63.4	75.4	90.3	37.9	55.6	100.7	67.9	79.0	91.9	110.0	103.9	91.9	96.0	94.7	89.8	110.0	103.2
0960	Flooring	104.5	64.7	93.1	108.4	64.7	95.9	101.2	46.6	85.5	95.8	58.4	85.1	93.8	64.9	85.5	92.0	58.4	82.3
0970, 0990	Wall Finishes & Painting/Coating	102.1	65.9	80.1	97.1	34.3	59.0	102.1	52.0	71.7	92.3	37.1	58.8	92.3	38.6	59.7	92.3	37.2	58.8
09	**FINISHES**	98.8	64.3	79.8	93.5	42.8	65.6	97.5	59.3	76.5	96.4	86.8	91.1	95.8	80.7	87.5	94.5	86.8	90.2
COVERS	**DIVS. 10 - 14, 25, 28, 41, 43, 44, 46**	100.0	82.5	96.4	100.0	67.2	93.2	100.0	79.7	95.8	100.0	96.7	99.3	100.0	106.3	101.3	100.0	96.7	99.3
21, 22, 23	**FIRE SUPPRESSION, PLUMBING & HVAC**	100.0	66.0	85.8	100.0	56.3	81.8	94.7	61.0	80.7	100.0	66.8	86.2	100.2	62.7	84.6	94.8	66.8	83.2
26, 27, 3370	**ELECTRICAL, COMMUNICATIONS & UTIL.**	101.7	72.1	86.0	103.0	68.0	84.4	98.7	70.8	83.9	99.0	79.9	88.8	98.6	79.9	88.7	96.4	79.9	87.6
MF2010	**WEIGHTED AVERAGE**	102.0	69.2	87.3	96.9	57.1	79.1	100.3	64.3	84.2	97.9	80.0	89.9	97.9	79.0	89.4	96.3	79.6	88.8

	DIVISION	MAINE																	
		HOULTON			KITTERY			LEWISTON			MACHIAS			PORTLAND			ROCKLAND		
		047			039			042			046			040 - 041			048		
		MAT.	INST.	TOTAL	MAT.	INST.	TOTAL	MAT.	INST.	TOTAL	MAT.	INST.	TOTAL	MAT.	INST.	TOTAL	MAT.	INST.	TOTAL
015433	CONTRACTOR EQUIPMENT		100.3	100.3		100.3	100.3		100.3	100.3		100.3	100.3		100.3	100.3		100.3	100.3
0241, 31 - 34	SITE & INFRASTRUCTURE, DEMOLITION	84.8	100.5	95.6	79.3	100.7	94.1	81.9	98.4	93.3	84.2	100.5	95.4	80.2	98.4	92.7	80.7	100.5	94.4
0310	Concrete Forming & Accessories	93.8	98.3	97.7	87.5	99.7	98.1	95.5	89.7	90.5	91.3	98.3	97.4	94.4	89.7	90.3	92.2	98.7	97.8
0320	Concrete Reinforcing	88.0	99.8	94.1	86.5	99.8	93.3	108.1	99.9	103.9	88.0	99.8	94.1	108.1	99.9	103.9	88.0	99.9	94.1
0330	Cast-in-Place Concrete	78.6	60.7	71.8	78.6	63.5	72.8	80.2	107.0	90.5	78.6	60.8	71.8	99.5	107.0	102.3	80.3	61.9	73.2
03	CONCRETE	99.8	85.2	92.8	94.0	86.7	90.5	98.5	96.8	97.7	99.3	85.2	92.5	108.1	96.8	102.7	96.1	85.7	91.1
04	MASONRY	97.0	59.2	74.1	110.3	60.8	80.3	97.6	61.0	75.4	97.0	59.5	74.3	97.7	61.0	75.5	91.4	61.0	73.0
05	METALS	85.6	83.8	85.0	85.3	83.8	84.8	90.1	80.0	86.6	85.6	83.7	85.0	91.4	80.0	87.5	85.5	83.9	84.9
06	WOOD, PLASTICS & COMPOSITES	95.5	110.4	104.0	87.5	110.4	100.6	97.4	96.8	97.1	92.5	110.4	102.8	96.6	96.8	96.7	93.4	110.4	103.1
07	THERMAL & MOISTURE PROTECTION	96.7	62.3	83.0	96.3	65.0	83.9	96.5	66.2	84.5	96.7	62.3	83.0	99.6	66.2	86.3	96.4	63.2	83.2
08	OPENINGS	102.3	88.0	98.7	102.2	89.3	98.9	105.6	82.4	99.7	102.3	88.0	98.7	105.6	82.4	99.7	102.3	88.0	98.7
0920	Plaster & Gypsum Board	96.8	110.0	106.1	91.5	110.0	104.6	100.2	96.0	97.3	95.0	110.0	105.6	101.0	96.0	97.5	95.0	110.0	105.6
0950, 0980	Ceilings & Acoustic Treatment	89.8	110.0	103.2	89.8	110.0	103.2	99.4	96.0	97.2	89.8	110.0	103.2	96.6	96.0	96.2	89.8	110.0	103.2
0960	Flooring	94.9	54.7	83.3	92.1	60.9	83.1	96.5	64.9	87.4	94.1	54.6	82.8	95.8	64.9	87.0	94.3	54.7	82.9
0970, 0990	Wall Finishes & Painting/Coating	92.3	38.6	59.7	92.3	37.2	58.8	92.3	38.6	59.7	92.3	38.6	59.7	92.3	38.6	59.7	92.3	38.6	59.7
09	FINISHES	96.0	85.8	90.4	94.1	87.5	90.5	98.5	80.7	88.7	95.5	85.8	90.2	97.4	80.7	88.2	95.2	86.0	90.1
COVERS	DIVS. 10 - 14, 25, 28, 41, 43, 44, 46	100.0	96.0	99.2	100.0	98.4	99.7	100.0	106.3	101.3	100.0	90.8	98.1	100.0	106.3	101.3	100.0	96.3	99.2
21, 22, 23	FIRE SUPPRESSION, PLUMBING & HVAC	94.8	65.7	82.7	94.8	67.4	83.4	100.2	62.7	84.6	94.8	65.8	82.7	100.0	62.7	84.5	94.8	66.2	82.9
26, 27, 3370	ELECTRICAL, COMMUNICATIONS & UTIL.	100.9	79.9	89.7	96.4	79.9	87.6	100.9	79.9	89.7	100.9	79.9	89.7	104.5	79.9	91.4	100.8	79.9	89.7
MF2010	WEIGHTED AVERAGE	95.9	79.1	88.4	95.0	80.2	88.4	98.2	79.0	89.6	95.8	78.9	88.2	99.8	79.0	90.5	94.9	79.5	88.0

	DIVISION	MAINE			MARYLAND														
		WATERVILLE			ANNAPOLIS			BALTIMORE			COLLEGE PARK			CUMBERLAND			EASTON		
		049			214			210 - 212			207 - 208			215			216		
		MAT.	INST.	TOTAL	MAT.	INST.	TOTAL	MAT.	INST.	TOTAL	MAT.	INST.	TOTAL	MAT.	INST.	TOTAL	MAT.	INST.	TOTAL
015433	CONTRACTOR EQUIPMENT		100.3	100.3		99.1	99.1		102.6	102.6		103.5	103.5		99.1	99.1		99.1	99.1
0241, 31 - 34	SITE & INFRASTRUCTURE, DEMOLITION	84.8	100.6	95.7	98.4	90.4	92.9	98.8	94.4	95.8	111.1	91.4	97.5	90.3	89.7	89.9	97.1	85.9	89.4
0310	Concrete Forming & Accessories	86.5	99.3	97.6	92.5	51.6	57.0	95.2	76.8	79.2	86.7	72.4	74.3	86.6	78.7	79.7	84.7	29.7	37.0
0320	Concrete Reinforcing	88.0	99.8	94.1	96.5	83.3	89.7	96.5	83.4	89.7	94.6	81.9	88.1	85.2	73.4	79.1	84.5	17.2	49.9
0330	Cast-in-Place Concrete	78.6	62.6	72.5	128.8	77.7	109.2	114.7	78.5	100.8	130.9	78.3	110.8	102.4	79.8	93.7	113.7	39.7	85.3
03	CONCRETE	100.6	86.2	93.7	115.6	68.1	92.9	108.6	79.6	94.7	116.2	77.4	97.6	97.0	79.0	88.4	105.6	32.8	70.7
04	MASONRY	107.4	61.2	79.4	96.1	69.8	80.2	97.3	69.8	80.7	116.4	70.8	88.8	95.0	78.8	85.2	109.1	31.3	61.9
05	METALS	85.5	83.9	85.0	92.3	95.0	93.2	93.4	95.8	94.2	81.3	96.1	86.4	90.0	90.9	90.3	90.2	59.0	79.5
06	WOOD, PLASTICS & COMPOSITES	86.6	110.4	100.2	93.5	44.3	65.3	96.3	78.2	85.9	82.3	72.7	76.8	86.7	78.0	81.7	84.3	29.7	53.0
07	THERMAL & MOISTURE PROTECTION	96.7	64.8	84.1	94.9	76.6	87.6	95.2	80.4	89.3	99.4	78.4	91.0	94.6	77.4	87.8	94.7	37.1	71.8
08	OPENINGS	102.3	90.0	99.2	95.4	64.7	87.6	97.7	83.0	94.0	98.3	76.5	92.8	96.6	78.4	92.0	94.6	26.8	77.4
0920	Plaster & Gypsum Board	91.5	110.0	104.6	96.9	43.0	59.0	98.5	77.8	83.9	95.4	71.9	78.9	93.2	77.7	82.2	91.8	27.9	46.8
0950, 0980	Ceilings & Acoustic Treatment	89.8	110.0	103.2	93.2	43.0	60.0	94.2	77.8	83.3	98.3	71.9	80.8	94.2	77.7	83.2	94.2	27.9	50.3
0960	Flooring	91.8	58.4	82.2	94.2	81.6	90.6	94.2	81.6	90.6	111.9	54.5	95.4	90.1	82.9	88.0	89.3	17.2	68.6
0970, 0990	Wall Finishes & Painting/Coating	92.3	37.1	58.8	98.5	85.3	90.5	98.5	85.3	90.5	121.0	80.4	96.4	98.5	79.0	86.6	98.5	21.9	52.0
09	FINISHES	94.6	86.8	90.3	94.9	58.1	74.6	95.5	78.1	85.9	102.8	69.3	84.3	93.0	79.4	85.5	93.0	27.1	56.7
COVERS	DIVS. 10 - 14, 25, 28, 41, 43, 44, 46	100.0	96.7	99.3	100.0	83.0	96.5	100.0	87.2	97.3	100.0	57.8	91.2	100.0	89.5	97.8	100.0	39.1	87.3
21, 22, 23	FIRE SUPPRESSION, PLUMBING & HVAC	94.8	66.8	83.2	100.0	85.7	94.1	100.1	85.7	94.1	94.9	88.1	92.0	94.6	71.3	84.9	94.6	27.8	66.8
26, 27, 3370	ELECTRICAL, COMMUNICATIONS & UTIL.	100.9	79.9	89.7	96.4	91.2	93.6	97.3	93.7	95.4	100.9	97.2	98.9	94.9	82.2	88.2	94.4	62.5	77.4
MF2010	WEIGHTED AVERAGE	96.3	80.0	89.0	99.1	78.5	89.9	98.9	84.8	92.6	99.2	83.0	92.0	94.7	80.1	88.1	96.4	41.9	72.0

	DIVISION	MARYLAND															MASSACHUSETTS		
		ELKTON			HAGERSTOWN			SALISBURY			SILVER SPRING			WALDORF			BOSTON		
		219			217			218			209			206			020 - 022, 024		
		MAT.	INST.	TOTAL	MAT.	INST.	TOTAL	MAT.	INST.	TOTAL	MAT.	INST.	TOTAL	MAT.	INST.	TOTAL	MAT.	INST.	TOTAL
015433	CONTRACTOR EQUIPMENT		99.1	99.1		99.1	99.1		99.1	99.1		96.2	96.2		96.2	96.2		106.1	106.1
0241, 31 - 34	SITE & INFRASTRUCTURE, DEMOLITION	84.6	86.3	85.8	88.3	90.8	90.0	97.1	85.5	89.1	97.6	84.0	88.2	105.0	83.1	89.9	90.3	107.9	102.4
0310	Concrete Forming & Accessories	90.3	79.8	81.2	85.8	78.5	79.4	97.6	28.2	37.4	95.6	52.3	58.0	102.4	45.7	53.2	99.3	140.8	135.3
0320	Concrete Reinforcing	84.5	50.4	67.0	85.2	73.3	79.1	84.5	50.4	67.0	93.6	76.9	85.0	94.2	50.4	71.7	107.0	147.5	127.8
0330	Cast-in-Place Concrete	92.0	45.0	74.0	97.5	79.7	90.7	113.7	38.5	84.9	134.0	60.5	105.9	150.2	53.9	113.3	101.4	151.4	120.6
03	CONCRETE	89.0	62.8	76.4	93.0	78.8	86.2	106.4	37.3	73.3	114.7	61.4	89.2	126.6	50.9	90.4	112.3	144.7	127.8
04	MASONRY	95.3	34.1	58.2	101.0	78.8	87.5	108.7	33.1	62.9	115.5	61.3	82.7	98.7	57.9	74.0	116.6	159.3	142.5
05	METALS	90.3	67.5	82.4	90.1	90.6	90.3	90.3	60.6	80.1	85.3	90.8	87.2	85.3	74.1	81.4	102.0	128.6	111.2
06	WOOD, PLASTICS & COMPOSITES	91.0	94.4	93.0	85.5	78.1	81.2	100.2	26.3	57.8	88.5	51.2	67.1	95.7	45.9	67.1	100.7	140.2	123.4
07	THERMAL & MOISTURE PROTECTION	94.3	46.2	75.2	94.3	71.2	85.1	94.9	54.2	78.7	101.7	74.2	90.8	102.2	69.9	89.4	96.8	150.0	117.9
08	OPENINGS	94.6	67.4	87.7	94.6	77.2	90.2	94.9	22.5	76.5	91.4	63.7	84.4	92.2	44.5	80.2	101.6	142.8	112.0
0920	Plaster & Gypsum Board	95.4	94.6	94.8	91.8	77.8	81.9	101.1	24.5	47.2	101.2	50.6	65.5	104.7	45.1	62.7	103.1	140.8	129.7
0950, 0980	Ceilings & Acoustic Treatment	94.2	94.6	94.4	95.2	77.8	83.7	94.2	24.5	48.0	107.6	50.6	69.8	107.6	45.1	66.2	98.4	140.8	126.5
0960	Flooring	91.9	51.7	80.4	89.8	82.9	87.8	95.7	51.7	83.1	119.0	54.5	100.5	123.2	51.7	102.7	96.3	176.2	119.3
0970, 0990	Wall Finishes & Painting/Coating	98.5	80.4	87.5	98.5	33.5	59.0	98.5	24.7	53.7	127.9	80.4	99.0	127.9	57.6	85.3	94.9	157.1	132.6
09	FINISHES	93.5	77.8	84.8	92.8	74.4	82.7	96.1	31.0	60.2	102.7	54.1	75.9	104.8	47.1	73.0	98.4	149.6	126.7
COVERS	DIVS. 10 - 14, 25, 28, 41, 43, 44, 46	100.0	47.5	89.1	100.0	89.6	97.8	100.0	38.3	87.2	100.0	50.1	89.6	100.0	41.0	87.7	100.0	120.3	104.2
21, 22, 23	FIRE SUPPRESSION, PLUMBING & HVAC	94.6	73.5	85.9	100.0	85.0	93.8	94.6	63.5	81.7	94.9	82.7	89.8	94.9	80.3	88.8	100.2	129.0	112.2
26, 27, 3370	ELECTRICAL, COMMUNICATIONS & UTIL.	96.3	53.5	73.5	94.7	82.2	88.1	93.1	62.5	76.8	98.1	97.2	97.6	95.3	97.2	96.3	100.0	134.8	118.5
MF2010	WEIGHTED AVERAGE	93.6	64.9	80.7	95.5	82.2	89.5	96.8	51.0	76.3	98.3	74.5	87.7	99.1	68.5	85.4	102.3	136.8	117.7

	DIVISION	MASSACHUSETTS																	
		BROCKTON			BUZZARDS BAY			FALL RIVER			FITCHBURG			FRAMINGHAM			GREENFIELD		
		023			025			027			014			017			013		
		MAT.	INST.	TOTAL	MAT.	INST.	TOTAL	MAT.	INST.	TOTAL	MAT.	INST.	TOTAL	MAT.	INST.	TOTAL	MAT.	INST.	TOTAL
015433	CONTRACTOR EQUIPMENT		102.2	102.2		102.2	102.2		103.2	103.2		100.3	100.3		101.4	101.4		100.3	100.3
0241, 31 - 34	SITE & INFRASTRUCTURE, DEMOLITION	87.5	104.3	99.1	79.0	104.2	96.4	86.6	104.4	98.8	80.5	104.2	96.8	77.3	103.8	95.6	83.7	102.7	96.8
0310	Concrete Forming & Accessories	99.7	129.4	125.5	97.5	128.7	124.6	99.7	129.0	125.2	91.9	120.5	116.7	98.5	129.2	125.1	90.4	109.8	107.2
0320	Concrete Reinforcing	108.1	147.4	128.3	86.7	123.1	105.4	108.1	123.1	115.8	86.7	142.7	115.6	86.7	147.0	117.8	90.3	121.9	106.6
0330	Cast-in-Place Concrete	96.6	151.3	117.5	80.2	151.0	107.3	93.4	151.6	115.7	84.1	145.3	107.5	84.1	139.1	105.2	86.5	123.2	100.5
03	CONCRETE	109.6	139.4	123.8	92.5	134.4	112.6	108.0	134.8	120.8	90.6	132.4	110.6	93.7	135.1	113.5	94.6	116.0	104.9
04	MASONRY	112.6	155.6	138.6	103.2	155.6	134.9	113.0	155.5	138.8	99.5	149.3	129.7	106.1	149.4	132.4	103.7	123.4	115.6
05	METALS	98.7	126.2	108.1	93.5	115.7	101.1	98.7	116.1	104.7	93.4	120.8	102.8	93.4	125.3	104.4	95.6	107.8	99.8
06	WOOD, PLASTICS & COMPOSITES	99.6	127.6	115.6	96.7	127.6	114.4	99.6	127.8	115.8	93.2	116.5	106.6	99.2	127.3	115.3	91.4	108.4	101.1
07	THERMAL & MOISTURE PROTECTION	96.7	145.8	116.2	96.1	139.7	113.4	96.7	138.9	113.5	96.2	136.6	112.3	96.2	141.7	114.3	96.2	117.8	104.8
08	OPENINGS	99.6	133.1	108.1	94.9	122.4	101.9	99.6	122.6	105.4	104.2	126.0	109.7	95.4	132.9	104.9	104.4	112.1	106.3
0920	Plaster & Gypsum Board	97.3	127.7	118.7	93.0	127.7	117.4	97.3	127.7	118.7	94.2	116.3	109.8	97.3	127.7	118.7	95.4	108.0	104.2
0950, 0980	Ceilings & Acoustic Treatment	100.5	127.7	118.5	90.9	127.7	115.3	100.5	127.7	118.5	89.8	116.3	107.4	89.8	127.7	114.9	97.3	108.0	104.4
0960	Flooring	96.8	176.2	119.7	95.5	176.2	118.7	96.6	176.2	119.5	93.8	176.2	117.5	95.3	176.2	118.6	93.0	135.4	105.2
0970, 0990	Wall Finishes & Painting/Coating	94.4	142.1	123.3	94.4	142.1	123.3	94.4	142.1	123.3	92.3	142.1	122.5	93.4	142.1	122.9	92.3	111.7	104.1
09	FINISHES	98.4	139.6	121.1	94.8	139.6	119.5	98.3	139.8	121.2	94.1	133.1	115.6	94.8	139.4	119.4	96.0	114.7	106.3
COVERS	DIVS. 10 - 14, 25, 28, 41, 43, 44, 46	100.0	117.5	103.6	100.0	117.5	103.6	100.0	118.1	103.8	100.0	92.9	98.5	100.0	117.0	103.5	100.0	105.6	101.2
21, 22, 23	FIRE SUPPRESSION, PLUMBING & HVAC	100.2	110.3	104.4	94.8	110.0	101.1	100.2	110.0	104.3	95.6	112.5	102.7	95.6	123.2	107.1	95.6	103.5	98.9
26, 27, 3370	ELECTRICAL, COMMUNICATIONS & UTIL.	99.1	101.0	100.1	96.0	101.0	98.6	99.1	101.0	100.1	101.4	104.0	102.8	97.5	126.9	113.1	101.4	95.0	98.0
MF2010	WEIGHTED AVERAGE	100.9	124.5	111.5	95.0	122.2	107.1	100.7	122.3	110.4	96.2	121.1	107.3	95.6	129.4	110.7	97.4	108.6	102.4

	DIVISION	MASSACHUSETTS																	
		HYANNIS			LAWRENCE			LOWELL			NEW BEDFORD			PITTSFIELD			SPRINGFIELD		
		026			019			018			027			012			010 - 011		
		MAT.	INST.	TOTAL	MAT.	INST.	TOTAL	MAT.	INST.	TOTAL	MAT.	INST.	TOTAL	MAT.	INST.	TOTAL	MAT.	INST.	TOTAL
015433	CONTRACTOR EQUIPMENT		102.2	102.2		102.2	102.2		100.3	100.3		103.2	103.2		100.3	100.3		100.3	100.3
0241, 31 - 34	SITE & INFRASTRUCTURE, DEMOLITION	84.4	104.2	98.0	88.2	104.3	99.3	87.3	104.2	99.0	85.2	104.4	98.4	88.3	102.6	98.2	87.8	102.9	98.2
0310	Concrete Forming & Accessories	92.1	128.7	123.9	99.5	129.5	125.5	96.5	129.6	125.2	99.7	129.0	125.2	96.5	108.4	106.8	96.7	110.4	108.6
0320	Concrete Reinforcing	86.7	123.1	105.4	107.2	137.3	122.7	108.1	137.0	123.0	108.1	123.1	115.8	89.6	106.2	98.2	108.1	121.9	115.2
0330	Cast-in-Place Concrete	88.1	151.0	112.2	97.4	148.2	116.8	88.5	148.2	111.4	82.3	151.6	108.8	96.6	121.4	106.1	92.1	124.2	104.4
03	CONCRETE	99.6	134.4	116.3	109.8	136.5	122.6	100.3	136.3	117.5	102.4	134.8	117.9	101.6	111.8	106.5	102.1	116.7	109.1
04	MASONRY	111.5	155.6	138.2	111.7	155.6	138.3	98.2	149.3	129.2	111.4	155.5	138.1	98.9	120.3	111.9	98.5	125.1	114.6
05	METALS	95.0	115.7	102.1	96.0	122.1	105.0	96.0	119.4	104.0	98.7	116.1	104.7	95.8	101.2	97.6	98.6	107.8	101.8
06	WOOD, PLASTICS & COMPOSITES	90.7	127.6	111.8	99.6	127.6	115.6	99.0	127.6	115.4	99.6	127.8	115.8	99.0	108.4	104.4	99.0	108.4	104.4
07	THERMAL & MOISTURE PROTECTION	96.4	139.7	113.6	96.7	145.4	116.0	96.5	143.4	115.1	96.7	138.9	113.5	96.5	116.5	104.5	96.5	118.6	105.3
08	OPENINGS	95.6	122.4	102.4	99.6	130.4	107.4	105.6	130.4	111.9	99.6	122.6	105.4	105.6	108.0	106.2	105.6	112.1	107.3
0920	Plaster & Gypsum Board	89.5	127.7	116.4	100.2	127.7	119.6	100.2	127.7	119.6	97.3	127.7	118.7	100.2	108.0	105.7	100.2	108.0	105.7
0950, 0980	Ceilings & Acoustic Treatment	93.0	127.7	116.0	99.4	127.7	118.2	99.4	127.7	118.2	100.5	127.7	118.5	99.4	108.0	105.1	99.4	108.0	105.1
0960	Flooring	93.4	176.2	117.2	95.8	176.2	118.9	95.8	176.2	118.9	96.6	176.2	119.5	96.0	135.4	107.4	95.7	135.4	107.2
0970, 0990	Wall Finishes & Painting/Coating	94.4	142.1	123.3	92.4	142.1	122.5	92.3	142.1	122.5	94.4	142.1	123.3	92.3	111.7	104.1	93.8	111.7	104.7
09	FINISHES	94.6	139.6	119.4	98.1	139.6	120.9	98.0	139.6	120.9	98.2	139.8	121.1	98.1	114.0	106.8	98.1	115.1	107.5
COVERS	DIVS. 10 - 14, 25, 28, 41, 43, 44, 46	100.0	117.5	103.6	100.0	117.6	103.7	100.0	117.6	103.7	100.0	118.1	103.8	100.0	104.5	100.9	100.0	106.2	101.3
21, 22, 23	FIRE SUPPRESSION, PLUMBING & HVAC	100.2	110.0	104.3	100.2	113.0	105.5	100.2	123.2	109.8	100.2	110.0	104.3	100.2	96.6	98.7	100.2	104.4	102.0
26, 27, 3370	ELECTRICAL, COMMUNICATIONS & UTIL.	96.5	101.0	98.9	100.3	126.9	114.4	100.9	126.8	114.7	99.9	101.0	100.5	100.9	95.0	97.7	100.9	96.9	98.8
MF2010	WEIGHTED AVERAGE	98.0	122.2	108.8	100.6	127.8	112.8	99.4	129.0	112.6	100.0	122.3	110.0	99.6	105.3	102.2	100.0	109.5	104.2

	DIVISION	MASSACHUSETTS			MICHIGAN														
		WORCESTER			ANN ARBOR			BATTLE CREEK			BAY CITY			DEARBORN			DETROIT		
		015 - 016			481			490			487			481			482		
		MAT.	INST.	TOTAL	MAT.	INST.	TOTAL	MAT.	INST.	TOTAL	MAT.	INST.	TOTAL	MAT.	INST.	TOTAL	MAT.	INST.	TOTAL
015433	CONTRACTOR EQUIPMENT		100.3	100.3		110.9	110.9		103.6	103.6		110.9	110.9		110.9	110.9		98.5	98.5
0241, 31 - 34	SITE & INFRASTRUCTURE, DEMOLITION	87.7	104.2	99.1	73.4	96.1	89.1	82.5	87.9	86.2	66.2	94.7	85.8	73.2	96.1	89.0	86.5	97.7	94.2
0310	Concrete Forming & Accessories	97.0	124.7	121.1	98.7	104.0	103.3	98.2	91.4	92.3	98.7	89.0	90.3	98.5	112.1	110.3	99.2	112.2	110.5
0320	Concrete Reinforcing	108.1	146.3	127.8	100.8	125.5	113.6	98.6	92.5	95.5	100.8	124.7	113.1	100.8	125.7	113.6	100.1	125.7	113.3
0330	Cast-in-Place Concrete	91.6	145.4	112.2	82.3	115.3	94.9	92.4	104.8	97.2	78.7	95.2	85.0	80.5	113.3	93.0	88.3	113.3	97.9
03	CONCRETE	101.9	135.0	117.7	90.3	112.3	100.9	95.0	95.5	95.3	88.5	98.6	93.4	89.4	115.3	101.8	93.3	114.2	103.3
04	MASONRY	98.0	149.3	129.1	96.8	110.4	105.0	95.3	88.1	90.9	96.4	89.6	92.3	96.6	113.7	107.0	95.8	113.7	106.7
05	METALS	98.7	122.4	106.8	93.1	120.2	102.4	92.9	87.7	91.1	93.7	116.9	101.7	93.1	120.6	102.6	97.0	102.8	99.0
06	WOOD, PLASTICS & COMPOSITES	99.4	122.0	112.4	100.4	101.7	101.2	100.9	91.6	95.6	100.4	88.3	93.5	100.4	112.2	107.2	100.8	112.2	107.3
07	THERMAL & MOISTURE PROTECTION	96.5	137.2	112.7	100.2	111.5	104.7	94.3	89.1	92.2	98.0	97.9	98.0	98.8	116.7	105.9	97.3	116.7	105.0
08	OPENINGS	105.6	129.9	111.8	92.1	105.6	95.5	92.0	85.5	90.4	92.1	95.9	93.1	92.1	110.6	96.8	93.6	111.3	98.1
0920	Plaster & Gypsum Board	100.2	122.0	115.5	99.1	100.7	100.3	90.1	87.4	88.2	99.1	86.9	90.5	99.1	111.5	107.8	99.1	111.5	107.8
0950, 0980	Ceilings & Acoustic Treatment	99.4	122.0	114.4	94.3	100.7	98.6	91.3	87.4	88.7	95.3	86.9	89.7	94.3	111.5	105.7	95.3	111.5	106.0
0960	Flooring	95.8	170.2	117.2	89.7	118.1	97.9	98.2	97.1	97.9	89.6	81.2	87.2	89.4	118.9	97.8	89.6	118.9	98.0
0970, 0990	Wall Finishes & Painting/Coating	92.3	142.1	122.5	93.0	104.2	99.8	97.6	82.8	88.6	93.0	85.1	88.2	93.0	104.3	99.9	94.5	104.3	100.5
09	FINISHES	98.0	135.1	118.4	92.6	106.2	100.1	96.1	91.6	93.6	92.3	86.4	89.1	92.4	112.7	103.6	93.8	112.7	104.2
COVERS	DIVS. 10 - 14, 25, 28, 41, 43, 44, 46	100.0	109.9	102.1	100.0	106.1	101.3	100.0	106.8	101.4	100.0	100.2	100.0	100.0	107.5	101.6	100.0	107.5	101.6
21, 22, 23	FIRE SUPPRESSION, PLUMBING & HVAC	100.2	112.6	105.3	100.0	102.1	100.9	99.9	92.0	96.6	100.0	87.5	94.8	100.0	107.8	103.3	100.1	111.2	104.7
26, 27, 3370	ELECTRICAL, COMMUNICATIONS & UTIL.	100.9	104.8	103.0	96.9	110.3	104.1	94.7	85.2	89.7	95.9	90.5	93.0	96.9	111.6	104.7	97.9	111.5	105.2
MF2010	WEIGHTED AVERAGE	100.0	122.7	110.1	95.2	107.7	100.8	95.8	90.4	93.4	94.6	93.9	94.3	95.0	111.2	102.2	96.7	110.2	102.8

DIVISION		MICHIGAN																	
		FLINT			GAYLORD			GRAND RAPIDS			IRON MOUNTAIN			JACKSON			KALAMAZOO		
		484 - 485			497			493,495			498 - 499			492			491		
		MAT.	INST.	TOTAL	MAT.	INST.	TOTAL	MAT.	INST.	TOTAL	MAT.	INST.	TOTAL	MAT.	INST.	TOTAL	MAT.	INST.	TOTAL
015433	CONTRACTOR EQUIPMENT		110.9	110.9		105.9	105.9		103.6	103.6		94.4	94.4		105.9	105.9		103.6	103.6
0241, 31 - 34	SITE & INFRASTRUCTURE, DEMOLITION	64.2	95.4	85.8	78.2	86.1	83.7	83.9	87.7	86.6	85.4	94.5	91.7	98.0	87.4	90.7	82.8	87.9	86.3
0310	Concrete Forming & Accessories	101.6	90.9	92.3	95.4	56.9	62.0	98.4	80.8	83.1	87.3	88.0	87.9	92.9	88.5	89.1	98.2	91.1	92.0
0320	Concrete Reinforcing	100.8	125.0	113.3	91.9	94.1	93.1	98.6	91.8	95.1	91.7	95.8	93.8	89.3	124.8	107.6	98.6	92.4	95.4
0330	Cast-in-Place Concrete	82.9	97.1	88.3	92.2	72.5	84.6	98.4	96.9	97.8	109.1	91.6	102.4	92.0	95.5	93.4	94.3	104.7	98.3
03	CONCRETE	90.8	100.2	95.3	91.4	70.8	81.6	98.1	87.9	93.2	101.1	90.8	96.2	86.5	98.4	92.2	98.3	95.3	96.9
04	MASONRY	96.8	98.0	97.5	105.6	84.1	92.6	91.2	81.8	85.5	92.1	90.8	91.3	86.3	94.7	91.4	94.0	88.1	90.4
05	METALS	93.1	117.7	101.6	94.0	101.7	96.6	94.0	85.1	90.9	93.4	91.5	92.7	94.2	115.2	101.4	92.9	87.3	91.0
06	WOOD, PLASTICS & COMPOSITES	103.6	88.6	95.0	93.3	49.8	68.4	98.1	80.1	87.8	87.9	88.0	88.0	92.1	86.9	89.1	100.9	91.6	95.6
07	THERMAL & MOISTURE PROTECTION	98.0	101.5	99.4	92.8	71.9	84.5	96.1	69.6	85.6	96.0	85.8	92.0	92.2	98.1	94.5	94.3	89.0	92.2
08	OPENINGS	92.1	96.1	93.1	94.0	55.0	84.2	95.7	70.7	89.4	99.3	74.9	93.1	93.0	88.3	91.8	92.0	85.5	90.4
0920	Plaster & Gypsum Board	100.9	87.2	91.2	89.3	47.0	59.5	90.7	75.6	80.0	59.5	88.1	79.6	87.8	85.2	85.9	90.1	87.4	88.2
0950, 0980	Ceilings & Acoustic Treatment	94.3	87.2	89.6	88.5	47.0	61.0	89.4	75.6	80.2	89.2	88.1	88.5	89.4	85.2	86.6	91.3	87.4	88.7
0960	Flooring	89.6	83.9	87.9	91.1	50.9	79.5	98.4	35.8	80.4	112.3	97.7	108.1	89.6	70.0	84.0	98.2	76.5	91.9
0970, 0990	Wall Finishes & Painting/Coating	93.0	92.0	92.4	93.2	46.5	64.9	97.6	34.7	59.5	114.3	58.3	80.3	93.2	77.3	83.5	97.6	82.8	88.6
09	FINISHES	91.9	88.8	90.2	96.1	52.6	72.1	95.9	67.0	80.0	99.6	86.7	92.5	97.2	83.0	89.4	96.1	87.5	91.4
COVERS	DIVS. 10 - 14, 25, 28, 41, 43, 44, 46	100.0	101.5	100.3	100.0	95.1	99.0	100.0	105.2	101.1	100.0	97.3	99.4	100.0	101.4	100.3	100.0	106.8	101.4
21, 22, 23	FIRE SUPPRESSION, PLUMBING & HVAC	100.0	96.1	98.4	94.9	82.6	89.7	99.9	80.6	91.8	94.8	87.4	91.7	94.9	93.1	94.1	99.9	83.0	92.9
26, 27, 3370	ELECTRICAL, COMMUNICATIONS & UTIL.	96.9	97.0	97.0	92.4	50.0	69.8	103.1	49.9	74.8	99.4	87.2	92.9	96.8	110.3	104.0	94.6	88.8	91.5
MF2010	WEIGHTED AVERAGE	94.9	98.2	96.4	94.5	73.4	85.1	97.3	76.8	88.1	96.7	88.8	93.2	94.0	96.8	95.2	96.1	88.4	92.7

DIVISION		MICHIGAN															MINNESOTA		
		LANSING			MUSKEGON			ROYAL OAK			SAGINAW			TRAVERSE CITY			BEMIDJI		
		488 - 489			494			480,483			486			496			566		
		MAT.	INST.	TOTAL	MAT.	INST.	TOTAL	MAT.	INST.	TOTAL	MAT.	INST.	TOTAL	MAT.	INST.	TOTAL	MAT.	INST.	TOTAL
015433	CONTRACTOR EQUIPMENT		110.9	110.9		103.6	103.6		95.8	95.8		110.9	110.9		94.4	94.4		98.6	98.6
0241, 31 - 34	SITE & INFRASTRUCTURE, DEMOLITION	81.0	94.8	90.5	80.5	87.8	85.5	77.0	95.5	89.8	67.0	94.7	86.1	72.8	94.0	87.4	95.7	98.4	97.6
0310	Concrete Forming & Accessories	102.2	89.9	91.6	98.6	89.6	90.8	94.9	101.2	100.3	98.7	88.9	90.2	87.3	72.3	74.3	87.4	96.1	95.0
0320	Concrete Reinforcing	100.8	124.8	113.2	99.3	91.8	95.4	91.4	122.4	107.3	100.8	124.7	113.1	93.1	91.6	92.3	98.7	110.4	104.7
0330	Cast-in-Place Concrete	87.4	94.9	90.3	92.1	97.8	94.3	72.0	113.0	87.7	81.3	95.2	86.6	85.1	64.5	77.3	103.6	109.5	105.9
03	CONCRETE	93.1	99.0	95.9	93.4	92.2	92.8	77.9	108.4	92.5	89.8	98.6	94.0	83.1	73.9	78.7	99.1	104.4	101.6
04	MASONRY	90.5	95.2	93.3	92.5	77.3	83.3	91.9	113.5	105.0	98.3	89.6	93.0	90.3	84.0	86.5	103.6	107.6	106.0
05	METALS	91.7	117.1	100.4	90.6	86.1	89.1	96.1	99.1	97.2	93.2	116.8	101.3	93.3	89.3	91.9	92.5	123.9	103.3
06	WOOD, PLASTICS & COMPOSITES	103.0	88.4	94.6	97.0	90.4	93.2	95.8	97.8	96.9	96.1	88.3	91.6	87.9	71.0	78.2	73.2	93.0	84.6
07	THERMAL & MOISTURE PROTECTION	99.2	94.1	97.2	93.5	78.9	87.7	96.7	114.6	103.8	98.8	97.9	98.5	95.2	72.5	86.2	107.4	99.7	104.3
08	OPENINGS	92.1	96.0	93.1	91.3	89.0	90.7	92.2	102.5	94.8	90.0	95.9	91.5	99.2	66.2	90.9	96.6	113.3	100.8
0920	Plaster & Gypsum Board	91.9	87.0	88.5	76.9	86.1	83.4	95.9	96.7	96.4	99.1	86.9	90.5	59.5	70.6	67.3	100.6	93.1	95.3
0950, 0980	Ceilings & Acoustic Treatment	94.3	87.0	89.4	92.2	86.1	88.2	93.5	96.7	95.6	94.3	86.9	89.4	89.2	70.6	76.8	132.5	93.1	106.3
0960	Flooring	99.1	83.9	94.8	97.4	65.9	88.3	87.5	117.4	96.1	89.7	81.2	87.3	112.2	50.9	94.6	112.6	127.6	116.9
0970, 0990	Wall Finishes & Painting/Coating	103.9	85.9	93.0	96.9	83.1	88.5	94.5	104.2	100.4	93.0	85.1	88.2	114.3	46.5	73.2	98.4	102.4	100.8
09	FINISHES	96.0	87.7	91.4	93.8	84.5	88.6	91.6	103.9	98.4	92.2	86.4	89.0	98.6	64.8	80.0	111.4	102.1	106.3
COVERS	DIVS. 10 - 14, 25, 28, 41, 43, 44, 46	100.0	101.1	100.2	100.0	106.4	101.3	100.0	104.5	100.9	100.0	100.2	100.0	100.0	93.6	98.7	100.0	99.4	99.9
21, 22, 23	FIRE SUPPRESSION, PLUMBING & HVAC	99.9	93.8	97.3	99.8	85.0	93.6	95.0	104.9	99.1	100.0	87.2	94.7	94.8	82.1	89.5	94.9	90.1	92.9
26, 27, 3370	ELECTRICAL, COMMUNICATIONS & UTIL.	99.5	95.0	97.1	95.0	78.8	86.4	99.3	96.1	97.6	94.5	92.1	93.2	94.1	78.3	85.7	102.4	101.4	101.9
MF2010	WEIGHTED AVERAGE	95.7	96.5	96.1	94.7	85.3	90.5	92.8	103.6	97.6	94.5	94.0	94.3	93.4	79.4	87.2	98.5	102.2	100.2

DIVISION		MINNESOTA																	
		BRAINERD			DETROIT LAKES			DULUTH			MANKATO			MINNEAPOLIS			ROCHESTER		
		564			565			556 - 558			560			553 - 555			559		
		MAT.	INST.	TOTAL	MAT.	INST.	TOTAL	MAT.	INST.	TOTAL	MAT.	INST.	TOTAL	MAT.	INST.	TOTAL	MAT.	INST.	TOTAL
015433	CONTRACTOR EQUIPMENT		101.5	101.5		98.6	98.6		101.0	101.0		101.5	101.5		104.7	104.7		101.0	101.0
0241, 31 - 34	SITE & INFRASTRUCTURE, DEMOLITION	93.8	103.5	100.5	93.7	98.9	97.3	96.5	101.9	100.2	90.1	104.2	99.8	99.1	107.5	104.9	98.0	101.8	100.6
0310	Concrete Forming & Accessories	90.7	99.1	98.0	84.8	99.0	97.1	100.6	118.1	115.8	97.4	107.0	105.7	102.6	132.1	128.2	103.3	113.0	111.7
0320	Concrete Reinforcing	97.6	110.6	104.3	98.7	110.4	104.7	93.1	110.8	102.2	97.4	121.2	109.7	93.3	121.6	107.9	93.1	121.3	107.6
0330	Cast-in-Place Concrete	112.6	114.6	113.4	100.6	113.9	105.7	105.0	109.4	106.7	103.7	113.7	107.5	117.5	122.1	119.3	115.2	105.7	111.6
03	CONCRETE	103.4	107.5	105.4	96.5	107.2	101.6	102.0	114.3	107.9	98.5	112.8	105.4	110.3	127.0	118.3	107.3	112.9	110.0
04	MASONRY	127.8	120.2	123.2	127.6	115.4	120.2	116.3	119.4	118.2	115.4	117.0	116.4	115.7	131.5	125.3	114.9	119.6	117.7
05	METALS	93.6	124.2	104.1	92.4	123.8	103.2	101.0	126.1	109.6	93.4	130.3	106.1	103.9	133.7	114.2	100.9	131.9	111.6
06	WOOD, PLASTICS & COMPOSITES	94.4	92.8	93.5	70.7	93.0	83.5	109.8	119.0	115.1	102.4	104.9	103.8	113.0	130.6	123.1	113.0	111.9	112.3
07	THERMAL & MOISTURE PROTECTION	107.0	113.6	109.6	107.2	113.9	109.9	107.9	119.4	112.5	107.3	104.8	106.3	107.7	133.6	118.0	108.0	106.5	107.4
08	OPENINGS	84.8	113.1	92.0	96.6	113.3	100.8	93.7	122.6	101.1	89.5	123.7	98.2	96.8	137.6	107.1	93.7	127.4	102.3
0920	Plaster & Gypsum Board	85.4	93.1	90.8	99.7	93.1	95.0	92.3	120.0	111.8	89.4	105.6	100.8	100.0	131.9	122.4	99.4	112.6	108.7
0950, 0980	Ceilings & Acoustic Treatment	65.5	93.1	83.8	132.5	93.1	106.3	88.3	120.0	109.3	65.5	105.6	92.0	90.2	131.9	117.8	88.3	112.6	104.5
0960	Flooring	111.4	127.6	116.0	111.2	127.6	115.9	108.3	129.0	114.2	113.5	127.6	117.6	106.1	127.6	112.3	108.5	127.6	114.0
0970, 0990	Wall Finishes & Painting/Coating	92.7	102.4	98.6	98.4	102.4	100.8	89.0	110.0	101.7	104.0	105.4	104.8	97.7	130.8	117.8	92.3	105.4	100.2
09	FINISHES	94.6	103.9	99.7	110.7	104.0	107.0	96.2	119.9	109.3	96.2	110.6	104.1	97.6	131.6	116.4	97.4	115.1	107.2
COVERS	DIVS. 10 - 14, 25, 28, 41, 43, 44, 46	100.0	101.7	100.4	100.0	102.1	100.4	100.0	101.0	100.2	100.0	101.9	100.4	100.0	111.1	102.3	100.0	104.9	101.0
21, 22, 23	FIRE SUPPRESSION, PLUMBING & HVAC	94.3	94.1	94.2	94.9	93.8	94.5	100.0	99.7	99.9	94.3	92.9	93.7	99.9	124.0	109.9	99.9	100.3	100.1
26, 27, 3370	ELECTRICAL, COMMUNICATIONS & UTIL.	99.8	111.2	105.9	102.2	70.7	85.5	100.0	111.2	106.0	106.9	91.0	98.4	98.9	115.6	107.8	98.0	91.0	94.2
MF2010	WEIGHTED AVERAGE	97.6	107.3	101.9	99.3	100.7	99.9	100.6	112.2	105.8	97.6	106.4	101.5	102.3	125.0	112.5	101.0	109.2	104.7

	DIVISION	MINNESOTA															MISSISSIPPI		
		SAINT PAUL			ST. CLOUD			THIEF RIVER FALLS			WILLMAR			WINDOM			BILOXI		
		550 - 551			563			567			562			561			395		
		MAT.	INST.	TOTAL	MAT.	INST.	TOTAL	MAT.	INST.	TOTAL	MAT.	INST.	TOTAL	MAT.	INST.	TOTAL	MAT.	INST.	TOTAL
015433	CONTRACTOR EQUIPMENT		101.0	101.0		101.5	101.5		98.6	98.6		101.5	101.5		101.5	101.5		98.3	98.3
0241, 31 - 34	SITE & INFRASTRUCTURE, DEMOLITION	98.8	102.6	101.5	88.1	105.5	100.1	94.4	98.4	97.1	88.0	103.2	98.5	81.3	101.9	95.5	100.9	88.5	92.4
0310	Concrete Forming & Accessories	97.4	131.8	127.3	88.3	127.6	122.4	87.9	95.5	94.5	88.1	98.3	97.0	91.9	89.8	90.1	95.5	57.3	62.3
0320	Concrete Reinforcing	89.9	121.6	106.2	97.6	121.4	109.9	99.1	110.3	104.9	97.2	121.1	109.5	97.2	120.1	109.0	91.8	60.6	75.7
0330	Cast-in-Place Concrete	117.5	121.5	119.0	99.3	121.1	107.7	102.7	93.6	99.2	100.9	91.4	97.3	87.3	68.3	80.0	106.2	56.1	87.0
03	CONCRETE	112.0	126.7	119.0	94.3	124.6	108.8	97.7	98.7	98.2	94.3	101.3	97.7	84.8	89.5	87.0	95.3	59.1	78.0
04	MASONRY	125.3	131.5	129.1	112.2	129.3	122.6	103.6	107.6	106.0	116.7	120.2	118.8	127.2	92.1	105.9	89.1	46.7	63.4
05	METALS	100.4	133.4	111.7	94.2	131.6	107.1	92.6	123.2	103.1	93.3	129.1	105.7	93.3	126.6	104.7	92.7	84.0	89.7
06	WOOD, PLASTICS & COMPOSITES	106.1	130.5	120.1	91.9	125.0	110.9	74.1	93.0	84.9	91.7	93.0	92.4	95.5	90.1	92.4	99.8	60.9	77.5
07	THERMAL & MOISTURE PROTECTION	109.2	133.4	118.8	107.1	124.9	114.2	108.3	102.6	106.0	107.0	112.0	109.0	106.9	87.2	99.1	91.6	58.0	78.2
08	OPENINGS	91.3	137.5	103.0	89.5	134.6	100.9	96.6	113.3	100.8	87.1	98.0	89.9	91.3	96.4	92.6	97.9	61.8	88.8
0920	Plaster & Gypsum Board	89.6	131.9	119.3	85.0	126.3	114.1	100.1	93.1	95.2	85.0	93.3	90.9	85.4	90.3	88.8	98.6	60.2	71.5
0950, 0980	Ceilings & Acoustic Treatment	86.5	131.9	116.6	65.5	126.3	105.8	132.5	93.1	106.3	65.5	93.3	83.9	65.5	90.3	81.9	96.5	60.2	72.4
0960	Flooring	102.4	127.6	109.7	107.2	127.6	113.0	112.2	127.6	116.6	109.2	127.6	114.5	112.0	127.6	116.5	119.4	55.8	101.1
0970, 0990	Wall Finishes & Painting/Coating	97.7	128.2	116.2	104.0	130.8	120.3	98.4	102.4	100.8	98.4	102.4	100.8	98.4	105.4	102.6	105.9	46.8	70.0
09	FINISHES	94.5	131.2	114.8	93.6	128.3	112.7	111.2	102.1	106.2	93.8	101.5	98.0	94.1	96.7	95.5	105.4	56.1	78.2
COVERS	DIVS. 10 - 14, 25, 28, 41, 43, 44, 46	100.0	110.8	102.2	100.0	108.7	101.8	100.0	99.3	99.9	100.0	101.7	100.3	100.0	96.5	99.3	100.0	62.0	92.1
21, 22, 23	FIRE SUPPRESSION, PLUMBING & HVAC	99.9	116.8	106.9	99.7	123.9	109.8	94.9	89.7	92.8	94.3	112.6	101.9	94.3	87.3	91.4	99.8	51.3	79.6
26, 27, 3370	ELECTRICAL, COMMUNICATIONS & UTIL.	98.5	115.6	107.6	99.8	115.6	108.2	99.3	70.7	84.1	99.8	92.5	95.9	106.9	90.9	98.4	98.7	59.4	77.8
MF2010	WEIGHTED AVERAGE	101.6	123.0	111.2	97.3	123.2	108.9	98.1	97.1	97.7	95.9	107.2	100.9	96.2	95.3	95.8	97.6	60.8	81.2

	DIVISION	MISSISSIPPI																	
		CLARKSDALE			COLUMBUS			GREENVILLE			GREENWOOD			JACKSON			LAUREL		
		386			397			387			389			390 - 392			394		
		MAT.	INST.	TOTAL	MAT.	INST.	TOTAL	MAT.	INST.	TOTAL	MAT.	INST.	TOTAL	MAT.	INST.	TOTAL	MAT.	INST.	TOTAL
015433	CONTRACTOR EQUIPMENT		98.3	98.3		98.3	98.3		98.3	98.3		98.3	98.3		98.3	98.3		98.3	98.3
0241, 31 - 34	SITE & INFRASTRUCTURE, DEMOLITION	98.6	88.3	91.5	99.7	88.6	92.1	104.6	88.6	93.6	101.6	88.0	92.3	96.2	88.6	90.9	106.0	88.4	93.9
0310	Concrete Forming & Accessories	86.4	53.4	57.8	84.5	55.6	59.4	83.0	69.2	71.0	95.4	53.4	59.0	94.8	62.3	66.6	84.5	57.4	61.0
0320	Concrete Reinforcing	98.9	44.8	71.0	98.3	45.9	71.3	99.5	65.0	81.7	98.9	56.6	77.1	91.8	58.6	74.7	98.9	42.1	69.7
0330	Cast-in-Place Concrete	103.6	57.0	85.7	108.3	62.5	90.8	106.7	59.3	88.5	111.4	56.6	90.4	90.4	60.1	78.8	105.9	58.8	87.9
03	CONCRETE	93.5	54.7	74.9	96.9	57.7	78.2	98.2	66.3	82.9	99.6	56.6	79.0	87.3	62.3	75.3	98.7	56.6	78.6
04	MASONRY	88.9	48.6	64.5	111.8	55.8	77.8	130.4	52.2	83.0	89.5	48.5	64.6	89.8	52.2	67.0	108.3	52.0	74.2
05	METALS	89.8	76.4	85.1	89.6	77.3	85.4	90.8	86.1	89.2	89.8	77.9	85.7	92.6	83.4	89.5	89.7	76.2	85.1
06	WOOD, PLASTICS & COMPOSITES	87.5	55.6	69.2	85.5	56.6	68.9	83.9	74.0	78.2	99.8	55.6	74.4	98.4	64.7	79.1	86.3	59.2	70.8
07	THERMAL & MOISTURE PROTECTION	91.4	50.6	75.2	91.5	55.2	77.1	91.7	64.2	80.8	91.7	50.5	75.4	93.4	63.2	81.4	91.7	57.0	77.9
08	OPENINGS	97.2	53.4	86.1	97.2	53.5	86.1	97.2	66.4	89.4	97.2	54.8	86.5	98.3	59.6	88.5	93.8	54.4	83.8
0920	Plaster & Gypsum Board	87.1	54.7	64.3	86.6	55.8	64.9	85.8	73.7	77.3	97.3	54.7	67.3	92.7	64.1	72.6	85.8	58.5	66.6
0950, 0980	Ceilings & Acoustic Treatment	89.7	54.7	66.5	89.7	55.8	67.2	91.8	73.7	79.8	89.7	54.7	66.5	93.7	64.1	74.1	89.7	58.5	69.0
0960	Flooring	113.0	55.8	96.6	111.8	62.6	97.7	110.9	55.8	95.1	119.4	55.8	101.1	119.8	55.8	101.4	109.9	55.8	94.4
0970, 0990	Wall Finishes & Painting/Coating	105.9	49.6	71.8	105.9	57.0	76.3	105.9	70.9	84.7	105.9	49.6	71.8	105.9	70.9	84.7	105.9	65.6	81.4
09	FINISHES	100.5	53.8	74.7	100.3	57.4	76.6	100.8	67.8	82.6	104.0	53.8	76.3	104.4	62.4	81.2	100.2	58.6	77.2
COVERS	DIVS. 10 - 14, 25, 28, 41, 43, 44, 46	100.0	62.4	92.2	100.0	63.6	92.4	100.0	65.5	92.8	100.0	62.3	92.2	100.0	64.4	92.6	100.0	41.6	87.9
21, 22, 23	FIRE SUPPRESSION, PLUMBING & HVAC	97.5	45.3	75.8	96.9	41.2	73.7	99.8	52.2	80.0	97.5	43.2	74.9	99.8	57.8	82.3	96.9	50.3	77.5
26, 27, 3370	ELECTRICAL, COMMUNICATIONS & UTIL.	98.0	50.3	72.6	95.9	53.3	73.3	98.0	68.0	82.1	98.0	47.2	70.9	101.9	68.0	83.9	97.6	58.9	77.0
MF2010	WEIGHTED AVERAGE	95.7	56.3	78.1	97.0	57.7	79.4	99.3	66.1	84.5	97.0	55.8	78.6	96.9	65.3	82.8	97.0	59.3	80.2

	DIVISION	MISSISSIPPI									MISSOURI								
		MCCOMB			MERIDIAN			TUPELO			BOWLING GREEN			CAPE GIRARDEAU			CHILLICOTHE		
		396			393			388			633			637			646		
		MAT.	INST.	TOTAL	MAT.	INST.	TOTAL	MAT.	INST.	TOTAL	MAT.	INST.	TOTAL	MAT.	INST.	TOTAL	MAT.	INST.	TOTAL
015433	CONTRACTOR EQUIPMENT		98.3	98.3		98.3	98.3		98.3	98.3		106.5	106.5		106.5	106.5		101.6	101.6
0241, 31 - 34	SITE & INFRASTRUCTURE, DEMOLITION	92.3	88.2	89.5	96.2	88.9	91.1	96.1	88.2	90.7	92.4	93.4	93.1	93.6	93.4	93.4	98.8	90.4	93.0
0310	Concrete Forming & Accessories	84.5	55.3	59.2	81.9	68.6	70.4	83.5	55.6	59.3	96.2	89.0	90.0	89.7	76.6	78.3	91.2	91.9	91.8
0320	Concrete Reinforcing	99.5	44.2	71.0	98.3	58.7	77.9	96.8	56.7	76.1	103.7	98.4	101.0	105.0	84.6	94.5	97.9	101.7	99.9
0330	Cast-in-Place Concrete	94.0	56.4	79.6	100.6	62.7	86.1	103.6	58.5	86.3	94.0	77.2	87.6	93.0	84.3	89.7	94.2	66.7	83.7
03	CONCRETE	87.1	55.1	71.8	91.2	66.0	79.2	93.0	58.2	76.3	94.5	87.9	91.4	93.7	82.3	88.2	99.7	85.5	92.9
04	MASONRY	113.3	47.9	73.7	88.7	56.5	69.2	119.9	51.7	78.6	126.0	98.0	109.1	123.1	79.0	96.4	99.7	92.8	95.5
05	METALS	89.9	74.5	84.6	90.8	83.6	88.3	89.7	78.2	85.7	96.5	112.5	102.0	97.6	106.0	100.5	92.0	97.2	93.8
06	WOOD, PLASTICS & COMPOSITES	85.5	58.6	70.1	82.8	70.7	75.9	84.5	56.6	68.5	93.5	89.3	91.1	86.5	73.8	79.2	85.7	98.1	92.9
07	THERMAL & MOISTURE PROTECTION	91.2	49.7	74.7	91.3	65.7	81.1	91.4	56.8	77.7	99.0	100.4	99.6	98.9	82.5	92.4	97.6	86.2	93.1
08	OPENINGS	97.2	55.2	86.6	97.2	68.9	90.0	97.2	54.7	86.4	99.6	84.8	95.8	99.5	75.8	93.5	89.6	86.2	88.7
0920	Plaster & Gypsum Board	86.6	57.8	66.4	85.8	70.3	74.9	85.8	55.8	64.7	94.8	89.0	90.7	92.1	73.0	78.7	98.0	97.9	97.9
0950, 0980	Ceilings & Acoustic Treatment	89.7	57.8	68.6	91.8	70.3	77.6	89.7	55.8	67.2	85.9	89.0	87.9	85.9	73.0	77.3	86.8	97.9	94.2
0960	Flooring	111.8	55.8	95.7	109.9	55.8	94.3	111.3	55.8	95.3	97.2	96.6	97.0	93.6	88.8	92.2	94.7	45.4	80.6
0970, 0990	Wall Finishes & Painting/Coating	105.9	45.5	69.3	105.9	82.0	91.4	105.9	63.2	80.0	100.9	105.6	103.8	100.9	74.3	84.8	95.1	48.3	66.7
09	FINISHES	99.7	54.9	75.0	99.7	68.3	82.4	99.8	56.7	76.0	95.8	91.1	93.2	94.4	76.7	84.6	96.0	81.0	87.7
COVERS	DIVS. 10 - 14, 25, 28, 41, 43, 44, 46	100.0	66.2	93.0	100.0	66.6	93.1	100.0	63.6	92.4	100.0	92.9	98.5	100.0	93.9	98.7	100.0	57.5	91.2
21, 22, 23	FIRE SUPPRESSION, PLUMBING & HVAC	96.9	37.9	72.3	99.8	60.1	83.3	97.8	47.5	76.9	94.5	88.7	92.1	99.9	97.1	98.7	94.9	45.0	74.1
26, 27, 3370	ELECTRICAL, COMMUNICATIONS & UTIL.	94.3	53.7	72.7	97.6	60.5	77.9	97.7	53.5	74.2	99.2	84.7	91.5	99.2	110.2	105.0	92.5	35.7	62.3
MF2010	WEIGHTED AVERAGE	95.4	55.3	77.5	95.9	67.0	83.0	97.1	58.8	80.0	98.1	92.2	95.5	99.2	91.3	95.7	95.2	71.1	84.4

	DIVISION	MISSOURI																	
		COLUMBIA			FLAT RIVER			HANNIBAL			HARRISONVILLE			JEFFERSON CITY			JOPLIN		
		652			636			634			647			650 - 651			648		
		MAT.	INST.	TOTAL	MAT.	INST.	TOTAL	MAT.	INST.	TOTAL	MAT.	INST.	TOTAL	MAT.	INST.	TOTAL	MAT.	INST.	TOTAL
015433	CONTRACTOR EQUIPMENT		107.4	107.4		106.5	106.5		106.5	106.5		101.6	101.6		107.4	107.4		105.0	105.0
0241, 31 - 34	SITE & INFRASTRUCTURE, DEMOLITION	107.4	95.7	99.3	95.2	93.5	94.0	90.0	93.2	92.2	90.6	95.5	94.0	103.5	95.6	98.0	98.0	97.8	97.8
0310	Concrete Forming & Accessories	89.4	81.1	82.2	101.9	89.0	90.7	94.6	79.0	81.0	88.4	106.1	103.7	99.3	81.0	83.4	102.6	75.0	78.7
0320	Concrete Reinforcing	100.3	113.4	107.0	105.0	107.4	106.2	103.1	98.3	100.6	97.5	113.2	105.6	99.0	108.4	103.9	101.0	73.4	86.8
0330	Cast-in-Place Concrete	91.2	84.3	88.6	97.1	96.5	96.9	89.0	76.7	84.3	96.6	103.4	99.2	91.8	84.3	88.9	102.1	67.0	88.6
03	CONCRETE	86.3	89.7	88.0	97.5	96.3	96.9	90.7	83.3	87.2	96.2	106.7	101.2	87.8	88.7	88.2	99.5	72.9	86.7
04	MASONRY	133.3	87.4	105.5	123.6	80.5	97.5	117.1	98.0	105.5	94.5	104.6	100.6	96.6	87.4	91.0	92.6	86.6	89.0
05	METALS	92.9	119.5	102.1	96.4	115.7	103.1	96.5	111.3	101.6	92.4	112.7	99.4	92.3	117.3	100.9	95.1	88.8	92.9
06	WOOD, PLASTICS & COMPOSITES	90.7	78.8	83.9	100.6	89.3	94.2	91.3	76.8	83.0	82.3	106.0	95.9	101.2	78.8	88.3	96.5	76.1	84.8
07	THERMAL & MOISTURE PROTECTION	93.8	87.6	91.4	99.2	95.7	97.8	98.9	94.1	97.0	96.9	106.5	100.7	93.9	87.6	91.4	97.3	73.2	87.7
08	OPENINGS	91.3	95.3	92.3	99.5	87.5	96.5	99.6	78.0	94.1	89.9	108.2	94.5	88.6	94.0	89.9	90.8	73.8	86.5
0920	Plaster & Gypsum Board	93.9	78.1	82.8	100.1	89.0	92.3	92.6	76.0	80.9	91.2	106.0	101.7	98.9	78.1	84.3	105.1	75.2	84.0
0950, 0980	Ceilings & Acoustic Treatment	89.5	78.1	81.9	85.9	89.0	87.9	85.9	76.0	79.3	86.8	106.0	99.6	88.6	78.1	81.6	88.7	75.2	79.7
0960	Flooring	109.4	96.7	105.8	100.4	88.8	97.0	96.3	96.6	96.4	90.3	104.9	94.5	117.0	96.7	111.2	119.9	42.6	97.7
0970, 0990	Wall Finishes & Painting/Coating	104.0	81.1	90.1	100.9	63.9	78.5	100.9	105.6	103.8	99.8	107.9	104.7	104.0	81.1	90.1	94.6	37.8	60.1
09	FINISHES	96.9	82.6	89.0	97.6	85.1	90.7	95.1	83.7	88.8	93.6	106.0	100.4	99.8	82.6	90.3	102.8	65.6	82.3
COVERS	DIVS. 10 - 14, 25, 28, 41, 43, 44, 46	100.0	95.6	99.1	100.0	96.2	99.2	100.0	56.6	91.0	100.0	99.9	100.0	100.0	95.6	99.1	100.0	64.1	92.5
21, 22, 23	FIRE SUPPRESSION, PLUMBING & HVAC	100.0	97.4	98.9	94.5	88.7	92.1	94.5	88.7	92.0	94.8	103.5	98.5	100.0	91.6	96.5	100.3	49.1	79.0
26, 27, 3370	ELECTRICAL, COMMUNICATIONS & UTIL.	94.3	89.9	91.9	104.2	110.2	107.4	97.8	84.7	90.8	99.7	103.8	101.9	99.1	89.9	94.2	90.2	70.5	79.8
MF2010	WEIGHTED AVERAGE	97.4	93.7	95.7	99.1	94.7	97.1	96.9	88.8	93.3	94.8	104.8	99.3	95.9	92.1	94.2	97.1	71.4	85.6

	DIVISION	MISSOURI																	
		KANSAS CITY			KIRKSVILLE			POPLAR BLUFF			ROLLA			SEDALIA			SIKESTON		
		640 - 641			635			639			654 - 655			653			638		
		MAT.	INST.	TOTAL	MAT.	INST.	TOTAL	MAT.	INST.	TOTAL	MAT.	INST.	TOTAL	MAT.	INST.	TOTAL	MAT.	INST.	TOTAL
015433	CONTRACTOR EQUIPMENT		102.8	102.8		98.9	98.9		101.1	101.1		107.4	107.4		99.5	99.5		101.1	101.1
0241, 31 - 34	SITE & INFRASTRUCTURE, DEMOLITION	92.0	97.7	96.0	91.6	89.8	90.4	78.1	93.0	88.4	107.0	93.8	97.9	98.3	92.1	94.0	81.6	93.2	89.6
0310	Concrete Forming & Accessories	101.3	106.9	106.2	87.2	79.6	80.6	87.6	74.1	75.9	96.4	92.4	92.9	93.6	81.5	83.1	88.5	75.2	76.9
0320	Concrete Reinforcing	96.0	114.1	105.3	103.9	87.3	95.4	107.1	84.6	95.5	100.8	68.9	84.4	99.3	106.4	102.9	106.4	84.5	95.1
0330	Cast-in-Place Concrete	94.9	107.8	99.8	97.0	76.6	89.2	74.6	81.5	77.3	93.3	87.5	91.1	97.5	85.1	92.7	79.8	83.9	81.3
03	CONCRETE	95.5	108.8	101.8	107.6	80.9	94.9	84.0	79.7	81.9	88.1	87.6	87.9	99.7	88.3	94.2	87.6	80.9	84.4
04	MASONRY	96.5	108.4	103.7	129.6	89.3	105.2	121.5	73.1	92.2	110.0	80.5	92.1	116.1	87.8	99.0	121.2	77.9	95.0
05	METALS	102.7	113.7	106.5	96.1	97.6	96.6	96.7	97.7	97.0	92.2	103.6	96.2	90.9	108.2	96.8	97.0	96.9	97.0
06	WOOD, PLASTICS & COMPOSITES	95.8	106.4	101.9	79.6	78.8	79.1	78.8	73.8	76.0	97.8	95.2	96.3	90.8	78.8	83.9	80.2	73.8	76.6
07	THERMAL & MOISTURE PROTECTION	97.0	109.5	101.9	104.9	87.3	97.9	103.6	82.3	95.1	94.0	92.8	93.5	99.2	94.8	97.4	103.8	80.7	94.6
08	OPENINGS	96.9	108.8	99.9	103.9	76.5	96.9	104.9	75.1	97.3	91.3	89.1	90.7	94.7	88.9	93.2	104.9	68.5	95.7
0920	Plaster & Gypsum Board	100.9	106.5	104.8	87.2	78.1	80.8	87.7	73.0	77.3	95.6	95.0	95.2	88.5	78.1	81.2	89.5	73.0	77.9
0950, 0980	Ceilings & Acoustic Treatment	94.3	106.5	102.4	84.0	78.1	80.1	85.9	73.0	77.3	89.5	95.0	93.2	89.5	78.1	81.9	85.9	73.0	77.3
0960	Flooring	95.9	108.8	99.6	74.8	96.7	81.1	88.7	82.2	86.8	113.3	96.7	108.6	86.8	97.9	90.0	89.1	82.2	87.1
0970, 0990	Wall Finishes & Painting/Coating	99.8	115.5	109.4	96.3	81.1	87.0	95.8	73.7	82.4	104.0	88.2	94.4	104.0	107.9	106.4	95.8	58.9	73.4
09	FINISHES	97.9	108.0	103.5	94.3	81.7	87.3	94.4	74.1	83.2	98.3	92.2	94.9	95.2	86.1	90.2	95.0	73.0	82.8
COVERS	DIVS. 10 - 14, 25, 28, 41, 43, 44, 46	100.0	101.3	100.3	100.0	54.9	90.6	100.0	91.8	98.3	100.0	59.1	91.5	100.0	93.1	98.6	100.0	93.5	98.7
21, 22, 23	FIRE SUPPRESSION, PLUMBING & HVAC	100.2	108.7	103.7	94.5	88.1	91.8	94.5	93.5	94.1	94.6	89.0	92.3	94.6	98.2	96.1	94.5	86.4	91.1
26, 27, 3370	ELECTRICAL, COMMUNICATIONS & UTIL.	101.3	104.2	102.8	98.0	84.6	90.9	98.3	110.2	104.6	92.6	89.8	91.1	94.0	103.8	99.2	97.3	110.2	104.1
MF2010	WEIGHTED AVERAGE	99.0	107.3	102.7	100.1	85.4	93.5	96.6	88.4	92.9	95.0	89.5	92.5	96.5	94.6	95.7	97.1	87.1	92.6

	DIVISION	MISSOURI									MONTANA								
		SPRINGFIELD			ST. JOSEPH			ST. LOUIS			BILLINGS			BUTTE			GREAT FALLS		
		656 - 658			644 - 645			630 - 631			590 - 591			597			594		
		MAT.	INST.	TOTAL	MAT.	INST.	TOTAL	MAT.	INST.	TOTAL	MAT.	INST.	TOTAL	MAT.	INST.	TOTAL	MAT.	INST.	TOTAL
015433	CONTRACTOR EQUIPMENT		101.9	101.9		101.6	101.6		107.6	107.6		98.9	98.9		98.6	98.6		98.6	98.6
0241, 31 - 34	SITE & INFRASTRUCTURE, DEMOLITION	100.1	93.7	95.6	93.5	93.7	93.6	93.4	97.2	96.1	95.9	96.8	96.5	106.4	96.1	99.3	110.4	96.8	101.0
0310	Concrete Forming & Accessories	101.7	75.7	79.1	101.2	89.4	91.0	101.9	102.7	102.6	98.0	67.3	71.3	87.2	65.3	68.2	97.8	68.8	72.6
0320	Concrete Reinforcing	96.5	112.9	104.9	94.8	112.8	104.1	96.0	109.2	102.8	96.6	81.5	88.8	104.9	81.5	92.8	96.7	81.5	88.8
0330	Cast-in-Place Concrete	99.1	73.9	89.4	94.8	104.5	98.5	93.0	108.3	98.9	113.0	67.6	95.6	124.3	69.3	103.3	131.4	65.6	106.2
03	CONCRETE	96.8	83.0	90.2	95.3	99.7	97.4	93.1	106.7	99.6	106.0	71.0	89.3	109.7	70.8	91.1	115.0	71.1	94.0
04	MASONRY	90.0	86.3	87.8	96.2	96.0	96.1	103.7	110.5	107.8	130.4	71.4	94.7	125.4	73.2	93.8	130.1	75.6	97.0
05	METALS	96.9	106.0	100.0	98.9	111.6	103.3	102.5	119.6	108.4	110.0	90.8	103.4	102.9	90.8	98.7	106.4	90.8	101.0
06	WOOD, PLASTICS & COMPOSITES	98.1	74.9	84.8	96.5	86.9	91.0	100.5	100.7	100.6	98.0	67.6	80.6	87.4	64.9	74.5	99.1	67.6	81.0
07	THERMAL & MOISTURE PROTECTION	98.0	81.6	91.5	97.4	97.0	97.3	99.0	107.1	102.2	106.3	70.2	92.0	106.2	70.1	91.9	106.7	69.0	91.7
08	OPENINGS	96.9	82.1	93.2	95.5	98.0	96.1	98.9	108.3	101.3	95.5	67.6	88.4	93.5	65.5	86.4	96.8	66.9	89.3
0920	Plaster & Gypsum Board	97.0	74.0	80.8	106.4	86.4	92.3	101.0	100.7	100.8	94.6	67.0	75.1	95.3	64.2	73.4	103.3	67.0	77.7
0950, 0980	Ceilings & Acoustic Treatment	89.5	74.0	79.2	93.4	86.4	88.7	90.5	100.7	97.3	102.1	67.0	78.8	106.5	64.2	78.5	108.4	67.0	80.9
0960	Flooring	111.2	42.6	91.5	99.1	97.9	98.7	100.1	100.1	100.1	111.9	74.2	101.1	109.1	78.4	100.3	117.8	78.4	106.4
0970, 0990	Wall Finishes & Painting/Coating	98.1	54.0	71.4	95.1	90.7	92.4	100.9	109.0	105.8	100.0	50.4	69.9	98.4	55.2	72.2	98.4	52.4	70.5
09	FINISHES	100.7	66.6	81.9	99.1	90.4	94.3	98.5	102.2	100.5	104.3	66.3	83.3	104.1	66.2	83.2	108.4	68.3	86.3
COVERS	DIVS. 10 - 14, 25, 28, 41, 43, 44, 46	100.0	93.7	98.7	100.0	97.1	99.4	100.0	102.2	100.5	100.0	77.7	95.4	100.0	77.5	95.3	100.0	79.2	95.7
21, 22, 23	FIRE SUPPRESSION, PLUMBING & HVAC	100.1	72.3	88.5	100.3	89.2	95.7	99.9	103.3	101.3	100.1	73.1	88.8	100.2	70.0	87.6	100.2	75.2	89.8
26, 27, 3370	ELECTRICAL, COMMUNICATIONS & UTIL.	98.7	74.0	85.5	99.7	78.1	88.2	101.9	110.6	106.5	94.7	78.3	86.0	102.7	74.1	87.5	94.8	75.1	84.3
MF2010	WEIGHTED AVERAGE	98.2	80.9	90.5	98.3	93.2	96.0	99.4	106.6	102.6	103.2	75.9	91.0	103.2	74.6	90.4	104.7	76.5	92.1

DIVISION		MONTANA																	
		HAVRE			HELENA			KALISPELL			MILES CITY			MISSOULA			WOLF POINT		
		595			596			599			593			598			592		
		MAT.	INST.	TOTAL	MAT.	INST.	TOTAL	MAT.	INST.	TOTAL	MAT.	INST.	TOTAL	MAT.	INST.	TOTAL	MAT.	INST.	TOTAL
015433	CONTRACTOR EQUIPMENT		98.6	98.6		98.6	98.6		98.6	98.6		98.6	98.6		98.6	98.6		98.6	98.6
0241, 31 - 34	SITE & INFRASTRUCTURE, DEMOLITION	115.9	96.5	102.5	99.1	96.0	96.9	95.6	95.7	95.7	103.2	96.1	98.3	86.4	95.9	93.0	124.4	96.2	105.0
0310	Concrete Forming & Accessories	81.3	58.4	61.4	98.0	68.0	72.0	89.8	56.3	60.7	96.5	56.8	62.1	89.8	63.6	67.1	90.3	51.8	56.8
0320	Concrete Reinforcing	105.7	65.3	84.9	100.0	65.0	82.0	107.7	81.2	94.0	105.2	65.3	84.7	106.6	81.5	93.7	106.8	65.3	85.4
0330	Cast-in-Place Concrete	133.9	64.2	107.2	102.0	64.6	87.6	107.9	61.7	90.2	118.2	63.8	97.4	91.6	63.8	80.9	132.4	64.2	106.3
03	CONCRETE	118.4	62.8	91.8	100.7	67.2	84.6	98.9	64.1	82.2	106.3	62.0	85.1	86.8	68.1	77.8	122.3	59.9	92.4
04	MASONRY	126.4	73.6	94.4	121.2	70.0	90.2	124.3	68.3	90.4	132.3	68.7	93.7	150.9	71.8	102.9	133.5	69.4	94.7
05	METALS	99.1	82.6	93.5	105.5	82.0	97.4	99.0	89.2	95.6	98.3	82.5	92.9	99.6	89.9	96.2	98.4	82.5	92.9
06	WOOD, PLASTICS & COMPOSITES	80.0	55.2	65.8	99.1	67.6	81.1	90.3	53.5	69.2	96.6	55.7	73.1	90.3	63.7	75.1	89.4	48.5	65.9
07	THERMAL & MOISTURE PROTECTION	106.5	66.8	90.7	106.5	71.0	92.4	105.8	67.7	90.7	106.1	64.9	89.8	105.5	69.5	91.2	107.0	63.9	89.9
08	OPENINGS	93.6	55.6	83.9	96.3	61.2	87.4	93.5	59.3	84.9	93.0	55.8	83.6	93.5	64.6	86.2	93.0	51.7	82.6
0920	Plaster & Gypsum Board	90.4	54.2	64.9	95.4	67.0	75.4	95.7	52.4	65.2	102.3	54.6	68.7	95.7	62.9	72.6	97.4	47.2	62.1
0950, 0980	Ceilings & Acoustic Treatment	106.5	54.2	71.8	105.6	67.0	80.0	106.5	52.4	70.6	104.7	54.6	71.5	106.5	62.9	77.6	104.7	47.2	66.6
0960	Flooring	105.7	78.4	97.8	117.8	62.4	101.9	111.3	78.4	101.8	117.5	74.2	105.1	111.3	78.4	101.8	113.0	74.2	101.9
0970, 0990	Wall Finishes & Painting/Coating	98.4	52.4	70.5	98.4	46.6	66.9	98.4	40.3	63.1	98.4	48.3	68.0	98.4	51.2	69.8	98.4	52.4	70.5
09	FINISHES	103.3	60.5	79.7	106.3	64.9	83.5	104.2	57.7	78.6	106.9	58.4	80.1	103.5	64.7	82.1	106.5	54.7	77.9
COVERS	DIVS. 10 - 14, 25, 28, 41, 43, 44, 46	100.0	77.1	95.2	100.0	79.2	95.7	100.0	76.1	95.0	100.0	75.4	94.9	100.0	76.9	95.2	100.0	74.8	94.8
21, 22, 23	FIRE SUPPRESSION, PLUMBING & HVAC	94.8	73.8	86.0	100.1	71.5	88.2	94.8	69.7	84.3	94.8	71.2	85.0	100.2	69.1	87.2	94.8	71.6	85.1
26, 27, 3370	ELECTRICAL, COMMUNICATIONS & UTIL.	94.8	67.4	80.2	101.0	75.1	87.2	99.3	73.5	85.6	94.8	70.2	81.7	100.4	73.5	86.1	94.8	65.4	79.2
MF2010	WEIGHTED AVERAGE	101.9	71.3	88.2	102.5	73.2	89.4	99.4	71.3	86.8	100.6	70.2	87.0	100.5	73.4	88.4	103.2	68.6	87.7

DIVISION		NEBRASKA																	
		ALLIANCE			COLUMBUS			GRAND ISLAND			HASTINGS			LINCOLN			MCCOOK		
		693			686			688			689			683 - 685			690		
		MAT.	INST.	TOTAL	MAT.	INST.	TOTAL	MAT.	INST.	TOTAL	MAT.	INST.	TOTAL	MAT.	INST.	TOTAL	MAT.	INST.	TOTAL
015433	CONTRACTOR EQUIPMENT		96.8	96.8		101.0	101.0		101.0	101.0		101.0	101.0		101.0	101.0		101.0	101.0
0241, 31 - 34	SITE & INFRASTRUCTURE, DEMOLITION	100.8	97.4	98.5	99.4	90.8	93.5	103.9	92.0	95.7	103.0	91.2	94.9	91.3	92.0	91.8	103.9	91.0	95.0
0310	Concrete Forming & Accessories	90.4	57.5	61.8	98.8	60.4	65.5	98.4	77.7	80.4	101.6	75.9	79.3	103.2	61.6	67.1	96.1	57.8	62.9
0320	Concrete Reinforcing	114.4	48.3	80.4	105.6	47.1	75.5	105.1	76.6	90.4	105.1	47.8	75.6	96.5	76.1	86.0	107.1	76.2	91.2
0330	Cast-in-Place Concrete	110.2	59.4	90.7	112.3	56.2	90.8	118.9	71.2	100.6	118.9	65.2	98.4	88.8	73.7	83.0	118.9	59.1	96.0
03	CONCRETE	118.3	56.9	88.9	104.6	57.5	82.1	109.3	75.9	93.3	109.5	67.6	89.5	91.7	69.5	81.1	109.4	62.8	87.1
04	MASONRY	110.8	68.3	85.1	115.9	73.2	90.0	108.9	80.8	91.8	117.8	90.5	101.3	97.2	71.2	81.4	106.0	68.4	83.2
05	METALS	98.5	65.2	87.0	91.1	74.9	85.5	92.9	87.0	90.9	93.4	76.5	87.6	97.9	86.1	93.8	93.7	84.9	90.6
06	WOOD, PLASTICS & COMPOSITES	84.3	56.6	68.4	95.5	61.6	76.0	94.7	78.5	85.4	98.4	78.5	87.0	97.6	57.0	74.3	92.7	56.7	72.1
07	THERMAL & MOISTURE PROTECTION	104.9	68.9	90.6	98.8	65.8	85.7	98.9	76.6	90.0	98.9	88.4	94.7	98.4	71.8	87.9	98.9	66.0	85.8
08	OPENINGS	90.9	52.7	81.2	90.6	57.3	82.2	90.7	74.2	86.5	90.7	66.5	84.5	96.1	58.4	86.6	90.7	59.9	82.9
0920	Plaster & Gypsum Board	86.4	55.3	64.5	95.2	60.4	70.7	94.3	77.8	82.7	96.5	77.8	83.3	96.4	55.6	67.7	94.3	55.3	66.8
0950, 0980	Ceilings & Acoustic Treatment	93.9	55.3	68.3	89.5	60.4	70.2	89.5	77.8	81.7	89.5	77.8	81.7	92.3	55.6	68.0	89.5	55.3	66.8
0960	Flooring	101.4	86.7	97.2	100.4	106.0	102.0	100.2	113.3	104.0	101.6	106.0	102.9	101.8	88.3	97.9	99.2	86.7	95.6
0970, 0990	Wall Finishes & Painting/Coating	172.0	59.0	103.4	96.2	68.6	79.4	96.2	72.8	82.0	96.2	68.6	79.4	96.2	83.8	88.7	96.2	51.4	69.0
09	FINISHES	100.6	62.2	79.4	97.6	68.6	81.6	97.7	82.9	89.6	98.4	80.1	88.3	98.5	68.0	81.7	97.4	61.3	77.5
COVERS	DIVS. 10 - 14, 25, 28, 41, 43, 44, 46	100.0	70.5	93.9	100.0	80.1	95.9	100.0	85.7	97.0	100.0	84.3	96.7	100.0	83.2	96.5	100.0	69.7	93.7
21, 22, 23	FIRE SUPPRESSION, PLUMBING & HVAC	94.8	60.8	80.7	94.7	74.0	86.1	100.1	80.1	91.7	94.7	77.2	87.4	99.9	80.0	91.7	94.7	74.5	86.3
26, 27, 3370	ELECTRICAL, COMMUNICATIONS & UTIL.	91.2	75.3	82.7	90.8	83.7	87.0	89.4	68.8	78.4	88.7	90.7	89.8	100.4	68.8	83.6	93.5	75.3	83.8
MF2010	WEIGHTED AVERAGE	99.6	66.8	85.0	96.7	72.9	86.1	98.5	79.8	90.1	97.7	80.7	90.1	97.7	74.9	87.5	97.5	71.9	86.0

DIVISION		NEBRASKA												NEVADA					
		NORFOLK			NORTH PLATTE			OMAHA			VALENTINE			CARSON CITY			ELKO		
		687			691			680 - 681			692			897			898		
		MAT.	INST.	TOTAL	MAT.	INST.	TOTAL	MAT.	INST.	TOTAL	MAT.	INST.	TOTAL	MAT.	INST.	TOTAL	MAT.	INST.	TOTAL
015433	CONTRACTOR EQUIPMENT		91.8	91.8		101.0	101.0		91.8	91.8		94.6	94.6		100.6	100.6		100.6	100.6
0241, 31 - 34	SITE & INFRASTRUCTURE, DEMOLITION	81.4	90.2	87.5	104.8	91.2	95.4	81.4	90.9	88.0	86.9	94.5	92.2	65.9	102.3	91.0	54.3	102.0	87.2
0310	Concrete Forming & Accessories	84.6	75.0	76.3	98.3	75.7	78.7	93.5	74.6	77.1	86.0	57.0	60.9	101.5	87.1	89.0	108.4	84.6	87.8
0320	Concrete Reinforcing	105.8	49.8	77.0	106.5	77.3	91.5	101.1	76.6	88.5	107.2	46.8	76.1	103.6	119.3	111.7	105.3	93.2	99.1
0330	Cast-in-Place Concrete	113.4	64.4	94.7	118.9	66.7	98.9	95.3	80.4	89.6	105.1	56.7	86.6	98.2	86.4	93.7	99.1	84.0	93.3
03	CONCRETE	103.5	66.9	86.0	109.5	73.5	92.3	94.3	77.2	86.1	105.9	55.7	81.9	102.3	93.1	97.9	100.0	86.2	93.4
04	MASONRY	122.7	71.0	91.4	93.1	90.5	91.6	102.7	81.6	89.9	105.8	68.4	83.1	116.9	75.8	92.0	119.3	68.5	88.5
05	METALS	94.3	68.1	85.3	93.1	86.9	91.0	97.2	79.1	91.0	104.3	66.3	91.2	94.7	104.2	98.0	93.7	91.4	92.9
06	WOOD, PLASTICS & COMPOSITES	79.1	78.0	78.4	94.6	78.5	85.4	89.2	74.1	80.5	79.0	56.2	65.9	90.2	87.0	88.3	99.6	86.3	92.0
07	THERMAL & MOISTURE PROTECTION	98.3	75.1	89.0	98.8	82.1	92.2	94.5	82.0	89.5	99.0	68.8	87.0	99.7	85.8	94.2	101.0	76.4	91.2
08	OPENINGS	92.3	67.0	85.9	89.9	73.5	85.8	99.0	73.9	92.6	92.4	53.9	82.7	95.2	106.4	98.0	98.0	83.9	94.4
0920	Plaster & Gypsum Board	94.7	77.8	82.8	94.3	77.8	82.7	99.3	73.8	81.4	96.7	55.3	67.5	94.0	86.4	88.7	98.9	85.8	89.6
0950, 0980	Ceilings & Acoustic Treatment	104.9	77.8	86.9	89.5	77.8	81.7	105.8	73.8	84.6	107.4	55.3	72.8	119.1	86.4	97.5	122.0	85.8	98.0
0960	Flooring	126.9	106.0	120.9	100.1	106.0	101.8	131.6	86.6	118.7	129.4	84.3	116.5	101.7	82.7	96.3	107.1	54.0	91.8
0970, 0990	Wall Finishes & Painting/Coating	170.4	68.6	108.6	96.2	68.6	79.4	170.4	70.4	109.7	170.4	71.1	110.2	99.5	85.0	90.7	98.5	85.0	90.3
09	FINISHES	118.9	80.1	97.5	97.7	80.1	88.0	121.1	76.1	96.3	120.9	62.6	88.8	103.2	85.3	93.3	104.6	79.1	90.5
COVERS	DIVS. 10 - 14, 25, 28, 41, 43, 44, 46	100.0	83.0	96.5	100.0	73.4	94.5	100.0	84.0	96.7	100.0	67.6	93.3	100.0	98.3	99.6	100.0	69.3	93.6
21, 22, 23	FIRE SUPPRESSION, PLUMBING & HVAC	94.5	77.1	87.2	100.1	75.4	89.8	99.9	79.9	91.5	94.4	58.8	79.6	99.9	79.9	91.5	97.4	76.5	88.7
26, 27, 3370	ELECTRICAL, COMMUNICATIONS & UTIL.	89.7	84.3	86.8	91.5	83.7	87.3	100.1	84.6	91.9	88.5	94.0	91.4	95.8	100.6	98.3	93.7	100.6	97.4
MF2010	WEIGHTED AVERAGE	98.4	76.4	88.6	97.8	80.9	90.3	99.8	80.6	91.2	99.5	68.7	85.7	98.6	90.8	95.1	97.7	84.4	91.8

DIVISION		NEVADA									NEW HAMPSHIRE								
		ELY			LAS VEGAS			RENO			CHARLESTON			CLAREMONT			CONCORD		
		893			889 - 891			894 - 895			036			037			032 - 033		
		MAT.	INST.	TOTAL	MAT.	INST.	TOTAL	MAT.	INST.	TOTAL	MAT.	INST.	TOTAL	MAT.	INST.	TOTAL	MAT.	INST.	TOTAL
015433	CONTRACTOR EQUIPMENT		100.6	100.6		100.6	100.6		100.6	100.6		100.3	100.3		100.3	100.3		100.3	100.3
0241, 31 - 34	SITE & INFRASTRUCTURE, DEMOLITION	58.9	102.5	89.0	60.7	104.9	91.2	58.8	102.3	88.8	81.6	96.4	91.8	75.8	96.4	90.0	86.8	98.3	94.8
0310	Concrete Forming & Accessories	102.0	106.9	106.2	103.7	117.3	115.5	98.8	87.0	88.6	85.3	44.2	49.7	90.9	44.2	50.4	92.4	86.2	87.0
0320	Concrete Reinforcing	104.1	86.3	94.9	96.0	127.3	112.1	98.4	126.0	112.6	86.5	47.4	66.4	86.5	47.4	66.4	86.5	90.8	88.7
0330	Cast-in-Place Concrete	106.4	106.2	106.3	103.0	114.1	107.3	112.5	86.3	102.5	94.1	50.6	77.4	86.5	50.6	72.7	102.4	88.0	96.9
03	CONCRETE	108.5	102.5	105.6	103.7	117.6	110.3	108.5	94.3	101.7	103.9	47.9	77.1	95.2	47.9	72.5	106.3	87.6	97.3
04	MASONRY	124.4	80.8	98.0	111.7	109.1	110.1	118.4	75.8	92.6	94.0	38.6	60.4	93.5	38.6	60.2	96.4	81.9	87.6
05	METALS	93.7	91.5	92.9	100.8	112.1	104.7	95.1	106.9	99.2	92.7	61.6	82.0	92.7	61.6	82.0	93.5	87.3	91.4
06	WOOD, PLASTICS & COMPOSITES	91.0	108.7	101.1	90.2	117.5	105.8	86.6	87.0	86.8	85.9	45.8	62.9	91.9	45.8	65.5	95.7	89.8	92.3
07	THERMAL & MOISTURE PROTECTION	101.7	88.7	96.5	115.2	102.8	110.3	101.2	85.8	95.1	95.9	41.8	74.4	95.8	41.8	74.3	94.7	84.9	90.8
08	OPENINGS	97.9	94.1	96.9	96.8	124.8	103.9	95.5	101.8	97.1	102.1	41.4	86.7	103.4	41.4	87.7	103.4	81.0	97.7
0920	Plaster & Gypsum Board	94.5	108.8	104.6	94.8	117.9	111.0	87.9	86.4	86.9	91.5	43.5	57.7	93.3	43.5	58.3	95.9	88.8	90.9
0950, 0980	Ceilings & Acoustic Treatment	122.0	108.8	113.3	132.3	117.9	122.7	128.5	86.4	100.6	89.8	43.5	59.1	89.8	43.5	59.1	88.9	88.8	88.8
0960	Flooring	104.6	60.4	91.9	95.7	104.3	98.2	101.3	82.7	96.0	90.7	33.9	74.4	93.1	33.9	76.1	90.8	94.6	91.9
0970, 0990	Wall Finishes & Painting/Coating	98.5	120.8	112.0	101.1	124.7	115.4	98.5	85.0	90.3	92.3	49.2	66.2	92.3	49.2	66.2	92.3	98.6	96.2
09	FINISHES	103.9	100.6	102.1	103.4	116.0	110.4	103.3	85.3	93.4	92.8	43.4	65.5	93.2	43.4	65.7	93.9	89.9	91.7
COVERS	DIVS. 10 - 14, 25, 28, 41, 43, 44, 46	100.0	69.4	93.6	100.0	96.9	99.4	100.0	98.3	99.6	100.0	39.7	87.5	100.0	39.7	87.5	100.0	63.8	92.5
21, 22, 23	FIRE SUPPRESSION, PLUMBING & HVAC	97.4	107.2	101.5	100.0	110.0	104.2	99.9	79.9	91.6	94.8	40.2	72.0	94.8	40.2	72.1	100.0	83.2	93.0
26, 27, 3370	ELECTRICAL, COMMUNICATIONS & UTIL.	94.0	111.3	103.2	98.1	119.1	109.2	94.4	100.6	97.7	98.0	34.9	64.4	98.0	34.9	64.4	98.9	60.7	78.6
MF2010	WEIGHTED AVERAGE	99.1	99.6	99.3	100.2	112.8	105.9	99.2	91.1	95.6	96.5	47.6	74.6	95.4	47.6	74.0	98.7	82.5	91.5

DIVISION		NEW HAMPSHIRE															NEW JERSEY		
		KEENE			LITTLETON			MANCHESTER			NASHUA			PORTSMOUTH			ATLANTIC CITY		
		034			035			031			030			038			082,084		
		MAT.	INST.	TOTAL	MAT.	INST.	TOTAL	MAT.	INST.	TOTAL	MAT.	INST.	TOTAL	MAT.	INST.	TOTAL	MAT.	INST.	TOTAL
015433	CONTRACTOR EQUIPMENT		100.3	100.3		100.3	100.3		100.3	100.3		100.3	100.3		100.3	100.3		98.4	98.4
0241, 31 - 34	SITE & INFRASTRUCTURE, DEMOLITION	88.6	96.5	94.1	75.9	96.1	89.9	85.1	98.3	94.3	89.3	98.3	95.6	83.5	99.7	94.7	92.1	103.6	100.0
0310	Concrete Forming & Accessories	89.6	42.3	48.5	99.7	47.2	54.1	94.8	86.8	87.8	96.7	86.8	88.1	86.6	90.7	90.2	104.5	126.8	123.8
0320	Concrete Reinforcing	86.5	56.4	71.0	87.3	34.3	60.0	108.1	90.9	99.3	108.1	90.9	99.3	86.5	91.0	88.8	82.4	116.4	99.9
0330	Cast-in-Place Concrete	94.5	52.1	78.3	84.8	52.8	72.6	105.3	110.8	107.4	89.5	110.8	97.6	84.9	117.0	97.2	80.2	130.2	99.4
03	CONCRETE	103.5	49.2	77.5	94.5	47.4	71.9	111.1	95.6	103.7	103.2	95.6	99.6	94.6	99.6	97.0	96.6	124.9	110.1
04	MASONRY	97.7	40.8	63.2	105.7	44.3	68.5	99.5	81.9	88.9	98.8	81.9	88.6	94.6	91.4	92.6	101.4	127.0	116.9
05	METALS	93.4	65.9	83.9	93.4	56.2	80.6	98.6	87.9	94.9	98.6	87.9	94.9	95.0	90.9	93.6	93.2	104.8	97.2
06	WOOD, PLASTICS & COMPOSITES	90.3	40.9	62.0	100.6	46.8	69.8	95.7	89.8	92.3	99.0	89.8	93.7	87.1	89.8	88.6	109.1	126.8	119.2
07	THERMAL & MOISTURE PROTECTION	96.3	43.8	75.5	95.9	47.7	76.7	95.9	88.4	92.9	96.7	88.4	93.4	96.3	112.0	102.5	96.4	126.6	108.4
08	OPENINGS	100.6	46.2	86.8	104.5	39.2	87.9	105.6	81.0	99.4	105.6	81.0	99.4	106.2	76.3	98.6	103.0	124.1	108.4
0920	Plaster & Gypsum Board	92.4	38.4	54.4	105.7	44.5	62.6	103.7	88.8	93.2	100.2	88.8	92.2	91.5	88.8	89.6	104.3	126.9	120.2
0950, 0980	Ceilings & Acoustic Treatment	89.8	38.4	55.8	89.8	44.5	59.8	95.7	88.8	91.1	99.4	88.8	92.4	91.9	88.8	89.9	89.8	126.9	114.4
0960	Flooring	92.6	55.6	82.0	101.6	33.9	82.2	96.9	94.6	96.2	95.8	94.6	95.5	91.2	94.6	92.2	99.5	134.3	109.5
0970, 0990	Wall Finishes & Painting/Coating	92.3	49.2	66.2	92.3	64.5	75.5	92.3	98.6	96.2	92.3	98.6	96.2	92.3	98.6	96.2	92.3	123.7	111.3
09	FINISHES	94.7	45.2	67.4	97.2	46.8	69.4	97.7	89.9	93.4	98.5	89.9	93.8	94.1	92.1	93.0	98.6	128.4	115.1
COVERS	DIVS. 10 - 14, 25, 28, 41, 43, 44, 46	100.0	54.5	90.6	100.0	41.5	87.8	100.0	63.9	92.5	100.0	63.9	92.5	100.0	67.1	93.2	100.0	112.7	102.6
21, 22, 23	FIRE SUPPRESSION, PLUMBING & HVAC	94.8	43.5	73.4	94.8	67.1	83.3	100.0	83.2	93.0	100.2	83.2	93.1	100.2	88.0	95.1	99.8	123.7	109.7
26, 27, 3370	ELECTRICAL, COMMUNICATIONS & UTIL.	98.0	34.9	64.4	99.3	60.6	78.7	99.6	77.7	87.9	100.8	77.7	88.5	98.6	77.7	87.5	95.5	135.1	116.6
MF2010	WEIGHTED AVERAGE	97.0	49.9	76.0	96.7	57.4	79.1	100.8	86.1	94.2	100.2	86.1	93.9	97.7	89.9	94.2	98.1	122.7	109.1

DIVISION		NEW JERSEY																	
		CAMDEN			DOVER			ELIZABETH			HACKENSACK			JERSEY CITY			LONG BRANCH		
		081			078			072			076			073			077		
		MAT.	INST.	TOTAL	MAT.	INST.	TOTAL	MAT.	INST.	TOTAL	MAT.	INST.	TOTAL	MAT.	INST.	TOTAL	MAT.	INST.	TOTAL
015433	CONTRACTOR EQUIPMENT		98.4	98.4		100.3	100.3		100.3	100.3		100.3	100.3		98.4	98.4		98.0	98.0
0241, 31 - 34	SITE & INFRASTRUCTURE, DEMOLITION	92.8	104.1	100.6	103.9	104.7	104.5	108.0	104.7	105.7	104.6	105.2	105.0	92.8	105.2	101.3	97.8	105.0	102.8
0310	Concrete Forming & Accessories	96.7	126.9	122.9	93.2	127.5	123.0	104.0	127.5	124.4	93.2	127.5	123.0	96.7	127.6	123.5	97.2	127.3	123.3
0320	Concrete Reinforcing	108.1	118.2	113.3	83.3	125.3	104.9	83.3	125.3	104.9	83.3	125.3	104.9	108.1	125.3	117.0	83.3	125.3	104.9
0330	Cast-in-Place Concrete	77.8	132.0	98.6	99.8	129.2	111.0	85.7	129.2	102.4	97.5	129.2	109.6	77.8	129.3	97.5	86.5	130.1	103.2
03	CONCRETE	97.4	125.9	111.0	104.6	126.7	115.2	99.9	126.8	112.8	102.6	126.8	114.2	97.4	126.6	111.4	102.3	126.8	114.0
04	MASONRY	92.5	130.3	115.4	96.9	127.5	115.5	112.5	127.5	121.6	100.7	127.5	117.0	89.7	127.5	112.6	105.2	127.0	118.4
05	METALS	98.5	105.2	100.8	93.2	112.3	99.8	94.7	112.3	100.8	93.2	112.4	99.8	98.5	110.0	102.5	93.3	109.8	99.0
06	WOOD, PLASTICS & COMPOSITES	99.0	126.8	114.9	98.4	126.8	114.7	112.3	126.8	120.6	98.4	126.8	114.7	99.0	126.8	114.9	100.1	126.7	115.4
07	THERMAL & MOISTURE PROTECTION	96.3	126.9	108.4	96.7	127.4	108.9	96.9	127.4	109.0	96.5	124.9	107.8	96.2	127.4	108.6	96.4	126.8	108.5
08	OPENINGS	105.6	124.1	110.3	108.6	125.0	112.8	106.6	125.0	111.2	105.8	125.0	110.7	105.6	125.0	110.5	101.4	125.0	107.4
0920	Plaster & Gypsum Board	100.2	126.9	119.0	96.8	126.9	118.0	104.8	126.9	120.3	96.8	126.9	118.0	100.2	126.9	119.0	99.0	126.9	118.6
0950, 0980	Ceilings & Acoustic Treatment	99.4	126.9	117.6	89.8	126.9	114.4	91.9	126.9	115.1	89.8	126.9	114.4	99.4	126.9	117.6	89.8	126.9	114.4
0960	Flooring	95.8	134.3	106.9	94.8	165.7	115.1	99.9	165.7	118.8	94.8	165.7	115.1	95.8	165.7	115.9	96.1	165.7	116.1
0970, 0990	Wall Finishes & Painting/Coating	92.3	123.7	111.3	92.2	125.7	112.5	92.2	125.7	112.5	92.2	125.7	112.5	92.3	125.7	112.6	92.3	123.7	111.3
09	FINISHES	99.0	128.4	115.2	96.5	134.1	117.2	99.9	134.1	118.8	96.3	134.1	117.2	98.9	134.1	118.4	97.4	133.7	117.5
COVERS	DIVS. 10 - 14, 25, 28, 41, 43, 44, 46	100.0	112.8	102.7	100.0	114.6	103.0	100.0	114.6	103.0	100.0	114.6	103.0	100.0	114.6	103.0	100.0	112.6	102.6
21, 22, 23	FIRE SUPPRESSION, PLUMBING & HVAC	100.2	123.7	110.0	99.8	127.1	111.2	100.2	127.6	111.6	99.8	127.1	111.2	100.2	127.1	111.4	99.8	127.5	111.4
26, 27, 3370	ELECTRICAL, COMMUNICATIONS & UTIL.	101.2	139.9	121.8	96.9	135.9	117.7	97.7	135.9	118.0	96.9	135.6	117.5	102.3	135.5	120.0	96.6	127.9	113.2
MF2010	WEIGHTED AVERAGE	99.4	124.0	110.4	99.6	125.5	111.2	100.6	125.6	111.8	99.3	125.4	111.0	99.3	125.3	110.9	98.9	124.1	110.2

	DIVISION	NEW JERSEY																	
		NEW BRUNSWICK			NEWARK			PATERSON			POINT PLEASANT			SUMMIT			TRENTON		
		088 - 089			070 - 071			074 - 075			087			079			085 - 086		
		MAT.	INST.	TOTAL	MAT.	INST.	TOTAL	MAT.	INST.	TOTAL	MAT.	INST.	TOTAL	MAT.	INST.	TOTAL	MAT.	INST.	TOTAL
015433	CONTRACTOR EQUIPMENT		98.0	98.0		100.3	100.3		100.3	100.3		98.0	98.0		100.3	100.3		98.0	98.0
0241, 31 - 34	SITE & INFRASTRUCTURE, DEMOLITION	105.4	105.1	105.2	108.2	104.7	105.8	106.1	105.2	105.5	107.0	105.0	105.6	105.5	104.7	105.0	90.3	105.0	100.5
0310	Concrete Forming & Accessories	99.5	127.5	123.8	93.2	127.5	123.0	95.3	127.5	123.3	94.6	127.1	122.8	95.9	127.5	123.3	94.0	116.5	113.5
0320	Concrete Reinforcing	83.3	125.3	104.9	108.1	125.3	117.0	108.1	125.3	117.0	83.3	125.2	104.9	83.3	125.3	104.9	108.1	120.6	114.6
0330	Cast-in-Place Concrete	99.1	130.4	111.1	103.4	129.2	113.3	99.2	129.2	110.7	99.1	130.0	111.0	82.8	129.2	100.6	95.5	130.1	108.8
03	CONCRETE	114.8	127.0	120.7	110.0	126.8	118.0	108.0	126.8	117.0	114.5	126.6	120.3	96.8	126.8	111.1	106.1	121.1	113.3
04	MASONRY	99.0	127.5	116.3	102.1	127.5	117.5	97.3	127.5	115.6	90.0	127.0	112.4	99.3	127.5	116.4	88.9	127.0	112.0
05	METALS	93.2	109.9	99.0	98.5	112.3	103.3	93.8	112.4	100.2	93.2	109.6	98.9	93.2	112.3	99.8	93.7	107.4	98.4
06	WOOD, PLASTICS & COMPOSITES	103.2	126.7	116.7	97.7	126.8	114.4	100.8	126.8	115.7	96.7	126.7	113.9	101.9	126.8	116.2	96.1	112.2	105.4
07	THERMAL & MOISTURE PROTECTION	96.5	127.0	108.6	96.4	127.4	108.8	96.8	124.9	107.9	96.6	126.8	108.6	97.1	127.4	109.2	92.6	124.9	105.4
08	OPENINGS	97.4	125.0	104.4	111.7	125.0	115.0	111.7	125.0	115.0	99.5	127.1	106.5	113.9	125.0	116.7	106.2	115.6	108.6
0920	Plaster & Gypsum Board	101.2	126.9	119.3	97.9	126.9	118.3	100.2	126.9	119.0	96.8	126.9	118.0	99.0	126.9	118.6	98.1	112.0	107.9
0950, 0980	Ceilings & Acoustic Treatment	89.8	126.9	114.4	95.7	126.9	116.4	99.4	126.9	117.6	89.8	126.9	114.4	89.8	126.9	114.4	96.6	112.0	106.8
0960	Flooring	97.4	165.7	117.0	96.0	165.7	116.1	95.8	165.7	115.9	94.8	134.3	106.1	96.1	165.7	116.1	96.2	165.7	116.2
0970, 0990	Wall Finishes & Painting/Coating	92.3	125.7	112.6	92.2	125.7	112.5	92.2	125.7	112.5	92.3	123.7	111.3	92.2	125.7	112.5	92.3	123.7	111.3
09	FINISHES	98.9	134.1	118.3	97.8	134.1	117.9	99.1	134.1	118.4	97.6	128.4	114.6	97.5	134.1	117.7	97.7	125.2	112.9
COVERS	DIVS. 10 - 14, 25, 28, 41, 43, 44, 46	100.0	114.5	103.0	100.0	114.6	103.0	100.0	114.6	103.0	100.0	108.1	101.7	100.0	114.6	103.0	100.0	111.0	102.3
21, 22, 23	FIRE SUPPRESSION, PLUMBING & HVAC	99.8	127.8	111.4	100.3	127.6	111.7	100.2	127.1	111.4	99.8	127.4	111.3	99.8	127.6	111.4	100.3	127.5	111.6
26, 27, 3370	ELECTRICAL, COMMUNICATIONS & UTIL.	96.3	132.8	115.7	100.6	135.6	119.2	102.3	135.9	120.2	95.5	127.9	112.7	97.7	135.9	118.0	102.4	137.7	121.2
MF2010	WEIGHTED AVERAGE	100.1	125.0	111.2	102.3	125.6	112.7	101.4	125.5	112.2	99.6	123.3	110.2	99.6	125.6	111.2	99.4	122.7	109.8

	DIVISION	NEW JERSEY			NEW MEXICO														
		VINELAND			ALBUQUERQUE			CARRIZOZO			CLOVIS			FARMINGTON			GALLUP		
		080,083			870 - 872			883			881			874			873		
		MAT.	INST.	TOTAL	MAT.	INST.	TOTAL	MAT.	INST.	TOTAL	MAT.	INST.	TOTAL	MAT.	INST.	TOTAL	MAT.	INST.	TOTAL
015433	CONTRACTOR EQUIPMENT		98.4	98.4		113.0	113.0		113.0	113.0		113.0	113.0		113.0	113.0		113.0	113.0
0241, 31 - 34	SITE & INFRASTRUCTURE, DEMOLITION	96.6	104.1	101.8	75.1	108.5	98.1	94.7	108.5	104.2	83.6	108.4	100.7	81.1	108.5	100.0	88.4	108.5	102.2
0310	Concrete Forming & Accessories	92.3	126.9	122.3	99.2	67.2	71.5	99.2	67.2	71.5	99.2	67.1	71.3	99.2	67.2	71.5	99.2	67.2	71.5
0320	Concrete Reinforcing	82.4	116.4	99.9	99.1	72.2	85.2	106.2	72.2	88.7	107.4	72.1	89.2	108.2	72.2	89.7	103.6	72.2	87.4
0330	Cast-in-Place Concrete	86.5	130.3	103.3	105.2	75.3	93.8	97.9	75.3	89.3	97.9	75.3	89.2	106.2	75.3	94.4	99.9	75.3	90.5
03	CONCRETE	101.8	124.9	112.9	105.0	72.0	89.2	119.2	72.0	96.6	107.9	71.9	90.7	108.6	72.0	91.1	114.5	72.0	94.2
04	MASONRY	89.2	127.0	112.1	107.9	64.3	81.5	104.8	64.3	80.3	104.8	64.3	80.3	114.8	64.3	84.2	104.9	64.3	80.3
05	METALS	93.1	104.9	97.2	101.0	89.4	97.0	97.8	89.4	94.9	97.5	89.2	94.6	98.8	89.4	95.5	97.8	89.4	94.9
06	WOOD, PLASTICS & COMPOSITES	93.9	126.8	112.7	90.3	67.6	77.3	90.4	67.6	77.3	90.4	67.6	77.3	90.4	67.6	77.3	90.4	67.6	77.3
07	THERMAL & MOISTURE PROTECTION	96.2	126.6	108.2	100.7	75.0	90.5	102.5	75.0	91.6	101.2	75.0	90.8	101.0	75.0	90.7	102.1	75.0	91.3
08	OPENINGS	99.0	124.1	105.4	95.3	70.2	88.9	94.1	70.2	88.0	94.3	70.2	88.2	98.3	70.2	91.2	98.3	70.2	91.2
0920	Plaster & Gypsum Board	95.0	126.9	117.5	93.2	66.2	74.2	85.0	66.2	71.8	85.0	66.2	71.8	85.0	66.2	71.8	85.0	66.2	71.8
0950, 0980	Ceilings & Acoustic Treatment	89.8	126.9	114.4	122.4	66.2	85.2	118.2	66.2	83.7	118.2	66.2	83.7	118.2	66.2	83.7	118.2	66.2	83.7
0960	Flooring	94.0	134.3	105.6	99.2	69.7	90.7	101.3	69.7	92.2	101.3	69.7	92.2	101.3	69.7	92.2	101.3	69.7	92.2
0970, 0990	Wall Finishes & Painting/Coating	92.3	123.7	111.3	103.1	52.8	72.6	98.5	52.8	70.8	98.5	52.8	70.8	98.5	52.8	70.8	98.5	52.8	70.8
09	FINISHES	96.3	128.4	114.0	102.8	65.9	82.5	103.2	65.9	82.6	102.0	65.9	82.1	101.6	65.9	81.9	102.7	65.9	82.4
COVERS	DIVS. 10 - 14, 25, 28, 41, 43, 44, 46	100.0	112.8	102.7	100.0	83.3	96.5	100.0	83.3	96.5	100.0	83.3	96.5	100.0	83.3	96.5	100.0	83.3	96.5
21, 22, 23	FIRE SUPPRESSION, PLUMBING & HVAC	99.8	123.7	109.7	100.0	69.8	87.4	96.9	69.8	85.6	96.9	69.4	85.4	99.9	69.8	87.4	96.7	69.8	85.5
26, 27, 3370	ELECTRICAL, COMMUNICATIONS & UTIL.	95.5	135.1	116.6	89.7	72.6	80.6	89.7	72.6	80.6	87.1	72.6	79.4	87.8	72.6	79.7	86.9	72.6	79.3
MF2010	WEIGHTED AVERAGE	97.5	122.8	108.8	99.1	75.0	88.3	100.0	75.0	88.8	97.9	74.9	87.6	99.8	75.0	88.7	99.3	75.0	88.5

	DIVISION	NEW MEXICO																	
		LAS CRUCES			LAS VEGAS			ROSWELL			SANTA FE			SOCORRO			TRUTH/CONSEQUENCES		
		880			877			882			875			878			879		
		MAT.	INST.	TOTAL	MAT.	INST.	TOTAL	MAT.	INST.	TOTAL	MAT.	INST.	TOTAL	MAT.	INST.	TOTAL	MAT.	INST.	TOTAL
015433	CONTRACTOR EQUIPMENT		87.9	87.9		113.0	113.0		113.0	113.0		113.0	113.0		113.0	113.0		87.9	87.9
0241, 31 - 34	SITE & INFRASTRUCTURE, DEMOLITION	85.8	86.7	86.4	80.7	108.5	99.9	85.4	108.5	101.3	79.0	108.5	99.3	77.3	108.5	98.8	96.6	86.7	89.8
0310	Concrete Forming & Accessories	96.1	66.0	70.0	99.2	67.2	71.5	99.2	67.2	71.5	98.4	67.2	71.4	99.2	67.2	71.5	96.1	66.0	70.0
0320	Concrete Reinforcing	102.4	72.0	86.7	105.3	72.2	88.2	107.4	72.2	89.3	106.2	72.2	88.7	107.4	72.2	89.3	101.2	72.0	86.2
0330	Cast-in-Place Concrete	92.4	67.4	82.8	103.3	75.3	92.6	97.9	75.3	89.2	107.6	75.3	95.2	101.2	75.3	91.3	111.0	67.5	94.3
03	CONCRETE	86.6	68.4	77.9	105.9	72.0	89.7	108.7	72.0	91.1	107.2	72.0	90.3	104.5	72.0	89.0	99.0	68.4	84.4
04	MASONRY	100.8	64.0	78.5	105.2	64.3	80.4	115.5	64.3	84.5	109.1	64.3	82.0	105.1	64.3	80.4	102.8	64.0	79.3
05	METALS	98.7	82.6	93.1	97.6	89.4	94.8	98.7	89.4	95.5	98.7	89.4	95.5	97.8	89.4	94.9	97.4	82.7	92.3
06	WOOD, PLASTICS & COMPOSITES	81.6	66.4	72.9	90.4	67.6	77.3	90.4	67.6	77.3	89.9	67.6	77.1	90.4	67.6	77.3	81.6	66.4	72.9
07	THERMAL & MOISTURE PROTECTION	87.1	70.3	80.4	100.6	75.0	90.5	101.2	75.0	90.8	100.3	75.0	90.3	100.6	75.0	90.4	87.7	70.3	80.8
08	OPENINGS	88.3	69.5	83.6	94.3	70.2	88.2	94.1	70.2	88.0	94.2	70.2	88.1	94.1	70.2	88.0	88.2	69.5	83.5
0920	Plaster & Gypsum Board	87.4	66.2	72.5	85.0	66.2	71.8	85.0	66.2	71.8	92.9	66.2	74.1	85.0	66.2	71.8	87.4	66.2	72.5
0950, 0980	Ceilings & Acoustic Treatment	108.8	66.2	80.6	118.2	66.2	83.7	118.2	66.2	83.7	119.1	66.2	84.0	118.2	66.2	83.7	108.8	66.2	80.6
0960	Flooring	133.7	69.7	115.3	101.3	69.7	92.2	101.3	69.7	92.2	101.3	69.7	92.2	101.3	69.7	92.2	133.7	69.7	115.3
0970, 0990	Wall Finishes & Painting/Coating	90.2	52.8	67.5	98.5	52.8	70.8	98.5	52.8	70.8	98.5	52.8	70.8	98.5	52.8	70.8	90.2	52.8	67.5
09	FINISHES	115.9	65.0	87.8	101.4	65.9	81.8	102.1	65.9	82.1	102.8	65.9	82.4	101.4	65.9	81.8	116.5	65.0	88.1
COVERS	DIVS. 10 - 14, 25, 28, 41, 43, 44, 46	100.0	80.4	95.9	100.0	83.3	96.5	100.0	83.3	96.5	100.0	83.3	96.5	100.0	83.3	96.5	100.0	80.4	95.9
21, 22, 23	FIRE SUPPRESSION, PLUMBING & HVAC	100.3	69.5	87.5	96.7	69.8	85.5	99.9	69.8	87.4	99.9	69.8	87.4	96.7	69.8	85.5	96.6	69.5	85.3
26, 27, 3370	ELECTRICAL, COMMUNICATIONS & UTIL.	88.3	67.8	77.4	89.7	72.6	80.6	88.6	72.6	80.1	100.0	72.6	85.4	87.5	72.6	79.6	91.4	72.6	81.4
MF2010	WEIGHTED AVERAGE	96.3	70.9	85.0	97.7	75.0	87.6	99.6	75.0	88.6	100.1	75.0	88.9	97.2	75.0	87.3	97.5	71.6	85.9

DIVISION		NEW MEXICO			NEW YORK															
		TUCUMCARI			ALBANY			BINGHAMTON			BRONX			BROOKLYN			BUFFALO			
		884			120 - 122			137 - 139			104			112			140 - 142			
		MAT.	INST.	TOTAL	MAT.	INST.	TOTAL	MAT.	INST.	TOTAL	MAT.	INST.	TOTAL	MAT.	INST.	TOTAL	MAT.	INST.	TOTAL	
015433	CONTRACTOR EQUIPMENT		113.0	113.0		114.2	114.2		114.5	114.5		114.5	114.5		115.4	115.4		96.1	96.1	
0241, 31 - 34	SITE & INFRASTRUCTURE, DEMOLITION	83.3	108.4	100.6	70.4	107.4	95.9	94.4	91.9	92.7	123.4	125.4	124.8	121.6	129.3	126.9	94.9	96.9	96.3	
0310	Concrete Forming & Accessories	99.2	67.1	71.3	99.0	97.4	97.6	103.6	84.4	86.9	104.2	171.6	162.7	110.2	171.2	163.1	100.2	112.4	110.8	
0320	Concrete Reinforcing	105.3	72.1	88.2	97.1	90.0	93.5	96.2	90.3	93.2	97.1	182.2	141.0	97.5	182.2	141.1	100.1	100.3	100.2	
0330	Cast-in-Place Concrete	97.9	75.3	89.2	76.6	104.5	87.3	98.3	96.7	97.7	91.8	159.3	117.7	100.3	157.6	122.3	104.3	117.7	109.5	
03	CONCRETE	107.1	71.9	90.3	96.0	99.1	97.5	98.0	91.4	94.8	99.8	167.6	132.3	110.7	166.8	137.6	109.1	111.2	110.1	
04	MASONRY	117.0	64.3	85.1	92.0	103.1	98.7	107.2	91.4	97.6	94.4	165.1	137.2	117.4	165.0	146.3	107.8	116.4	113.0	
05	METALS	97.5	89.2	94.6	100.0	104.1	101.4	96.6	113.6	102.5	102.2	143.1	116.2	104.5	140.8	117.0	101.7	94.2	99.1	
06	WOOD, PLASTICS & COMPOSITES	90.4	67.6	77.3	99.1	96.2	97.4	109.6	83.0	94.3	103.3	175.2	144.6	111.3	174.9	147.8	103.1	112.3	108.4	
07	THERMAL & MOISTURE PROTECTION	101.1	75.0	90.7	89.9	95.5	92.1	104.9	84.8	97.0	107.7	159.2	128.1	107.5	158.2	127.6	103.3	107.9	105.1	
08	OPENINGS	94.0	70.2	88.0	94.5	88.6	93.0	89.6	79.9	87.1	87.3	171.6	108.6	89.4	170.0	109.8	90.6	100.4	93.1	
0920	Plaster & Gypsum Board	85.0	66.2	71.8	103.6	95.8	98.1	114.4	82.0	91.6	109.4	177.1	157.1	112.0	177.1	157.8	100.8	112.4	109.0	
0950, 0980	Ceilings & Acoustic Treatment	118.2	66.2	83.7	97.8	95.8	96.5	100.6	82.0	88.3	88.4	177.1	147.2	97.0	177.1	150.1	101.1	112.4	108.6	
0960	Flooring	101.3	69.7	92.2	88.8	104.8	93.4	100.5	90.4	97.6	99.4	175.0	121.1	107.5	175.0	126.9	95.8	115.3	101.4	
0970, 0990	Wall Finishes & Painting/Coating	98.5	52.8	70.8	85.8	80.1	82.4	87.9	83.5	85.3	99.8	156.0	133.9	116.0	156.0	140.3	94.5	111.6	104.9	
09	FINISHES	101.9	65.9	82.0	95.7	97.1	96.4	98.0	85.1	90.9	104.9	171.5	141.6	114.4	171.2	145.7	97.3	113.7	106.3	
COVERS	DIVS. 10 - 14, 25, 28, 41, 43, 44, 46	100.0	83.3	96.5	100.0	93.9	98.7	100.0	94.6	98.9	100.0	127.5	105.7	100.0	126.6	105.5	100.0	105.9	101.2	
21, 22, 23	FIRE SUPPRESSION, PLUMBING & HVAC	96.9	69.4	85.4	100.1	94.2	97.6	100.5	82.5	93.0	100.3	161.5	125.8	99.8	161.4	125.4	99.9	92.6	96.9	
26, 27, 3370	ELECTRICAL, COMMUNICATIONS & UTIL.	89.7	72.6	80.6	102.9	93.6	98.0	101.4	92.2	96.5	97.0	172.8	137.3	100.8	172.8	139.1	99.9	98.7	99.3	
MF2010	WEIGHTED AVERAGE	98.6	74.9	88.0	97.2	97.9	97.5	98.8	90.2	95.0	99.9	160.3	126.9	104.0	160.1	129.1	100.5	102.9	101.6	

DIVISION		NEW YORK																	
		ELMIRA			FAR ROCKAWAY			FLUSHING			GLENS FALLS			HICKSVILLE			JAMAICA		
		148 - 149			116			113			128			115,117,118			114		
		MAT.	INST.	TOTAL	MAT.	INST.	TOTAL	MAT.	INST.	TOTAL	MAT.	INST.	TOTAL	MAT.	INST.	TOTAL	MAT.	INST.	TOTAL
015433	CONTRACTOR EQUIPMENT		115.9	115.9		115.4	115.4		115.4	115.4		114.2	114.2		115.4	115.4		115.4	115.4
0241, 31 - 34	SITE & INFRASTRUCTURE, DEMOLITION	90.8	93.3	92.5	125.7	129.3	128.2	125.7	129.3	128.2	62.6	106.7	93.0	114.0	128.1	123.8	119.5	129.3	126.2
0310	Concrete Forming & Accessories	85.2	87.1	86.8	96.6	171.2	161.3	100.5	171.2	161.8	87.5	88.6	88.5	93.3	142.7	136.2	100.5	171.2	161.8
0320	Concrete Reinforcing	101.4	87.2	94.1	97.5	182.2	141.1	99.2	182.2	141.9	96.8	90.0	93.3	97.5	182.0	141.0	97.5	182.2	141.1
0330	Cast-in-Place Concrete	97.5	96.7	97.2	108.5	157.6	127.3	108.5	157.6	127.3	74.8	102.1	85.2	92.1	153.1	115.5	100.3	157.6	122.3
03	CONCRETE	102.6	92.2	97.6	117.2	166.8	140.9	117.7	166.8	141.2	88.4	94.4	91.3	102.2	152.4	126.3	110.1	166.8	137.3
04	MASONRY	104.1	93.0	97.3	121.0	165.0	147.7	114.8	165.0	145.2	92.9	98.9	96.5	111.2	158.9	140.1	119.3	165.0	147.0
05	METALS	96.9	114.7	103.0	104.5	140.8	117.0	104.5	140.8	117.0	96.0	103.8	98.7	106.1	137.6	116.9	104.5	140.8	117.0
06	WOOD, PLASTICS & COMPOSITES	86.1	86.3	86.2	94.9	174.9	140.8	99.7	174.9	142.8	85.7	86.8	86.3	91.2	140.8	119.7	99.7	174.9	142.8
07	THERMAL & MOISTURE PROTECTION	103.9	86.1	96.9	107.4	158.2	127.6	107.5	158.2	127.6	89.4	92.0	90.4	107.1	150.0	124.1	107.3	158.2	127.5
08	OPENINGS	91.6	80.9	88.9	87.7	170.0	108.6	87.7	170.0	108.6	88.1	83.5	86.9	87.7	151.6	103.9	87.7	170.0	108.6
0920	Plaster & Gypsum Board	94.7	85.6	88.3	101.1	177.1	154.6	103.8	177.1	155.4	97.0	86.1	89.3	100.2	142.0	129.6	103.8	177.1	155.4
0950, 0980	Ceilings & Acoustic Treatment	97.3	85.6	89.6	86.7	177.1	146.6	86.7	177.1	146.6	95.1	86.1	89.1	85.6	142.0	123.0	86.7	177.1	146.6
0960	Flooring	85.7	90.4	87.1	102.3	175.0	123.2	103.8	175.0	124.3	80.8	104.8	87.7	101.2	150.7	115.4	103.8	175.0	124.3
0970, 0990	Wall Finishes & Painting/Coating	93.0	79.5	84.8	116.0	156.0	140.3	116.0	156.0	140.3	85.8	83.6	84.5	116.0	156.0	140.3	116.0	156.0	140.3
09	FINISHES	92.4	86.7	89.3	109.9	171.2	143.7	110.7	171.2	144.1	91.3	90.9	91.1	108.2	144.7	128.3	110.2	171.2	143.8
COVERS	DIVS. 10 - 14, 25, 28, 41, 43, 44, 46	100.0	91.1	98.2	100.0	126.6	105.5	100.0	126.6	105.5	100.0	85.6	97.0	100.0	120.5	104.3	100.0	126.6	105.5
21, 22, 23	FIRE SUPPRESSION, PLUMBING & HVAC	94.7	85.4	90.9	94.4	161.4	122.3	94.4	161.4	122.3	95.0	86.5	91.5	99.8	146.9	119.4	94.4	161.4	122.3
26, 27, 3370	ELECTRICAL, COMMUNICATIONS & UTIL.	98.1	95.5	96.7	109.0	172.8	142.9	109.0	172.8	142.9	96.3	90.4	93.2	100.1	144.0	123.5	99.0	172.8	138.2
MF2010	WEIGHTED AVERAGE	96.9	92.0	94.7	103.9	160.1	129.1	103.8	160.1	129.0	92.6	93.2	92.8	101.7	145.2	121.2	101.9	160.1	127.9

DIVISION		NEW YORK																	
		JAMESTOWN			KINGSTON			LONG ISLAND CITY			MONTICELLO			MOUNT VERNON			NEW ROCHELLE		
		147			124			111			127			105			108		
		MAT.	INST.	TOTAL	MAT.	INST.	TOTAL	MAT.	INST.	TOTAL	MAT.	INST.	TOTAL	MAT.	INST.	TOTAL	MAT.	INST.	TOTAL
015433	CONTRACTOR EQUIPMENT		92.9	92.9		115.4	115.4		115.4	115.4		115.4	115.4		114.5	114.5		114.5	114.5
0241, 31 - 34	SITE & INFRASTRUCTURE, DEMOLITION	92.1	92.6	92.4	121.3	125.7	124.3	122.9	129.3	127.3	116.5	125.7	122.9	133.0	120.1	124.1	131.5	120.1	123.6
0310	Concrete Forming & Accessories	85.2	80.0	80.7	89.3	109.8	107.1	104.7	171.2	162.4	96.2	109.8	108.0	94.8	131.9	127.0	109.4	131.9	128.9
0320	Concrete Reinforcing	101.5	74.6	87.7	97.1	133.3	115.8	97.5	182.2	141.1	96.4	133.3	115.4	95.9	181.1	139.8	96.0	181.1	139.8
0330	Cast-in-Place Concrete	101.2	95.6	99.1	101.0	133.0	113.3	103.5	157.6	124.2	94.6	133.0	109.3	102.6	132.9	114.2	102.6	132.9	114.2
03	CONCRETE	105.9	84.4	95.6	111.1	121.8	116.2	113.4	166.8	138.9	105.7	121.8	113.4	110.1	140.5	124.6	109.3	140.5	124.3
04	MASONRY	113.1	91.0	99.7	108.4	136.2	125.2	114.3	165.0	145.0	101.7	136.2	122.6	100.0	130.1	118.2	100.0	130.1	118.2
05	METALS	94.2	80.9	89.6	104.5	114.5	107.9	104.5	140.8	117.0	104.5	114.6	107.9	101.9	134.0	112.9	102.2	134.0	113.1
06	WOOD, PLASTICS & COMPOSITES	84.6	76.6	80.0	87.0	104.9	97.2	105.3	174.9	145.2	95.4	104.9	100.8	92.6	135.5	117.2	110.6	135.5	124.9
07	THERMAL & MOISTURE PROTECTION	103.5	84.9	96.1	109.1	135.7	119.7	107.4	158.2	127.6	108.8	135.7	119.5	108.6	137.5	120.1	108.6	137.5	120.1
08	OPENINGS	91.5	74.7	87.2	93.0	121.3	100.2	87.7	170.0	108.6	87.8	121.3	96.3	87.3	148.7	102.8	87.3	148.7	102.9
0920	Plaster & Gypsum Board	89.3	75.7	79.7	95.8	105.0	102.2	107.3	177.1	156.4	99.3	105.0	103.3	103.5	136.1	126.5	116.8	136.1	130.4
0950, 0980	Ceilings & Acoustic Treatment	94.5	75.7	82.0	80.1	105.0	96.6	86.7	177.1	146.6	80.1	105.0	96.6	86.5	136.1	119.4	86.5	136.1	119.4
0960	Flooring	88.6	93.1	89.9	94.7	76.6	89.5	105.4	175.0	125.4	97.4	76.6	91.4	91.3	173.8	115.0	98.4	173.8	120.1
0970, 0990	Wall Finishes & Painting/Coating	94.5	79.9	85.6	113.6	105.0	108.4	116.0	156.0	140.3	113.6	103.9	107.7	97.8	156.0	133.1	97.8	156.0	133.1
09	FINISHES	92.2	81.8	86.5	105.1	101.0	102.8	111.3	171.2	144.3	105.9	100.9	103.1	102.3	141.9	124.1	105.7	141.9	125.7
COVERS	DIVS. 10 - 14, 25, 28, 41, 43, 44, 46	100.0	93.2	98.6	100.0	109.4	101.9	100.0	126.6	105.5	100.0	109.4	101.9	100.0	115.3	103.2	100.0	115.3	103.2
21, 22, 23	FIRE SUPPRESSION, PLUMBING & HVAC	94.6	83.9	90.2	94.9	117.4	104.3	99.8	161.4	125.4	94.9	119.9	105.3	95.0	120.5	105.6	95.0	120.5	105.6
26, 27, 3370	ELECTRICAL, COMMUNICATIONS & UTIL.	97.0	81.8	88.9	97.9	118.2	108.7	99.5	172.8	138.5	97.9	118.2	108.7	95.2	153.6	126.3	95.2	153.6	126.3
MF2010	WEIGHTED AVERAGE	97.2	84.4	91.5	101.6	118.7	109.2	103.6	160.1	128.9	100.0	119.2	108.6	100.0	134.3	115.3	100.3	134.3	115.5

	DIVISION	NEW YORK																	
		NEW YORK			NIAGARA FALLS			PLATTSBURGH			POUGHKEEPSIE			QUEENS			RIVERHEAD		
		100 - 102			143			129			125 - 126			110			119		
		MAT.	INST.	TOTAL	MAT.	INST.	TOTAL	MAT.	INST.	TOTAL	MAT.	INST.	TOTAL	MAT.	INST.	TOTAL	MAT.	INST.	TOTAL
015433	CONTRACTOR EQUIPMENT		114.8	114.8		92.9	92.9		99.0	99.0		115.4	115.4		115.4	115.4		115.4	115.4
0241, 31 - 34	SITE & INFRASTRUCTURE, DEMOLITION	133.1	126.2	128.4	94.2	93.4	93.6	87.3	102.5	97.8	117.3	125.6	123.0	117.6	129.3	125.6	115.1	128.1	124.1
0310	Concrete Forming & Accessories	108.7	187.6	177.1	85.2	108.1	105.0	91.9	82.4	83.6	89.3	156.6	147.7	93.5	171.2	160.9	97.7	142.6	136.7
0320	Concrete Reinforcing	102.8	182.3	143.7	100.1	92.7	96.3	101.2	89.4	95.1	97.1	133.6	115.9	99.2	182.2	141.9	99.3	179.0	140.4
0330	Cast-in-Place Concrete	103.4	169.5	128.7	104.7	102.4	103.8	91.5	94.3	92.6	97.8	131.5	110.7	95.3	157.6	119.2	93.6	153.1	116.4
03	CONCRETE	110.6	178.2	142.9	108.4	102.5	105.6	102.3	87.6	95.3	108.2	142.2	124.4	105.6	166.8	134.9	102.9	151.9	126.3
04	MASONRY	105.7	173.9	147.0	121.1	111.7	115.4	88.9	88.3	88.5	101.7	132.2	120.2	108.6	165.0	142.8	116.8	158.9	142.3
05	METALS	116.5	143.3	125.7	96.9	88.7	94.1	100.5	83.4	94.6	104.5	116.5	108.6	104.5	140.8	117.0	106.5	136.5	116.8
06	WOOD, PLASTICS & COMPOSITES	107.8	192.2	156.2	84.6	107.4	97.7	93.5	80.5	86.1	87.0	168.4	133.7	91.4	174.9	139.3	96.7	140.8	122.0
07	THERMAL & MOISTURE PROTECTION	107.7	165.0	130.4	103.6	102.1	103.0	102.8	87.1	96.6	109.1	141.9	122.1	107.1	158.2	127.4	107.7	150.0	124.5
08	OPENINGS	92.6	180.7	114.9	91.5	95.4	92.5	97.3	79.4	92.8	93.0	152.3	108.0	87.7	170.0	108.6	87.7	150.9	103.7
0920	Plaster & Gypsum Board	117.5	194.6	171.8	89.3	107.4	102.0	119.9	79.2	91.2	95.8	170.4	148.3	100.2	177.1	154.4	102.3	142.0	130.2
0950, 0980	Ceilings & Acoustic Treatment	108.1	194.6	165.4	94.5	107.4	103.0	109.8	79.2	89.5	80.1	170.4	140.0	86.7	177.1	146.6	86.6	142.0	123.3
0960	Flooring	100.7	175.0	122.0	88.4	107.6	94.0	104.3	100.7	103.3	94.7	147.1	109.8	101.2	175.0	122.4	102.3	150.7	116.2
0970, 0990	Wall Finishes & Painting/Coating	99.8	156.0	133.9	94.5	102.5	99.3	110.6	77.0	90.2	113.6	103.9	107.7	116.0	156.0	140.3	116.0	156.0	140.3
09	FINISHES	111.2	183.6	151.2	92.3	108.2	101.1	101.9	85.2	92.7	104.9	151.7	130.7	108.6	171.2	143.1	108.9	144.7	128.6
COVERS	DIVS. 10 - 14, 25, 28, 41, 43, 44, 46	100.0	137.6	107.8	100.0	100.8	100.2	100.0	51.6	89.9	100.0	115.3	103.2	100.0	126.6	105.5	100.0	120.5	104.3
21, 22, 23	FIRE SUPPRESSION, PLUMBING & HVAC	100.2	166.0	127.6	94.6	92.4	93.7	94.9	81.4	89.2	94.9	119.7	105.2	99.8	161.4	125.4	99.9	145.6	119.0
26, 27, 3370	ELECTRICAL, COMMUNICATIONS & UTIL.	105.6	173.1	141.5	95.4	90.6	92.8	94.6	84.6	89.3	97.9	118.2	108.7	100.1	172.8	138.8	101.9	144.0	124.3
MF2010	WEIGHTED AVERAGE	106.0	166.3	133.0	98.3	98.0	98.2	97.5	85.0	91.9	100.7	130.5	114.0	101.9	160.1	127.9	102.5	144.7	121.4

	DIVISION	NEW YORK																	
		ROCHESTER			SCHENECTADY			STATEN ISLAND			SUFFERN			SYRACUSE			UTICA		
		144 - 146			123			103			109			130 - 132			133 - 135		
		MAT.	INST.	TOTAL	MAT.	INST.	TOTAL	MAT.	INST.	TOTAL	MAT.	INST.	TOTAL	MAT.	INST.	TOTAL	MAT.	INST.	TOTAL
015433	CONTRACTOR EQUIPMENT		116.6	116.6		114.2	114.2		114.5	114.5		114.5	114.5		114.2	114.2		114.2	114.2
0241, 31 - 34	SITE & INFRASTRUCTURE, DEMOLITION	71.8	108.5	97.1	72.0	107.3	96.4	138.0	125.4	129.3	127.8	121.1	123.2	93.4	106.7	102.6	70.2	104.4	93.8
0310	Concrete Forming & Accessories	101.5	93.7	94.7	103.4	97.0	97.8	94.3	171.6	161.4	103.2	125.5	122.6	102.4	89.1	90.9	104.0	82.2	85.1
0320	Concrete Reinforcing	101.0	90.6	95.7	95.8	90.0	92.8	97.1	182.2	141.0	96.0	133.6	115.3	97.1	90.7	93.8	97.1	88.3	92.6
0330	Cast-in-Place Concrete	97.9	99.0	98.3	90.3	103.9	95.5	102.6	159.3	124.3	99.5	129.8	111.1	91.3	96.9	93.4	83.5	89.3	85.7
03	CONCRETE	116.3	95.8	106.5	103.0	98.7	100.9	111.9	167.6	138.6	106.2	127.9	116.6	101.6	92.9	97.5	99.8	86.8	93.6
04	MASONRY	104.2	98.4	100.7	91.1	102.0	97.7	108.2	165.1	142.7	100.1	125.6	115.6	98.5	100.1	99.5	89.8	87.8	88.6
05	METALS	98.2	105.4	100.7	100.1	104.1	101.5	100.1	143.1	114.9	100.1	118.5	106.4	100.0	103.2	101.1	97.9	102.0	99.3
06	WOOD, PLASTICS & COMPOSITES	100.0	93.7	96.4	105.8	96.2	100.3	91.1	175.2	139.4	102.9	131.1	119.1	105.8	87.4	95.2	105.8	84.2	93.4
07	THERMAL & MOISTURE PROTECTION	98.7	94.6	97.1	90.5	95.0	92.3	108.0	159.2	128.3	108.5	134.2	118.7	101.1	96.2	99.2	90.3	90.7	90.5
08	OPENINGS	93.8	87.3	92.2	94.5	88.6	93.0	87.3	171.6	108.6	87.3	137.3	100.0	92.7	80.1	89.5	94.5	79.9	90.8
0920	Plaster & Gypsum Board	106.2	93.5	97.3	108.6	95.8	99.6	103.6	177.1	155.3	108.4	131.7	124.8	108.6	86.8	93.3	108.6	83.5	90.9
0950, 0980	Ceilings & Acoustic Treatment	102.1	93.5	96.4	100.6	95.8	97.4	88.4	177.1	147.2	86.5	131.7	116.5	100.6	86.8	91.5	100.6	83.5	89.3
0960	Flooring	84.6	104.0	90.2	88.2	104.8	93.0	95.4	175.0	118.3	94.6	150.7	110.7	90.1	90.5	90.2	88.2	91.2	89.1
0970, 0990	Wall Finishes & Painting/Coating	94.3	97.6	96.3	85.8	80.1	82.4	99.8	156.0	133.9	97.8	112.3	106.6	92.5	95.4	94.3	85.8	95.4	91.6
09	FINISHES	98.5	96.3	97.3	96.7	96.8	96.8	104.5	171.5	141.4	103.4	126.3	116.0	98.5	89.8	93.7	96.8	85.2	90.4
COVERS	DIVS. 10 - 14, 25, 28, 41, 43, 44, 46	100.0	94.5	98.8	100.0	93.5	98.7	100.0	127.5	105.7	100.0	113.2	102.7	100.0	95.7	99.1	100.0	87.2	97.3
21, 22, 23	FIRE SUPPRESSION, PLUMBING & HVAC	99.9	89.1	95.4	100.3	93.6	97.5	100.3	161.5	125.8	95.0	114.4	103.1	100.3	86.7	94.6	100.3	80.2	91.9
26, 27, 3370	ELECTRICAL, COMMUNICATIONS & UTIL.	100.1	91.6	95.6	101.5	93.6	97.3	97.0	172.8	137.3	103.7	126.4	115.8	101.5	92.4	96.6	99.6	92.4	95.8
MF2010	WEIGHTED AVERAGE	100.3	95.6	98.2	98.2	97.5	97.9	102.1	160.3	128.1	100.1	123.1	110.4	99.4	93.5	96.7	97.1	88.7	93.4

	DIVISION	NEW YORK									NORTH CAROLINA								
		WATERTOWN			WHITE PLAINS			YONKERS			ASHEVILLE			CHARLOTTE			DURHAM		
		136			106			107			287 - 288			281 - 282			277		
		MAT.	INST.	TOTAL	MAT.	INST.	TOTAL	MAT.	INST.	TOTAL	MAT.	INST.	TOTAL	MAT.	INST.	TOTAL	MAT.	INST.	TOTAL
015433	CONTRACTOR EQUIPMENT		114.2	114.2		114.5	114.5		114.5	114.5		94.9	94.9		94.9	94.9		100.4	100.4
0241, 31 - 34	SITE & INFRASTRUCTURE, DEMOLITION	78.2	106.8	97.9	123.0	120.1	121.0	132.5	119.9	123.8	109.5	74.8	85.6	110.7	74.9	86.0	109.8	83.6	91.7
0310	Concrete Forming & Accessories	89.2	92.0	91.6	108.1	131.9	128.8	108.4	131.9	128.8	92.0	38.1	45.3	96.5	40.0	47.4	94.6	38.7	46.1
0320	Concrete Reinforcing	97.7	90.7	94.1	96.0	181.1	139.8	99.8	181.2	141.7	90.2	41.6	65.2	90.6	38.8	63.9	91.6	48.7	69.5
0330	Cast-in-Place Concrete	97.2	96.7	97.0	91.1	132.9	107.1	101.9	132.9	113.8	113.2	45.8	87.4	119.5	47.7	92.0	115.1	46.0	88.6
03	CONCRETE	113.0	94.1	104.0	99.3	140.5	119.0	109.5	140.5	124.3	107.1	43.3	76.6	109.9	44.3	78.5	107.7	45.0	77.7
04	MASONRY	90.9	101.0	97.0	99.5	130.1	118.0	105.3	130.1	120.3	85.2	39.0	57.2	91.1	42.7	61.8	90.5	35.1	56.9
05	METALS	98.0	103.2	99.8	101.7	134.0	112.8	111.9	134.0	119.5	84.7	72.7	80.6	88.6	71.8	82.8	97.4	75.9	90.0
06	WOOD, PLASTICS & COMPOSITES	87.5	91.4	89.7	108.9	135.5	124.1	108.7	135.5	124.1	93.5	37.6	61.5	99.1	39.7	65.0	96.2	39.1	63.4
07	THERMAL & MOISTURE PROTECTION	90.6	97.0	93.1	108.3	137.5	119.9	108.6	137.5	120.1	98.2	42.1	75.9	95.7	43.6	75.0	98.9	44.0	77.1
08	OPENINGS	94.5	86.0	92.4	87.4	148.7	102.9	90.5	148.7	105.2	94.5	37.4	80.0	98.6	38.8	83.5	98.6	41.3	84.1
0920	Plaster & Gypsum Board	99.3	90.9	93.4	111.9	136.1	129.0	117.0	136.1	130.5	99.7	35.6	54.6	100.0	37.8	56.2	104.5	37.1	57.0
0950, 0980	Ceilings & Acoustic Treatment	100.6	90.9	94.2	86.5	136.1	119.4	106.2	136.1	126.0	92.0	35.6	54.6	94.8	37.8	57.0	95.7	37.1	56.8
0960	Flooring	81.3	92.0	84.4	96.7	173.8	118.8	96.3	173.8	118.5	106.2	44.0	88.3	109.3	44.2	90.6	109.9	44.0	91.0
0970, 0990	Wall Finishes & Painting/Coating	85.8	86.5	86.2	97.8	156.0	133.1	97.8	156.0	133.1	114.7	34.5	66.0	114.7	39.4	69.0	114.7	36.1	67.0
09	FINISHES	94.3	91.4	92.7	103.8	141.9	124.8	109.3	141.9	127.2	100.7	38.6	66.4	101.8	40.5	68.0	103.2	39.0	67.8
COVERS	DIVS. 10 - 14, 25, 28, 41, 43, 44, 46	100.0	96.1	99.2	100.0	115.3	103.2	100.0	120.4	104.2	100.0	66.9	93.1	100.0	67.4	93.2	100.0	71.6	94.1
21, 22, 23	FIRE SUPPRESSION, PLUMBING & HVAC	100.3	81.8	92.6	100.5	120.5	108.8	100.5	120.5	108.8	100.3	37.8	74.3	100.0	37.9	74.2	100.4	36.6	73.8
26, 27, 3370	ELECTRICAL, COMMUNICATIONS & UTIL.	101.5	87.5	94.1	95.2	153.6	126.3	103.8	153.6	130.3	97.1	37.4	65.3	99.5	48.6	72.4	97.8	47.3	71.0
MF2010	WEIGHTED AVERAGE	98.9	92.6	96.1	99.9	134.3	115.3	104.8	134.5	118.0	97.4	46.0	74.4	99.2	48.4	76.5	100.3	48.4	77.1

DIVISION		NORTH CAROLINA																	
		ELIZABETH CITY			FAYETTEVILLE			GASTONIA			GREENSBORO			HICKORY			KINSTON		
		279			283			280			270,272 - 274			286			285		
		MAT.	INST.	TOTAL	MAT.	INST.	TOTAL	MAT.	INST.	TOTAL	MAT.	INST.	TOTAL	MAT.	INST.	TOTAL	MAT.	INST.	TOTAL
015433	**CONTRACTOR EQUIPMENT**		104.7	104.7		100.4	100.4		94.9	94.9		100.4	100.4		100.4	100.4		100.4	100.4
0241, 31 - 34	**SITE & INFRASTRUCTURE, DEMOLITION**	115.3	84.8	94.2	108.5	83.7	91.4	109.2	74.8	85.4	109.7	83.7	91.8	108.9	83.7	91.5	107.6	83.5	91.0
0310	Concrete Forming & Accessories	81.4	33.6	39.9	91.7	39.4	46.4	98.3	39.0	46.9	94.3	39.4	46.6	88.4	37.2	44.0	85.0	31.5	38.6
0320	Concrete Reinforcing	89.7	40.5	64.4	93.7	48.8	70.6	90.6	41.6	65.3	90.6	57.1	73.3	90.2	18.9	53.5	89.7	18.3	53.0
0330	Cast-in-Place Concrete	115.3	37.3	85.4	118.7	49.0	92.0	110.7	53.1	88.6	114.2	47.3	88.6	113.2	46.9	87.8	109.2	37.5	81.8
03	**CONCRETE**	108.1	38.2	74.6	109.3	46.4	79.2	105.6	46.2	77.2	107.1	47.3	78.5	106.9	39.2	74.5	103.5	33.3	69.9
04	**MASONRY**	102.9	26.3	56.4	88.5	40.0	59.1	89.7	36.9	57.7	88.4	35.8	56.5	74.9	40.2	53.9	81.0	28.2	49.0
05	**METALS**	86.7	72.2	81.7	101.1	76.0	92.5	85.6	72.7	81.2	91.7	79.1	87.4	84.7	66.1	78.3	83.8	65.0	77.3
06	**WOOD, PLASTICS & COMPOSITES**	80.5	34.8	54.3	92.5	39.0	61.8	101.8	39.0	65.8	95.9	39.0	63.2	88.3	37.4	59.1	85.0	31.7	54.5
07	**THERMAL & MOISTURE PROTECTION**	98.2	33.1	72.3	98.1	42.6	76.0	98.4	42.3	76.1	98.7	41.7	76.0	98.5	41.0	75.6	98.4	31.8	71.9
08	**OPENINGS**	95.5	32.7	79.6	94.6	41.3	81.1	98.6	38.9	83.5	98.6	43.6	84.7	94.6	30.3	78.3	94.7	27.3	77.6
0920	Plaster & Gypsum Board	96.3	32.0	51.0	104.2	37.1	56.9	106.5	37.0	57.6	106.5	37.0	57.6	99.7	35.4	54.4	99.4	29.5	50.2
0950, 0980	Ceilings & Acoustic Treatment	95.7	32.0	53.5	92.9	37.1	55.9	95.7	37.0	56.8	95.7	37.0	56.8	92.0	35.4	54.4	95.7	29.5	51.8
0960	Flooring	100.3	23.9	78.3	106.3	44.0	88.4	109.9	44.0	91.0	109.9	40.7	90.0	106.0	33.9	85.3	103.1	19.4	79.0
0970, 0990	Wall Finishes & Painting/Coating	114.7	44.5	72.1	114.7	34.5	66.0	114.7	39.5	69.1	114.7	33.5	65.5	114.7	33.5	65.5	114.7	30.5	63.6
09	**FINISHES**	99.7	32.8	62.8	101.4	39.4	67.2	103.3	39.8	68.3	103.4	38.7	67.7	100.8	35.9	65.0	100.5	29.1	61.1
COVERS	**DIVS. 10 - 14, 25, 28, 41, 43, 44, 46**	100.0	75.5	94.9	100.0	72.4	94.3	100.0	66.8	93.1	100.0	67.1	93.2	100.0	66.8	93.1	100.0	70.5	93.9
21, 22, 23	**FIRE SUPPRESSION, PLUMBING & HVAC**	94.9	29.6	67.7	100.2	37.7	74.2	100.3	37.4	74.1	100.3	37.9	74.3	94.9	31.1	68.3	94.9	25.1	65.8
26, 27, 3370	**ELECTRICAL, COMMUNICATIONS & UTIL.**	97.6	29.7	61.5	95.7	51.6	72.2	96.4	34.7	63.6	96.8	37.6	65.3	94.4	31.4	60.9	94.1	49.0	70.1
MF2010	**WEIGHTED AVERAGE**	97.5	41.0	72.3	100.0	50.0	77.6	98.2	46.0	74.8	99.2	47.9	76.2	95.2	42.8	71.8	94.9	40.6	70.6

DIVISION		NORTH CAROLINA															NORTH DAKOTA		
		MURPHY			RALEIGH			ROCKY MOUNT			WILMINGTON			WINSTON-SALEM			BISMARCK		
		289			275 - 276			278			284			271			585		
		MAT.	INST.	TOTAL	MAT.	INST.	TOTAL	MAT.	INST.	TOTAL	MAT.	INST.	TOTAL	MAT.	INST.	TOTAL	MAT.	INST.	TOTAL
015433	**CONTRACTOR EQUIPMENT**		94.9	94.9		100.4	100.4		100.4	100.4		94.9	94.9		100.4	100.4		98.6	98.6
0241, 31 - 34	**SITE & INFRASTRUCTURE, DEMOLITION**	111.5	74.5	86.0	110.2	83.8	92.0	113.2	83.4	92.6	110.8	74.8	85.9	110.0	83.7	91.8	97.2	97.2	97.2
0310	Concrete Forming & Accessories	98.8	24.3	34.1	94.8	40.2	47.4	87.3	29.3	36.9	93.2	38.7	45.9	95.7	38.4	45.9	95.7	43.1	50.1
0320	Concrete Reinforcing	89.8	18.2	52.9	90.6	48.8	69.1	89.7	26.7	57.3	90.9	47.0	68.3	90.6	38.6	63.8	104.9	86.9	95.6
0330	Cast-in-Place Concrete	117.1	35.5	85.9	122.4	52.4	95.6	112.8	34.7	82.9	112.8	47.0	87.6	117.1	44.7	89.4	99.2	49.9	80.3
03	**CONCRETE**	110.4	29.4	71.6	111.3	47.9	80.9	108.7	32.8	72.3	106.9	45.0	77.3	108.6	42.5	77.0	99.9	55.1	78.4
04	**MASONRY**	77.5	26.0	46.3	89.6	40.9	60.1	81.3	26.3	48.0	75.2	37.0	52.1	88.6	36.8	57.2	107.5	60.9	79.3
05	**METALS**	82.8	64.5	76.5	89.6	75.9	84.9	85.9	65.3	78.8	84.6	75.0	81.3	89.6	71.3	83.3	96.3	86.7	93.0
06	**WOOD, PLASTICS & COMPOSITES**	102.5	22.4	56.5	96.2	39.7	63.8	87.3	29.4	54.1	95.3	38.4	62.6	95.9	38.3	62.9	82.3	37.6	56.7
07	**THERMAL & MOISTURE PROTECTION**	98.4	31.1	71.6	97.9	42.9	76.0	98.7	29.1	71.1	98.2	42.0	75.9	98.7	41.3	75.9	106.3	53.5	85.3
08	**OPENINGS**	94.5	24.9	76.9	95.4	41.4	81.7	94.7	26.0	77.3	94.7	39.9	80.8	98.6	38.2	83.3	96.9	49.4	84.8
0920	Plaster & Gypsum Board	105.4	19.9	45.2	100.0	37.7	56.1	98.6	27.1	48.3	101.3	36.4	55.6	106.5	36.3	57.1	95.9	36.0	53.7
0950, 0980	Ceilings & Acoustic Treatment	92.0	19.9	44.2	94.8	37.7	57.0	92.9	27.1	49.3	92.9	36.4	55.4	95.7	36.3	56.4	139.9	36.0	71.0
0960	Flooring	110.2	24.8	85.7	109.9	44.0	91.0	104.4	22.5	80.9	107.0	44.0	88.9	109.9	44.0	91.0	118.2	76.2	106.2
0970, 0990	Wall Finishes & Painting/Coating	114.7	18.5	56.3	114.7	38.0	68.2	114.7	32.4	64.8	114.7	33.4	65.4	114.7	35.2	66.5	98.4	31.7	57.9
09	**FINISHES**	102.8	23.1	58.8	102.2	40.4	68.1	100.8	28.0	60.6	101.3	38.8	66.8	103.4	38.9	67.9	114.3	46.1	76.7
COVERS	**DIVS. 10 - 14, 25, 28, 41, 43, 44, 46**	100.0	64.0	92.5	100.0	72.7	94.3	100.0	69.5	93.7	100.0	72.1	94.2	100.0	66.8	93.1	100.0	81.1	96.1
21, 22, 23	**FIRE SUPPRESSION, PLUMBING & HVAC**	94.9	18.8	63.2	100.1	36.1	73.5	94.9	32.7	69.0	100.3	37.5	74.2	100.3	36.5	73.7	100.2	56.4	82.0
26, 27, 3370	**ELECTRICAL, COMMUNICATIONS & UTIL.**	98.1	26.2	59.9	103.5	37.3	68.2	99.9	35.8	65.8	97.3	36.1	64.7	96.8	40.9	67.1	100.4	66.0	82.1
MF2010	**WEIGHTED AVERAGE**	96.2	33.4	68.1	99.6	48.1	76.6	96.6	39.8	71.2	96.9	46.3	74.3	99.1	46.5	75.6	100.8	63.0	83.9

DIVISION		NORTH DAKOTA																	
		DEVILS LAKE			DICKINSON			FARGO			GRAND FORKS			JAMESTOWN			MINOT		
		583			586			580 - 581			582			584			587		
		MAT.	INST.	TOTAL	MAT.	INST.	TOTAL	MAT.	INST.	TOTAL	MAT.	INST.	TOTAL	MAT.	INST.	TOTAL	MAT.	INST.	TOTAL
015433	**CONTRACTOR EQUIPMENT**		98.6	98.6		98.6	98.6		98.6	98.6		98.6	98.6		98.6	98.6		98.6	98.6
0241, 31 - 34	**SITE & INFRASTRUCTURE, DEMOLITION**	106.9	95.1	98.7	116.7	93.7	100.9	97.8	97.2	97.4	111.2	93.7	99.1	105.7	93.7	97.4	108.2	97.2	100.6
0310	Concrete Forming & Accessories	99.1	37.5	45.6	89.6	36.7	43.7	96.9	43.8	50.9	93.0	36.7	44.2	90.9	36.4	43.6	89.4	61.9	65.5
0320	Concrete Reinforcing	104.9	87.2	95.8	105.8	42.7	73.3	96.7	87.0	91.7	103.3	87.1	94.9	105.5	50.5	77.2	106.8	87.2	96.7
0330	Cast-in-Place Concrete	120.4	48.4	92.9	109.0	46.9	85.2	98.3	51.8	80.5	109.0	46.9	85.2	118.9	46.7	91.3	108.9	48.9	85.9
03	**CONCRETE**	113.7	52.1	84.2	113.4	42.7	79.5	108.5	56.1	83.4	110.2	51.0	81.9	112.3	44.0	79.6	108.7	63.2	86.9
04	**MASONRY**	116.3	66.7	86.2	119.4	60.3	83.6	109.4	60.9	80.0	110.6	66.2	83.7	130.1	35.4	72.7	110.0	65.6	83.1
05	**METALS**	96.3	85.8	92.7	96.2	64.7	85.4	98.7	87.1	94.7	96.3	81.1	91.1	96.3	66.0	85.8	96.6	87.4	93.4
06	**WOOD, PLASTICS & COMPOSITES**	85.7	34.6	56.4	75.6	34.6	52.1	82.2	38.1	56.9	79.2	34.6	53.6	77.2	34.6	52.8	75.3	62.7	68.1
07	**THERMAL & MOISTURE PROTECTION**	107.1	51.6	85.1	107.7	49.1	84.4	104.7	54.6	84.8	107.4	51.6	85.2	107.0	43.5	81.8	107.1	57.3	87.4
08	**OPENINGS**	96.9	43.1	83.3	96.9	32.5	80.6	96.8	49.7	84.9	96.9	43.1	83.3	96.9	34.7	81.1	97.0	62.9	88.4
0920	Plaster & Gypsum Board	113.6	32.9	56.8	104.7	32.9	54.2	95.9	36.6	54.1	106.5	32.9	54.7	106.0	32.9	54.6	104.7	61.8	74.5
0950, 0980	Ceilings & Acoustic Treatment	140.9	32.9	69.3	140.9	32.9	69.3	139.9	36.6	71.4	140.9	32.9	69.3	140.9	32.9	69.3	140.9	61.8	88.5
0960	Flooring	122.5	37.9	98.2	114.0	37.9	92.2	117.9	76.2	105.9	116.4	37.9	93.9	115.1	37.9	92.9	113.8	84.9	105.5
0970, 0990	Wall Finishes & Painting/Coating	98.4	24.2	53.4	98.4	24.2	53.4	98.4	75.3	84.4	98.4	24.2	53.4	98.4	24.2	53.4	98.4	26.5	54.8
09	**FINISHES**	118.6	34.6	72.3	116.0	34.6	71.1	114.3	51.3	79.6	116.3	34.6	71.3	115.4	34.6	70.8	115.1	62.1	85.8
COVERS	**DIVS. 10 - 14, 25, 28, 41, 43, 44, 46**	100.0	35.9	86.7	100.0	36.1	86.7	100.0	81.1	96.1	100.0	36.0	86.7	100.0	78.9	95.6	100.0	84.0	96.7
21, 22, 23	**FIRE SUPPRESSION, PLUMBING & HVAC**	94.9	56.1	78.7	94.9	69.6	84.4	100.2	59.4	83.2	100.3	36.4	73.7	94.9	38.4	71.4	100.3	56.4	82.0
26, 27, 3370	**ELECTRICAL, COMMUNICATIONS & UTIL.**	94.1	38.8	64.7	103.5	66.5	83.8	102.1	64.0	81.9	97.9	58.1	76.7	94.1	38.8	64.7	101.3	82.1	91.1
MF2010	**WEIGHTED AVERAGE**	101.7	56.0	81.3	102.8	58.2	82.9	102.4	64.3	85.4	102.6	53.8	80.8	101.9	46.8	77.3	102.5	70.0	88.0

DIVISION		NORTH DAKOTA			OHIO															
		WILLISTON			AKRON			ATHENS			CANTON			CHILLICOTHE			CINCINNATI			
		588			442 - 443			457			446 - 447			456			451 - 452			
		MAT.	INST.	TOTAL	MAT.	INST.	TOTAL	MAT.	INST.	TOTAL	MAT.	INST.	TOTAL	MAT.	INST.	TOTAL	MAT.	INST.	TOTAL	
015433	CONTRACTOR EQUIPMENT		98.6	98.6		96.8	96.8		91.3	91.3		96.8	96.8		101.6	101.6		101.4	101.4	
0241, 31 - 34	SITE & INFRASTRUCTURE, DEMOLITION	109.3	93.7	98.6	94.1	103.8	100.8	93.1	92.2	92.5	94.2	103.6	100.7	81.9	106.0	98.5	79.2	105.8	97.5	
0310	Concrete Forming & Accessories	94.5	36.7	44.4	101.2	93.9	94.9	93.2	90.1	90.5	101.2	82.3	84.8	94.9	94.7	94.7	96.6	83.7	85.4	
0320	Concrete Reinforcing	107.9	42.7	74.3	99.1	91.0	94.9	97.2	85.6	91.2	99.1	76.3	87.4	94.0	81.8	87.7	99.6	84.4	91.8	
0330	Cast-in-Place Concrete	109.0	46.9	85.2	100.5	100.4	100.5	98.8	96.7	98.0	101.5	96.0	99.4	89.7	90.0	89.8	82.6	88.3	84.8	
03	CONCRETE	110.1	42.7	77.9	98.8	94.8	96.9	98.4	90.5	94.6	99.3	85.3	92.6	93.0	90.5	91.8	88.0	85.5	86.8	
04	MASONRY	104.5	60.3	77.7	95.5	97.6	96.8	72.4	91.0	83.7	96.2	84.1	88.9	80.5	93.5	88.4	80.1	87.9	84.8	
05	METALS	96.4	64.7	85.5	93.0	80.2	88.6	100.8	70.2	90.3	93.0	73.5	86.3	93.3	84.8	90.4	95.7	86.0	92.4	
06	WOOD, PLASTICS & COMPOSITES	80.4	34.6	54.1	99.9	93.4	96.2	81.1	92.3	87.5	100.4	82.7	90.2	93.3	96.3	95.0	95.6	81.7	87.6	
07	THERMAL & MOISTURE PROTECTION	107.3	49.1	84.2	107.8	98.5	104.1	101.4	93.4	98.2	108.9	92.0	102.2	102.6	90.9	97.9	100.9	92.2	97.5	
08	OPENINGS	97.0	32.5	80.6	111.6	92.8	106.8	101.3	82.6	96.5	105.4	79.1	98.7	92.6	87.8	91.4	101.1	81.8	96.2	
0920	Plaster & Gypsum Board	106.5	32.9	54.7	92.0	92.9	92.6	91.8	91.7	91.8	93.4	81.8	85.2	93.3	96.4	95.5	95.3	81.4	85.5	
0950, 0980	Ceilings & Acoustic Treatment	140.9	32.9	69.3	91.5	92.9	92.4	97.3	91.7	93.6	91.5	81.8	85.1	91.3	96.4	94.7	92.2	81.4	85.0	
0960	Flooring	117.6	37.9	94.7	101.1	92.6	98.6	129.8	97.6	120.5	101.2	76.2	94.0	105.8	98.2	103.6	106.8	91.2	102.3	
0970, 0990	Wall Finishes & Painting/Coating	98.4	24.2	53.4	101.3	101.2	101.2	105.9	44.9	68.9	101.3	83.6	90.6	102.9	86.6	93.0	102.9	85.1	92.1	
09	FINISHES	116.6	34.6	71.4	99.5	94.5	96.8	102.0	87.9	94.2	99.8	81.3	89.6	99.0	95.5	97.0	99.4	84.8	91.4	
COVERS	DIVS. 10 - 14, 25, 28, 41, 43, 44, 46	100.0	36.1	86.7	100.0	81.4	96.1	100.0	58.0	91.3	100.0	61.3	91.9	100.0	77.7	95.4	100.0	92.5	98.5	
21, 22, 23	FIRE SUPPRESSION, PLUMBING & HVAC	94.9	69.6	84.4	99.9	89.5	95.5	94.6	56.0	78.5	99.9	74.7	89.4	95.0	82.2	89.7	99.9	82.6	92.7	
26, 27, 3370	ELECTRICAL, COMMUNICATIONS & UTIL.	98.3	66.5	81.4	99.9	92.5	96.0	98.3	98.0	98.1	99.1	88.8	93.6	98.4	79.5	88.3	96.9	80.6	88.2	
MF2010	WEIGHTED AVERAGE	101.0	58.2	81.9	99.8	92.6	96.6	97.0	81.0	89.8	99.2	82.5	91.8	94.5	88.5	91.8	96.0	86.3	91.7	

DIVISION		OHIO																	
		CLEVELAND			COLUMBUS			DAYTON			HAMILTON			LIMA			LORAIN		
		441			430 - 432			453 - 454			450			458			440		
		MAT.	INST.	TOTAL	MAT.	INST.	TOTAL	MAT.	INST.	TOTAL	MAT.	INST.	TOTAL	MAT.	INST.	TOTAL	MAT.	INST.	TOTAL
015433	CONTRACTOR EQUIPMENT		97.1	97.1		95.6	95.6		96.4	96.4		101.6	101.6		93.8	93.8		96.8	96.8
0241, 31 - 34	SITE & INFRASTRUCTURE, DEMOLITION	94.1	104.6	101.4	92.3	100.7	98.1	78.3	105.6	97.2	78.2	105.9	97.3	87.8	93.9	92.0	93.5	103.3	100.3
0310	Concrete Forming & Accessories	101.2	99.3	99.6	99.3	88.9	90.3	96.6	83.1	84.9	96.6	83.2	85.0	93.2	86.9	87.7	101.2	89.2	90.8
0320	Concrete Reinforcing	99.6	91.3	95.4	99.6	84.2	91.7	99.6	80.4	89.7	99.6	84.4	91.8	97.2	80.5	88.6	99.1	91.1	95.0
0330	Cast-in-Place Concrete	98.6	108.0	102.2	88.2	93.0	90.0	76.8	89.4	81.6	82.3	87.8	84.4	90.8	98.8	93.9	95.7	102.6	98.4
03	CONCRETE	97.9	100.0	98.9	93.4	89.0	91.3	85.1	84.4	84.8	87.9	85.1	86.5	91.9	89.5	90.7	96.3	93.5	95.0
04	MASONRY	100.3	104.6	102.9	91.0	97.1	94.7	79.7	85.0	82.9	80.0	86.9	84.2	98.8	84.5	90.1	91.8	97.5	95.3
05	METALS	94.7	83.9	91.0	94.7	81.2	90.1	94.8	77.3	88.8	95.0	85.9	91.8	100.8	80.4	93.8	93.6	81.8	89.6
06	WOOD, PLASTICS & COMPOSITES	98.9	97.0	97.8	103.4	86.7	93.8	96.9	82.1	88.4	95.6	81.7	87.6	81.0	87.0	84.5	99.9	87.2	92.6
07	THERMAL & MOISTURE PROTECTION	106.5	108.3	107.3	101.0	96.4	99.2	105.6	90.4	99.6	102.7	92.0	98.4	100.9	96.7	99.2	108.8	101.4	105.9
08	OPENINGS	101.3	94.8	99.6	103.5	83.2	98.4	101.1	81.0	96.0	98.5	81.8	94.2	101.3	81.1	96.2	105.4	89.4	101.4
0920	Plaster & Gypsum Board	91.1	96.6	95.0	97.1	86.3	89.5	95.3	81.8	85.8	95.3	81.4	85.5	91.8	86.3	87.9	92.0	86.5	88.1
0950, 0980	Ceilings & Acoustic Treatment	89.6	96.6	94.2	88.1	86.3	86.9	93.2	81.8	85.7	92.2	81.4	85.0	96.3	86.3	89.6	91.5	86.5	88.2
0960	Flooring	100.9	103.4	101.6	95.5	90.8	94.2	109.7	81.2	101.5	106.8	91.2	102.3	129.0	89.7	117.7	101.2	105.4	102.5
0970, 0990	Wall Finishes & Painting/Coating	101.3	106.6	104.5	97.8	93.3	95.1	102.9	80.1	89.1	102.9	81.4	89.9	105.9	85.9	93.8	101.3	106.6	104.5
09	FINISHES	99.1	100.7	100.0	97.5	89.5	93.1	100.5	82.2	90.4	99.4	84.2	91.0	101.2	87.2	93.5	99.6	93.8	96.4
COVERS	DIVS. 10 - 14, 25, 28, 41, 43, 44, 46	100.0	98.6	99.7	100.0	95.0	99.0	100.0	92.1	98.4	100.0	92.1	98.4	100.0	79.9	95.8	100.0	95.4	99.0
21, 22, 23	FIRE SUPPRESSION, PLUMBING & HVAC	99.8	99.3	99.6	99.9	90.7	96.1	100.6	78.9	91.6	100.4	82.1	92.8	94.6	85.2	90.7	99.9	84.8	93.6
26, 27, 3370	ELECTRICAL, COMMUNICATIONS & UTIL.	99.4	108.9	104.5	99.3	86.7	92.6	95.6	78.8	86.6	95.9	65.1	79.5	98.7	76.0	86.6	99.2	91.1	94.9
MF2010	WEIGHTED AVERAGE	99.0	100.5	99.7	97.9	90.3	94.5	95.8	83.6	90.3	95.7	83.8	90.4	97.4	85.0	91.9	98.7	91.5	95.5

DIVISION		OHIO																	
		MANSFIELD			MARION			SPRINGFIELD			STEUBENVILLE			TOLEDO			YOUNGSTOWN		
		448 - 449			433			455			439			434 - 436			444 - 445		
		MAT.	INST.	TOTAL	MAT.	INST.	TOTAL	MAT.	INST.	TOTAL	MAT.	INST.	TOTAL	MAT.	INST.	TOTAL	MAT.	INST.	TOTAL
015433	CONTRACTOR EQUIPMENT		96.8	96.8		95.3	95.3		96.4	96.4		100.0	100.0		98.2	98.2		96.8	96.8
0241, 31 - 34	SITE & INFRASTRUCTURE, DEMOLITION	90.5	104.3	100.0	88.6	99.3	96.0	78.6	104.2	96.3	134.3	109.0	116.8	91.5	100.2	97.5	94.0	104.2	101.0
0310	Concrete Forming & Accessories	91.3	87.9	88.4	96.1	75.4	78.2	96.6	86.0	87.4	98.7	87.8	89.2	99.3	96.8	97.2	101.2	84.1	86.4
0320	Concrete Reinforcing	90.2	76.9	83.3	91.8	84.1	87.9	99.6	81.4	90.3	89.4	86.6	87.9	99.6	91.0	95.2	99.1	86.4	92.5
0330	Cast-in-Place Concrete	93.1	91.8	92.6	80.6	89.5	84.0	78.9	89.3	82.9	87.5	94.6	90.2	88.2	104.7	94.5	99.6	98.6	99.2
03	CONCRETE	91.1	86.6	88.9	85.7	81.7	83.8	86.1	85.9	86.0	89.6	89.3	89.4	93.4	98.0	95.6	98.3	88.9	93.8
04	MASONRY	94.6	97.2	96.2	92.9	96.1	94.8	79.8	85.0	83.0	79.8	92.5	87.5	99.6	101.0	100.5	95.8	90.0	92.3
05	METALS	93.8	75.6	87.5	93.9	77.6	88.3	94.8	77.5	88.9	90.6	79.0	86.6	94.4	86.8	91.8	93.0	78.4	88.0
06	WOOD, PLASTICS & COMPOSITES	88.4	87.2	87.7	99.6	72.9	84.3	98.3	86.4	91.5	95.4	87.0	90.6	103.4	96.7	99.5	99.9	82.3	89.8
07	THERMAL & MOISTURE PROTECTION	108.3	96.8	103.7	100.6	84.4	94.2	105.5	90.8	99.7	112.4	94.5	105.3	102.7	105.4	103.8	109.0	91.5	102.1
08	OPENINGS	104.5	80.2	98.3	97.5	74.3	91.6	98.9	81.7	94.5	98.3	84.1	94.7	101.3	92.8	99.1	105.4	84.2	100.0
0920	Plaster & Gypsum Board	85.2	86.5	86.1	94.8	72.1	78.8	95.3	86.3	88.9	95.4	86.1	88.8	97.1	96.5	96.7	92.0	81.5	84.6
0950, 0980	Ceilings & Acoustic Treatment	92.4	86.5	88.5	88.1	72.1	77.5	93.2	86.3	88.6	84.9	86.1	85.7	88.1	96.5	93.7	91.5	81.5	84.8
0960	Flooring	96.6	104.2	98.8	94.0	73.7	88.1	109.7	81.2	101.5	122.5	84.1	111.4	94.8	100.2	96.4	101.2	87.8	97.4
0970, 0990	Wall Finishes & Painting/Coating	101.3	93.1	96.3	97.8	51.1	69.5	102.9	80.1	89.1	113.1	91.4	99.9	97.8	105.1	102.2	101.3	90.9	95.0
09	FINISHES	97.3	91.7	94.2	96.4	72.2	83.0	100.5	84.7	91.8	115.8	87.2	100.0	97.3	98.6	98.0	99.7	85.0	91.6
COVERS	DIVS. 10 - 14, 25, 28, 41, 43, 44, 46	100.0	83.3	96.5	100.0	58.3	91.3	100.0	92.6	98.5	100.0	83.5	96.6	100.0	82.4	96.3	100.0	78.7	95.6
21, 22, 23	FIRE SUPPRESSION, PLUMBING & HVAC	94.5	83.6	89.9	94.6	80.7	88.8	100.6	78.9	91.6	94.8	87.1	91.6	99.9	92.7	96.9	99.9	90.8	96.1
26, 27, 3370	ELECTRICAL, COMMUNICATIONS & UTIL.	96.5	78.9	87.1	92.7	74.3	82.9	95.6	76.5	85.4	87.3	98.2	93.1	99.6	97.6	98.5	99.2	80.0	89.0
MF2010	WEIGHTED AVERAGE	96.2	87.0	92.1	94.1	80.8	88.2	95.7	83.8	90.4	96.4	90.5	93.8	98.1	95.9	97.1	99.1	87.5	93.9

	DIVISION	OHIO			OKLAHOMA														
		ZANESVILLE			ARDMORE			CLINTON			DURANT			ENID			GUYMON		
		437 - 438			734			736			747			737			739		
		MAT.	INST.	TOTAL	MAT.	INST.	TOTAL	MAT.	INST.	TOTAL	MAT.	INST.	TOTAL	MAT.	INST.	TOTAL	MAT.	INST.	TOTAL
015433	CONTRACTOR EQUIPMENT		95.3	95.3		81.0	81.0		80.1	80.1		80.1	80.1		80.1	80.1		80.1	80.1
0241, 31 - 34	SITE & INFRASTRUCTURE, DEMOLITION	91.3	100.9	97.9	103.2	90.3	94.3	104.7	88.9	93.8	97.7	88.6	91.4	106.0	88.9	94.2	109.5	87.4	94.3
0310	Concrete Forming & Accessories	93.2	75.4	77.7	91.7	40.3	47.1	90.6	34.9	42.2	83.9	40.3	46.1	93.8	34.2	42.1	97.3	22.2	32.1
0320	Concrete Reinforcing	91.3	86.1	88.6	100.3	78.9	89.3	100.8	78.9	89.5	101.2	62.5	81.3	100.2	78.9	89.2	100.8	27.7	63.1
0330	Cast-in-Place Concrete	84.9	88.7	86.3	99.2	43.4	77.9	95.9	43.4	75.8	92.7	43.3	73.8	95.9	45.8	76.7	95.9	31.0	71.1
03	CONCRETE	89.1	81.8	85.6	95.7	49.2	73.5	95.0	46.8	71.9	90.4	46.0	69.1	95.5	47.3	72.4	98.1	27.0	64.1
04	MASONRY	89.9	84.0	86.3	96.8	57.0	72.7	125.2	57.0	83.9	95.4	65.0	77.0	106.0	57.0	76.3	99.9	22.6	53.1
05	METALS	95.2	79.9	89.9	91.5	69.8	84.1	91.6	69.7	84.1	91.6	63.1	81.8	93.1	69.7	85.1	92.1	39.8	74.1
06	WOOD, PLASTICS & COMPOSITES	95.8	72.9	82.7	97.2	38.9	63.8	96.2	31.6	59.1	88.1	38.9	59.9	99.9	30.7	60.2	103.6	20.3	55.8
07	THERMAL & MOISTURE PROTECTION	100.7	87.7	95.5	97.0	60.1	82.3	97.1	59.3	82.1	96.8	57.9	81.3	97.2	59.2	82.1	97.5	28.2	70.0
08	OPENINGS	97.5	76.3	92.1	95.1	49.6	83.5	95.1	45.6	82.6	95.1	45.5	82.5	95.1	45.0	82.4	95.2	21.7	76.6
0920	Plaster & Gypsum Board	92.2	72.1	78.0	78.4	37.8	49.8	77.9	30.2	44.3	73.5	37.8	48.4	79.3	29.3	44.1	79.8	18.6	36.7
0950, 0980	Ceilings & Acoustic Treatment	88.1	72.1	77.5	82.8	37.8	53.0	82.8	30.2	47.9	82.8	37.8	53.0	82.8	29.3	47.4	84.7	18.6	40.9
0960	Flooring	92.2	77.6	88.0	105.9	45.6	88.6	104.5	43.5	87.0	100.7	65.4	90.6	106.6	43.5	88.5	108.6	25.7	84.8
0970, 0990	Wall Finishes & Painting/Coating	97.8	72.8	82.7	97.1	55.2	71.7	97.1	55.2	71.7	97.1	55.2	71.7	97.1	55.2	71.7	97.1	19.7	50.1
09	FINISHES	95.8	74.7	84.2	91.3	41.0	63.6	91.0	36.3	60.9	88.8	45.2	64.7	91.9	35.8	60.9	93.4	22.2	54.1
COVERS	DIVS. 10 - 14, 25, 28, 41, 43, 44, 46	100.0	73.7	94.5	100.0	75.3	94.9	100.0	74.5	94.7	100.0	75.4	94.9	100.0	74.4	94.7	100.0	72.3	94.2
21, 22, 23	FIRE SUPPRESSION, PLUMBING & HVAC	94.6	84.9	90.6	94.8	62.7	81.4	94.8	62.7	81.4	94.8	62.8	81.5	100.2	62.7	84.6	94.8	23.6	65.1
26, 27, 3370	ELECTRICAL, COMMUNICATIONS & UTIL.	92.8	69.5	80.4	95.6	70.0	82.0	96.7	70.0	82.5	98.6	52.8	74.2	96.7	70.0	82.5	98.6	16.7	55.0
MF2010	WEIGHTED AVERAGE	94.6	81.1	88.6	95.2	61.0	79.9	96.7	59.6	80.2	94.3	58.5	78.3	97.4	59.6	80.5	96.5	31.1	67.3

	DIVISION	OKLAHOMA																	
		LAWTON			MCALESTER			MIAMI			MUSKOGEE			OKLAHOMA CITY			PONCA CITY		
		735			745			743			744			730 - 731			746		
		MAT.	INST.	TOTAL	MAT.	INST.	TOTAL	MAT.	INST.	TOTAL	MAT.	INST.	TOTAL	MAT.	INST.	TOTAL	MAT.	INST.	TOTAL
015433	CONTRACTOR EQUIPMENT		81.0	81.0		80.1	80.1		88.5	88.5		88.5	88.5		81.2	81.2		80.1	80.1
0241, 31 - 34	SITE & INFRASTRUCTURE, DEMOLITION	101.5	90.3	93.8	90.5	89.0	89.4	92.3	86.2	88.1	92.3	84.8	87.1	99.9	90.8	93.7	98.3	88.9	91.8
0310	Concrete Forming & Accessories	97.1	46.1	52.9	82.0	41.6	47.0	94.0	70.7	73.8	98.2	30.6	39.6	96.2	40.0	47.4	89.8	40.7	47.2
0320	Concrete Reinforcing	100.4	78.9	89.4	100.8	41.6	70.3	99.3	79.0	88.8	100.2	31.6	64.9	100.4	78.9	89.3	100.2	78.9	89.2
0330	Cast-in-Place Concrete	92.7	45.9	74.8	81.3	45.9	67.7	85.2	45.2	69.9	86.2	35.9	66.9	93.9	48.6	76.5	95.2	38.5	73.5
03	CONCRETE	91.8	52.6	73.1	81.3	43.5	63.2	85.6	64.1	75.3	87.3	33.9	61.7	92.4	50.8	72.5	92.4	47.7	71.0
04	MASONRY	99.8	57.0	73.9	114.6	57.6	80.1	98.1	58.2	73.9	117.4	46.2	74.3	99.3	58.8	74.8	90.7	57.6	70.6
05	METALS	96.7	69.8	87.5	91.5	53.5	78.4	91.5	82.3	88.3	93.1	55.9	80.3	98.6	69.8	88.7	91.5	69.5	83.9
06	WOOD, PLASTICS & COMPOSITES	102.6	46.7	70.6	85.7	42.2	60.7	99.8	79.2	88.0	104.0	30.4	61.8	101.9	37.6	65.0	95.5	39.5	63.4
07	THERMAL & MOISTURE PROTECTION	97.0	60.9	82.7	96.4	56.7	80.6	96.8	61.3	82.7	96.8	41.8	75.0	93.7	60.7	80.6	96.9	59.3	82.0
08	OPENINGS	96.7	53.7	85.8	95.1	40.9	81.4	95.1	71.9	89.2	95.0	29.5	78.4	96.7	48.7	84.6	95.1	49.9	83.6
0920	Plaster & Gypsum Board	81.4	45.8	56.3	72.2	41.2	50.3	79.3	79.1	79.2	81.4	28.8	44.4	83.4	36.4	50.3	77.9	38.4	50.1
0950, 0980	Ceilings & Acoustic Treatment	90.3	45.8	60.8	82.8	41.2	55.2	82.8	79.1	80.4	90.3	28.8	49.6	90.3	36.4	54.6	82.8	38.4	53.4
0960	Flooring	109.0	43.5	90.2	99.6	43.5	83.5	107.4	65.4	95.4	110.0	42.0	90.5	109.0	43.5	90.2	104.1	43.5	86.7
0970, 0990	Wall Finishes & Painting/Coating	97.1	55.2	71.7	97.1	40.7	62.9	97.1	83.4	88.8	97.1	29.6	56.1	97.1	55.2	71.7	97.1	55.2	71.7
09	FINISHES	93.9	45.2	67.0	87.7	41.1	62.0	90.7	72.2	80.5	93.4	32.5	59.8	93.8	40.2	64.3	90.4	41.1	63.3
COVERS	DIVS. 10 - 14, 25, 28, 41, 43, 44, 46	100.0	76.2	95.1	100.0	75.9	95.0	100.0	80.7	96.0	100.0	74.0	94.6	100.0	75.8	95.0	100.0	75.5	94.9
21, 22, 23	FIRE SUPPRESSION, PLUMBING & HVAC	100.2	62.7	84.6	94.8	30.8	68.2	94.8	61.6	81.0	100.2	23.9	68.4	100.0	63.6	84.9	94.8	61.2	80.8
26, 27, 3370	ELECTRICAL, COMMUNICATIONS & UTIL.	98.6	70.0	83.4	96.8	67.8	81.4	98.5	67.9	82.2	96.3	28.6	60.3	104.0	70.0	85.9	96.3	64.0	79.1
MF2010	WEIGHTED AVERAGE	97.6	62.3	81.8	93.7	51.2	74.7	94.0	69.0	82.8	96.7	39.6	71.2	98.2	61.5	81.8	94.3	59.6	78.8

	DIVISION	OKLAHOMA												OREGON					
		POTEAU			SHAWNEE			TULSA			WOODWARD			BEND			EUGENE		
		749			748			740 - 741			738			977			974		
		MAT.	INST.	TOTAL	MAT.	INST.	TOTAL	MAT.	INST.	TOTAL	MAT.	INST.	TOTAL	MAT.	INST.	TOTAL	MAT.	INST.	TOTAL
015433	CONTRACTOR EQUIPMENT		87.8	87.8		80.1	80.1		88.5	88.5		80.1	80.1		99.4	99.4		99.4	99.4
0241, 31 - 34	SITE & INFRASTRUCTURE, DEMOLITION	76.2	84.8	82.2	101.6	88.8	92.8	98.6	86.2	90.0	105.1	88.9	93.9	115.3	104.6	107.9	103.3	104.6	104.2
0310	Concrete Forming & Accessories	87.8	43.0	48.9	83.8	39.3	45.2	98.1	39.8	47.5	90.7	34.8	42.2	109.7	95.3	97.2	106.4	95.2	96.7
0320	Concrete Reinforcing	101.3	78.9	89.8	100.2	45.7	72.1	100.4	78.8	89.3	100.2	78.9	89.2	104.3	102.9	103.6	108.8	102.9	105.8
0330	Cast-in-Place Concrete	85.2	44.7	69.7	98.3	42.9	77.1	94.0	44.7	75.1	95.9	43.4	75.8	100.2	106.2	102.5	97.0	106.2	100.5
03	CONCRETE	87.6	51.6	70.4	94.2	42.3	69.3	92.5	50.2	72.2	95.3	46.7	72.0	111.8	100.3	106.3	102.8	100.2	101.6
04	MASONRY	99.1	57.6	74.0	115.9	56.4	79.8	98.9	58.8	74.6	93.8	57.0	71.5	109.9	101.5	104.8	107.4	101.5	103.8
05	METALS	91.5	81.9	88.2	91.4	55.3	79.0	96.4	80.9	91.1	91.7	69.7	84.1	88.7	97.9	91.9	89.4	97.8	92.3
06	WOOD, PLASTICS & COMPOSITES	92.6	42.4	63.8	87.9	38.7	59.7	103.2	38.7	66.2	96.4	31.5	59.1	98.0	93.1	95.1	93.9	93.1	93.4
07	THERMAL & MOISTURE PROTECTION	96.9	57.0	81.0	96.9	59.9	82.2	96.8	55.4	80.4	97.2	59.3	82.1	106.3	94.2	101.5	105.6	91.2	99.9
08	OPENINGS	95.1	52.0	84.2	95.0	38.1	80.6	96.7	48.9	84.6	95.1	45.6	82.6	97.5	100.6	98.3	97.9	100.6	98.6
0920	Plaster & Gypsum Board	76.6	41.2	51.7	73.5	37.5	48.2	81.4	37.4	50.4	78.5	30.1	44.4	106.9	92.7	96.9	104.6	92.7	96.2
0950, 0980	Ceilings & Acoustic Treatment	82.8	41.2	55.2	82.8	37.5	52.8	90.3	37.4	55.3	84.7	30.1	48.5	101.1	92.7	95.5	102.1	92.7	95.8
0960	Flooring	103.4	65.4	92.5	100.6	34.6	81.7	108.8	45.2	90.5	104.5	45.6	87.6	110.6	100.7	107.8	108.8	100.7	106.5
0970, 0990	Wall Finishes & Painting/Coating	97.1	55.2	71.7	97.1	36.7	60.5	97.1	43.6	64.7	97.1	55.2	71.7	108.1	70.5	85.3	108.1	70.5	85.3
09	FINISHES	88.3	47.2	65.6	89.0	36.7	60.2	93.2	40.0	63.9	91.6	36.6	61.3	107.4	93.1	99.5	105.5	93.1	98.7
COVERS	DIVS. 10 - 14, 25, 28, 41, 43, 44, 46	100.0	76.2	95.0	100.0	75.0	94.8	100.0	76.2	95.0	100.0	74.4	94.7	100.0	100.3	100.1	100.0	100.3	100.1
21, 22, 23	FIRE SUPPRESSION, PLUMBING & HVAC	94.8	61.3	80.8	94.8	62.0	81.1	100.2	48.4	78.6	94.8	62.7	81.4	94.7	98.7	96.4	100.1	98.7	99.5
26, 27, 3370	ELECTRICAL, COMMUNICATIONS & UTIL.	96.4	67.9	81.3	98.7	70.0	83.5	98.6	37.3	66.0	98.5	70.0	83.3	102.4	95.2	98.6	100.9	95.2	97.9
MF2010	WEIGHTED AVERAGE	93.3	62.3	79.5	96.0	57.3	78.7	97.4	54.2	78.1	95.4	59.7	79.4	100.3	98.4	99.5	99.8	98.3	99.1

	DIVISION	OREGON																	
		KLAMATH FALLS			MEDFORD			PENDLETON			PORTLAND			SALEM			VALE		
		976			975			978			970 - 972			973			979		
		MAT.	INST.	TOTAL	MAT.	INST.	TOTAL	MAT.	INST.	TOTAL	MAT.	INST.	TOTAL	MAT.	INST.	TOTAL	MAT.	INST.	TOTAL
015433	CONTRACTOR EQUIPMENT		99.4	99.4		99.4	99.4		96.7	96.7		99.4	99.4		99.4	99.4		96.7	96.7
0241, 31 - 34	SITE & INFRASTRUCTURE, DEMOLITION	120.0	104.6	109.3	112.1	104.6	106.9	110.5	98.0	101.8	106.2	104.6	105.1	98.4	104.6	102.7	97.0	97.9	97.6
0310	Concrete Forming & Accessories	103.2	95.1	96.2	102.4	95.1	96.1	102.9	95.5	96.5	107.5	95.4	97.0	105.2	95.3	96.6	108.9	94.5	96.4
0320	Concrete Reinforcing	104.3	102.9	103.6	106.1	102.9	104.4	103.6	103.1	103.3	109.6	103.1	106.2	109.7	103.1	106.3	101.0	102.9	102.0
0330	Cast-in-Place Concrete	100.2	106.2	102.5	100.2	106.2	102.5	100.9	107.6	103.5	99.7	106.3	102.2	88.9	106.2	95.6	79.5	107.2	90.1
03	CONCRETE	114.8	100.2	107.8	109.2	100.2	104.9	94.6	100.9	97.6	104.3	100.4	102.4	98.8	100.4	99.5	79.2	100.3	89.3
04	MASONRY	124.5	101.5	110.6	104.3	101.5	102.6	114.1	103.7	107.8	108.0	103.6	105.3	112.3	103.6	107.0	112.4	103.7	107.1
05	METALS	88.7	97.7	91.8	89.0	97.7	92.0	94.5	98.6	95.9	90.2	98.2	93.0	89.8	98.1	92.6	94.4	97.6	95.5
06	WOOD, PLASTICS & COMPOSITES	89.2	93.1	91.4	88.1	93.1	90.9	91.0	93.2	92.2	94.8	93.1	93.8	91.1	93.1	92.2	99.2	93.2	95.8
07	THERMAL & MOISTURE PROTECTION	106.5	92.7	101.0	106.2	92.7	100.8	98.0	93.5	96.2	105.6	97.3	102.3	103.5	94.8	100.1	97.4	91.1	94.9
08	OPENINGS	97.5	100.6	98.3	100.5	100.6	100.5	93.1	100.7	95.0	95.5	100.6	96.8	97.3	100.6	98.2	93.1	93.2	93.1
0920	Plaster & Gypsum Board	102.1	92.7	95.5	101.2	92.7	95.2	82.4	92.7	89.6	104.9	92.7	96.3	99.2	92.7	94.6	89.1	92.7	91.6
0950, 0980	Ceilings & Acoustic Treatment	109.5	92.7	98.3	114.7	92.7	100.1	69.1	92.7	84.7	104.2	92.7	96.6	104.4	92.7	96.6	69.1	92.7	84.7
0960	Flooring	107.3	100.7	105.4	106.6	100.7	104.9	74.9	100.7	82.3	106.3	100.7	104.7	109.0	100.7	106.6	76.8	100.7	83.7
0970, 0990	Wall Finishes & Painting/Coating	108.1	62.6	80.5	108.1	62.6	80.5	96.2	70.5	80.6	107.7	70.5	85.1	108.1	70.5	85.3	96.2	70.5	80.6
09	FINISHES	108.1	92.3	99.4	108.0	92.3	99.3	73.1	93.2	84.2	105.3	93.1	98.6	104.9	93.1	98.4	73.4	93.2	84.3
COVERS	DIVS. 10 - 14, 25, 28, 41, 43, 44, 46	100.0	100.3	100.1	100.0	100.3	100.1	100.0	98.8	99.8	100.0	100.4	100.1	100.0	100.3	100.1	100.0	100.6	100.1
21, 22, 23	FIRE SUPPRESSION, PLUMBING & HVAC	94.7	98.7	96.4	100.1	98.7	99.5	96.1	108.3	101.2	100.1	98.7	99.6	100.1	98.7	99.6	96.1	75.5	87.5
26, 27, 3370	ELECTRICAL, COMMUNICATIONS & UTIL.	101.0	84.7	92.4	104.8	84.7	94.1	92.6	96.4	94.6	101.1	102.4	101.8	106.3	95.2	100.4	92.6	78.4	85.0
MF2010	WEIGHTED AVERAGE	101.5	96.7	99.3	101.5	96.7	99.3	94.9	100.3	97.3	100.0	99.7	99.9	99.8	98.6	99.3	92.6	90.4	91.6

	DIVISION	PENNSYLVANIA																	
		ALLENTOWN			ALTOONA			BEDFORD			BRADFORD			BUTLER			CHAMBERSBURG		
		181			166			155			167			160			172		
		MAT.	INST.	TOTAL	MAT.	INST.	TOTAL	MAT.	INST.	TOTAL	MAT.	INST.	TOTAL	MAT.	INST.	TOTAL	MAT.	INST.	TOTAL
015433	CONTRACTOR EQUIPMENT		114.2	114.2		114.2	114.2		111.3	111.3		114.2	114.2		114.2	114.2		113.5	113.5
0241, 31 - 34	SITE & INFRASTRUCTURE, DEMOLITION	92.3	105.9	101.7	96.3	105.9	103.0	112.5	102.8	105.8	91.8	105.3	101.1	87.5	107.3	101.2	86.2	103.5	98.1
0310	Concrete Forming & Accessories	101.8	114.3	112.6	87.2	79.4	80.4	87.6	81.8	82.6	89.4	79.7	81.0	88.7	93.7	93.0	86.7	80.9	81.7
0320	Concrete Reinforcing	97.1	113.3	105.5	94.2	91.5	92.8	95.9	80.5	88.0	96.2	91.5	93.8	94.8	103.7	99.4	94.6	99.9	97.3
0330	Cast-in-Place Concrete	82.7	102.5	90.3	92.2	86.6	90.0	106.3	74.0	93.9	88.1	89.0	88.4	81.4	97.6	87.6	89.4	73.7	83.4
03	CONCRETE	96.2	110.8	103.2	91.2	85.6	88.6	101.9	80.3	91.5	98.1	86.6	92.6	82.9	98.1	90.2	104.2	83.6	94.3
04	MASONRY	95.3	101.1	98.8	96.9	69.6	80.3	100.6	85.7	91.6	94.1	83.1	87.4	99.0	96.9	97.7	95.2	89.5	91.7
05	METALS	100.2	124.7	108.6	94.1	111.9	100.2	97.1	104.7	99.7	98.0	111.3	102.6	93.8	119.1	102.5	98.0	114.3	103.6
06	WOOD, PLASTICS & COMPOSITES	105.1	118.0	112.5	83.3	81.3	82.2	87.8	81.2	84.0	90.2	77.5	82.9	84.9	91.2	88.5	93.8	81.3	86.6
07	THERMAL & MOISTURE PROTECTION	101.1	117.1	107.4	100.5	91.4	96.9	105.5	89.0	98.9	101.1	90.2	96.8	100.2	100.2	100.2	100.2	75.9	90.5
08	OPENINGS	92.7	116.9	98.8	87.3	85.9	86.9	93.9	82.9	91.1	93.1	87.6	91.7	87.3	101.4	90.8	89.5	82.8	87.8
0920	Plaster & Gypsum Board	106.5	118.3	114.8	97.8	80.5	85.6	98.3	80.5	85.7	99.2	76.6	83.3	98.3	90.7	92.9	103.2	80.5	87.2
0950, 0980	Ceilings & Acoustic Treatment	93.1	118.3	109.8	96.9	80.5	86.0	98.7	80.5	86.6	95.1	76.6	82.9	97.9	90.7	93.1	95.1	80.5	85.4
0960	Flooring	90.1	97.0	92.1	84.0	93.0	86.6	89.4	93.1	90.5	84.2	97.3	87.9	84.8	85.6	85.0	87.8	48.8	76.6
0970, 0990	Wall Finishes & Painting/Coating	92.5	76.0	82.5	88.6	105.7	99.0	95.0	72.8	81.5	92.5	94.2	93.6	88.6	105.7	99.0	92.5	62.2	74.1
09	FINISHES	96.7	107.6	102.7	95.1	83.9	88.9	96.9	81.5	88.4	94.5	82.7	88.0	95.0	92.2	93.5	95.1	72.6	82.6
COVERS	DIVS. 10 - 14, 25, 28, 41, 43, 44, 46	100.0	105.2	101.1	100.0	95.8	99.1	100.0	99.0	99.8	100.0	98.9	99.8	100.0	103.1	100.6	100.0	95.9	99.1
21, 22, 23	FIRE SUPPRESSION, PLUMBING & HVAC	100.3	110.3	104.5	99.8	82.9	92.8	94.7	87.6	91.8	94.9	89.2	92.5	94.5	94.5	94.5	95.0	89.0	92.5
26, 27, 3370	ELECTRICAL, COMMUNICATIONS & UTIL.	100.7	96.8	98.6	90.0	105.4	98.2	93.6	105.4	99.9	93.6	105.4	99.8	90.7	104.2	97.9	92.1	84.4	88.0
MF2010	WEIGHTED AVERAGE	98.3	108.6	102.9	95.0	90.5	93.0	97.6	91.1	94.7	95.9	93.1	94.7	92.6	100.3	96.0	96.1	88.6	92.7

	DIVISION	PENNSYLVANIA																	
		DOYLESTOWN			DUBOIS			ERIE			GREENSBURG			HARRISBURG			HAZLETON		
		189			158			164 - 165			156			170 - 171			182		
		MAT.	INST.	TOTAL	MAT.	INST.	TOTAL	MAT.	INST.	TOTAL	MAT.	INST.	TOTAL	MAT.	INST.	TOTAL	MAT.	INST.	TOTAL
015433	CONTRACTOR EQUIPMENT		94.4	94.4		111.3	111.3		114.2	114.2		111.3	111.3		113.5	113.5		114.2	114.2
0241, 31 - 34	SITE & INFRASTRUCTURE, DEMOLITION	107.2	91.6	96.4	118.5	102.9	107.8	92.8	106.5	102.3	107.9	104.9	105.8	82.2	104.8	97.8	85.6	106.1	99.8
0310	Concrete Forming & Accessories	86.1	131.0	125.1	87.1	82.9	83.4	101.3	87.3	89.1	93.7	93.6	93.6	94.7	86.0	87.1	83.9	87.2	86.7
0320	Concrete Reinforcing	94.0	141.0	118.2	95.4	91.8	93.5	96.2	93.4	94.8	95.4	103.7	99.6	97.1	102.0	99.6	94.3	108.0	101.3
0330	Cast-in-Place Concrete	78.2	91.1	83.1	102.4	92.3	98.5	90.6	85.5	88.6	98.6	97.2	98.1	87.1	94.3	89.9	78.2	90.9	83.1
03	CONCRETE	91.9	119.0	104.9	103.0	89.1	96.3	90.0	89.1	89.6	96.7	97.9	97.3	96.1	93.3	94.7	88.7	93.8	91.1
04	MASONRY	98.5	130.7	118.0	100.6	86.6	92.1	87.6	91.2	89.7	110.9	96.9	102.4	95.1	89.6	91.7	107.4	96.1	100.6
05	METALS	97.7	129.7	108.7	97.1	110.8	101.8	94.3	112.2	100.5	97.0	117.7	104.2	102.3	118.6	107.9	99.9	121.2	107.2
06	WOOD, PLASTICS & COMPOSITES	85.4	133.2	112.8	86.7	81.2	83.5	101.1	85.7	92.3	94.4	91.2	92.5	98.9	85.0	90.9	84.0	85.0	84.5
07	THERMAL & MOISTURE PROTECTION	99.4	134.3	113.3	105.7	94.7	101.3	100.8	93.2	97.8	105.3	99.8	103.1	107.0	108.0	107.4	100.7	106.4	102.9
08	OPENINGS	94.3	142.9	106.6	93.9	89.6	92.8	87.4	88.4	87.7	93.9	101.4	95.8	92.7	91.6	92.4	93.4	92.8	93.2
0920	Plaster & Gypsum Board	96.2	133.9	122.8	97.0	80.5	85.4	106.5	85.1	91.4	99.4	90.7	93.3	103.3	84.3	89.9	96.8	84.3	88.0
0950, 0980	Ceilings & Acoustic Treatment	92.3	133.9	119.9	98.7	80.5	86.6	93.1	85.1	87.8	97.8	90.7	93.1	96.9	84.3	88.5	94.2	84.3	87.6
0960	Flooring	74.4	134.5	91.7	89.1	99.2	92.0	89.5	97.0	91.6	92.6	71.9	86.6	90.7	92.6	91.3	81.6	95.7	85.6
0970, 0990	Wall Finishes & Painting/Coating	92.0	74.3	81.3	95.0	105.7	101.5	97.0	89.4	92.4	95.0	105.7	101.5	92.5	87.8	89.7	92.5	94.6	93.8
09	FINISHES	87.5	124.8	108.1	97.2	87.5	91.9	97.4	88.9	92.7	97.4	90.6	93.6	95.8	86.9	90.9	92.7	87.4	89.8
COVERS	DIVS. 10 - 14, 25, 28, 41, 43, 44, 46	100.0	76.9	95.2	100.0	99.0	99.8	100.0	100.9	100.2	100.0	103.0	100.6	100.0	96.3	99.2	100.0	99.8	100.0
21, 22, 23	FIRE SUPPRESSION, PLUMBING & HVAC	94.5	127.3	108.1	94.7	87.9	91.9	99.8	92.2	96.6	94.7	92.1	93.7	100.1	89.8	95.8	94.9	98.3	96.4
26, 27, 3370	ELECTRICAL, COMMUNICATIONS & UTIL.	93.0	120.5	107.6	94.3	105.4	100.2	92.0	86.0	88.8	94.3	105.4	100.2	97.3	84.5	90.5	94.5	87.6	90.8
MF2010	WEIGHTED AVERAGE	95.1	121.9	107.1	98.0	94.2	96.3	94.8	93.5	94.2	97.5	99.4	98.3	98.0	93.8	96.1	95.5	97.3	96.3

	DIVISION	PENNSYLVANIA																	
		INDIANA			JOHNSTOWN			KITTANNING			LANCASTER			LEHIGH VALLEY			MONTROSE		
		157			159			162			175 - 176			180			188		
		MAT.	INST.	TOTAL	MAT.	INST.	TOTAL	MAT.	INST.	TOTAL	MAT.	INST.	TOTAL	MAT.	INST.	TOTAL	MAT.	INST.	TOTAL
015433	CONTRACTOR EQUIPMENT		111.3	111.3		111.3	111.3		114.2	114.2		113.5	113.5		114.2	114.2		114.2	114.2
0241, 31 - 34	SITE & INFRASTRUCTURE, DEMOLITION	105.8	103.7	104.4	113.2	104.1	106.9	90.2	107.3	102.0	78.2	104.8	96.6	89.7	105.9	100.9	88.6	104.1	99.3
0310	Concrete Forming & Accessories	88.2	86.0	86.3	87.1	83.1	83.6	88.7	93.8	93.1	88.6	85.7	86.1	95.4	114.1	111.7	84.7	86.3	86.1
0320	Concrete Reinforcing	94.6	103.7	99.3	95.9	103.5	99.8	94.8	103.8	99.4	94.3	101.9	98.2	94.3	113.3	104.1	98.8	107.8	103.4
0330	Cast-in-Place Concrete	96.7	97.1	96.8	107.2	91.8	101.3	84.6	97.7	89.6	76.1	94.5	83.1	84.6	102.4	91.4	83.0	89.3	85.4
03	CONCRETE	94.4	94.4	94.4	102.7	91.3	97.2	85.5	98.2	91.6	91.4	93.2	92.3	95.2	110.7	102.6	93.8	92.6	93.2
04	MASONRY	96.8	98.5	97.9	97.7	86.1	90.7	103.1	98.5	100.3	100.7	90.2	94.3	95.3	101.1	98.8	95.3	95.8	95.6
05	METALS	97.2	117.5	104.2	97.1	116.3	103.7	93.9	119.3	102.7	98.0	118.1	104.9	99.8	124.3	108.3	98.1	116.3	104.4
06	WOOD, PLASTICS & COMPOSITES	88.6	81.2	84.4	86.7	81.2	83.5	84.9	91.2	88.5	96.4	85.0	89.9	97.2	118.0	109.1	84.8	85.3	85.1
07	THERMAL & MOISTURE PROTECTION	105.2	97.8	102.3	105.5	93.2	100.6	100.3	100.6	100.4	99.8	97.1	98.7	101.0	105.6	102.8	100.7	94.4	98.2
08	OPENINGS	93.9	93.2	93.7	93.9	89.5	92.8	87.3	98.7	90.2	89.5	95.6	91.1	93.4	116.9	99.4	89.7	93.0	90.5
0920	Plaster & Gypsum Board	98.8	80.5	85.9	96.7	80.5	85.3	98.3	90.7	92.9	105.4	84.3	90.5	101.2	118.3	113.3	97.0	84.6	88.3
0950, 0980	Ceilings & Acoustic Treatment	98.7	80.5	86.6	97.8	80.5	86.3	97.9	90.7	93.1	95.1	84.3	87.9	94.2	118.3	110.2	95.1	84.6	88.2
0960	Flooring	89.8	99.2	92.5	89.1	93.1	90.3	84.8	103.3	90.1	88.8	92.6	89.9	86.9	97.0	89.8	82.2	61.2	76.2
0970, 0990	Wall Finishes & Painting/Coating	95.0	105.7	101.5	95.0	105.7	101.5	88.6	105.7	99.0	92.5	61.5	73.7	92.5	71.5	79.8	92.5	98.4	96.1
09	FINISHES	96.6	88.8	92.3	96.5	86.4	90.9	95.1	95.4	95.3	95.0	84.1	89.0	95.2	106.0	101.1	93.4	82.5	87.4
COVERS	DIVS. 10 - 14, 25, 28, 41, 43, 44, 46	100.0	101.8	100.4	100.0	99.2	99.8	100.0	103.1	100.6	100.0	96.3	99.2	100.0	105.4	101.1	100.0	96.5	99.3
21, 22, 23	FIRE SUPPRESSION, PLUMBING & HVAC	94.7	92.1	93.6	94.7	88.2	92.0	94.5	96.7	95.4	95.0	90.0	92.9	94.9	110.3	101.3	94.9	97.8	96.1
26, 27, 3370	ELECTRICAL, COMMUNICATIONS & UTIL.	94.3	105.4	100.2	94.3	105.4	100.2	90.0	105.4	98.2	93.6	47.8	69.2	94.5	141.5	119.5	93.6	95.1	94.4
MF2010	WEIGHTED AVERAGE	96.3	98.2	97.1	97.6	94.9	96.4	93.2	101.4	96.9	94.7	88.3	91.8	96.0	114.2	104.2	94.9	96.3	95.5

	DIVISION	PENNSYLVANIA																	
		NEW CASTLE			NORRISTOWN			OIL CITY			PHILADELPHIA			PITTSBURGH			POTTSVILLE		
		161			194			163			190 - 191			150 - 152			179		
		MAT.	INST.	TOTAL	MAT.	INST.	TOTAL	MAT.	INST.	TOTAL	MAT.	INST.	TOTAL	MAT.	INST.	TOTAL	MAT.	INST.	TOTAL
015433	CONTRACTOR EQUIPMENT		114.2	114.2		97.9	97.9		114.2	114.2		97.3	97.3		112.6	112.6		113.5	113.5
0241, 31 - 34	SITE & INFRASTRUCTURE, DEMOLITION	88.0	107.3	101.3	88.0	98.9	95.6	86.4	105.5	99.6	94.4	98.2	97.0	111.4	107.1	108.4	81.1	104.9	97.5
0310	Concrete Forming & Accessories	88.7	94.0	93.3	88.6	133.4	127.5	88.7	81.0	82.0	104.8	137.6	133.3	101.1	94.9	95.7	80.6	86.7	85.9
0320	Concrete Reinforcing	93.6	93.4	93.5	92.2	148.5	121.2	94.8	93.2	94.0	95.1	148.5	122.6	96.5	104.0	100.3	93.6	101.9	97.8
0330	Cast-in-Place Concrete	82.2	96.7	87.8	78.1	131.3	98.5	79.9	94.6	85.5	95.8	131.3	109.4	102.4	97.4	100.5	81.0	78.8	80.2
03	CONCRETE	83.3	96.0	89.4	87.5	135.2	110.3	81.7	89.4	85.4	99.7	137.1	117.6	100.1	98.6	99.3	95.4	88.3	92.0
04	MASONRY	98.3	96.9	97.4	110.0	130.8	122.6	99.2	90.5	94.0	96.4	131.0	117.4	92.0	101.9	98.0	95.0	89.7	91.8
05	METALS	93.9	113.8	100.8	100.4	135.2	112.4	93.9	112.2	100.2	104.3	135.3	115.0	98.6	118.4	105.4	98.2	118.2	105.1
06	WOOD, PLASTICS & COMPOSITES	84.9	92.5	89.3	86.8	133.1	113.4	84.9	77.5	80.7	106.0	138.7	124.7	104.1	92.4	97.4	86.7	85.0	85.7
07	THERMAL & MOISTURE PROTECTION	100.2	97.3	99.0	97.1	134.8	112.1	100.1	93.4	97.4	97.9	135.4	112.8	105.5	101.3	103.8	99.9	105.6	102.2
08	OPENINGS	87.3	96.0	89.5	89.7	145.0	103.7	87.3	80.0	85.4	98.2	148.0	110.8	96.8	102.2	98.1	89.6	91.0	89.9
0920	Plaster & Gypsum Board	98.3	92.0	93.8	89.7	133.9	120.8	98.3	76.6	83.0	98.1	139.7	127.4	106.7	92.0	96.4	99.7	84.3	88.8
0950, 0980	Ceilings & Acoustic Treatment	97.9	92.0	94.0	104.3	133.9	123.9	97.9	76.6	83.8	104.3	139.7	127.7	98.7	92.0	94.3	95.1	84.3	87.9
0960	Flooring	84.8	60.7	77.9	85.2	140.6	101.1	84.8	99.2	88.9	92.8	140.6	106.5	96.2	101.0	97.6	85.0	92.6	87.2
0970, 0990	Wall Finishes & Painting/Coating	88.6	105.7	99.0	96.2	145.4	126.0	88.6	105.7	99.0	96.2	142.3	124.2	95.0	114.8	107.0	92.5	99.1	96.5
09	FINISHES	95.0	88.4	91.4	95.3	135.8	117.7	94.8	85.1	89.5	98.9	138.8	120.9	99.7	97.0	98.2	93.4	88.7	90.8
COVERS	DIVS. 10 - 14, 25, 28, 41, 43, 44, 46	100.0	103.2	100.7	100.0	119.3	104.0	100.0	100.1	100.0	100.0	121.5	104.5	100.0	103.1	100.7	100.0	97.1	99.4
21, 22, 23	FIRE SUPPRESSION, PLUMBING & HVAC	94.5	94.3	94.4	94.6	128.7	108.8	94.5	91.8	93.4	100.0	129.0	112.1	100.1	100.4	100.3	95.0	97.6	96.0
26, 27, 3370	ELECTRICAL, COMMUNICATIONS & UTIL.	90.7	93.3	92.1	93.4	141.8	119.1	92.7	104.2	98.8	98.3	141.9	121.5	96.9	105.4	101.4	91.6	89.6	90.5
MF2010	WEIGHTED AVERAGE	92.6	97.2	94.7	95.1	131.2	111.3	92.6	94.9	93.6	99.7	132.2	114.3	99.3	102.9	100.9	94.6	95.6	95.1

	DIVISION	PENNSYLVANIA																	
		READING			SCRANTON			STATE COLLEGE			STROUDSBURG			SUNBURY			UNIONTOWN		
		195 - 196			184 - 185			168			183			178			154		
		MAT.	INST.	TOTAL	MAT.	INST.	TOTAL	MAT.	INST.	TOTAL	MAT.	INST.	TOTAL	MAT.	INST.	TOTAL	MAT.	INST.	TOTAL
015433	CONTRACTOR EQUIPMENT		117.7	117.7		114.2	114.2		113.5	113.5		114.2	114.2		114.2	114.2		111.3	111.3
0241, 31 - 34	SITE & INFRASTRUCTURE, DEMOLITION	92.0	111.1	105.2	92.7	106.2	102.0	83.3	104.8	98.2	87.3	104.2	99.0	91.4	105.9	101.4	106.4	104.9	105.4
0310	Concrete Forming & Accessories	104.3	87.0	89.3	101.9	87.7	89.6	85.8	79.3	80.2	90.0	87.4	87.7	93.3	85.8	86.8	81.7	93.6	92.0
0320	Concrete Reinforcing	93.5	102.1	97.9	97.1	108.0	102.7	95.4	92.0	93.7	97.4	120.3	109.2	96.2	100.2	98.2	95.4	103.7	99.6
0330	Cast-in-Place Concrete	70.0	95.8	79.9	86.4	91.1	88.2	83.3	69.4	78.0	81.4	75.4	79.1	88.6	94.2	90.7	96.7	97.2	96.9
03	CONCRETE	86.4	94.3	90.2	98.0	94.1	96.2	98.4	79.9	89.5	92.6	90.8	91.7	98.8	92.8	95.9	94.1	97.9	95.9
04	MASONRY	98.3	92.9	95.0	95.6	96.2	96.0	100.8	80.6	88.5	91.7	100.7	97.1	94.1	87.3	90.0	113.1	96.9	103.2
05	METALS	100.7	119.3	107.1	102.4	121.5	108.9	97.8	112.7	102.9	99.9	121.8	107.5	97.9	117.4	104.6	96.9	117.7	104.0
06	WOOD, PLASTICS & COMPOSITES	106.1	84.6	93.8	105.1	85.3	93.8	92.5	81.3	86.1	91.1	85.3	87.8	94.7	85.0	89.1	80.2	91.2	86.5
07	THERMAL & MOISTURE PROTECTION	97.5	109.8	102.4	101.0	97.9	99.8	100.0	92.3	96.9	100.8	87.0	95.4	101.0	104.6	102.4	105.1	99.8	103.0
08	OPENINGS	95.6	95.4	95.5	92.7	93.0	92.8	89.5	85.9	88.6	93.4	89.7	92.5	89.7	90.4	89.9	93.8	101.4	95.8
0920	Plaster & Gypsum Board	102.6	84.0	89.5	108.6	84.6	91.7	101.4	80.5	86.6	98.9	84.6	88.9	100.0	84.3	88.9	93.6	90.7	91.6
0950, 0980	Ceilings & Acoustic Treatment	93.9	84.0	87.3	100.6	84.6	90.0	93.2	80.5	84.8	92.3	84.6	87.2	93.2	84.3	87.3	97.8	90.7	93.1
0960	Flooring	90.5	92.6	91.1	90.1	107.6	95.2	87.4	103.0	91.9	84.6	54.2	75.9	85.9	92.6	87.8	86.5	99.2	90.1
0970, 0990	Wall Finishes & Painting/Coating	94.7	99.1	97.4	92.5	98.4	96.1	92.5	105.7	100.5	92.5	69.5	78.6	92.5	99.1	96.5	95.0	105.7	101.5
09	FINISHES	96.5	88.0	91.8	98.5	91.8	94.8	94.1	85.9	89.6	93.7	78.9	85.5	94.7	88.1	91.1	94.7	94.4	94.6
COVERS	DIVS. 10 - 14, 25, 28, 41, 43, 44, 46	100.0	99.0	99.8	100.0	99.9	100.0	100.0	92.6	98.5	100.0	67.2	93.2	100.0	96.3	99.2	100.0	103.0	100.6
21, 22, 23	FIRE SUPPRESSION, PLUMBING & HVAC	100.1	106.9	102.9	100.3	98.4	99.5	95.0	84.7	90.7	94.9	100.1	97.1	94.9	89.6	92.7	94.7	92.1	93.7
26, 27, 3370	ELECTRICAL, COMMUNICATIONS & UTIL.	100.5	89.6	94.7	100.7	95.1	97.7	92.6	105.4	99.4	94.5	141.5	119.5	92.0	88.5	90.1	90.9	105.4	98.6
MF2010	WEIGHTED AVERAGE	97.4	99.6	98.4	99.0	98.7	98.9	95.5	91.4	93.7	95.3	102.3	98.4	95.5	94.0	94.8	96.6	99.9	98.0

	DIVISION	PENNSYLVANIA																	
		WASHINGTON			WELLSBORO			WESTCHESTER			WILKES-BARRE			WILLIAMSPORT			YORK		
		153			169			193			186 - 187			177			173 - 174		
		MAT.	INST.	TOTAL	MAT.	INST.	TOTAL	MAT.	INST.	TOTAL	MAT.	INST.	TOTAL	MAT.	INST.	TOTAL	MAT.	INST.	TOTAL
015433	CONTRACTOR EQUIPMENT		111.3	111.3		114.2	114.2		97.9	97.9		114.2	114.2		114.2	114.2		113.5	113.5
0241, 31 - 34	SITE & INFRASTRUCTURE, DEMOLITION	106.4	104.8	105.3	95.9	104.0	101.5	93.7	96.7	95.8	85.4	106.1	99.7	82.6	104.7	97.8	81.5	104.8	97.6
0310	Concrete Forming & Accessories	88.4	93.7	93.0	88.7	84.2	84.8	95.3	132.1	127.2	92.5	87.6	88.3	90.0	62.9	66.5	83.9	86.2	85.9
0320	Concrete Reinforcing	95.4	103.8	99.7	95.4	107.6	101.7	91.3	114.6	103.3	96.2	108.0	102.3	95.4	69.4	82.0	96.2	102.0	99.2
0330	Cast-in-Place Concrete	96.7	97.3	96.9	87.3	87.0	87.2	86.5	130.7	103.4	78.2	90.9	83.1	75.0	75.3	75.1	81.4	94.7	86.5
03	CONCRETE	94.5	98.0	96.2	100.8	90.9	96.1	95.1	127.9	110.8	89.5	94.0	91.7	86.3	70.3	78.6	96.6	93.5	95.1
04	MASONRY	97.3	98.5	98.0	100.8	89.9	94.2	103.7	130.8	120.1	108.1	95.8	100.7	86.2	91.5	89.4	95.1	90.2	92.1
05	METALS	96.8	117.9	104.1	97.9	116.4	104.3	100.4	118.6	106.7	98.1	121.6	106.2	98.0	101.9	99.3	99.7	118.6	106.2
06	WOOD, PLASTICS & COMPOSITES	88.6	91.2	90.1	89.6	85.0	86.9	94.5	133.1	116.6	93.7	85.2	88.8	91.1	55.3	70.6	90.5	85.0	87.3
07	THERMAL & MOISTURE PROTECTION	105.2	100.2	103.3	101.3	92.1	97.7	97.4	134.4	112.1	100.6	106.3	102.9	100.5	102.7	101.4	99.9	108.3	103.2
08	OPENINGS	93.8	101.4	95.8	93.1	92.8	93.0	89.7	129.4	99.7	89.7	97.5	91.7	89.7	57.9	81.6	89.5	91.6	90.1
0920	Plaster & Gypsum Board	98.5	90.7	93.0	98.7	84.3	88.5	91.9	133.9	121.5	100.1	84.5	89.1	99.7	53.8	67.3	101.5	84.3	89.3
0950, 0980	Ceilings & Acoustic Treatment	97.8	90.7	93.1	93.2	84.3	87.3	104.3	133.9	123.9	95.1	84.5	88.1	95.1	53.8	67.7	94.1	84.3	87.6
0960	Flooring	89.9	99.2	92.6	83.9	53.7	75.2	88.6	140.6	103.5	85.6	95.7	88.5	84.6	52.5	75.4	86.4	92.6	88.2
0970, 0990	Wall Finishes & Painting/Coating	95.0	105.7	101.5	92.5	98.4	96.1	96.2	145.4	126.0	92.5	98.4	96.1	92.5	98.4	96.1	92.5	87.8	89.7
09	FINISHES	96.4	94.8	95.5	94.4	79.7	86.3	97.1	133.7	117.3	94.5	89.3	91.6	94.0	63.9	77.4	93.9	87.0	90.1
COVERS	DIVS. 10 - 14, 25, 28, 41, 43, 44, 46	100.0	103.0	100.6	100.0	92.8	98.5	100.0	119.2	104.0	100.0	99.7	99.9	100.0	93.1	98.6	100.0	96.5	99.3
21, 22, 23	FIRE SUPPRESSION, PLUMBING & HVAC	94.7	95.5	95.1	94.9	89.0	92.5	94.6	127.2	108.1	94.9	98.2	96.3	94.9	89.7	92.8	100.3	90.1	96.0
26, 27, 3370	ELECTRICAL, COMMUNICATIONS & UTIL.	93.6	105.4	99.9	93.6	79.9	86.3	93.3	123.4	109.3	94.5	87.6	90.8	92.5	50.6	70.2	93.6	84.5	88.8
MF2010	WEIGHTED AVERAGE	96.2	100.8	98.3	96.7	91.0	94.2	96.1	124.7	108.9	95.2	97.7	96.3	93.3	79.7	87.2	96.5	94.0	95.4

	DIVISION	PUERTO RICO			RHODE ISLAND						SOUTH CAROLINA								
		SAN JUAN			NEWPORT			PROVIDENCE			AIKEN			BEAUFORT			CHARLESTON		
		009			028			029			298			299			294		
		MAT.	INST.	TOTAL	MAT.	INST.	TOTAL	MAT.	INST.	TOTAL	MAT.	INST.	TOTAL	MAT.	INST.	TOTAL	MAT.	INST.	TOTAL
015433	CONTRACTOR EQUIPMENT		89.6	89.6		101.9	101.9		101.9	101.9		100.0	100.0		100.0	100.0		100.0	100.0
0241, 31 - 34	SITE & INFRASTRUCTURE, DEMOLITION	125.4	90.5	101.3	82.5	103.5	97.0	83.8	103.5	97.4	127.9	86.0	99.0	122.3	84.2	96.0	103.5	83.8	89.9
0310	Concrete Forming & Accessories	92.1	17.6	27.4	99.7	115.6	113.5	98.4	115.6	113.3	96.0	71.0	74.3	95.0	40.1	47.4	94.1	41.9	48.8
0320	Concrete Reinforcing	194.4	12.4	100.7	108.1	114.4	111.3	108.1	114.4	111.3	91.7	68.8	79.9	90.8	27.7	58.3	90.6	59.8	74.7
0330	Cast-in-Place Concrete	101.6	30.6	74.4	78.6	120.3	94.5	96.5	120.3	105.6	89.3	72.9	83.0	89.3	49.5	74.1	104.8	50.2	83.9
03	CONCRETE	112.8	22.0	69.3	100.5	116.5	108.2	109.5	116.5	112.8	109.8	72.3	91.8	106.9	43.0	76.3	102.4	49.8	77.2
04	MASONRY	88.9	16.6	45.1	105.1	123.8	116.4	108.5	123.8	117.7	79.5	63.9	70.0	95.2	35.5	59.0	96.5	39.7	62.1
05	METALS	112.9	33.5	85.6	98.6	109.6	102.4	98.6	109.6	102.4	83.6	86.4	84.5	83.6	69.7	78.8	85.4	77.5	82.7
06	WOOD, PLASTICS & COMPOSITES	92.5	17.3	49.4	99.5	115.2	108.5	98.6	115.2	108.1	98.8	72.3	83.6	97.2	40.9	64.9	95.9	40.9	64.3
07	THERMAL & MOISTURE PROTECTION	125.6	21.1	84.0	96.6	115.4	104.1	94.3	115.4	102.7	99.5	69.8	87.7	99.2	38.5	75.1	98.3	43.1	76.4
08	OPENINGS	151.4	14.7	116.8	99.6	114.2	103.3	100.2	114.2	103.8	94.6	68.2	87.9	94.6	39.6	80.7	98.6	44.6	84.9
0920	Plaster & Gypsum Board	116.2	14.8	44.8	96.8	114.9	109.6	94.7	114.9	109.0	103.7	71.3	80.9	107.4	39.0	59.2	109.4	39.0	59.8
0950, 0980	Ceilings & Acoustic Treatment	234.7	14.8	88.9	96.5	114.9	108.7	95.8	114.9	108.5	92.0	71.3	78.3	95.7	39.0	58.1	95.7	39.0	58.1
0960	Flooring	200.5	13.9	146.9	96.6	131.6	106.7	96.3	131.6	106.4	108.7	69.9	97.5	110.4	54.1	94.2	109.9	61.5	96.0
0970, 0990	Wall Finishes & Painting/Coating	197.9	12.9	85.6	94.4	123.4	112.0	94.4	123.4	112.0	114.7	73.4	89.6	114.7	37.3	67.8	114.7	42.2	70.7
09	FINISHES	204.5	16.6	100.8	97.5	119.8	109.8	96.8	119.8	109.5	105.1	71.1	86.3	106.3	42.8	71.3	104.2	45.1	71.6
COVERS	DIVS. 10 - 14, 25, 28, 41, 43, 44, 46	100.0	17.1	82.8	100.0	103.2	100.7	100.0	103.2	100.7	100.0	79.4	95.7	100.0	48.2	89.2	100.0	72.5	94.3
21, 22, 23	FIRE SUPPRESSION, PLUMBING & HVAC	103.4	13.6	66.0	100.2	108.0	103.4	100.0	108.0	103.3	94.9	65.6	82.7	94.9	25.9	66.2	100.3	40.8	75.5
26, 27, 3370	ELECTRICAL, COMMUNICATIONS & UTIL.	128.5	12.9	67.0	99.9	101.2	100.6	98.3	101.2	99.8	94.4	69.4	81.1	98.2	32.9	63.5	96.6	96.4	96.5
MF2010	WEIGHTED AVERAGE	121.8	24.0	78.1	99.3	111.5	104.7	100.3	111.5	105.3	96.7	71.9	85.6	97.5	43.0	73.1	98.0	58.3	80.2

	DIVISION	SOUTH CAROLINA															SOUTH DAKOTA		
		COLUMBIA			FLORENCE			GREENVILLE			ROCK HILL			SPARTANBURG			ABERDEEN		
		290 - 292			295			296			297			293			574		
		MAT.	INST.	TOTAL	MAT.	INST.	TOTAL	MAT.	INST.	TOTAL	MAT.	INST.	TOTAL	MAT.	INST.	TOTAL	MAT.	INST.	TOTAL
015433	CONTRACTOR EQUIPMENT		100.0	100.0		100.0	100.0		100.0	100.0		100.0	100.0		100.0	100.0		98.6	98.6
0241, 31 - 34	SITE & INFRASTRUCTURE, DEMOLITION	101.3	83.7	89.2	115.0	83.8	93.4	109.3	83.4	91.4	107.1	83.3	90.6	109.1	83.4	91.4	96.1	94.6	95.0
0310	Concrete Forming & Accessories	95.8	41.0	48.2	82.3	41.1	46.6	93.5	41.0	47.9	92.1	38.6	45.7	96.6	41.0	48.3	96.3	39.7	47.2
0320	Concrete Reinforcing	90.6	59.6	74.6	90.3	59.7	74.5	90.2	47.6	68.3	91.0	47.4	68.6	90.2	47.6	68.3	103.4	41.0	71.3
0330	Cast-in-Place Concrete	107.6	50.8	85.8	89.3	50.0	74.2	89.3	49.9	74.2	89.3	44.8	72.2	89.3	49.9	74.2	103.8	46.2	81.8
03	CONCRETE	103.9	49.5	77.9	101.3	49.3	76.4	99.7	47.0	74.5	97.8	44.1	72.1	99.9	47.0	74.6	102.7	43.6	74.4
04	MASONRY	90.8	37.6	58.5	79.5	39.7	55.4	77.3	39.7	54.5	103.0	34.8	61.7	79.5	39.7	55.4	111.6	57.9	79.1
05	METALS	85.4	76.7	82.4	84.4	77.0	81.9	84.4	72.6	80.3	83.6	72.1	79.6	84.4	72.6	80.3	102.4	66.2	90.0
06	WOOD, PLASTICS & COMPOSITES	98.3	40.4	65.1	81.7	40.4	58.0	95.6	40.4	63.9	93.8	38.4	62.1	99.8	40.4	65.7	97.2	38.9	63.8
07	THERMAL & MOISTURE PROTECTION	97.2	41.8	75.2	98.6	43.1	76.6	98.5	43.1	76.5	98.4	40.6	75.5	98.6	43.1	76.5	106.1	51.0	84.2
08	OPENINGS	98.6	44.4	84.9	94.6	44.4	81.9	94.6	41.5	81.1	94.6	40.6	80.9	94.6	41.5	81.1	92.7	39.7	79.3
0920	Plaster & Gypsum Board	100.0	38.5	56.7	95.5	38.5	55.4	101.9	38.5	57.2	101.0	36.4	55.5	105.0	38.5	58.2	105.4	37.4	57.5
0950, 0980	Ceilings & Acoustic Treatment	94.8	38.5	57.5	92.9	38.5	56.8	92.0	38.5	56.5	92.0	36.4	55.2	92.0	38.5	56.5	108.1	37.4	61.2
0960	Flooring	106.5	46.1	89.1	100.8	46.1	85.1	107.4	60.4	93.9	106.4	47.0	89.3	108.9	60.4	95.0	118.6	53.2	99.8
0970, 0990	Wall Finishes & Painting/Coating	112.3	42.2	69.7	114.7	42.2	70.7	114.7	42.2	70.7	114.7	42.2	70.7	114.7	42.2	70.7	98.4	40.0	62.9
09	FINISHES	101.1	41.7	68.4	100.6	42.2	68.4	102.6	44.7	70.7	101.9	40.2	67.9	103.5	44.7	71.0	108.3	41.8	71.6
COVERS	DIVS. 10 - 14, 25, 28, 41, 43, 44, 46	100.0	72.2	94.2	100.0	72.2	94.2	100.0	72.2	94.2	100.0	71.4	94.0	100.0	72.2	94.2	100.0	45.2	88.6
21, 22, 23	FIRE SUPPRESSION, PLUMBING & HVAC	100.0	36.1	73.4	100.3	36.1	73.6	100.3	34.6	73.0	94.9	33.7	69.4	100.3	34.6	73.0	100.1	43.7	76.6
26, 27, 3370	ELECTRICAL, COMMUNICATIONS & UTIL.	98.2	49.3	72.2	94.4	46.9	69.1	96.6	50.7	72.2	96.6	50.7	72.2	96.6	50.7	72.2	98.4	55.5	75.6
MF2010	WEIGHTED AVERAGE	97.6	50.0	76.3	96.2	50.0	75.5	96.2	49.6	75.3	95.7	47.7	74.3	96.4	49.6	75.5	101.2	52.8	79.6

	DIVISION	SOUTH DAKOTA																	
		MITCHELL			MOBRIDGE			PIERRE			RAPID CITY			SIOUX FALLS			WATERTOWN		
		573			576			575			577			570 - 571			572		
		MAT.	INST.	TOTAL	MAT.	INST.	TOTAL	MAT.	INST.	TOTAL	MAT.	INST.	TOTAL	MAT.	INST.	TOTAL	MAT.	INST.	TOTAL
015433	CONTRACTOR EQUIPMENT		98.6	98.6		98.6	98.6		98.6	98.6		98.6	98.6		99.7	99.7		98.6	98.6
0241, 31 - 34	SITE & INFRASTRUCTURE, DEMOLITION	93.1	94.6	94.1	93.0	94.6	94.1	91.3	94.6	93.6	94.1	94.7	94.5	91.9	96.3	95.0	93.0	94.6	94.1
0310	Concrete Forming & Accessories	95.6	40.0	47.4	87.7	39.8	46.2	95.1	40.8	48.0	102.4	38.8	47.2	95.7	43.4	50.4	84.8	39.6	45.6
0320	Concrete Reinforcing	102.8	49.9	75.5	105.5	41.0	72.3	102.9	60.0	80.8	96.7	60.1	77.8	96.7	60.2	77.9	100.0	41.1	69.7
0330	Cast-in-Place Concrete	100.9	44.9	79.5	100.9	46.2	80.0	87.1	44.5	70.8	100.1	45.1	79.1	92.6	45.5	74.6	100.9	46.2	80.0
03	CONCRETE	100.4	44.8	73.8	100.3	43.6	73.2	93.5	47.1	71.3	99.6	46.5	74.2	94.2	48.7	72.4	99.4	43.6	72.7
04	MASONRY	99.8	55.7	73.1	109.0	57.9	78.1	107.1	56.0	76.1	109.0	57.1	77.5	104.6	57.8	76.2	136.1	60.2	90.1
05	METALS	101.4	66.3	89.4	101.5	66.1	89.3	102.4	75.0	93.0	104.3	75.5	94.4	104.5	75.8	94.6	101.4	66.7	89.5
06	WOOD, PLASTICS & COMPOSITES	96.2	39.5	63.7	86.9	39.2	59.6	95.1	39.8	63.4	100.3	36.0	63.4	95.9	42.5	65.3	84.0	38.9	58.1
07	THERMAL & MOISTURE PROTECTION	105.9	48.7	83.2	106.0	51.1	84.2	105.9	49.0	83.3	106.4	50.6	84.2	108.2	52.0	85.9	105.8	51.6	84.3
08	OPENINGS	91.2	40.3	78.3	94.1	39.3	80.2	95.8	45.8	83.2	96.8	43.8	83.4	97.1	47.3	84.5	91.2	39.4	78.1
0920	Plaster & Gypsum Board	103.8	38.0	57.5	97.9	37.7	55.5	95.9	38.3	55.3	104.6	34.4	55.2	102.5	41.1	59.3	96.2	37.4	54.8
0950, 0980	Ceilings & Acoustic Treatment	105.3	38.0	60.7	108.1	37.7	61.4	104.3	38.3	60.5	109.9	34.4	59.9	107.1	41.1	63.4	105.3	37.4	60.3
0960	Flooring	118.1	53.2	99.4	112.4	53.2	95.4	117.8	37.9	94.8	117.8	76.0	105.8	117.8	78.9	106.6	110.7	53.2	94.2
0970, 0990	Wall Finishes & Painting/Coating	98.4	43.7	65.2	98.4	44.8	65.9	98.4	48.0	67.8	98.4	48.0	67.8	98.4	48.0	67.8	98.4	40.0	62.9
09	FINISHES	107.2	42.5	71.5	105.4	42.5	70.7	105.8	40.2	69.6	108.3	45.8	73.9	107.3	50.0	75.7	104.1	41.8	69.7
COVERS	DIVS. 10 - 14, 25, 28, 41, 43, 44, 46	100.0	41.9	87.9	100.0	45.2	88.6	100.0	76.9	95.2	100.0	76.8	95.2	100.0	77.3	95.3	100.0	45.2	88.6
21, 22, 23	FIRE SUPPRESSION, PLUMBING & HVAC	94.7	43.3	73.3	94.7	43.8	73.5	100.0	43.4	76.4	100.1	43.9	76.7	100.1	42.2	76.0	94.7	43.8	73.5
26, 27, 3370	ELECTRICAL, COMMUNICATIONS & UTIL.	96.7	47.7	70.7	98.4	48.7	72.0	98.4	53.2	74.4	94.8	53.2	72.7	100.7	64.5	81.4	95.8	48.7	70.7
MF2010	WEIGHTED AVERAGE	98.3	51.6	77.4	99.1	52.0	78.0	99.7	54.5	79.5	101.0	55.3	80.6	100.6	57.9	81.5	99.7	52.2	78.4

	DIVISION	TENNESSEE																	
		CHATTANOOGA			COLUMBIA			COOKEVILLE			JACKSON			JOHNSON CITY			KNOXVILLE		
		373 - 374			384			385			383			376			377 - 379		
		MAT.	INST.	TOTAL	MAT.	INST.	TOTAL	MAT.	INST.	TOTAL	MAT.	INST.	TOTAL	MAT.	INST.	TOTAL	MAT.	INST.	TOTAL
015433	CONTRACTOR EQUIPMENT		103.8	103.8		97.8	97.8		97.8	97.8		104.0	104.0		97.7	97.7		97.7	97.7
0241, 31 - 34	SITE & INFRASTRUCTURE, DEMOLITION	101.5	98.4	99.3	89.5	86.9	87.7	95.5	86.3	89.2	98.0	96.9	97.2	110.0	87.1	94.2	87.7	87.6	87.6
0310	Concrete Forming & Accessories	97.2	58.2	63.4	83.5	62.3	65.1	83.7	38.0	44.0	90.2	48.3	53.8	84.5	41.5	47.1	96.0	44.6	51.4
0320	Concrete Reinforcing	83.5	63.1	73.0	86.6	63.1	74.5	86.6	63.0	74.5	86.6	59.9	72.8	84.1	61.9	72.7	83.5	62.0	72.4
0330	Cast-in-Place Concrete	97.6	63.3	84.5	92.3	54.0	77.7	104.6	45.6	82.1	102.2	52.2	83.0	78.5	64.0	72.9	91.7	50.8	76.0
03	CONCRETE	89.4	62.6	76.5	90.3	61.2	76.4	99.8	47.5	74.8	92.3	53.7	73.8	93.2	55.1	75.0	86.5	52.0	70.0
04	MASONRY	98.3	52.8	70.7	110.6	57.8	78.6	105.2	47.2	70.1	110.9	50.0	74.0	109.8	47.6	72.1	75.4	55.5	63.4
05	METALS	96.4	87.7	93.4	90.7	88.2	89.8	90.8	87.2	89.5	92.9	86.2	90.6	93.7	86.6	91.2	97.1	86.8	93.5
06	WOOD, PLASTICS & COMPOSITES	109.3	60.0	81.0	74.1	64.2	68.4	74.3	36.4	52.6	90.7	49.2	66.9	81.4	40.0	57.6	95.4	40.0	63.6
07	THERMAL & MOISTURE PROTECTION	93.7	60.5	80.5	89.2	60.0	77.6	89.6	49.0	73.5	91.4	52.0	75.7	89.3	56.4	76.2	87.5	60.0	76.6
08	OPENINGS	101.6	58.3	90.7	92.2	58.6	83.7	92.2	44.7	80.1	101.4	53.7	89.4	97.1	47.8	84.6	94.0	47.9	82.3
0920	Plaster & Gypsum Board	78.4	59.4	65.0	83.3	63.6	69.4	83.3	35.0	49.3	85.8	48.1	59.3	96.2	38.7	55.7	104.4	38.7	58.1
0950, 0980	Ceilings & Acoustic Treatment	97.4	59.4	72.2	84.1	63.6	70.5	84.1	35.0	51.5	93.3	48.1	63.4	93.3	38.7	57.1	94.2	38.7	57.4
0960	Flooring	100.0	59.8	88.5	94.0	21.7	73.2	94.1	61.9	84.9	93.0	43.1	78.7	96.2	42.9	80.9	101.2	57.7	88.7
0970, 0990	Wall Finishes & Painting/Coating	103.1	56.8	75.0	94.6	35.9	59.0	94.6	39.3	61.0	95.9	53.4	70.1	100.6	45.3	67.1	100.6	52.3	71.3
09	FINISHES	96.4	58.5	75.5	96.0	52.5	72.0	96.6	41.3	66.1	95.9	47.1	69.0	99.8	41.0	67.4	92.6	46.7	67.3
COVERS	DIVS. 10 - 14, 25, 28, 41, 43, 44, 46	100.0	45.8	88.7	100.0	52.4	90.1	100.0	45.6	88.7	100.0	47.9	89.2	100.0	48.1	89.2	100.0	50.9	89.8
21, 22, 23	FIRE SUPPRESSION, PLUMBING & HVAC	100.0	63.1	84.7	97.1	76.1	88.3	97.1	70.7	86.1	99.9	70.5	87.7	99.8	57.6	82.2	99.8	66.5	85.9
26, 27, 3370	ELECTRICAL, COMMUNICATIONS & UTIL.	103.3	70.8	86.0	94.8	62.3	77.5	96.6	62.3	78.4	101.7	63.5	81.4	93.0	50.3	70.3	99.3	58.7	77.7
MF2010	WEIGHTED AVERAGE	98.2	66.9	84.2	94.9	67.3	82.6	96.2	60.3	80.1	98.2	63.8	82.8	97.6	57.4	79.6	94.5	61.7	79.9

	DIVISION	TENNESSEE									TEXAS								
		MCKENZIE			MEMPHIS			NASHVILLE			ABILENE			AMARILLO			AUSTIN		
		382			375,380 - 381			370 - 372			795 - 796			790 - 791			786 - 787		
		MAT.	INST.	TOTAL	MAT.	INST.	TOTAL	MAT.	INST.	TOTAL	MAT.	INST.	TOTAL	MAT.	INST.	TOTAL	MAT.	INST.	TOTAL
015433	CONTRACTOR EQUIPMENT		97.8	97.8		101.7	101.7		104.8	104.8		88.5	88.5		88.5	88.5		87.9	87.9
0241, 31 - 34	SITE & INFRASTRUCTURE, DEMOLITION	95.1	86.6	89.3	89.6	92.6	91.7	97.2	99.9	99.0	99.0	85.8	89.9	99.1	86.9	90.7	95.2	86.0	88.8
0310	Concrete Forming & Accessories	91.3	41.0	47.6	96.2	61.5	66.1	97.4	65.2	69.5	95.6	40.9	48.2	96.2	55.2	60.7	89.7	54.4	59.1
0320	Concrete Reinforcing	86.7	63.8	74.9	89.7	64.5	76.7	89.7	63.5	76.2	97.5	52.6	74.3	97.5	50.8	73.4	97.7	48.6	72.4
0330	Cast-in-Place Concrete	102.4	61.0	86.6	88.3	63.7	78.9	89.2	67.9	81.1	97.5	49.1	79.0	96.6	56.1	81.1	84.0	49.7	70.8
03	CONCRETE	98.4	54.3	77.3	85.9	64.4	75.6	86.5	67.2	77.3	93.7	47.1	71.4	93.3	55.5	75.2	82.4	52.4	68.0
04	MASONRY	109.2	49.9	73.2	91.1	65.1	75.3	80.9	60.5	68.5	102.7	58.8	76.1	102.7	59.6	76.5	96.1	48.6	67.3
05	METALS	90.8	88.5	90.0	98.1	90.0	95.3	98.1	89.2	95.0	99.9	69.0	89.3	99.9	68.8	89.2	95.0	64.2	84.4
06	WOOD, PLASTICS & COMPOSITES	83.2	39.2	58.0	100.8	62.5	78.8	106.1	66.4	83.3	99.7	39.4	65.1	98.1	57.7	75.0	94.1	57.8	73.3
07	THERMAL & MOISTURE PROTECTION	89.6	50.2	74.0	88.4	65.6	79.3	92.2	63.3	80.7	96.9	51.6	78.9	97.0	54.8	80.2	87.2	51.6	73.0
08	OPENINGS	92.2	46.7	80.7	100.7	62.9	91.1	99.5	65.2	90.8	93.1	42.9	80.4	93.1	54.8	83.4	98.4	55.0	87.4
0920	Plaster & Gypsum Board	87.3	37.9	52.5	96.2	61.8	72.0	90.8	65.8	73.2	81.4	38.2	50.9	82.9	57.0	64.7	81.3	57.1	64.2
0950, 0980	Ceilings & Acoustic Treatment	84.1	37.9	53.5	95.5	61.8	73.1	92.3	65.8	74.7	90.3	38.2	55.7	90.3	57.0	68.2	84.3	57.1	66.2
0960	Flooring	97.3	41.6	81.3	96.5	38.1	79.7	102.3	67.9	92.4	109.0	77.9	100.1	108.8	77.9	99.9	90.8	41.8	76.7
0970, 0990	Wall Finishes & Painting/Coating	94.6	53.4	69.6	96.7	56.4	72.3	102.6	70.4	83.0	95.7	52.9	69.7	95.7	52.2	69.3	93.0	43.0	62.7
09	FINISHES	97.9	41.1	66.6	96.6	56.1	74.3	98.4	66.1	80.6	93.2	48.3	68.5	93.5	59.2	74.6	85.4	50.8	66.3
COVERS	DIVS. 10 - 14, 25, 28, 41, 43, 44, 46	100.0	30.3	85.5	100.0	80.8	96.0	100.0	81.6	96.2	100.0	75.7	94.9	100.0	73.9	94.6	100.0	77.6	95.3
21, 22, 23	FIRE SUPPRESSION, PLUMBING & HVAC	97.1	70.8	86.2	99.9	71.2	87.9	99.8	75.9	89.8	100.2	46.2	77.7	100.0	52.3	80.2	100.1	56.5	82.0
26, 27, 3370	ELECTRICAL, COMMUNICATIONS & UTIL.	96.4	68.3	81.4	100.3	66.4	82.3	100.9	63.3	80.9	98.8	51.6	73.7	101.5	68.5	84.0	100.7	65.9	82.2
MF2010	WEIGHTED AVERAGE	96.3	62.1	81.1	96.6	70.3	84.9	96.7	72.7	85.9	97.9	54.8	78.6	98.1	61.9	81.9	95.1	59.4	79.1

	DIVISION	TEXAS																	
		BEAUMONT			BROWNWOOD			BRYAN			CHILDRESS			CORPUS CHRISTI			DALLAS		
		776 - 777			768			778			792			783 - 784			752 - 753		
		MAT.	INST.	TOTAL	MAT.	INST.	TOTAL	MAT.	INST.	TOTAL	MAT.	INST.	TOTAL	MAT.	INST.	TOTAL	MAT.	INST.	TOTAL
015433	CONTRACTOR EQUIPMENT		89.5	89.5		88.5	88.5		89.5	89.5		88.5	88.5		95.4	95.4		98.1	98.1
0241, 31 - 34	SITE & INFRASTRUCTURE, DEMOLITION	93.2	86.0	88.2	105.6	85.2	91.5	84.2	87.6	86.6	110.7	85.7	93.4	131.5	81.7	97.1	116.2	87.3	96.2
0310	Concrete Forming & Accessories	102.8	51.1	57.9	94.1	34.7	42.6	81.7	42.3	47.5	94.1	62.8	66.9	96.2	38.7	46.3	94.8	65.1	69.1
0320	Concrete Reinforcing	97.4	43.3	69.5	97.8	49.8	73.1	99.9	48.7	73.5	97.6	50.0	73.1	97.0	48.0	71.8	99.1	52.2	75.0
0330	Cast-in-Place Concrete	92.9	53.8	78.0	105.1	44.1	81.8	74.4	67.2	71.7	100.0	50.7	81.1	93.1	46.4	75.3	101.5	57.0	84.4
03	CONCRETE	95.2	51.5	74.3	101.0	41.9	72.7	79.7	53.2	67.0	101.8	56.7	80.2	89.5	45.0	68.2	97.5	61.2	80.1
04	MASONRY	106.7	59.8	78.3	132.1	51.3	83.1	144.8	58.4	92.5	107.5	52.6	74.2	85.1	50.3	64.0	103.3	59.6	76.8
05	METALS	98.0	65.4	86.8	97.5	62.4	85.4	97.5	68.0	87.3	97.3	64.4	86.0	94.5	75.4	88.0	100.9	80.0	93.7
06	WOOD, PLASTICS & COMPOSITES	112.5	51.3	77.4	98.3	30.8	59.6	78.5	35.5	53.9	99.1	68.0	81.3	113.6	38.0	70.2	98.4	68.1	81.0
07	THERMAL & MOISTURE PROTECTION	103.9	57.6	85.5	97.1	46.1	76.8	96.4	57.6	81.0	97.5	50.8	79.0	96.2	46.4	76.4	90.2	64.4	80.0
08	OPENINGS	96.5	47.8	84.2	89.0	38.2	76.1	97.9	43.6	84.1	90.1	57.0	81.7	105.5	39.3	88.7	101.9	59.0	91.0
0920	Plaster & Gypsum Board	92.1	50.4	62.7	80.3	29.2	44.4	80.0	34.2	47.8	80.3	67.6	71.4	86.0	36.6	51.2	95.1	67.6	75.7
0950, 0980	Ceilings & Acoustic Treatment	102.1	50.4	67.8	86.6	29.2	48.6	95.2	34.2	54.8	86.6	67.6	74.0	84.3	36.6	52.7	93.1	67.6	76.2
0960	Flooring	107.2	75.3	98.0	107.9	47.2	90.5	80.0	55.7	73.0	106.9	43.2	88.6	103.3	66.6	92.8	99.8	50.4	85.6
0970, 0990	Wall Finishes & Painting/Coating	92.3	51.2	67.4	95.7	30.8	56.3	88.1	63.8	73.3	95.7	40.7	62.3	105.3	60.8	78.3	103.8	54.3	73.7
09	FINISHES	90.2	55.3	70.9	92.6	35.8	61.3	78.4	45.3	60.1	93.0	57.8	73.6	93.2	45.7	67.0	99.0	61.6	78.4
COVERS	DIVS. 10 - 14, 25, 28, 41, 43, 44, 46	100.0	79.2	95.7	100.0	40.4	87.6	100.0	78.4	95.5	100.0	76.1	95.0	100.0	76.5	95.1	100.0	81.8	96.2
21, 22, 23	FIRE SUPPRESSION, PLUMBING & HVAC	100.0	60.6	83.6	94.8	49.5	75.9	94.6	67.2	83.2	94.8	53.9	77.8	100.0	41.0	75.5	99.9	64.5	85.1
26, 27, 3370	ELECTRICAL, COMMUNICATIONS & UTIL.	97.7	74.2	85.2	94.2	46.4	68.8	95.2	67.0	80.2	98.7	46.0	70.7	95.0	55.9	74.2	97.4	69.1	82.3
MF2010	WEIGHTED AVERAGE	98.2	62.9	82.4	98.0	49.6	76.3	95.1	62.1	80.3	97.5	58.1	79.9	97.6	52.8	77.6	99.9	67.5	85.4

	DIVISION	TEXAS																	
		DEL RIO			DENTON			EASTLAND			EL PASO			FORT WORTH			GALVESTON		
		788			762			764			798 - 799,885			760 - 761			775		
		MAT.	INST.	TOTAL	MAT.	INST.	TOTAL	MAT.	INST.	TOTAL	MAT.	INST.	TOTAL	MAT.	INST.	TOTAL	MAT.	INST.	TOTAL
015433	CONTRACTOR EQUIPMENT		87.9	87.9		94.7	94.7		88.5	88.5		88.5	88.5		88.5	88.5		98.8	98.8
0241, 31 - 34	SITE & INFRASTRUCTURE, DEMOLITION	115.1	84.5	94.0	105.7	78.4	86.8	108.6	85.2	92.4	96.3	85.2	88.6	101.0	86.2	90.8	116.8	85.5	95.2
0310	Concrete Forming & Accessories	90.8	38.4	45.3	99.5	34.1	42.8	94.7	37.7	45.3	94.9	45.8	52.3	91.8	64.5	68.1	90.6	63.1	66.8
0320	Concrete Reinforcing	97.7	45.0	70.5	99.2	51.1	74.4	98.0	49.8	73.2	97.5	47.9	71.9	97.5	52.1	74.1	99.3	65.0	81.6
0330	Cast-in-Place Concrete	100.2	43.9	78.7	81.3	46.0	67.8	111.1	44.8	85.7	91.1	42.0	72.3	100.3	52.7	82.1	98.8	68.3	87.1
03	CONCRETE	105.3	42.5	75.3	79.8	43.4	62.4	105.6	43.5	75.9	90.4	45.8	69.1	94.9	58.7	77.6	96.8	66.8	82.4
04	MASONRY	100.0	53.0	71.5	140.7	49.7	85.5	101.2	51.1	70.9	100.4	48.9	69.2	96.8	59.6	74.2	101.4	58.5	75.4
05	METALS	93.4	59.3	81.7	97.1	76.0	89.9	97.3	62.8	85.4	99.7	63.3	87.2	96.9	68.4	87.1	99.0	88.0	95.2
06	WOOD, PLASTICS & COMPOSITES	93.5	36.5	60.8	108.1	33.7	65.4	105.2	34.6	64.7	99.2	48.5	70.1	100.7	68.0	81.9	95.8	63.1	77.1
07	THERMAL & MOISTURE PROTECTION	93.0	45.6	74.1	95.5	45.9	75.8	97.4	45.3	76.7	93.1	54.4	77.7	97.8	54.8	80.7	95.7	66.2	83.9
08	OPENINGS	99.2	34.8	82.9	103.5	39.0	87.2	78.0	40.3	68.5	93.1	44.5	80.8	87.9	58.9	80.6	102.8	62.9	92.7
0920	Plaster & Gypsum Board	80.9	35.2	48.7	83.8	32.1	47.4	80.3	33.2	47.1	85.9	47.4	58.8	83.9	67.6	72.4	87.0	62.5	69.8
0950, 0980	Ceilings & Acoustic Treatment	80.5	35.2	50.5	91.3	32.1	52.1	86.6	33.2	51.2	90.3	47.4	61.9	92.2	67.6	75.9	98.0	62.5	74.5
0960	Flooring	87.9	36.0	73.0	100.5	45.2	84.6	140.7	47.2	113.9	109.0	65.9	96.6	142.0	43.0	113.6	93.8	55.7	82.8
0970, 0990	Wall Finishes & Painting/Coating	93.0	27.1	53.0	105.5	31.2	60.4	97.1	32.2	57.7	95.7	36.8	60.0	97.1	53.8	70.9	99.5	49.5	69.2
09	FINISHES	85.6	36.4	58.5	89.1	35.2	59.4	102.9	38.2	67.2	93.4	48.8	68.8	104.2	60.0	79.8	88.3	59.9	72.7
COVERS	DIVS. 10 - 14, 25, 28, 41, 43, 44, 46	100.0	40.1	87.6	100.0	38.9	87.3	100.0	41.6	87.9	100.0	77.4	95.3	100.0	81.4	96.1	100.0	83.4	96.6
21, 22, 23	FIRE SUPPRESSION, PLUMBING & HVAC	94.6	47.3	74.9	94.8	40.7	72.3	94.8	49.8	76.1	100.0	36.2	73.5	100.0	58.9	82.9	94.6	67.2	83.2
26, 27, 3370	ELECTRICAL, COMMUNICATIONS & UTIL.	96.8	31.9	62.3	96.6	46.5	70.0	94.1	46.4	68.7	97.7	54.4	74.7	97.7	63.8	79.7	97.1	68.7	82.0
MF2010	WEIGHTED AVERAGE	97.0	47.0	74.6	97.3	48.4	75.4	96.8	50.3	76.0	97.0	51.7	76.8	97.7	63.6	82.4	97.6	69.2	84.9

	DIVISION	TEXAS																	
		GIDDINGS			GREENVILLE			HOUSTON			HUNTSVILLE			LAREDO			LONGVIEW		
		789			754			770 - 772			773			780			756		
		MAT.	INST.	TOTAL	MAT.	INST.	TOTAL	MAT.	INST.	TOTAL	MAT.	INST.	TOTAL	MAT.	INST.	TOTAL	MAT.	INST.	TOTAL
015433	CONTRACTOR EQUIPMENT		87.9	87.9		95.7	95.7		98.6	98.6		89.5	89.5		87.9	87.9		90.3	90.3
0241, 31 - 34	SITE & INFRASTRUCTURE, DEMOLITION	101.3	84.8	89.9	108.6	82.1	90.3	114.5	85.2	94.3	101.9	85.6	90.7	95.1	85.7	88.6	103.6	89.4	93.8
0310	Concrete Forming & Accessories	89.0	58.8	62.8	87.2	28.6	36.4	92.5	63.3	67.1	88.3	58.3	62.3	90.8	38.9	45.8	84.3	29.5	36.8
0320	Concrete Reinforcing	98.3	46.9	71.8	99.6	26.0	61.7	99.1	65.5	81.8	100.2	49.0	73.8	97.7	47.9	72.0	98.6	23.7	60.0
0330	Cast-in-Place Concrete	85.0	44.4	69.4	92.4	38.2	71.7	95.8	68.3	85.3	102.4	49.1	82.0	71.8	60.0	67.2	107.4	35.8	80.0
03	CONCRETE	85.8	52.2	69.7	90.1	33.5	63.0	94.5	66.9	81.3	102.8	54.1	79.5	80.9	48.9	65.6	105.6	31.9	70.3
04	MASONRY	108.1	53.1	74.7	164.1	48.4	94.0	101.2	68.7	81.5	144.2	44.8	83.9	93.8	48.5	66.3	159.5	44.6	89.9
05	METALS	92.9	61.7	82.1	98.1	63.5	86.2	102.0	88.9	97.5	97.4	63.9	85.9	96.0	63.5	84.8	91.8	50.6	77.6
06	WOOD, PLASTICS & COMPOSITES	92.5	63.1	75.6	89.7	27.7	54.2	98.1	63.1	78.0	87.0	61.7	72.5	93.5	37.9	61.6	83.9	30.0	53.0
07	THERMAL & MOISTURE PROTECTION	93.6	49.9	76.2	90.3	41.1	70.7	95.3	68.5	84.6	97.4	47.9	77.7	92.0	49.4	75.1	91.5	36.4	69.6
08	OPENINGS	98.4	51.2	86.5	100.7	27.3	82.1	105.0	62.9	94.3	97.8	50.3	85.8	99.0	40.0	84.1	91.6	27.3	75.4
0920	Plaster & Gypsum Board	79.6	62.5	67.6	89.6	26.1	44.9	89.6	62.5	70.5	84.4	61.1	68.0	82.0	36.6	50.0	87.3	28.5	45.9
0950, 0980	Ceilings & Acoustic Treatment	80.5	62.5	68.6	89.4	26.1	47.4	102.7	62.5	76.1	95.2	61.1	72.6	84.3	36.6	52.7	85.6	28.5	47.8
0960	Flooring	88.0	36.0	73.0	95.7	49.6	82.5	95.0	55.7	83.7	83.5	36.0	69.9	87.8	66.6	81.7	101.1	36.4	82.5
0970, 0990	Wall Finishes & Painting/Coating	93.0	32.2	56.1	103.8	26.5	56.9	99.5	63.8	77.8	88.1	33.0	54.7	93.0	57.3	71.3	95.6	30.9	56.3
09	FINISHES	84.4	52.9	67.0	95.7	31.9	60.5	95.5	61.5	76.8	81.1	52.6	65.4	84.8	45.2	63.0	101.2	30.9	62.4
COVERS	DIVS. 10 - 14, 25, 28, 41, 43, 44, 46	100.0	46.6	88.9	100.0	37.1	86.9	100.0	83.4	96.6	100.0	47.0	89.0	100.0	74.3	94.7	100.0	23.4	84.1
21, 22, 23	FIRE SUPPRESSION, PLUMBING & HVAC	94.6	61.5	80.9	94.6	46.7	74.6	99.9	67.4	86.4	94.6	62.4	81.2	100.0	39.7	74.9	94.5	29.5	67.4
26, 27, 3370	ELECTRICAL, COMMUNICATIONS & UTIL.	93.3	46.0	68.2	93.7	28.7	59.1	99.4	67.0	82.2	95.2	46.0	69.0	96.9	63.6	79.2	94.5	46.6	69.0
MF2010	WEIGHTED AVERAGE	94.1	57.0	77.5	99.6	43.7	74.6	100.0	70.4	86.8	98.6	56.7	79.9	94.7	53.1	76.1	99.7	40.7	73.4

DIVISION		TEXAS																	
		LUBBOCK			LUFKIN			MCALLEN			MCKINNEY			MIDLAND			ODESSA		
		793 - 794			759			785			750			797			797		
		MAT.	INST.	TOTAL	MAT.	INST.	TOTAL	MAT.	INST.	TOTAL	MAT.	INST.	TOTAL	MAT.	INST.	TOTAL	MAT.	INST.	TOTAL
015433	**CONTRACTOR EQUIPMENT**		97.2	97.2		90.3	90.3		95.4	95.4		95.7	95.7		97.2	97.2		88.5	88.5
0241, 31 - 34	**SITE & INFRASTRUCTURE, DEMOLITION**	125.6	84.1	97.0	98.0	90.5	92.8	136.2	81.6	98.5	104.2	82.7	89.4	129.4	84.1	98.1	99.2	86.1	90.2
0310	Concrete Forming & Accessories	93.6	53.6	58.9	87.3	59.9	63.5	95.6	36.3	44.1	86.4	38.3	44.7	97.0	40.3	47.8	95.6	40.2	47.5
0320	Concrete Reinforcing	98.7	52.6	74.9	100.2	61.7	80.4	97.2	47.9	71.8	99.6	51.1	74.6	99.7	51.9	75.1	97.5	51.8	74.0
0330	Cast-in-Place Concrete	97.6	49.4	79.2	96.1	45.6	76.8	101.2	44.7	79.6	86.6	46.8	71.4	103.8	46.6	81.9	97.5	45.6	77.6
03	**CONCRETE**	92.5	53.6	73.9	97.7	55.9	77.7	94.9	43.3	70.2	85.5	45.4	66.3	97.0	46.7	72.9	93.7	45.5	70.6
04	**MASONRY**	101.9	57.1	74.7	121.2	44.8	74.9	100.1	50.3	69.9	176.8	51.0	100.6	120.1	44.6	74.3	102.7	44.5	67.4
05	**METALS**	103.2	80.9	95.5	98.2	67.5	87.7	93.3	74.9	87.0	98.0	75.0	90.1	101.3	79.7	93.9	99.2	67.5	88.3
06	**WOOD, PLASTICS & COMPOSITES**	98.9	57.9	75.3	92.1	63.0	75.4	112.2	35.5	68.2	88.4	38.5	59.7	103.3	40.0	67.0	99.7	39.9	65.4
07	**THERMAL & MOISTURE PROTECTION**	87.7	53.8	74.2	91.3	47.6	73.9	96.2	44.7	75.7	90.2	47.9	73.4	87.9	45.2	70.9	96.9	44.4	76.1
08	**OPENINGS**	103.1	51.7	90.1	74.8	55.6	69.9	103.8	38.4	87.3	100.7	41.4	85.7	102.0	41.6	86.7	93.1	41.6	80.0
0920	Plaster & Gypsum Board	81.9	57.0	64.4	87.0	62.5	69.8	86.0	33.9	49.3	88.3	37.1	52.3	83.3	38.6	51.8	81.4	38.6	51.3
0950, 0980	Ceilings & Acoustic Treatment	92.2	57.0	68.9	80.0	62.5	68.4	84.3	33.9	50.9	89.4	37.1	54.7	89.4	38.6	55.7	90.3	38.6	56.1
0960	Flooring	101.5	40.8	84.1	135.2	39.6	107.7	102.8	66.6	92.4	95.4	36.9	78.6	103.0	50.4	87.9	109.0	50.4	92.2
0970, 0990	Wall Finishes & Painting/Coating	106.0	52.2	73.3	95.6	39.8	61.7	105.3	30.4	59.9	103.8	43.4	67.2	106.0	52.2	73.3	95.7	52.2	69.3
09	**FINISHES**	95.8	51.2	71.2	109.8	54.8	79.5	93.5	40.8	64.4	95.1	38.6	63.9	96.3	42.7	66.7	93.2	42.6	65.3
COVERS	**DIVS. 10 - 14, 25, 28, 41, 43, 44, 46**	100.0	77.5	95.3	100.0	77.4	95.3	100.0	74.2	94.6	100.0	43.0	88.2	100.0	77.1	95.2	100.0	76.8	95.2
21, 22, 23	**FIRE SUPPRESSION, PLUMBING & HVAC**	99.8	45.3	77.1	94.5	62.5	81.2	94.6	31.1	68.2	94.6	44.4	73.7	94.4	45.5	74.0	100.2	45.4	77.4
26, 27, 3370	**ELECTRICAL, COMMUNICATIONS & UTIL.**	97.4	52.6	73.5	96.0	46.0	69.4	94.1	36.3	63.3	93.8	48.0	69.4	97.4	42.8	68.3	98.9	42.7	69.0
MF2010	**WEIGHTED AVERAGE**	99.6	57.5	80.8	96.7	59.1	79.9	97.5	46.9	74.9	99.5	50.8	77.7	99.7	51.9	78.3	97.8	50.7	76.8

DIVISION		TEXAS																	
		PALESTINE			SAN ANGELO			SAN ANTONIO			TEMPLE			TEXARKANA			TYLER		
		758			769			781 - 782			765			755			757		
		MAT.	INST.	TOTAL	MAT.	INST.	TOTAL	MAT.	INST.	TOTAL	MAT.	INST.	TOTAL	MAT.	INST.	TOTAL	MAT.	INST.	TOTAL
015433	**CONTRACTOR EQUIPMENT**		90.3	90.3		88.5	88.5		90.4	90.4		88.5	88.5		90.3	90.3		90.3	90.3
0241, 31 - 34	**SITE & INFRASTRUCTURE, DEMOLITION**	105.7	90.3	95.1	101.6	86.2	91.0	94.8	90.3	91.7	89.3	84.9	86.3	91.5	90.2	90.6	103.2	90.3	94.3
0310	Concrete Forming & Accessories	79.5	59.0	61.8	94.4	38.0	45.5	90.9	57.0	61.5	97.3	35.8	43.9	93.5	35.8	43.4	88.8	34.6	41.8
0320	Concrete Reinforcing	97.8	49.8	73.1	97.7	51.8	74.0	104.6	47.4	75.2	97.8	49.5	72.9	97.7	51.8	74.1	98.6	52.0	74.6
0330	Cast-in-Place Concrete	87.8	44.1	71.1	99.2	43.7	77.9	70.4	68.1	69.5	81.3	49.3	69.1	88.5	44.9	71.8	105.5	43.8	81.9
03	**CONCRETE**	98.5	52.7	76.6	96.3	43.7	71.1	81.3	59.7	71.0	82.4	44.2	64.1	90.8	43.2	68.0	105.0	42.4	75.0
04	**MASONRY**	116.5	51.5	77.1	128.6	43.9	77.2	93.7	60.5	73.6	140.1	49.8	85.4	181.5	47.4	100.2	170.5	50.3	97.7
05	**METALS**	98.0	62.3	85.7	97.7	64.5	86.3	96.7	66.2	86.2	97.4	63.4	85.7	91.7	66.8	83.1	97.8	67.3	87.3
06	**WOOD, PLASTICS & COMPOSITES**	82.9	63.0	71.5	98.6	38.1	63.9	93.5	55.2	71.6	107.6	34.6	65.7	94.4	34.1	59.8	93.5	32.2	58.3
07	**THERMAL & MOISTURE PROTECTION**	91.7	48.7	74.6	96.9	44.0	75.9	91.7	64.1	80.7	96.5	45.7	76.3	91.1	49.4	74.5	91.6	53.9	76.6
08	**OPENINGS**	74.7	52.6	69.1	89.0	40.6	76.8	100.9	52.9	88.8	75.9	36.9	66.0	91.6	39.2	78.3	74.7	38.2	65.4
0920	Plaster & Gypsum Board	82.2	62.5	68.3	80.3	36.8	49.7	82.0	54.4	62.5	80.3	33.2	47.1	91.3	32.7	50.1	87.0	30.7	47.4
0950, 0980	Ceilings & Acoustic Treatment	80.0	62.5	68.4	86.6	36.8	53.6	84.3	54.4	64.5	86.6	33.2	51.2	85.6	32.7	50.6	80.0	30.7	47.3
0960	Flooring	126.9	61.2	108.0	108.0	36.0	87.3	87.8	66.8	81.7	143.3	43.3	114.6	108.7	61.6	95.2	136.9	38.4	108.6
0970, 0990	Wall Finishes & Painting/Coating	95.6	32.2	57.1	95.7	52.2	69.3	93.0	57.3	71.3	97.1	35.4	59.7	95.6	41.5	62.8	95.6	41.5	62.8
09	**FINISHES**	107.5	57.4	79.9	92.4	38.3	62.6	84.9	58.0	70.0	102.2	36.6	66.1	103.2	40.1	68.4	110.7	34.3	68.6
COVERS	**DIVS. 10 - 14, 25, 28, 41, 43, 44, 46**	100.0	44.2	88.4	100.0	74.2	94.6	100.0	80.6	96.0	100.0	38.4	87.2	100.0	74.3	94.7	100.0	74.0	94.6
21, 22, 23	**FIRE SUPPRESSION, PLUMBING & HVAC**	94.5	59.0	79.7	94.8	44.6	73.9	99.9	66.4	86.0	94.8	35.7	70.2	94.5	35.2	69.8	94.5	59.8	80.1
26, 27, 3370	**ELECTRICAL, COMMUNICATIONS & UTIL.**	91.6	46.0	67.4	98.7	48.6	72.1	97.3	63.6	79.4	95.5	68.8	81.3	95.8	64.8	79.3	94.5	55.1	73.5
MF2010	**WEIGHTED AVERAGE**	96.1	57.4	78.8	97.6	50.1	76.3	95.0	65.1	81.6	95.3	50.0	75.1	99.1	51.5	77.8	100.2	54.9	80.0

DIVISION		TEXAS															UTAH		
		VICTORIA			WACO			WAXAHACKIE			WHARTON			WICHITA FALLS			LOGAN		
		779			766 - 767			751			774			763			843		
		MAT.	INST.	TOTAL	MAT.	INST.	TOTAL	MAT.	INST.	TOTAL	MAT.	INST.	TOTAL	MAT.	INST.	TOTAL	MAT.	INST.	TOTAL
015433	**CONTRACTOR EQUIPMENT**		97.6	97.6		88.5	88.5		95.7	95.7		98.8	98.8		88.5	88.5		99.8	99.8
0241, 31 - 34	**SITE & INFRASTRUCTURE, DEMOLITION**	122.3	80.7	93.6	97.8	86.2	89.8	106.6	83.2	90.5	129.0	83.6	97.7	98.5	85.8	89.7	85.2	99.5	95.1
0310	Concrete Forming & Accessories	90.9	37.1	44.2	96.0	41.3	48.5	86.4	60.4	63.8	85.6	38.6	44.8	96.0	39.2	46.7	105.4	61.4	67.2
0320	Concrete Reinforcing	95.4	31.6	62.6	97.5	48.6	72.3	99.6	51.1	74.7	99.3	53.9	75.9	97.5	51.8	73.9	98.9	77.0	87.6
0330	Cast-in-Place Concrete	111.0	42.2	84.7	88.0	53.9	74.9	91.4	42.6	72.7	114.1	52.7	90.6	94.0	46.2	75.7	89.2	73.9	83.4
03	**CONCRETE**	104.5	40.0	73.6	88.9	48.1	69.4	89.3	53.9	72.4	108.6	48.3	79.7	91.9	45.2	69.6	108.0	69.1	89.4
04	**MASONRY**	118.0	36.2	68.4	99.4	57.0	73.7	164.7	51.0	95.8	102.6	47.2	69.0	99.9	58.8	75.0	107.3	63.4	80.7
05	**METALS**	97.6	71.3	88.5	99.8	66.4	88.3	98.1	77.0	90.9	99.0	78.6	92.0	99.7	69.0	89.2	100.4	77.6	92.6
06	**WOOD, PLASTICS & COMPOSITES**	99.7	37.9	64.2	106.0	37.1	66.5	88.4	68.1	76.7	89.7	34.4	58.0	106.0	37.0	66.4	82.2	59.7	69.3
07	**THERMAL & MOISTURE PROTECTION**	99.2	41.1	76.1	97.4	49.2	78.3	90.3	48.0	73.4	96.0	51.2	78.2	97.4	52.6	79.6	99.8	65.1	86.0
08	**OPENINGS**	102.6	37.1	86.0	87.9	37.9	75.3	100.7	57.4	89.8	102.8	39.0	86.7	87.9	42.1	76.3	90.0	58.7	82.1
0920	Plaster & Gypsum Board	85.0	36.5	50.9	81.4	35.8	49.2	88.6	67.6	73.8	82.6	32.9	47.6	81.4	35.6	49.2	85.3	58.3	66.3
0950, 0980	Ceilings & Acoustic Treatment	99.0	36.5	57.6	90.3	35.8	54.2	90.3	67.6	75.3	98.0	32.9	54.9	90.3	35.6	54.1	119.1	58.3	78.8
0960	Flooring	92.9	39.6	77.6	142.0	46.8	114.7	95.4	39.6	79.3	91.1	47.2	78.5	142.8	77.9	124.2	101.3	50.4	86.7
0970, 0990	Wall Finishes & Painting/Coating	99.6	39.8	63.3	97.1	35.4	59.7	103.8	39.8	65.0	99.5	33.0	59.2	99.6	52.2	70.8	98.5	63.3	77.1
09	**FINISHES**	86.4	37.9	59.7	103.3	40.5	68.7	95.6	55.7	73.6	87.9	38.6	60.7	103.7	46.8	72.3	102.8	58.9	78.6
COVERS	**DIVS. 10 - 14, 25, 28, 41, 43, 44, 46**	100.0	39.6	87.4	100.0	77.9	95.4	100.0	43.6	88.3	100.0	41.9	87.9	100.0	71.2	94.0	100.0	59.1	91.5
21, 22, 23	**FIRE SUPPRESSION, PLUMBING & HVAC**	94.6	33.5	69.2	100.2	52.4	80.3	94.6	43.4	73.3	94.6	62.5	81.2	100.2	47.7	78.3	99.9	70.2	87.6
26, 27, 3370	**ELECTRICAL, COMMUNICATIONS & UTIL.**	102.6	42.1	70.5	98.9	69.1	83.0	93.8	69.1	80.6	101.5	46.5	72.2	100.7	59.1	78.6	94.4	75.3	84.2
MF2010	**WEIGHTED AVERAGE**	99.9	44.4	75.1	97.5	57.1	79.4	99.4	58.2	81.0	99.8	54.8	79.7	98.1	55.5	79.1	99.5	70.7	86.6

DIVISION		UTAH												VERMONT					
		OGDEN			PRICE			PROVO			SALT LAKE CITY			BELLOWS FALLS			BENNINGTON		
		842,844			845			846 - 847			840 - 841			051			052		
		MAT.	INST.	TOTAL	MAT.	INST.	TOTAL	MAT.	INST.	TOTAL	MAT.	INST.	TOTAL	MAT.	INST.	TOTAL	MAT.	INST.	TOTAL
015433	CONTRACTOR EQUIPMENT		99.8	99.8		98.7	98.7		98.7	98.7		99.7	99.7		100.3	100.3		100.3	100.3
0241, 31 - 34	SITE & INFRASTRUCTURE, DEMOLITION	74.6	99.5	91.8	83.1	97.8	93.2	81.9	97.8	92.8	74.4	99.4	91.7	78.4	98.2	92.1	77.7	98.0	91.7
0310	Concrete Forming & Accessories	105.5	61.4	67.3	107.5	52.1	59.4	106.8	61.3	67.3	108.0	61.3	67.5	96.5	49.7	55.9	94.4	54.1	59.4
0320	Concrete Reinforcing	98.5	77.0	87.5	106.0	76.9	91.0	106.9	77.0	91.5	100.8	77.0	88.6	86.5	64.6	75.2	86.5	64.6	75.2
0330	Cast-in-Place Concrete	90.6	73.9	84.2	89.3	63.1	79.3	89.3	73.8	83.4	99.0	73.8	89.4	91.2	72.4	84.0	91.2	69.0	82.7
03	CONCRETE	97.9	69.1	84.2	109.2	61.3	86.3	107.7	69.1	89.2	117.2	69.1	94.2	99.9	60.9	81.2	99.7	61.7	81.5
04	MASONRY	100.9	63.4	78.1	112.7	63.4	82.8	112.8	63.4	82.8	114.3	63.4	83.4	97.5	61.9	75.9	106.6	61.8	79.4
05	METALS	100.9	77.6	92.9	97.7	76.9	90.5	98.7	77.4	91.4	105.6	77.4	95.9	93.0	69.4	84.8	92.9	68.4	84.5
06	WOOD, PLASTICS & COMPOSITES	82.2	59.7	69.3	84.9	48.1	63.8	83.5	59.7	69.8	83.8	59.7	70.0	99.0	39.1	64.7	96.2	45.7	67.2
07	THERMAL & MOISTURE PROTECTION	98.6	65.1	85.3	101.6	60.7	85.4	101.6	65.1	87.1	105.4	65.1	89.4	95.9	60.3	81.8	95.8	55.3	79.7
08	OPENINGS	90.0	58.7	82.1	94.6	51.3	83.6	94.6	58.7	85.5	91.9	58.7	83.5	102.2	41.9	86.9	102.2	44.2	87.5
0920	Plaster & Gypsum Board	85.3	58.3	66.3	87.5	46.4	58.5	85.7	58.3	66.4	88.2	58.3	67.2	98.1	36.6	54.8	95.9	43.5	59.0
0950, 0980	Ceilings & Acoustic Treatment	119.1	58.3	78.8	119.1	46.4	70.9	119.1	58.3	78.8	112.9	58.3	76.7	89.8	36.6	54.6	89.8	43.5	59.1
0960	Flooring	99.1	50.4	85.1	102.5	37.5	83.8	102.1	50.4	87.3	102.5	50.4	87.5	95.8	32.2	77.6	94.7	32.2	76.7
0970, 0990	Wall Finishes & Painting/Coating	98.5	63.3	77.1	98.5	40.4	63.2	98.5	51.7	70.1	101.4	51.7	71.2	92.3	28.7	53.7	92.3	28.7	53.7
09	FINISHES	100.9	58.9	77.7	103.9	46.9	72.5	103.5	57.6	78.2	101.7	57.6	77.4	94.8	42.8	66.1	94.2	46.7	68.0
COVERS	DIVS. 10 - 14, 25, 28, 41, 43, 44, 46	100.0	59.1	91.5	100.0	51.3	89.9	100.0	59.1	91.5	100.0	59.1	91.5	100.0	47.3	89.1	100.0	47.9	89.2
21, 22, 23	FIRE SUPPRESSION, PLUMBING & HVAC	99.9	70.2	87.6	97.2	70.4	86.1	99.9	70.2	87.5	100.1	70.2	87.6	94.8	96.9	95.7	94.8	91.5	93.4
26, 27, 3370	ELECTRICAL, COMMUNICATIONS & UTIL.	94.7	75.3	84.4	100.3	75.3	87.0	95.1	64.9	79.1	97.6	64.9	80.2	100.6	96.9	98.7	100.6	66.3	82.4
MF2010	WEIGHTED AVERAGE	97.5	70.7	85.5	100.0	67.1	85.3	100.0	68.9	86.1	102.0	69.1	87.3	96.6	73.7	86.4	97.0	68.9	84.4

DIVISION		VERMONT																	
		BRATTLEBORO			BURLINGTON			GUILDHALL			MONTPELIER			RUTLAND			ST. JOHNSBURY		
		053			054			059			056			057			058		
		MAT.	INST.	TOTAL	MAT.	INST.	TOTAL	MAT.	INST.	TOTAL	MAT.	INST.	TOTAL	MAT.	INST.	TOTAL	MAT.	INST.	TOTAL
015433	CONTRACTOR EQUIPMENT		100.3	100.3		100.3	100.3		100.3	100.3		100.3	100.3		100.3	100.3		100.3	100.3
0241, 31 - 34	SITE & INFRASTRUCTURE, DEMOLITION	79.1	98.2	92.3	80.2	96.9	91.7	77.5	96.2	90.4	81.2	96.9	92.0	80.9	96.9	91.9	77.6	96.4	90.6
0310	Concrete Forming & Accessories	96.7	49.5	55.8	93.9	52.3	57.8	95.0	44.6	51.3	99.1	52.3	58.5	97.0	52.3	58.2	94.0	44.7	51.2
0320	Concrete Reinforcing	85.6	64.6	74.8	108.1	81.2	94.3	87.3	38.2	62.0	86.1	81.2	83.6	108.1	81.2	94.3	85.6	43.6	64.0
0330	Cast-in-Place Concrete	94.1	69.3	84.6	99.5	101.2	100.1	88.4	55.5	75.8	105.3	101.2	103.7	89.4	101.2	93.9	88.4	55.8	75.9
03	CONCRETE	102.3	59.8	81.9	108.1	75.0	92.2	97.3	47.9	73.7	108.1	75.0	92.3	103.2	75.0	89.7	97.0	49.0	74.0
04	MASONRY	105.7	61.9	79.1	109.1	50.1	73.3	106.0	39.9	65.9	92.6	50.1	66.9	87.9	50.1	65.0	131.4	39.9	76.0
05	METALS	92.9	69.2	84.8	100.4	77.9	92.7	93.0	57.9	80.9	92.9	77.9	87.8	98.6	77.9	91.5	93.0	58.3	81.0
06	WOOD, PLASTICS & COMPOSITES	99.3	39.1	64.8	94.2	51.7	69.8	95.7	45.7	67.0	98.4	51.7	71.7	99.4	51.7	72.1	92.0	45.7	65.5
07	THERMAL & MOISTURE PROTECTION	95.9	54.2	79.3	95.8	60.1	81.6	95.7	41.7	74.3	95.3	60.6	81.5	96.0	60.6	81.9	95.7	41.7	74.2
08	OPENINGS	102.2	41.9	86.9	105.6	52.8	92.3	102.2	38.9	86.2	102.2	52.8	89.7	105.6	52.8	92.3	102.2	38.9	86.2
0920	Plaster & Gypsum Board	98.1	36.6	54.8	99.7	49.6	64.5	103.9	43.5	61.3	107.4	49.6	66.7	98.9	49.6	64.2	107.4	43.5	62.4
0950, 0980	Ceilings & Acoustic Treatment	89.8	36.6	54.6	91.9	49.6	63.9	89.8	43.5	59.1	88.9	49.6	62.9	94.7	49.6	64.8	89.8	43.5	59.1
0960	Flooring	96.0	32.2	77.6	96.0	58.3	85.2	99.0	32.2	79.8	102.5	58.3	89.8	95.8	58.3	85.0	102.5	32.2	82.3
0970, 0990	Wall Finishes & Painting/Coating	92.3	28.7	53.7	92.3	33.4	56.6	92.3	22.3	49.8	92.3	33.4	56.6	92.3	33.4	56.6	92.3	23.4	50.5
09	FINISHES	94.9	42.8	66.2	95.6	50.2	70.6	96.3	40.1	65.3	97.8	50.2	71.5	96.0	50.2	70.7	97.8	40.2	66.0
COVERS	DIVS. 10 - 14, 25, 28, 41, 43, 44, 46	100.0	47.3	89.1	100.0	73.1	94.4	100.0	39.7	87.5	100.0	73.1	94.4	100.0	73.1	94.4	100.0	39.7	87.5
21, 22, 23	FIRE SUPPRESSION, PLUMBING & HVAC	94.8	94.7	94.8	100.0	63.5	84.8	94.8	61.7	81.0	94.6	63.5	81.6	100.2	63.5	84.9	94.8	61.7	81.0
26, 27, 3370	ELECTRICAL, COMMUNICATIONS & UTIL.	100.6	96.9	98.7	99.8	66.3	82.0	100.6	66.3	82.4	101.2	66.3	82.6	100.7	66.3	82.4	100.6	66.3	82.4
MF2010	WEIGHTED AVERAGE	97.4	72.9	86.5	100.9	66.1	85.4	96.9	55.7	78.5	97.7	66.2	83.6	99.2	66.2	84.4	98.3	55.9	79.3

DIVISION		VERMONT			VIRGINIA														
		WHITE RIVER JCT.			ALEXANDRIA			ARLINGTON			BRISTOL			CHARLOTTESVILLE			CULPEPER		
		050			223			222			242			229			227		
		MAT.	INST.	TOTAL	MAT.	INST.	TOTAL	MAT.	INST.	TOTAL	MAT.	INST.	TOTAL	MAT.	INST.	TOTAL	MAT.	INST.	TOTAL
015433	CONTRACTOR EQUIPMENT		100.3	100.3		101.7	101.7		100.4	100.4		100.4	100.4		104.8	104.8		100.4	100.4
0241, 31 - 34	SITE & INFRASTRUCTURE, DEMOLITION	81.4	96.3	91.7	120.9	88.1	98.3	132.3	84.8	99.5	114.9	84.4	93.9	120.2	86.5	97.0	119.1	85.2	95.7
0310	Concrete Forming & Accessories	91.9	42.9	49.4	89.4	74.7	76.6	88.9	71.6	73.9	85.2	45.0	50.3	83.7	51.2	55.5	81.0	70.3	71.7
0320	Concrete Reinforcing	86.5	38.2	61.6	80.5	81.0	80.8	90.9	72.6	81.5	90.9	69.5	79.9	90.3	70.5	80.1	90.9	66.7	78.4
0330	Cast-in-Place Concrete	94.1	55.2	79.2	117.1	81.5	103.5	114.0	77.3	99.9	113.5	49.8	89.1	118.1	58.1	95.1	116.7	67.8	98.0
03	CONCRETE	104.3	47.1	76.9	110.0	79.2	95.3	114.4	74.4	95.3	110.6	53.1	83.1	111.6	58.9	86.4	108.3	69.6	89.8
04	MASONRY	118.4	38.6	70.0	91.8	71.6	79.5	105.7	66.3	81.8	93.9	42.5	62.7	119.7	49.1	76.9	107.6	63.2	80.7
05	METALS	93.0	58.0	80.9	93.0	95.2	93.7	91.7	82.3	88.5	90.5	86.2	89.0	90.8	88.6	90.1	90.9	81.2	87.5
06	WOOD, PLASTICS & COMPOSITES	93.2	44.0	65.0	95.0	73.8	82.8	91.9	73.8	81.5	84.6	43.2	60.9	83.0	49.3	63.7	81.9	73.8	77.3
07	THERMAL & MOISTURE PROTECTION	96.0	36.7	72.4	97.2	80.5	90.6	99.1	69.1	87.2	98.7	49.1	79.0	98.3	54.7	81.0	98.5	63.2	84.5
08	OPENINGS	102.2	40.2	86.5	98.6	75.7	92.8	96.5	69.6	89.7	99.9	49.4	87.1	97.9	54.3	86.9	98.3	65.1	89.9
0920	Plaster & Gypsum Board	94.2	41.6	57.2	106.5	72.8	82.8	103.0	72.8	81.8	98.4	41.4	58.2	97.9	46.9	62.0	98.6	72.8	80.5
0950, 0980	Ceilings & Acoustic Treatment	89.8	41.6	57.9	95.7	72.8	80.6	92.9	72.8	79.6	92.0	41.4	58.4	92.0	46.9	62.1	92.9	72.8	79.6
0960	Flooring	93.8	32.2	76.1	109.9	85.7	102.9	108.1	63.9	95.4	104.2	39.5	85.6	102.7	46.6	86.6	102.7	63.9	91.5
0970, 0990	Wall Finishes & Painting/Coating	92.3	19.5	48.2	129.1	108.3	116.5	129.1	82.8	101.0	114.7	44.8	72.3	114.7	63.3	83.5	129.1	56.3	84.9
09	FINISHES	94.0	38.6	63.4	104.4	79.8	90.8	104.2	71.9	86.4	100.5	43.1	68.9	100.1	50.8	72.9	100.8	68.8	83.2
COVERS	DIVS. 10 - 14, 25, 28, 41, 43, 44, 46	100.0	39.2	87.4	100.0	88.2	97.5	100.0	72.5	94.3	100.0	75.9	95.0	100.0	75.8	95.0	100.0	61.6	92.0
21, 22, 23	FIRE SUPPRESSION, PLUMBING & HVAC	94.8	61.3	80.9	100.3	89.6	95.9	100.3	86.6	94.6	94.9	55.0	78.3	94.9	58.2	79.6	94.9	64.3	82.2
26, 27, 3370	ELECTRICAL, COMMUNICATIONS & UTIL.	100.6	66.3	82.4	96.8	96.9	96.9	94.2	96.7	95.5	96.5	38.4	65.6	96.4	70.1	82.4	99.2	56.6	76.5
MF2010	WEIGHTED AVERAGE	98.3	55.1	79.0	100.2	85.5	93.6	101.2	80.1	91.8	98.4	55.1	79.0	99.8	63.4	83.5	99.1	67.8	85.1

	DIVISION	VIRGINIA																	
		FAIRFAX			FARMVILLE			FREDERICKSBURG			GRUNDY			HARRISONBURG			LYNCHBURG		
		220 - 221			239			224 - 225			246			228			245		
		MAT.	INST.	TOTAL	MAT.	INST.	TOTAL	MAT.	INST.	TOTAL	MAT.	INST.	TOTAL	MAT.	INST.	TOTAL	MAT.	INST.	TOTAL
015433	**CONTRACTOR EQUIPMENT**		100.4	100.4		104.8	104.8		100.4	100.4		100.4	100.4		100.4	100.4		100.4	100.4
0241, 31 - 34	**SITE & INFRASTRUCTURE, DEMOLITION**	131.4	85.7	99.8	115.9	85.8	95.1	118.5	85.4	95.6	112.5	83.2	92.3	128.0	84.5	98.0	113.4	84.6	93.5
0310	Concrete Forming & Accessories	83.6	72.1	73.7	95.0	46.6	53.0	83.6	50.9	55.2	88.1	38.8	45.3	79.9	45.1	49.7	85.2	64.5	67.2
0320	Concrete Reinforcing	90.9	80.9	85.7	90.9	53.7	71.7	91.6	80.0	85.6	89.7	54.3	71.4	90.9	48.9	69.2	90.3	70.2	79.9
0330	Cast-in-Place Concrete	114.0	78.3	100.3	118.1	56.1	94.3	115.7	54.7	92.3	113.5	50.1	89.3	114.0	61.3	93.8	113.5	61.2	93.5
03	**CONCRETE**	114.0	77.1	96.4	110.5	52.8	82.9	107.9	59.4	84.7	109.3	47.2	79.5	111.8	53.4	83.8	109.2	65.8	88.4
04	**MASONRY**	105.7	66.6	82.0	102.6	48.3	69.7	106.6	47.0	70.5	97.1	51.1	69.2	103.9	53.2	73.2	111.8	55.4	77.6
05	**METALS**	90.9	96.1	92.7	90.8	78.0	86.4	90.9	92.0	91.3	90.5	71.7	84.1	90.8	84.5	88.6	90.7	87.6	89.6
06	**WOOD, PLASTICS & COMPOSITES**	84.6	73.8	78.4	97.0	47.6	68.7	84.6	48.6	64.0	87.5	37.9	59.0	80.8	42.2	58.6	84.6	67.9	75.0
07	**THERMAL & MOISTURE PROTECTION**	99.1	69.5	87.3	98.3	49.5	78.9	98.5	52.7	80.3	98.6	48.8	78.9	98.9	62.8	84.6	98.5	56.9	82.0
08	**OPENINGS**	96.5	77.0	91.6	98.2	46.4	85.1	97.9	56.3	87.4	99.9	37.1	84.0	98.3	52.9	86.8	98.2	62.8	89.2
0920	Plaster & Gypsum Board	98.6	72.8	80.5	105.4	45.2	63.0	98.6	46.9	62.2	98.8	35.9	54.5	97.9	40.3	57.3	98.4	66.8	76.1
0950, 0980	Ceilings & Acoustic Treatment	92.9	72.8	79.6	92.0	45.2	61.0	92.9	46.9	62.4	92.0	35.9	54.8	92.0	40.3	57.7	92.0	66.8	75.3
0960	Flooring	104.5	66.8	93.7	109.9	63.7	96.6	104.5	78.2	96.9	105.7	35.6	85.6	102.3	78.2	95.4	104.2	44.1	86.9
0970, 0990	Wall Finishes & Painting/Coating	129.1	82.8	101.0	114.7	41.7	70.4	129.1	69.1	92.7	114.7	37.8	68.1	129.1	54.6	83.9	114.7	44.8	72.3
09	**FINISHES**	102.6	72.4	85.9	102.8	49.5	73.4	101.3	56.8	76.8	100.9	37.8	66.1	101.2	51.1	73.6	100.3	59.0	77.5
COVERS	**DIVS. 10 - 14, 25, 28, 41, 43, 44, 46**	100.0	72.5	94.3	100.0	50.8	89.8	100.0	76.7	95.2	100.0	48.6	89.3	100.0	75.6	94.9	100.0	79.5	95.7
21, 22, 23	**FIRE SUPPRESSION, PLUMBING & HVAC**	94.9	86.9	91.6	94.9	37.5	71.0	94.9	80.4	88.9	94.9	42.1	72.9	94.9	54.9	78.3	94.9	57.7	79.4
26, 27, 3370	**ELECTRICAL, COMMUNICATIONS & UTIL.**	97.7	96.7	97.2	93.4	50.9	70.8	94.4	93.2	93.8	96.5	48.4	70.9	96.7	84.8	90.4	97.6	57.2	76.1
MF2010	**WEIGHTED AVERAGE**	99.9	82.3	92.0	98.7	53.0	78.3	98.5	72.2	86.7	98.3	50.3	76.9	99.3	64.0	83.5	99.0	64.6	83.6

	DIVISION	VIRGINIA																	
		NEWPORT NEWS			NORFOLK			PETERSBURG			PORTSMOUTH			PULASKI			RICHMOND		
		236			233 - 235			238			237			243			230 - 232		
		MAT.	INST.	TOTAL	MAT.	INST.	TOTAL	MAT.	INST.	TOTAL	MAT.	INST.	TOTAL	MAT.	INST.	TOTAL	MAT.	INST.	TOTAL
015433	**CONTRACTOR EQUIPMENT**		104.8	104.8		105.4	105.4		104.8	104.8		104.7	104.7		100.4	100.4		104.8	104.8
0241, 31 - 34	**SITE & INFRASTRUCTURE, DEMOLITION**	114.0	87.2	95.5	112.1	88.2	95.6	118.7	87.5	97.2	112.3	86.6	94.5	111.8	83.2	92.1	111.3	87.5	94.9
0310	Concrete Forming & Accessories	94.3	67.4	71.0	95.8	67.5	71.3	88.1	60.8	64.4	84.3	55.4	59.3	88.1	39.0	45.5	93.8	60.8	65.2
0320	Concrete Reinforcing	90.6	70.9	80.4	90.6	70.9	80.4	90.3	71.2	80.4	90.3	70.8	80.3	89.7	54.3	71.5	90.6	71.2	80.6
0330	Cast-in-Place Concrete	114.7	61.4	94.3	125.4	61.5	100.9	121.9	56.2	96.7	113.6	61.0	93.5	113.5	50.2	89.3	110.6	56.2	89.8
03	**CONCRETE**	107.4	67.3	88.2	112.8	67.3	91.1	113.4	62.6	89.1	106.2	61.7	84.9	109.3	47.3	79.6	105.3	62.6	84.9
04	**MASONRY**	97.4	55.3	71.8	103.4	55.3	74.2	111.1	56.8	78.2	102.8	55.1	73.9	90.6	51.1	66.7	95.1	56.8	71.9
05	**METALS**	93.0	89.6	91.8	92.0	89.7	91.2	90.9	90.1	90.6	92.0	88.1	90.6	90.6	71.9	84.2	95.1	90.1	93.4
06	**WOOD, PLASTICS & COMPOSITES**	95.9	71.8	82.1	98.0	71.8	82.9	88.0	63.4	73.9	84.3	55.7	67.9	87.5	37.9	59.0	95.0	63.4	76.9
07	**THERMAL & MOISTURE PROTECTION**	98.3	57.3	82.0	95.9	57.3	80.6	98.3	57.6	82.2	98.3	55.0	81.1	98.6	48.8	78.9	97.1	57.6	81.4
08	**OPENINGS**	98.6	65.5	90.2	98.6	65.5	90.2	97.9	62.0	88.8	98.7	56.8	88.1	100.0	37.1	84.0	98.6	62.0	89.3
0920	Plaster & Gypsum Board	106.5	70.1	80.9	100.2	70.1	79.0	99.1	61.5	72.6	98.5	53.5	66.8	98.8	35.9	54.5	109.7	61.5	75.8
0950, 0980	Ceilings & Acoustic Treatment	95.7	70.1	78.7	95.7	70.1	78.7	92.9	61.5	72.1	95.7	53.5	67.7	92.0	35.9	54.8	98.5	61.5	74.0
0960	Flooring	109.9	53.4	93.7	109.3	53.4	93.2	105.2	75.2	96.6	101.8	67.5	92.0	105.7	35.6	85.6	109.3	75.2	99.5
0970, 0990	Wall Finishes & Painting/Coating	114.7	45.3	72.6	114.7	69.3	87.1	114.7	63.3	83.5	114.7	69.3	87.1	114.7	37.8	68.1	114.7	63.3	83.5
09	**FINISHES**	103.5	62.9	81.1	102.1	65.6	82.0	101.1	63.9	80.6	100.2	58.5	77.2	100.9	37.8	66.1	103.8	63.9	81.8
COVERS	**DIVS. 10 - 14, 25, 28, 41, 43, 44, 46**	100.0	81.3	96.1	100.0	81.3	96.1	100.0	76.1	95.0	100.0	79.5	95.7	100.0	48.6	89.3	100.0	76.1	95.0
21, 22, 23	**FIRE SUPPRESSION, PLUMBING & HVAC**	100.3	64.2	85.3	100.1	63.5	84.9	94.9	56.3	78.8	100.3	63.5	85.0	94.9	42.4	73.0	100.1	56.3	81.9
26, 27, 3370	**ELECTRICAL, COMMUNICATIONS & UTIL.**	96.5	59.3	76.7	101.7	61.2	80.1	96.7	70.1	82.6	94.6	61.1	76.8	96.5	52.3	73.0	101.3	70.1	84.7
MF2010	**WEIGHTED AVERAGE**	100.0	67.6	85.5	101.0	68.2	86.3	99.6	66.8	84.9	99.4	65.5	84.2	98.0	50.9	77.0	100.2	66.8	85.3

	DIVISION	VIRGINIA									WASHINGTON								
		ROANOKE			STAUNTON			WINCHESTER			CLARKSTON			EVERETT			OLYMPIA		
		240 - 241			244			226			994			982			985		
		MAT.	INST.	TOTAL	MAT.	INST.	TOTAL	MAT.	INST.	TOTAL	MAT.	INST.	TOTAL	MAT.	INST.	TOTAL	MAT.	INST.	TOTAL
015433	**CONTRACTOR EQUIPMENT**		100.4	100.4		104.8	104.8		100.4	100.4		91.3	91.3		103.9	103.9		103.9	103.9
0241, 31 - 34	**SITE & INFRASTRUCTURE, DEMOLITION**	111.5	84.5	92.9	116.5	86.3	95.7	126.5	84.6	97.6	99.2	90.6	93.3	94.5	113.8	107.8	95.8	113.8	108.2
0310	Concrete Forming & Accessories	94.1	64.2	68.2	87.7	53.5	58.0	82.2	50.2	54.4	122.4	71.5	78.2	109.1	101.4	102.4	99.0	101.6	101.2
0320	Concrete Reinforcing	90.6	70.2	80.1	90.3	52.7	70.9	90.3	85.1	87.6	110.2	85.8	97.6	115.1	102.0	108.3	115.7	102.0	108.7
0330	Cast-in-Place Concrete	129.0	61.1	103.0	118.1	54.3	93.6	114.0	57.4	92.3	102.6	83.7	95.4	103.8	108.3	105.6	94.5	108.5	99.9
03	**CONCRETE**	114.5	65.6	91.1	110.7	55.3	84.2	111.1	60.8	87.0	100.7	78.5	90.1	99.6	103.3	101.4	95.5	103.5	99.3
04	**MASONRY**	98.4	57.1	73.3	108.1	55.5	76.2	101.5	51.0	70.9	108.3	80.5	91.5	132.1	101.4	113.5	121.0	101.5	109.2
05	**METALS**	92.8	87.4	90.9	90.8	83.0	88.1	90.9	91.2	91.0	89.0	81.2	86.3	101.8	92.4	98.6	97.0	92.5	95.4
06	**WOOD, PLASTICS & COMPOSITES**	95.9	67.9	79.9	87.5	54.0	68.3	83.2	48.6	63.4	96.8	69.3	81.0	105.4	101.2	103.0	92.0	101.2	97.3
07	**THERMAL & MOISTURE PROTECTION**	98.3	56.0	81.5	98.3	53.4	80.5	99.0	60.1	83.5	152.5	77.5	122.7	101.8	99.6	100.9	101.4	98.5	100.3
08	**OPENINGS**	98.6	62.8	89.5	98.2	52.0	86.5	100.0	57.6	89.3	106.6	72.4	97.9	102.3	100.7	101.9	102.2	100.7	101.8
0920	Plaster & Gypsum Board	106.5	66.8	78.6	98.8	51.8	65.7	98.6	46.9	62.2	134.2	68.2	87.7	119.5	101.2	106.6	108.7	101.2	103.4
0950, 0980	Ceilings & Acoustic Treatment	95.7	66.8	76.6	92.0	51.8	65.3	92.9	46.9	62.4	103.9	68.2	80.2	113.3	101.2	105.3	107.5	101.2	103.3
0960	Flooring	109.9	47.6	92.0	105.1	41.7	86.9	103.8	78.2	96.4	104.4	49.8	88.7	117.4	96.7	111.4	110.2	96.9	106.4
0970, 0990	Wall Finishes & Painting/Coating	114.7	44.8	72.3	114.7	31.2	64.1	129.1	48.1	79.9	108.3	63.0	80.8	104.7	76.4	87.6	104.7	85.9	93.3
09	**FINISHES**	103.3	59.8	79.3	100.6	49.5	72.4	101.8	54.4	75.7	118.5	65.5	89.3	110.8	97.9	103.7	106.3	99.0	102.3
COVERS	**DIVS. 10 - 14, 25, 28, 41, 43, 44, 46**	100.0	79.4	95.7	100.0	79.1	95.7	100.0	57.1	91.1	100.0	83.8	96.6	100.0	100.3	100.1	100.0	101.7	100.3
21, 22, 23	**FIRE SUPPRESSION, PLUMBING & HVAC**	100.3	48.5	78.8	94.9	56.7	79.0	94.9	69.0	84.1	95.0	76.2	87.2	100.1	92.8	97.1	99.9	92.9	97.0
26, 27, 3370	**ELECTRICAL, COMMUNICATIONS & UTIL.**	96.5	41.3	67.1	95.2	44.9	68.4	94.9	46.6	69.2	101.8	94.2	97.8	104.4	84.1	93.6	105.8	95.8	100.5
MF2010	**WEIGHTED AVERAGE**	100.7	60.7	82.8	98.9	59.1	81.1	99.2	63.2	83.1	101.9	79.8	92.0	103.4	97.1	100.6	101.3	98.9	100.2

DIVISION		WASHINGTON																	
		RICHLAND			SEATTLE			SPOKANE			TACOMA			VANCOUVER			WENATCHEE		
		993			980 - 981,987			990 - 992			983 - 984			986			988		
		MAT.	INST.	TOTAL	MAT.	INST.	TOTAL	MAT.	INST.	TOTAL	MAT.	INST.	TOTAL	MAT.	INST.	TOTAL	MAT.	INST.	TOTAL
015433	CONTRACTOR EQUIPMENT		91.3	91.3		103.7	103.7		91.3	91.3		103.9	103.9		98.2	98.2		103.9	103.9
0241, 31 - 34	SITE & INFRASTRUCTURE, DEMOLITION	100.2	90.8	93.7	98.5	111.9	107.7	99.5	90.7	93.5	97.6	113.8	108.8	108.7	100.5	103.0	108.6	111.8	110.8
0310	Concrete Forming & Accessories	122.6	81.2	86.6	100.7	102.2	102.0	127.0	80.7	86.8	100.7	101.6	101.4	101.6	87.5	89.4	102.1	79.2	82.2
0320	Concrete Reinforcing	105.7	85.8	95.5	113.6	102.1	107.7	106.5	85.8	95.8	113.6	102.0	107.6	114.6	101.7	108.0	114.5	86.3	100.0
0330	Cast-in-Place Concrete	102.9	85.3	96.2	109.0	108.7	108.9	106.9	85.1	98.6	106.8	108.5	107.4	119.3	101.5	112.4	109.0	80.3	98.0
03	CONCRETE	100.3	83.3	92.2	102.6	104.0	103.2	102.7	83.1	93.3	101.5	103.5	102.4	111.8	94.9	103.7	109.0	80.9	95.6
04	MASONRY	108.1	83.0	92.9	125.5	101.5	111.0	108.8	83.0	93.1	125.4	101.5	110.9	126.5	96.1	108.1	128.5	82.9	100.9
05	METALS	89.4	81.7	86.8	103.7	94.5	100.5	91.8	81.3	88.2	103.7	92.5	99.8	101.0	93.0	98.3	101.0	82.8	94.7
06	WOOD, PLASTICS & COMPOSITES	97.0	80.4	87.5	96.5	101.2	99.2	105.6	80.4	91.2	95.2	101.2	98.7	87.5	85.2	86.2	96.3	78.2	85.9
07	THERMAL & MOISTURE PROTECTION	153.7	79.8	124.3	101.6	99.7	100.8	150.1	80.7	122.5	101.5	98.5	100.3	101.8	91.0	97.5	102.0	81.1	93.7
08	OPENINGS	109.1	74.0	100.2	104.6	100.7	103.6	109.7	74.0	100.6	102.9	100.7	102.3	100.5	91.1	98.1	102.5	72.8	94.9
0920	Plaster & Gypsum Board	134.2	79.7	95.8	112.8	101.2	104.6	134.5	79.7	95.9	115.6	101.2	105.5	113.6	85.0	93.5	117.4	77.4	89.3
0950, 0980	Ceilings & Acoustic Treatment	109.1	79.7	89.6	115.9	101.2	106.1	106.0	79.7	88.5	116.1	101.2	106.2	111.5	85.0	94.0	109.3	77.4	88.2
0960	Flooring	104.6	84.6	98.8	110.1	104.8	108.6	104.7	90.1	100.5	110.6	96.9	106.7	116.7	107.9	114.1	113.6	66.2	100.0
0970, 0990	Wall Finishes & Painting/Coating	108.3	63.0	80.8	104.7	85.9	93.3	108.3	63.0	80.8	104.7	85.9	93.3	109.6	66.6	83.6	104.7	63.0	79.4
09	FINISHES	119.9	79.7	97.7	108.7	100.6	104.2	119.4	80.8	98.1	109.1	99.0	103.5	107.3	88.7	97.0	109.8	74.6	90.4
COVERS	DIVS. 10 - 14, 25, 28, 41, 43, 44, 46	100.0	69.2	93.6	100.0	101.7	100.4	100.0	69.1	93.6	100.0	101.7	100.3	100.0	65.7	92.9	100.0	85.3	96.9
21, 22, 23	FIRE SUPPRESSION, PLUMBING & HVAC	100.5	102.2	101.2	100.0	118.3	107.6	100.4	77.5	90.8	100.1	92.9	97.1	100.2	94.1	97.7	94.7	84.4	90.4
26, 27, 3370	ELECTRICAL, COMMUNICATIONS & UTIL.	99.5	95.1	97.2	104.2	110.2	107.4	96.5	80.8	88.1	104.2	95.8	99.7	109.5	100.1	104.5	105.1	81.9	92.8
MF2010	WEIGHTED AVERAGE	103.4	88.0	96.5	103.8	106.6	105.0	103.7	80.8	93.5	103.5	98.9	101.4	104.8	93.8	99.9	103.3	83.7	94.6

DIVISION		WASHINGTON			WEST VIRGINIA														
		YAKIMA			BECKLEY			BLUEFIELD			BUCKHANNON			CHARLESTON			CLARKSBURG		
		989			258 - 259			247 - 248			262			250 - 253			263 - 264		
		MAT.	INST.	TOTAL	MAT.	INST.	TOTAL	MAT.	INST.	TOTAL	MAT.	INST.	TOTAL	MAT.	INST.	TOTAL	MAT.	INST.	TOTAL
015433	CONTRACTOR EQUIPMENT		103.9	103.9		100.4	100.4		100.4	100.4		100.4	100.4		100.4	100.4		100.4	100.4
0241, 31 - 34	SITE & INFRASTRUCTURE, DEMOLITION	100.4	112.4	108.7	106.1	86.7	92.7	105.8	86.7	92.6	112.6	86.4	94.5	103.7	88.3	93.1	113.2	86.4	94.7
0310	Concrete Forming & Accessories	101.0	96.7	97.2	83.8	88.3	87.7	85.1	87.4	87.1	84.5	85.5	85.4	99.4	90.2	91.4	82.3	85.5	85.1
0320	Concrete Reinforcing	114.1	86.4	99.8	89.3	85.3	87.2	89.3	67.1	77.9	89.8	84.3	87.0	90.6	85.5	88.0	89.8	84.9	87.3
0330	Cast-in-Place Concrete	114.1	84.9	102.9	111.0	102.7	107.9	111.0	102.4	107.7	110.7	101.1	107.0	102.7	102.9	102.8	121.3	101.1	113.6
03	CONCRETE	106.5	90.2	98.7	104.7	93.3	99.2	104.8	89.4	97.4	107.9	91.3	100.0	101.6	94.2	98.1	112.7	91.4	102.5
04	MASONRY	118.3	82.8	96.8	97.5	88.9	92.3	92.3	88.9	90.2	103.7	84.7	92.2	94.0	90.8	92.1	106.2	84.7	93.2
05	METALS	101.7	82.5	95.1	90.8	96.0	92.6	90.8	88.7	90.1	90.9	96.5	92.9	93.0	97.2	94.5	90.9	96.7	92.9
06	WOOD, PLASTICS & COMPOSITES	95.7	101.2	98.9	84.6	89.8	87.6	86.3	89.8	88.3	85.6	85.2	85.4	102.0	89.8	95.0	82.4	85.2	84.0
07	THERMAL & MOISTURE PROTECTION	101.6	82.7	94.1	98.4	87.1	93.9	98.4	87.1	93.9	98.7	86.9	94.0	95.7	87.9	92.6	98.6	86.9	94.0
08	OPENINGS	102.4	81.8	97.2	99.3	82.9	95.1	100.2	78.7	94.7	100.2	80.1	95.1	99.9	82.9	95.5	100.2	82.1	95.6
0920	Plaster & Gypsum Board	115.7	101.2	105.5	96.1	89.3	91.3	96.9	89.3	91.6	97.5	84.6	88.4	99.7	89.3	92.4	95.3	84.6	87.7
0950, 0980	Ceilings & Acoustic Treatment	111.4	101.2	104.6	90.1	89.3	89.6	90.1	89.3	89.6	92.0	84.6	87.1	93.8	89.3	90.8	92.0	84.6	87.1
0960	Flooring	111.8	55.0	95.4	101.2	106.0	102.6	101.8	106.0	103.0	101.5	100.2	101.1	109.9	106.0	108.8	100.3	100.2	100.2
0970, 0990	Wall Finishes & Painting/Coating	104.7	63.0	79.4	114.7	37.7	68.0	114.7	37.7	68.0	114.7	85.5	97.0	114.7	84.9	96.7	114.7	85.5	97.0
09	FINISHES	108.7	85.9	96.1	98.3	87.0	92.1	98.6	87.0	92.2	99.5	88.1	93.2	101.6	92.7	96.7	98.8	88.1	92.9
COVERS	DIVS. 10 - 14, 25, 28, 41, 43, 44, 46	100.0	98.7	99.7	100.0	59.3	91.5	100.0	59.2	91.5	100.0	97.7	99.5	100.0	99.3	99.9	100.0	97.7	99.5
21, 22, 23	FIRE SUPPRESSION, PLUMBING & HVAC	100.1	102.3	101.0	94.9	82.0	89.5	94.9	56.1	78.8	94.9	89.1	92.5	100.1	84.0	93.4	94.9	88.9	92.4
26, 27, 3370	ELECTRICAL, COMMUNICATIONS & UTIL.	107.5	95.1	100.9	92.4	89.4	90.8	95.3	70.4	82.0	96.8	93.5	95.0	99.8	89.5	94.3	96.8	93.5	95.0
MF2010	WEIGHTED AVERAGE	103.7	93.1	99.0	97.0	87.2	92.6	97.1	77.7	88.4	98.5	89.7	94.6	99.0	90.1	95.0	99.2	89.7	94.9

DIVISION		WEST VIRGINIA																	
		GASSAWAY			HUNTINGTON			LEWISBURG			MARTINSBURG			MORGANTOWN			PARKERSBURG		
		266			255 - 257			249			254			265			261		
		MAT.	INST.	TOTAL	MAT.	INST.	TOTAL	MAT.	INST.	TOTAL	MAT.	INST.	TOTAL	MAT.	INST.	TOTAL	MAT.	INST.	TOTAL
015433	CONTRACTOR EQUIPMENT		100.4	100.4		100.4	100.4		100.4	100.4		100.4	100.4		100.4	100.4		100.4	100.4
0241, 31 - 34	SITE & INFRASTRUCTURE, DEMOLITION	109.7	86.9	93.9	110.5	89.1	95.8	123.1	86.7	98.0	110.0	84.5	92.4	106.9	86.6	92.9	115.4	88.2	96.7
0310	Concrete Forming & Accessories	84.0	85.7	85.5	95.3	92.5	92.9	82.6	87.9	87.2	83.9	78.4	79.1	82.6	85.2	84.9	86.5	87.7	87.5
0320	Concrete Reinforcing	89.8	85.5	87.6	90.6	85.6	88.0	89.8	67.1	78.2	89.3	72.4	80.6	89.8	84.8	87.2	89.3	84.2	86.7
0330	Cast-in-Place Concrete	115.9	101.2	110.3	121.1	103.8	114.5	111.1	102.6	107.8	116.3	92.2	107.1	110.7	101.0	107.0	113.2	94.0	105.8
03	CONCRETE	108.5	91.6	100.5	110.6	95.6	103.4	114.7	89.7	102.7	108.7	82.6	96.2	104.5	91.3	98.2	110.1	89.9	100.4
04	MASONRY	107.3	87.7	95.4	97.8	92.4	94.5	96.5	88.9	91.9	99.3	72.1	82.8	126.4	84.7	101.1	82.4	86.6	84.9
05	METALS	90.9	97.2	93.1	93.1	97.4	94.5	90.9	89.2	90.3	91.1	84.5	88.8	91.0	96.4	92.9	91.7	96.5	93.3
06	WOOD, PLASTICS & COMPOSITES	84.7	85.2	85.0	95.9	91.9	93.6	82.7	89.8	86.8	84.6	79.9	81.9	82.7	85.2	84.1	85.9	86.8	86.4
07	THERMAL & MOISTURE PROTECTION	98.4	87.7	94.2	98.5	88.8	94.6	99.3	87.1	94.4	98.6	71.6	87.9	98.5	86.9	93.9	98.3	87.2	93.9
08	OPENINGS	98.1	80.4	93.6	98.6	84.1	94.9	100.2	78.7	94.7	101.6	67.7	93.0	101.6	82.1	96.7	99.2	81.0	94.6
0920	Plaster & Gypsum Board	96.6	84.6	88.1	105.4	91.5	95.7	95.3	89.3	91.1	96.6	79.1	84.3	95.3	84.6	87.7	98.4	86.2	89.8
0950, 0980	Ceilings & Acoustic Treatment	92.0	84.6	87.1	92.0	91.5	91.7	92.0	89.3	90.2	92.0	79.1	83.4	92.0	84.6	87.1	92.0	86.2	88.2
0960	Flooring	101.3	106.0	102.6	109.7	106.0	108.6	100.4	106.0	102.0	101.2	56.7	88.4	100.4	100.2	100.3	104.7	97.4	102.6
0970, 0990	Wall Finishes & Painting/Coating	114.7	84.9	96.7	114.7	84.9	96.7	114.7	37.7	68.0	114.7	44.5	72.1	114.7	85.5	97.0	114.7	85.0	96.7
09	FINISHES	99.0	89.2	93.6	102.3	94.4	97.9	100.0	87.0	92.8	99.0	71.4	83.8	98.4	88.1	92.7	100.7	89.1	94.3
COVERS	DIVS. 10 - 14, 25, 28, 41, 43, 44, 46	100.0	97.7	99.5	100.0	100.2	100.0	100.0	59.2	91.5	100.0	57.3	91.1	100.0	68.5	93.4	100.0	98.8	99.8
21, 22, 23	FIRE SUPPRESSION, PLUMBING & HVAC	94.9	89.9	92.8	100.3	83.3	93.3	94.9	81.8	89.5	94.9	84.5	90.6	94.9	88.8	92.4	100.2	86.8	94.6
26, 27, 3370	ELECTRICAL, COMMUNICATIONS & UTIL.	96.8	89.5	92.9	96.5	95.9	96.2	92.4	70.4	80.7	98.8	82.1	89.9	97.0	93.5	95.1	97.0	92.2	94.4
MF2010	WEIGHTED AVERAGE	98.4	89.9	94.6	100.2	91.6	96.3	98.9	83.2	91.9	98.6	79.1	89.9	99.2	88.8	94.6	99.1	89.4	94.8

DIVISION		WEST VIRGINIA									WISCONSIN								
		PETERSBURG			ROMNEY			WHEELING			BELOIT			EAU CLAIRE			GREEN BAY		
		268			267			260			535			547			541 - 543		
		MAT.	INST.	TOTAL	MAT.	INST.	TOTAL	MAT.	INST.	TOTAL	MAT.	INST.	TOTAL	MAT.	INST.	TOTAL	MAT.	INST.	TOTAL
015433	CONTRACTOR EQUIPMENT		100.4	100.4		100.4	100.4		100.4	100.4		101.0	101.0		100.7	100.7		98.6	98.6
0241, 31 - 34	SITE & INFRASTRUCTURE, DEMOLITION	106.1	87.0	92.9	109.1	87.0	93.9	116.2	87.3	96.2	94.2	106.3	102.5	95.9	103.1	100.9	99.9	99.4	99.5
0310	Concrete Forming & Accessories	85.4	80.6	81.2	82.0	80.8	80.9	88.1	85.7	86.0	96.4	96.7	96.7	98.3	98.5	98.4	102.9	97.8	98.4
0320	Concrete Reinforcing	89.3	73.3	81.0	89.8	70.6	79.9	88.7	84.9	86.7	98.3	118.7	108.8	95.2	103.3	99.4	93.5	89.7	91.5
0330	Cast-in-Place Concrete	110.7	95.7	104.9	115.9	95.7	108.2	113.2	101.2	108.6	101.9	95.2	99.3	101.6	96.0	99.5	105.1	100.7	103.4
03	CONCRETE	104.6	85.2	95.3	108.4	84.8	97.1	110.1	91.5	101.2	99.5	100.5	100.0	101.0	98.8	99.9	104.0	97.5	100.9
04	MASONRY	99.3	84.7	90.5	96.9	84.0	89.0	106.9	84.7	93.5	104.7	95.9	99.4	95.6	100.1	98.3	129.4	99.3	111.2
05	METALS	91.0	91.6	91.2	91.1	90.8	91.0	91.8	96.8	93.5	94.0	109.5	99.3	95.5	104.3	98.5	97.7	98.2	97.9
06	WOOD, PLASTICS & COMPOSITES	86.5	79.8	82.7	81.9	79.8	80.7	87.5	85.2	86.2	102.8	99.2	100.7	108.7	99.3	103.3	109.0	99.3	103.4
07	THERMAL & MOISTURE PROTECTION	98.5	71.7	87.9	98.6	72.0	88.0	98.7	87.4	94.2	105.2	93.1	100.4	105.2	86.7	97.9	106.7	87.0	98.9
08	OPENINGS	101.6	74.7	94.8	101.5	74.1	94.6	100.1	82.1	95.6	100.0	107.0	101.8	98.2	97.9	98.1	96.6	96.6	96.6
0920	Plaster & Gypsum Board	97.5	79.1	84.5	94.8	79.1	83.7	98.8	84.6	88.8	95.7	99.5	98.4	100.4	99.5	99.8	94.5	99.5	98.0
0950, 0980	Ceilings & Acoustic Treatment	92.0	79.1	83.4	92.0	79.1	83.4	92.0	84.6	87.1	83.7	99.5	94.2	97.2	99.5	98.7	89.8	99.5	96.3
0960	Flooring	102.3	100.2	101.7	100.1	98.7	99.7	105.7	100.2	104.1	107.3	118.6	110.5	98.7	114.3	103.2	118.8	114.3	117.5
0970, 0990	Wall Finishes & Painting/Coating	114.7	84.9	96.7	114.7	84.9	96.7	114.7	85.5	97.0	95.9	99.5	98.0	90.0	93.1	91.9	100.4	82.0	89.2
09	FINISHES	99.2	84.9	91.3	98.5	84.6	90.8	101.0	88.1	93.9	100.6	101.3	101.0	99.0	101.2	100.2	104.0	99.8	101.7
COVERS	DIVS. 10 - 14, 25, 28, 41, 43, 44, 46	100.0	67.8	93.3	100.0	97.1	99.4	100.0	99.4	99.9	100.0	92.9	98.5	100.0	95.3	99.0	100.0	95.0	99.0
21, 22, 23	FIRE SUPPRESSION, PLUMBING & HVAC	94.9	89.9	92.8	94.9	89.8	92.8	100.3	88.9	95.6	99.9	91.1	96.2	100.1	86.7	94.5	100.3	84.8	93.9
26, 27, 3370	ELECTRICAL, COMMUNICATIONS & UTIL.	100.5	82.2	90.8	99.7	82.2	90.4	93.9	93.5	93.7	99.1	89.4	94.0	102.1	92.2	96.8	96.1	84.6	90.0
MF2010	WEIGHTED AVERAGE	98.2	85.0	92.3	98.4	85.6	92.7	100.3	89.9	95.7	99.3	97.8	98.6	99.3	96.1	97.9	101.7	93.3	98.0

DIVISION		WISCONSIN																	
		KENOSHA			LA CROSSE			LANCASTER			MADISON			MILWAUKEE			NEW RICHMOND		
		531			546			538			537			530,532			540		
		MAT.	INST.	TOTAL	MAT.	INST.	TOTAL	MAT.	INST.	TOTAL	MAT.	INST.	TOTAL	MAT.	INST.	TOTAL	MAT.	INST.	TOTAL
015433	CONTRACTOR EQUIPMENT		99.1	99.1		100.7	100.7		101.0	101.0		101.0	101.0		89.0	89.0		101.5	101.5
0241, 31 - 34	SITE & INFRASTRUCTURE, DEMOLITION	100.6	103.8	102.8	89.1	103.2	98.8	93.9	106.6	102.7	89.9	106.7	101.5	95.0	96.7	96.2	90.9	103.8	99.8
0310	Concrete Forming & Accessories	99.8	105.5	104.8	87.1	98.6	97.1	95.8	97.6	97.4	95.8	99.4	98.9	98.3	117.3	114.8	94.5	105.4	104.0
0320	Concrete Reinforcing	98.1	101.9	100.1	95.0	92.0	93.4	99.6	91.8	95.6	98.3	92.2	95.2	98.3	102.3	100.4	92.5	103.2	98.0
0330	Cast-in-Place Concrete	110.9	99.8	106.7	91.4	93.4	92.1	101.3	99.0	100.4	93.2	99.7	95.7	99.6	111.8	104.2	105.8	82.9	97.0
03	CONCRETE	104.2	102.9	103.6	92.1	95.8	93.9	99.2	97.1	98.2	95.1	98.2	96.6	98.6	111.7	104.9	98.9	97.4	98.2
04	MASONRY	101.9	109.4	106.4	94.8	102.1	99.2	104.7	101.8	102.9	102.0	104.4	103.4	104.6	120.8	114.4	123.8	96.4	107.2
05	METALS	94.7	103.3	97.6	95.4	98.9	96.6	91.5	95.9	93.0	96.7	98.0	97.1	97.5	96.8	97.2	95.5	103.3	98.2
06	WOOD, PLASTICS & COMPOSITES	102.5	104.9	103.9	95.7	99.3	97.7	102.1	99.2	100.4	100.1	99.2	99.6	104.4	117.3	111.8	98.3	111.5	105.8
07	THERMAL & MOISTURE PROTECTION	105.0	98.1	102.3	104.8	86.3	97.4	105.0	81.2	95.6	101.8	92.5	98.1	103.2	113.8	107.4	107.2	93.9	101.9
08	OPENINGS	95.4	106.5	98.3	98.2	85.6	95.0	95.3	86.4	93.1	100.0	99.5	99.9	102.1	113.2	104.9	86.7	104.5	91.2
0920	Plaster & Gypsum Board	80.9	105.4	98.1	95.0	99.5	98.2	94.6	99.5	98.1	97.8	99.5	99.0	96.7	118.0	111.7	86.1	112.3	104.6
0950, 0980	Ceilings & Acoustic Treatment	83.7	105.4	98.1	95.3	99.5	98.1	79.9	99.5	92.9	87.6	99.5	95.5	87.4	118.0	107.7	64.5	112.3	96.2
0960	Flooring	125.9	113.2	122.2	91.3	114.3	97.9	106.4	107.1	106.6	101.3	107.1	103.0	110.3	120.4	113.2	111.2	114.3	112.1
0970, 0990	Wall Finishes & Painting/Coating	104.3	105.9	105.2	90.0	74.0	80.3	95.9	71.5	81.1	95.9	99.5	98.0	97.8	116.2	108.9	104.0	93.1	97.4
09	FINISHES	105.2	107.5	106.4	95.2	99.6	97.6	99.4	96.2	97.7	99.5	101.2	100.4	102.6	118.4	111.3	95.0	105.7	100.9
COVERS	DIVS. 10 - 14, 25, 28, 41, 43, 44, 46	100.0	100.3	100.1	100.0	96.0	99.2	100.0	52.8	90.2	100.0	95.9	99.1	100.0	103.9	100.8	100.0	93.0	98.5
21, 22, 23	FIRE SUPPRESSION, PLUMBING & HVAC	100.1	97.0	98.8	100.1	87.5	94.9	94.5	87.3	91.5	99.8	96.8	98.6	99.9	104.0	101.6	94.3	84.2	90.1
26, 27, 3370	ELECTRICAL, COMMUNICATIONS & UTIL.	99.6	101.5	100.6	102.4	92.4	97.1	98.9	92.4	95.4	102.1	96.2	98.9	100.6	100.3	100.4	100.0	92.4	96.0
MF2010	WEIGHTED AVERAGE	100.0	102.7	101.2	97.6	94.9	96.4	97.0	93.2	95.3	98.9	99.1	99.0	100.2	107.6	103.5	97.3	96.2	96.8

DIVISION		WISCONSIN																	
		OSHKOSH			PORTAGE			RACINE			RHINELANDER			SUPERIOR			WAUSAU		
		549			539			534			545			548			544		
		MAT.	INST.	TOTAL	MAT.	INST.	TOTAL	MAT.	INST.	TOTAL	MAT.	INST.	TOTAL	MAT.	INST.	TOTAL	MAT.	INST.	TOTAL
015433	CONTRACTOR EQUIPMENT		98.6	98.6		101.0	101.0		101.0	101.0		98.6	98.6		101.5	101.5		98.6	98.6
0241, 31 - 34	SITE & INFRASTRUCTURE, DEMOLITION	91.3	98.3	96.1	84.0	106.7	99.7	94.0	107.3	103.2	105.1	99.5	101.2	87.3	104.1	98.9	86.7	99.4	95.5
0310	Concrete Forming & Accessories	88.6	86.5	86.8	89.9	98.5	97.4	96.7	105.5	104.4	86.6	97.7	96.2	93.1	94.7	94.5	88.2	97.7	96.5
0320	Concrete Reinforcing	93.6	78.9	86.0	99.7	91.9	95.7	98.3	101.9	100.1	93.7	94.1	93.9	92.5	94.6	93.6	93.7	92.0	92.8
0330	Cast-in-Place Concrete	97.5	90.5	94.8	87.0	101.3	92.5	100.0	99.6	99.8	110.5	101.4	107.0	99.7	99.3	99.5	90.8	100.7	94.6
03	CONCRETE	93.6	86.9	90.4	87.1	98.3	92.5	98.5	102.8	100.6	106.2	98.5	102.5	93.2	96.6	94.8	88.4	97.9	93.0
04	MASONRY	109.1	89.9	97.4	103.6	101.7	102.5	104.6	109.3	107.5	126.6	101.3	111.3	123.1	106.3	112.9	108.6	99.3	103.0
05	METALS	95.7	90.8	94.1	92.2	96.8	93.8	95.7	103.3	98.3	95.6	99.2	96.9	96.5	100.8	98.0	95.5	99.1	96.7
06	WOOD, PLASTICS & COMPOSITES	92.1	88.1	89.8	94.5	99.2	97.2	103.0	104.9	104.1	89.8	99.3	95.2	96.8	93.3	94.8	91.6	99.3	96.0
07	THERMAL & MOISTURE PROTECTION	105.9	79.9	95.5	104.5	91.0	99.1	105.3	97.7	102.3	106.7	85.3	98.2	106.9	97.0	103.0	105.7	86.0	97.9
08	OPENINGS	92.0	84.9	90.2	95.5	93.5	95.0	100.0	106.5	101.6	92.0	88.6	91.2	86.7	93.3	88.3	92.2	97.2	93.5
0920	Plaster & Gypsum Board	84.4	88.0	86.9	89.1	99.5	96.4	95.7	105.4	102.5	83.9	99.5	94.9	85.9	93.6	91.3	84.4	99.5	95.0
0950, 0980	Ceilings & Acoustic Treatment	89.8	88.0	88.6	80.9	99.5	93.2	83.7	105.4	98.1	89.8	99.5	96.3	65.5	93.6	84.1	89.8	99.5	96.3
0960	Flooring	109.2	100.6	106.7	102.1	113.7	105.4	107.3	113.2	109.0	108.2	114.3	109.9	112.7	123.6	115.8	109.0	114.3	110.5
0970, 0990	Wall Finishes & Painting/Coating	98.4	69.8	81.0	95.9	71.5	81.1	95.8	111.0	105.0	98.4	65.0	78.1	92.7	102.5	98.6	98.4	82.0	88.4
09	FINISHES	99.2	86.9	92.4	96.9	98.3	97.7	100.6	108.0	104.7	100.0	97.4	98.6	94.5	100.8	98.0	98.8	99.8	99.3
COVERS	DIVS. 10 - 14, 25, 28, 41, 43, 44, 46	100.0	66.4	93.0	100.0	68.4	93.4	100.0	100.3	100.1	100.0	58.9	91.5	100.0	92.6	98.5	100.0	95.0	99.0
21, 22, 23	FIRE SUPPRESSION, PLUMBING & HVAC	94.9	76.3	87.2	94.5	92.3	93.6	99.9	97.0	98.7	94.9	86.8	91.5	94.3	91.8	93.2	94.9	85.5	91.0
26, 27, 3370	ELECTRICAL, COMMUNICATIONS & UTIL.	100.7	77.8	88.5	103.0	95.7	99.1	98.9	99.4	99.2	100.0	83.2	91.0	105.3	98.9	101.9	101.9	83.2	91.9
MF2010	WEIGHTED AVERAGE	97.0	84.1	91.2	95.4	96.3	95.8	99.4	102.8	100.9	99.8	92.2	96.4	97.1	98.1	97.6	96.2	93.4	95.0

	DIVISION	WYOMING																	
		CASPER			CHEYENNE			NEWCASTLE			RAWLINS			RIVERTON			ROCK SPRINGS		
		826			820			827			823			825			829 - 831		
		MAT.	INST.	TOTAL	MAT.	INST.	TOTAL	MAT.	INST.	TOTAL	MAT.	INST.	TOTAL	MAT.	INST.	TOTAL	MAT.	INST.	TOTAL
015433	CONTRACTOR EQUIPMENT		100.6	100.6		100.6	100.6		100.6	100.6		100.6	100.6		100.6	100.6		100.6	100.6
0241, 31 - 34	SITE & INFRASTRUCTURE, DEMOLITION	89.4	98.5	95.7	82.6	98.6	93.6	75.8	97.4	90.7	89.1	97.4	94.8	82.9	97.4	92.9	79.0	97.4	91.7
0310	Concrete Forming & Accessories	103.0	42.0	50.1	104.7	57.1	63.4	95.9	44.7	51.5	99.8	44.7	52.0	94.8	44.7	51.3	101.5	44.7	52.2
0320	Concrete Reinforcing	105.3	65.1	84.6	99.4	65.9	82.1	106.8	65.7	85.6	106.4	65.7	85.5	107.4	65.7	85.9	107.4	67.3	86.8
0330	Cast-in-Place Concrete	96.4	73.3	87.6	94.7	73.5	86.6	95.7	52.9	79.3	95.8	52.9	79.4	95.8	52.9	79.4	95.7	52.9	79.3
03	CONCRETE	101.7	58.0	80.8	100.6	64.9	83.5	101.1	52.3	77.7	114.2	52.3	84.6	109.0	52.3	81.8	101.6	52.6	78.1
04	MASONRY	101.5	36.6	62.2	98.3	50.5	69.4	95.1	45.3	64.9	95.1	45.3	64.9	95.1	45.3	64.9	153.6	45.3	88.0
05	METALS	98.1	68.7	88.0	99.3	70.3	89.3	94.4	68.1	85.3	94.4	68.1	85.4	94.5	68.1	85.4	95.3	68.7	86.1
06	WOOD, PLASTICS & COMPOSITES	94.3	39.2	62.7	96.0	59.0	74.8	86.3	45.7	63.0	90.3	45.7	64.7	85.3	45.7	62.6	93.7	45.7	66.2
07	THERMAL & MOISTURE PROTECTION	98.0	48.4	78.3	99.6	54.2	81.5	101.0	47.5	79.7	102.7	47.5	80.8	102.0	47.5	80.4	101.1	47.5	79.8
08	OPENINGS	94.7	44.1	81.9	96.8	54.9	86.2	101.1	47.6	87.5	100.7	47.6	87.3	100.9	47.6	87.4	101.5	48.0	87.9
0920	Plaster & Gypsum Board	101.1	37.2	56.1	92.6	57.7	68.0	86.6	43.9	56.6	87.9	43.9	57.0	86.6	43.9	56.6	96.3	43.9	59.4
0950, 0980	Ceilings & Acoustic Treatment	124.9	37.2	66.8	116.1	57.7	77.4	119.3	43.9	69.3	119.3	43.9	69.3	119.3	43.9	69.3	119.3	43.9	69.3
0960	Flooring	100.3	34.9	81.5	104.7	56.0	90.7	98.5	49.3	84.3	101.3	49.3	86.3	97.8	49.3	83.8	103.4	49.3	87.9
0970, 0990	Wall Finishes & Painting/Coating	98.5	46.3	66.8	101.8	46.3	68.2	97.7	29.7	56.4	97.7	29.7	56.4	97.7	29.7	56.4	97.7	29.7	56.4
09	FINISHES	106.9	39.8	69.9	106.2	55.8	78.4	101.0	43.2	69.1	103.6	43.2	70.3	101.8	43.2	69.5	103.7	43.2	70.3
COVERS	DIVS. 10 - 14, 25, 28, 41, 43, 44, 46	100.0	77.1	95.2	100.0	79.4	95.7	100.0	59.3	91.5	100.0	59.3	91.5	100.0	59.3	91.5	100.0	59.3	91.5
21, 22, 23	FIRE SUPPRESSION, PLUMBING & HVAC	100.0	56.2	81.8	100.0	56.3	81.8	97.2	53.9	79.2	97.2	53.9	79.2	97.2	53.9	79.2	99.9	53.9	80.8
26, 27, 3370	ELECTRICAL, COMMUNICATIONS & UTIL.	98.9	68.3	82.6	97.3	74.4	85.2	95.2	68.3	80.9	95.2	68.3	80.9	95.2	68.3	80.9	93.3	68.3	80.0
MF2010	WEIGHTED AVERAGE	99.5	58.5	81.2	99.2	64.7	83.8	97.3	58.0	79.7	99.6	58.0	81.0	98.6	58.0	80.4	101.5	58.1	82.1

	DIVISION	WYOMING												CANADA					
		SHERIDAN			WHEATLAND			WORLAND			YELLOWSTONE NAT'L PA			BARRIE, ONTARIO			BATHURST, NEW BRUNSWICK		
		828			822			824			821								
		MAT.	INST.	TOTAL	MAT.	INST.	TOTAL	MAT.	INST.	TOTAL	MAT.	INST.	TOTAL	MAT.	INST.	TOTAL	MAT.	INST.	TOTAL
015433	CONTRACTOR EQUIPMENT		100.6	100.6		100.6	100.6		100.6	100.6		100.6	100.6		101.6	101.6		101.4	101.4
0241, 31 - 34	SITE & INFRASTRUCTURE, DEMOLITION	83.0	98.6	93.7	79.8	97.5	92.0	77.5	97.4	91.2	77.5	97.8	91.5	115.0	101.5	105.7	94.2	97.3	96.3
0310	Concrete Forming & Accessories	102.6	47.1	54.4	97.8	45.1	52.1	97.9	44.7	51.7	97.9	47.0	53.7	123.1	95.7	99.3	102.2	68.0	72.5
0320	Concrete Reinforcing	107.4	65.8	86.0	106.8	65.7	85.6	107.4	65.7	85.9	109.3	66.8	87.4	180.1	89.0	133.2	148.4	60.0	102.9
0330	Cast-in-Place Concrete	99.0	73.4	89.2	99.9	53.5	82.1	95.7	52.9	79.3	95.7	56.4	80.6	167.6	94.5	139.6	138.2	65.1	110.2
03	CONCRETE	109.1	60.4	85.8	106.0	52.6	80.4	101.3	52.3	77.9	101.6	54.7	79.1	152.8	94.1	124.7	135.5	66.1	102.3
04	MASONRY	95.4	49.9	67.8	95.4	46.2	65.6	95.1	45.3	64.9	95.2	51.4	68.6	168.4	104.3	129.5	158.7	69.0	104.4
05	METALS	98.1	70.0	88.4	94.4	68.1	85.3	94.5	68.1	85.4	95.2	68.5	86.0	107.4	93.5	102.6	104.7	73.4	94.0
06	WOOD, PLASTICS & COMPOSITES	97.2	45.7	67.7	88.0	45.7	63.7	88.0	45.7	63.8	88.0	45.7	63.8	118.4	94.2	104.5	94.9	68.3	79.6
07	THERMAL & MOISTURE PROTECTION	102.1	52.6	82.4	101.3	48.5	80.3	101.0	47.5	79.7	100.6	50.7	80.8	111.4	94.9	104.8	104.8	66.1	89.4
08	OPENINGS	101.8	48.6	88.3	99.6	47.6	86.5	101.3	47.6	87.7	94.1	47.9	82.4	91.8	92.1	91.9	86.9	59.5	80.0
0920	Plaster & Gypsum Board	111.4	43.9	63.9	86.6	43.9	56.6	86.6	43.9	56.6	86.9	43.9	56.6	157.5	94.0	112.8	163.1	67.2	95.6
0950, 0980	Ceilings & Acoustic Treatment	122.4	43.9	70.4	119.3	43.9	69.3	119.3	43.9	69.3	120.2	43.9	69.6	95.2	94.0	94.4	96.2	67.2	77.0
0960	Flooring	101.3	49.3	86.4	99.8	49.3	85.3	99.8	49.3	85.3	99.8	49.3	85.3	132.0	102.7	123.5	115.2	49.0	96.2
0970, 0990	Wall Finishes & Painting/Coating	99.9	46.3	67.4	97.7	29.7	56.4	97.7	29.7	56.4	97.7	29.7	56.4	112.2	95.1	101.8	111.4	54.6	76.9
09	FINISHES	108.5	46.2	74.2	101.8	43.4	69.6	101.4	43.2	69.3	101.7	44.7	70.2	118.8	97.2	106.9	112.9	63.2	85.5
COVERS	DIVS. 10 - 14, 25, 28, 41, 43, 44, 46	100.0	77.9	95.4	100.0	59.6	91.6	100.0	59.3	91.5	100.0	61.4	92.0	140.0	82.1	128.0	140.0	71.9	125.9
21, 22, 23	FIRE SUPPRESSION, PLUMBING & HVAC	97.2	56.2	80.2	97.2	54.3	79.4	97.2	53.9	79.2	97.2	57.0	80.5	102.4	107.3	104.5	102.6	75.4	91.3
26, 27, 3370	ELECTRICAL, COMMUNICATIONS & UTIL.	97.7	68.3	82.1	95.2	74.4	84.1	95.2	68.3	80.9	94.1	68.3	80.4	120.1	96.6	107.6	121.9	66.0	92.2
MF2010	WEIGHTED AVERAGE	100.1	61.5	82.8	98.0	59.2	80.6	97.5	58.0	79.8	96.8	60.0	80.4	118.9	98.8	109.9	114.2	71.2	94.9

	DIVISION	CANADA																	
		BRANDON, MANITOBA			BRANTFORD, ONTARIO			BRIDGEWATER, NOVA SCOTIA			CALGARY, ALBERTA			CAP-DE-LA-MADELEINE, QUEBEC			CHARLESBOURG, QUEBEC		
		MAT.	INST.	TOTAL	MAT.	INST.	TOTAL	MAT.	INST.	TOTAL	MAT.	INST.	TOTAL	MAT.	INST.	TOTAL	MAT.	INST.	TOTAL
015433	CONTRACTOR EQUIPMENT		103.8	103.8		101.6	101.6		101.2	101.2		106.6	106.6		102.0	102.0		102.0	102.0
0241, 31 - 34	SITE & INFRASTRUCTURE, DEMOLITION	110.6	100.3	103.5	112.7	101.8	105.2	96.9	99.0	98.4	125.5	105.6	111.8	92.5	100.7	98.2	92.5	100.7	98.2
0310	Concrete Forming & Accessories	122.4	77.4	83.4	122.9	103.7	106.2	96.4	76.5	79.2	127.5	98.4	102.2	128.1	94.1	98.6	128.1	94.1	98.6
0320	Concrete Reinforcing	161.6	56.5	107.5	175.0	87.6	130.0	148.4	49.6	97.5	148.3	73.6	109.8	148.4	83.7	115.1	148.4	83.7	115.1
0330	Cast-in-Place Concrete	151.1	81.0	124.2	163.5	117.6	145.9	168.0	76.2	132.9	200.4	108.8	165.3	131.9	103.9	121.2	131.9	103.9	121.2
03	CONCRETE	135.2	75.3	106.5	150.0	105.2	128.6	148.0	72.0	111.6	167.1	97.3	133.7	130.6	95.5	113.8	130.6	95.5	113.8
04	MASONRY	162.8	71.9	107.7	164.7	109.1	131.0	159.7	76.5	109.3	215.0	92.7	140.9	159.7	92.6	119.0	159.7	92.6	119.0
05	METALS	107.6	79.0	97.8	106.8	94.2	102.5	105.9	76.3	95.7	148.0	89.8	128.0	105.0	90.5	100.0	105.0	90.5	100.0
06	WOOD, PLASTICS & COMPOSITES	114.0	78.4	93.6	116.3	102.3	108.3	87.1	75.9	80.7	114.5	98.1	105.1	127.7	94.0	108.4	127.7	94.0	108.4
07	THERMAL & MOISTURE PROTECTION	105.0	76.3	93.6	111.3	100.9	107.2	106.3	74.4	93.6	123.3	95.8	112.4	105.2	95.7	101.4	105.2	95.7	101.4
08	OPENINGS	92.9	69.0	86.8	91.7	98.7	93.4	83.5	69.0	79.8	92.9	87.5	91.5	92.9	86.3	91.2	92.9	86.3	91.2
0920	Plaster & Gypsum Board	130.2	77.4	93.0	157.5	102.4	118.7	156.2	75.1	99.1	173.7	97.6	120.1	188.5	93.7	121.7	188.5	93.7	121.7
0950, 0980	Ceilings & Acoustic Treatment	95.2	77.4	83.4	95.2	102.4	100.0	95.2	75.1	81.9	142.4	97.6	112.7	95.2	93.7	94.2	95.2	93.7	94.2
0960	Flooring	131.9	73.0	115.0	131.9	102.7	123.5	111.6	69.9	99.6	134.2	96.2	123.3	131.9	103.4	123.7	131.9	103.4	123.7
0970, 0990	Wall Finishes & Painting/Coating	111.4	62.1	81.5	111.4	104.4	107.2	111.4	67.9	85.0	111.3	112.0	111.7	111.4	97.1	102.7	111.4	97.1	102.7
09	FINISHES	114.6	75.3	92.9	117.5	104.0	110.1	110.5	74.6	90.7	131.9	99.9	114.3	120.3	96.4	107.2	120.3	96.4	107.2
COVERS	DIVS. 10 - 14, 25, 28, 41, 43, 44, 46	140.0	74.4	126.4	140.0	84.5	128.5	140.0	73.3	126.1	140.0	105.4	132.8	140.0	94.0	130.5	140.0	94.0	130.5
21, 22, 23	FIRE SUPPRESSION, PLUMBING & HVAC	102.6	90.9	97.7	102.6	110.6	105.9	102.6	91.2	97.9	100.4	94.7	98.0	103.2	98.4	101.2	103.2	98.4	101.2
26, 27, 3370	ELECTRICAL, COMMUNICATIONS & UTIL.	121.8	74.3	96.6	119.7	96.0	107.1	125.8	68.7	95.4	127.0	86.0	105.2	119.0	77.5	96.9	119.0	77.5	96.9
MF2010	WEIGHTED AVERAGE	116.1	80.3	100.1	118.1	103.0	111.3	115.8	79.1	99.4	130.9	94.8	114.8	114.9	93.0	105.1	114.9	93.0	105.1

DIVISION		CANADA																	
		CHARLOTTETOWN, PRINCE EDWARD ISLAND			CHICOUTIMI, QUEBEC			CORNER BROOK, NEWFOUNDLAND			CORNWALL, ONTARIO			DALHOUSIE, NEW BRUNSWICK			DARTMOUTH, NOVA SCOTIA		
		MAT.	INST.	TOTAL	MAT.	INST.	TOTAL	MAT.	INST.	TOTAL	MAT.	INST.	TOTAL	MAT.	INST.	TOTAL	MAT.	INST.	TOTAL
015433	CONTRACTOR EQUIPMENT		101.1	101.1		102.0	102.0		102.4	102.4		101.6	101.6		101.4	101.4		101.2	101.2
0241, 31 - 34	SITE & INFRASTRUCTURE, DEMOLITION	104.7	96.4	99.0	92.0	99.4	97.1	112.7	97.7	102.4	110.7	101.3	104.2	94.1	97.3	96.3	104.6	99.0	100.7
0310	Concrete Forming & Accessories	96.6	62.1	66.7	128.1	91.7	96.5	103.8	65.8	70.8	120.9	95.8	99.2	102.2	68.2	72.7	96.4	76.5	79.2
0320	Concrete Reinforcing	158.7	49.6	102.5	111.9	93.5	102.4	148.4	51.5	98.5	175.0	87.3	129.8	148.4	60.0	102.9	155.0	49.6	100.7
0330	Cast-in-Place Concrete	159.3	64.3	122.9	130.1	99.0	118.2	172.3	74.5	134.8	147.0	107.1	131.7	138.2	65.2	110.2	169.8	76.2	134.0
03	CONCRETE	153.7	61.2	109.4	124.3	94.5	110.1	162.4	66.9	116.7	141.6	98.1	120.8	135.5	66.2	102.3	149.9	72.0	112.6
04	MASONRY	160.1	66.6	103.4	160.2	92.6	119.2	160.0	68.2	104.4	163.4	100.0	125.0	156.6	69.0	103.5	171.2	76.5	113.8
05	METALS	105.9	69.1	93.2	104.5	91.6	100.1	108.1	74.7	96.6	106.8	92.8	101.9	104.7	73.6	94.0	108.0	76.3	97.1
06	WOOD, PLASTICS & COMPOSITES	87.4	61.6	72.6	127.7	91.7	107.1	95.9	64.8	78.1	115.1	95.1	103.7	94.9	68.3	79.6	87.1	75.9	80.7
07	THERMAL & MOISTURE PROTECTION	105.8	63.4	88.9	105.1	94.8	101.0	108.3	65.7	91.4	111.1	95.1	104.7	108.8	66.1	91.8	106.3	74.4	93.6
08	OPENINGS	95.2	53.1	84.5	92.4	79.4	89.1	99.2	60.8	89.5	92.9	91.8	92.6	86.9	59.5	80.0	83.5	69.0	79.8
0920	Plaster & Gypsum Board	160.3	60.4	90.0	188.5	91.3	120.1	163.1	63.6	93.0	228.3	95.0	134.4	163.1	67.2	95.6	158.8	75.1	99.8
0950, 0980	Ceilings & Acoustic Treatment	108.3	60.4	76.6	95.2	91.3	92.6	96.2	63.6	74.6	98.0	95.0	96.0	96.2	67.2	77.0	104.6	75.1	85.0
0960	Flooring	111.6	65.3	98.3	131.9	103.4	123.7	115.9	59.5	99.7	131.9	101.2	123.1	115.2	74.9	103.7	111.6	69.9	99.6
0970, 0990	Wall Finishes & Painting/Coating	111.4	45.5	71.4	111.4	103.2	106.4	111.4	65.4	83.5	111.4	97.3	102.8	111.4	54.6	76.9	111.4	67.9	85.0
09	FINISHES	114.7	61.3	85.3	120.3	95.5	106.6	114.1	64.6	86.8	126.5	97.3	110.4	112.9	68.4	88.4	112.8	74.6	91.7
COVERS	DIVS. 10 - 14, 25, 28, 41, 43, 44, 46	140.0	71.4	125.7	140.0	93.4	130.3	140.0	72.2	125.9	140.0	81.7	127.9	140.0	71.9	125.8	140.0	73.3	126.1
21, 22, 23	FIRE SUPPRESSION, PLUMBING & HVAC	102.6	68.9	88.6	102.6	97.8	100.6	102.6	77.6	92.2	103.2	108.2	105.3	102.6	75.4	91.3	102.6	91.2	97.9
26, 27, 3370	ELECTRICAL, COMMUNICATIONS & UTIL.	119.0	55.9	85.4	118.3	84.8	100.5	118.9	62.2	88.8	120.4	97.1	108.0	122.1	62.4	90.4	124.0	68.7	94.6
MF2010	WEIGHTED AVERAGE	117.6	66.4	94.7	113.9	93.2	104.7	119.6	71.5	98.1	118.0	99.1	109.5	114.2	71.4	95.1	117.2	79.1	100.1

DIVISION		CANADA																	
		EDMONTON, ALBERTA			FORT MCMURRAY, ALBERTA			FREDERICTON, NEW BRUNSWICK			GATINEAU, QUEBEC			GRANBY, QUEBEC			HALIFAX, NOVA SCOTIA		
		MAT.	INST.	TOTAL	MAT.	INST.	TOTAL	MAT.	INST.	TOTAL	MAT.	INST.	TOTAL	MAT.	INST.	TOTAL	MAT.	INST.	TOTAL
015433	CONTRACTOR EQUIPMENT		106.6	106.6		103.8	103.8		101.4	101.4		102.0	102.0		102.0	102.0		101.2	101.2
0241, 31 - 34	SITE & INFRASTRUCTURE, DEMOLITION	150.7	105.6	119.6	118.6	102.4	107.4	93.7	97.3	96.2	92.3	100.7	98.1	92.9	100.7	98.3	97.2	98.8	98.3
0310	Concrete Forming & Accessories	129.8	98.4	102.5	121.1	98.6	101.5	120.3	68.6	75.4	128.1	94.0	98.5	128.1	93.9	98.4	96.6	79.0	81.3
0320	Concrete Reinforcing	148.3	73.6	109.8	161.6	73.6	116.3	149.3	60.2	103.4	157.1	83.7	119.3	157.1	83.6	119.3	155.0	62.1	107.2
0330	Cast-in-Place Concrete	208.9	108.8	170.5	217.2	108.0	175.4	131.5	65.2	106.1	130.1	103.9	120.1	134.4	103.8	122.7	171.2	82.2	137.1
03	CONCRETE	171.5	97.3	136.0	174.9	97.2	137.7	133.6	66.4	101.4	130.9	95.4	113.9	133.1	95.4	115.1	150.6	77.5	115.6
04	MASONRY	202.5	92.7	136.0	204.3	92.7	136.6	177.3	70.7	112.7	159.6	92.6	119.0	160.0	92.6	119.1	172.8	84.2	119.1
05	METALS	148.6	89.8	128.4	132.1	89.9	117.6	104.6	74.2	94.2	105.0	90.3	100.0	105.0	90.3	100.0	119.0	81.8	106.2
06	WOOD, PLASTICS & COMPOSITES	111.0	98.1	103.6	110.8	98.1	103.5	116.9	68.3	89.0	127.7	94.0	108.4	127.7	94.0	108.4	87.3	77.9	81.9
07	THERMAL & MOISTURE PROTECTION	124.4	95.8	113.0	118.8	95.2	109.4	109.0	67.2	92.4	105.2	95.7	101.4	105.1	94.1	100.7	106.8	78.5	95.6
08	OPENINGS	92.9	87.5	91.5	92.9	87.5	91.5	86.8	58.3	79.6	92.9	81.3	89.9	92.9	81.3	89.9	83.5	72.8	80.8
0920	Plaster & Gypsum Board	164.5	97.6	117.4	156.8	97.7	115.2	181.0	67.2	100.9	147.7	93.7	109.7	150.6	93.7	110.5	159.2	77.1	101.4
0950, 0980	Ceilings & Acoustic Treatment	156.8	97.6	117.6	104.6	97.7	100.0	101.8	67.2	78.9	95.2	93.7	94.2	95.2	93.7	94.2	104.6	77.1	86.4
0960	Flooring	135.3	96.2	124.1	131.9	96.2	121.6	127.5	78.2	113.3	131.9	103.4	123.7	131.9	103.4	123.7	111.6	91.8	105.9
0970, 0990	Wall Finishes & Painting/Coating	111.4	112.0	111.8	111.4	101.1	105.1	111.4	69.6	86.0	111.4	97.1	102.7	111.4	97.1	102.7	111.4	81.5	93.2
09	FINISHES	136.0	99.9	116.1	120.9	98.7	108.6	119.9	70.7	92.8	115.5	96.4	105.0	115.8	96.4	105.1	112.8	81.9	95.8
COVERS	DIVS. 10 - 14, 25, 28, 41, 43, 44, 46	140.0	105.4	132.8	140.0	105.2	132.8	140.0	71.9	125.9	140.0	94.0	130.5	140.0	94.0	130.5	140.0	74.0	126.3
21, 22, 23	FIRE SUPPRESSION, PLUMBING & HVAC	100.4	94.7	98.0	103.0	106.8	104.6	102.6	85.7	95.6	103.2	98.4	101.2	102.6	98.4	100.8	99.7	79.8	91.4
26, 27, 3370	ELECTRICAL, COMMUNICATIONS & UTIL.	117.8	86.0	100.9	113.0	91.2	101.4	132.0	82.3	105.5	119.0	77.5	96.9	119.6	77.5	97.2	124.9	81.5	101.8
MF2010	WEIGHTED AVERAGE	131.1	94.8	114.9	127.0	97.7	113.9	116.8	76.8	98.9	114.6	92.7	104.8	114.8	92.7	104.9	118.1	81.7	101.8

DIVISION		CANADA																	
		HAMILTON, ONTARIO			HULL, QUEBEC			JOLIETTE, QUEBEC			KAMLOOPS, BRITISH COLUMBIA			KINGSTON, ONTARIO			KITCHENER, ONTARIO		
		MAT.	INST.	TOTAL	MAT.	INST.	TOTAL	MAT.	INST.	TOTAL	MAT.	INST.	TOTAL	MAT.	INST.	TOTAL	MAT.	INST.	TOTAL
015433	CONTRACTOR EQUIPMENT		108.5	108.5		102.0	102.0		102.0	102.0		105.3	105.3		103.9	103.9		103.8	103.8
0241, 31 - 34	SITE & INFRASTRUCTURE, DEMOLITION	113.4	113.3	113.3	92.3	100.7	98.1	93.0	100.7	98.3	116.6	104.3	108.1	110.7	105.1	106.8	100.3	105.3	103.7
0310	Concrete Forming & Accessories	130.9	96.4	100.9	128.1	94.0	98.5	128.1	94.1	98.6	123.3	97.1	100.6	121.0	95.9	99.2	117.3	87.5	91.4
0320	Concrete Reinforcing	167.9	87.2	126.4	157.1	83.7	119.3	148.4	83.7	115.1	116.6	81.5	98.5	175.0	87.3	129.8	110.5	87.1	98.5
0330	Cast-in-Place Concrete	151.9	98.1	131.3	130.1	103.9	120.1	135.5	103.9	123.4	117.3	107.9	113.7	147.0	107.0	131.7	140.7	88.8	120.9
03	CONCRETE	143.6	95.3	120.5	130.9	95.4	113.9	132.4	95.5	114.7	138.4	97.8	119.0	141.6	98.1	120.8	122.5	88.1	106.1
04	MASONRY	180.7	102.7	133.4	159.6	92.6	119.0	160.0	92.6	119.2	166.6	101.0	126.8	169.9	100.0	127.5	161.4	94.6	120.9
05	METALS	126.0	93.2	114.7	105.0	90.3	100.0	105.0	90.5	100.0	107.4	90.4	101.6	108.5	92.7	103.1	115.4	93.0	107.7
06	WOOD, PLASTICS & COMPOSITES	115.6	95.5	104.1	127.7	94.0	108.4	127.7	94.0	108.4	98.6	95.5	96.8	115.1	95.3	103.7	109.6	86.7	96.5
07	THERMAL & MOISTURE PROTECTION	121.8	97.6	112.2	105.1	95.7	101.4	105.1	95.7	101.4	120.5	92.4	109.3	111.1	96.3	105.2	110.7	92.6	103.5
08	OPENINGS	92.9	93.8	93.1	92.9	81.3	89.9	92.9	86.3	91.2	88.7	91.2	89.3	92.9	91.5	92.5	84.0	87.2	84.8
0920	Plaster & Gypsum Board	193.9	95.4	124.5	147.7	93.7	109.7	188.5	93.7	121.7	137.8	94.9	107.6	232.2	95.1	135.7	155.2	86.3	106.7
0950, 0980	Ceilings & Acoustic Treatment	124.1	95.4	105.0	95.2	93.7	94.2	95.2	93.7	94.2	95.2	94.9	95.0	112.1	95.1	100.8	104.6	86.3	92.5
0960	Flooring	134.2	106.2	126.2	131.9	103.4	123.7	131.9	103.4	123.7	129.8	57.8	109.1	131.9	101.2	123.1	127.9	106.2	121.7
0970, 0990	Wall Finishes & Painting/Coating	111.4	103.1	106.3	111.4	97.1	102.7	111.4	97.1	102.7	111.4	88.9	97.7	111.4	89.9	98.3	111.4	93.0	100.2
09	FINISHES	129.2	99.1	112.6	115.5	96.4	105.0	120.3	96.4	107.2	116.7	89.4	101.7	130.0	96.5	111.5	117.2	91.2	102.9
COVERS	DIVS. 10 - 14, 25, 28, 41, 43, 44, 46	140.0	108.0	133.4	140.0	94.0	130.5	140.0	94.0	130.5	140.0	105.5	132.8	140.0	81.7	127.9	140.0	104.8	132.7
21, 22, 23	FIRE SUPPRESSION, PLUMBING & HVAC	100.5	86.0	94.5	102.6	98.4	100.9	102.6	98.4	100.9	102.6	102.0	102.4	103.2	108.4	105.3	99.7	91.2	96.1
26, 27, 3370	ELECTRICAL, COMMUNICATIONS & UTIL.	127.5	89.4	107.2	121.1	77.5	97.9	119.6	77.5	97.2	123.5	87.5	104.3	120.4	95.7	107.2	120.9	86.9	102.8
MF2010	WEIGHTED AVERAGE	122.6	95.4	110.5	114.6	92.7	104.8	115.1	93.0	105.2	117.2	96.2	107.8	118.9	99.2	110.1	114.0	92.1	104.2

DIVISION		CANADA																	
		LAVAL, QUEBEC			LETHBRIDGE, ALBERTA			LLOYDMINSTER, ALBERTA			LONDON, ONTARIO			MEDICINE HAT, ALBERTA			MONCTON, NEW BRUNSWICK		
		MAT.	INST.	TOTAL	MAT.	INST.	TOTAL	MAT.	INST.	TOTAL	MAT.	INST.	TOTAL	MAT.	INST.	TOTAL	MAT.	INST.	TOTAL
015433	CONTRACTOR EQUIPMENT		102.0	102.0		103.8	103.8		103.8	103.8		104.0	104.0		103.8	103.8		101.4	101.4
0241, 31 - 34	SITE & INFRASTRUCTURE, DEMOLITION	92.8	100.7	98.3	110.6	103.1	105.4	110.5	102.4	104.9	113.3	105.5	107.9	109.2	102.5	104.6	93.6	97.5	96.3
0310	Concrete Forming & Accessories	128.2	94.0	98.5	122.3	98.6	101.8	120.6	88.1	92.4	131.0	90.2	95.6	122.2	88.1	92.6	102.2	68.5	73.0
0320	Concrete Reinforcing	157.1	83.7	119.3	161.6	73.6	116.3	161.6	73.6	116.3	146.7	86.0	115.4	161.6	73.6	116.3	148.4	63.2	104.5
0330	Cast-in-Place Concrete	134.4	103.9	122.7	162.9	108.0	141.9	151.1	104.0	133.1	149.2	95.9	128.8	151.1	104.0	133.0	133.1	65.4	107.2
03	CONCRETE	133.1	95.4	115.1	147.7	97.2	123.5	141.7	91.2	117.5	139.2	91.5	116.4	141.8	91.1	117.5	133.0	67.2	101.5
04	MASONRY	159.9	92.6	119.1	177.7	92.7	126.2	159.9	85.7	114.9	182.1	99.8	132.2	159.9	85.7	114.9	158.3	69.1	104.2
05	METALS	105.1	90.3	100.0	125.5	89.9	113.3	107.6	89.8	101.5	123.6	92.4	112.9	107.6	89.7	101.5	104.7	79.4	96.0
06	WOOD, PLASTICS & COMPOSITES	127.8	94.0	108.4	114.1	98.1	104.9	110.8	87.6	97.5	115.6	88.3	99.9	114.1	87.6	98.9	94.9	68.3	79.6
07	THERMAL & MOISTURE PROTECTION	105.7	95.7	101.7	116.2	95.2	107.9	113.4	90.5	104.3	122.6	95.3	111.8	119.4	90.5	107.9	108.8	67.6	92.4
08	OPENINGS	92.9	81.3	89.9	92.9	87.5	91.5	92.9	81.8	90.1	94.0	88.7	92.7	92.9	81.8	90.1	86.9	62.8	80.8
0920	Plaster & Gypsum Board	151.1	93.7	110.6	138.8	97.7	109.9	133.8	86.9	100.8	197.3	88.0	120.3	136.5	86.9	101.5	163.1	67.2	95.6
0950, 0980	Ceilings & Acoustic Treatment	95.2	93.7	94.2	103.7	97.7	99.7	95.2	86.9	89.7	136.3	88.0	104.3	95.2	86.9	89.7	96.2	67.2	77.0
0960	Flooring	131.9	103.4	123.7	131.9	96.2	121.6	131.9	96.2	121.6	134.4	106.2	126.3	131.9	96.2	121.6	115.2	74.9	103.7
0970, 0990	Wall Finishes & Painting/Coating	111.4	97.1	102.7	111.3	110.3	110.7	111.4	86.0	96.0	111.4	99.8	104.3	111.3	86.0	95.9	111.4	54.6	76.9
09	FINISHES	115.9	96.4	105.2	117.6	99.7	107.7	115.3	89.0	100.8	132.3	94.0	111.2	115.4	89.0	100.9	112.9	68.4	88.4
COVERS	DIVS. 10 - 14, 25, 28, 41, 43, 44, 46	140.0	94.0	130.5	140.0	105.2	132.8	140.0	101.5	132.0	140.0	106.6	133.1	140.0	101.5	132.0	140.0	71.9	125.9
21, 22, 23	FIRE SUPPRESSION, PLUMBING & HVAC	99.7	98.4	99.1	102.8	103.2	103.0	103.2	103.2	103.2	100.4	83.1	93.2	102.6	99.6	101.4	102.6	75.5	91.3
26, 27, 3370	ELECTRICAL, COMMUNICATIONS & UTIL.	120.8	77.5	97.8	114.5	91.2	102.1	112.0	91.2	100.9	119.5	86.9	102.2	112.0	91.2	100.9	127.2	66.0	94.7
MF2010	WEIGHTED AVERAGE	114.3	92.7	104.6	120.9	97.1	110.3	116.2	93.6	106.1	121.4	91.9	108.2	116.3	92.8	105.8	114.5	72.8	95.8

DIVISION		CANADA																	
		MONTREAL, QUEBEC			MOOSE JAW, SASKATCHEWAN			NEW GLASGOW, NOVA SCOTIA			NEWCASTLE, NEW BRUNSWICK			NORTH BAY, ONTARIO			OSHAWA, ONTARIO		
		MAT.	INST.	TOTAL	MAT.	INST.	TOTAL	MAT.	INST.	TOTAL	MAT.	INST.	TOTAL	MAT.	INST.	TOTAL	MAT.	INST.	TOTAL
015433	CONTRACTOR EQUIPMENT		103.6	103.6		100.0	100.0		101.2	101.2		101.4	101.4		101.6	101.6		103.8	103.8
0241, 31 - 34	SITE & INFRASTRUCTURE, DEMOLITION	99.6	99.7	99.7	110.3	96.7	100.9	97.1	99.0	98.4	94.2	97.3	96.3	112.7	100.8	104.5	112.6	105.1	107.4
0310	Concrete Forming & Accessories	136.0	92.0	97.8	106.1	64.7	70.1	96.4	76.5	79.2	102.2	68.3	72.7	122.9	93.1	97.0	123.1	95.3	99.0
0320	Concrete Reinforcing	153.4	93.6	122.6	114.0	65.7	89.1	148.4	49.6	97.5	148.4	60.0	102.9	175.0	86.8	129.6	175.0	89.6	131.0
0330	Cast-in-Place Concrete	180.9	100.4	150.1	146.6	75.1	119.2	169.8	76.2	134.0	138.2	65.2	110.2	163.5	92.4	136.3	162.8	93.3	136.2
03	CONCRETE	156.4	95.1	127.1	125.2	69.0	98.3	148.9	72.0	112.1	135.5	66.2	102.3	150.0	91.8	122.1	149.7	93.6	122.8
04	MASONRY	167.3	92.6	122.1	158.5	68.4	103.9	159.8	76.5	109.3	158.7	69.0	104.4	164.7	95.6	122.8	164.3	100.4	125.6
05	METALS	122.2	92.1	111.8	103.9	76.1	94.3	105.9	76.3	95.7	104.7	73.6	94.0	106.8	92.4	101.8	106.8	94.7	102.7
06	WOOD, PLASTICS & COMPOSITES	127.8	92.0	107.3	96.4	63.1	77.3	87.1	75.9	80.7	94.9	68.3	79.6	116.3	93.6	103.3	116.5	94.2	103.7
07	THERMAL & MOISTURE PROTECTION	118.6	95.3	109.3	104.6	68.2	90.1	106.3	74.4	93.6	108.8	66.1	91.8	111.3	91.2	103.3	111.7	93.0	104.3
08	OPENINGS	92.9	81.4	90.0	87.7	60.4	80.8	83.5	69.0	79.8	86.9	59.5	80.0	91.7	89.1	91.0	91.7	93.6	92.2
0920	Plaster & Gypsum Board	168.2	91.3	114.1	130.8	61.9	82.3	156.2	75.1	99.1	163.1	67.2	95.6	157.5	93.4	112.4	159.0	94.0	113.3
0950, 0980	Ceilings & Acoustic Treatment	113.8	91.3	98.9	95.2	61.9	73.2	95.2	75.1	81.9	96.2	67.2	77.0	95.2	93.4	94.0	99.0	94.0	95.7
0960	Flooring	133.6	103.4	124.9	120.6	65.7	104.9	111.6	69.9	99.6	115.2	74.9	103.7	131.9	101.2	123.1	131.9	109.0	125.3
0970, 0990	Wall Finishes & Painting/Coating	111.4	103.2	106.4	111.4	70.7	86.7	111.4	67.9	85.0	111.4	54.6	76.9	111.4	96.5	102.4	111.4	107.2	108.8
09	FINISHES	122.6	95.7	107.7	111.4	65.2	85.9	110.5	74.6	90.7	112.9	68.4	88.4	117.5	95.4	105.3	118.4	99.1	107.7
COVERS	DIVS. 10 - 14, 25, 28, 41, 43, 44, 46	140.0	94.0	130.4	140.0	71.1	125.7	140.0	73.3	126.1	140.0	71.9	125.9	140.0	80.2	127.6	140.0	107.5	133.3
21, 22, 23	FIRE SUPPRESSION, PLUMBING & HVAC	100.4	97.9	99.4	103.2	82.2	94.5	102.6	91.2	97.9	102.6	75.4	91.3	102.6	106.1	104.1	99.7	108.8	103.5
26, 27, 3370	ELECTRICAL, COMMUNICATIONS & UTIL.	128.3	84.8	105.1	122.1	67.0	92.8	119.3	68.7	92.4	121.1	66.0	91.8	119.7	97.0	107.6	122.2	96.6	108.6
MF2010	WEIGHTED AVERAGE	122.0	93.5	109.3	113.4	73.6	95.6	115.3	79.1	99.1	114.2	71.9	95.3	118.1	96.7	108.5	117.7	100.0	109.8

DIVISION		CANADA																	
		OTTAWA, ONTARIO			OWEN SOUND, ONTARIO			PETERBOROUGH, ONTARIO			PORTAGE LA PRAIRIE, MANITOBA			PRINCE ALBERT, SASKATCHEWAN			PRINCE GEORGE, BRITISH COLUMBIA		
		MAT.	INST.	TOTAL	MAT.	INST.	TOTAL	MAT.	INST.	TOTAL	MAT.	INST.	TOTAL	MAT.	INST.	TOTAL	MAT.	INST.	TOTAL
015433	CONTRACTOR EQUIPMENT		103.8	103.8		101.6	101.6		101.6	101.6		103.8	103.8		100.0	100.0		105.3	105.3
0241, 31 - 34	SITE & INFRASTRUCTURE, DEMOLITION	111.4	105.4	107.2	115.0	101.3	105.5	112.7	101.2	104.8	111.0	100.2	103.6	105.0	96.9	99.4	120.3	104.3	109.3
0310	Concrete Forming & Accessories	131.9	92.8	97.9	123.1	91.5	95.7	122.9	94.2	98.0	122.4	76.9	82.9	106.1	64.4	70.0	111.4	91.5	94.1
0320	Concrete Reinforcing	160.4	85.9	122.0	180.1	89.0	133.2	175.0	87.4	129.9	161.6	56.5	107.5	119.2	65.6	91.6	116.6	81.5	98.5
0330	Cast-in-Place Concrete	167.3	95.1	139.6	167.6	88.1	137.1	163.5	94.2	137.0	151.1	80.3	124.0	132.8	75.0	110.7	147.1	107.9	132.1
03	CONCRETE	150.3	92.4	122.6	152.8	90.0	122.7	150.0	93.0	122.7	135.2	74.8	106.3	119.0	68.9	95.0	152.6	95.3	125.2
04	MASONRY	181.1	99.8	131.8	168.4	101.7	127.9	164.7	102.8	127.2	162.8	70.8	107.0	157.5	68.4	103.5	169.0	101.0	127.8
05	METALS	124.4	92.3	113.4	107.4	93.3	102.5	106.8	92.9	102.0	107.6	79.0	97.8	103.9	75.9	94.3	107.4	90.5	101.6
06	WOOD, PLASTICS & COMPOSITES	115.4	92.0	102.0	118.4	90.1	102.2	116.3	92.2	102.5	114.0	78.4	93.6	96.4	63.1	77.3	98.6	87.7	92.3
07	THERMAL & MOISTURE PROTECTION	124.8	94.8	112.9	111.4	92.0	103.7	111.3	97.1	105.7	105.0	75.8	93.4	104.5	66.9	89.6	114.7	91.6	105.5
08	OPENINGS	92.9	90.7	92.3	91.8	88.4	91.0	91.7	91.0	91.5	92.9	69.0	86.8	86.3	60.4	79.7	88.7	87.0	88.3
0920	Plaster & Gypsum Board	234.1	91.8	133.9	157.5	89.8	109.8	157.5	91.9	111.3	130.2	77.4	93.0	130.8	61.9	82.3	134.8	86.9	101.1
0950, 0980	Ceilings & Acoustic Treatment	133.5	91.8	105.8	95.2	89.8	91.6	95.2	91.9	93.0	95.2	77.4	83.4	95.2	61.9	73.2	95.2	86.9	89.7
0960	Flooring	134.2	101.2	124.7	132.0	102.7	123.5	131.9	101.2	123.1	131.9	73.0	115.0	120.6	65.7	104.9	126.8	79.2	113.1
0970, 0990	Wall Finishes & Painting/Coating	111.4	94.6	101.2	112.2	95.1	101.8	111.4	99.0	103.8	111.4	62.1	81.5	111.4	60.3	80.4	111.4	88.9	97.7
09	FINISHES	135.4	94.7	113.0	118.8	94.1	105.2	117.5	96.1	105.7	114.6	75.1	92.8	111.4	64.0	85.2	115.5	88.5	100.6
COVERS	DIVS. 10 - 14, 25, 28, 41, 43, 44, 46	140.0	104.4	132.6	140.0	80.7	127.7	140.0	81.9	127.9	140.0	73.9	126.3	140.0	71.1	125.7	140.0	104.7	132.7
21, 22, 23	FIRE SUPPRESSION, PLUMBING & HVAC	100.5	83.6	93.4	102.4	106.0	103.9	102.6	109.9	105.7	102.6	90.3	97.5	103.2	74.1	91.1	102.6	102.1	102.4
26, 27, 3370	ELECTRICAL, COMMUNICATIONS & UTIL.	119.5	87.8	102.7	121.6	95.7	107.8	119.7	96.6	107.4	121.8	64.1	91.1	122.1	67.0	92.8	119.8	87.5	102.6
MF2010	WEIGHTED AVERAGE	123.0	92.3	109.3	119.1	96.8	109.1	118.1	98.8	109.5	116.1	78.5	99.3	112.3	71.6	94.1	118.5	95.4	108.2

	DIVISION	CANADA																	
		QUEBEC, QUEBEC			RED DEER, ALBERTA			REGINA, SASKATCHEWAN			RIMOUSKI, QUEBEC			ROUYN-NORANDA, QUEBEC			SAINT HYACINTHE, QUEBEC		
		MAT.	INST.	TOTAL	MAT.	INST.	TOTAL	MAT.	INST.	TOTAL	MAT.	INST.	TOTAL	MAT.	INST.	TOTAL	MAT.	INST.	TOTAL
015433	**CONTRACTOR EQUIPMENT**		104.0	104.0		103.8	103.8		100.0	100.0		102.0	102.0		102.0	102.0		102.0	102.0
0241, 31 - 34	**SITE & INFRASTRUCTURE, DEMOLITION**	97.4	99.8	99.0	109.2	102.5	104.6	111.6	98.4	102.5	92.8	99.4	97.4	92.3	100.7	98.1	92.8	100.7	98.3
0310	Concrete Forming & Accessories	137.5	92.2	98.2	139.4	88.1	94.9	106.3	90.0	92.2	128.1	91.7	96.5	128.1	94.0	98.5	128.1	94.0	98.5
0320	Concrete Reinforcing	145.4	93.6	118.7	161.6	73.6	116.3	126.2	79.0	101.8	111.9	93.5	102.4	157.1	83.7	119.3	157.1	83.7	119.3
0330	Cast-in-Place Concrete	148.2	100.8	130.0	151.1	104.0	133.0	156.2	95.4	132.9	136.9	99.0	122.4	130.1	103.9	120.1	134.4	103.9	122.7
03	**CONCRETE**	138.9	95.4	118.1	142.8	91.1	118.1	131.8	89.7	111.7	127.7	94.5	111.8	130.9	95.4	113.9	133.1	95.4	115.1
04	**MASONRY**	174.5	92.6	124.9	159.9	85.7	114.9	165.6	93.1	121.7	159.4	92.6	118.9	159.6	92.6	119.0	159.9	92.6	119.1
05	**METALS**	123.8	92.4	113.0	107.6	89.7	101.5	104.8	84.2	97.7	104.5	91.6	100.1	105.0	90.3	100.0	105.0	90.3	100.0
06	**WOOD, PLASTICS & COMPOSITES**	128.6	92.0	107.6	114.1	87.6	98.9	96.6	90.8	93.3	127.7	91.7	107.1	127.7	94.0	108.4	127.7	94.0	108.4
07	**THERMAL & MOISTURE PROTECTION**	120.8	95.5	110.7	129.1	90.5	113.8	106.1	85.9	98.1	105.1	94.8	101.0	105.1	95.7	101.4	105.4	95.7	101.6
08	**OPENINGS**	92.9	89.6	92.1	92.9	81.8	90.1	87.7	79.2	85.6	92.4	79.4	89.1	92.9	81.3	89.9	92.9	81.3	89.9
0920	Plaster & Gypsum Board	196.6	91.3	122.5	136.5	86.9	101.5	166.4	90.5	113.0	188.0	91.3	119.9	147.2	93.7	109.5	150.1	93.7	110.4
0950, 0980	Ceilings & Acoustic Treatment	119.1	91.3	100.7	95.2	86.9	89.7	116.8	90.5	99.4	93.4	91.3	92.0	93.4	93.7	93.6	93.4	93.7	93.6
0960	Flooring	134.2	103.4	125.3	134.2	96.2	123.3	120.6	68.0	105.5	131.9	103.4	123.7	131.9	103.4	123.7	131.9	103.4	123.7
0970, 0990	Wall Finishes & Painting/Coating	111.8	103.2	106.6	111.3	86.0	95.9	111.4	89.2	97.9	111.4	103.2	106.4	111.4	97.1	102.7	111.4	97.1	102.7
09	**FINISHES**	127.5	95.7	110.0	116.1	89.0	101.2	120.5	86.4	101.7	119.9	95.5	106.4	115.0	96.4	104.8	115.4	96.4	104.9
COVERS	**DIVS. 10 - 14, 25, 28, 41, 43, 44, 46**	140.0	94.1	130.5	140.0	101.5	132.0	140.0	77.9	127.1	140.0	93.4	130.3	140.0	94.0	130.5	140.0	94.0	130.5
21, 22, 23	**FIRE SUPPRESSION, PLUMBING & HVAC**	100.4	97.9	99.4	102.6	99.6	101.4	99.8	84.2	93.3	102.6	97.8	100.6	102.6	98.4	100.9	98.4	98.4	98.4
26, 27, 3370	**ELECTRICAL, COMMUNICATIONS & UTIL.**	119.6	84.8	101.1	112.0	91.2	100.9	123.6	89.2	105.3	119.6	84.8	101.1	119.6	77.5	97.2	120.4	77.5	97.6
MF2010	**WEIGHTED AVERAGE**	120.1	94.0	108.4	116.8	92.8	106.1	114.9	87.7	102.7	114.4	93.2	104.9	114.5	92.7	104.7	113.9	92.7	104.4

	DIVISION	CANADA																	
		SAINT JOHN, NEW BRUNSWICK			SARNIA, ONTARIO			SASKATOON, SASKATCHEWAN			SAULT STE MARIE, ONTARIO			SHERBROOKE, QUEBEC			SOREL, QUEBEC		
		MAT.	INST.	TOTAL	MAT.	INST.	TOTAL	MAT.	INST.	TOTAL	MAT.	INST.	TOTAL	MAT.	INST.	TOTAL	MAT.	INST.	TOTAL
015433	**CONTRACTOR EQUIPMENT**		101.4	101.4		101.6	101.6		100.0	100.0		101.6	101.6		102.0	102.0		102.0	102.0
0241, 31 - 34	**SITE & INFRASTRUCTURE, DEMOLITION**	94.4	98.3	97.1	111.2	101.3	104.4	106.1	98.4	100.8	101.2	100.9	101.0	92.9	100.7	98.3	93.0	100.7	98.3
0310	Concrete Forming & Accessories	120.3	65.2	72.5	121.7	101.9	104.6	106.3	90.0	92.2	111.4	89.9	92.8	128.1	94.0	98.5	128.1	94.1	98.6
0320	Concrete Reinforcing	148.4	63.4	104.6	123.5	88.8	105.6	119.2	79.0	98.5	112.4	87.0	99.3	157.1	83.7	119.3	148.4	83.7	115.1
0330	Cast-in-Place Concrete	136.1	65.2	109.0	150.8	108.7	134.7	142.0	95.4	124.1	135.5	92.5	119.0	134.4	103.9	122.7	135.5	103.9	123.4
03	**CONCRETE**	135.8	65.7	102.3	136.0	101.6	119.5	123.7	89.7	107.4	119.8	90.4	105.7	133.1	95.4	115.1	132.4	95.5	114.7
04	**MASONRY**	176.7	71.3	112.8	175.2	105.8	133.1	165.3	93.1	121.6	161.2	97.2	122.4	159.9	92.6	119.1	160.0	92.6	119.1
05	**METALS**	104.6	80.0	96.2	106.7	93.3	102.1	104.9	84.2	97.7	105.9	91.5	101.0	105.0	90.3	100.0	105.0	90.5	100.0
06	**WOOD, PLASTICS & COMPOSITES**	116.9	63.7	86.4	115.4	101.2	107.2	94.9	90.8	92.6	103.0	88.9	94.9	127.7	94.0	108.4	127.7	94.0	108.4
07	**THERMAL & MOISTURE PROTECTION**	109.0	66.5	92.1	111.3	100.9	107.2	105.2	85.9	97.5	110.2	92.7	103.2	105.2	95.7	101.4	105.1	95.7	101.4
08	**OPENINGS**	86.8	59.1	79.8	94.0	95.2	94.3	86.8	79.2	84.9	84.4	87.2	85.1	92.9	81.3	89.9	92.9	86.3	91.2
0920	Plaster & Gypsum Board	181.0	62.5	97.6	179.8	101.2	124.5	146.0	90.5	106.9	141.9	88.6	104.4	150.1	93.7	110.4	188.0	93.7	121.6
0950, 0980	Ceilings & Acoustic Treatment	101.8	62.5	75.7	99.0	101.2	100.5	116.8	90.5	99.4	95.2	88.6	90.8	93.4	93.7	93.6	93.4	93.7	93.6
0960	Flooring	127.5	74.9	112.4	131.9	110.7	125.8	120.6	68.0	105.5	123.8	104.7	118.3	131.9	103.4	123.7	131.9	103.4	123.7
0970, 0990	Wall Finishes & Painting/Coating	111.4	73.2	88.2	111.4	111.9	111.7	111.4	89.2	97.9	111.4	103.6	106.7	111.4	97.1	102.7	111.4	97.1	102.7
09	**FINISHES**	119.9	67.6	91.1	121.0	105.1	112.2	117.8	86.4	100.5	112.4	94.0	102.3	115.4	96.4	104.9	119.9	96.4	107.0
COVERS	**DIVS. 10 - 14, 25, 28, 41, 43, 44, 46**	140.0	71.3	125.7	140.0	83.5	128.3	140.0	77.9	127.1	140.0	106.0	132.9	140.0	94.0	130.5	140.0	94.0	130.5
21, 22, 23	**FIRE SUPPRESSION, PLUMBING & HVAC**	102.6	80.8	93.5	102.6	116.5	108.4	99.7	84.2	93.3	102.6	102.4	102.5	103.2	98.4	101.2	102.6	98.4	100.9
26, 27, 3370	**ELECTRICAL, COMMUNICATIONS & UTIL.**	130.6	86.4	107.1	122.9	99.0	110.2	123.9	89.2	105.4	121.5	97.0	108.4	119.7	77.5	97.2	119.6	77.5	97.2
MF2010	**WEIGHTED AVERAGE**	116.9	76.5	98.8	117.8	103.6	111.4	113.4	87.7	101.9	112.7	96.3	105.4	114.9	92.7	105.0	115.1	93.0	105.2

	DIVISION	CANADA																	
		ST CATHARINES, ONTARIO			ST JEROME, QUEBEC			ST JOHNS, NEWFOUNDLAND			SUDBURY, ONTARIO			SUMMERSIDE, PRINCE EDWARD ISLAND			SYDNEY, NOVA SCOTIA		
		MAT.	INST.	TOTAL	MAT.	INST.	TOTAL	MAT.	INST.	TOTAL	MAT.	INST.	TOTAL	MAT.	INST.	TOTAL	MAT.	INST.	TOTAL
015433	**CONTRACTOR EQUIPMENT**		101.6	101.6		102.0	102.0		102.4	102.4		101.6	101.6		101.1	101.1		101.2	101.2
0241, 31 - 34	**SITE & INFRASTRUCTURE, DEMOLITION**	101.1	102.0	101.8	92.3	100.7	98.1	113.1	99.2	103.5	101.2	101.5	101.4	105.0	96.4	99.1	92.3	99.0	96.9
0310	Concrete Forming & Accessories	115.2	96.1	98.6	128.1	94.0	98.5	103.8	74.0	78.0	111.4	89.9	92.8	96.6	62.2	66.8	96.4	76.5	79.2
0320	Concrete Reinforcing	111.5	87.2	98.9	157.1	83.7	119.3	164.6	65.0	113.3	112.4	85.6	98.6	148.4	49.7	97.5	148.4	49.6	97.5
0330	Cast-in-Place Concrete	134.4	97.9	120.4	130.1	103.9	120.1	172.3	86.9	139.6	135.5	94.2	119.7	161.1	64.3	124.1	130.9	76.2	109.9
03	**CONCRETE**	119.3	95.0	107.7	130.9	95.4	113.9	164.7	77.3	122.9	119.8	90.8	105.9	153.1	61.3	109.2	129.4	72.0	101.9
04	**MASONRY**	160.8	102.7	125.6	159.6	92.6	119.0	162.9	79.0	112.0	160.9	97.7	122.6	159.1	66.6	103.0	156.7	76.5	108.1
05	**METALS**	105.9	92.8	101.4	105.0	90.3	100.0	108.1	79.8	98.3	105.9	92.4	101.2	105.9	69.3	93.3	105.9	76.3	95.7
06	**WOOD, PLASTICS & COMPOSITES**	107.1	95.5	100.5	127.7	94.0	108.4	96.0	72.7	82.6	103.0	88.9	94.9	87.4	61.6	72.6	87.1	75.9	80.7
07	**THERMAL & MOISTURE PROTECTION**	110.7	98.3	105.8	105.1	95.7	101.4	109.7	76.8	96.6	110.2	92.7	103.2	105.7	64.9	89.5	106.3	74.4	93.6
08	**OPENINGS**	83.4	92.3	85.7	92.9	81.3	89.9	99.2	67.5	91.2	84.4	87.2	85.1	95.2	53.1	84.5	83.5	69.0	79.8
0920	Plaster & Gypsum Board	136.9	95.4	107.7	147.2	93.7	109.5	165.6	71.7	99.5	141.9	88.6	104.4	156.6	60.4	88.9	156.2	75.1	99.1
0950, 0980	Ceilings & Acoustic Treatment	99.0	95.4	96.6	93.4	93.7	93.6	103.7	71.7	82.5	95.2	88.6	90.8	95.2	60.4	72.2	95.2	75.1	81.9
0960	Flooring	126.4	102.7	119.6	131.9	103.4	123.7	115.9	61.5	100.3	123.8	104.7	118.3	111.6	65.3	98.3	111.6	69.9	99.6
0970, 0990	Wall Finishes & Painting/Coating	111.4	103.1	106.3	111.4	97.1	102.7	111.4	76.3	90.1	111.4	94.0	100.9	111.4	45.5	71.4	111.4	67.9	85.0
09	**FINISHES**	113.3	98.5	105.2	115.0	96.4	104.8	116.0	71.9	91.7	112.4	93.1	101.7	111.5	61.3	83.8	110.5	74.6	90.7
COVERS	**DIVS. 10 - 14, 25, 28, 41, 43, 44, 46**	140.0	81.8	127.9	140.0	94.0	130.5	140.0	74.7	126.4	140.0	106.1	133.0	140.0	71.4	125.7	140.0	73.3	126.1
21, 22, 23	**FIRE SUPPRESSION, PLUMBING & HVAC**	99.7	84.8	93.5	102.6	98.4	100.9	99.9	72.9	88.7	102.6	87.9	96.5	102.6	68.9	88.6	102.6	91.2	97.9
26, 27, 3370	**ELECTRICAL, COMMUNICATIONS & UTIL.**	122.7	87.9	104.2	120.3	77.5	97.6	119.7	73.8	95.3	119.7	87.9	102.8	118.2	55.9	85.1	119.3	68.7	92.4
MF2010	**WEIGHTED AVERAGE**	112.1	93.1	103.6	114.5	92.7	104.8	119.7	76.9	100.6	112.5	92.1	103.4	117.1	66.4	94.5	112.6	79.1	97.6

		CANADA																	
DIVISION		THUNDER BAY, ONTARIO			TIMMINS, ONTARIO			TORONTO, ONTARIO			TROIS RIVIERES, QUEBEC			TRURO, NOVA SCOTIA			VANCOUVER, BRITISH COLUMBIA		
		MAT.	INST.	TOTAL	MAT.	INST.	TOTAL	MAT.	INST.	TOTAL	MAT.	INST.	TOTAL	MAT.	INST.	TOTAL	MAT.	INST.	TOTAL
015433	CONTRACTOR EQUIPMENT		101.6	101.6		101.6	101.6		104.0	104.0		102.0	102.0		101.2	101.2		112.1	112.1
0241, 31 - 34	SITE & INFRASTRUCTURE, DEMOLITION	106.5	102.0	103.4	112.7	100.8	104.5	138.2	106.1	116.1	93.0	100.7	98.3	97.1	99.0	98.4	120.5	108.6	112.3
0310	Concrete Forming & Accessories	123.0	93.6	97.5	122.9	93.1	97.0	132.3	101.7	105.7	128.1	94.1	98.6	96.4	76.6	79.2	126.6	88.9	93.9
0320	Concrete Reinforcing	99.7	86.6	92.9	175.0	86.8	129.6	167.3	87.8	126.4	148.4	83.7	115.1	148.4	49.6	97.5	157.2	80.9	117.9
0330	Cast-in-Place Concrete	148.0	96.3	128.2	163.5	92.4	136.3	153.2	117.9	139.7	135.5	103.9	123.4	169.8	76.2	134.0	162.8	103.4	140.1
03	CONCRETE	128.3	93.3	111.5	150.0	91.8	122.1	144.3	104.4	125.2	132.4	95.5	114.7	148.9	72.0	112.1	153.2	92.6	124.2
04	MASONRY	161.7	101.9	125.4	164.7	95.6	122.8	194.1	107.3	141.5	160.0	92.6	119.1	159.8	76.5	109.3	176.7	95.9	127.7
05	METALS	105.7	91.7	100.9	106.8	92.4	101.8	126.1	93.8	115.0	105.0	90.5	100.0	105.9	76.3	95.7	162.8	91.8	138.3
06	WOOD, PLASTICS & COMPOSITES	116.5	92.3	102.6	116.3	93.6	103.3	116.5	100.2	107.2	127.7	94.0	108.4	87.1	75.9	80.7	114.9	87.0	98.9
07	THERMAL & MOISTURE PROTECTION	110.9	95.0	104.6	111.3	91.2	103.3	122.7	102.5	114.7	105.1	95.7	101.4	106.3	74.4	93.6	133.0	90.6	116.2
08	OPENINGS	82.4	89.9	84.3	91.7	89.1	91.0	91.7	98.4	93.4	92.9	86.3	91.2	83.5	69.0	79.8	94.6	86.0	92.4
0920	Plaster & Gypsum Board	172.5	92.0	115.8	157.5	93.4	112.4	180.3	100.2	123.9	188.0	93.7	121.6	156.2	75.1	99.1	155.1	86.1	106.5
0950, 0980	Ceilings & Acoustic Treatment	95.2	92.0	93.1	95.2	93.4	94.0	131.7	100.2	110.8	93.4	93.7	93.6	95.2	75.1	81.9	148.7	86.1	107.2
0960	Flooring	131.9	110.0	125.6	131.9	101.2	123.1	135.3	112.6	128.8	131.9	103.4	123.7	111.6	69.9	99.6	134.6	99.7	124.5
0970, 0990	Wall Finishes & Painting/Coating	111.4	91.4	99.3	111.4	96.5	102.4	113.4	107.2	109.6	111.4	97.1	102.7	111.4	67.9	85.0	111.3	83.1	94.2
09	FINISHES	118.9	96.7	106.6	117.5	95.4	105.3	131.6	104.4	116.6	119.9	96.4	107.0	110.5	74.6	90.7	130.8	89.9	108.2
COVERS	DIVS. 10 - 14, 25, 28, 41, 43, 44, 46	140.0	81.6	127.9	140.0	80.2	127.6	140.0	109.8	133.7	140.0	94.0	130.5	140.0	73.3	126.1	140.0	103.3	132.4
21, 22, 23	FIRE SUPPRESSION, PLUMBING & HVAC	99.7	84.6	93.4	102.6	106.1	104.1	100.4	99.9	100.2	102.6	98.4	100.9	102.6	91.2	97.9	100.4	86.6	94.7
26, 27, 3370	ELECTRICAL, COMMUNICATIONS & UTIL.	120.9	86.3	102.5	121.5	97.0	108.4	121.1	90.1	104.6	119.6	77.5	97.2	119.3	68.7	92.4	121.8	79.4	99.3
MF2010	WEIGHTED AVERAGE	113.6	91.9	103.9	118.3	96.7	108.6	123.6	100.7	113.4	115.1	93.0	105.2	115.3	79.1	99.1	129.1	90.7	111.9

		CANADA																	
DIVISION		VICTORIA, BRITISH COLUMBIA			WHITEHORSE, YUKON			WINDSOR, ONTARIO			WINNIPEG, MANITOBA			YARMOUTH, NOVA SCOTIA			YELLOWKNIFE, NWT		
		MAT.	INST.	TOTAL	MAT.	INST.	TOTAL	MAT.	INST.	TOTAL	MAT.	INST.	TOTAL	MAT.	INST.	TOTAL	MAT.	INST.	TOTAL
015433	CONTRACTOR EQUIPMENT		108.6	108.6		101.4	101.4		101.6	101.6		105.9	105.9		101.2	101.2		101.2	101.2
0241, 31 - 34	SITE & INFRASTRUCTURE, DEMOLITION	120.3	108.1	111.9	106.7	97.5	100.3	96.2	101.9	100.1	119.0	101.7	107.1	96.9	99.0	98.4	128.7	102.2	110.4
0310	Concrete Forming & Accessories	111.4	88.3	91.3	106.2	63.6	69.2	123.0	91.9	96.0	131.5	71.2	79.2	96.4	76.5	79.2	106.3	87.2	89.7
0320	Concrete Reinforcing	116.6	80.8	98.2	119.2	63.7	90.6	109.2	85.9	97.2	148.3	59.0	102.3	148.4	49.6	97.5	119.2	67.6	92.6
0330	Cast-in-Place Concrete	147.1	102.7	130.1	146.6	74.2	118.8	137.7	97.2	122.2	190.9	76.8	147.2	168.0	76.2	132.9	243.2	94.7	186.4
03	CONCRETE	152.6	91.9	123.6	126.0	67.9	98.2	121.1	92.7	107.5	162.5	71.3	118.9	148.0	72.0	111.6	174.5	86.2	132.3
04	MASONRY	168.1	95.9	124.4	157.9	66.7	102.6	161.0	100.7	124.4	180.3	72.4	114.9	159.7	76.5	109.3	166.4	80.9	114.6
05	METALS	107.4	87.6	100.6	105.1	75.6	95.0	105.8	92.2	101.1	156.2	75.6	128.5	105.9	76.3	95.7	106.0	81.8	97.7
06	WOOD, PLASTICS & COMPOSITES	98.6	86.9	91.9	96.4	62.2	76.8	116.5	90.8	101.7	114.5	72.1	90.1	87.1	75.9	80.7	96.4	88.9	92.1
07	THERMAL & MOISTURE PROTECTION	114.7	90.4	105.1	104.5	65.9	89.1	110.7	94.0	104.1	116.0	73.6	99.1	106.3	74.4	93.6	107.4	84.7	98.4
08	OPENINGS	88.7	82.0	87.0	86.3	59.0	79.4	82.1	89.6	84.0	92.9	63.9	85.5	83.5	69.0	79.8	86.3	74.6	83.3
0920	Plaster & Gypsum Board	134.8	86.1	100.5	119.2	60.9	78.1	160.8	90.5	111.3	164.3	70.7	98.4	156.2	75.1	99.1	169.1	88.5	112.3
0950, 0980	Ceilings & Acoustic Treatment	95.2	86.1	89.2	95.2	60.9	72.5	95.2	90.5	92.1	136.4	70.7	92.9	95.2	75.1	81.9	128.0	88.5	101.8
0960	Flooring	126.8	79.2	113.1	120.6	63.7	104.3	131.9	107.1	124.8	133.0	75.5	116.5	111.6	69.9	99.6	120.6	97.8	114.1
0970, 0990	Wall Finishes & Painting/Coating	111.4	83.1	94.2	111.4	59.3	79.8	111.4	95.9	102.0	111.4	52.7	75.8	111.4	67.9	85.0	111.4	86.4	96.2
09	FINISHES	115.5	86.3	99.4	110.0	62.9	84.0	117.1	95.4	105.1	128.0	70.4	96.3	110.5	74.6	90.7	122.9	88.5	103.9
COVERS	DIVS. 10 - 14, 25, 28, 41, 43, 44, 46	140.0	77.2	127.0	140.0	70.3	125.5	140.0	80.9	127.7	140.0	73.0	126.1	140.0	73.3	126.1	140.0	75.0	126.5
21, 22, 23	FIRE SUPPRESSION, PLUMBING & HVAC	102.6	86.6	95.9	102.6	80.8	93.5	99.7	84.2	93.3	100.5	69.5	87.6	102.6	91.2	97.9	102.9	103.2	103.1
26, 27, 3370	ELECTRICAL, COMMUNICATIONS & UTIL.	120.9	79.4	98.8	122.5	66.4	92.7	125.6	87.9	105.5	125.6	69.8	95.9	119.3	68.7	92.4	122.8	92.5	106.7
MF2010	WEIGHTED AVERAGE	118.5	88.7	105.2	113.2	72.4	95.0	112.7	91.6	103.3	128.8	73.5	104.1	115.1	79.1	99.0	121.5	90.5	107.7

Location Factors

Costs shown in RSMeans cost data publications are based on national averages for materials and installation. To adjust these costs to a specific location, simply multiply the base cost by the factor and divide by 100 for that city. The data is arranged alphabetically by state and postal zip code numbers. For a city not listed, use the factor for a nearby city with similar economic characteristics.

STATE/ZIP	CITY	MAT.	INST.	TOTAL
ALABAMA				
350-352	Birmingham	97.0	75.6	87.4
354	Tuscaloosa	95.8	59.2	79.4
355	Jasper	96.0	58.2	79.1
356	Decatur	95.8	60.8	80.1
357-358	Huntsville	95.8	69.0	83.8
359	Gadsden	95.6	58.2	78.9
360-361	Montgomery	96.8	57.2	79.1
362	Anniston	95.1	65.5	81.9
363	Dothan	95.8	52.6	76.5
364	Evergreen	95.3	54.1	76.9
365-366	Mobile	96.8	67.9	83.9
367	Selma	95.5	52.5	76.3
368	Phenix City	96.2	56.3	78.4
369	Butler	95.7	52.6	76.4
ALASKA				
995-996	Anchorage	127.2	114.5	121.5
997	Fairbanks	124.4	115.0	120.2
998	Juneau	122.8	112.6	118.3
999	Ketchikan	136.9	112.6	126.0
ARIZONA				
850,853	Phoenix	98.6	75.9	88.5
851,852	Mesa/Tempe	98.2	72.8	86.8
855	Globe	98.5	62.0	82.2
856-857	Tucson	96.9	72.0	85.8
859	Show Low	98.6	63.5	82.9
860	Flagstaff	100.7	74.4	88.9
863	Prescott	98.4	63.0	82.6
864	Kingman	96.7	74.2	86.7
865	Chambers	96.8	63.5	81.9
ARKANSAS				
716	Pine Bluff	96.7	64.8	82.4
717	Camden	94.2	50.7	74.8
718	Texarkana	95.9	51.9	76.3
719	Hot Springs	93.4	53.6	75.6
720-722	Little Rock	97.0	65.9	83.1
723	West Memphis	95.9	63.7	81.5
724	Jonesboro	96.7	61.6	81.0
725	Batesville	94.3	56.9	77.6
726	Harrison	95.7	53.0	76.6
727	Fayetteville	92.9	54.9	75.9
728	Russellville	94.2	56.1	77.2
729	Fort Smith	97.0	63.4	82.0
CALIFORNIA				
900-902	Los Angeles	100.6	115.4	107.2
903-905	Inglewood	96.1	112.6	103.5
906-908	Long Beach	97.8	112.6	104.5
910-912	Pasadena	96.5	112.7	103.7
913-916	Van Nuys	99.7	112.7	105.5
917-918	Alhambra	98.6	112.7	104.9
919-921	San Diego	100.6	108.2	104.0
922	Palm Springs	97.5	111.5	103.8
923-924	San Bernardino	95.2	111.2	102.3
925	Riverside	100.0	113.3	105.9
926-927	Santa Ana	97.2	111.4	103.6
928	Anaheim	99.9	113.7	106.1
930	Oxnard	101.4	113.1	106.6
931	Santa Barbara	100.6	113.1	106.2
932-933	Bakersfield	101.4	110.4	105.4
934	San Luis Obispo	101.2	109.8	105.0
935	Mojave	98.2	107.7	102.4
936-938	Fresno	101.7	115.3	107.8
939	Salinas	102.0	123.0	111.4
940-941	San Francisco	109.6	140.8	123.6
942,956-958	Sacramento	103.9	117.2	109.8
943	Palo Alto	102.3	129.4	114.4
944	San Mateo	105.1	131.4	116.9
945	Vallejo	102.9	127.8	114.0
946	Oakland	107.5	133.7	119.2
947	Berkeley	106.8	133.6	118.8
948	Richmond	106.1	128.9	116.3
949	San Rafael	107.4	130.3	117.6
950	Santa Cruz	106.4	123.1	113.9

STATE/ZIP	CITY	MAT.	INST.	TOTAL
CALIFORNIA (CONT'D)				
951	San Jose	104.6	134.3	117.9
952	Stockton	101.9	117.3	108.8
953	Modesto	101.9	116.6	108.5
954	Santa Rosa	101.7	134.9	116.5
955	Eureka	103.1	115.2	108.5
959	Marysville	102.4	113.8	107.5
960	Redding	104.1	113.8	108.4
961	Susanville	103.1	113.8	107.9
COLORADO				
800-802	Denver	101.4	84.9	94.0
803	Boulder	98.5	83.7	91.8
804	Golden	100.6	82.5	92.5
805	Fort Collins	102.3	77.1	91.0
806	Greeley	99.6	73.6	88.0
807	Fort Morgan	99.1	82.3	91.6
808-809	Colorado Springs	101.3	82.4	92.8
810	Pueblo	100.8	80.2	91.6
811	Alamosa	102.1	76.1	90.5
812	Salida	102.0	77.8	91.2
813	Durango	102.6	75.3	90.4
814	Montrose	101.2	74.7	89.4
815	Grand Junction	104.7	75.2	91.5
816	Glenwood Springs	102.1	78.0	91.3
CONNECTICUT				
060	New Britain	101.0	117.8	108.5
061	Hartford	102.1	117.8	109.1
062	Willimantic	101.6	117.8	108.9
063	New London	97.7	117.7	106.6
064	Meriden	99.7	117.7	107.8
065	New Haven	102.9	117.7	109.5
066	Bridgeport	102.2	116.4	108.6
067	Waterbury	101.7	117.7	108.9
068	Norwalk	101.6	116.5	108.3
069	Stamford	101.8	123.0	111.3
D.C.				
200-205	Washington	102.8	91.9	97.9
DELAWARE				
197	Newark	99.3	108.6	103.4
198	Wilmington	99.6	108.6	103.6
199	Dover	99.9	108.6	103.8
FLORIDA				
320,322	Jacksonville	97.8	69.7	85.2
321	Daytona Beach	98.0	76.4	88.3
323	Tallahassee	98.6	59.7	81.2
324	Panama City	99.2	57.5	80.6
325	Pensacola	101.9	65.1	85.4
326,344	Gainesville	99.5	70.6	86.6
327-328,347	Orlando	101.2	74.2	89.1
329	Melbourne	100.4	79.6	91.1
330-332,340	Miami	99.1	78.1	89.7
333	Fort Lauderdale	97.3	76.7	88.1
334,349	West Palm Beach	96.0	75.7	86.9
335-336,346	Tampa	99.0	83.6	92.1
337	St. Petersburg	101.3	66.1	85.6
338	Lakeland	98.0	83.5	91.5
339,341	Fort Myers	97.3	75.4	87.5
342	Sarasota	99.2	77.3	89.4
GEORGIA				
300-303,399	Atlanta	97.2	76.8	88.1
304	Statesboro	96.8	54.1	77.7
305	Gainesville	95.5	64.7	81.7
306	Athens	94.9	64.9	81.5
307	Dalton	96.8	60.5	80.6
308-309	Augusta	95.9	65.5	82.3
310-312	Macon	95.6	65.0	81.9
313-314	Savannah	97.7	60.8	81.2
315	Waycross	96.7	61.8	81.1
316	Valdosta	96.7	61.2	80.9
317,398	Albany	96.9	62.0	81.3
318-319	Columbus	96.8	65.3	82.7

STATE/ZIP	CITY	MAT.	INST.	TOTAL
HAWAII				
967	Hilo	111.9	120.6	115.8
968	Honolulu	115.1	120.6	117.5
STATES & POSS.				
969	Guam	133.3	61.7	101.3
IDAHO				
832	Pocatello	100.4	73.0	88.2
833	Twin Falls	101.7	53.1	80.0
834	Idaho Falls	99.0	59.0	81.1
835	Lewiston	109.3	78.3	95.4
836-837	Boise	99.5	73.3	87.8
838	Coeur d'Alene	108.5	75.4	93.7
ILLINOIS				
600-603	North Suburban	99.3	131.0	113.4
604	Joliet	99.2	137.7	116.4
605	South Suburban	99.3	131.0	113.4
606-608	Chicago	99.7	138.9	117.2
609	Kankakee	95.2	125.3	108.7
610-611	Rockford	98.1	122.7	109.1
612	Rock Island	95.7	99.0	97.2
613	La Salle	96.9	121.4	107.8
614	Galesburg	96.8	104.8	100.4
615-616	Peoria	99.2	107.3	102.8
617	Bloomington	96.1	108.7	101.7
618-619	Champaign	99.8	107.3	103.1
620-622	East St. Louis	94.6	107.5	100.4
623	Quincy	95.8	99.6	97.5
624	Effingham	95.1	103.6	98.9
625	Decatur	97.0	105.3	100.7
626-627	Springfield	97.5	106.2	101.4
628	Centralia	93.2	106.9	99.3
629	Carbondale	92.9	102.3	97.1
INDIANA				
460	Anderson	95.3	82.8	89.7
461-462	Indianapolis	98.3	85.7	92.7
463-464	Gary	96.7	108.1	101.8
465-466	South Bend	95.3	83.3	89.9
467-468	Fort Wayne	95.9	79.2	88.5
469	Kokomo	93.1	81.7	88.0
470	Lawrenceburg	92.5	78.9	86.4
471	New Albany	93.6	76.4	85.9
472	Columbus	95.9	81.6	89.5
473	Muncie	96.8	81.7	90.1
474	Bloomington	98.1	81.5	90.7
475	Washington	94.6	83.3	89.5
476-477	Evansville	95.9	86.0	91.5
478	Terre Haute	96.6	86.7	92.2
479	Lafayette	95.6	82.8	89.9
IOWA				
500-503,509	Des Moines	98.9	86.2	93.2
504	Mason City	97.5	68.9	84.7
505	Fort Dodge	97.7	60.6	81.1
506-507	Waterloo	99.4	73.4	87.8
508	Creston	98.0	80.2	90.0
510-511	Sioux City	100.1	71.9	87.5
512	Sibley	98.9	57.3	80.3
513	Spencer	100.6	57.3	81.3
514	Carroll	97.3	73.6	86.7
515	Council Bluffs	101.4	80.3	91.9
516	Shenandoah	97.9	73.6	87.1
520	Dubuque	99.7	81.2	91.4
521	Decorah	98.4	66.7	84.3
522-524	Cedar Rapids	100.5	85.5	93.8
525	Ottumwa	98.4	71.9	86.6
526	Burlington	97.5	78.6	89.1
527-528	Davenport	99.4	92.0	96.1
KANSAS				
660-662	Kansas City	99.2	97.3	98.3
664-666	Topeka	99.2	65.2	84.0
667	Fort Scott	97.6	75.2	87.6
668	Emporia	97.8	70.8	85.7
669	Belleville	99.6	66.5	84.8
670-672	Wichita	98.3	65.1	83.4
673	Independence	99.2	70.6	86.4
674	Salina	99.7	68.0	85.5
675	Hutchinson	94.5	66.4	82.0
676	Hays	98.7	66.8	84.4
677	Colby	99.5	66.8	84.9

STATE/ZIP	CITY	MAT.	INST.	TOTAL
KANSAS (CONT'D)				
678	Dodge City	101.0	67.8	86.2
679	Liberal	98.5	66.8	84.3
KENTUCKY				
400-402	Louisville	97.7	83.4	91.3
403-405	Lexington	96.3	75.3	86.9
406	Frankfort	97.1	74.3	86.9
407-409	Corbin	93.5	62.0	79.4
410	Covington	94.4	96.0	95.1
411-412	Ashland	93.2	98.5	95.6
413-414	Campton	94.4	62.6	80.2
415-416	Pikeville	95.7	80.3	88.8
417-418	Hazard	93.7	55.7	76.7
420	Paducah	92.6	86.0	89.7
421-422	Bowling Green	94.7	82.8	89.4
423	Owensboro	95.1	80.2	88.4
424	Henderson	92.3	86.3	89.6
425-426	Somerset	91.6	66.6	80.4
427	Elizabethtown	91.0	83.4	87.6
LOUISIANA				
700-701	New Orleans	102.0	69.2	87.3
703	Thibodaux	100.3	64.3	84.2
704	Hammond	97.4	56.1	78.9
705	Lafayette	99.7	60.5	82.2
706	Lake Charles	100.0	65.0	84.3
707-708	Baton Rouge	101.5	63.6	84.5
710-711	Shreveport	96.9	57.1	79.1
712	Monroe	97.2	54.3	78.0
713-714	Alexandria	97.2	53.8	77.8
MAINE				
039	Kittery	95.0	80.2	88.4
040-041	Portland	99.8	79.0	90.5
042	Lewiston	98.2	79.0	89.6
043	Augusta	97.9	80.0	89.9
044	Bangor	97.9	79.0	89.4
045	Bath	96.3	79.6	88.8
046	Machias	95.8	78.9	88.2
047	Houlton	95.9	79.1	88.4
048	Rockland	94.9	79.5	88.0
049	Waterville	96.3	80.0	89.0
MARYLAND				
206	Waldorf	99.1	68.5	85.4
207-208	College Park	99.2	83.0	92.0
209	Silver Spring	98.3	74.5	87.7
210-212	Baltimore	98.9	84.8	92.6
214	Annapolis	99.1	78.5	89.9
215	Cumberland	94.7	80.1	88.1
216	Easton	96.4	41.9	72.0
217	Hagerstown	95.5	82.2	89.5
218	Salisbury	96.8	51.0	76.3
219	Elkton	93.6	64.9	80.7
MASSACHUSETTS				
010-011	Springfield	100.0	109.5	104.2
012	Pittsfield	99.6	105.3	102.2
013	Greenfield	97.4	108.6	102.4
014	Fitchburg	96.2	121.1	107.3
015-016	Worcester	100.0	122.7	110.1
017	Framingham	95.6	129.4	110.7
018	Lowell	99.4	129.0	112.6
019	Lawrence	100.6	127.8	112.8
020-022, 024	Boston	102.3	136.8	117.7
023	Brockton	100.9	124.5	111.5
025	Buzzards Bay	95.0	122.2	107.1
026	Hyannis	98.0	122.2	108.8
027	New Bedford	100.0	122.3	110.0
MICHIGAN				
480,483	Royal Oak	92.8	103.6	97.6
481	Ann Arbor	95.2	107.7	100.8
482	Detroit	96.7	110.2	102.8
484-485	Flint	94.9	98.2	96.4
486	Saginaw	94.5	94.0	94.3
487	Bay City	94.6	93.9	94.3
488-489	Lansing	95.7	96.5	96.1
490	Battle Creek	95.8	90.4	93.4
491	Kalamazoo	96.1	88.4	92.7
492	Jackson	94.0	96.8	95.2
493,495	Grand Rapids	97.3	76.8	88.1
494	Muskegon	94.7	85.3	90.5

STATE/ZIP	CITY	MAT.	INST.	TOTAL
MICHIGAN (CONT'D)				
496	Traverse City	93.4	79.4	87.2
497	Gaylord	94.5	73.4	85.1
498-499	Iron mountain	96.7	88.8	93.2
MINNESOTA				
550-551	Saint Paul	101.6	123.0	111.2
553-555	Minneapolis	102.3	125.0	112.5
556-558	Duluth	100.6	112.2	105.8
559	Rochester	101.0	109.2	104.7
560	Mankato	97.6	106.4	101.5
561	Windom	96.2	95.3	95.8
562	Willmar	95.9	107.2	100.9
563	St. Cloud	97.3	123.2	108.9
564	Brainerd	97.6	107.3	101.9
565	Detroit Lakes	99.3	100.7	99.9
566	Bemidji	98.5	102.2	100.2
567	Thief River Falls	98.1	97.1	97.7
MISSISSIPPI				
386	Clarksdale	95.7	56.3	78.1
387	Greenville	99.3	66.1	84.5
388	Tupelo	97.1	58.8	80.0
389	Greenwood	97.0	55.8	78.6
390-392	Jackson	96.9	65.3	82.8
393	Meridian	95.9	67.0	83.0
394	Laurel	97.0	59.3	80.2
395	Biloxi	97.6	60.8	81.2
396	Mccomb	95.4	55.3	77.5
397	Columbus	97.0	57.7	79.4
MISSOURI				
630-631	St. Louis	99.4	106.6	102.6
633	Bowling Green	98.1	92.2	95.5
634	Hannibal	96.9	88.8	93.3
635	Kirksville	100.1	85.4	93.5
636	Flat River	99.1	94.7	97.1
637	Cape Girardeau	99.2	91.3	95.7
638	Sikeston	97.1	87.1	92.6
639	Poplar Bluff	96.6	88.4	92.9
640-641	Kansas City	99.0	107.3	102.7
644-645	St. Joseph	98.3	93.2	96.0
646	Chillicothe	95.2	71.1	84.4
647	Harrisonville	94.8	104.8	99.3
648	Joplin	97.1	71.4	85.6
650-651	Jefferson City	95.9	92.1	94.2
652	Columbia	97.4	93.7	95.7
653	Sedalia	96.5	94.6	95.7
654-655	Rolla	95.0	89.5	92.5
656-658	Springfield	98.2	80.9	90.5
MONTANA				
590-591	Billings	103.2	75.9	91.0
592	Wolf Point	103.2	68.6	87.7
593	Miles City	100.6	70.2	87.0
594	Great Falls	104.7	76.5	92.1
595	Havre	101.9	71.3	88.2
596	Helena	102.5	73.2	89.4
597	Butte	103.2	74.6	90.4
598	Missoula	100.5	73.4	88.4
599	Kalispell	99.4	71.3	86.8
NEBRASKA				
680-681	Omaha	99.8	80.6	91.2
683-685	Lincoln	97.7	74.9	87.5
686	Columbus	96.7	72.9	86.1
687	Norfolk	98.4	76.4	88.6
688	Grand Island	98.5	79.8	90.1
689	Hastings	97.7	80.7	90.1
690	Mccook	97.5	71.9	86.0
691	North Platte	97.8	80.9	90.3
692	Valentine	99.5	68.7	85.7
693	Alliance	99.6	66.8	85.0
NEVADA				
889-891	Las Vegas	100.2	112.8	105.9
893	Ely	99.1	99.6	99.3
894-895	Reno	99.2	91.1	95.6
897	Carson City	98.6	90.8	95.1
898	Elko	97.7	84.4	91.8
NEW HAMPSHIRE				
030	Nashua	100.2	86.1	93.9
031	Manchester	100.8	86.1	94.2

STATE/ZIP	CITY	MAT.	INST.	TOTAL
NEW HAMPSHIRE (CONT'D)				
032-033	Concord	98.7	82.5	91.5
034	Keene	97.0	49.9	76.0
035	Littleton	96.7	57.4	79.1
036	Charleston	96.5	47.6	74.6
037	Claremont	95.4	47.6	74.0
038	Portsmouth	97.7	89.9	94.2
NEW JERSEY				
070-071	Newark	102.3	125.6	112.7
072	Elizabeth	100.6	125.6	111.8
073	Jersey City	99.3	125.3	110.9
074-075	Paterson	101.4	125.5	112.2
076	Hackensack	99.3	125.4	111.0
077	Long Branch	98.9	124.1	110.2
078	Dover	99.6	125.5	111.2
079	Summit	99.6	125.6	111.2
080,083	Vineland	97.5	122.8	108.8
081	Camden	99.4	124.0	110.4
082,084	Atlantic City	98.1	122.7	109.1
085-086	Trenton	99.4	122.7	109.8
087	Point Pleasant	99.6	123.3	110.2
088-089	New Brunswick	100.1	125.0	111.2
NEW MEXICO				
870-872	Albuquerque	99.1	75.0	88.3
873	Gallup	99.3	75.0	88.5
874	Farmington	99.8	75.0	88.7
875	Santa Fe	100.1	75.0	88.9
877	Las Vegas	97.7	75.0	87.6
878	Socorro	97.2	75.0	87.3
879	Truth/Consequences	97.5	71.6	85.9
880	Las Cruces	96.3	70.9	85.0
881	Clovis	97.9	74.9	87.6
882	Roswell	99.6	75.0	88.6
883	Carrizozo	100.0	75.0	88.8
884	Tucumcari	98.6	74.9	88.0
NEW YORK				
100-102	New York	106.0	166.3	133.0
103	Staten Island	102.1	160.3	128.1
104	Bronx	99.9	160.3	126.9
105	Mount Vernon	100.0	134.3	115.3
106	White Plains	99.9	134.3	115.3
107	Yonkers	104.8	134.5	118.0
108	New Rochelle	100.3	134.3	115.5
109	Suffern	100.1	123.1	110.4
110	Queens	101.9	160.1	127.9
111	Long Island City	103.6	160.1	128.9
112	Brooklyn	104.0	160.1	129.1
113	Flushing	103.8	160.1	129.0
114	Jamaica	101.9	160.1	127.9
115,117,118	Hicksville	101.7	145.2	121.2
116	Far Rockaway	103.9	160.1	129.1
119	Riverhead	102.5	144.7	121.4
120-122	Albany	97.2	97.9	97.5
123	Schenectady	98.2	97.5	97.9
124	Kingston	101.6	118.7	109.2
125-126	Poughkeepsie	100.7	130.5	114.0
127	Monticello	100.0	119.2	108.6
128	Glens Falls	92.6	93.2	92.8
129	Plattsburgh	97.5	85.0	91.9
130-132	Syracuse	99.4	93.5	96.7
133-135	Utica	97.1	88.7	93.4
136	Watertown	98.9	92.6	96.1
137-139	Binghamton	98.8	90.2	95.0
140-142	Buffalo	100.5	102.9	101.6
143	Niagara Falls	98.3	98.0	98.2
144-146	Rochester	100.3	95.6	98.2
147	Jamestown	97.2	84.4	91.5
148-149	Elmira	96.9	92.0	94.7
NORTH CAROLINA				
270,272-274	Greensboro	99.2	47.9	76.2
271	Winston-Salem	99.1	46.5	75.6
275-276	Raleigh	99.6	48.1	76.6
277	Durham	100.3	48.4	77.1
278	Rocky Mount	96.6	39.8	71.2
279	Elizabeth City	97.5	41.0	72.3
280	Gastonia	98.2	46.0	74.8
281-282	Charlotte	99.2	48.4	76.5
283	Fayetteville	100.0	50.0	77.6
284	Wilmington	96.9	46.3	74.3
285	Kinston	94.9	40.6	70.6

STATE/ZIP	CITY	MAT.	INST.	TOTAL
NORTH CAROLINA (CONT'D)				
286	Hickory	95.2	42.8	71.8
287-288	Asheville	97.4	46.0	74.4
289	Murphy	96.2	33.4	68.1
NORTH DAKOTA				
580-581	Fargo	102.4	64.3	85.4
582	Grand Forks	102.6	53.8	80.8
583	Devils Lake	101.7	56.0	81.3
584	Jamestown	101.9	46.8	77.3
585	Bismarck	100.8	63.0	83.9
586	Dickinson	102.8	58.2	82.9
587	Minot	102.5	70.0	88.0
588	Williston	101.0	58.2	81.9
OHIO				
430-432	Columbus	97.9	90.3	94.5
433	Marion	94.1	80.8	88.2
434-436	Toledo	98.1	95.9	97.1
437-438	Zanesville	94.6	81.1	88.6
439	Steubenville	96.4	90.5	93.8
440	Lorain	98.7	91.5	95.5
441	Cleveland	99.0	100.5	99.7
442-443	Akron	99.8	92.6	96.6
444-445	Youngstown	99.1	87.5	93.9
446-447	Canton	99.2	82.5	91.8
448-449	Mansfield	96.2	87.0	92.1
450	Hamilton	95.7	83.8	90.4
451-452	Cincinnati	96.0	86.3	91.7
453-454	Dayton	95.8	83.6	90.3
455	Springfield	95.7	83.8	90.4
456	Chillicothe	94.5	88.5	91.8
457	Athens	97.0	81.0	89.8
458	Lima	97.4	85.0	91.9
OKLAHOMA				
730-731	Oklahoma City	98.2	61.5	81.8
734	Ardmore	95.2	61.0	79.9
735	Lawton	97.6	62.3	81.8
736	Clinton	96.7	59.6	80.2
737	Enid	97.4	59.6	80.5
738	Woodward	95.4	59.7	79.4
739	Guymon	96.5	31.1	67.3
740-741	Tulsa	97.4	54.2	78.1
743	Miami	94.0	69.0	82.8
744	Muskogee	96.7	39.6	71.2
745	Mcalester	93.7	51.2	74.7
746	Ponca City	94.3	59.6	78.8
747	Durant	94.3	58.5	78.3
748	Shawnee	96.0	57.3	78.7
749	Poteau	93.3	62.3	79.5
OREGON				
970-972	Portland	100.0	99.7	99.9
973	Salem	99.8	98.6	99.3
974	Eugene	99.8	98.3	99.1
975	Medford	101.5	96.7	99.3
976	Klamath Falls	101.5	96.7	99.3
977	Bend	100.3	98.4	99.5
978	Pendleton	94.9	100.3	97.3
979	Vale	92.6	90.4	91.6
PENNSYLVANIA				
150-152	Pittsburgh	99.3	102.9	100.9
153	Washington	96.2	100.8	98.3
154	Uniontown	96.6	99.9	98.0
155	Bedford	97.6	91.1	94.7
156	Greensburg	97.5	99.4	98.3
157	Indiana	96.3	98.2	97.1
158	Dubois	98.0	94.2	96.3
159	Johnstown	97.6	94.9	96.4
160	Butler	92.6	100.3	96.0
161	New Castle	92.6	97.2	94.7
162	Kittanning	93.2	101.4	96.9
163	Oil City	92.6	94.9	93.6
164-165	Erie	94.8	93.5	94.2
166	Altoona	95.0	90.5	93.0
167	Bradford	95.9	93.1	94.7
168	State College	95.5	91.4	93.7
169	Wellsboro	96.7	91.0	94.2
170-171	Harrisburg	98.0	93.8	96.1
172	Chambersburg	96.1	88.6	92.7
173-174	York	96.5	94.0	95.4
175-176	Lancaster	94.7	88.3	91.8

STATE/ZIP	CITY	MAT.	INST.	TOTAL
PENNSYLVANIA (CONT'D)				
177	Williamsport	93.3	79.7	87.2
178	Sunbury	95.5	94.0	94.8
179	Pottsville	94.6	95.6	95.1
180	Lehigh Valley	96.0	114.2	104.2
181	Allentown	98.3	108.6	102.9
182	Hazleton	95.5	97.3	96.3
183	Stroudsburg	95.3	102.3	98.4
184-185	Scranton	99.0	98.7	98.9
186-187	Wilkes-Barre	95.2	97.7	96.3
188	Montrose	94.9	96.3	95.5
189	Doylestown	95.1	121.9	107.1
190-191	Philadelphia	99.7	132.2	114.3
193	Westchester	96.1	124.7	108.9
194	Norristown	95.1	131.2	111.3
195-196	Reading	97.4	99.6	98.4
PUERTO RICO				
009	San Juan	121.8	24.0	78.1
RHODE ISLAND				
028	Newport	99.3	111.5	104.7
029	Providence	100.3	111.5	105.3
SOUTH CAROLINA				
290-292	Columbia	97.6	50.0	76.3
293	Spartanburg	96.4	49.6	75.5
294	Charleston	98.0	58.3	80.2
295	Florence	96.2	50.0	75.5
296	Greenville	96.2	49.6	75.3
297	Rock Hill	95.7	47.7	74.3
298	Aiken	96.7	71.9	85.6
299	Beaufort	97.5	43.0	73.1
SOUTH DAKOTA				
570-571	Sioux Falls	100.6	57.9	81.5
572	Watertown	99.7	52.2	78.4
573	Mitchell	98.3	51.6	77.4
574	Aberdeen	101.2	52.8	79.6
575	Pierre	99.7	54.5	79.5
576	Mobridge	99.1	52.0	78.0
577	Rapid City	101.0	55.3	80.6
TENNESSEE				
370-372	Nashville	96.7	72.7	85.9
373-374	Chattanooga	98.2	66.9	84.2
375,380-381	Memphis	96.6	70.3	84.9
376	Johnson City	97.6	57.4	79.6
377-379	Knoxville	94.5	61.7	79.9
382	Mckenzie	96.3	62.1	81.1
383	Jackson	98.2	63.8	82.8
384	Columbia	94.9	67.3	82.6
385	Cookeville	96.2	60.3	80.1
TEXAS				
750	Mckinney	99.5	50.8	77.7
751	Waxahackie	99.4	58.2	81.0
752-753	Dallas	99.9	67.5	85.4
754	Greenville	99.6	43.7	74.6
755	Texarkana	99.1	51.5	77.8
756	Longview	99.7	40.7	73.4
757	Tyler	100.2	54.9	80.0
758	Palestine	96.1	57.4	78.8
759	Lufkin	96.7	59.1	79.9
760-761	Fort Worth	97.7	63.6	82.4
762	Denton	97.3	48.4	75.4
763	Wichita Falls	98.1	55.5	79.1
764	Eastland	96.8	50.3	76.0
765	Temple	95.3	50.0	75.1
766-767	Waco	97.5	57.1	79.4
768	Brownwood	98.0	49.6	76.3
769	San Angelo	97.6	50.1	76.3
770-772	Houston	100.0	70.4	86.8
773	Huntsville	98.6	56.7	79.9
774	Wharton	99.8	54.8	79.7
775	Galveston	97.6	69.2	84.9
776-777	Beaumont	98.2	62.9	82.4
778	Bryan	95.1	62.1	80.3
779	Victoria	99.9	44.4	75.1
780	Laredo	94.7	53.1	76.1
781-782	San Antonio	95.0	65.1	81.6
783-784	Corpus Christi	97.6	52.8	77.6
785	Mc Allen	97.5	46.9	74.9
786-787	Austin	95.1	59.4	79.1

Location Factors

STATE/ZIP	CITY	MAT.	INST.	TOTAL
TEXAS (CONT'D)				
788	Del Rio	97.0	47.0	74.6
789	Giddings	94.1	57.0	77.5
790-791	Amarillo	98.1	61.9	81.9
792	Childress	97.5	58.1	79.9
793-794	Lubbock	99.6	57.5	80.8
795-796	Abilene	97.9	54.8	78.6
797	Midland	99.7	51.9	78.3
798-799,885	El Paso	97.0	51.7	76.8
UTAH				
840-841	Salt Lake City	102.0	69.1	87.3
842,844	Ogden	97.5	70.7	85.5
843	Logan	99.5	70.7	86.6
845	Price	100.0	67.1	85.3
846-847	Provo	100.0	68.9	86.1
VERMONT				
050	White River Jct.	98.3	55.1	79.0
051	Bellows Falls	96.6	73.7	86.4
052	Bennington	97.0	68.9	84.4
053	Brattleboro	97.4	72.9	86.5
054	Burlington	100.9	66.1	85.4
056	Montpelier	97.7	66.2	83.6
057	Rutland	99.2	66.2	84.4
058	St. Johnsbury	98.3	55.9	79.3
059	Guildhall	96.9	55.7	78.5
VIRGINIA				
220-221	Fairfax	99.9	82.3	92.0
222	Arlington	101.2	80.1	91.8
223	Alexandria	100.2	85.5	93.6
224-225	Fredericksburg	98.5	72.2	86.7
226	Winchester	99.2	63.2	83.1
227	Culpeper	99.1	67.8	85.1
228	Harrisonburg	99.3	64.0	83.5
229	Charlottesville	99.8	63.4	83.5
230-232	Richmond	100.2	66.8	85.3
233-235	Norfolk	101.0	68.2	86.3
236	Newport News	100.0	67.6	85.5
237	Portsmouth	99.4	65.5	84.2
238	Petersburg	99.6	66.8	84.9
239	Farmville	98.7	53.0	78.3
240-241	Roanoke	100.7	60.7	82.8
242	Bristol	98.4	55.1	79.0
243	Pulaski	98.0	50.9	77.0
244	Staunton	98.9	59.1	81.1
245	Lynchburg	99.0	64.6	83.6
246	Grundy	98.3	50.3	76.9
WASHINGTON				
980-981,987	Seattle	103.8	106.6	105.0
982	Everett	103.4	97.1	100.6
983-984	Tacoma	103.5	98.9	101.4
985	Olympia	101.3	98.9	100.2
986	Vancouver	104.8	93.8	99.9
988	Wenatchee	103.3	83.7	94.6
989	Yakima	103.7	93.1	99.0
990-992	Spokane	103.7	80.8	93.5
993	Richland	103.4	88.0	96.5
994	Clarkston	101.9	79.8	92.0
WEST VIRGINIA				
247-248	Bluefield	97.1	77.7	88.4
249	Lewisburg	98.9	83.2	91.9
250-253	Charleston	99.0	90.1	95.0
254	Martinsburg	98.6	79.1	89.9
255-257	Huntington	100.2	91.6	96.3
258-259	Beckley	97.0	87.2	92.6
260	Wheeling	100.3	89.9	95.7
261	Parkersburg	99.1	89.4	94.8
262	Buckhannon	98.5	89.7	94.6
263-264	Clarksburg	99.2	89.7	94.9
265	Morgantown	99.2	88.8	94.6
266	Gassaway	98.4	89.9	94.6
267	Romney	98.4	85.6	92.7
268	Petersburg	98.2	85.0	92.3
WISCONSIN				
530,532	Milwaukee	100.2	107.6	103.5
531	Kenosha	100.0	102.7	101.2
534	Racine	99.4	102.8	100.9
535	Beloit	99.3	97.8	98.6
537	Madison	98.9	99.1	99.0

STATE/ZIP	CITY	MAT.	INST.	TOTAL
WISCONSIN (CONT'D)				
538	Lancaster	97.0	93.2	95.3
539	Portage	95.4	96.3	95.8
540	New Richmond	97.3	96.2	96.8
541-543	Green Bay	101.7	93.3	98.0
544	Wausau	96.2	93.4	95.0
545	Rhinelander	99.8	92.2	96.4
546	La Crosse	97.6	94.9	96.4
547	Eau Claire	99.3	96.1	97.9
548	Superior	97.1	98.1	97.6
549	Oshkosh	97.0	84.1	91.2
WYOMING				
820	Cheyenne	99.2	64.7	83.8
821	Yellowstone Nat'l Park	96.8	60.0	80.4
822	Wheatland	98.0	59.2	80.6
823	Rawlins	99.6	58.0	81.0
824	Worland	97.5	58.0	79.8
825	Riverton	98.6	58.0	80.4
826	Casper	99.5	58.5	81.2
827	Newcastle	97.3	58.0	79.7
828	Sheridan	100.1	61.5	82.8
829-831	Rock Springs	101.5	58.1	82.1
CANADIAN FACTORS (reflect Canadian currency)				
ALBERTA				
	Calgary	130.9	94.8	114.8
	Edmonton	131.1	94.8	114.9
	Fort McMurray	127.0	97.7	113.9
	Lethbridge	120.9	97.1	110.3
	Lloydminster	116.2	93.6	106.1
	Medicine Hat	116.3	92.8	105.8
	Red Deer	116.8	92.8	106.1
BRITISH COLUMBIA				
	Kamloops	117.2	96.2	107.8
	Prince George	118.5	95.4	108.2
	Vancouver	129.1	90.7	111.9
	Victoria	118.5	88.7	105.2
MANITOBA				
	Brandon	116.1	80.3	100.1
	Portage la Prairie	116.1	78.5	99.3
	Winnipeg	128.8	73.5	104.1
NEW BRUNSWICK				
	Bathurst	114.2	71.2	94.9
	Dalhousie	114.2	71.4	95.1
	Fredericton	116.8	76.8	98.9
	Moncton	114.5	72.8	95.8
	Newcastle	114.2	71.9	95.3
	St. John	116.9	76.5	98.8
NEWFOUNDLAND				
	Corner Brook	119.6	71.5	98.1
	St Johns	119.7	76.9	100.6
NORTHWEST TERRITORIES				
	Yellowknife	121.5	90.5	107.7
NOVA SCOTIA				
	Bridgewater	115.8	79.1	99.4
	Dartmouth	117.2	79.1	100.1
	Halifax	118.1	81.7	101.8
	New Glasgow	115.3	79.1	99.1
	Sydney	112.6	79.1	97.6
	Truro	115.3	79.1	99.1
	Yarmouth	115.1	79.1	99.0
ONTARIO				
	Barrie	118.9	98.8	109.9
	Brantford	118.1	103.0	111.3
	Cornwall	118.0	99.1	109.5
	Hamilton	122.6	95.4	110.5
	Kingston	118.9	99.2	110.1
	Kitchener	114.0	92.1	104.2
	London	121.4	91.9	108.2
	North Bay	118.1	96.7	108.5
	Oshawa	117.7	100.0	109.8
	Ottawa	123.0	92.3	109.3
	Owen Sound	119.1	96.8	109.1
	Peterborough	118.1	98.8	109.5
	Sarnia	117.8	103.6	111.4

STATE/ZIP	CITY	MAT.	INST.	TOTAL
ONTARIO (CONT'D)	Sault Ste Marie	112.7	96.3	105.4
	St. Catharines	112.1	93.1	103.6
	Sudbury	112.5	92.1	103.4
	Thunder Bay	113.6	91.9	103.9
	Timmins	118.3	96.7	108.6
	Toronto	123.6	100.7	113.4
	Windsor	112.7	91.6	103.3
PRINCE EDWARD ISLAND				
	Charlottetown	117.6	66.4	94.7
	Summerside	117.1	66.4	94.5
QUEBEC				
	Cap-de-la-Madeleine	114.9	93.0	105.1
	Charlesbourg	114.9	93.0	105.1
	Chicoutimi	113.9	93.2	104.7
	Gatineau	114.6	92.7	104.8
	Granby	114.8	92.7	104.9
	Hull	114.6	92.7	104.8
	Joliette	115.1	93.0	105.2
	Laval	114.3	92.7	104.6
	Montreal	122.0	93.5	109.3
	Quebec	120.1	94.0	108.4
	Rimouski	114.4	93.2	104.9
	Rouyn-Noranda	114.5	92.7	104.7
	Saint Hyacinthe	113.9	92.7	104.4
	Sherbrooke	114.9	92.7	105.0
	Sorel	115.1	93.0	105.2
	St Jerome	114.5	92.7	104.8
	Trois Rivieres	115.1	93.0	105.2
SASKATCHEWAN				
	Moose Jaw	113.4	73.6	95.6
	Prince Albert	112.3	71.6	94.1
	Regina	114.9	87.7	102.7
	Saskatoon	113.4	87.7	101.9
YUKON				
	Whitehorse	113.2	72.4	95.0

R011105-05 Tips for Accurate Estimating

1. Use pre-printed or columnar forms for orderly sequence of dimensions and locations and for recording telephone quotations.
2. Use only the front side of each paper or form except for certain pre-printed summary forms.
3. Be consistent in listing dimensions: For example, length x width x height. This helps in rechecking to ensure that, the total length of partitions is appropriate for the building area.
4. Use printed (rather than measured) dimensions where given.
5. Add up multiple printed dimensions for a single entry where possible.
6. Measure all other dimensions carefully.
7. Use each set of dimensions to calculate multiple related quantities.
8. Convert foot and inch measurements to decimal feet when listing. Memorize decimal equivalents to .01 parts of a foot (1/8″ equals approximately .01′).
9. Do not "round off" quantities until the final summary.
10. Mark drawings with different colors as items are taken off.
11. Keep similar items together, different items separate.
12. Identify location and drawing numbers to aid in future checking for completeness.
13. Measure or list everything on the drawings or mentioned in the specifications.
14. It may be necessary to list items not called for to make the job complete.
15. Be alert for: Notes on plans such as N.T.S. (not to scale); changes in scale throughout the drawings; reduced size drawings; discrepancies between the specifications and the drawings.
16. Develop a consistent pattern of performing an estimate. For example:
    a. Start the quantity takeoff at the lower floor and move to the next higher floor.
    b. Proceed from the main section of the building to the wings.
    c. Proceed from south to north or vice versa, clockwise or counterclockwise.
    d. Take off floor plan quantities first, elevations next, then detail drawings.
17. List all gross dimensions that can be either used again for different quantities, or used as a rough check of other quantities for verification (exterior perimeter, gross floor area, individual floor areas, etc.).
18. Utilize design symmetry or repetition (repetitive floors, repetitive wings, symmetrical design around a center line, similar room layouts, etc.). Note: Extreme caution is needed here so as not to omit or duplicate an area.
19. Do not convert units until the final total is obtained. For instance, when estimating concrete work, keep all units to the nearest cubic foot, then summarize and convert to cubic yards.
20. When figuring alternatives, it is best to total all items involved in the basic system, then total all items involved in the alternates. Therefore you work with positive numbers in all cases. When adds and deducts are used, it is often confusing whether to add or subtract a portion of an item; especially on a complicated or involved alternate.

R011110-10 Architectural Fees

Tabulated below are typical percentage fees by project size, for good professional architectural service. Fees may vary from those listed depending upon degree of design difficulty and economic conditions in any particular area.

Rates can be interpolated horizontally and vertically. Various portions of the same project requiring different rates should be adjusted proportionately. For alterations, add 50% to the fee for the first $500,000 of project cost and add 25% to the fee for project cost over $500,000.

Architectural fees tabulated below include Structural, Mechanical and Electrical Engineering Fees. They do not include the fees for special consultants such as kitchen planning, security, acoustical, interior design, etc.

Civil Engineering fees are included in the Architectural fee for project sites requiring minimal design such as city sites. However, separate Civil Engineering fees must be added when utility connections require design, drainage calculations are needed, stepped foundations are required, or provisions are required to protect adjacent wetlands.

	Total Project Size in Thousands of Dollars						
Building Types	100	250	500	1,000	5,000	10,000	50,000
Factories, garages, warehouses, repetitive housing	9.0%	8.0%	7.0%	6.2%	5.3%	4.9%	4.5%
Apartments, banks, schools, libraries, offices, municipal buildings	12.2	12.3	9.2	8.0	7.0	6.6	6.2
Churches, hospitals, homes, laboratories, museums, research	15.0	13.6	12.7	11.9	9.5	8.8	8.0
Memorials, monumental work, decorative furnishings	—	16.0	14.5	13.1	10.0	9.0	8.3

R011110-30 Engineering Fees

Typical **Structural Engineering Fees** based on type of construction and total project size. These fees are included in Architectural Fees.

Type of Construction	Total Project Size (in thousands of dollars)			
	$500	$500-$1,000	$1,000-$5,000	Over $5000
Industrial buildings, factories & warehouses	Technical payroll times 2.0 to 2.5	1.60%	1.25%	1.00%
Hotels, apartments, offices, dormitories, hospitals, public buildings, food stores		2.00%	1.70%	1.20%
Museums, banks, churches and cathedrals		2.00%	1.75%	1.25%
Thin shells, prestressed concrete, earthquake resistive		2.00%	1.75%	1.50%
Parking ramps, auditoriums, stadiums, convention halls, hangars & boiler houses		2.50%	2.00%	1.75%
Special buildings, major alterations, underpinning & future expansion	▼	Add to above 0.5%	Add to above 0.5%	Add to above 0.5%

For complex reinforced concrete or unusually complicated structures, add 20% to 50%.

Typical **Mechanical and Electrical Engineering Fees** are based on the size of the subcontract. The fee structure for both are shown below. These fees are included in Architectural Fees.

Type of Construction	Subcontract Size							
	$25,000	$50,000	$100,000	$225,000	$350,000	$500,000	$750,000	$1,000,000
Simple structures	6.4%	5.7%	4.8%	4.5%	4.4%	4.3%	4.2%	4.1%
Intermediate structures	8.0	7.3	6.5	5.6	5.1	5.0	4.9	4.8
Complex structures	10.1	9.0	9.0	8.0	7.5	7.5	7.0	7.0

For renovations, add 15% to 25% to applicable fee.

General Requirements R0121 Allowances

R012157-20 Construction Time Requirements

Table at left is average construction time in months for different types of building projects. Table at right is the construction time in months for different size projects. Design time runs 25% to 40% of construction time.

Type Building	Construction Time	Project Value	Construction Time
Industrial Buildings	12 Months	Under $1,400,000	10 Months
Commercial Buildings	15 Months	Up to $3,800,000	15 Months
Research & Development	18 Months	Up to $19,000,000	21 Months
Institutional Buildings	20 Months	over $19,000,000	28 Months

R012909-80 Sales Tax by State

State sales tax on materials is tabulated below (5 states have no sales tax). Many states allow local jurisdictions, such as a county or city, to levy additional sales tax.

Some projects may be sales tax exempt, particularly those constructed with public funds.

State	Tax (%)	State	Tax (%)	State	Tax (%)	State	Tax (%)
Alabama	4	Illinois	6.25	Montana	0	Rhode Island	7
Alaska	0	Indiana	7	Nebraska	5.5	South Carolina	6
Arizona	5.6	Iowa	6	Nevada	6.85	South Dakota	4
Arkansas	6	Kansas	5.3	New Hampshire	0	Tennessee	7
California	8.25	Kentucky	6	New Jersey	7	Texas	6.25
Colorado	2.9	Louisiana	4	New Mexico	5	Utah	4.7
Connecticut	6	Maine	5	New York	4	Vermont	6
Delaware	0	Maryland	6	North Carolina	5.75	Virginia	5
District of Columbia	6	Massachusetts	6.25	North Dakota	5	Washington	6.5
Florida	6	Michigan	6	Ohio	5.5	West Virginia	6
Georgia	4	Minnesota	6.875	Oklahoma	4.5	Wisconsin	5
Hawaii	4	Mississippi	7	Oregon	0	Wyoming	4
Idaho	6	Missouri	4.225	Pennsylvania	6	Average	5.04 %

Sales Tax by Province (Canada)

GST - a value-added tax, which the government imposes on most goods and services provided in or imported into Canada. PST - a retail sales tax, which five of the provinces impose on the price of most goods and some services. QST - a value-added tax, similar to the federal GST, which Quebec imposes. HST - Three provinces have combined their retail sales tax with the federal GST into one harmonized tax.

Province	PST (%)	QST (%)	GST(%)	HST(%)
Alberta	0	0	5	0
British Columbia	0	0	0	12
Manitoba	7	0	5	0
New Brunswick	0	0	0	13
Newfoundland	0	0	0	13
Northwest Territories	0	0	5	0
Nova Scotia	0	0	0	15
Ontario	0	0	0	13
Prince Edward Island	10	0	5	0
Quebec	0	8.5	5	0
Saskatchewan	5	0	5	0
Yukon	0	0	5	0

R012909-85 Unemployment Taxes and Social Security Taxes

State Unemployment Tax rates vary not only from state to state, but also with the experience rating of the contractor. The Federal Unemployment Tax rate is 6.2% of the first $7,000 of wages. This is reduced by a credit of up to 5.4% for timely payment to the state. The minimum Federal Unemployment Tax is 0.8% after all credits.

Social Security (FICA) for 2011 is estimated at time of publication to be 7.65% of wages up to $106,800.

R012909-90 Overtime

One way to improve the completion date of a project or eliminate negative float from a schedule is to compress activity duration times. This can be achieved by increasing the crew size or working overtime with the proposed crew.

To determine the costs of working overtime to compress activity duration times, consider the following examples. Below is an overtime efficiency and cost chart based on a five, six, or seven day week with an eight through twelve hour day. Payroll percentage increases for time and one half and double time are shown for the various working days.

Days per Week	Hours per Day	Production Efficiency					Payroll Cost Factors	
		1st Week	2nd Week	3rd Week	4th Week	Average 4 Weeks	@ 1-1/2 Times	@ 2 Times
5	8	100%	100%	100%	100%	100 %	100 %	100 %
	9	100	100	95	90	96.25	105.6	111.1
	10	100	95	90	85	91.25	110.0	120.0
	11	95	90	75	65	81.25	113.6	127.3
	12	90	85	70	60	76.25	116.7	133.3
6	8	100	100	95	90	96.25	108.3	116.7
	9	100	95	90	85	92.50	113.0	125.9
	10	95	90	85	80	87.50	116.7	133.3
	11	95	85	70	65	78.75	119.7	139.4
	12	90	80	65	60	73.75	122.2	144.4
7	8	100	95	85	75	88.75	114.3	128.6
	9	95	90	80	70	83.75	118.3	136.5
	10	90	85	75	65	78.75	121.4	142.9
	11	85	80	65	60	72.50	124.0	148.1
	12	85	75	60	55	68.75	126.2	152.4

General Requirements | R0131 Project Management & Coordination

R013113-40 Builder's Risk Insurance

Builder's Risk Insurance is insurance on a building during construction. Premiums are paid by the owner or the contractor. Blasting, collapse and underground insurance would raise total insurance costs above those listed. Floater policy for materials delivered to the job runs $.75 to $1.25 per $100 value. Contractor equipment insurance runs $.50 to $1.50 per $100 value. Insurance for miscellaneous tools to $1,500 value runs from $3.00 to $7.50 per $100 value.

Tabulated below are New England Builder's Risk insurance rates in dollars per $100 value for $1,000 deductible. For $25,000 deductible, rates can be reduced 13% to 34%. On contracts over $1,000,000, rates may be lower than those tabulated. Policies are written annually for the total completed value in place. For "all risk" insurance (excluding flood, earthquake and certain other perils) add $.025 to total rates below.

Coverage	Frame Construction (Class 1)		Brick Construction (Class 4)		Fire Resistive (Class 6)	
	Range	Average	Range	Average	Range	Average
Fire Insurance	$.350 to $ .850	$.600	$.158 to $.189	$.174	$.052 to $.080	$.070
Extended Coverage	.115 to .200	.158	.080 to .105	.101	.081 to .105	.100
Vandalism	.012 to .016	.014	.008 to .011	.011	.008 to .011	.010
Total Annual Rate	$.477 to $1.066	$.772	$.246 to $.305	$.286	$.141 to $.196	$.180

R013113-50 General Contractor's Overhead

There are two distinct types of overhead on a construction project: Project Overhead and Main Office Overhead. Project Overhead includes those costs at a construction site not directly associated with the installation of construction materials. Examples of Project Overhead costs include the following:

1. Superintendent
2. Construction office and storage trailers
3. Temporary sanitary facilities
4. Temporary utilities
5. Security fencing
6. Photographs
7. Clean up
8. Performance and payment bonds

The above Project Overhead items are also referred to as General Requirements and therefore are estimated in Division 1. Division 1 is the first division listed in the CSI MasterFormat but it is usually the last division estimated. The sum of the costs in Divisions 1 through 49 is referred to as the sum of the direct costs.

All construction projects also include indirect costs. The primary components of indirect costs are the contractor's Main Office Overhead and profit. The amount of the Main Office Overhead expense varies depending on the the following:

1. Owner's compensation
2. Project managers and estimator's wages
3. Clerical support wages
4. Office rent and utilities
5. Corporate legal and accounting costs
6. Advertising
7. Automobile expenses
8. Association dues
9. Travel and entertainment expenses

These costs are usually calculated as a percentage of annual sales volume. This percentage can range from 35% for a small contractor doing less than $500,000 to 5% for a large contractor with sales in excess of $100 million.

R013113-60 Workers' Compensation Insurance Rates by Trade

The table below tabulates the national averages for Workers' Compensation insurance rates by trade and type of building. The average "Insurance Rate" is multiplied by the "% of Building Cost" for each trade. This produces the "Workers' Compensation Cost" by % of total labor cost, to be added for each trade by building type to determine the weighted average Workers' Compensation rate for the building types analyzed.

Trade	Insurance Rate (% Labor Cost) Range	Insurance Rate Average	% of Building Cost Office Bldgs.	% of Building Cost Schools & Apts.	% of Building Cost Mfg.	Workers' Compensation Office Bldgs.	Workers' Compensation Schools & Apts.	Workers' Compensation Mfg.
Excavation, Grading, etc.	3.5 % to 18.2%	9.1%	4.8%	4.9%	4.5%	0.44%	0.45%	0.41%
Piles & Foundations	4.8 to 51.8	18.0	7.1	5.2	8.7	1.28	0.94	1.57
Concrete	4.7 to 28.9	13.3	5.0	14.8	3.7	0.67	1.97	0.49
Masonry	4 to 28.5	12.7	6.9	7.5	1.9	0.88	0.95	0.24
Structural Steel	4.8 to 155.9	36.3	10.7	3.9	17.6	3.88	1.42	6.39
Miscellaneous & Ornamental Metals	4 to 21.2	9.9	2.8	4.0	3.6	0.28	0.40	0.36
Carpentry & Millwork	4.8 to 53.2	16.1	3.7	4.0	0.5	0.60	0.64	0.08
Metal or Composition Siding	4.8 to 40	14.5	2.3	0.3	4.3	0.33	0.04	0.62
Roofing	4.8 to 77.1	29.2	2.3	2.6	3.1	0.67	0.76	0.91
Doors & Hardware	2.5 to 40	10.9	0.9	1.4	0.4	0.10	0.15	0.04
Sash & Glazing	4.8 to 34.6	12.9	3.5	4.0	1.0	0.45	0.52	0.13
Lath & Plaster	3.1 to 29.9	11.4	3.3	6.9	0.8	0.38	0.79	0.09
Tile, Marble & Floors	2.8 to 18.9	8.5	2.6	3.0	0.5	0.22	0.26	0.04
Acoustical Ceilings	3.1 to 24.5	8.6	2.4	0.2	0.3	0.21	0.02	0.03
Painting	4.3 to 29.6	11.1	1.5	1.6	1.6	0.17	0.18	0.18
Interior Partitions	4.8 to 53.2	16.1	3.9	4.3	4.4	0.63	0.69	0.71
Miscellaneous Items	2.6 to 109.2	14.7	5.2	3.7	9.7	0.77	0.55	1.43
Elevators	1.9 to 11.9	6.1	2.1	1.1	2.2	0.13	0.07	0.13
Sprinklers	2.2 to 17.2	7.1	0.5	—	2.0	0.04	—	0.14
Plumbing	2.6 to 17.5	7.0	4.9	7.2	5.2	0.34	0.50	0.36
Heat., Vent., Air Conditioning	3.8 to 22.7	8.9	13.5	11.0	12.9	1.20	0.98	1.15
Electrical	2.4 to 11.7	5.8	10.1	8.4	11.1	0.59	0.49	0.64
Total	1.9 % to 155.9%	—	100.0%	100.0%	100.0%	14.26%	12.77%	16.14%
		Overall Weighted Average 14.39%						

Workers' Compensation Insurance Rates by States

The table below lists the weighted average Workers' Compensation base rate for each state with a factor comparing this with the national average of 14.1%.

State	Weighted Average	Factor	State	Weighted Average	Factor	State	Weighted Average	Factor
Alabama	20.0%	142	Kentucky	16.9%	120	North Dakota	11.2%	79
Alaska	16.7	118	Louisiana	28.5	202	Ohio	8.6	61
Arizona	8.9	63	Maine	13.3	94	Oklahoma	13.7	97
Arkansas	11.3	80	Maryland	15.1	107	Oregon	12.4	88
California	28.0	199	Massachusetts	12.7	90	Pennsylvania	16.1	114
Colorado	7.7	55	Michigan	17.7	126	Rhode Island	11.7	83
Connecticut	21.3	151	Minnesota	26.2	186	South Carolina	20.0	142
Delaware	9.6	68	Mississippi	14.9	106	South Dakota	17.3	123
District of Columbia	11.1	79	Missouri	18.4	130	Tennessee	13.5	96
Florida	10.7	76	Montana	14.9	106	Texas	11.3	80
Georgia	27.1	192	Nebraska	19.4	138	Utah	7.8	55
Hawaii	9.0	64	Nevada	10.2	72	Vermont	14.8	105
Idaho	10.1	72	New Hampshire	20.1	143	Virginia	10.3	73
Illinois	22.9	162	New Jersey	12.8	91	Washington	9.6	68
Indiana	5.7	40	New Mexico	17.7	126	West Virginia	13.0	92
Iowa	10.9	77	New York	6.1	43	Wisconsin	15.3	109
Kansas	8.4	60	North Carolina	17.5	124	Wyoming	5.3	38
			Weighted Average for U.S. is	14.4% of payroll = 100%				

Rates in the following table are the base or manual costs per $100 of payroll for Workers' Compensation in each state. Rates are usually applied to straight time wages only and not to premium time wages and bonuses.

The weighted average skilled worker rate for 35 trades is 14.1%. For bidding purposes, apply the full value of Workers' Compensation directly to total labor costs, or if labor is 38%, materials 42% and overhead and profit 20% of total cost, carry 38/80 x 14.1% = 6.7% of cost (before overhead and profit) into overhead. Rates vary not only from state to state but also with the experience rating of the contractor.

Rates are the most current available at the time of publication.

R013113-60 Workers' Compensation Insurance Rates by Trade and State (cont.)

State	Carpentry — 3 stories or less 5651	Carpentry — interior cab. work 5437	Carpentry — general 5403	Concrete Work — NOC 5213	Concrete Work — flat (flr., sdwk.) 5221	Electrical Wiring — inside 5190	Excavation — earth NOC 6217	Excavation — rock 6217	Glaziers 5462	Insulation Work 5479	Lathing 5443	Masonry 5022	Painting & Decorating 5474	Pile Driving 6003	Plastering 5480	Plumbing 5183	Roofing 5551	Sheet Metal Work (HVAC) 5538	Steel Erection — door & sash 5102	Steel Erection — inter., ornam. 5102	Steel Erection — structure 5040	Steel Erection — NOC 5057	Tile Work — (interior ceramic) 5348	Waterproofing 9014	Wrecking 5701
AL	25.95	16.64	27.92	13.20	9.70	9.29	11.19	11.19	22.26	17.03	10.11	25.50	17.74	21.90	15.41	8.77	47.76	9.10	9.40	9.40	45.83	21.35	15.04	7.80	45.83
AK	10.17	10.72	14.12	11.68	10.00	6.69	10.27	10.27	33.82	21.22	8.87	19.14	13.19	34.01	17.99	7.07	29.77	8.15	7.49	7.49	22.50	34.08	5.76	5.77	22.50
AZ	9.58	6.05	15.80	7.78	4.09	3.84	4.98	4.98	6.16	11.90	3.88	7.81	7.11	9.50	5.20	4.77	17.80	6.44	7.96	7.96	20.43	11.14	2.90	3.33	20.43
AR	10.25	6.32	13.03	7.90	6.10	5.13	7.72	7.72	8.76	9.71	5.62	8.38	8.83	12.60	11.45	5.58	21.23	5.31	6.19	6.19	27.01	32.27	6.75	3.52	27.01
CA	40.00	40.00	40.00	17.64	17.64	11.71	18.18	18.18	22.67	18.58	17.85	28.46	21.99	20.82	29.93	17.51	69.24	22.67	21.16	21.16	33.84	20.64	10.83	21.99	20.69
CO	7.64	4.80	6.96	6.20	4.69	2.91	5.75	5.75	5.42	9.91	3.93	7.92	5.68	8.44	5.60	4.11	14.33	3.85	5.41	5.41	24.62	9.37	4.39	3.11	9.37
CT	15.80	17.74	21.69	25.96	13.53	7.98	13.75	13.75	19.29	19.38	14.94	26.57	19.57	21.58	16.34	10.58	44.27	13.02	15.01	15.01	52.47	21.56	12.35	6.03	52.47
DE	9.85	9.85	7.39	7.74	6.33	3.39	5.87	5.87	8.38	7.39	8.38	8.64	10.01	11.74	8.38	5.02	17.66	6.50	8.74	8.74	17.91	8.74	5.92	8.64	17.91
DC	9.38	9.50	8.30	11.23	8.01	5.31	7.58	7.58	13.48	6.63	5.70	9.54	5.84	13.94	9.73	7.23	16.66	7.74	13.51	13.51	19.78	13.05	11.01	3.80	19.78
FL	9.80	6.83	10.73	10.69	5.31	4.72	5.78	5.78	8.37	7.60	4.04	8.78	7.50	38.24	12.58	4.75	17.85	6.49	8.06	8.06	20.73	10.19	4.07	4.15	20.73
GA	35.73	21.35	25.92	17.90	15.31	10.45	16.30	16.30	19.14	22.15	14.61	24.16	24.99	26.75	24.86	12.18	63.28	17.45	20.26	20.26	71.30	39.53	13.97	8.92	71.30
HI	7.56	7.43	13.92	8.59	4.90	4.26	5.38	5.38	8.46	6.34	6.06	9.00	7.28	11.84	10.47	4.00	12.79	4.33	6.28	6.28	29.40	10.64	5.80	4.77	29.40
ID	9.90	6.33	10.07	10.29	5.89	3.41	6.44	6.44	8.75	7.70	7.95	6.62	7.75	9.61	13.76	3.96	23.71	6.82	7.17	7.17	27.50	8.29	7.39	4.49	22.50
IL	23.76	14.07	22.20	28.86	11.72	9.02	10.65	10.65	18.69	19.31	15.58	20.36	12.84	27.34	20.79	11.11	35.16	13.05	18.56	18.56	84.61	24.43	16.31	5.51	84.61
IN	11.17	4.02	6.52	4.66	3.27	2.82	4.33	4.33	5.93	6.16	3.07	3.96	4.60	5.13	3.12	2.57	9.84	4.52	3.95	3.95	13.05	5.36	2.93	2.36	13.05
IA	13.13	9.09	11.58	14.33	7.45	4.42	7.61	7.61	8.93	6.14	5.23	8.37	6.58	9.12	8.83	5.86	20.63	4.44	5.40	5.40	33.37	19.27	7.35	3.98	33.37
KS	14.43	6.21	8.57	6.75	5.51	3.76	4.69	4.69	7.85	6.43	3.93	6.66	6.88	8.42	4.99	4.62	13.89	4.59	4.56	4.56	24.54	15.33	5.61	3.86	15.33
KY	12.63	12.30	24.88	12.96	7.51	5.36	11.65	11.65	16.68	13.04	8.83	7.70	11.00	21.98	12.55	5.87	34.41	11.42	10.07	10.07	60.00	17.23	18.88	4.22	60.00
LA	22.02	24.93	53.17	26.43	15.61	9.88	17.49	17.49	20.14	21.43	24.51	28.33	29.58	31.25	22.37	8.64	77.12	21.20	18.98	18.98	51.76	24.21	13.77	13.78	66.41
ME	14.70	10.85	21.69	14.37	9.79	5.55	9.56	9.56	13.86	11.69	6.39	11.50	12.62	13.12	11.21	8.38	21.49	8.22	6.97	6.97	30.84	16.36	5.67	4.87	30.84
MD	16.12	12.09	13.45	17.65	5.24	5.74	7.23	7.23	11.72	13.83	4.53	10.17	7.24	22.67	11.81	6.40	29.04	12.17	8.52	8.52	56.70	24.19	7.39	5.42	24.19
MA	7.50	5.93	11.92	21.45	6.62	3.17	4.17	4.17	8.93	8.94	6.53	11.62	5.01	14.99	4.89	3.88	30.35	5.74	6.76	6.76	44.61	39.97	6.45	2.50	29.46
MI	20.23	9.63	20.23	17.30	8.33	4.58	11.76	11.76	12.90	10.98	13.59	14.95	13.12	51.77	10.05	6.78	29.95	11.08	10.17	10.17	51.77	13.99	9.52	6.17	51.77
MN	21.58	21.93	31.55	10.75	14.50	5.25	11.40	11.40	34.58	20.50	12.03	17.40	14.38	28.68	12.03	9.30	59.58	18.78	11.58	11.58	155.90	9.80	13.50	6.93	9.80
MS	15.14	10.81	14.94	10.42	6.74	5.60	9.95	9.95	10.94	9.28	6.27	14.58	12.36	26.40	13.13	8.38	36.90	7.23	13.97	13.97	37.66	12.49	9.54	4.69	37.66
MO	18.48	13.84	14.93	21.56	11.45	8.02	12.16	12.16	11.41	15.50	10.00	16.97	12.30	21.47	15.98	11.55	34.76	9.09	12.35	12.35	34.33	46.40	12.73	6.91	34.33
MT	12.71	9.67	25.69	13.47	11.92	6.12	13.11	13.11	9.86	28.05	7.55	10.94	7.34	18.44	11.08	8.28	37.59	8.86	8.53	8.53	20.44	10.38	10.20	8.69	10.38
NE	22.48	16.10	22.18	21.90	12.55	8.83	14.33	14.33	13.63	21.55	9.85	18.65	13.40	17.78	15.03	10.45	37.18	11.60	13.93	13.93	45.53	36.08	9.05	6.43	64.50
NV	11.97	7.52	11.36	8.77	7.79	5.12	8.77	8.77	8.23	8.26	4.07	6.55	7.54	9.58	6.66	6.13	14.83	9.39	7.46	7.46	24.81	16.86	4.41	4.95	18.14
NH	19.09	11.63	18.72	25.24	11.74	6.65	11.22	11.22	10.69	13.49	8.48	21.03	25.47	16.39	10.11	9.96	43.20	9.01	12.97	12.97	79.22	23.21	11.47	6.64	79.22
NJ	14.86	9.38	14.81	14.44	9.32	4.32	8.00	8.00	7.69	10.21	10.65	12.88	10.30	16.15	10.65	5.67	38.90	6.72	9.23	9.23	16.02	16.09	7.69	5.87	16.52
NM	19.58	8.17	18.99	13.33	10.11	8.08	8.28	8.28	18.17	11.75	6.68	15.36	11.43	18.20	14.39	8.14	37.47	9.92	12.79	12.79	54.49	47.45	7.66	6.66	54.49
NY	5.97	2.54	5.59	6.71	5.39	2.37	3.49	3.49	15.45	2.98	5.00	6.56	4.25	6.03	3.16	3.16	12.06	4.13	5.04	5.04	10.02	7.90	2.79	2.53	4.13
NC	15.50	11.55	18.63	17.12	8.14	9.96	13.11	13.11	11.87	12.57	9.34	11.05	11.85	16.82	13.63	8.92	33.88	11.87	11.97	11.97	70.40	20.69	8.12	5.41	70.40
ND	10.56	10.56	10.56	5.17	5.17	3.66	4.44	4.44	10.56	10.56	7.94	6.35	6.94	14.58	7.94	5.56	20.76	5.56	14.58	14.58	14.58	19.14	10.56	20.76	9.31
OH	8.30	4.39	6.71	6.58	5.27	3.78	4.99	4.99	7.82	8.20	23.30	7.55	7.86	7.63	5.54	4.36	17.26	4.87	6.38	6.38	12.99	11.61	6.16	5.65	7.74
OK	11.94	8.70	12.16	13.07	6.43	6.00	10.98	10.98	12.76	12.68	8.07	9.29	7.99	15.81	16.64	7.06	21.53	9.05	11.30	11.30	36.18	23.97	6.37	6.21	36.18
OR	16.95	7.57	14.30	10.92	7.82	5.37	9.61	9.61	11.51	11.15	7.46	12.31	11.60	14.60	9.99	5.10	25.58	6.63	7.09	7.09	21.15	21.87	7.63	5.93	21.15
PA	16.22	16.22	13.12	16.61	12.01	6.79	9.95	9.95	12.89	13.12	12.89	14.10	14.78	17.69	12.89	8.58	32.97	8.95	16.30	16.30	24.57	15.11	9.36	14.00	24.57
RI	10.79	8.37	10.96	13.14	10.29	5.15	7.50	7.50	13.33	17.64	6.04	11.62	9.27	27.21	9.50	4.42	14.82	7.75	5.68	5.68	20.92	14.98	5.62	4.72	14.98
SC	21.43	18.13	22.49	16.89	9.59	11.45	11.78	11.78	16.99	18.58	11.06	13.93	15.50	20.61	15.75	12.36	50.08	13.06	13.91	13.91	41.22	32.53	10.50	6.33	41.22
SD	21.47	9.62	21.86	25.23	7.35	5.33	9.54	9.54	15.06	13.87	8.26	11.62	10.45	20.56	11.87	10.92	26.74	12.70	8.86	8.86	59.15	31.01	7.69	5.99	59.15
TN	14.09	10.66	11.66	12.50	6.18	6.09	12.14	12.14	12.52	9.27	8.02	12.10	9.94	24.60	10.80	6.06	25.65	8.80	8.58	8.58	25.35	25.53	8.11	4.08	25.35
TX	9.17	8.39	9.17	9.00	7.06	5.76	7.88	7.88	8.56	11.97	5.74	10.31	8.18	16.38	8.18	5.88	19.42	12.05	7.87	7.87	33.20	11.24	5.53	6.19	8.70
UT	11.11	6.27	8.58	8.13	6.04	2.96	6.70	6.70	7.35	5.44	3.32	7.07	6.19	8.25	4.53	3.79	21.11	5.03	5.21	5.21	14.21	11.11	4.24	3.59	18.87
VT	12.73	12.09	17.18	11.43	8.18	5.46	11.10	11.10	15.78	14.64	7.03	16.92	9.46	13.49	8.82	8.68	25.30	8.45	9.53	9.53	40.51	30.71	9.66	6.41	40.51
VA	10.44	8.34	8.71	9.83	5.37	4.39	7.46	7.46	8.80	7.36	8.95	7.50	8.31	10.07	7.92	4.93	23.55	6.37	6.66	6.66	27.15	20.56	4.98	3.09	27.15
WA	7.58	7.58	7.58	7.65	7.65	2.92	6.45	6.45	12.60	7.04	7.58	9.40	12.79	17.67	9.14	4.28	16.55	4.09	7.25	7.25	7.25	7.61	6.59	16.55	7.25
WV	14.41	10.17	14.80	12.02	6.15	6.42	8.69	8.69	11.48	9.73	8.06	10.16	10.75	17.16	10.53	6.79	24.97	8.18	7.97	7.97	34.22	7.91	8.31	3.83	17.17
WI	11.71	9.86	17.41	11.55	8.51	6.39	8.17	8.17	11.44	13.60	5.38	14.52	12.34	24.28	10.17	6.03	33.37	8.83	10.63	10.63	20.55	48.12	17.76	4.94	20.55
WY	4.77	4.77	4.77	4.77	4.77	4.77	4.77	4.77	4.77	4.77	4.77	4.77	4.77	4.77	4.77	4.77	4.77	4.77	4.77	4.77	4.77	4.77	4.77	4.77	4.77
AVG.	14.52	10.93	16.07	13.33	8.35	5.81	9.10	9.10	12.89	12.30	8.59	12.66	11.07	18.00	11.43	7.04	29.20	8.94	9.86	9.86	36.30	20.12	8.45	6.43	30.85

R013113-60 Workers' Compensation (cont.) (Canada in Canadian dollars)

Province		Alberta	British Columbia	Manitoba	Ontario	New Brunswick	Newfndld. & Labrador	Northwest Territories	Nova Scotia	Prince Edward Island	Quebec	Saskat-chewan	Yukon
Carpentry—3 stories or less	Rate	6.42	3.64	4.23	4.35	5.20	10.67	3.54	6.91	5.61	13.18	4.07	9.22
	Code	42143	721028	40102	723	238130	4226	4-41	4226	401	80110	1226	202
Carpentry—interior cab. work	Rate	2.43	2.62	4.23	4.35	4.64	4.29	3.54	4.73	3.44	13.18	2.41	9.22
	Code	42133	721021	40102	723	238350	4279	4-41	4274	402	80110	B1127	202
CARPENTRY—general	Rate	6.42	3.64	4.23	4.35	5.20	4.29	3.54	6.91	5.61	13.18	4.07	9.22
	Code	42143	721028	40102	723	238130	4299	4-41	4226	401	80110	12.02	202
CONCRETE WORK—NOC	Rate	4.11	3.87	6.79	16.50	5.20	10.67	3.54	4.85	5.61	14.82	4.80	7.12
	Code	42104	721010	40110	748	238110	4224	4-41	4224	401	80100	B1314	203
CONCRETE WORK—flat (flr. sidewalk)	Rate	4.11	3.87	6.79	16.50	5.20	10.67	3.54	4.85	5.61	14.82	4.80	7.12
	Code	42104	721010	40110	748	238110	4224	4-41	4224	401	80100	B1314	203
ELECTRICAL Wiring—inside	Rate	1.97	1.41	1.96	3.25	2.54	2.76	2.75	2.62	3.44	5.25	2.41	5.51
	Code	42124	721019	40203	704	238210	4261	4-46	4261	402	80170	B1105	206
EXCAVATION—earth NOC	Rate	2.36	3.25	4.52	4.68	5.20	3.67	3.79	2.85	2.82	8.43	2.42	5.51
	Code	40604	721031	40706	711	238190	4214	4-43	4214	404	80030	R1106	207
EXCAVATION—rock	Rate	2.36	3.25	4.52	4.68	5.20	3.67	3.79	2.85	2.82	8.43	2.42	5.51
	Code	40604	721031	40706	711	238190	4214	4-43	4214	404	80030	R1106	207
GLAZIERS	Rate	2.75	3.05	4.23	9.25	5.20	5.40	3.54	5.76	3.44	12.63	4.80	5.51
	Code	42121	715020	40109	751	238150	4233	4-41	4233	402	80150	B1304	212
INSULATION WORK	Rate	2.04	6.91	4.23	9.25	4.64	5.40	3.54	5.76	5.61	13.18	4.07	9.22
	Code	42184	721029	40102	751	238310	4234	4-41	4234	401	80110	B1207	202
LATHING	Rate	4.15	6.27	4.23	4.35	4.64	4.29	3.54	4.73	3.44	13.18	4.80	9.22
	Code	42135	721042	40102	723	238390	4279	4-41	4271	402	80110	B1316	202
MASONRY	Rate	4.11	3.50	4.23	11.15	5.20	5.40	3.54	5.76	5.61	14.82	4.80	9.22
	Code	42102	721037	40102	741	238140	4231	4-41	4231	401	80100	B1318	202
PAINTING & DECORATING	Rate	3.39	3.36	3.49	6.75	4.64	4.29	3.54	4.73	3.44	13.18	4.07	9.22
	Code	42111	721041	40105	719	238320	4275	4-41	4275	402	80110	B1201	202
PILE DRIVING	Rate	4.04	3.82	4.52	6.34	5.20	5.67	3.79	5.19	5.61	8.43	4.80	9.22
	Code	42159	722004	40706	732	238190	4129	4-43	4221	401	80030	B1310	202
PLASTERING	Rate	4.15	6.27	4.46	6.75	4.64	4.29	3.54	4.73	3.44	13.18	4.07	9.22
	Code	42135	721042	40108	719	238390	4271	4-41	4271	402	80110	B1221	202
PLUMBING	Rate	1.97	1.90	3.09	3.98	2.54	2.77	2.75	2.76	3.44	6.24	2.41	5.51
	Code	42122	721043	40204	707	238220	4241	4-46	4241	402	80160	B1101	214
ROOFING	Rate	6.59	6.18	7.23	13.30	5.20	10.67	3.54	7.45	5.61	19.78	4.80	9.22
	Code	42118	721036	40403	728	238160	4236	4-41	4236	401	80130	B1320	202
SHEET METAL WORK (HVAC)	Rate	1.97	1.90	7.23	3.98	2.54	2.77	2.75	2.76	3.44	6.24	2.41	5.51
	Code	42117	721043	40402	707	238220	4244	4-46	4244	402	80160	B1107	208
STEEL ERECTION—door & sash	Rate	2.04	12.66	10.32	16.50	5.20	10.67	3.54	5.76	5.61	22.38	4.80	9.22
	Code	42106	722005	40502	748	238120	4227	4-41	4227	401	80080	B1322	202
STEEL ERECTION—inter., ornam.	Rate	2.04	12.66	10.32	16.50	5.20	10.67	3.54	5.76	5.61	22.38	4.80	9.22
	Code	42106	722005	40502	748	238120	4227	4-41	4227	401	80080	B1322	202
STEEL ERECTION—structure	Rate	2.04	12.66	10.32	16.50	5.20	10.67	3.54	5.76	5.61	22.38	4.80	9.22
	Code	42106	722005	40502	748	238120	4227	4-41	4227	401	80080	B1322	202
STEEL ERECTION—NOC	Rate	2.04	12.66	10.32	16.50	5.20	10.67	3.54	5.76	5.61	22.38	4.80	9.22
	Code	42106	722005	40502	748	238120	4227	4-41	4227	401	80080	B1322	202
TILE WORK—inter. (ceramic)	Rate	3.24	3.62	2.00	6.75	4.64	4.29	3.54	4.73	3.44	13.18	4.80	9.22
	Code	42113	721054	40103	719	238340	4276	4-41	4276	402	80110	B1301	202
WATERPROOFING	Rate	3.39	3.35	4.23	4.35	5.20	4.29	3.54	5.76	3.44	19.78	4.07	9.22
	Code	42139	721016	40102	723	238190	4299	4-41	4239	402	80130	B1217	202
WRECKING	Rate	2.36	4.44	7.34	16.50	5.20	3.67	3.79	2.85	5.61	13.18	2.42	9.22
	Code	40604	721005	40106	748	238190	4211	4-43	4211	401	80110	R1125	202

General Requirements — R0131 Project Management & Coordination

R013113-80 Performance Bond

This table shows the cost of a Performance Bond for a construction job scheduled to be completed in 12 months. Add 1% of the premium cost per month for jobs requiring more than 12 months to complete. The rates are "standard" rates offered to contractors that the bonding company considers financially sound and capable of doing the work. Preferred rates are offered by some bonding companies based upon financial strength of the contractor. Actual rates vary from contractor to contractor and from bonding company to bonding company. Contractors should prequalify through a bonding agency before submitting a bid on a contract that requires a bond.

Contract Amount	Building Construction Class B Projects	Highways & Bridges: Class A New Construction	Highways & Bridges: Class A-1 Highway Resurfacing
First $ 100,000 bid	$25.00 per M	$15.00 per M	$9.40 per M
Next 400,000 bid	$ 2,500 plus $15.00 per M	$ 1,500 plus $10.00 per M	$ 940 plus $7.20 per M
Next 2,000,000 bid	8,500 plus 10.00 per M	5,500 plus 7.00 per M	3,820 plus 5.00 per M
Next 2,500,000 bid	28,500 plus 7.50 per M	19,500 plus 5.50 per M	15,820 plus 4.50 per M
Next 2,500,000 bid	47,250 plus 7.00 per M	33,250 plus 5.00 per M	28,320 plus 4.50 per M
Over 7,500,000 bid	64,750 plus 6.00 per M	45,750 plus 4.50 per M	39,570 plus 4.00 per M

General Requirements — R0151 Temporary Utilities

R015113-65 Temporary Power Equipment

Cost data for the temporary equipment was developed utilizing the following information.

1) Re-usable material-services, transformers, equipment and cords are based on new purchase and prorated to three projects.
2) PVC feeder includes trench and backfill.
3) Connections include disconnects and fuses.
4) Labor units include an allowance for removal.
5) No utility company charges or fees are included.
6) Concrete pads or vaults are not included.
7) Utility company conduits not included.

R015423-10 Steel Tubular Scaffolding

On new construction, tubular scaffolding is efficient up to 60′ high or five stories. Above this it is usually better to use a hung scaffolding if construction permits. Swing scaffolding operations may interfere with tenants. In this case, the tubular is more practical at all heights.

In repairing or cleaning the front of an existing building the cost of tubular scaffolding per S.F. of building front increases as the height increases above the first tier. The first tier cost is relatively high due to leveling and alignment.

The minimum efficient crew for erecting and dismantling is three workers. They can set up and remove 18 frame sections per day up to 5 stories high. For 6 to 12 stories high, a crew of four is most efficient. Use two or more on top and two on the bottom for handing up or hoisting. They can also set up and remove 18 frame sections per day. At 7' horizontal spacing, this will run about 800 S.F. per day of erecting and dismantling. Time for placing and removing planks must be added to the above. A crew of three can place and remove 72 planks per day up to 5 stories. For over 5 stories, a crew of four can place and remove 80 planks per day.

The table below shows the number of pieces required to erect tubular steel scaffolding for 1000 S.F. of building frontage. This area is made up of a scaffolding system that is 12 frames (11 bays) long by 2 frames high.

For jobs under twenty-five frames, add 50% to rental cost. Rental rates will be lower for jobs over three months duration. Large quantities for long periods can reduce rental rates by 20%.

Description of Component	Number of Pieces for 1000 S.F. of Building Front	Unit
5′ Wide Standard Frame, 6′-4″ High	24	Ea.
Leveling Jack & Plate	24	
Cross Brace	44	
Side Arm Bracket, 21″	12	
Guardrail Post	12	
Guardrail, 7′ section	22	
Stairway Section	2	
Stairway Starter Bar	1	
Stairway Inside Handrail	2	
Stairway Outside Handrail	2	
Walk-Thru Frame Guardrail	2	↓

Scaffolding is often used as falsework over 15′ high during construction of cast-in-place concrete beams and slabs. Two foot wide scaffolding is generally used for heavy beam construction. The span between frames depends upon the load to be carried with a maximum span of 5′.

Heavy duty shoring frames with a capacity of 10,000#/leg can be spaced up to 10′ O.C. depending upon form support design and loading.

Scaffolding used as horizontal shoring requires less than half the material required with conventional shoring.

On new construction, erection is done by carpenters.

Rolling towers supporting horizontal shores can reduce labor and speed the job. For maintenance work, catwalks with spans up to 70′ can be supported by the rolling towers.

R015423-20 Pump Staging

Pump staging is generally not available for rent. The table below shows the number of pieces required to erect pump staging for 2400 S.F. of building frontage. This area is made up of a pump jack system that is 3 poles (2 bays) wide by 2 poles high.

Item	Number of Pieces for 2400 S.F. of Building Front	Unit
Aluminum pole section, 24′ long	6	Ea.
Aluminum splice joint, 6′ long	3	
Aluminum foldable brace	3	
Aluminum pump jack	3	
Aluminum support for workbench/back safety rail	3	
Aluminum scaffold plank/workbench, 14″ wide x 24′ long	4	
Safety net, 22′ long	2	
Aluminum plank end safety rail	2	↓

The cost in place for this 2400 S.F. will depend on how many uses are realized during the life of the equipment.

R015433-10 Contractor Equipment

Rental Rates shown elsewhere in the book pertain to late model high quality machines in excellent working condition, rented from equipment dealers. Rental rates from contractors may be substantially lower than the rental rates from equipment dealers depending upon economic conditions; for older, less productive machines, reduce rates by a maximum of 15%. Any overtime must be added to the base rates. For shift work, rates are lower. Usual rule of thumb is 150% of one shift rate for two shifts; 200% for three shifts.

For periods of less than one week, operated equipment is usually more economical to rent than renting bare equipment and hiring an operator.

Costs to move equipment to a job site (mobilization) or from a job site (demobilization) are not included in rental rates, nor in any Equipment costs on any Unit Price line items or crew listings. These costs can be found elsewhere. If a piece of equipment is already at a job site, it is not appropriate to utilize mob/demob costs in an estimate again.

Rental rates vary throughout the country with larger cities generally having lower rates. Lease plans for new equipment are available for periods in excess of six months with a percentage of payments applying toward purchase.

Rental rates can also be treated as reimbursement costs for contractor-owned equipment. Owned equipment costs include depreciation, loan payments, interest, taxes, insurance, storage, and major repairs.

Monthly rental rates vary from 2% to 5% of the cost of the equipment depending on the anticipated life of the equipment and its wearing parts. Weekly rates are about 1/3 the monthly rates and daily rental rates about 1/3 the weekly rate.

The hourly operating costs for each piece of equipment include costs to the user such as fuel, oil, lubrication, normal expendables for the equipment, and a percentage of mechanic's wages chargeable to maintenance. The hourly operating costs listed do not include the operator's wages.

The daily cost for equipment used in the standard crews is figured by dividing the weekly rate by five, then adding eight times the hourly operating cost to give the total daily equipment cost, not including the operator. This figure is in the right hand column of the Equipment listings under Equipment Cost/Day.

Pile Driving rates shown for pile hammer and extractor do not include leads, crane, boiler or compressor. Vibratory pile driving requires an added field specialist during set-up and pile driving operation for the electric model. The hydraulic model requires a field specialist for set-up only. Up to 125 reuses of sheet piling are possible using vibratory drivers. For normal conditions, crane capacity for hammer type and size are as follows.

Crane Capacity	Hammer Type and Size		
	Air or Steam	Diesel	Vibratory
25 ton	to 8,750 ft.-lb.		70 H.P.
40 ton	15,000 ft.-lb.	to 32,000 ft.-lb.	170 H.P.
60 ton	25,000 ft.-lb.		300 H.P.
100 ton		112,000 ft.-lb.	

Cranes should be specified for the job by size, building and site characteristics, availability, performance characteristics, and duration of time required.

Backhoes & Shovels rent for about the same as equivalent size cranes but maintenance and operating expense is higher. Crane operators rate must be adjusted for high boom heights. Average adjustments: for 150′ boom add 2% per hour; over 185′, add 4% per hour; over 210′, add 6% per hour; over 250′, add 8% per hour and over 295′, add 12% per hour.

Tower Cranes of the climbing or static type have jibs from 50′ to 200′ and capacities at maximum reach range from 4,000 to 14,000 pounds. Lifting capacities increase up to maximum load as the hook radius decreases.

Typical rental rates, based on purchase price are about 2% to 3% per month.

Erection and dismantling runs between 500 and 2000 labor hours. Climbing operation takes 10 labor hours per 20′ climb. Crane dead time is about 5 hours per 40′ climb. If crane is bolted to side of the building add cost of ties and extra mast sections. Climbing cranes have from 80′ to 180′ of mast while static cranes have 80′ to 800′ of mast.

Truck Cranes can be converted to tower cranes by using tower attachments. Mast heights over 400′ have been used.

A single 100′ high material **Hoist and Tower** can be erected and dismantled in about 400 labor hours; a double 100′ high hoist and tower in about 600 labor hours. Erection times for additional heights are 3 and 4 labor hours per vertical foot respectively up to 150′, and 4 to 5 labor hours per vertical foot over 150′ high. A 40′ high portable Buck hoist takes about 160 labor hours to erect and dismantle. Additional heights take 2 labor hours per vertical foot to 80′ and 3 labor hours per vertical foot for the next 100′. Most material hoists do not meet local code requirements for carrying personnel.

A 150′ high **Personnel Hoist** requires about 500 to 800 labor hours to erect and dismantle. Budget erection time at 5 labor hours per vertical foot for all trades. Local code requirements or labor scarcity requiring overtime can add up to 50% to any of the above erection costs.

Earthmoving Equipment: The selection of earthmoving equipment depends upon the type and quantity of material, moisture content, haul distance, haul road, time available, and equipment available. Short haul cut and fill operations may require dozers only, while another operation may require excavators, a fleet of trucks, and spreading and compaction equipment. Stockpiled material and granular material are easily excavated with front end loaders. Scrapers are most economically used with hauls between 300′ and 1-1/2 miles if adequate haul roads can be maintained. Shovels are often used for blasted rock and any material where a vertical face of 8′ or more can be excavated. Special conditions may dictate the use of draglines, clamshells, or backhoes. Spreading and compaction equipment must be matched to the soil characteristics, the compaction required and the rate the fill is being supplied.

R015433-15 Heavy Lifting

Hydraulic Climbing Jacks

The use of hydraulic heavy lift systems is an alternative to conventional type crane equipment. The lifting, lowering, pushing, or pulling mechanism is a hydraulic climbing jack moving on a square steel jackrod from 1-5/8″ to 4″ square, or a steel cable. The jackrod or cable can be vertical or horizontal, stationary or movable, depending on the individual application. When the jackrod is stationary, the climbing jack will climb the rod and push or pull the load along with itself. When the climbing jack is stationary, the jackrod is movable with the load attached to the end and the climbing jack will lift or lower the jackrod with the attached load. The heavy lift system is normally operated by a single control lever located at the hydraulic pump.

The system is flexible in that one or more climbing jacks can be applied wherever a load support point is required, and the rate of lift synchronized.

Economic benefits have been demonstrated on projects such as: erection of ground assembled roofs and floors, complete bridge spans, girders and trusses, towers, chimney liners and steel vessels, storage tanks, and heavy machinery. Other uses are raising and lowering offshore work platforms, caissons, tunnel sections and pipelines.

R024119-10 Demolition Defined

Whole Building Demolition - Demolition of the whole building with no concern for any particular building element, component, or material type being demolished. This type of demolition is accomplished with large pieces of construction equipment that break up the structure, load it into trucks and haul it to a disposal site, but disposal or dump fees are not included. Demolition of below-grade foundation elements, such as footings, foundation walls, grade beams, slabs on grade, etc., is not included. Certain mechanical equipment containing flammable liquids or ozone-depleting refrigerants, electric lighting elements, communication equipment components, and other building elements may contain hazardous waste, and must be removed, either selectively or carefully, as hazardous waste before the building can be demolished.

Foundation Demolition - Demolition of below-grade foundation footings, foundation walls, grade beams, and slabs on grade. This type of demolition is accomplished by hand or pneumatic hand tools, and does not include saw cutting, or handling, loading, hauling, or disposal of the debris.

Gutting - Removal of building interior finishes and electrical/mechanical systems down to the load-bearing and sub-floor elements of the rough building frame, with no concern for any particular building element, component, or material type being demolished. This type of demolition is accomplished by hand or pneumatic hand tools, and includes loading into trucks, but not hauling, disposal or dump fees, scaffolding, or shoring. Certain mechanical equipment containing flammable liquids or ozone-depleting refrigerants, electric lighting elements, communication equipment components, and other building elements may contain hazardous waste, and must be removed, either selectively or carefully, as hazardous waste, before the building is gutted.

Selective Demolition - Demolition of a selected building element, component, or finish, with some concern for surrounding or adjacent elements, components, or finishes (see the first Subdivision(s) at the beginning of appropriate Divisions). This type of demolition is accomplished by hand or pneumatic hand tools, and does not include handling, loading, storing, hauling, or disposal of the debris, scaffolding, or shoring. "Gutting" methods may be used in order to save time, but damage that is caused to surrounding or adjacent elements, components, or finishes may have to be repaired at a later time.

Careful Removal - Removal of a piece of service equipment, building element or component, or material type, with great concern for both the removed item and surrounding or adjacent elements, components or finishes. The purpose of careful removal may be to protect the removed item for later re-use, preserve a higher salvage value of the removed item, or replace an item while taking care to protect surrounding or adjacent elements, components, connections, or finishes from cosmetic and/or structural damage. An approximation of the time required to perform this type of removal is 1/3 to 1/2 the time it would take to install a new item of like kind (see Reference Number R220105-10). This type of removal is accomplished by hand or pneumatic hand tools, and does not include loading, hauling, or storing the removed item, scaffolding, shoring, or lifting equipment.

Cutout Demolition - Demolition of a small quantity of floor, wall, roof, or other assembly, with concern for the appearance and structural integrity of the surrounding materials. This type of demolition is accomplished by hand or pneumatic hand tools, and does not include saw cutting, handling, loading, hauling, or disposal of debris, scaffolding, or shoring.

Rubbish Handling - Work activities that involve handling, loading or hauling of debris. Generally, the cost of rubbish handling must be added to the cost of all types of demolition, with the exception of whole building demolition.

Minor Site Demolition - Demolition of site elements outside the footprint of a building. This type of demolition is accomplished by hand or pneumatic hand tools, or with larger pieces of construction equipment, and may include loading a removed item onto a truck (check the Crew for equipment used). It does not include saw cutting, hauling or disposal of debris, and, sometimes, handling or loading.

R024119-20 Dumpsters

Dumpster rental costs on construction sites are presented in two ways.

The cost per week rental includes the delivery of the dumpster; its pulling or emptying once per week, and its final removal. The assumption is made that the dumpster contractor could choose to empty a dumpster by simply bringing in an empty unit and removing the full one. These costs also include the disposal of the materials in the dumpster.

The Alternate Pricing can be used when actual planned conditions are not approximated by the weekly numbers. For example, these lines can be used when a dumpster is needed for 4 weeks and will need to be emptied 2 or 3 times per week. Conversely the Alternate Pricing lines can be used when a dumpster will be rented for several weeks or months but needs to be emptied only a few times over this period.

Existing Conditions R0265 Underground Storage Tank Removal

R026510-20 Underground Storage Tank Removal

Underground Storage Tank Removal can be divided into two categories: Non-Leaking and Leaking. Prior to removing an underground storage tank, tests should be made, with the proper authorities present, to determine whether a tank has been leaking or the surrounding soil has been contaminated.

To safely remove Liquid Underground Storage Tanks:

1. Excavate to the top of the tank.
2. Disconnect all piping.
3. Open all tank vents and access ports.
4. Remove all liquids and/or sludge.
5. Purge the tank with an inert gas.
6. Provide access to the inside of the tank and clean out the interior using proper personal protective equipment (PPE).
7. Excavate soil surrounding the tank using proper PPE for on-site personnel.
8. Pull and properly dispose of the tank.
9. Clean up the site of all contaminated material.
10. Install new tanks or close the excavation.

Existing Conditions — R0282 Asbestos Remediation

R028213-20 Asbestos Removal Process

Asbestos removal is accomplished by a specialty contractor who understands the federal and state regulations regarding the handling and disposal of the material. The process of asbestos removal is divided into many individual steps. An accurate estimate can be calculated only after all the steps have been priced.

The steps are generally as follows:

1. Obtain an asbestos abatement plan from an industrial hygienist.
2. Monitor the air quality in and around the removal area and along the path of travel between the removal area and transport area. This establishes the background contamination.
3. Construct a two part decontamination chamber at entrance to removal area.
4. Install a HEPA filter to create a negative pressure in the removal area.
5. Install wall, floor and ceiling protection as required by the plan, usually 2 layers of fireproof 6 mil polyethylene.
6. Industrial hygienist visually inspects work area to verify compliance with plan.
7. Provide temporary supports for conduit and piping affected by the removal process.
8. Proceed with asbestos removal and bagging process. Monitor air quality as described in Step #2. Discontinue operations when contaminate levels exceed applicable standards.
9. Document the legal disposal of materials in accordance with EPA standards.
10. Thoroughly clean removal area including all ledges, crevices and surfaces.
11. Post abatement inspection by industrial hygienist to verify plan compliance.
12. Provide a certificate from a licensed industrial hygienist attesting that contaminate levels are within acceptable standards before returning area to regular use.

Existing Conditions — R0283 Lead Remediation

R028319-60 Lead Paint Remediation Methods

Lead paint remediation can be accomplished by the following methods.

1. Abrasive blast
2. Chemical stripping
3. Power tool cleaning with vacuum collection system
4. Encapsulation
5. Remove and replace
6. Enclosure

Each of these methods has strengths and weakness depending on the specific circumstances of the project. The following is an overview of each method.

1. **Abrasive blasting** is usually accomplished with sand or recyclable metallic blast. Before work can begin, the area must be contained to ensure the blast material with lead does not escape to the atmosphere. The use of vacuum blast greatly reduces the containment requirements. Lead abatement equipment that may be associated with this work includes a negative air machine. In addition, it is necessary to have an industrial hygienist monitor the project on a continual basis. When the work is complete, the spent blast sand with lead must be disposed of as a hazardous material. If metallic shot was used, the lead is separated from the shot and disposed of as hazardous material. Worker protection includes disposable clothing and respiratory protection.
2. **Chemical stripping** requires strong chemicals be applied to the surface to remove the lead paint. Before the work can begin, the area under/adjacent to the work area must be covered to catch the chemical and removed lead. After the chemical is applied to the painted surface it is usually covered with paper. The chemical is left in place for the specified period, then the paper with lead paint is pulled or scraped off. The process may require several chemical applications. The paper with chemicals and lead paint adhered to it, plus the containment and loose scrapings collected by a HEPA (High Efficiency Particulate Air Filter) vac, must be disposed of as a hazardous material. The chemical stripping process usually requires a neutralizing agent and several wash downs after the paint is removed. Worker protection includes a neoprene or other compatible protective clothing and respiratory protection with face shield. An industrial hygienist is required intermittently during the process.
3. **Power tool cleaning** is accomplished using shrouded needle blasting guns. The shrouding with different end configurations is held up against the surface to be cleaned. The area is blasted with hardened needles and the shroud captures the lead with a HEPA vac and deposits it in a holding tank. An industrial hygienist monitors the project, protective clothing and a respirator is required until air samples prove otherwise. When the work is complete the lead must be disposed of as a hazardous material.
4. **Encapsulation** is a method that leaves the well bonded lead paint in place after the peeling paint has been removed. Before the work can begin, the area under/adjacent to the work must be covered to catch the scrapings. The scraped surface is then washed with a detergent and rinsed. The prepared surface is covered with approximately 10 mils of paint. A reinforcing fabric can also be embedded in the paint covering. The scraped paint and containment must be disposed of as a hazardous material. Workers must wear protective clothing and respirators.
5. **Remove and replace** is an effective way to remove lead paint from windows, gypsum walls and concrete masonry surfaces. The painted materials are removed and new materials are installed. Workers should wear a respirator and tyvek suit. The demolished materials must be disposed of as hazardous waste if it fails the TCLP (Toxicity Characteristic Leachate Process) test.
6. **Enclosure** is the process that permanently seals lead painted materials in place. This process has many applications such as covering lead painted drywall with new drywall, covering exterior construction with tyvek paper then residing, or covering lead painted structural members with aluminum or plastic. The seams on all enclosing materials must be securely sealed. An industrial hygienist monitors the project, and protective clothing and a respirator is required until air samples prove otherwise.

All the processes require clearance monitoring and wipe testing as required by the hygienist.

R031113-10 Wall Form Materials

Aluminum Forms

Approximate weight is 3 lbs. per S.F.C.A. Standard widths are available from 4″ to 36″ with 36″ most common. Standard lengths of 2′, 4′, 6′ to 8′ are available. Forms are lightweight and fewer ties are needed with the wider widths. The form face is either smooth or textured.

Metal Framed Plywood Forms

Manufacturers claim over 75 reuses of plywood and over 300 reuses of steel frames. Many specials such as corners, fillers, pilasters, etc. are available. Monthly rental is generally about 15% of purchase price for first month and 9% per month thereafter with 90% of rental applied to purchase for the first month and decreasing percentages thereafter. Aluminum framed forms cost 25% to 30% more than steel framed.

After the first month, extra days may be prorated from the monthly charge. Rental rates do not include ties, accessories, cleaning, loss of hardware or freight in and out. Approximate weight is 5 lbs. per S.F. for steel; 3 lbs. per S.F. for aluminum.

Forms can be rented with option to buy.

Plywood Forms, Job Fabricated

There are two types of plywood used for concrete forms.

1. Exterior plyform which is completely waterproof. This is face oiled to facilitate stripping. Ten reuses can be expected with this type with 25 reuses possible.
2. An overlaid type consists of a resin fiber fused to exterior plyform. No oiling is required except to facilitate cleaning. This is available in both high density (HDO) and medium density overlaid (MDO). Using HDO, 50 reuses can be expected with 200 possible.

Plyform is available in 5/8″ and 3/4″ thickness. High density overlaid is available in 3/8″, 1/2″, 5/8″ and 3/4″ thickness.

5/8″ thick is sufficient for most building forms, while 3/4″ is best on heavy construction.

Plywood Forms, Modular, Prefabricated

There are many plywood forming systems without frames. Most of these are manufactured from 1-1/8″ (HDO) plywood and have some hardware attached. These are used principally for foundation walls 8′ or less high. With care and maintenance, 100 reuses can be attained with decreasing quality of surface finish.

Steel Forms

Approximate weight is 6-1/2 lbs. per S.F.C.A. including accessories. Standard widths are available from 2″ to 24″, with 24″ most common. Standard lengths are from 2′ to 8′, with 4′ the most common. Forms are easily ganged into modular units.

Forms are usually leased for 15% of the purchase price per month prorated daily over 30 days.

Rental may be applied to sale price, and usually rental forms are bought. With careful handling and cleaning 200 to 400 reuses are possible.

Straight wall gang forms up to 12′ x 20′ or 8′ x 30′ can be fabricated. These crane handled forms usually lease for approx. 9% per month.

Individual job analysis is available from the manufacturer at no charge.

R031113-30 Slipforms

The slipform method of forming may be used for forming circular silo and multi-celled storage bin type structures over 30′ high, and building core shear walls over eight stories high. The shear walls, usually enclose elevator shafts, stairwells, mechanical spaces, and toilet rooms. Reuse of the form on duplicate structures will reduce the height necessary and spread the cost of building the form. Slipform systems can be used to cast chimneys, towers, piers, dams, underground shafts or other structures capable of being extruded.

Slipforms are usually 4′ high and are raised semi-continuously by jacks climbing on rods which are embedded in the concrete. The jacks are powered by a hydraulic, pneumatic, or electric source and are available in 3, 6, and 22 ton capacities. Interior work decks and exterior scaffolds must be provided for placing inserts, embedded items, reinforcing steel, and concrete. Scaffolds below the form for finishers may be required. The interior work decks are often used as roof slab forms on silos and bin work. Form raising rates will range from 6″ to 20″ per hour for silos; 6″ to 30″ per hour for buildings; and 6″ to 48″ per hour for shaft work.

Reinforcing bars and stressing strands are usually hoisted by crane or gin pole, and the concrete material can be hoisted by crane, winch-powered skip, or pumps. The slipform system is operated on a continuous 24-hour day when a monolithic structure is desired. For least cost, the system is operated only during normal working hours.

Placing concrete will range from 0.5 to 1.5 labor-hours per C.Y. Bucks, blockouts, keyways, weldplates, etc. are extra.

R031113-40 Forms for Reinforced Concrete

Design Economy

Avoid many sizes in proportioning beams and columns.

From story to story avoid changing column dimensions. Gain strength by adding steel or using a richer mix. If a change in size of column is necessary, vary one dimension only to minimize form alterations. Keep beams and columns the same width.

From floor to floor in a multi-story building vary beam depth, not width, as that will leave slab panel form unchanged. It is cheaper to vary the strength of a beam from floor to floor by means of steel area than by 2″ changes in either width or depth.

Cost Factors

Material includes the cost of lumber, cost of rent for metal pans or forms if used, nails, form ties, form oil, bolts and accessories.

Labor includes the cost of carpenters to make up, erect, remove and repair, plus common labor to clean and move. Having carpenters remove forms minimizes repairs.

Improper alignment and condition of forms will increase finishing cost. When forms are heavily oiled, concrete surfaces must be neutralized before finishing. Special curing compounds will cause spillages to spall off in first frost. Gang forming methods will reduce costs on large projects.

Materials Used

Boards are seldom used unless their architectural finish is required. Generally, steel, fiberglass and plywood are used for contact surfaces. Labor on plywood is 10% less than with boards. The plywood is backed up with 2 x 4's at 12″ to 32″ O.C. Walers are generally 2 - 2 x 4's. Column forms are held together with steel yokes or bands. Shoring is with adjustable shoring or scaffolding for high ceilings.

Reuse

Floor and column forms can be reused four or possibly five times without excessive repair. Remember to allow for 10% waste on each reuse.

When modular sized wall forms are made, up to twenty uses can be expected with exterior plyform.

When forms are reused, the cost to erect, strip, clean and move will not be affected. 10% replacement of lumber should be included and about one hour of carpenter time for repairs on each reuse per 100 S.F.

The reuse cost for certain accessory items normally rented on a monthly basis will be lower than the cost for the first use.

After fifth use, new material required plus time needed for repair prevent form cost from dropping further and it may go up. Much depends on care in stripping, the number of special bays, changes in beam or column sizes and other factors.

Costs for multiple use of formwork may be developed as follows:

2 Uses

$$\frac{\text{(1st Use + Reuse)}}{2} = \text{avg. cost/2 uses}$$

3 Uses

$$\frac{\text{(1st Use + 2 Reuse)}}{3} = \text{avg. cost/3 uses}$$

4 Uses

$$\frac{\text{(1st use + 3 Reuse)}}{4} = \text{avg. cost/4 uses}$$

R031113-60 Formwork Labor-Hours

Item		Hours Required			Total Hours	Multiple Use		
	Unit	Fabricate	Erect & Strip	Clean & Move	1 Use	2 Use	3 Use	4 Use
Beam and Girder, interior beams, 12″ wide	100 S.F.	6.4	8.3	1.3	16.0	13.3	12.4	12.0
Hung from steel beams		5.8	7.7	1.3	14.8	12.4	11.6	11.2
Beam sides only, 36″ high		5.8	7.2	1.3	14.3	11.9	11.1	10.7
Beam bottoms only, 24″ wide		6.6	13.0	1.3	20.9	18.1	17.2	16.7
Box out for openings		9.9	10.0	1.1	21.0	16.6	15.1	14.3
Buttress forms, to 8′ high		6.0	6.5	1.2	13.7	11.2	10.4	10.0
Centering, steel, 3/4″ rib lath			1.0		1.0			
3/8″ rib lath or slab form			0.9		0.9			
Chamfer strip or keyway	100 L.F.		1.5		1.5	1.5	1.5	1.5
Columns, fiber tube 8″ diameter			20.6		20.6			
12″			21.3		21.3			
16″			22.9		22.9			
20″			23.7		23.7			
24″			24.6		24.6			
30″			25.6		25.6			
Columns, round steel, 12″ diameter			22.0		22.0	22.0	22.0	22.0
16″			25.6		25.6	25.6	25.6	25.6
20″			30.5		30.5	30.5	30.5	30.5
24″			37.7		37.7	37.7	37.7	37.7
Columns, plywood 8″ x 8″	100 S.F.	7.0	11.0	1.2	19.2	16.2	15.2	14.7
12″ x 12″		6.0	10.5	1.2	17.7	15.2	14.4	14.0
16″ x 16″		5.9	10.0	1.2	17.1	14.7	13.8	13.4
24″ x 24″		5.8	9.8	1.2	16.8	14.4	13.6	13.2
Columns, steel framed plywood 8″ x 8″			10.0	1.0	11.0	11.0	11.0	11.0
12″ x 12″			9.3	1.0	10.3	10.3	10.3	10.3
16″ x 16″			8.5	1.0	9.5	9.5	9.5	9.5
24″ x 24″			7.8	1.0	8.8	8.8	8.8	8.8
Drop head forms, plywood		9.0	12.5	1.5	23.0	19.0	17.7	17.0
Coping forms		8.5	15.0	1.5	25.0	21.3	20.0	19.4
Culvert, box			14.5	4.3	18.8	18.8	18.8	18.8
Curb forms, 6″ to 12″ high, on grade		5.0	8.5	1.2	14.7	12.7	12.1	11.7
On elevated slabs		6.0	10.8	1.2	18.0	15.5	14.7	14.3
Edge forms to 6″ high, on grade	100 L.F.	2.0	3.5	0.6	6.1	5.6	5.4	5.3
7″ to 12″ high	100 S.F.	2.5	5.0	1.0	8.5	7.8	7.5	7.4
Equipment foundations		10.0	18.0	2.0	30.0	25.5	24.0	23.3
Flat slabs, including drops		3.5	6.0	1.2	10.7	9.5	9.0	8.8
Hung from steel		3.0	5.5	1.2	9.7	8.7	8.4	8.2
Closed deck for domes		3.0	5.8	1.2	10.0	9.0	8.7	8.5
Open deck for pans		2.2	5.3	1.0	8.5	7.9	7.7	7.6
Footings, continuous, 12″ high		3.5	3.5	1.5	8.5	7.3	6.8	6.6
Spread, 12″ high		4.7	4.2	1.6	10.5	8.7	8.0	7.7
Pile caps, square or rectangular		4.5	5.0	1.5	11.0	9.3	8.7	8.4
Grade beams, 24″ deep		2.5	5.3	1.2	9.0	8.3	8.0	7.9
Lintel or Sill forms		8.0	17.0	2.0	27.0	23.5	22.3	21.8
Spandrel beams, 12″ wide		9.0	11.2	1.3	21.5	17.5	16.2	15.5
Stairs			25.0	4.0	29.0	29.0	29.0	29.0
Trench forms in floor		4.5	14.0	1.5	20.0	18.3	17.7	17.4
Walls, Plywood, at grade, to 8′ high		5.0	6.5	1.5	13.0	11.0	9.7	9.5
8′ to 16′		7.5	8.0	1.5	17.0	13.8	12.7	12.1
16′ to 20′		9.0	10.0	1.5	20.5	16.5	15.2	14.5
Foundation walls, to 8′ high		4.5	6.5	1.0	12.0	10.3	9.7	9.4
8′ to 16′ high		5.5	7.5	1.0	14.0	11.8	11.0	10.6
Retaining wall to 12′ high, battered		6.0	8.5	1.5	16.0	13.5	12.7	12.3
Radial walls to 12′ high, smooth		8.0	9.5	2.0	19.5	16.0	14.8	14.3
2′ chords		7.0	8.0	1.5	16.5	13.5	12.5	12.0
Prefabricated modular, to 8′ high		—	4.3	1.0	5.3	5.3	5.3	5.3
Steel, to 8′ high		—	6.8	1.2	8.0	8.0	8.0	8.0
8′ to 16′ high		—	9.1	1.5	10.6	10.3	10.2	10.2
Steel framed plywood to 8′ high		—	6.8	1.2	8.0	7.5	7.3	7.2
8′ to 16′ high		—	9.3	1.2	10.5	9.5	9.2	9.0

R032110-10 Reinforcing Steel Weights and Measures

Bar Designation No.**	Nominal Weight Lb./Ft.	U.S. Customary Units			SI Units			
		Nominal Dimensions*			Nominal Dimensions*			
		Diameter in.	Cross Sectional Area, in.^{2}	Perimeter in.	Nominal Weight kg/m	Diameter mm	Cross Sectional Area, cm^2	Perimeter mm
3	.376	.375	.11	1.178	.560	9.52	.71	29.9
4	.668	.500	.20	1.571	.994	12.70	1.29	39.9
5	1.043	.625	.31	1.963	1.552	15.88	2.00	49.9
6	1.502	.750	.44	2.356	2.235	19.05	2.84	59.8
7	2.044	.875	.60	2.749	3.042	22.22	3.87	69.8
8	2.670	1.000	.79	3.142	3.973	25.40	5.10	79.8
9	3.400	1.128	1.00	3.544	5.059	28.65	6.45	90.0
10	4.303	1.270	1.27	3.990	6.403	32.26	8.19	101.4
11	5.313	1.410	1.56	4.430	7.906	35.81	10.06	112.5
14	7.650	1.693	2.25	5.320	11.384	43.00	14.52	135.1
18	13.600	2.257	4.00	7.090	20.238	57.33	25.81	180.1

* The nominal dimensions of a deformed bar are equivalent to those of a plain round bar having the same weight per foot as the deformed bar.

** Bar numbers are based on the number of eighths of an inch included in the nominal diameter of the bars.

R032110-20 Metric Rebar Specification - ASTM A615-81

Grade 300 (300 MPa* = 43,560 psi; +8.7% vs. Grade 40)				
Grade 400 (400 MPa* = 58,000 psi; –3.4% vs. Grade 60)				
Bar No.	**Diameter mm**	**Area mm^2**	**Equivalent in.^{2}**	**Comparison with U.S. Customary Bars**
10M	11.3	100	.16	Between #3 & #4
15M	16.0	200	.31	#5 (.31 in.^{2})
20M	19.5	300	.47	#6 (.44 in.^{2})
25M	25.2	500	.78	#8 (.79 in.^{2})
30M	29.9	700	1.09	#9 (1.00 in.^{2})
35M	35.7	1000	1.55	#11 (1.56 in.^{2})
45M	43.7	1500	2.33	#14 (2.25 in.^{2})
55M	56.4	2500	3.88	#18 (4.00 in.^{2})

* MPa = megapascals

R032110-40 Weight of Steel Reinforcing Per Square Foot of Wall (PSF)

Reinforced Weights: The table below suggests the weights per square foot for reinforcing steel in walls. Weights are approximate and will be the same for all grades of steel bars. For bars in two directions, add weights for each size and spacing.

C/C Spacing in Inches	Bar Size								
	#3 Wt. (PSF)	#4 Wt. (PSF)	#5 Wt. (PSF)	#6 Wt. (PSF)	#7 Wt. (PSF)	#8 Wt. (PSF)	#9 Wt. (PSF)	#10 Wt. (PSF)	#11 Wt. (PSF)
2″	2.26	4.01	6.26	9.01	12.27				
3″	1.50	2.67	4.17	6.01	8.18	10.68	13.60	17.21	21.25
4″	1.13	2.01	3.13	4.51	6.13	8.10	10.20	12.91	15.94
5″	.90	1.60	2.50	3.60	4.91	6.41	8.16	10.33	12.75
6″	.752	1.34	2.09	3.00	4.09	5.34	6.80	8.61	10.63
8″	.564	1.00	1.57	2.25	3.07	4.01	5.10	6.46	7.97
10″	.451	.802	1.25	1.80	2.45	3.20	4.08	5.16	6.38
12″	.376	.668	1.04	1.50	2.04	2.67	3.40	4.30	5.31
18″	.251	.445	.695	1.00	1.32	1.78	2.27	2.86	3.54
24″	.188	.334	.522	.751	1.02	1.34	1.70	2.15	2.66
30″	.150	.267	.417	.600	.817	1.07	1.36	1.72	2.13
36″	.125	.223	.348	.501	.681	.890	1.13	1.43	1.77
42″	.107	.191	.298	.429	.584	.753	.97	1.17	1.52
48″	.094	.167	.261	.376	.511	.668	.85	1.08	1.33

R032110-50 Minimum Wall Reinforcement Weight (PSF)

This table lists the approximate minimum wall reinforcement weights per S.F. according to the specification of .12% of gross area for vertical bars and .20% of gross area for horizontal bars.

Location	Wall Thickness	Bar Size	Horizontal Steel Spacing C/C	Sq. In. Req'd per S.F.	Total Wt. per S.F.	Bar Size	Vertical Steel Spacing C/C	Sq. In. Req'd per S.F.	Total Wt. per S.F.	Horizontal & Vertical Steel Total Weight per S.F.
Both Faces	10″	#4	18″	.24	.89#	#3	18″	.14	.50#	1.39#
	12″	#4	16″	.29	1.00	#3	16″	.17	.60	1.60
	14″	#4	14″	.34	1.14	#3	13″	.20	.69	1.84
	16″	#4	12″	.38	1.34	#3	11″	.23	.82	2.16
	18″	#5	17″	.43	1.47	#4	18″	.26	.89	2.36
One Face	6″	#3	9″	.15	.50	#3	18″	.09	.25	.75
	8″	#4	12″	.19	.67	#3	11″	.12	.41	1.08
	10″	#5	15″	.24	.83	#4	16″	.14	.50	1.34

R032110-70 Bend, Place and Tie Reinforcing

Placing and tying by rodmen for footings and slabs runs from nine hrs. per ton for heavy bars to fifteen hrs. per ton for light bars. For beams, columns, and walls, production runs from eight hrs. per ton for heavy bars to twenty hrs. per ton for light bars. Overall average for typical reinforced concrete buildings is about fourteen hrs. per ton. These production figures include the time for placing of accessories and usual inserts, but not their material cost (allow 15% of the cost of delivered bent rods). Equipment handling is necessary for the larger-sized bars so that installation costs for the very heavy bars will not decrease proportionately.

Installation costs for splicing reinforcing bars include allowance for equipment to hold the bars in place while splicing as well as necessary scaffolding for iron workers.

R032110-80 Shop-Fabricated Reinforcing Steel

The material prices for reinforcing, shown in the unit cost sections of the book, are for 50 tons or more of shop-fabricated reinforcing steel and include:

1. Mill base price of reinforcing steel
2. Mill grade/size/length extras
3. Mill delivery to the fabrication shop
4. Shop storage and handling
5. Shop drafting/detailing
6. Shop shearing and bending
7. Shop listing
8. Shop delivery to the job site

Both material and installation costs can be considerably higher for small jobs consisting primarily of smaller bars, while material costs may be slightly lower for larger jobs.

Concrete — R0322 Welded Wire Fabric Reinforcing

R032205-30 Common Stock Styles of Welded Wire Fabric

This table provides some of the basic specifications, sizes, and weights of welded wire fabric used for reinforcing concrete.

	New Designation	Old Designation		Steel Area per Foot				Approximate Weight per 100 S.F.	
	Spacing — Cross Sectional Area (in.) — (Sq. in. 100)	Spacing — Wire Gauge (in.) — (AS & W)		Longitudinal		Transverse			
				in.	cm	in.	cm	lbs	kg
Rolls	6 x 6 — W1.4 x W1.4	6 x 6 — 10 x 10		.028	.071	.028	.071	21	9.53
	6 x 6 — W2.0 x W2.0	6 x 6 — 8 x 8	1	.040	.102	.040	.102	29	13.15
	6 x 6 — W2.9 x W2.9	6 x 6 — 6 x 6		.058	.147	.058	.147	42	19.05
	6 x 6 — W4.0 x W4.0	6 x 6 — 4 x 4		.080	.203	.080	.203	58	26.91
	4 x 4 — W1.4 x W1.4	4 x 4 — 10 x 10		.042	.107	.042	.107	31	14.06
	4 x 4 — W2.0 x W2.0	4 x 4 — 8 x 8	1	.060	.152	.060	.152	43	19.50
	4 x 4 — W2.9 x W2.9	4 x 4 — 6 x 6		.087	.227	.087	.227	62	28.12
	4 x 4 — W4.0 x W4.0	4 x 4 — 4 x 4		.120	.305	.120	.305	85	38.56
Sheets	6 x 6 — W2.9 x W2.9	6 x 6 — 6 x 6		.058	.147	.058	.147	42	19.05
	6 x 6 — W4.0 x W4.0	6 x 6 — 4 x 4		.080	.203	.080	.203	58	26.31
	6 x 6 — W5.5 x W5.5	6 x 6 — 2 x 2	2	.110	.279	.110	.279	80	36.29
	4 x 4 — W1.4 x W1.4	4 x 4 — 4 x 4		.120	.305	.120	.305	85	38.56

NOTES: 1. Exact W—number size for 8 gauge is W2.1
2. Exact W—number size for 2 gauge is W5.4

Concrete — R0330 Cast-In-Place Concrete

R033053-50 Industrial Chimneys

Foundation requirements in C.Y. of concrete for various sized chimneys.

Size Chimney	2 Ton Soil	3 Ton Soil	Size Chimney	2 Ton Soil	3 Ton Soil	Size Chimney	2 Ton Soil	3 Ton Soil
75′ x 3′-0″	13 C.Y.	11 C.Y.	160′ x 6′-6″	86 C.Y.	76 C.Y.	300′ x 10′-0″	325 C.Y.	245 C.Y.
85′ x 5′-6″	19	16	175′ x 7′-0″	108	95	350′ x 12′-0″	422	320
100′ x 5′-0″	24	20	200′ x 6′-0″	125	105	400′ x 14′-0″	520	400
125′ x 5′-6″	43	36	250′ x 8′-0″	230	175	500′ x 18′-0″	725	575

R033105-10 Proportionate Quantities

The tables below show both quantities per S.F. of floor areas as well as form and reinforcing quantities per C.Y. Unusual structural requirements would increase the ratios below. High strength reinforcing would reduce the steel weights. Figures are for 3000 psi concrete and 60,000 psi reinforcing unless specified otherwise.

Type of Construction	Live Load	Span	Per S.F. of Floor Area				Per C.Y. of Concrete		
			Concrete	Forms	Reinf.	Pans	Forms	Reinf.	Pans
Flat Plate	50 psf	15 Ft.	.46 C.F.	1.06 S.F.	1.71lb.		62 S.F.	101 lb.	
		20	.63	1.02	2.40		44	104	
		25	.79	1.02	3.03		35	104	
	100	15	.46	1.04	2.14		61	126	
		20	.71	1.02	2.72		39	104	
		25	.83	1.01	3.47		33	113	
Flat Plate (waffle construction) 20″ domes	50	20	.43	1.00	2.10	.84 S.F.	63	135	53 S.F.
		25	.52	1.00	2.90	.89	52	150	46
		30	.64	1.00	3.70	.87	42	155	37
	100	20	.51	1.00	2.30	.84	53	125	45
		25	.64	1.00	3.20	.83	42	135	35
		30	.76	1.00	4.40	.81	36	160	29
Waffle Construction 30″ domes	50	25	.69	1.06	1.83	.68	42	72	40
		30	.74	1.06	2.39	.69	39	87	39
		35	.86	1.05	2.71	.69	33	85	39
		40	.78	1.00	4.80	.68	35	165	40
Flat Slab (two way with drop panels)	50	20	.62	1.03	2.34		45	102	
		25	.77	1.03	2.99		36	105	
		30	.95	1.03	4.09		29	116	
	100	20	.64	1.03	2.83		43	119	
		25	.79	1.03	3.88		35	133	
		30	.96	1.03	4.66		29	131	
	200	20	.73	1.03	3.03		38	112	
		25	.86	1.03	4.23		32	133	
		30	1.06	1.03	5.30		26	135	
One Way Joists 20″ Pans	50	15	.36	1.04	1.40	.93	78	105	70
		20	.42	1.05	1.80	.94	67	120	60
		25	.47	1.05	2.60	.94	60	150	54
	100	15	.38	1.07	1.90	.93	77	140	66
		20	.44	1.08	2.40	.94	67	150	58
		25	.52	1.07	3.50	.94	55	185	49
One Way Joists 8″ x 16″ filler blocks	50	15	.34	1.06	1.80	.81 Ea.	84	145	64 Ea.
		20	.40	1.08	2.20	.82	73	145	55
		25	.46	1.07	3.20	.83	63	190	49
	100	15	.39	1.07	1.90	.81	74	130	56
		20	.46	1.09	2.80	.82	64	160	48
		25	.53	1.10	3.60	.83	56	190	42
One Way Beam & Slab	50	15	.42	1.30	1.73		84	111	
		20	.51	1.28	2.61		68	138	
		25	.64	1.25	2.78		53	117	
	100	15	.42	1.30	1.90		84	122	
		20	.54	1.35	2.69		68	154	
		25	.69	1.37	3.93		54	145	
	200	15	.44	1.31	2.24		80	137	
		20	.58	1.40	3.30		65	163	
		25	.69	1.42	4.89		53	183	
Two Way Beam & Slab	100	15	.47	1.20	2.26		69	130	
		20	.63	1.29	3.06		55	131	
		25	.83	1.33	3.79		43	123	
	200	15	.49	1.25	2.70		41	149	
		20	.66	1.32	4.04		54	165	
		25	.88	1.32	6.08		41	187	

R033105-10 Proportionate Quantities (cont.)

4000 psi Concrete and 60,000 psi Reinforcing—Form and Reinforcing Quantities per C.Y.					
Item	Size	Forms	Reinforcing	Minimum	Maximum
Columns (square tied)	10″ x 10″	130 S.F.C.A.	#5 to #11	220 lbs.	875 lbs.
	12″ x 12″	108	#6 to #14	200	955
	14″ x 14″	92	#7 to #14	190	900
	16″ x 16″	81	#6 to #14	187	1082
	18″ x 18″	72	#6 to #14	170	906
	20″ x 20″	65	#7 to #18	150	1080
	22″ x 22″	59	#8 to #18	153	902
	24″ x 24″	54	#8 to #18	164	884
	26″ x 26″	50	#9 to #18	169	994
	28″ x 28″	46	#9 to #18	147	864
	30″ x 30″	43	#10 to #18	146	983
	32″ x 32″	40	#10 to #18	175	866
	34″ x 34″	38	#10 to #18	157	772
	36″ x 36″	36	#10 to #18	175	852
	38″ x 38″	34	#10 to #18	158	765
	40″ x 40″	32	#10 to #18	143	692

Item	Size	Form	Spiral	Reinforcing	Minimum	Maximum
Columns (spirally reinforced)	12″ diameter	34.5 L.F.	190 lbs.	#4 to #11	165 lbs.	1505 lb.
		34.5	190	#14 & #18	—	1100
	14″	25	170	#4 to #11	150	970
		25	170	#14 & #18	800	1000
	16″	19	160	#4 to #11	160	950
		19	160	#14 & #18	605	1080
	18″	15	150	#4 to #11	160	915
		15	150	#14 & #18	480	1075
	20″	12	130	#4 to #11	155	865
		12	130	#14 & #18	385	1020
	22″	10	125	#4 to #11	165	775
		10	125	#14 & #18	320	995
	24″	9	120	#4 to #11	195	800
		9	120	#14 & #18	290	1150
	26″	7.3	100	#4 to #11	200	729
		7.3	100	#14 & #18	235	1035
	28″	6.3	95	#4 to #11	175	700
		6.3	95	#14 & #18	200	1075
	30″	5.5	90	#4 to #11	180	670
		5.5	90	#14 & #18	175	1015
	32″	4.8	85	#4 to #11	185	615
		4.8	85	#14 & #18	155	955
	34″	4.3	80	#4 to #11	180	600
		4.3	80	#14 & #18	170	855
	36″	3.8	75	#4 to #11	165	570
		3.8	75	#14 & #18	155	865
	40″	3.0	70	#4 to #11	165	500
		3.0	70	#14 & #18	145	765

R033105-10 Proportionate Quantities (cont.)

3000 psi Concrete and 60,000 psi Reinforcing—Form and Reinforcing Quantities per C.Y.						
Item	**Type**	**Loading**	**Height**	**C.Y./L.F.**	**Forms/C.Y.**	**Reinf./C.Y.**
Retaining Walls	Cantilever	Level Backfill	4 Ft.	0.2 C.Y.	49 S.F.	35 lbs.
			8	0.5	42	45
			12	0.8	35	70
			16	1.1	32	85
			20	1.6	28	105
		Highway Surcharge	4	0.3	41	35
			8	0.5	36	55
			12	0.8	33	90
			16	1.2	30	120
			20	1.7	27	155
		Railroad Surcharge	4	0.4	28	45
			8	0.8	25	65
			12	1.3	22	90
			16	1.9	20	100
			20	2.6	18	120
	Gravity, with Vertical Face	Level Backfill	4	0.4	37	None
			7	0.6	27	↓
			10	1.2	20	↓
		Sloping Backfill	4	0.3	31	↓
			7	0.8	21	↓
			10	1.6	15	↓

Item	Span	Live Load in Kips per Linear Foot							
		Under 1 Kip		2 to 3 Kips		4 to 5 Kips		6 to 7 Kips	
		Forms	Reinf.	Forms	Reinf.	Forms	Reinf.	Forms	Reinf.
Beams	10 Ft.	—	—	90 S.F.	170 #	85 S.F.	175 #	75 S.F.	185 #
	16	130 S.F.	165 #	85	180	75	180	65	225
	20	110	170	75	185	62	200	51	200
	26	90	170	65	215	62	215	—	—
	30	85	175	60	200	—	—	—	—

Item	Size	Type	Forms per C.Y.	Reinforcing per C.Y.
Spread Footings	Under 1 C.Y.	1,000 psf soil	24 S.F.	44 lbs.
		5,000	24	42
		10,000	24	52
	1 C.Y. to 5 C.Y.	1,000	14	49
		5,000	14	50
		10,000	14	50
	Over 5 C.Y.	1,000	9	54
		5,000	9	52
		10,000	9	56
Pile Caps (30 Ton Concrete Piles)	Under 5 C.Y.	shallow caps	20	65
		medium	20	50
		deep	20	40
	5 C.Y. to 10 C.Y.	shallow	14	55
		medium	15	45
		deep	15	40
	10 C.Y. to 20 C.Y.	shallow	11	60
		medium	11	45
		deep	12	35
	Over 20 C.Y.	shallow	9	60
		medium	9	45
		deep	10	40

R033105-10 Proportionate Quantities (cont.)

3000 psi Concrete and 60,000 psi Reinforcing — Form and Reinforcing Quantities per C.Y.						
Item	**Size**	**Pile Spacing**	**50 T Pile**	**100 T Pile**	**50 T Pile**	**100 T Pile**
Pile Caps (Steel H Piles)	Under 5 C.Y.	24″ O.C.	24 S.F.	24 S.F.	75 lbs.	90 lbs.
		30″	25	25	80	100
		36″	24	24	80	110
	5 C.Y. to 10 C.Y.	24″	15	15	80	110
		30″	15	15	85	110
		36″	15	15	75	90
	Over 10 C.Y.	24″	13	13	85	90
		30″	11	11	85	95
		36″	10	10	85	90

Item		8″ Thick		10″ Thick		12″ Thick		15″ Thick	
	Height	**Forms**	**Reinf.**	**Forms**	**Reinf.**	**Forms**	**Reinf.**	**Forms**	**Reinf.**
Basement Walls	7 Ft.	81 S.F.	44 lbs.	65 S.F.	45 lbs.	54 S.F.	44 lbs.	41 S.F.	43 lbs.
	8	↓	44	↓	45	↓	44	↓	43
	9	↓	46	↓	45	↓	44	↓	43
	10	↓	57	↓	45	↓	44	↓	43
	12	↓	83	↓	50	↓	52	↓	43
	14	↓	116	↓	65	↓	64	↓	51
	16	↓		↓	86	↓	90	↓	65
	18	↓		↓		↓	106	↓	70

R033105-20 Materials for One C.Y. of Concrete

This is an approximate method of figuring quantities of cement, sand and coarse aggregate for a field mix with waste allowance included.

With crushed gravel as coarse aggregate, to determine barrels of cement required, divide 10 by total mix; that is, for 1:2:4 mix, 10 divided by 7 = 1-3/7 barrels.

If the coarse aggregate is crushed stone, use 10-1/2 instead of 10 as given for gravel.

To determine tons of sand required, multiply barrels of cement by parts of sand and then by 0.2; that is, for the 1:2:4 mix, as above, 1-3/7 x 2 x .2 = .57 tons.

Tons of crushed gravel are in the same ratio to tons of sand as parts in the mix, or 4/2 x .57 = 1.14 tons.

1 bag cement = 94#
4 bags = 1 barrel

1 C.Y. sand or crushed gravel = 2700#
1 ton sand or crushed gravel = 20 C.F.

1 C.Y. crushed stone = 2575#
1 ton crushed stone = 21 C.F.

Average carload of cement is 692 bags; of sand or gravel is 56 tons.

Do not stack stored cement over 10 bags high.

R033105-30 Metric Equivalents of Cement Content for Concrete Mixes

94 Pound Bags per Cubic Yard	Kilograms per Cubic Meter	94 Pound Bags per Cubic Yard	Kilograms per Cubic Meter
1.0	55.77	7.0	390.4
1.5	83.65	7.5	418.3
2.0	111.5	8.0	446.2
2.5	139.4	8.5	474.0
3.0	167.3	9.0	501.9
3.5	195.2	9.5	529.8
4.0	223.1	10.0	557.7
4.5	251.0	10.5	585.6
5.0	278.8	11.0	613.5
5.5	306.7	11.5	641.3
6.0	334.6	12.0	669.2
6.5	362.5	12.5	697.1

a. If you know the cement content in pounds per cubic yard, multiply by .5933 to obtain kilograms per cubic meter.

b. If you know the cement content in 94 pound bags per cubic yard, multiply by 55.77 to obtain kilograms per cubic meter.

R033105-40 Metric Equivalents of Common Concrete Strengths (to convert other psi values to megapascals, multiply by 0.006895)

U.S. Values psi	SI Value Megapascals	Non-SI Metric Value kgf/cm^2*
2000	14	140
2500	17	175
3000	21	210
3500	24	245
4000	28	280
4500	31	315
5000	34	350
6000	41	420
7000	48	490
8000	55	560
9000	62	630
10,000	69	705

* kilograms force per square centimeter

R033105-50 Quantities of Cement, Sand and Stone for One C.Y. of Concrete per Various Mixes

This table can be used to determine the quantities of the ingredients for smaller quantities of site mixed concrete.

Concrete (C.Y.)	Mix = 1:1:1-3/4			Mix = 1:2:2.25			Mix = 1:2.25:3			Mix = 1:3:4		
	Cement (sacks)	Sand (C.Y.)	Stone (C.Y.)	Cement (sacks)	Sand (C.Y.)	Stone (C.Y.)	Cement (sacks)	Sand (C.Y.)	Stone (C.Y.)	Cement (sacks)	Sand (C.Y.)	Stone (C.Y.)
1	10	.37	.63	7.75	.56	.65	6.25	.52	.70	5	.56	.74
2	20	.74	1.26	15.50	1.12	1.30	12.50	1.04	1.40	10	1.12	1.48
3	30	1.11	1.89	23.25	1.68	1.95	18.75	1.56	2.10	15	1.68	2.22
4	40	1.48	2.52	31.00	2.24	2.60	25.00	2.08	2.80	20	2.24	2.96
5	50	1.85	3.15	38.75	2.80	3.25	31.25	2.60	3.50	25	2.80	3.70
6	60	2.22	3.78	46.50	3.36	3.90	37.50	3.12	4.20	30	3.36	4.44
7	70	2.59	4.41	54.25	3.92	4.55	43.75	3.64	4.90	35	3.92	5.18
8	80	2.96	5.04	62.00	4.48	5.20	50.00	4.16	5.60	40	4.48	5.92
9	90	3.33	5.67	69.75	5.04	5.85	56.25	4.68	6.30	45	5.04	6.66
10	100	3.70	6.30	77.50	5.60	6.50	62.50	5.20	7.00	50	5.60	7.40
11	110	4.07	6.93	85.25	6.16	7.15	68.75	5.72	7.70	55	6.16	8.14
12	120	4.44	7.56	93.00	6.72	7.80	75.00	6.24	8.40	60	6.72	8.88
13	130	4.82	8.20	100.76	7.28	8.46	81.26	6.76	9.10	65	7.28	9.62
14	140	5.18	8.82	108.50	7.84	9.10	87.50	7.28	9.80	70	7.84	10.36
15	150	5.56	9.46	116.26	8.40	9.76	93.76	7.80	10.50	75	8.40	11.10
16	160	5.92	10.08	124.00	8.96	10.40	100.00	8.32	11.20	80	8.96	11.84
17	170	6.30	10.72	131.76	9.52	11.06	106.26	8.84	11.90	85	9.52	12.58
18	180	6.66	11.34	139.50	10.08	11.70	112.50	9.36	12.60	90	10.08	13.32
19	190	7.04	11.98	147.26	10.64	12.36	118.76	9.84	13.30	95	10.64	14.06
20	200	7.40	12.60	155.00	11.20	13.00	125.00	10.40	14.00	100	11.20	14.80
21	210	7.77	13.23	162.75	11.76	13.65	131.25	10.92	14.70	105	11.76	15.54
22	220	8.14	13.86	170.05	12.32	14.30	137.50	11.44	15.40	110	12.32	16.28
23	230	8.51	14.49	178.25	12.88	14.95	143.75	11.96	16.10	115	12.88	17.02
24	240	8.88	15.12	186.00	13.44	15.60	150.00	12.48	16.80	120	13.44	17.76
25	250	9.25	15.75	193.75	14.00	16.25	156.25	13.00	17.50	125	14.00	18.50
26	260	9.64	16.40	201.52	14.56	16.92	162.52	13.52	18.20	130	14.56	19.24
27	270	10.00	17.00	209.26	15.12	17.56	168.76	14.04	18.90	135	15.02	20.00
28	280	10.36	17.64	217.00	15.68	18.20	175.00	14.56	19.60	140	15.68	20.72
29	290	10.74	18.28	224.76	16.24	18.86	181.26	15.08	20.30	145	16.24	21.46

R033105-65 Field-Mix Concrete

Presently most building jobs are built with ready-mixed concrete except at isolated locations and some larger jobs requiring over 10,000 C.Y. where land is readily available for setting up a temporary batch plant.

The most economical mix is a controlled mix using local aggregate proportioned by trial to give the required strength with the least cost of material.

R033105-70 Placing Ready-Mixed Concrete

For ground pours allow for 5% waste when figuring quantities.

Prices in the front of the book assume normal deliveries. If deliveries are made before 8 A.M. or after 5 P.M. or on Saturday afternoons add 30%. Negotiated discounts for large volumes are not included in prices in front of book.

For the lower floors without truck access, concrete may be wheeled in rubber-tired buggies, conveyer handled, crane handled or pumped. Pumping is economical if there is top steel. Conveyers are more efficient for thick slabs.

At higher floors the rubber-tired buggies may be hoisted by a hoisting tower and wheeled to location. Placement by a conveyer is limited to three floors and is best for high-volume pours. Pumped concrete is best when building has no crane access. Concrete may be pumped directly as high as thirty-six stories using special pumping techniques. Normal maximum height is about fifteen stories.

Best pumping aggregate is screened and graded bank gravel rather than crushed stone.

Pumping downward is more difficult than pumping upward. Horizontal distance from pump to pour may increase preparation time prior to pour. Placing by cranes, either mobile, climbing or tower types, continues as the most efficient method for high-rise concrete buildings.

R033105-85 Lift Slabs

The cost advantage of the lift slab method is due to placing all concrete, reinforcing steel, inserts and electrical conduit at ground level and in reduction of formwork. Minimum economical project size is about 30,000 S.F. Slabs may be tilted for parking garage ramps.

It is now used in all types of buildings and has gone up to 22 stories high in apartment buildings. Current trend is to use post-tensioned flat plate slabs with spans from 22′ to 35′. Cylindrical void forms are used when deep slabs are required. One pound of prestressing steel is about equal to seven pounds of conventional reinforcing.

To be considered cured for stressing and lifting, a slab must have attained 75% of design strength. Seven days are usually sufficient with four to five days possible if high early strength cement is used. Slabs can be stacked using two coats of a non-bonding agent to insure that slabs do not stick to each other. Lifting is done by companies specializing in this work. Lift rate is 5′ to 15′ per hour with an average of 10′ per hour. Total areas up to 33,000 S.F. have been lifted at one time. 24 to 36 jacking columns are common. Most economical bay sizes are 24′ to 28′ with four to fourteen stories most efficient. Continuous design reduces reinforcing steel cost. Use of post-tensioned slabs allows larger bay sizes.

R034105-30 Prestressed Precast Concrete Structural Units

Type	Location	Depth	Span in Ft.	Live Load Lb. per S.F.
Double Tee 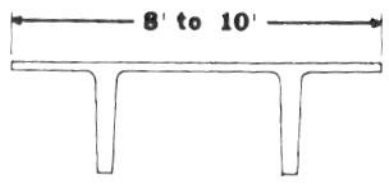 	Floor	28″ to 34″	60 to 80	50 to 80
	Roof	12″ to 24″	30 to 50	40
	Wall	Width 8′	Up to 55′ high	Wind
Multiple Tee 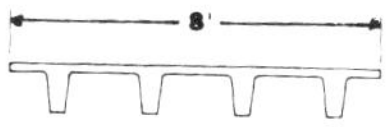 	Roof	8″ to 12″	15 to 40	40
	Floor	8″ to 12″	15 to 30	100
Plank  or 	Roof or Floor	4″ 6″ 8″ 10″ 12″	Roof / Floor 13 / 12 22 / 18 26 / 25 33 / 29 42 / 32	40 for Roof 100 for Floor
Single Tee 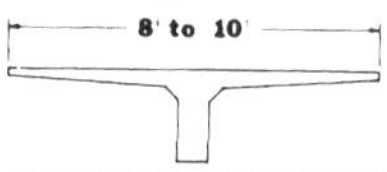 	Roof	28″ 32″ 36″ 48″	40 80 100 120	40
AASHO Girder 	Bridges	Type 4 5 6	100 110 125	Highway
Box Beam  	Bridges	15″ 27″ 33″	40 to 100	Highway

The majority of precast projects today utilize double tees rather than single tees because of speed and ease of installation. As a result casting beds at manufacturing plants are normally formed for double tees. Single tee projects will therefore require an initial set up charge to be spread over the individual single tee costs.

For floors, a 2″ to 3″ topping is field cast over the shapes. For roofs, insulating concrete or rigid insulation is placed over the shapes.

Member lengths up to 40′ are standard haul, 40′ to 60′ require special permits and lengths over 60′ must be escorted. Over width and/or over length can add up to 100% on hauling costs.

Large heavy members may require two cranes for lifting which would increase erection costs by about 45%. An eight man crew can install 12 to 20 double tees, or 45 to 70 quad tees or planks per day.

Grouting of connections must also be included.

Several system buildings utilizing precast members are available. Heights can go up to 22 stories for apartment buildings. Optimum design ratio is 3 S.F. of surface to 1 S.F. of floor area.

R034136-90 Prestressed Concrete, Post-Tensioned

In post-tensioned concrete the steel tendons are tensioned after the concrete has reached about 3/4 of its ultimate strength. The cableways are grouted after tensioning to provide bond between the steel and concrete. If bond is to be prevented, the tendons are coated with a corrosion-preventative grease and wrapped with waterproofed paper or plastic. Bonded tendons are usually used when ultimate strength (beams & girders) are controlling factors.

High strength concrete is used to fully utilize the steel, thereby reducing the size and weight of the member. A plasticizing agent may be added to reduce water content. Maximum size aggregate ranges from 1/2″ to 1-1/2″ depending on the spacing of the tendons.

The types of steel commonly used are bars and strands. Job conditions determine which is best suited. Bars are best for vertical prestresses since they are easy to support. The trend is for steel manufacturers to supply a finished package, cut to length, which reduces field preparation to a minimum.

Bars vary from 3/4″ to 1-3/8″ diameter. Table below gives time in labor-hours per tendon for placing, tensioning and grouting (if required) a 75′ beam. Tendons used in buildings are not usually grouted; tendons for bridges usually are grouted. For strands the table indicates the labor-hours per pound for typical prestressed units 100′ long. Simple span beams usually require one-end stressing regardless of lengths. Continuous beams are usually stressed from two ends. Long slabs are poured from the center outward and stressed in 75′ increments after the initial 150′ center pour.

Labor Hours per Tendon and per Pound of Prestressed Steel

Length	100′ Beam		75′ Beam		100′ Slab	
Type Steel	Strand		Bars		Strand	
Diameter	0.5″		3/4″	1-3/8″	0.5″	0.6″
Number	4	12	1	1	1	1
Force in Kips	100	300	42	143	25	35
Preparation & Placing Cables	3.6	7.4	0.9	2.9	0.9	1.1
Stressing Cables	2.0	2.4	0.8	1.6	0.5	0.5
Grouting, if required	2.5	3.0	0.6	1.3		
Total Labor Hours	8.1	12.8	2.3	5.8	1.4	1.6
Prestressing Steel Weights (Lbs.)	215	640	115	380	53	74
Labor-hours per Lb. Bonded	0.038	0.020	0.020	0.015		
Non-bonded					0.026	0.022

Flat Slab construction — 4000 psi concrete with span-to-depth ratio between 36 and 44. Two way post-tensioned steel averages 1.0 lb. per S.F. for 24′ to 28′ bays (usually strand) and additional reinforcing steel averages .5 lb. per S.F.

Pan and Joist construction — 4000 psi concrete with span-to-depth ratio 28 to 30. Post-tensioned steel averages .8 lb. per S.F. and reinforcing steel about 1.0 lb. per S.F. Placing and stressing averages 40 hours per ton of total material.

Beam construction — 4000 to 5000 psi concrete. Steel weights vary greatly.

Labor cost per pound goes down as the size and length of the tendon increase. The primary economic consideration is the cost per kip for the member.

Post-tensioning becomes feasible for beams and girders over 30′ long; for continuous two-way slabs over 20′ clear; also in transferring upper building loads over longer spans at lower levels. Post-tension suppliers will provide engineering services at no cost to the user. Substantial economies are possible by using post-tensioned Lift Slabs.

Concrete — R0345 Precast Architectural Concrete

R034513-10 Precast Concrete Wall Panels

Panels are either solid or insulated with plain, colored or textured finishes. Transportation is an important cost factor. Prices shown in the unit cost section of the book are based on delivery within 50 miles of a plant including fabricators' overhead and profit. Engineering data is available from fabricators to assist with construction details. Usual minimum job size for economical use of panels is about 5000 S.F. Small jobs can double the prices shown. For large, highly repetitive jobs, deduct up to 15% from the prices shown.

2″ thick panels cost about the same as 3″ thick panels, and maximum panel size is less. For building panels faced with granite, marble or stone, add the material prices from those unit cost sections to the plain panel price shown. There is a growing trend toward aggregate facings and broken rib finish rather than plain gray concrete panels.

No allowance has been made in the unit cost section for supporting steel framework. On one story buildings, panels may rest on grade beams and require only wind bracing and fasteners. On multi-story buildings panels can span from column to column and floor to floor. Plastic-designed steel-framed structures may have large deflections which slow down erection and raise costs.

Large panels are more economical than small panels on a S.F. basis. When figuring areas include all protrusions, returns, etc. Overhangs can triple erection costs. Panels over 45′ have been produced. Larger flat units should be prestressed. Vacuum lifting of smooth finish panels eliminates inserts and can speed erection.

Concrete — R0347 Site-Cast Concrete

R034713-20 Tilt Up Concrete Panels

The advantage of tilt up construction is in the low cost of forms and the placing of concrete and reinforcing. Panels up to 75′ high and 5-1/2″ thick have been tilted using strongbacks. Tilt up has been used for one to five story buildings and is well-suited for warehouses, stores, offices, schools and residences.

The panels are cast in forms on the floor slab. Most jobs use 5-1/2″ thick solid reinforced concrete panels. Sandwich panels with a layer of insulating materials are also used. Where dampness is a factor, lightweight aggregate is used. Optimum panel size is 300 to 500 S.F.

Slabs are usually poured with 3000 psi concrete which permits tilting seven days after pouring. Slabs may be stacked on top of each other and are separated from each other by either two coats of bond breaker or a film of polyethylene. Use of high early-strength cement allows tilting two days after a pour. Tilting up is done with a roller outrigger crane with a capacity of at least 1-1/2 times the weight of the panel at the required reach. Exterior precast columns can be set at the same time as the panels; interior precast columns can be set first and the panels clipped directly to them. The use of cast-in-place concrete columns is diminishing due to shrinkage problems. Structural steel columns are sometimes used if crane rails are planned. Panels can be clipped to the columns or lowered between the flanges. Steel channels with anchors may be used as edge forms for the slab. When the panels are lifted the channels form an integral steel column to take structural loads. Roof loads can be carried directly by the panels for wall heights to 14′.

Requirements of local building codes may be a limiting factor and should be checked. Building floor slabs should be poured first and should be a minimum of 5″ thick with 100% compaction of soil or 6″ thick with less than 100% compaction.

Setting times as fast as nine minutes per panel have been observed, but a safer expectation would be four panels per hour with a crane and a four-man setting crew. If crane erects from inside building, some provision must be made to get crane out after walls are erected. Good yarding procedure is important to minimize delays. Equalizing three-point lifting beams and self-releasing pick-up hooks speed erection. If panels must be carried to their final location, setting time per panel will be increased and erection costs may approach the erection cost range of architectural precast wall panels. Placing panels into slots formed in continuous footers will speed erection.

Reinforcing should be with #5 bars with vertical bars on the bottom. If surface is to be sandblasted, stainless steel chairs should be used to prevent rust staining.

Use of a broom finish is popular since the unavoidable surface blemishes are concealed.

Precast columns run from three to five times the C.Y. price of the panels only.

Concrete — R0352 Lightweight Concrete Roof Insulation

R035216-10 Lightweight Concrete

Lightweight aggregate concrete is usually purchased ready mixed, but it can also be field mixed.

Vermiculite or Perlite comes in bags of 4 C.F. under various trade names. Weight is about 8 lbs. per C.F. For insulating roof fill use 1:6 mix. For structural deck use 1:4 mix over gypsum boards, steeltex, steel centering, etc., supported by closely spaced joists or bulb trees. For structural slabs use 1:3:2 vermiculite sand concrete over steeltex, metal lath, steel centering, etc., on joists spaced 2′-0″ O.C. for maximum L.L. of 80 P.S.F. Use same mix for slab base fill over steel flooring or regular reinforced concrete slab when tile, terrazzo or other finish is to be laid over.

For slabs on grade use 1:3:2 mix when tile, etc., finish is to be laid over. If radiant heating units are installed use a 1:6 mix for a base. After coils are in place, cover with a regular granolithic finish (mix 1:3:2) to a minimum depth of 1-1/2″ over top of units.

Reinforce all slabs with 6 x 6 or 10 x 10 welded wire mesh.

Masonry — R0401 Maintenance of Masonry

R040130-10 Cleaning Face Brick

On smooth brick a person can clean 70 S.F. an hour; on rough brick 50 S.F. per hour. Use one gallon muriatic acid to 20 gallons of water for 1000 S.F. Do not use acid solution until wall is at least seven days old, but a mild soap solution may be used after two days.

Time has been allowed for clean-up in brick prices.

Masonry — R0405 Common Work Results for Masonry

R040513-10 Cement Mortar (material only)

Type N - 1:1:6 mix by volume. Use everywhere above grade except as noted below. - 1:3 mix using conventional masonry cement which saves handling two separate bagged materials.

Type M - 1:1/4:3 mix by volume, or 1 part cement, 1/4 (10% by wt.) lime, 3 parts sand. Use for heavy loads and where earthquakes or hurricanes may occur. Also for reinforced brick, sewers, manholes and everywhere below grade.

Mix Proportions by Volume and Compressive Strength of Mortar

Where Used	Allowable Proportions by Volume					Compressive Strength @ 28 days
	Mortar Type	Portland Cement	Masonry Cement	Hydrated Lime	Masonry Sand	
Plain Masonry	M	1	1	—	6	2500 psi
		1	—	1/4	3	
	S	1/2	1	—	4	1800 psi
		1	—	1/4 to 1/2	4	
	N	—	1	—	3	750 psi
		1	—	1/2 to 1-1/4	6	
	O	—	1	—	3	350 psi
		1	—	1-1/4 to 2-1/2	9	
	K	1	—	2-1/2 to 4	12	75 psi
Reinforced Masonry	PM	1	1	—	6	2500 psi
	PL	1	—	1/4 to 1/2	4	2500 psi

Note: The total aggregate should be between 2.25 to 3 times the sum of the cement and lime used.

The labor cost to mix the mortar is included in the productivity and labor cost of unit price lines in unit cost sections for brickwork, blockwork and stonework.

The material cost of mixed mortar is included in the material cost of those same unit price lines and includes the cost of renting and operating a 10 C.F. mixer at the rate of 200 C.F. per day.

There are two types of mortar color used. One type is the inert additive type with about 100 lbs. per M brick as the typical quantity required. These colors are also available in smaller-batch-sized bags (1 lb. to 15 lb.) which can be placed directly into the mixer without measuring. The other type is premixed and replaces the masonry cement. Dark green color has the highest cost.

R040519-50 Masonry Reinforcing

Horizontal joint reinforcing helps prevent wall cracks where wall movement may occur and in many locations is required by code. Horizontal joint reinforcing is generally not considered to be structural reinforcing and an unreinforced wall may still contain joint reinforcing.

Reinforcing strips come in 10′ and 12′ lengths and in truss and ladder shapes, with and without drips. Field labor runs between 2.7 to 5.3 hours per 1000 L.F. for wall thicknesses up to 12″.

The wire meets ASTM A82 for cold drawn steel wire and the typical size is 9 ga. sides and ties with 3/16″ diameter also available. Typical finish is mill galvanized with zinc coating at .10 oz. per S.F. Class I (.40 oz. per S.F.) and Class III (.80 oz per S.F.) are also available, as is hot dipped galvanizing at 1.50 oz. per S.F.

R042110-10 Economy in Bricklaying

Have adequate supervision. Be sure bricklayers are always supplied with materials so there is no waiting. Place best bricklayers at corners and openings.

Use only screened sand for mortar. Otherwise, labor time will be wasted picking out pebbles. Use seamless metal tubs for mortar as they do not leak or catch the trowel. Locate stack and mortar for easy wheeling.

Have brick delivered for stacking. This makes for faster handling, reduces chipping and breakage, and requires less storage space. Many dealers will deliver select common in 2′ x 3′ x 4′ pallets or face brick packaged. This affords quick handling with a crane or forklift and easy tonging in units of ten, which reduces waste.

Use wider bricks for one wythe wall construction. Keep scaffolding away from wall to allow mortar to fall clear and not stain wall.

On large jobs develop specialized crews for each type of masonry unit.

Consider designing for prefabricated panel construction on high rise projects.

Avoid excessive corners or openings. Each opening adds about 50% to labor cost for area of opening.

Bolting stone panels and using window frames as stops reduces labor costs and speeds up erection.

R042110-20 Common and Face Brick

Common building brick manufactured according to ASTM C62 and facing brick manufactured according to ASTM C216 are the two standard bricks available for general building use.

Building brick is made in three grades; SW, where high resistance to damage caused by cyclic freezing is required; MW, where moderate resistance to cyclic freezing is needed; and NW, where little resistance to cyclic freezing is needed. Facing brick is made in only the two grades SW and MW. Additionally, facing brick is available in three types; FBS, for general use; FBX, for general use where a higher degree of precision and lower permissible variation in size than FBS is needed; and FBA, for general use to produce characteristic architectural effects resulting from non-uniformity in size and texture of the units.

In figuring the material cost of brickwork, an allowance of 25% mortar waste and 3% brick breakage was included. If bricks are delivered palletized with 280 to 300 per pallet, or packaged, allow only 1-1/2% for breakage. Packaged or palletized delivery is practical when a job is big enough to have a crane or other equipment available to handle a package of brick. This is so on all industrial work but not always true on small commercial buildings.

The use of buff and gray face is increasing, and there is a continuing trend to the Norman, Roman, Jumbo and SCR brick.

Common red clay brick for backup is not used that often. Concrete block is the most usual backup material with occasional use of sand lime or cement brick. Building brick is commonly used in solid walls for strength and as a fire stop.

Brick panels built on the ground and then crane erected to the upper floors have proven to be economical. This allows the work to be done under cover and without scaffolding.

R042110-50 Brick, Block & Mortar Quantities

Running Bond					
Number of Brick per S.F. of Wall - Single Wythe with 3/8″ Joints				C.F. of Mortar per M Bricks, Waste Included	
Type Brick	Nominal Size (incl. mortar) L H W	Modular Coursing	Number of Brick per S.F.	3/8″ Joint	1/2″ Joint
Standard	8 x 2-2/3 x 4	3C=8″	6.75	10.3	12.9
Economy	8 x 4 x 4	1C=4″	4.50	11.4	14.6
Engineer	8 x 3-1/5 x 4	5C=16″	5.63	10.6	13.6
Fire	9 x 2-1/2 x 4-1/2	2C=5″	6.40	550 # Fireclay	—
Jumbo	12 x 4 x 6 or 8	1C=4″	3.00	23.8	30.8
Norman	12 x 2-2/3 x 4	3C=8″	4.50	14.0	17.9
Norwegian	12 x 3-1/5 x 4	5C=16″	3.75	14.6	18.6
Roman	12 x 2 x 4	2C=4″	6.00	13.4	17.0
SCR	12 x 2-2/3 x 6	3C=8″	4.50	21.8	28.0
Utility	12 x 4 x 4	1C=4″	3.00	15.4	19.6

For Other Bonds Standard Size Add to S.F. Quantities in Table to Left		
Bond Type	Description	Factor
Common	full header every fifth course	+20%
	full header every sixth course	+16.7%
English	full header every second course	+50%
Flemish	alternate headers every course	+33.3%
	every sixth course	+5.6%
Header = W x H exposed		+100%
Rowlock = H x W exposed		+100%
Rowlock stretcher = L x W exposed		+33.3%
Soldier = H x L exposed		—
Sailor = W x L exposed		-33.3%

Concrete Blocks Nominal Size	Approximate Weight per S.F.		Blocks per 100 S.F.	Mortar per M block, waste included	
	Standard	Lightweight		Partitions	Back up
2″ x 8″ x 16″	20 PSF	15 PSF	113	27 C.F.	36 C.F.
4″	30	20		41	51
6″	42	30		56	66
8″	55	38		72	82
10″	70	47		87	97
12″	85	55		102	112

Masonry — R0422 Concrete Unit Masonry

R042210-20 Concrete Block

The material cost of special block such as corner, jamb and head block can be figured at the same price as ordinary block of same size. Labor on specials is about the same as equal-sized regular block.

Bond beam and 16″ high lintel blocks are more expensive than regular units of equal size. Lintel blocks are 8″ long and either 8″ or 16″ high.

Use of motorized mortar spreader box will speed construction of continuous walls.

Hollow non-load-bearing units are made according to ASTM C129 and hollow load-bearing units according to ASTM C90.

Metals — R0505 Common Work Results for Metals

R050516-30 Coating Structural Steel

On field-welded jobs, the shop-applied primer coat is necessarily omitted. All painting must be done in the field and usually consists of red oxide rust inhibitive paint or an aluminum paint. The table below shows paint coverage and daily production for field painting.

See Division 09 97 13.23 for hot-dipped galvanizing and for field-applied cold galvanizing and other paints and protective coatings.

See Division 05 01 10.51 for steel surface preparation treatments such as wire brushing, pressure washing and sand blasting.

Type Construction	Surface Area per Ton	Coat	One Gallon Covers		In 8 Hrs. Person Covers		Average per Ton Spray	
			Brush	Spray	Brush	Spray	Gallons	Labor-hours
Light Structural	300 S.F. to 500 S.F.	1st	500 S.F.	455 S.F.	640 S.F.	2000 S.F.	0.9 gals.	1.6 L.H.
		2nd	450	410	800	2400	1.0	1.3
		3rd	450	410	960	3200	1.0	1.0
Medium	150 S.F. to 300 S.F.	All	400	365	1600	3200	0.6	0.6
Heavy Structural	50 S.F. to 150 S.F.	1st	400	365	1920	4000	0.2	0.2
		2nd	400	365	2000	4000	0.2	0.2
		3rd	400	365	2000	4000	0.2	0.2
Weighted Average	225 S.F.	All	400	365	1350	3000	0.6	0.6

R050521-20 Welded Structural Steel

Usual weight reductions with welded design run 10% to 20% compared with bolted or riveted connections. This amounts to about the same total cost compared with bolted structures since field welding is more expensive than bolts. For normal spans of 18′ to 24′ figure 6 to 7 connections per ton.

Trusses — For welded trusses add 4% to weight of main members for connections. Up to 15% less steel can be expected in a welded truss compared to one that is shop bolted. Cost of erection is the same whether shop bolted or welded.

General — Typical electrodes for structural steel welding are E6010, E6011, E60T and E70T. Typical buildings vary between 2# to 8# of weld rod per ton of steel. Buildings utilizing continuous design require about three times as much welding as conventional welded structures. In estimating field erection by welding, it is best to use the average linear feet of weld per ton to arrive at the welding cost per ton. The type, size and position of the weld will have a direct bearing on the cost per linear foot. A typical field welder will deposit 1.8# to 2# of weld rod per hour manually. Using semiautomatic methods can increase production by as much as 50% to 75%.

R050523-10 High Strength Bolts

Common bolts (A307) are usually used in secondary connections (see Division 05 05 23.10).

High strength bolts (A325 and A490) are usually specified for primary connections such as column splices, beam and girder connections to columns, column bracing, connections for supports of operating equipment or of other live loads which produce impact or reversal of stress, and in structures carrying cranes of over 5-ton capacity.

Allow 20 field bolts per ton of steel for a 6 story office building, apartment house or light industrial building. For 6 to 12 stories allow 25 bolts per ton, and above 12 stories, 30 bolts per ton. On power stations, 20 to 25 bolts per ton are needed.

R051223-10 Structural Steel

The bare material prices for structural steel, shown in the unit cost sections of the book, are for 100 tons of shop-fabricated structural steel and include:

1. Mill base price of structural steel
2. Mill scrap/grade/size/length extras
3. Mill delivery to a metals service center (warehouse)
4. Service center storage and handling
5. Service center delivery to a fabrication shop
6. Shop storage and handling
7. Shop drafting/detailing
8. Shop fabrication
9. Shop coat of primer paint
10. Shop listing
11. Shop delivery to the job site

In unit cost sections of the book that contain items for field fabrication of steel components, the bare material cost of steel includes:

1. Mill base price of structural steel
2. Mill scrap/grade/size/length extras
3. Mill delivery to a metals service center (warehouse)
4. Service center storage and handling
5. Service center delivery to the job site

R051223-20 Steel Estimating Quantities

One estimate on erection is that a crane can handle 35 to 60 pieces per day. Say the average is 45. With usual sizes of beams, girders, and columns, this would amount to about 20 tons per day. The type of connection greatly affects the speed of erection. Moment connections for continuous design slow down production and increase erection costs.

Short open web bar joists can be set at the rate of 75 to 80 per day, with 50 per day being the average for setting long span joists.

After main members are calculated, add the following for usual allowances: base plates 2% to 3%; column splices 4% to 5%; and miscellaneous details 4% to 5%, for a total of 10% to 13% in addition to main members.

The ratio of column to beam tonnage varies depending on type of steels used, typical spans, story heights and live loads.

It is more economical to keep the column size constant and to vary the strength of the column by using high strength steels. This also saves floor space. Buildings have recently gone as high as ten stories with 8″ high strength columns. For light columns under W8X31 lb. sections, concrete filled steel columns are economical.

High strength steels may be used in columns and beams to save floor space and to meet head room requirements. High strength steels in some sizes sometimes require long lead times.

Round, square and rectangular columns, both plain and concrete filled, are readily available and save floor area, but are higher in cost per pound than rolled columns. For high unbraced columns, tube columns may be less expensive.

Below are average minimum figures for the weights of the structural steel frame for different types of buildings using A36 steel, rolled shapes and simple joints. For economy in domes, rise to span ratio = .13. Open web joist framing systems will reduce weights by 10% to 40%. Composite design can reduce steel weight by up to 25% but additional concrete floor slab thickness may be required. Continuous design can reduce the weights up to 20%. There are many building codes with different live load requirements and different structural requirements, such as hurricane and earthquake loadings which can alter the figures.

Structural Steel Weights per S.F. of Floor Area

Type of Building	No. of Stories	Avg. Spans	L.L. #/S.F.	Lbs. Per S.F.	Type of Building	No. of Stories	Avg. Spans	L.L. #/S.F.	Lbs. Per S.F.
Steel Frame Mfg.	1	20'x20'	40	8	Apartments	2-8	20'x20'	40	8
		30'x30'		13		9-25			14
		40'x40'		18	Office	to 10	Various	80	10
Parking garage	4	Various	80	8.5		20			18
Domes (Schwedler)*	1	200'	30	10		30			26
		300'		15		over 50			35

R051223-25 Common Structural Steel Specifications

ASTM A992 (formerly A36, then A572 Grade 50) is the all-purpose carbon grade steel widely used in building and bridge construction.

The other high-strength steels listed below may each have certain advantages over ASTM A992 stuctural carbon steel, depending on the application. They have proven to be economical choices where, due to lighter members, the reduction of dead load and the associated savings in shipping cost can be significant.

ASTM A588 atmospheric weathering, high-strength low-alloy steels can be used in the bare (uncoated) condition, where exposure to normal atmosphere causes a tightly adherant oxide to form on the surface protecting the steel from further oxidation. ASTM A242 corrosion-resistant, high-strength low-alloy steels have enhanced atmospheric corrosion resistance of at least two times that of carbon structural steels with copper, or four times that of carbon structural steels without copper. The reduction or elimination of maintenance resulting from the use of these steels often offsets their higher initial cost.

Steel Type	ASTM Designation	Minimum Yield Stress in KSI	Shapes Available
Carbon	A36	36	All structural shape groups, and plates & bars up thru 8″ thick
	A529	50	Structural shape group 1, and plates & bars up thru 2″ thick
High-Strength Low-Alloy Quenched & Self-Tempered	A913	50	All structural shape groups
		60	
		65	
		70	
High-Strength Low-Alloy Columbium-Vanadium	A572	42	All structural shape groups, and plates & bars up thru 6″ thick
		50	All structural shape groups, and plates & bars up thru 4″ thick
		55	Structural shape groups 1 & 2, and plates & bars up thru 2″ thick
		60	Structural shape groups 1 & 2, and plates & bars up thru 1-1/4″ thick
		65	Structural shape group 1, and plates & bars up thru 1-1/4″ thick
High-Strength Low-Alloy Columbium-Vanadium	A992	50	All structural shape groups
Weathering High-Strength Low-Alloy	A242	42	Structural shape groups 4 & 5, and plates & bars over 1-1/2″ up thru 4″ thick
		46	Structural shape group 3, and plates & bars over 3/4″ up thru 1-1/2″ thick
		50	Structural shape groups 1 & 2, and plates & bars up thru 3/4″ thick
Weathering High-Strength Low-Alloy	A588	42	Plates & bars over 5″ up thru 8″ thick
		46	Plates & bars over 4″ up thru 5″ thick
		50	All structural shape groups, and plates & bars up thru 4″ thick
Quenched and Tempered	A852	70	Plates & bars up thru 4″ thick
Low-Alloy Quenched and Tempered Alloy	A514	90	Plates & bars over 2-1/2″ up thru 6″ thick
		100	Plates & bars up thru 2-1/2″ thick

R051223-30 High Strength Steels

The mill price of high strength steels may be higher than A992 carbon steel but their proper use can achieve overall savings thru total reduced weights. For columns with L/r over 100, A992 steel is best; under 100, high strength steels are economical. For heavy columns, high strength steels are economical when cover plates are eliminated. There is no economy using high strength steels for clip angles or supports or for beams where deflection governs. Thinner members are more economical than thick.

The per ton erection and fabricating costs of the high strength steels will be higher than for A992 since the same number of pieces, but less weight, will be installed.

R051223-35 Common Steel Sections

The upper portion of this table shows the name, shape, common designation and basic characteristics of commonly used steel sections. The lower portion explains how to read the designations used for the above illustrated common sections.

Shape & Designation	Name & Characteristics	Shape & Designation	Name & Characteristics
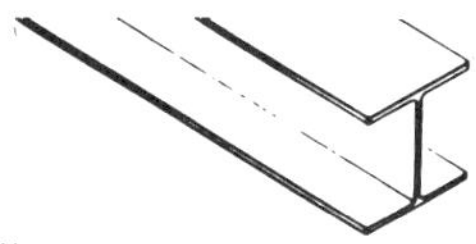 W	W Shape Parallel flange surfaces	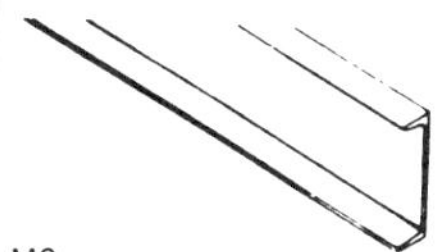 MC	Miscellaneous Channel Infrequently rolled by some producers
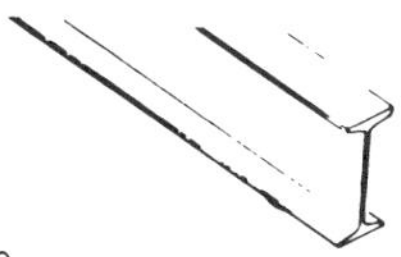 S	American Standard Beam (I Beam) Sloped inner flange	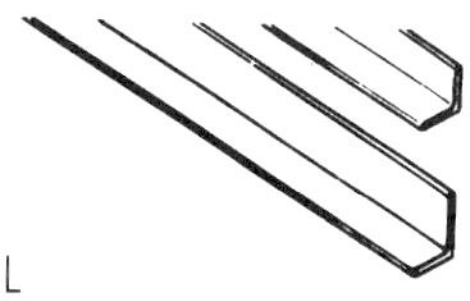 L	Angle Equal or unequal legs, constant thickness
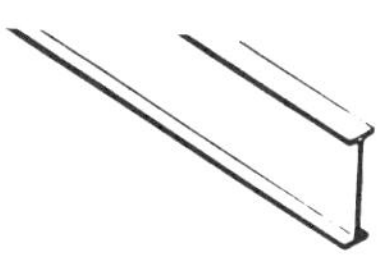 M	Miscellaneous Beams Cannot be classified as W, HP or S; infrequently rolled by some producers	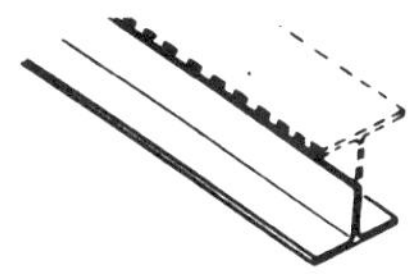 T	Structural Tee Cut from W, M or S on center of web
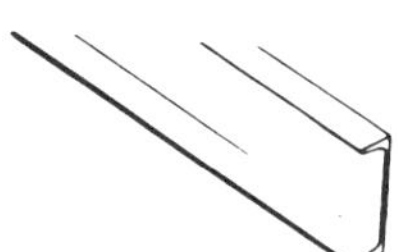 C	American Standard Channel Sloped inner flange	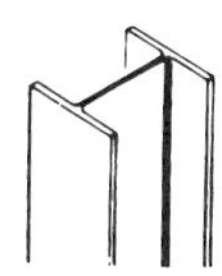 HP	Bearing Pile Parallel flanges and equal flange and web thickness

Common drawing designations follow:

W Shape
W 18 x 35
- 35: Weight in Pounds per Foot
- 18: Nominal Depth in Inches (Actual 17-3/4″)

American Standard Beam
S 12 x 31.8
- 31.8: Weight in Pounds per Foot
- 12: Depth in Inches

Miscellaneous Beam
M 8 x 6.5
- 6.5: Weight in Pounds per Foot
- 8: Depth in Inches

American Standard Channel
C 8 x 11.5
- 11.5: Weight in Pounds per Foot
- 8: Depth in Inches

Miscellaneous Channel
MC 8 x 22.8
- 22.8: Weight in Pounds per Foot
- 8: Depth in Inches

Angle
L 6 x 3-1/2 x 3/8
- 6: Length of Long Leg in Inches
- 3/8: Thickness of Each Leg in Inches
- 3-1/2: Length of Other Leg in Inches

Tee Cut from W16 x 100
WT 8 x 50
- 50: Weight in Pounds per Foot
- 8: Nominal Depth in Inches (Actual 8-1/2″)

Tee Cut from S12 x 35
ST 6 x 17.5
- 17.5: Weight in Pounds per Foot
- 6: Depth in Inches (Actual 6-1/4″)

Tee Cut from M10 x 9
MT 5 x 4.5
- 4.5: Weight in Pounds per Foot
- 5: Depth in Inches

Bearing Pile
HP 12 x 84
- 84: Weight in Pounds per Foot
- 12: Nominal Depth in Inches (Actual 12-1/4″)

R051223-45 Installation Time for Structural Steel Building Components

The following tables show the expected average installation times for various structural steel shapes. Table A presents installation times for columns, Table B for beams, Table C for light framing and bolts, and Table D for structural steel for various project types.

Table A

Description	Labor-Hours	Unit
Columns		
Steel, Concrete Filled		
3-1/2" Diameter	.933	Ea.
6-5/8" Diameter	1.120	Ea.
Steel Pipe		
3" Diameter	.933	Ea.
8" Diameter	1.120	Ea.
12" Diameter	1.244	Ea.
Structural Tubing		
4" x 4"	.966	Ea.
8" x 8"	1.120	Ea.
12" x 8"	1.167	Ea.
W Shape 2 Tier		
W8 x 31	.052	L.F.
W8 x 67	.057	L.F.
W10 x 45	.054	L.F.
W10 x 112	.058	L.F.
W12 x 50	.054	L.F.
W12 x 190	.061	L.F.
W14 x 74	.057	L.F.
W14 x 176	.061	L.F.

Table B

Description	Labor-Hours	Unit	Labor-Hours	Unit
Beams, W Shape				
W6 x 9	.949	Ea.	.093	L.F.
W10 x 22	1.037	Ea.	.085	L.F.
W12 x 26	1.037	Ea.	.064	L.F.
W14 x 34	1.333	Ea.	.069	L.F.
W16 x 31	1.333	Ea.	.062	L.F.
W18 x 50	2.162	Ea.	.088	L.F.
W21 x 62	2.222	Ea.	.077	L.F.
W24 x 76	2.353	Ea.	.072	L.F.
W27 x 94	2.581	Ea.	.067	L.F.
W30 x 108	2.857	Ea.	.067	L.F.
W33 x 130	3.200	Ea.	.071	L.F.
W36 x 300	3.810	Ea.	.077	L.F.

Table C

Description	Labor-Hours	Unit
Light Framing		
Angles 4" and Larger	.055	lbs.
Less than 4"	.091	lbs.
Channels 8" and Larger	.048	lbs.
Less than 8"	.072	lbs.
Cross Bracing Angles	.055	lbs.
Rods	.034	lbs.
Hanging Lintels	.069	lbs.
High Strength Bolts in Place		
3/4" Bolts	.070	Ea.
7/8" Bolts	.076	Ea.

Table D

Description	Labor-Hours	Unit	Labor-Hours	Unit
Apartments, Nursing Homes, etc.				
1-2 Stories	4.211	Piece	7.767	Ton
3-6 Stories	4.444	Piece	7.921	Ton
7-15 Stories	4.923	Piece	9.014	Ton
Over 15 Stories	5.333	Piece	9.209	Ton
Offices, Hospitals, etc.				
1-2 Stories	4.211	Piece	7.767	Ton
3-6 Stories	4.741	Piece	8.889	Ton
7-15 Stories	4.923	Piece	9.014	Ton
Over 15 Stories	5.120	Piece	9.209	Ton
Industrial Buildings				
1 Story	3.478	Piece	6.202	Ton

R051223-50 Subpurlins

Bulb tee subpurlins are structural members designed to support and reinforce a variety of roof deck systems such as precast cement fiber roof deck tiles, monolithic roof deck systems, and gypsum or lightweight concrete over formboard. Other uses include interstitial service ceiling systems, wall panel systems, and joist anchoring in bond beams. See Unit Price section for pricing on a square foot basis at 32-5/8" O.C. Maximum span is based on a 3-span condition with a total allowable vertical load of 40 psf.

R051223-80 Dimensions and Weights of Sheet Steel

Gauge No.	Approximate Thickness: Inches (in fractions) Wrought Iron	Inches (in decimal parts) Wrought Iron	Inches (in decimal parts) Steel	Millimeters Steel	Weight per S.F. in Ounces	Weight per S.F. in Lbs.	Weight per Square Meter in Kg.
0000000	1/2"	.5	.4782	12.146	320	20.000	97.650
000000	15/32"	.46875	.4484	11.389	300	18.750	91.550
00000	7/16"	.4375	.4185	10.630	280	17.500	85.440
0000	13/32"	.40625	.3886	9.870	260	16.250	79.330
000	3/8"	.375	.3587	9.111	240	15.000	73.240
00	11/32"	.34375	.3288	8.352	220	13.750	67.130
0	5/16"	.3125	.2989	7.592	200	12.500	61.030
1	9/32"	.28125	.2690	6.833	180	11.250	54.930
2	17/64"	.265625	.2541	6.454	170	10.625	51.880
3	1/4"	.25	.2391	6.073	160	10.000	48.820
4	15/64"	.234375	.2242	5.695	150	9.375	45.770
5	7/32"	.21875	.2092	5.314	140	8.750	42.720
6	13/64"	.203125	.1943	4.935	130	8.125	39.670
7	3/16"	.1875	.1793	4.554	120	7.500	36.320
8	11/64"	.171875	.1644	4.176	110	6.875	33.570
9	5/32"	.15625	.1495	3.797	100	6.250	30.520
10	9/64"	.140625	.1345	3.416	90	5.625	27.460
11	1/8"	.125	.1196	3.038	80	5.000	24.410
12	7/64"	.109375	.1046	2.657	70	4.375	21.360
13	3/32"	.09375	.0897	2.278	60	3.750	18.310
14	5/64"	.078125	.0747	1.897	50	3.125	15.260
15	9/128"	.0713125	.0673	1.709	45	2.813	13.730
16	1/16"	.0625	.0598	1.519	40	2.500	12.210
17	9/160"	.05625	.0538	1.367	36	2.250	10.990
18	1/20"	.05	.0478	1.214	32	2.000	9.765
19	7/160"	.04375	.0418	1.062	28	1.750	8.544
20	3/80"	.0375	.0359	.912	24	1.500	7.324
21	11/320"	.034375	.0329	.836	22	1.375	6.713
22	1/32"	.03125	.0299	.759	20	1.250	6.103
23	9/320"	.028125	.0269	.683	18	1.125	5.490
24	1/40"	.025	.0239	.607	16	1.000	4.882
25	7/320"	.021875	.0209	.531	14	.875	4.272
26	3/160"	.01875	.0179	.455	12	.750	3.662
27	11/640"	.0171875	.0164	.417	11	.688	3.357
28	1/64"	.015625	.0149	.378	10	.625	3.052

R053100-10 Decking Descriptions

General - All Deck Products

Steel deck is made by cold forming structural grade sheet steel into a repeating pattern of parallel ribs. The strength and stiffness of the panels are the result of the ribs and the material properties of the steel. Deck lengths can be varied to suit job conditions, but because of shipping considerations, are usually less than 40 feet. Standard deck width varies with the product used but full sheets are usually 12″, 18″, 24″, 30″, or 36″. Deck is typically furnished in a standard width with the ends cut square. Any cutting for width, such as at openings or for angular fit, is done at the job site.

Deck is typically attached to the building frame with arc puddle welds, self-drilling screws, or powder or pneumatically driven pins. Sheet to sheet fastening is done with screws, button punching (crimping), or welds.

Composite Floor Deck

After installation and adequate fastening, floor deck serves several purposes. It (a) acts as a working platform, (b) stabilizes the frame, (c) serves as a concrete form for the slab, and (d) reinforces the slab to carry the design loads applied during the life of the building. Composite decks are distinguished by the presence of shear connector devices as part of the deck. These devices are designed to mechanically lock the concrete and deck together so that the concrete and the deck work together to carry subsequent floor loads. These shear connector devices can be rolled-in embossments, lugs, holes, or wires welded to the panels. The deck profile can also be used to interlock concrete and steel.

Composite deck finishes are either galvanized (zinc coated) or phosphatized/painted. Galvanized deck has a zinc coating on both the top and bottom surfaces. The phosphatized/painted deck has a bare (phosphatized) top surface that will come into contact with the concrete. This bare top surface can be expected to develop rust before the concrete is placed. The bottom side of the deck has a primer coat of paint.

Composite floor deck is normally installed so the panel ends do not overlap on the supporting beams. Shear lugs or panel profile shape often prevent a tight metal to metal fit if the panel ends overlap; the air gap caused by overlapping will prevent proper fusion with the structural steel supports when the panel end laps are shear stud welded.

Adequate end bearing of the deck must be obtained as shown on the drawings. If bearing is actually less in the field than shown on the drawings, further investigation is required.

Roof Deck

Roof deck is not designed to act compositely with other materials. Roof deck acts alone in transferring horizontal and vertical loads into the building frame. Roof deck rib openings are usually narrower than floor deck rib openings. This provides adequate support of rigid thermal insulation board.

Roof deck is typically installed to endlap approximately 2″ over supports. However, it can be butted (or lapped more than 2″) to solve field fit problems. Since designers frequently use the installed deck system as part of the horizontal bracing system (the deck as a diaphragm), any fastening substitution or change should be approved by the designer. Continuous perimeter support of the deck is necessary to limit edge deflection in the finished roof and may be required for diaphragm shear transfer.

Standard roof deck finishes are galvanized or primer painted. The standard factory applied paint for roof deck is a primer paint and is not intended to weather for extended periods of time. Field painting or touching up of abrasions and deterioration of the primer coat or other protective finishes is the responsibility of the contractor.

Cellular Deck

Cellular deck is made by attaching a bottom steel sheet to a roof deck or composite floor deck panel. Cellular deck can be used in the same manner as floor deck. Electrical, telephone, and data wires are easily run through the chase created between the deck panel and the bottom sheet.

When used as part of the electrical distribution system, the cellular deck must be installed so that the ribs line up and create a smooth cell transition at abutting ends. The joint that occurs at butting cell ends must be taped or otherwise sealed to prevent wet concrete from seeping into the cell. Cell interiors must be free of welding burrs, or other sharp intrusions, to prevent damage to wires.

When used as a roof deck, the bottom flat plate is usually left exposed to view. Care must be maintained during erection to keep good alignment and prevent damage.

Cellular deck is sometimes used with the flat plate on the top side to provide a flat working surface. Installation of the deck for this purpose requires special methods for attachment to the frame because the flat plate, now on the top, can prevent direct access to the deck material that is bearing on the structural steel. It may be advisable to treat the flat top surface to prevent slipping.

Cellular deck is always furnished galvanized or painted over galvanized.

Form Deck

Form deck can be any floor or roof deck product used as a concrete form. Connections to the frame are by the same methods used to anchor floor and roof deck. Welding washers are recommended when welding deck that is less than 20 gauge thickness.

Form deck is furnished galvanized, prime painted, or uncoated. Galvanized deck must be used for those roof deck systems where form deck is used to carry a lightweight insulating concrete fill.

Wood, Plastics & Comp. R0611 Wood Framing

R061110-30 Lumber Product Material Prices

The price of forest products fluctuates widely from location to location and from season to season depending upon economic conditions. The bare material prices in the unit cost sections of the book show the National Average material prices in effect Jan. 1 of this book year. It must be noted that lumber prices in general may change significantly during the year.

Availability of certain items depends upon geographic location and must be checked prior to firm-price bidding.

Wood, Plastics & Comp. R0616 Sheathing

R061636-20 Plywood

There are two types of plywood used in construction: interior, which is moisture-resistant but not waterproofed, and exterior, which is waterproofed.

The grade of the exterior surface of the plywood sheets is designated by the first letter: A, for smooth surface with patches allowed; B, for solid surface with patches and plugs allowed; C, which may be surface plugged or may have knot holes up to 1″ wide; and D, which is used only for interior type plywood and may have knot holes up to 2-1/2″ wide. "Structural Grade" is specifically designed for engineered applications such as box beams. All CC & DD grades have roof and floor spans marked on them.

Underlayment-grade plywood runs from 1/4″ to 1-1/4″ thick. Thicknesses 5/8″ and over have optional tongue and groove joints which eliminate the need for blocking the edges. Underlayment 19/32″ and over may be referred to as Sturd-i-Floor.

The price of plywood can fluctuate widely due to geographic and economic conditions.

Typical uses for various plywood grades are as follows:

AA-AD Interior — cupboards, shelving, paneling, furniture

BB Plyform — concrete form plywood

CDX — wall and roof sheathing

Structural — box beams, girders, stressed skin panels

AA-AC Exterior — fences, signs, siding, soffits, etc.

Underlayment — base for resilient floor coverings

Overlaid HDO — high density for concrete forms & highway signs

Overlaid MDO — medium density for painting, siding, soffits & signs

303 Siding — exterior siding, textured, striated, embossed, etc.

Thermal & Moist. Protec. R0731 Shingles & Shakes

R073126-20 Roof Slate

16″, 18″ and 20″ are standard lengths, and slate usually comes in random widths. For standard 3/16″ thickness use 1-1/2″ copper nails. Allow for 3% breakage.

Thermal & Moist. Protec. R0751 Built-Up Bituminous Roofing

R075113-20 Built-Up Roofing

Asphalt is available in kegs of 100 lbs. each; coal tar pitch in 560 lb. kegs. Prepared roofing felts are available in a wide range of sizes, weights and characteristics. However, the most commonly used are #15 (432 S.F. per roll, 13 lbs. per square) and #30 (216 S.F. per roll, 27 lbs. per square).

Inter-ply bitumen varies from 24 lbs. per sq. (asphalt) to 30 lbs. per sq. (coal tar) per ply, MF4@ 25%. Flood coat bitumen also varies from 60 lbs. per sq. (asphalt) to 75 lbs. per sq. (coal tar), MF4@ 25%. Expendable equipment (mops, brooms, screeds, etc.) runs about 16% of the bitumen cost. For new, inexperienced crews this factor may be much higher.

Rigid insulation board is typically applied in two layers. The first is mechanically attached to nailable decks or spot or solid mopped to non-nailable decks; the second layer is then spot or solid mopped to the first layer. Membrane application follows the insulation, except in protected membrane roofs, where the membrane goes down first and the insulation on top, followed with ballast (stone or concrete pavers). Insulation and related labor costs are NOT included in prices for built-up roofing.

Thermal & Moist. Protec. R0752 Modified Bituminous Membrane Roofing

R075213-30 Modified Bitumen Roofing

The cost of modified bitumen roofing is highly dependent on the type of installation that is planned. Installation is based on the type of modifier used in the bitumen. The two most popular modifiers are atactic polypropylene (APP) and styrene butadiene styrene (SBS). The modifiers are added to heated bitumen during the manufacturing process to change its characteristics. A polyethylene, polyester or fiberglass reinforcing sheet is then sandwiched between layers of this bitumen. When completed, the result is a pre-assembled, built-up roof that has increased elasticity and weatherablility. Some manufacturers include a surfacing material such as ceramic or mineral granules, metal particles or sand.

The preferred method of adhering SBS-modified bitumen roofing to the substrate is with hot-mopped asphalt (much the same as built-up roofing). This installation method requires a tar kettle/pot to heat the asphalt, as well as the labor, tools and equipment necessary to distribute and spread the hot asphalt.

The alternative method for applying APP and SBS modified bitumen is as follows. A skilled installer uses a torch to melt a small pool of bitumen off the membrane. This pool must form across the entire roll for proper adhesion. The installer must unroll the roofing at a pace slow enough to melt the bitumen, but fast enough to prevent damage to the rest of the membrane.

Modified bitumen roofing provides the advantages of both built-up and single-ply roofing. Labor costs are reduced over those of built-up roofing because only a single ply is necessary. The elasticity of single-ply roofing is attained with the reinforcing sheet and polymer modifiers. Modifieds have some self-healing characteristics and because of their multi-layer construction, they offer the reliability and safety of built-up roofing.

Thermal & Moist. Protec. R0784 Firestopping

R078413-30 Firestopping

Firestopping is the sealing of structural, mechanical, electrical and other penetrations through fire-rated assemblies. The basic components of firestop systems are safing insulation and firestop sealant on both sides of wall penetrations and the top side of floor penetrations.

Pipe penetrations are assumed to be through concrete, grout, or joint compound and can be sleeved or unsleeved. Costs for the penetrations and sleeves are not included. An annular space of 1″ is assumed. Escutcheons are not included.

Metallic pipe is assumed to be copper, aluminum, cast iron or similar metallic material. Insulated metallic pipe is assumed to be covered with a thermal insulating jacket of varying thickness and materials.

Non-metallic pipe is assumed to be PVC, CPVC, FR Polypropylene or similar plastic piping material. Intumescent firestop sealant or wrap strips are included. Collars on both sides of wall penetrations and a sheet metal plate on the underside of floor penetrations are included.

Ductwork is assumed to be sheet metal, stainless steel or similar metallic material. Duct penetrations are assumed to be through concrete, grout or joint compound. Costs for penetrations and sleeves are not included. An annular space of 1/2″ is assumed.

Multi-trade openings include costs for sheet metal forms, firestop mortar, wrap strips, collars and sealants as necessary.

Structural penetrations joints are assumed to be 1/2″ or less. CMU walls are assumed to be within 1-1/2″ of metal deck. Drywall walls are assumed to be tight to the underside of metal decking.

Metal panel, glass or curtain wall systems include a spandrel area of 5′ filled with mineral wool foil-faced insulation. Fasteners and stiffeners are included.

R081313-20 Steel Door Selection Guide

Standard steel doors are classified into four levels, as recommended by the Steel Door Institute in the chart below. Each of the four levels offers a range of construction models and designs, to meet architectural requirements for preference and appearance, including full flush, seamless, and stile & rail. Recommended minimum gauge requirements are also included.

For complete standard steel door construction specifications and available sizes, refer to the Steel Door Institute Technical Data Series, ANSI A250.8-98 (SDI-100), and ANSI A250.4-94 Test Procedure and Acceptance Criteria for Physical Endurance of Steel Door and Hardware Reinforcements.

Level		Model	Construction	For Full Flush or Seamless		
				Min. Gauge	Thickness (in)	Thickness (mm)
I	Standard Duty	1 2	Full Flush Seamless	20	0.032	0.8
II	Heavy Duty	1 2	Full Flush Seamless	18	0.042	1.0
III	Extra Heavy Duty	1 2 3	Full Flush Seamless *Stile & Rail	16	0.053	1.3
IV	Maximum Duty	1 2	Full Flush Seamless	14	0.067	1.6

*Stiles & rails are 16 gauge; flush panels, when specified, are 18 gauge

Openings R0851 Metal Windows

R085123-10 Steel Sash

Ironworker crew will erect 25 S.F. or 1.3 sash unit per hour, whichever is less.

Mechanic will point 30 L.F. per hour.

Painter will paint 90 S.F. per coat per hour.

Glazier production depends on light size.

Allow 1 lb. special steel sash putty per 16″ x 20″ light.

Openings R0852 Wood Windows

R085216-10 Window Estimates

To ensure a complete window estimate, be sure to include the material and labor costs for each window, as well as the material and labor costs for an interior wood trim set.

Openings — R0871 Door Hardware

R087120-10 Hinges

All closer equipped doors should have ball bearing hinges. Lead lined or extremely heavy doors require special strength hinges. Usually 1-1/2 pair of hinges are used per door up to 7′-6″ high openings. Table below shows typical hinge requirements.

Use Frequency	Type Hinge Required	Type of Opening	Type of Structure
High	Heavy weight ball bearing	Entrances Toilet Rooms	Banks, Office buildings, Schools, Stores & Theaters Office buildings and Schools
Average	Standard weight ball bearing	Entrances Corridors Toilet Rooms	Dwellings Office buildings and Schools Stores
Low	Plain bearing	Interior	Dwellings

Door Thickness	Weight of Doors in Pounds per Square Foot				
	White Pine	Oak	Hollow Core	Solid Core	Hollow Metal
1-3/8″	3psf	6psf	1-1/2psf	3-1/2 — 4psf	6-1/2psf
1-3/4″	3-1/2	7	2-1/2	4-1/2 — 5-1/4	6-1/2
2-1/4″	4-1/2	9	—	5-1/2 — 6-3/4	6-1/2

Openings — R0881 Glass Glazing

R088110-10 Glazing Productivity

Some glass sizes are estimated by the "united inch" (height + width). The table below shows the number of lights glazed in an eight-hour period by the crew size indicated, for glass up to 1/4″ thick. Square or nearly square lights are more economical on a S.F. basis. Long slender lights will have a high S.F. installation cost. For insulated glass reduce production by 33%. For 1/2″ float glass reduce production by 50%. Production time for glazing with two glaziers per day averages: 1/4″ float glass 120 S.F.; 1/2″ float glass 55 S.F.; 1/2″ insulated glass 95 S.F.; 3/4″ insulated glass 75 S.F.

Glazing Method	United Inches per Light							
	40″	60″	80″	100″	135″	165″	200″	240″
Number of Men in Crew	**1**	**1**	**1**	**1**	**2**	**3**	**3**	**4**
Industrial sash, putty	60	45	24	15	18	—	—	—
With stops, putty bed	50	36	21	12	16	8	4	3
Wood stops, rubber	40	27	15	9	11	6	3	2
Metal stops, rubber	30	24	14	9	9	6	3	2
Structural glass	10	7	4	3	—	—	—	—
Corrugated glass	12	9	7	4	4	4	3	—
Storefronts	16	15	13	11	7	6	4	4
Skylights, putty glass	60	36	21	12	16	—	—	—
Thiokol set	15	15	11	9	9	6	3	2
Vinyl set, snap on	18	18	13	12	12	7	5	4
Maximum area per light	2.8 S.F.	6.3 S.F.	11.1 S.F.	17.4 S.F.	31.6 S.F.	47 S.F.	69 S.F.	100 S.F.

R092000-50 Lath, Plaster and Gypsum Board

Gypsum board lath is available in 3/8″ thick x 16″ wide x 4′ long sheets as a base material for multi-layer plaster applications. It is also available as a base for either multi-layer or veneer plaster applications in 1/2″ and 5/8″ thick–4′ wide x 8′, 10′ or 12′ long sheets. Fasteners are screws or blued ring shank nails for wood framing and screws for metal framing.

Metal lath is available in diamond mesh pattern with flat or self-furring profiles. Paper backing is available for applications where excessive plaster waste needs to be avoided. A slotted mesh ribbed lath should be used in areas where the span between structural supports is greater than normal. Most metal lath comes in 27″ x 96″ sheets. Diamond mesh weighs 1.75, 2.5 or 3.4 pounds per square yard, slotted mesh lath weighs 2.75 or 3.4 pounds per square yard. Metal lath can be nailed, screwed or tied in place.

Many **accessories** are available. Corner beads, flat reinforcing strips, casing beads, control and expansion joints, furring brackets and channels are some examples. Note that accessories are not included in plaster or stucco line items.

Plaster is defined as a material or combination of materials that when mixed with a suitable amount of water, forms a plastic mass or paste. When applied to a surface, the paste adheres to it and subsequently hardens, preserving in a rigid state the form or texture imposed during the period of elasticity.

Gypsum plaster is made from ground calcined gypsum. It is mixed with aggregates and water for use as a base coat plaster.

Vermiculite plaster is a fire-retardant plaster covering used on steel beams, concrete slabs and other heavy construction materials. Vermiculite is a group name for certain clay minerals, hydrous silicates or aluminum, magnesium and iron that have been expanded by heat.

Perlite plaster is a plaster using perlite as an aggregate instead of sand. Perlite is a volcanic glass that has been expanded by heat.

Gauging plaster is a mix of gypsum plaster and lime putty that when applied produces a quick drying finish coat.

Veneer plaster is a one or two component gypsum plaster used as a thin finish coat over special gypsum board.

Keenes cement is a white cementitious material manufactured from gypsum that has been burned at a high temperature and ground to a fine powder. Alum is added to accelerate the set. The resulting plaster is hard and strong and accepts and maintains a high polish, hence it is used as a finishing plaster.

Stucco is a Portland cement based plaster used primarily as an exterior finish.

Plaster is used on both interior and exterior surfaces. Generally it is applied in multiple-coat systems. A three-coat system uses the terms scratch, brown and finish to identify each coat. A two-coat system uses base and finish to describe each coat. Each type of plaster and application system has attributes that are chosen by the designer to best fit the intended use.

Gypsum Plaster Quantities for 100 S.Y.	2 Coat, 5/8″ Thick		3 Coat, 3/4″ Thick		
	Base	Finish	Scratch	Brown	Finish
	1:3 Mix	2:1 Mix	1:2 Mix	1:3 Mix	2:1 Mix
Gypsum plaster	1,300 lb.		1,350 lb.	650 lb.	
Sand	1.75 C.Y.		1.85 C.Y.	1.35 C.Y.	
Finish hydrated lime		340 lb.			340 lb.
Gauging plaster		170 lb.			170 lb.

Vermiculite or Perlite Plaster Quantities for 100 S.Y.	2 Coat, 5/8″ Thick		3 Coat, 3/4″ Thick		
	Base	Finish	Scratch	Brown	Finish
Gypsum plaster	1,250 lb.		1,450 lb.	800 lb.	
Vermiculite or perlite	7.8 bags		8.0 bags	3.3 bags	
Finish hydrated lime		340 lb.			340 lb.
Gauging plaster		170 lb.			170 lb.

Stucco-Three-Coat System Quantities for 100 S.Y.	On Wood Frame	On Masonry
Portland cement	29 bags	21 bags
Sand	2.6 C.Y.	2.0 C.Y.
Hydrated lime	180 lb.	120 lb.

R092910-10 Levels of Gypsum Drywall Finish

In the past, contract documents often used phrases such as "industry standard" and "workmanlike finish" to specify the expected quality of gypsum board wall and ceiling installations. The vagueness of these descriptions led to unacceptable work and disputes.

In order to resolve this problem, four major trade associations concerned with the manufacture, erection, finish, and decoration of gypsum board wall and ceiling systems have developed an industry-wide *Recommended Levels of Gypsum Board Finish.*

The finish of gypsum board walls and ceilings for specific final decoration is dependent on a number of factors. A primary consideration is the location of the surface and the degree of decorative treatment desired. Painted and unpainted surfaces in warehouses and other areas where appearance is normally not critical may simply require the taping of wallboard joints and 'spotting' of fastener heads. Blemish-free, smooth, monolithic surfaces often intended for painted and decorated walls and ceilings in habitated structures, ranging from single-family dwellings through monumental buildings, require additional finishing prior to the application of the final decoration.

Other factors to be considered in determining the level of finish of the gypsum board surface are (1) the type of angle of surface illumination (both natural and artificial lighting), and (2) the paint and method of application or the type and finish of wallcovering specified as the final decoration. Critical lighting conditions, gloss paints, and thin wall coverings require a higher level of gypsum board finish than do heavily textured surfaces which are subsequently painted or surfaces which are to be decorated with heavy grade wall coverings.

The following descriptions were developed jointly by the Association of the Wall and Ceiling Industries-International (AWCI), Ceiling & Interior Systems Construction Association (CISCA), Gypsum Association (GA), and Painting and Decorating Contractors of America (PDCA) as a guide.

Level 0: Used in temporary construction or wherever the final decoration has not been determined. Unfinished. No taping, finishing or corner beads are required. Also could be used where non-predecorated panels will be used in demountable-type partitions that are to be painted as a final finish.

Level 1: Frequently used in plenum areas above ceilings, in attics, in areas where the assembly would generally be concealed, or in building service corridors and other areas not normally open to public view. Some degree of sound and smoke control is provided; in some geographic areas, this level is referred to as "fire-taping," although this level of finish does not typically meet fire-resistant assembly requirements. Where a fire resistance rating is required for the gypsum board assembly, details of construction should be in accordance with reports of fire tests of assemblies that have met the requirements of the fire rating acceptable.

All joints and interior angles shall have tape embedded in joint compound. Accessories are optional at specifier discretion in corridors and other areas with pedestrian traffic. Tape and fastener heads need not be covered with joint compound. Surface shall be free of excess joint compound. Tool marks and ridges are acceptable.

Level 2: It may be specified for standard gypsum board surfaces in garages, warehouse storage, or other similar areas where surface appearance is not of primary importance.

All joints and interior angles shall have tape embedded in joint compound and shall be immediately wiped with a joint knife or trowel, leaving a thin coating of joint compound over all joints and interior angles. Fastener heads and accessories shall be covered with a coat of joint compound. Surface shall be free of excess joint compound. Tool marks and ridges are acceptable.

Level 3: Typically used in areas that are to receive heavy texture (spray or hand applied) finishes before final painting, or where commercial-grade (heavy duty) wall coverings are to be applied as the final decoration. This level of finish should not be used where smooth painted surfaces or where lighter weight wall coverings are specified. The prepared surface shall be coated with a drywall primer prior to the application of final finishes.

All joints and interior angles shall have tape embedded in joint compound and shall be immediately wiped with a joint knife or trowel, leaving a thin coating of joint compound over all joints and interior angles. One additional coat of joint compound shall be applied over all joints and interior angles. Fastener heads and accessories shall be covered with two separate coats of joint compound. All joint compounds shall be smooth and free of tool marks and ridges. The prepared surface shall be covered with a drywall primer prior to the application of the final decoration.

Level 4: This level should be used where residential grade (light duty) wall coverings, flat paints, or light textures are to be applied. The prepared surface shall be coated with a drywall primer prior to the application of final finishes. Release agents for wall coverings are specifically formulated to minimize damage if coverings are subsequently removed.

The weight, texture, and sheen level of the wall covering material selected should be taken into consideration when specifying wall coverings over this level of drywall treatment. Joints and fasteners must be sufficiently concealed if the wall covering material is lightweight, contains limited pattern, has a glossy finish, or has any combination of these features. In critical lighting areas, flat paints applied over light textures tend to reduce joint photographing. Gloss, semi-gloss, and enamel paints are not recommended over this level of finish.

All joints and interior angles shall have tape embedded in joint compound and shall be immediately wiped with a joint knife or trowel, leaving a thin coating of joint compound over all joints and interior angles. In addition, two separate coats of joint compound shall be applied over all flat joints and one separate coat of joint compound applied over interior angles. Fastener heads and accessories shall be covered with three separate coats of joint compound. All joint compounds shall be smooth and free of tool marks and ridges. The prepared surface shall be covered with a drywall primer like Sheetrock® First Coat prior to the application of the final decoration.

Level 5: The highest quality finish is the most effective method to provide a uniform surface and minimize the possibility of joint photographing and of fasteners showing through the final decoration. This level of finish is required where gloss, semi-gloss, or enamel is specified; when flat joints are specified over an untextured surface; or where critical lighting conditions occur. The prepared surface shall be coated with a drywall primer prior to the application of final decoration.

All joints and interior angles shall have tape embedded in joint compound and be immediately wiped with a joint knife or trowel, leaving a thin coating of joint compound over all joints and interior angles. Two separate coats of joint compound shall be applied over all flat joints and one separate coat of joint compound applied over interior angles. Fastener heads and accessories shall be covered with three separate coats of joint compound.

A thin skim coat of joint compound shall be trowel applied to the entire surface. Excess compound is immediately troweled off, leaving a film or skim coating of compound completely covering the paper. As an alternative to a skim coat, a material manufactured especially for this purpose may be applied such as Sheetrock® Tuff-Hide primer surfacer. The surface must be smooth and free of tool marks and ridges. The prepared surface shall be covered with a drywall primer prior to the application of the final decoration.

R096613-10 Terrazzo Floor

The table below lists quantities required for 100 S.F. of 5/8″ terrazzo topping, either bonded or not bonded.

Description	Bonded to Concrete 1-1/8″ Bed, 1:4 Mix	Not Bonded 2-1/8″ Bed and 1/4″ Sand
Portland cement, 94 lb. Bag	6 bags	8 bags
Sand	10 C.F.	20 C.F.
Divider strips, 4′ squares	50 L.F.	50 L.F.
Terrazzo fill, 50 lb. Bag	12 bags	12 bags
15 Lb. tarred felt		1 C.S.F.
Mesh 2 x 2 #14 galvanized		1 C.S.F.
Crew J-3	0.77 days	0.87 days

2′ x 2′ panels require 1.00 L.F. divider strip per S.F.

3′ x 3′ panels require 0.67 L.F. divider strip per S.F.

4′ x 4′ panels require 0.50 L.F. divider strip per S.F.

5′ x 5′ panels require 0.40 L.F. divider strip per S.F.

6′ x 6′ panels require 0.33 L.F. divider strip per S.F.

R097223-10 Wall Covering

The table below lists the quantities required for 100 S.F. of wall covering.

Description	Medium-Priced Paper	Expensive Paper
Paper	1.6 dbl. rolls	1.6 dbl. rolls
Wall sizing	0.25 gallon	0.25 gallon
Vinyl wall paste	0.6 gallon	0.6 gallon
Apply sizing	0.3 hour	0.3 hour
Apply paper	1.2 hours	1.5 hours

Most wallpapers now come in double rolls only.
To remove old paper, allow 1.3 hours per 100 S.F.

Finishes — R0991 Painting

R099100-10 Painting Estimating Techniques

Proper estimating methodology is needed to obtain an accurate painting estimate. There is no known reliable shortcut or square foot method. The following steps should be followed:

- List all surfaces to be painted, with an accurate quantity (area) of each. Items having similar surface condition, finish, application method and accessibility may be grouped together.
- List all the tasks required for each surface to be painted, including surface preparation, masking, and protection of adjacent surfaces. Surface preparation may include minor repairs, washing, sanding and puttying.
- Select the proper Means line for each task. Review and consider all adjustments to labor and materials for type of paint and location of work. Apply the height adjustment carefully. For instance, when applying the adjustment for work over 8' high to a wall that is 12' high, apply the adjustment only to the area between 8' and 12' high, and not to the entire wall.

When applying more than one percent (%) adjustment, apply each to the base cost of the data, rather than applying one percentage adjustment on top of the other.

When estimating the cost of painting walls and ceilings remember to add the brushwork for all cut-ins at inside corners and around windows and doors as a LF measure. One linear foot of cut-in with brush equals one square foot of painting.

All items for spray painting include the labor for roll-back.

Deduct for openings greater than 100 SF, or openings that extend from floor to ceiling and are greater than 5' wide. Do not deduct small openings.

The cost of brushes, rollers, ladders and spray equipment are considered to be part of a painting contractor's overhead, and should not be added to the estimate. The cost of rented equipment such as scaffolding and swing staging should be added to the estimate.

R099100-20 Painting

Item	Coat	One Gallon Covers			In 8 Hours a Laborer Covers			Labor-Hours per 100 S.F.		
		Brush	Roller	Spray	Brush	Roller	Spray	Brush	Roller	Spray
Paint wood siding	prime	250 S.F.	225 S.F.	290 S.F.	1150 S.F.	1300 S.F.	2275 S.F.	.695	.615	.351
	others	270	250	290	1300	1625	2600	.615	.492	.307
Paint exterior trim	prime	400	—	—	650	—	—	1.230	—	—
	1st	475	—	—	800	—	—	1.000	—	—
	2nd	520	—	—	975	—	—	.820	—	—
Paint shingle siding	prime	270	255	300	650	975	1950	1.230	.820	.410
	others	360	340	380	800	1150	2275	1.000	.695	.351
Stain shingle siding	1st	180	170	200	750	1125	2250	1.068	.711	.355
	2nd	270	250	290	900	1325	2600	.888	.603	.307
Paint brick masonry	prime	180	135	160	750	800	1800	1.066	1.000	.444
	1st	270	225	290	815	975	2275	.981	.820	.351
	2nd	340	305	360	815	1150	2925	.981	.695	.273
Paint interior plaster or drywall	prime	400	380	495	1150	2000	3250	.695	.400	.246
	others	450	425	495	1300	2300	4000	.615	.347	.200
Paint interior doors and windows	prime	400	—	—	650	—	—	1.230	—	—
	1st	425	—	—	800	—	—	1.000	—	—
	2nd	450	—	—	975	—	—	.820	—	—

Special Construction — R1311 Swimming Pools

R131113-20 Swimming Pools

Pool prices given per square foot of surface area include pool structure, filter and chlorination equipment, pumps, related piping, ladders/steps, maintenance kit, skimmer and vacuum system. Decks and electrical service to equipment are not included.

Residential in-ground pool construction can be divided into two categories: vinyl lined and gunite. Vinyl lined pool walls are constructed of different materials including wood, concrete, plastic or metal. The bottom is often graded with sand over which the vinyl liner is installed. Vermiculite or soil cement bottoms may be substituted for an added cost.

Gunite pool construction is used both in residential and municipal installations. These structures are steel reinforced for strength and finished with a white cement limestone plaster.

Municipal pools will have a higher cost because plumbing codes require more expensive materials, chlorination equipment and higher filtration rates.

Municipal pools greater than 1,800 S.F. require gutter systems to control waves. This gutter may be formed into the concrete wall. Often a vinyl/stainless steel gutter or gutter/wall system is specified, which will raise the pool cost.

Competition pools usually require tile bottoms and sides with contrasting lane striping, which will also raise the pool cost.

R133113-10 Air Supported Structures

Air supported structures are made from fabrics that can be classified into two groups: temporary and permanent. Temporary fabrics include nylon, woven polyethylene, vinyl film, and vinyl coated dacron. These have lifespans that range from five to fifteen plus years. The cost per square foot includes a fabric shell, tension cables, primary and back-up inflation systems and doors. The lower cost structures are used for construction shelters, bulk storage and pond covers. The more expensive are used for recreational structures and warehouses.

Permanent fabrics are teflon coated fiberglass. The life of this structure is twenty plus years. The high cost limits its application to architectural designed structures which call for a clear span covered area, such as stadiums and convention centers. Both temporary and permanent structures are available in translucent fabrics which eliminates the need for daytime lighting.

Areas to be covered vary from 10,000 S.F. to any area up to 1000 foot wide by any length. Height restrictions range from a maximum of 1/2 of width to a minimum of 1/6 of the width. Erection of even the largest of the temporary structures requires no more than a week.

Centrifugal fans provide the inflation necessary to support the structure during application of live loads. Airlocks are usually used at large entrances to prevent loss of static pressure. Some manufacturers employ propeller fans which generate sufficient airflow (30,000 CFM) to eliminate the need for airlocks. These fans may also be automatically controlled to resist high wind conditions, regulate humidity (air changes), and provide cooling and heat.

Insulation can be provided with the addition of a second or even third interior liner, creating a dead air space with an "R" value of four to nine. Some structures allow for the liner to be collapsed into the outer shell to enable the internal heat to melt accumulated snow. For cooling or air conditioning, the exterior face of the liner can be aluminized to reflect the sun's heat.

R133113-90 Seismic Bracing

Sometimes referred to as anti-sway bracing, this support system is required in earthquake areas. The individual components must be assembled to make a required system.

Example	C-Clamp 3/8″ rod	2 ea.
	Rod, continuous thread 3/8″	10 L.F.
	Field, weld 1″	2 ea.

Additionally, height factors must be taken into account. Add the following percentages to labor for elevated installations:

Height	Percentage
15′ to 20′ high	10%
21′ to 25′ high	20%
26′ to 30′ high	30%
31′ to 35′ high	40%
36′ to 40′ high	50%
41′ to 50′ high	60%

Special Construction R1334 Fabricated Engineered Structures

R133419-10 Pre-Engineered Steel Buildings

These buildings are manufactured by many companies and normally erected by franchised dealers throughout the U.S. The four basic types are: Rigid Frames, Truss type, Post and Beam and the Sloped Beam type. Most popular roof slope is low pitch of 1″ in 12″. The minimum economical area of these buildings is about 3000 S.F. of floor area. Bay sizes are usually 20′ to 24′ but can go as high as 30′ with heavier girts and purlins. Eave heights are usually 12′ to 24′ with 18′ to 20′ most typical.

Material prices shown in the Unit Price section are bare costs for the building shell only and do not include floors, foundations, anchor bolts, interior finishes or utilities. Costs assume at least three bays of 24′ each, a 1″ in 12″ roof slope, and they are based on 30 psf roof load and 20 psf wind load (wind load is a function of wind speed, building height, and terrain characteristics; this should be determined by a Registered Structural Engineer) and no unusual requirements. Costs include the structural frame, 26 ga. non-insulated colored corrugated or ribbed roofing and siding panels, fasteners, closures, trim and flashing but no allowance for insulation, doors, windows, skylights, gutters or downspouts. Very large projects would generally cost less for materials than the prices shown. For roof panel substitutions and wall panel substitutions, see appropriate Unit Price sections.

Conditions at the site, weather, shape and size of the building, and labor availability will affect the erection cost of the building.

R133423-30 Dome Structures

Steel — The four types are Lamella, Schwedler, Arch and Geodesic. For maximum economy, rise should be about 15 to 20% of diameter. Most common diameters are in the 200′ to 300′ range. Lamella domes weigh about 5 P.S.F. of floor area less than Schwedler domes. Schwedler dome weight in lbs. per S.F. approaches .046 times the diameter. Domes below 125′ diameter weigh .07 times diameter and the cost per ton of steel is higher. See R051223-20 for estimating weight.

Wood — Small domes are of sawn lumber, larger ones are laminated. In larger sizes, triaxial and triangular cost about the same; radial domes cost more. Radial domes are economical in the 60′ to 70′ diameter range. Most economical range of all types is 80′ to 200′ diameters. Diameters can run over 400′. All costs are quoted above the foundation. Prices include 2″ decking and a tension tie ring in place.

Plywood — Stock prefab geodesic domes are available with diameters from 24′ to 60′.

Fiberglass — Aluminum framed translucent sandwich panels with spans from 5′ to 45′ are commercially available.

Aluminum — Stressed skin aluminum panels form geodesic domes with spans ranging from 82′ to 232′. An aluminum space truss, triangulated or nontriangulated, with aluminum or clear acrylic closure panels can be used for clear spans of 40′ to 415′.

R142000-10 Freight Elevators

Capacities run from 2,000 lbs. to over 100,000 lbs. with 3,000 lbs. to 10,000 lbs. most common. Travel speeds are generally lower and control less intricate than on passenger elevators. Costs in the Unit Price Section are for hydraulic and geared elevators.

R142000-20 Elevator Selective Costs See R142000-40 for cost development.

	Passenger		Freight		Hospital	
A. Base Unit	Hydraulic	Electric	Hydraulic	Electric	Hydraulic	Electric
Capacity	1,500 lb.	2,000 lb.	2,000 lb.	4,000 lb.	4,000 lb.	4,000 lb.
Speed	100 F.P.M.	200 F.P.M.	100 F.P.M.	200 F.P.M.	100 F.P.M.	200 F.P.M.
#Stops/Travel Ft.	2/12	4/40	2/20	4/40	2/20	4/40
Push Button Oper.	Yes	Yes	Yes	Yes	Yes	Yes
Telephone Box & Wire	"	"	"	"	"	"
Emergency Lighting	"	"	No	No	"	"
Cab	Plastic Lam. Walls	Plastic Lam. Walls	Painted Steel	Painted Steel	Plastic Lam. Walls	Plastic Lam. Walls
Cove Lighting	Yes	Yes	No	No	Yes	Yes
Floor	V.C.T.	V.C.T.	Wood w/Safety Treads	Wood w/Safety Treads	V.C.T.	V.C.T.
Doors, & Speedside Slide	Yes	Yes	Yes	Yes	Yes	Yes
Gates, Manual	No	No	No	No	No	No
Signals, Lighted Buttons	Car and Hall	Car and Hall	Car and Hall	Car and Hall	Car and Hall	Car and Hall
O.H. Geared Machine	N.A.	Yes	N.A.	Yes	N.A.	Yes
Variable Voltage Contr.	"	"	N.A.	"	"	"
Emergency Alarm	Yes	"	Yes	"	Yes	"
Class "A" Loading	N.A.	N.A.	"	"	N.A.	N.A.

R142000-30 Passenger Elevators

Electric elevators are used generally but hydraulic elevators can be used for lifts up to 70′ and where large capacities are required. Hydraulic speeds are limited to 200 F.P.M. but cars are self leveling at the stops. On low rises, hydraulic installation runs about 15% less than standard electric types but on higher rises this installation cost advantage is reduced. Maintenance of hydraulic elevators is about the same as electric type but underground portion is not included in the maintenance contract.

In electric elevators there are several control systems available, the choice of which will be based upon elevator use, size, speed and cost criteria. The two types of drives are geared for low speeds and gearless for 450 F.P.M. and over.

The tables on the preceding pages illustrate typical installed costs of the various types of elevators available.

R142000-40 Elevator Cost Development

To price a new car or truck from the factory, you must start with the manufacturer's basic model, then add or exchange optional equipment and features. The same is true for pricing elevators.

Example:

Requirement: One-passenger elevator, five-story hydraulic, 2,500 lb. capacity, 12′ floor to floor, speed 150 F.P.M., emergency power switching and maintenance contract.

Description	Adjustment
A. Base Elevator: Hydraulic Passenger, 1500 lb. Capacity, 100 fpm, 2 Stops, Standard Finish	1 Ea.
B. Capacity Adjustment (2,500 lb.)	1 Ea.
C. Excess Travel Adjustment: 48′ Total Travel (4 x 12′) minus 12′ Base Unit Travel =	36 V.L.F.
D. Stops Adjustment: 5 Total Stops minus 2 Stops (Base Unit) =	3 Stops
E. Speed Adjustment (150 F.P.M.)	1 Ea.
F. Options:	
1. Intercom Service	1 Ea.
2. Emergency Power Switching, Automatic	1 Ea.
3. Stainless Steel Entrance Doors	5 Ea.
4. Maintenance Contract (12 Months)	1 Ea.
5. Position Indicator for main floor level (none indicated in Base Unit)	1 Ea.

Conveying Equipment — R1431 Escalators

R143110-10 Escalators

Moving stairs can be used for buildings where 600 or more people are to be carried to the second floor or beyond. Freight cannot be carried on escalators and at least one elevator must be available for this function.

Carrying capacity is 5,000 to 8,000 people per hour. Power requirement is 2 to 3 KW per hour and incline angle is 30°.

Conveying Equipment — R1432 Moving Walks

R143210-20 Moving Ramps and Walks

These are a specialized form of conveyor 3′ to 6′ wide with capacities of 3,600 to 18,000 persons per hour. Maximum speed is 140 F.P.M. and normal incline is 0° to 15°.

Local codes will determine the maximum angle. Outdoor units would require additional weather protection.

Plumbing — R2201 Operation & Maintenance of Plumbing

R220105-10 Demolition (Selective vs. Removal for Replacement)

Demolition can be divided into two basic categories.

One type of demolition involves the removal of material with no concern for its replacement. The labor-hours to estimate this work are found under "Selective Demolition" in the Fire Protection, Plumbing and HVAC Divisions. It is selective in that individual items or all the material installed as a system or trade grouping such as plumbing or heating systems are removed. This may be accomplished by the easiest way possible, such as sawing, torch cutting, or sledge hammer as well as simple unbolting.

The second type of demolition is the removal of some item for repair or replacement. This removal may involve careful draining, opening of unions, disconnecting and tagging of electrical connections, capping of pipes/ducts to prevent entry of debris or leakage of the material contained as well as transport of the item away from its in-place location to a truck/dumpster. An approximation of the time required to accomplish this type of demolition is to use half of the time indicated as necessary to install a new unit. For example; installation of a new pump might be listed as requiring 6 labor-hours so if we had to estimate the removal of the old pump we would allow an additional 3 hours for a total of 9 hours. That is, the complete replacement of a defective pump with a new pump would be estimated to take 9 labor-hours.

Plumbing — R2211 Facility Water Distribution

R221113-50 Pipe Material Considerations

1. Malleable fittings should be used for gas service.
2. Malleable fittings are used where there are stresses/strains due to expansion and vibration.
3. Cast fittings may be broken as an aid to disassembling of heating lines frozen by long use, temperature and minerals.
4. Cast iron pipe is extensively used for underground and submerged service.
5. Type M (light wall) copper tubing is available in hard temper only and is used for nonpressure and less severe applications than K and L.
6. Type L (medium wall) copper tubing, available hard or soft for interior service.
7. Type K (heavy wall) copper tubing, available in hard or soft temper for use where conditions are severe. For underground and interior service.
8. Hard drawn tubing requires fewer hangers or supports but should not be bent. Silver brazed fittings are recommended, however soft solder is normally used.
9. Type DMV (very light wall) copper tubing designed for drainage, waste and vent plus other non-critical pressure services.

Domestic/Imported Pipe and Fittings Cost

The prices shown in this publication for steel/cast iron pipe and steel, cast iron, malleable iron fittings are based on domestic production sold at the normal trade discounts. The above listed items of foreign manufacture may be available at prices of 1/3 to 1/2 those shown. Some imported items after minor machining or finishing operations are being sold as domestic to further complicate the system.

Caution: Most pipe prices in this book also include a coupling and pipe hangers which for the larger sizes can add significantly to the per foot cost and should be taken into account when comparing "book cost" with quoted supplier's cost.

R224000-40 Plumbing Fixture Installation Time

Item	Rough-In	Set	Total Hours	Item	Rough-In	Set	Total Hours
Bathtub	5	5	10	Shower head only	2	1	3
Bathtub and shower, cast iron	6	6	12	Shower drain	3	1	4
Fire hose reel and cabinet	4	2	6	Shower stall, slate		15	15
Floor drain to 4 inch diameter	3	1	4	Slop sink	5	3	8
Grease trap, single, cast iron	5	3	8	Test 6 fixtures			14
Kitchen gas range		4	4	Urinal, wall	6	2	8
Kitchen sink, single	4	4	8	Urinal, pedestal or floor	6	4	10
Kitchen sink, double	6	6	12	Water closet and tank	4	3	7
Laundry tubs	4	2	6	Water closet and tank, wall hung	5	3	8
Lavatory wall hung	5	3	8	Water heater, 45 gals. gas, automatic	5	2	7
Lavatory pedestal	5	3	8	Water heaters, 65 gals. gas, automatic	5	2	7
Shower and stall	6	4	10	Water heaters, electric, plumbing only	4	2	6

Fixture prices in front of book are based on the cost per fixture set in place. The rough-in cost, which must be added for each fixture, includes carrier, if required, some supply, waste and vent pipe connecting fittings and stops. The lengths of rough-in pipe are nominal runs which would connect to the larger runs and stacks. The supply runs and DWV runs and stacks must be accounted for in separate entries. In the eastern half of the United States it is common for the plumber to carry these to a point 5′ outside the building.

R224000-50 Water Cooler Application

Type of Service	Requirement
Office, School or Hospital	12 persons per gallon per hour
Office, Lobby or Department Store	4 or 5 gallons per hour per fountain
Light manufacturing	7 persons per gallon per hour
Heavy manufacturing	5 persons per gallon per hour
Hot heavy manufacturing	4 persons per gallon per hour
Hotel	.08 gallons per hour per room
Theatre	1 gallon per hour per 100 seats

Heating, Ventilating & A.C. R2331 HVAC Ducts & Casings

R233100-20 Ductwork

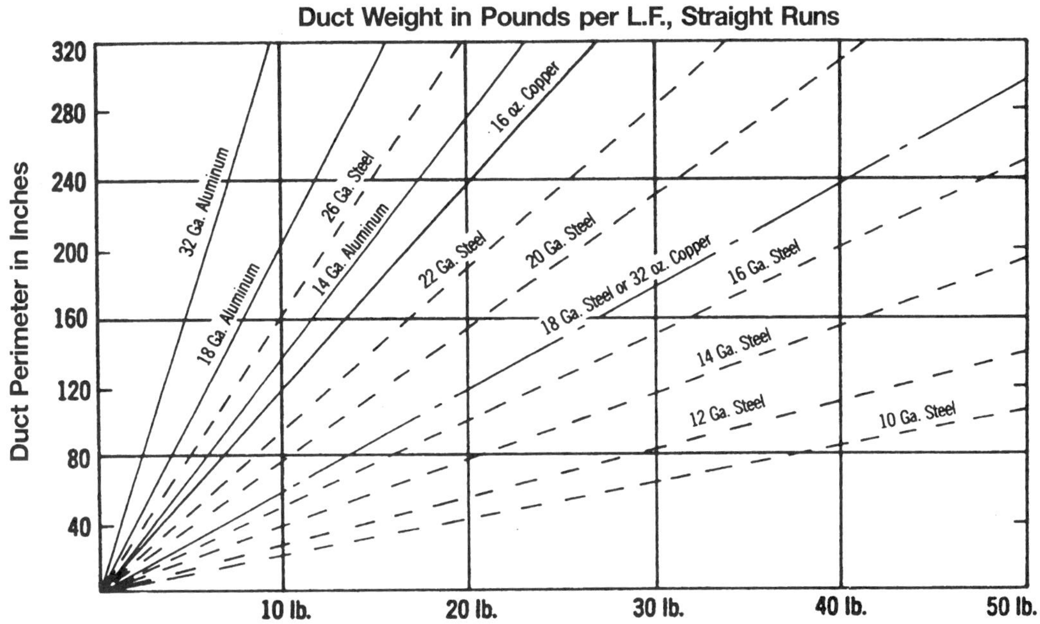

Add to the above for fittings; 90° elbow is 3 L.F.; 45° elbow is 2.5 L.F.; offset is 4 L.F.; transition offset is 6 L.F.; square-to-round transition is 4 L.F.; 90° reducing elbow is 5 L.F. For bracing and waste, add 20% to aluminum and copper, 15% to steel.

Heating, Ventilating & A.C. R2334 HVAC Fans

R233400-10 Recommended Ventilation Air Changes

Table below lists range of time in minutes per change for various types of facilities.

Facility	Minutes	Facility	Minutes	Facility	Minutes
Assembly Halls	2-10	Dance Halls	2-10	Laundries	1-3
Auditoriums	2-10	Dining Rooms	3-10	Markets	2-10
Bakeries	2-3	Dry Cleaners	1-5	Offices	2-10
Banks	3-10	Factories	2-5	Pool Rooms	2-5
Bars	2-5	Garages	2-10	Recreation Rooms	2-10
Beauty Parlors	2-5	Generator Rooms	2-5	Sales Rooms	2-10
Boiler Rooms	1-5	Gymnasiums	2-10	Theaters	2-8
Bowling Alleys	2-10	Kitchens-Hospitals	2-5	Toilets	2-5
Churches	5-10	Kitchens-Restaurant	1-3	Transformer Rooms	1-5

CFM air required for changes = Volume of room in cubic feet ÷ Minutes per change.

Heating, Ventilating & A.C. R2350 Central Heating Equipment

R235000-35 Heating (42° degrees latitude)

$$\text{Approximate S.F. radiation} = \frac{\text{S.F. sash}}{2} + \frac{\text{S.F. wall + roof — sash}}{20} + \frac{\text{C.F. building}}{200}$$

R235000-50 Factor for Determining Heat Loss for Various Types of Buildings

General: While the most accurate estimates of heating requirements would naturally be based on detailed information about the building being considered, it is possible to arrive at a reasonable approximation using the following procedure:

1. Calculate the cubic volume of the room or building.
2. Select the appropriate factor from Table 1 below. Note that the factors apply only to inside temperatures listed in the first column and to 0° F outside temperature.
3. If the building has bad north and west exposures, multiply the heat loss factor by 1.1.
4. If the outside design temperature is other than 0° F, multiply the factor from Table 1 by the factor from Table 2.
5. Multiply the cubic volume by the factor selected from Table 1. This will give the estimated BTUH heat loss which must be made up to maintain inside temperature.

Table 1 — Building Type	Conditions	Qualifications	Loss Factor*
Factories & Industrial Plants General Office Areas at 70° F	One Story	Skylight in Roof	6.2
		No Skylight in Roof	5.7
	Multiple Story	Two Story	4.6
		Three Story	4.3
		Four Story	4.1
		Five Story	3.9
		Six Story	3.6
	All Walls Exposed	Flat Roof	6.9
		Heated Space Above	5.2
	One Long Warm Common Wall	Flat Roof	6.3
		Heated Space Above	4.7
	Warm Common Walls on Both Long Sides	Flat Roof	5.8
		Heated Space Above	4.1
Warehouses at 60° F	All Walls Exposed	Skylights in Roof	5.5
		No Skylight in Roof	5.1
		Heated Space Above	4.0
	One Long Warm Common Wall	Skylight in Roof	5.0
		No Skylight in Roof	4.9
		Heated Space Above	3.4
	Warm Common Walls on Both Long Sides	Skylight in Roof	4.7
		No Skylight in Roof	4.4
		Heated Space Above	3.0

*Note: This table tends to be conservative particularly for new buildings designed for minimum energy consumption.

Table 2 — Outside Design Temperature Correction Factor (for Degrees Fahrenheit)									
Outside Design Temperature	50	40	30	20	10	0	-10	-20	-30
Correction Factor	.29	.43	.57	.72	.86	1.00	1.14	1.28	1.43

R235616-60 Solar Heating (Space and Hot Water)

Collectors should face as close to due South as possible, however, variations of up to 20 degrees on either side of true South are acceptable. Local climate and collector type may influence the choice between east or west deviations. Obviously they should be located so they are not shaded from the sun's rays. Incline collectors at a slope of latitude minus 5 degrees for domestic hot water and latitude plus 15 degrees for space heating.

Flat plate collectors consist of a number of components as follows: Insulation to reduce heat loss through the bottom and sides of the collector. The enclosure which contains all the components in this assembly is usually weatherproof and prevents dust, wind and water from coming in contact with the absorber plate. The cover plate usually consists of one or more layers of a variety of glass or plastic and reduces the reradiation by creating an air space which traps the heat between the cover and the absorber plates.

The absorber plate must have a good thermal bond with the fluid passages. The absorber plate is usually metallic and treated with a surface coating which improves absorptivity. Black or dark paints or selective coatings are used for this purpose, and the design of this passage and plate combination helps determine a solar system's effectiveness.

Heat transfer fluid passage tubes are attached above and below or integral with an absorber plate for the purpose of transferring thermal energy from the absorber plate to a heat transfer medium. The heat exchanger is a device for transferring thermal energy from one fluid to another.

Piping and storage tanks should be well insulated to minimize heat losses.

Size domestic water heating storage tanks to hold 20 gallons of water per user, minimum, plus 10 gallons per dishwasher or washing machine. For domestic water heating an optimum collector size is approximately 3/4 square foot of area per gallon of water storage. For space heating of residences and small commercial applications the collector is commonly sized between 30% and 50% of the internal floor area. For space heating of large commercial applications, collector areas less than 30% of the internal floor area can still provide significant heat reductions.

A supplementary heat source is recommended for Northern states for December through February.

The solar energy transmission per square foot of collector surface varies greatly with the material used. Initial cost, heat transmittance and useful life are obviously interrelated.

Heating, Ventilating & A.C. R2360 Central Cooling Equipment

R236000-20 Air Conditioning Requirements

BTU's per hour per S.F. of floor area and S.F. per ton of air conditioning.

Type of Building	BTU/Hr. per S.F.	S.F. per Ton	Type of Building	BTU/Hr per S.F.	S.F. per Ton	Type of Building	BTU/Hr per S.F.	S.F. per Ton
Apartments, Individual	26	450	Dormitory, Rooms	40	300	Libraries	50	240
Corridors	22	550	Corridors	30	400	Low Rise Office, Exterior	38	320
Auditoriums & Theaters	40	300/18*	Dress Shops	43	280	Interior	33	360
Banks	50	240	Drug Stores	80	150	Medical Centers	28	425
Barber Shops	48	250	Factories	40	300	Motels	28	425
Bars & Taverns	133	90	High Rise Office—Ext. Rms.	46	263	Office (small suite)	43	280
Beauty Parlors	66	180	Interior Rooms	37	325	Post Office, Individual Office	42	285
Bowling Alleys	68	175	Hospitals, Core	43	280	Central Area	46	260
Churches	36	330/20*	Perimeter	46	260	Residences	20	600
Cocktail Lounges	68	175	Hotel, Guest Rooms	44	275	Restaurants	60	200
Computer Rooms	141	85	Corridors	30	400	Schools & Colleges	46	260
Dental Offices	52	230	Public Spaces	55	220	Shoe Stores	55	220
Dept. Stores, Basement	34	350	Industrial Plants, Offices	38	320	Shop'g. Ctrs., Supermarkets	34	350
Main Floor	40	300	General Offices	34	350	Retail Stores	48	250
Upper Floor	30	400	Plant Areas	40	300	Specialty	60	200

*Persons per ton
12,000 BTU = 1 ton of air conditioning

R260519-92 Minimum Copper and Aluminum Wire Size Allowed for Various Types of Insulation

Minimum Wire Sizes										
	Copper		Aluminum			Copper		Aluminum		
Amperes	THW THWN or XHHW	THHN XHHW *	THW XHHW	THHN XHHW *	Amperes	THW THWN or XHHW	THHN XHHW *	THW XHHW	THHN XHHW *
15A	#14	#14	#12	#12	195	3/0	2/0	250kcmil	4/0
20	#12	#12	#10	#10	200	3/0	3/0	250kcmil	4/0
25	#10	#10	#10	#10	205	4/0	3/0	250kcmil	4/0
30	#10	#10	# 8	# 8	225	4/0	3/0	300kcmil	250kcmil
40	# 8	# 8	# 8	# 8	230	4/0	4/0	300kcmil	250kcmil
45	# 8	# 8	# 6	# 8	250	250kcmil	4/0	350kcmil	300kcmil
50	# 8	# 8	# 6	# 6	255	250kcmil	4/0	400kcmil	300kcmil
55	# 6	# 8	# 4	# 6	260	300kcmil	4/0	400kcmil	350kcmil
60	# 6	# 6	# 4	# 6	270	300kcmil	250kcmil	400kcmil	350kcmil
65	# 6	# 6	# 4	# 4	280	300kcmil	250kcmil	500kcmil	350kcmil
75	# 4	# 6	# 3	# 4	285	300kcmil	250kcmil	500kcmil	400kcmil
85	# 4	# 4	# 2	# 3	290	350kcmil	250kcmil	500kcmil	400kcmil
90	# 3	# 4	# 2	# 2	305	350kcmil	300kcmil	500kcmil	400kcmil
95	# 3	# 4	# 1	# 2	310	350kcmil	300kcmil	500kcmil	500kcmil
100	# 3	# 3	# 1	# 2	320	400kcmil	300kcmil	600kcmil	500kcmil
110	# 2	# 3	1/0	# 1	335	400kcmil	350kcmil	600kcmil	500kcmil
115	# 2	# 2	1/0	# 1	340	500kcmil	350kcmil	600kcmil	500kcmil
120	# 1	# 2	1/0	1/0	350	500kcmil	350kcmil	700kcmil	500kcmil
130	# 1	# 2	2/0	1/0	375	500kcmil	400kcmil	700kcmil	600kcmil
135	1/0	# 1	2/0	1/0	380	500kcmil	400kcmil	750kcmil	600kcmil
150	1/0	# 1	3/0	2/0	385	600kcmil	500kcmil	750kcmil	600kcmil
155	2/0	1/0	3/0	3/0	420	600kcmil	500kcmil		700kcmil
170	2/0	1/0	4/0	3/0	430		500kcmil		750kcmil
175	2/0	2/0	4/0	3/0	435		600kcmil		750kcmil
180	3/0	2/0	4/0	4/0	475		600kcmil		

*Dry Locations Only

Notes:

1. Size #14 to 4/0 is in AWG units (American Wire Gauge).
2. Size 250 to 750 is in kcmil units (Thousand Circular Mils).
3. Use next higher ampere value if exact value is not listed in table.
4. For loads that operate continuously increase ampere value by 25% to obtain proper wire size.
5. Refer to Table R260519-91 for the maximum circuit length for the various size wires.
6. Table R260519-92 has been written for estimating purpose only, based on ambient temperature of 30°C (86° F); for ambient temperature other than 30°C (86° F), ampacity correction factors will be applied.

R260533-22 Conductors in Conduit

Table below lists maximum number of conductors for various sized conduit using THW, TW or THWN insulations.

Copper Wire Size	Conduit Size																										
	1/2″			3/4″			1″			1-1/4″			1-1/2″			2″			2-1/2″			3″		3-1/2″		4″	
	TW	THW	THWN	TW	THW	THWN	TW	THW	THWN	TW	THW	THWN	TW	THW	THWN	TW	THW	THWN	TW	THW	THWN	THW	THWN	THW	THWN	THW	THWN
#14	9	6	13	15	10	24	25	16	39	44	29	69	60	40	94	99	65	154	142	93		143		192			
#12	7	4	10	12	8	18	19	13	29	35	24	51	47	32	70	78	53	114	111	76	164	117		157			
#10	5	4	6	9	6	11	15	11	18	26	19	32	36	26	44	60	43	73	85	61	104	95	160	127		163	
#8	2	1	3	4	3	5	7	5	9	12	10	16	17	13	22	28	22	36	40	32	51	49	79	66	106	85	136
#6		1	1		2	4		4	6		7	11		10	15		16	26		23	37	36	57	48	76	62	98
#4		1	1		1	2		3	4		5	7		7	9		12	16		17	22	27	35	36	47	47	60
#3		1	1		1	1		2	3		4	6		6	8		10	13		15	19	23	29	31	39	40	51
#2		1	1		1	1		2	3		4	5		5	7		9	11		13	16	20	25	27	33	34	43
#1					1	1		1	1		3	3		4	5		6	8		9	12	14	18	19	25	25	32
1/0					1	1		1	1		2	3		3	4		5	7		8	10	12	15	16	21	21	27
2/0					1	1		1	1		1	2		3	3		5	6		7	8	10	13	14	17	18	22
3/0					1	1		1	1		1	1		2	3		4	5		6	7	9	11	12	14	15	18
4/0						1		1	1		1	1		1	2		3	4		5	6	7	9	10	12	13	15
250 kcmil								1	1		1	1		1	1		2	3		4	4	6	7	8	10	10	12
300								1	1		1	1		1	1		2	3		3	4	5	6	7	8	9	11
350									1		1	1		1	1		1	2		3	3	4	5	6	7	8	9
400											1	1		1	1		1	1		2	3	4	5	5	6	7	8
500											1	1		1	1		1	1		1	2	3	4	4	5	6	7
600												1		1	1		1	1		1	1	3	3	4	4	5	5
700														1	1		1	1		1	1	2	3	3	4	4	5
750														1	1		1	1		1	1	2	2	3	3	4	4

R312316-40 Excavating

The selection of equipment used for structural excavation and bulk excavation or for grading is determined by the following factors.

1. Quantity of material
2. Type of material
3. Depth or height of cut
4. Length of haul
5. Condition of haul road
6. Accessibility of site
7. Moisture content and dewatering requirements
8. Availability of excavating and hauling equipment

Some additional costs must be allowed for hand trimming the sides and bottom of concrete pours and other excavation below the general excavation.

When planning excavation and fill, the following should also be considered.

1. Swell factor
2. Compaction factor
3. Moisture content
4. Density requirements

A typical example for scheduling and estimating the cost of excavation of a 15′ deep basement on a dry site when the material must be hauled off the site is outlined below.

Assumptions:

1. Swell factor, 18%
2. No mobilization or demobilization
3. Allowance included for idle time and moving on job
4. No dewatering, sheeting, or bracing
5. No truck spotter or hand trimming

Number of B.C.Y. per truck = 1.5 C.Y. bucket x 8 passes = 12 loose C.Y.

$$= 12 \text{ x } \frac{100}{118} = 10.2 \text{ B.C.Y. per truck}$$

Truck Haul Cycle:

Load truck, 8 passes	=	4 minutes
Haul distance, 1 mile	=	9 minutes
Dump time	=	2 minutes
Return, 1 mile	=	7 minutes
Spot under machine	=	1 minute
		23 minute cycle

Fleet Haul Production per day in B.C.Y.

$$4 \text{ trucks x } \frac{50 \text{ min. hour}}{23 \text{ min. haul cycle}} \text{ x 8 hrs. x 10.2 B.C.Y.}$$

$$= 4 \text{ x } 2.2 \text{ x } 8 \text{ x } 10.2 = 718 \text{ B.C.Y./day}$$

Add the mobilization and demobilization costs to the total excavation costs. When equipment is rented for more than three days, there is often no mobilization charge by the equipment dealer. On larger jobs outside of urban areas, scrapers can move earth economically provided a dump site or fill area and adequate haul roads are available. Excavation within sheeting bracing or cofferdam bracing is usually done with a clamshell and production is low, since the clamshell may have to be guided by hand between the bracing. When excavating or filling an area enclosed with a wellpoint system, add 10% to 15% to the cost to allow for restricted access. When estimating earth excavation quantities for structures, allow work space outside the building footprint for construction of the foundation and a slope of 1:1 unless sheeting is used.

R312316-45 Excavating Equipment

The table below lists THEORETICAL hourly production in C.Y./hr. bank measure for some typical excavation equipment. Figures assume 50 minute hours, 83% job efficiency, 100% operator efficiency, 90° swing and properly sized hauling units, which must be modified for adverse digging and loading conditions. Actual production costs in the front of the book average about 50% of the theoretical values listed here.

Equipment	Soil Type	B.C.Y. Weight	% Swell	1 C.Y.	1-1/2 C.Y.	2 C.Y.	2-1/2 C.Y.	3 C.Y.	3-1/2 C.Y.	4 C.Y.
Hydraulic Excavator "Backhoe" 15′ Deep Cut	Moist loam, sandy clay	3400 lb.	40%	85	125	175	220	275	330	380
	Sand and gravel	3100	18	80	120	160	205	260	310	365
	Common earth	2800	30	70	105	150	190	240	280	330
	Clay, hard, dense	3000	33	65	100	130	170	210	255	300
Power Shovel Optimum Cut (Ft.)	Moist loam, sandy clay	3400	40	170	245	295	335	385	435	475
				(6.0)	(7.0)	(7.8)	(8.4)	(8.8)	(9.1)	(9.4)
	Sand and gravel	3100	18	165	225	275	325	375	420	460
				(6.0)	(7.0)	(7.8)	(8.4)	(8.8)	(9.1)	(9.4)
	Common earth	2800	30	145	200	250	295	335	375	425
				(7.8)	(9.2)	(10.2)	(11.2)	(12.1)	(13.0)	(13.8)
	Clay, hard, dense	3000	33	120	175	220	255	300	335	375
				(9.0)	(10.7)	(12.2)	(13.3)	(14.2)	(15.1)	(16.0)
Drag Line Optimum Cut (Ft.)	Moist loam, sandy clay	3400	40	130	180	220	250	290	325	385
				(6.6)	(7.4)	(8.0)	(8.5)	(9.0)	(9.5)	(10.0)
	Sand and gravel	3100	18	130	175	210	245	280	315	375
				(6.6)	(7.4)	(8.0)	(8.5)	(9.0)	(9.5)	(10.0)
	Common earth	2800	30	110	160	190	220	250	280	310
				(8.0)	(9.0)	(9.9)	(10.5)	(11.0)	(11.5)	(12.0)
	Clay, hard, dense	3000	33	90	130	160	190	225	250	280
				(9.3)	(10.7)	(11.8)	(12.3)	(12.8)	(13.3)	(12.0)
				Wheel Loaders				Track Loaders		
				3 C.Y.	4 C.Y.	6 C.Y.	8 C.Y.	2-1/4 C.Y.	3 C.Y.	4 C.Y.
Loading Tractors	Moist loam, sandy clay	3400	40	260	340	510	690	135	180	250
	Sand and gravel	3100	18	245	320	480	650	130	170	235
	Common earth	2800	30	230	300	460	620	120	155	220
	Clay, hard, dense	3000	33	200	270	415	560	110	145	200
	Rock, well-blasted	4000	50	180	245	380	520	100	130	180

R312319-90 Wellpoints

A single stage wellpoint system is usually limited to dewatering an average 15′ depth below normal ground water level. Multi-stage systems are employed for greater depth with the pumping equipment installed only at the lowest header level. Ejectors with unlimited lift capacity can be economical when two or more stages of wellpoints can be replaced or when horizontal clearance is restricted, such as in deep trenches or tunneling projects, and where low water flows are expected. Wellpoints are usually spaced on 2-1/2′ to 10′ centers along a header pipe. Wellpoint spacing, header size, and pump size are all determined by the expected flow as dictated by soil conditions.

In almost all soils encountered in wellpoint dewatering, the wellpoints may be jetted into place. Cemented soils and stiff clays may require sand wicks about 12″ in diameter around each wellpoint to increase efficiency and eliminate weeping into the excavation. These sand wicks require 1/2 to 3 C.Y. of washed filter sand and are installed by using a 12″ diameter steel casing and hole puncher jetted into the ground 2′ deeper than the wellpoint. Rock may require predrilled holes.

Labor required for the complete installation and removal of a single stage wellpoint system is in the range of 3/4 to 2 labor-hours per linear foot of header, depending upon jetting conditions, wellpoint spacing, etc.

Continuous pumping is necessary except in some free draining soil where temporary flooding is permissible (as in trenches which are backfilled after each day's work). Good practice requires provision of a stand-by pump during the continuous pumping operation.

Systems for continuous trenching below the water table should be installed three to four times the length of expected daily progress to ensure uninterrupted digging, and header pipe size should not be changed during the job.

For pervious free draining soils, deep wells in place of wellpoints may be economical because of lower installation and maintenance costs. Daily production ranges between two to three wells per day, for 25′ to 40′ depths, to one well per day for depths over 50′.

Detailed analysis and estimating for any dewatering problem is available at no cost from wellpoint manufacturers. Major firms will quote "sufficient equipment" quotes or their affiliates offer lump sum proposals to cover complete dewatering responsibility.

Description for 200′ System with 8″ Header		Quantities
Equipment & Material	Wellpoints 25′ long, 2″ diameter @ 5′ O.C.	40 Each
	Header pipe, 8″ diameter	200 L.F.
	Discharge pipe, 8″ diameter	100 L.F.
	8″ valves	3 Each
	Combination jetting & wellpoint pump (standby)	1 Each
	Wellpoint pump, 8″ diameter	1 Each
	Transportation to and from site	1 Day
	Fuel for 30 days x 60 gal./day	1800 Gallons
	Lubricants for 30 days x 16 lbs./day	480 Lbs.
	Sand for points	40 C.Y.
Labor	Technician to supervise installation	1 Week
	Labor for installation and removal of system	300 Labor-hours
	4 Operators straight time 40 hrs./wk. for 4.33 wks.	693 Hrs.
	4 Operators overtime 2 hrs./wk. for 4.33 wks.	35 Hrs.

R312323-30 Compacting Backfill

Compaction of fill in embankments, around structures, in trenches, and under slabs is important to control settlement. Factors affecting compaction are:

1. Soil gradation
2. Moisture content
3. Equipment used
4. Depth of fill per lift
5. Density required

Production Rate:

$$\frac{1.75' \text{ plate width x 50 F.P.M. x 50 min./hr. x .67' lift}}{27 \text{ C.F. per C.Y.}} = 108.5 \text{ C.Y./hr.}$$

Production Rate for 4 Passes:

$$\frac{108.5 \text{ C.Y.}}{4 \text{ passes}} = 27.125 \text{ C.Y./hr. x 8 hrs.} = 217 \text{ C.Y./day}$$

Example:

Compact granular fill around a building foundation using a 21″ wide x 24″ vibratory plate in 8″ lifts. Operator moves at 50 F.P.M. working a 50 minute hour to develop 95% Modified Proctor Density with 4 passes.

R314116-40 Wood Sheet Piling

Wood sheet piling may be used for depths to 20′ where there is no ground water. If moderate ground water is encountered Tongue & Groove sheeting will help to keep it out. When considerable ground water is present, steel sheeting must be used.

For estimating purposes on trench excavation, sizes are as follows:

Depth	Sheeting	Wales	Braces	B.F. per S.F.
To 8′	3 x 12′s	6 x 8′s, 2 line	6 x 8′s, @ 10′	4.0 @ 8′
8′ x 12′	3 x 12′s	10 x 10′s, 2 line	10 x 10′s, @ 9′	5.0 average
12′ to 20′	3 x 12′s	12 x 12′s, 3 line	12 x 12′s, @ 8′	7.0 average

Sheeting to be toed in at least 2′ depending upon soil conditions. A five person crew with an air compressor and sheeting driver can drive and brace 440 SF/day at 8′ deep, 360 SF/day at 12′ deep, and 320 SF/day at 16′ deep. For normal soils, piling can be pulled in 1/3 the time to install. Pulling difficulty increases with the time in the ground. Production can be increased by high pressure jetting.

R314116-45 Steel Sheet Piling

Limiting weights are 22 to 38#/S.F. of wall surface with 27#/S.F. average for usual types and sizes. (Weights of piles themselves are from 30.7#/L.F. to 57#/L.F. but they are 15″ to 21″ wide.) Lightweight sections 12″ to 28″ wide from 3 ga. to 12 ga. thick are also available for shallow excavations. Piles may be driven two at a time with an impact or vibratory hammer (use vibratory to pull) hung from a crane without leads. A reasonable estimate of the life of steel sheet piling is 10 uses with up to 125 uses possible if a vibratory hammer is used. Used piling costs from 50% to 80% of new piling depending on location and market conditions. Sheet piling and H piles can be rented for about 30% of the delivered mill price for the first month and 5% per month thereafter. Allow 1 labor-hour per pile for cleaning and trimming after driving. These costs increase with depth and hydrostatic head. Vibratory drivers are faster in wet granular soils and are excellent for pile extraction. Pulling difficulty increases with the time in the ground and may cost more than driving. It is often economical to abandon the sheet piling, especially if it can be used as the outer wall form. Allow about 1/3 additional length or more for toeing into ground. Add bracing, waler and strut costs. Waler costs can equal the cost per ton of sheeting.

Earthwork R3145 Vibroflotation & Densification

R314513-90 Vibroflotation and Vibro Replacement Soil Compaction

Vibroflotation is a proprietary system of compacting sandy soils in place to increase relative density to about 70%. Typical bearing capacities attained will be 6000 psf for saturated sand and 12,000 psf for dry sand. Usual range is 4000 to 8000 psf capacity. Costs in the front of the book are for a vertical foot of compacted cylinder 6′ to 10′ in diameter.

Vibro replacement is a proprietary system of improving cohesive soils in place to increase bearing capacity. Most silts and clays above or below the water table can be strengthened by installation of stone columns.

The process consists of radial displacement of the soil by vibration. The created hole is then backfilled in stages with coarse granular fill which is thoroughly compacted and displaced into the surrounding soil in the form of a column.

The total project cost would depend on the number and depth of the compacted cylinders. The installing company guarantees relative soil density of the sand cylinders after compaction and the bearing capacity of the soil after the replacement process. Detailed estimating information is available from the installer at no cost.

Earthwork R3163 Drilled Caissons

R316326-60 Caissons

The three principal types of cassions are:

(1) **Belled Caissons,** which except for shallow depths and poor soil conditions, are generally recommended. They provide more bearing than shaft area. Because of its conical shape, no horizontal reinforcement of the bell is required.

(2) **Straight Shaft Caissons** are used where relatively light loads are to be supported by caissons that rest on high value bearing strata. While the shaft is larger in diameter than for belled types this is more than offset by the saving in time and labor.

(3) **Keyed Caissons** are used when extremely heavy loads are to be carried. A keyed or socketed caisson transfers its load into rock by a combination of end-bearing and shear reinforcing of the shaft. The most economical shaft often consists of a steel casing, a steel wide flange core and concrete. Allowable compressive stresses of .225 f′c for concrete, 16,000 psi for the wide flange core, and 9,000 psi for the steel casing are commonly used. The usual range of shaft diameter is 18″ to 84″. The number of sizes specified for any one project should be limited due to the problems of casing and auger storage. When hand work is to be performed, shaft diameters should not be less than 32″. When inspection of borings is required a minimum shaft diameter of 30″ is recommended. Concrete caissons are intended to be poured against earth excavation so permanent forms which add to cost should not be used if the excavation is clean and the earth sufficiently impervious to prevent excessive loss of concrete.

Soil Conditions for Belling		
Good	**Requires Handwork**	**Not Recommended**
Clay	Hard Shale	Silt
Sandy Clay	Limestone	Sand
Silty Clay	Sandstone	Gravel
Clayey Silt	Weathered Mica	Igneous Rock
Hard-pan		
Soft Shale		
Decomposed Rock		

Exterior Improvements R3292 Turf & Grasses

R329219-50 Seeding

The type of grass is determined by light, shade and moisture content of soil plus intended use. Fertilizer should be disked 4″ before seeding. For steep slopes disk five tons of mulch and lay two tons of hay or straw on surface per acre after seeding. Surface mulch can be staked, lightly disked or tar emulsion sprayed. Material for mulch can be wood chips, peat moss, partially rotted hay or straw, wood fibers and sprayed emulsions. Hemp seed blankets with fertilizer are also available. For spring seeding, watering is necessary. Late fall seeding may have to be reseeded in the spring. Hydraulic seeding, power mulching, and aerial seeding can be used on large areas.

R331113-80 Piping Designations

There are several systems currently in use to describe pipe and fittings. The following paragraphs will help to identify and clarify classifications of piping systems used for water distribution.

Piping may be classified by schedule. Piping schedules include 5S, 10S, 10, 20, 30, Standard, 40, 60, Extra Strong, 80, 100, 120, 140, 160 and Double Extra Strong. These schedules are dependent upon the pipe wall thickness. The wall thickness of a particular schedule may vary with pipe size.

Ductile iron pipe for water distribution is classified by Pressure Classes such as Class 150, 200, 250, 300 and 350. These classes are actually the rated water working pressure of the pipe in pounds per square inch (psi). The pipe in these pressure classes is designed to withstand the rated water working pressure plus a surge allowance of 100 psi.

The American Water Works Association (AWWA) provides standards for various types of **plastic pipe.** C-900 is the specification for polyvinyl chloride (PVC) piping used for water distribution in sizes ranging from 4″ through 12″. C-901 is the specification for polyethylene (PE) pressure pipe, tubing and fittings used for water distribution in sizes ranging from 1/2″ through 3″. C-905 is the specification for PVC piping sizes 14″ and greater.

PVC pressure-rated pipe is identified using the standard dimensional ratio (SDR) method. This method is defined by the American Society for Testing and Materials (ASTM) Standard D 2241. This pipe is available in SDR numbers 64, 41, 32.5, 26, 21, 17, and 13.5. Pipe with an SDR of 64 will have the thinnest wall while pipe with an SDR of 13.5 will have the thickest wall. When the pressure rating (PR) of a pipe is given in psi, it is based on a line supplying water at 73 degrees F.

The National Sanitation Foundation (NSF) seal of approval is applied to products that can be used with potable water. These products have been tested to ANSI/NSF Standard 14.

Valves and strainers are classified by American National Standards Institute (ANSI) Classes. These Classes are 125, 150, 200, 250, 300, 400, 600, 900, 1500 and 2500. Within each class there is an operating pressure range dependent upon temperature. Design parameters should be compared to the appropriate material dependent, pressure-temperature rating chart for accurate valve selection.

Transportation R3472 Railway Construction

R347216-10 Single Track R.R. Siding

The costs for a single track RR siding in the Unit Price section include the components shown in the table below.

Description of Component	Qty. per L.F. of Track	Unit
Ballast, 1-1/2″ crushed stone	.667	C.Y.
6″ x 8″ x 8′-6″ Treated timber ties, 22″ O.C.	.545	Ea.
Tie plates, 2 per tie	1.091	Ea.
Track rail	2.000	L.F.
Spikes, 6″, 4 per tie	2.182	Ea.
Splice bars w/ bolts, lock washers & nuts, @ 33′ O.C.	.061	Pair
Crew B-14 @ 57 L.F./Day	.018	Day

R347216-20 Single Track, Steel Ties, Concrete Bed

The costs for a R.R. siding with steel ties and a concrete bed in the Unit Price section include the components shown in the table below.

Description of Component	Qty. per L.F. of Track	Unit
Concrete bed, 9′ wide, 10″ thick	.278	C.Y.
Ties, W6x16 x 6′-6″ long, @ 30″ O.C.	.400	Ea.
Tie plates, 4 per tie	1.600	Ea.
Track rail	2.000	L.F.
Tie plate bolts, 1″, 8 per tie	3.200	Ea.
Splice bars w/bolts, lock washers & nuts, @ 33′ O.C.	.061	Pair
Crew B-14 @ 22 L.F./Day	.045	Day

Change Orders

Change Order Considerations

A Change Order is a written document, usually prepared by the design professional, and signed by the owner, the architect/engineer and the contractor. A change order states the agreement of the parties to: an addition, deletion, or revision in the work; an adjustment in the contract sum, if any; or an adjustment in the contract time, if any. Change orders, or "extras" in the construction process occur after execution of the construction contract and impact architects/engineers, contractors and owners.

Change orders that are properly recognized and managed can ensure orderly, professional and profitable progress for all who are involved in the project. There are many causes for change orders and change order requests. In all cases, change orders or change order requests should be addressed promptly and in a precise and prescribed manner. The following paragraphs include information regarding change order pricing and procedures.

The Causes of Change Orders

Reasons for issuing change orders include:

- Unforeseen field conditions that require a change in the work
- Correction of design discrepancies, errors or omissions in the contract documents
- Owner-requested changes, either by design criteria, scope of work, or project objectives
- Completion date changes for reasons unrelated to the construction process
- Changes in building code interpretations, or other public authority requirements that require a change in the work
- Changes in availability of existing or new materials and products

Procedures

Properly written contract documents must include the correct change order procedures for all parties—owners, design professionals and contractors—to follow in order to avoid costly delays and litigation.

Being "in the right" is not always a sufficient or acceptable defense. The contract provisions requiring notification and documentation must be adhered to within a defined or reasonable time frame.

The appropriate method of handling change orders is by a written proposal and acceptance by all parties involved. Prior to starting work on a project, all parties should identify their authorized agents who may sign and accept change orders, as well as any limits placed on their authority.

Time may be a critical factor when the need for a change arises. For such cases, the contractor might be directed to proceed on a "time and materials" basis, rather than wait for all paperwork to be processed—a delay that could impede progress. In this situation, the contractor must still follow the prescribed change order procedures, including but not limited to, notification and documentation.

All forms used for change orders should be dated and signed by the proper authority. Lack of documentation can be very costly, especially if legal judgments are to be made and if certain field personnel are no longer available. For time and material change orders, the contractor should keep accurate daily records of all labor and material allocated to the change. Forms that can be used to document change order work are available in *Means Forms for Building Construction Professionals.*

Owners or awarding authorities who do considerable and continual building construction (such as the federal government) realize the inevitability of change orders for numerous reasons, both predictable and unpredictable. As a result, the federal government, the American Institute of Architects (AIA), the Engineers Joint Contract Documents Committee (EJCDC) and other contractor, legal and technical organizations have developed standards and procedures to be followed by all parties to achieve contract continuance and timely completion, while being financially fair to all concerned.

In addition to the change order standards put forth by industry associations, there are also many books available on the subject.

Pricing Change Orders

When pricing change orders, regardless of their cause, the most significant factor is when the change occurs. The need for a change may be perceived in the field or requested by the architect/engineer *before* any of the actual installation has begun, or may evolve or appear *during* construction when the item of work in question is partially installed. In the latter cases, the original sequence of construction is disrupted, along with all contiguous and supporting systems. Change orders cause the greatest impact when they occur *after* the installation has been completed and must be uncovered, or even replaced. Post-completion changes may be caused by necessary design changes, product failure, or changes in the owner's requirements that are not discovered until the building or the systems begin to function.

Specified procedures of notification and record keeping must be adhered to and enforced regardless of the stage of construction: *before, during,* or *after* installation. Some bidding documents anticipate change orders by requiring that unit prices including overhead and profit percentages—for additional as well as deductible changes—be listed. Generally these unit prices do not fully take into account the ripple effect, or impact on other trades, and should be used for general guidance only.

When pricing change orders, it is important to classify the time frame in which the change occurs. There are two basic time frames for change orders: *pre-installation change orders,* which occur before the start of construction, and *post-installation change orders,* which involve reworking after the original installation. Change orders that occur between these stages may be priced according to the extent of work completed using a combination of techniques developed for pricing *pre-* and *post-installation* changes.

The following factors are the basis for a check list to use when preparing a change order estimate.

Factors To Consider When Pricing Change Orders

As an estimator begins to prepare a change order, the following questions should be reviewed to determine their impact on the final price.

General

- Is the change order work *pre-installation* or *post-installation?*

  Change order work costs vary according to how much of the installation has been completed. Once workers have the project scoped in their mind, even though they have not started, it can be difficult to refocus. Consequently they may spend more than the normal amount of time understanding the change. Also, modifications to work in place such as trimming or refitting usually take more time than was initially estimated. The greater the amount of work in place, the more reluctant workers are to change it. Psychologically they may resent the change and as a result the rework takes longer than normal. Post-installation change order estimates must include demolition of existing work as required to accomplish the change. If the work is performed at a later time, additional obstacles such as building finishes may be present which must be protected. Regardless of whether the change occurs pre-installation or post-installation, attempt to isolate the identifiable factors and price them separately. For example, add shipping costs that may be required pre-installation or any demolition required post-installation. Then analyze the potential impact on productivity of psychological and/or learning curve factors and adjust the output rates accordingly. One approach is to break down the typical workday into segments and quantify the impact on each segment. The following chart may be useful as a guide:

Activities (Productivity) Expressed as Percentages of a Workday

Task	Means Mechanical Cost Data (for New Construction)	Pre-Installation Change Orders	Post-Installation Change Orders
1. Study plans	3%	6%	6%
2. Material procurement	3%	3%	3%
3. Receiving and storing	3%	3%	3%
4. Mobilization	5%	5%	5%
5. Site movement	5%	5%	8%
6. Layout and marking	8%	10%	12%
7. Actual installation	64%	59%	54%
8. Clean-up	3%	3%	3%
9. Breaks—non-productive	6%	6%	6%
Total	100%	100%	100%

Change Order Installation Efficiency

The labor-hours expressed (for new construction) are based on average installation time, using an efficiency level of approximately 60-65%. For change order situations, adjustments to this efficiency level should reflect the daily labor-hour allocation for that particular occurrence.

If any of the specific percentages expressed in the above chart do not apply to a particular project situation, then those percentage points should be reallocated to the appropriate task(s). Example: Using data for new construction, assume there is no new material being utilized. The percentages for Tasks 2 and 3 would therefore be reallocated to other tasks. If the time required for Tasks 2 and 3 can now be applied to installation, we can add the time allocated for *Material Procurement* and *Receiving and Storing* to the *Actual Installation* time for new construction, thereby increasing the Actual Installation percentage.

This chart shows that, due to reduced productivity, labor costs will be higher than those for new construction by 5% to 15% for pre-installation change orders and by 15% to 25% for post-installation change orders. Each job and change order is unique and must be examined individually. Many factors, covered elsewhere in this section, can each have a significant impact on productivity and change order costs. All such factors should be considered in every case.

- Will the change substantially delay the original completion date?

  A significant change in the project may cause the original completion date to be extended. The extended schedule may subject the contractor to new wage rates dictated by relevant labor contracts. Project supervision and other project overhead must also be extended beyond the original completion date. The schedule extension may also put installation into a new weather season. For example, underground piping scheduled for October installation was delayed until January. As a result, frost penetrated the trench area, thereby changing the degree of difficulty of the task. Changes and delays may have a ripple effect throughout the project. This effect must be analyzed and negotiated with the owner.

- What is the net effect of a deduct change order?

  In most cases, change orders resulting in a deduction or credit reflect only bare costs. The contractor may retain the overhead and profit based on the original bid.

Materials

- Will you have to pay more or less for the new material, required by the change order, than you paid for the original purchase?

  The same material prices or discounts will usually apply to materials purchased for change orders as new construction. In some instances, however, the contractor may forfeit the advantages of competitive pricing for change orders. Consider the following example:

  A contractor purchased over $20,000 worth of fan coil units for an installation, and obtained the maximum discount. Some time later it was determined the project required an additional matching unit. The contractor has to purchase this unit from the original supplier to ensure a match. The supplier at this time may not discount the unit because of the small quantity, and the fact that he is no longer in a competitive situation. The impact of quantity on purchase can add between 0% and 25% to material prices and/or subcontractor quotes.

- If materials have been ordered or delivered to the job site, will they be subject to a cancellation charge or restocking fee?

  Check with the supplier to determine if ordered materials are subject to a cancellation charge. Delivered materials not used as result of a change order may be subject to a restocking fee if returned to the supplier. Common restocking charges run between 20% and 40%. Also, delivery charges to return the goods to the supplier must be added.

Labor

- How efficient is the existing crew at the actual installation?

  Is the same crew that performed the initial work going to do the change order? Possibly the change consists of the installation of a unit identical to one already installed; therefore the change should take less time. Be sure to consider this potential productivity increase and modify the productivity rates accordingly.

- If the crew size is increased, what impact will that have on supervision requirements?

  Under most bargaining agreements or management practices, there is a point at which a working foreman is replaced by a nonworking foreman. This replacement increases project overhead by adding a nonproductive worker. If additional workers are added to accelerate the project or to perform changes while maintaining the schedule, be sure to add additional supervision time if warranted. Calculate the hours involved and the additional cost directly if possible.

- What are the other impacts of increased crew size?

  The larger the crew, the greater the potential for productivity to decrease. Some of the factors that cause this productivity loss are: overcrowding (producing restrictive conditions in the working space), and possibly a shortage of any special tools and equipment required. Such factors affect not only the crew working on the elements directly involved in the change order, but other crews whose movement may also be hampered.

  As the crew increases, check its basic composition for changes by the addition or deletion of apprentices or nonworking foreman and quantify the potential effects of equipment shortages or other logistical factors.

- As new crews, unfamiliar with the project, are brought onto the site, how long will it take them to become oriented to the project requirements?

  The orientation time for a new crew to become 100% effective varies with the site and type of project. Orientation is easiest at a new construction site, and most difficult at existing, very restrictive renovation sites. The type of work also affects orientation time. When all elements of the work are exposed, such as concrete or masonry work, orientation is decreased. When the work is concealed or less visible, such as existing electrical systems, orientation takes longer. Usually orientation can be accomplished in one day or less. Costs for added orientation should be itemized and added to the total estimated cost.

- How much actual production can be gained by working overtime?

  Short term overtime can be used effectively to accomplish more work in a day. However, as overtime is scheduled to run beyond several weeks, studies have shown marked decreases in output. The following chart shows the effect of long term overtime on worker efficiency. If the anticipated change requires extended overtime to keep the job on schedule, these factors can be used as a guide to predict the impact on time and cost. Add project overhead, particularly supervision, that may also be incurred.

Days per Week	Hours per Day	Production Efficiency					Payroll Cost Factors	
		1 Week	2 Weeks	3 Weeks	4 Weeks	Average 4 Weeks	@ 1-1/2 Times	@ 2 Times
5	8	100%	100%	100%	100%	100%	100%	100%
	9	100	100	95	90	96.25	105.6	111.1
	10	100	95	90	85	91.25	110.0	120.0
	11	95	90	75	65	81.25	113.6	127.3
	12	90	85	70	60	76.25	116.7	133.3
6	8	100	100	95	90	96.25	108.3	116.7
	9	100	95	90	85	92.50	113.0	125.9
	10	95	90	85	80	87.50	116.7	133.3
	11	95	85	70	65	78.75	119.7	139.4
	12	90	80	65	60	73.75	122.2	144.4
7	8	100	95	85	75	88.75	114.3	128.6
	9	95	90	80	70	83.75	118.3	136.5
	10	90	85	75	65	78.75	121.4	142.9
	11	85	80	65	60	72.50	124.0	148.1
	12	85	75	60	55	68.75	126.2	152.4

Effects of Overtime

Caution: Under many labor agreements, Sundays and holidays are paid at a higher premium than the normal overtime rate.

The use of long-term overtime is counterproductive on almost any construction job; that is, the longer the period of overtime, the lower the actual production rate. Numerous studies have been conducted, and while they have resulted in slightly different numbers, all reach the same conclusion. The figure above tabulates the effects of overtime work on efficiency.

As illustrated, there can be a difference between the *actual* payroll cost per hour and the *effective* cost per hour for overtime work. This is due to the reduced production efficiency with the increase in weekly hours beyond 40. This difference between actual and effective cost results from overtime work over a prolonged period. Short-term overtime work does not result in as great a reduction in efficiency, and in such cases, effective cost may not vary significantly from the actual payroll cost. As the total hours per week are increased on a regular basis, more time is lost because of fatigue, lowered morale, and an increased accident rate.

As an example, assume a project where workers are working 6 days a week, 10 hours per day. From the figure above (based on productivity studies), the average effective productive hours over a four-week period are:

$$0.875 \times 60 = 52.5$$

Depending upon the locale and day of week, overtime hours may be paid at time and a half or double time. For time and a half, the overall (average) *actual* payroll cost (including regular and overtime hours) is determined as follows:

$$\frac{40 \text{ reg. hrs.} + (20 \text{ overtime hrs.} \times 1.5)}{60 \text{ hrs.}} = 1.167$$

Based on 60 hours, the payroll cost per hour will be 116.7% of the normal rate at 40 hours per week. However, because the effective production (efficiency) for 60 hours is reduced to the equivalent of 52.5 hours, the effective cost of overtime is calculated as follows:

For time and a half:

$$\frac{40 \text{ reg. hrs.} + (20 \text{ overtime hrs.} \times 1.5)}{52.5 \text{ hrs.}} = 1.33$$

Installed cost will be 133% of the normal rate (for labor).

Thus, when figuring overtime, the actual cost per unit of work will be higher than the apparent overtime payroll dollar increase, due to the reduced productivity of the longer workweek. These efficiency calculations are true only for those cost factors determined by hours worked. Costs that are applied weekly or monthly, such as equipment rentals, will not be similarly affected.

Equipment

- What equipment is required to complete the change order?

  Change orders may require extending the rental period of equipment already on the job site, or the addition of special equipment brought in to accomplish the change work. In either case, the additional rental charges and operator labor charges must be added.

Summary

The preceding considerations and others you deem appropriate should be analyzed and applied to a change order estimate. The impact of each should be quantified and listed on the estimate to form an audit trail.

Change orders that are properly identified, documented, and managed help to ensure the orderly, professional and profitable progress of the work. They also minimize potential claims or disputes at the end of the project.

Estimating Tips

- The cost figures in this Square Foot Cost section were derived from approximately 11,200 projects contained in the RSMeans database of completed construction projects. They include the contractor's overhead and profit, but do not generally include architectural fees or land costs. The figures have been adjusted to January of the current year. New projects are added to our files each year, and outdated projects are discarded. For this reason, certain costs may not show a uniform annual progression. In no case are all subdivisions of a project listed.
- These projects were located throughout the U.S. and reflect a tremendous variation in square foot (S.F.) and cubic foot (C.F.) costs. This is due to differences, not only in labor and material costs, but also in individual owners' requirements. For instance, a bank in a large city would have different features than one in a rural area. This is true of all the different types of buildings analyzed. Therefore, caution should be exercised when using these Square Foot costs. For example, for court houses, costs in the database are local court house costs and will not apply to the larger, more elaborate federal court houses. As a general rule, the projects in the 1/4 column do not include any site work or equipment, while the projects in the 3/4 column may include both equipment and site work. The median figures do not generally include site work.
- None of the figures "go with" any others. All individual cost items were computed and tabulated separately. Thus, the sum of the median figures for Plumbing, HVAC and Electrical will not normally total up to the total Mechanical and Electrical costs arrived at by separate analysis and tabulation of the projects.
- Each building was analyzed as to total and component costs and percentages. The figures were arranged in ascending order with the results tabulated as shown. The 1/4 column shows that 25% of the projects had lower costs and 75% had higher. The 3/4 column shows that 75% of the projects had lower costs and 25% had higher. The median column shows that 50% of the projects had lower costs and 50% had higher.
- There are two times when square foot costs are useful. The first is in the conceptual stage when no details are available. Then square foot costs make a useful starting point. The second is after the bids are in and the costs can be worked back into their appropriate units for information purposes. As soon as details become available in the project design, the square foot approach should be discontinued and the project priced as to its particular components. When more precision is required, or for estimating the replacement cost of specific buildings, the current edition of *RSMeans Square Foot Costs* should be used.
- In using the figures in this section, it is recommended that the median column be used for preliminary figures if no additional information is available. The median figures, when multiplied by the total city construction cost index figures (see City Cost Indexes) and then multiplied by the project size modifier at the end of this section, should present a fairly accurate base figure, which would then have to be adjusted in view of the estimator's experience, local economic conditions, code requirements, and the owner's particular requirements. There is no need to factor the percentage figures, as these should remain constant from city to city. All tabulations mentioning air conditioning had at least partial air conditioning.
- The editors of this book would greatly appreciate receiving cost figures on one or more of your recent projects, which would then be included in the averages for next year. All cost figures received will be kept confidential, except that they will be averaged with other similar projects to arrive at Square Foot cost figures for next year's book. See the last page of the book for details and the discount available for submitting one or more of your projects.

50 17 00 | S.F. Costs

		50 17 00 | S.F. Costs	UNIT	UNIT COSTS 1/4	UNIT COSTS MEDIAN	UNIT COSTS 3/4	% OF TOTAL 1/4	% OF TOTAL MEDIAN	% OF TOTAL 3/4	
01	0010	**APARTMENTS Low Rise (1 to 3 story)**	S.F.	67	84	112				01
	0020	Total project cost	C.F.	6	7.95	9.80				
	0100	Site work	S.F.	5.70	7.80	13.70	6.05%	10.55%	14.05%	
	0500	Masonry		1.31	3.05	5.30	1.54%	3.67%	6.35%	
	1500	Finishes		7.10	9.70	12	9.05%	10.75%	12.85%	
	1800	Equipment		2.18	3.31	4.92	2.73%	4.03%	5.95%	
	2720	Plumbing		5.20	6.70	8.50	6.65%	8.95%	10.05%	
	2770	Heating, ventilating, air conditioning		3.31	4.08	6	4.20%	5.60%	7.60%	
	2900	Electrical		3.86	5.15	6.90	5.20%	6.65%	8.40%	
	3100	Total: Mechanical & Electrical		13.40	17.05	21.50	15.90%	18.05%	23%	
	9000	Per apartment unit, total cost	Apt.	62,000	95,000	140,000				
	9500	Total: Mechanical & Electrical	"	11,800	18,500	24,200				
02	0010	**APARTMENTS Mid Rise (4 to 7 story)**	S.F.	88.50	107	132				02
	0020	Total project costs	C.F.	6.90	9.55	13.05				
	0100	Site work	S.F.	3.54	7	12.60	5.25%	6.70%	9.15%	
	0500	Masonry		5.90	8.10	11.10	5.10%	7.25%	10.50%	
	1500	Finishes		11.15	14.50	18.30	10.55%	13.45%	17.70%	
	1800	Equipment		2.57	3.85	5.05	2.54%	3.48%	4.31%	
	2500	Conveying equipment		1.91	2.43	2.94	1.94%	2.27%	2.69%	
	2720	Plumbing		5.20	8.30	8.80	5.70%	7.20%	8.95%	
	2900	Electrical		5.85	7.90	9.60	6.65%	7.20%	8.95%	
	3100	Total: Mechanical & Electrical		18.70	23.50	28.50	18.50%	21%	23%	
	9000	Per apartment unit, total cost	Apt.	100,000	118,000	195,500				
	9500	Total: Mechanical & Electrical	"	18,900	21,900	27,600				
03	0010	**APARTMENTS High Rise (8 to 24 story)**	S.F.	100	116	139				03
	0020	Total project costs	C.F.	9.75	11.35	14.50				
	0100	Site work	S.F.	3.64	5.90	8.25	2.58%	4.84%	6.15%	
	0500	Masonry		5.80	10.55	13.15	4.74%	9.65%	11.05%	
	1500	Finishes		11.15	13.90	16.45	9.75%	11.80%	13.70%	
	1800	Equipment		3.23	3.97	5.25	2.78%	3.49%	4.35%	
	2500	Conveying equipment		2.28	3.46	4.71	2.23%	2.78%	3.37%	
	2720	Plumbing		7.40	8.70	12.15	6.80%	7.20%	10.45%	
	2900	Electrical		6.90	8.70	11.75	6.45%	7.65%	8.80%	
	3100	Total: Mechanical & Electrical		20.50	26.50	31.50	17.95%	22.50%	24.50%	
	9000	Per apartment unit, total cost	Apt.	104,500	115,000	159,500				
	9500	Total: Mechanical & Electrical	"	22,600	25,800	27,300				
04	0010	**AUDITORIUMS**	S.F.	104	142	204				04
	0020	Total project costs	C.F.	6.55	9.10	13.05				
	2720	Plumbing	S.F.	6.30	9.15	11.10	5.85%	7.20%	8.70%	
	2900	Electrical		8.05	11.75	19.30	6.80%	9.05%	11.35%	
	3100	Total: Mechanical & Electrical		50.50	65.50	90.50	24.50%	29%	31.50%	
05	0010	**AUTOMOTIVE SALES**	S.F.	77	105	130				05
	0020	Total project costs	C.F.	5.10	6.10	7.90				
	2720	Plumbing	S.F.	3.52	6.10	6.65	2.89%	6.05%	6.50%	
	2770	Heating, ventilating, air conditioning		5.45	8.30	8.95	4.61%	10%	10.35%	
	2900	Electrical		6.20	9.75	13.25	7.25%	8.80%	12.15%	
	3100	Total: Mechanical & Electrical		19.50	27.50	33.50	18.90%	20.50%	22%	
06	0010	**BANKS**	S.F.	152	189	239				06
	0020	Total project costs	C.F.	10.80	14.70	19.40				
	0100	Site work	S.F.	17.35	26.50	38.50	7.85%	12.95%	16.95%	
	0500	Masonry		8.15	15.60	27.50	3.36%	6.95%	10.05%	
	1500	Finishes		14	20.50	25.50	5.85%	8.60%	11.55%	
	1800	Equipment		5.70	12.55	25.50	1%	5.55%	10.50%	
	2720	Plumbing		4.74	6.80	9.90	2.82%	3.90%	4.93%	
	2770	Heating, ventilating, air conditioning		9	12.05	16.05	4.86%	7.15%	8.50%	
	2900	Electrical		14.40	19.15	25	8.20%	10.20%	12.20%	
	3100	Total: Mechanical & Electrical		34.50	46.50	55.50	16.25%	19.45%	23%	
	3500	See also division 11 16 00 & 11 17 00								

50 17 00 | S.F. Costs

		50 17 00 | S.F. Costs	UNIT	UNIT COSTS 1/4	UNIT COSTS MEDIAN	UNIT COSTS 3/4	% OF TOTAL 1/4	% OF TOTAL MEDIAN	% OF TOTAL 3/4
13	0010	**CHURCHES**	S.F.	102	129	168			
	0020	Total project costs	C.F.	6.30	8	10.50			
	1800	Equipment	S.F.	1.22	2.91	6.20	.95%	2.11%	4.31%
	2720	Plumbing		3.97	5.55	8.15	3.51%	4.96%	6.25%
	2770	Heating, ventilating, air conditioning		9.30	12.10	17.10	7.50%	10%	12%
	2900	Electrical		8.60	11.80	15.85	7.30%	8.80%	10.95%
	3100	Total: Mechanical & Electrical		27	35.50	47.50	18.30%	22%	25%
	3500	See also division 11 91 00							
15	0010	**CLUBS, COUNTRY**	S.F.	109	132	166			
	0020	Total project costs	C.F.	8.80	10.75	14.85			
	2720	Plumbing	S.F.	6.60	9.80	22.50	5.60%	7.90%	10%
	2900	Electrical		8.60	11.80	15.35	7%	8.95%	11%
	3100	Total: Mechanical & Electrical		45.50	57	60.50	19%	26.50%	29.50%
17	0010	**CLUBS, SOCIAL Fraternal**	S.F.	87	125	168			
	0020	Total project costs	C.F.	5.45	8.25	9.85			
	2720	Plumbing	S.F.	5.50	6.85	10.35	5.60%	6.90%	8.55%
	2770	Heating, ventilating, air conditioning		7.90	9.55	12.30	8.20%	9.25%	14.40%
	2900	Electrical		6.55	10.80	12.35	6.50%	9.50%	10.55%
	3100	Total: Mechanical & Electrical		34	36.50	46.50	21%	23%	23.50%
18	0010	**CLUBS, Y.M.C.A.**	S.F.	114	146	181			
	0020	Total project costs	C.F.	5.05	8.45	12.60			
	2720	Plumbing	S.F.	6.90	13.80	15.45	5.65%	7.60%	10.85%
	2900	Electrical		8.95	10.95	16.40	6.25%	7.95%	9.25%
	3100	Total: Mechanical & Electrical		33.50	40.50	68	20.50%	22.50%	28.50%
19	0010	**COLLEGES Classrooms & Administration**	S.F.	111	151	201			
	0020	Total project costs	C.F.	8.10	11.75	18.10			
	0500	Masonry	S.F.	8.10	15.30	18.80	5.10%	7.45%	10.50%
	2720	Plumbing		5.65	11.55	21.50	5.10%	6.60%	8.95%
	2900	Electrical		9.20	14	19.45	7.80%	9.85%	12%
	3100	Total: Mechanical & Electrical		36.50	51.50	69	23.50%	28%	32%
21	0010	**COLLEGES Science, Engineering, Laboratories**	S.F.	206	241	279			
	0020	Total project costs	C.F.	11.80	17.20	19.55			
	1800	Equipment	S.F.	11.45	26	28.50	2%	6.45%	12.65%
	2900	Electrical		16.95	24	37	7.10%	9.40%	12.10%
	3100	Total: Mechanical & Electrical		63	74.50	116	28.50%	31.50%	41%
	3500	See also division 11 53 00							
23	0010	**COLLEGES Student Unions**	S.F.	131	178	215			
	0020	Total project costs	C.F.	7.30	9.60	11.85			
	3100	Total: Mechanical & Electrical	S.F.	49.50	53.50	63	23.50%	26%	29%
25	0010	**COMMUNITY CENTERS**	S.F.	108	134	181			
	0020	Total project costs	C.F.	7.05	10.10	13.10			
	1800	Equipment	S.F.	2.62	4.44	7.05	1.50%	3.04%	5.45%
	2720	Plumbing		5.15	9	12.25	4.85%	7%	8.95%
	2770	Heating, ventilating, air conditioning		8.20	12.05	17.15	6.80%	10.35%	12.90%
	2900	Electrical		8.85	11.80	17	7.20%	8.95%	10.45%
	3100	Total: Mechanical & Electrical		30.50	38	55	20%	25%	31%
28	0010	**COURT HOUSES**	S.F.	156	184	254			
	0020	Total project costs	C.F.	12	14.35	18.05			
	2720	Plumbing	S.F.	7.45	10.40	14.90	5.95%	7.45%	8.20%
	2900	Electrical		16.60	18.45	27	8.90%	10.45%	11.85%
	3100	Total: Mechanical & Electrical		46.50	61	66.50	22.50%	27.50%	30%
30	0010	**DEPARTMENT STORES**	S.F.	58	78	98.50			
	0020	Total project costs	C.F.	3.10	4.01	5.45			
	2720	Plumbing	S.F.	1.80	2.27	3.45	1.82%	4.21%	5.90%
	2770	Heating, ventilating, air conditioning		5.25	8.10	12.20	8.20%	9.10%	14.80%

50 17 00 | S.F. Costs

		50 17 00 | S.F. Costs	UNIT	UNIT COSTS 1/4	UNIT COSTS MEDIAN	UNIT COSTS 3/4	% OF TOTAL 1/4	% OF TOTAL MEDIAN	% OF TOTAL 3/4	
30	2900	Electrical	S.F.	6.65	9.10	10.75	9.05%	12.15%	14.95%	30
	3100	Total: Mechanical & Electrical		11.65	14.95	26	13.20%	21.50%	50%	
31	0010	**DORMITORIES Low Rise (1 to 3 story)**	S.F.	110	143	190				31
	0020	Total project costs	C.F.	6.20	10	15				
	2720	Plumbing	S.F.	6.60	8.80	11.15	8.05%	9%	9.65%	
	2770	Heating, ventilating, air conditioning		6.95	8.35	11.10	4.61%	8.05%	10%	
	2900	Electrical		7.25	11.05	15.10	6.40%	8.65%	9.50%	
	3100	Total: Mechanical & Electrical		38	40.50	63	22%	25%	27%	
	9000	Per bed, total cost	Bed	46,400	51,500	110,500				
32	0010	**DORMITORIES Mid Rise (4 to 8 story)**	S.F.	135	176	217				32
	0020	Total project costs	C.F.	14.85	16.30	19.50				
	2900	Electrical	S.F.	14.30	16.25	22	8.20%	10.20%	11.95%	
	3100	Total: Mechanical & Electrical	″	40	80	81	25.50%	34.50%	37.50%	
	9000	Per bed, total cost	Bed	19,200	43,700	253,000				
34	0010	**FACTORIES**	S.F.	51	76	117				34
	0020	Total project costs	C.F.	3.27	4.87	8.10				
	0100	Site work	S.F.	5.80	10.60	16.75	6.95%	11.45%	17.95%	
	2720	Plumbing		2.74	5.10	8.45	3.73%	6.05%	8.10%	
	2770	Heating, ventilating, air conditioning		5.35	7.65	10.35	5.25%	8.45%	11.35%	
	2900	Electrical		6.35	10	15.30	8.10%	10.50%	14.20%	
	3100	Total: Mechanical & Electrical		18.10	27.50	36.50	21%	28.50%	35.50%	
36	0010	**FIRE STATIONS**	S.F.	101	139	187				36
	0020	Total project costs	C.F.	5.90	8.10	10.80				
	0500	Masonry	S.F.	14.95	26.50	35	8.35%	11.55%	16.15%	
	1140	Roofing		3.29	8.90	10.15	1.90%	4.94%	5.05%	
	1580	Painting		2.56	3.82	3.91	1.37%	1.57%	2.07%	
	1800	Equipment		1.26	2.41	4.45	.62%	1.86%	3.54%	
	2720	Plumbing		5.65	9.10	13.15	5.85%	7.35%	9.45%	
	2770	Heating, ventilating, air conditioning		5.60	9.05	14	5.15%	7.40%	9.40%	
	2900	Electrical		7.20	12.65	17.15	6.80%	8.60%	10.60%	
	3100	Total: Mechanical & Electrical		37	47.50	53.50	18.40%	23%	26%	
37	0010	**FRATERNITY HOUSES & Sorority Houses**	S.F.	101	130	178				37
	0020	Total project costs	C.F.	10.05	10.50	12.65				
	2720	Plumbing	S.F.	7.65	8.75	16	6.80%	8%	10.85%	
	2900	Electrical		6.65	14.40	17.65	6.60%	9.90%	10.65%	
	3100	Total: Mechanical & Electrical		7.85	25.50	30.50		15.10%	15.90%	
38	0010	**FUNERAL HOMES**	S.F.	107	145	264				38
	0020	Total project costs	C.F.	10.90	12.10	23.50				
	2900	Electrical	S.F.	4.71	8.65	9.45	3.58%	4.44%	5.95%	
	3100	Total: Mechanical & Electrical		16.75	24.50	34	12.90%	12.90%	12.90%	
39	0010	**GARAGES, COMMERCIAL (Service)**	S.F.	60.50	93.50	129				39
	0020	Total project costs	C.F.	3.97	5.85	8.50				
	1800	Equipment	S.F.	3.40	7.65	11.90	2.21%	4.62%	6.80%	
	2720	Plumbing		4.18	6.45	11.75	5.45%	7.85%	10.65%	
	2730	Heating & ventilating		5.50	7.30	9.85	5.25%	6.85%	8.20%	
	2900	Electrical		5.75	8.75	12.60	7.15%	9.25%	10.85%	
	3100	Total: Mechanical & Electrical		12.65	24.50	36	12.35%	17.40%	26%	
40	0010	**GARAGES, MUNICIPAL (Repair)**	S.F.	94	119	169				40
	0020	Total project costs	C.F.	5.55	7.05	12.10				
	0500	Masonry	S.F.	8.30	16.25	25.50	5.60%	9.15%	12.50%	
	2720	Plumbing		3.98	7.65	14.40	3.59%	6.70%	7.95%	
	2730	Heating & ventilating		6.80	9.85	19	6.15%	7.45%	13.50%	
	2900	Electrical		6.55	10.50	14.95	6.65%	9.25%	11.15%	
	3100	Total: Mechanical & Electrical		30.50	42	61	21.50%	23%	28.50%	

	50 17 00 | S.F. Costs		UNIT	UNIT COSTS 1/4	UNIT COSTS MEDIAN	UNIT COSTS 3/4	% OF TOTAL 1/4	% OF TOTAL MEDIAN	% OF TOTAL 3/4	
41	0010	**GARAGES, PARKING**	S.F.	34.50	50.50	86.50				41
	0020	Total project costs	C.F.	3.24	4.40	6.40				
	2720	Plumbing	S.F.	.98	1.52	2.34	1.72%	2.70%	3.85%	
	2900	Electrical		1.89	2.32	3.65	4.33%	5.20%	6.30%	
	3100	Total: Mechanical & Electrical		3.87	5.40	6.70	7%	8.90%	11.05%	
	3200									
	9000	Per car, total cost	Car	14,600	18,300	23,300				
43	0010	**GYMNASIUMS**	S.F.	96	128	173				43
	0020	Total project costs	C.F.	4.79	6.50	8				
	1800	Equipment	S.F.	2.28	4.28	8.20	1.98%	3.30%	6.70%	
	2720	Plumbing		6.05	7.20	9.30	4.65%	6.40%	7.75%	
	2770	Heating, ventilating, air conditioning		6.55	9.95	19.95	5.15%	9.05%	11.10%	
	2900	Electrical		7.40	10.05	13.25	6.60%	8.30%	10.30%	
	3100	Total: Mechanical & Electrical		27	36.50	43.50	19.60%	23.50%	28%	
	3500	See also division 11 67 00								
46	0010	**HOSPITALS**	S.F.	184	230	315				46
	0020	Total project costs	C.F.	13.95	17.40	25				
	1800	Equipment	S.F.	4.68	9	15.50	.80%	2.53%	4.80%	
	2720	Plumbing		15.85	22.50	29	7.60%	9.10%	10.85%	
	2770	Heating, ventilating, air conditioning		23.50	30	41.50	7.80%	12.95%	16.65%	
	2900	Electrical		20.50	27.50	39	9.90%	11.75%	14.10%	
	3100	Total: Mechanical & Electrical		59.50	82.50	123	28%	33.50%	37%	
	9000	Per bed or person, total cost	Bed	213,500	294,000	338,500				
	9900	See also division 11 71 00								
48	0010	**HOUSING For the Elderly**	S.F.	90.50	115	141				48
	0020	Total project costs	C.F.	6.45	8.95	11.45				
	0100	Site work	S.F.	6.30	9.80	14.35	5.05%	7.90%	12.10%	
	0500	Masonry		2.76	10.30	15.05	1.30%	6.05%	11%	
	1800	Equipment		2.19	3.02	4.80	1.88%	3.23%	4.43%	
	2510	Conveying systems		2.20	2.96	4.02	1.78%	2.20%	2.81%	
	2720	Plumbing		6.75	8.60	10.80	8.15%	9.55%	10.50%	
	2730	Heating, ventilating, air conditioning		3.45	4.89	7.30	3.30%	5.60%	7.25%	
	2900	Electrical		6.75	9.15	11.75	7.30%	8.50%	10.25%	
	3100	Total: Mechanical & Electrical		24.50	29	37	18.10%	22.50%	29%	
	9000	Per rental unit, total cost	Unit	84,500	98,500	110,000				
	9500	Total: Mechanical & Electrical	"	18,800	21,600	25,200				
50	0010	**HOUSING Public (Low Rise)**	S.F.	76	106	138				50
	0020	Total project costs	C.F.	6.80	8.45	10.50				
	0100	Site work	S.F.	9.70	13.95	22.50	8.35%	11.75%	16.50%	
	1800	Equipment		2.07	3.38	5.15	2.26%	3.03%	4.24%	
	2720	Plumbing		5.50	7.25	9.20	7.15%	9.05%	11.60%	
	2730	Heating, ventilating, air conditioning		2.76	5.35	5.85	4.26%	6.05%	6.45%	
	2900	Electrical		4.61	6.85	9.50	5.10%	6.55%	8.25%	
	3100	Total: Mechanical & Electrical		22	28	31.50	14.50%	17.55%	26.50%	
	9000	Per apartment, total cost	Apt.	83,500	95,000	119,500				
	9500	Total: Mechanical & Electrical	"	17,800	22,000	24,400				
51	0010	**ICE SKATING RINKS**	S.F.	65	152	167				51
	0020	Total project costs	C.F.	4.79	4.90	5.65				
	2720	Plumbing	S.F.	2.43	4.56	4.66	3.12%	3.23%	5.65%	
	2900	Electrical		6.95	10.70	11.30	6.30%	10.15%	15.05%	
	3100	Total: Mechanical & Electrical		11.55	16.35	20	18.95%	18.95%	18.95%	
52	0010	**JAILS**	S.F.	198	256	330				52
	0020	Total project costs	C.F.	17.85	25	29				
	1800	Equipment	S.F.	7.75	23	39	2.77%	5.55%	10.35%	
	2720	Plumbing		19.10	25.50	34	7%	8.90%	13.35%	
	2770	Heating, ventilating, air conditioning		17.90	24	46	7.50%	9.45%	17.75%	
	2900	Electrical		21	28	35	8.80%	11.55%	14.95%	

	50 17 00 | S.F. Costs		Unit	Unit Costs 1/4	Unit Costs Median	Unit Costs 3/4	% of Total 1/4	% of Total Median	% of Total 3/4	
52	3100	Total: Mechanical & Electrical	S.F.	54	98.50	117	27.50%	30%	34%	52
53	0010	**LIBRARIES**	S.F.	127	162	209				53
	0020	Total project costs	C.F.	8.55	10.75	13.70				
	0500	Masonry	S.F.	10	17.50	29.50	5.80%	7.60%	11.60%	
	1800	Equipment		1.71	4.60	6.95	.37%	1.50%	4.07%	
	2720	Plumbing		4.63	6.75	9.10	3.38%	4.60%	5.70%	
	2770	Heating, ventilating, air conditioning		10.25	17.35	22.50	7.80%	10.95%	12.80%	
	2900	Electrical		12.85	16.65	21.50	8.35%	10.30%	11.95%	
	3100	Total: Mechanical & Electrical		37.50	48	59.50	20.50%	23%	26.50%	
54	0010	**LIVING, ASSISTED**	S.F.	116	137	161				54
	0020	Total project costs	C.F.	9.75	11.40	12.95				
	0500	Masonry	S.F.	3.41	4.06	4.77	2.36%	3.16%	3.86%	
	1800	Equipment		2.63	3.06	3.92	2.12%	2.47%	2.87%	
	2720	Plumbing		9.70	13	13.45	6.05%	8.15%	10.60%	
	2770	Heating, ventilating, air conditioning		11.50	12.05	13.20	7.95%	9.35%	9.70%	
	2900	Electrical		11.35	12.55	14.50	9%	10%	10.70%	
	3100	Total: Mechanical & Electrical		31.50	37	43	25%	28.50%	31.50%	
55	0010	**MEDICAL CLINICS**	S.F.	118	145	184				55
	0020	Total project costs	C.F.	8.65	11.20	14.85				
	1800	Equipment	S.F.	3.19	6.70	10.40	1.05%	2.94%	6.35%	
	2720	Plumbing		7.85	11.05	14.75	6.15%	8.40%	10.10%	
	2770	Heating, ventilating, air conditioning		9.35	12.25	18	6.65%	8.85%	11.35%	
	2900	Electrical		10.15	14.40	18.80	8.10%	10%	12.25%	
	3100	Total: Mechanical & Electrical		32.50	44	60	22.50%	27%	33.50%	
	3500	See also division 11 71 00								
57	0010	**MEDICAL OFFICES**	S.F.	111	137	168				57
	0020	Total project costs	C.F.	8.30	11.20	15.15				
	1800	Equipment	S.F.	3.66	7.25	10.30	.70%	5.10%	7.05%	
	2720	Plumbing		6.15	9.45	12.75	5.60%	6.80%	8.50%	
	2770	Heating, ventilating, air conditioning		7.40	10.70	14.10	6.10%	8%	9.70%	
	2900	Electrical		9.05	12.85	17.90	7.60%	9.80%	11.40%	
	3100	Total: Mechanical & Electrical		24	34.50	50	19.30%	22.50%	28.50%	
59	0010	**MOTELS**	S.F.	70	101	132				59
	0020	Total project costs	C.F.	6.25	8.35	13.65				
	2720	Plumbing	S.F.	7.10	9.05	10.75	9.45%	10.60%	12.55%	
	2770	Heating, ventilating, air conditioning		4.32	6.45	11.55	5.60%	5.60%	10%	
	2900	Electrical		6.60	8.35	10.40	7.45%	9.05%	10.45%	
	3100	Total: Mechanical & Electrical		21	28	48.50	18.50%	24%	25.50%	
	5000									
	9000	Per rental unit, total cost	Unit	35,600	67,500	73,000				
	9500	Total: Mechanical & Electrical	"	6,950	10,500	12,200				
60	0010	**NURSING HOMES**	S.F.	110	142	173				60
	0020	Total project costs	C.F.	8.60	10.75	14.70				
	1800	Equipment	S.F.	3.46	4.58	7.65	2.02%	3.62%	4.99%	
	2720	Plumbing		9.40	14.25	17.15	8.75%	10.10%	12.70%	
	2770	Heating, ventilating, air conditioning		9.90	15.05	19.90	9.70%	11.45%	11.80%	
	2900	Electrical		10.85	13.55	18.45	9.40%	10.55%	12.45%	
	3100	Total: Mechanical & Electrical		26	36.50	60.50	26%	29.50%	30.50%	
	9000	Per bed or person, total cost	Bed	48,700	61,000	78,500				
61	0010	**OFFICES Low Rise (1 to 4 story)**	S.F.	92.50	120	156				61
	0020	Total project costs	C.F.	6.60	9.10	12				
	0100	Site work	S.F.	7.40	13.15	19.35	6.20%	9.70%	13.55%	
	0500	Masonry		3.58	7.35	13.15	2.62%	5.45%	8.45%	
	1800	Equipment		.98	1.92	5.25	.60%	1.50%	3.50%	
	2720	Plumbing		3.31	5.10	7.50	3.66%	4.50%	6.10%	
	2770	Heating, ventilating, air conditioning		7.30	10.20	14.90	7.20%	10.30%	11.70%	
	2900	Electrical		7.55	10.75	15.25	7.45%	9.65%	11.35%	

50 17 00 | S.F. Costs

		50 17 00 | S.F. Costs	UNIT	UNIT COSTS 1/4	UNIT COSTS MEDIAN	UNIT COSTS 3/4	% OF TOTAL 1/4	% OF TOTAL MEDIAN	% OF TOTAL 3/4
61	3100	Total: Mechanical & Electrical	S.F.	21	28.50	42.50	18.20%	22%	27%
62	0010	**OFFICES Mid Rise (5 to 10 story)**	S.F.	98	119	161			
	0020	Total project costs	C.F.	6.95	8.85	12.55			
	2720	Plumbing	S.F.	2.96	4.58	6.60	2.83%	3.74%	4.50%
	2770	Heating, ventilating, air conditioning		7.45	10.60	16.95	7.65%	9.40%	11%
	2900	Electrical		7.25	9.40	14.05	6.35%	7.80%	10%
	3100	Total: Mechanical & Electrical		18.80	24.50	46	18.95%	21%	27.50%
63	0010	**OFFICES High Rise (11 to 20 story)**	S.F.	120	152	187			
	0020	Total project costs	C.F.	8.40	10.50	15.05			
	2900	Electrical	S.F.	7.30	8.90	13.20	5.80%	7.85%	10.50%
	3100	Total: Mechanical & Electrical		23.50	31.50	53	16.90%	23.50%	34%
64	0010	**POLICE STATIONS**	S.F.	144	189	239			
	0020	Total project costs	C.F.	11.50	14.05	19.25			
	0500	Masonry	S.F.	13.55	24	30	6.70%	9.10%	11.35%
	1800	Equipment		2.29	10.05	15.95	.98%	3.35%	6.70%
	2720	Plumbing		8.05	16.05	20	5.65%	6.90%	10.75%
	2770	Heating, ventilating, air conditioning		12.60	16.75	23.50	5.85%	10.55%	11.70%
	2900	Electrical		15.75	22.50	30	9.80%	11.85%	14.80%
	3100	Total: Mechanical & Electrical		51.50	62	84	25%	31.50%	32.50%
65	0010	**POST OFFICES**	S.F.	114	141	179			
	0020	Total project costs	C.F.	6.85	8.65	9.85			
	2720	Plumbing	S.F.	5.15	6.35	8	4.24%	5.30%	5.60%
	2770	Heating, ventilating, air conditioning		8	9.90	11	6.65%	7.15%	9.35%
	2900	Electrical		9.40	13.20	15.70	7.25%	9%	11%
	3100	Total: Mechanical & Electrical		27	35.50	40	16.25%	18.80%	22%
66	0010	**POWER PLANTS**	S.F.	790	1,050	1,925			
	0020	Total project costs	C.F.	21.50	47.50	101			
	2900	Electrical	S.F.	56	118	176	9.30%	12.75%	21.50%
	8100	Total: Mechanical & Electrical		139	450	1,025	32.50%	32.50%	52.50%
67	0010	**RELIGIOUS EDUCATION**	S.F.	91.50	120	147			
	0020	Total project costs	C.F.	5.10	7.30	9.10			
	2720	Plumbing	S.F.	3.83	5.45	7.70	4.40%	5.30%	7.10%
	2770	Heating, ventilating, air conditioning		9.70	10.95	15.50	10.05%	11.45%	12.35%
	2900	Electrical		7.25	10.25	13.55	7.60%	9.05%	10.35%
	3100	Total: Mechanical & Electrical		29.50	39	47.50	22%	23%	27%
69	0010	**RESEARCH Laboratories & Facilities**	S.F.	137	197	287			
	0020	Total project costs	C.F.	10.35	20	24			
	1800	Equipment	S.F.	6.10	12	29.50	.94%	4.58%	9.10%
	2720	Plumbing		13.75	17.60	28.50	6.15%	8.30%	10.80%
	2770	Heating, ventilating, air conditioning		12.35	41.50	49	7.25%	16.50%	17.50%
	2900	Electrical		16.05	26.50	43.50	9.45%	11.15%	15.40%
	3100	Total: Mechanical & Electrical		52.50	91.50	132	31.50%	37.50%	42%
70	0010	**RESTAURANTS**	S.F.	133	172	223			
	0020	Total project costs	C.F.	11.20	14.70	19.25			
	1800	Equipment	S.F.	8.60	21.50	32	6.10%	13%	15.65%
	2720	Plumbing		10.55	12.80	16.80	6.10%	8.15%	9%
	2770	Heating, ventilating, air conditioning		13.40	18.50	22.50	9.20%	12%	12.40%
	2900	Electrical		14.10	17.35	22.50	8.35%	10.55%	11.55%
	3100	Total: Mechanical & Electrical		43.50	46.50	60.50	21%	24.50%	29.50%
	9000	Per seat unit, total cost	Seat	4,875	6,500	7,700			
	9500	Total: Mechanical & Electrical	"	1,225	1,625	1,925			
72	0010	**RETAIL STORES**	S.F.	62	83.50	111			
	0020	Total project costs	C.F.	4.21	6	8.35			
	2720	Plumbing	S.F.	2.25	3.76	6.40	3.26%	4.60%	6.80%
	2770	Heating, ventilating, air conditioning		4.87	6.65	10	6.75%	8.75%	10.15%
	2900	Electrical		5.60	7.65	11.05	7.25%	9.90%	11.60%
	3100	Total: Mechanical & Electrical		14.90	19.05	26.50	17.05%	21%	23.50%

50 17 00 | S.F. Costs

			UNIT	UNIT COSTS 1/4	UNIT COSTS MEDIAN	UNIT COSTS 3/4	% OF TOTAL 1/4	% OF TOTAL MEDIAN	% OF TOTAL 3/4	
74	0010	**SCHOOLS Elementary**	S.F.	101	125	154				74
	0020	Total project costs	C.F.	6.65	8.50	11				
	0500	Masonry	S.F.	9.10	15.60	23	5.45%	10.65%	14.10%	
	1800	Equipment		2.77	4.73	8.70	1.90%	3.20%	4.70%	
	2720	Plumbing		5.85	8.25	11	5.70%	7.15%	9.35%	
	2730	Heating, ventilating, air conditioning		8.75	13.90	19.45	8.15%	10.80%	14.90%	
	2900	Electrical		9.55	12.65	16.10	8.40%	10.05%	11.70%	
	3100	Total: Mechanical & Electrical	↓	34.50	43	53	25%	27.50%	30%	
	9000	Per pupil, total cost	Ea.	11,600	17,300	38,700				
	9500	Total: Mechanical & Electrical	"	3,275	4,150	10,500				
76	0010	**SCHOOLS Junior High & Middle**	S.F.	105	128	157				76
	0020	Total project costs	C.F.	6.65	8.60	9.65				
	0500	Masonry	S.F.	13.40	17.60	20.50	8.60%	11.60%	14.35%	
	1800	Equipment		3.34	5.40	8.10	1.79%	3.09%	4.86%	
	2720	Plumbing		6.10	7.50	9.30	5.30%	6.80%	7.25%	
	2770	Heating, ventilating, air conditioning		12.15	14.80	26	8.90%	11.55%	14.20%	
	2900	Electrical		10.25	12.35	15.95	7.90%	9.35%	10.60%	
	3100	Total: Mechanical & Electrical	↓	33.50	43	53	23.50%	27%	29.50%	
	9000	Per pupil, total cost	Ea.	13,300	17,400	23,400				
78	0010	**SCHOOLS Senior High**	S.F.	108	133	167				78
	0020	Total project costs	C.F.	6.55	9.65	15.55				
	1800	Equipment	S.F.	2.86	6.75	9.50	1.88%	2.91%	4.56%	
	2720	Plumbing		6.10	9.15	16.75	5.60%	6.90%	8.30%	
	2770	Heating, ventilating, air conditioning		12.45	14.25	27	8.95%	11.60%	15%	
	2900	Electrical		11	14.25	21.50	8.65%	10.20%	12.25%	
	3100	Total: Mechanical & Electrical	↓	36	42.50	71	23.50%	26.50%	28.50%	
	9000	Per pupil, total cost	Ea.	10,300	20,900	26,100				
80	0010	**SCHOOLS Vocational**	S.F.	88	128	158				80
	0020	Total project costs	C.F.	5.45	7.85	10.85				
	0500	Masonry	S.F.	5.15	12.75	19.50	3.20%	4.61%	10.95%	
	1800	Equipment		2.75	6.85	9.55	1.24%	3.10%	4.26%	
	2720	Plumbing		5.65	8.40	12.35	5.40%	6.90%	8.55%	
	2770	Heating, ventilating, air conditioning		7.90	14.65	24.50	8.60%	11.90%	14.65%	
	2900	Electrical		9.15	12.55	17.15	8.45%	11%	13.20%	
	3100	Total: Mechanical & Electrical	↓	34.50	35	60.50	27.50%	29.50%	31%	
	9000	Per pupil, total cost	Ea.	12,300	32,800	49,000				
83	0010	**SPORTS ARENAS**	S.F.	77	103	158				83
	0020	Total project costs	C.F.	4.18	7.50	9.65				
	2720	Plumbing	S.F.	4.46	6.75	14.30	4.35%	6.35%	9.40%	
	2770	Heating, ventilating, air conditioning		9.60	11.35	15.75	8.80%	10.20%	13.55%	
	2900	Electrical		8	10.90	14.05	8.60%	9.90%	12.25%	
	3100	Total: Mechanical & Electrical	↓	19.95	35	47	21.50%	25%	27.50%	
85	0010	**SUPERMARKETS**	S.F.	71	82.50	96.50				85
	0020	Total project costs	C.F.	3.96	4.79	7.25				
	2720	Plumbing	S.F.	3.97	5	5.80	5.40%	6%	7.45%	
	2770	Heating, ventilating, air conditioning		5.85	7.75	9.45	8.60%	8.65%	9.60%	
	2900	Electrical		8.90	10.25	12.10	10.40%	12.45%	13.60%	
	3100	Total: Mechanical & Electrical	↓	23	25	32.50	20.50%	26.50%	31%	
86	0010	**SWIMMING POOLS**	S.F.	115	193	410				86
	0020	Total project costs	C.F.	9.25	11.50	12.55				
	2720	Plumbing	S.F.	10.65	12.15	16.45	4.80%	9.70%	20.50%	
	2900	Electrical		8.65	14	20.50	5.75%	6.95%	7.60%	
	3100	Total: Mechanical & Electrical	↓	21	54.50	73	11.15%	14.10%	23.50%	
87	0010	**TELEPHONE EXCHANGES**	S.F.	153	224	284				87
	0020	Total project costs	C.F.	9.50	15.25	21				
	2720	Plumbing	S.F.	6.45	9.95	14.60	4.52%	5.80%	6.90%	
	2770	Heating, ventilating, air conditioning	↓	14.95	30	37.50	11.80%	16.05%	18.40%	

	50 17 00 | S.F. Costs		UNIT	UNIT COSTS 1/4	UNIT COSTS MEDIAN	UNIT COSTS 3/4	% OF TOTAL 1/4	% OF TOTAL MEDIAN	% OF TOTAL 3/4	
87	2900	Electrical	S.F.	15.55	24.50	43.50	10.90%	14%	17.85%	87
	3100	Total: Mechanical & Electrical	↓	46	87	123	29.50%	33.50%	44.50%	
91	0010	**THEATERS**	S.F.	96	120	182				91
	0020	Total project costs	C.F.	4.44	6.55	9.65				
	2720	Plumbing	S.F.	3.20	3.47	14.20	2.92%	4.70%	6.80%	
	2770	Heating, ventilating, air conditioning		9.35	11.30	14	8%	12.25%	13.40%	
	2900	Electrical		8.40	11.35	23	8.05%	9.35%	12.25%	
	3100	Total: Mechanical & Electrical	↓	21.50	29	34.50	21.50%	25.50%	27.50%	
94	0010	**TOWN HALLS City Halls & Municipal Buildings**	S.F.	107	136	178				94
	0020	Total project costs	C.F.	9.80	11.75	16.50				
	2720	Plumbing	S.F.	4.48	8.40	15.40	4.31%	5.95%	7.95%	
	2770	Heating, ventilating, air conditioning		8.10	16.10	23.50	7.05%	9.05%	13.45%	
	2900	Electrical		10.20	14.55	19.85	8.05%	9.45%	11.65%	
	3100	Total: Mechanical & Electrical	↓	35.50	45	68.50	22%	26.50%	31%	
97	0010	**WAREHOUSES & Storage Buildings**	S.F.	40	59.50	85.50				97
	0020	Total project costs	C.F.	2.10	3.28	5.45				
	0100	Site work	S.F.	4.13	8.20	12.40	6.05%	12.95%	19.85%	
	0500	Masonry		2.41	5.70	12.30	3.73%	7.40%	12.30%	
	1800	Equipment		.64	1.39	7.80	.91%	1.82%	5.55%	
	2720	Plumbing		1.34	2.39	4.48	2.90%	4.80%	6.55%	
	2730	Heating, ventilating, air conditioning		1.53	4.30	5.75	2.41%	5%	8.90%	
	2900	Electrical		2.37	4.46	7.40	5.15%	7.20%	10.10%	
	3100	Total: Mechanical & Electrical	↓	6.60	10.15	20	12.75%	18.90%	26%	
99	0010	**WAREHOUSE & OFFICES Combination**	S.F.	49	65.50	90				99
	0020	Total project costs	C.F.	2.52	3.66	5.40				
	1800	Equipment	S.F.	.86	1.65	2.46	.52%	1.20%	2.40%	
	2720	Plumbing		1.90	3.37	4.93	3.74%	4.76%	6.30%	
	2770	Heating, ventilating, air conditioning		3	4.69	6.55	5%	5.65%	10.05%	
	2900	Electrical		3.30	4.91	7.75	5.75%	8%	10%	
	3100	Total: Mechanical & Electrical	↓	9.20	14.20	22.50	14.40%	19.95%	24.50%	

Square Foot Project Size Modifier

One factor that affects the S.F. cost of a particular building is the size. In general, for buildings built to the same specifications in the same locality, the larger building will have the lower S.F. cost. This is due mainly to the decreasing contribution of the exterior walls plus the economy of scale usually achievable in larger buildings. The Area Conversion Scale shown below will give a factor to convert costs for the typical size building to an adjusted cost for the particular project.

The Square Foot Base Size lists the median costs, most typical project size in our accumulated data, and the range in size of the projects.

The Size Factor for your project is determined by dividing your project area in S.F. by the typical project size for the particular Building Type. With this factor, enter the Area Conversion Scale at the appropriate Size Factor and determine the appropriate cost multiplier for your building size.

Example: Determine the cost per S.F. for a 100,000 S.F. Mid-rise apartment building.

$$\frac{\text{Proposed building area} = 100{,}000 \text{ S.F.}}{\text{Typical size from below} = 50{,}000 \text{ S.F.}} = 2.00$$

Enter Area Conversion scale at 2.0, intersect curve, read horizontally the appropriate cost multiplier of .94. Size adjusted cost becomes .94 x \$107.00 = \$101.00 based on national average costs.

Note: For Size Factors less than .50, the Cost Multiplier is 1.1
For Size Factors greater than 3.5, the Cost Multiplier is .90

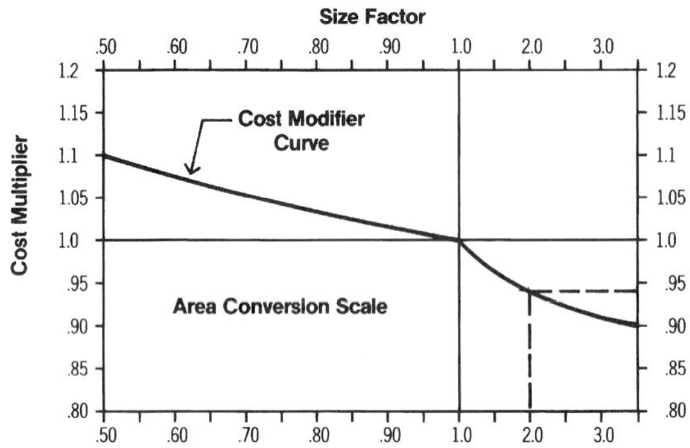

Square Foot Base Size							
Building Type	**Median Cost per S.F.**	**Typical Size Gross S.F.**	**Typical Range Gross S.F.**	**Building Type**	**Median Cost per S.F.**	**Typical Size Gross S.F.**	**Typical Range Gross S.F.**
Apartments, Low Rise	$ 84.00	21,000	9,700 - 37,200	Jails	$ 256.00	40,000	5,500 - 145,000
Apartments, Mid Rise	107.00	50,000	32,000 - 100,000	Libraries	162.00	12,000	7,000 - 31,000
Apartments, High Rise	116.00	145,000	95,000 - 600,000	Living, Assisted	137.00	32,300	23,500 - 50,300
Auditoriums	142.00	25,000	7,600 - 39,000	Medical Clinics	145.00	7,200	4,200 - 15,700
Auto Sales	105.00	20,000	10,800 - 28,600	Medical Offices	137.00	6,000	4,000 - 15,000
Banks	189.00	4,200	2,500 - 7,500	Motels	101.00	40,000	15,800 - 120,000
Churches	129.00	17,000	2,000 - 42,000	Nursing Homes	142.00	23,000	15,000 - 37,000
Clubs, Country	132.00	6,500	4,500 - 15,000	Offices, Low Rise	120.00	20,000	5,000 - 80,000
Clubs, Social	125.00	10,000	6,000 - 13,500	Offices, Mid Rise	119.00	120,000	20,000 - 300,000
Clubs, YMCA	146.00	28,300	12,800 - 39,400	Offices, High Rise	152.00	260,000	120,000 - 800,000
Colleges (Class)	151.00	50,000	15,000 - 150,000	Police Stations	189.00	10,500	4,000 - 19,000
Colleges (Science Lab)	241.00	45,600	16,600 - 80,000	Post Offices	141.00	12,400	6,800 - 30,000
College (Student Union)	178.00	33,400	16,000 - 85,000	Power Plants	1050.00	7,500	1,000 - 20,000
Community Center	134.00	9,400	5,300 - 16,700	Religious Education	120.00	9,000	6,000 - 12,000
Court Houses	184.00	32,400	17,800 - 106,000	Research	197.00	19,000	6,300 - 45,000
Dept. Stores	78.00	90,000	44,000 - 122,000	Restaurants	172.00	4,400	2,800 - 6,000
Dormitories, Low Rise	143.00	25,000	10,000 - 95,000	Retail Stores	83.50	7,200	4,000 - 17,600
Dormitories, Mid Rise	176.00	85,000	20,000 - 200,000	Schools, Elementary	125.00	41,000	24,500 - 55,000
Factories	76.00	26,400	12,900 - 50,000	Schools, Jr. High	128.00	92,000	52,000 - 119,000
Fire Stations	139.00	5,800	4,000 - 8,700	Schools, Sr. High	133.00	101,000	50,500 - 175,000
Fraternity Houses	130.00	12,500	8,200 - 14,800	Schools, Vocational	128.00	37,000	20,500 - 82,000
Funeral Homes	145.00	10,000	4,000 - 20,000	Sports Arenas	103.00	15,000	5,000 - 40,000
Garages, Commercial	93.50	9,300	5,000 - 13,600	Supermarkets	82.50	44,000	12,000 - 60,000
Garages, Municipal	119.00	8,300	4,500 - 12,600	Swimming Pools	193.00	20,000	10,000 - 32,000
Garages, Parking	50.50	163,000	76,400 - 225,300	Telephone Exchange	224.00	4,500	1,200 - 10,600
Gymnasiums	128.00	19,200	11,600 - 41,000	Theaters	120.00	10,500	8,800 - 17,500
Hospitals	230.00	55,000	27,200 - 125,000	Town Halls	136.00	10,800	4,800 - 23,400
House (Elderly)	115.00	37,000	21,000 - 66,000	Warehouses	59.50	25,000	8,000 - 72,000
Housing (Public)	106.00	36,000	14,400 - 74,400	Warehouse & Office	65.50	25,000	8,000 - 72,000
Ice Rinks	152.00	29,000	27,200 - 33,600				

Abbreviations

Abbreviation	Meaning
A	Area Square Feet; Ampere
ABS	Acrylonitrile Butadiene Stryrene; Asbestos Bonded Steel
A.C.	Alternating Current; Air-Conditioning; Asbestos Cement; Plywood Grade A & C
A.C.I.	American Concrete Institute
AD	Plywood, Grade A & D
Addit.	Additional
Adj.	Adjustable
af	Audio-frequency
A.G.A.	American Gas Association
Agg.	Aggregate
A.H.	Ampere Hours
A hr.	Ampere-hour
A.H.U.	Air Handling Unit
A.I.A.	American Institute of Architects
AIC	Ampere Interrupting Capacity
Allow.	Allowance
alt.	Altitude
Alum.	Aluminum
a.m.	Ante Meridiem
Amp.	Ampere
Anod.	Anodized
Approx.	Approximate
Apt.	Apartment
Asb.	Asbestos
A.S.B.C.	American Standard Building Code
Asbe.	Asbestos Worker
ASCE.	American Society of Civil Engineers
A.S.H.R.A.E.	American Society of Heating, Refrig. & AC Engineers
A.S.M.E.	American Society of Mechanical Engineers
A.S.T.M.	American Society for Testing and Materials
Attchmt.	Attachment
Avg., Ave.	Average
A.W.G.	American Wire Gauge
AWWA	American Water Works Assoc.
Bbl.	Barrel
B&B	Grade B and Better; Balled & Burlapped
B.&S.	Bell and Spigot
B.&W.	Black and White
b.c.c.	Body-centered Cubic
B.C.Y.	Bank Cubic Yards
BE	Bevel End
B.F.	Board Feet
Bg. cem.	Bag of Cement
BHP	Boiler Horsepower; Brake Horsepower
B.I.	Black Iron
Bit., Bitum.	Bituminous
Bit., Conc.	Bituminous Concrete
Bk.	Backed
Bkrs.	Breakers
Bldg.	Building
Blk.	Block
Bm.	Beam
Boil.	Boilermaker
B.P.M.	Blows per Minute
BR	Bedroom
Brg.	Bearing
Brhe.	Bricklayer Helper
Bric.	Bricklayer
Brk.	Brick
Brng.	Bearing
Brs.	Brass
Brz.	Bronze
Bsn.	Basin
Btr.	Better
Btu	British Thermal Unit
BTUH	BTU per Hour
B.U.R.	Built-up Roofing
BX	Interlocked Armored Cable
°C	degree centegrade
c	Conductivity, Copper Sweat
C	Hundred; Centigrade
C/C	Center to Center, Cedar on Cedar
C-C	Center to Center
Cab.	Cabinet
Cair.	Air Tool Laborer
Calc	Calculated
Cap.	Capacity
Carp.	Carpenter
C.B.	Circuit Breaker
C.C.A.	Chromate Copper Arsenate
C.C.F.	Hundred Cubic Feet
cd	Candela
cd/sf	Candela per Square Foot
CD	Grade of Plywood Face & Back
CDX	Plywood, Grade C & D, exterior glue
Cefi.	Cement Finisher
Cem.	Cement
CF	Hundred Feet
C.F.	Cubic Feet
CFM	Cubic Feet per Minute
c.g.	Center of Gravity
CHW	Chilled Water; Commercial Hot Water
C.I.	Cast Iron
C.I.P.	Cast in Place
Circ.	Circuit
C.L.	Carload Lot
Clab.	Common Laborer
Clam	Common maintenance laborer
C.L.F.	Hundred Linear Feet
CLF	Current Limiting Fuse
CLP	Cross Linked Polyethylene
cm	Centimeter
CMP	Corr. Metal Pipe
C.M.U.	Concrete Masonry Unit
CN	Change Notice
Col.	Column
CO_2	Carbon Dioxide
Comb.	Combination
Compr.	Compressor
Conc.	Concrete
Cont.	Continuous; Continued, Container
Corr.	Corrugated
Cos	Cosine
Cot	Cotangent
Cov.	Cover
C/P	Cedar on Paneling
CPA	Control Point Adjustment
Cplg.	Coupling
C.P.M.	Critical Path Method
CPVC	Chlorinated Polyvinyl Chloride
C.Pr.	Hundred Pair
CRC	Cold Rolled Channel
Creos.	Creosote
Crpt.	Carpet & Linoleum Layer
CRT	Cathode-ray Tube
CS	Carbon Steel, Constant Shear Bar Joist
Csc	Cosecant
C.S.F.	Hundred Square Feet
CSI	Construction Specifications Institute
C.T.	Current Transformer
CTS	Copper Tube Size
Cu	Copper, Cubic
Cu. Ft.	Cubic Foot
cw	Continuous Wave
C.W.	Cool White; Cold Water
Cwt.	100 Pounds
C.W.X.	Cool White Deluxe
C.Y.	Cubic Yard (27 cubic feet)
C.Y./Hr.	Cubic Yard per Hour
Cyl.	Cylinder
d	Penny (nail size)
D	Deep; Depth; Discharge
Dis., Disch.	Discharge
Db.	Decibel
Dbl.	Double
DC	Direct Current
DDC	Direct Digital Control
Demob.	Demobilization
d.f.u.	Drainage Fixture Units
D.H.	Double Hung
DHW	Domestic Hot Water
DI	Ductile Iron
Diag.	Diagonal
Diam., Dia	Diameter
Distrib.	Distribution
Div.	Division
Dk.	Deck
D.L.	Dead Load; Diesel
DLH	Deep Long Span Bar Joist
Do.	Ditto
Dp.	Depth
D.P.S.T.	Double Pole, Single Throw
Dr.	Drive
Drink.	Drinking
D.S.	Double Strength
D.S.A.	Double Strength A Grade
D.S.B.	Double Strength B Grade
Dty.	Duty
DWV	Drain Waste Vent
DX	Deluxe White, Direct Expansion
dyn	Dyne
e	Eccentricity
E	Equipment Only; East
Ea.	Each
E.B.	Encased Burial
Econ.	Economy
E.C.Y	Embankment Cubic Yards
EDP	Electronic Data Processing
EIFS	Exterior Insulation Finish System
E.D.R.	Equiv. Direct Radiation
Eq.	Equation
EL	elevation
Elec.	Electrician; Electrical
Elev.	Elevator; Elevating
EMT	Electrical Metallic Conduit; Thin Wall Conduit
Eng.	Engine, Engineered
EPDM	Ethylene Propylene Diene Monomer
EPS	Expanded Polystyrene
Eqhv.	Equip. Oper., Heavy
Eqlt.	Equip. Oper., Light
Eqmd.	Equip. Oper., Medium
Eqmm.	Equip. Oper., Master Mechanic
Eqol.	Equip. Oper., Oilers
Equip.	Equipment
ERW	Electric Resistance Welded
E.S.	Energy Saver
Est.	Estimated
esu	Electrostatic Units
E.W.	Each Way
EWT	Entering Water Temperature
Excav.	Excavation
Exp.	Expansion, Exposure
Ext.	Exterior
Extru.	Extrusion
f.	Fiber stress
F	Fahrenheit; Female; Fill
Fab.	Fabricated

Abbreviations

FBGS	Fiberglass
F.C.	Footcandles
f.c.c.	Face-centered Cubic
f'c.	Compressive Stress in Concrete; Extreme Compressive Stress
F.E.	Front End
FEP	Fluorinated Ethylene Propylene (Teflon)
F.G.	Flat Grain
F.H.A.	Federal Housing Administration
Fig.	Figure
Fin.	Finished
Fixt.	Fixture
Fl. Oz.	Fluid Ounces
Flr.	Floor
F.M.	Frequency Modulation; Factory Mutual
Fmg.	Framing
Fdn.	Foundation
Fori.	Foreman, Inside
Foro.	Foreman, Outside
Fount.	Fountain
fpm	Feet per Minute
FPT	Female Pipe Thread
Fr.	Frame
F.R.	Fire Rating
FRK	Foil Reinforced Kraft
FRP	Fiberglass Reinforced Plastic
FS	Forged Steel
FSC	Cast Body; Cast Switch Box
Ft.	Foot; Feet
Ftng.	Fitting
Ftg.	Footing
Ft lb.	Foot Pound
Furn.	Furniture
FVNR	Full Voltage Non-Reversing
FXM	Female by Male
Fy.	Minimum Yield Stress of Steel
g	Gram
G	Gauss
Ga.	Gauge
Gal., gal.	Gallon
gpm, GPM	Gallon per Minute
Galv.	Galvanized
Gen.	General
G.F.I.	Ground Fault Interrupter
Glaz.	Glazier
GPD	Gallons per Day
GPH	Gallons per Hour
GPM	Gallons per Minute
GR	Grade
Gran.	Granular
Grnd.	Ground
H	High Henry
H.C.	High Capacity
H.D.	Heavy Duty; High Density
H.D.O.	High Density Overlaid
H.D.P.E.	high density polyethelene
Hdr.	Header
Hdwe.	Hardware
Help.	Helper Average
HEPA	High Efficiency Particulate Air Filter
Hg	Mercury
HIC	High Interrupting Capacity
HM	Hollow Metal
HMWPE	high molecular weight polyethylene
H.O.	High Output
Horiz.	Horizontal
H.P.	Horsepower; High Pressure
H.P.F.	High Power Factor
Hr.	Hour
Hrs./Day	Hours per Day
HSC	High Short Circuit
Ht.	Height
Htg.	Heating
Htrs.	Heaters
HVAC	Heating, Ventilation & Air-Conditioning
Hvy.	Heavy
HW	Hot Water
Hyd.;Hydr.	Hydraulic
Hz.	Hertz (cycles)
I.	Moment of Inertia
IBC	International Building Code
I.C.	Interrupting Capacity
ID	Inside Diameter
I.D.	Inside Dimension; Identification
I.F.	Inside Frosted
I.M.C.	Intermediate Metal Conduit
In.	Inch
Incan.	Incandescent
Incl.	Included; Including
Int.	Interior
Inst.	Installation
Insul.	Insulation/Insulated
I.P.	Iron Pipe
I.P.S.	Iron Pipe Size
I.P.T.	Iron Pipe Threaded
I.W.	Indirect Waste
J	Joule
J.I.C.	Joint Industrial Council
K	Thousand; Thousand Pounds; Heavy Wall Copper Tubing, Kelvin
K.A.H.	Thousand Amp. Hours
KCMIL	Thousand Circular Mils
KD	Knock Down
K.D.A.T.	Kiln Dried After Treatment
kg	Kilogram
kG	Kilogauss
kgf	Kilogram Force
kHz	Kilohertz
Kip.	1000 Pounds
KJ	Kiljoule
K.L.	Effective Length Factor
K.L.F.	Kips per Linear Foot
Km	Kilometer
K.S.F.	Kips per Square Foot
K.S.I.	Kips per Square Inch
kV	Kilovolt
kVA	Kilovolt Ampere
K.V.A.R.	Kilovar (Reactance)
KW	Kilowatt
KWh	Kilowatt-hour
L	Labor Only; Length; Long; Medium Wall Copper Tubing
Lab.	Labor
lat	Latitude
Lath.	Lather
Lav.	Lavatory
lb.; #	Pound
L.B.	Load Bearing; L Conduit Body
L. & E.	Labor & Equipment
lb./hr.	Pounds per Hour
lb./L.F.	Pounds per Linear Foot
lbf/sq.in.	Pound-force per Square Inch
L.C.L.	Less than Carload Lot
L.C.Y.	Loose Cubic Yard
Ld.	Load
LE	Lead Equivalent
LED	Light Emitting Diode
L.F.	Linear Foot
L.F. Nose	Linear Foot of Stair Nosing
L.F. Rsr	Linear Foot of Stair Riser
Lg.	Long; Length; Large
L & H	Light and Heat
LH	Long Span Bar Joist
L.H.	Labor Hours
L.L.	Live Load
L.L.D.	Lamp Lumen Depreciation
lm	Lumen
lm/sf	Lumen per Square Foot
lm/W	Lumen per Watt
L.O.A.	Length Over All
log	Logarithm
L-O-L	Lateralolet
long.	longitude
L.P.	Liquefied Petroleum; Low Pressure
L.P.F.	Low Power Factor
LR	Long Radius
L.S.	Lump Sum
Lt.	Light
Lt. Ga.	Light Gauge
L.T.L.	Less than Truckload Lot
Lt. Wt.	Lightweight
L.V.	Low Voltage
M	Thousand; Material; Male; Light Wall Copper Tubing
M^2CA	Meters Squared Contact Area
m/hr.; M.H.	Man-hour
mA	Milliampere
Mach.	Machine
Mag. Str.	Magnetic Starter
Maint.	Maintenance
Marb.	Marble Setter
Mat; Mat'l.	Material
Max.	Maximum
MBF	Thousand Board Feet
MBH	Thousand BTU's per hr.
MC	Metal Clad Cable
M.C.F.	Thousand Cubic Feet
M.C.F.M.	Thousand Cubic Feet per Minute
M.C.M.	Thousand Circular Mils
M.C.P.	Motor Circuit Protector
MD	Medium Duty
M.D.O.	Medium Density Overlaid
Med.	Medium
MF	Thousand Feet
M.F.B.M.	Thousand Feet Board Measure
Mfg.	Manufacturing
Mfrs.	Manufacturers
mg	Milligram
MGD	Million Gallons per Day
MGPH	Thousand Gallons per Hour
MH, M.H.	Manhole; Metal Halide; Man-Hour
MHz	Megahertz
Mi.	Mile
MI	Malleable Iron; Mineral Insulated
mm	Millimeter
Mill.	Millwright
Min., min.	Minimum, minute
Misc.	Miscellaneous
ml	Milliliter, Mainline
M.L.F.	Thousand Linear Feet
Mo.	Month
Mobil.	Mobilization
Mog.	Mogul Base
MPH	Miles per Hour
MPT	Male Pipe Thread

Abbreviations

Abbreviation	Meaning
MRT	Mile Round Trip
ms	Millisecond
M.S.F.	Thousand Square Feet
Mstz.	Mosaic & Terrazzo Worker
M.S.Y.	Thousand Square Yards
Mtd., mtd.	Mounted
Mthe.	Mosaic & Terrazzo Helper
Mtng.	Mounting
Mult.	Multi; Multiply
M.V.A.	Million Volt Amperes
M.V.A.R.	Million Volt Amperes Reactance
MV	Megavolt
MW	Megawatt
MXM	Male by Male
MYD	Thousand Yards
N	Natural; North
nA	Nanoampere
NA	Not Available; Not Applicable
N.B.C.	National Building Code
NC	Normally Closed
N.E.M.A.	National Electrical Manufacturers Assoc.
NEHB	Bolted Circuit Breaker to 600V.
N.L.B.	Non-Load-Bearing
NM	Non-Metallic Cable
nm	Nanometer
No.	Number
NO	Normally Open
N.O.C.	Not Otherwise Classified
Nose.	Nosing
N.P.T.	National Pipe Thread
NQOD	Combination Plug-on/Bolt on Circuit Breaker to 240V.
N.R.C.	Noise Reduction Coefficient/ Nuclear Regulator Commission
N.R.S.	Non Rising Stem
ns	Nanosecond
nW	Nanowatt
OB	Opposing Blade
OC	On Center
OD	Outside Diameter
O.D.	Outside Dimension
ODS	Overhead Distribution System
O.G.	Ogee
O.H.	Overhead
O&P	Overhead and Profit
Oper.	Operator
Opng.	Opening
Orna.	Ornamental
OSB	Oriented Strand Board
O.S.&Y.	Outside Screw and Yoke
Ovhd.	Overhead
OWG	Oil, Water or Gas
Oz.	Ounce
P.	Pole; Applied Load; Projection
p.	Page
Pape.	Paperhanger
P.A.P.R.	Powered Air Purifying Respirator
PAR	Parabolic Reflector
Pc., Pcs.	Piece, Pieces
P.C.	Portland Cement; Power Connector
P.C.F.	Pounds per Cubic Foot
P.C.M.	Phase Contrast Microscopy
P.E.	Professional Engineer; Porcelain Enamel; Polyethylene; Plain End
Perf.	Perforated
PEX	Cross linked polyethylene
Ph.	Phase
P.I.	Pressure Injected
Pile.	Pile Driver
Pkg.	Package
Pl.	Plate
Plah.	Plasterer Helper
Plas.	Plasterer
Pluh.	Plumbers Helper
Plum.	Plumber
Ply.	Plywood
p.m.	Post Meridiem
Pntd.	Painted
Pord.	Painter, Ordinary
pp	Pages
PP, PPL	Polypropylene
P.P.M.	Parts per Million
Pr.	Pair
P.E.S.B.	Pre-engineered Steel Building
Prefab.	Prefabricated
Prefin.	Prefinished
Prop.	Propelled
PSF, psf	Pounds per Square Foot
PSI, psi	Pounds per Square Inch
PSIG	Pounds per Square Inch Gauge
PSP	Plastic Sewer Pipe
Pspr.	Painter, Spray
Psst.	Painter, Structural Steel
P.T.	Potential Transformer
P. & T.	Pressure & Temperature
Ptd.	Painted
Ptns.	Partitions
Pu	Ultimate Load
PVC	Polyvinyl Chloride
Pvmt.	Pavement
Pwr.	Power
Q	Quantity Heat Flow
Qt.	Quart
Quan., Qty.	Quantity
Q.C.	Quick Coupling
r	Radius of Gyration
R	Resistance
R.C.P.	Reinforced Concrete Pipe
Rect.	Rectangle
Reg.	Regular
Reinf.	Reinforced
Req'd.	Required
Res.	Resistant
Resi.	Residential
Rgh.	Rough
RGS	Rigid Galvanized Steel
R.H.W.	Rubber, Heat & Water Resistant; Residential Hot Water
rms	Root Mean Square
Rnd.	Round
Rodm.	Rodman
Rofc.	Roofer, Composition
Rofp.	Roofer, Precast
Rohe.	Roofer Helpers (Composition)
Rots.	Roofer, Tile & Slate
R.O.W.	Right of Way
RPM	Revolutions per Minute
R.S.	Rapid Start
Rsr	Riser
RT	Round Trip
S.	Suction; Single Entrance; South
SC	Screw Cover
SCFM	Standard Cubic Feet per Minute
Scaf.	Scaffold
Sch., Sched.	Schedule
S.C.R.	Modular Brick
S.D.	Sound Deadening
S.D.R.	Standard Dimension Ratio
S.E.	Surfaced Edge
Sel.	Select
S.E.R., S.E.U.	Service Entrance Cable
S.F.	Square Foot
S.F.C.A.	Square Foot Contact Area
S.F. Flr.	Square Foot of Floor
S.F.G.	Square Foot of Ground
S.F. Hor.	Square Foot Horizontal
S.F.R.	Square Feet of Radiation
S.F. Shlf.	Square Foot of Shelf
S4S	Surface 4 Sides
Shee.	Sheet Metal Worker
Sin.	Sine
Skwk.	Skilled Worker
SL	Saran Lined
S.L.	Slimline
Sldr.	Solder
SLH	Super Long Span Bar Joist
S.N.	Solid Neutral
S-O-L	Socketolet
sp	Standpipe
S.P.	Static Pressure; Single Pole; Self-Propelled
Spri.	Sprinkler Installer
spwg	Static Pressure Water Gauge
S.P.D.T.	Single Pole, Double Throw
SPF	Spruce Pine Fir
S.P.S.T.	Single Pole, Single Throw
SPT	Standard Pipe Thread
Sq.	Square; 100 Square Feet
Sq. Hd.	Square Head
Sq. In.	Square Inch
S.S.	Single Strength; Stainless Steel
S.S.B.	Single Strength B Grade
sst, ss	Stainless Steel
Sswk.	Structural Steel Worker
Sswl.	Structural Steel Welder
St.;Stl.	Steel
S.T.C.	Sound Transmission Coefficient
Std.	Standard
Stg.	Staging
STK	Select Tight Knot
STP	Standard Temperature & Pressure
Stpi.	Steamfitter, Pipefitter
Str.	Strength; Starter; Straight
Strd.	Stranded
Struct.	Structural
Sty.	Story
Subj.	Subject
Subs.	Subcontractors
Surf.	Surface
Sw.	Switch
Swbd.	Switchboard
S.Y.	Square Yard
Syn.	Synthetic
S.Y.P.	Southern Yellow Pine
Sys.	System
t.	Thickness
T	Temperature; Ton
Tan	Tangent
T.C.	Terra Cotta
T & C	Threaded and Coupled
T.D.	Temperature Difference
Tdd	Telecommunications Device for the Deaf
T.E.M.	Transmission Electron Microscopy
TFE	Tetrafluoroethylene (Teflon)
T. & G.	Tongue & Groove; Tar & Gravel
Th., Thk.	Thick
Thn.	Thin
Thrded	Threaded
Tilf.	Tile Layer, Floor
Tilh.	Tile Layer, Helper
THHN	Nylon Jacketed Wire
THW.	Insulated Strand Wire

Abbreviations

THWN	Nylon Jacketed Wire
T.L.	Truckload
T.M.	Track Mounted
Tot.	Total
T-O-L	Threadolet
T.S.	Trigger Start
Tr.	Trade
Transf.	Transformer
Trhv.	Truck Driver, Heavy
Trlr	Trailer
Trlt.	Truck Driver, Light
TTY	Teletypewriter
TV	Television
T.W.	Thermoplastic Water Resistant Wire
UCI	Uniform Construction Index
UF	Underground Feeder
UGND	Underground Feeder
U.H.F.	Ultra High Frequency
U.I.	United Inch
U.L.	Underwriters Laboratory
Uld.	unloading
Unfin.	Unfinished
URD	Underground Residential Distribution
US	United States
USP	United States Primed
UTP	Unshielded Twisted Pair
V	Volt
V.A.	Volt Amperes
V.C.T.	Vinyl Composition Tile
VAV	Variable Air Volume
VC	Veneer Core
Vent.	Ventilation
Vert.	Vertical
V.F.	Vinyl Faced
V.G.	Vertical Grain
V.H.F.	Very High Frequency
VHO	Very High Output
Vib.	Vibrating
V.L.F.	Vertical Linear Foot
Vol.	Volume
VRP	Vinyl Reinforced Polyester
W	Wire; Watt; Wide; West
w/	With
W.C.	Water Column; Water Closet
W.F.	Wide Flange
W.G.	Water Gauge
Wldg.	Welding
W. Mile	Wire Mile
W-O-L	Weldolet
W.R.	Water Resistant
Wrck.	Wrecker
W.S.P.	Water, Steam, Petroleum
WT., Wt.	Weight
WWF	Welded Wire Fabric
XFER	Transfer
XFMR	Transformer
XHD	Extra Heavy Duty
XHHW,	Cross-Linked Polyethylene Wire
XLPE	Insulation
XLP	Cross-linked Polyethylene
Y	Wye
yd	Yard
yr	Year
Δ	Delta
%	Percent
~	Approximately
Ø	Phase; diameter
@	At
#	Pound; Number
<	Less Than
>	Greater Than
Z	zone

A

B

F

G

H

I

M

Q

R

S

U

X

Y

Z

Notes

Notes

Notes

Notes

Division Notes

	CREW	DAILY OUTPUT	LABOR-HOURS	UNIT	2011 BARE COSTS MAT.	LABOR	EQUIP.	TOTAL	TOTAL INCL O&P

Division Notes

	CREW	DAILY OUTPUT	LABOR-HOURS	UNIT	2011 BARE COSTS MAT.	LABOR	EQUIP.	TOTAL	TOTAL INCL O&P

Division Notes

	CREW	DAILY OUTPUT	LABOR-HOURS	UNIT	2011 BARE COSTS MAT.	LABOR	EQUIP.	TOTAL	TOTAL INCL O&P

Division Notes

	CREW	DAILY OUTPUT	LABOR-HOURS	UNIT	2011 BARE COSTS MAT.	LABOR	EQUIP.	TOTAL	TOTAL INCL O&P

Reed Construction Data/RSMeans— a tradition of excellence in construction cost information and services since 1942

Table of Contents

For more information visit the RSMeans website at www.rsmeans.com

Book Selection Guide The following table provides definitive information on the content of each cost data publication. The number of lines of data provided in each unit price or assemblies division, as well as the number of reference tables and crews, is listed for each book. The presence of other elements such as equipment rental costs, historical cost indexes, city cost indexes, square foot models, or cross-referenced indexes is also indicated. You can use the table to help select the RSMeans book that has the quantity and type of information you most need in your work.

Unit Cost Divisions	Building Construction	Mechanical	Electrical	Commercial Renovation	Square Foot	Site Work Landsc.	Green Building	Interior	Concrete Masonry	Open Shop	Heavy Construction	Light Commercial	Facilities Construction	Plumbing	Residential
1	571	392	401	492		509	39	317	455	570	499	235	985	402	172
2	746	266	85	685		967	168	364	221	743	706	440	1163	290	242
3	1526	216	206	887		1329	859	304	1833	1526	1468	375	1552	183	359
4	898	24	0	698		700	182	573	1133	874	593	490	1145	0	405
5	1916	263	167	1024		870	1803	1025	833	1907	1117	895	1910	339	727
6	1855	67	69	1442		450	537	1161	300	1834	381	1544	1466	22	2056
7	1437	209	125	1465		534	766	544	485	1435	26	1143	1497	218	896
8	2093	61	10	2233		297	1146	1731	676	2018	0	1622	2450	0	1397
9	1812	72	26	1627		278	441	1910	386	1774	15	1542	2064	54	1354
10	988	17	10	608		221	22	803	159	988	29	490	1074	232	221
11	1012	208	165	484		126	54	862	28	997	0	231	1026	169	110
12	562	0	2	318		248	150	1706	12	551	19	364	1756	23	317
13	738	119	111	247		349	128	255	69	737	261	80	748	74	79
14	280	36	0	229		28	0	259	0	279	0	12	299	21	6
21	78	0	16	30		0	0	250	0	78	0	60	368	373	0
22	1141	7183	155	1105		1510	1015	621	20	1103	1676	834	7114	8929	632
23	1186	7077	608	799		157	894	677	38	1106	110	661	5154	1872	416
26	1249	496	9916	1006		746	555	1068	55	1237	523	1095	9852	441	560
27	67	0	261	34		9	0	59	0	67	35	48	263	0	4
28	96	58	124	71		0	21	78	0	99	0	40	139	44	25
31	1477	735	608	798		2794	38	7	1208	1421	2865	600	1540	660	606
32	759	54	0	840		4195	343	359	274	730	1710	380	1611	166	429
33	502	1010	465	201		1950	41	0	225	501	1873	113	1536	1181	111
34	109	0	20	4		144	0	0	31	64	122	0	129	0	0
35	18	0	0	0		166	0	0	0	18	281	0	83	0	0
41	52	0	0	24		7	0	22	0	52	30	0	59	14	0
44	75	83	0	0		0	0	0	0	0	0	0	75	75	0
46	23	6	0	0		112	99	0		23	112	0	23	23	0
48	0	0	0	0		0	19	0		0	0	0	0	0	0
Totals	23266	18652	13550	17351		18696	9320	14955	8441	22732	14451	13294	47081	15805	11094

Assem Div	Building Construction	Mechanical	Elec-trical	Commercial Renovation	Square Foot	Site Work Landscape	Assemblies	Green Building	Interior	Concrete Masonry	Heavy Construction	Light Commercial	Facilities Construction	Plumbing	Asm Div	Residential
A		15	0	188	150	575	596	0	0	534	569	154	24	0	1	374
B		0	0	825	2465	0	5562	56	329	1911	368	2025	143	0	2	217
C		0	0	647	866	0	1236	0	1561	146	0	766	248	0	3	588
D		1064	811	712	1793	0	2423	233	765	0	0	1270	1026	1083	4	853
E		0	0	86	260	0	297	0	5	0	0	257	5	0	5	392
F		0	0	0	123	0	124	0	0	0	0	123	3	0	6	358
G		527	294	317	202	2673	668	0	0	535	1189	200	293	685	7	300
															8	760
															9	80
															10	0
															11	0
															12	0
Totals		1606	1105	2775	5859	3248	10906	289	2660	3126	2126	4795	1742	1768		3922

Reference Section	Building Construction Costs	Mechanical	Electrical	Commercial Renovation	Square Foot	Site Work Landscape	Assem.	Green Building	Interior	Concrete Masonry	Open Shop	Heavy Construction	Light Commercial	Facilities Construction	Plumbing	Resi.
Reference Tables	yes	yes	yes	yes	no	yes	yes	yes	yes	yes	yes	yes	yes	yes	yes	yes
Models					107								46			28
Crews	505	505	505	483		505		505	505	505	481	505	481	483	505	481
Equipment Rental Costs	yes	yes	yes	yes		yes		yes	yes	yes	yes	yes	yes	yes	yes	yes
Historical Cost Indexes	yes	yes	yes	yes	yes	yes	yes	yes	yes	yes	yes	yes	yes	yes	yes	no
City Cost Indexes	yes	yes	yes	yes	yes	yes	yes	yes	yes	yes	yes	yes	yes	yes	yes	yes

For more information visit the RSMeans website at www.rsmeans.com

RSMeans Building Construction Cost Data 2011

Offers you unchallenged unit price reliability in an easy-to-use format. Whether used for verifying complete, finished estimates or for periodic checks, it supplies more cost facts better and faster than any comparable source. Over 21,000 unit prices have been updated for 2011. The City Cost Indexes and Location Factors cover over 930 areas, for indexing to any project location in North America. Order and get *RSMeans Quarterly Update Service* FREE. You'll have year-long access to the RSMeans Estimating **HOTLINE** FREE with your subscription. Expert assistance when using RSMeans data is just a phone call away.

$174.95 per copy
Available Sept. 2010
Catalog no. 60011

Unit prices organized according to the latest MasterFormat!

RSMeans Green Building Cost Data 2011

Estimate, plan, and budget the costs of green building for both new commercial construction and renovation work with this first edition of RSMeans Green Building Cost Data. More than 9,000 unit costs for a wide array of green building products plus assemblies costs. Easily identified cross references to LEED and Green Globes building rating systems criteria.

NEW!!!!

$144.95 per copy
Available Nov. 2010
Catalog no. 60551

RSMeans Mechanical Cost Data 2011

- **HVAC**
- **Controls**

Total unit and systems price guidance for mechanical construction. . . materials, parts, fittings, and complete labor cost information. Includes prices for piping, heating, air conditioning, ventilation, and all related construction.

Plus new 2011 unit costs for:

- Thousands of installed HVAC/controls assemblies components
- "On-site" Location Factors for over 930 cities and towns in the U.S. and Canada
- Crews, labor, and equipment

$174.95 per copy
Available Oct. 2010
Catalog no. 60021

Unit prices organized according to the latest MasterFormat!

RSMeans Facilities Construction Cost Data 2011

For the maintenance and construction of commercial, industrial, municipal, and institutional properties. Costs are shown for new and remodeling construction and are broken down into materials, labor, equipment, and overhead and profit. Special emphasis is given to sections on mechanical, electrical, furnishings, site work, building maintenance, finish work, and demolition.

More than 45,000 unit costs, plus assemblies costs and a comprehensive Reference Section are included.

$439.95 per copy
Available Nov. 2010
Catalog no. 60201

For more information
visit the RSMeans website
at www.rsmeans.com

Annual Cost Guides

RSMeans Square Foot Costs 2011

Accurate and Easy To Use!

- **Updated price information** based on nationwide figures from suppliers, estimators, labor experts, and contractors
- "How-to-Use" sections, with **clear examples** of commercial, residential, industrial, and institutional structures
- Realistic graphics, offering true-to-life illustrations of building projects
- Extensive information on using square foot cost data, including sample estimates and alternate pricing methods

$184.95 per copy
Available Oct. 2010
Catalog no. 60051

RSMeans Commercial Renovation Cost Data 2011

Commercial/Multifamily Residential

Use this valuable tool to estimate commercial and multifamily residential renovation and remodeling.

Includes: New costs for hundreds of unique methods, materials, and conditions that only come up in repair and remodeling, PLUS:

- Unit costs for more than 15,000 construction components
- Installed costs for more than 90 assemblies
- More than 930 "on-site" localization factors for the U.S. and Canada

Unit prices organized according to the latest MasterFormat!

$144.95 per copy
Available Oct. 2010
Catalog no. 60041

RSMeans Electrical Cost Data 2011

Pricing information for every part of electrical cost planning. More than 13,000 unit and systems costs with design tables; clear specifications and drawings; engineering guides; illustrated estimating procedures; complete labor-hour and materials costs for better scheduling and procurement; and the latest electrical products and construction methods.

- A variety of special electrical systems including cathodic protection
- Costs for maintenance, demolition, HVAC/mechanical, specialties, equipment, and more

$174.95 per copy
Available Oct. 2010
Catalog no. 60031

Unit prices organized according to the latest MasterFormat!

RSMeans Electrical Change Order Cost Data 2011

RSMeans Electrical Change Order Cost Data provides you with electrical unit prices exclusively for pricing change orders—based on the recent, direct experience of contractors and suppliers. Analyze and check your own change order estimates against the experience others have had doing the same work. It also covers productivity analysis and change order cost justifications. With useful information for calculating the effects of change orders and dealing with their administration.

$169.95 per copy
Available Dec. 2010
Catalog no. 60231

Annual Cost Guides

For more information
visit the RSMeans website
at www.rsmeans.com

RSMeans Assemblies Cost Data 2011

RSMeans Assemblies Cost Data takes the guesswork out of preliminary or conceptual estimates. Now you don't have to try to calculate the assembled cost by working up individual component costs. We've done all the work for you.

Presents detailed illustrations, descriptions, specifications, and costs for every conceivable building assembly—240 types in all—arranged in the easy-to-use UNIFORMAT II system. Each illustrated "assembled" cost includes a complete grouping of materials and associated installation costs, including the installing contractor's overhead and profit.

$285.95 per copy
Available Sept. 2010
Catalog no. 60061

RSMeans Open Shop Building Construction Cost Data 2011

The latest costs for accurate budgeting and estimating of new commercial and residential construction. . . renovation work. . . change orders. . . cost engineering.

RSMeans Open Shop "BCCD" will assist you to:

- Develop benchmark prices for change orders
- Plug gaps in preliminary estimates and budgets
- Estimate complex projects
- Substantiate invoices on contracts
- Price ADA-related renovations

Unit prices organized according to the latest MasterFormat!

$134.95 per copy
Available Dec. 2010
Catalog no. 60151

RSMeans Residential Cost Data 2011

Contains square foot costs for 30 basic home models with the look of today, plus hundreds of custom additions and modifications you can quote right off the page. With costs for the 100 residential systems you're most likely to use in the year ahead. Complete with blank estimating forms, sample estimates, and step-by-step instructions.

Now contains line items for cultured stone and brick, PVC trim lumber, and TPO roofing.

$109.95 per copy
Available Oct. 2010
Catalog no. 60171

Unit prices organized according to the latest MasterFormat!

RSMeans Site Work & Landscape Cost Data 2011

Includes unit and assemblies costs for earthwork, sewerage, piped utilities, site improvements, drainage, paving, trees & shrubs, street openings/repairs, underground tanks, and more. Contains 57 tables of assemblies costs for accurate conceptual estimates.

Includes:

- Estimating for infrastructure improvements
- Environmentally-oriented construction
- ADA-mandated handicapped access
- Hazardous waste line items

$169.95 per copy
Available Dec. 2010
Catalog no. 60281

RSMeans Facilities Maintenance & Repair Cost Data 2011

RSMeans Facilities Maintenance & Repair Cost Data gives you a complete system to manage and plan your facility repair and maintenance costs and budget efficiently. Guidelines for auditing a facility and developing an annual maintenance plan. Budgeting is included, along with reference tables on cost and management, and information on frequency and productivity of maintenance operations.

The only nationally recognized source of maintenance and repair costs. Developed in cooperation with the Civil Engineering Research Laboratory (CERL) of the Army Corps of Engineers.

$389.95 per copy
Available Nov. 2010
Catalog no. 60301

RSMeans Light Commercial Cost Data 2011

Specifically addresses the light commercial market, which is a specialized niche in the construction industry. Aids you, the owner/designer/contractor, in preparing all types of estimates—from budgets to detailed bids. Includes new advances in methods and materials.

Assemblies Section allows you to evaluate alternatives in the early stages of design/planning.

Over 11,000 unit costs ensure that you have the prices you need. . . when you need them.

Unit prices organized according to the latest MasterFormat!

$119.95 per copy
Available Nov. 2010
Catalog no. 60181

For more information
visit the RSMeans website
at www.rsmeans.com

Annual Cost Guides

RSMeans Concrete & Masonry Cost Data 2011

Provides you with cost facts for virtually all concrete/masonry estimating needs, from complicated formwork to various sizes and face finishes of brick and block—all in great detail. The comprehensive Unit Price Section contains more than 8,000 selected entries. Also contains an Assemblies [Cost] Section, and a detailed Reference Section that supplements the cost data.

Unit prices organized according to the latest MasterFormat!

$159.95 per copy
Available Dec. 2010
Catalog no. 60111

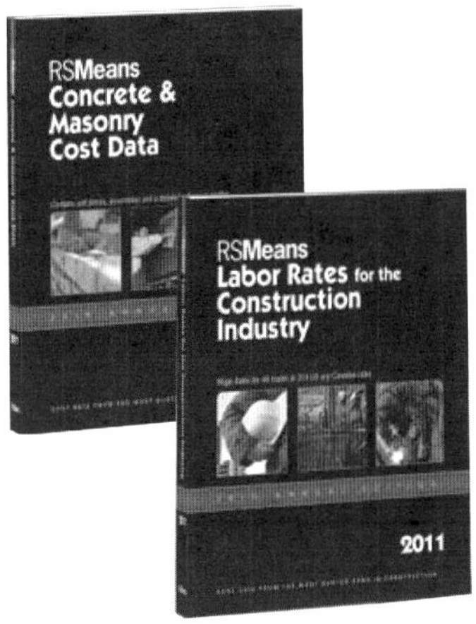

RSMeans Labor Rates for the Construction Industry 2011

Complete information for estimating labor costs, making comparisons, and negotiating wage rates by trade for more than 300 U.S. and Canadian cities. With 46 construction trades listed by local union number in each city, and historical wage rates included for comparison. Each city chart lists the county and is alphabetically arranged with handy visual flip tabs for quick reference.

$389.95 per copy
Available Dec. 2010
Catalog no. 60121

RSMeans Construction Cost Indexes 2011

What materials and labor costs will change unexpectedly this year? By how much?

- Breakdowns for 318 major cities
- National averages for 30 key cities
- Expanded five major city indexes
- Historical construction cost indexes

$362.00 per year (subscription)
Catalog no. 50141

$90.50 individual quarters
Catalog no. 60141 A,B,C,D

RSMeans Interior Cost Data 2011

Provides you with prices and guidance needed to make accurate interior work estimates. Contains costs on materials, equipment, hardware, custom installations, furnishings, and labor costs. . . for new and remodel commercial and industrial interior construction, including updated information on office furnishings, and reference information.

Unit prices organized according to the latest MasterFormat!

$179.95 per copy
Available Nov. 2010
Catalog no. 60091

RSMeans Heavy Construction Cost Data 2011

A comprehensive guide to heavy construction costs. Includes costs for highly specialized projects such as tunnels, dams, highways, airports, and waterways. Information on labor rates, equipment, and materials costs is included. Features unit price costs, systems costs, and numerous reference tables for costs and design.

$174.95 per copy
Available Dec. 2010
Catalog no. 60161

Unit prices organized according to the latest MasterFormat!

RSMeans Plumbing Cost Data 2011

Comprehensive unit prices and assemblies for plumbing, irrigation systems, commercial and residential fire protection, point-of-use water heaters, and the latest approved materials. This publication and its companion, *RSMeans Mechanical Cost Data*, provide full-range cost estimating coverage for all the mechanical trades.

Now contains updated costs for potable water and radiant heat systems and high efficiency fixtures.

$174.95 per copy
Available Oct. 2010
Catalog no. 60211

Value Engineering: Practical Applications

For Design, Construction, Maintenance & Operations

by Alphonse Dell'Isola, PE

A tool for immediate application—for engineers, architects, facility managers, owners, and contractors. Includes making the case for VE—the management briefing; integrating VE into planning, budgeting, and design; conducting life cycle costing; using VE methodology in design review and consultant selection; case studies; VE workbook; and a life cycle costing program on disk.

$79.95 per copy
Over 450 pages, illustrated, softcover
Catalog no. 67319A

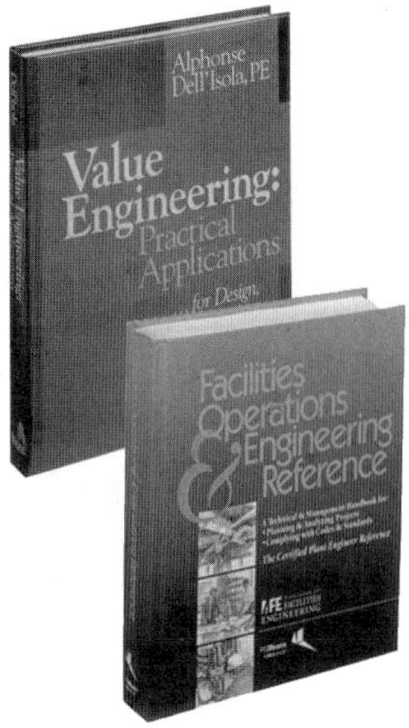

Facilities Operations & Engineering Reference

by the Association for Facilities Engineering and RSMeans

An all-in-one technical reference for planning and managing facility projects and solving day-to-day operations problems. Selected as the official certified plant engineer reference, this handbook covers financial analysis, maintenance, HVAC and energy efficiency, and more.

$54.98 per copy
Over 700 pages, illustrated, hardcover
Catalog no. 67318

The Building Professional's Guide to Contract Documents

3rd Edition

by Waller S. Poage, AIA, CSI, CVS

A comprehensive reference for owners, design professionals, contractors, and students

- Structure your documents for maximum efficiency.
- Effectively communicate construction requirements.
- Understand the roles and responsibilities of construction professionals.
- Improve methods of project delivery.

$32.48 per copy, 400 pages
Diagrams and construction forms, hardcover
Catalog no. 67261A

Complete Book of Framing

by Scot Simpson

This straightforward, easy-to-learn method will help framers, carpenters, and handy homeowners build their skills in rough carpentry and framing. Shows how to frame all the parts of a house: floors, walls, roofs, door & window openings, and stairs—with hundreds of color photographs and drawings that show every detail.

The book gives beginners all the basics they need to go from zero framing knowledge to a journeyman level.

$29.95 per copy
352 pages, softcover
Catalog no. 67353

Cost Planning & Estimating for Facilities Maintenance

In this unique book, a team of facilities management authorities shares their expertise on:

- Evaluating and budgeting maintenance operations
- Maintaining and repairing key building components
- Applying *RSMeans Facilities Maintenance & Repair Cost Data* to your estimating

Covers special maintenance requirements of the ten major building types.

$89.95 per copy
Over 475 pages, hardcover
Catalog no. 67314

Life Cycle Costing for Facilities

by Alphonse Dell'Isola and Dr. Steven Kirk

Guidance for achieving higher quality design and construction projects at lower costs! Cost-cutting efforts often sacrifice quality to yield the cheapest product. Life cycle costing enables building designers and owners to achieve both. The authors of this book show how LCC can work for a variety of projects — from roads to HVAC upgrades to different types of buildings.

$99.95 per copy
450 pages, hardcover
Catalog no. 67341

Interior Home Improvement Costs 9th Edition

Updated estimates for the most popular remodeling and repair projects—from small, do-it-yourself jobs to major renovations and new construction. Includes: kitchens & baths; new living space from your attic, basement, or garage; new floors, paint, and wallpaper; tearing out or building new walls; closets, stairs, and fireplaces; new energy-saving improvements, home theaters, and more!

$24.95 per copy
250 pages, illustrated, softcover
Catalog no. 67308E

Exterior Home Improvement Costs 9th Edition

Updated estimates for the most popular remodeling and repair projects—from small, do-it-yourself jobs, to major renovations and new construction. Includes: curb appeal projects—landscaping, patios, porches, driveways, and walkways; new windows and doors; decks, greenhouses, and sunrooms; room additions and garages; roofing, siding, and painting; "green" improvements to save energy & water.

$24.95 per copy
Over 275 pages, illustrated, softcover
Catalog no. 67309E

Builder's Essentials: Plan Reading & Material Takeoff

For Residential and Light Commercial Construction

by Wayne J. DelPico

A valuable tool for understanding plans and specs, and accurately calculating material quantities. Step-by-step instructions and takeoff procedures based on a full set of working drawings.

$35.95 per copy
Over 420 pages, softcover
Catalog no. 67307

Means Unit Price Estimating Methods

New 4th Edition

This new edition includes up-to-date cost data and estimating examples, updated to reflect changes to the CSI numbering system and new features of RSMeans cost data. It describes the most productive, universally accepted ways to estimate, and uses checklists and forms to illustrate shortcuts and timesavers. A model estimate demonstrates procedures. A new chapter explores computer estimating alternatives.

$59.95 per copy
Over 350 pages, illustrated, softcover
Catalog no. 67303B

Concrete Repair and Maintenance Illustrated

by Peter Emmons

Hundreds of illustrations show users how to analyze, repair, clean, and maintain concrete structures for optimal performance and cost effectiveness. From parking garages to roads and bridges to structural concrete, this comprehensive book describes the causes, effects, and remedies for concrete wear and failure. Invaluable for planning jobs, selecting materials, and training employees, this book is a must-have for concrete specialists, general contractors, facility managers, civil and structural engineers, and architects.

$34.98 per copy
300 pages, illustrated, softcover
Catalog no. 67146

Total Productive Facilities Management

by Richard W. Sievert, Jr.

Today, facilities are viewed as strategic resources. . . elevating the facility manager to the role of asset manager supporting the organization's overall business goals. Now, Richard Sievert Jr., in this well-articulated guidebook, sets forth a new operational standard for the facility manager's emerging role. . . a comprehensive program for managing facilities as a true profit center.

$29.98 per copy
275 pages, softcover
Catalog no. 67321

Reference Books

For more information visit the RSMeans website at www.rsmeans.com

Means Illustrated Construction Dictionary Condensed, 2nd Edition

The best portable dictionary for office or field use—an essential tool for contractors, architects, insurance and real estate personnel, facility managers, homeowners, and anyone who needs quick, clear definitions for construction terms. The second edition has been further enhanced with updates and hundreds of new terms and illustrations . . . in keeping with the most recent developments in the construction industry.

Now with a quick-reference Spanish section. Includes tools and equipment, materials, tasks, and more!

$59.95 per copy
Over 500 pages, softcover
Catalog no. 67282A

Repair & Remodeling Estimating 4th Edition

By Edward B. Wetherill and RSMeans

This important reference focuses on the unique problems of estimating renovations of existing structures, and helps you determine the true costs of remodeling through careful evaluation of architectural details and a site visit.

New section on disaster restoration costs.

$69.95 per copy
Over 450 pages, illustrated, hardcover
Catalog no. 67265B

Facilities Planning & Relocation

by David D. Owen

A complete system for planning space needs and managing relocations. Includes step-by-step manual, over 50 forms, and extensive reference section on materials and furnishings.

New lower price and user-friendly format.

$49.98 per copy
Over 450 pages, softcover
Catalog no. 67301

Square Foot & UNIFORMAT Assemblies Estimating Methods 3rd Edition

Develop realistic square foot and assemblies costs for budgeting and construction funding. The new edition features updated guidance on square foot and assemblies estimating using UNIFORMAT II. An essential reference for anyone who performs conceptual estimates.

$69.95 per copy
Over 300 pages, illustrated, softcover
Catalog no. 67145B

Electrical Estimating Methods 3rd Edition

Expanded edition includes sample estimates and cost information in keeping with the latest version of the CSI MasterFormat and UNIFORMAT II. Complete coverage of fiber optic and uninterruptible power supply electrical systems, broken down by components, and explained in detail. Includes a new chapter on computerized estimating methods. A practical companion to *RSMeans Electrical Cost Data.*

$64.95 per copy
Over 325 pages, hardcover
Catalog no. 67230B

Mechanical Estimating Methods 4th Edition

Completely updated, this guide assists you in making a review of plans, specs, and bid packages, with suggestions for takeoff procedures, listings, substitutions, and pre-bid scheduling for all components of HVAC. Includes suggestions for budgeting labor and equipment usage. Compares materials and construction methods to allow you to select the best option.

$64.95 per copy
Over 350 pages, illustrated, softcover
Catalog no. 67294B

Construction Business Management

by Nick Ganaway

Only 43% of construction firms stay in business after four years. Make sure your company thrives with valuable guidance from a pro with 25 years of success as a commercial contractor. Find out what it takes to build all aspects of a business that is profitable, enjoyable, and enduring. With a bonus chapter on retail construction.

$49.95 per copy
200 pages, softcover
Catalog no. 67352

Project Scheduling & Management for Construction 3rd Edition

by David R. Pierce, Jr.

A comprehensive yet easy-to-follow guide to construction project scheduling and control—from vital project management principles through the latest scheduling, tracking, and controlling techniques. The author is a leading authority on scheduling, with years of field and teaching experience at leading academic institutions. Spend a few hours with this book and come away with a solid understanding of this essential management topic.

$64.95 per copy
Over 300 pages, illustrated, hardcover
Catalog no. 67247B

The Practice of Cost Segregation Analysis

by Bruce A. Desrosiers and Wayne J. DelPico

This expert guide walks you through the practice of cost segregation analysis, which enables property owners to defer taxes and benefit from "accelerated cost recovery" through depreciation deductions on assets that are properly identified and classified.

With a glossary of terms, sample cost segregation estimates for various building types, key information resources, and updates via a dedicated website, this book is a critical resource for anyone involved in cost segregation analysis.

$99.95 per copy
Over 225 pages
Catalog no. 67345

Preventive Maintenance for Multi-Family Housing

by John C. Maciha

Prepared by one of the nation's leading experts on multi-family housing.

This complete PM system for apartment and condominium communities features expert guidance, checklists for buildings and grounds maintenance tasks and their frequencies, a reusable wall chart to track maintenance, and a dedicated website featuring customizable electronic forms. A must-have for anyone involved with multi-family housing maintenance and upkeep.

$89.95 per copy
225 pages
Catalog no. 67346

How to Estimate with Means Data & CostWorks

New 3rd Edition

by RSMeans and Saleh A. Mubarak, Ph.D.

New 3rd Edition—fully updated with new chapters, plus new CD with updated *CostWorks* cost data and MasterFormat organization. Includes all major construction items—with more than 300 exercises and two sets of plans that show how to estimate for a broad range of construction items and systems—including general conditions and equipment costs.

$59.95 per copy
272 pages, softcover
Includes CostWorks CD
Catalog no. 67324B

Job Order Contracting

Expediting Construction Project Delivery

by Allen Henderson

Expert guidance to help you implement JOC—fast becoming the preferred project delivery method for repair and renovation, minor new construction, and maintenance projects in the public sector and in many states and municipalities. The author, a leading JOC expert and practitioner, shows how to:

- Establish a JOC program
- Evaluate proposals and award contracts
- Handle general requirements and estimating
- Partner for maximum benefits

$89.95 per copy
192 pages, illustrated, hardcover
Catalog no. 67348

Estimating Building Costs

For the Residential & Light Commercial Contractor

by Wayne J. DelPico

Step-by-step estimating methods for residential and light commercial contractors. Includes a detailed look at every construction specialty—explaining all the components, takeoff units, and labor needed for well-organized, complete estimates. Covers correctly interpreting plans and specifications, and developing accurate and complete labor and material costs.

$29.95 per copy
Over 400 pages, illustrated, softcover
Catalog no. 67343

Building & Renovating Schools

This all-inclusive guide covers every step of the school construction process—from initial planning, needs assessment, and design, right through moving into the new facility. A must-have resource for anyone concerned with new school construction or renovation. With square foot cost models for elementary, middle, and high school facilities, and real-life case studies of recently completed school projects.

The contributors to this book—architects, construction project managers, contractors, and estimators who specialize in school construction—provide start-to-finish, expert guidance on the process.

$69.98 per copy
Over 425 pages, hardcover
Catalog no. 67342

Reference Books

For more information visit the RSMeans website at www.rsmeans.com

The Homeowner's Guide to Mold By Michael Pugliese

Expert guidance to protect your health and your home.

Mold, whether caused by leaks, humidity or flooding, is a real health and financial issue—for homeowners and contractors. This full-color book explains:

- Construction and maintenance practices to prevent mold
- How to inspect for and remove mold
- Mold remediation procedures and costs
- What to do after a flood
- How to deal with insurance companies

$21.95 per copy
144 pages, softcover
Catalog no. 67344

Universal Design Ideas for Style, Comfort & Safety

by RSMeans and Lexicon Consulting, Inc.

Incorporating universal design when building or remodeling helps people of any age and physical ability more fully and safely enjoy their living spaces. This book shows how universal design can be artfully blended into the most attractive homes. It discusses specialized products like adjustable countertops and chair lifts, as well as simple ways to enhance a home's safety and comfort. With color photos and expert guidance, every area of the home is covered. Includes budget estimates that give an idea how much projects will cost.

$21.95
160 pages, illustrated, softcover
Catalog no. 67354

Landscape Estimating Methods

5th Edition

Answers questions about preparing competitive landscape construction estimates, with up-to-date cost estimates and the new MasterFormat classification system. Expanded and revised to address the latest materials and methods, including new coverage on approaches to green building. Includes:

- Step-by-step explanation of the estimating process
- Sample forms and worksheets that save time and prevent errors

$64.95 per copy
Over 350 pages, softcover
Catalog no. 67295C

Plumbing Estimating Methods 3rd Edition

by Joseph Galeno and Sheldon Greene

Updated and revised! This practical guide walks you through a plumbing estimate, from basic materials and installation methods through change order analysis. *Plumbing Estimating Methods* covers residential, commercial, industrial, and medical systems, and features sample takeoff and estimate forms and detailed illustrations of systems and components.

$29.98 per copy
330+ pages, softcover
Catalog no. 67283B

Understanding & Negotiating Construction Contracts

by Kit Werremeyer

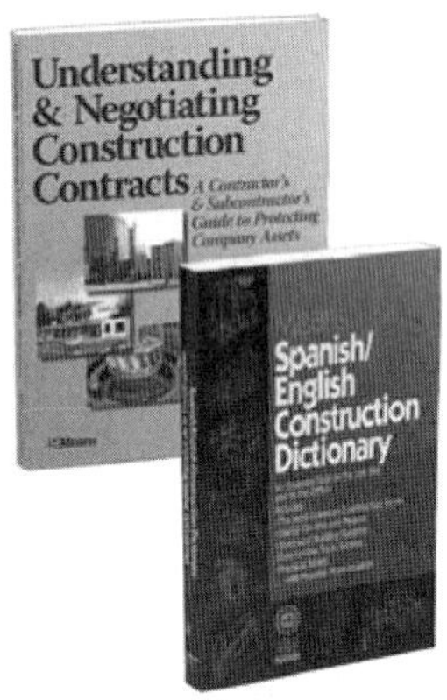

Take advantage of the author's 30 years' experience in small-to-large (including international) construction projects. Learn how to identify, understand, and evaluate high risk terms and conditions typically found in all construction contracts—then negotiate to lower or eliminate the risk, improve terms of payment, and reduce exposure to claims and disputes. The author avoids "legalese" and gives real-life examples from actual projects.

$69.95 per copy
300 pages, softcover
Catalog no. 67350

Spanish/English Construction Dictionary 2nd Edition

by RSMeans and the International Code Council

This expanded edition features thousands of the most common words and useful phrases in the construction industry with easy-to-follow pronunciations in both Spanish and English. Over 800 new terms, phrases, and illustrations have been added. It also features a new stand-alone "Safety & Emergencies" section, with colored pages for quick access.

Unique to this dictionary are the systems illustrations showing the relationship of components in the most common building systems for all major trades.

$23.95 per copy
Over 400 pages
Catalog no. 67327A

How Your House Works

by Charlie Wing

A must-have reference for every homeowner, handyman, and contractor—for repair, remodeling, and new construction. This book uncovers the mysteries behind just about every major appliance and building element in your house—from electrical, heating and AC, to plumbing, framing, foundations and appliances. Clear, "exploded" drawings show exactly how things should be put together and how they function—what to check if they don't work, and what you can do that might save you having to call in a professional.

$21.95 per copy
160 pages, softcover
Catalog no. 67351

Professional Development

Means CostWorks® CD Training

This one-day course helps users become more familiar with the functionality of *Means CostWorks* program. Each menu, icon, screen, and function found in the program is explained in depth. Time is devoted to hands-on estimating exercises.

Some of what you'll learn:

- Searching the database utilizing all navigation methods
- Exporting RSMeans Data to your preferred spreadsheet format
- Viewing crews, assembly components, and much more!
- Automatically regionalizing the database

You are required to bring your own laptop computer for this course.

When you register for this course you will receive an outline for your laptop requirements.

Also offering web training for CostWorks CD!

Unit Price Estimating

This interactive *two-day* seminar teaches attendees how to interpret project information and process it into final, detailed estimates with the greatest accuracy level.

The most important credential an estimator can take to the job is the ability to visualize construction, and estimate accurately.

Some of what you'll learn:

- Interpreting the design in terms of cost
- The most detailed, time-tested methodology for accurate pricing
- Key cost drivers—material, labor, equipment, staging, and subcontracts
- Understanding direct and indirect costs for accurate job cost accounting and change order management

Who should attend: Corporate and government estimators and purchasers, architects, engineers... and others needing to produce accurate project estimates.

Square Foot & Assemblies Estimating

This *two-day* course teaches attendees how to quickly deliver accurate square foot estimates using limited budget and design information.

Some of what you'll learn:

- How square foot costing gets the estimate done faster
- Taking advantage of a "systems" or "assemblies" format
- The RSMeans "building assemblies/square foot cost approach"
- How to create a reliable preliminary systems estimate using bare-bones design information

Who should attend: Facilities managers, facilities engineers, estimators, planners, developers, construction finance professionals... and others needing to make quick, accurate construction cost estimates at commercial, government, educational, and medical facilities.

Facilities Maintenance & Repair Estimating

This *two-day* course teaches attendees how to plan, budget, and estimate the cost of ongoing and preventive maintenance and repair for existing buildings and grounds.

Some of what you'll learn:

- The most financially favorable maintenance, repair, and replacement scheduling and estimating
- Auditing and value engineering facilities
- Preventive planning and facilities upgrading
- Determining both in-house and contract-out service costs
- Annual, asset-protecting M&R plan

Who should attend: Facility managers, maintenance supervisors, buildings and grounds superintendents, plant managers, planners, estimators... and others involved in facilities planning and budgeting.

Mechanical & Electrical Estimating

This *two-day* course teaches attendees how to prepare more accurate and complete mechanical/electrical estimates, avoiding the pitfalls of omission and double-counting, while understanding the composition and rationale within the RSMeans Mechanical/Electrical database.

Some of what you'll learn:

- The unique way mechanical and electrical systems are interrelated
- M&E estimates–conceptual, planning, budgeting, and bidding stages
- Order of magnitude, square foot, assemblies, and unit price estimating
- Comparative cost analysis of equipment and design alternatives

Who should attend: Architects, engineers, facilities managers, mechanical and electrical contractors... and others needing a highly reliable method for developing, understanding, and evaluating mechanical and electrical contracts.

Green Building: Planning & Construction

In this *two-day* course, learn about tailoring a building and its placement on the site to the local climate, site conditions, culture, and community. Includes information to reduce resource consumption and augment resource supply.

Some of what you'll learn:

- Green technologies, materials, systems, and standards
- Modeling tools and how to use them
- Cost vs. value of green products over their life cycle
- Low-cost green strategies, and economic incentives and funding
- Health, comfort, and productivity

Who should attend: Contractors, project managers, building owners, building officials, healthcare and insurance professionals.

Practical Project Management for Construction Professionals

In this *two-day* course, acquire the essential knowledge and develop the skills to effectively and efficiently execute the day-to-day responsibilities of the construction project manager.

Covers:

- General conditions of the construction contract
- Contract modifications: change orders and construction change directives
- Negotiations with subcontractors and vendors
- Effective writing: notification and communications
- Dispute resolution: claims and liens

Who should attend: Architects, engineers, owner's representatives, project managers.

Facilities Construction Estimating

In this *two-day* course, professionals working in facilities management can get help with their daily challenges to establish budgets for all phase of a project

Some of what you'll learn:

- Determine the full scope of a project
- Identify the scope of risks & opportunities
- Creative solutions to estimating issues
- Organizing estimates for presentation & discussion
- Special techniques for repair/remodel and maintenance projects
- Negotiating project change orders

Who should attend: Facility managers, engineers, contractors, facility trades-people, planners & project managers.

Professional Development

For more information visit the RSMeans website at www.rsmeans.com

Assessing Scope of Work for Facility Construction Estimating

This *two-day* practical training program addresses the vital importance of understanding the SCOPE of projects in order to produce accurate cost estimates for facility repair and remodeling.

Some of what you'll learn:

- Discussions of site visits, plans/specs, record drawings of facilities, and site-specific lists
- Review of CSI divisions, including means, methods, materials, and the challenges of scoping each topic
- Exercises in SCOPE identification and SCOPE writing for accurate estimating of projects
- Hands-on exercises that require SCOPE, take-off, and pricing

Who should attend: Corporate and government estimators, planners, facility managers, and others who need to produce accurate project estimates.

Site Work & Heavy Construction Estimating

This *two-day* course teaches attendees how to estimate earthwork, site utilities, foundations, and site improvements, using the assemblies and the unit price methods.

Some of what you'll learn:

- Basic site work and heavy construction estimating skills
- Estimating foundations, utilities, earthwork, and site improvements
- Correct equipment usage, quality control, and site investigation for estimating purposes

Who should attend: Project managers, design engineers, estimators, and contractors doing site work and heavy construction.

2011 RSMeans Seminar Schedule

Note: call for exact dates and details.

Location	Dates
Las Vegas, NV	March
Washington, DC	April
Denver, CO	May
San Francisco, CA	June
Washington, DC	September
Dallas, TX	September
Las Vegas, NV	October
Atlantic City, NJ	October
Orlando, FL	November
San Diego, CA	December

2011 Regional Seminar Schedule

Location	Dates
Seattle, WA	January and August
Dallas/Ft. Worth, TX	January
Austin, TX	February
Anchorage, AK	March and September
Oklahoma City, OK	March
Phoenix, AZ	March
Aiea, HI	May
Kansas City, MO	May
Bethesda, MD	June
Columbus, GA	June
El Segundo, CA	August
St. Louis, MO	August
Houston, TX	October
Colorado Springs, CO	October
Baltimore, MD	November
Sacramento, CA	November
San Antonio, TX	December
Raleigh, NC	December

1-800-334-3509, ext. 5115

Registration Information

Register early... Save up to $100! Register 30 days before the start date of a seminar and save $100 off your total fee. *Note: This discount can be applied only once per order. It cannot be applied to team discount registrations or any other special offer.*

How to register Register by phone today! The RSMeans toll-free number for making reservations is **1-800-334-3509, ext. 5115.**

Individual seminar registration fee - $935. *Means CostWorks*® training registration fee - $375. To register by mail, complete the registration form and return, with your full fee, to: RSMeans Seminars, 700 Longwater Drive, Norwell, MA 02061.

Government pricing All federal government employees save off the regular seminar price. Other promotional discounts cannot be combined with the government discount.

Team discount program for two to four seminar registrations. Call for pricing: 1-800-334-3509, ext. 5115

Multiple course discounts When signing up for two or more courses, call for pricing.

Refund policy Cancellations will be accepted up to ten business days prior to the seminar start. There are no refunds for cancellations received later than ten working days prior to the first day of the seminar. A $150 processing fee will be applied for all cancellations. Written notice of cancellation is required. Substitutions can be made at any time before the session starts. **No-shows are subject to the full seminar fee.**

AACE approved courses Many seminars described and offered here have been approved for 14 hours (1.4 recertification credits) of credit by the AACE International Certification Board toward meeting the continuing education requirements for recertification as a Certified Cost Engineer/Certified Cost Consultant.

AIA Continuing Education We are registered with the AIA Continuing Education System (AIA/CES) and are committed to developing quality learning activities in accordance with the CES criteria. Many seminars meet the AIA/CES criteria for Quality Level 2. AIA members may receive (14) learning units (LUs) for each two-day RSMeans course.

Daily course schedule The first day of each seminar session begins at 8:30 a.m. and ends at 4:30 p.m. The second day begins at 8:00 a.m. and ends at 4:00 p.m. Participants are urged to bring a hand-held calculator, since many actual problems will be worked out in each session.

Continental breakfast Your registration includes the cost of a continental breakfast, and a morning and afternoon refreshment break. These informal segments allow you to discuss topics of mutual interest with other seminar attendees. (You are free to make your own lunch and dinner arrangements.)

Hotel/transportation arrangements RSMeans arranges to hold a block of rooms at most host hotels. To take advantage of special group rates when making your reservation, be sure to mention that you are attending the RSMeans seminar. You are, of course, free to stay at the lodging place of your choice. **(Hotel reservations and transportation arrangements should be made directly by seminar attendees.)**

Important Class sizes are limited, so please register as soon as possible.

Note: *Pricing subject to change.*

Registration Form

ADDS-1000

Call 1-800-334-3509, ext. 5115 to register or FAX this form 1-800-632-6732. Visit our website: www.rsmeans.com

Please register the following people for the RSMeans construction seminars as shown here. We understand that we must make our own hotel reservations if overnight stays are necessary.

☐ Full payment of $__________________enclosed.

☐ Bill me

Please print name of registrant(s)

(To appear on certificate of completion)

____

____

____

____

P.O. #: ________________________

GOVERNMENT AGENCIES MUST SUPPLY PURCHASE ORDER NUMBER OR TRAINING FORM.

Firm name________________________

Address________________________

City/State/Zip________________________

Telephone no.______________ Fax no.______________

E-mail address________________________

Charge registration(s) to: ☐ MasterCard ☐ VISA ☐ American Express

Account no.______________Exp. date______________

Cardholder's signature________________________

Seminar name________________________

Seminar City ________________________

Please mail check to: RSMeans Seminars, 700 Longwater Drive, Norwell, MA 02061 USA

For more information
visit the RSMeans website
at www.rsmeans.com

New Titles

The Gypsum Construction Handbook by USG

An invaluable reference of construction procedures for gypsum drywall, cement board, veneer plaster, and conventional plaster. This new edition includes the newest product developments, installation methods, fire- and sound-rated construction information, illustrated framing-to-finish application instructions, estimating and planning information, and more. Great for architects and engineers; contractors, builders, and dealers; apprentices and training programs; building inspectors and code officials; and anyone interested in gypsum construction. Features information on tools and safety practices, a glossary of construction terms, and a list of important agencies and associations. Also available in Spanish.

$29.95
Over 575 pages, illustrated, softcover
Catalog no. 67357A

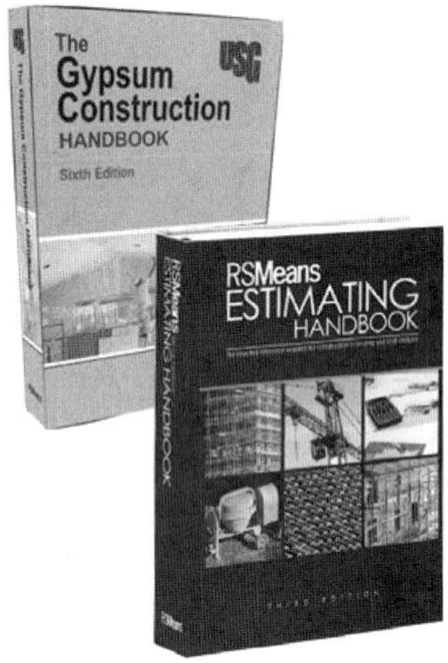

RSMeans Estimating Handbook 3rd Edition

Widely used in the industry for tasks ranging from routine estimates to special cost analysis projects, this handbook has been completely updated and reorganized with new and expanded technical information.

RSMeans Estimating Handbook will help construction professionals:

- Evaluate architectural plans and specifications
- Prepare accurate quantity takeoffs
- Compare design alternatives and costs
- Perform value engineering
- Double-check estimates and quotes
- Estimate change orders

$99.95 per copy
Over 900 pages, hardcover
Catalog No. 67276B

Means Illustrated Construction Dictionary

Unabridged 4th Edition, with CD-ROM

Long regarded as the industry's finest, *Means Illustrated Construction Dictionary* is now even better. With nearly 20,000 terms and more than 1,400 illustrations and photos, it is the clear choice for the most comprehensive and current information. The companion CD-ROM that comes with this new edition adds extra features such as: larger graphics and expanded definitions.

$99.95 per copy
Over 790 pages, illustrated, hardcover
Catalog no. 67292B

Green Building: Project Planning & Cost Estimating 3rd Edition

Since the widely read first edition of this book, green building has gone from a growing trend to a major force in design and construction.

This new edition has been updated with the latest in green building technologies, design concepts, standards, and costs. Full-color with all new case studies—plus a new chapter on commercial real estate.

$99.95 per copy
Over 450 pages, softcover
Catalog no. 67338B

Green Home Improvement by Daniel D. Chiras, PhD

With energy costs rising and environmental awareness increasing, many people are looking to make their homes greener. This book makes it easy, with 65 projects and actual costs and projected savings that will help homeowners prioritize their green improvements.

Projects range from simple water savers that cost only a few dollars, to bigger-ticket items such as efficient appliances and HVAC systems, green roofing, flooring, landscaping, and much more. With color photos and cost estimates, each project compares options and describes the work involved, the benefits, and the savings.

$24.95
320 pages, illustrated, softcover
Catalog no. 67355

Residential & Light Commercial Construction Standards 3rd Edition, Updated by RSMeans and contributing authors

This book provides authoritative requirements and recommendations compiled from leading professional associations, industry publications, and building code organizations. This all-in-one reference helps establish a standard for workmanship, quickly resolve disputes, and avoid defect claims. Updated third edition includes new coverage of green building, seismic, hurricane, and mold-resistant construction.

$59.95
Over 550 pages, illustrated, softcover
Catalog no. 67322B

ORDER TOLL FREE 1-800-334-3509
OR FAX 1-800-632-6732

2011 Order Form

Qty.	Book no.	COST ESTIMATING BOOKS	Unit Price	Total
	60061	Assemblies Cost Data 2011	$285.95	
	60011	Building Construction Cost Data 2011	174.95	
	60111	Concrete & Masonry Cost Data 2011	154.95	
	50141	Construction Cost Indexes 2011 (subscription)	362.00	
	60141A	Construction Cost Index–January 2011	90.50	
	60141B	Construction Cost Index–April 2011	90.50	
	60141C	Construction Cost Index–July 2011	90.50	
	60141D	Construction Cost Index–October 2011	90.50	
	60341	Contr. Pricing Guide: Resid. R & R Costs 2011	39.95	
	60231	Electrical Change Order Cost Data 2011	169.95	
	60031	Electrical Cost Data 2011	174.95	
	60201	Facilities Construction Cost Data 2011	439.95	
	60301	Facilities Maintenance & Repair Cost Data 2011	389.95	
	60551	Green Building Cost Data 2011	144.95	
	60161	Heavy Construction Cost Data 2011	174.95	
	60091	Interior Cost Data 2011	179.95	
	60121	Labor Rates for the Const. Industry 2011	389.95	
	60181	Light Commercial Cost Data 2011	119.95	
	60021	Mechanical Cost Data 2011	174.95	
	60151	Open Shop Building Const. Cost Data 2011	134.95	
	60211	Plumbing Cost Data 2011	174.95	
	60041	Commercial Renovation Cost Data 2011	144.95	
	60171	Residential Cost Data 2011	109.95	
	60281	Site Work & Landscape Cost Data 2011	169.95	
	60051	Square Foot Costs 2011	184.95	
	62010	Yardsticks for Costing (2010)	174.95	
	62011	Yardsticks for Costing (2011)	174.95	
		REFERENCE BOOKS		
	67329	Bldrs Essentials: Best Bus. Practices for Bldrs	29.95	
	67298A	Bldrs Essentials: Framing/Carpentry 2nd Ed.	15.98	
	67298AS	Bldrs Essentials: Framing/Carpentry Spanish	15.98	
	67307	Bldrs Essentials: Plan Reading & Takeoff	35.95	
	67342	Building & Renovating Schools	69.98	
	67339	Building Security: Strategies & Costs	44.98	
	67353	Complete Book of Framing	29.95	
	67146	Concrete Repair & Maintenance Illustrated	34.98	
	67352	Construction Business Management	49.95	
	67314	Cost Planning & Est. for Facil. Maint.	89.95	
	67328A	Designing & Building with the IBC, 2nd Ed.	99.95	
	67230B	Electrical Estimating Methods, 3rd Ed.	64.95	
	64777A	Environmental Remediation Est. Methods, 2nd Ed.	49.98	
	67343	Estimating Bldg. Costs for Resi. & Lt. Comm.	29.95	
	67276B	Estimating Handbook, 3rd Ed.	$99.95	
	67318	Facilities Operations & Engineering Reference	54.98	
	67301	Facilities Planning & Relocation	49.98	
	67338B	Green Building: Proj. Planning & Cost Est., 3rd Ed.	99.95	
	67355	Green Home Improvement	24.95	
	67357A	The Gypsum Construction Handbook	29.95	
	67357S	The Gypsum Construction Handbook (Spanish)	29.95	
	67323	Historic Preservation: Proj. Planning & Est.	49.98	
	67308E	Home Improvement Costs–Int. Projects, 9th Ed.	24.95	

Qty.	Book no.	REFERENCE BOOKS (Cont.)	Unit Price	Total
	67309E	Home Improvement Costs–Ext. Projects, 9th Ed.	24.95	
	67344	Homeowner's Guide to Mold	21.95	
	67324B	How to Est.w/Means Data & CostWorks, 3rd Ed.	59.95	
	67351	How Your House Works	21.95	
	67282A	Illustrated Const. Dictionary, Condensed, 2nd Ed.	59.95	
	67292B	Illustrated Const. Dictionary, w/CD-ROM, 4th Ed.	99.95	
	67348	Job Order Contracting	89.95	
	67295C	Landscape Estimating Methods, 5th Ed.	64.95	
	67341	Life Cycle Costing for Facilities	99.95	
	67294B	Mechanical Estimating Methods, 4th Ed.	64.95	
	67283B	Plumbing Estimating Methods, 3rd Ed.	29.98	
	67345	Practice of Cost Segregation Analysis	99.95	
	67337	Preventive Maint. for Higher Education Facilities	149.95	
	67346	Preventive Maint. for Multi-Family Housing	89.95	
	67326	Preventive Maint. Guidelines for School Facil.	149.95	
	67247B	Project Scheduling & Management for Constr. 3rd Ed.	64.95	
	67265B	Repair & Remodeling Estimating Methods, 4th Ed.	69.95	
	67322B	Resi. & Light Commercial Const. Stds., 3rd Ed.	59.95	
	67327A	Spanish/English Construction Dictionary, 2nd Ed.	23.95	
	67145B	Sq. Ft. & Assem. Estimating Methods, 3rd Ed.	69.95	
	67321	Total Productive Facilities Management	29.98	
	67350	Understanding and Negotiating Const. Contracts	69.95	
	67303B	Unit Price Estimating Methods, 4th Ed.	59.95	
	67354	Universal Design	21.95	
	67319A	Value Engineering: Practical Applications	79.95	

MA residents add 6.25% state sales tax	
Shipping & Handling**	
Total (U.S. Funds)*	

Prices are subject to change and are for U.S. delivery only. *Canadian customers may call for current prices. **Shipping & handling charges: Add 7% of total order for check and credit card payments. Add 9% of total order for invoiced orders.

Send order to: **ADDV-1000**

Name (please print) ______________________

Company ______________________

☐ **Company**

☐ **Home** **Address** ______________________

____

City/State/Zip ______________________

Phone # ______________ P.O. # ______________

(Must accompany all orders being billed)

Mail to: RSMeans, 700 Longwater Drive, Norwell, MA 02061

RSMeans Project Cost Report

By filling out this report, your project data will contribute to the database that supports the RSMeans Project Cost Square Foot Data. When you fill out this form, RSMeans will provide a $30 discount off one of the RSMeans products advertised in the preceding pages. Please complete the form including all items where you have cost data, and all the items marked (✔).

$30.00 Discount per product for each report you submit.

Project Description (No remodeling projects, please.)

✔ Building Use (Office, School...) ____________

✔ Address (City, State) ____________

____

✔ Total Building Area (SF) ____________

✔ Ground Floor (SF) ____________

✔ Frame (Wood, Steel...) ____________

✔ Exterior Wall (Brick, Tilt-up...) ____________

✔ Basement: (check one) ☐ Full ☐ Partial ☐ None

✔ Number Stories ____________

✔ Floor-to-Floor Height ____________

✔ Volume (C.F.) ____________

% Air Conditioned ____________ Tons ____________

Total Project Cost $ ____________

Owner ____________

Architect ____________

General Contractor ____________

✔ Bid Date ____________

✔ Typical Bay Size ____________

✔ Capacity ____________

✔ Labor Force: _____ % Union _____ % Non-Union

✔ Project Description (Circle one number in each line)

1. Economy 2. Average 3. Custom 4. Luxury
1. Square 2. Rectangular 3. Irregular 4. Very Irregular

Comments ____________

____

A	✔	General Conditions	$
B	✔	Site Work	$
C	✔	Concrete	$
D	✔	Masonry	$
E	✔	Metals	$
F	✔	Wood & Plastics	$
G	✔	Thermal & Moisture Protection	$
GR		Roofing & Flashing	$
H	✔	Doors and Windows	$
J	✔	Finishes	$
JP		Painting & Wall Covering	$

K	✔	Specialties	$
L	✔	Equipment	$
M	✔	Furnishings	$
N	✔	Special Construction	$
P	✔	Conveying Systems	$
Q	✔	Mechanical	$
QP		Plumbing	$
QB		HVAC	$
R	✔	Electrical	$
S	✔	Mech./Elec. Combined	$

Please specify the RSMeans product you wish to receive.
Complete the address information.

Product Name ____________

Product Number ____________

Your Name ____________

Title ____________

Company ____________

☐ Company

☐ Home Street Address ____________

City, State, Zip ____________

Return by mail or Fax 888-492-6770.

Method of Payment:

Credit Card # ____________

Expiration Date ____________

Check ____________

Purchase Order ____________

RSMeans
a division of Reed Construction Data
Square Foot Costs Department
700 Longwater Drive
Norwell, MA 02061